ABOUT THE AUTHORS

Gerard J. Tortora Jerry Tortora is a professor of biology and teaches microbiology, human anatomy, and physiology at Bergen Community College in Paramus, New Jersey. He received his M.A. in Biology from Montclair State College in 1965. He belongs to a number of biology/microbiology organizations, such as the American Society of Microbiology (ASM), Human Anatomy and Physiology Society (HAPS), American Association for the Advancement of Science (AAAS), National Education Association (NEA), New Jersey Educational Association (NJEA), and the Metropolitan Association of College and University Biologists (MACUB). Jerry is the author of numerous biological science textbooks. In 1995, he was selected as one of the finest faculty scholars of Bergen Community College and was named Distinguished Faculty Scholar. In 1996, Jerry received a National Institute for Staff and Organizational Development (NISOD) excellence award from the University of Texas and was selected to represent Bergen Community College in a campaign to increase awareness of the contributions of community colleges to higher education.

Berdell R. Funke Bert Funke received his Ph.D., M.S., and B.S. in microbiology from Kansas State University. He has spent his professional years as a professor of microbiology at North Dakota State University. He taught introductory microbiology, including laboratory sections, general microbiology, food microbiology, soil microbiology, clinical parasitology, and pathogenic microbiology. As a research scientist in the Experiment Station at North Dakota State, he has published numerous papers in soil microbiology and food microbiology.

Christine L. Case Chris Case is a registered microbiologist and a professor of microbiology at Skyline College in San Bruno, California, where she has taught for the past 38 years. She received her Ed.D. in curriculum and instruction from Nova Southeastern University and her M.A. in microbiology from San Francisco State University. She was Director for the Society for Industrial Microbiology (SIM) and is an active member of the ASM and Northern California SIM. She received the ASM and California Hayward outstanding educator awards. In 2008, Chris received the SACNAS Distinguished Community/Tribal College Mentor Award for her commitment to her students, several of whom have presented at undergraduate research conferences and won awards. In addition to teaching, Chris contributes regularly to the professional literature, develops innovative educational methodologies, and maintains a personal and professional commitment to conservation and the importance of science in society. Chris is also an avid photographer, and many of her photographs appear in this book.

PREFACE

Since the publication of the first edition nearly 30 years ago, well over one million students have used *Microbiology: An Introduction* at colleges and universities around the world, making it the leading textbook for non-majors microbiology. The tenth edition continues to be a comprehensive beginning text, assuming no previous study of biology or chemistry. The text is appropriate for students in a wide variety of programs, including the allied health sciences, biological science, environmental sciences, animal science, forestry, agriculture, home economics, and the liberal arts.

HALLMARKS OF *MICROBIOLOGY: AN INTRODUCTION*

We have retained in this new edition features that made the previous editions so popular:

- **An appropriate balance between microbiological fundamentals and applications, and between medical applications and other applied areas of microbiology.** As in previous editions, basic microbiological principles are given greater emphasis than are applications, and health-related applications are featured.
- **Straightforward presentation of complex topics.** Each section of the text has been written with the student in mind. Our book is known for its clear explanations and consistent pedagogy.
- **Clear, accurate, and pedagogically effective illustrations and photos.** Step-by-step diagrams that closely coordinate with narrative descriptions aid student comprehension of concepts. Clear and accurate renderings of processes and structures focus students on what they need to learn. The quantity and quality of electron and light micrographs is unmatched in the market.
- **Flexible organization.** We have organized the book in what we think is a useful fashion while recognizing that the material might be effectively presented in a number of other sequences. For instructors who wish to use a different order, we have made each chapter as independent as possible and have included numerous cross-references. The Instructor's Guide, written by Christine Case, provides detailed guidelines for organizing the material in several other ways.

NEW TO THE TENTH EDITION

Please see pages x–xviii for a visual introduction to the new tenth edition.

The changes in this edition address instructors' biggest challenge in the introductory microbiology course: the wide range of student levels, including student under-preparedness. The tenth edition meets all students at their respective levels of skill and understanding.

The highlights of the tenth edition can be seen in the new Foundation Figures, the new features that help students check their understanding as they progress through each chapter, and the improved boxes that prepare students to start thinking like a clinician. Content and currency have also been substantially updated.

Foundation Figures

In order to help students focus on and master the core concepts of microbiology, the authors have integrated text and visuals into twenty specially designed Foundation Figures. These Foundation Figures include both a key concept statement that ensures students understand the central concept of the figure and an explanation of how each figure is foundational to further learning in the course. In addition, throughout the entire book the illustration program has been dramatically revised and updated with new art styles and a brighter color palette that has more contrasting colors and more dimensionality.

Features that Help Students Check Their Understanding

New Check Your Understanding questions encourage students to engage interactively with the material and self-assess their understanding of the Learning Objectives as they progress through each chapter. New Draw It questions are now included in the end-of-chapter Review Questions, asking students to sketch a rough diagram or fill in missing parts of a figure or graph. The popular Figure Legend Questions have been retained and improved.

Support for Students to Start Thinking like a Clinician

Revised and redesigned Applications of Microbiology boxes describe current and practical uses of microbiology. New and updated Clinical Focus boxes contain data from the *Morbidity and Mortality Weekly Report* modified into clinical problem-solving questions that help students develop their critical thinking skills and give them an active role while reading. Diseases in Focus boxes replace disease comparison tables, organizing comparative information about similar diseases in a discovery-oriented and visually interesting format that provides a helpful study tool for students.

Updates in Content and Currency

Antimicrobial resistance, biofilms, bioterrorism, and evolution receive special attention and increased emphasis. The immunity chapters—Chapters 16 and 17—have received a careful and sig-

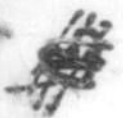

nificant update for increased currency, clarity, and accuracy, without an increase in the level of detail. Taxonomy, nomenclature, and disease incidence data are current through August 2008.

Cutting-Edge Media Supplements

Turn to the inside front cover of this book for an overview of the new MyMicrobiologyPlace website with its simple three-step approach to learning. Pages xiv–xv provide more details about the exciting new student and instructor media, including the MP3 Tutor Sessions, the 3D MicroFlix animations, and the Instructor's Resource DVD/CD-ROM.

CHAPTER-BY-CHAPTER REVISIONS

Every chapter in this edition has been thoroughly revised, and data in the text, tables, and figures have been updated through August 2008 where possible. The main changes for each chapter are summarized below.

Part One
Fundamentals of Microbiology

Chapter 1: The Microbial World and You

- The table "Making Scientific Names Familiar" has been moved to this chapter from Chapter 10.
- Biofilms are introduced.
- Discussion of emerging infectious diseases has been updated, including a section on antibiotic-resistant bacteria.

Chapter 2: Chemical Principles

- Definitions have been expanded, including definitions of *cis* and *trans fatty acids*.
- Figure 2.16 is now a Foundation Figure.

Chapter 3: Observing Microorganisms through a Microscope

- Two-photon microscopy is included.
- Several new photos illustrate microscopic images.
- Figure 3.2 is now a Foundation Figure.

Chapter 4: Functional Anatomy of Prokaryotic and Eukaryotic Cells

- Figure 4.6 is now a Foundation Figure.
- The discussion of flagella, fimbriae, and pili has been revised, as has the discussion of the lipopolysaccharide.
- The discussion of facilitated diffusion has been revised, and a new figure compares types of diffusion across membranes, including aquaporins.

Chapter 5: Microbial Metabolism

- The section on biochemical tests has been expanded.
- New renditions of enzymes are more realistic.
- Figure 5.11 is now a Foundation Figure.
- A new Clinical Focus box illustrates the use of biochemical tests to identify slow-growing mycobacteria.

Chapter 6: Microbial Growth

- The discussion of biofilms previously appearing in Chapter 27 has been moved to this chapter and been significantly updated and expanded.
- Discussion of anaerobic growth media and methods has been updated.
- A discussion of Biosafety Levels has been added, including a figure illustrating Biosafety Level 4.
- A new figure showing differential medium is included.
- Figure 6.15 is now a Foundation Figure.
- A new Clinical Focus box illustrates the role of biofilms in causing nosocomial infections.

Chapter 7: The Control of Microbial Growth

- The definition of *sterilization* has been updated and qualified in consideration of the existence of prions.
- Figure 7.1 is now a Foundation Figure.
- Discussion of ultra-high temperature (UHT) has been clarified.
- New products and newly approved uses are included.
- A new Clinical Focus box illustrates the relationship between improper disinfection and nosocomial infection.

Chapter 8: Microbial Genetics

- Figure 8.2 is now a Foundation Figure.
- Discussion of genetic combination by crossing over has been revised for clarity.
- snRNPs are defined.
- Inducible and repressible operons are explained and compared in separate figures.

Chapter 9: Biotechnology and Recombinant DNA

- Figure 9.1 is now a Foundation Figure.
- Gene silencing, reverse genetics, and real-time PCR are discussed.
- A new Clincal Focus box describes using reverse-transcription PCR to track a norovirus outbreak.

Part Two
A Survey of the Microbial World

Chapter 10: Classification of Microorganisms

- Figure 10.1 is now a Foundation Figure.
- Photos of fossil and living stromatolites are included.
- The use of transport media is explained.

Chapter 11: The Prokaryotes: Domains Bacteria and Archaea

- Several new bacterial groups are discussed: *Pelagibacter, Acinetobacter baumanii,* Planctomycetes, *Gemmata obscuriglobus.*
- Discussion of the theoretical minimal size of a bacterium and its genetic requirements has been revised.

Chapter 12: The Eukaryotes: Fungi, Algae, Protozoa, and Helminths

- Examples of new uses of fungi as pesticides are listed.
- Discussion of the oomycotes is expanded to include introduction of *Phytophthora* into the United States. The oomycote life cycle is illustrated in a new figure.
- Heartworm is included.
- A new Clinical Focus box highlights cryptosporidial diarrhea, the most common pathogen associated with swimming.

Chapter 13: Viruses, Viroids, and Prions

- The chapter begins with the use of retroviridae to genetically modify cells.
- Figure 13.15 is now a Foundation Figure.
- Bee colony collapse is mentioned.
- The Clinical Focus box on the evolution and occurrence of avian flu has been updated.

Part Three
Interaction Between Microbe and Host

Chapter 14: Principles of Disease and Epidemiology

- Figure 14.3 is now a Foundation Figure.
- Statistics on notifiable infectious diseases have been updated.
- A new Clinical Focus box illustrates the emergence of hospital-acquired and community-acquired MRSA.

Chapter 15: Microbial Mechanisms of Pathogenicity

- Discussion of A-B toxins has been expanded and clarified.
- Figure 15.5, action of an exotoxin, has been revised and expanded.
- Figures 15.4 and 15.9 are now Foundation Figures.
- A new Clinical Focus box illustrates role of biofilms and endotoxins in postoperative infections.

Chapter 16: Innate Immunity: Nonspecific Defenses of the Host

- Treatment of several topics has been expanded and/or reorganized and clarified: physical and chemical factors in the first line of defense; formed elements in blood; the lymphatic system (including additional illustrations); adherence, acute-phase proteins, complement, iron-binding proteins, and antimicrobial peptides.
- The role of biofilms in evading phagocytosis is included.
- Figures 16.7 and Figure 16.9 are now Foundation Figures.
- The Applications of Microbiology box on serum collection has been revised to include testing for complement to monitor immune complex diseases in patients.

Chapter 17: Adaptive Immunity: Specific Defenses of the Host

- A new photo depicts actual antibody morphology shown by atomic force microscopy.
- Several important figures have been extensively revised for accuracy and clarity:
  - Figure 17.5. Clonal selection and differentiation of B cells
  - Figure 17.10. Activation of $CD4^+$ T cells
  - Figure 17.11. Killing of virus-infected target cell by cytotoxic T lymphocyte
  - Figure 17.19. The dual nature of the immune system (now a Foundation Figure)
- A new photo and illustration (Figure 17.9) show M cells found within Peyer's patches.
- Discussion of the major histocompatibility complex (MHC) has been revised and improved.
- Nomenclature conventions have been updated for T cells (for example, T helper cell, $CD4^+$ T cell).
- The discussions of T cells, dendritic cells, and cytokines have been completely revised.
- A new Applications of Microbiology box describes the possible use of IL-12 to treat psoriasis.

Chapter 18: Practical Applications of Immunology

- Figure 18.2 is now a Foundation Figure.
- Discussions of DNA vaccines and adjuvants have been updated and revised.
- The tables of vaccine schedules have been updated.
- A new Clinical Focus box illustrates the success of vaccination in eliminating measles in the U.S. and highlights the importance of measles as a cause of death in developing countries.

Chapter 19: Disorders Associated with the Immune System

- Coverage of blood groups includes a discussion of the relationship between certain blood groups and their relative resistance or susceptibility to certain diseases.
- A discussion of the autoimmune disease psoriasis and its associated arthritis has been introduced, along with the current treatments with monoclonal antibodies.
- The discussion of stem cells has been updated, and a new figure (Figure 19.10) the derivation of stem cells and stem cell lines.
- The discussion of HIV and AIDS has been revised and updated. Especially important is the complete revision of Figure 19.13, which shows the sequence of attachment, fusion, and entry of the virus into the target $CD4^+$ T cell.
- Figure 19.16 is now a Foundation Figure.

Chapter 20: Antimicrobial Drugs

- Figure 20.2 is now a Foundation Figure.
- The historical importance of the sulfa drugs is given more prominence.
- The current methods used for the discovery of new antibiotics are discussed, including rapid throughput methods.
- The discussion of antibiotics has been updated to admit new antibiotics. The discussion of antivirals for the treatment of HIV/AIDS has been especially updated and revised to include the latest developments in this constantly changing area.
- The discussion of resistance to antibiotics has been completely revised and expanded, and a new Foundation Figure (Figure 20.20) illustrates the most important target areas for resistance.
- The concluding discussion on the future of antibiotic development and the prospect for unconventional antibiotics has been completely revised and updated.

Part Four
Microorganisms and Human Disease

Chapter 21: Microbial Diseases of the Skin and Eyes

- The discussion of *Staphylococcus aureus* has been completely rewritten to emphasize the importance of MRSA.

- The discussion of impetigo and scalded skin syndrome has been revised, and discussion of a new disease, Buruli ulcer, has been added.
- Some of the newer treatments for acne now have an expanded discussion.

Chapter 22: Microbial Diseases of the Nervous System

- A new figure (Figure 22.4) illustrates a spinal tap.
- The discussion of cryptococcosis has been revised to include a newer pathogen.
- A brief description of prions has been included to supplement that given in Chapter 13.
- The discussion of chronic disease syndrome has been completely revised and now includes the CDC's diagnostic definition and the alternative name of myalgic encephalomyelitis.

Chapter 23: Microbial Diseases of the Cardiovascular and Lymphatic Systems

- The definitions of the similar terms *septicemia* and *sepsis* have been revised.
- Discussions of brucellosis and rat-bite fever have been completely rewritten.
- The discussion of ehrlichiosis has been revised to include the new terminology of anaplasmosis.
- Discussion of the disease chikungunya fever has been added because of its current spread into temperate climates.
- The discussion of malaria has been revised completely to better differentiate between prophylaxis and therapy.

Chapter 24: Microbial Diseases of the Respiratory System

- The discussion of pertussis has been revised to better describe some of the latest developments, especially the recent increase in cases.
- The discussion of tuberculosis has been updated and revised to include more on extensively resistant strains of the pathogen and some of the more recent testing methods.
- Discussion of influenza has been thoroughly revised and updated, especially the means by which mutants arise and the infectiveness of the avian flu virus.

Chapter 25: Microbial Diseases of the Digestive System

- The discussion of traveler's diarrhea has been rewritten to include the important pathogen enteroaggregative *E. coli.*
- Recent therapeutic drugs for HBV have been included.
- The discussion of noroviruses has been updated with special attention to decontamination methods available to deal with outbreaks.

Chapter 26: Microbial Diseases of the Urinary and Reproductive Systems

- The discussion of vaginal microbiota has been extensively revised.
- The introductory discussion of syphilis, especially relating to recent genetic analysis on its probable origin in the New World, has been revised.
- The discussion of testing for syphilis has been revised.
- The TORCH panel of tests is included.

Part Five
Environmental and Applied Microbiology

Chapter 27: Environmental Microbiology

- The sulfur cycle figure (Figure 27.7)has been completely redrawn.
- The discussion of biodegradable plastics has been revised and updated.

Chapter 28: Applied and Industrial Microbiology

- The discussion of biofuels has been expanded.

ACKNOWLEDGMENTS

In preparing for this textbook, we have benefited from the guidance and advice of a large number of microbiology instructors across the country. The reviewers listed provided constructive criticism and valuable suggestions at various stages of the revision. We gratefully acknowledge our debt to these individuals.

TENTH EDITION REVIEWERS

Cynthia Anderson
Mt. San Antonio College

Rod Anderson
Ohio Northern University

Terry Austin
Temple College

Joan Baird
Rose State College

Archna Bhasin
Valdosta State University

Victoria Bingham
Daytona Beach College

Phyllis Braun
Fairfield University

Donald P. Breakwell
Brigham Young University

Sandra Burnett
Brigham Young University

Susan Capasso
St. Vincent's College

Carol Castaneda
Indiana University Northwest

James K. Collins
University of Arizona

Lee Couch
University of New Mexico

Ellen C. Cover
Lamar University

Jean Cremins
Middlesex Community College

Melissa A. Deadmond
Truckee Meadows Community College

Janet M. Decker
University of Arizona

Vivian Elder
Ozarks Technical Community College

Christina Gan
Highline Community College

Pete Haddix
Auburn University Montgomery

Rachel Hirst
Massasoit Community College

Dawn Janich
Community College of Philadelphia

Judy Kaufman
Monroe Community College

Malda Kocache
George Mason University

John M. Lammert
Gustavus Adolphus College

Paul A. LeBlanc
University of Alabama

Michael W. Lema
Midlands Technical College

John Lennox
The Pennsylvania State University

Shawn Lester
Montgomery College

Leslie Lichtenstein
Massasoit Community College

Eric Lifson
Bucks County Community College

Suzanne Long
Monroe Community College

William C. Matthai
Tarrant County College Northeast

Philip Mixter
Washington State University

Rita B. Moyes
Texas A&M University

Ellyn R. Mulcahy
Johnson County Community College

Tim R. Mullican
Dakota Wesleyan University

Richard L. Myers
Missouri State University

Kabi Neupane
Leeward Community College

Lourdes P. Norman
Florida Community College, Jacksonville

Eric R. Paul
Southwestern Oklahoma State University

Judy L. Penn
Shoreline Community College

Jack Pennington
St. Louis Community College, Forest Park

Indiren Pillay
Culver-Stockton College

Ronny Priefer
Niagara University

Todd P. Primm
Sam Houston State University

Mary L. Puglia
Central Arizona College

Amy J. Reese
Cedar Crest College

Lois Sealy
Valencia Community College West

Heather Smith
Riverside City College

Kate Sutton
Clark College

Paul H. Tomasek
California State University, Northridge

David J. Wartell
Harrisburg Area Community College

MJ Weintraub
Raymond Walters College

Ruth Wrightsman
Saddleback College

Anne Zayaitz
Kutztown University

Michele Zwolinski
Weber State University

We also thank the staff at Benjamin Cummings for their dedication to excellence. Leslie Berriman, our executive editor, successfully kept us all focused on where we wanted this revision to go. Robin Pille, project editor, masterfully managed the development of the book, keeping communication lines open and ensuring the highest quality at every stage. Sally Peyrefitte's careful attention to continuity and detail in her copyedit of both text and art served to keep concepts and information clear throughout.

Janet Vail and Wendy Earl expertly guided the text through the production process. Lisa Torri and the talented staff at Precision Graphics effectively managed the ambitious overhaul of our large and complex beautiful new art program. Jean Lake coordinated the many complex stages of the art development and rendering. The photo researcher, Maureen Spuhler, working closely with Senior Photo and Art Manager Travis Amos, made sure we had clear and striking images throughout the book. tani hasegawa created the elegant interior design, and Yvo Riezebos did a wonderful job with the cover. The skilled team at Progressive Information Technologies, led by Michelle Jones, did an outstanding job moving this book quickly and beautifully through composition. Stacey Weinberger guided the book through the manufacturing process.

Katie Heimsoth impeccably handled the instructor supplements and also was the editor for the new edition of Johnson/Case *Laboratory Experiments in Microbiology.* Kelly Reed brought her creativity and teaching experience to bear on the development of the student supplements including the new *Get Ready for Microbiology* workbook. Lucinda Bingham managed the media program, working many miracles to produce the impressive array of resources on the website and Instructor's Resource DVD/CD-ROM including the new MicroFlix. Leslie Austin and James Bruce managed the print and media supplements through the complex production stages.

Neena Bali, Senior Marketing Manager, and the entire Pearson Science sales force do a stellar job presenting this book to instructors and students and ensuring its unwavering status as the best-selling microbiology textbook.

We would all like to acknowledge our spouses and families, who have provided invaluable support throughout the writing process.

Finally, we have an enduring appreciation for our students, whose comments and suggestions provide insight and remind us of their needs. This text is for them.

Gerard J. Tortora ***Berdell R. Funke*** ***Christine L. Case***

A visual approach to teaching...

NEW Foundation Figures emphasize core concepts in microbiology and give students the foundation they need to succeed in the course.

The introductory text explains how the figure is foundational to other concepts students will learn later.

In its Tenth Edition, this best-selling textbook addresses the #1 challenge of the microbiology course: the wide variance in student levels, including the under-preparedness of many students. New and highly visual **Foundation Figures** get students to focus on and engage with core microbiology content.

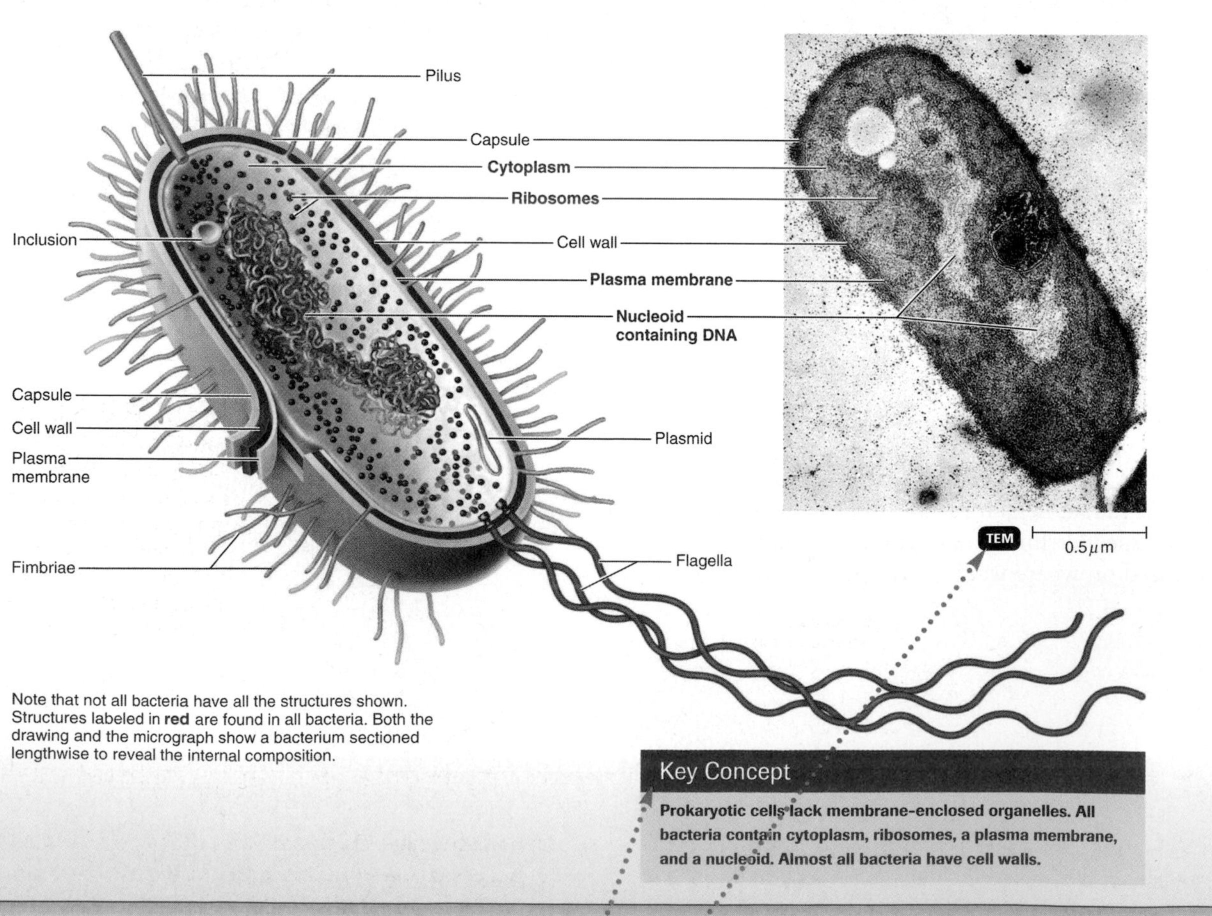

Figure 4.6
FOUNDATION FIGURE The Structure of a Prokaryotic Cell
This prokaryotic cell shows typical structures that may be found in bacteria. Each of the labeled structures will be discussed individually in this chapter. As you will see in later chapters, some of these structures contribute to bacterial virulence, play a role in bacterial identification, and are targets of antimicrobial agents.
Pilus
Capsule
Cytoplasm
Ribosomes
Inclusion
Cell wall
Plasma membrane
Nucleoid containing DNA
Capsule
Cell wall
Plasma membrane
Plasmid
Fimbriae
Flagella
TEM
0.5 μm
Note that not all bacteria have all the structures shown. Structures labeled in red are found in all bacteria. Both the drawing and the micrograph show a bacterium sectioned lengthwise to reveal the internal composition.
Key Concept
Prokaryotic cells lack membrane-enclosed organelles. All bacteria contain cytoplasm, ribosomes, a plasma membrane, and a nucleoid. Almost all bacteria have cell walls.

The Key Concept box presents the big picture, helping to ensure students understand the central concept presented by the figure.

Clear and consistent TEM/SEM/LM icons appear for all micrograph images, showing at a glance which type of microscope was used.

...the foundations of microbiology

Figure 5.11

FOUNDATION FIGURE An Overview of Respiration and Fermentation

A small version of this overview figure will be included in other figures throughout the chapter to indicate the relationships of different reactions to the overall processes of respiration and fermentation.

1 Glycolysis produces ATP and reduces NAD^+ to NADH while oxidizing glucose to pyruvic acid. In respiration, the pyruvic acid is converted to the first reactant in the Krebs cycle.

2 The Krebs cycle produces ATP and reduces NAD^+ (and another electron carrier called $FADH_2$) while giving off CO_2. The NADH and $FADH_2$ from both processes carry electrons to the electron transport chain.

3 In the electron transport chain, the energy of the electrons is used to produce a great deal of ATP.

In fermentation, the pyruvic acid and the electrons carried by NADH from glycolysis are incorporated into fermentation end-products.

Key Concept

To produce energy from glucose, microbes use two general processes: respiration and fermentation. Both usually start with glycolysis but follow different subsequent pathways.

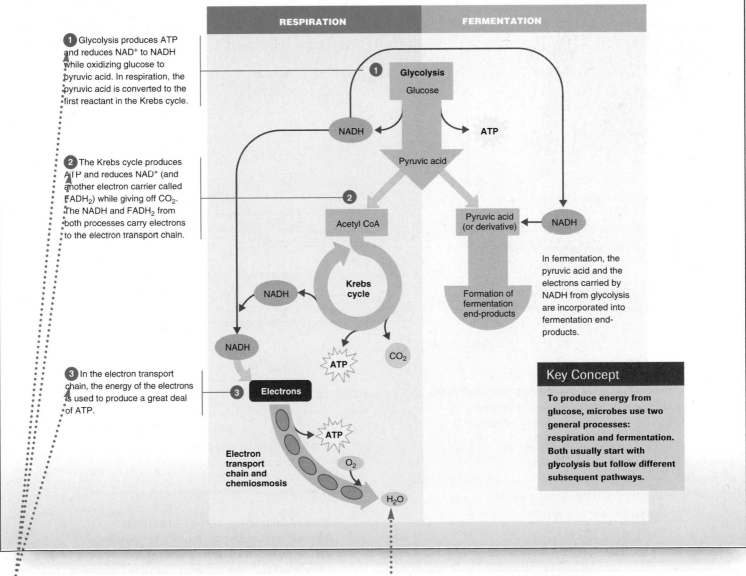

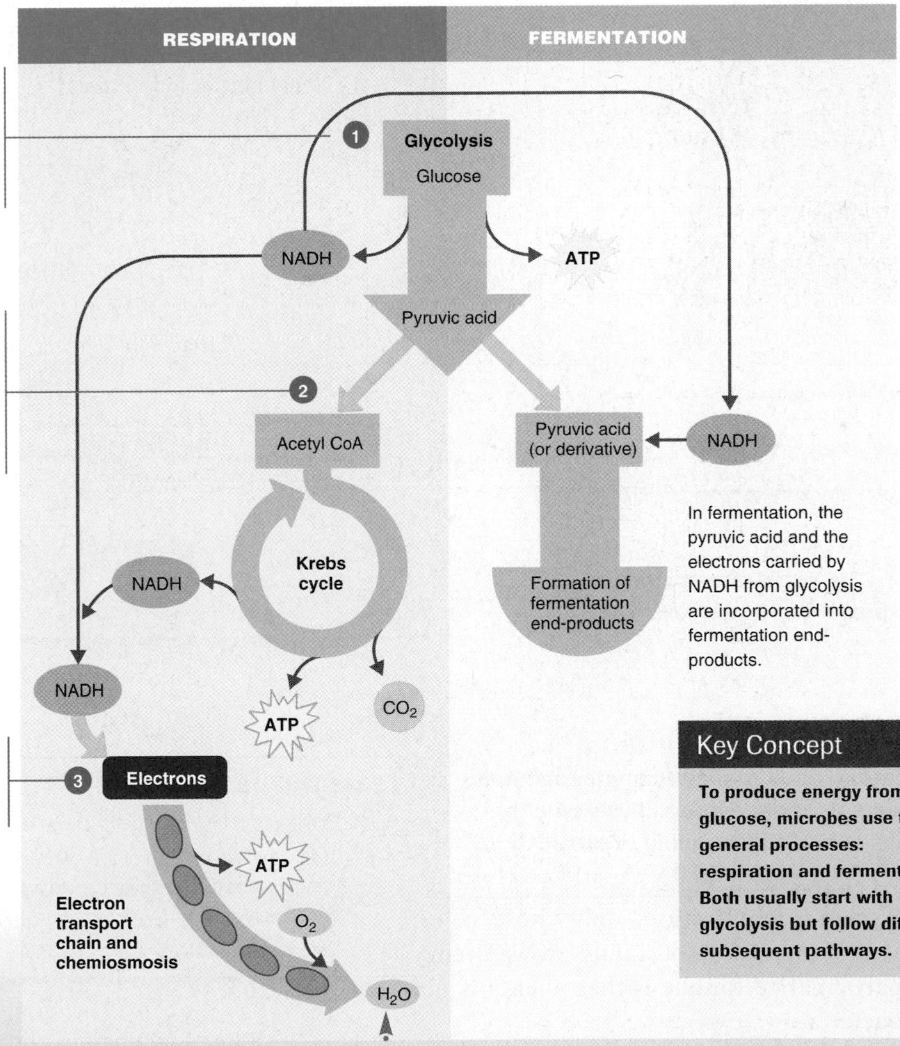

Easy-to-find blue step numbers guide the eye through complex processes, breaking them down into clear, manageable pieces that make concepts easier to teach and understand.

Consistent use of symbols and colors enables students to progress from familiar parts of illustrated processes to unfamiliar ones with confidence. Molecules such as ATP are the same color and shape throughout the book.

For a complete list of Foundation Figures, turn to page xxxi.

Frequent opportunities for students to check their understanding...

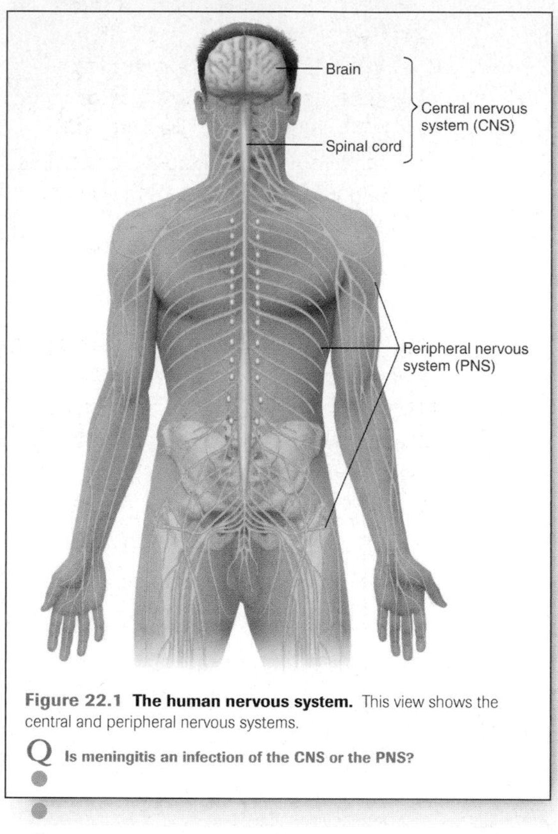

Figure 22.1 The human nervous system. This view shows the central and peripheral nervous systems.

Q Is meningitis an infection of the CNS or the PNS?

Figure-legend questions ask students to apply concepts presented in the text.

NEW Draw It questions give students an opportunity to interact with figures and develop a deeper understanding of the content. Suggested answers to the Draw It questions are provided in the Answers section at the back of the book and replicate how an actual student's work might look.

3. **DRAW IT** Label the parts of the compound light microscope in the figure below, and then draw the path of light from the illuminator to your eye.

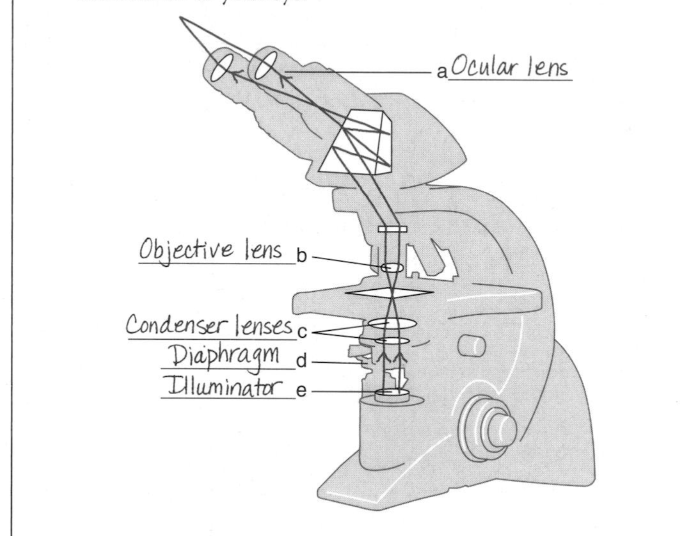

NEW Check Your Understanding questions appear at the ends of major chapter sections, encouraging students to engage interactively with the text and self-assess their understanding of the section.

LEARNING OBJECTIVES

6-8 Distinguish chemically defined and complex media.

6-9 Justify the use of each of the following: anaerobic techniques, living host cells, candle jars, selective and differential media, enrichment medium.

6-10 Differentiate biosafety levels 1, 2, 3, and 4.

medium of the same composition, only those colo[...] organisms capable of using phenol should grow. A rema[...] aspect of this particular technique is that phenol is normally lethal to most bacteria.

CHECK YOUR UNDERSTANDING

- ✓ Could humans exist on chemically defined media, at least under laboratory conditions? **6-8**
- ✓ Could Louis Pasteur, in the 1800s, have grown rabies viruses in cell culture instead of in living animals? **6-9**
- ✓ What BSL is your laboratory? **6-10**

Shared numbering between Learning Objectives and Check Your Understanding questions helps students determine which objectives they have achieved and which require further study.

...and think like a clinician

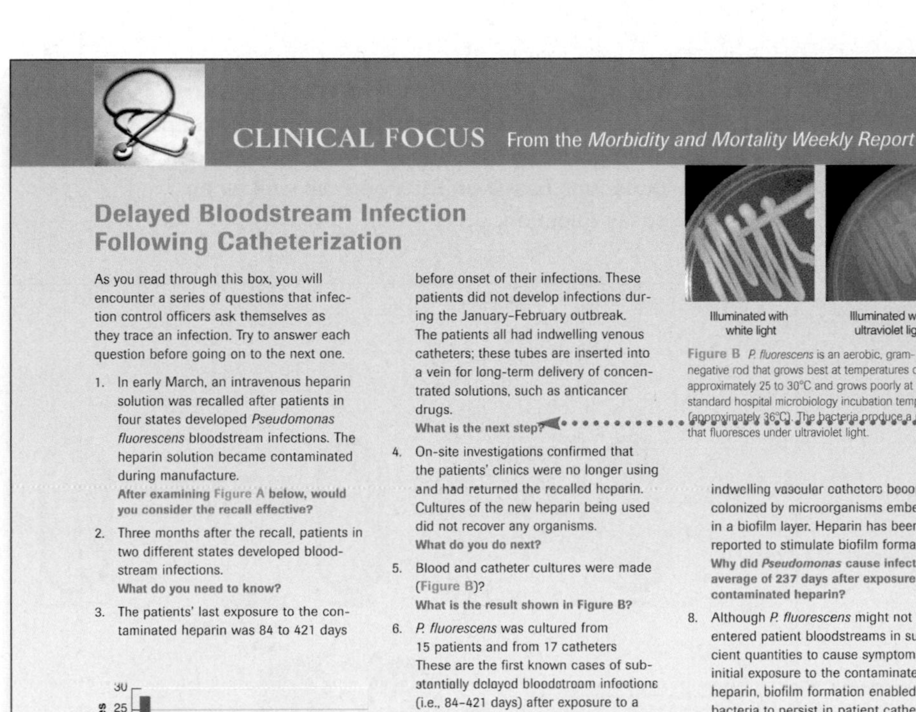

CLINICAL FOCUS From the *Morbidity and Mortality Weekly Report*

Delayed Bloodstream Infection Following Catheterization

As you read through this box, you will encounter a series of questions that infection control officers ask themselves as they trace an infection. Try to answer each question before going on to the next one.

1. In early March, an intravenous heparin solution was recalled after patients in four states developed *Pseudomonas fluorescens* bloodstream infections. The heparin solution became contaminated during manufacture.
After examining Figure A below, would you consider the recall effective?
2. Three months after the recall, patients in two different states developed bloodstream infections.
What do you need to know?
3. The patients' last exposure to the contaminated heparin was 84 to 421 days before onset of their infections. These patients did not develop infections during the January–February outbreak. The patients all had indwelling venous catheters; these tubes are inserted into a vein for long-term delivery of concentrated solutions, such as anticancer drugs.
What is the next step?
4. On-site investigations confirmed that the patients' clinics were no longer using and had returned the recalled heparin. Cultures of the new heparin being used did not recover any organisms.
What do you do next?
5. Blood and catheter cultures were made (Figure B)?
What is the result shown in Figure B?
6. *P. fluorescens* was cultured from 15 patients and from 17 catheters. These are the first known cases of substantially delayed bloodstream infections (i.e., 84–421 days) after exposure to a contaminated intravenous solution.
What was the source of these infections?
7. Scanning electron microscopy at the Centers for Disease Control and Prevention showed that *P. fluorescens* colonized the inside of the catheters by forming biofilms; previous electron microscopy studies have indicated that nearly all indwelling vascular catheters become colonized by microorganisms embedded in a biofilm layer. Heparin has been reported to stimulate biofilm formation.
Why did *Pseudomonas* cause infection an average of 237 days after exposure to the contaminated heparin?
8. Although *P. fluorescens* might not have entered patient bloodstreams in sufficient quantities to cause symptoms on initial exposure to the contaminated heparin, biofilm formation enabled the bacteria to persist in patient catheters. The bacteria might have proliferated in the biofilm, from which they were disrupted by subsequent, uncontaminated intravenous solutions and released into the bloodstream, finally causing symptoms.

Figure A Occurrence of *P. fluorescens* bloodstream infections in patients with intravenous catheters.

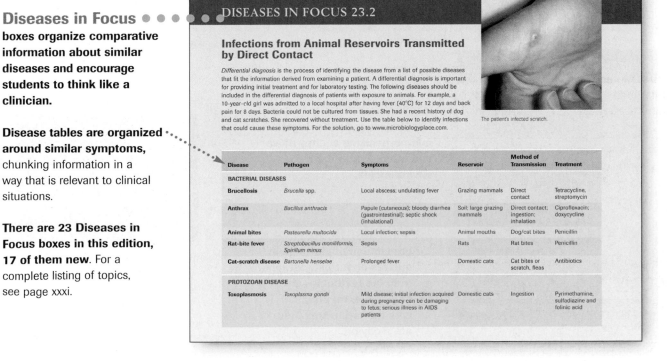

Figure B *P. fluorescens* is an aerobic, gram-negative rod that grows best at temperatures of approximately 25 to 30°C and grows poorly at the standard hospital microbiology incubation temperature (approximately 36°C). The bacteria produce a pigment that fluoresces under ultraviolet light.

Source: Adapted from *MMWR* 55(35): 961–963 (9/8/06).

Clinical Focus boxes contain ***Morbidity and Mortality Weekly Report*** **data from the Centers for Disease Control and Prevention (CDC) modified into clinical problem-solving scenarios and questions to help students develop their critical thinking skills.**

Students are asked to stop and try to answer questions as they walk through each clinical case, putting themselves in the place of the health care professional.

Diseases in Focus boxes organize comparative information about similar diseases and encourage students to think like a clinician.

Disease tables are organized around similar symptoms, chunking information in a way that is relevant to clinical situations.

There are 23 Diseases in Focus boxes in this edition, 17 of them new. For a complete listing of topics, see page xxxi.

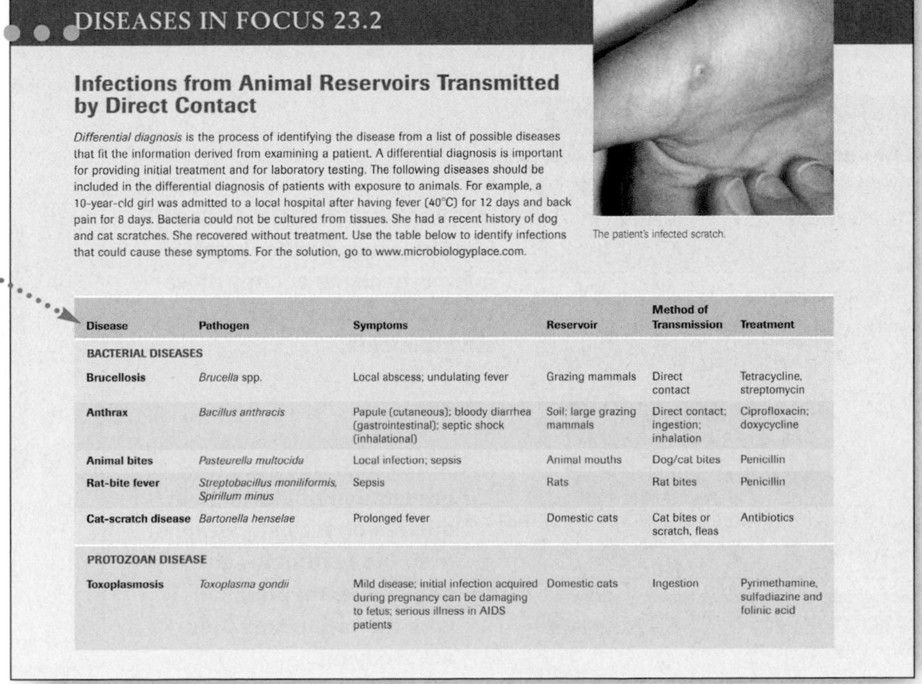

DISEASES IN FOCUS 23.2

Infections from Animal Reservoirs Transmitted by Direct Contact

Differential diagnosis is the process of identifying the disease from a list of possible diseases that fit the information derived from examining a patient. A differential diagnosis is important for providing initial treatment and for laboratory testing. The following diseases should be included in the differential diagnosis of patients with exposure to animals. For example, a 10-year-old girl was admitted to a local hospital after having fever (40°C) for 12 days and back pain for 8 days. Bacteria could not be cultured from tissues. She had a recent history of dog and cat scratches. She recovered without treatment. Use the table below to identify infections that could cause these symptoms. For the solution, go to www.microbiologyplace.com.

The patient's infected scratch.

Disease	Pathogen	Symptoms	Reservoir	Method of Transmission	Treatment
BACTERIAL DISEASES					
Brucellosis	*Brucella* spp.	Local abscess; undulating fever	Grazing mammals	Direct contact	Tetracycline, streptomycin
Anthrax	*Bacillus anthracis*	Papule (cutaneous); bloody diarrhea (gastrointestinal); septic shock (inhalational)	Soil; large grazing mammals	Direct contact; ingestion; inhalation	Ciprofloxacin; doxycycline
Animal bites	*Pasteurella multocida*	Local infection; sepsis	Animal mouths	Dog/cat bites	Penicillin
Rat-bite fever	*Streptobacillus moniliformis, Spirillum minus*	Sepsis	Rats	Rat bites	Penicillin
Cat-scratch disease	*Bartonella henselae*	Prolonged fever	Domestic cats	Cat bites or scratch, fleas	Antibiotics
PROTOZOAN DISEASE					
Toxoplasmosis	*Toxoplasma gondii*	Mild disease; initial infection acquired during pregnancy can be damaging to fetus; serious illness in AIDS patients	Domestic cats	Ingestion	Pyrimethamine, sulfadiazine and folinic acid

Unsurpassed online resources...

PEARSON
mymicrobiology place™

will help students get ready for tests with its simple three-step approach:

1. **Take a Pre-Test** and obtain a personalized study plan.
2. **Learn and Practice** with animations, tutorials, and MP3 Tutor Sessions.
3. **Test Yourself** with quizzes and a chapter post-test.

See the inside front cover of this book for details.

MP3
tutor sessions

NEW MP3 Tutor Sessions are downloadable audio study guides for each chapter of the textbook, allowing students to study on the go. They include mini-lectures about the toughest topics together with audio quizzes so students can self-assess their understanding.

NEW Foundation Figure quizzes
give students extra practice with core concepts. Each gradable quiz includes multiple choice questions based on the figure as well as an essay question.

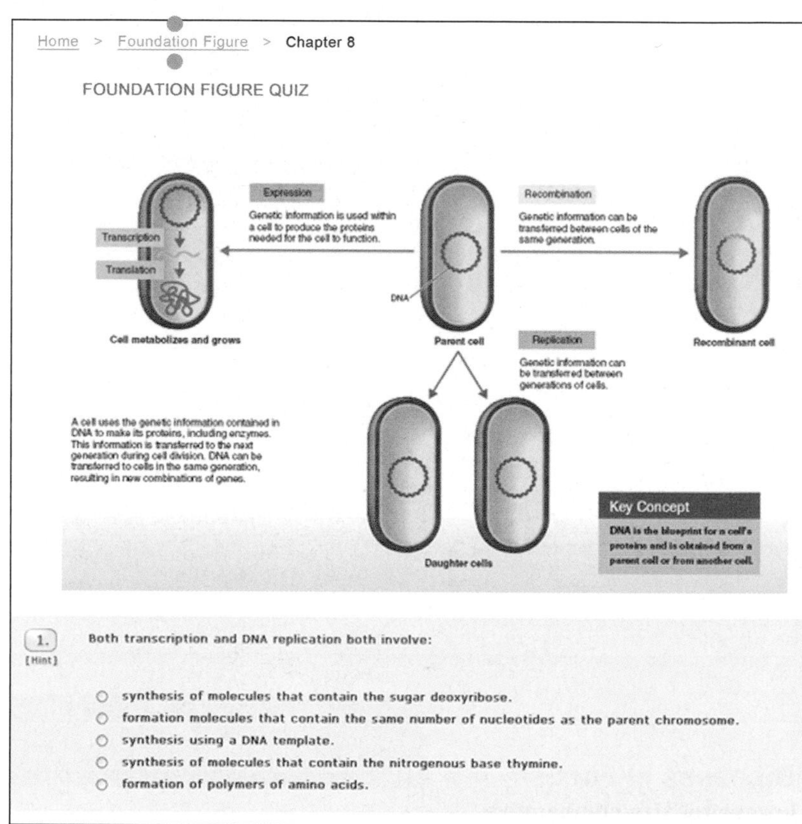
Home > Foundation Figure > Chapter 8
FOUNDATION FIGURE QUIZ
Expression
Genetic information is used within a cell to produce the proteins needed for the cell to function.
Recombination
Genetic information can be transferred between cells of the same generation.
Transcription
Translation
DNA
Cell metabolizes and grows
Parent cell
Recombinant cell
Replication
Genetic information can be transferred between generations of cells.
A cell uses the genetic information contained in DNA to make its proteins, including enzymes. This information is transferred to the next generation during cell division. DNA can be transferred to cells in the same generation, resulting in new combinations of genes.
Daughter cells
Key Concept
DNA is the blueprint for a cell's proteins and is obtained from a parent cell or from another cell.
1.
[Hint]
Both transcription and DNA replication both involve:
synthesis of molecules that contain the sugar deoxyribose.
formation molecules that contain the same number of nucleotides as the parent chromosome.
synthesis using a DNA template.
synthesis of molecules that contain the nitrogenous base thymine.
formation of polymers of amino acids.

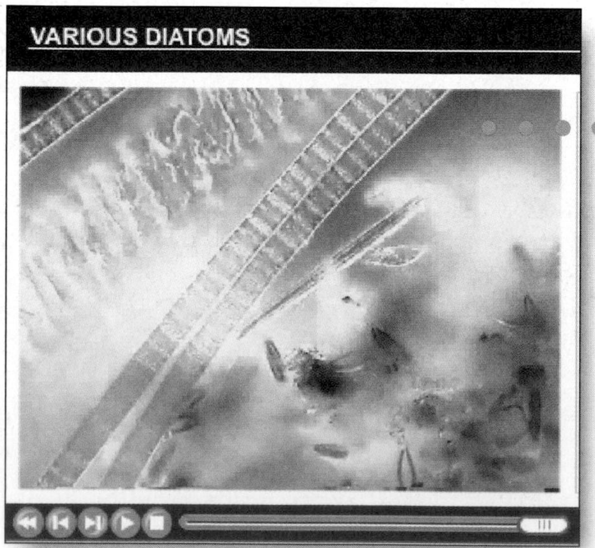
VARIOUS DIATOMS

25 Microbiology Videos
feature live-action footage of microorganisms as they move and interact with their environments.

Microbe Reviews—
gradable multiple-choice quizzes about micrographs from the textbook—give students extra practice with the organisms they are studying.

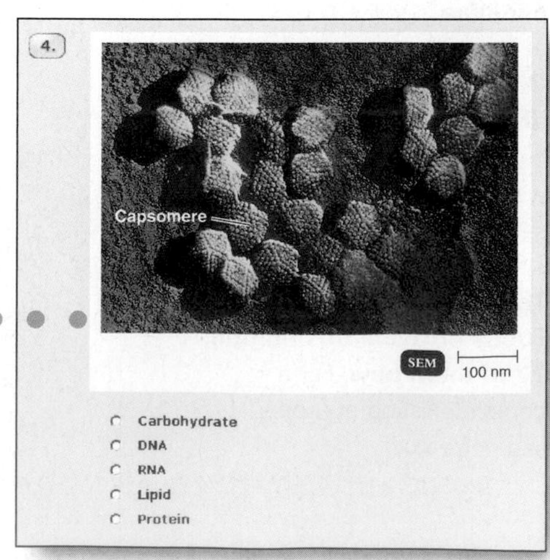
4.
Capsomere
SEM
100 nm
Carbohydrate
DNA
RNA
Lipid
Protein

...for additional practice and assessment

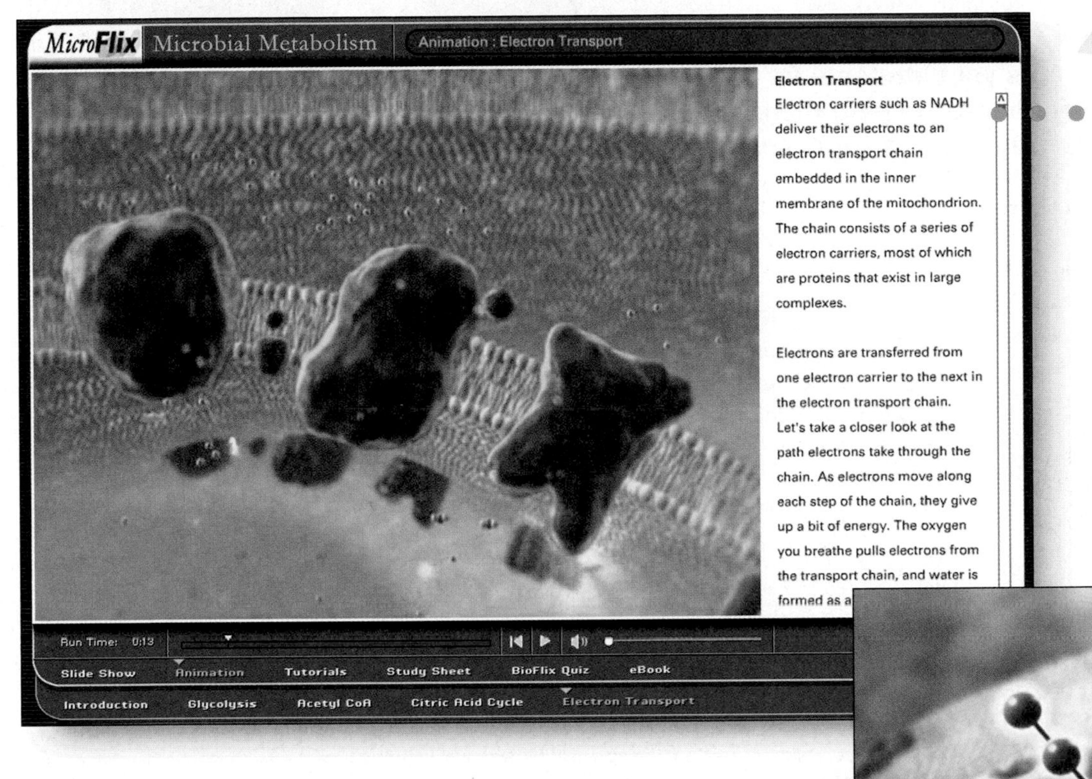

MicroFlix

NEW MicroFlix are 3D, movie-quality animations with self-paced tutorials and gradable quizzes that help students master the three toughest topics in microbiology: metabolism, DNA replication, and immunology. Students can review the fundamentals by viewing the animations, completing the tutorial, printing a personal review sheet, and taking the quiz. Students also have access to BioFlix animationsthat help them review relevant concepts from general biology.

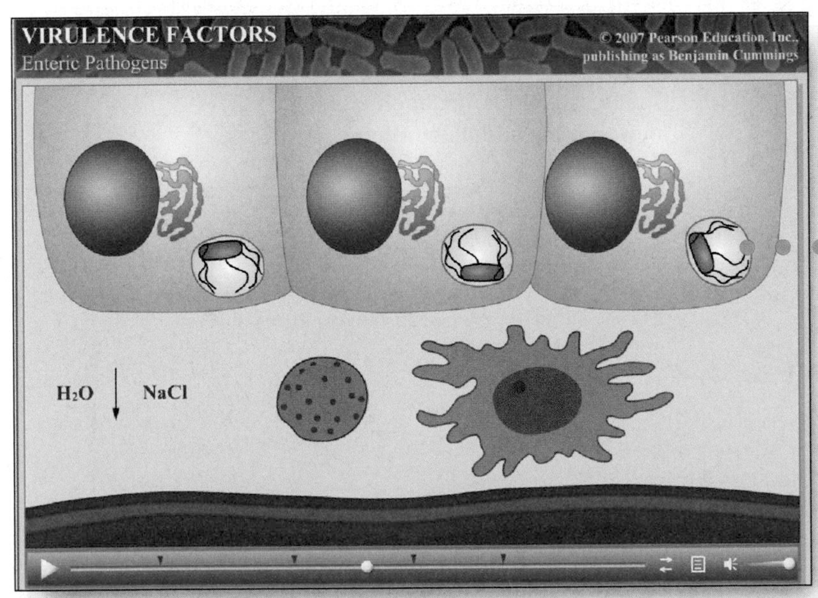

115 multi-step Microbiology Animations explain and visually demonstrate core concepts, providing an additional chance for students to learn. They are accompanied by gradable quizzes. References to the Microbiology Animations appear throughout the chapters of the book.

A gradebook feature allows instructors to track and record their students' performance on website quizzes and tests.

Are your students ready?

NEW ***Get Ready for Microbiology***
quickly prepares your students for the microbiology course, helping them brush up on the skills they need to succeed.

Sharpen Your Study Skills
Brush up on Math & Science
Quiz Yourself

Get Ready for Microbiology
LORI K. GARRETT · JUDY MEIER PENN

CONTENTS

Chapter 1 Study Skills
Chapter 2 Math Basics
Chapter 3 Terminology
Chapter 4 Chemistry Basics
Chapter 5 Biology Basics
Chapter 6 Cell Biology
Chapter 7 Microbiology Basics

Your Starting Point
pre-tests students' grasp of chapter content before they start the chapter.

Your Starting Point

Answer the following questions to assess your chemistry knowledge.

1. The most basic unit of a chemical substance is the ____________
2. What are the three states of matter? ____________
3. An atom is made of what three subatomic particles? ____________

Domain Bacteria

Domain Eukarya

Protists (multiple kingdoms)

Kingdom Fungi

Domain Archaea

Kingdom Animalia

Kingdom Fungi

FIGURE 7.1 Types of microorganisms

Chapters include textbook quality photographs and illustrations.

TIME TO TRY

Examine the three drawings below. Circle the ones that you think are cells.

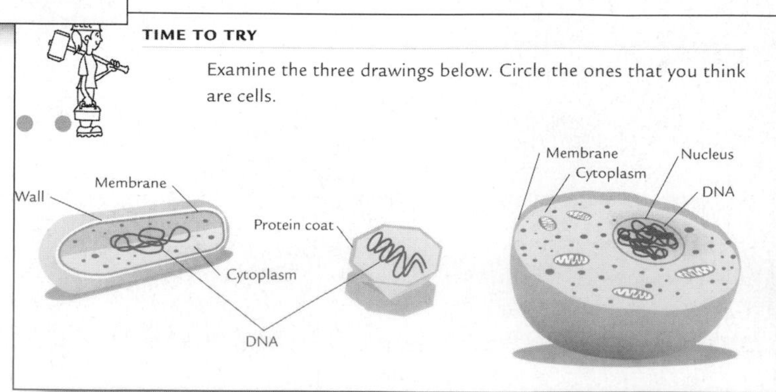

Engaging features like Time to Try provide a simple experiment or quick question that gives students a chance to practice what they just learned.

The MyMicrobiologyPlace website includes additional practice and assessment material for *Get Ready for Microbiology,* including a Diagnostic Test to get students started.

And for the microbiology lab...

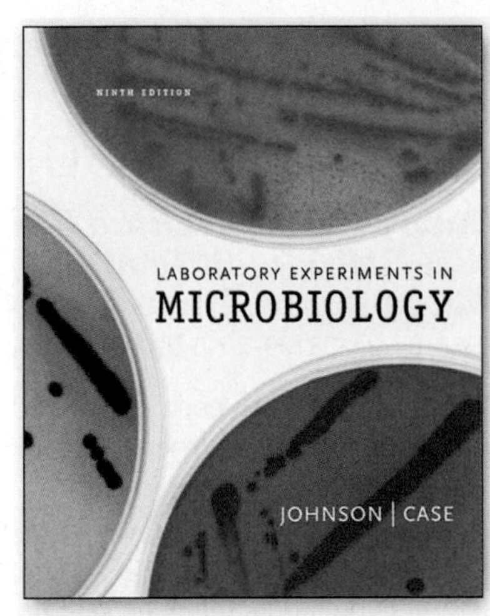

NOW IN FULL COLOR

***Laboratory Experiments in Microbiology*, Ninth Edition**
by Ted R. Johnson | Christine L. Case

The new full-color design makes each lab exercise easier to navigate. The quick reference sections (Objectives, Materials, Cultures, Techniques Required) are clearly distinguished from the instructive and procedural sections (Background, Procedure) through colored headings.

EXERCISE 4

Transfer of Bacteria: Aseptic Technique

Study without thinking is worthless; thinking without study is DANGEROUS.
—CONFUCIUS

Objectives

After completing this exercise, you should be able to:

1. Provide the rationale for aseptic technique.
2. Differentiate among the following: broth culture, agar slant, and agar deep.
3. Aseptically transfer bacteria from one form of culture medium to another.

BACKGROUND

In the laboratory, bacteria must be cultured to facilitate identification and to examine their growth and metabolism. Bacteria are **inoculated,** or introduced, into various forms of culture media to keep them alive and to study their growth—without introducing unwanted microbes, or **contaminants,** into the media. **Aseptic technique** is used in microbiology to exclude contaminants.

All culture media are **sterilized,** or rendered free of all life, prior to use. Sterilization is usually accomplished using an autoclave. Containers of culture media, such as test tubes or Petri plates, should not be opened until you are ready to work with them, and even then, you should not leave them open.

Culture Media

Broth cultures provide large numbers of bacteria in a small space and are easily transported. **Agar slants** are test tubes containing solid culture media that were left at an angle while the agar solidified. Agar slants, like Petri plates, provide a solid growth surface, but slants are easier to store and transport than Petri plates. Agar is allowed to solidify in the bottom of a test tube to make an **agar deep.** Deeps are often used to grow bacteria that require less oxygen than is present on the surface of the medium. Semisolid agar deeps containing 0.5–0.7% agar instead of the usual 1.5% agar can be used to determine whether a bacterium is motile. Motile bacteria will move away from the point of inoculation, giving the appearance of an inverted pine tree.

Inoculation

Aseptic transfer and inoculation are usually performed with a sterile, heat-resistant, noncorroding Nichrome wire attached to an insulated handle. When the end of the wire is bent into a loop, it is called an **inoculating loop;** when straight, it is an **inoculating needle** (FIGURE 4.1). For special purposes, cultures may also be transferred with sterile cotton swabs, pipettes, glass rods, or syringes. You will learn these techniques in later exercises.

Whether to use an inoculating loop or a needle depends on the form of the medium; after completing this exercise, you will be able to decide which instrument to use.

MATERIALS

FIRST PERIOD

Tubes containing nutrient broth (3)
Tubes containing nutrient agar slants (3)
Tubes containing nutrient semisolid agar deeps (3)
Inoculating loop

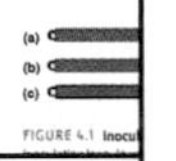

FIGURE 4.1 Inocul

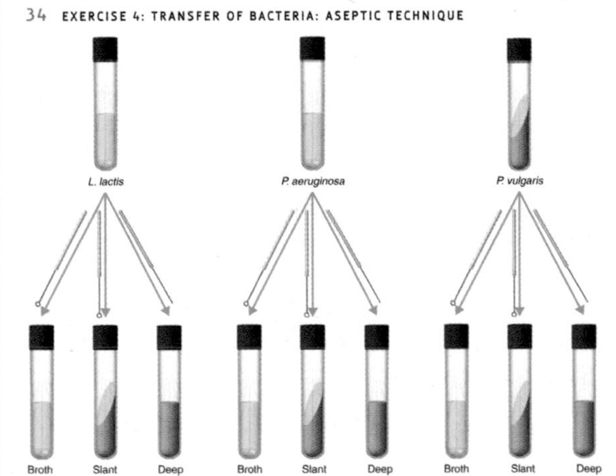

FIGURE 4.2 **The procedure.** Inoculate one tube of each medium with each of the cultures using a loop or needle as indicated. Work with only one culture at a time to avoid contamination.

Inoculating needle
Test-tube rack

SECOND PERIOD

Methylene blue

CULTURES

Lactococcus lactis broth
Pseudomonas aeruginosa broth
Proteus vulgaris slant

TECHNIQUES REQUIRED

Compound light microscopy, Exercise 1
Wet mount, Exercise 2

PROCEDURE First Period

2. Work with only one of the bacterial cultures at a time to prevent any mix-ups or cross-contamination (FIGURE 4.2). Label one tube of each medium with the name of the first broth culture, your name, the date, and your lab section. Inoculate each tube as described below (FIGURE 4.2), and then work with the next culture. Begin with one of the broth cultures, and gently tap the bottom of it to resuspend the sediment.
3. To inoculate nutrient broth, hold the inoculating loop in your dominant hand and one of the broth cultures of bacteria in the other hand.
   a. Sterilize your inoculating loop by holding it in the hottest part of the flame (at the edge of the inner blue area) or the electric incinerator until it is red hot (FIGURE 4.3a). The entire wire should get red. Why? ______
   b. Holding the loop like a pencil or paintbrush, curl the little finger of the same hand around

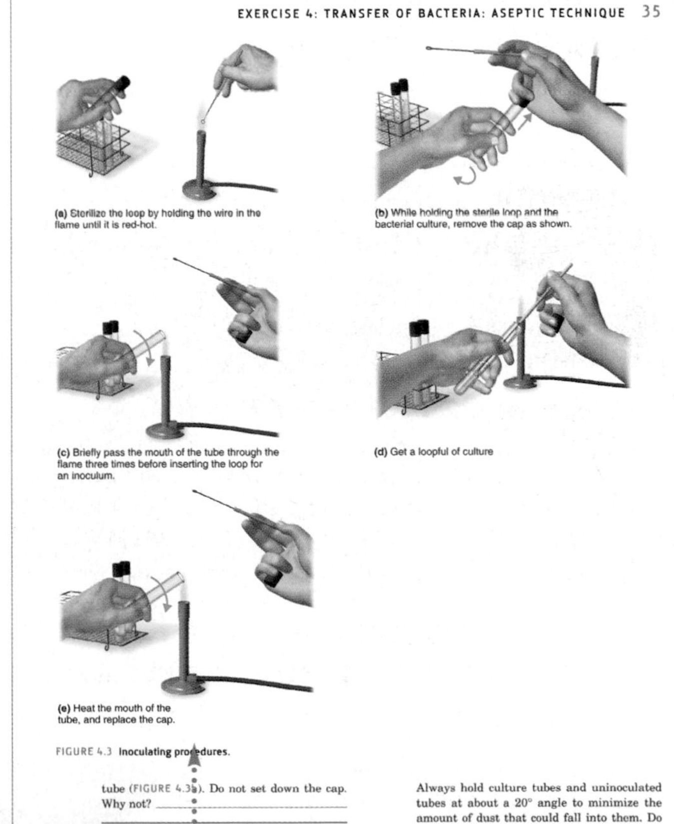

(a) Sterilize the loop by holding the wire in the flame until it is red-hot.

(b) While holding the sterile loop and the bacterial culture, remove the cap as shown.

(c) Briefly pass the mouth of the tube through the flame three times before inserting the loop for an inoculum.

(d) Get a loopful of culture

(e) Heat the mouth of the tube, and replace the cap.

FIGURE 4.3 **Inoculating procedures.**

tube (FIGURE 4.3b). Do not set down the cap. Why not? ______

Always hold culture tubes and uninoculated tubes at about a 20° angle to minimize the amount of dust that could fall into them. Do

Rich, vibrant micrographs and other photographs now appear within the relevant exercise, right next to the narrative, allowing students to easily view the images in the context of the exercise and to better interpret their results.

Clear, colorful, and realistically rendered step-by-step diagrams walk students through each procedure, providing visual instructions in addition to narrative ones.

The best support for instructors and students

INSTRUCTOR SUPPLEMENTS

Instructors Resource DVD/ CD-ROM (with TestGen® Computerized Test Bank CD)

978-0-321-58190-7 / 0-321-58190-3

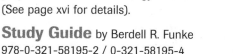

This media tool has been overhauled to make planning and presenting media easier. It includes:

- All figures from the book with and without labels in JPEG format
- All figures from the book with the Label Edit feature in PowerPoint® format
- Select "process" figures from the book with the Step Edit feature in PowerPoint format
- All tables from the book
- Multimedia, including the Microbiology Animations, Microbiology Videos, and MicroFlix Animations
- PowerPoint lecture outlines, including figures and tables from the book and links to multimedia
- PRS-enabled Active Lecture Clicker Questions
- PRS-enabled Quiz Show Clicker Questions
- The Instructor's Guide and Testbank as editable Microsoft® Word files
- A CD-ROM of the Test Bank in TestGen® format

Instructor's Visual Guide

978-0-321-58191-4 / 0-321-58191-1

Instructor's Guide/Test Bank by Christine L. Case

978-0-321-58187-7 / 0-321-58187-3

Transparency Acetates

978-0-321-58189-1 / 0-321-58189-X

CourseCompass™/ WebCT / Blackboard

Pre-loaded book-specific content and test item files accompanying the text are available in several course management formats. Contact your local Benjamin Cummings sales representative for more information. To locate your rep, use the "Find Your Rep" search feature at: www.pearsonhighered.com/educator

STUDENT SUPPLEMENTS

MyMicrobiologyPlace Website

www.microbiologyplace.com

For a full description of the website, see the inside front cover of this book.

Get Ready for Microbiology

by Lori K. Garrett and Judy Meier Penn

978-0-321-59250-7 / 0-321-59250-6

This new brief primer saves classroom time and frustration by helping students quickly prepare for their microbiology course. (See page xvi for details).

Study Guide by Berdell R. Funke

978-0-321-58195-2 / 0-321-58195-4

Students can master key concepts and earn a better grade with the help of the clear writing and thought-provoking exercises found in this Study Guide, which includes concise explanations of key conceps, art labeling exercises, and a variety of self-test questions with answers.

Study Card

978-0-321-58196-9 / 0-321-58196-2

This six-panel, full-color study card provides students with a quick reference to the three most challenging topics in microbiology: metabolism, genetics, and immunology.

The Microbe Files: Cases in Microbiology for the Undergraduate by Marjorie K. Cowan

With answers: 978-0-8053-4927-6 / 0-8053-4927-8

Without answers: 978-0-8053-4928-3 / 0-8053-4928-6

The Microbe Files provides microbiology students with a fascinating series of short cases that help them apply what they have learned in the course.

Scientific American: *Current Issues in Microbiology*

Vol.1: 978-0-8053-4623-7 / 0-8053-4623-6

Vol.2: 978-0-3215-3816-1/ 0-3215-3816-1

Accessible, dynamic, and relevant articles from *Scientific American* magazine present key issues in microbiology, and end-of-article questions help students check their comprehension and make connections between science and society.

FOR THE LAB

Laboratory Experiments in Microbiology, Ninth Edition

by Ted R. Johnson and Christine L. Case

978-0-321-56028-5 / 0-321-56028-0

Containing 57 thoroughly class-tested exercises, the ninth edition features a new full-color design and art program. (See previous page for details.)

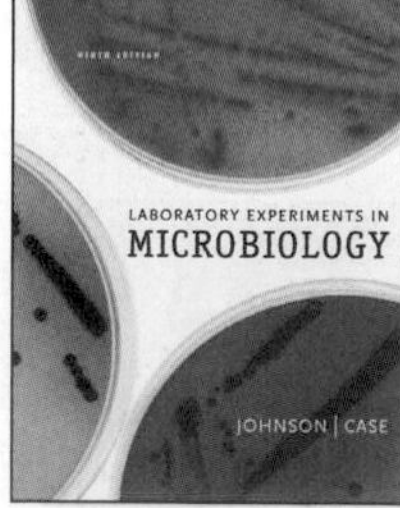

Techniques in Microbiology: A Student Handbook

by John M. Lammert

978-0-13-224011-6 / 0-13-224011-4

This vivid, full-color handbook guides students in manipulations and preparations needed in the microbiology laboratory. The techniques are the ones that are used frequently for studying microbes in the laboratory and include those identified by the American Society for Microbiology (ASM) in its recommendations for the Microbiology Lab Core Curriculum.

BRIEF CONTENTS

PART ONE Fundamentals of Microbiology

1 The Microbial World and You 1
2 Chemical Principles 26
3 Observing Microorganisms through a Microscope 54
4 Functional Anatomy of Prokaryotic and Eukaryotic Cells 76
5 Microbial Metabolism 113
6 Microbial Growth 156
7 The Control of Microbial Growth 184
8 Microbial Genetics 210
9 Biotechnology and Recombinant DNA 246

PART TWO A Survey of the Microbial World

10 Classification of Microorganisms 273
11 The Prokaryotes: Domains Bacteria and Archaea 299
12 The Eukaryotes: Fungi, Algae, Protozoa, and Helminths 329
13 Viruses, Viroids, and Prions 367

PART THREE Interaction between Microbe and Host

14 Principles of Disease and Epidemiology 399
15 Microbial Mechanisms of Pathogenicity 428
16 Innate Immunity: Nonspecific Defenses of the Host 449
17 Adaptive Immunity: Specific Defenses of the Host 476
18 Practical Applications of Immunology 500
19 Disorders Associated with the Immune System 522
20 Antimicrobial Drugs 553

PART FOUR Microorganisms and Human Disease

21 Microbial Diseases of the Skin and Eyes 584
22 Microbial Diseases of the Nervous System 610
23 Microbial Diseases of the Cardiovascular and Lymphatic Systems 637
24 Microbial Diseases of the Respiratory System 674
25 Microbial Diseases of the Digestive System 705
26 Microbial Diseases of the Urinary and Reproductive Systems 743

PART FIVE Environmental and Applied Microbiology

27 Environmental Microbiology 766
28 Applied and Industrial Microbiology 793

Answers to Review and Multiple Choice Study Questions AN-1

Appendix A Metabolic Pathways AP-1
Appendix B Exponents, Exponential Notation, Logarithms, and Generation Time AP-7
Appendix C Methods for Taking Clinical Samples AP-8
Appendix D Pronunciation of Scientific Names AP-9
Appendix E Word Roots Used in Microbiology AP-13
Appendix F Classification of Bacteria According to *Bergey's Manual* AP-16

Glossary G-1

Credits C-1

Index I-1

CONTENTS

PART ONE Fundamentals of Microbiology

1 The Microbial World and You 1

Microbes in Our Lives 2

Naming and Classifying Microorganisms 2
- Nomenclature 2
- Types of Microorganisms 3
- Classification of Microorganisms 6

A Brief History of Microbiology 6
- The First Observations 7
- The Debate Over Spontaneous Generation 8
- The Golden Age of Microbiology 9
- The Birth of Modern Chemotherapy: Dreams of a "Magic Bullet" 12
- Modern Developments in Microbiology 13

Microbes and Human Welfare 16
- Recycling Vital Elements 16
- Sewage Treatment: Using Microbes to Recycle Water 17
- Bioremediation: Using Microbes to Clean Up Pollutants 17
- Insect Pest Control by Microorganisms 17
- Modern Biotechnology and Recombinant DNA Technology 17

Microbes and Human Disease 18
- Normal Microbiota 18
- Biofilms 18
- Infectious Diseases 19
- Emerging Infectious Diseases 19

Study Outline 22

Study Questions 23

2 Chemical Principles 26

The Structure of Atoms 27
- Chemical Elements 27
- Electronic Configurations 28

How Atoms Form Molecules: Chemical Bonds 28
- Ionic Bonds 28
- Covalent Bonds 30
- Hydrogen Bonds 31
- Molecular Weight and Moles 32

Chemical Reactions 32
- Energy in Chemical Reactions 32
- Synthesis Reactions 32
- Decomposition Reactions 32
- Exchange Reactions 33
- The Reversibility of Reactions 34

IMPORTANT BIOLOGICAL MOLECULES 34

Inorganic Compounds 34
- Water 34
- Acids, Bases, and Salts 35
- Acid–Base Balance: The Concept of pH 35

Organic Compounds 37
- Structure and Chemistry 37
- Carbohydrates 39
- Lipids 40
- Proteins 42
- Nucleic Acids 47
- Adenosine Triphosphate (ATP) 47

Study Outline 49

Study Questions 51

3 Observing Microorganisms through a Microscope 54

Units of Measurement 55

Microscopy: The Instruments 55
- Light Microscopy 56
- Two-Photon Microscopy 62
- Scanning Acoustic Microscopy 63
- Electron Microscopy 63
- Scanned-Probe Microscopy 65

Preparation of Specimens for Light Microscopy 68
Preparing Smears for Staining 68
Simple Stains 69
Differential Stains 69
Special Stains 71
Study Outline 73
Study Questions 74

4 Functional Anatomy of Prokaryotic and Eukaryotic Cells 76

Comparing Prokaryotic and Eukaryotic Cells: An Overview 77
THE PROKARYOTIC CELL 77
The Size, Shape, and Arrangement of Bacterial Cells 77
Structures External to the Cell Wall 79
Glycocalyx 79
Flagella 81
Axial Filaments 82
Fimbriae and Pili 83
The Cell Wall 84
Composition and Characteristics 85
Cell Walls and The Gram Stain Mechanism 87
Atypical Cell Walls 87
Damage to the Cell Wall 88
Structures Internal to the Cell Wall 89
The Plasma (Cytoplasmic) Membrane 89
The Movement of Materials across Membranes 91
Cytoplasm 94
The Nucleoid 94
Ribosomes 95
Inclusions 95
Endospores 96
THE EUKARYOTIC CELL 98
Flagella and Cilia 98
The Cell Wall and Glycocalyx 98
The Plasma (Cytoplasmic) Membrane 100
Cytoplasm 100
Ribosomes 101
Organelles 102
The Nucleus 102
Endoplasmic Reticulum 103
Golgi Complex 104
Lysosomes 104
Vacuoles 104
Mitochondria 104
Chloroplasts 105
Peroxisomes 105
Centrosome 105
The Evolution of Eukaryotes 106
Study Outline 108
Study Questions 110

5 Microbial Metabolism 113

Catabolic and Anabolic Reactions 114
Enzymes 115
Collision Theory 115
Enzymes and Chemical Reactions 115
Enzyme Specificity and Efficiency 116
Naming Enzymes 116
Enzyme Components 116
The Mechanism of Enzymatic Action 117
Factors Influencing Enzymatic Activity 118
Feedback Inhibition 120
Ribozymes 121
Energy Production 121
Oxidation-Reduction Reactions 122
The Generation of ATP 122
Metabolic Pathways of Energy Production 123
Carbohydrate Metabolism 124
Glycolysis 124
Alternatives to Glycolysis 125
Cellular Respiration 127
Fermentation 132
Lipid and Protein Catabolism 136
Biochemical Tests and Bacterial Identification 137
Photosynthesis 140
The Light-Dependent Reactions: Photophosphorylation 140
The Light-Independent Reactions: The Calvin-Benson Cycle 140

Chapter 5 continues

Chapter 5, continued

A Summary of Energy Production Mechanisms 141

Metabolic Diversity among Organisms 142
- Photoautotrophs 143
- Photoheterotrophs 145
- Chemoautotrophs 145
- Chemoheterotrophs 145

Metabolic Pathways of Energy Use 146
- Polysaccharide Biosynthesis 146
- Lipid Biosynthesis 146
- Amino Acid and Protein Biosynthesis 146
- Purine and Pyrimidine Biosynthesis 147

The Integration of Metabolism 147

Study Outline 150

Study Questions 153

6 Microbial Growth 156

The Requirements for Growth 157
- Physical Requirements 157
- Chemical Requirements 160
- Biofilms 162

Culture Media 164
- Chemically Defined Media 165
- Complex Media 165
- Anaerobic Growth Media and Methods 166
- Special Culture Techniques 167
- Selective and Differential Media 168
- Enrichment Culture 169

Obtaining Pure Cultures 170

Preserving Bacterial Cultures 170

The Growth of Bacterial Cultures 171
- Bacterial Division 171
- Generation Time 171
- Logarithmic Representation of Bacterial Populations 171
- Phases of Growth 172
- Direct Measurement of Microbial Growth 174
- Estimating Bacterial Numbers by Indirect Methods 178

Study Outline 180

Study Questions 181

7 The Control of Microbial Growth 184

The Terminology of Microbial Control 185

The Rate of Microbial Death 186

Actions of Microbial Control Agents 186
- Alteration of Membrane Permeability 186
- Damage to Proteins and Nucleic Acids 187

Physical Methods of Microbial Control 187
- Heat 188
- Filtration 191
- Low Temperatures 191
- High Pressure 192
- Desiccation 192
- Osmotic Pressure 192
- Radiation 192

Chemical Methods of Microbial Control 195
- Principles of Effective Disinfection 195
- Evaluating a Disinfectant 195
- Types of Disinfectants 195

Microbial Characteristics and Microbial Control 202

Study Outline 205

Study Questions 207

8 Microbial Genetics 210

Structure and Function of the Genetic Material 211
- Genotype and Phenotype 211
- DNA and Chromosomes 211
- The Flow of Genetic Information 212
- DNA Replication 212
- RNA and Protein Synthesis 216

The Regulation of Bacterial Gene Expression 221
- Repression and Induction 224
- The Operon Model of Gene Expression 224
- Positive Regulation 225

Mutation: Change in the Genetic Material 226
- Types of Mutations 227
- Mutagens 229
- The Frequency of Mutation 231
- Identifying Mutants 231
- Identifying Chemical Carcinogens 232

Genetic Transfer and Recombination 233
Transformation in Bacteria 234
Conjugation in Bacteria 236
Transduction in Bacteria 237
Plasmids and Transposons 237
Genes and Evolution 241
Study Outline 242
Study Questions 244

9 Biotechnology and Recombinant DNA 246

Introduction to Biotechnology 247
Recombinant DNA Technology 247
An Overview of Recombinant DNA Procedures 247
Tools of Biotechnology 247
Selection 249
Mutation 249
Restriction Enzymes 249
Vectors 250
Polymerase Chain Reaction 251
Techniques of Genetic Modification 253
Inserting Foreign DNA into Cells 253
Obtaining DNA 254
Selecting a Clone 256
Making a Gene Product 257
Applications of rDNA 258
Therapeutic Applications 258
The Human Genome Project 261
Scientific Applications 261
Agricultural Applications 264
Safety Issues and the Ethics of Using rDNA 268
Study Outline 269
Study Questions 271

PART TWO A Survey of the Microbial World

10 Classification of Microorganisms 273

The Study of Phylogenetic Relationships 274
The Three Domains 274
A Phylogenetic Hierarchy 277
Classification of Organisms 278
Scientific Nomenclature 278
The Taxonomic Hierarchy 279
Classification of Prokaryotes 279
Classification of Eukaryotes 281
Classification of Viruses 282
Methods of Classifying and Identifying Microorganisms 282
Morphological Characteristics 284
Differential Staining 285
Biochemical Tests 285
Serology 287
Phage Typing 288
Fatty Acid Profiles 288
Flow Cytometry 288
DNA Base Composition 288
DNA Fingerprinting 289
The Polymerase Chain Reaction 290
Nucleic Acid Hybridization 291
Putting Classification Methods Together 293
Study Outline 295
Study Questions 296

11 The Prokaryotes: Domains Bacteria and Archaea 299

The Prokaryotic Groups 300
DOMAIN BACTERIA 302
The Proteobacteria 302
The Alphaproteobacteria 303
The Betaproteobacteria 305
The Gammaproteobacteria 306
The Deltaproteobacteria 312
The Epsilonproteobacteria 312
The Nonproteobacteria Gram-Negative Bacteria 313
Cyanobacteria (The Oxygenic Photosynthetic Bacteria) 313
Purple and Green Photosynthetic Bacteria (The Anoxygenic Photosynthetic Bacteria) 315
The Gram-Positive Bacteria 315
Firmicutes (Low G + C Gram-Positive Bacteria) 316
Actinobacteria (High G + C Gram-Positive Bacteria) 320

Chapter 11 continues

Chapter 11, continued

Planctomycetes 322

Chlamydiae 322

Spirochaetes 322

Bacteriodetes 324

Fusobacteria 324

DOMAIN ARCHAEA 325

Diversity within the Archaea 325

MICROBIAL DIVERSITY 325

Discoveries Illustrating the Range of Diversity 326

Study Outline 327

Study Questions 327

12 The Eukaryotes: Fungi, Algae, Protozoa, and Helminths 329

Fungi 330
- Characteristics of Fungi 331
- Medically Important Phyla of Fungi 333
- Fungal Diseases 335
- Economic Effects of Fungi 339

Lichens 339

Algae 340
- Characteristics of Algae 341
- Selected Phyla of Algae 342
- Roles of Algae in Nature 344

Protozoa 345
- Characteristics of Protozoa 346
- Medically Important Phyla of Protozoa 346

Slime Molds 351

Helminths 352
- Characteristics of Helminths 353
- Platyhelminths 356
- Nematodes 358

Arthropods as Vectors 361

Study Outline 363

Study Questions 365

13 Viruses, Viroids, and Prions 367

General Characteristics of Viruses 368
- Host Range 368
- Viral Size 369

Viral Structure 370
- Nucleic Acid 371
- Capsid and Envelope 372
- General Morphology 373

Taxonomy of Viruses 373

Isolation, Cultivation, and Identification of Viruses 374
- Growing Bacteriophages in the Laboratory 374
- Growing Animal Viruses in the Laboratory 377
- Viral Identification 379

Viral Multiplication 379
- Multiplication of Bacteriophages 379
- Multiplication of Animal Viruses 382

Viruses and Cancer 389
- The Transformation of Normal Cells into Tumor Cells 390
- DNA Oncogenic Viruses 391
- RNA Oncogenic Viruses 391

Latent Viral Infections 392

Persistent Viral Infections 392

Prions 392

Plant Viruses and Viroids 393

Study Outline 395

Study Questions 397

PART THREE Interaction between Microbe and Host

14 Principles of Disease and Epidemiology 399

Pathology, Infection, and Disease 400

Normal Microbiota 400
- Relationships between the Normal Microbiota and the Host 401
- Opportunistic Microorganisms 403
- Cooperation among Microorganisms 404

The Etiology of Infectious Diseases 404
- Koch's Postulates 404
- Exceptions to Koch's Postulates 404

Classifying Infectious Diseases 406
- Occurrence of a Disease 406
- Severity or Duration of a Disease 406
- Extent of Host Involvement 407

Patterns of Disease 408
- Predisposing Factors 408
- Development of Disease 408

The Spread of Infection 409
- Reservoirs of Infection 409
- Transmission of Disease 409

Nosocomial (Hospital-Acquired) Infections 413
- Microorganisms in the Hospital 414
- Compromised Host 415
- Chain of Transmission 415
- Control of Nosocomial Infections 416

Emerging Infectious Diseases 416

Epidemiology 418
- Descriptive Epidemiology 419
- Analytical Epidemiology 419
- Experimental Epidemiology 420
- Case Reporting 420
- The Centers for Disease Control and Prevention 420

Study Outline 423

Study Questions 425

15 Microbial Mechanisms of Pathogenicity 428

How Microorganisms Enter a Host 429
- Portals of Entry 429
- The Preferred Portal of Entry 429
- Numbers of Invading Microbes 429
- Adherence 431

How Bacterial Pathogens Penetrate Host Defenses 432
- Capsules 432
- Cell Wall Components 432
- Enzymes 432
- Antigenic Variation 433
- Penetration into the Host Cell Cytoskeleton 433

How Bacterial Pathogens Damage Host Cells 434
- Using the Host's Nutrients: Siderophores 434
- Direct Damage 434
- The Production of Toxins 434
- Plasmids, Lysogeny, and Pathogenicity 439

Pathogenic Properties of Viruses 441
- Viral Mechanisms for Evading Host Defenses 441
- Cytopathic Effects of Viruses 441

Pathogenic Properties of Fungi, Protozoa, Helminths, and Algae 442
- Fungi 443
- Protozoa 443
- Helminths 444
- Algae 444

Portals of Exit 444

Study Outline 445

Study Questions 447

16 Innate Immunity: Nonspecific Defenses of the Host 449

The Concept of Immunity 450

FIRST LINE OF DEFENSE: SKIN AND MUCOUS MEMBRANES 450

Physical Factors 451

Chemical Factors 453

Normal Microbiota and Innate Immunity 453

SECOND LINE OF DEFENSE 454

Formed Elements in Blood 454

The Lymphatic System 456

Phagocytes 457
- Actions of Phagocytic Cells 457
- The Mechanism of Phagocytosis 458
- Microbial Evasion of Phagocytosis 459

Inflammation 460
- Vasodilation and Increased Permeability of Blood Vessels 460
- Phagocyte Migration and Phagocytosis 462
- Tissue Repair 462

Fever 463

Chapter 16 continues

Chapter 16, continued

Antimicrobial Substances 463
- The Complement System 463
- Interferons 468
- Iron-Binding Proteins 470
- Antimicrobial Peptides 470

Study Outline 472

Study Questions 474

17 Adaptive Immunity: Specific Defenses of the Host 476

The Adaptive Immune System 477

Dual Nature of the Adaptive Immune System 477
- Humoral Immunity 477
- Cellular Immunity 477

Antigens and Antibodies 478
- The Nature of Antigens 478
- The Nature of Antibodies 479

B Cells and Humoral Immunity 482
- Clonal Selection of Antibody-Producing Cells 482
- The Diversity of Antibodies 484

Antigen–Antibody Binding and Its Results 484

T Cells and Cellular Immunity 486
- Classes of T Cells 487
- T Helper Cells ($CD4^+$ T Cells) 487
- T Cytotoxic Cells ($CD8^+$ T Cells) 488
- T Regulatory Cells 489

Antigen-Presenting Cells (APCs) 489
- Dendritic Cells 490
- Macrophages 490

Extracellular Killing by the Immune System 491

Antibody-Dependent Cell-Mediated Cytotoxicity 491

Cytokines: Chemical Messengers of Immune Cells 491

Immunological Memory 493

Types of Adaptive Immunity 494

Study Outline 497

Study Questions 498

18 Practical Applications of Immunology 500

Vaccines 501
- Principles and Effects of Vaccination 501
- Types of Vaccines and Their Characteristics 501
- The Development of New Vaccines 504
- Safety of Vaccines 506

Diagnostic Immunology 507
- Immunologic-Based Diagnostic Tests 507
- Monoclonal Antibodies 507
- Precipitation Reactions 509
- Agglutination Reactions 510
- Neutralization Reactions 512
- Complement-Fixation Reactions 512
- Fluorescent-Antibody Techniques 513
- Enzyme-Linked Immunosorbent Assay (ELISA) 514
- Western Blotting (Immunoblotting) 516
- The Future of Diagnostic Immunology 516

Study Outline 519

Study Questions 520

19 Disorders Associated with the Immune System 522

Hypersensitivity 523
- Type I (Anaphylactic) Reactions 523
- Type II (Cytotoxic) Reactions 526
- Type III (Immune Complex) Reactions 528
- Type IV (Delayed Cell-Mediated) Reactions 529

Autoimmune Diseases 532
- Cytotoxic Autoimmune Reactions 532
- Immune Complex Autoimmune Reactions 532
- Cell-Mediated Autoimmune Reactions 532

Reactions Related to the Human Leukocyte Antigen (HLA) Complex 533
- Reactions to Transplantation 534
- Immunosuppression 536

The Immune System and Cancer 537
- Immunotherapy for Cancer 538

Immunodeficiencies 538
- Congenital Immunodeficiencies 538
- Acquired Immunodeficiencies 538

Acquired Immunodeficiency Syndrome (AIDS) 539
- The Origin of AIDS 540
- HIV Infection 540
- Diagnostic Methods 545
- HIV Transmission 545
- AIDS Worldwide 546
- Preventing and Treating AIDS 547
- The AIDS Epidemic and the Importance of the Scientific Research 548

Study Outline 549

Study Questions 551

20 Antimicrobial Drugs 553

The History of Chemotherapy 554
- Antibiotic Discovery Today 554

The Spectrum of Antimicrobial Activity 555

The Action of Antimicrobial Drugs 555
- Inhibiting Cell Wall Synthesis 556
- Inhibiting Protein Synthesis 556
- Injuring the Plasma Membrane 558
- Inhibiting Nucleic Acid Synthesis 558
- Inhibiting the Synthesis of Essential Metabolites 558

A Survey of Commonly Used Antimicrobial Drugs 559
- Antibacterial Antibiotics: Inhibitors of Cell Wall Synthesis 559
- Antimycobacterial Antibiotics 563
- Inhibitors of Protein Synthesis 563
- Injury to the Plasma Membrane 566
- Inhibitors of Nucleic Acid (DNA/RNA) Synthesis 567
- Competitive Inhibitors of the Synthesis of Essential Metabolites 567
- Antifungal Drugs 567
- Antiviral Drugs 569
- Antiprotozoan and Antihelminthic Drugs 571

Tests to Guide Chemotherapy 572
- The Diffusion Methods 572
- Broth Dilution Tests 572

Resistance to Antimicrobial Drugs 573
- Mechanisms of Resistance 574
- Antibiotic Misuse 575
- Cost and Prevention of Resistance 576

Antibiotic Safety 576

Effects of Combinations of Drugs 578

The Future of Chemotherapeutic Agents 578
- Antimicrobial Peptides 578
- Antisense Agents 579

Study Outline 580

Study Questions 582

PART FOUR Microorganisms and Human Disease

21 Microbial Diseases of the Skin and Eyes 584

Structure and Function of the Skin 585
- Mucous Membranes 585

Normal Microbiota of the Skin 585

Microbial Diseases of the Skin 586
- Bacterial Diseases of the Skin 586
- Viral Diseases of the Skin 595
- Fungal Diseases of the Skin and Nails 600
- Parasitic Infestation of the Skin 602

Microbial Diseases of the Eye 603
- Inflammation of the Eye Membranes: Conjunctivitis 603
- Bacterial Diseases of the Eye 603
- Other Infectious Diseases of the Eye 605

Study Outline 606

Study Questions 608

22 Microbial Diseases of the Nervous System 610

Structure and Function of the Nervous System 611

Bacterial Diseases of the Nervous System 611
- Bacterial Meningitis 612
- Tetanus 615
- Botulism 616
- Leprosy 619

Viral Diseases of the Nervous System 620

Poliomyelitis 620
- Rabies 622
- Arboviral Encephalitis 624

Chapter 22 continues

Chapter 22, continued

Fungal Disease of the Nervous System 626

Cryptococcus neoformans Meningitis (Cryptococcosis) 626

Protozoan Diseases of the Nervous System 627

African Trypanosomiasis 627

Amebic Meningoencephalitis 629

Nervous System Diseases Caused by Prions 629

Bovine Spongiform Encephalopathy and Variant Creutzfeldt-Jakob Disease 631

Disease Caused by Unidentified Agents 633

Chronic Fatigue Syndrome 633

Study Outline 633

Study Questions 635

23 Microbial Diseases of the Cardiovascular and Lymphatic Systems 637

Structure and Function of the Cardiovascular and Lymphatic Systems 638

Bacterial Diseases of the Cardiovascular and Lymphatic Systems 638

Sepsis and Septic Shock 639

Bacterial Infections of the Heart 641

Rheumatic Fever 641

Tularemia 642

Brucellosis (Undulant Fever) 643

Anthrax 645

Gangrene 646

Systemic Diseases Caused by Bites and Scratches 647

Vector-Transmitted Diseases 648

Viral Diseases of the Cardiovascular and Lymphatic Systems 655

Burkitt's Lymphoma 655

Infectious Mononucleosis 656

Other Diseases and Epstein-Barr Virus 657

Cytomegalovirus Infections 658

Chikungunya Fever 658

Classic Viral Hemorrhagic Fevers 658

Emerging Viral Hemorrhagic Fevers 659

Protozoan Diseases of the Cardiovascular and Lymphatic Systems 660

Chagas' Disease (American Trypanosomiasis) 661

Toxoplasmosis 661

Malaria 663

Leishmaniasis 665

Babesiosis 666

Helminthic Diseases of the Cardiovascular and Lymphatic Systems 666

Schistosomiasis 666

Swimmer's Itch 667

Study Outline 669

Study Questions 671

24 Microbial Diseases of the Respiratory System 674

Structure and Function of the Respiratory System 675

Normal Microbiota of the Respiratory System 675

MICROBIAL DISEASES OF THE UPPER RESPIRATORY SYSTEM 676

Bacterial Diseases of the Upper Respiratory System 677

Streptococcal Pharyngitis (Strep Throat) 677

Scarlet Fever 677

Diphtheria 677

Otitis Media 679

Viral Diseases of the Upper Respiratory System 679

The Common Cold 679

MICROBIAL DISEASES OF THE LOWER RESPIRATORY SYSTEM 680

Bacterial Diseases of the Lower Respiratory System 680

Pertussis (Whooping Cough) 680

Tuberculosis 682

Bacterial Pneumonias 684

Meliodosis 690

Viral Diseases of the Lower Respiratory System 692

Viral Pneumonia 692

Respiratory Syncytial Virus (RSV) 692

Influenza (Flu) 692

Fungal Diseases of the Lower Respiratory System 695

Histoplasmosis 695

Coccidiomycosis 696

Pneumocystis Pneumonia 697
Blastomycosis (North American Blastomycosis) 697
Other Fungi Involved in Respiratory Disease 697

Study Outline 700

Study Questions 702

25 Microbial Diseases of the Digestive System 705

Structure and Function of the Digestive System 706

Normal Microbiota of the Digestive System 706

Bacterial Diseases of the Mouth 707
Dental Caries (Tooth Decay) 707
Periodontal Disease 709

Bacterial Diseases of the Lower Digestive System 710
Staphylococcal Food Poisoning (Staphylococcal Enterotoxicosis) 711
Shigellosis (Bacillary Dysentery) 712
Salmonellosis (*Salmonella* Gastroenteritis) 712
Typhoid Fever 714
Cholera 716
Noncholera Vibrios 717
Escherichia coli Gastroenteritis 717
Campylobacter Gastroenteritis 718
Helicobacter Peptic Ulcer Disease 718
Yersinia Gastroenteritis 720
Clostridium perfringens Gastroenteritis 720
Clostridium difficile–Associated Diarrhea 720
Bacillus cereus Gastroenteritis 720

Viral Diseases of the Digestive System 721
Mumps 721
Hepatitis 721
Viral Gastroenteritis 728

Fungal Diseases of the Digestive System 729
Ergot Poisoning 729
Aflatoxin Poisoning 730

Protozoan Diseases of the Digestive System 730
Giardiasis 730
Cryptosporidiosis 731
Cyclospora Diarrheal Infection 731
Amoebic Dysentery (Amoebiasis) 731

Helminthic Diseases of the Digestive System 732
Tapeworms 732
Hydatid Disease 733
Nematodes 734

Study Outline 738

Study Questions 741

26 Microbial Diseases of the Urinary and Reproductive Systems 743

Structure and Function of the Urinary System 744

Structure and Function of the Reproductive Systems 744

Normal Microbiota of the Urinary and Reproductive Systems 745

DISEASES OF THE URINARY SYSTEM 746

Bacterial Diseases of the Urinary System 746
Cystitis 746
Pyelonephritis 746
Leptospirosis 746

DISEASES OF THE REPRODUCTIVE SYSTEMS 747

Bacterial Diseases of the Reproductive Systems 747
Gonorrhea 747
Nongonococcal Urethritis (NGU) 750
Pelvic Inflammatory Disease (PID) 751
Syphilis 752
Lymphogranuloma Venereum 755
Chancroid (Soft Chancre) 756
Bacterial Vaginosis 756

Viral Diseases of the Reproductive System 757
Genital Herpes 757
Genital Warts 758
AIDS 758

Fungal Disease of the Reproductive Systems 758
Candidiasis 758

Protozoan Disease of the Reproductive Systems 760
Trichomoniasis 760
The TORCH Panel of Tests 760

Study Outline 762

Study Questions 764

PART FIVE Environmental and Applied Microbiology

27 Environmental Microbiology 766

Microbial Diversity and Habitats 767
Symbiosis 767

Soil Microbiology and Biogeochemical Cycles 768
The Carbon Cycle 768
The Nitrogen Cycle 770
The Sulfur Cycle 772
Life without Sunshine 773
The Phosphorus Cycle 774
The Degradation of Synthetic Chemicals in Soils and Water 775

Aquatic Microbiology and Sewage Treatment 776
Aquatic Microorganisms 776
The Role of Microorganisms in Water Quality 778
Water Treatment 782
Sewage (Wastewater) Treatment 783

Study Outline 789

Study Questions 791

28 Applied and Industrial Microbiology 793

Food Microbiology 794
Foods and Disease 794
Industrial Food Canning 794
Aseptic Packaging 795
Radiation and Industrial Food Preservation 796
High-Pressure Food Preservation 797
The Role of Microorganisms in Food Production 797

Industrial Microbiology 800
Fermentation Technology 801
Industrial Products 804
Alternative Energy Sources Using Microorganisms 806
Biofuels 807
Industrial Microbiology and the Future 808

Study Outline 809

Study Questions 810

Answers to Review and Multiple Choice Study Questions AN-1

Appendix A Metabolic Pathways AP-1

Appendix B Exponents, Exponential Notation, Logarithms, and Generation Time AP-7

Appendix C Methods for Taking Clinical Samples AP-8

Appendix D Pronunciation of Scientific Names AP-9

Appendix E Word Roots Used in Microbiology AP-13

Appendix F Classification of Bacteria According to *Bergey's Manual* AP-16

Glossary G-1

Credits C-1

Index I-1

FEATURES

FOUNDATION FIGURES

Figure 2.16 The Structure of DNA 48
Figure 3.2 Sizes and Resolutions 58
Figure 4.6 The Structure of a Prokaryotic Cell 80
Figure 5.11 An Overview of Respiration and Fermentation 125
Figure 6.15 The Bacterial Growth Curve 173
Figure 7.1 A Microbial Death Curve 187
Figure 8.2 The Flow of Genetic Information 213
Figure 9.1 A Typical Genetic Modification Procedure 248
Figure 10.1 The Three-Domain System 275
Figure 13.15 Replication of a DNA-Containing Animal Virus 386
Figure 14.3 Koch's Postulates 405
Figure 15.4 Toxins 435
Figure 15.9 Mechanisms of Pathogenicity 445
Figure 16.7 The Mechanism of Phagocytosis 458
Figure 16.9 Outcomes of Complement Activation 464
Figure 17.19 The Dual Nature of the Adaptive Immune System 496
Figure 18.2 The Production of Monoclonal Antibodies 508
Figure 19.16 The Progression of HIV Infection 543
Figure 20.2 Major Modes of Action of Antimicrobial Drugs 556
Figure 20.20 Resistance to Antibiotics 575

CLINICAL FOCUS

Human Tuberculosis—New York City 144
Delayed Bloodstream Infection Following Catheterization 164
Infection Following Steroid Injection 201
Tracking West Nile Virus 223
Norovirus—Who Is Responsible for the Outbreak? 266
The Most Frequent Cause of Recreational Waterborne Diarrhea 355
Influenza: Crossing the Species Barrier 370
Nosocomial Infections 422
Inflammation of the Eye 440
A World Health Problem 505
A Delayed Rash 531
Antibiotics in Animal Feed Linked to Human Disease 577
Infections in the Gym 593
A Neurological Disease 625
A Sick Child 644
Outbreak 691
A Foodborne Infection 715
Survival of the Fittest 751

APPLICATIONS OF MICROBIOLOGY

Designer Jeans: Are They Really Stone Washed? 3
Bioremediation—Bacteria Clean Up Pollution 33
What Is That Slime? 57
Why Microbiologists Study Termites 107
What Is Fermentation? 135
Mass Deaths of Marine Mammals Spur Veterinary Microbiology 283
Bacteria and Insect Sex 307
Serum Collection 467
Interleukin-12: The Next "Magic Bullet"? 493
Protection against Bioterrorism 649
A Safe Blood Supply 727
Biosensors: Bacteria That Detect Pollutants and Pathogens 780
From Plant Disease to Shampoo and Salad Dressing 801

DISEASES IN FOCUS

21.1: Macular Rashes 589
21.2: Vesicular and Pustular Rashes 590
21.3: Patchy Redness and Pimple-Like Conditions 592
21.4: Microbial Diseases of the Eye 604
22.1: Meningitis and Encephalitis 617
22.2: Types of Arboviral Encephalitis 628
22.3: Microbial Diseases with Neurological Symptoms or Paralysis 632
23.1: Infections from Human Reservoirs 643
23.2: Infections from Animal Reservoirs Transmitted by Direct Contact 650
23.3: Infections Transmitted by Vectors 651
23.4: Viral Hemorrhagic Fevers 660
23.5: Infections Transmitted by Soil and Water 668
24.1: Microbial Diseases of the Upper Respiratory System 681
24.2: Common Bacterial Pneumonias 687
24.3: Microbial Diseases of the Lower Respiratory System 699
25.1: Diseases of the Mouth 710
25.2: Bacterial Diseases of the Lower Digestive System 722
25.3: Characteristics of Viral Hepatitis 724
25.4: Viral Diseases of the Digestive System 729
25.5: Fungal, Protozoan, and Helminthic Diseases of the Lower Digestive System 734
26.1: Bacterial Diseases of the Urinary System 748
26.2: Characteristics of the Most Common Types of Vaginitis and Vaginosis 759
26.3: Microbial Diseases of the Reproductive Systems 761

1 The Microbial World and You

The overall theme of this textbook is the relationship between microbes (very small organisms that usually require a microscope to be seen) and our lives. This relationship involves not only the familiar harmful effects of certain microorganisms, such as disease and food spoilage, but also their many beneficial effects. In this chapter we introduce you to some of the many ways microbes affect our lives. They have been fruitful subjects of study for a number of years, as you will see in the short history of microbiology that opens the chapter. We then discuss the incredible diversity of microorganisms and their ecological importance in maintaining balance in the environment by recycling chemical elements such as carbon and nitrogen between the soil, organisms, and the atmosphere. We will also examine how microbes are used in commercial and industrial applications to produce foods, chemicals, and drugs (such as antibiotics) and to treat sewage, control pests, and clean up pollutants. Finally, we will discuss microbes as the cause of such diseases as avian (bird) flu, West Nile encephalitis, mad cow disease, diarrhea, hemorrhagic fever, and AIDS.

UNDER THE MICROSCOPE

Bacteria on the human tongue. Most of the bacteria found in the mouth are harmless.

Q&A

Advertisements tell you that bacteria and viruses are all over your home and that you need to buy antibacterial cleaning products. Should you?

Look for the answer in the chapter.

Microbes in Our Lives

LEARNING OBJECTIVE

1-1 List several ways in which microbes affect our lives.

For many people, the words *germ* and *microbe* bring to mind a group of tiny creatures that do not quite fit into any of the categories in that old question, "Is it animal, vegetable, or mineral?" **Microbes,** also called **microorganisms,** are minute living things that individually are usually too small to be seen with the unaided eye. The group includes bacteria (Chapter 11), fungi (yeasts and molds), protozoa, and microscopic algae (Chapter 12). It also includes viruses, those noncellular entities sometimes regarded as straddling the border between life and nonlife (Chapter 13). You will be introduced to each of these groups of microbes shortly.

We tend to associate these small organisms only with major diseases such as AIDS, uncomfortable infections, or such common inconveniences as spoiled food. However, the majority of microorganisms make crucial contributions by helping to maintain the balance of living organisms and chemicals in our environment. Marine and freshwater microorganisms form the basis of the food chain in oceans, lakes, and rivers. Soil microbes help break down wastes and incorporate nitrogen gas from the air into organic compounds, thereby recycling chemical elements between the soil, water, life, and air. Certain microbes play important roles in *photosynthesis,* a food- and oxygen-generating process that is critical to life on Earth. Humans and many other animals depend on the microbes in their intestines for digestion and the synthesis of some vitamins that their bodies require, including some B vitamins for metabolism and vitamin K for blood clotting.

Microorganisms also have many commercial applications. They are used in the synthesis of such chemical products as vitamins, organic acids, enzymes, alcohols, and many drugs. The process by which microbes produce acetone and butanol was discovered in 1914 by Chaim Weizmann, a Russian-born chemist working in England. With the outbreak of World War I in August of that year, the production of acetone became very important for making cordite (a smokeless form of gunpowder used in munitions). Weizmann's discovery played a significant role in determining the outcome of the war.

The food industry also uses microbes in producing vinegar, sauerkraut, pickles, alcoholic beverages, green olives, soy sauce, buttermilk, cheese, yogurt, and bread. In addition, enzymes from microbes can now be manipulated to cause the microbes to produce substances they normally do not synthesize. These substances include cellulose, digestive aids, and drain cleaner, plus important therapeutic substances such as insulin. Microbial enzymes may even have helped produce your favorite pair of jeans (see the box on the facing page).

Though only a minority of microorganisms are **pathogenic** (disease-producing), practical knowledge of microbes is necessary for medicine and the related health sciences. For example, hospital workers must be able to protect patients from common microbes that are normally harmless but pose a threat to the sick and injured.

Today we understand that microorganisms are found almost everywhere. Yet not long ago, before the invention of the microscope, microbes were unknown to scientists. Thousands of people died in devastating epidemics, the causes of which were not understood. Entire families died because vaccinations and antibiotics were not available to fight infections.

We can get an idea of how our current concepts of microbiology developed by looking at a few historic milestones in microbiology that have changed our lives. First, however, we will look at the major groups of microbes and how they are named and classified.

CHECK YOUR UNDERSTANDING

✓ Describe some of the destructive and beneficial actions of microbes. **1-1***

Naming and Classifying Microorganisms

LEARNING OBJECTIVES

1-2 Recognize the system of scientific nomenclature that uses two names: a genus and a specific epithet.

1-3 Differentiate the major characteristics of each group of microorganisms.

1-4 List the three domains.

Nomenclature

The system of nomenclature (naming) for organisms in use today was established in 1735 by Carolus Linnaeus. Scientific names are latinized because Latin was the language traditionally used by scholars. Scientific nomenclature assigns each organism two names—the **genus** (plural: *genera*) is the first name and is always capitalized; the **specific epithet** (**species** name) follows and is not capitalized. The organism is referred to by both the genus and the specific epithet, and both names are underlined or italicized. By custom, after a scientific name has been mentioned once, it can be abbreviated with the initial of the genus followed by the specific epithet.

Scientific names can, among other things, describe an organism, honor a researcher, or identify the habitat of a species. For example, consider *Staphylococcus aureus* (staf-i-lō-kok′kus ô′rē-us), a bacterium commonly found on human skin. *Staphylo-* describes the clustered arrangement of the cells; *coccus* indicates that they are shaped like spheres. The specific epithet, *aureus,* is Latin for golden, the color of many colonies of this

*Orange numbers following Check Your Understanding questions refer back to the corresponding Learning Objectives.

Designer Jeans: Made by Microbes?

Denim blue jeans have become increasingly popular ever since Levi Strauss and Jacob Davis first made them for California gold miners in 1873. Now, companies that manufacture blue jeans are turning to microbiology to develop environmentally sound production methods that minimize toxic wastes and the costs associated with treating toxic wastes. Moreover, microbiological methods can provide abundant, renewable raw materials.

Stone Washing?

A softer denim, called "stone-washed," was introduced in the 1980s. The fabric is not really washed with rocks. Enzymes, called cellulases, from *Trichoderma* fungus are used to digest some of the cellulose in the cotton, thereby softening it. Unlike many chemical reactions, enzymes usually operate at safe temperatures and pH. Moreover, enzymes are proteins, so they are readily degraded for removal from wastewater.

Fabric

Cotton production requires large tracts of land, pesticides, and fertilizer, and the crop yield depends on the weather. However, bacteria can produce both cotton and polyester with less environmental impact. *Gluconacetobacter xylinus* bacteria make cellulose by attaching glucose units to simple chains in the outer membrane of the bacterial cell wall. The cellulose microfibrils are extruded through pores in the outer membrane, and bundles of microfibrils then twist into ribbons,

Bleaching

Peroxide is a safer bleaching agent than chlorine and can be easily removed from fabric and wastewater by enzymes. Researchers at Novo Nordisk Biotech cloned a mushroom peroxidase gene in yeast and grew the yeasts in washing machine conditions. The yeast that survived the washing machine were selected as the peroxidase producers.

Indigo

Chemical synthesis of indigo requires a high pH and produces waste that explodes in contact with air. However, a California biotechnology company, Genencor, has developed a method to produce indigo by using bacteria. In the Genencor labs, researchers put the gene for conversion of the bacterial by-product indole to indigo from a soil bacterium, *Pseudomonas putida*, into *Escherichia coli* bacteria, which then turned blue.

Plastic

Microbes can even make plastic zippers and packaging material for the jeans. Over 25 bacteria make polyhydroxyalkanoate (PHA) inclusion granules as a food reserve. PHAs are similar to common plastics, and because they are made by bacteria, they are also readily degraded by many bacteria. PHAs could provide a biodegradable alternative to conventional plastic, which is made from petroleum.

E. coli bacteria produce indigo from trytophan.

Indigo-producing *E. coli* bacteria.

bacterium. **Table 1.1** contains more examples. The genus of the bacterium *Escherichia coli* (esh-ë-rik′-ē-ä kō′lī or kō′lē) is named for a scientist, Theodor Escherich, whereas its specific epithet, *coli*, reminds us that *E. coli* live in the colon, or large intestine.

CHECK YOUR UNDERSTANDING

✓ Distinguish a genus from a specific epithet. **1-2**

Types of Microorganisms

The classification and identification of microorganisms is discussed in Chapter 10. Here is an overview of the major groups.

Bacteria

Bacteria (singular: **bacterium**) are relatively simple, single-celled (unicellular) organisms. Because their genetic material is not enclosed in a special nuclear membrane, bacterial cells are

Table 1.1 Making Scientific Names Familiar

Use the word roots guide in Appendix E to find out what the name means. The name will not seem so strange if you translate it. When you encounter a new name, practice saying it out loud. The exact pronunciation is not as important as the familiarity you will gain. Guidelines for pronunciation are given in Appendix D.

Following are some examples of microbial names you may encounter in the popular press as well as in the lab.

	Pronunciation	Source of Genus Name	Source of Specific Epithet
Salmonella typhimurium (bacterium)	sal-mōn-el′lä tī-fi-mur′ē-um	Honors public health microbiologist Daniel Salmon	Causes stupor (*typh-*) in mice (*muri-*)
Streptococcus pyogenes (bacterium)	strep-tō-kok′kus pī-äj′en-ēz	Appearance of cells in chains (*strepto-*)	Forms pus (*pyo-*)
Saccharomyces cerevisiae (yeast)	sak-ä-rō-mī′ses se-ri-vis′ē-ī	Fungus (*-myces*) that uses sugar (*saccharo-*)	Makes beer (*cerevisia*)
Penicillium chrysogenum (fungus)	pen-i-sil′lē-um krī-so′jen-um	Tuftlike or paintbrush (*penicill-*) appearance microscopically	Produces a yellow (*chryso-*) pigment
Trypanosoma cruzi (protozoan)	tri-pa-nō-sō′mä krūz′ē	Corkscrew- (*trypano-*, borer; soma-, body)	Honors epidemiologist Oswaldo Cruz

called **prokaryotes** (prō-kar′e-ōts), from Greek words meaning prenucleus. Prokaryotes include both bacteria and archaea.

Bacterial cells generally appear in one of several shapes. *Bacillus* (bä-sil′lus) (rodlike), illustrated in Figure 1.1a, *coccus* (kok′kus)(spherical or ovoid), and *spiral* (corkscrew or curved) are among the most common shapes, but some bacteria are star-shaped or square (see Figures 4.1 through 4.5, pages 78–79). Individual bacteria may form pairs, chains, clusters, or other groupings; such formations are usually characteristic of a particular genus or species of bacteria.

Bacteria are enclosed in cell walls that are largely composed of a carbohydrate and protein complex called *peptidoglycan.* (By contrast, cellulose is the main substance of plant and algal cell walls.) Bacteria generally reproduce by dividing into two equal cells; this process is called *binary fission.* For nutrition, most bacteria use organic chemicals, which in nature can be derived from either dead or living organisms. Some bacteria can manufacture their own food by photosynthesis, and some can derive nutrition from inorganic substances. Many bacteria can "swim" by using moving appendages called *flagella.* (For a complete discussion of bacteria, see Chapter 11.)

Archaea

Like bacteria, **archaea** (är′kē-ä) consist of prokaryotic cells, but if they have cell walls, the walls lack peptidoglycan. Archaea, often found in extreme environments, are divided into three main groups. The *methanogens* produce methane as a waste product from respiration. The *extreme halophiles* (*halo* = salt; *philic* = loving) live in extremely salty environments such as the Great Salt Lake and the Dead Sea. The *extreme thermophiles* (*therm* = heat) live in hot sulfurous water, such as hot springs at Yellowstone National Park. Archaea are not known to cause disease in humans.

Fungi

Fungi (singular: **fungus**) are **eukaryotes** (yū-kar′ē-ōts), organisms whose cells have a distinct nucleus containing the cell's genetic material (DNA), surrounded by a special envelope called the nuclear membrane. Organisms in the Kingdom Fungi may be unicellular or multicellular (see Chapter 12, page 330). Large multicellular fungi, such as mushrooms, may look somewhat like plants, but they cannot carry out photosynthesis, as most plants can. True fungi have cell walls composed primarily of a substance called *chitin.* The unicellular forms of fungi, *yeasts,* are oval microorganisms that are larger than bacteria. The most typical fungi are *molds* (Figure 1.1b). Molds form visible masses called *mycelia,* which are composed of long filaments *(hyphae)* that branch and intertwine. The cottony growths sometimes found on bread and fruit are mold mycelia. Fungi can reproduce sexually or asexually. They obtain nourishment by absorbing solutions of organic material from their environment—whether soil, seawater, fresh water, or an animal or plant host. Organisms called *slime molds* have characteristics of both fungi and amoebas. They are discussed in detail in Chapter 12.

Protozoa

Protozoa (singular: **protozoan**) are unicellular eukaryotic microbes (see Chapter 12, page 345). Protozoa move by

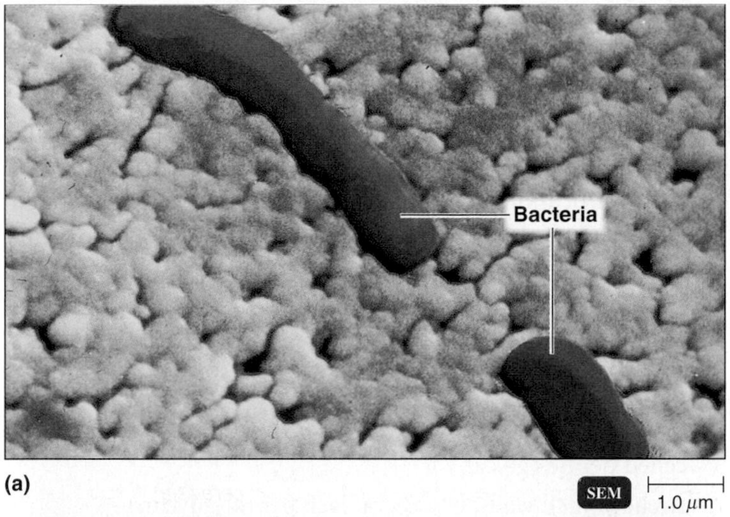

Bacteria

(a) SEM 1.0 μm

Sporangia

(b) SEM 400 μm

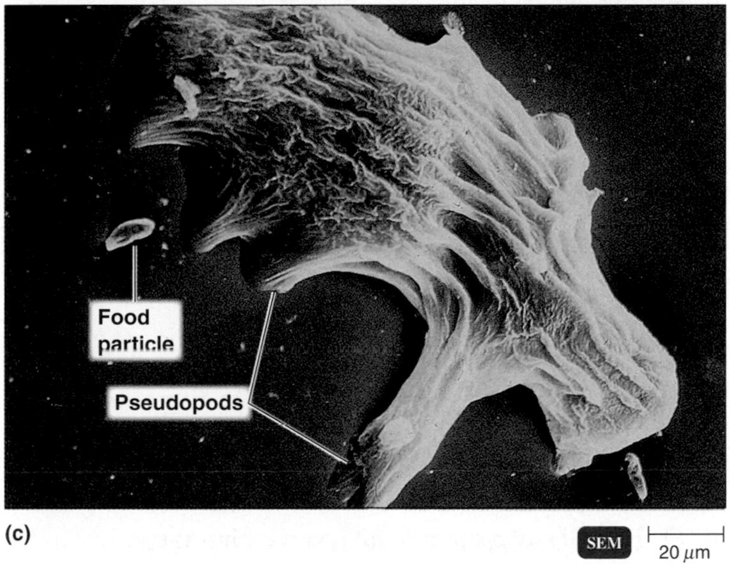

Food particle

Pseudopods

(c) SEM 20 μm

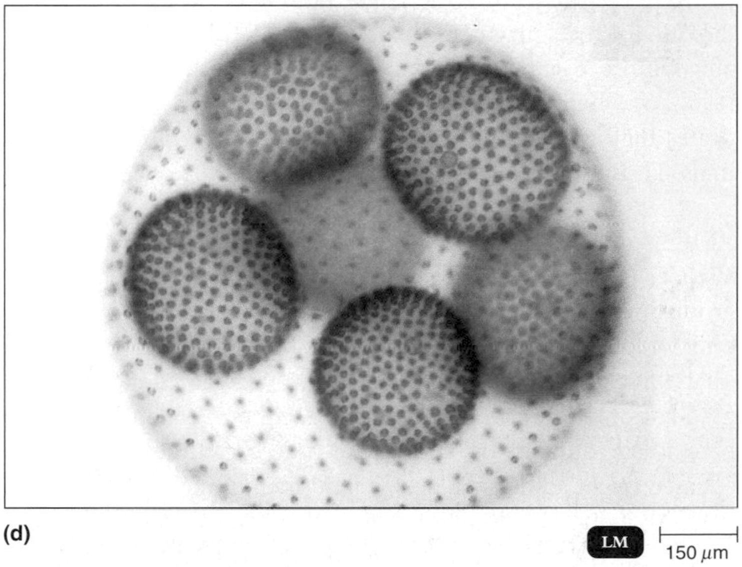

(d) LM 150 μm

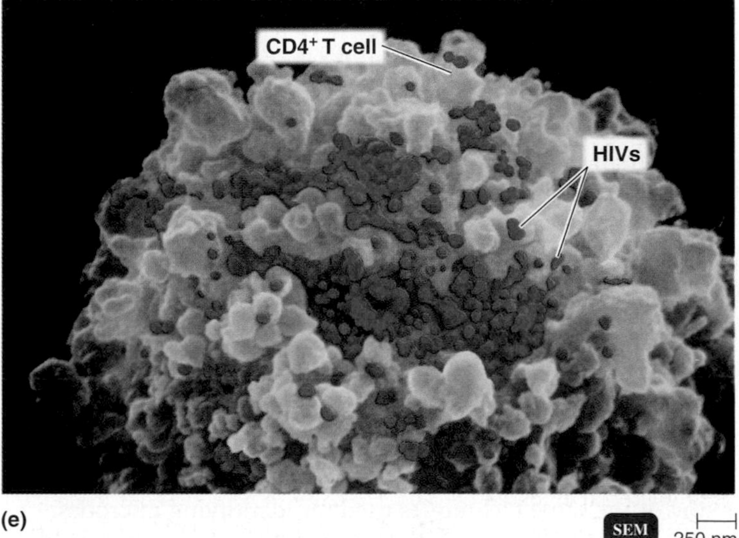

(e) SEM 250 nm

Figure 1.1 Types of microorganisms. *Note:* Throughout the book, a red icon under a micrograph indicates that the micrograph has been artificially colored. (**a**) The rod-shaped bacterium *Haemophilus influenzae,* one of the bacterial causes of pneumonia. (**b**) *Mucor,* a common bread mold, is a type of fungus. When released from sporangia, spores that land on a favorable surface germinate into a network of hyphae (filaments) that absorb nutrients. (**c**) An amoeba, a protozoan, approaching a food particle. (**d**) The pond alga, *Volvox.* (**e**) Several human immunodeficiency viruses (HIVs), the causative agent of AIDS, budding from a $CD4^+$ T cell.

Q How are bacteria, archaea, fungi, protozoa, algae, and viruses distinguished on the basis of cellular structure?

pseudopods, flagella, or cilia. Amoebas (**Figure 1.1c**) move by using extensions of their cytoplasm called *pseudopods* (false feet). Other protozoa have long *flagella* or numerous shorter appendages for locomotion called *cilia*. Protozoa have a variety of shapes and live either as free entities or as *parasites* (organisms that derive nutrients from living hosts) that absorb or ingest organic compounds from their environment. Protozoa can reproduce sexually or asexually.

Algae

Algae (singular: **alga**) are photosynthetic eukaryotes with a wide variety of shapes and both sexual and asexual reproductive forms (**Figure 1.1d**). The algae of interest to microbiologists are usually unicellular (see Chapter 12, page 340). The cell walls of many algae, are composed of a carbohydrate called *cellulose*. Algae are abundant in fresh and salt water, in soil, and in association with plants. As photosynthesizers, algae need light, water, and carbon dioxide for food production and growth, but they do not generally require organic compounds from the environment. As a result of photosynthesis, algae produce oxygen and carbohydrates that are then utilized by other organisms, including animals. Thus, they play an important role in the balance of nature.

Viruses

Viruses (**Figure 1.1e**) are very different from the other microbial groups mentioned here. They are so small that most can be seen only with an electron microscope, and they are acellular (not cellular). Structurally very simple, a virus particle contains a core made of only one type of nucleic acid, either DNA or RNA. This core is surrounded by a protein coat. Sometimes the coat is encased by an additional layer, a lipid membrane called an envelope. All living cells have RNA *and* DNA, can carry out chemical reactions, and can reproduce as self-sufficient units. Viruses can reproduce only by using the cellular machinery of other organisms. Thus, viruses are considered to be living when they multiply within host cells they infect. In this sense, viruses are parasites of other forms of life. On the other hand, viruses are not considered to be living because outside living hosts, they are inert. (Viruses will be discussed in detail in Chapter 13.)

Multicellular Animal Parasites

Although multicellular animal parasites are not strictly microorganisms, they are of medical importance and therefore will be discussed in this text. Animals are eukaryotes. The two major groups of parasitic worms are the flatworms and the roundworms, collectively called **helminths** (see Chapter 12, page 352). During some stages of their life cycle, helminths are microscopic in size. Laboratory identification of these organisms includes many of the same techniques used for identifying microbes.

CHECK YOUR UNDERSTANDING

✓ Which groups of microbes are prokaryotes? Which are eukaryotes? **1-3**

Classification of Microorganisms

Before the existence of microbes was known, all organisms were grouped into either the animal kingdom or the plant kingdom. When microscopic organisms with characteristics of animals and plants were discovered late in the seventeenth century, a new system of classification was needed. Still, biologists could not agree on the criteria for classifying the new organisms they were seeing until the late 1970s.

In 1978, Carl Woese devised a system of classification based on the cellular organization of organisms. It groups all organisms in three domains as follows:

1. Bacteria (cell walls contain a protein-carbohydrate complex called peptidoglycan)
2. Archaea (cell walls, if present, lack peptidoglycan)
3. Eukarya, which includes the following:
   - Protists (slime molds, protozoa, and algae)
   - Fungi (unicellular yeasts, multicellular molds, and mushrooms)
   - Plants (includes mosses, ferns, conifers, and flowering plants)
   - Animals (includes sponges, worms, insects, and vertebrates)

Classification will be discussed in more detail in Chapters 10 through 12.

CHECK YOUR UNDERSTANDING

✓ What are the three domains? **1-4**

A Brief History of Microbiology

LEARNING OBJECTIVES

1-5 Explain the importance of observations made by Hooke and van Leeuwenhoek.

1-6 Compare spontaneous generation and biogenesis.

1-7 Identify the contributions to microbiology made by Needham, Spallanzani, Virchow, and Pasteur.

1-8 Explain how Pasteur's work influenced Lister and Koch.

1-9 Identify the importance of Koch's postulates.

1-10 Identify the importance of Jenner's work.

1-11 Identify the contributions to microbiology made by Ehrlich and Fleming.

1-12 Define *bacteriology*, *mycology*, *parasitology*, *immunology*, and *virology*.

1-13 Explain the importance of microbial genetics and molecular biology.

The science of microbiology dates back only 200 years, yet the recent discovery of *Mycobacterium tuberculosis* (mī-kō-bak-ti'rē-um tü-bėr-ku-lō'sis) DNA in 3000-year-old Egyptian mummies reminds us that microorganisms have been around

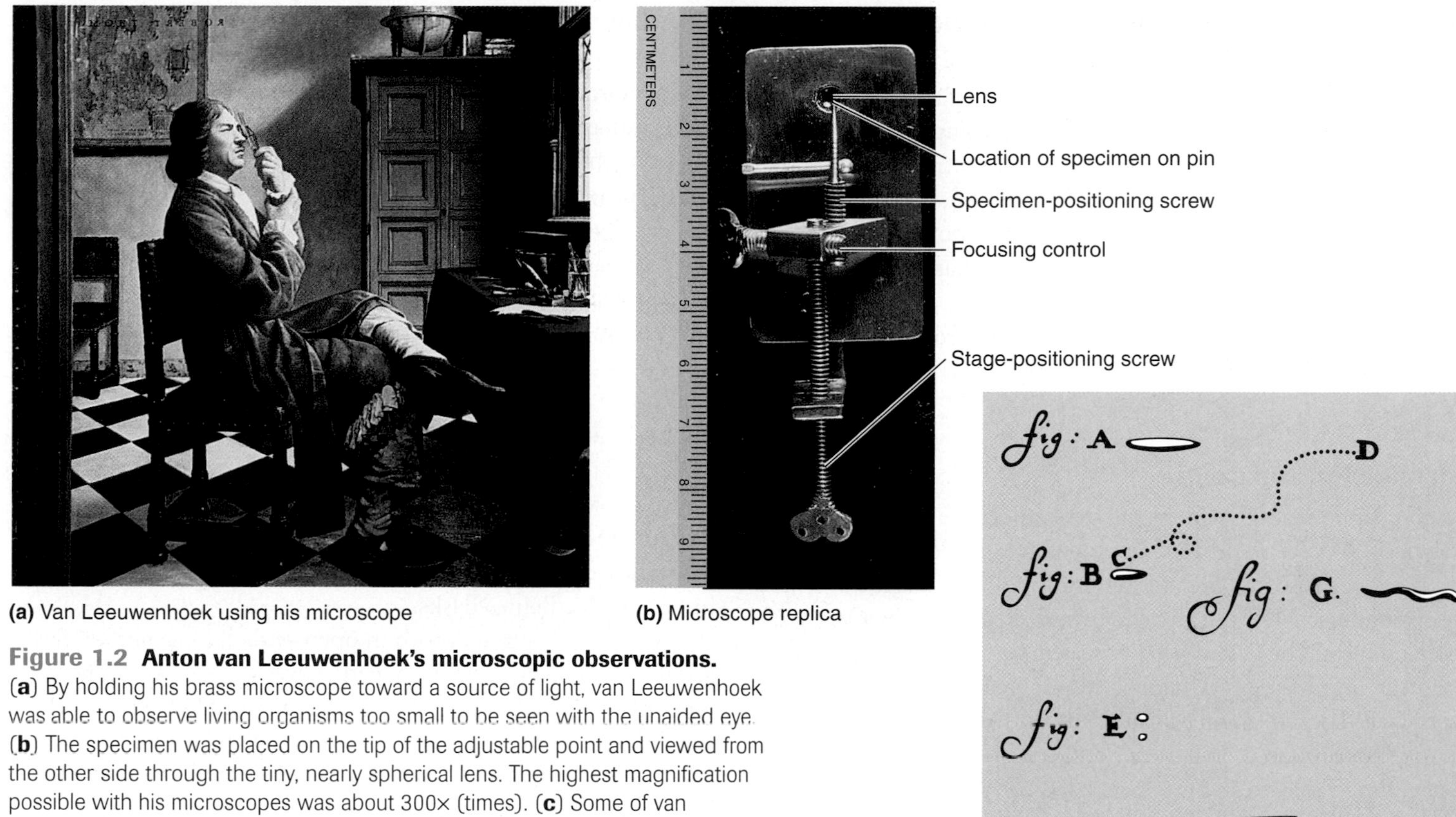

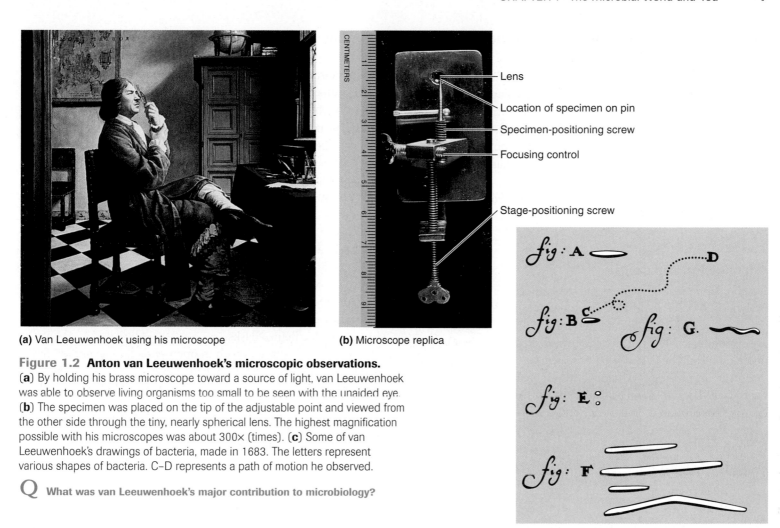

(a) Van Leeuwenhoek using his microscope

(b) Microscope replica

(c) Drawings of bacteria

Figure 1.2 Anton van Leeuwenhoek's microscopic observations. (**a**) By holding his brass microscope toward a source of light, van Leeuwenhoek was able to observe living organisms too small to be seen with the unaided eye. (**b**) The specimen was placed on the tip of the adjustable point and viewed from the other side through the tiny, nearly spherical lens. The highest magnification possible with his microscopes was about 300× (times). (**c**) Some of van Leeuwenhoek's drawings of bacteria, made in 1683. The letters represent various shapes of bacteria. C–D represents a path of motion he observed.

Q What was van Leeuwenhoek's major contribution to microbiology?

for much longer. In fact, bacterial ancestors were the first living cells to appear on Earth. Although we know relatively little about what earlier people thought about the causes, transmission, and treatment of disease, the history of the past few hundred years is better known. Let's look now at some key developments in microbiology that have spurred the field to its current high-technology state.

The First Observations

One of the most important discoveries in the history of biology occurred in 1665 with the help of a relatively crude microscope. After observing a thin slice of cork, an Englishman, Robert Hooke, reported to the world that life's smallest structural units were "little boxes," or "cells," as he called them. Using his improved version of a compound microscope (one that uses two sets of lenses), Hooke was able to see individual cells. Hooke's discovery marked the beginning of the **cell theory**—the theory that *all living things are composed of cells.* Subsequent investigations into the structure and functions of cells were based on this theory.

Though Hooke's microscope was capable of showing large cells, he lacked the resolution that would have allowed him to see microbes clearly. The Dutch merchant and amateur scientist Anton van Leeuwenhoek was probably the first actually to observe live microorganisms through the magnifying lenses of more than 400 microscopes he constructed. Between 1673 and 1723, he wrote a series of letters to the Royal Society of London describing the "animalcules" he saw through his simple, single-lens microscope. Van Leeuwenhoek made detailed drawings of "animalcules" in rainwater, in his own feces, and in material scraped from his teeth. These drawings have since been identified as representations of bacteria and protozoa (**Figure 1.2**).

CHECK YOUR UNDERSTANDING

✓ What is the cell theory? **1-5**

The Debate over Spontaneous Generation

After van Leeuwenhoek discovered the previously "invisible" world of microorganisms, the scientific community of the time became interested in the origins of these tiny living things. Until the second half of the nineteenth century, many scientists and philosophers believed that some forms of life could arise spontaneously from nonliving matter; they called this hypothetical process **spontaneous generation.** Not much more than 100 years ago, people commonly believed that toads, snakes, and mice could be born of moist soil; that flies could emerge from manure; and that maggots, the larvae of flies, could arise from decaying corpses.

Evidence Pro and Con

A strong opponent of spontaneous generation, the Italian physician Francesco Redi set out in 1668 (even before van Leeuwenhoek's discovery of microscopic life) to demonstrate that maggots did not arise spontaneously from decaying meat. Redi filled two jars with decaying meat. The first was left unsealed; the flies laid their eggs on the meat, and the eggs developed into larvae. The second jar was sealed, and because the flies could not lay their eggs on the meat, no maggots appeared. Still, Redi's antagonists were not convinced; they claimed that fresh air was needed for spontaneous generation. So Redi set up a second experiment, in which he covered a jar with a fine net instead of sealing it. No larvae appeared in the gauze-covered jar, even though air was present. Maggots appeared only when flies were allowed to leave their eggs on the meat.

Redi's results were a serious blow to the long-held belief that large forms of life could arise from nonlife. However, many scientists still believed that small organisms, such as van Leeuwenhoek's "animalcules," were simple enough to be generated from nonliving materials.

The case for spontaneous generation of microorganisms seemed to be strengthened in 1745, when John Needham, an Englishman, found that even after he heated nutrient fluids (chicken broth and corn broth) before pouring them into covered flasks, the cooled solutions were soon teeming with microorganisms. Needham claimed that microbes developed spontaneously from the fluids. Twenty years later, Lazzaro Spallanzani, an Italian scientist, suggested that microorganisms from the air probably had entered Needham's solutions after they were boiled. Spallanzani showed that nutrient fluids heated *after* being sealed in a flask did not develop microbial growth. Needham responded by claiming the "vital force" necessary for spontaneous generation had been destroyed by the heat and was kept out of the flasks by the seals.

This intangible "vital force" was given all the more credence shortly after Spallanzani's experiment, when Anton Laurent Lavoisier showed the importance of oxygen to life. Spallanzani's observations were criticized on the grounds that there was not enough oxygen in the sealed flasks to support microbial life.

The Theory of Biogenesis

The issue was still unresolved in 1858, when the German scientist Rudolf Virchow challenged the case for spontaneous generation with the concept of **biogenesis,** the claim that living cells can arise only from preexisting living cells. Arguments about spontaneous generation continued until 1861, when the issue was resolved by the French scientist Louis Pasteur.

With a series of ingenious and persuasive experiments, Pasteur demonstrated that microorganisms are present in the air and can contaminate sterile solutions, but that air itself does not create microbes. He filled several short-necked flasks with beef broth and then boiled their contents. Some were then left open and allowed to cool. In a few days, these flasks were found to be contaminated with microbes. The other flasks, sealed after boiling, were free of microorganisms. From these results, Pasteur reasoned that microbes in the air were the agents responsible for contaminating nonliving matter such as the broths in Needham's flasks.

Pasteur next placed broth in open-ended, long-necked flasks and bent the necks into S-shaped curves (**Figure 1.3**). The contents of these flasks were then boiled and cooled. The broth in the flasks did not decay and showed no signs of life, even after months. Pasteur's unique design allowed air to pass into the flask, but the curved neck trapped any airborne microorganisms that might contaminate the broth. (Some of these original vessels are still on display at the Pasteur Institute in Paris. They have been sealed but, like the flask shown in Figure 1.3, show no sign of contamination more than 100 years later.)

Pasteur showed that microorganisms can be present in nonliving matter—on solids, in liquids, and in the air. Furthermore, he demonstrated conclusively that microbial life can be destroyed by heat and that methods can be devised to block the access of airborne microorganisms to nutrient environments. These discoveries form the basis of **aseptic techniques,** techniques that prevent contamination by unwanted microorganisms, which are now the standard practice in laboratory and many medical procedures. Modern aseptic techniques are among the first and most important things that a beginning microbiologist learns.

Pasteur's work provided evidence that microorganisms cannot originate from mystical forces present in nonliving materials. Rather, any appearance of "spontaneous" life in nonliving solutions can be attributed to microorganisms that were already present in the air or in the fluids themselves. Scientists now believe that a form of spontaneous generation probably did occur on the primitive Earth when life first began, but they agree that this does not happen under today's environmental conditions.

CHECK YOUR UNDERSTANDING

- ✓ What evidence supported spontaneous generation? **1-6**
- ✓ How was spontaneous generation disproved? **1-7**

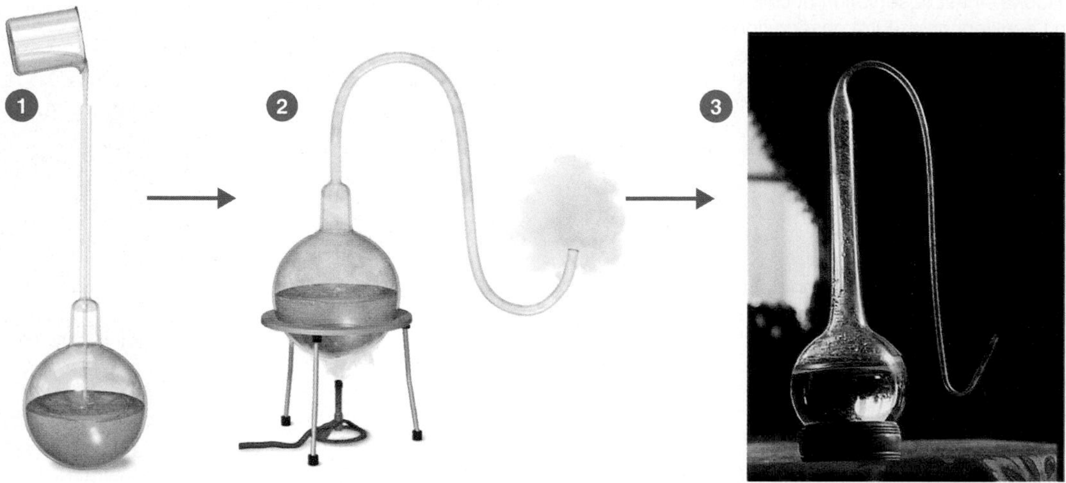

Figure 1.3 **Pasteur's experiment disproving the theory of spontaneous generation.** (1) Pasteur first poured beef broth into a long-necked flask. (2) Next he heated the neck of the flask and bent it into an S-shaped curve; then he boiled the broth for several minutes. (3) Microorganisms did not appear in the cooled solution, even after long periods, as you can see in this recent photograph of an actual flask Pasteur used in a similar experiment.

Q **What are aseptic techniques, and how did Pasteur contribute to their development?**

The Golden Age of Microbiology

For about 60 years, beginning with the work of Pasteur, there was an explosion of discoveries in microbiology. The period from 1857 to 1914 has been appropriately named the Golden Age of Microbiology. During this period, rapid advances, spearheaded mainly by Pasteur and Robert Koch, led to the establishment of microbiology as a science. Discoveries during these years included both the agents of many diseases and the role of immunity in preventing and curing disease. During this productive period, microbiologists studied the chemical activities of microorganisms, improved the techniques for performing microscopy and culturing microorganisms, and developed vaccines and surgical techniques. Some of the major events that occurred during the Golden Age of Microbiology are listed in Figure 1.4.

Fermentation and Pasteurization

One of the key steps that established the relationship between microorganisms and disease occurred when a group of French merchants asked Pasteur to find out why wine and beer soured. They hoped to develop a method that would prevent spoilage when those beverages were shipped long distances. At the time, many scientists believed that air converted the sugars in these fluids into alcohol. Pasteur found instead that microorganisms called yeasts convert the sugars to alcohol in the absence of air. This process, called **fermentation** (see Chapter 5, page 132), is used to make wine and beer. Souring and spoilage are caused by different microorganisms called bacteria. In the presence of air, bacteria change the alcohol in the beverage into vinegar (acetic acid).

Pasteur's solution to the spoilage problem was to heat the beer and wine just enough to kill most of the bacteria that caused the spoilage. The process, called **pasteurization,** is now commonly used to reduce spoilage and kill potentially harmful bacteria in milk as well as in some alcoholic drinks. Showing the connection between spoilage of food and microorganisms was a major step toward establishing the relationship between disease and microbes.

The Germ Theory of Disease

As we have seen, the fact that many kinds of diseases are related to microorganisms was unknown until relatively recently. Before the time of Pasteur, effective treatments for many diseases were discovered by trial and error, but the causes of the diseases were unknown.

The realization that yeasts play a crucial role in fermentation was the first link between the activity of a microorganism and physical and chemical changes in organic materials. This discovery alerted scientists to the possibility that microorganisms might have similar relationships with plants and animals—specifically, that microorganisms might cause disease. This idea was known as the **germ theory of disease.**

The germ theory was a difficult concept for many people to accept at that time because for centuries disease was believed to be punishment for an individual's crimes or misdeeds. When the inhabitants of an entire village became ill, people often blamed the disease on demons appearing as foul odors from sewage or on poisonous vapors from swamps. Most people born in Pasteur's time found it inconceivable that "invisible" microbes

1665 Hooke—First observation of cells
1673 van Leeuwenhoek—First observation of live microorganisms
1735 Linnaeus—Nomenclature for organisms
1798 Jenner—First vaccine
1835 Bassi—Silkworm fungus
1840 Semmelweis—Childbirth fever
1853 DeBary—Fungal plant disease

GOLDEN AGE OF MICROBIOLOGY

1857 Pasteur—Fermentation
1861 Pasteur—Disproved spontaneous generation
1864 Pasteur—Pasteurization
1867 Lister—Aseptic surgery
1876 *Koch—Germ theory of disease
1879 Neisser—*Neisseria gonorrhoeae*
1881 *Koch—Pure cultures
Finley—Yellow fever
1882 *Koch—*Mycobacterium tuberculosis*
Hess—Agar (solid) media
1883 *Koch—*Vibrio cholerae*
1884 *Metchnikoff—Phagocytosis
Gram—Gram-staining procedure
Escherich—*Escherichia coli*
1887 Petri—Petri dish
1889 Kitasato—*Clostridium tetani*
1890 *von Behring—Diphtheria antitoxin
*Ehrlich—Theory of immunity
1892 Winogradsky—Sulfur cycle
1898 Shiga—*Shigella dysenteriae*
1908 *Ehrlich—Syphilis
1910 Chagas—*Trypanosoma cruzi*
1911 * Rous—Tumor-causing virus (1966 Nobel Prize)

1928 *Fleming, Chain, Florey—Penicillin
Griffith—Transformation in bacteria
1934 Lancefield—Streptococcal antigens
1935 *Stanley, Northrup, Sumner—Crystallized virus
1941 Beadle and Tatum—Relationship between genes and enzymes
1943 *Delbrück and Luria—Viral infection of bacteria
1944 Avery, MacLeod, McCarty—Genetic material is DNA
1946 Lederberg and Tatum—Bacterial conjugation
1953 *Watson and Crick—DNA structure
1957 *Jacob and Monod—Protein synthesis regulation
1959 Stewart—Viral cause of human cancer
1962 *Edelman and Porter—Antibodies
1964 Epstein, Achong, Barr—Epstein-Barr virus as cause of human cancer
1971 *Nathans, Smith, Arber—Restriction enzymes (used for recombinant DNA technology)
1973 Berg, Boyer, Cohen—Genetic engineering
1975 Dulbecco, Temin, Baltimore—Reverse transcriptase
1978 Woese—Archaea
*Mitchell—Chemiosmotic mechanism
1981 Margulis—Origin of eukaryotic cells
1982 *Klug—Structure of tobacco mosaic virus
1983 *McClintock—Transposons

1988 *Deisenhofer, Huber, Michel—Bacterial photosynthesis pigments
1994 Cano—Reported to have cultured 40-million-year-old bacteria
1997 *Prusiner—Prions

Louis Pasteur (1822–1895)
Demonstrated that life did not arise spontaneously from nonliving matter.

Robert Koch (1843–1910)
Established experimental steps for directly linking a specific microbe to a specific disease.

Rebecca C. Lancefield (1895–1981)
Classified streptococci according to serotypes (variants within a species)

Figure 1.4 Milestones in microbiology, highlighting those that occurred during the Golden Age of Microbiology. An asterisk (*) indicates a Nobel laureate.

Why was the Golden Age of Microbiology so named?

could travel through the air to infect plants and animals or remain on clothing and bedding to be transmitted from one person to another. But gradually scientists accumulated the information needed to support the new germ theory.

In 1865, Pasteur was called upon to help fight silkworm disease, which was ruining the silk industry throughout Europe. Years earlier, in 1835, Agostino Bassi, an amateur microscopist, had proved that another silkworm disease was caused by a fungus. Using data provided by Bassi, Pasteur found that the more recent infection was caused by a protozoan, and he developed a method for recognizing afflicted silkworm moths.

In the 1860s, Joseph Lister, an English surgeon, applied the germ theory to medical procedures. Lister was aware that in the 1840s, the Hungarian physician Ignaz Semmelweis had demonstrated that physicians, who at the time did not disinfect their hands, routinely transmitted infections (puerperal, or childbirth, fever) from one obstetrical patient to another. Lister had also heard of Pasteur's work connecting microbes to animal diseases. Disinfectants were not used at the time, but Lister knew that phenol (carbolic acid) kills bacteria, so he began treating surgical wounds with a phenol solution. The practice so reduced the incidence of infections and deaths that other surgeons quickly adopted it. Lister's technique was one of the earliest medical attempts to control infections caused by microorganisms. In fact, his findings proved that microorganisms cause surgical wound infections.

The first proof that bacteria actually cause disease came from Robert Koch in 1876. Koch, a German physician, was Pasteur's young rival in the race to discover the cause of anthrax, a disease that was destroying cattle and sheep in Europe. Koch discovered rod-shaped bacteria now known as *Bacillus anthracis* (bä-sil'lus an-thrā'sis) in the blood of cattle that had died of anthrax. He cultured the bacteria on nutrients and then injected samples of the culture into healthy animals. When these animals became sick and died, Koch isolated the bacteria in their blood and compared them with the bacteria originally isolated. He found that the two sets of blood cultures contained the same bacteria.

Koch thus established a sequence of experimental steps for directly relating a specific microbe to a specific disease. These steps are known today as **Koch's postulates** (see Figure 14.3, page 405). During the past 100 years, these same criteria have been invaluable in investigations proving that specific microorganisms cause many diseases. Koch's postulates, their limitations, and their application to disease will be discussed in greater detail in Chapter 14.

Vaccination

Often a treatment or preventive procedure is developed before scientists know why it works. The smallpox vaccine is an example. On May 4, 1796, almost 70 years before Koch established that a specific microorganism causes anthrax, Edward Jenner, a young British physician, embarked on an experiment to find a way to protect people from smallpox.

Smallpox epidemics were greatly feared. The disease periodically swept through Europe, killing thousands, and it wiped out 90% of the American Indians on the East Coast when European settlers first brought the infection to the New World.

When a young milkmaid informed Jenner that she couldn't get smallpox because she already had been sick from cowpox—a much milder disease—he decided to put the girl's story to the test. First Jenner collected scrapings from cowpox blisters. Then he inoculated a healthy 8-year-old volunteer with the cowpox material by scratching the person's arm with a pox-contaminated needle. The scratch turned into a raised bump. In a few days, the volunteer became mildly sick but recovered and never again contracted either cowpox or smallpox. The process was called *vaccination,* from the Latin word *vacca,* meaning cow. Pasteur gave it this name in honor of Jenner's work. The protection from disease provided by vaccination (or by recovery from the disease itself) is called **immunity.** We will discuss the mechanisms of immunity in Chapter 17.

Years after Jenner's experiment, in about 1880, Pasteur discovered why vaccinations work. He found that the bacterium that causes fowl cholera lost its ability to cause disease (lost its *virulence,* or became *avirulent*) after it was grown in the laboratory for long periods. However, it—and other microorganisms with decreased virulence—was able to induce immunity against subsequent infections by its virulent counterparts. The discovery of this phenomenon provided a clue to Jenner's successful experiment with cowpox. Both cowpox and smallpox are caused by viruses. Even though cowpox virus is not a laboratory-produced derivative of smallpox virus, it is so closely related to the smallpox virus that it can induce immunity to both viruses. Pasteur used the term *vaccine* for cultures of avirulent microorganisms used for preventive inoculation.

Jenner's experiment marked the first time in a Western culture that a living viral agent—the cowpox virus—was used to produce immunity. Physicians in China had immunized patients by removing scales from drying pustules of a person suffering from a mild case of smallpox, grinding the scales to a fine powder, and inserting the powder into the nose of the person to be protected.

Some vaccines are still produced from avirulent microbial strains that stimulate immunity to the related virulent strain. Other vaccines are made from killed virulent microbes, from isolated components of virulent microorganisms, or by genetic engineering techniques.

CHECK YOUR UNDERSTANDING

- ✓ Summarize in your own words the germ theory of disease. **1-8**
- ✓ What is the importance of Koch's postulates? **1-9**
- ✓ What is the significance of Jenner's discovery? **1-10**

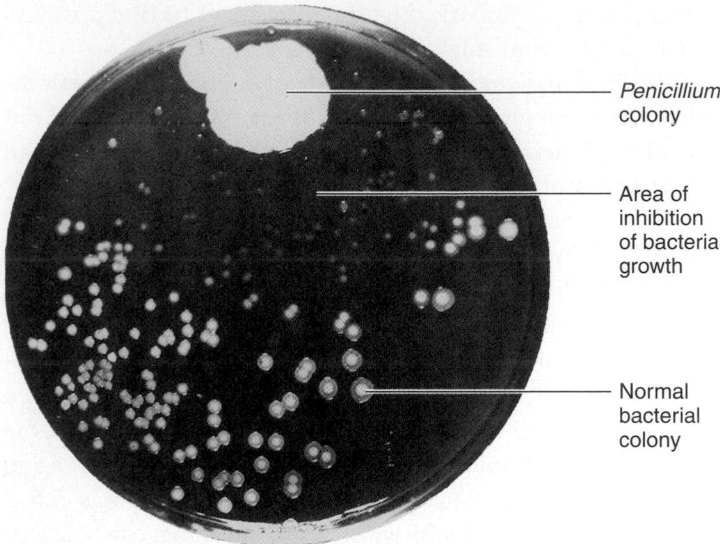

Figure 1.5 The discovery of penicillin. Alexander Fleming took this photograph in 1928. The colony of *Penicillium* mold accidentally contaminated the plate and inhibited nearby bacterial growth.

Q What are some of the problems associated with antibiotics?

The Birth of Modern Chemotherapy: Dreams of a "Magic Bullet"

After the relationship between microorganisms and disease was established, medical microbiologists next focused on the search for substances that could destroy pathogenic microorganisms without damaging the infected animal or human. Treatment of disease by using chemical substances is called **chemotherapy.** (The term also commonly refers to chemical treatment of non-infectious diseases, such as cancer.) Chemicals produced naturally by bacteria and fungi to act against other microorganisms are called **antibiotics.** Chemotherapeutic agents prepared from chemicals in the laboratory are called **synthetic drugs.** The success of chemotherapy is based on the fact that some chemicals are more poisonous to microorganisms than to the hosts infected by the microbes. Antimicrobial therapy will be discussed in further detail in Chapter 20.

The First Synthetic Drugs

Paul Ehrlich, a German physician, was the imaginative thinker who fired the first shot in the chemotherapy revolution. As a medical student, Ehrlich speculated about a "magic bullet" that could hunt down and destroy a pathogen without harming the infected host. Ehrlich launched a search for such a bullet. In 1910, after testing hundreds of substances, he found a chemotherapeutic agent called *salvarsan,* an arsenic derivative effective against syphilis. The agent was named salvarsan because it was considered to offer salvation from syphilis and it contained arsenic. Before this discovery, the only known chemical in Europe's medical arsenal was an extract from the bark of a South American tree, *quinine,* which had been used by Spanish conquistadors to treat malaria.

By the late 1930s, researchers had developed several other synthetic drugs that could destroy microorganisms. Most of these drugs were derivatives of dyes. This came about because the dyes synthesized and manufactured for fabrics were routinely tested for antimicrobial qualities by microbiologists looking for a "magic bullet." In addition, *sulfonamides* (sulfa drugs) were synthesized at about the same time.

A Fortunate Accident—Antibiotics

In contrast to the sulfa drugs, which were deliberately developed from a series of industrial chemicals, the first antibiotic was discovered by accident. Alexander Fleming, a Scottish physician and bacteriologist, almost tossed out some culture plates that had been contaminated by mold. Fortunately, he took a second look at the curious pattern of growth on the contaminated plates. Around the mold was a clear area where bacterial growth had been inhibited (**Figure 1.5**). Fleming was looking at a mold that could inhibit the growth of a bacterium. The mold was later identified as *Penicillium notatum* (pen-i-sil'lē-um nō-tā'tum), later renamed *Penicillium chrysogenum* (krī-so'jen-um), and in 1928 Fleming named the mold's active inhibitor *penicillin.* Thus, penicillin is an antibiotic produced by a fungus. The enormous usefulness of penicillin was not apparent until the 1940s, when it was finally tested clinically and mass produced.

Since these early discoveries, thousands of other antibiotics have been discovered. Unfortunately, antibiotics and other chemotherapeutic drugs are not without problems. Many antimicrobial chemicals are too toxic to humans for practical use; they kill the pathogenic microbes, but they also damage the infected host. For reasons we will discuss later, toxicity to humans is a particular problem in the development of drugs for treating viral diseases. Viral growth depends on life processes of normal host cells. Thus, there are very few successful antiviral drugs, because a drug that would interfere with viral reproduction would also likely affect uninfected cells of the body.

Another major problem associated with antimicrobial drugs is the emergence and spread of new strains of microorganisms that are resistant to antibiotics. Over the years, more and more microbes have developed resistance to antibiotics that at one time were very effective against them. Drug resistance results from genetic changes in microbes that enables them to tolerate a certain amount of an antibiotic that would normally inhibit them (see the box in Chapter 26, page 751). These changes might include the production by microbes of chemicals (enzymes) that inactivate antibiotics, changes in the surface of a microbe that prevent an antibiotic from attaching to it, and prevention of an antibiotic from entering the microbe.

The recent appearance of vancomycin-resistant *Staphylococcus aureus* and *Enterococcus faecalis* (en-te-rō-kok'kus fe-kā'lis) has alarmed health care professionals because it indicates that some previously treatable bacterial infections may soon be impossible to treat with antibiotics.

CHECK YOUR UNDERSTANDING

✓ What was Ehrlich's "magic bullet"? **1-11**

Modern Developments in Microbiology

The quest to solve drug resistance, identify viruses, and develop vaccines requires sophisticated research techniques and correlated studies that were never dreamed of in the days of Koch and Pasteur.

The groundwork laid during the Golden Age of Microbiology provided the basis for several monumental achievements during the twentieth century (Table 1.2). New branches of microbiology were developed, including immunology and virology. Most recently, the development of a set of new methods called recombinant DNA technology has revolutionized research and practical applications in all areas of microbiology.

Bacteriology, Mycology, and Parasitology

Bacteriology, the study of bacteria, began with van Leeuwenhoek's first examination of tooth scrapings. New pathogenic bacteria are still discovered regularly. Many bacteriologists, like their predecessor Pasteur, look at the roles of bacteria in food and the environment. One intriguing discovery came in 1997, when Heide Schulz discovered a bacterium large enough to be seen with the unaided eye (0.2 mm wide). This bacterium, which she named *Thiomargarita namibiensis* (thī′o-mä-gär-e-tä na′mib-ē-en-sis), lives in the mud on the African coast. *Thiomargarita* is unusual because of its size and its ecological niche. The bacterium consumes hydrogen sulfide, which would be toxic to mud-dwelling animals (Figure 11.28, page 326).

Mycology, the study of fungi, includes medical, agricultural, and ecological branches. Recall that Bassi's work leading up to the germ theory of disease was on a fungal pathogen. Fungal infection rates have been rising during the past decade, accounting for 10% of hospital-acquired infections. Climatic and environmental changes (severe drought) are thought to account for the tenfold increase in *Coccidioides immitis* (kok-sid-ē-oi′dēz im′mi-tis) infections in California. New techniques for diagnosing and treating fungal infections are currently being investigated.

Parasitology is the study of protozoa and parasitic worms. Because many parasitic worms are large enough to be seen with the unaided eye, they have been known for thousands of years. It has been speculated that the medical symbol, the caduceus, represents the removal of parasitic guinea worms (Figure 1.6).

The clearing of rain forests has exposed laborers to previously undiscovered parasites. Previously unknown parasitic diseases are also being found in patients whose immune systems have been suppressed by organ transplants, cancer chemotherapy, and AIDS.

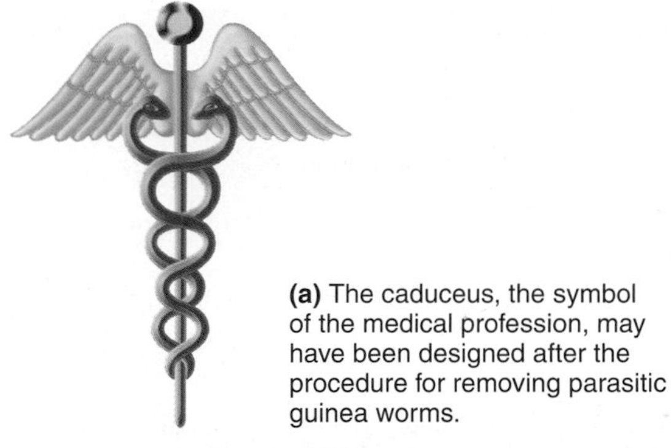

(a) The caduceus, the symbol of the medical profession, may have been designed after the procedure for removing parasitic guinea worms.

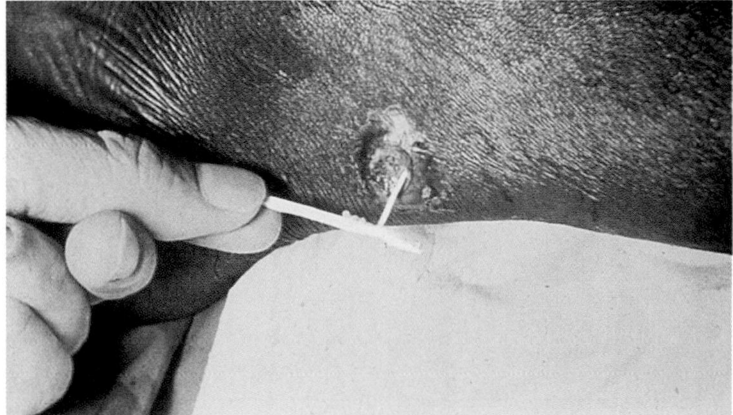

(b) A doctor removes a guinea worm (*Dracunculus medinensis*) from the subcutaneous tissue of a patient by winding it onto a stick.

Figure 1.6 Parasitology: the study of protozoa and parasitic worms.

Q How do bacteriology, mycology, and parasitology differ?

Bacteriology, mycology, and parasitology are currently going through a "golden age of classification." Recent advances in **genomics,** the study of all of an organism's genes, have allowed scientists to classify bacteria and fungi according to their genetic relationships with other bacteria, fungi, and protozoa. Previously these microorganisms were classified according to a limited number of visible characteristics.

Immunology

Immunology, the study of immunity, actually dates back in Western culture to Jenner's first vaccine in 1796. Since then, knowledge about the immune system has accumulated steadily and expanded rapidly during the twentieth century. Vaccines are now available for numerous diseases, including measles, rubella (German measles), mumps, chickenpox, pneumococcal pneumonia, tetanus, tuberculosis, influenza, whooping cough, polio, and hepatitis B. The smallpox vaccine was so effective that the disease has been eliminated. Public health officials estimate that polio will be eradicated within a few years because of the polio vaccine. In 1960, interferons, substances generated by the body's own immune system, were discovered. Interferons inhibit

Table 1.2 Selected Nobel Prizes Awarded for Research in Microbiology

Nobel Laureates	Year of Presentation	Country of Birth	Contribution
Emil A. von Behring	1901	Germany	Developed a diphtheria antitoxin
Ronald Ross	1902	England	Discovered how malaria is transmitted
Robert Koch	1905	Germany	Cultured tuberculosis bacteria
Paul Ehrlich	1908	Germany	Developed theories on immunity
Elie Metchnikoff	1908	Russia	Described phagocytosis, the intake of solid materials by cells
Alexander Fleming, Ernst Chain, and Howard Florey	1945	Scotland England England	Discovered penicillin
Selman A. Waksman	1952	Ukraine	Discovered streptomycin
Hans A. Krebs	1953	Germany	Discovered chemical steps of the Krebs cycle in carbohydrate metabolism
John F. Enders, Thomas H. Weller, and Frederick C. Robbins	1954	United States	Cultured poliovirus in cell cultures
Joshua Lederberg, George Beadle, and Edward Tatum	1958	United States	Described genetic control of biochemical reactions
Frank Macfarlane Burnet and Peter Brian Medawar	1960	Australia Great Britain	Discovered acquired immune tolerance
James D. Watson, Frances H. C. Crick, and Maurice A. F. Wilkins	1962	United States England New Zealand	Identified the physical structure of DNA
François Jacob, Jacques Monod, and André Lwoff	1965	France	Described how protein synthesis is regulated in bacteria
Peyton Rous	1966	United States	Discovered cancer-causing viruses
Max Delbrück, Alfred D. Hershey, and Salvador E. Luria	1969	Germany United States Italy	Described the mechanism of viral infection of bacterial cells
Gerald M. Edelman and Rodney R. Porter	1972	United States England	Described the nature and structure of antibodies
Renato Dulbecco, Howard Temin, and David Baltimore	1975	United States	Discovered reverse transcriptase and described how RNA viruses could cause cancer
Daniel Nathans, Hamilton Smith, and Werner Arber	1978	United States United States Switzerland	Described the action of restriction enzymes (now used in recombinant DNA technology)
Peter Mitchell	1978	England	Described the chemiosmotic mechanism for ATP synthesis
Paul Berg	1980	United States	Performed experiments in gene splicing

Table 1.2 (continued)

Nobel Laureates	Year of Presentation	Country of Birth	Contribution
Aaron Klug	1982	South Africa	Described the structure of tobacco mosaic virus (TMV)
Barbara McClintock	1983	United States	Discovered transposons (small segments of DNA that can move from one region of a DNA molecule to another)
César Milstein, Georges J.F. Köhler, and Niels Kai Jerne	1984	Argentina Germany Denmark	Developed a technique for producing monoclonal antibodies (single pure antibodies)
Susumu Tonegawa	1987	Japan	Described the genetics of antibody production
Johann Deisenhofer, Robert Huber, and Hartmut Michel	1988	Germany	Described the structure of bacterial photosynthetic pigments
J. Michael Bishop and Harold E. Varmus	1989	United States	Discovered cancer-causing genes called oncogenes
Joseph E. Murray and E. Donnall Thomas	1990	United States	Performed the first successful organ transplants by using immuno-suppressive agents
Edmond H. Fisher and Edwin G. Krebs	1992	United States	Discovered protein kinases, enzymes that regulate cell growth
Richard J. Roberts and Phillip A. Sharp	1993	Great Britain United States	Discovered that a gene can be separated onto different segments of DNA
Kary B. Mullis	1993	United States	Discovered the polymerase chain reaction to amplify (make multiple copies of) DNA
Peter C. Doherty and Rolf M. Zinkernagel	1996	Australia Switzerland	Discovered how cytotoxic T cells recognize virus-infected cells prior to destroying them
Stanley B. Prusiner	1997	United States	Discovered and named proteinaceous infectious particles (prions) and demonstrated a relationship between prions and deadly neurological diseases in humans and animals
Peter Agre and Roderick MacKirron	2003	United States	Discovered water and ion channels in plasma membranes
Aaron Ciechanover, Avram Hershko, and Irwin Rose	2004	Israel Israel United States	Discovered how cells dispose of unwanted proteins in proteasomes
Barry Marshall and J. Robin Warren	2005	Australia	Discovered that *Helicobacter pylori* causes peptic ulcers
Andrew Fire and Craig Mello	2006	United States	Discovered RNA interference (RNAi), or gene silencing, by double-stranded RNA
Harald zur Hausen	2008	Germany	Discovered that human papilloma viruses cause cervical cancer
Françoise Barré-Sinoussi and Luc Montagnier	2008	France	Discovered human immunodeficiency virus (HIV)

replication of viruses and have triggered considerable research related to the treatment of viral diseases and cancer. One of today's biggest challenges for immunologists is learning how the immune system might be stimulated to ward off the virus responsible for AIDS, a disease that destroys the immune system.

A major advance in immunology occurred in 1933, when Rebecca Lancefield proposed that streptococci be classified according to serotypes (variants within a species) based on certain components in the cell walls of the bacteria. Streptococci are responsible for a variety of diseases, such as sore throat (strep throat), streptococcal toxic shock, and septicemia (blood poisoning). Her research permits the rapid identification of specific pathogenic streptococci based on immunological techniques.

Virology

The study of viruses, **virology,** actually originated during the Golden Age of Microbiology. In 1892, Dmitri Iwanowski reported that the organism that caused mosaic disease of tobacco was so small that it passed through filters fine enough to stop all known bacteria. At the time, Iwanowski was not aware that the organism in question was a virus in the sense that we now understand the term. In 1935, Wendell Stanley demonstrated that the organism, called tobacco mosaic virus (TMV), was fundamentally different from other microbes and so simple and homogeneous that it could be crystallized like a chemical compound. Stanley's work facilitated the study of viral structure and chemistry. Since the development of the electron microscope in the 1940s, microbiologists have been able to observe the structure of viruses in detail, and today much is known about their structure and activity.

Recombinant DNA Technology

Microorganisms can now be genetically modified to manufacture large amounts of human hormones and other urgently needed medical substances. In the late 1960s, Paul Berg showed that fragments of human or animal DNA (genes) that code for important proteins can be attached to bacterial DNA. The resulting hybrid was the first example of **recombinant DNA.** When recombinant DNA is inserted into bacteria (and other microbes), it can be used to make large quantities of the desired protein. The technology that developed from this technique is called **recombinant DNA technology,** and it had its origins in two related fields. The first, **microbial genetics,** studies the mechanisms by which microorganisms inherit traits. The second, **molecular biology,** specifically studies how genetic information is carried in molecules of DNA and how DNA directs the synthesis of proteins.

Although molecular biology encompasses all organisms, much of our knowledge of how genes determine specific traits has been revealed through experiments with bacteria. Until the 1930s, all genetic research was based on the study of plant and animal cells. But in the 1940s, scientists turned to unicellular organisms, primarily bacteria, which have several advantages for genetic and biochemical research. For one thing, bacteria are less complex than plants and animals. For another, the life cycles of many bacteria require less than an hour, so scientists can cultivate very large numbers of individuals for study in a relatively short time.

Once science turned to the study of unicellular life, progress in genetics began to occur rapidly. In 1941, George W. Beadle and Edward L. Tatum demonstrated the relationship between genes and enzymes. DNA was established as the hereditary material in 1944 by Oswald Avery, Colin MacLeod, and Maclyn McCarty. In 1946, Joshua Lederberg and Edward L. Tatum discovered that genetic material could be transferred from one bacterium to another by a process called conjugation. Then, in 1953, James Watson and Francis Crick proposed a model for the structure and replication of DNA. The early 1960s witnessed a further explosion of discoveries relating to the way DNA controls protein synthesis. In 1961, François Jacob and Jacques Monod discovered messenger RNA (ribonucleic acid), a chemical involved in protein synthesis, and later they made the first major discoveries about the regulation of gene function in bacteria. During the same period, scientists were able to break the genetic code and thus understand how the information for protein synthesis in messenger RNA is translated into the amino acid sequence for making proteins.

CHECK YOUR UNDERSTANDING

✓ Define *bacteriology, mycology, parasitology, immunology,* and *virology.* **1-12**

✓ Differentiate microbial genetics from molecular biology. **1-13**

Microbes and Human Welfare

LEARNING OBJECTIVES

1-14 List at least four beneficial activities of microorganisms.

1-15 Name two examples of biotechnology that use recombinant DNA technology and two examples that do not.

As mentioned earlier, only a minority of all microorganisms are pathogenic. Microbes that cause food spoilage, such as soft spots on fruits and vegetables, decomposition of meats, and rancidity of fats and oils, are also a minority. The vast majority of microbes benefit humans, other animals, and plants in many ways. The following sections outline some of these beneficial activities. In later chapters, we will discuss these activities in greater detail.

Recycling Vital Elements

Discoveries made by two microbiologists in the 1880s have formed the basis for today's understanding of the biogeochemical cycles that support life on Earth. Martinus Beijerinck and Sergei Winogradsky were the first to show how bacteria help recycle vital elements between the soil and the atmosphere. **Microbial ecology,** the study of the relationship between microorganisms and their environment, originated with the work of Beijerinck and Winogradsky. Today, microbial ecology has branched out and includes the study of how microbial populations interact with plants and animals in various environments. Among the concerns of microbial ecologists are water pollution and toxic chemicals in the environment.

The chemical elements carbon, nitrogen, oxygen, sulfur, and phosphorus are essential for life and abundant, but not necessarily in forms that organisms can use. Microorganisms are primarily responsible for converting these elements into forms that plants and animals can use. Microorganisms, primarily bacteria and fungi, play a key role in returning carbon dioxide to the atmosphere when they decompose organic wastes and dead plants and animals. Algae, cyanobacteria, and higher plants use the carbon dioxide during photosynthesis to produce carbohydrates for animals, fungi, and bacteria. Nitrogen is abundant in the atmosphere but in that form is not usable by plants and animals. Only bacteria can naturally convert atmospheric nitrogen to a form available to plants and animals.

Sewage Treatment: Using Microbes to Recycle Water

Our society's growing awareness of the need to preserve the environment has made people more conscious of the responsibility to recycle precious water and prevent the pollution of rivers and oceans. One major pollutant is sewage, which consists of human excrement, waste water, industrial wastes, and surface runoff. Sewage is about 99.9% water, with a few hundredths of 1% suspended solids. The remainder is a variety of dissolved materials.

Sewage treatment plants remove the undesirable materials and harmful microorganisms. Treatments combine various physical processes with the action of beneficial microbes. Large solids such as paper, wood, glass, gravel, and plastic are removed from sewage; left behind are liquid and organic materials that bacteria convert into such by-products as carbon dioxide, nitrates, phosphates, sulfates, ammonia, hydrogen sulfide, and methane. (We will discuss sewage treatment in detail in Chapter 27.)

Bioremediation: Using Microbes to Clean Up Pollutants

In 1988, scientists began using microbes to clean up pollutants and toxic wastes produced by various industrial processes. For example, some bacteria can actually use pollutants as energy sources; others produce enzymes that break down toxins into less harmful substances. By using bacteria in these ways—a process known as **bioremediation**—toxins can be removed from underground wells, chemical spills, toxic waste sites, and oil spills, such as the *Exxon Valdez* disaster of 1989 (see the box in Chapter 2, page 33). In addition, bacterial enzymes are used in drain cleaners to remove clogs without adding harmful chemicals to the environment. In some cases, microorganisms indigenous to the environment are used; in others, genetically modified microbes are used. Among the most commonly used microbes are certain species of bacteria of the genera *Pseudomonas* (sū-dō-mō'nas) and *Bacillus* (bä-sil'lus). *Bacillus* enzymes are also used in household detergents to remove spots from clothing.

Insect Pest Control by Microorganisms

Besides spreading diseases, insects can cause devastating crop damage. Insect pest control is therefore important for both agriculture and the prevention of human disease.

The bacterium *Bacillus thuringiensis* (thůr-in-jē-en'sis) has been used extensively in the United States to control such pests as alfalfa caterpillars, bollworms, corn borers, cabbageworms, tobacco budworms, and fruit tree leaf rollers. It is incorporated into a dusting powder that is applied to the crops these insects eat. The bacteria produce protein crystals that are toxic to the digestive systems of the insects. The toxin gene has been inserted into some plants to make them insect resistant.

By using microbial rather than chemical insect control, farmers can avoid harming the environment. Many chemical insecticides, such as DDT, remain in the soil as toxic pollutants and are eventually incorporated into the food chain.

Modern Biotechnology and Recombinant DNA Technology

Earlier, we touched on the commercial use of microorganisms to produce some common foods and chemicals. Such practical applications of microbiology are called **biotechnology.** Although biotechnology has been used in some form for centuries, techniques have become much more sophisticated in the past few decades. In the last several years, biotechnology has undergone a revolution through the advent of recombinant DNA technology to expand the potential of bacteria, viruses, and yeast cells and other fungi as miniature biochemical factories. Cultured plant and animal cells, as well as intact plants and animals, are also used as recombinant cells and organisms.

The applications of recombinant DNA technology are increasing with each passing year. Recombinant DNA techniques have been used thus far to produce a number of natural proteins, vaccines, and enzymes. Such substances have great potential for medical use; some of them are described in Table 9.1 on page 249.

A very exciting and important outcome of recombinant DNA techniques is **gene therapy**—inserting a missing gene or replacing a defective one in human cells. This technique uses a harmless virus to carry the missing or new gene into certain host cells, where the gene is picked up and inserted into the appropriate chromosome. Since 1990, gene therapy has been used to treat patients with adenosine deaminase (ADA) deficiency, a cause of severe combined immunodeficiency disease (SCID), in which cells of the immune system are inactive or missing; Duchenne's muscular dystrophy, a muscle-destroying disease; cystic fibrosis, a disease of the secreting portions of the respiratory passages, pancreas, salivary glands, and sweat glands; and LDL-receptor deficiency, a condition in which low-density lipoprotein (LDL) receptors are defective and LDL cannot enter cells. The LDL remains in the blood in high concentrations and increases the risk of atherosclerosis and coronary artery disease because it leads to fatty plaque formation in blood vessels. Results are still being evaluated. Other genetic diseases may also be treatable by gene therapy in the future, including hemophilia, an inability of the blood to clot normally; diabetes, elevated blood sugar levels; sickle cell disease, an abnormal kind of hemoglobin; and one type of hypercholesterolemia, high blood cholesterol.

Beyond medical applications, recombinant DNA techniques have also been applied to agriculture. For example, genetically altered strains of bacteria have been developed to protect fruit against frost damage, and bacteria are being modified to control insects that damage crops. Recombinant DNA has also been used to improve the appearance, flavor, and shelf life of fruits and vegetables. Potential agricultural uses of recombinant DNA include drought resistance, resistance to insects and microbial diseases, and increased temperature tolerance in crops.

CHECK YOUR UNDERSTANDING

- ✓ Name two beneficial uses of bacteria. **1-14**
- ✓ Differentiate biotechnology from recombinant DNA technology. **1-15**

Microbes and Human Disease

LEARNING OBJECTIVES

1-16 Define *normal microbiota* and *resistance.*

1-17 Define *biofilm.*

1-18 Define *emerging infectious disease.*

Normal Microbiota

We all live from birth until death in a world filled with microbes, and we all have a variety of microorganisms on and inside our bodies. These microorganisms make up our **normal microbiota,** or *flora** (**Figure 1.7**). The normal microbiota not only do us no harm, but also in some cases can actually benefit us. For example, some normal microbiota protect us against disease by preventing the overgrowth of harmful microbes, and others produce useful substances such as vitamin K and some B vitamins. Unfortunately, under some circumstances normal microbiota can make us sick or infect people we contact. For instance, when some normal microbiota leave their habitat, they can cause disease.

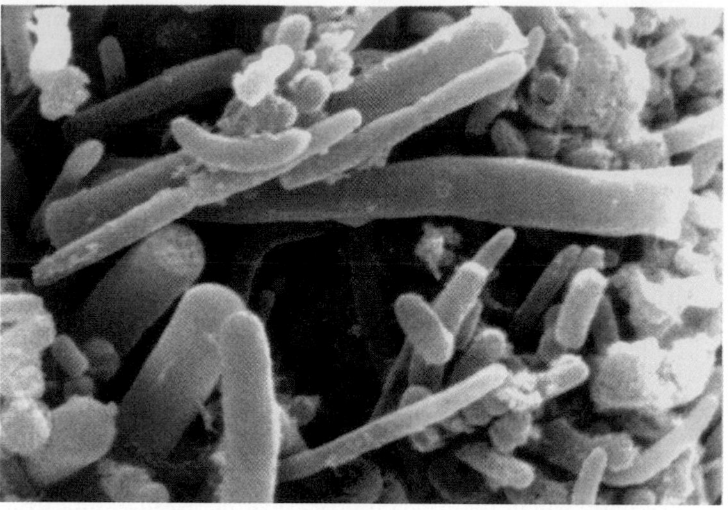

Figure 1.7 Several types of bacteria found as part of the normal microbiota on the surface of the human tongue.

Q How are normal microbiota beneficial?

When is a microbe a welcome part of a healthy human, and when is it a harbinger of disease? The distinction between health and disease is in large part a balance between the natural defenses of the body and the disease-producing properties of microorganisms. Whether our bodies overcome the offensive tactics of a particular microbe depends on our **resistance**—the ability to ward off diseases. Important resistance is provided by the barrier of the skin, mucous membranes, cilia, stomach acid, and antimicrobial chemicals such as interferons. Microbes can be destroyed by white blood cells, by the inflammatory response, by fever, and by specific responses of our immune system. Sometimes, when our natural defenses are not strong enough to overcome an invader, they have to be supplemented by antibiotics or other drugs.

Biofilms

In nature, microorganisms may exist as single cells that float or swim independently in a liquid, or they may attach to each other and/or some usually solid surface. This latter mode of behavior is called a **biofilm,** a complex aggregation of microbes. The slime covering a rock in a lake is a biofilm. Use your tongue to feel the biofilm on your teeth. Biofilms can be beneficial. They protect your mucous membranes from harmful microbes, and biofilms in lakes are an important food for aquatic animals. Biofilms can also be harmful. They can clog water pipes, and on medical implants such as joint prostheses and catheters (**Figure 1.8**), they can cause such infections as endocarditis (inflammation of

*At one time, bacteria and fungi were thought to be plants, and thus the term *flora* was used.

the heart.) Bacteria in biofilms are often resistant to antibiotics because the biofilm offers a protective barrier. See the box in Chapter 3 on page 57. Biofilms will be discussed in Chapter 6.

Infectious Diseases

An **infectious disease** is a disease in which pathogens invade a susceptible host, such as a human or an animal. In the process, the pathogen carries out at least part of its life cycle inside the host, and disease frequently results. By the end of World War II, many people believed that infectious diseases were under control. They thought malaria would be eradicated through the use of the insecticide DDT to kill mosquitoes, that a vaccine would prevent diphtheria, and that improved sanitation measures would help prevent cholera transmission. Malaria is far from eliminated. Since 1986, local outbreaks have been identified in New Jersey, California, Florida, New York, and Texas, and the disease infects 300 million people worldwide. In 1994, diphtheria appeared in the United States, brought by travelers from the newly independent states of the former Soviet Union, which were experiencing a massive diphtheria epidemic. The epidemic was brought under control in 1998. Cholera outbreaks still occur in less-developed parts of the world.

Emerging Infectious Diseases

These recent outbreaks point to the fact that infectious diseases not only are not disappearing, but seem to be reemerging and increasing. In addition, a number of new diseases—**emerging infectious diseases (EIDs)**—have cropped up in recent years. These are diseases that are new or changing and are increasing or have the potential to increase in incidence in the near future. Some of the factors that have contributed to the development of EIDs are evolutionary changes in existing organisms (*Vibrio cholerae* O139; vib′rē-ō kol′-er-ī); the spread of known diseases to new geographic regions or populations by modern transportation (West Nile virus); and increased human exposure to new, unusual infectious agents in areas that are undergoing ecologic changes such as deforestation and construction (e.g., Venezuelan hemorrhagic virus). EIDs also develop as a result of antimicrobial resistance (vancomycin-resistant *S. aureus*). An increasing number of incidents in recent years highlights the extent of the problem.

Avian influenza A (H5N1), or **bird flu,** caught the attention of the public in 2003, when it killed millions of poultry and 24 people in eight countries in southeast Asia. Avian influenza viruses occur in birds worldwide. Certain wild birds, particularly waterfowl, do not get sick but carry the virus in their intestines and shed it in saliva, nasal secretions, and feces. Most often, the wild birds spread influenza to domesticated birds, in which the virus causes death.

Influenza A viruses are found in many different animals, including ducks, chickens, pigs, whales, horses, and seals.

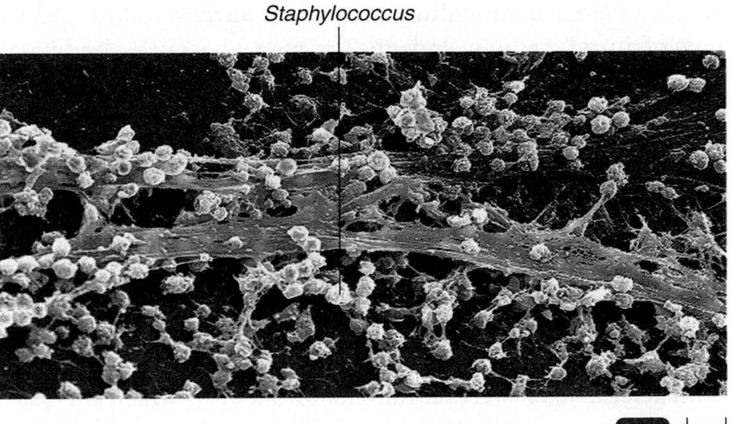

Figure 1.8 Biofilm on a catheter. *Staphylococcus* bacteria stick to solid surfaces, forming a slimy layer. Bacteria that break away from this biofilm can cause infections.

Q Why don't antibiotics kill these bacteria?

Normally, each subtype of influenza A virus is specific to certain species. However, influenza A viruses normally seen in one species sometimes can cross over and cause illness in another species, and all subtypes of influenza A virus can infect pigs. Although it is unusual for people to get influenza infections directly from animals, sporadic human infections and outbreaks caused by certain avian influenza A viruses and pig influenza viruses have been reported. As of 2008, avian influenza had sickened 242 people, and about half of them died. Fortunately, the virus has not yet evolved to be transmitted successfully among humans.

Human infections with avian influenza viruses detected since 1997 have not resulted in sustained human-to-human transmission. However, because influenza viruses have the potential to change and gain the ability to spread easily between people, monitoring for human infection and person-to-person transmission is important (see the box in Chapter 13 on page 370). The U.S. Food and Drug Administration (FDA) approved a human vaccine against the avian influenza virus in April 2007.

Antibiotics are critical in treating bacterial infections. However, years of overuse and misuse of these drugs have created environments in which antibiotic-resistant bacteria thrive. Random mutations in bacterial genes can make a bacterium resistant to an antibiotic. In the presence of that antibiotic, this bacterium has an advantage over other, susceptible bacteria and is able to proliferate. Antibiotic-resistant bacteria have become a global health crisis.

Staphylococcus aureus causes a wide range of human infections from pimples and boils to pneumonia, food poisoning, and surgical wound infections, and it is a significant cause of hospital-associated infections. After penicillin's initial success in treating *S. aureus* infection, penicillin-resistant *S. aureus* became a major threat in hospitals in the 1950s, requiring the use of methicillin.

In the 1980s, **methicillin-resistant *S. aureus,*** called **MRSA,** emerged and became endemic in many hospitals, leading to increasing use of vancomycin. In the late 1990s, *S. aureus* infections that were less sensitive to vancomycin (**vancomycin-intermediate *S. aureus,*** or **VISA**) were reported. In 2002, an infection caused by **vancomycin-resistant *S. aureus* (VRSA)** in a patient in the United States was reported.

Q&A The antibacterial substances added to various household cleaning products are similar to antibiotics in many ways. When used correctly, they inhibit bacterial growth. However, wiping every household surface with these antibacterial agents creates an environment in which the resistant bacteria survive. Unfortunately, when you really need to disinfect your homes and hands—for example, when a family member comes home from a hospital and is still vulnerable to infection—you may encounter mainly resistant bacteria.

Routine housecleaning and hand-washing are necessary, but standard soaps and detergents (without added antibacterials) are fine for these tasks. In addition, quickly evaporating chemicals, such as chlorine bleach, alcohol, ammonia, and hydrogen peroxide, remove potentially pathogenic bacteria but do not leave residues that select for the growth of resistant bacteria.

West Nile encephalitis (WNE) is inflammation of the brain caused by West Nile virus. WNE was first diagnosed in the West Nile region of Uganda in 1937. In 1999 the virus made its first North American appearance in humans in New York City. In 2007, West Nile virus infected over 3600 people in 43 states. West Nile virus is now established in nonmigratory birds in 47 states. The virus, which is carried by birds, is transmitted between birds—and to horses and humans—by mosquitoes. West Nile virus may have arrived in the United States in an infected traveler or in migratory birds.

In 1996, countries worldwide were refusing to import beef from the United Kingdom, where hundreds of thousands of cattle born after 1988 had to be killed because of an epidemic of **bovine spongiform encephalopathy** (en-sef-a-lop′a-thē), also called **BSE** or **mad cow disease.** BSE first came to the attention of microbiologists in 1986 as one of a handful of diseases caused by an infectious protein called a *prion.* Studies suggest that the source of disease was cattle feed prepared from sheep infected with their own version of the disease. Cattle are herbivores (plant-eaters), but adding protein to their feed improves their growth and health. **Creutzfeldt-Jakob disease** (kroits′felt yä′kôb), or **CJD,** is a human disease also caused by a prion. The incidence of CJD in the United Kingdom is similar to the incidence in other countries. However, by 2005 the United Kingdom reported 154 human cases of CJD caused by a new variant related to the bovine disease (see Chapter 22).

Escherichia coli is a normal inhabitant of the large intestine of vertebrates, including humans, and its presence is beneficial because it helps produce certain vitamins and breaks down otherwise undigestible foodstuffs (see Chapter 25). However, a strain called ***E. coli* O157:H7** causes bloody diarrhea when it grows in the intestines. This strain was first recognized in 1982 and since then has emerged as a public health problem. It is now one of the leading causes of diarrhea worldwide. In 1996, some 9000 people in Japan became ill, and 7 died, as a result of infection by *E. coli* O157:H7. The recent outbreaks of *E. coli* O157:H7 in the United States, associated with contamination of undercooked meat and unpasteurized beverages, have led public health officials to call for the development of new methods of testing for bacteria in food.

In 1995, infections of so-called **flesh-eating bacteria** were reported on the front pages of major newspapers. The bacteria are more correctly named invasive group A *Streptococcus* (strep-tō-kok′kus), or IGAS. Rates of IGAS in the United States, Scandinavia, England, and Wales have been increasing.

In 1995, a hospital laboratory technician in Democratic Republic of Congo (DROC) who had fever and bloody diarrhea underwent surgery for a suspected perforated bowel. Afterward he started hemorrhaging, and his blood began clotting in his blood vessels. A few days later, health care workers in the hospital where he was staying developed similar symptoms. One of them was transferred to a hospital in a different city; personnel in the second hospital who cared for this patient also developed symptoms. By the time the epidemic was over, 315 people had contracted **Ebola hemorrhagic fever** (hem-ôr-raj′ik), or **EHF,** and over 75% of them died. The epidemic was controlled when microbiologists instituted training on the use of protective equipment and educational measures in the community. Close personal contact with infectious blood or other body fluids or tissue (see Chapter 23) leads to human-to-human transmission.

Microbiologists first isolated Ebola viruses from humans during earlier outbreaks in DROC in 1976. (The virus is named after Congo's Ebola River.) In 2008, an Ebola virus outbreak occurred in Uganda with 149 cases. In 1989 and 1996, outbreaks among monkeys imported into the United States from the Philippines were caused by another Ebola virus but were not associated with human disease.

Recorded cases of **Marburg virus,** another hemorrhagic fever virus, are rare. The first cases were laboratory workers in Europe who handled African green monkeys from Uganda. Four outbreaks were identified in Africa between 1975 and 1998, involving 2 to 154 people with 56% mortality. In 2004, an outbreak killed 227 people. Microbiologists have been studying many animals but have not yet discovered the natural reservoir (source) of EHF and Marburg viruses.

In 1993, an outbreak of **cryptosporidiosis** (krip-tō-spô-rid-ē-ō′sis) transmitted through the public water supply in Milwaukee, Wisconsin, resulted in diarrheal illness in an estimated 403,000 persons. The microorganism responsible for this outbreak was the protozoan *Cryptosporidium* (krip-tō-spô-ri′dē-um). First reported as a cause of human disease in 1976, it is responsible for up to 30% of the diarrheal illness in developing countries. In the United States, transmission has occurred via drinking water, swimming pools, and contaminated hospital supplies.

AIDS (acquired immunodeficiency syndrome) first came to public attention in 1981 with reports from Los Angeles that a few young homosexual men had died of a previously rare type of pneumonia known as *Pneumocystis* (nü-mō-sis′tis) pneumonia. These men had experienced a severe weakening of the immune system, which normally fights infectious diseases. Soon these cases were correlated with an unusual number of occurrences of a rare form of cancer, Kaposi's sarcoma, among young homosexual men. Similar increases in such rare diseases were found among hemophiliacs and intravenous drug users.

Researchers quickly discovered that the cause of AIDS was a previously unknown virus (see Figure 1.1e). The virus, now called **human immunodeficiency virus (HIV),** destroys $CD4^+$ T cells, one type of white blood cell important to immune system defenses. Sickness and death result from microorganisms or cancerous cells that might otherwise have been defeated by the body's natural defenses. So far, the disease has been inevitably fatal once symptoms develop.

By studying disease patterns, medical researchers found that HIV could be spread through sexual intercourse, by contaminated needles, from infected mothers to their newborns via breast milk, and by blood transfusions—in short, by the transmission of body fluids from one person to another. Since 1985, blood used for transfusions has been carefully checked for the presence of HIV, and it is now quite unlikely that the virus can be spread by this means.

By the end of 2007, over 1 million people in the United States had been diagnosed with AIDS, and over 50% of them had died as a result of the disease. A great many more people had tested positive for the presence of HIV in their blood. As of 2007, health officials estimated that 1.2 million Americans have HIV infection. In 2007, the World Health Organization (WHO) estimated that over 30 million people worldwide are living with HIV/AIDS and that 14,000 new infections occur every day.

Since 1994, new treatments have extended the life span of people with AIDS; however, approximately 40,000 new cases occur annually in the United States. The majority of individuals with AIDS are in the sexually active age group. Because heterosexual partners of AIDS sufferers are at high risk of infection, public health officials are concerned that even more women and minorities will contract AIDS. In 1997, HIV diagnoses began increasing among women and minorities. Among the AIDS cases reported in 2005, 26% were women, and 49% were African American.

In the months and years to come, scientists will continue to apply microbiological techniques to help them learn more about the structure of the deadly HIV, how it is transmitted, how it grows in cells and causes disease, how drugs can be directed against it, and whether an effective vaccine can be developed. Public health officials have also focused on prevention through education.

AIDS poses one of this century's most formidable health threats, but it is not the first serious epidemic of a sexually transmitted disease. Syphilis was also once a fatal epidemic disease. As recently as 1941, syphilis caused an estimated 14,000 deaths per year in the United States. With few drugs available for treatment and no vaccines to prevent it, efforts to control the disease focused mainly on altering sexual behavior and on the use of condoms. The eventual development of drugs to treat syphilis contributed significantly to preventing the spread of the disease. According to the Centers for Disease Control and Prevention (CDC), reported cases of syphilis dropped from a record high of 575,000 in 1943 to an all-time low of 5979 cases in 2004. Since then, however, the number of cases has been increasing.

Just as microbiological techniques helped researchers in the fight against syphilis and smallpox, they will help scientists discover the causes of new emerging infectious diseases in the twenty-first century. Undoubtedly there will be new diseases. Ebola virus and *Influenzavirus* are examples of viruses that may be changing their abilities to infect different host species. Emerging infectious diseases will be discussed further in Chapter 14 on page 416.

Infectious diseases may reemerge because of antibiotic resistance (see the box in Chapter 26 on page 751) and through the use of microorganisms as weapons. (See the box in Chapter 23 on page 644.) The breakdown of public health measures for previously controlled infections has resulted in unexpected cases of tuberculosis, whooping cough, and diphtheria (see Chapter 24).

CHECK YOUR UNDERSTANDING

- ✓ Differentiate normal microbiota and infectious disease. **1-16**
- ✓ Why are biofilms important? **1-17**
- ✓ What factors contribute to the emergence of an infectious disease? **1-18**

* * *

The diseases we have mentioned are caused by viruses, bacteria, protozoa, and prions—types of microorganisms. This book introduces you to the enormous variety of microscopic organisms. It shows you how microbiologists use specific techniques and procedures to study the microbes that cause such diseases as AIDS and diarrhea—and diseases that have yet to be discovered. You will also learn how the body responds to microbial infection and how certain drugs combat microbial diseases. Finally, you will learn about the many beneficial roles that microbes play in the world around us.

STUDY OUTLINE

The **MyMicrobiologyPlace** website (**www.microbiologyplace.com**) will help you get ready for tests with its simple three-step approach: ❶ **take a pre-test** and obtain a personalized study plan, ❷ **learn and practice** with animations, tutorials, and MP3 tutor sessions, and ❸ **test yourself** with quizzes and a chapter post-test.

Microbes in Our Lives (p. 2)

1. Living things too small to be seen with the unaided eye are called microorganisms.
2. Microorganisms are important in maintaining Earth's ecological balance.
3. Some microorganisms live in humans and other animals and are needed to maintain good health.
4. Some microorganisms are used to produce foods and chemicals.
5. Some microorganisms cause disease.

Naming and Classifying Microorganisms (pp. 2–6)

Nomenclature (p. 2)

1. In a nomenclature system designed by Carolus Linnaeus (1735), each living organism is assigned two names.
2. The two names consist of a genus and a specific epithet, both of which are underlined or italicized.

Types of Microorganisms (pp. 3–6)

Bacteria (pp. 3–4)

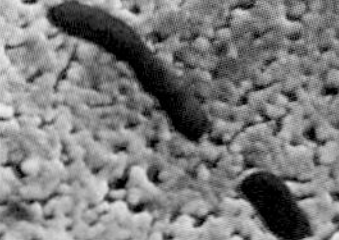

3. Bacteria are unicellular organisms. Because they have no nucleus, the cells are described as prokaryotic.
4. The three major basic shapes of bacteria are bacillus, coccus, and spiral.
5. Most bacteria have a peptidoglycan cell wall; they divide by binary fission, and they may possess flagella.
6. Bacteria can use a wide range of chemical substances for their nutrition.

Archaea (p. 4)

7. Archaea consist of prokaryotic cells; they lack peptidoglycan in their cell walls.
8. Archaea include methanogens, extreme halophiles, and extreme thermophiles.

Fungi (p. 4)

9. Fungi (mushrooms, molds, and yeasts) have eukaryotic cells (cells with a true nucleus). Most fungi are multicellular.
10. Fungi obtain nutrients by absorbing organic material from their environment.

Protozoa (pp. 4, 6)

11. Protozoa are unicellular eukaryotes.
12. Protozoa obtain nourishment by absorption or ingestion through specialized structures.

Algae (p. 6)

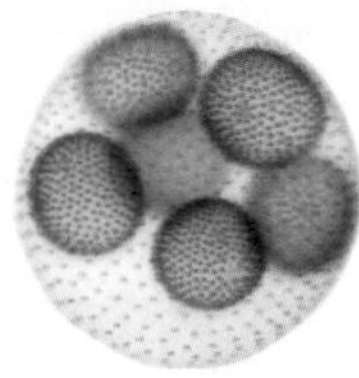

13. Algae are unicellular or multicellular eukaryotes that obtain nourishment by photosynthesis.
14. Algae produce oxygen and carbohydrates that are used by other organisms.

Viruses (p. 6)

15. Viruses are noncellular entities that are parasites of cells.
16. Viruses consist of a nucleic acid core (DNA or RNA) surrounded by a protein coat. An envelope may surround the coat.

Multicellular Animal Parasites (p. 6)

17. The principal groups of multicellular animal parasites are flatworms and roundworms, collectively called helminths.
18. The microscopic stages in the life cycle of helminths are identified by traditional microbiological procedures.

Classification of Microorganisms (p. 6)

19. All organisms are classified into Bacteria, Archaea, and Eukarya. Eukarya include protists, fungi, plants, and animals.

A Brief History of Microbiology (pp. 6–16)

The First Observations (p. 7)

1. Robert Hooke observed that cork was composed of "little boxes"; he introduced the term *cell* (1665).
2. Hooke's observations laid the groundwork for development of the cell theory, the concept that all living things are composed of cells.
3. Anton van Leeuwenhoek, using a simple microscope, was the first to observe microorganisms (1673).

The Debate over Spontaneous Generation (p. 8)

4. Until the mid-1880s, many people believed in spontaneous generation, the idea that living organisms could arise from nonliving matter.
5. Francesco Redi demonstrated that maggots appear on decaying meat only when flies are able to lay eggs on the meat (1668).
6. John Needham claimed that microorganisms could arise spontaneously from heated nutrient broth (1745).
7. Lazzaro Spallanzani repeated Needham's experiments and suggested that Needham's results were due to microorganisms in the air entering his broth (1765).
8. Rudolf Virchow introduced the concept of biogenesis: living cells can arise only from preexisting cells (1858).
9. Louis Pasteur demonstrated that microorganisms are in the air everywhere and offered proof of biogenesis (1861).
10. Pasteur's discoveries led to the development of aseptic techniques used in laboratory and medical procedures to prevent contamination by microorganisms.

The Golden Age of Microbiology (pp. 9–11)

11. The science of microbiology advanced rapidly between 1857 and 1914.

Fermentation and Pasteurization (p. 9)

12. Pasteur found that yeast ferment sugars to alcohol and that bacteria can oxidize the alcohol to acetic acid.
13. A heating process called pasteurization is used to kill bacteria in some alcoholic beverages and milk.

The Germ Theory of Disease (pp. 9, 11)

14. Agostino Bassi (1835) and Pasteur (1865) showed a causal relationship between microorganisms and disease.
15. Joseph Lister introduced the use of a disinfectant to clean surgical wounds in order to control infections in humans (1860s).
16. Robert Koch proved that microorganisms cause disease. He used a sequence of procedures, now called Koch's postulates (1876), that are used today to prove that a particular microorganism causes a particular disease.

Vaccination (p. 11)

17. In a vaccination, immunity (resistance to a particular disease) is conferred by inoculation with a vaccine.
18. In 1798, Edward Jenner demonstrated that inoculation with cowpox material provides humans with immunity to smallpox.
19. About 1880, Pasteur discovered that avirulent bacteria could be used as a vaccine for fowl cholera; he coined the word *vaccine.*
20. Modern vaccines are prepared from living avirulent microorganisms or killed pathogens, from isolated components of pathogens, and by recombinant DNA techniques.

The Birth of Modern Chemotherapy: Dreams of a "Magic Bullet" (pp. 12–13)

21. Chemotherapy is the chemical treatment of a disease.
22. Two types of chemotherapeutic agents are synthetic drugs (chemically prepared in the laboratory) and antibiotics (substances produced naturally by bacteria and fungi to inhibit the growth of other microorganisms).
23. Paul Ehrlich introduced an arsenic-containing chemical called salvarsan to treat syphilis (1910).
24. Alexander Fleming observed that the *Penicillium* fungus inhibited the growth of a bacterial culture. He named the active ingredient penicillin (1928).
25. Penicillin has been used clinically as an antibiotic since the 1940s.
26. Researchers are tackling the problem of drug-resistant microbes.

Modern Developments in Microbiology (pp. 13–16)

27. Bacteriology is the study of bacteria, mycology is the study of fungi, and parasitology is the study of parasitic protozoa and worms.
28. Microbiologists are using genomics, the study of all of an organism's genes, to classify bacteria, fungi, and protozoa.
29. The study of AIDS, analysis of the action of interferons, and the development of new vaccines are among the current research interests in immunology.
30. New techniques in molecular biology and electron microscopy have provided tools for advancing our knowledge of virology.
31. The development of recombinant DNA technology has helped advance all areas of microbiology.

Microbes and Human Welfare (pp. 16–18)

1. Microorganisms degrade dead plants and animals and recycle chemical elements to be used by living plants and animals.
2. Bacteria are used to decompose organic matter in sewage.
3. Bioremediation processes use bacteria to clean up toxic wastes.
4. Bacteria that cause diseases in insects are being used as biological controls of insect pests. Biological controls are specific for the pest and do not harm the environment.
5. Using microbes to make products such as foods and chemicals is called biotechnology.
6. Using recombinant DNA, bacteria can produce important substances such as proteins, vaccines, and enzymes.
7. In gene therapy, viruses are used to carry replacements for defective or missing genes into human cells.
8. Genetically modified bacteria are used in agriculture to protect plants from frost and insects and to improve the shelf life of produce.

Microbes and Human Disease (pp. 18–21)

1. Everyone has microorganisms in and on the body; these make up the normal microbiota, or flora.
2. The disease-producing properties of a species of microbe and the host's resistance are important factors in determining whether a person will contract a disease.
3. Bacterial communities that form slimy layers on surfaces are called biofilms.
4. An infectious disease is one in which pathogens invade a susceptible host.
5. An emerging infectious disease (EID) is a new or changing disease showing an increase in incidence in the recent past or a potential to increase in the near future.

STUDY QUESTIONS

Answers to the Review and Multiple Choice questions can be found by turning to the blue Answers tab at the back of the textbook.

Review

1. How did the idea of spontaneous generation come about?
2. Briefly state the role microorganisms play in each of the following:
   a. biological control of pests
   b. recycling of elements
   c. normal microbiota
   d. sewage treatment
   e. human insulin production
   f. vaccine production
   g. biofilms

3. Into which field of microbiology would the following scientists best fit?

Researcher Who	Field
_____ **a.** Studies biodegradation of toxic wastes	**1.** Biotechnology
_____ **b.** Studies the causative agent of Ebola hemorrhagic fever	**2.** Immunology
_____ **c.** Studies the production of human proteins by bacteria	**3.** Microbial ecology
_____ **d.** Studies the symptoms of AIDS	**4.** Microbial genetics
_____ **e.** Studies the production of toxin by *E. coli*	**5.** Microbial physiology
_____ **f.** Studies the life cycle of *Cryptosporidium*	**6.** Molecular biology
_____ **g.** Develops gene therapy for a disease	**7.** Mycology
_____ **h.** Studies the fungus *Candida albicans*	**8.** Virology

4. Match the microorganisms in column A to their descriptions in column B.

Column A	Column B
_____ **a.** Archaea	**1.** Not composed of cells
_____ **b.** Algae	**2.** Cell wall made of chitin
_____ **c.** Bacteria	**3.** Cell wall made of peptidoglycan
_____ **d.** Fungi	**4.** Cell wall made of cellulose; photosynthetic
_____ **e.** Helminths	**5.** Unicellular, complex cell structure lacking a cell wall
_____ **f.** Protozoa	**6.** Multicellular animals
_____ **g.** Viruses	**7.** Prokaryote without peptidoglycan cell wall

5. Match the people in column A to their contribution toward the advancement of microbiology, in column B.

Column A	Column B
_____ **a.** Avery, MacLeod, and McCarty	**1.** Developed vaccine against smallpox
_____ **b.** Beadle and Tatum	**2.** Discovered how DNA controls protein synthesis in a cell
_____ **c.** Berg	**3.** Discovered penicillin
_____ **d.** Ehrlich	**4.** Discovered that DNA can be transferred from one bacterium to another
_____ **e.** Fleming	**5.** Disproved spontaneous generation
_____ **f.** Hooke	**6.** First to characterize a virus
_____ **g.** Iwanowski	**7.** First to use disinfectants in surgical procedures
_____ **h.** Jacob and Monod	**8.** First to observe bacteria
_____ **i.** Jenner	**9.** First to observe cells in plant material and name them
_____ **j.** Koch	**10.** Observed that viruses are filterable
_____ **k.** Lancefield	**11.** Proved that DNA is the hereditary material
_____ **l.** Lederberg and Tatum	**12.** Proved that microorganisms can cause disease
_____ **m.** Lister	**13.** Said living cells arise from preexisting living cells
_____ **n.** Pasteur	**14.** Showed that genes code for enzymes
_____ **o.** Stanley	**15.** Spliced animal DNA to bacterial DNA
_____ **p.** van Leeuwenhoek	**16.** Used bacteria to produce acetone
_____ **q.** Virchow	**17.** Used the first synthetic chemotherapeutic agent
_____ **r.** Weizmann	**18.** Proposed a classification system for streptococci based on antigens in their cell walls

6. The genus name of a bacterium is "erwinia," and the specific epithet is "amylovora." Write the scientific name of this organism correctly. Using this name as an example, explain how scientific names are chosen.
7. It is possible to purchase the following microorganisms in a retail store. Provide a reason for buying each.
   a. *Bacillus thuringiensis*
   b. *Saccharomyces*
8. **DRAW IT** Show where airborne microbes ended up in Pasteur's experiment.

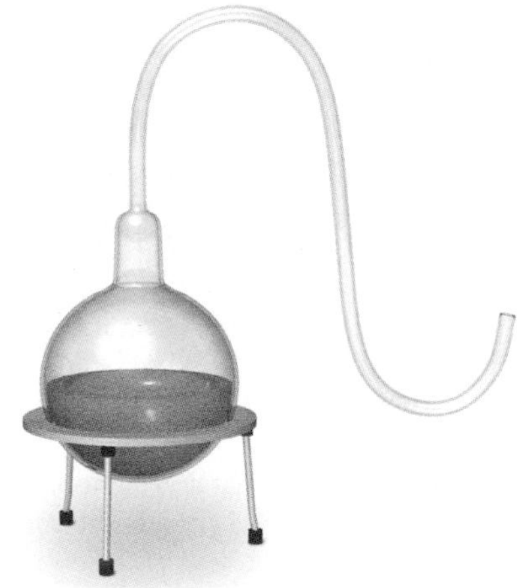

Multiple Choice

1. Which of the following is a scientific name?
   a. *Mycobacterium tuberculosis*
   b. Tubercle bacillus
2. Which of the following is *not* a characteristic of bacteria?
   a. are prokaryotic
   b. have peptidoglycan cell walls
   c. have the same shape
   d. grow by binary fission
   e. have the ability to move
3. Which of the following is the most important element of Koch's germ theory of disease? The animal shows disease symptoms when
   a. the animal has been in contact with a sick animal.
   b. the animal has a lowered resistance.
   c. a microorganism is observed in the animal.
   d. a microorganism is inoculated into the animal.
   e. microorganisms can be cultured from the animal.
4. Recombinant DNA is
   a. DNA in bacteria.
   b. the study of how genes work.
   c. the DNA resulting when genes of two different organisms are mixed.
   d. the use of bacteria in the production of foods.
   e. the production of proteins by genes.
5. Which of the following statements is the best definition of biogenesis?
   a. Nonliving matter gives rise to living organisms.
   b. Living cells can only arise from preexisting cells.
   c. A vital force is necessary for life.
   d. Air is necessary for living organisms.
   e. Microorganisms can be generated from nonliving matter.

6. Which of the following is a beneficial activity of microorganisms?
   a. Some microorganisms are used as food for humans.
   b. Some microorganisms use carbon dioxide.
   c. Some microorganisms provide nitrogen for plant growth.
   d. Some microorganisms are used in sewage treatment processes.
   e. all of the above
7. It has been said that bacteria are essential for the existence of life on Earth. Which of the following would be the essential function performed by bacteria?
   a. control insect populations
   b. directly provide food for humans
   c. decompose organic material and recycle elements
   d. cause disease
   e. produce human growth hormones such as insulin
8. Which of the following is an example of bioremediation?
   a. application of oil-degrading bacteria to an oil spill
   b. application of bacteria to a crop to prevent frost damage
   c. fixation of gaseous nitrogen into usable nitrogen
   d. production by bacteria of a human protein such as interferon
   e. all of the above
9. Spallanzani's conclusion about spontaneous generation was challenged because Lavoisier had just shown that oxygen was the vital component of air. Which of the following statements is true?
   a. All life requires air.
   b. Only disease-causing organisms require air.
   c. Some microbes do not require air.
   d. Pasteur kept air out of his biogenesis experiments.
   e. Lavoisier was mistaken.
10. Which of the following statements about *E. coli* is *not* true?
    a. *E. coli* was the first disease-causing bacterium identified by Koch.
    b. *E. coli* is part of the normal microbiota of humans.
    c. *E. coli* is beneficial in human intestines.
    d. A disease-causing strain of *E. coli* causes bloody diarrhea.
    e. none of the above

Critical Thinking

1. How did the theory of biogenesis lead the way for the germ theory of disease?
2. Even though the germ theory of disease was not demonstrated until 1876, why did Semmelweis (1840) and Lister (1867) argue for the use of aseptic techniques?
3. Find at least three supermarket products made by microorganisms. (*Hint:* The label will state the scientific name of the organism or include the word *culture, fermented,* or *brewed.*)
4. People once believed all microbial diseases would be controlled by the twenty-first century. Name one emerging infectious disease. List three reasons why we are identifying new diseases now.

Clinical Applications

1. The prevalence of arthritis in the United States is 1 in 100,000 children. However, 1 in 10 children in Lyme, Connecticut, developed arthritis between June and September in 1973. Allen Steere, a rheumatologist at Yale University, investigated the cases in Lyme and found that 25% of the patients remembered having a skin rash during their arthritic episode and that the disease was treatable with penicillin. Steere concluded that this was a new infectious disease and did not have an environmental, genetic, or immunologic cause.
   a. What was the factor that caused Steere to reach his conclusion?
   b. What is the disease?
   c. Why was the disease more prevalent between June and September?
2. In 1864, Lister observed that patients recovered completely from simple fractures, but that compound fractures had "disastrous consequences." He knew that the application of phenol (carbolic acid) to fields in the town of Carlisle prevented cattle disease. In 1864, Lister treated compound fractures with phenol, and his patients recovered without complications. How was Lister influenced by Pasteur's work? Why was Koch's work still needed?

2 Chemical Principles

We can see a tree rot and smell milk going sour, but we might not realize what is happening on a microscopic level. In both cases, microbes are conducting chemical operations. The tree rots when microorganisms decompose the wood. Milk turns sour from the production of lactic acid by bacteria. Most of the activities of microorganisms are the result of a series of chemical reactions.

Like all organisms, microorganisms use nutrients to make chemical building blocks for growth and other functions essential to life. For most microorganisms, synthesizing these building blocks requires them to break down nutrient substances and use the energy released to assemble the resulting molecular fragments into new substances.

The chemistry of microbes is one of the most important concerns of microbiologists. Knowledge of chemistry is essential to understanding what roles microorganisms play in nature, how they cause disease, how methods for diagnosing disease are developed, how the body's defenses combat infection, and how antibiotics and vaccines are produced to combat the harmful effects of microbes. To understand the changes that occur in microorganisms and the changes microbes make in the world around us, we need to know how molecules are formed and how they interact.

UNDER THE MICROSCOPE

Salmonella typhimurium. This *Salmonella* bacterium is entering a human epithelial cell.

Q&A

Salmonella release a regulatory molecule containing amino acids and phosphate that causes a human cell's cytoskeleton to change shape, thereby allowing the bacterium to enter the cell. What type of chemical is this regulatory molecule?

Look for the answer in the chapter.

The Structure of Atoms

LEARNING OBJECTIVE

2-1 Describe the structure of an atom and its relation to the physical properties of elements.

All matter—whether air, rock, or a living organism—is made up of small units called atoms. An **atom** is the smallest component of a pure substance that exhibits physical and chemical properties of that substance; an atom cannot be subdivided into smaller substances without losing its properties. Atoms interact with each other in certain combinations to form **molecules.** Living cells are made up of molecules, some of which are very complex. The science of the interaction between atoms and molecules is called **chemistry.**

Atoms are the smallest units of matter that enter into chemical reactions. Every atom has a centrally located **nucleus** and particles called **electrons** that move around the nucleus in patterns known as electronic configurations (Figure 2.1). The nuclei of most atoms are stable—that is, they do not change spontaneously—and nuclei do not participate in chemical reactions. The nucleus is made up of positively (+) charged particles called **protons** and uncharged (neutral) particles called **neutrons.** The nucleus, therefore, bears a net positive charge. A **charge** is a property of some subatomic particles that produces an attractive or repulsive force between them; particles of opposite charge attract each other, and particles of the same charge repel each other. Neutrons and protons have approximately the same weight, which is about 1840 times that of an electron. The charge on electrons is negative (−), and in all atoms the number of electrons is equal to the number of protons. Because the total positive charge of the nucleus equals the total negative charge of the electrons, each atom is electrically neutral.

The number of protons in an atomic nucleus ranges from one (in a hydrogen atom) to more than 100 (in the largest atoms known). Atoms are often listed by their **atomic number,** the number of protons in the nucleus. The total number of protons and neutrons in an atom is its approximate **atomic weight.**

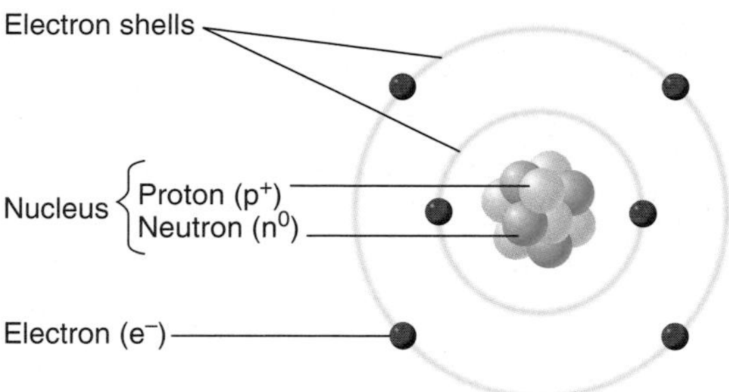

Figure 2.1 The structure of an atom. In this simplified diagram of a carbon atom, note the central location of the nucleus. The nucleus contains six neutrons and six protons, although not all the protons are visible in this view. The six electrons move about the nucleus in regions called electron shells, shown here as circles.

Q What is the atomic number of this atom?

Chemical Elements

All atoms with the same number of protons behave the same way chemically and are classified as the same **chemical element.** Each element has its own name and a one- or two-letter symbol, usually derived from the English or Latin name for the element. For example, the symbol for the element hydrogen is H, and the symbol for carbon is C. The symbol for sodium is Na—the first two letters of its Latin name, *natrium*—to distinguish it from nitrogen, N, and from sulfur, S. There are 92 naturally occurring elements. However, only about 26 elements are commonly found in living things. Table 2.1 lists some of the chemical elements found in living organisms, including their atomic numbers and weights. The elements most abundant in living matter are hydrogen, carbon, nitrogen, and oxygen.

Most elements have several **isotopes**—atoms with different numbers of neutrons in their nuclei. All isotopes of an element have the same number of protons in their nuclei, but their atomic weights differ because of the difference in the number of neutrons. For example, in a natural sample of oxygen, all the atoms contain eight protons. However, 99.76% of the atoms have eight neutrons, 0.04% contain nine neutrons, and the

Table 2.1 The Elements of Life*

Element	Symbol	Atomic Number	Approximate Atomic Weight
Hydrogen	H	1	1
Carbon	C	6	12
Nitrogen	N	7	14
Oxygen	O	8	16
Sodium	Na	11	23
Magnesium	Mg	12	24
Phosphorus	P	15	31
Sulfur	S	16	32
Chlorine	Cl	17	35
Potassium	K	19	39
Calcium	Ca	20	40
Iron	Fe	26	56
Iodine	I	53	127

*Hydrogen, carbon, nitrogen, and oxygen are the most abundant chemical elements in living organisms.

remaining 0.2% contain ten neutrons. Therefore, the three isotopes composing a natural sample of oxygen have atomic weights of 16, 17, and 18, although all will have the atomic number 8. Atomic numbers are written as a subscript to the left of an element's chemical symbol. Atomic weights are written as a superscript above the atomic number. Thus, natural oxygen isotopes are represented as ${}^{16}_{8}O$, ${}^{17}_{8}O$, and ${}^{18}_{8}O$. Isotopes of certain elements are extremely useful in biological research, medical diagnosis, the treatment of some disorders, and some forms of sterilization.

Electronic Configurations

In an atom, electrons are arranged in **electron shells,** which are regions corresponding to different **energy levels.** The arrangement is called an **electronic configuration.** Shells are layered outward from the nucleus, and each shell can hold a characteristic maximum number of electrons—two electrons in the innermost shell (lowest energy level), eight electrons in the second shell, and eight electrons in the third shell, if it is the atom's outermost (valence) shell. The fourth, fifth, and sixth electron shells can each accommodate 18 electrons, although there are some exceptions to this generalization. Table 2.2 shows the electronic configurations for atoms of some elements found in living organisms.

The outermost shell tends to be filled with the maximum number of electrons. An atom can give up, accept, or share electrons with other atoms to fill this shell. The chemical properties of atoms are largely a function of the number of electrons in the outermost electron shell. When its outer shell is filled, the atom is chemically stable, or inert: it does not tend to react with other atoms. Helium (atomic number 2) and neon (atomic number 10) are examples of atoms of inert gases whose outer shells are filled.

When an atom's outer electron shell is only partially filled, the atom is chemically unstable. Such an atom reacts with other atoms, and this reaction depends, in part, on the degree to which the outer energy levels are filled. Notice the number of electrons in the outer energy levels of the atoms in Table 2.2. We will see later how the number correlates with the chemical reactivity of the elements.

CHECK YOUR UNDERSTANDING

✓ How does ${}^{14}_{6}C$ differ from ${}^{12}_{6}C$? What is the atomic number of each carbon atom? The atomic weight? **2-1**

How Atoms Form Molecules: Chemical Bonds

LEARNING OBJECTIVE

2-2 Define *ionic bond, covalent bond, hydrogen bond, molecular weight,* and *mole.*

When the outermost energy level of an atom is not completely filled by electrons, you can think of it as having either unfilled spaces or extra electrons in that energy level, depending on whether it is easier for the atom to gain or lose electrons. For example, an atom of oxygen, with two electrons in the first energy level and six in the second, has two unfilled spaces in the second electron shell; an atom of magnesium has two extra electrons in its outermost shell. The most chemically stable configuration for any atom is to have its outermost shell filled, as do the inert gases. Therefore, for these two atoms to attain that state, oxygen must gain two electrons, and magnesium must lose two electrons. All atoms tend to combine so that the extra electrons in the outermost shell of one atom fill the spaces of the outermost shell of the other atom; for example, oxygen and magnesium combine so that the outermost shell of each atom has the full complement of eight electrons.

The **valence,** or combining capacity, of an atom is the number of extra or missing electrons in its outermost electron shell. For example, hydrogen has a valence of 1 (one unfilled space, or one extra electron), oxygen has a valence of 2 (two unfilled spaces), carbon has a valence of 4 (four unfilled spaces, or four extra electrons), and magnesium has a valence of 2 (two extra electrons).

Basically, atoms achieve the full complement of electrons in their outermost energy shells by combining to form molecules, which are made up of atoms of one or more elements. A molecule that contains at least two different kinds of atoms, such as H_2O (the water molecule), is called a **compound.** In H_2O, the subscript 2 indicates that there are two atoms of hydrogen; the absence of a subscript indicates that there is only one atom of oxygen. Molecules hold together because the valence electrons of the combining atoms form attractive forces, called **chemical bonds,** between the atomic nuclei. Therefore, valence may also be viewed as the bonding capacity of an element. Because energy is required for chemical bond formation, each chemical bond possesses a certain amount of potential chemical energy.

In general, atoms form bonds in one of two ways: by either gaining or losing electrons from their outer electron shell, or by sharing outer electrons. When atoms have gained or lost outer electrons, the chemical bond is called an ionic bond. When outer electrons are shared, the bond is called a covalent bond. Although we will discuss ionic and covalent bonds separately, the kinds of bonds actually found in molecules do not belong entirely to either category. Instead, bonds range from the highly ionic to the highly covalent.

Ionic Bonds

Atoms are electrically neutral when the number of positive charges (protons) equals the number of negative charges (electrons). But when an isolated atom gains or loses electrons, this balance is upset. If the atom gains electrons, it acquires an overall negative charge; if the atom loses electrons, it acquires an overall positive charge. Such a negatively or positively charged atom (or group of atoms) is called an **ion.**

Consider the following examples. Sodium (Na) has 11 protons and 11 electrons, with one electron in its outer electron shell.

Table 2.2 Electronic Configurations for the Atoms of Some Elements Found in Living Organisms

Element	First Electron Shell (2)*	Second Electron Shell (8)*	Third Electron Shell (8)*	Diagram	Number of Valence (Outermost) Shell Electrons	Number of Unfilled Spaces	Maximum Number of Bonds Formed
Hydrogen	1	—	—	H	1	1	1
Carbon	2	4	—	C	4	4	4
Nitrogen	2	5	—	N	5	3	5
Oxygen	2	6	—	O	6	2	2
Magnesium	2	8	2	Mg	2	6	2
Phosphorus	2	8	5	P	5	3	5
Sulfur	2	8	6	S	6	2	6

*Numbers in parentheses indicate the maximum number of electrons in their respective shells.

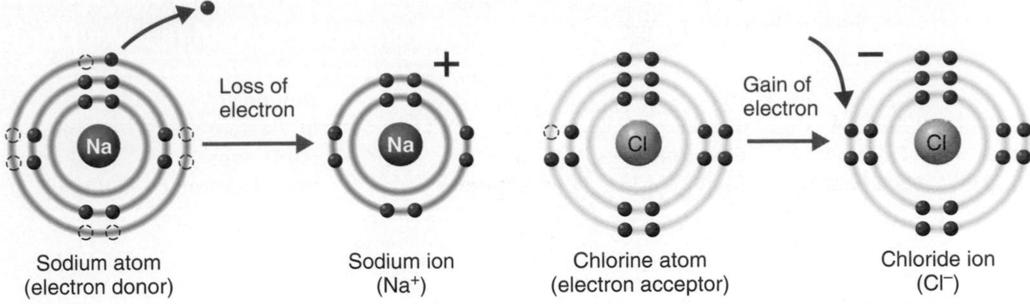

(a) A sodium atom (Na) loses one electron to an electron acceptor and forms a sodium ion (Na^+). A chlorine atom (Cl) accepts one electron from an electron donor to become a chloride ion (Cl^-).

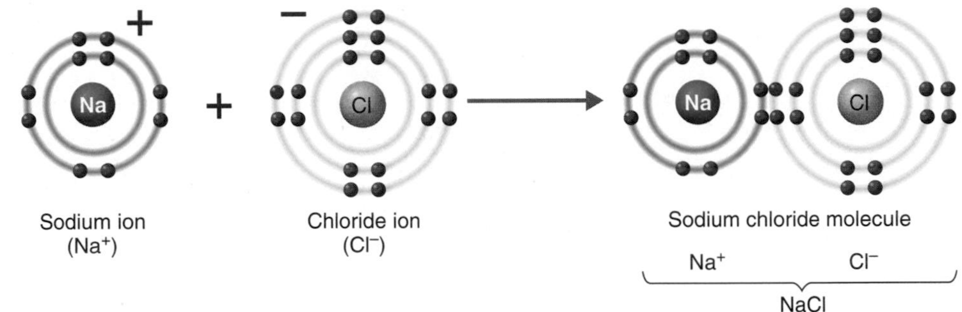

(b) The sodium and chloride ions are attracted because of their opposite charges and are held together by an ionic bond to form a molecule of sodium chloride.

Figure 2.2 Ionic bond formation.

 **What is an ionic bond?**

Sodium tends to lose the single outer electron; it is an *electron donor* (Figure 2.2a). When sodium donates an electron to another atom, it is left with 11 protons and only 10 electrons and so has an overall charge of $+1$. This positively charged sodium atom is called a sodium ion and is written as Na^+. Chlorine (Cl) has a total of 17 electrons, seven of them in the outer electron shell. Because this outer shell can hold eight electrons, chlorine tends to pick up an electron that has been lost by another atom; it is an *electron acceptor* (see Figure 2.2a). By accepting an electron, chlorine totals 18 electrons. However, it still has only 17 protons in its nucleus. The chloride ion therefore has a charge of -1 and is written as Cl^-.

The opposite charges of the sodium ion (Na^+) and chloride ion (Cl^-) attract each other. The attraction, an ionic bond, holds the two atoms together, and a molecule is formed (Figure 2.2b). The formation of this molecule, called sodium chloride (NaCl) or table salt, is a common example of ionic bonding. Thus, an **ionic bond** is an attraction between ions of opposite charge that holds them together to form a stable molecule. Put another way, an ionic bond is an attraction between atoms in which one atom loses electrons and another atom gains electrons. Strong ionic bonds, such as those that hold Na^+ and Cl^- together in salt crystals, have limited importance in living cells. But the weaker ionic bonds formed in aqueous (water) solutions are important in biochemical reactions in microbes and other organisms. For example, weaker ionic bonds assume a role in certain antigen–antibody reactions—that is, reactions in which molecules produced by the immune system (antibodies) combine with foreign substances (antigens) to combat infection.

In general, an atom whose outer electron shell is less than half-filled will lose electrons and form positively charged ions, called **cations.** Examples of cations are the potassium ion (K^+), calcium ion (Ca^{2+}), and sodium ion (Na^+). When an atom's outer electron shell is more than half-filled, the atom will gain electrons and form negatively charged ions, called **anions.** Examples are the iodide ion (I^-), chloride ion (Cl^-), and sulfide ion (S^{2-}).

Covalent Bonds

A **covalent bond** is a chemical bond formed by two atoms sharing one or more pairs of electrons. Covalent bonds are stronger and far more common in organisms than are true ionic bonds. In the hydrogen molecule, H_2, two hydrogen atoms share a pair of electrons. Each hydrogen atom has its own electron plus one electron from the other atom (Figure 2.3a). The shared pair of electrons actually orbits the nuclei of both atoms. Therefore, the outer electron shells of both atoms are filled. Atoms that

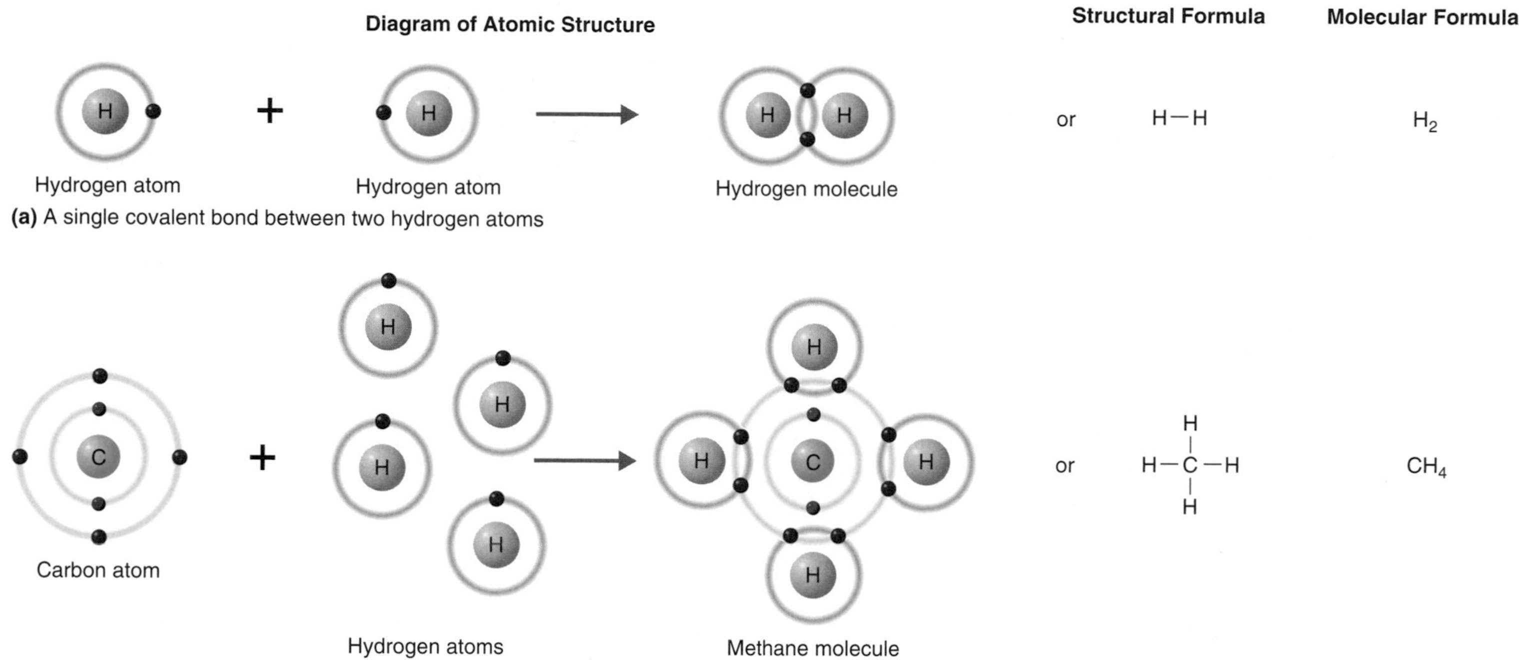

(a) A single covalent bond between two hydrogen atoms

(b) Single covalent bonds between four hydrogen atoms and a carbon atom, forming a methane molecule

Figure 2.3 Covalent bond formation. On the right are simpler ways to represent molecules. In structural formulas, each covalent bond is written as a straight line between the symbols for two atoms. In molecular formulas, the number of atoms in each molecule is noted by subscripts.

Q What is a covalent bond?

share only one pair of electrons form a *single covalent bond.* For simplicity, a single covalent bond is expressed as a single line between the atoms (H—H). Atoms that share two pairs of electrons form a *double covalent bond,* expressed as two single lines (=). A *triple covalent bond,* expressed as three single lines (≡), occurs when atoms share three pairs of electrons.

The principles of covalent bonding that apply to atoms of the same element also apply to atoms of different elements. Methane (CH_4) is an example of covalent bonding between atoms of different elements (Figure 2.3b). The outer electron shell of the carbon atom can hold eight electrons but has only four; each hydrogen atom can hold two electrons but has only one. Consequently, in the methane molecule the carbon atom gains four hydrogen electrons to complete its outer shell, and each hydrogen atom completes its pair by sharing one electron from the carbon atom. Each outer electron of the carbon atom orbits both the carbon nucleus and a hydrogen nucleus. Each hydrogen electron orbits both its own nucleus and the carbon nucleus.

Elements such as hydrogen and carbon, whose outer electron shells are half-filled, form covalent bonds quite easily. In fact, in living organisms, carbon almost always forms covalent bonds; it almost never becomes an ion. *Remember:* Covalent bonds are formed by the *sharing* of electrons between atoms. Ionic bonds are formed by *attraction* between atoms that have lost or gained electrons and are therefore positively or negatively charged.

Hydrogen Bonds

Another chemical bond of special importance to all organisms is the **hydrogen bond,** in which a hydrogen atom that is covalently bonded to one oxygen or nitrogen atom is attracted to another oxygen or nitrogen atom. Such bonds are weak and do not bind atoms into molecules. However, they do serve as bridges between different molecules or between various portions of the same molecule.

When hydrogen combines with atoms of oxygen or nitrogen, the relatively large nucleus of these larger oxygen or nitrogen atoms has more protons and attracts the hydrogen electron more strongly than does the small hydrogen nucleus. Thus, in a molecule of water (H_2O), all the electrons tend to be closer to the oxygen nucleus than to the hydrogen nuclei. As a result, the oxygen portion of the molecule has a slightly negative charge, and the hydrogen portion of the molecule has a slightly positive charge (Figure 2.4a). When the positively charged end of one molecule is attracted to the negatively charged end of another molecule, a hydrogen bond is formed (Figure 2.4b). This attraction can also occur between hydrogen and other atoms of the same molecule, especially in large molecules. Oxygen and nitrogen are the elements most frequently involved in hydrogen bonding.

Hydrogen bonds are considerably weaker than either ionic or covalent bonds; they have only about 5% of the strength of covalent bonds. Consequently, hydrogen bonds are formed and

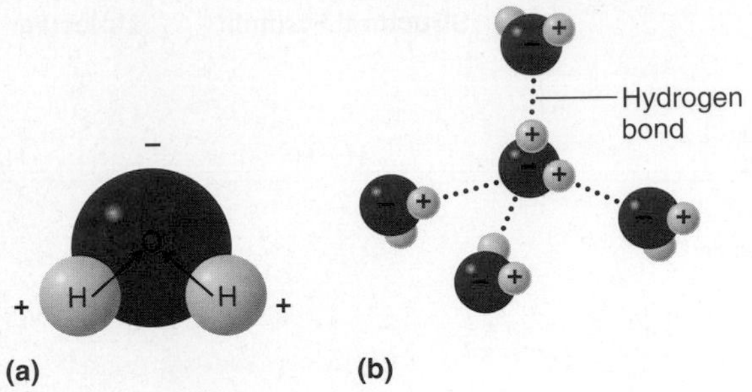

Figure 2.4 Hydrogen bond formation in water. (**a**) In a water molecule, the electrons of the hydrogen atoms are strongly attracted to the oxygen atom. Therefore, the part of the water molecule containing the oxygen atom has a slightly negative charge, and the part containing hydrogen atoms has a slightly positive charge. (**b**) In a hydrogen bond between water molecules, the hydrogen of one water molecule is attracted to the oxygen of another water molecule. Many water molecules may be attracted to each other by hydrogen bonds (black dots).

Q Which chemical elements are usually involved in hydrogen bonding?

broken relatively easily. This property accounts for the temporary bonding that occurs between certain atoms of large and complex molecules, such as proteins and nucleic acids. Even though hydrogen bonds are relatively weak, large molecules containing several hundred of these bonds have considerable strength and stability.

Molecular Weight and Moles

You have seen that bond formation results in the creation of molecules. Molecules are often discussed in terms of units of measure called molecular weight and moles. The **molecular weight** of a molecule is the sum of the atomic weights of all its atoms. To relate the molecular level to the laboratory level, we use a unit called the mole. One **mole** of a substance is its molecular weight expressed in grams. For example, 1 mole of water weighs 18 grams because the molecular weight of H_2O is 18, or $[(2 \times 1) + 16]$.

CHECK YOUR UNDERSTANDING

✓ Differentiate an ionic bond from a covalent bond. 2-2

Chemical Reactions

LEARNING OBJECTIVE

2-3 Diagram three basic types of chemical reactions.

As we said earlier, **chemical reactions** involve the making or breaking of bonds between atoms. After a chemical reaction, the total number of atoms remains the same, but there are new molecules with new properties because the atoms have been rearranged.

Energy in Chemical Reactions

Some change of energy occurs whenever bonds between atoms are formed or broken during chemical reactions. This energy is called **chemical energy.** All chemical bonds require energy when they are broken and release chemical energy when they are formed. A chemical reaction that absorbs more energy than it releases is called an **endergonic reaction** (*endo* = within), meaning that energy is directed inward. A chemical reaction that releases more energy than it absorbs is called an **exergonic reaction** (*exo* = out), meaning that energy is directed outward.

In this section we will look at three basic types of chemical reactions common to all living cells. By becoming familiar with these reactions, you will be able to understand the specific chemical reactions we will discuss later, particularly in Chapter 5.

Synthesis Reactions

When two or more atoms, ions, or molecules combine to form new and larger molecules, the reaction is called a **synthesis reaction.** To synthesize means to put together, and a synthesis reaction *forms new bonds.* Synthesis reactions can be expressed in the following way:

A	+	B	Combine to form →	AB
Atom, ion, or molecule A		Atom, ion, or molecule B		New molecule AB

The combining substances, A and B, are called the *reactants;* the substance formed by the combination, AB, is the *product.* The arrow indicates the direction in which the reaction proceeds.

Pathways of synthesis reactions in living organisms are collectively called anabolic reactions, or simply **anabolism** (an-ab′ō-lizm). The combining of sugar molecules to form starch and of amino acids to form proteins are two examples of anabolism.

Decomposition Reactions

The reverse of a synthesis reaction is a **decomposition reaction.** To decompose means to break down into smaller parts, and in a decomposition reaction *bonds are broken.* Typically, decomposition reactions split large molecules into smaller molecules, ions, or atoms. A decomposition reaction occurs in the following way:

AB	Breaks down into →	A	+	B
Molecule AB		Atom, ion, or molecule A		Atom, ion, or molecule B

Decomposition reactions that occur in living organisms are collectively called catabolic reactions, or simply **catabolism** (ka-tab′ō-lizm). An example of catabolism is the breakdown of sucrose (table sugar) into simpler sugars, glucose and fructose, during digestion. Bacterial decomposition of petroleum is discussed in the box on the facing page.

Bioremediation—Bacteria Clean Up Pollution

Although many bacteria have dietary requirements similar to ours—that's why they cause food spoilage—others metabolize (or chemically process) substances that are toxic to most plants and animals: heavy metals, sulfur, nitrogen gas, petroleum, and mercury.

Bacteria that can degrade many pollutants are naturally present in soil and water but in such small numbers that they cannot deal with large-scale contamination efficiently. Scientists are now working to improve the efficiency of natural pollution fighters. Using bacteria to degrade pollutants is called *bioremediation.*

One of the most promising successes for bioremediation occurred on an Alaskan beach following the *Exxon Valdez* oil spill. Several naturally occurring *Pseudomonas* bacteria are able to degrade oil for their carbon and energy requirements. In the presence of air, they remove two carbon atoms at a time from a large petroleum molecule (see the figure).

The bacteria degrade the oil too slowly to clean up an oil spill. However, scientists hit on a very simple way to speed up the process—with no need for recombinant DNA. They simply dumped ordinary nitrogen and phosphorus plant fertilizers (bioenhancers) onto a test beach. The number of oil-degrading bacteria increased compared with that on unfertilized control beaches, and oil was quickly cleared from the test beach.

Another group of bacteria is being investigated for its ability to clean up mercury contamination. Mercury is present in such common substances as discarded paint and fluorescent bulbs and can leach into soil and water from garbage dumps. *Desulfovibrio desulfuricans* bacteria actually make the mercury more dangerous by adding a methyl group, converting it into highly toxic methyl mercury. Methyl mercury in ponds or marshes sticks to small organisms such as plankton, which are eaten by larger organisms, which in turn are eaten by fish. Fish and human poisonings have been attributed to the ingestion of methyl mercury.

However, other bacteria, such as species of *Pseudomonas,* may offer the solution. To avoid mercury poisoning, these bacteria first convert methyl mercury to mercuric ion:

$$\underset{\text{Methyl mercury}}{CH_3Hg} \longrightarrow \underset{\text{Methane}}{CH_4} + \underset{\text{Mercuric ion}}{Hg^{2+}}$$

Many bacteria can then convert the positively charged mercuric ion to the relatively harmless elemental form by adding electrons, which they take from hydrogen atoms:

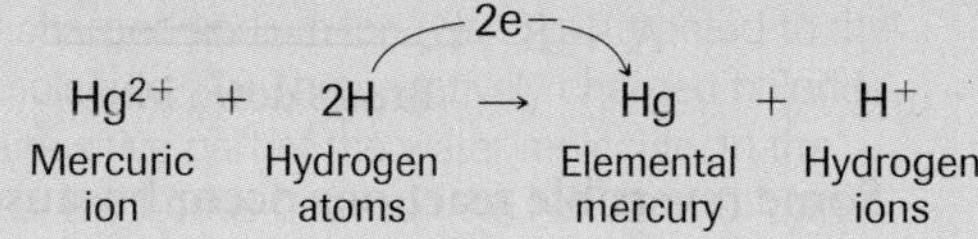

Unlike some forms of environmental cleanup, in which dangerous substances are removed from one place only to be dumped in another, bacterial cleanup eliminates the toxic substance and often returns a harmless or useful substance to the environment.

Typical saturated hydrocarbon found in petroleum

Two-carbon unit can be metabolized in cell

Exchange Reactions

All chemical reactions are based on synthesis and decomposition. Many reactions, such as **exchange reactions,** are actually part synthesis and part decomposition. An exchange reaction works in the following way:

$$AB + CD \xrightarrow{\text{Recombine to form}} AD + BC$$

First, the bonds between A and B and between C and D are broken in a decomposition process. New bonds are then formed between A and D and between B and C in a synthesis process. For

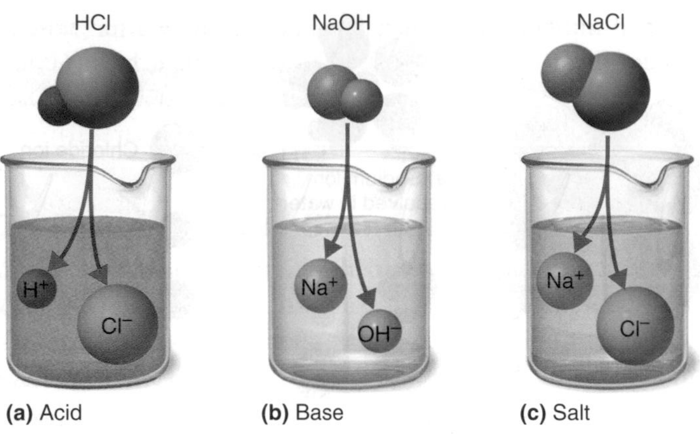

Figure 2.6 Acids, bases, and salts. (**a**) In water, hydrochloric acid (HCl) dissociates into H^+ and Cl^-. (**b**) Sodium hydroxide (NaOH), a base, dissociates into OH^- and Na^+ in water. (**c**) In water, table salt (NaCl) dissociates into positive ions (Na^+) and negative ions (Cl^-), neither of which are H^+ or OH^-.

Q How do acids and bases differ?

Biochemical reactions—that is, chemical reactions in living systems—are extremely sensitive to even small changes in the acidity or alkalinity of the environments in which they occur. In fact, H^+ and OH^- are involved in almost all biochemical processes, and any deviation from a cell's narrow band of normal H^+ and OH^- concentrations can dramatically modify the cell's functions. For this reason, the acids and bases that are continually formed in an organism must be kept in balance.

It is convenient to express the amount of H^+ in a solution by a logarithmic **pH** scale, which ranges from 0 to 14 (Figure 2.7). The term *pH* means potential of hydrogen. On a logarithmic scale, a change of one whole number represents a *tenfold* change from the previous concentration. Thus, a solution of pH 1 has ten times more hydrogen ions than a solution of pH 2 and has 100 times more hydrogen ions than a solution of pH 3.

A solution's pH is calculated as $-\log_{10}[H^+]$, the negative logarithm to the base 10 of the hydrogen ion concentration (denoted by brackets), determined in moles per liter $[H^+]$. For example, if the H^+ concentration of a solution is 1.0×10^{-4} moles/liter, or 10^{-4}, its pH equals $-\log_{10}10^{-4} = -(-4) = 4$; this is about the pH value of wine (see Appendix B). The pH values of some human body fluids and other common substances are also shown in Figure 2.7. In the laboratory, you will usually measure the pH of a solution with a pH meter or with chemical test papers.

Acidic solutions contain more H^+ than OH^- and have a pH lower than 7. If a solution has more OH^- than H^+, it is a basic, or alkaline, solution. In pure water, a small percentage of the molecules are dissociated into H^+ and OH^-, so it has a pH of 7. Because the concentrations of H^+ and OH^- are equal, this pH is said to be the pH of a neutral solution.

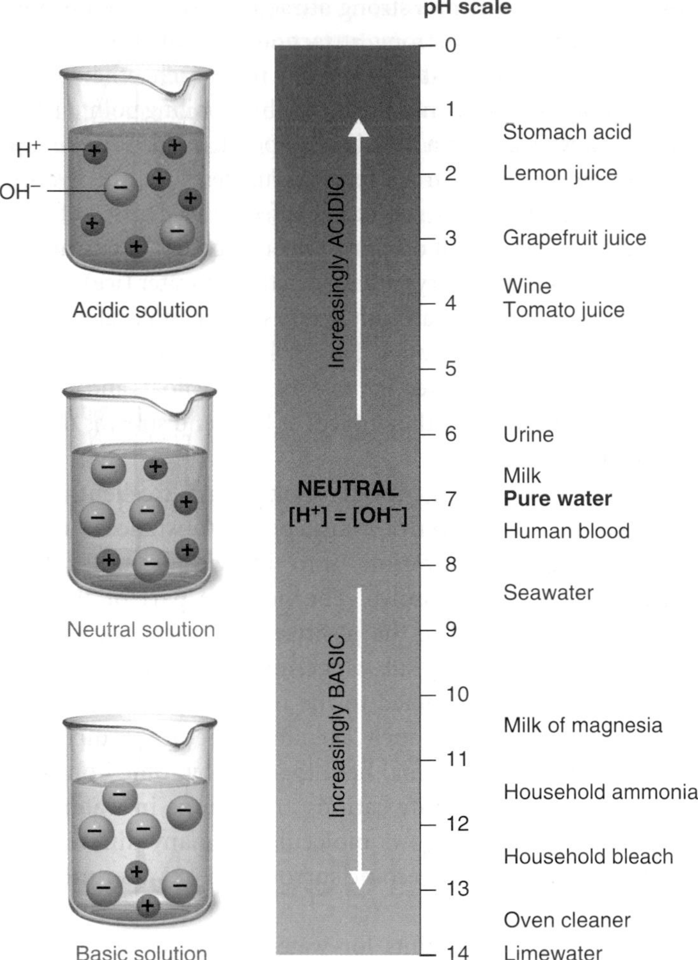

Figure 2.7 The pH scale.
As pH values decrease from 14 to 0, the H^+ concentration increases. Thus, the lower the pH, the more acidic the solution; the higher the pH, the more basic the solution. If the pH value of a solution is below 7, the solution is acidic; if the pH is above 7, the solution is basic (alkaline). The approximate pH values of some human body fluids and common substances are shown next to the pH scale.

Q At what pH are the concentrations of H^+ and OH^- equal?

Keep in mind that the pH of a solution can be changed. We can increase its acidity by adding substances that will increase the concentration of hydrogen ions. As a living organism takes up nutrients, carries out chemical reactions, and excretes wastes, its balance of acids and bases tends to change, and the pH fluctuates. Fortunately, organisms possess natural pH **buffers,** compounds that help keep the pH from changing drastically. But the pH in our environment's water and soil can be altered by waste products from organisms, pollutants from industry, or fertilizers used in agricultural fields or gardens. When bacteria are grown in a laboratory medium, they excrete waste products such as acids that can alter the pH of the medium. If this effect were to continue, the medium would become acidic enough to inhibit bacterial enzymes and kill the bacteria. To prevent this problem, pH buffers are added to the culture medium. One very effective pH buffer for some culture media uses a mixture of K_2HPO_4 and KH_2PO_4 (see Table 6.3, page 166).

Different microbes function best within different pH ranges, but most organisms grow best in environments with a pH value between 6.5 and 8.5. Among microbes, fungi are best able to tolerate acidic conditions, whereas the prokaryotes called cyanobacteria tend to do well in alkaline habitats. *Propionibacterium acnes* (prō-pē-on-ē-bak-ti′rē-um ak′nēz), a bacterium that causes acne, has as its natural environment human skin, which tends to be slightly acidic, with a pH of about 4. *Thiobacillus ferrooxidans* (thī-ō-bä-sil′lus fer-rō-oks′i-danz) is a bacterium that metabolizes elemental sulfur and produces sulfuric acid (H_2SO_4). Its pH range for optimum growth is from 1 to 3.5. The sulfuric acid produced by this bacterium in mine water is important in dissolving uranium and copper from low-grade ore (see Chapter 28).

CHECK YOUR UNDERSTANDING

✓ Why is the polarity of a water molecule important? **2-4**

✓ Antacids neutralize acid by the following reaction.
$Mg(OH)_2 + 2HCl \rightarrow MgCl_2 + H_2O$
Identify the acid, base, and salt. **2-5**

Organic Compounds

LEARNING OBJECTIVES

2-6 Distinguish organic and inorganic compounds.

2-7 Define *functional group*.

2-8 Identify the building blocks of carbohydrates.

2-9 Differentiate simple lipids, complex lipids, and steroids.

2-10 Identify the building blocks and structure of proteins.

2-11 Identify the building blocks of nucleic acids.

2-12 Describe the role of ATP in cellular activities.

Inorganic compounds, excluding water, constitute about 1–1.5% of living cells. These relatively simple components, whose molecules have only a few atoms, cannot be used by cells to perform complex biological functions. Organic molecules, whose carbon atoms can combine in an enormous variety of ways with other carbon atoms and with atoms of other elements, are relatively complex and thus are capable of more complex biological functions.

Structure and Chemistry

In the formation of organic molecules, carbon's four outer electrons can participate in up to four covalent bonds, and carbon atoms can bond to each other to form straight-chain, branched-chain, or ring structures.

In addition to carbon, the most common elements in organic compounds are hydrogen (which can form one bond), oxygen (two bonds), and nitrogen (three bonds). Sulfur (two bonds) and phosphorus (five bonds) appear less often. Other elements are found, but only in a relatively few organic compounds. The elements that are most abundant in living organisms are the same as those that are most abundant in organic compounds (see Table 2.1).

The chain of carbon atoms in an organic molecule is called the **carbon skeleton;** a huge number of combinations is possible for carbon skeletons. Most of these carbons are bonded to hydrogen atoms. The bonding of other elements with carbon and hydrogen forms characteristic **functional groups,** specific groups of atoms that are most commonly involved in chemical reactions and are responsible for most of the characteristic chemical properties and many of the physical properties of a particular organic compound (**Table 2.3**).

Different functional groups confer different properties on organic molecules. For example, the hydroxyl group of alcohols is hydrophilic (water-loving) and thus attracts water molecules to it. This attraction helps dissolve organic molecules containing hydroxyl groups. Because the carboxyl group is a source of hydrogen ions, molecules containing it have acidic properties. Amino groups, by contrast, function as bases because they readily accept hydrogen ions. The sulfhydryl group helps stabilize the intricate structure of many proteins.

Functional groups help us classify organic compounds. For example, the —OH group is present in each of the following molecules:

```
    H               H   H
    |               |   |
H—C—OH          H—C—C—OH
    |               |   |
    H               H   H
 Methanol          Ethanol

        H   H   H
        |   |   |
    H—C—C—C—H
        |   |   |
        H  OH   H
      Isopropanol
```

Because the characteristic reactivity of the molecules is based on the —OH group, they are grouped together in a class called alcohols. The —OH group is called the *hydroxyl group* and is not to be confused with the *hydroxide ion* (OH^-) of bases. The hydroxyl group of alcohols does not ionize at neutral pH; it is covalently bonded to a carbon atom.

When a class of compounds is characterized by a certain functional group, the letter *R* can be used to stand for the remainder of the molecule. For example, alcohols in general may be written R—OH.

Like disaccharides, polysaccharides can be split apart into their constituent sugars through hydrolysis. Unlike monosaccharides and disaccharides, however, they usually lack the characteristic sweetness of sugars such as fructose and sucrose and usually are not soluble in water.

One important polysaccharide is *glycogen,* which is composed of glucose subunits and is synthesized as a storage material by animals and some bacteria. *Cellulose,* another important glucose polymer, is the main component of the cell walls of plants and most algae. Although cellulose is the most abundant carbohydrate on Earth, it can be digested by only a few organisms that have the appropriate enzyme. The polysaccharide *dextran,* which is produced as a sugary slime by certain bacteria, is used in a blood plasma substitute. *Chitin* is a polysaccharide that makes up part of the cell wall of most fungi and the exoskeletons of lobsters, crabs, and insects. *Starch* is a polymer of glucose produced by plants and used as food by humans.

Many animals, including humans, produce enzymes called *amylases* that can break the bonds between the glucose molecules in glycogen. However, this enzyme cannot break the bonds in cellulose. Bacteria and fungi that produce enzymes called *cellulases* can digest cellulose. Cellulases from the fungus *Trichoderma* (trik′ō-dėr-mä) are used for a variety of industrial purposes. One of the more unusual uses is producing stone-washed denim. Because washing the fabric with rocks would damage washing machines, cellulase is used to digest, and therefore soften, the cotton. (See the box in Chapter 1, page 3.)

CHECK YOUR UNDERSTANDING

✓ Give an example of a monosaccharide, a disaccharide, and a polysaccharide. **2-8**

Lipids

If lipids were suddenly to disappear from the Earth, all living cells would collapse in a pool of fluid, because lipids are essential to the structure and function of membranes that separate living cells from their environment. **Lipids** (*lip* = fat) are a second major group of organic compounds found in living matter. Like carbohydrates, they are composed of atoms of carbon, hydrogen, and oxygen, but lipids lack the 2:1 ratio between hydrogen and oxygen atoms. Even though lipids are a very diverse group of compounds, they share one common characteristic: they are *nonpolar* molecules so, unlike water, do not have a positive and a negative end (pole). Therefore, most lipids are insoluble in water but dissolve readily in nonpolar solvents, such as ether and chloroform. Lipids provide the structure of membranes and some cell walls and function in energy storage.

Simple Lipids

Simple lipids, called *fats* or *triglycerides,* contain an alcohol called *glycerol* and a group of compounds known as *fatty acids.* Glycerol molecules have three carbon atoms to which are attached three hydroxyl (—OH) groups (**Figure 2.9a**). Fatty acids consist of long hydrocarbon chains (composed only of carbon and hydrogen atoms) ending in a carboxyl (—COOH, organic acid) group (**Figure 2.9b**). Most common fatty acids contain an even number of carbon atoms.

A molecule of fat is formed when a molecule of glycerol combines with one to three fatty acid molecules. The number of fatty acid molecules determines whether the fat molecule is a monoglyceride, diglyceride, or triglyceride (**Figure 2.9c**). In the reaction, one to three molecules of water are formed (dehydration), depending on the number of fatty acid molecules reacting. The chemical bond formed where the water molecule is removed is called an *ester linkage.* In the reverse reaction, hydrolysis, a fat molecule is broken down into its component fatty acid and glycerol molecules.

Because the fatty acids that form lipids have different structures, there is a wide variety of lipids. For example, three molecules of fatty acid A might combine with a glycerol molecule. Or one molecule each of fatty acids A, B, and C might unite with a glycerol molecule (see Figure 2.9c).

The primary function of lipids is to form plasma membranes that enclose cells. A plasma membrane supports the cell and allows nutrients and wastes to pass in and out; therefore, the lipids must maintain the same viscosity, regardless of the temperature of the surroundings. The membrane must be about as viscous as olive oil, without getting too fluid when warmed or too thick when cooled. As everyone who has ever cooked a meal knows, animal fats (such as butter) are usually solid at room temperature, whereas vegetable oils are usually liquid at room temperature. The difference in their respective melting points is due to the degrees of saturation of the fatty acid chains. A fatty acid is said to be *saturated* when it has no double bonds, in which case the carbon skeleton contains the maximum number of hydrogen atoms (see Figure 2.9c and **Figure 2.10a**). Saturated chains become solid more easily because they are relatively straight and are thus able to pack together more closely than unsaturated chains. The double bonds of *unsaturated* chains create kinks in the chain, which keep the chains apart from one another (**Figure 2.10b**). Note in Figure 2.9c that the H atoms on either side of the double-bond in oleic acid are on the same side of the unsaturated fatty acid. Such an unsaturated fatty acid is called a *cis* fatty acid. If, instead, the H atoms are on opposite sides of the double bond, the unsaturated acid is called a *trans* fatty acid.

Complex Lipids

Complex lipids contain such elements as phosphorus, nitrogen, and sulfur, in addition to the carbon, hydrogen, and oxygen found in simple lipids. The complex lipids called *phospholipids* are made up of glycerol, two fatty acids, and, in place of a third fatty acid, a phosphate group bonded to one of several organic groups (see Figure 2.10a). Phospholipids are the lipids that build

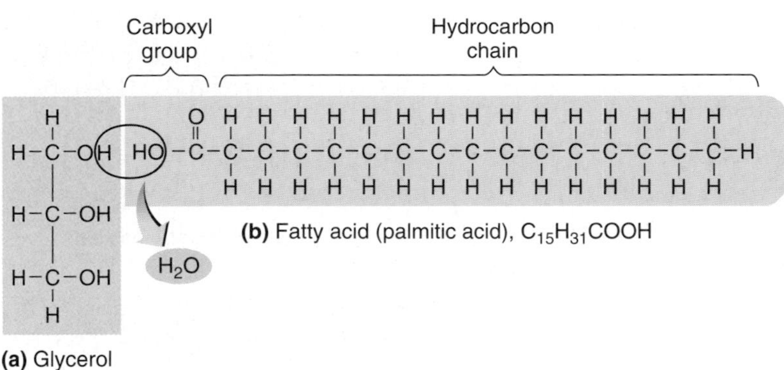

Figure 2.9 Structural formulas of simple lipids. **(a)** Glycerol. **(b)** Palmitic acid, a saturated fatty acid. **(c)** The chemical combination of a molecule of glycerol and three fatty acid molecules (palmitic, stearic, and oleic in this example) forms one molecule of fat (triglyceride) and three molecules of water in a dehydration synthesis reaction. Oleic acid is a *cis* fatty acid. The bond between glycerol and each fatty acid is called an ester linkage. The addition of three water molecules to a fat forms glycerol and three fatty acid molecules in a hydrolysis reaction.

Q How do saturated and unsaturated fatty acids differ?

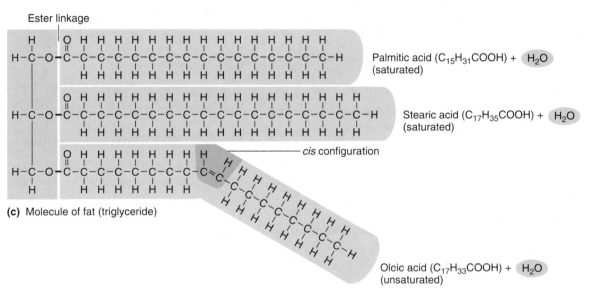

Ester linkage

Palmitic acid ($C_{15}H_{31}COOH$) + H_2O (saturated)

Stearic acid ($C_{17}H_{35}COOH$) + H_2O (saturated)

cis configuration

Oleic acid ($C_{17}H_{33}COOH$) + H_2O (unsaturated)

(c) Molecule of fat (triglyceride)

membranes; they are essential to a cell's survival. Phospholipids have polar as well as nonpolar regions (Figure 2.10a and b; see also Figure 4.13, page 86). When placed in water, phospholipid molecules twist themselves in such a way that all polar (hydrophilic) portions orient themselves toward the polar water molecules, with which they then form hydrogen bonds. (Recall that *hydrophilic* means water-loving.) This forms the basic structure of a plasma membrane (**Figure 2.10c**). Polar portions consist of a phosphate group and glycerol. In contrast to the polar regions, all nonpolar (hydrophobic) parts of the phospholipid make contact only with the nonpolar portions of neighboring molecules. (*Hydrophobic* means water-fearing.) Nonpolar portions consist of fatty acids. This characteristic behavior makes phospholipids particularly suitable for their role as a major component of the membranes that enclose cells. Phospholipids enable the membrane to act as a barrier that separates the contents of the cell from the water-based environment in which it lives.

Some complex lipids are useful in identifying certain bacteria. For example, the cell wall of *Mycobacterium tuberculosis* (mī-kō-bak-ti′rē-um tü-bėr-kū-lō′sis), the bacterium that causes tuberculosis, is distinguished by its lipid-rich content. The cell wall contains complex lipids such as waxes and glycolipids (lipids with carbohydrates attached) that give the bacterium distinctive staining characteristics. Cell walls rich in such complex lipids are characteristic of all members of the genus *Mycobacterium.*

Steroids

Steroids are structurally very different from lipids. **Figure 2.11** shows the structure of the steroid cholesterol, with the four interconnected carbon rings that are characteristic of steroids. When an —OH group is attached to one of the rings, the steroid is called a *sterol* (an alcohol). Sterols are important constituents of the plasma membranes of animal cells and of one group of bacteria (mycoplasmas), and they are also found in fungi and plants. The sterols separate the fatty acid chains and thus prevent the packing that would harden the plasma membrane at low temperatures (see Figure 2.10c).

CHECK YOUR UNDERSTANDING

✓ How do simple lipids differ from complex lipids? 2-9

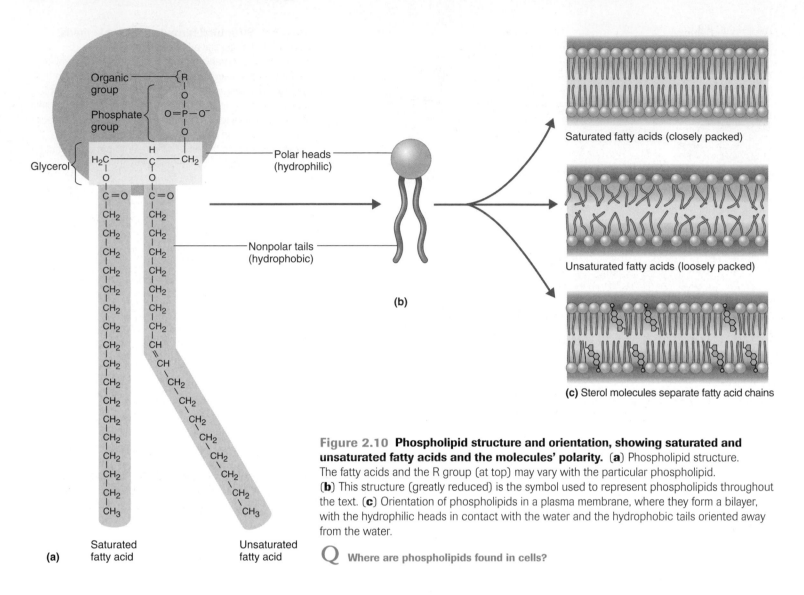

Figure 2.10 Phospholipid structure and orientation, showing saturated and unsaturated fatty acids and the molecules' polarity. (**a**) Phospholipid structure. The fatty acids and the R group (at top) may vary with the particular phospholipid. (**b**) This structure (greatly reduced) is the symbol used to represent phospholipids throughout the text. (**c**) Orientation of phospholipids in a plasma membrane, where they form a bilayer, with the hydrophilic heads in contact with the water and the hydrophobic tails oriented away from the water.

Q Where are phospholipids found in cells?

Proteins

Proteins are organic molecules that contain carbon, hydrogen, oxygen, and nitrogen. Some also contain sulfur. If you were to separate and weigh all the groups of organic compounds in a living cell, the proteins would tip the scale. Hundreds of different proteins can be found in any single cell, and together they make up 50% or more of a cell's dry weight.

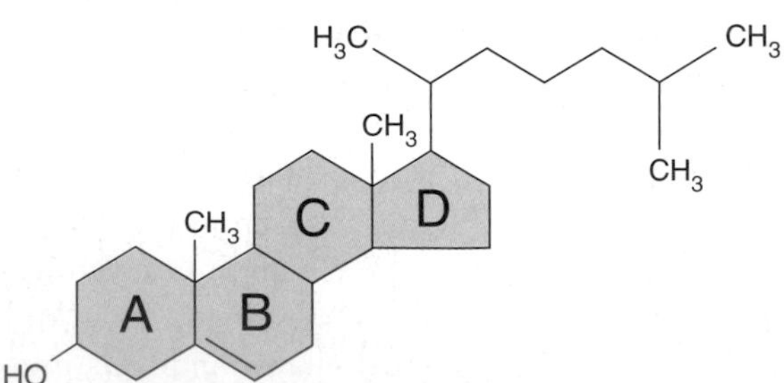

Figure 2.11 Cholesterol, a steroid. Note the four "fused" carbon rings (labeled A–D), which are characteristic of steroid molecules. The hydrogen atoms attached to the carbons at the corners of the rings have been omitted. The —OH group (colored red) makes this molecule a sterol.

Q Where are sterols found in cells?

Proteins are essential ingredients in all aspects of cell structure and function. *Enzymes* are the proteins that speed up biochemical reactions. But proteins have other functions as well. *Transporter proteins* help transport certain chemicals into and out of cells. Other proteins, such as the *bacteriocins* produced by many bacteria, kill other bacteria. Certain *toxins,* called exotoxins, produced by some disease-causing microorganisms are also proteins. Some proteins play a role in the *contraction* of animal muscle cells and the *movement* of microbial and other types of cells. Other proteins are integral parts of *cell structures* such as walls, membranes, and cytoplasmic components. Still others, such as the *hormones* of certain organisms, have regulatory functions. As we will see in Chapter 17, proteins called *antibodies* play a role in vertebrate immune systems.

Amino Acids

Just as monosaccharides are the building blocks of larger carbohydrate molecules, and fatty acids and glycerol are the building blocks of fats, **amino acids** are the building blocks of proteins. Amino acids contain at least one carboxyl (—COOH) group and one amino (—NH_2) group attached to the same carbon atom, called an alpha-carbon (written C_α) (**Figure 2.12a**). Such amino acids are called *alpha-amino acids.* Also attached to the alpha-carbon is a side group (R group), which is the amino acid's distinguishing feature. The side group can be a hydrogen atom, an unbranched or branched chain of atoms, or a ring structure that is cyclic (all carbon) or heterocyclic (when an atom other than carbon is included in the ring). **Figure 2.12b** shows the structural formula of tyrosine, an amino acid that has a cyclic side group. The side group can contain functional groups, such as the sulfhydryl group (—SH), the hydroxyl group (—OH), or additional carboxyl or amino groups. These side groups and the carboxyl and alpha-amino groups affect the total structure of a protein, described later. The structures and standard abbreviations of the 20 amino acids found in proteins are shown in **Table 2.4**.

Most amino acids exist in either of two configurations called **stereoisomers,** designated by D and L. These configurations are mirror images, corresponding to "right-handed" (D) and "left-handed" (L) three-dimensional shapes (**Figure 2.13**). The amino acids found in proteins are always the L-isomers (except for glycine, the simplest amino acid, which does not have stereoisomers). However, D-amino acids occasionally occur in nature—for example, in certain bacterial cell walls and antibiotics. (Many other kinds of organic molecules also can exist in D and L forms. One example is the sugar glucose, which occurs in nature as D-glucose.)

Although only 20 different amino acids occur naturally in proteins, a single protein molecule can contain from 50 to hundreds of amino acid molecules, which can be arranged in an almost infinite number of ways to make proteins of different lengths, compositions, and structures. The number of proteins is practically endless, and every living cell produces many different proteins.

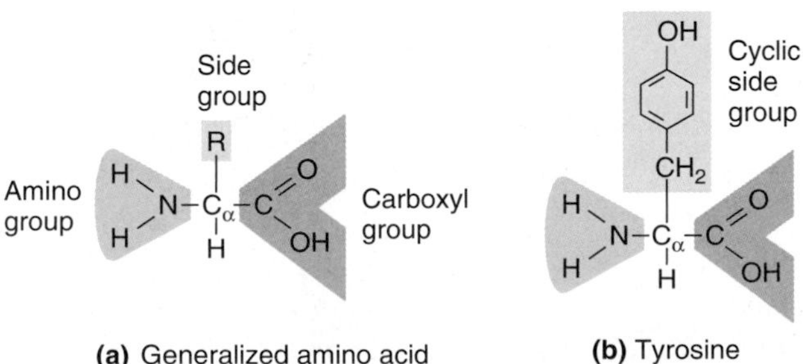

Figure 2.12 Amino acid structure. (**a**) The general structural formula for an amino acid. The alpha-carbon (C_α) is shown in the center. Different amino acids have different R groups, also called side groups. (**b**) Structural formula for the amino acid tyrosine, which has a cyclic side group.

Q What distinguishes one amino acid from another?

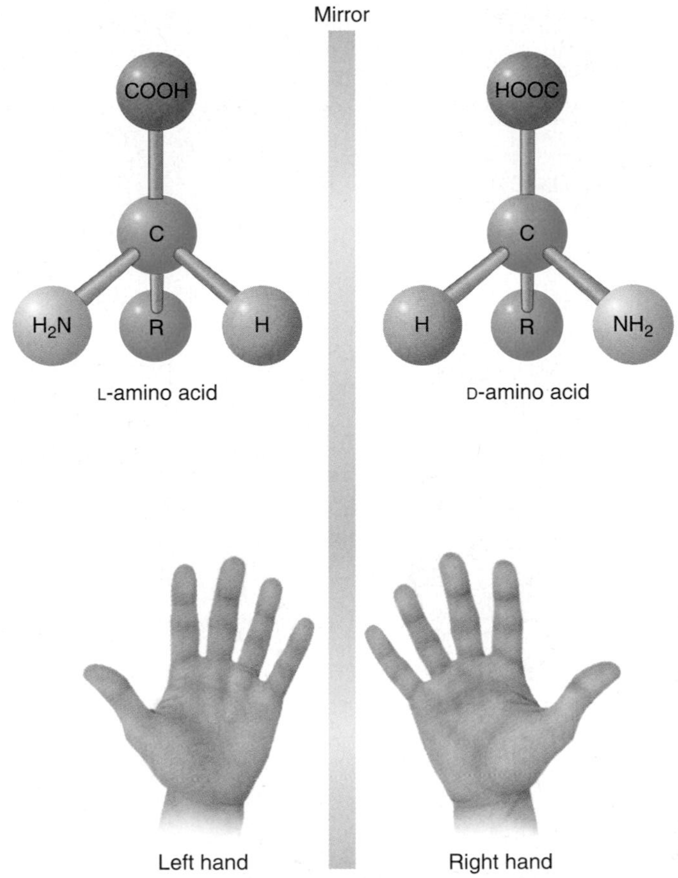

Figure 2.13 The L- and D-isomers of an amino acid, shown with ball-and-stick models. The two isomers, like left and right hands, are mirror images of each other and cannot be superimposed on one another. (Try it!)

Q Which isomer is always found in proteins?

Peptide Bonds

Amino acids bond between the carbon atom of the carboxyl (—COOH) group of one amino acid and the nitrogen atom of the amino (—NH_2) group of another (**Figure 2.14**). The bonds between amino acids are called **peptide bonds.** For every peptide bond formed between two amino acids, one water molecule is released; thus, peptide bonds are formed by dehydration synthesis. In Figure 2.14 the resulting compound is called a *dipeptide* because it consists of two amino acids joined by a peptide bond. Adding another amino acid to a dipeptide would form a *tripeptide.* Further additions of amino acids would produce a long, chainlike molecule called a *peptide* (4–9 amino acids) or *polypeptide* (10–2000 or more amino acids).

Table 2.4 The 20 Amino Acids Found in Proteins*

Amino acid	Structural formula	Characteristic R group
Glycine (Gly)	$H_2N-CH(\mathbf{H})-COOH$	Hydrogen atom
Alanine (Ala)	$H_2N-CH(\mathbf{CH_3})-COOH$	Unbranched chain
Valine (Val)	$H_2N-CH(\mathbf{CH(CH_3)_2})-COOH$	Branched chain
Leucine (Leu)	$H_2N-CH(\mathbf{CH_2-CH(CH_3)_2})-COOH$	Branched chain
Isoleucine (Ile)	$H_2N-CH(\mathbf{CH(CH_3)-CH_2-CH_3})-COOH$	Branched chain
Serine (Ser)	$H_2N-CH(\mathbf{CH_2-OH})-COOH$	Hydroxyl (—OH) group
Threonine (Thr)	$H_2N-CH(\mathbf{CH(OH)CH_3})-COOH$	Hydroxyl (—OH) group
Cysteine (Cys)	$H_2N-CH(\mathbf{CH_2-SH})-COOH$	Sulphur-containing (—SH) group
Methionine (Met)	$H_2N-CH(\mathbf{CH_2-CH_2-S-CH_3})-COOH$	Thioether (SC) group
Glutamic acid (Glu)	$H_2N-CH(\mathbf{CH_2-CH_2-COOH})-COOH$	Additional carboxyl (—COOH) group, acidic
Aspartic acid (Asp)	$H_2N-CH(\mathbf{CH_2-COOH})-COOH$	Additional carboxyl (—COOH) group, acidic
Lysine (Lys)	$H_2N-CH(\mathbf{CH_2-CH_2-CH_2-CH_2-NH_2})-COOH$	Additional amino (—NH_2) group, basic
Arginine (Arg)	$H_2N-CH(\mathbf{CH_2-CH_2-CH_2-NH-C(=NH)-NH_2})-COOH$	Additional amino (—NH_2) group, basic
Asparagine (Asn)	$H_2N-CH(\mathbf{CH_2-C(=O)NH_2})-COOH$	Additional amino (—NH_2) group, basic
Glutamine (Gln)	$H_2N-CH(\mathbf{CH_2-CH_2-C(=O)NH_2})-COOH$	Additional amino (—NH_2) group, basic
Phenylalanine (Phe)	$H_2N-CH(\mathbf{CH_2-C_6H_5})-COOH$ (benzene ring)	Cyclic
Tyrosine (Tyr)	$H_2N-CH(\mathbf{CH_2-C_6H_4-OH})-COOH$	Cyclic
Histidine (His)	$H_2N-CH(\mathbf{CH_2}$–imidazole ring with $\mathbf{HN^+}$ and $\mathbf{NH})-COOH$	Heterocyclic
Tryptophan (Trp)	$H_2N-CH(\mathbf{CH_2}$–indole ring with $\mathbf{NH})-COOH$	Heterocyclic
Proline (Pro)	$HN-CH-COOH$, ring closed by $\mathbf{H_2C-CH_2-CH_2}$	Heterocyclic

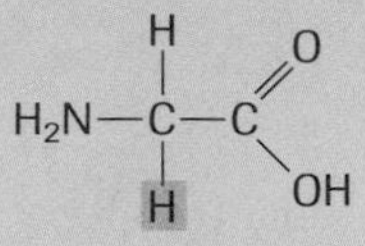

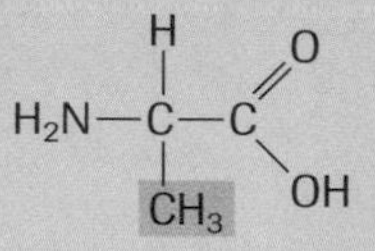

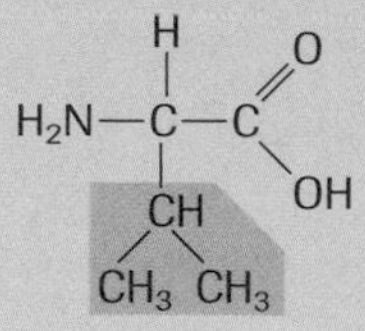

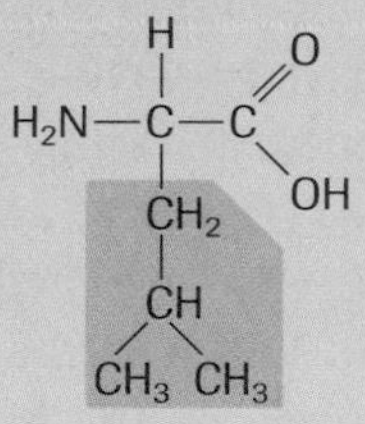

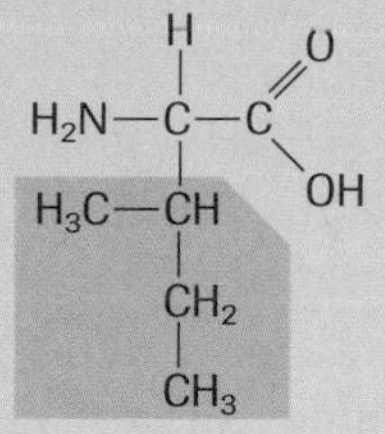

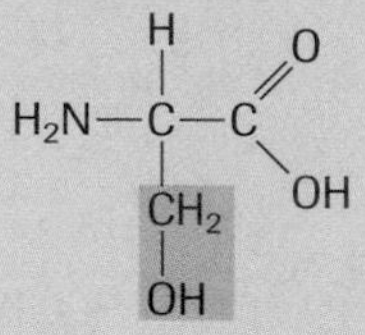

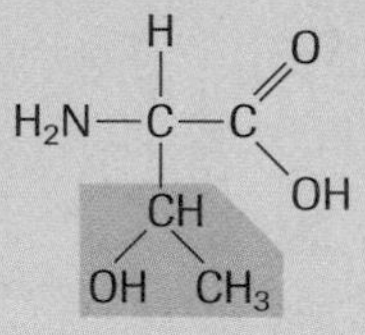

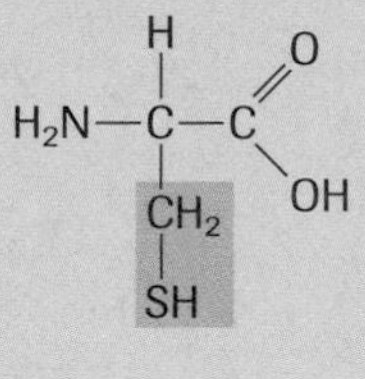

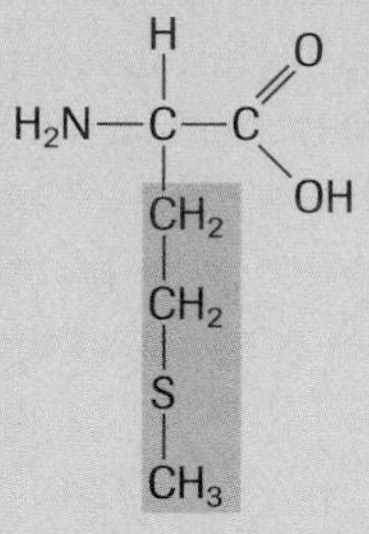

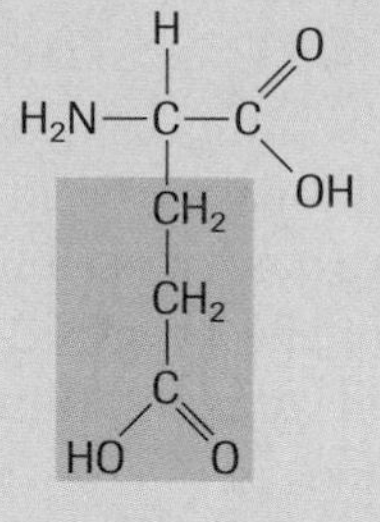

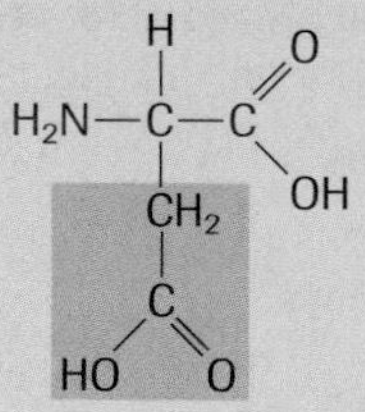

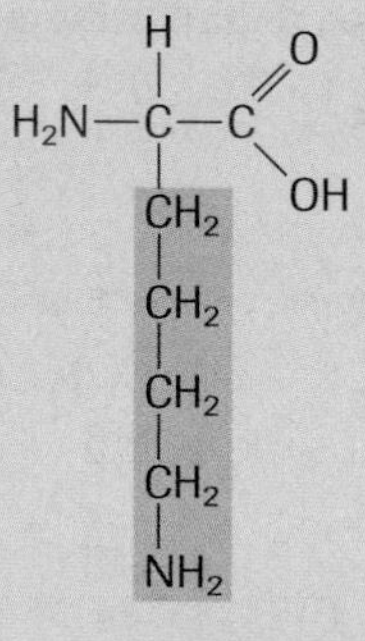

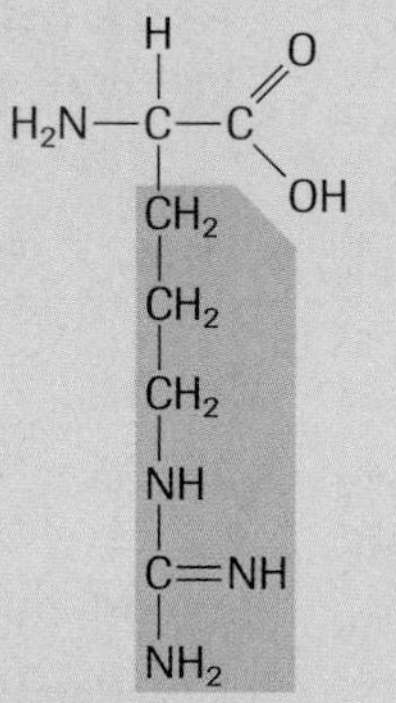

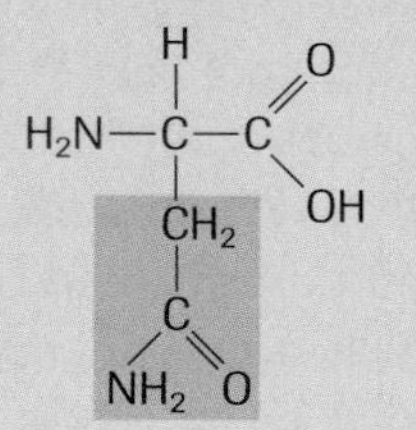

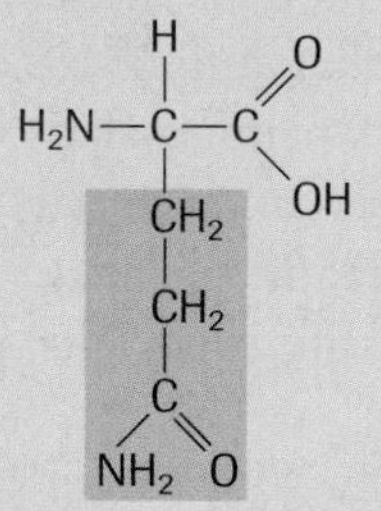

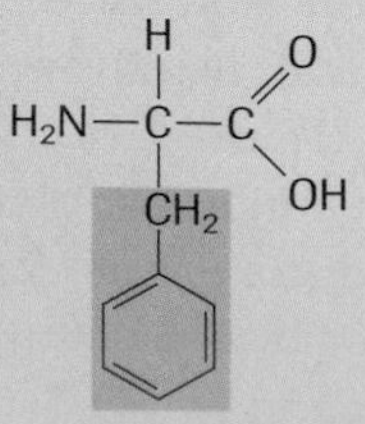

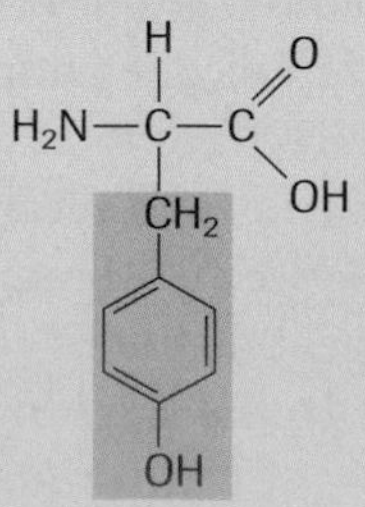

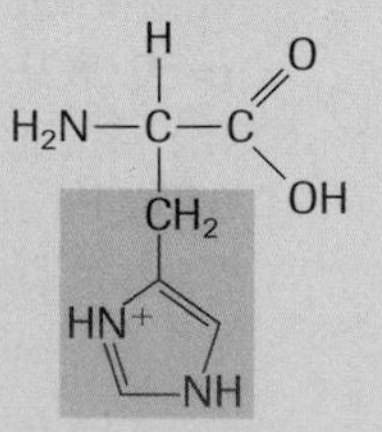

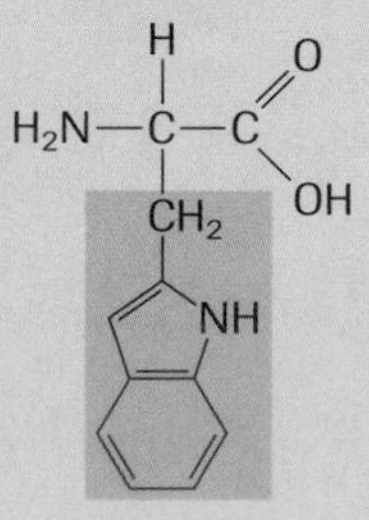

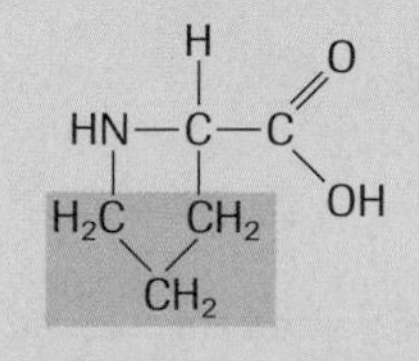

*Shown are the amino acid names, including the three-letter abbreviation in parentheses (above), their structural formulas (center), and characteristic R group (below). Note that cysteine and methionine are the only amino acids that contain sulfur.

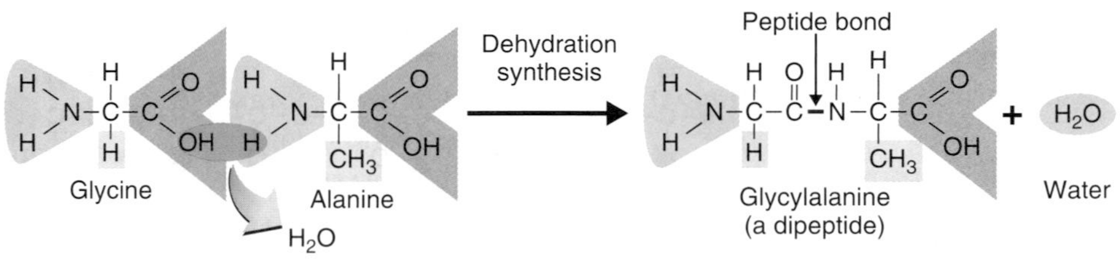

Figure 2.14 Peptide bond formation by dehydration synthesis. The amino acids glycine and alanine combine to form a dipeptide. The newly formed bond between the carbon atom of glycine and the nitrogen atom of alanine is called a peptide bond.

Q How are amino acids related to proteins?

Levels of Protein Structure

Proteins vary tremendously in structure. Different proteins have different architectures and different three-dimensional shapes. This variation in structure is directly related to their diverse functions.

When a cell makes a protein, the polypeptide chain folds spontaneously to assume a certain shape. One reason for folding of the polypeptide is that some parts of a protein are attracted to water and other parts are repelled by it. In practically every case, the function of a protein depends on its ability to recognize and bind to some other molecule. For example, an enzyme binds specifically with its substrate. A hormonal protein binds to a receptor on a cell whose function it will alter. An antibody binds to an antigen (foreign substance) that has invaded the body. The unique shape of each protein permits it to interact with specific other molecules in order to carry out specific functions.

Proteins are described in terms of four levels of organization: primary, secondary, tertiary, and quaternary. The *primary structure* is the unique sequence in which the amino acids are linked together to form a polypeptide chain (Figure 2.15a). This sequence is genetically determined. Alterations in sequence can have profound metabolic effects. For example, a single incorrect amino acid in a blood protein can produce the deformed hemoglobin molecule characteristic of sickle cell disease. But proteins do not exist as long, straight chains. Each polypeptide chain folds and coils in specific ways into a relatively compact structure with a characteristic three-dimensional shape.

A protein's *secondary structure* is the localized, repetitious twisting or folding of the polypeptide chain. This aspect of a protein's shape results from hydrogen bonds joining the atoms of peptide bonds at different locations along the polypeptide chain. The two types of secondary protein structures are clockwise spirals called *helices* (singular: *helix*) and pleated sheets, which form from roughly parallel portions of the chain (Figure 2.15b). Both structures are held together by hydrogen bonds between oxygen or nitrogen atoms that are part of the polypeptide's backbone.

Tertiary structure refers to the overall three-dimensional structure of a polypeptide chain (Figure 2.15c). The folding is not repetitive or predictable, as in secondary structure. Whereas secondary structure involves hydrogen bonding between atoms of the amino and carboxyl groups involved in the peptide bonds, tertiary structure involves several interactions between various amino acid side groups in the polypeptide chain. For example, amino acids with nonpolar (hydrophobic) side groups usually interact at the core of the protein, out of contact with water. This *hydrophobic interaction* helps contribute to tertiary structure. Hydrogen bonds between side groups, and ionic bonds between oppositely charged side groups, also contribute to tertiary structure. Proteins that contain the amino acid cysteine form strong covalent bonds called *disulfide bridges.* These bridges form when two cysteine molecules are brought close together by the folding of the protein. Cysteine molecules contain sulfhydryl groups (—SH), and the sulfur of one cysteine molecule bonds to the sulfur on another, forming (by the removal of hydrogen atoms) a disulfide bridge (S—S) that holds parts of the protein together.

Some proteins have a *quaternary structure,* which consists of an aggregation of two or more individual polypeptide chains (subunits) that operate as a single functional unit. Figure 2.15d shows a hypothetical protein consisting of two polypeptide chains. More commonly, proteins have two or more kinds of polypeptide subunits. The bonds that hold a quaternary structure together are basically the same as those that maintain tertiary structure. The overall shape of a protein may be globular (compact and roughly spherical) or fibrous (threadlike).

If a protein encounters a hostile environment in terms of temperature, pH, or salt concentrations, it may unravel and lose its characteristic shape. This process is called **denaturation** (see Figure 5.6, page 119). As a result of denaturation, the protein is no longer functional. This process will be discussed in more detail in Chapter 5 with regard to denaturation of enzymes.

Q&A The proteins we have been discussing are *simple proteins,* which contain only amino acids. *Conjugated proteins* are combinations of amino acids with other organic or inorganic components. Conjugated proteins are named by their non–amino acid component. Thus, glycoproteins contain sugars, nucleoproteins contain nucleic acids, metalloproteins contain metal atoms, lipoproteins contain lipids, and phosphoproteins contain phosphate groups. Phosphoproteins are important regulators of activity in eukaryotic cells. Bacterial synthesis of phosphoproteins may be important for the survival of bacteria such as *Legionella pneumophila* that grow inside host cells.

CHECK YOUR UNDERSTANDING

✓ What two functional groups are in all amino acids? 2-10

Figure 2.16

FOUNDATION FIGURE The Structure of DNA

This figure provides an overview of the structure of DNA, a double-stranded molecule that stores genetic information in all cells. Familiarity with DNA's structure and function is essential for understanding genetics, recombinant DNA techniques, and the emergence of antibiotic resistance and new diseases.

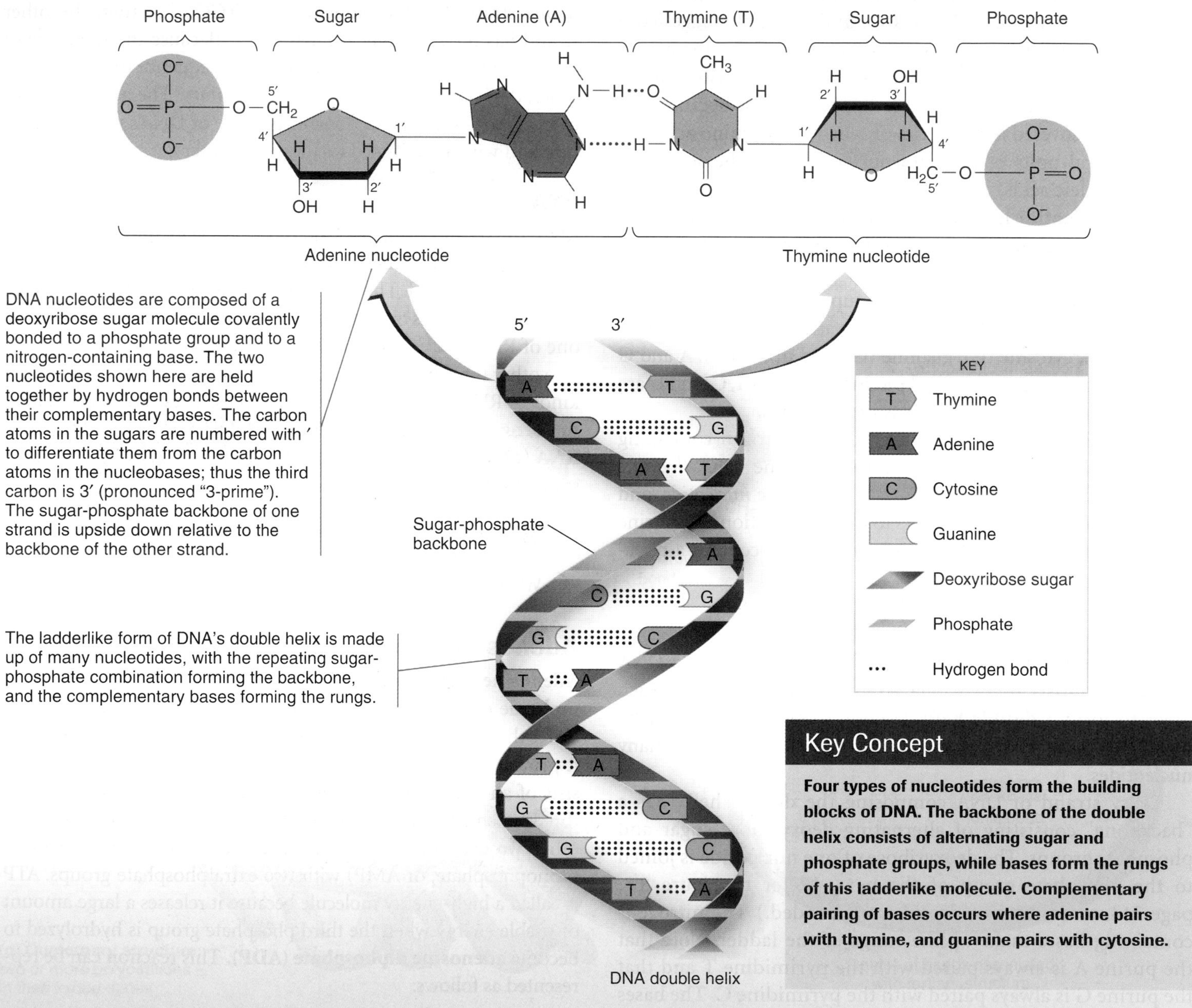

Phosphate
Sugar
Adenine (A)
Thymine (T)
Sugar
Phosphate
Adenine nucleotide
Thymine nucleotide
DNA nucleotides are composed of a deoxyribose sugar molecule covalently bonded to a phosphate group and to a nitrogen-containing base. The two nucleotides shown here are held together by hydrogen bonds between their complementary bases. The carbon atoms in the sugars are numbered with ′ to differentiate them from the carbon atoms in the nucleobases; thus the third carbon is 3′ (pronounced "3-prime"). The sugar-phosphate backbone of one strand is upside down relative to the backbone of the other strand.
5′
3′
Sugar-phosphate backbone
The ladderlike form of DNA's double helix is made up of many nucleotides, with the repeating sugar-phosphate combination forming the backbone, and the complementary bases forming the rungs.
KEY
Thymine
Adenine
Cytosine
Guanine
Deoxyribose sugar
Phosphate
Hydrogen bond
DNA double helix

Key Concept

Four types of nucleotides form the building blocks of DNA. The backbone of the double helix consists of alternating sugar and phosphate groups, while bases form the rungs of this ladderlike molecule. Complementary pairing of bases occurs where adenine pairs with thymine, and guanine pairs with cytosine.

A cell's supply of ATP at any particular time is limited. Whenever the supply needs replenishing, the reaction goes in the reverse direction; the addition of a phosphate group to ADP and the input of energy produces more ATP. The energy required to attach the terminal phosphate group to ADP is supplied by the cell's various oxidation reactions, particularly the oxidation of glucose. ATP can be stored in every cell, where its potential energy is not released until needed.

CHECK YOUR UNDERSTANDING

✓ Which can provide more energy for a cell and why: ATP or ADP? **2-12**

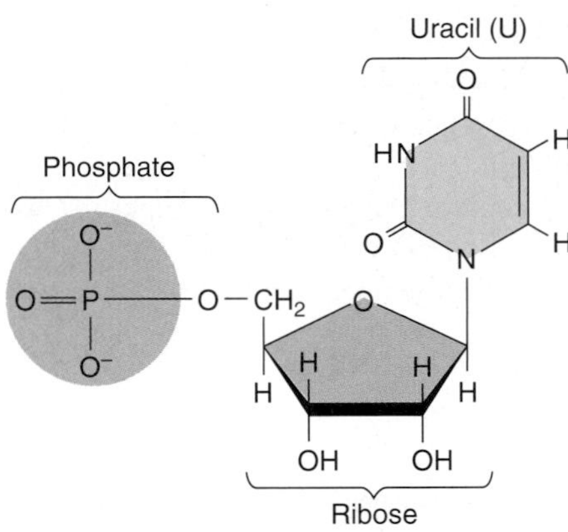

Figure 2.17 A uracil nucleotide of RNA.

Q How do DNA and RNA differ in structure?

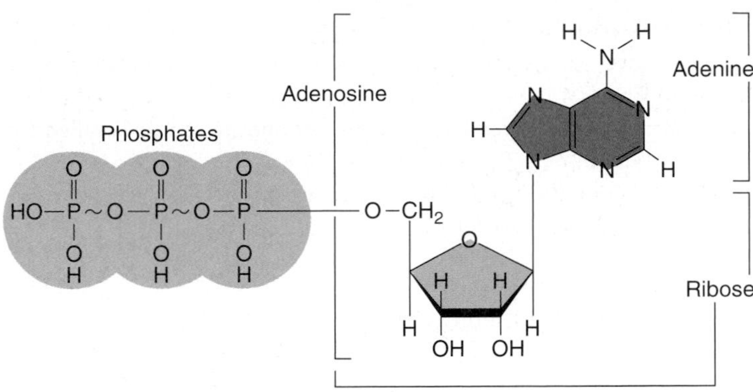

Figure 2.18 The structure of ATP. High-energy phosphate bonds are indicated by wavy lines. When ATP breaks down to ADP and inorganic phosphate, a large amount of chemical energy is released for use in other chemical reactions.

STUDY OUTLINE

The **MyMicrobiologyPlace** website (**www.microbiologyplace.com**) will help you get ready for tests with its simple three-step approach: ❶ **take a pre-test** and obtain a personalized study plan, ❷ **learn and practice** with animations, tutorials, and MP3 tutor sessions, and ❸ **test yourself** with quizzes and a chapter post-test.

Introduction (p. 26)

1. The science of the interaction between atoms and molecules is called chemistry.
2. The metabolic activities of microorganisms involve complex chemical reactions.
3. Microbes break down nutrients to obtain energy and to make new cells.

The Structure of Atoms (pp. 27–28)

1. An atom is the smallest unit of a chemical element that exhibits the properties of that element.
2. Atoms consist of a nucleus, which contains protons and neutrons, and electrons, which move around the nucleus.
3. The atomic number is the number of protons in the nucleus; the total number of protons and neutrons is the atomic weight.

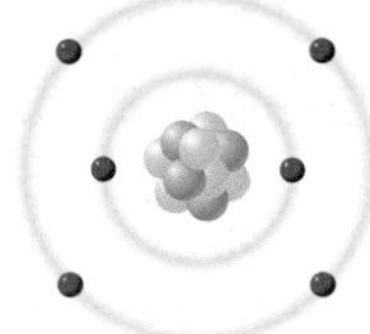

Chemical Elements (pp. 27–28)

4. Atoms with the same number of protons and the same chemical behavior are classified as the same chemical element.
5. Chemical elements are designated by abbreviations called chemical symbols.
6. About 26 elements are commonly found in living cells.
7. Atoms that have the same atomic number (are of the same element) but different atomic weights are called isotopes.

Electronic Configurations (p. 28)

8. In an atom, electrons are arranged around the nucleus in electron shells.
9. Each shell can hold a characteristic maximum number of electrons.
10. The chemical properties of an atom are due largely to the number of electrons in its outermost shell.

How Atoms Form Molecules: Chemical Bonds (pp. 28–32)

1. Molecules are made up of two or more atoms; molecules consisting of at least two different kinds of atoms are called compounds.
2. Atoms form molecules in order to fill their outermost electron shells.
3. Attractive forces that bind the atomic nuclei of two atoms together are called chemical bonds.
4. The combining capacity of an atom—the number of chemical bonds the atom can form with other atoms—is its valence.

Ionic Bonds (pp. 28, 30)

5. A positively or negatively charged atom or group of atoms is called an ion.
6. A chemical attraction between ions of opposite charge is called an ionic bond.
7. To form an ionic bond, one ion is an electron donor, and the other ion is an electron acceptor.

Covalent Bonds (pp. 30–31)

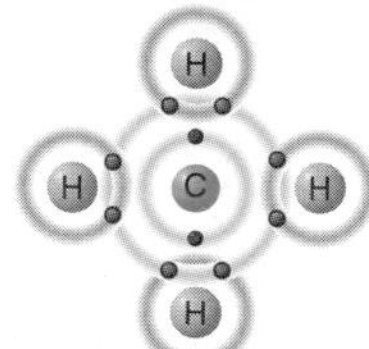

8. In a covalent bond, atoms share pairs of electrons.
9. Covalent bonds are stronger than ionic bonds and are far more common in organisms.

Hydrogen Bonds (pp. 31–32)

10. A hydrogen bond exists when a hydrogen atom covalently bonded to one oxygen or nitrogen atom is attracted to another oxygen or nitrogen atom.
11. Hydrogen bonds form weak links between different molecules or between parts of the same large molecule.

Molecular Weight and Moles (p. 32)

12. The molecular weight is the sum of the atomic weights of all the atoms in a molecule.
13. A mole of an atom, ion, or molecule is equal to its atomic or molecular weight expressed in grams.

Chemical Reactions (pp. 32–34)

1. Chemical reactions are the making or breaking of chemical bonds between atoms.
2. A change of energy occurs during chemical reactions.
3. Endergonic reactions require more energy than they release; exergonic reactions release more energy.
4. In a synthesis reaction, atoms, ions, or molecules are combined to form a larger molecule.
5. In a decomposition reaction, a larger molecule is broken down into its component molecules, ions, or atoms.
6. In an exchange reaction, two molecules are decomposed, and their subunits are used to synthesize two new molecules.
7. The products of reversible reactions can readily revert to form the original reactants.

IMPORTANT BIOLOGICAL MOLECULES (pp. 34–49)

Inorganic Compounds (pp. 34–37)

1. Inorganic compounds are usually small, ionically bonded molecules.
2. Water and many common acids, bases, and salts are examples of inorganic compounds.

Water (pp. 34–35)

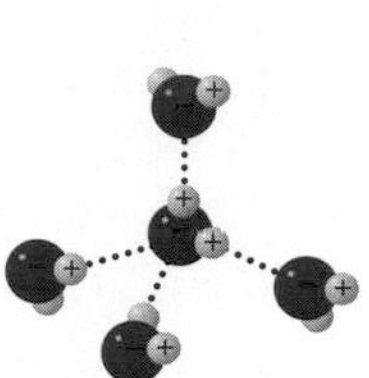

3. Water is the most abundant substance in cells.
4. Because water is a polar molecule, it is an excellent solvent.
5. Water is a reactant in many of the decomposition reactions of digestion.
6. Water is an excellent temperature buffer.

Acids, Bases, and Salts (p. 35)

7. An acid dissociates into H^+ and anions.
8. A base dissociates into OH^- and cations.
9. A salt dissociates into negative and positive ions, neither of which is H^+ or OH^-.

Acid–Base Balance: The Concept of pH (pp. 35–37)

10. The term *pH* refers to the concentration of H^+ in a solution.
11. A solution of pH 7 is neutral; a pH value below 7 indicates acidity; pH above 7 indicates alkalinity.
12. The pH inside a cell and in culture media is stabilized with pH buffers.

Organic Compounds (pp. 37–49)

1. Organic compounds always contain carbon and hydrogen.
2. Carbon atoms form up to four bonds with other atoms.
3. Organic compounds are mostly or entirely covalently bonded, and many of them are large molecules.

Structure and Chemistry (pp. 37–38)

4. A chain of carbon atoms forms a carbon skeleton.
5. Functional groups of atoms are responsible for most of the properties of organic molecules.
6. The letter *R* may be used to denote the remainder of an organic molecule.
7. Frequently encountered classes of molecules are R—OH (alcohols) and R—COOH (organic acids).
8. Small organic molecules may combine into very large molecules called macromolecules.
9. Monomers usually bond together by dehydration synthesis, or condensation reactions, that form water and a polymer.
10. Organic molecules may be broken down by hydrolysis, a reaction involving the splitting of water molecules.

Carbohydrates (pp. 39–40)

11. Carbohydrates are compounds consisting of atoms of carbon, hydrogen, and oxygen, with hydrogen and oxygen in a 2:1 ratio.
12. Carbohydrates include sugars and starches.
13. Carbohydrates can be classified as monosaccharides, disaccharides, and polysaccharides.
14. Monosaccharides contain from three to seven carbon atoms.
15. Isomers are two molecules with the same chemical formula but different structures and properties—for example, glucose ($C_6H_{12}O_6$) and fructose ($C_6H_{12}O_6$).
16. Monosaccharides may form disaccharides and polysaccharides by dehydration synthesis.

Lipids (pp. 40–41)

17. Lipids are a diverse group of compounds distinguished by their insolubility in water.
18. Simple lipids (fats) consist of a molecule of glycerol and three molecules of fatty acids.
19. A saturated lipid has no double bonds between carbon atoms in the fatty acids; an unsaturated lipid has one or more double bonds. Saturated lipids have higher melting points than unsaturated lipids.
20. Phospholipids are complex lipids consisting of glycerol, two fatty acids, and a phosphate group.
21. Steroids have carbon ring structures; sterols have a functional hydroxyl group.

Proteins (pp. 42–46)

22. Amino acids are the building blocks of proteins.
23. Amino acids consist of carbon, hydrogen, oxygen, nitrogen, and sometimes sulfur.
24. Twenty amino acids occur naturally in proteins.
25. By linking amino acids, peptide bonds (formed by dehydration synthesis) allow the formation of polypeptide chains.
26. Proteins have four levels of structure: primary (sequence of amino acids), secondary (helices or pleats), tertiary (overall three-dimensional structure of a polypeptide), and quaternary (two or more polypeptide chains).
27. Conjugated proteins consist of amino acids combined with other organic or inorganic compounds.

Nucleic Acids (p. 47)

28. Nucleic acids—DNA and RNA—are macromolecules consisting of repeating nucleotides.
29. A nucleotide is composed of a pentose, a phosphate group, and a nitrogen-containing base. A nucleoside is composed of a pentose and a nitrogen-containing base.
30. A DNA nucleotide consists of deoxyribose (a pentose) and one of the following nitrogen-containing bases: thymine or cytosine (pyrimidines) or adenine or guanine (purines).
31. DNA consists of two strands of nucleotides wound in a double helix. The strands are held together by hydrogen bonds between purine and pyrimidine nucleotides: AT and GC.
32. Genes consist of sequences of nucleotides.
33. An RNA nucleotide consists of ribose (a pentose) and one of the following nitrogen-containing bases: cytosine, guanine, adenine, or uracil.

Adenosine Triphosphate (ATP) (pp. 47–49)

34. ATP stores chemical energy for various cellular activities.
35. When the bond to ATP's terminal phosphate group is hydrolyzed, energy is released.
36. The energy from oxidation reactions is used to regenerate ATP from ADP and inorganic phosphate.

STUDY QUESTIONS

Answers to the Review and Multiple Choice questions can be found by turning to the blue Answers tab at the back of the textbook.

Review

1. What is a chemical element?
2. **DRAW IT** Diagram the electronic configuration of a carbon atom.
3. What type of bond holds the following atoms together?
   a. Li^+ and Cl^- in LiCl
   b. carbon and oxygen atoms in methanol
   c. oxygen atoms in O_2
   d. a hydrogen atom of one nucleotide to a nitrogen or oxygen atom of another nucleotide in:

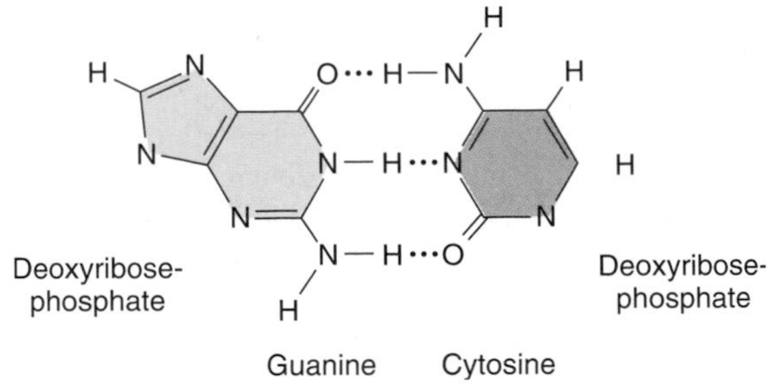

Deoxyribose-phosphate
Guanine
Cytosine
Deoxyribose-phosphate

4. Classify the following types of chemical reactions.
   a. glucose + fructose $\rightarrow$ sucrose + H_2O
   b. lactose $\rightarrow$ glucose + galactose
   c. $NH_4Cl + H_2O \rightarrow NH_4OH + HCl$
   d. ATP $\rightleftharpoons$ ADP + Ⓟ$_i$
5. Bacteria use the enzyme urease to obtain nitrogen in a form they can use from urea in the following reaction:

$$\underset{\text{Urea}}{CO(NH_2)_2} + H_2O \longrightarrow \underset{\text{Ammonia}}{2NH_3} + \underset{\text{Carbon dioxide}}{CO_2}$$

   What purpose does the enzyme serve in this reaction? What type of reaction is this?
6. Classify the following as subunits of either a carbohydrate, lipid, protein, or nucleic acid.
   a. $CH_3—(CH_2)_7—CH=CH—(CH_2)_7—COOH$
   Oleic acid
   b. $H—C(NH_2)(CH_2OH)—COOH$
   Serine
   c. $C_6H_{12}O_6$
   d. Thymine nucleotide

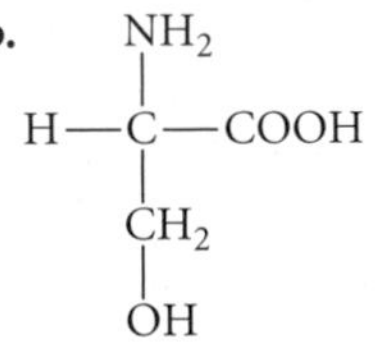

7. **DRAW IT** The artificial sweetener aspartame, or NutraSweet, is made by joining aspartic acid to methylated phenylalanine, as shown below.

$$H_2N-CH(CH_2-COOH)-C(=O)-OH \;+\; H_2N-CH(CH_2-C_6H_5)-C(=O)-O-CH_3 \;\rightleftharpoons$$

$$H_2N-CH(CH_2-COOH)-C(=O)-N(H)-CH(CH_2-C_6H_5)-C(=O)-O-CH_3 + H_2O$$

a. What types of molecules are aspartic acid and phenylalanine?
b. What direction is the hydrolysis reaction (left to right or right to left)?
c. What direction is the dehydration synthesis reaction?
d. Circle the atoms involved in the formation of water.
e. Identify the peptide bond.

8. **DRAW IT** The following diagram shows the bacteriorhodopsin protein. Indicate the regions of primary, secondary, and tertiary structure. Does this protein have quaternary structure?

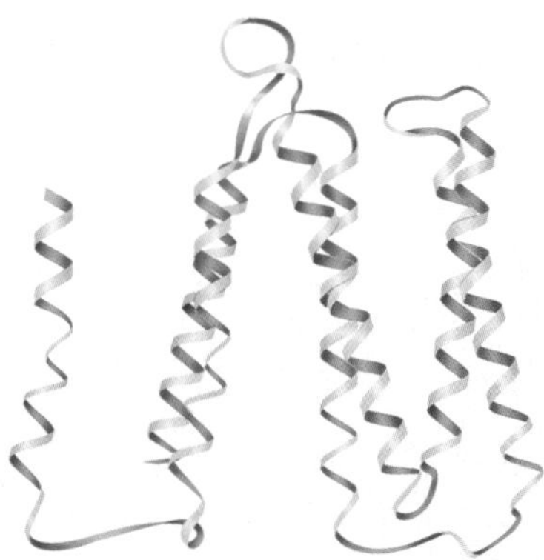

9. **DRAW IT** Draw a simple lipid, and show how it could be modified to a phospholipid.

Multiple Choice

Radioisotopes are frequently used to label molecules in a cell. The fate of atoms and molecules in a cell can then be followed. This process is the basis for questions 1–3.

1. Assume *E. coli* are grown in a nutrient medium containing the radioisotope ^{16}N. After a 48-hour incubation period, the ^{16}N would most likely be found in the *E. coli's*
   a. carbohydrates. b. lipids. c. proteins. d. water. e. none of the above

2. If *Pseudomonas* bacteria are supplied with radioactively labeled cytosine, after a 24-hour incubation period this cytosine would most likely be found in the cells'
   a. carbohydrates. b. DNA. c. lipids. d. water. e. proteins.

3. If *E. coli* were grown in a medium containing the radioactive isotope ^{32}P, the ^{32}P would be found in all of the following molecules of the cell *except*
   a. ATP. b. carbohydrates. c. DNA. d. plasma membrane. e. none of the above

4. A carbonated drink, pH 3, is __________ times more acid than distilled water.
   a. 4 b. 10 c. 100 d. 1000 e. 10,000

5. The best definition of ATP is that it is:
   a. a molecule stored for food use.
   b. a molecule that supplies energy to do work.
   c. a molecule stored for an energy reserve.
   d. a molecule used as a source of phosphate.

6. Which of the following is an organic molecule?
   a. H_2O (water)
   b. O_2 (oxygen)
   c. $C_{18}H_{29}SO_3$ (Styrofoam)
   d. FeO (iron oxide)
   e. $F_2C{=}CF_2$ (Teflon)

Classify each of the molecules on the left as an acid, base, or salt. The dissociation products of the molecules are shown to help you.

7. $HNO_3 \rightarrow H^+ + NO_3^-$
8. $H_2SO_4 \rightarrow 2H^+ + SO_4^{2-}$
9. $NaOH \rightarrow Na^+ + OH^-$
10. $MgSO_4 \rightarrow Mg^{2+} + SO_4^{2-}$

a. acid
b. base
c. salt

Critical Thinking

1. When you blow bubbles into a glass of water, the following reactions take place:

$$H_2O + CO_2 \xrightarrow{A} H_2CO_3 \xrightarrow{B} H^+ + HCO_3^-$$

   a. What type of reaction is *A*?
   b. What does reaction *B* tell you about the type of molecule H_2CO_3 is?

2. What are the common structural characteristics of ATP and DNA molecules?

3. What happens to the relative amount of unsaturated lipids in the plasma membrane when *E. coli* bacteria grown at 25°C are then grown at 37°C?

4. Giraffes, termites, and koalas eat only plant matter. Because animals cannot digest cellulose, how do you suppose these animals get nutrition from the leaves and wood they eat?

Clinical Applications

1. *Ralstonia* bacteria make poly-β-hydroxybutyrate (PHB), which is used to make a biodegradable plastic. PHB consists of many of the monomers shown below. What type of molecule is PHB? What is the most likely reason a cell would store this molecule?

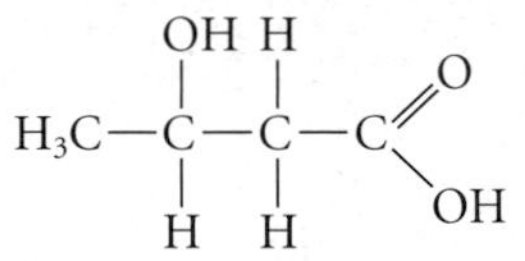

$$H_3C-\underset{H}{\overset{OH}{C}}-\underset{H}{\overset{H}{C}}-C{\scriptstyle\begin{matrix}\nearrow O\\ \searrow OH\end{matrix}}$$

2. *Thiobacillus ferrooxidans* was responsible for destroying buildings in the Midwest by causing changes in the earth. The original rock, which contained lime ($CaCO_3$) and pyrite (FeS_2), expanded as bacterial metabolism caused gypsum ($CaSO_4$) crystals to form. How did *T. ferrooxidans* bring about the change from lime to gypsum?
3. When growing in an animal, *Bacillus anthracis* produces a capsule that is resistant to phagocytosis. The capsule is composed of D-glutamic acid. Why is this capsule resistant to digestion by the host's phagocytes? (Phagocytes are white blood cells that destroy bacteria.)
4. The antibiotic amphotericin B causes leaks in cells by combining with sterols in the plasma membrane. Would you expect to use amphotericin B against a bacterial infection? A fungal infection? Offer a reason why amphotericin B has severe side effects in humans.
5. You can smell sulfur when boiling eggs. What amino acids do you expect in the egg?

3 Observing Microorganisms Through a Microscope

Microorganisms are much too small to be seen with the unaided eye; they must be observed with a microscope. The word *microscope* is derived from the Latin word *micro*, which means small, and the Greek word *skopos*, to look at. Modern microbiologists use microscopes that produce, with great clarity, magnifications that range from ten to thousands of times greater than those of van Leeuwenhoek's single lens (see Figure 1.2b on page 7). This chapter describes how different types of microscopes function and why one type might be used in preference to another.

Some microbes are more readily visible than others because of their larger size or more easily observable features. Many microbes, however, must undergo several staining procedures before their cell walls, capsules, and other structures lose their colorless natural state. The last part of this chapter explains some of the more commonly used methods of preparing specimens for examination through a light microscope.

You may wonder how we are going to sort, count, and measure the specimens we will study. To answer these questions, this chapter opens with a discussion of how to use the metric system for measuring microbes.

UNDER THE MICROSCOPE

Mycobacterium tuberculosis is the causative agent of tuberculosis.

Q&A

Acid-fast staining of a patient's sputum is a rapid, reliable, and inexpensive method to diagnose tuberculosis. What color would bacterial cells appear if the patient has tuberculosis?

Look for the answer in the chapter.

Units of Measurement

LEARNING OBJECTIVE

3-1 List the metric units of measurement that are used for microorganisms.

Because microorganisms and their component parts are so very small, they are measured in units that are unfamiliar to many of us in everyday life. When measuring microorganisms, we use the metric system. The standard unit of length in the metric system is the meter (m). A major advantage of the metric system is that the units are related to each other by factors of 10. Thus, 1 m equals 10 decimeters (dm) or 100 centimeters (cm) or 1000 millimeters (mm). Units in the U.S. system of measure do not have the advantage of easy conversion by a single factor of 10. For example, we use 3 feet or 36 inches to equal 1 yard.

Microorganisms and their structural components are measured in even smaller units, such as micrometers and nanometers. A **micrometer (μm)** is equal to 0.000001 m (10^{-6} m). The prefix *micro* indicates that the unit following it should be divided by 1 million, or 10^6 (see the "Exponential Notation" section in Appendix B). A **nanometer (nm)** is equal to 0.000000001 m (10^{-9} m). Angstrom (Å) was previously used for 10^{-10} m, or 0.1 nm.

Table 3.1 presents the basic metric units of length and some of their U.S. equivalents. In Table 3.1, you can compare the microscopic units of measurement with the commonly known macroscopic units of measurement, such as centimeters, meters, and kilometers. If you look ahead to Figure 3.2 on page 58, you will see the relative sizes of various organisms on the metric scale.

CHECK YOUR UNDERSTANDING

✓ If a microbe measures 10 μm in length, how long is it in nanometers? **3-1**

Microscopy: The Instruments

LEARNING OBJECTIVES

3-2 Diagram the path of light through a compound microscope.

3-3 Define *total magnification* and *resolution.*

3-4 Identify a use for darkfield, phase-contrast, differential interference contrast, fluorescence, confocal, two-photon, and scanning acoustic microscopy, and compare each with brightfield illumination.

3-5 Explain how electron microscopy differs from light microscopy.

3-6 Identify one use for the TEM, SEM, and scanned-probe microscopes.

The simple microscope used by van Leeuwenhoek in the seventeenth century had only one lens and was similar to a magnifying glass. However, van Leeuwenhoek was the best lens grinder in the world in his day. His lenses were ground with such precision that a single lens could magnify a microbe 300×. His simple microscopes enabled him to be the first person to see bacteria (see Figure 1.2, page 7).

Contemporaries of van Leeuwenhoek, such as Robert Hooke, built compound microscopes, which have multiple lenses. In fact, a Dutch spectacle maker, Zaccharias Janssen, is credited with making the first compound microscope around 1600. However, these early compound microscopes were of poor quality and could not be used to see bacteria. It was not until about 1830 that a significantly better microscope was developed by Joseph Jackson Lister (the father of Joseph Lister). Various improvements to Lister's microscope resulted in the development of the modern compound microscope, the kind used in microbiology laboratories today. Microscopic studies of live specimens have revealed dramatic interactions between microbes (see the Applications of Microbiology box on page 57.) **Animation** Microscopy and Staining: Overview. **www.microbiologyplace.com**

Table 3.1 Metric Units of Length and U.S. Equivalents

Metric Unit	Meaning of Prefix	Metric Equivalent	U.S. Equivalent
1 kilometer (km)	*kilo* = 1000	1000 m = 10^3 m	3280.84 ft or 0.62 mi; 1 mi = 1.61 km
1 meter (m)		Standard unit of length	39.37 in or 3.28 ft or 1.09 yd
1 decimeter (dm)	*deci* = 1/10	0.1 m = 10^{-1} m	3.94 in
1 centimeter (cm)	*centi* = 1/100	0.01 m = 10^{-2} m	0.394 in; 1 in = 2.54 cm
1 millimeter (mm)	*milli* = 1/1000	0.001 m = 10^{-3} m	
1 micrometer (μm)	*micro* = 1/1,000,000	0.000001 m = 10^{-6} m	
1 nanometer (nm)	*nano* = 1/1,000,000,000	0.000000001 m = 10^{-9} m	
1 picometer (pm)	*pico* = 1/1,000,000,000,000	0.000000000001 m = 10^{-12} m	

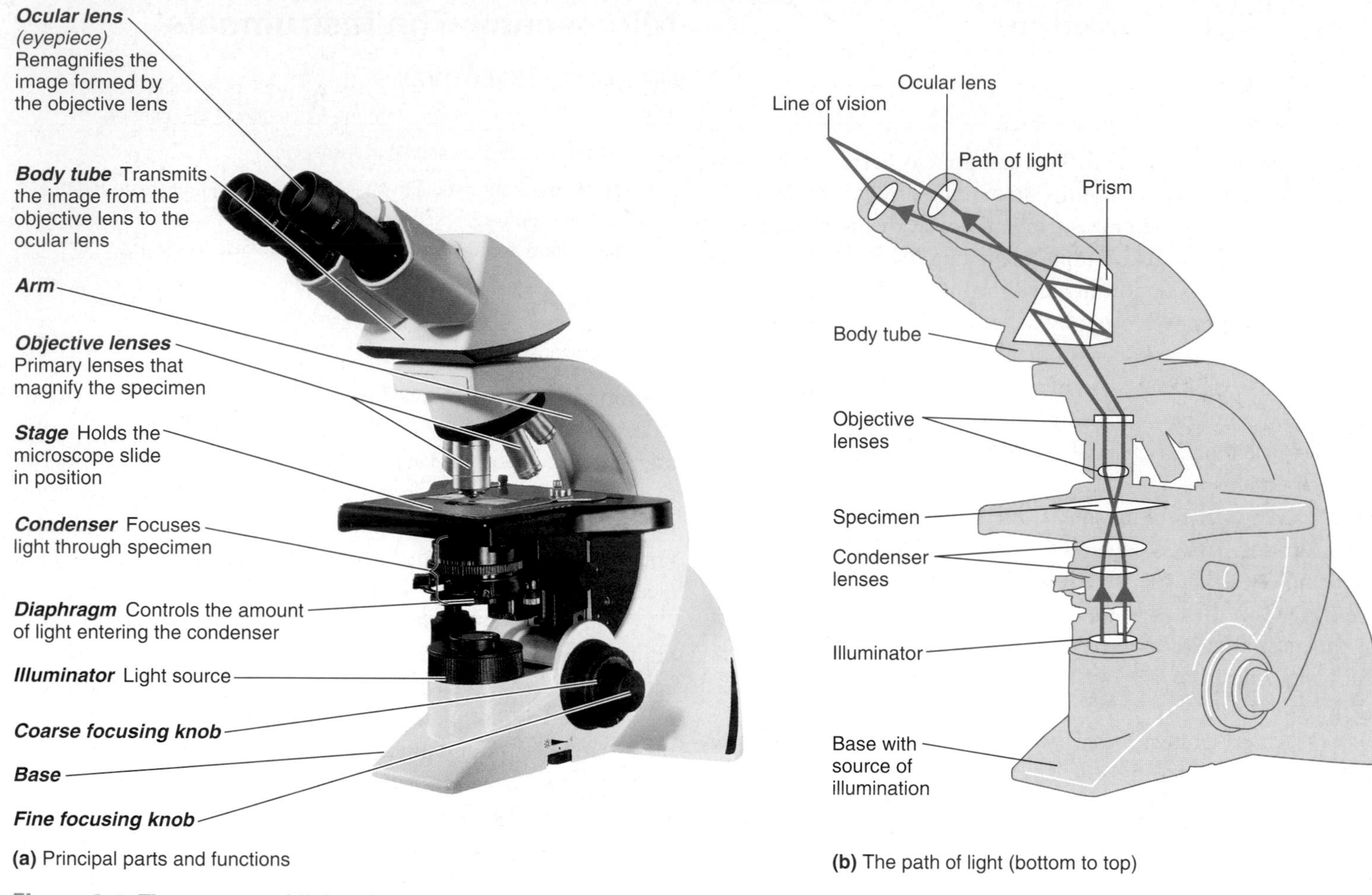

Figure 3.1 The compound light microscope.

Q How is the total magnification of a compound light microscope calculated?

Light Microscopy

Light microscopy refers to the use of any kind of microscope that uses visible light to observe specimens. Here we examine several types of light microscopy.

Compound Light Microscopy

A modern **compound light microscope** has a series of lenses and uses visible light as its source of illumination (Figure 3.1a). With a compound light microscope, we can examine very small specimens as well as some of their fine detail. A series of finely ground lenses (Figure 3.1b) forms a clearly focused image that is many times larger than the specimen itself. This magnification is achieved when light rays from an **illuminator,** the light source, pass through a **condenser,** which has lenses that direct the light rays through the specimen. From here, light rays pass into the **objective lenses,** the lenses closest to the specimen. The image of the specimen is magnified again by the **ocular lens,** or *eyepiece.*

We can calculate the **total magnification** of a specimen by multiplying the objective lens magnification (power) by the ocular lens magnification (power). Most microscopes used in microbiology have several objective lenses, including 10× (low power), 40× (high power), and 100× (oil immersion, which is described shortly). Most ocular lenses magnify specimens by a factor of 10. Multiplying the magnification of a specific objective lens with that of the ocular, we see that the total magnifications would be 100× for low power, 400× for high power, and 1000× for oil immersion. Some compound light microscopes can achieve a magnification of 2000× with the oil immersion lens.

Resolution (also called *resolving power*) is the ability of the lenses to distinguish fine detail and structure. Specifically, it refers to the ability of the lenses to distinguish two points a specified distance apart. For example, if a microscope has a resolving power of 0.4 nm, it can distinguish two points if they are at least 0.4 nm apart. A general principle of microscopy is that the shorter the wavelength of light used in the instrument, the greater the resolution. The white light used in a compound light microscope has a relatively long wavelength and cannot resolve structures smaller than about 0.2 μm. This fact and other practical

What Is That Slime?

When bacteria grow, they often stay together in packs called biofilms. This can result in a slimy film on rocks, on food, inside pipes, and on implanted medical devices. Bacterial cells interact and exhibit multicellular organization (**Figure A**).

Pseudomonas aeruginosa can grow within a human without causing disease until they form a biofilm that overcomes the host's immune system. Biofilm-forming *P. aeruginosa* bacteria colonize the lungs of cystic fibrosis patients and are a leading cause of death in these patients (**Figure B**). Perhaps biofilms that lead to disease can be prevented by new drugs that destroy the inducer (discussed shortly).

Figure A *Paenibacillus.* As one small colony moves away from the parent colony, other groups of cells follow the first colony. Soon, all of the other bacteria join the relocation to form this spiraling colony.

Myxobacteria

Myxobacteria are found in decaying organic material and fresh water throughout the world. Although they are bacteria, many myxobacteria never exist as individual cells. *Myxococcus xanthus* cells appear to hunt in packs. In their natural aqueous habitat, *M. xanthus* cells form spherical colonies that surround prey bacteria, where they can secrete digestive enzymes and absorb the nutrients. On solid substrates, other myxobacterial cells glide over a solid surface, leaving slime trails that are followed by other cells. When food is scarce, the cells aggregate to form a mass. Cells within the mass differentiate into a fruiting body that consists of a slime stalk and clusters of spores, as shown in **Figure C**.

Vibrio

Vibrio fischeri is a bioluminescent bacterium that lives as a symbiont in the light-producing organ of squid and certain fish. When free-living, the bacteria are at a low concentration and do not give off light. However, when they grow in their host, they are highly concentrated, and each cell is induced to produce the enzyme luciferase, which is used in the chemical pathway of bioluminescence.

How Bacterial Group Behavior Works

Cell density alters gene expression in bacterial cells in a process called quorum sensing. In law, a quorum is the minimum number of members necessary to conduct business. *Quorum sensing* is the ability of bacteria to communicate and coordinate behavior. Bacteria that use quorum sensing produce and secrete a signaling chemical called an *inducer.* As the inducer diffuses into the surrounding medium, other bacterial cells move toward the source and begin producing inducer. The concentration of inducer increases with increasing cell numbers. This, in turn, attracts more cells and initiates synthesis of more inducer.

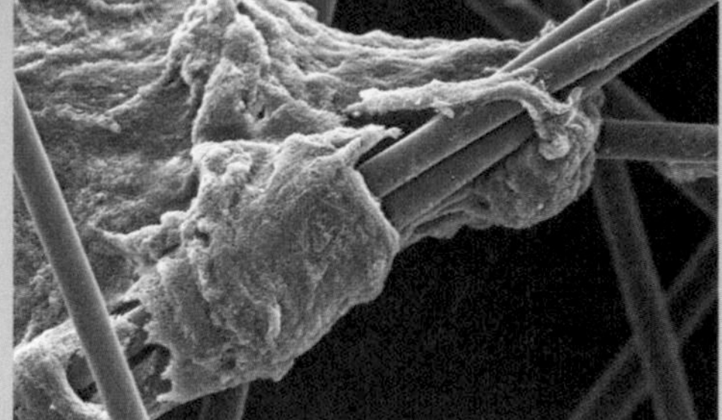

Figure B *Pseudomonas aeruginosa* biofilm. 5 µm LM

Figure C A fruiting body of a myxobacterium.

Figure 3.2

FOUNDATION FIGURE Sizes and Resolutions

This figure illustrates two concepts: 1) the relative sizes of various specimens, and 2) the resolution of the human eye and several microscopes. The micrographs included throughout this textbook (like the ones below) will have size bars and symbols to help you identify the actual size of the specimen and the type of microscope used for that image. A red symbol indicates that a micrograph has been artificially colorized. See Table 3.2 on pages 66–68 for more examples.

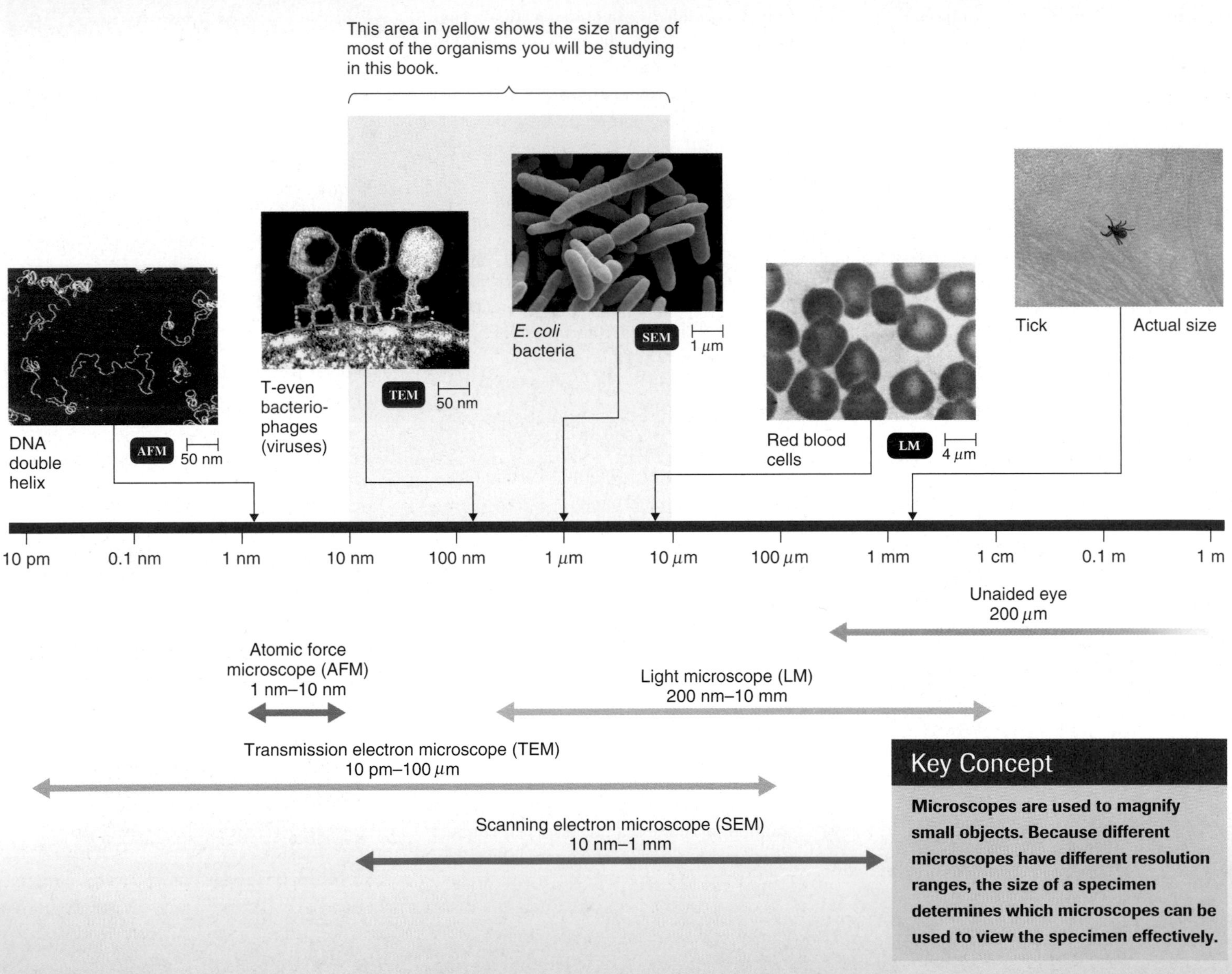

Key Concept

Microscopes are used to magnify small objects. Because different microscopes have different resolution ranges, the size of a specimen determines which microscopes can be used to view the specimen effectively.

considerations limit the magnification achieved by even the best compound light microscopes to about 2000×. By comparison, van Leeuwenhoek's microscopes had a resolution of 1 μm.

Figure 3.2 shows various specimens that can be resolved by the human eye, light microscope, and electron microscope.

To obtain a clear, finely detailed image under a compound light microscope, specimens must be made to contrast sharply with their *medium* (substance in which they are suspended). To attain such contrast, we must change the refractive index of specimens from that of their medium. The **refractive index** is a measure of the light-bending ability of a medium. We change the refractive index of specimens by staining them, a procedure we will discuss shortly. Light rays move in a straight line through a single medium. After the specimen is stained, when light rays pass through the two materials (the specimen and its medium) with different refractive indexes, the rays change direction (refract) from a straight path by bending or changing angle at the boundary between the materials and increase the image's contrast between the specimen and the medium. As the light rays travel away from the specimen, they spread out and enter the objective lens, and the image is thereby magnified.

To achieve high magnification (1000×) with good resolution, the objective lens must be small. Although we want light traveling through the specimen and medium to refract differently, we do not want to lose light rays after they have passed through the stained specimen. To preserve the direction of light rays at the highest magnification, immersion oil is placed between the glass slide and the oil immersion objective lens (**Figure 3.3**). The immersion oil has the same refractive index as glass, so the oil becomes part of the optics of the glass of the microscope. Unless immersion oil is used, light rays are refracted as they enter the air from the slide, and the objective lens would have to be increased in diameter to capture most of them. The oil has the same effect as increasing the objective lens diameter; therefore, it improves the resolving power of the lenses. If oil is not used with an oil immersion objective lens, the image becomes fuzzy, with poor resolution.

Under usual operating conditions, the field of vision in a compound light microscope is brightly illuminated. By focusing the light, the condenser produces a **brightfield illumination** (**Figure 3.4a**).

It is not always desirable to stain a specimen. However, an unstained cell has little contrast with its surroundings and is therefore difficult to see. Unstained cells are more easily observed with the modified compound microscopes described in the next section.

CHECK YOUR UNDERSTANDING

- ✓ Through what lenses does light pass in a compound microscope? **3-2**
- ✓ What does it mean when a microscope has a resolution of 0.2 nm? **3-3**

Darkfield Microscopy

A **darkfield microscope** is used to examine live microorganisms that either are invisible in the ordinary light microscope, cannot be stained by standard methods, or are so distorted by staining that their characteristics then cannot be identified. Instead of the normal condenser, a darkfield microscope uses a darkfield condenser that contains an opaque disk. The disk blocks light that would enter the objective lens directly. Only light that is reflected off (turned away from) the specimen enters the objective lens. Because there is no direct background light, the specimen appears light against a black background—the dark field (**Figure 3.4b**). This technique is frequently used to examine unstained microorganisms suspended in liquid. One use for darkfield microscopy is the examination of very thin spirochetes, such as *Treponema pallidum* (tre-pō-nē′mä pal′li-dum), the causative agent of syphilis.

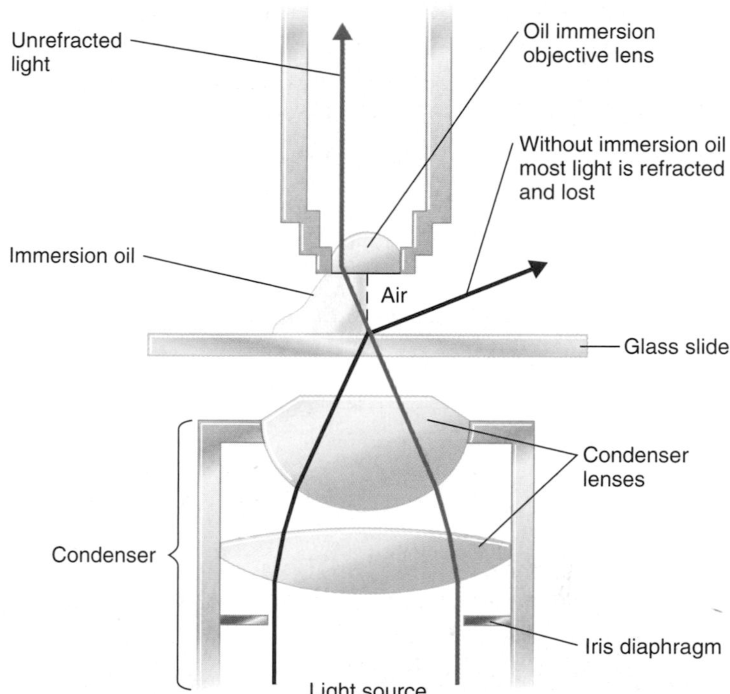

Figure 3.3 Refraction in the compound microscope using an oil immersion objective lens. Because the refractive indexes of the glass microscope slide and immersion oil are the same, the light rays do not refract when passing from one to the other when an oil immersion objective lens is used. This method produces images with better resolution at magnifications greater than 900×.

Q What is meant by *resolution*?

Phase-Contrast Microscopy

Another way to observe microorganisms is with a **phase-contrast microscope.** Phase-contrast microscopy is especially useful because it permits detailed examination of internal structures in *living* microorganisms. In addition, it is not necessary to fix (attach the microbes to the microscope slide) or stain the specimen—procedures that could distort or kill the microorganisms.

The principle of phase-contrast microscopy is based on the wave nature of light rays and the fact that light rays can be *in phase* (their peaks and valleys match) or *out of phase.* If the wave peak of light rays from one source coincides with the wave peak of light rays from another source, the rays interact to produce *reinforcement* (relative brightness). However, if the wave peak from one light source coincides with the wave trough from another light source, the rays interact to produce *interference* (relative darkness). In a phase-contrast microscope, one set of light rays comes directly from the light source. The other set

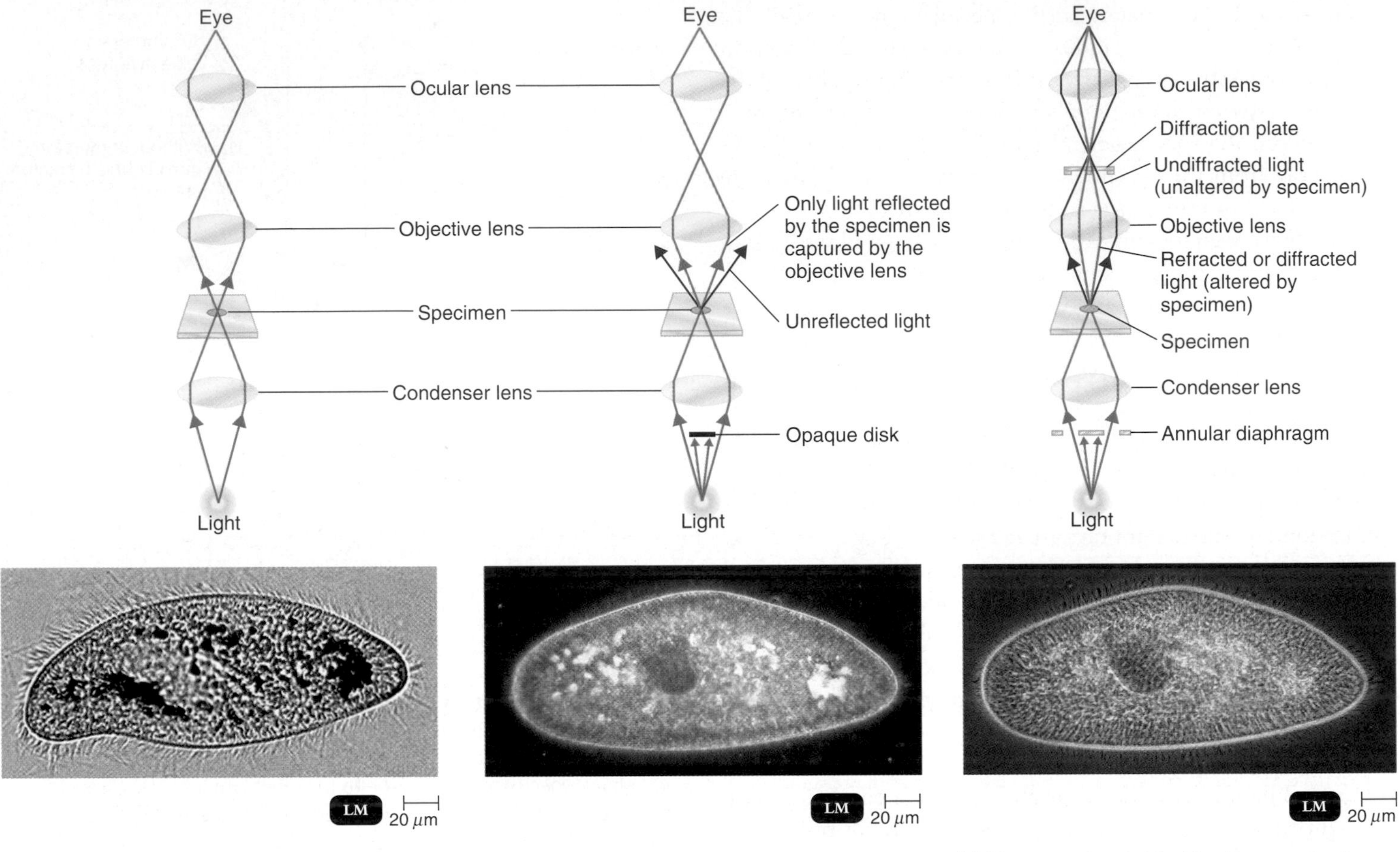

(a) Brightfield. (Top) The path of light in brightfield microscopy, the type of illumination produced by regular compound light microscopes. (Bottom) Brightfield illumination shows internal structures and the outline of the transparent pellicle (external covering).

(b) Darkfield. (Top) The darkfield microscope uses a special condenser with an opaque disk that eliminates all light in the center of the beam. The only light that reaches the specimen comes in at an angle; thus, only light reflected by the specimen (blue rays) reaches the objective lens. (Bottom) Against the black background seen with darkfield microscopy, edges of the cell are bright, some internal structures seem to sparkle, and the pellicle is almost visible.

(c) Phase-contrast. (Top) In phase-contrast microscopy, the specimen is illuminated by light passing through an annular (ring shaped) diaphragm. Direct light rays (unaltered by the specimen) travel a different path than light rays that are reflected or diffracted as they pass through the specimen. These two sets of rays are combined at the eye. Reflected or diffracted light rays are indicated in blue; direct rays are red. (Bottom) Phase-contrast microscopy shows greater differentiation of internal structures and clearly shows the pellicle.

Figure 3.4 Brightfield, darkfield, and phase-contrast microscopy. The illustrations show the contrasting light pathways of each of these types of microscopy. The photographs compare the protozoan *Paramecium* using these three different microscopy techniques.

Q What are the advantages of brightfield, darkfield, and phase-contrast microscopy?

comes from light that is reflected or diffracted from a particular structure in the specimen. (*Diffraction* is the scattering of light rays as they "touch" a specimen's edge. The diffracted rays are bent away from the parallel light rays that pass farther from the specimen.) When the two sets of light rays—direct rays and reflected or diffracted rays—are brought together, they form an image of the specimen on the ocular lens, containing areas that are relatively light (in phase), through shades of gray, to black (out of phase; Figure 3.4c). In phase-contrast microscopy, the internal structures of a cell become more sharply defined.

Differential Interference Contrast (DIC) Microscopy

Differential interference contrast (DIC) microscopy is similar to phase-contrast microscopy in that it uses differences in refractive indexes. However, a DIC microscope uses two beams of light instead of one. In addition, prisms split each light beam, adding contrasting colors to the specimen. Therefore, the resolution of a DIC microscope is higher than that of a standard phase-contrast microscope. Also, the image is brightly colored and appears nearly three-dimensional (Figure 3.5).

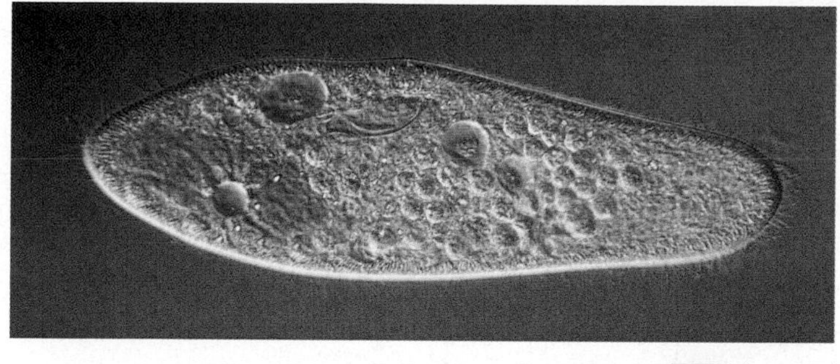

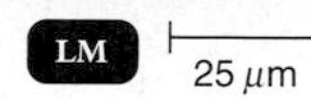

Figure 3.5 Differential interference contrast (DIC) microscopy. Like phase-contrast, DIC uses differences in refractive indexes to produce an image, in this case a *Paramecium.* The colors in the image are produced by prisms that split the two light beams used in this process.

Q Why is the resolution of a DIC microscope higher than that of a phase-contrast microscope?

Fluorescence Microscopy

Fluorescence microscopy takes advantage of **fluorescence,** the ability of substances to absorb short wavelengths of light (ultraviolet) and give off light at a longer wavelength (visible). Some organisms fluoresce naturally under ultraviolet light; if the specimen to be viewed does not naturally fluoresce, it is stained with one of a group of fluorescent dyes called *fluorochromes.* When microorganisms stained with a fluorochrome are examined under a fluorescence microscope with an ultraviolet or near-ultraviolet light source, they appear as luminescent, bright objects against a dark background.

Fluorochromes have special attractions for different microorganisms. For example, the fluorochrome auramine O, which glows yellow when exposed to ultraviolet light, is strongly absorbed by *Mycobacterium tuberculosis,* the bacterium that causes tuberculosis. When the dye is applied to a sample of material suspected of containing the bacterium, the bacterium can be detected by the appearance of bright yellow organisms against a dark background. *Bacillus anthracis,* the causative agent of anthrax, appears apple green when stained with another fluorochrome, fluorescein isothiocyanate (FITC).

The principal use of fluorescence microscopy is a diagnostic technique called the **fluorescent-antibody (FA) technique,** or **immunofluorescence. Antibodies** are natural defense molecules that are produced by humans and many animals in reaction to a foreign substance, or **antigen.** Fluorescent antibodies for a particular antigen are obtained as follows: an animal is injected with a specific antigen, such as a bacterium, and the animal then begins to produce antibodies against that antigen. After a sufficient time, the antibodies are removed from the serum of the animal. Next, as shown in Figure 3.6a, a fluorochrome is chemically combined with the antibodies. These fluorescent antibodies

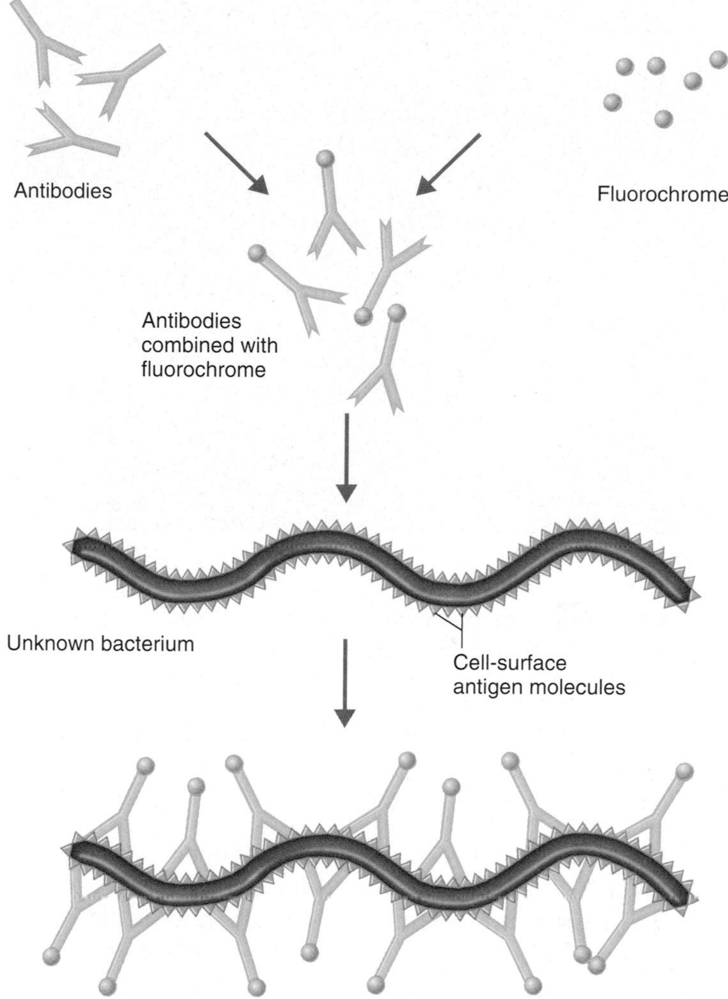

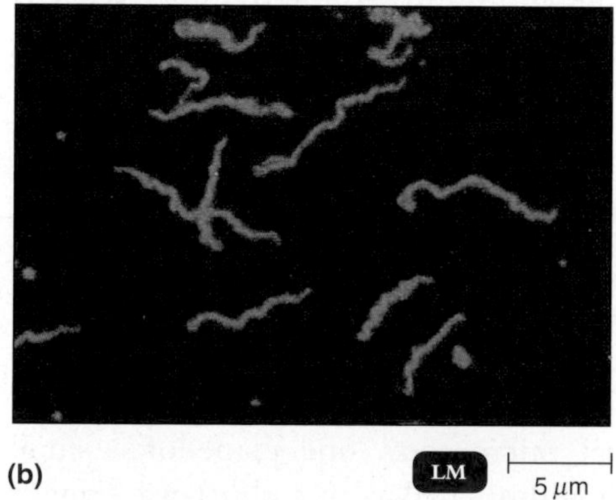

Figure 3.6 The principle of immunofluorescence. (**a**) A type of fluorochrome is combined with antibodies against a specific type of bacterium. When the preparation is added to bacterial cells on a microscope slide, the antibodies attach to the bacterial cells, and the cells fluoresce when illuminated with ultraviolet light. (**b**) In the fluorescent treponemal antibody absorption (FTA-ABS) test for syphilis shown here, *Treponema pallidum* shows up as green cells against a darker background.

Q Why won't other bacteria fluoresce in the FTA-ABS test?

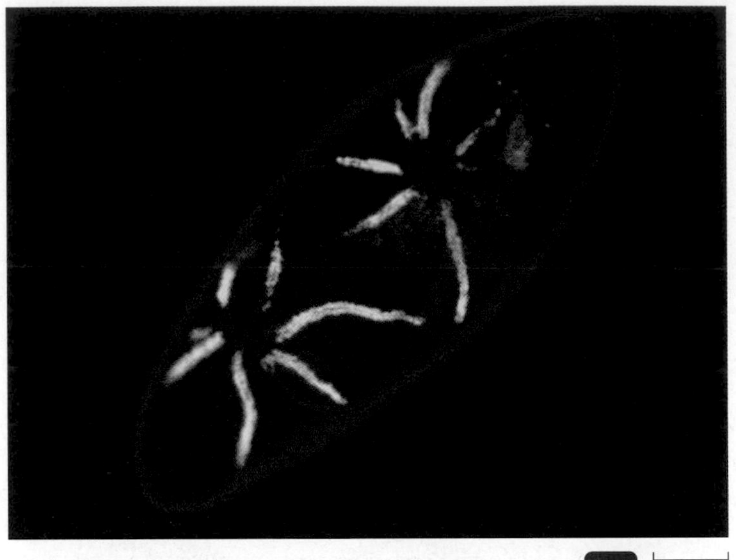

Figure 3.7 Confocal microscopy. Confocal microscopy produces three-dimensional images and can be used to look inside cells. Shown here are contractile vacuoles in *Paramecium multimicronucleatum.*

Q What feature of confocal microscopy eliminates the blurring that occurs with other microscopes?

are then added to a microscope slide containing an unknown bacterium. If this unknown bacterium is the same bacterium that was injected into the animal, the fluorescent antibodies bind to antigens on the surface of the bacterium, causing it to fluoresce.

This technique can detect bacteria or other pathogenic microorganisms, even within cells, tissues, or other clinical specimens (**Figure 3.6b**). Of paramount importance, it can be used to identify a microbe in minutes. Immunofluorescence is especially useful in diagnosing syphilis and rabies. We will say more about antigen-antibody reactions and immunofluorescence in Chapter 18.

Confocal Microscopy

Confocal microscopy is a technique in light microscopy used to reconstruct three-dimensional images. Like fluorescent microscopy, specimens are stained with fluorochromes so they will emit, or return, light. But instead of illuminating the entire field, in confocal microscopy, one plane of a small region of a specimen is illuminated with a short-wavelength (blue) light which passes the returned light through an aperture aligned with the illuminated region. Each plane corresponds to an image of a fine slice that has been physically cut from a specimen. Successive planes and regions are illuminated until the entire specimen has been scanned. Because confocal microscopy uses a pinhole aperture, it eliminates the blurring that occurs with other microscopes. As a result, exceptionally clear two-dimensional images can be obtained, with improved resolution of up to 40% over that of other microscopes.

Most confocal microscopes are used in conjunction with computers to construct three-dimensional images. The scanned planes of a specimen, which resemble a stack of images, are converted to a digital form that can be used by a computer to construct a three-dimensional representation. The reconstructed images can be rotated and viewed in any orientation. This technique has been used to obtain three-dimensional images of entire cells and cellular components (**Figure 3.7**). In addition, confocal microscopy can be used to evaluate cellular physiology by monitoring the distributions and concentrations of substances such as ATP and calcium ions.

Two-Photon Microscopy

As in confocal microscopy, specimens are stained with a fluorochrome for **two-photon microscopy (TPM).** Two-photon microscopy uses long-wavelength (red) light, and therefore two photons, instead of one, are needed to excite the fluorochrome to emit light. The longer wavelength allows imaging of living cells in tissues up to 1 mm deep (**Figure 3.8**). Confocal microscopy can image cells in detail only to a depth of less than 100 µm. Additionally, the longer wavelength is less likely to generate singlet oxygen, which damages cells (see page 161). Another advantage of TPM is that it can track the activity of cells in real time. For example, cells of the immune system have been observed responding to an antigen. **Animation** Light Microscopy. **www.microbiologyplace.com**

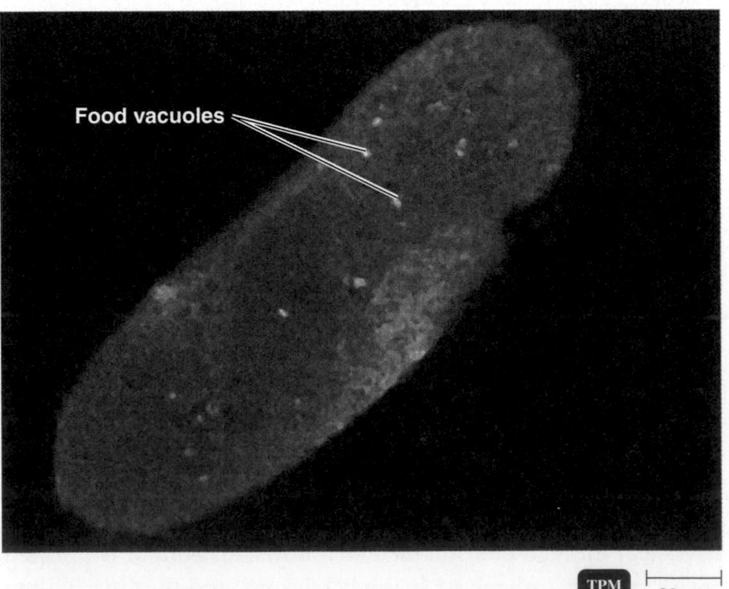

Figure 3.8 Two-photon microscopy (TPM). This procedure makes it possible to image living cells up to 1 mm deep in detail. This image shows movement of food vacuoles in a living *Paramecium.*

What are the advantages of TPM?

Scanning Acoustic Microscopy

Scanning acoustic microscopy (SAM) basically consists of interpreting the action of a sound wave sent through a specimen. A sound wave of a specific frequency travels through the specimen, and a portion of it is reflected back every time it hits an interface within the material. The resolution is about 1 μm. SAM is used to study living cells attached to another surface, such as cancer cells, artery plaque, and bacterial biofilms that foul equipment (Figure 3.9).

CHECK YOUR UNDERSTANDING

✓ How are brightfield, darkfield, phase-contrast, and fluorescence microscopy similar? **3-4**

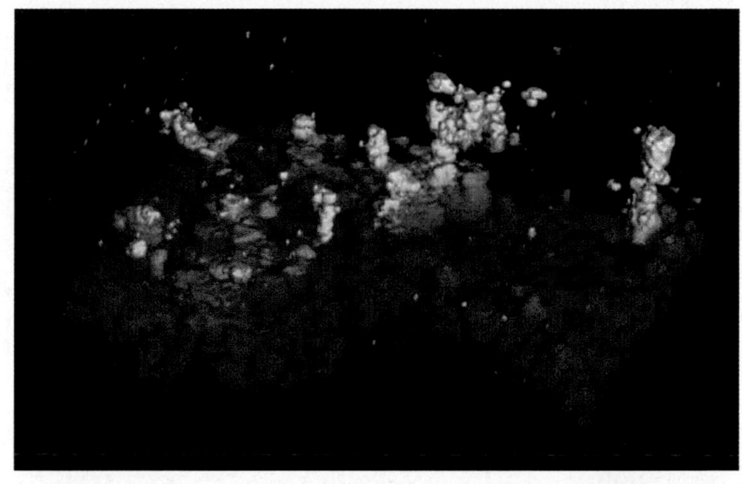

Figure 3.9 Scanning acoustic microscopy (SAM) of a bacterial biofilm on glass. Scanning acoustic microscopy essentially consists of interpreting the action of sound waves through a specimen. © 2006 IEEE.

Q What is the principal use of SAM?

Electron Microscopy

Objects smaller than about 0.2 μm, such as viruses or the internal structures of cells, must be examined with an **electron microscope.** In electron microscopy, a beam of electrons is used instead of light. Like light, free electrons travel in waves. The resolving power of the electron microscope is far greater than that of the other microscopes described here so far. The better resolution of electron microscopes is due to the shorter wavelengths of electrons; the wavelengths of electrons are about 100,000 times smaller than the wavelengths of visible light. Thus, electron microscopes are used to examine structures too small to be resolved with light microscopes. Images produced by electron microscopes are always black and white, but they may be colored artificially to accentuate certain details.

Instead of using glass lenses, an electron microscope uses electromagnetic lenses to focus a beam of electrons onto a specimen. There are two types of electron microscopes: the transmission electron microscope and the scanning electron microscope.

Transmission Electron Microscopy

In the **transmission electron microscope (TEM),** a finely focused beam of electrons from an electron gun passes through a specially prepared, ultrathin section of the specimen (Figure 3.10a). The beam is focused on a small area of the specimen by an electromagnetic condenser lens that performs roughly the same function as the condenser of a light microscope—directing the beam of electrons in a straight line to illuminate the specimen.

Electron microscopes use electromagnetic lenses to control illumination, focus, and magnification. Instead of being placed on a glass slide, as in light microscopes, the specimen is usually placed on a copper mesh grid. The beam of electrons passes through the specimen and then through an electromagnetic objective lens, which magnifies the image. Finally, the electrons are focused by an electromagnetic projector lens (rather than by an ocular lens as in a light microscope) onto a fluorescent screen or photographic plate. The final image, called a *transmission electron micrograph,* appears as many light and dark areas, depending on the number of electrons absorbed by different areas of the specimen.

In practice, the transmission electron microscope can resolve objects as close together as 2.5 nm, and objects are generally magnified 10,000 to 100,000×. Because most microscopic specimens are so thin, the contrast between their ultrastructures and the background is weak. Contrast can be greatly enhanced by using a "stain" that absorbs electrons and produces a darker image in the stained region. Salts of various heavy metals, such as lead, osmium, tungsten, and uranium, are commonly used as stains. These metals can be fixed onto the specimen (*positive staining*) or used to increase the electron opacity of the surrounding field (*negative staining*). Negative staining is useful for the study of the very smallest specimens, such as virus particles, bacterial flagella, and protein molecules.

In addition to positive and negative staining, a microbe can be viewed by a technique called *shadow casting.* In this procedure, a heavy metal such as platinum or gold is sprayed at an angle of about 45° so that it strikes the microbe from only one side. The metal piles up on one side of the specimen, and the uncoated area on the opposite side of the specimen leaves a clear area behind it as a shadow. This gives a three-dimensional effect to the specimen and provides a general idea of the size and shape of the specimen (see the TEM in Figure 4.6, page 80).

Transmission electron microscopy has high resolution and is extremely valuable for examining different layers of specimens. However, it does have certain disadvantages. Because

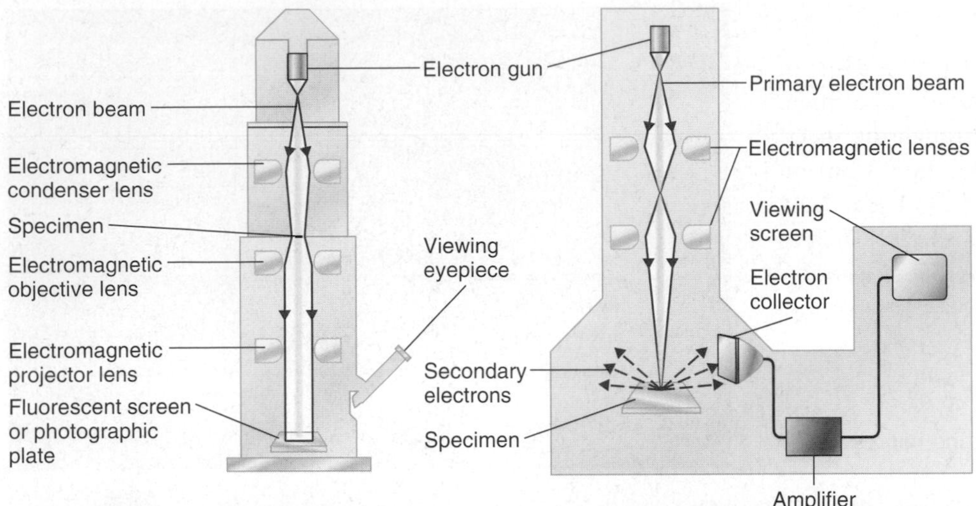

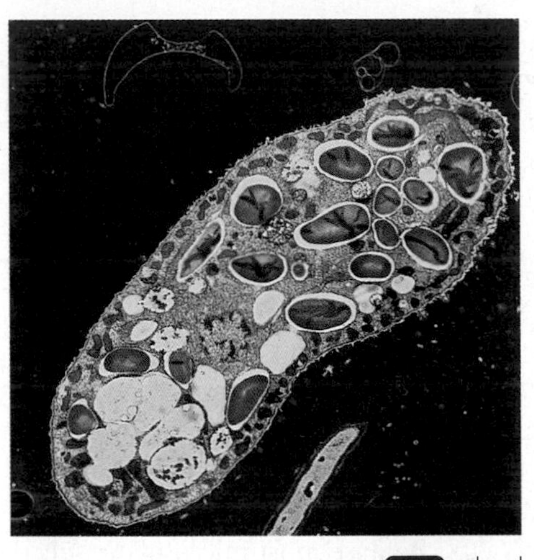

(a) Transmission. (Top) In a transmission electron microscope, electrons pass through the specimen and are scattered. Magnetic lenses focus the image onto a fluorescent screen or photographic plate. (Bottom) This colorized transmission electron micrograph (TEM) shows a thin slice of *Paramecium.* In this type of microscopy, the internal structures present in the slice can be seen.

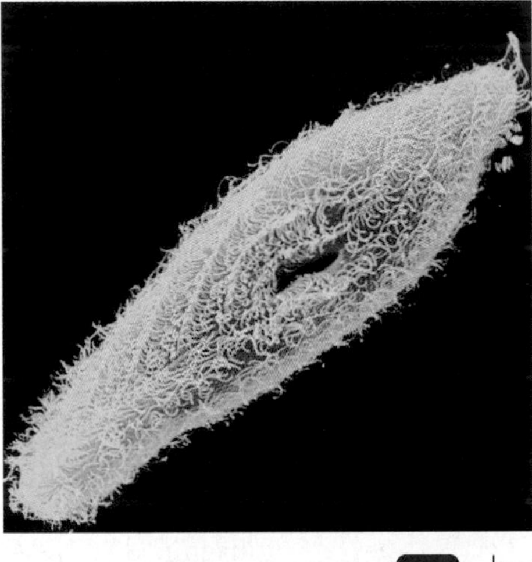

(b) Scanning. (Top) In a scanning electron microscope, primary electrons sweep across the specimen and knock electrons from its surface. These secondary electrons are picked up by a collector, amplified, and transmitted onto a viewing screen or photographic plate. (Bottom) In this colorized scanning electron micrograph (SEM), the surface structures of *Paramecium* can be seen. Note the three-dimensional appearance of this cell, in contrast to the two-dimensional appearance of the transmission electron micrograph in part (a).

Figure 3.10 Transmission and scanning electron microscopy. The illustrations show the pathways of electron beams used to create images of the specimens. The photographs show a *Paramecium* viewed with both of these types of electron microscopes. Although electron micrographs are normally black and white, these and other electron micrographs in this book have been artifically colorized for emphasis.

Q How do TEM and SEM images of the same organism differ?

electrons have limited penetrating power, only a very thin section of a specimen (about 100 nm) can be studied effectively. Thus, the specimen has no three-dimensional aspect. In addition, specimens must be fixed, dehydrated, and viewed under a high vacuum to prevent electron scattering. These treatments not only kill the specimen, but also cause some shrinkage and distortion, sometimes to the extent that there may appear to be additional structures in a prepared cell. Structures that appear as a result of the method of preparation are called *artifacts.*

Scanning Electron Microscopy

The **scanning electron microscope (SEM)** overcomes the problem of sectioning associated with a transmission electron microscope. A scanning electron microscope provides striking three-dimensional views of specimens (Figure 3.10b). In scanning electron microscopy, an electron gun produces a finely focused beam of electrons called the primary electron beam. These electrons pass through electromagnetic lenses and are directed over the surface of the specimen. The primary electron beam knocks electrons out of the surface of the specimen,

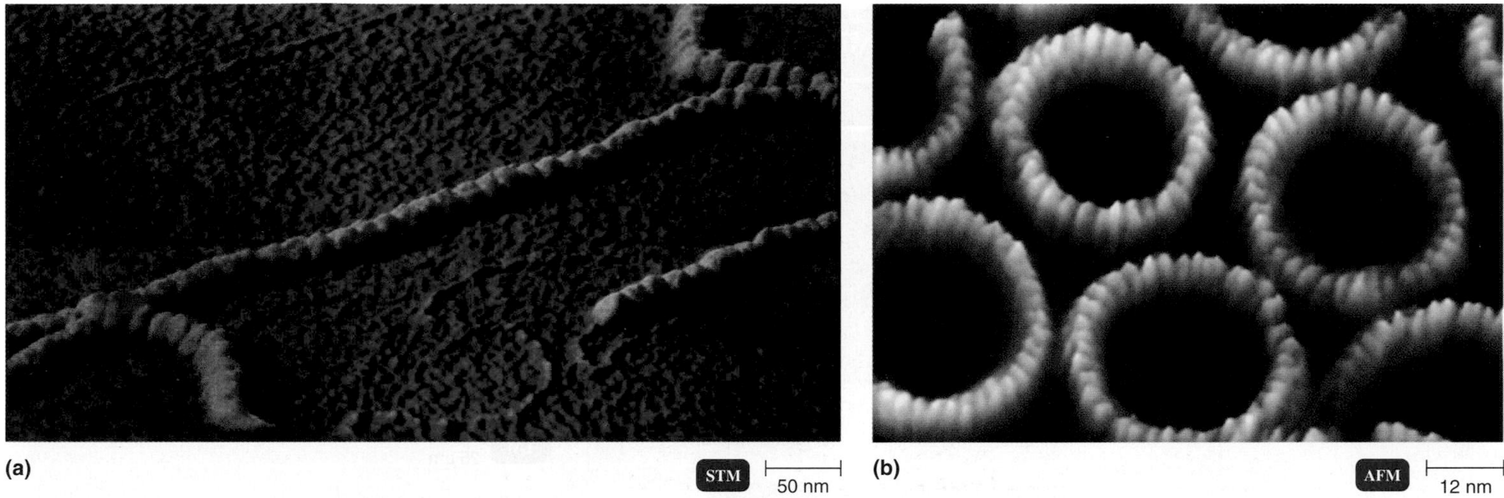

Figure 3.11 Scanned-probe microscopy. (**a**) Scanning tunneling microscopy (STM) image of RecA protein from *E.coli*. This protein is involved in repair of DNA. (**b**) Atomic force microscopy (AFM) image of perfringoglysin O toxin from *Clostridium perfringens*. This toxin makes holes in human plasma membranes.

Q What is the principle employed in scanned-probe microscopy?

and the secondary electrons thus produced are transmitted to an electron collector, amplified, and used to produce an image on a viewing screen or photographic plate. The image is called a *scanning electron micrograph*. This microscope is especially useful in studying the surface structures of intact cells and viruses. In practice, it can resolve objects as close together as 10 nm, and objects are generally magnified 1000 to 10,000×.

CHECK YOUR UNDERSTANDING

✓ Why do electron microscopes have greater resolution than light microscopes? **3-5**

Scanned-Probe Microscopy

Since the early 1980s, several new types of microscopes, called **scanned-probe microscopes,** have been developed. They use various kinds of probes to examine the surface of a specimen at very close range, and they do so without modifying the specimen or exposing it to damaging, high-energy radiation. Such microscopes can be used to map atomic and molecular shapes, to characterize magnetic and chemical properties, and to determine temperature variations inside cells. Among the new scanned-probe microscopes are the scanning tunneling microscope and the atomic force microscope, discussed next.

Scanning Tunneling Microscopy

Scanning tunneling microscopy (STM) uses a thin metal (tungsten) probe that scans a specimen and produces an image revealing the bumps and depressions of the atoms on the surface of the specimen (**Figure 3.11a**). The resolving power of an STM is much greater than that of an electron microscope; it can resolve features that are only about 1/100 the size of an atom. Moreover, special preparation of the specimen for observation is not needed. STMs are used to provide incredibly detailed views of molecules such as DNA.

Atomic Force Microscopy

In **atomic force microscopy (AFM),** a metal-and-diamond probe is gently forced down onto a specimen. As the probe moves along the surface of the specimen, its movements are recorded, and a three-dimensional image is produced (**Figure 3.11b**). As with STM, AFM does not require special specimen preparation. AFM is used to image both biological substances (in nearly atomic detail) (See also Figure 17.3b on page 480.) and molecular processes (such as the assembly of fibrin, a component of a blood clot).

The various types of microscopy just described are summarized in **Table 3.2** (pp. 66–68). **Animation** Electron Microscopy. **www.microbiologyplace.com**

CHECK YOUR UNDERSTANDING

✓ For what is TEM used? SEM? Scanned-probe microscopy? **3-6**

Table 3.2 A Summary of Various Types of Microscopes (continued)

Microscope Type	Distinguishing Features	Typical Image	Principal Uses
Scanned-Probe			
Scanning tunneling	Uses a thin metal probe that scans a specimen and produces an image revealing the bumps and depressions of the atoms on the surface of the specimen. Resolving power is much greater than that of an electron microscope. No special preparation required.	RecA protein from *E. coli* STM 5 nm	Provides very detailed views of molecules inside cells.
Atomic force	Uses a metal-and-diamond probe gently forced down along the surface of the specimen. Produces a three-dimensional image. No special preparation required.	Penfringoglycin O toxin from *Clostridium perfringens* AFM 11 nm	Provides three-dimensional images of biological specimens at high resolution in nearly atomic detail and can measure physical properties of biological specimens and molecular processes.

Preparation of Specimens for Light Microscopy

LEARNING OBJECTIVES

3-7 Differentiate an acidic dye from a basic dye.

3-8 Explain the purpose of simple staining.

3-9 List the steps in preparing a Gram stain, and describe the appearance of gram-positive and gram-negative cells after each step.

3-10 Compare and contrast the Gram stain and the acid-fast stain.

3-11 Explain why each of the following is used: capsule stain, endospore stain, flagella stain.

Because most microorganisms appear almost colorless when viewed through a standard light microscope, we often must prepare them for observation. One way to do this is to stain (color) the specimen. Next we will discuss several different staining procedures.

Preparing Smears for Staining

Most initial observations of microorganisms are made with stained preparations. **Staining** simply means coloring the microorganisms with a dye that emphasizes certain structures. Before the microorganisms can be stained, however, they must be **fixed** (attached) to the microscope slide. Fixing simultaneously kills the microorganisms and fixes them to the slide. It also preserves various parts of microbes in their natural state with only minimal distortion.

When a specimen is to be fixed, a thin film of material containing the microorganisms is spread over the surface of the slide. This film, called a **smear,** is allowed to air dry. In most staining procedures the slide is then fixed by passing it through the flame of a Bunsen burner several times, smear side up, or by covering the slide with methyl alcohol for 1 minute. Stain is applied and then washed off with water; then the slide is blotted with absorbent paper. Without fixing, the stain might wash the microbes off the slide. The stained microorganisms are now ready for microscopic examination.

Stains are salts composed of a positive and a negative ion, one of which is colored and is known as the *chromophore.* The color of so-called **basic dyes** is in the positive ion; in **acidic dyes,** it is in the negative ion. Bacteria are slightly negatively charged at pH 7. Thus, the colored positive ion in a basic dye is attracted to the negatively charged bacterial cell. Basic dyes, which include crystal violet, methylene blue, malachite green, and safranin, are more commonly used than acidic dyes. Acidic dyes are not attracted to most types of bacteria because the dye's negative ions are repelled by the negatively charged bacterial surface, so the stain colors the background instead. Preparing colorless bacteria against a colored background is called **negative staining.** It is valuable for observing overall cell shapes, sizes, and capsules because the cells are made highly visible against a contrasting dark background (see Figure 3.14a on page 72). Distortions of cell size and shape are minimized because fixing is not necessary and the cells do not pick up the stain. Examples of acidic dyes are eosin, acid fuchsin, and nigrosin.

To apply acidic or basic dyes, microbiologists use three kinds of staining techniques: simple, differential, and special.

Simple Stains

A **simple stain** is an aqueous or alcohol solution of a single basic dye. Although different dyes bind specifically to different parts of cells, the primary purpose of a simple stain is to highlight the entire microorganism so that cellular shapes and basic structures are visible. The stain is applied to the fixed smear for a certain length of time and then washed off, and the slide is dried and examined. Occasionally, a chemical is added to the solution to intensify the stain; such an additive is called a **mordant.** One function of a mordant is to increase the affinity of a stain for a biological specimen; another is to coat a structure (such as a flagellum) to make it thicker and easier to see after it is stained with a dye. Some of the simple stains commonly used in the laboratory are methylene blue, carbolfuchsin, crystal violet, and safranin.

CHECK YOUR UNDERSTANDING

✓ Why doesn't a negative stain color a cell? **3-7**

✓ Why is fixing necessary for most staining procedures? **3-8**

Differential Stains

Unlike simple stains, **differential stains** react differently with different kinds of bacteria and thus can be used to distinguish them. The differential stains most frequently used for bacteria are the Gram stain and the acid-fast stain.

Gram Stain

The **Gram stain** was developed in 1884 by the Danish bacteriologist Hans Christian Gram. It is one of the most useful staining procedures because it classifies bacteria into two large groups: gram-positive and gram-negative.

In this procedure (**Figure 3.12a**),

1. A heat-fixed smear is covered with a basic purple dye, usually crystal violet. Because the purple stain imparts its color to all cells, it is referred to as a **primary stain.**
2. After a short time, the purple dye is washed off, and the smear is covered with iodine, a mordant. When the iodine is washed off, both gram-positive and gram-negative bacteria appear dark violet or purple.
3. Next, the slide is washed with alcohol or an alcohol-acetone solution. This solution is a **decolorizing agent,** which removes the purple from the cells of some species but not from others.
4. The alcohol is rinsed off, and the slide is then stained with safranin, a basic red dye. The smear is washed again, blotted dry, and examined microscopically.

The purple dye and the iodine combine in the cytoplasm of each bacterium and color it dark violet or purple. Bacteria that retain this color after the alcohol has attempted to decolorize them are classified as **gram-positive;** bacteria that lose the dark violet or purple color after decolorization are classified as **gram-negative** (**Figure 3.12b**). Because gram-negative bacteria are colorless after the alcohol wash, they are no longer visible. This is why the basic dye safranin is applied; it turns the gram-negative bacteria pink. Stains such as safranin that have a contrasting color to the primary stain are called **counterstains.** Because gram-positive bacteria retain the original purple stain, they are not affected by the safranin counterstain.

As you will see in Chapter 4, different kinds of bacteria react differently to the Gram stain because structural differences in their cell walls affect the retention or escape of a combination of crystal violet and iodine, called the crystal violet–iodine (CV–I) complex. Among other differences, gram-positive bacteria have a thicker peptidoglycan (disaccharides and amino acids) cell wall than gram-negative bacteria. In addition, gram-negative bacteria contain a layer of lipopolysaccharide (lipids and polysaccharides) as part of their cell wall (see Figure 4.13, page 86). When applied to both gram-positive and gram-negative cells, crystal violet and then iodine readily enter the cells. Inside the cells, the crystal violet and iodine combine to form CV–I. This complex is larger than the crystal violet molecule that entered the cells, and, because of its size, it cannot be washed out of the intact peptidoglycan layer of gram-positive cells by alcohol. Consequently, gram-positive cells retain the color of the crystal violet dye. In gram-negative cells, however, the alcohol wash disrupts the outer lipopolysaccharide layer, and the CV–I complex is washed out through the thin layer of peptidoglycan. As a result, gram-negative cells are colorless until counterstained with safranin, after which they are pink.

In summary, gram-positive cells retain the dye and remain purple. Gram-negative cells do not retain the dye; they are colorless until counterstained with a red dye.

The Gram method is one of the most important staining techniques in medical microbiology. But Gram staining results are not universally applicable, because some bacterial cells stain poorly or not at all. The Gram reaction is most consistent when it is used on young, growing bacteria.

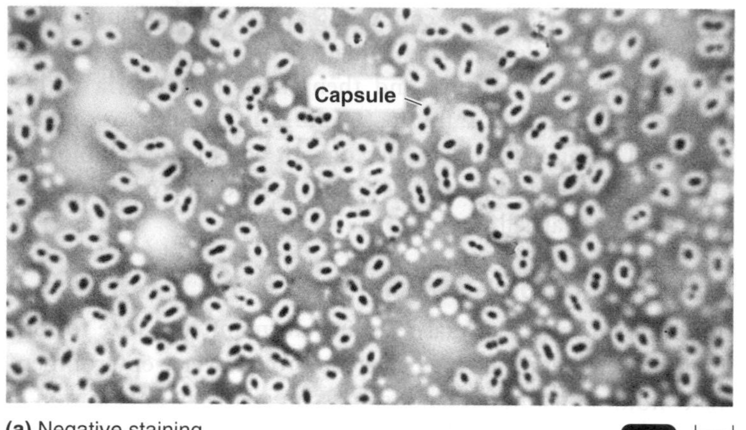

(a) Negative staining LM 5 μm

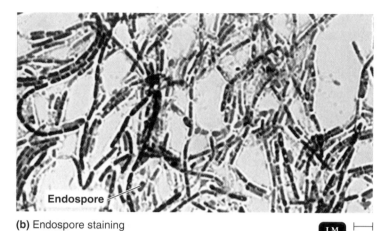

(b) Endospore staining LM 5 μm

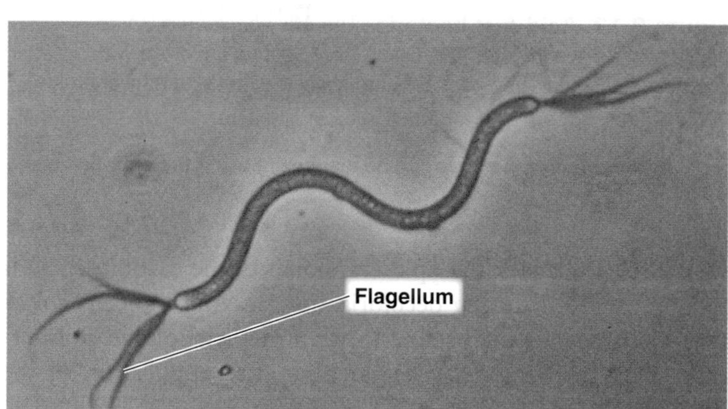

(c) Flagella staining LM 5 μm

Figure 3.14 Special staining. (**a**) Capsule staining provides a contrasting background, so the capsules of these bacteria, *Klebsiella pneumoniae,* show up as light areas surrounding the stained cells. (**b**) Endospores are seen as green ovals in these rod-shaped cells of the bacterium *Bacillus cereus,* using the Schaeffer-Fulton endospore stain. (**c**) Flagella appear as wavy extensions from the ends of these cells of the bacterium *Spirillum volutans.* In relation to the body of the cell, the flagella are much thicker than normal because layers of the stain have accumulated from treatment of the specimen with a mordant.

Q Of what value are capsules, endospores, and flagella to bacteria?

Table 3.3 A Summary of Various Stains and Their Uses

Stain	Principal Uses
Simple (methylene blue, carbolfuchsin, crystal violet, safranin)	Used to highlight microorganisms to determine cellular shapes and arrangements. Aqueous or alcohol solution of a single basic dye stains cells. (Sometimes a mordant is added to intensify the stain.)
Differential	Used to distinguish different kinds of bacteria.
Gram	Classifies bacteria into two large groups: gram-positive and gram-negative. Gram-positive bacteria retain the crystal violet stain and appear purple. Gram-negative bacteria do not retain the crystal violet stain; they remain colorless until counterstained with safranin and then appear pink.
Acid-fast	Used to distinguish *Mycobacterium* species and some species of *Nocardia.* Acid-fast bacteria, once stained with carbolfuchsin and treated with acid-alcohol, remain red because they retain the carbolfuchsin stain. Non–acid-fast bacteria, when stained and treated the same way and then stained with methylene blue, appear blue because they lose the carbolfuchsin stain and are then able to accept the methylene blue stain.
Special	Used to color and isolate various structures, such as capsules, endospores, and flagella; sometimes used as a diagnostic aid.
Negative	Used to demonstrate the presence of capsules. Because capsules do not accept most stains, the capsules appear as unstained halos around bacterial cells and stand out against a contrasting background.
Endospore	Used to detect the presence of endospores in bacteria. When malachite green is applied to a heat-fixed smear of bacterial cells, the stain penetrates the endospores and stains them green. When safranin (red) is then applied, it stains the remainder of the cells red or pink.
Flagella	Used to demonstrate the presence of flagella. A mordant is used to build up the diameters of flagella until they become visible microscopically when stained with carbolfuchsin.

STUDY OUTLINE

The **MyMicrobiologyPlace** website (**www.microbiologyplace.com**) will help you get ready for tests with its simple three-step approach: ❶ **take a pre-test** and obtain a personalized study plan, ❷ **learn and practice** with animations, tutorials, and MP3 tutor sessions, and ❸ **test yourself** with quizzes and a chapter post-test.

Units of Measurement (p. 55)

1. The standard unit of length is the meter (m).
2. Microorganisms are measured in micrometers, μm (10^{-6} m), and in nanometers, nm (10^{-9} m).

Microscopy: The Instruments (p. 55)

1. A simple microscope consists of one lens; a compound microscope has multiple lenses.

Light Microscopy (pp. 56, 58–62)

Compound Light Microscopy (pp. 56, 58–59)

2. The most common microscope used in microbiology is the compound light microscope (LM).
3. The total magnification of an object is calculated by multiplying the magnification of the objective lens by the magnification of the ocular lens.
4. The compound light microscope uses visible light.
5. The maximum resolution, or resolving power (the ability to distinguish two points) of a compound light microscope is 0.2 μm; maximum magnification is 2000×.
6. Specimens are stained to increase the difference between the refractive indexes of the specimen and the medium.
7. Immersion oil is used with the oil immersion lens to reduce light loss between the slide and the lens.
8. Brightfield illumination is used for stained smears.
9. Unstained cells are more productively observed using darkfield, phase-contrast, or DIC microscopy.

Darkfield Microscopy (p. 59)

10. The darkfield microscope shows a light silhouette of an organism against a dark background.
11. It is most useful for detecting the presence of extremely small organisms.

Phase-Contrast Microscopy (pp. 59–60)

12. A phase-contrast microscope brings direct and reflected or diffracted light rays together (in phase) to form an image of the specimen on the ocular lens.
13. It allows the detailed observation of living organisms.

Differential Interference Contrast (DIC) Microscopy (p. 60)

14. The DIC microscope provides a colored, three-dimensional image of the object being observed.
15. It allows detailed observations of living cells.

Fluorescence Microscopy (pp. 61–62)

16. In fluorescence microscopy, specimens are first stained with fluorochromes and then viewed through a compound microscope by using an ultraviolet light source.
17. The microorganisms appear as bright objects against a dark background.
18. Fluorescence microscopy is used primarily in a diagnostic procedure called fluorescent-antibody (FA) technique, or immunofluorescence.

Confocal Microscopy (p. 62)

19. In confocal microscopy, a specimen is stained with a fluorescent dye and illuminated with short-wavelength light.
20. Using a computer to process the images, two-dimensional and three-dimensional images of cells can be produced.

Two-Photon Microscopy (p. 62)

21. In TPM, a live specimen is stained with a fluorescent dye and illuminated with long-wavelength light.

Scanning Acoustic Microscopy (p. 63)

22. Scanning acoustic microscopy (SAM) is based on the interpretation of sound waves through a specimen.
23. It is used to study living cells attached to surfaces such as cancer cells, artery plaque, and biofilms.

Electron Microscopy (pp. 63–65)

24. Instead of light, a beam of electrons is used with an electron microscope.
25. Instead of glass lenses, electromagnets control focus, illumination, and magnification.
26. Thin sections of organisms can be seen in an electron micrograph produced using a transmission electron microscope (TEM). Magnification: 10,000–100,000×. Resolving power: 2.5 nm.
27. Three-dimensional views of the surfaces of whole microorganisms can be obtained with a scanning electron microscope (SEM). Magnification: 1000–10,000×. Resolving power: 20 nm.

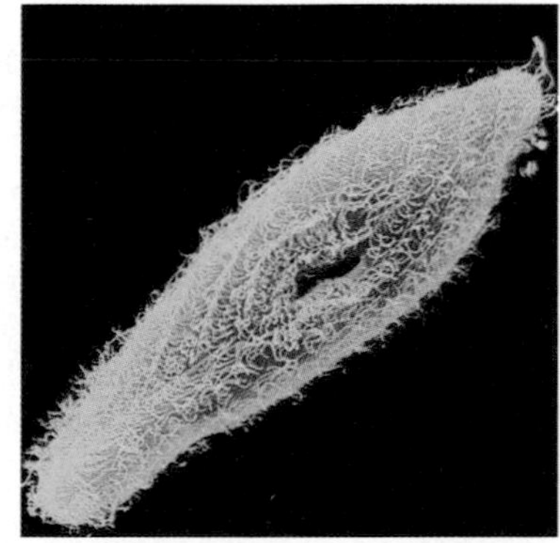

Scanned-Probe Microscopy (p. 65)

28. Scanning tunneling microscopy (STM) and atomic force microscopy (AFM) produce three-dimensional images of the surface of a molecule.

Preparation of Specimens for Light Microscopy (pp. 68–72)

Preparing Smears for Staining (pp. 68–69)

1. Staining means coloring a microorganism with a dye to make some structures more visible.
2. Fixing uses heat or alcohol to kill and attach microorganisms to a slide.

3. A smear is a thin film of material used for microscopic examination.
4. Bacteria are negatively charged, and the colored positive ion of a basic dye will stain bacterial cells.
5. The colored negative ion of an acidic dye will stain the background of a bacterial smear; a negative stain is produced.

Simple Stains (p. 69)

6. A simple stain is an aqueous or alcohol solution of a single basic dye.
7. It is used to make cellular shapes and arrangements visible.
8. A mordant may be used to improve bonding between the stain and the specimen.

Differential Stains (pp. 69–71)

9. Differential stains, such as the Gram stain and acid-fast stain, differentiate bacteria according to their reactions to the stains.
10. The Gram stain procedure uses a purple stain (crystal violet), iodine as a mordant, an alcohol decolorizer, and a red counterstain.
11. Gram-positive bacteria retain the purple stain after the decolorization step; gram-negative bacteria do not and thus appear pink from the counterstain.
12. Acid-fast microbes, such as members of the genera *Mycobacterium* and *Nocardia,* retain carbolfuchsin after acid-alcohol decolorization and appear red; non–acid-fast microbes take up the methylene blue counterstain and appear blue.

Special Stains (pp. 71–72)

13. Negative staining is used to make microbial capsules visible.
14. The endospore stain and flagella stain are special stains that color only certain parts of bacteria.

STUDY QUESTIONS

Answers to the Review and Multiple Choice questions can be found by turning to the blue Answers tab at the back of the textbook.

Review

1. Fill in the following blanks.
   a. 1 μm = ________ m
   b. 1 ________ = 10^{-9} m
   c. 1 μm = ________ nm
2. Which type of microscope would be best to use to observe each of the following?
   a. a stained bacterial smear
   b. unstained bacterial cells: the cells are small, and no detail is needed
   c. unstained live tissue when it is desirable to see some intracellular detail
   d. a sample that emits light when illuminated with ultraviolet light
   e. intracellular detail of a cell that is 1 μm long
   f. unstained live cells in which intracellular structures are shown in color
3. **DRAW IT** Label the parts of the compound light microscope in the figure below, and then draw the path of light from the illuminator to your eye.

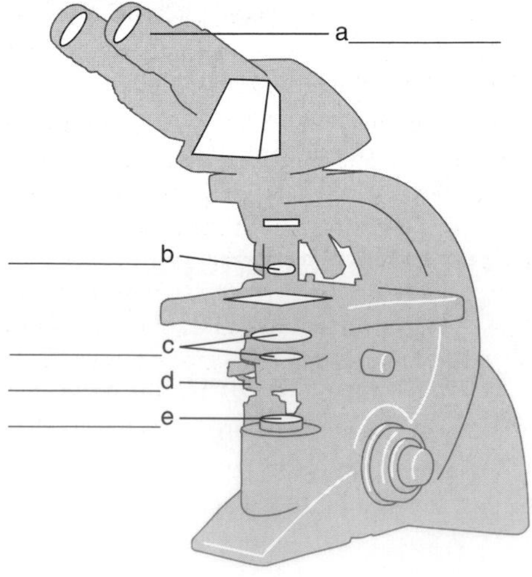

4. Calculate the total magnification of the nucleus of a cell being observed through a compound light microscope with a 10× ocular lens and an oil immersion lens.
5. The maximum magnification of a compound microscope is (a) ________; that of an electron microscope, (b) ________. The maximum resolution of a compound microscope is (c) ________; that of an electron microscope, (d) ________. One advantage of a scanning electron microscope over a transmission electron microscope is (e) ________.
6. Why is a mordant used in the Gram stain? In the flagella stain?
7. What is the purpose of a counterstain in the acid-fast stain?
8. What is the purpose of a decolorizer in the Gram stain? In the acid-fast stain?
9. Fill in the following table regarding the Gram stain:

	Appearance After This Step of	
Steps	**Gram-Positive Cells**	**Gram-Negative Cells**
Crystal violet	a. ________	e. ________
Iodine	b. ________	f. ________
Alcohol-acetone	c. ________	g. ________
Safranin	d. ________	h. ________

Multiple Choice

1. Assume you stain *Bacillus* by applying malachite green with heat and then counterstain with safranin. Through the microscope, the green structures are
   a. cell walls.
   b. capsules.
   c. endospores.
   d. flagella.
   e. impossible to identify.
2. Three-dimensional images of live cells can be produced with
   a. darkfield microscopy.
   b. fluorescence microscopy.
   c. scanning electron microscopy.
   d. two-photon microscopy.
   e. all of the above.

3. Carbolfuchsin can be used as a simple stain and a negative stain. As a simple stain, the pH is
 a. 2.
 b. higher than the negative stain.
 c. lower than the negative stain.
 d. the same as the negative stain.
4. Looking at the cell of a photosynthetic microorganism, you observe that the chloroplasts are green in brightfield microscopy and red in fluorescence microscopy. You conclude that
 a. chlorophyll is fluorescent.
 b. the magnification has distorted the image.
 c. you're not looking at the same structure in both microscopes.
 d. the stain masked the green color.
 e. none of the above
5. Which of the following is *not* a functionally analogous pair of stains?
 a. nigrosin and malachite green
 b. crystal violet and carbolfuchsin
 c. safranin and methylene blue
 d. ethanol-acetone and acid-alcohol
 e. none of the above
6. Which of the following pairs is mismatched?
 a. capsule—negative stain
 b. cell arrangement—simple stain
 c. cell size—negative stain
 d. Gram stain—bacterial identification
 e. none of the above
7. Assume you stain *Clostridium* by applying a basic stain, carbolfuchsin, with heat, decolorizing with acid-alcohol, and counterstaining with an acidic stain, nigrosin. Through the microscope, the endospores are ___1___, and the cells are stained ___2___.
 a. 1—red; 2—black
 b. 1—black; 2—colorless
 c. 1—colorless; 2—black
 d. 1—red; 2—colorless
 e. 1—black; 2—red
8. Assume that you are viewing a Gram-stained field of red cocci and blue bacilli through the microscope. You can safely conclude that you have
 a. made a mistake in staining.
 b. two different species.
 c. old bacterial cells.
 d. young bacterial cells.
 e. none of the above
9. In 1996, scientists described a new tapeworm parasite that had killed at least one person. The initial examination of the patient's abdominal mass was most likely made using
 a. brightfield microscopy.
 b. darkfield microscopy.
 c. electron microscopy.
 d. phase-contrast microscopy.
 e. fluorescence microscopy.
10. Which of the following is *not* a modification of a compound light microscope?
 a. brightfield microscopy
 b. darkfield microscopy
 c. electron microscopy
 d. phase-contrast microscopy
 e. fluorescence microscopy

Critical Thinking

1. In a Gram stain, one step could be omitted and still allow differentiation between gram-positive and gram-negative cells. What is that one step?
2. Using a good compound light microscope with a resolving power of 0.3 μm, a 10× ocular lens, and a 100× oil immersion lens, would you be able to discern two objects separated by 3 μm? 0.3 μm? 300 nm?
3. Why isn't the Gram stain used on acid-fast bacteria? If you did Gram stain acid-fast bacteria, what would their Gram reaction be? What is the Gram reaction of non–acid-fast bacteria?
4. Endospores can be seen as refractile structures in unstained cells and as colorless areas in Gram-stained cells. Why is it necessary to do an endospore stain to verify the presence of endospores?

Clinical Applications

1. In 1882, German bacteriologist Paul Erhlich described a method for staining *Mycobacterium* and noted, "It may be that all disinfecting agents which are acidic will be without effect on this [tubercle] bacillus, and one will have to be limited to alkaline agents." How did he reach this conclusion without testing disinfectants?
2. Laboratory diagnosis of *Neisseria gonorrhoeae* infection is based on microscopic examination of Gram-stained pus. Locate the bacteria in this light micrograph. What is the disease?

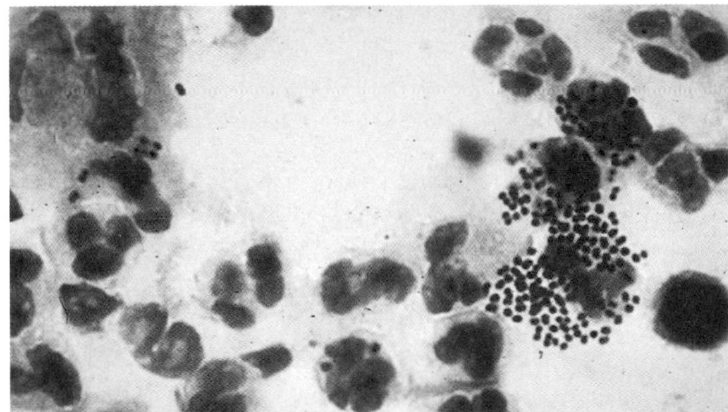

3. Assume that you are viewing a Gram-stained sample of vaginal discharge. Large (10 μm) nucleated red cells are coated with small (0.5 μm × 1.5 μm) blue cells on their surfaces. What is the most likely explanation for the red and blue cells?
4. A sputum sample from Calle, a 30-year-old Asian elephant, was smeared onto a slide and air dried. The smear was fixed, covered with carbolfuchsin, and heated for 5 minutes. After washing with water, acid-alcohol was placed on the smear for 30 seconds. Finally, the smear was stained with methylene blue for 30 seconds, washed with water, and dried. On examination at 1000×, the zoo veterinarian saw red rods on the slide. What infections do the results suggest? (Calle was treated and recovered.)

4 Functional Anatomy of Prokaryotic and Eukaryotic Cells

Despite their complexity and variety, all living cells can be classified into two groups, prokaryotes and eukaryotes, based on certain structural and functional characteristics. In general, prokaryotes are structurally simpler and smaller than eukaryotes. The DNA (genetic material) of prokaryotes is usually arranged in a single, circularly arranged chromosome and is not surrounded by a membrane; the DNA of eukaryotes is found in multiple chromosomes in a membrane-enclosed nucleus. Prokaryotes lack membrane-enclosed organelles, specialized structures that carry on various activities. Additional differences are discussed shortly.

Plants and animals are entirely composed of eukaryotic cells. In the microbial world, bacteria and archaea are prokaryotes. Other cellular microbes—fungi (yeasts and molds), protozoa, and algae—are eukaryotes. Humans exploit the differences between bacterial (prokaryotes) and human cells (eukaryotes) to protect themselves from disease. For example, certain drugs kill or inhibit bacteria while not harming human cells, and chemicals on the surface of bacteria stimulate the body to mount a defensive response to eliminate them.

Viruses, as noncellular elements, do not fit into any organizational scheme of living cells. They are genetic particles that replicate but are unable to perform the usual chemical activities of living cells. Viruses will be discussed in Chapter 13. In this chapter we will concentrate on prokaryotic and eukaryotic cells.

UNDER THE MICROSCOPE

Staphylococcus aureus destroying a phagocyte. *S. aureus* bacteria produce a leukocidin toxin that destroys the host's white blood cells.

Q&A

Penicillin was called a "miracle drug" because it doesn't harm human cells. Why doesn't it?

Look for the answer in the chapter.

Comparing Prokaryotic and Eukaryotic Cells: An Overview

LEARNING OBJECTIVE

4-1 Compare and contrast the overall cell structure of prokaryotes and eukaryotes.

Prokaryotes and eukaryotes are chemically similar, in the sense that they both contain nucleic acids, proteins, lipids, and carbohydrates. They use the same kinds of chemical reactions to metabolize food, build proteins, and store energy. It is primarily the structure of cell walls and membranes, and the absence of *organelles* (specialized cellular structures that have specific functions), that distinguish prokaryotes from eukaryotes.

The chief distinguishing characteristics of **prokaryotes** (from the Greek words meaning prenucleus) are as follows:

1. Their DNA is not enclosed within a membrane and is usually a singular circularly arranged chromosome. (Some bacteria, such as *Vibrio cholerae,* have two chromosomes, and some bacteria have a linearly arranged chromosome.)
2. Their DNA is not associated with histones (special chromosomal proteins found in eukaryotes); other proteins are associated with the DNA.
3. They lack membrane-enclosed organelles.
4. Their cell walls almost always contain the complex polysaccharide peptidoglycan.
5. They usually divide by **binary fission.** During this process, the DNA is copied, and the cell splits into two cells. Binary fission involves fewer structures and processes than eukaryotic cell division.

Eukaryotes (from the Greek words meaning true nucleus) have the following distinguishing characteristics:

1. Their DNA is found in the cell's nucleus, which is separated from the cytoplasm by a nuclear membrane, and the DNA is found in multiple chromosomes.
2. Their DNA is consistently associated with chromosomal proteins called histones and with nonhistones.
3. They have a number of membrane-enclosed organelles, including mitochondria, endoplasmic reticulum, Golgi complex, lysosomes, and sometimes chloroplasts.
4. Their cell walls, when present, are chemically simple.
5. Cell division usually involves mitosis, in which chromosomes replicate and an identical set is distributed into each of two nuclei. This process is guided by the mitotic spindle, a football-shaped assembly of microtubules. Division of the cytoplasm and other organelles follows so that the two cells produced are identical to each other.

Additional differences between prokaryotic and eukaryotic cells are listed in Table 4.2, page 101. Next we describe, in detail, the parts of the prokaryotic cell.

CHECK YOUR UNDERSTANDING

✓ What is the main feature that distinguishes prokaryotes from eukaryotes? **4-1**

THE PROKARYOTIC CELL

The members of the prokaryotic world make up a vast heterogeneous group of very small unicellular organisms. Prokaryotes include bacteria and archaea. The majority of prokaryotes, including the photosynthesizing cyanobacteria, are bacteria. Although bacteria and archaea look similar, their chemical composition is different, as will be described later. The thousands of species of bacteria are differentiated by many factors, including morphology (shape), chemical composition (often detected by staining reactions), nutritional requirements, biochemical activities, and sources of energy (sunlight or chemicals). It is estimated that 99% of the bacteria in nature exist in biofilms (see pages 57 and 162).

The Size, Shape, and Arrangement of Bacterial Cells

LEARNING OBJECTIVE

4-2 Identify the three basic shapes of bacteria.

Bacteria come in a great many sizes and several shapes. Most bacteria range from 0.2 to 2.0 μm in diameter and from 2 to 8 μm in length. They have a few basic shapes: spherical **coccus** (plural: **cocci,** meaning berries), rod-shaped **bacillus** (plural: **bacilli,** meaning little staffs), and **spiral.**

Cocci are usually round but can be oval, elongated, or flattened on one side. When cocci divide to reproduce, the cells can remain attached to one another. Cocci that remain in pairs after dividing are called **diplococci;** those that divide and remain attached in chainlike patterns are called **streptococci**

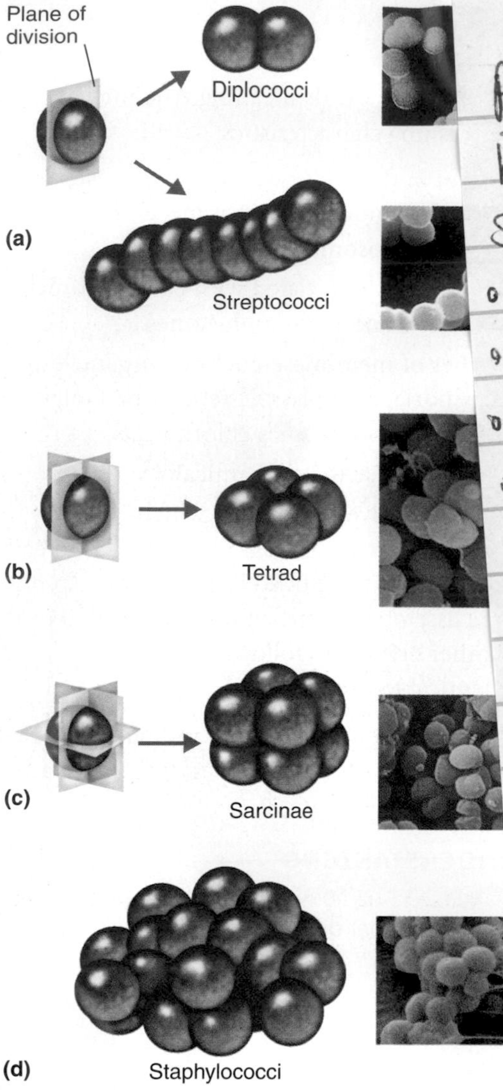

Plane of division

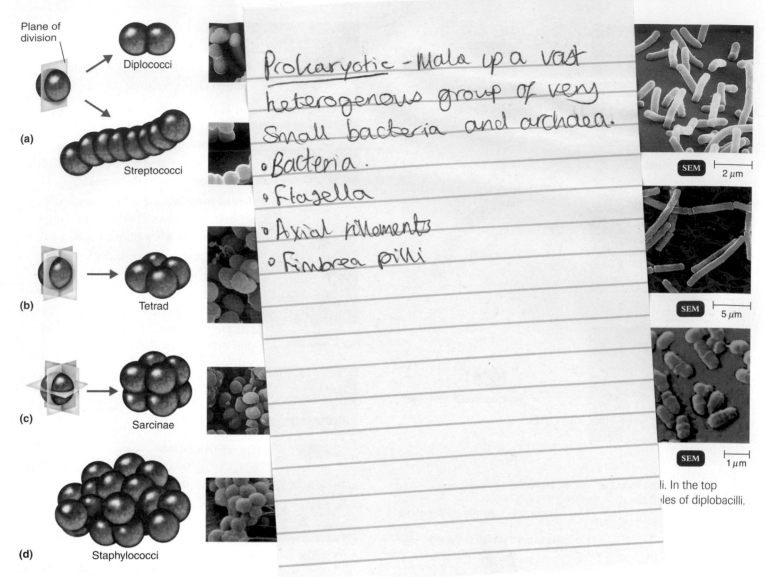

Diplococci

(a)

Streptococci

(b)

Tetrad

(c)

Sarcinae

(d)

Staphylococci

Prokaryotic - Make up a vast heterogenous group of very Small bacteria and archaea.

- Bacteria.
- Flagella
- Axial filaments
- Fimbrea pilli

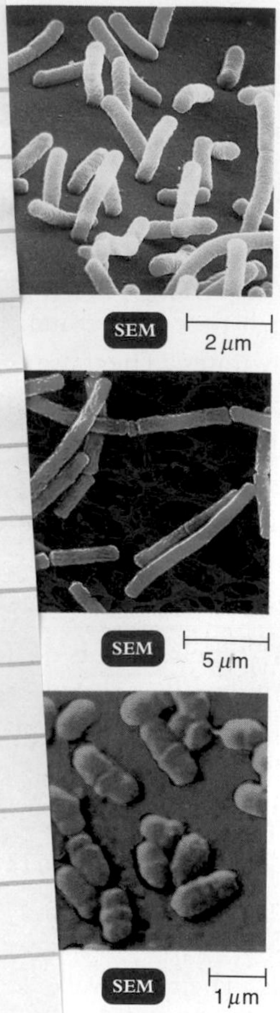

SEM 2 μm

SEM 5 μm

SEM 1 μm

...li. In the top ...les of diplobacilli.

Figure 4.1 Arrangements of cocci. (**a**) Division in one plane produces diplococci and streptococci. (**b**) Division in two planes produces tetrads. (**c**) Division in three planes produces sarcinae, and (**d**) division in multiple planes produces staphylococci.

Q How do the planes of division determine the arrangement of cells?

(Figure 4.1a). Those that divide in two planes and remain in groups of four are known as **tetrads** (Figure 4.1b). Those that divide in three planes and remain attached in cubelike groups of eight are called **sarcinae** (Figure 4.1c). Those that divide in multiple planes and form grapelike clusters or broad sheets are called **staphylococci** (Figure 4.1d). These group characteristics are frequently helpful in identifying certain cocci.

Bacilli divide only across their short axis, so there are fewer groupings of bacilli than of cocci. Most bacilli appear as single rods (Figure 4.2a). **Diplobacilli** appear in pairs after division (Figure 4.2b), and **streptobacilli** occur in chains (Figure 4.2c). Some bacilli look like straws. Others have tapered ends, like cigars. Still others are oval and look so much like cocci that they are called **coccobacilli** (Figure 4.2d).

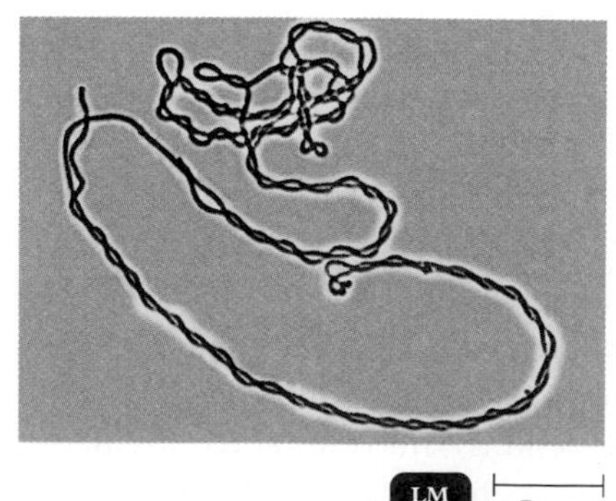

LM 5 μm

Figure 4.3 A double-stranded helix formed by *Bacillus subtilis*.

Q What is the difference between the term bacillus and *Bacillus*?

"Bacillus" has two meanings in microbiology. As we have just used it, bacillus refers to a bacterial shape. When capitalized and italicized, it refers to a specific genus. For example, the bacterium *Bacillus anthracis* is the causative agent of anthrax. Bacillus cells often form long, twisted chains of cells (Figure 4.3).

Spiral bacteria have one or more twists; they are never straight. Bacteria that look like curved rods are called **vibrios** (Figure 4.4a). Others, called **spirilla,** have a helical shape, like a corkscrew, and fairly rigid bodies (Figure 4.4b). Yet another group of spirals are helical and flexible; they are called **spirochetes** (Figure 4.4c). Unlike the spirilla, which use propeller-like external appendages called flagella to move, spirochetes move by means of axial filaments, which resemble flagella but are contained within a flexible external sheath.

In addition to the three basic shapes, there are star-shaped cells (genus *Stella*; Figure 4.5a); rectangular, flat cells (halophilic archaea) of the genus *Haloarcula* (Figure 4.5b); and triangular cells.

The shape of a bacterium is determined by heredity. Genetically, most bacteria are **monomorphic;** that is, they maintain a single shape. However, a number of environmental conditions can alter that shape. If the shape is altered, identification becomes difficult. Moreover, some bacteria, such as *Rhizobium* (rī-zō′bē-um) and *Corynebacterium* (kô-rī-nē-bak-ti′rē-um), are genetically **pleomorphic,** which means they can have many shapes, not just one.

The structure of a typical prokaryotic cell is shown in Figure 4.6. We will discuss its components according to the following organization: (1) structures external to the cell wall, (2) the cell wall itself, and (3) structures internal to the cell wall.

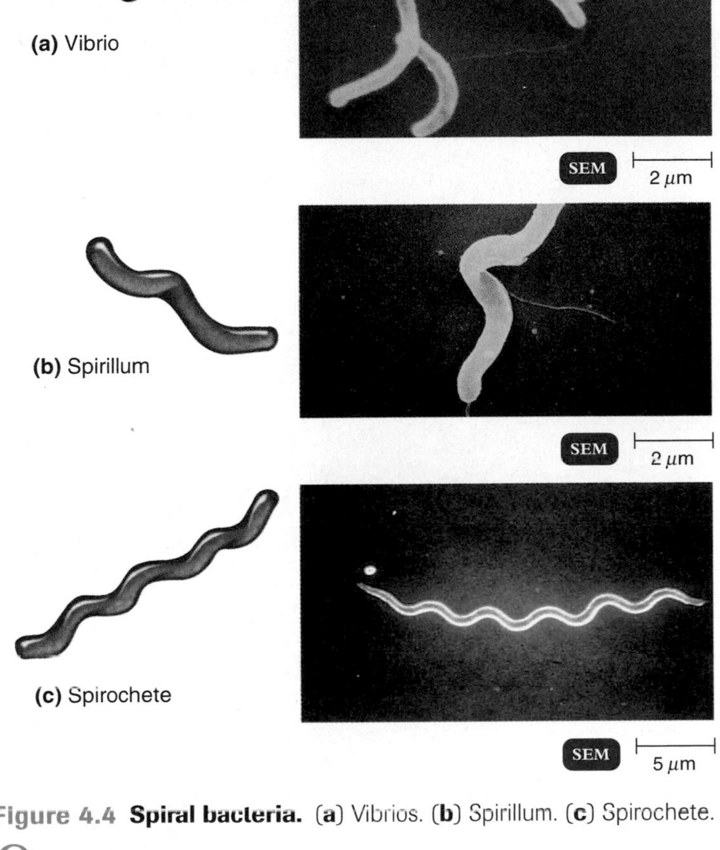

Figure 4.4 Spiral bacteria. (a) Vibrios. (b) Spirillum. (c) Spirochete.

Q **What is the distinguishing feature of spirochete bacteria?**

CHECK YOUR UNDERSTANDING

✓ How would you be able to identify streptococci through a microscope? **4-2**

Structures External to the Cell Wall

LEARNING OBJECTIVES

4-3 Describe the structure and function of the glycocalyx

4-4 Differentiate flagella, axial filaments, fimbriae, and pili.

Among the possible structures external to the prokaryotic cell wall are the glycocalyx, flagella, axial filaments, fimbriae, and pili.

Glycocalyx

Many prokaryotes secrete on their surface a substance called glycocalyx. **Glycocalyx** (meaning sugar coat) is the general term used for substances that surround cells. The bacterial glycocalyx is a viscous (sticky), gelatinous polymer that is external to the cell wall and composed of polysaccharide, polypeptide, or both. Its chemical composition varies widely with the species. For the most part, it is made inside the cell and secreted to the cell surface. If the substance is organized and is firmly attached to the cell wall, the glycocalyx is described as a **capsule.** The presence of a capsule can be determined by using negative staining,

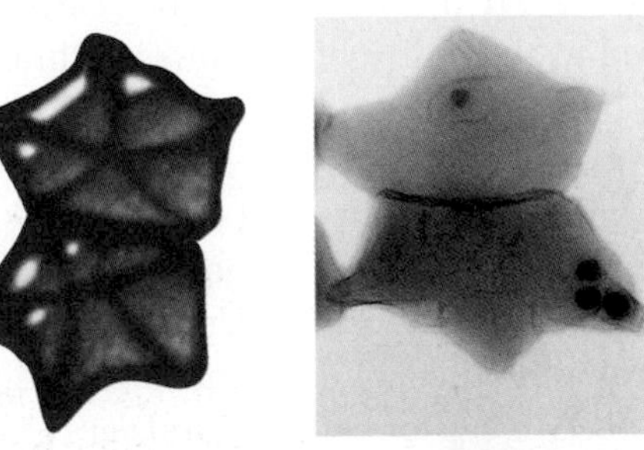
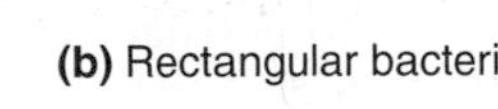

(a) Star-shaped bacteria

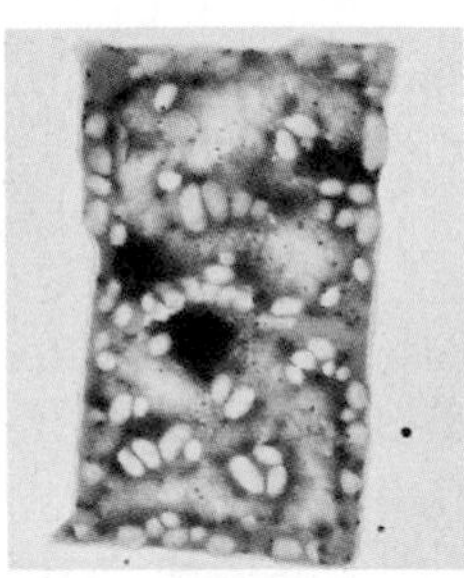
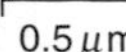

(b) Rectangular bacteria

Figure 4.5 Star-shaped and rectangular prokaryotes. (a) *Stella* (star-shaped). (b) *Haloarcula,* a genus of halophilic archaea (rectangular cells).

Q **What are the common bacterial shapes?**

Figure 4.6

FOUNDATION FIGURE The Structure of a Prokaryotic Cell

This prokaryotic cell shows typical structures that may be found in bacteria. Each of the labeled structures will be discussed individually in this chapter. As you will see in later chapters, some of these structures contribute to bacterial virulence, play a role in bacterial identification, and are targets of antimicrobial agents.

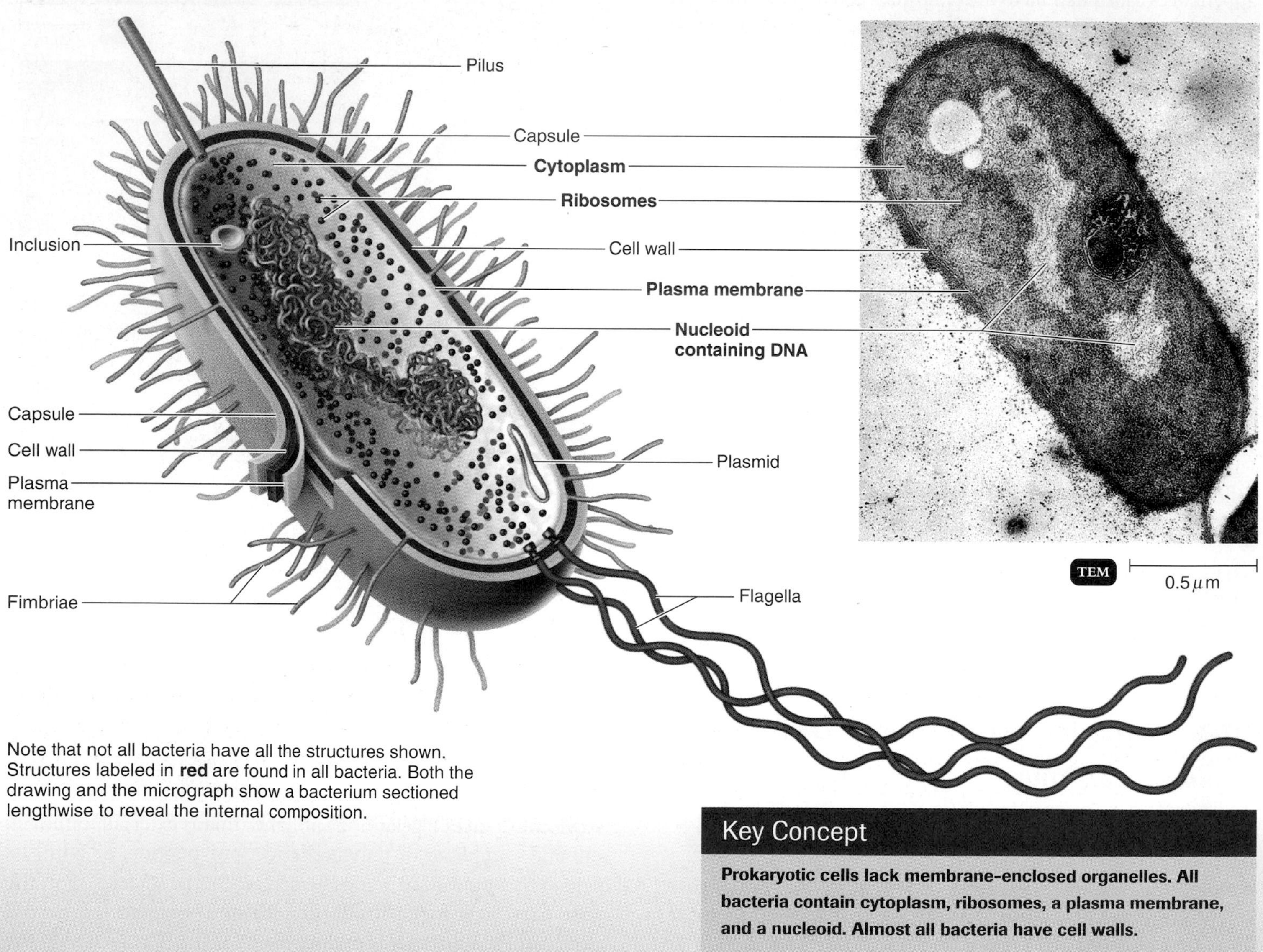

Note that not all bacteria have all the structures shown. Structures labeled in **red** are found in all bacteria. Both the drawing and the micrograph show a bacterium sectioned lengthwise to reveal the internal composition.

Key Concept

Prokaryotic cells lack membrane-enclosed organelles. All bacteria contain cytoplasm, ribosomes, a plasma membrane, and a nucleoid. Almost all bacteria have cell walls.

described in Chapter 3 (see Figure 3.14a, page 72). If the substance is unorganized and only loosely attached to the cell wall, the glycocalyx is described as a **slime layer.**

In certain species, capsules are important in contributing to bacterial virulence (the degree to which a pathogen causes disease). Capsules often protect pathogenic bacteria from phagocytosis by the cells of the host. (As you will see later, phagocytosis is the ingestion and digestion of microorganisms and other solid particles.) For example, *Bacillus anthracis* produces a capsule of D-glutamic acid. (Recall from Chapter 2 that the D forms of amino acids are unusual.) Because only encapsulated *B. anthracis* causes anthrax, it is speculated that the capsule may prevent its being destroyed by phagocytosis.

Another example involves *Streptococcus pneumoniae* (strep-tō-kok'kus nü-mō'nē-ī), which causes pneumonia only when the cells are protected by a polysaccharide capsule. Unencapsulated *S. pneumoniae* cells cannot cause pneumonia and are readily phagocytized. The polysaccharide capsule of *Klebsiella* (kleb-sē-el'lä) also prevents phagocytosis and allows the bacterium to adhere to and colonize the respiratory tract.

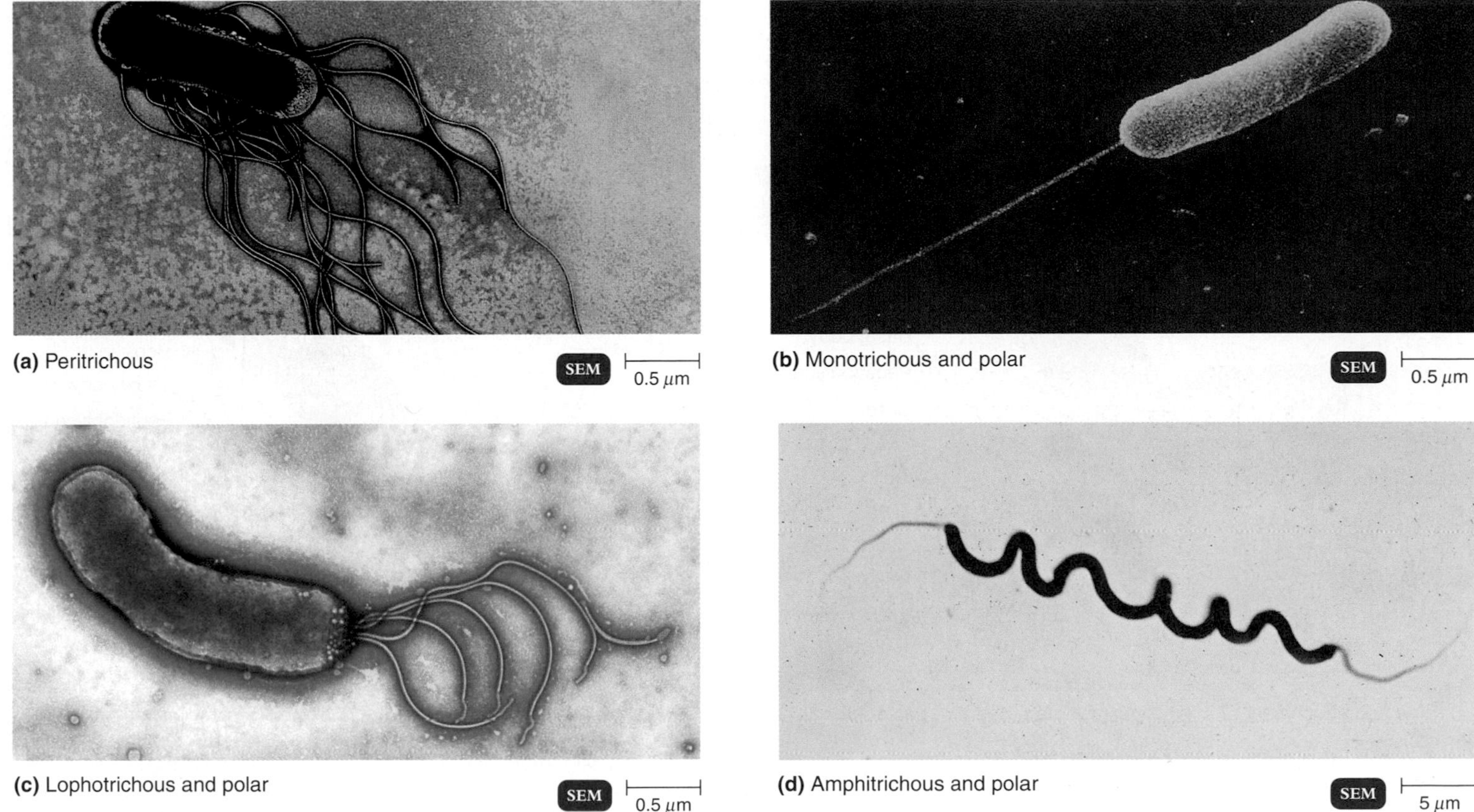

Figure 4.7 Arrangements of bacterial flagella. (**a**) Peritrichous. (**b**)–(**d**) Polar.

Q **What are some of the key differences and similarities between flagella and endoflagella?**

The glycocalyx is a very important component of biofilms (see page 162). A glycocalyx that helps cells in a biofilm attach to their target environment and to each other is called an **extracellular polymeric substance (EPS).** The EPS protects the cells within it, facilitates communication among them, and enables the cells to survive by attaching to various surfaces in their natural environment.

Through attachment, bacteria can grow on diverse surfaces such as rocks in fast-moving streams, plant roots, human teeth, medical implants, water pipes, and even other bacteria. *Streptococcus mutans* (mū′tans), an important cause of dental caries, attaches itself to the surface of teeth by a glycocalyx. *S. mutans* may use its capsule as a source of nutrition by breaking it down and utilizing the sugars when energy stores are low. *Vibrio cholerae* (vib′-rē-o kol′-er-ī), the cause of cholera, produces a glycocalyx that helps it attach to the cells of the small intestine. A glycocalyx also can protect a cell against dehydration, and its viscosity may inhibit the movement of nutrients out of the cell.

Flagella

Some prokaryotic cells have **flagella** (singular: **flagellum**), which are long filamentous appendages that propel bacteria. Bacteria that lack flagella are referred to as **atrichous** (without projections). Flagella may be **peritrichous** (distributed over the entire cell; Figure 4.7a) or **polar** (at one or both poles or ends of the cell). If polar, flagella may be **monotrichous** (a single flagellum at one pole; Figure 4.7b), **lophotrichous** (a tuft of flagella coming from one pole; Figure 4.7c), or **amphitrichous** (flagella at both poles of the cell; Figure 4.7d).

A flagellum has three basic parts (Figure 4.8). The long outermost region, the *filament,* is constant in diameter and contains the globular (roughly spherical) protein *flagellin* arranged in several chains that intertwine and form a helix around a hollow core. In most bacteria, filaments are not covered by a membrane or sheath, as in eukaryotic cells. The filament is attached to a slightly wider *hook,* consisting of a different protein. The third portion of a flagellum is the *basal body,* which anchors the flagellum to the cell wall and plasma membrane.

The basal body is composed of a small central rod inserted into a series of rings. Gram-negative bacteria contain two pairs of rings; the outer pair of rings is anchored to various portions of the cell wall, and the inner pair of rings is anchored to the plasma membrane. In gram-positive bacteria, only the inner pair is present. As you will see later, the flagella (and cilia) of eukaryotic cells are more complex than those of prokaryotic cells.

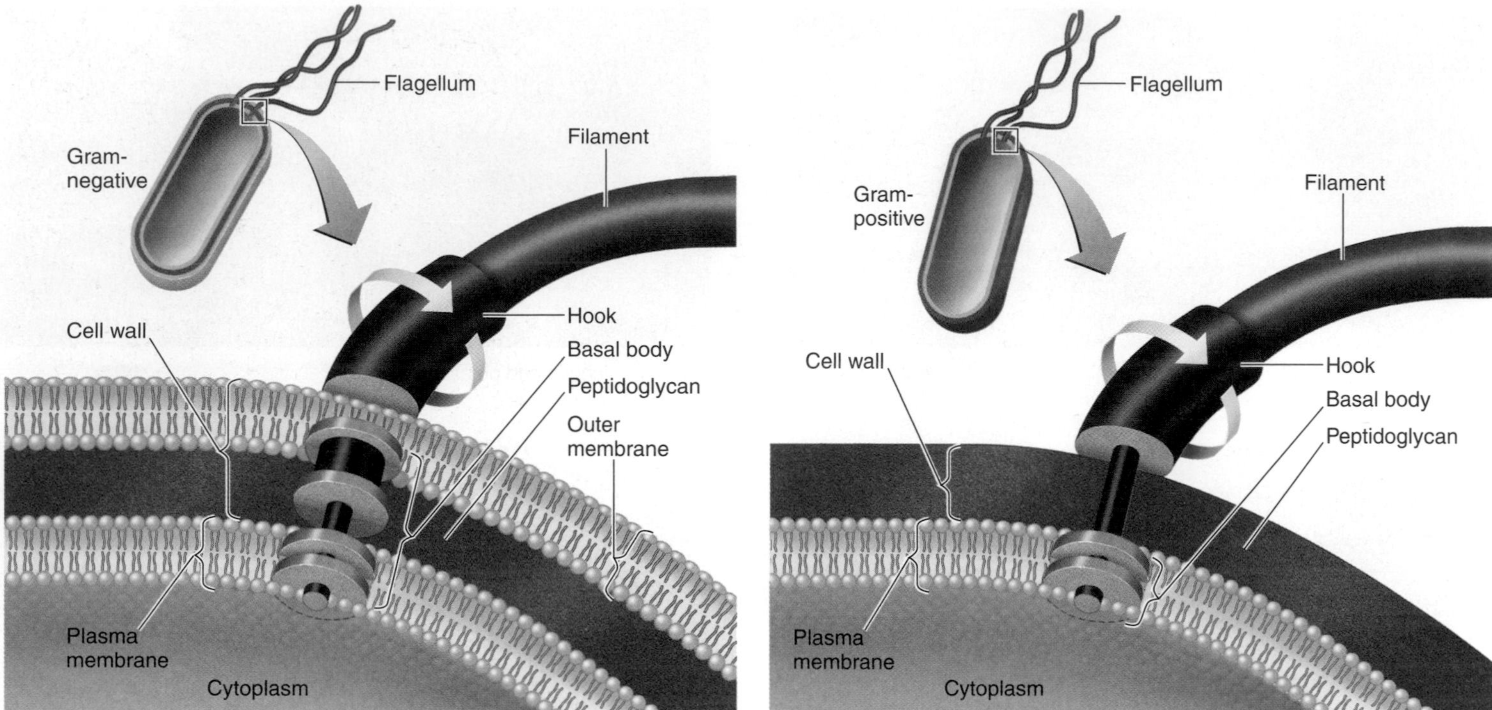

(a) Parts and attachment of a flagellum of a gram-negative bacterium

(b) Parts and attachment of a flagellum of a gram-positive bacterium

Figure 4.8 The structure of a prokaryotic flagellum. The parts and attachment of a flagellum of a gram-negative bacterium and gram-positive bacterium are shown in these highly schematic diagrams.

How do the basal bodies of gram-negative and gram-positive bacteria differ?

Each prokaryotic flagellum is a semirigid, helical structure that moves the cell by rotating from the basal body. The rotation of a flagellum is either clockwise or counterclockwise around its long axis. (Eukaryotic flagella, by contrast, undulate in a wavelike motion.) The movement of a prokaryotic flagellum results from rotation of its basal body and is similar to the movement of the shaft of an electric motor. As the flagella rotate, they form a bundle that pushes against the surrounding liquid and propels the bacterium. Flagellar rotation depends on the cell's continuous generation of energy.

Bacterial cells can alter the speed and direction of rotation of flagella and thus are capable of various patterns of **motility,** the ability of an organism to move by itself. When a bacterium moves in one direction for a length of time, the movement is called a "run" or "swim." "Runs" are interrupted by periodic, abrupt, random changes in direction called "tumbles." Then, a "run" resumes. "Tumbles" are caused by a reversal of flagellar rotation (**Figure 4.9a**). Some species of bacteria endowed with many flagella—*Proteus* (prō′tē-us), for example (**Figure 4.9b**)—can "swarm," or show rapid wavelike movement across a solid culture medium.

One advantage of motility is that it enables a bacterium to move toward a favorable environment or away from an adverse one. The movement of a bacterium toward or away from a particular stimulus is called **taxis.** Such stimuli include chemicals **(chemotaxis)** and light **(phototaxis)**. Motile bacteria contain receptors in various locations, such as in or just under the cell wall. These receptors pick up chemical stimuli, such as oxygen, ribose, and galactose. In response to the stimuli, information is passed to the flagella. If the chemotactic signal is positive, called an *attractant,* the bacteria move toward the stimulus with many runs and few tumbles. If the chemotactic signal is negative, called a *repellent,* the frequency of tumbles increases as the bacteria move away from the stimulus.

The flagellar protein called **H antigen** is useful for distinguishing among **serovars,** or variations within a species, of gram-negative bacteria (see page 310). For example, there are at least 50 different H antigens for *E. coli.* Those serovars identified as *E. coli* O157:H7 are associated with foodborne epidemics (see Chapter 1, page 20). **Animations** Motility; Flagella: Structure, Movement, Arrangement. **www.microbiologyplace.com**

Axial Filaments

Spirochetes are a group of bacteria that have unique structure and motility. One of the best-known spirochetes is *Treponema pallidum* (tre-pō-nē′mä pal′li-dum), the causative agent of

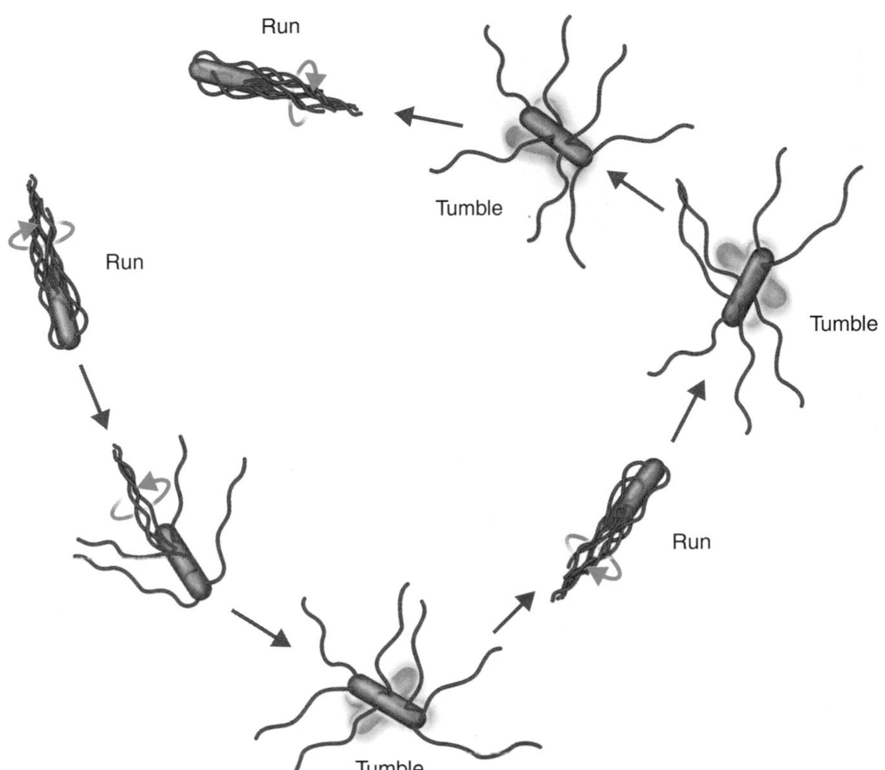

(a) A bacterium running and tumbling. Notice that the direction of flagellar rotation (blue arrows) determines which of these movements occurs. Gray arrows indicate direction of movement of the microbe.

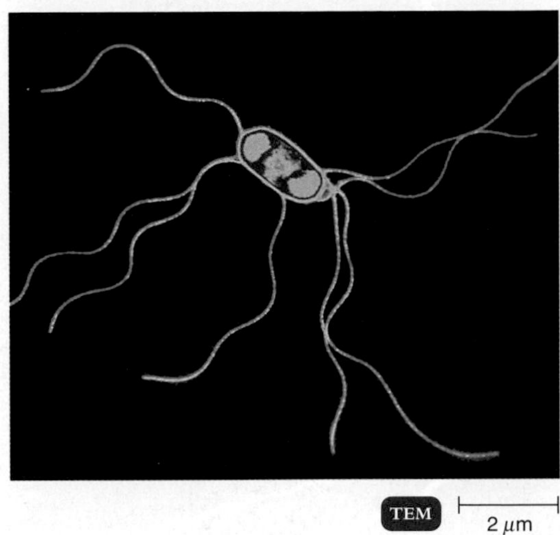

(b) A *Proteus* cell in the swarming stage may have more than 1000 peritrichous flagella.

Figure 4.9 Flagella and bacterial motility.

Q Do bacterial flagella push or pull a cell?

syphilis. Another spirochete is *Borrelia burgdorferi* (bôr′-rel-ē-a burg-dor′fer-ē), the causative agent of Lyme disease. Spirochetes move by means of **axial filaments,** or **endoflagella,** bundles of fibrils that arise at the ends of the cell beneath an outer sheath and spiral around the cell (Figure 4.10).

Axial filaments, which are anchored at one end of the spirochete, have a structure similar to that of flagella. The rotation of the filaments produces a movement of the outer sheath that propels the spirochetes in a spiral motion. This type of movement is similar to the way a corkscrew moves through a cork. This corkscrew motion probably enables a bacterium such as *T. pallidum* to move effectively through body fluids. **Animation** Spirochetes. **www.microbiologyplace.com**

Fimbriae and Pili

Many gram-negative bacteria contain hairlike appendages that are shorter, straighter, and thinner than flagella and are used for attachment and transfer of DNA rather than for motility. These structures, which consist of a protein called *pilin* arranged helically around a central core, are divided into two types, fimbriae and pili, having very different functions. (Some microbiologists use the two terms interchangeably to refer to all such structures, but we distinguish between them.)

Fimbriae (singular: **fimbria**) can occur at the poles of the bacterial cell or can be evenly distributed over the entire surface of the cell. They can number anywhere from a few to several hundred per cell (Figure 4.11). Fimbriae have a tendency to adhere to each other and to surfaces. As a result, they are involved in forming biofilms and other aggregations on the surfaces of liquids, glass, and rocks. Fimbriae can also help bacteria adhere to epithelial surfaces in the body. For example, fimbriae on the bacterium *Neisseria gonorrhoeae* (nī-se′rē-ä go-nôr-rē′ī), the causative agent of gonorrhea, help the microbe colonize mucous membranes. Once colonization occurs, the bacteria can cause disease. The fimbriae of *E. coli* O157 enable this bacterium to adhere to the lining of the small intestine, where it causes a severe watery diarrhea. When fimbriae are absent (because of genetic mutation), colonization cannot happen, and no disease ensues.

Pili (singular: **pilus**) are usually longer than fimbriae and number only one or two per cell. Pili are involved in motility and DNA transfer. In one type of motility, called **twitching motility,** a pilus extends by the addition of subunits of pilin, makes contact with a surface or another cell, and then retracts (powerstroke) as the pilin subunits are disassembled. This is called the *grappling hook model* of twitching motility and results in short, jerky, intermittent movements. Twitching motility has been observed in *Pseudomonas aeruginosa*, *Neisseria gonorrhoeae*, and some strains of *E. coli.* The other type of motility associated with pili is **gliding motility,** the smooth gliding movement of myxobacteria. Although the exact mechanism is

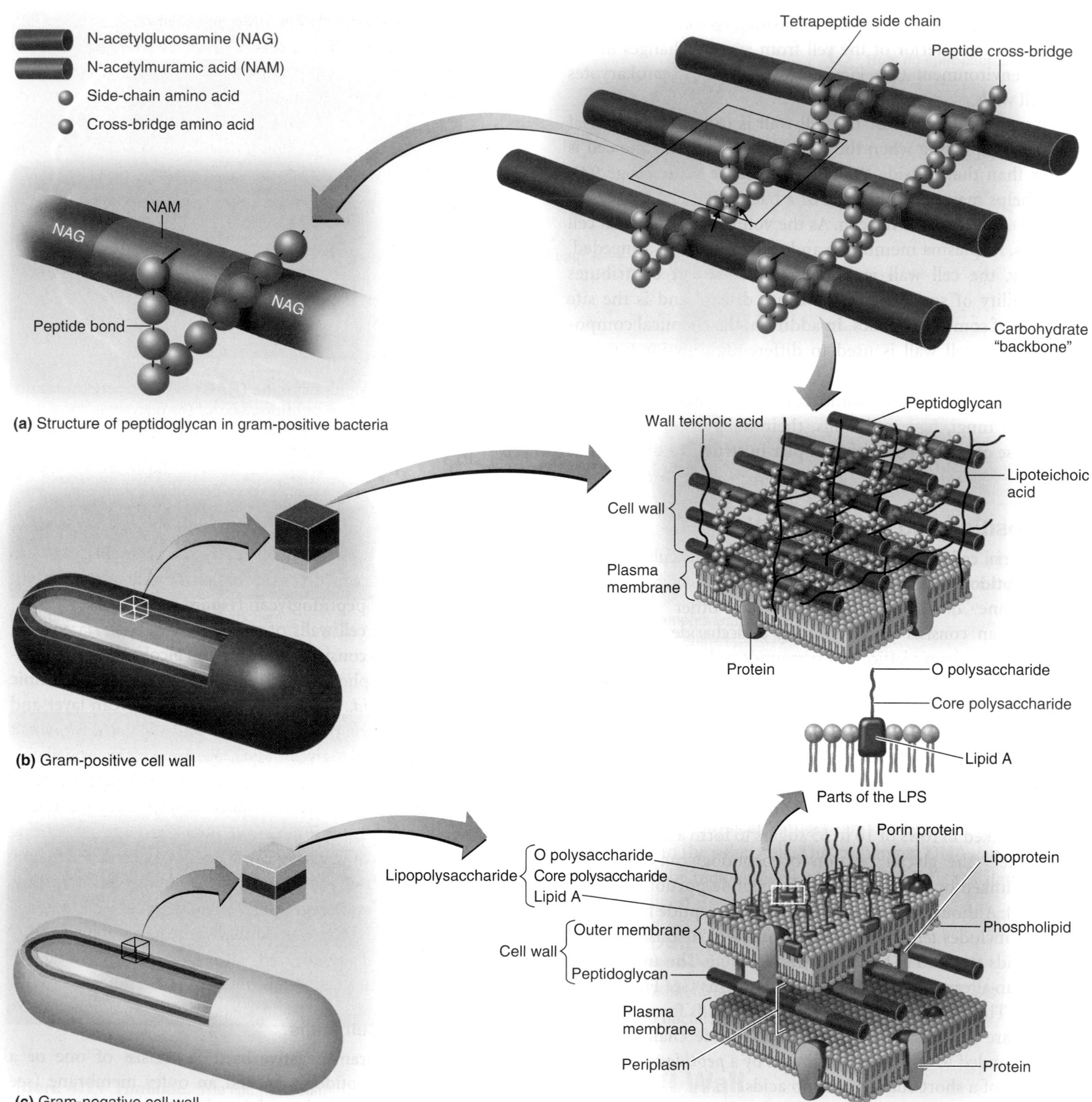

Figure 4.13 Bacterial cell walls. (**a**) The structure of peptidoglycan in gram-positive bacteria. Together the carbohydrate backbone (glycan portion) and tetrapeptide side chains (peptide portion) make up peptidoglycan. The frequency of peptide cross-bridges and the number of amino acids in these bridges vary with species of bacteria. The small arrows indicate where penicillin interferes with the linkage of peptidoglycan rows by peptide cross-bridges. (**b**) A gram-positive cell wall. (**c**) A gram-negative cell wall.

Q **What are the major structural differences between gram-positive and gram-negative cell walls?**

The *outer membrane* of the gram-negative cell consists of lipopolysaccharides (LPS), lipoproteins, and phospholipids (see Figure 4.13c). The outer membrane has several specialized functions. Its strong negative charge is an important factor in evading phagocytosis and the actions of complement (lyses cells and promotes phagocytosis), two components of the defenses of the host (discussed in detail in Chapter 16). The outer membrane also provides a barrier to certain antibiotics (for example, penicillin), digestive enzymes such as lysozyme, detergents, heavy metals, bile salts, and certain dyes.

However, the outer membrane does not provide a barrier to all substances in the environment because nutrients must pass through to sustain the metabolism of the cell. Part of the permeability of the outer membrane is due to proteins in the membrane, called **porins,** that form channels. Porins permit the passage of molecules such as nucleotides, disaccharides, peptides, amino acids, vitamin B_{12}, and iron.

The **lipopolysaccharide (LPS)** of the outer membrane is a large complex molecule that contains lipids and carbohydrates and consists of three components: (1) lipid A, (2) a core polysaccharide, and (3) an O polysaccharide. **Lipid A** is the lipid portion of the LPS and is embedded in the top layer of the outer membrane. When gram-negative bacteria die, they release lipid A, which functions as an endotoxin (Chapter 15). Lipid A is responsible for the symptoms associated with infections by gram-negative bacteria such as fever, dilation of blood vessels, shock, and blood clotting. The **core polysaccharide** is attached to lipid A and contains unusual sugars. Its role is structural—to provide stability. The **O polysaccharide** extends outward from the core polysaccharide and is composed of sugar molecules. The O polysaccharide functions as an antigen and is useful for distinguishing species of gram-negative bacteria. For example, the foodborne pathogen *E. coli* O157:H7 is distinguished from other serovars by certain laboratory tests that test for these specific antigens. This role is comparable to that of teichoic acids in gram-positive cells.

Cell Walls and the Gram Stain Mechanism

Now that you have studied the Gram stain (in Chapter 3, page 69) and the chemistry of the bacterial cell wall (in the previous section), it is easier to understand the mechanism of the Gram stain. The mechanism is based on differences in the structure of the cell walls of gram-positive and gram-negative bacteria and how each reacts to the various reagents (substances used for producing a chemical reaction). Crystal violet, the primary stain, stains both gram-positive and gram-negative cells purple because the dye enters the cytoplasm of both types of cells. When iodine (the mordant) is applied, it forms large crystals with the dye that are too large to escape through the cell wall. The application of alcohol dehydrates the peptidoglycan of gram-positive cells to make it more impermeable to the crystal violet-iodine. The effect on gram-negative cells is quite different; alcohol dissolves the outer membrane of gram-negative cells and even leaves small holes in the thin peptidoglycan layer through which crystal violet-iodine diffuse. Because gram-negative bacteria are colorless after the alcohol wash, the addition of safranin (the counterstain) turns the cells pink. Safranin provides a contrasting color to the primary stain (crystal violet). Although gram-positive and gram-negative cells both absorb safranin, the pink color of safranin is masked by the darker purple dye previously absorbed by gram-positive cells.

In any population of cells, some gram-positive cells will give a gram-negative response. These cells are usually dead. However, there are a few gram-positive genera that show an increasing number of gram-negative cells as the culture ages. *Bacillus* and *Clostridium* are examples and are often described as *gram-variable.*

A comparison of some of the characteristics of gram-positive and gram-negative bacteria is presented in **Table 4.1**.

Atypical Cell Walls

Among prokaryotes, certain types of cells have no walls or have very little wall material. These include members of the genus *Mycoplasma* (mī-kō-plaz′mä) and related organisms (see Figure 11.20, page 319). Mycoplasmas are the smallest known bacteria that can grow and reproduce outside living host cells. Because of their size and because they have no cell walls, they pass through most bacterial filters and were first mistaken for viruses. Their plasma membranes are unique among bacteria in having lipids called *sterols,* which are thought to help protect them from lysis (rupture).

Archaea may lack walls or may have unusual walls composed of polysaccharides and proteins but not peptidoglycan. These walls do, however, contain a substance similar to peptidoglycan called *pseudomurein.* Pseudomurein contains N-acetyltalosaminuronic acid instead of NAM and lacks the D-amino acids found in bacterial cell walls. Archaea generally cannot be Gram-stained but appear gram-negative because they do not contain peptidoglycan.

Acid-Fast Cell Walls

Recall from Chapter 3 that the acid-fast stain is used to identify all bacteria of the genus *Mycobacterium* and pathogenic species of *Nocardia.* These bacteria contain high concentrations (60%) of a hydrophobic waxy lipid (**mycolic acid**) in their cell wall that prevents the uptake of dyes, including those used in the Gram stain. The mycolic acid forms a layer outside of a thin layer of peptidoglycan. The mycolic acid and peptidoglycan are held together by a polysaccharide. The hydrophobic waxy cell wall causes both cultures of *Mycobacterium* to clump and to stick to the walls of the flask. Acid-fast bacteria can be stained with carbolfuchsin; heating enhances penetration of the stain. The carbolfuchsin penetrates the cell wall, binds to cytoplasm, and resists removal by washing with acid-alcohol. Acid-fast bacteria

Table 4.1 Some Comparative Characteristics of Gram-Positive and Gram-Negative Bacteria

Characteristic	Gram-Positive	Gram-Negative
	LM 4 um	LM 4 um
Gram Reaction	Retain crystal violet dye and stain blue or purple	Can be decolorized to accept counterstain (safranin) and stain pink or red
Peptidoglycan Layer	Thick (multilayered)	Thin (single-layered)
Teichoic Acids	Present in many	Absent
Periplasmic Space	Absent	Present
Outer Membrane	Absent	Present
Lipopolysaccharide (LPS) Content	Virtually none	High
Lipid and Lipoprotein Content	Low (acid-fast bacteria have lipids linked to peptidoglycan)	High (because of presence of outer membrane)
Flagellar Structure	2 rings in basal body	4 rings in basal body
Toxins Produced	Exotoxins	Endotoxins and exotoxins
Resistance to Physical Disruption	High	Low
Cell Wall Disruption by Lysozyme	High	Low (requires pretreatment to destabilize outer membrane)
Susceptibility to Penicillin and Sulfonamide	High	Low
Susceptibility to Streptomycin, Chloramphenicol, and Tetracycline	Low	High
Inhibition by Basic Dyes	High	Low
Susceptibility to Anionic Detergents	High	Low
Resistance to Sodium Azide	High	Low
Resistance to Drying	High	Low

retain the red color of carbolfuchsin because it is more soluble in the cell wall mycolic acid than in the acid-alcohol. If the mycolic acid layer is removed from the cell wall of acid-fast bacteria, they will stain gram-positive with the Gram stain.

Damage to the Cell Wall

Chemicals that damage bacterial cell walls, or interfere with their synthesis, often do not harm the cells of an animal host because the bacterial cell wall is made of chemicals unlike those in eukaryotic cells. Thus, cell wall synthesis is the target for some antimicrobial drugs. One way the cell wall can be damaged is by exposure to the digestive enzyme *lysozyme*. This enzyme occurs naturally in some eukaryotic cells and is a constituent of tears, mucus, and saliva. Lysozyme is particularly active on the major cell wall components of most gram-positive bacteria, making them vulnerable to lysis. Lysozyme catalyzes hydrolysis of the bonds between the sugars in the repeating disaccharide "backbone" of peptidoglycan. This act is analogous to cutting the steel supports of a bridge with a cutting torch: the gram-positive cell wall is almost completely destroyed by lysozyme. The cellular contents that remain surrounded by the plasma membrane may remain intact if lysis does not occur; this wall-less cell is termed a **protoplast.** Typically, a protoplast is spherical and is still capable of carrying on metabolism.

Some members of the genus *Proteus,* as well as other genera, can lose their cell walls and swell into irregularly shaped cells called **L forms,** named for the Lister Institute, where they were discovered. They may form spontaneously or develop in response to penicillin (which inhibits cell wall formation) or lysozyme (which removes the cell wall). L forms can live and divide repeatedly or return to the walled state.

When lysozyme is applied to gram-negative cells, usually the wall is not destroyed to the same extent as in gram-positive cells; some of the outer membrane also remains. In this case, the cellular contents, plasma membrane, and remaining outer wall layer are called a **spheroplast,** also a spherical structure. For lysozyme to exert its effect on gram-negative cells, the cells are first treated with EDTA (ethylenediaminetetraacetic acid). EDTA weakens ionic bonds in the outer membrane and thereby damages it, giving the lysozyme access to the peptidoglycan layer.

Protoplasts and spheroplasts burst in pure water or very dilute salt or sugar solutions because the water molecules from the surrounding fluid rapidly move into and enlarge the cell, which has a much lower internal concentration of water. This rupturing, called **osmotic lysis,** will be discussed in detail shortly.

As noted earlier, certain antibiotics, such as penicillin, destroy bacteria by interfering with the formation of the peptide cross-bridges of peptidoglycan, thus preventing the formation of a functional cell wall. Most gram-negative bacteria are not as susceptible to penicillin as gram-positive bacteria are because the outer membrane of gram-negative bacteria forms a barrier that inhibits the entry of this and other substances, and gram-negative bacteria have fewer peptide cross-bridges. However, gram-negative bacteria are quite susceptible to some β-lactam antibiotics that penetrate the outer membrane better than penicillin. Antibiotics will be discussed in more detail in Chapter 20.

CHECK YOUR UNDERSTANDING

- ✓ Why are drugs that target cell wall synthesis useful? **4-5**
- ✓ Why are mycoplasmas resistant to antibiotics that interfere with cell wall synthesis? **4-6**
- ✓ How do protoplasts differ from L forms? **4-7**

Structures Internal to the Cell Wall

LEARNING OBJECTIVES

4-8 Describe the structure, chemistry, and functions of the prokaryotic plasma membrane.

4-9 Define *simple diffusion, facilitated diffusion, osmosis, active transport,* and *group translocation.*

4-10 Identify the functions of the nucleoid and ribosomes.

4-11 Identify the functions of four inclusions.

4-12 Describe the functions of endospores, sporulation, and endospore germination.

Thus far, we have discussed the prokaryotic cell wall and structures external to it. We will now look inside the prokaryotic cell and discuss the structures and functions of the plasma membrane and components within the cytoplasm of the cell.

The Plasma (Cytoplasmic) Membrane

The **plasma (cytoplasmic) membrane** (or *inner membrane*) is a thin structure lying inside the cell wall and enclosing the cytoplasm of the cell (see Figure 4.6). The plasma membrane of prokaryotes consists primarily of phospholipids (see Figure 2.10, page 42), which are the most abundant chemicals in the membrane, and proteins. Eukaryotic plasma membranes also contain carbohydrates and sterols, such as cholesterol. Because they lack sterols, prokaryotic plasma membranes are less rigid than eukaryotic membranes. One exception is the wall-less prokaryote *Mycoplasma,* which contains membrane sterols.

Structure

In electron micrographs, prokaryotic and eukaryotic plasma membranes (and the outer membranes of gram-negative bacteria) look like two-layered structures; there are two dark lines with a light space between the lines (**Figure 4.14a**). The phospholipid molecules are arranged in two parallel rows, called a *lipid bilayer* (**Figure 4.14b**). As introduced in Chapter 2, each phospholipid molecule contains a polar head, composed of a phosphate group and glycerol that is hydrophilic (water-loving) and soluble in water, and nonpolar tails, composed of fatty acids that are hydrophobic (water-fearing) and insoluble in water (**Figure 4.14c**). The polar heads are on the two surfaces of the lipid bilayer, and the nonpolar tails are in the interior of the bilayer.

The protein molecules in the membrane can be arranged in a variety of ways. Some, called *peripheral proteins,* are easily removed from the membrane by mild treatments and lie at the inner or outer surface of the membrane. They may function as enzymes that catalyze chemical reactions, as a "scaffold" for support, and as mediators of changes in membrane shape during movement. Other proteins, called *integral proteins,* can be removed from the membrane only after disrupting the lipid bilayer (by using detergents, for example). Most integral proteins penetrate the membrane completely and are called

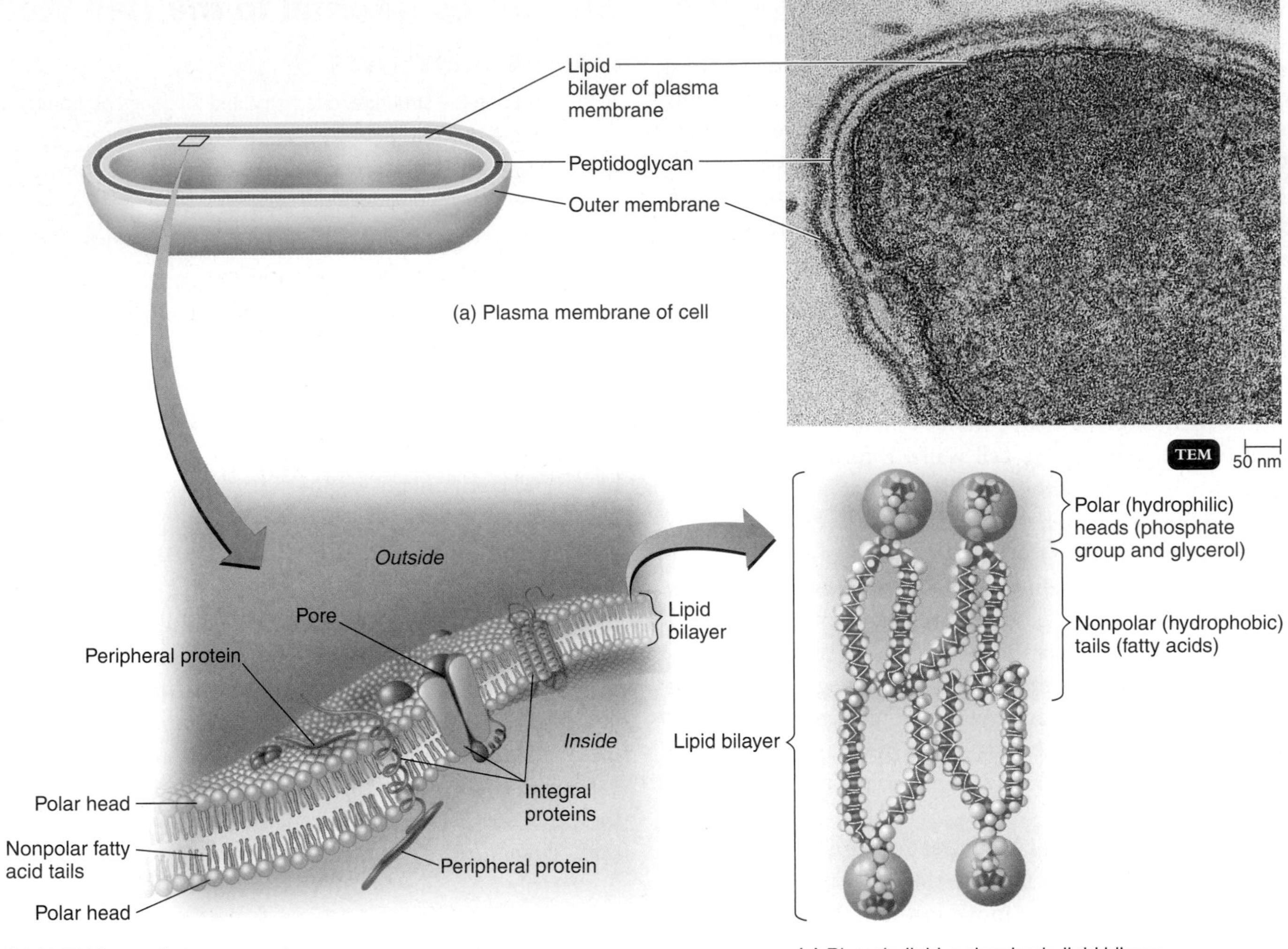

Figure 4.14 Plasma membrane. **(a)** A diagram and micrograph showing the lipid bilayer forming the inner plasma membrane of the gram-negative bacterium *Aquaspirillum serpens.* Layers of the cell wall, including the outer membrane, can be seen outside the inner membrane. **(b)** A portion of the inner membrane showing the lipid bilayer and proteins. The outer membrane of gram-negative bacteria is also a lipid bilayer. **(c)** Space-filling models of several molecules as they are arranged in the lipid bilayer.

Q What is the action of polymyxins on plasma membranes?

transmembrane proteins. Some integral proteins are channels that have a pore, or hole, through which substances enter and exit the cell.

Many of the proteins and some of the lipids on the outer surface of the plasma membrane have carbohydrates attached to them. Proteins attached to carbohydrates are called **glycoproteins;** lipids attached to carbohydrates are called **glycolipids.** Both glycoproteins and glycolipids help protect and lubricate the cell and are involved in cell-to-cell interactions. For example, glycoproteins play a role in certain infectious diseases. The influenza virus and the toxins that cause cholera and botulism enter their target cells by first binding to glycoproteins on their plasma membranes.

Studies have demonstrated that the phospholipid and protein molecules in membranes are not static but move quite freely within the membrane surface. This movement is most probably associated with the many functions performed by the plasma membrane. Because the fatty acid tails cling together, phospholipids in the presence of water form a self-sealing bilayer; as a result, breaks and tears in the membrane heal themselves. The membrane must be about as viscous as olive oil, which allows membrane proteins to move freely enough to perform their functions without destroying the structure of the membrane. This dynamic arrangement of phospholipids and proteins is referred to as the **fluid mosaic model.**

Functions

The most important function of the plasma membrane is to serve as a selective barrier through which materials enter and exit the cell. In this function, plasma membranes have

selective permeability (sometimes called *semipermeability*). This term indicates that certain molecules and ions pass through the membrane, but that others are prevented from passing through it. The permeability of the membrane depends on several factors. Large molecules (such as proteins) cannot pass through the plasma membrane, possibly because these molecules are larger than the pores in integral proteins that function as channels. But smaller molecules (such as water, oxygen, carbon dioxide, and some simple sugars) usually pass through easily. Ions penetrate the membrane very slowly. Substances that dissolve easily in lipids (such as oxygen, carbon dioxide, and nonpolar organic molecules) enter and exit more easily than other substances because the membrane consists mostly of phospholipids. The movement of materials across plasma membranes also depends on transporter molecules, which will be described shortly.

Plasma membranes are also important to the breakdown of nutrients and the production of energy. The plasma membranes of bacteria contain enzymes capable of catalyzing the chemical reactions that break down nutrients and produce ATP. In some bacteria, pigments and enzymes involved in photosynthesis are found in infoldings of the plasma membrane that extend into the cytoplasm. These membranous structures are called **chromatophores** or **thylakoids** (Figure 4.15).

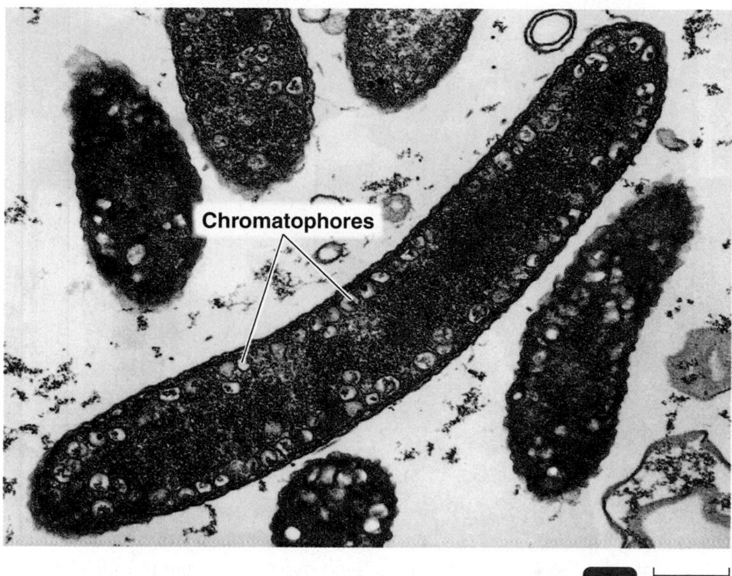

Figure 4.15 Chromatophores. In this micrograph of *Rhodospirillum rubrum,* a purple (nonsulfur) bacterium, the chromatophores are clearly visible.

Q What is the function of chromatophores?

When viewed with an electron microscope, bacterial plasma membranes often appear to contain one or more large, irregular folds called **mesosomes.** Many functions have been proposed for mesosomes. However, it is now known that they are artifacts, not true cell structures. Mesosomes are believed to be folds in the plasma membrane that develop by the process used for preparing specimens for electron microscopy. **Animations** Membrane Structure; Membrane Permeability. **www.microbiologyplace.com**

Destruction of the Plasma Membrane by Antimicrobial Agents

Because the plasma membrane is vital to the bacterial cell, it is not surprising that several antimicrobial agents exert their effects at this site. In addition to the chemicals that damage the cell wall and thereby indirectly expose the membrane to injury, many compounds specifically damage plasma membranes. These compounds include certain alcohols and quaternary ammonium compounds, which are used as disinfectants. By disrupting the membrane's phospholipids, a group of antibiotics known as the *polymyxins* cause leakage of intracellular contents and subsequent cell death. This mechanism will be discussed in Chapter 20.

The Movement of Materials across Membranes

Materials move across plasma membranes of both prokaryotic and eukaryotic cells by two kinds of processes: passive and active. In *passive processes,* substances cross the membrane from an area of high concentration to an area of low concentration (move with the concentration gradient, or difference), without any expenditure of energy (ATP) by the cell. In *active processes,* the cell must use energy (ATP) to move substances from areas of low concentration to areas of high concentration (against the concentration gradient).

Passive Processes Passive processes include simple diffusion, facilitated diffusion, and osmosis.

Simple diffusion is the net (overall) movement of molecules or ions from an area of high concentration to an area of low concentration (Figure 4.16 and Figure 4.17a). The movement continues until the molecules or ions are evenly distributed. The point of even distribution is called *equilibrium.* Cells rely on simple diffusion to transport certain small molecules, such as oxygen and carbon dioxide, across their cell membranes.

In **facilitated diffusion,** integral membrane proteins function as channels or carriers that facilitate the movement of ions or large molecules across the plasma membrane. Such integral proteins are called *transporters* or *permeases.* Facilitated diffusion is similar to simple diffusion in that the cell *does not* expend energy, because the substance moves from a high to a low concentration. The process differs from simple diffusion in its use of transporters. Some transporters permit the passage of mostly small, inorganic ions that are too hydrophilic to penetrate the nonpolar interior of the lipid bilayer (Figure 4.17b). These transporters, which are common in prokaryotes, are nonspecific and allow the passage of a wide variety of ions (or even small molecules). Other transporters, which are common in eukaryotes, are specific and transport only specific, usually larger, molecules, such as simple

(a) (b)

Figure 4.16 The principle of simple diffusion. (**a**) After a dye pellet is put into a beaker of water, the molecules of dye in the pellet diffuse into the water from an area of high dye concentration to areas of low dye concentration. (**b**) The dye potassium permanganate in the process of diffusing.

Q What is the distinguishing feature of a passive process?

sugars (glucose, fructose, and galactose) and vitamins. In this process, the transported substance binds to a specific transporter on the outer surface of the plasma membrane, which undergoes a change of shape; then the transporter releases the substance on the other side of the membrane (Figure 4.17c).

In some cases, molecules that bacteria need are too large to be transported into the cells by these methods. Most bacteria, however, produce enzymes that can break down large molecules into simpler ones (such as proteins into amino acids, or polysaccharides into simple sugars). Such enzymes, which are released by the bacteria into the surrounding medium, are appropriately called *extracellular enzymes.* Once the enzymes degrade the large molecules, the subunits move into the cell with the help of transporters. For example, specific carriers retrieve DNA bases, such as the purine guanine, from extracellular media and bring them into the cell's cytoplasm.

Osmosis is the net movement of solvent molecules across a selectively permeable membrane from an area with a high concentration of solvent molecules (low concentration of solute molecules) to an area of low concentration of solvent molecules (high concentration of solute molecules). In living systems, the chief solvent is water. Water molecules may pass through plasma membranes by moving through the lipid bilayer by simple diffusion or through integral membrane proteins, called *aquaporins,* that function as water channels (Figure 4.17d).

Osmosis may be demonstrated with the apparatus shown in Figure 4.18a. A sack constructed from cellophane, which is a selectively permeable membrane, is filled with a solution of 20% sucrose (table sugar). The cellophane sack is placed into a beaker containing distilled water. Initially, the concentrations of water on either side of the membrane are different. Because of the sucrose molecules, the concentration of water is lower inside the cellophane sack. Therefore, water moves from the beaker (where its concentration is higher) into the cellophane sack (where its concentration is lower).

There is no movement of sugar out of the cellophane sack into the beaker, however, because the cellophane is impermeable to molecules of sugar—the sugar molecules are too large to go

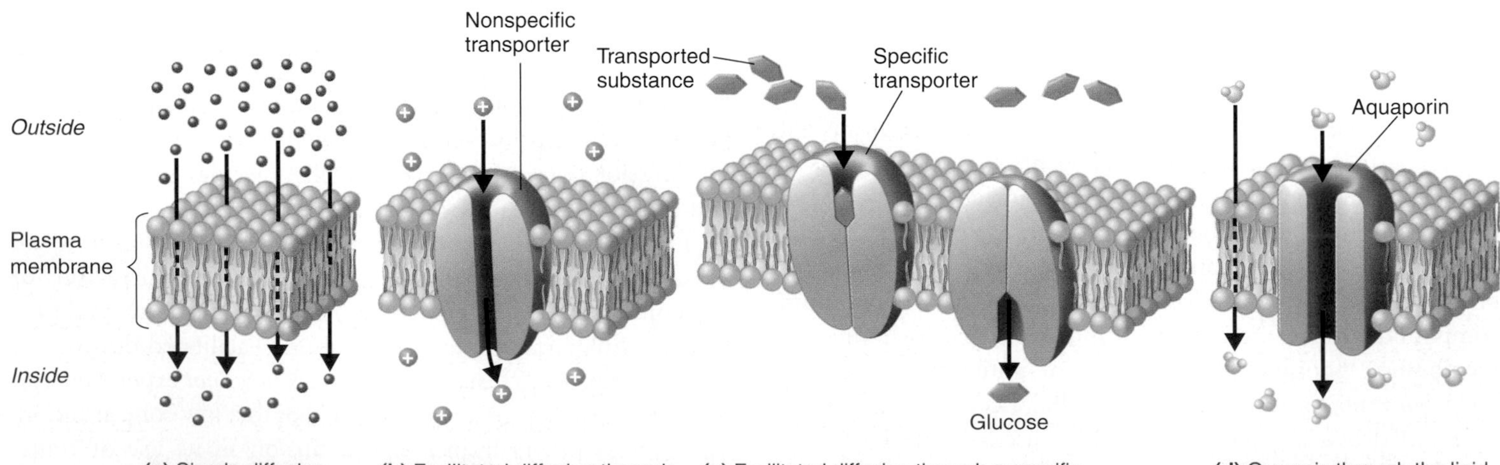

Figure 4.17 Facilitated diffusion. Transporter proteins in the plasma membrane transport molecules across the membrane from an area of high concentration to one of low concentration (with the concentration gradient). The transporter undergoes a change in shape to transport the substance. The process does not require ATP.

Q How does simple diffusion differ from facilitated diffusion?

(a) At beginning of osmotic pressure experiment

(b) At equilibrium

(c) Isotonic solution. No net movement of water

(d) Hypotonic solution. Water moves into the cell. If the cell wall is strong, it contains the swelling. If the cell wall is weak or damaged, the cell bursts (osmotic lysis).

(e) Hypertonic solution. Water moves out of the cell, causing its cytoplasm to shrink (plasmolysis).

Figure 4.18 The principle of osmosis. (**a**) Setup at the beginning of an osmotic pressure experiment. Water molecules start to move from the beaker into the sack along the concentration gradient. (**b**) Setup at equilibrium. The osmotic pressure exerted by the solution in the sack pushes water molecules from the sack back into the beaker to balance the rate of water entry into the sack. The height of the solution in the glass tube at equilibrium is a measure of the osmotic pressure. (**c**)–(**e**) The effects of various solutions on bacterial cells.

through the pores of the membrane. As water moves into the cellophane sack, the sugar solution becomes increasingly dilute, and, because the cellophane sack has expanded to its limit as a result of an increased volume of water, water begins to move up the glass tube. In time, the water that has accumulated in the cellophane sack and the glass tube exerts a downward pressure that forces water molecules out of the cellophane sack and back into the beaker. This movement of water through a selectively permeable membrane produces osmotic pressure. **Osmotic pressure** is the pressure required to prevent the movement of pure water (water with no solutes) into a solution containing some solutes. In other words, osmotic pressure is the pressure needed to stop the flow of water across the selectively permeable membrane (cellophane). When water molecules leave and enter the cellophane sack at the same rate, equilibrium is reached (Figure 4.18b).

A bacterial cell may be subjected to any of three kinds of osmotic solutions: isotonic, hypotonic, or hypertonic. An **isotonic solution** is a medium in which the overall concentration of solutes equals that found inside a cell (*iso* means equal). Water leaves and enters the cell at the same rate (no net change); the cell's contents are in equilibrium with the solution outside the cell wall (Figure 4.18c).

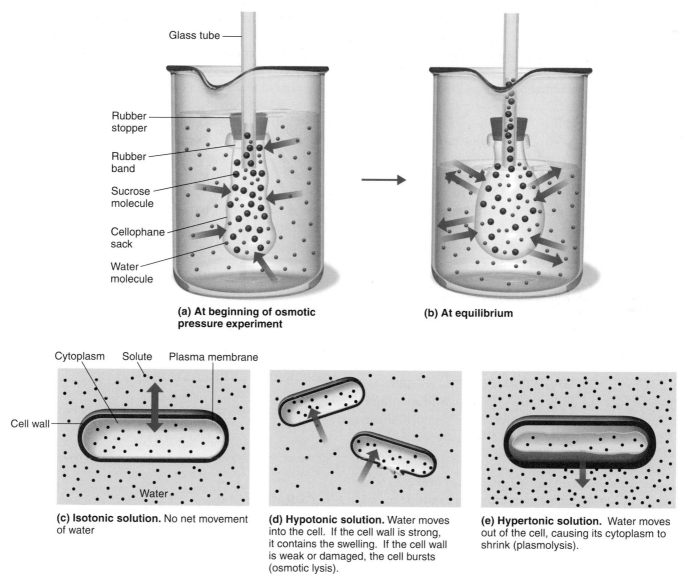

Earlier we mentioned that lysozyme and certain antibiotics (such as penicillin) damage bacterial cell walls, causing the cells to rupture, or lyse. Such rupturing occurs because bacterial cytoplasm usually contains such a high concentration of solutes that, when the wall is weakened or removed, additional water enters the cell by osmosis. The damaged (or removed) cell wall cannot constrain the swelling of the cytoplasmic membrane, and the membrane bursts. This is an example of osmotic lysis caused by immersion in a hypotonic solution. A **hypotonic solution** outside the cell is a medium whose concentration of solutes is lower than that inside the cell (*hypo* means under or less). Most bacteria live in hypotonic solutions, and the cell wall resists further osmosis and protects cells from lysis. Cells with weak cell walls, such as gram-negative bacteria, may burst or undergo osmotic lysis as a result of excessive water intake (**Figure 4.18d**).

A **hypertonic solution** is a medium having a higher concentration of solutes than inside the cell has (*hyper* means above or more). Most bacterial cells placed in a hypertonic solution shrink and collapse or *plasmolyze* because water leaves the cells by osmosis (**Figure 4.18e**). Keep in mind that the terms *isotonic, hypotonic,* and *hypertonic* describe the concentration of solutions outside the cell *relative to* the concentration inside the cell. **Animations** Passive Transport: Principles of Diffusion, Special Types of Diffusion. **www.microbiologyplace.com**

Active Processes Simple diffusion and facilitated diffusion are useful mechanisms for transporting substances into cells when the concentrations of the substances are greater outside the cell. However, when a bacterial cell is in an environment in which nutrients are in low concentration, the cell must use active processes, such as active transport and group translocation, to accumulate the needed substances.

In performing **active transport,** the cell *uses energy* in the form of ATP to move substances across the plasma membrane. Among the substances actively transported are ions (for example Na^+, K^+, H^+, Ca^{2+}, and Cl^-), amino acids, and simple sugars. Although these substances can also be moved into cells by passive processes, their movement by active processes can go against the concentration gradient, allowing a cell to accumulate needed materials. The movement of a substance in active transport is usually from outside to inside, even though the concentration might be much higher inside the cell. Like facilitated diffusion, active transport depends on transporter proteins in the plasma membrane (see Figure 4.17b,c). There appears to be a different transporter for each transported substance or group of closely related transported substances. Active transport enables microbes to move substances across the plasma membrane at a constant rate, even if they are in short supply.

In active transport, the substance that crosses the membrane is not altered by transport across the membrane. In **group translocation,** a special form of active transport that occurs exclusively in prokaryotes, the substance is chemically altered during transport across the membrane. Once the substance is altered and inside the cell, the plasma membrane is impermeable to it, so it remains inside the cell. This important mechanism enables a cell to accumulate various substances even though they may be in low concentrations outside the cell. Group translocation requires energy supplied by high-energy phosphate compounds, such as phosphoenolpyruvic acid (PEP).

One example of group translocation is the transport of the sugar glucose, which is often used in growth media for bacteria. While a specific carrier protein is transporting the glucose molecule across the membrane, a phosphate group is added to the sugar. This phosphorylated form of glucose, which cannot be transported out, can then be used in the cell's metabolic pathways.

Some eukaryotic cells (those without cell walls) can use two additional active transport processes called phagocytosis and pinocytosis. These processes, which do not occur in bacteria, are explained on page 100. **Animations** Active Transport: Types, Overview. **www.microbiologyplace.com**

CHECK YOUR UNDERSTANDING

✓ Which agents can cause injury to the bacterial plasma membrane? **4-8**

✓ How are simple diffusion and facilitated diffusion similar? How are they different? **4-9**

Cytoplasm

For a prokaryotic cell, the term **cytoplasm** refers to the substance of the cell inside the plasma membrane (see Figure 4.6). Cytoplasm is about 80% water and contains primarily proteins (enzymes), carbohydrates, lipids, inorganic ions, and many low-molecular-weight compounds. Inorganic ions are present in much higher concentrations in cytoplasm than in most media. Cytoplasm is thick, aqueous, semitransparent, and elastic. The major structures in the cytoplasm of prokaryotes are a nucleoid (containing DNA), particles called ribosomes, and reserve deposits called inclusions. Protein filaments in the cytoplasm are most likely responsible for the rod and helical cell shapes of bacteria.

Prokaryotic cytoplasm lacks certain features of eukaryotic cytoplasm, such as a cytoskeleton and cytoplasmic streaming. These features will be described later.

The Nucleoid

The **nucleoid** of a bacterial cell (see Figure 4.6) usually contains a single long, continuous, and frequently circularly arranged thread of double-stranded DNA called the **bacterial chromosome.** This is the cell's genetic information, which carries all the information required for the cell's structures and functions.

Unlike the chromosomes of eukaryotic cells, bacterial chromosomes are not surrounded by a nuclear envelope (membrane) and do not include histones. The nucleoid can be spherical, elongated, or dumbbell-shaped. In actively growing bacteria, as much as 20% of the cell volume is occupied by DNA because such cells presynthesize nuclear material for future cells. The chromosome is attached to the plasma membrane. Proteins in the plasma membrane are believed to be responsible for replication of the DNA and segregation of the new chromosomes to daughter cells during cell division.

In addition to the bacterial chromosome, bacteria often contain small usually circular, double-stranded DNA molecules called **plasmids** (see the F factor in Figure 8.27a, page 238). These molecules are extrachromosomal genetic elements; that is, they are not connected to the main bacterial chromosome, and they replicate independently of chromosomal DNA. Research indicates that plasmids are associated with plasma membrane proteins. Plasmids usually contain from 5 to 100 genes that are generally not crucial for the survival of the bacterium under normal environmental conditions; plasmids may be gained or lost without harming the cell. Under certain conditions, however, plasmids are an advantage to cells. Plasmids may carry genes for such activities as antibiotic resistance, tolerance to toxic metals, the production of toxins, and the synthesis of enzymes. Plasmids can be transferred from one bacterium to another. In fact, plasmid DNA is used for gene manipulation in biotechnology.

Ribosomes

All eukaryotic and prokaryotic cells contain **ribosomes,** which function as the sites of protein synthesis. Cells that have high rates of protein synthesis, such as those that are actively growing, have a large number of ribosomes. The cytoplasm of a prokaryotic cell contains tens of thousands of these very small structures, which give the cytoplasm a granular appearance (see Figure 4.6).

Ribosomes are composed of two subunits, each of which consists of protein and a type of RNA called *ribosomal RNA (rRNA)*. Prokaryotic ribosomes differ from eukaryotic ribosomes in the number of proteins and rRNA molecules they contain; they are also somewhat smaller and less dense than ribosomes of eukaryotic cells. Accordingly, prokaryotic ribosomes are called 70S ribosomes (**Figure 4.19**), and those of eukaryotic cells are known as 80S ribosomes. The letter S refers to Svedberg units, which indicate the relative rate of sedimentation during ultra-high-speed centrifugation. Sedimentation rate is a function of the size, weight, and shape of a particle. The subunits of a 70S ribosome are a small 30S subunit containing one molecule of rRNA and a larger 50S subunit containing two molecules of rRNA.

Several antibiotics work by inhibiting protein synthesis on prokaryotic ribosomes. Antibiotics such as streptomycin and gentamicin attach to the 30S subunit and interfere with protein synthesis. Other antibiotics, such as erythromycin and chloramphenicol, interfere with protein synthesis by attaching to the 50S subunit. Because of differences in prokaryotic and eukaryotic ribosomes, the microbial cell can be killed by the antibiotic while the eukaryotic host cell remains unaffected.

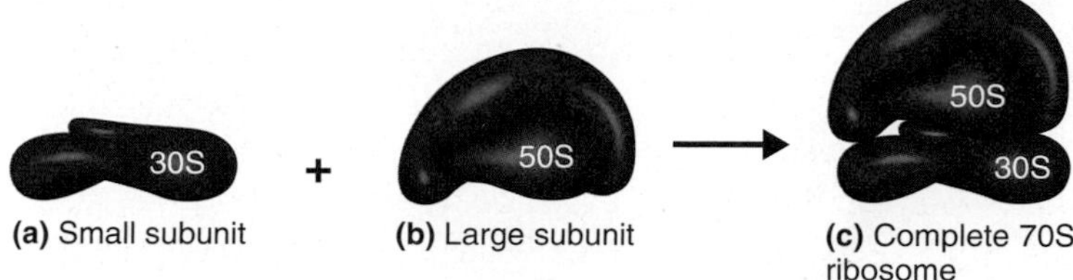

Figure 4.19 The prokaryotic ribosome. (**a**) A small 30S subunit and (**b**) a large 50S subunit make up (**c**) the complete 70S prokaryotic ribosome.

Q What is the importance of the differences between prokaryotic and eukaryotic ribosomes with regard to antibiotic therapy?

Inclusions

Within the cytoplasm of prokaryotic cells are several kinds of reserve deposits, known as **inclusions.** Cells may accumulate certain nutrients when they are plentiful and use them when the environment is deficient. Evidence suggests that macromolecules concentrated in inclusions avoid the increase in osmotic pressure that would result if the molecules were dispersed in the cytoplasm. Some inclusions are common to a wide variety of bacteria, whereas others are limited to a small number of species and therefore serve as a basis for identification.

Metachromatic Granules

Metachromatic granules are large inclusions that take their name from the fact that they sometimes stain red with certain blue dyes such as methylene blue. Collectively they are known as **volutin.** Volutin represents a reserve of inorganic phosphate (polyphosphate) that can be used in the synthesis of ATP. It is generally formed by cells that grow in phosphate-rich environments. Metachromatic granules are found in algae, fungi, and protozoa, as well as in bacteria. These granules are characteristic of *Corynebacterium diphtheriae* (kô-rī-nē-bak-ti′rē-um dif-thi′rē-ī), the causative agent of diphtheria; thus, they have diagnostic significance.

Polysaccharide Granules

Inclusions known as **polysaccharide granules** typically consist of glycogen and starch, and their presence can be demonstrated when iodine is applied to the cells. In the presence of iodine, glycogen granules appear reddish brown and starch granules appear blue.

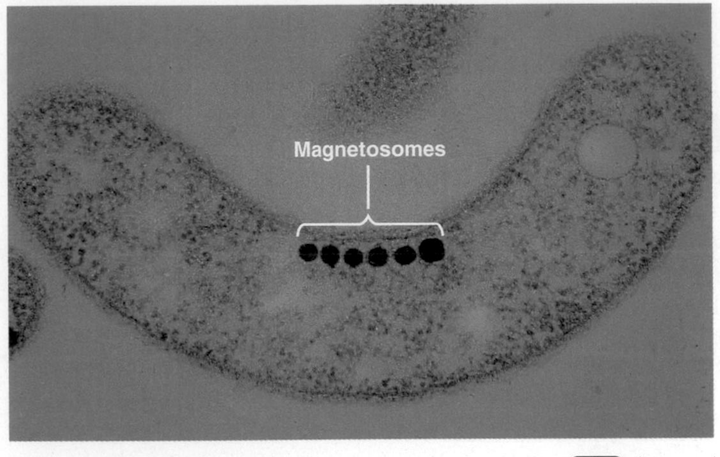

Figure 4.20 Magnetosomes. This micrograph of *Magnetospirillum magnetotacticum* shows a chain of magnetosomes. The outer membrane of the gram-negative wall is also visible.

Q What is the function of magnetosomes?

Lipid Inclusions

Lipid inclusions appear in various species of *Mycobacterium, Bacillus, Azotobacter* (ä-zō-tō-bak'tėr), *Spirillum* (spī-ril'lum), and other genera. A common lipid-storage material, one unique to bacteria, is the polymer *poly-β-hydroxybutyric acid.* Lipid inclusions are revealed by staining cells with fat-soluble dyes, such as Sudan dyes.

Sulfur Granules

Certain bacteria—for example, the "sulfur bacteria" that belong to the genus *Thiobacillus*—derive energy by oxidizing sulfur and sulfur-containing compounds. These bacteria may deposit **sulfur granules** in the cell, where they serve as an energy reserve.

Carboxysomes

Carboxysomes are inclusions that contain the enzyme ribulose 1,5-diphosphate carboxylase. Photosynthetic bacteria use carbon dioxide as their sole source of carbon and require this enzyme for carbon dioxide fixation. Among the bacteria containing carboxysomes are nitrifying bacteria, cyanobacteria, and thiobacilli.

Gas Vacuoles

Hollow cavities found in many aquatic prokaryotes, including cyanobacteria, anoxygenic photosynthetic bacteria, and halobacteria are called **gas vacuoles.** Each vacuole consists of rows of several individual *gas vesicles,* which are hollow cylinders covered by protein. Gas vacuoles maintain buoyancy so that the cells can remain at the depth in the water appropriate for them to receive sufficient amounts of oxygen, light, and nutrients.

Magnetosomes

Magnetosomes are inclusions of iron oxide (Fe_3O_4), formed by several gram-negative bacteria such as *Magnetospirillum magnetotacticum,* that act like magnets (Figure 4.20). Bacteria may use magnetosomes to move downward until they reach a suitable attachment site. In vitro, magnetosomes can decompose hydrogen peroxide, which forms in cells in the presence of oxygen. Researchers speculate that magnetosomes may protect the cell against hydrogen peroxide accumulation.

Endospores

When essential nutrients are depleted, certain gram-positive bacteria, such as those of the genera *Clostridium* and *Bacillus,* form specialized "resting" cells called **endospores** (Figure 4.21). As you will see later, some members of the genus *Clostridium* cause diseases such as gangrene, tetanus, botulism, and food poisoning. Some members of the genus *Bacillus* cause anthrax and food poisoning. Unique to bacteria, endospores are highly durable dehydrated cells with thick walls and additional layers. They are formed internal to the bacterial cell membrane.

When released into the environment, they can survive extreme heat, lack of water, and exposure to many toxic chemicals and radiation. For example, 7500-year-old endospores of *Thermoactinomyces vulgaris* (thėr-mō-ak-tin-ō-mī'sēs vul-ga'ris) from the freezing muds of Elk Lake in Minnesota have germinated when rewarmed and placed in a nutrient medium, and 25- to 40-million-year-old endospores found in the gut of a stingless bee entombed in amber (hardened tree resin) in the Dominican Republic are reported to have germinated when placed in nutrient media. Although true endospores are found in gram-positive bacteria, one gram-negative species, *Coxiella burnetii* (käks-ē-el'lä bėr-ne'tē-ē), the cause of Q fever, forms endosporelike structures that resist heat and chemicals and can be stained with endospore stains (see Figure 24.14, page 690).

The process of endospore formation within a vegetative cell takes several hours and is known as **sporulation** or **sporogenesis** (Figure 4.21a). Vegetative cells of endospore-forming bacteria begin sporulation when a key nutrient, such as the carbon or nitrogen source, becomes scarce or unavailable. In the first observable stage of sporulation, a newly replicated bacterial chromosome and a small portion of cytoplasm are isolated by an ingrowth of the plasma membrane called a *spore septum.* The spore septum becomes a double-layered membrane that surrounds the chromosome and cytoplasm. This structure, entirely enclosed within the original cell, is called a *forespore.* Thick layers of peptidoglycan are laid down between the two membrane layers. Then a thick *spore coat* of protein forms around the outside membrane; this coat is responsible for the resistance of endospores to many harsh chemicals. The original cell is degraded, and the endospore is released.

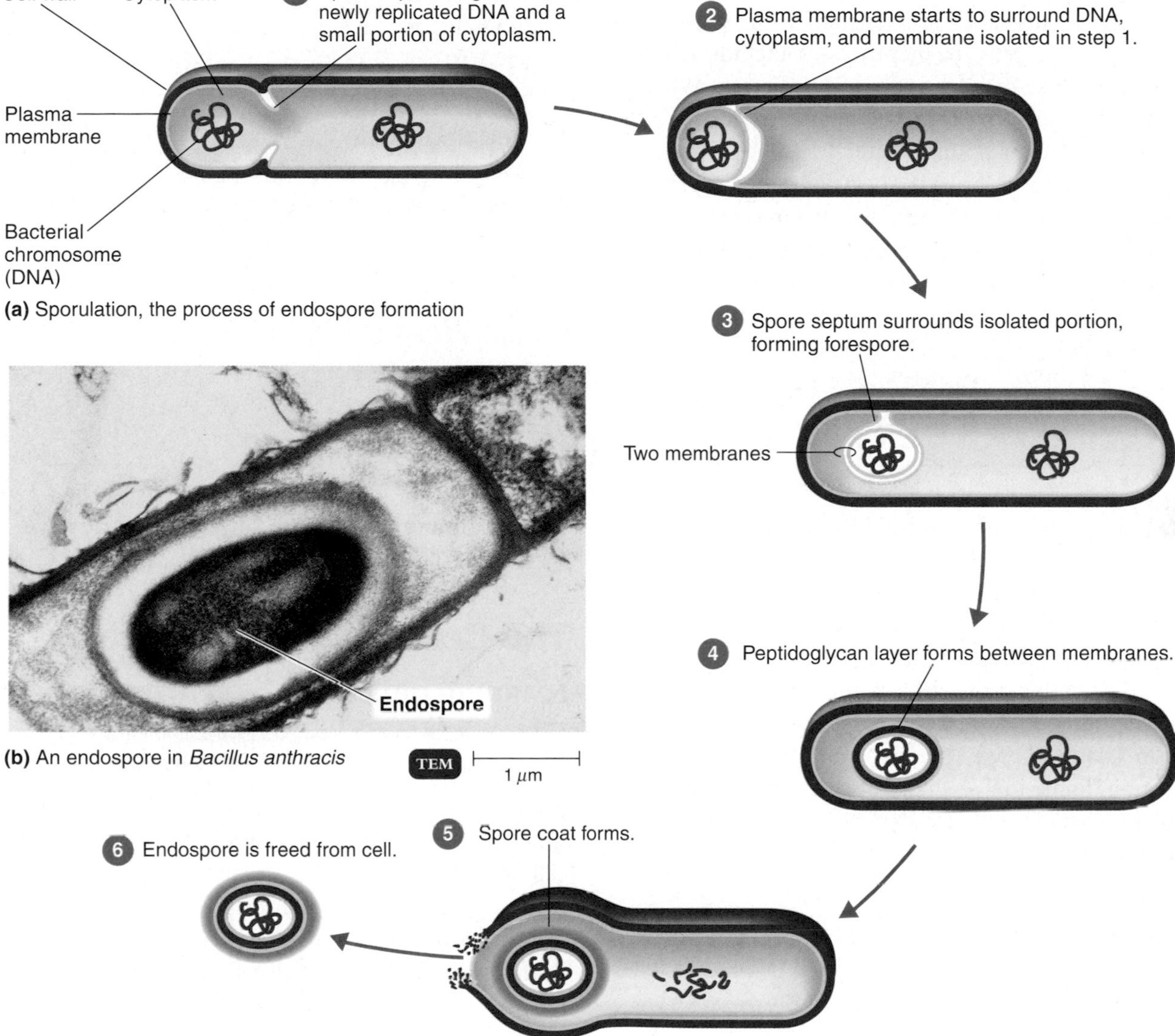
Cell wall
Cytoplasm
1 Spore septum begins to isolate newly replicated DNA and a small portion of cytoplasm.
Plasma membrane
Bacterial chromosome (DNA)
(a) Sporulation, the process of endospore formation
2 Plasma membrane starts to surround DNA, cytoplasm, and membrane isolated in step 1.
3 Spore septum surrounds isolated portion, forming forespore.
Two membranes
4 Peptidoglycan layer forms between membranes.
Endospore
(b) An endospore in Bacillus anthracis
TEM
1 μm
5 Spore coat forms.
6 Endospore is freed from cell.

Figure 4.21 Formation of endospores by sporulation.

Q Under what conditions are endospores formed by bacteria?

The diameter of the endospore may be the same as, smaller than, or larger than the diameter of the vegetative cell. Depending on the species, the endospore might be located *terminally* (at one end), *subterminally* (near one end; **Figure 4.21b**), or *centrally* inside the vegetative cell. When the endospore matures, the vegetative cell wall ruptures (lyses), killing the cell, and the endospore is freed.

Most of the water present in the forespore cytoplasm is eliminated by the time sporulation is complete, and endospores do not carry out metabolic reactions. The highly dehydrated endospore core contains only DNA, small amounts of RNA, ribosomes, enzymes, and a few important small molecules. The latter include a strikingly large amount of an organic acid called *dipicolinic acid* (found in the cytoplasm), which is accompanied by a large number of calcium ions. These cellular components are essential for resuming metabolism later.

Endospores can remain dormant for thousands of years. An endospore returns to its vegetative state by a process called **germination.** Germination is triggered by physical or chemical damage to the endospore's coat. The endospore's enzymes then break down the extra layers surrounding the endospore, water enters, and metabolism resumes. Because one vegetative cell forms a single endospore, which, after germination, remains one cell, sporulation in bacteria is *not* a means of reproduction. This process does not increase the number of cells. Bacterial endospores differ from spores formed by (prokaryotic) actinomycetes and the eukaryotic fungi and algae, which detach from the parent and develop into another organism and, therefore, represent reproduction.

Endospores are important from a clinical viewpoint and in the food industry because they are resistant to processes that normally kill vegetative cells. Such processes include heating, freezing, desiccation, use of chemicals, and radiation. Whereas most vegetative cells are killed by temperatures above 70°C, endospores can survive in boiling water for several hours or more. Endospores of thermophilic (heat-loving) bacteria can survive in boiling water for 19 hours. Endospore-forming bacteria are a problem in the food industry because they are likely to survive underprocessing, and, if conditions for growth occur, some species produce toxins and disease. Special methods for controlling organisms that produce endospores are discussed in Chapter 7.

CHECK YOUR UNDERSTANDING

- ✓ Where is the DNA located in a prokaryotic cell? **4-10**
- ✓ What is the general function of inclusions? **4-11**
- ✓ Under what conditions do endospores form? **4-12**

* * *

Having examined the functional anatomy of the prokaryotic cell, we will now look at the functional anatomy of the eukaryotic cell.

THE EUKARYOTIC CELL

As mentioned earlier, eukaryotic organisms include algae, protozoa, fungi, plants, and animals. The eukaryotic cell is typically larger and structurally more complex than the prokaryotic cell (**Figure 4.22**). When the structure of the prokaryotic cell in Figure 4.6 is compared with that of the eukaryotic cell, the differences between the two types of cells become apparent. The principal differences between prokaryotic and eukaryotic cells are summarized in **Table 4.2,** page 101.

The following discussion of eukaryotic cells will parallel our discussion of prokaryotic cells by starting with structures that extend to the outside of the cell.

Flagella and Cilia

LEARNING OBJECTIVE

4-13 Differentiate prokaryotic and eukaryotic flagella.

Many types of eukaryotic cells have projections that are used for cellular locomotion or for moving substances along the surface of the cell. These projections contain cytoplasm and are enclosed by the plasma membrane. If the projections are few and are long in relation to the size of the cell, they are called **flagella.** If the projections are numerous and short, they are called **cilia** (singular: **cilium**).

Algae of the genus *Euglena* (ū-glē′na) use a flagellum for locomotion, whereas protozoa, such as *Tetrahymena* (tet-rä-hī′me-nä), use cilia for locomotion (**Figure 4.23a** and **Figure 4.23b**). Both flagella and cilia are anchored to the plasma membrane by a basal body, and both consist of nine pairs of microtubules (doublets) arranged in a ring, plus another two microtubules in the center of the ring, an arrangement called a *9 + 2 array* (**Figure 4.23c**). **Microtubules** are long, hollow tubes made up of a protein called *tubulin.* A prokaryotic flagellum rotates, but a eukaryotic flagellum moves in a wavelike manner (**Figure 4.23d**). To help keep foreign material out of the lungs, ciliated cells of the human respiratory system move the material along the surface of the cells in the bronchial tubes and trachea toward the throat and mouth (see Figure 16.4, page 452).

The Cell Wall and Glycocalyx

LEARNING OBJECTIVE

4-14 Compare and contrast prokaryotic and eukaryotic cell walls and glycocalyxes.

Most eukaryotic cells have cell walls, although they are generally much simpler than those of prokaryotic cells. Many algae have cell walls consisting of the polysaccharide *cellulose* (as do all plants); other chemicals may be present as well. Cell walls of some fungi also contain cellulose, but in most fungi the principal structural component of the cell wall is the polysaccharide *chitin,* a polymer of N-acetylglucosamine (NAG) units. (Chitin is also the main structural component of the exoskeleton of crustaceans and insects.) The cell walls of yeasts contain the polysaccharides *glucan* and *mannan.* In eukaryotes that lack a cell wall, the plasma membrane may be the outer covering; however, cells that have direct contact with the environment may have coatings outside the plasma membrane. Protozoa do not have a typical cell wall; instead, they have a flexible outer protein covering called a *pellicle.*

In other eukaryotic cells, including animal cells, the plasma membrane is covered by a **glycocalyx,** a layer of material containing substantial amounts of sticky carbohydrates. Some of these carbohydrates are covalently bonded to proteins and lipids in the plasma membrane, forming glycoproteins and glycolipids that anchor the glycocalyx to the cell. The glycocalyx strengthens the cell surface, helps attach cells together, and may contribute to cell–cell recognition.

Q&A Eukaryotic cells do not contain peptidoglycan, the framework of the prokaryotic cell wall. This is significant medically because antibiotics, such as penicillins and cephalosporins, act against peptidoglycan and therefore do not affect human eukaryotic cells.

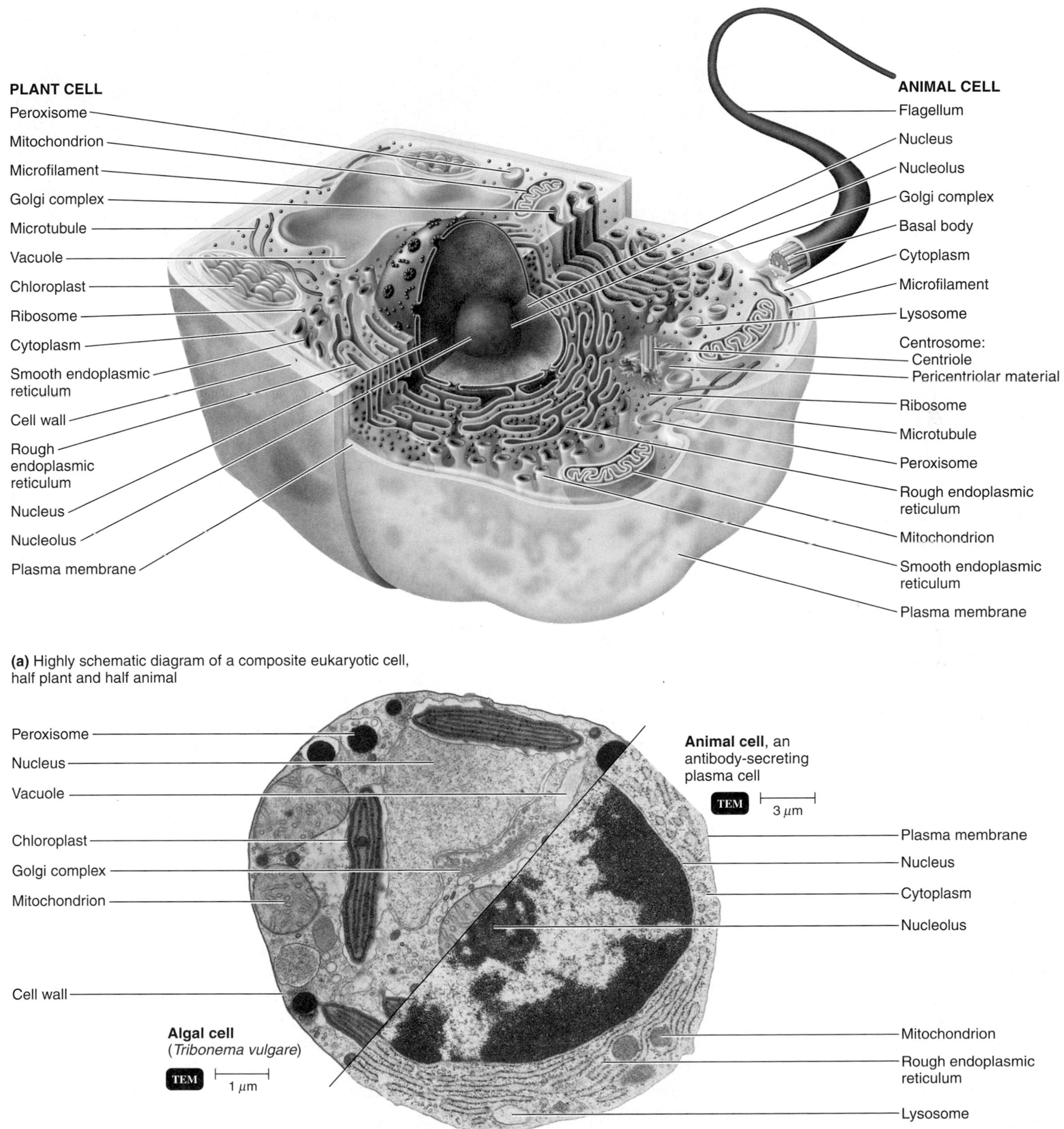

(a) Highly schematic diagram of a composite eukaryotic cell, half plant and half animal

(b) Transmission electron micrographs of plant and animal cells.

Figure 4.22 Eukaryotic cells showing typical structures.

Q What kingdoms contain eukaryotic organisms?

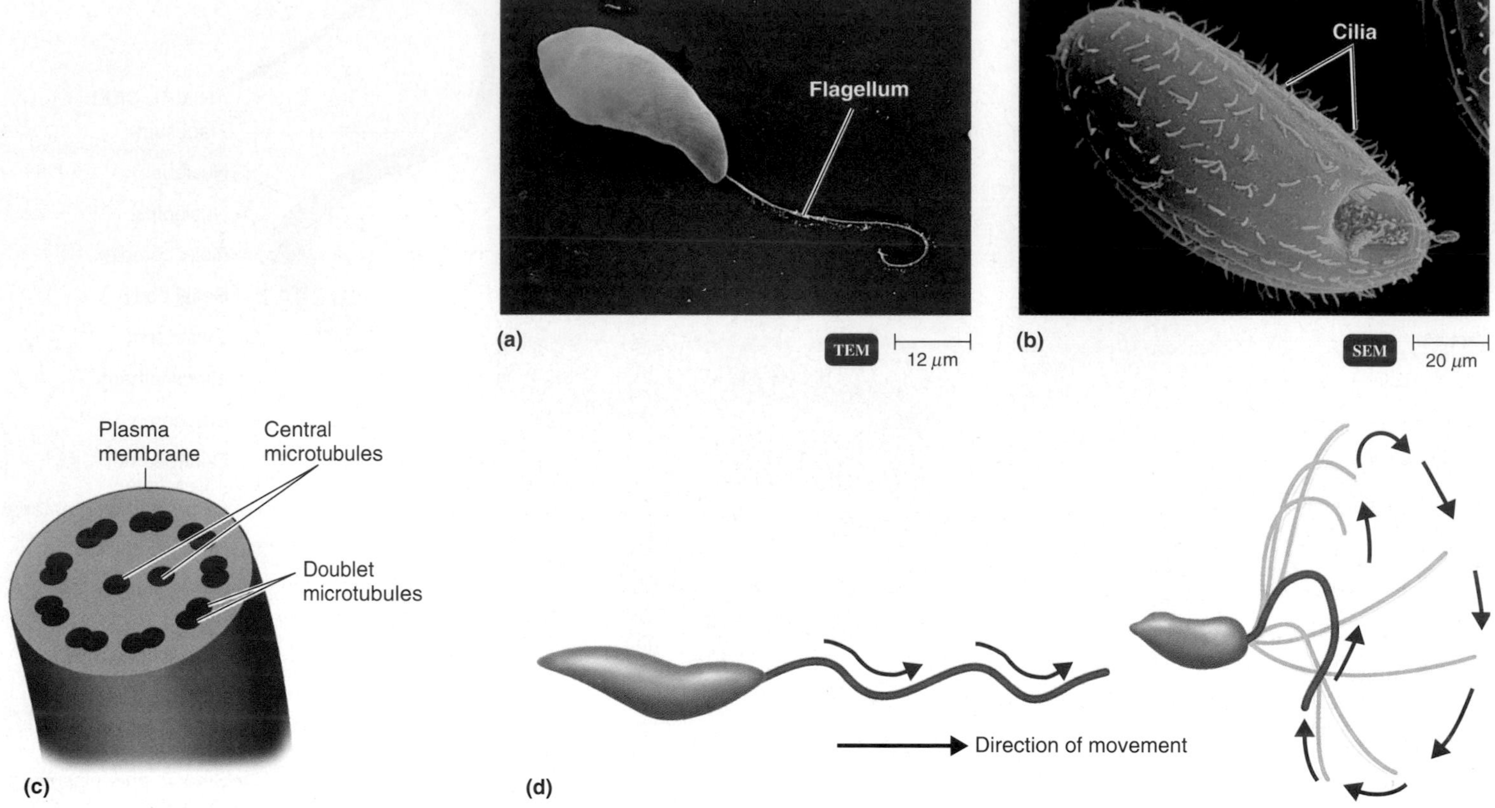

Figure 4.23 Eukaryotic flagella and cilia. (**a**) A micrograph of *Euglena,* a chlorophyll-containing alga, with its flagellum. (**b**) A micrograph of *Tetrahymena,* a common freshwater protozoan, with cilia. (**c**) The internal structure of a flagellum (or cilium), showing the 9 + 2 arrangement of microtubules. (**d**) The pattern of movement of a eukaryotic flagellum.

Q **How do prokaryotic and eukaryotic flagella differ?**

The Plasma (Cytoplasmic) Membrane

LEARNING OBJECTIVE

4-15 Compare and contrast prokaryotic and eukaryotic plasma membranes.

The **plasma (cytoplasmic) membrane** of eukaryotic and prokaryotic cells is very similar in function and basic structure. There are, however, differences in the types of proteins found in the membranes. Eukaryotic membranes also contain carbohydrates, which serve as attachment sites for bacteria and as receptor sites that assume a role in such functions as cell–cell recognition. Eukaryotic plasma membranes also contain *sterols,* complex lipids not found in prokaryotic plasma membranes (with the exception of *Mycoplasma* cells). Sterols seem to be associated with the ability of the membranes to resist lysis resulting from increased osmotic pressure.

Substances can cross eukaryotic and prokaryotic plasma membranes by simple diffusion, facilitated diffusion, osmosis, or active transport. Group translocation does not occur in eukaryotic cells. However, eukaryotic cells can use a mechanism called **endocytosis.** This occurs when a segment of the plasma membrane surrounds a particle or large molecule, encloses it, and brings it into the cell.

Two very important types of endocytosis are phagocytosis and pinocytosis. During *phagocytosis,* cellular projections called pseudopods engulf particles and bring them into the cell. Phagocytosis is used by white blood cells to destroy bacteria and foreign substances (see Figure 16.8, page 461, and further discussion in Chapter 16). In *pinocytosis,* the plasma membrane folds inward, bringing extracellular fluid into the cell, along with whatever substances are dissolved in the fluid. Pinocytosis is one of the ways viruses can enter animal cells (see Figure 13.14a, page 384).

Cytoplasm

LEARNING OBJECTIVE

4-16 Compare and contrast prokaryotic and eukaryotic cytoplasms.

The **cytoplasm** of eukaryotic cells encompasses the substance inside the plasma membrane and outside the nucleus (see Figure 4.22). The cytoplasm is the substance in which various

Table 4.2 Principal Differences between Prokaryotic and Eukaryotic Cells

Eukaryotes

Prokaryotes

10 μm

LM

Characteristic	Prokaryotic	Eukaryotic
Size of Cell	Typically 0.2–2.0 μm in diameter	Typically 10–100 μm in diameter
Nucleus	No nuclear membrane or nucleoli	True nucleus, consisting of nuclear membrane and nucleoli
Membrane-Enclosed Organelles	Absent	Present; examples include lysosomes, Golgi complex, endoplasmic reticulum, mitochondria, and chloroplasts
Flagella	Consist of two protein building blocks	Complex; consist of multiple microtubules
Glycocalyx	Present as a capsule or slime layer	Present in some cells that lack a cell wall
Cell Wall	Usually present; chemically complex (typical bacterial cell wall includes peptidoglycan)	When present, chemically simple (includes cellulose and chitin)
Plasma Membrane	No carbohydrates and generally lacks sterols	Sterols and carbohydrates that serve as receptors
Cytoplasm	No cytoskeleton or cytoplasmic streaming	Cytoskeleton; cytoplasmic streaming
Ribosomes	Smaller size (70S)	Larger size (80S); smaller size (70S) in organelles
Chromosome (DNA)	Usually single circular chromosome; typically lacks histones	Multiple linear chromosomes with histones
Cell Division	Binary fission	Involves mitosis
Sexual Recombination	None; transfer of DNA only	Involves meiosis

cellular components are found. (The term **cytosol** refers to the fluid portion of cytoplasm.) A major difference between eukaryotic and prokaryotic cytoplasm is that eukaryotic cytoplasm has a complex internal structure, consisting of exceedingly small rods (*microfilaments* and *intermediate filaments*) and cylinders (*microtubules*). Together, they form the **cytoskeleton.** The cytoskeleton provides support and shape and assists in transporting substances through the cell (and even in moving the entire cell, as in phagocytosis). The movement of eukaryotic cytoplasm from one part of the cell to another, which helps distribute nutrients and move the cell over a surface, is called **cytoplasmic streaming.** Another difference between prokaryotic and eukaryotic cytoplasm is that many of the important enzymes found in the cytoplasmic fluid of prokaryotes are sequestered in the organelles of eukaryotes.

Ribosomes

LEARNING OBJECTIVE

4-17 Compare the structure and function of eukaryotic and prokaryotic ribosomes.

Attached to the outer surface of rough endoplasmic reticulum (discussed on page 103) are **ribosomes** (see Figure 4.25), which are also found free in the cytoplasm. As in prokaryotes, ribosomes are the sites of protein synthesis in the cell.

The ribosomes of eukaryotic endoplasmic reticulum and cytoplasm are somewhat larger and denser than those of prokaryotic cells. These eukaryotic ribosomes are 80S ribosomes, each of which consists of a large 60S subunit containing three molecules of rRNA and a smaller 40S subunit with one molecule

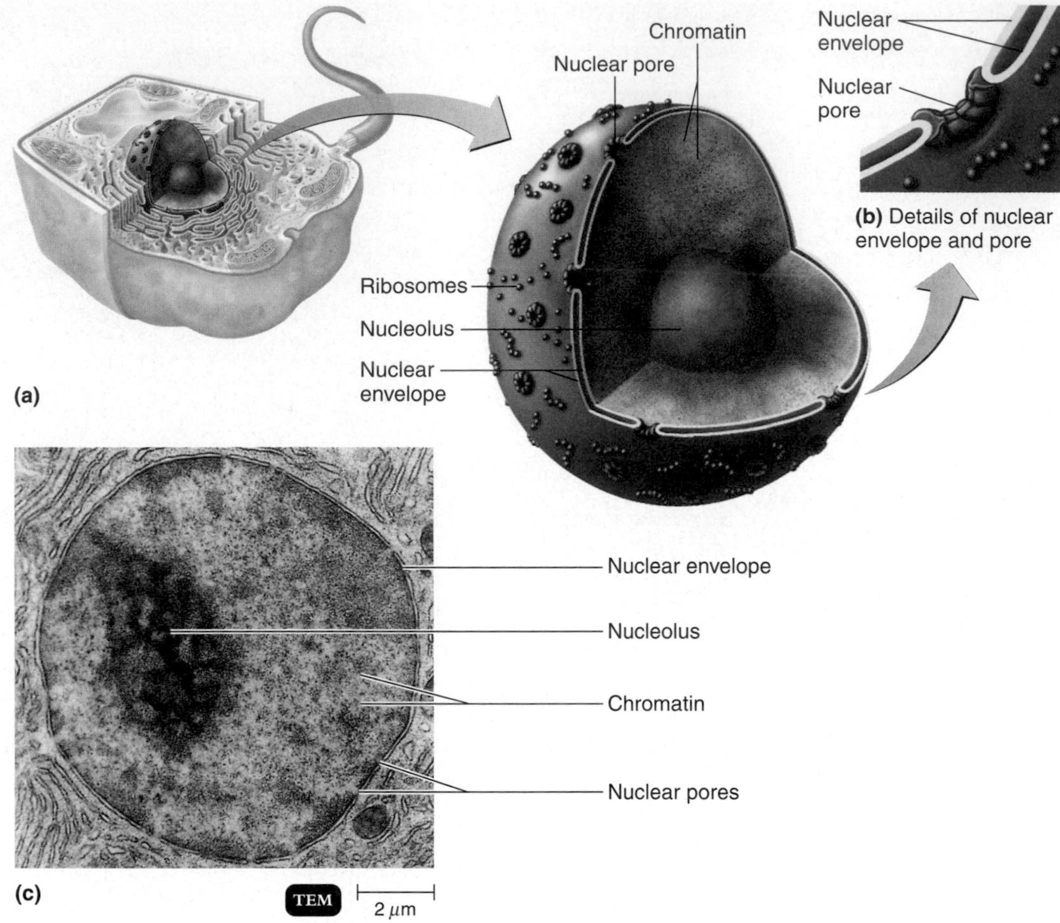

Figure 4.24 The eukaryotic nucleus. (**a, b**) Drawings of details of a nucleus. (**c**) A micrograph of a nucleus.

Q What keeps the nucleus suspended in the cell?

of rRNA. The subunits are made separately in the nucleolus and, once produced, exit the nucleus and join together in the cytosol. Chloroplasts and mitochondria contain 70S ribosomes, which may indicate their evolution from prokaryotes. (This theory is discussed on page 106) The role of ribosomes in protein synthesis will be discussed in more detail in Chapter 8.

Some ribosomes, called *free ribosomes,* are unattached to any structure in the cytoplasm. Primarily, free ribosomes synthesize proteins used *inside* the cell. Other ribosomes, called *membrane-bound ribosomes,* attach to the nuclear membrane and the endoplasmic reticulum. These ribosomes synthesize proteins destined for insertion in the plasma membrane or for export from the cell. Ribosomes located within mitochondria synthesize mitochondrial proteins. Sometimes 10 to 20 ribosomes join together in a stringlike arrangement called a *polyribosome.*

CHECK YOUR UNDERSTANDING

- ✓ Identify at least one significant difference between eukaryotic and prokaryotic flagella and cilia, cell walls, plasma membranes, and cytoplasm. **4-13–4-16**
- ✓ The antibiotic erythromycin binds with the 50S portion of a ribosome. What effect does this have on a prokaryotic cell? On a eukaryotic cell? **4-17**

Organelles

LEARNING OBJECTIVES

4-18 Define *organelle.*

4-19 Describe the functions of the nucleus, endoplasmic reticulum, Golgi complex, lysosomes, vacuoles, mitochondria, chloroplasts, peroxisomes, and centrosomes.

Organelles are structures with specific shapes and specialized functions and are characteristic of eukaryotic cells. They include the nucleus, endoplasmic reticulum, Golgi complex, lysosomes, vacuoles, mitochondria, chloroplasts, peroxisomes, and centrosomes. Not all of the organelles described are found in all cells. Certain cells have their own type and distribution of organelles based on specialization, age, and level of activity.

The Nucleus

The most characteristic eukaryotic organelle is the nucleus (see Figure 4.22). The **nucleus** (Figure 4.24) is usually spherical or oval, is frequently the largest structure in the cell, and contains almost all of the cell's hereditary information (DNA). Some DNA is also found in mitochondria and in the chloroplasts of photosynthetic organisms.

The nucleus is surrounded by a double membrane called the **nuclear envelope.** Both membranes resemble the plasma membrane in structure. Tiny channels in the membrane called **nuclear pores** allow the nucleus to communicate with the cytoplasm (Figure 4.24b). Nuclear pores control the movement of substances between the nucleus and cytoplasm. Within the nuclear envelope are one or more spherical bodies called **nucleoli** (singular: **nucleolus**). Nucleoli are actually condensed regions of chromosomes where ribosomal RNA is being synthesized. Ribosomal RNA is an essential component of ribosomes.

The nucleus also contains most of the cell's DNA, which is combined with several proteins, including some basic proteins called **histones** and nonhistones. The combination of about 165 base pairs of DNA and 9 molecules of histones is referred to as a *nucleosome.* When the cell is not reproducing, the DNA and its associated proteins appear as a threadlike mass called **chromatin.** During nuclear division, the chromatin coils into shorter and thicker rodlike bodies called **chromosomes.** Prokaryotic chromosomes do not undergo this process, do not have histones, and are not enclosed in a nuclear envelope.

Eukaryotic cells require two elaborate mechanisms: mitosis and meiosis to segregate chromosomes prior to cell division. Neither process occurs in prokaryotic cells.

Endoplasmic Reticulum

Within the cytoplasm of eukaryotic cells is the **endoplasmic reticulum,** or **ER,** an extensive network of flattened membranous sacs or tubules called **cisterns** (Figure 4.25). The ER network is continuous with the nuclear envelope (see Figure 4.22a).

Most eukaryotic cells contain two distinct, but interrelated, forms of ER that differ in structure and function. The membrane of **rough ER** is continuous with the nuclear membrane and usually unfolds into a series of flattened sacs. The outer surface of rough ER is studded with ribosomes, the sites of protein synthesis. Proteins synthesized by ribosomes that are attached to rough ER enter cisterns within the ER for processing and sorting. In some cases, enzymes within the cisterns attach the proteins to carbohydrates to form glycoproteins. In other cases, enzymes attach the proteins to phospholipids, also synthesized by rough ER. These molecules may be incorporated into organelle membranes or the plasma membrane. Thus, rough ER is a factory for synthesizing secretory proteins and membrane molecules.

Smooth ER extends from the rough ER to form a network of membrane tubules (see Figure 4.25). Unlike rough ER, smooth ER does not have ribosomes on the outer surface of its membrane. However, smooth ER contains unique enzymes that make it functionally more diverse than rough ER. Although it does not synthesize proteins, smooth ER does synthesize phospholipids, as does rough ER. Smooth ER also synthesizes fats and steroids, such as estrogens and testosterone. In liver cells, enzymes of the smooth ER help release glucose into the bloodstream and inactivate or detoxify drugs and other potentially harmful substances (for example, alcohol). In muscle cells, calcium ions released from the sarcoplasmic reticulum, a form of smooth ER, trigger the contraction process.

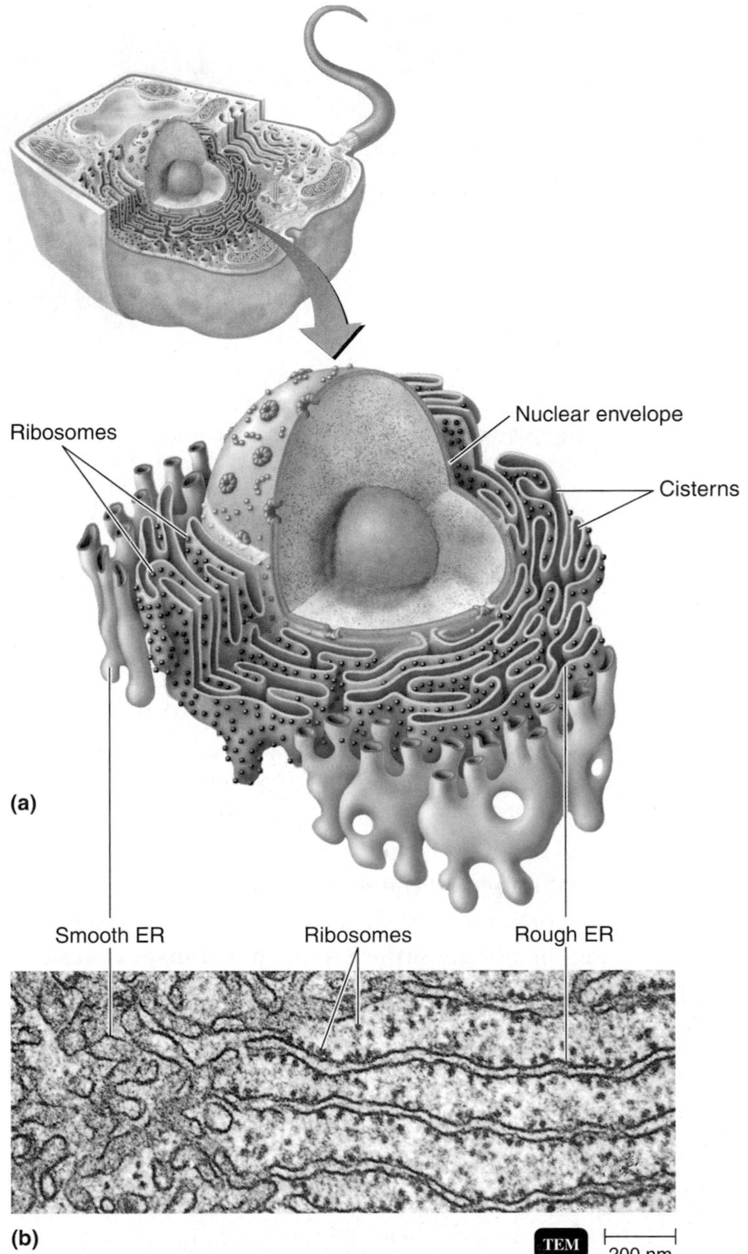

Figure 4.25 Rough endoplasmic reticulum and ribosomes. (**a**) A drawing of details of the endoplasmic reticulum. (**b**) A micrograph of the endoplasmic reticulum and ribosomes.

Q What is the difference between rough ER and smooth ER?

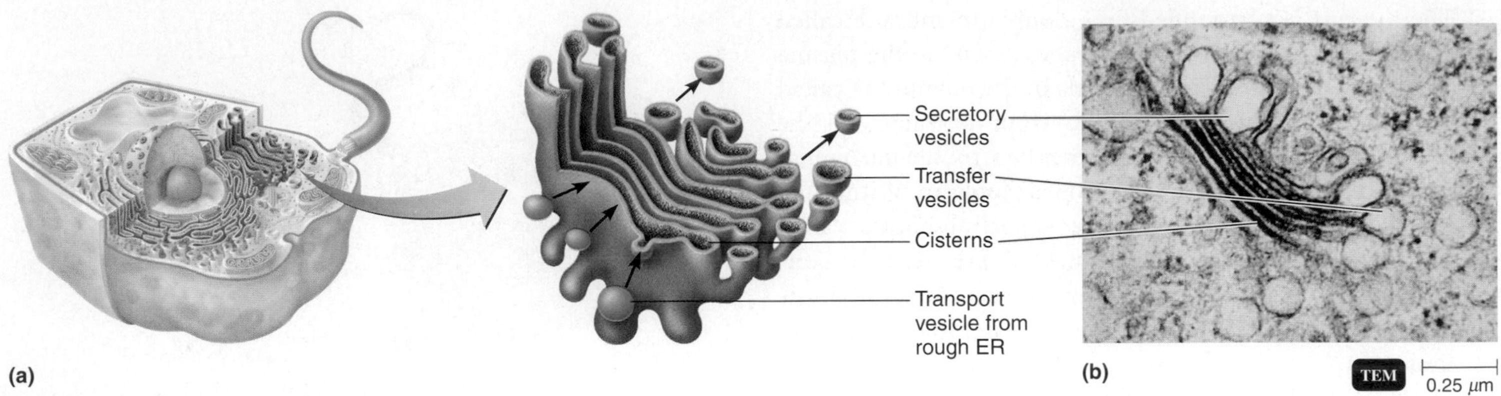

Figure 4.26 Golgi complex. (**a**) A drawing of details of a Golgi complex. (**b**) A micrograph of a Golgi complex.

Q What are the functions of the Golgi complex?

Golgi Complex

Most of the proteins synthesized by ribosomes attached to rough ER are ultimately transported to other regions of the cell. The first step in the transport pathway is through an organelle called the **Golgi complex.** It consists of 3 to 20 cisterns that resemble a stack of pita bread (Figure 4.26). The cisterns are often curved, giving the Golgi complex a cuplike shape.

Proteins synthesized by ribosomes on the rough ER are surrounded by a portion of the ER membrane, which eventually buds from the membrane surface to form a **transport vesicle.** The transport vesicle fuses with a cistern of the Golgi complex, releasing proteins into the cistern. The proteins are modified and move from one cistern to another via **transfer vesicles** that bud from the cisterns' edges. Enzymes in the cisterns modify the proteins to form glycoproteins, glycolipids, and lipoproteins. Some of the processed proteins leave the cisterns in **secretory vesicles,** which detach from the cistern and deliver the proteins to the plasma membrane, where they are discharged by exocytosis. Other processed proteins leave the cisterns in vesicles that deliver their contents to the plasma membrane for incorporation into the membrane. Finally, some processed proteins leave the cisterns in vesicles that are called **storage vesicles.** The major storage vesicle is a lysosome, whose structure and functions are discussed next.

Lysosomes

Lysosomes are formed from Golgi complexes and look like membrane-enclosed spheres. Unlike mitochondria, lysosomes have only a single membrane and lack internal structure (see Figure 4.22). But they contain as many as 40 different kinds of powerful digestive enzymes capable of breaking down various molecules. Moreover, these enzymes can also digest bacteria that enter the cell. Human white blood cells, which use phagocytosis to ingest bacteria, contain large numbers of lysosomes.

Vacuoles

A **vacuole** (see Figure 4.22) is a space or cavity in the cytoplasm of a cell that is enclosed by a membrane called a *tonoplast.* In plant cells, vacuoles may occupy 5–90% of the cell volume, depending on the type of cell. Vacuoles are derived from the Golgi complex and have several diverse functions. Some vacuoles serve as temporary storage organelles for substances such as proteins, sugars, organic acids, and inorganic ions. Other vacuoles form during endocytosis to help bring food into the cell. Many plant cells also store metabolic wastes and poisons that would otherwise be injurious if they accumulated in the cytoplasm. Finally, vacuoles may take up water, enabling plant cells to increase in size and also providing rigidity to leaves and stems.

Mitochondria

Spherical or rod-shaped organelles called **mitochondria** (singular: **mitochondrion**) appear throughout the cytoplasm of most eukaryotic cells (see Figure 4.22). The number of mitochondria per cell varies greatly among different types of cells. For example, the protozoan *Giardia* has no mitochondria, whereas liver cells contain 1000 to 2000 per cell. A mitochondrion consists of a double membrane similar in structure to the plasma membrane (Figure 4.27). The outer mitochondrial membrane is smooth, but the inner mitochondrial membrane is arranged in a series of folds called **cristae** (singular: **crista**). The center of the mitochondrion is a semifluid substance called the **matrix.** Because of the nature and arrangement of the cristae, the inner membrane provides an enormous surface area on which chemical reactions can occur. Some proteins that function in cellular respiration, including the enzyme that makes ATP, are located on the cristae of the inner mitochondrial membrane, and many of the metabolic steps involved in cellular respiration are concentrated in the matrix (see Chapter 5). Mitochondria are often called the "powerhouses of the cell" because of their central role in ATP production.

Mitochondria contain 70S ribosomes and some DNA of their own, as well as the machinery necessary to replicate, transcribe, and translate the information encoded by their DNA. In addition, mitochondria can reproduce more or less on their own by growing and dividing in two.

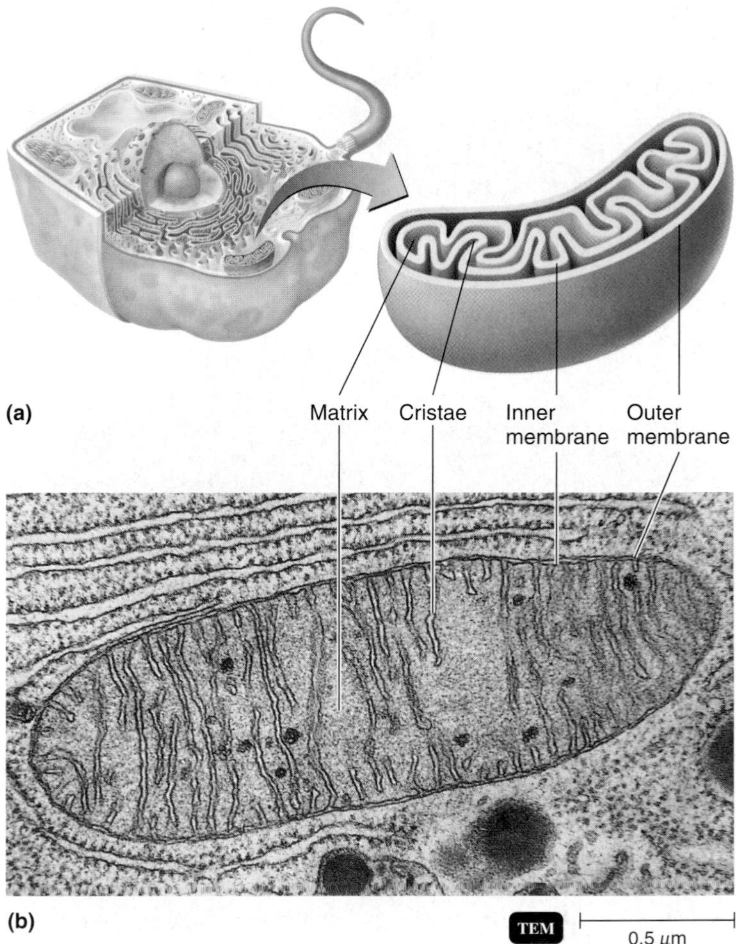

Figure 4.27 Mitochondria. (**a**) A drawing of details of a mitochondrion. (**b**) A micrograph of a mitochondrion from a rat pancreas cell.

Q How are mitochondria similar to prokaryotic cells?

Chloroplasts

Algae and green plants contain a unique organelle called a **chloroplast** (Figure 4.28), a membrane-enclosed structure that contains both the pigment chlorophyll and the enzymes required for the light-gathering phases of photosynthesis (see Chapter 5). The chlorophyll is contained in flattened membrane sacs called **thylakoids;** stacks of thylakoids are called *grana* (singular: **granum**) (see Figure 4.28).

Like mitochondria, chloroplasts contain 70S ribosomes, DNA, and enzymes involved in protein synthesis. They are capable of multiplying on their own within the cell. The way both chloroplasts and mitochondria multiply—by increasing in size and then dividing in two—is strikingly reminiscent of bacterial multiplication.

Peroxisomes

Organelles similar in structure to lysosomes, but smaller, are called **peroxisomes** (see Figure 4.22). Although peroxisomes were once thought to form by budding off the ER, it is now generally agreed that they form by the division of preexisting peroxisomes.

Peroxisomes contain one or more enzymes that can oxidize various organic substances. For example, substances such as amino acids and fatty acids are oxidized in peroxisomes as part of normal metabolism. In addition, enzymes in peroxisomes oxidize toxic substances, such as alcohol. A by-product of the oxidation reactions is hydrogen peroxide (H_2O_2), a potentially toxic compound. However, peroxisomes also contain the enzyme *catalase,* which decomposes H_2O_2 (see Chapter 6, page 162). Because the generation and degradation of H_2O_2 occurs within the same organelle, peroxisomes protect other parts of the cell from the toxic effects of H_2O_2.

Centrosome

The **centrosome,** located near the nucleus, consists of two components: the pericentriolar area and centrioles (see Figure 4.22). The *pericentriolar material* is a region of the cytosol composed of a dense network of small protein fibers. This area is the organizing center for the mitotic spindle, which plays a critical role in cell division, and for microtubule formation in nondividing cells. Within the pericentriolar material is a pair of cylindrical structures called *centrioles,* each of which is composed of nine clusters of three microtubules (triplets) arranged in a circular pattern, an arrangement called a *9 + 0 array.* The 9 refers to the nine clusters of microtubules, and the 0 refers to the absence of microtubules in the center. The long axis of one centriole is at a right angle to the long axis of the other.

CHECK YOUR UNDERSTANDING

✓ Compare the structure of the nucleus of a eukaryote and the nucleoid of a prokaryote. **4-18**

✓ How do rough and smooth ER compare structurally and functionally? **4-19**

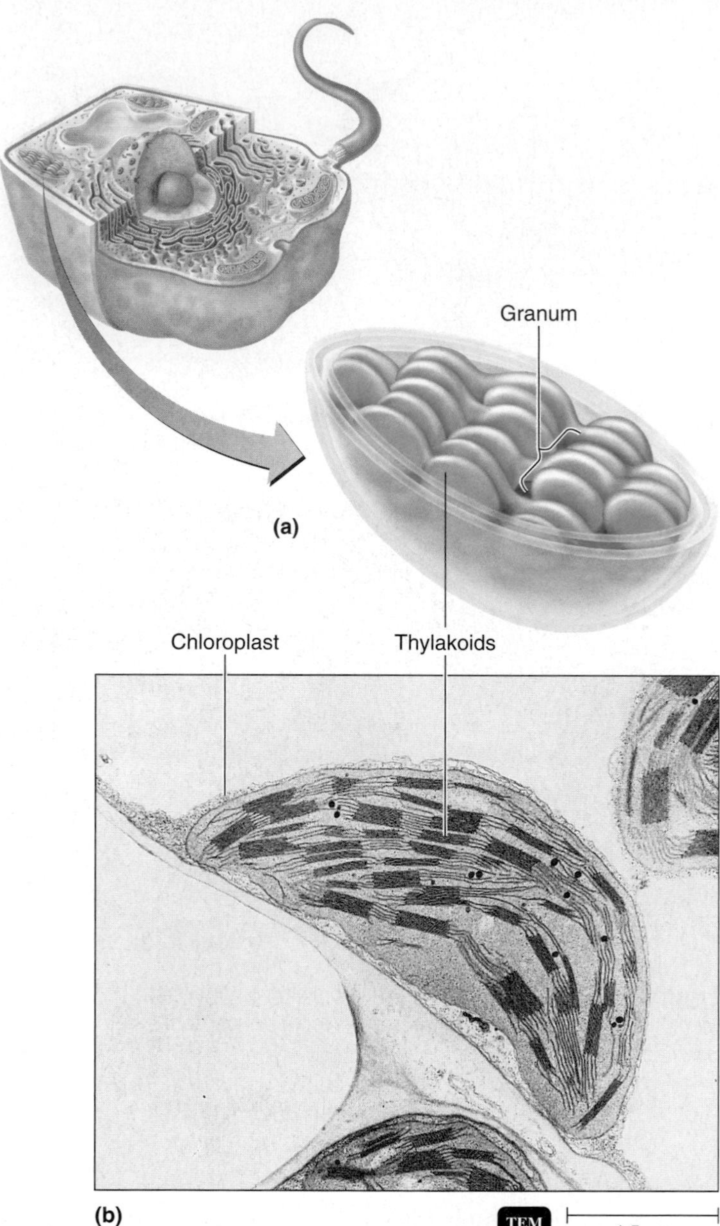

Figure 4.28 Chloroplasts. Photosynthesis occurs in chloroplasts; the light-trapping pigments are located on the thylakoids. **(a)** A drawing of details of a chloroplast, showing grana. **(b)** A micrograph of chloroplasts in a plant cell.

Q What are the similarities between chloroplasts and prokaryotic cells?

The Evolution of Eukaryotes

LEARNING OBJECTIVE

4-20 Discuss evidence that supports the endosymbiotic theory of eukaryotic evolution.

Biologists generally believe that life arose on Earth in the form of very simple organisms, similar to prokaryotic cells, about 3.5 to 4 billion years ago. About 2.5 billion years ago, the first eukaryotic cells evolved from prokaryotic cells. Recall that prokaryotes and eukaryotes differ mainly in that eukaryotes contain highly specialized organelles. The theory explaining the origin of eukaryotes from prokaryotes, pioneered by Lynn Margulis, is the **endosymbiotic theory.** According to this theory, larger bacterial cells lost their cell walls and engulfed smaller bacterial cells. This relationship, in which one organism lives within another, is called *endosymbiosis* (*symbiosis* = living together).

According to the endosymbiotic theory, the ancestral eukaryote developed a rudimentary nucleus when the plasma membrane folded around the chromosome (see Figure 10.2, page 277). This cell, called a nucleoplasm, may have ingested aerobic bacteria. Some ingested bacteria lived inside the host nucleoplasm. This arrangement evolved into a symbiotic relationship in which the host nucleoplasm supplied nutrients and the endosymbiotic bacterium produced energy that could be used by the nucleoplasm. Similarly, chloroplasts may be descendants of photosynthetic prokaryotes ingested by this early nucleoplasm. Eukaryotic flagella and cilia are believed to have originated from symbiotic associations between the plasma membrane of early eukaryotes and motile spiral bacteria called spirochetes. A living example that suggests how flagella developed is described in the box on facing page.

Studies comparing prokaryotic and eukaryotic cells provide evidence for the endosymbiotic theory. For example, both mitochondria and chloroplasts resemble bacteria in size and shape. Further, these organelles contain circular DNA, which is typical of prokaryotes, and the organelles can reproduce independently of their host cell. Moreover, mitochondrial and chloroplast ribosomes resemble those of prokaryotes, and their mechanism of protein synthesis is more similar to that found in bacteria than eukaryotes. Also, the same antibiotics that inhibit protein synthesis on ribosomes in bacteria also inhibit protein synthesis on ribosomes in mitochondria and chloroplasts.

CHECK YOUR UNDERSTANDING

✓ Which three organelles are not associated with the Golgi complex? What does this suggest about their origin? **4-20**

* * *

Our next concern is to examine microbial metabolism. In Chapter 5, you will learn about the importance of enzymes to microorganisms and the ways microbes produce and use energy.

Why Microbiologists Study Termites

Although termites are famous for their ability to eat wood, causing damage to wooden structures and recycling cellulose in the soil, they are unable to digest the wood that they eat. To break down the cellulose, termites enlist the help of a variety of microorganisms. Some termites, for example, dig tunnels in the wood, then inoculate the tunnels with fungi that grow on the wood. These termites then eat the fungi, not the wood itself.

What microbiologists find more interesting are the termites that contain, within their digestive tracts, symbiotic microorganisms that digest the cellulose that the termites chew and swallow. Even more fascinating to microbiologists is the fact that these microorganisms themselves can survive only because of even smaller symbionts that live on and within them, without which they would not even be able to move. By studying how a single termite survives, microbiologists have begun to gain an entirely new understanding of symbiosis.

The termite's dependence on nitrogen-fixing bacteria to supply its nitrogen and on protozoans such as *Trichonympha sphaerica* to digest cellulose is an example of endosymbiosis, a symbiotic relationship with an organism that lives inside the body of the host organism (in this case, within the hindgut of the termite).

The picture is more complicated than this, however, for *T. sphaerica* itself is unable to digest cellulose without the aid of bacteria that live within its body: in other words, the protozoan has its own endosymbionts.

Certain hindgut flagellates such as *T. sphaerica* also demonstrate another form of symbiosis—ectosymbiosis, a symbiotic relationship with organisms that live outside its body. Recent advances in microscopy have shown that these flagellates are covered by precise rows consisting of thousands of bacteria, either rods or spirochetes. If these bacteria are killed, the protozoan is unable to move. Instead of using its own flagella, the protozoan relies on the rows of bacteria to row it about like oarsmen in a boat.

The protozoan *Mixotricha,* for example, has rows of spirochetes on its surface. As shown in part (a) of the figure, the end of each spirochete abuts against a swelling known as a bracket. The spirochetes undulate in unison, thereby creating waves of motion along *Mixotricha*'s surface.

Rod-shaped bacteria align in grooves that cover the surface of devescovinids, another group of termite-hindgut protozoans. Each rod has twelve flagella that overlap the flagella of the adjacent bacteria to form a continuous filament along the groove; see part (b) of the figure. The bacteria rotate their flagella, thus creating coordinated waves along all these rows of filaments, which propel the protozoan.

Sid Tamm and his colleagues at Boston University have found that the protozoa cannot control motility of the ectosymbiotics. *Mixotricha* uses its flagella to steer, and the bacteria push the protozoa forward—shoving and being shoved by its neighbors, like "bumper cars."

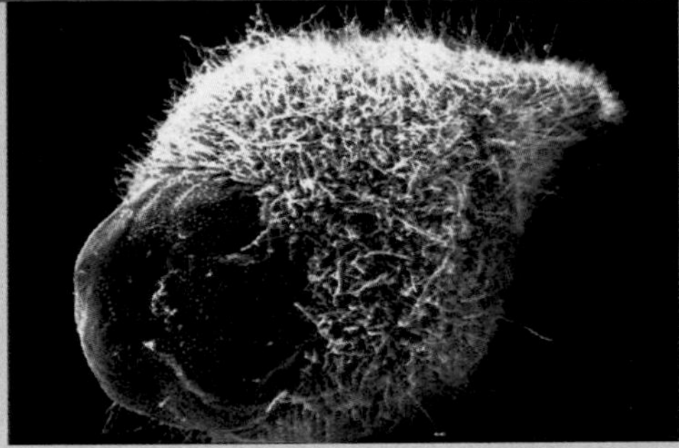

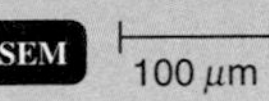

Mixotricha, a protozoan that lives in the termite gut.

Arrangements of bacteria on the surfaces of two protozoans.

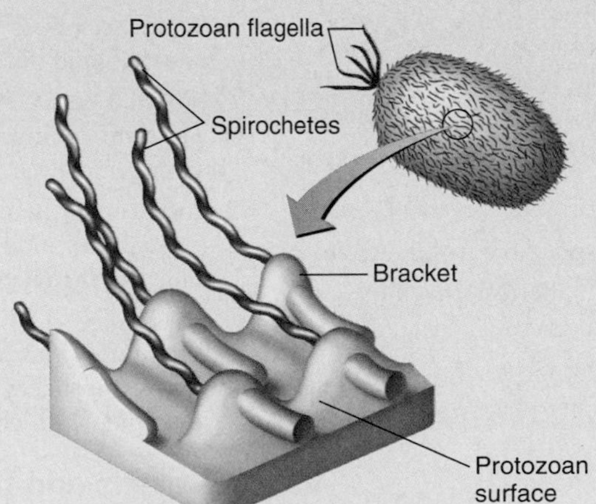

(a) Spirochetes attached to brackets over the surface of a *Mixotricha* protozoan align themselves and move in unison.

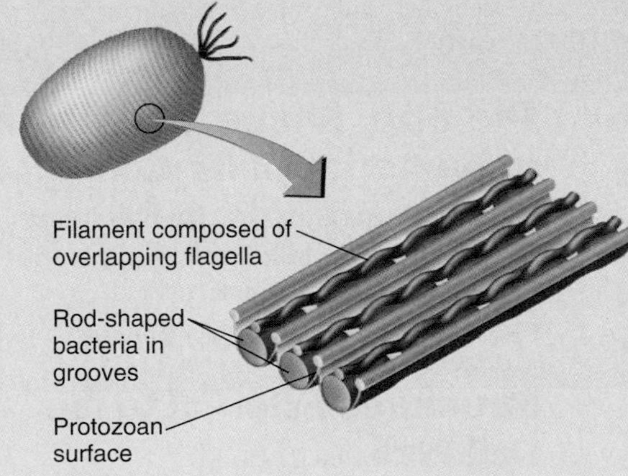

(b) On this devescovinid protozoan, the flagella from one rod-shaped bacterium overlap the next to form a continuous filament.

STUDY OUTLINE

Comparing Prokaryotic and Eukaryotic Cells: An Overview (p. 77)

1. Prokaryotic and eukaryotic cells are similar in their chemical composition and chemical reactions.
2. Prokaryotic cells lack membrane-enclosed organelles (including a nucleus).
3. Peptidoglycan is found in prokaryotic cell walls but not in eukaryotic cell walls.
4. Eukaryotic cells have a membrane-bound nucleus and other organelles.

THE PROKARYOTIC CELL (pp. 77–98)

1. Bacteria are unicellular, and most of them multiply by binary fission.
2. Bacterial species are differentiated by morphology, chemical composition, nutritional requirements, biochemical activities, and source of energy.

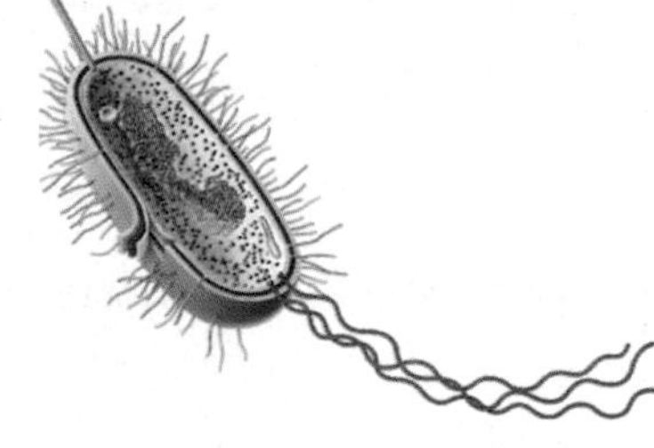

The Size, Shape, and Arrangement of Bacterial Cells (pp. 77–79)

1. Most bacteria are 0.2 to 2.0 μm in diameter and 2 to 8 μm in length.
2. The three basic bacterial shapes are coccus (spherical), bacillus (rod-shaped), and spiral (twisted).
3. Pleomorphic bacteria can assume several shapes.

Structures External to the Cell Wall (pp. 79–84)

Glycocalyx (pp. 79–81)

1. The glycocalyx (capsule, slime layer, or extracellular polysaccharide) is a gelatinous polysaccharide and/or polypeptide covering.
2. Capsules may protect pathogens from phagocytosis.
3. Capsules enable adherence to surfaces, prevent desiccation, and may provide nutrients.

Flagella (pp. 81–82)

4. Flagella are relatively long filamentous appendages consisting of a filament, hook, and basal body.
5. Prokaryotic flagella rotate to push the cell.
6. Motile bacteria exhibit taxis; positive taxis is movement toward an attractant, and negative taxis is movement away from a repellent.
7. Flagellar (H) protein is an antigen.

Axial Filaments (pp. 82–83)

8. Spiral cells that move by means of an axial filament (endoflagellum) are called spirochetes.
9. Axial filaments are similar to flagella, except that they wrap around the cell.

Fimbriae and Pili (pp. 83–84)

10. Fimbriae help cells adhere to surfaces.
11. Pili are involved in twitching motility and DNA transfer.

The Cell Wall (pp. 84-89)

Composition and Characteristics (pp. 85-87)

1. The cell wall surrounds the plasma membrane and protects the cell from changes in water pressure.
2. The bacterial cell wall consists of peptidoglycan, a polymer consisting of NAG and NAM and short chains of amino acids.
3. Penicillin interferes with peptidoglycan synthesis.
4. Gram-positive cell walls consist of many layers of peptidoglycan and also contain teichoic acids.
5. Gram-negative bacteria have a lipopolysaccharide-lipoprotein-phospholipid outer membrane surrounding a thin peptidoglycan layer.
6. The outer membrane protects the cell from phagocytosis and from penicillin, lysozyme, and other chemicals.
7. Porins are proteins that permit small molecules to pass through the outer membrane; specific channel proteins allow other molecules to move through the outer membrane.
8. The lipopolysaccharide component of the outer membrane consists of sugars (O polysaccharides), which function as antigens, and lipid A, which is an endotoxin.

Cell Walls and the Gram Stain Mechanism (p. 87)

9. The crystal violet–iodine complex combines with peptidoglycan.
10. The decolorizer removes the lipid outer membrane of gram-negative bacteria and washes out the crystal violet.

Atypical Cell Walls (pp. 87–88)

11. *Mycoplasma* is a bacterial genus that naturally lacks cell walls.
12. Archaea have pseudomurein; they lack peptidoglycan.
13. Acid-fast cell walls have a layer of mycolic acid outside a thin peptidoglycan layer.

Damage to the Cell Wall (pp. 88–89)

14. In the presence of lysozyme, gram-positive cell walls are destroyed, and the remaining cellular contents are referred to as a protoplast.
15. In the presence of lysozyme, gram-negative cell walls are not completely destroyed, and the remaining cellular contents are referred to as a spheroplast.
16. L forms are gram-positive or gram-negative bacteria that do not make a cell wall.
17. Antibiotics such as penicillin interfere with cell wall synthesis.

Structures Internal to the Cell Wall (pp. 89–98)

The Plasma (Cytoplasmic) Membrane (pp. 89–91)

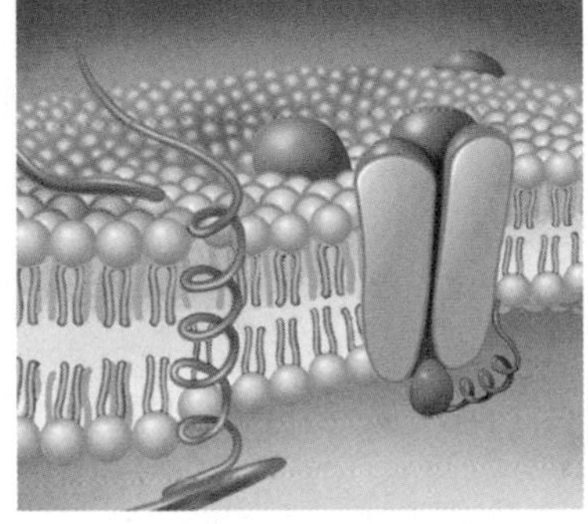

1. The plasma membrane encloses the cytoplasm and is a lipid bilayer with peripheral and integral proteins (the fluid mosaic model).
2. The plasma membrane is selectively permeable.
3. Plasma membranes contain enzymes for metabolic reactions, such as nutrient breakdown, energy production, and photosynthesis.
4. Mesosomes, irregular infoldings of the plasma membrane, are artifacts, not true cell structures.
5. Plasma membranes can be destroyed by alcohols and polymyxins.

The Movement of Materials across Membranes (pp. 91–94)

6. Movement across the membrane may be by passive processes, in which materials move from areas of higher to lower concentration and no energy is expended by the cell.
7. In simple diffusion, molecules and ions move until equilibrium is reached.
8. In facilitated diffusion, substances are transported by transporter proteins across membranes from areas of high to low concentration.
9. Osmosis is the movement of water from areas of high to low concentration across a selectively permeable membrane until equilibrium is reached.
10. In active transport, materials move from areas of low to high concentration by transporter proteins, and the cell must expend energy.
11. In group translocation, energy is expended to modify chemicals and transport them across the membrane.

Cytoplasm (p. 94)

12. Cytoplasm is the fluid component inside the plasma membrane.
13. The cytoplasm is mostly water, with inorganic and organic molecules, DNA, ribosomes, and inclusions.

The Nucleoid (pp. 94–95)

14. The nucleoid contains the DNA of the bacterial chromosome.
15. Bacteria can also contain plasmids, which are circular, extrachromosomal DNA molecules.

Ribosomes (p. 95)

16. The cytoplasm of a prokaryote contains numerous 70S ribosomes; ribosomes consist of rRNA and protein.
17. Protein synthesis occurs at ribosomes; it can be inhibited by certain antibiotics.

Inclusions (pp. 95–96)

18. Inclusions are reserve deposits found in prokaryotic and eukaryotic cells.
19. Among the inclusions found in bacteria are metachromatic granules (inorganic phosphate), polysaccharide granules (usually glycogen or starch), lipid inclusions, sulfur granules, carboxysomes (ribulose 1,5-diphosphate carboxylase), magnetosomes (Fe_3O_4), and gas vacuoles.

Endospores (pp. 96–98)

20. Endospores are resting structures formed by some bacteria; they allow survival during adverse environmental conditions.
21. The process of endospore formation is called sporulation; the return of an endospore to its vegetative state is called germination.

THE EUKARYOTIC CELL (pp. 98–106)

Flagella and Cilia (p. 98)

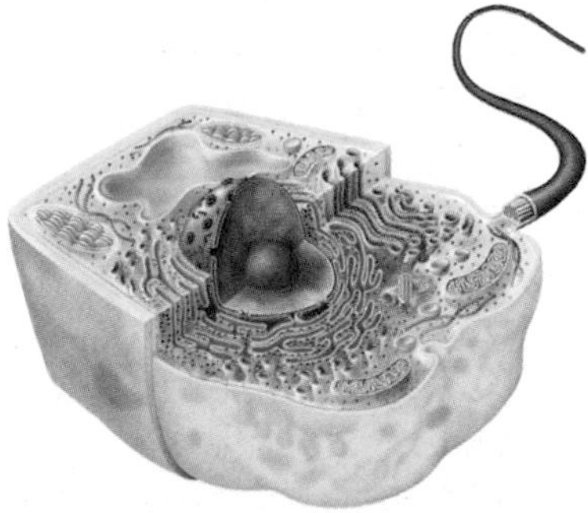

1. Flagella are few and long in relation to cell size; cilia are numerous and short.
2. Flagella and cilia are used for motility, and cilia also move substances along the surface of the cells.
3. Both flagella and cilia consist of an arrangement of nine pairs and two single microtubules.

The Cell Wall and Glycocalyx (p. 98)

1. The cell walls of many algae and some fungi contain cellulose.
2. The main material of fungal cell walls is chitin.
3. Yeast cell walls consist of glucan and mannan.
4. Animal cells are surrounded by a glycocalyx, which strengthens the cell and provides a means of attachment to other cells.

The Plasma (Cytoplasmic) Membrane (p. 100)

1. Like the prokaryotic plasma membrane, the eukaryotic plasma membrane is a phospholipid bilayer containing proteins.
2. Eukaryotic plasma membranes contain carbohydrates attached to the proteins and sterols not found in prokaryotic cells (except *Mycoplasma* bacteria).

3. Eukaryotic cells can move materials across the plasma membrane by the passive processes used by prokaryotes and by active transport and endocytosis (phagocytosis and pinocytosis).

Cytoplasm (pp. 100–101)

1. The cytoplasm of eukaryotic cells includes everything inside the plasma membrane and external to the nucleus.
2. The chemical characteristics of the cytoplasm of eukaryotic cells resemble those of the cytoplasm of prokaryotic cells.
3. Eukaryotic cytoplasm has a cytoskeleton and exhibits cytoplasmic streaming.

Ribosomes (pp. 101–102)

1. 80S ribosomes are found in the cytoplasm or attached to the rough endoplasmic reticulum.

Organelles (pp. 102–106)

1. Organelles are specialized membrane-enclosed structures in the cytoplasm of eukaryotic cells.
2. The nucleus, which contains DNA in the form of chromosomes, is the most characteristic eukaryotic organelle.
3. The nuclear envelope is connected to a system of membranes in the cytoplasm called the endoplasmic reticulum (ER).
4. The ER provides a surface for chemical reactions, serves as a transporting network, and stores synthesized molecules. Protein synthesis and transport occur on rough ER; lipid synthesis occurs on smooth ER.
5. The Golgi complex consists of flattened sacs called cisterns. It functions in membrane formation and protein secretion.
6. Lysosomes are formed from Golgi complexes. They store digestive enzymes.
7. Vacuoles are membrane-enclosed cavities derived from the Golgi complex or endocytosis. They are usually found in plant cells that store various substances, increase cell size, and provide rigidity to leaves and stems.
8. Mitochondria are the primary sites of ATP production. They contain 70S ribosomes and DNA, and they multiply by binary fission.
9. Chloroplasts contain chlorophyll and enzymes for photosynthesis. Like mitochondria, they contain 70S ribosomes and DNA and multiply by binary fission.

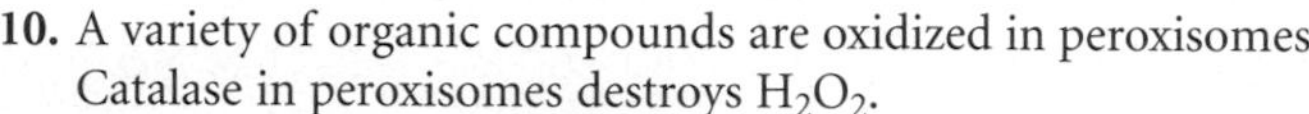

10. A variety of organic compounds are oxidized in peroxisomes. Catalase in peroxisomes destroys H_2O_2.
11. The centrosome consists of the pericentriolar material and centrioles. Centrioles are 9 triplet microtubules involved in formation of the mitotic spindle and microtubules.

The Evolution of Eukaryotes (p. 106)

1. According to the endosymbiotic theory, eukaryotic cells evolved from symbiotic prokaryotes living inside other prokaryotic cells.

STUDY QUESTIONS

Answers to the Review and Multiple Choice questions can be found by turning to the blue Answers tab at the back of the textbook.

Review

1. **DRAW IT** Diagram each of the following flagellar arrangements:
   a. lophotrichous
   b. monotrichous
   c. peritrichous
   d. amphitrichous
   e. polar
2. Endospore formation is called (a) ________. It is initiated by (b) ________. Formation of a new cell from an endospore is called (c) ________. This process is triggered by (d) ________.
3. **DRAW IT** Draw the bacterial shapes listed in (a), (b), and (c). Then draw the shapes in (d), (e), and (f), showing how they are special conditions of a, b, and c, respectively.
   a. spiral
   b. bacillus
   c. coccus
   d. spirochetes
   e. streptobacilli
   f. staphylococci
4. Match the structures in column A to their functions in column B.

Column A	Column B
_____ a. Cell wall	1. Attachment to surfaces
_____ b. Endospore	2. Cell wall formation
_____ c. Fimbriae	3. Motility
_____ d. Flagella	4. Protection from osmotic lysis
_____ e. Glycocalyx	5. Protection from phagocytes
_____ f. Pili	6. Resting
_____ g. Plasma membrane	7. Protein synthesis
_____ h. Ribosomes	8. Selective permeability
	9. Transfer of genetic material

5. Why is an endospore called a resting structure? Of what advantage is an endospore to a bacterial cell?

6. Compare and contrast the following:
   a. simple diffusion and facilitated diffusion
   b. active transport and facilitated diffusion
   c. active transport and group translocation
7. Answer the following questions using the diagrams provided, which represent cross sections of bacterial cell walls.
   a. Which diagram represents a gram-positive bacterium? How can you tell?

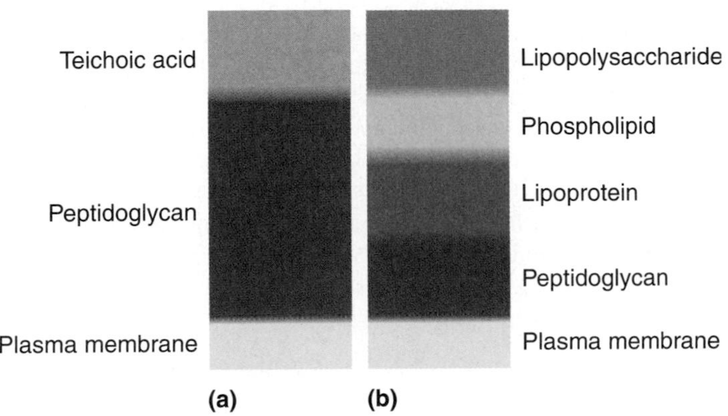

   b. Explain how the Gram stain works to distinguish these two types of cell walls.
   c. Why does penicillin have no effect on most gram-negative cells?
   d. How do essential molecules enter cells through each wall?
   e. Which cell wall is toxic to humans?
8. Starch is readily metabolized by many cells, but a starch molecule is too large to cross the plasma membrane. How does a cell obtain the glucose molecules from a starch polymer? How does the cell transport these glucose molecules across the plasma membrane?
9. Match the characteristics of eukaryotic cells in column A with their functions in column B.

Column A	Column B
______ a. Pericentriolar material	1. Digestive enzyme storage
______ b. Chloroplasts	2. Oxidation of fatty acids
______ c. Golgi complex	3. Microtubule formation
______ d. Lysosomes	4. Photosynthesis
______ e. Mitochondria	5. Protein synthesis
______ f. Peroxisomes	6. Respiration
______ g. Rough ER	7. Secretion

Multiple Choice

1. Which of the following is *not* a distinguishing characteristic of prokaryotic cells?
   a. They usually have a single, circular chromosome.
   b. They lack membrane-enclosed organelles.
   c. They have cell walls containing peptidoglycan.
   d. Their DNA is not associated with histones.
   e. They lack a plasma membrane.

Use the following choices to answer questions 2–4.
   a. No change will result; the solution is isotonic.
   b. Water will move into the cell.
   c. Water will move out of the cell.
   d. The cell will undergo osmotic lysis.
   e. Sucrose will move into the cell from an area of higher concentration to one of lower concentration.

2. Which statement best describes what happens when a gram-positive bacterium is placed in distilled water and penicillin?
3. Which statement best describes what happens when a gram-negative bacterium is placed in distilled water and penicillin?
4. Which statement best describes what happens when a gram-positive bacterium is placed in an aqueous solution of lysozyme and 10% sucrose?
5. Which of the following statements best describes what happens to a cell exposed to polymyxins that destroy phospholipids?
   a. In an isotonic solution, nothing will happen.
   b. In a hypotonic solution, the cell will lyse.
   c. Water will move into the cell.
   d. Intracellular contents will leak from the cell.
   e. Any of the above might happen.
6. Which of the following is *not* true about fimbriae?
   a. They are composed of protein.
   b. They may be used for attachment.
   c. They are found on gram-negative cells.
   d. They are composed of pilin.
   e. They may be used for motility.
7. Which of the following pairs is mismatched?
   a. glycocalyx—adherence
   b. pili—reproduction
   c. cell wall—toxin
   d. cell wall—protection
   e. plasma membrane—transport
8. Which of the following pairs is mismatched?
   a. metachromatic granules—stored phosphates
   b. polysaccharide granules—stored starch
   c. lipid inclusions—poly-β-hydroxybutyric acid
   d. sulfur granules—energy reserve
   e. ribosomes—protein storage
9. You have isolated a motile, gram-positive cell with no visible nucleus. You can assume this cell has
   a. ribosomes.
   b. mitochondria.
   c. an endoplasmic reticulum.
   d. a Golgi complex.
   e. all of the above
10. The antibiotic amphothericin B disrupts plasma membranes by combining with sterols; it will affect all of the following cells *except*
   a. animal cells.
   b. bacterial cells.
   c. fungal cells.
   d. *Mycoplasma* cells.
   e. plant cells.

Critical Thinking

1. How can prokaryotic cells be smaller than eukaryotic cells and still carry on all the functions of life?
2. The smallest eukaryotic cell is the motile alga *Micromonas*. What is the minimum number of organelles this alga must have?
3. Two types of prokaryotic cells have been distinguished: bacteria and archaea. How do these cells differ from each other? How are they similar?

4. In 1985, a 0.5-mm cell was discovered in surgeonfish and named *Epulopiscium fishelsoni* (see Figure 11.16 page 316). It was presumed to be a protozoan. In 1993, researchers determined that *Epulopiscium* was actually a gram-positive bacterium. Why do you suppose this organism was initially identified as a protozoan? What evidence would change the classification to bacterium?
5. When *E. coli* cells are exposed to a hypertonic solution, the bacteria produce a transporter protein that can move K^+ (potassium ions) into the cell. Of what value is the active transport of K^+, which requires ATP?

Clinical Applications

1. A child with a bloodborne *Neisseria* infection was treated with gentamicin. After treatment, *Neisseria* could not be cultured from her blood, indicating that the bacteria were killed. However, her symptoms became worse. Annually, nearly half of similar patients die. Explain why antibiotic treatment made her symptoms increase.
2. *Clostridium botulinum* is a strict anaerobe; that is, it is killed by the molecular oxygen (O_2) present in air. Humans can die of botulism from eating foods in which *C. botulinum* is growing. How does this bacterium survive on plants picked for human consumption? Why are home-canned foods most often the source of botulism?
3. Within a 3-day period at a large hospital, five patients undergoing hemodialysis developed fever and chills. *Pseudomonas aeruginosa* and *Klebsiella pneumoniae* were isolated from three of the patients. *P. aeruginosa, K. pneumoniae,* and *Pantoea agglomerans* were isolated from the dialysis system. Why do all three bacteria cause similar symptoms?
4. A South San Francisco child enjoyed bath time at his home because of the colorful orange and red water. The water did not have this rusty color at its source, and the water department could not culture the *Thiobacillus* bacteria responsible for the rusty color from the source. How were the bacteria getting into the household water? What bacterial structures make this possible?
5. Live cultures of *Bacillus thuringiensis* (Dipel) and *B. subtilis* (Kodiak) are sold as pesticides. What bacterial structures make it possible to package and sell these bacteria? For what purpose is each product used? (*Hint:* Refer to Chapter 11.)

5 Microbial Metabolism

Now that you are familiar with the structure of prokaryotic cells, we can discuss the activities that enable these microbes to thrive. The life-support processes of even the most structurally simple organism involve a large number of complex biochemical reactions. Most, although not all, of the biochemical processes of bacteria also occur in eukaryotic microbes and in the cells of multicellular organisms, including humans. However, the reactions that are unique to bacteria are fascinating because they allow microorganisms to do things we cannot do. For example, some bacteria can live on cellulose, whereas others can live on petroleum. Through their metabolism, bacteria recycle elements after other organisms have used them. Still other bacteria can live on diets of such inorganic substances as carbon dioxide, iron, sulfur, hydrogen gas, and ammonia.

This chapter examines some representative chemical reactions that either produce energy (the catabolic reactions) or use energy (the anabolic reactions) in microorganisms. We will also look at how these various reactions are integrated within the cell.

UNDER THE MICROSCOPE

E. coli O157:H7. This bacterium causes one of the most severe foodborne diseases called hemolytic uremic syndrome (HUS), a complication in which the kidneys fail.

Q&A

E. coli O157. *Escherichia coli* is an important member of the microbiota of the large intestine, but *E. coli* O157 causes severe diarrhea and hemorrhagic colitis. It is therefore important to diagnose *E. coli* O157, both to treat the patient and to trace the source of infection. All *E. coli* look alike through a microscope, however; so how can *E. coli* O157 be differentiated?

Look for the answer in the chapter.

Catabolic and Anabolic Reactions

LEARNING OBJECTIVES

5-1 Define *metabolism,* and describe the fundamental differences between anabolism and catabolism.

5-2 Identify the role of ATP as an intermediate between catabolism and anabolism.

We use the term **metabolism** to refer to the sum of all chemical reactions within a living organism. Because chemical reactions either release or require energy, metabolism can be viewed as an energy-balancing act. Accordingly, metabolism can be divided into two classes of chemical reactions: those that release energy and those that require energy.

In living cells, the enzyme-regulated chemical reactions that release energy are generally the ones involved in **catabolism,** the breakdown of complex organic compounds into simpler ones. These reactions are called *catabolic,* or *degradative,* reactions. Catabolic reactions are generally *hydrolytic reactions* (reactions which use water and in which chemical bonds are broken), and they are *exergonic* (produce more energy than they consume). An example of catabolism occurs when cells break down sugars into carbon dioxide and water.

The enzyme-regulated energy-requiring reactions are mostly involved in **anabolism,** the building of complex organic molecules from simpler ones. These reactions are called *anabolic,* or *biosynthetic,* reactions. Anabolic processes often involve *dehydration synthesis* reactions (reactions that release water), and they are *endergonic* (consume more energy than they produce). Examples of anabolic processes are the formation of proteins from amino acids, nucleic acids from nucleotides, and polysaccharides from simple sugars. These biosynthetic reactions generate the materials for cell growth.

Catabolic reactions provide building blocks for anabolic reactions and furnish the energy needed to drive anabolic reactions. This coupling of energy-requiring and energy-releasing reactions is made possible through the molecule adenosine triphosphate (ATP). (You can review its structure in Figure 2.18, page 49.) ATP stores energy derived from catabolic reactions and releases it later to drive anabolic reactions and perform other cellular work. Recall from Chapter 2 that a molecule of ATP consists of an adenine, a ribose, and three phosphate groups. When the terminal phosphate group is split from ATP, adenosine diphosphate (ADP) is formed, and energy is released to drive anabolic reactions. Using Ⓟ to represent a phosphate group (Ⓟ$_i$ represents inorganic phosphate, which is not bound to any other molecule), we write this reaction as follows:

$$\text{ATP} \rightarrow \text{ADP} + \text{Ⓟ}_i + \text{energy}$$

Then, the energy from catabolic reactions is used to combine ADP and a Ⓟ to resynthesize ATP:

$$\text{ADP} + \text{Ⓟ}_i + \text{energy} \rightarrow \text{ATP}$$

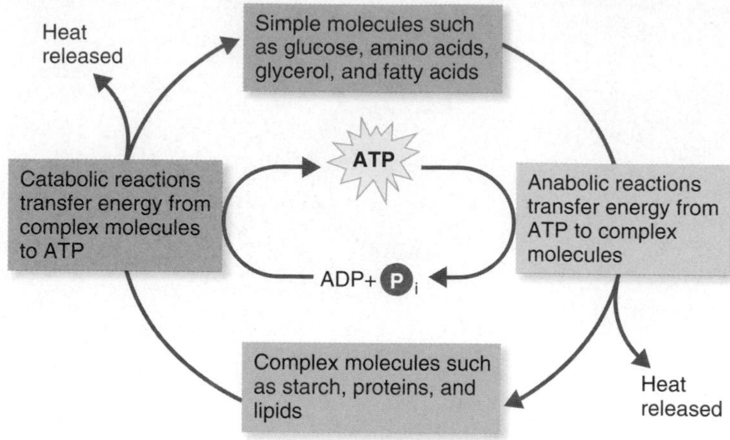

Figure 5.1 The role of ATP in coupling anabolic and catabolic reactions. When complex molecules are split apart (catabolism), some of the energy is transferred to and trapped in ATP, and the rest is given off as heat. When simple molecules are combined to form complex molecules (anabolism), ATP provides the energy for synthesis, and again some energy is given off as heat.

Q Which molecule facilitates the coupling of anabolic and catabolic reactions?

Thus, anabolic reactions are coupled to ATP breakdown, and catabolic reactions are coupled to ATP synthesis. This concept of coupled reactions is very important; you will see why by the end of this chapter. For now, you should know that the chemical composition of a living cell is constantly changing: some molecules are broken down while others are being synthesized. This balanced flow of chemicals and energy maintains the life of a cell.

The role of ATP in coupling anabolic and catabolic reactions is shown in Figure 5.1. Only part of the energy released in catabolism is actually available for cellular functions because part of the energy is lost to the environment as heat. Because the cell must use energy to maintain life, it has a continuous need for new external sources of energy.

Before we discuss how cells produce energy, let's first consider the principal properties of a group of proteins involved in almost all biologically important chemical reactions: enzymes. A cell's **metabolic pathways** (sequences of chemical reactions) are determined by its enzymes, which are in turn determined by the cell's genetic makeup. **Animation** Metabolism: Overview. **www.microbiologyplace.com**

CHECK YOUR UNDERSTANDING

- ✓ Distinguish catabolism from anabolism. **5-1**
- ✓ How is ATP an intermediate between catabolism and anabolism? **5-2**

Enzymes

LEARNING OBJECTIVES

5-3 Identify the components of an enzyme.

5-4 Describe the mechanism of enzymatic action.

5-5 List the factors that influence enzymatic activity.

5-6 Distinguish competitive and noncompetitive inhibition.

5-7 Define *ribozyme.*

Collision Theory

We indicated in Chapter 2 that chemical reactions occur when chemical bonds are formed or broken. For reactions to take place, atoms, ions, or molecules must collide. The **collision theory** explains how chemical reactions occur and how certain factors affect the rates of those reactions. The basis of the collision theory is that all atoms, ions, and molecules are continuously moving and are thus continuously colliding with one another. The energy transferred by the particles in the collision can disrupt their electron structures enough to break chemical bonds or form new bonds.

Several factors determine whether a collision will cause a chemical reaction: the velocities of the colliding particles, their energy, and their specific chemical configurations. Up to a point, the higher the particles' velocities, the more probable that their collision will cause a reaction. Also, each chemical reaction requires a specific level of energy. But even if colliding particles possess the minimum energy needed for reaction, no reaction will take place unless the particles are properly oriented toward each other.

Let's assume that molecules of substance AB (the reactant) are to be converted to molecules of substances A and B (the products). In a given population of molecules of substance AB, at a specific temperature, some molecules possess relatively little energy; the majority of the population possesses an average amount of energy; and a small portion of the population has high energy. If only the high-energy AB molecules are able to react and be converted to A and B molecules, then only relatively few molecules at any one time possess enough energy to react in a collision. The collision energy required for a chemical reaction is its **activation energy,** which is the amount of energy needed to disrupt the stable electronic configuration of any specific molecule so that the electrons can be rearranged.

The **reaction rate**—the frequency of collisions containing sufficient energy to bring about a reaction—depends on the number of reactant molecules at or above the activation energy level. One way to increase the reaction rate of a substance is to raise its temperature. By causing the molecules to move faster, heat increases both the frequency of collisions and the number of molecules that attain activation energy. The number of collisions also increases when pressure is increased or when the reactants are more concentrated (because the distance between molecules is thereby decreased). In living systems, enzymes increase the reaction rate without raising the temperature.

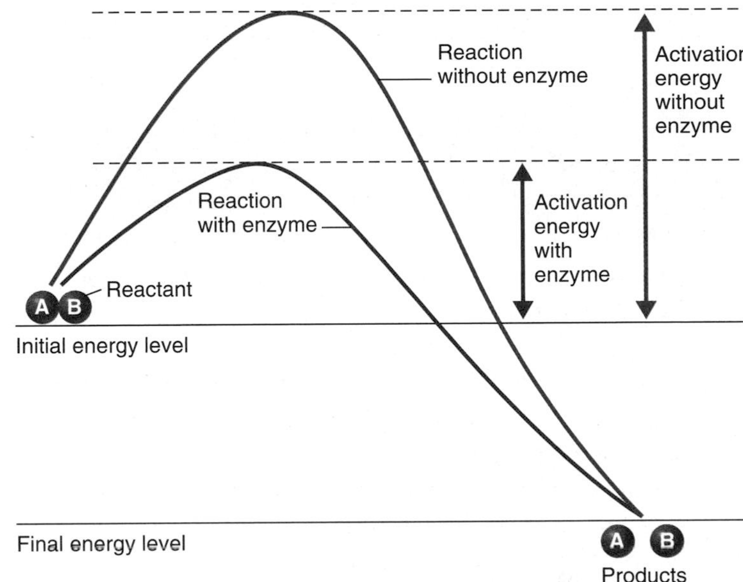

Figure 5.2 Energy requirements of a chemical reaction. This graph shows the progress of the reaction AB → A + B both without (blue line) and with (red line) an enzyme. The presence of an enzyme lowers the activation energy of the reaction (see arrows). Thus, more molecules of reactant AB are converted to products A and B because more molecules of reactant AB possess the activation energy needed for the reaction.

Q How do enzymes speed up chemical reactions?

Enzymes and Chemical Reactions

Substances that can speed up a chemical reaction without being permanently altered themselves are called **catalysts.** In living cells, **enzymes** serve as biological catalysts. As catalysts, enzymes are specific. Each acts on a specific substance, called the enzyme's **substrate** (or substrates, when there are two or more reactants), and each catalyzes only one reaction. For example, sucrose (table sugar) is the substrate of the enzyme sucrase, which catalyzes the hydrolysis of sucrose to glucose and fructose.

As catalysts, enzymes typically accelerate chemical reactions. The three-dimensional enzyme molecule has an *active site,* a region that interacts with a specific chemical substance (see Figure 5.4 on page 118).

The enzyme orients the substrate into a position that increases the probability of a reaction. The **enzyme–substrate complex** formed by the temporary binding of enzyme and reactants enables the collisions to be more effective and lowers the activation energy of the reaction (Figure 5.2). The enzyme therefore speeds up the reaction by increasing the number of AB molecules that attain sufficient activation energy to react.

An enzyme's ability to accelerate a reaction without the need for an increase in temperature is crucial to living systems because a significant temperature increase would destroy cellular proteins. The crucial function of enzymes, therefore, is to speed up biochemical reactions at a temperature that is compatible with the normal functioning of the cell.

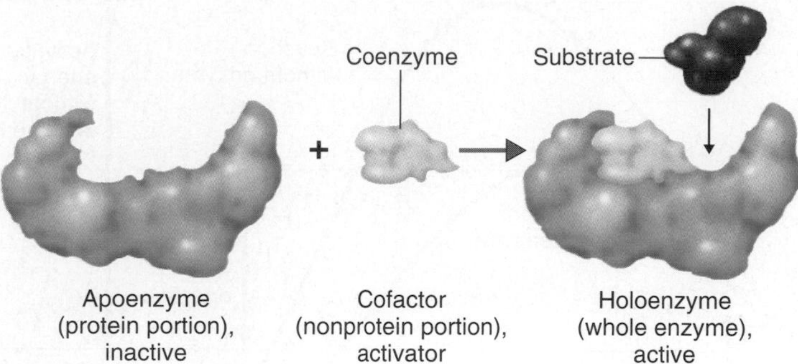

Figure 5.3 Components of a holoenzyme. Many enzymes require both an apoenzyme (protein portion) and a cofactor (nonprotein portion) to become active. The cofactor can be a metal ion, or if it is an organic molecule, it is called a coenzyme (as shown here). The apoenzyme and cofactor together make up the holoenzyme, or whole enzyme. The substrate is the reactant acted upon by the enzyme.

Q Which substances usually function as coenzymes?

Enzyme Specificity and Efficiency

The specificity of enzymes is made possible by their structures. Enzymes are generally large globular proteins that range in molecular weight from about 10,000 to several million. Each of the thousands of known enzymes has a characteristic three-dimensional shape with a specific surface configuration as a result of its primary, secondary, and tertiary structures (see Figure 2.15, page 46). The unique configuration of each enzyme enables it to "find" the correct substrate from among the large number of diverse molecules in the cell.

Enzymes are extremely efficient. Under optimum conditions, they can catalyze reactions at rates 10^8 to 10^{10} times (up to 10 billion times) higher than those of comparable reactions without enzymes. The **turnover number** (maximum number of substrate molecules an enzyme molecule converts to product each second) is generally between 1 and 10,000 and can be as high as 500,000. For example, the enzyme DNA polymerase I, which participates in the synthesis of DNA, has a turnover number of 15, whereas the enzyme lactate dehydrogenase, which removes hydrogen atoms from lactic acid, has a turnover number of 1000.

Many enzymes exist in the cell in both active and inactive forms. The rate at which enzymes switch between these two forms is determined by the cellular environment.

Naming Enzymes

The names of enzymes usually end in *-ase.* All enzymes can be grouped into six classes, according to the type of chemical reaction they catalyze (Table 5.1). Enzymes within each of the major classes are named according to the more specific types of reactions they assist. For example, the class called *oxidoreductases* is involved with oxidation-reduction reactions (described shortly). Enzymes in the oxidoreductase class that remove hydrogen from a substrate are called *dehydrogenases;* those that add molecular oxygen (O_2) are called *oxidases.* As you will see later, dehydrogenase and oxidase enzymes have even more specific names, such as lactate dehydrogenase and cytochrome oxidase, depending on the specific substrates on which they act.

Enzyme Components

Although some enzymes consist entirely of proteins, most consist of both a protein portion, called an **apoenzyme,** and a nonprotein component, called a **cofactor.** Ions of iron, zinc, magnesium, or calcium are examples of cofactors. If the cofactor is an organic molecule, it is called a **coenzyme.** Apoenzymes are inactive by themselves; they must be activated by cofactors. Together, the apoenzyme and cofactor form a **holoenzyme,** or whole, active enzyme (Figure 5.3). If the cofactor is removed, the apoenzyme will not function.

Table 5.1 Enzyme Classification Based on Type of Chemical Reaction Catalyzed

Class	Type of Chemical Reaction Catalyzed	Examples
Oxidoreductase	Oxidation-reduction, in which oxygen and hydrogen are gained or lost	Cytochrome oxidase, lactate dehydrogenase
Transferase	Transfer of functional groups, such as an amino group, acetyl group, or phosphate group	Acetate kinase, alanine deaminase
Hydrolase	Hydrolysis (addition of water)	Lipase, sucrase
Lyase	Removal of groups of atoms without hydrolysis	Oxalate decarboxylase, isocitrate lyase
Isomerase	Rearrangement of atoms within a molecule	Glucose-phosphate isomerase, alanine racemase
Ligase	Joining of two molecules (using energy usually derived from the breakdown of ATP)	Acetyl-CoA synthetase, DNA ligase

Table 5.2 Selected Vitamins and Their Coenzymatic Functions

Vitamin	Function
Vitamin B_1 (Thiamine)	Part of coenzyme cocarboxylase; has many functions, including the metabolism of pyruvic acid
Vitamin B_2 (Riboflavin)	Coenzyme in flavoproteins; active in electron transfers
Niacin (Nicotinic Acid)	Part of NAD molecule*; active in electron transfers
Vitamin B_6 (Pyridoxine)	Coenzyme in amino acid metabolism
Vitamin B_{12} (Cyanocobalamin)	Coenzyme (methyl cyanocobalamide) involved in the transfer of methyl groups; active in amino acid metabolism
Pantothenic Acid	Part of coenzyme A molecule; involved in the metabolism of pyruvic acid and lipids
Biotin	Involved in carbon dioxide fixation reactions and fatty acid synthesis
Folic Acid	Coenzyme used in the synthesis of purines and pyrimidines
Vitamin E	Needed for cellular and macromolecular syntheses
Vitamin K	Coenzyme used in electron transport (naphthoquinones and quinones)

*NAD = nicotinamide adenine dinucleotide

Coenzymes may assist the enzyme by accepting atoms removed from the substrate or by donating atoms required by the substrate. Some coenzymes act as electron carriers, removing electrons from the substrate and donating them to other molecules in subsequent reactions. Many coenzymes are derived from vitamins (Table 5.2). Two of the most important coenzymes in cellular metabolism are **nicotinamide adenine dinucleotide (NAD^+)** and **nicotinamide adenine dinucleotide phosphate ($NADP^+$).** Both compounds contain derivatives of the B vitamin niacin (nicotinic acid), and both function as electron carriers. Whereas NAD^+ is primarily involved in catabolic (energy-yielding) reactions, $NADP^+$ is primarily involved in anabolic (energy-requiring) reactions. The flavin coenzymes, such as **flavin mononucleotide (FMN)** and **flavin adenine dinucleotide (FAD),** contain derivatives of the B vitamin riboflavin and are also electron carriers. Another important coenzyme, **coenzyme A (CoA),** contains a derivative of pantothenic acid, another B vitamin. This coenzyme plays an important role in the synthesis and breakdown of fats and in a series of oxidizing reactions called the Krebs cycle. We will come across all of these coenzymes in our discussion of metabolism later in the chapter.

As noted earlier, some cofactors are metal ions, including iron, copper, magnesium, manganese, zinc, calcium, and cobalt. Such cofactors may help catalyze a reaction by forming a bridge between the enzyme and a substrate. For example, magnesium (Mg^{2+}) is required by many phosphorylating enzymes (enzymes that transfer a phosphate group from ATP to another substrate). The Mg^{2+} can form a link between the enzyme and the ATP molecule. Most trace elements required by living cells are probably used in some such way to activate cellular enzymes.

The Mechanism of Enzymatic Action

Enzymes lower the activation energy of chemical reactions. The general sequence of events in enzyme action is as follows (Figure 5.4a):

1. The surface of the substrate contacts a specific region of the surface of the enzyme molecule, called the **active site.**
2. A temporary intermediate compound forms, called an **enzyme–substrate complex.**
3. The substrate molecule is transformed by the rearrangement of existing atoms, the breakdown of the substrate molecule, or in combination with another substrate molecule.
4. The transformed substrate molecules—the products of the reaction—are released from the enzyme molecule because they no longer fit in the active site of the enzyme.
5. The unchanged enzyme is now free to react with other substrate molecules.

As a result of these events, an enzyme speeds up a chemical reaction.

As mentioned earlier, enzymes have *specificity* for particular substrates. For example, a specific enzyme may be able to hydrolyze a peptide bond only between two specific amino acids.

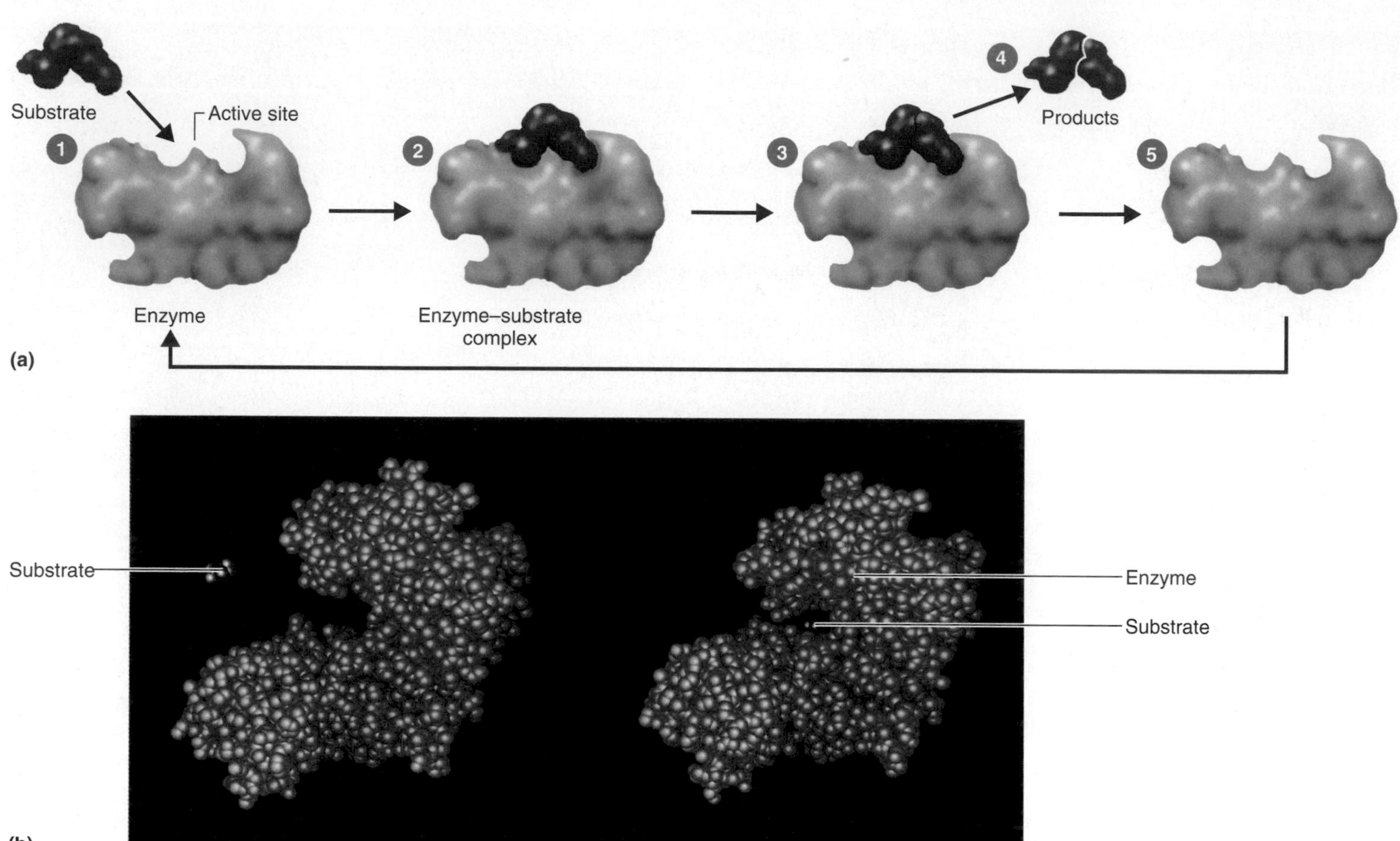

Figure 5.4 The mechanism of enzymatic action. (**a**) 1 The substrate contacts the active site on the enzyme to form 2 an enzyme–substrate complex. 3 The substrate is then transformed into products, 4 the products are released, and 5 the enzyme is recovered unchanged. In the example shown, the transformation into products involves a breakdown of the substrate into two products. Other transformations, however, may occur. (**b**) Left: A molecular model of the enzyme in step 1 of part (a). The active site of the enzyme can be seen here as a groove on the surface of the protein. Right: As the enzyme and substrate meet in step 2 of part (a), the enzyme changes shape slightly to fit together more tightly with the substrate.

Q **What is the function of enzymes in living organisms?**

Other enzymes can hydrolyze starch but not cellulose; even though both starch and cellulose are polysaccharides composed of glucose subunits, the orientations of the subunits in the two polysaccharides differ. Enzymes have this specificity because the three-dimensional shape of the active site fits the substrate somewhat as a lock fits with its key (**Figure 5.4b**). However, the active site and substrate are flexible, and they change shape somewhat as they meet to fit together more tightly. The substrate is usually much smaller than the enzyme, and relatively few of the enzyme's amino acids make up the active site.

A certain compound can be a substrate for several different enzymes that catalyze different reactions, so the fate of a compound depends on the enzyme that acts on it. At least four different enzymes can act on glucose 6-phosphate, a molecule important in cell metabolism, and each reaction will yield a different product. **Animations** Enzymes: Overview, Steps in a Reaction. **www.microbiologyplace.com**

Factors Influencing Enzymatic Activity

Enzymes are subject to various cellular controls. Two primary types are the control of enzyme *synthesis* (see Chapter 8) and the control of enzyme *activity* (how much enzyme is present versus how active it is).

Several factors influence the activity of an enzyme. Among the more important are temperature, pH, substrate concentration, and the presence or absence of inhibitors.

Temperature

The rate of most chemical reactions increases as the temperature increases. Molecules move more slowly at lower temperatures

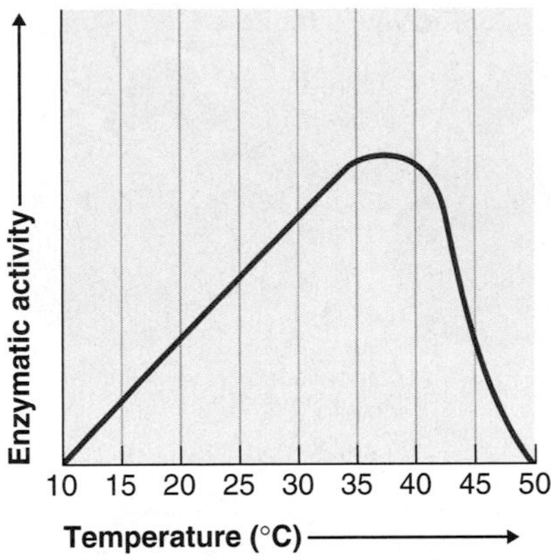

(a) **Temperature.** The enzymatic activity (rate of reaction catalyzed by the enzyme) increases with increasing temperature until the enzyme, a protein, is denatured by heat and inactivated. At this point, the reaction rate falls steeply.

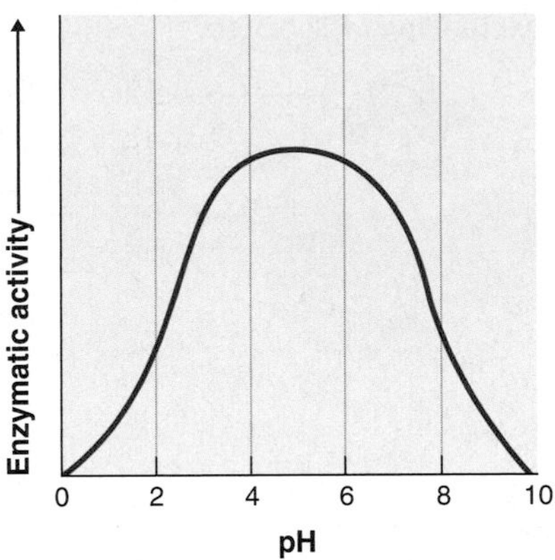

(b) **pH.** The enzyme illustrated is most active at about pH 5.0.

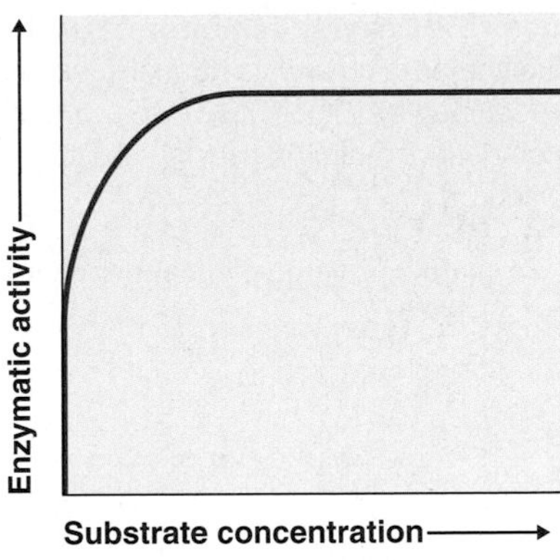

(c) **Substrate concentration.** With increasing concentration of substrate molecules, the rate of reaction increases until the active sites on all the enzyme molecules are filled, at which point the maximum rate of reaction is reached.

Figure 5.5 Factors that influence enzymatic activity, plotted for a hypothetical enzyme.

Q How will this enzyme act at 25°C? At 45°C? At pH 7?

than at higher temperatures and so may not have enough energy to cause a chemical reaction. For enzymatic reactions, however, elevation beyond a certain temperature (the optimal temperature) drastically reduces the rate of reaction (Figure 5.5a). The optimal temperature for most disease-producing bacteria in the human body is between 35°C and 40°C. The rate of reaction declines beyond the optimal temperature because of the enzyme's **denaturation,** the loss of its characteristic three-dimensional structure (tertiary configuration) (Figure 5.6). Denaturation of a protein involves the breakage of hydrogen bonds and other noncovalent bonds; a common example is the transformation of uncooked egg white (a protein called albumin) to a hardened state by heat.

Denaturation of an enzyme changes the arrangement of the amino acids in the active site, altering its shape and causing the enzyme to lose its catalytic ability. In some cases, denaturation is partially or fully reversible. However, if denaturation continues until the enzyme has lost its solubility and coagulates, the enzyme cannot regain its original properties. Enzymes can also be denatured by concentrated acids, bases, heavy-metal ions (such as lead, arsenic, or mercury), alcohol, and ultraviolet radiation.

pH

Most enzymes have an optimum pH at which their activity is characteristically maximal. Above or below this pH value, enzyme activity, and therefore the reaction rate, decline (Figure 5.5b). When the H^+ concentration (pH) in the medium is changed drastically, the protein's three-dimensional structure is altered. Extreme changes in pH can cause denaturation. Acids (and bases) alter a protein's three-dimensional structure because the H^+ (and OH^-) compete with hydrogen and ionic bonds in an enzyme, resulting in the enzyme's denaturation.

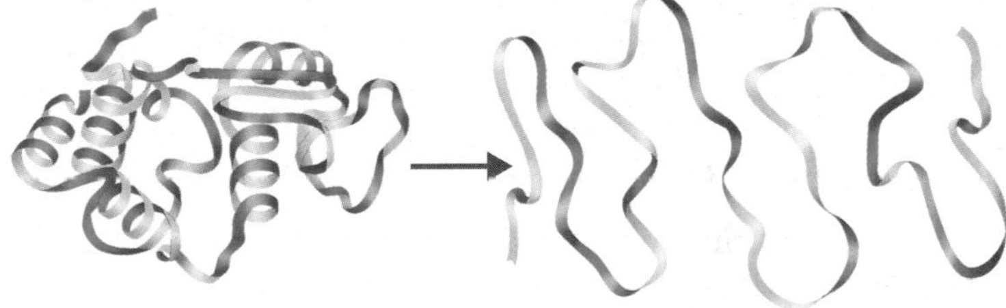

Figure 5.6 Denaturation of a protein. Breakage of the noncovalent bonds (such as hydrogen bonds) that hold the active protein in its three-dimensional shape renders the denatured protein nonfunctional.

Q What factors may cause denaturation?

Substrate Concentration

There is a maximum rate at which a certain amount of enzyme can catalyze a specific reaction. Only when the concentration of substrate(s) is extremely high can this maximum rate be attained. Under conditions of high substrate concentration, the enzyme is said to be in **saturation;** that is, its active site is always occupied by substrate or product molecules. In this condition, a further increase in substrate concentration will not affect the reaction rate because all active sites are already in use (Figure 5.5c). Under normal cellular conditions, enzymes are not saturated with substrate(s). At any given time, many of the enzyme molecules are inactive for lack of substrate; thus, the substrate concentration is likely to influence the rate of reaction.

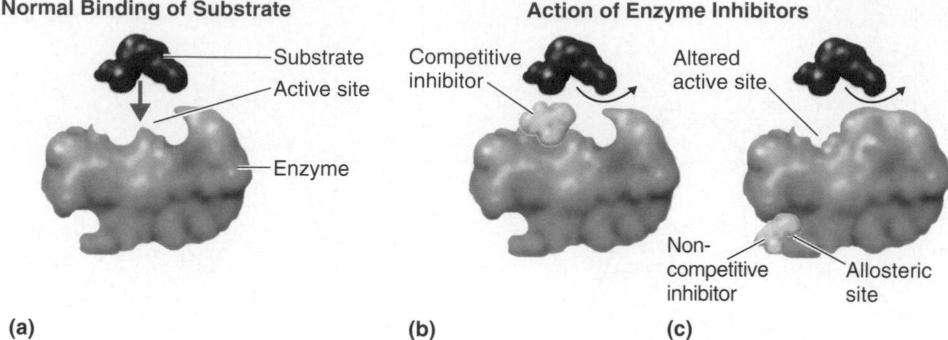

Figure 5.7 Enzyme inhibitors. (**a**) An uninhibited enzyme and its normal substrate. (**b**) A competitive inhibitor. (**c**) One type of noncompetitive inhibitor, causing allosteric inhibition.

Q How do competitive inhibitors operate?

Inhibitors

An effective way to control the growth of bacteria is to control their enzymes. Certain poisons, such as cyanide, arsenic, and mercury, combine with enzymes and prevent them from functioning. As a result, the cells stop functioning and die.

Enzyme inhibitors are classified as either competitive or noncompetitive inhibitors (Figure 5.7). **Competitive inhibitors** fill the active site of an enzyme and compete with the normal substrate for the active site. A competitive inhibitor can do this because its shape and chemical structure are similar to those of the normal substrate (Figure 5.7b). However, unlike the substrate, it does not undergo any reaction to form products. Some competitive inhibitors bind irreversibly to amino acids in the active site, preventing any further interactions with the substrate. Others bind reversibly, alternately occupying and leaving the active site; these slow the enzyme's interaction with the substrate. Increasing the substrate concentration can overcome reversible competitive inhibition. As active sites become available, more substrate molecules than competitive inhibitor molecules are available to attach to the active sites of enzymes.

One good example of a competitive inhibitor is sulfanilamide (a sulfa drug), which inhibits the enzyme whose normal substrate is *para*-aminobenzoic acid (PABA):

NH_2 — $O{=}S{=}O$ — (benzene ring) — NH_2
Sulfanilamide

HO — C = O — (benzene ring) — NH_2
PABA

PABA is an essential nutrient used by many bacteria in the synthesis of folic acid, a vitamin that functions as a coenzyme. When sulfanilamide is administered to bacteria, the enzyme that normally converts PABA to folic acid combines instead with the sulfanilamide. Folic acid is not synthesized, and the bacteria cannot grow. Because human cells do not use PABA to make their folic acid, sulfanilamide kills bacteria but does not harm human cells.

Noncompetitive inhibitors do not compete with the substrate for the enzyme's active site; instead, they interact with another part of the enzyme (Figure 5.7c). In this process, called **allosteric** ("other space") **inhibition,** the inhibitor binds to a site on the enzyme other than the substrate's binding site, called the **allosteric site.** This binding causes the active site to change its shape, making it nonfunctional. As a result, the enzyme's activity is reduced. This effect can be either reversible or irreversible, depending on whether the active site can return to its original shape. In some cases, allosteric interactions can activate an enzyme rather than inhibit it. Another type of noncompetitive inhibition can operate on enzymes that require metal ions for their activity. Certain chemicals can bind or tie up the metal ion activators and thus prevent an enzymatic reaction. Cyanide can bind the iron in iron-containing enzymes, and fluoride can bind calcium or magnesium. Substances such as cyanide and fluoride are sometimes called *enzyme poisons* because they permanently inactivate enzymes. **Animations** Enzymes: Competitive Inhibition, Non-competitive Inhibition. **www.microbiologyplace.com**

Feedback Inhibition

Allosteric inhibitors play a role in a kind of biochemical control called **feedback inhibition,** or **end-product inhibition.** This control mechanism stops the cell from making more of a substance than it needs and thereby wasting chemical resources. In some metabolic reactions, several steps are required for the synthesis of a particular chemical compound, called the *end-product.* The process is similar to an assembly line, with each step catalyzed by a separate enzyme (Figure 5.8). In many anabolic pathways, the final product can allosterically inhibit the activity of one of the enzymes earlier in the pathway. This phenomenon is feedback inhibition.

Feedback inhibition generally acts on the first enzyme in a metabolic pathway (similar to shutting down an assembly line by stopping the first worker). Because the enzyme is inhibited, the product of the first enzymatic reaction in the pathway is not synthesized. Because that unsynthesized product would normally be

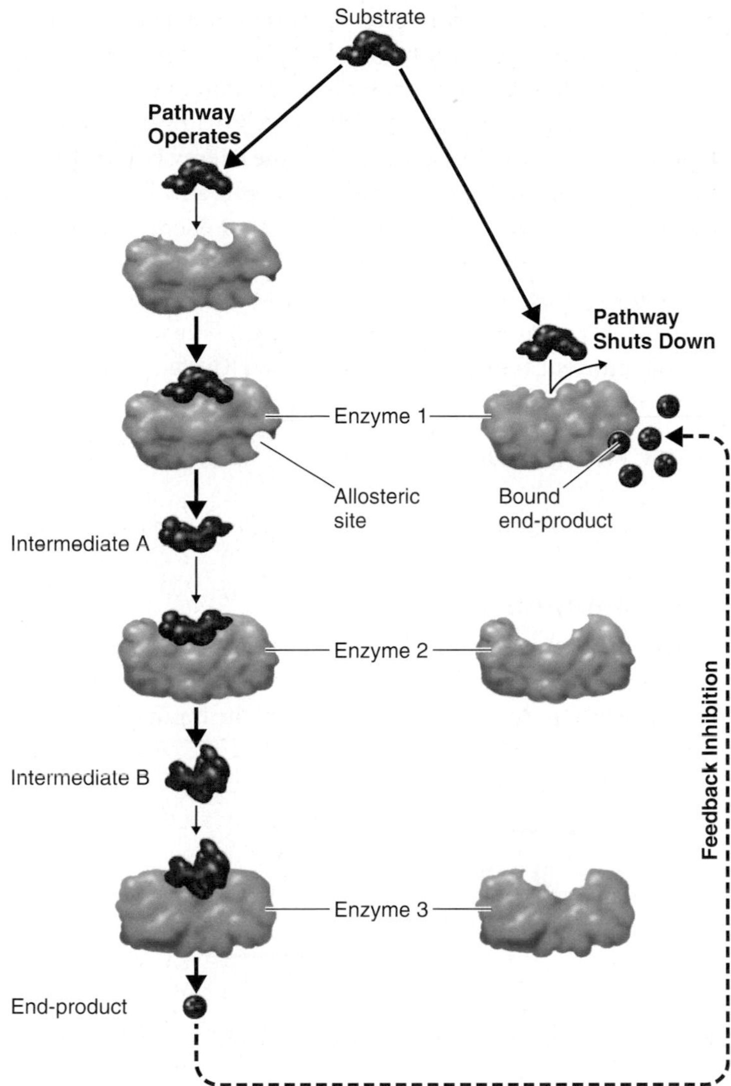

Figure 5.8 Feedback inhibition.

Q What is meant by *feedback inhibition*?

the substrate for the second enzyme in the pathway, the second reaction stops immediately as well. Thus, even though only the first enzyme in the pathway is inhibited, the entire pathway shuts down, and no new end-product is formed. By inhibiting the first enzyme in the pathway, the cell also keeps metabolic intermediates from accumulating. As the cell uses up the existing end-product, the first enzyme's allosteric site more often remains unbound, and the pathway resumes activity.

The bacterium *E. coli* can be used to demonstrate feedback inhibition in the synthesis of the amino acid isoleucine, which is required for the cell's growth. In this metabolic pathway, the amino acid threonine is enzymatically converted to isoleucine in five steps. If isoleucine is added to the growth medium for *E. coli,* it inhibits the first enzyme in the pathway, and the bacteria stop synthesizing isoleucine. This condition is maintained until the supply of isoleucine is depleted. This type of feedback inhibition is also involved in regulating the cells' production of other amino acids, as well as vitamins, purines, and pyrimidines.

Ribozymes

Prior to 1982, it was believed that only protein molecules had enzymatic activity. Researchers working on microbes discovered a unique type of RNA called a **ribozyme.** Like protein enzymes, ribozymes function as catalysts, have active sites that bind to substrates, and are not used up in a chemical reaction. Ribozymes specifically act on strands of RNA by removing sections and splicing together the remaining pieces. In this respect, ribozymes are more restricted than protein enzymes in terms of the diversity of substrates with which they interact.

CHECK YOUR UNDERSTANDING

- ✓ What is a coenzyme? **5-3**
- ✓ Why is enzyme specificity important? **5-4**
- ✓ What happens to an enzyme below its optimal temperature? Above its optimal temperature? **5-5**
- ✓ Why is feedback inhibition noncompetitive inhibition? **5-6**
- ✓ What is a ribozyme? **5-7**

Energy Production

LEARNING OBJECTIVES

5-8 Explain the term *oxidation-reduction.*

5-9 List and provide examples of three types of phosphorylation reactions that generate ATP.

5-10 Explain the overall function of metabolic pathways.

Nutrient molecules, like all molecules, have energy associated with the electrons that form bonds between their atoms. When it is spread throughout the molecule, this energy is difficult for the cell to use. Various reactions in catabolic pathways, however, concentrate the energy into the bonds of ATP, which serves as a convenient energy carrier. ATP is generally referred to as having "high-energy" bonds. Actually, a better term is probably *unstable bonds.* Although the amount of energy in these bonds is not exceptionally large, it can be released quickly and easily. In a sense, ATP is similar to a highly flammable liquid such as kerosene. Although a large log might eventually burn to produce more heat than a cup of kerosene, the kerosene is easier to ignite and provides heat more quickly and conveniently. In a similar way, the "high-energy" unstable bonds of ATP provide the cell with readily available energy for anabolic reactions.

Before discussing the catabolic pathways, we will consider two general aspects of energy production: the concept of oxidation-reduction and the mechanisms of ATP generation.

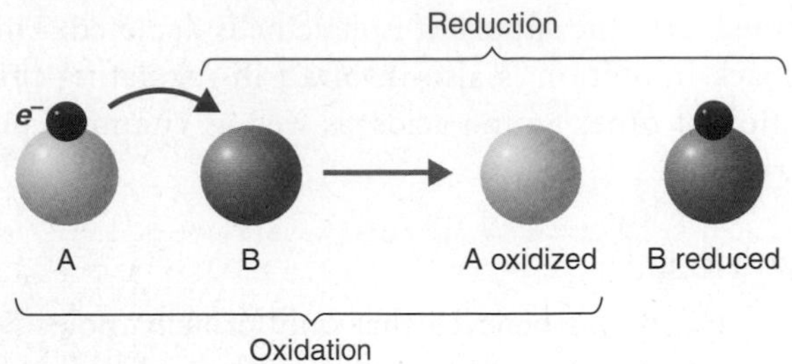

Figure 5.9 Oxidation-reduction. An electron is transferred from molecule A to molecule B. In the process, molecule A is oxidized and molecule B is reduced.

Q How do oxidation and reduction differ?

Oxidation-Reduction Reactions

Oxidation is the removal of electrons (e^-) from an atom or molecule, a reaction that often produces energy. Figure 5.9 shows an example of an oxidation in which molecule A loses an electron to molecule B. Molecule A has undergone oxidation (meaning that it has lost one or more electrons), whereas molecule B has undergone **reduction** (meaning that it has gained one or more electrons).* Oxidation and reduction reactions are always coupled; in other words, each time one substance is oxidized, another is simultaneously reduced. The pairing of these reactions is called **oxidation-reduction** or a **redox reaction.**

In many cellular oxidations, electrons and protons (hydrogen ions, H^+) are removed at the same time; this is equivalent to the removal of hydrogen atoms, because a hydrogen atom is made up of one proton and one electron (see Table 2.2, page 29). Because most biological oxidations involve the loss of hydrogen atoms, they are also called **dehydrogenation** reactions. Figure 5.10 shows an example of a biological oxidation. An organic molecule is oxidized by the loss of two hydrogen atoms, and a molecule of NAD^+ is reduced. Recall from our earlier discussion of coenzymes that NAD^+ assists enzymes by accepting hydrogen atoms removed from the substrate, in this case the organic molecule. As shown in Figure 5.10, NAD^+ accepts two electrons and one proton. One proton (H^+) is left over and is released into the surrounding medium. The reduced coenzyme, NADH, contains more energy than NAD^+. This energy can be used to generate ATP in later reactions.

An important point to remember about biological oxidation-reduction reactions is that cells use them in catabolism to extract energy from nutrient molecules. Cells take nutrients, some of which serve as energy sources, and degrade them from highly reduced compounds (with many hydrogen atoms) to highly oxidized compounds. For example, when a cell oxidizes a molecule of glucose ($C_6H_{12}O_6$) to CO_2 and H_2O, the energy in the glucose molecule is removed in a stepwise manner and ultimately is trapped by ATP, which can then serve as an energy source for energy-requiring reactions. Compounds such as glucose that have many hydrogen atoms are highly reduced compounds, containing a large amount of potential energy. Thus, glucose is a valuable nutrient for organisms. **Animation** Oxidation-Reduction Reactions. **www.microbiologyplace.com**

*The terms do not seem logical until one considers the history of the discovery of these reactions. When mercury is roasted, it gains weight as mercuric oxide is formed; this was called *oxidation*. Later it was determined that the mercury actually *lost* electrons, and the observed *gain* in oxygen was a direct result of this. Oxidation, therefore, is a *loss* of electrons, and reduction is a *gain* of electrons, but the gain and loss of electrons is not usually apparent as chemical-reaction equations are usually written. For example, in the equations for aerobic respiration on page 132, notice that each carbon in glucose had only one oxygen originally, and later, as carbon dioxide, each carbon now has two oxygens. However, the gain or loss of electrons actually responsible for this is not apparent.

CHECK YOUR UNDERSTANDING

✓ Why is glucose such an important molecule for organisms? **5-8**

The Generation of ATP

Much of the energy released during oxidation-reduction reactions is trapped within the cell by the formation of ATP. Specifically, a phosphate group, Ⓟ, is added to ADP with the input of energy to form ATP:

$$\overbrace{\text{Adenosine—Ⓟ}\sim\text{Ⓟ}}^{\text{ADP}} + \text{Energy} + \text{Ⓟ} \rightarrow \underbrace{\text{Adenosine—Ⓟ}\sim\text{Ⓟ}\sim\text{Ⓟ}}_{\text{ATP}}$$

The symbol ~ designates a "high-energy" bond—that is, one that can readily be broken to release usable energy. The high-energy bond that attaches the third Ⓟ in a sense contains the energy stored in this reaction. When this Ⓟ is removed, usable energy is released. The addition of Ⓟ to a chemical compound is called **phosphorylation.** Organisms use three mechanisms of phosphorylation to generate ATP from ADP.

Substrate-Level Phosphorylation

In **substrate-level phosphorylation,** ATP is usually generated when a high-energy Ⓟ is directly transferred from a phosphorylated compound (a substrate) to ADP. Generally, the Ⓟ has acquired its energy during an earlier reaction in which the substrate itself was oxidized. The following example shows only the carbon skeleton and the Ⓟ of a typical substrate:

$$\text{C—C—C} \sim \text{Ⓟ} + \text{ADP} \rightarrow \text{C—C—C} + \text{ATP}$$

Oxidative Phosphorylation

In **oxidative phosphorylation,** electrons are transferred from organic compounds to one group of electron carriers (usually to NAD^+ and FAD). Then, the electrons are passed through a series

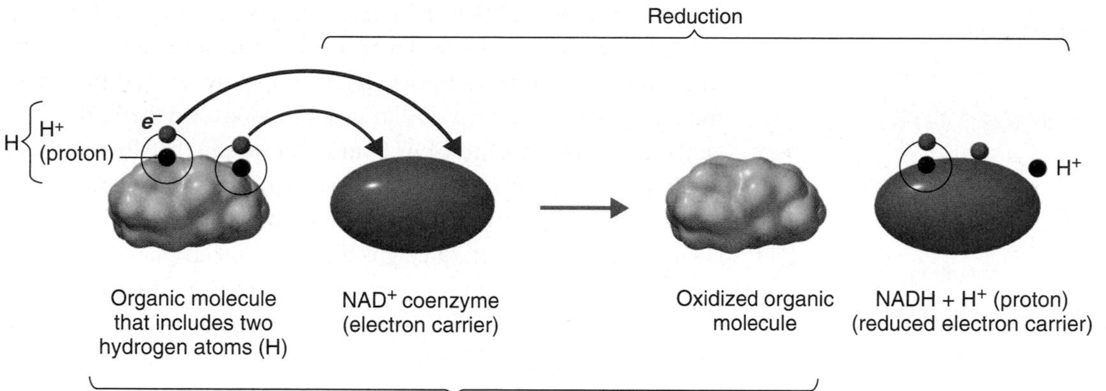

Figure 5.10 Representative biological oxidation. Two electrons and two protons (altogether equivalent to two hydrogen atoms) are transferred from an organic substrate molecule to a coenzyme, NAD^+. NAD^+ actually receives one hydrogen atom and one electron, and one proton is released into the medium. NAD^+ is reduced to NADH, which is a more energy-rich molecule.

Q How do organisms use oxidation-reduction reactions?

of different electron carriers to molecules of oxygen (O_2) or other oxidized inorganic and organic molecules. This process occurs in the plasma membrane of prokaryotes and in the inner mitochondrial membrane of eukaryotes. The sequence of electron carriers used in oxidative phosphorylation is called an **electron transport chain (system)** (see Figure 5.14, page 129). The transfer of electrons from one electron carrier to the next releases energy, some of which is used to generate ATP from ADP through a process called *chemiosmosis,* to be described on page 130.

Photophosphorylation

The third mechanism of phosphorylation, **photophosphorylation,** occurs only in photosynthetic cells, which contain light-trapping pigments such as chlorophylls. In photosynthesis, organic molecules, especially sugars, are synthesized with the energy of light from the energy-poor building blocks carbon dioxide and water. Photophosphorylation starts this process by converting light energy to the chemical energy of ATP and NADPH, which, in turn, are used to synthesize organic molecules. As in oxidative phosphorylation, an electron transport chain is involved.

CHECK YOUR UNDERSTANDING

✓ Outline the three ways that ATP is generated. 5-9

Metabolic Pathways of Energy Production

Organisms release and store energy from organic molecules by a series of controlled reactions rather than in a single burst. If the energy were released all at once, as a large amount of heat, it could not be readily used to drive chemical reactions and would, in fact, damage the cell. To extract energy from organic compounds and store it in chemical form, organisms pass electrons from one compound to another through a series of oxidation-reduction reactions.

As noted earlier, a sequence of enzymatically catalyzed chemical reactions occurring in a cell is called a metabolic pathway. Below is a hypothetical metabolic pathway that converts starting material A to end-product F in a series of five steps:

NAD^+ $NADH + H^+$
A (Starting material) —(1)→ B —(2)→

ADP + Ⓟ ATP
C —(3)→ D ⇌(4) E —(5)→ F
O_2 (reactant in step 5); CO_2 H_2O End-product

The first step is the conversion of molecule A to molecule B. The curved arrow indicates that the reduction of coenzyme NAD^+ to NADH is coupled to that reaction; the electrons and protons come from molecule A. Similarly, the two arrows in (3) show a coupling of two reactions. As C is converted to D, ADP is converted to ATP; the energy needed comes from C as it transforms into D. The reaction converting D to E is readily reversible, as indicated by the double arrow. In the fifth step, the curved arrow leading from O_2 indicates that O_2 is a reactant in the reaction. The curved arrows leading to CO_2 and H_2O indicate that these substances are secondary products produced in the reaction, in addition to F, the end-product that (presumably) interests us the most. Secondary products such as CO_2 and H_2O shown here are sometimes called *by-products* or *waste products.* Keep in mind that almost every reaction in a metabolic pathway is catalyzed by a specific enzyme; sometimes the name of the enzyme is printed near the arrow.

CHECK YOUR UNDERSTANDING

✓ What is the purpose of metabolic pathways? 5-10

Carbohydrate Catabolism

LEARNING OBJECTIVES

5-11 Describe the chemical reactions of glycolysis.

5-12 Identify the functions of the pentose phosphate and Entner-Doudoroff pathways.

5-13 Explain the products of the Krebs cycle.

5-14 Describe the chemiosmotic model for ATP generation.

5-15 Compare and contrast aerobic and anaerobic respiration.

5-16 Describe the chemical reactions of, and list some products of, fermentation.

Most microorganisms oxidize carbohydrates as their primary source of cellular energy. **Carbohydrate catabolism,** the breakdown of carbohydrate molecules to produce energy, is therefore of great importance in cell metabolism. Glucose is the most common carbohydrate energy source used by cells. Microorganisms can also catabolize various lipids and proteins for energy production (page 136).

To produce energy from glucose, microorganisms use two general processes: *cellular respiration* and *fermentation.* (In discussing cellular respiration, we frequently refer to the process simply as respiration, but it should not be confused with breathing.) Both cellular respiration and fermentation usually start with the same first step, glycolysis, but follow different subsequent pathways (**Figure 5.11**). Before examining the details of glycolysis, respiration, and fermentation, we will first look at a general overview of the processes.

As shown in Figure 5.11, the respiration of glucose typically occurs in three principal stages: glycolysis, the Krebs cycle, and the electron transport chain (system).

1. Glycolysis is the oxidation of glucose to pyruvic acid with the production of some ATP and energy-containing NADH.
2. The Krebs cycle is the oxidation of acetyl CoA (a derivative of pyruvic acid) to carbon dioxide, with the production of some ATP, energy-containing NADH, and another reduced electron carrier, $FADH_2$ (the reduced form of flavin adenine dinucleotide).
3. In the electron transport chain (system), NADH and $FADH_2$ are oxidized, contributing the electrons they have carried from the substrates to a "cascade" of oxidation-reduction reactions involving a series of additional electron carriers. Energy from these reactions is used to generate a considerable amount of ATP. In respiration, most of the ATP is generated in the third step.

Because respiration involves a long series of oxidation-reduction reactions, the entire process can be thought of as involving a flow of electrons from the energy-rich glucose molecule to the relatively energy-poor CO_2 and H_2O molecules. The coupling of ATP production to this flow is somewhat analogous to producing electrical power by using energy from a flowing stream. Carrying the analogy further, you could imagine a stream flowing down a gentle slope during glycolysis and the Krebs cycle, supplying energy to turn two old-fashioned waterwheels. Then the stream rushes down a steep slope in the electron transport chain, supplying energy for a large modern power plant. In a similar way, glycolysis and the Krebs cycle generate a small amount of ATP and also supply the electrons that generate a great deal of ATP at the electron transport chain stage.

Typically, the initial stage of fermentation is also glycolysis (Figure 5.11). However, once glycolysis has taken place, the pyruvic acid is converted into one or more different products, depending on the type of cell. These products might include alcohol (ethanol) and lactic acid. Unlike respiration, there is no Krebs cycle or electron transport chain in fermentation. Accordingly, the ATP yield, which comes only from glycolysis, is much lower.

Glycolysis

Glycolysis, the oxidation of glucose to pyruvic acid, is usually the first stage in carbohydrate catabolism. Most microorganisms use this pathway; in fact, it occurs in most living cells.

Glycolysis is also called the *Embden-Meyerhof pathway.* The word *glycolysis* means splitting of sugar, and this is exactly what happens. The enzymes of glycolysis catalyze the splitting of glucose, a six-carbon sugar, into two three-carbon sugars. These sugars are then oxidized, releasing energy, and their atoms are rearranged to form two molecules of pyruvic acid. During glycolysis NAD^+ is reduced to NADH, and there is a net production of two ATP molecules by substrate-level phosphorylation. Glycolysis does not require oxygen; it can occur whether oxygen is present or not. This pathway is a series of ten chemical reactions, each catalyzed by a different enzyme. The steps are outlined in **Figure 5.12**; see also Figure A.2 in Appendix A for a more detailed representation of glycolysis.

To summarize the process, glycolysis consists of two basic stages, a preparatory stage and an energy-conserving stage:

1. First, in the preparatory stage (steps 1–4 in Figure 5.12), two molecules of ATP are used as a six-carbon glucose molecule is phosphorylated, restructured, and split into two three-carbon compounds: glyceraldehyde 3-phosphate (GP) and dihydroxyacetone phosphate (DHAP). 5 DHAP is readily converted to GP. (The reverse reaction may also occur.) The conversion of DHAP into GP means that from this point on in glycolysis, two molecules of GP are fed into the remaining chemical reactions.
2. In the energy-conserving stage (steps 6–10 in Figure 5.12), the two three-carbon molecules are oxidized in several steps to two molecules of pyruvic acid. In these reactions, two molecules of NAD^+ are reduced to NADH, and four molecules of ATP are formed by substrate-level phosphorylation.

Because two molecules of ATP were needed to get glycolysis started and four molecules of ATP are generated by the process, *there is a net gain of two molecules of ATP for each molecule of glucose that is oxidized.* **Animations** Glycolysis: Overview, Steps. **www.microbiologyplace.com**

Figure 5.11

FOUNDATION FIGURE An Overview of Respiration and Fermentation

A small version of this overview figure will be included in other figures throughout the chapter to indicate the relationships of different reactions to the overall processes of respiration and fermentation.

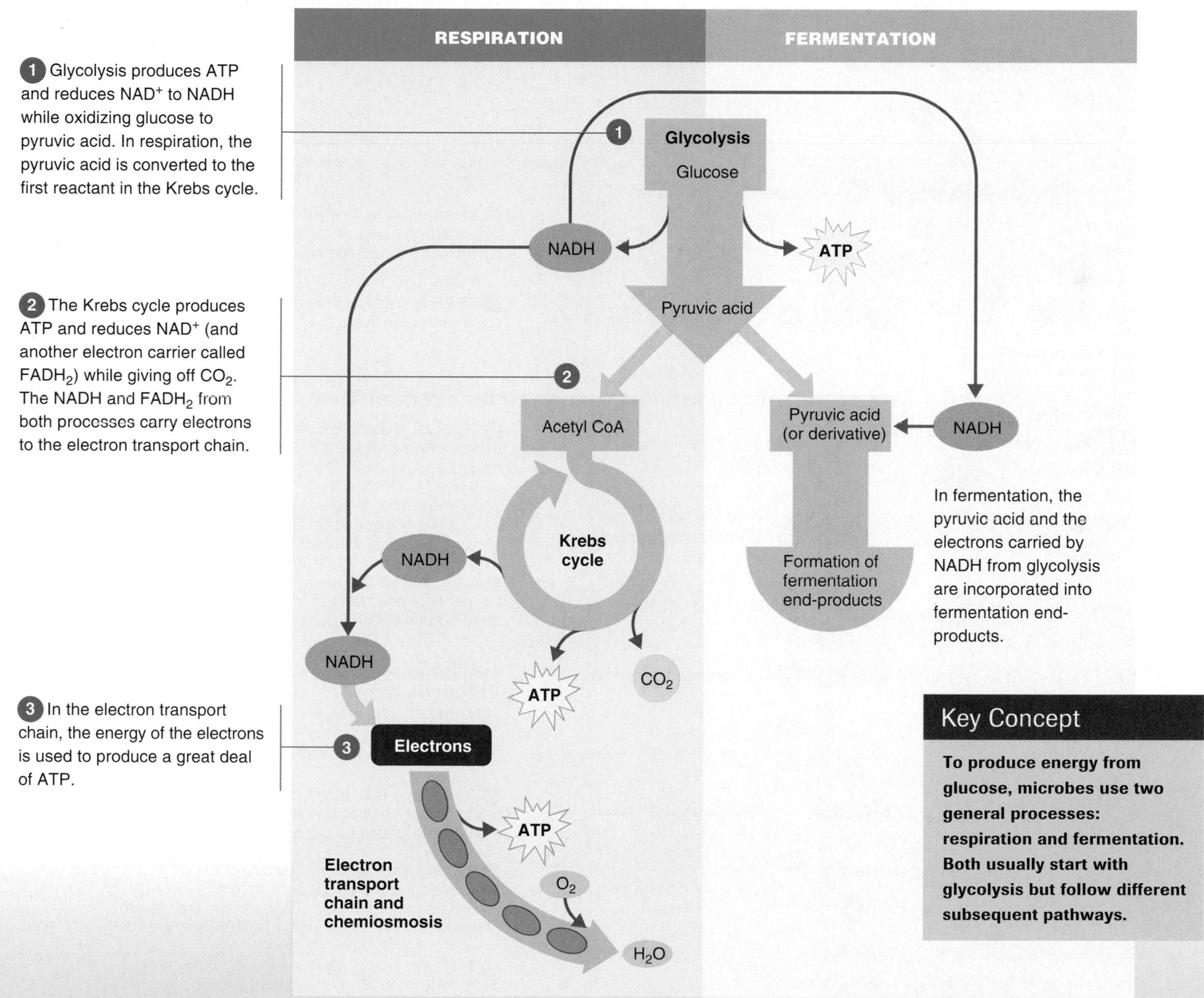

1 Glycolysis produces ATP and reduces NAD^+ to NADH while oxidizing glucose to pyruvic acid. In respiration, the pyruvic acid is converted to the first reactant in the Krebs cycle.
2 The Krebs cycle produces ATP and reduces NAD^+ (and another electron carrier called $FADH_2$) while giving off CO_2. The NADH and $FADH_2$ from both processes carry electrons to the electron transport chain.
3 In the electron transport chain, the energy of the electrons is used to produce a great deal of ATP.
RESPIRATION
FERMENTATION
Glycolysis
Glucose
NADH
ATP
Pyruvic acid
Acetyl CoA
Pyruvic acid (or derivative)
NADH
Krebs cycle
NADH
Formation of fermentation end-products
In fermentation, the pyruvic acid and the electrons carried by NADH from glycolysis are incorporated into fermentation end-products.
NADH
ATP
CO_2
Electrons
ATP
Electron transport chain and chemiosmosis
O_2
H_2O

Key Concept

To produce energy from glucose, microbes use two general processes: respiration and fermentation. Both usually start with glycolysis but follow different subsequent pathways.

Alternatives to Glycolysis

Many bacteria have another pathway in addition to glycolysis for the oxidation of glucose. The most common alternative is the pentose phosphate pathway; another alternative is the Entner-Doudoroff pathway.

The Pentose Phosphate Pathway

The **pentose phosphate pathway** (or *hexose monophosphate shunt*) operates simultaneously with glycolysis and provides a means for the breakdown of five-carbon sugars (pentoses) as well as glucose (see Figure A.3 in Appendix A for a more detailed representation of

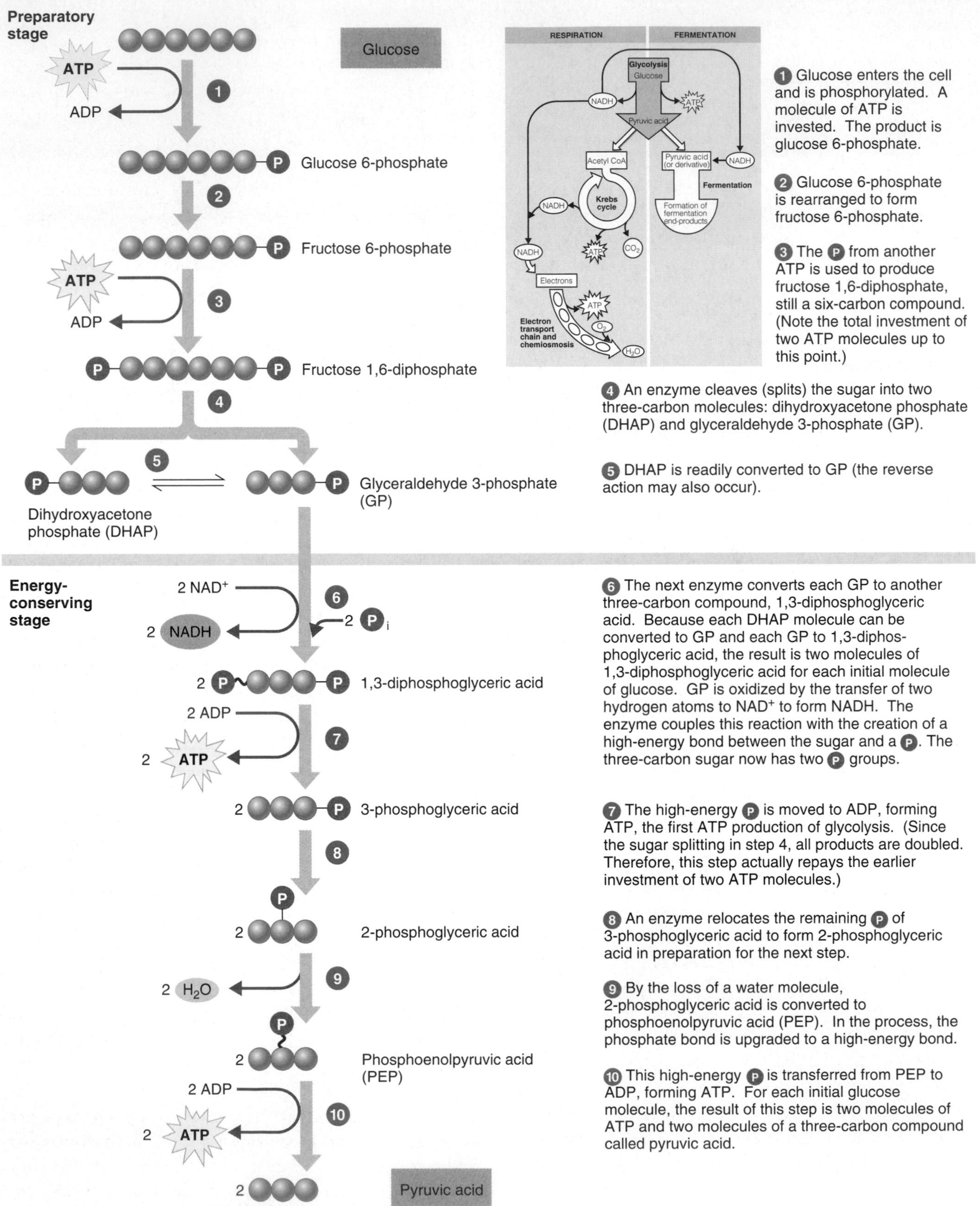

Preparatory stage
Glucose
ATP
ADP
1
Glucose 6-phosphate
2
Fructose 6-phosphate
ATP
ADP
3
Fructose 1,6-diphosphate
4
5
Glyceraldehyde 3-phosphate (GP)
Dihydroxyacetone phosphate (DHAP)
RESPIRATION
FERMENTATION
Glycolysis
Glucose
NADH
ATP
Pyruvic acid
Acetyl CoA
Pyruvic acid (or derivative)
NADH
Fermentation
Krebs cycle
Formation of fermentation end-products
NADH
NADH
ATP
CO_2
Electrons
ATP
Electron transport chain and chemiosmosis
O_2
H_2O
1 Glucose enters the cell and is phosphorylated. A molecule of ATP is invested. The product is glucose 6-phosphate.
2 Glucose 6-phosphate is rearranged to form fructose 6-phosphate.
3 The P from another ATP is used to produce fructose 1,6-diphosphate, still a six-carbon compound. (Note the total investment of two ATP molecules up to this point.)
4 An enzyme cleaves (splits) the sugar into two three-carbon molecules: dihydroxyacetone phosphate (DHAP) and glyceraldehyde 3-phosphate (GP).
5 DHAP is readily converted to GP (the reverse action may also occur).
Energy-conserving stage
2 NAD^+
2 NADH
6
2 P_i
2 1,3-diphosphoglyceric acid
2 ADP
2 ATP
7
2 3-phosphoglyceric acid
8
2 2-phosphoglyceric acid
2 H_2O
9
2 Phosphoenolpyruvic acid (PEP)
2 ADP
2 ATP
10
2 Pyruvic acid
6 The next enzyme converts each GP to another three-carbon compound, 1,3-diphosphoglyceric acid. Because each DHAP molecule can be converted to GP and each GP to 1,3-diphosphoglyceric acid, the result is two molecules of 1,3-diphosphoglyceric acid for each initial molecule of glucose. GP is oxidized by the transfer of two hydrogen atoms to NAD^+ to form NADH. The enzyme couples this reaction with the creation of a high-energy bond between the sugar and a P. The three-carbon sugar now has two P groups.
7 The high-energy P is moved to ADP, forming ATP, the first ATP production of glycolysis. (Since the sugar splitting in step 4, all products are doubled. Therefore, this step actually repays the earlier investment of two ATP molecules.)
8 An enzyme relocates the remaining P of 3-phosphoglyceric acid to form 2-phosphoglyceric acid in preparation for the next step.
9 By the loss of a water molecule, 2-phosphoglyceric acid is converted to phosphoenolpyruvic acid (PEP). In the process, the phosphate bond is upgraded to a high-energy bond.
10 This high-energy P is transferred from PEP to ADP, forming ATP. For each initial glucose molecule, the result of this step is two molecules of ATP and two molecules of a three-carbon compound called pyruvic acid.

Figure 5.12 An outline of the reactions of glycolysis (Embden-Meyerhof pathway). The inset indicates the relationship of glycolysis to the overall processes of respiration and fermentation. A more detailed version of glycolysis is presented in Figure A.2 in Appendix A.

Q What is glycolysis?

the pentose phosphate pathway). A key feature of this pathway is that it produces important intermediate pentoses used in the synthesis of (1) nucleic acids, (2) glucose from carbon dioxide in photosynthesis, and (3) certain amino acids. The pathway is an important producer of the reduced coenzyme NADPH from $NADP^+$. The pentose phosphate pathway yields a net gain of only one molecule of ATP for each molecule of glucose oxidized. Bacteria that use the pentose phosphate pathway include *Bacillus subtilis* (sub'til-us), *E. coli, Leuconostoc mesenteroides* (lü-kō-nos'tok mes-en-ter-oi'dēz), and *Enterococcus faecalis* (fē-kāl'is).

The Entner-Doudoroff Pathway

From each molecule of glucose, the **Entner-Doudoroff pathway** produces two molecules of NADPH and one molecule of ATP for use in cellular biosynthetic reactions (see Figure A.4 in Appendix A for a more detailed representation). Bacteria that have the enzymes for the Entner-Doudoroff pathway can metabolize glucose without either glycolysis or the pentose phosphate pathway. The Entner-Doudoroff pathway is found in some gram-negative bacteria, including *Rhizobium, Pseudomonas* (sū-dō-mō'nas), and *Agrobacterium* (ag-rō-bak-ti'rē-um); it is generally not found among gram-positive bacteria. Tests for the ability to oxidize glucose by this pathway are sometimes used to identify *Pseudomonas* in the clinical laboratory.

CHECK YOUR UNDERSTANDING

- ✓ What happens during the preparatory and energy-conserving stages of glycolysis? **5-11**
- ✓ What is the value of the pentose phosphate and Entner-Doudoroff pathways if they produce only one ATP molecule? **5-12**

Cellular Respiration

After glucose has been broken down to pyruvic acid, the pyruvic acid can be channeled into the next step of either fermentation (page 132) or cellular respiration (see Figure 5.11). **Cellular respiration,** or simply **respiration,** is defined as an ATP-generating process in which molecules are oxidized and the final electron acceptor is (almost always) an inorganic molecule. An essential feature of respiration is the operation of an electron transport chain.

There are two types of respiration, depending on whether an organism is an **aerobe,** which uses oxygen, or an **anaerobe,** which does not use oxygen and may even be killed by it. In **aerobic respiration,** the final electron acceptor is O_2; in **anaerobic respiration,** the final electron acceptor is an inorganic molecule other than O_2 or, rarely, an organic molecule. First we will describe respiration as it typically occurs in an aerobic cell.

Aerobic Respiration

The Krebs Cycle The **Krebs cycle,** also called the *tricarboxylic acid (TCA) cycle* or *citric acid cycle,* is a series of biochemical reactions in which the large amount of potential chemical energy stored in acetyl CoA is released step by step (see Figure 5.11). In this cycle, a series of oxidations and reductions transfer that potential energy, in the form of electrons, to electron carrier coenzymes, chiefly NAD^+. The pyruvic acid derivatives are oxidized; the coenzymes are reduced.

Pyruvic acid, the product of glycolysis, cannot enter the Krebs cycle directly. In a preparatory step, it must lose one molecule of CO_2 and become a two-carbon compound (**Figure 5.13**, at top). This process is called **decarboxylation.** The two-carbon compound, called an *acetyl group,* attaches to coenzyme A through a high-energy bond; the resulting complex is known as *acetyl coenzyme A (acetyl CoA).* During this reaction, pyruvic acid is also oxidized and NAD^+ is reduced to NADH.

Remember that the oxidation of one glucose molecule produces two molecules of pyruvic acid, so for each molecule of glucose, two molecules of CO_2 are released in this preparatory step, two molecules of NADH are produced, and two molecules of acetyl CoA are formed. Once the pyruvic acid has undergone decarboxylation and its derivative (the acetyl group) has attached to CoA, the resulting acetyl CoA is ready to enter the Krebs cycle.

As acetyl CoA enters the Krebs cycle, CoA detaches from the acetyl group. The two-carbon acetyl group combines with a four-carbon compound called oxaloacetic acid to form the six-carbon citric acid. This synthesis reaction requires energy, which is provided by the cleavage of the high-energy bond between the acetyl group and CoA. The formation of citric acid is thus the first step in the Krebs cycle. The major chemical reactions of this cycle are outlined in Figure 5.13; a more detailed representation of the Krebs cycle is provided in Figure A.5 in Appendix A. Keep in mind that each reaction is catalyzed by a specific enzyme.

The chemical reactions of the Krebs cycle fall into several general categories; one of these is decarboxylation. For example, in step ❸ isocitric acid, a six-carbon compound, is decarboxylated to the five-carbon compound called α-ketoglutaric acid. Another decarboxylation takes place in step ❹. Because one decarboxylation has taken place in the preparatory step and two in the Krebs cycle, all three carbon atoms in pyruvic acid are eventually released as CO_2 by the Krebs cycle. This represents the conversion to CO_2 of all six carbon atoms contained in the original glucose molecule.

Another general category of Krebs cycle chemical reactions is oxidation-reduction. For example, in step ❸, two hydrogen atoms are lost during the conversion of the six-carbon isocitric acid to a five-carbon compound. In other words, the six-carbon compound is oxidized. Hydrogen atoms are also released in the Krebs cycle in steps ❹, ❻, and ❽ and are picked up by the coenzymes NAD^+ and FAD. Because NAD^+ picks up two electrons but only one additional proton, its reduced form is represented as NADH; however, FAD picks up two complete hydrogen atoms and is reduced to $FADH_2$.

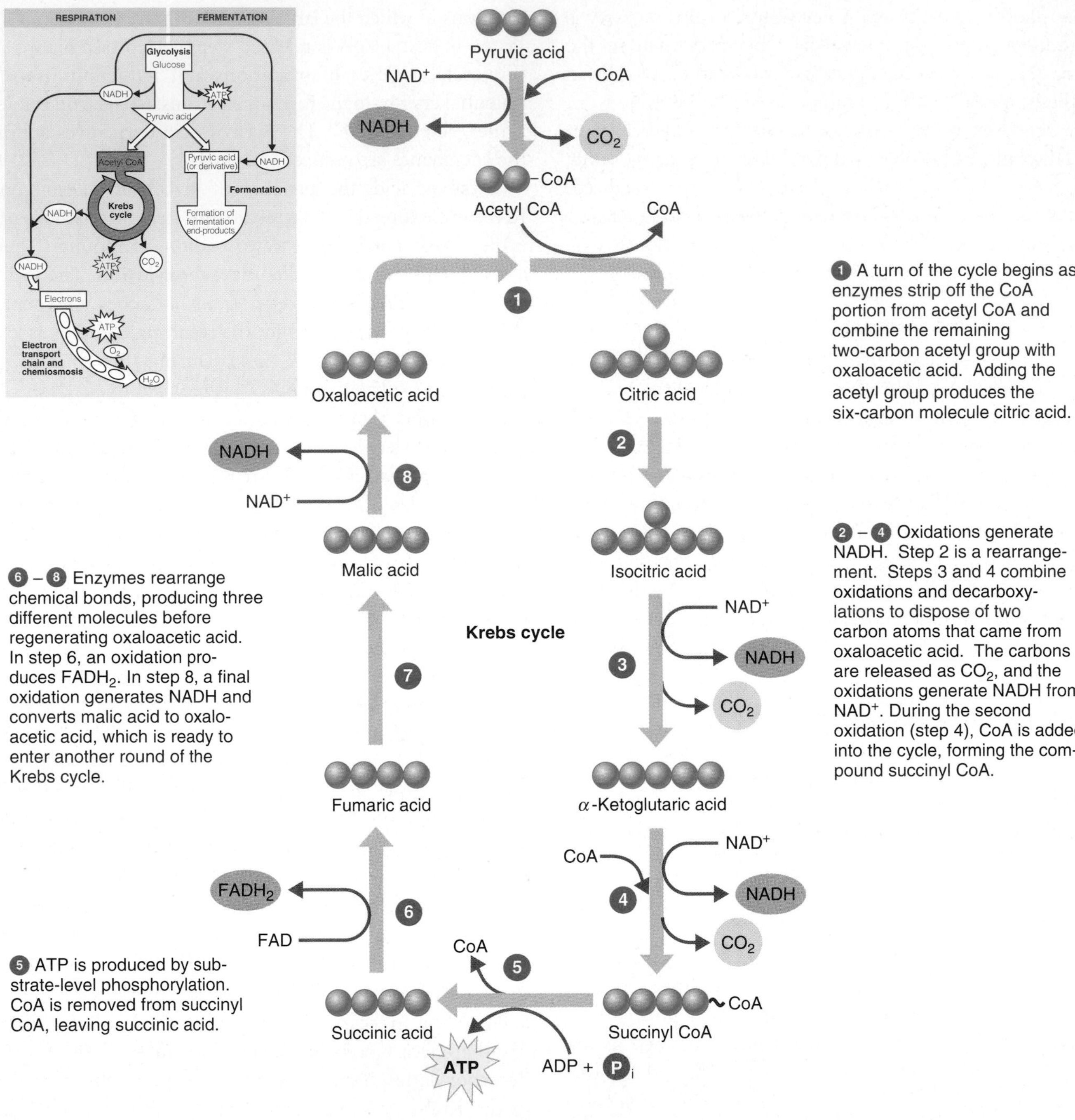
RESPIRATION
FERMENTATION
Glycolysis
Glucose
NADH
ATP
Pyruvic acid
Acetyl CoA
Pyruvic acid (or derivative)
NADH
Fermentation
Krebs cycle
Formation of fermentation end-products
NADH
ATP
CO2
NADH
Electrons
ATP
Electron transport chain and chemiosmosis
O2
H2O
Pyruvic acid
NAD+
CoA
NADH
CO2
CoA
Acetyl CoA
CoA
1
Oxaloacetic acid
Citric acid
1 A turn of the cycle begins as enzymes strip off the CoA portion from acetyl CoA and combine the remaining two-carbon acetyl group with oxaloacetic acid. Adding the acetyl group produces the six-carbon molecule citric acid.
NADH
8
NAD+
2
Malic acid
Isocitric acid
2 – 4 Oxidations generate NADH. Step 2 is a rearrangement. Steps 3 and 4 combine oxidations and decarboxylations to dispose of two carbon atoms that came from oxaloacetic acid. The carbons are released as CO2, and the oxidations generate NADH from NAD+. During the second oxidation (step 4), CoA is added into the cycle, forming the compound succinyl CoA.
6 – 8 Enzymes rearrange chemical bonds, producing three different molecules before regenerating oxaloacetic acid. In step 6, an oxidation produces FADH2. In step 8, a final oxidation generates NADH and converts malic acid to oxaloacetic acid, which is ready to enter another round of the Krebs cycle.
Krebs cycle
NAD+
NADH
3
7
CO2
Fumaric acid
α-Ketoglutaric acid
NAD+
CoA
FADH2
NADH
4
6
FAD
CO2
CoA
5
5 ATP is produced by substrate-level phosphorylation. CoA is removed from succinyl CoA, leaving succinic acid.
CoA
Succinic acid
Succinyl CoA
ATP
ADP + Pi

Figure 5.13 The Krebs cycle. The inset indicates the relationship of the Krebs cycle to the overall process of respiration. A more detailed version of the Krebs cycle is presented in Figure A.5 in Appendix A.

Q What are the products of the Krebs cycle?

If we look at the Krebs cycle as a whole, we see that for every two molecules of acetyl CoA that enter the cycle, four molecules of CO_2 are liberated by decarboxylation, six molecules of NADH and two molecules of $FADH_2$ are produced by oxidation-reduction reactions, and two molecules of ATP are generated by substrate-level phosphorylation. Many of the intermediates in the Krebs cycle also play a role in other pathways, especially in amino acid biosynthesis (page 146).

The CO_2 produced in the Krebs cycle is ultimately liberated into the atmosphere as a gaseous by-product of aerobic respiration. (Humans produce CO_2 from the Krebs cycle in most cells of the body and discharge it through the lungs during exhalation.) The reduced coenzymes NADH and $FADH_2$ are the most important products of the Krebs cycle because they contain most of the energy originally stored in glucose. During the next phase of respiration, a series of reductions indirectly transfers the

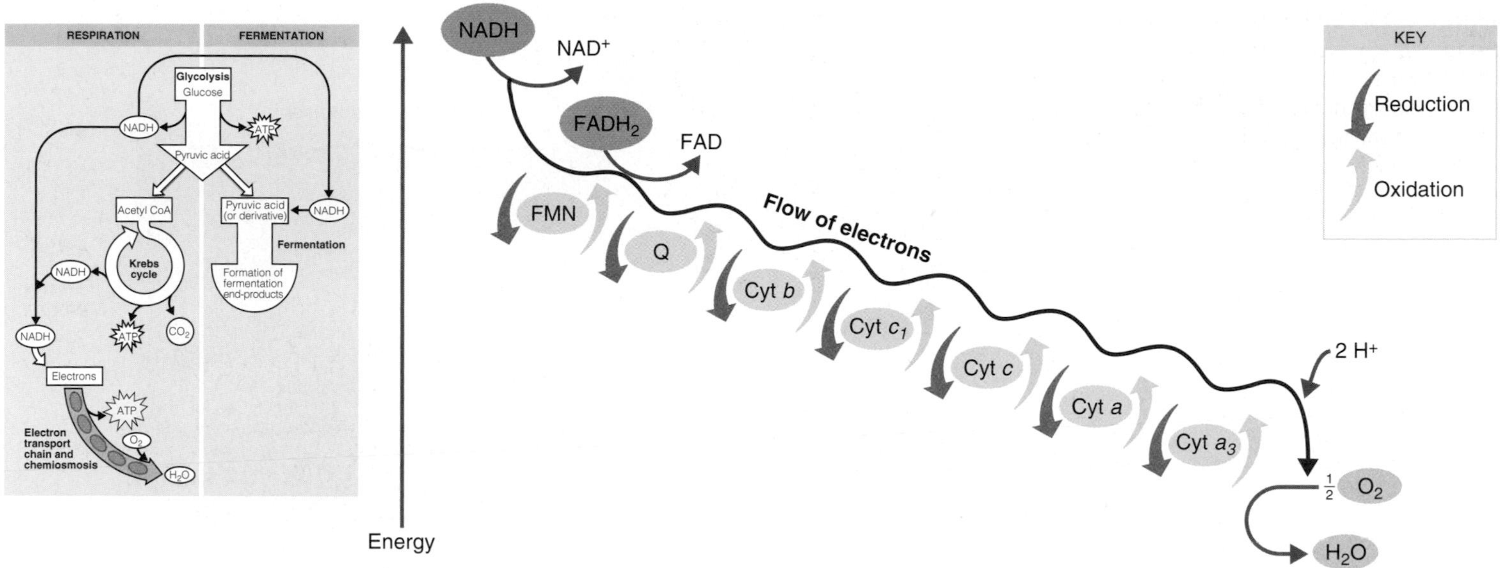

Figure 5.14 An electron transport chain (system). The inset indicates the relationship of the electron transport chain to the overall process of respiration. In the mitochondrial electron transport chain shown, the electrons pass along the chain in a gradual and stepwise fashion, so energy is released in manageable quantities. To learn where ATP is formed, see Figure 5.16.

Q What are the functions of the electron transport chain?

energy stored in those coenzymes to ATP. These reactions are collectively called the electron transport chain. **Animations** Krebs Cycle: Overview, Steps. **www.microbiologyplace.com**

The Electron Transport Chain (System) An **electron transport chain (system)** consists of a sequence of carrier molecules that are capable of oxidation and reduction. As electrons are passed through the chain, there occurs a stepwise release of energy, which is used to drive the chemiosmotic generation of ATP, to be described shortly. The final oxidation is irreversible. In eukaryotic cells, the electron transport chain is contained in the inner membrane of mitochondria; in prokaryotic cells, it is found in the plasma membrane.

There are three classes of carrier molecules in electron transport chains. The first are **flavoproteins.** These proteins contain flavin, a coenzyme derived from riboflavin (vitamin B_2), and are capable of performing alternating oxidations and reductions. One important flavin coenzyme is flavin mononucleotide (FMN). The second class of carrier molecules are **cytochromes,** proteins with an iron-containing group (heme) capable of existing alternately as a reduced form (Fe^{2+}) and an oxidized form (Fe^{3+}). The cytochromes involved in electron transport chains include cytochrome *b* (cyt *b*), cytochrome c_1 (cyt c_1), cytochrome *c* (cyt *c*), cytochrome *a* (cyt *a*), and cytochrome a_3 (cyt a_3). The third class is known as **ubiquinones,** or **coenzyme Q,** symbolized Q; these are small nonprotein carriers.

The electron transport chains of bacteria are somewhat diverse, in that the particular carriers used by a bacterium and the order in which they function may differ from those of other bacteria and from those of eukaryotic mitochondrial systems. Even a single bacterium may have several types of electron transport chains. However, keep in mind that all electron transport chains achieve the same basic goal: to release energy as electrons are transferred from higher-energy compounds to lower-energy compounds. Much is known about the electron transport chain in the mitochondria of eukaryotic cells, so this is the chain we will describe.

The first step in the mitochondrial electron transport chain involves the transfer of high-energy electrons from NADH to FMN, the first carrier in the chain (**Figure 5.14**). This transfer actually involves the passage of a hydrogen atom with two electrons to FMN, which then picks up an additional H^+ from the surrounding aqueous medium. As a result of the first transfer, NADH is oxidized to NAD^+, and FMN is reduced to $FMNH_2$. In the second step in the electron transport chain, $FMNH_2$ passes $2H^+$ to the other side of the mitochondrial membrane (see Figure 5.16) and passes two electrons to Q. As a result, $FMNH_2$ is oxidized to FMN. Q also picks up an additional $2H^+$ from the surrounding aqueous medium and releases it on the other side of the membrane.

The next part of the electron transport chain involves the cytochromes. Electrons are passed successively from Q to cyt *b*, cyt c_1, cyt *c*, cyt *a*, and cyt a_3. Each cytochrome in the chain is reduced as it picks up electrons and is oxidized as it gives up electrons. The last cytochrome, cyt a_3, passes its electrons to molecular

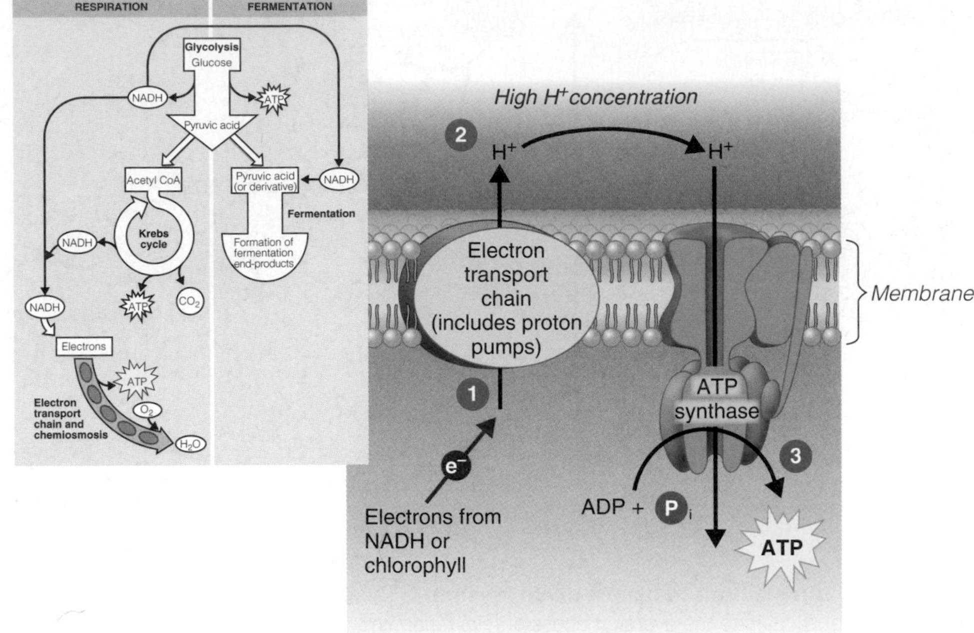

Figure 5.15 Chemiosmosis. An overview of the mechanism of chemiosmosis. The membrane shown could be a prokaryotic plasma membrane, a eukaryotic mitochondrial membrane, or a photosynthetic thylakoid. The numbered steps are described in the text.

Q What is the proton motive force?

oxygen (O_2), which becomes negatively charged and then picks up protons from the surrounding medium to form H_2O.

Notice that Figure 5.14 shows $FADH_2$, which is derived from the Krebs cycle, as another source of electrons. However, $FADH_2$ adds its electrons to the electron transport chain at a lower level than NADH. Because of this, the electron transport chain produces about one-third less energy for ATP generation when $FADH_2$ donates electrons than when NADH is involved.

An important feature of the electron transport chain is the presence of some carriers, such as FMN and Q, that accept and release protons as well as electrons, and other carriers, such as cytochromes, that transfer electrons only. Electron flow down the chain is accompanied at several points by the active transport (pumping) of protons from the matrix side of the inner mitochondrial membrane to the opposite side of the membrane. The result is a buildup of protons on one side of the membrane. Just as water behind a dam stores energy that can be used to generate electricity, this buildup of protons provides energy for the generation of ATP by the chemiosmotic mechanism.

The Chemiosmotic Mechanism of ATP Generation The mechanism of ATP synthesis using the electron transport chain is called **chemiosmosis.** To understand chemiosmosis, we need to recall several concepts that were introduced in Chapter 4 as part of the section on the movement of materials across membranes (page 91). Recall that substances diffuse passively across membranes from areas of high concentration to areas of low concentration; this diffusion yields energy. Recall also that the movement of substances *against* such a concentration gradient *requires* energy and that, in such an active transport of molecules or ions across biological membranes, the required energy is usually provided by ATP. In chemiosmosis, the energy released when a substance moves along a gradient is used to *synthesize* ATP. The "substance" in this case refers to protons. In respiration, chemiosmosis is responsible for most of the ATP that is generated. The steps of chemiosmosis are as follows (**Figure 5.15** and **Figure 5.16**):

1. As energetic electrons from NADH (or chlorophyll) pass down the electron transport chain, some of the carriers in the chain pump—actively transport—protons across the membrane. Such carrier molecules are called *proton pumps.*
2. The phospholipid membrane is normally impermeable to protons, so this one-directional pumping establishes a proton gradient (a difference in the concentrations of protons on the two sides of the membrane). In addition to a concentration gradient, there is an electrical charge gradient. The excess H^+ on one side of the membrane makes that side positively charged compared with the other side. The resulting electrochemical gradient has potential energy, called the *proton motive force.*
3. The protons on the side of the membrane with the higher proton concentration can diffuse across the membrane only through special protein channels that contain an enzyme called *ATP synthase.* When this flow occurs, energy is released and is used by the enzyme to synthesize ATP from ADP and Ⓟi.

Figure 5.16 shows in detail how the electron transport chain operates in eukaryotes to drive the chemiosmotic mechanism. ❶ Energetic electrons from NADH pass down the electron transport chains. Within the inner mitochondrial membrane, the carriers of the electron transport chain are organized into three complexes, with Q transporting electrons between the first and

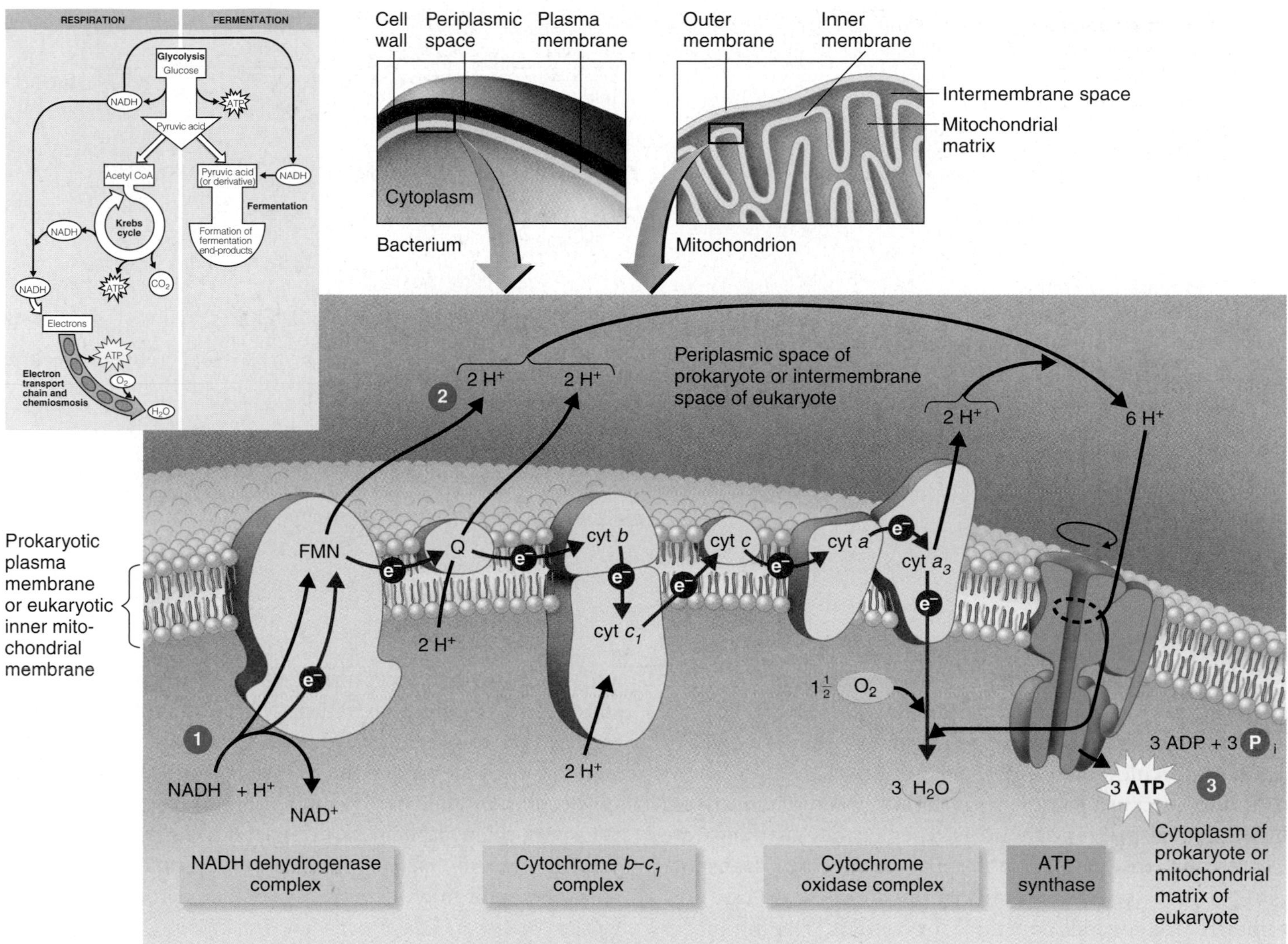

Figure 5.16 Electron transport and the chemiosmotic generation of ATP. Electron carriers are organized into three complexes, and protons (H^+) are pumped across the membrane at three points. In a prokaryotic cell, protons are pumped across the plasma membrane from the cytoplasmic side. In a eukaryotic cell, they are pumped from the matrix side of the mitochondrial membrane to the opposite side. The flow of electrons is indicated with red arrows.

Q Where does chemiosmosis occur in eukaryotes? In prokaryotes?

second complexes, and cyt *c* transporting them between the second and third complexes. ❷ Three components of the system pump protons: the first and third complexes and Q. At the end of the chain, electrons join with protons and oxygen (O_2) in the matrix fluid to form water (H_2O). Thus, O_2 is the final electron acceptor.

Both prokaryotic and eukaryotic cells use the chemiosmotic mechanism to generate energy for ATP production. However, in eukaryotic cells, ❸ the inner mitochondrial membrane contains the electron transport carriers and ATP synthase, whereas in most prokaryotic cells, the plasma membrane does so. An electron transport chain also operates in photophosphorylation and is located in the thylakoid membrane of cyanobacteria and eukaryotic chloroplasts.

A Summary of Aerobic Respiration The electron transport chain regenerates NAD^+ and FAD, which can be used again in glycolysis and the Krebs cycle. The various electron transfers in the electron transport chain generate about 34 molecules of ATP from each molecule of glucose oxidized: approximately three from each of the ten molecules of NADH (a total of 30), and approximately two from each of the two molecules of $FADH_2$ (a total of four). To arrive at the total number of ATP molecules generated for each molecule of glucose, the 34 from chemiosmosis are added to those generated by oxidation in glycolysis and the Krebs cycle. In aerobic respiration among prokaryotes, a total of 38 molecules of ATP can be generated from one molecule of glucose. Note that four of those ATPs come from substrate-level phosphorylation in glycolysis and

Table 5.3 ATP Yield during Prokaryotic Aerobic Respiration of One Glucose Molecule

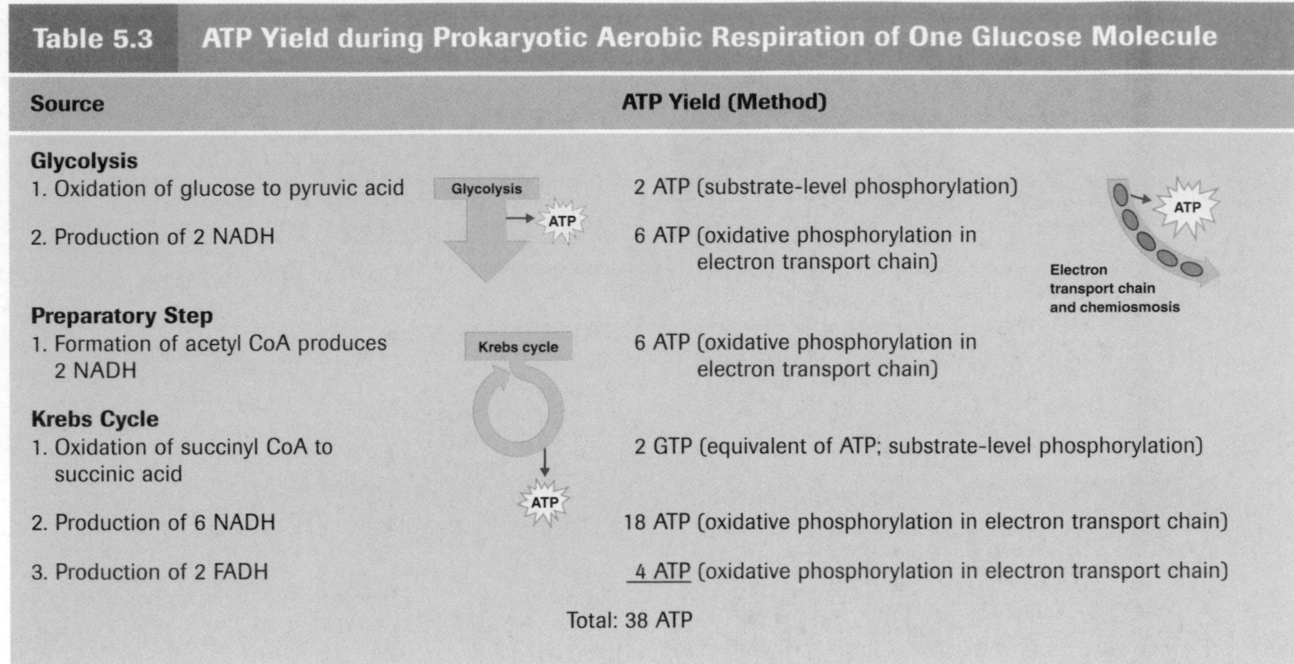

Source	ATP Yield (Method)
Glycolysis	
1. Oxidation of glucose to pyruvic acid	2 ATP (substrate-level phosphorylation)
2. Production of 2 NADH	6 ATP (oxidative phosphorylation in electron transport chain)
Preparatory Step	
1. Formation of acetyl CoA produces 2 NADH	6 ATP (oxidative phosphorylation in electron transport chain)
Krebs Cycle	
1. Oxidation of succinyl CoA to succinic acid	2 GTP (equivalent of ATP; substrate-level phosphorylation)
2. Production of 6 NADH	18 ATP (oxidative phosphorylation in electron transport chain)
3. Production of 2 FADH	4 ATP (oxidative phosphorylation in electron transport chain)
	Total: 38 ATP

the Krebs cycle. **Table 5.3** provides a detailed accounting of the ATP yield during prokaryotic aerobic respiration.

Aerobic respiration among eukaryotes produces a total of only 36 molecules of ATP. There are fewer ATPs than in prokaryotes because some energy is lost when electrons are shuttled across the mitochondrial membranes that separate glycolysis (in the cytoplasm) from the electron transport chain. No such separation exists in prokaryotes. We can now summarize the overall reaction for aerobic respiration in prokaryotes as follows:

$$\underset{\text{Glucose}}{C_6H_{12}O_6} + \underset{\text{Oxygen}}{6\ O_2} + 38\ ADP + 38\ \text{Ⓟ}_i \longrightarrow \underset{\text{Carbon dioxide}}{6\ CO_2} + \underset{\text{Water}}{6\ H_2O} + 38\ ATP$$

A summary of the various stages of aerobic respiration in prokaryotes is presented in **Figure 5.17**.

Anaerobic Respiration

In anaerobic respiration, the final electron acceptor is an inorganic substance other than oxygen (O_2). Some bacteria, such as *Pseudomonas* and *Bacillus,* can use a nitrate ion (NO_3^-) as a final electron acceptor; the nitrate ion is reduced to a nitrite ion (NO_2^-), nitrous oxide (N_2O), or nitrogen gas (N_2). Other bacteria, such as *Desulfovibrio* (dē-sul-fō-vib′rē-ō), use sulfate (SO_4^{2-}) as the final electron acceptor to form hydrogen sulfide (H_2S). Still other bacteria use carbonate (CO_3^{2-}) to form methane (CH_4). Anaerobic respiration by bacteria using nitrate and sulfate as final acceptors is essential for the nitrogen and sulfur cycles that occur in nature. The amount of ATP generated in anaerobic respiration varies with the organism and the pathway. Because only part of the Krebs cycle operates under anaerobic conditions, and because not all the carriers in the electron transport chain participate in anaerobic respiration, the ATP yield is never as high as in aerobic respiration. Accordingly, anaerobes tend to grow more slowly than aerobes. **Animations** Electron Transport Chain: Overview, The Process, Factors Affecting ATP Yield. **www.microbiologyplace.com**

CHECK YOUR UNDERSTANDING

- ✓ What are the principal products of the Krebs cycle? **5-13**
- ✓ How do carrier molecules function in the electron transport chain? **5-14**
- ✓ Compare the energy yield (ATP) of aerobic and anaerobic respiration. **5-15**

Fermentation

After glucose has been broken down into pyruvic acid, the pyruvic acid can be completely broken down in respiration, as previously described, or it can be converted to an organic product in fermentation, whereupon NAD^+ and $NADP^+$ are regenerated and can enter another round of glycolysis (see Figure 5.11). **Fermentation** can be defined in several ways (see the box, page 135), but we define it here as a process that

1. releases energy from sugars or other organic molecules, such as amino acids, organic acids, purines, and pyrimidines;

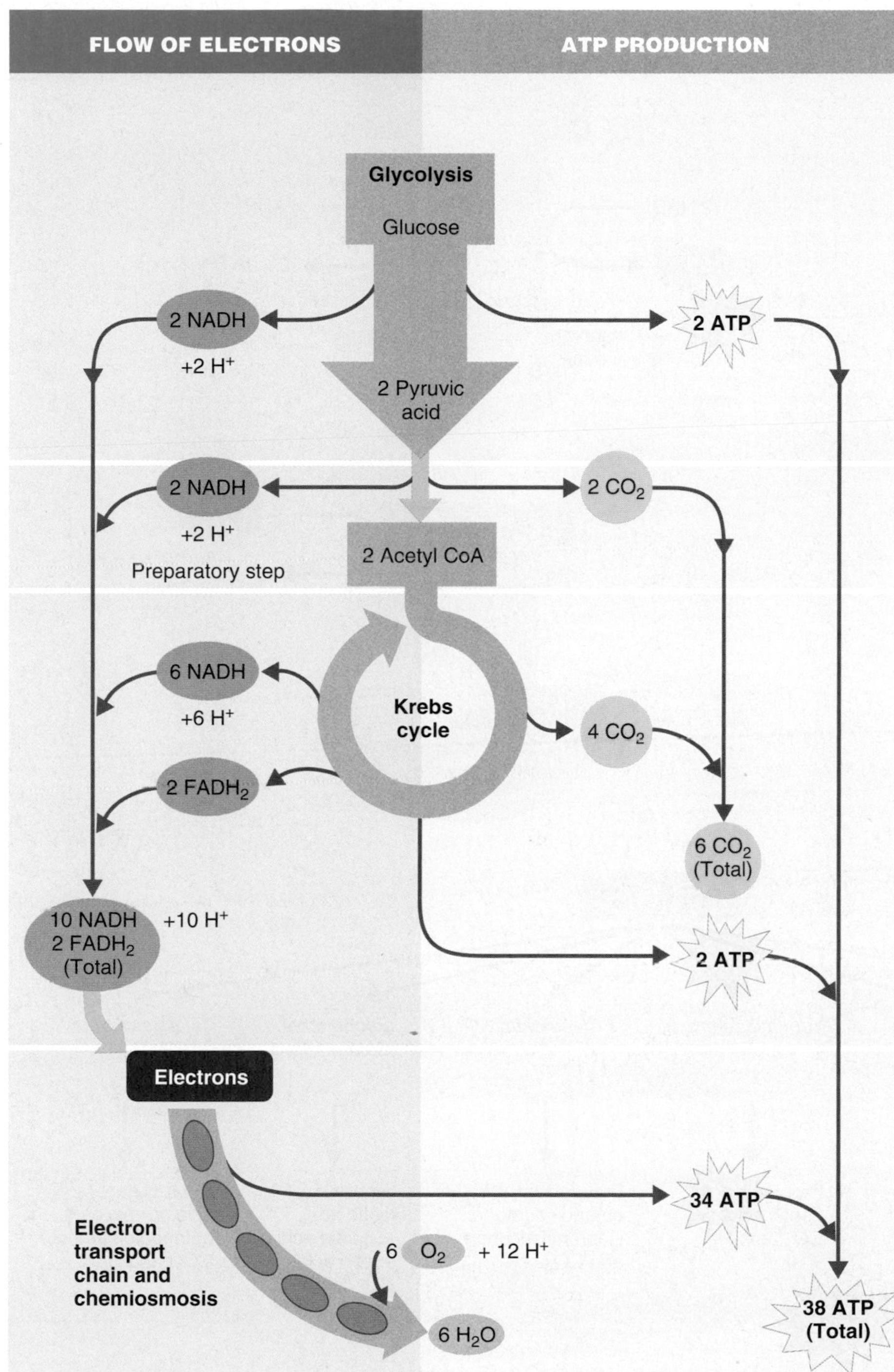

Figure 5.17 A summary of aerobic respiration in prokaryotes. Glucose is broken down completely to carbon dioxide and water, and ATP is generated. This process has three major phases: glycolysis, the Krebs cycle, and the electron transport chain. The preparatory step is between glycolysis and the Krebs cycle. The key event in aerobic respiration is that electrons are picked up from intermediates of glycolysis and the Krebs cycle by NAD^+ or FAD and are carried by NADH or $FADH_2$ to the electron transport chain. NADH is also produced during the conversion of pyruvic acid to acetyl CoA. Most of the ATP generated by aerobic respiration is made by the chemiosmotic mechanism during the electron transport chain phase; this is called oxidative phosphorylation.

Q How do aerobic and anaerobic respiration differ?

2. does not require oxygen (but sometimes can occur in its presence);
3. does not require the use of the Krebs cycle or an electron transport chain;
4. uses an organic molecule as the final electron acceptor;
5. produces only small amounts of ATP (only one or two ATP molecules for each molecule of starting material) because much of the original energy in glucose remains in the chemical bonds of the organic end-products, such as lactic acid or ethanol.

During fermentation, electrons are transferred (along with protons) from reduced coenzymes (NADH, NADPH) to pyruvic acid or its derivatives (**Figure 5.18a**). Those final electron acceptors are reduced to the end-products shown in **Figure 5.18b**. An essential function of the second stage of

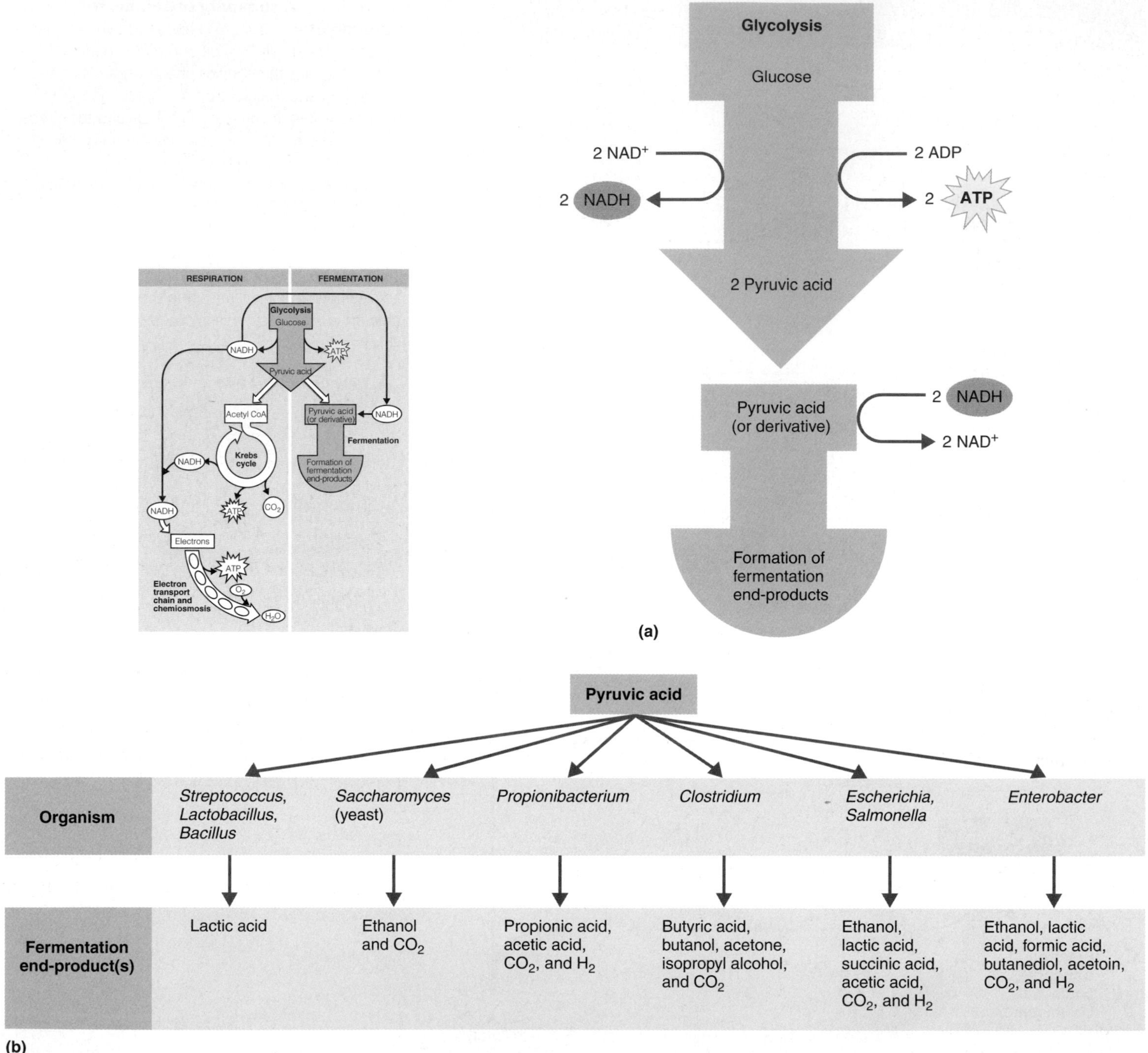

Organism	*Streptococcus, Lactobacillus, Bacillus*	*Saccharomyces* (yeast)	*Propionibacterium*	*Clostridium*	*Escherichia, Salmonella*	*Enterobacter*
Fermentation end-product(s)	Lactic acid	Ethanol and CO_2	Propionic acid, acetic acid, CO_2, and H_2	Butyric acid, butanol, acetone, isopropyl alcohol, and CO_2	Ethanol, lactic acid, succinic acid, acetic acid, CO_2, and H_2	Ethanol, lactic acid, formic acid, butanediol, acetoin, CO_2, and H_2

(b)

Figure 5.18 Fermentation. The inset indicates the relationship of fermentation to the overall energy-producing processes. (**a**) An overview of fermentation. The first step is glycolysis, the conversion of glucose to pyruvic acid. In the second step, the reduced coenzymes from glycolysis or its alternatives (NADH, NADPH) donate their electrons and hydrogen ions to pyruvic acid or a derivative to form a fermentation end-product. (**b**) End-products of various microbial fermentations.

Q During which phase of fermentation is ATP generated?

fermentation is to ensure a steady supply of NAD^+ and $NADP^+$ so that glycolysis can continue. In fermentation, ATP is generated only during glycolysis.

Microorganisms can ferment various substrates; the end-products depend on the particular microorganism, the substrate, and the enzymes that are present and active. Chemical analyses of these end-products are useful in identifying microorganisms. We next consider two of the more important processes: lactic acid fermentation and alcohol fermentation.

What Is Fermentation?

To many people, *fermentation* simply means the production of alcohol: grains and fruits are fermented to produce beer and wine. If a food soured, you might say it was "off" or fermented. Here are some definitions of *fermentation.* They range from informal, general usage to more scientific definitions.

1. Any spoilage of food by microorganisms (general use)
2. Any process that produces alcoholic beverages or acidic dairy products (general use)
3. Any large-scale microbial process occurring with or without air (common definition used in industry)
4. Any energy-releasing metabolic process that takes place only under anaerobic conditions (becoming more scientific)
5. Any metabolic process that releases energy from a sugar or other organic molecule, does not require oxygen or an electron transport system, and uses an organic molecule as the final electron acceptor (this is the definition we use in this book)

Lactic Acid Fermentation

During glycolysis, which is the first phase of **lactic acid fermentation,** a molecule of glucose is oxidized to two molecules of pyruvic acid (Figure 5.19; see also Figure 5.10). This oxidation generates the energy that is used to form the two molecules of ATP. In the next step, the two molecules of pyruvic acid are reduced by two molecules of NADH to form two molecules of lactic acid (Figure 5.19a). Because lactic acid is the end-product of the reaction, it undergoes no further oxidation, and most of the energy produced by the reaction remains stored in the lactic acid. Thus, this fermentation yields only a small amount of energy.

Two important genera of lactic acid bacteria are *Streptococcus* and *Lactobacillus* (lak-tō-bä-sil′lus). Because these microbes produce only lactic acid, they are referred to as **homolactic** (or *homofermentative*). Lactic acid fermentation can result in food spoilage. However, the process can also produce yogurt from milk, sauerkraut from fresh cabbage, and pickles from cucumbers.

Alcohol Fermentation

Alcohol fermentation also begins with the glycolysis of a molecule of glucose to yield two molecules of pyruvic acid and two molecules of ATP. In the next reaction, the two molecules of pyruvic acid are converted to two molecules of acetaldehyde and two molecules of CO_2 (Figure 5.19b). The two molecules of acetaldehyde are next reduced by two molecules of NADH to form two molecules of ethanol. Again, alcohol fermentation is a low-energy-yield process because most of the energy contained in the initial glucose molecule remains in the ethanol, the end-product.

Alcohol fermentation is carried out by a number of bacteria and yeasts. The ethanol and carbon dioxide produced by the yeast *Saccharomyces* (sak-ä-rō-mī′sēs) are waste products for yeast cells but are useful to humans. Ethanol made by yeasts is the alcohol in alcoholic beverages, and carbon dioxide made by yeasts causes bread dough to rise.

Organisms that produce lactic acid as well as other acids or alcohols are known as **heterolactic** (or *heterofermentative*) and often use the pentose phosphate pathway.

Table 5.4 lists some of the various microbial fermentations used by industry to convert inexpensive raw materials into useful end-products. Table 5.5 provides a summary comparison of aerobic respiration, anaerobic respiration, and fermentation.
Animation Fermentation. **www.microbiologyplace.com**

CHECK YOUR UNDERSTANDING

✓ List four compounds that can be made from pyruvic acid by an organism that uses fermentation. **5-16**

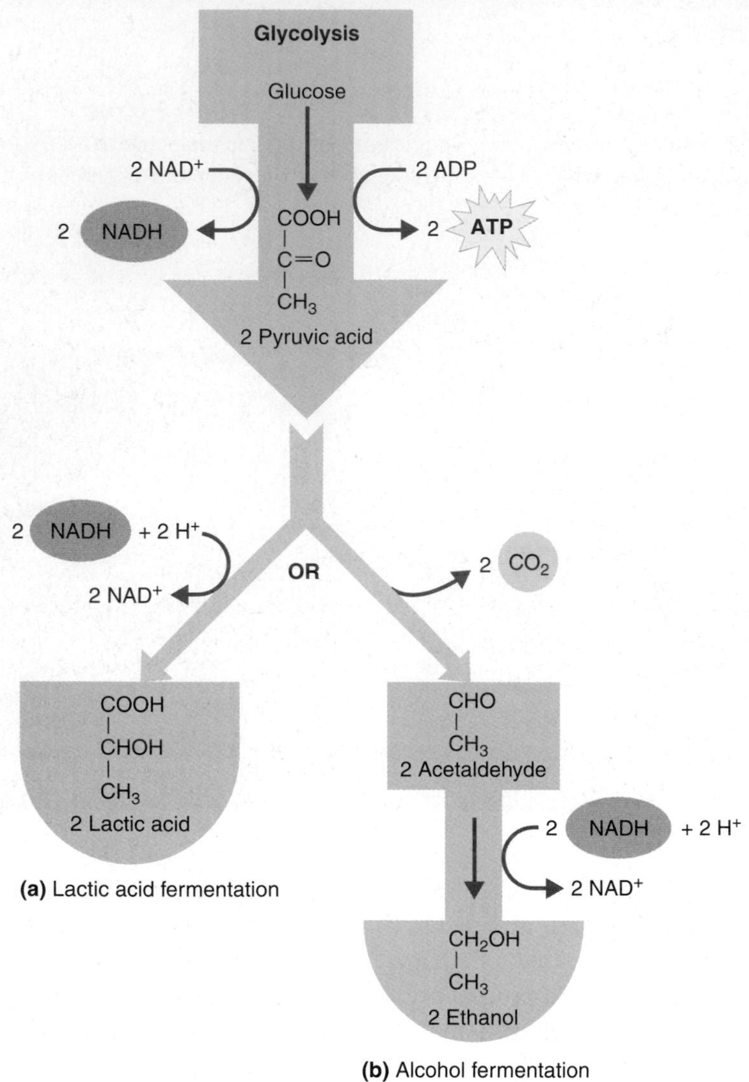

Figure 5.19 Types of fermentation.

Q What is the difference between homolactic and heterolactic fermentation?

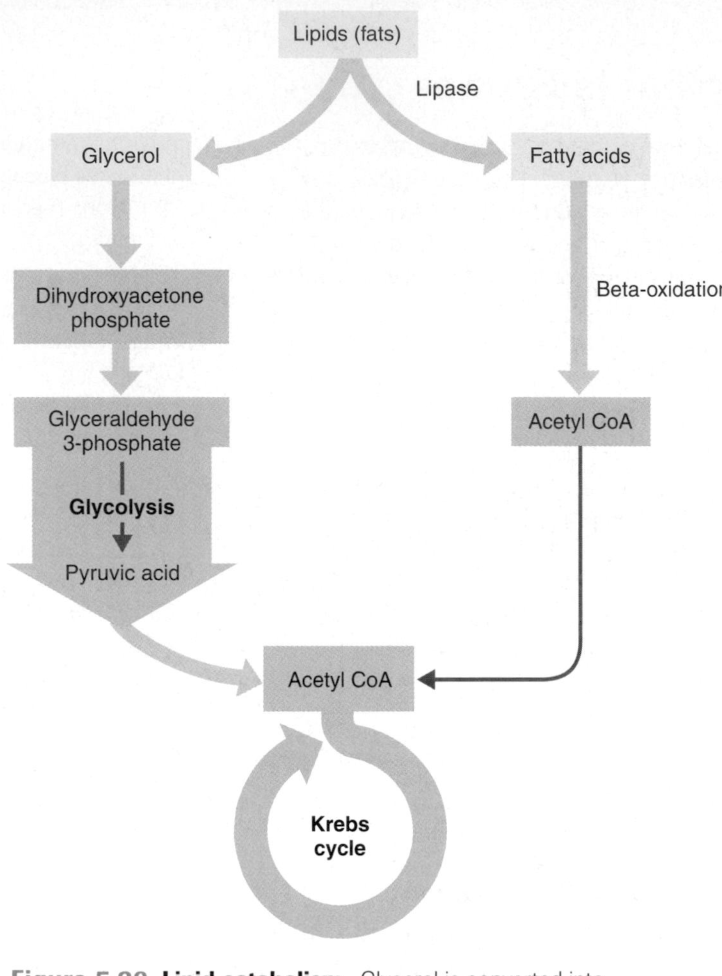

Figure 5.20 Lipid catabolism. Glycerol is converted into dihydroxyacetone phosphate (DHAP) and catabolized via glycolysis and the Krebs cycle. Fatty acids undergo beta-oxidation, in which carbon fragments are split off two at a time to form acetyl CoA, which is catabolized via the Krebs cycle.

Q What is the role of lipases?

Lipid and Protein Catabolism

LEARNING OBJECTIVE

5-17 Describe how lipids and proteins undergo catabolism.

Our discussion of energy production has emphasized the oxidation of glucose, the main energy-supplying carbohydrate. However, microbes also oxidize lipids and proteins, and the oxidations of all these nutrients are related.

Recall that fats are lipids consisting of fatty acids and glycerol. Microbes produce extracellular enzymes called *lipases* that break fats down into their fatty acid and glycerol components. Each component is then metabolized separately (Figure 5.20). The Krebs cycle functions in the oxidation of glycerol and fatty acids. Many bacteria that hydrolyze fatty acids can use the same enzymes to degrade petroleum products. Although these bacteria are a nuisance when they grow in a fuel storage tank, they are beneficial when they grow in oil spills. Beta-oxidation (the oxidation of fatty acids) of petroleum is illustrated in the box in Chapter 2 (page 33).

Proteins are too large to pass unaided through plasma membranes. Microbes produce extracellular *proteases* and *peptidases,* enzymes that break down proteins into their component amino acids, which can cross the membranes. However, before amino acids can be catabolized, they must be enzymatically converted to other substances that can enter the Krebs cycle. In one such conversion, called **deamination,** the amino group of an amino acid is removed and converted to an ammonium ion (NH_4^+), which can be excreted from the cell. The remaining organic acid can enter the Krebs cycle. Other conversions involve **decarboxylation** (the removal of —COOH) and **dehydrogenation.**

Table 5.4 Some Industrial Uses for Different Types of Fermentations*

Fermentation End-Product(s)	Industrial or Commercial Use	Starting Material	Microorganism
Ethanol	Beer	Malt extract	*Saccharomyces cerevisiae* (yeast, a fungus)
	Wine	Grape or other fruit juices	*Saccharomyces cerevisiae* (yeast)
	Fuel	Agricultural wastes	*Saccharomyces cerevisiae* (yeast)
Acetic Acid	Vinegar	Ethanol	*Acetobacter*
Lactic Acid	Cheese, yogurt	Milk	*Lactobacillus, Streptococcus*
	Rye bread	Grain, sugar	*Lactobacillus delbruckii*
	Sauerkraut	Cabbage	*Lactobacillus plantarum*
	Summer sausage	Meat	*Pediococcus*
Propionic Acid and Carbon Dioxide	Swiss cheese	Lactic acid	*Propionibacterium freudenreichii*
Acetone and Butanol	Pharmaceutical, industrial uses	Molasses	*Clostridium acetobutylicum*
Glycerol	Pharmaceutical, industrial uses	Molasses	*Saccharomyces cerevisiae* (yeast)
Citric Acid	Flavoring	Molasses	*Aspergillus* (fungus)
Methane	Fuel	Acetic acid	*Methanosarcina*
Sorbose	Vitamin C (ascorbic acid)	Sorbitol	*Gluconobacter*

*Unless otherwise noted, the microorganisms listed are bacteria.

Table 5.5 Aerobic Respiration, Anaerobic Respiration, and Fermentation Compared

Energy-Producing Process	Growth Conditions	Final Hydrogen (Electron) Acceptor	Type of Phosphorylation Used to Generate ATP	ATP Molecules Produced per Glucose Molecule
Aerobic Respiration	Aerobic	Molecular oxygen (O_2)	Substrate-level and oxidative	36 (eukaryotes) 38 (prokaryotes)
Anaerobic Respiration	Anaerobic	Usually an inorganic substance (such as NO_3^-, SO_4^{2-}, or CO_3^{2-}) but not molecular oxygen (O_2)	Substrate-level and oxidative	Variable (fewer than 38 but more than 2)
Fermentation	Aerobic or anaerobic	An organic molecule	Substrate-level	2

A summary of the interrelationships of carbohydrate, lipid, and protein catabolism is shown in Figure 5.21.

CHECK YOUR UNDERSTANDING

✓ What are the end-products of lipid and protein catabolism? **5-17**

Biochemical Tests and Bacterial Identification

LEARNING OBJECTIVE

5-18 Provide two examples of the use of biochemical tests to identify bacteria in the laboratory.

Photosynthesis

LEARNING OBJECTIVES

5-19 Compare and contrast cyclic and noncyclic photophosphorylation.

5-20 Compare and contrast the light-dependent and light-independent reactions of photosynthesis.

5-21 Compare and contrast oxidative phosphorylation and photophosphorylation.

In all of the metabolic pathways just discussed, organisms obtain energy for cellular work by oxidizing organic compounds. But where do organisms obtain these organic compounds? Some, including animals and many microbes, feed on matter produced by other organisms. For example, bacteria may catabolize compounds from dead plants and animals, or may obtain nourishment from a living host.

Other organisms synthesize complex organic compounds from simple inorganic substances. The major mechanism for such synthesis is a process called **photosynthesis,** which is used by plants and many microbes. Essentially, photosynthesis is the conversion of light energy from the sun into chemical energy. The chemical energy is then used to convert CO_2 from the atmosphere to more reduced carbon compounds, primarily sugars. The word *photosynthesis* summarizes the process: *photo* means light, and *synthesis* refers to the assembly of organic compounds. This synthesis of sugars by using carbon atoms from CO_2 gas is also called **carbon fixation.** Continuation of life as we know it on Earth depends on the recycling of carbon in this way (see Figure 27.3 on page 769). Cyanobacteria, algae, and green plants all contribute to this vital recycling with photosynthesis.

Photosynthesis can be summarized with the following equations:

1. Plants, algae, and cyanobacteria use water as a hydrogen donor, releasing O_2.

$$6\,CO_2 + 12\,H_2O + \text{Light energy} \rightarrow C_6H_{12}O_6 + 6\,H_2O + 6\,O_2$$

2. Purple sulfur and green sulfur bacteria use H_2S as a hydrogen donor, producing sulfur granules.

$$6\,CO_2 + 12\,H_2S + \text{Light energy} \rightarrow C_6H_{12}O_6 + 6\,H_2O + 12\,S$$

In the course of photosynthesis, electrons are taken from hydrogen atoms, an energy-poor molecule, and incorporated into sugar, an energy-rich molecule. The energy boost is supplied by light energy, although indirectly.

Photosynthesis takes place in two stages. In the first stage, called the **light-dependent (light) reactions,** light energy is used to convert ADP and Ⓟ to ATP. In addition, in the predominant form of the light-dependent reactions, the electron carrier $NADP^+$ is reduced to NADPH. The coenzyme NADPH, like NADH, is an energy-rich carrier of electrons. In the second stage, the **light-independent (dark) reactions,** these electrons are used along with energy from ATP to reduce CO_2 to sugar.

The Light-Dependent Reactions: Photophosphorylation

Photophosphorylation is one of the three ways ATP is formed, and it occurs only in photosynthetic cells. In this mechanism, light energy is absorbed by chlorophyll molecules in the photosynthetic cell, exciting some of the molecules' electrons. The chlorophyll principally used by green plants, algae, and cyanobacteria is *chlorophyll a.* It is located in the membranous thylakoids of chloroplasts in algae and green plants (see Figure 4.28, page 106) and in the thylakoids found in the photosynthetic structures of cyanobacteria. Other bacteria use *bacteriochlorophylls.*

The excited electrons jump from the chlorophyll to the first of a series of carrier molecules, an electron transport chain similar to that used in respiration. As electrons are passed along the series of carriers, protons are pumped across the membrane, and ADP is converted to ATP by chemiosmosis. In **cyclic photophosphorylation,** the electrons eventually return to chlorophyll (**Figure 5.25a**). In **noncyclic photophosphorylation,** which is used in oxygenic organisms, the electrons released from chlorophyll do not return to chlorophyll but become incorporated into NADPH (**Figure 5.25b**). The electrons lost from chlorophyll are replaced by electrons from H_2O. To summarize: the products of noncyclic photophosphorylation are ATP (formed by chemiosmosis using energy released in an electron transport chain), O_2 (from water molecules), and NADPH (in which the hydrogen electrons and protons were derived ultimately from water).

The Light-Independent Reactions: The Calvin-Benson Cycle

The light-independent (dark) reactions are so named because they require no light directly. They include a complex cyclic pathway called the **Calvin-Benson cycle,** in which CO_2 is "fixed"—that is, used to synthesize sugars (**Figure 5.26**, see also Figure A.1 in Appendix A). **Animations** Photosythesis: Overview; Comparing Prokaryotes and Eukaryotes; Light Reaction: Cyclic Photophosphorylation; Light Reaction: Noncyclic Photophosphorylation; Light Independent Reactions. **www.microbiologyplace.com**

CHECK YOUR UNDERSTANDING

- ✓ How is photosynthesis important to catabolism? **5-19**
- ✓ What is made during the light-dependent reactions? **5-20**
- ✓ How are oxidative phosphorylation and photophosphorylation similar? **5-21**

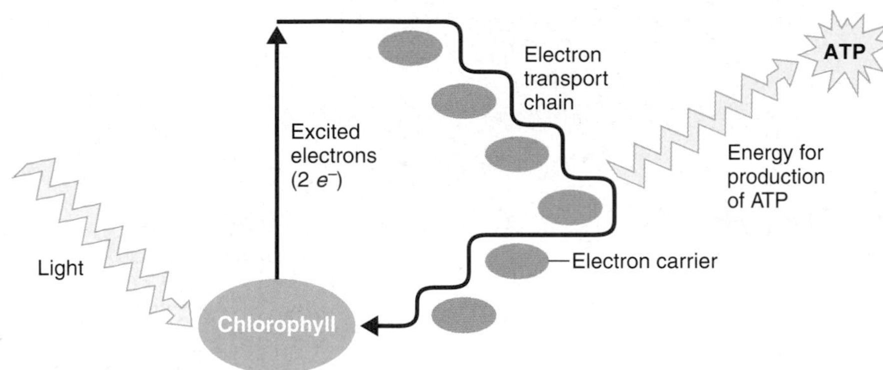

(a) Cyclic photophosphorylation

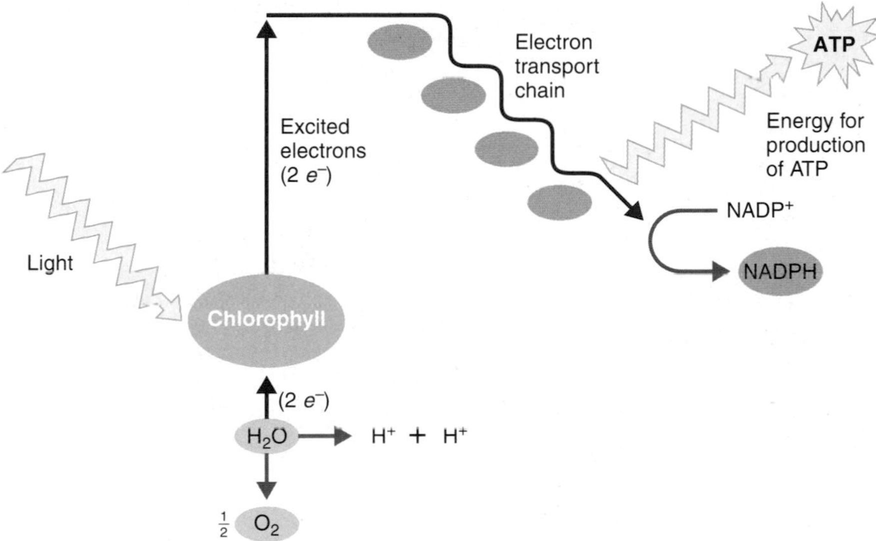

(b) Noncyclic photophosphorylation

Figure 5.25 Photophosphorylation. (**a**) In cyclic photophosphorylation, electrons released from chlorophyll by light return to chlorophyll after passage along the electron transport chain. The energy from electron transfer is converted to ATP. (**b**) In noncyclic photophosphorylation, electrons released from chlorophyll are replaced by electrons from water. The chlorophyll electrons are passed along the electron transport chain to the electron acceptor $NADP^+$. $NADP^+$ combines with electrons and with hydrogen ions from water, forming NADPH.

Q How are oxidative phosphorylation and photophosphorylation similar?

A Summary of Energy Production Mechanisms

LEARNING OBJECTIVE

5-22 Write a sentence to summarize energy production in cells.

In the living world, energy passes from one organism to another in the form of the potential energy contained in the bonds of chemical compounds. Organisms obtain the energy from oxidation reactions. To obtain energy in a usable form, a cell must have an electron (or hydrogen) donor, which serves as an initial energy source within the cell. Electron donors can be as diverse as photosynthetic pigments, glucose or other organic compounds, elemental sulfur, ammonia, or hydrogen gas (Figure 5.27). Next, electrons removed from the chemical energy sources are transferred to electron carriers, such as the coenzymes NAD^+, $NADP^+$, and FAD. This transfer is an oxidation-reduction reaction; the initial energy source is oxidized as this first electron carrier is reduced. During this phase, some ATP is produced. In the third stage, electrons are transferred from electron carriers to their final electron acceptors in further oxidation-reduction reactions, producing more ATP.

In aerobic respiration, oxygen (O_2) serves as the final electron acceptor. In anaerobic respiration, inorganic substances other than oxygen, such as nitrate ions (NO_3^-) or sulfate ions (SO_4^{2-}), serve as the final electron acceptors. In fermentation, organic compounds serve as the final electron acceptors. In aerobic and anaerobic respiration, a series of electron carriers called an electron transport chain releases energy that is used by the mechanism of chemiosmosis to synthesize ATP. Regardless of their energy sources, all organisms use similar oxidation-reduction reactions to transfer electrons and similar mechanisms to use the energy released to produce ATP.

CHECK YOUR UNDERSTANDING

✓ Summarize how oxidation enables organisms to get energy from glucose, sulfur, or sunlight. **5-22**

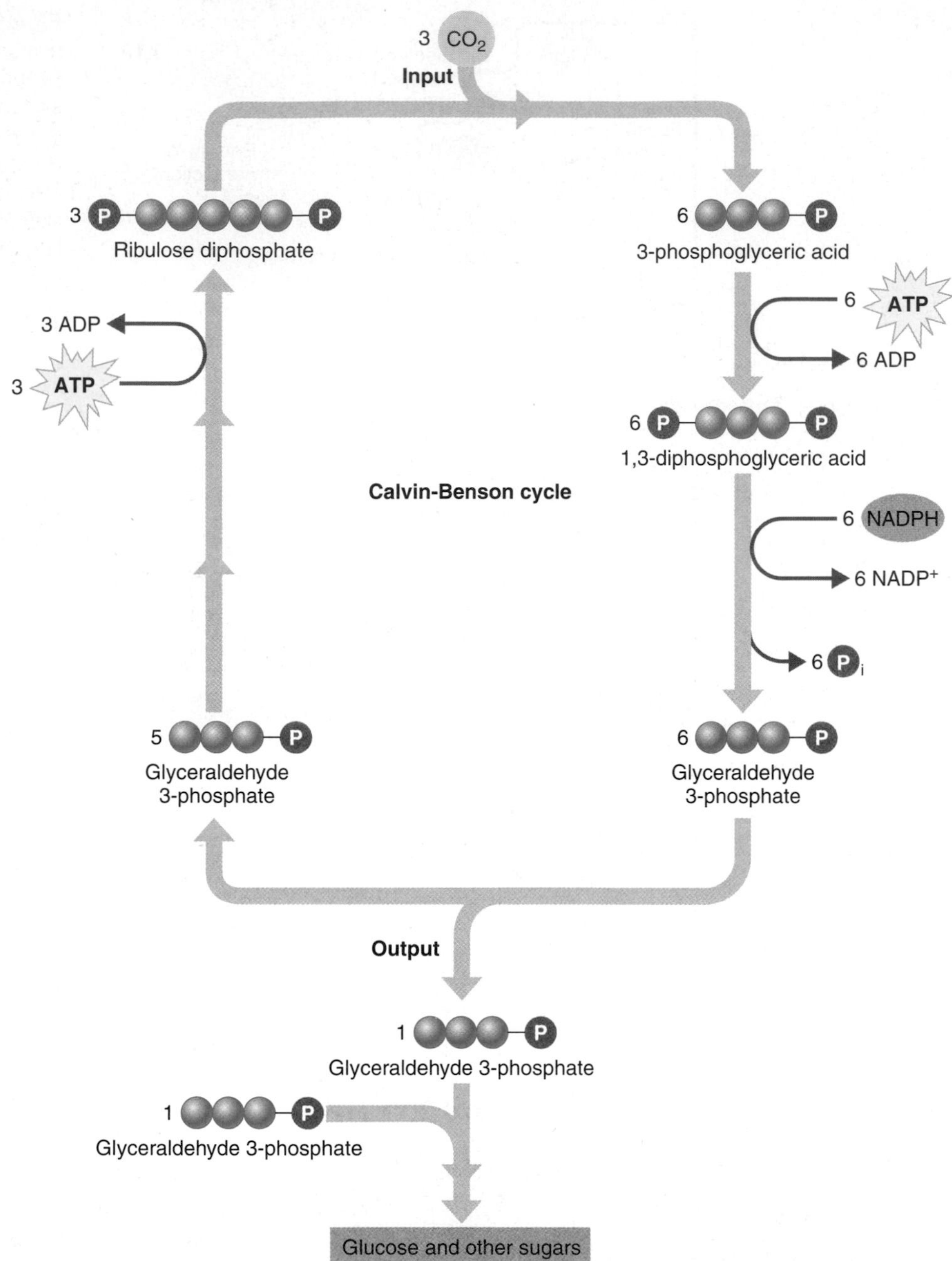

Figure 5.26 A simplified version of the Calvin-Benson cycle. This diagram shows three turns of the cycle, in which three molecules of CO_2 are fixed and one molecule of glyceraldehyde 3-phosphate is produced and leaves the cycle. Two molecules of glyceraldehyde 3-phosphate are needed to make one molecule of glucose. Therefore, the cycle must turn six times for each glucose molecule produced, requiring a total investment of 6 molecules of CO_2, 18 molecules of ATP, and 12 molecules of NADPH. A more detailed version of this cycle is presented in Figure A.1 in Appendix A.

Q In the Calvin-Benson cycle, which molecule is used to synthesize sugars?

Metabolic Diversity among Organisms

LEARNING OBJECTIVE

5-23 Categorize the various nutritional patterns among organisms according to carbon source and mechanisms of carbohydrate catabolism and ATP generation.

We have looked in detail at some of the energy-generating metabolic pathways that are used by animals and plants, as well as by many microbes. Microbes are distinguished by their great metabolic diversity, however, and some can sustain themselves on inorganic substances by using pathways that are unavailable to either plants or animals. All organisms, including microbes, can be classified metabolically according to their *nutritional pattern*—their source of energy and their source of carbon.

First considering the energy source, we can generally classify organisms as phototrophs or chemotrophs. **Phototrophs** use light as their primary energy source, whereas **chemotrophs** depend on oxidation-reduction reactions of inorganic or organic compounds for energy. For their principal carbon source, **autotrophs** (self-feeders) use carbon dioxide, and **heterotrophs** (feeders on others) require an organic carbon source. Autotrophs

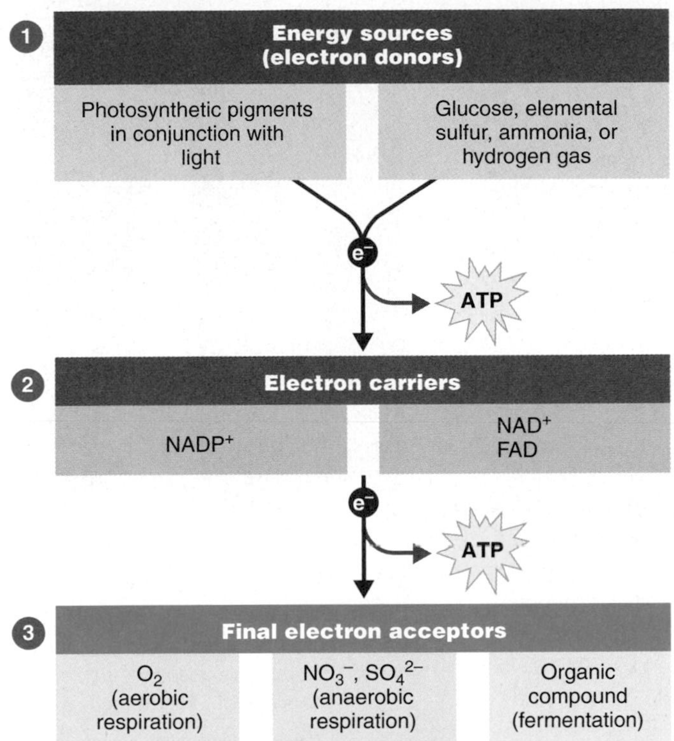

Figure 5.27 Requirements of ATP production. The production of ATP requires ❶ an energy source (electron donor), ❷ the transfer of electrons to an electron carrier during an oxidation-reduction reaction, and ❸ the transfer of electrons to a final electron acceptor.

Q Are energy-generating reactions oxidations or reductions?

are also referred to as *lithotrophs* (rock eating), and heterotrophs are also referred to as *organotrophs.*

If we combine the energy and carbon sources, we derive the following nutritional classifications for organisms: *photoautotrophs, photoheterotrophs, chemoautotrophs,* and *chemoheterotrophs* (Figure 5.28). Almost all of the medically important microorganisms discussed in this book are chemoheterotrophs. Typically, infectious organisms catabolize substances obtained from the host.

Photoautotrophs

Photoautotrophs use light as a source of energy and carbon dioxide as their chief source of carbon. They include photosynthetic bacteria (green and purple bacteria and cyanobacteria), algae, and green plants. In the photosynthetic reactions of cyanobacteria, algae, and green plants, the hydrogen atoms of water are used to reduce carbon dioxide, and oxygen gas is given off. Because this photosynthetic process produces O_2, it is sometimes called **oxygenic.**

In addition to the cyanobacteria (see Figure 11.13, page 314), there are several other families of photosynthetic prokaryotes. Each is classified according to the way it reduces CO_2. These bacteria cannot use H_2O to reduce CO_2 and cannot carry on photosynthesis when oxygen is present (they must have an anaerobic environment). Consequently, their photosynthetic process does not produce O_2 and is called

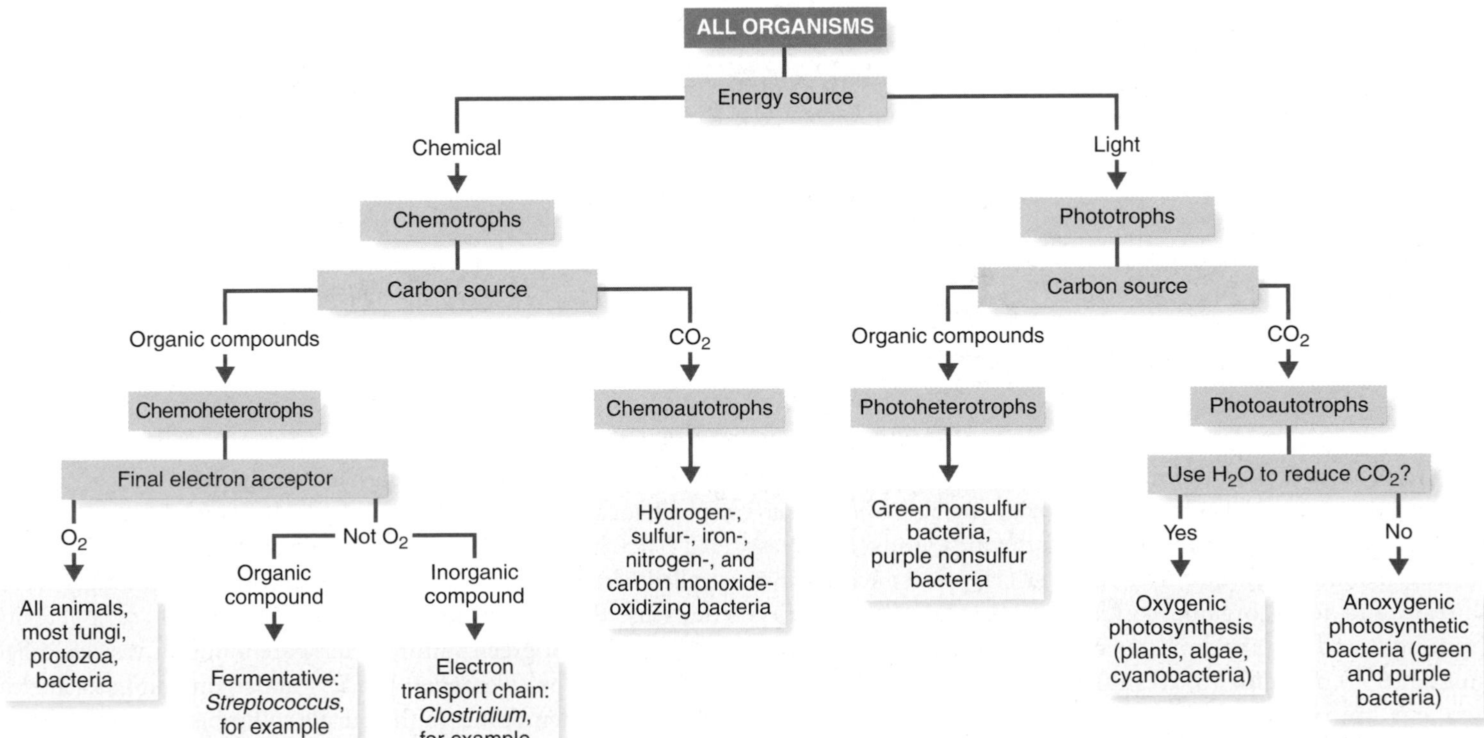

Figure 5.28 A nutritional classification of organisms.

Q What is the basic difference between chemotrophs and phototrophs?

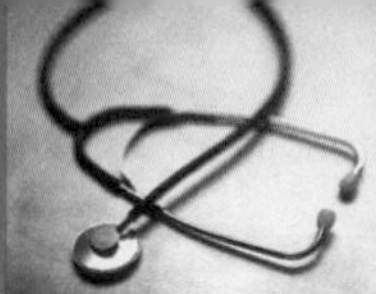

CLINICAL FOCUS From the *Morbidity and Mortality Weekly Report*

Human Tuberculosis—New York City

As you read through this box, you will encounter a series of questions that laboratory technicians ask themselves as they identify bacteria. Try to answer each question before going on to the next one.

1. A 15-month old, U.S.-born boy in New York City died of peritoneal tuberculosis (TB). Caused by one of several closely related species in the *Mycobacterium tuberculosis* complex, TB is a reportable condition in the United States. Peritoneal TB is a disease of the intestines and abdominal cavity.
   What organ is usually associated with tuberculosis? How might someone get peritoneal TB?
2. Pulmonary TB is contracted by inhaling the bacteria; ingesting the bacteria can result in peritoneal TB. The first step is observing acid-fast bacteria in nodules on the boy's organs.
   What is the next step?
3. Speciation of the *M. tuberculosis* complex is done by biochemical testing in reference laboratories (**Figure A**). The bacteria need to be grown in culture media. Slow-growing mycobacteria may take up to 6 weeks to form colonies.
   After colonies have been isolated, what is the next step?
4. In this case, the bacteria are slow-growing. According to the identification scheme, the urease test should be performed.
   What is the result shown in Figure B?
5. The urease test is positive.
   What test is done next?
6. The nitrate reduction test is performed. It shows that the bacteria do not produce the enzyme nitrate reductase.
   What is the bacterium?
7. *M. bovis* is a pathogen that primarily infects cattle. However, humans also can become infected, most commonly by consuming unpasteurized milk products from infected cows. In industrialized nations, human TB caused by *M. bovis* is rare because of milk pasteurization and culling of infected cattle herds. This investigation identified 35 cases of human *M. bovis* infection in New York City. Fresh cheese brought from Mexico was a likely source of infection. No evidence of human-to-human transmission has been found. Products from unpasteurized cow's milk have been associated with certain infectious diseases and carry the risk of transmitting *M. bovis* if imported from countries where the bacterium is common in cattle. Everyone should avoid consuming products from unpasteurized cow's milk.

Test Control

Figure B The urease test. In a positive test, bacterial urease hydrolyzes urea, producing ammonia. The ammonia raises the pH, and the indicator in the medium turns to fuchsia.

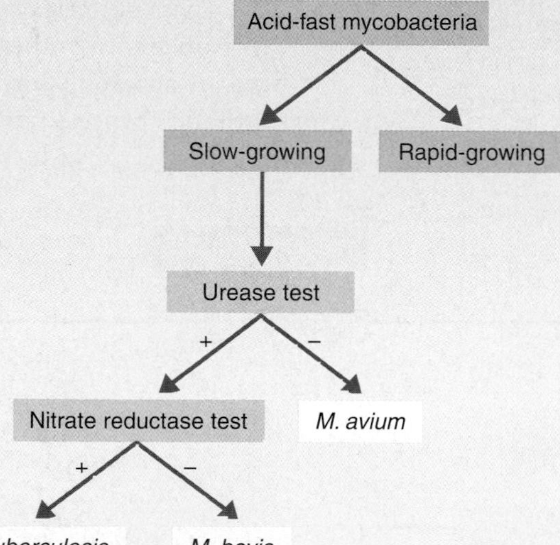

Figure A An identification scheme for selected species of slow-growing mycobacteria.

Source: Adapted from *MMWR* 54(24): 605-608, June 24, 2005.

anoxygenic. The anoxygenic photoautotrophs are the green and purple bacteria. The **green bacteria,** such as *Chlorobium* (klô-rō′bē-um), use sulfur (S), sulfur compounds (such as hydrogen sulfide, H_2S), or hydrogen gas (H_2) to reduce carbon dioxide and form organic compounds. Applying the energy from light and the appropriate enzymes, these bacteria oxidize sulfide (S^{2-}) or sulfur (S) to sulfate (SO_4^{2-}) or oxidize hydrogen gas to water (H_2O). The **purple bacteria,** such as *Chromatium* (krō-mā′tē-um), also use sulfur, sulfur compounds, or hydrogen gas to reduce carbon dioxide. They are distinguished from the green bacteria by their type of chlorophyll, location of stored sulfur, and ribosomal RNA.

The chlorophylls used by these photosynthetic bacteria are called *bacteriochlorophylls,* and they absorb light at longer wavelengths than that absorbed by chlorophyll *a.* Bacteriochlorophylls of green sulfur bacteria are found in vesicles called *chlorosomes* (or *chlorobium vesicles*) underlying and attached to the plasma membrane. In the purple sulfur bacteria, the bacteriochlorophylls are located in invaginations of the plasma membrane (*chromatophores*).

Table 5.6 Photosynthesis Compared in Selected Eukaryotes and Prokaryotes

Characteristic	*Eukaryotes*	*Prokaryotes*		
	Algae, Plants	**Cyanobacteria**	**Green Bacteria**	**Purple Bacteria**
Substance That Reduces CO_2	H atoms of H_2O	H atoms of H_2O	Sulfur, sulfur compounds, H_2 gas	Sulfur, sulfur compounds, H_2 gas
Oxygen Production	Oxygenic	Oxygenic (and anoxygenic)	Anoxygenic	Anoxygenic
Type of Chlorophyll	Chlorophyll *a*	Chlorophyll *a*	Bacteriochlorophyll *a*	Bacteriochlorophyll *a* or *b*
Site of Photosynthesis	Chloroplasts with thylakoids	Thylakoids	Chlorosomes	Chromatophores
Environment	Aerobic	Aerobic (and anaerobic)	Anaerobic	Anaerobic

Table 5.6 summarizes several characteristics that distinguish eukaryotic photosynthesis from prokaryotic photosynthesis.

Photoheterotrophs

Photoheterotrophs use light as a source of energy but cannot convert carbon dioxide to sugar; rather, they use as sources of carbon organic compounds, such as alcohols, fatty acids, other organic acids, and carbohydrates. Photoheterotrophs are anoxygenic. The **green nonsulfur bacteria,** such as *Chloroflexus* (klô-rō-flex'us), and **purple nonsulfur bacteria,** such as *Rhodopseudomonas* (rō-dō-sū-dō-mō'nas), are photoheterotrophs.

Chemoautotrophs

Chemoautotrophs use the electrons from reduced inorganic compounds as a source of energy, and they use CO_2 as their principal source of carbon. They fix CO_2 in the Calvin-Benson Cycle (see Figure 5.26). Inorganic sources of energy for these organisms include hydrogen sulfide (H_2S) for *Beggiatoa* (bej-jē-ä-tō'ä); elemental sulfur (S) for *Thiobacillus thiooxidans;* ammonia (NH_3) for *Nitrosomonas* (nī-trō-sō-mō'näs); nitrite ions (NO_2^-) for *Nitrobacter* (nī-trō-bak'tėr); hydrogen gas (H_2) for *Hydrogenomonas* (hī-drō-je-nō-mō'näs); ferrous iron (Fe^{2+}) for *Thiobacillus ferrooxidans;* and carbon monoxide (CO) for *Pseudomonas carboxydohydrogena* (kär'boks-i-dō-hi-drō-je-nä). The energy derived from the oxidation of these inorganic compounds is eventually stored in ATP, which is produced by oxidative phosphorylation.

Chemoheterotrophs

When we discuss photoautotrophs, photoheterotrophs, and chemoautotrophs, it is easy to categorize the energy source and carbon source because they occur as separate entities. However, in chemoheterotrophs, the distinction is not as clear because the energy source and carbon source are usually the same organic compound—glucose, for example. **Chemoheterotrophs** specifically use the electrons from hydrogen atoms in organic compounds as their energy source.

Heterotrophs are further classified according to their source of organic molecules. **Saprophytes** live on dead organic matter, and **parasites** derive nutrients from a living host. Most bacteria, and all fungi, protozoa, and animals, are chemoheterotrophs.

Bacteria and fungi can use a wide variety of organic compounds for carbon and energy sources. This is why they can live in diverse environments. Understanding microbial diversity is scientifically interesting and economically important. In some situations microbial growth is undesirable, such as when rubber-degrading bacteria destroy a gasket or shoe sole. However, these same bacteria might be beneficial if they decomposed discarded rubber products, such as used tires. *Rhodococcus erythropolis* (rō-dō-kok'kus er-i-throp'ō-lis) is widely distributed in soil and can cause disease in humans and other animals. However, this same species is able to replace sulfur atoms in petroleum with atoms of oxygen. A Texas company is currently using *R. erythropolis* to produce desulfurized oil.

CHECK YOUR UNDERSTANDING

✓ Almost all medically important microbes belong to which of the four aforementioned groups? **5-23**

* * *

We will next consider how cells use ATP pathways for the synthesis of organic compounds such as carbohydrates, lipids, proteins, and nucleic acids.

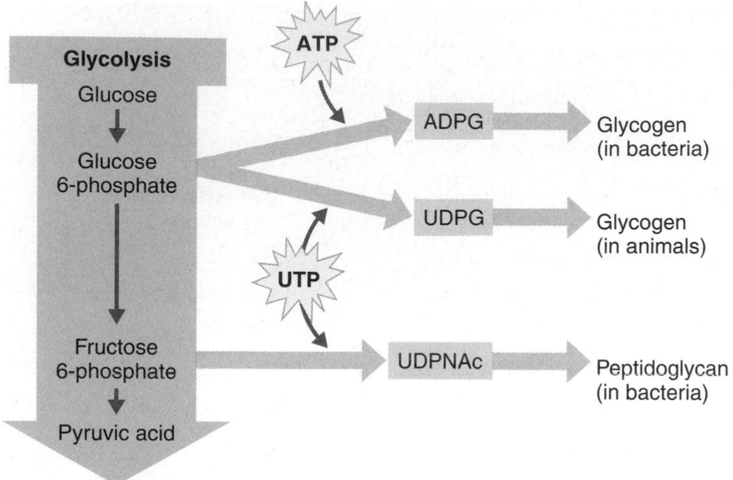

Figure 5.29 The biosynthesis of polysaccharides.

Q How are polysaccharides used in cells?

Metabolic Pathways of Energy Use

LEARNING OBJECTIVE

5-24 Describe the major types of anabolism and their relationship to catabolism.

Up to now we have been considering energy production. Through the oxidation of organic molecules, organisms produce energy by aerobic respiration, anaerobic respiration, and fermentation. Much of this energy is given off as heat. The complete metabolic oxidation of glucose to carbon dioxide and water is considered a very efficient process, but about 45% of the energy of glucose is lost as heat. Cells use the remaining energy, which is trapped in the bonds of ATP, in a variety of ways. Microbes use ATP to provide energy for the transport of substances across plasma membranes—the process called active transport that we discussed in Chapter 4. Microbes also use some of their energy for flagellar motion (also discussed in Chapter 4). Most of the ATP, however, is used in the production of new cellular components. This production is a continuous process in cells and, in general, is faster in prokaryotic cells than in eukaryotic cells.

Autotrophs build their organic compounds by fixing carbon dioxide in the Calvin-Benson cycle (see Figure 5.26). This requires both energy (ATP) and electrons (from the oxidation of NADPH). Heterotrophs, by contrast, must have a ready source of organic compounds for biosynthesis—the production of needed cellular components, usually from simpler molecules. The cells use these compounds as both the carbon source and the energy source. We will next consider the biosynthesis of a few representative classes of biological molecules: carbohydrates, lipids, amino acids, purines, and pyrimidines. As we do so, keep in mind that synthesis reactions require a net input of energy.

Polysaccharide Biosynthesis

Microorganisms synthesize sugars and polysaccharides. The carbon atoms required to synthesize glucose are derived from the intermediates produced during processes such as glycolysis and the Krebs cycle and from lipids or amino acids. After synthesizing glucose (or other simple sugars), bacteria may assemble it into more complex polysaccharides, such as glycogen. For bacteria to build glucose into glycogen, glucose units must be phosphorylated and linked. The product of glucose phosphorylation is glucose 6-phosphate. Such a process involves the expenditure of energy, usually in the form of ATP. In order for bacteria to synthesize glycogen, a molecule of ATP is added to glucose 6-phosphate to form *adenosine diphosphoglucose (ADPG)* (**Figure 5.29**). Once ADPG is synthesized, it is linked with similar units to form glycogen.

Using a nucleotide called uridine triphosphate (UTP) as a source of energy and glucose 6-phosphate, animals synthesize glycogen (and many other carbohydrates) from *uridine diphosphoglucose, UDPG* (see Figure 5.29). A compound related to UDPG, called *UDP-N-acetylglucosamine (UDPNAc),* is a key starting material in the biosynthesis of peptidoglycan, the substance that forms bacterial cell walls. UDPNAc is formed from fructose 6-phosphate, and the reaction also uses UTP.

Lipid Biosynthesis

Because lipids vary considerably in chemical composition, they are synthesized by a variety of routes. Cells synthesize fats by joining glycerol and fatty acids. The glycerol portion of the fat is derived from dihydroxyacetone phosphate, an intermediate formed during glycolysis. Fatty acids, which are long-chain hydrocarbons (hydrogen linked to carbon), are built up when two-carbon fragments of acetyl CoA are successively added to each other (**Figure 5.30**). As with polysaccharide synthesis, the building units of fats and other lipids are linked via dehydration synthesis reactions that require energy, not always in the form of ATP.

The most important role of lipids is to serve as structural components of biological membranes, and most membrane lipids are phospholipids. A lipid of a very different structure, cholesterol, is also found in plasma membranes of eukaryotic cells. Waxes are lipids that are important components of the cell wall of acid-fast bacteria. Other lipids, such as carotenoids, provide the red, orange, and yellow pigments of some microorganisms. Some lipids form portions of chlorophyll molecules. Lipids also function in energy storage. Recall that the breakdown products of lipids after biological oxidation feed into the Krebs cycle.

Amino Acid and Protein Biosynthesis

Amino acids are required for protein biosynthesis. Some microbes, such as *E. coli,* contain the enzymes necessary to use starting materials, such as glucose and inorganic salts, for the

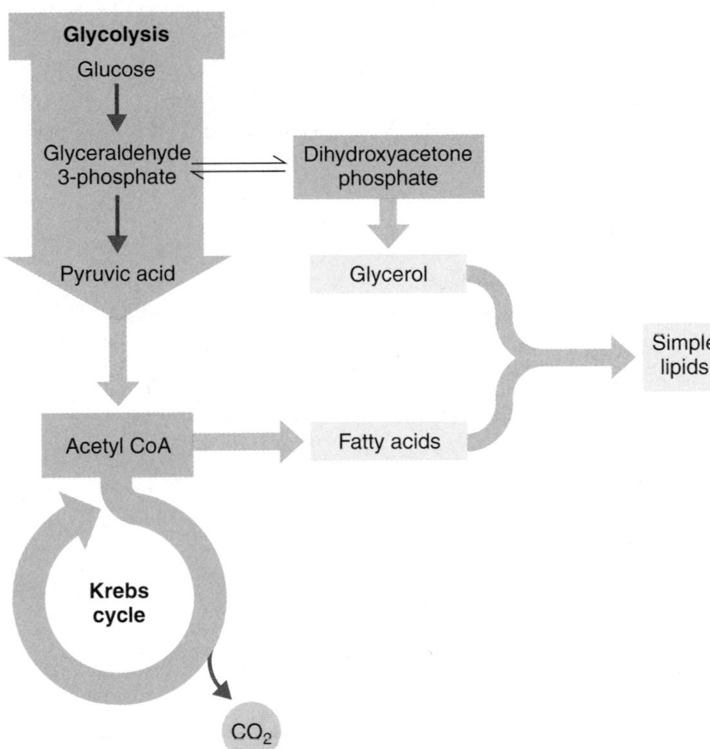

Figure 5.30 The biosynthesis of simple lipids.

Q What is the primary use of lipids in cells?

synthesis of all the amino acids they need. Organisms with the necessary enzymes can synthesize all amino acids directly or indirectly from intermediates of carbohydrate metabolism (Figure 5.31a). Other microbes require that the environment provide some preformed amino acids.

One important source of the *precursors* (intermediates) used in amino acid synthesis is the Krebs cycle. Adding an amine group to pyruvic acid or to an appropriate organic acid of the Krebs cycle converts the acid into an amino acid. This process is called **amination.** If the amine group comes from a preexisting amino acid, the process is called **transamination** (Figure 5.31b).

Most amino acids within cells are destined to be building blocks for protein synthesis. Proteins play major roles in the cell as enzymes, structural components, and toxins, to name just a few uses. The joining of amino acids to form proteins involves dehydration synthesis and requires energy in the form of ATP. The mechanism of protein synthesis involves genes and is discussed in Chapter 8.

Purine and Pyrimidine Biosynthesis

Recall from Chapter 2 that the informational molecules DNA and RNA consist of repeating units called *nucleotides,* each of which consists of a purine or pyrimidine, a pentose (five-carbon sugar), and a phosphate group. The five-carbon sugars of nucleotides are derived from either the pentose phosphate pathway or the Entner-Doudoroff pathway. Certain amino acids—aspartic acid, glycine, and glutamine—made from intermediates produced during glycolysis and in the Krebs cycle participate in the biosyntheses of purines and pyrimidines (Figure 5.32). The carbon and nitrogen atoms derived from these amino acids form the purine and pyrimidine rings, and the energy for synthesis is provided by ATP. DNA contains all the information necessary to determine the specific structures and functions of cells. Both RNA and DNA are required for protein synthesis. In addition, such nucleotides as ATP, NAD^+, and $NADP^+$ assume roles in stimulating and inhibiting the rate of cellular metabolism. The synthesis of DNA and RNA from nucleotides will be discussed in Chapter 8.

CHECK YOUR UNDERSTANDING

✓ Where do amino acids required for protein synthesis come from? 5-24

The Integration of Metabolism

LEARNING OBJECTIVE

5-25 Define *amphibolic pathways.*

We have seen thus far that the metabolic processes of microbes produce energy from light, inorganic compounds, and organic compounds. Reactions also occur in which energy is used for biosynthesis. With such a variety of activity, you might imagine that anabolic and catabolic reactions occur independently of each other in space and time. Actually, anabolic and catabolic reactions are joined through a group of common intermediates (identified as key intermediates in Figure 5.33). Both anabolic and catabolic reactions also share some metabolic pathways, such as the Krebs cycle. For example, reactions in the Krebs cycle not only participate in the oxidation of glucose but also produce intermediates that can be converted to amino acids. Metabolic pathways that function in both anabolism and catabolism are called **amphibolic pathways,** meaning that they are dual-purpose.

Amphibolic pathways bridge the reactions that lead to the breakdown and synthesis of carbohydrates, lipids, proteins, and nucleotides. Such pathways enable simultaneous reactions to occur in which the breakdown product formed in one reaction is used in another reaction to synthesize a different compound, and vice versa. Because various intermediates are common to both anabolic and catabolic reactions, mechanisms exist that regulate synthesis and breakdown pathways and allow these reactions to occur simultaneously. One such mechanism involves the use of different coenzymes for opposite pathways. For example, NAD^+ is involved in catabolic reactions, whereas $NADP^+$ is involved in anabolic reactions. Enzymes can also coordinate anabolic and catabolic reactions by accelerating or inhibiting the rates of biochemical reactions.

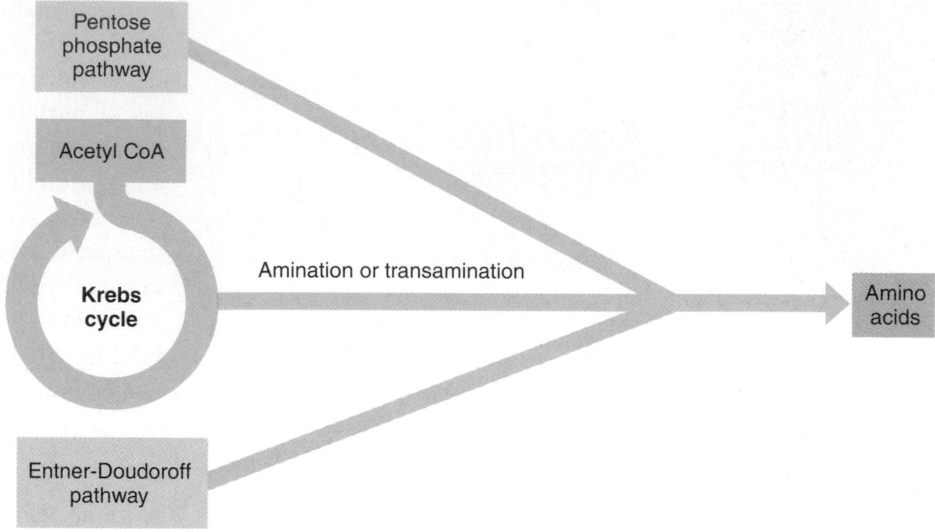

(a) Amino acid biosynthesis

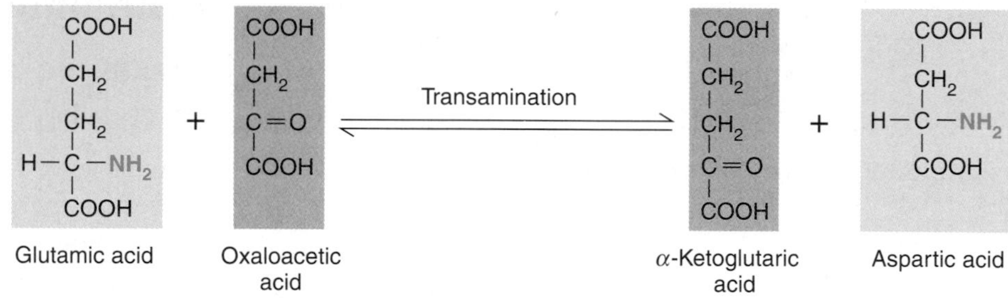

(b) Process of transamination

Figure 5.31 The biosynthesis of amino acids. (**a**) Pathways of amino acid biosynthesis through amination or transamination of intermediates of carbohydrate metabolism from the Krebs cycle, pentose phosphate pathway, and Entner-Doudoroff pathway. (**b**) Transamination, a process by which new amino acids are made with the amine groups from old amino acids. Glutamic acid and aspartic acid are both amino acids; the other two compounds are intermediates in the Krebs cycle.

Q What is the function of amino acids in cells?

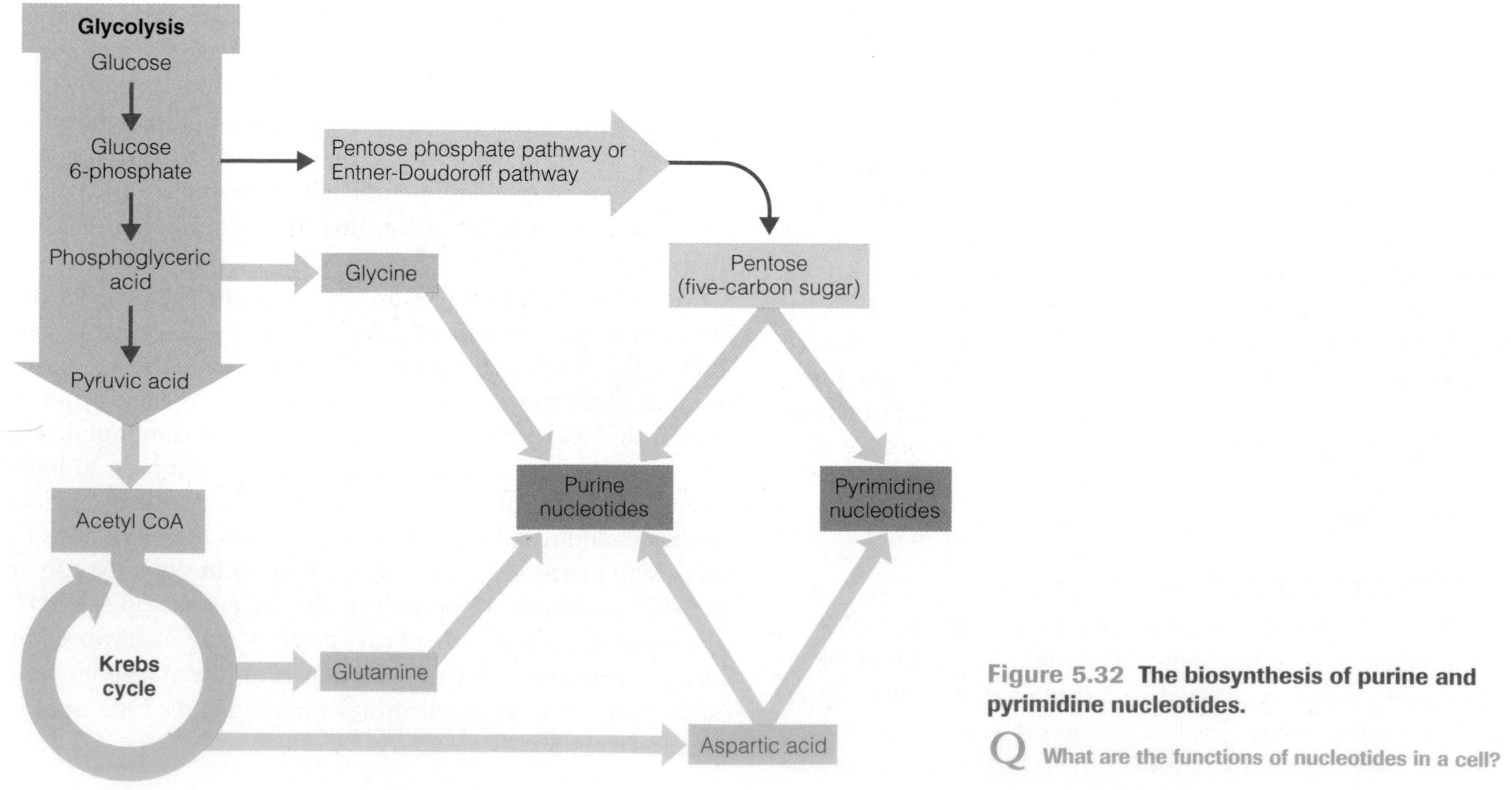

Figure 5.32 The biosynthesis of purine and pyrimidine nucleotides.

Q What are the functions of nucleotides in a cell?

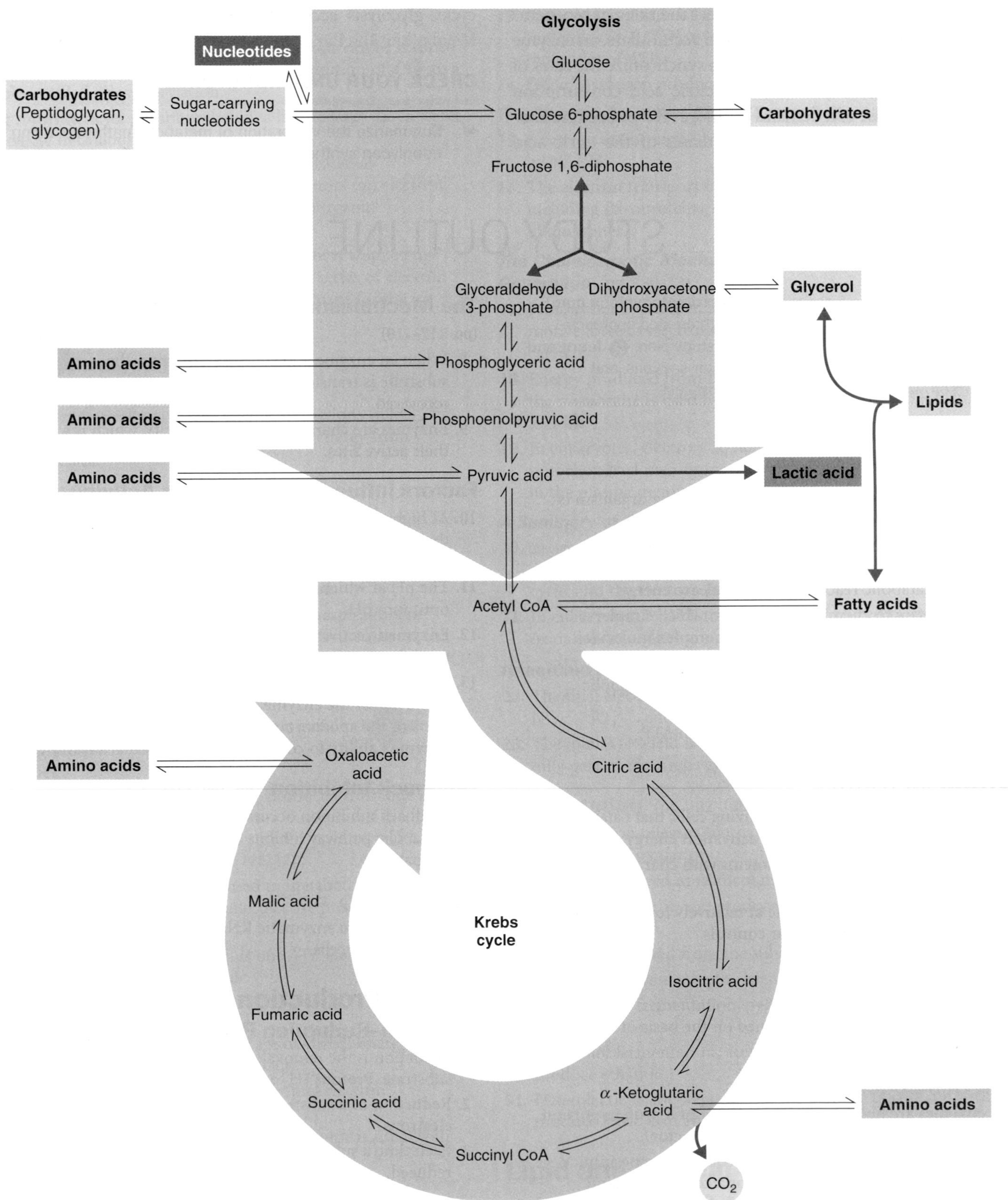

Figure 5.33 The integration of metabolism. Key intermediates are shown. Although not indicated in the figure, amino acids and ribose are used in the synthesis of purine and pyrimidine nucleotides (see Figure 5.32). The double arrows indicate amphibolic pathways.

Q What is an amphibolic pathway?

6 Microbial Growth

When we talk about microbial growth, we are really referring to the *number* of cells, not the *size* of the cells. Microbes that are "growing" are increasing in number, accumulating into *colonies* (groups of cells large enough to be seen without a microscope) of hundreds of thousands of cells or *populations* of billions of cells. Although individual cells approximately double in size during their lifetime, this change is not very significant compared with the size increases observed during the lifetime of plants and animals.

Microbial populations can become incredibly large in a very short time, as we will see later in this chapter. By understanding the conditions necessary for microbial growth, we can determine how to control the growth of microbes that cause diseases and food spoilage. We can also learn how to encourage the growth of helpful microbes and those we wish to study.

In this chapter we will examine the physical and chemical requirements for microbial growth, the various kinds of culture media, bacterial cell division, the phases of microbial growth, and the methods of measuring microbial growth.

UNDER THE MICROSCOPE

Escherichia coli. *E. coli*, a common inhabitant of the human intestine, can live with or without oxygen.

Q&A

Oxygen in the atmosphere is essential for human life. How can some bacteria grow in the absence of oxygen?

Look for the answer in the chapter.

The Requirements for Growth

LEARNING OBJECTIVES

6-1 Classify microbes into five groups on the basis of preferred temperature range.

6-2 Identify how and why the pH of culture media is controlled.

6-3 Explain the importance of osmotic pressure to microbial growth.

6-4 Name a use for each of the four elements (carbon, nitrogen, sulfur, and phosphorus) needed in large amounts for microbial growth.

6-5 Explain how microbes are classified on the basis of oxygen requirements.

6-6 Identify ways in which aerobes avoid damage by toxic forms of oxygen.

The requirements for microbial growth can be divided into two main categories: physical and chemical. Physical aspects include temperature, pH, and osmotic pressure. Chemical requirements include sources of carbon, nitrogen, sulfur, phosphorus, oxygen, trace elements, and organic growth factors.

Physical Requirements

Temperature

Most microorganisms grow well at the temperatures that humans favor. However, certain bacteria are capable of growing at extremes of temperature that would certainly hinder the survival of almost all eukaryotic organisms.

Microorganisms are classified into three primary groups on the basis of their preferred range of temperature: **psychrophiles** (cold-loving microbes), **mesophiles** (moderate-temperature–loving microbes), and **thermophiles** (heat-loving microbes). Most bacteria grow only within a limited range of temperatures, and their maximum and minimum growth temperatures are only about 30°C apart. They grow poorly at the high and low temperature extremes within their range.

Each bacterial species grows at particular minimum, optimum, and maximum temperatures. The **minimum growth temperature** is the lowest temperature at which the species will grow. The **optimum growth temperature** is the temperature at which the species grows best. The **maximum growth temperature** is the highest temperature at which growth is possible. By graphing the growth response over a temperature range, we can see that the optimum growth temperature is usually near the top of the range; above that temperature the rate of growth drops off rapidly (Figure 6.1). This happens presumably because the high temperature has inactivated necessary enzymatic systems of the cell.

The ranges and maximum growth temperatures that define bacteria as psychrophiles, mesophiles, or thermophiles are not rigidly defined. Psychrophiles, for example, were originally considered simply to be organisms capable of growing at 0°C. However, there seem to be two fairly distinct groups capable of growth at that temperature. One group, composed of psychrophiles in the strictest sense, can grow at 0°C but has an optimum growth temperature of about 15°C. Most of these organisms are so sensitive to higher temperatures that they will not even grow in a reasonably warm room (25°C). Found mostly in the oceans' depths or in certain polar regions, such organisms seldom cause problems in food preservation. The other group that can grow at 0°C has higher optimum temperatures, usually 20–30°C and cannot grow above about 40°C. Organisms of this type are much more common than psychrophiles and are the most likely to be encountered in low-temperature food spoilage because they grow fairly well at refrigerator temperatures. We will use the term **psychrotrophs,** which food microbiologists favor, for this group of spoilage microorganisms.

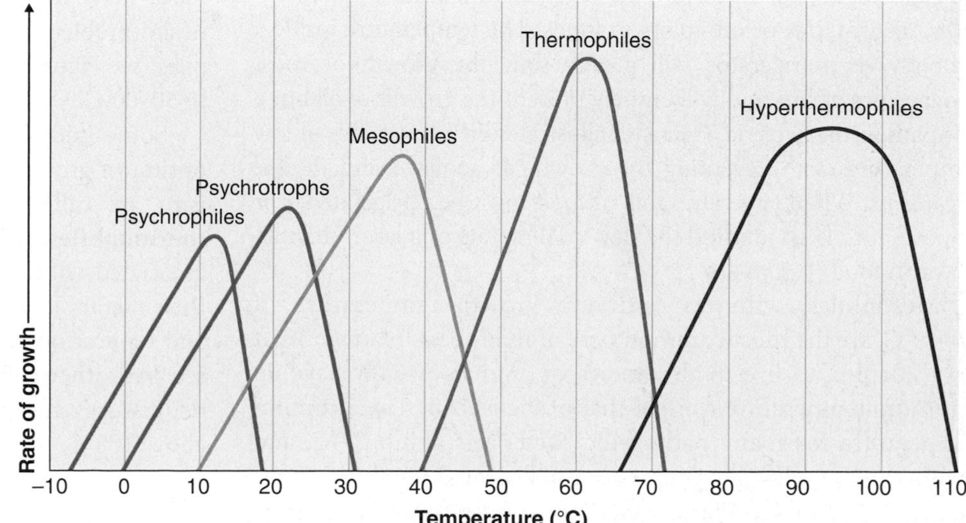

Figure 6.1 Typical growth rates of different types of microorganisms in response to temperature. The peak of the curve represents optimum growth (fastest reproduction). Notice that the reproductive rate drops off very quickly at temperatures only a little above the optimum. At either extreme of the temperature range, the reproductive rate is much lower than the rate at the optimum temperature.

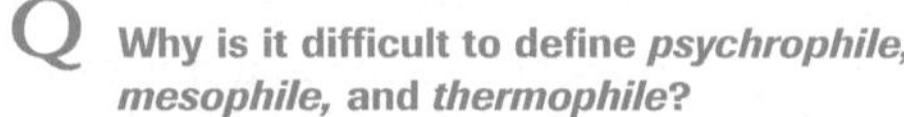
Q Why is it difficult to define *psychrophile, mesophile,* and *thermophile*?

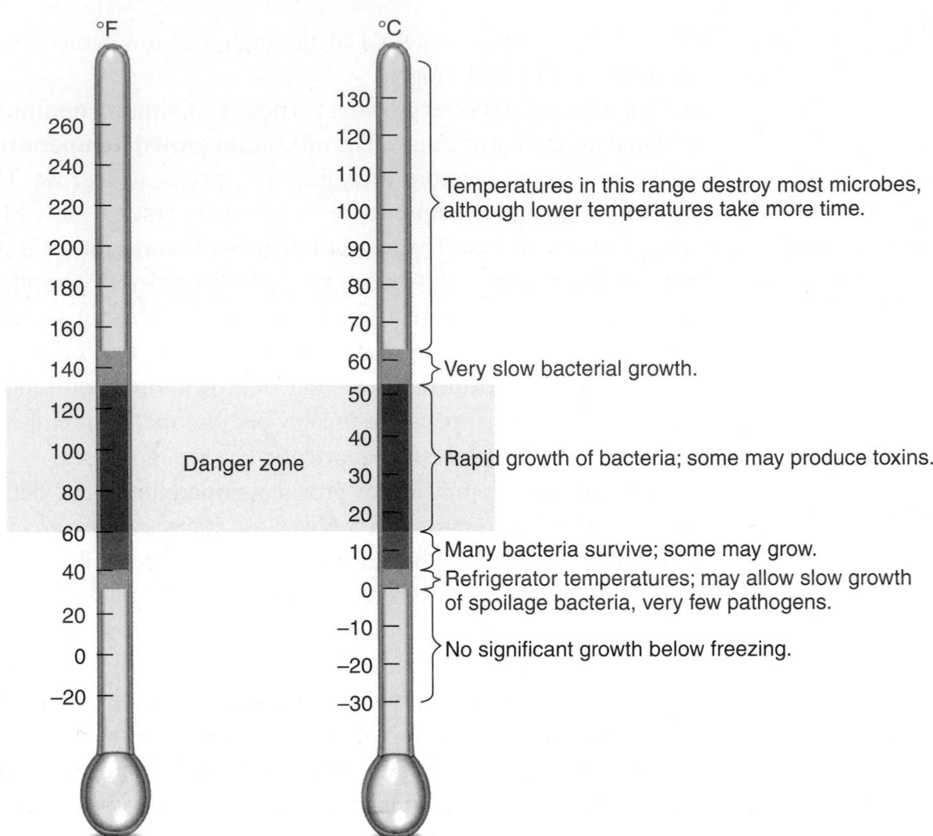
°F
°C
260
240
220
200
180
160
140
120
100
80
60
40
20
0
–20
130
120
110
100
90
80
70
60
50
40
30
20
10
0
–10
–20
–30
Danger zone
Temperatures in this range destroy most microbes, although lower temperatures take more time.
Very slow bacterial growth.
Rapid growth of bacteria; some may produce toxins.
Many bacteria survive; some may grow.
Refrigerator temperatures; may allow slow growth of spoilage bacteria, very few pathogens.
No significant growth below freezing.

Figure 6.2 Food preservation temperatures. Low temperatures decrease microbial reproduction rates, which is the basic principle of refrigeration. There are always some exceptions to the temperature responses shown here; for example, certain bacteria grow well at temperatures that would kill most bacteria, and a few bacteria can actually grow at temperatures well below freezing.

Q Which bacterium would theoretically be more likely to grow at refrigerator temperatures: a human intestinal pathogen or a soilborne plant pathogen?

Refrigeration is the most common method of preserving household food supplies. It is based on the principle that microbial reproductive rates decrease at low temperatures. Although microbes usually survive even subfreezing temperatures (they might become entirely dormant), they gradually decline in number. Some species decline faster than others. Psychrotrophs actually do not grow well at low temperatures, except in comparison with other organisms; given time, however, they are able to slowly degrade food. Such spoilage might take the form of mold mycelium, slime on food surfaces, or off-tastes or off-colors in foods. The temperature inside a properly set refrigerator will greatly slow the growth of most spoilage organisms and will entirely prevent the growth of all but a few pathogenic bacteria. Figure 6.2 illustrates the importance of low temperatures for preventing the growth of spoilage and disease organisms. When large amounts of food must be refrigerated, it is important to keep in mind the slow cooling rate of a large quantity of warm food (Figure 6.3).

Mesophiles, with an optimum growth temperature of 25–40°C, are the most common type of microbe. Organisms that have adapted to live in the bodies of animals usually have an optimum temperature close to that of their hosts. The optimum temperature for many pathogenic bacteria is about 37°C, and incubators for clinical cultures are usually set at about this temperature. The mesophiles include most of the common spoilage and disease organisms.

Thermophiles are microorganisms capable of growth at high temperatures. Many of these organisms have an optimum growth temperature of 50–60°C, about the temperature of water from a hot water tap. Such temperatures can also be reached in sunlit soil and in thermal waters such as hot springs. Remarkably, many thermophiles cannot grow at temperatures below about 45°C. Endospores formed by thermophilic bacteria are unusually heat resistant and may survive the usual heat treatment given canned goods. Although elevated storage temperatures may cause surviving endospores to germinate and grow, thereby spoiling the food, these thermophilic bacteria are not considered a public health problem. Thermophiles are important in organic compost piles (see Figure 27.10), in which the temperature can rise rapidly to 50–60°C.

Some microbes, members of the Archaea (page 4), have an optimum growth temperature of 80°C or higher. These organisms are called **hyperthermophiles,** or sometimes **extreme thermophiles.** Most of these organisms live in hot springs associated with volcanic activity; sulfur is usually important in their metabolic activity. The known record for bacterial growth and replication at high temperatures is about 121°C near deep-sea hydrothermal vents. The immense pressure in the ocean depths prevents water from boiling even at temperatures well above 100°C.

pH

Recall from Chapter 2 (page 35) that pH refers to the acidity or alkalinity of a solution. Most bacteria grow best in a narrow pH range near neutrality, between pH 6.5 and 7.5. Very few bacteria

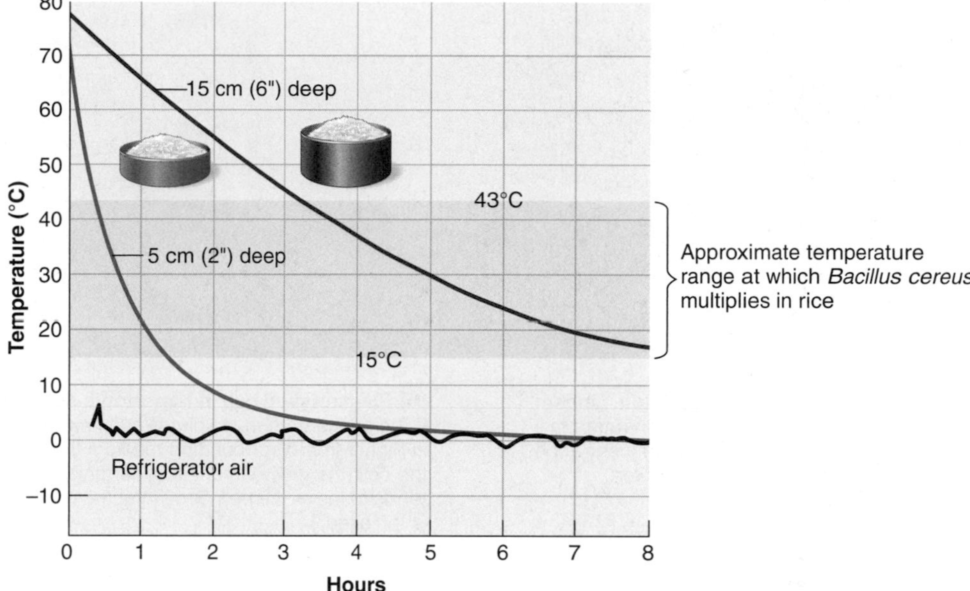

Figure 6.3 The effect of the amount of food on its cooling rate in a refrigerator and its chance of spoilage. Notice that in this example, the pan of rice with a depth of 5 cm (2 in) cooled through the incubation temperature range of the *Bacillus cereus* in about 1 hour, whereas the pan of rice with a depth of 15 cm (6 in) remained in this temperature range for about 5 hours.

Q Given a shallow pan and a deep pot with the same volume, which would cool faster?

grow at an acidic pH below about pH 4. This is why a number of foods, such as sauerkraut, pickles, and many cheeses, are preserved from spoilage by acids produced by bacterial fermentation. Nonetheless, some bacteria, called **acidophiles,** are remarkably tolerant of acidity. One type of chemoautotrophic bacteria, which is found in the drainage water from coal mines and oxidizes sulfur to form sulfuric acid, can survive at a pH value of 1. Molds and yeasts will grow over a greater pH range than bacteria will, but the optimum pH of molds and yeasts is generally below that of bacteria, usually about pH 5 to 6. Alkalinity also inhibits microbial growth but is rarely used to preserve foods.

When bacteria are cultured in the laboratory, they often produce acids that eventually interfere with their own growth. To neutralize the acids and maintain the proper pH, chemical buffers are included in the growth medium. The peptones and amino acids in some media act as buffers, and many media also contain phosphate salts. Phosphate salts have the advantage of exhibiting their buffering effect in the pH growth range of most bacteria. They are also nontoxic; in fact, they provide phosphorus, an essential nutrient.

Osmotic Pressure

Microorganisms obtain almost all their nutrients in solution from the surrounding water. Thus, they require water for growth, and their composition is 80–90% water. High osmotic pressures have the effect of removing necessary water from a cell. When a microbial cell is in a solution whose concentration of solutes is higher than in the cell (the environment is *hypertonic* to the cell), the cellular water passes out through the plasma membrane to the high solute concentration. (See the discussion of osmosis in Chapter 4, pages 92–94, and review Figure 4.18 for the three types of solution environments a cell may encounter.) This osmotic loss of water causes **plasmolysis,** or shrinkage of the cell's cytoplasm (Figure 6.4).

The importance of this phenomenon is that the growth of the cell is inhibited as the plasma membrane pulls away from the cell wall. Thus, the addition of salts (or other solutes) to a solution, and the resulting increase in osmotic pressure, can be used to preserve foods. Salted fish, honey, and sweetened condensed milk are preserved largely by this mechanism; the high salt or sugar concentrations draw water out of any microbial cells that are present and thus prevent their growth. These effects of osmotic pressure are roughly related to the *number* of dissolved molecules and ions in a volume of solution.

Some organisms, called **extreme halophiles,** have adapted so well to high salt concentrations that they actually require them for growth. In this case, they may be termed **obligate halophiles.** Organisms from such saline waters as the Dead Sea often require nearly 30% salt, and the inoculating loop (a device for handling bacteria in the laboratory) used to transfer them must first be dipped into a saturated salt solution. More common are **facultative halophiles,** which do not require high salt concentrations but are able to grow at salt concentrations up to 2%, a concentration that inhibits the growth of many other organisms. A few species of facultative halophiles can tolerate even 15% salt.

Most microorganisms, however, must be grown in a medium that is nearly all water. For example, the concentration of agar (a complex polysaccharide isolated from marine algae) used to solidify microbial growth media is usually about 1.5%. If markedly higher concentrations are used, the increased osmotic pressure can inhibit the growth of some bacteria.

If the osmotic pressure is unusually low (the environment is *hypotonic*)—such as in distilled water, for example—water tends to enter the cell rather than leave it. Some microbes that have a relatively weak cell wall may be lysed by such treatment.

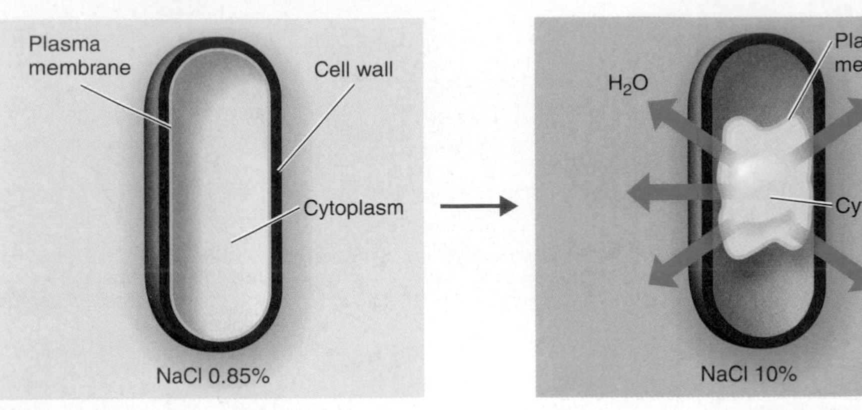

(a) Normal cell in isotonic solution. Under these conditions, the solute concentration in the cell is equivalent to a solute concentration of 0.85% sodium chloride (NaCl). See Figure 4.18.

(b) Plasmolyzed cell in hypertonic solution. If the concentration of solutes such as NaCl is higher in the surrounding medium than in the cell (the environment is hypertonic), water tends to leave the cell. Growth of the cell is inhibited.

Figure 6.4 Plasmolysis.

Q Name a food preserved by high osmotic pressure.

CHECK YOUR UNDERSTANDING

- ✓ Why are hyperthermophiles that grow at temperatures above 100°C seemingly limited to oceanic depths? **6-1**
- ✓ Other than controlling acidity, what is an advantage of using phosphate salts as buffers in growth media? **6-2**
- ✓ Why might primitive civilizations have used food preservation techniques that rely on osmotic pressure? **6-3**

Chemical Requirements

Carbon

Besides water, one of the most important requirements for microbial growth is carbon. Carbon is the structural backbone of living matter; it is needed for all the organic compounds that make up a living cell. Half the dry weight of a typical bacterial cell is carbon. Chemoheterotrophs get most of their carbon from the source of their energy—organic materials such as proteins, carbohydrates, and lipids. Chemoautotrophs and photoautotrophs derive their carbon from carbon dioxide.

Nitrogen, Sulfur, and Phosphorus

In addition to carbon, microorganisms need other elements to synthesize cellular material. For example, protein synthesis requires considerable amounts of nitrogen as well as some sulfur. The syntheses of DNA and RNA also require nitrogen and some phosphorus, as does the synthesis of ATP, the molecule so important for the storage and transfer of chemical energy within the cell. Nitrogen makes up about 14% of the dry weight of a bacterial cell, and sulfur and phosphorus together constitute about another 4%.

Organisms use nitrogen primarily to form the amino group of the amino acids of proteins. Many bacteria meet this requirement by decomposing protein-containing material and reincorporating the amino acids into newly synthesized proteins and other nitrogen-containing compounds. Other bacteria use nitrogen from ammonium ions (NH_4^+), which are already in the reduced form and are usually found in organic cellular material. Still other bacteria are able to derive nitrogen from nitrates (compounds that dissociate to give the nitrate ion, NO_3^-, in solution).

Some important bacteria, including many of the photosynthesizing cyanobacteria (page 140), use gaseous nitrogen (N_2) directly from the atmosphere. This process is called **nitrogen fixation.** Some organisms that can use this method are free-living, mostly in the soil, but others live cooperatively in symbiosis with the roots of legumes such as clover, soybeans, alfalfa, beans, and peas. The nitrogen fixed in the symbiosis is used by both the plant and the bacterium (see Chapter 27).

Sulfur is used to synthesize sulfur-containing amino acids and vitamins such as thiamine and biotin. Important natural sources of sulfur include the sulfate ion (SO_4^{2-}), hydrogen sulfide, and the sulfur-containing amino acids.

Phosphorus is essential for the synthesis of nucleic acids and the phospholipids of cell membranes. Among other places, it is also found in the energy bonds of ATP. A source of phosphorus is the phosphate ion (PO_4^{3-}). Potassium, magnesium, and calcium are also elements that microorganisms require, often as cofactors for enzymes (see Chapter 5, pages 116–117).

Trace Elements

Microbes require very small amounts of other mineral elements, such as iron, copper, molybdenum, and zinc; these are referred to as **trace elements.** Most are essential for the functions of certain enzymes, usually as cofactors. Although these elements are sometimes added to a laboratory medium, they are usually assumed to be naturally present in tap water and other components of media. Even most distilled waters contain adequate amounts, but tap water is sometimes specified to ensure that these trace minerals will be present in culture media.

Table 6.1 The Effect of Oxygen on the Growth of Various Types of Bacteria

	a. Obligate Aerobes	b. Facultative Anaerobes	c. Obligate Anaerobes	d. Aerotolerant Anaerobes	e. Microaerophiles
Effect of Oxygen on Growth	Only aerobic growth; oxygen required	Both aerobic and anaerobic growth; greater growth in presence of oxygen	Only anaerobic growth; ceases in presence of oxygen	Only anaerobic growth; but continues in presence of oxygen	Only aerobic growth; oxygen required in low concentration
Bacterial Growth in Tube of Solid Growth Medium					
Explanation of Growth Patterns	Growth occurs only where high concentrations of oxygen have diffused into the medium	Growth is best where most oxygen is present, but occurs throughout tube	Growth occurs only where there is no oxygen	Growth occurs evenly; oxygen has no effect	Growth occurs only where a low concentration of oxygen has diffused into medium
Explanation of Oxygen's Effects	Presence of enzymes catalase and superoxide dismutase (SOD) allows toxic forms of oxygen to be neutralized; can use oxygen	Presence of enzymes catalase and SOD allows toxic forms of oxygen to be neutralized; can use oxygen	Lacks enzymes to neutralize harmful forms of oxygen; cannot tolerate oxygen	Presence of one enzyme, SOD, allows harmful forms of oxygen to be partially neutralized; tolerates oxygen	Produce lethal amounts of toxic forms of oxygen if exposed to normal atmospheric oxygen

Oxygen

We are accustomed to thinking of molecular oxygen (O_2) as a necessity of life, but it is actually in a sense a poisonous gas. Very little molecular oxygen existed in the atmosphere during most of Earth's history—in fact, it is possible that life could not have arisen had oxygen been present. However, many current forms of life have metabolic systems that require oxygen for aerobic respiration. As we have seen, hydrogen atoms that have been stripped from organic compounds combine with oxygen to form water, as shown in Figure 5.14 (page 129). This process yields a great deal of energy while neutralizing a potentially toxic gas—a very neat solution, all in all.

Microbes that use molecular oxygen (aerobes) produce more energy from nutrients than microbes that do not use oxygen (anaerobes). Organisms that require oxygen to live are called **obligate aerobes** (Table 6.1a).

Q&A Obligate aerobes are at a disadvantage because oxygen is poorly soluble in the water of their environment. Therefore, many of the aerobic bacteria have developed, or retained, the ability to continue growing in the absence of oxygen. Such organisms are called **facultative anaerobes** (Table 6.1b). In other words, facultative anaerobes can use oxygen when it is present but are able to continue growth by using fermentation or anaerobic respiration when oxygen is not available. However, their efficiency in producing energy decreases in the absence of oxygen. Examples of facultative anaerobes are the familiar *Escherichia coli* that are found in the human intestinal tract. Many yeasts are also facultative anaerobes. Recall from the discussion of anaerobic respiration in Chapter 5 (page 132) that many microbes are able to substitute other electron acceptors, such as nitrate ions, for oxygen, which is something humans are unable to do.

Obligate anaerobes (Table 6.1c) are bacteria that are unable to use molecular oxygen for energy-yielding reactions. In fact, most are harmed by it. The genus *Clostridium* (klôs-tri′dē-um), which contains the species that cause tetanus and botulism, is the most familiar example. These bacteria do use oxygen atoms present in cellular materials; the atoms are usually obtained from water.

Understanding how organisms can be harmed by oxygen requires a brief discussion of the toxic forms of oxygen:

1. **Singlet oxygen** ($^1O_2^-$) is normal molecular oxygen (O_2) that has been boosted into a higher-energy state and is extremely reactive.
2. **Superoxide radicals** ($O_2^{\cdot-}$), or **superoxide anions,** are formed in small amounts during the normal respiration of organisms that use oxygen as a final electron acceptor, forming water. In the presence of oxygen, obligate anaerobes also appear to form some superoxide radicals, which are so

toxic to cellular components that all organisms attempting to grow in atmospheric oxygen must produce an enzyme, **superoxide dismutase (SOD),** to neutralize them. Their toxicity is caused by their great instability, which leads them to steal an electron from a neighboring molecule, which in turn becomes a radical and steals an electron, and so on. Aerobic bacteria, facultative anaerobes growing aerobically, and aerotolerant anaerobes (discussed shortly) produce SOD, with which they convert the superoxide radical into molecular oxygen (O_2) and hydrogen peroxide (H_2O_2):

$$O_2^{\bar{\cdot}} + O_2^{\bar{\cdot}} + 2\,H^+ \longrightarrow H_2O_2 + O_2$$

3. The hydrogen peroxide produced in this reaction contains the **peroxide anion** O_2^{2-} and is also toxic. In Chapter 7 (page 202) we will encounter it as the active principle in the antimicrobial agents hydrogen peroxide and benzoyl peroxide. Because the hydrogen peroxide produced during normal aerobic respiration is toxic, microbes have developed enzymes to neutralize it. The most familiar of these is **catalase,** which converts it into water and oxygen:

$$2\,H_2O_2 \longrightarrow 2\,H_2O + O_2$$

Catalase is easily detected by its action on hydrogen peroxide. When a drop of hydrogen peroxide is added to a colony of bacterial cells producing catalase, oxygen bubbles are released. Anyone who has put hydrogen peroxide on a wound will recognize that cells in human tissue also contain catalase. The other enzyme that breaks down hydrogen peroxide is **peroxidase,** which differs from catalase in that its reaction does not produce oxygen:

$$H_2O_2 + 2\,H^+ \longrightarrow 2\,H_2O$$

Another important form of reactive oxygen, **ozone (O_3),** is also discussed on page 202.

4. The **hydroxyl radical** ($OH\cdot$) is another intermediate form of oxygen and probably the most reactive. It is formed in the cellular cytoplasm by ionizing radiation. Most aerobic respiration produces traces of hydroxyl radicals, but they are transient.

These toxic forms of oxygen are an essential component of one of the body's most important defenses against pathogens, phagocytosis (see page 457 and Figure 16.7). In the phagolysosome of the phagocytic cell, ingested pathogens are killed by exposure to singlet oxygen, superoxide radicals, peroxide anions of hydrogen peroxide, and hydroxyl radicals and other oxidative compounds.

Obligate anaerobes usually produce neither superoxide dismutase nor catalase. Because aerobic conditions probably lead to an accumulation of superoxide radicals in their cytoplasm, obligate anaerobes are extremely sensitive to oxygen.

Aerotolerant anaerobes (Table 6.1d) cannot use oxygen for growth, but they tolerate it fairly well. On the surface of a solid medium, they will grow without the use of special techniques (discussed later) required for obligate anaerobes. Many of the aerotolerant bacteria characteristically ferment carbohydrates to lactic acid. As lactic acid accumulates, it inhibits the growth of aerobic competitors and establishes a favorable ecological niche for lactic acid producers. A common example of lactic acid–producing aerotolerant anaerobes is the lactobacilli used in the production of many acidic fermented foods, such as pickles and cheese. In the laboratory, they are handled and grown much like any other bacteria, but they make no use of the oxygen in the air. These bacteria can tolerate oxygen because they possess SOD or an equivalent system that neutralizes the toxic forms of oxygen previously discussed.

A few bacteria are **microaerophiles** (Table 6.1e). They are aerobic; they do require oxygen. However, they grow only in oxygen concentrations lower than those in air. In a test tube of solid nutrient medium, they grow only at a depth where small amounts of oxygen have diffused into the medium; they do not grow near the oxygen-rich surface or below the narrow zone of adequate oxygen. This limited tolerance is probably due to their sensitivity to superoxide radicals and peroxides, which they produce in lethal concentrations under oxygen-rich conditions.

Organic Growth Factors

Essential organic compounds an organism is unable to synthesize are known as **organic growth factors;** they must be directly obtained from the environment. One group of organic growth factors for humans is vitamins. Most vitamins function as coenzymes, the organic cofactors required by certain enzymes in order to function. Many bacteria can synthesize all their own vitamins and do not depend on outside sources. However, some bacteria lack the enzymes needed for the synthesis of certain vitamins, and for them those vitamins are organic growth factors. Other organic growth factors required by some bacteria are amino acids, purines, and pyrimidines.

CHECK YOUR UNDERSTANDING

- ✓ If bacterial cells were given a sulfur source containing radioactive sulfur (^{35}S) in their culture media, in what molecules would the ^{35}S be found in the cells? **6-4**
- ✓ How would one determine whether a microbe is a strict anaerobe? **6-5**
- ✓ Oxygen is so pervasive in the environment that it would be very difficult for a microbe to always avoid physical contact with it. What, therefore, is the most obvious way for a microbe to avoid damage? **6-6**

Biofilms

LEARNING OBJECTIVE

6-7 Describe the formation of biofilms and their potential for causing infection.

In nature, microorganisms seldom live in the isolated single-species colonies that we see on laboratory plates. They more typically live in communities called **biofilms.** This fact was not well appreciated until the development of confocal microscopy (see page 62) made the three-dimensional structure of biofilms more visible. Biofilms reside in a matrix made up primarily of polysaccharides, but also containing DNA and proteins, that is often informally called *slime.* A biofilm also can be considered a *hydrogel,* which is a complex polymer containing many times its dry weight in water. Cell-to-cell chemical communication, or *quorum sensing,* allows bacteria to coordinate their activity and group together into communities that provide benefits not unlike those of multicellular organisms (see the box in Chapter 3, page 57). Therefore, biofilms are not just bacterial slime layers but biological systems; the bacteria are organized into a coordinated, functional community. Biofilms are usually attached to a surface, such as a rock in a pond, a human tooth (plaque; see Figure 25.3 on page 708), or a mucous membrane. This community might be of a single species or of a diverse group of microorganisms. Biofilms also might take other, more varied forms. The floc that forms in certain types of sewage treatment (see Figure 27.20, page 786) is an example. In fast-flowing streams, the biofilm might be in the form of filamentous streamers. Within a biofilm community, the bacteria are able to share nutrients and are sheltered from harmful factors in the environment, such as desiccation, antibiotics, and the body's immune system. The close proximity of microorganisms within a biofilm might also have the advantage of facilitating the transfer of genetic information by, for example, conjugation.

A biofilm usually begins to form when a free-swimming (*planktonic*) bacterium attaches to a surface. If these bacteria grew in a uniformly thick monolayer, they would become overcrowded, nutrients would not be available in lower depths, and toxic wastes could accumulate. Microorganisms in biofilm communities sometimes avoid these problems by forming pillar-like structures (**Figure 6.5**) with channels between them, through which water can carry incoming nutrients and outgoing wastes. This constitutes a primitive circulatory system. Individual microbes and clumps of slime occasionally leave the established biofilm and move to a new location where the biofilm becomes extended. Such a biofilm is generally composed of a surface layer about 10 μm thick, with pillars that extend up to 200 μm above it.

The microorganisms in biofilms can work cooperatively to carry out complex tasks. For example, the digestive systems of ruminant animals, such as cattle, require many different microbial species to break down cellulose. The microbes in a ruminant's digestive system are located mostly within biofilm communities. Biofilms are also essential elements in the proper functioning of sewage treatment systems, which we will discuss in Chapter 27. They can also, however, be a problem in pipes and tubing, where their accumulations impede circulation.

Biofims are an important factor in human health. For example, microbes in biofilms are probably 1000 times more resistant to microbicides. Experts at the Centers for Disease Control and Prevention (CDC) estimate that 70% of human bacterial infections involve biofilms. Most nosocomial infections (infections

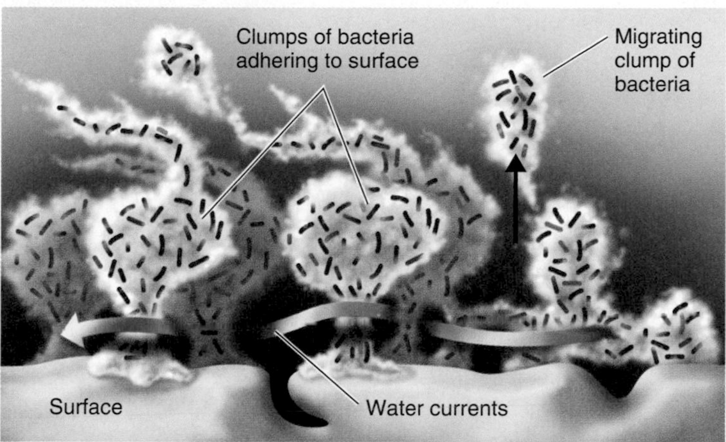

Water currents move, as shown by the blue arrow, among pillars of slime formed by the growth of bacteria attached to solid surfaces. This allows efficient access to nutrients and removal of bacterial waste products. Individual slime-forming bacteria or bacteria in clumps of slime detach and move to new locations. See Figure 1.8.

Figure 6.5 Biofilms.

Q What is the biofilm that forms on teeth called?

acquired in health care facilities) are probably related to biofilms on medical catheters (see Figure 1.8 on page 19 and Figure 21.3 on page 587). In fact, biofilms form on almost all indwelling medical devices, including mechanical heart valves. Biofilms, which also can include those formed by fungi such as *Candida,* are encountered in many disease conditions, such as infections related to the use of contact lenses, dental caries (see page 707), and infections by pseudomonad bacteria (see page 308). See the box on page 164.

One approach to preventing biofilm formation is to incorporate antimicrobials into surfaces on which biofilms might form (see page 57). Because the chemical signals that allow quorum sensing are essential to biofilm formation, research is underway to determine the makeup of these chemical signals and perhaps block them. Another approach involves the discovery that lactoferrin (see page 470), which is abundant in many human secretions, can inhibit biofilm formation. Lactoferrin binds iron, especially among the pseudomonads that are responsible for cystic fibrosis biofilms, the cause of the pathology of this hereditary disease. The lack of iron inhibits the surface motility essential for the aggregation of the bacteria into biofilms.

Most laboratory methods in microbiology today use organisms being cultured in their planktonic mode. However, microbiologists now predict that there will be an increasing focus on how microorganisms actually live in relation with one another and that this will be considered in industrial and medical research.

CHECK YOUR UNDERSTANDING

✓ Identify a way in which pathogens find it advantageous to form biofilms. **6-7**

CLINICAL FOCUS From the *Morbidity and Mortality Weekly Report*

Delayed Bloodstream Infection Following Catheterization

As you read through this box, you will encounter a series of questions that infection control officers ask themselves as they trace an infection. Try to answer each question before going on to the next one.

1. In early March, an intravenous heparin solution was recalled after patients in four states developed *Pseudomonas fluorescens* bloodstream infections. The heparin solution became contaminated during manufacture.

   After examining Figure A below, would you consider the recall effective?

2. Three months after the recall, patients in two different states developed bloodstream infections.

   What do you need to know?

3. The patients' last exposure to the contaminated heparin was 84 to 421 days before onset of their infections. These patients did not develop infections during the January–February outbreak. The patients all had indwelling venous catheters; these tubes are inserted into a vein for long-term delivery of concentrated solutions, such as anticancer drugs.

   What is the next step?

4. On-site investigations confirmed that the patients' clinics were no longer using and had returned the recalled heparin. Cultures of the new heparin being used did not recover any organisms.

   What do you do next?

5. Blood and catheter cultures were made (**Figure B**)?

   What is the result shown in Figure B?

6. *P. fluorescens* was cultured from 15 patients and from 17 catheters These are the first known cases of substantially delayed bloodstream infections (i.e., 84–421 days) after exposure to a contaminated intravenous solution.

   What was the source of these infections?

7. Scanning electron microscopy at the Centers for Disease Control and Prevention showed that *P. fluorescens* colonized the inside of the catheters by forming biofilms; previous electron microscopy studies have indicated that nearly all indwelling vascular catheters become colonized by microorganisms embedded in a biofilm layer. Heparin has been reported to stimulate biofilm formation.

   Why did *Pseudomonas* cause infection an average of 237 days after exposure to the contaminated heparin?

8. Although *P. fluorescens* might not have entered patient bloodstreams in sufficient quantities to cause symptoms on initial exposure to the contaminated heparin, biofilm formation enabled the bacteria to persist in patient catheters. The bacteria might have proliferated in the biofilm, from which they were disrupted by subsequent, uncontaminated intravenous solutions and released into the bloodstream, finally causing symptoms.

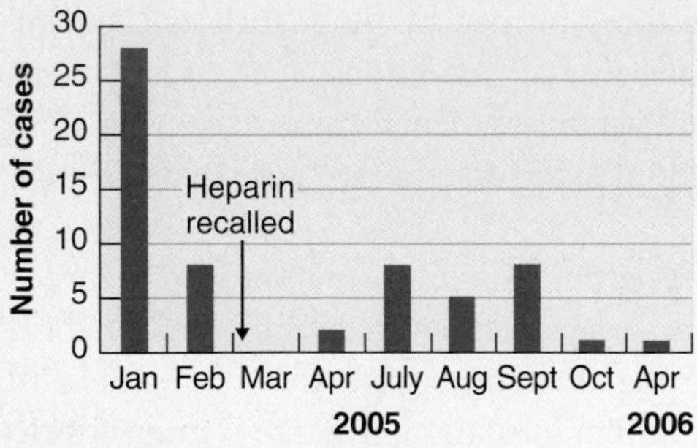

Figure A Occurrence of *P. fluorescens* bloodstream infections in patients with intravenous catheters.

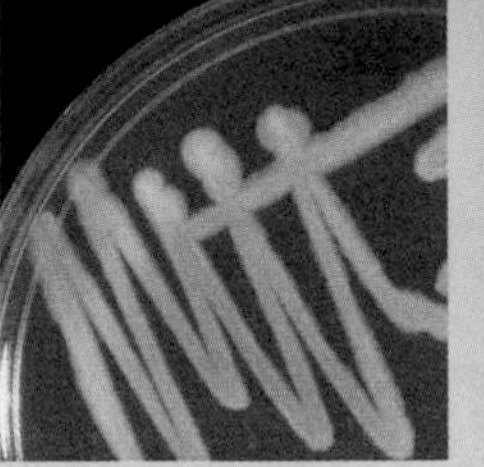
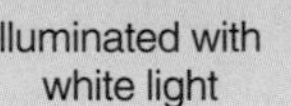
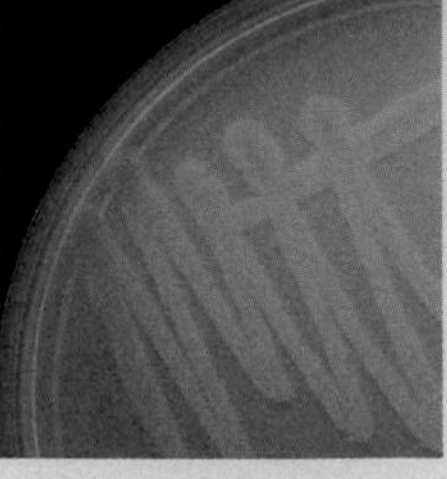

Figure B *P. fluorescens* is an aerobic, gram-negative rod that grows best at temperatures of approximately 25 to 30°C and grows poorly at the standard hospital microbiology incubation temperature (approximately 36°C). The bacteria produce a pigment that fluoresces under ultraviolet light.

Source: Adapted from *MMWR* 55(35): 961–963 (9/8/06).

Culture Media

LEARNING OBJECTIVES

6-8 Distinguish chemically defined and complex media.

6-9 Justify the use of each of the following: anaerobic techniques, living host cells, candle jars, selective and differential media, enrichment medium.

6-10 Differentiate biosafety levels 1, 2, 3, and 4.

A nutrient material prepared for the growth of microorganisms in a laboratory is called a **culture medium.** Some bacteria can grow well on just about any culture medium; others require special media, and still others cannot grow on any nonliving medium yet developed. Microbes that are introduced into a culture medium to initiate growth are called an **inoculum.** The microbes that grow and multiply in or on a culture medium are referred to as a **culture.**

Suppose we want to grow a culture of a certain microorganism, perhaps the microbes from a particular clinical specimen. What criteria must the culture medium meet? First, it must contain the right nutrients for the specific microorganism we want to grow. It should also contain sufficient moisture, a properly adjusted pH, and a suitable level of oxygen, perhaps none at all. The medium must initially be **sterile**—that is, it must initially contain no living microorganisms—so that the culture will contain only the microbes (and their offspring) we add to the medium. Finally, the growing culture should be incubated at the proper temperature.

A wide variety of media are available for the growth of microorganisms in the laboratory. Most of these media, which are available from commercial sources, have premixed components and require only the addition of water and then sterilization. Media are constantly being developed or revised for use in the isolation and identification of bacteria that are of interest to researchers in such fields as food, water, and clinical microbiology.

When it is desirable to grow bacteria on a solid medium, a solidifying agent such as agar is added to the medium. A complex polysaccharide derived from a marine alga, **agar** has long been used as a thickener in foods such as jellies and ice cream.

Agar has some very important properties that make it valuable to microbiology, and no satisfactory substitute has ever been found. Few microbes can degrade agar, so it remains solid. Also, agar liquefies at about 100°C (the boiling point of water) and at sea level remains liquid until the temperature drops to about 40°C. For laboratory use, agar is held in water baths at about 50°C. At this temperature, it does not injure most bacteria when it is poured over them (as shown in Figure 6.17a, page 176). Once the agar has solidified, it can be incubated at temperatures approaching 100°C before it again liquefies; this property is particularly useful when thermophilic bacteria are being grown.

Agar media are usually contained in test tubes or *Petri dishes.* The test tubes are called *slants* when they are allowed to solidify with the tube held at an angle so that a large surface area for growth is available. When the agar solidifies in a vertical tube, it is called a *deep.* Petri dishes, named for their inventor, are shallow dishes with a lid that nests over the bottom to prevent contamination; when filled, they are called *Petri* (or culture) *plates.*

Chemically Defined Media

To support microbial growth, a medium must provide an energy source, as well as sources of carbon, nitrogen, sulfur, phosphorus, and any organic growth factors the organism is unable to synthesize. A **chemically defined medium** is one whose exact chemical composition is known. For a chemoheterotroph, the chemically defined medium must contain organic growth factors that serve as a source of carbon and energy. For example, as shown in **Table 6.2**, glucose is included in the medium for growing the chemoheterotroph *E. coli.*

As **Table 6.3** shows, many organic growth factors must be provided in the chemically defined medium used to cultivate a species of *Neisseria* (page 306). Organisms that require many growth factors are described as *fastidious.* Organisms of this type, such as *Lactobacillus* (page 318), are sometimes used in tests that determine the concentration of a particular vitamin in a substance. To perform such a *microbiological assay,* a growth medium is prepared that contains all the growth requirements of the bacterium except the vitamin being assayed. Then the medium, test substance, and bacterium are combined, and the growth of bacteria is measured. This bacterial growth, which is reflected by the amount of lactic acid produced, will be proportional to the amount of vitamin in the test substance. The more lactic acid, the more the *Lactobacillus* cells have been able to grow, so the more vitamin is present.

Table 6.2 A Chemically Defined Medium for Growing a Typical Chemoheterotroph, Such as *Escherichia coli*

Constituent	Amount
Glucose	5.0 g
Ammonium phosphate, monobasic ($NH_4H_2PO_4$)	1.0 g
Sodium chloride (NaCl)	5.0 g
Magnesium sulfate ($MgSO_4 \cdot 7H_2O$)	0.2 g
Potassium phosphate, dibasic (K_2HPO_4)	1.0 g
Water	1 liter

Complex Media

Chemically defined media are usually reserved for laboratory experimental work or for the growth of autotrophic bacteria. Most heterotrophic bacteria and fungi, such as you would work with in an introductory lab course, are routinely grown on **complex media** made up of nutrients including extracts from yeasts, meat, or plants, or digests of proteins from these and other sources. The exact chemical composition varies slightly from batch to batch. **Table 6.4** shows one widely used recipe.

In complex media, the energy, carbon, nitrogen, and sulfur requirements of the growing microorganisms are provided primarily by protein. Protein is a large, relatively insoluble molecule that a minority of microorganisms can utilize directly, but a partial digestion by acids or enzymes reduces protein to shorter chains of amino acids called *peptones.* These small, soluble fragments can be digested by most bacteria.

Vitamins and other organic growth factors are provided by meat extracts or yeast extracts. The soluble vitamins and minerals from the meats or yeasts are dissolved in the extracting water, which is then evaporated so that these factors are concentrated. (These extracts also supplement the organic nitrogen and carbon compounds.) Yeast extracts are particularly rich in the B vitamins. If a complex medium is in liquid form, it is called **nutrient broth.** When agar is added, it is called **nutrient agar.** (This terminology can be confusing; just remember that agar itself is not a nutrient.)

Table 6.3 A Chemically Defined Medium for Growing a Fastidious Chemoheterotrophic Bacterium, Such as *Neisseria gonorrhoeae*

Constituent	Amount	Constituent	Amount
Carbon and Energy Sources		**Amino Acids**	
Glucose	9.1 g	Cysteine	1.5 g
Starch	9.1 g	Arginine, proline (each)	0.3 g
Sodium acetate	1.8 g	Glutamic acid, methionine (each)	0.2 g
Sodium citrate	1.4 g	Asparagine, isoleucine, serine (each)	0.2 g
Oxaloacetate	0.3 g	Cystine	0.06 g
Salts		**Organic Growth Factors**	
Potassium phosphate, dibasic (K_2HPO_4)	12.7 g	Calcium pantothenate	0.02 g
Sodium chloride (NaCl)	6.4 g	Thiamine	0.02 g
Potassium phosphate, monobasic (KH_2PO_4)	5.5 g	Nicotinamide adenine dinucleotide	0.01 g
Sodium bicarbonate ($NaHCO_3$)	1.2 g	Uracil	0.006 g
Potassium sulfate (K_2SO_4)	1.1 g	Biotin	0.005 g
Sodium sulfate (Na_2SO_4)	0.9 g	Hypoxanthine	0.003 g
Magnesium chloride ($MgCl_2$)	0.5 g	**Reducing Agent**	
Ammonium chloride (NH_4Cl)	0.4 g	Sodium thioglycolate	0.00003 g
Potassium chloride (KCl)	0.4 g	**Water**	1 liter
Calcium chloride ($CaCl_2$)	0.006 g		
Ferric nitrate [$Fe(NO_3)_3$]	0.006 g		

Source: R. M. Atlas, *Handbook of Microbiological Media,* Ann Arbor, MI: CRC Press, 1993.

Table 6.4 Composition of Nutrient Agar, a Complex Medium for the Growth of Heterotrophic Bacteria

Constituent	Amount
Peptone (partially digested protein)	5.0 g
Beef extract	3.0 g
Sodium chloride	8.0 g
Agar	15.0 g
Water	1 liter

Anaerobic Growth Media and Methods

The cultivation of anaerobic bacteria poses a special problem. Because anaerobes might be killed by exposure to oxygen, special media called **reducing media** must be used. These media contain ingredients, such as sodium thioglycolate, that chemically combine with dissolved oxygen and deplete the oxygen in the culture medium. To routinely grow and maintain pure cultures of obligate anaerobes, microbiologists use reducing media stored in ordinary, tightly capped test tubes. These media are heated shortly before use, to drive off absorbed oxygen.

When the culture must be grown in Petri plates to observe individual colonies, several methods are available. Laboratories that work with relatively few culture plates at a time can use systems that can incubate the microorganisms in sealed boxes and jars in which the oxygen is chemically removed after the culture plates have been introduced and the container sealed. Some systems require that water be added to an envelope of chemicals before the container is closed, as shown in **Figure 6.6**, and require a catalyst. The chemicals produce hydrogen and carbon dioxide (about 4–10%) and remove the oxygen in the container by combining it, in the presence of the catalyst, with hydrogen to form water. In another commercially available system, the envelope of chemicals (the active ingredient is ascorbic acid) is simply opened to expose it to oxygen in the container's atmosphere. No water or catalyst is needed. The atmosphere in such containers usually has less than 5% oxygen, about 18% CO_2, and no hydrogen. In a

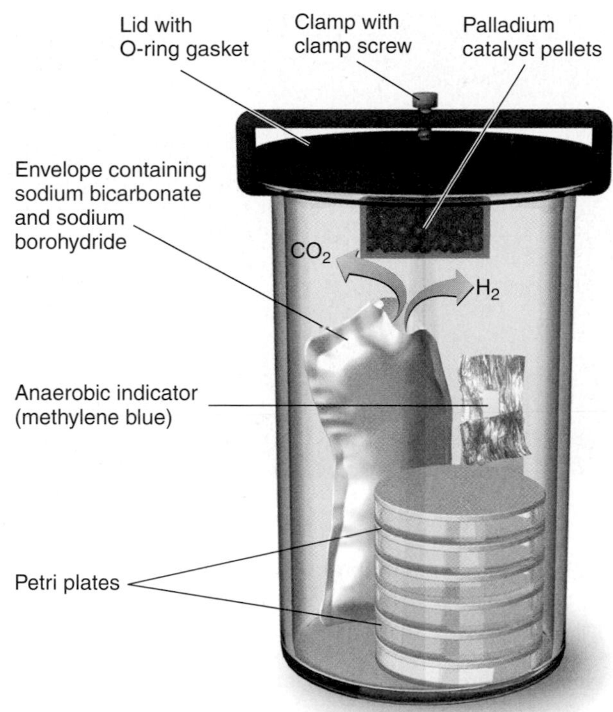

Figure 6.6 A jar for cultivating anaerobic bacteria on Petri plates. When water is mixed with the chemical packet containing sodium bicarbonate and sodium borohydride, hydrogen and carbon dioxide are generated. Reacting on the surface of a palladium catalyst in a screened reaction chamber, which may also be incorporated into the chemical packet, the hydrogen and atmospheric oxygen in the jar combine to form water. The oxygen is thus removed. Also in the jar is an anaerobic indicator containing methylene blue, which is blue when oxidized and turns colorless when the oxygen is removed (as shown here).

Q What is the technical name for bacteria that require a higher-than-atmospheric-concentration of CO_2 for growth?

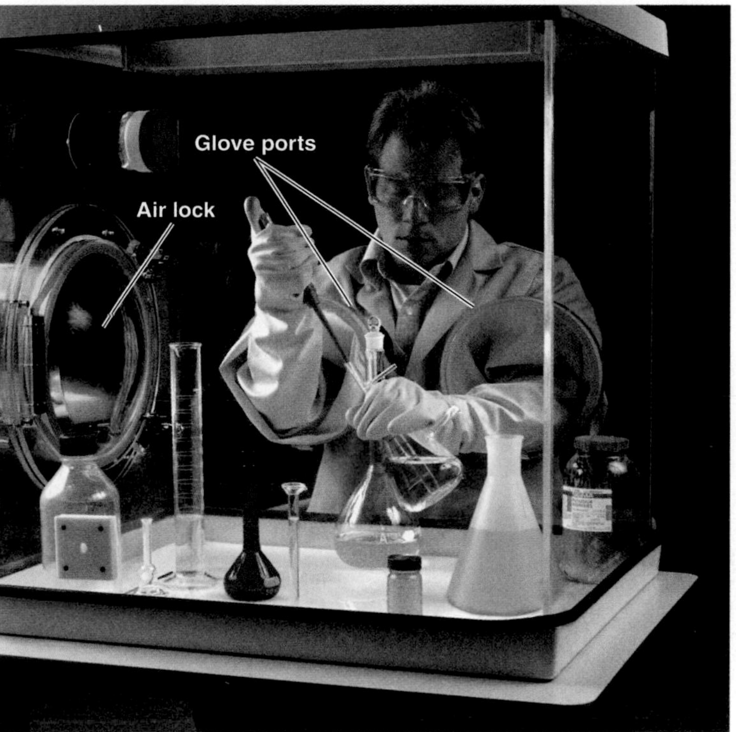

Figure 6.7 An anaerobic chamber. The technician is pipetting a bacterial suspension into a flask inside an anaerobic chamber filled with an inert, oxygen-free gas. His arms and hands are encased in glove ports. Organisms and materials enter and leave through the air-lock opening that is visible to the left.

Q In what way would an anaerobic chamber resemble the Space Laboratory orbiting in the vacuum of space?

recently introduced system, each individual Petri plate (OxyPlate) becomes an anaerobic chamber. The medium in the plate contains an enzyme, oxyrase, which combines oxygen with hydrogen, removing oxygen as water is formed.

Laboratories that have a large volume of work with anaerobes often use an anaerobic chamber, such as that shown in Figure 6.7. The chamber is filled with inert gases (typically about 85% N_2, 10% H_2, and 5% CO_2) and is equipped with air locks to introduce cultures and materials.

Special Culture Techniques

Many bacteria have never been successfully grown on artificial laboratory media. *Mycobacterium leprae,* the leprosy bacillus, is now usually grown in armadillos, which have a relatively low body temperature that matches the requirements of the microbe. Another example is the syphilis spirochete, although certain nonpathogenic strains of this microbe have been grown on laboratory media. With few exceptions, the obligate intracellular bacteria, such as the rickettsias and the chlamydias, do not grow on artificial media. Like viruses, they can reproduce only in a living host cell. See the discussion of cell culture, page 378.

Many clinical laboratories have special *carbon dioxide incubators* in which to grow aerobic bacteria that require concentrations of CO_2 higher or lower than that found in the atmosphere. Desired CO_2 levels are maintained by electronic controls. High CO_2 levels are also obtained with simple *candle jars.* Cultures are placed in a large sealed jar containing a lighted candle, which consumes oxygen. The candle stops burning when the air in the jar has a lowered concentration of oxygen (but one still adequate for the growth of aerobic bacteria). An elevated concentration of CO_2 is also present. Microbes that grow better at high CO_2 concentrations are called **capnophiles.** The low-oxygen, high-CO_2 conditions resemble those found in the intestinal tract, respiratory tract, and other body tissues where pathogenic bacteria grow.

Candle jars are still used occasionally, but more often commercially available chemical packets are used to generate carbon dioxide atmospheres in containers. When only one or two Petri plates of cultures are to be incubated, clinical laboratory investigators often use small plastic bags with self-contained chemical

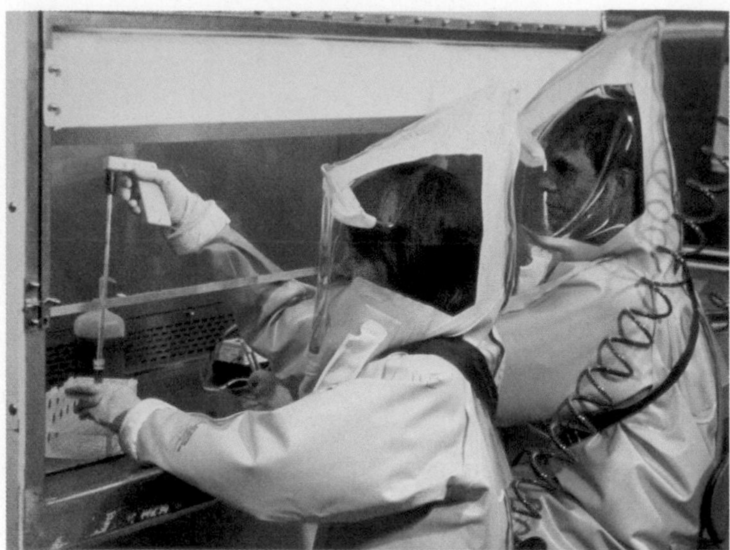

Figure 6.8 Technicians in a biosafety level 4 (BSL-4) laboratory. Personnel working in a BSL-4 facility wear a "space suit" that is connected to an outside air supply.

If a technician were working with pathogenic prions, how would material leaving the lab be rendered noninfectious? (*Hint*: See Chapter 7.)

gas generators that are activated by crushing the packet or moistening it with a few milliliters of water. These packets are sometimes specially designed to provide precise concentrations of carbon dioxide (usually higher than can be obtained in candle jars) and oxygen for culturing organisms such as the microaerophilic *Campylobacter* bacteria (page 312).

Some microorganisms are so dangerous that they can be handled only under extraordinary systems of containment called *biosafety level 4 (BSL-4)*. Level 4 labs are popularly known as "the hot zone." Only a handful of such labs exists in the United States. The lab is a sealed environment within a larger building and has an atmosphere under negative pressure, so that aerosols containing pathogens will not escape. Both intake and exhaust air is filtered through high-efficiency particulate air filters (see HEPA filters, page 191); the exhaust air is filtered twice. All waste materials leaving the lab are rendered noninfectious. The personnel wear "space suits" that are connected to an air supply (**Figure 6.8**).

Less dangerous organisms are handled at lower levels of biosafety. For example, a basic microbiology teaching laboratory would be BSL-1. Organisms that present a moderate risk of infection can be handled at BSL-2 levels, that is, on open laboratory benchtops with appropriate gloves, lab coats, or possibly face and eye protection. BSL-3 labs are intended for highly infectious airborne pathogens such as the tuberculosis agent. Biological safety cabinets similar in appearance to the anaerobic chamber shown in Figure 6.7 are used. The laboratory itself should be negatively pressurized and equipped with air filters to prevent release of the pathogen from the laboratory.

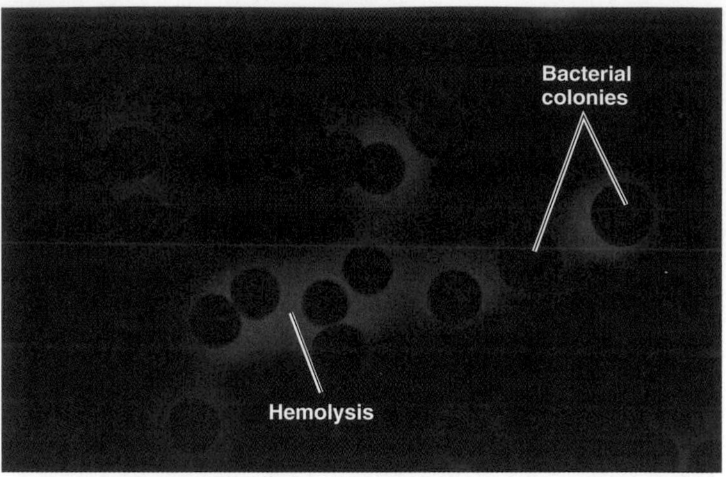

Figure 6.9 Blood agar, a differential medium containing red blood cells. The bacteria have lysed the red blood cells (beta-hemolysis), causing the clear areas around the colonies.

Q Of what value are hemolysins to pathogens?

Selective and Differential Media

In clinical and public health microbiology, it is frequently necessary to detect the presence of specific microorganisms associated with disease or poor sanitation. For this task, selective and differential media are used. **Selective media** are designed to suppress the growth of unwanted bacteria and encourage the growth of the desired microbes. For example, bismuth sulfite agar is one medium used to isolate the typhoid bacterium, the gram-negative *Salmonella typhi* (tī′fē), from feces. Bismuth sulfite inhibits gram-positive bacteria and most gram-negative intestinal bacteria (other than *S. typhi*), as well. Sabouraud's dextrose agar, which has a pH of 5.6, is used to isolate fungi that outgrow most bacteria at this pH.

Differential media make it easier to distinguish colonies of the desired organism from other colonies growing on the same plate. Similarly, pure cultures of microorganisms have identifiable reactions with differential media in tubes or plates. Blood agar (which contains red blood cells) is a medium that microbiologists often use to identify bacterial species that destroy red blood cells. These species, such as *Streptococcus pyogenes* (pī-äj′en-ēz), the bacterium that causes strep throat, show a clear ring around their colonies (beta-hemolysis, page 319) where they have lysed the surrounding blood cells (**Figure 6.9**).

Sometimes, selective and differential characteristics are combined in a single medium. Suppose we want to isolate the common bacterium *Staphylococcus aureus,* found in the nasal passages. This organism has a tolerance for high concentrations of sodium chloride; it can also ferment the carbohydrate mannitol to form acid. Mannitol salt agar contains 7.5% sodium

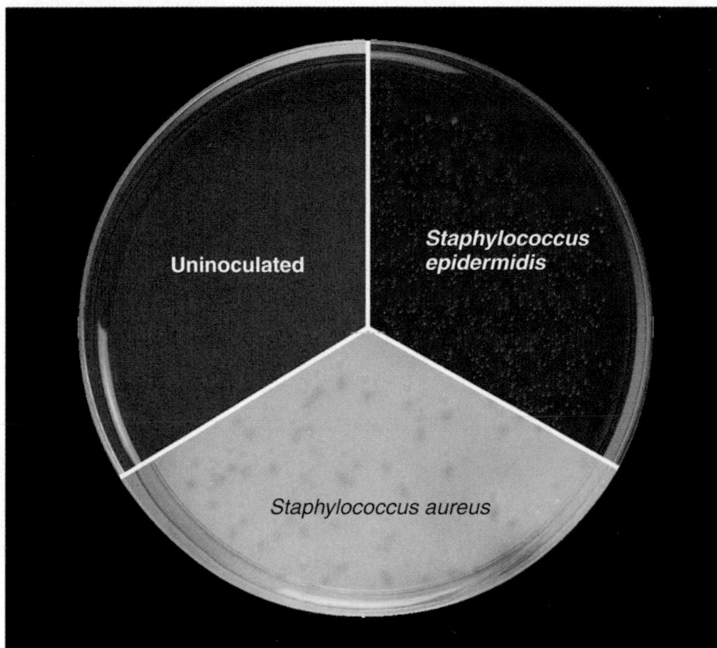

Figure 6.10 Differential medium. Bacterial colonies on differential media have a distinctive appearance. This medium is mannitol salt agar, and the bacteria in the colonies capable of fermenting the mannitol in the medium to acid, cause a change of color. Actually, this medium is also *selective* because the high salt concentration prevents the growth of most bacteria except *Staphlylococcus* spp.

Q Are bacteria capable of growing at a high osmotic pressure likely to be capable of growing in the mucus found in nostrils?

chloride, which will discourage the growth of competing organisms and thus *select for* (favor the growth of) *S. aureus.* This salty medium also contains a pH indicator that changes color if the mannitol in the medium is fermented to acid; the mannitol-fermenting colonies of *S. aureus* are thus *differentiated from* colonies of bacteria that do not ferment mannitol. Bacteria that grow at the high salt concentration *and* ferment mannitol to acid can be readily identified by the color change (Figure 6.10). These are probably colonies of *S. aureus,* and their identification can be confirmed by additional tests. The use of differential media to identify toxin-producing *E. coli* is discussed in Chapter 5, page 139.

Enrichment Culture

Because bacteria present in small numbers can be missed, especially if other bacteria are present in much larger numbers, it is sometimes necessary to use an **enrichment culture.** This is often the case for soil or fecal samples. The medium (enrichment medium) for an enrichment culture is usually liquid and provides nutrients and environmental conditions that favor the growth of a particular microbe but not others. In this sense, it is also a selective medium, but it is designed to increase very small numbers of the desired type of organism to detectable levels.

Suppose we want to isolate from a soil sample a microbe that can grow on phenol and is present in much smaller numbers than other species. If the soil sample is placed in a liquid enrichment medium in which phenol is the only source of carbon and energy, microbes unable to metabolize phenol will not grow. The culture medium is allowed to incubate for a few days, and then a small amount of it is transferred into another flask of the same medium. After a series of such transfers, the surviving population will consist of bacteria capable of metabolizing phenol. The bacteria are given time to grow in the medium between transfers; this is the enrichment stage. (See the box in Chapter 28, page 801.) Any nutrients in the original inoculum are rapidly diluted out with the successive transfers. When the last dilution is streaked onto a solid medium of the same composition, only those colonies of organisms capable of using phenol should grow. A remarkable aspect of this particular technique is that phenol is normally lethal to most bacteria.

CHECK YOUR UNDERSTANDING

- ✓ Could humans exist on chemically defined media, at least under laboratory conditions? **6-8**
- ✓ Could Louis Pasteur, in the 1800s, have grown rabies viruses in cell culture instead of in living animals? **6-9**
- ✓ What BSL is your laboratory? **6-10**

* * *

Table 6.5 summarizes the purposes of the main types of culture media.

Table 6.5 Culture Media

Type	Purpose
Chemically Defined	Growth of chemoautotrophs and photoautotrophs; microbiological assays
Complex	Growth of most chemoheterotrophic organisms
Reducing	Growth of obligate anaerobes
Selective	Suppression of unwanted microbes; encouraging desired microbes
Differential	Differentiation of colonies of desired microbes from others
Enrichment	Similar to selective media but designed to increase numbers of desired microbes to detectable levels

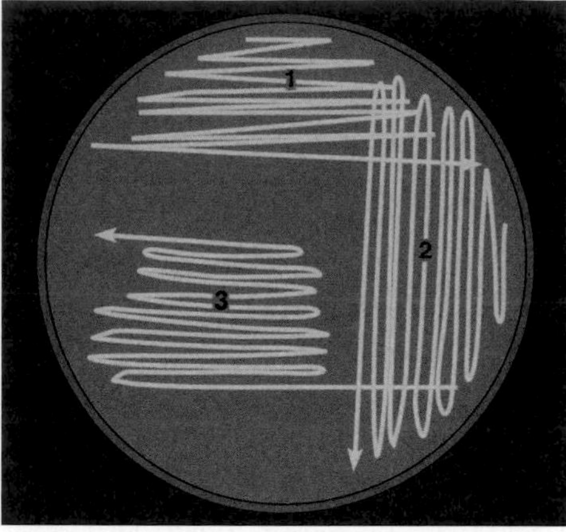

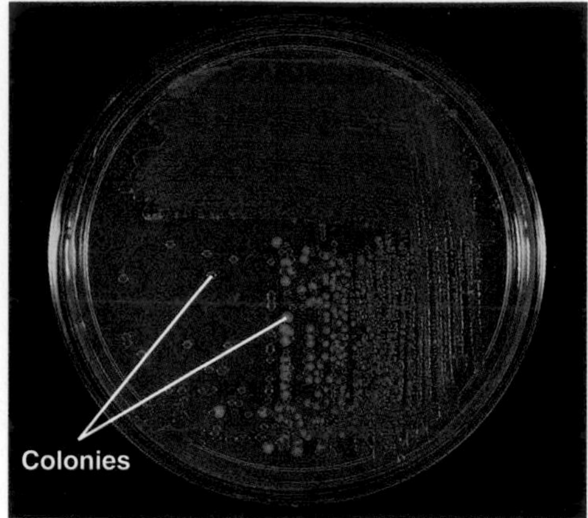

Figure 6.11 The streak plate method for isolating pure bacterial cultures. (**a**) Arrows indicate the direction of streaking. Streak series 1 is made from the original bacterial culture. The incoculating loop is sterilized following each streak series. In series 2 and 3, the loop picks up bacteria from the previous series, diluting the number of cells each time. There are numerous variants of such patterns. (**b**) In series 3 of this example, notice that well-isolated colonies of bacteria of two different types, red and yellow, have been obtained.

Q **Is a colony formed as a result of streaking a plate always derived from a single bacterium?**

Obtaining Pure Cultures

LEARNING OBJECTIVES

6-11 Define *colony.*

6-12 Describe how pure cultures can be isolated by using the streak plate method.

Most infectious materials, such as pus, sputum, and urine, contain several different kinds of bacteria; so do samples of soil, water, or food. If these materials are plated out onto the surface of a solid medium, colonies will form that are exact copies of the original organism. A visible **colony** theoretically arises from a single spore or vegetative cell or from a group of the same microorganisms attached to one another in clumps or chains. Microbial colonies often have a distinctive appearance that distinguishes one microbe from another (see Figure 6.10). The bacteria must be distributed widely enough so that the colonies are visibly separated from each other.

Most bacteriological work requires pure cultures, or clones, of bacteria. The isolation method most commonly used to get pure cultures is the **streak plate method** (Figure 6.11). A sterile inoculating loop is dipped into a mixed culture that contains more than one type of microbe and is streaked in a pattern over the surface of the nutrient medium. As the pattern is traced, bacteria are rubbed off the loop onto the medium. The last cells to be rubbed off the loop are far enough apart to grow into isolated colonies. These colonies can be picked up with an inoculating loop and transferred to a test tube of nutrient medium to form a pure culture containing only one type of bacterium.

The streak plate method works well when the organism to be isolated is present in large numbers relative to the total population. However, when the microbe to be isolated is present only in very small numbers, its numbers must be greatly increased by selective enrichment before it can be isolated with the streak plate method.

CHECK YOUR UNDERSTANDING

✓ Can you think of any reason why a colony does not grow to an infinite size, or at least fill the confines of the Petri plate? **6-11**

✓ Could a pure culture of bacteria be obtained by the streak plate method if there were only one desired microbe in a bacterial suspension of billions? **6-12**

Preserving Bacterial Cultures

LEARNING OBJECTIVE

6-13 Explain how microorganisms are preserved by deep-freezing and lyophilization (freeze-drying).

Refrigeration can be used for the short-term storage of bacterial cultures. Two common methods of preserving microbial cultures for long periods are deep-freezing and lyophilization. **Deep-freezing** is a process in which a pure culture of microbes is placed in a suspending liquid and quick-frozen at temperatures ranging from $-50°C$ to $-95°C$. The culture can usually be thawed and cultured even several years later. During **lyophilization (freeze-drying),** a suspension of microbes is quickly frozen at temperatures ranging from $-54°C$ to $-72°C$, and the water is removed by a high vacuum (sublimation). While under vacuum, the container is sealed by melting the glass with a high-temperature torch. The remaining powderlike residue that contains the surviving microbes can be stored for years. The organisms can be revived at any time by hydration with a suitable liquid nutrient medium.

CHECK YOUR UNDERSTANDING

✓ If the Space Station in Earth orbit suddenly ruptured, the humans on board would die instantly from cold and the vacuum of space. Would all the bacteria in the capsule also be killed? **6-13**

The Growth of Bacterial Cultures

LEARNING OBJECTIVES

6-14 Define *bacterial growth,* including *binary fission.*

6-15 Compare the phases of microbial growth, and describe their relation to generation time.

6-16 Explain four direct methods of measuring cell growth.

6-17 Differentiate direct and indirect methods of measuring cell growth.

6-18 Explain three indirect methods of measuring cell growth.

Being able to represent graphically the enormous populations resulting from the growth of bacterial cultures is an essential part of microbiology. It is also necessary to be able to determine microbial numbers, either directly, by counting, or indirectly, by measuring their metabolic activity.

Bacterial Division

As we mentioned at the beginning of the chapter, bacterial growth refers to an increase in bacterial numbers, not an increase in the size of the individual cells. Bacteria normally reproduce by **binary fission** (Figure 6.12).

A few bacterial species reproduce by **budding;** they form a small initial outgrowth (a bud) that enlarges until its size approaches that of the parent cell, and then it separates. Some filamentous bacteria (certain actinomycetes) reproduce by producing chains of conidiospores carried externally at the tips of the filaments. A few filamentous species simply fragment, and the fragments initiate the growth of new cells. **Animations** Binary Fission; Bacterial Growth: Overview. **www.microbiologyplace.com**

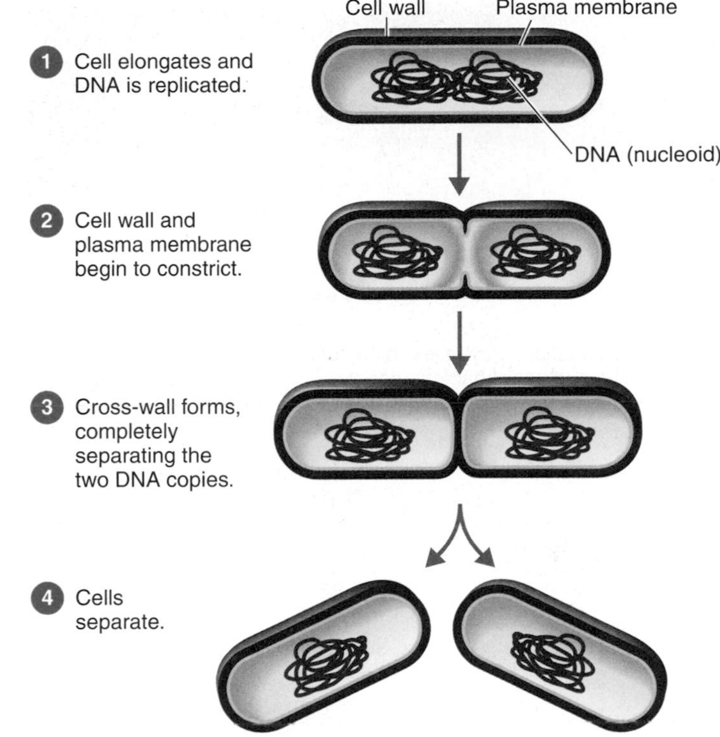

Cell wall
Plasma membrane
1 Cell elongates and DNA is replicated.
DNA (nucleoid)
2 Cell wall and plasma membrane begin to constrict.
3 Cross-wall forms, completely separating the two DNA copies.
4 Cells separate.

(a) A diagram of the sequence of cell division

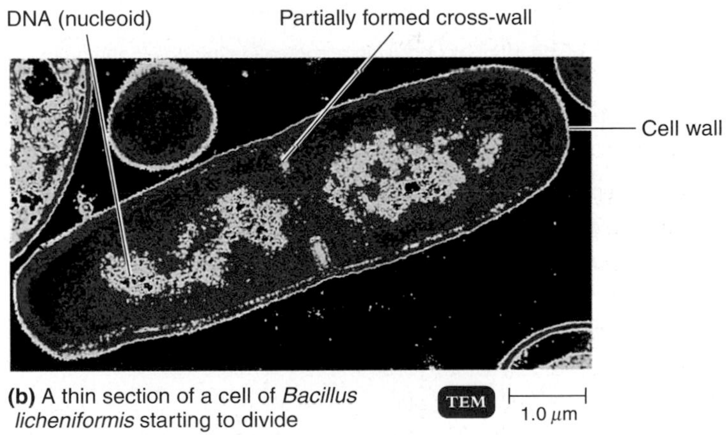

(b) A thin section of a cell of *Bacillus licheniformis* starting to divide

Figure 6.12 Binary fission in bacteria.

Q Do all bacteria reproduce by binary fission?

Generation Time

For purposes of calculating the generation time of bacteria, we will consider only reproduction by binary fission, which is by far the most common method. As you can see in Figure 6.13, one cell's division produces two cells, two cells' divisions produce four cells, and so on. When the number of cells in each generation is expressed as a power of 2, the exponent tells the number of doublings (generations) that have occurred.

The time required for a cell to divide (and its population to double) is called the **generation time.** It varies considerably among organisms and with environmental conditions, such as temperature. Most bacteria have a generation time of 1 to 3 hours; others require more than 24 hours per generation. (The math required to calculate generation times is presented in Appendix B.) If binary fission continues unchecked, an enormous number of cells will be produced. If a doubling occurred every 20 minutes—which is the case for *E. coli* under favorable conditions—after 20 generations a single initial cell would increase to over 1 million cells. This would require a little less than 7 hours. In 30 generations, or 10 hours, the population would be 1 billion, and in 24 hours it would be a number trailed by 21 zeros. It is difficult to graph population changes of such enormous magnitude by using arithmetic numbers. This is why logarithmic scales are generally used to graph bacterial growth. Understanding logarithmic representations of bacterial populations requires some use of mathematics and is necessary for anyone studying microbiology. (See Appendix B.)

Logarithmic Representation of Bacterial Populations

To illustrate the difference between logarithmic and arithmetic graphing of bacterial populations, let's express 20 bacterial generations both logarithmically and arithmetically. In five

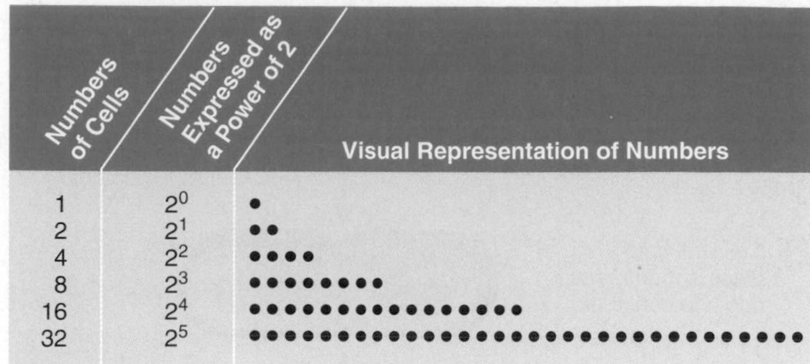

(a) Visual representation of increase in bacterial number over five generations. The number of bacteria doubles in each generation. The superscript indicates the generation; that is, 2^5 = 5 generations.

Generation Number	Number of Cells	Log_{10} of Number of Cells
0	2^0 = 1	0
5	2^5 = 32	1.51
10	2^{10} = 1,024	3.01
15	2^{15} = 32,768	4.52
16	2^{16} = 65,536	4.82
17	2^{17} = 131,072	5.12
18	2^{18} = 262,144	5.42
19	2^{19} = 524,288	5.72
20	2^{20} = 1,048,576	6.02

(b) Conversion of the number of cells in a population into the logarithmic expression of this number. To arrive at the numbers in the center column, use the y^x key on your calculator. Enter 2 on the calculator; press y^x; enter 5; then press the = sign. The calculator will show the number 32. Thus, the fifth-generation population of bacteria will total 32 cells. To arrive at the numbers in the right-hand column, use the log key on your calculator. Enter the number 32; then press the log key. The calculator will show, rounded off, that the log_{10} of 32 is 1.51.

Figure 6.13 Cell division.

Q If a single bacterium reproduced every 20 minutes, how many would there be in 2 hours?

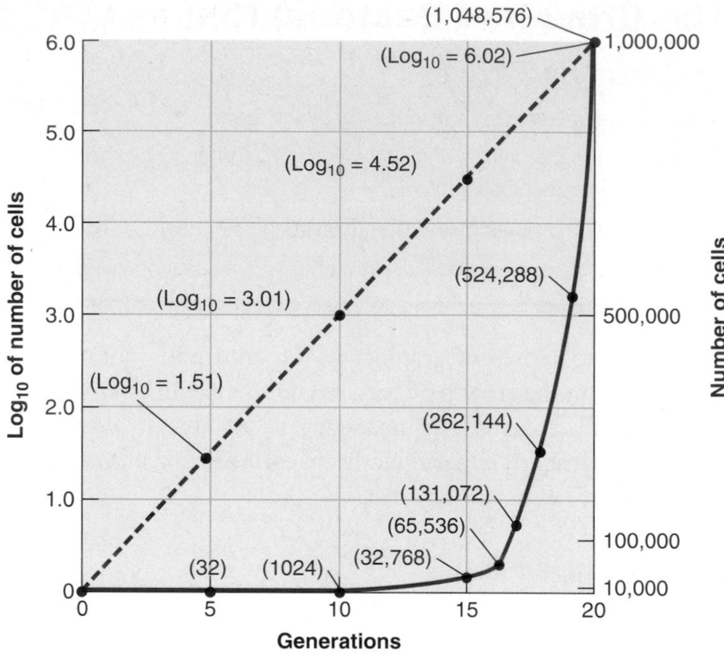

Figure 6.14 A growth curve for an exponentially increasing population, plotted logarithmically (dashed line) and arithmetically (solid line). This figure demonstrates why it is necessary to graph changes in the immense numbers of bacterial populations by logarithmic plots rather than by arithmetic numbers. For example, note that at ten generations the line representing arithmetic numbers has not even perceptibly left the baseline, whereas the logarithmic plot point for the tenth generation (3.01) is halfway up the graph.

Q If the arithmetic numbers (solid line) were plotted for two more generations, would the line still be on the page?

generations (2^5), there would be 32 cells; in ten generations (2^{10}), there would be 1024 cells, and so on. (If your calculator has a y^x key and a log key, you can duplicate the numbers in the third column of Figure 6.13.)

In **Figure 6.14**, notice that the arithmetically plotted line (solid) does not clearly show the population changes in the early stages of the growth curve at this scale. In fact, the first ten generations do not even appear to leave the baseline. Furthermore, another one or two arithmetic generations graphed to the same scale would greatly increase the height of the graph and take the line off the page.

The dashed line in Figure 6.14 shows how these plotting problems can be avoided by graphing the log_{10} of the population numbers. The log_{10} of the population is plotted at 5, 10, 15, and 20 generations. Notice that a straight line is formed and that a thousand times this population (1,000,000,000, or log_{10} 9.0) could be accommodated in relatively little extra space. However, this advantage is obtained at the cost of distorting our "common sense" perception of the actual situation. We are not accustomed to thinking in logarithmic relationships, but it is necessary for a proper understanding of graphs of microbial populations.

CHECK YOUR UNDERSTANDING

✓ Can a complex organism, such as a beetle, divide by binary fission? **6-14**

Phases of Growth

When a few bacteria are inoculated into a liquid growth medium and the population is counted at intervals, it is possible to plot a **bacterial growth curve** that shows the growth of cells over time (**Figure 6.15**). There are four basic phases of growth: the lag, log, stationary, and death phases. **Animation** Bacterial Growth Curve. **www.microbiologyplace.com**

Figure 6.15

FOUNDATION FIGURE The Bacterial Growth Curve

The concept of the bacterial growth curve is fundamental to understanding population dynamics and control in, for example, food preservation and spoilage; industrial microbiology, such as ethanol production; and the course and treatment of infectious disease.

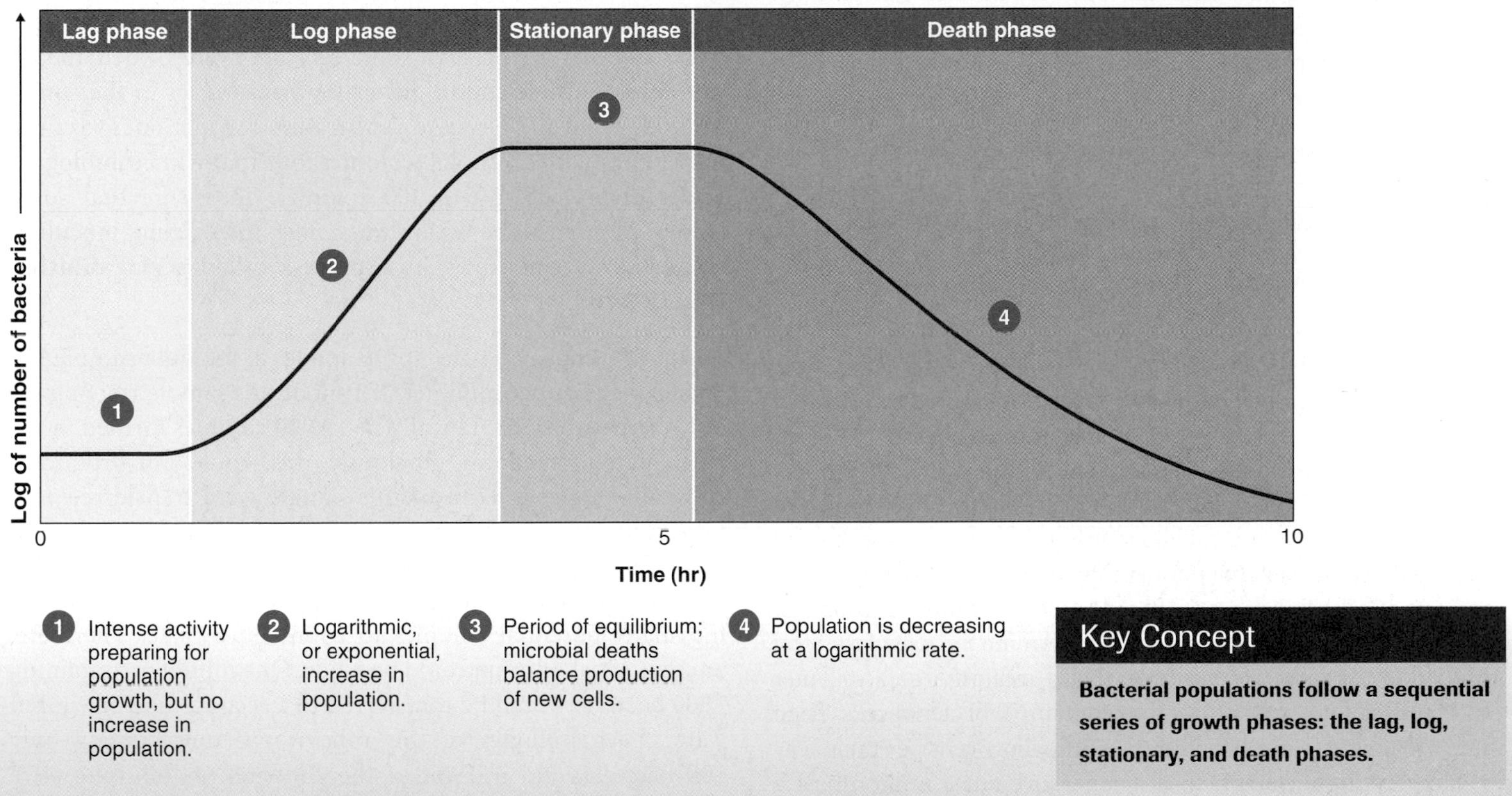

1 Intense activity preparing for population growth, but no increase in population.

2 Logarithmic, or exponential, increase in population.

3 Period of equilibrium; microbial deaths balance production of new cells.

4 Population is decreasing at a logarithmic rate.

Key Concept

Bacterial populations follow a sequential series of growth phases: the lag, log, stationary, and death phases.

The Lag Phase

For a while, the number of cells changes very little because the cells do not immediately reproduce in a new medium. This period of little or no cell division is called the **lag phase,** and it can last for 1 hour or several days. During this time, however, the cells are not dormant. The microbial population is undergoing a period of intense metabolic activity involving, in particular, synthesis of enzymes and various molecules. (The situation is analogous to a factory being equipped to produce automobiles; there is considerable tooling-up activity but no immediate increase in the automobile population.)

The Log Phase

Eventually, the cells begin to divide and enter a period of growth, or logarithmic increase, called the **log phase,** or **exponential growth phase.** Cellular reproduction is most active during this period, and generation time reaches a constant minimum. Because the generation time is constant, a logarithmic plot of growth during the log phase is a straight line. The log phase is the time when cells are most active metabolically and is preferred for industrial purposes where, for example, a product needs to be produced efficiently.

The Stationary Phase

If exponential growth continues unchecked, startlingly large numbers of cells could arise. For example, a single bacterium (at a weight of 9.5×10^{-13} g per cell) dividing every 20 minutes for only 25.5 hours can theoretically produce a population equivalent in weight to that of an 80,000-ton aircraft carrier. In reality, this does not happen. Eventually, the growth rate slows, the number of microbial deaths balances the number of new cells, and the population stabilizes. This period of equilibrium is called the **stationary phase.**

What causes exponential growth to stop is not always clear. The exhaustion of nutrients, accumulation of waste products, and harmful changes in pH may all play a role.

The Death Phase

The number of deaths eventually exceeds the number of new cells formed, and the population enters the **death phase,** or **logarithmic decline phase.** This phase continues until the population is diminished to a tiny fraction of the number of cells in the previous phase or until the population dies out entirely. Some species pass through the entire series of phases in only a few days; others retain some surviving cells almost indefinitely. Microbial death will be discussed further in Chapter 7.

CHECK YOUR UNDERSTANDING

✓ If two mice started a family within a fixed enclosure, with a fixed food supply, would the population curve be the same as a bacterial growth curve? **6-15**

Direct Measurement of Microbial Growth

The growth of microbial populations can be measured in a number of ways. Some methods measure cell numbers; other methods measure the population's total mass, which is often directly proportional to cell numbers. Population numbers are usually recorded as the number of cells in a milliliter of liquid or in a gram of solid material. Because bacterial populations are usually very large, most methods of counting them are based on direct or indirect counts of very small samples; calculations then determine the size of the total population. Assume, for example, that a millionth of a milliliter (10^{-6} ml) of sour milk is found to contain 70 bacterial cells. Then there must be 70 times 1 million, or 70 million, cells per milliliter.

However, it is not practical to measure out a millionth of a milliliter of liquid or a millionth of a gram of food. Therefore, the procedure is done indirectly, in a series of dilutions. For example, if we add 1 ml of milk to 99 ml of water, each milliliter of this dilution now has one-hundredth as many bacteria as each milliliter of the original sample had. By making a series of such dilutions, we can readily estimate the number of bacteria in our original sample. To count microbial populations in solid foods (such as hamburger), an homogenate of one part food to nine parts water is finely ground in a food blender. Samples of this initial one-tenth dilution can then be transferred with a pipette for further dilutions or cell counts.

Plate Counts

The most frequently used method of measuring bacterial populations is the **plate count.** An important advantage of this method is that it measures the number of viable cells. One disadvantage may be that it takes some time, usually 24 hours or more, for visible colonies to form. This can be a serious problem in some applications, such as quality control of milk, when it is not possible to hold a particular lot for this length of time.

Plate counts assume that each live bacterium grows and divides to produce a single colony. This is not always true, because bacteria frequently grow linked in chains or as clumps (see Figure 4.1, page 78). Therefore, a colony often results, not from a single bacterium, but from short segments of a chain or from a bacterial clump. To reflect this reality, plate counts are often reported as **colony-forming units (CFU).**

When a plate count is performed, it is important that only a limited number of colonies develop in the plate. When too many colonies are present, some cells are overcrowded and do not develop; these conditions cause inaccuracies in the count. The U.S. Food and Drug Administration convention is to count only plates with 25 to 250 colonies, but many microbiologists prefer plates with 30 to 300 colonies. To ensure that some colony counts will be within this range, the original inoculum is diluted several times in a process called **serial dilution** (**Figure 6.16**).

Serial Dilutions Let's say, for example, that a milk sample has 10,000 bacteria per milliliter. If 1 ml of this sample were plated out, there would theoretically be 10,000 colonies formed in the Petri plate of medium. Obviously, this would not produce a countable plate. If 1 ml of this sample were transferred to a tube containing 9 ml of sterile water, each milliliter of fluid in this tube would now contain 1000 bacteria. If 1 ml of this sample were inoculated into a Petri plate, there would still be too many potential colonies to count on a plate. Therefore, another serial dilution could be made. One milliliter containing 1000 bacteria would be transferred to a second tube of 9 ml of water. Each milliliter of this tube would now contain only 100 bacteria, and if 1 ml of the contents of this tube were plated out, potentially 100 colonies would be formed—an easily countable number.

Pour Plates and Spread Plates A plate count is done by either the pour plate method or the spread plate method. The **pour plate method** follows the procedure shown in **Figure 6.17a**. Either 1.0 ml or 0.1 ml of dilutions of the bacterial suspension is introduced into a Petri dish. The nutrient medium, in which the agar is kept liquid by holding it in a water bath at about 50°C, is poured over the sample, which is then mixed into the medium by gentle agitation of the plate. When the agar solidifies, the plate is incubated. With the pour plate technique, colonies will grow within the nutrient agar (from cells suspended in the nutrient medium as the agar solidifies) as well as on the surface of the agar plate.

This technique has some drawbacks because some relatively heat-sensitive microorganisms may be damaged by the melted agar and will therefore be unable to form colonies. Also, when certain differential media are used, the distinctive appearance of the colony on the surface is essential for diagnostic purposes. Colonies that form beneath the surface of a pour plate are not satisfactory for such tests. To avoid these problems, the

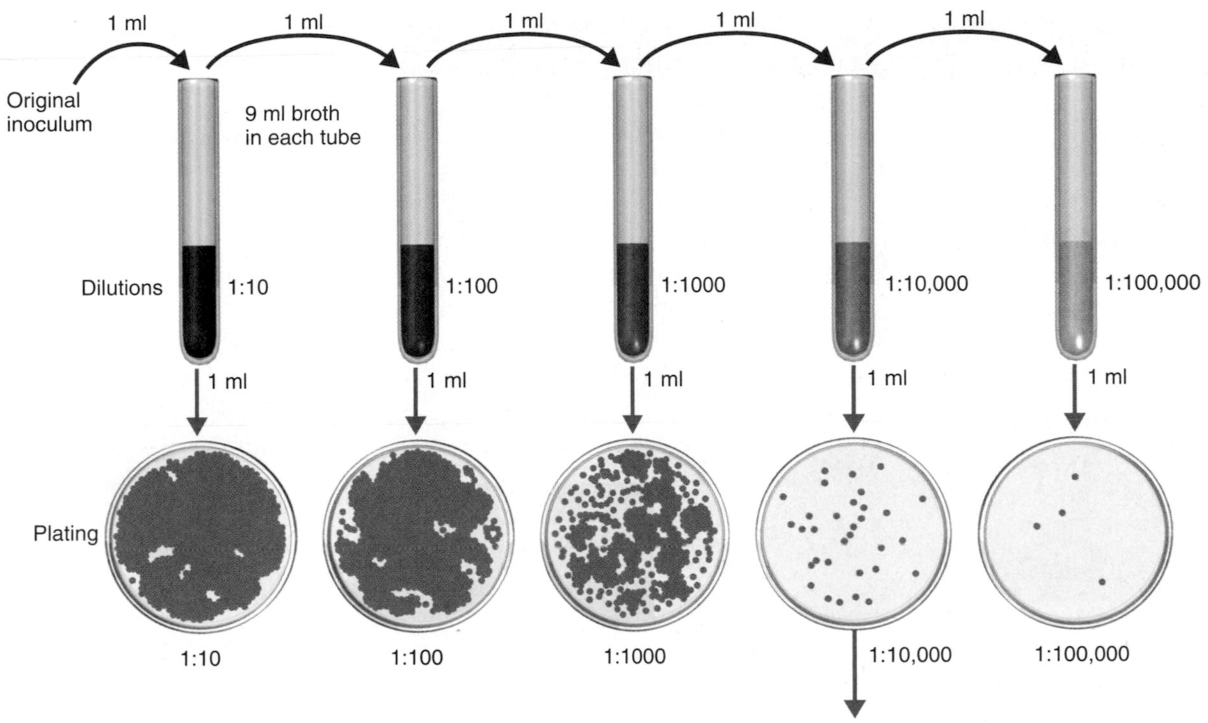

Calculation: Number of colonies on plate × reciprocal of dilution of sample = number of bacteria/ml
(For example, if 32 colonies are on a plate of 1:10,000 dilution, then the count is 32 × 10,000 = 320,000 bacteria/ml in sample.)

Figure 6.16 Serial dilutions and plate counts. In serial dilutions, the original inoculum is diluted in a series of dilution tubes. In our example, each succeeding dilution tube will have only one-tenth the number of microbial cells as the preceding tube. Then, samples of the dilution are used to inoculate Petri plates, on which colonies grow and can be counted. This count is then used to estimate the number of bacteria in the original sample.

Why were the dilutions of 1:1000 and 1:100,000 not counted? Theoretically, how many colonies should appear on the 1:1000 plate?

spread plate method is frequently used instead (**Figure 6.17b**). A 0.1-ml inoculum is added to the surface of a prepoured, solidified agar medium. The inoculum is then spread uniformly over the surface of the medium with a specially shaped, sterilized glass or metal rod. This method positions all the colonies on the surface and avoids contact between the cells and melted agar.

Filtration

When the quantity of bacteria is very small, as in lakes or relatively pure streams, bacteria can be counted by **filtration** methods (**Figure 6.18**). In this technique, at least 100 ml of water are passed through a thin membrane filter whose pores are too small to allow bacteria to pass. Thus, the bacteria are filtered out and retained on the surface of the filter. This filter is then transferred to a Petri dish containing a pad soaked in liquid nutrient medium, where colonies arise from the bacteria on the filter's surface. This method is applied frequently to detection and enumeration of coliform bacteria, which are indicators of fecal contamination of food or water (see Chapter 27). The colonies formed by these bacteria are distinctive when a differential nutrient medium is used. (The colonies shown in Figure 6.18b are examples of coliforms.)

The Most Probable Number (MPN) Method

Another method for determining the number of bacteria in a sample is the **most probable number (MPN) method,** illustrated in **Figure 6.19.** This statistical estimating technique is based on the fact that the greater the number of bacteria in a sample, the more dilution is needed to reduce the density to the point at which no bacteria are left to grow in the tubes in a dilution series. The MPN method is most useful when the microbes being counted will not grow on solid media (such as the chemoautotrophic nitrifying bacteria). It is also useful when the growth of bacteria in a liquid differential medium is used to identify the microbes (such as coliform bacteria, which selectively ferment lactose to acid, in water testing). The MPN is only a statement that there is a 95% chance that the bacterial population falls within a certain range and that the MPN is statistically the most probable number.

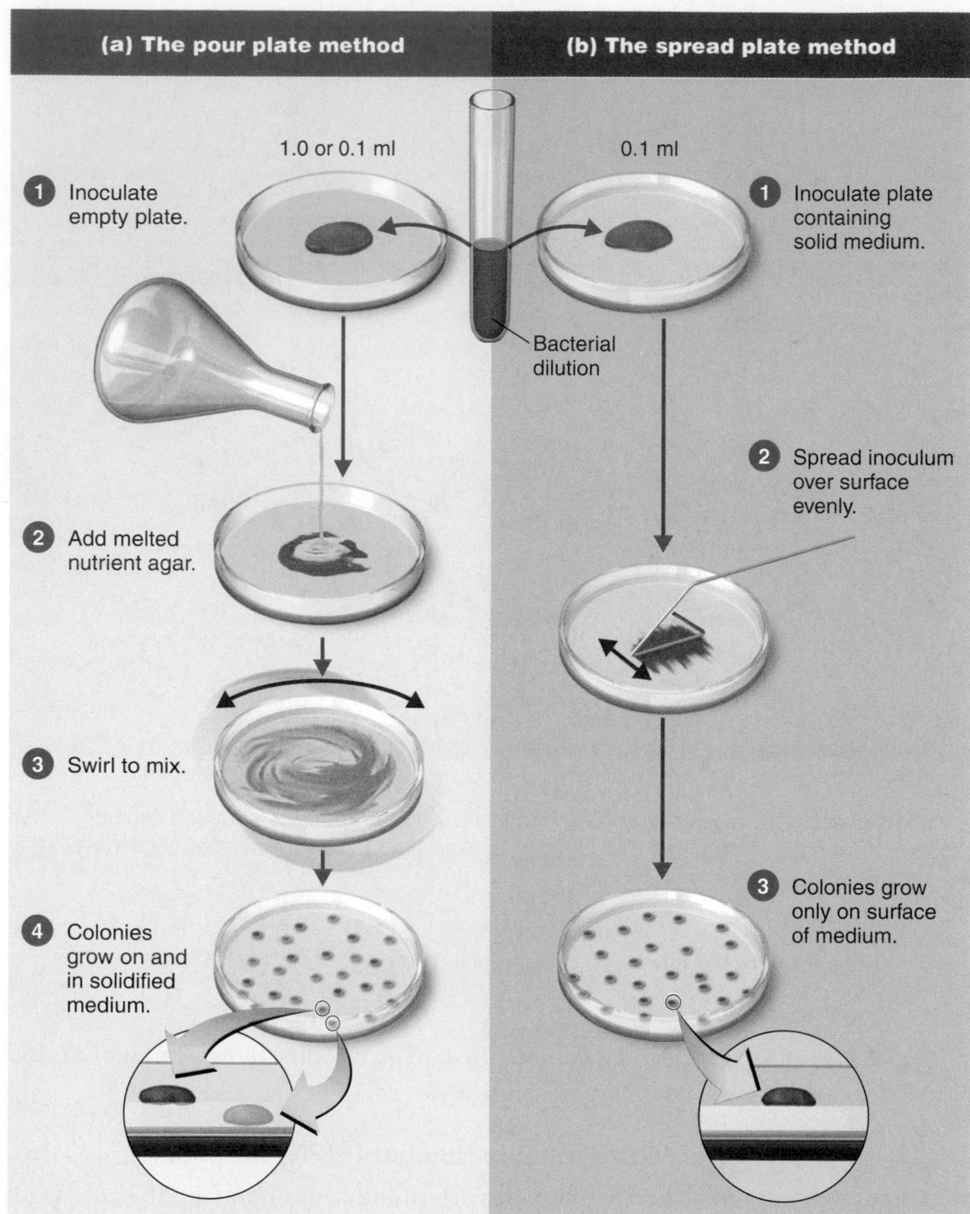
(a) The pour plate method
(b) The spread plate method
1.0 or 0.1 ml
0.1 ml
1 Inoculate empty plate.
1 Inoculate plate containing solid medium.
Bacterial dilution
2 Spread inoculum over surface evenly.
2 Add melted nutrient agar.
3 Swirl to mix.
3 Colonies grow only on surface of medium.
4 Colonies grow on and in solidified medium.

Figure 6.17 Methods of preparing plates for plate counts. (**a**) The pour plate method. (**b**) The spread plate method.

Q What are the advantages of each method?

Direct Microscopic Count

In the method known as the **direct microscopic count,** a measured volume of a bacterial suspension is placed within a defined area on a microscope slide. Because of time considerations, this method is often used to count the number of bacteria in milk. A 0.01-ml sample is spread over a marked square centimeter of slide, stain is added so that the bacteria can be seen, and the sample is viewed under the oil immersion objective lens. The area of the viewing field of this objective can be determined. Once the number of bacteria has been counted in several different fields, the average number of bacteria per viewing field can be calculated. From these data, the number of bacteria in the square centimeter over which the sample was spread can also be calculated. Because this area on the slide contained 0.01 ml of sample, the number of bacteria in each milliliter of the suspension is the number of bacteria in the sample times 100.

A specially designed slide called a *Petroff-Hausser cell counter* is also used in direct microscopic counts (**Figure 6.20**).

Motile bacteria are difficult to count by this method, and, as happens with other microscopic methods, dead cells are about as likely to be counted as live ones. In addition to these disadvantages, a rather high concentration of cells is required to be countable—about 10 million bacteria per milliliter. The chief advantage of microscopic counts is that no incubation time is required, and they are usually reserved for applications in which time is the primary consideration. This advantage also holds for *electronic cell counters,* sometimes known as *Coulter counters,* which automatically count the number of cells in a measured

Figure 6.18 Counting bacteria by filtration.

Q Could you make a pour plate in the usual Petri dish with a 10-ml inoculum? If not, why not?

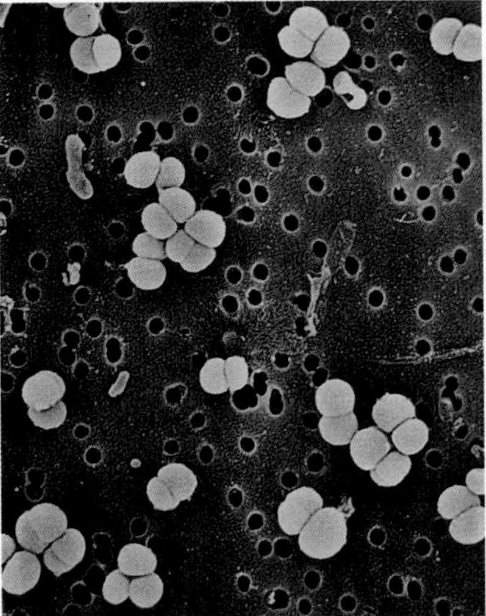

(a) The bacteria in 100 ml of water were sieved out onto the surface of a membrane filter. Note that the holes in the filter are smaller than the bacteria. SEM 1 μm

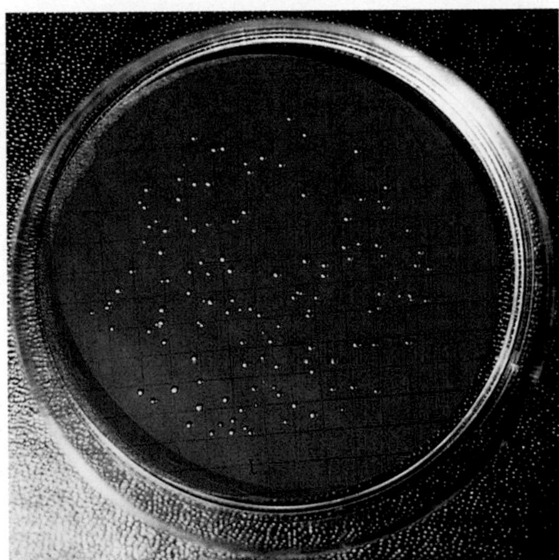

(b) A filter such as shown in photo (a), with the bacteria much more widely spaced, was placed on a pad saturated with liquid Endo medium, which is used for enumerating coliforms. The individual bacteria grew into visible colonies. There are 124 colonies visible, so we would record 124 bacteria per 100 ml of water sample.

Volume of Inoculum for Each Set of Five Tubes	Tubes of Nutrient Medium (Sets of Five Tubes)	Number of Positive Tubes in Set
10 ml		5
1 ml		3
0.1 ml		1

(a) Most probable number (MPN) dilution series. In this example, there are three sets of tubes and five tubes in each set. Each tube in the first set of five tubes receives 10 ml of the inoculum, such as a sample of water. Each tube in the second set of five tubes receives 1 ml of the sample, and the third set, 0.1 ml each. There were enough bacteria in the sample so that all five tubes in the first set showed bacterial growth and were recorded as positive. In the second set, which received only one-tenth as much inoculum, only three tubes were positive. In the third set, which received one-hundredth as much inoculum, only one tube was positive.

Combination of Positives	MPN Index/ 100 ml	95% Confidence Limits	
		Lower	Upper
4-2-0	22	6.8	50
4-2-1	26	9.8	70
4-3-0	27	9.9	70
4-3-1	33	10	70
4-4-0	34	14	100
5-0-0	23	6.8	70
5-0-1	31	10	70
5-0-2	43	14	100
5-1-0	33	10	100
5-1-1	46	14	120
5-1-2	63	22	150
5-2-0	49	15	150
5-2-1	70	22	170
5-2-2	94	34	230
5-3-0	79	22	220
5-3-1	110	34	250
5-3-2	140	52	400

(b) MPN table. MPN tables enable us to calculate for a sample the microbial numbers that are statistically likely to lead to such a result. The number of positive tubes is recorded for each set: in the shaded example, 5, 3, and 1. If we look up this combination in an MPN table, we find that the MPN index per 100 ml is 110. Statistically, this means that 95% of the water samples that give this result contain 34–250 bacteria, with 110 being the most probable number.

Figure 6.19 The most probable number (MPN) method.

Q Under what circumstances is the MPN method used to determine the number of bacteria in a sample?

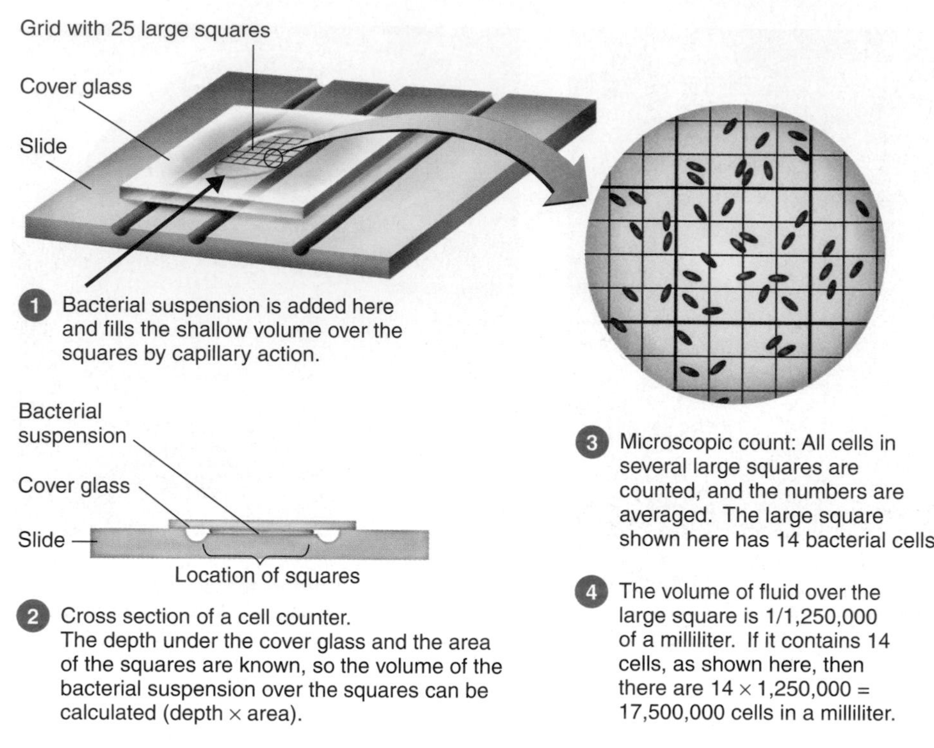

Grid with 25 large squares
Cover glass
Slide
1 Bacterial suspension is added here and fills the shallow volume over the squares by capillary action.
Bacterial suspension
Cover glass
Slide
Location of squares
2 Cross section of a cell counter. The depth under the cover glass and the area of the squares are known, so the volume of the bacterial suspension over the squares can be calculated (depth × area).
3 Microscopic count: All cells in several large squares are counted, and the numbers are averaged. The large square shown here has 14 bacterial cells.
4 The volume of fluid over the large square is 1/1,250,000 of a milliliter. If it contains 14 cells, as shown here, then there are 14 × 1,250,000 = 17,500,000 cells in a milliliter.

Figure 6.20 Direct microscopic count of bacteria with a Petroff-Hausser cell counter. The average number of cells within a large square multiplied by a factor of 1,250,000 gives the number of bacteria per milliliter.

Q This type of counting, despite its obvious disadvantages, is often used in estimating the bacterial population in dairy products. Why?

volume of liquid. These instruments are used in some research laboratories and hospitals.

CHECK YOUR UNDERSTANDING

✓ Why is it difficult to measure realistically the growth of a filamentous mold isolate by the plate count method? **6-16**

Estimating Bacterial Numbers by Indirect Methods

It is not always necessary to count microbial cells to estimate their numbers. In science and industry, microbial numbers and activity are determined by some of the following indirect means as well.

Turbidity

For some types of experimental work, estimating **turbidity** is a practical way of monitoring bacterial growth. As bacteria multiply in a liquid medium, the medium becomes turbid, or cloudy with cells.

The instrument used to measure turbidity is a *spectrophotometer* (or colorimeter). In the spectrophotometer, a beam of light is transmitted through a bacterial suspension to a light-sensitive detector (Figure 6.21). As bacterial numbers increase, less light will reach the detector. This change of light will register on the instrument's scale as the *percentage of transmission.* Also printed on the instrument's scale is a logarithmic expression called the *absorbance* (sometimes called *optical density,* or *OD,* which is calculated as Abs = 2 − log of % transmittance.) The absorbance is used to plot bacterial growth. When the bacteria are in logarithmic growth or decline, a graph of absorbance versus time will form an approximately straight line. If absorbance readings are matched with plate counts of the same culture, this correlation can be used in future estimations of bacterial numbers obtained by measuring turbidity.

More than a million cells per milliliter must be present for the first traces of turbidity to be visible. About 10 million to 100 million cells per milliliter are needed to make a suspension turbid enough to be read on a spectrophotometer. Therefore, turbidity is not a useful measure of contamination of liquids by relatively small numbers of bacteria.

Metabolic Activity

Another indirect way to estimate bacterial numbers is to measure a population's *metabolic activity.* This method assumes that the amount of a certain metabolic product, such as acid or CO_2, is in direct proportion to the number of bacteria present. An example of a practical application of a metabolic test is the microbiological assay in which acid production is used to determine amounts of vitamins.

Dry Weight

For filamentous bacteria and molds, the usual measuring methods are less satisfactory. A plate count would not measure this increase in filamentous mass. In plate counts of actinomycetes (see Figure 11.22, page 321) and molds, it is mostly the number of asexual spores that is counted instead. This is not a good measure of growth. One of the better ways to measure the growth of filamentous organisms is by *dry weight.* In this

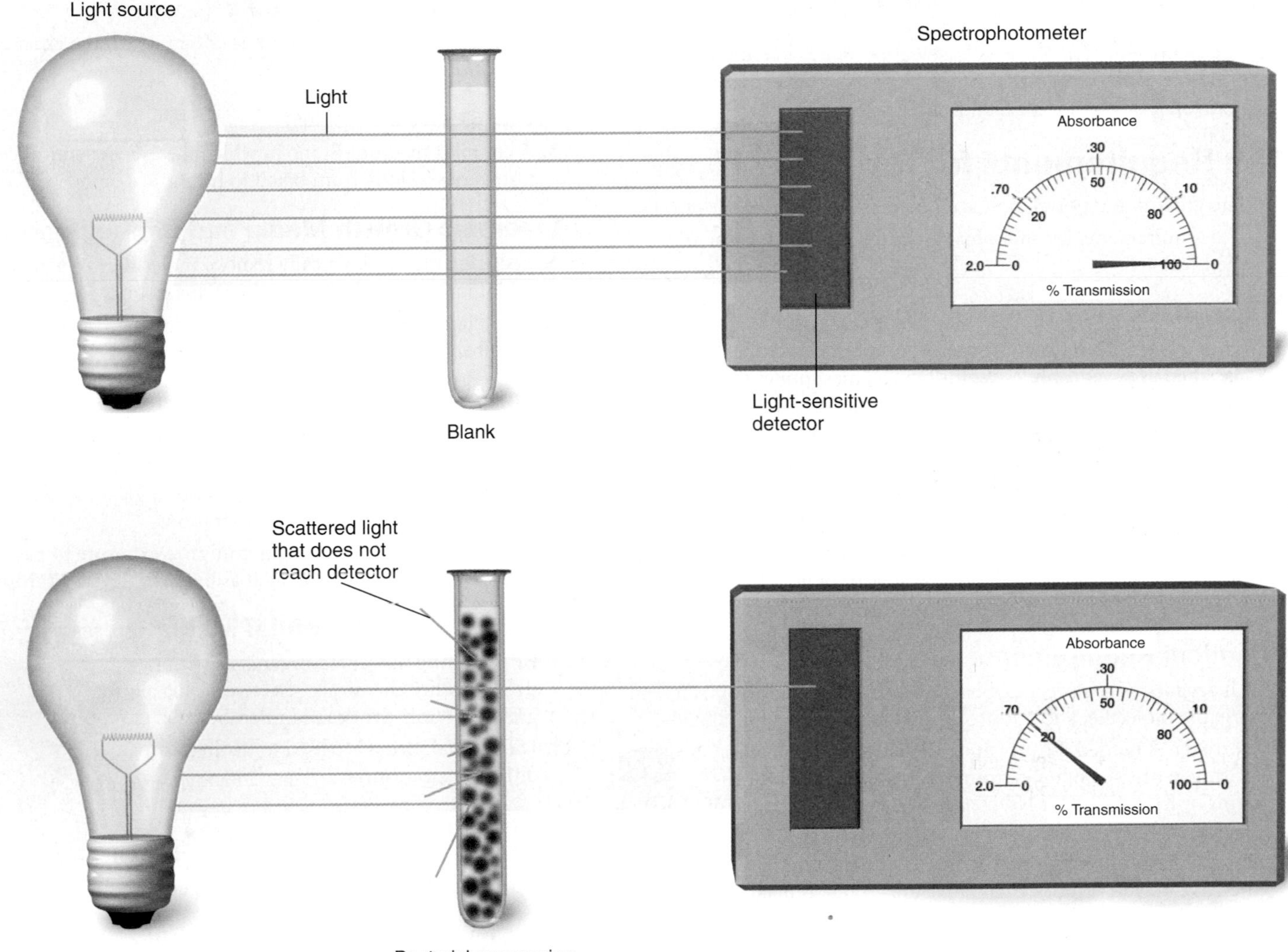

Figure 6.21 Turbidity estimation of bacterial numbers. The amount of light striking the light-sensitive detector on the spectrophotometer is inversely proportional to the number of bacteria under standardized conditions. The less light transmitted, the more bacteria in the sample. The turbidity of the sample could be reported as either 20% transmittance or 0.7 absorbance. Readings in absorbance are a logarithmic function and are sometimes useful in plotting data.

Q Is turbidity a direct or an indirect method of measuring bacterial growth?

procedure, the fungus is removed from the growth medium, filtered to remove extraneous material, and dried in a desiccator. It is then weighed. For bacteria, the same basic procedure is followed.

CHECK YOUR UNDERSTANDING

✓ Direct methods usually require an incubation time for a colony. Why is this not always feasible for analysis of foods? **6-17**

✓ If there is no good method for analyzing a product for its vitamin content, what is a feasible method of determining the vitamin content? **6-18**

* * *

You now have a basic understanding of the requirements for, and measurements of, microbial growth. In Chapter 7, we will look at how this growth is controlled in laboratories, hospitals, industry, and our homes.

STUDY OUTLINE

The **MyMicrobiologyPlace** website (**www.microbiologyplace.com**) will help you get ready for tests with its simple three-step approach: ❶ **take a pre-test** and obtain a personalized study plan, ❷ **learn and practice** with animations, tutorials, and MP3 tutor sessions, and ❸ **test yourself** with quizzes and a chapter post-test.

The Requirements for Growth (pp. 157–163)

1. The growth of a population is an increase in the number of cells.
2. The requirements for microbial growth are both physical and chemical.

Physical Requirements (pp. 157–160)

3. On the basis of preferred temperature ranges, microbes are classified as psychrophiles (cold-loving), mesophiles (moderate-temperature–loving), and thermophiles (heat-loving).
4. The minimum growth temperature is the lowest temperature at which a species will grow, the optimum growth temperature is the temperature at which it grows best, and the maximum growth temperature is the highest temperature at which growth is possible.
5. Most bacteria grow best at a pH value between 6.5 and 7.5.
6. In a hypertonic solution, most microbes undergo plasmolysis; halophiles can tolerate high salt concentrations.

Chemical Requirements (pp. 160–162)

7. All organisms require a carbon source; chemoheterotrophs use an organic molecule, and autotrophs typically use carbon dioxide.
8. Nitrogen is needed for protein and nucleic acid synthesis. Nitrogen can be obtained from the decomposition of proteins or from NH_4^+ or NO_3^-; a few bacteria are capable of nitrogen (N_2) fixation.
9. On the basis of oxygen requirements, organisms are classified as obligate aerobes, facultative anaerobes, obligate anaerobes, aerotolerant anaerobes, and microaerophiles.
10. Aerobes, facultative anaerobes, and aerotolerant anaerobes must have the enzymes superoxide dismutase ($2\ O_2^- + 2\ H^+ \longrightarrow O_2 + H_2O_2$) and either catalase ($2\ H_2O_2 \longrightarrow 2\ H_2O + O_2$) or peroxidase ($H_2O_2 + 2\ H^+ \longrightarrow 2\ H_2O$).
11. Other chemicals required for microbial growth include sulfur, phosphorus, trace elements, and, for some microorganisms, organic growth factors.

Biofilms (pp. 162–163)

1. Microbes adhere to surfaces and accumulate as biofilms on solid surfaces in contact with water.
2. Biofilms form on teeth, contact lenses, and catheters.
3. Microbes in biofilms are more resistant to antibiotics than are free swimming microbes.

Culture Media (pp. 164–169)

1. A culture medium is any material prepared for the growth of bacteria in a laboratory.
2. Microbes that grow and multiply in or on a culture medium are known as a culture.
3. Agar is a common solidifying agent for a culture medium.

Chemically Defined Media (p. 165)

4. A chemically defined medium is one in which the exact chemical composition is known.

Complex Media (p. 165)

5. A complex medium is one in which the exact chemical composition varies slightly from batch to batch.

Anaerobic Growth Media and Methods (pp. 166–167)

6. Reducing media chemically remove molecular oxygen (O_2) that might interfere with the growth of anaerobes.
7. Petri plates can be incubated in an anaerobic jar, anaerobic chamber, or OxyPlate.

Special Culture Techniques (pp. 167–168)

8. Some parasitic and fastidious bacteria must be cultured in living animals or in cell cultures.
9. CO_2 incubators or candle jars are used to grow bacteria that require an increased CO_2 concentration.
10. Procedures and equipment to minimize exposure to pathogenic microorganisms are designated as biosafety levels 1 through 4.

Selective and Differential Media (pp. 168–169)

11. By inhibiting unwanted organisms with salts, dyes, or other chemicals, selective media allow growth of only the desired microbes.
12. Differential media are used to distinguish different organisms.

Enrichment Culture (p. 169)

13. An enrichment culture is used to encourage the growth of a particular microorganism in a mixed culture.

Obtaining Pure Cultures (pp. 169–170)

1. A colony is a visible mass of microbial cells that theoretically arose from one cell.
2. Pure cultures are usually obtained by the streak plate method.

Preserving Bacterial Cultures (p. 170)

1. Microbes can be preserved for long periods of time by deep-freezing or lyophilization (freeze-drying).

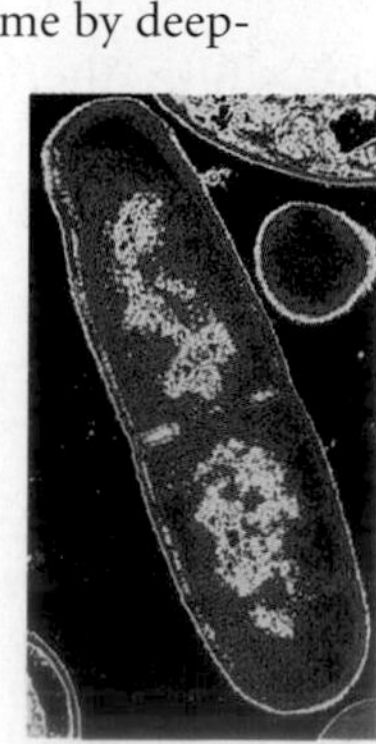

The Growth of Bacterial Cultures (pp. 171–179)

Bacterial Division (p. 171)

1. The normal reproductive method of bacteria is binary fission, in which a single cell divides into two identical cells.
2. Some bacteria reproduce by budding, aerial spore formation, or fragmentation.

Generation Time (p. 171)

3. The time required for a cell to divide or a population to double is known as the generation time.

Logarithmic Representation of Bacterial Populations (pp. 171–172)

4. Bacterial division occurs according to a logarithmic progression (two cells, four cells, eight cells, and so on).

Phases of Growth (pp. 172–174)

5. During the lag phase, there is little or no change in the number of cells, but metabolic activity is high.
6. During the log phase, the bacteria multiply at the fastest rate possible under the conditions provided.
7. During the stationary phase, there is an equilibrium between cell division and death.
8. During the death phase, the number of deaths exceeds the number of new cells formed.

Direct Measurement of Microbial Growth (pp. 174–178)

9. A standard plate count reflects the number of viable microbes and assumes that each bacterium grows into a single colony; plate counts are reported as number of colony-forming units (CFU).
10. A plate count may be done by either the pour plate method or the spread plate method.
11. In filtration, bacteria are retained on the surface of a membrane filter and then transferred to a culture medium to grow and subsequently be counted.
12. The most probable number (MPN) method can be used for microbes that will grow in a liquid medium; it is a statistical estimation.
13. In a direct microscopic count, the microbes in a measured volume of a bacterial suspension are counted with the use of a specially designed slide.

Estimating Bacterial Numbers by Indirect Methods (pp. 178–179)

14. A spectrophotometer is used to determine turbidity by measuring the amount of light that passes through a suspension of cells.
15. An indirect way of estimating bacterial numbers is measuring the metabolic activity of the population (for example, acid production or oxygen consumption).
16. For filamentous organisms such as fungi, measuring dry weight is a convenient method of growth measurement.

STUDY QUESTIONS

Answers to the Review and Multiple Choice questions can be found by turning to the blue Answers tab at the back of the textbook.

Review

1. Describe binary fission.
2. Macronutrients (needed in relatively large amounts) are often listed as CHONPS. What does each of these letters indicate, and why are they needed by the cell?
3. Define and explain the importance of each of the following:
   a. catalase
   b. hydrogen peroxide
   c. peroxidase
   d. superoxide radical
   e. superoxide dismutase
4. Seven methods of measuring microbial growth were explained in this chapter. Categorize each as either a direct or an indirect method.
5. By deep-freezing, bacteria can be stored without harm for extended periods. Why do refrigeration and freezing preserve foods?
6. A pastry chef accidentally inoculated a cream pie with six *S. aureus* cells. If *S. aureus* has a generation time of 60 minutes, how many cells would be in the cream pie after 7 hours?
7. Nitrogen and phosphorus added to beaches following an oil spill encourage the growth of natural oil-degrading bacteria. Explain why the bacteria do not grow if nitrogen and phosphorus are not added.
8. Differentiate complex and chemically defined media.
9. **DRAW IT** Draw the following growth curves for *E. coli*, starting with 100 cells with a generation time of 30 minutes at 35°C, 60 minutes at 20°C, and 3 hours at 5°C.
   a. The cells are incubated for 5 hours at 35°C.
   b. After 5 hours, the temperature is changed to 20°C for 2 hours.
   c. After 5 hours at 35°C, the temperature is changed to 5°C for 2 hours followed by 35°C for 5 hours.

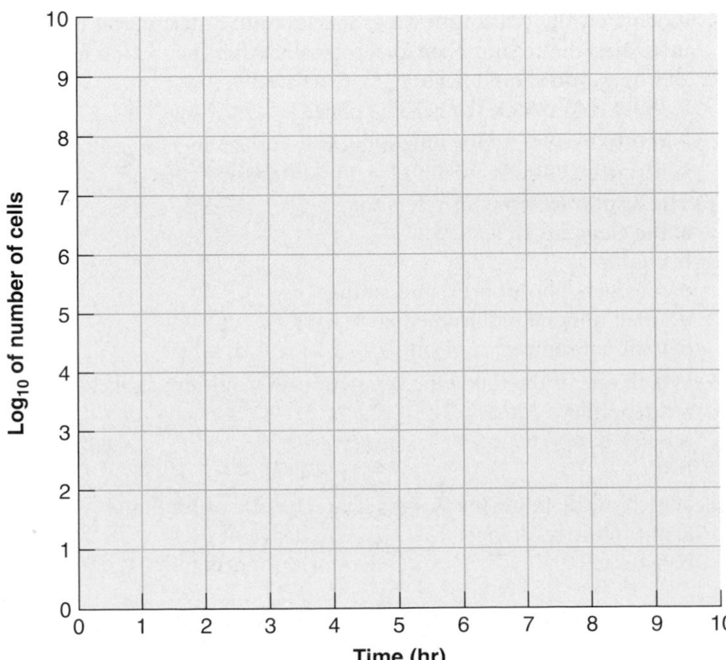

Multiple Choice

Use the following information to answer questions 1 and 2. Two culture media were inoculated with four different bacteria. After incubation, the following results were obtained:

Organism	Medium 1	Medium 2
Escherichia coli	Red colonies	No growth
Staphylococcus aureus	No growth	Growth
Staphylococcus epidermidis	No growth	Growth
Salmonella enterica	Colorless colonies	No growth

1. Medium 1 is
   a. selective.
   b. differential.
   c. both selective and differential.
2. Medium 2 is
   a. selective.
   b. differential.
   c. both selective and differential.

Use the following graph to answer questions 3 and 4.

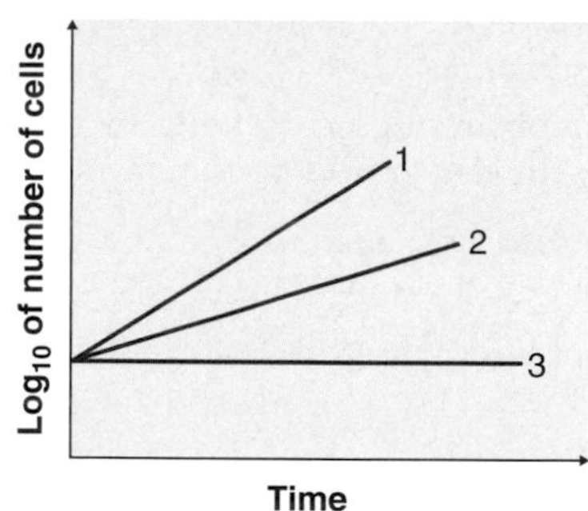

3. Which of the lines best depicts the log phase of a thermophile incubated at room temperature?
4. Which of the lines best depicts the log phase of *Listeria monocytogenes* growing in a human?
5. Assume you inoculated 100 facultatively anaerobic cells onto nutrient agar and incubated the plate aerobically. You then inoculated 100 cells of the same species onto nutrient agar and incubated the second plate anaerobically. After incubation for 24 hours, you should have
   a. more colonies on the aerobic plate.
   b. more colonies on the anaerobic plate.
   c. the same number of colonies on both plates.
6. The term *trace elements* refers to
   a. the elements CHONPS.
   b. vitamins.
   c. nitrogen, phosphorus, and sulfur.
   d. small mineral requirements.
   e. toxic substances.
7. Which one of the following temperatures would most likely kill a mesophile?
   a. −50°C
   b. 0°C
   c. 9°C
   d. 37°C
   e. 60°C
8. Which of the following is *not* a characteristic of biofilms?
   a. antibiotic resistance
   b. hydrogel
   c. iron deficiency
   d. quorum sensing
9. Which of the following types of media would *not* be used to culture aerobes?
   a. selective media
   b. reducing media
   c. enrichment media
   d. differential media
   e. complex media
10. An organism that has peroxidase and superoxide dismutase but lacks catalase is most likely an
    a. aerobe.
    b. aerotolerant anaerobe.
    c. obligate anaerobe.

Critical Thinking

1. *E. coli* was incubated with aeration in a nutrient medium containing two carbon sources, and the following growth curve was made from this culture.
   a. Explain what happened at the time marked x.
   b. Which substrate provided "better" growth conditions for the bacteria? How can you tell?

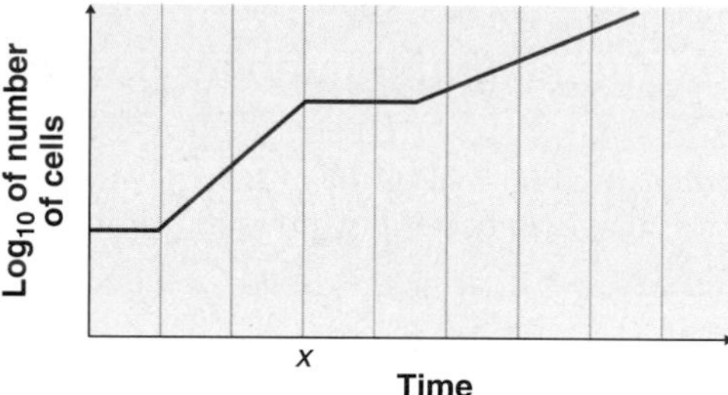

2. *Clostridium* and *Streptococcus* are both catalase-negative. *Streptococcus* grows by fermentation. Why is *Clostridium* killed by oxygen, whereas *Streptococcus* is not?
3. Most laboratory media contain a fermentable carbohydrate and peptone because the majority of bacteria require carbon, nitrogen, and energy sources in these forms. How are these three needs met by glucose–minimal salts medium? (*Hint:* See Table 6.2.)
4. Flask A contains yeast cells in glucose–minimal salts broth incubated at 30°C with aeration. Flask B contains yeast cells in glucose–minimal salts broth incubated at 30°C in an anaerobic jar. The yeasts are facultative anaerobes.
   a. Which culture produced more ATP?
   b. Which culture produced more alcohol?
   c. Which culture had the shorter generation time?
   d. Which culture had the greater cell mass?
   e. Which culture had the higher absorbance?

Clinical Applications

1. Assume that after washing your hands, you leave ten bacterial cells on a new bar of soap. You then decide to do a plate count of the soap after it was left in the soap dish for 24 hours. You dilute 1 g of the soap 1:10^6 and plate it on standard plate count agar. After 24 hours of incubation, there are 168 colonies. How many bacteria were on the soap? How did they get there?
2. Heat lamps are commonly used to maintain foods at about 50°C for as long as 12 hours in cafeteria serving lines. The following experiment was conducted to determine whether this practice poses a potential health hazard.

Beef cubes were surface-inoculated with 500,000 bacterial cells and incubated at 43–53°C to establish temperature limits for bacterial growth. The following results were obtained from standard plate counts performed on beef cubes at 6 and 12 hours after inoculation:

		Bacteria per Gram of Beef After	
	Temp. (°C)	**6 hr**	**12 hr**
Staphylococcus aureus	43	140,000,000	740,000,000
	51	810,000	59,000
	53	650	300
Salmonella typhimurium	43	3,200,000	10,000,000
	51	950,000	83,000
	53	1,200	300
Clostridium perfringens	43	1,200,000	3,600,000
	51	120,000	3,800
	53	300	300

Draw the growth curves for each organism. What holding temperature would you recommend? Assuming that cooking kills bacteria in foods, how could these bacteria contaminate the cooked foods? What disease does each organism cause? (*Hint:* See Chapter 25.)

3. The number of bacteria in saliva samples was determined by collecting the saliva, making serial dilutions, and inoculating nutrient agar by the pour plate method. The plates were incubated aerobically for 48 hours at 37°C.

	Bacteria per ml Saliva	
	Before Using Mouthwash	**After Using Mouthwash**
Mouthwash 1	13.1×10^6	10.9×10^6
Mouthwash 2	11.7×10^6	14.2×10^5
Mouthwash 3	9.3×10^5	7.7×10^5

What can you conclude from these data? Did all the bacteria present in each saliva sample grow?

7 The Control of Microbial Growth

The scientific control of microbial growth began only about 100 years ago. Recall from Chapter 1 that Pasteur's work on microorganisms led scientists to believe that microbes were a possible cause of disease. In the mid-1800s, the Hungarian physician Ignaz Semmelweis and English physician Joseph Lister used this thinking to develop some of the first microbial control practices for medical procedures. These practices included washing hands with microbe-killing chloride of lime and using the techniques of **aseptic surgery** to prevent microbial contamination of surgical wounds. Until that time, hospital-acquired infections, or *nosocomial infections,* were the cause of death in at least 10% of surgical cases, and deaths of delivering mothers were as high as 25%. Ignorance of microbes was such that during the American Civil War, a surgeon might have cleaned his scalpel on his bootsole between incisions.

Over the last century, scientists have continued to develop a variety of physical methods and chemical agents to control microbial growth. In Chapter 20 we will discuss methods for controlling microbes once infection has occurred, mainly antibiotic chemotherapy.

UNDER THE MICROSCOPE

Bacteria trapped on a membrane filter (note the two holes shown).

Q&A

Filtration can be used to remove microorganisms from water and solutions. In what situation is filtration the only practical way to eliminate undesirable microbes?

Look for the answer in the chapter.

The Terminology of Microbial Control

LEARNING OBJECTIVE

7-1 Define the following key terms related to microbial control: *sterilization, disinfection, antisepsis, degerming, sanitization, biocide, germicide, bacteriostasis,* and *asepsis.*

A word frequently used, and misused, in discussing the control of microbial growth is *sterilization.* **Sterilization** is the removal or destruction of *all forms* of microbial life. Prions, however, have a high resistance to all forms of sterilization (see page 203), and for practical purposes this requires a modification of the term (even if unstated). Therefore, the definition of *sterilization* usually implies the absence of prions. Heating is the most common method used for killing microbes, including the most resistant forms, such as endospores. A sterilizing agent is called a **sterilant.** Liquids or gases can be sterilized by filtration.

One would think that canned food in the supermarket is completely sterile. In reality, the heat treatment required to ensure absolute sterility would unnecessarily degrade the quality of the food. Instead, food is subjected only to enough heat to destroy the endospores of *Clostridium botulinum* (bo-tū-lī′num), which can produce a deadly toxin. This limited heat treatment is termed **commercial sterilization.** The endospores of a number of thermophilic bacteria, capable of causing food spoilage but not human disease, are considerably more resistant to heat than *C. botulinum.* If present, they will survive, but their survival is usually of no practical consequence; they will not grow at normal food storage temperatures. If canned foods in a supermarket were incubated at temperatures in the growth range of these thermophiles (above about 45°C), significant food spoilage would occur.

Complete sterilization is often not required in other settings. For example, the body's normal defenses can cope with a few microbes entering a surgical wound. A drinking glass or a fork in a restaurant requires only enough microbial control to prevent the transmission of possibly pathogenic microbes from one person to another.

Control directed at destroying harmful microorganisms is called **disinfection.** It usually refers to the destruction of vegetative (non–endospore-forming) pathogens, which is not the same thing as complete sterility. Disinfection might make use of chemicals, ultraviolet radiation, boiling water, or steam. In practice, the term is most commonly applied to the use of a chemical (a *disinfectant*) to treat an inert surface or substance. When this treatment is directed at living tissue, it is called **antisepsis,** and the chemical is then called an *antiseptic.* Therefore, in practice the same chemical might be called a disinfectant for one use and an antiseptic for another. Of course, many chemicals suitable for swabbing a tabletop would be too harsh to use on living tissue.

There are modifications of disinfection and antisepsis. For example, when someone is about to receive an injection, the skin is swabbed with alcohol—the process of **degerming** (or *degermation*), which mostly results in the mechanical removal, rather than the killing, of most of the microbes in a limited area. Restaurant glassware, china, and tableware are subjected to **sanitization,** which is intended to lower microbial counts to safe public health levels and minimize the chances of disease transmission from one user to another. This is usually accomplished by high-temperature washing or, in the case of glassware in a bar, washing in a sink followed by a dip in a chemical disinfectant.

Table 7.1 summarizes the terminology relating to the control of microbial growth.

Table 7.1 Terminology Relating to the Control of Microbial Growth

	Definition	Comments
Sterilization	Destruction or removal of all forms of microbial life, including endospores but with the possible exception of prions.	Usually done by steam under pressure or a sterilizing gas, such as ethylene oxide.
Commercial Sterilization	Sufficient heat treatment to kill endospores of *Clostridium botulinum* in canned food.	More-resistant endospores of thermophilic bacteria may survive, but they will not germinate and grow under normal storage conditions.
Disinfection	Destruction of vegetative pathogens.	May make use of physical or chemical methods.
Antisepsis	Destruction of vegetative pathogens on living tissue.	Treatment is almost always by chemical antimicrobials.
Degerming	Removal of microbes from a limited area, such as the skin around an injection site.	Mostly a mechanical removal by an alcohol-soaked swab.
Sanitization	Treatment intended to lower microbial counts on eating and drinking utensils to safe public health levels.	May be done with high-temperature washing or by dipping into a chemical disinfectant.

Names of treatments that cause the outright death of microbes have the suffix *-cide,* meaning kill. A **biocide,** or **germicide,** kills microorganisms (usually with certain exceptions, such as endospores); a *fungicide* kills fungi; a *virucide* inactivates viruses; and so on. Other treatments only inhibit the growth and multiplication of bacteria; their names have the suffix *-stat* or *-stasis,* meaning to stop or to steady, as in **bacteriostasis.** Once a bacteriostatic agent is removed, growth might resume.

Sepsis, from the Greek for decay or putrid, indicates bacterial contamination, as in septic tanks for sewage treatment. (The term is also used to describe a disease condition; see Chapter 23, page 639.) *Aseptic* means that an object or area is free of pathogens. Recall from Chapter 1 that **asepsis** is the absence of significant contamination. Aseptic techniques are important in surgery to minimize contamination from the instruments, operating personnel, and the patient.

CHECK YOUR UNDERSTANDING

✓ The usual definition of *sterilization* is the removal or destruction of all forms of microbial life; how could there be practical exceptions to this simple definition? **7-1**

The Rate of Microbial Death

LEARNING OBJECTIVE

7-2 Describe the patterns of microbial death caused by treatments with microbial control agents.

When bacterial populations are heated or treated with antimicrobial chemicals, they usually die at a constant rate. For example, suppose a population of 1 million microbes has been treated for 1 minute, and 90% of the population has died. We are now left with 100,000 microbes. If the population is treated for another minute, 90% of *those* microbes die, and we are left with 10,000 survivors. In other words, for each minute the treatment is applied, 90% of the remaining population is killed (**Table 7.2**).

Table 7.2 Microbial Exponential Death Rate: An Example

Time (min)	Deaths per Minute	Number of Survivors
0	0	1,000,000
1	900,000	100,000
2	90,000	10,000
3	9000	1000
4	900	100
5	90	10
6	9	1

If the death curve is plotted logarithmically, the death rate is constant, as shown by the straight line in **Figure 7.1a**.

Several factors influence the effectiveness of antimicrobial treatments:

- *The number of microbes.* The more microbes there are to begin with, the longer it takes to eliminate the entire population (**Figure 7.1b**).
- *Environmental influences.* The presence of organic matter often inhibits the action of chemical antimicrobials. In hospitals, the presence of organic matter in blood, vomit, or feces influences the selection of disinfectants. Microbes in surface biofilms (see page 162) are difficult for biocides to reach effectively. Because their activity is due to temperature-dependent chemical reactions, disinfectants work somewhat better under warm conditions.

  The nature of the suspending medium is also a factor in heat treatment. Fats and proteins are especially protective, and a medium rich in these substances protects microbes, which will then have a higher survival rate. Heat is also measurably more effective under acidic conditions.
- *Time of exposure.* Chemical antimicrobials often require extended exposure to affect more-resistant microbes or endospores. See the discussion of equivalent treatments on page 191.
- *Microbial characteristics.* The concluding section of this chapter discusses how microbial characteristics affect the choice of chemical and physical control methods.

CHECK YOUR UNDERSTANDING

✓ How is it possible that a solution containing a million bacteria would take longer to sterilize than one containing a half-million bacteria? **7-2**

Actions of Microbial Control Agents

LEARNING OBJECTIVE

7-3 Describe the effects of microbial control agents on cellular structures.

In this section, we examine the ways various agents actually kill or inhibit microbes.

Alteration of Membrane Permeability

A microorganism's plasma membrane (see Figure 4.14, page 90), located just inside the cell wall, is the target of many microbial control agents. This membrane actively regulates the passage of nutrients into the cell and the elimination of wastes from the cell. Damage to the lipids or proteins of the plasma membrane by antimicrobial agents causes cellular contents to leak into the surrounding medium and interferes with the growth of the cell.

Figure 7.1

FOUNDATION FIGURE A Microbial Death Curve

The concept of a killing curve for microbial populations, including the elements of time and the size of the initial population, is especially useful in food preservation and in the sterilization of media or medical supplies.

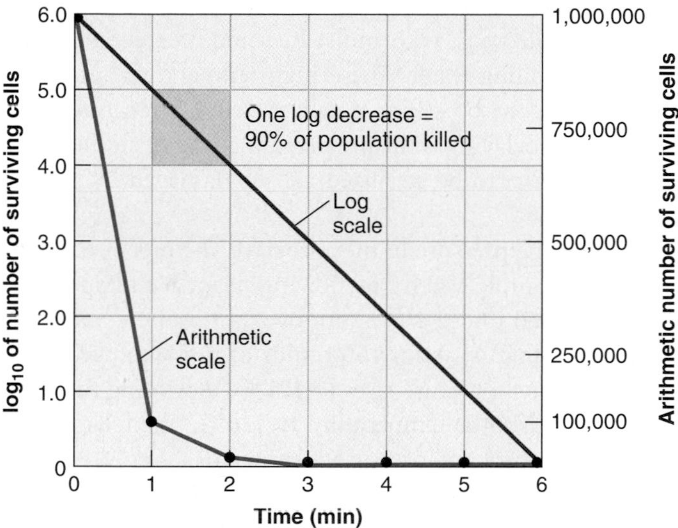

(a) The data are plotted logarithmically (red line) and arithmetically (blue line). This graph has been constructed so that the log and arithmetic scales coincide at two points: at 1 cell and 1 million cells. In this example, the cells are dying at a constant rate of 90% each minute. Note that the attempt to plot the population arithmetically is not practical; at 3 minutes the population of 1000 cells would be only a hundredth of the graphed distance between 100,000 and the baseline. Logarithmic numbers are needed to graph this properly, although at the cost of our easy perception of the situation.

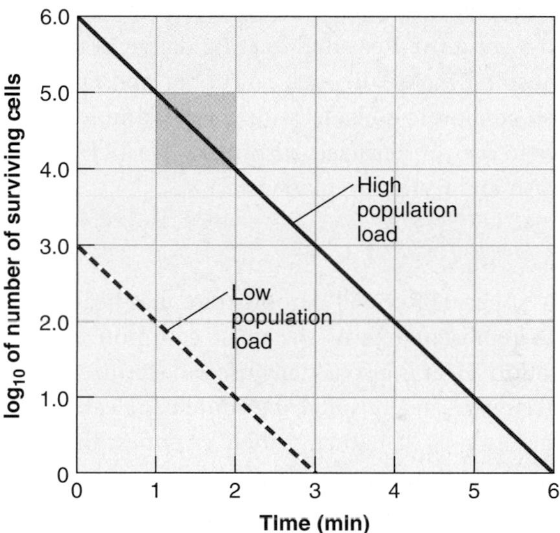

(b) The effect of high or low initial load of microbes. If the rate of killing is the same, it will take longer to kill all members of a larger population than a smaller one. This is true for both heat and chemical treatments.

Key Concept

It is necessary to use logarithmic numbers to graph bacterial populations effectively. Logarithmic death curves can demonstrate, for instance, the effects of initial population size on the time necessary to achieve sterilization.

Damage to Proteins and Nucleic Acids

Bacteria are sometimes thought of as "little bags of enzymes." Enzymes, which are primarily protein, are vital to all cellular activities. Recall that the functional properties of proteins are the result of their three-dimensional shape (see Figure 2.15, page 46). This shape is maintained by chemical bonds that link adjoining portions of the amino acid chain as it folds back and forth upon itself. Some of those bonds are hydrogen bonds, which are susceptible to breakage by heat or certain chemicals; breakage results in denaturation of the protein. Covalent bonds are stronger but are also subject to attack. For example, disulfide bridges, which play an important role in protein structure by joining amino acids with exposed sulfhydryl (—SH) groups, can be broken by certain chemicals or sufficient heat.

The nucleic acids DNA and RNA are the carriers of the cell's genetic information. Damage to these nucleic acids by heat, radiation, or chemicals is frequently lethal to the cell; the cell can no longer replicate, nor can it carry out normal metabolic functions such as the synthesis of enzymes.

CHECK YOUR UNDERSTANDING

✓ Would a chemical microbial control agent that affects plasma membranes affect humans? **7-3**

Physical Methods of Microbial Control

LEARNING OBJECTIVES

7-4 Compare the effectiveness of moist heat (boiling, autoclaving, pasteurization) and dry heat.

7-5 Describe how filtration, low temperatures, high pressure, desiccation, and osmotic pressure suppress microbial growth.

7-6 Explain how radiation kills cells.

As early as the Stone Age, humans likely were already using some physical methods of microbial control to preserve foods. Drying (desiccation) and salting (osmotic pressure) were probably among the earliest techniques.

When selecting methods of microbial control, one must consider what else, besides the microbes, a particular method will affect. For example, heat might inactivate certain vitamins or antibiotics in a solution. Repeated heating damages many laboratory and hospital materials, such as rubber and latex tubing. There are also economic considerations; for example, it may be less expensive to use presterilized, disposable plasticware than to repeatedly wash and resterilize glassware.

Heat

A visit to any supermarket will demonstrate that heat-preserved canned goods represent one of the most common methods of food preservation. Heat is also usually used to sterilize laboratory media and glassware and hospital instruments. Heat appears to kill microorganisms by denaturing their enzymes; the resultant changes to the three-dimensional shapes of these proteins inactivate them (see Figure 5.6, page 119).

Heat resistance varies among different microbes; these differences can be expressed through the concept of thermal death point. **Thermal death point (TDP)** is the lowest temperature at which all the microorganisms in a particular liquid suspension will be killed in 10 minutes.

Another factor to be considered in sterilization is the length of time required. This is expressed as **thermal death time (TDT),** the minimal length of time for all bacteria in a particular liquid culture to be killed at a given temperature. Both TDP and TDT are useful guidelines that indicate the severity of treatment required to kill a given population of bacteria.

Decimal reduction time (DRT, or *D value*) is a third concept related to bacterial heat resistance. DRT is the time, in minutes, in which 90% of a population of bacteria at a given temperature will be killed (in Table 7.2 and Figure 7.1a, DRT is 1 minute). In Chapter 28 you can find an important application of DRT in the canning industry; see the discussion of the 12D treatment of canned goods in Chapter 28.

Moist Heat Sterilization

Moist heat kills microorganisms primarily by coagulating proteins (denaturation), which is caused by breakage of the hydrogen bonds that hold the proteins in their three-dimensional structure. This coagulation process is familiar to anyone who has watched an egg white frying.

One type of moist heat "sterilization" is boiling, which kills vegetative forms of bacterial pathogens, almost all viruses, and fungi and their spores within about 10 minutes, usually much faster. Free-flowing (unpressurized) steam is essentially the same temperature as boiling water. Endospores and some viruses, however, are not destroyed this quickly. Some hepatitis viruses, for example, can survive up to 30 minutes of boiling, and some bacterial endospores can resist boiling for more than 20 hours. Boiling is therefore not always a reliable sterilization procedure. However, brief boiling, even at high altitudes, will kill most pathogens. The use of boiling to sanitize glass baby bottles is a familiar example.

Reliable sterilization with moist heat requires temperatures above that of boiling water. These high temperatures are most commonly achieved by steam under pressure in an **autoclave** (**Figure 7.2**). Autoclaving is the preferred method of sterilization, unless the material to be sterilized can be damaged by heat or moisture.

The higher the pressure in the autoclave, the higher the temperature. For example, when free-flowing steam at a temperature of 100°C is placed under a pressure of 1 atmosphere above sea level pressure—that is, about 15 pounds of pressure per square inch (psi)—the temperature rises to 121°C. Increasing the pressure to 20 psi raises the temperature to 126°C. The relationship between temperature and pressure is shown in **Table 7.3.**

Sterilization in an autoclave is most effective when the organisms either are contacted by the steam directly or are contained in a small volume of aqueous (primarily water) liquid. Under these conditions, steam at a pressure of about 15 psi (121°C) will kill *all* organisms (but not prions, see page 392) and their endospores in about 15 minutes.

Autoclaving is used to sterilize culture media, instruments, dressings, intravenous equipment, applicators, solutions, syringes, transfusion equipment, and numerous other items that can withstand high temperatures and pressures. Large industrial autoclaves are called *retorts* (see Figure 28.2 on page 795), but the same principle applies for the common household pressure cooker used in the home canning of foods.

Heat requires extra time to reach the center of solid materials, such as canned meats, because such materials do not develop the efficient heat-distributing convection currents that occur in liquids. Heating large containers also requires extra time. **Table 7.4** shows the different time requirements for sterilizing liquids in various container sizes.

Unlike sterilizing aqueous solutions, sterilizing the surface of a solid requires that steam actually contact it. To sterilize dry glassware, bandages, and the like, care must be taken to ensure that steam contacts all surfaces. For example, aluminum foil is impervious to steam and should not be used to wrap dry materials that are to be sterilized; paper should be used instead. Care should also be taken to avoid trapping air in the bottom of a dry container: trapped air will not be replaced by steam, because steam is lighter than air. The trapped air is the equivalent of a small hot-air oven, which, as we will see shortly, requires a higher temperature and longer time to sterilize materials. Containers that can trap air should be placed in a tipped position so that

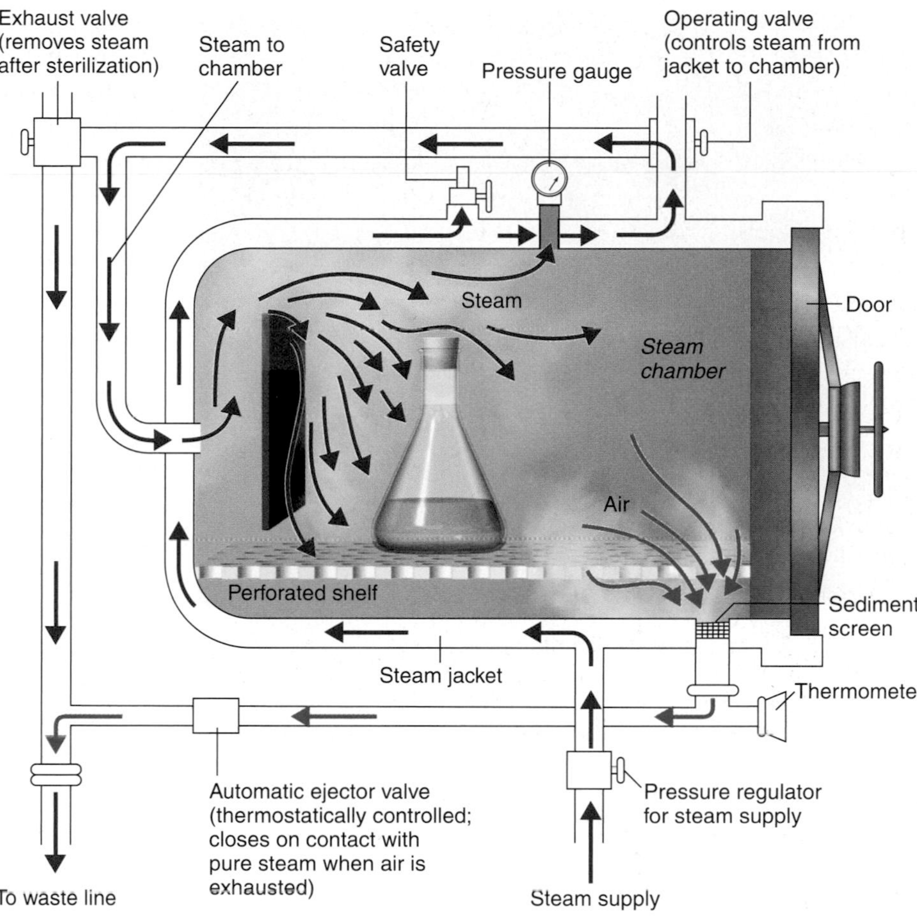

Figure 7.2 An autoclave. The entering steam forces the air out of the bottom (blue arrows). The automatic ejector valve remains open as long as an air-steam mixture is passing out of the waste line. When all the air has been ejected, the higher temperature of the pure steam closes the valve, and the pressure in the chamber increases.

Q How would an empty, uncapped flask be positioned for sterilization in an autoclave?

Table 7.3 The Relationship between the Pressure and Temperature of Steam at Sea Level*

Pressure (psi in Excess of Atmospheric Pressure)	Temperature (°C)
0	100
5	110
10	116
15	121
20	126
30	135

*At higher altitudes, the atmospheric pressure is less, a phenomenon that must be taken into account in operating an autoclave. For example, to reach sterilizing temperatures (121°C) in Denver, Colorado, whose altitude is 5280 feet (1600 meters), the pressure shown on the autoclave gauge would need to be higher than the 15 psi shown in the table.

Table 7.4 The Effect of Container Size on Autoclave Sterilization Times for Liquid Solutions*

Container Size	Liquid Volume	Sterilization Time (min)
Test tube: 18 × 150 mm	10 ml	15
Erlenmeyer flask: 125 ml	95 ml	15
Erlenmeyer flask: 2000 ml	1500 ml	30
Fermentation bottle: 9000 ml	6750 ml	70

*Sterilization times in the autoclave include the time for the contents of the containers to reach sterilization temperatures. For smaller containers, this is only 5 min or less, but for a 9000-ml bottle it might be as much as 70 min. A container is usually not filled past 75% of its capacity.

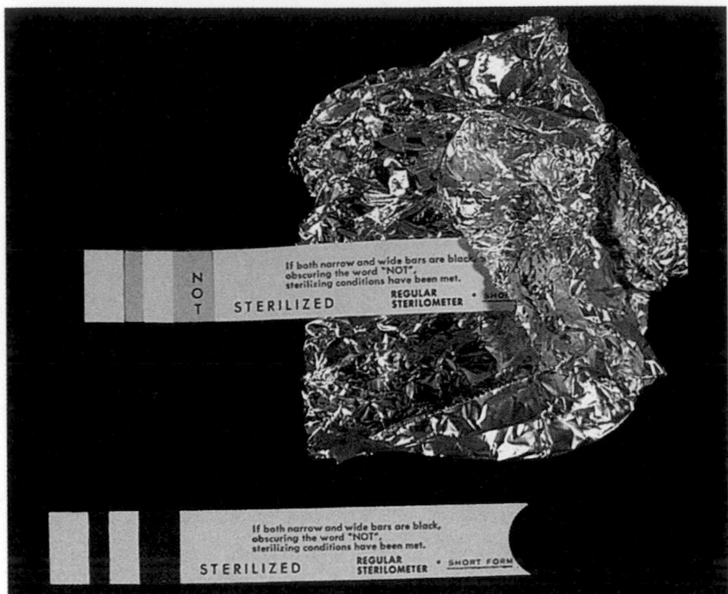

Figure 7.3 Examples of sterilization indicators. The strips indicate whether the item has been properly sterilized. The word *NOT* appears if heating has been inadequate. In the illustration, the indicator that was wrapped with aluminum foil was not sterilized because steam couldn't penetrate the foil.

Q **What should have been used instead of aluminum foil to wrap the items?**

the steam will force out the air. Products that do not permit penetration by moisture, such as mineral oil or petroleum jelly, are not sterilized by the same methods used to sterilize aqueous solutions.

Several commercially available methods can indicate whether heat treatment has achieved sterilization. Some of these are chemical reactions in which an indicator changes color when the proper times and temperatures have been reached (Figure 7.3). In some designs, the word *sterile* or *autoclaved* appears on wrappings or tapes. In another method, a pellet contained within a glass vial melts. A widely used test consists of preparations of specified species of bacterial endospores impregnated into paper strips. After the strips are autoclaved, they can then be aseptically inoculated into culture media. Growth in the culture media indicates survival of the endospores and therefore inadequate processing. Other designs use endospore suspensions that can be released, after heating, into a surrounding culture medium within the same vial.

Steam under pressure fails to sterilize when the air is not completely exhausted. This can happen with the premature closing of the autoclave's automatic ejector valve (see Figure 7.2). The principles of heat sterilization have a direct bearing on home canning. As anyone familiar with home canning knows, the steam must flow vigorously out of the valve in the lid for several minutes to carry with it all the air before the pressure cooker is sealed. If the air is not completely exhausted, the container will not reach the temperature expected for a given pressure. Because of the possibility of botulism, a kind of food poisoning resulting from improper canning methods (see Chapter 22, page 616), anyone doing home canning should obtain reliable directions and follow them exactly.

Pasteurization

Recall from Chapter 1 that in the early days of microbiology, Louis Pasteur found a practical method of preventing the spoilage of beer and wine. Pasteur used mild heating, which was sufficient to kill the organisms that caused the particular spoilage problem without seriously damaging the taste of the product. The same principle was later applied to milk to produce what we now call pasteurized milk. The intent of **pasteurization** of milk was to eliminate pathogenic microbes. It also lowers microbial numbers, which prolongs milk's good quality under refrigeration. Many relatively heat-resistant (**thermoduric**) bacteria survive pasteurization, but these are unlikely to cause disease or cause refrigerated milk to spoil.

Products other than milk, such as ice cream, yogurt, and beer, all have their own pasteurization times and temperatures, which often differ considerably. There are several reasons for these variations. For example, heating is less efficient in foods that are more viscous, and fats in food can have a protective effect on microorganisms. The dairy industry routinely uses a test to determine whether products have been pasteurized: the *phosphatase test* (phosphatase is an enzyme naturally present in milk). If the product has been pasteurized, phosphatase will have been inactivated.

Most milk pasteurization today uses temperatures of at least 72°C, but for only 15 seconds. This treatment, known as **high-temperature short-time (HTST) pasteurization,** is applied as the milk flows continuously past a heat exchanger. In addition to killing pathogens, HTST pasteurization lowers total bacterial counts, so the milk keeps well under refrigeration.

Milk can also be sterilized—something quite different from pasteurization—by **ultra-high-temperature (UHT)** treatments. It can then be stored for several months without refrigeration (also see *commercial sterilization,* page 794). UHT-treated milk is widely sold in Europe and is especially necessary in less developed parts of the world where refrigeration facilities are not always available. In the United States, UHT is sometimes used on the small containers of coffee creamers found in restaurants. To avoid giving the milk a cooked taste, the process avoids having the milk touch a surface hotter than the milk itself. Usually, the liquid milk (or juice) is sprayed through a nozzle into a chamber filled with high-temperature steam under pressure. A small volume of fluid sprayed into an atmosphere of high-temperature steam exposes a relatively large surface area on the fluid droplets to heating by the steam;

sterilizing temperatures are reached almost instantaneously. After reaching a temperature of 140°C for 4 seconds, the fluid is rapidly cooled in a vacuum chamber. The milk or juice is then packaged in a presterilized, airtight container.

The heat treatments we have just discussed illustrate the concept of **equivalent treatments:** as the temperature is increased, much less time is needed to kill the same number of microbes. For example, the destruction of highly resistant endospores might take 70 minutes at 115°C, whereas only 7 minutes might be needed at 125°C. Both treatments yield the same result.

Dry Heat Sterilization

Dry heat kills by oxidation effects. A simple analogy is the slow charring of paper in a heated oven, even when the temperature remains below the ignition point of paper. One of the simplest methods of dry heat sterilization is direct **flaming.** You will use this procedure many times in the microbiology laboratory when you sterilize inoculating loops. To effectively sterilize the inoculating loop, you heat the wire to a red glow. A similar principle is used in *incineration,* an effective way to sterilize and dispose of contaminated paper cups, bags, and dressings.

Another form of dry heat sterilization is **hot-air sterilization.** Items to be sterilized by this procedure are placed in an oven. Generally, a temperature of about 170°C maintained for nearly 2 hours ensures sterilization. The longer period and higher temperature (relative to moist heat) are required because the heat in water is more readily transferred to a cool body than is the heat in air. For example, imagine the different effects of immersing your hand in boiling water at 100°C (212°F) and of holding it in a hot-air oven at the same temperature for the same amount of time.

Filtration

Q&A Recall from Chapter 6 that *filtration* is the passage of a liquid or gas through a screenlike material with pores small enough to retain microorganisms (often the same apparatus used for counting; see Figure 6.18, page 177). A vacuum is created in the receiving flask; air pressure then forces the liquid through the filter. Filtration is used to sterilize heat-sensitive materials, such as some culture media, enzymes, vaccines, and antibiotic solutions.

Some operating theaters and rooms occupied by burn patients receive filtered air to lower the numbers of airborne microbes. **High-efficiency particulate air (HEPA) filters** remove almost all microorganisms larger than about 0.3 μm in diameter.

In the early days of microbiology, hollow candle-shaped filters of unglazed porcelain were used to filter liquids. The long and indirect passageways through the walls of the filter adsorbed the bacteria. Unseen pathogens that passed through the filters (causing such diseases as rabies) were called *filterable viruses.*

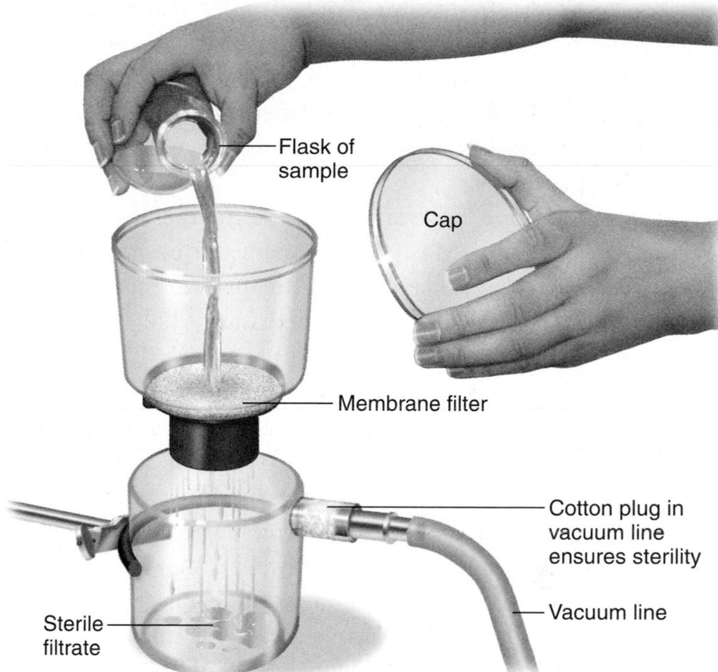

Figure 7.4 Filter sterilization with a disposable, presterilized plastic unit. The sample is placed into the upper chamber and forced through the membrane filter by a vacuum in the lower chamber. Pores in the membrane filter are smaller than the bacteria, so bacteria are retained on the filter. The sterilized sample can then be decanted from the lower chamber. Similar equipment with removable filter disks is used to count bacteria in samples (see Figure 6.18).

Q How is a plastic filtration apparatus presterilized? (Assume the plastic cannot be heat sterilized.)

See the discussion of filtration in modern water treatment on page 782.

In recent years, **membrane filters,** composed of such substances as cellulose esters or plastic polymers, have become popular for industrial and laboratory use (**Figure 7.4**). These filters are only 0.1 mm thick. The pores of membrane filters include, for example, 0.22-μm and 0.45-μm sizes, which are intended for bacteria. Some very flexible bacteria, such as spirochetes, or the wall-less mycoplasma, will sometimes pass through such filters, however. Filters are available with pores as small as 0.01 μm, a size that will retain viruses and even some large protein molecules.

Low Temperatures

The effect of low temperatures on microorganisms depends on the particular microbe and the intensity of the application. For example, at temperatures of ordinary refrigerators (0–7°C), the metabolic rate of most microbes is so reduced that they cannot reproduce or synthesize toxins. In other words, ordinary refrigeration has a bacteriostatic effect. Yet psychrotrophs do grow slowly at refrigerator temperatures and will alter the appearance and taste of foods after a time. For example, a single microbe reproducing only three times a day would reach a population of more than 2 million within a week. Pathogenic bacteria generally will not grow at refrigerator temperatures, but for at least one important exception, see the discussion of listeriosis in Chapter 22 (page 614).

Surprisingly, some bacteria can grow at temperatures several degrees below freezing. Most foods remain unfrozen until −2°C or lower. Rapidly attained subfreezing temperatures tend to render microbes dormant but do not necessarily kill them. Slow freezing is more harmful to bacteria; the ice crystals that form and grow disrupt the cellular and molecular structure of the bacteria. Thawing, being inherently slower, is actually the more damaging part of a freeze-thaw cycle. Once frozen, one-third of the population of some vegetative bacteria might survive a year, whereas other species might have very few survivors after this time. Many eukaryotic parasites, such as the roundworms that cause human trichinellosis, are killed by several days of freezing temperatures. Some important temperatures associated with microorganisms and food spoilage are shown in Figure 6.2 (page 158).

High Pressure

High pressure applied to liquid suspensions is transferred instantly and evenly throughout the sample. If the pressure is high enough, it alters the molecular structures of proteins and carbohydrates, resulting in the rapid inactivation of vegetative bacterial cells. Endospores are relatively resistant to high pressure. They can, however, be killed by other techniques, such as combining high pressure with elevated temperatures or by alternating pressure cycles that cause spore germination, followed by pressure-caused death of the resulting vegetative cells. Fruit juices preserved by high-pressure treatments have been marketed in Japan and the United States. An advantage is that these treatments preserve the flavors, colors, and nutrient values of the products.

Desiccation

In the absence of water, known as **desiccation,** microorganisms cannot grow or reproduce but can remain viable for years. Then, when water is made available to them, they can resume their growth and division. This is the principle that underlies lyophilization, or freeze-drying, a laboratory process for preserving microbes described in Chapter 6 (page 170). Certain foods are also freeze-dried (for example, coffee and some fruit additives for dry cereals).

The resistance of vegetative cells to desiccation varies with the species and the organism's environment. For example, the gonorrhea bacterium can withstand desiccation for only about an hour, but the tuberculosis bacterium can remain viable for months. Viruses are generally resistant to desiccation, but they are not as resistant as bacterial endospores, some of which have survived for centuries. This ability of certain dried microbes and endospores to remain viable is important in a hospital setting. Dust, clothing, bedding, and dressings might contain infectious microbes in dried mucus, urine, pus, and feces.

Osmotic Pressure

The use of high concentrations of salts and sugars to preserve food is based on the effects of *osmotic pressure.* High concentrations of these substances create a hypertonic environment that causes water to leave the microbial cell (see Figure 6.4, page 160). This process resembles preservation by desiccation, in that both methods deny the cell the moisture it needs for growth. The principle of osmotic pressure is used in the preservation of foods. For example, concentrated salt solutions are used to cure meats, and thick sugar solutions are used to preserve fruits.

As a general rule, molds and yeasts are much more capable than bacteria of growing in materials with low moisture or high osmotic pressures. This property of molds, sometimes combined with their ability to grow under acidic conditions, is the reason that molds, rather than bacteria, cause spoilage of fruits and grains. It is also part of the reason molds are able to form mildew on a damp wall or a shower curtain.

Radiation

Radiation has various effects on cells, depending on its wavelength, intensity, and duration. Radiation that kills microorganisms (sterilizing radiation) is of two types: ionizing and nonionizing.

Ionizing radiation—gamma rays, X rays, or high-energy electron beams—has a wavelength shorter than that of nonionizing radiation, less than about 1 nm. Therefore, it carries much more energy (**Figure 7.5**). *Gamma rays* are emitted by certain radioactive elements such as cobalt, and electron beams are produced by accelerating electrons to high energies in special machines. *X rays,* which are produced by machines in a manner similar to the production of electron beams, are similar to gamma rays. Gamma rays penetrate deeply but may require hours to sterilize large masses; *high-energy electron beams* have much lower penetrating power but usually require only a few seconds of exposure. The principal effect of ionizing radiation is the ionization of water, which forms highly reactive hydroxyl radicals (see the discussion of toxic forms of oxygen in Chapter 6, pages 161–162). These radicals react with organic cellular components, especially DNA.

The so-called target theory of damage by radiation supposes that ionizing particles, or packets of energy, pass through or close to vital portions of the cell; these constitute "hits." One, or a few, hits may only cause nonlethal mutations, some of them conceivably useful. More hits are likely to cause sufficient mutations to kill the microbe.

The food industry has recently renewed its interest in the use of radiation for food preservation (discussed more fully in Chapter 28). Low-level ionizing radiation, used for years in

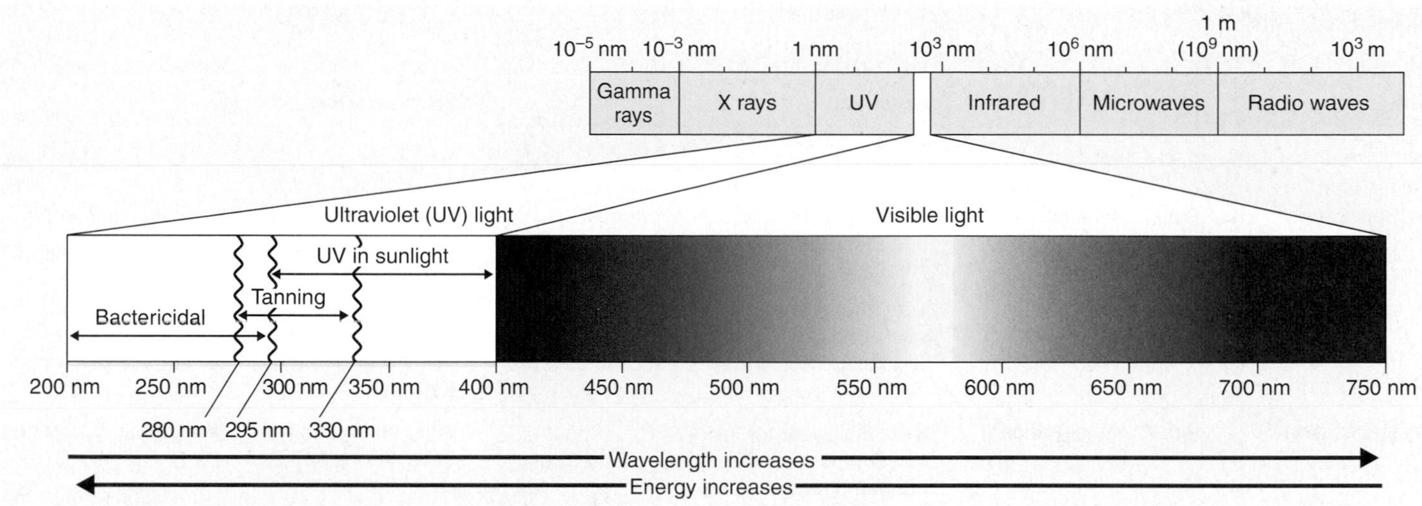

Figure 7.5 The radiant energy spectrum. Visible light and other forms of radiant energy radiate through space as waves of various lengths. Ionizing radiation, such as gamma rays and X rays, has a wavelength shorter than 1 nm. Nonionizing radiation, such as ultraviolet (UV) light, has a wavelength between 1 nm and about 380 nm, where the visible spectrum begins.

Q How might increased UV radiation (due to decrease in the ozone layer) affect the Earth's ecosystems?

many countries, has been approved in the United States for processing spices and certain meats and vegetables. Ionizing radiation, especially high-energy electron beams, is used to sterilize pharmaceuticals and disposable dental and medical supplies, such as plastic syringes, surgical gloves, suturing materials, and catheters. As a protection against bioterrorism, the postal service often uses electron beam radiation to sterilize certain classes of mail.

Nonionizing radiation has a wavelength longer than that of ionizing radiation, usually greater than about 1 nm. The best example of nonionizing radiation is ultraviolet (UV) light. UV light damages the DNA of exposed cells by causing bonds to form between adjacent pyrimidine bases, usually thymines, in DNA chains (see Figure 8.20, page 230). These *thymine dimers* inhibit correct replication of the DNA during reproduction of the cell. The UV wavelengths most effective for killing microorganisms are about 260 nm; these wavelengths are specifically absorbed by cellular DNA. UV radiation is also used to control microbes in the air. A UV, or "germicidal," lamp is commonly found in hospital rooms, nurseries, operating rooms, and cafeterias. UV light is also used to disinfect vaccines and other medical products. A major disadvantage of UV light as a disinfectant is that the radiation is not very penetrating, so the organisms to be killed must be directly exposed to the rays. Organisms protected by solids and such coverings as paper, glass, and textiles are not affected. Another potential problem is that UV light can damage human eyes, and prolonged exposure can cause burns and skin cancer in humans.

Sunlight contains some UV radiation, but the shorter wavelengths—those most effective against bacteria—are screened out by the ozone layer of the atmosphere. The antimicrobial effect of sunlight is due almost entirely to the formation of singlet oxygen in the cytoplasm (see Chapter 6, page 161). Many pigments produced by bacteria provide protection from sunlight.

Microwaves do not have much direct effect on microorganisms, and bacteria can readily be isolated from the interior of recently operated microwave ovens. Moisture-containing foods are heated by microwave action, and the heat will kill most vegetative pathogens. Solid foods heat unevenly because of the uneven distribution of moisture. For this reason, pork cooked in a microwave oven has been responsible for outbreaks of trichinellosis.

CHECK YOUR UNDERSTANDING

✓ How is microbial growth in canned foods prevented? **7-4**

✓ Why would a can of pork take longer to sterilize at a given temperature than a can of soup that also contained pieces of pork? **7-5**

✓ What is the connection between the killing effect of radiation and hydroxyl radical forms of oxygen? **7-6**

* * *

Table 7.5 summarizes the physical methods of microbial control.

Table 7.5 Physical Methods Used to Control Microbial Growth

Methods	Mechanism of Action	Comment	Preferred Use
Heat			
1. Moist heat			
a. Boiling or flowing steam	Protein denaturation	Kills vegetative bacterial and fungal pathogens and almost all viruses within 10 min; less effective on endospores	Dishes, basins, pitchers, various equipment
b. Autoclaving	Protein denaturation	Very effective method of sterilization; at about 15 psi of pressure (121°C), all vegetative cells and their endospores are killed in about 15 min	Microbiological media, solutions, linens, utensils, dressings, equipment, and other items that can withstand temperature and pressure
2. Pasteurization	Protein denaturation	Heat treatment for milk (72°C for about 15 sec) that kills all pathogens and most nonpathogens	Milk, cream, and certain alcoholic beverages (beer and wine)
3. Dry heat			
a. Direct flaming	Burning contaminants to ashes	Very effective method of sterilization	Inoculating loops
b. Incineration	Burning to ashes	Very effective method of sterilization	Paper cups, contaminated dressings, animal carcasses, bags, and wipes
c. Hot-air sterilization	Oxidation	Very effective method of sterilization but requires temperature of 170°C for about 2 hr	Empty glassware, instruments, needles, and glass syringes
Filtration	Separation of bacteria from suspending liquid	Removes microbes by passage of a liquid or gas through a screenlike material; most filters in use consist of cellulose acetate or nitrocellulose	Useful for sterilizing liquids (enzymes, vaccines) that are destroyed by heat
Cold			
1. Refrigeration	Decreased chemical reactions and possible changes in proteins	Has a bacteriostatic effect	Food, drug, and culture preservation
2. Deep-freezing (see Chapter 6, page 170)	Decreased chemical reactions and possible changes in proteins	An effective method for preserving microbial cultures, in which cultures are quick-frozen between −50° and −95°C	Food, drug, and culture preservation
3. Lyophilization (see Chapter 6, page 170)	Decreased chemical reactions and possible changes in proteins	Most effective method for long-term preservation of microbial cultures; water removed by high vacuum at low temperature	Food, drug, and culture preservation
High Pressure	Alteration of molecular structure of proteins and carbohydrates	Preservation of colors, flavors, nutrient values	Fruit juices
Desiccation	Disruption of metabolism	Involves removing water from microbes; primarily bacteriostatic	Food preservation
Osmotic Pressure	Plasmolysis	Results in loss of water from microbial cells	Food preservation
Radiation			
1. Ionizing	Destruction of DNA	Not widespread in routine sterilization	Sterilizing pharmaceuticals and medical and dental supplies
2. Nonionizing	Damage to DNA	Radiation not very penetrating	Control of closed environment with UV (germicidal) lamp

Chemical Methods of Microbial Control

LEARNING OBJECTIVES

7-7 List the factors related to effective disinfection.

7-8 Interpret the results of use-dilution tests and the disk-diffusion method.

7-9 Identify the methods of action and preferred uses of chemical disinfectants.

7-10 Differentiate halogens used as antiseptics from halogens used as disinfectants.

7-11 Identify the appropriate uses for surface-active agents.

7-12 List the advantages of glutaraldehyde over other chemical disinfectants.

7-13 Identify chemical sterilizers.

Chemical agents are used to control the growth of microbes on both living tissue and inanimate objects. Unfortunately, few chemical agents achieve sterility; most of them merely reduce microbial populations to safe levels or remove vegetative forms of pathogens from objects. A common problem in disinfection is the selection of an agent. No single disinfectant is appropriate for all circumstances.

Principles of Effective Disinfection

By reading the label, we can learn a great deal about a disinfectant's properties. Usually the label indicates what groups of organisms the disinfectant is effective against. Remember that the concentration of a disinfectant affects its action, so it should always be diluted exactly as specified by the manufacturer.

Also consider the nature of the material being disinfected. For example, are organic materials present that might interfere with the action of the disinfectant? Similarly, the pH of the medium often has a great effect on a disinfectant's activity.

Another very important consideration is whether the disinfectant will easily make contact with the microbes. An area might need to be scrubbed and rinsed before the disinfectant is applied. In general, disinfection is a gradual process. Thus, to be effective, a disinfectant might need to be left on a surface for several hours.

Evaluating a Disinfectant

Use-Dilution Tests

There is a need to evaluate the effectiveness of disinfectants and antiseptics. The current standard is the American Official Analytical Chemist's **use-dilution test.** Metal or glass cylinders (8 mm × 10 mm) are dipped into standardized cultures of the test bacteria grown in liquid media, removed, and dried at 37°C for a short time. The dried cultures are then placed into a solution of the disinfectant at the concentration recommended by the manufacturer and left there for 10 minutes at 20°C. Following this exposure, the cylinders are transferred to a medium that permits the growth of any surviving bacteria. The effectiveness of the disinfectant can then be determined by the number of cultures that grow.

Variations of this method are used for testing the effectiveness of antimicrobial agents against endospores, mycobacteria that cause tuberculosis, viruses, and fungi, because they are difficult to control with chemicals. Also, tests of antimicrobials intended for special purposes, such as dairy utensil disinfection, may substitute other test bacteria.

The Disk-Diffusion Method

The **disk-diffusion method** is used in teaching laboratories to evaluate the efficacy of a chemical agent. A disk of filter paper is soaked with a chemical and placed on an agar plate that has been previously inoculated and incubated with the test organism. After incubation, if the chemical is effective, a clear zone representing inhibition of growth can be seen around the disk (Figure 7.6).

Disks containing antibiotics are commercially available and used to determine microbial susceptibility to antibiotics (see Figure 20.17, page 572).

Types of Disinfectants

Phenol and Phenolics

Lister was the first to use **phenol** (carbolic acid) to control surgical infections in the operating room. Its use had been suggested by its effectiveness in controlling odor in sewage. It is now rarely used as an antiseptic or disinfectant because it irritates the skin and has a disagreeable odor. It is often used in throat lozenges for its local anesthetic effect but has little antimicrobial effect at the low concentrations used. At concentrations above 1% (such as in some throat sprays), however, phenol has a significant antibacterial effect. The structure of a phenol molecule is shown in Figure 7.7a.

Derivatives of phenol, called **phenolics,** contain a molecule of phenol that has been chemically altered to reduce its irritating qualities or increase its antibacterial activity in combination with a soap or detergent. Phenolics exert antimicrobial activity by injuring lipid-containing plasma membranes, which results in leakage of cellular contents. The cell wall of mycobacteria, the causes of tuberculosis and leprosy, are rich in lipids, which make them susceptible to phenol derivatives. A useful property of phenolics as disinfectants is that they remain active in the presence of organic compounds, are stable, and persist for long periods after application. For these reasons, phenolics are suitable agents for disinfecting pus, saliva, and feces.

One of the most frequently used phenolics is derived from coal tar, a group of chemicals called *cresols.* A very important cresol is *O-phenylphenol* (see Figure 7.6 and Figure 7.7b), the main ingredient in most formulations of Lysol. Cresols are very good surface disinfectants.

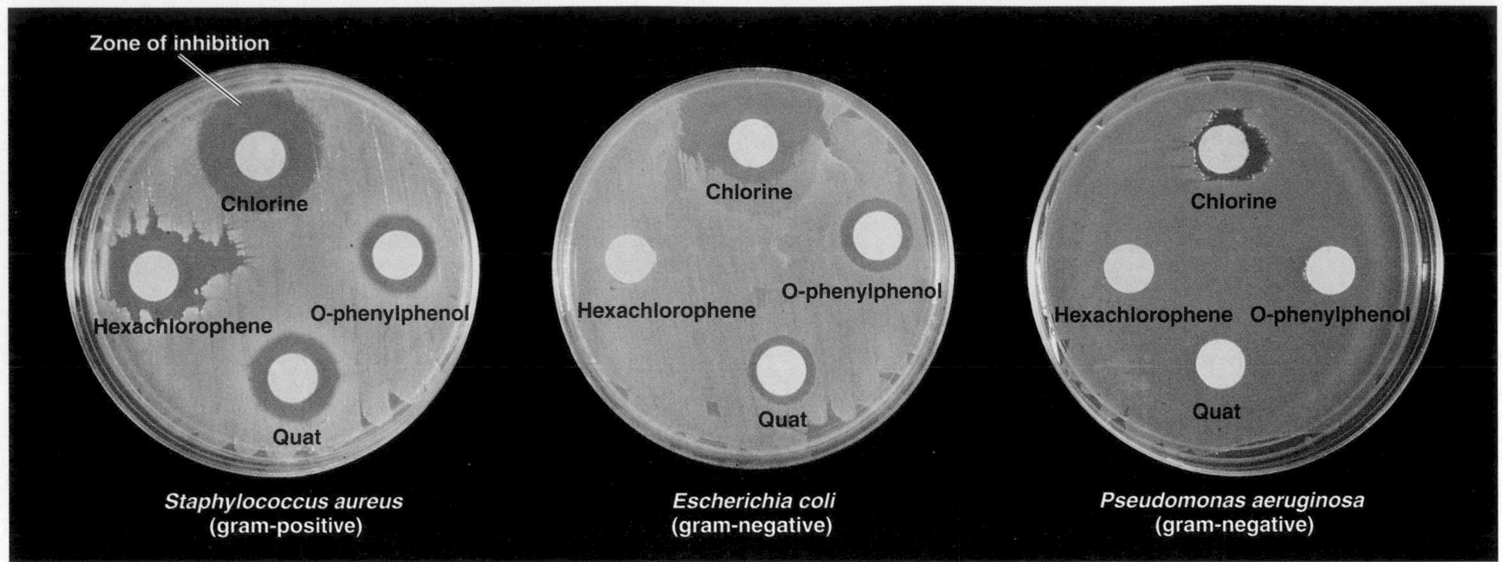

Figure 7.6 Evaluation of disinfectants by the disk-diffusion method. In this experiment, paper disks are soaked in a solution of disinfectant and placed on the surface of a nutrient medium on which a culture of test bacteria has been spread to produce uniform growth.

At the top of each plate, the tests show that chlorine (as sodium hypochlorite) was effective against all the test bacteria but was more effective against gram-positive bacteria.

At the bottom row of each plate, the tests show that the quaternary ammonium compound ("quat") was also more effective against the gram-positive bacteria, but it did not affect the pseudomonads at all.

At the left side of each plate, the tests show that hexachlorophene was effective against gram-positive bacteria only.

At the right sides, O-phenylphenol was ineffective against pseudomonads but was almost equally effective against the gram-positive bacteria and the gram-negative bacteria.

All four chemicals worked against the gram-positive test bacteria, but only one of the four chemicals affected pseudomonads.

Q Which group of bacteria is most resistant to the disinfectants tested?

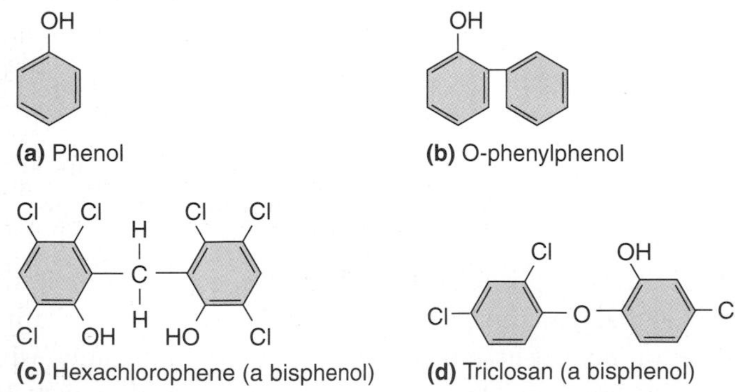

Figure 7.7 The structure of phenolics and bisphenols.

Q Some lozenges intended to alleviate the symptoms of a sore throat contain phenol. Why include this ingredient?

Bisphenols

Bisphenols are derivatives of phenol that contain two phenolic groups connected by a bridge (*bis* indicates two). One bisphenol, *hexachlorophene* (Figure 7.6 and **Figure 7.7c**), is an ingredient of a prescription lotion, pHisoHex, used for surgical and hospital microbial control procedures. Gram-positive staphylococci and streptococci, which can cause skin infections in newborns, are particularly susceptible to hexachlorophene, so it is often used to control such infections in nurseries. However, excessive use of this bisphenol, such as bathing infants with it several times a day, can lead to neurological damage.

Another widely used bisphenol is *triclosan* (**Figure 7.7d**), an ingredient in antibacterial soaps and at least one toothpaste. Triclosan has even been incorporated into kitchen cutting boards and the handles of knives and other plastic kitchenware. Its use is now so widespread that resistant bacteria have been reported, and concerns about its effect on microbes' resistance to certain antibiotics have been raised. Triclosan inhibits an enzyme needed for the biosynthesis of fatty acids (lipids), which mainly affects the integrity of the plasma membrane. It is especially effective against gram-positive bacteria but also works well against yeasts and gram-negative bacteria. There are certain exceptions, such as *Pseudomonas aeruginosa,* a gram-negative bacterium that is very resistant to triclosan, as well as to many other antibiotics and disinfectants (see the discussions on pages 308, 414, and 591).

Biguanides

Biguanides have a broad spectrum of activity, with a mode of action primarily affecting bacterial cell membranes. They are especially effective against gram-positive bacteria. Biguanides are also effective against gram-negative bacteria, with the significant exception of most pseudomonads. Biguanides are not sporicidal but have some activity against enveloped viruses. The best known biguanide is *chlorhexidine,* which is frequently used for microbial control on skin and mucous membranes. Combined with a detergent or alcohol, chlorhexidine is much used for surgical hand scrubs and preoperative skin preparation in patients. *Alexidine* is a similar biguanide and is more rapid in its action.

Halogens

The **halogens,** particularly iodine and chlorine, are effective antimicrobial agents, both alone and as constituents of inorganic or organic compounds. *Iodine* (I_2) is one of the oldest and most effective antiseptics. It is active against all kinds of bacteria, many endospores, various fungi, and some viruses. Iodine impairs protein synthesis and alters cell membranes, apparently by forming complexes with amino acids and unsaturated fatty acids.

Iodine is available as a **tincture**—that is, in solution in aqueous alcohol—and as an iodophor. An **iodophor** is a combination of iodine and an organic molecule, from which the iodine is released slowly. Iodophors have the antimicrobial activity of iodine, but they do not stain and are less irritating. The most common commercial preparation is Betadine, which is a *povidone-iodine.* Povidone is a surface-active iodophor that improves the wetting action and serves as a reservoir of free iodine. Iodines are used mainly for skin disinfection and wound treatment. Many campers are familiar with using iodine for water treatment.

Chlorine (Cl_2), as a gas or in combination with other chemicals, is another widely used disinfectant. Its germicidal action is caused by the hypochlorous acid (HOCl) that forms when chlorine is added to water:

(1)

$$\underset{\text{Chlorine}}{Cl_2} + \underset{\text{Water}}{H_2O} \rightleftharpoons \underset{\text{Hydrogen ion}}{H^+} + \underset{\text{Chloride ion}}{Cl^-} + \underset{\text{Hypochlorous acid}}{HOCl}$$

(2)

$$\underset{\text{Hypochlorous acid}}{HOCl} \rightleftharpoons \underset{\text{Hydrogen ion}}{H^+} + \underset{\text{Hypochlorite ion}}{OCl^-}$$

Hypochlorous acid is a strong oxidizing agent that prevents much of the cellular enzyme system from functioning. Hypochlorous acid is the most effective form of chlorine because it is neutral in electrical charge and diffuses as rapidly as water through the cell wall. Because of its negative charge, the hypochlorite ion (OCl^-) cannot enter the cell freely.

A liquid form of compressed chlorine gas is used extensively for disinfecting municipal drinking water, water in swimming pools, and sewage. Several compounds of chlorine are also effective disinfectants. For example, solutions of *calcium hypochlorite* [$Ca(OCl)_2$] are used to disinfect dairy equipment and restaurant eating utensils. This compound, once called chloride of lime, was used as early as 1825, long before the concept of a germ theory for disease, to soak hospital dressings in Paris hospitals. It was also the disinfectant used in the 1840s by Semmelweis to control hospital infections during childbirth, as mentioned in Chapter 1, page 11. Another chlorine compound, *sodium hypochlorite* (NaOCl; see Figure 7.6), is used as a household disinfectant and bleach (Clorox) and as a disinfectant in dairies, food-processing establishments, and hemodialysis systems. When the quality of drinking water is in question, household bleach can provide a rough equivalent of municipal chlorination. After two drops of bleach are added to a liter of water (four drops if the water is cloudy) and the mixture has sat for 30 minutes, the water is considered safe for drinking under emergency conditions.

The food-processing industry makes wide use of chlorine dioxide solution as a surface disinfectant because it does not leave residual tastes or odors. As a disinfectant, it has a broad spectrum of activity against bacteria and viruses and at high concentrations is even effective against cysts and endospores. At low concentrations, chlorine dioxide can be used as an antiseptic. (Also, see page 201 for the use of chlorine dioxide as a sterilant and disinfectant).

An important group of chlorine compounds are the *chloramines,* combinations of chlorine and ammonia. Most municipal water-treatment systems mix ammonia with chlorine to form chloramines. (Chloramines are toxic to aquarium fish, but pet shops sell chemicals to neutralize them.) U.S. military forces in the field are issued tablets (Chlor-Floc) that contain *sodium dichloroisocyanurate,* a chloramine combined with an agent that flocculates (coagulates) suspended materials in a water sample, causing them to settle out, clarifying the water. Chloramines are also used to sanitize glassware and eating utensils and to treat dairy and food-manufacturing equipment. They are relatively stable compounds that release chlorine over long periods. Chloramines are relatively effective in organic matter but have the disadvantages of acting more slowly and being less effective than hypochlorite.

Alcohols

Alcohols effectively kill bacteria and fungi but not endospores and nonenveloped viruses. The mechanism of action of alcohol is usually protein denaturation, but alcohol can also disrupt membranes and dissolve many lipids, including the lipid component of enveloped viruses. Alcohols have the advantage of acting and then evaporating rapidly and leaving no residue. When the skin is swabbed (degermed) before an injection, most of the microbial control activity comes from simply wiping away dirt and microorganisms, along with skin oils. However, alcohols are unsatisfactory antiseptics when applied to wounds. They cause coagulation of a layer of protein under which bacteria continue to grow.

Table 7.6 Biocidal Action of Various Concentrations of Ethanol in Aqueous Solution against *Streptococcus pyogenes*

Concentration of Ethanol (%)	*Time of Exposure (sec)* 10	20	30	40	50
100	G	G	G	G	G
95	NG	NG	NG	NG	NG
90	NG	NG	NG	NG	NG
80	NG	NG	NG	NG	NG
70	NG	NG	NG	NG	NG
60	NG	NG	NG	NG	NG
50	G	G	NG	NG	NG
40	G	G	G	G	G

NOTE:
G = growth
NG = no growth

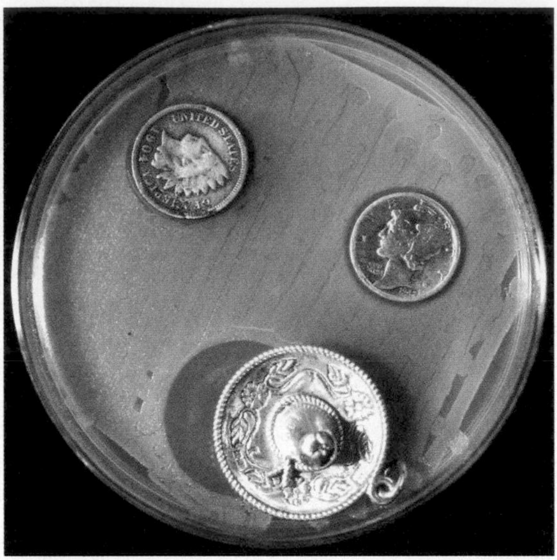

Figure 7.8 Oligodynamic action of heavy metals. Clear zones where bacterial growth has been inhibited are seen around the sombrero charm (pushed aside), the dime, and the penny. The charm and the dime contain silver; the penny contains copper.

Q The coins used in this demonstration were minted many years ago; why were current coins not used?

Two of the most commonly used alcohols are ethanol and isopropanol. The recommended optimum concentration of *ethanol* is 70%, but concentrations between 60% and 95% seem to kill as well (Table 7.6). Pure ethanol is less effective than aqueous solutions (ethanol mixed with water) because denaturation requires water. *Isopropanol,* often sold as rubbing alcohol, is slightly superior to ethanol as an antiseptic and disinfectant. Moreover, it is less volatile, less expensive, and more easily obtained than ethanol. Purell, a very popular commercial preparation used as a hand cleaner, contains 62–65% ethanol combined with skin moisturizers.

Ethanol and isopropanol are often used to enhance the effectiveness of other chemical agents. For example, an aqueous solution of Zephiran (described on page 199) kills about 40% of the population of a test organism in 2 minutes, whereas a tincture of Zephiran kills about 85% in the same period. To compare the effectiveness of tinctures and aqueous solutions, see Figure 7.10 on page 200.

Heavy Metals and Their Compounds

Several heavy metals can be biocidal or antiseptic, including silver, mercury, and copper. The ability of very small amounts of heavy metals, especially silver and copper, to exert antimicrobial activity is referred to as **oligodynamic action** (*oligo* means few). Centuries ago, Egyptians found that putting silver coins in water barrels served to keep the water clean of unwanted organic growths. This action can be seen when we place a coin or other clean piece of metal containing silver or copper on a culture on an inoculated Petri plate. Extremely small amounts of metal diffuse from the coin and inhibit the growth of bacteria for some distance around the coin (Figure 7.8). This effect is produced by the action of heavy metal ions on microbes. When the metal ions combine with the sulfhydryl groups on cellular proteins, denaturation results.

Silver is used as an antiseptic in a 1% *silver nitrate* solution. At one time, many states required that the eyes of newborns be treated with a few drops of silver nitrate to guard against an infection of the eyes called gonorrheal neonatal ophthalmia, which the infants might have contracted as they passed through the birth canal. In recent years, antibiotics have replaced silver nitrate for this purpose.

Recently, there has been renewed interest in the use of silver as an antimicrobial agent. Silver-impregnated dressings that slowly release silver ions have proven especially useful against antibiotic-resistant bacteria. The enthusiasm for incorporating silver in all manner of consumer products is increasing. Among the newer products being sold are plastic food containers infused with silver nanoparticles, which are intended to keep food fresher, and silver-infused athletic shirts and socks, which are claimed to minimize odors.

A combination of silver and the drug sulfadiazine, *silver-sulfadiazine,* is the most common formulation. It is available as a topical cream for use on burns. Silver can also be incorporated into indwelling catheters, which are a common source of hospital infections, and in wound dressings. *Surfacine* is a relatively new antimicrobial for application to surfaces, either animate or inanimate. It contains water-insoluble silver iodide in a polymer carrier and is very persistent, lasting at least 13 days. When a bacterium contacts the surface, the cell's outer membrane is recognized, and a lethal amount of silver ions is released.

Inorganic mercury compounds, such as *mercuric chloride*, have a long history of use as disinfectants. They have a very broad spectrum of activity; their effect is primarily bacteriostatic. However, their use is now limited because of their toxicity, corrosiveness, and ineffectiveness in organic matter. At present, the primary use of mercurials is to control mildew in paints.

Copper in the form of *copper sulfate* or other copper-containing additives is used chiefly to destroy green algae (algicide) that grow in reservoirs, stock ponds, swimming pools, and fish tanks. If the water does not contain excessive organic matter, copper compounds are effective in concentrations of one part per million of water. To prevent mildew, copper compounds such as *copper 8-hydroxyquinoline* are sometimes included in paint.

Another metal used as an antimicrobial is zinc. The effect of trace amounts of zinc can be seen on weathered roofs of buildings down-slope from galvanized (zinc-coated) fittings. The roof is lighter-colored where biological growth, mostly algae, is impeded. Copper- and zinc-treated shingles are available. *Zinc chloride* is a common ingredient in mouthwashes, and *zinc pyrithione* is an ingredient in antidandruff shampoos.

Surface-Active Agents

Surface-active agents, or **surfactants,** can decrease surface tension among molecules of a liquid. Such agents include soaps and detergents.

Soaps and Detergents Soap has little value as an antiseptic, but it does have an important function in the mechanical removal of microbes through scrubbing. The skin normally contains dead cells, dust, dried sweat, microbes, and oily secretions from oil glands. Soap breaks the oily film into tiny droplets, a process called *emulsification,* and the water and soap together lift up the emulsified oil and debris and float them away as the lather is washed off. In this sense, soaps are good degerming agents.

Acid-Anionic Sanitizers *Acid-anionic* surface-active sanitizers are very important in cleaning dairy utensils and equipment. Their sanitizing ability is related to the negatively charged portion (anion) of the molecule, which reacts with the plasma membrane. These sanitizers, which act on a wide spectrum of microbes, including troublesome thermoduric bacteria, are nontoxic, noncorrosive, and fast acting.

Quaternary Ammonium Compounds (Quats) The most widely used surface-active agents are the cationic detergents, especially the **quaternary ammonium compounds (quats).** Their cleansing ability is related to the positively charged portion—the cation—of the molecule. Their name is derived from the fact that they are modifications of the four-valence ammonium ion, NH_4^+ (Figure 7.9). Quaternary ammonium compounds are strongly bactericidal against gram-positive bacteria and less active against gram-negative bacteria (see Figure 7.6).

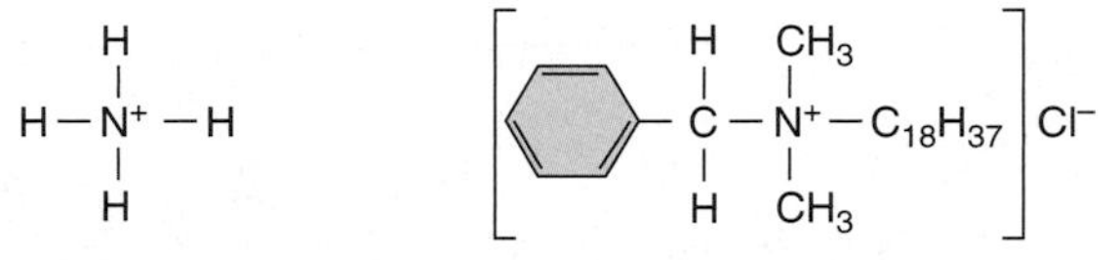

Figure 7.9 The ammonium ion and a quaternary ammonium compound, benzalkonium chloride (Zephiran). Notice how other groups replace the hydrogens of the ammonium ion.

Q Are quats most effective against gram-positive or gram-negative bacteria?

Quats are also fungicidal, amoebicidal, and virucidal against enveloped viruses. They do not kill endospores or mycobacteria. (See the box on page 201.) Their chemical mode of action is unknown, but they probably affect the plasma membrane. They change the cell's permeability and cause the loss of essential cytoplasmic constituents, such as potassium.

Two popular quats are Zephiran, a brand name of *benzalkonium chloride* (see Figure 7.9), and Cepacol, a brand name of *cetylpyridinium chloride.* They are strongly antimicrobial, colorless, odorless, tasteless, stable, easily diluted, and nontoxic, except at high concentrations. If your mouthwash bottle fills with foam when shaken, the mouthwash probably contains a quat. However, organic matter interferes with their activity, and they are rapidly neutralized by soaps and anionic detergents.

Anyone involved in medical applications of quats should remember that certain bacteria, such as some species of *Pseudomonas,* not only survive in quaternary ammonium compounds but actively grow in them. These microbes are resistant not only to the disinfectant solution but also to gauze and bandages moistened with it, because the fibers tend to neutralize the quats.

Before we move on to the next group of chemical agents, refer to Figure 7.10, which compares the effectiveness of some of the antiseptics we have discussed so far.

Chemical Food Preservatives

Chemical preservatives are frequently added to foods to retard spoilage. *Sulfur dioxide* (SO_2) has long been used as a disinfectant, especially in wine-making. Homer's *Odyssey,* written nearly 2800 years ago, mentions its use. Among the more common additives are sodium benzoate, sorbic acid, and calcium propionate. These chemicals are simple organic acids, or salts of organic acids, which the body readily metabolizes and which are generally judged to be safe in foods. *Sorbic acid,* or its more soluble salt *potassium sorbate,* and *sodium benzoate* prevent

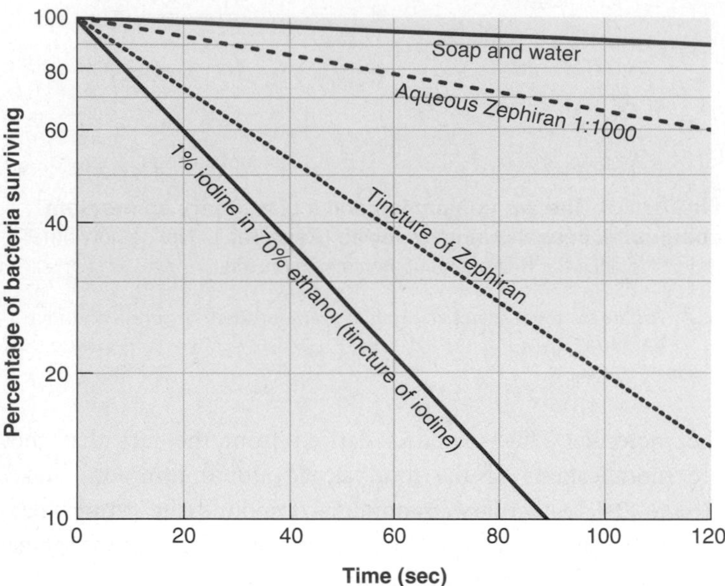

Figure 7.10 A comparison of the effectiveness of various antiseptics. The steeper the downward slope of the killing curve, the more effective the antiseptic is. A 1% iodine in 70% ethanol solution is the most effective; soap and water are the least effective. Notice that a tincture of Zephiran is more effective than an aqueous solution of the same antiseptic.

Q Why is the tincture of Zephiran more effective than the aqueous solution?

molds from growing in certain acidic foods, such as cheese and soft drinks. Such foods, usually with a pH of 5.5 or lower, are most susceptible to spoilage by molds. *Calcium propionate,* an effective fungistat used in bread, prevents the growth of surface molds and the *Bacillus* bacterium that causes ropy bread. These organic acids inhibit mold growth, not by affecting the pH but by interfering with the mold's metabolism or the integrity of the plasma membrane.

Sodium nitrate and *sodium nitrite* are added to many meat products, such as ham, bacon, hot dogs, and sausage. The active ingredient is sodium nitrite, which certain bacteria in the meats can also produce from sodium nitrate. These bacteria use nitrate as a substitute for oxygen under anaerobic conditions. The nitrite has two main functions: to preserve the pleasing red color of the meat by reacting with blood components in the meat, and to prevent the germination and growth of any botulism endospores that might be present. Nitrite selectively inhibits certain iron-containing enzymes of *Clostridium botulinum.* There has been some concern that the reaction of nitrites with amino acids can form certain carcinogenic products known as **nitrosamines,** and the amount of nitrites added to foods has generally been reduced recently for this reason. However, the use of nitrites continues because of their established value in preventing botulism. Because nitrosamines are formed in the body from other sources, the added risk posed by a limited use of nitrates and nitrites in meats is lower than was once thought.

Antibiotics

The antimicrobials discussed in this chapter are not useful for ingestion or injection to treat disease. Antibiotics are used for this purpose. The use of antibiotics is highly restricted; however, at least two have considerable use in food preservation. Neither is of value for clinical purposes. *Nisin,* (see page 578) is often added to cheese to inhibit the growth of certain endospore-forming spoilage bacteria. It is an example of a bacteriocin, a protein that is produced by one bacterium and inhibits another (see Chapter 8, page 239). Nisin is present naturally in small amounts in many dairy products. It is tasteless, readily digested, and nontoxic. *Natamycin* (pimaricin) is an antifungal antibiotic approved for use in foods, mostly cheese.

Aldehydes

Aldehydes are among the most effective antimicrobials. Two examples are formaldehyde and glutaraldehyde. They inactivate proteins by forming covalent cross-links with several organic functional groups on proteins ($—NH_2$, —OH, —COOH, and —SH). *Formaldehyde gas* is an excellent disinfectant. However, it is more commonly available as *formalin,* a 37% aqueous solution of formaldehyde gas. Formalin was once used extensively to preserve biological specimens and inactivate bacteria and viruses in vaccines.

Glutaraldehyde is a chemical relative of formaldehyde that is less irritating and more effective than formaldehyde. Glutaraldehyde is used to disinfect hospital instruments, including endoscopes and respiratory therapy equipment, but they must be carefully cleaned first. When used in a 2% solution (Cidex), it is bactericidal, tuberculocidal, and virucidal in 10 minutes and sporicidal in 3 to 10 hours. Glutaraldehyde is one of the few liquid chemical disinfectants that can be considered a sterilizing agent. However, 30 minutes is often considered the maximum time allowed for a sporicide to act, which is a criterion glutaraldehyde cannot meet. Both glutaraldehyde and formalin are used by morticians for embalming.

A possible replacement for glutaraldehyde for many uses is *ortho-phthalaldehyde* (OPA), which is more effective against many microbes and has fewer irritating properties.

Chemical Sterilization

Sterilization with liquid chemicals is possible, but even sporicidal chemicals such as glutaraldehyde are usually not considered to be practical sterilants. However, the gaseous chemosterilants are frequently used as substitutes for physical sterilization processes. Their application requires a closed chamber similar to a steam autoclave. Probably the most familiar example is *ethylene oxide:*

$$H_2C—CH_2$$
$$\ \backslash\ /$$
$$O$$

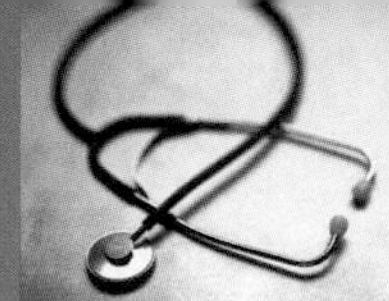

CLINICAL FOCUS

Infection Following Steroid Injection

As you read through this box, you will encounter a series of questions that infection control officers ask themselves as they track the source of infection. Try to answer each question before going on to the next one.

1. During a 3-month period, an infectious diseases physician reported to the department of health 12 patients with *Mycobacterium abscessus* joint and soft-tissue infections. Slow-growing mycobacteria, including *M. tuberculosis* and *M. leprae*, are human pathogens. These infections, however, were caused by rapidly growing mycobacteria (RGM).
   Where are these RGMs normally found? (*Hint:* Read page 320.)
2. All 12 patients received injections for arthritis from the same physician. The injection procedure consisted of cleaning the skin with cotton balls soaked in diluted (1:10) Zephiran, painting the skin with commercially prepared iodine swabs, anesthetizing the area with 0.5 ml of 1% lidocaine in a sterile 22-gauge needle and syringe, and injecting 0.5–1.0 ml of betamethasone, a cortisone-like steroid, into the joint with a sterile 20-gauge needle and syringe.
   What do you need to do to determine the source of the infection?
3. Cultures were made from swabs of the inside surface of an open metal container of forceps and the inside surface of a metal container of cotton balls. Cultures were also made of the iodine prep swabs, Zephiran-soaked cotton balls, solutions of lidocaine and betamethasone, diluted and undiluted Zephiran, and a sealed bottle of distilled water used to dilute the Zephiran.
   What type of disinfectant is Zephiran?
4. *M. abscessus* was grown only from cotton balls soaked in Zephiran. A disk-diffusion assay was performed (see the figure above).
   What two problems occurred, and how would you prevent future infections?
5. Dilute Zephiran is not effective against *M. abscessus*, and the disinfecting ability of Zephiran and other quats is reduced by the presence of organic material, such as cotton balls. To avoid inoculating patients with bacteria, cotton balls should not be stored in the disinfectant.

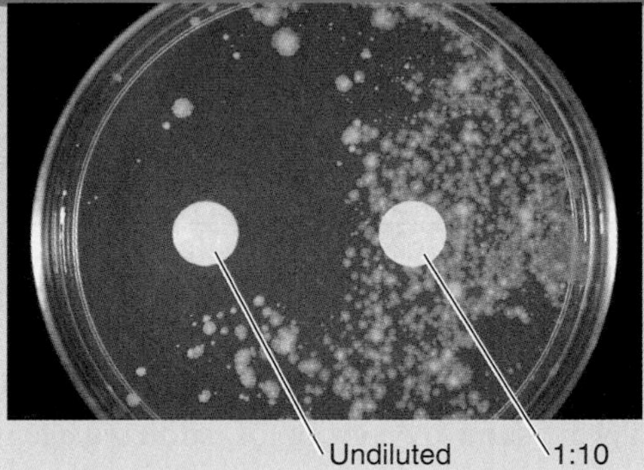

Disk-diffusion test of Zephiran against *M. abscessus.*

Source: Adapted from *Clinical Infectious Diseases* 36:954–962 (2003).

Its activity depends on *alkylation,* that is, replacing the proteins' labile hydrogen atoms in a chemical group (such as —SH, —COOH, or —CH_2CH_2OH) with a chemical radical. This leads to cross-linking of nucleic acids and proteins and inhibits vital cellular functions. Ethylene oxide kills all microbes and endospores but requires a lengthy exposure period of several hours. It is toxic and explosive in its pure form, so it is usually mixed with a nonflammable gas, such as carbon dioxide. Among its advantages is that it carries out sterilization at ambient temperatures and it is highly penetrating. Larger hospitals often are able to sterilize even mattresses in special ethylene oxide sterilizers.

Chlorine dioxide is a short-lived gas that is usually manufactured at the place of use. Notably, it has been used to fumigate enclosed building areas contaminated with endospores of anthrax. It is much more stable in aqueous solution. Its most common use is in water treatment prior to chlorination, where its purpose is to remove, or reduce the formation of, certain carcinogenic compounds sometimes formed in the chlorination of water.

Plasmas

In addition to the traditional three states of matter—liquid, gas, and solid—there might be considered to exist a fourth state of matter, plasma. **Plasma** is a state of matter in which a gas is excited, in this case by an electromagnetic field, to make a mixture of nuclei with assorted electrical charges and free electrons. Health care facilities are increasingly facing the challenge of sterilizing metal or plastic surgical instruments used for many newer procedures in arthroscopic or laparoscopic surgery. Such devices have long, hollow tubes, many with an interior diameter of only a few millimeters, and are difficult to sterilize. *Plasma sterilization* is a reliable method for this. The instruments are placed in a container in which a combination of a vacuum, electromagnetic field, and chemicals such as hydrogen peroxide (sometimes with peracetic acid, as well) form the plasma. Such plasmas have many free radicals that quickly destroy even endospore-forming microbes. The advantage of plasma sterilization, which has elements of both physical and chemical sterilization, is that it requires only low temperatures, but it is relatively expensive.

Supercritical Fluids

The use of supercritical fluids in sterilization combines chemical and physical methods. When carbon dioxide is compressed into a "supercritical" state, it has properties of both a liquid (with increased solubility) and a gas (with a lowered surface tension).

Organisms exposed to *supercritical carbon dioxide* are inactivated, including most vegetative organisms that cause spoilage and foodborne pathogens. Even endospore inactivation requires a temperature of only about 45°C. Used for a number of years in treating certain foods, supercritical carbon dioxide has more recently been used to decontaminate medical implants, such as bone, tendons, or ligaments taken from donor patients.

Peroxygens and Other Forms of Oxygen

Peroxygens are a group of oxidizing agents that includes hydrogen peroxide and peracetic acid.

Hydrogen peroxide is an antiseptic found in many household medicine cabinets and in hospital supply rooms. It is not a good antiseptic for open wounds. It is quickly broken down to water and gaseous oxygen by the action of the enzyme catalase, which is present in human cells (see Chapter 6, page 162). However, hydrogen peroxide does effectively disinfect inanimate objects; in such applications, it is even sporicidal at high concentrations. On a nonliving surface, the normally protective enzymes of aerobic bacteria and facultative anaerobes are overwhelmed by high concentrations of peroxide. Because of these factors, and its rapid degradation into harmless water and oxygen, the food industry is increasing its use of hydrogen peroxide for aseptic packaging (see Figure 28.4). The packaging material passes through a hot solution of the chemical before being assembled into a container. In addition, many wearers of contact lenses are familiar with hydrogen peroxide's use as a disinfectant. After the lens is disinfected, a platinum catalyst in the lens-disinfecting kit destroys residual hydrogen peroxide so that it does not persist on the lens, where it might cause eye irritation.

Heated *hydrogen peroxide* can be used as a gaseous sterilant. It is not as penetrating as ethylene oxide and also cannot be used to sterilize textiles or liquids.

Peracetic acid (peroxyactic acid, or *PAA)* is one of the most effective liquid chemical sporicides available and can be used as a sterilant. Its mode of action is similar to that of hydrogen peroxide. It is generally effective on endospores and viruses within 30 minutes and kills vegetative bacteria and fungi in less than 5 minutes. PAA has many applications in the disinfection of food-processing and medical equipment, especially endoscopes, because it leaves no toxic residues (only water and small amounts of acetic acid) and is minimally affected by the presence of organic matter. The FDA has approved use of PAA for the washing of fruits and vegetables.

Other oxidizing agents include *benzoyl peroxide*, which is probably most familiar as the main ingredient in over-the-counter medications for acne. *Ozone* (O_3) is a highly reactive form of oxygen that is generated by passing oxygen through high-voltage electrical discharges (see Figure 27.16, page 783). It is responsible for the air's rather fresh odor after a lightning storm, in the vicinity of electrical sparking, or around an ultraviolet light. Ozone is often used to supplement chlorine in the disinfection of water because it helps neutralize tastes and odors. Although ozone is a more effective killing agent than chlorine, its residual activity is difficult to maintain in water.

CHECK YOUR UNDERSTANDING

- ✓ If you wanted to disinfect a surface contaminated by vomit and a surface contaminated by a sneeze, why would your choice of disinfectant make a difference? **7-7**
- ✓ Which is more likely to be used in a medical clinic laboratory, a use-dilution test or a disk-diffusion test? **7-8**
- ✓ Why is alcohol effective against some viruses and not others? **7-9**
- ✓ Is Betadine an antiseptic or a disinfectant when it is used on skin? **7-10**
- ✓ What characteristics make surface-active agents attractive to the dairy industry? **7-11**
- ✓ What chemical disinfectants can be considered sporicides? **7-12**
- ✓ What chemicals are used to sterilize? **7-13**

Microbial Characteristics and Microbial Control

LEARNING OBJECTIVE

7-14 Explain how the type of microbe affects the control of microbial growth.

Many biocides tend to be more effective against gram-positive bacteria, as a group, than against gram-negative bacteria. This principle is illustrated in **Figure 7.11**, which presents a simplified hierarchy of relative resistance of major microbial groups to biocides. A principal factor in this relative resistance to biocides is the external lipopolysaccharide layer of gram-negative bacteria. Within gram-negative bacteria, members of the genera *Pseudomonas* and *Burkholderia* are of special interest. These closely related bacteria are unusually resistant to biocides (see Figure 7.6) and will even grow actively in some disinfectants and antiseptics, most notably the quaternary ammonium compounds. In Chapter 20, you will see that these bacteria are also resistant to many antibiotics. This resistance to chemical antimicrobials is related mostly to the characteristics of their *porins* (structural openings in the wall of gram-negative bacteria; see Figure 4.13c, page 86). Porins are highly selective of molecules that they permit to enter the cell.

The mycobacteria are another group of non–endospore-forming bacteria that exhibit greater than normal resistance to chemical biocides. (See the box on page 201.) This group includes *Mycobacterium tuberculosis,* the pathogen that causes tuberculosis. The cell wall of this organism and other members of this genus have a waxy, lipid-rich component. Instruction labels on disinfectants often state whether they are tuberculocidal, indicating that they are effective against mycobacteria.

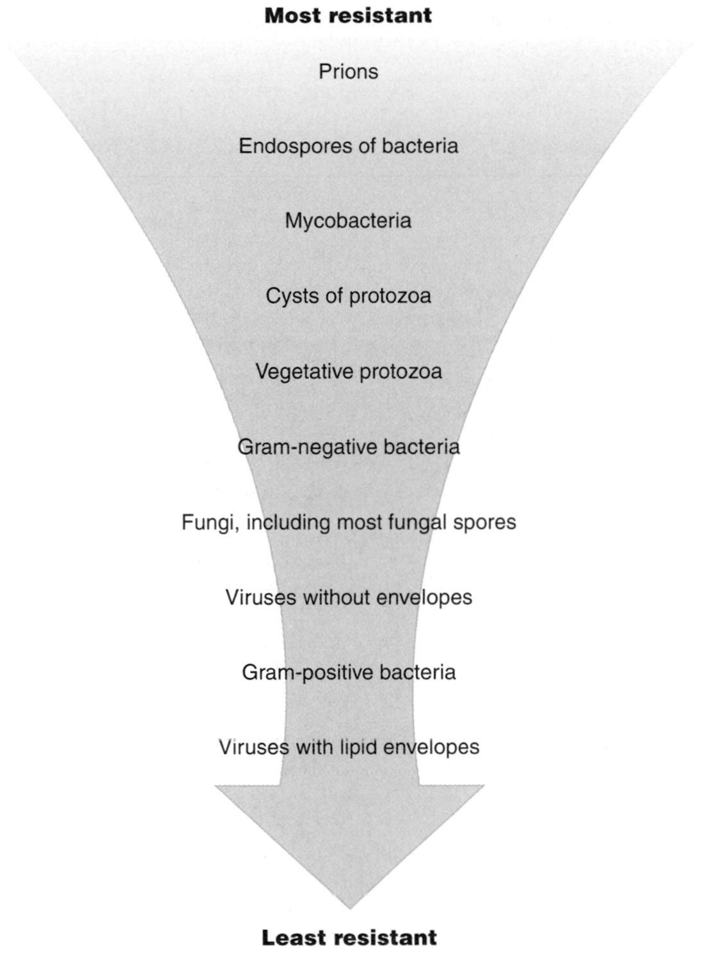

Most resistant

Prions

Endospores of bacteria

Mycobacteria

Cysts of protozoa

Vegetative protozoa

Gram-negative bacteria

Fungi, including most fungal spores

Viruses without envelopes

Gram-positive bacteria

Viruses with lipid envelopes

Least resistant

Figure 7.11 **Decreasing order of resistance of microorganisms to chemical biocides.**

Q **Why are viruses with lipid-containing envelopes relatively susceptible to certain biocides?**

Special tuberculocidal tests have been developed to evaluate the effectiveness of biocides against this bacterial group.

Bacterial endospores are affected by relatively few biocides. (The activity of the major chemical antimicrobial groups against mycobacteria and endospores is summarized in Table 7.7.) The cysts and oocysts of protozoa are also relatively resistant to chemical disinfection.

Viruses are not especially resistant to biocides, with the exception of viruses that possess a lipid-containing envelope. Antimicrobials that are lipid-soluble are more likely to be effective against enveloped viruses. The label of such an agent will indicate that it is effective against lipophilic viruses. Nonenveloped viruses, which have only a protein coat, are more resistant—fewer biocides are active against them.

A special problem, not yet completely solved, is the reliable killing of prions. Prions are infectious proteins that are the cause of neurological diseases known as spongiform encephalopathies, such as the popularly named mad cow disease (see Chapter 22, page 631). To destroy prions, infected animal carcasses are incinerated. A major problem is the disinfection of surgical instruments exposed to prion contamination. Normal autoclaving has proven to be inadequate. The World Health Organization (WHO) and the Centers for Disease Control and Prevention (CDC) have recommended the combined use of a solution of sodium hydroxide and autoclaving at 134°C. Recent reports indicate that surgical instruments have been successfully treated to inactivate prions, which are proteins, by addition of protease enzymes to the cleaning solution. Surgeons sometimes resort to using disposable instruments.

In summary, it is important to remember that microbial control methods, especially biocides, are not uniformly effective against all microbes.

Table 7.7 The Effectiveness of Chemical Antimicrobials against Endospores and Mycobacteria

Chemical Agent	Endospores	Mycobacteria
Mercury	No activity	No activity
Phenolics	Poor	Good
Bisphenols	No activity	No activity
Quaternary Ammonium Compounds	No activity	No activity
Chlorines	Fair	Fair
Iodine	Poor	Good
Alcohols	Poor	Good
Glutaraldehyde	Fair	Good
Chlorhexidine	No activity	Fair

CHECK YOUR UNDERSTANDING

✓ The presence or absence of endospores has an obvious effect on microbial control, but why are gram-negative bacteria more resistant to chemical biocides than gram-positive bacteria? **7-14**

* * *

Table 7.8 summarizes chemical agents used to control microbial growth.

The compounds discussed in this chapter are not generally useful in the treatment of diseases. Antibiotics and the pathogens against which they are active will be discussed in Chapter 20.

Table 7.8 Chemical Agents Used to Control Microbial Growth

Chemical Agent	Mechanism of Action	Preferred Use	Comment
Phenol and Phenolics			
1. Phenol	Disruption of plasma membrane, denaturation of enzymes.	Rarely used, except as a standard of comparison.	Seldom used as a disinfectant or antiseptic because of its irritating qualities and disagreeable odor.
2. Phenolics	Disruption of plasma membrane, denaturation of enzymes.	Environmental surfaces, instruments, skin surfaces, and mucous membranes.	Derivatives of phenol that are reactive even in the presence of organic material; O-phenylphenol is an example.
3. Bisphenols	Probably disruption of plasma membrane.	Disinfectant hand soaps and skin lotions.	Triclosan is an especially common example of a bisphenol. Broad spectrum, but most effective against gram-positives.
Biguanides (Chlorhexidine)	Disruption of plasma membrane.	Skin disinfection, especially for surgical scrubs.	Bactericidal to gram-positives and gram-negatives; nontoxic, persistent.
Halogens	Iodine inhibits protein function and is a strong oxidizing agent; chlorine forms the strong oxidizing agent hypochlorous acid, which alters cellular components.	Iodine is an effective antiseptic available as a tincture and an iodophor; chlorine gas is used to disinfect water; chlorine compounds are used to disinfect dairy equipment, eating utensils, household items, and glassware.	Iodine and chlorine may act alone or as components of inorganic and organic compounds.
Alcohols	Protein denaturation and lipid dissolution.	Thermometers and other instruments; in swabbing the skin with alcohol before an injection, most of the disinfecting action probably comes from a simple wiping away (degerming) of dirt and some microbes.	Bactericidal and fungicidal, but not effective against endospores or nonenveloped viruses; commonly used alcohols are ethanol and isopropanol.
Heavy Metals and Their Compounds	Denaturation of enzymes and other essential proteins.	Silver nitrate may be used to prevent gonorrheal neonatal ophthalmia; silver-sulfadiazine used as a topical cream on burns; copper sulfate is an algicide.	Heavy metals such as silver and mercury are biocidal.
Surface-Active Agents			
Soaps and detergents	Mechanical removal of microbes through scrubbing.	Skin degerming and removal of debris.	Many antibacterial soaps contain antimicrobials.
Acid-anionic sanitizers	Not certain; may involve enzyme inactivation or disruption.	Sanitizers in dairy and food-processing industries.	Wide spectrum of activity; nontoxic, noncorrosive, fast-acting.
Quaternary ammonium compounds (cationic detergents)	Enzyme inhibition, protein denaturation, and disruption of plasma membranes.	Antiseptic for skin, instruments, utensils, rubber goods.	Bactericidal, bacteriostatic, fungicidal, and virucidal against enveloped viruses; examples of quats are Zephiran and Cepacol.
Chemical Food Preservatives			
Organic acids	Metabolic inhibition, mostly affecting molds; action not related to their acidity.	Sorbic acid and benzoic acid effective at low pH; parabens much used in cosmetics, shampoos; calcium propionate used in bread.	Widely used to control mold and some bacteria in foods and cosmetics.
Nitrates/nitrites	Active ingredient is nitrite, which is produced by bacterial action on nitrate. Nitrite inhibits certain iron-containing enzymes of anaerobes.	Meat products such as ham, bacon, hot dogs, sausage.	Prevents growth of *Clostridium botulinum* in food; also imparts a red color.

Table 7.8 (continued)

Chemical Agent	Mechanism of Action	Preferred Use	Comment
Aldehydes	Protein denaturation.	Glutaraldehyde (Cidex) is less irritating than formaldehyde and is used for disinfecting medical equipment.	Very effective antimicrobials.
Chemical Sterilization			
Ethylene oxide and other gaseous sterilants	Inhibits vital cellular functions.	Mainly for sterilization of materials that would be damaged by heat.	Ethylene oxide is the most commonly used. Heated hydrogen peroxide and chlorine dioxide have special uses.
Plasma sterilization	Inhibits vital cellular functions.	Especially useful for tubular medical instruments.	Usually hydrogen peroxide excited in a vacuum by an electromagnetic field.
Supercritical fluids	Inhibits vital cellular functions.	Especially useful for sterilizing organic medical implants.	Carbon dioxide compressed to a supercritical state.
Peroxygens and Other Forms of Oxygen	Oxidation.	Contaminated surfaces; some deep wounds, in which they are very effective against oxygen-sensitive anaerobes.	Ozone is widely used as a supplement for chlorination; hydrogen peroxide is a poor antiseptic but a good disinfectant. Peracetic acid is especially effective.

STUDY OUTLINE

The **MyMicrobiologyPlace** website (**www.microbiologyplace.com**) will help you get ready for tests with its simple three-step approach: ❶ **take a pre-test** and obtain a personalized study plan, ❷ **learn and practice** with animations, tutorials, and MP3 tutor sessions, and ❸ **test yourself** with quizzes and a chapter post-test.

The Terminology of Microbial Control (pp. 185–186)

1. The control of microbial growth can prevent infections and food spoilage.
2. Sterilization is the process of removing or destroying all microbial life on an object.
3. Commercial sterilization is heat treatment of canned foods to destroy *C. botulinum* endospores.
4. Disinfection is the process of reducing or inhibiting microbial growth on a nonliving surface.
5. Antisepsis is the process of reducing or inhibiting microorganisms on living tissue.
6. The suffix *-cide* means to kill; the suffix *-stat* means to inhibit.
7. Sepsis is bacterial contamination.

The Rate of Microbial Death (p. 186)

1. Bacterial populations subjected to heat or antimicrobial chemicals usually die at a constant rate.
2. Such a death curve, when plotted logarithmically, shows this constant death rate as a straight line.
3. The time it takes to kill a microbial population is proportional to the number of microbes.
4. Microbial species and life cycle phases (e.g., endospores) have different susceptibilities to physical and chemical controls.
5. Organic matter may interfere with heat treatments and chemical control agents.
6. Longer exposure to lower heat can produce the same effect as shorter time at higher heat.

Actions of Microbial Control Agents (pp. 186–187)

Alteration of Membrane Permeability (p. 186)

1. The susceptibility of the plasma membrane is due to its lipid and protein components.
2. Certain chemical control agents damage the plasma membrane by altering its permeability.

Damage to Proteins and Nucleic Acids (p. 187)

3. Some microbial control agents damage cellular proteins by breaking hydrogen bonds and covalent bonds.
4. Other agents interfere with DNA and RNA and protein synthesis.

Physical Methods of Microbial Control (pp. 187–194)

Heat (pp. 188–191)

1. Heat is frequently used to kill microorganisms.
2. Moist heat kills microbes by denaturing enzymes.
3. Thermal death point (TDP) is the lowest temperature at which all the microbes in a liquid culture will be killed in 10 minutes.
4. Thermal death time (TDT) is the length of time required to kill all bacteria in a liquid culture at a given temperature.
5. Decimal reduction time (DRT) is the length of time in which 90% of a bacterial population will be killed at a given temperature.
6. Boiling (100°C) kills many vegetative cells and viruses within 10 minutes.
7. Autoclaving (steam under pressure) is the most effective method of moist heat sterilization. The steam must directly contact the material to be sterilized.
8. In HTST pasteurization, a high temperature is used for a short time (72°C for 15 seconds) to destroy pathogens without altering the flavor of the food. Ultra-high-temperature (UHT) treatment (140°C for 4 seconds) is used to sterilize dairy products.
9. Methods of dry heat sterilization include direct flaming, incineration, and hot-air sterilization. Dry heat kills by oxidation.
10. Different methods that produce the same effect (reduction in microbial growth) are called equivalent treatments.

Filtration (p. 191)

11. Filtration is the passage of a liquid or gas through a filter with pores small enough to retain microbes.
12. Microbes can be removed from air by high-efficiency particulate air (HEPA) filters.
13. Membrane filters composed of cellulose esters are commonly used to filter out bacteria, viruses, and even large proteins.

Low Temperatures (pp. 191–192)

14. The effectiveness of low temperatures depends on the particular microorganism and the intensity of the application.
15. Most microorganisms do not reproduce at ordinary refrigerator temperatures (0–7°C).
16. Many microbes survive (but do not grow) at the subzero temperatures used to store foods.

High Pressure (p. 192)

17. High pressure denatures proteins in vegetative cells.

Desiccation (p. 192)

18. In the absence of water, microorganisms cannot grow but can remain viable.
19. Viruses and endospores can resist desiccation.

Osmotic Pressure (p. 192)

20. Microorganisms in high concentrations of salts and sugars undergo plasmolysis.
21. Molds and yeasts are more capable than bacteria of growing in materials with low moisture or high osmotic pressure.

Radiation (pp. 192–194)

22. The effects of radiation depend on its wavelength, intensity, and duration.
23. Ionizing radiation (gamma rays, X rays, and high-energy electron beams) has a high degree of penetration and exerts its effect primarily by ionizing water and forming highly reactive hydroxyl radicals.
24. Ultraviolet (UV) radiation, a form of nonionizing radiation, has a low degree of penetration and causes cell damage by making thymine dimers in DNA that interfere with DNA replication; the most effective germicidal wavelength is 260 nm.
25. Microwaves can kill microbes indirectly as materials get hot.

Chemical Methods of Microbial Control (pp. 195–202)

1. Chemical agents are used on living tissue (as antiseptics) and on inanimate objects (as disinfectants).
2. Few chemical agents achieve sterility.

Principles of Effective Disinfection (p. 195)

3. Careful attention should be paid to the properties and concentration of the disinfectant to be used.
4. The presence of organic matter, degree of contact with microorganisms, and temperature should also be considered.

Evaluating a Disinfectant (p. 195)

5. In the use-dilution test, bacterial survival in the manufacturer's recommended dilution of a disinfectant is determined.
6. Viruses, endospore-forming bacteria, mycobacteria, and fungi can also be used in the use-dilution test.
7. In the disk-diffusion method, a disk of filter paper is soaked with a chemical and placed on an inoculated agar plate; a zone of inhibition indicates effectiveness.

Types of Disinfectants (pp. 195–202)

Phenol and Phenolics (p. 195)

OH

8. Phenolics exert their action by injuring plasma membranes.

Bisphenols (p. 196)

9. Bisphenols such as triclosan (over the counter) and hexachlorophene (prescription) are widely used in household products.

Biguanides (pp. 196–197)

10. Biguanides damage plasma membranes of vegetative cells.

Halogens (p. 197)

11. Some halogens (iodine and chlorine) are used alone or as components of inorganic or organic solutions.
12. Iodine may combine with certain amino acids to inactivate enzymes and other cellular proteins.
13. Iodine is available as a tincture (in solution with alcohol) or as an iodophor (combined with an organic molecule).
14. The germicidal action of chlorine is based on the formation of hypochlorous acid when chlorine is added to water.

Alcohols (pp. 197–198)

15. Alcohols exert their action by denaturing proteins and dissolving lipids.
16. In tinctures, they enhance the effectiveness of other antimicrobial chemicals.
17. Aqueous ethanol (60–95%) and isopropanol are used as disinfectants.

Heavy Metals and Their Compounds (pp. 198–199)

18. Silver, mercury, copper, and zinc are used as germicides.
19. They exert their antimicrobial action through oligodynamic action. When heavy metal ions combine with sulfhydryl (—SH) groups, proteins are denatured.

Surface-Active Agents (p. 199)

20. Surface-active agents decrease the surface tension among molecules of a liquid; soaps and detergents are examples.
21. Soaps have limited germicidal action but assist in removing microorganisms.
22. Acid-anionic detergents are used to clean dairy equipment.
23. Quats are cationic detergents attached to NH_4^+.
24. By disrupting plasma membranes, quats allow cytoplasmic constituents to leak out of the cell.
25. Quats are most effective against gram-positive bacteria.

Chemical Food Preservatives (pp. 199–200)

26. SO_2, sorbic acid, benzoic acid, and propionic acid inhibit fungal metabolism and are used as food preservatives.
27. Nitrate and nitrite salts prevent germination of *C. botulinum* endospores in meats.

Antibiotics (p. 200)

28. Nisin and natamycin are anibiotics used to preserve foods, especially cheese.

Aldehydes (p. 200)

29. Aldehydes such as formaldehyde and glutaraldehyde exert their antimicrobial effect by inactivating proteins.
30. They are among the most effective chemical disinfectants.

Chemical Sterilization (pp. 200–201)

31. Ethylene oxide is the gas most frequently used for sterilization.
32. It penetrates most materials and kills all microorganisms by protein denaturation.

Plasmas (p. 201)

33. Free radicals in plasma gases are used to sterilize plastic instruments.

Supercritical Fluids (pp. 201–202)

34. Supercritical fluids, which have properties of liquid and gas, can sterilize at low temperatures.

Peroxygens and Other Forms of Oxygen (p. 202)

35. Hydrogen peroxide, peracetic acid, benzoyl peroxide, and ozone exert their antimicrobial effect by oxidizing molecules inside cells.

Microbial Characteristics and Microbial Control (pp. 202–205)

1. Gram-negative bacteria are generally more resistant than gram-positive bacteria to disinfectants and antiseptics.
2. Mycobacteria, endospores, and protozoan cysts and oocysts are very resistant to disinfectants and antiseptics.
3. Nonenveloped viruses are generally more resistant than enveloped viruses to disinfectants and antiseptics.
4. Prions are resistant to disinfection and autoclaving.

STUDY QUESTIONS

Answers to the Review and Multiple Choice questions can be found by turning to the blue Answers tab at the back of the textbook.

Review

1. The thermal death time for a suspension of *Bacillus subtilis* endospores is 30 minutes in dry heat and less than 10 minutes in an autoclave. Which type of heat is more effective? Why?
2. If pasteurization does not achieve sterilization, why is pasteurization used to treat food?
3. Thermal death point is not considered an accurate measure of the effectiveness of heat sterilization. List three factors that can alter thermal death point.
4. The antimicrobial effect of gamma radiation is due to (a) ___________. The antimicrobial effect of ultraviolet radiation is due to (b) ___________.
5. **DRAW IT** A bacterial culture was in log phase in the following figure. At time *x*, an antibacterial compound was added to the culture. Draw the lines indicating addition of a bactericidal compound and a bacteriostatic compound. Explain why the viable count does not immediately drop to zero at *x*.

6. How do autoclaving, hot air, and pasteurization illustrate the concept of equivalent treatments?
7. How do salts and sugars preserve foods? Why are these considered physical rather than chemical methods of microbial control? Name one food that is preserved with sugar and one preserved with salt. How do you account for the occasional growth of *Penicillium* mold in jelly, which is 50% sucrose?
8. The use-dilution values for two disinfectants tested under the same conditions are as follows: Disinfectant A—1:2; Disinfectant B—1:10,000. If both disinfectants are designed for the same purpose, which would you select?
9. A large hospital washes burn patients in a stainless steel tub. After each patient, the tub is cleaned with a quat. It was noticed that 14 of 20 burn patients acquired *Pseudomonas* infections after being bathed. Provide an explanation for this high rate of infection.

Multiple Choice

1. Which of the following does *not* kill endospores?
   a. autoclaving
   b. incineration
   c. hot-air sterilization
   d. pasteurization
   e. All of the above kill endospores.
2. Which of the following is most effective for sterilizing mattresses and plastic Petri dishes?
   a. chlorine
   b. ethylene oxide
   c. glutaraldehyde
   d. autoclaving
   e. nonionizing radiation
3. Which of these disinfectants does *not* act by disrupting the plasma membrane?
   a. phenolics
   b. phenol
   c. quaternary ammonium compounds
   d. halogens
   e. biguanides
4. Which of the following *cannot* be used to sterilize a heat-labile solution stored in a plastic container?
   a. gamma radiation
   b. ethylene oxide
   c. nonionizing radiation
   d. autoclaving
   e. short-wavelength radiation
5. Which of the following is *not* a characteristic of quaternary ammonium compounds?
   a. bactericidal against gram-positive bacteria
   b. sporicidal
   c. amoebicidal
   d. fungicidal
   e. kills enveloped viruses
6. A classmate is trying to determine how a disinfectant might kill cells. You observed that when he spilled the disinfectant in your reduced litmus milk, the litmus turned blue again. You suggest to your classmate that
   a. the disinfectant might inhibit cell wall synthesis.
   b. the disinfectant might oxidize molecules.
   c. the disinfectant might inhibit protein synthesis.
   d. the disinfectant might denature proteins.
   e. he take his work away from yours.
7. Which of the following is most likely to be bactericidal?
   a. membrane filtration
   b. ionizing radiation
   c. lyophilization (freeze-drying)
   d. deep-freezing
   e. all of the above
8. Which of the following is used to control microbial growth in foods?
   a. organic acids
   b. alcohols
   c. aldehydes
   d. heavy metals
   e. all of the above

Use the following information to answer questions 9 and 10. The data were obtained from a use-dilution test comparing four disinfectants against *Salmonella choleraesuis*. G = growth, NG = no growth

	Bacterial Growth after Exposure to			
Dilution	Disinfectant A	Disinfectant B	Disinfectant C	Disinfectant D
1:2	NG	G	NG	NG
1:4	NG	G	NG	G
1:8	NG	G	G	G
1:16	G	G	G	G

9. Which disinfectant is the most effective?
10. Which disinfectant(s) is (are) bactericidal?
    a. A, B, C, and D
    b. A, C, and D
    c. A only
    d. B only
    e. none of the above

Critical Thinking

1. The disk-diffusion method was used to evaluate three disinfectants. The results were as follows:

Disinfectant	Zone of Inhibition
X	0 mm
Y	5 mm
Z	10 mm

   a. Which disinfectant was the most effective against the organism?
   b. Can you determine whether compound Y was bactericidal or bacteriostatic?
2. Why is each of the following bacteria often resistant to disinfectants?
   a. *Mycobacterium*
   b. *Pseudomonas*
   c. *Bacillus*
3. A use-dilution test was used to evaluate two disinfectants against *Salmonella choleraesuis*. The results were as follows:

	Bacterial Growth after Exposures		
Time of Exposure (min)	Disinfectant A	Disinfectant B Diluted with Distilled Water	Disinfectant B Diluted with Tap Water
10	G	NG	G
20	G	NG	NG
30	NG	NG	NG

   a. Which disinfectant was the most effective?
   b. Which disinfectant should be used against *Staphylococcus*?

4. To determine the lethal action of microwave radiation, two 10^5 suspensions of *E. coli* were prepared. One cell suspension was exposed to microwave radiation while wet, whereas the other was lyophilized (freeze-dried) and then exposed to radiation. The results are shown in the following figure. Dashed lines indicate the temperature of the samples. What is the most likely method of lethal action of microwave radiation? How do you suppose these data might differ for *Clostridium*?

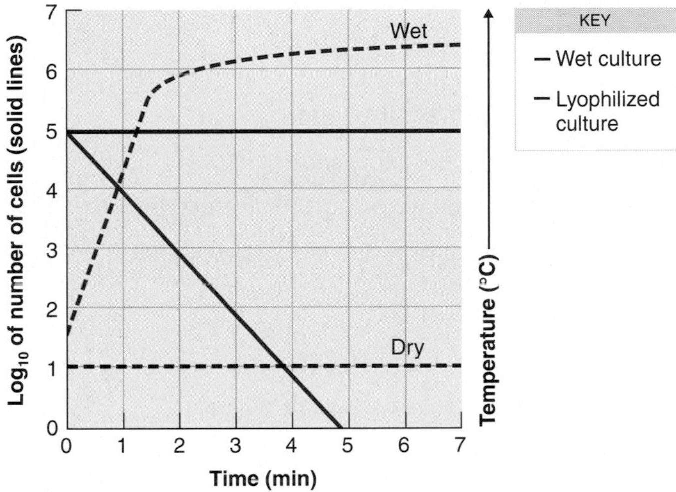

Clinical Applications

1. *Entamoeba histolytica* and *Giardia lamblia* were isolated from the stool sample of a 45-year-old man, and *Shigella sonnei* was isolated from the stool sample of an 18-year-old woman. Both patients experienced diarrhea and severe abdominal cramps, and prior to onset of digestive symptoms both had been treated by the same chiropractor. The chiropractor had administered colonic irrigations (enemas) to these patients. The device used for this treatment was a gravity-dependent apparatus using 12 liters of tap water. There were no check valves to prevent backflow, so all parts of the apparatus could have become contaminated with feces during each colonic treatment. The chiropractor provided colonic treatment to four or five patients per day. Between patients, the adaptor piece that is inserted into the rectum was placed in a "hot-water sterilizer."

   What two errors were made by the chiropractor?

2. Between March 9 and April 12, five chronic peritoneal dialysis patients at one hospital became infected with *Pseudomonas aeruginosa*. Four patients developed peritonitis (inflammation of the abdominal cavity), and one developed a skin infection at the catheter insertion site. All patients with peritonitis had low-grade fever, cloudy peritoneal fluid, and abdominal pain. All patients had permanent indwelling peritoneal catheters, which the nurse wiped with gauze that had been soaked with an iodophor solution each time the catheter was connected to or disconnected from the machine tubing. Aliquots of the iodophor were transferred from stock bottles to small in-use bottles. Cultures from the dialysate concentrate and the internal areas of the dialysis machines were negative; iodophor from a small in-use plastic container yielded a pure culture of *P. aeruginosa*.

   What improper technique led to this infection?

3. Eleven patients received injections of methylprednisolone and lidocaine to relieve the pain and inflammation of arthritis at the same orthopedic surgery office. All of them developed septic arthritis caused by *Serratia marcescens*. Unopened bottles of methylprednisolone from the same lot numbers tested sterile; the methylprednisolone was preserved with a quat. Cotton balls were used to wipe multiple-use injection vials before the medication was drawn into a disposable syringe. The site of injection on each patient was also wiped with a cotton ball. The cotton balls were soaked in benzalkonium chloride, and fresh cotton balls were added as the jar was emptied. Opened methylprednisolone containers and the jar of cotton balls contained *S. marcescens*.

   How was the infection transmitted? What part of the routine procedure caused the contamination?

8 Microbial Genetics

Virtually all the microbial traits you have read about in earlier chapters are controlled or influenced by heredity. The inherited traits of microbes include their shape and structural features, their metabolism, their ability to move or behave in various ways, and their ability to interact with other organisms—perhaps causing disease. Individual organisms transmit these characteristics to their offspring through genes.

Researchers are trying to solve the difficult medical problem of antibiotic resistance. Microorganisms can become resistant to antibiotics in any of several ways, all of which depend on genetics. The emergence of vancomycin-resistant *Staphylococcus aureus* (VRSA) poses a serious threat to patient care. In this chapter you will see how VRSA acquired this trait.

Emerging diseases provide another reason why it is important to understand genetics. New diseases are the results of genetic changes in some existing organism; for example, *E. coli* O157:H7 acquired the genes for Shiga toxin from *Shigella.*

Currently, microbiologists are using genetics to discover relatedness among organisms, to explore the origins of organisms such as HIV and West Nile virus, and to study the potential for avian influenza viruses to infect humans.

UNDER THE MICROSCOPE

Bacterial chromosome. The single chromosome, normally tightly packed inside a bacterial cell, has burst from an *E. coli* cell after the cell wall and plasma membrane were damaged.

Q&A

A 2-year-old girl was brought to the emergency department with bloody diarrhea, vomiting, fever, and renal failure. She was diagnosed with hemolytic uremic syndrome caused by *E. coli* O157:H7. *E. coli* is found naturally in the human large intestine, and there it is beneficial. However, the strain designated *E. coli* O157:H7 produces Shiga toxin, which causes severe illness and has emerged as a leading cause of foodborne illness. How did *E. coli* acquire this gene from *Shigella*?

Look for the answer in the chapter.

Structure and Function of the Genetic Material

LEARNING OBJECTIVES

8-1 Define *genetics, genome, chromosome, gene, genetic code, genotype, phenotype,* and *genomics.*

8-2 Describe how DNA serves as genetic information.

8-3 Describe the process of DNA replication.

8-4 Describe protein synthesis, including transcription, RNA processing, and translation.

8-5 Compare protein synthesis in prokaryotes and eukaryotes.

Genetics is the science of heredity; it includes the study of what genes are, how they carry information, how they are replicated and passed to subsequent generations of cells or passed between organisms, and how the expression of their information within an organism determines the particular characteristics of that organism. The genetic information in a cell is called the **genome.** A cell's genome includes its chromosomes and plasmids. **Chromosomes** are structures containing DNA that physically carry hereditary information; the chromosomes contain the genes. **Genes** are segments of DNA (except in some viruses, in which they are made of RNA) that code for functional products. We saw in Chapter 2 that DNA is a macromolecule composed of repeating units called *nucleotides.* Recall that each nucleotide consists of a nucleobase (adenine, thymine, cytosine, or guanine), deoxyribose (a pentose sugar), and a phosphate group (see Figure 2.16, page 48). The DNA within a cell exists as long strands of nucleotides twisted together in pairs to form a double helix. Each strand has a string of alternating sugar and phosphate groups (its *sugar-phosphate backbone*), and a nitrogenous base is attached to each sugar in the backbone. The two strands are held together by hydrogen bonds between their nitrogenous bases. The **base pairs** always occur in a specific way: adenine always pairs with thymine, and cytosine always pairs with guanine. Because of this specific base pairing, the base sequence of one DNA strand determines the base sequence of the other strand. The two strands of DNA are thus *complementary.*

The structure of DNA helps explain two primary features of biological information storage. First, the linear sequence of bases provides the actual information. Genetic information is encoded by the sequence of bases along a strand of DNA, in much the same way as our written language uses a linear sequence of letters to form words and sentences. The genetic language, however, uses an alphabet with only four letters—the four kinds of nucleobases in DNA (or RNA). But 1000 of these four bases, the number contained in an average-sized gene, can be arranged in 4^{1000} different ways. This astronomically large number explains how genes can be varied enough to provide all the information a cell needs to grow and perform its functions. The **genetic code,** the set of rules that determines how a nucleotide sequence is converted into the amino acid sequence of a protein, is discussed in more detail later in the chapter.

Second, the complementary structure allows for the precise duplication of DNA during cell division. Each daughter cell receives one of the original strands from the parent; thus ensuring one strand that functions correctly.

Much of cellular metabolism is concerned with translating the genetic message of genes into specific proteins. A gene usually codes for a messenger RNA (mRNA) molecule, which ultimately results in the formation of a protein. Alternatively, the gene product can be a ribosomal RNA (rRNA) or a transfer RNA (tRNA). As we will see, all of these types of RNA are involved in the process of protein synthesis. When the ultimate molecule for which a gene codes (a protein, for example) has been produced, we say that the gene has been *expressed.*

Genotype and Phenotype

The **genotype** of an organism is its genetic makeup, the information that codes for all the particular characteristics of the organism. The genotype represents *potential* properties, but not the properties themselves. **Phenotype** refers to *actual, expressed* properties, such as the organism's ability to perform a particular chemical reaction. Phenotype, then, is the manifestation of genotype.

In molecular terms, an organism's genotype is its collection of genes, its entire DNA. What constitutes the organism's phenotype in molecular terms? In a sense, an organism's phenotype is its collection of proteins. Most of a cell's properties derive from the structures and functions of its proteins. In microbes, most proteins are either *enzymatic* (catalyze particular reactions) or *structural* (participate in large functional complexes such as membranes or flagella). Even phenotypes that depend on structural macromolecules other than proteins (such as lipids or polysaccharides) rely indirectly on proteins. For instance, the structure of a complex lipid or polysaccharide molecule results from the catalytic activities of enzymes that synthesize, process, and degrade those molecules. Thus, although it is not completely accurate to say that phenotypes are due only to proteins, it is a useful simplification.

DNA and Chromosomes

Bacteria typically have a single circular chromosome consisting of a single circular molecule of DNA with associated proteins. The chromosome is looped and folded and attached at one or several points to the plasma membrane. The DNA of *E. coli,* the most-studied bacterial species, has about 4.6 million base pairs and is about 1 mm long—1000 times longer than the entire cell

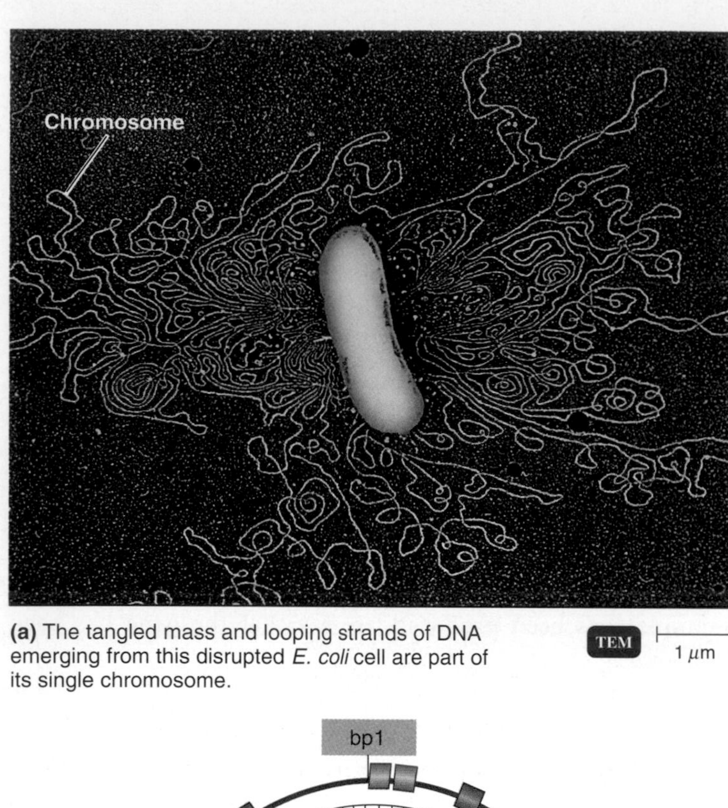

(a) The tangled mass and looping strands of DNA emerging from this disrupted *E. coli* cell are part of its single chromosome.

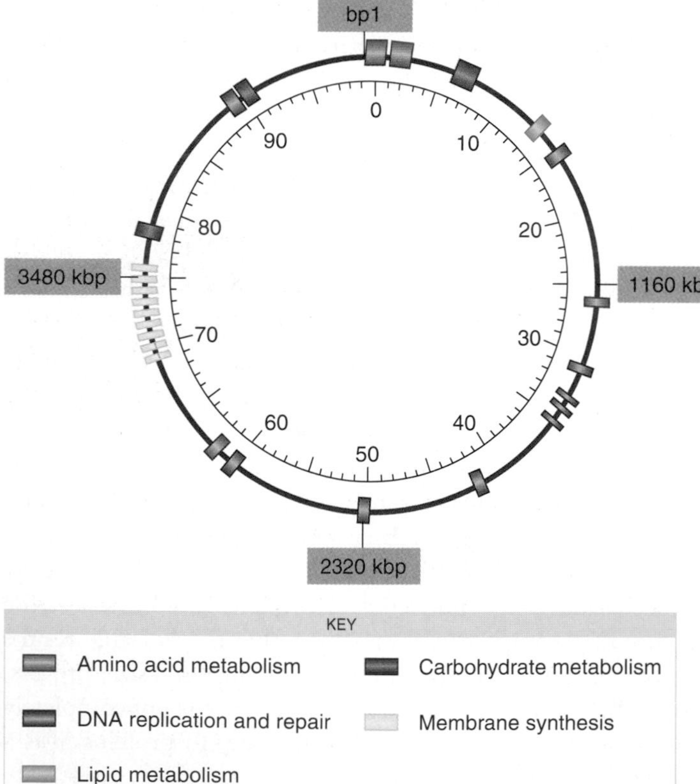

(b) A genetic map of the chromosome of *E. coli*. The numbers inside the circle indicate the number of minutes it takes to transfer the genes during mating between two cells; the numbers in colored boxes indicate the number of base pairs.

Figure 8.1 A prokaryotic chromosome.

Q What is a gene? What is an open-reading frame?

(**Figure 8.1a**). However, the chromosome takes up only about 10% of the cell's volume because the DNA is twisted, or *supercoiled*—much like a telephone cord when you put the handset back on the receiver.

The location of genes on a bacterial chromosome can be determined by experiments on the transfer of genes from one cell to another. These processes will be discussed later in this chapter. The bacterial chromosome map that results is marked in minutes corresponding to when the genes are transferred from a donor cell to a recipient cell (**Figure 8.1b**).

In recent years, the complete base sequences of several bacterial chromosomes have been determined. Computers are used to search for *open-reading frames,* that is, regions of DNA that are likely to encode a protein. As you will see later, these are base sequences between start and stop codons. The sequencing and molecular characterization of genomes is called **genomics.** The use of genomics to track West Nile virus is described in the box on page 223.

The Flow of Genetic Information

DNA replication makes possible the flow of genetic information from one generation to the next. As shown in **Figure 8.2**, the DNA of a cell replicates before cell division so that each offspring cell receives a chromosome identical to the parent's. Within each metabolizing cell, the genetic information contained in DNA also flows in another way: it is transcribed into mRNA and then translated into protein. We describe the processes of transcription and translation later in this chapter.

CHECK YOUR UNDERSTANDING

- ✓ Give a clinical application of genomics. **8-1**
- ✓ Why is the base pairing in DNA important? **8-2**

DNA Replication

In DNA replication, one "parental" double-stranded DNA molecule is converted to two identical "daughter" molecules. The complementary structure of the nitrogenous base sequences in the DNA molecule is the key to understanding DNA replication. Because the bases along the two strands of double-helical DNA are complementary, one strand can act as a template for the production of the other strand (**Figure 8.3a**).

DNA replication requires the presence of several cellular proteins that direct a particular sequence of events. Enzymes involved in DNA replication and other processes are listed in **Table 8.1** on page 215. When replication begins, the supercoiling is relaxed by *topoisomerase* or *gyrase*, and the two strands of parental DNA are unwound by *helicase* and separated from

Figure 8.2

FOUNDATION FIGURE The Flow of Genetic Information

Using the example of a bacterium with a single circular chromosome, this figure summarizes how genetic information is used within and passed between cells. Small versions of the relevant portions of this overview figure will appear in other figures throughout the chapter to indicate the relationships of different processes.

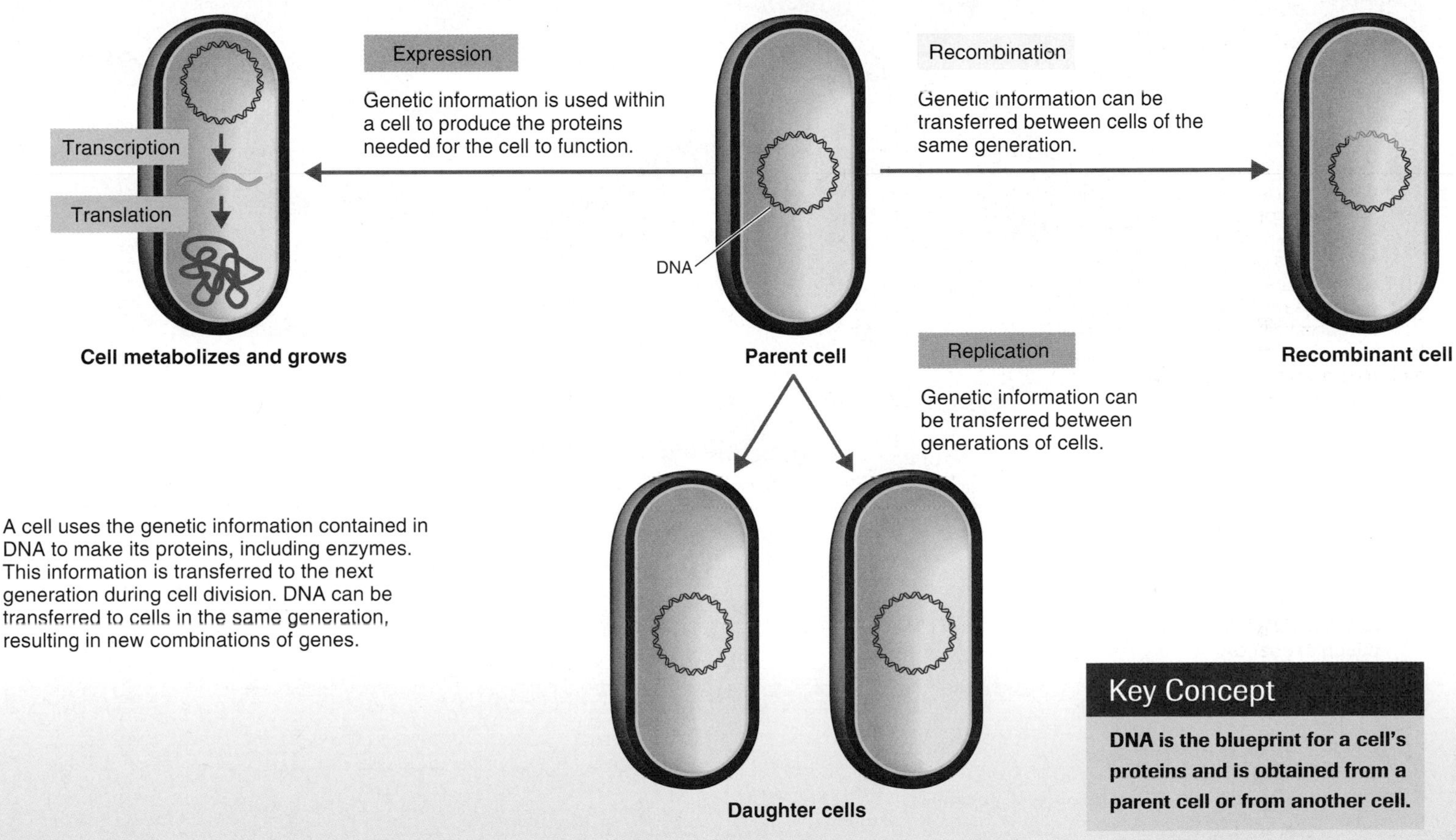

Expression
Genetic information is used within a cell to produce the proteins needed for the cell to function.
Transcription
Translation
Cell metabolizes and grows
DNA
Parent cell
Recombination
Genetic information can be transferred between cells of the same generation.
Recombinant cell
Replication
Genetic information can be transferred between generations of cells.
Daughter cells

A cell uses the genetic information contained in DNA to make its proteins, including enzymes. This information is transferred to the next generation during cell division. DNA can be transferred to cells in the same generation, resulting in new combinations of genes.

Key Concept

DNA is the blueprint for a cell's proteins and is obtained from a parent cell or from another cell.

each other in one small DNA segment after another. Free nucleotides present in the cytoplasm of the cell are matched up to the exposed bases of the single-stranded parental DNA. Where thymine is present on the original strand, only adenine can fit into place on the new strand; where guanine is present on the original strand, only cytosine can fit into place, and so on. Any bases that are improperly base-paired are removed and replaced by replication enzymes. Once aligned, the newly added nucleotide is joined to the growing DNA strand by an enzyme called **DNA polymerase.** Then the parental DNA is unwound a bit further to allow the addition of the next nucleotides. The point at which replication occurs is called the *replication fork*.

As the replication fork moves along the parental DNA, each of the unwound single strands combines with new nucleotides. The original strand and this newly synthesized daughter strand then rewind. Because each new double-stranded DNA molecule contains one original (conserved) strand and one new strand, the process of replication is referred to as **semiconservative replication.**

Before looking at DNA replication in more detail, let's take a closer look at the structure of DNA (see Figure 2.16, on page 48). It is important to understand the concept that the paired DNA strands are oriented in opposite directions relative to each other. Notice in Figure 2.16 that the carbon atoms of the sugar component of each nucleotide are numbered 1′ (pronounced "one prime") to 5′. For the paired bases to be next to each other, the sugar components in one strand are upside down relative to the other. The end with the hydroxyl attached to the 3′ carbon is called the 3′ end of the DNA strand; the end having a phosphate attached

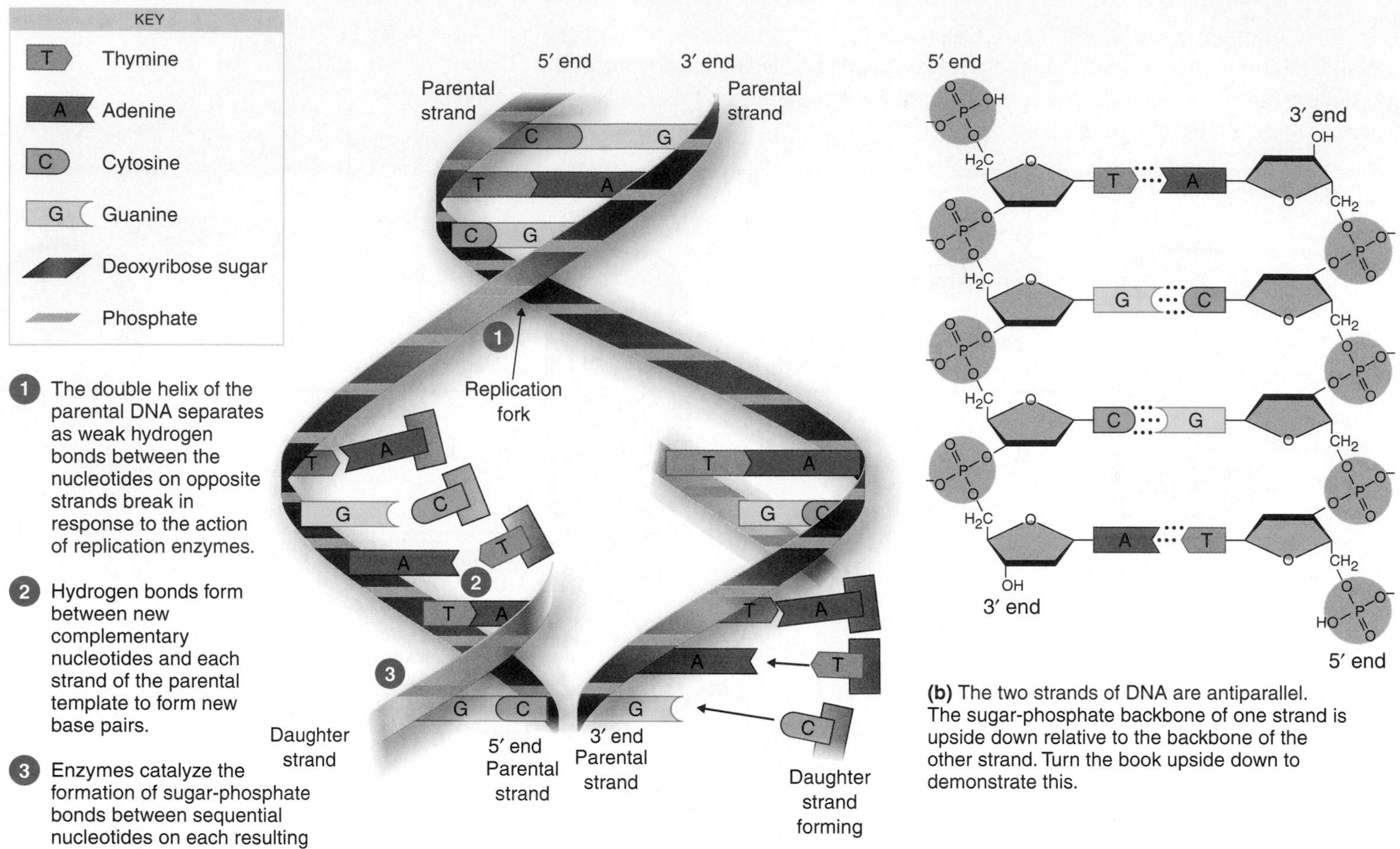
KEY
Thymine
Adenine
Cytosine
Guanine
Deoxyribose sugar
Phosphate
1 The double helix of the parental DNA separates as weak hydrogen bonds between the nucleotides on opposite strands break in response to the action of replication enzymes.
2 Hydrogen bonds form between new complementary nucleotides and each strand of the parental template to form new base pairs.
3 Enzymes catalyze the formation of sugar-phosphate bonds between sequential nucleotides on each resulting daughter strand.
5′ end
3′ end
Parental strand
Parental strand
Replication fork
Daughter strand
5′ end Parental strand
3′ end Parental strand
Daughter strand forming
5′ end
3′ end
OH
3′ end
5′ end

(a) The replication fork.

(b) The two strands of DNA are antiparallel. The sugar-phosphate backbone of one strand is upside down relative to the backbone of the other strand. Turn the book upside down to demonstrate this.

Figure 8.3 DNA replication.

Q What is meant by semiconservative replication?

to the 5′ carbon is called the 5′ end. The way in which the two strands fit together dictates that the 5′ → 3′ direction of one strand runs counter to the 5′ → 3′ direction of the other strand (Figure 8.3b). This structure of DNA affects the replication process because DNA polymerases can add new nucleotides to the 3′ end only. Therefore, as the replication fork moves along the parental DNA, the two new strands must grow in different directions.

DNA replication requires a great deal of energy. The energy is supplied from the nucleotides, which are actually nucleoside triphosphates. You already know about ATP; the only difference between ATP and the adenine nucleotide in DNA is the sugar component. Deoxyribose is the sugar in the nucleosides used to synthesize DNA, and nucleoside triphosphates with ribose are used to synthesize RNA. Two phosphate groups are removed to add the nucleotide to a growing strand of DNA; hydrolysis of the nucleoside is exergonic and provides energy to make the new bonds in the DNA strand (Figure 8.4).

Figure 8.5 provides more detail about the many steps that go into this complex process.

DNA replication by some bacteria, such as *E. coli*, goes *bidirectionally* around the chromosome (Figure 8.6). Two replication forks move in opposite directions away from the origin of replication. Because the bacterial chromosome is a closed loop, the replication forks eventually meet when replication is completed. The two loops must be separated by a topoisomerase. Much evidence shows an association between the bacterial plasma membrane and the origin of replication. After duplication, if each copy of the origin binds to the membrane at opposite poles, then each daughter cell receives one copy of the DNA molecule—that is, one complete chromosome.

DNA replication is an amazingly accurate process. Typically, mistakes are made at a rate of only 1 in every 10^{10} bases incorporated. Such accuracy is largely due to the *proofreading* capability of DNA polymerase. As each new base is added, the enzyme evaluates whether it forms the proper complementary base-pairing structure. If not, the enzyme excises the improper

Table 8.1	Important Enzymes in DNA Replication, Expression, and Repair
DNA Gyrase	Relaxes supercoiling ahead of the replication fork
DNA Ligase	Makes covalent bonds to join DNA strands; joins Okazaki fragments and new segments in excision repair
DNA Polymerase	Synthesizes DNA; proofreads and repairs DNA
Endonucleases	Cut DNA backbone in a strand of DNA; facilitate repair and insertions
Exonucleases	Cut DNA from an exposed end of DNA; facilitate repair
Helicase	Unwinds double-stranded DNA
Methylase	Adds methyl group to selected bases in newly made DNA
Photolyase	Uses visible light energy to separate UV-induced pyrimidine dimers
Primase	Makes RNA primers from a DNA template
Ribozyme	RNA enzyme that removes introns and splices exons together
RNA Polymerase	Copies RNA from a DNA template
snRNP	RNA-protein complex that removes introns and splices exons together
Topoisomerase	Relaxes supercoiling ahead of the replication fork; separates DNA circles at the end of DNA replication
Transposase	Cuts DNA backbone leaving single-stranded "sticky ends"

base and replaces it with the correct one. In this way, DNA can be replicated very accurately, allowing each daughter chromosome to be virtually identical to the parental DNA. **Animations** DNA Replication: Overview, Forming the Replication Fork, Replication Proteins, Synthesis. **www.microbiologyplace.com**

CHECK YOUR UNDERSTANDING

✓ Describe DNA replication, including the functions of DNA gyrase, DNA ligase, and DNA polymerase. **8-3**

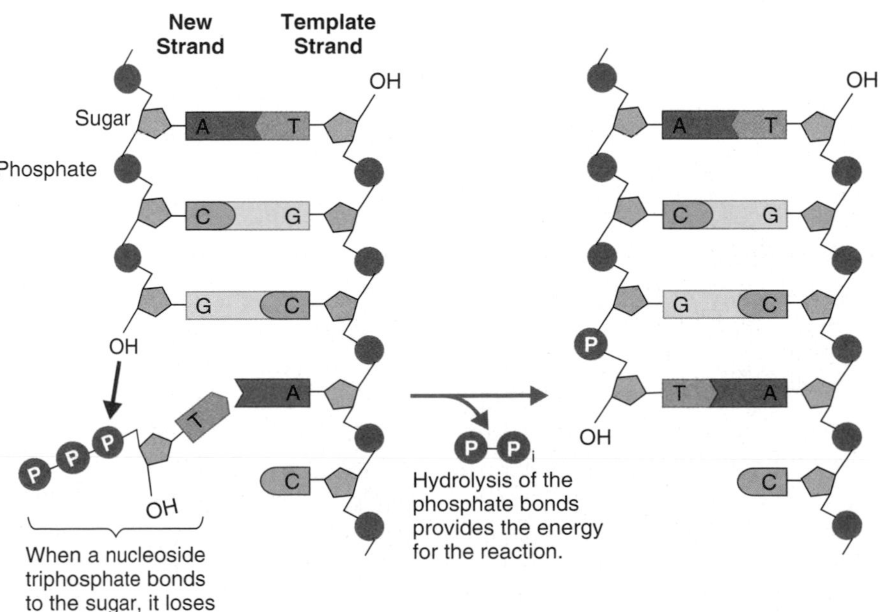

Figure 8.4 Adding a nucleotide to DNA.

Q **Why is one strand "upside down" relative to the other strand? Why can't both strands "face" the same way?**

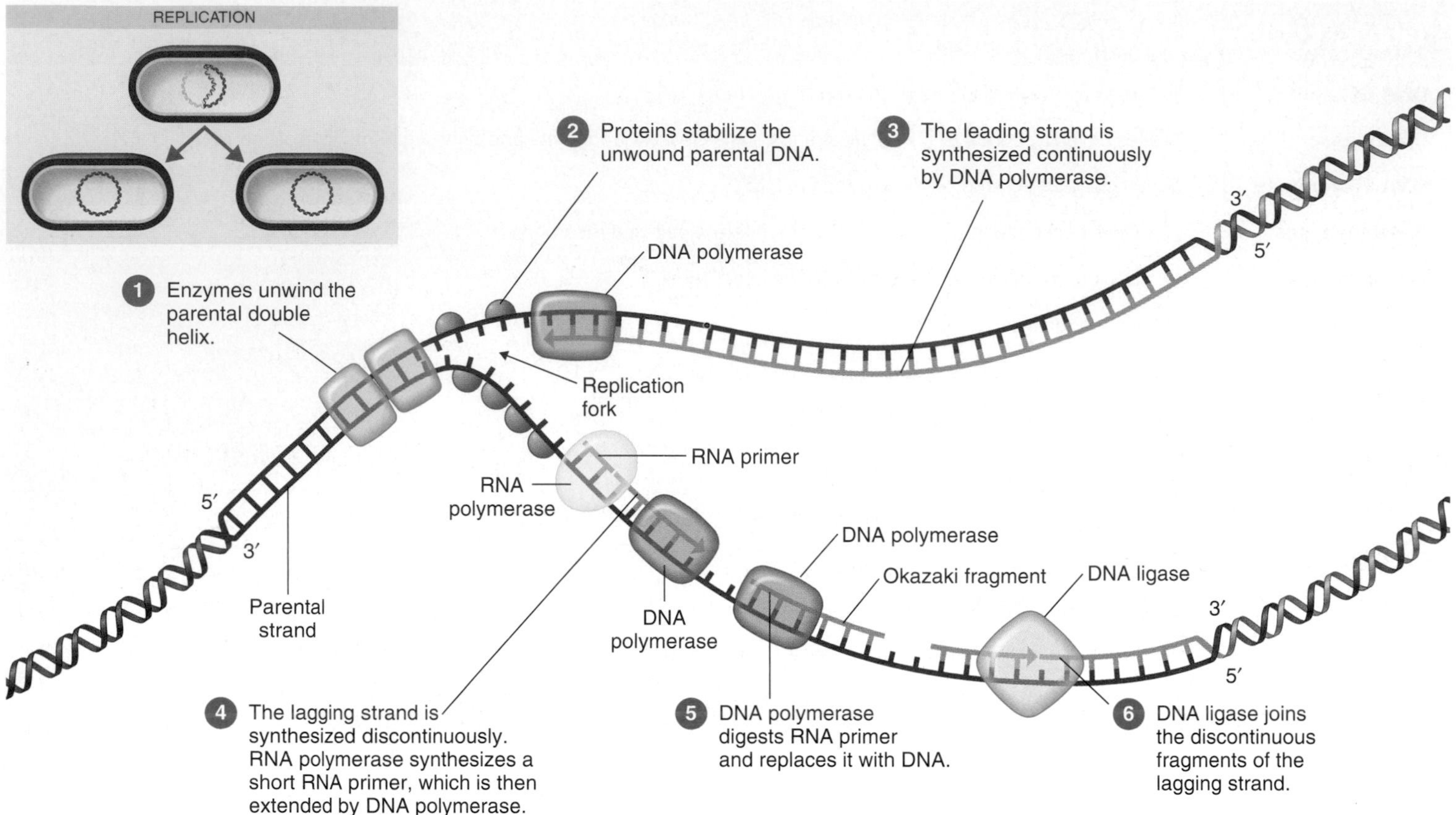

Figure 8.5 A summary of events at the DNA replication fork.

Q **Why is one strand of DNA synthesized discontinuously?**

RNA and Protein Synthesis

How is the information in DNA used to make the proteins that control cell activities? In the process of *transcription,* genetic information in DNA is copied, or transcribed, into a complementary base sequence of RNA. The cell then uses the information encoded in this RNA to synthesize specific proteins through the process of *translation.* We now take a closer look at these two processes as they occur in a bacterial cell.

Transcription

Transcription is the synthesis of a complementary strand of RNA from a DNA template. We will discuss transcription in prokaryotic cells here. Transcription in eukaryotes is discussed on page 220. As mentioned earlier, there are three kinds of RNA in bacterial cells: messenger RNA, ribosomal RNA, and transfer RNA. Ribosomal RNA forms an integral part of ribosomes, the cellular machinery for protein synthesis. Transfer RNA is also involved in protein synthesis, as we will see. **Messenger RNA (mRNA)** carries the coded information for making specific proteins from DNA to ribosomes, where proteins are synthesized.

During transcription, a strand of mRNA is synthesized using a specific portion of the cell's DNA as a template. In other words, the genetic information stored in the sequence of nitrogenous bases of DNA is rewritten so that the same information appears in the base sequence of mRNA. As in DNA replication, a G in the DNA template dictates a C in the mRNA being made, a C in the DNA template dictates a G in the mRNA, and a T in the DNA template dictates an A in the mRNA. However, an A in the DNA template dictates a uracil (U) in the mRNA, because RNA contains U instead of T. (U has a chemical structure slightly different from T, but it base-pairs in the same way.) If, for example, the template portion of DNA has the base sequence 3′-ATGCAT, the newly synthesized mRNA strand will have the complementary base sequence 5′-UACGUA.

The process of transcription requires both an enzyme called *RNA polymerase* and a supply of RNA nucleotides (**Figure 8.7**). Transcription begins when RNA polymerase binds to the DNA at a site called the **promoter.** Only one of the two DNA strands serves as the template for RNA synthesis for a given gene. Like DNA, RNA is synthesized in the $5' \rightarrow 3'$ direction. RNA

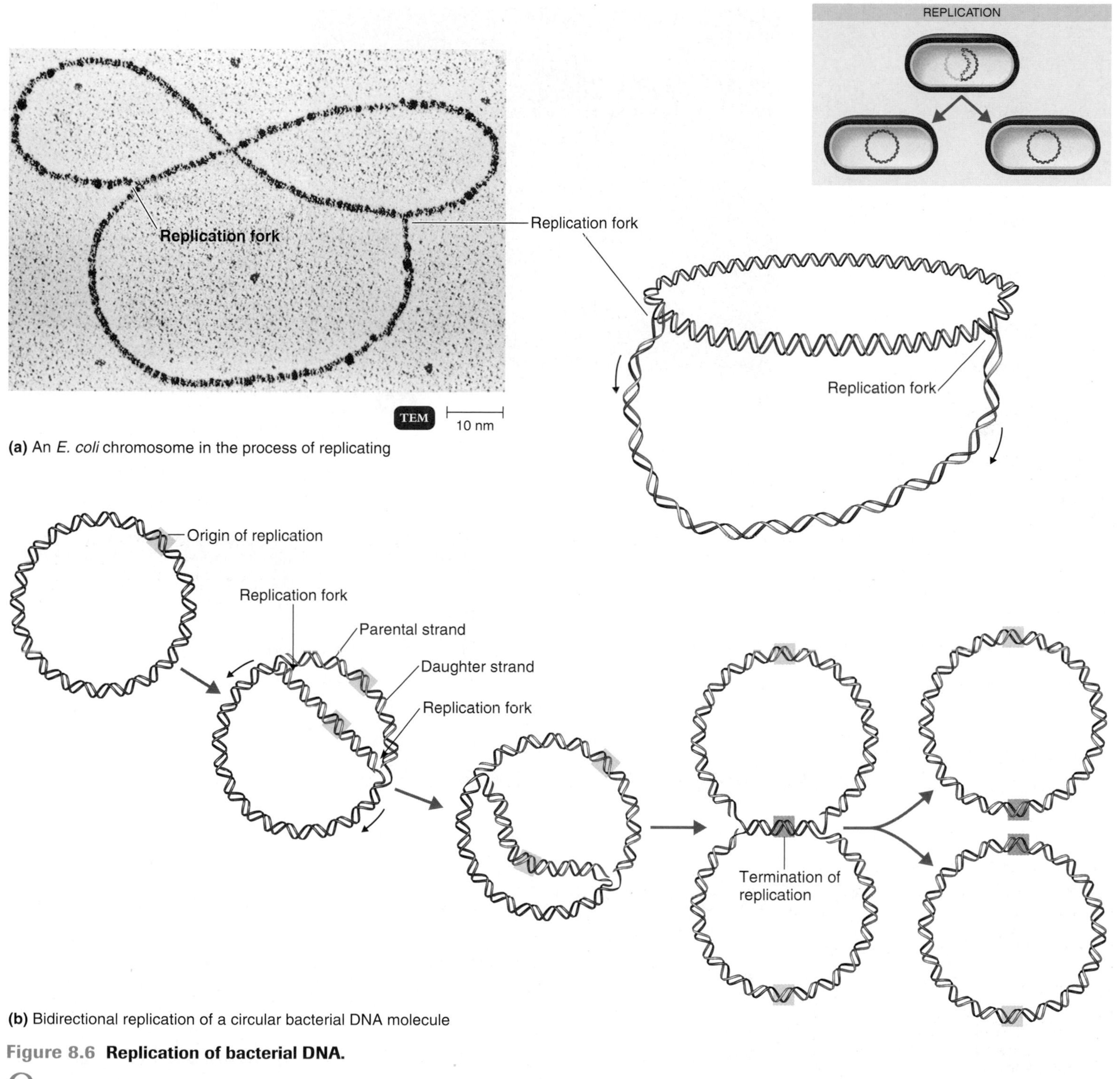

(a) An *E. coli* chromosome in the process of replicating

(b) Bidirectional replication of a circular bacterial DNA molecule

Figure 8.6 Replication of bacterial DNA.

Q **What is the origin of replication?**

synthesis continues until RNA polymerase reaches a site on the DNA called the **terminator.**

The process of transcription allows the cell to produce short-term copies of genes that can be used as the direct source of information for protein synthesis. Messenger RNA acts as an intermediate between the permanent storage form, DNA, and the process that uses the information, translation. **Animations** Transcription: Overview, Process. **www.microbiologyplace.com**

Translation

We have seen how the genetic information in DNA is transferred to mRNA during transcription. Now we will see how mRNA

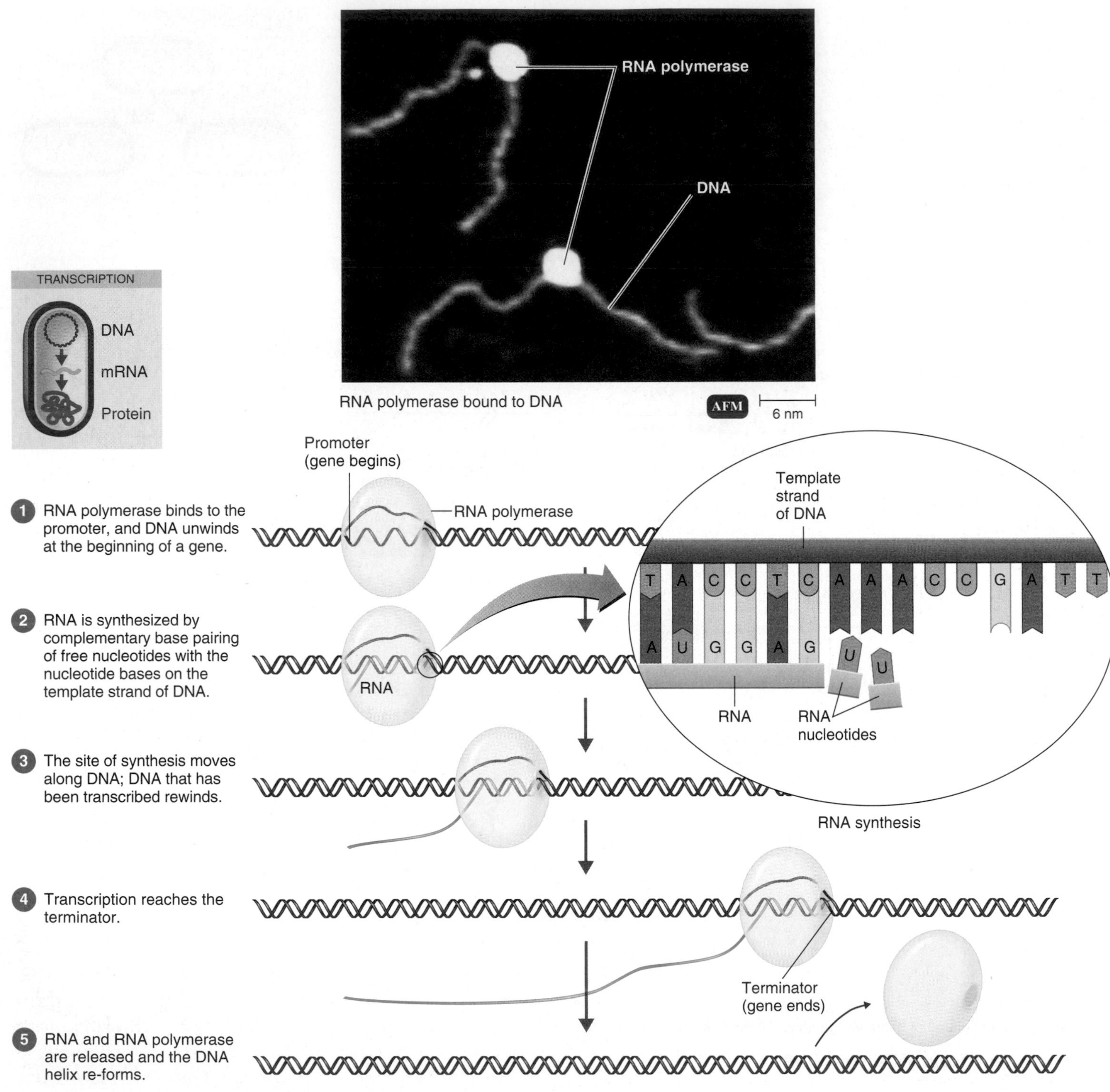
RNA polymerase
DNA
RNA polymerase bound to DNA
AFM
6 nm
TRANSCRIPTION
DNA
mRNA
Protein
Promoter
(gene begins)
1 RNA polymerase binds to the promoter, and DNA unwinds at the beginning of a gene.
RNA polymerase
Template strand of DNA
T A C C T C A A A C C G A T T
A U G G A G
U U
RNA
RNA nucleotides
2 RNA is synthesized by complementary base pairing of free nucleotides with the nucleotide bases on the template strand of DNA.
RNA
3 The site of synthesis moves along DNA; DNA that has been transcribed rewinds.
RNA synthesis
4 Transcription reaches the terminator.
Terminator
(gene ends)
5 RNA and RNA polymerase are released and the DNA helix re-forms.

Figure 8.7 The process of transcription. The orienting diagram indicates the relationship of transcription to the overall flow of genetic information within a cell.

Q In transcription, what is copied, and what is made?

serves as the source of information for the synthesis of proteins. Protein synthesis is called **translation** because it involves decoding the "language" of nucleic acids and converting that information into the "language" of proteins.

The language of mRNA is in the form of **codons,** groups of three nucleotides, such as AUG, GGC, or AAA. The sequence of codons on an mRNA molecule determines the sequence of amino acids that will be in the protein being synthesized. Each codon "codes" for a particular amino acid. This is the genetic code (Figure 8.8).

Codons are written in terms of their base sequence in mRNA. Notice that there are 64 possible codons but only 20 amino acids. This means that most amino acids are signaled by several alternative codons, a situation referred to as the **degeneracy** of the code. For example, leucine has six codons, and alanine has four codons. Degeneracy allows for a certain amount of change, or mutation, in the DNA without affecting the protein ultimately produced.

Of the 64 codons, 61 are sense codons, and 3 are nonsense codons. **Sense codons** code for amino acids, and **nonsense codons** (also called *stop codons*) do not. Rather, the nonsense codons—UAA, UAG, and UGA—signal the end of the protein molecule's synthesis. The start codon that initiates the synthesis of the protein molecule is AUG, which is also the codon for methionine. In bacteria, the start AUG codes for formylmethionine rather than the methionine found in other parts of the protein. The initiating methionine is often removed later, so not all proteins begin with methionine.

The codons of mRNA are converted into protein through the process of translation. The codons of an mRNA are "read" sequentially; and, in response to each codon, the appropriate amino acid is assembled into a growing chain. The site of translation is the ribosome, and **transfer RNA (tRNA)** molecules both recognize the specific codons and transport the required amino acids.

Each tRNA molecule has an **anticodon,** a sequence of three bases that is complementary to a codon. In this way, a tRNA molecule can base-pair with its associated codon. Each tRNA can also carry on its other end the amino acid encoded by the codon that the tRNA recognizes. The functions of the ribosome are to direct the orderly binding of tRNAs to codons and to assemble the amino acids brought there into a chain, ultimately producing a protein.

Figure 8.9 shows the details of translation. The necessary components assemble: the two ribosomal subunits, a tRNA with the anticodon UAC, and the mRNA molecule to be translated, along with several additional protein factors. This sets up the start codon (AUG) in the proper position to allow translation to begin. After the ribosome joins the first two amino acids with a peptide bond, the first tRNA molecule leaves the ribosome. The ribosome then moves along the mRNA to the next

First position	Second position: U	C	A	G	Third position
U	UUU Phe	UCU Ser	UAU Tyr	UGU Cys	U
	UUC	UCC	UAC	UGC	C
	UUA Leu	UCA	UAA Stop	UGA Stop	A
	UUG	UCG	UAG Stop	UGG Trp	G
C	CUU Leu	CCU Pro	CAU His	CGU Arg	U
	CUC	CCC	CAC	CGC	C
	CUA	CCA	CAA Gln	CGA	A
	CUG	CCG	CAG	CGG	G
A	AUU Ile	ACU Thr	AAU Asn	AGU Ser	U
	AUC	ACC	AAC	AGC	C
	AUA	ACA	AAA Lys	AGA Arg	A
	AUG Met/start	ACG	AAG	AGG	G
G	GUU Val	GCU Ala	GAU Asp	GGU Gly	U
	GUC	GCC	GAC	GGC	C
	GUA	GCA	GAA Glu	GGA	A
	GUG	GCG	GAG	GGG	G

Figure 8.8 The genetic code. The three nucleotides in an mRNA codon are designated, respectively, as the first position, second position, and third position of the codon on the mRNA. Each set of three nucleotides specifies a particular amino acid, represented by a three-letter abbreviation (see Table 2.4, page 44). The codon AUG, which specifies the amino acid methionine, is also the start of protein synthesis. The word *Stop* identifies the nonsense codons that signal the termination of protein synthesis.

Q Why is the genetic code described as degenerate?

codon. As the proper amino acids are brought into line one by one, peptide bonds are formed between them, and a polypeptide chain results. (See Figure 2.14, page 45.) Translation ends when one of the three nonsense codons in the mRNA is reached. The ribosome then comes apart into its two subunits, and the mRNA and newly synthesized polypeptide chain are released. The ribosome, the mRNA, and the tRNAs are then available to be used again.

The ribosome moves along the mRNA in the $5' \rightarrow 3'$ direction. As a ribosome moves along the mRNA, it will soon allow the start codon to be exposed. Additional ribosomes can then assemble and begin synthesizing protein. In this way, there are usually a number of ribosomes attached to a single mRNA, all at various stages of protein synthesis. In prokaryotic cells, the

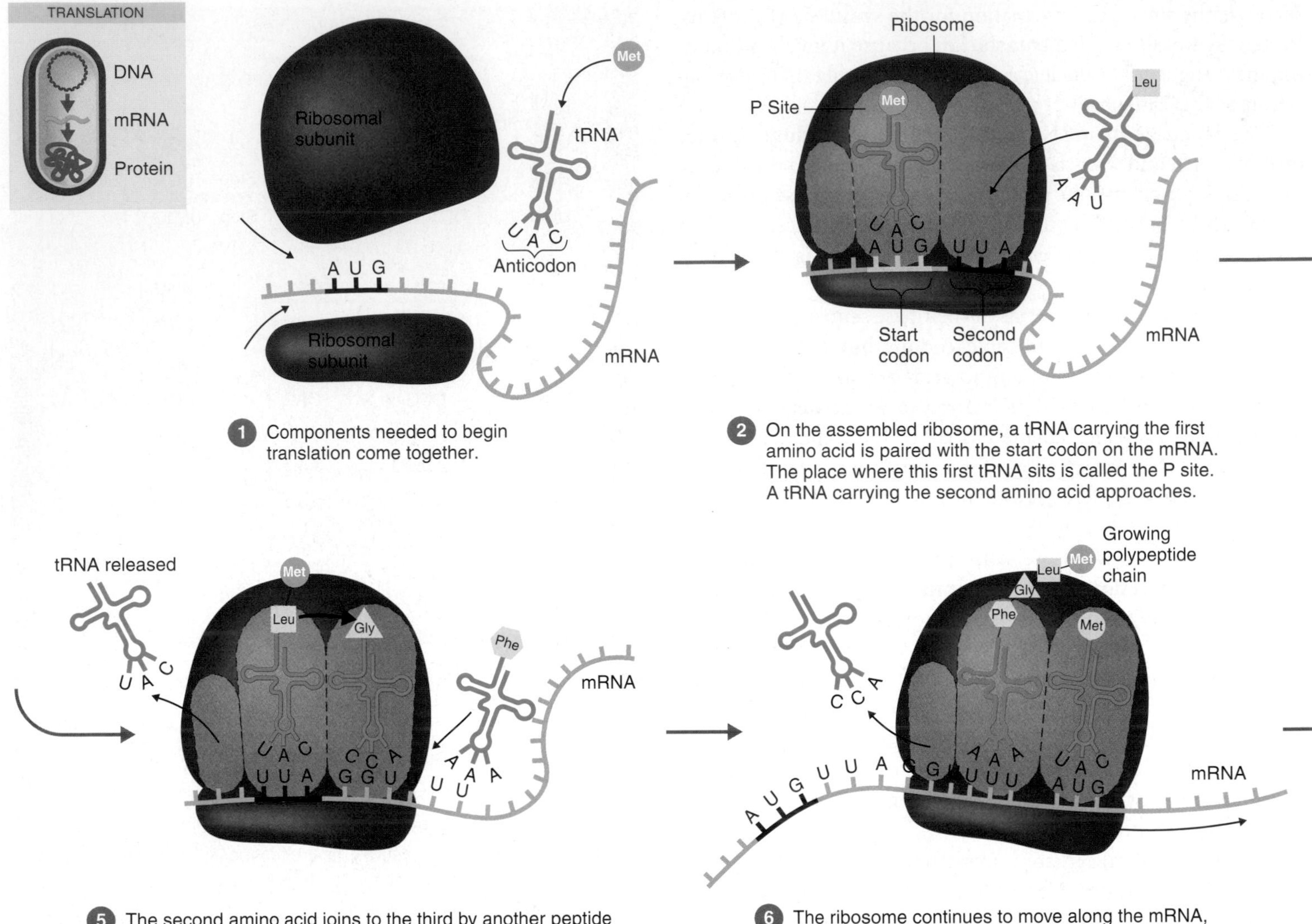

1 Components needed to begin translation come together.

2 On the assembled ribosome, a tRNA carrying the first amino acid is paired with the start codon on the mRNA. The place where this first tRNA sits is called the P site. A tRNA carrying the second amino acid approaches.

5 The second amino acid joins to the third by another peptide bond, and the first tRNA is released from the E site.

6 The ribosome continues to move along the mRNA, and new amino acids are added to the polypeptide.

Figure 8.9 The process of translation. The overall goal of translation is to produce proteins using mRNAs as the source of biological information. The complex cycle of events illustrated here shows the primary role of tRNA and ribosomes in the decoding of this information. The ribosome acts as the site where the mRNA-encoded information is decoded, as well as the site where individual amino acids are connected into polypeptide chains. The tRNA molecules act as the actual "translators"—one end of each tRNA recognizes a specific mRNA codon, while the other end carries the amino acid coded for by that codon.

Q Why is this process called translation?

translation of mRNA into protein can begin even before transcription is complete (Figure 8.10). Because mRNA is produced in the cytoplasm, the start codons of an mRNA being transcribed are available to ribosomes before the entire mRNA molecule is even made.

In eukaryotic cells, transcription takes place in the nucleus. The mRNA must be completely synthesized and moved through the nuclear membrane to the cytoplasm before translation can begin. In addition, the RNA undergoes processing before it leaves the nucleus. In eukaryotic cells, the regions of genes that code for proteins are often interrupted by noncoding DNA. Thus, eukaryotic genes are composed of **exons,** the regions of DNA *expressed,* and **introns,** the *intervening* regions of DNA that do not encode protein. In the nucleus, RNA polymerase synthesizes a molecule called an RNA transcript that contains copies of the introns. Particles called **small nuclear ribonucleoproteins,** abbreviated **snRNPs** and pronounced "snurps," remove the introns and splice the exons together. In some organisms, the introns act as ribozymes to catalyze their own removal (Figure 8.11).

* * *

To summarize, genes are the units of biological information encoded by the sequence of nucleotide bases in DNA. A gene is

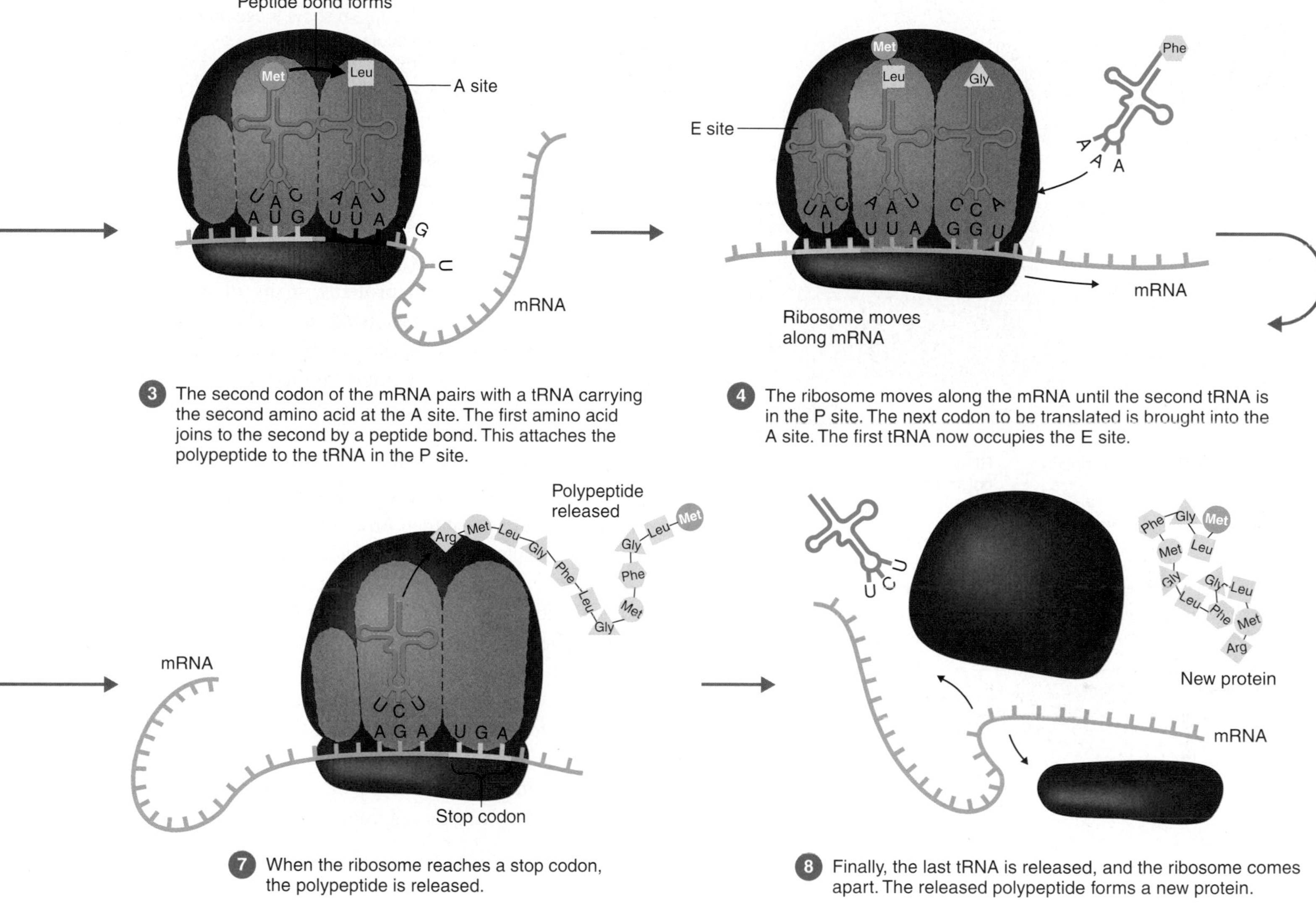

3 The second codon of the mRNA pairs with a tRNA carrying the second amino acid at the A site. The first amino acid joins to the second by a peptide bond. This attaches the polypeptide to the tRNA in the P site.

4 The ribosome moves along the mRNA until the second tRNA is in the P site. The next codon to be translated is brought into the A site. The first tRNA now occupies the E site.

7 When the ribosome reaches a stop codon, the polypeptide is released.

8 Finally, the last tRNA is released, and the ribosome comes apart. The released polypeptide forms a new protein.

Figure 8.9 The process of translation. (continued)

expressed, or turned into a product within the cell, through the processes of transcription and translation. The genetic information carried in DNA is transferred to a temporary mRNA molecule by transcription. Then, during translation, the mRNA directs the assembly of amino acids into a polypeptide chain: a ribosome attaches to mRNA, tRNAs deliver the amino acids to the ribosome as directed by the mRNA codon sequence, and the ribosome assembles the amino acids into the chain that will be the newly synthesized protein. **Animations** Translation: Overview, Genetic Code, Process. **www.microbiologyplace.com**

CHECK YOUR UNDERSTANDING

- ✓ What is the role of the promoter, terminator, and mRNA in transcription? **8-4**
- ✓ How does mRNA production in eukaryotes differ from the process in prokaryotes? **8-5**

The Regulation of Bacterial Gene Expression

LEARNING OBJECTIVE

8-6 Define *operon.*

8-7 Explain the regulation of gene expression in bacteria by induction, repression, and catabolite repression.

A cell's genetic machinery and its metabolic machinery are integrated and interdependent. Recall from Chapter 5 that the bacterial cell carries out an enormous number of metabolic reactions. The common feature of all metabolic reactions is that they are catalyzed by enzymes. Also recall from Chapter 5 (page 120) that feedback inhibition stops a cell from performing unneeded chemical reactions. Feedback inhibition stops enzymes that have already been synthesized. We will now look at mechanisms to prevent synthesis of enzymes that are not needed.

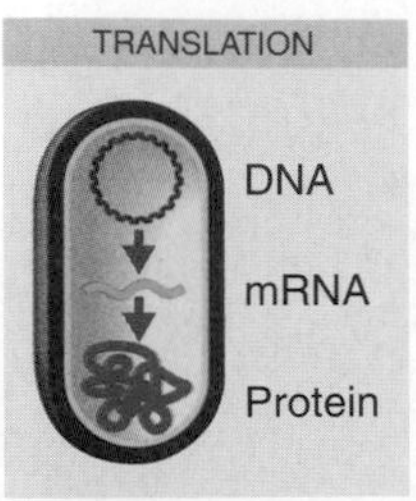

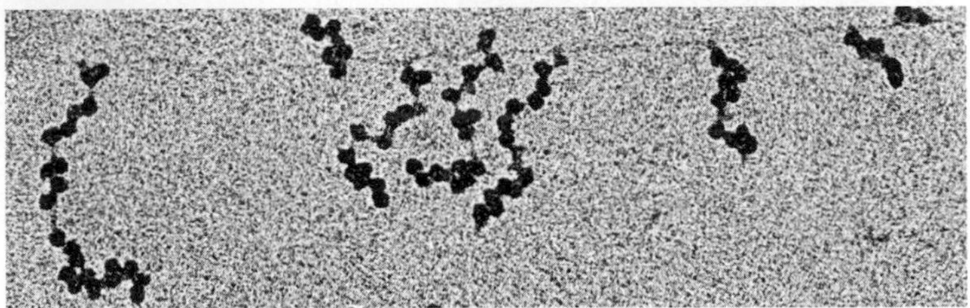

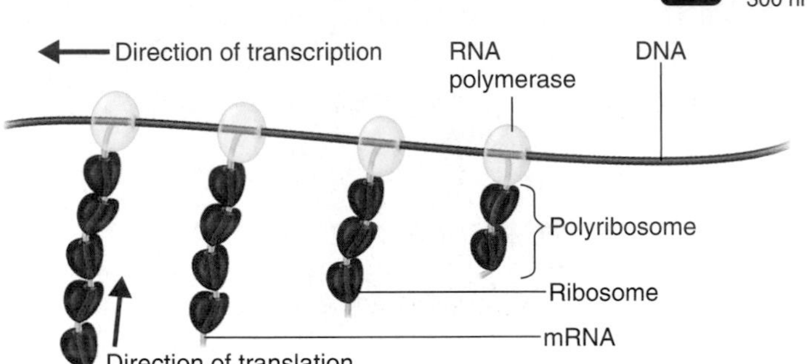

Figure 8.10 Simultaneous transcription and translation in bacteria. The micrograph and diagram show these processes in a single bacterial gene. Many molecules of mRNA are being synthesized simultaneously. The longest mRNA molecules were the first to be transcribed at the promoter. Note the ribosomes attached to the newly forming mRNA. The newly synthesized polypeptides are not shown.

Q Why can translation begin before transcription is complete in prokaryotes but not in eukaryotes?

We have seen that genes, through transcription and translation, direct the synthesis of proteins, many of which serve as enzymes—the very enzymes used for cellular metabolism. Because protein synthesis requires a huge amount of energy, regulation of protein synthesis is important to the cell's energy economy. Cells save energy by making only those proteins needed at a particular time. Next we look at how chemical reactions are regulated by controlling the synthesis of the enzymes.

Many genes, perhaps 60–80%, are not regulated but are instead *constitutive,* meaning that their products are constantly produced at a fixed rate. Usually these genes, which are effectively turned on all the time, code for enzymes that the cell needs in fairly large amounts for its major life processes; the enzymes of glycolysis are examples. The production of other enzymes is regulated so that they are present only when needed. *Trypanosoma,* the protozoan parasite that causes African sleeping sickness, has hundreds of genes coding for surface glycoproteins. Each protozoan cell turns on only one glycoprotein gene at a time. As the host's immune system kills parasites with one type of surface molecule, parasites expressing a different surface glycoprotein can continue to grow.

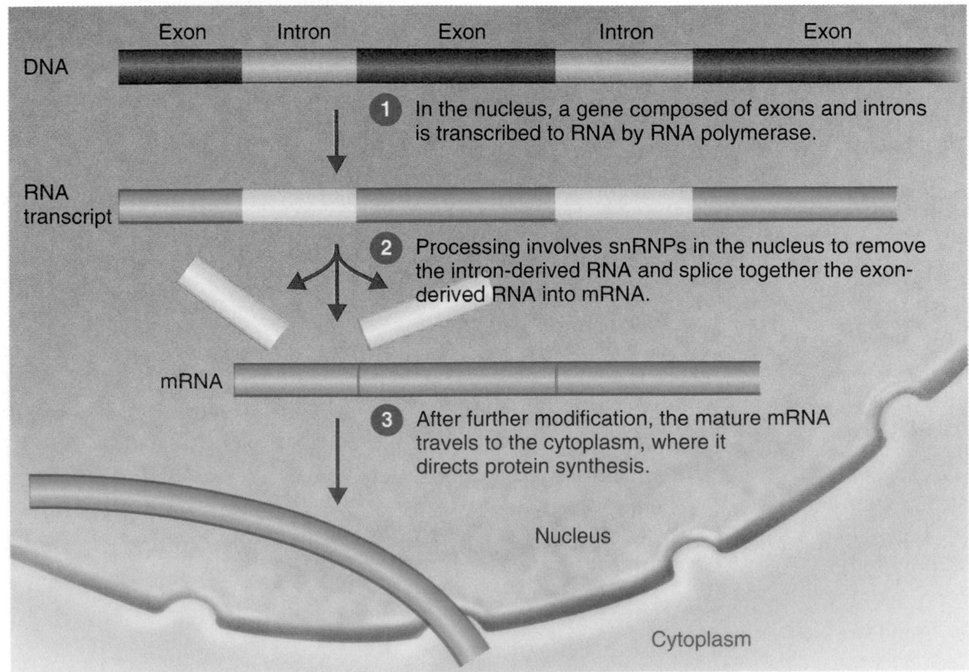

Exon
Intron
Exon
Intron
Exon
DNA
1 In the nucleus, a gene composed of exons and introns is transcribed to RNA by RNA polymerase.
RNA transcript
2 Processing involves snRNPs in the nucleus to remove the intron-derived RNA and splice together the exon-derived RNA into mRNA.
mRNA
3 After further modification, the mature mRNA travels to the cytoplasm, where it directs protein synthesis.
Nucleus
Cytoplasm

Figure 8.11 RNA processing in eukaryotic cells.

Q Why can't the RNA transcript be used for translation?

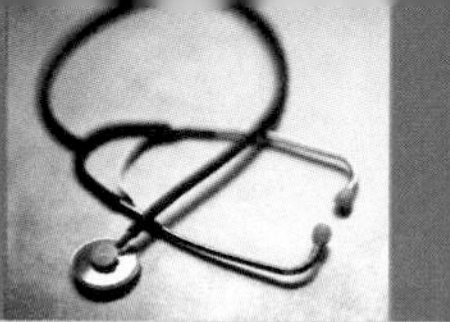

CLINICAL FOCUS

Tracking West Nile Virus

On August 23, 1999, an infectious disease physician from a hospital in northern Queens contacted the New York City Department of Health (NYCDOH) to report two patients with encephalitis. On investigation, NYCDOH initially identified a cluster of six patients with encephalitis, five of whom had profound muscle weakness and required respiratory support. No bacteria were cultured from the patients' blood or cerebrospinal fluid. Viruses transmitted by mosquitoes are a likely cause of aseptic encephalitis during the summer months. These viruses are called arboviruses. Arboviruses, *ar*thropod-*bo*rne, are viruses that are maintained in nature through biological transmission between susceptible vertebrate hosts by blood-feeding arthropods such as mosquitoes.

Nucleic acid sequencing of these isolates was performed at the CDC on September 23. Comparison of the nucleic acid sequences to databases indicated that the viruses were closely related to West Nile virus (WNV, see the photo), which had never been isolated in the Western Hemisphere.

By 2007, WNV had been found in birds in all states except Alaska and Hawaii. The recognition of WNV in the Western Hemisphere in the summer of 1999 marked the first introduction in recent history of an Old World flavivirus into the New World. The United States is not alone, however, in reporting new or heightened activity in humans and other animals. As of 2007, WNV caused human encephalitis in Mexico and in Canada, and incursions of flaviviruses into new areas are likely to continue through increasing global commerce and travel.

West Nile virus was first isolated in 1937 in the West Nile district of Uganda. In the early 1950s, scientists recognized WNV encephalitis outbreaks in humans in Egypt and Israel. Initially considered a minor arbovirus, WNV has emerged as a major public health and veterinary concern in southern Europe, the Mediterranean basin, and North America.

Currently, researchers are looking at the virus's genome for clues about its path around the world. The flavivirus genome consists of a positive, single-stranded RNA 11,000 to 12,000 nucleotides long. (Positive RNA can act as mRNA and be translated.) The virus has acquired several mutations, and researchers are looking for clues in these mutations to determine the virus's journey.

1. Using the portions of the genomes (shown below) that encode viral proteins, can you determine how similar are these viruses? Can you figure out its movement around the world?
   Determine the amino acids encoded, and divide the viruses based on percentage of similarity to the Uganda strain.
2. Based on amino acids, there are two groups called clades.
   Which group is the older?
3. The North American and Australian strains have accumulated more mutations, so these should be more recent.
   Calculate the percentage of difference between nucleotides to see how the viruses are related within their clade.
4. Although genetically related groups or clades can be seen, the actual journey of the virus remains elusive.

Source: Adapted from CDC data.

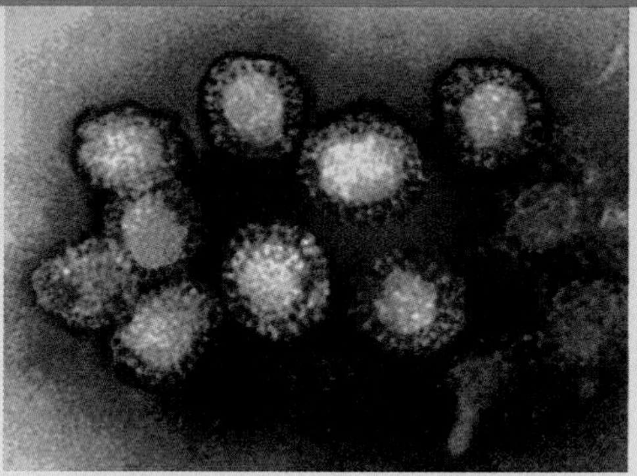

West Nile virus

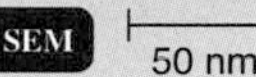

Australia	A	C	C	C	C	G	T	C	C	A	C	C	C	T	T	T	C	A	A	T	T
Egypt	A	A	T	C	G	A	T	C	A	T	C	T	T	C	G	T	C	G	A	T	C
France	A	A	T	C	G	A	T	C	A	T	C	G	T	C	G	T	C	G	A	T	C
Israel	A	T	C	C	A	T	T	C	A	T	C	C	T	C	A	T	C	G	A	T	T
Italy	A	T	C	C	A	C	T	C	A	T	C	C	T	C	G	T	C	G	A	T	T
Kenya	A	T	C	C	A	C	T	C	A	T	C	C	T	C	G	T	C	G	A	T	T
Mexico	A	A	C	C	C	T	T	C	C	T	C	C	C	C	T	T	C	G	A	T	T
United States	A	A	C	C	C	C	T	C	C	T	C	C	C	C	T	T	C	G	A	T	T
Uganda	A	T	A	C	G	A	T	C	A	T	G	C	T	C	G	T	C	C	A	T	C

Repression and Induction

Two genetic control mechanisms known as repression and induction regulate the transcription of mRNA and consequently the synthesis of enzymes from them. These mechanisms control the formation and amounts of enzymes in the cell, not the activities of the enzymes.

Repression

The regulatory mechanism that inhibits gene expression and decreases the synthesis of enzymes is called **repression.** Repression is usually a response to the overabundance of an end-product of a metabolic pathway; it causes a decrease in the rate of synthesis of the enzymes leading to the formation of that product. Repression is mediated by regulatory proteins called **repressors,** which block the ability of RNA polymerase to initiate transcription from the repressed genes. The default position of a repressible gene is *on.*

Induction

The process that turns on the transcription of a gene or genes is **induction.** A substance that acts to induce transcription of a gene is called an **inducer,** and enzymes that are synthesized in the presence of inducers are *inducible enzymes.* The genes required for lactose metabolism in *E. coli* are a well-known example of an inducible system. One of these genes codes for the enzyme β-galactosidase, which splits the substrate lactose into two simple sugars, glucose and galactose. (β refers to the type of linkage that joins the glucose and galactose.) If *E. coli* is placed into a medium in which no lactose is present, the organisms contain almost no β-galactosidase; however, when lactose is added to the medium, the bacterial cells produce a large quantity of the enzyme. Lactose is converted in the cell to the related compound allolactose, which is the inducer for these genes; the presence of lactose thus indirectly induces the cells to synthesize more enzyme. The default position of an inducible gene is *off.* **Animations** Operons: Induction, Repression. **www.microbiologyplace.com**

The Operon Model of Gene Expression

Details of the control of gene expression by induction and repression are described by the operon model. François Jacob and Jacques Monod formulated this general model in 1961 to account for the regulation of protein synthesis. They based their model on studies of the induction of the enzymes of lactose catabolism in *E. coli.* In addition to β-galactosidase, these enzymes include lac permease, which is involved in the transport of lactose into the cell, and transacetylase, which metabolizes certain disaccharides other than lactose.

The genes for the three enzymes involved in lactose uptake and utilization are next to each other on the bacterial chromosome and are regulated together (**Figure 8.12**). These genes,

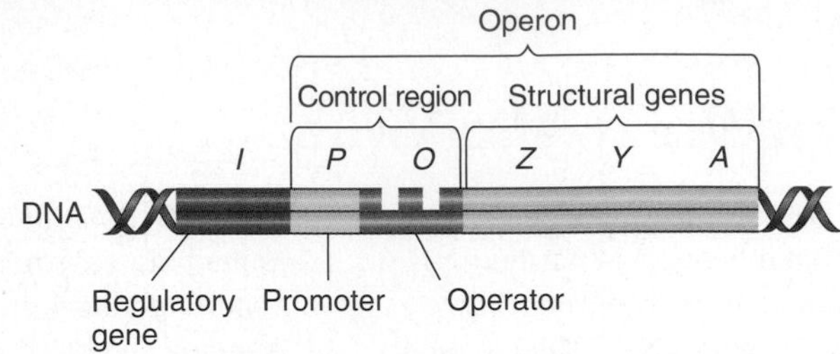

1 **Structure of the operon.** The operon consists of the promoter (*P*) and operator (*O*) sites and structural genes that code for the protein. The operon is regulated by the product of the regulatory gene (*I*).

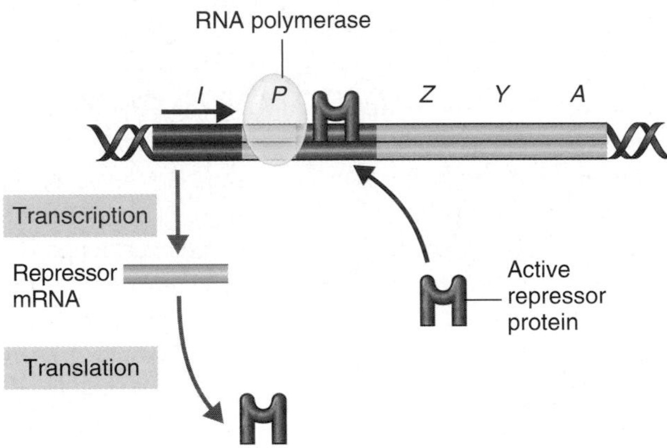

2 **Repressor active, operon off.** The repressor protein binds with the operator, preventing transcription from the operon.

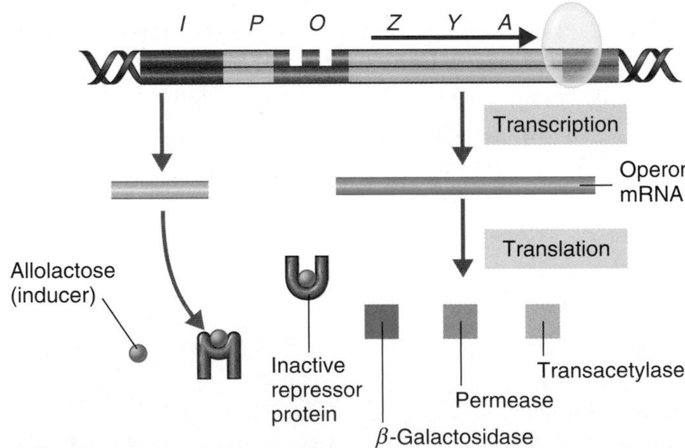

3 **Repressor inactive, operon on.** When the inducer allolactose binds to the repressor protein, the inactivated repressor can no longer block transcription. The structural genes are transcribed, ultimately resulting in the production of the enzymes needed for lactose catabolism.

Figure 8.12 An inducible operon. Lactose-digesting enzymes are produced in the presence of lactose. In *E. coli*, the genes for the three enzymes are in the *lac* operon. β-galactosidase is encoded by *lacZ*. The *lacY* gene encodes the lac permease, and *lacA* encodes transacetylase, whose function in lactose metabolism is still unclear.

Q **What causes transcription of an inducible enzyme?**

which determine the structures of proteins, are called *structural genes* to distinguish them from an adjoining control region on the DNA. When lactose is introduced into the culture medium, the *lac* structural genes are all transcribed and translated rapidly and simultaneously. We will now see how this regulation occurs.

In the control region of the *lac* operon are two relatively short segments of DNA. One, the *promoter,* is the region of DNA where RNA polymerase initiates transcription. The other is the **operator,** which is like a traffic light that acts as a go or stop signal for transcription of the structural genes. A set of operator and promoter sites and the structural genes they control define an **operon;** thus, the combination of the three *lac* structural genes and the adjoining control regions is called the *lac* operon.

A regulatory gene called the *I gene* encodes a **repressor** protein that switches inducible and repressible operons on or off. The *lac* operon is an **inducible operon** (see Figure 8.12). In the absence of lactose, the repressor binds to the operator site, thus preventing transcription. If lactose is present, the repressor binds to a metabolite of lactose instead of to the operator, and lactose-digesting enzymes are transcribed.

In **repressible operons,** the structural genes are transcribed until they are turned off, or *repressed* (**Figure 8.13**). The genes for the enzymes involved in the synthesis of tryptophan are regulated in this manner. The structural genes are transcribed and translated, leading to tryptophan synthesis. When excess tryptophan is present, the tryptophan acts as a **corepressor** binding to the repressor protein. The repressor protein can now bind to the operator, stopping further tryptophan synthesis.

Animation Operons: Overview. **www.microbiologyplace.com**

CHECK YOUR UNDERSTANDING

✓ What is an operon? **8-6**

Positive Regulation

Regulation of the lactose operon also depends on the level of glucose in the medium, which in turn controls the intracellular level of the small molecule **cyclic AMP (cAMP),** a substance derived from ATP that serves as a cellular alarm signal. Enzymes that metabolize glucose are constitutive, and cells grow at their maximal rate with glucose as their carbon source because they can use it most efficiently (**Figure 8.14**). When glucose is no longer available, cAMP accumulates in the cell. The cAMP binds to the allosteric site of *catabolic activator protein (CAP).* CAP then binds to the *lac* promoter, which initiates transcription by making it easier for RNA polymerase to bind to the promoter. Thus transcription of the *lac* operon requires both the presence of lactose and the absence of glucose (**Figure 8.15**).

Cyclic AMP is an example of an *alarmone,* a chemical alarm signal that promotes a cell's response to environmental or nutritional stress. (In this case, the stress is the lack of glucose.) The

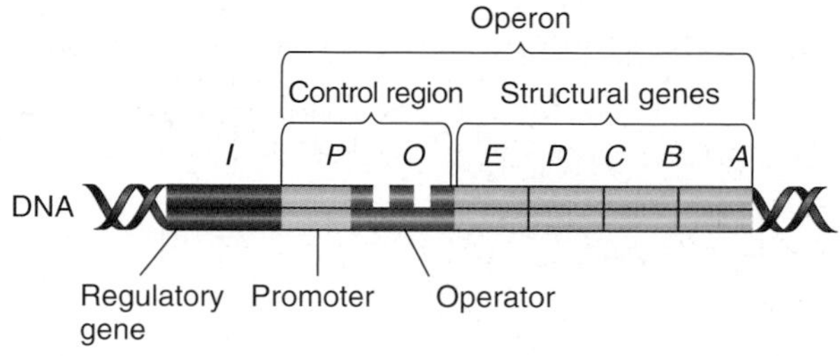

1 **Structure of the operon.** The operon consists of the promoter (*P*) and operator (*O*) sites and structural genes that code for the protein. The operon is regulated by the product of the regulatory gene (*I*).

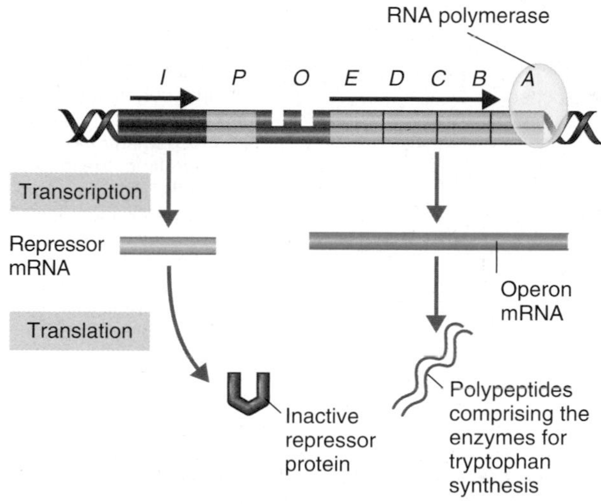

2 **Repressor inactive, operon on.** The repressor is inactive, and transcription and translation proceed, leading to the synthesis of tryptophan.

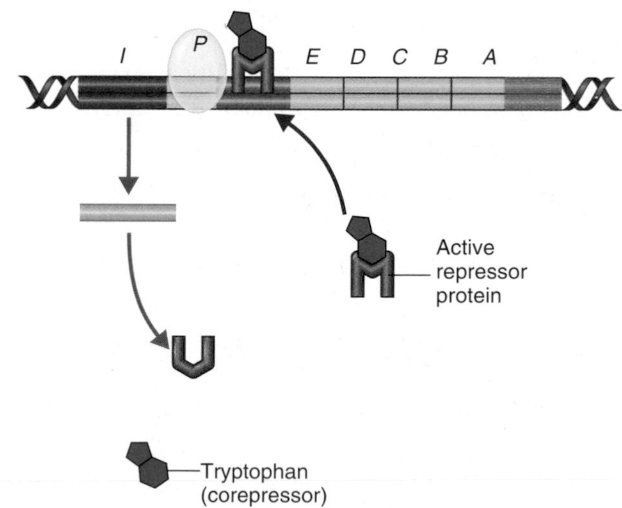

3 **Repressor active, operon off.** When the corepressor tryptophan binds to the repressor protein, the activated repressor binds with the operator, preventing transcription from the operon.

Figure 8.13 A repressible operon. Tryptophan, an amino acid, is produced by anabolic enzymes encoded by five structural genes. Accumulation of tryptophan represses transcription of these genes, preventing further synthesis of tryptophan. The *E. coli trp* operon is shown here.

Q What causes transcription of a repressible enzyme?

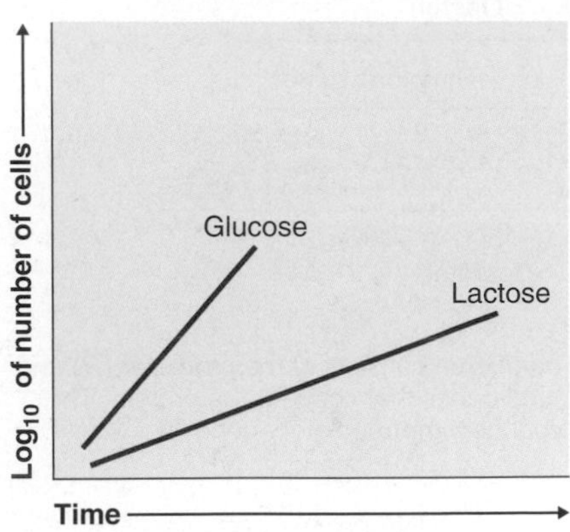

(a) Bacteria growing on glucose as the sole carbon source grow faster than on lactose.

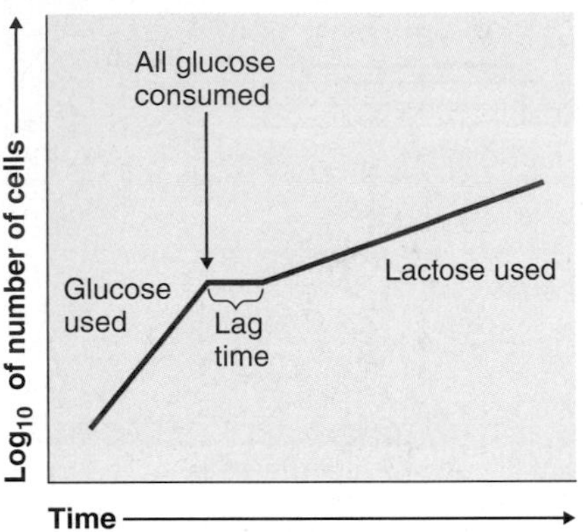

(b) Bacteria growing in a medium containing glucose and lactose first consume the glucose and then, after a short lag time, the lactose. During the lag time, intracellular cAMP increases, the *lac* operon is transcribed, more lactose is transported into the cell, and β-galactosidase is synthesized to break down lactose.

Figure 8.14 The growth rate of *E. coli* on glucose and lactose.

Q **When both glucose and lactose are present, why will cells use glucose first?**

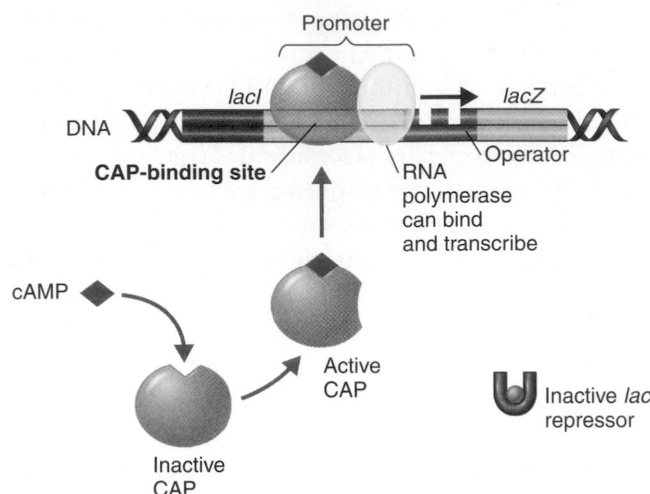

(a) Lactose present, glucose scarce (cAMP level high). If glucose is scarce, the high level of cAMP activates CAP, and the *lac* operon produces large amounts of mRNA for lactose digestion.

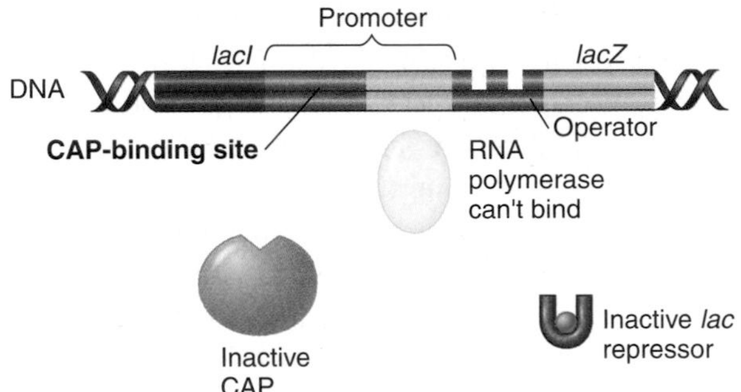

(b) Lactose present, glucose present (cAMP level low). When glucose is present, cAMP is scarce, and CAP is unable to stimulate transcription.

Figure 8.15 Positive regulation of the *lac* operon.

Will transcription of the *lac* operon occur in the presence of lactose and glucose? In the presence of lactose and the absence of glucose? In the presence of glucose and the absence of lactose?

same mechanism involving cAMP allows the cell to grow on other sugars. Inhibition of the metabolism of alternative carbon sources by glucose is termed **catabolite repression** (or the *glucose effect*). When glucose is available, the level of cAMP in the cell is low, and consequently CAP is not bound.

CHECK YOUR UNDERSTANDING

✓ What is the role of cAMP in catabolite repression? **8-7**

Mutation: Change in the Genetic Material

LEARNING OBJECTIVES

8-8 Classify mutations by type.

8-9 Define *mutagen.*

8-10 Describe two ways mutations can be repaired.

8-11 Describe the effect of mutagens on the mutation rate.

8-12 Outline the methods of direct and indirect selection of mutants.

8-13 Identify the purpose of and outline the procedure for the Ames test.

A **mutation** is a change in the base sequence of DNA. Such a change in the base sequence of a gene will sometimes cause a change in the product encoded by that gene. For example, when the gene for an enzyme mutates, the enzyme encoded by the gene may become inactive or less active because its amino acid sequence has changed. Such a change in genotype may be disadvantageous, or even lethal, if the cell loses a phenotypic trait it needs. However, a mutation can be beneficial if, for instance, the altered enzyme encoded by the mutant gene has a new or enhanced activity that benefits the cell.

Many simple mutations are silent (neutral); the change in DNA base sequence causes no change in the activity of the product encoded by the gene. Silent mutations commonly occur when one nucleotide is substituted for another in the DNA, especially at a location corresponding to the third position of the mRNA codon. Because of the degeneracy of the genetic code, the resulting new codon might still code for the

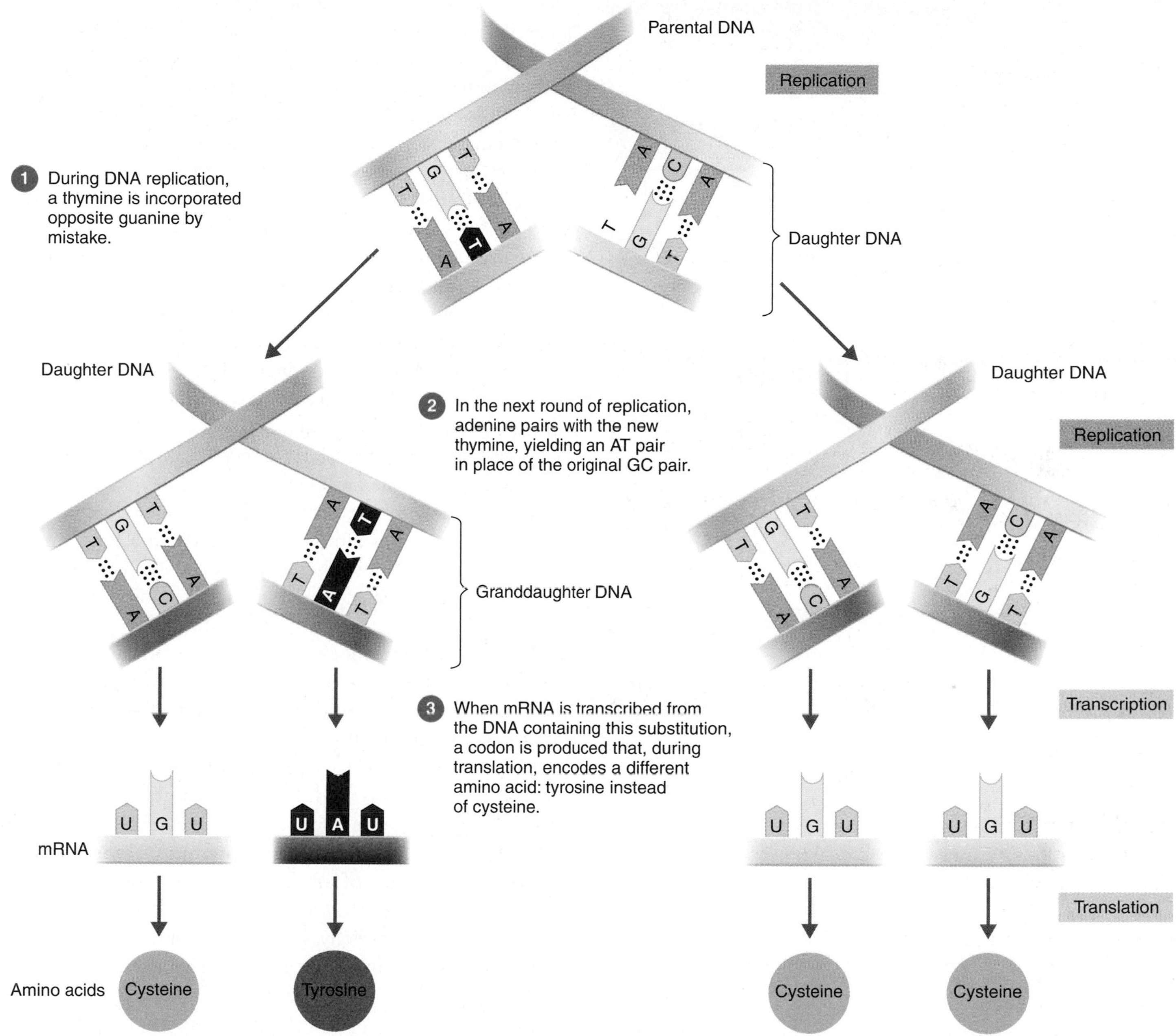

Figure 8.16 Base substitution. This mutation leads to an altered protein in a granddaughter cell.

Q Does a base substitution always result in a different amino acid?

same amino acid. Even if the amino acid is changed, the function of the protein may not change if the amino acid is in a nonvital portion of the protein, or is chemically very similar to the original amino acid.

Types of Mutations

The most common type of mutation involving single base pairs is **base substitution** (or *point mutation*), in which a single base at one point in the DNA sequence is replaced with a different base. When the DNA replicates, the result is a substituted base pair (Figure 8.16). For example, AT might be substituted for GC, or CG for GC. If a base substitution occurs within a gene that codes for a protein, the mRNA transcribed from the gene will carry an incorrect base at that position. When the mRNA is translated into protein, the incorrect base may cause the insertion of an incorrect amino acid in the protein. If the base substitution results in an amino acid substitution in the synthesized protein, this change in the DNA is known as a **missense mutation** (Figure 8.17a and Figure 8.17b).

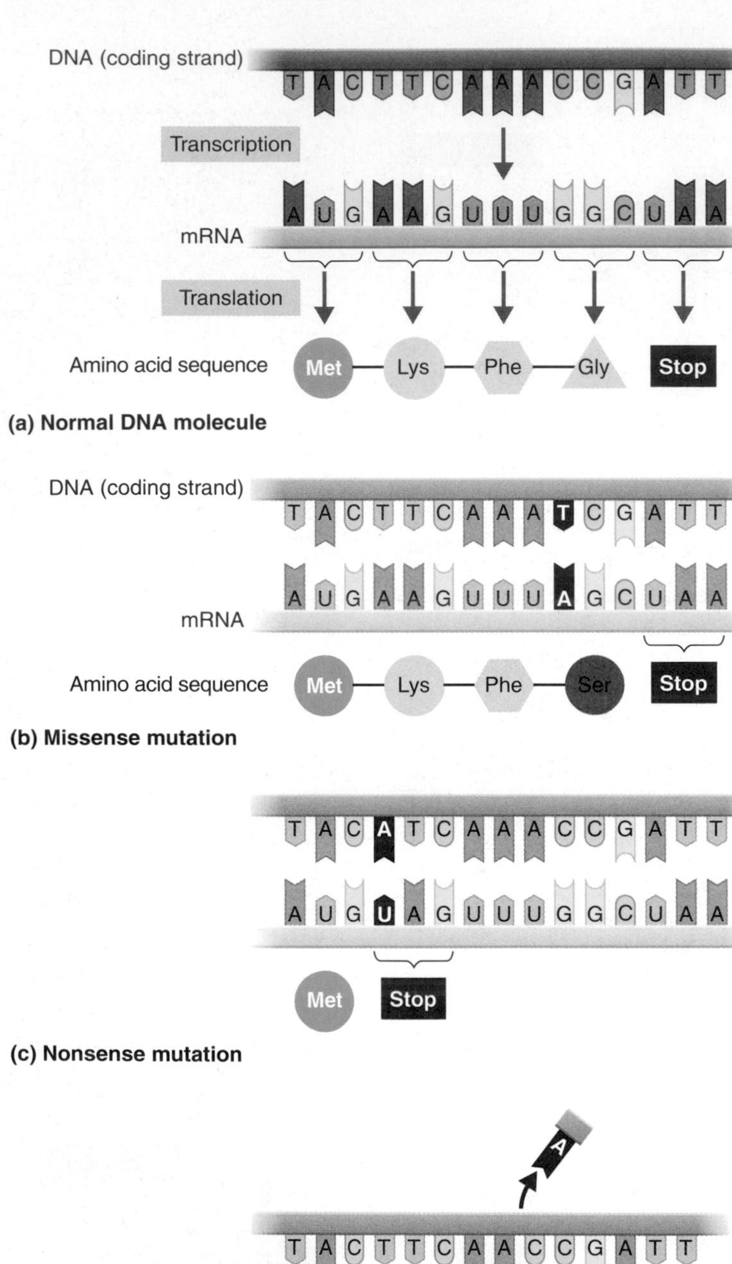

Figure 8.17 Types of mutations and their effects on the amino acid sequences of proteins.

Q On what basis are missense, nonsense, and frameshift mutations distinguished?

The effects of such mutations can be dramatic. For example, sickle cell disease is caused by a single change in the gene for globin, the protein component of hemoglobin. Hemoglobin is primarily responsible for transporting oxygen from the lungs to the tissues. A single missense mutation, a change from an A to a T at a specific site, results in the change from glutamic acid to valine in the protein. The effect of this change is that the shape of the hemoglobin molecule changes under conditions of low oxygen, altering the shape of the red blood cells such that movement of the cells through small capillaries is greatly impeded.

By creating a nonsense (stop) codon in the middle of an mRNA molecule, some base substitutions effectively prevent the synthesis of a complete functional protein; only a fragment is synthesized. A base substitution resulting in a nonsense codon is thus called a **nonsense mutation** (Figure 8.17c).

Besides base-pair mutations, there are also changes in DNA called **frameshift mutations,** in which one or a few nucleotide pairs are deleted or inserted in the DNA (Figure 8.17d). This mutation can shift the "translational reading frame"—that is, the three-by-three grouping of nucleotides recognized as codons by the tRNAs during translation. For example, deleting one nucleotide pair in the middle of a gene causes changes in many amino acids downstream from the site of the original mutation. Frameshift mutations almost always result in a long stretch of altered amino acids and the production of an inactive protein from the mutated gene. In most cases, a nonsense codon will eventually be encountered and thereby terminate translation.

Occasionally, mutations occur where significant numbers of bases are added to (inserted into) a gene. Huntington's disease, for example, is a progressive neurological disorder caused by extra bases inserted into a particular gene. The reason these insertions occur in this particular gene is still being studied.

Base substitutions and frameshift mutations may occur spontaneously because of occasional mistakes made during DNA replication. These **spontaneous mutations** apparently occur in the absence of any mutation-causing agents. Agents in the environment, such as certain chemicals and radiation, that directly or indirectly bring about mutations are called **mutagens.** Almost any agent that can chemically or physically react with DNA can potentially cause mutations. A wide variety of chemicals, many of which are common in nature or in households, are known to be mutagens. Many forms of radiation, including X rays and ultraviolet light, are also mutagenic, as discussed shortly.

In the microbial world, certain mutations result in resistance to antibiotics (see the box in Chapter 26, page 751) or altered pathogenicity. A mutation in a gene encoding the outer membrane may increase pathogenicity; for example, *Salmonella enterica* with an altered outer membrane can survive in phagocytes. A mutation in a capsule-encoding gene may result in decreased pathogenicity because phagocytes can destroy the bacteria, as in the cases of *Streptococcus pneumoniae, Haemophilus influenzae,* and *Neisseria meningitidis.*

CHECK YOUR UNDERSTANDING

✓ How can a mutation be beneficial? **8-8**

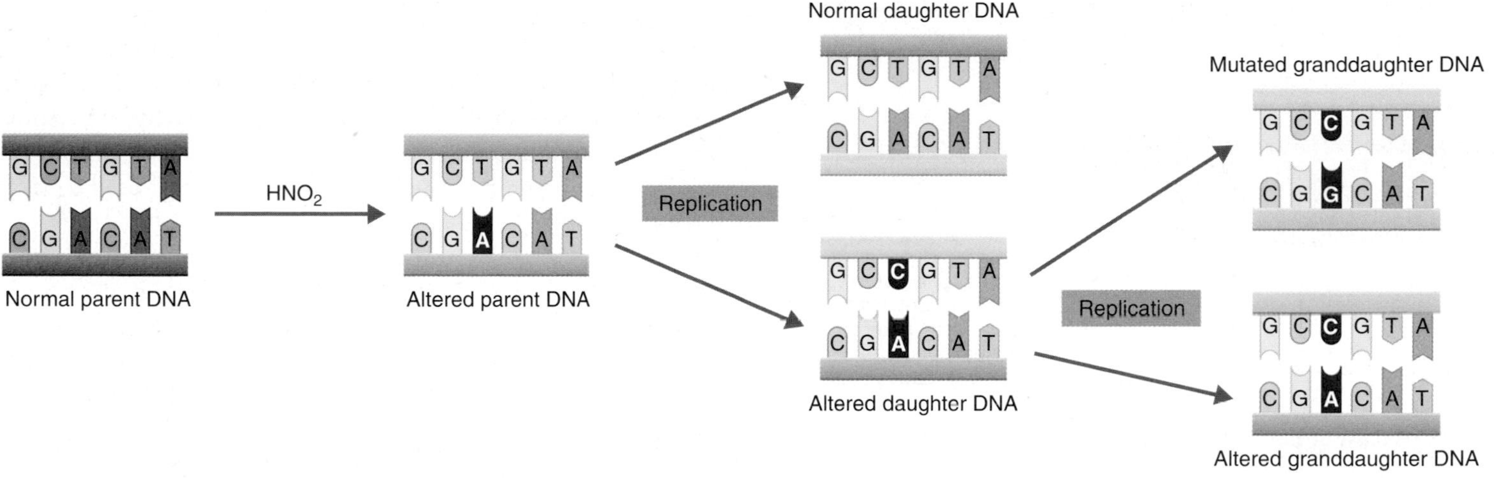

Figure 8.18 Nitrous acid (HNO_2) as a mutagen. The nitrous acid alters an adenine in such a way that it pairs with cytosine instead of thymine.

Q What is a mutagen?

Mutagens

Chemical Mutagens

One of the many chemicals known to be a mutagen is nitrous acid. Figure 8.18 shows how exposure of DNA to nitrous acid can convert the base adenine (A) to a form that no longer pairs with thymine (T) but instead pairs with cytosine (C). When DNA containing such modified adenines replicates, one daughter DNA molecule will have a base-pair sequence different from that of the parent DNA. Eventually, some AT base pairs of the parent will have been changed to GC base pairs in a granddaughter cell. Nitrous acid makes a specific base-pair change in DNA. Like all mutagens, it alters DNA at random locations.

Another type of chemical mutagen is the **nucleoside analog.** These molecules are structurally similar to normal nitrogenous bases, but they have slightly altered base-pairing properties. Examples, 2-aminopurine and 5-bromouracil, are shown in Figure 8.19. When nucleoside analogs are given to growing cells, the analogs are randomly incorporated into cellular DNA in

Figure 8.19 Nucleoside analogs and the nitrogenous bases they replace. A nucleoside is phosphorylated and the resulting nucleotide used to synthesize DNA.

Q Why do these drugs kill cells?

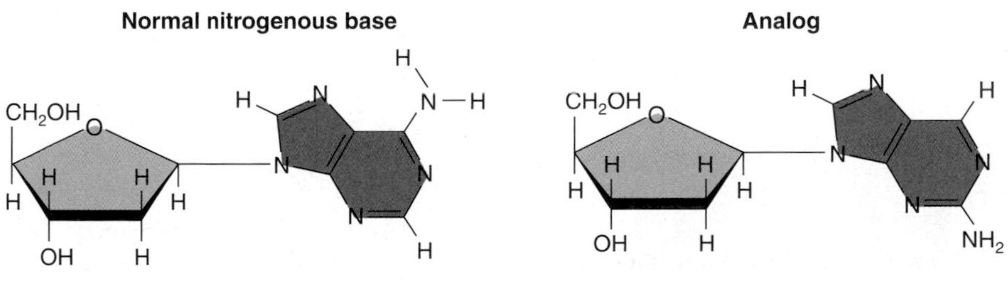

(a) The 2-aminopurine is incorporated into DNA in place of adenine but can pair with cytosine, so an AT pair becomes a CG pair.

Thymine nucleoside

5-Bromouracil nucleoside

(b) The 5-bromouracil is used as an anticancer drug because it is mistaken for thymine by cellular enzymes but pairs with cytosine. In the next DNA replication, an AT pair becomes a GC pair.

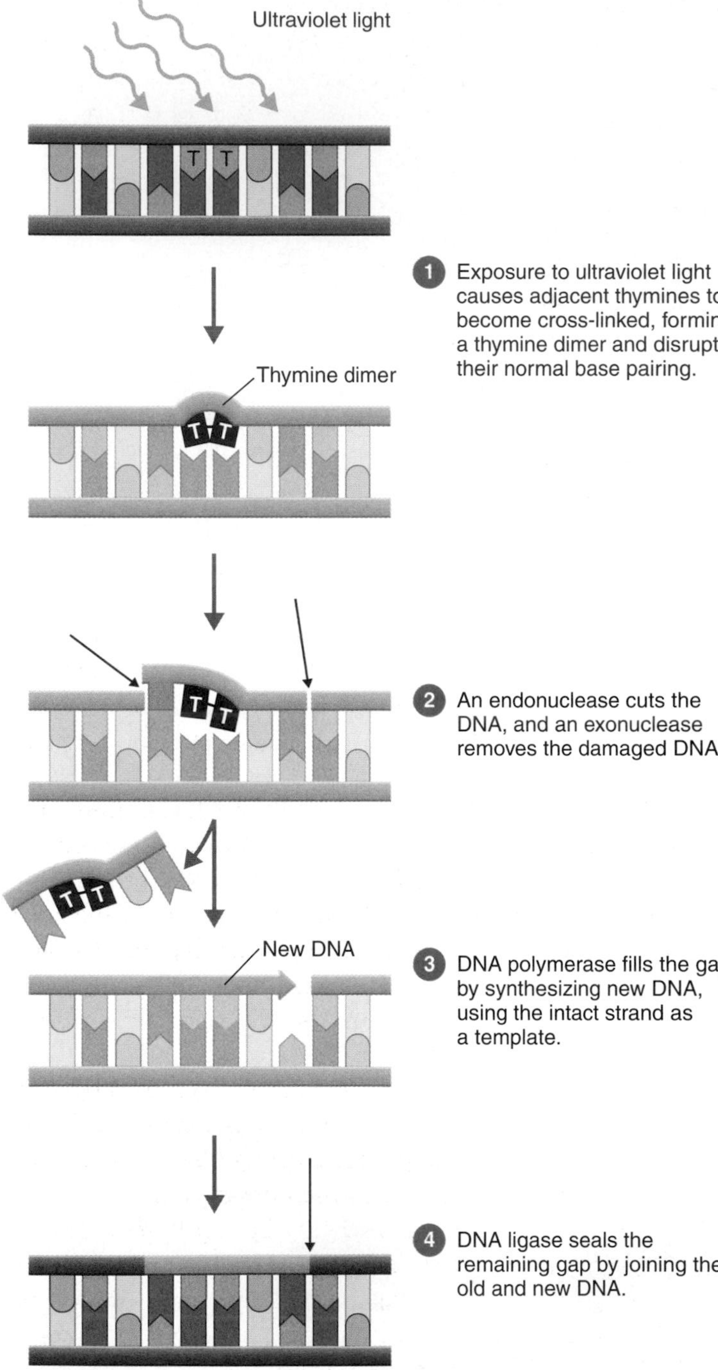
Ultraviolet light
T T
1 Exposure to ultraviolet light causes adjacent thymines to become cross-linked, forming a thymine dimer and disrupting their normal base pairing.
Thymine dimer
2 An endonuclease cuts the DNA, and an exonuclease removes the damaged DNA.
New DNA
3 DNA polymerase fills the gap by synthesizing new DNA, using the intact strand as a template.
4 DNA ligase seals the remaining gap by joining the old and new DNA.

Figure 8.20 The creation and repair of a thymine dimer caused by ultraviolet light. After exposure to UV light, adjacent thymines can become cross-linked, forming a thymine dimer. In the absence of visible light, the nucleotide excision repair mechanism is used in a cell to repair the damage.

Q How do excision repair enzymes "know" which strand is incorrect?

place of the normal bases. Then, during DNA replication, the analogs cause mistakes in base pairing. The incorrectly paired bases will be copied during subsequent replication of the DNA, resulting in base-pair substitutions in the progeny cells. Some antiviral and antitumor drugs are nucleoside analogs, including AZT (azidothymidine), one of the primary drugs used to treat HIV infection.

Still other chemical mutagens cause small deletions or insertions, which can result in frameshifts. For instance, under certain conditions, benzopyrene, which is present in smoke and soot, is an effective *frameshift mutagen.* Aflatoxin—produced by *Aspergillus flavus* (a-spėr-jil′lus flā′vus), a mold that grows on peanuts and grain—is a frameshift mutagen, as are the acridine dyes used experimentally against herpesvirus infections. Frameshift mutagens usually have the right size and chemical properties to slip between the stacked base pairs of the DNA double helix. They may work by slightly offsetting the two strands of DNA, leaving a gap or bulge in one strand or the other. When the staggered DNA strands are copied during DNA synthesis, one or more base pairs can be inserted or deleted in the new double-stranded DNA. Interestingly, frameshift mutagens are often potent carcinogens.

Radiation

X rays and gamma rays are forms of radiation that are potent mutagens because of their ability to ionize atoms and molecules. The penetrating rays of ionizing radiation cause electrons to pop out of their usual shells (see Chapter 2). These electrons bombard other molecules and cause more damage, and many of the resulting ions and free radicals (molecular fragments with unpaired electrons) are very reactive. Some of these ions can combine with bases in DNA, resulting in errors in DNA replication and repair that produce mutations. An even more serious outcome is the breakage of covalent bonds in the sugar-phosphate backbone of DNA, which causes physical breaks in chromosomes.

Another form of mutagenic radiation is ultraviolet (UV) light, a nonionizing component of ordinary sunlight. However, the most mutagenic component of UV light (wavelength 260 nm) is screened out by the ozone layer of the atmosphere. The most important effect of direct UV light on DNA is the formation of harmful covalent bonds between certain bases. Adjacent thymines in a DNA strand can cross-link to form thymine dimers. Such dimers, unless repaired, may cause serious damage or death to the cell because it cannot properly transcribe or replicate such DNA.

Bacteria and other organisms have enzymes that can repair UV-induced damage. **Photolyases,** also known as *light-repair enzymes,* use visible light energy to separate the dimer back to the original two thymines. **Nucleotide excision repair,** shown in Figure 8.20, is not restricted to UV-induced damage; it can repair mutations from other causes as well. Enzymes cut out the incorrect base and fill in the gap with newly synthesized DNA that is complementary to the correct strand. For many years biologists questioned how the incorrect base could be distinguished from the correct base if it was not physically distorted like a thymine dimer. In 1970, Hamilton Smith provided the answer with the discovery of **methylases.** These enzymes add a methyl group to selected bases soon after a DNA strand is made. A repair endonuclease then cuts the nonmethylated strand.

Exposure to UV light in humans, such as by excessive suntanning, causes a large number of thymine dimers in skin cells. Unrepaired dimers may result in skin cancers. Humans with xeroderma pigmentosum, an inherited condition that results in increased sensitivity to UV light, have a defect in nucleotide excision repair; consequently, they have an increased risk of skin cancer.

The Frequency of Mutation

The **mutation rate** is the probability that a gene will mutate when a cell divides. The rate is usually stated as a power of 10, and because mutations are very rare, the exponent is always a negative number. For example, if there is one chance in 10,000 that a gene will mutate when the cell divides, the mutation rate is 1/10,000, which is expressed as 10^{-4}. Spontaneous mistakes in DNA replication occur at a very low rate, perhaps only once in 10^9 replicated base pairs (a mutation rate of 10^{-9}). Because the average gene has about 10^3 base pairs, the spontaneous rate of mutation is about one in 10^6 (a million) replicated genes.

Mutations usually occur more or less randomly along a chromosome. The occurrence of random mutations at low frequency is an essential aspect of the adaptation of species to their environment, for evolution requires that genetic diversity be generated randomly and at a low rate. For example, in a bacterial population of significant size—say, greater than 10^7 cells—a few new mutant cells will always be produced in every generation. Most mutations either are harmful and likely to be removed from the gene pool when the individual cell dies or are neutral. However, a few mutations may be beneficial. For example, a mutation that confers antibiotic resistance is beneficial to a population of bacteria that is regularly exposed to antibiotics. Once such a trait has appeared through mutation, cells carrying the mutated gene are more likely than other cells to survive and reproduce as long as the environment stays the same. Soon most of the cells in the population will have the gene; an evolutionary change will have occurred, although on a small scale.

A mutagen usually increases the spontaneous rate of mutation, which is about one in 10^6 replicated genes, by a factor of 10 to 1000 times. In other words, in the presence of a mutagen, the normal rate of 10^{-6} mutations per replicated gene becomes a rate of 10^{-5} to 10^{-3} per replicated gene. Mutagens are used experimentally to enhance the production of mutant cells for research on the genetic properties of microorganisms and for commercial purposes. **Animations** Mutations: Types, Repair; Mutagens. **www.microbiologyplace.com**

CHECK YOUR UNDERSTANDING

- ✓ How are mutations caused by chemicals? By radiation? **8-9**
- ✓ How can mutations be repaired? **8-10**
- ✓ How do mutagens affect the mutation rate? **8-11**

Identifying Mutants

Mutants can be detected by selecting or testing for an altered phenotype. Whether or not a mutagen is used, mutant cells with specific mutations are always rare compared with other cells in the population. The problem is detecting such a rare event.

Experiments are usually performed with bacteria because they reproduce rapidly, so large numbers of organisms (more than 10^9 per milliliter of nutrient broth) can easily be used. Furthermore, because bacteria generally have only one copy of each gene per cell, the effects of a mutated gene are not masked by the presence of a normal version of the gene, as in many eukaryotic organisms.

Positive (direct) selection involves the detection of mutant cells by rejection of the unmutated parent cells. For example, suppose we were trying to find mutant bacteria that are resistant to penicillin. When the bacterial cells are plated on a medium containing penicillin, the mutant can be identified directly. The few cells in the population that are resistant (mutants) will grow and form colonies, whereas the normal, penicillin-sensitive parental cells cannot grow.

To identify mutations in other kinds of genes, **negative (indirect) selection** can be used. This process selects a cell that cannot perform a certain function, using the technique of **replica plating.** For example, suppose we wanted to use replica plating to identify a bacterial cell that has lost the ability to synthesize the amino acid histidine (**Figure 8.21**). First, about 100 bacterial cells are inoculated onto an agar plate. This plate, called the master plate, contains a medium with histidine on which all cells will grow. After 18 to 24 hours of incubation, each cell reproduces to form a colony. Then a pad of sterile material, such as latex, filter paper, or velvet, is pressed over the master plate, and some of the cells from each colony adhere to the velvet. Next, the velvet is pressed down onto two (or more) sterile plates. One plate contains a medium without histidine, and one contains a medium with histidine on which the original, nonmutant bacteria can grow. Any colony that grows on the medium with histidine on the master plate but that cannot synthesize its own histidine will not be able to grow on the medium without histidine. The mutant colony can then be identified on

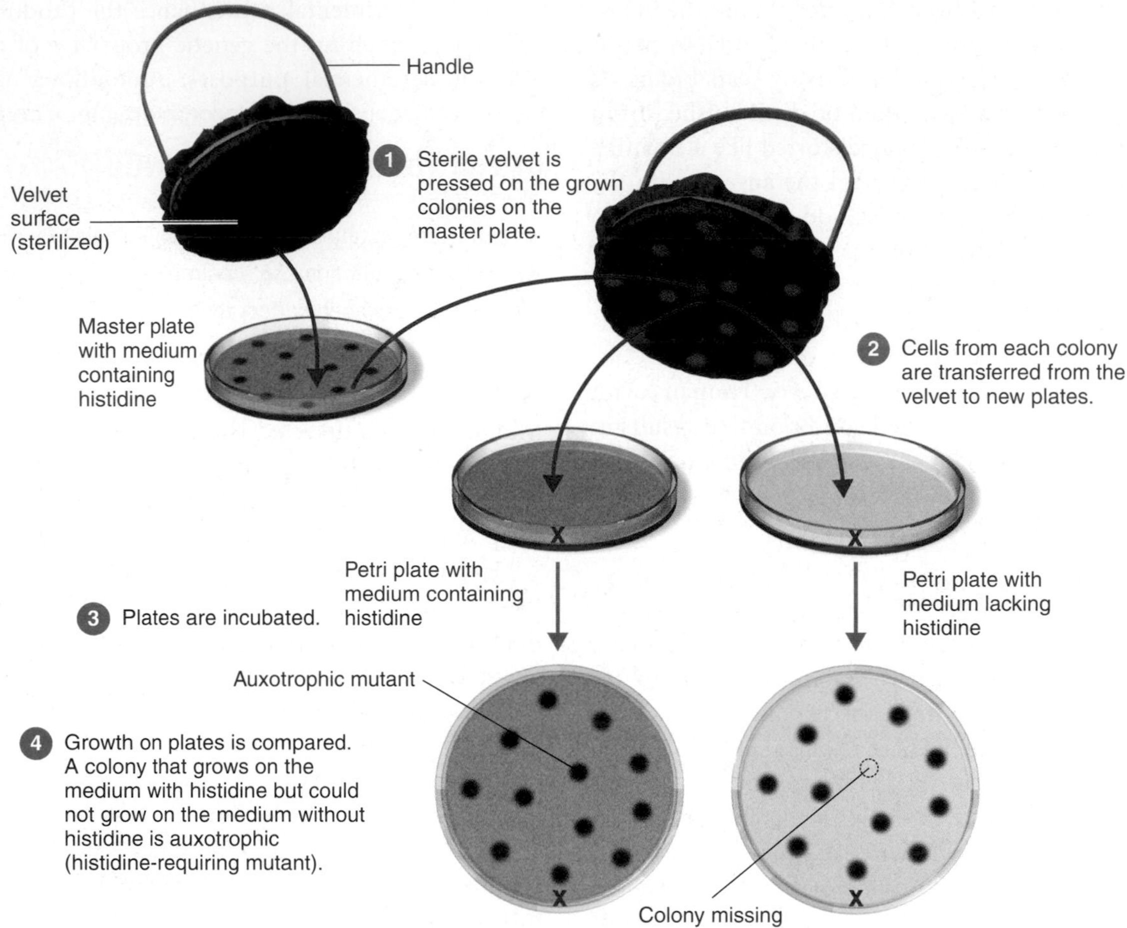
Handle
Velvet surface (sterilized)
1 Sterile velvet is pressed on the grown colonies on the master plate.
Master plate with medium containing histidine
2 Cells from each colony are transferred from the velvet to new plates.
3 Plates are incubated.
Petri plate with medium containing histidine
Petri plate with medium lacking histidine
Auxotrophic mutant
4 Growth on plates is compared. A colony that grows on the medium with histidine but could not grow on the medium without histidine is auxotrophic (histidine-requiring mutant).
Colony missing
X
X

Figure 8.21 Replica plating. In this example, the auxotrophic mutant cannot synthesize histidine. The plates must be carefully marked (with an X here) to maintain orientation so that colony positions are known in relation to the original master plate.

Q What is an auxotroph?

the master plate. Of course, because mutants are so rare (even those induced by mutagens), many plates must be screened with this technique to isolate a specific mutant.

Replica plating is a very effective means of isolating mutants that require one or more new growth factors. Any mutant microorganism having a nutritional requirement that is absent in the parent is known as an **auxotroph.** For example, an auxotroph may lack an enzyme needed to synthesize a particular amino acid and will therefore require that amino acid as a growth factor in its nutrient medium.

Identifying Chemical Carcinogens

Many known mutagens have been found to be **carcinogens,** substances that cause cancer in animals, including humans. In recent years, chemicals in the environment, the workplace, and the diet have been implicated as causes of cancer in humans. The usual subjects of tests to determine potential carcinogens are animals, and the testing procedures are time-consuming and expensive. Now there are faster and less expensive procedures for the preliminary screening of potential carcinogens. One of these, called the **Ames test,** uses bacteria as carcinogen indicators.

The Ames test is based on the observation that exposure of mutant bacteria to mutagenic substances may cause new mutations that reverse the effect (the change in phenotype) of the original mutation. These are called *reversions.* Specifically, the test measures the reversion of histidine auxotrophs of *Salmonella* (his^- cells, mutants that have lost the ability to synthesize histidine) to histidine-synthesizing cells (his^+) after treatment with a mutagen (**Figure 8.22**). Bacteria are incubated in both the presence and absence of the substance being tested. Because animal enzymes must activate many chemicals into forms that are chemically reactive for mutagenic or carcinogenic activity to appear, the chemical to be tested and the mutant bacteria are incubated together with rat liver extract, a rich source of activation enzymes. If the substance being tested is mutagenic, it will cause the reversion of his^- bacteria to his^+ bacteria at a rate higher than the spontaneous reversion rate. The number of

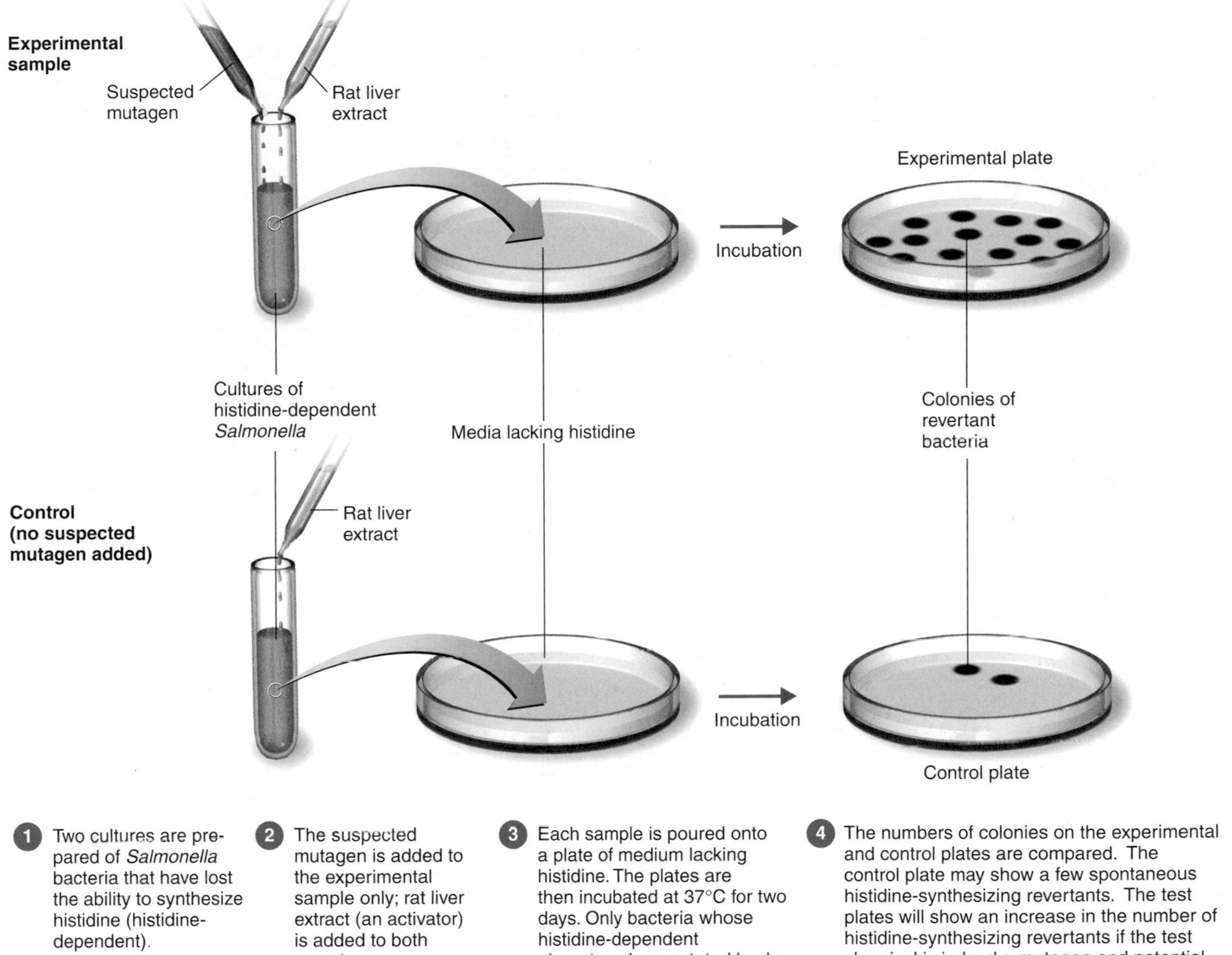

Experimental sample
Suspected mutagen
Rat liver extract
Experimental plate
Incubation
Cultures of histidine-dependent *Salmonella*
Media lacking histidine
Colonies of revertant bacteria
Control (no suspected mutagen added)
Rat liver extract
Incubation
Control plate
1 Two cultures are prepared of *Salmonella* bacteria that have lost the ability to synthesize histidine (histidine-dependent).
2 The suspected mutagen is added to the experimental sample only; rat liver extract (an activator) is added to both samples.
3 Each sample is poured onto a plate of medium lacking histidine. The plates are then incubated at 37°C for two days. Only bacteria whose histidine-dependent phenotype has mutated back (reverted) to histidine-synthesizing will grow into colonies.
4 The numbers of colonies on the experimental and control plates are compared. The control plate may show a few spontaneous histidine-synthesizing revertants. The test plates will show an increase in the number of histidine-synthesizing revertants if the test chemical is indeed a mutagen and potential carcinogen. The higher the concentration of mutagen used, the more revertant colonies will result.

Figure 8.22 The Ames reverse gene mutation test.

Q Do all mutagens cause cancer?

observed revertants indicates the degree to which a substance is mutagenic and therefore possibly carcinogenic.

The test can be used in many ways. Several potential mutagens can be qualitatively tested by spotting the individual chemicals on small paper disks on a single plate inoculated with bacteria. In addition, mixtures such as wine, blood, smoke condensates, and extracts of foods can also be tested to see whether they contain mutagenic substances.

About 90% of the substances found by the Ames test to be mutagenic have also been shown to be carcinogenic in animals. By the same token, the more mutagenic substances have generally been found to be more carcinogenic.

CHECK YOUR UNDERSTANDING

- ✓ How would you isolate an antibiotic-resistant bacterium? An antibiotic-sensitive bacterium? **8-12**
- ✓ What is the principle behind the Ames test? **8-13**

Genetic Transfer and Recombination

LEARNING OBJECTIVES

8-14 Differentiate horizontal and vertical gene transfer.

8-15 Compare the mechanisms of genetic recombination in bacteria.

8-16 Describe the functions of plasmids and transposons.

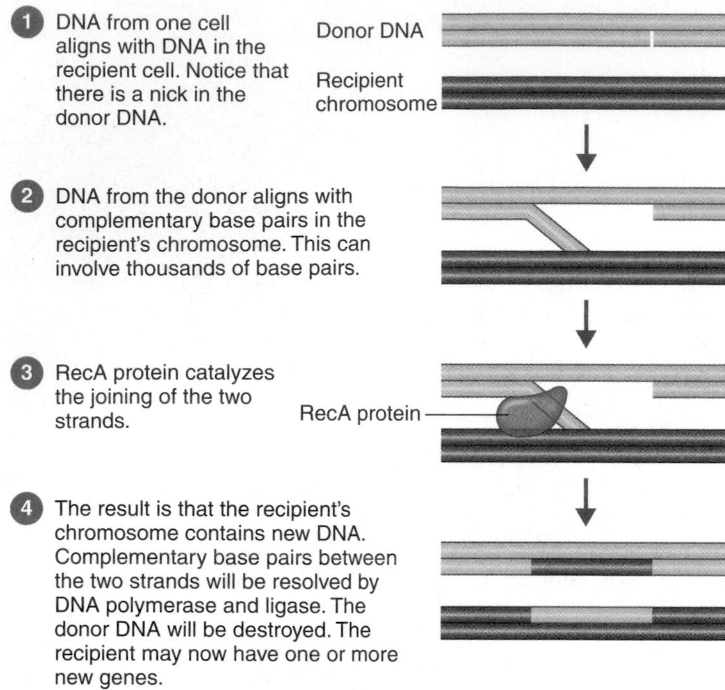

1 DNA from one cell aligns with DNA in the recipient cell. Notice that there is a nick in the donor DNA.
Donor DNA
Recipient chromosome
2 DNA from the donor aligns with complementary base pairs in the recipient's chromosome. This can involve thousands of base pairs.
3 RecA protein catalyzes the joining of the two strands.
RecA protein
4 The result is that the recipient's chromosome contains new DNA. Complementary base pairs between the two strands will be resolved by DNA polymerase and ligase. The donor DNA will be destroyed. The recipient may now have one or more new genes.

Figure 8.23 Genetic recombination by crossing over. Foreign DNA can be inserted into a chromosome by breaking and rejoining the chromosome. This can insert one or more new genes into the chromosome. A photograph of RecA protein is shown in Figure 3.11a.

Q What type of enzyme breaks the DNA? What enzyme rejoins the pieces of DNA?

Genetic recombination refers to the exchange of genes between two DNA molecules to form new combinations of genes on a chromosome. Figure 8.23 shows one mechanism for genetic recombination. If a cell picks up a foreign DNA (called donor DNA in the figure), some of it could insert into the cell's chromosome—a process called **crossing over**—and some of the genes carried by the chromosomes are shuffled. The DNA has recombined, so that the chromosome now carries a portion of the donor's DNA.

If A and B represent DNA from different individuals, how are they brought close enough together to recombine? In eukaryotes, genetic recombination is an ordered process that usually occurs as part of the sexual cycle of the organism. Crossing over generally takes place during the formation of reproductive cells, such that these cells contain recombinant DNA. In bacteria, genetic recombination can happen in a number of ways, which we will discuss in the following sections.

Like mutation, genetic recombination contributes to a population's genetic diversity, which is the source of variation in evolution. In highly evolved organisms such as present-day microbes, recombination is more likely than mutation to be beneficial because recombination will less likely destroy a gene's function and may bring together combinations of genes that enable the organism to carry out a valuable new function.

The major protein that constitutes the flagella of *Salmonella* is also one of the primary proteins that causes our immune systems to respond. However, these bacteria have the capability of producing two different flagellar proteins. As our immune system mounts a response against those cells containing one form of the flagellar protein, those organisms producing the second are not affected. Which flagellar protein is produced is determined by a recombination event that apparently occurs somewhat randomly within the chromosomal DNA. Thus, by altering the flagellar protein produced, *Salmonella* can better avoid the defenses of the host.

Vertical gene transfer occurs when genes are passed from an organism to its offspring. Plants and animals transmit their genes by vertical transmission. Bacteria can pass their genes not only to their offspring, but also laterally, to other microbes of the same generation. This is known as **horizontal gene transfer** (see Figure 8.2). Horizontal gene transfer between bacteria occurs in several ways. In all of the mechanisms, the transfer involves a **donor cell** that gives a portion of its total DNA to a **recipient cell.** Once transferred, part of the donor's DNA is usually incorporated into the recipient's DNA; the remainder is degraded by cellular enzymes. The recipient cell that incorporates donor DNA into its own DNA is called a *recombinant.* The transfer of genetic material between bacteria is by no means a frequent event; it may occur in only 1% or less of an entire population. Let's examine in detail the specific types of genetic transfer.

Transformation in Bacteria

During the process of **transformation,** genes are transferred from one bacterium to another as "naked" DNA in solution. This process was first demonstrated over 70 years ago, although it was not understood at the time. Not only did transformation show that genetic material could be transferred from one bacterial cell to another, but study of this phenomenon eventually led to the conclusion that DNA is the genetic material. The initial experiment on transformation was performed by Frederick Griffith in England in 1928 while he was working with two strains of *Streptococcus pneumoniae.* One, a virulent (pathogenic) strain, has a polysaccharide capsule that prevents phagocytosis. The bacteria grow and cause pneumonia. The other, an avirulent strain, lacks the capsule and does not cause disease.

Griffith was interested in determining whether injections of heat-killed bacteria of the encapsulated strain could be used to vaccinate mice against pneumonia. As he expected, injections of living encapsulated bacteria killed the mouse (Figure 8.24a); injections of live nonencapsulated bacteria (Figure 8.24b) or dead encapsulated bacteria (Figure 8.24c) did not kill the mouse. However, when the dead encapsulated bacteria were mixed with live nonencapsulated bacteria and injected into the mice, many

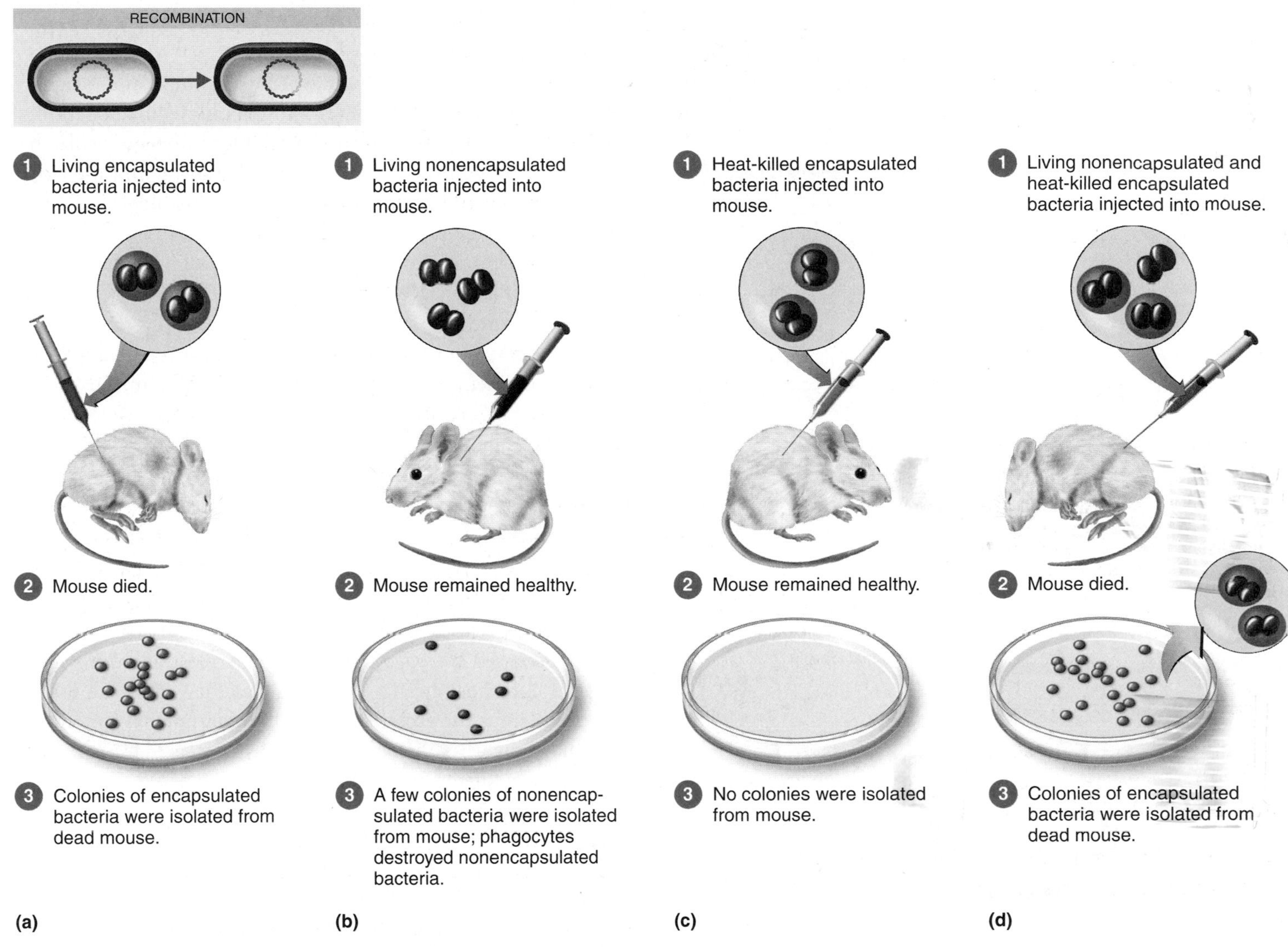

Figure 8.24 Griffith's experiment demonstrating genetic transformation. (**a**) Living encapsulated bacteria caused disease and death when injected into a mouse. (**b**) Living nonencapsulated bacteria are readily destroyed by the phagocytic defenses of the host, so the mouse remained healthy after injection. (**c**) After being killed by heat, encapsulated bacteria lost the ability to cause disease. (**d**) However, the combination of living nonencapsulated bacteria and heat-killed encapsulated bacteria (neither of which alone cause disease) did cause disease. Somehow, the live nonencapsulated bacteria were transformed by the dead encapsulated bacteria so that they acquired the ability to form a capsule and therefore cause disease. Subsequent experiments proved the transforming factor to be DNA.

Q Why did encapsulated bacteria kill the mouse while nonencapsulated bacteria did not? What killed the mouse in (d)?

of the mice died. In the blood of the dead mice, Griffith found living, encapsulated bacteria. Hereditary material (genes) from the dead bacteria had entered the live cells and changed them genetically so that their progeny were encapsulated and therefore virulent (Figure 8.24d).

Subsequent investigations based on Griffith's research revealed that bacterial transformation could be carried out without mice. A broth was inoculated with live nonencapsulated bacteria. Dead encapsulated bacteria were then added to the broth. After incubation, the culture was found to contain living bacteria that were encapsulated and virulent. The nonencapsulated bacteria had been transformed; they had acquired a new hereditary trait by incorporating genes from the killed encapsulated bacteria.

The next step was to extract various chemical components from the killed cells to determine which component caused the transformation. These crucial experiments were performed in the United States by Oswald T. Avery and his associates Colin M. MacLeod and Maclyn McCarty. After years of research, they announced in 1944 that the component responsible for transforming harmless *S. pneumoniae* into virulent strains was DNA. Their results provided one of the conclusive indications that DNA was indeed the carrier of genetic information.

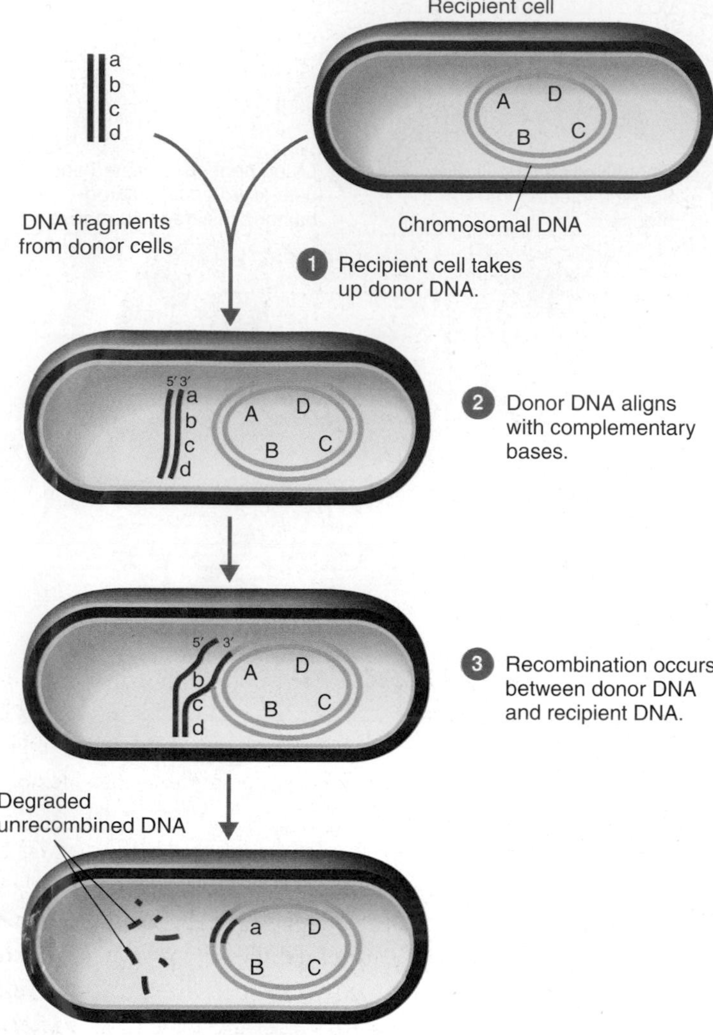

Figure 8.25 The mechanism of genetic transformation in bacteria. Some similarity is needed for the donor and recipient to align. Genes *a, b, c,* and *d* may be mutations of genes *A, B, C,* and *D.*

Q What type of enzyme cuts the donor DNA?

Since the time of Griffith's experiment, considerable information has been gathered about transformation. In nature, some bacteria, perhaps after death and cell lysis, release their DNA into the environment. Other bacteria can then encounter the DNA and, depending on the particular species and growth conditions, take up fragments of DNA and integrate them into their own chromosomes by recombination. A protein called RecA (see Figure 3.11a, page 65) binds to the cell's DNA and then to donor DNA causing the exchange of strands. A recipient cell with this new combination of genes is a kind of hybrid, or recombinant cell (**Figure 8.25**). All the descendants of such a recombinant cell will be identical to it. Transformation occurs naturally among very few genera of bacteria, including *Bacillus, Haemophilus* (hē-mä′fi-lus), *Neisseria, Acinetobacter* (a-si-ne′tō-bak-tėr), and certain strains of the genera *Streptococcus* and *Staphylococcus.*

Transformation works best when the donor and recipient cells are very closely related. Even though only a small portion of a cell's DNA is transferred to the recipient, the molecule that must pass through the recipient cell wall and membrane is still very large. When a recipient cell is in a physiological state in which it can take up the donor DNA, it is said to be competent. **Competence** results from alterations in the cell wall that make it permeable to large DNA molecules.

The well-understood and widely used bacterium *E. coli* is not naturally competent for transformation. However, a simple laboratory treatment enables *E. coli* to readily take up DNA. The discovery of this treatment has enabled researchers to use *E. coli* for genetic engineering, discussed in Chapter 9.

Conjugation in Bacteria

Another mechanism by which genetic material is transferred from one bacterium to another is known as **conjugation.** Conjugation is mediated by one kind of *plasmid,* a circular piece of DNA that replicates independently from the cell's chromosome (discussed on page 238). However, plasmids differ from bacterial chromosomes in that the genes they carry are usually not essential for the growth of the cell under normal conditions. The plasmids responsible for conjugation are transmissible between cells during conjugation.

Conjugation differs from transformation in two major ways. First, conjugation requires direct cell-to-cell contact. Second, the conjugating cells must generally be of opposite mating type; donor cells must carry the plasmid, and recipient cells usually do not. In gram-negative bacteria, the plasmid carries genes that code for the synthesis of *sex pili,* projections from the donor's cell surface that contact the recipient and help bring the two cells into direct contact (**Figure 8.26a**). Gram-positive bacterial cells produce sticky surface molecules that cause cells to come into direct contact with each other. In the process of conjugation, the plasmid is replicated during the transfer of a single-stranded copy of the plasmid DNA to the recipient, where the complementary strand is synthesized (**Figure 8.26b**).

Because most experimental work on conjugation has been done with *E. coli,* we will describe the process in this organism. In *E. coli,* the **F factor (fertility factor)** was the first plasmid observed to be transferred between cells during conjugation. Donors carrying F factors (F^+ cells) transfer the plasmid to recipients (F^- cells), which become F^+ cells as a result (**Figure 8.27a**). In some cells carrying F factors, the factor integrates into the chromosome, converting the F^+ cell to an **Hfr cell** (high frequency of recombination) (**Figure 8.27b**). When conjugation occurs between an Hfr cell and an F^- cell, the Hfr cell's chromosome (with its integrated F factor) replicates, and a parental strand of the chromosome is transferred to the recipient

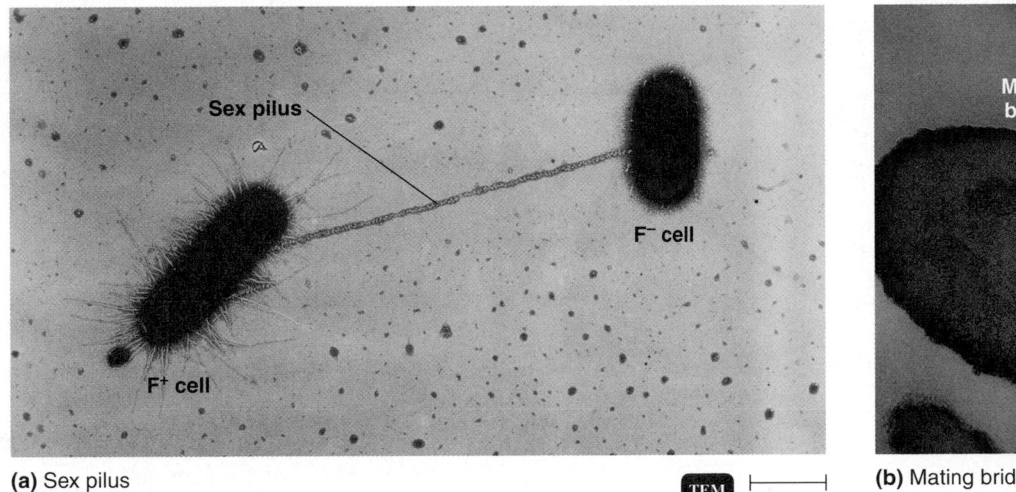

(a) Sex pilus

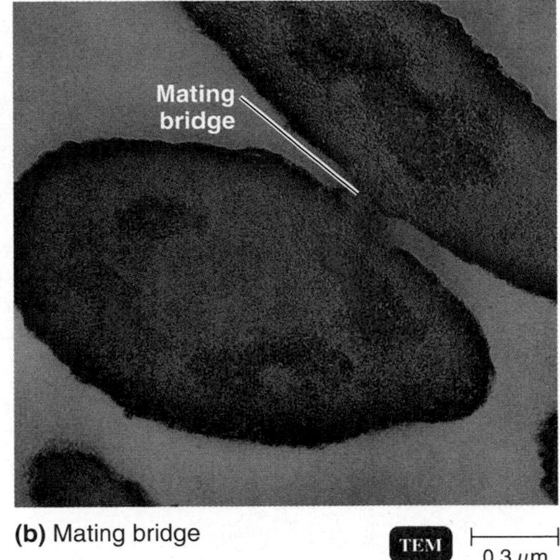

(b) Mating bridge

Figure 8.26 Bacterial conjugation.

Q What is an F^+ cell?

cell (Figure 8.27c). Replication of the Hfr chromosome begins in the middle of the integrated F factor, and a small piece of the F factor leads the chromosomal genes into the F^- cell. Usually, the chromosome breaks before it is completely transferred. Once within the recipient cell, donor DNA can recombine with the recipient's DNA. (Donor DNA that is not integrated is degraded.) Therefore, by conjugation with an Hfr cell, an F^- cell may acquire new versions of chromosomal genes (just as in transformation). However, it remains an F^- cell because it did not receive a complete F factor during conjugation.

Conjugation is used to map the location of genes on a bacterial chromosome (see Figure 8.1b). The genes for the synthesis of threonine *(thr)* and leucine *(leu)* are first, reading clockwise from 0. Their locations were determined by conjugation experiments. Assume that conjugation is allowed for only 1 minute between an Hfr strain that is his^+, pro^+, thr^+, and leu^+, and an F^- strain that is his^-, pro^-, thr^-, and leu^-. If the F^- acquired the ability to synthesize threonine, then the *thr* gene is located early in the chromosome, between 0 and 1 minute. If after 2 minutes the F^- cell now becomes thr^+ and leu^+, the order of these two genes on the chromosome must be *thr, leu.*

Transduction in Bacteria

A third mechanism of genetic transfer between bacteria is **transduction.** In this process, bacterial DNA is transferred from a donor cell to a recipient cell inside a virus that infects bacteria, called a **bacteriophage,** or **phage.** (Phages will be discussed further in Chapter 13.)

To understand how transduction works, we will consider the life cycle of one type of transducing phage of *E. coli;* this phage carries out **generalized transduction** (Figure 8.28).

During phage reproduction, phage DNA and proteins are synthesized by the host bacterial cell. The phage DNA should be packaged inside the phage protein coat. However, bacterial DNA, plasmid DNA, or even DNA of another virus may be packaged inside a phage protein coat.

Q&A All genes contained within a bacterium infected by a generalized transducing phage are equally likely to be packaged in a phage coat and transferred. In another type of transduction, called **specialized transduction,** only certain bacterial genes are transferred. In one type of specialized transduction, the phage codes for certain toxins produced by their bacterial hosts, such as diphtheria toxin for *Corynebacterium diphtheriae* (kôr′-i-nē-bak-ti-rē-um dif-thi′-re-ī), erythrogenic toxin for *Streptococcus pyogenes*, and Shiga toxin for *E. coli* O157:H7. Specialized transduction will be discussed in Chapter 13 (page 382). In addition to mutation, transformation, and conjugation, transduction is another way bacteria acquire new genotypes. **Animations** Horizontal Gene Transfer: Overview; Transformation; Transduction: Generalized Transduction; Conjugation: Overview, F Factor, Hfr Conjugation, Chromosome Mapping. **www.microbiologyplace.com**

CHECK YOUR UNDERSTANDING

✓ Differentiate horizontal and vertical gene transfer. **8-14**

✓ Compare conjugation between the following pairs: $F^+ \times F^-$, $Hfr \times F^-$. **8-15**

Plasmids and Transposons

Plasmids and transposons are genetic elements that provide additional mechanisms for genetic change. They occur in both

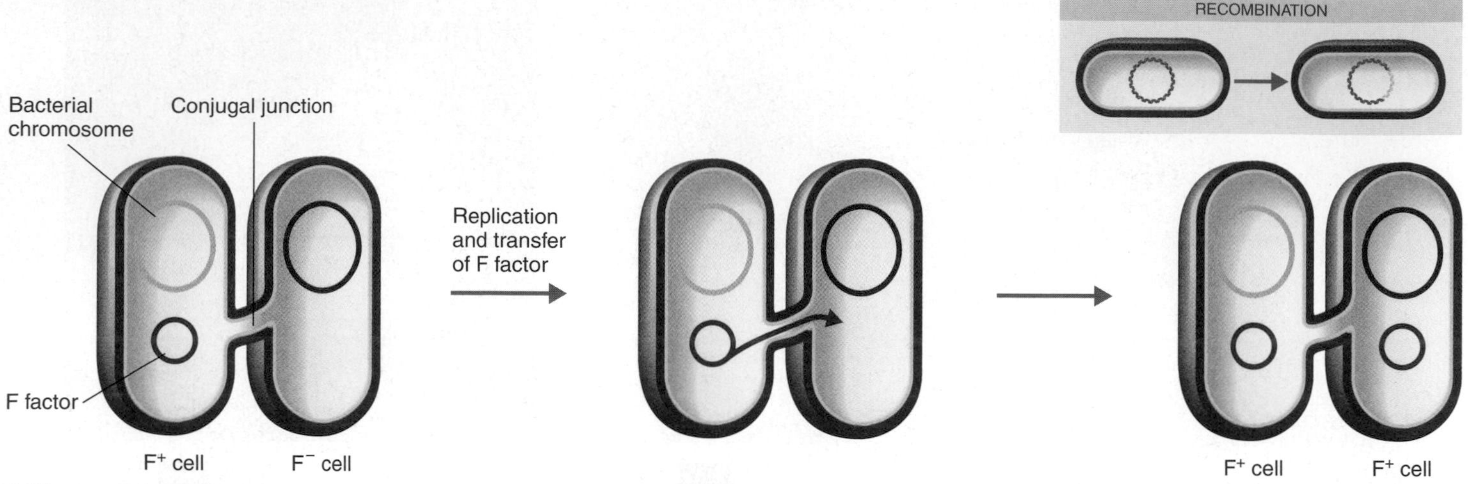

(a) When an F factor (a plasmid) is transferred from a donor (F^+) to a recipient (F^-), the F^- cell is converted to an F^+ cell.

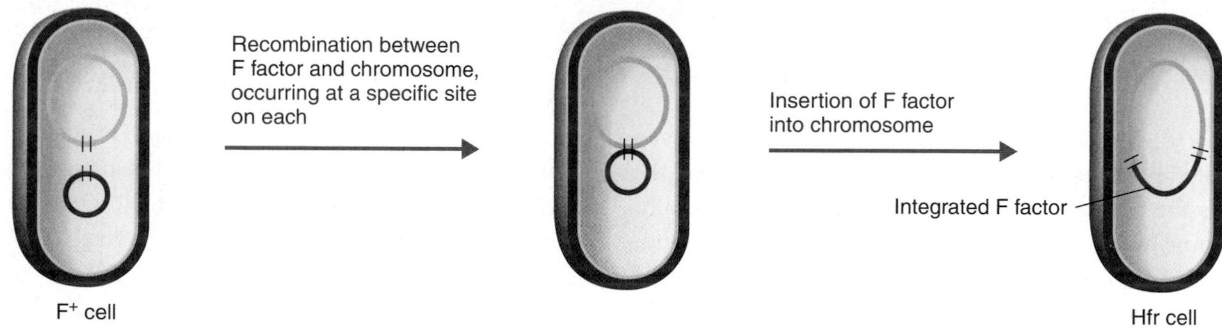

(b) When an F factor becomes integrated into the chromosome of an F^+ cell, it makes the cell a high frequency of recombination (Hfr) cell.

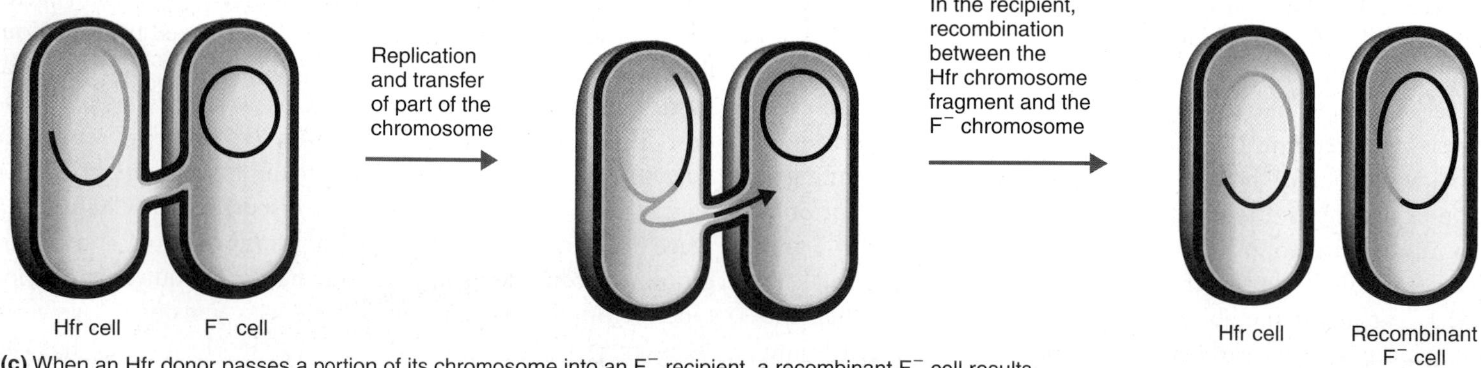

(c) When an Hfr donor passes a portion of its chromosome into an F^- recipient, a recombinant F^- cell results.

Figure 8.27 Conjugation in *E. coli.*

Q How does conjugation differ from transformation?

prokaryotic and eukaryotic organisms, but this discussion focuses on their role in genetic change in prokaryotes.

Plasmids

Recall from Chapter 4 (page 85) that plasmids are self-replicating, gene-containing circular pieces of DNA about 1–5% the size of the bacterial chromosome (**Figure 8.29a**). They are found mainly in bacteria but also in some eukaryotic microorganisms, such as *Saccharomyces cerevisiae.* The F factor is a **conjugative plasmid** that carries genes for sex pili and for the transfer of the plasmid to another cell. Although plasmids are usually dispensable, under certain conditions genes carried by plasmids can be crucial to the survival and growth of the cell. For example, **dissimilation plasmids** code for enzymes that trigger the catabolism of certain unusual sugars and hydrocarbons. Some species of *Pseudomonas* can actually use such exotic substances as toluene, camphor, and hydrocarbons of petroleum as primary carbon and energy sources because they have catabolic enzymes encoded by genes carried on plasmids. Such specialized capabilities permit the survival of those microorganisms in very diverse and challenging environments. Because of their ability to degrade and detoxify a variety of unusual compounds, many of them are being investigated for possible use in the cleanup of environmental wastes. (See the box in Chapter 2, page 33.)

Other plasmids code for proteins that enhance the pathogenicity of a bacterium. The strain of *E. coli* that causes infant diarrhea and traveler's diarrhea carries plasmids that code for toxin production and for bacterial attachment to intestinal cells. Without these plasmids, *E. coli* is a harmless resident of the large intestine; with them, it is pathogenic. Other plasmid-encoded toxins include the exfoliative toxin of *Staphylococcus aureus, Clostridium tetani* neurotoxin, and toxins of *Bacillus anthracis.* Still other plasmids contain genes for the synthesis of **bacteriocins,** toxic proteins that kill other bacteria. These plasmids have been found in many bacterial genera, and they are useful markers for the identification of certain bacteria in clinical laboratories.

Resistance factors (R factors) are plasmids that have significant medical importance. They were first discovered in Japan in the late 1950s after several dysentery epidemics. In some of these epidemics, the infectious agent was resistant to the usual antibiotic. Following isolation, the pathogen was also found to be resistant to a number of different antibiotics. In addition, other normal bacteria from the patients (such as *E. coli*) proved to be resistant as well. Researchers soon discovered that these bacteria acquired resistance through the spread of genes from one organism to another. The plasmids that mediated this transfer are R factors.

R factors carry genes that confer upon their host cell resistance to antibiotics, heavy metals, or cellular toxins. Many R factors contain two groups of genes. One group is called the **resistance transfer factor (RTF)** and includes genes for plasmid replication and conjugation. The other group, the **r-determinant,** has the resistance genes; it codes for the production of enzymes that inactivate certain drugs or toxic substances (Figure 8.29b). Different R factors, when present in the same cell, can recombine to produce R factors with new combinations of genes in their r-determinants.

In some cases, the accumulation of resistance genes within a single plasmid is quite remarkable. For example, Figure 8.29b shows a genetic map of resistance plasmid R100. Carried on this plasmid are resistance genes for sulfonamides, streptomycin, chloramphenicol, and tetracycline, as well as genes for resistance to mercury. This particular plasmid can be transferred between a number of enteric species, including *Escherichia, Klebsiella,* and *Salmonella.*

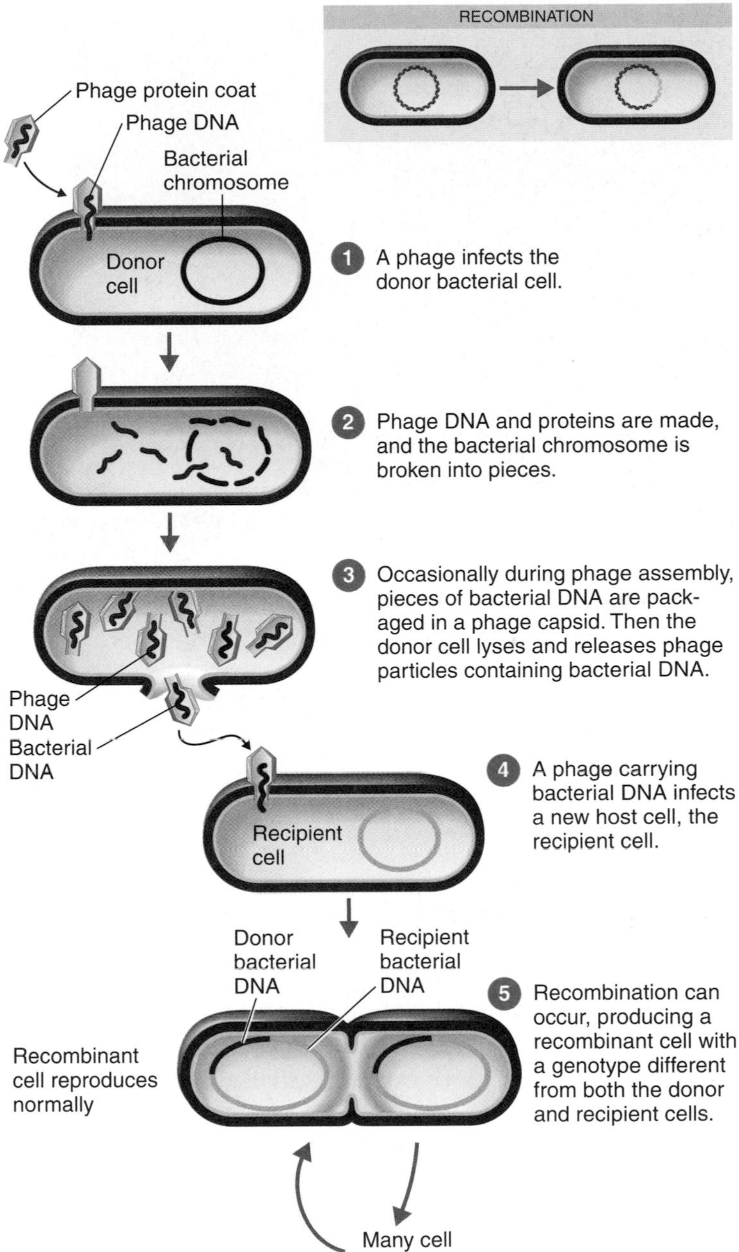

Figure 8.28 Transduction by a bacteriophage. Shown here is generalized transduction, in which any bacterial DNA can be transferred from one cell to another.

Q What is transduction?

R factors present very serious problems for treating infectious diseases with antibiotics. The widespread use of antibiotics in medicine and agriculture (see the box in Chapter 20 on page 577) has led to the preferential survival (selection) of bacteria that have R factors, so populations of resistant bacteria grow larger and larger. The transfer of resistance between bacterial cells of a population, and even between bacteria of different genera, also contributes to the problem. The ability to reproduce sexually with

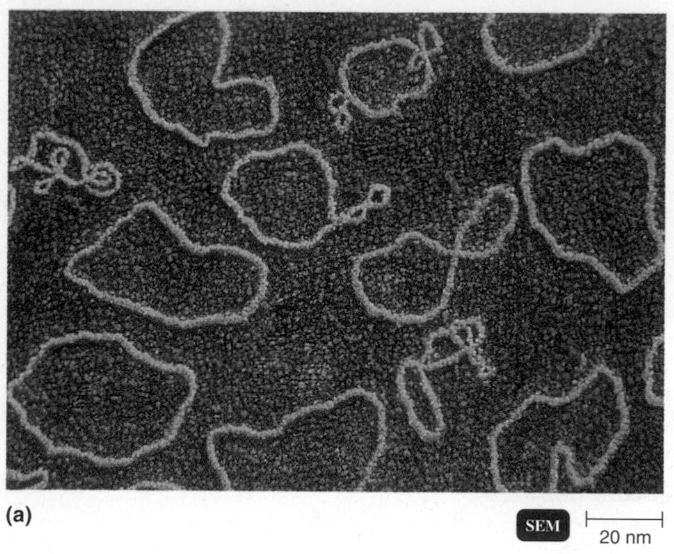

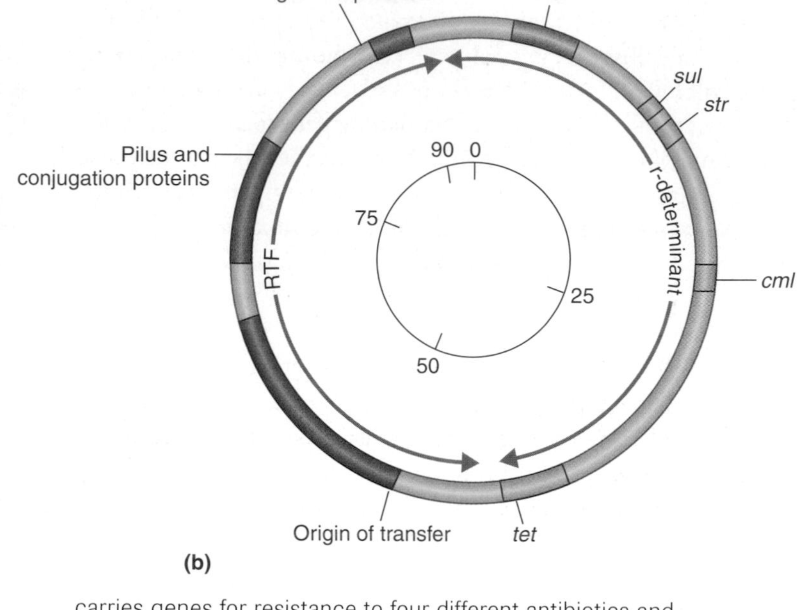

Figure 8.29 R factor, a type of plasmid. (**a**) Plasmids from *Bacteroides fragilis* bacteria that encode resistance to the antibiotic clindamycin. (**b**) A diagram of an R factor, which has two parts: the RTF contains genes needed for plasmid replication and transfer of the plasmid by conjugation, and the r-determinant carries genes for resistance to four different antibiotics and mercury (*sul* = sulfonamide resistance, *str* = streptomycin resistance, *cml* = chloramphenicol resistance, *tet* = tetracycline resistance, *mer* = mercury resistance); numbers are base pairs × 1000.

Q Why are R factors important in the treatment of infectious diseases?

members of its own species defines a eukaryotic species. However, a bacterial species can conjugate and transfer plasmids to other species. *Neisseria* may have acquired its penicillinase-producing plasmid from *Streptococcus,* and *Agrobacterium* can transfer plasmids to plant cells (see Figure 9.20, page 265). Nonconjugative plasmids may be transferred from one cell to another by inserting themselves into a conjugative plasmid or a chromosome or by transformation when released from a dead cell. Insertion is made possible by an insertion sequence, which will be discussed shortly.

Plasmids are an important tool for genetic engineering, discussed in Chapter 9 (page 250).

Transposons

Transposons are small segments of DNA that can move (be "transposed") from one region of a DNA molecule to another. These pieces of DNA are 700 to 40,000 base pairs long.

In the 1950s, American geneticist Barbara McClintock discovered transposons in corn, but they occur in all organisms and have been studied most thoroughly in microorganisms. They may move from one site to another site on the same chromosome or to another chromosome or plasmid. As you might imagine, the frequent movement of transposons could wreak havoc inside a cell. For example, as transposons move about on chromosomes, they may insert themselves *within* genes, inactivating them. Fortunately, transposition occurs relatively rarely. The frequency of transposition is comparable to the spontaneous mutation rate that occurs in bacteria—that is, from 10^{-5} to 10^{-7} per generation.

All transposons contain the information for their own transposition. As shown in **Figure 8.30a**, the simplest transposons, also called **insertion sequences (IS),** contain only a gene that codes for an enzyme (*transposase,* which catalyzes the cutting and resealing of DNA that occurs in transposition) and recognition sites. *Recognition sites* are short inverted repeat sequences of DNA that the enzyme recognizes as recombination sites between the transposon and the chromosome.

Complex transposons also carry other genes not connected with the transposition process. For example, bacterial transposons may contain genes for enterotoxin or for antibiotic resistance (**Figure 8.30b**). Plasmids such as R factors are frequently made up of a collection of transposons (**Figure 8.30c**).

Transposons with antibiotic resistance genes are of practical interest, but there is no limitation on the kinds of genes that transposons can have. Thus, transposons provide a natural

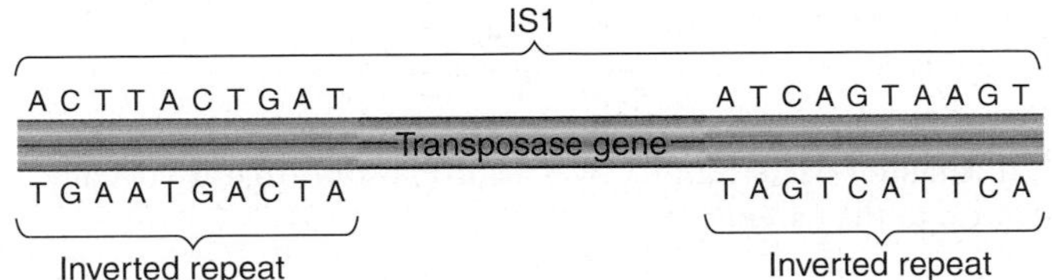

(a) An insertion sequence (IS), the simplest transposon, contains a gene for transposase, the enzyme that catalyzes transposition. The tranposase gene is bounded at each end by inverted repeat sequences that function as recognition sites for the transposon. IS1 is one example of an insertion sequence, shown here with simplified IR sequences.

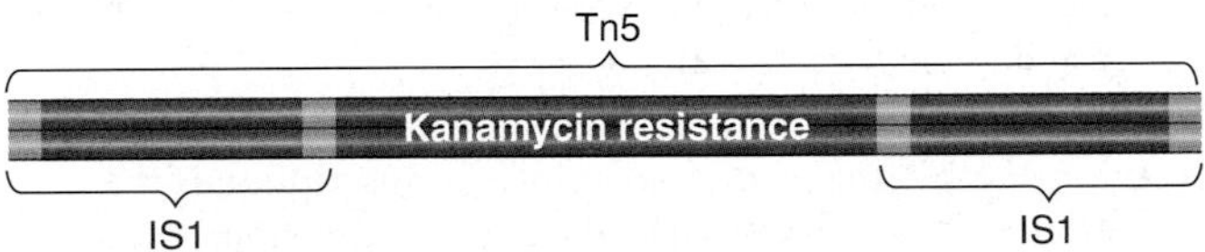

(b) Complex transposons carry other genetic material in addition to transposase genes. The example shown here, Tn5, carries the gene for kanamycin resistance and has complete copies of the insertion sequence IS1 at each end.

1 Transposase cuts DNA, leaving sticky ends.

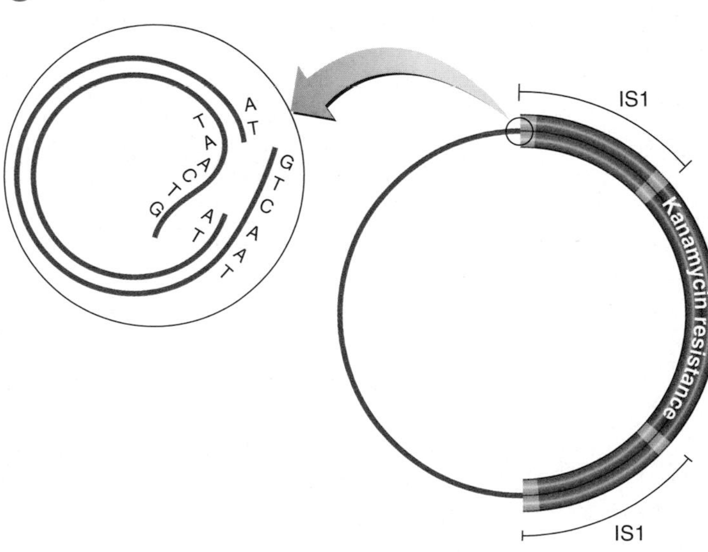

2 Sticky ends of transposon and target DNA anneal.

(c) Insertion of the transposon Tn5 into R100 plasmid.

Figure 8.30 **Transposons and insertion.**

Q **Why are transposons sometimes referred to as "jumping genes"?**

mechanism for the movement of genes from one chromosome to another. Furthermore, because they may be carried between cells on plasmids or viruses, they can also spread from one organism—or even species—to another. For example, vancomycin resistance was transferred from *Enterococcus faecalis* to *Staphylococcus aureus* via a transposon called Tn1546. Transposons are thus a potentially powerful mediator of evolution in organisms. **Animations** Transposons: Overview, Insertion Sequences, Complex Transposons **www.microbiologyplace.com**

CHECK YOUR UNDERSTANDING

✓ What types of genes do plasmids carry? **8-16**

Genes and Evolution

LEARNING OBJECTIVE

8-17 Discuss how genetic mutation and recombination provide material for natural selection to act upon.

We have now seen how gene activity can be controlled by the cell's internal regulatory mechanisms and how genes themselves can be altered or rearranged by mutation, transposition, and recombination. All these processes provide diversity in the descendants of cells. Diversity provides the raw material for evolution, and natural selection provides its driving force. Natural selection will act on diverse populations to ensure the survival of those fit for that particular environment. The different kinds of microorganisms that exist today are the result of a long history of evolution. Microorganisms have continually changed by alterations in their genetic properties and acquisition of adaptations to many different habitats. See the box on antibiotic resistance in Chapter 26, page 751, for an example of natural selection.

CHECK YOUR UNDERSTANDING

✓ Natural selection means that the environment favors survival of some genotypes. From where does diversity in genotypes come? **8-17**

STUDY OUTLINE

The **MyMicrobiologyPlace** website (**www.microbiologyplace.com**) will help you get ready for tests with its simple three-step approach: ❶ **take a pre-test** and obtain a personalized study plan, ❷ **learn and practice** with animations, tutorials, and MP3 tutor sessions, and ❸ **test yourself** with quizzes and a chapter post-test.

Structure and Function of the Genetic Material (pp. 211–221)

1. Genetics is the study of what genes are, how they carry information, how their information is expressed, and how they are replicated and passed to subsequent generations or other organisms.
2. DNA in cells exists as a double-stranded helix; the two strands are held together by hydrogen bonds between specific nitrogenous base pairs: AT and CG.
3. A gene is a segment of DNA, a sequence of nucleotides, that codes for a functional product, usually a protein.
4. The DNA in a cell is duplicated before the cell divides, so each daughter cell receives the same genetic information.

Genotype and Phenotype (p. 211)

5. Genotype is the genetic composition of an organism, its entire complement of DNA.
6. Phenotype is the expression of the genes: the proteins of the cell and the properties they confer on the organism.

DNA and Chromosomes (pp. 211–212)

7. The DNA in a chromosome exists as one long double helix associated with various proteins that regulate genetic activity.
8. Bacterial DNA is circular; the chromosome of *E. coli*, for example, contains about 4 million base pairs and is approximately 1000 times longer than the cell.
9. Genomics is the molecular characterization of genomes.

The Flow of Genetic Information (p. 212)

10. Information contained in the DNA is transcribed into RNA and translated into proteins.

DNA Replication (pp. 212–215)

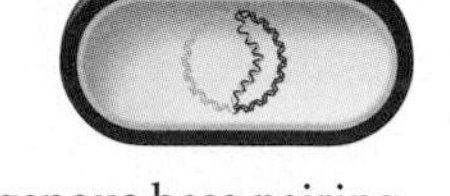

11. During DNA replication, the two strands of the double helix separate at the replication fork, and each strand is used as a template by DNA polymerases to synthesize two new strands of DNA according to the rules of nitrogenous base pairing.
12. The result of DNA replication is two new strands of DNA, each having a base sequence complementary to one of the original strands.
13. Because each double-stranded DNA molecule contains one original and one new strand, the replication process is called semiconservative.
14. DNA is synthesized in one direction designated 5′ → 3′. At the replication fork, the leading strand is synthesized continuously and the lagging strand discontinuously.
15. DNA polymerase proofreads new molecules of DNA and removes mismatched bases before continuing DNA synthesis.
16. Each daughter bacterium receives a chromosome that is virtually identical to the parent's.

RNA and Protein Synthesis (pp. 216–221)

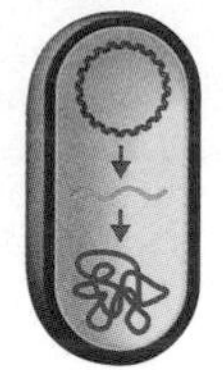

17. During transcription, the enzyme RNA polymerase synthesizes a strand of RNA from one strand of double-stranded DNA, which serves as a template.
18. RNA is synthesized from nucleotides containing the bases A, C, G, and U, which pair with the bases of the DNA strand being transcribed.
19. RNA polymerase binds the promoter; transcription begins at AUG; the region of DNA that is the end point of transcription is the terminator; RNA is synthesized in the 5′ → 3′ direction.
20. Translation is the process in which the information in the nucleotide base sequence of mRNA is used to dictate the amino acid sequence of a protein.
21. The mRNA associates with ribosomes, which consist of rRNA and protein.
22. Three-base segments of mRNA that specify amino acids are called codons.
23. The genetic code refers to the relationship among the nucleotide base sequence of DNA, the corresponding codons of mRNA, and the amino acids for which the codons code.
24. The genetic code is degenerate; that is, most amino acids are coded for by more than one codon.
25. Of the 64 codons, 61 are sense codons (which code for amino acids), and 3 are nonsense codons (which do not code for amino acids and are stop signals for translation).
26. The start codon, AUG, codes for methionine.
27. Specific amino acids are attached to molecules of tRNA. Another portion of the tRNA has a base triplet called an anticodon.
28. The base pairing of codon and anticodon at the ribosome results in specific amino acids being brought to the site of protein synthesis.
29. The ribosome moves along the mRNA strand as amino acids are joined to form a growing polypeptide; mRNA is read in the 5′ → 3′ direction.
30. Translation ends when the ribosome reaches a stop codon on the mRNA.

The Regulation of Bacterial Gene Expression (pp. 221–226)

1. Regulating protein synthesis at the gene level is energy-efficient because proteins are synthesized only as they are needed.
2. Constitutive enzymes produce products at a fixed rate. Examples are genes for the enzymes in glycolysis.
3. For these gene regulatory mechanisms, the control is aimed at mRNA synthesis.

Repression and Induction (p. 224)

4. Repression controls the synthesis of one or several (repressible) enzymes.
5. When cells are exposed to a particular end-product, the synthesis of enzymes related to that product decreases.

6. In the presence of certain chemicals (inducers), cells synthesize more enzymes. This process is called induction.
7. An example of induction is the production of β-galactosidase by *E. coli* in the presence of lactose; lactose can then be metabolized.

The Operon Model of Gene Expression (p. 224)

8. The formation of enzymes is determined by structural genes.
9. In bacteria, a group of coordinately regulated structural genes with related metabolic functions, plus the promoter and operator sites that control their transcription, are called an operon.
10. In the operon model for an inducible system, a regulatory gene codes for the repressor protein.
11. When the inducer is absent, the repressor binds to the operator, and no mRNA is synthesized.
12. When the inducer is present, it binds to the repressor so that it cannot bind to the operator; thus, mRNA is made, and enzyme synthesis is induced.
13. In repressible systems, the repressor requires a corepressor in order to bind to the operator site; thus, the corepressor controls enzyme synthesis.

Positive Regulation (pp. 224–226)

14. Transcription of structural genes for catabolic enzymes (such as β-galactosidase) is induced by the absence of glucose. Cyclic AMP and CRP must bind to a promoter in the presence of an alternative carbohydrate.
15. The presence of glucose inhibits the metabolism of alternative carbon sources by catabolite repression.

Mutation: Change in the Genetic Material (pp. 226–233)

1. A mutation is a change in the nitrogenous base sequence of DNA; that change causes a change in the product coded for by the mutated gene.
2. Many mutations are neutral, some are disadvantageous, and others are beneficial.

Types of Mutations (pp. 227–229)

3. A base substitution occurs when one base pair in DNA is replaced with a different base pair.
4. Alterations in DNA can result in missense mutations (which cause amino acid substitutions) or nonsense mutations (which create stop codons).
5. In a frameshift mutation, one or a few base pairs are deleted or added to DNA.
6. Mutagens are agents in the environment that cause permanent changes in DNA.
7. Spontaneous mutations occur without the presence of any mutagen.

Mutagens (pp. 229–231)

8. Chemical mutagens include base-pair mutagens, nucleoside analogs, and frameshift mutagens.
9. Ionizing radiation causes the formation of ions and free radicals that react with DNA; base substitutions or breakage of the sugar-phosphate backbone results.
10. Ultraviolet (UV) radiation is nonionizing; it causes bonding between adjacent thymines.
11. Damage to DNA caused by UV radiation can be repaired by enzymes that cut out and replace the damaged portion of DNA.
12. Light-repair enzymes repair thymine dimers in the presence of visible light.

The Frequency of Mutation (p. 231)

13. Mutation rate is the probability that a gene will mutate when a cell divides; the rate is expressed as 10 to a negative power.
14. Mutations usually occur randomly along a chromosome.
15. A low rate of spontaneous mutations is beneficial in providing the genetic diversity needed for evolution.

Identifying Mutants (pp. 231–232)

16. Mutants can be detected by selecting or testing for an altered phenotype.
17. Positive selection involves the selection of mutant cells and the rejection of nonmutated cells.
18. Replica plating is used for negative selection—to detect, for example, auxotrophs that have nutritional requirements not possessed by the parent (nonmutated) cell.

Identifying Chemical Carcinogens (pp. 232–233)

19. The Ames test is a relatively inexpensive and rapid test for identifying possible chemical carcinogens.
20. The test assumes that a mutant cell can revert to a normal cell in the presence of a mutagen and that many mutagens are carcinogens.

Genetic Transfer and Recombination (pp. 233–241)

1. Genetic recombination, the rearrangement of genes from separate groups of genes, usually involves DNA from different organisms; it contributes to genetic diversity.
2. In crossing over, genes from two chromosomes are recombined into one chromosome containing some genes from each original chromosome.
3. Vertical gene transfer occurs during reproduction when genes are passed from an organism to its offspring.
4. Horizontal gene transfer in bacteria involves a portion of the cell's DNA being transferred from donor to recipient.
5. When some of the donor's DNA has been integrated into the recipient's DNA, the resultant cell is called a recombinant.

Transformation in Bacteria (pp. 234–236)

6. During this process, genes are transferred from one bacterium to another as "naked" DNA in solution.
7. This process occurs naturally among a few genera of bacteria.

Conjugation in Bacteria (pp. 236–237)

8. This process requires contact between living cells.
9. One type of genetic donor cell is an F^+; recipient cells are F^-. F cells contain plasmids called F factors; these are transferred to the F^- cells during conjugation.
10. When the plasmid becomes incorporated into the chromosome, the cell is called an Hfr (high frequency of recombination) cell.
11. During conjugation, an Hfr cell can transfer chromosomal DNA to an F^- cell. Usually, the Hfr chromosome breaks before it is fully transferred.

Transduction in Bacteria (p. 237)

12. In this process, DNA is passed from one bacterium to another in a bacteriophage and is then incorporated into the recipient's DNA.
13. In generalized transduction, any bacterial genes can be transferred.

Plasmids and Transposons (pp. 237–241)

14. Plasmids are self-replicating circular molecules of DNA carrying genes that are not usually essential for the cell's survival.
15. There are several types of plasmids, including conjugative plasmids, dissimilation plasmids, plasmids carrying genes for toxins or bacteriocins, and resistance factors.
16. Transposons are small segments of DNA that can move from one region to another region of the same chromosome or to a different chromosome or a plasmid.
17. Transposons are found in chromosomes, in plasmids, and in the genetic material of viruses. They vary from simple (insertion sequences) to complex.
18. Complex transposons can carry any type of gene, including antibiotic-resistance genes, and are thus a natural mechanism for moving genes from one chromosome to another.

Genes and Evolution (p. 241)

1. Diversity is the precondition for evolution.
2. Genetic mutation and recombination provide a diversity of organisms, and the process of natural selection allows the growth of those best adapted to a given environment.

STUDY QUESTIONS

Answers to the Review and Multiple Choice questions can be found by turning to the blue Answers tab at the back of the textbook.

Review

1. Briefly describe the components of DNA, and explain its functional relationship to RNA and protein.
2. **DRAW IT** Identify and mark each of the following on the portion of DNA undergoing replication: replication fork, DNA polymerase, RNA primer, parent strands, leading strand, lagging strand, the direction of replication on each strand, and the 5′ end of each strand.

3. The following is a code for a strand of DNA.

		1	2	3	4	5	6	7	8	9	10	11	12	13	14	15	16	17	18	19
DNA	3′	A	T	A	T	_	_	_	T	T	T	_	_	_	_	_	_	_	_	_
mRNA												C	G	U				U	G	A
tRNA															U	G	G			
Amino Acid									Met			______			______			______		

ATAT = Promoter sequence

a. Using the genetic code provided in Figure 8.8, fill in the blanks to complete the segment of DNA shown.
b. Fill in the blanks to complete the sequence of amino acids coded for by this strand of DNA.
c. Write the code for the complementary strand of DNA completed in part (a).
d. What would be the effect if C were substituted for T at base 10?
e. What would be the effect if A were substituted for G at base 11?
f. What would be the effect if G were substituted for T at base 14?
g. What would be the effect if C were inserted between bases 9 and 10?
h. How would UV radiation affect this strand of DNA?
i. Identify a nonsense sequence in this strand of DNA.

4. Match the following examples of mutagens.

Column A	Column B
_____ **a.** A mutagen that is incorporated into DNA in place of a normal base	**1.** Frameshift mutagen
_____ **b.** A mutagen that causes the formation of highly reactive ions	**2.** Nucleoside analog
_____ **c.** A mutagen that alters adenine so that it base-pairs with cytosine	**3.** Base-pair mutagen
_____ **d.** A mutagen that causes insertions	**4.** Ionizing radiation
_____ **e.** A mutagen that causes the formation of pyrimidine dimers	**5.** Nonionizing radiation

5. Use the following metabolic pathway to answer the questions that follow it.

$$\text{Substrate } A \xrightarrow{\text{enzyme } a} \text{Intermediate } B \xrightarrow{\text{enzyme } b} \text{End-product } C$$

a. If enzyme *a* is inducible and is not being synthesized at present, a (1) __________ protein must be bound tightly to the (2) __________ site. When the inducer is present, it will bind to the (3) __________ so that (4) __________ can occur.
b. If enzyme *a* is repressible, end-product *C*, called a (1) __________, causes the (2) __________ to bind to the (3) __________. What causes derepression?
c. If enzyme *a* is constitutive, what effect, if any, will the presence of *A* or *C* have on it?

6. Which sequence is the best target for damage by UV radiation: AGGCAA, CTTTGA, or GUAAAU? Why aren't all bacteria killed when they are exposed to sunlight?
7. You are provided with cultures with the following characteristics:
   Culture 1: F^+, genotype $A^+ B^+ C^+$
   Culture 2: F^-, genotype $A^- B^- C^-$
   a. Indicate the possible genotypes of a recombinant cell resulting from the conjugation of cultures 1 and 2.
   b. Indicate the possible genotypes of a recombinant cell resulting from conjugation of the two cultures after the F^+ has become an Hfr cell.
8. Why are semiconservative replication and degeneracy of the genetic code advantageous to the survival of species?
9. Why are mutation and recombination important in the process of natural selection and the evolution of organisms?

Multiple Choice

Match the following terms to the definitions in questions 1and 2.

a. conjugation
c. transduction
e. translation
b. transcription
d. transformation

1. Transfer of DNA from a donor to a recipient cell by a bacteriophage.
2. Transfer of DNA from a donor to a recipient as naked DNA in solution.
3. Feedback inhibition differs from repression because feedback inhibition
   a. is less precise.
   b. is slower acting.
   c. stops the action of preexisting enzymes.
   d. stops the synthesis of new enzymes.
   e. all of the above
4. Bacteria can acquire antibiotic resistance by all of the following *except*
   a. mutation.
   b. insertion of transposons.
   c. conjugation.
   d. snRNPs.
   e. transformation.
5. Suppose you inoculate three flasks of minimal salts broth with *E. coli.* Flask A contains glucose. Flask B contains glucose and lactose. Flask C contains lactose. After a few hours of incubation, you test the flasks for the presence of β-galactosidase. Which flask(s) do you predict will have this enzyme?
   a. A
   b. B
   c. C
   d. A and B
   e. B and C
6. Plasmids differ from transposons because plasmids
   a. become inserted into chromosomes.
   b. are self-replicated outside the chromosome.
   c. move from chromosome to chromosome.
   d. carry genes for antibiotic resistance.
   e. none of the above

Use the following choices to answer questions 7 and 8.

a. catabolite repression
c. induction
e. translation
b. DNA polymerase
d. repression

7. Mechanism by which the presence of glucose inhibits the *lac* operon.
8. The mechanism by which lactose controls the *lac* operon.
9. Two daughter cells are most likely to inherit which one of the following from the parent cell?
   a. a change in a nucleotide in mRNA
   b. a change in a nucleotide in tRNA
   c. a change in a nucleotide in rRNA
   d. a change in a nucleotide in DNA
   e. a change in a protein
10. Which of the following is *not* a method of horizontal gene transfer?
   a. binary fission
   b. conjugation
   c. integration of a transposon
   d. transduction
   e. transformation

Critical Thinking

1. Nucleoside analogs and ionizing radiation are used in treating cancer. These mutagens can cause cancer, so why do you suppose they are used to treat the disease?
2. Replication of the *E. coli* chromosome takes 40 to 45 minutes, but the organism has a generation time of 26 minutes. How does the cell have time to make complete chromosomes for each daughter cell? For each granddaughter cell?
3. *Pseudomonas* has a plasmid containing the *mer* operon, which includes the gene for mercuric reductase. This enzyme catalyzes the reduction of the mercuric ion Hg^{2+} to the uncharged form of mercury, Hg^0. Hg^{2+} is quite toxic to cells; Hg^0 is not.
   a. What do you suppose is the inducer for this operon?
   b. The protein encoded by one of the *mer* genes binds Hg^{2+} in the periplasm and brings it into the cell. Why would a cell bring in a toxin?
   c. What is the value of the *mer* operon to *Pseudomonas*?

Clinical Applications

1. Ciprofloxacin, erythromycin, and acyclovir are used to treat microbial infections. Ciprofloxacin inhibits DNA gyrase. Erythromycin binds in front of the A site on the 50S subunit of a ribosome. Acyclovir is a guanine analog.
   a. What steps in protein synthesis are inhibited by each drug?
   b. Which drug is more effective against bacteria? Why?
   c. Which drugs will have effects on the host's cells? Why?
   d. Use the index to identify the disease for which acyclovir is primarily used. Why is it more effective than erythromycin for treating this disease?
2. HIV, the virus that causes AIDS, was isolated from three individuals, and the amino acid sequences for the viral coat were determined. Of the amino acid sequences shown below, which two of the viruses are most closely related? How can these amino acid sequences be used to identify the source of a virus?

Patient	Viral Amino Acid Sequence											
A	Asn	Gln	Thr	Ala	Ala	Ser	Lys	Asn	Ile	Asp	Ala	Leu
B	Asn	Leu	His	Ser	Asp	Lys	Ile	Asn	Ile	Ile	Leu	Leu
C	Asn	Gln	Thr	Ala	Asp	Ser	Ile	Val	Ile	Asp	Ala	Leu

3. Human herpesvirus-8 (HHV-8) is common in parts of Africa, the Middle East, and the Mediterranean, but is rare elsewhere except in AIDS patients. Genetic analyses indicate that the African strain is not changing, whereas the Western strain is accumulating changes. Using the portions of the HHV-8 genomes (shown below) that encode one of the viral proteins, how similar are these two viruses? What mechanism can account for the changes? What disease does HHV-8 cause?

Western	3′-ATGGAGTTCTTCTGGACAAGA
African	3′-ATAAACTTTTTCTTGACAACG

9 Biotechnology and Recombinant DNA

For thousands of years, people have been consuming foods that are produced by the action of microorganisms. Bread, chocolate, and soy sauce are some of the best-known examples. But it was only just over 100 years ago that scientists showed that microorganisms are responsible for these products. This knowledge opened the way for using microorganisms to produce other important products. Since World War I, microbes have been used to produce a variety of chemicals, such as ethanol, acetone, and citric acid. Since World War II, microorganisms have been grown on a large scale to produce antibiotics. More recently, microbes and their enzymes are replacing a variety of chemical processes involved in manufacturing such products as paper, textiles, and fructose. Using microbes or their enzymes instead of chemical syntheses offers several advantages: microbes may use inexpensive, abundant raw materials, such as starch; microbes work at normal temperatures and pressure, thereby avoiding the need for expensive and dangerous pressurized systems; and microbes don't produce toxic, hard-to-treat wastes.

In this chapter you will learn the tools and techniques that are used to research and develop a product. You will also learn how recombinant DNA technology is used to track outbreaks of infectious disease and to provide evidence for courts of law in forensic microbiology.

UNDER THE MICROSCOPE

Escherichia coli. This bacterium has been genetically modified to produce a human protein, gamma interferon. Unlike human cells, it doesn't secrete the protein, so the cells will be lysed to harvest the protein.

Q&A

Thirty years ago, researchers knew that interferons are an effective antiviral agent. However, they had discovered that they are species specific, so that interferons to be used in humans must be produced in human cells. Researchers feared that interferons would thus always be in limited supply. Can you think of a way to increase the supply of interferons so that they can be used to treat diseases?

Look for the answer in the chapter.

Introduction to Biotechnology

LEARNING OBJECTIVES

9-1 Compare and contrast biotechnology, genetic modification, and recombinant DNA technology.

9-2 Identify the roles of a clone and a vector in making recombinant DNA.

Biotechnology is the use of microorganisms, cells, or cell components to make a product. Microbes have been used in the commercial production of foods, vaccines, antibiotics, and vitamins for years. Bacteria are also used in mining to extract valuable elements from ore (see Figure 28.14, page 807). Additionally, animal cells have been used to produce viral vaccines since the 1950s. Until the 1980s, products made by living cells were all made by naturally occurring cells; the role of scientists was to find the appropriate cell and develop a method for large-scale cultivation of the cells.

Now, microorganisms as well as entire plants are being used as "factories" to produce chemicals that the organisms don't naturally make. The latter is made possible by inserting genes into cells by **recombinant DNA (rDNA) technology,** which is sometimes called *genetic engineering*. The development of rDNA technology is expanding the practical applications of biotechnology almost beyond imagination.

Recombinant DNA Technology

Recall from Chapter 8 that recombination of DNA occurs naturally in microbes. In the 1970s and 1980s, scientists developed artificial techniques for making recombinant DNA.

A gene from a vertebrate animal, including a human, can be inserted into the DNA of a bacterium, or a gene from a virus into a yeast may be used. In many cases, the recipient can then be made to express the gene, which may code for a commercially useful product. Thus, bacteria with genes for human insulin are now being used to produce insulin for treating diabetes, and a vaccine for hepatitis B is being made by yeast carrying a gene for part of the hepatitis virus (the yeast produces a viral coat protein). Scientists hope that such an approach may prove useful in producing vaccines against other infectious agents, thus eliminating the need to use whole organisms, as in conventional vaccines.

The rDNA techniques can also be used to make thousands of copies of the same DNA molecule—to *amplify* DNA, thus generating sufficient DNA for various kinds of experimentation and analysis. This technique has practical application for identifying microbes, such as viruses, that can't be cultured.

An Overview of Recombinant DNA Procedures

Figure 9.1 presents an overview of some of the procedures typically used for making rDNA, along with some promising applications. The gene of interest is inserted into the vector DNA in vitro. In this example, the vector is a plasmid. The DNA molecule chosen as a vector must be a self-replicating type, such as a plasmid or a viral genome. This recombinant vector DNA is taken up by a cell such as a bacterium, where it can multiply. The cell containing the recombinant vector is then grown in culture to form a **clone** of many genetically identical cells, each of which carries copies of the vector. This cell clone therefore contains many copies of the gene of interest. This is why DNA vectors are often called *gene-cloning vectors,* or simply *cloning vectors.* (In addition to referring to a culture of identical cells, the word *clone* is also routinely used as a verb, to describe the entire process, as in "to clone a gene.")

The final step varies according to whether the gene itself or the product of the gene is of interest. From the cell clone, the researcher may isolate ("harvest") large quantities of the gene of interest, which may then be used for a variety of purposes. The gene may even be inserted into another vector for introduction into another kind of cell (such as a plant or animal cell). Alternatively, if the gene of interest is expressed (transcribed and translated) in the cell clone, its protein product can be harvested and used for a variety of purposes.

The advantages of using recombinant DNA for obtaining such proteins is illustrated by one of its early successes, human growth hormone (hGH). Some individuals do not produce adequate amounts of hGH, so their growth is stunted. In the past, hGH needed to correct this deficiency had to be obtained from human pituitary glands at autopsy. (hGH from other animals is not effective in humans.) This practice was not only expensive but also dangerous because on several occasions neurological diseases were transmitted with the hormone. Human growth hormone produced by genetically modified *E. coli* is a pure and cost-effective product. Recombinant DNA techniques also result in faster production of the hormone than traditional methods might allow.

CHECK YOUR UNDERSTANDING

- ✓ Differentiate biotechnology and recombinant DNA technology. **9-1**
- ✓ In one sentence, describe how a vector and clone are used. **9-2**

Tools of Biotechnology

LEARNING OBJECTIVES

9-3 Compare selection and mutation.

9-4 Define *restriction enzymes*, and outline how they are used to make recombinant DNA.

9-5 List the four properties of vectors.

9-6 Describe the use of plasmid and viral vectors.

9-7 Outline the steps in PCR, and provide an example of its use.

Research scientists and technicians isolate bacteria and fungi from natural environments such as soil and water to find, or *select,* the

Figure 9.1

FOUNDATION FIGURE A Typical Genetic Modification Procedure

This figure provides an overview of making a recombinant cell and presents examples of applications. Each step of genetic modification introduced in this figure will be discussed in greater detail later in the chapter.

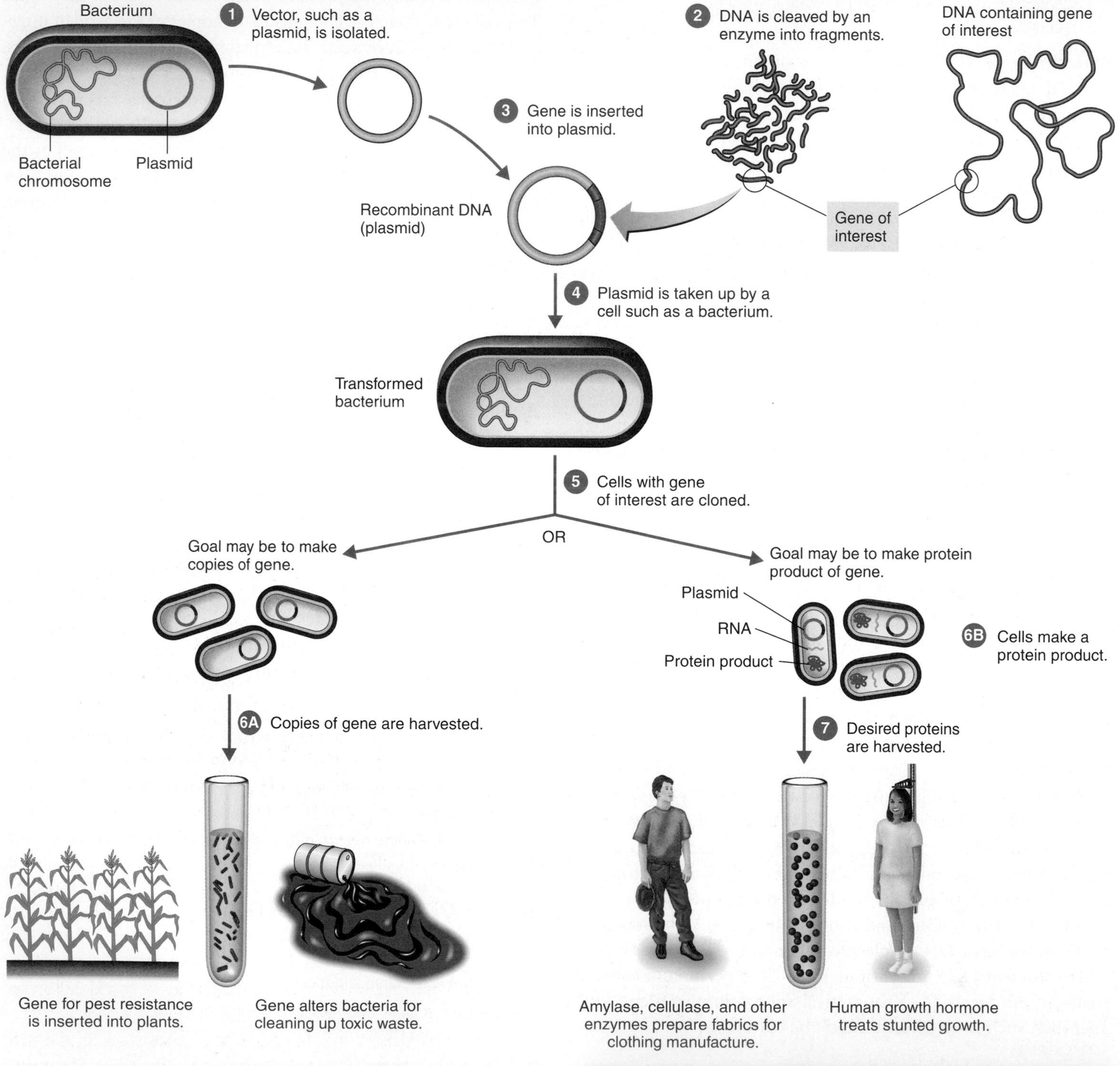

Bacterium
Bacterial chromosome
Plasmid
1 Vector, such as a plasmid, is isolated.
2 DNA is cleaved by an enzyme into fragments.
DNA containing gene of interest
3 Gene is inserted into plasmid.
Recombinant DNA (plasmid)
Gene of interest
4 Plasmid is taken up by a cell such as a bacterium.
Transformed bacterium
5 Cells with gene of interest are cloned.
OR
Goal may be to make copies of gene.
Goal may be to make protein product of gene.
Plasmid
RNA
Protein product
6B Cells make a protein product.
6A Copies of gene are harvested.
7 Desired proteins are harvested.
Gene for pest resistance is inserted into plants.
Gene alters bacteria for cleaning up toxic waste.
Amylase, cellulase, and other enzymes prepare fabrics for clothing manufacture.
Human growth hormone treats stunted growth.

Key Concept

Genes from the cell of one organism can be inserted into the cell of another organism and expressed in that cell. Genetically modified cells can be used to produce a wide variety of useful products.

organisms that produce a desired product. The selected organism can be mutated to make more product or to make a better product.

Selection

In nature, organisms with characteristics that enhance survival are more likely to survive and reproduce than are variants that lack the desirable traits. This is called *natural selection.* Humans use **artificial selection** to select desirable breeds of animals or strains of plants to cultivate. As microbiologists learned how to isolate and grow microorganisms in pure culture, they were able to select the ones that could accomplish the desired objective, such as brewing beer more efficiently, for example, or producing a new antibiotic. Over 2000 strains of antibiotic-producing bacteria have been discovered by testing soil bacteria and selecting the strains that produce an antibiotic. The box in Chapter 28 describes the selection of a bacterium that converts a waste product into a valuable product.

Mutation

As we saw in Chapter 8, mutations are responsible for much of the diversity of life. A bacterium with a mutation that confers resistance to an antibiotic will survive and reproduce in the presence of that antibiotic. Biologists working with antibiotic-producing microbes discovered that they could create new strains by exposing microbes to mutagens. After random mutations were created in penicillin-producing *Penicillium* by exposing fungal cultures to radiation, the highest-yielding variant among the survivors was selected for another exposure to a mutagen. Using mutations, biologists increased the amount of penicillin produced by the fungus over 1000 times.

Screening each mutant for penicillin production is a tedious process. **Site-directed mutagenesis** can be used to make a specific change in a gene. Suppose you determine that changing one amino acid will make a laundry enzyme work better in cold water. Using the genetic code (see Figure 8.8, page 219), you could, using the techniques described next, produce the sequence of DNA that encodes that amino acid and insert it into the gene for that enzyme.

The science of molecular genetics has advanced to such a degree that many routine cloning procedures are performed using prepackaged materials and procedures that are very much like cookbook recipes. Scientists have a grab bag of methods from which to choose, depending on the ultimate application of their experiments. Next we describe some of the most important tools and techniques, and later we will consider some specific applications.

Restriction Enzymes

Recombinant DNA technology has its technical roots in the discovery of **restriction enzymes,** a special class of DNA-cutting enzymes that exist in many bacteria. First isolated in 1970, restriction enzymes in nature had actually been observed earlier, when certain bacteriophages were found to have a restricted host range. If these phages were used to infect bacteria other than their usual hosts, restriction enzymes in the new host destroyed almost all the phage DNA. Restriction enzymes protect a bacterial cell by hydrolyzing phage DNA. The bacterial DNA is protected from digestion because the cell **methylates** (adds methyl groups to) some of the cytosines in its DNA. The purified forms of these bacterial enzymes are used in today's laboratories.

What is important for rDNA techniques is that a restriction enzyme recognizes and cuts, or *digests,* only one particular sequence of nucleotide bases in DNA, and it cuts this sequence in the same way each time. Typical restriction enzymes used in cloning experiments recognize four-, six-, or eight-base sequences. Hundreds of restriction enzymes are known, each producing DNA fragments with characteristic ends. A few restriction enzymes are listed in **Table 9.1**. You can see they are named for their bacterial source. Some of these enzymes (e.g., *Hae*III) cut both strands of DNA in the same place, producing **blunt ends,** and others make staggered cuts in the two strands—cuts that are not directly opposite each other (**Figure 9.2**). These staggered ends, or **sticky ends,** are most useful in rDNA because they can be used to join two different pieces of DNA that were cut by the same restriction enzyme. The sticky ends "stick" to stretches of single-stranded DNA by complementary base pairing.

Notice in Figure 9.2 that the darker base sequences on the two strands are the same, but they run in opposite directions. Staggered cuts leave stretches of single-stranded DNA at the ends of the DNA fragments. If two fragments of DNA from different sources have been produced by the action of the same restriction enzyme, the two pieces will have identical sets of sticky ends and can be spliced (recombined) in vitro. The sticky ends join spontaneously by

Table 9.1 Selected Restriction Enzymes Used in rDNA Technology

Enzyme	Bacterial Source	Recognition Sequence
*Bam*HI	*Bacillus amyloliquefaciens*	G↓G A T C C G C T A G↑G
*Eco*RI	*Escherichia coli*	G↓A A T T C C T T A A↑G
*Hae*III	*Haemophilus aegyptius*	G G↓C C C C↑G G
*Hin*dIII	*Haemophilus influenzae*	A↓A G C T T T T C G A↑A

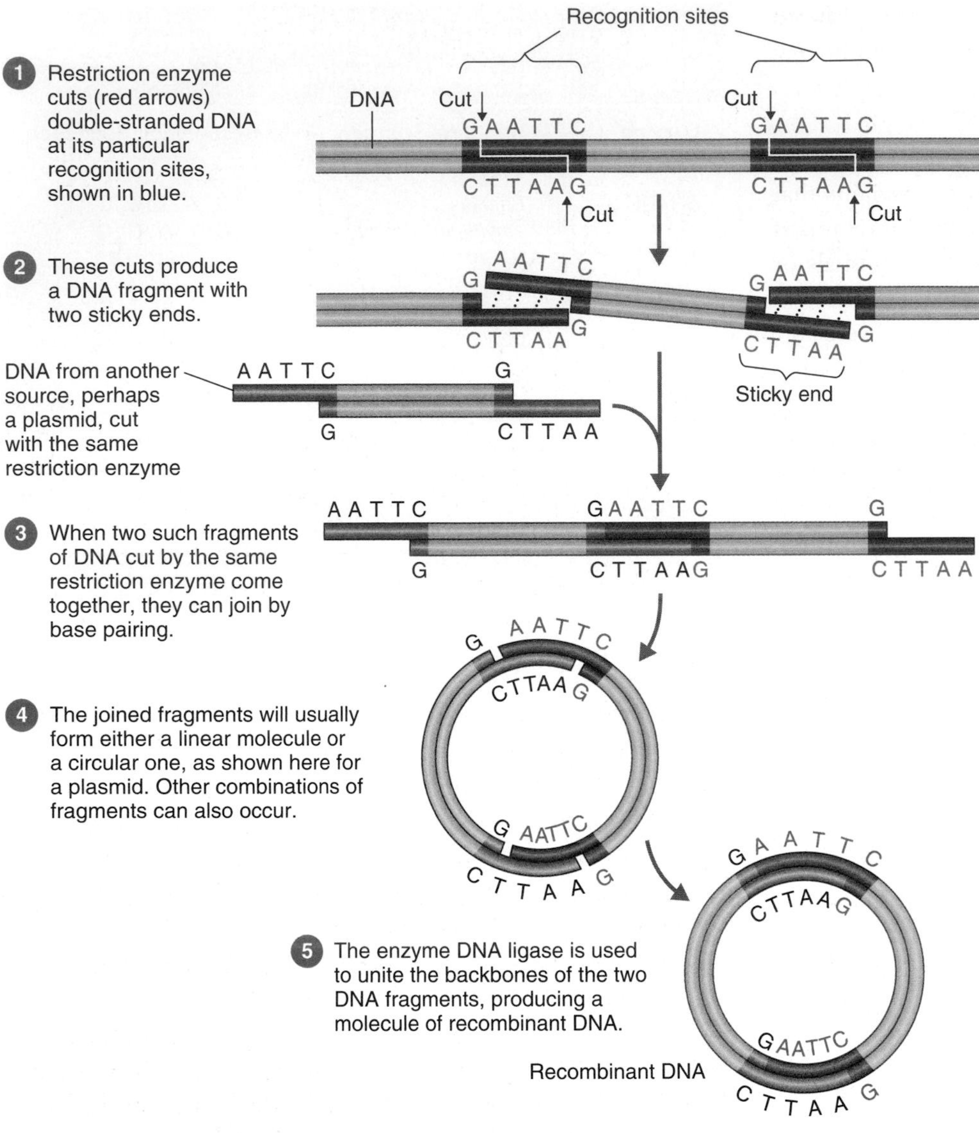

Recognition sites
1 Restriction enzyme cuts (red arrows) double-stranded DNA at its particular recognition sites, shown in blue.
DNA
Cut
GAATTC
CTTAAG
Cut
2 These cuts produce a DNA fragment with two sticky ends.
Sticky end
DNA from another source, perhaps a plasmid, cut with the same restriction enzyme
AATTC
G
CTTAA
3 When two such fragments of DNA cut by the same restriction enzyme come together, they can join by base pairing.
4 The joined fragments will usually form either a linear molecule or a circular one, as shown here for a plasmid. Other combinations of fragments can also occur.
5 The enzyme DNA ligase is used to unite the backbones of the two DNA fragments, producing a molecule of recombinant DNA.
Recombinant DNA

Figure 9.2 The role of a restriction enzyme in making recombinant DNA.

Q **Why are restriction enzymes used to make recombinant DNA?**

hydrogen bonding (base pairing). The enzyme DNA ligase is used to covalently link the backbones of the DNA pieces, producing an rDNA molecule. **Animation** Recombinant DNA Technology. **www.microbiologyplace.com**

Vectors

A great variety of different types of DNA molecules can serve as vectors, provided that they have certain properties. The most important property is self-replication; once in a cell, a vector must be capable of replicating. Any DNA that is inserted in the vector will be replicated in the process. Thus, vectors serve as vehicles for the replication of desired DNA sequences.

Vectors also need to be of a size that allows them to be manipulated outside the cell during recombinant DNA procedures. Smaller vectors are more easily manipulated than larger DNA molecules, which tend to be more fragile. Preservation is another important property of vectors. The circular form of DNA molecules is important in protecting the DNA of the vector from destruction by the recipient of the vector. Notice in **Figure 9.3** that the DNA of a plasmid is circular. Another preservation mechanism occurs when the DNA of a virus inserts itself quickly into the chromosome of the host (see Chapter 13, page 380).

When it is necessary to retrieve cells containing the vector, a marker gene contained within the vector can often help make selection easy. Common selectable marker genes are for antibiotic resistance or for an enzyme that carries out an easily identified reaction.

Plasmids are one of the primary vectors in use, particularly variants of R factor plasmids. Plasmid DNA can be cut with the same restriction enzymes as the DNA to be cloned, so that all pieces of the DNA will have the same sticky ends. When the pieces are mixed, the DNA to be cloned will become inserted into the plasmid (see Figure 9.2). Note that other possible combinations of fragments can occur as well, including the plasmid re-forming a circle with no DNA inserted.

Some plasmids are capable of existing in several different species. They are called **shuttle vectors** and can be used to move cloned DNA sequences among organisms, such as among bacterial, yeast, and mammalian cells, or among bacterial, fungal, and plant cells. Shuttle vectors can be very useful in the process of genetically modifying multicellular organisms—for example, by trying to insert herbicide resistance genes into plants.

A different kind of vector is viral DNA. This type of vector can usually accept much larger pieces of foreign DNA than plasmids can. After the DNA has been inserted into the viral vector, it can be cloned in the virus's host cells. The choice of a suitable vector depends on many factors, including the organism that will receive the new gene and the size of the DNA to be cloned. Retroviruses, adenoviruses, and herpesviruses are being used to insert corrective genes into human cells that have defective genes. Gene therapy is discussed on page 259.

CHECK YOUR UNDERSTANDING

- ✓ How are selection and mutation used in biotechnology? **9-3**
- ✓ What is the value of restriction enzymes in recombinant DNA technology? **9-4**
- ✓ What criteria must a vector meet? **9-5**
- ✓ Why is a vector used in recombinant DNA technology? **9-6**

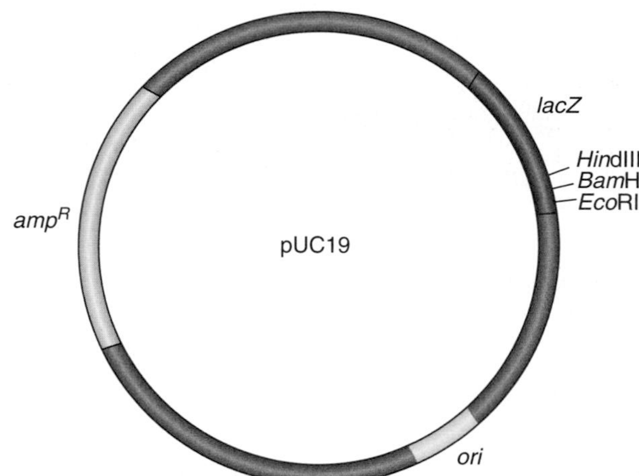

Figure 9.3 A plasmid used for cloning. A plasmid vector used for cloning in the bacterium *E. coli* is pUC19. An origin of replication *(ori)* allows the plasmid to be self-replicating. Two genes, one encoding resistance to the antibiotic ampicillin *(amp^R)* and one encoding the enzyme β-galactosidase *(lacZ)*, serve as marker genes. Foreign DNA can be inserted at the restriction enzyme sites.

Q **What is a vector in recombinant DNA technology?**

Polymerase Chain Reaction

The **polymerase chain reaction (PCR)** is a technique by which small samples of DNA can be quickly amplified, that is, increased to quantities that are large enough for analysis.

Starting with just one gene-sized piece of DNA, PCR can be used to make literally billions of copies in only a few hours. The PCR process is shown in Figure 9.4.

1. Each strand of the target DNA will serve as a template for DNA synthesis.
2. To this DNA is added a supply of the four nucleotides (for assembly into new DNA) and the enzyme for catalyzing the synthesis, DNA polymerase (see Chapter 8, page 213). Short pieces of nucleic acid called primers are also added to help start the reaction. The primers are complementary to the ends of the target DNA and
3. will hybridize to the fragments to be amplified.
4. Then, the polymerase synthesizes new complementary strands.
5. After each cycle of synthesis, the DNA is heated to convert all the new DNA into single strands. Each newly synthesized DNA strand serves in turn as a template for more new DNA.

As a result, the process proceeds exponentially. All of the necessary reagents are added to a tube, which is placed in a *thermalcycler.* The thermalcycler can be set for the desired temperatures, times, and number of cycles. Use of an automated thermalcycler is made possible by the use of DNA polymerase taken from a thermophilic bacterium such as *Thermus aquaticus;* the enzyme from such organisms can survive the heating phase without being destroyed. Thirty cycles, completed in just a few hours, will increase the amount of target DNA by more than a billion times.

The amplified DNA can be seen by gel electrophoresis. In *real-time PCR*, the newly made DNA is tagged with a fluorescent dye, so that the levels of fluorescence can be measured after every PCR cycle (that's the *real time* aspect). Another PCR procedure called *reverse-transcription PCR* uses viral RNA or a cell's mRNA as the template. The enzyme, reverse transcriptase, makes DNA from the RNA template, and the DNA is then amplified.

Note that PCR can only be used to amplify relatively small, specific sequences of DNA as determined by the choice of primers. It cannot be used to amplify an entire genome.

PCR can be applied to any situation that requires the amplification of DNA. Especially noteworthy are diagnostic tests that use PCR to detect the presence of infectious agents in situations in which they would otherwise be undetectable. **Animations** PCR: Overview, Components, Process. **www.microbiologyplace.com**

CHECK YOUR UNDERSTANDING

- ✓ For what is each of the following used in PCR: primer, DNA polymerase, 94°C? **9-7**

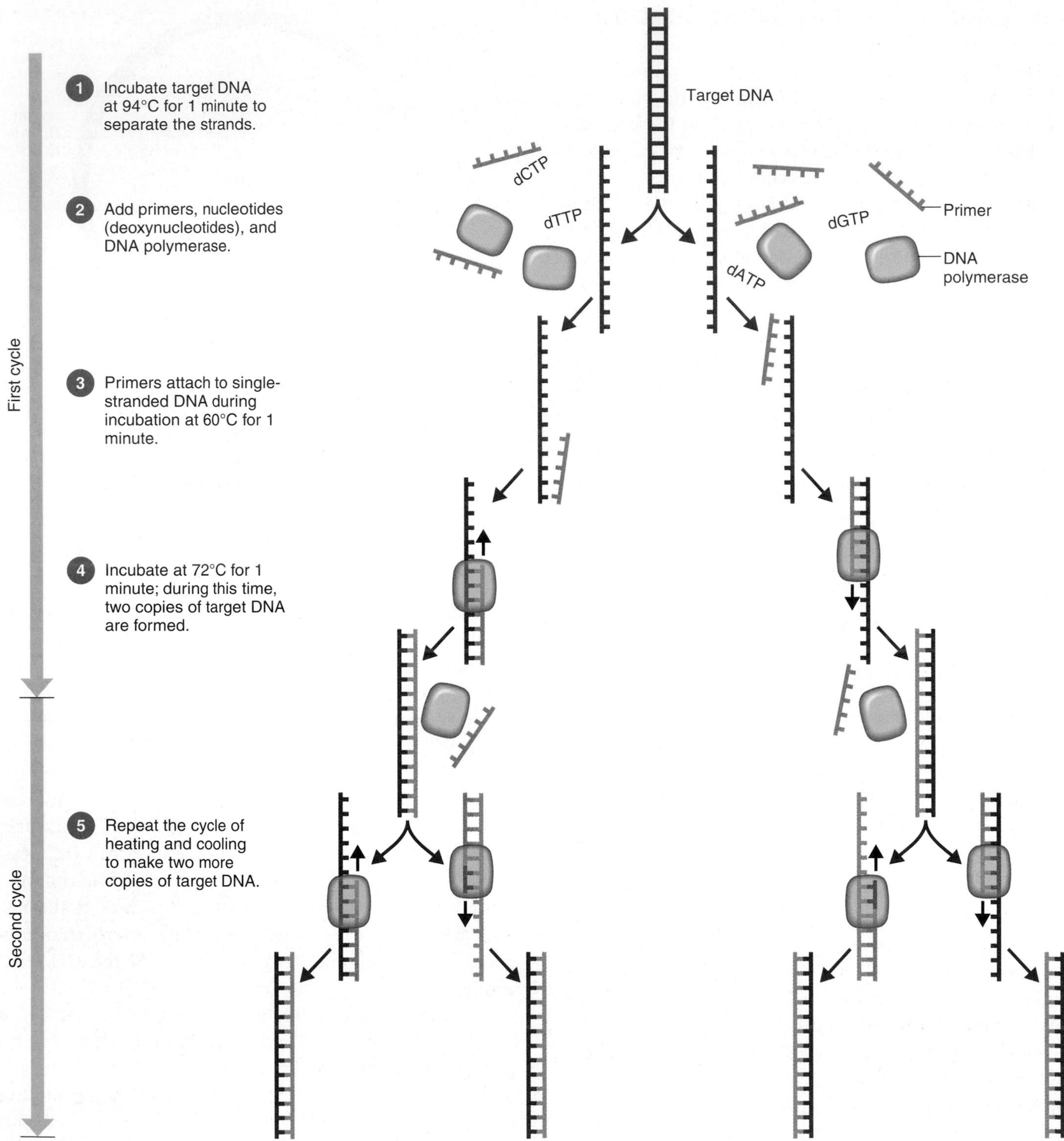

Figure 9.4 The polymerase chain reaction.

Q What enzyme copies DNA in PCR?

Techniques of Genetic Modification

LEARNING OBJECTIVES

9-8 Describe five ways of getting DNA into a cell.

9-9 Describe how a genomic library is made.

9-10 Differentiate cDNA from synthetic DNA.

9-11 Explain how each of the following is used to locate a clone: antibiotic-resistance genes, DNA probes, gene products.

9-12 List one advantage of modifying each of the following: *E. coli*, *Saccharomyces cerevisiae*, mammalian cells, plant cells.

Inserting Foreign DNA into Cells

Recombinant DNA procedures require that DNA molecules be manipulated outside the cell and then returned to living cells. There are several ways to introduce DNA into cells. The choice of method is usually determined by the type of vector and host cell being used.

In nature, plasmids are usually transferred between closely related microbes by cell-to-cell contact, such as in conjugation. In genetic engineering, a plasmid must be inserted into a cell by **transformation,** a procedure during which cells can take up DNA from the surrounding environment (see Chapter 8, page 234). Many cell types, including *E. coli,* yeast, and mammalian cells, do not naturally transform; however, simple chemical treatments can make all of these cell types *competent,* or able to take up external DNA. For *E. coli,* the procedure for making cells competent is to soak them in a solution of calcium chloride for a brief period. Following this treatment, the now-competent cells are mixed with the cloned DNA and given a mild heat shock. Some of these cells will then take up the DNA.

There are other ways to transfer DNA to cells. A process called **electroporation** uses an electrical current to form microscopic pores in the membranes of cells; the DNA then enters the cells through the pores. Electroporation is generally applicable to all cells; those with cell walls often must be converted to protoplasts first (see Chapter 4, page 89). **Protoplasts** are produced by enzymatically removing the cell wall, thereby allowing more direct access to the plasma membrane.

The process of **protoplast fusion** also takes advantage of the properties of protoplasts. Protoplasts in solution fuse at a low but significant rate; the addition of polyethylene glycol increases the frequency of fusion (Figure 9.5a). In the new hybrid cell, the DNA derived from the two "parent" cells may undergo natural recombination. This method is especially valuable in the genetic manipulation of plant and algal cells (Figure 9.5b).

A remarkable way of introducing foreign DNA into plant cells is to literally shoot it directly through the thick cellulose walls using a gene gun (Figure 9.6). Microscopic particles of tungsten or gold are coated with DNA and propelled by a burst of helium through the plant cell walls. Some of the cells express the introduced DNA as though it were their own.

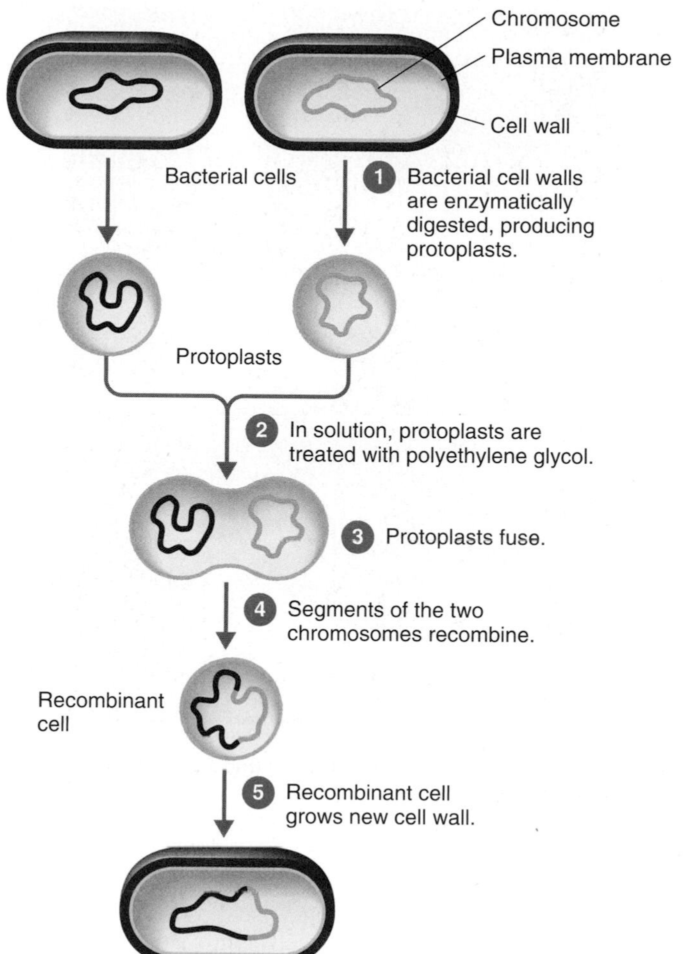
Chromosome
Plasma membrane
Cell wall
Bacterial cells
1 Bacterial cell walls are enzymatically digested, producing protoplasts.
Protoplasts
2 In solution, protoplasts are treated with polyethylene glycol.
3 Protoplasts fuse.
4 Segments of the two chromosomes recombine.
Recombinant cell
5 Recombinant cell grows new cell wall.

(a) Process of protoplast fusion

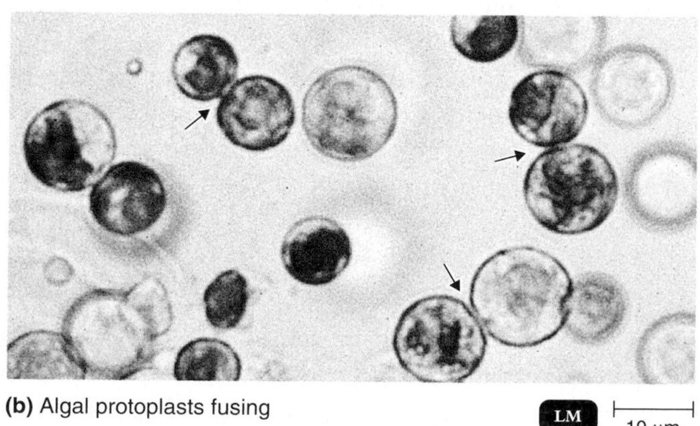
LM
10 μm

(b) Algal protoplasts fusing

Figure 9.5 Protoplast fusion. (**a**) A diagram of protoplast fusion with bacterial cells. (**b**) Protoplast algal cells are shown fusing at the arrows. Removal of the cell wall left only the delicate plasma membrane to bind the cell contents together, allowing the exchange of DNA.

Q What is a protoplast?

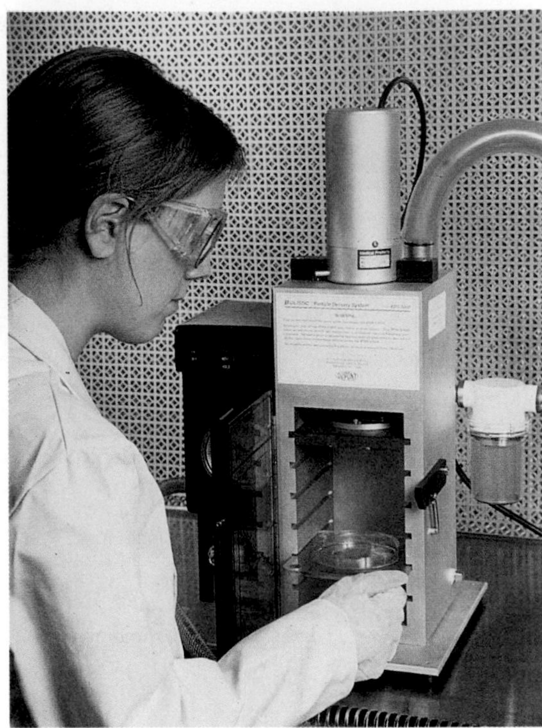

Figure 9.6 A gene gun, which can be used to insert DNA-coated "bullets" into a cell.

Q Name four other methods of inserting DNA into a cell.

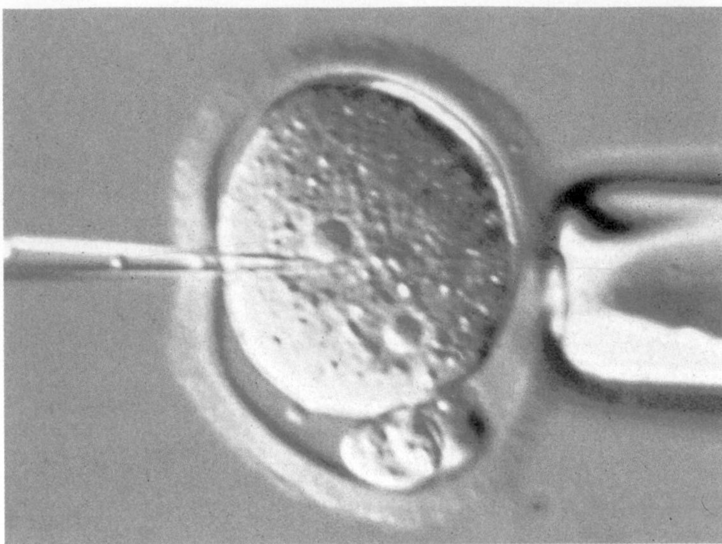

Figure 9.7 The microinjection of foreign DNA into a fertilized mouse egg. The egg is first immobilized by applying mild suction to the large, blunt, holding pipette (right). Several hundred copies of the gene of interest are then injected into the nucleus of the cell through the tiny end of the micropipette (left).

Q Why is microinjection impractical for bacterial and fungal cells?

DNA can be introduced directly into an animal cell by **microinjection.** This technique requires the use of a glass micropipette with a diameter that is much smaller than the cell. The micropipette punctures the plasma membrane, and DNA can be injected through it (Figure 9.7).

Thus, there is a great variety of different restriction enzymes, vectors, and methods of inserting DNA into cells. But foreign DNA will survive only if it is either present on a self-replicating vector or incorporated into one of the cell's chromosomes by recombination.

Obtaining DNA

We have seen how genes can be cloned into vectors by using restriction enzymes and how genes can be transformed or transferred into a variety of cell types. But how do biologists obtain the genes they are interested in? There are two main sources of genes: (1) genomic libraries containing either natural copies of genes or cDNA copies of genes made from mRNA, and (2) synthetic DNA.

Genomic Libraries

Isolating specific genes as individual pieces of DNA is seldom practical. Therefore, researchers interested in genes from a particular organism start by extracting the organism's DNA, which can be obtained from cells of any organism, whether plant, animal, or microbe, by lysing the cells and precipitating the DNA. This process results in a DNA mass that includes the organism's entire genome. After the DNA is digested by restriction enzymes, the restriction fragments are then spliced into plasmid or phage vectors, and the recombinant vectors are introduced into bacterial cells. The goal is to make a collection of clones large enough to ensure that at least one clone exists for every gene in the organism. This collection of clones containing different DNA fragments is called a **genomic library;** each "book" is a bacterial or phage strain that contains a fragment of the genome (Figure 9.8). Such libraries are essential for maintaining and retrieving DNA clones; they can even be purchased commercially.

Cloning genes from eukaryotic organisms presents a specific problem. Genes of eukaryotic cells generally contain both **exons,** stretches of DNA that code for protein, and **introns,** intervening stretches of DNA that do not code for protein. When the RNA transcript of such a gene is converted to mRNA, the introns are removed (see Figure 8.11 on page 222). In cloning genes of eukaryotic cells, it is desirable to use a version of the gene that lacks introns because a gene that includes introns may be too large to work with easily. In addition, if such a gene is put into a bacterial cell, the bacterium will not usually be able to remove the introns from the RNA transcript, and therefore it will not be able to make the correct protein product. However, an artificial gene that contains only exons can be produced by using an enzyme called **reverse transcriptase** to synthesize **complementary DNA (cDNA)** from an mRNA template (Figure 9.9). This synthesis is the reverse of the normal DNA-to-RNA transcription process. A DNA copy of mRNA is produced by reverse transcriptase.

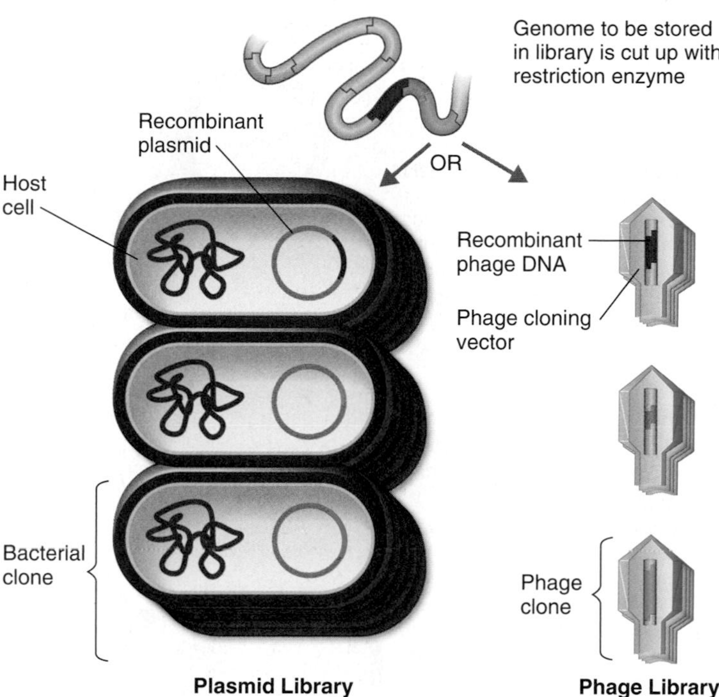

Figure 9.8 Genomic libraries. Each fragment of DNA, containing about one gene, is carried by a vector, either a plasmid within a bacterial cell or a phage.

Q Differentiate an RFLP from a gene.

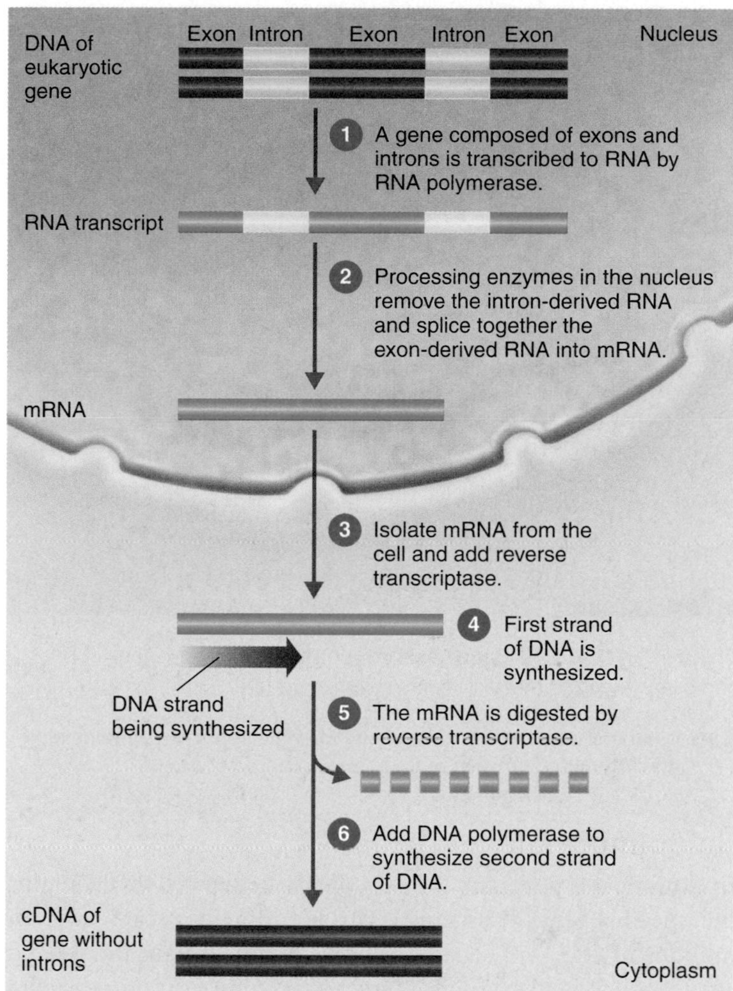

Figure 9.9 Making complementary DNA (cDNA) for a eukaryotic gene. Reverse transcriptase catalyzes the synthesis of double-stranded DNA from an RNA template.

Q How does reverse transcriptase differ from DNA polymerase?

Following this, the mRNA is enzymatically digested away. DNA polymerase then synthesizes a complementary strand of DNA, creating a double-stranded piece of DNA containing the information from the mRNA. Molecules of cDNA produced from a mixture of all the mRNAs from a tissue or cell type can then be cloned to form a cDNA library.

The cDNA method is the most common method of obtaining eukaryotic genes. A difficulty with this method is that long molecules of mRNA may not be completely reverse-transcribed into DNA; the reverse transcription often aborts, forming only parts of the desired gene.

Synthetic DNA

Under certain circumstances, genes can be made in vitro with the help of DNA synthesis machines (Figure 9.10). A keyboard on the machine is used to enter the desired sequence of nucleotides, much as letters are entered into a word processor to compose a sentence. A microprocessor controls the synthesis of the DNA from stored supplies of nucleotides and the other necessary reagents. A chain of over 120 nucleotides can be synthesized by this method. Unless the gene is very small, at least several chains must be synthesized separately and linked together to form an entire gene.

The difficulty of this approach, of course, is that the sequence of the gene must be known before it can be synthesized. If the gene has not already been isolated, then the only way to predict the DNA sequence is by knowing the amino acid sequence of the protein product of the gene. If this amino acid sequence is known, in principle one can work backward through the genetic code to obtain the DNA sequence. Unfortunately, the degeneracy of the code prevents an unambiguous determination; thus, if the protein contains a leucine, for example, which of the six codons for leucine is the one in the gene?

For these reasons, it is rare to clone a gene by synthesizing it directly, although some commercial products such as insulin, interferon, and somatostatin are produced from chemically synthesized genes. Desired restriction sites were added to the synthetic genes so that the genes could be inserted into plasmid vectors for cloning in *E. coli*. Synthetic DNA plays a much more useful role in selection procedures, as we will see.

CHECK YOUR UNDERSTANDING

- ✓ Contrast the five ways of putting DNA into a cell. **9-8**
- ✓ What is the purpose of a genomic library? **9-9**
- ✓ Why isn't cDNA synthetic? **9-10**

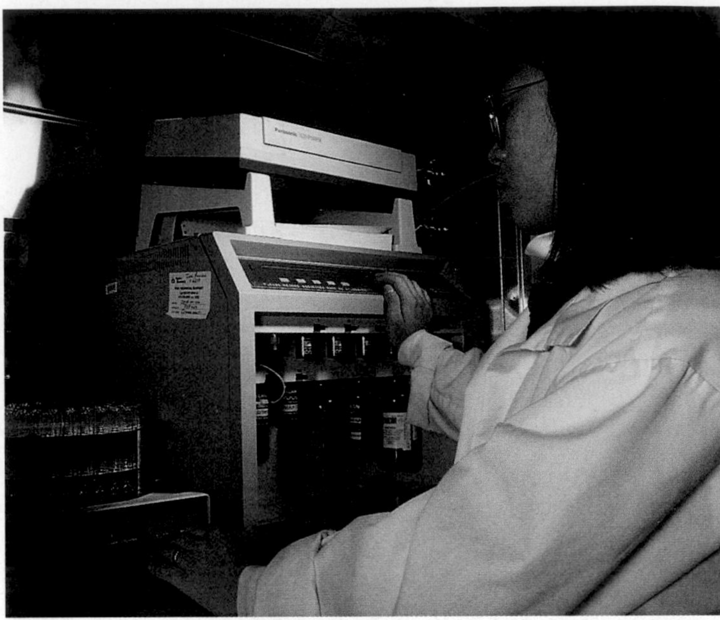

Figure 9.10 A DNA synthesis machine. Short sequences of DNA can be synthesized by instruments such as this one.

Q What are some of the disadvantages of using a DNA synthesis machine?

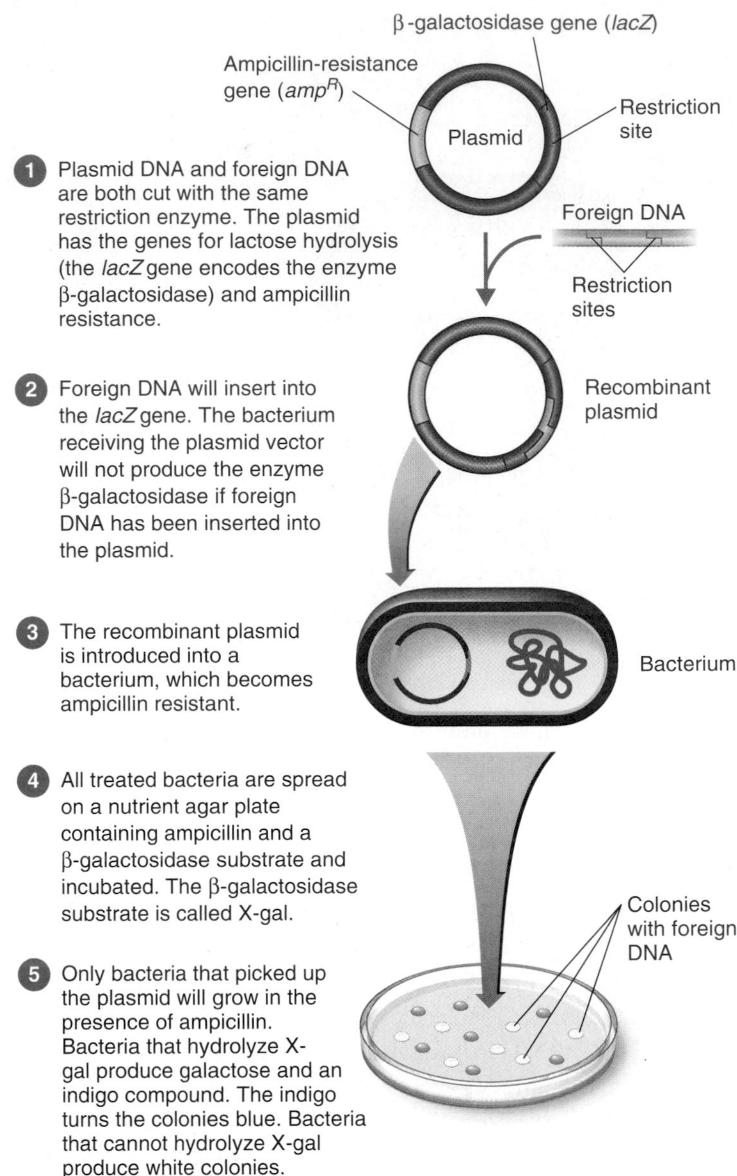

β-galactosidase gene (*lacZ*)
Ampicillin-resistance gene (amp^R)
Restriction site
Plasmid
1 Plasmid DNA and foreign DNA are both cut with the same restriction enzyme. The plasmid has the genes for lactose hydrolysis (the *lacZ* gene encodes the enzyme β-galactosidase) and ampicillin resistance.
Foreign DNA
Restriction sites
2 Foreign DNA will insert into the *lacZ* gene. The bacterium receiving the plasmid vector will not produce the enzyme β-galactosidase if foreign DNA has been inserted into the plasmid.
Recombinant plasmid
3 The recombinant plasmid is introduced into a bacterium, which becomes ampicillin resistant.
Bacterium
4 All treated bacteria are spread on a nutrient agar plate containing ampicillin and a β-galactosidase substrate and incubated. The β-galactosidase substrate is called X-gal.
Colonies with foreign DNA
5 Only bacteria that picked up the plasmid will grow in the presence of ampicillin. Bacteria that hydrolyze X-gal produce galactose and an indigo compound. The indigo turns the colonies blue. Bacteria that cannot hydrolyze X-gal produce white colonies.

Figure 9.11 Blue-white screening, one method of selecting recombinant bacteria.

Q Why are some colonies blue and others white?

Selecting a Clone

In cloning, it is necessary to select the particular cell that contains the specific gene of interest. This is difficult because out of millions of cells, only a very few cells might contain the desired gene. Here we will examine a typical screening procedure known as *blue-white screening,* from the color of the bacterial colonies formed at the end of the screening process.

The plasmid vector used contains a gene (amp^R) coding for resistance to the antibiotic ampicillin. The host bacterium will not be able to grow on the test medium, which contains ampicillin, unless the vector has transferred the ampicillin-resistance gene. The plasmid vector also contains a second gene, this one for the enzyme β-galactosidase (*lacZ*). Notice in Figure 9.3 that there are several sites in *lacZ* that can be cut by restriction enzymes.

The procedure is shown in **Figure 9.11**. The two genes, called marker genes, are used so that the insertion of plasmid DNA into the host bacterium can be determined. In the blue-white screening procedure, a library of bacteria is cultured in a medium called X-gal. X-gal contains two essential components other than those necessary to support normal bacterial growth. One is the antibiotic ampicillin, which prevents the growth of any bacterium that has not successfully received the ampicillin-resistance gene from the plasmid. The other, called X-gal, is a substrate for β-galactosidase.

Only bacteria that picked up the plasmid will grow—because they are now ampicillin resistant. Bacteria that picked up the recombinant plasmid—in which the new gene was inserted into the *lacZ* gene—will not hydrolyze lactose and will produce white colonies. If a bacterium received the original plasmid containing the intact *lacZ* gene, the cells will hydrolyze X-gal to produce a blue-colored compound; the colony will be blue.

What remains to be done can still be difficult. The above procedure has isolated white colonies known to contain foreign DNA, but it is still not known whether this is the desired fragment of foreign DNA. A second procedure is needed to identify these bacteria. If the foreign DNA in the plasmid codes for the production of an identifiable product, the bacterial isolate only needs to be grown in culture and tested. However, in some cases the gene itself must be identified in the host bacterium.

Colony hybridization is a common method of identifying cells that carry a specific cloned gene. **DNA probes,** short seg-

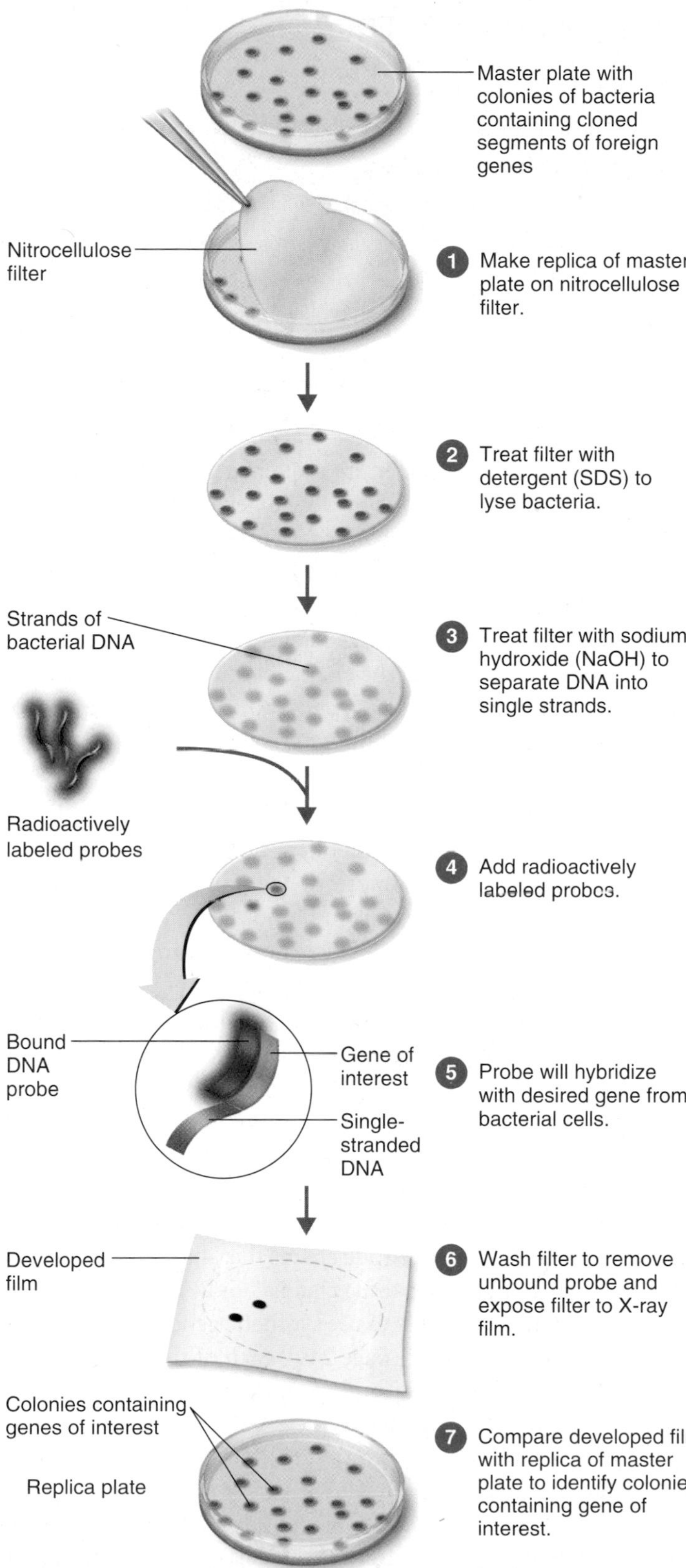

Master plate with colonies of bacteria containing cloned segments of foreign genes
Nitrocellulose filter
1 Make replica of master plate on nitrocellulose filter.
2 Treat filter with detergent (SDS) to lyse bacteria.
Strands of bacterial DNA
3 Treat filter with sodium hydroxide (NaOH) to separate DNA into single strands.
Radioactively labeled probes
4 Add radioactively labeled probes.
Bound DNA probe
Gene of interest
Single-stranded DNA
5 Probe will hybridize with desired gene from bacterial cells.
Developed film
6 Wash filter to remove unbound probe and expose filter to X-ray film.
Colonies containing genes of interest
Replica plate
7 Compare developed film with replica of master plate to identify colonies containing gene of interest.

Figure 9.12 Colony hybridization: using a DNA probe to identify a cloned gene of interest.

Q What is a DNA probe?

ments of single-stranded DNA that are complementary to the desired gene, are synthesized. If the DNA probe finds a match, it will adhere to the target gene. The DNA probe is labeled with a radioactive element or fluorescent dye so its presence can be determined. A typical colony hybridization experiment is shown in Figure 9.12. An array of DNA probes arranged in a DNA chip can be used to identify pathogens (see Figure 10.17, page 293).

Making a Gene Product

Q&A We have just seen how to identify cells carrying a particular gene. The gene products are frequently the objective of genetic modification. Most of the earliest work in genetic modification used *E. coli* to synthesize the gene products. *E. coli* is easily grown, and researchers are very familiar with this bacterium and its genetics. For example, some inducible promoters, such as that of the *lac* operon, have been cloned, and cloned genes can be attached to such promoters. The synthesis of great amounts of the cloned gene product can then be directed by the addition of an inducer. Such a method has been used to produce gamma interferon in *E. coli* (Figure 9.13). However, *E. coli* also has several disadvantages. Like other gram-negative bacteria, it produces endotoxins as part of the outer layer of its cell wall. Because endotoxins cause fever and shock in animals, their accidental presence in products intended for human use would be a serious problem.

Another disadvantage of *E. coli* is that it does not usually secrete protein products. To obtain a product, cells must usually be broken open and the product purified from the resulting "soup" of cell components. Recovering the product from such a

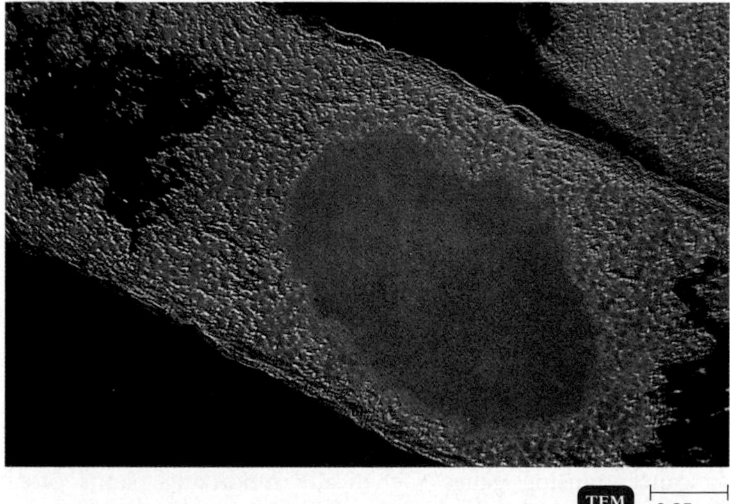

TEM
0.25 μm

Figure 9.13 *E. coli* genetically engineered to produce gamma interferon, a human protein that promotes an immune response. The product, visible here as an orange-colored substance, can be released by lysis of the cell.

Q What is one advantage of using *E. coli* for genetic engineering? One disadvantage?

mixture is expensive when done on an industrial scale. It is more economical to have an organism secrete the product so that it can be recovered continuously from the growth medium. One approach has been to link the product to a natural *E. coli* protein that the bacterium does secrete. However, gram-positive bacteria, such as *Bacillus subtilis,* are more likely to secrete their products and are often preferred industrially for that reason.

Another microbe being used as a vehicle for expressing genetically engineered genes is baker's yeast, *Saccharomyces cerevisiae.* Its genome is only about four times larger than that of *E. coli* and is probably the best understood eukaryotic genome. Yeasts may carry plasmids, and the plasmids are easily transferred into yeast cells after their cell walls have been removed. As eukaryotic cells, yeasts may be more successful in expressing foreign eukaryotic genes than bacteria. Furthermore, yeasts are likely to continuously secrete the product. Because of all these factors, yeasts have become the eukaryotic workhorse of biotechnology.

Mammalian cells in culture, even human cells, can be genetically modified much like bacteria to produce various products. Scientists have developed effective methods of growing certain mammalian cells in culture as hosts for growing viruses (see Chapter 13, page 378). Mammalian cells are often the best suited to making protein products for medical use because the cells secrete their products and there is a low risk of toxins or allergens. Using mammalian cells to make foreign gene products on an industrial scale often requires a preliminary step of cloning the gene in bacteria. Consider the example of colony-stimulating factor (CSF). A protein produced naturally in tiny amounts by white blood cells, CSF is valuable because it stimulates the growth of certain cells that protect against infection. To produce huge amounts of CSF industrially, the gene is first inserted into a plasmid, and bacteria are used to make multiple copies of the plasmid (see Figure 9.1). The recombinant plasmids are inserted into mammalian cells that are grown in bottles.

Plant cells can also be grown in culture, altered by recombinant DNA techniques, and then used to generate genetically modified plants. Such plants may prove useful as sources of valuable products, such as plant alkaloids (the painkiller codeine, for example), the isoprenoids that are the basis of synthetic rubber, and melanin (the animal skin pigment) for use in sunscreens. Genetically modified plants have many advantages for the production of human therapeutic agents, including vaccines and antibodies. The advantages include large-scale, low-cost production using agriculture and low risk of product contamination by mammalian pathogens or cancer-causing genes. Genetically modifying plants often requires use of a bacterium. We will return to the topic of genetically modified plants later in the chapter (page 264).

CHECK YOUR UNDERSTANDING

✓ How are recombinant clones identified? **9-11**

✓ What types of cells are used for cloning rDNA? **9-12**

Applications of rDNA

LEARNING OBJECTIVES

9-13 List at least five applications of rDNA technology.

9-14 Define RNAi.

9-15 Discuss the value of the Human Genome Project.

9-16 Define the following terms: *random shotgun sequencing, bioinformatics, proteomics.*

9-17 Diagram the Southern blotting procedure, and provide an example of its use.

9-18 Diagram DNA fingerprinting, and provide an example of its use.

9-19 Outline genetic engineering with *Agrobacterium.*

We have now described the entire sequence of events in cloning a gene. As indicated earlier, such cloned genes can be applied in a variety of ways. One is to produce useful substances more efficiently and less expensively (see the box in Chapter 1, page 3). Another is to obtain information from the cloned DNA that is useful for either basic research, medicine, or forensics. A third is to use cloned genes to alter the characteristics of cells or organisms. The box in Chapter 27, page 780, describes the use of recombinant cells to detect pollutants.

Therapeutic Applications

An extremely valuable pharmaceutical product is the hormone insulin, a small protein produced by the pancreas that controls the body's uptake of glucose from blood. For many years, people with insulin-dependent diabetes have controlled their disease by injecting insulin obtained from the pancreases of slaughtered animals. Obtaining this insulin is an expensive process, and the insulin from animals is not as effective as human insulin.

Because of the value of human insulin and the small size of the protein, producing human insulin by recombinant DNA techniques was an early goal for the pharmaceutical industry. To produce the hormone, synthetic genes were first constructed for each of the two short polypeptide chains that make up the insulin molecule. The small size of these chains—only 21 and 30 amino acids long—made it possible to use synthetic genes. Following the procedure described earlier (page 249), each of the two synthetic genes was inserted into a plasmid vector and linked to the end of a gene coding for the bacterial enzyme β-galactosidase, so that the insulin polypeptide was coproduced with the enzyme. Two different *E. coli* bacterial cultures were used, one to produce each of the insulin polypeptide chains. The polypeptides were then recovered from the bacteria, separated from the β-galactosidase, and chemically joined to make human insulin. This accomplishment was one of the early commercial successes of DNA technology, and it illustrates a number of the principles and procedures discussed in this chapter.

Another human hormone that is now being produced commercially by genetic modification of *E. coli* is somatostatin. At one time 500,000 sheep brains were needed to produce 5 mg of animal somatostatin for experimental purposes. By contrast, only 8 liters of a genetically modified bacterial culture are now required to obtain the equivalent amount of the human hormone.

Subunit vaccines, consisting only of a protein portion of a pathogen, are being made by genetically modifying yeasts. Subunit vaccines have been produced for a number of diseases, notably hepatitis B. One of the advantages of a subunit vaccine is that there is no chance of becoming infected from the vaccine. The protein is harvested from genetically modified cells and purified for use as a vaccine. Animal viruses such as vaccinia virus can be genetically modified to carry a gene for another microbe's surface protein. When injected, the virus acts as a vaccine against the other microbe.

DNA vaccines are usually circular plasmids that include a gene encoding a viral protein under the transcriptional control of a promoter region active in human cells. The plasmids are cloned in bacteria. Several trial vaccines against HIV, SARS, influenza, and malaria are being tested. Vaccines are discussed in Chapter 18 (page 501). Table 9.2 lists some other important rDNA products used in medical therapy.

The importance of recombinant DNA technology to medical research cannot be emphasized enough. Artificial blood for use in transfusions can now be prepared using human hemoglobin produced in genetically modified pigs. Sheep have also been genetically modified to produce a number of drugs in their milk. This procedure has no apparent effect upon the sheep, and they provide a ready source of raw material for the product that does not require sacrificing animals.

Gene therapy may eventually provide cures for some genetic diseases. It is possible to imagine removing some cells from a person and transforming them with a normal gene to replace a defective or mutated gene. When these cells are returned to the person, they should function normally. For example, gene therapy has been used to treat hemophilia B and severe combined immunodeficiency. Adenoviruses and retroviruses are used most often to deliver genes; however, some researchers are working with plasmid vectors. The first gene therapy to treat hemophilia in humans was done in 1999. An attenuated retrovirus was used as the vector. Several gene therapy trials are in progress using genetically modified adenovirus carrying human gene *p53* to treat a variety of cancers. The *p53* gene, which encodes a tumor-suppressing protein, is the most frequently mutated gene in cancer cells.

The number of gene therapy trials will increase as technical improvements are made and initial attempts are successful. However, there is a great deal of preliminary work to do, and cures may not be possible for all genetic diseases. Antisense DNA (see page 267) introduced into cells is also being explored to treat hepatitis, cancers, and one type of coronary artery disease.

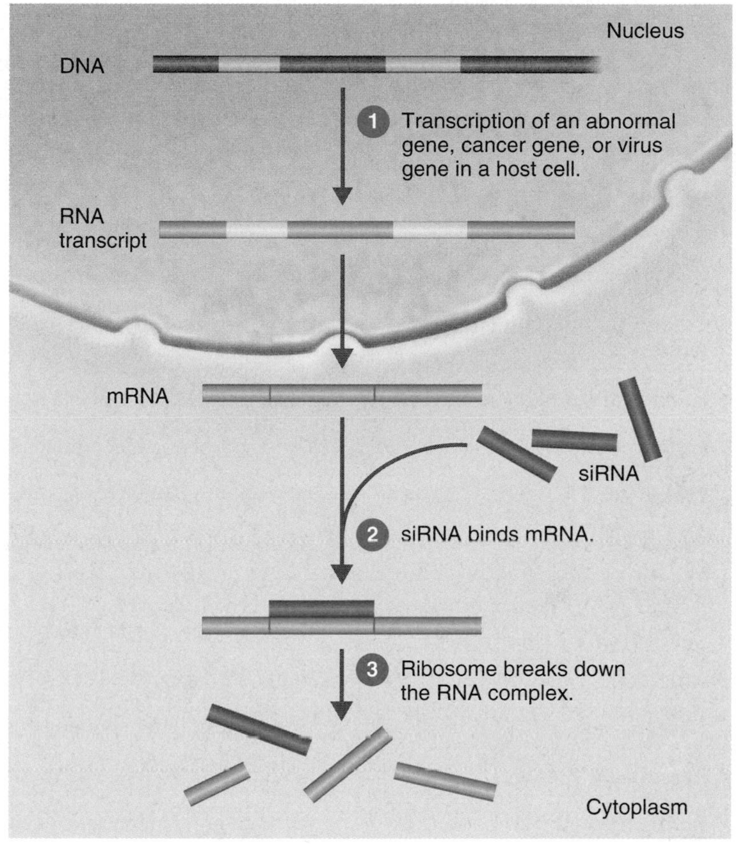

Figure 9.14 Gene silencing could provide treatments for a wide range of diseases.

Q Does RNAi act during or after transcription?

Gene silencing is a natural process that occurs in a wide variety of organisms and is apparently a defense against viruses and transposons. A new technology called **RNA interference (RNAi)** holds promise for gene therapy and treating cancer and viral infections. Double-stranded RNAs called **small interfering RNAs (siRNA)** that target a particular gene, such as a virus gene, can be introduced into a cell (Figure 9.14). The siRNA molecules bind to mRNA causing its enzymatic destruction, thus *silencing* the expression of a gene. In mice, RNAi has been shown to inhibit hepatitis B virus. The siRNA can be injected into a cell or introduced in a DNA vector. A small DNA insert encoding siRNA against the gene of interest could be cloned into a DNA vector. When transferred into a cell, the cell would produce the desired siRNA.

CHECK YOUR UNDERSTANDING

- ✓ Explain how rDNA technology can be used to treat disease and to prevent disease. 9-13
- ✓ What is gene silencing? 9-14

Table 9.2 Some Pharmaceutical Products of Genetic Engineering

Product	Comments
Antitrypsin	Assists emphysema patients; produced by genetically modified sheep
Bone Morphogenic Proteins	Induces new bone formation; useful in healing fractures and reconstructive surgery; produced by mammalian cell culture
Cervical Cancer Vaccine	Consists of viral proteins produced by *S. cerevisiae*
Colony-Stimulating Factor (CSF)	Counteracts effects of chemotherapy; improves resistance to infectious disease such as AIDS; treatment of leukemia; produced by *E. coli* and *S. cerevisiae*
Epidermal Growth Factor (EGF)	Heals wounds, burns, ulcers; produced by *E. coli*
Erythropoietin (EPO)	Treatment of anemia; produced by mammalian cell culture
Factor VII	Treatment of hemorrhagic strokes; produced by mammalian cell culture
Factor VIII	Treatment of hemophilia; improves clotting; produced by mammalian cell culture
Interferon	
IFN–α	Therapy for leukemia, melanoma, and hepatitis; produced by *E. coli* and *S. cerevisiae* (yeast)
IFN–β	Treatment for multiple sclerosis; produced by mammalian cell culture
IFN–γ	Treatment of chronic granulomatous disease; produced by *E. coli*
Hepatitis B Vaccine	Produced by *S. cerevisiae* that carries hepatitis-virus gene on a plasmid
Human Growth Hormone (hGH)	Corrects growth deficiencies in children; produced by *E. coli*
Human Insulin	Therapy for diabetes; better tolerated than insulin extracted from animals; produced by *E. coli*
Influenza Vaccine	Trial vaccine made from *E. coli* or *S. cerevisiae* carrying virus genes
Interleukins	Regulate the immune system; possible treatment for cancer; produced by *E. coli*
Monoclonal Antibodies	Possible therapy for cancer and transplant rejection; used in diagnostic tests; produced by mammalian cell culture (from fusion of cancer cell and antibody-producing cell)
Orthoclone OKT3 Muromonab-CD3	Monoclonal antibody used in transplant patients to help suppress the immune system, reducing the chance of tissue rejection; produced by mouse cells
Prourokinase	Anticoagulant; therapy for heart attacks; produced by *E. coli* and yeast
Pulmozyme (rhDNase)	Enzyme used to break down mucous secretions in cystic fibrosis patients; produced by mammalian cell culture
Relaxin	Used to ease childbirth; produced by *E. coli*
Superoxide Dismutase (SOD)	Minimizes damage caused by oxygen free radicals when blood is resupplied to oxygen-deprived tissues; produced by *S. cerevisiae* and *Komagataella pastoris* (yeast)
Taxol	Plant product used for treatment for ovarian cancer; produced in *E. coli*
Tissue Plasminogen Activator (Activase)	Dissolves the fibrin of blood clots; therapy for heart attacks; produced by mammalian cell culture
Tumor Necrosis Factor (TNF)	Causes disintegration of tumor cells; produced by *E. coli*

The Human Genome Project

One monumental project involving rDNA technology is the Human Genome Project. The Human Genome Project was an international 13-year effort, formally begun in October 1990 and completed in 2003. The goal of this project was to sequence the entire human genome, approximately 3 billion nucleotide pairs, comprising 20,000 to 25,000 genes. Thousands of people in 18 countries participated in this project. Researchers collected blood (female) or sperm (male) samples from a large number of donors. Only a few samples were processed as DNA resources, and the source names are protected so that neither donors nor scientists know whose samples were used. Development of shotgun sequencing (discussed shortly) greatly speeded the process, and the genome is nearly complete.

One surprising finding was that less than 2% of the genome encodes a functional product—the other 98% is being called "junk DNA." This includes introns, the chromosome ends (called *telomeres*), and the transposons (repeating sequences that make up more than half of the human genome; see Chapter 8, page 240). Currently, researchers are locating specific genes and determining their functions.

The next goal of researchers is the Human Proteome Project, which will map all the proteins expressed in human cells. Even before it is completed, however, it is yielding data that are of immense value to our understanding of biology. It will also eventually be of great medical benefit, especially for the diagnosis and treatment of genetic diseases.

Scientific Applications

Recombinant DNA technology can be used to make products, but this is not its only important application. Because of its ability to produce many copies of DNA, it can serve as a sort of DNA "printing press." Once a large amount of a particular piece of DNA is available, various analytic techniques, discussed in this section, can be used to "read" the information contained in the DNA.

What kind of information can be obtained from cloned DNA? One kind is provided by the process of **DNA sequencing**—the determination of the exact sequence of nucleotide bases in DNA.

One technique for genome sequencing is **random shotgun sequencing.** Small pieces of a genome are sequenced, and the sequences are then assembled using a computer. Any gaps between the pieces then have to be found and sequenced (Figure 9.15). The sequences of entire viral genomes are now relatively easy to obtain. The genomes of *Saccharomyces cerevisiae* yeast, *E. coli,* and over 70 other microbes have been mapped; another 100 are in progress. The Human Genome Project was a massive undertaking resulting in sequencing the human genome. The published maps are between 70% and 99% complete, with some gaps remaining to be filled. Most of the gaps are repeated sequences that do not encode genes. For example, the *S. cerevisiae* genome is 93% complete, but the gene encoding regions are 100% complete. Computer applications can also be used to look for protein encoding regions, which can then be "translated" by computer software.

DNA sequencing has produced an enormous amount of information that has spawned the new field of **bioinformatics,**

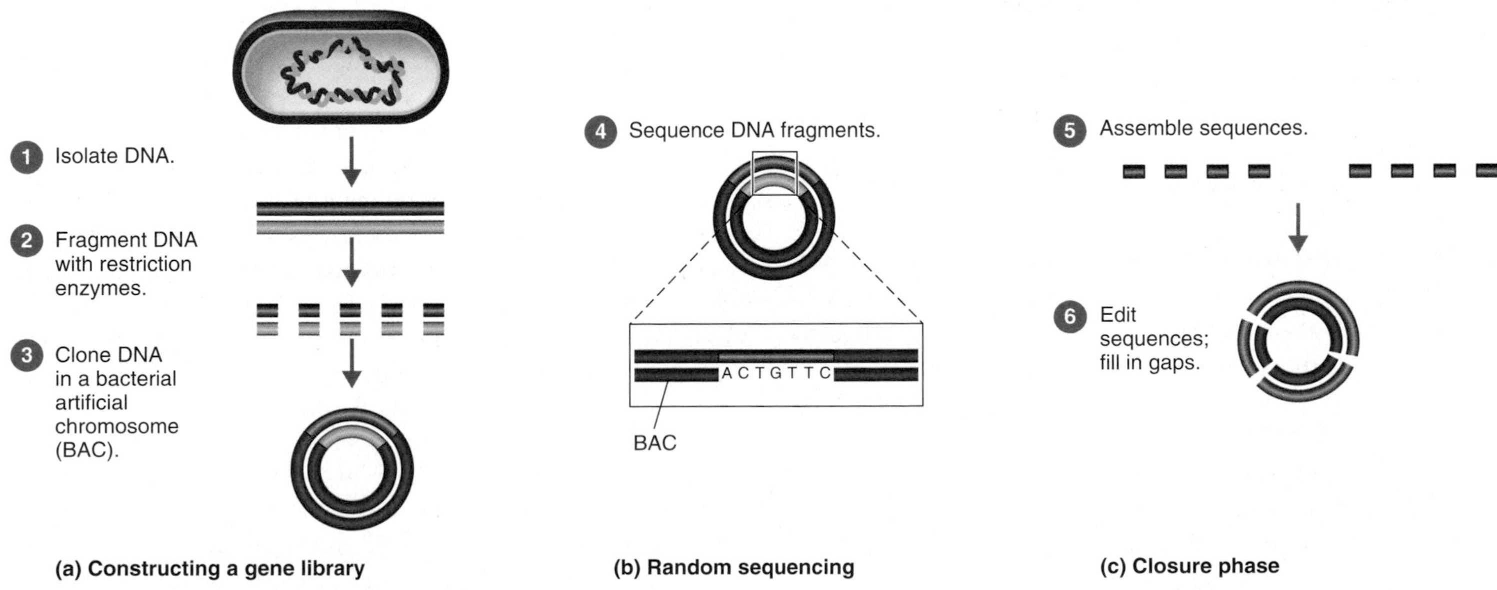

Figure 9.15 Random shotgun sequencing. In this technique a genome is cut into pieces, and each piece is sequenced. Then the pieces are fit together. There may be gaps if a specific DNA fragment was not sequenced.

Q **Does this technique identify genes and their locations?**

the science of understanding the function of genes through computer-assisted analysis. DNA sequences are stored in web-based databases referred to as GenBank. Genomic information can be searched with computer programs to find specific sequences or to look for similar patterns in the genomes of different organisms. Microbial genes are now being searched to identify molecules that are the virulence factors of pathogens. By comparing genomes, researchers discovered that *Chlamydia trachomatis* (tra-kō′mä-tis) produces a toxin similar to that of *Clostridium difficile* (dif′fi-sē-il).

The next goal is to identify proteins encoded by these genes. **Proteomics** is the science of determining all of the proteins expressed in a cell.

Reverse genetics is an approach to discovering the function of a gene from a genetic sequence. Reverse genetics attempts to connect a given genetic sequence with specific effects on the organism. For example, if you mutate or block a gene (see the discussion of gene silencing on page 259), you can then look for a characteristic the organism lost.

An example of the use of human DNA sequencing is the identification and cloning of the mutant gene that causes cystic fibrosis (CF). CF is characterized by the oversecretion of mucus, leading to blocked respiratory passageways. The sequence of the mutated gene can be used as a diagnostic tool in a hybridization technique called **Southern blotting** (Figure 9.16), named for Ed Southern, who developed the technique in 1975.

In this technique,

1. human DNA is first digested with a restriction enzyme, yielding thousands of fragments of various sizes. The different fragments are then separated by **gel electrophoresis.**
2. The fragments are put in a well at one end of a layer of agarose gel. Then an electrical current is passed through the gel. While the charge is applied, the different-sized pieces of DNA migrate through the gel at different rates. The fragments are called **RFLPs,** for *restriction fragment length polymorphisms.*

3–4. The separated fragments are transferred onto a filter by blotting.

5. The fragments on the filter are then exposed to a radioactive probe made from the cloned gene of interest, in this case the CF gene. The probe will hybridize to this mutant gene but not to the normal gene.
6. Fragments to which the probe binds are identified by exposing the filter to X-ray film. With this method, any person's DNA can be tested for the presence of the mutated gene.

This process, called **genetic screening,** can now be used to screen for several hundred genetic diseases. Such screening procedures can be performed on prospective parents and also on fetal tissue. Two of the more commonly screened genes are those associated with inherited forms of breast cancer and the gene responsible for Huntington's disease.

Forensic Microbiology

Several important diagnostic tools, many of which rely on the technique of hybridization, are now available as a result of rDNA technology. Recall that hybridization enables the identification of a particular DNA sequence among many others. This is exactly what is necessary in most diagnostic situations—identifying a particular pathogen among many others.

DNA probes such as the ones used to screen gene libraries are promising tools for the rapid identification of microorganisms. For use in medical diagnosis, these probes are derived from the DNA of a pathogenic microbe and are labeled (with a radioactive tag, for example). The probe then serves a diagnostic function by combining with the DNA of the pathogen to reveal its location in body tissue (or perhaps its presence in food). Probes are also being used in nonmedical aspects of microbiology—for example, for locating and identifying specific microbes in soil. We will discuss PCR and DNA probes further in Chapter 10.

For several years, microbiologists have used RFLPs in a method of identification known as **DNA fingerprinting** to identify bacterial or viral pathogens (Figure 9.17). DNA fingerprinting is also used in forensic medicine to determine paternity or to prove that blood on a murder suspect's clothes came from the murder victim. Southern blotting requires substantial amounts of DNA. As mentioned earlier, small samples of DNA can be quickly amplified, or increased to quantities large enough for analysis, by PCR.

DNA chips (see Figure 10.17, page 293) and *PCR microarrays* that can screen a sample for multiple pathogens at once are being developed. In a PCR microarray, up to 22 primers from different microorganisms can be used to initiate the PCR. The microorganism is identified if DNA is copied from one of the primers.

The genomics of pathogens has become a mainstay of monitoring, preventing, and controlling infectious disease. The use of genomics to trace a disease outbreak is described in the box on page 266. The new field of **forensic microbiology** developed because hospitals and food manufacturers can be sued in courts of law and because microorganisms can be used as weapons. The requirements to prove in a court of law the source of a microbe are stricter than for the medical community. For example, to prove intent to commit harm requires collecting evidence properly and establishing a chain of custody of that evidence. Microbial properties that are unimportant in public health may be important clues in forensic investigations. The American Academy of Microbiology recently proposed professional certification in forensic microbiology.

DNA can often be extracted from preserved and fossilized materials, including mummies and extinct plants and animals. Although such material is very rare, and usually partially degraded,

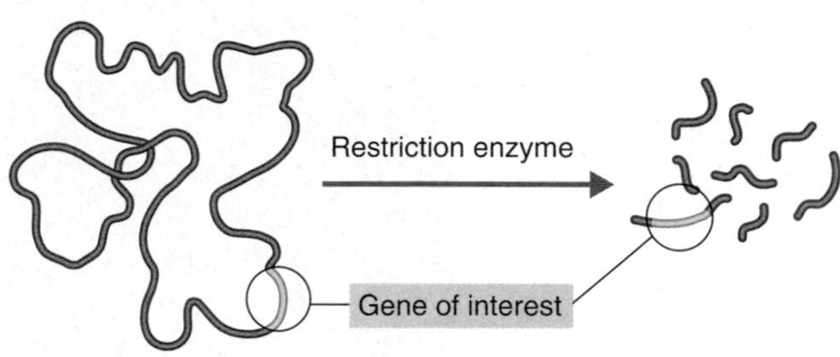

1 DNA containing the gene of interest is extracted from human cells and cut into fragments by restriction enzymes.

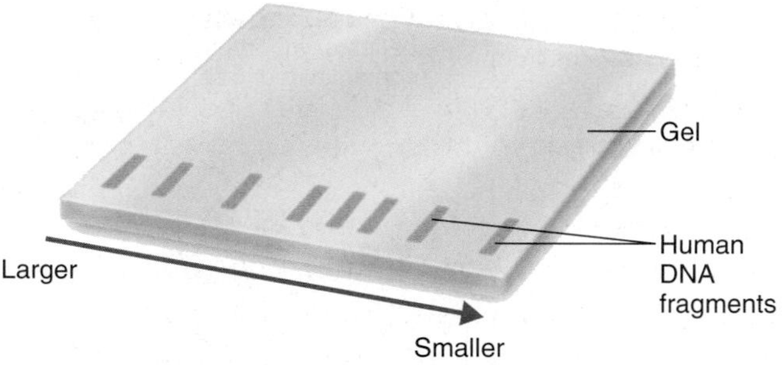

2 The fragments are separated according to size by gel electrophoresis. Each band consists of many copies of a particular DNA fragment. The bands are invisible but can be made visible by staining.

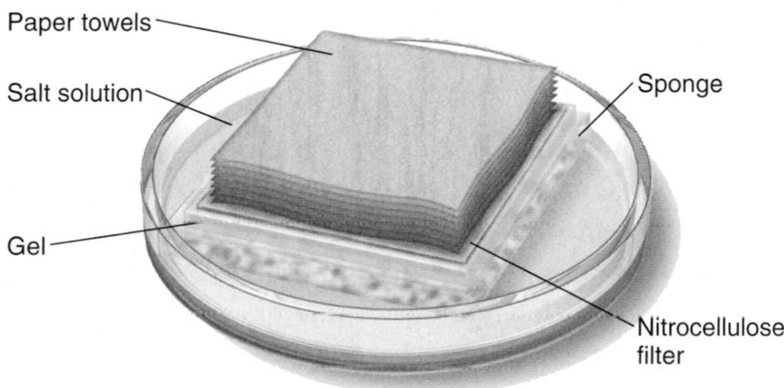

3 The DNA bands are transferred to a nitrocellulose filter by blotting. The solution passes through the gel and filter to the paper towels.

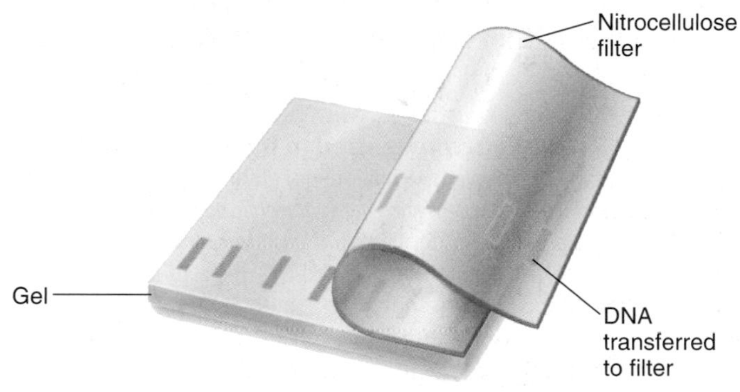

4 This produces a nitrocellulose filter with DNA fragments positioned exactly as on the gel.

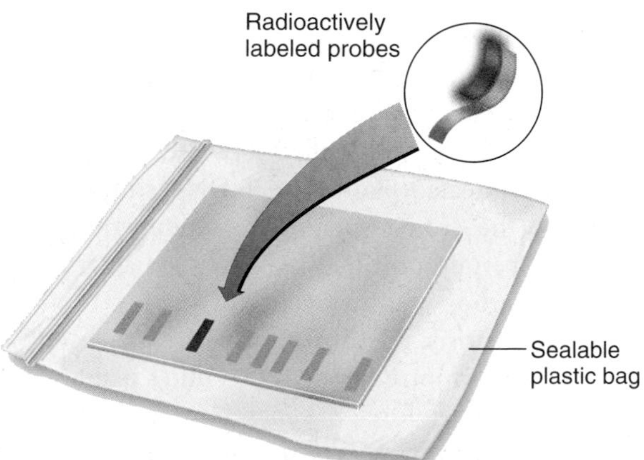

5 The filter is exposed to a radioactively labeled probe for a specific gene. The probe will base-pair (hybridize) with a short sequence present on the gene.

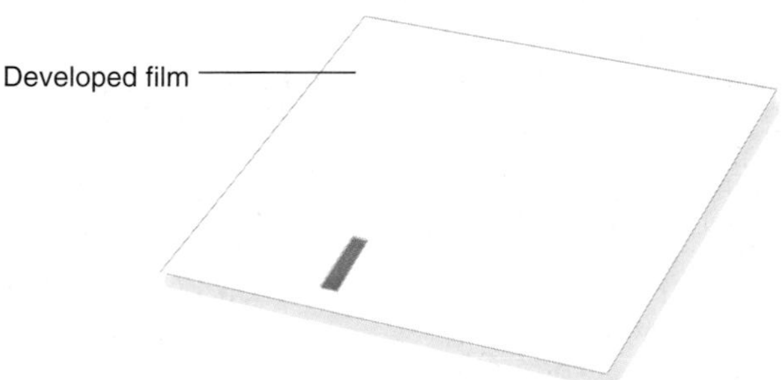

6 The filter is then exposed to X-ray film. The fragment containing the gene of interest is identified by a band on the developed film.

Figure 9.16 Southern blotting.

Q What is the purpose of Southern blotting?

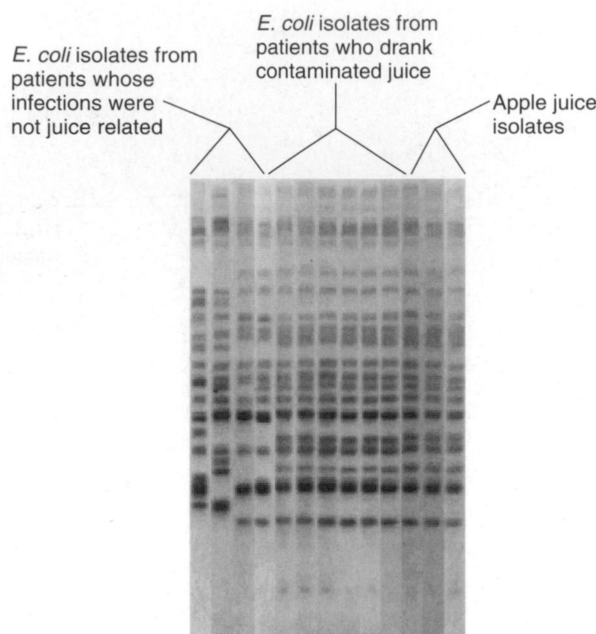

Figure 9.17 DNA fingerprints used to track an infectious disease. This figure shows the relationship among DNA patterns of bacterial isolates from an outbreak of *Escherichia coli* O157:H7. The isolates from apple juice are identical to the patterns of isolates from patients who drank the contaminated juice but different from those from patients whose infections were not juice related.

Q What is forensic microbiology?

PCR enables researchers to study this genetic material that no longer exists in its natural form. The study of unusual organisms has also led to advances in basic taxonomy; this will be discussed in Chapter 10.

Nanotechnology

Nanotechnology deals with the design and manufacture of extremely small electronic circuits and mechanical devices built at the molecular level of matter. Molecule-sized robots or computers can be used to detect contamination in food, diseases in plants, or biological weapons. However, the small machines require small (a nanometer is 10^{-9} meters; 1000 nm fit in 1 μm) wires and components. Bacteria may provide the needed small metals. Researchers at the U.S. Geological Survey have cultured several anaerobic bacteria that reduce toxic selenium, Se^{4+}, to nontoxic elemental Se^0 which forms into nanospheres (Figure 9.18).

CHECK YOUR UNDERSTANDING

- ✓ How are shotgun sequencing, bioinformatics, and proteomics related to the Human Genome Project? **9-15, 9-16**
- ✓ What is Southern blotting? **9-17**
- ✓ Why do RFLPs result in a DNA fingerprint? **9-18**

Figure 9.18 ***Bacillus*** **cells growing on selenium form chains of elemental selenium.**

Q What might bacteria provide for nanotechnology?

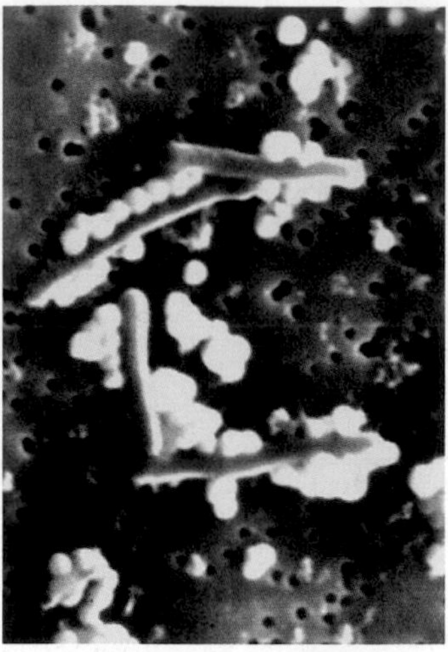

Agricultural Applications

The process of selecting for genetically desirable plants has always been a time-consuming one. Performing conventional plant crosses is laborious and involves waiting for the planted seed to germinate and for the plant to mature. Plant breeding has been revolutionized by the use of plant cells grown in culture. Clones of plant cells, including cells that have been genetically altered by recombinant DNA techniques, can be grown in large numbers. These cells can then be induced to regenerate whole plants, from which seeds can be harvested.

Recombinant DNA can be introduced into plant cells in several ways. Previously we mentioned protoplast fusion and the use of DNA-coated "bullets." The most elegant method, however, makes use of a plasmid called the **Ti plasmid** (*Ti* stands for tumor-inducing), which occurs naturally in the bacterium *Agrobacterium tumefaciens* (tu′me-fash-enz). This bacterium infects certain plants, in which the Ti plasmid causes the formation of a tumorlike growth called a crown gall (Figure 9.19). A part of the Ti plasmid, called T-DNA, integrates into the genome of the infected plant. The T-DNA stimulates local cellular growth (the crown gall) and simultaneously causes the production of certain products used by the bacteria as a source of nutritional carbon and nitrogen.

For plant scientists, the attraction of the Ti plasmid is that it provides a vehicle for introducing rDNA into a plant (Figure 9.20). A scientist can insert foreign genes into the T-DNA, put the recombinant plasmid back into the *Agrobacterium* cell, and use the bacterium to insert the recombinant Ti plasmid into a plant cell. The plant cell with the foreign gene can then be used to generate a new plant. With luck, the new plant will express the foreign gene. Unfortunately, *Agrobacterium* does not naturally infect grasses, so it cannot be used to improve grains such as wheat, rice, or corn.

Noteworthy accomplishments of this approach are the introduction into plants of resistance to the herbicide glyphosate, and a *Bacillus thuringiensis*–derived insecticidal toxin (Bt). Normally, the herbicide kills both weeds and useful plants by inhibiting an enzyme necessary for making certain essential amino acids. *Salmonella* bacteria happen to have this enzyme, and some salmonellae have a mutant enzyme that is resistant to the herbicide. When the DNA for this enzyme is introduced into a crop plant, the crop becomes resistant to the herbicide, which then kills only the weeds. There is now a variety of plants in which different herbicide and pesticide resistances have been engineered. Resistance to drought, viral infection, and several other environmental stresses has also been engineered into crop plants. *Bacillus thuringiensis* bacteria are pathogenic to some insects because they produce a protein called Bt toxin that interferes with the insect digestive tract. The Bt gene has been inserted into a variety of crop plants, including cotton and potatoes, so insects that eat the plants will be killed.

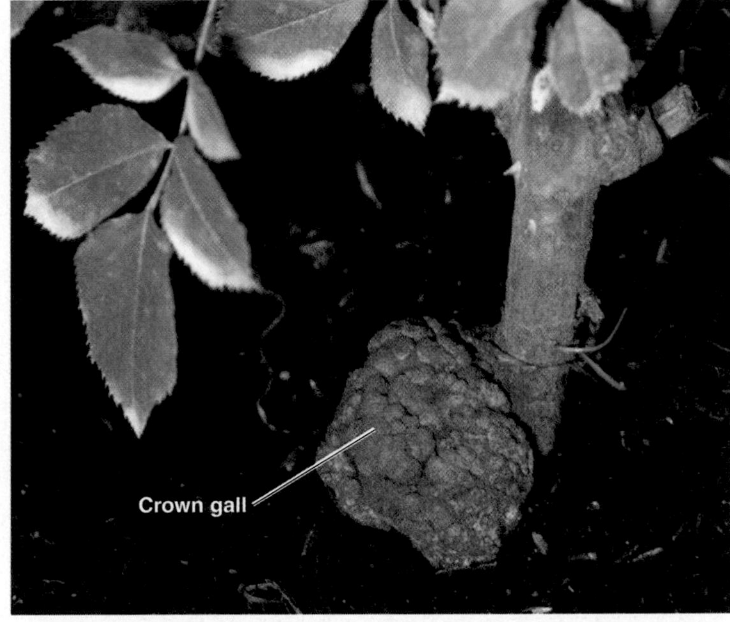

Figure 9.19 Crown gall disease on a rose plant. The tumorlike growth is stimulated by a gene on the Ti plasmid carried by the bacterium *Agrobacterium tumefaciens,* which has infected the plant.

Q What are some of the agricultural applications of recombinant DNA technology?

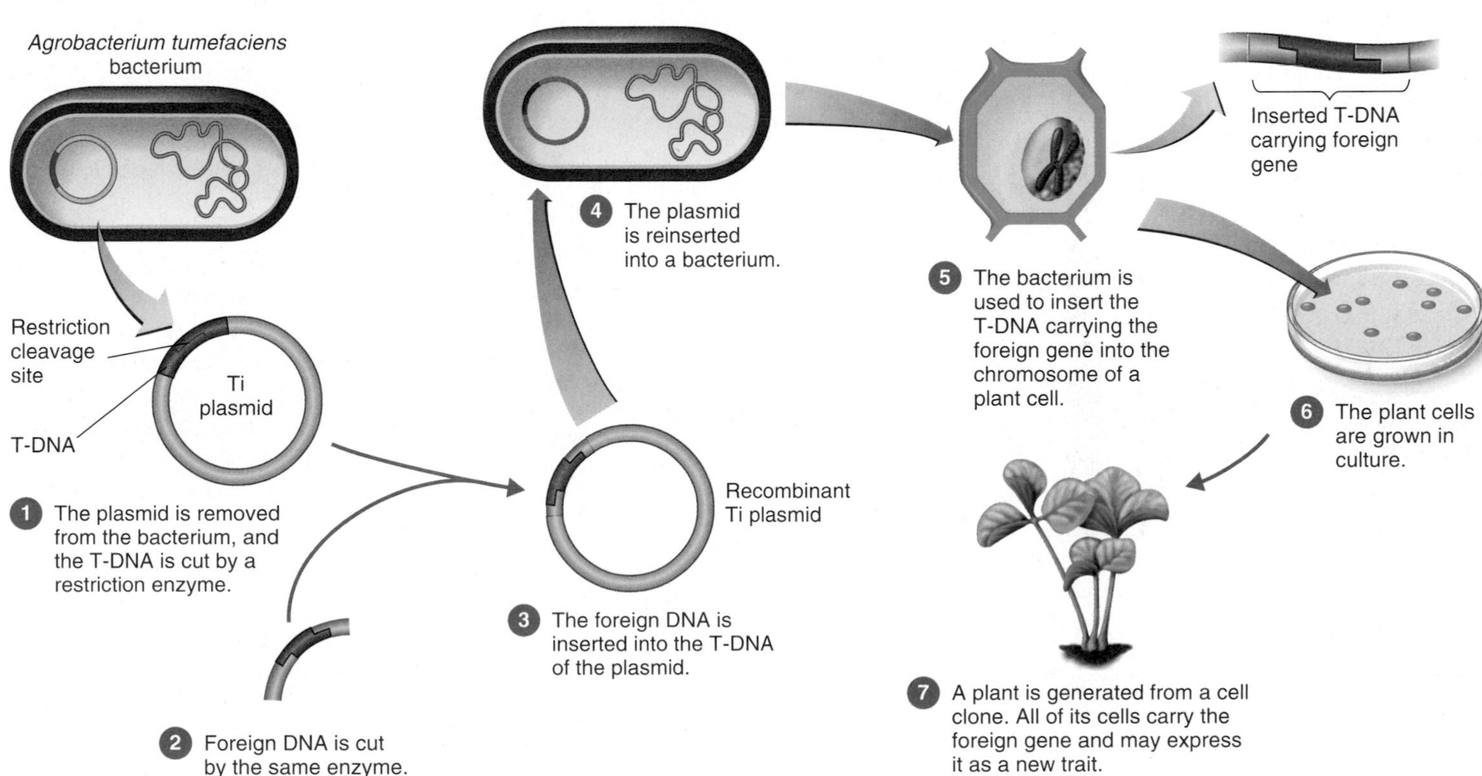

Agrobacterium tumefaciens bacterium
Restriction cleavage site
T-DNA
Ti plasmid
1 The plasmid is removed from the bacterium, and the T-DNA is cut by a restriction enzyme.
2 Foreign DNA is cut by the same enzyme.
Recombinant Ti plasmid
3 The foreign DNA is inserted into the T-DNA of the plasmid.
4 The plasmid is reinserted into a bacterium.
5 The bacterium is used to insert the T-DNA carrying the foreign gene into the chromosome of a plant cell.
Inserted T-DNA carrying foreign gene
6 The plant cells are grown in culture.
7 A plant is generated from a cell clone. All of its cells carry the foreign gene and may express it as a new trait.

Figure 9.20 Using the Ti plasmid as a vector for genetic modification in plants.

Q Why is the Ti plasmid important to biotechnology?

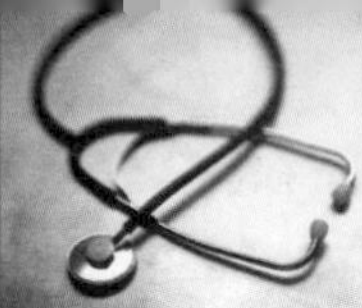

CLINICAL FOCUS From the *Morbidity and Mortality Weekly Report*

Norovirus—Who Is Responsible for the Outbreak?

As you read through this box, you will encounter a series of questions that microbiologists ask themselves as they trace a disease outbreak. Whether the microbiologist is called as expert witness in court will depend on whether a lawsuit is filed. Try to answer each question before going on to the next one.

1. On May 7, the Kent County (Michigan) Health Department was notified of a gastroenteritis outbreak among 115 people. A case was defined as vomiting and diarrhea and fever, cramps, or nausea.
   What do you need to know?
2. Ill people included 23 school employees, 55 publishing company employees, 9 employees of a social service organization, and 28 other people (see **Figure A**).
   What is the next step?

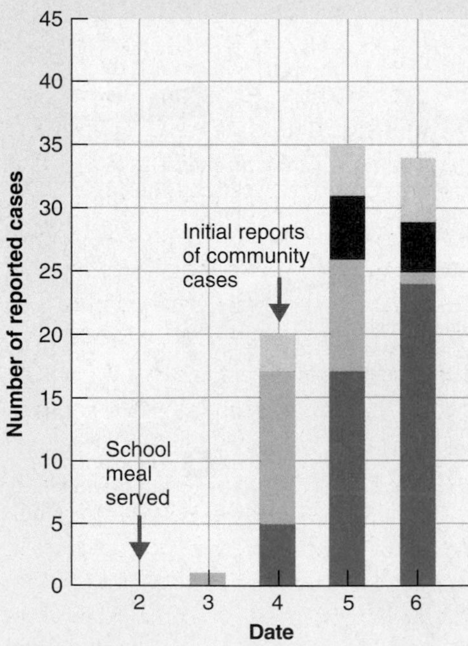

Figure A Number of cases reported.

3. On May 2, the school staff was served a party-sized sandwich catered by a national franchise restaurant. On May 3, the publishing company and social service staff luncheons were catered by the same restaurant. The remaining 28 people ate sandwiches at the same restaurant.
   What do you need to know?
4. Exposures to 16 food items were analyzed. Results indicated that eating lettuce was significantly associated with illness.
   What should you do now?
5. Reverse transcription PCR (RT-PCR) using a norovirus primer was done on stool samples (**Figure B**).
   What can you conclude?
6. RT-PCR confirmed norovirus infection. Sequence analysis performed on 21 stool specimens demonstrated 100% sequence homology for the 21 specimens.
   What would you do next?
7. Investigators learned that a food handler employed by the restaurant had experienced vomiting and diarrhea on May 1. The food handler believed he had acquired illness from his child. The child's illness was traced to an ill cousin who had been exposed to norovirus at a child-care center. The food handler's vomiting ended by the early morning of May 2, and he retuned to work at the restaurant later that morning.
   What should you look for now?

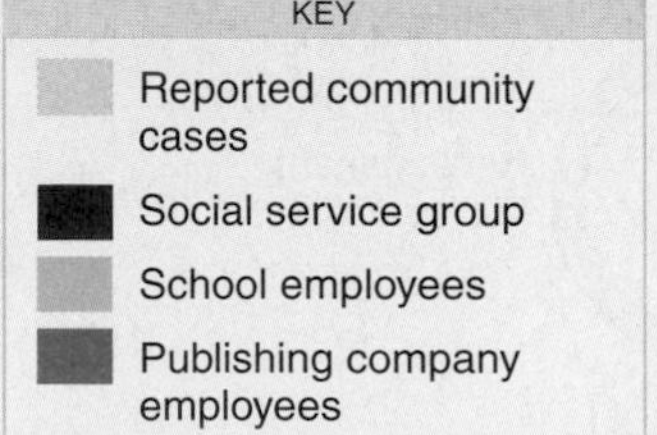

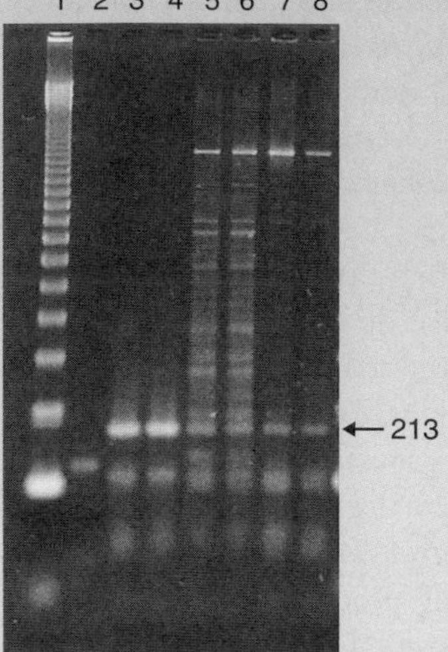

Figure B Results of PCR of patient samples. Lane 1, 123-bp size ladders. Lane 2, negative RT-PCR control; Lanes 3–8, patient samples. Norovirus is identified by the 213-bp band of DNA.

8. Sequence analysis on viruses from the food handler and eight ill customers were identical to the strains identified in step 7.
   Where do you look next?
9. The lettuce was sliced each morning by the food handler who had been sick. Inspection revealed that the food preparation sink was being used for handwashing. The sink was not sanitized before and after the lettuce was washed. State guidelines specify cleaning procedures and the bleach concentrations required. The restaurant was not cleaned appropriately until after it had been closed for nearly 1 week by the health department.

Source: Adapted from *MMWR* 55(14): 395–397, April 14, 2006.

Table 9.3 Some Agriculturally Important Products of Genetic Engineering

Product	Comments
Agricultural Products	
Bt cotton and Bt corn	Plants have toxin-producing gene from *Bacillus thuringiensis;* toxin kills insects that eat plants
MacGregor tomatoes	Antisense gene blocks pectin degradation, so fruits have longer shelf life
Pseudomonas fluorescens bacterium	Has toxin-producing gene from insect pathogen *B. thuringiensis;* toxin kills root-eating insects that ingest bacteria
Pseudomonas syringae, ice-minus bacterium	Lacks normal protein product that initiates undesirable ice formation on plants
Rhizobium meliloti bacterium	Modified for enhanced nitrogen fixation
Round up (glyphosate)-resistant crops	Plants have bacterial gene; allows use of herbicide on weeds without damaging crops
Animal Husbandry Products	
Bovine growth hormone (bGH)	Improves weight gain and milk production in cattle; produced by *E. coli*
Porcine growth hormone (pGH)	Improves weight gain in swine; produced by *E. coli*
Transgenic animals	Genetic modification of animals to produce medically useful products in their milk
Other Food Production Products	
Cellulase	Enzyme that degrades cellulose to make animal feedstocks; produced by *E. coli*
Rennin	Causes formation of milk curds for dairy products; produced by *Aspergillus niger*

Another example involves MacGregor tomatoes, which stay firm after harvest because the gene for polygalacturonase (PG), the enzyme that breaks down pectin, is suppressed. The suppression was accomplished by **antisense DNA technology.** First, a length of DNA complementary to the PG mRNA is synthesized. This antisense DNA is taken up by the cell and binds to the mRNA to inhibit translation. The DNA-RNA hybrid is broken down by the cell's enzymes, freeing the antisense DNA to disable another mRNA.

Perhaps the most exciting potential use of genetically modified plants concerns nitrogen fixation, the ability to convert the nitrogen gas in the air to compounds that living cells can use (see page 771). The availability of such nitrogen-containing nutrients is usually the main factor limiting crop growth. But in nature, only certain bacteria have genes for carrying out this process. Some plants, such as alfalfa, benefit from a symbiotic relationship with these microbes. Species of the symbiotic bacterium *Rhizobium* have already been genetically modified for enhanced nitrogen fixation. In the future, *Rhizobium* strains may be designed that can colonize such crop plants as corn and wheat, perhaps eliminating their requirement for nitrogen fertilizer. The ultimate goal would be to introduce functioning nitrogen-fixation genes directly into the plants. Although this goal cannot be achieved with our current knowledge, work toward it will continue because of its potential for dramatically increasing the world's food supply.

An example of a genetically modified bacterium now in agricultural use is *Pseudomonas fluorescens* that has been engineered to produce a toxin normally produced by *Bacillus thuringiensis.* This toxin kills certain plant pathogens, such as the European corn borer. The genetically altered *Pseudomonas,* which produces much more toxin than *B. thuringiensis,* can be added to plant seeds and in time will enter the vascular system of the growing plant. Its toxin is ingested by the feeding borer larvae and kills them (but is harmless to humans and other warm-blooded animals).

Animal husbandry has also benefited from rDNA technology. We have seen how one of the early commercial products of rDNA was human growth hormone. By similar methods it is possible to manufacture bovine growth hormone (bGH). When bGH is injected into beef cattle, it increases their weight gain; in dairy cows, it also causes a 10% increase in milk production. Such procedures have met with resistance from consumers, especially in Europe, primarily as a result of as-yet unsubstantiated fears that some of the bGH would be present in the milk or meat of these cattle and might be harmful to humans.

Table 9.3 lists these and several other genetically engineered products used in agriculture and animal husbandry.

CHECK YOUR UNDERSTANDING

✓ Of what value is the plant pathogen *Agrobacterium*? **9-19**

Safety Issues and the Ethics of Using rDNA

LEARNING OBJECTIVE

9-20 List the advantages of, and problems associated with, the use of genetic modification techniques.

There will always be concern about the safety of any new technology, and genetic modification and biotechnology are certainly no exceptions. One reason for this concern is that it is nearly impossible to prove that something is entirely safe under all conceivable conditions. People worry that the same techniques that can alter a microbe or plant to make them useful to humans could also inadvertently make them pathogenic to humans or otherwise dangerous to living organisms or could create an ecological nightmare. Therefore, laboratories engaged in rDNA research must meet rigorous standards of control to avoid either accidentally releasing genetically modified organisms into the environment or exposing humans to any risk of infection. To reduce risk further, microbiologists engaged in genetic modification often delete from the microbes' genomes certain genes that are essential for growth in environments outside the laboratory. Genetically modified organisms intended for use in the environment (in agriculture, for example) may be engineered to contain "suicide genes"—genes that eventually turn on to produce a toxin that kills the microbes, thus ensuring that they will not survive in the environment for very long after they have accomplished their task.

The safety issues in agricultural biotechnology are similar to those concerning chemical pesticides: toxicity to humans and to nonpest species. Although not shown to be harmful, genetically modified foods have not been popular with consumers. In 1999, researchers in Ohio noticed that humans may develop allergies to *Bacillus thuringiensis* (Bt) toxin after working in fields sprayed with the insecticide. And an Iowa study showed that the caterpillar stage of monarch butterflies could be killed by ingesting windblown Bt-carrying pollen that landed on milkweed, the caterpillars' normal food. Crop plants can be genetically modified for herbicide resistance so that fields can be sprayed to eliminate weeds without killing the desired crop. However, if the modified plants pollinate related weed species, weeds could become resistant to herbicides, making it more difficult to control unwanted plants. An unanswered question is whether releasing genetically modified organisms will alter evolution as genes move to wild species.

These developing technologies also raise a variety of moral and ethical issues. If genetic screening for diseases becomes routine, who should have access to this information? Should employers or insurance companies have the right to know the results of such tests? Restricting access to such information will be very difficult, which raises questions concerning the right to privacy. How can we be assured that such information will not be used to discriminate against certain groups?

Applications of genetic screening techniques are not limited to adults. The ability to diagnose a genetic disease in a fetus adds even more controversy to the abortion debate. Genetic counseling, which provides advice and counseling to prospective parents with family histories of genetic disease, is becoming more important in considerations about whether or not to have children. As more is learned about genetic causes for various diseases, such as cancer or Huntington's disease, reproductive choices might become more difficult for some families.

What extra burdens does molecular genetics place on our already overtaxed health care system? Genetic screening and gene therapy are expensive procedures, and we need to consider how they will be delivered to the public as the technology develops. Will a sufficient number of genetic counselors be available to work with people? Will expensive medical cures and treatments be available only to those who can afford them?

There are probably just as many harmful applications of a new technology as there are helpful ones. It is particularly easy to imagine rDNA technology being used to develop new and powerful biological weapons. In addition, because such research efforts are performed under top-secret conditions, it is virtually impossible for the general public to learn of them.

Perhaps more than most new technologies, molecular genetics holds the promise of affecting human life in previously unimaginable ways. It is important that society and individuals be given every opportunity to understand the potential impact of these new developments.

Like the invention of the microscope, the development of rDNA techniques is causing profound changes in science, agriculture, and human health care. With this technology only slightly more than 30 years old, it is difficult to predict exactly what changes will occur. However, it is likely that within another 30 years, many of the treatments and diagnostic methods discussed in this book will have been replaced by far more powerful techniques based on the unprecedented ability to manipulate DNA precisely.

CHECK YOUR UNDERSTANDING

✓ Identify two advantages and two problems associated with genetically modified organisms. **9-20**

STUDY OUTLINE

The **MyMicrobiologyPlace** website (**www.microbiologyplace.com**) will help you get ready for tests with its simple three-step approach: ❶ **take a pre-test** and obtain a personalized study plan, ❷ **learn and practice** with animations, tutorials, and MP3 tutor sessions, and ❸ **test yourself** with quizzes and a chapter post-test.

Introduction to Biotechnology (p. 247)

1. Biotechnology is the use of microorganisms, cells, or cell components to make a product.

Recombinant DNA Technology (p. 247)

2. Closely related organisms can exchange genes in natural recombination.
3. Genes can be transferred among unrelated species via laboratory manipulation, called recombinant DNA technology.
4. Recombinant DNA is DNA that has been artificially manipulated to combine genes from two different sources.

An Overview of Recombinant DNA Procedures (p. 247)

5. A desired gene is inserted into a DNA vector, such as a plasmid or a viral genome.
6. The vector inserts the DNA into a new cell, which is grown to form a clone.
7. Large quantities of the gene product can be harvested from the clone.

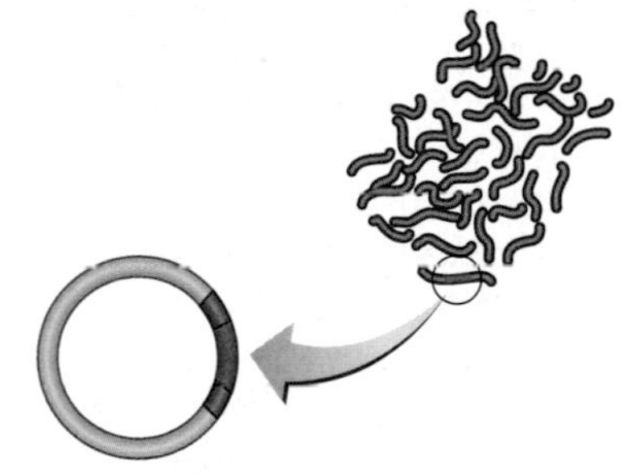

Tools of Biotechnology (pp. 247–252)

Selection (p. 249)

1. Microbes with desirable traits are selected for culturing by artificial selection.

Mutation (p. 249)

2. Mutagens are used to cause mutations that might result in a microbe with desirable traits.
3. Site-directed mutagenesis is used to change a specific codon in a gene.

Restriction Enzymes (pp. 249–250)

4. Prepackaged kits are available for rDNA techniques.
5. A restriction enzyme recognizes and cuts only one particular nucleotide sequence in DNA.
6. Some restriction enzymes produce sticky ends, short stretches of single-stranded DNA at the ends of the DNA fragments.
7. Fragments of DNA produced by the same restriction enzyme will spontaneously join by base pairing. DNA ligase can covalently link the DNA backbones.

Vectors (pp. 250–251)

8. Shuttle vectors are plasmids that can exist in several different species.
9. A plasmid containing a new gene can be inserted into a cell by transformation.
10. A virus containing a new gene can insert the gene into a cell.

Polymerase Chain Reaction (pp. 251–252)

11. The polymerase chain reaction (PCR) is used to make multiple copies of a desired piece of DNA enzymatically.
12. PCR can be used to increase the amounts of DNA in samples to detectable levels. This may allow sequencing of genes, the diagnosis of genetic diseases, or the detection of viruses.

Techniques of Genetic Modification (pp. 253–258)

Inserting Foreign DNA into Cells (pp. 253–254)

1. Cells can take up naked DNA by transformation. Chemical treatments are used to make cells that are not naturally competent take up DNA.
2. Pores made in protoplasts and animal cells by electric current in the process of electroporation can provide entrance for new pieces of DNA.
3. Protoplast fusion is the joining of cells whose cell walls have been removed.
4. Foreign DNA can be introduced into plant cells by shooting DNA-coated particles into the cells.
5. Foreign DNA can be injected into animal cells by using a fine glass micropipette.

Obtaining DNA (pp. 254–256)

6. Genomic libraries can be made by cutting up an entire genome with restriction enzymes and inserting the fragments into bacterial plasmids or phages.

7. Complementary DNA (cDNA) made from mRNA by reverse transcription can be cloned in genomic libraries.
8. Synthetic DNA can be made in vitro by a DNA synthesis machine.

Selecting a Clone (pp. 256–257)

9. Antibiotic-resistance markers on plasmid vectors are used to identify cells containing the engineered vector by direct selection.
10. In blue-white screening, the vector contains the genes for *amp*^{R} and β-galactosidase.
11. The desired gene is inserted into the β-galactosidase gene site, destroying the gene.
12. Clones containing the recombinant vector will be resistant to ampicillin and unable to hydrolyze X-gal (white colonies). Clones containing the vector without the new gene will be blue. Clones lacking the vector will not grow.
13. Clones containing foreign DNA can be tested for the desired gene product.
14. A short piece of labeled DNA called a DNA probe can be used to identify clones carrying the desired gene.

Making a Gene Product (pp. 257–258)

15. *E. coli* is used to produce proteins using rDNA because *E. coli* is easily grown and its genomics are well understood.
16. Efforts must be made to ensure that *E. coli*'s endotoxin does not contaminate a product intended for human use.
17. To recover the product, *E. coli* must be lysed, or the gene must be linked to a gene that produces a naturally secreted protein.
18. Yeasts can be genetically modified and are likely to secrete a gene product continuously.
19. Genetically modified mammalian cells can be grown to produce proteins such as hormones for medical use.
20. Genetically modified plant cells can be grown and used to produce plants with new properties.

Applications of rDNA (pp. 258–267)

1. Cloned DNA is used to produce products, study the cloned DNA, and alter the phenotype of an organism.

Therapeutic Applications (pp. 258–261)

2. Synthetic genes linked to the β-galactosidase gene (*lacZ*) in a plasmid vector were inserted into *E. coli,* allowing *E. coli* to produce and secrete the two polypeptides used to make human insulin.
3. Cells and viruses can be modified to produce a pathogen's surface protein, which can be used as a vaccine.
4. DNA vaccines consist of rDNA cloned in bacteria.
5. Gene therapy can be used to cure genetic diseases by replacing the defective or missing gene.

The Human Genome Project (p. 261)

6. Recombinant DNA techniques were used to map the human genome through the Human Genome Project.
7. This will provide tools for diagnosis and possibly the repair of genetic diseases.

Scientific Applications (pp. 261–264)

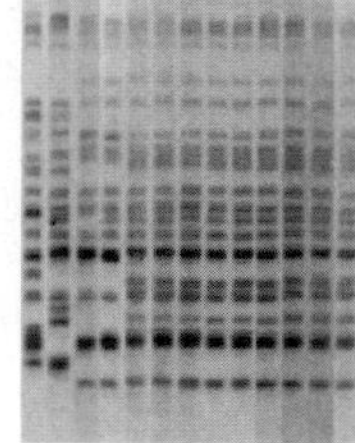

8. Recombinant DNA techniques can be used to increase understanding of DNA, for genetic fingerprinting, and for gene therapy.
9. DNA sequencing machines are used to determine the nucleotide base sequence of restriction fragments in random shotgun sequencing.
10. Bioinformatics is the use of computer applications to study genetic data; proteomics is the study of a cell's proteins.
11. Southern blotting can be used to locate a gene in a cell.
12. DNA probes can be used to quickly identify a pathogen in body tissue or food.
13. Forensic microbiologists use DNA fingerprinting to identify the source of bacterial or viral pathogens.

Agricultural Applications (pp. 264–267)

14. Cells from plants with desirable characteristics can be cloned to produce many identical cells. These cells can then be used to produce whole plants from which seeds can be harvested.
15. Plant cells can be modified by using the Ti plasmid vector. The tumor-producing T genes are replaced with desired genes, and the recombinant DNA is inserted into *Agrobacterium.* The bacterium naturally transforms its plant hosts.
16. Genes for glyphosate resistance, Bt toxin, and pectinase suppression have been engineered into crop plants.
17. Genetically modified *Rhizobium* has enhanced nitrogen fixation.
18. Genetically modified *Pseudomonas* is a biological insecticide that produces *Bacillus thuringiensis* toxin.
19. Bovine growth hormone is being produced by *E. coli.*

Safety Issues and the Ethics of Using rDNA (p. 268)

1. Strict safety standards are used to avoid the accidental release of genetically modified microorganisms.
2. Some microbes used in rDNA cloning have been altered so that they cannot survive outside the laboratory.
3. Microorganisms intended for use in the environment may be modified to contain suicide genes so that the organisms do not persist in the environment.
4. Genetic technology raises a number of ethical questions: Should employers and insurance companies have access to a person's genetic records? Will some people be targeted for either breeding or sterilization? Will genetic counseling be available to everyone?
5. Genetically modified crops must be safe for consumption and for release in the environment.

STUDY QUESTIONS

Answers to the Review and Multiple Choice questions can be found by turning to the blue Answers tab at the back of the textbook.

Review

1. Compare and contrast the following terms:
   a. *cDNA* and *gene*
   b. *restriction fragment* and *gene*
   c. *DNA probe* and *gene*
   d. *DNA polymerase* and *DNA ligase*
   e. *rDNA* and *cDNA*
   f. *genome* and *proteome*
2. Differentiate the following terms. Which one is "hit and miss"—that is, does not add a specific gene to a cell?
   a. protoplast fusion
   b. gene gun
   c. microinjection
   d. electroporation
3. **DRAW IT** Using the following map of plasmid pMICRO, diagram the locations of the restriction fragments that result from digesting pMICRO with *Eco*RI, *Hin*dIII, and both enzymes together following electrophoresis. Which enzyme makes the smallest fragment containing the tetracycline-resistance gene?

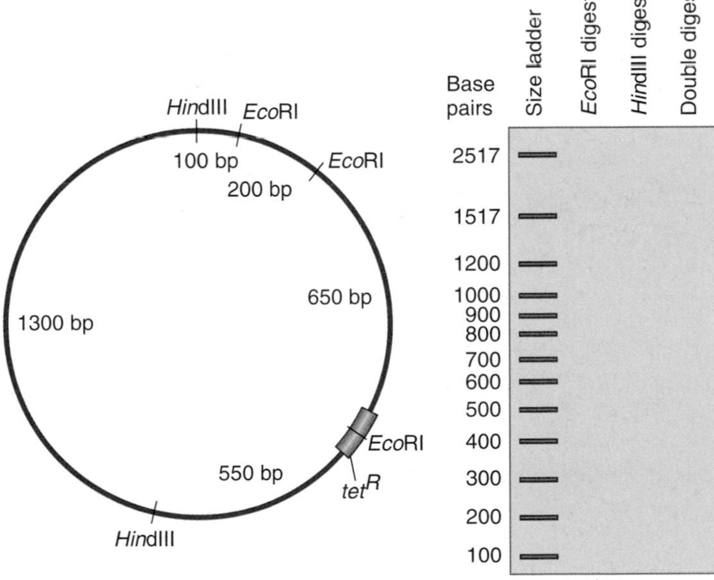

4. Some commonly used restriction enzymes are listed in Table 9.1 on page 249.
   a. Indicate which enzymes produce sticky ends.
   b. Of what value are sticky ends in making recombinant DNA?
5. Suppose you want multiple copies of a gene you have synthesized. How would you obtain the necessary copies by cloning? By PCR?
6. Describe a recombinant DNA experiment in two or three sentences. Use the following terms: intron, exon, DNA, mRNA, cDNA, RNA polymerase, reverse transcriptase.
7. List at least two examples of the use of rDNA in medicine and in agriculture.
8. You are attempting to insert a gene for saltwater tolerance into a plant by using the Ti plasmid. In addition to the desired gene, you add a gene for tetracycline resistance (*tet*^{R}) to the plasmid. What is the purpose of the *tet*^{R} gene?
9. How does RNAi "silence" a gene?

Multiple Choice

1. Restriction enzymes were first discovered with the observation that
   a. DNA is restricted to the nucleus.
   b. phage DNA is destroyed in a host cell.
   c. foreign DNA is kept out of a cell.
   d. foreign DNA is restricted to the cytoplasm.
   e. all of the above
2. The DNA probe, 3′-GGCTTA, will hybridize with which of the following?
   a. 5′-CCGUUA
   b. 5′-CCGAAT
   c. 5′-GGCTTA
   d. 3′-CCGAAT
   e. 3′-GGCAAU
3. Which of the following is the fourth basic step to genetically modify a cell?
   a. transformation
   b. ligation
   c. plasmid cleavage
   d. restriction-enzyme digestion of gene
   e. isolation of gene
4. The following enzymes are used to make cDNA. What is the second enzyme used to make cDNA?
   a. reverse transcriptase
   b. ribozyme
   c. RNA polymerase
   d. DNA polymerase
5. If you put a gene in a virus, the next step in genetic modification would be
   a. insertion of a plasmid.
   b. transformation.
   c. transduction.
   d. PCR.
   e. Southern blotting.
6. You have a small gene that you want replicated by PCR. You add radioactively labeled nucleotides to the PCR thermalcycler. After three replication cycles, what percentage of the DNA single-strands are radioactively labeled?
   a. 0%
   b. 12.5%
   c. 50%
   d. 87.5%
   e. 100%

Match the following choices to the statements in questions 7 through 10.

a. antisense
b. clone
c. library
d. Southern blot
e. vector

7. Pieces of human DNA stored in yeast cells.
8. A population of cells carrying a desired plasmid.
9. Self-replicating DNA for transmitting a gene from one organism to another.
10. A gene that hybridizes with mRNA.

Critical Thinking

1. Design an experiment using vaccinia virus to make a vaccine against the AIDS virus (HIV).
2. Why did the use of DNA polymerase from the bacterium *Thermus aquaticus* allow researchers to add the necessary reagents to tubes in a preprogrammed heating block?
3. The following picture shows bacterial colonies growing on X-gal plus ampicillin in a blue-white screening test. Which colonies have the recombinant plasmid? The small satellite colonies do not have the plasmid. Why did they start growing on the medium 48 hours after the larger colonies?

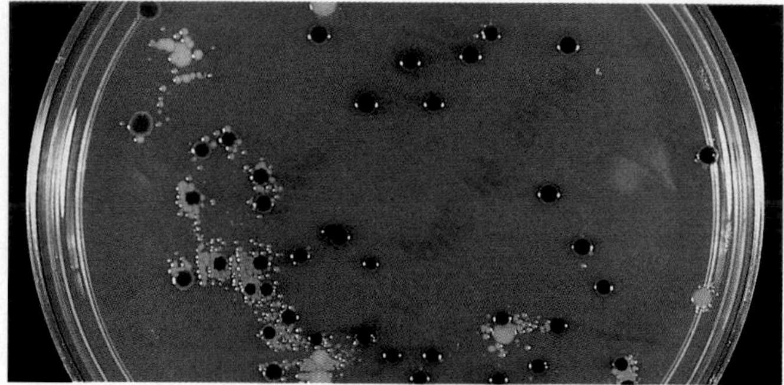

4. Using the restriction enzyme *Eco*RI, the following gel electrophoresis patterns were obtained from digests of various DNA molecules from a transformation experiment. Can you conclude from these data that transformation occurred? Explain why or why not.

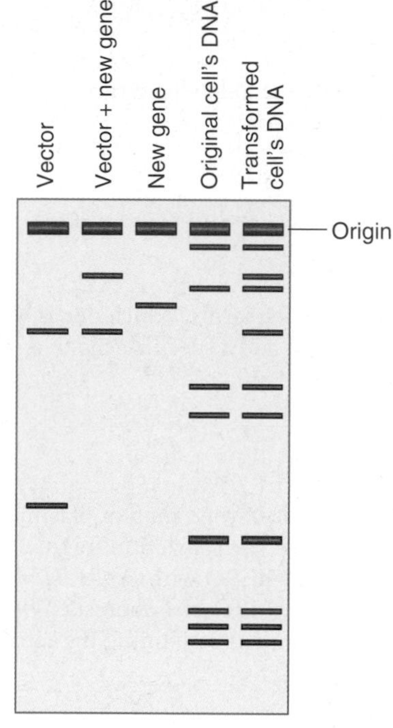

Clinical Applications

1. PCR has been used to examine oysters for the presence of *Vibrio cholerae*. Oysters from different areas were homogenized, and DNA was extracted from the homogenates. The DNA was digested by the restriction enzyme *Hinc*II. A primer for the hemolysin gene of *V. cholerae* was used for the PCR reaction. After PCR, each sample was electrophoresed and stained with a probe for the hemolysin gene. Which of the oyster samples were (was) positive for *V. cholerae*? How can you tell? Why look for *V. cholerae* in oysters? What is the advantage of PCR over conventional biochemical tests to identify the bacteria?

10 Classification of Microorganisms

The science of classification, especially the classification of living forms, is called *taxonomy* (from the Greek for orderly arrangement). The objective of taxonomy is to classify living organisms—that is, to establish the relationships between one group of organisms and another and to differentiate them. There may be as many as 100 million different living organisms, but fewer than 10% have been discovered, much less classified and identified.

Taxonomy also provides a common reference for identifying organisms already classified. For example, when a bacterium suspected of causing a specific disease is isolated from a patient, characteristics of that isolate are matched to lists of characteristics of previously classified bacteria to identify the isolate (see the box on page 283). Finally, taxonomy is a basic and necessary tool for scientists, providing a universal language of communication.

Modern taxonomy is an exciting and dynamic field. New techniques in molecular biology and genetics are providing new insights into classification and evolution. In this chapter, you will learn the various classification systems, the different criteria used for classification, and tests that are used to identify microorganisms that have already been classified.

UNDER THE MICROSCOPE

Pneumocystis jirovecii. This was thought to be a protozoan until DNA analysis showed it is a fungus. *P. jirovecii* causes pneumonia in immunocompromised people.

Q&A

Why does it matter whether an organism is classified as a protozoan or a fungus?

Look for the answer in the chapter.

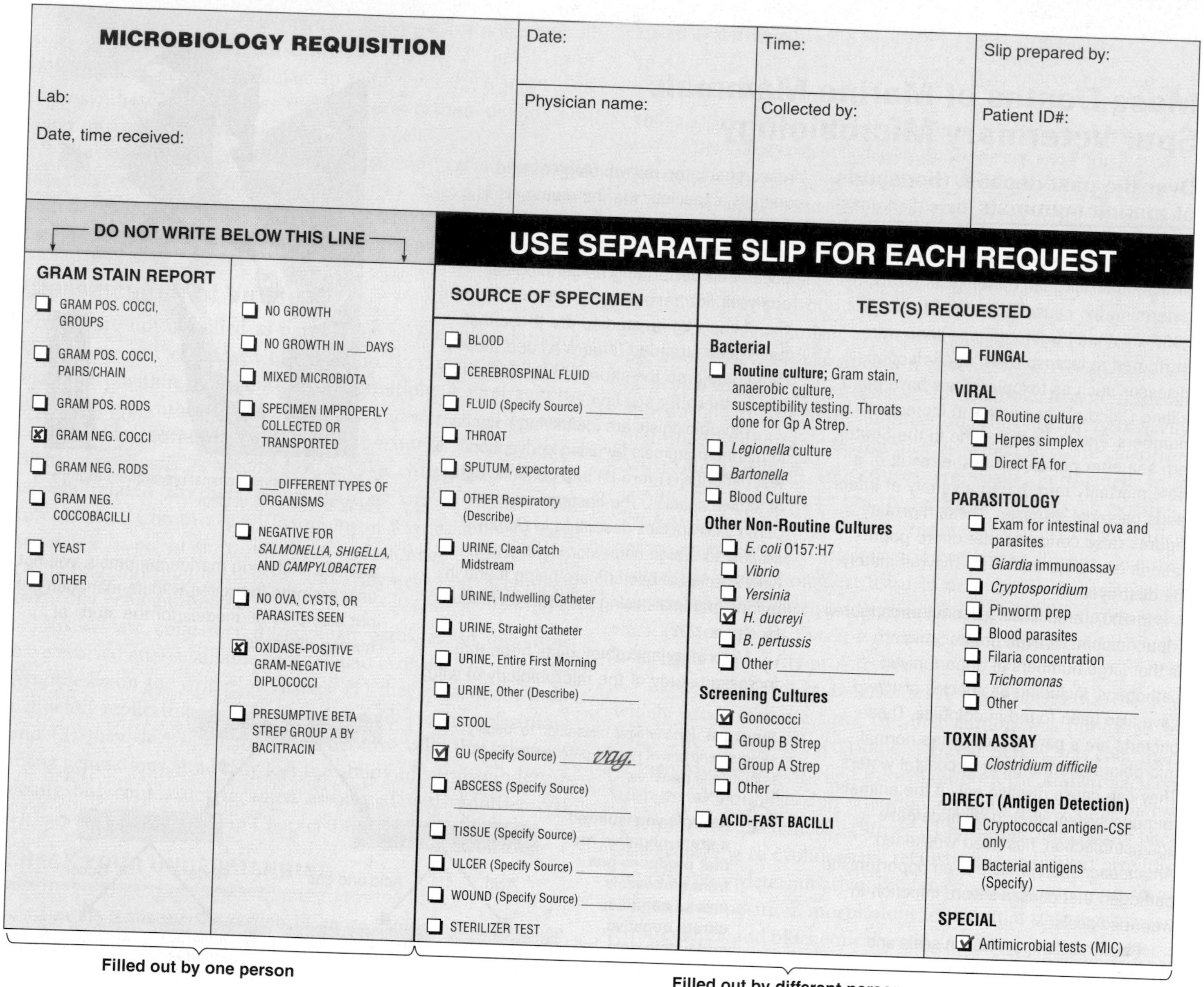

MICROBIOLOGY REQUISITION

Lab:

Date, time received:

Date:

Physician name:

Time:

Collected by:

Slip prepared by:

Patient ID#:

DO NOT WRITE BELOW THIS LINE

USE SEPARATE SLIP FOR EACH REQUEST

GRAM STAIN REPORT

- ☐ GRAM POS. COCCI, GROUPS
- ☐ GRAM POS. COCCI, PAIRS/CHAIN
- ☐ GRAM POS. RODS
- ☒ GRAM NEG. COCCI
- ☐ GRAM NEG. RODS
- ☐ GRAM NEG. COCCOBACILLI
- ☐ YEAST
- ☐ OTHER

- ☐ NO GROWTH
- ☐ NO GROWTH IN ___DAYS
- ☐ MIXED MICROBIOTA
- ☐ SPECIMEN IMPROPERLY COLLECTED OR TRANSPORTED
- ☐ ___DIFFERENT TYPES OF ORGANISMS
- ☐ NEGATIVE FOR *SALMONELLA, SHIGELLA,* AND *CAMPYLOBACTER*
- ☐ NO OVA, CYSTS, OR PARASITES SEEN
- ☒ OXIDASE-POSITIVE GRAM-NEGATIVE DIPLOCOCCI
- ☐ PRESUMPTIVE BETA STREP GROUP A BY BACITRACIN

SOURCE OF SPECIMEN

- ☐ BLOOD
- ☐ CEREBROSPINAL FLUID
- ☐ FLUID (Specify Source) ______
- ☐ THROAT
- ☐ SPUTUM, expectorated
- ☐ OTHER Respiratory (Describe) ______
- ☐ URINE, Clean Catch Midstream
- ☐ URINE, Indwelling Catheter
- ☐ URINE, Straight Catheter
- ☐ URINE, Entire First Morning
- ☐ URINE, Other (Describe) ______
- ☐ STOOL
- ☑ GU (Specify Source) vag.
- ☐ ABSCESS (Specify Source) ______
- ☐ TISSUE (Specify Source) ______
- ☐ ULCER (Specify Source) ______
- ☐ WOUND (Specify Source) ______
- ☐ STERILIZER TEST

TEST(S) REQUESTED

Bacterial

- ☐ **Routine culture;** Gram stain, anaerobic culture, susceptibility testing. Throats done for Gp A Strep.
- ☐ *Legionella* culture
- ☐ *Bartonella*
- ☐ Blood Culture

Other Non-Routine Cultures

- ☐ *E. coli* O157:H7
- ☐ *Vibrio*
- ☐ *Yersinia*
- ☑ *H. ducreyi*
- ☐ *B. pertussis*
- ☐ Other ______

Screening Cultures

- ☑ Gonococci
- ☐ Group B Strep
- ☐ Group A Strep
- ☐ Other ______

☐ **ACID-FAST BACILLI**

☐ **FUNGAL**

VIRAL

- ☐ Routine culture
- ☐ Herpes simplex
- ☐ Direct FA for ______

PARASITOLOGY

- ☐ Exam for intestinal ova and parasites
- ☐ *Giardia* immunoassay
- ☐ *Cryptosporidium*
- ☐ Pinworm prep
- ☐ Blood parasites
- ☐ Filaria concentration
- ☐ *Trichomonas*
- ☐ Other ______

TOXIN ASSAY

- ☐ *Clostridium difficile*

DIRECT (Antigen Detection)

- ☐ Cryptococcal antigen-CSF only
- ☐ Bacterial antigens (Specify) ______

SPECIAL

- ☑ Antimicrobial tests (MIC)

Filled out by one person

Filled out by different person

Figure 10.7 A clinical microbiology lab report form. In health care, morphology and differential staining are important in determining the proper treatment for microbial diseases. A clinician completes the form to identify the sample and specific tests. In this case, a genitourinary sample will be examined for sexually transmitted infections. The red notations are the lab technician's report of the Gram stain and culture results. [Minimal inhibitory concentration (MIC) of antibiotics will be discussed in Chapter 20, page 572.]

Q **What diseases are suspected if the "acid-fast bacilli" box is checked?**

processes. In clinical microbiology, a physician will swab a patient's pus or tissue surface. The swab is inserted into a tube of transport medium. **Transport media** are usually not nutritive and are designed to prolong viability of fastidious pathogens. The physician will note the type of specimen and test(s) requested on a lab requisition form (**Figure 10.7**). The information returned by the lab technician will help the physician begin treatment (see the box in Chapter 5, page 144).

Morphological Characteristics

Morphological (structural) characteristics have helped taxonomists classify organisms for 200 years. Higher organisms are frequently classified according to observed anatomical detail. But many microorganisms look too similar to be classified by their structures. Through a microscope, organisms that might differ in metabolic or physiological properties may look alike. Literally hundreds of bacterial species are small rods or small cocci.

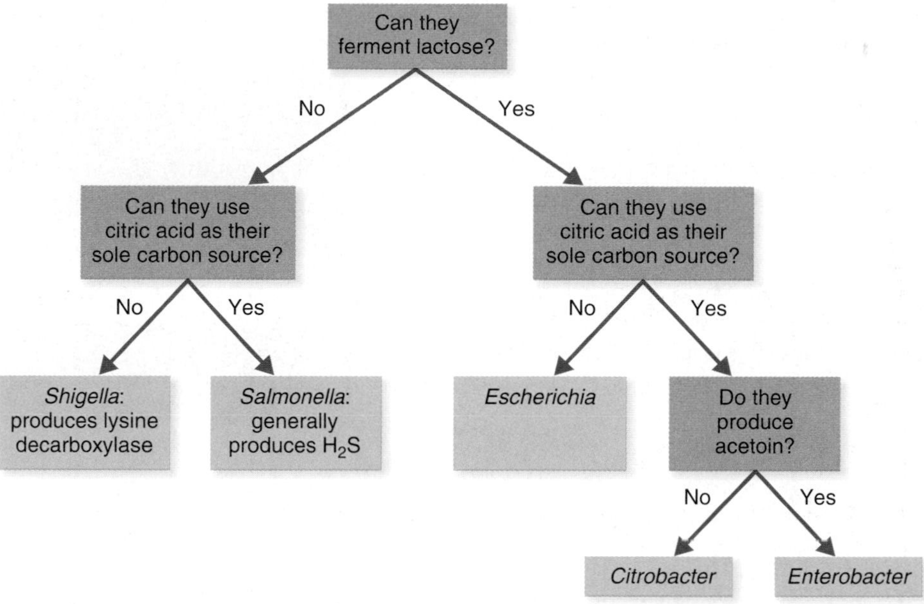

Figure 10.8 The use of metabolic characteristics to identify selected genera of enteric bacteria.

Q Assume you have a gram-negative bacterium that produces acid from lactose and cannot use citric acid as its sole carbon source. What is the bacterium?

Q&A Larger size and the presence of intracellular structures does not always mean easy classification, however. *Pneumocystis* (nü-mō-sis'tis) pneumonia is the most common opportunistic infection in AIDS patients. Until the AIDS epidemic, the causative agent of this infection, *P. jirovecii* (ye-rō'vet-zē̄ē) [formerly *P. carinii* (kär-i'nē-ē)] was rarely seen in humans. *Pneumocystis* lacks structures that can be easily used for identification (see Figure 24.20, page 698), and its taxonomic position has been uncertain since its discovery in 1909 by Carlos Chagas in mice. Although it was classified as a protozoan, recent studies comparing its rRNA sequence with those of protozoa, *Euglena,* cellular slime molds, plants, mammals, and fungi have shown that *Pneumocystis* is actually a member of the Kingdom Fungi. Researchers are culturing *Pneumocystis,* and they have developed some useful treatments for *Pneumocystis* pneumonia. Perhaps as researchers take into account this organism's relatedness to fungi, appropriate treatments will result.

Cell morphology tells us little about phylogenetic relationships. However, morphological characteristics are still useful in identifying bacteria. For example, differences in such structures as endospores or flagella can be helpful.

Differential Staining

Recall from Chapter 3 that one of the first steps in identifying bacteria is differential staining. Most bacteria are either gram-positive or gram-negative. Other differential stains, such as the acid-fast stain, can be useful for a more limited group of microorganisms. Recall that these stains are based on the chemical composition of cell walls and therefore are not useful in identifying either the wall-less bacteria or the archaea with unusual walls. Microscopic examination of a Gram stain or an acid-fast stain is used to obtain information quickly in the clinical environment.

Biochemical Tests

Enzymatic activities are widely used to differentiate bacteria. Even closely related bacteria can usually be separated into distinct species by subjecting them to biochemical tests, such as one to determine their ability to ferment an assortment of selected carbohydrates. For one example of the use of biochemical tests to identify bacteria (in this instance, in marine mammals), see the box on page 283. Moreover, biochemical tests can provide insight into a species' niche in the ecosystem. For example, a bacterium that can fix nitrogen gas or oxidize elemental sulfur will provide important nutrients for plants and animals. This will be discussed in Chapter 27.

Enteric, gram-negative bacteria are a large heterogeneous group of microbes whose natural habitat is the intestinal tract of humans and other animals. This family contains several pathogens that cause diarrheal illness. A number of tests have been developed so that technicians can quickly identify the pathogens, a clinician can then provide appropriate treatment, and epidemiologists can locate the source of an illness. All members of the family Enterobacteriaceae are oxidase-negative. Among the enteric bacteria are members of the genera *Escherichia*, *Enterobacter*, *Shigella*, *Citrobacter*, and *Salmonella*. *Escherichia*, *Enterobacter*, and *Citrobacter*, which ferment lactose to produce acid and gas, can be distinguished from *Salmonella* and *Shigella,* which do not. Further biochemical testing, as represented in **Figure 10.8**, can differentiate among the genera.

The time needed to identify bacteria can be reduced considerably by the use of selective and differential media or by rapid identification methods. Recall from Chapter 6 (page 168) that selective media contain ingredients that suppress the growth of

1 One tube containing media for 15 biochemical tests is inoculated with an unknown enteric bacterium.

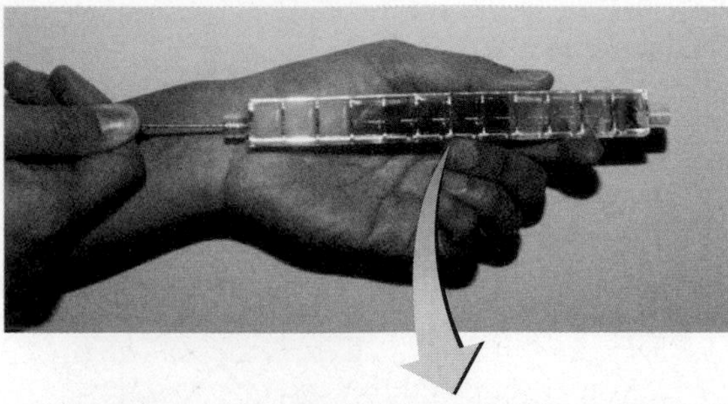

2 After incubation, the tube is observed for results.

3 The value for each positive test is circled, and the numbers from each group of tests are added to give the ID value.

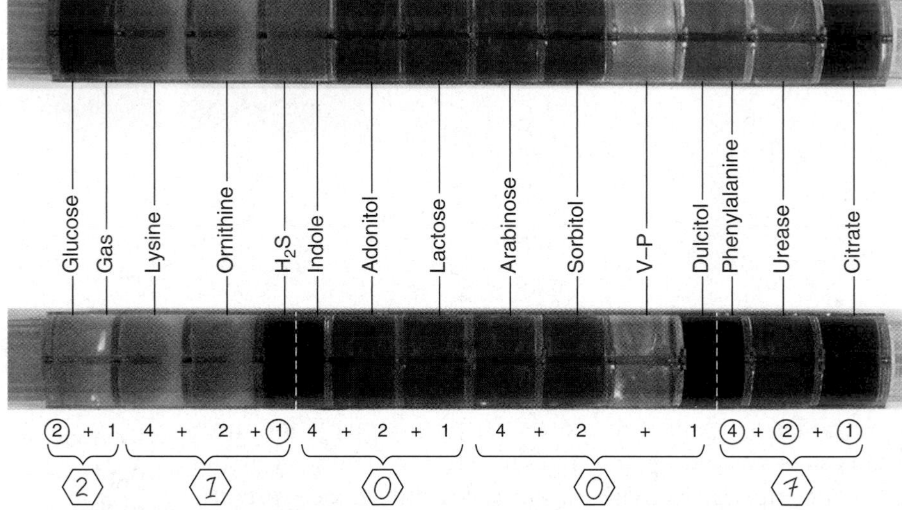

4 Comparing the resultant ID value with a computerized listing shows that the organism in the tube is *Proteus mirabilis.*

ID Value	Organism	Atypical Test Results	Confirmatory Test
21006	*Proteus mirabilis*	Ornithine$^-$	Sucrose
21007	*Proteus mirabilis*	Ornithine$^-$	
21020	*Salmonella choleraesuis*	Lysine$^-$	

Figure 10.9 One type of rapid identification method for bacteria: Enterotube II from Becton Dickinson. This example shows results for a typical strain of *P. mirabilis;* however, other strains may produce different test results, which are listed in the Atypical Test Results column. The V–P test is used to confirm an identification.

Q How can one species have two different ID values?

competing organisms and encourage the growth of desired ones, and that differential media allow the desired organism to form a colony that is somehow distinctive.

Bergey's Manual does not evaluate the relative importance of each biochemical test and does not always describe strains. In diagnosing an infection, clinicians must identify a particular species and even a particular strain to proceed with proper treatment. To this end, specific series of biochemical tests have been developed for fast identification in hospital laboratories. Rapid test systems have been developed for yeasts and other fungi, as well as bacteria.

Rapid identification methods are manufactured for groups of medically important bacteria, such as the enterics. Such tools are designed to perform several biochemical tests simultaneously and can identify bacteria within 4 to 24 hours. This is sometimes called **numerical identification** because the results of each test are assigned a number. In the simplest form, a positive test would be assigned a value of 1, and a negative is assigned a value of 0. In most commercial testing kits, test results are assigned numbers ranging from 1 to 4 that are based on the relative reliability and importance of each test, and the resulting total is compared to a database of known organisms.

In the example shown in Figure 10.9, an unknown enteric bacterium is inoculated into a tube designed to perform 15 biochemical tests. After incubation, results in each compartment are recorded. Notice that each test is assigned a value; the number

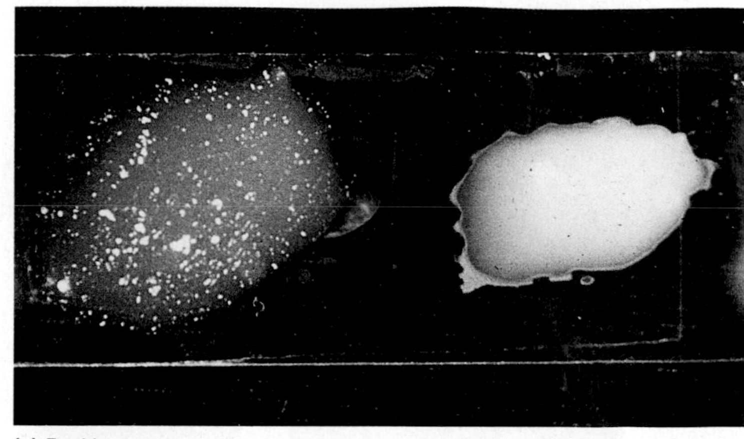

(a) Positive test (b) Negative test

Figure 10.10 A slide agglutination test. (**a**) In a positive test, the grainy appearance is due to the clumping (agglutination) of the bacteria. (**b**) In a negative test, the bacteria are still evenly distributed in the saline and antiserum.

Q Agglutination results when the bacteria are mixed with ________.

derived from scoring all the tests is called the ID value. Fermentation of glucose is important, and a positive reaction is valued at 2, compared with the production of acetoin (V–P test, or the Voges–Proskauer test), which has no value.

A computerized interpretation of the simultaneous test results is essential and is provided by the manufacturer. A limitation of biochemical testing is that mutations and plasmid acquisition can result in strains with different characteristics. Unless a large number of tests is used, an organism could be incorrectly identified.

Serology

Serology is the science that studies serum and immune responses that are evident in serum (see Chapter 18). Microorganisms are antigenic; that is, microorganisms that enter an animal's body stimulate it to form antibodies. Antibodies are proteins that circulate in the blood and combine in a highly specific way with the bacteria that caused their production. For example, the immune system of a rabbit injected with killed typhoid bacteria (antigens) responds by producing antibodies against typhoid bacteria. Solutions of such antibodies used in the identification of many medically important microorganisms are commercially available; such a solution is called an **antiserum** (plural: *antisera*). If an unknown bacterium is isolated from a patient, it can be tested against known antisera and often identified quickly.

In a procedure called a **slide agglutination test,** samples of an unknown bacterium are placed in a drop of saline on each of several slides. Then a different known antiserum is added to each sample. The bacteria agglutinate (clump) when mixed with antibodies that were produced in response to that species or strain of bacterium; a positive test is indicated by the presence of agglutination. Positive and negative slide agglutination tests are shown in Figure 10.10.

Serological testing can differentiate not only among microbial species, but also among strains within species. Strains with different antigens are called **serotypes, serovars,** or **biovars.** See the discussion of *Escherichia* and *Salmonella* serovars on page 310. As mentioned in Chapter 1, Rebecca Lancefield was able to classify serotypes of streptococci by studying serological reactions. She found that the different antigens in the cell walls of various serotypes of streptococci stimulate the formation of different antibodies. In contrast, because closely related bacteria also produce some of the same antigens, serological testing can be used to screen bacterial isolates for possible similarities. If an antiserum reacts with proteins from different bacterial species or strains, these bacteria can be tested further for relatedness.

Serological testing was used to determine whether the increase in number of cases of necrotizing fasciitis in the United States and England since 1987 was due to a common source of the infections. No common source was located, but there has been an increase in two serotypes of *Streptococcus pyogenes* that have been dubbed the "flesh-eating" bacteria.

A test called the **enzyme-linked immunosorbent assay (ELISA)** is widely used because it is fast and can be read by a computer scanner (Figure 10.11; see also Figure 18.14, page 518). In a direct ELISA, known antibodies are placed in (and adhere to) the wells of a microplate, and an unknown type of bacterium is added to each well. A reaction between the known antibodies and the bacteria provides identification of the bacteria. An ELISA is used in AIDS testing to detect the presence of antibodies against human immunodeficiency virus (HIV), the virus that causes AIDS (see Figure 19.13, page 540).

Another serological test, **Western blotting,** is also used to identify antibodies in a patient's serum (Figure 10.12). HIV infection is confirmed by Western blotting, and Lyme disease, caused by *Borrelia burgdorferi,* is often diagnosed by the Western blot.

1. Proteins from a known bacterium or virus are separated by an electric current in electrophoresis.
2. The proteins are then transferred to a filter by blotting.
3. Patient's serum is washed over the filter. If the patient has antibodies to one of the proteins in the filter (in this case, *Borrelia* proteins), the antibodies and protein will combine. Anti-human serum linked to an enzyme is then washed over the filter.
4. This will be made visible as a colored band on the filter after addition of the enzyme's substrate.

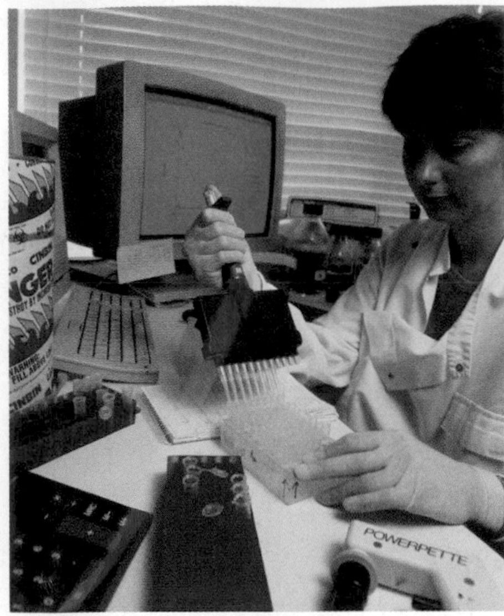

(a)

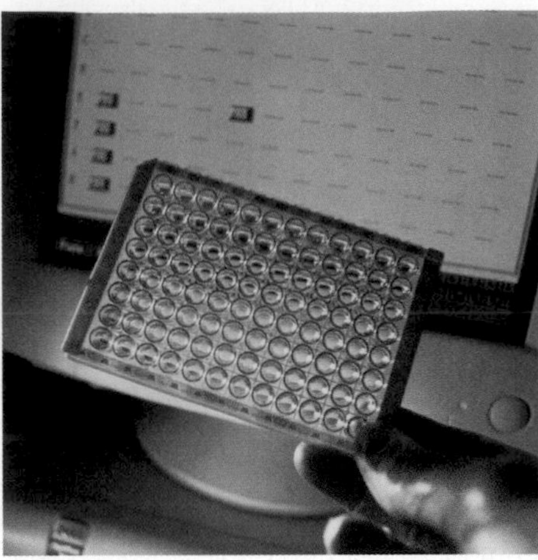

(b)

Figure 10.11 An ELISA test. (**a**) A technician uses a micropipette to add samples to a microplate for an ELISA. (**b**) ELISA results are then read by the computer scanner.

Q What are the similarities between the slide agglutination test and the ELISA test?

Phage Typing

Like serological testing, phage typing looks for similarities among bacteria. Both techniques are useful in tracing the origin and course of a disease outbreak. **Phage typing** is a test for determining which phages a bacterium is susceptible to. Recall from Chapter 8 (page 237) that bacteriophages (phages) are bacterial viruses and that they usually cause lysis of the bacterial cells they infect. They are highly specialized, in that they usually infect only members of a particular species, or even particular strains within a species. One bacterial strain might be susceptible to two different phages, whereas another strain of the same species might be susceptible to those two phages plus a third phage. Bacteriophages will be discussed further in Chapter 13.

The sources of food-associated infections can be traced by phage typing. One version of this procedure starts with a plate totally covered with bacteria growing on agar. A drop of each different phage type to be used in the test is then placed on the bacteria. Wherever the phages are able to infect and lyse the bacterial cells, clearings in the bacterial growth (called plaques) appear (**Figure 10.13**). Such a test might show, for instance, that bacteria isolated from a surgical wound have the same pattern of phage sensitivity as those isolated from the operating surgeon or surgical nurses. This result establishes that the surgeon or a nurse is the source of infection.

Fatty Acid Profiles

Bacteria synthesize a wide variety of fatty acids, and in general, these fatty acids are constant for a particular species. Commercial systems have been designed to separate cellular fatty acids to compare them to fatty acid profiles of known organisms. Fatty acid profiles, called **FAME** (*f*atty *a*cid *m*ethyl *e*ster), are widely used in clinical and public health laboratories.

Flow Cytometry

Flow cytometry can be used to identify bacteria in a sample without culturing the bacteria. In a *flow cytometer,* a moving fluid containing bacteria is forced through a small opening (see Figure 18.12, page 516). The simplest method detects the presence of bacteria by detecting the difference in electrical conductivity between cells and the surrounding medium. If the fluid passing through the opening is illuminated by a laser, the scattering of light provides information about the cell size, shape, density, and surface, which is analyzed by a computer. Fluorescence can be used to detect naturally fluorescent cells, such as *Pseudomonas,* or cells tagged with fluorescent dyes.

Milk can be a vehicle for disease transmission. A proposed test that uses flow cytometry to detect *Listeria* in milk could save time because the bacteria would not need to be cultured for identification. Antibodies against *Listeria* can be labeled with a fluorescent dye and added to the milk to be tested. The milk is passed through the flow cytometer, which records the fluorescence of the antibody-labeled cells.

DNA Base Composition

Taxonomists can use an organism's **DNA base composition** to draw conclusions about relatedness. This base composition is usually expressed as the percentage of guanine plus cytosine (G + C). The base composition of a single species is theoretically a fixed property; thus, a comparison of the G + C content in different species can reveal the degree of species relatedness. As we saw in Chapter 8, each guanine (G) in DNA has a complementary cytosine (C). Similarly, each adenine (A) in the DNA has a complementary thymine (T). Therefore, the percentage of DNA bases that are GC pairs also tells us the percentage that are AT pairs

1 If Lyme disease is suspected in a patient: Electrophoresis is used to separate *Borrelia burgdorferi* proteins in the serum. Proteins move at different rates based on their charge and size when the gel is exposed to an electric current.

2 The bands are transferred to a nitrocellulose filter by blotting. Each band consists of many molecules of a particular protein (antigen). The bands are not visible at this point.

3 The proteins (antigens) are positioned on the filter exactly as they were on the gel. The filter is then washed with patient's serum followed by antihuman antibodies tagged with an enzyme. The patient antibodies that combine with their specific antigen are visible (shown here in red) when the enzyme's substrate is added.

4 The test is read. If the tagged antibodies stick to the filter, evidence of the presence of the microorganism in question—in this case, *B. burgdorferi*—has been found in the patient's serum.

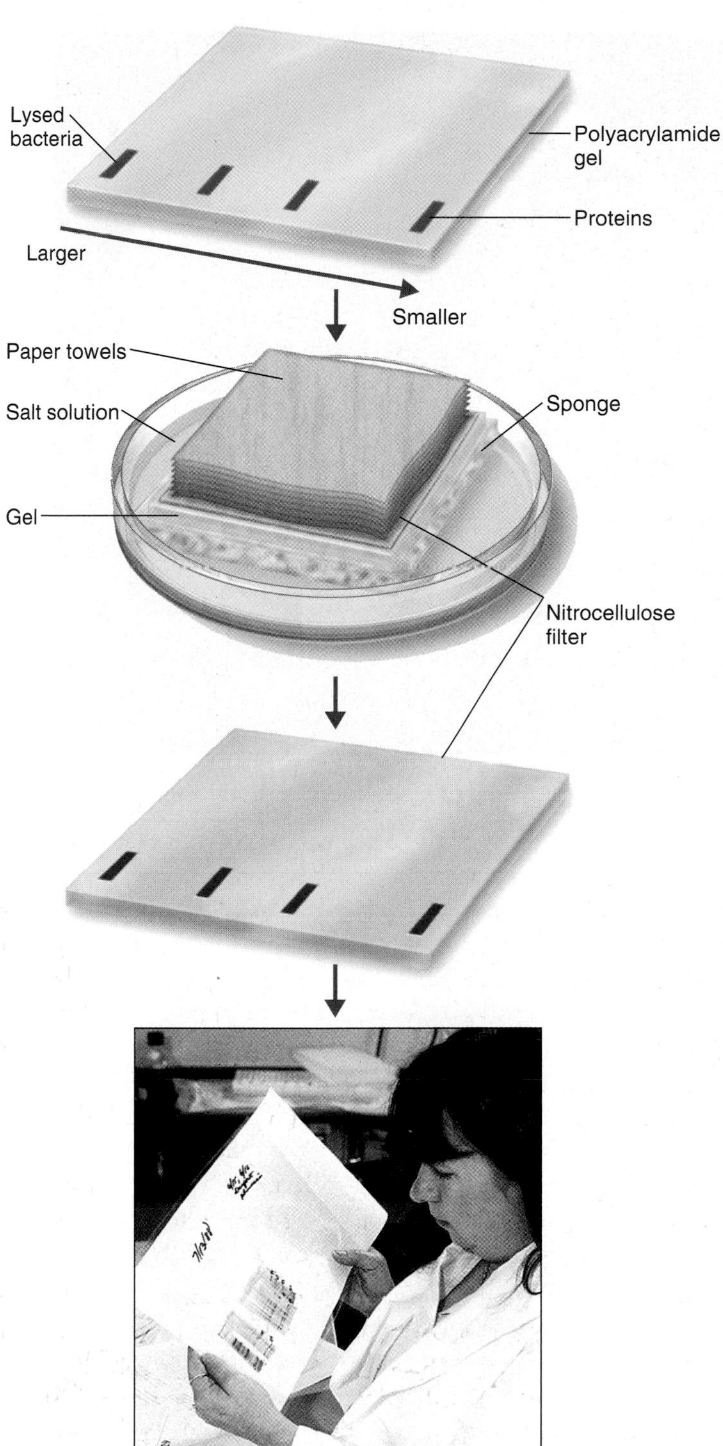

Figure 10.12 The Western blot. Proteins separated by electrophoresis can be detected by their reactions with antibodies.

Q Name two diseases that may be diagnosed by Western blotting.

(GC + AT = 100%). Two organisms that are closely related and hence have many identical or similar genes will have similar amounts of the various bases in their DNA. However, if there is a difference of more than 10% in their percentage of GC pairs (for example, if one bacterium's DNA contains 40% GC and another bacterium has 60% GC), then these two organisms are probably not related. Of course, two organisms that have the same percentage of GC are not necessarily closely related; other supporting data are needed to draw conclusions about their phylogenetic relationship.

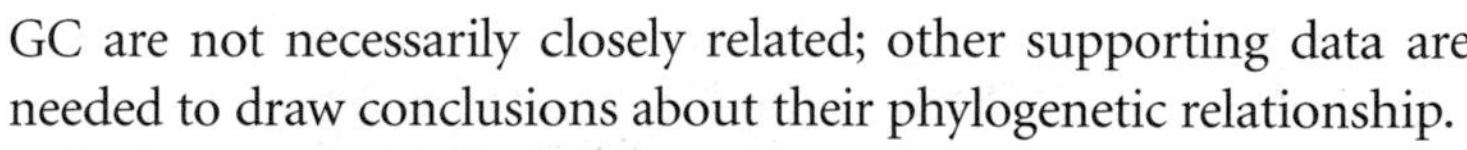

DNA Fingerprinting

Actually determining the entire sequence of bases in an organism's DNA is now possible with modern biochemical methods, but this

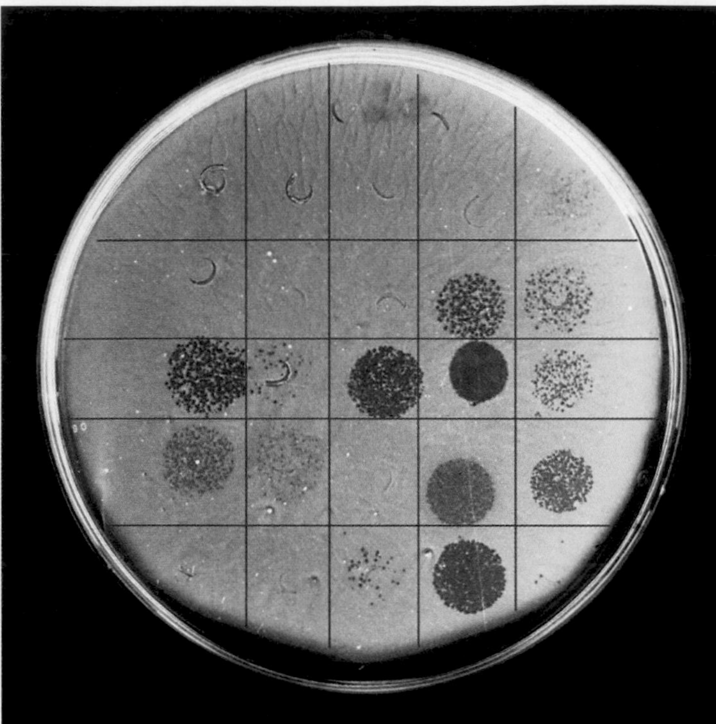

Figure 10.13 Phage typing of a strain of *Salmonella enterica.* The tested strain was grown over the entire plate. Plaques, or areas of lysis, were produced by bacteriophages, indicating that the strain was sensitive to infection by these phages. Phage typing is used to distinguish *S. enterica* serotypes and *Staphylococcus aureus* types.

Q What is being identified in phage typing?

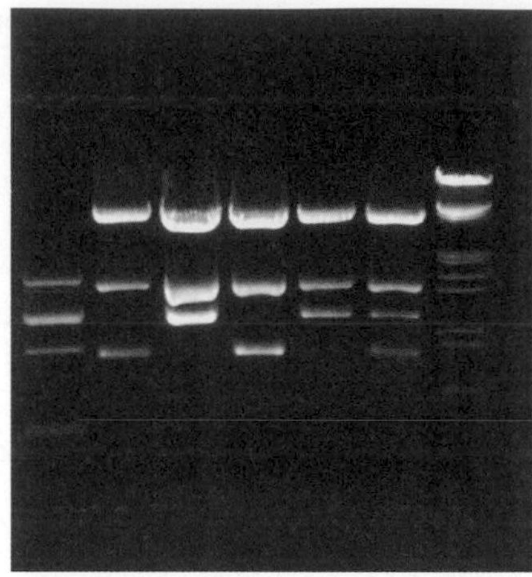

Figure 10.14 DNA fingerprints. Plasmids from seven different bacteria were digested with the same restriction enzyme. Each digest was put in a different well (origin) in the agarose gel. An electrical current was then applied to the gel to separate the fragments by size and electrical charge. The DNA was made visible by staining with ethidium bromide, which fluoresces under ultraviolet light. Comparison of the lanes shows that none of the DNA samples (and therefore none of the bacteria) is identical.

Q What is an RFLP?

is currently impractical for laboratory identification because of the great amount of time required. However, the use of restriction enzymes enables researchers to compare the base sequences of different organisms. Restriction enzymes cut a molecule of DNA everywhere a specific base sequence occurs, producing restriction fragments (as discussed in Chapter 9, page 249). For example, the enzyme *Eco*RI cuts DNA at the arrows in the sequence

$$\ldots G^{\downarrow}AATTC\ldots$$
$$\ldots CTTAA_{\uparrow}G\ldots$$

In this technique, the DNA from two microorganisms is treated with the same restriction enzyme, and the restriction fragments (RFLPs) produced are separated by electrophoresis (see Figure 9.17, page 264). A comparison of the number and sizes of restriction fragments that are produced from different organisms provides information about their genetic similarities and differences; the more similar the patterns, or *DNA fingerprints,* the more closely related the organisms are expected to be (**Figure 10.14**).

DNA fingerprinting is used to determine the source of hospital-acquired infections. In one hospital, patients undergoing coronary-bypass surgery developed infections caused by *Rhodococcus bronchialis* (rō-dō-kok′kus bron-kē′al-is). The DNA fingerprints of the patients' bacteria and the bacteria of one nurse were identical. The hospital was thus able to break the chain of transmission of this infection by encouraging this nurse to use aseptic technique. DNA fingerprinting to locate the source of tomato-associated diarrhea is described in the box in Chapter 25 on page 715.

The Polymerase Chain Reaction

When a microorganism cannot be cultured by conventional methods, the causative agent of an infectious disease might not be recognized. However, a technique called the **polymerase chain reaction (PCR)** can be used to increase the amount of microbial DNA to levels that can be tested by gel electrophoresis (see Chapter 9, page 251). If a primer for a specific microorganism is used, the presence of amplified DNA indicates that microorganism is present.

In 1992, researchers used PCR to determine the causative agent of Whipple's disease, which was previously an unknown bacterium now named *Tropheryma whipplei* (trō′fėr-ē-mä whip′plē-ē). Whipple's disease was first described in 1907 by George Whipple as a gastrointestinal and nervous system disorder caused by an unknown bacillus. No one has been able to culture the bacterium to identify it, and thus PCR provides the only reliable methods of diagnosing and treating the disease.

In recent years, PCR made possible several discoveries. For example, in 1992, Raul Cano used PCR to amplify DNA from

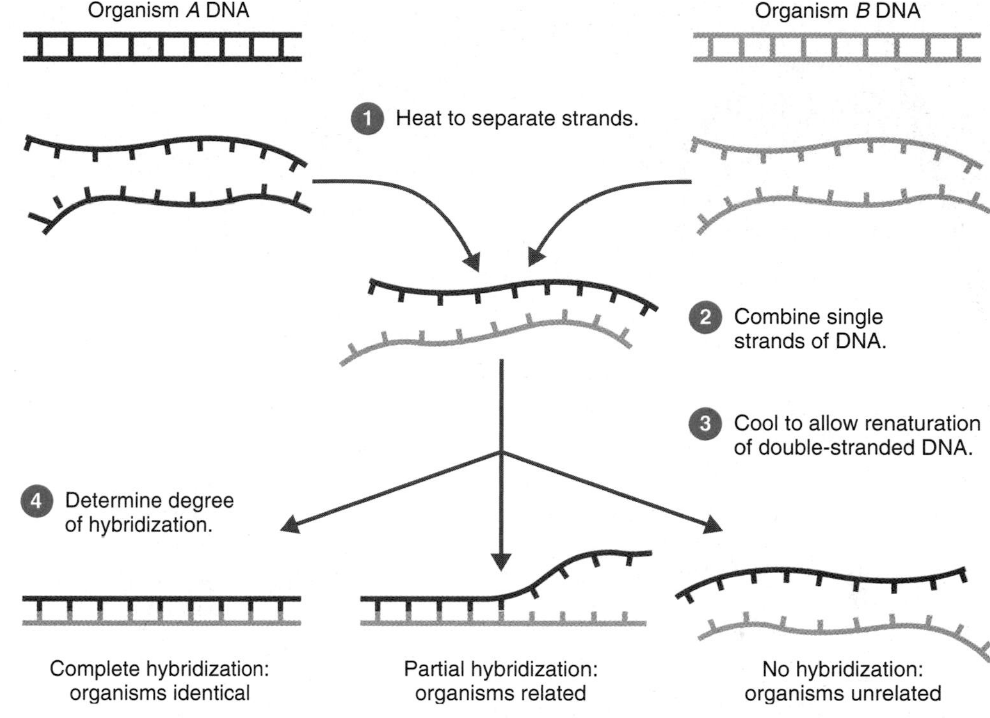

Figure 10.15 DNA-DNA hybridization. The greater the amount of pairing between DNA strands from different organisms (hybridization), the more closely the organisms are related.

Q What is the principle involved in DNA probes?

Bacillus bacteria in amber that was 25 to 40 million years old. These primers were made from rRNA sequences in living *B. circulans* to amplify DNA coding for rRNA in the amber. These primers will cause amplification of DNA from other *Bacillus* species but do not cause amplification of DNA from other bacteria that might have been present, such as *Escherichia* or *Pseudomonas.* The DNA was sequenced after amplification. This information was used to determine the relationships between the ancient bacteria and modern bacteria.

In 1993, microbiologists identified a *Hantavirus* as the cause of an outbreak of hemorrhagic fever in the American Southwest using PCR. The identification was made in record time—less than 2 weeks. PCR was used in 1994 to identify the causative agent of a new tickborne disease (human granulocytic ehrlichiosis) as the bacterium *Ehrlichia chaffeensis* (ėr′lik-ē-ä chaf′ē-en-sis)(page 654). PCR is used to identify the source of rabies viruses; see the box in Chapter 22 (page 625).

TaqMan is a commercial system that uses PCR to identify pathogenic *E. coli* in food and water. With this system, the newly amplified *E. coli* DNA fluoresces and can be detected using gel electrophoresis.

Nucleic Acid Hybridization

If a double-stranded molecule of DNA is subjected to heat, the complementary strands will separate as the hydrogen bonds between the bases break. If the single strands are then cooled slowly, they will reunite to form a double-stranded molecule identical to the original double strand. (This reunion occurs because the single strands have complementary sequences.) When this technique is applied to separated DNA strands from two different organisms, it is possible to determine the extent of similarity between the base sequences of the two organisms. This method is known as **nucleic acid hybridization.** The procedure assumes that if two species are similar or related, a major portion of their nucleic acid sequences will also be similar. The procedure measures the ability of DNA strands from one organism to hybridize (bind through complementary base pairing) with the DNA strands of another organism (**Figure 10.15**). The greater the degree of hybridization, the greater the degree of relatedness.

Similar hybridization reactions can occur between any single-stranded nucleic acid chain: DNA-DNA, RNA-RNA, DNA-RNA. An RNA transcript will hybridize with the separated template DNA to form a DNA-RNA hybrid molecule. Nucleic acid hybridization reactions are the basis of several techniques (described below) that are used to detect the presence of microorganisms and to identify unknown organisms.

Southern Blotting

Nucleic acid hybridization can be used to identify unknown microorganisms by **Southern blotting** (see Figure 9.16, page 263). In addition, rapid identification methods using **DNA probes** are being developed. One method involves breaking DNA extracted from *Salmonella* into fragments with a restriction enzyme, then selecting a specific fragment as the probe for *Salmonella* (**Figure 10.16**). This fragment must be able to hybridize with the DNA of all *Salmonella* strains, but not with the DNA of closely related enteric bacteria.

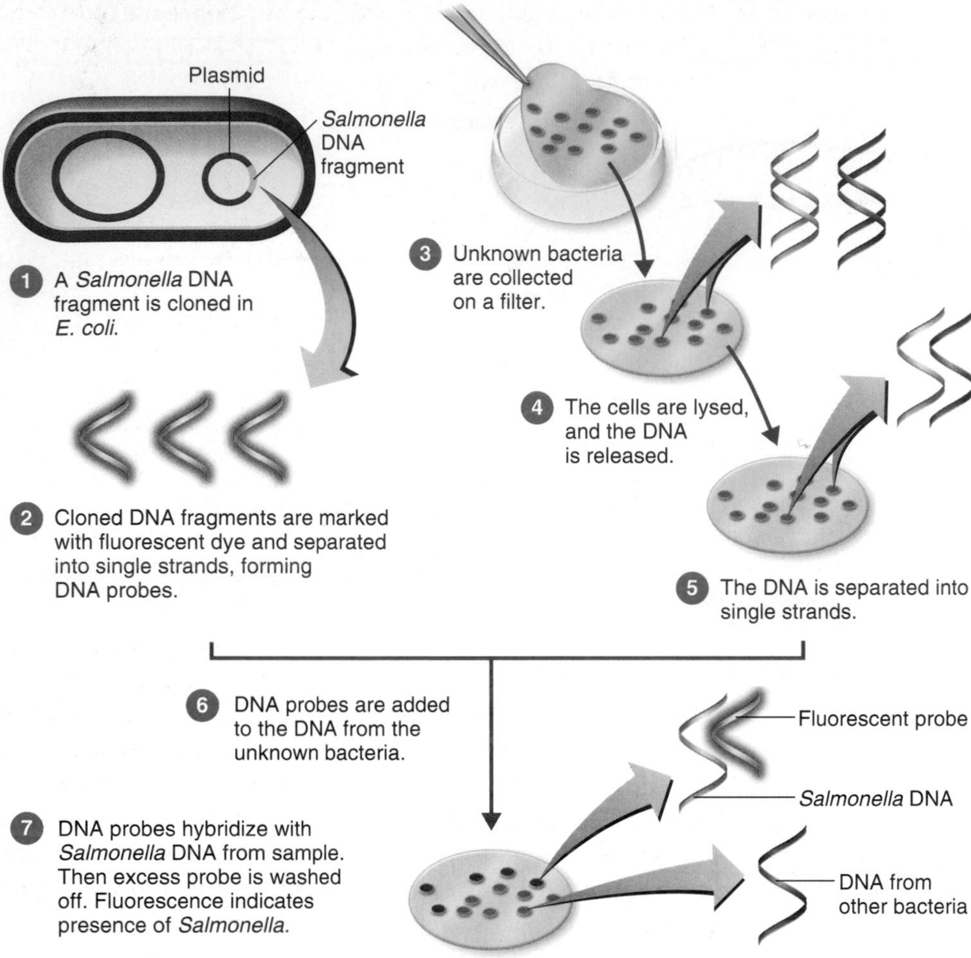
Plasmid
Salmonella DNA fragment
1 A Salmonella DNA fragment is cloned in E. coli.
2 Cloned DNA fragments are marked with fluorescent dye and separated into single strands, forming DNA probes.
3 Unknown bacteria are collected on a filter.
4 The cells are lysed, and the DNA is released.
5 The DNA is separated into single strands.
6 DNA probes are added to the DNA from the unknown bacteria.
7 DNA probes hybridize with Salmonella DNA from sample. Then excess probe is washed off. Fluorescence indicates presence of Salmonella.
Fluorescent probe
Salmonella DNA
DNA from other bacteria

Figure 10.16 A DNA probe used to identify bacteria. Southern blotting is used to detect specific DNA. This modification of the Southern blot is used to detect *Salmonella*.

Q Why do the DNA probe and cellular DNA hybridize?

DNA Chips

An exciting new technology is the **DNA chip**, which will make it possible to quickly detect a pathogen in a host by identifying a gene that is unique to that pathogen (**Figure 10.17**). The DNA chip is composed of DNA probes. A sample containing DNA from an unknown organism is labeled with a fluorescent dye and added to the chip. Hybridization between the probe DNA and DNA in the sample is detected by fluoresence.

Ribotyping and Ribosomal RNA Sequencing

Ribotyping is currently being used to determine the phylogenetic relationships among organisms. There are several advantages to using rRNA. First, all cells contain ribosomes. Second, RNA genes have undergone few changes over time so all members of a domain, phylum, and, in some cases, a genus, have the same "signature" sequences in their rRNA. The rRNA used most often is a component of the smaller portion of ribosomes. A third advantage of rRNA sequencing is that cells do not have to be cultured in the laboratory.

DNA can be amplified by PCR using an rRNA primer for specific signature sequences. The amplified fragments are subsequently cut with one or more restriction enzymes and separated by electrophoresis. The resulting band patterns can then be compared. Then the rRNA genes in the amplified fragments can be sequenced to determine evolutionary relationships between organisms. This technique is useful for classifying a newly discovered organism to domain or phylum or to determine the general types of organisms present in one environment. More specific probes (see page 291) are needed to identify individual species, however.

Fluorescent In Situ Hybridization (FISH)

Fluorescent dye-labeled RNA or DNA probes are used to specifically stain microorganisms in place, or in situ. This technique is called **fluorescent in situ hybridization,** or **FISH.** Cells are treated so the probe enters the cells and reacts with target ribosome in the cell (in situ). FISH is used to determine the identity, abundance, and relative activity of microorganisms in an environment and can be used to detect bacteria that have not yet been cultured. Using FISH, a tiny bacterium, *Pelagibacter* (pel-aj'ē-bak-tėr), was discovered in the ocean and determined to be related to the rickettsias (page 303). As probes are developed,

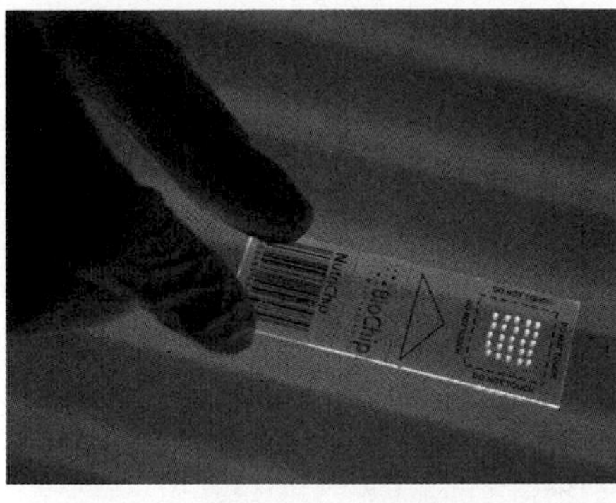

(a) A DNA chip can be manufactured to contain hundreds of thousands of synthetic single-stranded DNA sequences. Assume that each DNA sequence was unique to a different gene.

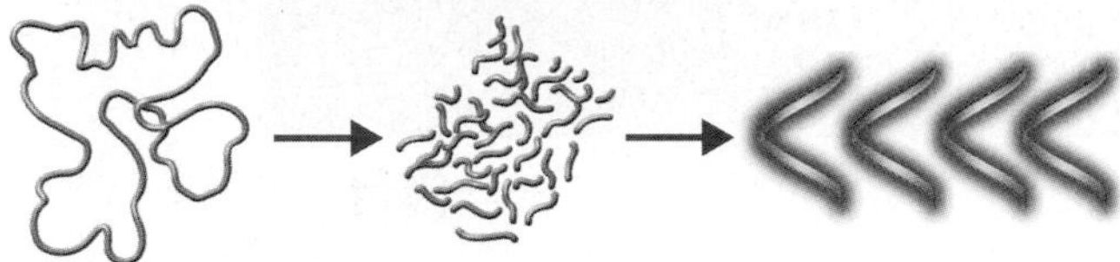

(b) Unknown DNA from a patient is separated into single strands, enzymatically cut, and labeled with a fluorescent dye.

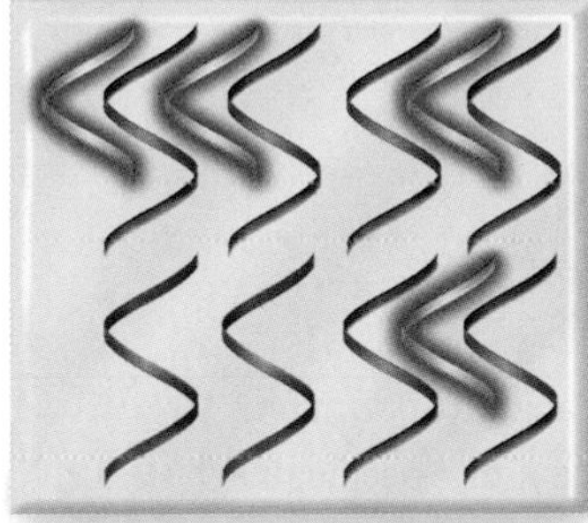

(c) The unknown DNA is inserted into the chip and allowed to hybridize with the DNA on the chip.

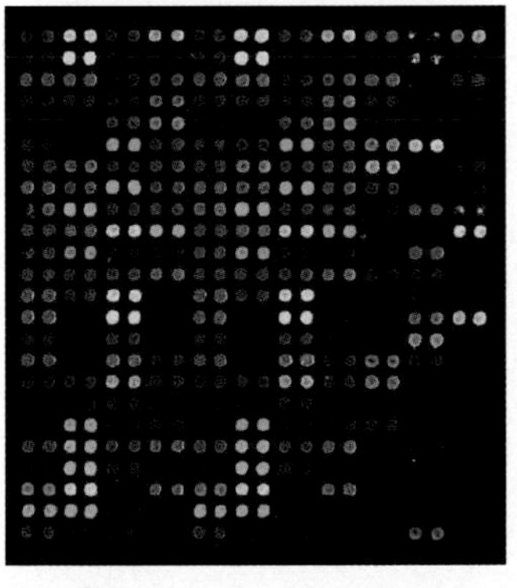

(d) The tagged DNA will bind only to the complementary DNA on the chip. The bound DNA will be detected by its fluorescent dye and analyzed by a computer. The red light is a gene expressed in normal cells; green is a mutated gene expressed in tumor cells; and yellow, in both cells.

Figure 10.17 DNA chip technology. DNA chip technology will enable much more rapid, inexpensive analysis using many more probes. Chip technology is new; however, scientists soon hope to use DNA chips to detect microbes in a human or environmental sample, to identify cancer genes, to identify potential suspects in a crime or victims of a crime or catastrophe, to identify endangered species, and to match organ donors. If the DNA chip in (**d**) was constructed from different bacterial DNA, each spot would represent a bacterium, and the spot giving off fluorescent light would identify the bacterium in the sample.

Q **What is on the chip to make it specific for a particular microorganism?**

FISH can be used to detect bacteria in drinking water or bacteria in a patient without the normal 24-hour or longer wait required for culturing the bacteria (Figure 10.18).

Putting Classification Methods Together

Morphological characteristics, differential staining, and biochemical testing were the only identification tools available just a few years ago. Technological advancements are making it possible to use nucleic acid analysis techniques, once reserved for classification, for routine identification. Information obtained about microbes is used to identify and classify the organisms. Two methods of using the information are described below.

Dichotomous Keys

Dichotomous keys are widely used for identification. In a dichotomous key, identification is based on successive questions, and each question has two possible answers (*dichotomous* means cut in two). After answering one question, the investigator is directed to another question until an organism is identified. Although these keys often have little to do with phylogenetic relationships, they are invaluable for identification. For example, a dichotomous key for bacteria could begin with an easily determined characteristic, such as cell shape, and move on to the ability to ferment a sugar. Dichotomous keys are shown in Figure 10.8 and in the box on page 283. **Animations** Dichotomous Keys: Overview, Sample with Flowchart, Practice. **www.microbiologyplace.com**

Cladograms

Cladograms are maps that show evolutionary relationships among organisms (*clado-* means branch). Cladograms are shown in Figures 10.1 and 10.6. Each branch point on the cladogram is defined by a feature shared by various species on that branch. Historically, cladograms for vertebrates were made using fossil evidence; however, rRNA sequences are now being used to confirm assumptions based on fossils. As we said earlier, most microorganisms do not leave fossils; therefore, rRNA sequencing is primarily used to make cladograms for microorganisms. The small rRNA subunit used has 1500 bases, and computer programs do the calculations. The steps for constructing a cladogram are shown in Figure 10.19.

1. Two rRNA sequences are aligned, and
2. the percentage of similarity between the sequences is calculated.
3. Then the horizontal branches are drawn in a length proportional to the calculated percent similarity. All species beyond a node (branch point) have similar rRNA sequences, suggesting that they arose from an ancestor at that node.

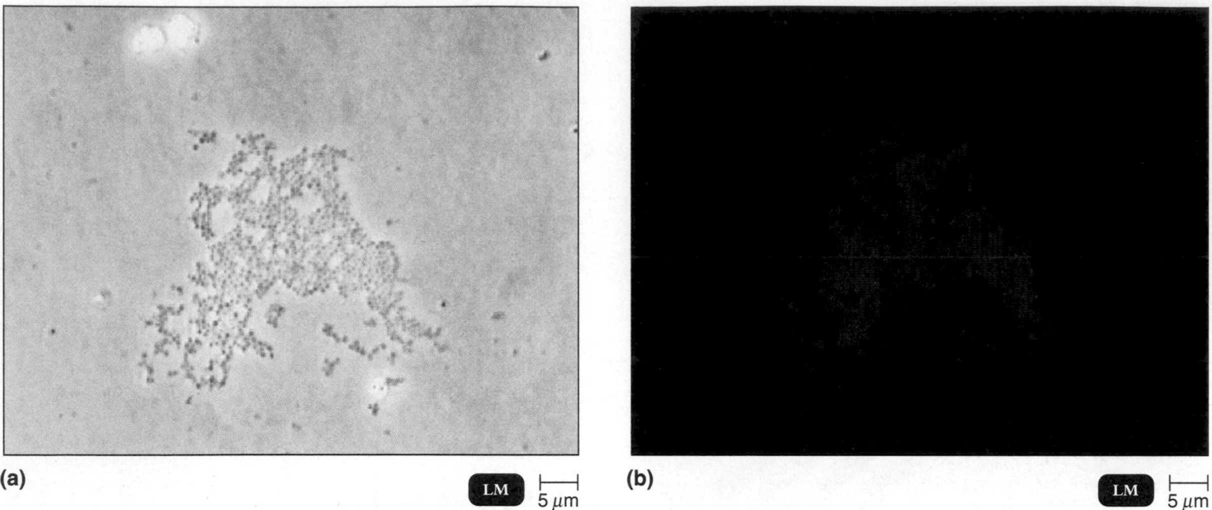

Figure 10.18 FISH, or fluorescent in situ hybridization. A DNA or RNA probe attached to fluorescent dyes is used to identify chromosomes. Bacteria seen with phase-contrast microscopy (**a**) are identified with a fluorescent-labeled probe that hybridizes with a specific sequence of DNA in *Staphylococcus aureus* (**b**).

Q What is stained using the FISH technique?

CHECK YOUR UNDERSTANDING

- ✓ What is in *Bergey's Manual*? **10-13**
- ✓ Design a rapid test for a *Staphylococcus aureus*. (*Hint:* See Figure 6.10, page 169.) **10-14**
- ✓ What is tested in Western blotting and Southern blotting? **10-15**
- ✓ What is identified by phage typing? **10-16**
- ✓ Why does PCR identify a microbe? **10-17**
- ✓ Which techniques involve nucleic acid hybridization? **10-18**
- ✓ Is a cladogram used for identification or classification? **10-12, 10-19**

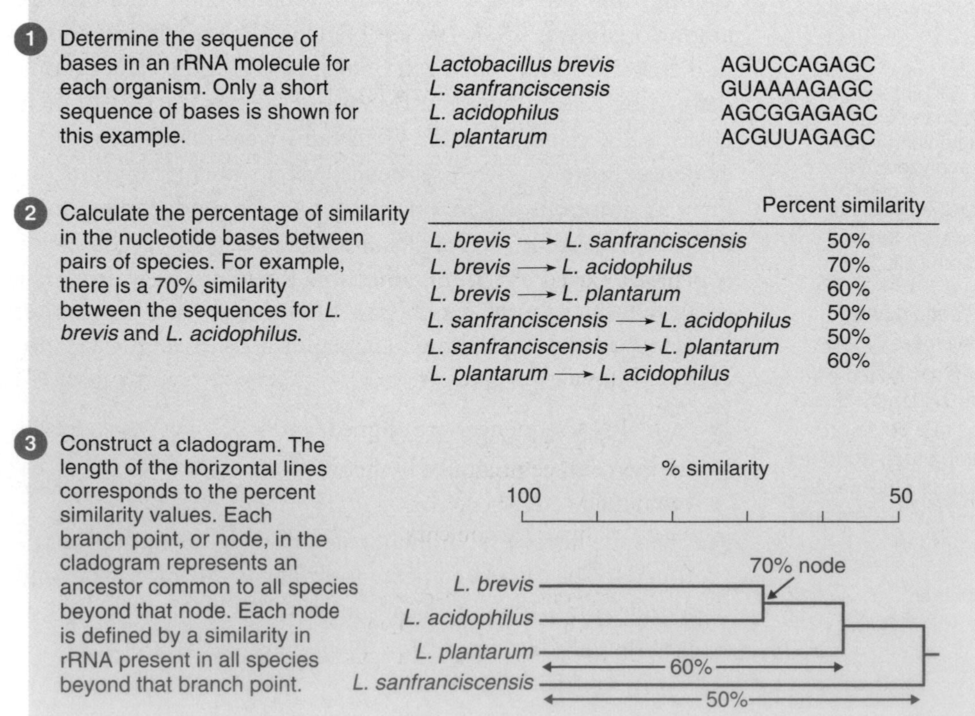

1 Determine the sequence of bases in an rRNA molecule for each organism. Only a short sequence of bases is shown for this example.
Lactobacillus brevis AGUCCAGAGC
L. sanfranciscensis GUAAAAGAGC
L. acidophilus AGCGGAGAGC
L. plantarum ACGUUAGAGC
2 Calculate the percentage of similarity in the nucleotide bases between pairs of species. For example, there is a 70% similarity between the sequences for L. brevis and L. acidophilus.
Percent similarity
L. brevis → L. sanfranciscensis 50%
L. brevis → L. acidophilus 70%
L. brevis → L. plantarum 60%
L. sanfranciscensis → L. acidophilus 50%
L. sanfranciscensis → L. plantarum 50%
L. plantarum → L. acidophilus 60%
3 Construct a cladogram. The length of the horizontal lines corresponds to the percent similarity values. Each branch point, or node, in the cladogram represents an ancestor common to all species beyond that node. Each node is defined by a similarity in rRNA present in all species beyond that branch point.
% similarity
100
50
70% node
L. brevis
L. acidophilus
L. plantarum
L. sanfranciscensis
60%
50%

Figure 10.19 Building a cladogram.

Q Why do *L. brevis* and *L. acidophilus* branch from the same node?

STUDY OUTLINE

The **MyMicrobiologyPlace** website (**www.microbiologyplace.com**) will help you get ready for tests with its simple three-step approach: ❶ **take a pre-test** and obtain a personalized study plan, ❷ **learn and practice** with animations, tutorials, and MP3 tutor sessions, and ❸ **test yourself** with quizzes and a chapter post-test.

Introduction (p. 273)

1. Taxonomy is the science of the classification of organisms. Its goal is to show relationships among organisms.
2. Taxonomy also provides a means of identifying organisms.

The Study of Phylogenetic Relationships (pp. 274–278)

1. Phylogeny is the evolutionary history of a group of organisms.
2. The taxonomic hierarchy shows evolutionary, or phylogenetic, relationships among organisms.
3. Bacteria were separated into the Kingdom Prokaryotae in 1968.
4. Living organisms were divided into five kingdoms in 1969.

The Three Domains (pp. 274–277)

5. Living organisms are currently classified into three domains. A domain can be divided into kingdoms.
6. In this system, plants, animals, fungi, and protists belong to the Domain Eukarya.
7. Bacteria (with peptidoglycan) form a second domain.
8. Archaea (with unusual cell walls) are placed in the Domain Archaea.

A Phylogenetic Hierarchy (pp. 277–278)

9. Organisms are grouped into taxa according to phylogenetic relationships (from a common ancestor).
10. Some of the information for eukaryotic relationships is obtained from the fossil record.
11. Prokaryotic relationships are determined by rRNA sequencing.

Classification of Organisms (pp. 278–282)

Scientific Nomenclature (pp. 278–279)

1. According to scientific nomenclature, each organism is assigned two names, or a binomial: a genus and a specific epithet, or species.
2. Rules for the assignment of names to bacteria are established by the International Committee on Systematics of Prokaryotes.
3. Rules for naming fungi and algae are published in the *International Code of Botanical Nomenclature.*
4. Rules for naming protozoa are found in the *International Code of Zoological Nomenclature.*

The Taxonomic Hierarchy (p. 279)

5. A eukaryotic species is a group of organisms that interbreeds with each other but does not breed with individuals of another species.
6. Similar species are grouped into a genus; similar genera are grouped into a family; families, into an order; orders, into a class; classes, into a phylum; phyla, into a kingdom; and kingdoms, into a domain.

Classification of Prokaryotes (pp. 279–281)

7. *Bergey's Manual of Systematic Bacteriology* is the standard reference on bacterial classification.
8. A group of bacteria derived from a single cell is called a strain.
9. Closely related strains constitute a bacterial species.

Classification of Eukaryotes (pp. 281–282)

10. Eukaryotic organisms may be classified into the Kingdom Fungi, Plantae, or Animalia.
11. Protists are mostly unicellular organisms; these organisms are currently being assigned to kingdoms.
12. Fungi are absorptive chemoheterotrophs that develop from spores.
13. Multicellular photoautotrophs are placed in the Kingdom Plantae.
14. Multicellular ingestive heterotrophs are classified as Animalia.

Classification of Viruses (p. 282)

15. Viruses are not placed in a kingdom. They are not composed of cells and cannot grow without a host cell.
16. A viral species is a population of viruses with similar characteristics that occupies a particular ecological niche.

Methods of Classifying and Identifying Microorganisms (pp. 282–294)

1. *Bergey's Manual of Determinative Bacteriology* is the standard reference for laboratory identification of bacteria.
2. Morphological characteristics are useful in identifying microorganisms, especially when aided by differential staining techniques.
3. The presence of various enzymes, as determined by biochemical tests, is used in identifying microorganisms.
4. Serological tests, involving the reactions of microorganisms with specific antibodies, are useful in determining the identity of strains and species, as well as relationships among organisms. ELISA and Western blotting are examples of serological tests.
5. Phage typing is the identification of bacterial species and strains by determining their susceptibility to various phages.
6. Fatty acid profiles can be used to identify some organisms.
7. Flow cytometry measures physical and chemical characteristics of cells.
8. The percentage of GC base pairs in the nucleic acid of cells can be used in the classification of organisms.
9. The number and sizes of DNA fragments, or DNA fingerprints, produced by restriction enzymes are used to determine genetic similarities.
10. The polymerase chain reaction (PCR) can be used to amplify a small amount of microbial DNA in a sample. The presence or identification of an organism is indicated by amplified DNA.

11. Single strands of DNA, or of DNA and RNA, from related organisms will hydrogen-bond to form a double-stranded molecule; this bonding is called nucleic acid hybridization.
12. Southern blotting, DNA chips, and FISH are examples of nucleic acid hybridization techniques.

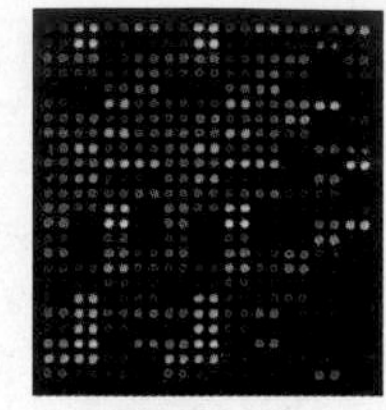

13. The sequence of bases in ribosomal RNA can be used in the classification of organisms.
14. Dichotomous keys are used for the identification of organisms. Cladograms show phylogenetic relationships among organisms.

STUDY QUESTIONS

Answers to the Review and Multiple Choice questions can be found by turning to the blue Answers tab at the back of the textbook.

Review

1. Which of the following organisms are most closely related? Are any two the same species? On what did you base your answer?

Characteristic	A	B	C	D
Morphology	Rod	Coccus	Rod	Rod
Gram Reaction	+	–	–	+
Glucose Utilization	Fermentative	Oxidative	Fermentative	Fermentative
Cytochrome Oxidase	Present	Present	Absent	Absent
GC Moles %	48–52	23–40	50–54	49–53

2. Here is some additional information on the organisms in question 1:

Organism	% DNA Hybridization
A and B	5–15
A and C	5–15
A and D	70–90
B and C	10–20
B and D	2–5

Which of these organisms are most closely related? Compare this answer with your response to review question 1.

3. DRAW IT Use the information in the table below to complete the dichotomous key to these organisms. What is the purpose of a dichotomous key? Look up each genus in Chapter 11, and provide an example of why this organism is of interest to humans.

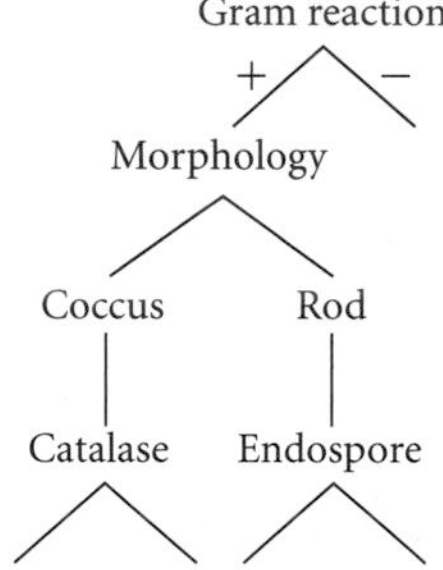

	Morphology	Gram Reaction	Acid from Glucose	Growth in Air (21% O_2)	Motile by Peritrichous Flagella	Presence of Cytochrome Oxidase	Produce Catalase
Staphylococcus aureus	Coccus	+	+	+	–	–	+
Streptococcus pyogenes	Coccus	+	+	+	–	–	–
Mycoplasma pneumoniae	Coccus	–	+	+ (Colonies < 1 mm)	–	–	+
Clostridium botulinum	Rod	+	+	–	+	–	–
Escherichia coli	Rod	–	+	+	+	–	+
Pseudomonas aeruginosa	Rod	–	+	+	–	+	+
Campylobacter fetus	Vibrio	–	–	–	–	+	+
Listeria monocytogenes	Rod	+	+	+	+	–	+

4. **DRAW IT** Use the additional information shown below to construct a cladogram for some of the organisms used in question 3. What is the purpose of a cladogram? How does your cladogram differ from a dichotomous key for these organisms?

	Similarity in rRNA Bases
P. aeruginosa—M. pneumoniae	52%
P. aeruginosa—C. botulinum	52%
P. aeruginosa—E. coli	79%
M. pneumoniae—C. botulinum	65%
M. pneumoniae—E. coli	52%
E. coli—C. botulinum	52%

% similarity

100 50

Multiple Choice

1. *Bergey's Manual of Systematic Bacteriology* differs from *Bergey's Manual of Determinative Bacteriology* in that the former
   a. groups bacteria into species.
   b. groups bacteria according to phylogenetic relationships.
   c. groups bacteria according to pathogenic properties.
   d. groups bacteria into 19 species.
   e. all of the above
2. *Bacillus* and *Lactobacillus* are not in the same order. This indicates that which one of the following is *not* sufficient to assign an organism to a taxon?
   a. biochemical characteristics
   b. amino acid sequencing
   c. phage typing
   d. serology
   e. morphological characteristics
3. Which of the following is used to classify organisms into the Kingdom Fungi?
   a. ability to photosynthesize; possess a cell wall
   b. unicellular; possess cell wall; prokaryotic
   c. unicellular; lacking cell wall; eukaryotic
   d. absorptive; possess cell wall; eukaryotic
   e. ingestive; lacking cell wall; multicellular; prokaryotic
4. Which of the following is *not* true about scientific nomenclature?
   a. Each name is specific.
   b. Names vary with geographical location.
   c. The names are standardized.
   d. Each name consists of a genus and specific epithet.
   e. It was first designed by Linnaeus.
5. You could identify an unknown bacterium by all of the following *except*
   a. hybridizing a DNA probe from a known bacterium with the unknown's DNA.
   b. making a fatty acid profile of the unknown.
   c. specific antiserum agglutinating the unknown.
   d. ribosomal RNA sequencing.
   e. percentage of guanine + cytosine.
6. The wall-less mycoplasmas are considered to be related to gram-positive bacteria. Which of the following would provide the most compelling evidence for this?
   a. They share common rRNA sequences.
   b. Some gram-positive bacteria and some mycoplasmas produce catalase.
   c. Both groups are prokaryotic.
   d. Some gram-positive bacteria and some mycoplasmas have coccus-shaped cells.
   e. Both groups contain human pathogens.

Use the following choices to answer questions 7 and 8.
   a. Animalia
   b. Fungi
   c. Plantae
   d. Firmicutes (gram-positive bacteria)
   e. Proteobacteria (gram-negative bacteria)

7. Into which group would you place a multicellular organism that has a mouth and lives inside the human liver?
8. Into which group would you place a photosynthetic organism that lacks a nucleus and has a thin peptidoglycan wall surrounded by an outer membrane?

Use the following choices to answer questions 9 and 10.
   1. 9 + 2 flagella
   2. 70S ribosome
   3. fimbria
   4. nucleus
   5. peptidoglycan
   6. plasma membrane

9. Which is (are) found in all three domains?
   a. 2, 6
   b. 5
   c. 2, 4, 6
   d. 1, 3, 5
   e. all six
10. Which is (are) found *only* in prokaryotes?
   a. 1, 4, 6
   b. 3, 5
   c. 1, 2
   d. 4
   e. 2, 4, 5

Critical Thinking

1. The GC content of *Micrococcus* is 66–75 moles %, and of *Staphylococcus,* 30–40 moles %. According to this information, would you conclude that these two genera are closely related?
2. Describe the use of a DNA probe and PCR for:
   a. rapid identification of an unknown bacterium.
   b. determining which of a group of bacteria are most closely related.
3. SF medium is a selective medium, developed in the 1940s, to test for fecal contamination of milk and water. Only certain gram-positive cocci can grow in this medium. Why is it named SF? Using this medium, which genus will you culture? (*Hint:* Refer to page 279.)

Clinical Applications

1. A 55-year-old veterinarian was admitted to a hospital with a 2-day history of fever, chest pain, and cough. Gram-positive cocci were detected in his sputum, and he was treated for lobar pneumonia with penicillin. The next day, another Gram stain of his sputum revealed gram-negative rods, and he was switched to ampicillin and gentamicin. A sputum culture showed

biochemically inactive gram-negative rods identified as *Pantoea (Enterobacter) agglomerans.* After fluorescent-antibody staining and phage typing, *Yersinia pestis* was identified in the patient's sputum and blood, and chloramphenicol and tetracycline were administered. The patient died 3 days after admission to the hospital. Tetracycline was given to his 220 contacts (hospital personnel, family, and co-workers). What disease did the patient have? Discuss what went wrong in the diagnosis and how his death might have been prevented. Why were the 220 other people treated? (*Hint:* Refer to Chapter 23.)

2. A 6-year-old girl was admitted to a hospital with endocarditis. Blood cultures showed a gram-positive, aerobic rod identified by the hospital laboratory as *Corynebacterium xerosis.* The girl died after 6 weeks of treatment with intravenous penicillin and chloramphenicol. The bacterium was tested by another laboratory and identified as *C. diphtheriae.* The following test results were obtained by each laboratory:

	Hospital Lab	Other lab
Catalase	+	+
Nitrate reduction	+	+
Urea	–	–
Esculin hydrolysis	–	–
Glucose fermentation	+	+
Sucrose fermentation	–	+
Serological test for toxin production	Not done	+

Provide a possible explanation for the incorrect identification. What are the potential public health consequences of misidentifying *C. diphtheriae*? (*Hint:* Refer to Chapter 24.)

3. Using the following information, create a dichotomous key for distinguishing these unicellular organisms. Which cause human disease?

	Mitochondria?	Chlorophyll?	Nutritional Type?	Motile?
Euglena	+	+	Both	+
Giardia	–	–	Heterotroph	+
Nosema	–	–	Heterotroph	–
Pfiesteria	+	+	Autotroph	+
Trichomonas	–	–	Heterotroph	+
Trypanosoma	+	–	Heterotroph	+

Using the additional information shown below, create a dichotomous key for these organisms. Do your two keys differ? Explain why. Which key is more useful for laboratory identification? For classification?

	rRNA base #																			
	1	2	3	4	5	6	7	8	9	10	11	12	13	14	15	16	17	18	19	20
Euglena	C	C	A	G	G	U	U	G	U	U	C	C	A	G	U	U	U	U	A	A
Giardia	C	C	A	U	A	U	U	U	U	U	G	A	C	G	A	A	G	G	U	C
Nosema	C	C	A	U	A	U	U	U	U	U	A	A	C	G	A	A	G	G	C	C
Pfiesteria	C	C	A	A	C	U	U	A	U	U	C	C	A	G	U	U	U	C	A	G
Trichomonas	C	C	A	U	A	U	U	U	U	U	G	A	C	G	A	A	G	G	G	C
Trypanosoma	C	C	A	C	G	U	U	G	U	U	C	C	A	G	U	U	U	A	A	A

11 The Prokaryotes: Domains Bacteria and Archaea

When biologists first encountered microscopic bacteria, they were puzzled as to how to classify them. Bacteria were clearly not animals or rooted plants. Attempts to build a taxonomic system for bacteria based on the phylogenetic system developed for plants and animals failed (see page 274). In the earlier editions of *Bergey's Manual* bacteria were grouped by morphology (rod, coccus), staining reactions, presence of endospores, and other obvious features. Although this system had its practical uses, microbiologists were aware that it had many limitations, a bit as though bats and birds were being grouped together as being winged. The knowledge of bacteria at the molecular level has now expanded to such a degree that it is possible to base the latest edition of *Bergey's Manual* on a phylogenetic system. For example, the genera *Rickettsia* and *Chlamydia* are no longer grouped together by their common requirement for intracellular growth. While members of the genus, *Chlamydia*, are now found in a phylum familiarly named, Chlamydia, the rickettsias are now grouped in a distant phylum, Proteobacteria, in the strangely named class, Alphaproteobacteria. Some microbiologists find such changes upsetting, but they reflect important differences. These differences are primarily in the ribosomal RNA (rRNA) of the microbes, a genetic component which is slow to change (see page 292) and performs the same functions in all organisms.

UNDER THE MICROSCOPE

Thiomargarita namibiensis. This is a rare giant of a bacterium.

Q&A

Bacteria are single-celled organisms that must absorb their nutrients by simple diffusion. The dimensions of *T. namibiensis* are several hundred times larger than most bacteria, much too large for simple diffusion to operate. How does the bacterium solve this problem?

Look for the answer in the chapter.

The Prokaryotic Groups

In the second edition of *Bergey's Manual,* the prokaryotes are grouped into two **domains,** the **Archaea** and the **Bacteria.** Both domains consist of prokaryotic cells. Spelled without capitalization, that is, archaea and bacteria, these terms denote organisms that fit into these domains. Each domain is divided into phyla, each phylum into classes, and so on. The phyla discussed in this chapter are summarized in **Table 11.1** (see also Appendix F).

Table 11.1 Selected Prokaryotes from *Bergey's Manual of Systematic Bacteriology,* Second Edition*

Phylum Class	Order	Important Genera	Special Features
DOMAIN BACTERIA			
Proteobacteria			
Alphaproteobacteria	Caulobacterales	*Caulobacter*	Stalked
	Rickettsiales	*Ehrlichia*	Obligately intracellular human pathogens
		Rickettsia	Obligately intracellular human pathogens
		Wolbachia	Symbionts of insects
	Rhizobiales	*Agrobacterium*	Plant pathogens
		Bartonella	Human pathogens
		Beijerinckia	Free-living nitrogen fixers
		Bradyrhizobium	Symbiotic nitrogen fixers
		Brucella	Human pathogens
		Hyphomicrobium	Budding
		Nitrobacter	Nitrifying
		Rhizobium	Symbiotic nitrogen fixers
	Rhodospirillales	*Acetobacter*	Acetic acid producers
		Azospirillum	Nitrogen fixers
		Gluconobacter	Acetic acid producers
		Rhodospirillum	Photosynthetic, anoxygenic
Betaproteobacteria	Burkholderiales	*Burkholderia*	Opportunistic pathogens
		Bordetella	Human pathogens
		Sphaerotilus	Sheathed
	Hydrogenophilales	*Thiobacillus*	Sulfur oxidizers
	Neisseriales	*Neisseria*	Human pathogens
	Nitrosomonadales	*Nitrosomonas*	Nitrifying
		Spirillum	Found in stagnant fresh water
	Rhodocyclales	*Zoogloea*	Sewage treatment
Gammaproteobacteria	Chromatiales	*Chromatium*	Photosynthetic, anoxygenic
	Thiotrichales	*Beggiatoa*	Sulfur oxiders
		Thiomargarita	Giant bacterium
		Francisella	Human pathogens
	Legionellales	*Legionella*	Human pathogens
		Coxiella	Obligately intracellular human pathogens
	Pseudomonadales	*Azomonas*	Free-living nitrogen fixers
		Azotobacter	Free-living nitrogen fixers
		Moraxella	Human pathogens
		Pseudomonas	Opportunistic pathogens
	Vibrionales	*Vibrio*	Human pathogens
	Enterobacteriales	*Citrobacter*	Opportunistic pathogens
		Enterobacter	Opportunistic pathogens
		Erwinia	Plant pathogens
		Escherichia	Normal intestinal bacteria, some pathogens
		Klebsiella	Opportunistic pathogens
		Proteus	Human intestinal bacteria, occasional pathogens

*See Appendix F for a complete taxonomic listing. This table includes prokaryotes mentioned in this text. Descriptions such as *pathogenic* mean that this trait is common in the genus but not that all members of the genus may have this trait.

Table 11.1 (continued)

Phylum Class	**Order**	**Important Genera**	**Special Features**
		Salmonella	Human pathogens
		Serratia	Red pigment, opportunistic pathogens
		Shigella	Human pathogens
		Yersinia	Human pathogens
	Pasteurellales	*Haemophilus*	Human pathogens
		Pasteurella	Human pathogens
Deltaproteobacteria	Bdellovibrionales	*Bdellovibrio*	Parasites of bacteria
	Desulfovibrionales	*Desulfovibrio*	Sulfate reducers
	Myxococcales	*Myxococcus*	Gliding, fruiting
		Stigmatella	Gliding, fruiting
Epsilonproteobacteria	Campylobacterales	*Campylobacter*	Human pathogens
		Helicobacter	Human pathogens, carcinogenic
Nonproteobacteria, Gram-Negative Bacteria			
Cyanobacteria			
		Anabaena	Photosynthetic, oxygenic
		Gloeocapsa	Photosynthetic, oxygenic
Chlorobi			
		Chlorobium	Photosynthetic, anoxygenic
Chloroflexi			
		Chloroflexus	Photosynthetic, anoxygenic
Firmicutes (The Low G + C Gram-Positive Bacteria)			
	Clostridiales	*Clostridium*	Anaerobes, endospores, some human pathogens
		Epulopiscium	Giant bacterium
		Sarcina	Occur in cubical packets
	Mycoplasmatales†	*Mycoplasma*	No cell wall, human pathogens
		Spiroplasma	No cell wall, pleomorphic, plant pathogens
		Ureaplasma	No cell wall, ammonia from urea
	Bacillales	*Bacillus*	Endospores, some pathogens
		Listeria	Human pathogens
		Staphylococcus	Some human pathogens
	Lactobacillales	*Enterococcus*	Opportunistic pathogens
		Lactobacillus	Lactic acid producers
		Streptococcus	Many human pathogens
Actinobacteria (The High G + C Gram-Positive Bacteria)			
	Actinomycetales	*Actinomyces*	Filamentous, branching, some human pathogens
		Corynebacterium	Human pathogens
		Frankia	Symbiotic nitrogen-fixers
		Gardnerella	Human pathogens
		Mycobacterium	Acid fast, human pathogens
		Nocardia	Filamentous, branching, opportunistic pathogens
		Propionibacterium	Propionic acid producers
		Streptomyces	Filamentous branching, many produce antibiotics

†The bacteria in the order Mycoplasmatales are genetically related to the low G + C gram-positive bacteria, but they lack a cell wall and stain gram-negative.

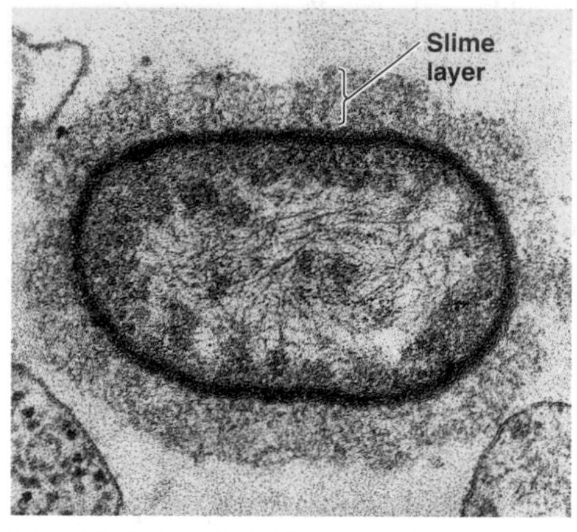
Slime layer

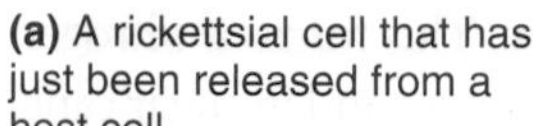

(a) A rickettsial cell that has just been released from a host cell

TEM .04 μm

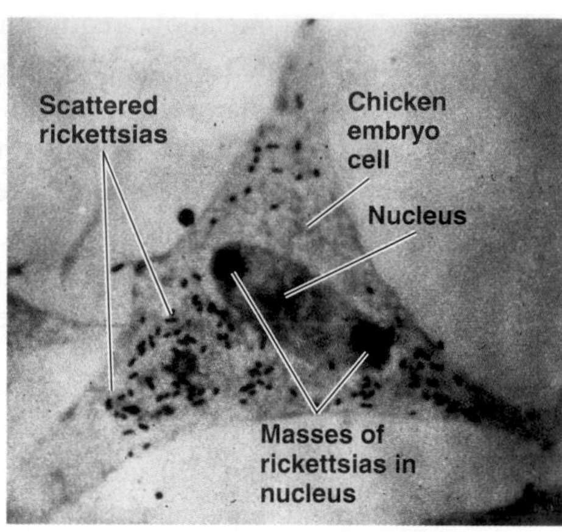
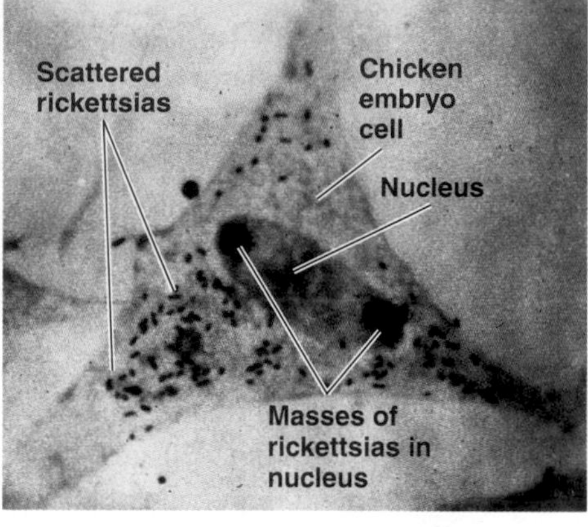
Scattered rickettsias
Chicken embryo cell
Nucleus
Masses of rickettsias in nucleus

LM 5 μm

Figure 11.1 Rickettsias.

Q How are rickettsias transmitted from one host to another?

(b) Rickettsias grow only within a host cell, such as the chicken embryo cell shown here. Note the scattered rickettsias within the cell and the compact masses of rickettsias in the cell nucleus.

Caulobacter* and *Hyphomicrobium Members of the genus *Caulobacter* (kô-lō-bak′tėr) are found in low-nutrient aquatic environments, such as lakes. They feature stalks that anchor the organisms to surfaces (Figure 11.2). This arrangement increases their nutrient uptake because they are exposed to a continuously changing flow of water and because the stalk increases the surface-to-volume ratio of the cell. Also, if the surface to which they anchor is a living host, these bacteria can use the host's excretions as nutrients. When the nutrient concentration is exceptionally low, the size of the stalk increases, evidently to provide an even greater surface area for nutrient absorption.

Budding bacteria do not divide by binary fission into two nearly identical cells. The budding process resembles the asexual reproductive processes of many yeasts (Figure 12.3, page 332). The parent cell retains its identity while the bud increases in size until it separates as a complete new cell. An example is the genus *Hyphomicrobium* (hī-fō-mī-krō′bē-um), as shown in Figure 11.3. These bacteria, like the caulobacteria, are found in low-nutrient aquatic environments and have even been found growing in laboratory water baths. Both *Caulobacter* and *Hyphomicrobium* produce prominent prosthecae.

Rhizobium, Bradyrhizobium,* and *Agrobacterium The *Rhizobium* (rī-zō′bē-um) and *Bradyrhizobium* (brăd-ē-rī-zō′bē-um) are two of the more important genera of a group of agriculturally important bacteria that specifically infect the roots of leguminous plants, such as beans, peas, or clover. For simplicity these bacteria are known by the common name of **rhizobia.** The presence of rhizobia in the roots leads to formation of nodules in which the rhizobia and plant form a symbiotic relationship, resulting in the fixation of nitrogen from the air for use by the plant (see Figure 27.5 on page 773).

Like rhizobia, the genus *Agrobacterium* (ag′rō-bak-ti′rē-um) has the ability to invade plants. However, they do not induce root nodules or fix nitrogen. Of particular interest is *Agrobacterium tumefaciens.* This is a plant pathogen that causes a disease called crown gall; the crown is the area of the plant where the roots and stem merge. The tumorlike gall is induced when *A. tumefaciens* inserts a plasmid containing bacterial genetic information into the plant's chromosomal DNA (see Figure 9.19, page 265). For this reason, microbial geneticists are very interested in this

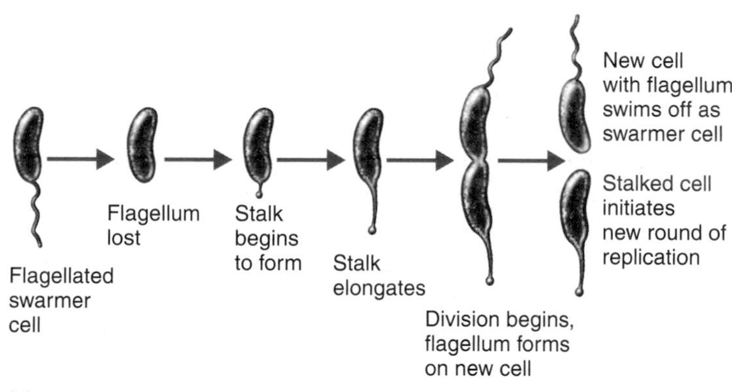
Flagellated swarmer cell
Flagellum lost
Stalk begins to form
Stalk elongates
Division begins, flagellum forms on new cell
New cell with flagellum swims off as swarmer cell
Stalked cell initiates new round of replication

(a)

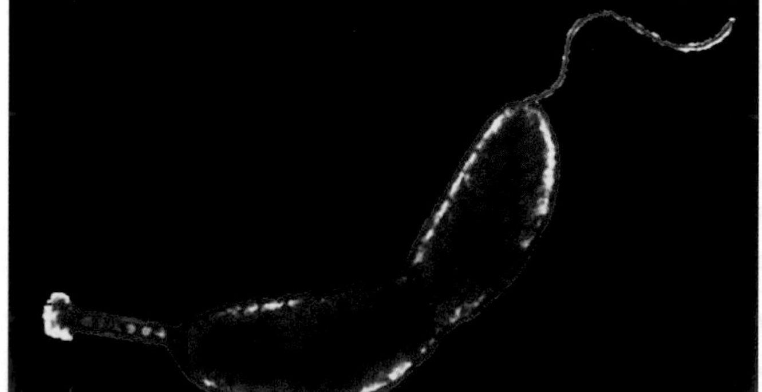

(b) TEM 0.4 μm

Figure 11.2 *Caulobacter*.

Q What is the competitive advantage provided by attaching to a surface?

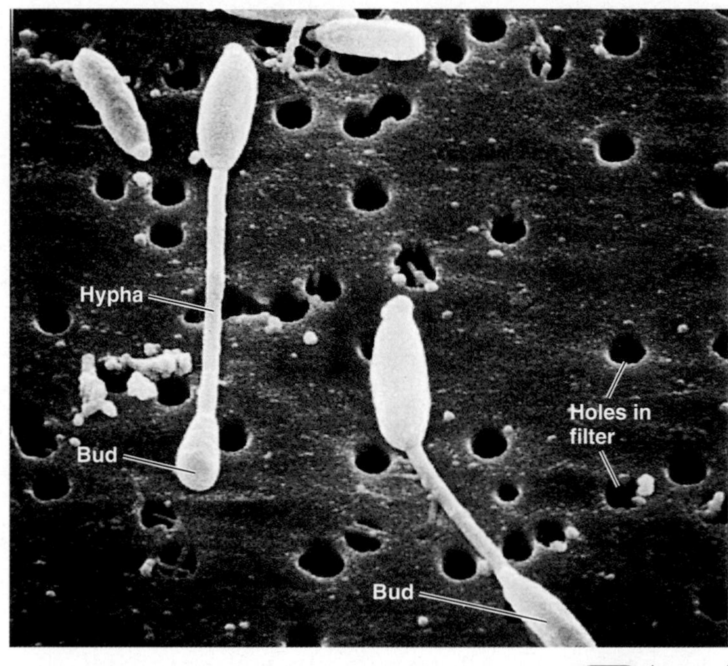

Figure 11.3 ***Hyphomicrobium*, a type of budding bacterium.**

Q **Most bacteria do not reproduce by budding; what method do they use?**

organism. Plasmids are the most common vector that scientists use to carry new genes into a cell, and the thick wall of plants is especially difficult to penetrate (see Figure 9.20, page 265).

Bartonella The genus *Bartonella* (bär′tō-nel-la) contains several members that are human pathogens. The best known is *Bartonella henselae,* a gram-negative bacillus that causes cat-scratch disease.

Brucella *Brucella* (brü′sel-la) bacteria are small nonmotile coccobacilli. All species of *Brucella* are obligate parasites of mammals and cause the disease brucellosis. Of medical interest is the ability of *Brucella* to survive phagocytosis, an important element of the body's defense against bacteria (see Chapter 16, page 457).

Nitrobacter* and *Nitrosomonas *Nitrobacter* (nī-trō-bak′tėr) and *Nitrosomonas* (nī-trō-sō-mō′nas) are genera of nitrifying bacteria that are of great importance to the environment and to agriculture. They are chemoautotrophs capable of using inorganic chemicals as energy sources and carbon dioxide as the only source of carbon, from which they synthesize all of their complex chemical makeup. The energy sources of the genera *Nitrobacter* and *Nitrosomonas* (the latter is a member of the betaproteobacteria) are reduced nitrogenous compounds. *Nitrobacter* species oxidize ammonium (NH_4^+) to nitrite (NO_2^-), which is in turn oxidized by *Nitrosomonas* species to nitrates (NO_3^-) in the process of *nitrification.* Nitrate is important to agriculture; it is a nitrogen form that is highly mobile in soil and therefore likely to be encountered and used by plants.

Wolbachia *Wolbachia* (wol-ba′kē-ä) are probably the most common infectious bacterial genus in the world. Even so, little is known about *Wolbachia;* they live only inside the cells of their hosts, usually insects (a relationship known as *endosymbiosis*). Therefore, *Wolbachia* escape detection by the usual culture methods. This fascinating group of bacteria is described further in the box on page 307.

CHECK YOUR UNDERSTANDING

✓ Make a dichotomous key to distinguish the alphaproteobacteria described in this chapter. (*Hint:* See page 303 for a completed example.) **11-1**

The Betaproteobacteria

There is considerable overlap between the betaproteobacteria and the alphaproteobacteria, for example, among the nitrifying bacteria discussed earlier. The betaproteobacteria often use nutrient substances that diffuse away from areas of anaerobic decomposition of organic matter, such as hydrogen gas, ammonia, and methane. Several important pathogenic bacteria are found in this group.

Thiobacillus *Thiobacillus* (thī-ō-bä-sil′lus) species and other sulfur-oxidizing bacteria are important in the sulfur cycle (see Figure 27.7, page 774). These chemoautotrophic bacteria are capable of obtaining energy by oxidizing the reduced forms of sulfur, such as hydrogen sulfide (H_2S), or elemental sulfur (S^0), into sulfates (SO_4^{2-}).

Spirillum The habitat of the genus *Spirillum* (spī-ril′lum) is mainly fresh water. An important morphological difference from the helical spirochetes (discussed on page 322) is that *Spirillum* bacteria are motile by conventional polar flagella, rather than axial filaments. The spirilla are relatively large, gram-negative, aerobic bacteria. *Spirillum volutans* (vō-lū-tans) is often used as a demonstration slide when microbiology students are first introduced to the operation of the microscope (**Figure 11.4**).

Sphaerotilus Sheathed bacteria, which include *Sphaerotilus natans* (sfe-rä′ti-lus na′tans), are found in fresh water and in sewage. These gram-negative bacteria with polar flagella form a hollow, filamentous sheath in which to live (**Figure 11.5**). Sheaths are protective and also aid in nutrient accumulation. *Sphaerotilus* probably contributes to bulking, an important problem in sewage treatment (see Chapter 27).

Burkholderia The genus *Burkholderia* was formerly grouped with the genus *Pseudomonas,* which is now classified under the gammaproteobacteria. Like the pseudomonads, almost all *Burkholderia* species are motile by a single polar flagellum or tuft of flagella. The best known species is the aerobic, gram-negative rod *Burkholderia cepacia* (berk′hōld-ėr-ē-ä se-pā′se-ä). It has an extraordinary nutritional spectrum and is capable of degrading

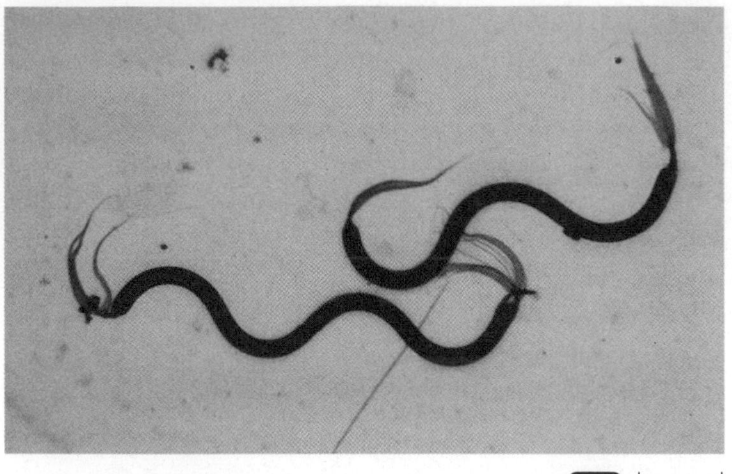

Figure 11.4 ***Spirillum volutans.*** These large helical bacteria are found in aquatic environments. Note the polar flagella.

Q Is this bacterium motile? How can you tell?

more than 100 different organic molecules. This capability is often a factor in the contamination of equipment and drugs in hospitals; these bacteria may actually grow in disinfectant solutions (see the box in Chapter 15, page 440). This bacterium is also a problem for persons with the genetic lung disease cystic fibrosis, in whom it metabolizes accumulated respiratory secretions. *Burkholderia pseudomallei* (sūdo-mal′lē-ī) is a resident in moist soils and is the cause of a severe disease (melioidosis) endemic in southeast Asia and northern Australia.

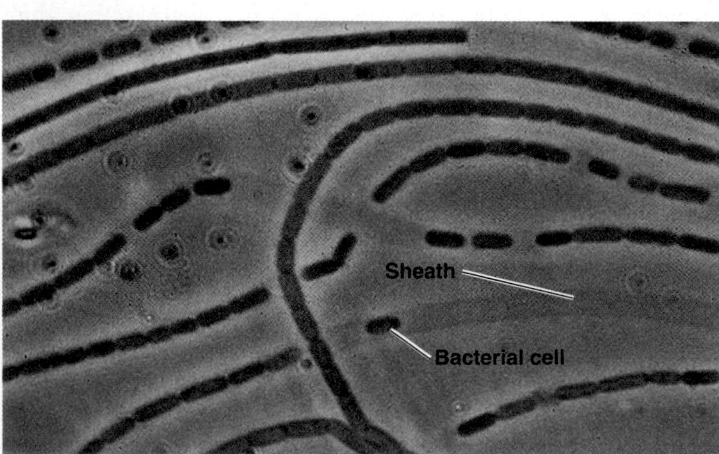

Figure 11.5 ***Sphaerotilus natans.*** These sheathed bacteria are found in dilute sewage and aquatic environments. They form elongated sheaths in which the bacteria live. The bacteria have flagella (not visible here) and can eventually swim free of the sheath.

Q How does the sheath help the cell?

Bordetella Of special importance is the nonmotile, aerobic, gram-negative rod *Bordetella pertussis* (bôr′de-tel-lä pėr-tus′sis). This serious pathogen is the cause of pertussis, or whooping cough.

Neisseria Bacteria of the genus *Neisseria* (nī-se′rē-ä) are aerobic, gram-negative cocci that usually inhabit the mucous membranes of mammals. Pathogenic species include the gonococcus bacterium *Neisseria gonorrhoeae* (go-nôr-rē′ī), the causative agent of gonorrhoea (**Figure 11.6** and the box in Chapter 26, page 751), and *N. meningitidis* (men-nin-ji′ti-dis), the agent of meningococcal meningitis.

Zoogloea The genus *Zoogloea* (zō′ō-glē-ä) is important in the context of aerobic sewage-treatment processes, such as the activated sludge system (see Figure 27.22, page 787). As they grow, *Zoogloea* bacteria form fluffy, slimy masses that are essential to the proper operation of such systems.

CHECK YOUR UNDERSTANDING

✓ Make a dichotomous key to distinguish the betaproteobacteria described in this chapter. **11-2**

The Gammaproteobacteria

The gammaproteobacteria constitute the largest subgroup of the proteobacteria and include a great variety of physiological types. One species that is used in industrial microbiology is described in the box in Chapter 28 on page 801.

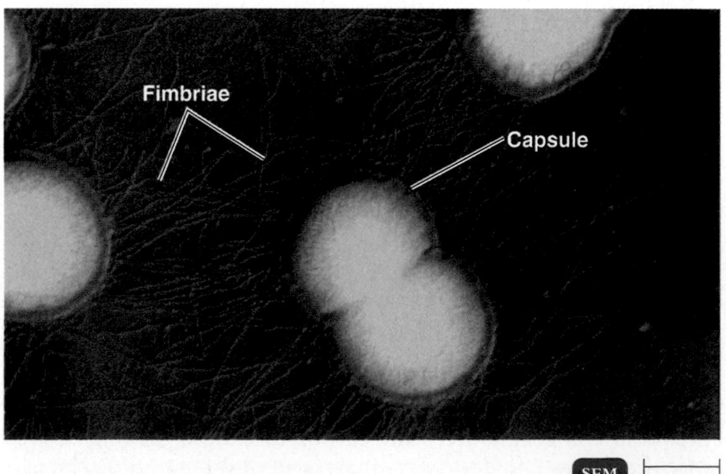

Figure 11.6 **The gram-negative coccus *Neisseria gonorrhoeae.*** Notice the paired arrangement (diplococci). The fimbriae enable the organism to attach to mucous membranes and thus contribute to its pathogenicity. *N. gonorrhoeae* causes gonorrhea.

Q For what are fimbriae used?

Bacteria and Insect Sex

***Wolbachia* is quite possibly the most common** infectious bacterial genus on Earth. Although these bacteria were first discovered in 1924, little had been known about them until the 1990s. They escape detection by the usual culture methods because they live as endosymbionts in the cells of insects and other invertebrates (**Figure A**).

Wolbachia infect over a million species of insects and other invertebrates. In all, as many as 75% of species of animals surveyed carry this bacterium. *Wolbachia* is essential to nematodes. If the bacterium is killed with antibiotics, the host worm dies. Pea aphids infected with *Wolbachia* are not killed by a normally lethal parasitic wasp larva; the bacterium is harmless in the aphid but kills the wasp.

In some insects, *Wolbachia* destroys males of its host species. *Wolbachia* can turn males into females by interfering with the male hormone. As shown in **Figure B**, if a male and female insect are uninfected with *Wolbachia*, they produce offspring normally. If only the male is infected, the insects fail to reproduce. If one or both insects of a mating pair are infected, only the infected females reproduce—and transmit *Wolbachia* in the cytoplasm of their eggs. Offspring produced without fertilization are female. The result is that the bacteria are transmitted to the next generation. This type of reproduction, called parthenogenesis, has been seen in a variety of insects and in some amphibians and reptiles. Thus a question arises: Is *Wolbachia* always responsible?

Eukaryotic species are defined as organisms that reproduce only with members of their own species. This reproductive isolation prevents the production of hybrids and thus maintains the uniqueness of each species. In the laboratory, researchers have found that after antibiotic treatment, wasps of one species will produce hybrid offspring with another species. This raises the question about the influence *Wolbachia* has had on the evolution of insects. Did insects that were not infected reproduce successfully outside their species?

A virulent strain of *Wolbachia* called "popcorn" causes host cells to lyse, or "pop," which eventually kills the host insect. On the one hand, the popcorn strain might be used to kill mosquitoes. On the other hand, eliminating *Wolbachia* from pest insects could result in a lower number of females, thus reducing population growth.

The unique biology of *Wolbachia* has attracted researchers interested in questions that range from the evolutionary implications of infection to commercial uses of *Wolbachia*.

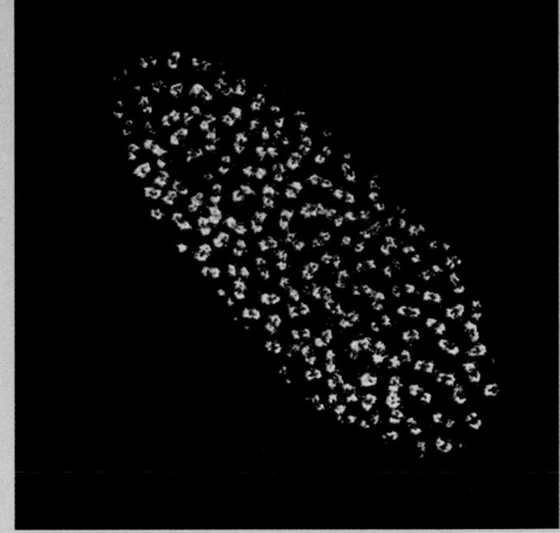

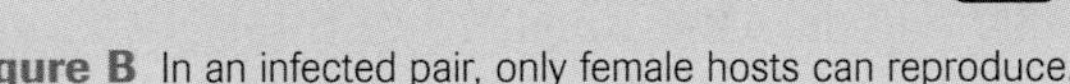

Figure A *Wolbachia* are red inside the cells of this fruit fly embryo.

Figure B In an infected pair, only female hosts can reproduce.

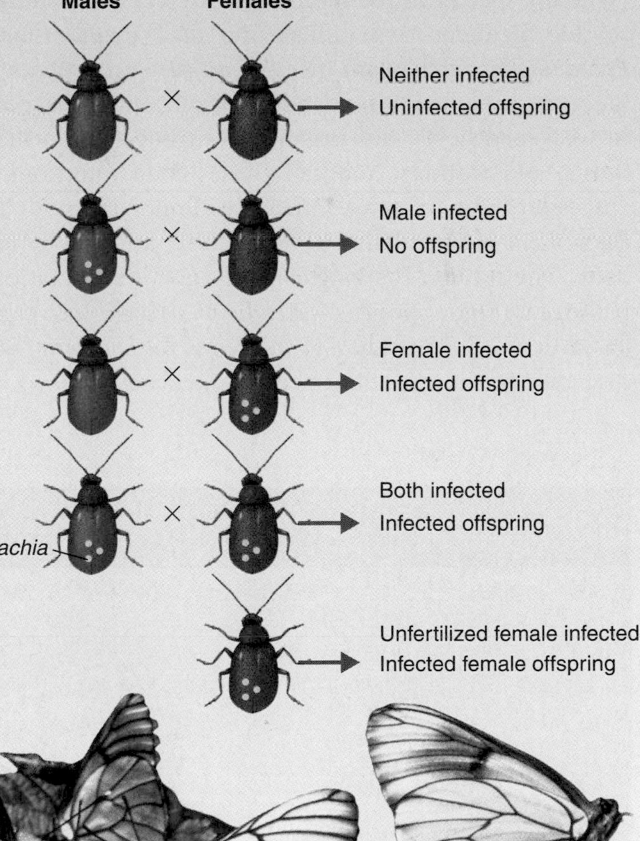

Beggiatoa *Beggiatoa alba* (bej′jē-ä-tō-ä al′ba), the only species of this unusual genus, grows in aquatic sediments at the interface between the aerobic and anaerobic layers. Morphologically, it resembles certain filamentous cyanobacteria (page 313), but it is not photosynthetic. Motility is by gliding. The mechanism is the production of slime, which attaches to the surface on which movement occurs and also provides lubrication allowing the organism to glide.

Nutritionally *B. alba* uses hydrogen sulfide (H_2S) as an energy source and accumulates internal granules of sulfur. The ability of this organism to obtain energy from an inorganic compound was an important factor in the discovery of autotrophic metabolism.

Francisella *Francisella* (fran′sis-el′lä) is a genus of small, pleomorphic bacteria that grow only on complex media enriched with blood or tissue extracts. *Francisella tularensis* (tü′lär-en-sis) causes the disease tularemia. (See the box in Chapter 23, page 644.)

Pseudomonadales

Members of the order Pseudomonadales are gram-negative aerobic rods or cocci. The most important genus in this group is *Pseudomonas.*

Pseudomonas A very important genus, *Pseudomonas* (sū-dō-mō′nas) consists of aerobic, gram-negative rods that are motile by polar flagella, either single or tufts (**Figure 11.7**). Pseudomonads are very common in soil and other natural environments.

Many species of pseudomonads excrete extracellular, water-soluble pigments that diffuse into their media. One species, *Pseudomonas aeruginosa* (ā-rü-ji-nō′sä), produces a soluble, blue-green pigmentation. Under certain conditions, particularly in weakened hosts, this organism can infect the urinary tract, burns, and wounds, and can cause blood infections (sepsis), abscesses, and meningitis. Other pseudomonads produce soluble fluorescent pigments that glow when illuminated by ultraviolet light. One species, *P. syringae* (sėr′in-gī), is an occasional plant pathogen. (Some species of *Pseudomonas* have been transferred, based upon rRNA studies, to the genus *Burkholderia,* which was discussed previously with the betaproteobacteria.)

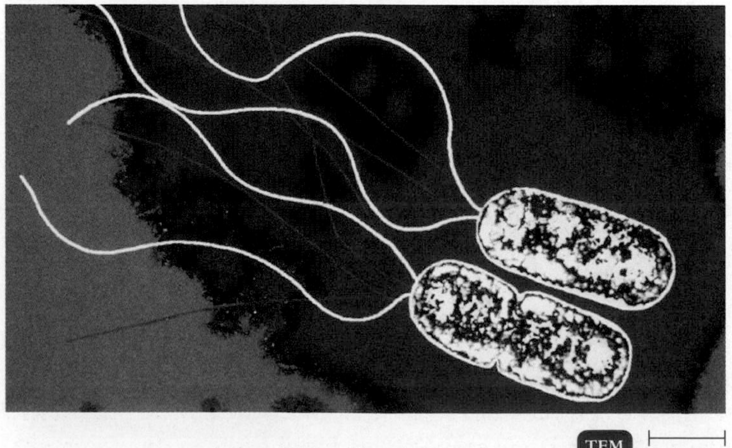

Figure 11.7 *Pseudomonas.* This photo of a pair of *Pseudomonas* bacteria shows polar flagella that are a characteristic of the genus. In some species only a single flagellum is present (see Figure 4.7b, page 81). Note that one cell (on the bottom) is beginning to divide.

Q How does the nutritional diversity of these bacteria make them a problem in hospitals?

Pseudomonads have almost as much genetic capacity as the eukaryotic yeasts and almost half as much as a fruit fly. Although these bacteria are less efficient than some other heterotrophic bacteria in utilizing many of the more common nutrients, they make use of their genetic capacity by compensating for this in other ways. For example, pseudomonads synthesize an unusually large number of enzymes and can metabolize a wide variety of substrates. Therefore, they probably contribute significantly to the decomposition of uncommon chemicals, such as pesticides, that are added to soil.

In hospitals and other places where pharmaceutical agents are prepared, the ability of pseudomonads to grow on minute traces of unusual carbon sources, such as soap residues or cap-liner adhesives found in a solution, has been unexpectedly troublesome. (See the box on page 164.) Pseudomonads are even capable of growth in some antiseptics, such as quaternary ammonium compounds. Their resistance to most antibiotics has also been a source of medical concern. This resistance is probably related to the characteristics of the cell wall porins, which control the entrance of molecules through the cell wall (see Chapter 4, page 87). The large genome of pseudomonads also codes for several very efficient efflux pump systems (page 575) that eject antibiotics from the cell before they can function. Pseudomonads are responsible for about one in ten nosocomial infections (hospital-acquired infections; see page 413), especially among infections in burn units. Persons with cystic fibrosis are also especially prone to infections by *Pseudomonas* and the closely related *Burkholderia.*

Although pseudomonads are classified as aerobic, some are capable of substituting nitrate for oxygen as a terminal electron acceptor. This process, anaerobic respiration, yields almost as much energy as aerobic respiration (see page 132). In this way, pseudomonads cause important losses of valuable nitrogen in fertilizer and soil. Nitrate (NO_3^-) is the form of fertilizer nitrogen most easily used by plants. Under anaerobic conditions, as in water-logged soil, pseudomonads eventually convert this valuable nitrate into nitrogen gas (N_2), which is lost to the atmosphere (see Figure 27.4, page 770).

Many pseudomonads can grow at refrigerator temperatures. This characteristic, combined with their ability to utilize proteins and lipids, makes them an important contributor to food spoilage.

Azotobacter* and *Azomonas Some nitrogen-fixing bacteria, such as *Azotobacter* (ā-zō-tō-bak′tėr) and *Azomonas* (ā-zō-mō′nas), are free-living in soil. These large, ovoid, heavily capsulated bacteria are frequently used in laboratory demonstrations of nitrogen fixation. However, to fix agriculturally significant amounts of nitrogen, they would require energy sources, such as carbohydrates, that are in limited supply in soil.

Moraxella Members of the genus *Moraxella* (mô-raks-el′lä) are strictly aerobic coccobacilli—that is, intermediate in shape between cocci and rods. *Moraxella lacunata* (la-kü-nä′tä) is implicated in conjunctivitis, an inflammation of the conjunctiva, the membrane that covers the eye and lines the eyelids.

Acinetobacter The genus *Acinetobacter* (a-si-nē′tō-bak-tėr) is aerobic and in stained preparations typically forms pairs. The bacteria occur naturally in soil and water. A member of this genus, *Acinetobacter baumanii* (bou′man-ē-ē), is an increasing concern to the medical community because of the rapidity with which it becomes resistant to antibiotics. Some strains are resistant to most available antibiotics. Not yet widespread in the United States, *A. baumanii* is an opportunistic pathogen primarily found in a hospital setting. The antibiotic resistance of the pathogen, combined with the weakened health of infected hospital patients, has resulted in an unusually high mortality rate. *A. baumanii* is primarily a respiratory pathogen, but it also infects skin and soft tissues and wounds and occasionally invades the bloodstream. It is more environmentally hardy than most gram-negative bacteria, and, once established in a hospital, it becomes difficult to eliminate.

Legionellales

The genera *Legionella* and *Coxiella* are closely associated in the second edition of *Bergey's Manual,* where both are placed in the same order, Legionellales. Because the *Coxiella* share an intracellular lifestyle with the rickettsial bacteria, they were previously considered rickettsial in nature and grouped with them. *Legionella* bacteria grow readily on suitable artificial media.

Legionella *Legionella* (lē-jä-nel′lä) bacteria were originally isolated during a search for the cause of an outbreak of pneumonia now known as legionellosis. The search was difficult because these bacteria did not grow on the usual laboratory isolation media then available. After intensive effort, special media were developed that enabled researchers to isolate and culture the first *Legionella.* Microbes of this genus are now known to be relatively common in streams, and they colonize such habitats as warm-water supply lines in hospitals and water in the cooling towers of air conditioning systems. (See the box in Chapter 24, page 691.) An ability to survive and reproduce within aquatic amoebae often makes them difficult to eradicate in water systems.

Coxiella *Coxiella burnetii* (käks-ē-el′lä bėr-ne′tē-ē), which causes Q fever, was formerly grouped with the rickettsia. Like them, *Coxiella* bacteria require a mammalian host cell to reproduce. Unlike rickettsias, *Coxiella* are not transmitted among humans by insect or tick bites. Although cattle ticks harbor the organism, it is most commonly transmitted by aerosols or contaminated milk. A sporelike body is present in *C. burnetii* (see Figure 24.14b, page 690). This might explain the bacterium's relatively high resistance to the stresses of airborne transmission and heat treatment.

Vibrionales

Members of the order Vibrionales are facultatively anaerobic gram-negative rods. Many are slightly curved. They are found mostly in aquatic habitats.

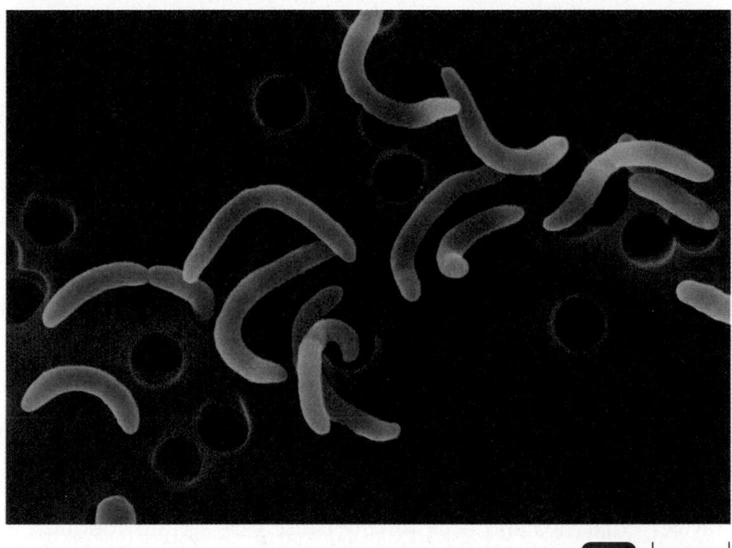

Figure 11.8 ***Vibrio cholerae.*** Notice the slight curvature of these rods, which is a characteristic of the genus. The circles that appear in the photo are holes in the membrane filter on which the bacteria are resting.

Q **What disease does *Vibrio cholerae* cause?**

Vibrio Members of the genus *Vibrio* (vib′rē-ō) are rods that are often slightly curved (**Figure 11.8**). One important pathogen is *Vibrio cholerae* (kol′er-ī), the causative agent of cholera. The disease is characterized by a profuse and watery diarrhea. *V. parahaemolyticus* (pa-rä-hē-mō-li′ti-kus) causes a less serious form of gastroenteritis. Usually inhabiting coastal salt waters, it is transmitted to humans mostly by raw or undercooked shellfish.

Enterobacteriales

The members of the order Enterobacteriales are facultatively anaerobic, gram-negative rods that are, if motile, peritrichously flagellated. Morphologically, the rods are straight. This is an important bacterial group, often commonly called **enterics.** This reflects the fact that they inhabit the intestinal tracts of humans and other animals. Most enterics are active fermenters of glucose and other carbohydrates.

Because of the clinical importance of enterics, there are many techniques to isolate and identify them. An identification method for some enterics is shown in Figure 10.9 (page 286), which incorporates a modern tool using 15 biochemical tests. Biochemical tests are especially important in clinical laboratory work and in food and water microbiology.

Enterics have fimbriae that help them adhere to surfaces or mucous membranes. Specialized sex pili aid in the exchange of genetic information between cells, which often includes antibiotic resistance (see Figures 8.26 and 8.27, pages 237 and 238).

Enterics, like many bacteria, produce proteins called bacteriocins that cause the lysis of closely related species of bacteria. Bacteriocins may help maintain the ecological balance of various enterics in the intestines.

Escherichia The bacterial species *Escherichia coli* is one of the most common inhabitants of the human intestinal tract and is probably the most familiar organism in microbiology. Recall from previous chapters that a great deal is known about the biochemistry and genetics of *E. coli,* and it continues to be an important tool for basic biological research—many researchers consider it almost a laboratory pet. Its presence in water or food is an indication of fecal contamination (see Chapter 27, page 781). *E. coli* is not usually pathogenic. However, it can be a cause of urinary tract infections, and certain strains produce enterotoxins that cause traveler's diarrhea and occasionally cause very serious foodborne disease (see *E. coli* O157:H7 in Chapter 25, page 714).

Salmonella Almost all members of the genus *Salmonella* (sal′mōn-el-lä) are potentially pathogenic. Accordingly, there are extensive biochemical and serological tests to clinically isolate and identify salmonellae. Salmonellae are common inhabitants of the intestinal tracts of many animals, especially poultry and cattle. Under unsanitary conditions, they can contaminate food.

The nomenclature of the genus *Salmonella* is unusual. Instead of multiple species, members of the genus *Salmonella* that are infectious to warm-blooded animals can be considered for practical purposes to be a single species, *Salmonella enterica* (en-ter′i-kä). This species is divided into more than 2400 **serovars,** that is, *serological varieties.* The term **serotype** is often used to mean the same thing. By way of explanation of these terms, when salmonellae are injected into appropriate animals, their flagella, capsules, and cell walls serve as *antigens* that cause the animals to form *antibodies* in their blood that are specific for each of these structures. Thus, *serological* means are used to differentiate the microorganisms. Serology is discussed more fully in Chapter 18, but for now it will be sufficient to state that it can be used to differentiate and identify bacteria.

A serovar such as *Salmonella typhimurium* (tī-fi-mur′ē-um) is not a species and should be more properly written as "*Salmonella enterica* serovar Typhimurium." The convention now used by the Centers for Disease Control and Prevention (CDC) is to spell out the entire name at the first mention and then abbreviate it as, for example, *Salmonella* Typhimurium. For simplicity, we will identify serovars of salmonellae in this text as we would species, that is, *S. typhimurium,* etc.

Specific antibodies, which are available commercially, can be used to differentiate *Salmonella* serovars by a system known as the Kauffmann-White scheme. This scheme designates an organism by numbers and letters that correspond to specific antigens on the organism's capsule, cell wall, and flagella, which are identified by the letters K, O, and H, respectively. For example, the antigenic formula for the bacterium *S. typhimurium* is O$\underline{1}$,4,[5],12:H,i,1,2.* Many salmonellae are named only by their antigenic formulas. Serovars can be further differentiated by special biochemical or physiological properties into **biovars,** or **biotypes.**

A recent taxonomic arrangement based upon the latest molecular technology adds another species, *Salmonella bongori* (bon′gôr-ē). This is a resident of "cold-blooded" animals—it was originally isolated from a lizard in the town of Bongor in the African desert nation of Chad—and is rarely found in humans.

Typhoid fever, caused by *Salmonella typhi* (tī′fē), is the most severe illness caused by any member of the genus *Salmonella.* A less severe gastrointestinal disease caused by other salmonellae is called salmonellosis. Salmonellosis is one of the most common forms of foodborne illness. (See the box in Chapter 25 on page 715.)

Shigella Species of *Shigella* (shi-gel′lä) are responsible for a disease called bacillary dysentery, or shigellosis. Unlike salmonellae, they are found only in humans. Some strains of *Shigella* can cause life-threatening dysentery (see Chapter 25, page 712).

Klebsiella Members of the genus *Klebsiella* (kleb-sē-el′lä) are commonly found in soil or water. Many isolates are capable of fixing nitrogen from the atmosphere, which has been proposed as being a nutritional advantage in isolated populations with little protein nitrogen in their diet. The species *Klebsiella pneumoniae* (nü-mō′nē-ī) occasionally causes a serious form of pneumonia in humans.

Serratia *Serratia marcescens* (ser-rä′tē-ä mär-ses′sens) is a bacterial species distinguished by its production of red pigment. In hospital situations, the organism can be found on catheters, in saline irrigation solutions, and in other supposedly sterile solutions. Such contamination is probably the cause of many urinary and respiratory tract infections in hospitals.

Proteus Colonies of *Proteus* (prō′tē-us) bacteria growing on agar exhibit a swarming type of growth. Swarmer cells with many flagella (**Figure 11.9a**) move outward on the edges of the colony and then revert to normal cells with only a few flagella and reduced motility. Periodically, new generations of highly motile swarmer cells develop, and the process is repeated. As a result, a *Proteus* colony has the distinctive appearance of a series of concentric rings (**Figure 11.9b**). This genus of bacteria is implicated in many infections of the urinary tract and in wounds.

*The letters derive from the original German usage: K represents the German for capsule. (Salmonellae with capsules are identified serologically by a particular capsular antigen named Vi, for virulence.) Colonies that spread in a thin film over the agar surface were described by the German word for film, *hauch.* The motility needed to form a film implied the presence of flagella, and the letter H came to be assigned to the antigens of flagella. Nonmotile bacteria were described as *ohne hauch,* without film, and the O came to be assigned to the cell surface or body antigens. This terminology is also used in the naming of *E. coli* O157:H7, *Vibrio cholerae* O:1, and others.

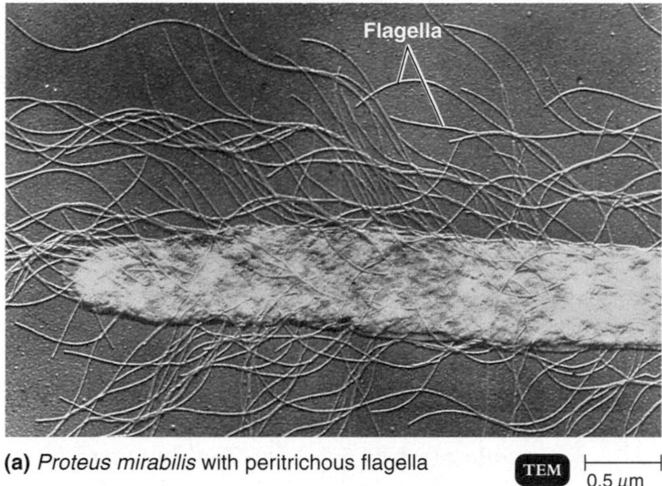

(a) *Proteus mirabilis* with peritrichous flagella

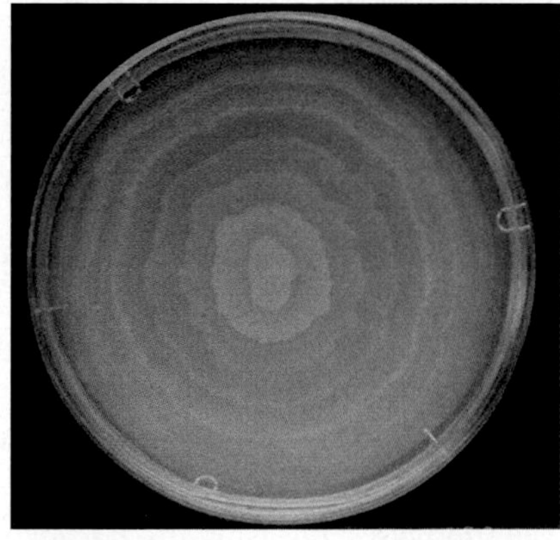

(b) A swarming colony of *Proteus mirabilis*, showing concentric rings of growth

Figure 11.9 ***Proteus mirabilis.*** Chemical communication between bacterial cells causes changes from cells adapted to swimming in fluid (few flagella) to cells that are able to move on surfaces (numerous flagella). The concentric growth (**b**) results from periodic synchronized conversion to the highly flagellated form capable of movement on surfaces.

Q **The photo of the *Proteus* cell is probably a swarmer cell. How would you know?**

Yersinia *Yersinia pestis* (yėr-sin′ē-ä pes′tis) causes plague, the Black Death of medieval Europe. Urban rats in some parts of the world and ground squirrels in the American Southwest carry these bacteria. Fleas usually transmit the organisms among animals and to humans, although contact with respiratory droplets from infected animals and people can be involved in transmission.

Erwinia *Erwinia* (ėr-wi′nē-ä) species are primarily plant pathogens; some cause plant soft-rot diseases. These species produce enzymes that hydrolyze the pectin between individual plant cells. This causes the plant cells to separate from each other, a disease that plant pathologists term *plant rot.*

Enterobacter Two *Enterobacter* (en-te-rō-bak′tėr) species, *E. cloacae* (klō-ā′kī), and *E. aerogenes* (ā-rä′jen-ēz), can cause urinary tract infections and hospital-acquired infections. They are widely distributed in humans and animals, as well as in water, sewage, and soil.

Pasteurellales

The bacteria in the Order Pasteurellales are nonmotile; they are best known as human and animal pathogens.

Pasteurella The genus *Pasteurella* (pas-tyėr-el′lä) is primarily known as a pathogen of domestic animals. It causes sepsis in cattle, fowl cholera in chickens and other fowl, and pneumonia in several types of animals. The best-known species is *Pasteurella multocida* (mul-tō′si-dä), which can be transmitted to humans by dog and cat bites. It is also a prominent member of the microbiota of saliva of the relatively slow-moving Komodo dragon, a large reptile found on an Indonesian island, that bites more mobile prey and waits several days for their death. The Komodo dragon is not venomous, but its prey dies from an especially virulent strain of *P. multocida* introduced by its bite.

Haemophilus *Haemophilus* (hē-mä′fil-us) is a very important genus of pathogenic bacteria. These organisms inhabit the mucous membranes of the upper respiratory tract, mouth, vagina, and intestinal tract. The best-known species that affects humans is *Haemophilus influenzae* (in-flü-en′zī), named long ago because of the erroneous belief that it was responsible for influenza.

The name *Haemophilus* is derived from the bacteria's requirement for blood in their culture medium (*hemo* = blood). They are unable to synthesize important parts of the cytochrome system needed for respiration, and they obtain these substances from the heme fraction, known as the **X factor,** of blood hemoglobin. The culture medium must also supply the cofactor nicotinamide adenine dinucleotide (from either NAD^+ or $NADP^+$), which is known as **V factor.** Clinical laboratories use tests for the requirement of X and V factors to identify isolates as *Haemophilus* species.

Haemophilus influenzae is responsible for several important diseases. It has been a common cause of meningitis in young children and is a frequent cause of earaches. Other clinical conditions caused by *H. influenzae* include epiglotitis (a life-threatening condition in which the epiglottis becomes infected and inflamed), septic arthritis in children, bronchitis, and pneumonia. *Haemophilus ducreyi* (dü-krā′ē) is the cause of the sexually transmitted disease chancroid.

CHECK YOUR UNDERSTANDING

✓ Make a dichotomous key to distinguish the orders of gammaproteobacteria described in this chapter. **11-3**

The Deltaproteobacteria

The deltaproteobacteria are distinctive in that they include some bacteria that are predators on other bacteria. Bacteria in this group are also important contributors to the sulfur cycle.

Bdellovibrio *Bdellovibrio* (del-lō-vib'rē-ō) is a particularly interesting genus. It attacks other gram-negative bacteria. It attaches tightly (*bdella* = leech; **Figure 11.10**), and after penetrating the outer layer of gram-negative bacteria, it reproduces within the periplasm. There, the cell elongates into a tight spiral, which then fragments almost simultaneously into several individual flagellated cells. The host cell then lyses, releasing the *Bdellovibrio* cells.

Desulfovibrionales

Members of the order Desulfovibrionales are sulfur reducing bacteria. They are obligately anaerobic bacteria that use oxidized forms of sulfur, such as sulfates (SO_4^{2-}) or elemental sulfur (S^0) rather than oxygen as electron acceptors. The product of this reduction is hydrogen sulfide (H_2S). (Because the H_2S is not assimilated as a nutrient, this type of metabolism is termed *dissimilatory.*) The activity of these bacteria releases millions of tons of H_2S into the atmosphere every year and plays a key part in the sulfur cycle (see Figure 27.7 on page 774). Sulfur-oxidizing bacteria such as *Beggiatoa* are able to use H_2S either as part of photosynthesis or as an autotrophic energy source.

Desulfovibrio The best studied sulfur-reducing genus is *Desulfovibrio* (dē'sul-fō-vib'rē-ō), which is found in anaerobic sediments and in the intestinal tracts of humans and animals. Sulfur-reducing and sulfate-reducing bacteria use organic compounds such as lactate, ethanol, or fatty acids as electron donors. This reduces sulfur or sulfate to H_2S. When H_2S reacts with iron it forms insoluble FeS, which is responsible for the black color of many sediments.

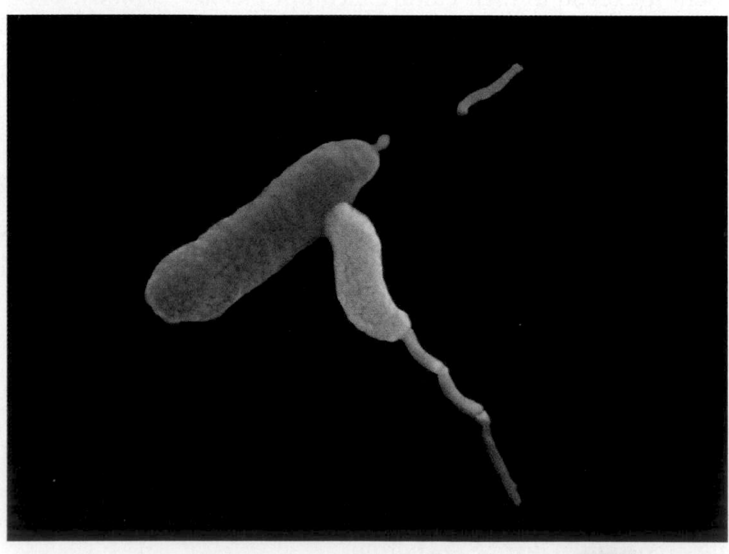

Figure 11.10 ***Bdellovibrio bacteriovorus.*** The yellow bacterium is *B. bacteriovorus.* It is attacking a bacterial cell shown in blue.

Q Would this bacterium attack *Staphylococcus aureus*?

Myxococcales

In the first edition of *Bergey's Manual,* the Myxococcales were classified among the fruiting and gliding bacteria. They illustrate the most complex life cycle of all bacteria, part of which is predatory upon other bacteria.

Myxococcus Vegetative cells of the myxobacteria (*myxo* = nasal mucus) move by gliding and leave behind a slime trail. *Myxococcus xanthus* (micks-ō-kok'kus zan'thus) and *M. fulvus* (ful'vus) are well-studied representatives of the genus. As they move, their source of nutrition is the bacteria they encounter, enzymatically lyse, and digest. Large numbers of these gram-negative microbes eventually aggregate (**Figure 11.11a**). Where the moving cells aggregate, they differentiate and form a macroscopic stalked fruiting body that contains large numbers of resting cells called *myxospores* (**Figure 11.11b**). Differentiation is usually triggered by low nutrients. Under proper conditions, usually a change in nutrients, the myxospores germinate and form new vegetative gliding cells. You might note the resemblance to the life cycle of the eukaryotic cellular slime molds in Figure 12.22 (page 353).

CHECK YOUR UNDERSTANDING

✓ Make a dichotomous key to distinguish the deltaproteobacteria described in this chapter. **11-4**

The Epsilonproteobacteria

The epsilonproteobacteria are slender gram-negative rods that are helical or curved. We will discuss the two important genera, both of which are motile by means of flagella and are microaerophilic.

Campylobacter Members of the genus *Campylobacter* (kam'pi-lō-bak-tėr) are microaerophilic vibrios; each cell has one polar flagellum. One species of *Campylobacter, C. fetus* (fē'tus), causes spontaneous abortion in domestic animals. Another species, *C. jejuni* (je-ju'ni), is a leading cause of outbreaks of foodborne intestinal disease.

Helicobacter Members of the genus *Helicobacter* are microaerophilic curved rods with multiple flagella. The species *Helicobacter pylori* (hē'lik-ō-bak-tėr pī-lōr'ē) has been identified as the most common cause of peptic ulcers in humans and a cause of stomach cancer (**Figure 11.12**; see also Figure 25.13 on page 719).

CHECK YOUR UNDERSTANDING

✓ Make a dichotomous key to distinguish the epsilonproteobacteria described in this chapter. **11-5**

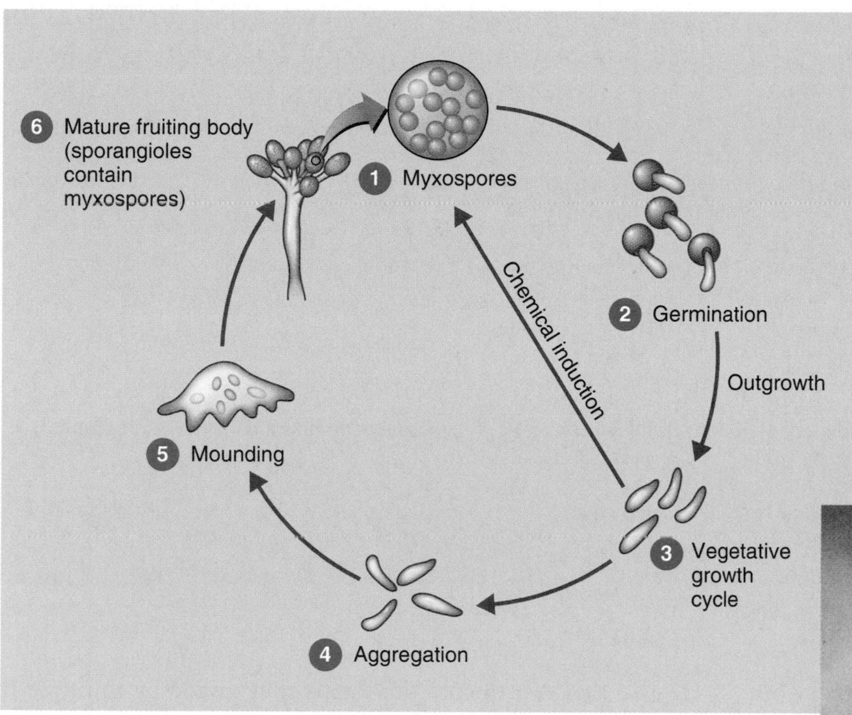

(a) Life cycle of Myxococcales

1 Myxospores are produced by a fruiting body or by chemical induction of vegetative cells.

2 Myxospores germinate and form gram-negative vegetative cells, which divide to reproduce.

3 Vegetative myxobacteria are motile by gliding, forming visible slime trails.

4 Under certain conditions, the vegetative cells swarm to central locations, forming an aggregation.

5 Aggregations of cells heap up into a mound, an early fruiting body.

6 Mounds of myxobacteria differentiate into a fruiting body, which produces myxospores packed within sporangioles.

(b) A myxobacterium fruiting body; the sporangioles contain myxospores

Figure 11.11 Myxococcales.

Q What is the feeding stage of this organism?

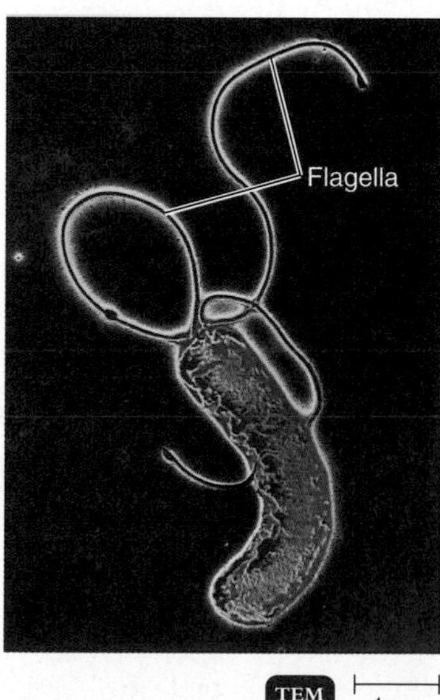

Figure 11.12 *Helicobacter pylori.* *H. pylori,* a curved rod, is an example of a helical bacterium that does not make a complete twist.

Q How do helical bacteria differ from spirochetes?

The Nonproteobacteria Gram-Negative Bacteria

LEARNING OBJECTIVES

11-6 Differentiate the groups of nonproteobacteria gram-negative bacteria described in this chapter by drawing a dichotomous key.

11-7 Compare and contrast purple and green photosynthetic bacteria with the cyanobacteria.

There are a number of important gram-negative bacteria that are not closely related to the gram-negative proteobacteria. They include several physiologically and morphologically distinctive photosynthesizing bacteria, such as those included in the phyla Cyanobacteria (cyanobacteria), Chlorobi (green sulfur bacteria), and Chloroflexi (green nonsulfur bacteria). The cyanobacteria produce oxygen during photosynthesis (are *oxygenic*), and the green sulfur and green nonsulfur bacteria do not produce oxygen (are *anoxygenic*). These groups are summarized in **Table 11.2**.

Cyanobacteria (The Oxygenic Photosynthetic Bacteria)

The cyanobacteria, named for their characteristic blue-green (*cyan*) pigmentation, were once called blue-green algae. Although they resemble the eukaryotic algae and often occupy the same environmental niches, this is a misnomer because they

Table 11.2 Selected Characteristics of Photosynthesizing Bacteria

Common Name	Example	Phylum	Comments	Electron Donor for CO_2 Reduction	Oxygenic or Anoxygenic
Cyanobacteria	*Anabaena*	Cyanobacteria	Plantlike photosynthesis; some use bacterial photosynthesis under anaerobic conditions	Usually H_2O	Usually oxygenic
Green nonsulfur bacteria	*Chloroflexus*	Chloroflexi	Grow chemoheterotrophically in aerobic environments	Organic compounds	Anoxygenic
Green sulfur bacteria	*Chlorobium*	Chlorobi	Deposit sulfur granules inside cells	Usually H_2S	Anoxygenic
Purple nonsulfur bacteria	*Rhodospirillum*	Proteobacteria	Can grow chemoheterotrophically as well	Organic compounds	Anoxygenic
Purple sulfur bacteria	*Chromatium*	Proteobacteria	Deposit sulfur granules inside cells	Usually H_2S	Anoxygenic

are bacteria; algae are not. However, cyanobacteria do carry out oxygenic photosynthesis, as do the eukaryotic plants and algae (see Chapter 12). Many of the cyanobacteria are capable of fixing nitrogen from the atmosphere. In most cases, this activity is located in specialized cells called **heterocysts,** which contain enzymes that fix nitrogen gas (N_2) into ammonium (NH_4^+) that can be used by the growing cell (**Figure 11.13a**). Species that grow in water usually have gas vacuoles that provide buoyancy, helping the cell float at a favorable environment. Cyanobacteria that move about on solid surfaces use gliding motility.

Cyanobacteria are morphologically varied. They range from unicellular forms that divide by simple binary fission (**Figure 11.13b**), to colonial forms that divide by multiple fission, to filamentous forms that reproduce by fragmentation of the filaments. The filamentous forms usually exhibit some differentiation of cells that are often bound together within an envelope or sheath.

Evidence indicates that oxygenic cyanobacteria played an important part in the development of life on Earth, which originally had very little free oxygen that would support life as we are familiar with it. Fossil evidence indicates that when cyanobacteria first appeared, the atmosphere contained only about 0.1% free oxygen. When oxygen-producing eukaryotic plants appeared

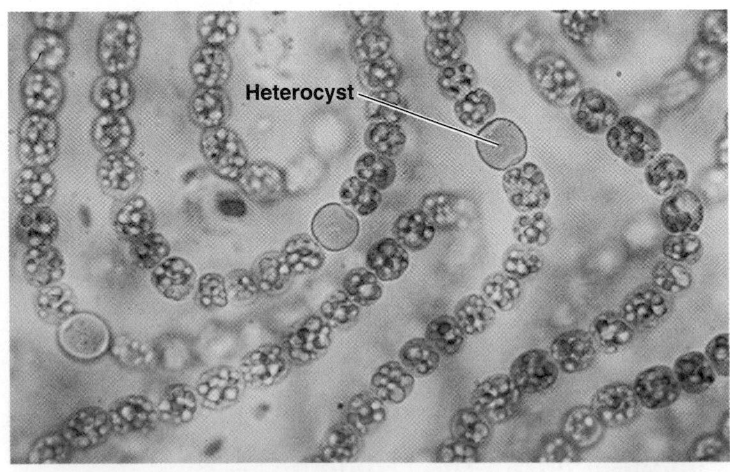

(a) Filamentous cyanobacterium showing heterocysts, in which nitrogen-fixing activity is located LM 10 μm

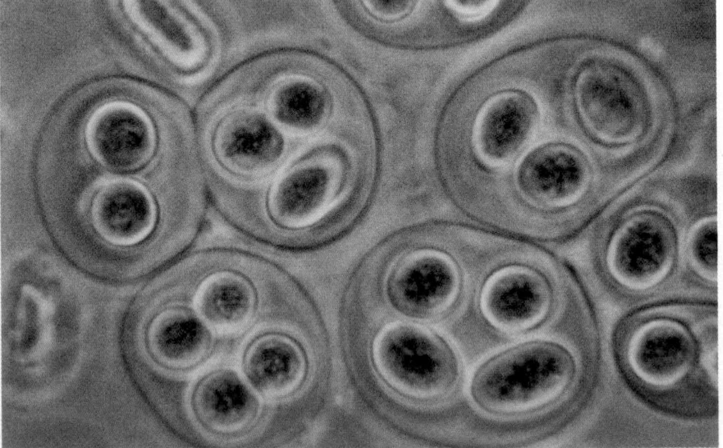

(b) A unicellular, nonfilamentous cyanobacterium, *Gloeocapsa*. Groups of these cells, which divide by binary fission, are held together by the surrounding glycocalyx.

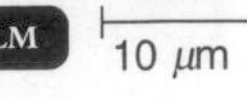

10 μm

Figure 11.13 Cyanobacteria.

How does the photosynthesis of the cyanobacteria differ from that of the purple sulfur bacteria?

millions of years later, the concentration of oxygen was more than 10%. The increase presumably was a result of photosynthetic activity by cyanobacteria. The atmosphere we breathe today contains about 20% oxygen.

Cyanobacteria, especially those that fix nitrogen, are extremely important to the environment. They occupy environmental niches similar to those occupied by the eukaryotic algae (see Figure 12.10, page 341), but the ability of many of the cyanobacteria to fix nitrogen makes them even more adaptable in nutritionally poor environments. The environmental role of cyanobacteria is presented more fully in Chapter 27, in the discussion of eutrophication (the nutritional overenrichment of bodies of water).

Purple and Green Photosynthetic Bacteria (The Anoxygenic Photosynthetic Bacteria)

The photosynthetic bacteria are taxonomically confusing. The phyla Cyanobacteria, Chlorobi, and Chloroflexi are gram-negative, but they are not genetically included in the proteobacteria. The photosynthetic purple sulfur bacteria and purple nonsulfur bacteria are genetically included in, respectively, the alphaproteobacteria and gammaproteobacteria, although for simplicity we will discuss them at this point. These photosynthetic bacteria, which are not necessarily colored purple or green, are generally anaerobic. Their habitat is usually the deep sediments of lakes and ponds. Like plants, algae, and the cyanobacteria, purple and green bacteria carry out photosynthesis to make carbohydrates (CH_2O). Growing as they do in aquatic depths, these bacteria possess bacteriochlorophyll that makes use of parts of the visible spectrum not intercepted by photosynthetic organisms located at higher levels. Also, unlike plantlike photosynthesis, the photosynthesis of purple or green bacteria is anoxygenic.

Cyanobacteria, as well as eukaryotic plants and algae produce oxygen (O_2) from water (H_2O) as they carry out photosynthesis:

$$(1)\ 2H_2O + CO_2 \xrightarrow{\text{light}} (CH_2O) + H_2O + O_2$$

The *purple sulfur* and *green sulfur bacteria* use reduced sulfur compounds, such as hydrogen sulfide (H_2S), instead of water, and they produce granules of sulfur (S^0) rather than oxygen, as follows:

$$(2)\ 2H_2S + CO_2 \xrightarrow{\text{light}} (CH_2O) + H_2O + 2S^0$$

Chromatium (krō-mā′tē-um), shown in **Figure 11.14**, is a representative genus. At one time, an important question in biology concerned the source of the oxygen produced by plant photosynthesis: was it from CO_2 or from H_2O? Until the introduction of radioisotope tracers, which traced the oxygen in water and carbon dioxide and finally settled the question, comparison of equations 1 and 2 was the best evidence that the oxygen source was from H_2O. It is important, also, to compare these two equations for an understanding of how reduced sulfur compounds, such as H_2S, can substitute for H_2O in photosynthesis. See "Life Without Sunshine" on page 773.

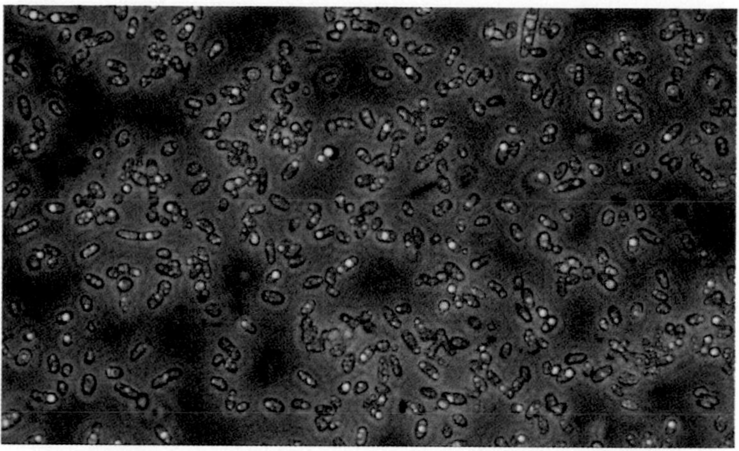

Figure 11.14 Purple sulfur bacteria. This photomicrograph of cells of the genus *Chromatium* shows the intracellular sulfur granules as multicolored refractile objects. The reason the sulfur accumulates can be surmised from inspection of equation 2 in the discussion.

Q What does *anoxygenic* mean?

Other photoautotrophs, the *purple nonsulfur* and *green nonsulfur bacteria,* use organic compounds, such as acids and carbohydrates, for the photosynthetic reduction of carbon dioxide.

Morphologically, the photosynthetic bacteria are very diverse, with spirals, rods, cocci, and even budding forms.

CHECK YOUR UNDERSTANDING

- ✓ Make a dichotomous key to distinguish the gram-negative nonproteobacteria described in this chapter. **11-6**
- ✓ Both the purple and green photosynthetic bacteria and the photosynthetic cyanobacteria use plantlike photosynthesis to make carbohydrates. In what way does the photosynthesis carried out by these two groups differ from plant photosynthesis? **11-7**

The Gram-Positive Bacteria

LEARNING OBJECTIVES

11-8 Differentiate the genera of firmicutes described in this chapter by drawing a dichotomous key.

11-9 Differentiate the actinobacteria described in this chapter by drawing a dichotomous key.

The gram-positive bacteria can be divided into two groups: those that have a high G + C ratio, and those that have a low G + C ratio (see "Nucleic Acids," page 47). To illustrate the variations in

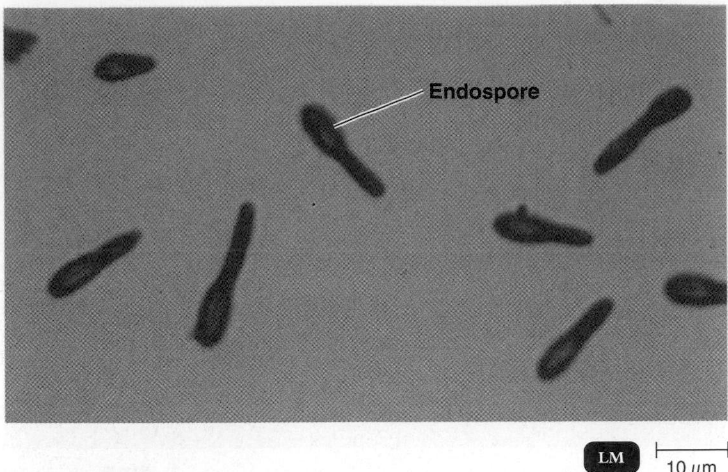

Figure 11.15 ***Clostridium tetani.*** The endospores of clostridia usually distend the cell wall as shown here.

Q What physiological characteristic of *Clostridium* makes it a problem in contamination of deep wounds?

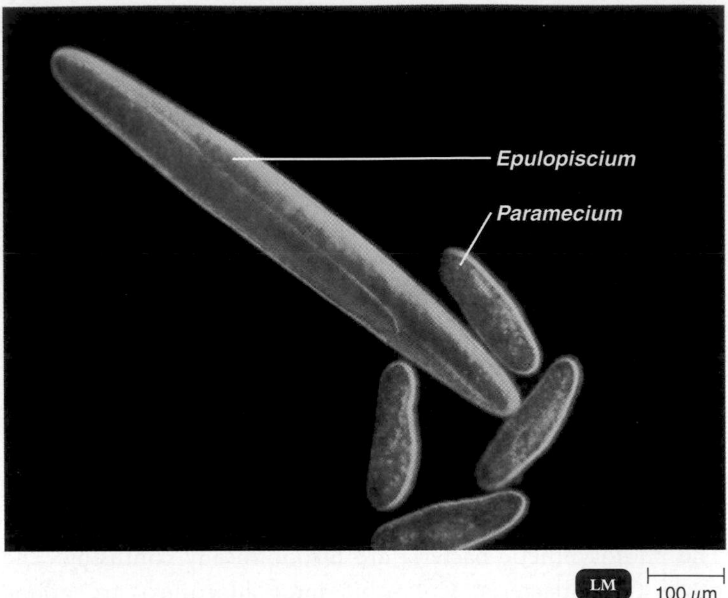

Figure 11.16 A giant prokaryote, *Epulopiscium fishelsoni.*

Q Why is *Epulopiscium* not in the same domain as *Paramecium*?

G + C ratio, the genus *Streptococcus* has a low G + C content of 33 to 44%; and the genus *Clostridium* has a low content of 21 to 54%. Included with the gram-positive, low G + C bacteria are the mycoplasmas, even though they lack a cell wall and therefore do not have a Gram reaction. Their G + C ratio is 23 to 40%.

By contrast, filamentous actinomycetes of the genus *Streptomyces* have a high G + C content of 69 to 73%. Gram-positive bacteria of a more conventional morphology, such as the genera *Corynebacterium* and *Mycobacterium,* have a G + C content of 51 to 63% and 62 to 70%, respectively.

These bacterial groups are placed into separate phyla, the **Firmicutes (low G + C ratios)** and **Actinobacteria (high G + C ratios).**

Firmicutes (Low G + C Gram-Positive Bacteria)

Low G + C gram-positive bacteria are assigned to the phylum Firmicutes. This group includes important endospore-forming bacteria such as the genera *Clostridium* and *Bacillus.* Also of extreme importance in medical microbiology are the genera *Staphylococcus, Enterococcus,* and *Streptococcus.* In industrial microbiology, the genus *Lactobacillus,* which produces lactic acid, is well known. The mycoplasma, which do not possess a cell wall, are also found in this phylum.

Clostridiales

Clostridium Members of the genus *Clostridium* (klôs-tri′dē-um) are obligate anaerobes. The rod-shaped cells contain endospores that usually distend the cell (**Figure 11.15**). The formation of endospores by bacteria is important to both medicine and the food industry because of the endospore's resistance to heat and many chemicals. Diseases associated with clostridia include tetanus, caused by *C. tetani* (te′tan-e); botulism, caused by *C. botulinum* (bo-tū-lī′num); and gas gangrene, caused by *C. perfringens* (per-frin′jens) and other clostridia. *C. perfringens* is also the cause of a common form of foodborne diarrhea. *C. difficile* (dif′fi-sē-il) is an inhabitant of the intestinal tract that may cause a serious diarrhea. This occurs only when antibiotic therapy alters the normal intestinal microbiota, allowing overgrowth by toxin-producing *C. difficile.*

Epulopiscium Biologists have long considered bacteria to be small by necessity because they lack the nutrient transport systems used by higher, eukaryotic organisms and because they depend on simple diffusion to obtain nutrients. These characteristics would seem to critically limit size. So, when a cigar-shaped organism living symbiotically in the gut of the Red Sea surgeonfish was first observed in 1985, it was considered to be a protozoan. Certainly, its size suggested this: the organism was as large as 80 μm × 600 μm —over half a millimeter in length—large enough to be seen with the unaided eye (**Figure 11.16**). Compared to the familiar bacterium *E. coli,* which is about 1 μm × 2 μm, this organism would be about a million times larger in volume.

Further investigation of the new organism showed that certain external structures thought to resemble the cilia of protozoa were actually similar to bacterial flagella, and it did not have a membrane-enclosed nucleus. Ribosomal RNA analysis conclusively placed *Epulopiscium* (ep′ū-lō-pis-ē-um) with the

Figure 11.17 *Bacillus.*

Q What structure is made by both *Clostridium* and *Bacillus*?

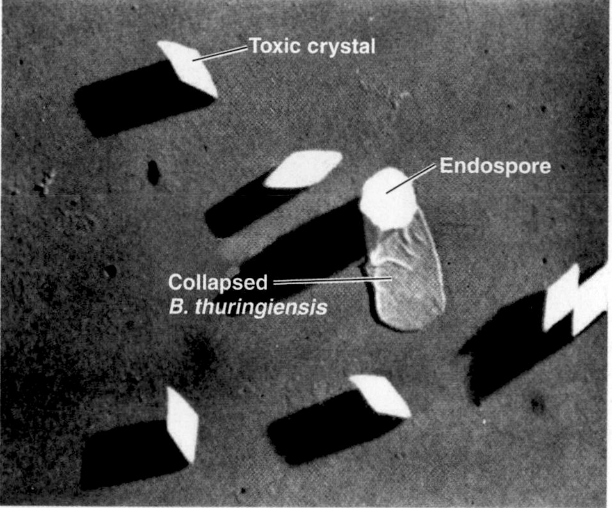

(a) *Bacillus thuringiensis.* The diamond-shaped crystals shown next to the endospore are toxic to insects that ingest them. This electron micrograph was made using the technique of shadow casting described on page 63.

TEM 2.5 μm

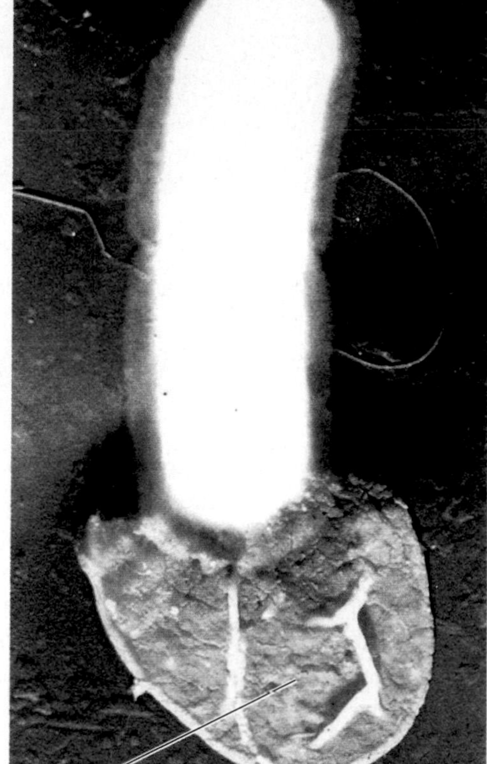

(b) *Bacillus cereus.* This cell of *B. cereus* is shown emerging from the endospore.

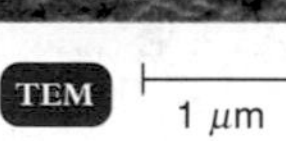

prokaryotes. (The name means "guest at the banquet of a fish." It is literally bathed in semidigested food.) It most closely resembles gram-positive bacteria of the genus *Clostridium.* Strangely, the species *Epulopiscium fishelsoni* (fish-el-sō′nē) does not reproduce by binary fission. Daughter cells formed within the cell are released through a slit opening in the parent cell. This may be related to the evolutionary development of sporulation.

Recently it was discovered that this bacterium does not rely on diffusion to distribute nutrients. Instead, it makes use of its larger genetic capacity—it has 25 times as much DNA as a human cell and as many as 85,000 copies of at least one gene—to manufacture proteins at internal sites where they are needed. (We will describe another, more recently discovered giant bacterium, *Thiomargarita,* on page 326.)

Bacillales

The order Bacillales includes several important genera of gram-positive rods and cocci.

Bacillus Bacteria of the genus *Bacillus* are typically rods that produce endospores. They are common in soil, and only a few are pathogenic to humans. Several species produce antibiotics.

Bacillus anthracis (bä-sil′lus an-thrā′sis) causes anthrax, a disease of cattle, sheep, and horses that can be transmitted to humans. It is often mentioned as a possible agent of biological warfare. (See the box in Chapter 23 on page 644.) The anthrax bacillus is a nonmotile facultative anaerobe, often forming chains in culture. The centrally located endospore does not distend the walls. *Bacillus thuringiensis* (thur-in-jē-en′sis) is probably the best-known microbial insect pathogen (**Figure 11.17a**). It produces intracellular crystals when it sporulates. Commercial preparations containing endospores and crystalline toxin (Bt) of this bacterium are sold in gardening supply shops to be sprayed on plants. *Bacillus cereus* (se′rē-us) (**Figure 11.17b**) is a common bacterium in the environment and occasionally is identified as a cause of food poisoning, especially in starchy foods such as rice.

The three species of the genus *Bacillus* that we have just described are dramatically different in important ways, especially their disease-causing properties. However, they are so closely related that taxonomists consider them to be variants of a single species, differing almost entirely in genes carried on plasmids, which are easily transferred from one bacterium to another.

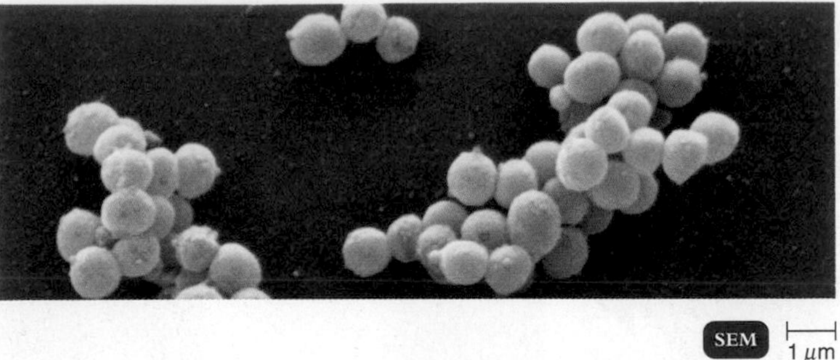

Figure 11.18 ***Staphylococcus aureus.*** Notice the grapelike clusters of these gram-positive cocci.

Q **What is an environmental advantage of a pigment?**

Staphylococcus Staphylococci typically occur in grapelike clusters (**Figure 11.18**). The most important staphylococcal species is *Staphylococcus aureus* (staf-i-lō-kok'kus ô'rē-us), which is named for its yellow-pigmented colonies (*aureus* = golden). Members of this species are facultative anaerobes.

Some characteristics of the staphylococci account for their pathogenicity, which takes many forms. They grow comparatively well under conditions of high osmotic pressure and low moisture, which partially explains why they can grow and survive in nasal secretions (many of us carry the bacteria in our nostrils) and on the skin. This also explains how *S. aureus* can grow in some foods with high osmotic pressure (such as ham and other cured meats) or in low-moisture foods that tend to inhibit the growth of other organisms. The yellow pigment probably confers some protection from the antimicrobial effects of sunlight.

S. aureus produces many toxins that contribute to the bacterium's pathogenicity by increasing its ability to invade the body or damage tissue. The infection of surgical wounds by *S. aureus* is a common problem in hospitals. And its ability to develop resistance quickly to such antibiotics as penicillin contributes to its danger to patients in hospital environments. (See the box in Chapter 14, page 422.) *S. aureus* produces the toxin responsible for toxic shock syndrome, a severe infection characterized by high fever and vomiting, sometimes even death. *S. aureus* also produces an **enterotoxin** that causes vomiting and nausea when ingested; it is one of the most common causes of food poisoning.

Lactobacillales

Several important genera are found in the order Lactobacillales. The genus *Lactobacillus* is a representative of the industrially important lactic acid–producing bacteria. Most lack a cytochrome system and are unable to use oxygen as an electron acceptor. Unlike most obligate anaerobes, though, they are aerotolerant and capable of growth in the presence of oxygen. But compared to oxygen-utilizing microbes, they grow poorly. However, the production of lactic acid from simple carbohydrates inhibits the growth of competing organisms and allows them to grow competitively in spite of their inefficient metabolism. The genus *Streptococcus* shares the metabolic characteristics of the genus *Lactobacillus.* There are several industrially important species, but the streptococci are best known for their pathogenicity. The genera *Enterococcus* and *Listeria* are more conventional metabolically. Both are facultative anaerobes and several species are important pathogens.

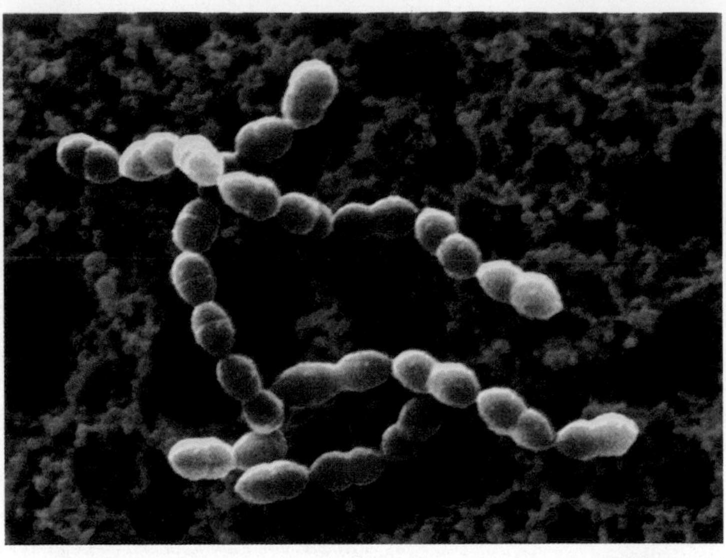

Figure 11.19 ***Streptococcus.*** Notice the chains of cells characteristic of most streptococci. Many of the spherical cells are dividing and are somewhat oval in appearance—especially when viewed with a light microscope, which has lower magnification than this electron micrograph.

 **How does the arrangement of *Streptococcus* differ from *Staphylococcus*?**

Lactobacillus In humans, bacteria of the genus *Lactobacillus* (lak-tō-bä-sil'lus) are located in the vagina, intestinal tract, and oral cavity. Lactobacilli are used commercially in the production of sauerkraut, pickles, buttermilk, and yogurt. Typically, a succession of lactobacilli, each more acid tolerant than its predecessor, participates in these lactic acid fermentations.

Streptococcus Members of the genus *Streptococcus* (strep-tō-kok'kus) are spherical, gram-positive bacteria that typically appear in chains (**Figure 11.19**). They are a taxonomically complex group, probably responsible for more illnesses and causing a greater variety of diseases than any other group of bacteria.

Pathogenic streptococci produce several extracellular substances that contribute to their pathogenicity. Among them are products that destroy phagocytic cells that ingest them. Enzymes produced by some streptococci spread infections by digesting connective tissue of the host, which may also result in extensive tissue destruction. (See the discussion on necrotizing fasciitis on page 591). Infections

(a) Individual cells of *M. pneumoniae.* Arrowheads indicate terminal structures that probably aid in attachment to eukaryotic cells, which then become infected.

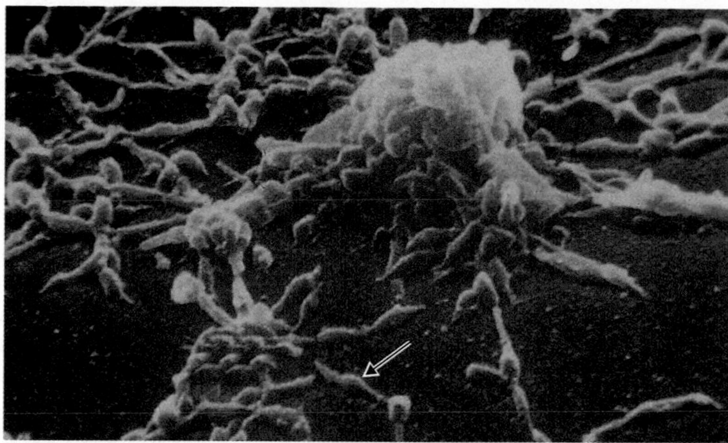
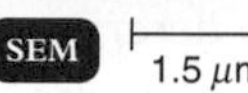

(b) This micrograph shows the filamentous growth of *M. pneumoniae.* Some individual cells can also be seen (arrow). The organism reproduces by fragmentation of the filaments at the bulges.

Figure 11.20 ***Mycoplasma pneumoniae.*** Bacteria such as *M. pneumoniae* have no cell walls, and their morphology is irregular (pleomorphic)

Q How can the cell structure of mycoplasmas account for their pleomorphism?

are also allowed to spread from sites of injury by enzymes that lyse the fibrin (a threadlike protein) of blood clots.

A few nonpathogenic species of streptococci are important in the production of dairy products (see Chapter 28, page 798).

Beta-hemolytic streptococci. A useful basis for the classification of some streptococci is their colonial appearance when grown on blood agar. The *beta-hemolytic* species produce a hemolysin that forms a clear zone of hemolysis on blood agar (see Figure 6.9 on page 168). This group includes the principal pathogen of the streptococci, *Streptococcus pyogenes* (pī-äj′en-ēz), also known as the beta-hemolytic group A streptococcus. Group A represents one of an antigenic group (A through G) within the hemolytic streptococci. Among the diseases caused by *S. pyogenes* are scarlet fever, pharyngitis (sore throat), erysipelas, impetigo, and rheumatic fever. The most important virulence factor is the M protein on the bacterial surface (see Figure 21.6, page 591) by which the bacteria avoid phagocytosis. Another member of the beta-hemolytic streptococci is *Streptococcus agalactiae* (ā′gal-acī-ē-ī), in the beta-hemolytic group B. It is the only species with the group B antigen and is the cause of an important disease of the newborn, neonatal sepsis.

Non-beta-hemolytic streptococci. Certain streptococci are not beta-hemolytic, but when grown on blood agar, their colonies are surrounded by a distinctive greening. These are the *alpha-hemolytic* streptococci. The greening represents a partial destruction of the red blood cells caused mostly by the action of bacteria-produced hydrogen peroxide, but it appears only when the bacteria grow in the presence of oxygen. The most important pathogen in this group is *Streptococcus pneumoniae,* the cause of pneumococcal pneumonia. Also included among the alpha-hemolytic streptococci are species of streptococci called *viridans streptococci.* However, not all species form the alpha-hemolytic greening (*virescent* = green), so this is not really a satisfactory group name. Probably the most significant pathogen of the group is *Streptococcus mutans* (mū′tans), the primary cause of dental caries.

Enterococcus The enterococci are adapted to areas of the body that are rich in nutrients but low in oxygen, such as the gastrointestinal tract, vagina, and oral cavity. They are also found in large numbers in human stool. Because they are relatively hardy microbes, they persist as contaminants in a hospital environment, on hands, bedding, and even as a fecal aerosol. In recent years they have become a leading cause of nosocomial infections, especially because of their high resistance to most antibiotics. Two species, *Enterococcus faecalis* (en-te-rō-kok′kus fe-kā′lis) and *Enterococcus faecium* (fē′sē-um), are responsible for much of the infections of surgical wounds and the urinary tract. In medical settings they frequently enter the bloodstream through invasive procedures, such as indwelling catheters.

Listeria The pathogenic species of the genus *Listeria, Listeria monocytogenes* (lis-te′rē-ä mo-nō-sī-to′je-nēz), can contaminate food, especially dairy products. Important characteristics of *L. monocytogenes* are that it survives within phagocytic cells and is capable of growth at refrigeration temperatures. If it infects a pregnant woman, the organism poses the threat of stillbirth or serious damage to the fetus.

Mycoplasmatales

The mycoplasmas are highly pleomorphic because they lack a cell wall (**Figure 11.20**) and can produce filaments that resemble fungi, hence their name (*mykes* = fungus, and *plasma* = formed). Cells of the genus *Mycoplasma* (mī-ko-plaz'ma) are very small, ranging in size from 0.1 to 0.25 μm, with a cell volume that is only about 5% of that of a typical bacillus. Because their size and plasticity allowed them to pass through filters that retained bacteria, they were originally considered to be viruses. Mycoplasmas may represent the smallest self-replicating organisms that are capable of a free-living existence. One species has only 517 genes; the minimum necessary is between 265 and 350. Studies of their DNA suggest that they are genetically related to the gram-positive bacterial group that includes the genera *Bacillus, Streptococcus,* and *Lactobacillus* but have gradually lost genetic material. The term *degenerative evolution* has been used to describe this process.

The most significant human pathogen among the mycoplasmas is *M. pneumoniae* (nu-mō'nē-ī), which is the cause of a common form of mild pneumonia. Other genera in the order Mycoplasmatales are *Spiroplasma* (spī-rō-plaz'mä), cells with a tight corkscrew morphology that are serious plant pathogens and common parasites of plant-feeding insects, and *Ureaplasma* (ū-rē-ä-plaz'mä), so named because they can enzymatically hydrolyze the urea in urine and are occasionally associated with urinary tract infections.

Mycoplasmas can be grown on artificial media that provide them with sterols (if necessary) and other special nutritional or physical requirements. Colonies are less than 1 mm in diameter and have a characteristic "fried egg" appearance when viewed under magnification (see Figure 24.13, page 688). For many purposes, cell culture methods are often more satisfactory. In fact, mycoplasmas grow so well by this method that they are a frequent contamination problem in cell culture laboratories.

CHECK YOUR UNDERSTANDING

✔ Make a dichotomous key to distinguish the low G + C gram-positive bacteria described in this chapter. **11-8**

Actinobacteria (High G + C Gram-Positive Bacteria)

High G + C gram-positive bacteria are in the phylum Actinobacteria. Many bacteria in this phylum are highly pleomorphic in their morphology; the genera *Corynebacterium* and *Gardnerella,* for example, and several genera such as *Streptomyces* grow only as extended, often branching filaments. Several important pathogenic genera are found in the Actinobacteria, such as the *Mycobacterium* species causing tuberculosis and leprosy. The genera *Streptomyces, Frankia, Actinomyces,* and *Nocardia* are often informally called actinomycetes (from the Greek *actino* = ray) because they have a radiate, or starlike, form of growth by reason of their often-branching filaments. Superficially, their morphology resembles that of filamentous fungi; however, the actinomycetes are prokaryotic cells, and their filaments have a diameter much smaller than that of the eukaryotic molds. Some actinomycetes further resemble molds by their possession of externally carried asexual spores that are used for reproduction. Filamentous bacteria, like filamentous fungi, are very common inhabitants in soil, where a filamentous pattern of growth has advantages. The filamentous organism can bridge water-free gaps between soil particles to move to a new nutritional site. This morphology also gives the organism a much higher surface-to-volume ratio and improves its ability to absorb nutrients in the highly competitive soil environment.

Mycobacterium The mycobacteria are aerobic, non–endospore-forming rods. The name *myco,* meaning funguslike, was derived from their occasional exhibition of filamentous growth (see Figure 24.8, page 682). Many of the characteristics of mycobacteria, such as acid-fast staining, drug resistance, and pathogenicity, are related to their distinctive cell wall, which is structurally similar to gram-negative bacteria (see Figure 4.13c, page 86). However, the outermost lipopolysaccharide layer in mycobacteria is replaced by mycolic acids, which form a waxy, water-resistant layer. This makes the bacteria resistant to stresses such as drying. Also, few antimicrobial drugs are able to enter the cell. (See the box in Chapter 7 on page 201.) Nutrients enter the cell through this layer very slowly, which is a factor in the slow growth rate of mycobacteria; it sometimes takes weeks for visible colonies to appear. The mycobacteria include the important pathogens *Mycobacterium tuberculosis* (mī-kō-bak-ti'rē-um tü-ber-kū-lō'sis), which causes tuberculosis, and *M. leprae* (lep'rī), which causes leprosy. A number of other mycobacteria species are found in soil and water and are occasional pathogens.

Corynebacterium The corynebacteria (*coryne* = club-shaped) tend to be pleomorphic, and their morphology often varies with the age of the cells. The best-known species is *Corynebacterium diphtheriae* (kôr'i-nē-bak-ti-rē-um dif-thi'rē-ī), the causative agent of diphtheria.

Propionibacterium The name of the genus *Propionibacterium* (prō-pē-on'ē-bak-ti-rē-um) is derived from the organism's ability to form propionic acid; some species are important in the fermentation of Swiss cheese. *Propionibacterium acnes* (ak'nēz) are bacteria that are commonly found on human skin and are implicated as the primary bacterial cause of acne.

Gardnerella *Gardnerella vaginalis* (gard-ne-rel'la va-jin-al'is) is a bacterium that causes one of the most common forms of vaginitis. There has always been some difficulty in assigning a taxonomic position in this species, which is gram-variable, and which exhibits a highly pleomorphic morphology.

Frankia The genus *Frankia* (frank'ē-ä) causes nitrogen-fixing nodules to form in alder tree roots, much as rhizobia cause nodules on the roots of legumes (see Figure 27.5, page 773).

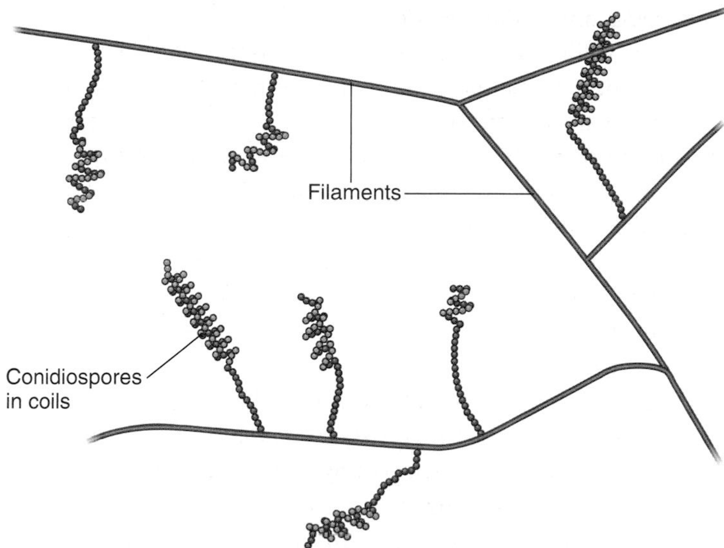

(a) Drawing of a typical streptomycete showing filamentous, branching growth with asexual reproductive conidiospores at the filament tips

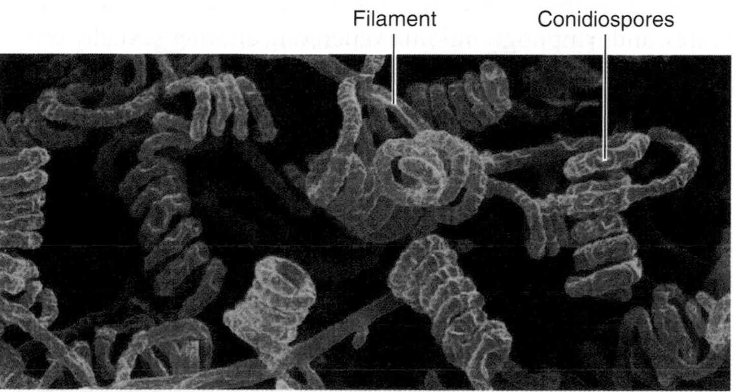

(b) Coils of conidiospores supported by filaments of the streptomycete

Figure 11.21 ***Streptomyces.***

Q **Why is *Streptomyces* not classified with fungi?**

Streptomyces The genus *Streptomyces* (strep-tō-mī′ses) is the best known of the actinomycetes and is one of the bacteria most commonly isolated from soil (**Figure 11.21**). The reproductive asexual spores of *Streptomyces* are formed at the ends of aerial filaments. If each spore lands on a suitable substrate, it is capable of germinating into a new colony. These organisms are strict aerobes. They often produce extracellular enzymes that enable them to utilize proteins, polysaccharides (such as starch and cellulose), and many other organic materials found in soil. *Streptomyces* characteristically produce a gaseous compound called *geosmin,* which gives fresh soil its typical musty odor. Species of *Streptomyces* are valuable because they produce most of our commercial antibiotics (see Table 20.1, page 555). This has led to intensive study of the genus—there are nearly 500 described species.

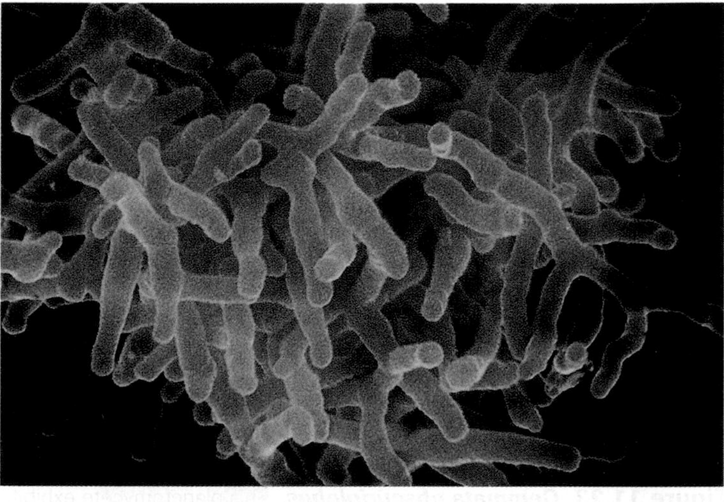

Figure 11.22 ***Actinomyces.*** Notice the branched filamentous morphology.

Q **Why are these bacteria not classified as fungi?**

Actinomyces The genus *Actinomyces* (ak-tin-ō-mī′sēs) consists of facultative anaerobes that are found in the mouth and throat of humans and animals. They occasionally form filaments that can fragment (**Figure 11.22**). One species, *Actinomyces israelii* (is-rā′lē-ē), causes actinomycosis, a tissue-destroying disease usually affecting the head, neck, or lungs.

Nocardia The genus *Nocardia* (nō-kär′dē-ä) morphologically resembles *Actinomyces;* however, these bacteria are aerobic. To reproduce, they form rudimentary filaments, which fragment into short rods. The structure of their cell wall resembles that of the mycobacteria; therefore, they are often acid-fast. *Nocardia* species are common in soil. Some species, such as *Nocardia asteroides* (as′ter-oi-dēz), occasionally cause a chronic, difficult-to-treat pulmonary infection. *N. asteroides* is also one of the causative agents of mycetoma, a localized destructive infection of the feet or hands.

CHECK YOUR UNDERSTANDING

✓ Make a dichotomous key to distinguish the high G + C gram-positive bacteria described in this chapter. **11-9**

* * *

The fifth and final volume of the second edition of *Bergey's Manual of Systematic Bacteriology* is scheduled to contain a varied assortment of phyla such as Planctomycetes Chlamydiae, Spirochaetes, Bacteroidetes, and Fusobacteria. Several important pathogens, such as the genera *Chlamydia, Borrelia,* and *Treponema,* are included. Members of the genera *Bacteroides* and *Fusobacterium* are profuse and important inhabitants of the human intestinal tract.

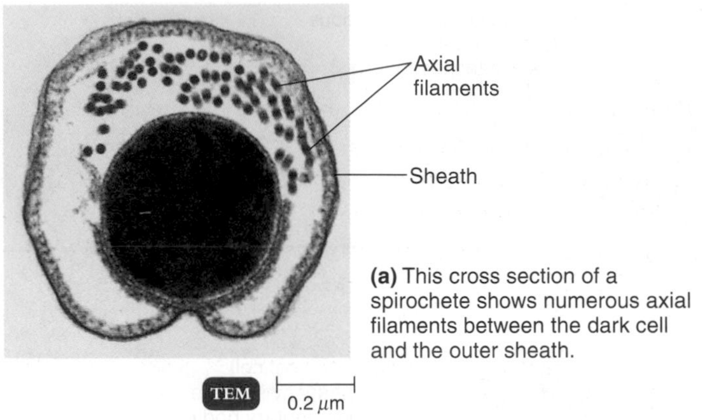

(a) This cross section of a spirochete shows numerous axial filaments between the dark cell and the outer sheath.

TEM 0.2 μm

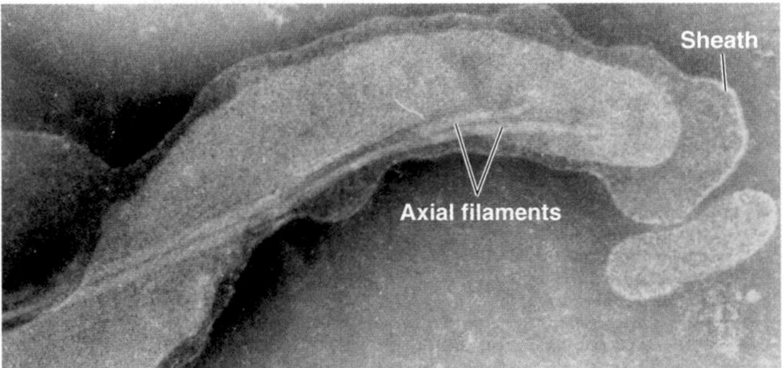

(b) This micrograph of a portion of *Treponema pallidum* shows the sheath, which has shrunk away from the cell, and two axial filaments attached near one end of the cell under the sheath.

Figure 11.25 Spirochetes. Spirochetes are helical and have axial filaments under an outer sheath that enables them to move by a corkscrewlike rotation.

Q How does a spirochete's motility differ from that of *Spirillum* (see Figure 11.4)?

and swine, so domestic dogs and cats are routinely immunized against leptospirosis. The tightly coiled cells of *Leptospira* are shown in Figure 26.4 on page 747.

Bacteroidetes

The phylum Bacteroidetes includes several genera of anaerobic bacteria. Included are the genus *Bacteroides,* a common inhabitant of the human intestinal tract, and the genus *Prevotella,* found in the human mouth. Also included in the phylum Bacteroidetes are the important soil bacteria with gliding motility of the genus *Cytophaga.*

Bacteroides Bacteria of the genus *Bacteroides* (bak-tė-roi′dēz) live in the human intestinal tract in numbers approaching 1 billion per gram of feces. Some *Bacteroides* species also reside in anaerobic habitats such as the gingival crevice (see Figure 25.2 on page 707) and are also frequently recovered from deep tissue infections. *Bacteroides* organisms are nonmotile and do not form endospores. Infections caused by *Bacteroides* often result from puncture wounds or surgery and are a frequent cause of peritonitis, an inflammation resulting from a perforated bowel.

Cytophaga Members of the genus *Cytophaga* (sī-täf′ag-a) are important in the degradation of cellulose and chitin, which are both abundant in soil. Gliding motility places the microbe in close contact with these substrates so that enzymatic action is very efficient.

Fusobacteria

The fusiform bacteria comprise another phylum of anaerobes. These bacteria are often pleomorphic but, as their name suggests, may be spindle-shaped (*fuso* = spindle).

Fusobacterium Members of the genus *Fusobacterium* (fū-sō-bak-ti′rē-um) are long and slender, with pointed rather than blunt ends (Figure 11.26). In humans, they are found most often in the gingival crevice of the gums and may be responsible for some dental abscesses.

CHECK YOUR UNDERSTANDING

✓ Make a dichotomous key to distinguish planctomeycetes, chlamydias, spirochetes, Bacteroidetes, *Cytophaga*, and *Fusobacterium*. **11-10**

SEM 1 μm

Figure 11.26 *Fusobacterium.* This is a common anaerobic rod found in the human intestine. Notice the characteristic pointed ends.

Q In what other place in the human body do you often find *Fusobacterium*?

DOMAIN ARCHAEA

In the late 1970s, a distinctive type of prokaryotic cell was discovered. Most strikingly, their cell walls lacked the peptidoglycan common to most bacteria. It soon became clear that they also shared many rRNA sequences, and the sequences were different from either those of the Domain Bacteria or the eukaryotic organisms. These differences were so significant that these organisms now constitute a new taxonomic grouping, the Domain Archaea.

Diversity within the Archaea

LEARNING OBJECTIVE

11-11 Name a habitat for each group of archaea.

This exceptionally interesting group of prokaryotes is highly diverse. Most archaea are of conventional morphology, that is, rods, cocci, and helixes, but some are of very unusual morphology, as illustrated in Figure 11.27. Some are gram-positive, others gram-negative; some may divide by binary fission, others by fragmentation or budding; a few lack cell walls. Cultivated members of the archaea (singular: *archaeon*) can be placed into five physiological or nutritional groups.

Physiologically, archaea are found under extreme environmental conditions. **Extremophiles,** as they are known, include halophiles, thermophiles, and acidophiles (also see pages 158–159). There are no known pathogenic archaea. *Halophiles* thrive in salt concentrations of more than 25%, such as found in the Great Salt Lake and solar evaporating ponds. Examples of these are found in the genus *Halobacterium* (ha-lo-bak-ti′re-um), some of which may even require such salt concentrations in order to grow. The optimal growth temperatures of extremely *thermophilic* archaea is 80°C or higher. The present record growth temperature is 121°C, established by archaea growing near a hydrothermal vent at 2000 meters deep in the ocean. *Acidophilic* archaea can be found growing at pH values below zero, and frequently at elevated temperatures, as well. An example is *Sulfolobus* (sul′fo-lo-bus), whose optimal pH is about 2 and optimal temperature is more than 70°C.

Nutritionally, the ocean contains numerous *nitrifying* archaea that oxidize ammonia for energy. Some might also be found in soils. *Methanogens* are strictly anaerobic archaea that produce methane as an end-product by combining hydrogen (H_2) with carbon dioxide (CO_2). There are no known bacterial methanogens. These archaea are of considerable economic importance when they are used in sewage treatment (see the discussion of sludge digestion in Chapter 27 on pages 785–787). Methanogens are also part of the microbiota of the human colon, vagina, and mouth.

CHECK YOUR UNDERSTANDING

✓ What kind of archaea would populate solar evaporating ponds? **11-11**

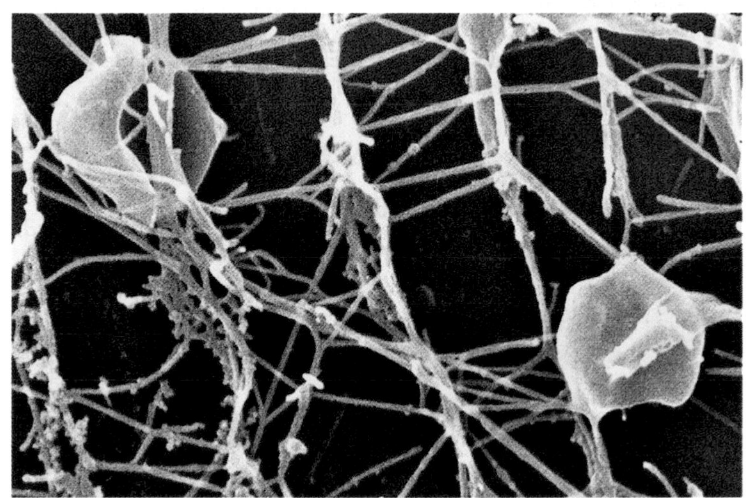

Figure 11.27 Archaea. *Pyrodictium abyssi,* an unusual member of the archaea found growing in deep ocean sediment at a temperature of 110°C. The cells are disk-shaped with a network of tubules (cannulae). Most archaea are more conventional in their morphology.

Q Do the terms included in the name, *pyro* and *abyssi,* suggest a basis for the naming of this bacterium?

MICROBIAL DIVERSITY

The Earth provides a seemingly infinite number of environmental niches, and novel life forms have evolved to fill them. Many of the microbes that exist in these niches cannot be cultivated by conventional methods on conventional growth media and have remained unknown. In recent years, however, isolation and identification methods have become much more sophisticated, and microbes that fill these niches are being identified—many without being cultivated. Particularly interesting are bacteria that test the theoretical limits of size for prokaryotes.

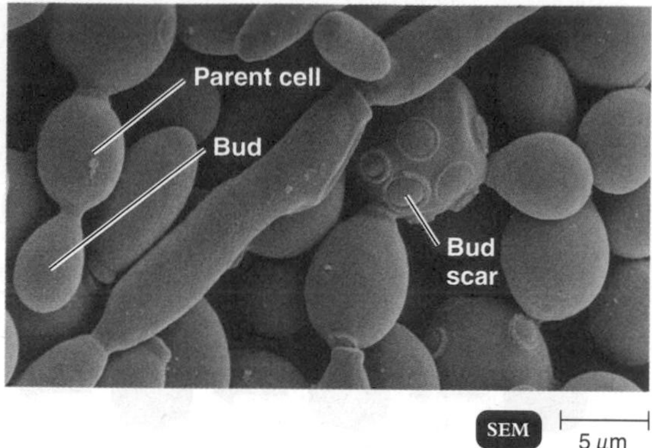

Figure 12.3 A budding yeast. A micrograph of *Saccharomyces cerevisiae* in various stages of budding.

Q How does a bud differ from a spore?

hyphae grow to form a filamentous mass called a **mycelium,** which is visible to the unaided eye (Figure 12.2b).

Yeasts Yeasts are nonfilamentous, unicellular fungi that are typically spherical or oval. Like molds, yeasts are widely distributed in nature; they are frequently found as a white powdery coating on fruits and leaves. **Budding yeasts,** such as *Saccharomyces* (sak-ä-rō-mi′sēs), divide unevenly.

In budding (Figure 12.3), the parent cell forms a protuberance (bud) on its outer surface. As the bud elongates, the parent cell's nucleus divides, and one nucleus migrates into the bud. Cell wall material is then laid down between the bud and parent cell, and the bud eventually breaks away.

One yeast cell can in time produce up to 24 daughter cells by budding. Some yeasts produce buds that fail to detach themselves; these buds form a short chain of cells called a **pseudohypha.** *Candida albicans* (kan′did-ä al′bi-kanz) attaches to human epithelial cells as a yeast but usually requires pseudohyphae to invade deeper tissues (see Figure 21.17a, page 601).

Fission yeasts, such as *Schizosaccharomyces* (skiz-ō-sak-ä-rō-mī′sēs), divide evenly to produce two new cells. During fission, the parent cell elongates, its nucleus divides, and two daughter cells are produced. Increases in the number of yeast cells on a solid medium produce a colony similar to a bacterial colony.

Yeasts are capable of facultative anaerobic growth. Yeasts can use oxygen or an organic compound as the final electron acceptor; this is a valuable attribute because it allows these fungi to survive in various environments. If given access to oxygen, yeasts perform aerobic respiration to metabolize carbohydrates to carbon dioxide and water; denied oxygen, they ferment carbohydrates and produce ethanol and carbon dioxide. This fermentation is used in the brewing, wine-making, and baking industries. *Saccharomyces* species produce ethanol in brewed beverages and carbon dioxide for leavening bread dough.

Dimorphic Fungi Some fungi, most notably the pathogenic species, exhibit **dimorphism**—two forms of growth. Such fungi can grow either as a mold or as a yeast. The moldlike forms produce vegetative and aerial hyphae; the yeastlike forms reproduce by budding. Dimorphism in pathogenic fungi is temperature-dependent: at 37°C, the fungus is yeastlike, and at 25°C, it is moldlike. (See Figure 24.16, page 695.) However, the appearance of the dimorphic (in this instance, nonpathogenic) fungus shown in Figure 12.4 changes with CO_2 concentration.

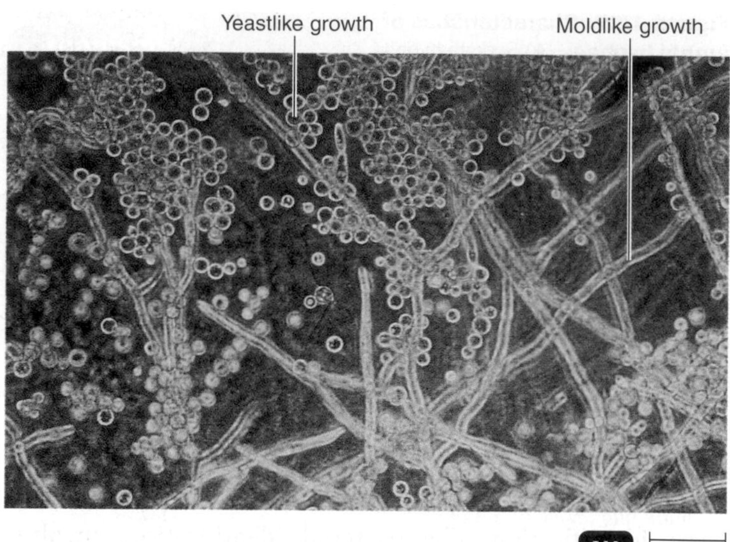

Figure 12.4 Fungal dimorphism. Dimorphism in the fungus *Mucor indicus* depends on CO_2 concentration. On the agar surface, *Mucor* exhibits yeastlike growth, but in the agar it is moldlike.

Q What is fungal dimorphism?

Life Cycle

Filamentous fungi can reproduce asexually by fragmentation of their hyphae. In addition, both sexual and asexual reproduction in fungi occurs by the formation of **spores.** In fact, fungi are usually identified by spore type.

Fungal spores, however, are quite different from bacterial endospores. Bacterial endospores allow a bacterial cell to survive adverse environmental conditions (see Chapter 4). A single vegetative bacterial cell forms one endospore, which eventually germinates to produce a single vegetative bacterial cell. This process is not reproduction because it does not increase the total number of bacterial cells. But after a mold forms a spore, the spore detaches from the parent and germinates into a new mold (see Figure 12.1c). Unlike the bacterial endospore, this is a true reproductive spore; a second organism grows from the spore. Although fungal spores can survive for extended periods in dry or hot environments, most do not exhibit the extreme tolerance and longevity of bacterial endospores.

Spores are formed from aerial hyphae in a number of different ways, depending on the species. Fungal spores can be either asexual or sexual. **Asexual spores** are formed by the hyphae of one organism. When these spores germinate, they become organisms that are genetically identical to the parent. **Sexual spores** result from the fusion of nuclei from two opposite mating strains of the same species of fungus. Fungi produce sexual spores less frequently than asexual spores. Organisms that grow from sexual spores will have genetic characteristics of both parental strains. Because spores are of considerable importance in identifying fungi, we will next look at some of the various types of asexual and sexual spores.

Asexual Spores Asexual spores are produced by an individual fungus through mitosis and subsequent cell division; there is no fusion of the nuclei of cells. Two types of asexual spores are produced by fungi. One type is a **conidiospore,** or **conidium** (plural: *conidia*), a unicellular or multicellular spore that is not enclosed in a sac (Figure 12.5a). Conidia are produced in a chain at the end of a **conidiophore.** Such spores are produced by *Aspergillus* (a-spėr-jil'lus). Conidia formed by the fragmentation of a septate hypha into single, slightly thickened cells are called **arthroconidia** (Figure 12.5b). One species that produces such spores is *Coccidioides immitis* (kok-sid-ē-oi'dēz im'mi-tis) (see Figure 24.18, page 696). Another type of conidium, **blastoconidia,** consists of buds coming off the parent cell (Figure 12.5c). Such spores are found in some yeasts, such as *Candida albicans* and *Cryptococcus.* A **chlamydoconidium** is a thick-walled spore formed by rounding and enlargement within a hyphal segment (Figure 12.5d). A fungus that produces chlamydoconidia is the yeast *C. albicans.*

The other type of asexual spore is a **sporangiospore,** formed within a **sporangium,** or sac, at the end of an aerial hypha called a **sporangiophore.** The sporangium can contain hundreds of sporangiospores (Figure 12.5e). Such spores are produced by *Rhizopus.*

Sexual Spores A fungal sexual spore results from sexual reproduction, which consists of three phases:

1. **Plasmogamy.** A haploid nucleus of a donor cell (+) penetrates the cytoplasm of a recipient cell (−).
2. **Karyogamy.** The (+) and (−) nuclei fuse to form a diploid zygote nucleus.
3. **Meiosis.** The diploid nucleus gives rise to haploid nuclei (sexual spores), some of which may be genetic recombinants.

The sexual spores produced by fungi characterize the phyla. In laboratory settings, most fungi exhibit only asexual spores. Consequently, clinical identification is based on microscopic examination of asexual spores.

Nutritional Adaptations

Fungi are generally adapted to environments that would be hostile to bacteria. Fungi are chemoheterotrophs, and, like bacteria, they absorb nutrients rather than ingesting them as animals do. However, fungi differ from bacteria in certain environmental requirements and in the following nutritional characteristics:

- Fungi usually grow better in an environment with a pH of about 5, which is too acidic for the growth of most common bacteria.
- Almost all molds are aerobic. Most yeasts are facultative anaerobes.
- Most fungi are more resistant to osmotic pressure than bacteria; most can therefore grow in relatively high sugar or salt concentrations.
- Fungi can grow on substances with a very low moisture content, generally too low to support the growth of bacteria.
- Fungi require somewhat less nitrogen than bacteria for an equivalent amount of growth.
- Fungi are often capable of metabolizing complex carbohydrates, such as lignin (a component of wood), that most bacteria cannot use for nutrients.

These characteristics enable fungi to grow on such unlikely substrates as bathroom walls, shoe leather, and discarded newspapers.

CHECK YOUR UNDERSTANDING

- ✓ Assume you isolated a single-celled organism that has a cell wall. How would you determine that it is a fungus and not a bacterium? **12-1**
- ✓ Contrast the mechanism of conidiospore and ascospore formation. **12-2**

Medically Important Phyla of Fungi

This section provides an overview of medically important phyla of fungi. The actual diseases they cause will be studied in Chapters 21 through 26. Note that not all fungi cause disease.

The genera named in the following phyla include many that are readily found as contaminants in foods and in laboratory bacterial cultures. Although these genera are not all of primary medical importance, they are typical examples of their respective groups.

Zygomycota

The Zygomycota, or conjugation fungi, are saprophytic molds that have coenocytic hyphae. An example is *Rhizopus stolonifer,* the common black bread mold. The asexual spores of *Rhizopus* are sporangiospores (Figure 12.6, lower right). The dark sporangiospores inside the sporangium give *Rhizopus* its descriptive common name. When the sporangium breaks open, the sporangiospores are dispersed. If they fall on a suitable medium, they will germinate into a new mold thallus.

The sexual spores are zygospores. A **zygospore** is a large spore enclosed in a thick wall (Figure 12.6, lower left). This type of spore results from the fusion of the nuclei of two cells that are morphologically similar to each other.

(a) Conidia are arranged in chains at the end of *Aspergillus flavus* conidiophore. SEM 5 μm

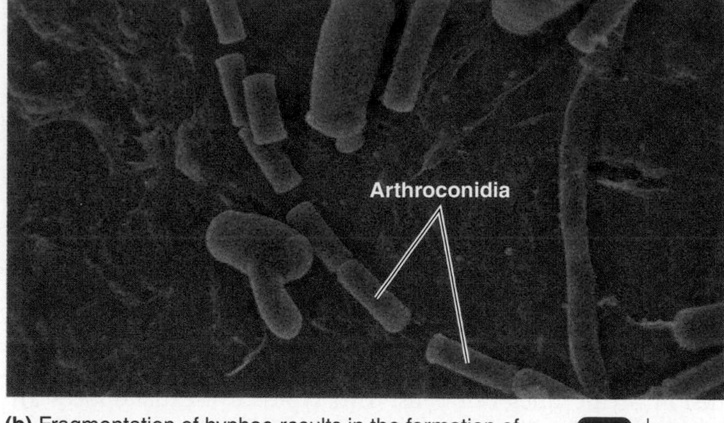

(b) Fragmentation of hyphae results in the formation of arthroconidia in *Coccidioides immitis.*

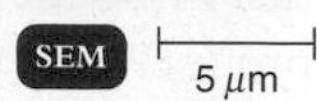

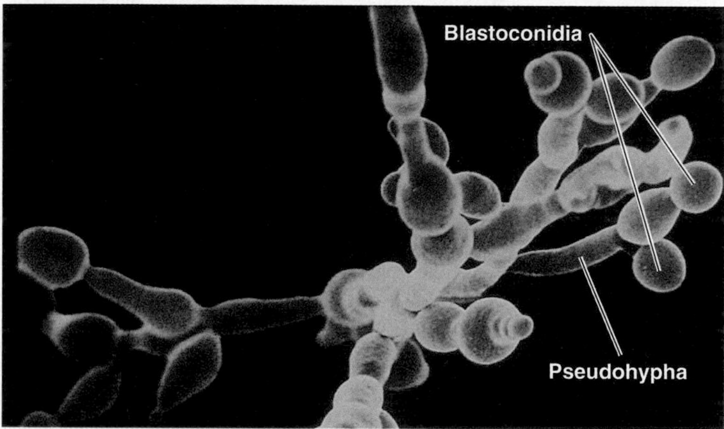

(c) Blastoconidia are formed from the buds of a parent cell of *Candida albicans.* SEM 10 μm

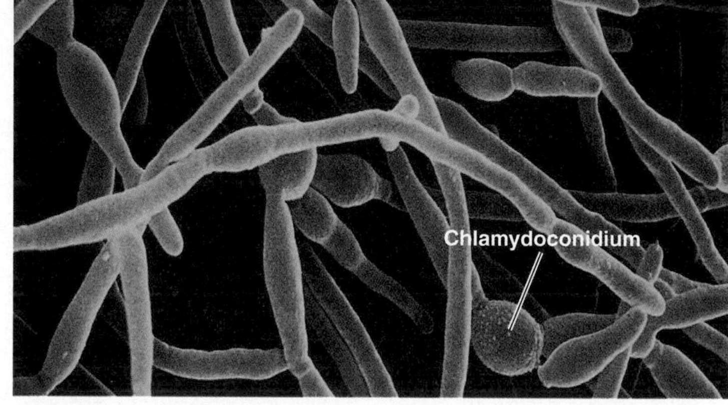

(d) Chlamydoconidia are thick-walled cells within hyphae of this *Candida albicans.*

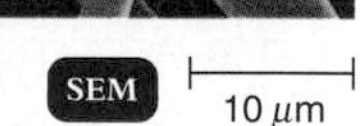

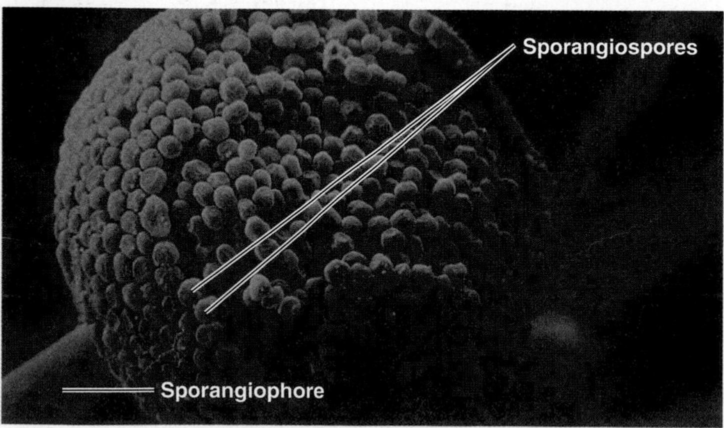

(e) Sporangiospores are formed within a sporangium (spore sac) of *Rhizopus.*

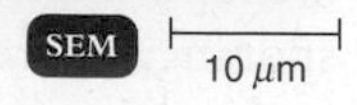

Figure 12.5 Representative asexual spores.

Q **What are the green powdery structures on moldy food?**

Ascomycota

The Ascomycota, or sac fungi, include molds with septate hyphae and some yeasts. Their asexual spores are usually conidia produced in long chains from the conidiophore. The term *conidia* means dust, and these spores freely detach from the chain at the slightest disturbance and float in the air like dust.

An **ascospore** results from the fusion of the nuclei of two cells that can be either morphologically similar or dissimilar. These spores are produced in a saclike structure called an **ascus** (Figure 12.7, lower left). The members of this phylum are called sac fungi because of the ascus.

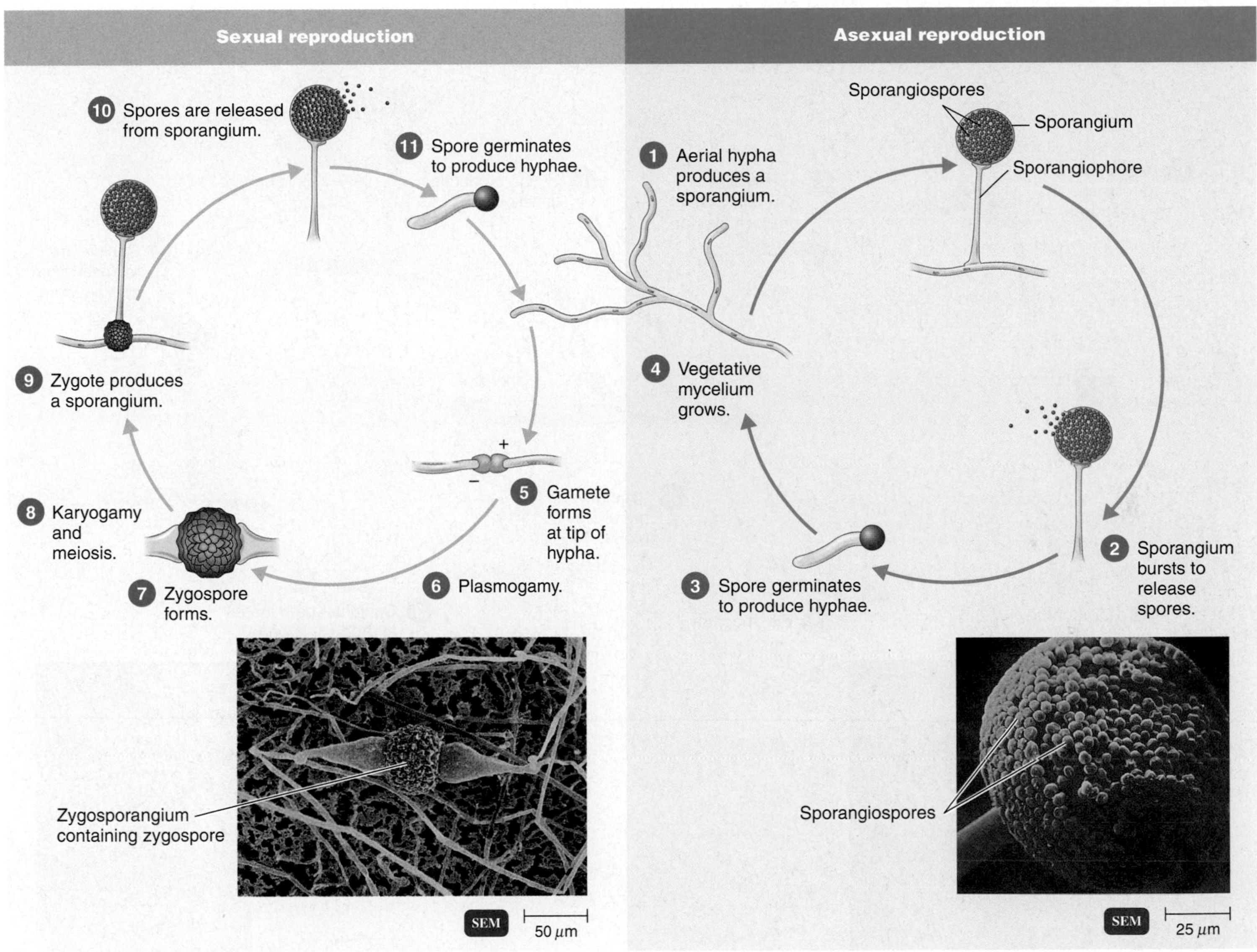

Figure 12.6 The life cycle of *Rhizopus,* a zygomycete. This fungus will reproduce asexually most of the time. Two opposite mating strains (designated + and –) are necessary for sexual reproduction.

Q What is an opportunistic mycosis?

Basidiomycota

The Basidiomycota, or club fungi, also possess septate hyphae. This phylum includes fungi that produce mushrooms. **Basidiospores** are formed externally on a base pedestal called a **basidium** (Figure 12.8). (The common name of the fungus is derived from the club shape of the basidium.) There are usually four basidiospores per basidium. Some of the basidiomycota produce asexual conidiospores.

The fungi we have looked at thus far are **teleomorphs;** that is, they produce both sexual and asexual spores. Some ascomycetes have lost the ability to reproduce sexually. These asexual fungi are called **anamorphs.** *Penicillium* is an example of an anamorph that arose from a mutation in a teleomorph. Historically, fungi whose sexual cycle had not been observed were put in a "holding category" called *Deuteromycota.* Now, mycologists are using rRNA sequencing to classify these organisms. Most of these previously unclassified deuteromycetes are anamorph phases of Ascomycota, and a few are basidiomycetes.

Table 12.3 on page 338 lists some fungi that cause human diseases. Two generic names are given for some of the fungi because medically important fungi that are well known by their anamorph, or asexual, name are often referred to by that name.

Fungal Diseases

Any fungal infection is called a **mycosis.** Mycoses are generally chronic (long-lasting) infections because fungi grow slowly.

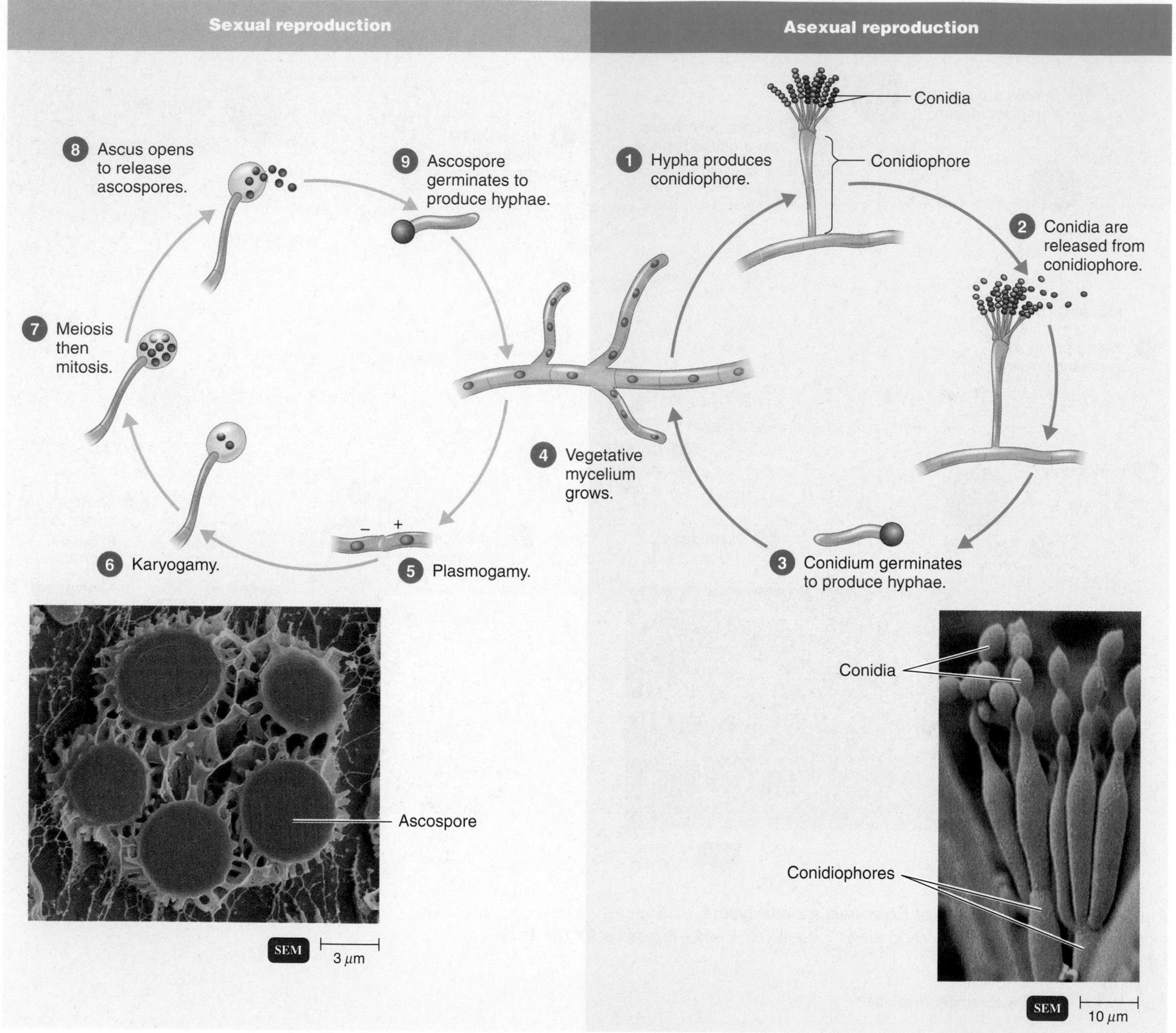

Figure 12.7 The life cycle of *Talaromyces*, an ascomycete. Occasionally, when two opposite mating cells from two different strains (+ and –) fuse, sexual reproduction occurs.

Q Name one ascomycete that can infect humans.

Mycoses are classified into five groups according to the degree of tissue involvement and mode of entry into the host: systemic, subcutaneous, cutaneous, superficial, or opportunistic. In Chapter 10, we saw that fungi are related to animals. Consequently, drugs that affect fungal cells may also affect animal cells. This fact makes fungal infections of humans and other animals often difficult to treat.

Systemic mycoses are fungal infections deep within the body. They are not restricted to any particular region of the body but can affect a number of tissues and organs. Systemic mycoses are usually caused by fungi that live in the soil. Inhalation of spores is the route of transmission; these infections typically begin in the lungs and then spread to other body tissues. They are not contagious from animal to human or from human to human. Two systemic mycoses, histoplasmosis and coccidioidomycosis, are discussed in Chapter 24.

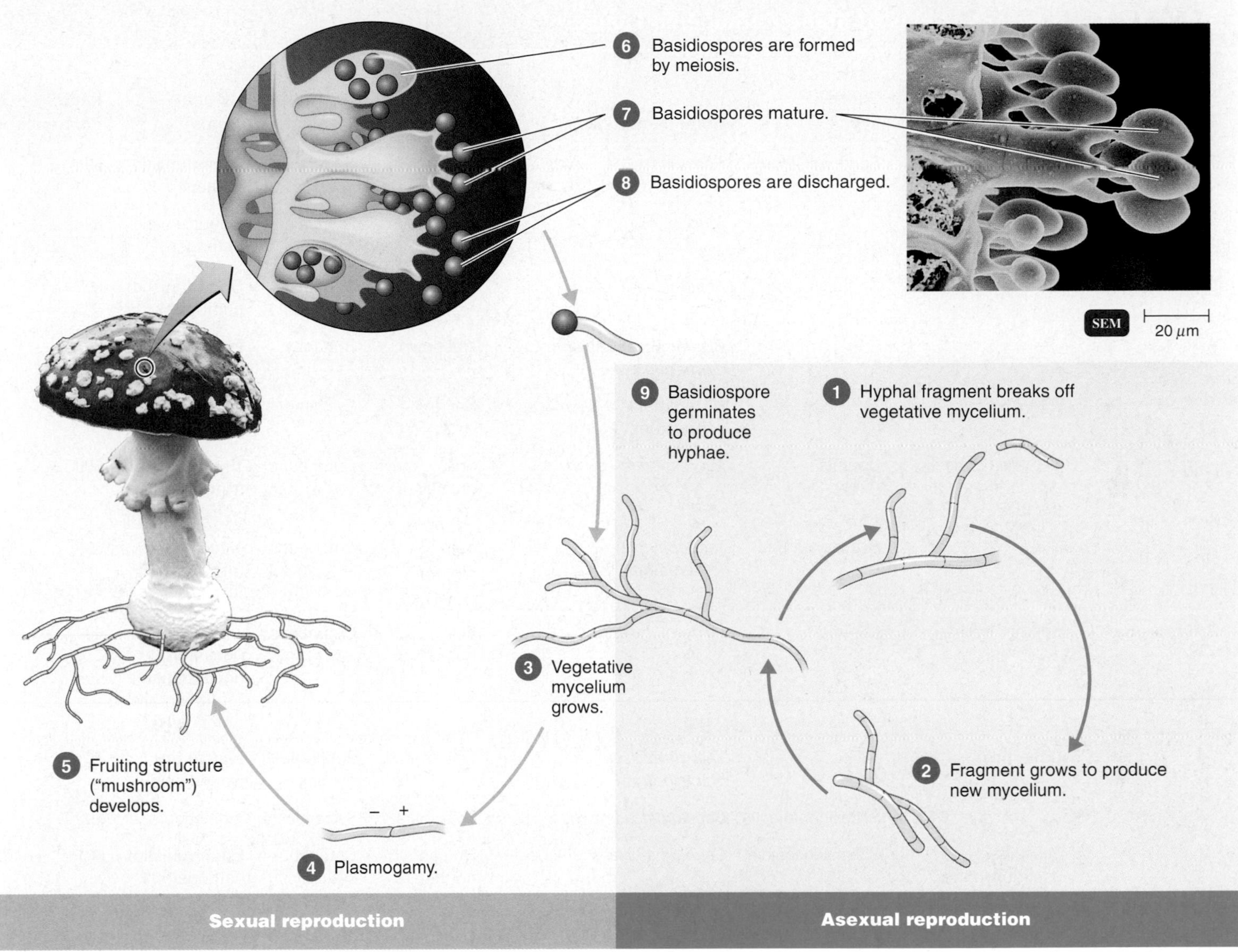
6 Basidiospores are formed by meiosis.
7 Basidiospores mature.
8 Basidiospores are discharged.
SEM
20 μm
9 Basidiospore germinates to produce hyphae.
1 Hyphal fragment breaks off vegetative mycelium.
3 Vegetative mycelium grows.
2 Fragment grows to produce new mycelium.
5 Fruiting structure ("mushroom") develops.
– +
4 Plasmogamy.
Sexual reproduction
Asexual reproduction

Figure 12.8 A generalized life cycle of a basidiomycete. Mushrooms appear after cells from two mating strains (+ and –) have fused.

Q **On what basis are fungi classified into phyla?**

Subcutaneous mycoses are fungal infections beneath the skin caused by saprophytic fungi that live in soil and on vegetation. Sporotrichosis is a subcutaneous infection acquired by gardeners and farmers (Chapter 21, page 601). Infection occurs by direct implantation of spores or mycelial fragments into a puncture wound in the skin.

Fungi that infect only the epidermis, hair, and nails are called **dermatophytes,** and their infections are called *dermatomycoses* or **cutaneous mycoses** (see Figure 21.16, page 600). Dermatophytes secrete keratinase, an enzyme that degrades **keratin,** a protein found in hair, skin, and nails. Infection is transmitted from human to human or from animal to human by direct contact or by contact with infected hairs and epidermal cells (as from barber shop clippers or shower room floors).

The fungi that cause **superficial mycoses** are localized along hair shafts and in superficial (surface) epidermal cells. These infections are prevalent in tropical climates.

An **opportunistic pathogen** is generally harmless in its normal habitat but can become pathogenic in a host who is seriously debilitated or traumatized, who is under treatment with broad-spectrum antibiotics, whose immune system is suppressed by drugs or by an immune disorder, or who has a lung disease.

Pneumocystis is an opportunistic pathogen in individuals with compromised immune systems and is the most common life-threatening infection in AIDS patients (see Figure 24.20, page 698). It was first classified as a protozoan, but recent studies of its RNA indicate it is a unicellular anamorphic fungus. Another example of an opportunistic pathogen is the fungus *Stachybotrys*

Table 12.3 Characteristics of Some Pathogenic Fungi

Phylum	Growth Characteristics	Asexual Spore Types	Human Pathogens	Habitat	Type of Mycosis	Clinical Notes	Page Reference
Zygomycota	Nonseptate hyphae	Sporangiospores	*Rhizopus*	Ubiquitous	Systemic	Opportunistic pathogen	698
			Mucor	Ubiquitous	Systemic	Opportunistic pathogen	698
Ascomycota	Dimorphic	Conidia	*Aspergillus*	Ubiquitous	Systemic	Opportunistic pathogen	697
			Blastomyces (Ajellomyces†) dermatitidis*	Unknown	Systemic	Inhalation	697
			Histoplasma(Ajellomyces†) capsulatum*	Soil	Systemic	Inhalation	695
	Septate hyphae, strong affinity for keratin	Conidia	*Microsporum*	Soil, animals	Cutaneous	Tinea capitis (ringworm)	600
		Arthroconidia	*Trichophyton* (Arthroderma†)*	Soil, animals	Cutaneous	Tinea pedis (athlete's foot)	600
Anamorphs	Septate hyphae	Conidia	*Epidermophyton*	Soil, humans	Cutaneous	Tinea cruris (jock itch), tinea unguium (of fingernails or toenails)	600
	Dimorphic		*Sporothrix schenckii, Stachybotrys*	Soil	Subcutaneous	Puncture wound	601
		Arthroconidia	*Coccidioides immitis*	Soil	Systemic	Inhalation	696
	Yeastlike, pseudohyphae	Chlamydoconidia	*Candida albicans*	Human normal microbiota	Cutaneous, systemic, mucocutaneous	Opportunistic pathogen	601
	Unicellular	None	*Pneumocystis*	Ubiquitous	Systemic	Opportunistic pathogen	697
Basidiomycota	Septate hyphae; includes rusts and smuts, and plant pathogens; yeastlike encapsulated cells	Conidia	*Cryptococcus neoformans* (Filobasidiella)†*	Soil, bird feces	Systemic	Inhalation	626
			Malassezia	Human skin	Cutaneous	Dandruff; dermatitis	586

*Anamorph name.
†Teleomorph name.

(sta′ke-botris), which normally grows on cellulose found in dead plants but in recent years has been found growing on water-damaged walls of homes.

Mucormycosis is an opportunistic mycosis caused by *Rhizopus* and *Mucor* (mū′kôr); the infection occurs mostly in patients with diabetes mellitus, with leukemia, or undergoing treatment with immunosuppressive drugs. Aspergillosis is also an opportunistic mycosis; it is caused by *Aspergillus* (see Figure 12.2). This disease occurs in people who have debilitating lung diseases or cancer and have inhaled *Aspergillus* spores.

Opportunistic infections by *Cryptococcus* and *Penicillium* can cause fatal diseases in AIDS patients. These opportunistic fungi may be transmitted from one person to an uninfected person but do not usually infect immunocompetent people. **Yeast infection,** or candidiasis, is most frequently caused by *Candida albicans* and may occur as vulvovaginal candidiasis or thrush, a mucocutaneous candidiasis. Candidiasis frequently occurs in newborns, in people with AIDS, and in people being treated with broad-spectrum antibiotics (see Figure 21.17, page 601).

Some fungi cause disease by producing toxins. These toxins are discussed in Chapter 15.

Economic Effects of Fungi

Fungi have been used in biotechnology for many years. *Aspergillus niger* (nī-jėr), for example, has been used to produce citric acid for foods and beverages since 1914. The yeast *Saccharomyces cerevisiae* is used to make bread and wine. It is also genetically modified to produce a variety of proteins, including hepatitis B vaccine. *Trichoderma* is used commercially to produce the enzyme cellulase, which is used to remove plant cell walls to produce a clear fruit juice. When the anticancer drug taxol, which is produced by yew trees, was discovered, there was concern that the yew forests of the U.S. Northwest coast would be decimated to harvest the drug. However, the fungus *Taxomyces* (tax'ō-mī'sēs) also produces taxol.

Fungi are used as biological controls of pests. In 1990, the fungus *Entomophaga* (en'tō mo fäg ä) unexpectedly proliferated and killed gypsy moths that were destroying trees in the eastern United States. Scientists are investigating the use of several fungi to kill pests:

- *Metarrhizium* grows on plant roots, and parasitic weevils die after eating the roots.
- A fungus that was first spotted on insects feeding on eggplants in Texas may become a new biocontrol for the widespread and costly agricultural pests known as whiteflies. The fungus *Coniothyrium minitans* (kon'ē-oth-rē-um mi'ni-tanz) feeds on fungi that destroy soybeans and other bean crops.
- A foam filled with *Paecilomyces fumosoroseus* is being used as a biological alternative to chemicals to kill termites hiding inside tree trunks and other hard-to-reach places.

In contrast to these beneficial effects, fungi can have undesirable effects for agriculture because of their nutritional adaptations. As most of us have observed, mold spoilage of fruits, grains, and vegetables is relatively common, but bacterial spoilage of such foods is not. There is little moisture on the unbroken surfaces of such foods, and the interiors of fruits are too acidic for many bacteria to grow there. Jams and jellies also tend to be acidic, and they have a high osmotic pressure from the sugars they contain. These factors all discourage bacterial growth but readily support the growth of molds. A paraffin layer on top of a jar of homemade jelly helps deter mold growth because molds are aerobic and the paraffin layer keeps out the oxygen. However, fresh meats and certain other foods are such good substrates for bacterial growth that bacteria not only will outgrow molds but also will actively suppress mold growth in these foods.

The spreading chestnut tree, of which Longfellow wrote, no longer grows in the United States except in a few widely isolated locations; a fungal blight killed virtually all of them. This blight was caused by the ascomycete *Cryphonectria parasitica* (kri-fō-nek'trē-ä par-ä-si'ti-kä), which was introduced from China around 1904. The fungus allows the tree roots to live and put forth shoots regularly, but then it kills the shoots just as regularly. *Cryphonectria*-resistant chestnuts are being developed. Another imported fungal plant disease is Dutch elm disease, caused by *Ceratocystis ulmi* (sē-rä-tō-sis'tis ul'me). Carried from tree to tree by a bark beetle, the fungus blocks the afflicted tree's circulation. The disease has devastated the American elm population.

CHECK YOUR UNDERSTANDING

- ✓ List the asexual and sexual spores made by Zygomycetes, Ascomycetes, and Basidiomycetes. **12-3**
- ✓ Are yeasts beneficial or harmful? **12-4**

Lichens

LEARNING OBJECTIVES

12-5 List the distinguishing characteristics of lichens, and describe their nutritional needs.

12-6 Describe the roles of the fungus and the alga in a lichen.

A **lichen** is a combination of a green alga (or a cyanobacterium) and a fungus. Lichens are placed in the Kingdom Fungi and are classified according to the fungal partner, most often an ascomycete. The two organisms exist in a *mutualistic* relationship, in which each partner benefits. The lichen is very different from either the alga or fungus growing alone, and if the partners are separated, the lichen no longer exists. Approximately 13,500 species of lichens occupy quite diverse habitats. Because they can inhabit areas in which neither fungi nor algae could survive alone, lichens are often the first life forms to colonize newly exposed soil or rock. Lichens secrete organic acids that chemically weather rock, and they accumulate nutrients needed for plant growth. Also found on trees, concrete structures, and rooftops, lichens are some of the slowest-growing organisms on Earth.

Lichens can be grouped into three morphologic categories (**Figure 12.9a**). *Crustose lichens* grow flush or encrusting onto the substratum, *foliose lichens* are more leaflike, and *fruticose lichens* have fingerlike projections. The lichen's thallus, or body, forms when fungal hyphae grow around algal cells to become the **medulla** (**Figure 12.9b**). Fungal hyphae project below the lichen body to form **rhizines,** or holdfasts. Fungal hyphae also form a **cortex,** or protective covering, over the algal layer and sometimes under it as well. After incorporation into a lichen thallus, the alga continues to grow, and the growing hyphae can incorporate new algal cells.

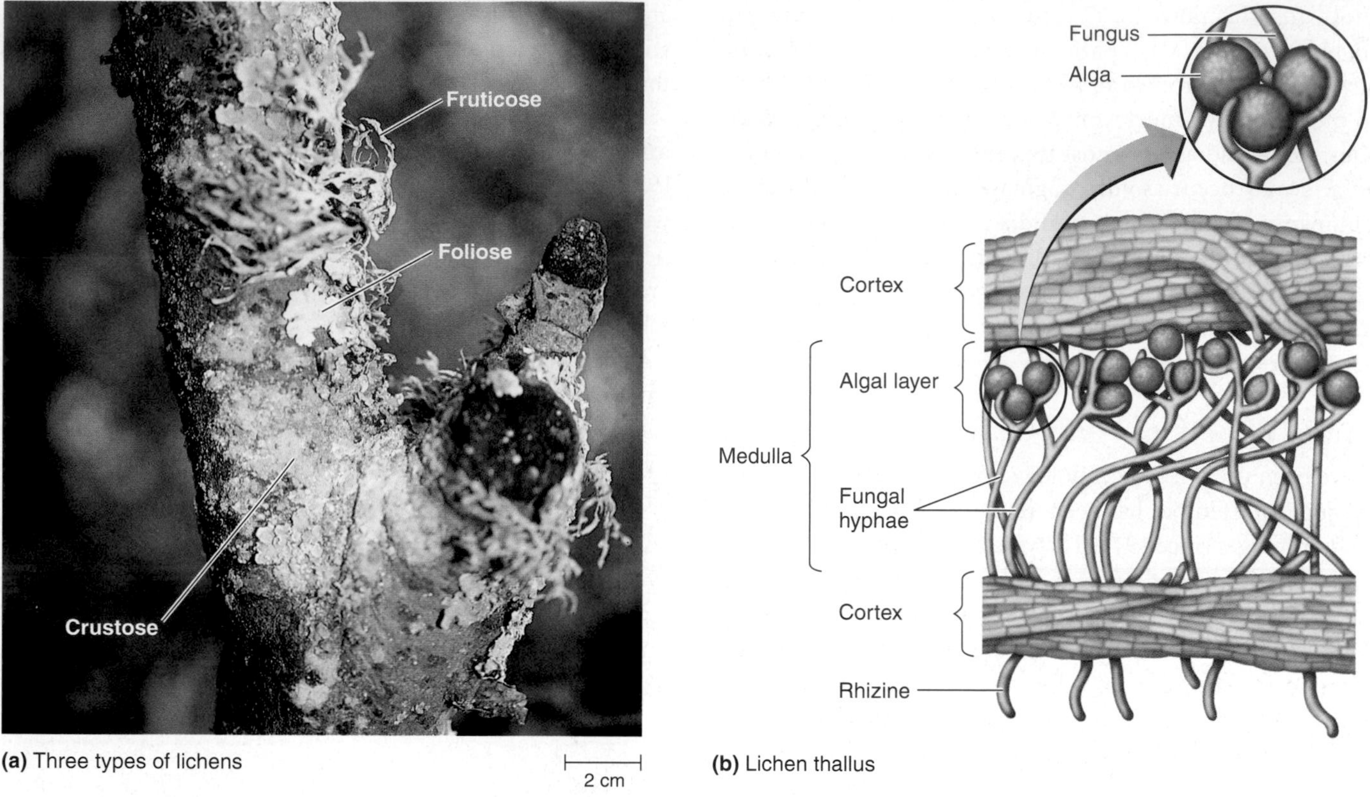

Figure 12.9 Lichens. The lichen medulla is composed of fungal hyphae surrounding the algal layer. The protective cortex is a layer of fungal hyphae that covers the surface and sometimes the bottom of the lichen.

Q In what ways are lichens unique?

When the algal partner is cultured separately in vitro, about 1% of the carbohydrates produced during photosynthesis are released into the culture medium; however, when the alga is associated with a fungus, the algal plasma membrane is more permeable, and up to 60% of the products of photosynthesis are released to the fungus or are found as end-products of fungal metabolism. The fungus clearly benefits from this association. The alga, while giving up valuable nutrients, is in turn compensated; it receives from the fungus both protection from desiccation (cortex) and attachment (rhizines).

Lichens had considerable economic importance in ancient Greece and other parts of Europe as dyes for clothing. Usnic acid from *Usnea* is used as an antimicrobial agent in China. Erythrolitmin, the dye used in litmus paper to indicate changes in pH, is extracted from a variety of lichens. Some lichens or their acids can cause allergic contact dermatitis in humans.

Populations of lichens readily incorporate cations (positively charged ions) into their thalli. Therefore, the concentrations and types of cations in the atmosphere can be determined by chemical analyses of lichen thalli. In addition, the presence or absence of species that are quite sensitive to pollutants can be used to ascertain air quality. A 1985 study in the Cuyahoga Valley in Ohio revealed that 81% of the 172 lichen species that were present in 1917 were gone. Because this area is severely affected by air pollution, the inference is that air pollutants, primarily sulfur dioxide (the major contributor to acid precipitation), caused the death of sensitive species.

Lichens are the major food for tundra herbivores such as caribou and reindeer. After the 1986 Chernobyl nuclear disaster, 70,000 reindeer in Lapland that had been raised for food had to be destroyed because of high levels of radiation. The lichens on which the reindeer fed had absorbed radioactive cesium-137, which had spread in the air.

CHECK YOUR UNDERSTANDING

- ✓ What is the role of lichens in nature? **12-5**
- ✓ What is the role of the fungus in a lichen? **12-6**

Algae

LEARNING OBJECTIVES

12-7 List the defining characteristics of algae.

12-8 List the outstanding characteristics of the five phyla of algae discussed in this chapter.

12-9 Identify two beneficial and two harmful effects of algae.

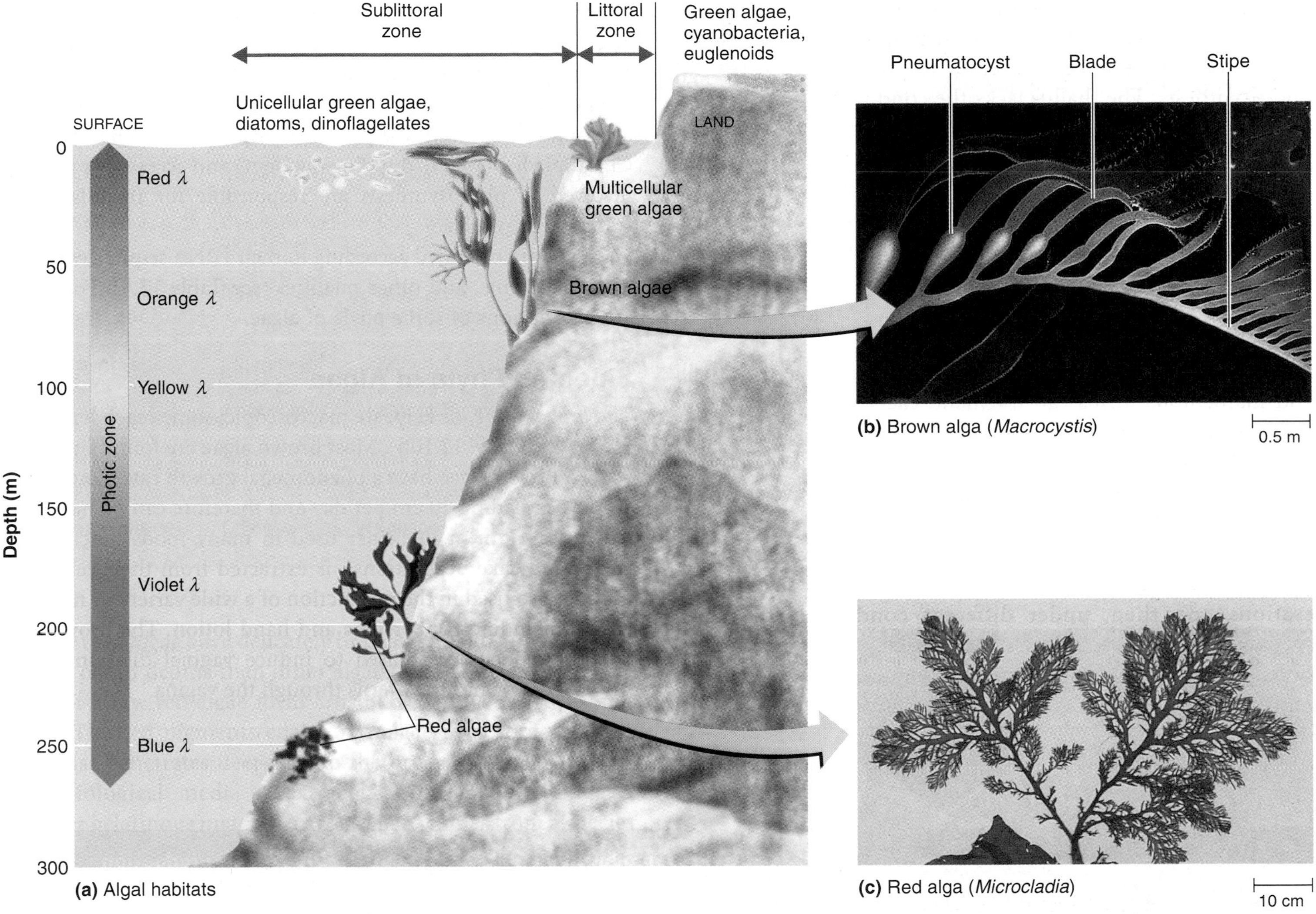

(a) Algal habitats

(b) Brown alga (*Macrocystis*)

(c) Red alga (*Microcladia*)

Figure 12.10 Algae and their habitats. (**a**) Although unicellular and filamentous algae can be found on land, they frequently exist in marine and freshwater environments as plankton. Multicellular green, brown, and red algae require a suitable attachment site, adequate water for support, and light of the appropriate wavelengths. (**b**) *Macrocystis porifera,* a brown alga. The hollow stipe and gas-filled pneumatocysts hold the thallus upright ensuring that sufficient sunlight is received for growth. (**c**) *Microcladia,* a red alga. The delicately branched red algae get their color from phycobiliprotein accessory pigments.

Q What red alga is toxic for humans?

Algae are familiar as the large brown kelp in coastal waters, the green scum in a puddle, and the green stains on soil or on rocks. A few algae are responsible for food poisonings. Some algae are unicellular; others form chains of cells (are filamentous); and a few have thalli.

Algae are mostly aquatic, although some are found in soil or on trees when sufficient moisture is available there. Unusual algal habitats include the hair of both the sedentary South American sloth and the polar bear. Water is necessary for physical support, reproduction, and the diffusion of nutrients. Generally, algae are found in cool temperate waters, although the large floating mats of the brown alga *Sargassum* (sär-gas′sum) are found in the subtropical Sargasso Sea. Some species of brown algae grow in antarctic waters.

Characteristics of Algae

Algae are relatively simple eukaryotic photoautotrophs that lack the tissues (roots, stem, and leaves) of plants. The identification of unicellular and filamentous algae requires microscopic examination. Most algae are found in the ocean. Their locations depend on the availability of appropriate nutrients, wavelengths of light, and surfaces on which to grow. Probable locations for representative algae are shown in Figure 12.10a.

Vegetative Structures

The body of a multicellular alga is called a thallus. Thalli of the larger multicellular algae, those commonly called seaweeds,

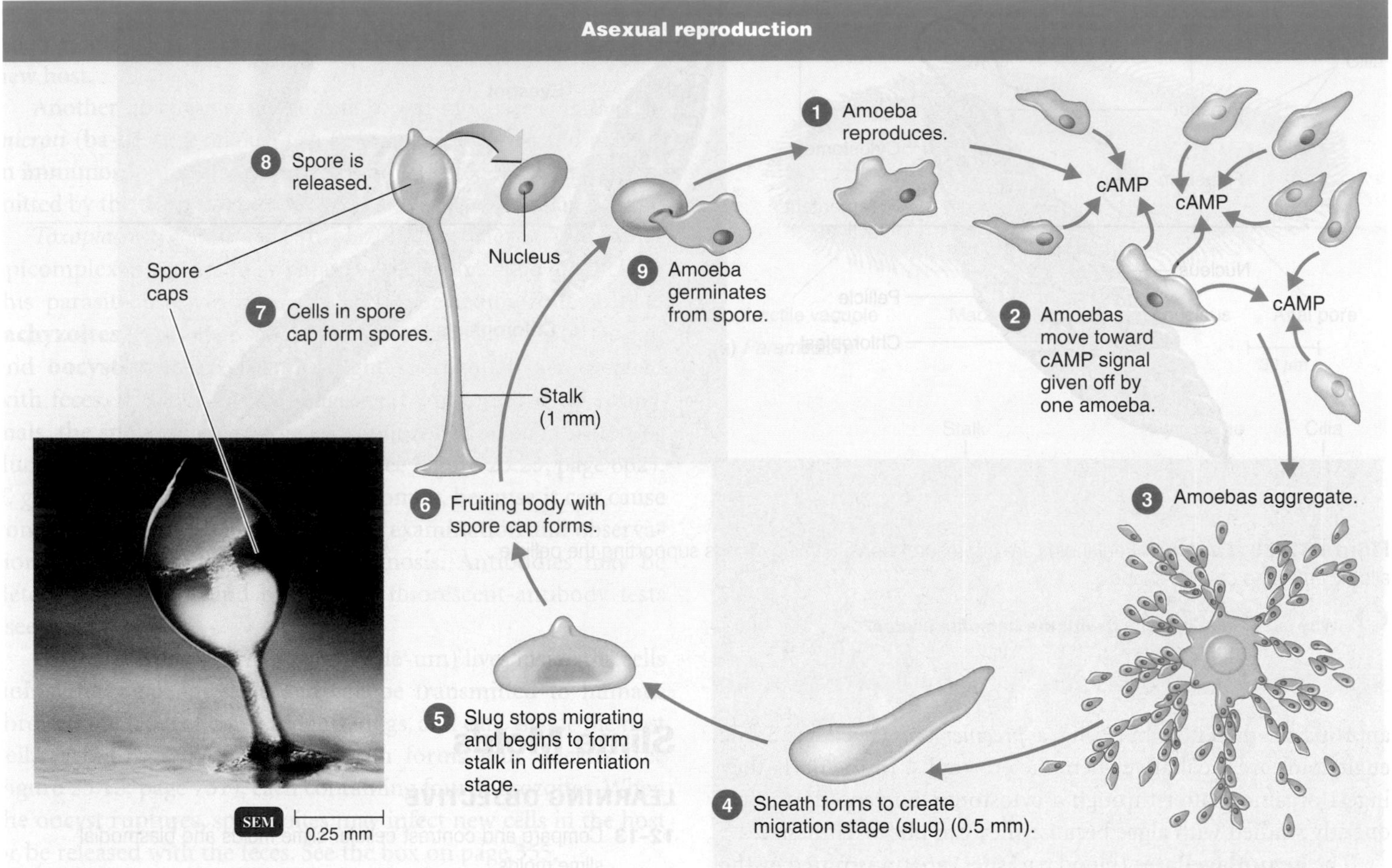
Asexual reproduction
1 Amoeba reproduces.
cAMP
cAMP
cAMP
2 Amoebas move toward cAMP signal given off by one amoeba.
3 Amoebas aggregate.
4 Sheath forms to create migration stage (slug) (0.5 mm).
5 Slug stops migrating and begins to form stalk in differentiation stage.
6 Fruiting body with spore cap forms.
7 Cells in spore cap form spores.
8 Spore is released.
9 Amoeba germinates from spore.
Nucleus
Spore caps
Stalk (1 mm)
SEM
0.25 mm

Figure 12.21 **The generalized life cycle of a cellular slime mold.** The micrograph shows a spore cap of *Dictyostelium.*

Q **What characteristics do slime molds share with protozoa? With fungi?**

Plasmodial slime molds were first scientifically reported in 1729. They belong to a separate phylum. A plasmodial slime mold exists as a mass of protoplasm with many nuclei (it is multinucleated). This mass of protoplasm is called a **plasmodium** (Figure 12.22). The entire plasmodium moves as a giant amoeba; it engulfs organic debris and bacteria. Biologists have found that musclelike proteins forming microfilaments account for the movement of the plasmodium.

When plasmodial slime molds are grown in laboratories, a phenomenon called **cytoplasmic streaming** is observed, during which the protoplasm within the plasmodium moves and changes both its speed and direction so that the oxygen and nutrients are evenly distributed. The plasmodium continues to grow as long as there is enough food and moisture for it to thrive.

When either is in short supply, the plasmodium separates into many groups of protoplasm; each of these groups forms a stalked sporangium, in which haploid spores (a resistant, resting form of the slime mold) develop. When conditions improve, these spores germinate, fuse to form diploid cells, and develop into a multinucleated plasmodium.

CHECK YOUR UNDERSTANDING

✓ Why are slime molds classified with amoeba and not fungi? **12-13**

Helminths

LEARNING OBJECTIVES

12-14 List the distinguishing characteristics of parasitic helminths.

12-15 Provide a rationale for the elaborate life cycle of parasitic worms.

12-16 List the characteristics of the two classes of parasitic platyhelminths, and give an example of each.

12-17 Describe a parasitic infection in which humans serve as a definitive host, as an intermediate host, and as both.

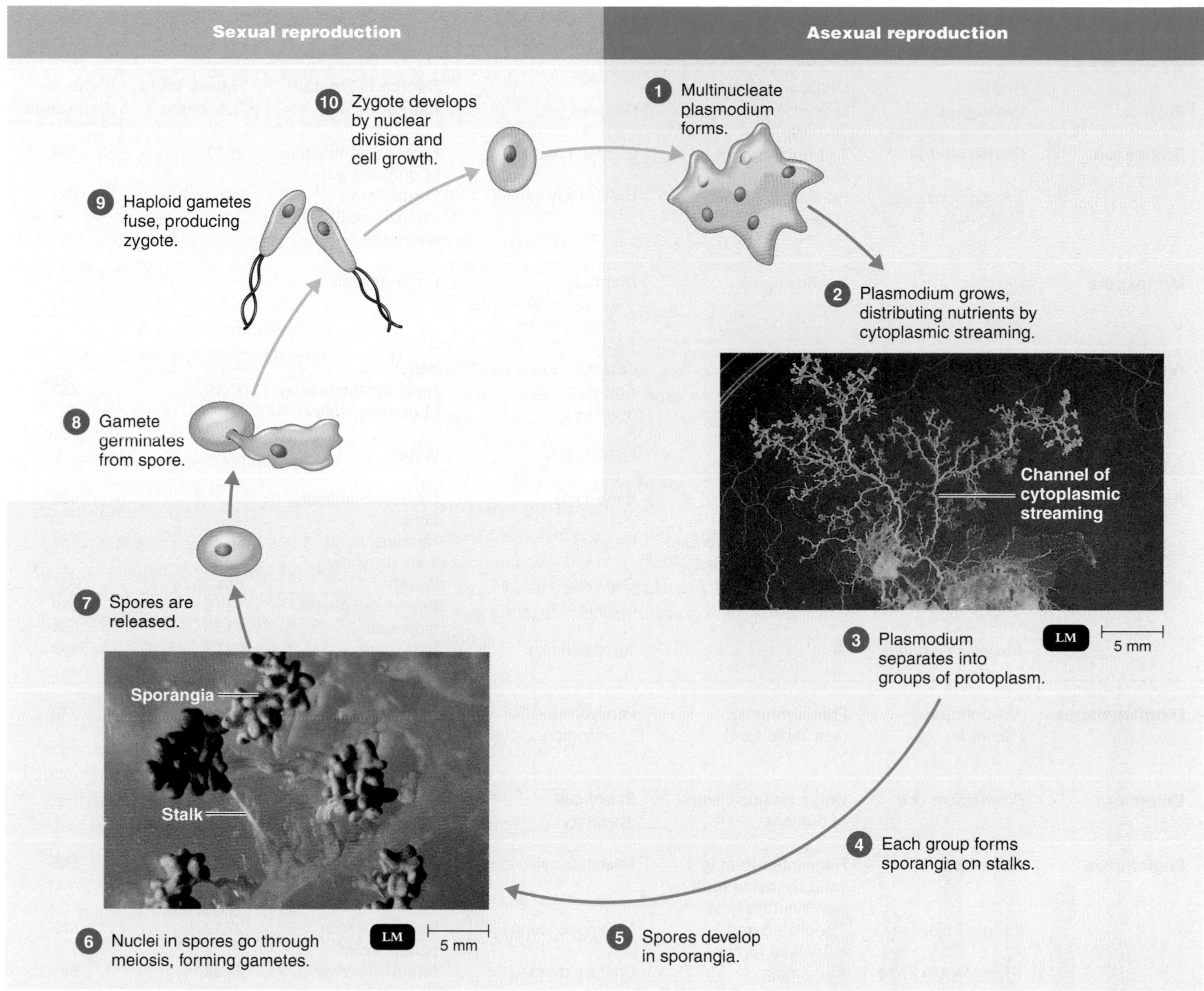
Sexual reproduction
Asexual reproduction
1 Multinucleate plasmodium forms.
2 Plasmodium grows, distributing nutrients by cytoplasmic streaming.
Channel of cytoplasmic streaming
3 Plasmodium separates into groups of protoplasm.
LM
5 mm
4 Each group forms sporangia on stalks.
5 Spores develop in sporangia.
Sporangia
Stalk
6 Nuclei in spores go through meiosis, forming gametes.
LM
5 mm
7 Spores are released.
8 Gamete germinates from spore.
9 Haploid gametes fuse, producing zygote.
10 Zygote develops by nuclear division and cell growth.

Figure 12.22 The life cycle of a plasmodial slime mold. *Physarum* is pictured in the photomicrographs.

Q How do cellular and acellular slime molds differ?

12-18 List the characteristics of parasitic nematodes, and give an example of infective eggs and infective larvae.

12-19 Compare and contrast platyhelminths and nematodes.

A number of parasitic animals spend part or all of their lives in humans. Most of these animals belong to two phyla: Platyhelminthes (flatworms) and Nematoda (roundworms). These worms are commonly called **helminths.** There are also free-living species in these phyla, but we will limit our discussion to the parasitic species. Diseases caused by parasitic worms are discussed in Part Four.

Characteristics of Helminths

Helminths are multicellular eukaryotic animals that generally possess digestive, circulatory, nervous, excretory, and reproductive systems. Parasitic helminths must be highly specialized to

Figure 12.23 Infection by a parasitic platyhelminth. An increase in the trematode *Ribeiroia* in recent years has caused deformed frogs. Frogs with multiple limbs have been found from Minnesota to California. Cercaria of the trematode infect tadpoles. The encysted metacercariae displace developing limb buds, causing abnormal limb development. The increase in the parasite may be due to fertilizer runoff that increases algae, which are food for the parasite's intermediate host snail.

Q What tailed stage of the parasite lives in a snail?

Platyhelminths

Members of the Phylum Platyhelminthes, the **flatworms,** are dorsoventrally flattened. The classes of parasitic flatworms include the trematodes and cestodes. These parasites cause disease or developmental disturbances in a wide variety of animals (Figure 12.23).

Trematodes

Trematodes, or **flukes,** often have flat, leaf-shaped bodies with a ventral sucker and an oral sucker (Figure 12.24). The suckers hold the organism in place. Flukes obtain food by absorbing it through their nonliving outer covering, called the **cuticle.** Flukes are given common names according to the tissue of the definitive host in which the adults live (for example, lung fluke, liver fluke, blood fluke). The Asian liver fluke *Clonorchis sinensis* (klo-nôr'kis si-nen'sis) is occasionally seen in immigrants in the United States, but it cannot be transmitted because its intermediate hosts are not in the United States.

To exemplify a fluke's life cycle, let's look at the lung fluke, *Paragonimus westermani* (păr-ä-gōn'e-mus we-ster-ma'nē). The intermediate hosts for this fluke, and therefore the fluke itself, occur throughout the world, including the United States and Canada. The adult lung fluke lives in the bronchioles of humans and other mammals and is approximately 6 mm wide and 12 mm long. The hermaphroditic adults liberate eggs into the bronchi. Because sputum that contains eggs is frequently swallowed, the eggs are usually excreted in feces of the definitive host. If the life cycle is to continue, the eggs must reach a body of water. A series of steps occurs that ensure adult flukes can mature in the lungs of a new host. The life cycle is shown in Figure 12.25.

In a laboratory diagnosis, sputum and feces are examined microscopically for fluke eggs. Infection results from eating undercooked crayfish, and the disease can be prevented by thoroughly cooking crayfish.

The cercariae of the blood fluke *Schistosoma* (shis-tō-sō'ma) are not ingested. Instead, they burrow through the skin of the human host and enter the circulatory system. The adults are found in certain abdominal and pelvic veins. The disease schistosomiasis is a major world health problem; it will be discussed further in Chapter 23 (page 666).

Cestodes

Cestodes, or **tapeworms,** are intestinal parasites. Their structure is shown in Figure 12.26. The head, or **scolex** (plural: *scoleces*), has suckers for attaching to the intestinal mucosa of the definitive host; some species also have small hooks for attachment. Tapeworms do not ingest the tissues of their hosts; in fact, they completely lack a digestive system. To obtain nutrients from the small intestine, they absorb food through their cuticle. The body consists of segments called **proglottids.** Proglottids are continu-

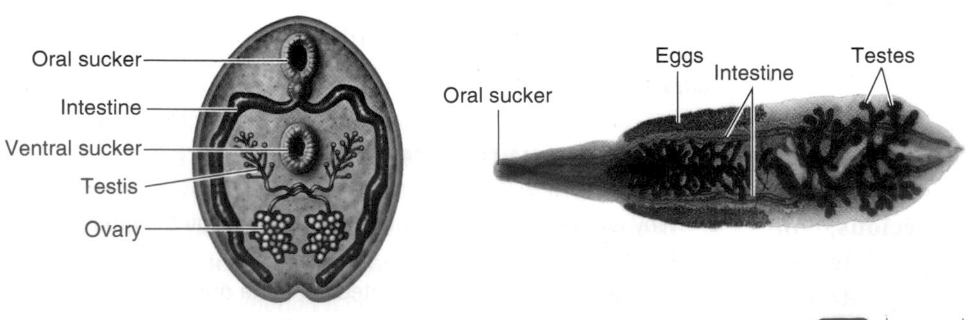

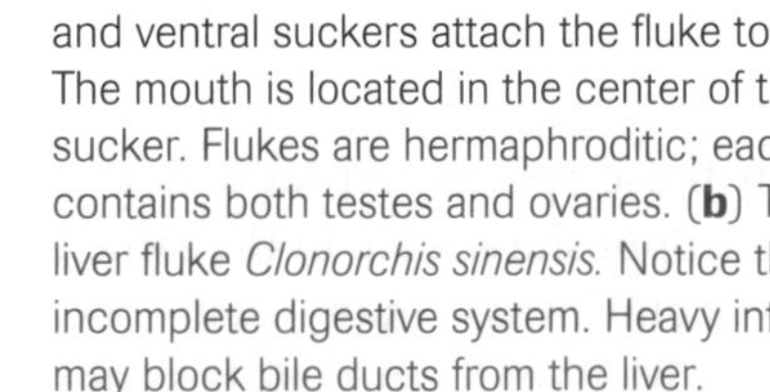

Figure 12.24 Flukes. (**a**) General anatomy of an adult fluke, shown in cross section. The oral and ventral suckers attach the fluke to the host. The mouth is located in the center of the oral sucker. Flukes are hermaphroditic; each animal contains both testes and ovaries. (**b**) The Asian liver fluke *Clonorchis sinensis.* Notice the incomplete digestive system. Heavy infestations may block bile ducts from the liver.

Q Why is the flatworm digestive system called "incomplete"?

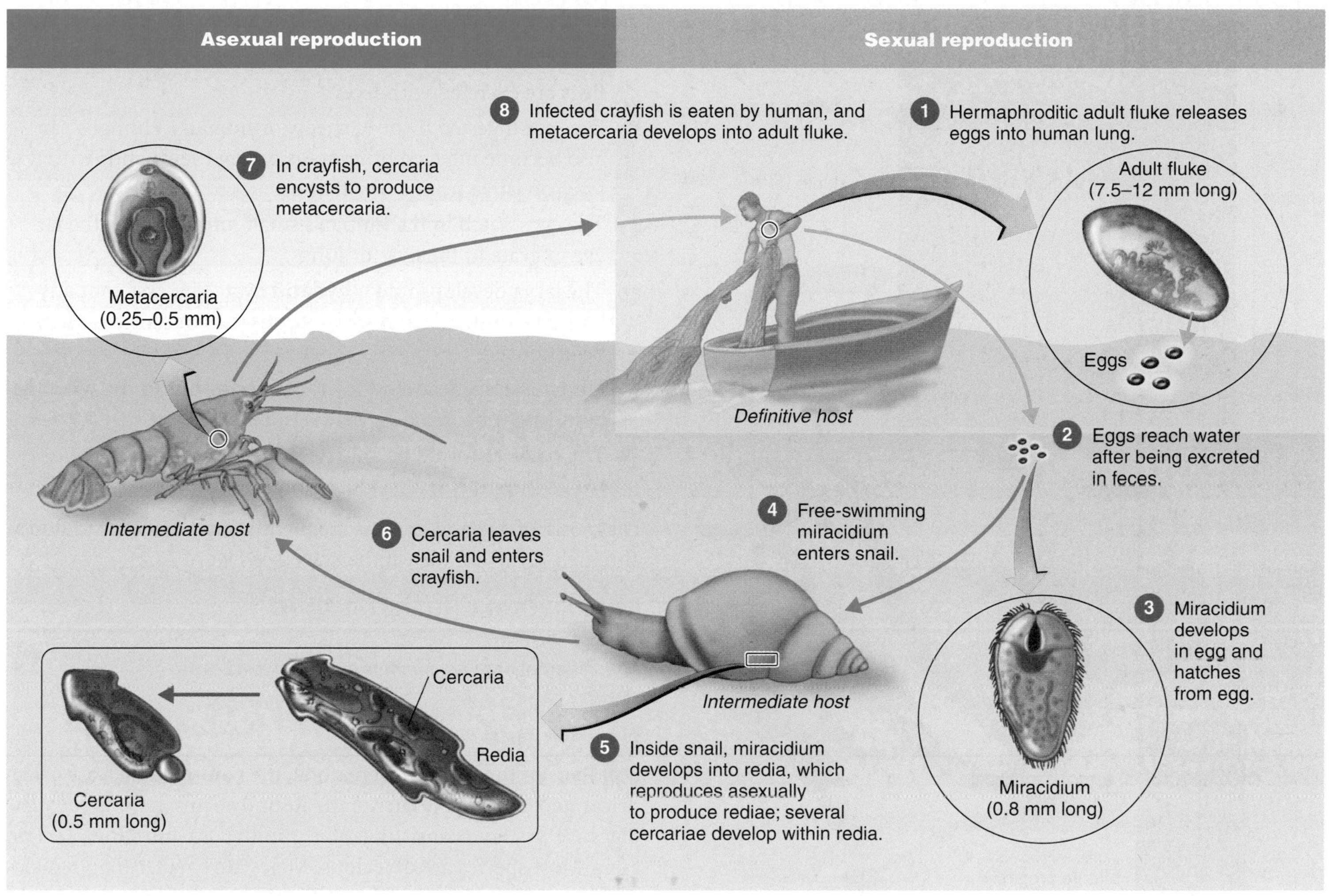

Figure 12.25 The life cycle of the lung fluke *Paragonimus westermani*. The trematode reproduces sexually in a human and asexually in a snail, its first intermediate host. Larvae encysted in the second intermediate host, a crayfish, infect humans when ingested. Also see the *Schistosoma* life cycle in Figure 23.27 (page 667).

Q Of what value is this complex life cycle to *Paragonimus*?

ally produced by the neck region of the scolex, as long as the scolex is attached and alive. Each mature proglottid contains both male and female reproductive organs. The proglottids farthest away from the scolex are the mature ones containing eggs. Mature proglottids are essentially bags of eggs, each of which is infective to the proper intermediate host.

Humans as Definitive Hosts The adults of *Taenia saginata* (te′nē-ä sa-ji-nä′tä), the beef tapeworm, live in humans and can reach a length of 6 meters. The scolex is about 2 millimeters long and is followed by a thousand or more proglottids. The feces of an infected human contain mature proglottids, each of which contains thousands of eggs. As the proglottids wriggle away from the fecal material, they increase their chances of being ingested by an animal that is grazing. Upon ingestion by cattle, the larvae hatch from the eggs and bore through the intestinal wall. The larvae migrate to muscle (meat), in which they encyst as **cysticerci.** When the cysticerci are ingested by humans, all but the scolex is digested. The scolex anchors itself in the small intestine and begins producing proglottids.

Diagnosis of tapeworm infection in humans is based on the presence of mature proglottids and eggs in feces. Cysticerci can be seen macroscopically in meat; their presence is referred to as "measly beef." Inspecting beef that is intended for human consumption for "measly" appearance is one way to prevent infections by beef tapeworm. Another method of prevention is to avoid the use of untreated human sewage as fertilizer in grazing pastures.

Humans are the only known definitive host of the pork tapeworm, *Taenia solium.* Adult worms living in the human intestine produce eggs, which are passed out in feces. When eggs are eaten by pigs, the larval helminth encysts in the pig's muscles; humans

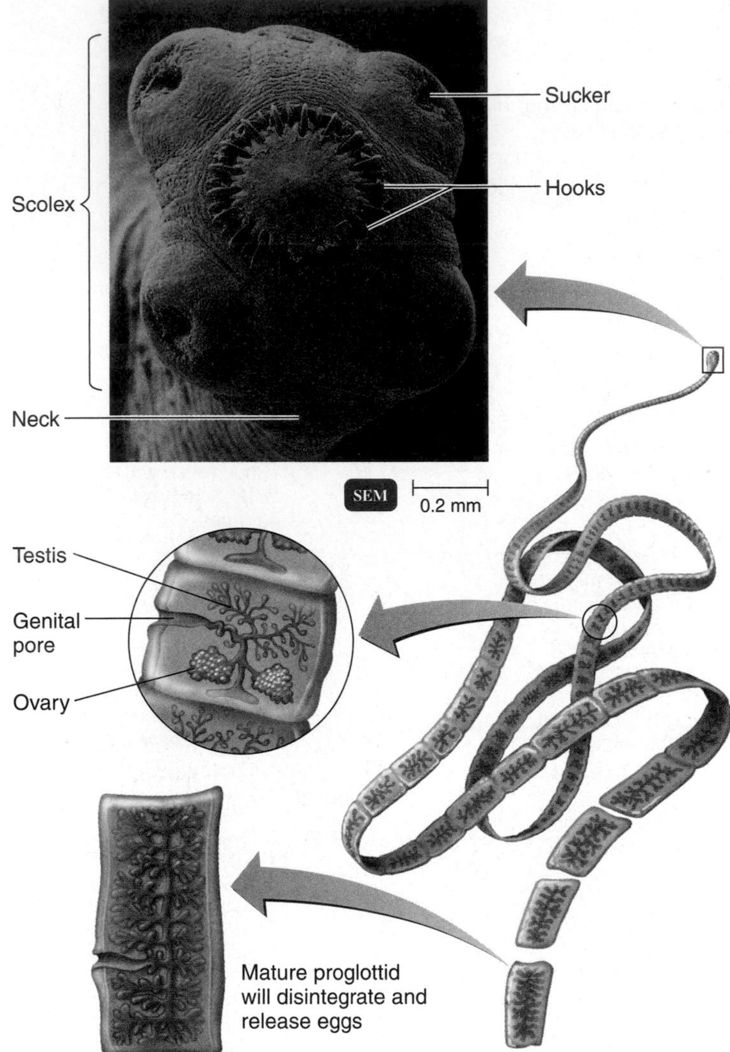

Figure 12.26 General anatomy of an adult tapeworm. The scolex, shown in the micrograph, consists of suckers and hooks that attach to the host's tissues. The body lengthens as new proglottids form at the neck. Each mature proglottid contains both testes and ovaries.

Q What are the similarities between tapeworms and flukes?

become infected when they eat undercooked pork. The human-pig-human cycle of *T. solium* is common in Latin America, Asia, and Africa. In the United States, however, *T. solium* is virtually nonexistent in pigs; the parasite is transmitted from human to human. Eggs shed by one person and ingested by another person hatch, and the larvae encyst in the brain and other parts of the body, causing cysticercosis (see Figure 25.22, page 733). The human hosting *T. solium*'s larvae is serving as an intermediate host. Approximately 7% of the few hundred cases reported in recent years were acquired by people who had never been outside the United States. They may have become infected through household contact with people who were born in or had traveled in other countries.

Humans as Intermediate Hosts Humans are the intermediate hosts for *Echinococcus granulosus* (ē-kīn-ō-kok'kus gra-nū-lō'sus), shown in Figure 12.27. Dogs and coyotes are the definitive hosts for this minute (2–8 mm) tapeworm.

1. Eggs are excreted with feces.
2. Eggs are ingested by deer, sheep, or humans. Humans can also become infected by contaminating their hands with dog feces or saliva from a dog that has licked itself.
3. The eggs hatch in the human's small intestine, and the larvae migrate to the liver or lungs.
4. The larva develops into a **hydatid cyst.** The cyst contains "brood capsules," from which thousands of scoleces might be produced.
5. Humans are a dead-end for the parasite, but in the wild, the cysts might be in a deer that is eaten by a wolf.
6. The scoleces would be able to attach themselves in the wolf's intestine and produce proglottids.

Diagnosis of hydatid cysts is frequently made only on autopsy, although X rays can detect the cysts (see Figure 25.21, page 733).

CHECK YOUR UNDERSTANDING

✓ Differentiate *Paragonimus* and *Taenia.* **12-16**

Nematodes

Members of the Phylum Nematoda, the **roundworms,** are cylindrical and tapered at each end. Roundworms have a *complete* digestive system, consisting of a mouth, an intestine, and an anus. Most species are dioecious. Males are smaller than females and have one or two hardened **spicules** on their posterior ends. Spicules are used to guide sperm to the female's genital pore.

Some species of nematodes are free-living in soil and water, and others are parasites on plants and animals. Some nematodes pass their entire life cycle, from egg to mature adult, in a single host.

Nematode infections of humans can be divided into two categories: those in which the egg is infective, and those in which the larva is infective.

Eggs Infective for Humans

The pinworm *Enterobius vermicularis* (en-te-rō'bē-us ver-mi-kū-lar'is) spends its entire life in a human host (Figure 12.28). Adult pinworms are found in the large intestine. From there, the female pinworm migrates to the anus to deposit her eggs on the perianal skin. The eggs can be ingested by the host or by another person exposed through contaminated clothing or bedding. Pinworm infections are diagnosed by the Graham sticky-tape method. A piece of transparent tape is placed on the perianal skin in such a way that the sticky side picks up eggs that were deposited earlier. The tape is then microscopically examined for the presence of eggs.

Ascaris lumbricoides (as'kar-is lum-bri-koi'dēz) is a large nematode (30 cm in length) that infects over one billion people worldwide (Figure 25.24, page 736). It is dioecious with **sexual dimorphism;** that is, the male and female worms look distinctly

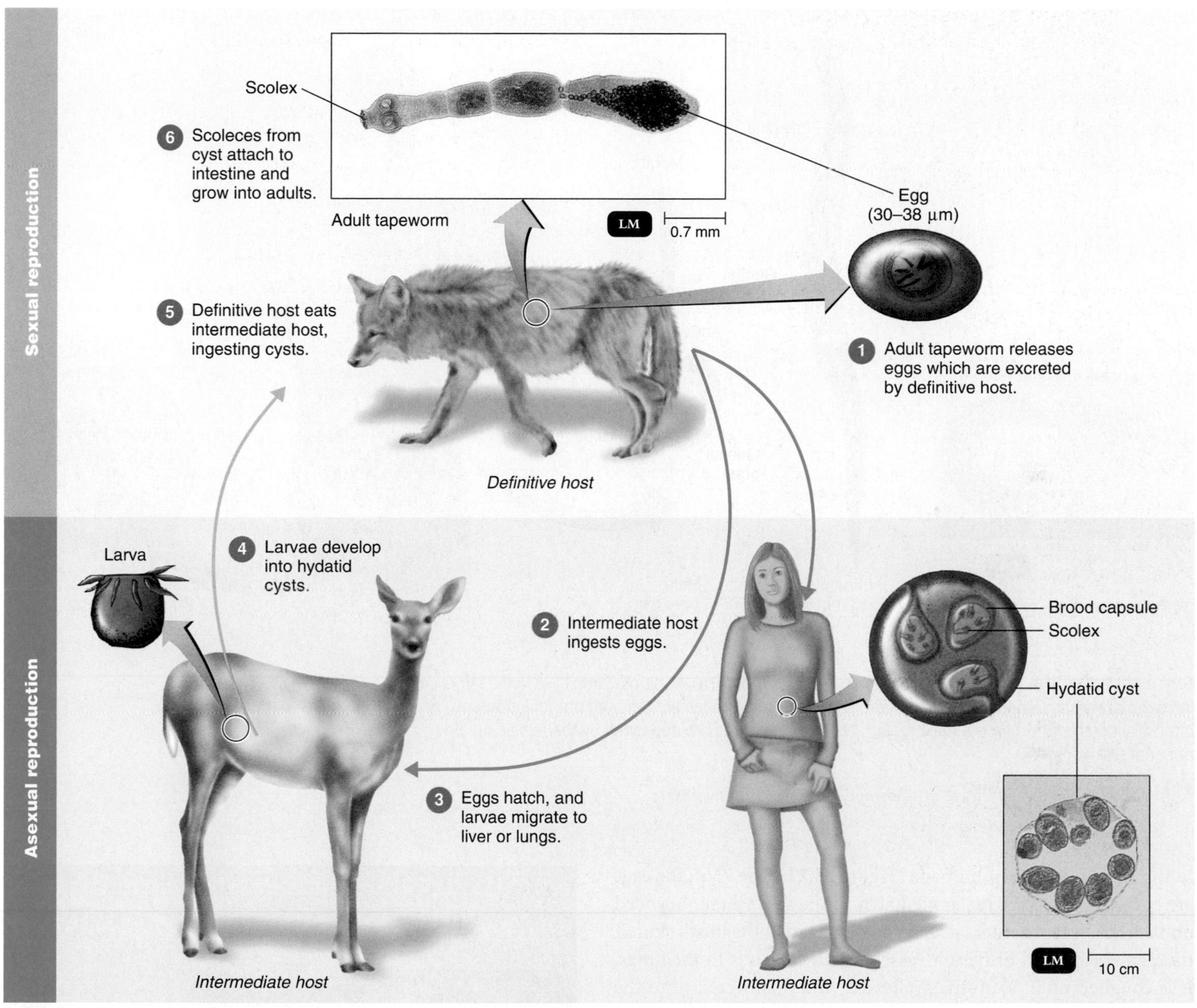

Sexual reproduction
Asexual reproduction
Scolex
6 Scoleces from cyst attach to intestine and grow into adults.
Adult tapeworm
LM
0.7 mm
Egg (30–38 µm)
5 Definitive host eats intermediate host, ingesting cysts.
1 Adult tapeworm releases eggs which are excreted by definitive host.
Definitive host
Larva
4 Larvae develop into hydatid cysts.
2 Intermediate host ingests eggs.
Brood capsule
Scolex
Hydatid cyst
3 Eggs hatch, and larvae migrate to liver or lungs.
LM
10 cm
Intermediate host
Intermediate host

Figure 12.27 Life cycle of the tapeworm *Echinococcus granulosus*. This tiny tapeworm is found in the intestines of dogs, wolves, and foxes. The adult of the closely related *Echinococcus multilocularis.* The photomicrograph shows a hydatid cyst. The parasite can complete its life cycle only if the cysts are ingested by a definitive host that eats the intermediate host.

Q Why isn't being in a human of benefit to *Echinococcus*?

different, the male being smaller with a curled tail. The adult *Ascaris* lives in the small intestines of humans exclusively; it feeds primarily on semidigested food. Eggs, excreted with feces, can survive in the soil for long periods until accidentally ingested by another host. The eggs hatch in the small intestine of the host. The larvae then burrow out of the intestine and enter the blood. They are carried to the lungs, where they grow. The larvae will then be coughed up, swallowed, and returned to the small intestine, where they mature into adults.

Diagnosis is frequently made when the adult worms are excreted with feces. *Ascaris* eggs remain in soil for 10 years. Children become infected by putting their hands and toys in their mouths. Preventing infection in humans is managed by proper sanitary habits.

Larvae Infective for Humans

Adult hookworms, *Necator americanus* (ne-kā′tôr ä-me-ri-ka′nus) and *Ancylostoma duodenale* (an-sil-os′toma dü′o-den-al-ē), live

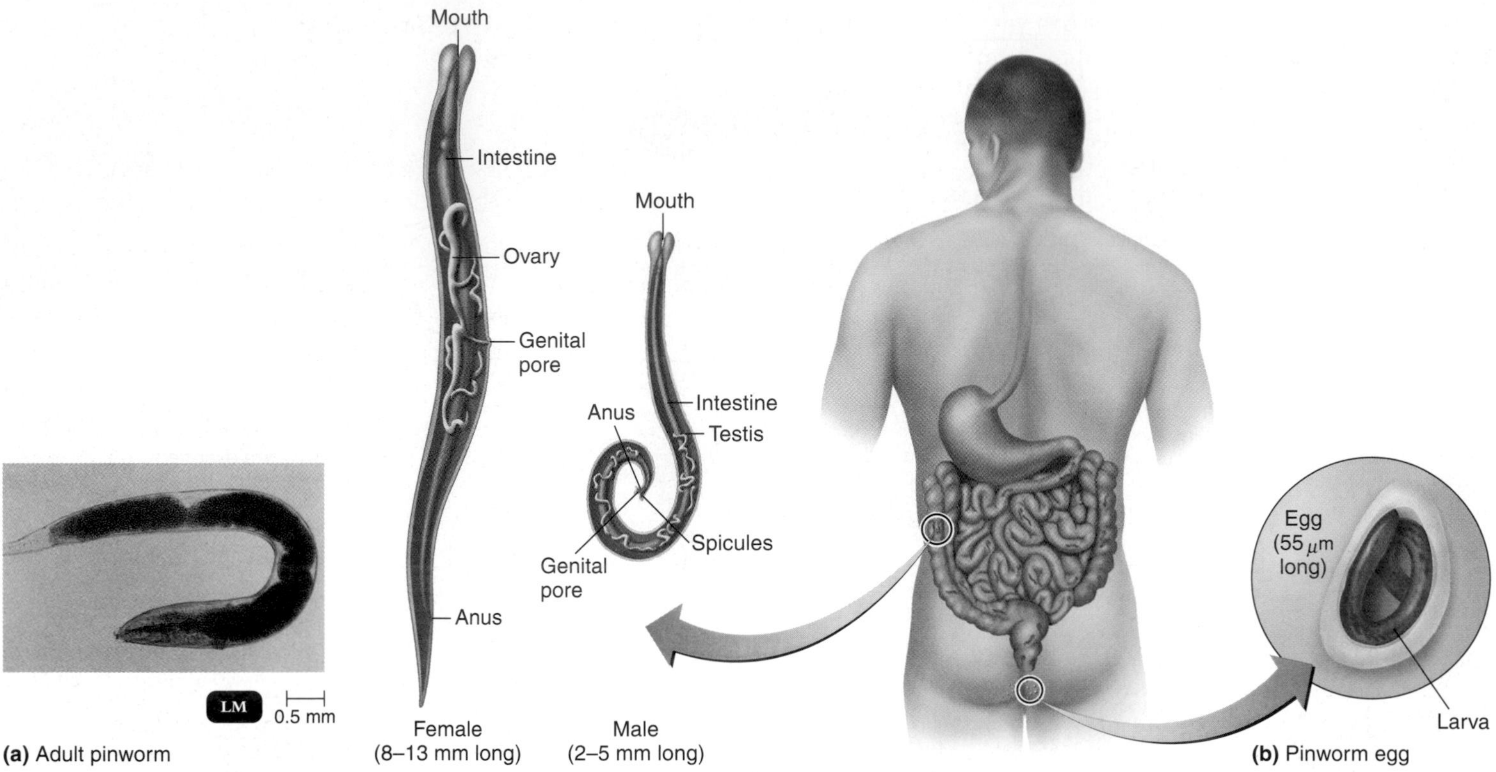

Figure 12.28 The pinworm *Enterobius vermicularis.* (**a**) Adult pinworms live in the large intestine of humans. Most roundworms are dioecious, and the female (left and photomicrograph) is often distinctly larger than the male (right). (**b**) Pinworm eggs are deposited by the female on the perianal skin at night.

Q Are humans the definitive or intermediate host for pinworms?

in the small intestine of humans (Figure 25.23, page 736); the eggs are excreted in feces. The larvae hatch in the soil, where they feed on bacteria. A larva enters its host by penetrating the host's skin. It then enters a blood or lymph vessel, which carries it to the lungs. It is coughed up in sputum, swallowed, and finally carried to the small intestine. Diagnosis is based on the presence of eggs in feces. People can avoid hookworm infections by wearing shoes.

Trichinellosis is caused by a nematode that the host acquires by eating encysted larvae in undercooked meat of infected animals (see pages 736–737). The nematode, *Dirofilaria immitis* (dir′ō-fi-lãr-ē-ä im′mi-tis), is spread from host to host through the bites of *Aedes* mosquitoes. It primarily affects dogs and cats, but it can infest human lungs. Larvae injected by the mosquito migrate to various organs, where they mature into adults. The parasitic worm is called a **heartworm** because the adult stage is often in the animal host's heart, where it can kill its host through congestive heart failure (Figure 12.29). The disease occurs on every continent except Antarctica. *Wolbachia* bacteria appear to be essential to development of the worm embryos (see the box in Chapter 11 on page 307.)

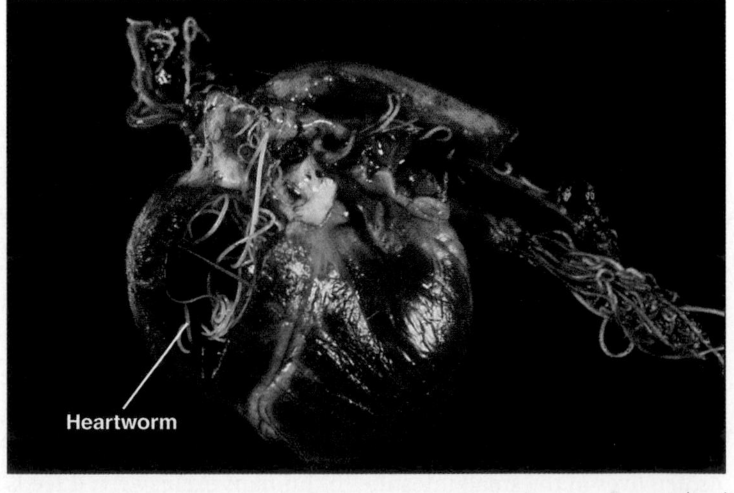

Figure 12.29 The heartworm *Dirofilaria immitis.* Four adult *D. immitis* in the right ventricle of a dog's heart. Each worm is 12–30 centimeters long.

Q How do roundworms and flatworms differ?

Four genera of roundworms called *anisakines*, or wriggly worms, can be transmitted to humans from infected fish and squid. Anisakine larvae are in the fish's intestinal mesenteries and migrate to the muscle when the fish dies. Freezing or thorough cooking will kill the larvae.

Table 12.6 lists representative parasitic helminths of each phylum and class and the diseases they cause.

CHECK YOUR UNDERSTANDING

- ✓ What is the definitive host for *Enterobius*? **12-17**
- ✓ What stage of *Dirofilaria immitis* is infectious for dogs and cats? **12-18**
- ✓ You find a parasitic worm in a baby's diapers. How would you know whether it's a *Taenia* or a *Necator*? **12-19**

Arthropods as Vectors

LEARNING OBJECTIVES

12-20 Define *arthropod vector*.

12-21 Differentiate a tick from a mosquito, and name a disease transmitted by each.

Arthropods are animals characterized by segmented bodies, hard external skeletons, and jointed legs. With nearly 1 million species, this is the largest phylum in the animal kingdom. Although arthropods are not microbes themselves, we will briefly describe them here because a few suck the blood of humans and other animals and can transmit microbial diseases while doing so. Arthropods that carry pathogenic

Table 12.6 Representative Parasitic Helminths

Phylum	Class	Human Parasites	Intermediate Host	Definitive Host Site	Stage Passed to Humans; Method	Disease	Location in Humans	Figure Reference
Platyhelminthes	**Trematodes**	*Paragonimus westermani*	Freshwater snails and crayfish	Humans; lungs	Metacercaria in crayfish; ingested	Paragonimiasis (lung fluke)	Lungs	12.25
		Schistosoma	Freshwater snails	Humans	Cercariae; through skin	Schistosomiasis	Veins	23.27 23.28
	Cestodes	*Taenia saginata*	Cattle	Humans; small intestine	Cysticerci in beef; ingested	Tapeworm	Small intestine	—
		Taenia solium	Humans; pigs	Humans	Eggs; ingested	Neurocysti-cercosis	Brain; any tissue	25.22
		Echinococcus granulosus	Humans	Dogs and other animals; intestines	Eggs from other animals; ingested	Hydatidosis	Lungs, liver brain	12.27, 25.23
Nematoda		*Ascaris lumbricoides*	—	Humans; small intestine	Eggs; ingested	Ascariasis	Small intestine	25.25
		Enterobius vermicularis	—	Humans; large intestine	Eggs; ingested	Pinworm	Large intestine	12.28
		Necator americanus	—	Humans; small intestine	Larvae; through skin	Hookworm	Small intestine	25.24
		Ancylostoma duodenale	—	Humans; small intestine	Larvae; through skin	Hookworm	Small Intestine	25.24
		Trichinella spiralis	—	Humans, pigs, and other mammals; small intestine	Larvae; ingested	Trichinellosis	Muscles	25.26
		Anisakines	Marine fish and squid	Marine mammals	Larvae in fish; ingested	Anisakiasis (sashimi worms)	Gastrointestinal tract	—

Table 12.7 Important Arthropod Vectors of Human Diseases

Class	Order	Vector	Disease	Figure Reference
Arachnida	Mites and ticks	*Dermacentor* (tick)	Rocky Mountain spotted fever	—
		Ixodes (tick)	Lyme disease, babesiosis, ehrlichiosis	12.31
		Ornithodorus (tick)	Relapsing fever	—
Insecta	Sucking lice	*Pediculus* (human louse)	Epidemic typhus, relapsing fever	12.32a
	Fleas	*Xenopsylla* (rat flea)	Endemic murine typhus, plague	12.32b
	True flies	*Chrysops* (deer fly)	Tularemia	12.32c
		Aedes (mosquito)	Dengue fever, yellow fever, heartworm	12.29
		Anopheles (mosquito)	Malaria	12.30
		Culex (mosquito)	Arboviral encephalitis	—
		Glossina (tsetse fly)	African trypanosomiasis	—
	True bugs	*Triatoma* (kissing bug)	Chagas' disease	12.32d

microorganisms are called **vectors.** Scabies and pediculosis are diseases that are caused by arthropods (see Chapter 21, pages 602–603).

Representative classes of arthropods include the following:

- Arachnida (eight legs): spiders, mites, ticks
- Crustacea (four antennae): crabs, crayfish
- Insecta (six legs): bees, flies, lice

Table 12.7 lists those arthropods that are important vectors, and Figures 12.30, 12.31, and 12.32 illustrate some of them. These insects and ticks reside on an animal only when they are feeding. An exception to this is the louse, which spends its entire life on its host and cannot survive for long away from a host.

Some vectors are just a mechanical means of transport for a pathogen. For example, houseflies lay their eggs on decaying organic matter, such as feces. While doing so, a housefly can pick up a pathogen on its feet or body and transport the pathogen to our food.

Some parasites multiply in their vectors. When this happens, the parasites can accumulate in the vector's feces or saliva. Large numbers of parasites can then be deposited on or in the host while the vector is feeding there. The spirochete that causes Lyme disease is transmitted by ticks in this manner (see Chapter 23, page 651), and the West Nile virus is transmitted in the same way by mosquitoes (see Chapter 22, page 626).

As discussed earlier, *Plasmodium* is an example of a parasite that requires that its vector also be the definitive host. *Plasmodium* can sexually reproduce only in the gut of an

Figure 12.30 Mosquito. A female mosquito sucking blood from human skin. Mosquitoes transmit several pathogens from person to person, including the yellow fever and West Nile viruses.

Q When is a vector also a definitive host?

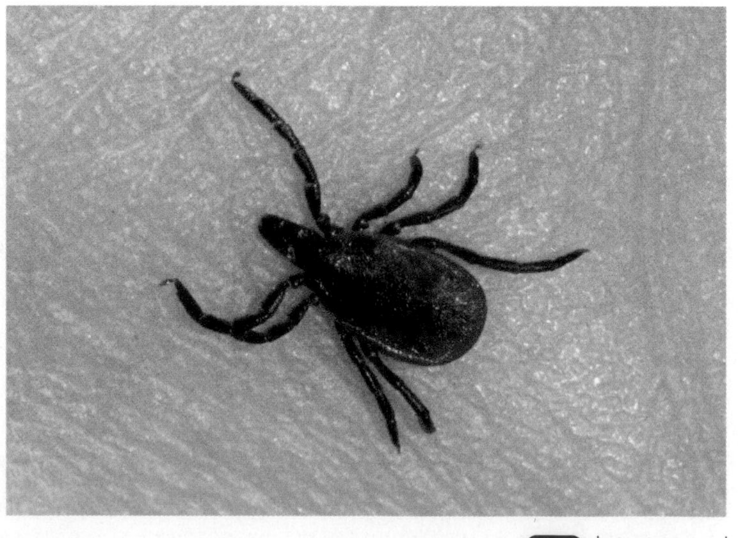

Figure 12.31 Tick. *Ixodes pacificus* is the Lyme disease vector on the West Coast.

Q Why aren't ticks classified as insects?

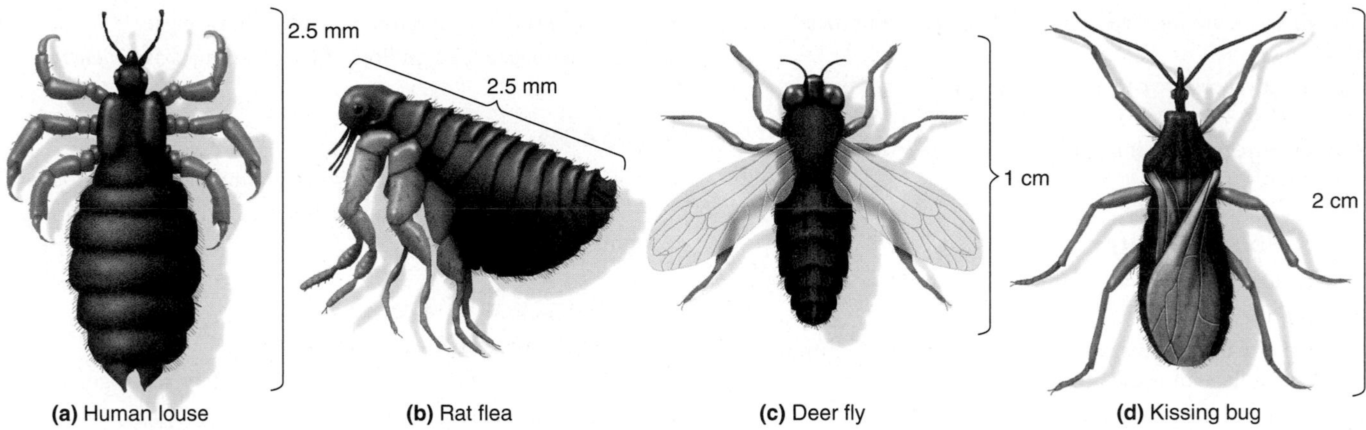

Figure 12.32 Arthropod vectors. (**a**) The human louse, *Pediculus.* (**b**) The rat flea, *Xenopsylla.* (**c**) The deer fly, *Chrysops.* (**d**) The kissing bug, *Triatoma.*

Q **Name one pathogen carried by each of these vectors.**

Anopheles mosquito. *Plasmodium* is introduced into a human host with the mosquito's saliva, which acts as an anticoagulant that keeps blood flowing.

To eliminate vectorborne diseases (such as African sleeping sickness), health workers focus on eradicating the vectors.

CHECK YOUR UNDERSTANDING

✓ Vectors can be divided into three major types, according to the roles they play for the parasite. List the three types of vectors and a disease transmitted by each. **12-20**

✓ Assume you see an arthropod on your arm. How will you determine whether it is a tick or a flea? **12-21**

STUDY OUTLINE

The **MyMicrobiologyPlace** website (**www.microbiologyplace.com**) will help you get ready for tests with its simple three-step approach: **1** **take a pre-test** and obtain a personalized study plan, **2** **learn and practice** with animations, tutorials, and MP3 tutor sessions, and **3** **test yourself** with quizzes and a chapter post-test.

Fungi (pp. 330–339)

1. Mycology is the study of fungi.
2. The number of serious fungal infections is increasing.
3. Fungi are aerobic or facultatively anaerobic chemoheterotrophs.
4. Most fungi are decomposers, and a few are parasites of plants and animals.

Characteristics of Fungi (pp. 331–333)

5. A fungal thallus consists of filaments of cells called hyphae; a mass of hyphae is called a mycelium.
6. Yeasts are unicellular fungi. To reproduce, fission yeasts divide symmetrically, whereas budding yeasts divide asymmetrically.
7. Buds that do not separate from the mother cell form pseudohyphae.
8. Pathogenic dimorphic fungi are yeastlike at 37°C and moldlike at 25°C.
9. Fungi are classified according to rRNA.
10. Sporangiospores and conidiospores are produced asexually.
11. Sexual spores are usually produced in response to special circumstances, often changes in the environment.
12. Fungi can grow in acidic, low-moisture, aerobic environments.
13. They are able to metabolize complex carbohydrates.

Medically Important Phyla of Fungi (pp. 333–335)

14. The Zygomycota have coenocytic hyphae and produce sporangiospores and zygospores.
15. The Ascomycota have septate hyphae and produce ascospores and frequently conidiospores.
16. Basidiomycota have septate hyphae and produce basidiospores; some produce conidiospores.
17. Teleomorphic fungi produce sexual and asexual spores; anamorphic fungi produce asexual spores only.

Fungal Diseases (pp. 335–339)

18. Systemic mycoses are fungal infections deep within the body that affect many tissues and organs.
19. Subcutaneous mycoses are fungal infections beneath the skin.
20. Cutaneous mycoses affect keratin-containing tissues such as hair, nails, and skin.

21. Superficial mycoses are localized on hair shafts and superficial skin cells.
22. Opportunistic mycoses are caused by fungi that are not usually pathogenic.
23. Opportunistic mycoses can infect any tissues. However, they are usually systemic.

Economic Effects of Fungi (p. 339)

24. *Saccharomyces* and *Trichoderma* are used in the production of foods.
25. Fungi are used for the biological control of pests.
26. Mold spoilage of fruits, grains, and vegetables is more common than bacterial spoilage of these products.
27. Many fungi cause diseases in plants.

Lichens (pp. 339–340)

1. A lichen is a mutualistic combination of an alga (or a cyanobacterium) and a fungus.
2. The alga photosynthesizes, providing carbohydrates for the lichen; the fungus provides a holdfast.
3. Lichens colonize habitats that are unsuitable for either the alga or the fungus alone.
4. Lichens may be classified on the basis of morphology as crustose, foliose, or fruticose.

Algae (pp. 340–345)

1. Algae are unicellular, filamentous, or multicellular (thallic).
2. Most algae live in aquatic environments.

Characteristics of Algae (pp. 341–342)

3. Algae are eukaryotic; most are photoautotrophs.
4. The thallus of multicellular algae usually consists of a stipe, a holdfast, and blades.
5. Algae reproduce asexually by cell division and fragmentation.
6. Many algae reproduce sexually.
7. Photoautotrophic algae produce oxygen.
8. Algae are classified according to their structures and pigments.

Selected Phyla of Algae (pp. 342–344)

9. Brown algae (kelp) may be harvested for algin.
10. Red algae grow deeper in the ocean than other algae.
11. Green algae have cellulose and chlorophyll *a* and *b* and store starch.
12. Diatoms are unicellular and have pectin and silica cell walls; some produce a neurotoxin.
13. Dinoflagellates produce neurotoxins that cause paralytic shellfish poisoning and ciguatera.
14. The oomycotes are heterotrophic; they include decomposers and pathogens.

Roles of Algae in Nature (p. 344–345)

15. Algae are the primary producers in aquatic food chains.
16. Planktonic algae produce most of the molecular oxygen in the Earth's atmosphere.
17. Petroleum is the fossil remains of planktonic algae.
18. Unicellular algae are symbionts in such animals as *Tridacna*.

Protozoa (pp. 345–351)

1. Protozoa are unicellular, eukaryotic chemoheterotrophs.
2. Protozoa are found in soil and water and as normal microbiota in animals.

Characteristics of Protozoa (p. 346)

3. The vegetative form is called a trophozoite.
4. Asexual reproduction is by fission, budding, or schizogony.
5. Sexual reproduction is by conjugation.
6. During ciliate conjugation, two haploid nuclei fuse to produce a zygote.
7. Some protozoa can produce a cyst that provides protection during adverse environmental conditions.
8. Protozoa have complex cells with a pellicle, a cytostome, and an anal pore.

Medically Important Phyla of Protozoa (pp. 346–351)

9. Archaezoa lack mitochondria and have flagella; they include *Trichomonas* and *Giardia.*
10. Microsporidia lack mitochondria and microtubules; microsporans cause diarrhea in AIDS patients.
11. Amoebozoa are amoeba; they include *Entamoeba* and *Acanthamoeba.*
12. Apicomplexa have apical organelles for penetrating host tissue; they include *Plasmodium* and *Cryptosporidium.*
13. Ciliophora move by means of cilia; *Balantidium coli* is the human parasitic ciliate.
14. Euglenozoa move by means of flagella and lack sexual reproduction; they include *Trypanosoma.*

Slime Molds (pp. 351–352)

1. Cellular slime molds resemble amoebas and ingest bacteria by phagocytosis.
2. Plasmodial slime molds consist of a multinucleated mass of protoplasm that engulfs organic debris and bacteria as it moves.

Helminths (pp. 352–361)

1. Parasitic flatworms belong to the Phylum Platyhelminthes.
2. Parasitic roundworms belong to the Phylum Nematoda.

Characteristics of Helminths (pp. 353–355)

3. Helminths are multicellular animals; a few are parasites of humans.
4. The anatomy and life cycle of parasitic helminths are modified for parasitism.
5. The adult stage of a parasitic helminth is found in the definitive host.

6. Each larval stage of a parasitic helminth requires an intermediate host.
7. Helminths can be monoecious or dioecious.

Platyhelminths (pp. 356–358)

8. Flatworms are dorsoventrally flattened animals; parasitic flatworms may lack a digestive system.
9. Adult trematodes, or flukes, have an oral and ventral sucker with which they attach to host tissue.
10. Eggs of trematodes hatch into free-swimming miracidia that enter the first intermediate host; two generations of rediae develop; the rediae become cercariae that bore out of the first intermediate host and penetrate the second intermediate host; cercariae encyst as metacercariae; the metacercariae develop into adults in the definitive host.
11. A cestode, or tapeworm, consists of a scolex (head) and proglottids.
12. Humans serve as the definitive host for the beef tapeworm, and cattle are the intermediate host.
13. Humans serve as the definitive host and can be an intermediate host for the pork tapeworm.
14. Humans serve as the intermediate host for *Echinococcus granulosus;* the definitive hosts are dogs, wolves, and foxes.

Nematodes (pp. 358–361)

15. Roundworms have a complete digestive system.
16. The nematodes that infect humans with their eggs include pinworm and *Ascaris.*
17. The nematodes that infect humans with their larvae include hookworm and *Trichinella.*

Arthropods as Vectors (pp. 361–363)

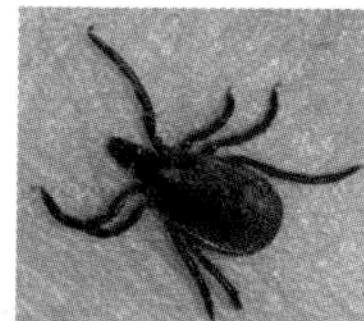

1. Jointed-legged animals, including ticks and insects, belong to the Phylum Arthropoda.
2. Arthropods that carry diseases are called vectors.
3. Vectorborne diseases are most effectively eliminated by controlling or eradicating the vectors.

STUDY QUESTIONS

Answers to the Review and Multiple Choice questions can be found by turning to the blue Answers tab at the back of the textbook.

Review

1. Following is a list of fungi, their methods of entry into the body, and sites of infections they cause. Categorize each type of mycosis as cutaneous, opportunistic, subcutaneous, superficial, or systemic.

Genus	Method of Entry	Site of Infection	Mycosis
Blastomyces	Inhalation	Lungs	(a) ________
Sporothrix	Puncture	Ulcerative lesions	(b) ________
Microsporum	Contact	Fingernails	(c) ________
Trichosporon	Contact	Hair shafts	(d) ________
Aspergillus	Inhalation	Lungs	(e) ________

2. A mixed culture of *Escherichia coli* and *Penicillium chrysogenum* is inoculated onto the following culture media. On which medium would you expect each to grow? Why?
   a. 0.5% peptone in tap water **b.** 10% glucose in tap water
3. Briefly discuss the importance of lichens in nature. Briefly discuss the importance of algae in nature.
4. Differentiate cellular and plasmodial slime molds. How does each survive adverse environmental conditions?
5. Complete the following table.

Phylum	Method of Motility	One Human Parasite
Archaezoa	(a) ________	(b) ________
Microspora	(c) ________	(d) ________
Amoebozoa	(e) ________	(f) ________
Apicomplexa	(g) ________	(h) ________
Ciliophora	(i) ________	(j) ________
Euglenozoa	(k) ________	(l) ________

6. Why is it significant that *Trichomonas* does not have a cyst stage? Name a protozoan parasite that does have a cyst stage.
7. By what means are helminthic parasites transmitted to humans?
8. **DRAW IT** A generalized life cycle of the liver fluke *Clonorchis sinensis* is shown below. Label the fluke's stages. Identify the intermediate host(s). Identify the definitive host(s). To what phylum and class does this animal belong?

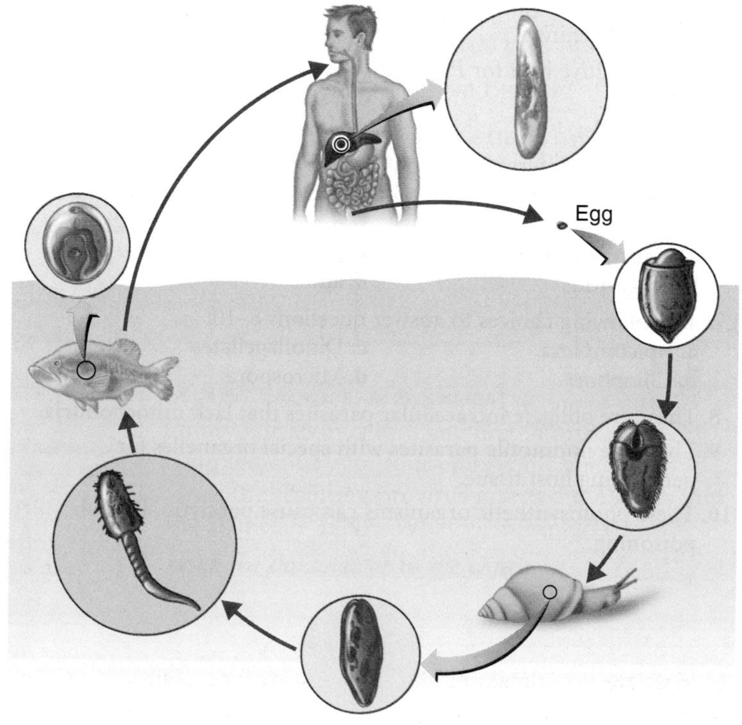

General Characteristics of Viruses

LEARNING OBJECTIVE

13-1 Differentiate a virus from a bacterium.

One hundred years ago, researchers could not imagine submicroscopic particles, and thus they described the infectious agent as *contagium vivum fluidum*—a contagious fluid. By the 1930s, scientists had begun using the word *virus*, the Latin word for poison, to describe these filterable agents. The nature of viruses, however, remained elusive until 1935, when Wendell Stanley, an American chemist, isolated tobacco mosaic virus, making it possible for the first time to carry out chemical and structural studies on a purified virus. At about the same time, the invention of the electron microscope made it possible to see viruses.

The question of whether viruses are living organisms has an ambiguous answer. Life can be defined as a complex set of processes resulting from the actions of proteins specified by nucleic acids. The nucleic acids of living cells are in action all the time. Because viruses are inert outside living host cells, in this sense they are not considered to be living organisms. However, once viruses enter a host cell, the viral nucleic acids become active, and viral multiplication results. In this sense, viruses are alive when they multiply in the host cells they infect. From a clinical point of view, viruses can be considered alive because they cause infection and disease, just as pathogenic bacteria, fungi, and protozoa do. Depending on one's viewpoint, a virus may be regarded as an exceptionally complex aggregation of nonliving chemicals, or as an exceptionally simple living microorganism.

How, then, do we define *virus*? Viruses were originally distinguished from other infectious agents because they are especially small (filterable) and because they are **obligatory intracellular parasites**—that is, they absolutely require living host cells in order to multiply. However, both of these properties are shared by certain small bacteria, such as some rickettsias. Viruses and bacteria are compared in **Table 13.1**.

The truly distinctive features of viruses are now known to relate to their simple structural organization and their mechanism of multiplication. Accordingly, **viruses** are entities that

- Contain a single type of nucleic acid, either DNA or RNA.
- Contain a protein coat (sometimes itself enclosed by an envelope of lipids, proteins, and carbohydrates) that surrounds the nucleic acid.
- Multiply inside living cells by using the synthesizing machinery of the cell.
- Cause the synthesis of specialized structures that can transfer the viral nucleic acid to other cells.

Viruses have few or no enzymes of their own for metabolism; for example, they lack enzymes for protein synthesis and ATP generation. To multiply, viruses must take over the metabolic machinery of the host cell. This fact has considerable medical significance for the development of antiviral drugs, because most drugs that would interfere with viral multiplication would also interfere with the functioning of the host cell and therefore are too toxic for clinical use. (Antiviral drugs are discussed in Chapter 20.)

Table 13.1 Viruses and Bacteria Compared

	Bacteria		Viruses
	Typical Bacteria	Rickettsias/ Chlamydias	
Intracellular Parasite	No	Yes	Yes
Plasma Membrane	Yes	Yes	No
Binary Fission	Yes	Yes	No
Pass through Bacteriological Filters	No	No/Yes	Yes
Possess Both DNA and RNA	Yes	Yes	No
ATP-Generating Metabolism	Yes	Yes/No	No
Ribosomes	Yes	Yes	No
Sensitive to Antibiotics	Yes	Yes	No
Sensitive to Interferon	No	No	Yes

Host Range

The **host range** of a virus is the spectrum of host cells the virus can infect. There are viruses that infect invertebrates, vertebrates, plants, protists, fungi, and bacteria. However, most viruses are able to infect specific types of cells of only one host species. In rare cases, viruses cross the host-range barrier, thus expanding their host range. An example is described in the box on page 370. In this chapter, we are concerned mainly with viruses that infect either humans or bacteria. Viruses that infect bacteria are called **bacteriophages,** or **phages.**

The particular host range of a virus is determined by the virus's requirements for its specific attachment to the host cell and the availability within the potential host of cellular factors required for viral multiplication. For the virus to infect the host cell, the outer surface of the virus must chemically interact with specific receptor sites on the surface of the cell. The two complementary components are held together by weak bonds, such as hydrogen bonds. The combination of many attachment and receptor sites leads to a strong association between host cell and virus. For some bacteriophages, the receptor site is part of the cell wall of the host; in other cases, it is part of the fimbriae or flagella. For animal viruses, the receptor sites are on the plasma membranes of the host cells.

The potential to use viruses to treat diseases is intriguing because of their narrow host range and their ability to kill their

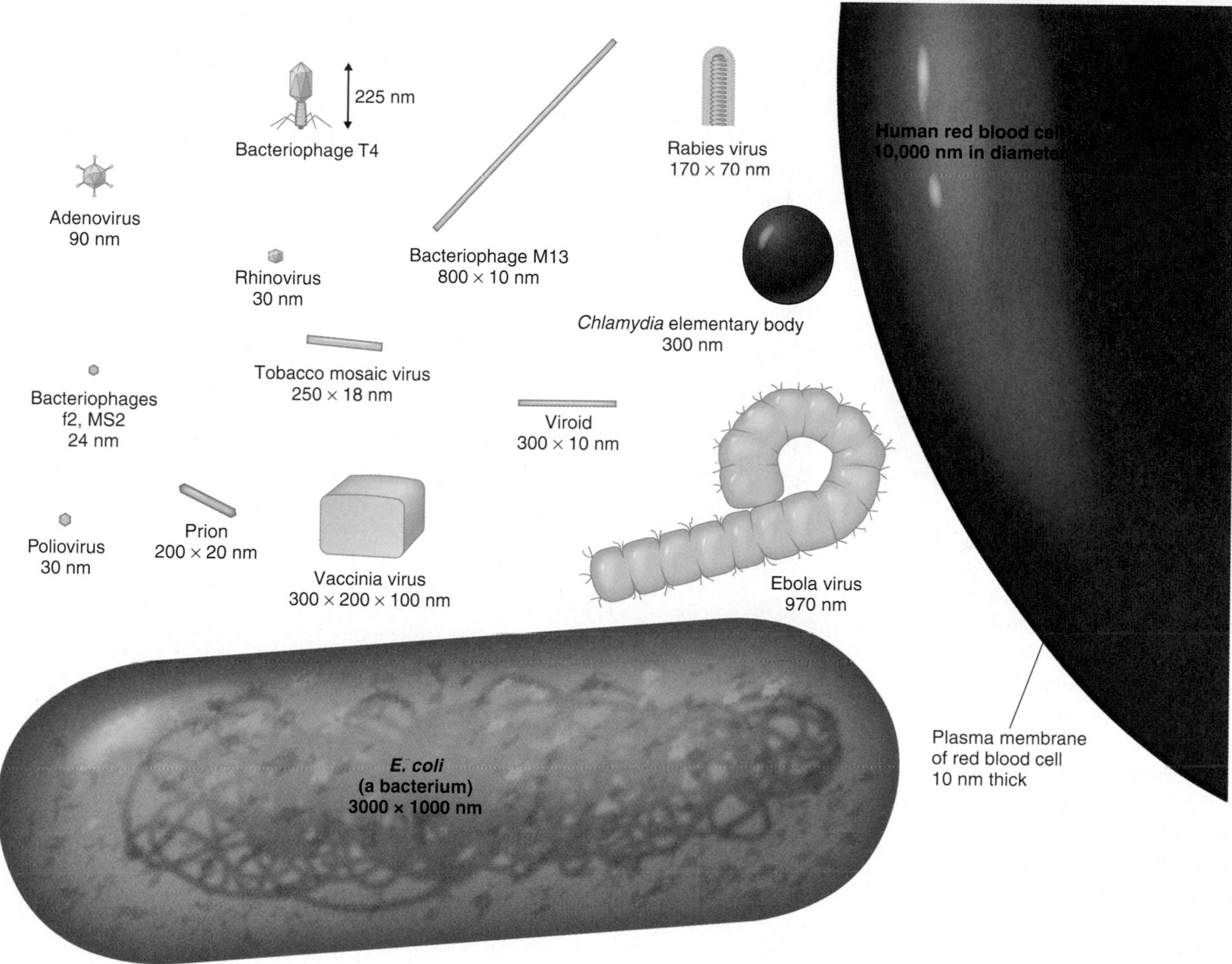

Figure 13.1 Virus sizes. The sizes of several viruses (teal blue) and bacteria (brown) are compared with a human red blood cell, shown to the right of the microbes. Dimensions are given in nanometers (nm) and are either diameters or length by width.

Q How do viruses differ from bacteria?

host cells. The idea of *phage therapy*—using bacteriophage to treat bacterial infections, has been around for 100 years. Recent advances in our understanding of virus-host interactions have fueled new studies in the field of phage therapy.

Experimentally induced viral infections in cancer patients during the 1920s suggested that viruses might have antitumor activity. These tumor-destroying, or *oncolytic,* viruses may selectively infect and kill tumor cells or cause an immune response against tumor cells. Some viruses naturally infect tumor cells, and other viruses can be genetically modified to infect tumor cells. At present several studies are underway to determine the killing mechanism of oncolytic viruses and the safety of using viral therapy.

Viral Size

Viral sizes are determined with the aid of electron microscopy. Different viruses vary considerably in size. Although most are quite a bit smaller than bacteria, some of the larger viruses (such as the vaccinia virus) are about the same size as some very small bacteria (such as the mycoplasmas, rickettsias, and chlamydias). Viruses range from 20 to 1000 nm in length. The comparative sizes of several viruses and bacteria are shown in Figure 13.1.

CHECK YOUR UNDERSTANDING

✓ How could the small size of viruses have helped researchers detect viruses before the invention of the electron microscope? **13-1**

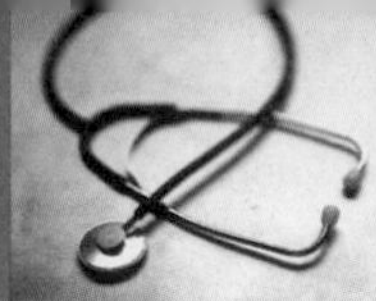

CLINICAL FOCUS From the *Morbidity and Mortality Weekly Report*

Influenza: Crossing the Species Barrier

Influenza A viruses are found in many different animals, including birds, pigs, whales, horses, and seals. Sometimes influenza A viruses seen in one species can cross over and cause illness in another species. For example, up until 1998, only H1N1 viruses circulated widely in the U.S. pig population. In 1998, H3N2 viruses from humans were introduced into the pig population and caused widespread disease among pigs. The subtypes differ because of certain proteins on the surface of the virus (hemagglutinin [HA] and neuraminidase [NA] proteins). There are 16 different HA subtypes and 9 different NA subtypes of influenza A viruses.

Now many different combinations of H and N proteins are possible?

Each combination is a different subtype. When we talk about "human flu viruses," we are referring to those subtypes that occur widely in humans. There are only three known subtypes of human influenza viruses (H1N1, H1N2, and H3N2).

What's different about the Table A viruses?

These subtypes occur mainly in birds. Avian influenza (bird flu) viruses do not usually infect humans. Avian influenza viruses may be transmitted to humans: (1) directly from birds or from avian-virus-contaminated environments or (2) through an intermediate host, such as a pig. Pigs are an important carrier because they can be infected with both human and avian flu. The influenza virus genome is composed of eight separate segments. A segmented genome allows virus genes to mix and create a new influenza A virus if viruses from two different species infect the same person or animal (see the figure). This is known as *antigenic shift.*

Why have there been so few human cases of avian influenza?

Thus far, avian flu viruses have not caused outbreaks in the human population because they are not infective via human-to-human transmission. All cases in Table A can be attributed to outbreaks in poultry except one noteworthy probable transmission from a daughter to her mother.

On what continents have H5N1 human infections occurred?

During the twentieth century, the emergence of new influenza A virus subtypes caused three pandemics, all of which spread around the world within one year of being detected (see Table B). It is likely that some genetic parts of these influenza A strains originally came from birds. Many scientists believe it is only a matter of time until the next influenza pandemic occurs.

Source: Adapted from *MMWR* sources.

Table A Recent Cases of Avian Influenza Virus Infections in Humans

Influenza virus	Location	Year	Human cases
H5N1	Southeast asia	2005–2007	350 sporadic
H5N1	Egypt	2006–2007	38
H5N1	China	2006–2007	25
H7N2	United Kingdom	2007	4
H5N1	Turkey	2006	2
H5N1	Iraq	2006	3
H5N1	Azerbaijan	2006	1
H5N1	Djibouti	2006	1
H5N1	Nigeria	2007	1
H5N1	Southeast Asia	2005	130 sporadic
H5N1	Southeast Asia	2004	35 sporadic
H7N3	Canada	2004	Eye infections among poultry workers
H7N2	New York	2003	1
H7N7	Netherlands	2003	89
H5N1	China	2003	2
H7N2	Virginia	2002	1
H9N2	China	1999	2
H5N1	China	1997	18

Viral Structure

LEARNING OBJECTIVE

13-2 Describe the chemical and physical structure of both an enveloped and a nonenveloped virus.

A **virion** is a complete, fully developed, infectious viral particle composed of nucleic acid and surrounded by a protein coat that protects it from the environment and is a vehicle of transmission from one host cell to another. Viruses are classified by differences in the structures of these coats.

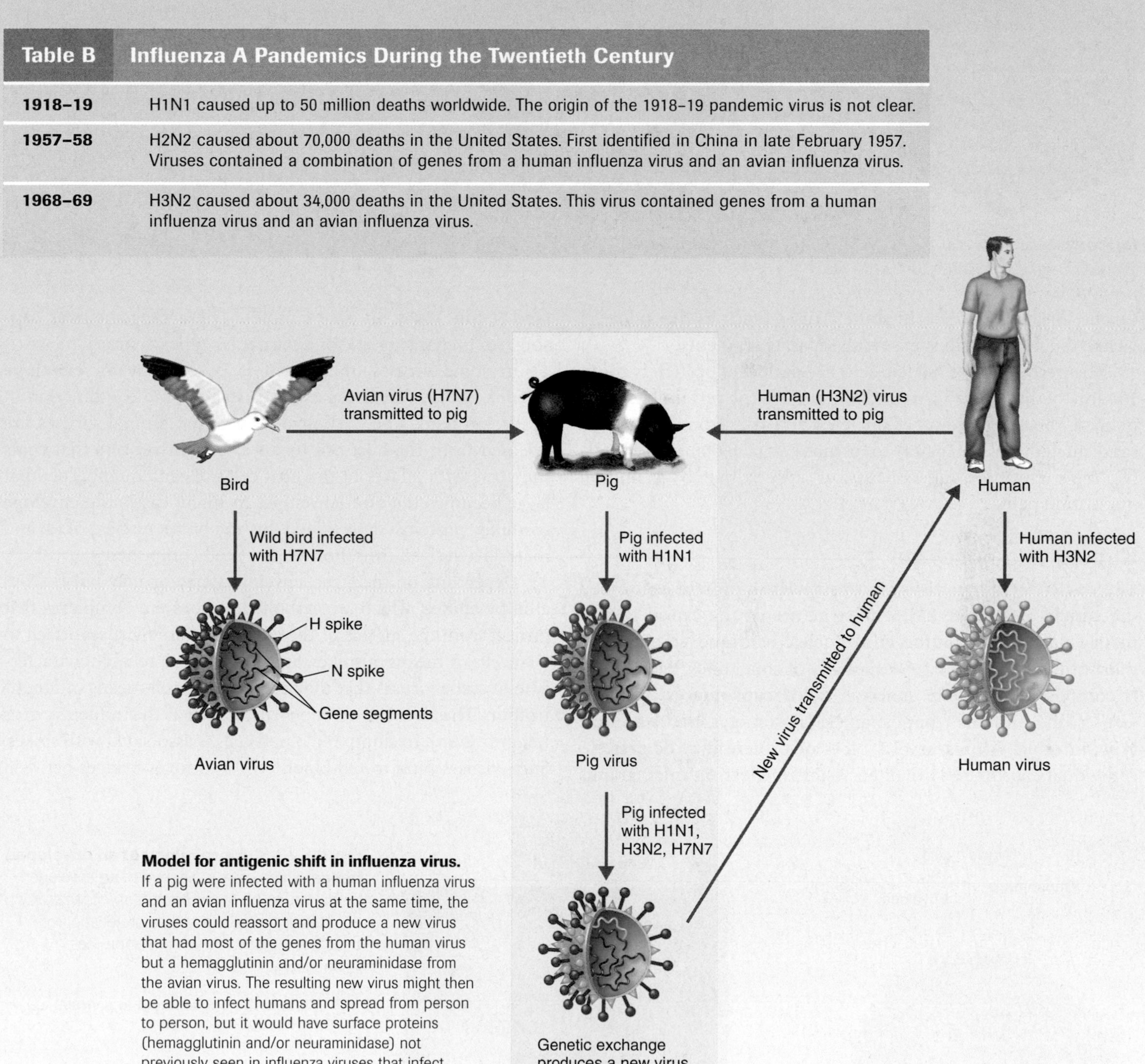

Table B	Influenza A Pandemics During the Twentieth Century
1918–19	H1N1 caused up to 50 million deaths worldwide. The origin of the 1918–19 pandemic virus is not clear.
1957–58	H2N2 caused about 70,000 deaths in the United States. First identified in China in late February 1957. Viruses contained a combination of genes from a human influenza virus and an avian influenza virus.
1968–69	H3N2 caused about 34,000 deaths in the United States. This virus contained genes from a human influenza virus and an avian influenza virus.

Model for antigenic shift in influenza virus. If a pig were infected with a human influenza virus and an avian influenza virus at the same time, the viruses could reassort and produce a new virus that had most of the genes from the human virus but a hemagglutinin and/or neuraminidase from the avian virus. The resulting new virus might then be able to infect humans and spread from person to person, but it would have surface proteins (hemagglutinin and/or neuraminidase) not previously seen in influenza viruses that infect humans.

Nucleic Acid

In contrast to prokaryotic and eukaryotic cells, in which DNA is always the primary genetic material (and RNA plays an auxiliary role), a virus can have either DNA or RNA—but never both. The nucleic acid of a virus can be single-stranded or double-stranded. Thus, there are viruses with the familiar double-stranded DNA, with single-stranded DNA, with double-stranded RNA, and with single-stranded RNA. Depending on the virus, the nucleic acid

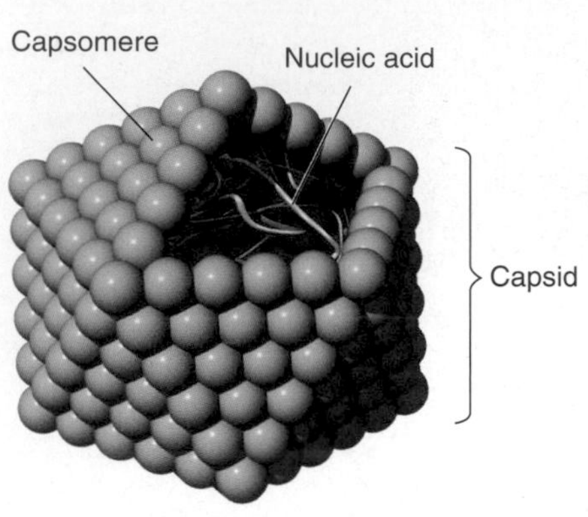

(a) A polyhedral virus

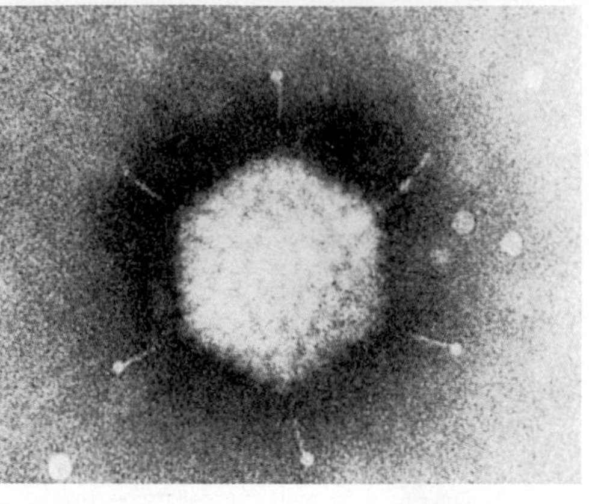

(b) *Mastadenovirus*

Figure 13.2 Morphology of a nonenveloped polyhedral virus. (a) A diagram of a polyhedral (icosahedral) virus. (b) A micrograph of the adenovirus *Mastadenovirus.* Individual capsomeres are visible.

Q **What is the chemical composition of a capsid?**

can be linear or circular. In some viruses (such as the influenza virus), the nucleic acid is in several separate segments.

The percentage of nucleic acid in relation to protein is about 1% for the influenza virus and about 50% for certain bacteriophages. The total amount of nucleic acid varies from a few thousand nucleotides (or pairs) to as many as 250,000 nucleotides. (*E. coli*'s chromosome consists of approximately 4 million nucleotide pairs.)

Capsid and Envelope

The nucleic acid of a virus is protected by a protein coat called the **capsid** (Figure 13.2a). The structure of the capsid is ultimately determined by the viral nucleic acid and accounts for most of the mass of a virus, especially of small ones. Each capsid is composed of protein subunits called **capsomeres.** In some viruses, the proteins composing the capsomeres are of a single type; in other viruses, several types of protein may be present. Individual capsomeres are often visible in electron micrographs (see Figure 13.2b for an example). The arrangement of capsomeres is characteristic of a particular type of virus.

In some viruses, the capsid is covered by an **envelope** (Figure 13.3a), which usually consists of some combination of lipids, proteins, and carbohydrates. Some animal viruses are released from the host cell by an extrusion process that coats the virus with a layer of the host cell's plasma membrane; that layer becomes the viral envelope. In many cases, the envelope contains proteins determined by the viral nucleic acid and materials derived from normal host cell components.

Depending on the virus, envelopes may or may not be covered by **spikes,** which are carbohydrate-protein complexes that project from the surface of the envelope. Some viruses attach to host cells by means of spikes. Spikes are such a reliable characteristic of some viruses that they can be used as a means of identification. The ability of certain viruses, such as the influenza virus (Figure 13.3b), to clump red blood cells is associated with spikes. Such viruses bind to red blood cells and form bridges between

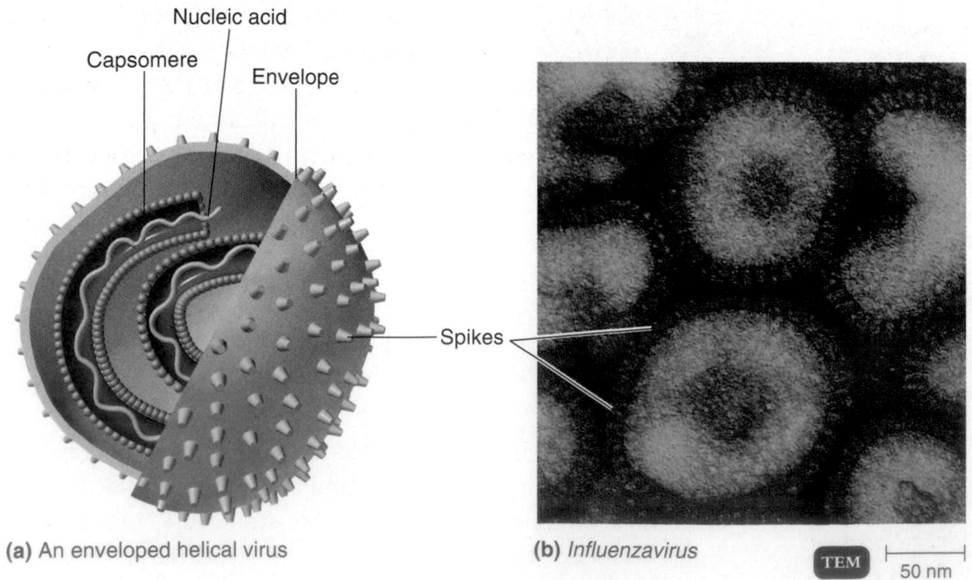

(a) An enveloped helical virus

(b) *Influenzavirus*

Figure 13.3 Morphology of an enveloped helical virus. (a) A diagram of an enveloped helical virus. (b) A micrograph of *Influenzavirus* A2. Notice the halo of spikes projecting from the outer surface of each envelope (see Chapter 24).

Q **What is the nucleic acid in a virus?**

Figure 13.4 Morphology of a helical virus. (**a**) A diagram of a portion of a helical virus. A row of capsomeres has been removed to reveal the nucleic acid. (**b**) A micrograph of Ebola virus, a filovirus showing a helical rod.

What is the chemical composition of a capsomere?

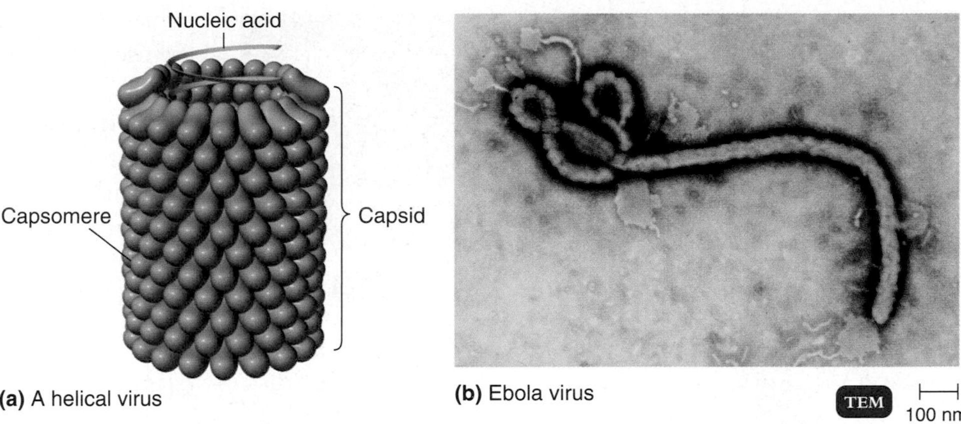

(**a**) A helical virus

(**b**) Ebola virus

them. The resulting clumping is called *hemagglutination* and is the basis for several useful laboratory tests. (See Figure 18.7, page 511.)

Viruses whose capsids are not covered by an envelope are known as **nonenveloped viruses** (see Figure 13.2). The capsid of a nonenveloped virus protects the nucleic acid from nuclease enzymes in biological fluids and promotes the virus's attachment to susceptible host cells.

When the host has been infected by a virus, the host immune system is stimulated to produce antibodies (proteins that react with the surface proteins of the virus). This interaction between host antibodies and virus proteins should inactivate the virus and stop the infection. However, some viruses can escape antibodies because regions of the genes that code for these viruses' surface proteins are susceptible to mutations. The progeny of mutant viruses have altered surface proteins, such that the antibodies are not able to react with them. Influenza virus frequently undergoes such changes in its spikes. This is why you can get influenza more than once. Although you may have produced antibodies to one influenza virus, the virus can mutate and infect you again.

General Morphology

Viruses may be classified into several different morphological types on the basis of their capsid architecture. The structure of these capsids has been revealed by electron microscopy and a technique called X-ray crystallography.

Helical Viruses

Helical viruses resemble long rods that may be rigid or flexible. The viral nucleic acid is found within a hollow, cylindrical capsid that has a helical structure (**Figure 13.4**). The viruses that cause rabies and Ebola hemorrhagic fever are helical viruses.

Polyhedral Viruses

Many animal, plant, and bacterial viruses are polyhedral, or many-sided, viruses. The capsid of most polyhedral viruses is in the shape of an *icosahedron,* a regular polyhedron with 20 triangular faces and 12 corners (see Figure 13.2a). The capsomeres of each face form an equilateral triangle. An example of a polyhedral virus in the shape of an icosahedron is the adenovirus (shown in Figure 13.2b). Another icosahedral virus is the poliovirus.

Enveloped Viruses

As noted earlier, the capsid of some viruses is covered by an envelope. Enveloped viruses are roughly spherical. When helical or polyhedral viruses are enclosed by envelopes, they are called *enveloped helical* or *enveloped polyhedral viruses.* An example of an enveloped helical virus is the influenza virus (see Figure 13.3b). An example of an enveloped polyhedral (icosahedral) virus is the herpes simplex virus (see Figure 13.16b).

Complex Viruses

Some viruses, particularly bacterial viruses, have complicated structures and are called **complex viruses.** One example of a complex virus is a bacteriophage. Some bacteriophages have capsids to which additional structures are attached (**Figure 13.5a**). In this figure, notice that the capsid (head) is polyhedral and that the tail sheath is helical. The head contains the nucleic acid. Later in the chapter, we will discuss the functions of the other structures, such as the tail sheath, tail fibers, plate, and pin. Another example of complex viruses are poxviruses, which do not contain clearly identifiable capsids but have several coats around the nucleic acid (**Figure 13.5b**).

CHECK YOUR UNDERSTANDING

✓ Diagram a nonenveloped polyhedral virus that has spikes. **13-2**

Taxonomy of Viruses

LEARNING OBJECTIVES

13-3 Define *viral species.*

13-4 Give an example of a family, genus, and common name for a virus.

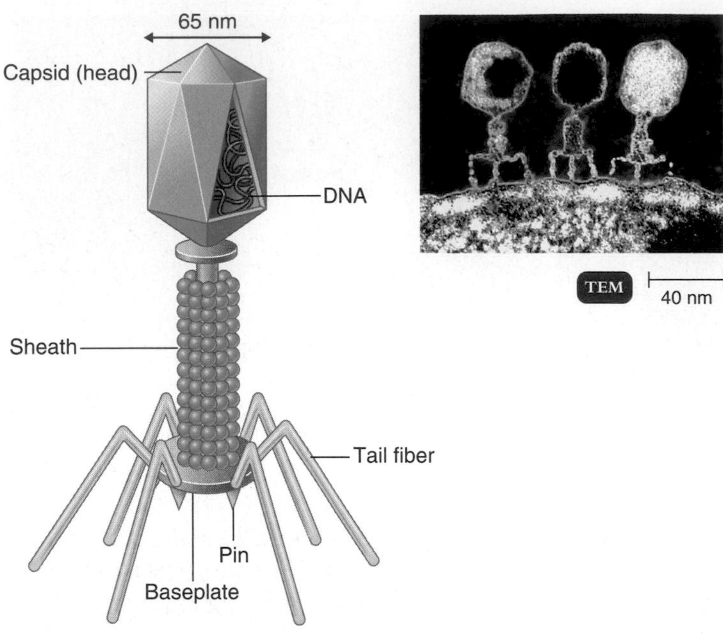

(a) A T-even bacteriophage

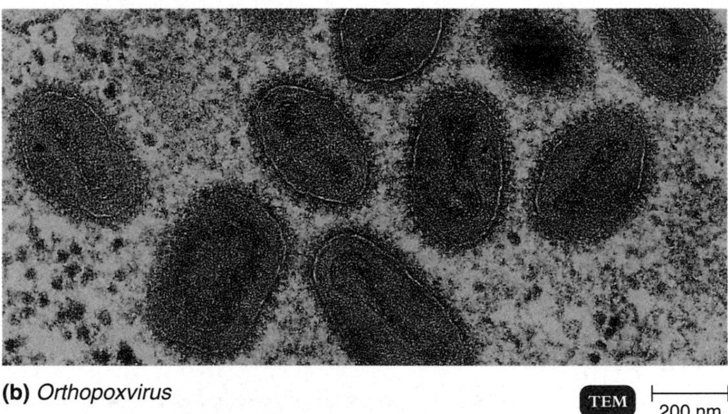

(b) *Orthopoxvirus*

Figure 13.5 Morphology of complex viruses. (**a**) A diagram and micrograph of a T-even bacteriophage. (**b**) A micrograph of variola virus, a species in the genus *Orthopoxvirus,* which causes smallpox.

Q **What is the value of a capsid to a virus?**

Just as we need taxonomic categories of plants, animals, and bacteria, we need viral taxonomy to help us organize and understand newly discovered organisms. The oldest classification of viruses is based on symptomatology, such as for diseases that affect the respiratory system. This system was convenient but not scientifically acceptable because the same virus may cause more than one disease, depending on the tissue affected. In addition, this system artificially grouped viruses that do not infect humans.

Virologists began addressing the problem of viral taxonomy in 1966 with the formation of the International Committee on Taxonomy of Viruses (ICTV). Since then, the ICTV has been grouping viruses into families based on (1) nucleic acid type, (2) strategy for replication, and (3) morphology. The suffix *-virus* is used for genus names; family names end in *-viridae;* and order names end in *-ales.* In formal usage, the family and genus names are used in the following manner: Family Herpesviridae, genus *Simplexvirus,* human herpesvirus 2.

A **viral species** is a group of viruses sharing the same genetic information and ecological niche (host range). Specific epithets for viruses are not used. Thus, viral species are designated by descriptive common names, such as human immunodeficiency virus (HIV), with subspecies (if any) designated by a number (HIV-1). Table 13.2 presents a summary of the classification of viruses that infect humans.

CHECK YOUR UNDERSTANDING

✓ How does a virus species differ from a bacterial species? **13-3**

✓ Attach the proper endings to *Papilloma-* to show the family and genus that includes HPV, the cause of cervical cancer. **13-4**

Isolation, Cultivation, and Identification of Viruses

LEARNING OBJECTIVES

13-5 Describe how bacteriophages are cultured.

13-6 Describe how animal viruses are cultured.

13-7 List three techniques used to identify viruses.

The fact that viruses cannot multiply outside a living host cell complicates their detection, enumeration, and identification. Viruses must be provided with living cells instead of a fairly simple chemical medium. Living plants and animals are difficult and expensive to maintain, and pathogenic viruses that grow only in higher primates and human hosts cause additional complications. However, viruses that use bacterial cells as a host (bacteriophages) are rather easily grown on bacterial cultures. This is one reason so much of our understanding of viral multiplication has come from bacteriophages.

Growing Bacteriophages in the Laboratory

Bacteriophages can be grown either in suspensions of bacteria in liquid media or in bacterial cultures on solid media. The use of solid media makes possible the *plaque method* for detecting and counting viruses. A sample of bacteriophage is mixed with host bacteria and melted agar. The agar containing the bacteriophages and host bacteria is then poured into a Petri plate containing a hardened layer of agar growth medium. The virus-bacteria mixture solidifies into a thin top layer that contains a layer of bacteria approximately one cell thick. Each virus infects a bacterium, multiplies, and releases several hundred new viruses. These newly produced viruses infect other bacteria in the immediate vicinity, and more new viruses are produced. Following several viral multiplication cycles, all the bacteria in the area surrounding the original virus are destroyed. This produces a number of clearings, or **plaques,** visible against a lawn of bacterial growth

Table 13.2 Families of Viruses That Affect Humans

Characteristics/ Dimensions	Viral Family	Important Genera	Clinical or Special Features
Single-Stranded DNA Nonenveloped			
18–25 nm	Parvoviridae	Human parvovirus B19	Fifth disease; anemia in immunocompromised patients. Refer to Chapter 21.
Double-Stranded DNA Nonenveloped			
70–90 nm	Adenoviridae	*Mastadenovirus*	Medium-sized viruses that cause various respiratory infections in humans; some cause tumors in animals.
40–57 nm	Papovaviridae	*Papillomavirus* (human wart virus) *Polyomavirus*	Small viruses that induce tumors; the human wart virus (papilloma) and certain viruses that produce cancer in animals (polyoma and simian) belong to this family. Refer to Chapters 21 and 26.
Double-Stranded DNA Enveloped			
200–350 nm	Poxviridae	*Orthopoxvirus* (vaccinia and smallpox viruses) *Molluscipoxvirus*	Very large, complex, brick-shaped viruses that cause diseases such as smallpox (variola), molluscum contagiosum (wartlike skin lesion), and cowpox. Refer to Chapter 21.
150–200 nm	Herpesviridae	*Simplexvirus* (HHV-1 and 2) *Varicellovirus* (HHV-3) *Lymphocryptovirus* (HHV-4) *Cytomegalovirus* (HHV-5) *Roseolovirus* (HHV-6) HHV-7 Kaposi's sarcoma (HHV-8)	Medium-sized viruses that cause various human diseases, such as fever blisters, chickenpox, shingles, and infectious mononucleosis; causes a type of human cancer called Burkitt's lymphoma. Refer to Chapters 21, 23, and 26.
42 nm	Hepadnaviridae	*Hepadnavirus* (hepatitis B virus)	After protein synthesis, hepatitis B virus uses reverse transcriptase to produce its DNA from mRNA; causes hepatitis B and liver tumors. Refer to Chapter 25.
Single-Stranded RNA, + Strand Nonenveloped			
28–30 nm	Picornaviridae	*Enterovirus* *Rhinovirus* (common cold virus) Hepatitis A virus	At least 70 human enteroviruses are known, including the polio-, coxsackie-, and echoviruses; more than 100 rhinoviruses exist and are the most common cause of colds. Refer to Chapters 22, 24, and 25.
35–40 nm	Caliciviridae	Hepatitis E virus *Norovirus*	Includes causes of gastroenteritis and one cause of human hepatitis. Refer to Chapter 25.
Single-Stranded RNA, + Strand Enveloped			
60–70 nm	Togaviridae	*Alphavirus* *Rubivirus* (rubella virus)	Included are many viruses transmitted by arthropods (*Alphavirus*); diseases include eastern equine encephalitis (EEE) and western equine encephalitis (WEE). Rubella virus is transmitted by the respiratory route. Refer to Chapters 21, 22, and 23.

➤

Table 13.2 Families of Viruses That Affect Humans (continued)

Characteristics/ Dimensions	Viral Family	Important Genera	Clinical or Special Features
40–50 nm	Flaviviridae	*Flavivirus* *Pestivirus* Hepatitis C virus	Can replicate in arthropods that transmit them; diseases include yellow fever, dengue and St. Louis and West Nile encephalitis. Refer to Chapters 22, 23, and 25.
80–160 nm	Coronaviridae	*Coronavirus*	Associated with upper respiratory tract infections and the common cold; SARS virus. Refer to Chapter 24.
– Strand, One Strand of RNA			
70–180 nm	Rhabdoviridae	*Vesiculovirus* (vesicular stomatatis virus) *Lyssavirus* (rabies virus)	Bullet-shaped viruses with a spiked envelope; cause rabies and numerous animal diseases. Refer to Chapter 22.
80–14,000 nm	Filoviridae	*Filovirus*	Enveloped, helical viruses; Ebola and Marburg viruses are filoviruses. Refer to Chapter 23.
150–300 nm	Paramyxoviridae	*Paramyxovirus* *Morbillivirus* (measles virus)	Paramyxoviruses cause parainfluenza, mumps, and Newcastle disease in chickens. Refer to Chapters 21, 24, and 25.
32 nm	Deltaviridae	Hepatitis D	Depend on coinfection with hepadnavirus. Refer to Chapter 25.
– Strand, Multiple Strands of RNA			
80–200 nm	Orthomyxoviridae	Influenza virus A, B, and C	Envelope spikes can agglutinate red blood cells. Refer to Chapter 24.
90–120 nm	Bunyaviridae	*Bunyavirus* (California encephalitis virus) *Hantavirus*	Hantaviruses cause hemorrhagic fevers such as Korean hemorrhagic fever and *Hantavirus* pulmonary syndrome; associated with rodents. Refer to Chapters 22, 23.
110–130 nm	Arenaviridae	*Arenavirus*	Helical capsids contain RNA-containing granules; cause lymphocytic choriomeningitis, Venezuelan hemorrhagic fever, and Lassa fever. Refer to Chapter 23.
Produce DNA			
100–120 nm	Retroviridae	Oncoviruses *Lentivirus* (HIV)	Includes all RNA tumor viruses. Oncoviruses cause leukemia and tumors in animals; the *Lentivirus* HIV causes AIDS. Refer to Chapter 19.
Double-Stranded RNA Nonenveloped			
60–80 nm	Reoviridae	*Reovirus* *Rotavirus*	Involved in mild respiratory infections and gastroenteritis; an unclassified species causes Colorado tick fever. Refer to Chapter 25.

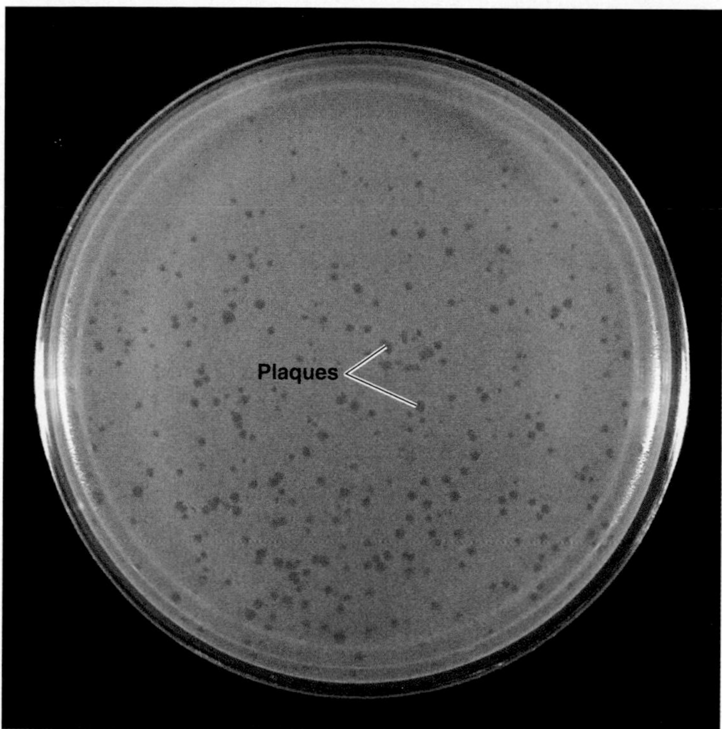

Figure 13.6 Viral plaques formed by bacteriophages. Clear viral plaques of various sizes have been formed by bacteriophage λ (lambda) on a lawn of *E. coli.*

Q What is a plaque-forming unit?

on the surface of the agar (**Figure 13.6**). While the plaques form, uninfected bacteria elsewhere in the Petri plate multiply rapidly and produce a turbid background.

Each plaque theoretically corresponds to a single virus in the initial suspension. Therefore, the concentrations of viral suspensions measured by the number of plaques are usually given in terms of **plaque-forming units (PFU).**

Growing Animal Viruses in the Laboratory

In the laboratory, three methods are commonly used for culturing animal viruses. These methods involve using living animals, embryonated eggs, or cell cultures.

In Living Animals

Some animal viruses can be cultured only in living animals, such as mice, rabbits, and guinea pigs. Most experiments to study the immune system's response to viral infections must also be performed in virally infected live animals. Animal inoculation may be used as a diagnostic procedure for identifying and isolating a virus from a clinical specimen. After the animal is inoculated with the specimen, the animal is observed for signs of disease or is killed so that infected tissues can be examined for the virus.

Some human viruses cannot be grown in animals or can be grown but do not cause disease. The lack of natural animal models for AIDS has slowed our understanding of its disease process and prevented experimentation with drugs that inhibit growth of the virus in vivo. Chimpanzees can be infected with one subspecies of human immunodeficiency virus (HIV-1, genus *Lentivirus*), but because they do not show symptoms of the disease, they cannot be used to study the effects of viral growth and disease treatments. AIDS vaccines are presently being tested in humans, but the disease progresses so slowly in humans that it can take years to determine the effectiveness of these vaccines. In 1986, simian AIDS (an immunodeficiency disease of green monkeys) was reported, followed in 1987 by feline AIDS (an immunodeficiency disease of domestic cats). These diseases are caused by lentiviruses, which are closely related to HIV, and the diseases develop within a few months, thus providing a model for studying viral growth in different tissues. In 1990, a way to infect mice with human AIDS was found when immunodeficient mice were grafted to produce human T cells and human gamma globulin. The mice provide a reliable model for studying viral replication, although they do not provide models for vaccine development.

In Embryonated Eggs

If the virus will grow in an *embryonated egg*, this can be a fairly convenient and inexpensive form of host for many animal viruses. A hole is drilled in the shell of the embryonated egg, and a viral suspension or suspected virus-containing tissue is injected into the fluid of the egg. There are several membranes in an egg, and the virus is injected near the one most appropriate for its growth (**Figure 13.7**). Viral growth is signaled by the death of the embryo,

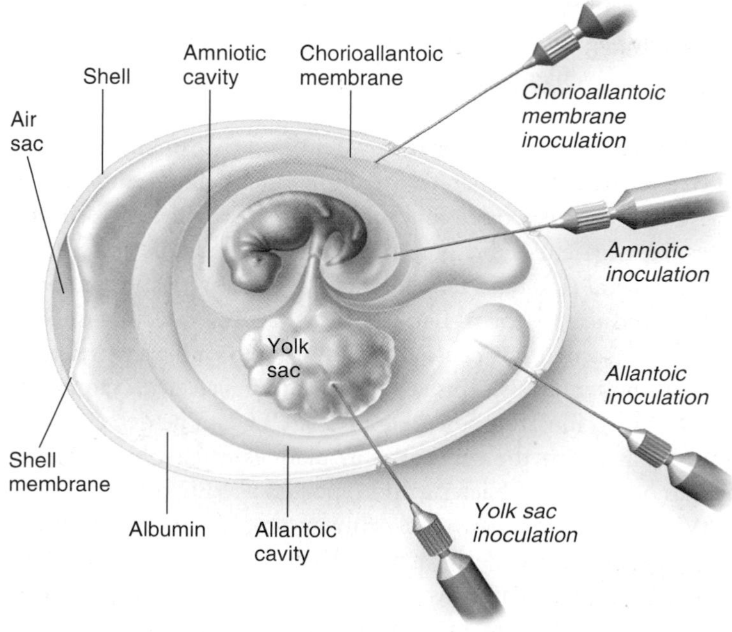

Figure 13.7 Inoculation of an embryonated egg. The injection site determines the membrane on which the viruses will grow.

Q Why are viruses grown in eggs and not in culture media?

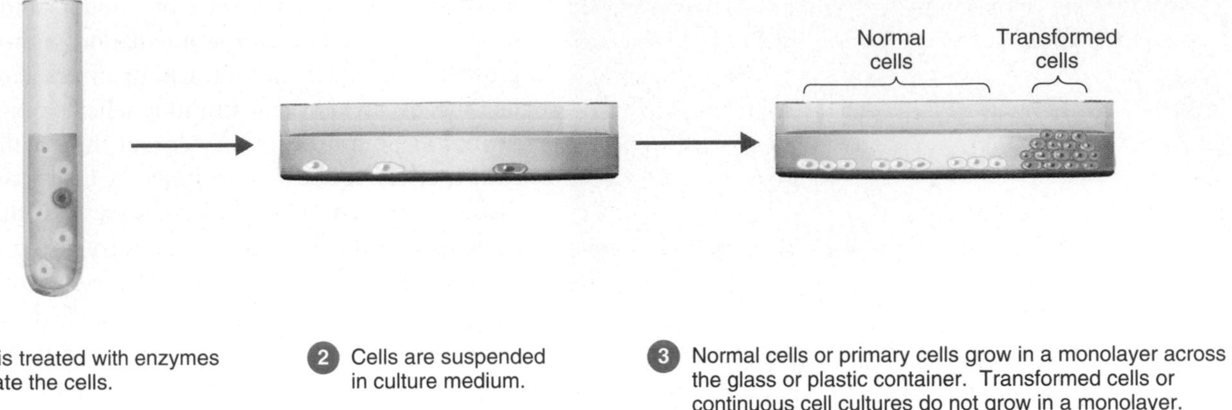
Normal cells
Transformed cells
1 A tissue is treated with enzymes to separate the cells.
2 Cells are suspended in culture medium.
3 Normal cells or primary cells grow in a monolayer across the glass or plastic container. Transformed cells or continuous cell cultures do not grow in a monolayer.

Figure 13.8 Cell cultures. Transformed cells can be grown indefinitely in laboratory culture.

Q Why are transformed cells referred to as "immortal"?

by embryo cell damage, or by the formation of typical pocks or lesions on the egg membranes. This method was once the most widely used method of viral isolation and growth, and it is still used to grow viruses for some vaccines. For this reason, you may be asked if you are allergic to eggs before receiving a vaccination, because egg proteins may be present in the viral vaccine preparations. (Allergic reactions will be discussed in Chapter 19.)

In Cell Cultures

Cell cultures have replaced embryonated eggs as the preferred type of growth medium for many viruses. Cell cultures consist of cells grown in culture media in the laboratory. Because these cultures are generally rather homogeneous collections of cells and can be propagated and handled much like bacterial cultures, they are more convenient to work with than whole animals or embryonated eggs.

Cell culture lines are started by treating a slice of animal tissue with enzymes that separate the individual cells (Figure 13.8). These cells are suspended in a solution that provides the osmotic pressure, nutrients, and growth factors needed for the cells to grow. Normal cells tend to adhere to the glass or plastic container and reproduce to form a monolayer. Viruses infecting such a monolayer sometimes cause the cells of the monolayer to deteriorate as they multiply. This cell deterioration is called **cytopathic effect (CPE)**, illustrated in Figure 13.9. CPE can be detected and counted in much the same way as plaques caused by bacteriophages on a lawn of bacteria and reported as PFU/ml.

Viruses may be grown in primary or continuous cell lines. **Primary cell lines,** derived from tissue slices, tend to die out after only a few generations. Certain cell lines, called **diploid cell lines,** developed from human embryos can be maintained for about 100 generations and are widely used for culturing viruses that require a human host. Cell lines developed from embryonic human cells are used to culture rabies virus for a rabies vaccine called human diploid culture vaccine (see Chapter 22).

When viruses are routinely grown in a laboratory, **continuous cell lines** are used. These are transformed (cancerous) cells that can be maintained through an indefinite number of generations, and they are sometimes called immortal cell lines (see the discussion of transformation on page 390). One of these, the HeLa cell line, was isolated from the cancer of a woman (**He**nrietta **La**cks) who died in 1951. After years of laboratory cultivation, many such cell lines have lost almost all the original characteristics of the cell, but these changes have not interfered with the use of the cells for viral propagation. In spite of the success of cell culture in viral isolation and growth, there are still some viruses that have never been successfully cultivated in cell culture.

The idea of cell culture dates back to the end of the nineteenth century, but it was not a practical laboratory technique

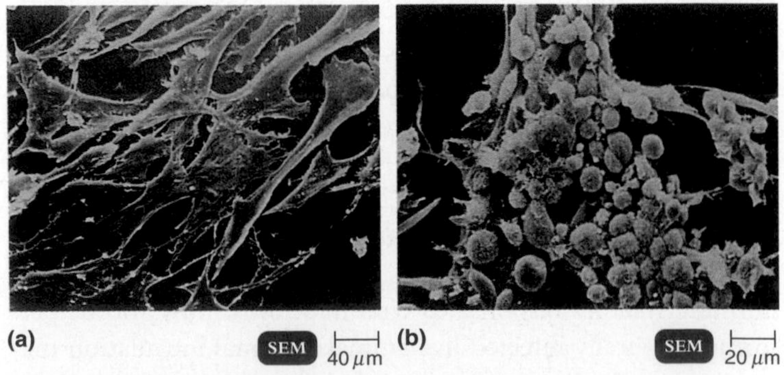
(a)
SEM
40 μm
(b)
SEM
20 μm

Figure 13.9 The cytopathic effect of viruses. (**a**) Uninfected mouse cells align next to each other, forming a monolayer. (**b**) The same cells 24 hours after infection with vesicular stomatitis virus (VSV) (see Figure 13.18a). Notice the cells pile up and "round up."

Q How did VSV infection affect the cells?

until the development of antibiotics in the years following World War II. A major problem with cell culture is that the cell lines must be kept free of microbial contamination. The maintenance of cell culture lines requires trained technicians with considerable experience working on a full-time basis. Because of these difficulties, most hospital laboratories and many state health laboratories do not isolate and identify viruses in clinical work. Instead, the tissue or serum samples are sent to central laboratories that specialize in such work.

Viral Identification

Identifying viral isolates is not an easy task. For one thing, viruses cannot be seen at all without the use of an electron microscope. Serological methods, such as Western blotting, are the most commonly used means of identification (see Figure 10.12, page 289). In these tests, the virus is detected and identified by its reaction with antibodies. We will discuss antibodies in detail in Chapter 17 and a number of immunological tests for identifying viruses in Chapter 18. Observation of cytopathic effects, described in Chapter 15 (page 441), is also useful for the identification of a virus.

Virologists can identify and characterize viruses by using such modern molecular methods as use of restriction fragment length polymorphisms (RFLPs) and the polymerase chain reaction (PCR) (Chapter 9, page 251). PCR was used to amplify viral RNA to identify the West Nile virus in 1999 in the United States and the SARS-associated coronavirus in China in 2002.

CHECK YOUR UNDERSTANDING

- ✓ What is the plaque method? **13-5**
- ✓ Why are continuous cell lines of more practical use than primary cell lines for culturing viruses? **13-6**
- ✓ What tests could you use to identify influenza virus in a patient? **13-7**

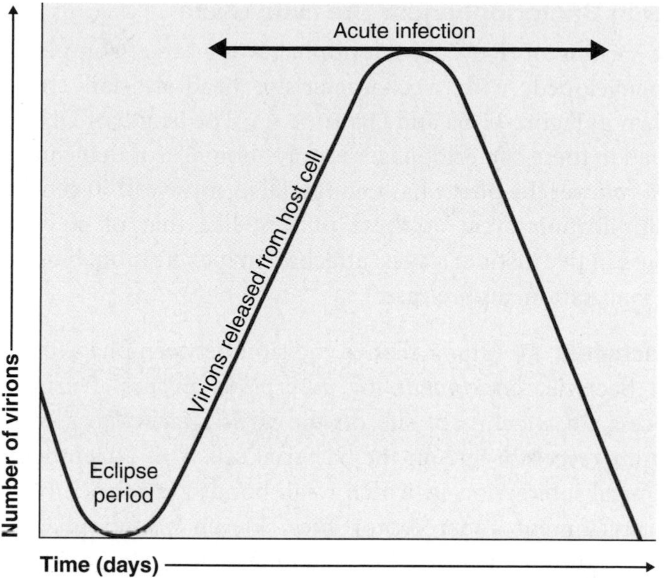

Figure 13.10 A viral one-step growth curve. No new infective virions are found in a culture until after biosynthesis and maturation have taken place. Most infected cells die as a result of infection; consequently, new virions will not be produced.

Q What can be found in the cell during biosynthesis and maturation?

Viral Multiplication

LEARNING OBJECTIVES

13-8 Describe the lytic cycle of T-even bacteriophages.

13-9 Describe the lysogenic cycle of bacteriophage lambda.

13-10 Compare and contrast the multiplication cycle of DNA and RNA-containing animal viruses.

The nucleic acid in a virion contains only a few of the genes needed for the synthesis of new viruses. These include genes for the virion's structural components, such as the capsid proteins, and genes for a few of the enzymes used in the viral life cycle. These enzymes are synthesized and functional only when the virus is within the host cell. Viral enzymes are almost entirely concerned with replicating or processing viral nucleic acid. Enzymes needed for protein synthesis, ribosomes, tRNA, and energy production are supplied by the host cell and are used for synthesizing viral proteins, including viral enzymes. Although the smallest nonenveloped virions do not contain any preformed enzymes, the larger virions may contain one or a few enzymes, which usually function in helping the virus penetrate the host cell or replicate its own nucleic acid.

Thus, for a virus to multiply, it must invade a host cell and take over the host's metabolic machinery. A single virion can give rise to several or even thousands of similar viruses in a single host cell. This process can drastically change the host cell and usually causes its death. In a few viral infections, cells survive and continue to produce viruses indefinitely.

The multiplication of viruses can be demonstrated with a **one-step growth curve** (Figure 13.10). The data are obtained by infecting every cell in a culture and then testing the culture medium and cells for virions and viral proteins and nucleic acids.

Multiplication of Bacteriophages

Although the means by which a virus enters and exits a host cell may vary, the basic mechanism of viral multiplication is similar for all viruses. Bacteriophages can multiply by two alternative mechanisms: the lytic cycle or the lysogenic cycle. The **lytic cycle** ends with the lysis and death of the host cell, whereas the host cell remains alive in the **lysogenic cycle.** Because the *T-even bacteriophages* (T2, T4, and T6) have been studied most extensively, we will describe the multiplication of T-even bacteriophages in their host, *E. coli,* as an example of the lytic cycle.

T-Even Bacteriophages: The Lytic Cycle

The virions of T-even bacteriophages are large, complex, and nonenveloped, with a characteristic head-and-tail structure shown in Figure 13.5a and Figure 13.11. The length of DNA contained in these bacteriophages is only about 6% of that contained in *E. coli,* yet the phage has enough DNA for over 100 genes. The multiplication cycle of these phages, like that of all viruses, occurs in five distinct stages: attachment, penetration, biosynthesis, maturation, and release.

Attachment 1 After a chance collision between phage particles and bacteria, *attachment,* or *adsorption,* occurs. During this process, an attachment site on the virus attaches to a complementary receptor site on the bacterial cell. This attachment is a chemical interaction in which weak bonds are formed between the attachment and receptor sites. T-even bacteriophages use fibers at the end of the tail as attachment sites. The complementary receptor sites are on the bacterial cell wall.

Penetration 2 After attachment, the T-even bacteriophage injects its DNA (nucleic acid) into the bacterium. To do this, the bacteriophage's tail releases an enzyme, **phage lysozyme,** which breaks down a portion of the bacterial cell wall. During the process of *penetration,* the tail sheath of the phage contracts, and the tail core is driven through the cell wall. When the tip of the core reaches the plasma membrane, the DNA from the bacteriophage's head passes through the tail core, through the plasma membrane, and enters the bacterial cell. The capsid remains outside the bacterial cell. Therefore, the phage particle functions like a hypodermic syringe to inject its DNA into the bacterial cell.

Biosynthesis 3 Once the bacteriophage DNA has reached the cytoplasm of the host cell, the biosynthesis of viral nucleic acid and protein occurs. Host protein synthesis is stopped by virus-induced degradation of the host DNA, viral proteins that interfere with transcription, or the repression of translation.

Initially, the phage uses the host cell's nucleotides and several of its enzymes to synthesize many copies of phage DNA. Soon after, the biosynthesis of viral proteins begins. Any RNA transcribed in the cell is mRNA transcribed from phage DNA for the biosynthesis of phage enzymes and capsid proteins. The host cell's ribosomes, enzymes, and amino acids are used for translation. Genetic controls regulate when different regions of phage DNA are transcribed into mRNA during the multiplication cycle. For example, early messages are translated into early phage proteins, the enzymes used in the synthesis of phage DNA. Also, late messages are translated into late phage proteins for the synthesis of capsid proteins.

For several minutes following infection, complete phages cannot be found in the host cell. Only separate components—DNA and protein—can be detected. The period during viral multiplication when complete, infective virions are not yet present is called the **eclipse period.**

Maturation 4 In the next sequence of events, *maturation* occurs. In this process, bacteriophage DNA and capsids are assembled into complete virions. The viral components essentially assemble into a viral particle spontaneously, eliminating the need for many nonstructural genes and gene products. The phage heads and tails are separately assembled from protein subunits, and the head is filled with phage DNA and attached to the tail.

Release 5 The final stage of viral multiplication is the *release* of virions from the host cell. The term **lysis** is generally used for this stage in the multiplication of T-even phages because in this case, the plasma membrane actually breaks open (lyses). Lysozyme, which is encoded by a phage gene, is synthesized within the cell. This enzyme causes the bacterial cell wall to break down, and the newly produced bacteriophages are released from the host cell. The released bacteriophages infect other susceptible cells in the vicinity, and the viral multiplication cycle is repeated within those cells.

Bacteriophage Lambda (λ): The Lysogenic Cycle

In contrast to T-even bacteriophages, some viruses do not cause lysis and death of the host cell when they multiply. These *lysogenic phages* (also called *temperate phages*) may indeed proceed through a lytic cycle, but they are also capable of incorporating their DNA into the host cell's DNA to begin a lysogenic cycle. In **lysogeny,** the phage remains latent (inactive). The participating bacterial host cells are known as *lysogenic cells.*

We will use the bacteriophage λ (lambda), a well-studied lysogenic phage, as an example of the lysogenic cycle (Figure 13.12).

1 Upon penetration into an *E. coli* cell,

2 the originally linear phage DNA forms a circle.

3A This circle can multiply and be transcribed,

4A leading to the production of new phage and to cell lysis (the lytic cycle).

3B Alternatively, the circle can recombine with and become part of the circular bacterial DNA (the lysogenic cycle). The inserted phage DNA is now called a **prophage.** Most of the prophage genes are repressed by two repressor proteins that are the products of phage genes. These repressors stop transcription of all the other phage genes by binding to operators. Thus, the phage genes that would otherwise direct the synthesis and release of new virions are turned off, in much the same way that the genes of the *E. coli lac* operon are turned off by the *lac* repressor (Figure 8.12, page 225).

Every time the host cell's machinery replicates the bacterial chromosome,

4B it also replicates the prophage DNA. The prophage remains latent within the progeny cells.

5 However, a rare spontaneous event, or the action of UV light or certain chemicals, can lead to the excision (popping-out) of the phage DNA, and to initiation of the lytic cycle.

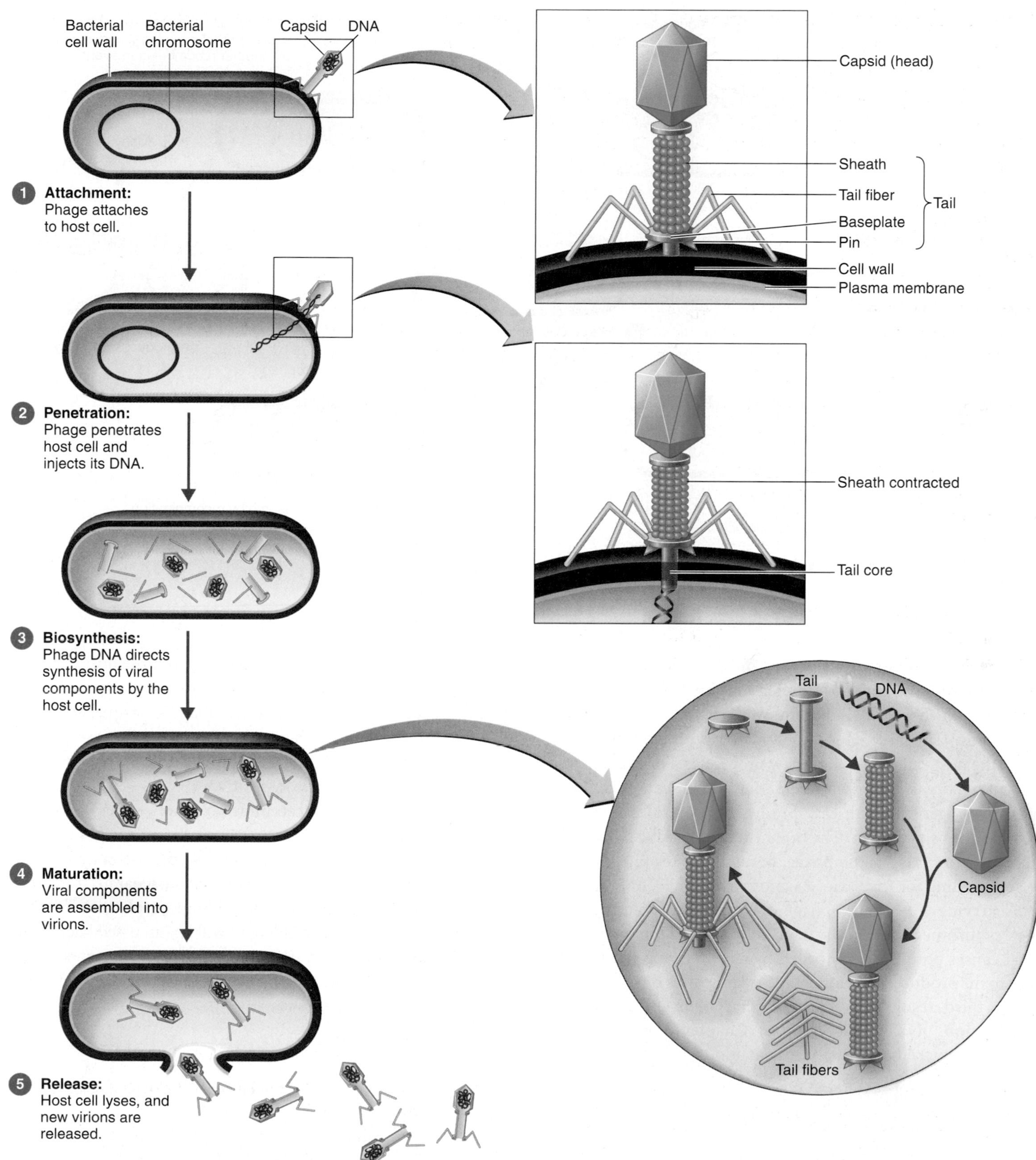

Figure 13.11 The lytic cycle of a T-even bacteriophage.

Q What is the result of the lytic cycle?

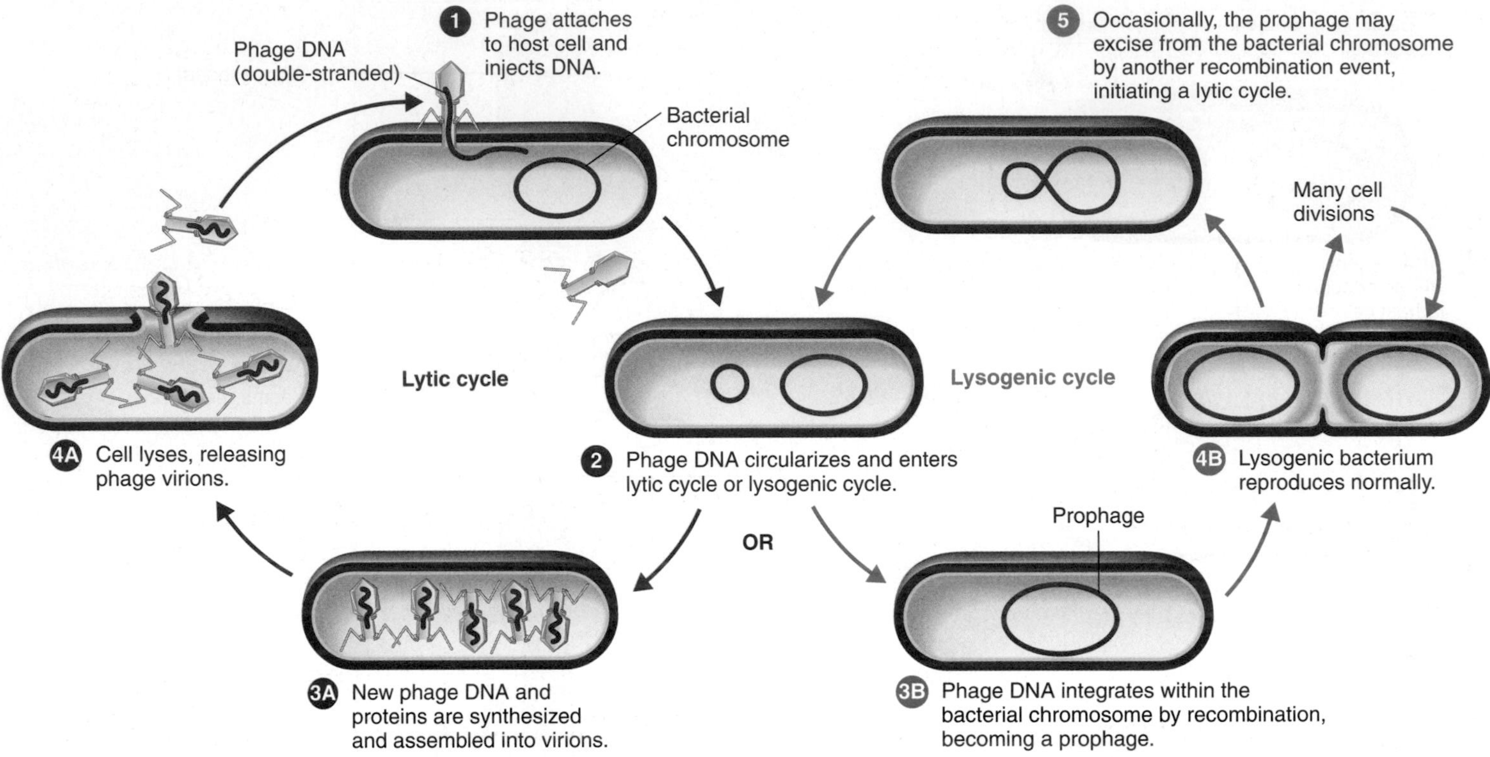

Figure 13.12 The lysogenic cycle of bacteriophage λ in *E. coli.*

Q How does lysogeny differ from the lytic cycle?

There are three important results of lysogeny. First, the lysogenic cells are immune to reinfection by the same phage. (However, the host cell is not immune to infection by other phage types.) The second result of lysogeny is **phage conversion;** that is, the host cell may exhibit new properties. For example, the bacterium *Corynebacterium diphtheriae,* which causes diphtheria, is a pathogen whose disease-producing properties are related to the synthesis of a toxin. The organism can produce toxin only when it carries a lysogenic phage, because the prophage carries the gene coding for the toxin. As another example, only streptococci carrying a lysogenic phage are capable of causing toxic shock syndrome. The toxin produced by *Clostridium botulinum,* which causes botulism, is encoded by a prophage gene, as is the Shiga toxin produced by pathogenic strains of *E. coli.*

The third result of lysogeny is that it makes **specialized transduction** possible. Recall from Chapter 8 that bacterial genes can be picked up in a phage coat and transferred to another bacterium in a process called generalized transduction (see Figure 8.28, page 239). Any bacterial genes can be transferred by generalized transduction because the host chromosome is broken down into fragments, any of which can be packaged into a phage coat. In specialized transduction, however, only certain bacterial genes can be transferred.

Specialized transduction is mediated by a lysogenic phage, which packages bacterial DNA *along with* its own DNA in the same capsid. When a prophage is excised from the host chromosome, adjacent genes from either side may remain attached to the phage DNA. In Figure 13.13, bacteriophage λ has picked up the *gal* gene for galactose fermentation from its galactose-positive host. The phage carries this gene to a galactose-negative cell, which then becomes galactose-positive.

Certain animal viruses can undergo processes very similar to lysogeny. Animal viruses that can remain latent in cells for long periods without multiplying or causing disease may become inserted into a host chromosome or remain separate from host DNA in a repressed state (as some lysogenic phages). Cancer-causing viruses may also be latent, as will be discussed later in the chapter. **Animations** Viral Replication: Virulent Bacteriophages, Temperate Bacteriophages; Transduction: Specialized Transduction. **www.microbiologyplace.com**

CHECK YOUR UNDERSTANDING

- ✓ How do bacteriophages get nucleotides and amino acids if they don't have any metabolic enzymes? **13-8**
- ✓ *Vibrio cholerae* produces toxin and is capable of causing cholera only when it is lysogenic. What does this mean? **13-9**

Multiplication of Animal Viruses

The multiplication of animal viruses follows the basic pattern of bacteriophage multiplication but has several differences, summarized in Table 13.3. Animal viruses differ from phages in their mechanism of entering the host cell. Also, once the virus is inside, the synthesis and assembly of the new viral components are somewhat different, partly because of the differences between prokaryotic cells and eukaryotic cells. Animal viruses may have certain types of enzymes not found in phages. Finally, the mechanisms of maturation and release, and the effects on the host cell, differ in animal viruses and phages.

In the following discussion of the multiplication of animal viruses, we will consider the processes that are shared by both DNA- and RNA-containing animal viruses. These processes are attachment, entry, uncoating, and release. We will also examine how DNA- and RNA-containing viruses differ with respect to their processes of biosynthesis.

Attachment

Like bacteriophages, animal viruses have attachment sites that attach to complementary receptor sites on the host cell's surface. However, the receptor sites of animal cells are proteins and glycoproteins of the plasma membrane. Moreover, animal viruses do not possess appendages like the tail fibers of some bacteriophages. The attachment sites of animal viruses are distributed over the surface of the virus. The sites themselves vary from one group of viruses to another. In adenoviruses, which are icosahedral viruses, the attachment sites are small fibers at the corners of the icosahedron (see Figure 13.2b). In many of the enveloped viruses, such as influenza virus, the attachment sites are spikes located on the surface of the envelope (see Figure 13.3b). As soon as one spike attaches to a host receptor, additional receptor sites on the same cell migrate to the virus. Attachment is completed when many sites are bound.

Receptor sites are inherited characteristics of the host. Consequently, the receptor for a particular virus can vary from person to person. This could account for the individual differences in susceptibility to a particular virus. For example, people who lack the cellular receptor (called P antigen) for parvovirus B19, are naturally resistant to infection and do not get fifth disease (see page 600). Understanding the nature of attachment can lead to the development of drugs that prevent viral infections. Monoclonal antibodies (discussed in Chapter 17) that combine with a virus's attachment site or the cell's receptor site may soon be used to treat some viral infections.

Entry

Following attachment, entry occurs. Viruses enter into eukaryotic cells by **pinocytosis,** an active cellular process by which nutrients and other molecules are brought into a cell (Chapter 4, page 100). A cell's plasma membrane continuously folds inward to form vesicles. These vesicles contain elements that originate outside the cell and are brought into the interior of the cell to be digested. If a virion attaches to the plasma membrane of a potential host cell, the host cell will enfold the virion into a fold of plasma membrane, forming a vesicle (Figure 13.14a).

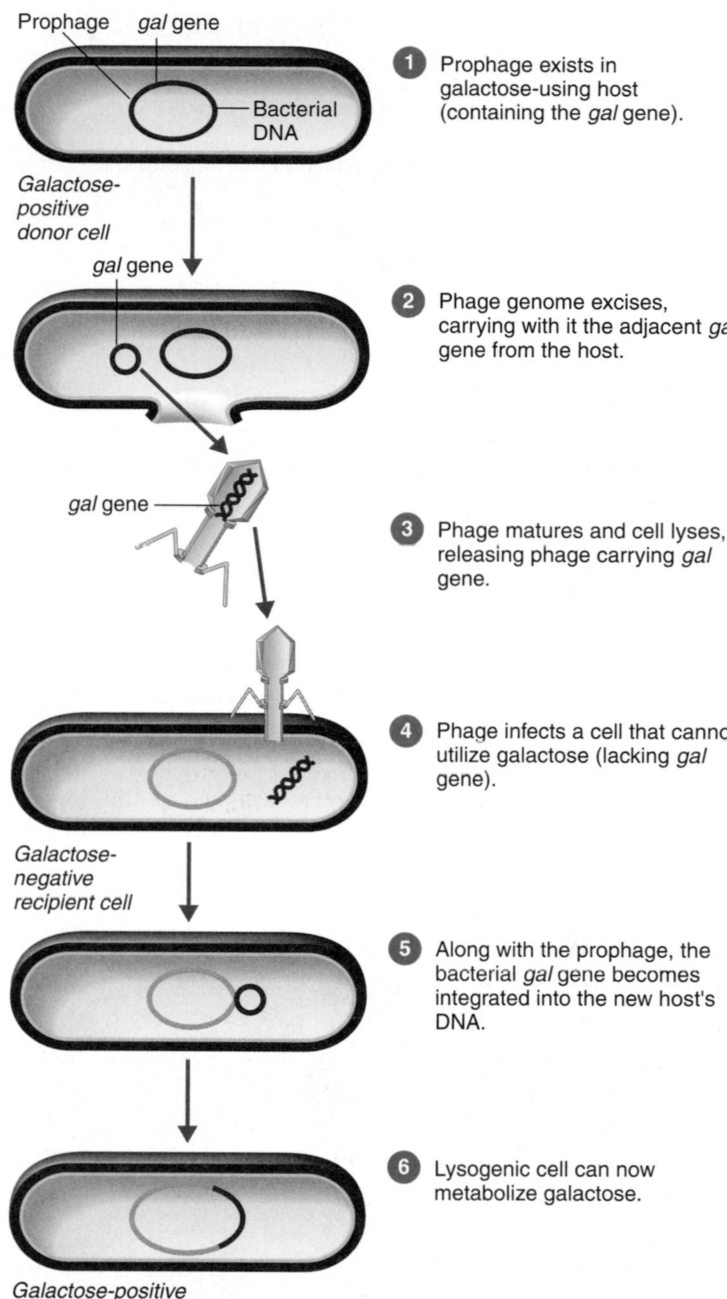
Prophage
gal gene
Bacterial DNA
Galactose-positive donor cell
1 Prophage exists in galactose-using host (containing the *gal* gene).
gal gene
2 Phage genome excises, carrying with it the adjacent *gal* gene from the host.
gal gene
3 Phage matures and cell lyses, releasing phage carrying *gal* gene.
4 Phage infects a cell that cannot utilize galactose (lacking *gal* gene).
Galactose-negative recipient cell
5 Along with the prophage, the bacterial *gal* gene becomes integrated into the new host's DNA.
6 Lysogenic cell can now metabolize galactose.
Galactose-positive recombinant cell

Figure 13.13 Specialized transduction. When a prophage is excised from its host chromosome, it can take with it a bit of the adjacent DNA from the bacterial chromosome.

Q How does specialized transduction differ from the lytic cycle?

Enveloped viruses can enter by an alternative method called **fusion,** in which the viral envelope fuses with the plasma membrane and releases the capsid into the cell's cytoplasm. For example, HIV penetrates cells by this method (Figure 13.14b).

Table 13.3 Bacteriophage and Animal Viral Multiplication Compared

Stage	Bacteriophages	Animal Viruses
Attachment	Tail fibers attach to cell wall proteins	Attachment sites are plasma membrane proteins and glycoproteins
Entry	Viral DNA injected into host cell	Capsid enters by endocytosis or fusion
Uncoating	Not required	Enzymatic removal of capsid proteins
Biosynthesis	In cytoplasm	In nucleus (DNA viruses) or cytoplasm (RNA viruses)
Chronic infection	Lysogeny	Latency; slow viral infections; cancer
Release	Host cell lysed	Enveloped viruses bud out; nonenveloped viruses rupture plasma membrane

Uncoating

Viruses disappear during the eclipse period of an infection because they are taken apart inside the cell. **Uncoating** is the separation of the viral nucleic acid from its protein coat once the virion is enclosed within the vesicle. The capsid is digested when the cell attempts to digest the vesicle's contents, or the nonenveloped capsid may be released into the cytoplasm of the host cell. This process varies with the type of virus. Some animal viruses accomplish uncoating by the action of lysosomal enzymes of the host cell. These enzymes degrade the proteins of the viral capsid. The uncoating of poxviruses is completed by a specific enzyme encoded by the viral DNA and synthesized soon after infection. For other viruses, uncoating appears to be exclusively caused by enzymes in the host cell cytoplasm. For at least one virus, the

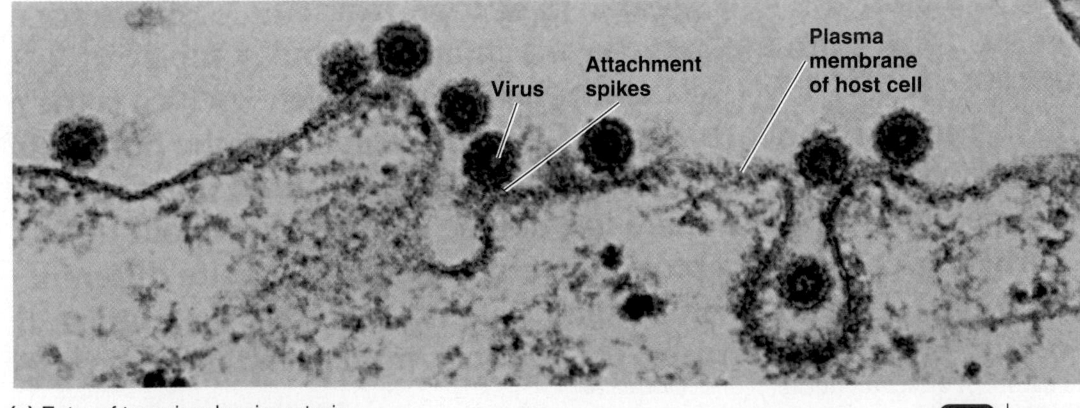

(a) Entry of togavirus by pinocytosis

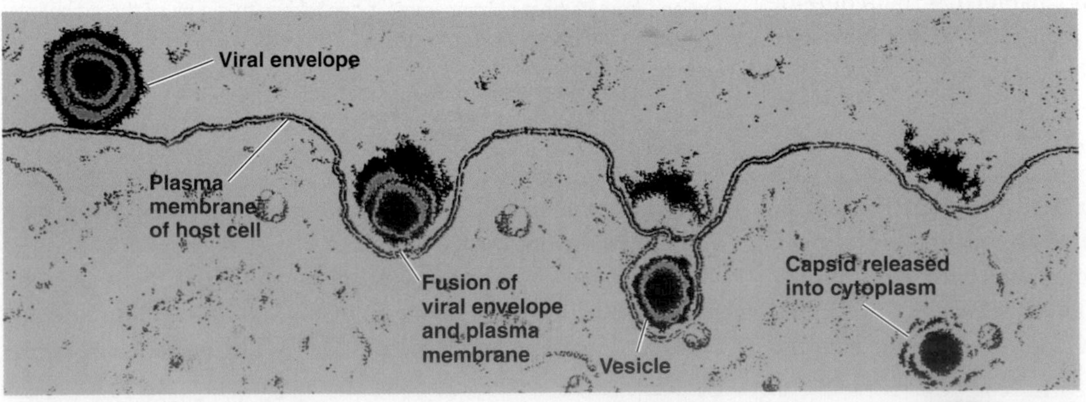

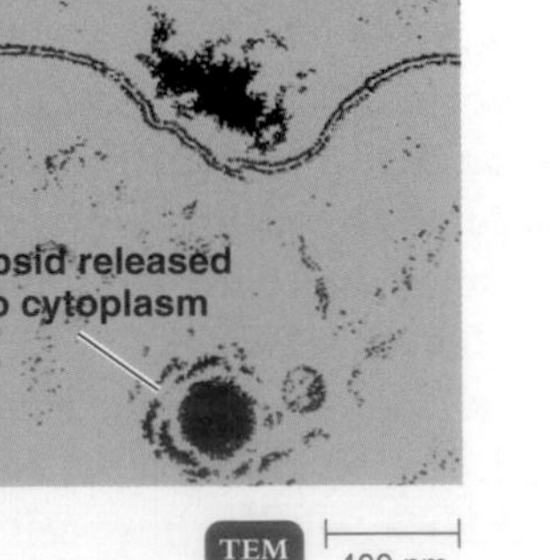

(b) Entry of herpesvirus by fusion

Figure 13.14 The entry of viruses into host cells. After attachment, viruses enter host cells by **(a)** pinocytosis or **(b)** fusion.

Q In which process is the cell actively taking in the virus?

poliovirus, uncoating seems to begin while the virus is still attached to the host cell's plasma membrane.

The Biosynthesis of DNA Viruses

Generally, DNA-containing viruses replicate their DNA in the nucleus of the host cell by using viral enzymes, and they synthesize their capsid and other proteins in the cytoplasm by using host cell enzymes. Then the proteins migrate into the nucleus and are joined with the newly synthesized DNA to form virions. These virions are transported along the endoplasmic reticulum to the host cell's membrane for release. Herpesviruses, papovaviruses, adenoviruses, and hepadnaviruses all follow this pattern of biosynthesis (**Table 13.4**). Poxviruses are an exception because all of their components are synthesized in the cytoplasm.

As an example of the multiplication of a DNA virus, the sequence of events in papovavirus is shown in **Figure 13.15**.

1–2 Following attachment, entry, and uncoating, the viral DNA is released into the nucleus of the host cell.

3 Transcription of a portion of the viral DNA—the "early" genes—occurs next. Translation follows. The products of these genes are enzymes that are required for the multiplication of viral DNA. In most DNA viruses, early transcription is carried out with the host's transcriptase (RNA polymerase); poxviruses, however, contain their own transcriptase.

4 Sometime after the initiation of DNA replication, transcription and translation of the remaining "late" viral genes occur. Late proteins include capsid and other structural proteins.

5 This leads to the synthesis of capsid proteins, which occurs in the cytoplasm of the host cell.

6 After the capsid proteins migrate into the nucleus of the host cell, maturation occurs; the viral DNA and capsid proteins assemble to form complete viruses.

7 Complete viruses are then released from the host cell.

Some DNA viruses are described below.

Adenoviridae Named after adenoids, from which they were first isolated, adenoviruses cause acute respiratory diseases—the common cold (**Figure 13.16a**).

Poxviridae All diseases caused by poxviruses, including smallpox and cowpox, include skin lesions (see Figure 21.10, page 595). *Pox* refers to pus-filled lesions. Viral multiplication is started by viral transcriptase; the viral components are synthesized and assembled in the cytoplasm of the host cell.

Herpesviridae Nearly 100 herpesviruses are known (**Figure 13.16b**). They are named after the spreading (*herpetic*) appearance of cold sores. Species of human herpesviruses (HHV) include HHV-1 and HHV-2, both in the genus *Simplexvirus,* which cause cold sores; HHV-3, genus *Varicellovirus,* which cause chickenpox; HHV-4, genus *Lymphocryptovirus,* which causes infectious mononucleosis; HHV-5, genus *Cytomegalovirus,* which causes CMV inclusion disease; HHV-6, genus *Roseolovirus,* which causes roseola; HHV-7, which infects most infants, causing measleslike rashes; and HHV-8, which causes Kaposi's sarcoma, primarily in AIDS patients.

Papovaviridae Papovaviruses are named for *pa*pillomas (warts), *po*lyomas (tumors), and *va*cuolation (cytoplasmic vacuoles produced by some of these viruses). Warts are caused by members of the genus *Papillomavirus.* Some *Papillomavirus* species

Table 13.4 The Biosynthesis of DNA and RNA Viruses Compared

Viral Nucleic Acid	Virus Family	Special Features of Biosynthesis
DNA, single-stranded	Parvoviridae	Cellular enzyme transcribes viral DNA in nucleus
DNA, double-stranded	Herpesviridae Papovaviridae Poxviridae	Cellular enzyme transcribes viral DNA in nucleus Viral enzyme transcribes viral DNA in virion, in cytoplasm
DNA, reverse transcriptase	Hepadnaviridae	Cellular enzyme transcribes viral DNA in nucleus; reverse transcriptase copies mRNA to make viral DNA
RNA, + strand	Picornaviridae Togaviridae	Viral RNA functions as a template for synthesis of RNA polymerase which copies − strand RNA to make mRNA in cytoplasm
RNA, − strand	Rhabdoviridae	Viral enzyme copies viral RNA to make mRNA in cytoplasm
RNA, double-stranded	Reoviridae	Viral enzyme copies − strand RNA to make mRNA in cytoplasm
RNA, reverse transcriptase	Retroviridae	Viral enzyme copies viral RNA to make DNA in cytoplasm; DNA moves to nucleus

Figure 13.15

FOUNDATION FIGURE Replication of a DNA-Containing Animal Virus

This figure shows the general replication of DNA-containing viruses, using a papovavirus as an example. Understanding the phases of viral replication is important in drug development strategies and disease pathology, discussed in later chapters.

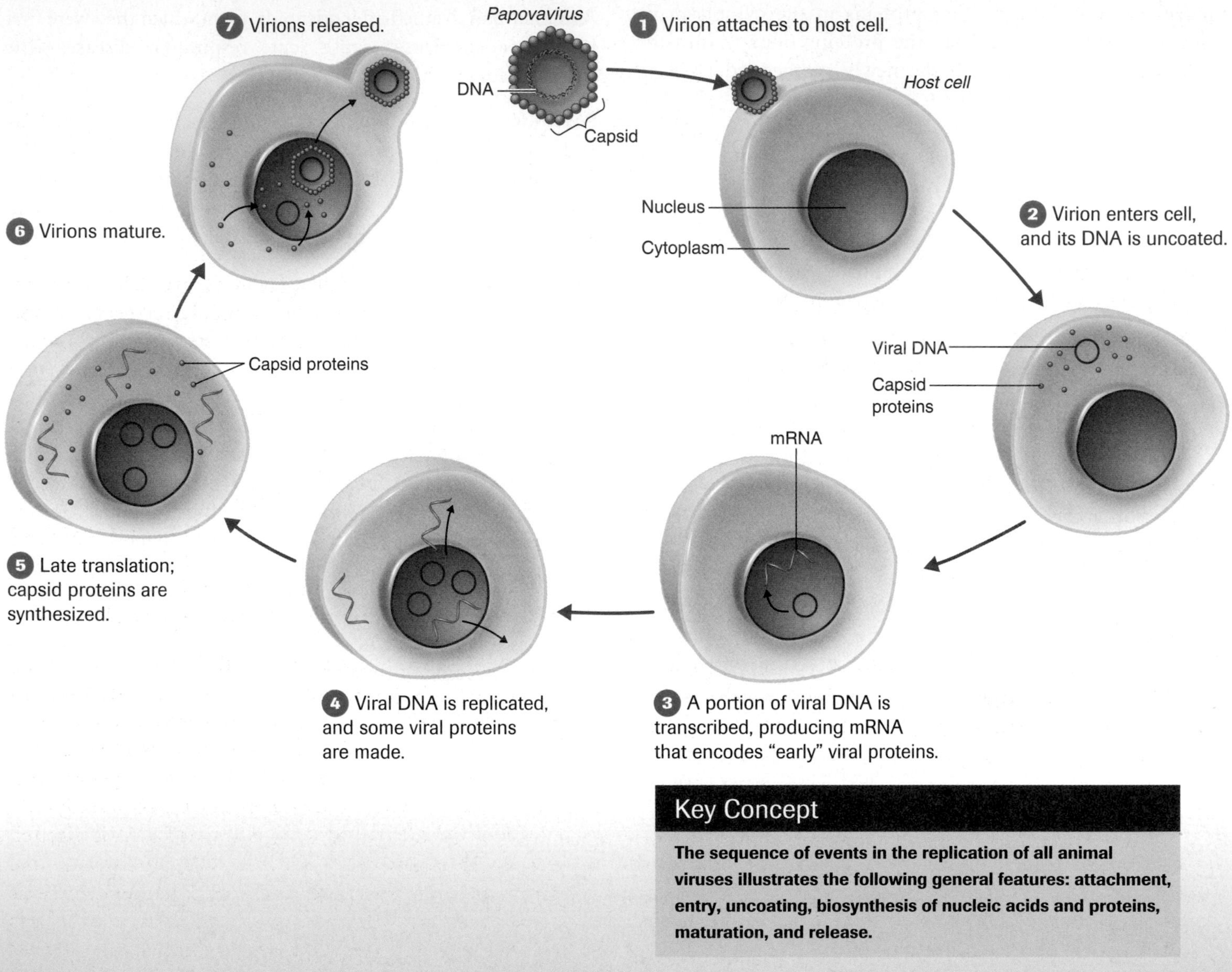

Papovavirus
DNA
Capsid
1 Virion attaches to host cell.
Host cell
Nucleus
Cytoplasm
2 Virion enters cell, and its DNA is uncoated.
Viral DNA
Capsid proteins
mRNA
3 A portion of viral DNA is transcribed, producing mRNA that encodes "early" viral proteins.
4 Viral DNA is replicated, and some viral proteins are made.
5 Late translation; capsid proteins are synthesized.
Capsid proteins
6 Virions mature.
7 Virions released.

Key Concept

The sequence of events in the replication of all animal viruses illustrates the following general features: attachment, entry, uncoating, biosynthesis of nucleic acids and proteins, maturation, and release.

are capable of transforming cells and causing cancer. Viral DNA is replicated in the host cell's nucleus along with host cell chromosomes. Host cells may proliferate, resulting in a tumor.

Hepadnaviridae Hepadnaviridae are so named because they cause *hepa*titis and contain *DNA* (Figure 25.15, page 725). The only genus in this family causes hepatitis B. (Hepatitis A, C, D, E, F, and G viruses, although not related to each other, are RNA viruses. Hepatitis is discussed in Chapter 25.) Hepadnaviruses differ from other DNA viruses because they synthesize DNA by copying RNA, using viral reverse transcriptase. This enzyme is discussed later with the retroviruses, the only other family with reverse transcriptase.

Figure 13.16 DNA-containing animal viruses. (**a**) Negatively stained adenoviruses that have been concentrated in a centrifuge gradient. The individual capsomeres are clearly visible. (**b**) The envelope around this herpes simplex virus capsid has broken, giving a "fried egg" appearance.

What is the morphology of these viruses?

(**a**) *Mastadenovirus* SEM 100 nm

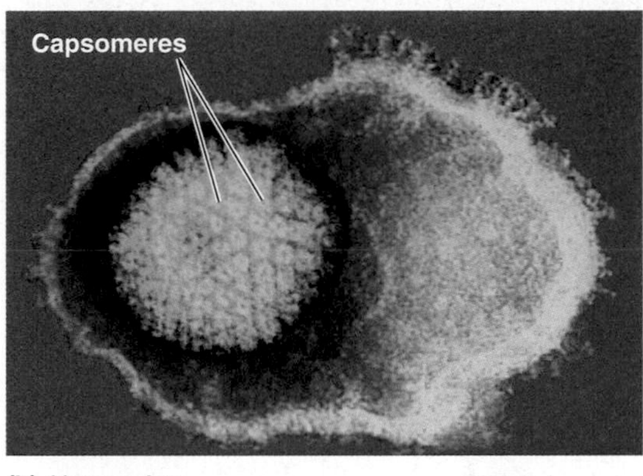

(**b**) *Herpesvirus* TEM 80 nm

The Biosynthesis of RNA Viruses

The multiplication of RNA viruses is essentially the same as that of DNA viruses, except that several different mechanisms of mRNA formation occur among different groups of RNA viruses (see Table 13.4). Although the details of these mechanisms are beyond the scope of this text, for comparative purposes we will trace the multiplication cycles of the four nucleic acid types of RNA viruses (three of which are shown in Figure 13.17). RNA viruses multiply in the host cell's cytoplasm. The major differences among the multiplication processes of these viruses lie in how mRNA and viral RNA are produced. Once viral RNA and viral proteins are synthesized, maturation occurs by similar means among all animal viruses, as will be discussed shortly.

Picornaviridae Picornaviruses, such as poliovirus (see Chapter 22, page 620), are single-stranded RNA viruses. They are the smallest viruses; and the prefix *pico-* (small) plus *RNA* gives these viruses their name. The RNA within the virion is called a **sense strand** (or **+ strand**), because it can act as mRNA. After attachment, penetration, and uncoating are completed, the single-stranded viral RNA (Figure 13.17a) is translated into two principal proteins, which inhibit the host cell's synthesis of RNA and protein and which form an enzyme called *RNA-dependent RNA polymerase.* This enzyme catalyzes the synthesis of another strand of RNA, which is complementary in base sequence to the original infecting strand. This new strand, called an **antisense strand** (or **− strand**), serves as a template to produce additional + strands. The + strands may serve as mRNA for the translation of capsid proteins, may become incorporated into capsid proteins to form a new virus, or may serve as a template for continued RNA multiplication. Once viral RNA and viral protein are synthesized, maturation occurs.

Togaviridae Togaviruses, which include *ar*thropod-*bo*rne *arbo*viruses or alphaviruses (see Chapter 22, page 624), also contain a single + strand of RNA. Togaviruses are enveloped viruses; their name is from the Latin word for covering, *toga*. Keep in mind that these are not the only enveloped viruses. After a − strand is made from the + strand, two types of mRNA are transcribed from the − strand. One type of mRNA is a short strand that codes for envelope proteins; the other, longer strand serves as mRNA for capsid proteins and can become incorporated into a capsid.

Rhabdoviridae Rhabdoviruses, such as rabiesvirus (genus *Lyssavirus*; see Chapter 22, page 622), are usually bullet-shaped (Figure 13.18a). *Rhabdo-* is from the Greek word for rod, which is not really an accurate description of their morphology. They contain a single − strand of RNA (Figure 13.17b). They also contain an *RNA-dependent RNA polymerase* that uses the − strand as a template from which to produce a + strand. The + strand serves as mRNA and as a template for synthesis of new viral RNA.

Reoviridae Reoviruses were named for their habitats: the respiratory and enteric (digestive) systems of humans. They were not associated with any diseases when first discovered, so they were considered orphan viruses. Their name comes from the first letters of *r*espiratory, *e*nteric, and *o*rphan. Three serotypes are now known to cause respiratory tract and intestinal tract infections.

The capsid containing the double-stranded RNA is digested upon entering a host cell. Viral mRNA is produced in the cytoplasm, where it is used to synthesize more viral proteins (Figure 13.17c). One of the newly synthesized viral proteins acts as *RNA-dependent RNA polymerase* to produce more − strands of RNA. The mRNA + and − strands form the double-stranded RNA that is then surrounded by capsid proteins.

Retroviridae Many retroviruses infect vertebrates (Figure 13.18b). One genus of retrovirus, *Lentivirus,* includes the subspecies HIV-1 and HIV-2, which cause AIDS (see Chapter 19, pages 539–548). The retroviruses that cause cancer will be discussed later in this chapter.

The formation of mRNA and RNA for new retrovirus virions is shown in Figure 13.19 on page 390. These viruses carry **reverse transcriptase,** which uses the viral RNA as a template to produce complementary double-stranded DNA. This enzyme also degrades the original viral RNA. The name *retrovirus* is derived from the first letters of *re*verse *tr*anscriptase. The viral DNA is then

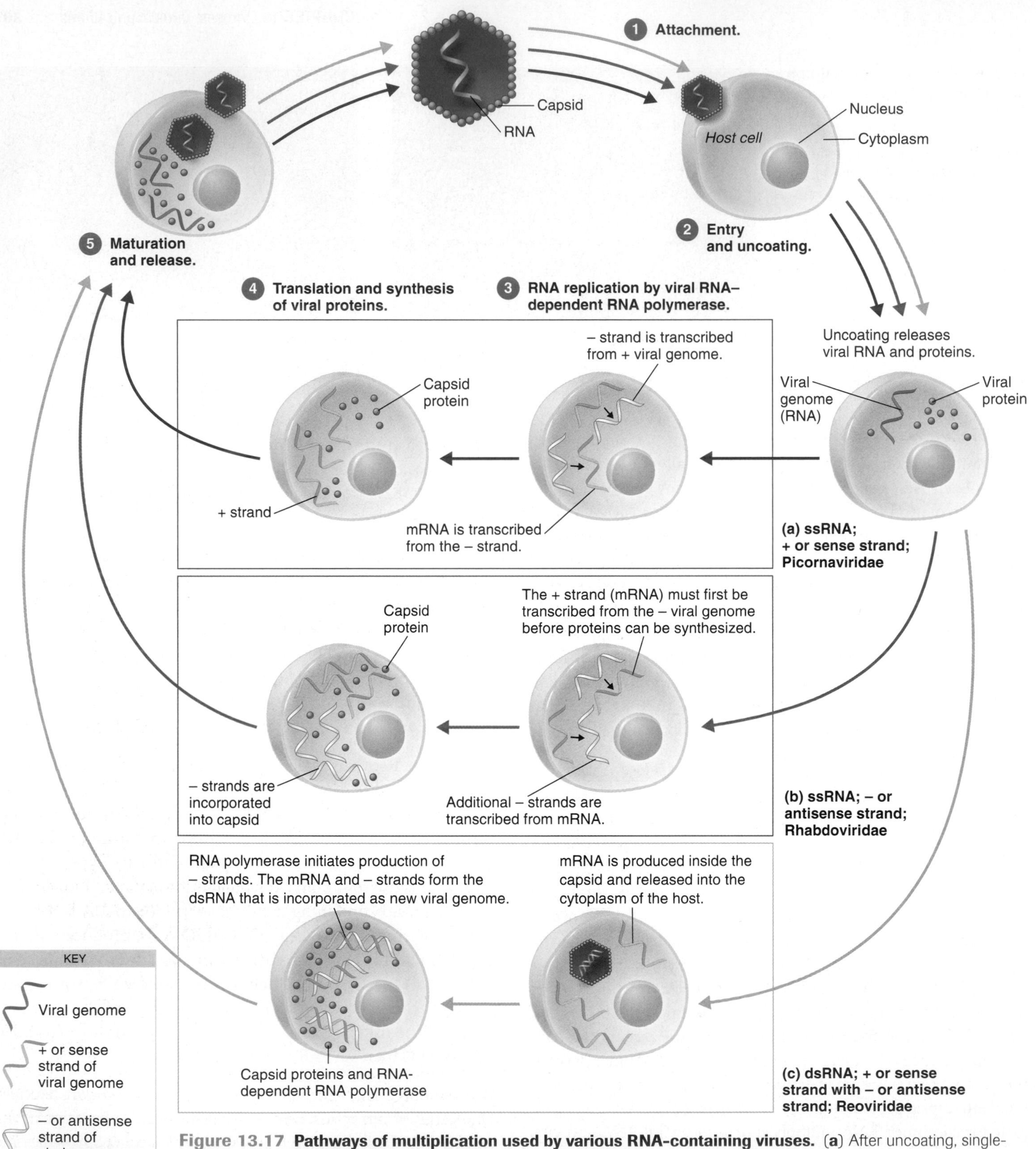

1 Attachment.
Capsid
RNA
Nucleus
Host cell
Cytoplasm
2 Entry and uncoating.
5 Maturation and release.
4 Translation and synthesis of viral proteins.
3 RNA replication by viral RNA–dependent RNA polymerase.
Uncoating releases viral RNA and proteins.
Viral genome (RNA)
Viral protein
– strand is transcribed from + viral genome.
Capsid protein
+ strand
mRNA is transcribed from the – strand.
(a) ssRNA; + or sense strand; Picornaviridae
The + strand (mRNA) must first be transcribed from the – viral genome before proteins can be synthesized.
Capsid protein
– strands are incorporated into capsid
Additional – strands are transcribed from mRNA.
(b) ssRNA; – or antisense strand; Rhabdoviridae
RNA polymerase initiates production of – strands. The mRNA and – strands form the dsRNA that is incorporated as new viral genome.
mRNA is produced inside the capsid and released into the cytoplasm of the host.
Capsid proteins and RNA-dependent RNA polymerase
(c) dsRNA; + or sense strand with – or antisense strand; Reoviridae
KEY
Viral genome
+ or sense strand of viral genome
– or antisense strand of viral genome
ss = single-stranded
ds = double-stranded

Figure 13.17 Pathways of multiplication used by various RNA-containing viruses. (**a**) After uncoating, single-stranded RNA (ssRNA) viruses with a + strand genome are able to synthesize proteins directly from their + strand. Using the + strand as a template, they transcribe − strands to produce additional + strands to serve as mRNA and be incorporated into capsid proteins as the viral genome. (**b**) The ssRNA viruses with a − strand genome must transcribe a + strand to serve as mRNA before they begin synthesizing proteins. The mRNA transcribes additional − strands for incorporation into capsid protein. Both ssRNA and (**c**) dsRNA viruses must use mRNA (+ strand) to code for proteins, including capsid proteins.

Q Why is − strand RNA made by picornaviruses and reoviruses? By rhabdoviruses?

integrated into a host cell chromosome as a **provirus.** Unlike a prophage, the provirus never comes out of the chromosome. As a provirus, HIV is protected from the host's immune system and antiviral drugs.

Sometimes the provirus simply remains in a latent state and replicates when the DNA of the host cell replicates. In other cases, the provirus is expressed and produces new viruses, which may infect adjacent cells. Mutagens such as gamma radiation can induce expression of a provirus. In oncogenic retroviruses, the provirus can also convert the host cell into a tumor cell; possible mechanisms for this phenomenon will be discussed later.

Maturation and Release

The first step in viral maturation is the assembly of the protein capsid; this assembly is usually a spontaneous process. The capsids of many animal viruses are enclosed by an envelope consisting of protein, lipid, and carbohydrate, as noted earlier. Examples of such viruses include orthomyxoviruses and paramyxoviruses. The envelope protein is encoded by the viral genes and is incorporated into the plasma membrane of the host cell. The envelope lipid and carbohydrate are encoded by host cell genes and are present in the plasma membrane. The envelope actually develops around the capsid by a process called **budding** (Figure 13.20).

After the sequence of attachment, entry, uncoating, and biosynthesis of viral nucleic acid and protein, the assembled capsid containing nucleic acid pushes through the plasma membrane. As a result, a portion of the plasma membrane, now the envelope, adheres to the virus. This extrusion of a virus from a host cell is one method of release. Budding does not immediately kill the host cell, and in some cases the host cell survives.

Nonenveloped viruses are released through ruptures in the host cell plasma membrane. In contrast to budding, this type of release usually results in the death of the host cell. **Animations** Viral Replication: Overview; Animal Viruses. **www.microbiologyplace.com**

CHECK YOUR UNDERSTANDING

✓ Describe the principal events of attachment, entry, uncoating, biosynthesis, maturation, and release of an enveloped DNA-containing virus. **13-10**

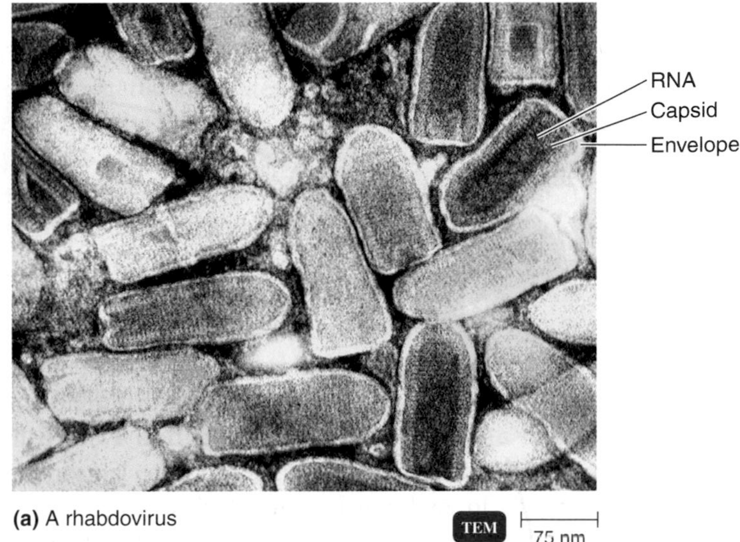

(a) A rhabdovirus

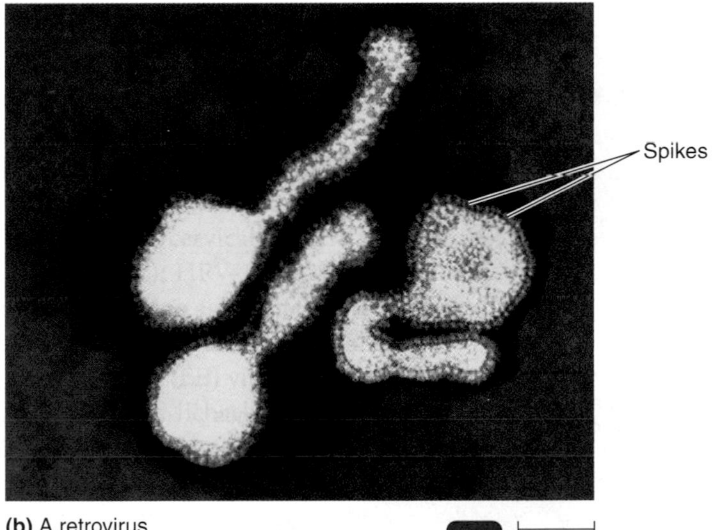

(b) A retrovirus

Figure 13.18 RNA-containing animal viruses. (**a**) Vesicular stomatitis viruses, a member of the family Rhabdoviridae. (**b**) Mouse mammary tumor virus, a Retroviridae, causes tumors in mice.

Q Why do viruses with a + strand of RNA make a − strand of RNA?

Viruses and Cancer

LEARNING OBJECTIVES

13-11 Define *oncogene* and *transformed cell.*

13-12 Discuss the relationship between DNA- and RNA-containing viruses and cancer.

Several types of cancer are now known to be caused by viruses. Molecular biological research shows that the mechanisms of the diseases are similar, even when a virus does not cause the cancer.

The relationship between cancers and viruses was first demonstrated in 1908, when virologists Wilhelm Ellerman and Olaf Bang, working in Denmark, were trying to isolate the causative agent of chicken leukemia. They found that leukemia could be transferred to healthy chickens by cell-free filtrates that contained viruses. Three years later, F. Peyton Rous, working at the Rockefeller Institute in New York, found that a chicken **sarcoma** (cancer of connective tissue) can be similarly transmitted. Virus-induced **adenocarcinomas** (cancers of glandular epithelial tissue) in mice were discovered in 1936. At that time, it was clearly shown that mouse mammary gland tumors are transmitted from mother to offspring through the mother's milk.

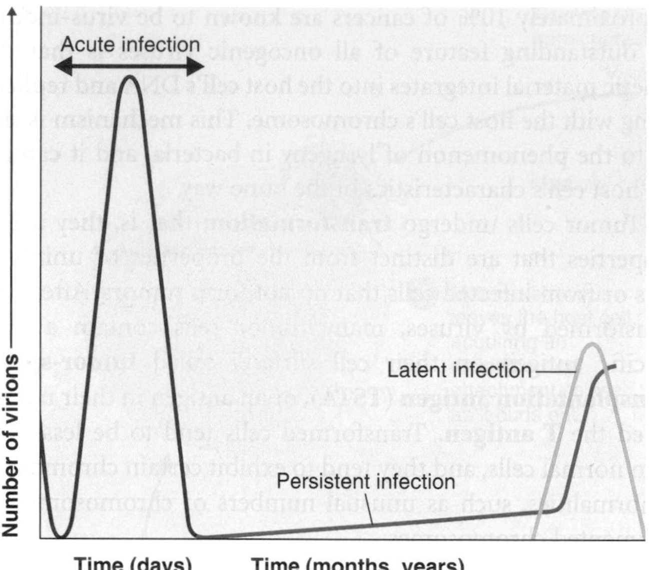

Figure 13.21 Latent and persistent viral infections.

Q How do latent and persistent infections differ?

Q&A The ability of retroviruses to induce tumors is related to their production of a reverse transcriptase by the mechanism described earlier (see Figure 13.19). The provirus, which is the double-stranded DNA molecule synthesized from the viral RNA, becomes integrated into the host cell's DNA; new genetic material is thereby introduced into the host's genome, and this is the key reason retroviruses can contribute to cancer. Some retroviruses contain oncogenes; others contain promoters that turn on oncogenes or other cancer-causing factors.

CHECK YOUR UNDERSTANDING

✓ What is a provirus? **13-11**

✓ How can an RNA virus cause cancer if it doesn't have DNA to insert into a cell's genome? **13-12**

Latent Viral Infections

LEARNING OBJECTIVE

13-13 Provide an example of a latent viral infection.

A virus can remain in equilibrium with the host and not actually produce disease for a long period, often many years. The oncogenic viruses just discussed are examples of such latent infections. All of the human herpesviruses can remain in host cells throughout the life of an individual. When herpesviruses are reactivated by immunosuppression (for example, AIDS), the resulting infection may be fatal. The classic example of such a **latent infection** in viruses is the infection of the skin by herpes simplex virus, which produces cold sores. This virus inhabits the host's nerve cells but causes no damage until it is activated by a stimulus such as fever or sunburn—hence the term *fever blister.*

In some individuals, viruses are produced, but symptoms never appear. Even though a large percentage of the human population carries the herpes simplex virus, only 10 to 15% of people carrying the virus exhibit the disease. The virus of some latent infections can exist in a lysogenic state within host cells.

The chickenpox virus (*Varicellovirus*) can also exist in a latent state. Chickenpox (varicella) is a skin disease that is usually acquired in childhood. The virus gains access to the skin via the blood. From the blood, some viruses may enter nerves, where they remain latent. Later, changes in the immune (T-cell) response can activate these latent viruses, causing shingles (zoster). The shingles rash appears on the skin along the nerve in which the virus was latent. Shingles occurs in 10 to 20% of people who have had chickenpox.

Persistent Viral Infections

LEARNING OBJECTIVE

13-14 Differentiate persistent viral infections from latent viral infections.

A **persistent** or **chronic viral infection** occurs gradually over a long period. Typically, persistent viral infections are fatal.

A number of persistent viral infections have in fact been shown to be caused by conventional viruses. For example, several years after causing measles, the measles virus can be responsible for a rare form of encephalitis called subacute sclerosing panencephalitis (SSPE). A persistent viral infection is apparently different from a latent viral infection in that, in most persistent viral infections, detectable infectious virus gradually builds up over a long period, rather than appearing suddenly (**Figure 13.21**).

Several examples of latent and persistent viral infections are listed in **Table 13.5**.

CHECK YOUR UNDERSTANDING

✓ Is shingles a persistent or latent infection? **13-13, 13-14**

Prions

LEARNING OBJECTIVE

13-15 Discuss how a protein can be infectious.

A few infectious diseases are caused by prions. In 1982, American neurobiologist Stanley Prusiner proposed that infectious proteins caused a neurological disease in sheep called scrapie. The infectivity of scrapie-infected brain tissue is reduced by treatment with proteases but not by treatment with radiation, suggesting that the infectious agent is pure protein. Prusiner coined the name **prion** for *pro*teinaceous *in*fectious particle.

Nine animal diseases now fall into this category, including the "mad cow disease" that emerged in cattle in Great Britain in 1987. All nine are neurological diseases called spongiform encephalopathies because large vacuoles develop in the brain (Figure 22.18a, page 630). The human diseases are kuru, Creutzfeldt-Jakob disease (CJD), Gerstmann-Sträussler-Scheinker syndrome, and fatal familial insomnia. (Neurological diseases are discussed in Chapter 22.) These diseases run in families, which indicates a possible genetic cause. However, they cannot be purely inherited, because mad cow disease arose from feeding scrapie-infected sheep meat to cattle, and the new (bovine) variant was transmitted to humans who ate undercooked beef from infected cattle (see Chapter 1, page 20). Additionally, CJD has been transmitted with transplanted nerve tissue and contaminated surgical instruments.

These diseases are caused by the conversion of a normal host glycoprotein called PrP^{C} (for cellular prion protein) into an infectious form called PrP^{Sc} (for scrapie protein). The gene for PrP^{C} is located on chromosome 20 in humans. Recent evidence suggests that PrP^{C} is involved in regulating cell death. (See the discussion of apoptosis on page 489.) One hypothesis for how an infectious agent that lacks any nucleic acid can reproduce is shown in **Figure 13.22**.

The actual cause of cell damage is not known. Fragments of PrP^{Sc} molecules accumulate in the brain, forming plaques; these plaques are used for postmortem diagnosis, but they do not appear to be the cause of cell damage. **Animations** Prion Reproduction: Overview, Characteristics, Diseases. **www.microbiologyplace.com**

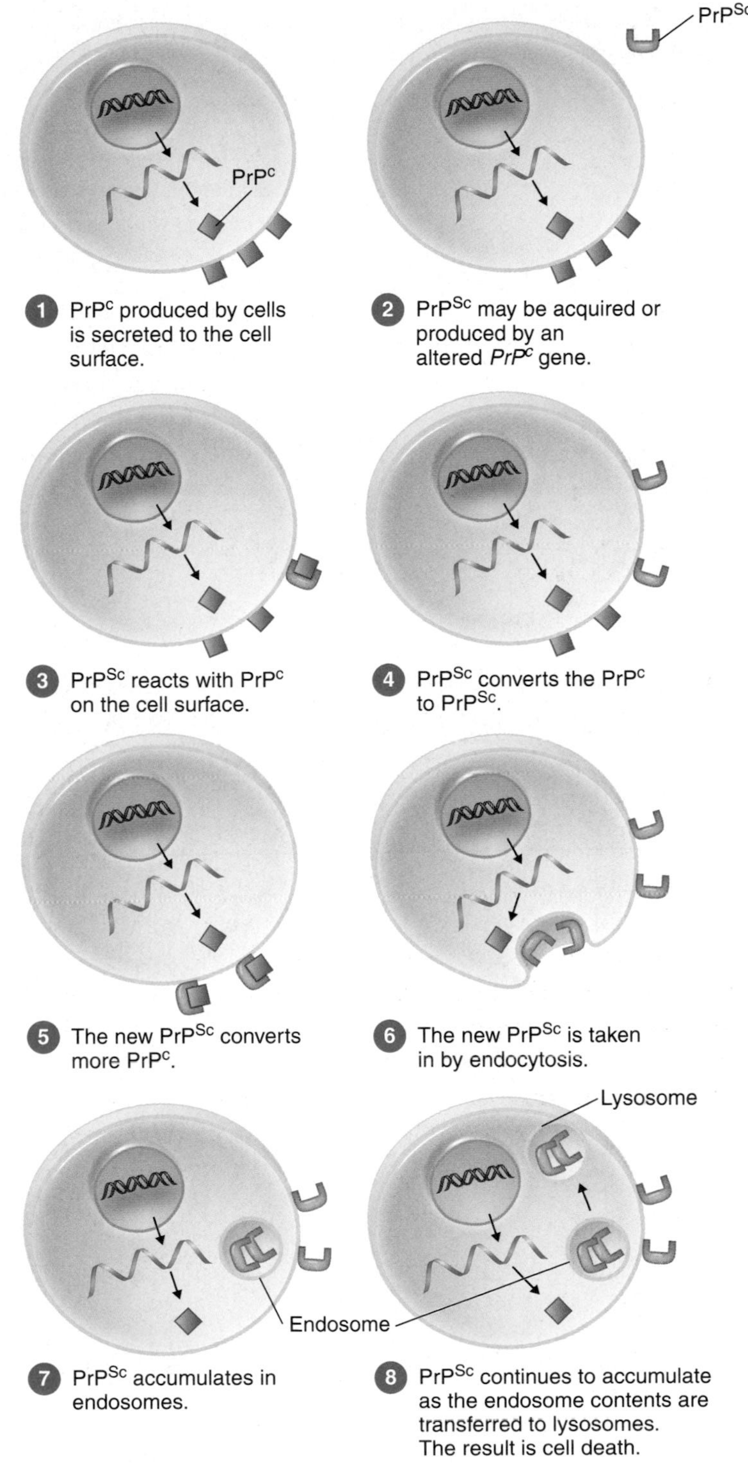

PrP^Sc
PrP^c
1 PrP^c produced by cells is secreted to the cell surface.
2 PrP^Sc may be acquired or produced by an altered PrP^c gene.
3 PrP^Sc reacts with PrP^c on the cell surface.
4 PrP^Sc converts the PrP^c to PrP^Sc.
5 The new PrP^Sc converts more PrP^c.
6 The new PrP^Sc is taken in by endocytosis.
Lysosome
Endosome
7 PrP^Sc accumulates in endosomes.
8 PrP^Sc continues to accumulate as the endosome contents are transferred to lysosomes. The result is cell death.

Figure 13.22 How a protein can be infectious. If an abnormal prion protein (PrP^{Sc}) enters a cell, it changes a normal prion protein to PrP^{Sc}, which now can change another normal PrP^{C}, resulting in an accumulation of the abnormal PrP^{Sc}.

Q How do prions differ from viruses?

Plant Viruses and Viroids

LEARNING OBJECTIVES

13-16 Differentiate virus, viroid, and prion.

13-17 Describe the lytic cycle for a plant virus.

Plant viruses resemble animal viruses in many respects: plant viruses are morphologically similar to animal viruses, and they have similar types of nucleic acid (**Table 13.6**). In fact, some plant viruses can multiply inside insect cells. Plant viruses cause many diseases of economically important crops, including beans (bean mosaic virus), corn and sugarcane (wound tumor virus), and potatoes (potato yellow dwarf virus). Viruses can cause color change, deformed growth, wilting, and stunted growth in their plant hosts. Some hosts, however, remain symptomless and only serve as reservoirs of infection.

Plant cells are generally protected from disease by an impermeable cell wall. Viruses must enter through wounds or be assisted by other plant parasites, including nematodes, fungi, and, most often, insects that suck the plant's sap. Once one plant is infected, it can spread infection to other plants in its pollen and seeds.

In laboratories, plant viruses are cultured in protoplasts (plant cells with the cell walls removed) and in insect cell cultures.

Table 13.5 Examples of Latent and Persistent Viral Infections in Humans

Disease	Primary Effect	Causative Virus
Latent	**No symptoms during latency, viruses not usually released.**	
Cold sores	Skin and mucous membrane lesions; genital lesions	Herpes simplex 1 and 2
Leukemia	Increased white blood cell growth	HTLV-1 and -2
Shingles	Skin lesions	*Varicellovirus* (Herpesvirus)
Persistent	**Viruses continuously released**	
Cervical cancer	Increased cell growth	Human papillomavirus
HIV/AIDS	Decreased CD_4^+T cells	HIV-1 and -2 (*Lentivirus*)
Liver cancer	Increased cell growth	Hepatitis B virus
Persistent enterovirus infection	Mental deterioration associated with AIDS	Echoviruses
Progressive encephalitis	Rapid mental deterioration	Rubella virus
Subacute sclerosing panencephalitis (SSPE)	Mental deterioration	Measles virus

Table 13.6 Classification of Some Major Plant Viruses

Characteristic	Viral Family	Viral Genus or Unclassified Members	Morphology	Method of Transmission
Double-stranded DNA, nonenveloped	Papovaviridae	Cauliflower mosaic virus		Aphids
Single-stranded RNA, + strand, nonenveloped	Potyviridae	Watermelon wilt		Whiteflies
	Tetraviridae	*Tobamovirus*		Wounds
Single-stranded RNA, − strand, enveloped	Rhabdoviridae	Potato yellow dwarf virus		Leafhoppers and aphids
Double-stranded RNA, nonenveloped	Reovirus	Wound tumor virus		Leafhoppers

Some plant diseases are caused by **viroids,** short pieces of naked RNA, only 300 to 400 nucleotides long, with no protein coat. The nucleotides are often internally paired, so the molecule has a closed, folded, three-dimensional structure that presumably helps protect it from attack by cellular enzymes. The RNA does not code for any proteins. Thus far, viroids have been conclusively identified as pathogens only of plants. Annually, infections by viroids, such as potato spindle tuber viroid, result in losses of millions of dollars from crop damage (**Figure 13.23**).

Current research on viroids has revealed similarities between the base sequences of viroids and introns. Recall from Chapter 8 (page 220) that introns are sequences of genetic material that do not code for polypeptides. This observation has led to

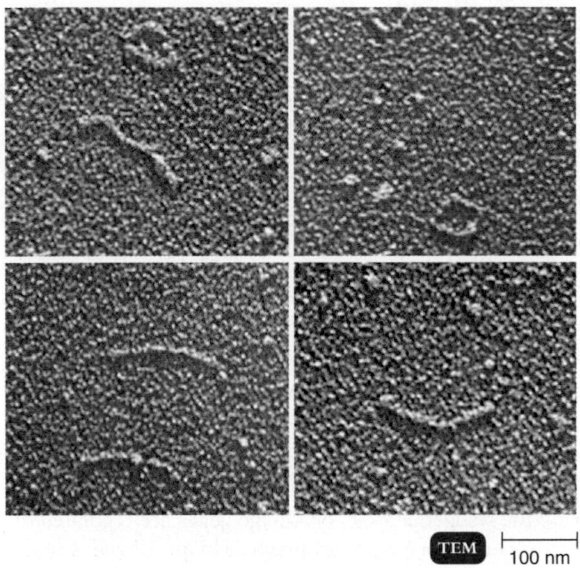

Figure 13.23 Linear and circular potato spindle tuber viroid (PSTV).

Q How do viroids differ from prions?

the hypothesis that viroids evolved from introns, leading to speculation that future researchers may discover animal viroids.

CHECK YOUR UNDERSTANDING

✓ Contrast viroids and prions, and for each name a disease it causes. **13-15, 13-16**

✓ How do plant viruses enter host cells? **13-17**

STUDY OUTLINE

The **MyMicrobiology Place** wesbite (**www.microbiologyplace.com**) will help you get ready for tests with its simple three-step approach: ❶ **take a pre-test** and obtain a personalized study plan, ❷ **learn and practice** with animations, tutorials, and MP3 tutor sessions, and ❸ **test yourself** with quizzes and a chapter post-test.

General Characteristics of Viruses (pp. 368–369)

1. Depending on one's viewpoint, viruses may be regarded as exceptionally complex aggregations of nonliving chemicals or as exceptionally simple living microbes.
2. Viruses contain a single type of nucleic acid (DNA or RNA) and a protein coat, sometimes enclosed by an envelope composed of lipids, proteins, and carbohydrates.
3. Viruses are obligatory intracellular parasites. They multiply by using the host cell's synthesizing machinery to cause the synthesis of specialized elements that can transfer the viral nucleic acid to other cells.

Host Range (pp. 368–369)

4. *Host range* refers to the spectrum of host cells in which a virus can multiply.
5. Most viruses infect only specific types of cells in one host species.
6. Host range is determined by the specific attachment site on the host cell's surface and the availability of host cellular factors.

Viral Size (p. 369)

7. Viral size is ascertained by electron microscopy.
8. Viruses range from 20 to 1000 nm in length.

Viral Structure (pp. 370–373)

1. A virion is a complete, fully developed viral particle composed of nucleic acid surrounded by a coat.

Nucleic Acid (pp. 371–372)

2. Viruses contain either DNA or RNA, never both, and the nucleic acid may be single- or double-stranded, linear or circular, or divided into several separate molecules.
3. The proportion of nucleic acid in relation to protein in viruses ranges from about 1% to about 50%.

Capsid and Envelope (pp. 372–373)

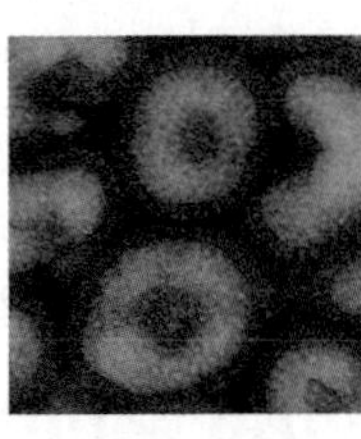

4. The protein coat surrounding the nucleic acid of a virus is called the capsid.
5. The capsid is composed of subunits, capsomeres, which can be a single type of protein or several types.
6. The capsid of some viruses is enclosed by an envelope consisting of lipids, proteins, and carbohydrates.
7. Some envelopes are covered with carbohydrate-protein complexes called spikes.

General Morphology (p. 373)

8. Helical viruses (for example, Ebola virus) resemble long rods, and their capsids are hollow cylinders surrounding the nucleic acid.
9. Polyhedral viruses (for example, adenovirus) are many-sided. Usually the capsid is an icosahedron.

10. Enveloped viruses are covered by an envelope and are roughly spherical but highly pleomorphic. There are also enveloped helical viruses (for example, influenza virus) and enveloped polyhedral viruses (for example, *Simplexvirus*).
11. Complex viruses have complex structures. For example, many bacteriophages have a polyhedral capsid with a helical tail attached.

Taxonomy of Viruses (pp. 373–374)

1. Classification of viruses is based on type of nucleic acid, strategy for replication, and morphology.
2. Virus family names end in *-viridae;* genus names end in *-virus.*
3. A viral species is a group of viruses sharing the same genetic information and ecological niche.

Isolation, Cultivation, and Identification of Viruses (pp. 374–379)

1. Viruses must be grown in living cells.
2. The easiest viruses to grow are bacteriophages.

Growing Bacteriophages in the Laboratory (pp. 374, 377)

3. The plaque method mixes bacteriophages with host bacteria and nutrient agar.
4. After several viral multiplication cycles, the bacteria in the area surrounding the original virus are destroyed; the area of lysis is called a plaque.
5. Each plaque originates with a single viral particle; the concentration of viruses is given as plaque-forming units.

Growing Animal Viruses in the Laboratory (pp. 377–379)

6. Cultivation of some animal viruses requires whole animals.
7. Simian AIDS and feline AIDS provide models for studying human AIDS.
8. Some animal viruses can be cultivated in embryonated eggs.
9. Cell cultures are cells growing in culture media in the laboratory.
10. Primary cell lines and embryonic diploid cell lines grow for a short time in vitro.
11. Continuous cell lines can be maintained in vitro indefinitely.
12. Viral growth can cause cytopathic effects in the cell culture.

Viral Identification (p. 379)

13. Serological tests are used most often to identify viruses.
14. Viruses may be identified by RFLPs and PCR.

Viral Multiplication (pp. 379–389)

1. Viruses do not contain enzymes for energy production or protein synthesis.
2. For a virus to multiply, it must invade a host cell and direct the host's metabolic machinery to produce viral enzymes and components.

Multiplication of Bacteriophages (pp. 379–382)

3. During the lytic cycle, a phage causes the lysis and death of a host cell.
4. Some viruses can either cause lysis or have their DNA incorporated as a prophage into the DNA of the host cell. The latter situation is called lysogeny.
5. During the attachment phase of the lytic cycle, sites on the phage's tail fibers attach to complementary receptor sites on the bacterial cell.
6. In penetration, phage lysozyme opens a portion of the bacterial cell wall, the tail sheath contracts to force the tail core through the cell wall, and phage DNA enters the bacterial cell. The capsid remains outside.

7. In biosynthesis, transcription of phage DNA produces mRNA coding for proteins necessary for phage multiplication. Phage DNA is replicated, and capsid proteins are produced. During the eclipse period, separate phage DNA and protein can be found.
8. During maturation, phage DNA and capsids are assembled into complete viruses.
9. During release, phage lysozyme breaks down the bacterial cell wall, and the new phages are released.
10. During the lysogenic cycle, prophage genes are regulated by a repressor coded for by the prophage. The prophage is replicated each time the cell divides.
11. Exposure to certain mutagens can lead to excision of the prophage and initiation of the lytic cycle.
12. Because of lysogeny, lysogenic cells become immune to reinfection with the same phage and may undergo phage conversion.
13. A lysogenic phage can transfer bacterial genes from one cell to another through transduction. Any genes can be transferred in generalized transduction, and specific genes can be transferred in specialized transduction.

Multiplication of Animal Viruses (pp. 382–389)

14. Animal viruses attach to the plasma membrane of the host cell.
15. Entry occurs by endocytosis or fusion.
16. Animal viruses are uncoated by viral or host cell enzymes.
17. The DNA of most DNA viruses is released into the nucleus of the host cell. Transcription of viral DNA and translation produce viral DNA and, later, capsid proteins. Capsid proteins are synthesized in the cytoplasm of the host cell.
18. DNA viruses include members of the families Adenoviridae, Poxviridae, Herpesviridae, Papovaviridae, and Hepadnaviridae.
19. Multiplication of RNA viruses occurs in the cytoplasm of the host cell. RNA-dependent RNA polymerase synthesizes a double-stranded RNA.
20. Picornaviridae + strand RNA acts as mRNA and directs the synthesis of RNA-dependent RNA polymerase.
21. Togaviridae + strand RNA acts as a template for RNA-dependent RNA polymerase, and mRNA is transcribed from a new − RNA strand.
22. Rhabdoviridae − strand RNA is a template for viral RNA-dependent RNA polymerase, which transcribes mRNA.
23. Reoviridae are digested in host cell cytoplasm to release mRNA for viral biosynthesis.
24. Retroviridae reverse transcriptase (RNA-dependent DNA polymerase) transcribes DNA from RNA.
25. After maturation, viruses are released. One method of release (and envelope formation) is budding. Nonenveloped viruses are released through ruptures in the host cell membrane.

Viruses and Cancer (pp. 389)

1. The earliest relationship between cancer and viruses was demonstrated in the early 1900s, when chicken leukemia and chicken sarcoma were transferred to healthy animals by cell-free filtrates.

The Transformation of Normal Cells Into Tumor Cells (pp. 390–391)

2. When activated, oncogenes transform normal cells into cancerous cells.
3. Viruses capable of producing tumors are called oncogenic viruses.
4. Several DNA viruses and retroviruses are oncogenic.
5. The genetic material of oncogenic viruses becomes integrated into the host cell's DNA.
6. Transformed cells lose contact inhibition, contain virus-specific antigens (TSTA and T antigen), exhibit chromosome abnormalities, and can produce tumors when injected into susceptible animals.

DNA Oncogenic Viruses (p. 391)

7. Oncogenic viruses are found among the Adenoviridae, Herpesviridae, Poxviridae, and Papovaviridae.
8. The EB virus, a herpesvirus, causes Burkitt's lymphoma and nasopharyngeal carcinoma. *Hepadnavirus* causes liver cancer.

RNA Oncogenic Viruses (p. 391)

9. Among the RNA viruses, only retroviruses seem to be oncogenic.
10. HTLV-1 and HTLV-2 have been associated with human leukemia and lymphoma.
11. The virus's ability to produce tumors is related to the production of reverse transcriptase. The DNA synthesized from the viral RNA becomes incorporated as a provirus into the host cell's DNA.
12. A provirus can remain latent, can produce viruses, or can transform the host cell.

Latent Viral Infections (p. 392)

1. A latent viral infection is one in which the virus remains in the host cell for long periods without producing an infection.
2. Examples are cold sores and shingles.

Persistent Viral Infections (p. 392)

1. Persistent viral infections are disease processes that occur over a long period and are generally fatal.
2. Persistent viral infections are caused by conventional viruses; viruses accumulate over a long period.

Prions (pp. 392–393)

1. Prions are infectious proteins first discovered in the 1980s.
2. Prion diseases, such as CJD and mad cow disease, all involve the degeneration of brain tissue.
3. Prion diseases are the result of an altered protein; the cause can be a mutation in the normal gene for PrP^{C} or contact with an altered protein (PrP^{Sc}).

Plant Viruses and Viroids (pp. 393–395)

1. Plant viruses must enter plant hosts through wounds or with invasive parasites, such as insects.
2. Some plant viruses also multiply in insect (vector) cells.
3. Viroids are infectious pieces of RNA that cause some plant diseases, such as potato spindle tuber disease.

STUDY QUESTIONS

Answers to the Review and Multiple Choice questions can be found by turning to the blue Answers tab at the back of the textbook.

Review

1. Why do we classify viruses as obligatory intracellular parasites?
2. List the four properties that define a virus. What is a virion?
3. Describe the four morphological classes of viruses, then diagram and give an example of each.
4. **DRAW IT** Label the principal events of attachment, biosynthesis, entry, and maturation of a + stranded RNA virus. Draw in uncoating.

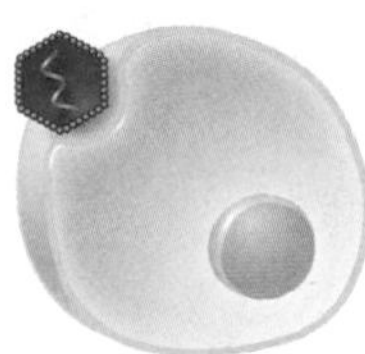
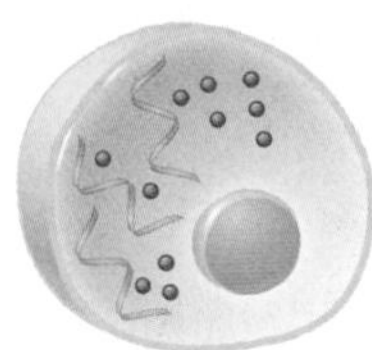
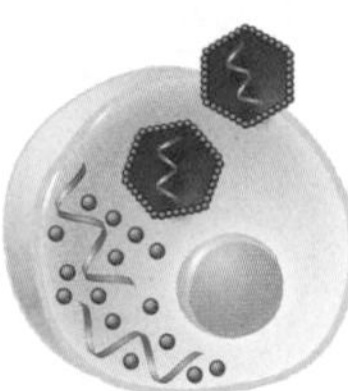

5. Compare biosynthesis of a + stranded RNA and a – stranded RNA virus.
6. Some antibiotics activate phage genes. MRSA releasing Panton-Valentine leukocidin causes a life-threatening disease. Why can this happen following antibiotic treatment?
7. Recall from Chapter 1 that Koch's postulates are used to determine the etiology of a disease. Why is it difficult to determine the etiology of
   a. a viral infection, such as influenza?
   b. cancer?
8. Persistent viral infections such as (a) ___________ might be caused by (b) ___________ that are (c) ___________.
9. Plant viruses cannot penetrate intact plant cells because (a) ___________; therefore, they enter cells by (b) ___________. Plant viruses can be cultured in (c) ___________.

Multiple Choice

1. Place the following in the most likely order for biosynthesis of a bacteriophage: (1) phage lysozyme; (2) mRNA; (3) DNA; (4) viral proteins; (5) DNA polymerase.
   a. 5, 4, 3, 2, 1
   b. 1, 2, 3, 4, 5
   c. 5, 3, 4, 2, 1
   d. 3, 5, 2, 4, 1
   e. 2, 5, 3, 4, 1
2. The molecule serving as mRNA can be incorporated in the newly synthesized virus capsids of all of the following *except*
   a. + strand RNA picornaviruses.
   b. + strand RNA togaviruses.
   c. − strand RNA rhabdoviruses.
   d. double-stranded RNA reoviruses.
   e. double-stranded DNA herpesviruses.
3. A virus with RNA-dependent RNA polymerase
   a. synthesizes DNA from an RNA template.
   b. synthesizes double-stranded RNA from an RNA template.
   c. synthesizes double-stranded RNA from a DNA template.
   d. transcribes mRNA from DNA.
   e. none of the above
4. Which of the following would be the first step in the biosynthesis of a virus with reverse transcriptase?
   a. A complementary strand of RNA must be synthesized.
   b. Double-stranded RNA must be synthesized.
   c. A complementary strand of DNA must be synthesized from an RNA template.
   d. A complementary strand of DNA must be synthesized from a DNA template.
   e. none of the above
5. An example of lysogeny in animals could be
   a. slow viral infections.
   b. latent viral infections.
   c. T-even bacteriophages.
   d. infections resulting in cell death.
   e. none of the above
6. The ability of a virus to infect an organism is regulated by
   a. the host species.
   b. the type of cells.
   c. the availability of an attachment site.
   d. cell factors necessary for viral replication.
   e. all of the above
7. Which of the following statements is *not* true?
   a. Viruses contain DNA or RNA.
   b. The nucleic acid of a virus is surrounded by a protein coat.
   c. Viruses multiply inside living cells using viral mRNA, tRNA, and ribosomes.
   d. Viruses cause the synthesis of specialized infectious elements.
   e. Viruses multiply inside living cells.
8. Place the following in the order in which they are found in a host cell: (1) capsid proteins; (2) infective phage particles; (3) phage nucleic acid.
   a. 1, 2, 3
   b. 3, 2, 1
   c. 2, 1, 3
   d. 3, 1, 2
   e. 1, 3, 2
9. Which of the following does *not* initiate DNA synthesis?
   a. a double-stranded DNA virus (Poxviridae)
   b. a DNA virus with reverse transcriptase (Hepadnaviridae)
   c. an RNA virus with reverse transcriptase (Retroviridae)
   d. a single-stranded RNA virus (Togaviridae)
   e. none of the above
10. A viral species is not defined on the basis of the disease symptoms it causes. The best example of this is
    a. polio.
    b. rabies.
    c. hepatitis.
    d. chickenpox and shingles.
    e. measles.

Critical Thinking

1. Discuss the arguments for and against the classification of viruses as living organisms.
2. In some viruses, capsomeres function as enzymes as well as structural supports. Of what advantage is this to the virus?
3. Why was the discovery of simian AIDS and feline AIDS important?
4. Prophages and proviruses have been described as being similar to bacterial plasmids. What similar properties do they exhibit? How are they different?

Clinical Applications

1. A 40-year-old man who was seropositive for HIV experienced abdominal pain, fatigue, and low-grade fever (38°C) for 2 weeks. A chest X-ray examination revealed lung infiltrates. Gram and acid-fast stains were negative. A viral culture revealed the cause of his symptoms: a large, enveloped polyhedral virus with double-stranded DNA. What is the disease? Which virus causes it? Why was a viral culture done after the Gram and acid-fast stain results were obtained?
2. A newborn female developed extensive vesicular and ulcerative lesions over her face and chest. What is the most likely cause of her symptoms? How would you determine the viral cause of this disease without doing a viral culture?
3. Thirty-two people in the same town reported to their physicians with fever (40°C), jaundice, and tender abdomen. All 32 had eaten an ice-slush beverage purchased from a local convenience store. Liver function tests were abnormal. Over the next several months, the symptoms subsided, and liver function returned to normal. What is the disease? This disease could be caused by a member of the Picornaviridae, Hepadnaviridae, or Flaviviridae. Differentiate among these families by method of transmission, morphology, nucleic acid, and type of replication.

14 Principles of Disease and Epidemiology

Now that you have a basic understanding of the structures and functions of microorganisms and some idea of the variety of microorganisms that exist, we can consider how the human body and various microorganisms interact in terms of health and disease.

We all have defenses to keep us healthy. In spite of these, however, we are still susceptible to **pathogens** (disease-causing microorganisms). A rather delicate balance exists between our defenses and the pathogenic mechanisms of microorganisms. When our defenses resist these pathogenic capabilities, we maintain our health—when the pathogen's capability overcomes our defenses, disease results. After the disease has become established, an infected person may recover completely, suffer temporary or permanent damage, or die.

In Part Three we examine some of the principles of infection and disease, the mechanisms by which pathogens cause disease, the body's defenses against disease, and the ways that microbial diseases can be prevented by immunization and controlled by drugs. This first chapter discusses the general principles of disease, starting with a discussion of the meaning and scope of pathology. In the last section of this chapter, "Epidemiology," you will learn how these principles are useful in studying and controlling disease.

UNDER THE MICROSCOPE

Staphylococcus aureus. These pathogenic bacteria are often found on the skin of healthy people.

Q&A

A patient entered the hospital to have torn cartilage removed from her right knee. The surgery was scheduled as a same-day procedure. Unfortunately, she subsequently developed pneumonia and wasn't released until 10 days later. How would you account for these events?

Look for the answer in the chapter.

Pathology, Infection, and Disease

LEARNING OBJECTIVE

14-1 Define *pathology, etiology, infection,* and *disease.*

Pathology is the scientific study of disease (*pathos* = suffering; *logos* = science). Pathology is first concerned with the cause, or **etiology,** of disease. Second, it deals with **pathogenesis,** the manner in which a disease develops. Third, pathology is concerned with the *structural* and *functional changes* brought about by disease and with their final effects on the body.

Although the terms *infection* and *disease* are sometimes used interchangeably, they differ somewhat in meaning. **Infection** is the invasion or colonization of the body by pathogenic microorganisms; **disease** occurs when an infection results in any change from a state of health. Disease is an abnormal state in which part or all of the body is not properly adjusted or incapable of performing its normal functions. An infection may exist in the absence of detectable disease. For example, the body may be infected with the virus that causes AIDS, but there may be no symptoms of the disease.

The presence of a particular type of microorganism in a part of the body where it is not normally found is also called an infection—and may lead to disease. For example, although large numbers of *E. coli* are normally present in the healthy intestine, their infection of the urinary tract usually results in disease.

Few microorganisms are pathogenic. In fact, the presence of some microorganisms can even benefit the host. Therefore, before we discuss the role of microorganisms in causing disease, let's examine the relationship of the microorganisms to the healthy human body.

CHECK YOUR UNDERSTANDING

✓ What are the objectives of pathology? **14-1**

Normal Microbiota

LEARNING OBJECTIVES

14-2 Define *normal* and *transient microbiota.*

14-3 Compare commensalism, mutualism, and parasitism, and give an example of each.

14-4 Contrast normal microbiota and transient microbiota with opportunistic microorganisms.

Animals, including humans, are generally free of microbes in utero. At birth, however, normal and characteristic microbial populations begin to establish themselves. Just before a woman gives birth, lactobacilli in her vagina multiply rapidly. The newborn's first contact with microorganisms is usually with these lactobacilli, and they become the predominant organisms in the newborn's intestine. More microorganisms are introduced to the newborn's body from the environment when breathing begins and feeding starts. After birth, *E. coli* and other bacteria acquired from foods begin to inhabit the large intestine. These microorganisms remain there throughout life and, in response to altered environmental conditions, may increase or decrease in number and contribute to disease.

Many other usually harmless microorganisms establish themselves inside other parts of the normal adult body and on its surface. A typical human body contains 1×10^{13} body cells, yet harbors an estimated 1×10^{14} bacterial cells (10 times more bacterial cells than human cells). This gives you an idea of the abundance of microorganisms that normally reside in the human body. The microorganisms that establish more or less permanent residence (colonize) but that do not produce disease under normal conditions are members of the body's **normal microbiota,** or **normal flora** (**Figure 14.1**). Others, called **transient microbiota,** may be present for several days, weeks, or months and then disappear. Microorganisms are not found throughout the entire human body but are localized in certain regions, as shown in **Table 14.1** on page 402.

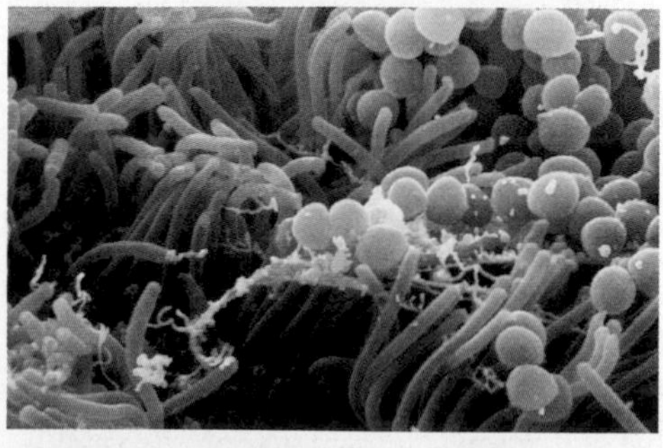

(a) Bacteria (orange spheres) on the surface of the nasal epithelium SEM 2 μm

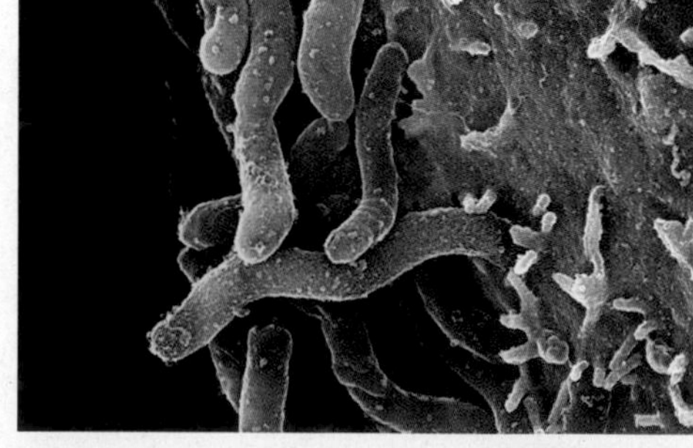

(b) Bacteria on the lining of the stomach SEM 2 μm

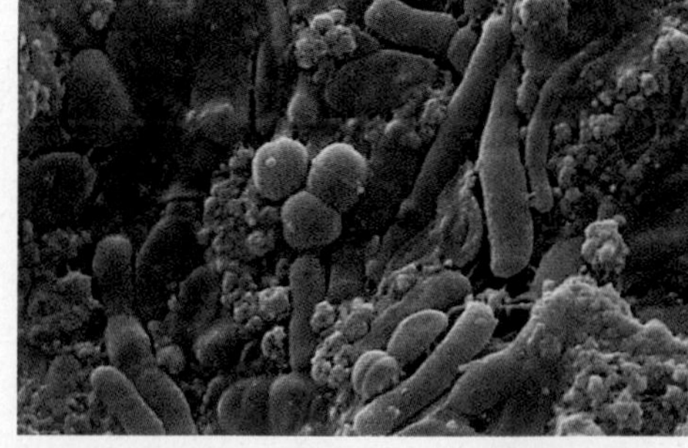

(c) Bacteria in the large intestine

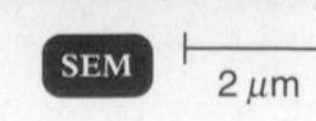

Figure 14.1 Representative normal microbiota for different regions of the body.

Of what value are normal microbiota?

Many factors determine the distribution and composition of the normal microbiota. Among these are nutrients, physical and chemical factors, defenses of the host, and mechanical factors. Microbes vary with respect to the types of nutrients that they can use as an energy source. Accordingly, microbes can colonize only those body sites that can supply the appropriate nutrients. These nutrients may be derived from secretory and excretory products of cells, substances in body fluids, dead cells, and foods in the gastrointestinal tract.

A number of physical and chemical factors affect the growth of microbes and thus the growth and composition of the normal microbiota. Among these are temperature, pH, available oxygen and carbon dioxide, salinity, and sunlight.

You will learn in Chapters 16 and 17 that the human body has certain defenses against microbes. These defenses include a variety of molecules and activated cells that kill microbes, inhibit their growth, prevent their adhesion to host cell surfaces, and neutralize toxins that microbes produce. Although these defenses are extremely important against pathogens, their role in determining and regulating the normal microbiota is unclear.

Certain regions of the body are subjected to mechanical forces that may affect colonization by the normal microbiota. For example, the chewing actions of the teeth and tongue movements can dislodge microbes attached to tooth and mucosal surfaces. In the gastrointestinal tract, the flow of saliva and digestive secretions and the various muscular movements of the throat, esophagus, stomach, and intestines can remove unattached microbes. The flushing action of urine also removes unattached microbes. In the respiratory system, mucus traps microbes, which cilia then propel toward the throat for elimination.

The conditions provided by the host at a particular body site vary from one person to another. Among the factors that also affect the normal microbiota are age, nutritional status, diet, health status, disability, hospitalization, emotional state, stress, climate, geography, personal hygiene, living conditions, occupation, and lifestyle.

The principal normal microbiota in different regions of the body and some distinctive features of each region are listed in Table 14.1. Normal microbiota are also discussed more specifically in Part Four.

Animals with no microbiota whatsoever can be reared in the laboratory. Most germfree mammals used in research are obtained by breeding them in a sterile environment. On the one hand, research with germfree animals has shown that microbes are not absolutely essential to animal life. On the other hand, this research has shown that germfree animals have undeveloped immune systems and are unusually susceptible to infection and serious disease. Germfree animals also require more calories and vitamins than do normal animals.

Relationships between the Normal Microbiota and the Host

Once established, the normal microbiota can benefit the host by preventing the overgrowth of harmful microorganisms. This phenomenon is called **microbial antagonism,** or **competitive exclusion.** Microbial antagonism involves competition among microbes. One consequence of this competition is that the normal microbiota protect the host against colonization by potentially pathogenic microbes by competing for nutrients, producing substances harmful to the invading microbes, and affecting conditions such as pH and available oxygen. When this balance between normal microbiota and pathogenic microbes is upset, disease can result. For example, the normal bacterial microbiota of the adult human vagina maintains a local pH of about 4. The presence of normal microbiota inhibits the overgrowth of the yeast *Candida albicans,* which can grow when the balance between normal microbiota and pathogens is upset and when pH is altered. If the bacterial population is eliminated by antibiotics, excessive douching, or deodorants, the pH of the vagina reverts to nearly neutral, and *C. albicans* can flourish and become the dominant microorganism there. This condition can lead to a form of vaginitis (vaginal infection).

Another example of microbial antagonism occurs in the large intestine. *E. coli* cells produce *bacteriocins,* proteins that inhibit the growth of other bacteria of the same or closely related species, such as pathogenic *Salmonella* and *Shigella.* A bacterium that makes a particular bacteriocin is not killed by that bacteriocin but may be killed by other ones. Bacteriocins are used in medical microbiology to help identify different strains of bacteria. Such identification helps determine whether several outbreaks of an infectious disease are caused by one or more strains of a bacterium.

A final example involves another bacterium, *Clostridium difficile* (dif′-fi-sē-il), also in the large intestine. The normal microbiota of the large intestine effectively inhibit *C. difficile,* possibly by making host receptors unavailable, competing for available nutrients, or producing bacteriocins. However, if the normal microbiota are eliminated (for example, by antibiotics), *C. difficile* can become a problem. This microbe is responsible for nearly all gastrointestinal infections that follow antibiotic therapy, from mild diarrhea to severe or even fatal colitis (inflammation of the colon).

The relationship between the normal microbiota and the host is called **symbiosis,** a relationship between two organisms in which at least one organism is dependent on the other (**Figure 14.2**). In the symbiotic relationship called **commensalism,** one of the organisms benefits, and the other is unaffected. Many of the microorganisms that make up our normal microbiota are commensals; these include the corynebacteria that inhabit the surface of the eye and certain saprophytic mycobacteria that inhabit the ear and external genitals. These bacteria live on secretions and sloughed-off cells, and they bring no apparent benefit or harm to the host.

Table 14.1 Representative Normal Microbiota by Body Region

Region	Principal Components	Comments
Skin	*Propionibacterium, Staphylococcus, Corynebacterium, Micrococcus, Acinetobacter, Brevibacterium; Pityrosporum* (fungus), *Candida* (fungus), *Malassezia* (fungus)	• Most of the microbes in direct contact with skin do not become residents because secretions from sweat and oil glands have antimicrobial properties. • Keratin is a resistant barrier, and the low pH of the skin inhibits many microbes. • The skin also has a relatively low moisture content.
Eyes (Conjunctiva)	*Staphylococcus epidermidis, S. aureus,* diphtheroids, *Propionibacterium, Corynebacterium,* streptococci, *Micrococcus*	• The conjunctiva, a continuation of the skin or mucous membrane, contains basically the same microbiota found on the skin. • Tears and blinking also eliminate some microbes or inhibit others from colonizing.

Nose and throat (upper respiratory system)
Eyes (conjunctiva)
Mouth
Skin
Large intestine
Urinary and reproductive systems (lower urethra in both sexes and vagina in females)

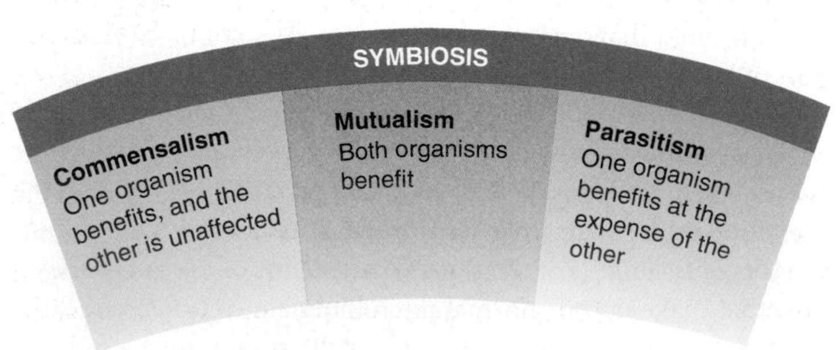
SYMBIOSIS
Commensalism
One organism benefits, and the other is unaffected
Mutualism
Both organisms benefit
Parasitism
One organism benefits at the expense of the other

Figure 14.2 Symbiosis.

Q Which type of symbiosis is best represented by the relationship between humans and *E. coli*?

Mutualism is a type of symbiosis that benefits both organisms. For example, the large intestine contains bacteria, such as *E. coli,* that synthesize vitamin K and some B vitamins. These vitamins are absorbed into the bloodstream and distributed for use by body cells. In exchange, the large intestine provides nutrients used by the bacteria, resulting in their survival.

Recent interest in the importance of bacteria to human health has led to the study of probiotics. **Probiotics** (*pro* = for, *bios* = life) are live microbial cultures applied to or ingested that are intended to exert a beneficial effect. Probiotics may be administered with *prebiotics,* which are chemicals that selectively promote the growth of beneficial bacteria. Several studies have shown that ingesting certain lactic acid bacteria (LAB) can alleviate diarrhea and prevent colonization by *Salmonella enterica* during antibiotic therapy. If these LAB colonize the large intestine, the lactic acid and bacteriocins they produce can inhibit the growth of certain pathogens. Researchers are also testing the use of LAB to prevent surgical wound infections caused by

Table 14.1 (continued)

Region	Principal Components	Comments
Nose and Throat (Upper Respiratory System)	*Staphylococcus aureus, S. epidermidis,* and aerobic diphtheroids in the nose; *S. epidermidis, S. aureus,* diphtheroids, *Streptococcus pneumoniae, Haemophilus,* and *Neisseria* in the throat	• Although some normal microbiota are potential pathogens, their ability to cause disease is reduced by microbial antagonism. • Nasal secretions kill or inhibit many microbes, and mucus and ciliary action remove many microbes.
Mouth	*Streptococcus, Lactobacillus, Actinomyces, Bacteroides, Veillonella, Neisseria, Haemophilis, Fusobacterium, Treponema, Staphylococcus, Corynebacterium,* and *Candida* (fungus)	• Abundant moisture, warmth, and the constant presence of food make the mouth an ideal environment that supports very large and diverse microbial populations on the tongue, cheeks, teeth, and gums. • However, biting, chewing, tongue movements, and salivary flow dislodge microbes. Saliva contains several antimicrobial substances.
Large Intestine	*Escherichia coli, Bacteroides, Fusobacterium, Lactobacillus, Enterococcus, Bifidobacterium, Enterobacter, Citrobacter, Proteus, Klebsiella, Candida* (fungus)	• The large intestine contains the largest numbers of resident microbiota in the body because of its available moisture and nutrients. • Mucus and periodic shedding of the lining prevent many microbes from attaching to the lining of the gastrointestinal tract, and the mucosa produces several antimicrobiol chemicals. • Diarrhea also flushes out some of the normal microbiota.
Urinary and Reproductive Systems	*Staphylococcus, Micrococcus, Enterococcus, Lactobacillus, Bacteroides,* aerobic diphtheroids, *Pseudomonas, Klebsiella,* and *Proteus* in urethra; lactobacilli, *Streptococcus, Clostridium, Candida albicans* (fungus), and *Trichomonas vaginalis* (protozoan) in vagina	• The lower urethra in both sexes has a resident population; the vagina has its acid-tolerant population of microbes because of the nature of its secretions. • Mucus and periodic shedding of the lining prevent microbes from attaching to the lining; urine flow mechanically removes microbes, and the pH of urine and urea are antimicrobial. • Cilia and mucus expel microbes from the cervix of the uterus into the vagina, and the acidity of the vagina inhibits or kills microbes.

Staphylococcus aureus and vaginal infections caused by *E. coli.* In a Stanford University study, HIV infection was reduced in women treated with a LAB that was genetically modified to produce CD4 protein that binds to HIV.

In still another kind of symbiosis, one organism benefits by deriving nutrients at the expense of the other; this relationship is called **parasitism.** Many disease-causing bacteria are parasites.

Opportunistic Microorganisms

Although categorizing symbiotic relationships by type is convenient, keep in mind that under certain conditions the relationship can change. For example, given the proper circumstances, a mutualistic organism, such as *E. coli,* can become harmful. *E. coli* is generally harmless as long as it remains in the large intestine; but if it gains access to other body sites, such as the urinary bladder, lungs, spinal cord, or wounds, it may cause urinary tract infections, pulmonary infections, meningitis, or abscesses, respectively. Microbes such as *E. coli* are called **opportunistic pathogens.** They ordinarily do not cause disease in their normal habitat in a healthy person but may do so in a different environment. For example, microbes that gain access through broken skin or mucous membranes can cause opportunistic infections. Or, if the host is already weakened or compromised by infection, microbes that are usually harmless can cause disease. AIDS is often accompanied by a common opportunistic infection, *Pneumocystis* pneumonia, caused by the opportunistic organism *Pneumocystis jirovecii* (see Figure 24.20, page 698). This secondary infection can develop in AIDS patients because their immune systems are suppressed. Before the AIDS epidemic, this type of pneumonia was rare. Opportunistic pathogens possess other features that contribute to their ability to cause disease. For example, they are present in or on the body or in the external environment in relatively large numbers. Some opportunistic pathogens may be found in locations in or on the body that are somewhat protected from the body's defenses, and some are resistant to antibiotics.

In addition to the usual symbionts, many people carry other microorganisms that are generally regarded as pathogenic but that may not cause disease in those people. Among the pathogens that are frequently carried in healthy individuals are echoviruses (*echo* comes from *e*nteric *c*ytopathogenic *h*uman *o*rphan), which can cause intestinal diseases, and adenoviruses, which can cause respiratory diseases. *Neisseria meningitidis,* which often resides benignly in the respiratory tract, can cause meningitis, a disease that inflames the coverings of the brain and spinal cord. *Streptococcus pneumoniae,* a normal resident of the nose and throat, can cause a type of pneumonia.

Cooperation among Microorganisms

It is not only competition among microbes that can cause disease; cooperation among microbes can also be a factor in causing disease. For example, pathogens that cause periodontal disease and gingivitis have been found to have receptors, not for the teeth, but for the oral streptococci that colonize the teeth.

CHECK YOUR UNDERSTANDING

✓ How do normal microbiota differ from transient microbiota? **14-2**

✓ Give several examples of microbial antagonism. **14-3**

✓ How can opportunistic pathogens cause infections? **14-4**

The Etiology of Infectious Diseases

LEARNING OBJECTIVE

14-5 List Koch's postulates.

Some diseases—such as polio, Lyme disease, and tuberculosis—have a well-known etiology. Some have an etiology that is not completely understood; for example, the relationship between certain viruses and cancer. For still others, such as Alzheimer disease, the etiology is unknown. Of course, not all diseases are caused by microorganisms. For example, the disease hemophilia is an *inherited (genetic) disease;* osteoarthritis and cirrhosis are considered *degenerative diseases.* There are several other categories of disease, but here we will discuss only *infectious diseases,* those caused by microorganisms. To see how microbiologists determine the etiology of an infectious disease, we will discuss in greater detail the work of Robert Koch, which was introduced in Chapter 1 (pages 9–11).

Koch's Postulates

In the historical overview of microbiology presented in Chapter 1, we briefly discussed Koch's famous postulates. Recall that Koch was a German physician who played a major role in establishing that microorganisms cause specific diseases. In 1877, he published some early papers on anthrax, a disease of cattle that can also occur in humans. Koch demonstrated that certain bacteria, today known as *Bacillus anthracis,* were always present in the blood of animals that had the disease and were not present in healthy animals. He knew that the mere presence of the bacteria did not prove that they had caused the disease; the bacteria could have been there as a result of the disease. Thus, he experimented further.

He took a sample of blood from a sick animal and injected it into a healthy one. The second animal developed the same disease and died. He repeated this procedure many times, always with the same results. (A key criterion in the validity of any scientific proof is that experimental results be repeatable.) Koch also cultivated the microorganism in fluids outside the animal's body, and he demonstrated that the bacterium would cause anthrax even after many culture transfers.

Koch showed that a specific infectious disease (anthrax) is caused by a specific microorganism (*B. anthracis*) that can be isolated and cultured on artificial media. He later used the same methods to show that the bacterium *Mycobacterium tuberculosis* is the causative agent of tuberculosis.

Koch's research provides a framework for the study of the etiology of any infectious disease. Today, we refer to Koch's experimental requirements as **Koch's postulates** (**Figure 14.3**). They are summarized as follows:

1. The same pathogen must be present in every case of the disease.
2. The pathogen must be isolated from the diseased host and grown in pure culture.
3. The pathogen from the pure culture must cause the disease when it is inoculated into a healthy, susceptible laboratory animal.
4. The pathogen must be isolated from the inoculated animal and must be shown to be the original organism.

Exceptions to Koch's Postulates

Although Koch's postulates are useful in determining the causative agent of most bacterial diseases, there are some exceptions. For example, some microbes have unique culture requirements. The bacterium *Treponema pallidum* is known to cause syphilis, but virulent strains have never been cultured on artificial media. The causative agent of leprosy, *Mycobacterium leprae,* has also never been grown on artificial media. Moreover, many rickettsial and viral pathogens cannot be cultured on artificial media because they multiply only within cells.

The discovery of microorganisms that cannot grow on artificial media has necessitated some modifications of Koch's postulates and the use of alternative methods of culturing and detecting certain microbes. For example, when researchers looking for the microbial cause of legionellosis (Legionnaires' disease) were unable to isolate the microbe directly from a victim, they took the alternative step of inoculating a victim's lung tissue into guinea pigs. These guinea pigs developed the

Figure 14.3

FOUNDATION FIGURE Koch's Postulates

Koch's postulates are used to determine the etiology of a disease, which is the beginning of treatment and prevention, as will be seen in Part Four of this book. Microbiologists use these steps to identify the cause of emerging diseases.

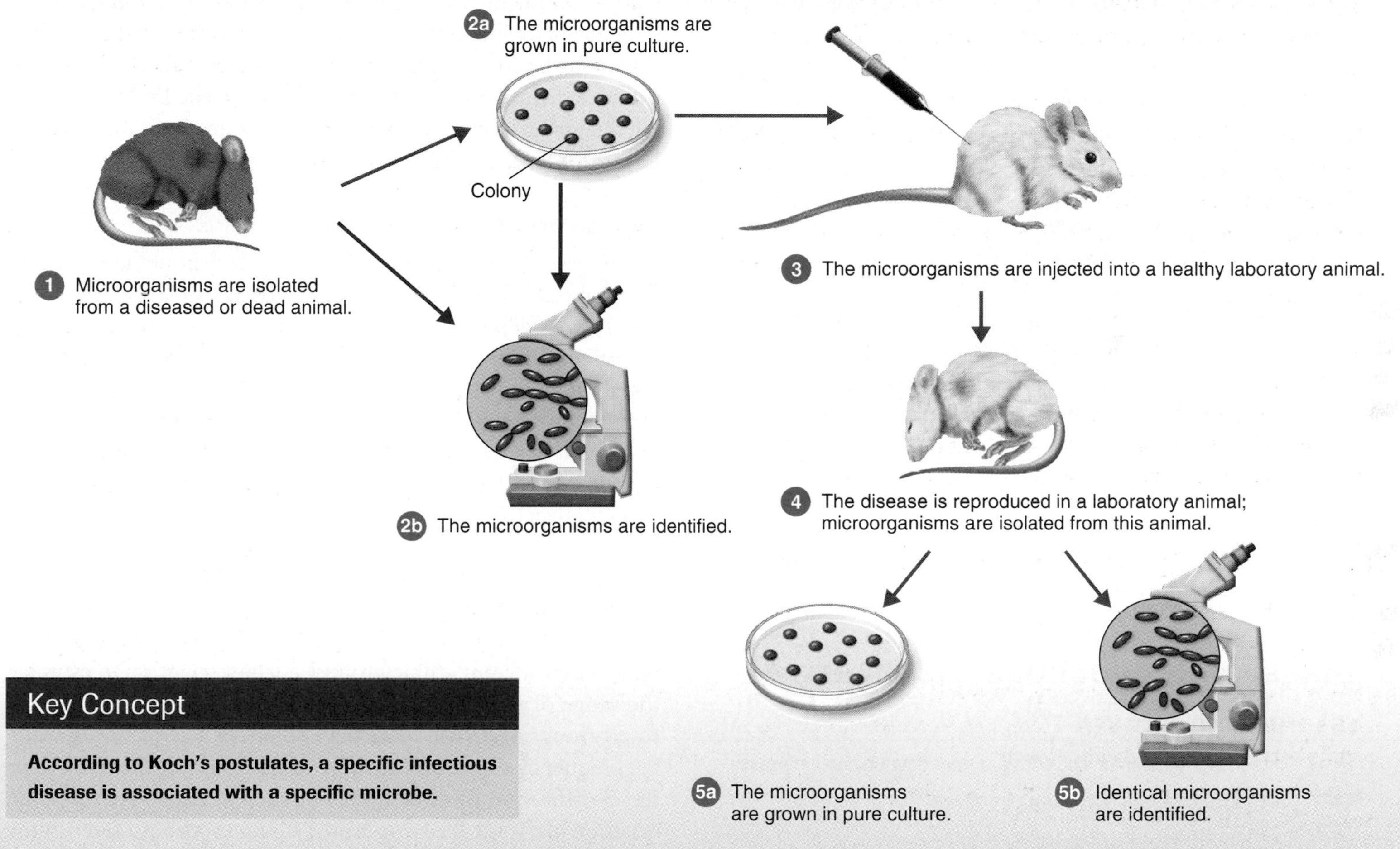

1 Microorganisms are isolated from a diseased or dead animal.
2a The microorganisms are grown in pure culture.
Colony
2b The microorganisms are identified.
3 The microorganisms are injected into a healthy laboratory animal.
4 The disease is reproduced in a laboratory animal; microorganisms are isolated from this animal.
5a The microorganisms are grown in pure culture.
5b Identical microorganisms are identified.

Key Concept

According to Koch's postulates, a specific infectious disease is associated with a specific microbe.

disease's pneumonia-like symptoms, whereas guinea pigs inoculated with tissue from an unafflicted person did not. Then tissue samples from the diseased guinea pigs were cultured in yolk sacs of chick embryos, a method (see Figure 13.7, page 377) that reveals the growth of extremely small microbes. After the embryos were incubated, electron microscopy revealed rod-shaped bacteria in the chick embryos. Finally, modern immunological techniques (discussed in Chapter 18) were used to show that the bacteria in the chick embryos were the same bacteria as those in the guinea pigs and in afflicted humans.

In a number of situations, a human host exhibits certain signs and symptoms that are associated only with a certain pathogen and its disease. For example, the pathogens responsible for diphtheria and tetanus cause distinguishing signs and symptoms that no other microbe can produce. They are unequivocally the only organisms that produce their respective diseases. But some infectious diseases are not as clear-cut and provide another exception to Koch's postulates. For example, nephritis (inflammation of the kidneys) can involve any of several different pathogens, all of which cause the same signs and symptoms . Thus, it is often difficult to know which particular microorganism is causing a disease. Other infectious diseases that sometimes have poorly defined etiologies are pneumonia, meningitis, and peritonitis (inflammation of the peritoneum, the membrane that lines the abdomen and covers the organs within them).

Still another exception to Koch's postulates results because some pathogens can cause several disease conditions. *Mycobacterium tuberculosis,* for example, is implicated in diseases of the lungs, skin, bones, and internal organs. *Streptococcus pyogenes* can cause sore throat, scarlet fever, skin infections (such as erysipelas), and osteomyelitis (inflammation of bone), among other diseases. When clinical signs and symptoms are used together with laboratory methods, these infections can usually be distinguished from infections of the same organs by other pathogens.

Ethical considerations may also impose an exception to Koch's postulates. For example, some agents that cause disease in humans have no other known host. An example is human immunodeficiency virus (HIV), the cause of AIDS. This poses the ethical question of whether humans can be intentionally inoculated with infectious agents. In 1721, King George I told condemned prisoners they could be inoculated with smallpox to test a smallpox vaccine (see Chapter 18). He promised their freedom if they lived. Human experiments with untreatable diseases are not acceptable today. Sometimes accidental inoculation does occur. A contaminated red bone marrow transplant satisfied the third Koch's postulate to prove that a herpesvirus caused cancer (see page 375).

CHECK YOUR UNDERSTANDING

✔ Explain some exceptions to Koch's postulates. **14-5**

Classifying Infectious Diseases

LEARNING OBJECTIVES

14-6 Differentiate a communicable from a noncommunicable disease.

14-7 Categorize diseases according to frequency of occurrence.

14-8 Categorize diseases according to severity.

14-9 Define *herd immunity.*

Every disease that affects the body alters body structures and functions in particular ways, and these alterations are usually indicated by several kinds of evidence. For example, the patient may experience certain **symptoms,** or changes in body function, such as pain and *malaise* (a vague feeling of body discomfort). These *subjective* changes are not apparent to an observer. The patient can also exhibit **signs,** which are *objective* changes the physician can observe and measure. Frequently evaluated signs include lesions (changes produced in tissues by disease), swelling, fever, and paralysis. A specific group of symptoms or signs may always accompany a particular disease; such a group is called a **syndrome.** The diagnosis of a disease is made by evaluation of the signs and symptoms, together with the results of laboratory tests.

Diseases are often classified in terms of how they behave within a host and within a given population. Any disease that spreads from one host to another, either directly or indirectly, is said to be a **communicable disease.** Chickenpox, measles, genital herpes, typhoid fever, and tuberculosis are examples. Chickenpox and measles are also examples of **contagious diseases,** that is, diseases that are *easily* spread from one person to another. A **noncommunicable disease** is not spread from one host to another. These diseases are caused by microorganisms that normally inhabit the body and only occasionally produce disease or by microorganisms that reside outside the body and produce disease only when introduced into the body. An example is tetanus: *Clostridium tetani* produces disease only when it is introduced into the body via abrasions or wounds.

Occurrence of a Disease

To understand the full scope of a disease, we should know something about its occurrence. The **incidence** of a disease is the number of people in a population who develop a disease during a particular time period. It is an indicator of the spread of the disease. The **prevalence** of a disease is the number of people in a population who develop a disease at a specified time, regardless of when it first appeared. Prevalence takes into account both old and new cases. It's an indicator of how seriously and how long a disease affects a population. For example, the incidence of AIDS in the United States in 2007 was 56,300 whereas the prevalence in that same year was estimated to be about 1,185,000. Knowing the incidence and the prevalence of a disease in different populations (for example, in populations representing different geographic regions or different ethnic groups) enables scientists to estimate the range of the disease's occurrence and its tendency to affect some groups of people more than others.

Frequency of occurrence is another criterion that is used in the classification of diseases. If a particular disease occurs only occasionally, it is called a **sporadic disease;** typhoid fever in the United States is such a disease. A disease constantly present in a population is called an **endemic disease;** an example of such a disease is the common cold. If many people in a given area acquire a certain disease in a relatively short period, it is called an **epidemic disease;** influenza is an example of a disease that often achieves epidemic status. **Figure 14.4** shows the epidemic incidence of AIDS in the United States. Some authorities consider gonorrhea and certain other sexually transmitted infections to be epidemic at this time as well (see Figures 26.5 and 26.6, page 749). An epidemic disease that occurs worldwide is called a **pandemic disease.** We experience pandemics of influenza from time to time. Some authorities also consider AIDS to be pandemic.

Severity or Duration of a Disease

Another useful way of defining the scope of a disease is in terms of its severity or duration. An **acute disease** is one that develops rapidly but lasts only a short time; a good example is influenza.

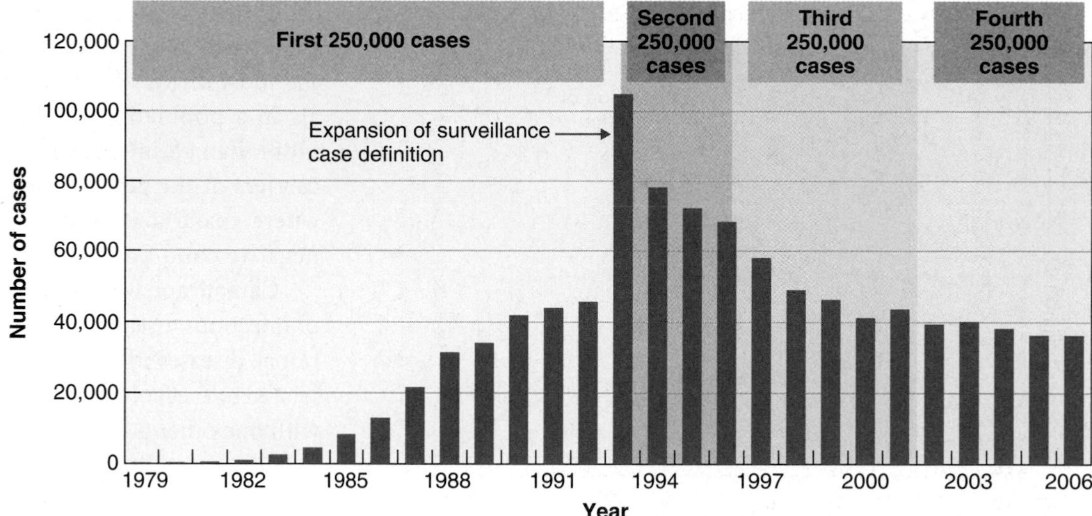

Figure 14.4 Reported AIDS cases in the United States. Notice that the first 250,000 cases occurred over a 12-year period, whereas the second through fourth 250,000 cases in this epidemic occurred in just 3 to 6 years. Much of the increase shown for 1993 is due to an expanded definition of AIDS cases adopted in that year.
Source: CDC.

Q What was the incidence of AIDS in 2004?

A **chronic disease** develops more slowly, and the body's reactions may be less severe, but the disease is likely to continue or recur for long periods. Infectious mononucleosis, tuberculosis, and hepatitis B fall into this category. A disease that is intermediate between acute and chronic is described as a **subacute disease;** an example is subacute sclerosing panencephalitis, a rare brain disease characterized by diminished intellectual function and loss of nervous function. A **latent disease** is one in which the causative agent remains inactive for a time but then becomes active to produce symptoms of the disease; an example is shingles, one of the diseases caused by varicella virus.

The rate at which a disease or an epidemic spreads and the number of individuals involved are determined in part by the immunity of the population. Vaccination can provide long-lasting and sometimes lifelong protection of an individual against certain diseases. People who are immune to an infectious disease will not be carriers, thereby reducing the occurrence of the disease. Immune individuals act as a barrier to the spread of infectious agents. Even though a highly communicable disease may cause an epidemic, many nonimmune people will be protected because of the unlikelihood of their coming into contact with an infected person. A great advantage of vaccination is that enough individuals in a population will be protected from a disease to prevent its rapid spread to those in the population who are not vaccinated. When many immune people are present in a community, **herd immunity** exists.

Extent of Host Involvement

Infections can also be classified according to the extent to which the host's body is affected. A **local infection** is one in which the invading microorganisms are limited to a relatively small area of the body. Some examples of local infections are boils and abscesses. In a **systemic (generalized) infection,** microorganisms or their products are spread throughout the body by the blood or lymph. Measles is an example of a systemic infection. Very often, agents of a local infection enter a blood or lymphatic vessel and spread to other specific parts of the body, where they are confined to specific areas of the body. This condition is called a **focal infection.** Focal infections can arise from infections in areas such as the teeth, tonsils, or sinuses.

Sepsis is a toxic inflammatory condition arising from the spread of microbes, especially bacteria or their toxins, from a focus of infection. **Septicemia,** also called blood poisoning, is a systemic infection arising from the multiplication of pathogens in the blood. Septicemia is a common example of sepsis. The presence of bacteria in the blood is known as **bacteremia. Toxemia** refers to the presence of toxins in the blood (as occurs in tetanus), and **viremia** refers to the presence of viruses in blood.

The state of host resistance also determines the extent of infections. A **primary infection** is an acute infection that causes the initial illness. A **secondary infection** is one caused by an opportunistic pathogen after the primary infection has weakened the body's defenses. Secondary infections of the skin and respiratory tract are common and are sometimes more dangerous than the primary infections. *Pneumocystis* pneumonia as a consequence of AIDS is an example of a secondary infection; streptococcal bronchopneumonia following influenza is an example of a secondary infection that is more serious than the primary infection. A **subclinical (inapparent) infection** is one that does not cause any noticeable illness. Poliovirus and hepatitis A virus, for example, can be carried by people who never develop the illness.

CHECK YOUR UNDERSTANDING

- ✓ Does *Clostridium perfringens* (page 646) cause a communicable disease? **14-6**
- ✓ Distinguish the incidence from the prevalence of a disease. **14-7**
- ✓ List two examples of acute and chronic diseases. **14-8**
- ✓ How does herd immunity develop? **14-9**

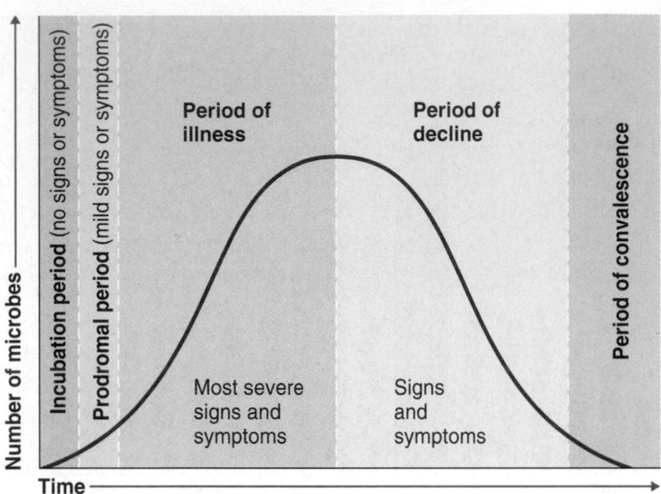

Figure 14.5 The stages of a disease.

Q During which periods can a disease be transmitted?

Patterns of Disease

LEARNING OBJECTIVES

14-10 Identify four predisposing factors for disease.

14-11 Put the following in proper sequence, according to the pattern of disease: period of decline, period of convalescence, period of illness, prodromal period, incubation period.

A definite sequence of events usually occurs during infection and disease. As you will learn shortly, for an infectious disease to occur, there must be a reservoir of infection as a source of pathogens. Next, the pathogen must be transmitted to a susceptible host by direct contact, by indirect contact, or by vectors. Transmission is followed by invasion, in which the microorganism enters the host and multiplies. Following invasion, the microorganism injures the host through a process called pathogenesis (discussed further in the next chapter). The extent of injury depends on the degree to which host cells are damaged, either directly or by toxins. Despite the effects of all these factors, the occurrence of disease ultimately depends on the resistance of the host to the activities of the pathogen.

Predisposing Factors

Certain predisposing factors also affect the occurrence of disease. A **predisposing factor** makes the body more susceptible to a disease and may alter the course of the disease. Gender is sometimes a predisposing factor; for example, females have a higher incidence of urinary tract infections than males, whereas males have higher rates of pneumonia and meningitis. Other aspects of genetic background may play a role as well. For example, sickle cell disease is a severe, life-threatening form of anemia that occurs when the genes for the disease are inherited from both parents. Individuals who carry only one sickle cell gene have a condition called sickle cell trait and are normal unless specially tested. However, they are relatively resistant to the most serious form of malaria. The potential that individuals in a population might inherit a life-threatening disease is more than counterbalanced by protection from malaria among carriers of the gene for sickle cell trait. Of course, in countries where malaria is not present, sickle cell trait is an entirely negative condition.

Climate and weather seem to have some effect on the incidence of infectious diseases. In temperate regions, the incidence of respiratory diseases increases during the winter. This increase may be related to the fact that when people stay indoors, the closer contact with one other facilitates the spread of respiratory pathogens.

Other predisposing factors include inadequate nutrition, fatigue, age, environment, habits, lifestyle, occupation, preexisting illness, chemotherapy, and emotional disturbances. It is often difficult to know the exact relative importance of the various predisposing factors.

Development of Disease

Once a microorganism overcomes the defenses of the host, development of the disease follows a certain sequence that tends to be similar whether the disease is acute or chronic (**Figure 14.5**).

Incubation Period

The **incubation period** is the interval between the initial infection and the first appearance of any signs or symptoms. In some diseases, the incubation period is always the same; in others, it is quite variable. The time of incubation depends on the specific microorganism involved, its virulence (degree of pathogenicity), the number of infecting microorganisms, and the resistance of the host. (See Table 15.1, page 430, for the incubation periods of a number of microbial diseases.)

Prodromal Period

The **prodromal period** is a relatively short period that follows the period of incubation in some diseases. The prodromal period is characterized by early, mild symptoms of disease, such as general aches and malaise.

Period of Illness

During the **period of illness,** the disease is most severe. The person exhibits overt signs and symptoms of disease, such as fever, chills, muscle pain (myalgia), sensitivity to light (photophobia), sore throat (pharyngitis), lymph node enlargement (lymphadenopathy), and gastrointestinal disturbances. During the period of illness, the number of white blood cells may increase or decrease. Generally, the patient's immune response and other defense mechanisms overcome the pathogen, and the period of illness ends. When the disease is not successfully overcome (or successfully treated), the patient dies during this period.

Period of Decline

During the **period of decline,** the signs and symptoms subside. The fever decreases, and the feeling of malaise diminishes. During this phase, which may take from less than 24 hours to several days, the patient is vulnerable to secondary infections.

Period of Convalescence

During the **period of convalescence,** the person regains strength and the body returns to its prediseased state. Recovery has occurred.

We all know that during the period of illness, people can serve as reservoirs of disease and can easily spread infections to other people. However, you should also know that people can spread infection during incubation and convalescence as well. This is especially true of diseases such as typhoid fever and cholera, in which the convalescing person carries the pathogenic microorganism for months or even years.

CHECK YOUR UNDERSTANDING

- ✓ What is a predisposing factor? **14-10**
- ✓ The incubation period for a cold is 3 days, and the period of disease is usually 5 days. If the person next to you has a cold, when will you know whether you contracted it? **14-11**

The Spread of Infection

LEARNING OBJECTIVES

14-12 Define *reservoir of infection.*

14-13 Contrast human, animal, and nonliving reservoirs, and give one example of each.

14-14 Explain three methods of disease transmission.

Now that you have an understanding of normal microbiota, the etiology of infectious diseases, and the types of infectious diseases, we will examine the sources of pathogens and how diseases are transmitted.

Reservoirs of Infection

For a disease to perpetuate itself, there must be a continual source of the disease organisms. This source can be either a living organism or an inanimate object that provides a pathogen with adequate conditions for survival and multiplication and an opportunity for transmission. Such a source is called a **reservoir of infection.** These reservoirs may be human, animal, or nonliving.

Human Reservoirs

The principal living reservoir of human disease is the human body itself. Many people harbor pathogens and transmit them directly or indirectly to others. People with signs and symptoms of a disease may transmit the disease; in addition, some people can harbor pathogens and transmit them to others without exhibiting any signs of illness. These people, called **carriers,** are important living reservoirs of infection. Some carriers have inapparent infections for which no signs or symptoms are ever exhibited. Other people, such as those with latent diseases, carry a disease during its symptom-free stages—during the incubation period (before symptoms appear) or during the convalescent period (recovery). Typhoid Mary is an example of a carrier (see page 714). Human carriers play an important role in the spread of such diseases as AIDS, diphtheria, typhoid fever, hepatitis, gonorrhea, amoebic dysentery, and streptococcal infections.

Animal Reservoirs

Both wild and domestic animals are living reservoirs of microorganisms that can cause human diseases. Diseases that occur primarily in wild and domestic animals and can be transmitted to humans are called **zoonoses** (zō-ō-no′-sēz) (singular: *zoonosis*). Rabies (found in bats, skunks, foxes, dogs, and coyotes), and Lyme disease (found in field mice) are examples of zoonoses. Other representative zoonoses are presented in **Table 14.2**.

About 150 zoonoses are known. The transmission of zoonoses to humans can occur via one of many routes: by direct contact with infected animals; by direct contact with domestic pet waste (such as cleaning a litter box or bird cage); by contamination of food and water; by air from contaminated hides, fur, or feathers; by consuming infected animal products; or by insect vectors (insects that transmit pathogens).

Nonliving Reservoirs

The two major nonliving reservoirs of infectious disease are soil and water. Soil harbors such pathogens as fungi, which cause mycoses such as ringworm and systemic infections; *Clostridium botulinum,* the bacterium that causes botulism; and *C. tetani,* the bacterium that causes tetanus. Because both species of clostridia are part of the normal intestinal microbiota of horses and cattle, the bacteria are found especially in soil where animal feces are used as fertilizer.

Water that has been contaminated by the feces of humans and other animals is a reservoir for several pathogens, notably those responsible for gastrointestinal diseases. These include *Vibrio cholerae,* which causes cholera, and *Salmonella typhi,* which causes typhoid fever. Other nonliving reservoirs include foods that are improperly prepared or stored. They may be sources of diseases such as trichinellosis and salmonellosis.

Transmission of Disease

The causative agents of disease can be transmitted from the reservoir of infection to a susceptible host by three principal routes: contact, vehicles, and vectors.

Table 14.2 Selected Zoonoses

Disease	Causative Agent	Reservoir	Transmission Due To	Chapter Reference
Viral				
Influenza (some types)	*Influenzavirus*	Swine, birds	Direct contact	24
Rabies	*Lyssavirus*	Bats, skunks, foxes, dogs, raccoons	Direct contact (bite)	22
West Nile encephalitis	*Flavivirus*	Horses, birds	*Aedes* and *Culex* mosquito bite	22
Hantavirus pulmonary syndrome	*Hantavirus*	Rodents (primarily deer mice)	Direct contact with rodent saliva, feces, or urine	23
Bacterial				
Anthrax	*Bacillus anthracis*	Domestic livestock	Direct contact with contaminated hides or animals; air; food	23
Brucellosis	*Brucella* spp.	Domestic livestock	Direct contact with contaminated milk, meat, or animals	23
Plague	*Yersinia pestis*	Rodents	Flea bites	23
Cat-scratch disease	*Bartonella henselae*	Domestic cats	Direct contact	23
Ehrlichiosis	*Ehrlichia* spp.	Deer, rodents	Tick bites	23
Leptospirosis	*Leptospira*	Wild mammals, domestic dogs and cats	Direct contact with urine, soil, water	26
Lyme disease	*Borrelia burgdorferi*	Field mice	Tick bites	23
Psittacosis (ornithosis)	*Chlamydophila psittaci*	Birds, especially parrots	Direct contact	24
Rocky Mountain spotted fever	*Rickettsia rickettsii*	Rodents	Tick bites	23
Salmonellosis	*Salmonella enterica*	Poultry, reptiles	Ingestion of contaminated food and water and putting hands in mouth	25
Endemic typhus	*Rickettsia typhi*	Rodents	Flea bites	23
Fungal				
Ringworm	*Trichophyton* *Microsporum* *Epidermophyton*	Domestic mammals	Direct contact; fomites (nonliving objects)	21
Protozoan				
Malaria	*Plasmodium* spp.	Monkeys	*Anopheles* mosquito bite	23
Toxoplasmosis	*Toxoplasma gondii*	Cats and other mammals	Ingestion of contaminated meat or by direct contact with infected tissues or fecal matter	23
Helminthic				
Tapeworm (pork)	*Taenia solium*	Pigs	Ingestion of undercooked contaminated pork	25
Trichinellosis	*Trichinella spiralis*	Pigs, bears	Ingestion of undercooked contaminated pork	25

Contact Transmission

Contact transmission is the spread of an agent of disease by direct contact, indirect contact, or droplet transmission. **Direct contact transmission,** also known as *person-to-person transmission,* is the direct transmission of an agent by physical contact between its source and a susceptible host; no intermediate object is involved (**Figure 14.6a**). The most common forms of direct contact transmission are touching, kissing, and sexual intercourse. Among the diseases that can be transmitted by direct contact are viral respiratory tract diseases (the common cold and influenza), staphylococcal infections, hepatitis A, measles, scarlet fever, and sexually transmitted infections (syphilis, gonorrhea, and genital herpes). Direct contact is also one way to spread AIDS and infectious mononucleosis. To guard against person-to-person transmission, health care workers use gloves and other protective measures (**Figure 14.6b**). Potential pathogens can also be transmitted by direct contact from animals (or animal products) to humans. Examples are the pathogens causing rabies and anthrax.

Indirect contact transmission occurs when the agent of disease is transmitted from its reservoir to a susceptible host by means of a nonliving object. The general term for any nonliving

(a) Direct contact transmission

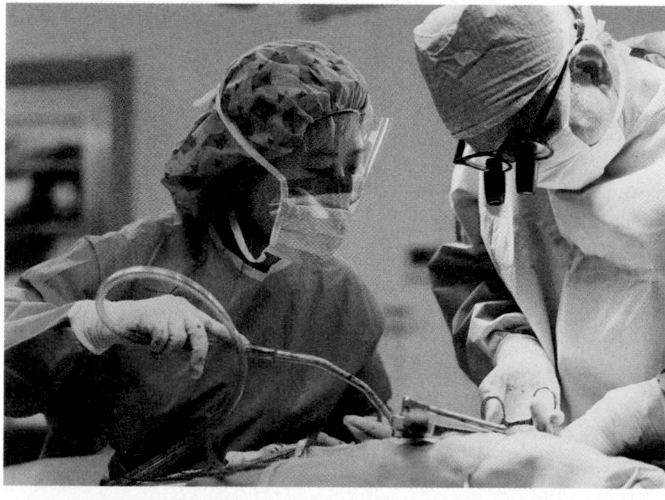

(b) Preventing direct contact transmission through the use of gloves, masks, and face shields

(c) Indirect contact transmission

(d) Droplet transmission

Figure 14.6 Contact transmission.

Q **Can you name a disease transmitted by direct contact, a disease transmitted by indirect contact, and a disease transmitted by droplet transmission?**

object involved in the spread of an infection is a **fomite.** Examples of fomites are tissues, handkerchiefs, towels, bedding, diapers, drinking cups, eating utensils, toys, money, and thermometers (**Figure 14.6c**). Contaminated syringes serve as fomites in transmitting AIDS and hepatitis B. Other fomites may transmit diseases such as tetanus.

Droplet transmission is a third type of contact transmission in which microbes are spread in *droplet nuclei* (mucus droplets) that travel only short distances (**Figure 14.6d**). These droplets are discharged into the air by coughing, sneezing, laughing, or talking and travel less than 1 meter from the reservoir to the host. One sneeze may produce 20,000 droplets. Disease agents that travel such short distances are not regarded as airborne (airborne transmission is discussed shortly). Examples of diseases spread by droplet transmission are influenza, pneumonia, and pertussis (whooping cough).

Vehicle Transmission

Vehicle transmission is the transmission of disease agents by a medium, such as water, food, or air (**Figure 14.7**). Other media include blood and other body fluids, drugs, and intravenous fluids. An outbreak of *Salmonella* infections caused by vehicle transmission is described in the box in Chapter 25 (page 715). Here we will discuss water, food, and air as vehicles of transmission.

In *waterborne transmission,* pathogens are usually spread by water contaminated with untreated or poorly treated sewage. Diseases transmitted via this route include cholera, waterborne shigellosis, and leptospirosis. In *foodborne transmission,*

(a) Water

(b) Food

(c) Air

Figure 14.7 Vehicle transmission.

Q How does vehicle transmission differ from contact transmission?

pathogens are generally transmitted in foods that are incompletely cooked, poorly refrigerated, or prepared under unsanitary conditions. Foodborne pathogens cause diseases such as food poisoning and tapeworm infestation.

Airborne transmission refers to the spread of agents of infection by droplet nuclei in dust that travel more than 1 meter from the reservoir to the host. For example, microbes are spread by droplets, which may be discharged in a fine spray from the mouth and nose during coughing and sneezing (see Figure 14.6d). These droplets are small enough to remain airborne for prolonged periods. The virus that causes measles and the bacterium that causes tuberculosis can be transmitted via airborne droplets. Dust particles can harbor various pathogens. Staphylococci and streptococci can survive on dust and be transmitted by the airborne route. Spores produced by certain fungi are also transmitted by the airborne route and can cause such diseases as histoplasmosis, coccidioidomycosis, and blastomycosis (see Chapter 24).

Vectors

Arthropods are the most important group of disease **vectors**—animals that carry pathogens from one host to another. (Insects and other arthropod vectors are discussed in Chapter 12, page 361.) Arthropod vectors transmit disease by two general methods. **Mechanical transmission** is the passive transport of the pathogens on the insect's feet or other body parts (**Figure 14.8**). If the insect makes contact with a host's food, pathogens can be transferred to the food and later swallowed by the host. Houseflies, for instance, can transfer the pathogens of typhoid fever and bacillary dysentery (shigellosis) from the feces of infected people to food.

Biological transmission is an active process and is more complex. The arthropod bites an infected person or animal and ingests some of the infected blood (see Figure 12.30, page 362). The pathogens then reproduce in the vector, and the increase in the number of pathogens increases the possibility that they will be transmitted to another host. Some parasites reproduce in the gut of the arthropod; these can be passed

Figure 14.8 Mechanical transmission.

Q How do mechanical transmission and biological transmission by vectors differ?

Table 14.3 Representative Arthropod Vectors and the Diseases They Transmit

Disease	Causative Agent	Arthropod Vector	Chapter Reference
Malaria	*Plasmodium* spp.	*Anopheles* (mosquito)	23
African trypanosomiasis	*Trypanosoma brucei gambiense* and *T. b. rhodesiense*	*Glossina* (tsetse fly)	22
Chagas' disease	*T. cruzi*	*Triatoma* (kissing bug)	23
Yellow fever	*Alphavirus* (yellow fever virus)	*Aedes* (mosquito)	23
Dengue	*Alphavirus* (dengue fever virus)	*A. aegypti* (mosquito)	23
Arthropod-borne encephalitis	*Alphavirus* (encephalitis virus)	*Culex* (mosquito)	22
Ehrlichiosis	*Ehrlichia* spp.	*Ixodes* spp. (tick)	23
Epidemic typhus	*Rickettsia prowazekii*	*Pediculus humanus* (louse)	23
Endemic murine typhus	*R. typhi*	*Xenopsylla cheopis* (rat flea)	23
Rocky Mountain spotted fever	*R. rickettsii*	*Dermacentor andersoni* and other species (tick)	23
Plague	*Yersinia pestis*	*X. cheopis* (rat flea)	23
Relapsing fever	*Borrelia* spp.	*Ornithodorus* spp. (soft tick)	23
Lyme disease	*B. burgdorferi*	*Ixodes* spp. (tick)	23

with feces. If the arthropod defecates or vomits while biting a potential host, the parasite can enter the wound. Other parasites reproduce in the vector's gut and migrate to the salivary gland; these are directly injected into a bite. Some protozoan and helminthic parasites use the vector as a host for a developmental stage in their life cycle.

Table 14.3 lists a few important arthropod vectors and the diseases they transmit.

CHECK YOUR UNDERSTANDING

- ✓ Why are carriers important reservoirs of infection? **14-12**
- ✓ How are zoonoses transmitted to humans? **14-13**
- ✓ Give an example of contact transmission, vehicle transmission, mechanical transmission, and biological transmission. **14-14**

Nosocomial (Hospital-Acquired) Infections

LEARNING OBJECTIVES

14-15 Define *nosocomial infections,* and explain their importance.

14-16 Define *compromised host.*

14-17 List several methods of disease transmission in hospitals.

14-18 Explain how nosocomial infections can be prevented.

Q&A A **nosocomial** (nōs-ō-kō′mē-al) **infection** does not show any evidence of being present or incubating at the time of admission to a hospital; it is acquired as a result of a hospital stay. (The word *nosocomial* is derived from the Greek word for hospital; the term also includes infections acquired in nursing homes and other health care facilities.)

The Centers for Disease Control and Prevention (CDC) estimates that 5 to 15% of all hospital patients acquire some type of nosocomial infection. The work of pioneers in aseptic techniques such as Lister and Semmelweis (Chapter 1, page 11) decreased the rate of nosocomial infections considerably. However, despite modern advances in sterilization techniques and disposable materials, the rate of nosocomial infections has increased 36% during the last 20 years. In the United States, about 2 million people per year contract nosocomial infections, and nearly 20,000 die as a result. Nosocomial infections represent the eighth leading cause of death in the United States (the top three are heart disease, cancer, and strokes).

Nosocomial infections result from the interaction of several factors: (1) microorganisms in the hospital environment, (2) the compromised (or weakened) status of the host, and (3) the chain of transmission in the hospital. **Figure 14.9** illustrates that the presence of any one of these factors alone is generally not enough to cause infection; it is the interaction of all three factors that poses a significant risk of nosocomial infection.

Severity or Duration of a Disease (pp. 406–407)

8. The scope of a disease can be defined as acute, chronic, subacute, or latent.
9. Herd immunity is the presence of immunity to a disease in most of the population.

Extent of Host Involvement (p. 407)

10. A local infection affects a small area of the body; a systemic infection is spread throughout the body via the circulatory system.
11. A primary infection is an acute infection that causes the initial illness.
12. A secondary infection can occur after the host is weakened from a primary infection.
13. An inapparent, or subclinical, infection does not cause any signs of disease in the host.

Patterns of Disease (pp. 408–409)

Predisposing Factors (p. 408)

1. A predisposing factor is one that makes the body more susceptible to disease or alters the course of a disease.
2. Examples include gender, climate, age, fatigue, and inadequate nutrition.

Development of Disease (pp. 408–409)

3. The incubation period is the interval between the initial infection and the first appearance of signs and symptoms.
4. The prodromal period is characterized by the appearance of the first mild signs and symptoms.
5. During the period of illness, the disease is at its height, and all disease signs and symptoms are apparent.
6. During the period of decline, the signs and symptoms subside.
7. During the period of convalescence, the body returns to its prediseased state, and health is restored.

The Spread of Infection (pp. 409–413)

Reservoirs of Infection (p. 409)

1. A continual source of infection is called a reservoir of infection.
2. People who have a disease or are carriers of pathogenic microorganisms are human reservoirs of infection.
3. Zoonoses are diseases that affect wild and domestic animals and can be transmitted to humans.
4. Some pathogenic microorganisms grow in nonliving reservoirs, such as soil and water.

Transmission of Disease (pp. 409–413)

5. Transmission by direct contact involves close physical contact between the source of the disease and a susceptible host.
6. Transmission by fomites (inanimate objects) constitutes indirect contact.
7. Transmission via saliva or mucus in coughing or sneezing is called droplet transmission.
8. Transmission by a medium such as water, food, or air is called vehicle transmission.
9. Airborne transmission refers to pathogens carried on water droplets or dust for a distance greater than 1 meter.
10. Arthropod vectors carry pathogens from one host to another by both mechanical and biological transmission.

Nosocomial (Hospital-Acquired) Infections (pp. 413–416)

1. A nosocomial infection is any infection that is acquired during the course of stay in a hospital, nursing home, or other health care facility.
2. About 5 to 15% of all hospitalized patients acquire nosocomial infections.

Microorganisms in the Hospital (p. 414)

3. Certain normal microbiota are often responsible for nosocomial infections when they are introduced into the body through such medical procedures as surgery and catheterization.
4. Opportunistic, drug-resistant gram-negative bacteria are the most frequent causes of nosocomial infections.

Compromised Host (p. 415)

5. Patients with burns, surgical wounds, and suppressed immune systems are the most susceptible to nosocomial infections.

Chain of Transmission (pp. 415–416)

6. Nosocomial infections are transmitted by direct contact between staff members and patients and between patients.
7. Fomites such as catheters, syringes, and respiratory devices can transmit nosocomial infections.

Control of Nosocomial Infections (p. 416)

8. Aseptic techniques can prevent nosocomial infections.
9. Hospital infection control staff members are responsible for overseeing the proper cleaning, storage, and handling of equipment and supplies.

Emerging Infectious Diseases (pp. 416–418)

1. New diseases and diseases with increasing incidences are called emerging infectious diseases (EIDs).
2. EIDs can result from the use of antibiotics and pesticides, climatic changes, travel, the lack of vaccinations, and improved case reporting.
3. The CDC, NIH, and WHO are responsible for surveillance and responses to emerging infectious diseases.

Epidemiology (pp. 418–422)

1. The science of epidemiology is the study of the transmission, incidence, and frequency of disease.
2. Modern epidemiology began in the mid-1800s with the works of Snow, Semmelweis, and Nightingale.
3. In descriptive epidemiology, data about infected people are collected and analyzed.
4. In analytical epidemiology, a group of infected people is compared with an uninfected group.

5. In experimental epidemiology, controlled experiments designed to test hypotheses are performed.
6. Case reporting provides data on incidence and prevalence to local, state, and national health officials.
7. The Centers for Disease Control and Prevention (CDC) is the main source of epidemiologic information in the United States.
8. The CDC publishes the *Morbidity and Mortality Weekly Report* to provide information on morbidity (incidence) and mortality (deaths).

STUDY QUESTIONS

Answers to the Review and Multiple Choice questions can be found by turning to the blue Answers tab in the back of the textbook.

Review

1. Differentiate the terms in each of the following pairs:
   a. etiology and pathogenesis
   b. infection and disease
   c. communicable disease and noncommunicable disease
2. Define *symbiosis*. Differentiate commensalism, mutualism, and parasitism, and give an example of each.
3. **DRAW IT** Using the data below, draw a graph showing the incidence of influenza during a typical year. Indicate the endemic and epidemic levels.

Month	Percentage of Physician Visits for Influenza-like Symptoms
Jan	2.33
Feb	3.21
Mar	2.08
Apr	1.47
May	0.97
Jun	0.30
Jul	0.30
Aug	0.20
Sep	0.20
Oct	1.18
Nov	1.54
Dec	2.39

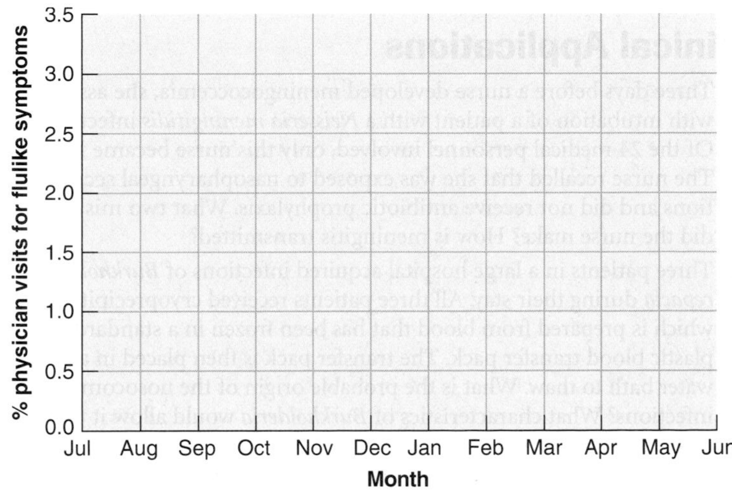

4. Indicate whether each of the following conditions is typical of subacute, chronic, or acute infections.
   a. The patient experiences a rapid onset of malaise; symptoms last 5 days.
   b. The patient experiences cough and breathing difficulty for months.
   c. The patient has no apparent symptoms and is a known carrier.
5. Of all the hospital patients with infections, one-third do not enter the hospital with an infection. How do they acquire these infections? What is the method of transmission of these infections? What is the reservoir of infection?
6. Distinguish symptoms from signs as signals of disease.
7. How can a local infection become a systemic infection?
8. Why are some organisms that constitute the normal microbiota described as commensals, whereas others are described as mutualistic?
9. Put the following in the correct order to describe the pattern of disease: period of convalescence, prodromal period, period of decline, incubation period, period of illness.

Multiple Choice

1. The emergence of new infectious diseases is probably due to all of the following *except*
   a. the need of bacteria to cause disease.
   b. the ability of humans to travel by air.
   c. changing environments (e.g., flood, drought, pollution).
   d. a pathogen crossing the species barrier.
   e. the increasing human population.
2. All members of a group of ornithologists studying barn owls in the wild have had salmonellosis (*Salmonella* gastroenteritis). One birder is experiencing her third infection. What is the most likely source of their infections?
   a. The ornithologists are eating the same food.
   b. They are contaminating their hands while handling the owls and nests.
   c. One of the workers is a *Salmonella* carrier.
   d. Their drinking water is contaminated.
3. Which of the following statements is *not* true?
   a. *E. coli* never causes disease.
   b. *E. coli* provides vitamin K for its host.
   c. *E. coli* often exists in a mutualistic relationship with humans.
   d. *E. coli* gets nutrients from intestinal contents.
4. Which of the following is *not* one of Koch's postulates?
   a. The same pathogen must be present in every case of the disease.
   b. The pathogen must be isolated and grown in pure culture from the diseased host.
   c. The pathogen from pure culture must cause the disease when inoculated into a healthy, susceptible laboratory animal.

15 Microbial Mechanisms of Pathogenicity

Now that you have a basic understanding of how microorganisms cause disease, we will take a look at some of the specific properties of microorganisms that contribute to **pathogenicity,** the ability to cause disease by overcoming the defenses of a host, and **virulence,** the degree or extent of pathogenicity. (As discussed throughout the chapter, the term *host* usually refers to humans.) Microbes don't try to cause disease; the microbial cells are getting food and defending themselves.

To humans, it doesn't make sense for a parasite to kill its host. However, nature does not have a plan for evolution; the genetic variations that give rise to evolution are due to random mutations, not to logic. According to natural selection, organisms best adapted to their environments will reproduce. Coevolution between a parasite and its host seems to occur: the behavior of one influences that of the other. For example, the cholera pathogen, *Vibrio cholerae,* quickly induces diarrhea, threatening the host's life from a loss of fluids and salts but providing a way to transmit the pathogen to another person by contaminating the water supply.

Keep in mind that many of the properties contributing to microbial pathogenicity and virulence are unclear or unknown. We do know, however, that if the microbe overpowers the host defenses, disease results.

UNDER THE MICROSCOPE

Bacteria (shown here in purple) have the ability to attach to host tissues such as the human skin, as shown here.

Q&A

Almost every pathogen has a mechanism for attaching to host tissues at their portal of entry. What is this attachment called, and how does it occur?

Look for the answer in the chapter.

How Microorganisms Enter a Host

LEARNING OBJECTIVES

15-1 Identify the principal portals of entry.

15-2 Define ID_{50} and LD_{50}.

15-3 Using examples, explain how microbes adhere to host cells.

To cause disease, most pathogens must gain access to the host, adhere to host tissues, penetrate or evade host defenses, and damage the host tissues. However, some microbes do not cause disease by directly damaging host tissue. Instead, disease is due to the accumulation of microbial waste products. Some microbes, such as those that cause dental caries and acne, can cause disease without penetrating the body. Pathogens can gain entrance to the human body and other hosts through several avenues, which are called **portals of entry.**

Portals of Entry

The portals of entry for pathogens are mucous membranes, skin, and direct deposition beneath the skin or membranes (the parenteral route).

Mucous Membranes

Many bacteria and viruses gain access to the body by penetrating mucous membranes lining the respiratory tract, gastrointestinal tract, genitourinary tract, and conjunctiva, a delicate membrane that covers the eyeballs and lines the eyelids. Most pathogens enter through the mucous membranes of the gastrointestinal and respiratory tracts.

The respiratory tract is the easiest and most frequently traveled portal of entry for infectious microorganisms. Microbes are inhaled into the nose or mouth in drops of moisture and dust particles. Diseases that are commonly contracted via the respiratory tract include the common cold, pneumonia, tuberculosis, influenza, measles, and smallpox.

Microorganisms can gain access to the gastrointestinal tract in food and water and via contaminated fingers. Most microbes that enter the body in these ways are destroyed by hydrochloric acid (HCl) and enzymes in the stomach or by bile and enzymes in the small intestine. Those that survive can cause disease. Microbes in the gastrointestinal tract can cause poliomyelitis, hepatitis A, typhoid fever, amoebic dysentery, giardiasis, shigellosis (bacillary dysentery), and cholera. These pathogens are then eliminated with feces and can be transmitted to other hosts via contaminated water, food, or fingers.

The genitourinary tract is a portal of entry for pathogens that are contracted sexually. Some microbes that cause sexually transmitted infections (STIs) may penetrate an unbroken mucous membrane. Others require a cut or abrasion of some type. Examples of STIs are HIV infection, genital warts, chlamydia, herpes, syphilis, and gonorrhea.

Skin

The skin is the largest organ of the body in terms of surface area and weight and is an important defense against disease. Unbroken skin is impenetrable by most microorganisms. Some microbes gain access to the body through openings in the skin, such as hair follicles and sweat gland ducts. Larvae of the hookworm actually bore through intact skin, and some fungi grow on the keratin in skin or infect the skin itself.

The conjunctiva is a delicate mucous membrane that lines the eyelids and covers the white of the eyeballs. Although it is a relatively effective barrier against infection, certain diseases such as conjunctivitis, trachoma, and ophthalmia neonatorium are acquired through the conjunctiva.

The Parenteral Route

Other microorganisms gain access to the body when they are deposited directly into the tissues beneath the skin or into mucous membranes when these barriers are penetrated or injured. This route is called the **parenteral route.** Punctures, injections, bites, cuts, wounds, surgery, and splitting of the skin or mucous membrane due to swelling or drying can all establish parenteral routes. HIV, the hepatitis viruses, and bacteria that cause tetanus and gangrene can be transmitted parenterally.

The Preferred Portal of Entry

Even after microorganisms have entered the body, they do not necessarily cause disease. The occurrence of disease depends on several factors, only one of which is the portal of entry. Many pathogens have a preferred portal of entry that is a prerequisite to their being able to cause disease. If they gain access to the body by another portal, disease might not occur. For example, the bacteria of typhoid fever, *Salmonella typhi,* produce all the signs and symptoms of the disease when swallowed (preferred route), but if the same bacteria are rubbed on the skin, no reaction (or only a slight inflammation) occurs. Streptococci that are inhaled (preferred route) can cause pneumonia; those that are swallowed generally do not produce signs or symptoms. Some pathogens, such as *Yersinia pestis,* the microorganism that causes plague, and *Bacillus anthracis,* the causative agent of anthrax, can initiate disease from more than one portal of entry. The preferred portals of entry for some common pathogens are listed in **Table 15.1**.

Numbers of Invading Microbes

If only a few microbes enter the body, they will probably be overcome by the host's defenses. However, if large numbers of microbes gain entry, the stage is probably set for disease. Thus, the likelihood of disease increases as the number of pathogens increases.

The virulence of a microbe is often expressed as the $\mathbf{ID_{50}}$ (infectious dose for 50% of a sample population). The 50 is not

Table 15.1 Portals of Entry for the Pathogens of Some Common Diseases

Portal of Entry	Pathogen*	Disease	Incubation Period
Mucous Membranes			
Respiratory tract	*Streptococcus pneumoniae*	Pneumococcal pneumonia	Variable
	Mycobacterium tuberculosis[†]	Tuberculosis	Variable
	Bordetella pertussis	Whooping cough (pertussis)	12–20 days
	Influenza virus *(Influenzavirus)*	Influenza	18–36 hours
	Measles virus *(Morbillivirus)*	Measles (rubeola)	11–14 days
	Rubella virus *(Rubivirus)*	German measles (rubella)	2–3 weeks
	Epstein-Barr virus *(Lymphocryptovirus)*	Infectious mononucleosis	2–6 weeks
	Varicella-zoster virus *(Varicellovirus)*	Chickenpox (varicella) (primary infection)	14–16 days
	Histoplasma capsulatum (fungus)	Histoplasmosis	5–18 days
Gastrointestinal tract	*Shigella* spp.	Shigellosis (bacillary dysentery)	1–2 days
	Brucella spp.	Brucellosis (undulant fever)	6–14 days
	Vibrio cholerae	Cholera	1–3 days
	Salmonella enterica	Salmonellosis	7–22 hours
	Salmonella typhi	Typhoid fever	14 days
	Hepatitis A virus *(Hepatovirus)*	Hepatitis A	15–50 days
	Mumps virus *(Rubulavirus)*	Mumps	2–3 weeks
	Trichinella spiralis (helminth)	Trichinellosis	2–28 days
Genitourinary tract	*Neisseria gonorrhoeae*	Gonorrhea	3–8 days
	Treponema pallidum	Syphilis	9–90 days
	Chlamydia trachomatis	Nongonococcal urethritis	1–3 weeks
	Herpes simplex virus type 2	Herpes virus infections	4–10 days
	Human immunodeficiency virus (HIV)[‡]	AIDS	10 years
	Candida albicans (fungus)	Candidiasis	2–5 days
Skin or Parenteral Route			
	Clostridium perfringens	Gas gangrene	1–5 days
	Clostridium tetani	Tetanus	3–21 days
	Rickettsia rickettsii	Rocky Mountain spotted fever	3–12 days
	Hepatitis B virus *(Hepadnavirus)*[‡]	Hepatitis B	6 weeks–6 months
	Rabiesvirus *(Lyssavirus)*	Rabies	10 days–1 year
	Plasmodium spp. (protozoan)	Malaria	2 weeks

*All pathogens are bacteria, unless indicated otherwise. For viruses, the viral species and/or genus name is given.
[†]These pathogens can also cause disease after entering the body via the gastrointestinal tract.
[‡]These pathogens can also cause disease after entering the body via the parenteral route. Hepatitis B virus and HIV can also cause disease after entering the body via the genitourinary tract.

an absolute value; rather, it is used to compare relative virulence under experimental conditions. *Bacillus anthracis* can cause infection via three different portals of entry. The ID_{50} through the skin (cutaneous anthrax) is 10 to 50 endospores; the ID_{50} for inhalation anthrax is inhalation of 10,000 to 20,000 endospores; and the ID_{50} for gastrointestinal anthrax is ingestion of 250,000 to 1,000,000 endospores. These data show that cutaneous anthrax is significantly easier to acquire than either the inhalation or the gastrointestinal forms. A study of *Vibrio cholerae* showed that the ID_{50} is 10^8 cells; but if stomach acid is neutralized with bicarbonate, the number of cells required to cause an infection decreases significantly.

The potency of a toxin is often expressed as the $\mathbf{LD_{50}}$ (lethal dose for 50% of a sample population). For example, the LD_{50} for botulinum toxin in mice is 0.03 ng/kg; for Shiga toxin, 250 ng/kg; and staphylococcal enterotoxin, 1350 ng/kg. In other words,

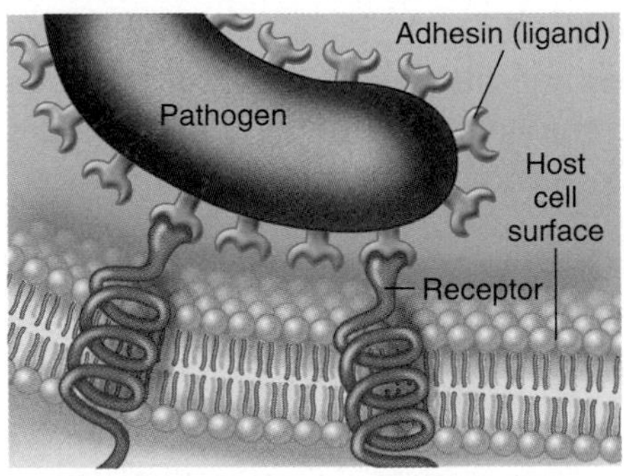

(a) Surface molecules on a pathogen, called adhesins or ligands, bind specifically to complementary surface receptors on cells of certain host tissues.

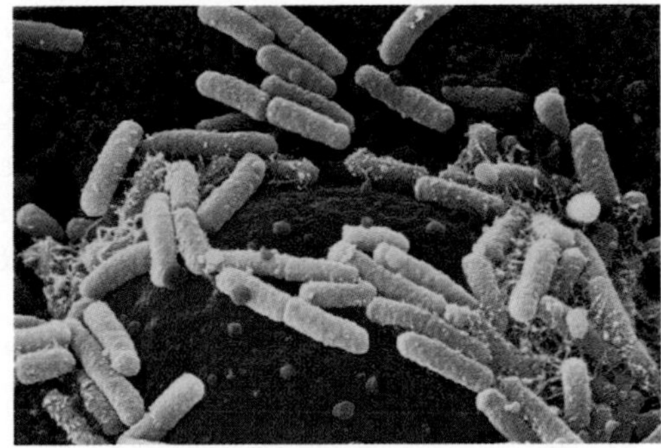

(b) *E. coli* bacteria (yellow-green) on human urinary bladder cells. SEM 1 μm

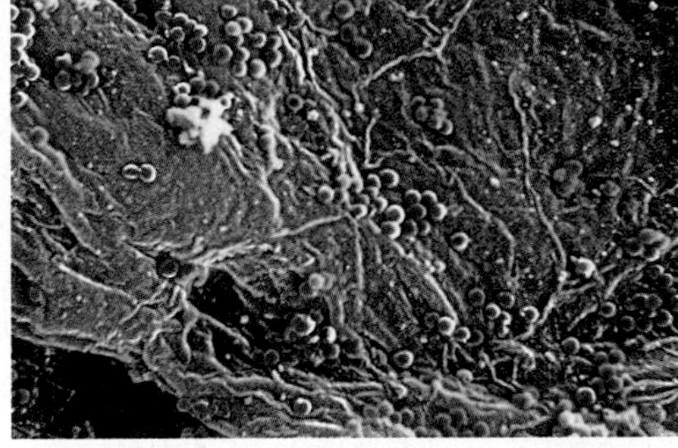

(c) Bacteria (purple) adhering to human skin.

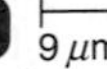

Figure 15.1 Adherence.

Q How are most adhesins classified chemically?

compared to the other two toxins, a much smaller dose of botulinum toxin is needed to cause symptoms.

Adherence

Almost all pathogens have some means of attaching themselves to host tissues at their portal of entry. For most pathogens, this attachment, called **adherence** (or **adhesion**), is a necessary step in pathogenicity. (Of course, nonpathogens also have structures for attachment.) The attachment between pathogen and host is accomplished by means of surface molecules on the pathogen called **adhesins** or **ligands** that bind specifically to complementary surface **receptors** on the cells of certain host tissues (Figure 15.1). Adhesins may be located on a microbe's glycocalyx or on other microbial surface structures, such as pili, fimbriae, and flagella (see Chapter 4).

The majority of adhesins on the microorganisms studied so far are glycoproteins or lipoproteins. The receptors on host cells are typically sugars, such as mannose. Adhesins on different strains of the same species of pathogen can vary in structure. Different cells of the same host can also have different receptors that vary in structure. If adhesins, receptors, or both can be altered to interfere with adherence, infection can often be prevented (or at least controlled).

The following examples illustrate the diversity of adhesins. *Streptococcus mutans,* a bacterium that plays a key role in tooth decay, attaches to the surface of teeth by its glycocalyx. An enzyme produced by *S. mutans,* called glucosyltransferase, converts glucose (derived from sucrose or table sugar) into a sticky polysaccharide called dextran, which forms the glycocalyx. *Actinomyces* bacterial cells have fimbriae that adhere to the glycocalyx of *S. mutans.* The combination of *S. mutans, Actinomyces,* and dextran make up dental plaque and contribute to dental caries (tooth decay; see Chapter 25, page 707).

Microbes have the ability to come together in masses, cling to surfaces, and take in and share available nutrients. These communities, which constitute masses of microbes and their extracellular products that can attach to living and nonliving surfaces, are called **biofilms** (discussed in more detail in Chapter 6, page 162). Examples of biofilms include the dental plaque on teeth, the algae on the walls of swimming pools, and the scum that accumulates on shower doors. A biofilm forms when microbes adhere to a particular surface that is typically moist and contains organic matter. The first microbes to attach are usually bacteria. Once they adhere to the surface, they multiply and secrete a glycocalyx that further attaches the bacteria to each other and to the surface (see Figure 6.5, page 163). In some cases, biofilms can be several layers thick and may contain several types of microbes. Biofilms represent another method of adherence and are important because they resist disinfectants and antibiotics. This characteristic is significant, especially when biofilms colonize structures such as teeth, medical catheters, stents, heart valves, hip replacement components, and contact lenses. Dental plaque is actually a biofilm that mineralizes over time. It is estimated that biofilms are involved in 65% of all human bacterial infections. See the box in Chapter 6 on page 164.

Enteropathogenic strains of *E. coli* (those responsible for gastrointestinal disease) have adhesins on fimbriae that adhere only to specific kinds of cells in certain regions of the small intestine. After adhering, *Shigella* and *E. coli* induce endocytosis as a vehicle to enter host cells and then multiply within them (see Figure 25.7, page 712). *Treponema pallidum,* the causative agent of syphilis, uses its tapered end as a hook to attach to host cells. *Listeria monocytogenes,* which causes meningitis, spontaneous abortions, and stillbirths, produces an adhesin for a specific receptor on host cells. *Neisseria gonorrhoeae,* the causative agent of gonorrhea, also has fimbriae containing adhesins, which in this case permit attachment to cells with appropriate receptors in the genitourinary tract, eyes, and pharynx. *Staphylococcus aureus,* which can cause skin infections, binds to skin by a mechanism of adherence that resembles viral attachment (see Chapter 13).

CHECK YOUR UNDERSTANDING

- ✓ List three portals of entry, and describe how microorganisms gain access through each. **15-1**
- ✓ The LD_{50} of botulinum toxin is 0.03 ng/kg; the LD_{50} of *Salmonella* toxin is 12 mg/kg. Which is the more potent toxin? **15-2**
- ✓ How would a drug that binds mannose on human cells affect a pathogenic bacterium? **15-3**

How Bacterial Pathogens Penetrate Host Defenses

LEARNING OBJECTIVES

15-4 Explain how capsules and cell wall components contribute to pathogenicity.

15-5 Compare the effects of coagulases, kinases, hyaluronidase, and collagenase.

15-6 Define and give an example of *antigenic variation.*

15-7 Describe how bacteria use the host cell's cytoskeleton to enter the cell.

Although some pathogens can cause damage on the surface of tissues, most must penetrate tissues to cause disease. Here we will consider several factors that contribute to the ability of bacteria to invade a host.

Capsules

Recall from Chapter 4 that some bacteria make glycocalyx material that forms capsules around their cell walls; this property increases the virulence of the species. The capsule resists the host's defenses by impairing phagocytosis, the process by which certain cells of the body engulf and destroy microbes (see Chapter 16, page 458). The chemical nature of the capsule appears to prevent the phagocytic cell from adhering to the bacterium. However, the human body can produce antibodies against the capsule, and when these antibodies are present on the capsule surface, the encapsulated bacteria are easily destroyed by phagocytosis.

One bacterium that owes its virulence to the presence of a polysaccharide capsule is *Streptococcus pneumoniae,* the causative agent of pneumococcal pneumonia (see Figure 24.12, page 686). Some strains of this organism have capsules, and others do not. Strains with capsules are virulent, but strains without capsules are avirulent because they are susceptible to phagocytosis. Other bacteria that produce capsules related to virulence are *Klebsiella pneumoniae,* a causative agent of bacterial pneumonia; *Haemophilus influenzae,* a cause of pneumonia and meningitis in children; *Bacillus anthracis,* the cause of anthrax; and *Yersinia pestis,* the causative agent of plague. Keep in mind that capsules are not the only cause of virulence. Many nonpathogenic bacteria produce capsules, and the virulence of some pathogens is not related to the presence of a capsule.

Cell Wall Components

The cell walls of certain bacteria contain chemical substances that contribute to virulence. For example, *Streptococcus pyogenes* produces a heat-resistant and acid-resistant protein called **M protein** (see Figure 21.6, page 591). This protein is found on both the cell surface and fimbriae. The M protein mediates attachment of the bacterium to epithelial cells of the host and helps the bacterium resist phagocytosis by white blood cells. The protein thereby increases the virulence of the microorganism. Immunity to *S. pyogenes* depends on the body's production of an antibody specific to M protein. *Neisseria gonorrhoeae* grows inside human epithelial cells and leukocytes. These bacteria use **fimbriae** and an outer membrane protein called **Opa** to attach to host cells. Following attachment by both Opa and fimbriae, the host cells take in the bacteria. (Bacteria that produce Opa form *opa*que colonies on culture media.) The **waxy lipid** (mycolic acid) that makes up the cell wall of *Mycobacterium tuberculosis* also increases virulence by resisting digestion by phagocytes, and can even multiply inside phagocytes.

Enzymes

The virulence of some bacteria is thought to be aided by the production of extracellular enzymes (*exoenzymes*) and related substances. These chemicals can digest materials between cells and form or digest blood clots, among other functions.

Coagulases are bacterial enzymes that coagulate (clot) the fibrinogen in blood. Fibrinogen, a plasma protein produced by the liver, is converted by coagulases into fibrin, the threads that form a blood clot. The fibrin clot may protect the bacterium from phagocytosis and isolate it from other defenses of the host. Coagulases are produced by some members of the genus *Staphylococcus;* they may be involved in the walling-off process in boils produced by staphylococci. However, some staphylococci that do not produce coagulases are still virulent. (Capsules may be more important to their virulence.)

Bacterial **kinases** are bacterial enzymes that break down fibrin and thus digest clots formed by the body to isolate the infection. One of the better-known kinases is *fibrinolysin (streptokinase),* which is produced by such streptococci as *Streptococcus pyogenes.* Another kinase, *staphylokinase,* is produced by *Staphylococcus aureus.* Injected into the blood, streptokinase has been used to remove some types of blood clots in cases of heart attacks due to obstructed coronary arteries.

Hyaluronidase is another enzyme secreted by certain bacteria, such as streptococci. It hydrolyzes hyaluronic acid, a type of polysaccharide that holds together certain cells of the body, particularly cells in connective tissue. This digesting action is thought to be involved in the tissue blackening of infected wounds and to help the microorganism spread from its initial site of infection. Hyaluronidase is also produced by some clostridia that cause gas gangrene. For therapeutic use, hyaluronidase may be mixed with a drug to promote the spread of the drug through a body tissue.

Another enzyme, **collagenase,** produced by several species of *Clostridium,* facilitates the spread of gas gangrene. Collagenase breaks down the protein collagen, which forms the connective tissue of muscles and other body organs and tissues.

As a defense against adherence of pathogens to mucosal surfaces, the body produces a class of antibodies called IgA antibodies. There are some pathogens with the ability to produce enzymes, called **IgA proteases,** that can destroy these antibodies. *N. gonorrhoeae* has this ability, as do *N. meningitidis,* the causative agent of meningococcal meningitis, and other microbes that infect the central nervous system.

Antigenic Variation

In Chapter 17 you will learn that *adaptive (acquired) immunity* refers to a specific defensive response of the body to an infection or to antigens. In the presence of antigens the body produces proteins called antibodies, which bind to the antigens and inactivate or destroy them. However, some pathogens can alter their surface antigens, by a process called **antigenic variation.** Thus, by the time the body mounts an immune response against a pathogen, the pathogen has already altered its antigens and is unaffected by the antibodies. Some microbes can activate alternative genes, resulting in antigenic changes. For example, *N. gonorrhoeae* has several copies of the Opa-encoding gene, resulting in cells with different antigens and in cells that express different antigens over time.

A wide range of microbes is capable of antigenic variation. Examples include *Influenzavirus,* the causative agent of influenza (flu); *Neisseria gonorrhoeae,* the causative agent of gonorrhea; and *Trypanosoma brucei gambiense* (tri-pa'nō-sō-mä brüs'ē gam-bē-ens'), the causative agent of African trypanosomiasis (sleeping sickness). See Figure 22.16, page 629.

Penetration into the Host Cell Cytoskeleton

As previously noted, microbes attach to host cells by adhesins. The interaction triggers signals in the host cell that activate factors that can result in the entrance of some bacteria. The actual mechanism is provided by the host cell cytoskeleton. Recall from Chapter 4 that eukaryotic cytoplasm has a complex internal structure (the cytoskeleton), consisting of protein filaments called microfilaments, intermediate filaments, and microtubules. A major component of the cytoskeleton is a protein called actin, which is used by some microbes to penetrate host cells and by others to move through and between host cells.

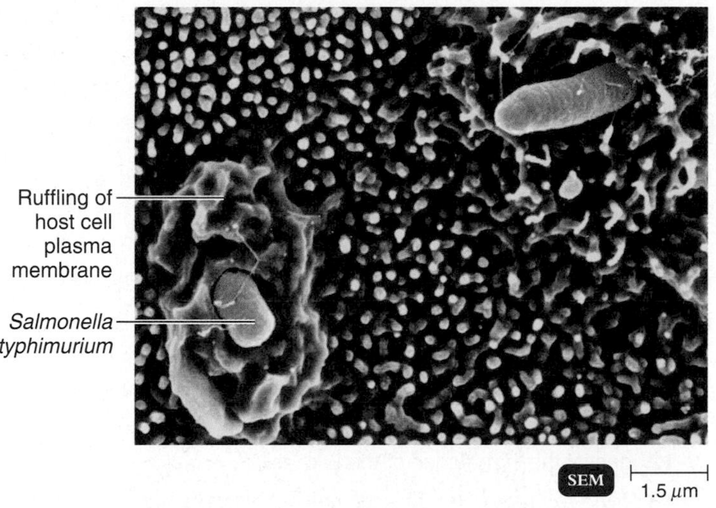

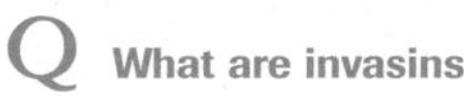

Figure 15.2 ***Salmonella*** **entering intestinal epithelial cells as a result of ruffling.**

Q What are invasins?

Salmonella strains and *E. coli* make contact with the host cell plasma membrane. This leads to dramatic changes in the membrane at the point of contact. The microbes produce surface proteins called **invasins** that rearrange nearby actin filaments of the cytoskeleton. For example, when *S. typhimurium* makes contact with a host cell, invasins of the microbe cause the appearance of the host cell plasma membrane to resemble the splash of a drop of a liquid hitting a solid surface. This effect, called *membrane ruffling,* is the result of disruption in the cytoskeleton of the host cell (Figure 15.2). The microbe sinks into the ruffle and is engulfed by the host cell.

Once inside the host cell, certain bacteria such as *Shigella* species and *Listeria* species can actually use actin to propel themselves through the host cell cytoplasm and from one host cell to another. The condensation of actin on one end of the bacteria propels them through the cytoplasm. The bacteria also make contact with membrane junctions that form part of a transport network between host cells. The bacteria use a glycoprotein called *cadherin,* which bridges the junctions, to move from cell to cell.

The study of the numerous interactions between microbes and host cell cytoskeleton is a very intense area of investigation on virulence mechanisms.

CHECK YOUR UNDERSTANDING

- ✓ What function do capsules and M proteins have in common? **15-4**
- ✓ Would you expect a bacterium to make coagulase and kinase simultaneously? **15-5**
- ✓ Many vaccines provide years of protection against a disease. Why doesn't the influenza vaccine offer more than a few months of protection? **15-6**
- ✓ How does *E. coli* cause membrane ruffling? **15-7**

How Bacterial Pathogens Damage Host Cells

LEARNING OBJECTIVES

15-8 Describe the function of siderophores.

15-9 Provide an example of direct damage, and compare this to toxin production.

15-10 Contrast the nature and effects of exotoxins and endotoxins.

15-11 Outline the mechanisms of action of A-B toxins, membrane-disrupting toxins, and superantigens. Classify diphtheria toxin, erythrogenic toxin, botulinum toxin, tetanus toxin, *Vibrio* enterotoxin, and staphylococcal enterotoxin.

15-12 Identify the importance of the LAL assay.

15-13 Using examples, describe the roles of plasmids and lysogeny in pathogenicity.

When a microorganism invades a body tissue, it initially encounters phagocytes of the host. If the phagocytes are successful in destroying the invader, no further damage is done to the host. But if the pathogen overcomes the host's defense, then the microorganism can damage host cells in four basic ways: (1) by using the host's nutrients; (2) by causing direct damage in the immediate vicinity of the invasion; (3) by producing toxins, transported by blood and lymph, that damage sites far removed from the original site of invasion; and (4) by inducing hypersensitivity reactions. This fourth mechanism is considered in detail in Chapter 19. For now, we will discuss only the first three mechanisms.

Using the Host's Nutrients: Siderophores

Iron is required for the growth of most pathogenic bacteria. However, the concentration of free iron in the human body is fairly low because most of the iron is tightly bound to iron-transport proteins, such as lactoferrin, transferrin, and ferritin, as well as hemoglobin. These are discussed in more detail in Chapter 16. To obtain free iron, some pathogens secrete proteins called **siderophores** (Figure 15.3). When a pathogen needs iron, siderophores are released into the medium where they take the iron away from iron-transport proteins by binding the iron even more tightly. Once the iron-siderophore complex is formed, it is taken up by siderophore receptors on the bacterial surface. Then the iron is brought into the bacterium. In some cases, the iron is released from the complex to enter the bacterium; in other cases, the iron enters as part of the complex.

As an alternative to iron acquisition by siderophores, some pathogens have receptors that bind directly to iron-transport proteins and hemoglobin. Then these are taken into the bacterium directly along with the iron. Also, it is possible that some bacteria produce toxins (described shortly) when iron levels are low. The toxins kill host cells, releasing their iron and thereby making it available to the bacteria.

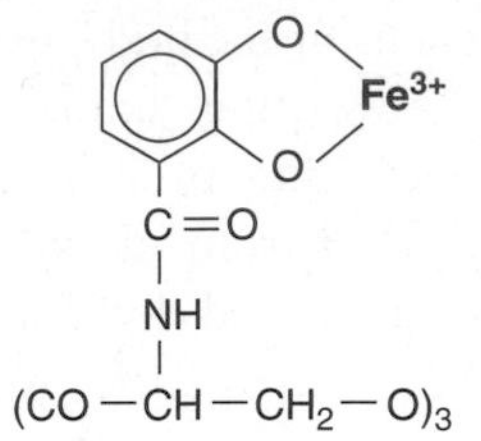

Fe^{3+}
C=O
NH
$(CO-CH-CH_2-O)_3$

Figure 15.3 Structure of enterobactin, one type of bacterial siderophore. Note where the iron (Fe^{3+}) is attached to the siderophore.

Q Of what value are siderophores?

Direct Damage

Once pathogens attach to host cells, they can cause direct damage as the pathogens use the host cell for nutrients and produce waste products. As pathogens metabolize and multiply in cells, the cells usually rupture. Many viruses and some intracellular bacteria and protozoa that grow in host cells are released when the host cell ruptures. Following their release, pathogens that rupture cells can spread to other tissues in even greater numbers. Some bacteria, such as *E. coli, Shigella, Salmonella,* and *Neisseria gonorrhoeae,* can induce host epithelial cells to engulf them by a process that resembles phagocytosis. These pathogens can disrupt host cells as they pass through and can then be extruded from the host cells by a reverse phagocytosis process, enabling them to enter other host cells. Some bacteria can also penetrate host cells by excreting enzymes and by their own motility; such penetration can itself damage the host cell. Most damage by bacteria, however, is done by toxins. **Animations** Virulence Factors: Penetrating Host Tissues, Hiding from Host Defenses, Enteric Pathogens. **www.microbiologyplace.com**

The Production of Toxins

Toxins are poisonous substances that are produced by certain microorganisms. They are often the primary factor contributing to the pathogenic properties of those microbes. The capacity of microorganisms to produce toxins is called **toxigenicity.** Toxins transported by the blood or lymph can cause serious, and sometimes fatal, effects. Some toxins produce fever, cardiovascular disturbances, diarrhea, and shock. Toxins can also inhibit protein synthesis, destroy blood cells and blood vessels, and disrupt the nervous system by causing spasms. Of the 220 or so known bacterial toxins, nearly 40% cause disease by damaging eukaryotic cell membranes. The term **toxemia** refers to the presence of toxins in the blood. Toxins are of two general types, based on their position relative to the microbial cell: exotoxins and endotoxins.

Exotoxins

Exotoxins are produced inside some bacteria as part of their growth and metabolism and are secreted by the bacterium into the surrounding medium or released following lysis (Figure 15.4a). Exotoxins are proteins, and many are enzymes that catalyze only certain biochemical reactions. Because of the enzymatic nature of most exotoxins, even small amounts are quite harmful because they can act over and over again. Bacteria that produce exotoxins may be gram-positive or gram-negative. The genes for most (perhaps all) exotoxins are carried on bacterial plasmids or phages.

Figure 15.4

FOUNDATION FIGURE Exotoxins and Endotoxins

Damage to the host by bacteria is often due to toxins. Many of the diseases discussed in Part Four of the book will involve the activity of these toxins.

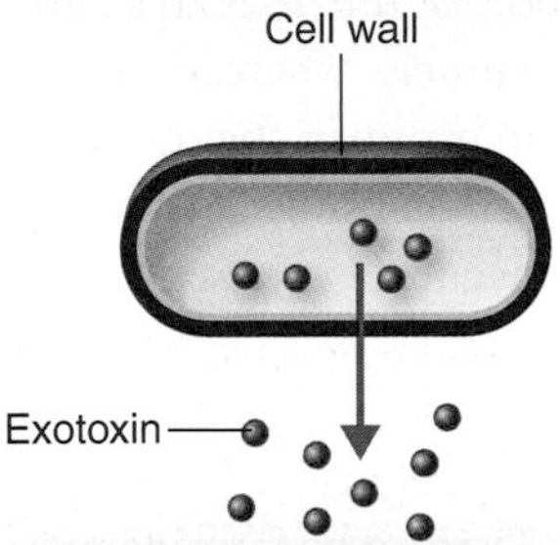

(a) Exotoxins are proteins produced inside pathogenic bacteria, most commonly gram-positive bacteria, as part of their growth and metabolism. The exotoxins are then secreted or released into the surrounding medium following lysis.

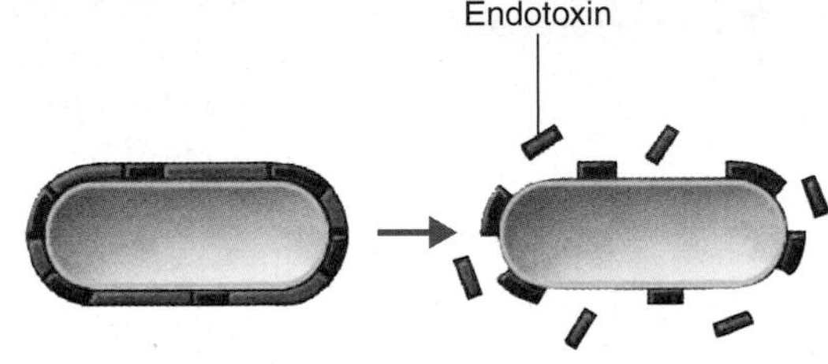

(b) Endotoxins are the lipid portions of lipopolysaccharides (LPSs) that are part of the outer membrane of the cell wall of gram-negative bacteria (lipid A; see Figure 4.13c). The endotoxins are liberated when the bacteria die and the cell wall breaks apart.

Key Concept

Toxins are of two general types: exotoxins and endotoxins.

Because exotoxins are soluble in body fluids, they can easily diffuse into the blood and are rapidly transported throughout the body.

Exotoxins work by destroying particular parts of the host's cells or by inhibiting certain metabolic functions. They are highly specific in their effects on body tissues. Exotoxins are among the most lethal substances known. Only 1 mg of the botulinum exotoxin is enough to kill 1 million guinea pigs. Fortunately, only a few bacterial species produce such potent exotoxins.

Diseases caused by bacteria that produce exotoxins are often caused by minute amounts of exotoxins, not by the bacteria themselves. It is the exotoxins that produce the specific signs and symptoms of the disease. Thus, exotoxins are disease-specific. For example, botulism is usually due to ingesting the exotoxin, not a bacterial infection. Likewise, staphylococcal food poisoning is an *intoxication,* not an infection.

The body produces antibodies called **antitoxins** that provide immunity to exotoxins. When exotoxins are inactivated by heat or by formaldehyde, iodine, or other chemicals, they no longer cause the disease but can still stimulate the body to produce antitoxins. Such altered exotoxins are called **toxoids.** When toxoids are injected into the body as a vaccine, they stimulate antitoxin production so that immunity is produced. Diphtheria and tetanus can be prevented by toxoid vaccination.

Naming Exotoxins Exotoxins are named on the basis of several characteristics. One is the type of host cell that is attacked. For example, *neurotoxins* attack nerve cells, *cardiotoxins* attack heart cells, *hepatotoxins* attack liver cells, *leukotoxins* attack leukocytes, *enterotoxins* attack the lining of the gastrointestinal tract, and *cytotoxins* attack a wide variety of cells. Some exotoxins are named for the diseases with which they are associated. Examples include *diphtheria toxin* (cause of diphtheria) and *tetanus toxin* (cause of tetanus). Other exotoxins are named for the specific bacterium that produces them, for example, *botulinum toxin (Clostridium botulinum)* and *Vibrio enterotoxin (Vibrio cholerae).*

Types of Exotoxins Exotoxins are divided into three principal types on the basis of their structure and function: (1) A-B toxins, (2) membrane-disrupting toxins, and (3) superantigens.

A-B Toxins **A-B toxins** were the first toxins to be studied intensively and are so named because they consist of two parts designated A and B, both of which are polypeptides. Most exotoxins are A-B toxins. The A part is the active (enzyme) component, and the B part is the binding component. An example of an A-B toxin is the diphtheria toxin, which is illustrated in Figure 15.5.

1. In the first step, the A-B toxin is released from the bacterium.
2. The B component attaches to a host cell receptor.
3. The plasma membrane of the host cell invaginates (folds inward) at the point where the A-B exotoxin and plasma receptor make contact, and the exotoxin enters the cell by endocytosis.
4. The A-B exotoxin and receptor are enclosed by a pinched-off portion of the membrane.
5. The A-B components of the exotoxin separate. The A component alters the function of the host cell, often by inhibiting protein synthesis. The B component is released from the host cell, and the receptor is inserted into the plasma membrane for reuse.

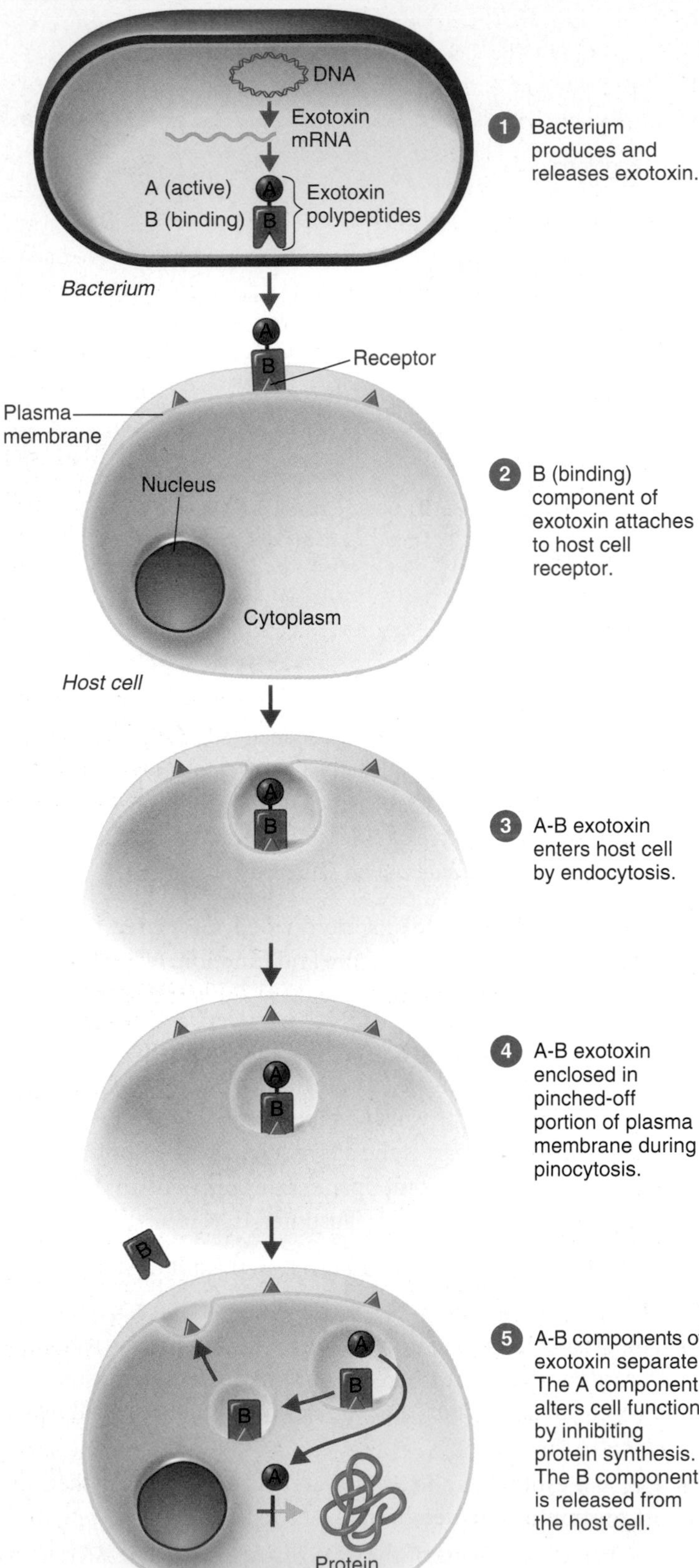
DNA
Exotoxin mRNA
A (active)
B (binding)
Exotoxin polypeptides
Bacterium
1 Bacterium produces and releases exotoxin.
Receptor
Plasma membrane
Nucleus
Cytoplasm
Host cell
2 B (binding) component of exotoxin attaches to host cell receptor.
3 A-B exotoxin enters host cell by endocytosis.
4 A-B exotoxin enclosed in pinched-off portion of plasma membrane during pinocytosis.
5 A-B components of exotoxin separate. The A component alters cell function by inhibiting protein synthesis. The B component is released from the host cell.
Protein

Figure 15.5 The action of an A-B exotoxin. A proposed model for the mechanism of action of diphtheria toxin.

Q **Why is this called an A-B toxin?**

Membrane-Disrupting Toxins **Membrane-disrupting toxins** cause lysis of host cells by disrupting their plasma membranes. Some do this by forming protein channels in the plasma membrane; others disrupt the phospholipid portion of the membrane. The cell-lysing exotoxin of *Staphylococcus aureus* is an example of an exotoxin that forms protein channels, whereas that of *Clostridium perfringens* is an example of an exotoxin that disrupts the phospholipids. Membrane-disrupting toxins contribute to virulence by killing host cells, especially phagocytes, and by aiding the escape of bacteria from sacs within phagocytes (phagosomes) into the host cell's cytoplasm.

Membrane-disrupting toxins that kill phagocytic leukocytes (white blood cells) are called **leukocidins.** They act by forming protein channels. Leukocidins are also active against macrophages, phagocytes present in tissues. Most leukocidins are produced by staphylococci and streptococci. The damage to phagocytes decreases host resistance. Membrane-disrupting toxins that destroy erythrocytes (red blood cells), also by forming protein channels, are called **hemolysins.** Important producers of hemolysins include staphylococci and streptococci. Hemolysins produced by streptococci are called **streptolysins.** One kind, called *streptolysin O (SLO)*, is so named because it is inactivated by atmospheric oxygen. Another kind of streptolysin is called *streptolysin S (SLS)* because it is stable in an oxygen environment. Both streptolysins can cause lysis not only of red blood cells, but also of white blood cells (whose function is to kill the streptococci) and other body cells.

Superantigens **Superantigens** are antigens that provoke a very intense immune response. They are bacterial proteins. Through a series of interactions with various cells of the immune system, superantigens nonspecifically stimulate the proliferation of immune cells called T cells. These cells are types of white blood cells (lymphocytes) that act against foreign organisms and tissues (transplants) and regulate the activation and proliferation of other cells of the immune system. In response to superantigens, T cells are stimulated to release enormous amounts of chemicals called cytokines. *Cytokines* are small protein molecules produced by various body cells, especially T cells, that regulate immune responses and mediate cell-to-cell communication (see Chapter 17, page 492). The excessively high levels of cytokines released by T cells enter the bloodstream and give rise to a number of symptoms, including fever, nausea, vomiting, diarrhea, and sometimes shock and even death. Bacterial superantigens include the staphylococcal toxins that cause food poisoning and toxic shock syndrome.

Representative Exotoxins Next we briefly describe a few of the more notable exotoxins (antitoxins will be discussed further in Chapter 18).

Diphtheria Toxin *Corynebacterium diphtheriae* produces the *diphtheria toxin* only when it is infected by a lysogenic phage carrying the *tox* gene. This cytotoxin inhibits protein synthesis in eukaryotic cells. It does this using an A-B toxin mechanism, which is illustrated in Figure 15.5.

Erythrogenic Toxins *Streptococcus pyogenes* has the genetic material to synthesize three types of cytotoxins, designated A, B, and C. These *erythrogenic* (*erythro* = red; *gen* = producing) *toxins* are superantigens that damage the plasma membranes of blood capillaries under the skin and produce a red skin rash. Scarlet fever, caused by *S. pyogenes* exotoxins, is named for this characteristic rash.

Botulinum Toxin *Botulinum toxin* is produced by *Clostridium botulinum.* Although toxin production is associated with the germination of endospores and the growth of vegetative cells, little of the toxin appears in the medium until it is released by lysis late in growth. Botulinum toxin is an A-B neurotoxin; it acts at the neuromuscular junction (the junction between nerve cells and muscle cells) and prevents the transmission of impulses from the nerve cell to the muscle. The toxin accomplishes this by binding to nerve cells and inhibiting the release of a neurotransmitter called acetylcholine. As a result, botulinum toxin causes paralysis in which muscle tone is lacking (flaccid paralysis). *C. botulinum* produces several different types of botulinum toxin, and each possesses a different potency.

Tetanus Toxin *Clostridium tetani* produces tetanus neurotoxin, also known as *tetanospasmin.* This A-B toxin reaches the central nervous system and binds to nerve cells that control the contraction of various skeletal muscles. These nerve cells normally send inhibiting impulses that prevent random contractions and terminate completed contractions. The binding of tetanospasmin blocks this relaxation pathway (see Chapter 22, page 615). The result is uncontrollable muscle contractions, producing the convulsive symptoms (spasmodic contractions) of tetanus, or "lockjaw."

Vibrio *Enterotoxin* *Vibrio cholerae* produces an A-B enterotoxin called *cholera toxin.* Subunit B binds to epithelial cells, and subunit A causes cells to secrete large amounts of fluids and electrolytes (ions). Normal muscular contractions are disturbed, leading to severe diarrhea that may be accompanied by vomiting. *Heat-labile enterotoxin* (so named because it is more sensitive to heat than are most toxins), produced by some strains of *E. coli,* has an action identical to that of *Vibrio* enterotoxin.

Staphylococcal Enterotoxin *Staphylococcus aureus* produces a superantigen that affects the intestines in the same way as *Vibrio* enterotoxin. A strain of *S. aureus* also produces a superantigen that results in the symptoms associated with toxic shock syndrome (see Chapter 21, page 588). A summary of diseases produced by exotoxins is shown in **Table 15.2**.

Endotoxins

Endotoxins differ from exotoxins in several ways. Endotoxins are part of the outer portion of the cell wall of gram-negative bacteria (**Figure 15.4b**). Recall from Chapter 4 that gram-negative bacteria have an outer membrane surrounding the peptidoglycan layer of the cell wall. This outer membrane consists of lipoproteins, phospholipids, and lipopolysaccharides (LPSs) (see Figure 4.13c, page 86). The lipid portion of LPS, called **lipid A,** is the endotoxin. Thus, endotoxins are lipopolysaccharides, whereas exotoxins are proteins.

Endotoxins are released when gram-negative bacteria die and their cell walls undergo lysis, thus liberating the endotoxin. (Endotoxins are also released during bacterial multiplication.) Antibiotics used to treat diseases caused by gram-negative bacteria can lyse the bacterial cells; this reaction releases endotoxin and may lead to an immediate worsening of the symptoms, but the condition usually improves as the endotoxin breaks down. Endotoxins exert their effects by stimulating macrophages to release cytokines in very high concentrations. At these levels, cytokines are toxic. All endotoxins produce the same signs and symptoms, regardless of the species of microorganism, although not to the same degree. These include chills, fever, weakness, generalized aches, and, in some cases, shock and even death. Endotoxins can also induce miscarriage.

Another consequence of endotoxins is the activation of blood-clotting proteins, causing the formation of small blood clots. These blood clots obstruct capillaries, and the resulting decreased blood supply induces the death of tissues. This condition is referred to as *disseminated intravascular coagulation (DIC).*

The fever (pyrogenic response) caused by endotoxins is believed to occur as depicted in **Figure 15.6.**

1. Gram-negative bacteria are ingested by phagocytes.
2. As the bacteria are degraded in vacuoles, the LPSs of the bacterial cell wall are released. These endotoxins cause macrophages to produce cytokines called **interleukin-1 (IL-1),** formerly called *endogenous pyrogen,* and **tumor necrosis factor alpha (TNF-α).**
3. The cytokines are carried via the blood to the hypothalamus, a temperature control center in the brain.
4. The cytokines induce the hypothalamus to release lipids called prostaglandins, which reset the thermostat in the hypothalamus at a higher temperature. The result is a fever.

Bacterial cell death caused by lysis or antibiotics can also produce fever by this mechanism. Both aspirin and acetaminophen reduce fever by inhibiting the synthesis of prostaglandins. (The function of fever in the body is discussed in Chapter 16, page 463.)

Shock refers to any life-threatening decrease in blood pressure. Shock caused by bacteria is called **septic shock.** Gram-negative bacteria cause *endotoxic shock.* Like fever, the shock produced by endotoxins is related to the secretion of a cytokine by macrophages. Phagocytosis of gram-negative bacteria causes the phagocytes to secrete tumor necrosis factor (TNF), sometimes called *cachectin.* TNF binds to many tissues in the body and alters their metabolism in a number of ways. One effect of TNF is damage to blood capillaries; their permeability is increased, and they lose large amounts of fluid. The result is a drop in blood pressure that results in shock.

Table 15.2 Diseases Caused by Exotoxins

Disease	Bacterium	Type of Exotoxin	Mechanism
Botulism	*Clostridium botulinum*	A-B	Neurotoxin prevents the transmission of nerve impulses; flaccid paralysis results.
Tetanus	*Clostridium tetani*	A-B	Neurotoxin blocks nerve impulses to muscle relaxation pathway; results in uncontrollable muscle contractions.
Diphtheria	*Corynebacterium diphtheriae*	A-B	Cytotoxin inhibits protein synthesis, especially in nerve, heart, and kidney cells.
Scalded skin syndrome	*Staphylococcus aureus*	A-B	One exotoxin causes skin layers to separate and slough off (scalded skin).
Cholera	*Vibrio cholerae*	A-B	Enterotoxin causes secretion of large amounts of fluids and electrolytes that result in diarrhea.
Traveler's diarrhea	Enterotoxigenic *Escherichia coli* and *Shigella* spp.	A-B	Enterotoxin causes secretion of large amounts of fluids and electrolytes that result in diarrhea.
Anthrax	*Bacillus anthracis*	A-B	Two A components enter the cell via the same B. The A proteins cause shock and reduce the immune response.
Gas gangrene and food poisoning	*Clostridium perfringens* and other species of *Clostridium*	Membrane-disrupting	One exotoxin (cytotoxin) causes massive red blood cell destruction (hemolysis); another exotoxin (enterotoxin) is related to food poisoning and causes diarrhea.
Antibiotic-associated diarrhea	*Clostridium difficile*	Membrane-disrupting	Enterotoxin causes secretion of fluids and electrolytes that results in diarrhea; cytotoxin disrupts host cytoskeleton.
Food poisoning	*Staphylococcus aureus*	Superantigen	Enterotoxin causes secretion of fluids and electrolytes that results in diarrhea.
Toxic shock syndrome (TSS)	*Staphylococcus aureus*	Superantigen	Toxin causes secretion of fluids and electrolytes from capillaries that decreases blood volume and lowers blood pressure.

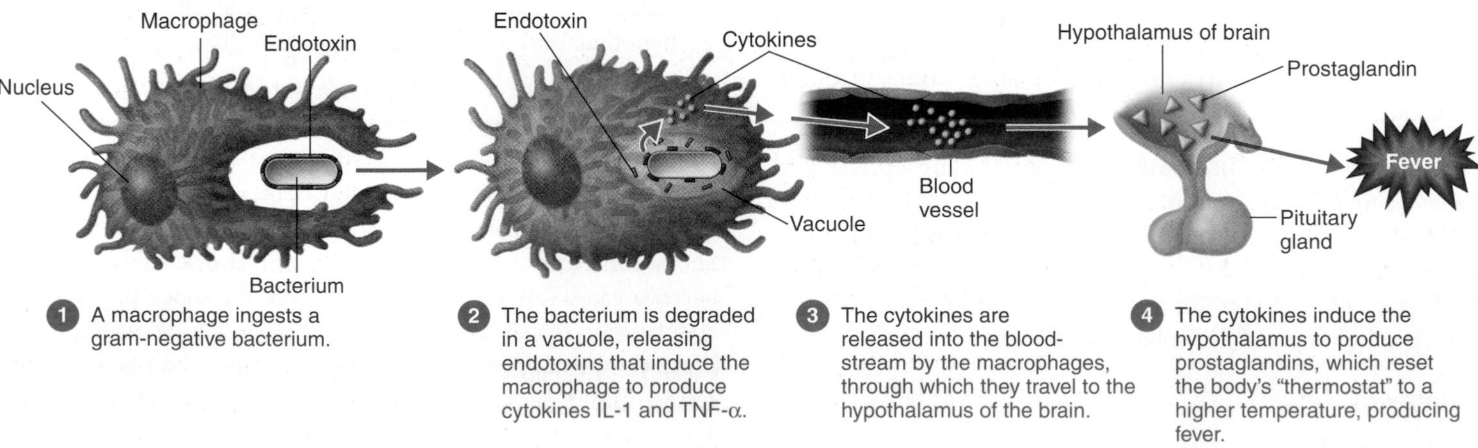

Macrophage
Endotoxin
Nucleus
Bacterium
Endotoxin
Cytokines
Vacuole
Blood vessel
Hypothalamus of brain
Prostaglandin
Fever
Pituitary gland
1 A macrophage ingests a gram-negative bacterium.
2 The bacterium is degraded in a vacuole, releasing endotoxins that induce the macrophage to produce cytokines IL-1 and TNF-α.
3 The cytokines are released into the blood-stream by the macrophages, through which they travel to the hypothalamus of the brain.
4 The cytokines induce the hypothalamus to produce prostaglandins, which reset the body's "thermostat" to a higher temperature, producing fever.

Figure 15.6 Endotoxins and the pyrogenic response. The proposed mechanism by which endotoxins cause fever.

Q What is an endotoxin?

Table 15.3 Exotoxins and Endotoxins

Property	Exotoxin	Endotoxin
Bacterial Source	Mostly from gram-positive bacteria	Gram-negative bacteria
Relation to Microorganism	Metabolic product of growing cell	Present in LPS of outer membrane of cell wall and released with destruction of cell or during cell division
Chemistry	Proteins, usually with two parts (A-B)	Lipid portion (lipid A) of LPS of outer membrane (lipopolysaccharide).
Pharmacology (Effect on Body)	Specific for a particular cell structure or function in the host (mainly affects cell functions, nerves, and gastrointestinal tract)	General, such as fever, weaknesses, aches, and shock; all produce the same effects
Heat Stability	Unstable; can usually be destroyed at 60–80°C (except staphylococcal enterotoxin)	Stable; can withstand autoclaving (121°C for 1 hour)
Toxicity (Ability to Cause Disease)	High	Low
Fever-Producing	No	Yes
Immunology (Relation to Antibodies)	Can be converted to toxoids to immunize against toxin; neutralized by antitoxin	Not easily neutralized by antitoxin; therefore, effective toxoids cannot be made to immunize against toxin
Lethal Dose	Small	Considerably larger
Representative Diseases	Gas gangrene, tetanus, botulism, diphtheria, scarlet fever	Typhoid fever, urinary tract infections, and meningococcal meningitis

Low blood pressure has serious effects on the kidneys, lungs, and gastrointestinal tract. In addition, the presence of gram-negative bacteria such as *Haemophilus influenzae* type b in cerebrospinal fluid causes the release of IL-1 and TNF. These, in turn, cause a weakening of the blood–brain barrier that normally protects the central nervous system from infection. The weakened barrier lets phagocytes in, but this also lets more bacteria enter from the bloodstream. In the United States, 750,000 cases of septic shock occur each year. One-third of the patients die within a month, and nearly half die within 6 months.

Endotoxins do not promote the formation of effective antitoxins against the carbohydrate component of an endotoxin. Antibodies are produced, but they tend not to counter the effect of the toxin; sometimes, in fact, they actually enhance its effect.

Representative microorganisms that produce endotoxins are *Salmonella typhi* (the causative agent of typhoid fever), *Proteus* spp. (frequently the causative agents of urinary tract infections), and *Neisseria meningitidis* (the causative agent of meningococcal meningitis).

It is important to have a sensitive test to identify the presence of endotoxins in drugs, medical devices, and body fluids. Materials that have been sterilized may contain endotoxins, even though no bacteria can be cultured from them. An example is discussed in the box on page 440. One such laboratory test is called the **Limulus amoebocyte lysate (LAL) assay,** which can detect even minute amounts of endotoxin. The hemolymph (blood) of the Atlantic coast horseshoe crab, *Limulus polyphemus,* contains white blood cells called amoebocytes, which have large amounts of a protein (lysate) that causes clotting. In the presence of endotoxin, amoebocytes in the crab hemolymph lyse and liberate their clotting protein. The resulting gel-clot (precipitate) is a positive test for the presence of endotoxin. The degree of the reaction is measured using a spectrophotometer (see Figure 6.21, page 179).

Table 15.3 compares exotoxins and endotoxins. **Animations** Virulence Factors: Exotoxins, Endotoxins. **www.microbiologyplace.com**

Plasmids, Lysogeny, and Pathogenicity

Recall from Chapters 4 (page 95) and 8 (page 237) that plasmids are small, circular DNA molecules that are not connected to the main bacterial chromosome and are capable of independent replication. One group of plasmids, called R (resistance) factors, is responsible for the resistance of some microorganisms to antibiotics. In addition, a plasmid may carry the information that determines a microbe's pathogenicity. Examples of virulence

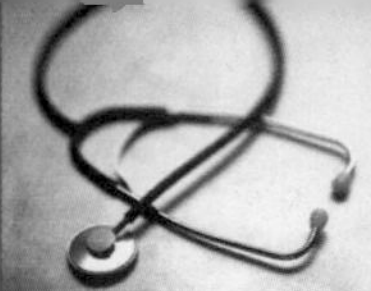

CLINICAL FOCUS From the *Morbidity and Mortality Weekly Report*

Inflammation of the Eye

As you read through this box, you will encounter a series of questions that infection control officers ask themselves as they trace an infection. Try to answer each question before going on to the next one.

1. On October 11, an ophthalmologist performed outpatient cataract surgery on ten patients (**Figure A**). Later that day, he noted that eight of the ten patients had an unusual degree of inflammation and that their pupils were fixed and did not respond to light. Lens replacements were performed on the left eye of five patients and on the right eye of three patients.
   What can you conclude?
2. Infections take 3 to 4 days to show symptoms. The same-day onset of symptoms in these cases suggests a reaction to a toxin or other chemical called *toxic anterior segment syndrome (TASS)*. TASS is caused by (1) chemicals on surgical instruments, resulting from improper or insufficient cleaning; (2) products introduced into the eye during surgery, such as washing solutions or medications; or (3) other substances that enter the eye during or after surgery, such as topical ointments or talc from surgical gloves.
   What do you need to know now?
3. The epinephrine used during surgery and the enzymatic solution for the ultrasonic bath used to clean surgical instruments were sterile. Medications with different lot numbers were used in each case. The surgeon is board certified and has 20 years of experience. The autoclave used to sterilize ophthalmic equipment was functioning normally. Single-use topical iodine antiseptic was used, and a new sterile tip for the corneal extraction was used for each patient.
   What is the next step?
4. An LAL assay of the solution from the ultrasonic bath was performed and was positive.
   What does this indicate?
5. Endotoxin was present in the solution in the ultrasonic bath.
   What is the source of the endotoxin? Is this condition treated with antibiotics?
6. Gram-negative bacteria such as *Burkholderia* are commonly found in liquid reservoirs and moist environments. Gram-negative bacteria can colonize water pipes and laboratory containers used to hold water (**Figure B**). Bacteria from these biofilms could be washed into solutions. The bacteria, killed by autoclaving, release endotoxin. Most patients recover nicely after treatment with a topical anti-inflammatory drug, such as prednisone. Antibiotics should not be used, because TASS is not an infection.

Cataract extraction is one of the most common surgeries in the United States, with approximately 2 million procedures performed each year. A nationwide outbreak of TASS in 2005 was attributed to a commercially distributed irrigating solution contaminated with endotoxin. Preventing TASS depends primarily on using appropriate protocols for cleaning and sterilizing surgical equipment and paying careful attention to all solutions, medications, and ophthalmic devices used during surgery.

Source: Adapted from *MMWR* 56(25):629–930, June 28, 2007.

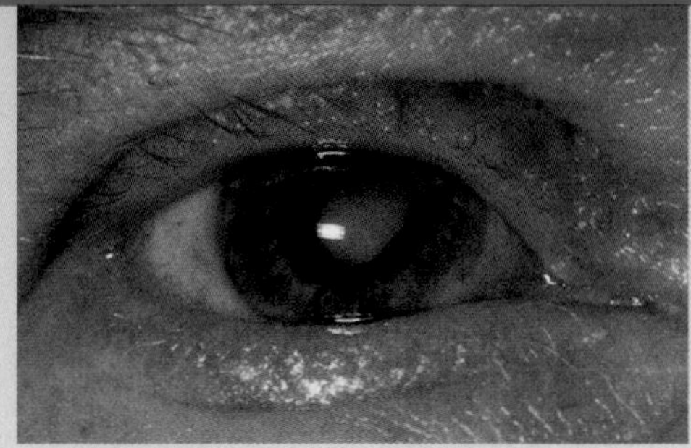

Figure A Cataract

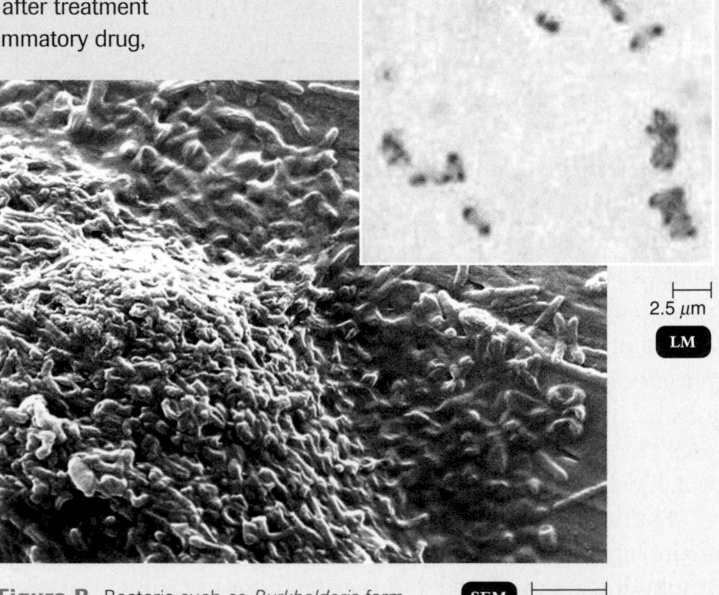

Figure B Bacteria such as *Burkholderia* form biofilms in water pipes and storage containers. Inset: Gram-stained *Burkholderia* showing typical bipolar staining.

factors that are encoded by plasmid genes are tetanus neurotoxin, heat-labile enterotoxin, and staphylococcal enterotoxin. Other examples are dextransucrase, an enzyme produced by *Streptococcus mutans* that is involved in tooth decay; adhesins and coagulase produced by *Staphylococcus aureus;* and a type of fimbria specific to enteropathogenic strains of *E. coli.*

In Chapter 13, we noted that some bacteriophages (viruses that infect bacteria) can incorporate their DNA into the bacterial chromosome, becoming a prophage, and thus remain latent (do not cause lysis of the bacterium). Such a state is called *lysogeny,* and cells containing a prophage are said to be lysogenic. One outcome of lysogeny is that the host bacterial cell and its progeny may exhibit new properties encoded by the bacteriophage DNA. Such a change in the characteristics of a microbe due to a prophage is called **lysogenic conversion.** As a result of lysogenic conversion, the bacterial cell is immune to infection by the same type of phage. In addition, lysogenic cells are of medical importance because some bacterial pathogenesis is caused by the prophages they contain.

Among the bacteriophage genes that contribute to pathogenicity are the genes for diphtheria toxin, erythrogenic toxins, staphylococcal enterotoxin and pyrogenic toxin, botulinum neurotoxin, and the capsule produced by *Streptococcus pneumoniae.* The gene for Shiga toxin in *E. coli* O157 is encoded by phage genes. Pathogenic strains of *Vibrio cholerae* carry lysogenic phages. These phages can transmit the cholera toxin gene to nonpathogenic *V. cholerae* strains, increasing the number of pathogenic bacteria.

CHECK YOUR UNDERSTANDING

- ✓ Of what value are siderophores? **15-8**
- ✓ How does toxigenicity differ from direct damage? **15-9**
- ✓ Differentiate an exotoxin from an endotoxin. **15-10**
- ✓ Food poisoning can be divided into two categories: food infection and food intoxication. On the basis of toxin production by bacteria, explain the difference between these two categories. **15-11**
- ✓ Washwater containing *Pseudomonas* was sterilized and used to wash cardiac catheters. Three patients developed fever, chills, and hypotension following cardiac catheterization. The water and catheters were sterile. Why did the patients show these reactions? How should the water have been tested? **15-12**
- ✓ How can lysogeny turn the normally harmless *E. coli* into a pathogen? **15-13**

Pathogenic Properties of Viruses

LEARNING OBJECTIVE

15-14 List nine cytopathic effects of viral infections.

The pathogenic properties of viruses depend on their gaining access to a host, evading the host's defenses, and then causing damage to or death of the host cell while reproducing themselves.

Viral Mechanisms for Evading Host Defenses

Viruses have a variety of mechanisms that enable them to evade destruction by the host's immune response (see Chapter 17, page 487). For example, viruses can penetrate and grow inside host cells, where components of the immune system cannot reach them. Viruses gain access to cells because they have attachment sites for receptors on their target cells. When such an attachment site is brought together with an appropriate receptor, the virus can bind to and penetrate the cell. Some viruses gain access to host cells because their attachment sites mimic substances useful to those cells. For example, the attachment sites of rabies virus can mimic the neurotransmitter acetylcholine. As a result, the virus can enter the host cell along with the neurotransmitter.

The AIDS virus (HIV) goes further by hiding its attachment sites from the immune response and by attacking components of the immune system directly. Like most viruses, HIV is cell-specific; that is, it attacks only particular body cells. HIV attacks only those cells that have a surface marker called the CD4 protein, most of which are cells of the immune system called T cells (T lymphocytes). Binding sites on HIV are complementary to the CD4 protein. The surface of the virus is folded to form ridges and valleys, and the HIV binding sites are located on the floors of the valleys. CD4 proteins are long enough and slender enough to reach these binding sites, whereas antibody molecules made against HIV are too large to make contact with the sites. As a result, it is difficult for these antibodies to destroy HIV.

Cytopathic Effects of Viruses

Infection of a host cell by an animal virus usually kills the host cell (see Chapter 13, page 378). Death can be caused by the accumulation of large numbers of multiplying viruses, by the effects of viral proteins on the permeability of the host cell's plasma membrane, or by inhibition of host DNA, RNA, or protein synthesis. The visible effects of viral infection are known as **cytopathic effects (CPE).** Those cytopathic effects that result in cell death are called *cytocidal effects;* those that result in cell damage but not cell death are called *noncytocidal effects.* CPEs are used to diagnose many viral infections.

Cytopathic effects vary with the virus. One difference is the point in the viral infection cycle at which the effects occur. Some viral infections result in early changes in the host cell; in other infections, changes are not seen until a much later stage. A virus can produce one or more of the following cytopathic effects:

1. At some stage in their multiplication, cytocidal viruses cause the macromolecular synthesis within the host cell to stop. Some viruses, such as herpes simplex virus, irreversibly stop mitosis.
2. When a cytocidal virus infects a cell, it causes the cell's lysosomes to release their enzymes, resulting in destruction of intracellular contents and host cell death.
3. **Inclusion bodies** are granules found in the cytoplasm or nucleus of some infected cells (Figure 15.7a). These granules are sometimes viral parts—nucleic acids or proteins in the process of being assembled into virions. The granules vary in size, shape, and staining properties, according to the virus.

3. Most microorganisms cannot penetrate intact skin; they enter hair follicles and sweat ducts.
4. Some microorganisms can gain access to tissues by inoculation through the skin and mucous membranes in bites, injections, and other wounds. This route of penetration is called the parenteral route.

The Preferred Portal of Entry (p. 429)

5. Many microorganisms can cause infections only when they gain access through their specific portal of entry.

Numbers of Invading Microbes (pp. 429–431)

6. Virulence can be expressed as LD_{50} (lethal dose for 50% of the inoculated hosts) or ID_{50} (infectious dose for 50% of the inoculated hosts).

Adherence (pp. 431–432)

7. Surface projections on a pathogen called adhesins (ligands) adhere to complementary receptors on the host cells.
8. Adhesins can be glycoproteins or lipoproteins and are frequently associated with fimbriae.
9. Mannose is the most common receptor.
10. Biofilms provide attachment and resistance to antimicrobial agents.

How Bacterial Pathogens Penetrate Host Defenses (pp. 432–434)

Capsules (p. 432)

1. Some pathogens have capsules that prevent them from being phagocytized.

Cell Wall Components (p. 432)

2. Proteins in the cell wall can facilitate adherence or prevent a pathogen from being phagocytized.

Enzymes (pp. 432–433)

3. Local infections can be protected in a fibrin clot caused by the bacterial enzyme coagulase.
4. Bacteria can spread from a focal infection by means of kinases (which destroy blood clots), hyaluronidase (which destroys a mucopolysaccharide that holds cells together), and collagenase (which hydrolyzes connective tissue collagen).
5. IgA proteases destroy IgA antibodies.

Antigenic Variation (p. 433)

6. Some microbes vary expression of antigens, thus avoiding the host's antibodies.

Penetration into the Host Cell Cytoskeleton (p. 433)

7. Bacteria may produce proteins that alter the actin of the host cell's cytoskeleton allowing bacteria into the cell.

How Bacterial Pathogens Damage Host Cells (pp. 434–441)

Using the Host's Nutrients: Siderophores (p. 434)

1. Bacteria get iron from the host using siderophores.

Direct Damage (p. 434)

2. Host cells can be destroyed when pathogens metabolize and multiply inside the host cells.

The Production of Toxins (pp. 434–439)

3. Poisonous substances produced by microorganisms are called toxins; toxemia refers to the presence of toxins in the blood. The ability to produce toxins is called toxigenicity.
4. Exotoxins are produced by bacteria and released into the surrounding medium. Exotoxins, not the bacteria, produce the disease symptoms.
5. Antibodies produced against exotoxins are called antitoxins.
6. A-B toxins consist of an active component that inhibits a cellular process and a binding component that attaches the two portions to the target cell, e.g., diphtheria toxin.
7. Membrane-disrupting toxins cause cell lysis, e.g., hemolysins.
8. Superantigens cause release of cytokines, which cause fever, nausea, and other symptoms; e.g., toxic shock syndrome toxin.
9. Endotoxins are lipopolysaccharides (LPS), the lipid A component of the cell wall of gram-negative bacteria.
10. Bacterial cell death, antibiotics, and antibodies may cause the release of endotoxins.
11. Endotoxins cause fever (by inducing the release of interleukin-1) and shock (because of a TNF-induced decrease in blood pressure).
12. Endotoxins allow bacteria to cross the blood–brain barrier.
13. The *Limulus* amoebocyte lysate (LAL) assay is used to detect endotoxins in drugs and on medical devices.

Plasmids, Lysogeny, and Pathogenicity (pp. 440–441)

14. Plasmids may carry genes for antibiotic resistance, toxins, capsules, and fimbriae.
15. Lysogenic conversion can result in bacteria with virulence factors, such as toxins or capsules.

Pathogenic Properties of Viruses (pp. 441–443)

1. Viruses avoid the host's immune response by growing inside cells.
2. Viruses gain access to host cells because they have attachment sites for receptors on the host cell.
3. Visible signs of viral infections are called cytopathic effects (CPE).
4. Some viruses cause cytocidal effects (cell death), and others cause noncytocidal effects (damage but not death).
5. Cytopathic effects include stopping mitosis, lysis, formation of inclusion bodies, cell fusion, antigenic changes, chromosomal changes, and transformation.

Pathogenic Properties of Fungi, Protozoa, Helminths, and Algae (pp. 443–444)

1. Symptoms of fungal infections can be caused by capsules, toxins, and allergic responses.

2. Symptoms of protozoan and helminthic diseases can be caused by damage to host tissue or by the metabolic waste products of the parasite.
3. Some protozoa change their surface antigens while growing in a host, thus avoiding destruction by the host's antibodies.
4. Some algae produce neurotoxins that cause paralysis when ingested by humans.

Portals of Exit (pp. 444–445)

1. Pathogens have definite portals of exit.
2. Three common portals of exit are the respiratory tract via coughing or sneezing, the gastrointestinal tract via saliva or feces, and the genitourinary tract via secretions from the vagina or penis.
3. Arthropods and syringes provide a portal of exit for microbes in blood.

STUDY QUESTIONS

Answers to the Review and Multiple Choice questions can be found by turning to the blue Answers tab at the back of the textbook.

Review

1. Compare pathogenicity with virulence.
2. How are capsules and cell wall components related to pathogenicity? Give specific examples.
3. Describe how hemolysins, leukocidins, coagulase, kinases, hyaluronidase, siderophores, and IgA proteases might contribute to pathogenicity.
4. Explain how drugs that bind each of the following would affect pathogenicity:
   a. iron in the host's blood
   b. *Neisseria gonorrhoeae* fimbriae
   c. *Streptococcus pyogenes* M protein
5. Compare and contrast the following aspects of endotoxins and exotoxins: bacterial source, chemistry, toxicity, and pharmacology. Give an example of each toxin.
6. **DRAW IT** Label this diagram to show how the Shiga toxin enters and inhibits protein synthesis in a human cell.

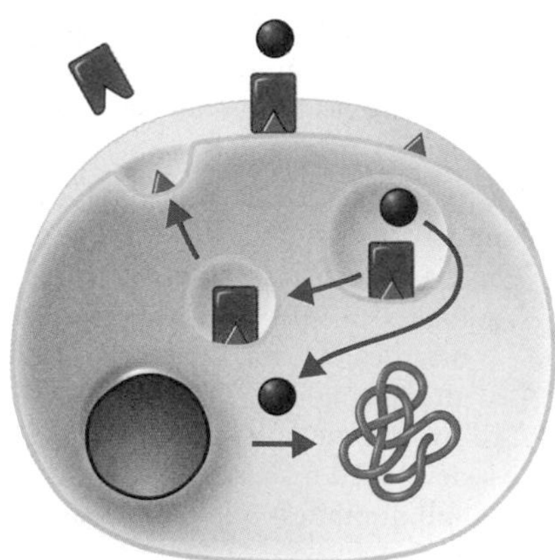

7. Describe the factors contributing to the pathogenicity of fungi, protozoa, and helminths.
8. Which of the following genera is the most infectious?

Genus	ID_{50}	Genus	ID_{50}
Legionella	1 cell	*Shigella*	200 cells
Salmonella	10^5 cells	*Treponema*	52 cells

9. How can viruses and protozoa avoid being killed by the host's immune response?

Multiple Choice

1. The removal of plasmids reduces virulence in which of the following organisms?
   a. *Clostridium tetani*
   b. *Escherichia coli*
   c. *Staphylococcus aureus*
   d. *Streptococcus mutans*
   e. *Clostridium botulinum*
2. What is the LD_{50} for the bacterial toxin tested in the example below?

Dilution (μg/kg)	No. of Animals Died	No. of Animals Survived
a. 6	0	6
b. 12.5	0	6
c. 25	3	3
d. 50	4	2
e. 100	6	0

3. Which of the following is *not* a portal of entry for pathogens?
   a. mucous membranes of the respiratory tract
   b. mucous membranes of the gastrointestinal tract
   c. skin
   d. blood
   e. parenteral route
4. All of the following can occur during bacterial infection. Which would prevent all of the others?
   a. vaccination against fimbriae
   b. phagocytosis
   c. inhibition of phagocytic digestion
   d. destruction of adhesins
   e. alteration of cytoskeleton
5. The ID_{50} for *Campylobacter* sp. is 500 cells; the ID_{50} for *Cryptosporidium* sp. is 100 cells. Which of the following statements is *not* true?
   a. Both microbes are pathogens.
   b. Both microbes produce infections in 50% of the inoculated hosts.
   c. *Cryptosporidium* is more virulent than *Campylobacter.*
   d. *Campylobacter* and *Cryptosporidium* are equally virulent; they cause infections in the same number of test animals.
   e. The severity of infections caused by *Campylobacter* and *Cryptosporidium* cannot be determined by the information provided.

6. An encapsulated bacterium can be virulent because the capsule
   a. resists phagocytosis.
   b. is an endotoxin.
   c. destroys host tissues.
   d. interferes with physiological processes.
   e. has no effect; because many pathogens do not have capsules, capsules do not contribute to virulence.
7. A drug that binds to mannose on human cells would prevent
   a. the entrance of *Vibrio* enterotoxin.
   b. the attachment of pathogenic *E. coli*.
   c. the action of botulinum toxin.
   d. streptococcal pneumonia.
   e. the action of diphtheria toxin.
8. The earliest smallpox vaccines were infected tissue rubbed into the skin of a healthy person. The recipient of such a vaccine usually developed a mild case of smallpox, recovered, and was immune thereafter. What is the most likely reason this vaccine did not kill more people?
   a. Skin is the wrong portal of entry for smallpox.
   b. The vaccine consisted of a mild form of the virus.
   c. Smallpox is normally transmitted by skin-to-skin contact.
   d. Smallpox is a virus.
   e. The virus mutated.
9. Which of the following does *not* represent the same mechanism for avoiding host defenses as the others?
   a. Rabies virus attaches to the receptor for the neurotransmitter acetylcholine.
   b. *Salmonella* attaches to the receptor for epidermal growth factor.
   c. Epstein-Barr (EB) virus binds to the host receptor for complement.
   d. Surface protein genes in *Neisseria gonorrhoeae* mutate frequently.
   e. none of the above
10. Which of the following statements is true?
    a. The primary goal of a pathogen is to kill its host.
    b. Evolution selects for the most virulent pathogens.
    c. A successful pathogen doesn't kill its host before it is transmitted.
    d. A successful pathogen never kills its host.

Critical Thinking

1. The graph below shows confirmed cases of enteropathogenic *E. coli*. Why is the incidence seasonal?

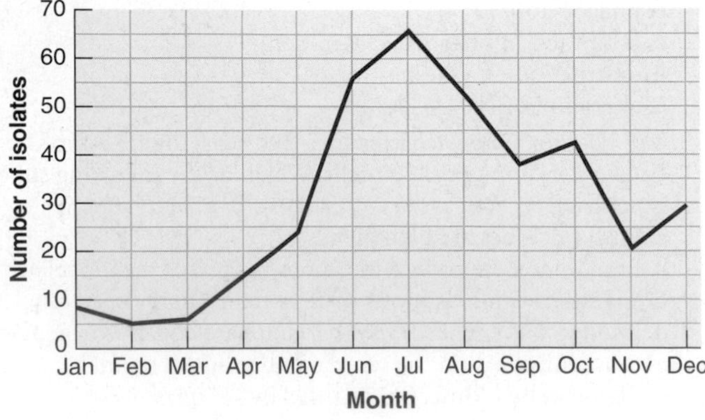

2. The cyanobacterium *Microcystis aeruginosa* produces a peptide that is toxic to humans. According to the graph below, when is this bacterium most toxic?

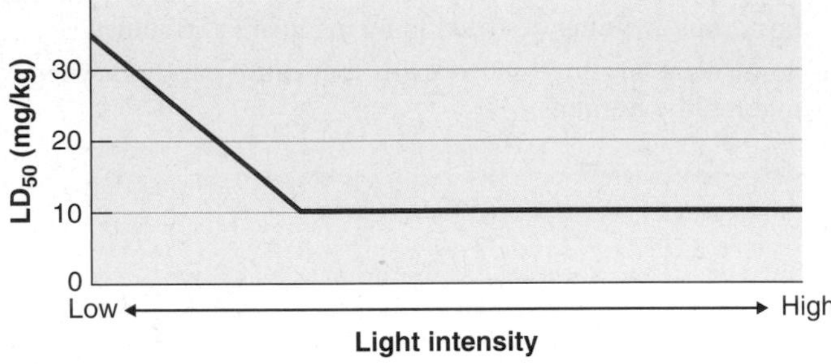

3. When injected into rats, the ID_{50} for *Salmonella typhimurium* is 10^6 cells. If sulfonamides are injected with the salmonellae, the ID_{50} is 35 cells. Explain the change in ID_{50} value.
4. How do each of the following strategies contribute to the virulence of the pathogen? What disease does each organism cause?

Strategy	Pathogen
Changes its cell wall after entry into host	*Yersinia pestis*
Uses urea to produce ammonia	*Helicobacter pylori*
Causes host to make more receptors	*Rhinovirus*

Clinical Applications

1. On July 8, a woman was given an antibiotic for presumptive sinusitis. However, her condition worsened, and she was unable to eat for 4 days because of severe pain and tightness of the jaw. On July 12, she was admitted to a hospital with severe facial spasms. She reported that on July 5 she had incurred a puncture wound at the base of her big toe; she cleaned the wound but did not seek medical attention. What caused her symptoms? Was her condition due to an infection or an intoxication? Can she transmit this condition to another person?
2. Explain whether each of the following examples is a food infection or intoxication. What is the probable etiological agent in each case?
   a. Eighty-two people who ate shrimp in Louisiana, developed diarrhea, nausea, headache, and fever from 4 hours to 2 days after eating.
   b. Two people in Vermont who ate barracuda caught in Florida developed malaise, nausea, blurred vision, breathing difficulty, and numbness 3 to 6 hours after eating.
3. Cancer patients undergoing chemotherapy are normally *more* susceptible to infections. However, a patient receiving an antitumor drug that inhibited cell division was resistant to *Salmonella*. Provide a possible mechanism for the resistance.

16 Innate Immunity: Nonspecific Defenses of the Host

From our discussion to this point, you can see that pathogenic microorganisms are endowed with special properties that enable them to cause disease if given the right opportunity. If microorganisms never encountered resistance from the host, we would constantly be ill and would eventually die of various diseases. In most cases, however, our body's defenses prevent this from happening. Some of these defenses are designed to keep out microorganisms altogether, other defenses remove the microorganisms if they do get in, and still others combat them if they remain inside.

Our ability to ward off disease caused by microbes or their products and to protect against environmental agents such as pollen, drugs, foods, chemicals, and animal dander is called **immunity,** or **resistance.** Vulnerability or lack of immunity is referred to as **susceptibility.** We have two lines of defense against pathogens. The first line of defense is our skin and mucous membranes. The second line of defense consists of various defensive cells, inflammation, fever, and antimicrobial substances produced by the body.

UNDER THE MICROSCOPE

Burkholderia cepacia. These gram-negative bacteria can degrade a wide variety of organic molecules, including some disinfectants.

Q&A

During one year, nosocomial infections occurred in 74 patients in one hospital. All 74 patients were intubated and mechanically ventilated. The infections were caused by *Burkholderia cepacia* transmitted in nonsterile mouthwash. Why did these patients develop infections while others who used the mouthwash were not infected?

Look for the answer in the chapter.

Innate Immunity		Adaptive Immunity (Chapter 17)
First line of defense	**Second line of defense**	**Third line of defense**
• Intact skin • Mucous membranes and their secretions • Normal microbiota	• Phagocytes, such as neutrophils, eosinophils, dendritic cells, and macrophages • Inflammation • Fever • Antimicrobial substances	• Specialized lymphocytes: T cells and B cells • Antibodies

Figure 16.1 An overview of the body's defenses. Innate immunity involves defenses against any pathogen, regardless of species; adaptive immunity involves defenses against a specific pathogen.

Q What is the difference between immunity and susceptibility?

The Concept of Immunity

LEARNING OBJECTIVES

16-1 Differentiate innate and adaptive immunity.

16-2 Define *Toll-like receptors.*

When microbes attack our bodies, we defend ourselves by utilizing our various mechanisms of immunity. In general, there are two types of immunity: innate and adaptive (Figure 16.1). **Innate immunity** refers to defenses that are present at birth. They are always present and available to provide rapid responses to protect us against disease. Innate immunity does not involve specific recognition of a microbe. Further, innate immunity does not have a memory response, that is, a more rapid and stronger immune reaction to the same microbe at a later date. Among the components of innate immunity are the first line of defense (skin and mucous membranes) and the second line of defense (natural killer cells and phagocytes, inflammation, fever, and antimicrobial substances). Innate immune responses represent immunity's early-warning system and are designed to prevent microbes from gaining access into the body and to help eliminate those that do gain access.

Adaptive immunity is based on a specific response to a specific microbe once a microbe has breached the innate immunity defenses. It adapts or adjusts to handle a particular microbe. Unlike innate immunity, adaptive immunity is slower to respond, but it does have a memory component. Adaptive immunity involves lymphocytes (a type of white blood cell) called T cells (T lymphocytes) and B cells (B lymphocytes) and will be discussed in detail in Chapter 17. Here, we concentrate on innate immunity.

As noted previously, the innate immune system responds rapidly to invaders by detecting them and then attempting to eliminate them. It has recently been learned that the responses of the innate system are activated by protein receptors in the plasma membranes of defensive cells; among these activators are **Toll-like receptors (TLRs).** These TLRs attach to various components commonly found on pathogens that are called **pathogen-associated molecular patterns (PAMPs).** See Figure 16.7. Examples include the lipopolysaccharide (LPS) of the outer membrane of gram-negative bacteria, the flagellin in the flagella of motile bacteria, the peptidoglycan in the cell wall of gram-positive bacteria, the DNA of bacteria, and the DNA and RNA of viruses. TLRs also attach to components of fungi and parasites. You will learn later in this chapter that two of the defensive cells involved in innate immunity are called macrophages and dendritic cells. When the TLRs on these cells encounter the PAMPs of microbes, such as the LPS of gram-negative bacteria, the TLRs induce the defensive cells to release chemicals called cytokines. **Cytokines** (*cyto-* = cell; *-kinesis* = motion) are proteins that regulate the intensity and duration of immune responses. One role of cytokines is to recruit other macrophages and dendritic cells, as well as other defensive cells, to isolate and destroy the microbes as part of the inflammatory response. Cytokines can also activate the T cells and B cells involved in adaptive immunity. You will learn more about the different cytokines and their functions in Chapter 17. **Animation** Host Defenses: The Big Picture. **www.microbiologyplace.com**

CHECK YOUR UNDERSTANDING

✓ Which defense system, innate or adaptive immunity, prevents entry of microbes into the body? **16-1**

✓ What relationship do Toll-like receptors have to pathogen-associated molecular patterns? **16-2**

FIRST LINE OF DEFENSE: SKIN AND MUCOUS MEMBRANES

LEARNING OBJECTIVES

16-3 Describe the role of the skin and mucous membranes in innate immunity.

16-4 Differentiate physical from chemical factors, and list five examples of each.

16-5 Describe the role of normal microbiota in innate immunity.

The skin and mucous membranes are the body's first line of defense against environmental pathogens. This function results from both physical and chemical factors. Whereas physical factors include barriers to entry or processes that remove microbes from the body's surface, chemical factors include substances made by the body that inhibit microbial growth or destroy them.

Physical Factors

The intact **skin** is the human body's largest organ in terms of surface area and weight and is an extremely important component of the first line of defense (see Figure 16.1). It consists of two distinct portions: the dermis and the epidermis (**Figure 16.2**). The **dermis,** the skin's inner, thicker portion, is composed of connective tissue. The **epidermis,** the outer, thinner portion, is in direct contact with the external environment. The epidermis consists of many layers of continuous sheets of tightly packed epithelial cells with little or no material between the cells. The top layer of epidermal cells is dead and contains a protective protein called **keratin.** The periodic shedding of the top layer helps remove microbes at the surface. In addition, the dryness of the skin is a major factor in inhibiting microbial growth on the skin. Although normal microbiota and other microbes are present on the entire skin, they are most numerous on moist areas of the skin. When the skin is moist, as in hot, humid climates, skin infections are quite common, especially fungal infections such as athlete's foot. These fungi hydrolyze keratin when water is available.

If we consider the closely packed cells, continuous layering, the presence of keratin, and the dryness and shedding of the skin, we can see why the intact skin provides such a formidable barrier to the entrance of microorganisms. Microorganisms rarely, if ever, penetrate the intact surface of healthy epidermis. However, when the epithelial surface is broken, a subcutaneous (below-the-skin) infection often develops. The bacteria most likely to cause infection are the staphylococci that normally inhabit the epidermis, hair follicles, and sweat and oil glands of the skin. Infections of the skin and underlying tissues frequently result from burns, cuts, stab wounds, or other conditions that break the skin.

Epithelial cells called *endothelial cells* that line blood and lymphatic vessels are not closely packed like those of the epidermis. Although this arrangement permits defensive cells to move from blood into tissues during inflammation, it also permits microbes to move into and out of blood and lymph.

Mucous membranes also consist of an epithelial layer and an underlying connective tissue layer. Mucous membranes are an important component of the first line of defense (see Figure 16.1) and inhibit the entrance of many microorganisms. Mucous membranes line the entire gastrointestinal, respiratory, and genitourinary tracts. The epithelial layer of a mucous membrane secretes a fluid called **mucus,** a slightly viscous (thick) glycoprotein produced by goblet cells of a mucous membrane. Among other functions, mucus prevents the tracts from drying out. Some pathogens that can thrive on the moist secretions of a mucous membrane are able to penetrate the membrane if the microorganism is present in sufficient numbers. *Treponema pallidum* is such a pathogen. This penetration may be facilitated by toxic substances produced by the microorganism, prior injury by viral infection, or mucosal irritation.

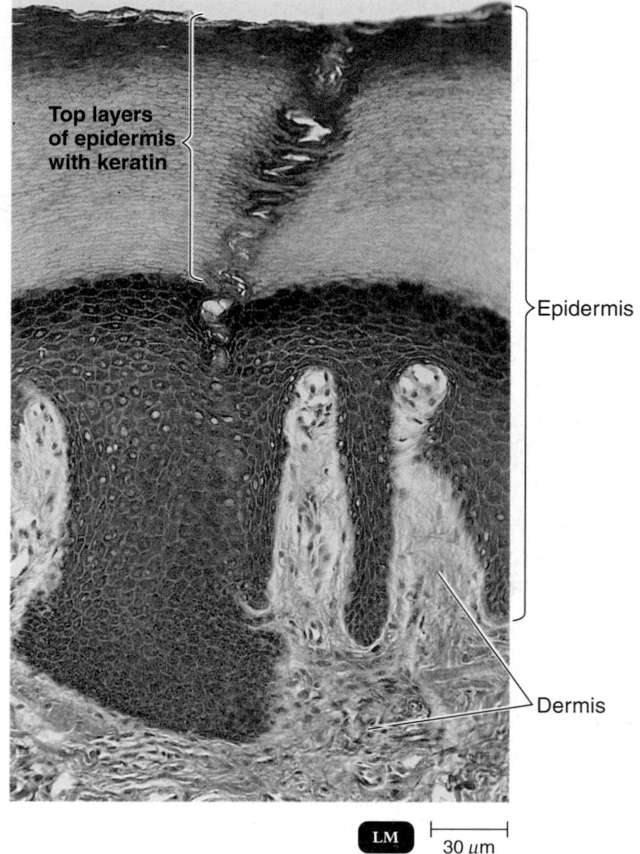

Figure 16.2 A section through human skin. The thin layers at the top of this photomicrograph contain keratin. These layers and the darker purple cells beneath them make up the epidermis. The lighter purple material below the epidermis is the dermis.

Q The intact skin and mucous membranes are components of which line of defense against pathogens?

Besides the physical barrier presented by the skin and mucous membranes, several other physical factors help protect certain epithelial surfaces. One such mechanism that protects the eyes is the **lacrimal apparatus,** a group of structures that manufactures and drains away tears (**Figure 16.3**). The lacrimal glands, located toward the upper, outermost portion of each eye socket, produce the tears and pass them under the upper eyelid. From here, tears pass toward the corner of the eye near the nose and into two small holes that lead through tubes (lacrimal canals) to the nose. The tears are spread over the surface of the eyeball by blinking. Normally, the tears evaporate or pass into the nose as fast as they are produced. This continual washing action helps keep microorganisms from settling on the surface of the eye. If an irritating substance or large numbers of microorganisms come in contact with the eye, the lacrimal glands start to

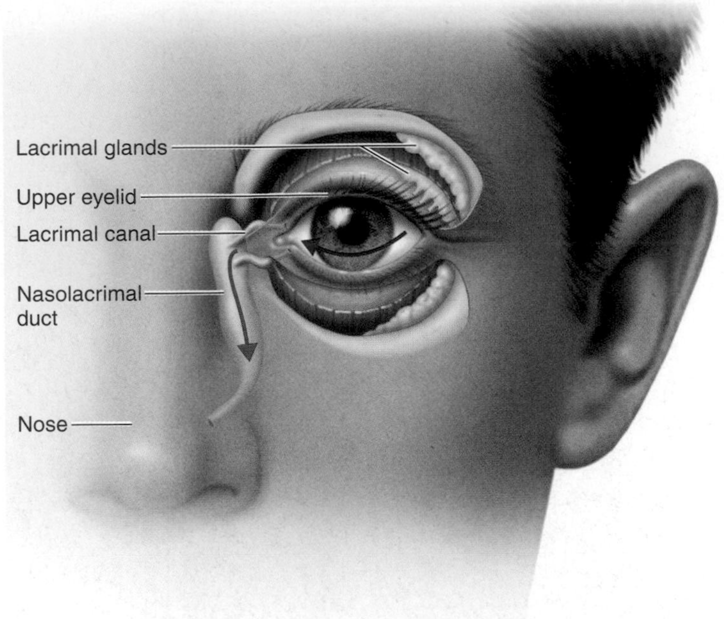

Figure 16.3 The lacrimal apparatus. The washing action of the tears is shown by the red arrow passing over the surface of the eyeball. Tears produced by the lacrimal glands pass across the surface of the eyeball into two small holes that convey the tears into the lacrimal canals and the nasolacrimal duct.

Q How does the lacrimal apparatus protect the eyes against infections?

secrete heavily, and the tears accumulate more rapidly than they can be carried away. This excessive production is a protective mechanism because the excess tears dilute and wash away the irritating substance or microorganisms.

In a cleansing action very similar to that of tears, **saliva,** produced by the salivary glands, helps dilute the numbers of microorganisms and wash them from both the surface of the teeth and the mucous membrane of the mouth. This helps prevent colonization by microbes.

Q&A The respiratory and gastrointestinal tracts have many physical forms of defense. **Mucus,** produced by mucous membranes, traps many of the microorganisms that enter the respiratory and gastrointestinal tracts. The mucous membrane of the nose also has mucus-coated **hairs** that filter inhaled air and trap microorganisms, dust, and pollutants. The cells of the mucous membrane of the lower respiratory tract are covered with **cilia.** By moving synchronously, these cilia propel inhaled dust and microorganisms that have become trapped in mucus upward toward the throat. This so-called **ciliary escalator** (Figure 16.4) keeps the mucus blanket moving toward the throat at a rate of 1 to 3 cm per hour; coughing and sneezing speed up the escalator. Some substances in cigarette smoke are toxic to cilia and can seriously impair the functioning of the ciliary escalator by inhibiting or destroying the cilia. Mechanically ventilated patients are vulnerable to respiratory tract infections because the ciliary escalator mechanism is inhibited. Microorganisms are also prevented from entering the lower respiratory tract by a small lid of cartilage called the **epiglottis,** which covers the larynx (voicebox) during swallowing.

The cleansing of the urethra by the flow of **urine** is another physical factor that prevents microbial colonization in the genitourinary tract. As you will see later, when infections or urinary catheters alter urine flow, urinary tract infections may develop. **Vaginal secretions** likewise move microorganisms out of the female body.

Peristalsis, defecation, and **vomiting** also expel microbes. Peristalsis is a series of coordinated contractions that propel food along the gastrointestinal tract. Mass peristalsis of large intestinal contents into the rectum results in defecation. In response to microbial toxins, the muscles of the gastrointestinal tract contract vigorously, resulting in vomiting and/or diarrhea, which may also rid the body of microbes.

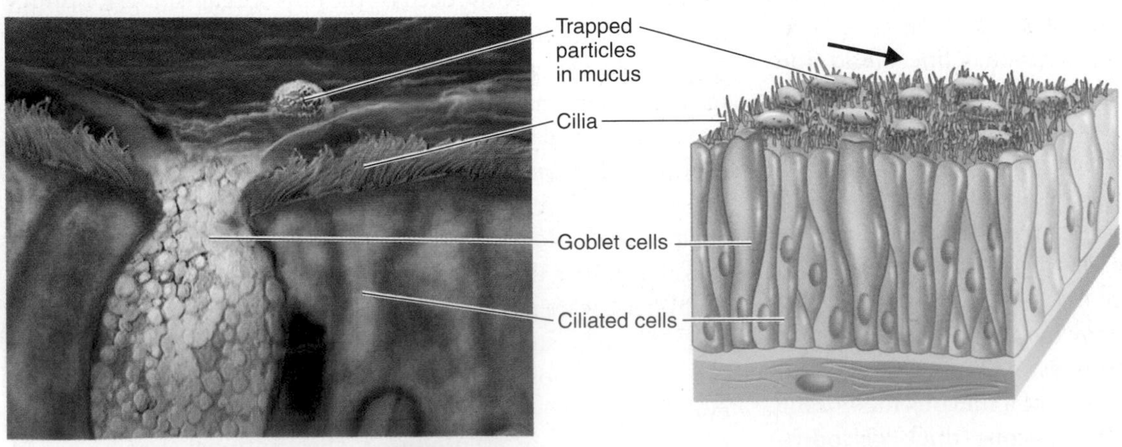

Figure 16.4 The ciliary escalator.

Q What is the function of the ciliary escalator?

Chemical Factors

Physical factors alone do not account for the high degree of resistance of skin and mucous membranes to microbial invasion. Certain chemical factors also play important roles.

Sebaceous (oil) glands of the skin produce an oily substance called **sebum** that prevents hair from drying and becoming brittle. Sebum also forms a protective film over the surface of the skin. One of the components of sebum is unsaturated fatty acids, which inhibit the growth of certain pathogenic bacteria and fungi. The low pH of the skin, between pH 3 and 5, is caused in part by the secretion of fatty acids and lactic acid. The skin's acidity probably discourages the growth of many other microorganisms.

Bacteria that live commensally on the skin decompose sloughed-off skin cells, and the resultant organic molecules and the end-products of their metabolism produce body odor. As we will see in Chapter 21, certain bacteria commonly found on the skin metabolize sebum, and this metabolism forms free fatty acids that cause the inflammatory response associated with acne. Isotretinoin (Accutane), a derivative of vitamin A that prevents sebum formation, is a treatment for a very severe type of acne called cystic acne.

The sweat glands of the skin produce **perspiration,** which helps maintain body temperature, eliminate certain wastes, and flush microorganisms from the surface of the skin. Perspiration also contains **lysozyme,** an enzyme capable of breaking down cell walls of gram-positive bacteria and, to a lesser extent, gram-negative bacteria (see Figure 4.13, page 86). Specifically, lysozyme breaks chemical bonds on peptidoglycan, which destroys the cell walls. Lysozyme is also found in tears, saliva, nasal secretions, tissue fluids, and urine, where it exhibits its antimicrobial activity. Alexander Fleming was studying lysozyme in 1929 when he accidentally discovered the antimicrobial effects of penicillin (see Figure 1.5, page 12).

Saliva contains not only an enzyme (salivary amylase) that digests starch, but also a number of substances that inhibit microbial growth. These include lysozyme, urea, and uric acid. The pH acid of saliva (6.55–6.85) also inhibits some microbes. Saliva also contains an antibody (immunoglobulin A) that prevents attachment of microbes so that they cannot penetrate mucous membranes. **Gastric juice** is produced by the glands of the stomach. It is a mixture of hydrochloric acid, enzymes, and mucus. The very high acidity of gastric juice (pH 1.2–3.0) is sufficient to destroy bacteria and most bacterial toxins, except those of *Clostridium botulinum* and *Staphylococcus aureus.* However, many enteric pathogens are protected by food particles and can enter the intestines via the gastrointestinal tract. In contrast, the bacterium *Helicobacter pylori* neutralizes stomach acid, thereby allowing the bacterium to grow in the stomach. Its growth initiates an immune response that results in gastritis and ulcers.

Vaginal secretions play a role in antibacterial activity in two ways. Glycogen produced by vaginal epithelial cells is broken down into lactic acid by *Lactobacillus acidophilus.* This creates an acid pH (3–5) that inhibits microbes. Cervical mucus also has some antimicrobial activity.

Urine, in addition to containing lysozyme, has an acid pH (average 6) that inhibits microbes. Also, urine contains urea and other metabolic by-products, such as uric acid, hippuric acid, and indicant, which inhibit microbes.

Later in the chapter, we will discuss another group of chemicals, the antimicrobial peptides, which play a very important role in innate immunity.

Normal Microbiota and Innate Immunity

Technically speaking, the **normal microbiota** are not usually considered part of the first line of defense of the innate immune system, but they are discussed here because of the considerable protection they afford (see Figure 16.1). Chapter 14 described several relationships between normal microbiota and host cells. Some of these relationships help prevent the overgrowth of pathogens and thus may be considered components of innate immunity. For example, in microbial antagonism, the normal microbiota prevent pathogens from colonizing the host by competing with them for nutrients (competitive exclusion), by producing substances that are harmful to the pathogens, and by altering conditions that affect the survival of the pathogens, such as pH and oxygen availability. The presence of normal microbiota in the vagina, for example, alters pH, thus preventing overpopulation by *Candida albicans,* a pathogenic yeast that causes vaginitis. In the large intestine, *E. coli* bacteria produce bacteriocins that inhibit the growth of *Salmonella* and *Shigella.*

In commensalism, one organism uses the body of a larger organism as its physical environment and may make use of the body to obtain nutrients. Thus in commensalism, one organism benefits while the other is unaffected. Most microbes that are part of the commensal microbiota are found on the skin and in the gastrointestinal tract. The majority of such microbes are bacteria that have highly specialized attachment mechanisms and precise environmental requirements for survival. Normally, such microbes are harmless, but they may cause disease if their environmental conditions change. These opportunistic pathogens include *E. coli, Staphylococcus aureus, S. epidermidis, Enterococcus faecalis, Pseudomonas aeruginosa,* and oral streptococci.

CHECK YOUR UNDERSTANDING

- ✓ Identify one physical factor and one chemical factor that prevent microbes from entering the body through skin and mucous membranes. **16-3**
- ✓ Identify one physical factor and one chemical factor that prevent microbes from entering or colonizing the body through the eyes, digestive tract, and respiratory tract. **16-4**
- ✓ Distinguish microbial antagonism from commensalism. **16-5**

SECOND LINE OF DEFENSE

When microbes penetrate the first line of defense, they encounter a second line of defense that includes defensive cells, such as phagocytic cells; inflammation; fever; and antimicrobial substances.

Before we look at the phagocytic cells, it will be helpful to first have an understanding of the cellular components of blood.

Formed Elements in Blood

LEARNING OBJECTIVES

16-6 Classify leukocytes, and describe the roles of granulocytes and monocytes.

16-7 Define *differential white blood cell count.*

Blood consists of fluid, called **plasma,** and **formed elements**—that is, cells and cell fragments suspended in plasma (Table 16.1). Of the formed elements listed in Table 16.1, those that concern us at present are the **leukocytes,** or white blood cells.

Leukocytes are divided into two main categories based on their appearance under a light microscope: granulocytes and agranulocytes. **Granulocytes** owe their name to the presence of large granules in their cytoplasm that can be seen under a light microscope after staining. They are differentiated into three types of cells on the basis of how the granules stain: neutrophils, basophils, and eosinophils. The granules of **neutrophils** stain pale lilac with a mixture of acidic and basic dyes. Neutrophils are also commonly called *polymorphonuclear leukocytes (PMNs)*, or *polymorphs.* (The term *polymorphonuclear* refers to the fact that the nuclei of neutrophils contain two to five lobes.) Neutrophils, which are highly phagocytic and motile, are active in the initial stages of an infection (see Figure 16.1). They have the ability to leave the blood, enter an infected tissue, and destroy microbes and foreign particles. **Basophils** stain blue-purple with the basic dye methylene blue. Basophils release substances, such as histamine, that are important in inflammation and allergic responses. **Eosinophils** stain red or orange with the acidic dye eosin. Eosinophils are somewhat phagocytic and also have the ability to leave the blood. Their major function is to produce toxic proteins against certain parasites, such as helminths. Although eosinophils are physically too small to ingest and destroy helminths, they can attach to the outer surface of the parasites and discharge peroxide ions that destroy them (see Figure 17.14, page 490). Their number increases significantly during certain parasitic worm infections and hypersensitivity (allergy) reactions.

Agranulocytes also have granules in their cytoplasm, but the granules are not visible under the light microscope after staining. There are three different types of granulocytes: monocytes, dendritic cells, and lymphocytes. **Monocytes** are not actively phagocytic until they leave circulating blood, enter body tissues, and mature into **macrophages.** In fact, the proliferation of lymphocytes is one factor responsible for the swelling of lymph nodes during an infection. As blood and lymph that contain microorganisms pass through organs with macrophages, the microorganisms are removed by phagocytosis. Macrophages also dispose of worn out blood cells.

Dendritic cells (see Figure 16.1) are believed to be derived from monocytes. They have long extensions that resemble the dendrites of nerve cells, thus their name. Dendritic cells are especially abundant in the epidermis of the skin, mucous membranes, the thymus, and lymph nodes. The function of dendritic cells is to destroy microbes by phagocytosis and to initiate adaptive immunity responses (see Chapter 17, page 490).

Lymphocytes include natural killer cells, T cells, and B cells. **Natural killer (NK) cells** are found in blood and in the spleen, lymph nodes, and red bone marrow. NK cells have the ability to kill a wide variety of infected body cells and certain tumor cells. NK cells attack any body cells that display abnormal or unusual plasma membrane proteins. The binding of NK cells to a target cell, such as an infected human cell, causes the release of vesicles containing toxic substances from NK cells. Some granules contain a protein called **perforin,** which inserts into the plasma membrane of the target cell and creates channels (perforations) in the membrane. As a result, extracellular fluid flows into the target cell and the cell bursts, a process called **cytolysis** (sī-tol′-i-sis; *cyto-* = cell; *-lysis* = loosening). Other granules of NK cells release **granzymes,** which are protein-digesting enzymes that induce the target cell to undergo apoptosis, or self-destruction. This type of attack kills infected cells but not the microbes inside the cells; the released microbes, which may or may not be intact, can be destroyed by phagocytes.

T cells and **B cells** are not usually phagocytic but play a key role in adaptive immunity (see Chapter 17). They occur in lymphoid tissues of the lymphatic system and also circulate in blood.

During many kinds of infections, especially bacterial infections, the total number of white blood cells increases as a protective response to combat the microbes; this increase is called *leukocytosis.* During the active stage of infection, the leukocyte count might double, triple, or quadruple, depending on the severity of the infection. Diseases that might cause such an elevation in the leukocyte count include meningitis, infectious mononucleosis, appendicitis, pneumococcal pneumonia, and gonorrhea. Other diseases, such as salmonellosis and brucellosis, and some viral and rickettsial infections may cause a *decrease* in the leukocyte count, called *leukopenia.* Leukopenia may be related to either impaired white blood cell production or the effect of increased sensitivity of white blood cell membranes to damage by complement, antimicrobial serum proteins discussed later in the chapter. Leukocyte increase or decrease can be detected by a **differential white blood cell count,** which is a calculation of the percentage of each kind of

Table 16.1 Formed Elements in Blood

I. Erythrocytes (Red Blood Cells)

4.8–5.4 million per μL or mm^3

Function: Transport of O_2 and CO_2

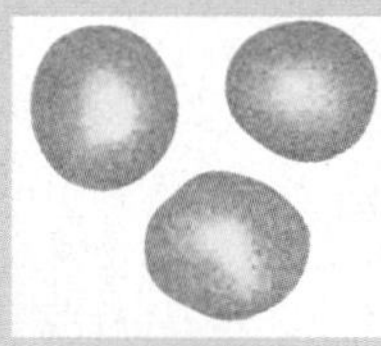

LM 4 μm

II. Leukocytes (White Blood Cells)

5000–10,000 per μL or mm^3

A. Granulocytes (stained)

1. Neutrophils (PMNs) (60–70% of leukocytes)
   Function: Phagocytosis

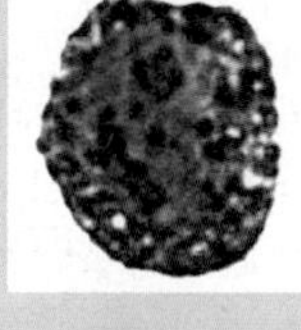

LM 4 μm

2. Basophils (0.5–1%)
   Function: Production of histamine

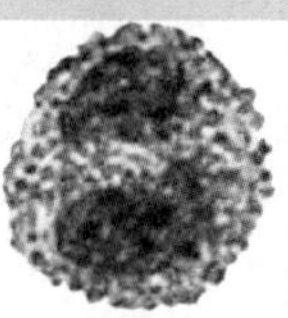

LM 3 μm

3. Eosinophils (2–4%)
   Functions: Production of toxic proteins against certain parasites; some phagocytosis

LM 4 μm

B. Agranulocytes (stained)

1. Monocytes (3–8%)
   Function: Phagocytosis (when they mature into macrophages)

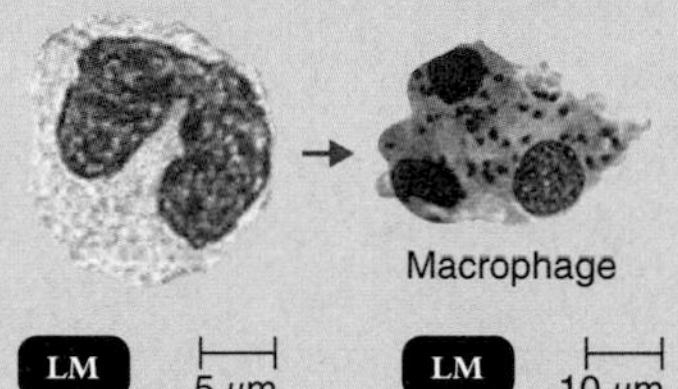

LM 5 μm LM 10 μm

2. Dendritic cells
   Functions: Derived from monocytes; phagocytosis and initiation of adaptive immune responses

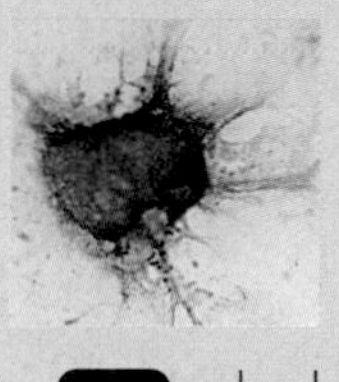

LM 10 μm

3. Lymphocytes (20–25%)
   - Natural killer (NK) cells
     Function: Destroy target cells by cytolysis and apoptosis

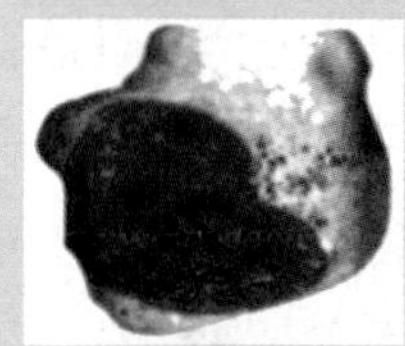

LM 2.5 μm

   - T cells
     Function: Cell-mediated immunity (discussed in Chapter 17)

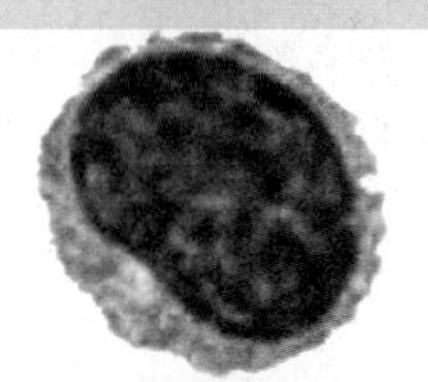

LM 15 μm

   - B cells
     Function: Descendants of B cells (plasma cells) produce antibodies

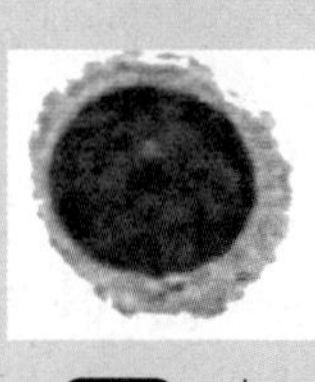

LM 8 μm

III. Platelets

150,000–400,000 per μL or mm^3

Function: Blood clotting

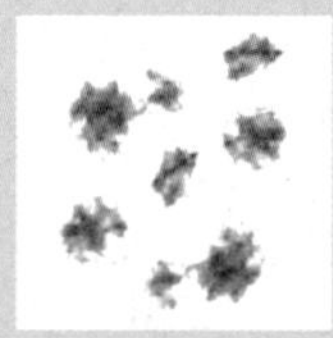

LM 2.5 μm

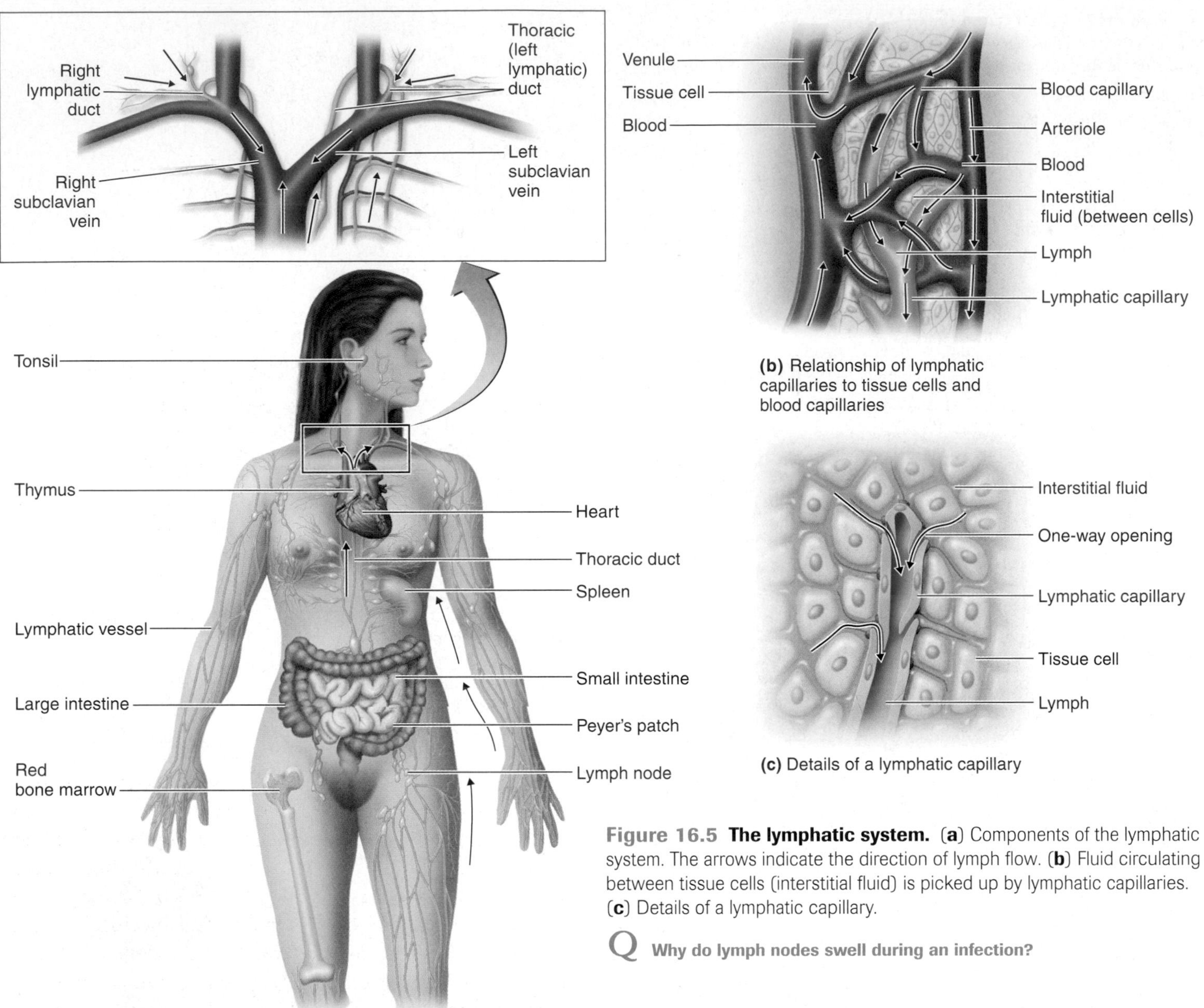

(a) Components of lymphatic system

(b) Relationship of lymphatic capillaries to tissue cells and blood capillaries

(c) Details of a lymphatic capillary

Figure 16.5 The lymphatic system. (**a**) Components of the lymphatic system. The arrows indicate the direction of lymph flow. (**b**) Fluid circulating between tissue cells (interstitial fluid) is picked up by lymphatic capillaries. (**c**) Details of a lymphatic capillary.

Q Why do lymph nodes swell during an infection?

white cell in a sample of 100 white blood cells. The percentages in a normal differential white blood cell count are shown in parentheses in Table 16.1.

The Lymphatic System

LEARNING OBJECTIVE

16-8 Differentiate the lymphatic and blood circulatory systems.

The **lymphatic system** consists of a fluid called *lymph*, vessels called *lymphatic vessels*, a number of structures and organs containing *lymphoid tissue*, and *red bone marrow*, where stem cells develop into blood cells, including lymphocytes (**Figure 16.5a**). Lymphoid tissue contains large numbers of lymphocytes, including T cells, B cells, and phagocytic cells that participate in immune responses.

Lymphatic vessels begin as microscopic *lymphatic capillaries* located in spaces between cells (**Figure 16.5b** and **Figure 16.5c**). The lymphatic capillaries permit interstitial fluid derived from blood plasma to flow into them, but not out. Within the lymphatic capillaries, the fluid is called lymph. Lymphatic capillaries converge to form larger lymphatic vessels. These vessels, like veins, have one-way valves to keep lymph flowing in one direction only. At intervals along the lymphatic vessels, lymph flows through bean-shaped *lymph nodes* (Figure 16.5a). Lymph nodes are the sites of activation of T cells and B cells, which destroy microbes by immune responses (Chapter 17). Also within lymph nodes are reticular fibers, which trap microbes, and macrophages and dendritic cells, which destroy microbes by phagocytosis. All lymph eventually passes into the *thoracic (left lymphatic) duct* and *right lymphatic duct* and then into their respective subclavian veins, where the fluid is now called blood plasma. The blood plasma moves through the cardiovascular system and ultimately becomes interstitial fluid between tissue cells, and another cycle begins.

Lymphoid tissues and organs are scattered throughout the mucous membranes that line the gastrointestinal, respiratory, urinary, and reproductive tracts. They protect against microbes that are ingested or inhaled. Multiple large aggregations of lymphoid tissues are located in specific parts of the body. These include the *tonsils* in the throat and *Peyer's patches* in the small intestine. See Figure 17.9, page 488.

The *spleen* contains lymphocytes and macrophages that monitor the blood for microbes and secreted products such as toxins, much like lymph nodes monitor lymph. The *thymus* serves as a site for T cell maturation. It also contains dendritic cells and macrophages. **Animation** Host Defenses: Overview. **www.microbiologyplace.com**

CHECK YOUR UNDERSTANDING

- ✓ Compare the structures and function of monocytes and neutrophils. **16-6**
- ✓ Describe the six different types of white blood cells, and name a function for each type. **16-7**
- ✓ What is the function of lymph nodes? **16-8**

Phagocytes

LEARNING OBJECTIVES

16-9 Define *phagocyte* and *phagocytosis.*

16-10 Describe the process of phagocytosis, and include the stages of adherence and ingestion.

16-11 Identify six mechanisms of avoiding destruction by phagocytosis.

Phagocytosis (from Greek words meaning eat and cell) is the ingestion of a microorganism or other substances (such as debris) by a cell. We have previously mentioned phagocytosis as the method of nutrition of certain protozoa. Phagocytosis is also involved in clearing away debris such as dead body cells and denatured proteins. In this chapter, phagocytosis is discussed as a means by which cells in the human body counter infection as part of the second line of defense. The cells that perform this function are collectively called **phagocytes,** all of which are types of white blood cells or derivatives of white blood cells.

Actions of Phagocytic Cells

When an infection occurs, both granulocytes (especially neutrophils, but also eosinophils and dendritic cells) and monocytes migrate to the infected area. During this migration, monocytes enlarge and develop into actively phagocytic macrophages (Figure 16.6). These cells leave the blood and migrate into tissues where they enlarge and develop into macrophages. Some macrophages, called **fixed macrophages,** or *histiocytes,* are resident in certain tissues and organs of the body. Fixed macrophages are found in the liver (Kupffer's cells), lungs (alveolar macrophages), nervous system (microglial cells), bronchial tubes, spleen (splenic macrophages), lymph nodes, red bone marrow, and the peritoneal cavity surrounding abdominal organs (peritoneal macrophages). Other macrophages are motile and are called **free (wandering) macrophages,** which roam the tissues and gather at sites of infection or inflammation. The various macrophages of the body constitute the **mononuclear phagocytic (reticuloendothelial) system.**

During the course of an infection, a shift occurs in the type of white blood cell that predominates in the bloodstream. Granulocytes, especially neutrophils, dominate during the initial phase of bacterial infection, at which time they are actively phagocytic; this dominance is indicated by their increased number in a differential white blood cell count. However, as the infection progresses, the macrophages dominate; they scavenge and phagocytize remaining living bacteria and dead or dying bacteria. The increased number of monocytes (which develop into macrophages) is also reflected in a differential white blood cell count.

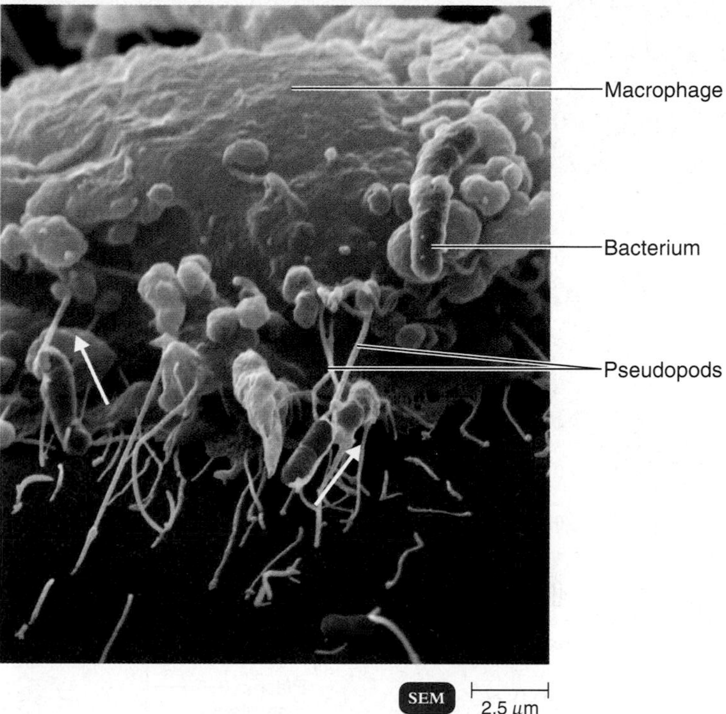

Figure 16.6 A macrophage engulfing rod-shaped bacteria. Macrophages in the mononuclear phagocytic system remove microorganisms after the initial phase of infection.

Q What are monocytes?

Figure 16.7

FOUNDATION FIGURE The Mechanism of Phagocytosis

Phagocytosis is an important second line of defense that is activated when the first line fails. Phagocytes also play an important role in supporting adaptive immunity by stimulating T and B cells, as will be seen in the next chapter.

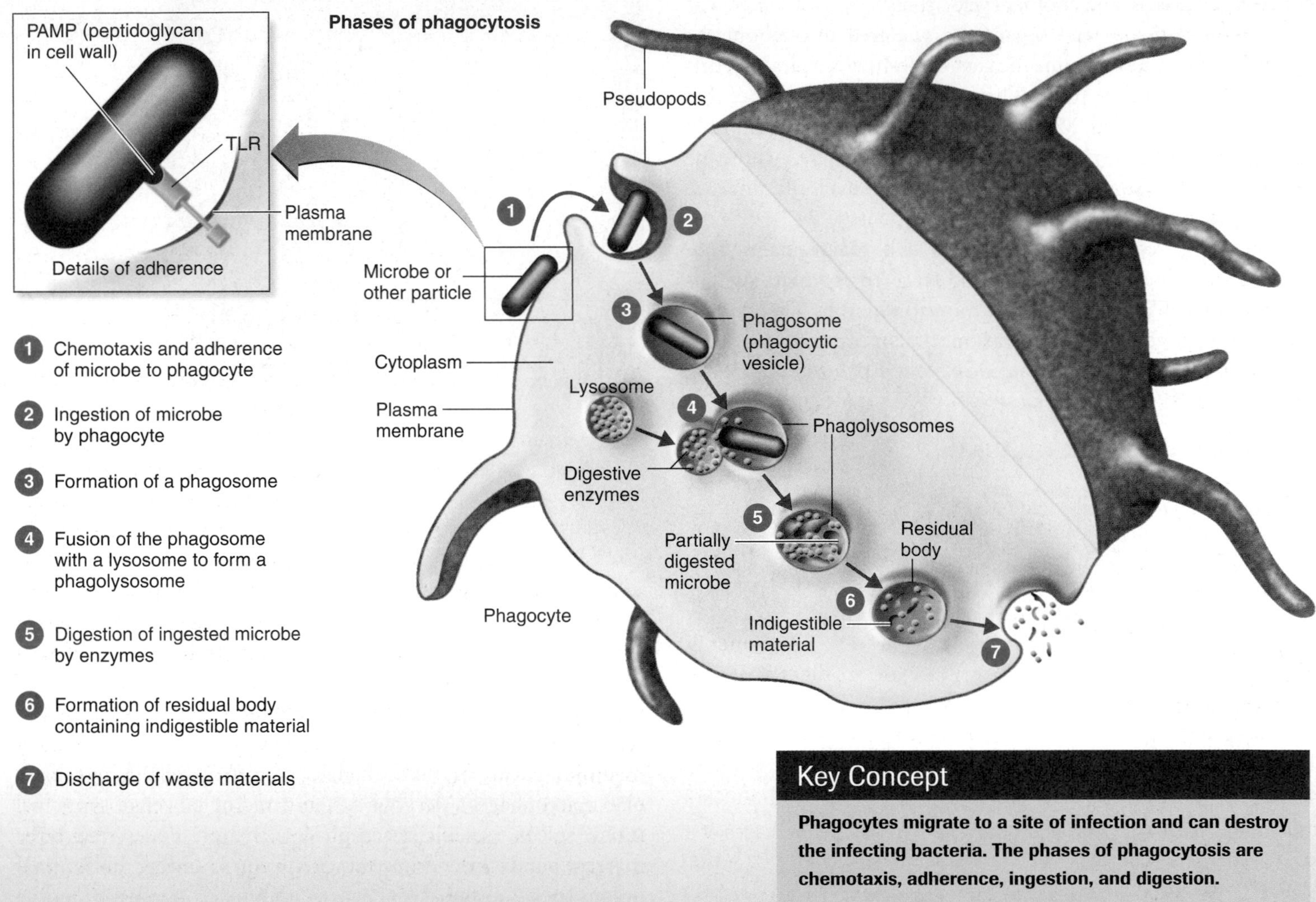

Key Concept

Phagocytes migrate to a site of infection and can destroy the infecting bacteria. The phases of phagocytosis are chemotaxis, adherence, ingestion, and digestion.

The Mechanism of Phagocytosis

How does phagocytosis occur? For the convenience of study, we will divide phagocytosis into four main phases: chemotaxis, adherence, ingestion, and digestion (Figure 16.7).

Chemotaxis

1 **Chemotaxis** is the chemical attraction of phagocytes to microorganisms. (The mechanism of chemotaxis is discussed in Chapter 4, page 82.) Among the chemotactic chemicals that attract phagocytes are microbial products, components of white blood cells and damaged tissue cells, cytokines released by other white blood cells, and peptides derived from complement, a system of host defense discussed later in the chapter.

Adherence

As it pertains to phagocytosis, **adherence** is the attachment of the phagocyte's plasma membrane to the surface of the microorganism or other foreign material. Adherence is facilitated by the attachment of pathogen-associated molecular patterns (PAMPs) of microbes to receptors, such as Toll-like receptors (TLRs), on the surface of phagocytes. The binding of PAMPs to TLRs not only initiates phagocytosis, but also induces the phagocyte to release specific cytokines that recruit additional phagocytes.

In some instances, adherence occurs easily, and the microorganism is readily phagocytized. Microorganisms can be more readily phagocytized if they are first coated with certain serum proteins that promote attachment of the microorganisms to the phagocyte. This coating process is called **opsonization.** The proteins that act as *opsonins* include some components of the complement system and antibody molecules (described later in this chapter and in Chapter 17).

Ingestion

❷ Following adherence, **ingestion** occurs. During this process, the plasma membrane of the phagocyte extends projections called **pseudopods** that engulf the microorganism. (See also Figure 16.6.) ❸ Once the microorganism is surrounded, the pseudopods meet and fuse, surrounding the microorganism with a sac called a **phagosome,** or *phagocytic vesicle.* The membrane of a phagosome has enzymes that pump protons (H^+) into the phagosome, reducing the pH to about 4. At this pH, hydrolytic enzymes are activated.

Digestion

In this phase of phagocytosis, the phagosome pinches off from the plasma membrane and enters the cytoplasm. Within the cytoplasm, it contacts lysosomes that contain digestive enzymes and bactericidal substances (see Chapter 4, page 104). ❹ On contact, the phagosome and lysosome membranes fuse to form a single, larger structure called a **phagolysosome.** ❺ The contents of the phagolysosome brought in by ingestion are digested in the phagolysosome.

Lysosomal enzymes that attack microbial cells directly include lysozyme, which hydrolyzes peptidoglycan in bacterial cell walls. A variety of other enzymes, such as lipases, proteases, ribonuclease, and deoxyribonuclease, hydrolyze other macromolecular components of microorganisms. Lysosomes also contain enzymes that can produce toxic oxygen products such as superoxide radical ($O_2^{\overline{\cdot}}$), hydrogen peroxide (H_2O_2), nitric oxide (NO), singlet oxygen ($^1O_2^-$), and hydroxyl radical (OH·) (see Chapter 6, pages 161–162). Toxic oxygen products are produced by a process called an *oxidative burst.* Other enzymes can make use of these toxic oxygen products in killing ingested microorganisms. For example, the enzyme myeloperoxidase converts chloride (Cl^-) ions and hydrogen peroxide into highly toxic hypochlorous acid (HOCl). The acid contains hypochlorous ions, which are found in household bleach and account for its antimicrobial activity (see Chapter 7, page 197).

❻ After enzymes have digested the contents of the phagolysosome brought into the cell by ingestion, the phagolysosome contains indigestible material and is called a *residual body.* ❼ This residual body then moves toward the cell boundary and discharges its wastes outside the cell.

Microbial Evasion of Phagocytosis

The ability of a pathogen to cause disease is related to its ability to evade phagocytosis. Some bacteria have structures that inhibit adherence, such as the M protein and capsules. As mentioned in Chapter 15 (page 432), the M protein of *Streptococcus pyogenes* inhibits the attachment of phagocytes to their surfaces and makes adherence more difficult. Organisms with large capsules include *Streptococcus pneumoniae* and *Haemophilus influenzae* type b. Heavily encapsulated microorganisms like these can be phagocytized only if the phagocyte traps the microorganism against a rough surface, such as a blood vessel, blood clot, or connective tissue fiber, from which the microbe cannot slide away.

Other microbes may be ingested but not killed. For example, *Staphylococcus* produces leukocidins that may kill phagocytes by causing the release of the phagocyte's own lysosomal enzymes into its cytoplasm. Streptolysin released by streptococci has a similar mechanism.

A number of intracellular pathogens secrete pore-forming toxins that lyse phagocyte cell membranes once inside the phagocyte. For example, *Trypanosoma cruzi* (the causative agent of American trypanosomiasis), and *Listeria monocytogenes* (the causative agent of listeriosis), produce membrane attack complexes that lyse phagolysosome membranes and release the microbes into the cytoplasm of the phagocyte, where they propagate. Later the microbes secrete more membrane attack complexes that lyse the plasma membrane (see page 465) and release the microbes from the phagocyte, resulting in lysis of the phagocyte and infection of neighboring cells by the microbe.

Still other microbes have the ability to survive inside phagocytes. *Coxiella burnetii,* the causative agent of Q fever, actually requires the low pH inside a phagolysosome to replicate. *Listeria monocytogenes, Shigella* (the causative agent of shigellosis), and *Rickettsia* (the causative agent of Rocky Mountain spotted fever and typhus) have the ability to escape from a phagosome before it fuses with a lysosome. *Mycobacterium tuberculosis* (the causative agent of tuberculosis), HIV (the causative agent of AIDS), *Chlamydia* (the causative agent of trachoma, nongonococcal urethritis, and lymphogranuloma venereum), *Leishmania* (the causative agent of leishmaniasis), and *Plasmodium* (malarial parasites) can prevent both the fusion of a phagosome with a lysosome and the proper acidification of digestive enzymes. The microbes then multiply within the phagocyte, almost completely filling it. In most cases, the phagocyte dies and the microbes are released by autolysis to infect other cells. Still other microbes, such as the causative agents of tularemia and brucellosis, can remain dormant within phagocytes for months or years at a time.

Biofilms also play a role in evading phagocytes. Bacteria that are part of biofilms are much more resistant to phagocytosis because the phagocytes cannot detach bacteria from the biofilm prior to phagocytosis. Moreover, the neutrophil response against *P. aeruginosa* in a biofilm is slower than against free-floating bacteria. In addition, although some bacteria in a biofilm, such as *Pseudomonas aeruginosa*, can activate the oxidative burst response, it is weaker than in free-floating bacteria. **Animations** Virulence Factors: Hiding from Host Defenses, Inactivating Host Defenses; Phagocytosis: Overview, Mechanism, Microbes That Evade It. **www.microbiologyplace.com**

CHECK YOUR UNDERSTANDING

- ✓ What do fixed and wandering macrophages do? **16-9**
- ✓ What is the role of TLRs in phagocytosis? **16-10**
- ✓ How does each of these bacteria avoid destruction by phagocytes? *Streptococcus pneumoniae, Staphylococcus aureus, Listeria monocytogenes, Mycobacterium tuberculosis, Rickettsia* **16-11**

* * *

In addition to providing innate (nonspecific) resistance for the host, phagocytosis plays a role in adaptive immunity. Macrophages help T and B cells perform vital adaptive immune functions. In Chapter 17, we will discuss in more detail how phagocytosis supports adaptive immunity.

In the next section, we will see how phagocytosis often occurs as part of another innate mechanism of resistance: inflammation.

Inflammation

LEARNING OBJECTIVES

16-12 List the stages of inflammation.

16-13 Describe the roles of vasodilation, kinins, prostaglandins, and leukotrienes in inflammation.

16-14 Describe phagocyte migration.

Damage to the body's tissues triggers a local defensive response called **inflammation,** another component of the second line of defense (see Figure 16.1). The damage can be caused by microbial infection, physical agents (such as heat, radiant energy, electricity, or sharp objects), or chemical agents (acids, bases, and gases). Inflammation is usually characterized by four signs and symptoms: *redness, pain, heat,* and *swelling.* Sometimes a fifth, *loss of function,* is present; its occurrence depends on the site and extent of damage.

If the cause of an inflammation is removed in a relatively short period of time, the inflammatory response is intense and is referred to as an *acute inflammation.* An example is the response to a boil caused by *S. aureus.* If, instead, the cause of an inflammation is difficult or impossible to remove, the inflammatory response is longer lasting but less intense (although overall more destructive). This type of inflammation is referred to as a *chronic inflammation.* An example is the response to a chronic infection such as tuberculosis, caused by *M. tuberculosis.*

Inflammation has the following functions: (1) to destroy the injurious agent, if possible, and to remove it and its by-products from the body; (2) if destruction is not possible, to limit the effects on the body by confining or walling off the injurious agent and its by-products; and (3) to repair or replace tissue damaged by the injurious agent or its by-products.

During the early stages of inflammation, microbial structures, such as flagellin, lipopolysaccharides (LPS), and bacterial DNA stimulate the Toll-like receptors of macrophages to produce cytokines, such as tumor necrosis factor alpha (TNF-α). In response to TNF-α in the blood, the liver synthesizes a group of proteins called **acute-phase proteins;** other acute-phase proteins are present in the blood in an inactive form and are converted to an active form during inflammation. Acute-phase proteins induce both local and systemic responses and include proteins such as C-reactive protein, mannose-binding lectin (page 467), and several specialized proteins such as fibrinogen for blood clotting and kinins for vasodilation.

All of the cells involved in inflammation have receptors for TNF-α and are activated by it to produce more of their own TNF-α. This amplifies the inflammatory response. Unfortunately, excessive production of TNF-α may lead to disorders such as rheumatoid arthritis and Crohn's disease. In Chapter 18 (page 507), you will learn that monoclonal antibodies are used therapeutically to treat such inflammatory disorders.

For purposes of our discussion, we will divide the process of inflammation into three stages: vasodilation and increased permeability of blood vessels, phagocyte migration and phagocytosis, and tissue repair.

Vasodilation and Increased Permeability of Blood Vessels

Immediately following tissue damage, blood vessels dilate (increase in diameter) in the area of damage, and their permeability increases (**Figure 16.8a** and **Figure 16.8b**). Dilation of blood vessels, called **vasodilation,** increases blood flow to the damaged area and is responsible for the redness (erythema) and heat associated with inflammation.

Increased permeability permits defensive substances normally retained in the blood to pass through the walls of the blood vessels and enter the injured area. The increase in permeability, which permits fluid to move from the blood into tissue spaces, is responsible for the **edema** (accumulation of fluid) of inflammation. The pain of inflammation can be caused by nerve damage, irritation by toxins, or the pressure of edema.

❶ Vasodilation and the increase in permeability of blood vessels are caused by a number of chemicals released by damaged cells in response to injury. One such substance is **histamine,** a chemical present in many cells of the body, especially in mast cells in connective tissue, circulating basophils, and blood platelets. Histamine is released in direct response to the injury of cells that contain it; it is also released in response to stimulation by certain components of the complement system (discussed later). Phagocytic granulocytes attracted to the site of injury can also produce chemicals that cause the release of histamine.

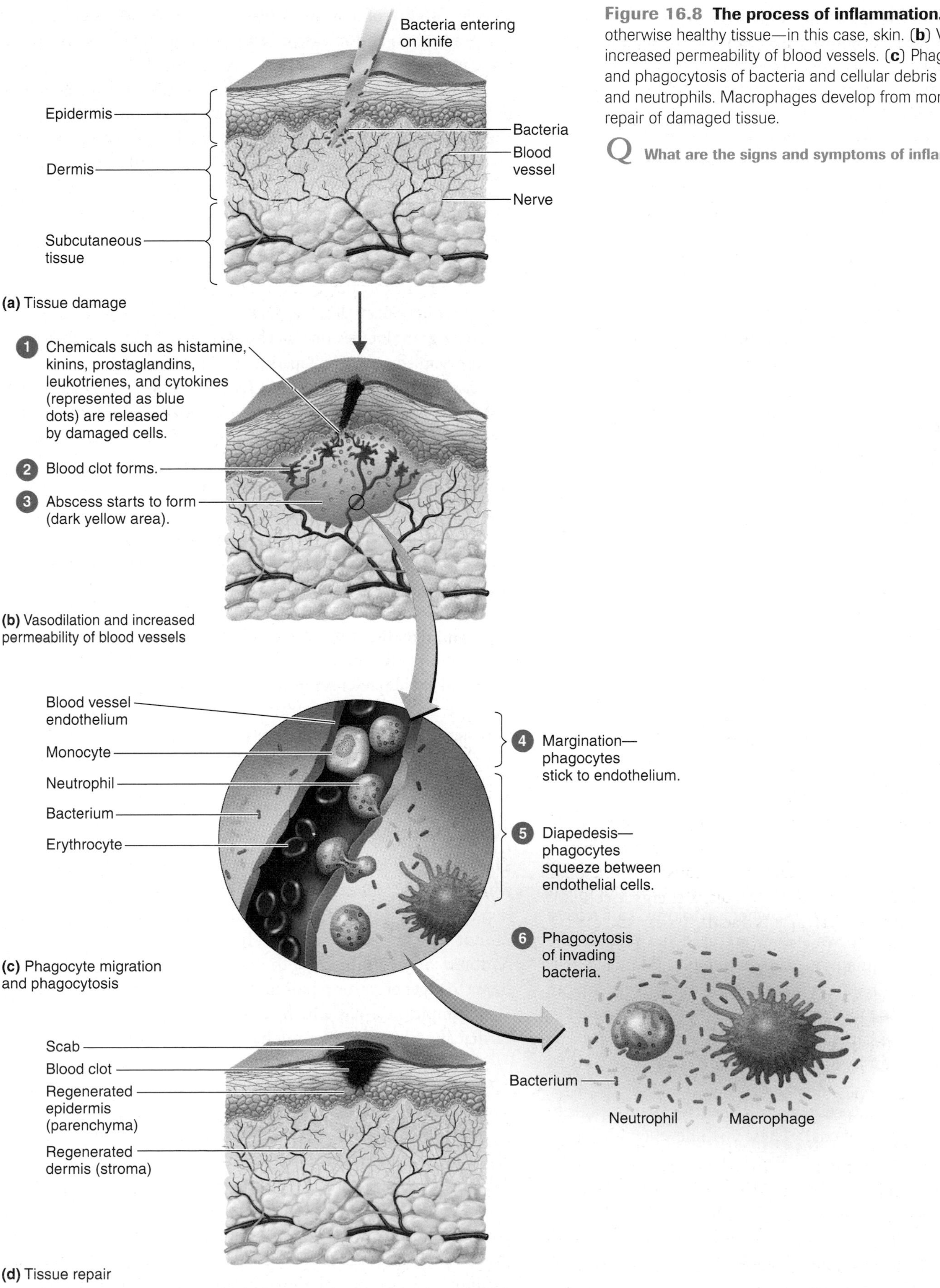

Figure 16.8 The process of inflammation. (**a**) Damage to otherwise healthy tissue—in this case, skin. (**b**) Vasodilation and increased permeability of blood vessels. (**c**) Phagocyte migration and phagocytosis of bacteria and cellular debris by macrophages and neutrophils. Macrophages develop from monocytes. (**d**) The repair of damaged tissue.

Q What are the signs and symptoms of inflammation?

Kinins are another group of substances that cause vasodilation and increased permeability of blood vessels. Kinins are present in blood plasma, and once activated, they play a role in chemotaxis by attracting phagocytic granulocytes, chiefly neutrophils, to the injured area.

Prostaglandins, substances released by damaged cells, intensify the effects of histamine and kinins and help phagocytes move through capillary walls. **Leukotrienes** are substances produced by mast cells (cells especially numerous in the connective tissue of the skin and respiratory system, and in blood vessels) and basophils. Leukotrienes cause increased permeability of blood vessels and help attach phagocytes to pathogens. Various components of the complement system stimulate the release of histamine, attract phagocytes, and promote phagocytosis.

Activated fixed macrophages also secrete **cytokines,** which bring about vasodilation and increased permeability. Vasodilation and the increased permeability of blood vessels also help deliver clotting elements of blood into the injured area. ❷ The blood clots that form around the site of activity prevent the microbe (or its toxins) from spreading to other parts of the body. ❸ As a result, there may be a localized collection of **pus,** a mixture of dead cells and body fluids, in a cavity formed by the breakdown of body tissues. This focus of infection is called an **abscess.** Common abscesses include pustules and boils.

The next stage in inflammation involves the migration of phagocytes to the injured area.

Phagocyte Migration and Phagocytosis

Generally, within an hour after the process of inflammation is initiated, phagocytes appear on the scene (Figure 16.8c). ❹ As the flow of blood gradually decreases, phagocytes (both neutrophils and monocytes) begin to stick to the inner surface of the endothelium (lining) of blood vessels. This sticking process in response to local cytokines is called **margination.** The cytokines alter cellular adhesion molecules (CAMs) on cells lining blood vessels, causing the phagocytes to stick at the site of inflammation. (Margination is also involved in red bone marrow, where cytokines can release phagocytes into circulation when they are needed.) ❺ Then the collected phagocytes begin to squeeze between the endothelial cells of the blood vessel to reach the damaged area. This migration, which resembles amoeboid movement, is called **diapedesis;** the migratory process can take as little as 2 minutes. ❻ The phagocytes then begin to destroy invading microorganisms by phagocytosis.

As mentioned earlier, certain chemicals attract neutrophils to the site of injury (chemotaxis). These include chemicals produced by microorganisms and even other neutrophils; other chemicals are kinins, leukotrienes, chemokines, and components of the complement system. Chemokines are cytokines that are chemotactic for phagocytes and T cells and thus stimulate both the inflammatory response and an adaptive immune response. The availability of a steady stream of neutrophils is ensured by the production and release of additional granulocytes from red bone marrow.

As the inflammatory response continues, monocytes follow the granulocytes into the infected area. Once the monocytes are contained in the tissue, they undergo changes in biological properties and become free macrophages. The granulocytes predominate in the early stages of infection but tend to die off rapidly. Macrophages enter the picture during a later stage of the infection, once granulocytes have accomplished their function. They are several times more phagocytic than granulocytes and are large enough to phagocytize tissue that has been destroyed, granulocytes that have been destroyed, and invading microorganisms.

After granulocytes or macrophages engulf large numbers of microorganisms and damaged tissue, they themselves eventually die. As a result, pus forms, and its formation usually continues until the infection subsides. At times, the pus pushes to the surface of the body or into an internal cavity for dispersal. On other occasions the pus remains even after the infection is terminated. In this case, the pus is gradually destroyed over a period of days and is absorbed by the body.

As effective as phagocytosis is in contributing to innate resistance, there are times when the mechanism becomes less functional in response to certain conditions. For example, some individuals are born with an inability to produce phagocytes. And with age, there is a progressive decline in the efficiency of phagocytosis. Recipients of heart or kidney transplants have impaired innate defenses as a result of receiving drugs that prevent the rejection of the transplant. Radiation treatments can also depress innate immune responses by damaging red bone marrow. Even certain diseases such as AIDS and cancer can cause defective functioning of innate defenses.

Tissue Repair

The final stage of inflammation is tissue repair, the process by which tissues replace dead or damaged cells (Figure 16.8d). Repair begins during the active phase of inflammation, but it cannot be completed until all harmful substances have been removed or neutralized at the site of injury. The ability of a tissue to regenerate, or repair itself, depends on the type of tissue. For example, skin has a high capacity for regeneration, whereas cardiac muscle tissue does not have a high capacity to regenerate.

A tissue is repaired when its stroma or parenchyma produces new cells. The *stroma* is the supporting connective tissue, and the *parenchyma* is the functioning part of the tissue. For example, the capsule around the liver that encloses and protects it is part of the stroma because it is not involved in the functions of the liver; liver cells (hepatocytes) that perform the functions of the liver are part of the parenchyma. If only parenchymal cells are active in repair, a perfect or near-perfect reconstruction of the tissue occurs. A familiar example of perfect reconstruction is a minor skin cut, in which parenchymal cells are more active in repair. However, if repair cells of the stroma of the skin are more active, scar tissue is formed.

As noted earlier, some microbes have various mechanisms that enable them to evade phagocytosis. Such microbes often induce a chronic inflammatory response, which can result in significant damage to body tissues. The most significant feature of chronic inflammation is the accumulation and activation of macrophages in the infected area. Cytokines released by activated macrophages induce fibroblasts in the tissue stroma to synthesize collagen fibers. These fibers aggregate to form scar tissue, a process called *fibrosis*. Because scar tissue is not specialized to perform the functions of the previously healthy tissue, fibrosis can interfere with the normal function of the tissue. **Animation** Inflammation: Overview, Steps. **www.microbiologyplace.com**

CHECK YOUR UNDERSTANDING

- ✓ What purposes does inflammation serve? **16-12**
- ✓ What causes the redness, swelling, and pain associated with inflammation? **16-13**
- ✓ What is margination? **16-14**

Fever

LEARNING OBJECTIVE

16-15 Describe the cause and effects of fever.

Inflammation is a local response of the body to injury. There are also systemic, or overall, responses; one of the most important is **fever,** an abnormally high body temperature, a third component of the second line of defense (see Figure 16.1). The most frequent cause of fever is infection from bacteria (and their toxins) or viruses.

Body temperature is controlled by a part of the brain called the hypothalamus. The hypothalamus is sometimes called the body's thermostat, and it is normally set at 37°C (98.6°F). It is believed that certain substances affect the hypothalamus by setting it at a higher temperature. Recall from Chapter 15 that when phagocytes ingest gram-negative bacteria, the lipopolysaccharides (LPS) of the cell wall (endotoxins) are released, causing the phagocytes to release the cytokines interleukin-1 (formerly called endogenous pyrogen), along with TNF-α. These cytokines cause the hypothalamus to release prostaglandins that reset the hypothalamic thermostat at a higher temperature, thereby causing fever (see Figure 15.6, page 438).

Assume that the body is invaded by pathogens and that the thermostat setting is increased to 39°C (102.2°F). To adjust to the new thermostat setting, the body responds by constricting blood vessels, increasing the rate of metabolism, and **shivering,** all of which raise body temperature. Even though body temperature is climbing higher than normal, the skin remains cold, and shivering occurs. This condition, called a *chill,* is a definite sign that body temperature is rising. When body temperature reaches the setting of the thermostat, the chill disappears. The body will continue to maintain its temperature at 39°C until the cytokines are eliminated. The thermostat is then reset to 37°C. As the infection subsides, heat-losing mechanisms such as vasodilation and sweating go into operation. The skin becomes warm, and the person begins to sweat. This phase of the fever, called the **crisis,** indicates that body temperature is falling.

Up to a certain point, fever is considered a defense against disease. Interleukin-1 helps step up the production of T cells. High body temperature intensifies the effect of antiviral interferons (page 468) and increases production of transferrins that decrease the iron available to microbes (page 470). Also, because the high temperature speeds up the body's reactions, it may help body tissues repair themselves more quickly.

Among the complications of fever are tachycardia (rapid heart rate), which may compromise older persons with cardiopulmonary disease; increased metabolic rate, which may produce acidosis; dehydration; electrolyte imbalances; seizures in young children; and delirium and coma. As a rule, death results if body temperature rises above 44 to 46°C (112–114°F).

CHECK YOUR UNDERSTANDING

- ✓ Why does a chill indicate that a fever is about to occur? **16-15**

Antimicrobial Substances

LEARNING OBJECTIVES

16-16 List the major components of the complement system.

16-17 Describe three pathways of activating complement.

16-18 Describe three consequences of complement activation.

16-19 Define *interferons.*

16-20 Compare and contrast the actions of IFN-α and IFN-β with IFN-γ.

16-21 Describe the role of iron-binding proteins in innate immunity.

16-22 Describe the role of antimicrobial peptides in innate immunity.

The body produces certain antimicrobial substances, a final component of the second line of defense (see Figure 16.1), in addition to the chemical factors mentioned earlier. Among the most important of these are the proteins of the complement system: interferons, iron-binding proteins, and antimicrobial peptides.

The Complement System

The **complement system** is a defensive system consisting of over 30 proteins produced by the liver and found circulating in blood serum (see the box on page 467) and within tissues throughout the body. The complement system is so-named because it "complements" the cells of the immune system in destroying microbes. The complement system is not adaptable and does not change

Figure 16.9

FOUNDATION FIGURE Outcomes of Complement Activation

The complement system is another way the body fights infection and destroys pathogens. This component of innate immunity "complements" other immune reactions described in this and the following chapters.

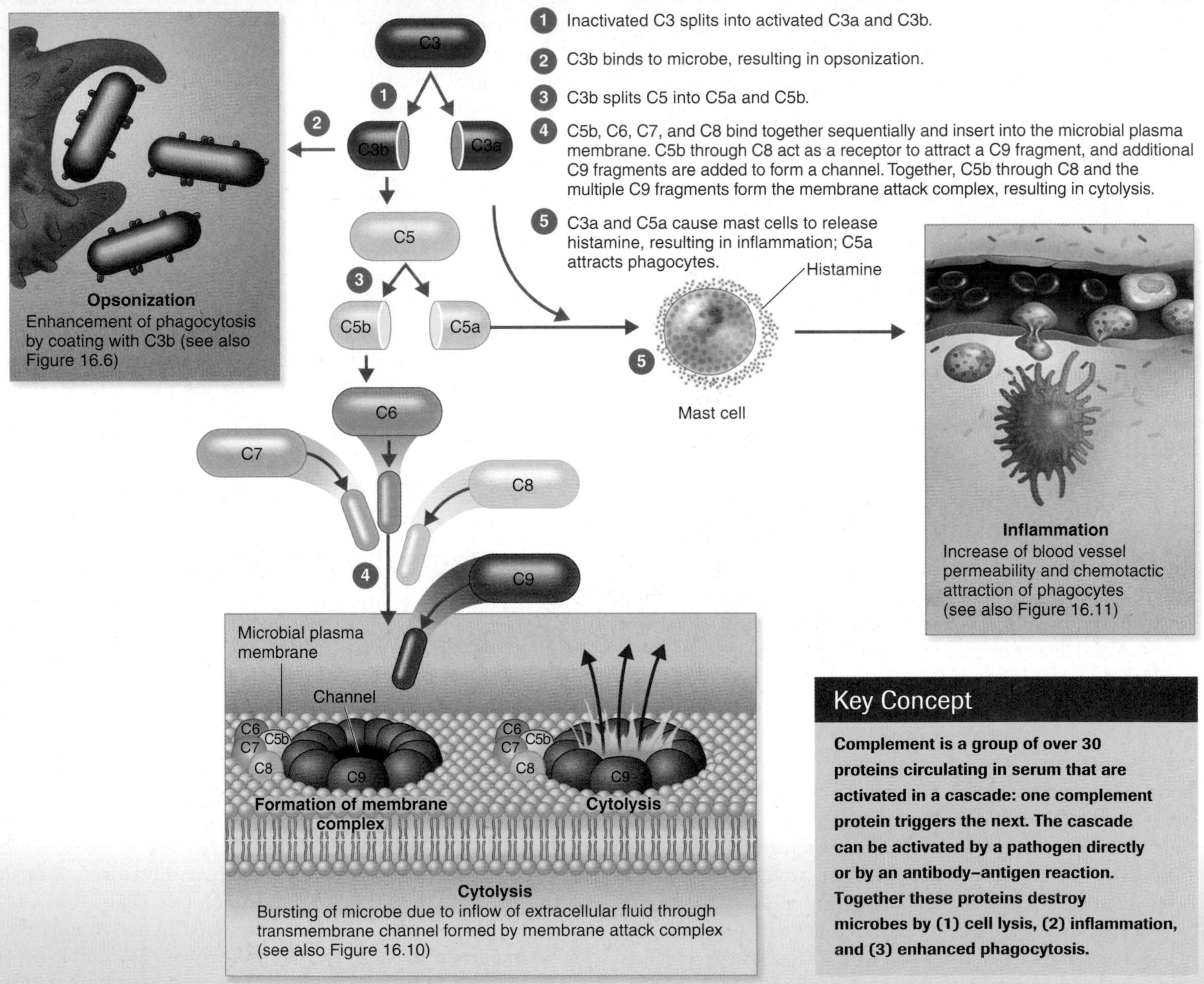

1 Inactivated C3 splits into activated C3a and C3b.
2 C3b binds to microbe, resulting in opsonization.
3 C3b splits C5 into C5a and C5b.
4 C5b, C6, C7, and C8 bind together sequentially and insert into the microbial plasma membrane. C5b through C8 act as a receptor to attract a C9 fragment, and additional C9 fragments are added to form a channel. Together, C5b through C8 and the multiple C9 fragments form the membrane attack complex, resulting in cytolysis.
5 C3a and C5a cause mast cells to release histamine, resulting in inflammation; C5a attracts phagocytes.
C3
C3b
C3a
C5
C5b
C5a
C6
C7
C8
C9
Histamine
Mast cell
Opsonization
Enhancement of phagocytosis by coating with C3b (see also Figure 16.6)
Inflammation
Increase of blood vessel permeability and chemotactic attraction of phagocytes (see also Figure 16.11)
Microbial plasma membrane
Channel
Formation of membrane complex
Cytolysis
Cytolysis
Bursting of microbe due to inflow of extracellular fluid through transmembrane channel formed by membrane attack complex (see also Figure 16.10)

Key Concept

Complement is a group of over 30 proteins circulating in serum that are activated in a cascade: one complement protein triggers the next. The cascade can be activated by a pathogen directly or by an antibody–antigen reaction. Together these proteins destroy microbes by (1) cell lysis, (2) inflammation, and (3) enhanced phagocytosis.

over the course of a person's lifetime; for these reasons, it belongs to the innate immune system. However, it can be recruited and brought into action by the adaptive immune system. Together, proteins of the complement system destroy microbes by (1) cytolysis, (2) inflammation, and (3) phagocytosis and also prevent excessive damage to host tissues. Complement proteins are usually designated by an uppercase letter C and are inactive until they are split into fragments (products). The proteins are numbered C1 through C9, named for the order in which they were discovered. The fragments are activated proteins and are indicated by the lowercase letters *a* and *b*. For example, inactive complement protein C3 is split into two activated fragments, C3a and

C3b. The activated fragments carry out the destructive actions of the C1 through C9 complement proteins.

Complement proteins act in a *cascade*; that is, one reaction triggers another, which in turn triggers another, and so on. Also, as part of the cascade, more product is formed with each succeeding reaction so that the effect is amplified many times as the reactions continue.

The Result of Complement Activation

The C3 protein can be activated by three mechanisms that will be described shortly. Activation of C3 (Figure 16.9) is very important because it starts a cascade that results in cytolysis, inflammation, and phagocytosis.

1. Inactive C3 splits into activated C3a and C3b.
2. C3b binds to the surface of a microbe, and receptors on phagocytes attach to the C3b. Thus C3b enhances *phagocytosis* by coating a microbe, a process called *opsonization*, or *immune adherence*. Opsonization promotes attachment of a phagocyte to a microbe.
3. C3b also initiates a series of reactions that result in cytolysis. First, C3b splits C5 into C5b and C5a. Fragments C5b, C6, C7, and C8 bind together sequentially and insert into the plasma membrane of the invading cell. C5b through C8 act as a receptor that attracts a C9 fragment. Additional C9 fragments are added to form a transmembrane channel. Together, C5b through C8 and the multiple C9 fragments form the **membrane attack complex (MAC).**
4. The transmembrane channels (holes) of the MAC result in *cytolysis*, the bursting of the microbial cell due to the inflow of extracellular fluid through the channels (Figure 16.10).
5. C3a and C5a bind to mast cells and cause them to release histamine and other chemicals that increase blood vessel permeability during *inflammation* (Figure 16.11). C5a also functions as a very powerful chemotactic factor that attracts phagocytes to the site of an infection.

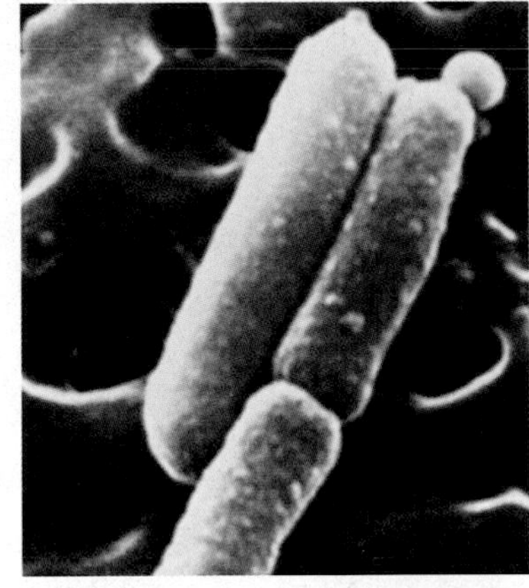
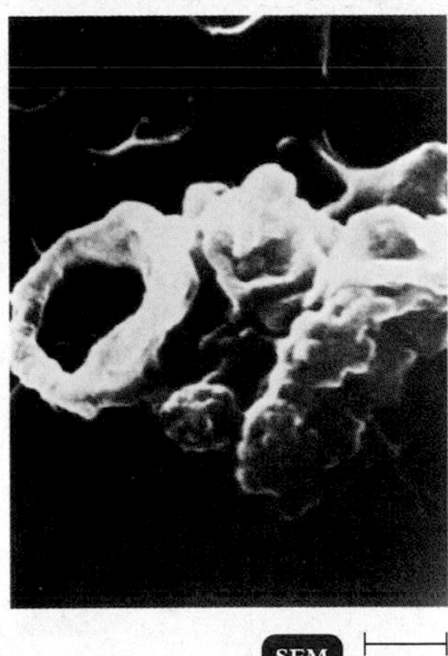

Figure 16.10 Cytolysis caused by complement. Micrographs of a rod-shaped bacterium before cytolysis (left) and after cytolysis (right). *Source:* Reprinted from Schreiber, R. D. et al. "Bactericidal Activity of the Alternative Complement Pathway Generated from 11 Isolated Plasma Proteins." *Journal of Experimental Medicine*, 149:870–882, 1979.

Q How does complement aid in fighting infections?

Bacteria that are not killed by the MAC are said to be *MAC-resistant.*

Host cell plasma membranes contain proteins that protect against lysis by preventing attachment of the MAC proteins to their surfaces. Also, the MAC forms the basis for the complement fixation test used to diagnose some diseases. This is explained in the box on page 467 and in Chapter 18 (see Figure 18.10, page 514). Gram-negative bacteria are more susceptible to cytolysis because they have only one or very few layers of peptidoglycan to protect the plasma membrane from the effects of complement. The many layers of peptidoglycan of gram-positive bacteria limit access of complement to the plasma membrane and thus interfere with cytolysis.

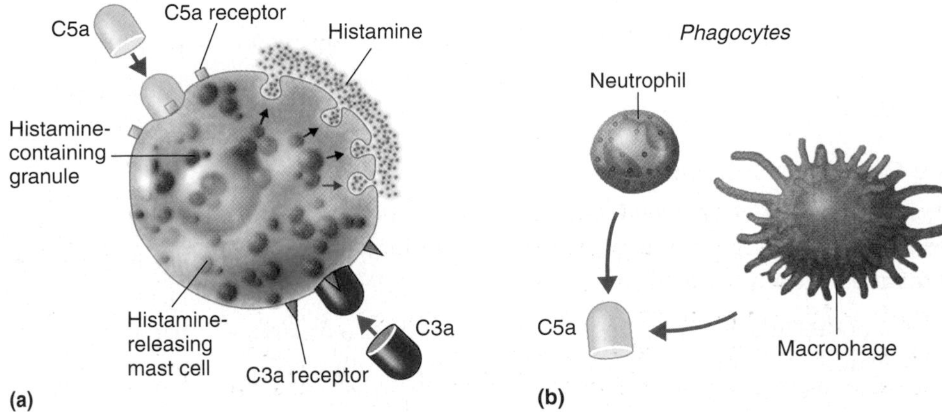

Figure 16.11 Inflammation stimulated by complement. (**a**) C3a and C5a bound to mast cells, basophils, and platelets trigger the release of histamine, which increases blood vessel permeability. (**b**) C5a functions as a chemotactic factor that attracts phagocytes to the site of complement activation.

Q How is complement inactivated?

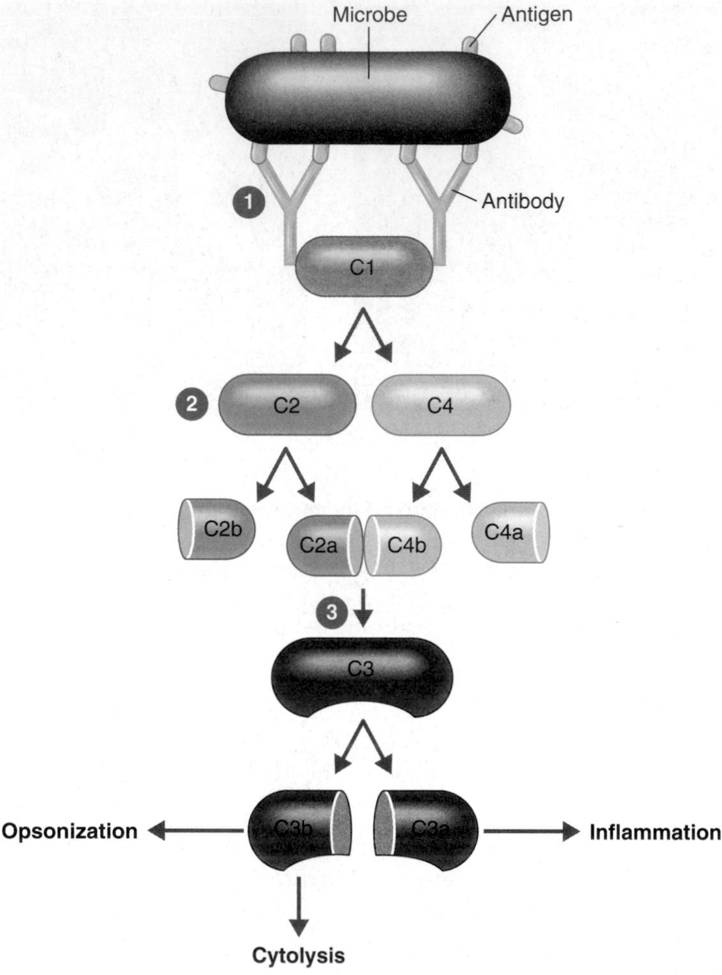

1. C1 is activated by binding to antigen–antibody complexes.
2. Activated C1 splits C2 into C2a and C2b, and C4 into C4a and C4b.
3. C2a and C4b combine and activate C3, splitting it into C3a and C3b (see also Figure 16.9).

Figure 16.12 Classical pathway of complement activation. This pathway is initiated by an antigen–antibody reaction. The splitting of C3 into C3a and C3b starts a cascade that results in cytolysis, inflammation, and opsonization (see also Figure 16.9).

Q What happens after C3 is cleaved into C3a and C3b?

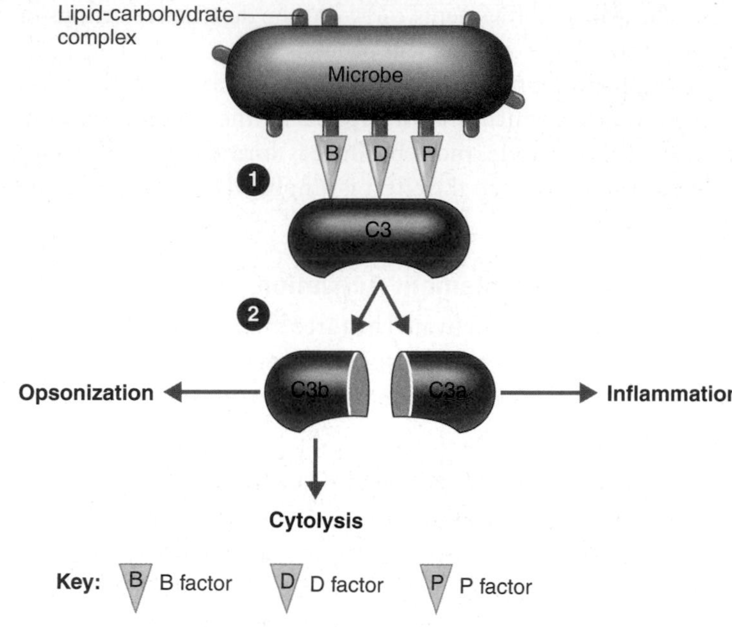

1. C3 combines with factors B, D, and P on the surface of a microbe.
2. This causes C3 to split into fragments C3a and C3b.

Figure 16.13 Alternative pathway of complement activation. This pathway is initiated by contact between certain complement proteins (C3 and factors B, D, and P) and a pathogen. There are no antibodies involved. Once C3a and C3b are formed, they participate in cytolysis, inflammation, and opsonization per the classical pathway (see also Figure 16.9).

Q How is the alternative pathway similar to the classical pathway?

The cascade of complement proteins that occurs during an infection is called **complement activation** and may occur in three pathways.

The Classical Pathway

The **classical pathway** (Figure 16.12), so named because it was the first to be discovered, is initiated when antibodies bind to antigens (microbes) and occurs as follows:

1. Antibodies attach to antigens (for example, proteins or large polysaccharides on the surface of a bacterium or other cell), forming antigen–antibody complexes. The antigen–antibody complexes bind and activate C1.
2. Next, activated C1 activates C2 and C4 by splitting them. C2 is split into fragments called C2a and C2b, and C4 is split into fragments called C4a and C4b.
3. C2a and C4b combine and together they activate C3 by splitting it into C3a and C3b. The C3 fragments then initiate cytolysis, inflammation, and opsonization (see Figure 16.9).

The Alternative Pathway

The **alternative pathway** is so named because it was discovered after the classical pathway. Unlike the classical pathway, the alternative pathway does not involve antibodies. The alternative pathway is activated by contact between certain complement proteins and a pathogen (Figure 16.13). ❶ C3 is constantly present in the blood. It combines with complement proteins called factor B, factor D, and factor P (properdin) on the surface of a pathogenic microbe. The complement proteins are attracted to microbial cell surface material (mostly lipid–carbohydrate

Serum Collection

It is common to draw more than one blood sample for laboratory tests. The blood is collected in tubes with caps of different colors (**Figure A**). Whole blood may be needed to culture microbes or to type the blood. Serum may be needed to test for enzymes or other chemicals in the blood. Serum is the straw-colored liquid remaining after blood is allow to clot. Blood plasma is the liquid remaining after formed elements are removed from unclotted blood, for example, by centrifugation.

Which sample would you use to count blood cells? To test for complement?
Complement activity is measured because complement deficiency may be associated with recurrent bacterial infections. Moreover, complement is a key component in immune complex diseases. A decrease in serum complement, which occurs as complement is used in immune complexes, can be used to monitor progress and treatment of immune complex diseases such as systemic lupus erythematosus and rheumatoid arthritis.

Total complement activity is measured as shown in **Figure B**. Dilutions of the patient's serum are mixed with sheep red blood cells (RBCs) and antibodies against sheep RBCs. Following incubation at 37°C for 20 minutes, the degree of hemolysis is determined.

What is the purpose of the RBCs and antiRBCs?
The antibodies will react with the antigen (RBCs). This will activate complement in the patient's serum. The degree of lysis is relative to the amount of complement present and is expressed as a percentage of the hemolysis produced by a serum pool collected from 50 healthy blood donors.

Figure A Collecting blood cells and serum.

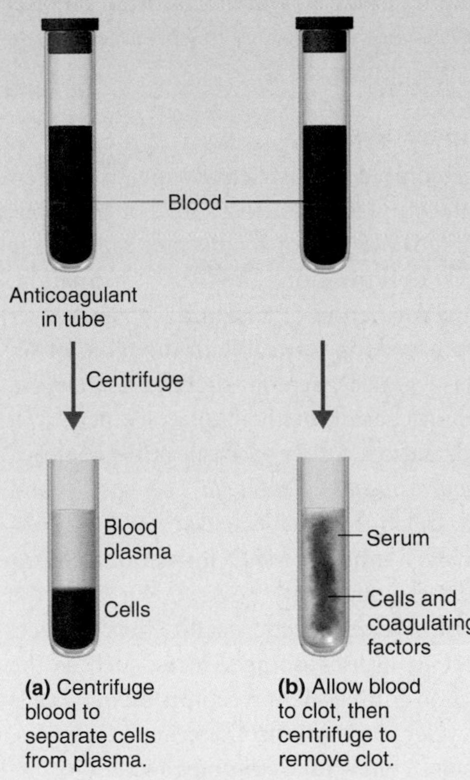

Figure B Testing for complement.

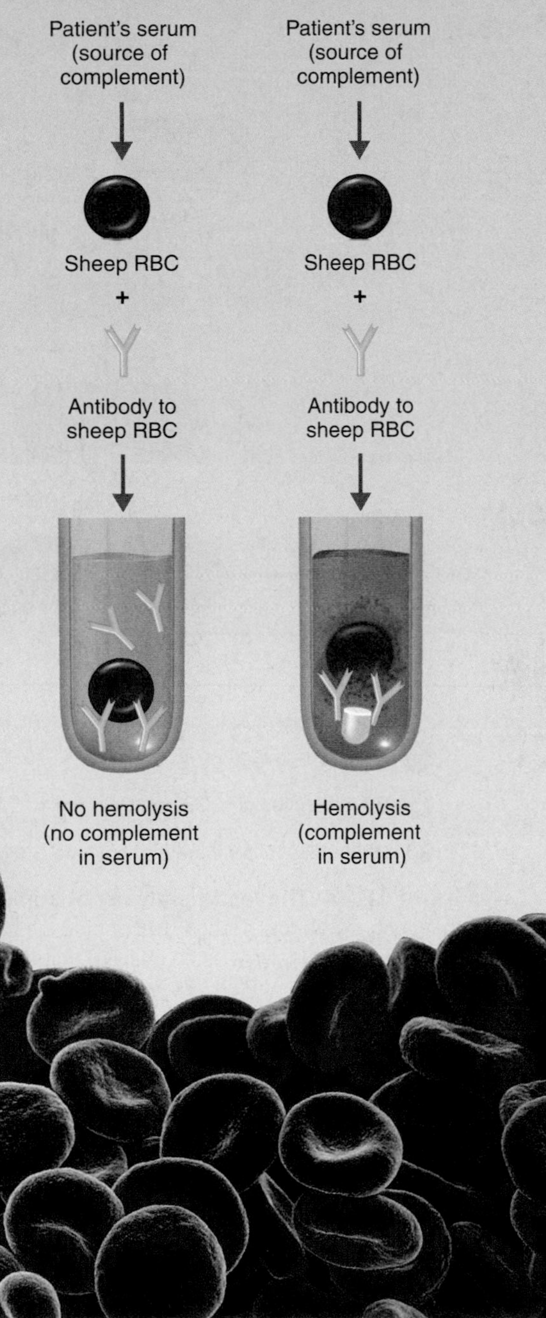

complexes of certain bacteria and fungi). ❷ Once the complement proteins combine and interact, C3 is split into fragments C3a and C3b. As in the classical pathway, C3a participates in inflammation, and C3b functions in cytolysis and opsonization (see Figure 16.9).

The Lectin Pathway

The **lectin pathway** is the most recently discovered mechanism for complement activation. When macrophages ingest bacteria, viruses, and other foreign matter by phagocytosis, they release cytokines that stimulate the liver to produce **lectins,** proteins that bind to carbohydrates (**Figure 16.14**).

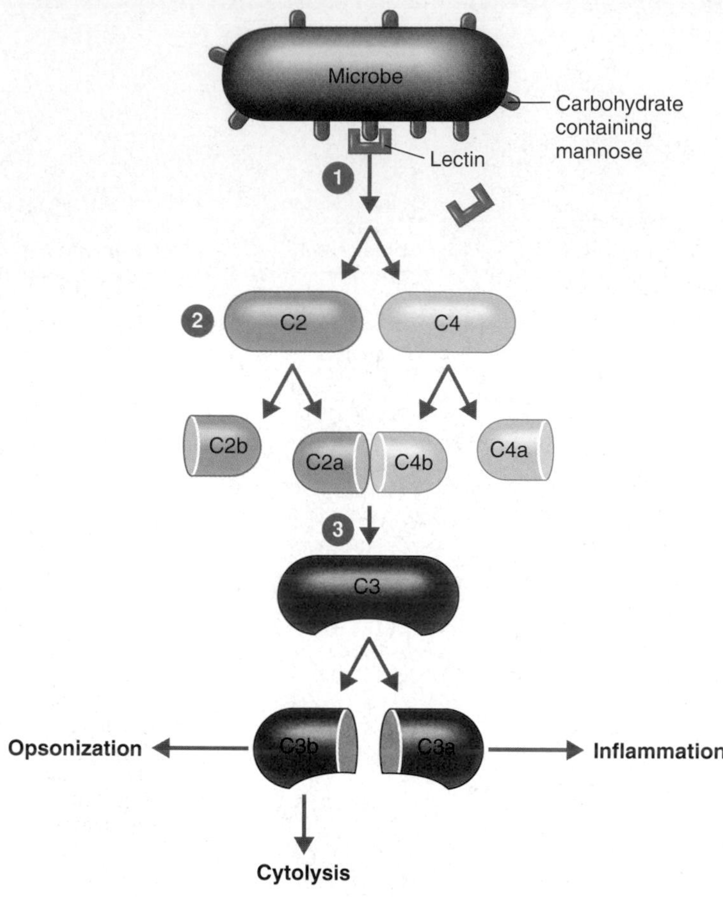

1 Lectin binds to an invading cell.

2 Bound lectin splits C2 and C4.

3 C2a and C4b combine and activate C3 (see also Figure 16.9).

Figure 16.14 The lectin pathway of complement activation. When mannose-binding lectin (MBL) binds to mannose on the surface of microbes, MBL functions as an opsonin that enhances phagocytosis and activates complement (see Figure 16.9).

Q How do the lectin and alternative pathways differ from the classical pathway?

1 One such lectin, **mannose-binding lectin (MBL),** binds to the carbohydrate mannose. MBL binds to many pathogens because the MBL molecules recognize a distinctive pattern of carbohydrates that includes mannose, which is found in bacterial cell walls and on some viruses. As a result of binding, MBL functions as an opsonin to enhance phagocytosis and

2 activates C2 and C4;

3 C2a and C4b activate C3 (see Figure 16.9).

Regulation of Complement

Once complement is activated, its destructive capabilities usually cease very quickly to minimize the destruction of host cells. This is accomplished by various regulatory proteins in the host's blood and on certain cells, such as blood cells. The regulatory proteins are present at higher concentrations than the complement proteins. One example of a regulatory protein is *CD59*, which prevents the assembly of C9 molecule to form the MAC. The proteins bring about the breakdown of activated complement and function as inhibitors and destructive enzymes.

Complement and Disease

In addition to its importance in defense, the complement system assumes a role in causing disease as a result of inherited deficiencies. Deficiencies of C1, C2, or C4 cause collagen vascular disorders that result in hypersensitivity (anaphylaxis); deficiency of C3, though rare, results in increased susceptibility to recurrent infections with pyogenic microbes; and C5 through C9 defects result in increased susceptibility to *Neisseria meningitidis* and *N. gonorrhoeae* infections. Complement may play a role in diseases with an immune component, such as asthma, systemic lupus erythematosus, various forms of arthritis, multiple sclerosis, and inflammatory bowel disease. Complement has also been implicated in Alzheimer disease and other neurodegenerative disorders.

Evading the Complement System

Some bacteria evade the complement system by means of their capsules, which prevent complement activation. For example, some capsules contain large amounts of a substance called sialic acid, which discourages opsonization and MAC formation. Other capsules inhibit the formation of C3b and C4b and cover the C3b to prevent it from making contact with the receptor on phagocytes. Some gram-negative bacteria, such as *Salmonella*, can lengthen the O polysaccharide of their LPS (see page 87), which prevents MAC formation. Other gram-negative bacteria, such as *Neisseria gonorrhoeae, Bordetella pertussis,* and *Haemophilus influenzae,* attach their sialic acid to the lipid A of their LPS, which ultimately inhibits MAC formation. Gram-positive cocci release an enzyme that breaks down C5a, the fragment that serves as a chemotactic factor that attracts phagocytes. With respect to viruses, some viruses, such as the Epstein-Barr virus, attach to complement receptors on body cells to initiate their life cycle. **Animations** Complement System: Overview, Activation, Results. **www.microbiologyplace.com**

CHECK YOUR UNDERSTANDING

✓ What is complement? **16-16**

✓ List the steps of complementation activation via (1) the classical pathway, (2) the alternative pathway, and (3) the lectin pathway. **16-17**

✓ Summarize the major outcomes of complement activation. **16-18**

Interferons

Because viruses depend on their host cells to provide many functions of viral multiplication, it is difficult to inhibit viral multiplication without affecting the host cell itself. One way the infected host counters viral infections is with a family of

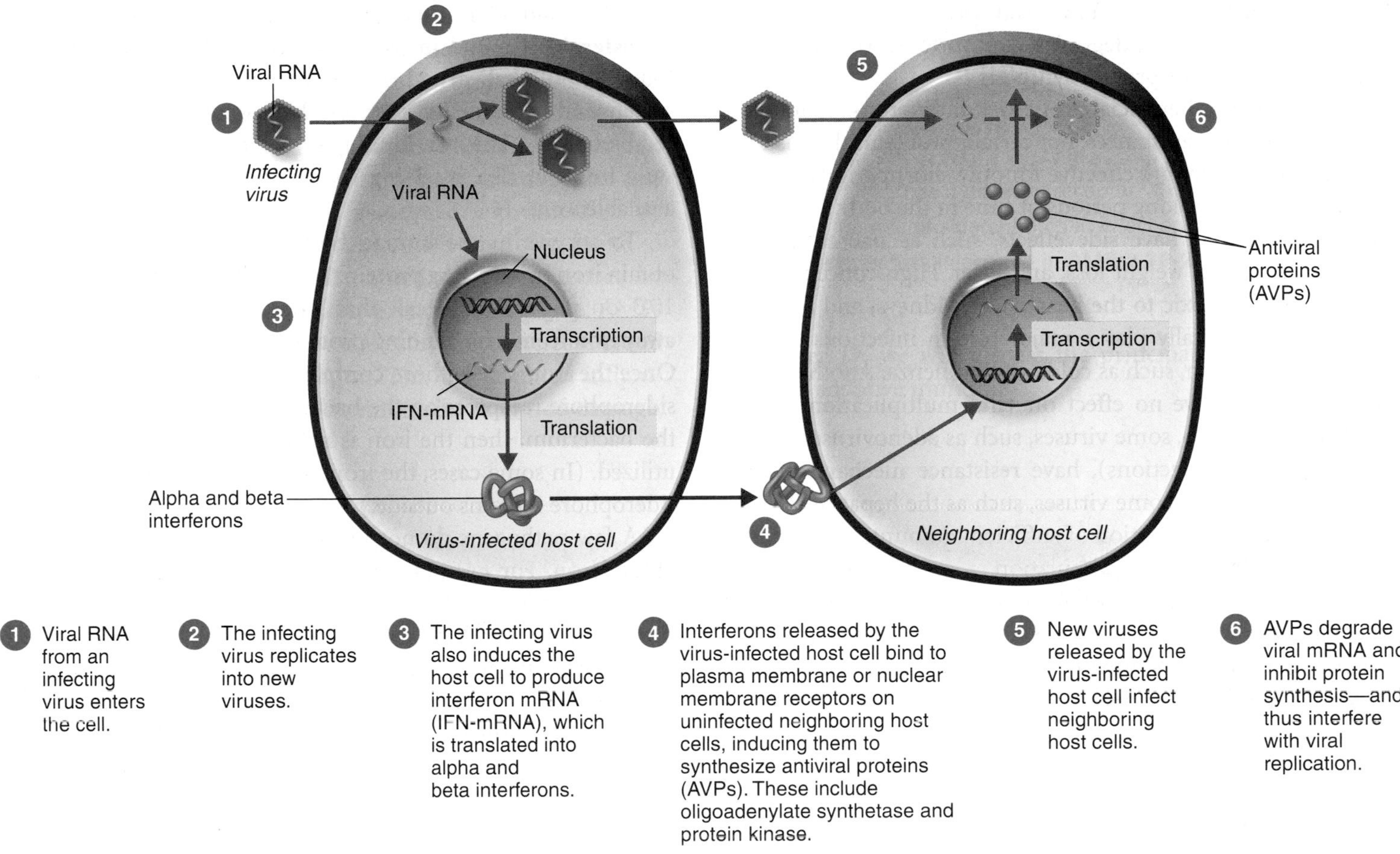

1 Viral RNA from an infecting virus enters the cell.

2 The infecting virus replicates into new viruses.

3 The infecting virus also induces the host cell to produce interferon mRNA (IFN-mRNA), which is translated into alpha and beta interferons.

4 Interferons released by the virus-infected host cell bind to plasma membrane or nuclear membrane receptors on uninfected neighboring host cells, inducing them to synthesize antiviral proteins (AVPs). These include oligoadenylate synthetase and protein kinase.

5 New viruses released by the virus-infected host cell infect neighboring host cells.

6 AVPs degrade viral mRNA and inhibit protein synthesis—and thus interfere with viral replication.

Figure 16.15 Antiviral action of alpha and beta interferons (IFNs). Interferons are host-cell-specific but not virus-specific.

Q **How does interferon stop viruses?**

cytokines called interferons. **Interferons (IFNs)** are a class of similar antiviral proteins produced by certain animal cells, such as lymphocytes and macrophages, after viral stimulation. One of the principal functions of interferons is to interfere with viral multiplication.

An interesting feature of interferons is that they are host-cell–specific but not virus-specific. Interferons produced by human cells protect human cells but produce little antiviral activity for cells of other species, such as mice or chickens. However, the interferons of a species are active against a number of different viruses.

Just as different animal species produce different interferons, different types of cells in the same animal also produce different interferons. Human interferons are of three principal types: *alpha interferon* (IFN-α), *beta interferon* (IFN-β), and *gamma interferon* (IFN-γ). There are also various subtypes of interferons within each of the principal groups. In the human body, interferons are produced by fibroblasts in connective tissue and by lymphocytes and other leukocytes. Each of the three types of interferons produced by these cells can have a slightly different effect on the body.

All interferons are small proteins, with molecular weights between 15,000 and 30,000. They are quite stable at low pH and are fairly resistant to heat.

Gamma interferon is produced by lymphocytes; it induces neutrophils and macrophages to kill bacteria. IFN-γ causes macrophages to produce nitric oxide that appears to kill bacteria as well as tumor cells by inhibiting ATP production. Neutrophils and macrophages in individuals with an inherited condition called chronic granulomatous disease (CGD) do not kill bacteria. When these people take a recombinant IFN-γ, their neutrophils and macrophages kill bacteria. Gamma interferon is not a cure and must be taken for the life of the CGD individual. As you will see in Chapter 17, IFN-γ increases the expression of class I and class II molecules and increases antigen presentation.

Both IFN-α and IFN-β are produced by virus-infected host cells only in very small quantities and diffuse to uninfected neighboring cells (Figure 16.15). They react with plasma or nuclear membrane receptors, inducing the uninfected cells to manufacture mRNA for the synthesis of **antiviral proteins (AVPs).** These proteins are enzymes that disrupt various stages

17 Adaptive Immunity: Specific Defenses of the Host

In Chapter 16, we discussed innate defenses of the body against microorganisms in the environment. These defenses include the skin and mucous membranes, phagocytosis, and inflammation. Humans, and even the simplest animals, have some defenses that are always present to provide instant protection against infection. Taken altogether, these defenses are termed **innate immunity.** Immunity is a term derived from the Latin word *immunis,* meaning to exempt. (For a review of the body's innate defense systems, see Figure 16.1, page 450).

Higher animals also are protected by the more specialized adaptive immunity. **Adaptive immunity** is induced; that is, it adapts to a specific microbial invader or a foreign substance. Adaptive immunity will be the focus of this chapter.

UNDER THE MICROSCOPE

Dendritic cell. Dendritic cells are antigen-presenting cells that play a key role in helping your immune system differentiate self from nonself.

Q&A

Why doesn't the adaptive immune system normally attack your own body tissues?

Look for the answer in the chapter.

The Adaptive Immune System

LEARNING OBJECTIVE

17-1 Differentiate innate from adaptive immunity.

It was recognized long ago that immunity to certain infectious diseases can be acquired during an individual's life. If a person recovered from smallpox or measles, that person was almost always immune to that disease when exposed to it again. In some way they had acquired a memory of the infection, an important factor in adaptive immunity. Eventually, as medicine developed over the centuries, methods were found to mimic the adaptive immunity to disease by deliberately exposing people to harmless versions of pathogens that caused certain diseases, thus rendering them immune; we now call this *vaccination* (see Chapter 18, pages 501–506). Vaccination against smallpox, the first disease for which vaccination was developed, predated by nearly a hundred years any knowledge of microscopic pathogens. However, the systematic advancement of the science of adaptive immunity required the concept of the germ theory of diseases; that is, that specific pathogenic microbes are responsible for specific diseases (see page 404). **Animation** Host Defenses: The Big Picture. **www.microbiologyplace.com**

CHECK YOUR UNDERSTANDING

✓ Is vaccination an example of innate or adaptive immunity? **17-1**

Dual Nature of the Adaptive Immune System

LEARNING OBJECTIVE

17-2 Differentiate humoral from cellular immunity.

To introduce many of the terms and concepts important to understanding adaptive immunity, let's first briefly review the development of the theoretical basis. It will also serve to emphasize the dual nature of adaptive immunity—that it consists of a humoral and a cellular component.

In 1887, when Louis Pasteur first observed the immunity that developed in chickens when he accidentally injected them with weakened pathogens, he erroneously hypothesized that this was due to the depletion of some essential nutrient that was required by the pathogens to multiply. Events moved rapidly in the next few years. Emil von Behring, working with diphtheria and tetanus bacteria, showed that the culture medium in which they were grown contained an apparent *toxin* that was fatal to animals when injected into them. When rabbits were injected with very small amounts of toxin, however, they would often survive. Blood serum from these surviving animals was found to have a surprising property. When it was injected into other animals after they had received a typically lethal dose of toxin, they showed no ill effects. Apparently, some factor in the serum from the surviving rabbits had neutralized the lethal toxins. They named this factor *antitoxin*. For this work, in 1901, von Behring received the first Nobel Prize awarded in Physiology or Medicine.

Paul Ehrlich, a German physician, found that a measured amount of antitoxin would neutralize an equivalent amount of toxin. He also determined that the protective nature of antitoxin was highly specific—it neutralized only the toxin that had given rise to it. The scientific community quickly focused attention on this new concept of immunity that clearly showed immense promise in medicine. It was found that similarly derived protective serum factors (now called by the more general name of *antibodies*) were also produced by exposure to bacteria and other pathogens. It was soon learned that antibodies arose against not only microbial pathogens and toxins, but also many other particulate substances, such as plant pollen and red blood cells, that the body recognized as alien, or *nonself.* Such substances that caused the production of antibodies were called *antigens*—from *anti*body *gen*erators. The combination of an antibody and a particulate antigen caused them to visibly clump together or, in some cases, lyse. It was found that the lysis of a cell bearing an antigen, in response to an antibody against it, required another element naturally found in the blood. This was *complement* (see Chapter 16, page 463), so called because it complements the action of antibodies.

Humoral Immunity

From ancient times until well into the nineteenth century, the medical community believed fervently that health depended on four different body fluids, or *humors*: blood, phlegm, black bile, and yellow bile. The new science of immunology adopted the term **humoral immunity** when describing immunity brought about by antibodies. **Animation** Humoral Immunity: Overview. **www.microbiologyplace.com**

Cellular Immunity

As recently as the mid-1950s, immunology was a relatively simple science. Medical science was familiar with innate immunity largely based on nonspecific phagocytic cells and a humoral immunity based on antibodies that specifically targeted certain toxins or particulate antigens. Up to that time the **thymus,** a lymphoid organ found in the upper chest, and which slowly atrophies after puberty, had no known function. Similarly, in birds the **bursa of Fabricius,** a structure resembling a lymph node, also had no known function. Experiments designed to determine the function of the thymus or the bursa by removing them had no apparent effect. A breakthrough in the science of immunology occurred in 1956, when an experimenter happened to remove the bursa from very young, rather than mature, birds. There were, as usual, no discernible effects, and the birds were set aside. In an unrelated experiment, nearly a year later, these same birds were selected to produce antibodies against pathogenic *Salmonella* bacteria. This was done by inoculating them with old, presumably weakened, cultures of the bacteria and, in effect, vaccinating them. To the surprise of the experimenter, the birds produced no antibodies, and some of them became ill and died. Following this lead, it was found that birds that had their bursa removed as adults had a normal immune response—they produced antibodies when injected with bacteria. However, when the bursa was removed from very young birds, they developed into adult birds with defective immune systems. This clearly indicated that the bursa was needed for *maturation* of the immune system and the eventual ability to produce antibodies.

This important work was obscurely published as a two-page note in the journal, *Poultry Science*. There it might have remained, unnoticed by researchers interested in the immunology of humans. Someone, however, happened to call it to the attention of another immunologist who was also attempting to determine the function of the thymus by removing it from his experimental mice. Following the direction suggested by the article in *Poultry Science*, this researcher then tried removing the thymus from newborn rather than mature mice. These mice, like the chickens with their bursa removed, matured into mice that produced no antibodies and, importantly, were also very slow in rejecting skin transplants from other mice.

This was a revolutionary development, showing that there was more to the immune system than the antibody-based immunity that had been taught for the past 50 years. Ehrlich had proposed that there were certain specialized cells that made an antibody in response to contact with an antigen—which then served as a pattern. This new work seemed to point at the need for at least two cells to produce an antibody. One type of cell recognized the antigen as being foreign and passed this information on to a second type of cell that actually produced the antibodies. The identity of these cells was unknown, but it was thought that they probably were white blood cells called lymphocytes (see Table 16.1, page 455), which were plentiful in lymphoid tissue such as the thymus and the bursa of Fabricius.

Before this time, the function of lymphocytes had been mysterious, but by the 1960s the technology required to isolate and culture lymphocytes in the laboratory was in place. The source of lymphocytes was determined to be in the liver during the first few weeks of development. By about the third month, the bone marrow replaces the liver as the source of lymphocytes. From the red bone marrow, stem cells produce lymphocytes that begin their "schooling"—that is, they differentiate. Some mature in the bone marrow and become **B cells** (named for the bursa of Fabricius) that recognize antigens and make specific antibodies against them. The recognition of different antigens depends on receptors to the antigens that coat the surface of the B cell. Other lymphocytes mature under the influence of the thymus and are therefore called T lymphocytes, or **T cells,** the basis of **cellular immunity.** Both T cells and B cells are found primarily in blood and lymphoid organs. The T cells, like B cells, respond to antigens by means of receptors on their surface—**T-cell receptors (TCRs).** Contact with an antigen complementary to a TCR can cause certain types of T cells to proliferate and secrete *cytokines* rather than antibodies. These are chemical messengers that impart instructions to other cells to perform certain functions (see page 492).

CHECK YOUR UNDERSTANDING

✓ How was basic research on chicken diseases related to the discoveries of both humoral and cellular immunity? **17-2**

Antigens and Antibodies

LEARNING OBJECTIVES

17-3 Define *antigen, epitope,* and *hapten.*

17-4 Explain the function of antibodies, and describe their structural and chemical characteristics.

17-5 Name one function for each of the five classes of antibodies.

Antigens and antibodies play key roles in the response of the immune system. Antigens provoke a highly specific immune response that, in humoral immunity, results in the production of antibodies that are capable of recognizing the antigen that gave rise to them. Antigens that cause such a response are, therefore, often more descriptively known as *immunogens.*

The Nature of Antigens

Most **antigens** are either proteins or large polysaccharides. Lipids and nucleic acids are usually antigenic only when combined with proteins and polysaccharides. Antigenic compounds are often components of invading microbes, such as capsules, cell walls, flagella, fimbriae, and toxins of bacteria; the coats of viruses; or the surfaces of other types of microbes. Nonmicrobial antigens include pollen, egg white, blood cell surface molecules, serum proteins from other individuals or species, and surface molecules of transplanted tissues and organs.

Generally, antibodies recognize and interact with specific regions on antigens called **epitopes** or **antigenic determinants** (**Figure 17.1**). The nature of this interaction depends on the size, shape, and chemical structure of the binding site on the antibody molecule.

Most antigens have a molecular weight of 10,000 or higher. A foreign substance that has a low molecular weight is often not antigenic unless it is attached to a carrier molecule. These low molecular weight compounds are called **haptens** (from the Greek *hapto*, to grasp; **Figure 17.2**). Once an antibody against the hapten has been formed, the antibody will react with the hapten independent of the carrier molecule. Penicillin is a good example of a hapten. This drug is not antigenic by itself, but some people develop an allergic reaction to it. (Allergic reactions are a type of

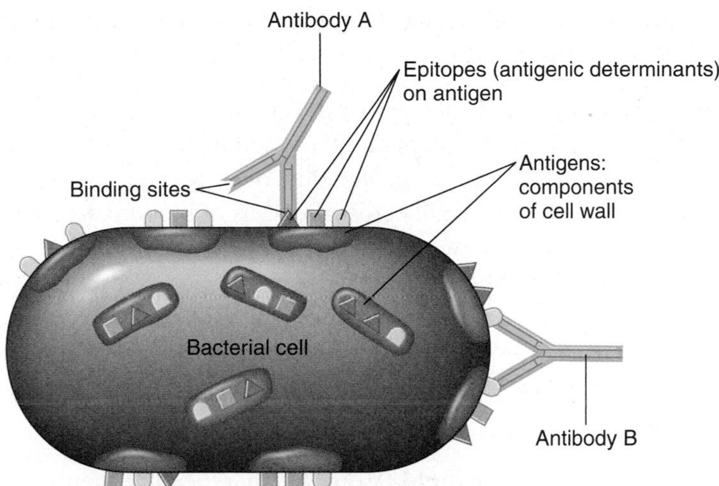

Figure 17.1 Epitopes (antigenic determinants). In this illustration the epitopes are components of the antigen—the bacterial cell wall. Each antigen carries more than one epitope. Each Y-shaped antibody molecule has two binding sites that can attach to a specific epitope on an antigen. An antibody can also bind to identical epitopes on two different cells at the same time (see Figure 18.5, page 510), which can cause neighboring cells to aggregate.

Q **Which antibody type (IgG or IgM) should be the more efficient in aggregating cells?**

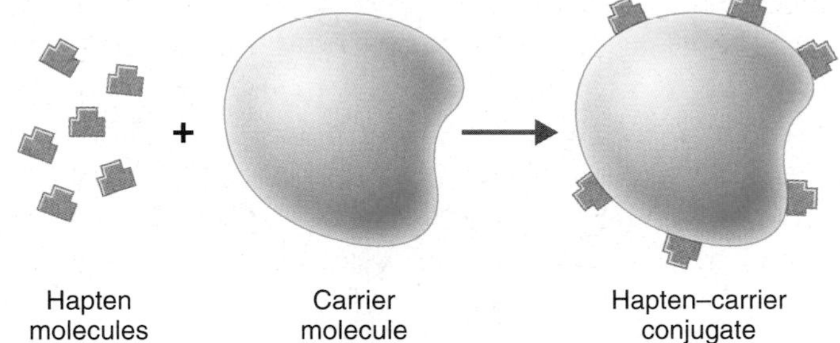

Figure 17.2 Haptens. A hapten is a molecule too small to stimulate antibody formation by itself. However, when the hapten is combined with a larger carrier molecule, usually a serum protein, the hapten and its carrier together form a conjugate that can stimulate an immune response.

Q **How does a hapten differ from an antigen?**

immune response). In these people, when penicillin combines with host proteins, the resulting combined molecule initiates an immune response.

CHECK YOUR UNDERSTANDING

✓ Does an antibody necessarily react with a bacterium as an antigen or as an epitope? **17-3**

The Nature of Antibodies

Antibodies are globulin proteins (a protein family with a compact, globular form)—therefore, we have come to use the term **immunoglobulins (Ig)** for antibodies. Globulin proteins are relatively soluble. Antibodies are made in response to an antigen and can recognize and bind to the antigen. As was seen in Figure 17.1, a bacterium or virus may have several epitopes that cause the production of different antibodies.

Each antibody has at least two identical sites that bind to epitopes. These sites are known as **antigen-binding sites.** The number of antigen-binding sites on an antibody is called the **valence** of that antibody. For example, most human antibodies have two binding sites; therefore, they are bivalent.

Antibody Structure

Because a bivalent antibody has the simplest molecular structure, it is called a **monomer.** A typical antibody monomer has four protein chains: two identical *light chains* and two identical *heavy chains.* ("Light" and "heavy" refer to the relative molecular weights.) The chains are joined by disulfide links (see Figure 2.15c, page 46) and other bonds to form a Y-shaped molecule (Figure 17.3). The Y-shaped molecule is flexible and can assume a T shape (notice the hinge region in Figure 17.3a).

The two sections located at the ends of the Y's arms are called *variable (V) regions.* These bind to the epitopes (Figure 17.3b). The amino acid sequences and, therefore, the three-dimensional structure of these two variable regions are identical on any one antibody. Their structure reflects the nature of the antigen for which they are specific—they are specific to the two antigen-binding sites found on each antibody monomer. The stem of the antibody monomer and the lower parts of the arms of the Y are called the *constant (C) regions.* They are the same for a particular class of immunoglobulin. There are five major types of C regions, which account for the five major classes of immunoglobulins (described shortly).

The stem of the Y-shaped antibody monomer is called the *Fc region,* so named because when antibody structure was first being identified, it was a fragment (F) that crystallized (c) in cold storage.

These Fc regions are often important in immunological reactions. If left exposed after both antigen-binding sites attach to an antigen such as a bacterium, the Fc regions of adjacent antibodies can bind complement. This leads to the destruction of the bacterium (see Figure 16.10, page 465). Conversely, the Fc region may bind to a cell, leaving the antigen-binding sites of adjacent antibodies free to react with antigens (see Figure 19.1a, page 524).

Immunoglobulin Classes

The simplest and most abundant immunoglobulins are monomers, but they can also assume some differences in size and arrangement.

The five classes of Igs are designated IgG, IgM, IgA, IgD, and IgE. Each class has a different role in the immune response. The structures of IgG, IgD, and IgE molecules resemble the structure shown in Figure 17.3a. Molecules of IgA and IgM are aggregates of two or five monomers, respectively, that are joined together.

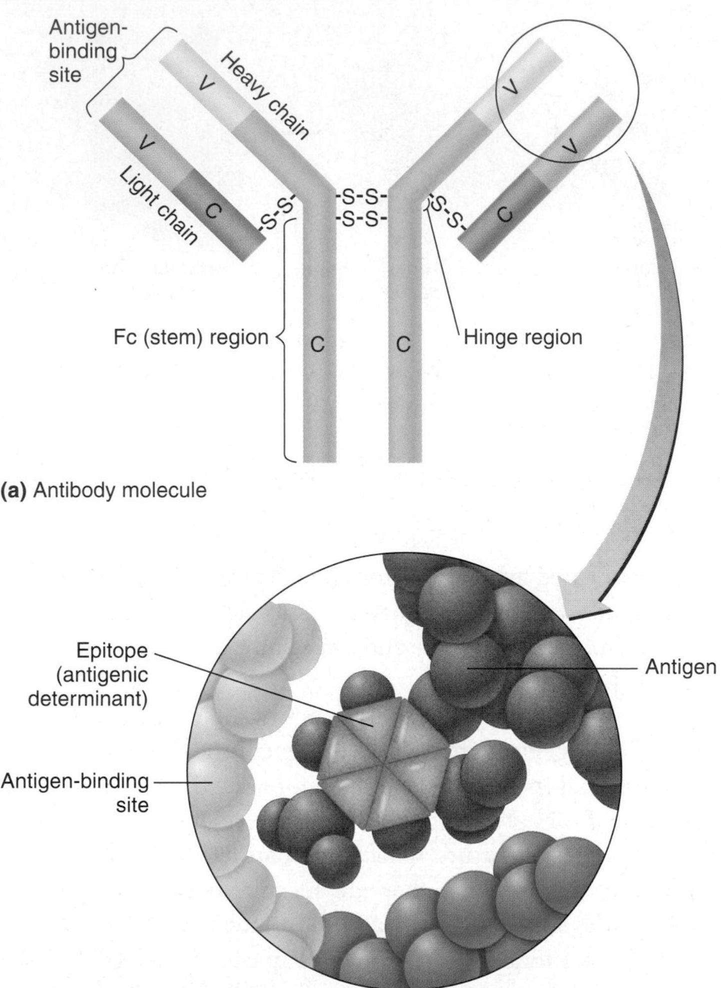

(a) Antibody molecule

(b) Enlarged antigen-binding site bound to an epitope

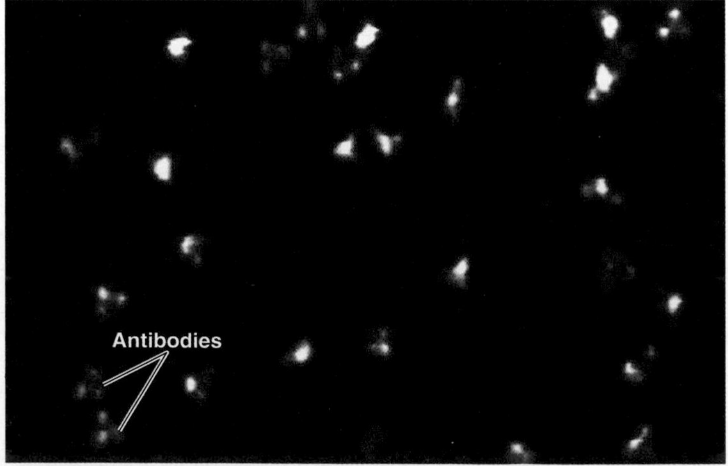

(c) Antibody molecules shown by atomic force microscopy (see page 65) AFM 5 nm

Figure 17.3 The structure of a typical antibody molecule. The Y-shaped molecule is composed of two light chains and two heavy chains linked by disulfide bridges (S—S). Most of the molecule is made up of constant regions (C), which are the same for all antibodies of the same class. The amino acid sequences of the variable regions (V), which form the two antigen-binding sites, differ from molecule to molecule.

Q What is responsible for the specificity of each different antibody?

The structures and characteristics of the immunoglobulin classes are summarized in **Table 17.1**.

IgG **IgG** (the name is derived from the blood fraction gamma globulin; see Figure 17.18, page 495) accounts for about 80% of all antibodies in serum. In regions of inflammation these monomer antibodies readily cross the walls of blood vessels and enter tissue fluids. Maternal IgG antibodies, for example, can cross the placenta and confer passive immunity to a fetus (see page 494). IgG antibodies protect against circulating bacteria and viruses, neutralize bacterial toxins, trigger the complement system, and, when bound to antigens, enhance the effectiveness of phagocytic cells.

IgM Antibodies of the **IgM** (from *macro,* reflecting their large size) class make up 5–10% of the antibodies in serum. IgM has a pentamer structure consisting of five monomers held together by a polypeptide called a *J (joining) chain* (see Table 17.1). The large size of the molecule prevents IgM from moving about as freely as IgG does, so IgM antibodies generally remain in blood vessels without entering the surrounding tissues.

IgM is the predominant type of antibody involved in the response to the ABO blood group antigens on the surface of red blood cells (see Table 19.2, page 527). It is much more effective than IgG in causing the clumping of cells and viruses (see the discussion of agglutination on page 484) and in reactions involving the activation of complement (see Figure 16.9 on page 464).

The fact that IgM appears first in response to a primary infection and is relatively short-lived makes it uniquely valuable in the diagnosis of disease. If high concentrations of IgM against a pathogen are detected in a patient, it is likely that the disease observed is caused by that pathogen. The detection of IgG, which is relatively long-lived, may indicate only that immunity against a particular pathogen was acquired in the more distant past.

IgA **IgA** accounts for only about 10–15% of the antibodies in serum, but it is by far the most common form in mucous membranes and in body secretions such as mucus, saliva, tears, and breast milk. If we take this into consideration, IgA is the most abundant immunoglobulin in the body (IgG is the most abundant in serum). The form of IgA that circulates in serum, *serum IgA,* is usually in the form of a monomer. The most effective form of IgA, however, consists of two connected monomers that form a *dimer* called *secretory IgA.* It is produced in this form by plasma cells in the mucous membranes. Each dimer then enters and passes through a mucosal cell, where it acquires a polypeptide called a *secretory component* that protects it from enzymatic degradation. The main function of secretory IgA is probably to prevent the attachment of microbial pathogens to mucosal surfaces. This is especially important in resistance to intestinal

Table 17.1 A Summary of Immunoglobulin Classes

Characteristics	IgG	IgM	IgA	IgD	IgE
		Disulfide bond; J chain	J chain; Secretory component		
Structure	Monomer	Pentamer	Dimer (with secretory component)	Monomer	Monomer
Percentage of Total Serum Antibody	80%	5–10%	10–15%*	0.2%	0.002%
Location	Blood, lymph, intestine	Blood, lymph, B cell surface (as monomer)	Secretions (tears, saliva, mucus, intestine, milk), blood, lymph	B cell surface, blood, lymph	Bound to mast and basophil cells throughout body, blood
Molecular Weight	150,000	970,000	405,000	175,000	190,000
Half-Life in Serum	23 days	5 days	6 days	3 days	2 days
Complement Fixation	Yes	Yes	No†	No	No
Placental Transfer	Yes	No	No	No	No
Known Functions	Enhances phagocytosis; neutralizes toxins and viruses; protects fetus and newborn	Especially effective against microorganisms and agglutinating antigens; first antibodies produced in response to initial infection	Localized protection on mucosal surfaces	Serum function not known; presence on B cells functions in initiation of immune response	Allergic reactions; possibly lysis of parasitic worms

*Percentage in serum only; if mucous membranes and body secretions are included, percentage is much higher.
†May be yes via alternative pathway.

and respiratory pathogens. Because IgA immunity is relatively short-lived, the length of immunity to many respiratory infections is correspondingly short. IgA's presence in a mother's milk, especially the colostrum (see page 494) probably helps protect infants from gastrointestinal infections.

IgD IgD antibodies make up only about 0.2% of the total serum antibodies. Their structure resembles that of IgG molecules. IgD antibodies are found in blood, lymph, and particularly on the surfaces of B cells (Figure 17.4). IgD has no well-defined function.

IgE Antibodies of the **IgE** class are slightly larger than IgG molecules; they constitute only 0.0002% of the total serum antibodies. IgE molecules bind tightly by their Fc (stem) regions to receptors on mast cells and basophils, specialized cells that participate in allergic reactions (see Chapter 19). When an antigen such as pollen cross-links with the IgE antibodies attached to a mast cell or basophil (see Figure 19.1a, page 524), that cell releases histamine and other chemical mediators. These chemicals provoke a response—for example, an allergic reaction such as hay fever. However, the response can be protective as well, for it attracts complement and phagocytic cells. This is especially useful when the antibodies bind to parasitic worms. The concentration of IgE is greatly increased during some allergic reactions and parasitic infections, which is often diagnostically useful.

CHECK YOUR UNDERSTANDING

- ✓ The original theoretical concepts of an antibody called for a rod with antigentic determinants at each end. What is the primary advantage of the Y-shaped structure that eventually emerged? **17-4**
- ✓ Which class of antibody is most likely to protect you from a common cold? **17-5**

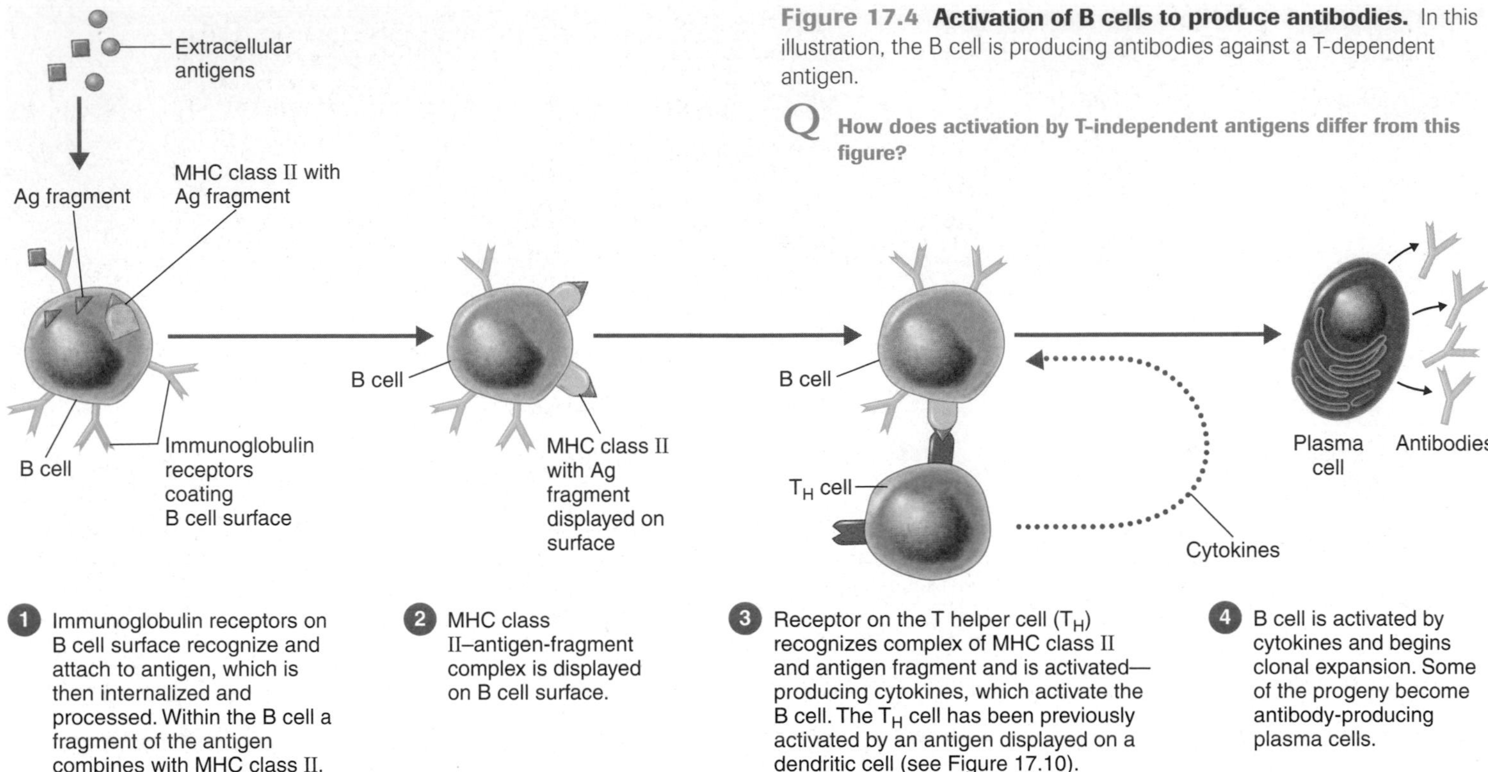

Extracellular antigens
Ag fragment
MHC class II with Ag fragment
B cell
Immunoglobulin receptors coating B cell surface
B cell
MHC class II with Ag fragment displayed on surface
B cell
T_H cell
Cytokines
Plasma cell
Antibodies
1 Immunoglobulin receptors on B cell surface recognize and attach to antigen, which is then internalized and processed. Within the B cell a fragment of the antigen combines with MHC class II.
2 MHC class II–antigen-fragment complex is displayed on B cell surface.
3 Receptor on the T helper cell (T_H) recognizes complex of MHC class II and antigen fragment and is activated—producing cytokines, which activate the B cell. The T_H cell has been previously activated by an antigen displayed on a dendritic cell (see Figure 17.10).
4 B cell is activated by cytokines and begins clonal expansion. Some of the progeny become antibody-producing plasma cells.

Figure 17.4 Activation of B cells to produce antibodies. In this illustration, the B cell is producing antibodies against a T-dependent antigen.

Q How does activation by T-independent antigens differ from this figure?

B Cells and Humoral Immunity

LEARNING OBJECTIVES

17-6 Compare and contrast T-dependent and T-independent antigens.

17-7 Differentiate plasma cell from memory cell.

17-8 Describe clonal selection.

17-9 Describe how a human can produce different antibodies.

As we have seen, the humoral (antibody-mediated) response is carried out by antibodies. Antibodies are produced by a special group of lymphocytes called B cells. The process that leads to the production of antibodies starts when B cells are exposed to *free*, or *extracellular, antigens.*

Clonal Selection of Antibody-Producing Cells

Each B cell carries immunoglobulins on its surface that are part of its makeup. The majority of the B cell's surface immunoglobulins are IgM and IgD—all of which are specific for the recognition of the same epitope. Ten percent or fewer of B cells carry other classes of immunoglobulins, but in certain locations their numbers may be high; for example, B cells in the intestinal mucosa are rich in IgA. B cells may carry at least 100,000 identical immunoglobulin molecules embedded in their surface membranes.

When a B cell's immunoglobulins bind to the epitope for which they become specific, the B cell is *activated.* An activated B cell undergoes *clonal expansion,* or proliferation. B cells usually require the assistance of a *T helper cell* (T_H) as shown in Figure 17.4. (T cells will be the subject of a detailed discussion later in this chapter). An antigen that requires a T_H cell for antibody production is known as a **T-dependent antigen.** T-dependent antigens are mainly proteins, such as those found on viruses, bacteria, foreign red blood cells, and haptens with their carrier molecules. For antibodies to be produced in response to a T-dependent antigen, it is necessary that both B and T cells be activated and interact. The process is initiated when the B cell contacts an antigen. It is important to note that the antigen contacts the surface immunoglobulins on the B cell and is enzymatically processed within the B cell and that fragments of it are combined with the **major histocompatibility complex (MHC).** The MHC is a collection of genes that encode molecules of genetically diverse glycoproteins (that is, part carbohydrate and part protein) that are found on the plasma membranes of mammalian nucleated cells. The study of the MHC originated when it was found that tissue rejection reflected the presence of cell surface molecules called histocompatibility antigens. Therefore, in humans the MHC is also called the human leuckocyte antigen (HLA) system, which is discussed further in Chapter 19, page 533. The combination of the antigenic fragments and the MHC are then displayed on the B cell's surface for the receptors on the T helper cells to identify. The MHC molecule identifies the host, and its use here prevents the immune system from making antibodies that would be harmful to the host. In this instance, the MHC is of class II, which is found only on the surface of *antigen-presenting cells (APCs)*—in this case, a B cell. We will encounter other APCs later in this chapter.

As shown in Figure 17.4, the T_H cell in contact with the antigenic fragment presented on the surface of the B cell becomes

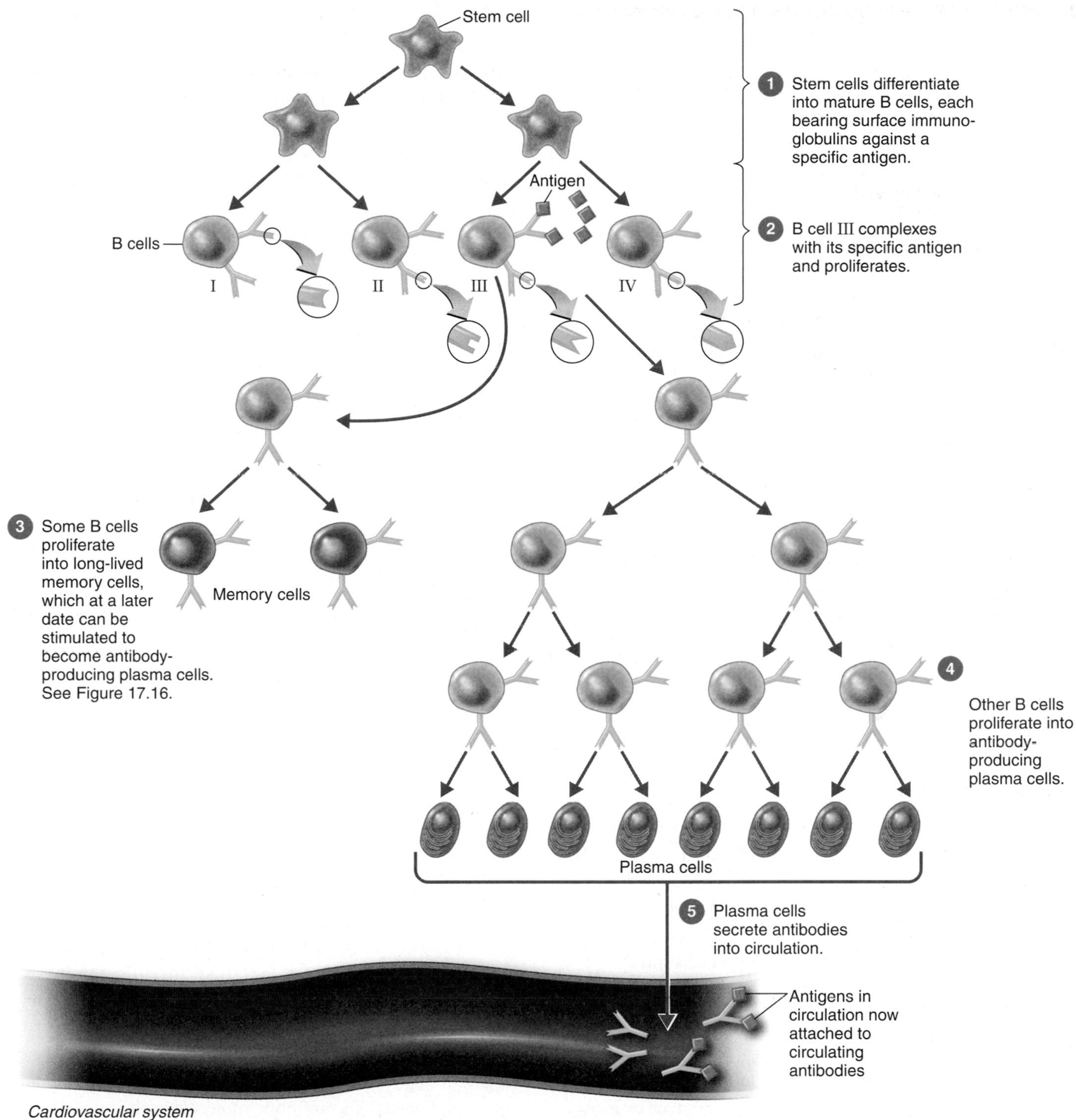
Stem cell
1 Stem cells differentiate into mature B cells, each bearing surface immunoglobulins against a specific antigen.
Antigen
B cells
I
II
III
IV
2 B cell III complexes with its specific antigen and proliferates.
3 Some B cells proliferate into long-lived memory cells, which at a later date can be stimulated to become antibody-producing plasma cells. See Figure 17.16.
Memory cells
4
Other B cells proliferate into antibody-producing plasma cells.
Plasma cells
5 Plasma cells secrete antibodies into circulation.
Antigens in circulation now attached to circulating antibodies
Cardiovascular system

Figure 17.5 Clonal selection and differentiation of B cells. B cells can recognize an almost infinite number of antigens, but each particular cell recognizes only one type of antigen. An encounter with a particular antigen triggers the proliferation of a cell that is specific for that antigen (here, B cell "III") into a clone of cells with the same specificity, hence the term *clonal selection.*

Q What caused cell "III" to respond?

activated and begins producing cytokines. These deliver a message that causes the activation of the B cell. An activated B cell proliferates into a large clone of cells, some of which will differentiate into antibody-producing **plasma cells.** Other clones of the activated B cell become long-lived **memory cells** that are responsible for the enhanced secondary response to an antigen (see Figure 17.16, page 494). This phenomenon, as shown in Figure 17.5, is called **clonal selection.** (A similar process occurs with T cells, as we will see later in the chapter.) The pool of B cells does not contain many that are harmfully reactive against

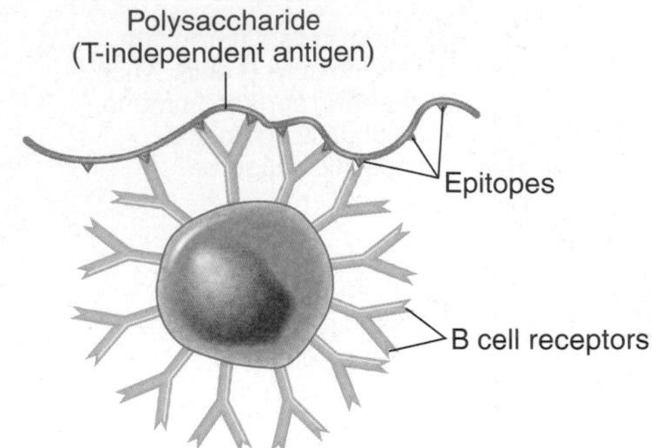

Figure 17.6 T-independent antigens. T-independent antigens have repeating units (epitopes) that can cross-link several antigen receptors on the some B cell. These antigens stimulate the B cell to make antibodies without the aid of T helper cells. The polysaccharides of bacterial capsules are examples of this type of antigen.

Q How can you differentiate T-dependent from T-independent antigens?

host tissue, or self. These are usually eliminated at the immature lymphocyte stage by the process of **clonal deletion.**

Antigens that stimulate B cells directly without the help of T cells are called **T-independent antigens.** Such antigens are characterized by repeating subunits such as are found in polysaccharides or lipopolysaccharides. Bacterial capsules are often good examples of T-independent antigens. The repeating subunits, as shown in Figure 17.6, can bind to multiple B cell receptors, which is probably why they do not require T cell assistance. T-independent antigens generally provoke a weaker immune response than do T-dependent antigens. This response is composed primarily of IgM, and no memory cells are generated. The immune system of infants may not be stimulated by T-independent antigens until about age 2. **Animations** Antigen Processing and Presentation: Overview; Humoral Immunity: Clonal Selection and Expansion. **www.microbiologyplace.com**

CHECK YOUR UNDERSTANDING

- ✓ Would pneumococcal pneumonia (see Figure 24.12, page 686) require a T_H cell to stimulate a B cell to form antibodies? **17-6**
- ✓ Plasma cells produce antibodies; do they also produce memory cells? **17-7**
- ✓ In what way does a B cell that encounters an antigen function as an antigen-presenting cell? **17-8**

The Diversity of Antibodies

The human immune system is capable of recognizing a mind-boggling number of different antigens—estimates are of a minimum of 10^{15} antigens. The number of genes required for this amount of diversity would seem to require a major part of an individual's inherited DNA. The work of the Japanese immunologist Susumu Tonegawa, for which he received a Nobel prize in 1987, showed how this diversity could be obtained by a set of just hundreds, not billions, of genes. Simplistically, the mechanism is analogous to the generation of huge numbers of words from a limited alphabet. This "alphabet" is found in the genetic makeup of the variable (V) region amino acid sequence of the immunoglobulin molecule, which can be linked to various amino acid sequences in the antibody's constant (C) region (see Figure 17.3). These combinations greatly reduce the amount of genetic information needed, so that a different gene is not needed to respond to each antigen.

CHECK YOUR UNDERSTANDING

- ✓ On what part of the antibody molecule do we find the amino acid sequence that makes the huge genetic diversity of antibody production possible? **17-9**

Antigen–Antibody Binding and Its Results

LEARNING OBJECTIVE

17-10 Describe four outcomes of an antigen–antibody reaction.

When an antibody encounters an antigen for which it is specific, an **antigen–antibody complex** rapidly forms. An antibody binds to an antigen such as a bacterium at a specific portion called the *epitope*, or *antigenic determinant* (see Figure 17.1).

The strength of the bond between an antigen and an antibody is called **affinity.** In general, the closer the physical fit between antigen and antibody, the higher the affinity. Antibodies tend to recognize the shape of the antigen's epitope. They also exhibit a capability for **specificity** that is remarkable. They can distinguish between minor differences in the amino acid sequence of a protein and even between two isomers (see Figure 2.13, page 43). Therefore, antibodies can be used to differentiate between the viruses of chickenpox and measles and between bacteria of different species, for example.

The binding of an antibody to an antigen protects the host by tagging foreign cells and molecules for destruction by phagocytes and complement. The antibody molecule itself is not damaging to the antigen. Foreign organisms and toxins are rendered harmless by only a few mechanisms, as summarized in Figure 17.7. These are agglutination, opsonization, neutralization, antibody-dependent cell-mediated cytotoxicity, and the activation of complement leading to inflammation and cell lysis (see Figure 16.10, page 465).

In **agglutination,** antibodies cause antigens to clump together. For example, the two antigen-binding sites of an IgG antibody can combine with epitopes on two different foreign cells, aggregating the cells into clumps that are more easily ingested by phagocytes. Because of its more numerous binding sites, IgM is more effective at cross-linking and aggregating particulate antigens (see Figure 18.5, page 510). IgG requires 100 to 1000 times

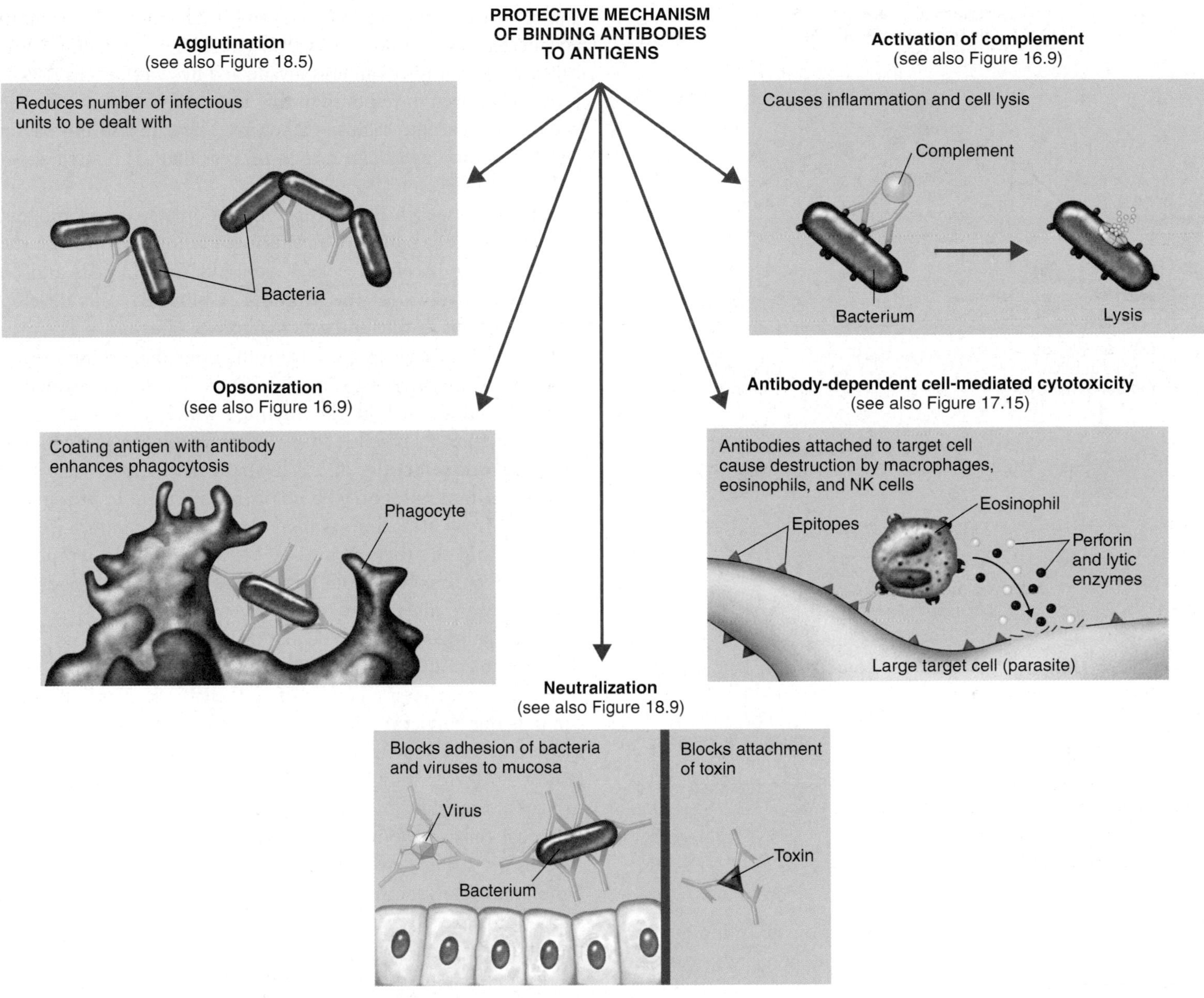

Figure 17.7 The results of antigen–antibody binding. The binding of antibodies to antigens to form antigen–antibody complexes tags foreign cells and molecules for destruction by phagocytes and complement.

Q What are some of the possible outcomes of an antigen–antibody reaction?

as many molecules for the same results. (In Chapter 18, we will see how agglutination is important in the diagnosis of some diseases.)

For **opsonization** (from the Greek *opsonare,* meaning to cater), the antigen, such as a bacterium, is coated with antibodies that enhance its ingestion and lysis by phagocytic cells. **Antibody-dependent cell-mediated cytotoxicity** (see page 491 and Figure 17.15) resembles opsonization in that the target organism becomes coated with antibodies; however, destruction of the target cell is by immune system cells that remain external to the target cell.

In **neutralization,** IgG antibodies inactivate microbes by blocking their attachment to host cells, and they neutralize toxins in a similar manner.

Finally, either IgG or IgM antibodies may trigger **activation of the complement system.** For example, inflammation is caused by infection or tissue injury (see Figure 16.8, page 461). One aspect of inflammation is that it will often cause microbes in the inflamed area to become coated with certain proteins. This, in turn, leads to the attachment to the microbe of an antibody–complement complex. This complex lyses the microbe, which then attracts phagocytes and other defensive immune system cells to the area.

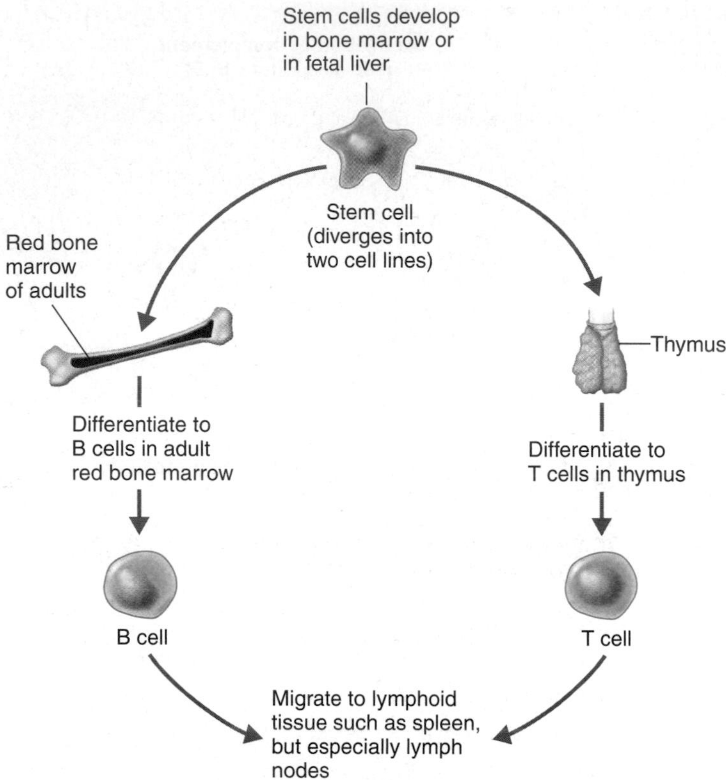
Stem cells develop in bone marrow or in fetal liver
Stem cell (diverges into two cell lines)
Red bone marrow of adults
Thymus
Differentiate to B cells in adult red bone marrow
Differentiate to T cells in thymus
B cell
T cell
Migrate to lymphoid tissue such as spleen, but especially lymph nodes

Figure 17.8 Differentiation of T cells and B cells. Both B cells and T cells originate from stem cells in adult red bone marrow or in the fetal liver. (Red blood cells, macrophages, neutrophils, and other white blood cells also originate from these same stem cells.) Some cells pass through the thymus and emerge as mature T cells. Other cells probably remain in the red bone marrow and become B cells. Both types of cells then migrate to lymphoid tissues, such as the lymph nodes or spleen.

Q Which cells, T or B, make antibodies?

As you will see in Chapter 19, the action of antibodies can be harmful. **Animation** Humoral Immunity: Antibody Function. **www.microbiologyplace.com**

CHECK YOUR UNDERSTANDING

✓ Antibodies and what other component of the immune system are required for the lysis of a target antigenic cell? **17-10**

T Cells and Cellular Immunity

LEARNING OBJECTIVES

17-11 Describe at least one function of each of the following: M cells, T_H1 cells, T_H2 cells, T_C cells, T_{reg} cells, CTLs, NK cells.

17-12 Differentiate T helper, T cytotoxic, and T regulatory cells.

17-13 Differentiate T_H1 and T_H2 cells.

17-14 Define *apoptosis.*

Humoral antibodies are effective against pathogens such as viruses and bacteria that are circulating freely, where the antibodies can contact them. Intracellular antigens, such as a virus within an infected cell, are not exposed to circulating antibodies. Some bacteria and parasites can also invade and live within cells. T cells probably evolved in response to this aspect of pathogenicity—the need to combat intracellular pathogens. They are also the way in which the immune system recognizes cells that are nonself, especially cancer cells.

Like B cells, each T cell is specific for only a certain antigen. Rather than the coating of immunoglobulins that provide the specificity for B cells, T cells have TCRs. Like B cells and all other cells involved in the immune response, T cells develop from stem cells in the red bone marrow (**Figure 17.8**). The precursors of the T cells migrate from the bone marrow and reach maturity in the thymus. Most immature T cells, an estimated 98%, are eliminated in the thymus, which is analogous to clonal deletion in B cells. This reflects a weeding-out process, called **thymic selection,** of T cells that will not specifically recognize self-molecules of MHC. This is important in preventing the body from attacking its own tissues. Next, mature T cells migrate from the thymus by way of the blood and lymphatic system to various lymphoid tissues (see Figure 16.5, page 456) where they are most likely to encounter antigens.

Most pathogens of the type that the cellular immune system is designed to combat first enter the gastrointestinal tract or lungs, where they encounter a barrier of epithelial cells. Normally, they can pass this barrier in the gastrointestinal tract only by way of a scattered array of gateway cells called **microfold cells,** or **M cells** (**Figure 17.9**; see also Figure 25.7, page 712). (Instead of the myriad of fingerlike microvilli found on the surface of absorptive epithelial cells of the intestinal tract, M cells have microfolds). M cells are located over **Peyer's patches,** which are secondary lymphoid organs located on the intestinal wall. M cells are well adapted to take up antigens from the intestinal tract and allow their transfer to the lymphocytes and antigen-presenting cells of the immune system found throughout the intestinal tract, just under the epithelial-cell layer but especially in the Peyer's patches. It is also here that antibodies, mostly IgA essential for mucosal immunity, are formed and migrate to the mucosal lining.

The recognition of antigens by a T cell requires that they be first processed by specialized **antigen-presenting cells (APCs).** This resembles the situation previously discussed in humoral immunity in which a B cell served as the APC (see Figure 17.4). After processing, an antigenic fragment is presented on the APC surface together with a molecule of the MHC. APCs are described fully on page 489; they include activated macrophages and, most important, dendritic cells.

The body's ability to make new T cells decreases with age, beginning in late adolescence. Eventually, the T-cell-producing thymus becomes less active, and red bone marrow produces fewer B cells. As a result, the immune system is relatively weak in older adults. However, sufficient long-lived T and B memory cells survive to make immunization of older adults effective for such diseases as influenza and pneumococcal pneumonia. **Animations** Cell-Mediated Immunity: Overview, Helper T Cells. **www.microbiologyplace.com**

CHECK YOUR UNDERSTANDING

✓ What antibody is the primary one produced when an antigen is taken up by an M cell? **17-11**

Classes of T Cells

There are classes of T cells that have different functions, rather like the classes of immunoglobulins. For example, T helper cells cooperate with B cells in the production of antibodies, mainly through cytokine signaling (see Figure 17.4). Therefore, T helper cells are an important part of humoral immunity; they are an even more essential element of cellular immunity. In their contributions to cellular immunity, T cells do not contribute to the production of antibodies but interact more directly with antigens. Primarily, the two populations of T cells that concern us here are **T helper cells (T_H)** cells and **T cytotoxic cells (T_C).** A T_C cell can differentiate into an effector cell called a **cytotoxic T lymphocyte (CTL).**

T cells are also classified by certain glycoproteins on their surface called **clusters of differentiation,** or **CD.** These are membrane molecules that are especially important for adhesion to receptors. The CDs of greatest interest are CD4 and CD8; cells that carry these molecules are named **$CD4^+$** and **$CD8^+$** cells, respectively. (For the importance of these molecules in HIV infection, see Figure 19.13 on page 540). T_H cells are classified as $CD4^+$, which bind to MHC class II molecules on B cells and APCs (Figures 17.4 and 17.10). T_C cells are classified as $CD8^+$, which bind to MHC class I molecules (Figure 17.11 and page 488).

T Helper Cells ($CD4^+$ T Cells)

We have seen that an essential part of the body's innate defenses is phagocytosis by cells such as macrophages. Macrophages, when functioning as APCs, also are important in adaptive cellular immunity. T_H cells can recognize an antigen presented on the surface of a macrophage and activate the macrophage, making it more effective in both phagocytosis and in antigen presentation. Even more important as APCs are dendritic cells. (See page 490.) Dendritic cells are especially important in the activation of $CD4^+$ T cells and in developing their effector functions (**Figure 17.10**).

For a $CD4^+$ T cell to become activated, its TCR recognizes an antigen that has been processed and is presented as fragments held in a complex with proteins of MHC class II on the surface of the APC. This is the initial signal for activation; a second signal, the costimulatory signal, which is present on the APC and the T_H cell, is also required. Because activation should be directed against harmful pathogens, the displayed antigenic fragments should include *Toll-like receptors* (see Chapter 16, page 450), which signal a dangerous microbe. The activated T_H cell begins to proliferate at the rate of two to three cell cycles a day and to secrete cytokines, which are essential for its effector functions. The proliferating T_H cell differentiates into populations of T_H1 and T_H2 cells; it also forms a population of memory cells.

The cytokines produced by **T_H1 cells,** especially IFN-γ, activate mostly those cells related to important elements of cellular immunity, such as delayed-type hypersensitivity (see page 529), and are also responsible for activation of macrophages (see page 490). They also stimulate the production of antibodies that

(a) M cell on Peyer's patch. Note the tips of the closely packed microvilli on the surrounding epithelial cells.

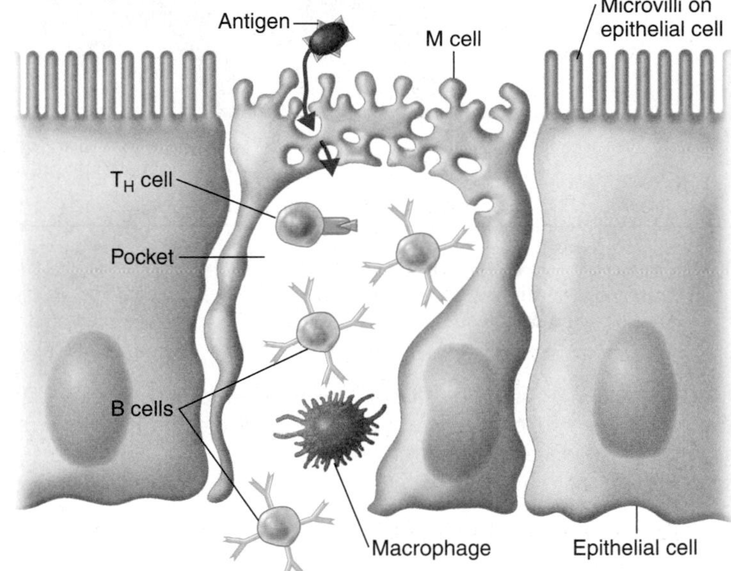

(b) M cells facilitate contact between antigens passing through the intestinal tract and cells of the body's immune system.

Figure 17.9 M cells. M cells are located within Peyer's patches (see Figure 16.5, page 456), which are located on the intestinal wall. Their function is to transport antigens encountered in the digestive tract to contact lymphocytes and antigen-presenting cells (see page 488) of the immune system.

Q Why are M cells especially important for immune defenses against diseases affecting the digestive system?

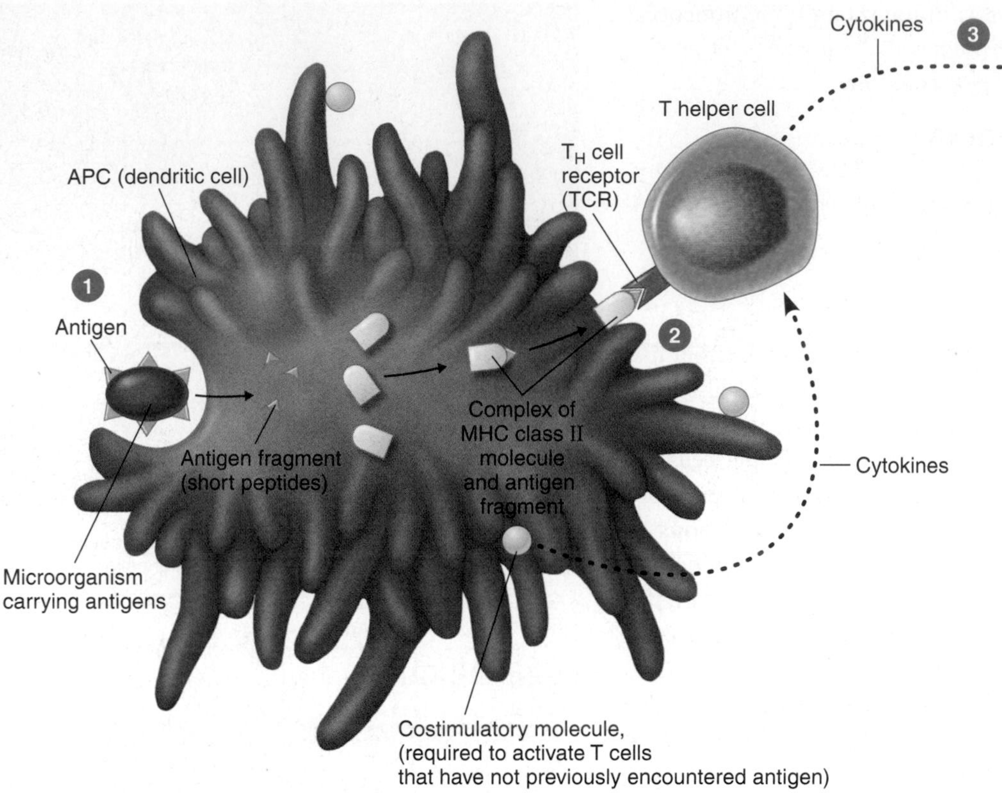

1 An APC encounters and ingests a microorganism. The antigen is enzymatically processed into short peptides, which combine with MHC class II molecules and are displayed on the surface of the APC.

2 A receptor (TCR) on the surface of the $CD4^+$ T helper cell (T_H cell) binds to the MHC–antigen complex. If this includes a Toll-like receptor, the APC is stimulated to secrete a costimulatory molecule. These two signals activate the T_H cell, which produces cytokines.

3 The cytokines cause the T_H cell to proliferate and to develop its effector functions.

Figure 17.10 Activation of $CD4^+$ T helper cells. To activate a $CD4^+$ T helper cell, at least two signals are required: the first signal is the binding of the TCR to the processed antigen, and the second signal requires a costimulating cytokine, such as IL-2 and others. Once activated, the T_H cell secretes cytokines that affect the effector functions of multiple cell types of the immune system.

Q What is the role of the macrophage?

promote phagocytosis and are especially effective in enhancing the activity of complement, such as opsonization and inflammation (see Figure 16.9, page 464). As shown in **Figure 17.11**, the generation of cytotoxic T lymphocytes also requires action by a T_H1 cell.

T_H2 cells produce cytokines that are associated primarily with the production of antibodies, especially IgE, that are important in allergic reactions (see the discussion of hypersensitivity on page 523). They are also important in the activation of the eosinophils that defend against infections by extracellular parasites such as helminths (see Figure 17.15 on page 491). **Animation:** Antigen Processing and Presentation: Steps. **www.microbiologyplace.com**

T Cytotoxic Cells ($CD8^+$ T cells)

T cytotoxic cells, despite their name, are not capable of attacking any target cell as they emerge from the thymus; rather, they are precursors to CTLs, which do have this capability. This differentiation requires sequential, and complex, activation of the precursor T_C by an antigen processed by a dendritic cell and interaction with a T_H cell and costimulatory signals. The resulting CTL is an effector cell that has the ability to recognize and kill target cells that are considered nonself (see Figure 17.11). Primarily, these target cells are self-cells that have been altered by infection with a pathogen, especially viruses. On their surface they carry fragments of **endogenous antigens** that are generally synthesized within the cell and are mostly of viral or parasitic origin. Other important target cells are tumor cells (see Figure 19.10, page 535) and transplanted foreign tissue. Rather than reacting with antigenic fragments presented by an APC in complex with MHC class II molecules, the $CD8^+$ T cell recognizes endogenous antigens on the target cell's surface that are in combination with an MHC class I molecule. MHC class I molecules are found on nucleated cells; therefore, a CTL can attack almost any cell of the host that has been altered.

In its attack, a CTL attaches to the target cell and releases a pore-forming protein, **perforin.** Pore formation contributes to the subsequent death of the cell and is similar to the action of the complement membrane attack complex described in Chapter 16

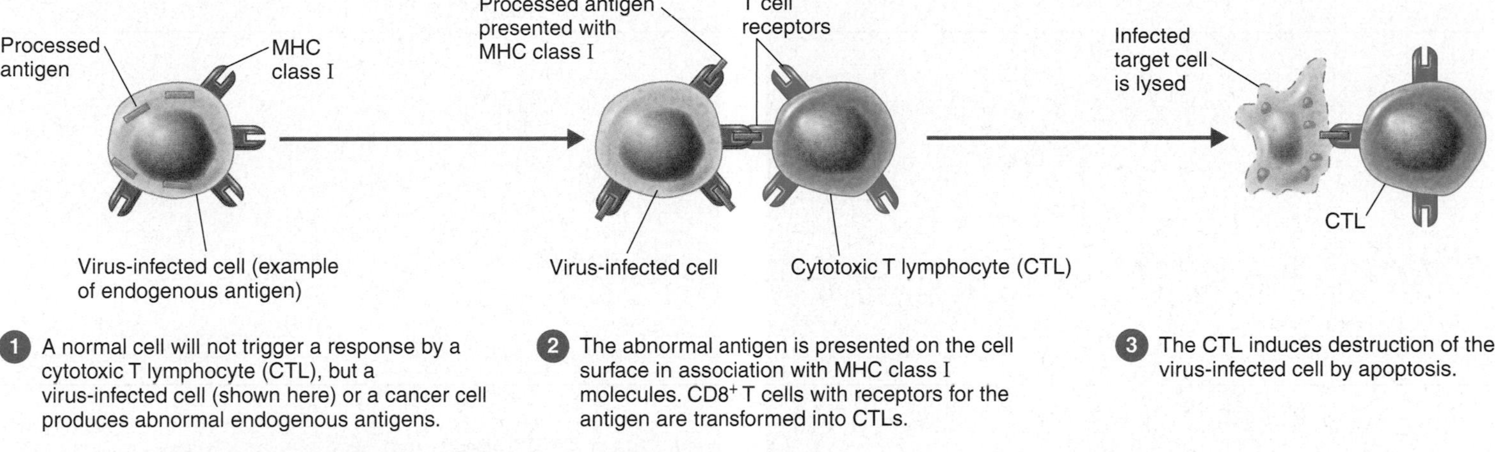
Processed antigen
MHC class I
Virus-infected cell (example of endogenous antigen)
Processed antigen presented with MHC class I
T cell receptors
Virus-infected cell
Cytotoxic T lymphocyte (CTL)
Infected target cell is lysed
CTL
1 A normal cell will not trigger a response by a cytotoxic T lymphocyte (CTL), but a virus-infected cell (shown here) or a cancer cell produces abnormal endogenous antigens.
2 The abnormal antigen is presented on the cell surface in association with MHC class I molecules. CD8+ T cells with receptors for the antigen are transformed into CTLs.
3 The CTL induces destruction of the virus-infected cell by apoptosis.

Figure 17.11 Killing of virus-infected target cell by cytotoxic T lymphocyte.

Q Differentiate a $CD8^+$ T cell from a $CD4^+$ T cell.

(see page 465). **Granzymes,** proteases that induce apoptosis, are then able to enter through the pore. **Apoptosis** (from the Greek for falling away like leaves) is also called *programmed cell death.* Cells that die from apoptosis first cut their genome into fragments, and the external membranes bulge outward in a manner called *blebbing* (Figure 17.12). Signals are displayed on the cell's surface that attract circulating phagocytes to digest the remains before any significant leakage of contents occurs. This has the advantage of preventing the spread of infectious viruses into other cells. Also, after apoptosis the immune response is rapidly ended, which lessens nonspecific damage to tissue.

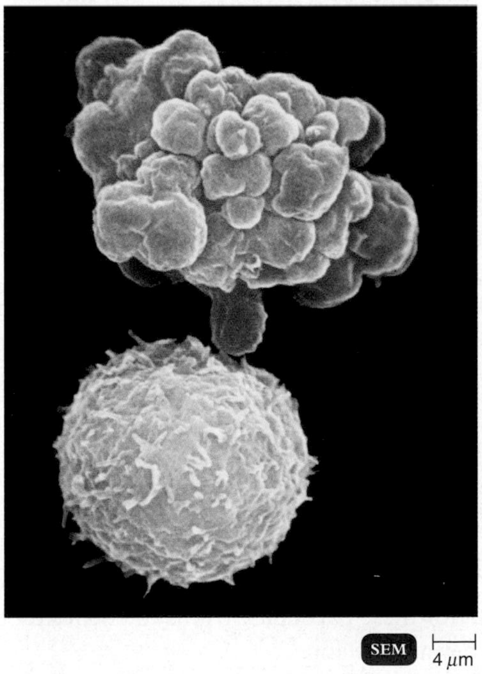

SEM 4 μm

Figure 17.12 Apoptosis. A normal B cell is shown at the bottom of the photo. A B cell undergoing apoptosis is at the top. Notice the bubble-like blebs.

Q What is apoptosis?

T Regulatory Cells

T regulatory cells (T_{reg}), formerly called *T suppressor cells,* make up about 5–10% of the T cell population. They are a subset of the $CD4^+$ T helper cells and are distinguished by carrying an additional CD25 molecule. Their primary function is to combat autoimmunity by suppressing T cells that escape deletion in the thymus without the necessary "education" to avoid reacting against the body's self. They are also useful in protecting, from the immune system, the intestinal bacteria required for digestion and other useful functions. Similarly, in pregnancy they may play a role in protecting the fetus from rejection as nonself. **Animation** Cell-Mediated Immunity: Cytotoxic T Cells. **www.microbiologyplace.com**

CHECK YOUR UNDERSTANDING

- ✓ Which T cell type is generally involved when a B cell reacts with an antigen and produces antibodies against the antigen? **17-12**
- ✓ Which is the T cell type that is generally involved in allergic reactions? **17-13**
- ✓ What is another name for apoptosis, one that describes its function? **17-14**

Antigen-Presenting Cells (APCs)

LEARNING OBJECTIVE

17-15 Define *antigen-presenting cell.*

Although B cells are a form of antigen-presenting cell (APC) that we have already discussed with humoral immunity, we will now consider other APCs associated with cellular immunity.

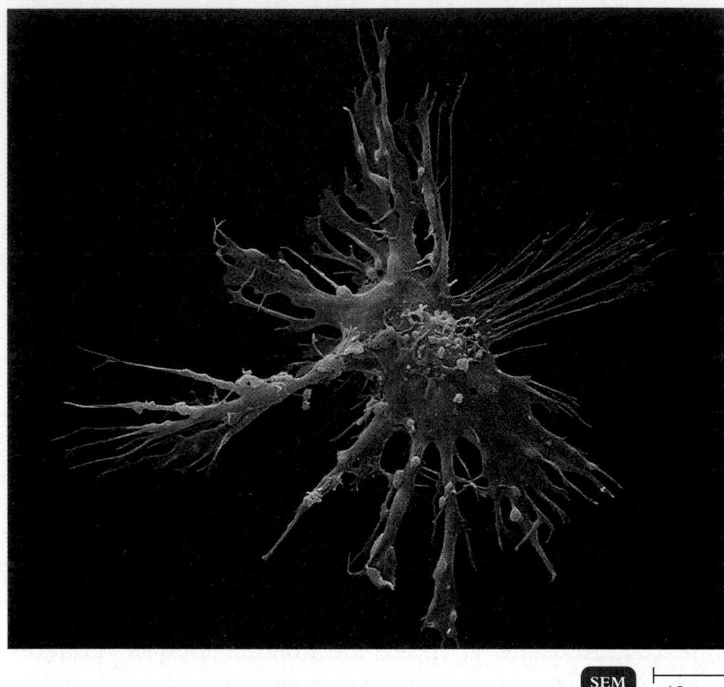

Figure 17.13 A dendritic cell. These antigen-presenting cells are named for their long arms, or dendrites. They are especially plentiful in the skin and mucous membranes.

Q What is the role of dendritic cells in immunity?

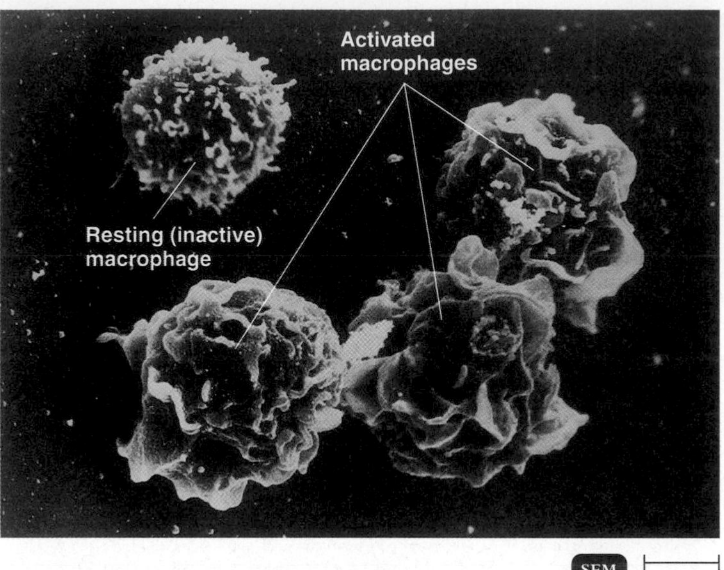

Figure 17.14 Activated macrophages. When activated, macrophages become larger and ruffled.

Q How do macrophages become activated?

These APC's are the dendritic cells and the activated macrophages. **Animation** Antigen Processing and Presentation: MHC. **www.microbiologyplace.com**

Dendritic Cells

Dendritic cells (DCs) are characterized by long extensions called dendrites (Figure 17.13) because they resemble the dendrites of nerve cells. They were first identified in anatomical studies of the skin by Langerhans in 1868, and the dendritic cells in the skin and genital tract are still called *Langerhans cells*, or *Langerhans DC*. (Vaccines injected between skin layers, where there are more such dendritic cells, are often more effective than injections into muscle). This represents only one of at least four populations of DCs named for their derivation or location. Other populations are found in the lymph nodes, spleen, thymus, blood, and various tissues, except the brain. Dendritic cells that act as sentinels in these tissues engulf invading microbes, degrade them, and transfer them to lymph nodes for display to T cells located there. Dendritic cells are the principal APCs to induce immune responses by T cells. Macrophages, although more efficient for phagocytosis, are less efficient in antigen presentation for T cells in cellular immunity, but they play a key role in the later stages of the adaptive response.

Macrophages

Macrophages (from the Greek for large eaters) are cells usually found in a resting state. We have already discussed the function of macrophages in phagocytosis. They are important for innate immunity and for ridding the body of worn out blood cells and other debris, such as cellular remnants from apoptosis. Their phagocytic capabilities are greatly increased when they are stimulated to become **activated macrophages** (Figure 17.14). This activation can be initiated by ingestion of antigenic material. Other stimuli, such as cytokines produced by an activated helper T cell, can further enhance their capabilities. Once activated, macrophages are more effective as phagocytes and as APCs. Activated macrophages are important factors in the control of cancer cells and such intracellular pathogens as the tubercle bacillus and virus-infected cells. Their appearance becomes recognizably different as well—they are larger and become ruffled.

After taking up an antigen, APCs tend to migrate to lymph nodes or other lymphoid centers on the mucosa, where they present the antigen to T cells located there. T cells carrying receptors that are capable of binding with any specific antigen are present in relatively limited numbers. Migration increases the opportunity for these particular T cells to encounter the antigen for which they are specific.

CHECK YOUR UNDERSTANDING

✓ Are dendritic cells considered primarily part of the humoral or the cellular immune system? **17-15**

Extracellular Killing by the Immune System

LEARNING OBJECTIVE

17-16 Describe the function of natural killer cells.

We have seen how the action of a CTL can lead to the destruction of a target cell. A component of the innate immune system that has not yet been discussed can also destroy certain virus-infected cells and tumor cells. These are granular leukocytes (10–15% of circulating lymphocytes) called **natural killer (NK) cells.** They can also attack parasites, which are normally much larger than bacteria, as illustrated in Figure 17.15. In contrast to CTLs, NK cells are not immunologically specific; that is, they do not need to be stimulated by an antigen. They first contact the target cell and determine if it expresses MHC class I self-antigens. If it does not—which is often true in the early stages of viral infection and with some infecting viruses that have developed a system of interfering with the usual presentation of antigens on an APC—they kill the target cell by mechanisms similar to that of a CTL. Tumor cells also have a reduced number of MHC class I molecules on their surfaces. NK cells cause pores to form in the target cell, which leads to either lysis or apoptosis.

The functions of NK cells and the other principal cells involved in cellular immunity are briefly summarized in Table 17.2.

CHECK YOUR UNDERSTANDING

✓ How does the natural killer cell respond if the target cell does not have MHC class I molecules on its surface? **17-16**

Antibody-Dependent Cell-Mediated Cytotoxicity

LEARNING OBJECTIVE

17-17 Describe the role of antibodies and natural killer cells in antibody-dependent cell-mediated cytotoxicity.

With the help of antibodies produced by the humoral immune system, the cell-mediated immune system can stimulate natural killer cells (see page 454) and cells of the innate defense system, such as macrophages, to kill targeted cells. In this way, an organism such as a protozoan or a helminth that is too large to be phagocytized can be attacked by immune system cells. This is referred to as **antibody-dependent cell-mediated cytotoxicity (ADCC).** As is illustrated in Figure 17.15, the target cell is first coated with antibodies. A variety of cells of the immune system bind to the F_C regions of these antibodies, and thus to the target cell. The target cell is then lysed by substances secreted by the attacking cells.

Table 17.2 Principal Cells That Function in Cell-Mediated Immunity

Cell	Function
T Helper (T_H1) Cell	Activates cells related to cell-mediated immunity: macrophages, T_c cells, and natural killer cells
T Helper (T_H2) Cell	Stimulates production of eosinophils, IgM, and IgE
Cytotoxic T Lymphocyte (CTL)	Destroys target cells on contact; generated from T cytotoxic (T_C) cell
T Regulatory (T_{reg}) cell	Regulates immune response and helps maintain tolerance
Activated Macrophage	Enhanced phagocytic activity; attacks cancer cells
Natural Killer (NK) Cell	Attacks and destroys target cells; participates in antibody-dependent cell-mediated cytotoxicity

CHECK YOUR UNDERSTANDING

✓ What makes a natural killer cell, which is not immunologically specific, attack a particular target cell? **17-17**

Cytokines: Chemical Messengers of Immune Cells

LEARNING OBJECTIVE

17-18 Identify at least one function of each of the following: cytokines, interleukins, chemokines, interferons, TNF, and hematopoietic cytokines.

The immune response requires complex interactions between different cells. The communication required for this is mediated by chemical messengers called **cytokines** (from the Greek for cell and to move). These are soluble proteins or glycoproteins that are produced by practically all cells of the immune system in response to a stimulus. Many cytokines—there are probably more than 200—have common names that reflect their functions known at the time of their discovery; some are now known to have multiple functions. A cytokine acts only on a cell that has a receptor for it.

Cytokines that serve as communicators between leukocytes (white blood cells) are now known as **interleukins** (between leukocytes). When enough information, including the amino acid sequence, is known, these cytokines are assigned an interleukin number, such as IL-1, and so on, by an international committee.

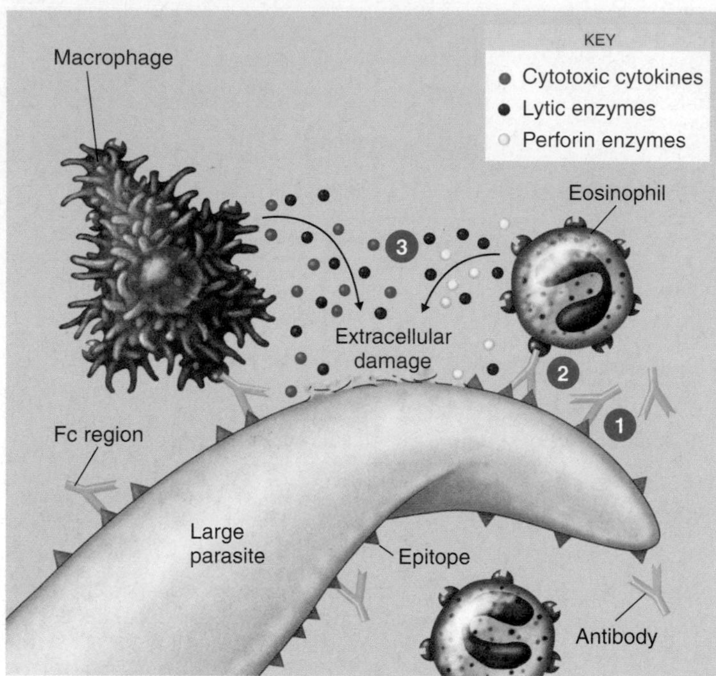

(a) Organisms, such as many parasites, that are too large for ingestion by phagocytic cells must be attacked externally.

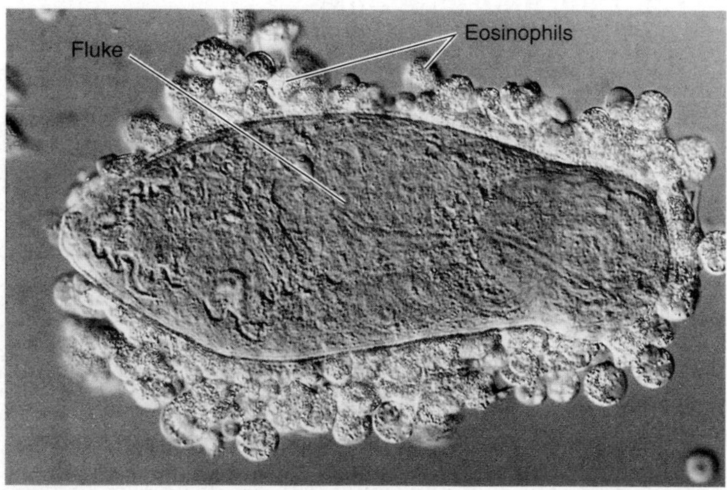

(b) Eosinophils adhering to the larval stage of a parasitic fluke SEM 20 μm

Figure 17.15 Antibody-dependent cell-mediated cytotoxicity (ADCC). If an organism, such as a parasitic worm, is too large for ingestion and destruction by phagocytosis, it can be attacked by immune system cells that remain external to it. 1 The target cell is first coated with antibodies. 2 Cells of the immune system, such as eosinophils, macrophages, and NK cells (not shown), bind to the Fc regions of the attached antibodies. 3 The target cell is then lysed by substances secreted by the cells of the immune system.

Q Why is ADCC important protection against parasitic protozoa and helminths?

The role of cytokines in stimulating the immune system has suggested their use as therapeutic agents (see the box on the facing page.)

A family of small cytokines that induces the migration of leukocytes into areas of infection or tissue damage is called **chemokines,** from *chemotaxis.* They are especially important in inflammation. Certain chemokine receptors are important for infection by HIV (see Chapter 19, page 541).

Another family of cytokines is the **interferons** (see page 468), originally named for one of their functions, protecting cells from viral infection. They are assigned designations such as IFN-α, and so on (see page 469). A number of these are available as commercial products in treating disease conditions such as hepatitis and some cancers.

A very important cytokine family is that of **tumor necrosis factor,** known by abbreviations such as TNF-α, and so on. TNF was originally named because tumor cells were observed to be one of its targets. These cytokines are a strong factor in inflammatory reactions of autoimmune diseases such as rheumatoid arthritis. Monoclonal antibodies (see page 507) that block the action of TNF are an available therapy for some of these conditions.

A family of cytokines, **hematopoietic cytokines,** function in controlling the pathways by which stem cells develop into different red or white blood cells (see page 486). Some of these are interleukins with designations such as IL-3, and so on (some are used therapeutically, as described in the box on page 493); others are termed *colony stimulating factors (CSF).* An example is G-CSF (granulocyte-colony stimulating factor). This particular CSF stimulates the production of neutrophils from their granulocyte/monocyte precursors. Another, GM-CSF, is used therapeutically to increase the numbers of protective macrophages and granulocytes in patients undergoing red bone marrow transplants.

Cytokines, among other things, may stimulate cells to produce more cytokines. This feedback loop occasionally gets out of control, resulting in a harmful overproduction of cytokines—a **cytokine storm.** These can do significant damage to tissues, which appears to be a factor in the pathology of certain diseases and conditions such as influenza, graft-versus-host disease (page 536), and sepsis. Also, see the discussion of superantigens on page 436.

Q&A You have now learned that there are several mechanisms at work to prevent the adaptive immune system from attacking so-called self. These include the deletion of immune system cells that fail to recognize the host's own tissues. They also include the major histocompatibility complex system, such as found on antigen-presenting cells (see Figure 17.11), that must combine with foreign antigens before they stimulate humoral or cellular immune reactions. There are still questions

Interleukin-12: The Next "Magic Bullet"?

Worldwide, HIV/AIDS-related illnesses kill 3 million people, and measles kills 1 million people annually. If laboratory tests are any indication of the future, the cytokine IL-12 (interleukin-12) could be the "magic bullet" against cancer and a host of other diseases.

Since its discovery in the 1980s, IL-12 proved to be different from the other cytokines. It inhibits the humoral response and activates T_H1 cellular immunity (see Figure A).

Scientists at the National Institute of Allergy and Infectious Diseases (NIAID) have found that treating mice with IL-12 can help activate phagocytes to kill *Histoplasma* fungi, *Leishmania*, *Cryptosporidium*, and *Toxoplasma gondii* protozoa, as well as *Mycobacterium avium.* The last three are common opportunistic infections in people with late-stage AIDS. The NIAID recently began recruiting people with both AIDS and *Mycobacterium avium* complex into a clinical trial to test IL-12.

Researchers have shown that HIV and measles virus decrease the production of IL-12, which may make the patient more susceptible to secondary infections. When treated with IL-12, however, T_H cells taken from HIV-positive people responded to viruses, including HIV.

Interleukin-12 is known to inhibit about 20 kinds of tumors in mice by inhibiting blood vessel growth to tumors. Several clinical trials are in progress to test the effectiveness of IL-12 in patients with advanced kidney cancer.

Because IL-12 activates the T_H1 pathway, it can cause the symptoms associated with chronic inflammatory diseases, including Crohn's disease, psoriasis, rheumatoid arthritis, and multiple sclerosis. NIAID researchers developed the concept of blocking IL-12 in patients with these diseases. There are two approaches to blocking IL-12. One approach uses RNAi to block transcription, thus preventing production and secretion of IL-12. The other approach, which uses monoclonal antibodies that bind secreted IL-12, was tested on more than 300 psoriasis patients in 2007. Patients showed dramatic improvements after treatment with IL-12 monoclonal antibodies compared to the placebo-treated patients.

Will IL-12 be a panacea? Further studies are needed to determine whether treatment with IL-12 could have adverse effects, such as autoimmune disease, or whether blocking IL-12 could allow growth of cancer cells.

Figure A Interleukin-12 leads to a T_H1 cell response.

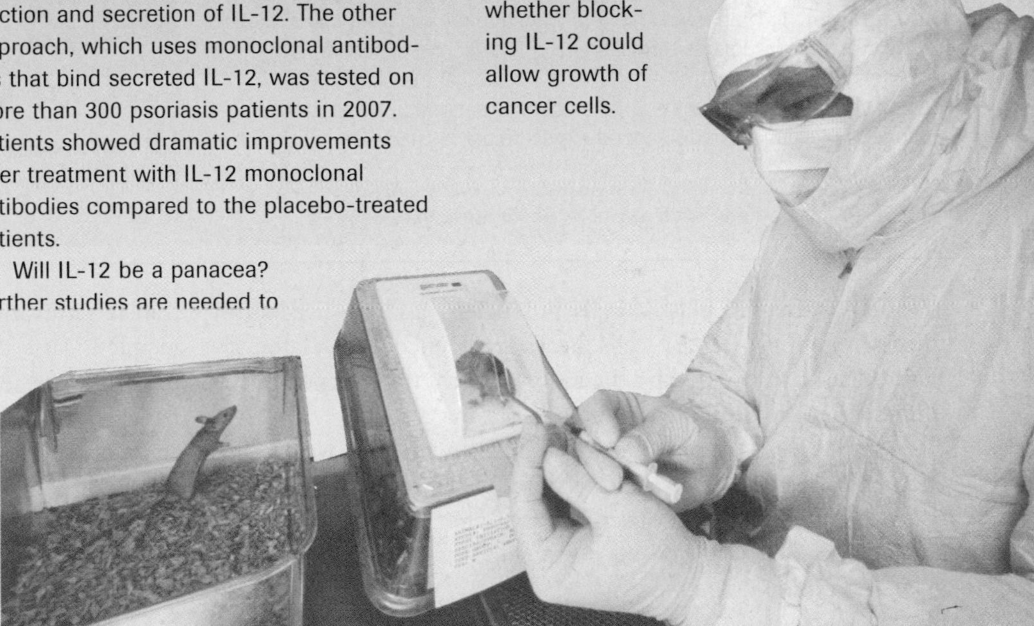

that have not been totally answered, such as why the body does not reject the fetus, which is nonself.

CHECK YOUR UNDERSTANDING

✓ What is the function of cytokines? **17-18**

Immunological Memory

LEARNING OBJECTIVE

17-19 Distinguish a primary from a secondary immune response.

The intensity of the antibody-mediated humoral response can be reflected by the **antibody titer,** the relative amount of antibody in the serum. After the initial contact with an antigen, the exposed person's serum contains no detectable antibodies for 4 to 7 days. Then there is a slow rise in antibody titer: first, IgM class antibodies are produced, followed by IgG peaking in about 10 to 17 days, after which antibody titer gradually declines. This pattern is characteristic of a **primary response** to an antigen.

The antibody-mediated immune responses of the host intensify after a second exposure to an antigen. This **secondary response** is also called the **memory** (or **anamnestic**) **response.** As shown in Figure 17.16 this response is comparatively more rapid, reaching a peak in only 2 to 7 days, lasts many days, and is considerably greater in magnitude. By way of explanation, as is shown in Figure 17.5, some activated B cells do not become

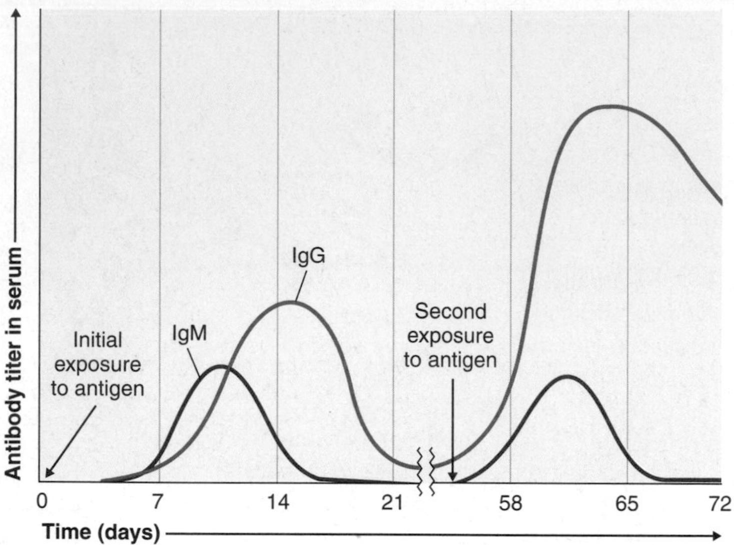

Figure 17.16 The primary and secondary immune responses to an antigen. IgM appears first in response to the initial exposure. IgG follows and provides longer-term immunity. The second exposure to the same antigen stimulates the memory cells (formed at the time of initial exposure) to rapidly produce a large amount of antibody. The antibodies produced in response to this second exposure are mostly IgG.

Q **Why do many diseases, such as measles, occur only once in a person, yet others, such as colds, occur more than once?**

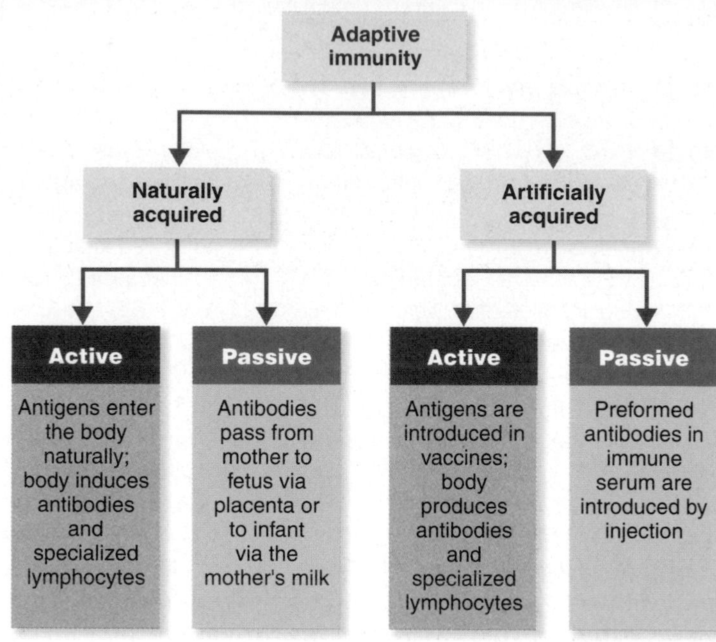

Figure 17.17 Types of adaptive immunity.

Q **Which type of immunity, active or passive, lasts longer?**

antibody-producing plasma cells but persist as long-lived but nonproliferating memory cells. Years, or even decades later, if these cells are stimulated by the same antigen, they very rapidly differentiate into antibody-producing plasma cells.

A similar response occurs with T cells, which, as we see in Chapter 19, is necessary for establishing the lifelong memory for distinguishing self from nonself. **Animations** Humoral Immunity: Primary Immune Response, Secondary Immune Response. **www.microbiologyplace.com**

CHECK YOUR UNDERSTANDING

✔ Is the anamnestic response primary or secondary? **17-19**

Types of Adaptive Immunity

LEARNING OBJECTIVE

17-20 Contrast the four types of adaptive immunity.

Adaptive immunity refers to the protection an animal develops against certain specific microbes or foreign substances. The various manifestations of adaptive immunity are summarized in **Figure 17.17**.

Immunity can be acquired either actively or passively. Immunity is acquired *actively* when a person is exposed to microorganisms or foreign substances and the immune system responds. Immunity is acquired *passively* when antibodies are transferred from one person to another. Passive immunity in the recipient lasts only as long as the antibodies are present—in most cases, a few weeks. Both actively acquired immunity and passively acquired immunity can be obtained by natural or artificial means.

- **Naturally acquired active immunity** develops when a person is exposed to antigens, becomes ill, and then recovers. Once acquired, immunity is lifelong for some diseases, such as measles. For certain other diseases, especially intestinal diseases, the immunity may last for only a few years. *Subclinical infections* or *inapparent infections* (those that produce no noticeable symptoms or signs of illness) can also confer immunity.
- **Naturally acquired passive immunity** involves the natural transfer of antibodies from a mother to her infant. Antibodies in a pregnant woman cross the placenta to her fetus—*transplacental transfer.* If the mother is immune to diphtheria, rubella, or polio, for example, the newborn will be temporarily immune to these diseases as well. Certain antibodies are also passed from the mother to her nursing infant in breast milk, especially in the first secretions, called *colostrum.* In the infant, this passive immunity generally lasts only as long as the transmitted antibodies persist—usually a few weeks or months. These maternal antibodies are essential for providing immunity to the infant until its own immune system matures. Colostrum is even more important to some other mammals; calves, for example, do not have

antibodies that cross the placenta and rely on colostrum ingested during the first day of life. Researchers often specify fetal calf serum for certain experimental uses because it does not contain maternal antibodies.

- **Artificially acquired active immunity** is the result of vaccination—which will be discussed in Chapter 18. **Vaccination,** also called **immunization,** introduces **vaccines** into the body. These are antigens such as killed or living microorganisms or inactivated bacterial toxins.
- **Artificially acquired passive immunity** involves the injection of antibodies (rather than antigens) into the body. These antibodies come from an animal or person who is already immune to the disease.

Because blood serum is easily obtained (see the box on page 467) and contains a considerable concentration of antibodies, **antiserum** has become a generic term for blood-derived fluids containing antibodies. Hence, the study of reactions between antibodies and antigens is called **serology.** As shown in Figure 17.18, when a sample of serum is subjected to an electrical current in the laboratory during gel electrophoresis (see Chapter 9), the proteins dissolved within it move at different rates. The globulin proteins separate into fractions that are termed alpha (α), beta (β), and gamma (γ) for their relative motility. Because the gamma fraction, called **gamma globulin,** contains most of the antibodies, it is often used to transfer passive immunity.

When immune serum globulin from an individual who is immune to a disease is injected into another individual, it confers an immediate passive protection against the disease. However, although artificially acquired passive immunity is immediate, it is short-lived because antibodies are degraded by the recipient. The half-life of an injected antibody (the time required for half of the antibodies to disappear) is typically about 3 weeks.

CHECK YOUR UNDERSTANDING

✓ What type of adaptive immunity is involved when gamma globulin is injected into a person? **17-20**

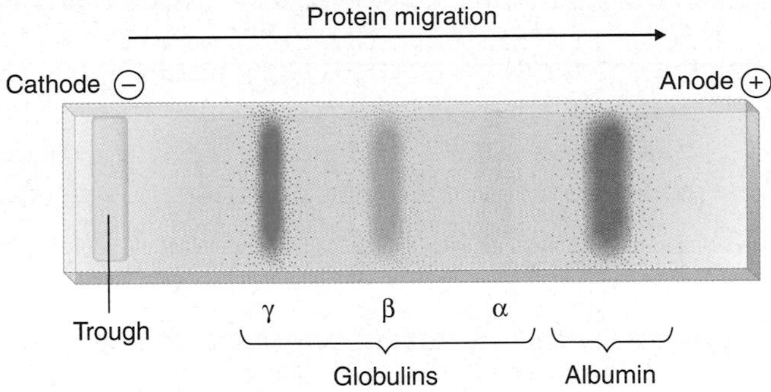

Figure 17.18 The separation of serum proteins by gel electrophoresis. In this procedure, serum is placed in a trough cut into a gel. In response to an electrical current, the negatively charged proteins of the serum migrate through the gel from the negatively charged end (cathode) to the positively charged end (anode).

Q What serum fraction contains the most antibodies?

* * *

This chapter on immunology is intended to provide you with the general concepts on the subject. It should give you the information you need to understand the chapters that follow, which emphasize some of the more practical and clinical aspects of immunology. Immunology can be a very complex subject; some of you, in pursuit of your academic major, will be taking a full course in immunology later and will study the subject in much more detail. The presentation here, you will then find, has been considerably simplified—although you might not have realized it. Figure 17.19 summarizes the material covered in this chapter, especially emphasizing the dual nature of immunology.

Figure 17.19

FOUNDATION FIGURE The Dual Nature of the Adaptive Immune System

This figure summarizes the functions of the adaptive immune system covered in Chapter 17. Refer to Figure 16.1, on page 450, to review how adaptive immunity fits into the bigger picture of the body's defenses. The basic concepts of immunity summarized here are necessary for understanding the practical and clinical aspects of immunology covered in subsequent chapters.

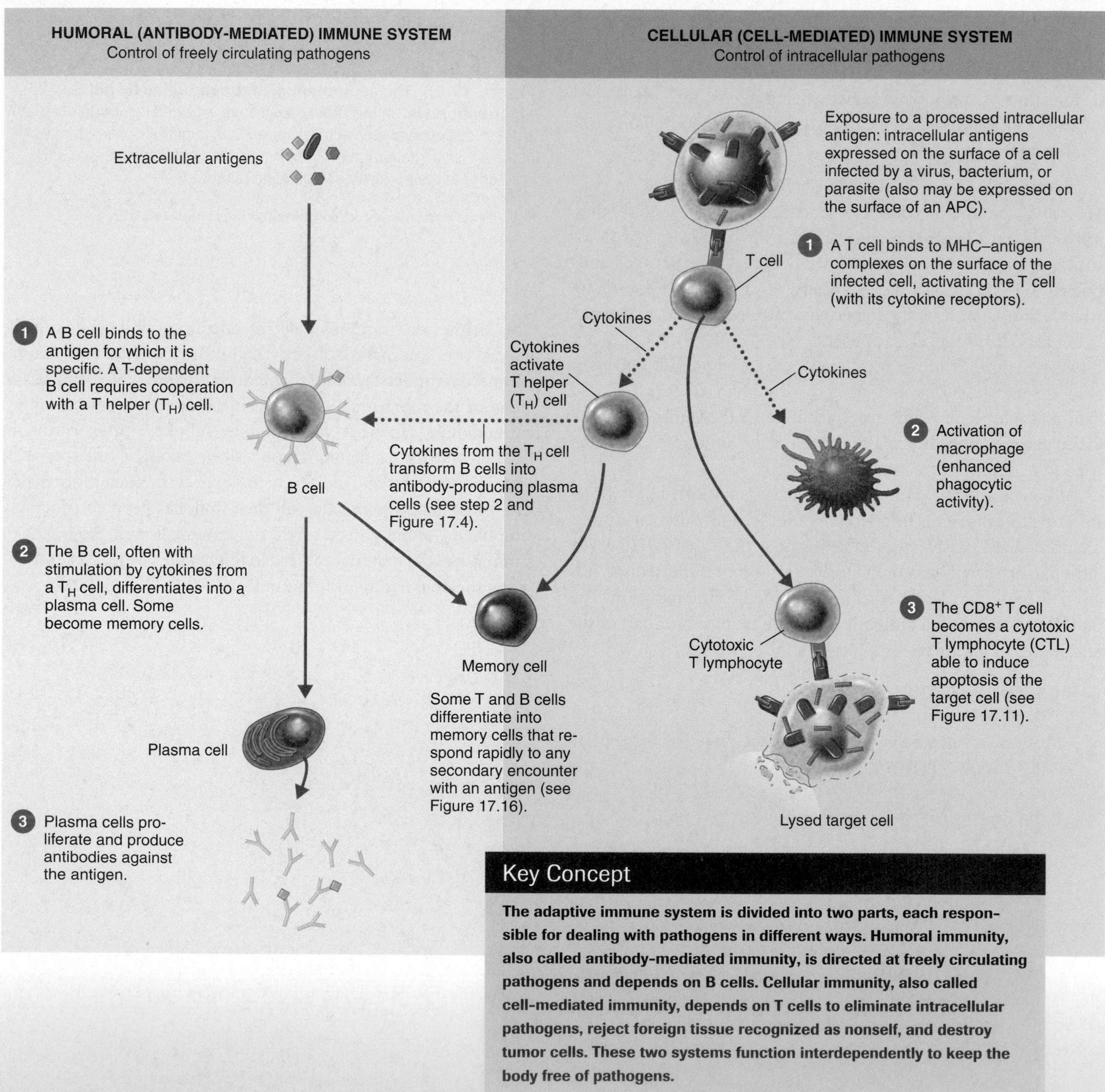

HUMORAL (ANTIBODY-MEDIATED) IMMUNE SYSTEM
Control of freely circulating pathogens
CELLULAR (CELL-MEDIATED) IMMUNE SYSTEM
Control of intracellular pathogens
Extracellular antigens
1 A B cell binds to the antigen for which it is specific. A T-dependent B cell requires cooperation with a T helper (T_H) cell.
B cell
2 The B cell, often with stimulation by cytokines from a T_H cell, differentiates into a plasma cell. Some become memory cells.
Plasma cell
3 Plasma cells proliferate and produce antibodies against the antigen.
Exposure to a processed intracellular antigen: intracellular antigens expressed on the surface of a cell infected by a virus, bacterium, or parasite (also may be expressed on the surface of an APC).
T cell
1 A T cell binds to MHC–antigen complexes on the surface of the infected cell, activating the T cell (with its cytokine receptors).
Cytokines
Cytokines activate T helper (T_H) cell
Cytokines
Cytokines from the T_H cell transform B cells into antibody-producing plasma cells (see step 2 and Figure 17.4).
2 Activation of macrophage (enhanced phagocytic activity).
Memory cell
Some T and B cells differentiate into memory cells that respond rapidly to any secondary encounter with an antigen (see Figure 17.16).
Cytotoxic T lymphocyte
3 The CD8$^+$ T cell becomes a cytotoxic T lymphocyte (CTL) able to induce apoptosis of the target cell (see Figure 17.11).
Lysed target cell

Key Concept

The adaptive immune system is divided into two parts, each responsible for dealing with pathogens in different ways. Humoral immunity, also called antibody-mediated immunity, is directed at freely circulating pathogens and depends on B cells. Cellular immunity, also called cell-mediated immunity, depends on T cells to eliminate intracellular pathogens, reject foreign tissue recognized as nonself, and destroy tumor cells. These two systems function interdependently to keep the body free of pathogens.

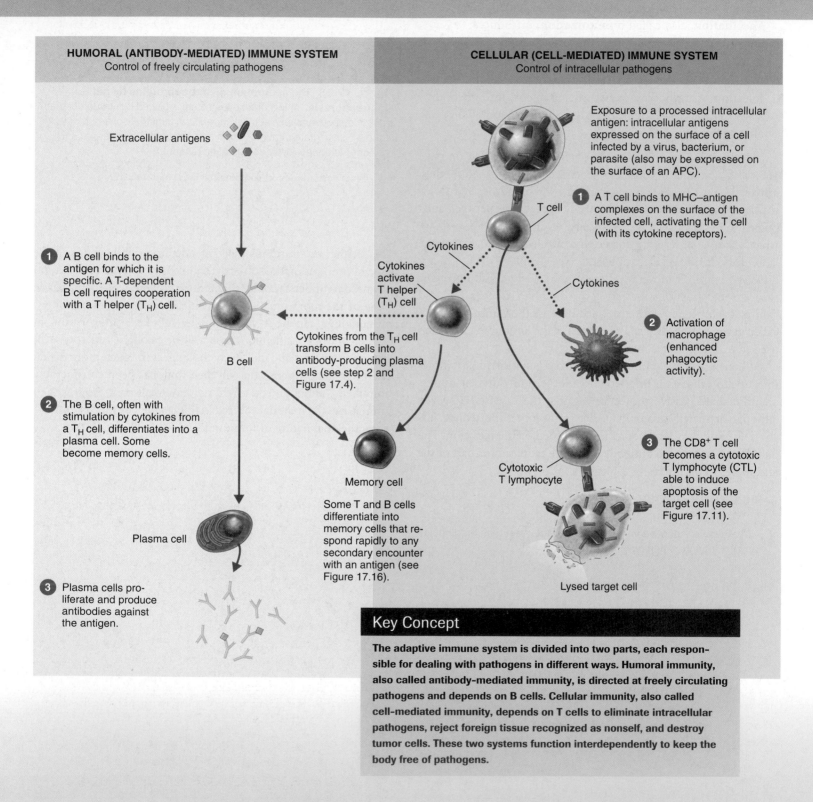

STUDY OUTLINE

The **MyMicrobiologyPlace** website (**www.microbiologyplace.com**) will help you get ready for tests with its simple three-step approach: ❶ **take a pre-test** and obtain a personalized study plan, ❷ **learn and practice** with animations, tutorials, and MP3 tutor sessions, and ❸ **test yourself** with quizzes and a chapter post-test.

The Adaptive Immune System (p. 477)

1. An individual's genetically predetermined resistance to certain diseases is called innate immunity.
2. Adaptive immunity is the ability of the body to specifically react to a microbial infection.

Dual Nature of the Adaptive Immune System (pp. 477–478)

1. Red bone marrow stem cells produce lymphocytes. Lymphocytes that mature in bone marrow become B cells.
2. Humoral immunity involves antibodies, which are found in serum and lymph and are produced by B cells.
3. Lymphocytes that migrate through the thymus become T cells. Cellular immunity involves T cells.
4. T cell receptors recognize antigens.

Antigens and Antibodies (pp. 478–482)

The Nature of Antigens (pp. 478–479)

1. An antigen (or immunogen) is a chemical substance that causes the body to produce specific antibodies.
2. As a rule, antigens are proteins or large polysaccharides. Antibodies are formed against specific regions on antigens called epitopes, or antigenic determinants.

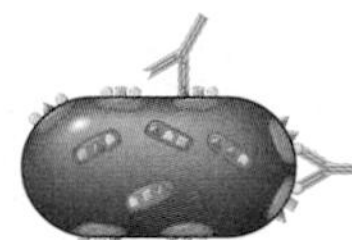

3. A hapten is a low-molecular-weight substance that cannot cause the formation of antibodies unless combined with a carrier molecule; haptens react with their antibodies independently of the carrier molecule.

The Nature of Antibodies (pp. 479–482)

4. An antibody, or immunoglobulin, is a protein produced by B cells in response to an antigen and is capable of combining specifically with that antigen.
5. Typical monomers consist of four polypeptide chains: two heavy chains and two light chains.
6. Within each chain is a variable (V) region that binds the epitope and a constant (C) region that distinguishes the different classes of antibodies.
7. An antibody monomer is Y-shaped or T-shaped: the V regions form the tips, the C regions form the base and F_C (stem) region.
8. The F_C region can attach to a host cell or to complement.
9. IgG antibodies are the most prevalent in serum; they provide naturally acquired passive immunity, neutralize bacterial toxins, participate in complement fixation, and enhance phagocytosis.
10. IgM antibodies consist of five monomers held by a joining chain; they are involved in agglutination and complement fixation.
11. Serum IgA antibodies are monomers; secretory IgA antibodies are dimers that protect mucosal surfaces from invasion by pathogens.
12. IgD antibodies are on B cells; they may delete B cells that produce antibodies against self.
13. IgE antibodies bind to mast cells and basophils and are involved in allergic reactions.

B Cells and Humoral Immunity (pp. 482–486)

Clonal Selection of Antibody-Producing Cells (pp. 482–488)

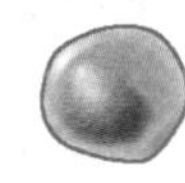

1. Red bone marrow stem cells give rise to B cells with IgM and IgD on their surfaces, which recognize specific epitopes.
2. For T-dependent antigens, B cells are selected by antigens with repeating epitopes.
3. For T-independent antigens, B cells are activated by a T cell. The T cell was activated by an antigenic fragment presented with host MHC II.
4. Activated B cells differentiate into plasma cells and memory cells.
5. B cells that recognize self are eliminated by clonal deletion.

The Diversity of Antibodies (p. 484)

6. During development, the genes in embryonic B cells recombine so that mature B cells each have different genes for the V region of their antibodies.

Antigen–Antibody Binding and Its Results (pp. 484–486)

1. An antigen–antibody complex forms when an antibody binds to its specific epitopes on an antigen.
2. Agglutination results when an antibody combines with epitopes on two different cells.
3. Opsonization enhances phagocytosis of the antigen.
4. Antibodies that attach to microbes or toxins cause neutralization.
5. Complement activation results in cell lysis.

T Cells and Cellular Immunity (pp. 486–489)

1. Red bone marrow stem cells give rise to T cells, which mature in the thymus gland. Thymic selection removes T cells that don't recognize MHC-self molecules.
2. T-cell receptors on T cells recognize antigens.
3. T cells recognize antigens processed by antigen-presenting cells.
4. T cells recognize antigens in association with MHC on an APC.

Classes of T Cells (p. 487)

5. T cells are classified according to their functions and cell-surface glycoproteins called CDs.

T Helper Cells ($CD4^+$ T Cells) (pp. 487–488)

6. T_H1 cells activate cells involved in cellular immunity.
7. T_H2 cells are associated with allergic reactions and parasitic infections.

8. T helper cells, or $CD4^+$ T cells, are activated by MHC class II on APCs. After binding an APC, $CD4^+$ T cells secrete cytokines that activate other T cells and B cells.

T Cytotoxic Cells ($CD8^+$ T Cells) (pp. 488–489)

9. T cytotoxic cells (T_C), or $CD8^+$ T cells, are activated by endogenous antigens and MHC class I on a target cell and are transformed into a CTL.
10. CTLs lyse or induce apoptosis in the target cell.

T Regulatory Cells (p. 489)

11. T regulatory cells (T_{reg}) suppress T cells against self.

Antigen-Presenting Cells (APCs) (pp. 489–490)

1. APCs include B cells, dendritic cells, and macrophages.
2. Dendritic cells are the primary APCs.
3. Activated macrophages are effective phagocytes and APCs.
4. APCs carry antigens to lymphoid tissues where T cells that recognize the antigen are located.

Extracellular Killing by the Immune System (p. 491)

1. Natural killer (NK) cells lyse virus-infected cells, tumor cells, and parasites. They kill cells that do not express MHC class I antigens.

Antibody-Dependent Cell-Mediated Cytotoxicity (pp. 491–492)

1. In ADCC, NK cells and macrophages lyse antibody-coated cells.

Cytokines: Chemical Messengers of Immune Cells (pp. 492–493)

1. Cells of the immune system communicate with each other by means of chemicals called cytokines.
2. Interleukins (IL) are cytokines that serve as communicators between leukocytes.
3. Chemokines cause leukocytes to migrate to an infection.
4. Alpha interferon and IFN-β protect cells against viruses. Gamma interferon increases phagocytosis.
5. Tumor necrosis factor promotes the inflammatory reaction.
6. Hematopoietic cytokines promote development of white blood cells.
7. Overproduction of cytokines leads to a cytokine storm, which results in tissue damage.

Immunological Memory (pp. 493–494)

1. The relative amount of antibody in serum is called the antibody titer.
2. The response of the body to the first contact with an antigen is called the primary response. It is characterized by the appearance of IgM followed by IgG.
3. Subsequent contact with the same antigen results in a very high antibody titer and is called the secondary, anamnestic, or memory response. The antibodies are primarily IgG.

Types of Adaptive Immunity (pp. 494–496)

1. Immunity resulting from infection is called naturally acquired active immunity; this type of immunity may be long-lasting.
2. Antibodies transferred from a mother to a fetus (transplacental transfer) or to a newborn in colostrum results in naturally acquired passive immunity in the newborn; this type of immunity can last up to a few months.
3. Immunity resulting from vaccination is called artificially acquired active immunity and can be long-lasting.
4. Artificially acquired passive immunity refers to humoral antibodies acquired by injection; this type of immunity can last for a few weeks.
5. Serum containing antibodies is often called antiserum.
6. When serum is separated by gel electrophoresis, antibodies are found in the gamma fraction of the serum and are termed immune serum globulin, or gamma globulin.

STUDY QUESTIONS

Answers to the Review and Multiple choice questions can be found by turning to the blue Answers tab at the back of the textbook.

Review

1. Contrast the terms in the following pairs:
   a. innate and adaptive immunity
   b. humoral and cellular immunity
   c. active and passive immunity
   d. T_H1 and T_H2 cells
   e. natural and artificial immunity
   f. T-dependent and T-independent antigens
   g. $CD8^+$ T cell and CTL
   h. immunoglobulin and TCR
2. What does MHC stand for? What is the function of MHC? What types of T cells interact with MHC class I? With MHC class II?
3. **DRAW IT** Label the heavy chains, light chains, and variable and F_C regions of this typical antibody. Indicate where the antibody binds to antigen. Sketch an IgM antibody.

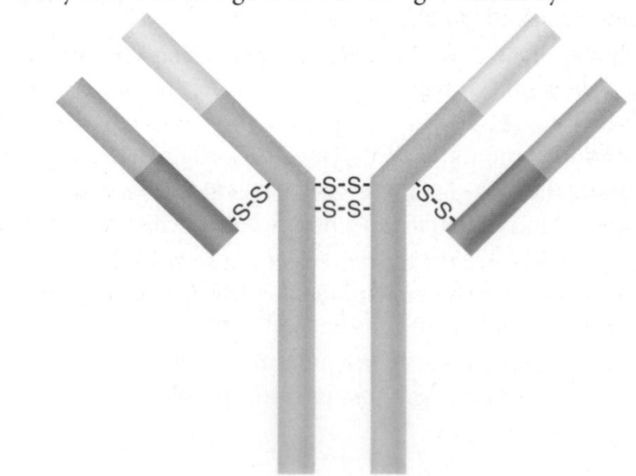

4. Diagram the roles that T cells and B cells play in immunity.
5. Explain a function for the following types of cells: T_C, T_H, and T_{reg}. What is a cytokine?
6. **DRAW IT**
   a. In the graph below, at time *A* the host was injected with tetanus toxoid. Show the response to a booster close at time B.
   b. Draw the antibody response of this same individual to exposure to a new antigen indicated at time *B*.

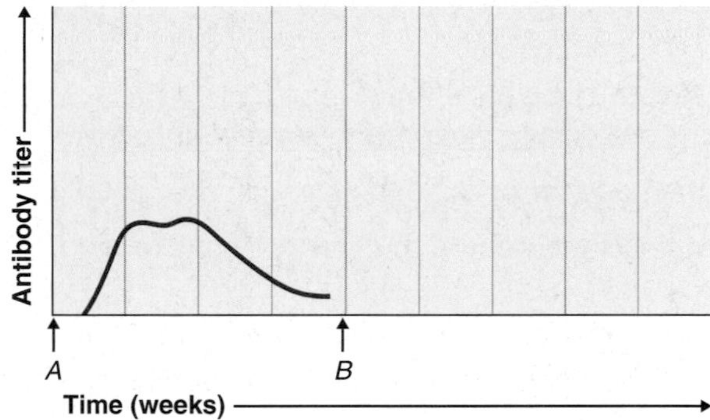

7. How would each of the following prevent infection?
   a. antibodies against *Neisseria gonorrhoeae* fimbriae
   b. antibodies against host cell mannose
8. How can a human make over ten billion different antibodies with only 35,000 different genes?
9. Explain why a person who recovers from a disease can attend others with the disease without fear of contracting it.

Multiple Choice

Match the following choices to questions 1–4:
- **a.** innate resistance
- **b.** naturally acquired active immunity
- **c.** naturally acquired passive immunity
- **d.** artificially acquired active immunity
- **e.** artificially acquired passive immunity

1. The type of protection provided by the injection of diphtheria toxoid.
2. The type of protection provided by the injection of antirabies serum.
3. The type of protection resulting from recovery from an infection.
4. A newborn's immunity to yellow fever.

Match the following choices to the statements in questions 5–7:
- **a.** IgA
- **b.** IdD
- **c.** IgE
- **d.** IgG
- **e.** IgM

5. Antibodies that protect the fetus and newborn.
6. The first antibodies synthesized; especially effective against microorganisms.
7. Antibodies that are bound to mast cells and involved in allergic reactions.
8. Put the following in the correct sequence to elicit an antibody response: (1) T_H cell recognizes B cell; (2) APC contacts antigen; (3) antigen fragment goes to surface of APC; (4) T_H recognizes antigen digest and MHC; (5) B cell proliferates.
   a. 1, 2, 3, 4, 5
   b. 5, 4, 3, 2, 1
   c. 3, 4, 5, 1, 2
   d. 2, 3, 4, 1, 5
   e. 4, 5, 3, 1, 2
9. A kidney-transplant patient experienced a cytotoxic rejection of his new kidney. Place the following in order for that rejection: (1) apoptosis occurs; (2) $CD8^+$ T cell becomes CTL; (3) granzymes released; (4) MHC class I activates $CD8^+$ T cell; (5) perforin released.
   a. 1, 2, 3, 4, 5
   b. 5, 4, 3, 2, 1
   c. 4, 2, 5, 3,1
   d. 3, 4, 5, 1, 2
   e. 2, 3, 4, 1, 5
10. Patients with Chédiak-Higashi syndrome suffer from various types of cancer. These patients are most likely lacking which of the following:
   a. T_R cells
   b. T_H1 cells
   c. B cells
   d. NK cells
   e. T_H2 cells

Critical Thinking

1. Injections of T_C cells completely removed all hepatitis B viruses from infected mice, but they killed only 5% of the infected liver cells. Explain how T_C cells cured the mice.
2. Why is dietary protein deficiency associated with increased susceptibility to infections?
3. A positive tuberculin skin test shows cellular immunity to *Mycobacterium tuberculosis.* How could a person acquire this immunity?
4. On her vacation to Australia, Janet was bitten by a poisonous sea snake. She survived because the emergency room physician injected her with antivenin to neutralize the toxin. What is antivenin? How is it obtained?

Clinical Applications

1. A woman had life-threatening salmonellosis that was successfully treated with anti-*Salmonella.* Why did this treatment work, when antibiotics and her own immune system failed?
2. A patient with AIDS has a low T_H cell count. Why does this patient have trouble making antibodies? How does this patient make *any* antibodies?
3. A patient with chronic diarrhea was found to lack IgA in his secretions, although he had a normal level of serum IgA. What was this patient found to be unable to produce?
4. Newborns (under 1 year) who contract dengue fever have a higher chance of dying from it if their mothers had dengue fever prior to pregnancy. Explain why.
5. A woman died from a *Capnocytophaga* bacterial infection introduced by a dog bite. *Capnocytophaga* kills only people who lack a spleen. What is the relationship between infection and the spleen?

18 Practical Applications of Immunology

In Chapter 17, we learned the basics of the immune system, in which the body recognizes foreign microbes, toxins, or tissues. In response, the body forms antibodies and activates other immune system cells that are programmed to recognize and neutralize or destroy this foreign material if the body encounters it again.

In this chapter, we will discuss some useful tools that have been developed from knowledge of the basics of the immune system. Vaccines were briefly mentioned in the previous chapter; in this chapter we will expand our discussion of this important field of immunology. The diagnosis of disease frequently depends on tests that make use of antibodies and the specificity of the immune system.

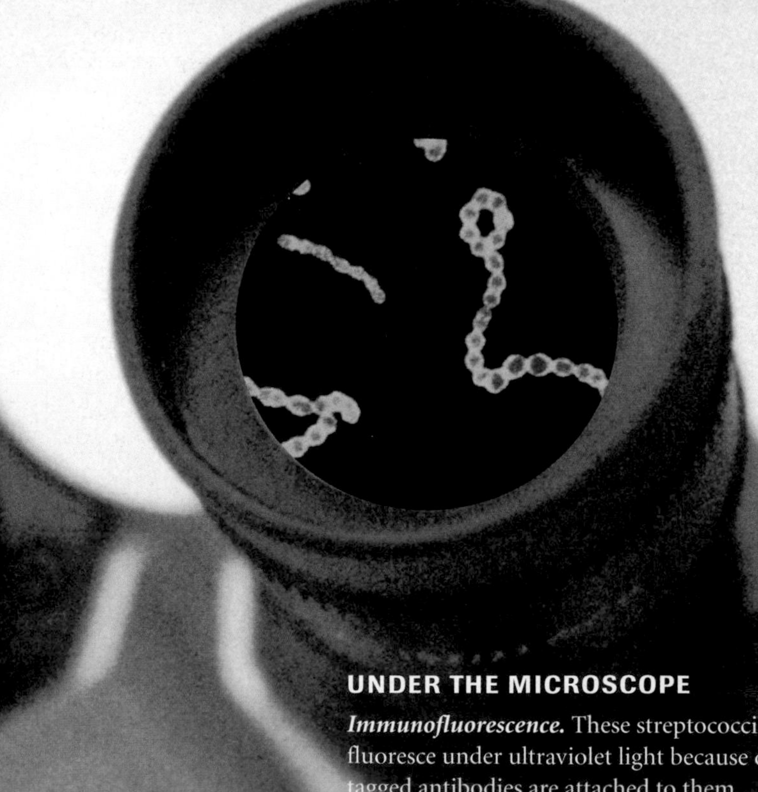

UNDER THE MICROSCOPE

Immunofluorescence. These streptococci fluoresce under ultraviolet light because dye-tagged antibodies are attached to them.

Q&A

The bacteria shown in the microscope here would be visible even if fluorescent antibodies were not attached to them. Why would this test be even more valuable for testing for such organisms as the rabies virus?

Look for the answer in the chapter.

Vaccines

LEARNING OBJECTIVES

18-1 Define *vaccine.*

18-2 Explain why vaccination works.

18-3 Differentiate the following, and provide an example of each: attenuated, inactivated, toxoid, subunit, and conjugated vaccines.

18-4 Contrast subunit vaccines and nucleic acid vaccines.

18-5 Compare and contrast the production of whole-agent vaccines, recombinant vaccines, and DNA vaccines.

18-6 Define *adjuvant.*

18-7 Explain the value of vaccines, and discuss acceptable risks for vaccines.

Long before the invention of vaccines, it was known that people who recovered from certain diseases, such as smallpox, were immune to the disease thereafter. Chinese physicians may have been the first to try to exploit this phenomenon to prevent disease when they had children inhale dried smallpox scabs.

In 1717, Lady Mary Montagu reported from her travels in Turkey that an "old woman comes with a nutshell full of the matter of the best sort of smallpox and asks what veins you please to have opened, and puts into the vein as much venom as can lie upon the head of her needle." This practice usually led to a week of mild illness, and the person was subsequently protected from smallpox. Called **variolation,** this procedure became commonplace in England. Unfortunately, however, it sometimes resulted in a serious case of smallpox. In eighteenth-century England, the mortality rate associated with variolation was about 1%, still a significant improvement over the 50% mortality rate that could be expected from smallpox.

One person who received this treatment, at the age of 8, was Edward Jenner. As a physician, Jenner subsequently encountered patients who did not respond with the usual reactions to variolation. Many of them, especially dairymaids, told him that they had no fear of smallpox because they had already had cowpox. Cowpox is a mild disease that causes lesions on cows' udders; dairymaids' hands often became infected during milking. Motivated by his childhood memory of variolation, Jenner began a series of experiments in 1798 in which he deliberately inoculated people with cowpox in an attempt to prevent smallpox. To honor Jenner's work, the term *vaccination* (from the Latin *vacca*, meaning cow) was coined. A **vaccine** is a suspension of organisms or fractions of organisms that is used to induce immunity. Two centuries later, the disease of smallpox has been eliminated worldwide by vaccination, and two other viral diseases, measles and polio, are also targeted for elimination. See the box on page 505. **Animation** Vaccines: Function. **www.microbiologyplace.com**

CHECK YOUR UNDERSTANDING

✓ What is the etymology (origin) of the word *vaccine*? **18-1**

Principles and Effects of Vaccination

We now know that Jenner's inoculations worked because the cowpox virus, which is not a serious pathogen, is closely related to the smallpox virus. The injection, by skin scratches, provoked a primary immune response in the recipients, leading to the formation of antibodies and long-term memory cells. Later, when the recipient encountered the smallpox virus, the memory cells were stimulated, producing a rapid, intense secondary immune response (see Figure 17.16, page 494). This response mimics the immunity gained by recovering from the disease. The cowpox vaccine was soon replaced by a vaccinia virus vaccine. The vaccinia virus also confers immunity to smallpox, although, strangely, little is known with certainty about the origin of this important virus. It is genetically distinct from cowpox virus and may be a hybrid of an accidental mixing of cowpox and smallpox viruses or perhaps once may have been the cause of a now-extinct disease, horsepox. The development of vaccines based on the model of the smallpox vaccine is the single most important application of immunology.

Many communicable diseases can be controlled by behavioral and environmental methods. For example, proper sanitation can prevent the spread of cholera, and the use of condoms can slow the spread of sexually transmitted infections. If prevention fails, bacterial diseases can often be treated with antibiotics. Viral diseases, however, often cannot be effectively treated once contracted. Therefore, vaccination is frequently the only feasible method of controlling viral disease. Controlling a disease does not necessarily require that everyone be immune to it. If most of the population is immune, a phenomenon called *herd immunity,* outbreaks are limited to sporadic cases because there are not enough susceptible individuals to support the spread of epidemics.

The principal vaccines used to prevent bacterial and viral diseases in the United States are listed in **Table 18.1** and **Table 18.2**. Recommendations for childhood immunizations against some of these diseases are given in **Table 18.3**. American travelers who might be exposed to cholera, yellow fever, or other diseases not endemic in the United States can obtain current immunization recommendations from the U.S. Public Health Service and local public health agencies.

CHECK YOUR UNDERSTANDING

✓ Vaccination is often the only feasible way to control most viral diseases; why is this? **18-2**

Types of Vaccines and Their Characteristics

There are now several basic types of vaccine. Some of the newer vaccines take full advantage of knowledge and technology developed in recent years.

Attenuated whole-agent vaccines use living but attenuated (weakened) microbes. Live vaccines more closely mimic an actual infection. Lifelong immunity, especially with viruses, is often achieved without booster immunizations, and an effectiveness

Table 18.3 Recommended Immunization Schedule for Persons Aged 0–6 Years—United States, 2008 (CDC)

Age ▶ / Vaccine ▼	Birth	1 month	2 months	4 months	6 months	12 months	15 months	18 months	19–23 months	2–3 years	4–6 years
Hepatitis B	HepB	HepB			HepB						
Rotavirus			Rota	Rota	Rota						
Diphtheria, Tetanus, Pertussis			DTaP	DTaP	DTaP		DTaP				DTaP
***Haemophilus influenzae* type b**			Hib	Hib	Hib	Hib					
Pneumococcal*			PCV	PCV	PCV	PCV				PPV	
Inactivated Poliovirus			IPV	IPV	IPV						IPV
Influenza					Influenza (Yearly)						
Measles, Mumps, Rubella						MMR					MMR
Varicella						Varicella					Varicella
Hepatitis A†						HepA (2 doses)					
Meningococcal‡										MCV4	

Note: Vaccines are listed under routinely recommended ages. Bars indicate range of recommended ages for immunization. For those who fall behind or start late, see the catch-up schedule. Additional information at www.cdc.gov/vaccines/rec/schedules/

* PCV = Pneumococcal conjugate vaccine, PPV = Pneumococcal polysaccharide vaccine.

† The two doses at least 6 mo. apart.

‡ Meningococcal conjugate vaccine (MCV4) for children aged 2–10 years with defective immune systems and certain other high risk situations.

diseases, which should lower costs. **Animation** Vaccines: Types. **www.microbiology place.com**

CHECK YOUR UNDERSTANDING

- Experience has shown that attenuated vaccines tend to be more effective than inactivated vaccines. Why? **18-3**
- Which is more likely to be useful in preventing a disease caused by an encapsulated bacterium such as the pneumococcus: a subunit vaccine or a nucleic acid vaccine? **18-4**

The Development of New Vaccines

An effective vaccine is the most desirable method of disease control. It prevents the targeted disease from ever occurring at all in an individual, and it is generally the most economical. This is especially important in developing parts of the world.

Although interest in vaccine development declined with the introduction of antibiotics, it has intensified in recent years. Fear of litigation contributed to the decrease in the development of new vaccines in the United States. However, passage of the National Childhood Vaccine Injury Act in 1986, which limits the liability of vaccine manufacturers, has now helped reverse this trend. Even so, pharmaceutical companies find that the most profitable drugs are those that must be taken daily for extended periods, for example, drugs taken for high blood pressure. In contrast, a vaccine that is required for a few injections or even only once in a lifetime is inherently less attractive.

Historically, vaccines could be developed only by growing the pathogen in usefully large amounts. The early successful viral vaccines were developed by animal cultivation. The vaccinia virus for smallpox was grown on the shaved bellies of calves, for example.

The introduction of vaccines against polio, measles, mumps, and a number of other viral diseases that would not grow in anything but a living human awaited the development of cell culture techniques. Cell cultures from human sources or, more often, from animals such as monkeys that are closely related to humans enabled growth of these viruses on a large scale. A convenient animal that will grow many viruses is the chick embryo (see Figure 13.7, page 377). Viruses for several vaccines (influenza, for example) are grown this way (see Figure 18.1). Interestingly, the first vaccine against hepatitis B virus used viral antigens extracted from the blood of chronically infected humans because no other source was available.

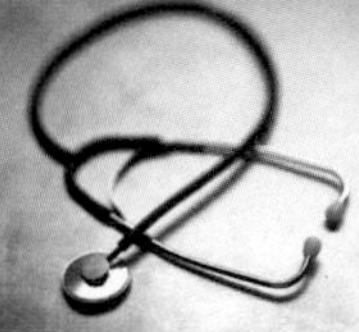

CLINICAL FOCUS From the *Morbidity and Mortality Weekly Report*

A World Health Problem

As you read through this box, you will encounter a series of questions that public health scientists ask themselves as they try to reduce the occurrence of diseases. Try to answer each question before going to the next one.

1. On May 14, a 17-year-old girl developed a fever and tiny red spots with blue-white centers inside her mouth (**Figure A**). She developed a rash on her face on May 16; the rash spread over her trunk and extremities. Subsequently, a 2-year-old developed a fever and pneumonia. In all, 34 people from her church developed a maculopapular rash and fever (≥38°C), and at least one of the following: fever, cough, conjunctivitis, coldlike symptoms.

   What is the disease? *(Hint:* See Table 21.1 on page 589.)

2. Measles was confirmed by testing for IgM measles antibodies. Measles is a highly contagious viral illness that can cause pneumonia, diarrhea, encephalitis, and death.

   What do you need to know?

3. The index patient had traveled to Romania for 2 weeks. She and the other infected people had not been vaccinated against measles.

   Why didn't more people get measles?

4. In 1920, prior to development of the measles vaccine, nearly 500,000 cases of measles occurred, with over 7500 deaths, in the United States. In 2007, only 30 cases of measles were reported in the United States (**Figure B**). However, measles is still endemic in many countries (**Figure C**). Worldwide, there are 70 million cases each year. Measles is still one of the top 20 causes of death, killing 600 children per day.

   What would happen if we stopped vaccinating against measles?

5. If there were no vaccines, there would be many more cases of disease. Along with more disease, there would be serious sequelae and more deaths. Some vaccine-preventable diseases are still quite prevalent in other parts of the world. As occurred in this case, travelers can unknowingly bring these diseases into the United States, and if we were not protected by vaccinations, these diseases could quickly spread throughout the population, causing epidemics here.

   The Measles Initiative is a partnership—led by the American Red Cross, the United Nations Foundation, UNICEF, the U.S. Centers for Disease Control and Prevention, and the World Health Organization—committed to reducing measles deaths worldwide. The Measles Initiative has supported the vaccination of more than 400 million children in over 50 countries. In 2000, measles caused approximately 757,000 deaths, mostly in children under 5. By 2006, measles deaths were reduced to 242,000 people worldwide.

Source: Adapted from *MMWR* 54(42):1073–1075, October 28, 2005.

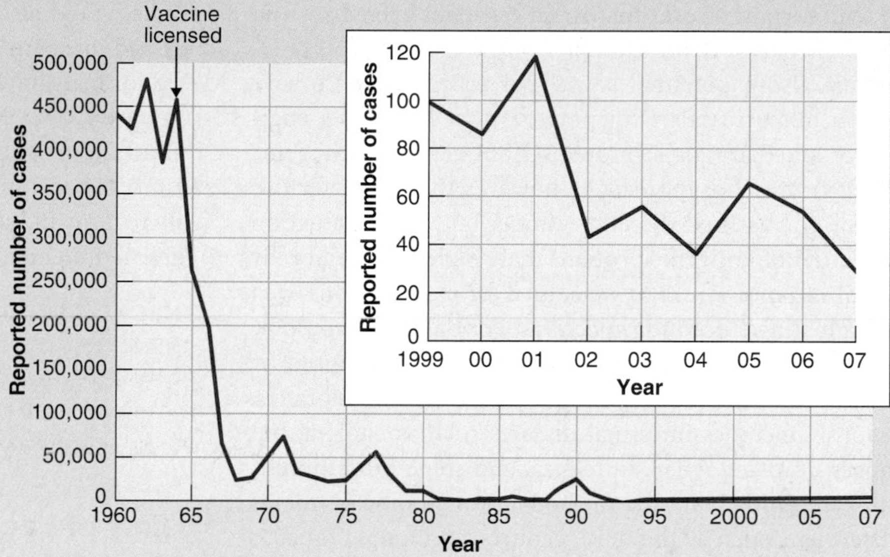

Figure B Reported numbers of measles cases in the United States, 1960–2007. (CDC, 2008)

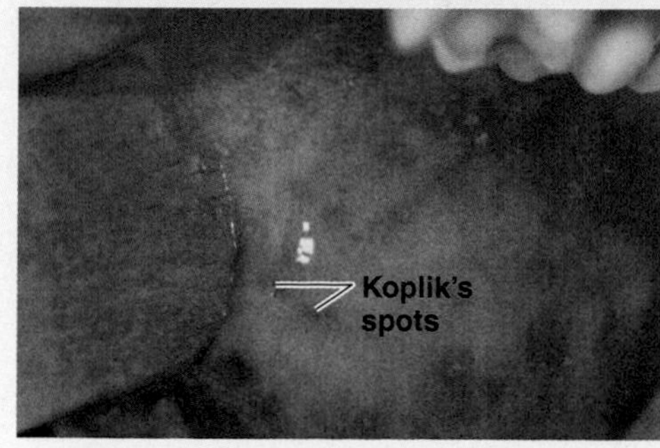

Figure A Koplik's spots inside the cheeks.

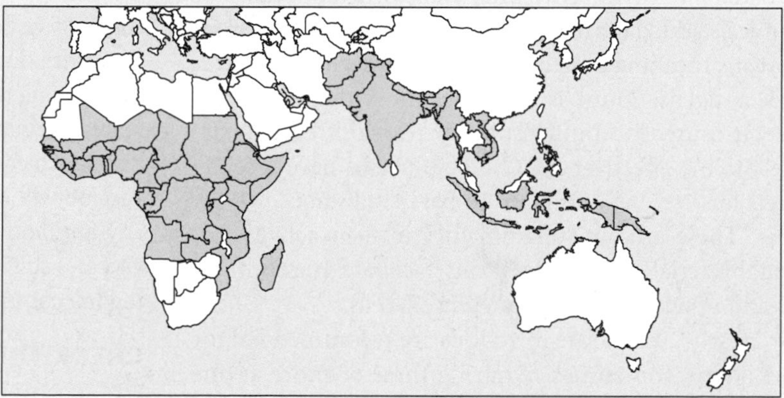

Figure C Countries with the highest measles mortality.

Recombinant vaccines and DNA vaccines do not need a cell or animal host to grow the vaccine's microbe. This avoids a major problem with certain viruses that so far have not been grown in cell culture—hepatitis B, for example.

Plants are also a potential source for vaccines. There have already been human trials with potatoes that have been engineered to produce antigenic proteins from certain pathogenic bacteria and viruses. It is more likely, however, that plants for this purpose will not be used directly as food, but as a production system for doses of antigenic proteins that would be taken orally as pills. Oral vaccines would be welcomed for many reasons even beyond eliminating a need for injections. For one, they would be especially effective in protecting against the diseases caused by pathogens invading the body through the mucous membranes. This obviously includes intestinal diseases such as cholera, but the pathogens causing AIDS, influenza, and some other nonintestinal diseases initially invade the body through mucous membranes elsewhere, such as the nose, genitals, and lungs. Tobacco plants are a leading candidate for this use because they are unlikely to contaminate the food chain.

The so-called golden age of immunology occurred from about 1870 to 1910, when most of the basic elements of immunology were discovered and several important vaccines were developed. We may soon be entering another golden age, in which new technologies are brought to bear on emerging infectious diseases and problems arising from the decreasing effectiveness of antibiotics. It is remarkable that there are no useful vaccines against chlamydias, fungi, protozoa, or helminthic parasites of humans. Moreover, vaccines for some diseases, such as cholera and tuberculosis, are not reliably protective. At present, vaccines for at least 75 diseases are under development, ranging from those for prominent deadly diseases such as AIDS and malaria to such commonplace conditions as earaches. But we will probably find that the easy vaccines have already been made.

Infectious diseases are not the only possible target of vaccines. Researchers are investigating vaccines' potential for treating and preventing cocaine addiction, Alzheimer disease, and cancer and for contraception.

Work is underway to improve effectiveness of antigens. For example, chemicals added for this purpose, called **adjuvants** (from the Latin *adjuvare*, meaning to help), greatly improve the effectiveness of many antigens. Alum is the adjuvant with the longest history of use. It causes local inflammatory reactions that apparently increase vaccine effectiveness. Other adjuvants have recently been registered for use, including an oil-based substance, MF59, and virosomes. These are not inflammatory in their activity but mimic certain bacterial components and facilitate transport to lymph nodes and uptake by antigen-presenting cells.

Currently, nearly 20 separate injections are recommended for infants and children, sometimes requiring three or more at one appointment. Developing additional multiple combinations of vaccines would be of some help. The U.S. Food and Drug Administration (FDA) has recently approved such a combination for five childhood diseases. Delivery in ways other than by needle would also be a desirable advance. Many have already received injection by high-pressure "guns," which are commonly used for mass inoculations, and an intranasal spray for influenza is now available. Most current vaccines have primarily induced humoral, or antibody-based, immunity. Vaccines effective against diseases such as HIV infection, tuberculosis, and malaria will require induction of effective cellular immunity as well. These requirements are not necessarily mutually exclusive.

CHECK YOUR UNDERSTANDING

✓ Which type of vaccine did Louis Pasteur develop, whole-agent, recombinant, or DNA? **18-5**

✓ What is the derivation of the word *adjuvant*? **18-6**

Safety of Vaccines

We have seen how variolation, the first attempt to provide immunity to smallpox, sometimes caused the disease it was intended to prevent. At the time, however, the risk was considered very worthwhile. As you will see later in this book, the oral polio vaccine on rare occasions may *cause* the disease. In 1999, a vaccine to prevent infant diarrhea caused by rotaviruses was withdrawn from the market because several recipients developed a life-threatening intestinal obstruction. However, public reaction to such risks has changed; most parents have never seen a case of polio or measles and therefore tend to view the risk of these diseases as a remote abstraction. Moreover, reports or rumors of harmful effects often lead people to avoid certain vaccines for themselves or their children. In particular, a possible connection between the MMR vaccine and autism has received widespread publicity. Autism is a poorly understood developmental condition that causes a child to withdraw from reality. Because autism is usually diagnosed at the age of 18 to 30 months, about the time vaccine immunization schedules are nearing completion, some people have attempted to make a cause-and-effect connection. Medically, however, most experts agree that autism is a condition with a major genetic component and begins before birth. Extensive scientific surveys have provided no evidence to support a connection between the usual childhood vaccines and autism or any other disease condition. Some experts even recommend again introducing the rotavirus vaccine that was withdrawn in the United States, maintaining that it would be well justified on a risk-versus-benefit calculation in much of the underdeveloped world. No vaccine will ever be perfectly safe or perfectly effective—neither is any antibiotic or most other drugs, for that matter. Nevertheless, vaccines still remain the safest and most effective means of preventing infectious disease in children.

CHECK YOUR UNDERSTANDING

✓ What is the name of a currently used oral vaccine that occasionally causes the disease it is intended to prevent? **18-7**

Diagnostic Immunology

LEARNING OBJECTIVES

18-8 Differentiate sensitivity from specificity in a diagnostic test.

18-9 Define *monoclonal antibodies,* and identify their advantage over conventional antibody production.

18-10 Explain how precipitation reactions and immunodiffusion tests work.

18-11 Differentiate direct from indirect agglutination tests.

18-12 Differentiate agglutination from precipitation tests.

18-13 Define *hemagglutination.*

18-14 Explain how a neutralization test works.

18-15 Differentiate precipitation from neutralization tests.

18-16 Explain the basis for the complement-fixation test.

18-17 Compare and contrast direct and indirect fluorescent-antibody tests.

18-18 Explain how direct and indirect ELISA tests work.

18-19 Explain how Western blotting works.

18-20 Explain the importance of monoclonal antibodies.

Throughout most of history, diagnosing a disease was essentially a matter of observing a patient's signs and symptoms. The writings of ancient and medieval physicians left descriptions of many diseases that are recognizable even today. Essential elements of diagnostic tests are sensitivity and specificity. **Sensitivity** is the probability that the test is reactive if the specimen is a true positive. **Specificity** is the probability that a positive test will *not* be reactive if a specimen is a true negative.

Immunologic-Based Diagnostic Tests

Knowledge of the high specificity of the immune system soon suggested that this might be used in diagnosing diseases. In fact, it was an accidental observation that led to one of the first diagnostic tests for an infectious disease. More than 100 years ago, Robert Koch was trying to develop a vaccine against tuberculosis. He observed that when guinea pigs with the disease were injected with a suspension of *Mycobacterium tuberculosis,* the site of the injection became red and slightly swollen a day or two later. You may recognize this symptom as a positive result for the widely used tuberculin skin test (see Figure 24.10, page 684)—many colleges and universities require the test as part of admission procedures. Koch, of course, had no idea of the mechanism of cell-mediated immunity that caused this phenomenon, nor did he know of the existence of antibodies.

Since the time of Robert Koch, immunology has given us many other invaluable diagnostic tools, most of which are based on interactions of humoral antibodies with antigens. A known antibody can be used to identify an *unknown* pathogen (antigen) by its reaction with it. This reaction can be reversed, and a *known* pathogen can be used, for example, to determine the presence of an unknown antibody in a person's blood—which would determine whether he or she had immunity to the pathogen. One problem that must be overcome in antibody-based diagnostic tests is that antibodies cannot be seen directly. Even at magnifications of well over 100,000×, they appear only as fuzzy, ill-defined particles (see Figure 17.3c on page 480). Therefore, their presence must be established indirectly. We will describe a number of ingenious solutions to this problem.

Other problems that had to be overcome were that antibodies produced in an animal were mixed with numerous other antibodies produced in that animal and the quantities of any particular antibody were severely limited.

CHECK YOUR UNDERSTANDING

✓ What property of the immune system suggested its use as an aid for diagnosing disease: specificity or sensitivity? **18-8**

Monoclonal Antibodies

As soon as it was determined that antibodies were produced by specialized cells (B cells), it was understood that these were a potential source of a single type of antibody. If such a B cell producing a single type of antibody could be isolated and cultivated, it would be able to produce the desired antibody in nearly unlimited quantities and without contamination by other antibodies. Unfortunately, a B cell reproduces only a few times under the usual cell culture conditions. This problem was largely solved with the discovery of a method to isolate and indefinitely cultivate B cells capable of producing a single type of antibody. Neils Jerne, Georges Köchler, and César Milstein made this discovery in 1975, for which they were awarded a Nobel prize.

Scientists have long observed that antibody-producing B cells may become cancerous. In this case, their proliferation is unchecked, and they are called *myelomas.* These cancerous B cells can be isolated and propagated indefinitely in cell culture. Cancer cells, in this sense, are "immortal." The breakthrough came in combining an "immortal" cancerous B cell with an antibody-producing normal B cell. When fused, this combination is termed a **hybridoma.**

When a hybridoma is grown in culture, its genetically identical cells continue to produce the type of antibody characteristic of the ancestral B cell. The importance of the technique is that clones of the antibody-secreting cells now can be maintained indefinitely in cell culture and can produce immense quantities of identical antibody molecules. Because all of these antibody molecules are produced by a single hybridoma clone, they are called **monoclonal antibodies,** or **Mabs** (Figure 18.2).

Monoclonal antibodies are useful for three reasons: they are uniform; they are highly specific; and they can be produced readily in large quantities. Because of these qualities, Mabs have assumed enormous importance as diagnostic tools. For instance, commercial kits use Mabs to recognize several bacterial pathogens, and nonprescription pregnancy tests use Mabs to indicate the presence of a hormone excreted only in the urine of a pregnant woman (see Figure 18.13, page 517).

Figure 18.2

FOUNDATION FIGURE The Production of Monoclonal Antibodies

An appreciation of how monoclonal antibodies, an important advance in medicine, are produced is helpful for understanding the functions and applications of many common diagnostic and therapeutic tools that will be discussed in this and subsequent chapters.

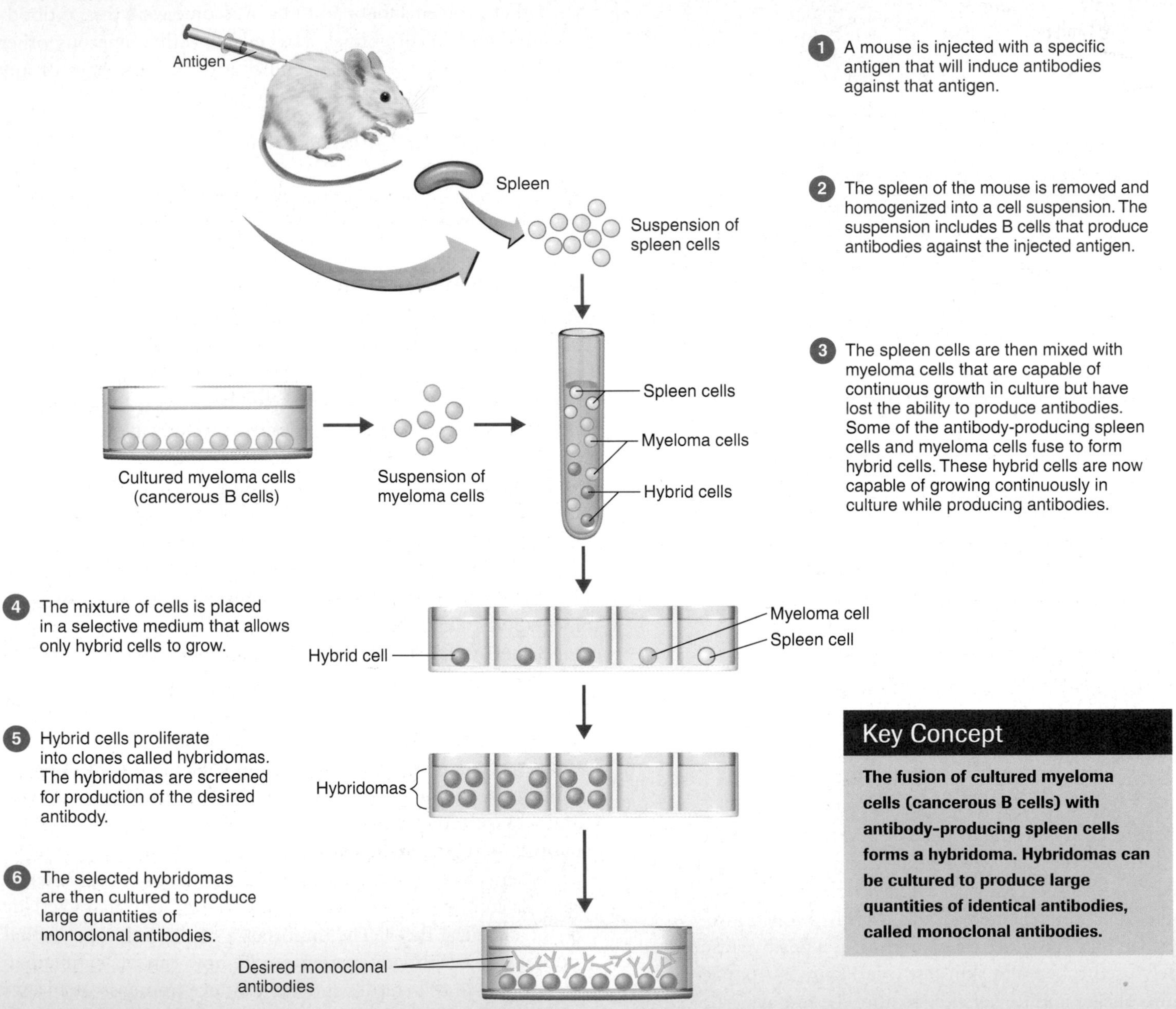

Antigen
1 A mouse is injected with a specific antigen that will induce antibodies against that antigen.
Spleen
Suspension of spleen cells
2 The spleen of the mouse is removed and homogenized into a cell suspension. The suspension includes B cells that produce antibodies against the injected antigen.
Spleen cells
Myeloma cells
Hybrid cells
Cultured myeloma cells (cancerous B cells)
Suspension of myeloma cells
3 The spleen cells are then mixed with myeloma cells that are capable of continuous growth in culture but have lost the ability to produce antibodies. Some of the antibody-producing spleen cells and myeloma cells fuse to form hybrid cells. These hybrid cells are now capable of growing continuously in culture while producing antibodies.
4 The mixture of cells is placed in a selective medium that allows only hybrid cells to grow.
Hybrid cell
Myeloma cell
Spleen cell
5 Hybrid cells proliferate into clones called hybridomas. The hybridomas are screened for production of the desired antibody.
Hybridomas
6 The selected hybridomas are then cultured to produce large quantities of monoclonal antibodies.
Desired monoclonal antibodies

Key Concept

The fusion of cultured myeloma cells (cancerous B cells) with antibody-producing spleen cells forms a hybridoma. Hybridomas can be cultured to produce large quantities of identical antibodies, called monoclonal antibodies.

Monoclonal antibodies are also being used therapeutically to overcome unwanted effects of the immune system. For example, *muromonab-CD3* has been used since 1986 to minimize rejection of kidney transplants. For these purposes, monoclonal antibodies are prepared that react with the T cells that are responsible for rejecting the transplanted tissue. The Mabs suppress the T cell activity.

The use of Mabs is revolutionizing the treatment of many illnesses. The FDA has approved several for the treatment of specific diseases. The inflammation of rheumatoid arthritis and the intestinal inflammatory condition Crohn's disease can be treated with *infliximab (Remicade)* or *entanercept (Enbrel)*, which target and block the action of a cause of the inflammation, tumor necrosis factor alpha (see page 492). A cancer of the lymphatic system, non-Hodgkin's lymphoma, can be treated with a combination of Mabs, *ibritumonab (Zevalin)* and *rituximab (Rituxan)*. These target and destroy cancer cells but are used only when other treatments have proved unsuccessful. Breast cancer can be treated with some success with *trastuzumab (Herceptin)*. This Mab binds to a specific site called the HER-2 receptor, which occurs in about 30% of women; this limits the spread of the cancer.

The therapeutic use of Mabs had been limited because these antibodies once were produced only by mouse (murine) cells. The immune systems of patients reacted against the foreign mouse proteins, leading to rashes, swelling, and even occasional kidney failure, plus the destruction of the Mabs. For example, the success of *muromonab-CD3* in minimizing tissue rejection was severely limited by side effects related to the administration of the foreign (murine) fraction of the monoclonal antibody.

In recognition of this problem, researchers are developing new generations of Mabs that are less likely to cause side effects due to their "foreignness." Essentially, the more human the antibody, the more successful it is likely to be. Researchers are exploring several approaches.

Chimeric monoclonal antibodies use genetically modified mice to make a human–murine hybrid. The variable part of the antibody molecule, including the antigen-binding sites (see Figure 17.3a), is murine. The remainder of the antibody molecule, the constant region, has been derived from a human source. These Mabs are about 66% human. An example is rituximab.

Humanized antibodies are constructed so that the murine portion is limited to the antigen-binding sites. The balance of the variable region and all of the constant region are derived from human sources. Such Mabs are about 90% human. Examples are alemtuzumab and trastuzumab.

The eventual goal is to develop **fully human antibodies.** One approach is to genetically modify mice to contain human antibody genes. The mice would produce antibodies that are fully human; in some cases, it might even be possible to produce an antibody that is an exact match to the patient.

It is also possible that Mab therapies may succeed so well that it would be difficult to produce them in sufficient volumes. Several potential solutions to this problem are under investigation. For example, the use of mice could be avoided entirely by using bacteriophages to insert desired genes into bacteria, which would be able to produce the desired Mabs on an industrial scale. Another approach to the problem is to genetically modify animals that can secrete the Mabs in their milk. Genetic alteration of plants to produce Mabs is another possible avenue to large-scale production.

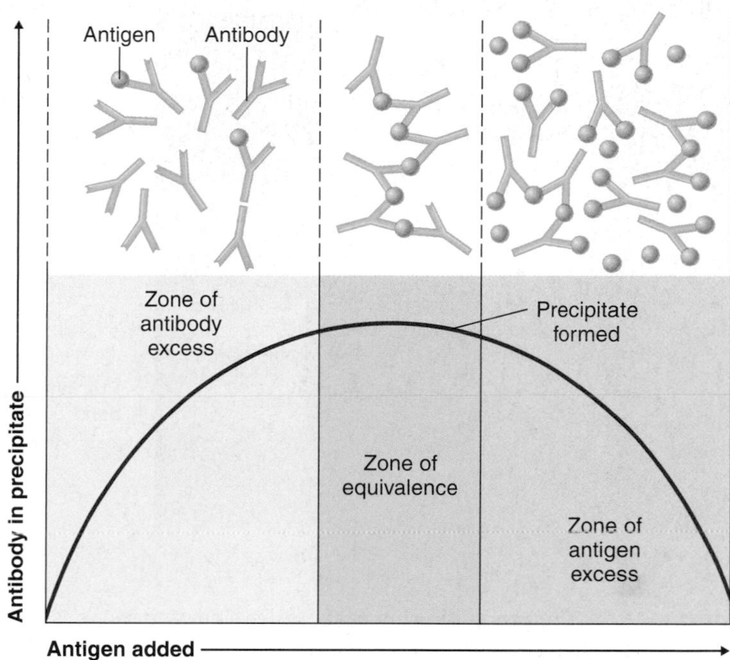

Figure 18.3 A precipitation curve. The curve is based on the ratio of antigen to antibody. The maximum amount of precipitate forms in the zone of equivalence, where the ratio is roughly equivalent.

Q How does precipitation differ from agglutination?

CHECK YOUR UNDERSTANDING

✓ The blood of an infected cow would have a considerable amount of antibodies against the infectious pathogen in its blood. How would an equivalent amount of monoclonal antibodies be more useful? **18-9**

Precipitation Reactions

Precipitation reactions involve the reaction of soluble antigens with IgG or IgM antibodies to form large, interlocking molecular aggregates called *lattices.*

Precipitation reactions occur in two distinct stages. First, the antigens and antibodies rapidly form small antigen–antibody complexes. This interaction occurs within seconds and is followed by a slower reaction, which may take minutes to hours, in which the antigen–antibody complexes form lattices that precipitate from solution. Precipitation reactions normally occur only when the ratio of antigen to antibody is optimal. Figure 18.3 shows that no visible precipitate forms when either component is in excess. The optimal ratio is produced when separate solutions of antigen and antibody are placed adjacent to each other and allowed to diffuse together. In a **precipitin ring test** (Figure 18.4), a cloudy line of precipitation (ring) appears in the area in which the optimal ratio has been reached (the *zone of equivalence*).

Immunodiffusion tests are precipitation reactions carried out in an agar gel medium, on either a Petri plate or a

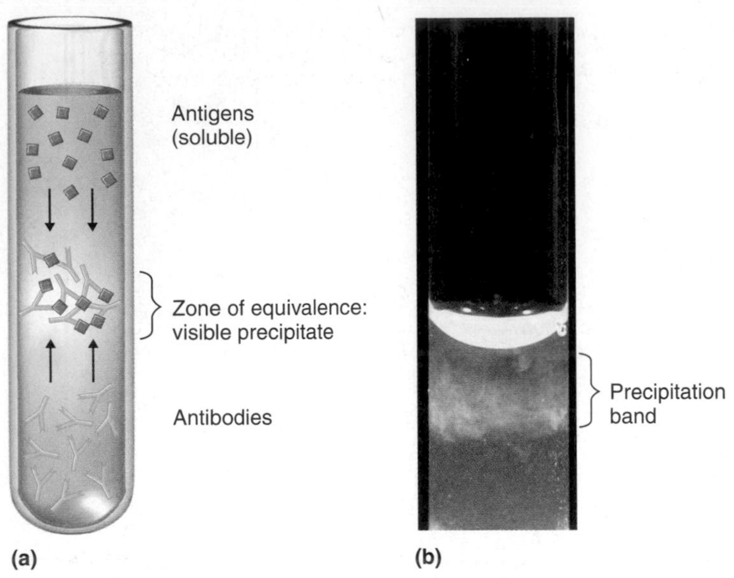

Figure 18.4 The precipitin ring test. (**a**) This drawing shows the diffusion of antigens and antibodies toward each other in a small-diameter test tube. Where they reach equal proportions, in the zone of equivalence, a visible line or ring of precipitate is formed. (**b**) A photograph of a precipitin band.

Q What causes the visible line?

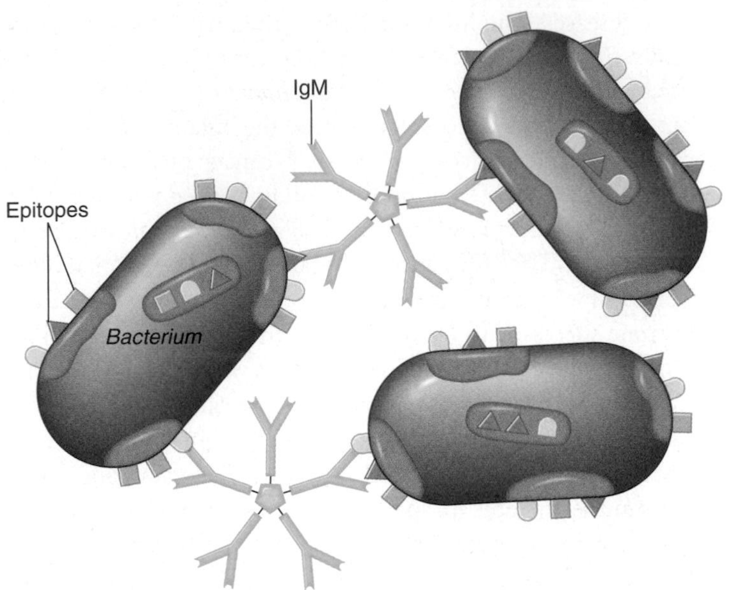

Figure 18.5 An agglutination reaction. When antibodies react with epitopes on antigens carried on neighboring cells, such as these bacteria (or red blood cells), the particulate antigens (cells) agglutinate. IgM, the most efficient immunoglobulin for agglutination, is shown here, but IgG also participates in agglutination reactions.

Q Draw an agglutination reaction involving IgG.

microscope slide. A line of visible precipitate develops between the wells at the point where the optimal antigen–antibody ratio is reached.

Other tests use electrophoresis to speed up the movement of antigen and antibody in a gel, sometimes in less than an hour, with this method. The techniques of immunodiffusion and electrophoresis can be combined in a procedure called **immunoelectrophoresis.** The procedure is used in research to separate proteins in human serum and is the basis of certain diagnostic tests. It is an essential part of the Western blot test used in AIDS testing (see Figure 10.12, page 289 and page 516).

CHECK YOUR UNDERSTANDING

✓ Why does the reaction of a precipitation test become visible only in a narrow range? **18-10**

Agglutination Reactions

Whereas precipitation reactions involve *soluble* antigens, agglutination reactions involve either *particulate* antigens (particles such as cells that carry antigenic molecules) or soluble antigens adhering to particles. These antigens can be linked together by antibodies to form visible aggregates, a reaction called **agglutination** (Figure 18.5). Agglutination reactions are very sensitive, relatively easy to read (see Figure 10.10, page 287), and available in great variety. Agglutination tests are classified as either direct or indirect.

Direct Agglutination Tests

Direct agglutination tests detect antibodies against relatively large cellular antigens, such as those on red blood cells, bacteria, and fungi. At one time they were carried out in a series of test tubes, but now they are usually done in plastic *microtiter plates,* which have many shallow wells that take the place of the individual test tubes. The amount of particulate antigen in each well is the same, but the amount of serum that contains antibodies is diluted, so that each successive well has half the antibodies of the previous well. These tests are used, for example, to test for brucellosis and to separate *Salmonella* isolates into serovars, types defined by serological means.

Clearly, the more antibody we start with, the more dilutions it will take to lower the amount to the point where there is not enough antibody for the antigen to react with. This is the measure of **titer,** or concentration of serum antibody (Figure 18.6). For infectious diseases in general, the higher the serum antibody titer, the greater the immunity to the disease. However, the titer alone is of limited use in diagnosing an existing illness. There is no way to know whether the measured antibodies were generated in response to the immediate infection or to an earlier illness. For diagnostic purposes, *a rise in titer* is significant; that is, the titer is higher later in the course of the disease than at its outset. Also, if it can be demonstrated that the person's blood had no antibody titer before the illness but has a significant titer while the disease is progressing, this change, called **seroconversion,** is also diagnostic. This situation is frequently encountered with HIV infections.

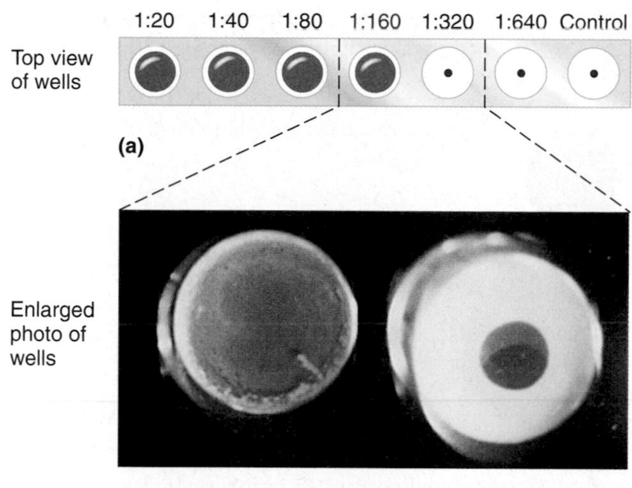

(a) Each well in this microtiter plate contains, from left to right, only half the concentration of serum that is contained in the preceding well. Each well contains the same concentration of particulate antigens, in this instance red blood cells.

Side view of wells

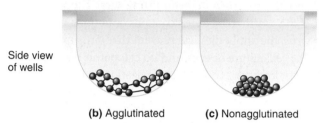

(b) Agglutinated **(c)** Nonagglutinated

(b) In a positive (agglutinated) reaction, sufficient antibodies are present in the serum to link the antigens together, forming a mat of antigen–antibody complexes on the bottom of the well.

(c) In a negative (nonagglutinated) reaction, not enough antibodies are present to cause the linking of antigens. The particulate antigens roll down the sloping sides of the well, forming a pellet at the bottom. In this example, the antibody titer is 160 because the well with a 1:160 concentration is the most dilute concentration that produces a positive reaction.

Figure 18.6 Measuring antibody titer with the direct agglutination test.

Q What is meant by the term *antibody titer*?

Some diagnostic tests specifically identify IgM antibodies. As discussed in Chapter 17, short-lived IgM is more likely to reflect a response to a current disease condition.

Indirect (Passive) Agglutination Tests

Antibodies against soluble antigens can be detected by agglutination tests if the antigens are adsorbed onto particles such as bentonite clay or, most often, minute latex spheres, each about one-tenth of the diameter of a bacterium. Such tests, known as *latex agglutination tests,* are commonly used for the rapid detection of serum antibodies against many bacterial and viral diseases. In such **indirect (passive) agglutination tests,** the antibody reacts with the soluble antigen adhering to the particles (Figure 18.7). The particles then agglutinate with one another, much as particles do in the direct agglutination tests. The same principle can be applied in reverse by using particles coated with antibodies to

Figure 18.7 Reactions in indirect agglutination tests. These tests are performed using antigens or antibodies coated onto particles such as minute latex spheres.

Q Differentiate direct from indirect agglutination tests.

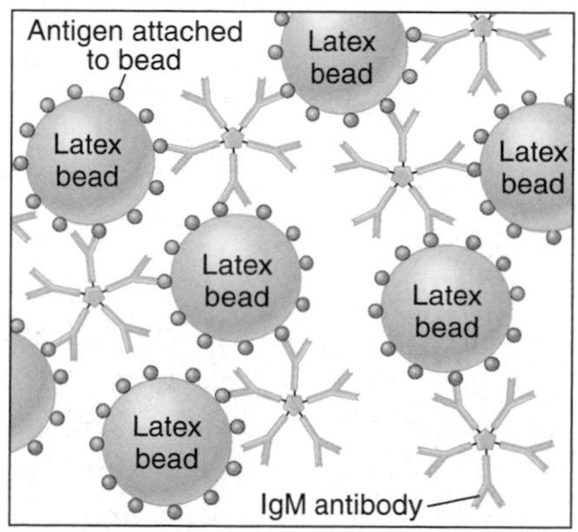

(a) Reaction in a positive indirect test for antibodies. When particles (latex beads here) are coated with antigens, agglutination indicates the presence of antibodies, such as the IgM shown here.

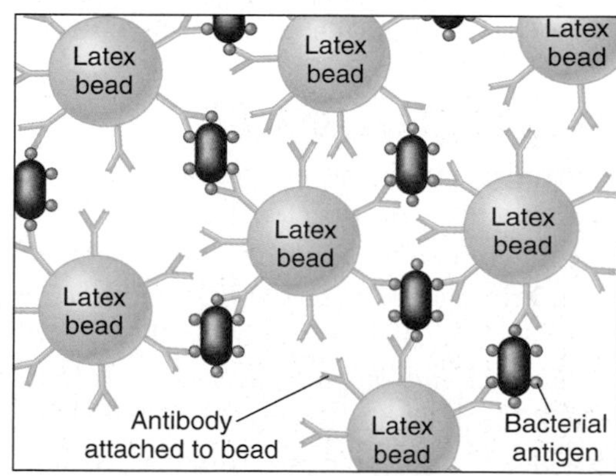

(b) Reaction in a positive indirect test for antigens. When particles are coated with monoclonal antibodies, agglutination indicates the presence of antigens.

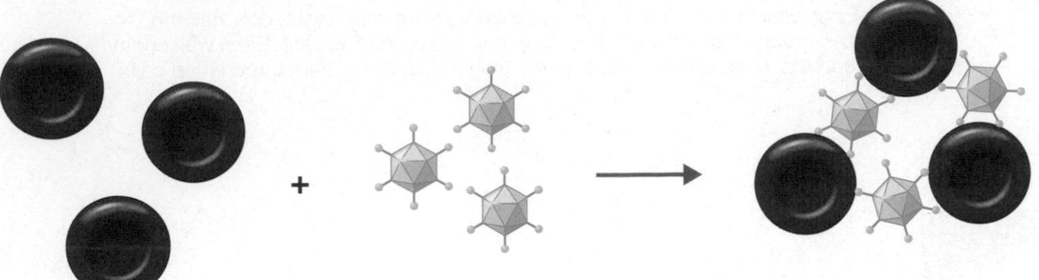

Figure 18.8 Viral hemagglutination. Viral hemagglutination is not an antigen–antibody reaction.

Q **What causes agglutination in viral hemagglutination?**

detect the antigens against which they are specific. This approach is especially common in tests for the streptococci that cause sore throats. A diagnosis can be completed in about 10 minutes.

Hemagglutination

When agglutination reactions involve the clumping of red blood cells, the reaction is called **hemagglutination.** These reactions, which involve red blood cell surface antigens and their complementary antibodies, are used routinely in blood typing (see Table 19.2, page 527) and in the diagnosis of infectious mononucleosis.

Certain viruses, such as those causing mumps, measles, and influenza, have the ability to agglutinate red blood cells without an antigen–antibody reaction; this process is called **viral hemagglutination** (Figure 18.8). This type of hemagglutination can be inhibited by antibodies that neutralize the agglutinating virus. Diagnostic tests based on such neutralization reactions are discussed in the next section.

CHECK YOUR UNDERSTANDING

- ✓ Why wouldn't a direct agglutination test work very well with viruses? **18-11**
- ✓ Which test detects soluble antigens, agglutination or precipitation? **18-12**
- ✓ Certain diagnostic tests require red blood cells that clump visibly. What are these tests called? **18-13**

Neutralization Reactions

Neutralization is an antigen–antibody reaction in which the harmful effects of a bacterial exotoxin or a virus are blocked by specific antibodies. These reactions were first described in 1890, when investigators observed that immune serum could neutralize the toxic substances produced by the diphtheria pathogen, *Corynebacterium diphtheriae.* Such a neutralizing substance, which is called an antitoxin, is a specific antibody produced by a host as it responds to a bacterial exotoxin or its corresponding toxoid (inactivated toxin). The antitoxin combines with the exotoxin to neutralize it (Figure 18.9a). Antitoxins produced in an animal can be injected into humans to provide passive immunity against a toxin. Antitoxins from horses are routinely used to prevent or treat diphtheria and botulism; tetanus antitoxin is usually of human origin.

These therapeutic uses of neutralization reactions have led to their use as diagnostic tests. Viruses that exhibit their cytopathic (cell-damaging) effects in cell culture or embryonated eggs can be used to detect the presence of neutralizing viral antibodies (see page 441). If the serum to be tested contains antibodies against the particular virus, the antibodies will prevent that virus from infecting cells in the cell culture or eggs, and no cytopathic effects will be seen. Such tests, known as in vitro neutralization tests, can thus be used both to identify a virus and to ascertain the viral antibody titer. In vitro neutralization tests are comparatively complex to carry out and are becoming less common in modern clinical laboratories.

A neutralization test used mostly for the serological typing of viruses is the **viral hemagglutination inhibition test.** Certain viruses such as those causing influenza, mumps, and measles have surface proteins that will cause the agglutination of red blood cells. This test finds its most common use in the subtyping of influenza viruses, although more laboratories are likely to be familiar with ELISA tests for this purpose. If a person's serum contains antibodies against these viruses, these antibodies will react with the viruses and neutralize them (Figure 18.9b). For example, if hemagglutination occurs in a mixture of measles virus and red blood cells but does not occur when the patient's serum is added to the mixture, this result indicates that the serum contains antibodies that have bound to and neutralized the measles virus.

CHECK YOUR UNDERSTANDING

- ✓ In what way is there a connection between hemagglutination and certain viruses? **18-14**
- ✓ Which of these tests is an antigen–antibody reaction: precipitation or viral hemagglutination inhibition? **18-15**

Complement-Fixation Reactions

In Chapter 16 (pages 463–468), we discussed a group of serum proteins collectively called complement. During most antigen–antibody reactions, complement binds to the antigen–antibody complex and is used up, or fixed. This process of **complement fixation** can be used to detect very small amounts of antibody. Antibodies that do not produce a visible reaction, such as precipitation or agglutination, can be demonstrated by the fixing of complement during the

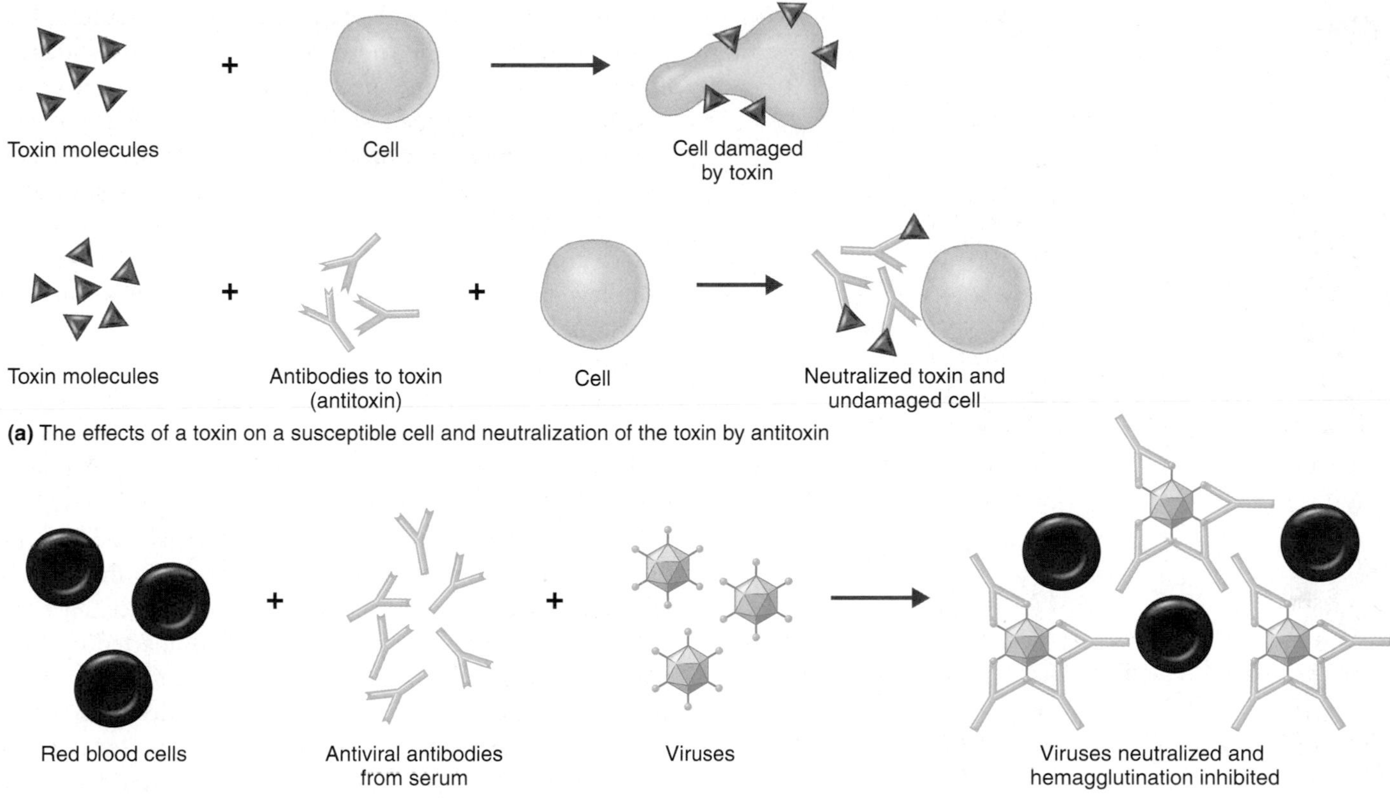

(a) The effects of a toxin on a susceptible cell and neutralization of the toxin by antitoxin

(b) Viral hemagglutination test to detect antibodies to a virus. These viruses will normally cause hemagglutination when mixed with red blood cells. If antibodies to the virus are present, as shown here, they neutralize and inhibit hemagglutination.

Figure 18.9 **Reactions in neutralization tests.**

Q **Why does hemagglutination indicate that a patient does not have a specific disease?**

antigen–antibody reaction. Complement fixation was once used in the diagnosis of syphilis (Wassermann test) and is still used to diagnose certain viral, fungal, and rickettsial diseases. The complement-fixation test requires great care and good controls, one reason the trend is to replace it with newer, simpler tests. The test is performed in two stages: complement fixation and indicator (Figure 18.10).

CHECK YOUR UNDERSTANDING

✓ Why is complement given its name? **18-16**

Fluorescent-Antibody Techniques

Fluorescent-antibody (FA) techniques can identify microorganisms in clinical specimens and can detect the presence of a specific antibody in serum (Figure 18.11). These techniques combine fluorescent dyes such as fluorescein isothiocyanate (FITC) with antibodies to make them fluoresce when exposed to ultraviolet light (see Figure 3.6, page 61). These procedures are quick, sensitive, and very specific; the FA test for rabies can be performed in a few hours and has an accuracy rate close to 100%.

Q&A Fluorescent-antibody tests are of two types, direct and indirect. **Direct FA tests** are usually used to identify a microorganism in a clinical specimen (Figure 18.11a). During this procedure, the specimen containing the antigen to be identified is fixed onto a slide. Fluorescein-labeled antibodies are then added, and the slide is incubated briefly. Next the slide is washed to remove any antibody not bound to antigen and is then examined under the fluorescence microscope for yellow-green fluorescence. The residual antibody will be visible even if the antigen, such as a virus, is submicroscopic in size.

Indirect FA tests are used to detect the presence of a specific antibody in serum following exposure to a microorganism (Figure 18.11b). They are often more sensitive than direct tests. During this procedure, a known antigen is fixed onto a slide. The test serum is then added, and, if antibody that is specific to that microbe is present, it reacts with the antigen to form a bound complex. For the antigen–antibody complex to be seen, fluorescein-labeled **antihuman immune serum globulin (anti-HISG),** an antibody that reacts specifically with *any* human antibody, is added to the slide. Anti-HISG will be present only if the specific antibody has reacted with its antigen and is therefore present as well. After the slide has been incubated and washed

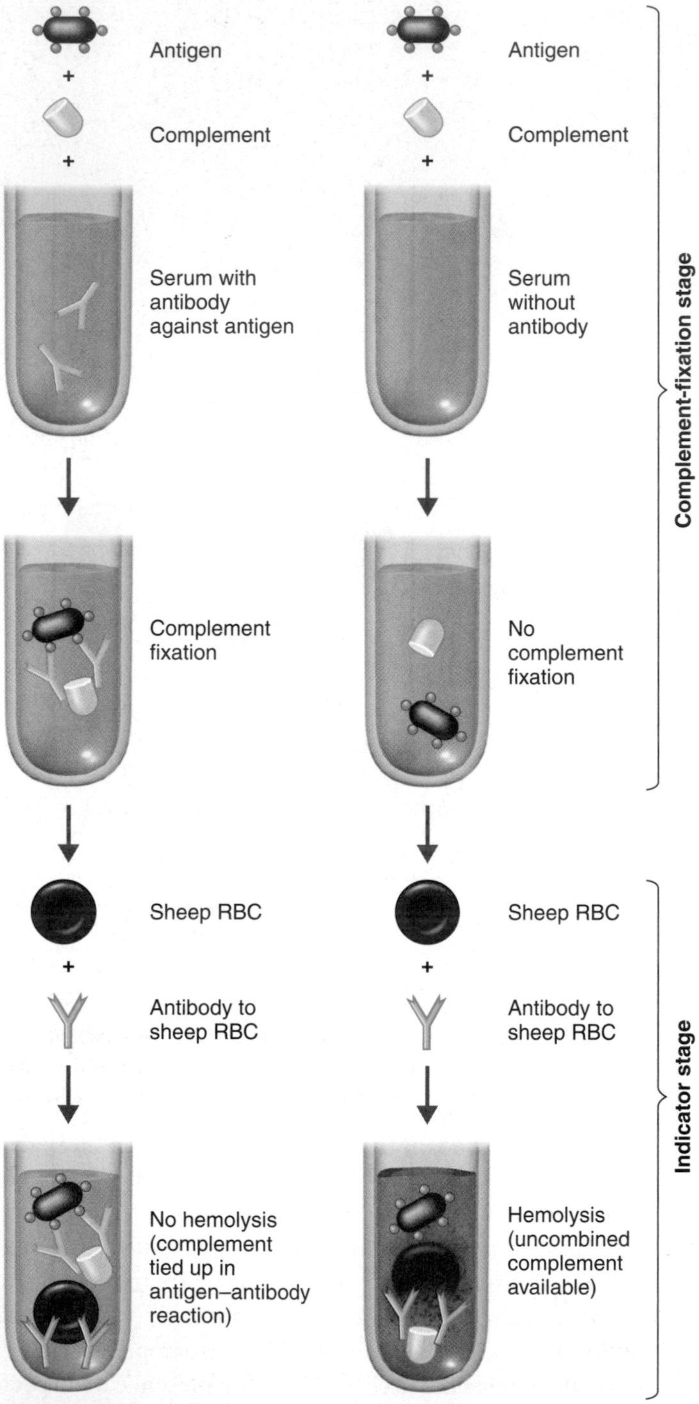

(a) Positive test. All available complement is fixed by the antigen–antibody reaction; no hemolysis occurs, so the test is positive for the presence of antibodies.

(b) Negative test. No antigen–antibody reaction occurs. The complement remains, and the red blood cells are lysed in the indicator stage, so the test is negative.

Figure 18.10 The complement-fixation test. This test is used to indicate the presence of antibodies to a known antigen. Complement will combine (be fixed) with an antibody that is reacting with an antigen. If all the complement is fixed in the complement-fixation stage, then none will remain to cause hemolysis of the red blood cells in the indicator stage.

Q Why does red blood cell lysis indicate that the patient does not have a specific disease?

(to remove unbound antibody), it is examined under a fluorescence microscope. If the known antigen fixed to the slide appears fluorescent, the antibody specific to the test antigen is present.

An especially interesting adaptation of fluorescent antibodies is the **fluorescence-activated cell sorter (FACS).** In Chapter 17, we learned that T cells carry antigenically specific molecules such as CD4 and CD8 on their surface, and these are characteristic of certain groups of T cells. The depletion of $CD4^+$ T cells is used to follow the progression of AIDS; their populations can be determined with a FACS.

The FACS is a modification of a *flow cytometer,* in which a suspension of cells leaves a nozzle as droplets containing no more than one cell each (see page 288). A laser beam strikes each cell-containing droplet and is then received by a detector that determines certain characteristics such as size (Figure 18.12). If the cells carry FA markers to identify them as $CD4^+$ or $CD8^+$ T cells, the detector can measure this fluorescence. As the laser beam detects a cell of a preselected size or fluorescence, an electrical charge, either positive or negative, can be imparted to it. As the charged droplet falls between electrically charged plates, it is attracted to one receiving tube or another, effectively separating cells of different types. Millions of cells can be separated in an hour with this process, all under sterile conditions, which allows them to be used in experimental work.

An interesting application of the flow cytometer is sorting sperm cells to separate male (Y-carrying) and female (X-carrying) sperm. The female sperm (meaning that it will result in a female embryo when it fertilizes the egg) contains more DNA, 2.8% more in humans, 4% in animals. When the sperm is stained with a fluorescent dye specific for DNA, the female sperm glows more brightly when illuminated by the laser beam because it has more DNA and therefore can be separated out. The technique was developed for agricultural purposes. However, it has received medical approval for use in humans where couples carry genes for inherited diseases that affect only boys.

CHECK YOUR UNDERSTANDING

✓ Which test is used to detect antibodies against a pathogen: the direct or the indirect fluorescent-antibody test? **18-17**

Enzyme-Linked Immunosorbent Assay (ELISA)

The **enzyme-linked immunosorbent assay (ELISA)** is the most widely used of a group of tests known as *enzyme immunoassay (EIA).* There are two basic methods. The *direct ELISA* detects antigens, and the *indirect ELISA* detects antibodies. A microtiter plate with numerous shallow wells is used in both procedures (see Figure 10.11a, page 288). Variations of the test exist; for example, the reagents can be bound to tiny latex particles rather than to the surfaces of the microtiter plates. ELISA procedures are popular primarily because they require little interpretive skill to read; the results tend to be clearly positive or clearly negative.

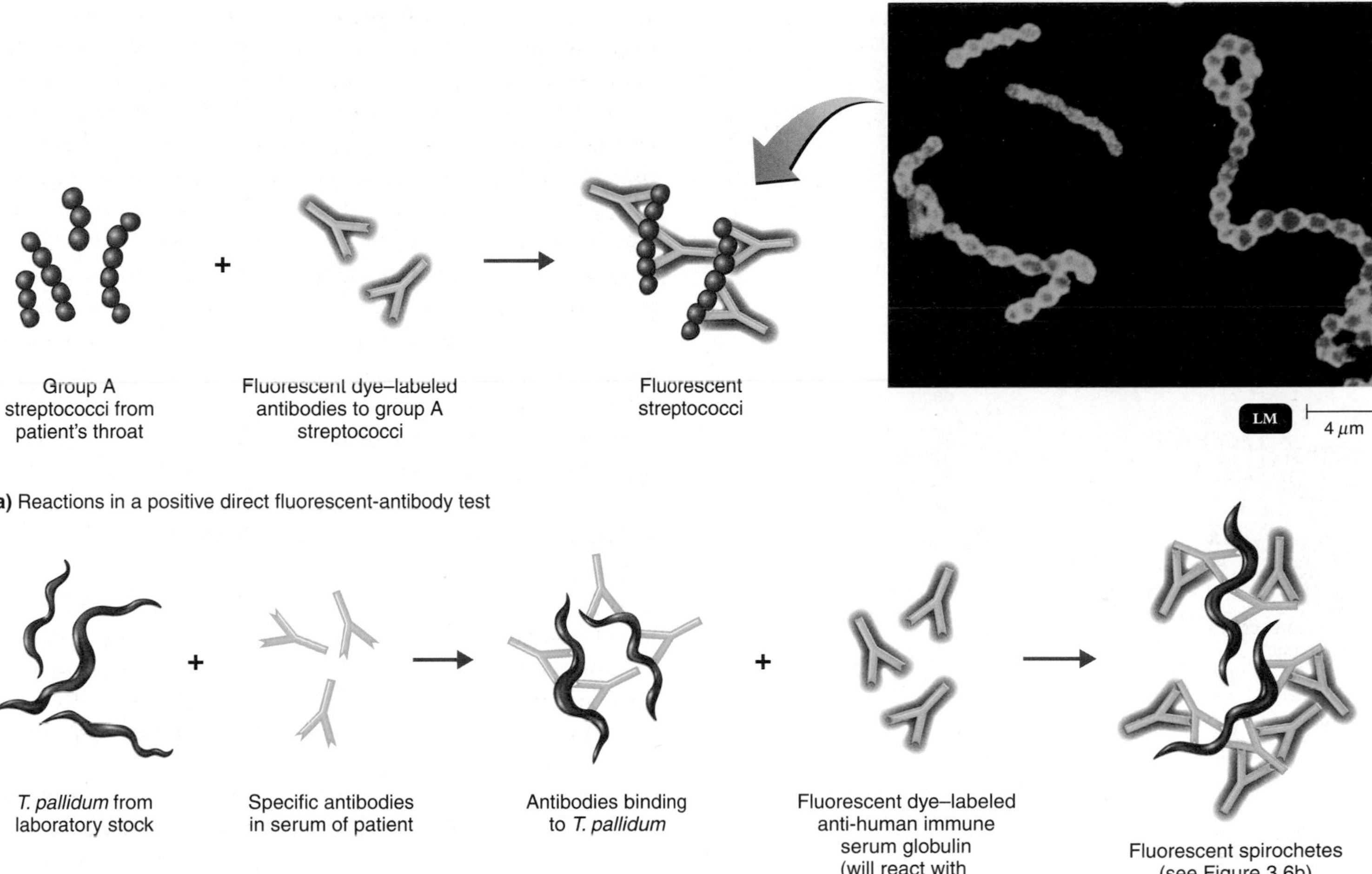

Figure 18.11 Fluorescent-antibody (FA) techniques. (**a**) A direct FA test to identify group A streptococci. (**b**) In an indirect FA test such as that used in the diagnosis of syphilis, the fluorescent dye is attached to antihuman immune serum globulin, which reacts with any human immunoglobulin (such as the *Treponema pallidum*-specific antibody) that has previously reacted with the antigen. The reaction is viewed through a fluorescence microscope, and the antigen with which the dye-tagged antibody has reacted fluoresces (glows) in the ultraviolet illumination.

Q Differentiate a direct from an indirect FA test.

Many ELISA tests are available for clinical use in the form of commercially prepared kits. Procedures are often highly automated, with the results read by a scanner and printed out by computer (see Figure 10.11, page 288). Some tests based on this principle are also available for use by the public; one example is a commonly available home pregnancy test (**Figure 18.13**).*

*An Egyptian papyrus dating to 1350 B.C. described a pregnancy test in which a woman would urinate on wheat and barley seeds for several days. Growth of the seeds was an indication of pregnancy. In 1963 this theory was tested, and 70% of the time the urine of a pregnant woman did indeed promote growth whereas the urine of men or nonpregnant women did not. Elevated levels of estrogens were considered a likely cause. (The use of two seed species was intended to determine gender: barley, male; wheat, female).

Direct ELISA

The direct ELISA method is shown in **Figure 18.14a** (page 518). A common use of the direct ELISA test is to detect the presence of drugs in urine. For these tests, antibodies specific for the drug are adsorbed to the well on the microtiter plate. (The availability of monoclonal antibodies has been essential to the widespread use of the ELISA test.) When the patient's urine sample is added to the well, any of the drug that it contained would bind to the antibody and is captured. The well is rinsed to remove any unbound drug. To make a visible test, more antibodies specific to the drug are now added (these antibodies have an enzyme attached to them—therefore, the term *enzyme-linked*) and will react with the already-captured drug, forming a "sandwich" of antibody/drug/enzyme-linked antibody. This positive test can be

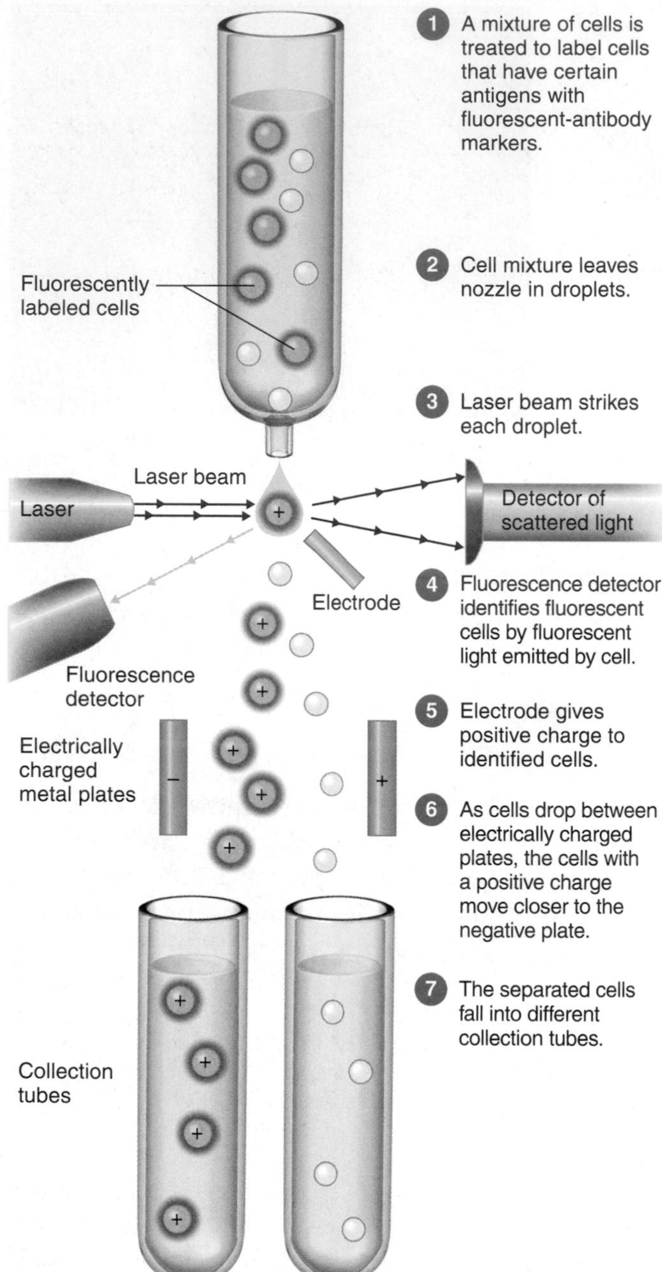

1 A mixture of cells is treated to label cells that have certain antigens with fluorescent-antibody markers.
2 Cell mixture leaves nozzle in droplets.
3 Laser beam strikes each droplet.
4 Fluorescence detector identifies fluorescent cells by fluorescent light emitted by cell.
5 Electrode gives positive charge to identified cells.
6 As cells drop between electrically charged plates, the cells with a positive charge move closer to the negative plate.
7 The separated cells fall into different collection tubes.
Fluorescently labeled cells
Laser beam
Laser
Detector of scattered light
Electrode
Fluorescence detector
Electrically charged metal plates
Collection tubes

Figure 18.12 The fluorescence-activated cell sorter (FACS). This technique can be used to separate different classes of T cells. A fluorescence-labeled antibody reacts with, for example, the CD4 molecule on a T cell.

Q Provide an application of FACS to follow the progress of HIV infection.

detected by adding a substrate for the linked enzyme; a visible color is produced by the enzyme reacting with its substrate.

Indirect ELISA

The indirect ELISA text, illustrated in Figure 18.14b, detects antibodies in a patient's sample rather than an antigen such as a drug. Indirect ELISA tests are used, for example, to screen blood for antibodies to HIV (see page 545). For such a purpose, the microtiter well contains an antigen, such as the inactivated virus that causes the disease the test is designed to diagnose. A sample of the patient's blood is added to the well; if it contains antibodies against the virus, they will react with the virus. The well is rinsed to remove unbound antibodies. If antibodies in the blood and the virus in the well have attached to each other, they will remain in the well—a positive test. To make a positive test visible, some anti-HISG (an immunoglobulin that will attach to *any* antibody, including the one in the patient's serum that has attached to the virus in the well; see page 513) is added. The anti-HISG is linked to an enzyme. A positive test consists of a "sandwich" or a virus/antibody/enzyme-linked-anti-HISG. At this point, the substrate for the enzyme is added, and a positive test is detected by the color change caused by the enzyme linked to the anti-HISG.

Western Blotting (Immunoblotting)

Western blotting, often simply called *immunoblotting*, can be used to identify a specific protein in a mixture. When this specific protein is an antibody, the technique is valuable in diagnosing disease. The components of the mixture are separated by electrophoresis in a gel and then transferred to a protein-binding sheet (blotter). There the protein/antigen is flooded with an enzyme-linked antibody. The location of the antigen and the enzyme-linked antibody reactant can be visualized, usually with a color-reacting label similar to an ELISA test reaction (see Figure 18.14). The most frequent application is in a confirmatory test for HIV infection (see page 545). Figure 10.12 on page 289 illustrates the procedure.*

CHECK YOUR UNDERSTANDING

✓ Which test is used to detect antibodies against a pathogen, the direct or the indirect ELISA test? **18-18**

✓ How are antibodies detected in Western blotting? **18-19**

The Future of Diagnostic and Therapeutic Immunology

The introduction of monoclonal antibodies has revolutionized diagnostic immunology by making available large, economical amounts of specific antibodies. This has led to many newer diagnostic tests that are more sensitive, specific, rapid, and simpler to use. For example, tests to diagnose sexually transmitted chlamydial infections and certain protozoan-caused intestinal parasitic diseases are coming into common use. These tests had previously required relatively difficult culture or microscopic methods for diagnosis. At the same time, the use of many of the classic serological tests, such as complement-fixation tests, is declining.

*The naming of Western blotting is a bit of scientific whimsy. The Southern blotting procedure used for detecting DNA fragments (page 262), which was named for the inventor Ed Southern, led to a similar procedure for detecting mRNA fragments being named Northern blotting. This "directional system" was continued when a new blotting procedure for identifying proteins was developed—hence, Western blotting.

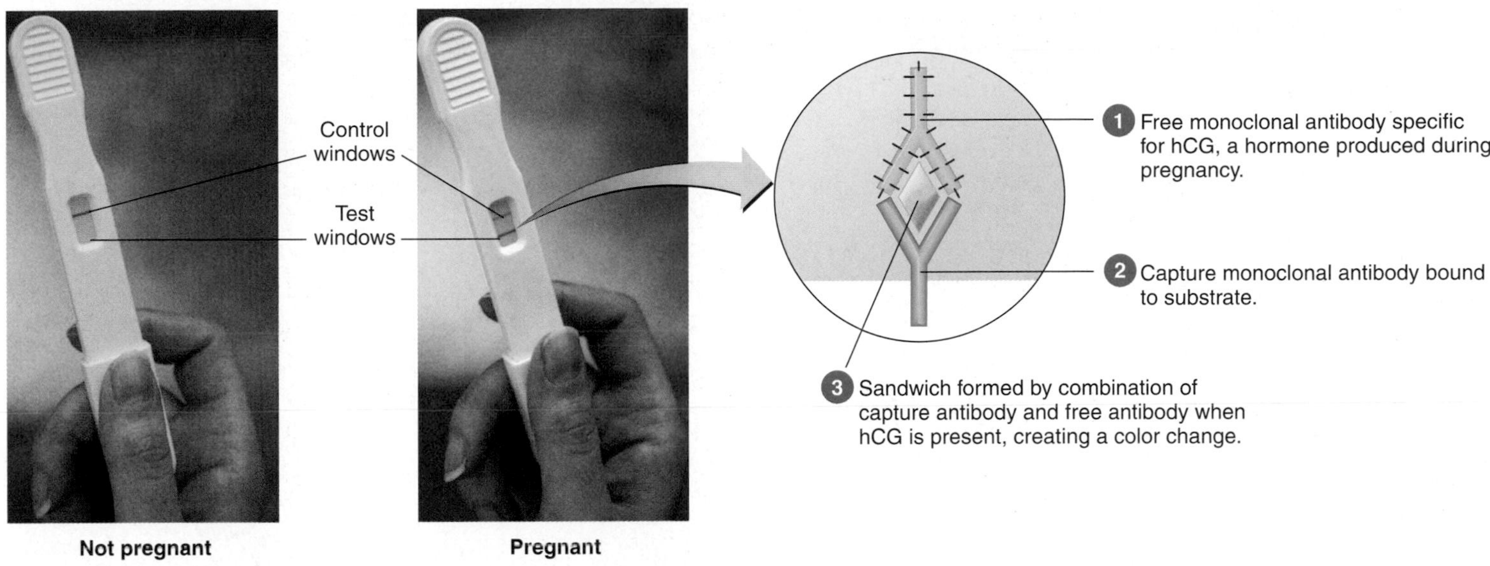
Control windows
Test windows
1 Free monoclonal antibody specific for hCG, a hormone produced during pregnancy.
2 Capture monoclonal antibody bound to substrate.
3 Sandwich formed by combination of capture antibody and free antibody when hCG is present, creating a color change.
Not pregnant
Pregnant

Figure 18.13 The use of monoclonal antibodies in a home pregnancy test. Home pregnancy tests detect a hormone called human chorionic gonadotropin (hCG) that is excreted only in the urine of a pregnant woman.

Q What is the antigen in the home pregnancy test?

Most newer tests will require less human judgment to read, and they require fewer highly trained personnel.

The use of certain *nonimmunological* tests, such as the PCR and DNA probes that were discussed in Chapter 10 (page 292), is increasing. Some of these tests will become automated to a significant degree. For example, a DNA chip (see Figure 10.17, page 293) containing over 50,000 DNA probes for genetic information expected in possible pathogens can be exposed to a test sample. This chip is scanned and its data automatically analyzed. PCR tests are also becoming highly automated.

Most of the diagnostic tests described in this chapter are those used in the developed world, where laboratory budgets are lavish compared to the funds available in much of the world. In many countries the money available for all medical expenditures, for diagnosis and treatment alike, is tragically small.

The diseases that most of these diagnostic methods target are also those that are more likely to be found in developed countries. In many parts of the world, especially tropical Africa and tropical Asia, there is an urgent need for diagnostic tests for diseases endemic in those areas, such as malaria, leishmaniasis, AIDS, Chagas' disease, and tuberculosis. These tests will need to be inexpensive and simple enough to be carried out by personnel with minimal training.

The tests described in this chapter are most often used to detect existing disease. In the future, diagnostic testing will probably also be directed at *preventing* disease. In the United States we regularly see reports of outbreaks of foodborne disease. Sampling methods that would allow complete identification, (including specific pathogenic serovars), within a few hours, or even minutes, would be a vast saving in time. Such rapid diagnostic tests would be especially valuable for tracking outbreaks of infectious disease carried by fruits and vegetables, such as the Salmonella outbreak in the summer of 2008. These products do not have the identifying marks that facilitate the tracking of packaged food products. This saving in time would be translated into immense economic savings for growers and retailers. It would also lead to less human illness and, perhaps, a saving of lives.

Not every topic discussed in this chapter is necessarily directed at the detection and prevention of disease. As was mentioned on page 509, Mabs have applications in the therapy of disease as well. These are already in use to treat certain cancers such as breast cancer and non-Hodgkin's lymphoma, as well as inflammatory diseases such as rheumatoid arthritis. Currently, Mabs are being tested for many disease conditions; these include asthma, sepsis, coronary artery disease, and several viral infections. They are also being studied as a way to treat the immune-caused neurological disease, multiple sclerosis.

CHECK YOUR UNDERSTANDING

✓ How has the development of monoclonal antibodies revolutionized diagnostic immunology? **18-20**

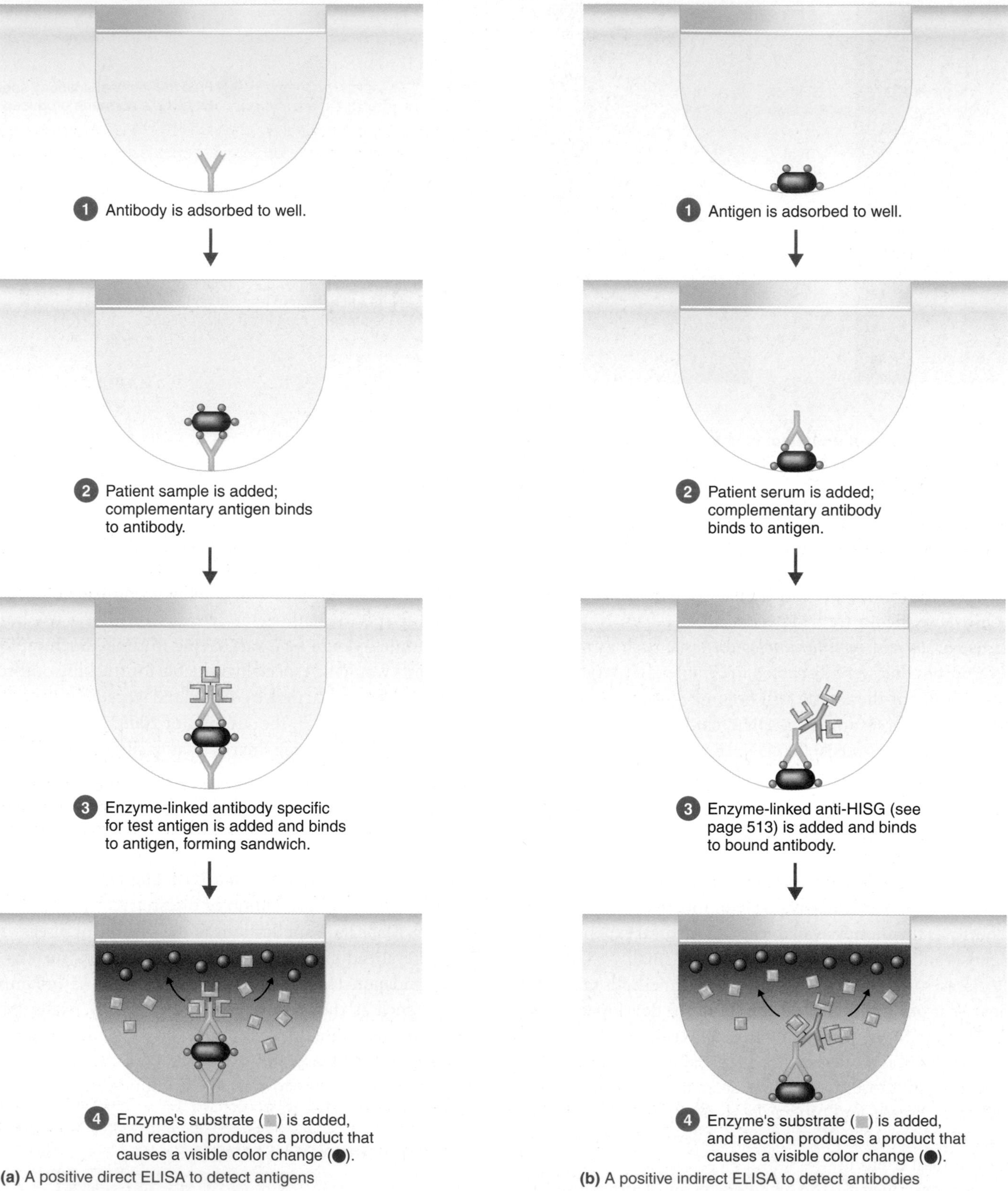

1 Antibody is adsorbed to well.
2 Patient sample is added; complementary antigen binds to antibody.
3 Enzyme-linked antibody specific for test antigen is added and binds to antigen, forming sandwich.
4 Enzyme's substrate (■) is added, and reaction produces a product that causes a visible color change (●).
(a) A positive direct ELISA to detect antigens
1 Antigen is adsorbed to well.
2 Patient serum is added; complementary antibody binds to antigen.
3 Enzyme-linked anti-HISG (see page 513) is added and binds to bound antibody.
4 Enzyme's substrate (■) is added, and reaction produces a product that causes a visible color change (●).
(b) A positive indirect ELISA to detect antibodies

Figure 18.14 The ELISA method. The components are usually contained in small wells of a microtiter plate. For an illustration of a technician carrying out an ELISA test on such a microtiter plate and the use of a computer to read the results, see Figure 10.11 on page 288.

Q Differentiate a direct from an indirect ELISA test.

STUDY OUTLINE

The **MyMicrobiologyPlace** website **(www.microbiologyplace.com)** will help you get ready for tests with its simple three-step approach: ❶ **take a pre-test** and obtain a personalized study plan, ❷ **learn and practice** with animations, tutorials, and MP3 tutor sessions, and ❸ **test yourself** with quizzes and a chapter post-test.

Vaccines (pp. 501–506)

1. Edward Jenner developed the modern practice of vaccination when he inoculated people with cowpox virus to protect them against smallpox.

Principles and Effects of Vaccination (p. 501)

2. Herd immunity results when most of a population is immune to a disease.

Types of Vaccines and Their Characteristics (pp. 501–505)

3. Attenuated whole-agent vaccines consist of attenuated (weakened) microorganisms; attenuated virus vaccines generally provide lifelong immunity.
4. Inactivated whole-agent vaccines consist of killed bacteria or viruses.
5. Toxoids are inactivated toxins.
6. Subunit vaccines consist of antigenic fragments of a microorganism; these include recombinant vaccines and acellular vaccines.
7. Conjugated vaccines combine the desired antigen with a protein that boosts the immune response.
8. Nucleic acid vaccines, or DNA vaccines, cause the recipient to make the antigenic protein.

The Development of New Vaccines (pp. 505–506)

9. Viruses for vaccines may be grown in animals, cell cultures, or chick embryos.
10. Recombinant vaccines and nucleic acid vaccines do not need to be grown in cells or animals.
11. Genetically modified plants may someday provide edible vaccines.
12. Adjuvants improve the effectiveness of some antigens.

Safety of Vaccines (p. 506)

13. Vaccines are the safest and most effective means of controlling infectious diseases.

Diagnostic Immunology (p. 507)

Immunologic-Based Diagnostic Tests (p. 507)

1. Many tests based on the interactions of antibodies and antigens have been developed to determine the presence of antibodies or antigens in a patient.
2. The sensitivity of a diagnostic test is determined by the percentage of positive samples it correctly detects; and its specificity is determined by the percentage of false positive results it gives.

Monoclonal Antibodies (pp. 507–509)

3. Hybridomas are produced in the laboratory by fusing a cancerous cell with an antibody-secreting plasma cell.
4. A hybridoma cell culture produces large quantities of the plasma cell's antibodies, called monoclonal antibodies.
5. Monoclonal antibodies are used in serological identification tests, to prevent tissue rejections, and to make immunotoxins to treat cancer.

Precipitation Reactions (pp. 509–510)

6. The interaction of soluble antigens with IgG or IgM antibodies leads to precipitation reactions.
7. Precipitation reactions depend on the formation of lattices and occur best when antigen and antibody are present in optimal proportions. Excesses of either component decrease lattice formation and subsequent precipitation.
8. The precipitin ring test is performed in a small tube.
9. Immunodiffusion procedures are precipitation reactions carried out in an agar gel medium.
10. Immunoelectrophoresis combines electrophoresis with immunodiffusion for the analysis of serum proteins.

Agglutination Reactions (pp. 510–512)

11. The interaction of particulate antigens (cells that carry antigens) with antibodies leads to agglutination reactions.
12. Diseases may be diagnosed by combining the patient's serum with a known antigen.
13. Diseases can be diagnosed by a rising titer or seroconversion (from no antibodies to the presence of antibodies).
14. Direct agglutination reactions can be used to determine antibody titer.
15. Antibodies cause visible agglutination of soluble antigens affixed to latex spheres in indirect or passive agglutination tests.
16. Hemagglutination reactions involve agglutination reactions using red blood cells. Hemagglutination reactions are used in blood typing, the diagnosis of certain diseases, and the identification of viruses.

Neutralization Reactions (p. 512)

17. In neutralization reactions, the harmful effects of a bacterial exotoxin or virus are eliminated by a specific antibody.
18. An antitoxin is an antibody produced in response to a bacterial exotoxin or a toxoid that neutralizes the exotoxin.
19. In a virus neutralization test, the presence of antibodies against a virus can be detected by the antibodies' ability to prevent cytopathic effects of viruses in cell cultures.

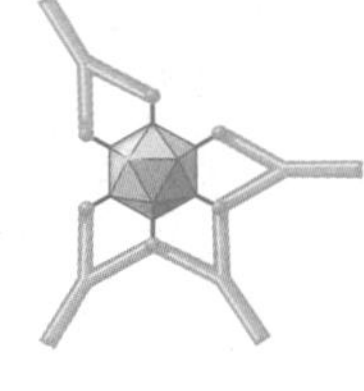

20. Antibodies against certain viruses can be detected by their ability to interfere with viral hemagglutination in viral hemagglutination inhibition tests.

Complement-Fixation Reactions (pp. 512–513)

21. Complement-fixation reactions are serological tests based on the depletion of a fixed amount of complement in the presence of an antigen–antibody reaction.

Fluorescent-Antibody Techniques (pp. 513–514)

22. Fluorescent-antibody techniques use antibodies labeled with fluorescent dyes.

23. Direct fluorescent-antibody tests are used to identify specific microorganisms.
24. Indirect fluorescent-antibody tests are used to demonstrate the presence of antibody in serum.
25. A fluorescence-activated cell sorter can be used to detect and count cells labeled with fluorescent antibodies.

Enzyme-Linked Immunosorbent Assay (ELISA) (pp. 514–516)

26. ELISA techniques use antibodies linked to an enzyme.
27. Antigen–antibody reactions are detected by enzyme activity. If the indicator enzyme is present in the test well, an antigen–antibody reaction has occurred.

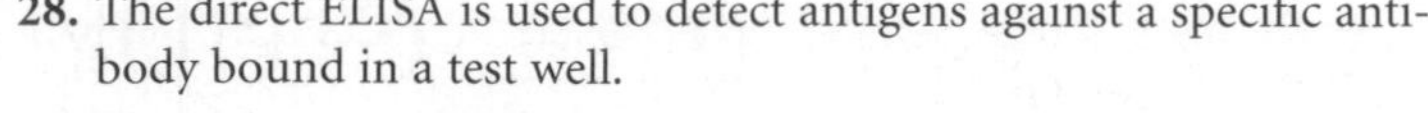

28. The direct ELISA is used to detect antigens against a specific antibody bound in a test well.
29. The indirect ELISA is used to detect antibodies against an antigen bound in a test well.

Western Blotting (Immunoblotting) (p. 516)

30. Serum antibodies separated by electrophoresis are identified with an enzyme-linked antibody.

The Future of Diagnostic Immunology (p. 516-518)

31. The use of monoclonal antibodies will continue to make new diagnostic tests possible.

STUDY QUESTIONS

Answers to the Review and Multiple Choice questions can be found by turning to the blue Answers at the back of the textbook.

Review

1. Classify the following vaccines by type. Which could cause the disease it is supposed to prevent?
   a. attenuated measles virus
   b. dead *Rickettsia prowazekii*
   c. *Vibrio cholerae* toxoid
   d. hepatitis B antigen produced in yeast cells
   e. purified polysaccharides from *Streptococcus pyogenes*
   f. *Haemophilus influenzae* polysaccharide bound to diphtheria toxoid
   g. a plasmid containing genes for influenza A protein
2. Define the following terms, and give an example of how each reaction is used diagnostically:
   a. viral hemagglutination
   b. hemagglutination inhibition
   c. passive agglutination
3. **DRAW IT** Label the components of the direct and indirect FA tests in the following situations. Which test is direct? Which test provides definitive proof of disease?

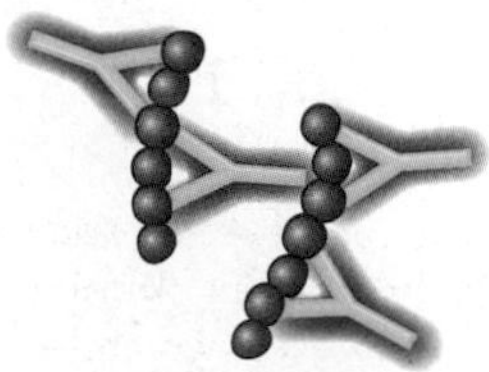

(a) Rabies can be diagnosed postmortem by mixing fluorescent-labeled antibodies with brain tissue.

(b) Syphilis can be diagnosed by adding the patient's serum to a slide fixed with *Treponema pallidum*. Anti-human immune serum globulin tagged with a fluorescent dye is added.

4. **DRAW IT** Label components of the direct and indirect ELISA tests in the following situations. Which test is direct? Which test provides definitive proof of disease?

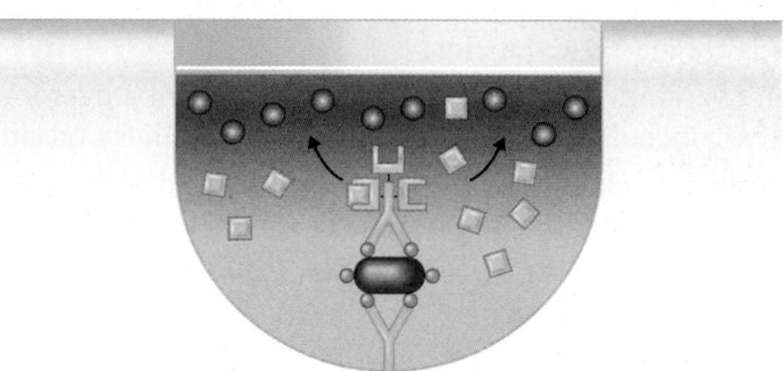

(a) Respiratory secretions to detect respiratory syncytial virus.

(b) Blood to detect human immunodeficiency virus antibodies.

5. How are monoclonal antibodies produced? What is their advantage over horse serum?
6. Explain the effects of excess antigen and antibody on the precipitation reaction. How is the precipitin ring test different from an immunodiffusion test?
7. How does the antigen in an agglutination reaction differ from that in a precipitation reaction?

8. Match the following serological tests in column A to the descriptions in column B.

Column A	Column B
________ a. Precipitation	1. Occurs with particulate antigens
________ b. Western blotting	2. Uses an enzyme for the indicator
________ c. Agglutination	3. Uses red blood cells for the indicator
________ d. Complement fixation	4. Uses antihuman immune serum globulin
________ e. Neutralization	5. Occurs with a free soluble antigen
________ f. ELISA	6. Used to determine the presence of antitoxin

9. Match each of the following tests in column A to its positive reaction in column B.

Column A	Column B
________ a. Agglutination	1. Peroxidase activity
________ b. Complement fixation	2. Harmful effects of agents not seen
________ c. ELISA	3. No hemolysis
________ d. FA test	4. Cloudy white line
________ e. Neutralization	5. Cell clumping
________ f. Precipitation	6. Fluorescence

Multiple Choice

Use the following choices to answer questions 1 and 2:

a. hemolysis
b. hemagglutination
c. hemagglutination-inhibition
d. no hemolysis
e. precipitin ring forms

1. Patient's serum, influenza virus, sheep red blood cells, and anti-sheep red blood cells are mixed in a tube. What happens if the patient has antibodies against influenza?
2. Patient's serum, *Chlamydia,* guinea pig complement, sheep red blood cells, and anti-sheep red blood cells are mixed in a tube. What happens if the patient has antibodies against *Chlamydia*?
3. The examples in questions 1 and 2 are
   a. direct tests. **b.** indirect tests.

Use the following choices to answer questions 4 and 5:

a. anti-*Brucella* **c.** substrate for the enzyme
b. *Brucella*

4. Which is the third step in a direct ELISA test?
5. Which item is from the patient in an indirect ELISA test?
6. In an immunodiffusion test, a strip of filter paper containing diphtheria antitoxin is placed on a solid culture medium. Then bacteria are streaked perpendicular to the filter paper. If the bacteria are toxigenic,
   a. the filter paper will turn red.
   b. a line of antigen–antibody precipitate will form.
   c. the cells will lyse.
   d. the cells will fluoresce.
   e. none of the above

Use the following choices to answer questions 7–9.

a. direct fluorescent antibody **d.** killed rabies virus
b. indirect fluorescent antibody **e.** none of the above
c. rabies immune globulin

7. Treatment given to a person bitten by a rabid bat.
8. Test used to identify rabies virus in the brain of a dog.
9. Test used to detect the presence of antibodies in a patient's serum.
10. In an agglutination test, eight serial dilutions to determine antibody titer were set up: Tube 1 contained a 1:2 dilution; tube 2, a 1:4, and so on. If tube 5 is the last tube showing agglutination, what is the antibody titer?
    a. 5 **b.** 1:5 **c.** 32 **d.** 1:32

Critical Thinking

1. What problems are associated with the use of attenuated whole-agent vaccines?
2. Many of the serological tests require a supply of antibodies against pathogens. For example, to test for *Salmonella,* anti-*Salmonella* antibodies are mixed with the unknown bacterium. How are these antibodies obtained?
3. A test for antibodies against *Treponema pallidum* uses the antigen cardiolipin and the patient's serum (suspected of having antibodies). Why do the antibodies react with cardiolipin? What is the disease?

Clinical Applications

1. Which of the following is proof of a disease state? Why doesn't the other situation confirm a disease state? What is the disease?
   a. *Mycobacterium tuberculosis* is isolated from a patient.
   b. Antibodies against *M. tuberculosis* are found in a patient.
2. Streptococcal erythrogenic toxin is injected into a person's skin in the Dick test. What results are expected if a person has antibodies against this toxin? What type of immunological reaction is this? What is the disease?
3. The following data were obtained from FA tests for anti-*Legionella* in four people. What conclusions can you draw? What is the disease?

	Antibody Titer			
	Day 1	Day 7	Day 14	Day 21
Patient A	128	256	512	1024
Patient B	0	0	0	0
Patient C	256	256	256	256
Patient D	0	0	128	512

4. Maria decided against the relatively new chickenpox vaccine and used her parents' method: she wanted her children to get chickenpox so that they would develop natural immunity. Her two children did get chickenpox. Her son had slight itching and skin vesicles, but her daughter was hospitalized for months with streptococcal cellulitis and underwent several skin grafts before recovering. Maria's housekeeper contracted chickenpox from the children and subsequently died. Almost half of the deaths due to chickenpox occur in adults.
   a. What responsibilities do parents have for their children's health?
   b. What rights do individuals have? Should vaccination be required by law?
   c. What responsibilities do individuals (e.g., parents) have for the health of society?
   d. Vaccines are given to healthy people, so what risks are acceptable?

19 Disorders Associated with the Immune System

In this chapter, we will see that not all immune system responses produce a desirable result. A familiar example is hay fever, which results from repeated exposure to plant pollen. Most of us also know that a blood transfusion will be rejected if the blood of the donor and the blood of the recipient are not compatible and that rejection is also a potential problem with transplanted organs. One's own tissue may be mistakenly attacked by the immune system, causing diseases we classify as autoimmune. Certain antigens, called superantigens, indiscriminately activate many T-cell receptors at once (see cytokine storm in Chapter 17, page 493), resulting in damage to tissue (see also Chapter 15, page 436).

Some people are born with a defective immune system, and in all of us the effectiveness of our immune system declines with age. Our immune systems can be deliberately crippled (*immunosuppressed*) to prevent the rejection of transplanted organs. Disease can also impair the immune system, especially infection by HIV, a virus that specifically attacks the immune system.

UNDER THE MICROSCOPE

Pollen grains. At certain times of year plants distribute pollen grains such as these, causing the sneezing, teary eyes, and runny noses of "hay fever."

Q&A

How can microscopic pollen from plants cause acute discomfort to many people?

Look for the answer in the chapter.

Table 19.1 Types of Hypersensitivity

Type of Reaction	Time Before Clinical Signs	Characteristics	Examples
Type I (Anaphylactic)	<30 min	IgE binds to mast cells or basophils; causes degranulation of mast cell or basophil and release of reactive substances such as histamine	Anaphylactic shock from drug injections and insect venom; common allergic conditions, such as hay fever, asthma
Type II (Cytotoxic)	5–12 hours	Antigen causes formation of IgM and IgG antibodies that bind to target cell; when combined with action of complement, destroys target cell	Transfusion reactions, Rh incompatibility
Type III (Immune Complex)	3–8 hours	Antibodies and antigens form complexes that cause damaging inflammation	Arthus reactions, serum sickness
Type IV (Delayed Cell-Mediated, or Delayed Hypersensitivity)	24–48 hours	Antigens activate T_C that kill target cell	Rejection of transplanted tissues; contact dermatitis, such as poison ivy; certain chronic diseases, such as tuberculosis

Hypersensitivity

LEARNING OBJECTIVES

19-1 Define *hypersensitivity.*

19-2 Describe the mechanism of anaphylaxis.

19-3 Compare and contrast systemic and localized anaphylaxis.

19-4 Explain how allergy skin tests work.

19-5 Define *desensitization* and *blocking antibody.*

19-6 Describe the mechanism of cytotoxic reactions and how drugs can induce them.

19-7 Describe the basis of the ABO and Rh blood group systems.

19-8 Explain the relationships among blood groups, blood, transfusions, and hemolytic disease of the newborn.

19-9 Describe the mechanism of immune complex reactions.

19-10 Describe the mechanism of delayed cell-mediated reactions, and name two examples.

The term *hypersensitivity* refers to an antigenic response beyond that which is considered normal; the term *allergy* is more familiar and is essentially synonymous. Hypersensitivity responses occur in individuals who have been *sensitized* by previous exposure to an antigen, which in this context is sometimes called an **allergen.** When an individual who was previously sensitized is exposed to that antigen again, his or her immune system reacts to it in a damaging manner. The four principal types of hypersensitivity reactions, summarized in Table 19.1, are anaphylactic, cytotoxic, immune complex, and cell-mediated (or delayed-type) reactions.

CHECK YOUR UNDERSTANDING

✓ Are all immune responses beneficial? **19-1**

Type I (Anaphylactic) Reactions

Type I, or anaphylactic, reactions often occur within 2 to 30 minutes after a person sensitized to an antigen is reexposed to that antigen. *Anaphylaxis* means "the opposite of protected," from the prefix *ana-*, meaning against, and the Greek *phylaxis,* meaning protection. **Anaphylaxis** is an inclusive term for the reactions caused when certain antigens combine with IgE antibodies. Anaphylactic responses can be *systemic reactions,* which produce shock and breathing difficulties and are sometimes fatal, or *localized reactions,* which include common allergic conditions such as hay fever, asthma, and hives (slightly raised, often itchy and reddened areas of the skin).

The IgE antibodies produced in response to an antigen, such as insect venom or plant pollen, bind to the surfaces of cells such as mast cells and basophils. These two cell types are similar in morphology and in their contribution to allergic reactions. **Mast cells** are especially prevalent in the connective tissue of the skin and respiratory tract and in surrounding blood vessels. The name is from the German word *mastzellen,* meaning "well fed"; they are packed with granules that at one time were mistakenly thought to have been ingested (Figure 19.1a). **Basophils** circulate in the bloodstream, where they constitute fewer than 1% of the leukocytes. Both are filled with granules containing a variety of chemicals called *mediators.*

Mast cells and basophils can have as many as 500,000 sites for IgE attachment. The Fc (stem) region of an IgE antibody (see Figure 17.3, page 481) can attach to one of these specific receptor sites on such a cell, leaving two antigen-binding sites free. Of course, the attached IgE monomers will not all be specific for the same antigen. But when an antigen such as plant pollen encounters

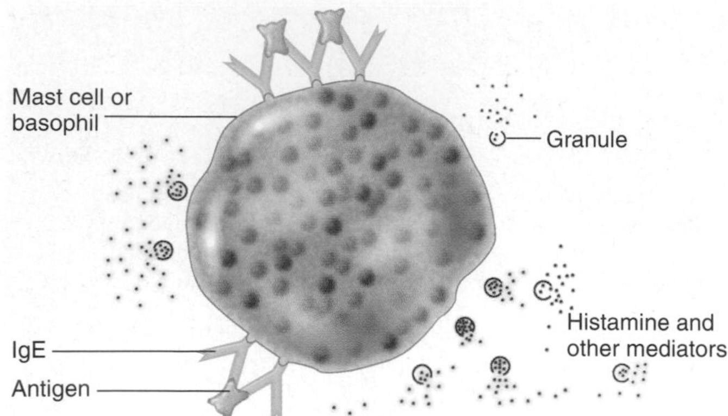

(a) IgE antibodies, produced in response to an antigen, coat mast cells and basophils. When an antigen bridges the gap between two adjacent antibody molecules of the same specificity, the cell undergoes degranulation and releases histamine and other mediators.

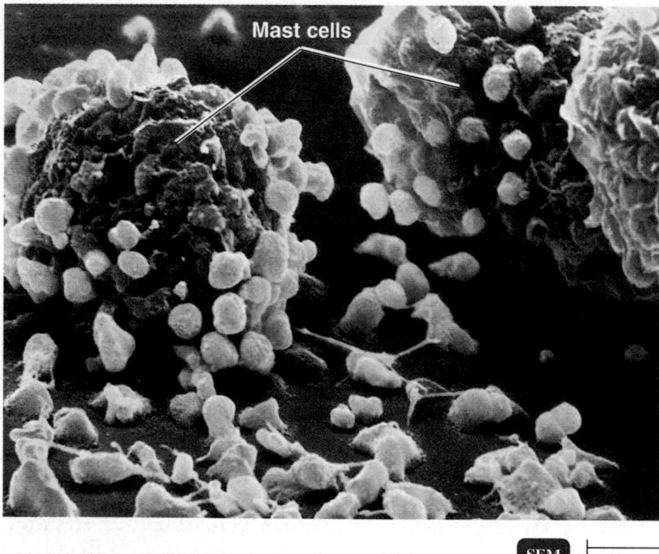

(b) A degranulated mast cell that has reacted with an antigen and released granules of histamine and other reactive mediators.

Figure 19.1 The mechanism of anaphylaxis.

Q What type of cells do IgE antibodies bind to?

two adjacent antibodies of the same appropriate specificity, it can bind to one antigen-binding site on each antibody, bridging the space between them. This bridge triggers the mast cell or basophil to undergo **degranulation,** which releases the granules inside these cells and also the mediators they contain (Figure 19.1b).

These mediators cause the unpleasant and damaging effects of an allergic reaction. The best-known mediator is **histamine.** The release of histamine increases the permeability and distension of blood capillaries, resulting in edema (swelling) and erythema (redness). Other effects include increased mucus secretion (a runny nose, for example) and smooth muscle contraction, which in the respiratory bronchi results in breathing difficulty.

Other mediators include **leukotrienes** of various types and **prostaglandins.** These mediators are not preformed and stored in the granules but are synthesized by the antigen-triggered cell. Because leukotrienes tend to cause prolonged contractions of certain smooth muscles, their action contributes to the spasms of the bronchial tubes that occur during asthmatic attacks. Prostaglandins affect smooth muscles of the respiratory system and increase mucus secretion.

Collectively, all these mediators serve as chemotactic agents that, in a few hours, attract neutrophils and eosinophils to the site of the degranulated cell. They then activate various factors that cause inflammatory symptoms, such as distension of the capillaries, swelling, increased secretion of mucus, and involuntary contractions of smooth muscles.

Systemic Anaphylaxis

At the turn of the twentieth century, two French biologists studied the responses of dogs to the venom of stinging jellyfish. Large doses of venom usually killed the dogs, but sometimes a few survived the injections. These surviving dogs were used for repeat experiments with the venom, and the results were surprising. Even a very tiny dose of the venom, one that should have been almost harmless, killed the dogs. They suffered difficulty in respiration, entered shock as their cardiovascular systems collapsed, and quickly died. This phenomenon was called *anaphylactic shock.*

Systemic anaphylaxis (or *anaphylactic shock*) can result when an individual sensitized to an antigen is exposed to it again. Injected antigens are more likely to cause a dramatic response than antigens introduced via other portals of entry. The release of mediators causes peripheral blood vessels throughout the body to enlarge, resulting in a drop in blood pressure (shock). This reaction can be fatal within a few minutes. There is very little time to act once someone develops systemic anaphylaxis. Treatment usually involves self-administration with a preloaded syringe of epinephrine, a drug that constricts blood vessels and raises the blood pressure. In the United States, 50 to 60 people die each year from anaphylactic shock caused by insect stings.

You may be acquainted with someone who reacts to penicillin in this way. In these individuals, the penicillin, which is a hapten (it cannot induce antibody formation by itself; see Chapter 17), combines with a carrier serum protein. Only then is penicillin immunogenic. Penicillin allergy probably occurs in about 2% of the population. Skin tests for penicillin sensitivity are available. Patients who have a positive skin test can be desensitized (see page 526) by an orally administered series of increasing doses of penicillin V. The interval between doses is only 15 minutes and is completed within 4 hours. The desensitization is valid only for an uninterrupted penicillin series immediately following the procedure. Allergy to penicillin also includes risk from exposure to some related drugs, such as carbapenem (page 561).

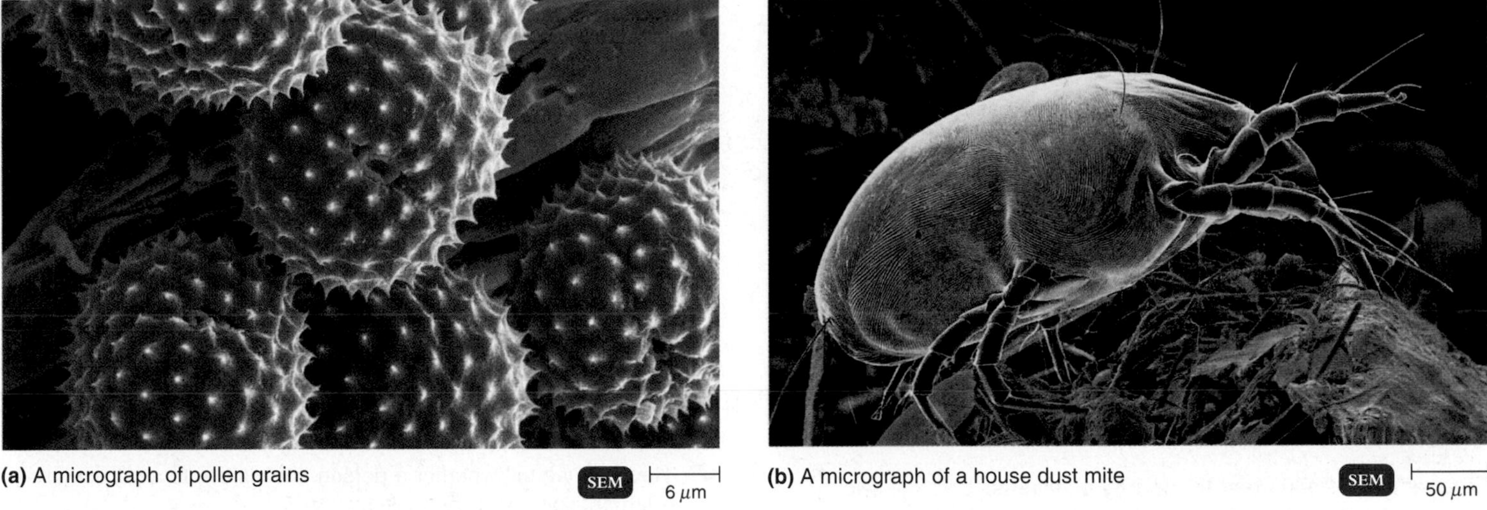

(a) A micrograph of pollen grains

(b) A micrograph of a house dust mite

Figure 19.2 Localized anaphylaxis. Inhaled antigens such as these are a common cause of localized anaphylaxis.

Q Compare localized and systemic anaphylaxis.

Localized Anaphylaxis

Whereas sensitization to injected antigens is a common cause of systemic anaphylaxis, **localized anaphylaxis** is usually associated with antigens that are ingested (foods) or inhaled (pollen)(Figure 19.2a). The symptoms depend primarily on the route by which the antigen enters the body.

Q&A In allergies involving the upper respiratory system, such as hay fever, sensitization usually involves mast cells in the mucous membranes of the upper respiratory tract. The airborne antigen might be a common environmental material such as plant pollen, fungal spores, feces of house dust mites (Figure 19.2b), or animal dander.* The typical symptoms are itchy and teary eyes, congested nasal passages, coughing, and sneezing. Antihistamine drugs, which compete for histamine receptor sites, are often used to treat these symptoms.

Asthma is an allergic reaction that mainly affects the lower respiratory system. Symptoms such as wheezing and shortness of breath are caused by the constriction of smooth muscles in the bronchial tubes.

For unknown reasons, asthma is becoming a near epidemic, affecting about 10% of children in Western society, although they often outgrow it in time. It is speculated that lack of childhood exposure in the developed world to many infections, the so-called *hygiene hypothesis,* is a factor in the increase in the incidence of asthma. Mental or emotional stress can also be a contributing factor in precipitating an attack. Symptoms of asthma are usually controlled by aerosol inhalants, which, unfortunately, may be difficult for very small children to use. Xolair (omalizumab) is a newly available drug but is a very expensive treatment for severe allergic asthma. It blocks IgE.

Antigens that enter the body via the gastrointestinal tract can also sensitize an individual. Many of us may know someone who is allergic to a particular food. Frequently, so-called food allergies may not be related to hypersensitivity at all and are more accurately described as *food intolerances.* For example, many people are unable to digest the lactose in milk because they lack the enzyme that breaks down this disaccharide milk sugar. The lactose enters the intestine, where it osmotically retains fluid, causing diarrhea.

Gastrointestinal upset is a common symptom of food allergies, but it can also result from many other factors. Hives are more characteristic of a true food allergy, and ingestion of the antigen may result in systemic anaphylaxis. Death has even resulted when a person sensitive to fish ate french fries that had been prepared in oil previously used to fry fish. Skin tests are not reliable indicators for the diagnosis of food-related allergies, and completely controlled tests for hypersensitivity to ingested foods are very difficult to perform. Only eight foods are responsible for 97% of food-related allergies: eggs, peanuts, tree-grown nuts, milk, soy, fish, wheat, and peas. Most children exhibiting allergies to milk, egg, wheat, and soy develop tolerance as they age, but reactions to peanuts, tree nuts, and seafood tend to persist.

An estimated 1.5 million Americans are allergic to peanuts, and as many as 100 deaths occur annually. Therefore, considerable research is underway on the problem, ranging from drugs and vaccines to developing less allergenic peanuts. It is interesting that China has a relatively low incidence of peanut allergy, although peanuts are common in Chinese foods. This may be because Chinese cookery involves boiling and lower-temperature

**Dander* is a general term for microscopic particles from the fur or skin of animals. A cat, for example, carries about 100 mg of dander on its coat and sheds about 0.1 mg a day. This accumulates in upholstery and carpeting. People with allergies to mice, gerbils, and similar small animals are more likely to be allergic to components of urine accumulating in cages.

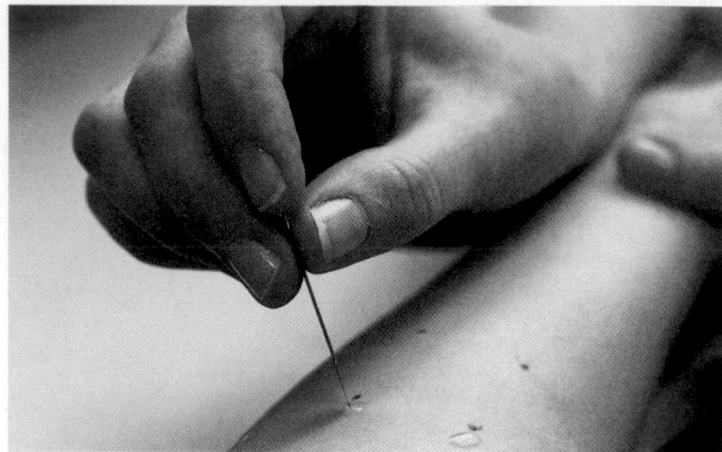

Figure 19.3 A skin test to identify allergens. Drops of fluid containing test substances are placed on the skin. A light scratch is made with a needle to allow the substances to penetrate the skin. Reddening and swelling at the site identify the substance as a probable cause of an allergic reaction.

Q What is scratched into the skin in a skin test?

frying, whereas dry roasting of peanuts concentrates allergic properties. Some children who have a relatively low level of peanut-specific IgE may outgrow their peanut allergy.

Sulfites, to which many people have an allergy, are a frequent problem. Their use is widespread in foods and beverages; and, although food labels should indicate their presence, they may be difficult to avoid in practice. A food product may have come into contact with a food allergen through processing machinery or cookware previously used for other foods. In one U.S. Food and Drug Administration report, 25% of bakery, ice cream, and candy products tested positive for peanut allergens even though peanuts were not listed on the required product labels. In the United States it is estimated that 200 persons a year die of severe allergic reactions to foods.

Preventing Anaphylactic Reactions

Avoiding contact with the sensitizing antigen is the most obvious way to prevent allergic reactions. Unfortunately, avoidance is not always possible. Some individuals experience an allergic reaction after eating an assortment of foods. In such cases, they may not know exactly what antigen they are sensitive to. In some cases, skin tests might be of use in diagnosis (Figure 19.3). These tests involve inoculating small amounts of the suspected antigen just beneath the epidermis of the skin. Sensitivity to the antigen is indicated by a rapid inflammatory reaction that produces redness, swelling, and itching at the inoculation site. This small affected area is called a *wheal.*

Once the responsible antigen has been identified, the person can either try to avoid contact with it or undergo **desensitization.** This procedure usually consists of a series of gradually increasing dosages of the antigen carefully injected beneath the skin. The objective is to cause the production of IgG rather than IgE antibodies in the hope that the circulating IgG antibodies will act as *blocking antibodies* to intercept and neutralize the antigens before they can react with cell-bound IgE. Desensitization is not a routinely successful procedure, but it is effective in 65–75% of individuals whose allergies are induced by inhaled antigens and in a reported 97% of people allergic to insect venom.

CHECK YOUR UNDERSTANDING

- ✓ In what tissues do we find the mast cells that are major contributors to allergic reactions such as hay fever? **19-2**
- ✓ Which is the more dangerous to life, systemic or localized anaphylaxis? **19-3**
- ✓ How can we tell whether a person is sensitive to a particular allergen, such as a tree pollen? **19-4**
- ✓ Which antibody types need to be blocked to desensitize a person subject to allergies? **19-5**

Type II (Cytotoxic) Reactions

Type II (cytotoxic) reactions generally involve the activation of complement by the combination of IgG or IgM antibodies with an antigenic cell. This activation stimulates complement to lyse the affected cell, which might be either a foreign cell or a host cell that carries a foreign antigenic determinant (such as a drug) on its surface. Additional cellular damage may be caused within 5 to 8 hours by the action of macrophages and other cells that attack antibody-coated cells.

The most familiar cytotoxic hypersensitivity reactions are *transfusion reactions,* in which red blood cells are destroyed as a result of reacting with circulating antibodies. These involve blood group systems that include the ABO and Rh antigens.

The ABO Blood Group System

In 1901, Karl Landsteiner discovered that human blood could be grouped into four principal types, which were designated A, B, AB, and O. This method of classification is called the **ABO blood group system.** Since then, other blood group systems, such as the Lewis system and the MN system, have been discovered, but our discussion will be limited to two of the best known, the ABO and the Rh systems. The main features of the ABO blood group system are summarized in Table 19.2.

A person's ABO blood type depends on the presence or absence of carbohydrate antigens located on the cell membranes of red blood cells (RBCs). Cells of blood type O lack both A and B antigens. Table 19.2 shows that the plasma of individuals with a given blood type, such as A, have antibodies against the alternative blood type, anti-B antibody. These antibodies are presumed to arise in response to microorganisms and ingested foodstuffs that have antigenic determinants very similar to blood group antigens. Individuals with type AB cells have plasma with no antibodies to either A or B antigens. Type O individuals have antibodies against both A and B antigens.

Table 19.2 The ABO Blood Group System

Blood Group	Erythrocyte or Red Blood Cell Antigens	Illustration	Plasma Antibodies	Blood That Can Be Received	Frequency (% U.S. Population)		
					White	Black	Asian
AB	A and B	A, B	Neither anti-A nor anti-B antibodies	A, B, AB, O	3	4	5
B	B		Anti-A	B, O	9	20	27
A	A		Anti-B	A, O	41	27	28
O	Neither A nor B		Anti-A and Anti-B	O	47	49	40

When a transfusion is incompatible, as when type B blood is transfused into a person with type A blood, the antigens on the type B blood cells will react with anti-B antibodies in the recipient's serum. This antigen–antibody reaction activates complement, which in turn causes lysis of the donor's RBCs as they enter the recipient's system.

A relationship between blood types and certain diseases has been observed, which may be related to the skewing of blood types in certain geographical areas. For example, individuals with type O blood are more susceptible to the incidence and severity of cholera and other diarrheas, whereas individuals with type B are much less affected. This tendency seems to be reflected in the blood types found in the Indian subcontinent, where type B is common and type O less so. Iceland has a relatively low percentage of blood types A and AB, possibly caused by the greater susceptibility of these blood types to a succession of smallpox epidemics within this geographically restricted population. More than half of the population of Africa is of type O, which tends to be less severely affected by malaria.

The Rh Blood Group System

In the 1930s, researchers discovered the presence of a different surface antigen on human red blood cells. Soon after they injected rabbits with RBCs from rhesus monkeys, the rabbit serum contained antibodies that were directed against the monkey blood cells but that would also agglutinate some human RBCs. This indicated that a common antigen was present on both human and monkey red blood cells. The antigen was named the **Rh factor** (*Rh* for rhesus monkey). The roughly 85% of the population whose cells possess this antigen are called Rh^+; those lacking this RBC antigen (about 15%) are Rh^-. Antibodies that react with the Rh antigen do not occur naturally in the serum of Rh^- individuals, but exposure to this antigen can sensitize their immune systems to produce anti-Rh antibodies.

Blood Transfusions and Rh Incompatibility If blood from an Rh^+ donor is given to an Rh^- recipient, the donor's RBCs stimulate the production of anti-Rh antibodies in the recipient. If the recipient then receives Rh^+ RBCs in a subsequent transfusion, a rapid, serious hemolytic reaction will develop.

Hemolytic Disease of the Newborn Blood transfusions are not the only way in which an Rh^- person can become sensitized to Rh^+ blood. When an Rh^- woman and an Rh^+ man produce a child, there is a 50% chance that the child will be Rh^+ (**Figure 19.4**). If the child is Rh^+, the Rh^- mother can become sensitized to this antigen during birth when the placental membranes tear and the fetal Rh^+ RBCs enter the maternal circulation, causing the mother's body to produce anti-Rh antibodies of the IgG type. If the fetus in a subsequent pregnancy is Rh^+, her anti-Rh antibodies will cross the placenta and destroy the fetal RBCs. The fetal body responds to this immune attack by producing large numbers of immature RBCs called erythroblasts. Thus, the term *erythroblastosis fetalis* was once used to describe what is now called **hemolytic disease of the newborn (HDNB).** Before the birth of a fetus with this condition, the maternal circulation removes most of the toxic by-products of fetal RBC disintegration. After birth, however, the fetal blood is no longer purified by the mother, and the newborn develops jaundice and severe anemia.

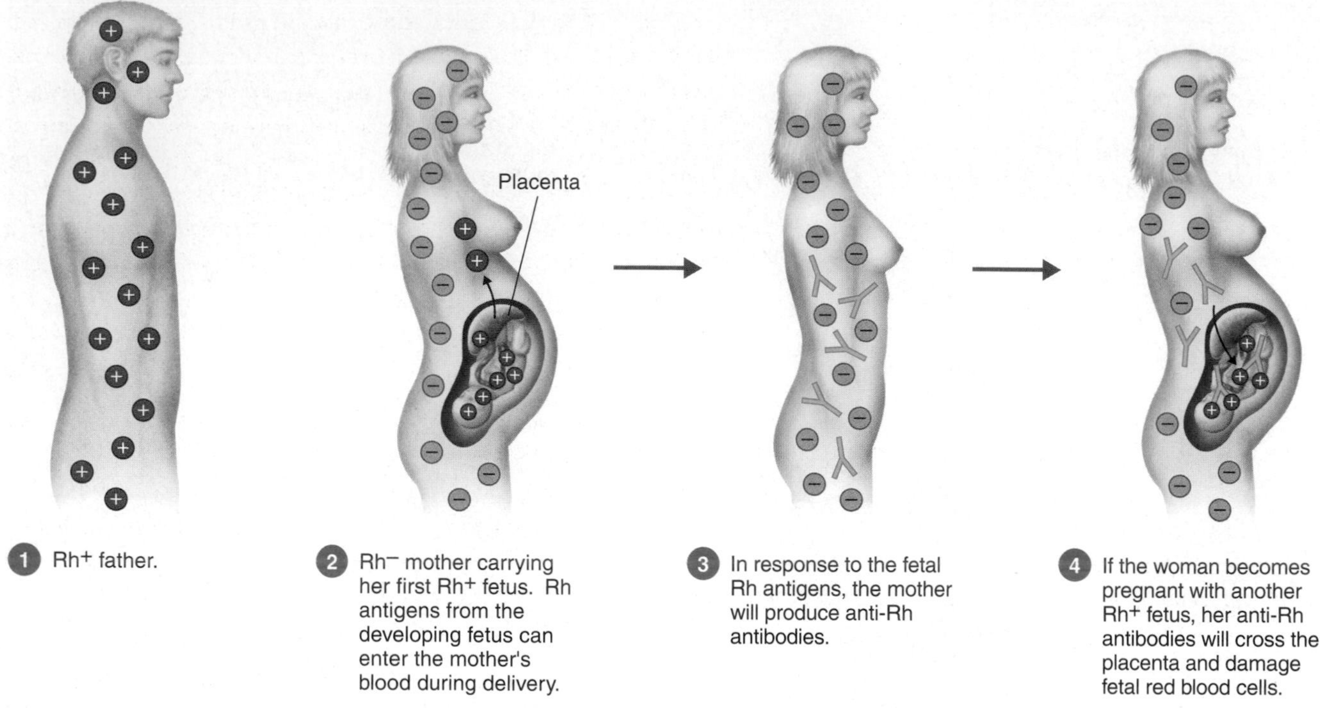

Placenta
1 Rh+ father.
2 Rh− mother carrying her first Rh+ fetus. Rh antigens from the developing fetus can enter the mother's blood during delivery.
3 In response to the fetal Rh antigens, the mother will produce anti-Rh antibodies.
4 If the woman becomes pregnant with another Rh+ fetus, her anti-Rh antibodies will cross the placenta and damage fetal red blood cells.

Figure 19.4 Hemolytic disease of the newborn.

Q What type of antibodies cross the placenta?

HDNB is usually prevented today by passive immunization of the Rh^- mother at the time of delivery of any Rh^+ infant with anti-Rh antibodies, which are available commercially (RhoGAM). These anti-Rh antibodies combine with any fetal Rh^+ RBCs that have entered the mother's circulation, so it is much less likely that she will become sensitized to the Rh antigen. If the disease is not prevented, the newborn's Rh^+ blood, contaminated with maternal antibodies, may have to be replaced by transfusion of uncontaminated blood.

Drug-Induced Cytotoxic Reactions

Blood platelets (thrombocytes) are minute cell-like bodies that are destroyed by drug-induced cytotoxic reactions in the disease called **thrombocytopenic purpura.** The drug molecules are usually haptens because they are too small to be antigenic by themselves; but, in the situation illustrated in Figure 19.5, a platelet has become coated with molecules of a drug (quinine is a familiar example), and the combination is antigenic. Both antibody and complement are needed for lysis of the platelet. Because platelets are necessary for blood clotting, their loss results in hemorrhages that appear on the skin as purple spots (purpura).

Drugs may bind similarly to white or red blood cells, causing local hemorrhaging and yielding symptoms described as "blueberry muffin" skin mottling. Immune-caused destruction of granulocytic white cells is called **agranulocytosis,** and it affects the body's phagocytic defenses. When RBCs are destroyed in the same manner, the condition is termed **hemolytic anemia.**

CHECK YOUR UNDERSTANDING

- ✓ What, besides an allergen and an antibody, is required to precipitate a cytotoxic reaction? **19-6**
- ✓ What are the antigens located on the cell membranes of type O blood? **19-7**
- ✓ If a fetus that is Rh^+ can be damaged by anti-Rh antibodies of the mother, why does such damage never happen during the first such pregnancy? **19-8**

Type III (Immune Complex) Reactions

Type III reactions involve antibodies against soluble antigens circulating in the serum. (In contrast, type II immune reactions are directed against antigens located on cell or tissue surfaces.) The antigen–antibody complexes are deposited in organs and cause inflammatory damage.

Immune complexes form only when certain ratios of antigen and antibody occur. The antibodies involved are usually IgG. A significant excess of antibody leads to the formation of complement-fixing complexes that are rapidly removed from the body by phagocytosis. When there is a significant excess of antigen, soluble complexes form that do not fix complement and do not cause inflammation. However, when a certain antigen–antibody ratio exists, usually with a slight excess of antigen, the soluble complexes that form are small and escape phagocytosis.

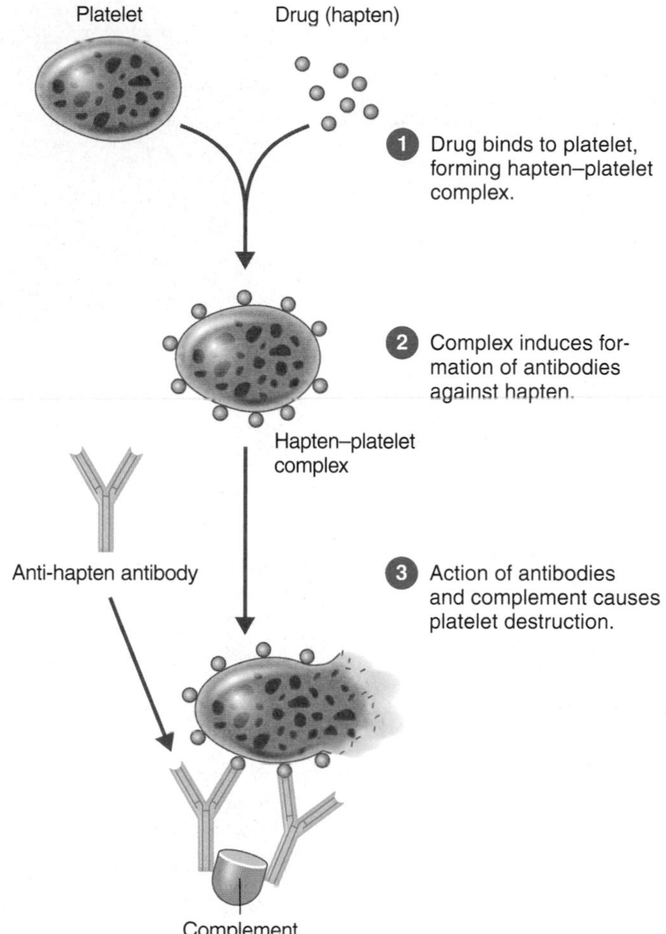

Figure 19.5 Drug-induced thrombocytopenic purpura. Molecules of a drug such as quinine accumulate on the surface of a platelet and stimulate an immune response that destroys the platelet.

Q What actually destroys the platelets in thrombocytopenic purpura?

Figure 19.6 illustrates the consequences. These complexes circulate in the blood, pass between endothelial cells of the blood vessels, and become trapped in the basement membrane beneath the cells. In this location, they may activate complement and cause a transient inflammatory reaction: attracting neutrophils that release enzymes. Repeated introduction of the same antigen can lead to more serious inflammatory reactions, causing damage to the basement membrane's endothelial cells within 2 to 8 hours.

Glomerulonephritis is an immune complex condition, usually resulting from an infection, that causes inflammatory damage to the kidney glomeruli, which are sites of blood filtration.

CHECK YOUR UNDERSTANDING

✓ Are the antigens causing immune complex reactions soluble or insoluble? **19-9**

Type IV (Delayed Cell-Mediated) Reactions

Up to this point we have discussed humoral immune responses involving IgE, IgG, or IgM. Type IV reactions involve cell-mediated immune responses and are caused mainly by T cells. Instead of occurring within a few minutes or hours after a sensitized individual is again exposed to an antigen, these **delayed cell-mediated reactions,** (or **delayed hypersensitivity**) are not apparent for a day or more. A major factor in the delay is the time required for the participating T cells and macrophages to migrate to and accumulate near the foreign antigens. Transplant rejection is most commonly mediated by cytotoxic T lymphocytes (CTLs; page 487), but other mechanisms are by antibody-dependent cell-mediated cytotoxicity (ADCC; page 492) or complement-mediated lysis (page 464). Another example is described in the box on page 531.

Figure 19.6 Immune complex–mediated hypersensitivity.

Q Name one immune complex disease.

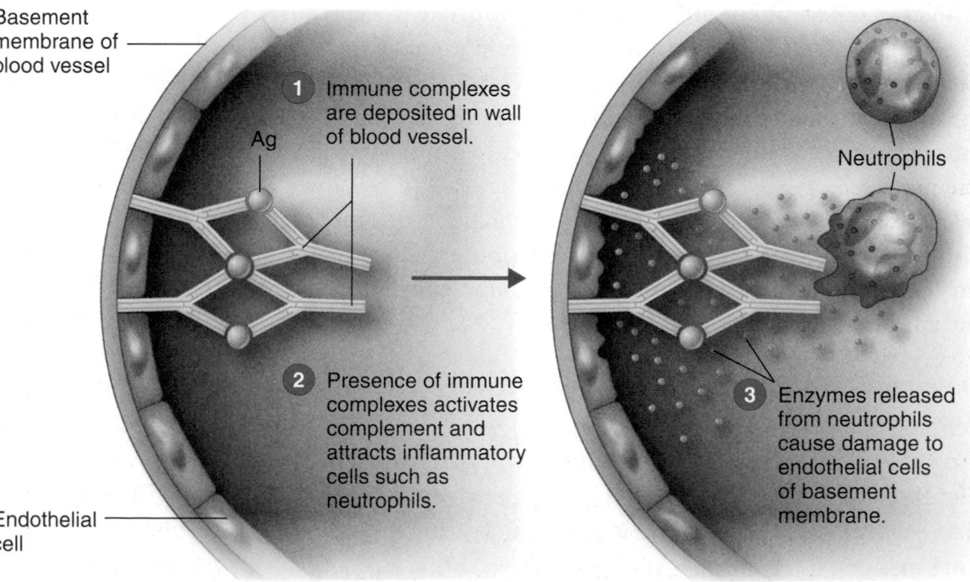

Basement membrane of blood vessel
1 Immune complexes are deposited in wall of blood vessel.
Ag
Neutrophils
2 Presence of immune complexes activates complement and attracts inflammatory cells such as neutrophils.
3 Enzymes released from neutrophils cause damage to endothelial cells of basement membrane.
Endothelial cell

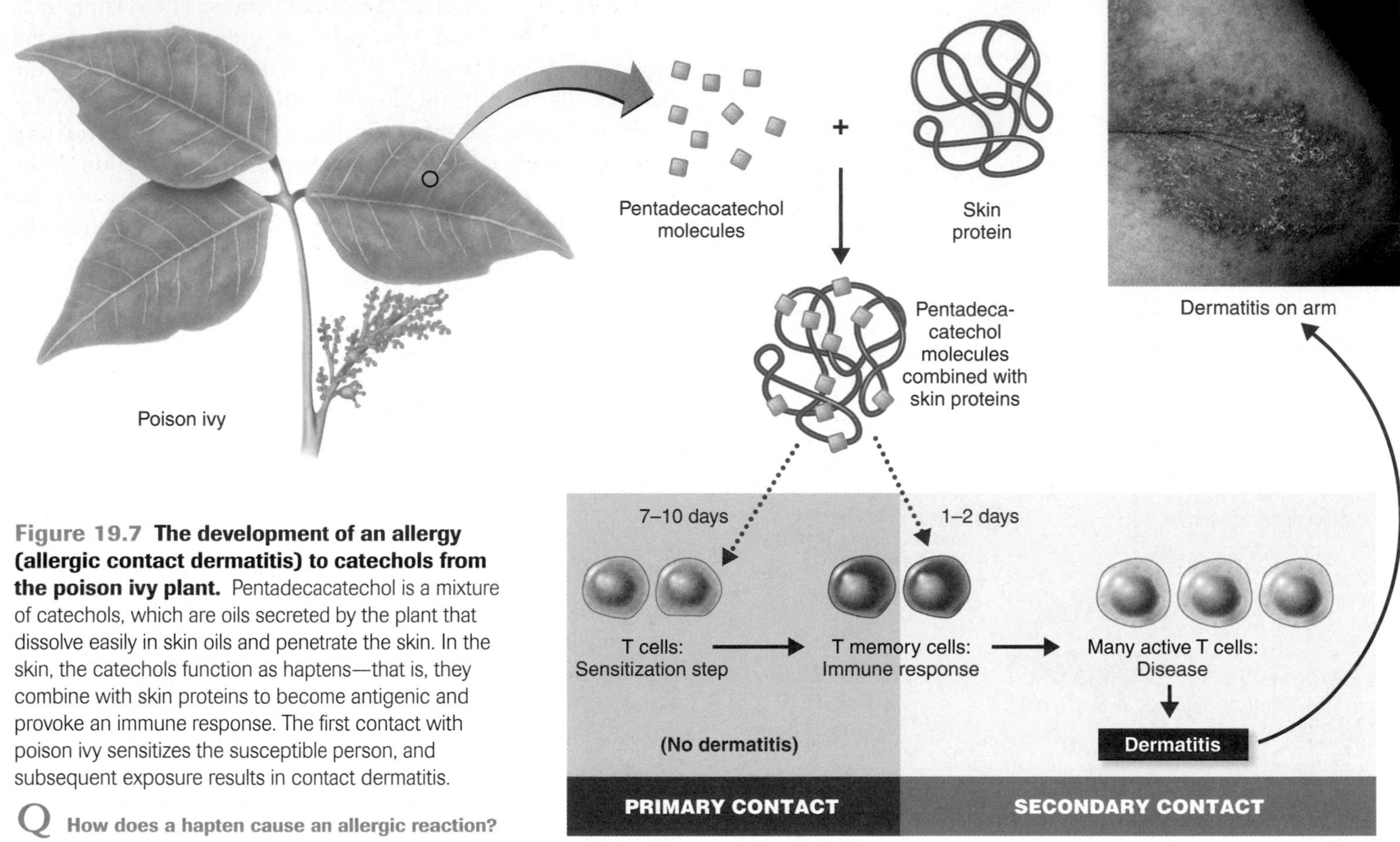

Figure 19.7 The development of an allergy (allergic contact dermatitis) to catechols from the poison ivy plant. Pentadecacatechol is a mixture of catechols, which are oils secreted by the plant that dissolve easily in skin oils and penetrate the skin. In the skin, the catechols function as haptens—that is, they combine with skin proteins to become antigenic and provoke an immune response. The first contact with poison ivy sensitizes the susceptible person, and subsequent exposure results in contact dermatitis.

Q How does a hapten cause an allergic reaction?

Causes of Delayed Cell-Mediated Reactions

Sensitization for delayed hypersensitivity reactions occurs when certain foreign antigens, particularly those that bind to tissue cells, are phagocytized by macrophages and then presented to receptors on the T-cell surface. Contact between the antigenic determinant sites and the appropriate T cell causes the T cell to proliferate into mature differentiated T cells and memory cells.

When a person sensitized in this way is reexposed to the same antigen, a delayed hypersensitivity reaction might result. Memory cells from the initial exposure activate T cells, which release destructive cytokines in their interaction with the target antigen. In addition, some cytokines contribute to the inflammatory reaction to the foreign antigen by attracting macrophages to the site and activating them.

Delayed Cell-Mediated Hypersensitivity Reactions of the Skin

We have seen that hypersensitivity symptoms are frequently displayed on the skin. One delayed hypersensitivity reaction that involves the skin is the familiar skin test for tuberculosis. Because *Mycobacterium tuberculosis* is often located within macrophages, this organism can stimulate a delayed cell-mediated immune response. As a screening test, protein components of the bacteria are injected into the skin. If the recipient has (or has had) a previous infection by tuberculosis bacteria, an inflammatory reaction to the injection of these antigens will appear on the skin in 1 to 2 days (see Figure 24.11, page 686); this interval is typical of delayed hypersensitivity reactions.

Allergic contact dermatitis, another common manifestation of delayed cell-mediated hypersensitivity, is usually caused by haptens that combine with proteins (particularly the amino acid lysine) in the skin of some people to produce an immune response. Reactions to poison ivy (Figure 19.7), cosmetics, and the metals in jewelry (especially nickel) are familiar examples of these allergies.

The increasing exposure to latex in condoms, in certain catheters, and in gloves used by health care workers has led to a greater awareness of hypersensitivity to latex. Death from anaphylactic shock can also occur. During a recent 4-year period, 15 fatalities among patients were caused by exposure to latex

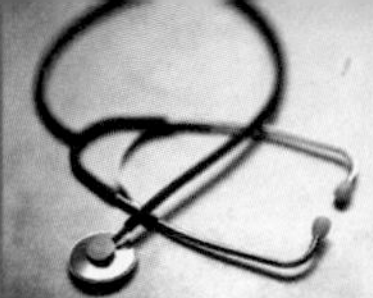

CLINICAL FOCUS

A Delayed Rash

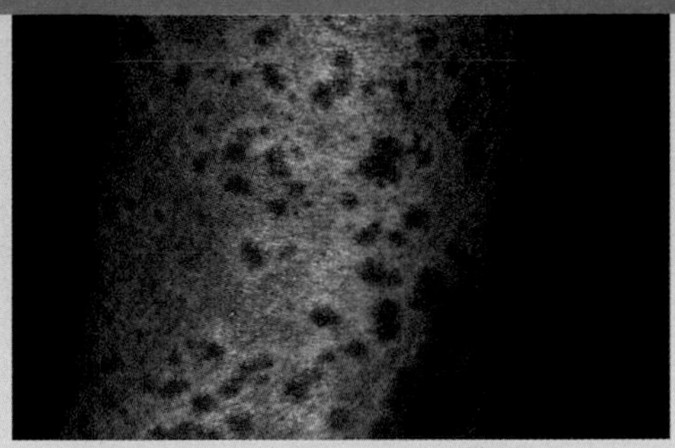

As you read through this box, you will encounter a series of questions that health care professionals ask as they determine the cause of a patient's symptoms. Try to answer each question before going on to the next one.

1. A 65-year-old woman with hip and shoulder replacements made a routine dental appointment. She asked for her usual prescription of cephalothin. The nurse practitioner prescribed penicillin, saying it is less expensive. Because of her hip and shoulder implants, antibiotics were prescribed for 2 days after any dental work.

   Why are patients with medical implants more susceptible to infection from dental work?

2. Oral bacteria introduced to the bloodstream during dental work can colonize on medical implants. The resulting biofilm may be a source of serious systemic infections. The dental cleaning went well. Seven days later, the woman developed a maculopapular rash over her legs and torso (see the figure).

   What are the most likely causes of a rash, in the absence of fever or other signs of infection?

3. A rash is likely due to an allergic reaction.

   What questions would you ask the patient?

4. The patient had not tried any new foods, cleaning agents, or clothing. She said the only thing different during the past 10 days was her taking the penicillin. The nurse practitioner said penicillin couldn't be the cause because responses to penicillin occur within minutes to hours after exposure.

   Was the nurse practitioner correct?

5. Immediate reactions that occur within minutes to hours suggest an antibody-mediated allergy. Delayed reactions, such as in this patient, that occur after days to weeks suggest a type IV cell-mediated reaction.

   What cells are responsible for a type IV hypersensitivity? What antibodies are involved in a type I hypersensitivity?

6. Sensitized T cells are involved in delayed hypersensitivity reactions, including antibiotic-induced rashes. Drug-specific IgE antibodies are responsible for type I immediate hypersensitivity reactions.

   What should the nurse practitioner have asked?

7. The nurse practitioner should have asked whether the patient had any drug allergies. However, in this case, the patient had no prior drug-induced allergy.

   Was this the patient's first exposure to penicillin?

8. Allergic reactions do not occur on the first exposure to an antigen. The prior exposure could have occurred once during the patient's life. Many immunologists feel that the overuse of penicillin 40 years ago for bacterial infections resulted in an increased frequency of allergic reactions.

   Most patients who have a history of penicillin allergy, however, will tolerate cephalosporins.

tubing used for enemas or, during abdominal surgery, to the latex gloves of the surgeon. Many hospitals now even restrict entry of latex balloons.

Among physicians and nurses, 5–12% report this type of hypersensitivity to latex surgical gloves (**Figure 19.8**). Several different proteins in natural rubber may be involved, and different proteins may cause different immune reactions. Low-protein latex gloves, which are also powder-free to minimize aerosols, elicit fewer adverse reactions. Synthetic polymers such as vinyl and, especially, nitrile are alternatives to latex, but even nitrile gloves occasionally cause allergic reactions. Most gloves that are made of natural latex, as well as those made of nitrile and neoprene, contain chemical additives called accelerators. Accelerators promote cross-linkage, which adds strength and resiliency, but they have been implicated in allergic reactions. One type of nitrile glove that does not contain accelerators has been developed and has received U.S. Food and Drug Administration (FDA) registration as a Class II medical device that can be labeled as nonallergenic.

Many persons who develop an allergy to latex for some reason also have evidence of allergies to certain fruits, most commonly avocado, chestnut, banana, and kiwi. Latex paint, however, does not pose a threat of hypersensitivity reactions. Despite its name, latex paint contains no natural latex, but only synthetic nonallergenic chemical polymers.

The identity of the environmental factor causing the dermatitis can usually be determined by a *patch test.* Samples of suspected materials are taped to the skin; after 48 hours, the area is examined for inflammation.

CHECK YOUR UNDERSTANDING

✓ What is the primary reason for the delay in a delayed cell-mediated reaction? **19-10**

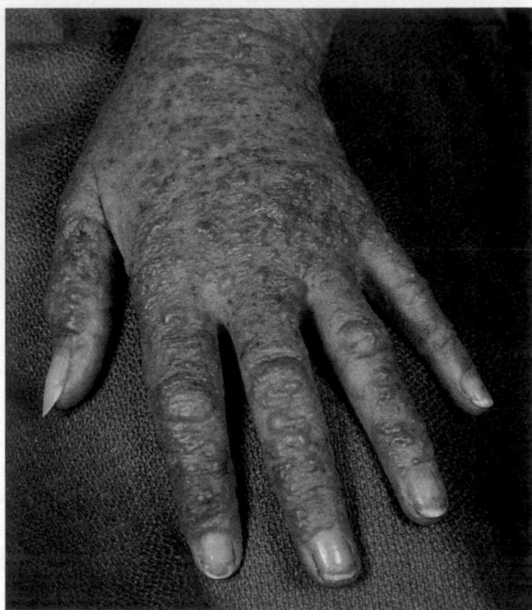

Figure 19.8 Allergic contact dermatitis. This person's hand exhibits a severe case of delayed contact dermatitis from wearing latex surgical gloves.

Q What is allergic contact dermatitis?

Autoimmune Diseases

LEARNING OBJECTIVES

19-11 Describe a mechanism for self-tolerance.

19-12 Give an example of immune complex, cytotoxic, and cell-mediated autoimmune diseases.

When the action of the immune system is in response to self-antigens and causes damage to one's own organs, the result is an **autoimmune disease.** More than 40 autoimmune diseases have been identified. Although relatively rare, they affect about 5% of the population in the developed world. About 75% of the cases of autoimmune disease selectively affect women. Treatments for autoimmune diseases are improving as knowledge of the mechanisms controlling immune reactions improves.

Autoimmune diseases occur when there is a loss of **self-tolerance,** the immune system's ability to discriminate self from nonself. In the generally accepted model by which T cells become capable of distinguishing self from nonself, the cells acquire this ability during their passage through the thymus. As we saw in Chapter 17 (page 485), any T cells that will target host cells are eliminated by *thymic selection* during this period. This step makes it unlikely that the T cell will attack its own tissue cells.

In autoimmune diseases, the loss of self-tolerance leads to the production of antibodies or a response by sensitized T cells against a person's own tissue antigens. Autoimmune reactions, and the diseases they cause, can be cytotoxic, immune complex, or cell-mediated in nature.

Autoimmunity involves antibodies that attack self. These antibodies may be made in response to an infectious agent such as a virus, but sequence similarities between viral and self-proteins may cause the antibodies to attack self cells. Hepatitis C virus may be responsible for autoimmune hepatitis by this mechanism.

Cytotoxic Autoimmune Reactions

Graves' disease and myasthenia gravis are two examples of disorders caused by cytotoxic autoimmune reactions. Both diseases involve antibody reactions to cell-surface antigens, although there is no cytotoxic destruction of the cells.

Graves' disease is caused by antibodies called long-acting thyroid stimulators. These antibodies attach to receptors on thyroid gland cells that are the normal target cells of the thyroid-stimulating hormone produced by the pituitary gland. The result is that the thyroid gland is stimulated to produce increased amounts of thyroid hormones and becomes greatly enlarged. The most striking signs of the disease are goiter (a disfiguring swelling of the thyroid gland) and markedly bulging, staring eyes.

Myasthenia gravis is a disease in which muscles become progressively weaker. It is caused by antibodies that coat the acetylcholine receptors at the junctions at which nerve impulses reach the muscles. Eventually, the muscles controlling the diaphragm and the rib cage may fail to receive the necessary nerve signals, and respiratory arrest and death result.

Immune Complex Autoimmune Reactions

Systemic lupus erythematosus is a systemic autoimmune disease, involving immune complex reactions, that mainly affects women. The etiology of the disease is not completely understood, but afflicted individuals produce antibodies directed at components of their own cells, including DNA, which is probably released during the normal breakdown of tissues, especially the skin. The most damaging effects of the disease result from deposits of immune complexes in the kidney glomeruli.

Crippling **rheumatoid arthritis** is a disease in which immune complexes of IgM, IgG, and complement are deposited in the joints. In fact, immune complexes called *rheumatoid factors* may be formed by IgM binding to the Fc region of normal IgG. These factors are found in 70% of individuals suffering from rheumatoid arthritis. The chronic inflammation caused by this deposition eventually leads to severe damage to the cartilage and bone of the joint.

Cell-Mediated Autoimmune Reactions

Multiple sclerosis is one of the more common autoimmune diseases, affecting mostly younger adults. Most individuals with multiple sclerosis are whites living in northern latitudes; women are twice as likely to have the disease. It is a neurological disease in which T cells and macrophages attack the myelin sheath of nerves. Symptoms range from only fatigue and weakness to, in some cases, eventual severe paralysis. The disease progresses slowly, over many years. New attacks that worsen the condition are often separated by long periods of remission. There is considerable evidence of genetic susceptibility, probably not from a single gene, but from several genes that interact. The etiology of multiple sclerosis is unknown, but epidemiological evidence indicates that it probably involves some infective agent or agents acquired during early adolescence. The Epstein-Barr virus (page 391) is frequently mentioned as a prime suspect. No cure exists, but treatments with interferons and several drugs that interfere with immune processes can significantly slow progression of symptoms.

Insulin-dependent diabetes mellitus is a familiar condition caused by immunological destruction of insulin-secreting cells of the pancreas. T cells are clearly implicated in this disease; animals that are genetically likely to develop diabetes fail to do so when their thymus is removed in infancy.

The fairly common skin condition **psoriasis** is an autoimmune disorder characterized by itchy-red patches of thickened skin. As many as 25% of patients develop **psoriatic arthritis.** Several topical and systemic therapies such as corticosteroids and methotrexate are available to help control psoriasis of the skin. Psoriasis is considered to be a T_H1 disease and can be effectively treated with immunosuppressants that target T cells and especially the cytokine, TNF-α (see page 460), an important factor in inflammation. For psoriatic arthritis, as well as rheumatoid arthritis, the most effective treatments are injections of monoclonal antibodies that inhibit TNF-α. See the box in Chapter 17 on page 494 for another new treatment.

CHECK YOUR UNDERSTANDING

- ✓ What is the importance of clonal deletion in the thymus? **19-11**
- ✓ What organ is affected in Graves' disease? **19-12**

Reactions Related to the Human Leukocyte Antigen (HLA) Complex

LEARNING OBJECTIVES

- **19-13** Define *HLA complex,* and explain its importance in disease susceptibility and tissue transplants.
- **19-14** Explain how a transplant is rejected.
- **19-15** Define *privileged site.*
- **19-16** Discuss the role of stem cells in transplantation.
- **19-17** Define *autograft, isograft, allograft,* and *xenotransplant.*
- **19-18** Explain how graft-versus-host disease occurs.
- **19-19** Explain how rejection of a transplant is prevented.

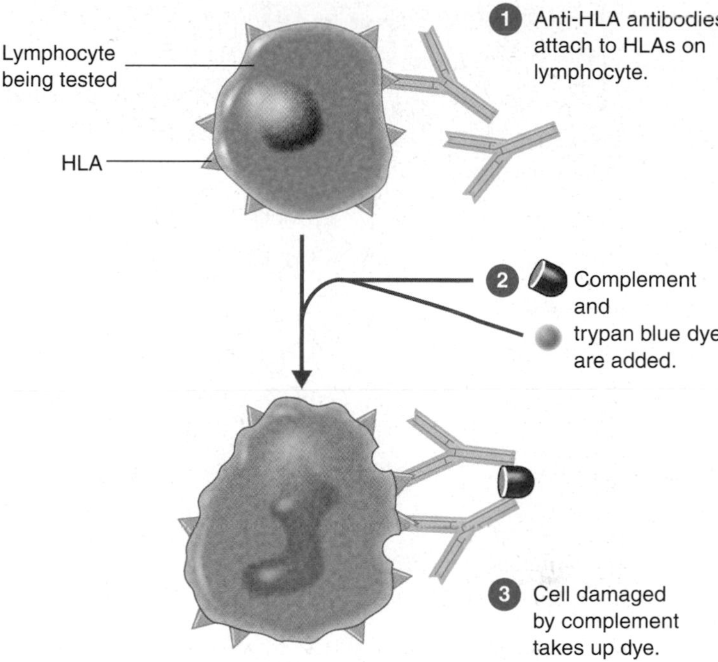
Lymphocyte being tested
HLA
1 Anti-HLA antibodies attach to HLAs on lymphocyte.
2 Complement and trypan blue dye are added.
3 Cell damaged by complement takes up dye.

Figure 19.9 Tissue typing, a serological method. Lymphocytes from the person being tested are incubated with laboratory test stocks of anti-HLA antibodies specific for a particular HLA. If the antibodies react with the antigens on a lymphocyte, then complement damages the lymphocyte and dye can enter the cell. Such a positive test result indicates that the person has the particular HLA being tested for.

Q **Why is tissue typing done?**

The inherited genetic characteristics of individuals are expressed not only in the color of their eyes and the curl of their hair but also in the composition of the self molecules on their cell surfaces. Some of these are called **histocompatibility antigens.** The genes controlling the production of the most important of these self molecules are known as the **major histocompatibility complex (MHC).** In humans, these genes are called the **human leukocyte antigen (HLA) complex.** We encountered these self molecules in Chapter 17 (page 482), where we saw that most antigens can stimulate an immune reaction only if they are associated with an MHC molecule.

A process called *HLA typing* is used to identify and compare HLAs. Certain HLAs are related to an increased susceptibility to specific diseases; one medical application of HLA typing is to identify such susceptibility. A few of these relationships are summarized in **Table 19.3**.

Another important medical application of HLA typing is in transplant surgery, in which the donor and the recipient must be matched by *tissue typing.* The serological technique shown in **Figure 19.9** is the one most often used. In serological tissue typing, the laboratory uses standardized antisera or monoclonal antibodies that are specific for particular HLAs.

Table 19.3 Diseases Related to Specific Human Leukocyte Antigens (HLAs)

Disease	Increased Risk of Occurrence with Specific HLA*	Description
Inflammatory Diseases		
Multiple sclerosis	5 times	Progressive inflammatory disease affecting nervous system
Rheumatic fever	4–5 times	Cross-reaction with antibodies against streptococcal antigen
Endocrine Diseases		
Addison's disease	4–10 times	Deficiency in production of hormones by adrenal gland
Graves' disease	10–12 times	Antibodies attached to certain receptors in the thyroid gland cause it to enlarge and produce excessive hormones
Malignant Disease		
Hodgkin's disease	1.4–1.8 times	Cancer of lymph nodes

*Compared to the general population.

A promising new technique for analyzing HLA is the use of *PCR,* the *polymerase chain reaction,* to amplify the cell's DNA (see Figure 9.4, page 252). If this is done for both donor and recipient, a match between donor DNA and recipient DNA can then be made. Having such a DNA match and matching ABO blood type between the donor and the recipient should result in a much higher success rate in transplant surgery.

Other factors may be involved in the success of a transplant, however. In Chapter 17, we briefly introduced a hypothesis that the body's reaction to transplanted foreign tissue may be a response to surgery-damaged cells. In other words, tissue rejection may result from a learned reaction to the danger signal posed by damaged cells, rather than a learned reaction to nonself.

CHECK YOUR UNDERSTANDING

✓ What is the relationship between the major histocompatibility complex in humans and the human leukocyte antigen complex? **19-13**

Reactions to Transplantation

In sixteenth-century Italy, crimes were often punished by cutting off the offender's nose. A surgeon of the time, in his attempts to repair this mutilation, observed that if skin was taken from the patient, it healed properly, but if it was taken from another person, it did not. He called this a manifestation of "the force and power of individuality."

We now know the principles behind this phenomenon. Transplants recognized as nonself are rejected—attacked by T cells that directly lyse the grafted cells, by macrophages activated by T cells, and, in certain cases, by antibodies, which activate the complement system and injure blood vessels supplying the transplanted tissue. However, transplants that are not rejected can add many healthy years to a person's life.

Since the first kidney transplant was performed in 1954, this particular type of transplant has become a nearly routine medical procedure. Other types of transplants that are now feasible include bone marrow, lungs, heart, liver, and cornea. Tissues and organs for transplant are usually taken from recently deceased individuals, although one of a pair of organs, such as a kidney, occasionally comes from a living donor. A donor can also donate as much as half of a healthy liver.

Privileged Sites and Privileged Tissue

Some transplants or grafts do not stimulate an immune response. A transplanted cornea, for example, is rarely rejected, mainly because antibodies usually do not circulate into that portion of the eye, which is therefore considered an immunologically **privileged site.** (However, rejections do occur, especially when the cornea has developed many blood vessels from corneal infections or damage.) The brain is also an immunologically privileged site, probably because it does not have lymphatic vessels and because the walls of the blood vessels in the brain differ from blood vessel walls elsewhere in the body (the blood–brain barrier is discussed in Chapter 22). Someday it may even be possible to graft foreign nerves to replace damaged nerves in the brain and spinal cord.

How animals tolerate pregnancy without rejecting the fetus is only partially understood. During pregnancy, the tissues of two genetically different individuals are in direct contact. An important factor seems to be that the MHC classes I and II in the cells that form the outer layer of the placenta, and come into contact with maternal tissue, are not of the specific types that stimulate a cellular immune response. The fetus is also protected by certain proteins it synthesizes, which have immunosuppressive properties. But there is no single, simple explanation.

It is possible to transplant **privileged tissue** that does not stimulate an immune rejection. An example is replacing a person's damaged heart valve with a valve from a pig's heart. However, privileged sites and tissues are more the exception than the rule.

Stem Cells

A development that promises to transform transplantation medicine is the use of **stem cells** (see Figure 17.8, page 487), master cells that are capable of generating any of the myriad cell types that make up the body. The most interest is centered on **embryonic stem cells (ESCs).** These cells can be isolated from the very earliest stage of an embryo, usually from discarded embryos created for attempts at in vitro fertilization. The ESCs are *pluripotent,* meaning they are capable of generating many different types of tissue cells. Figure 19.10 shows how these cells are harvested from the blastocyst stage, a hollow ball of 100 to 150 undifferentiated cells that is reached a few days after the egg is fertilized. When cultured, ESCs can be trained to produce different cell lines, such as muscle, nerve, or blood cells.

In the medical community there is great interest in using ESCs in therapy. For example, theoretically these cells could be used to regenerate damaged heart tissue or the failing insulin-producing cells in the pancreas that lead to diabetes. Damaged cartilage in the joints of rheumatoid arthritis patients might be replaced. Neurological conditions such as Parkinson disease or trauma-caused paralysis might also be candidates for treatment. There is even the prospect of growing complete new organs. In some instances the original donor might be the recipient, ensuring a genetic match of the tissues. Fortunately, human ESCs seem to express few MHC class I antigens and no class II antigens. This eases the problem of immune rejection but does not solve it. In any case, researchers want to create pluripotent cells that genetically match the patient or otherwise evade immune rejection.

Because ESCs are derived from embryos, even at their microscopic stage, many people object to their use. Possible alternatives are *adult stem cells (ASCs)*, which exist in some tissues such as the blood or skin. These produce only a very few different cell types, mostly of the tissue type of origin, and are difficult to cultivate. A promising new avenue of research is to genetically reprogram ASCs by using viruses to insert genes into skin cells, or other adult cells, to convert them into *induced pluripotent stem cells* (*iPS*). Other nonembryonic sources of stem cells are cord blood cells, considered ASCs, harvested from umbilical cords (see page 536). These are primarily *hematopoietic stem cells (HSCs)*, which are progenitors of blood and lymphatic (immune system) cells. Bone marrow transplants (see page 536) are a form of stem cell transplantation, mostly HSCs.

1 (1 day) Embryo, usually a fertilized egg discarded from attempt at in vitro fertilization.

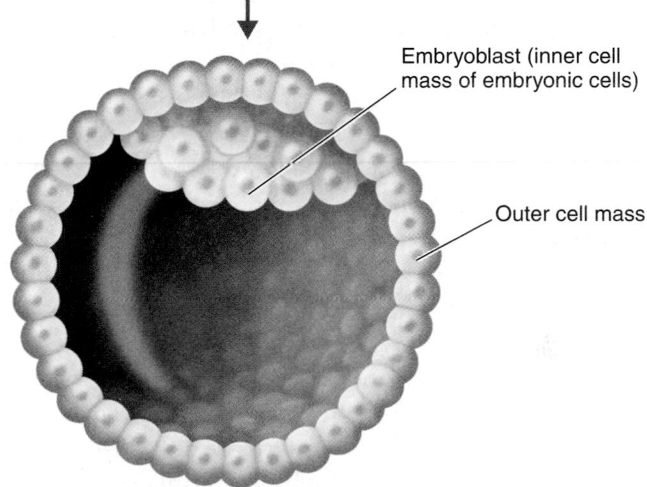

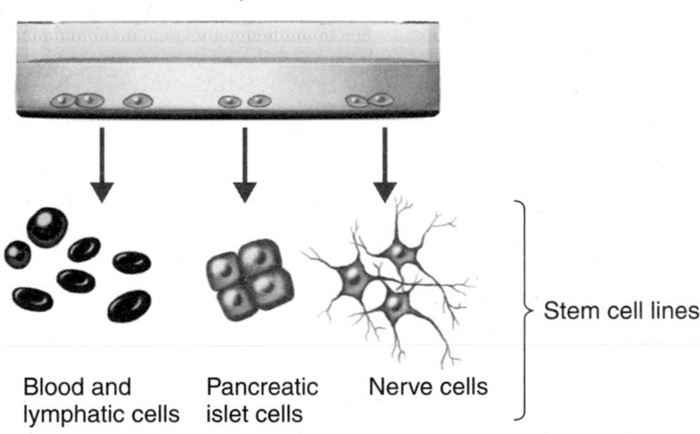

Embryoblast (inner cell mass of embryonic cells)
Outer cell mass
2 (1–5 days) Blastocyst stage; the embryo divides repeatedly and forms a hollow ball of cells about the size of the period at the end of a sentence.

Pancreatic islet cells

3 Embryonic stem cells from embryoblast are grown on feeder cells in culture medium. Stem cell lines and groups of stem cells form colonies in culture medium. Different conditions, as well as growth factors added to culture medium, direct stem cells to become stem cell lines for various tissues of the body (e.g., blood and lymphatic cells, pancreatic islet cells, nerve cells).

Figure 19.10 Derivation of embryonic stem cells.

Q What does *pluripotent* mean?

Grafts

When one's own tissue is grafted to another part of the body, as is done in burn treatment or in plastic surgery, the graft is not rejected. Recent technology has made it possible to use a few cells of a burn patient's uninjured skin to culture extensive sheets of new skin. This new skin is an example of an **autograft.** Identical twins have the same genetic makeup; therefore, skin or organs such as kidneys may be transplanted between them without provoking an immune response. Such a transplant is called an **isograft.**

Most transplants, however, are made between people who are not identical twins, and these transplants do trigger an immune response. Attempts are made to match the HLAs of the donor and recipient as closely as possible to reduce the chances of rejection. Because HLAs of close relatives are most likely to match, blood relatives, especially siblings, are the preferred donors. Grafts between people who are not identical twins are called **allografts.**

Because of the shortage of available organs, medical researchers hope to increase the success of **xenotransplantation products** (formerly called **xenografts**), which are tissues or organs that have been transplanted from animals. However, the body tends to mount an especially severe immune assault on such transplants. Unsatisfactory attempts have been made to use organs from baboons and other nonhuman primates. Research interest is high in genetically modifying pigs—an animal that is in plentiful supply, is of the right size, and generates relatively little public sympathy—to make acceptable donors of organs. The primary concern about xenotransplantation products is the possibility of transferring harmful animal viruses.

Preliminary research is underway that may eventually allow some bones and organs to be grown from the host's own tissue cells.

To be successful, xenotransplantation products must overcome **hyperacute rejection,** caused by the development in early infancy of antibodies against all distantly related animals such as pigs. With the aid of complement, these antibodies attack the transplanted animal tissue and destroy it within an hour. Hyperacute rejection occurs in human-to-human transplants only when antibodies have been preformed because of previous transfusions, transplantations, or pregnancies. Liver transplantation among humans is unusual in one respect; this organ usually resists hyperacute rejection, and HLA typing is not as important as in other types of tissue.

Bone Marrow Transplants

Transplants of bone marrow are frequently in the news. The recipients are usually individuals who lack the capacity to produce B cells and T cells vital for immunity or who are suffering from leukemia. Recall from Chapter 17 that bone marrow stem cells give rise to red blood cells and immune system lymphocytes. The goal of bone marrow transplants is to enable the recipient to produce healthy red blood, or immune system, cells. However, such transplants can result in **graft-versus-host (GVH) disease.** The transplanted bone marrow contains immunocompetent cells that mount primarily a cell-mediated immune response against the tissue into which they have been transplanted. Because the recipients lack effective immunity, GVH disease is a serious complication and can even be fatal.

An extremely promising technique for avoiding this problem is the use of *umbilical cord blood* instead of bone marrow. This blood is harvested from the placenta and umbilical cords of newborns, material that would otherwise be discarded. It is very rich in the stem cells found in bone marrow. Not only do these cells proliferate into the variety of cells required by the recipient, but also, because stem cells from this source are younger and less mature, the "matching" requirements are also less stringent than with bone marrow. As a result, GVH disease is less likely to occur.

CHECK YOUR UNDERSTANDING

- ✓ What immune system cells are involved in the rejection of nonself transplants? **19-14**
- ✓ Why is a transplanted cornea usually not rejected as nonself? **19-15**
- ✓ Differentiate an embryonic stem cell from an adult stem cell. **19-16**
- ✓ Which type of transplant is most subject to hyperacute rejection? **19-17**
- ✓ When red bone marrow is transplanted, many immunocompetent cells are included. How can this be bad? **19-18**

Immunosuppression

To keep the problem of transplant rejection in perspective, it is useful to remember that the immune system is simply doing its job and has no way of recognizing that its attack against the transplant is not helpful. In an attempt to prevent rejection, the recipient of an allograft usually receives treatment to suppress this normal immune response against the graft.

In transplantation surgery, it is generally desirable to suppress cell-mediated immunity, the most important factor in transplant rejection. If humoral (antibody-based) immunity is not suppressed, much of the ability to resist microbial infection will remain. In 1976, the drug *cyclosporine* was isolated from a mold. The successful transplantation of organs such as hearts and livers generally dates from the discovery of cyclosporine. Cyclosporine suppresses the secretion of interleukin-2 (IL-2), disrupting cell-mediated immunity by cytotoxic T cells. Following the success of this drug, other immunosuppressant drugs soon followed. *Tacrolimus* (FK506) has a mechanism similar to that of cyclosporine and is a frequent alternative, although both have many serious side effects. Neither cyclosporine nor tacrolimus has much effect on antibody production by the humoral immune system. Both of these drugs remain the mainstay for most regimens to prevent rejection of transplants. Some newer drugs, such as *sirolimus* (Rapamune), are among those that inhibit both cell-mediated and humoral immunity. This can be an advantage if chronic or hyperacute rejection by antibodies

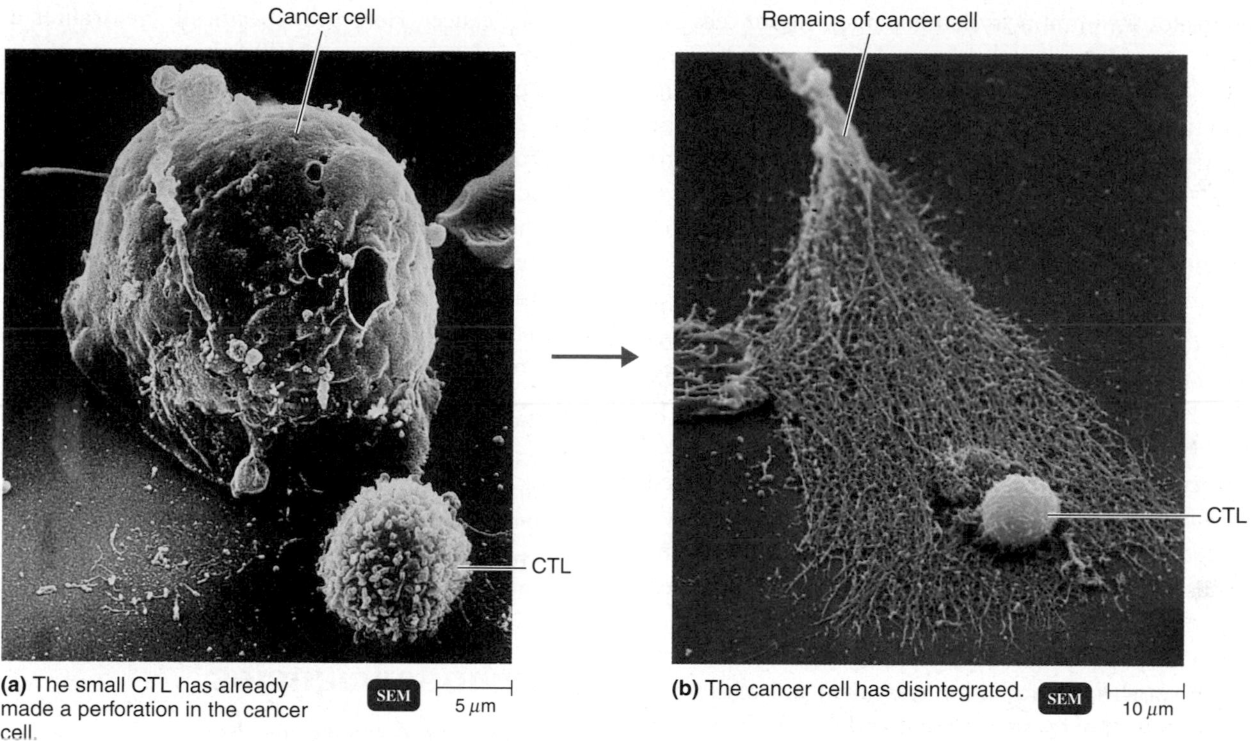

(a) The small CTL has already made a perforation in the cancer cell.

(b) The cancer cell has disintegrated.

Figure 19.11 The interaction between a cytotoxic T lymphocyte (CTL) and a cancer cell.

Q **CTLs can lyse cancer cells. How do they do this? (*Hint:* See Figure 17.10.)**

are a consideration. Sirolimus is best known for its use in stents, cylindrical meshes designed to keep blood vessels open after removal of blockages. Drugs such as *mycophenolate mofetil* inhibit the proliferation of T cells and B cells. Some biological agents such as the chimeric monoclonal antibodies (page 509) *basiliximab* and *daclizumab* also block IL-2 and are useful immunosuppressives. Immunosuppressive agents are usually administered in combinations.

An interesting observation is that occasionally a transplant recipient discontinues using immunosuppressant drugs but, surprisingly, does not reject the transplant. The reasons for this are uncertain.

CHECK YOUR UNDERSTANDING

✓ What cytokine is usually the target of immunosuppressant drugs intended to block transplant rejection? **19-19**

The Immune System and Cancer

LEARNING OBJECTIVES

19-20 Describe how the immune system responds to cancer and how cells evade immune responses.

19-21 Give two examples of immunotherapy.

Like an infectious disease, cancer represents a failure of the body's defenses, including the immune system. Some of the most promising avenues for effective cancer therapy make use of immunological techniques.

Nearly 100 years ago it was recognized that cancer cells arise frequently in the body and that they are usually eliminated by the immune system much like any other invading cell—the concept of **immune surveillance.** It was postulated that the cell-mediated immune system probably arose to combat cancer cells and that the appearance of a cancerous growth represented a failure of the immune system. This concept has been supported by the observation that cancers occur most often in older adults, whose immune systems are becoming less efficient, or in the very young, whose immune systems may not have developed fully or properly. Also, individuals who are immunosuppressed by either natural or artificial means are more susceptible to certain cancers.

A cell becomes cancerous when it undergoes transformation and begins to proliferate without control (see Chapter 13, page 389). The surfaces of tumor cells acquire tumor-associated antigens that mark them as nonself to the immune system. Figure 19.11 illustrates the attack on such a cancer cell by activated T_C cells (cytotoxic T lymphocytes, or CTLs). Activated macrophages can also destroy cancer cells. Although a healthy immune system serves to prevent most cancers, it has limitations. In some cases there is no antigenic epitope for the immune system to target. Tumor cells can even reproduce so rapidly that they exceed the capacity of the immune system to deal with them. Finally, if the tumor cell begins reproducing in tissues and becomes vascularized (connected to the body's blood supply), it usually becomes invisible to the immune system.

Immunotherapy for Cancer

The hypothesis that cancer represents a failure of the immune system has led to the thought that the immune system might be used to prevent or cure cancer—that is, **immunotherapy.**

At the turn of the twentieth century, William B. Coley, a physician at a New York City hospital, observed that if cancer patients contracted typhoid fever, their cancers often diminished noticeably. Following this lead, Coley made mixtures of killed streptococci (gram-positive) and *Serratia marcescens* (gram-negative) bacteria. These so-called Coley's toxins were injected into cancer patients to simulate a bacterial infection. Some of this work was very promising, but its results were inconsistent, and advances in surgery and radiation treatment caused it to be nearly forgotten. It is now recognized that endotoxins from such bacteria are powerful stimulants for the production of tumor necrosis factor (TNF) by macrophages. TNF is a small protein that interferes with the blood supply of cancers in animals.

Other research determined many years ago that if animals were injected with dead tumor cells, as with a vaccine, they did not develop tumors when injected with live cells from these tumors. Similarly, cancers sometimes undergo spontaneous remission that is probably related to the immune system gaining the advantage. Treating or preventing cancer by immunological means will probably be an increasingly important approach. An attractive aspect is that this approach avoids the damage to healthy cells caused by chemotherapy and radiation treatments. Already, a vaccine for Marek's disease, a cancer of chickens, has been successful. Vaccines to protect cats from feline leukemia have also been shown to provide considerable protection.

Cancer vaccines might be either *therapeutic* (used to treat existing cancers) or *prophylactic* (to prevent the development of cancer). Actually, prophylactic vaccines already exist. Hepatitis B virus is a common cause of liver cancer, and a vaccine against infection by this virus is widely used. A vaccine recommended for young girls, Gardasil, minimizes the chance of later development of cervical cancer caused by strains of a virus that also causes genital warts.

Although work with potential cancer vaccines has been in progress for almost a century, it is an area that is only in an early stage. Currently, a cancer vaccine would probably be most useful at preventing recurrences after other forms of treatment because the immune system would have to deal with a smaller number of cells.

Monoclonal antibodies are a promising tool for delivering cancer treatment. A humanized monoclonal antibody, *Herceptin* (see Chapter 18, page 509), is currently being used to treat a form of breast cancer. Herceptin specifically neutralizes a genetically determined growth factor, HER2, that promotes the proliferation of the cancer cells. It is expressed in relatively high quantities in about 25–30% of breast cancer patients.

Another approach is to combine a monoclonal antibody with a toxic agent, forming an **immunotoxin.** Theoretically, an immunotoxin might be used to specifically target and kill cells of a tumor with little damage to healthy cells. There have been some promising results in clinical trials, but not with large tumor masses in which many of the tumor cells cannot be reached by the immunotoxin.

CHECK YOUR UNDERSTANDING

- ✓ What is the function of tumor-associated antigens in the development of cancer? **19-20**
- ✓ Give an example of a prophylactic cancer vaccine that is in current use. **19-21**

Immunodeficiencies

LEARNING OBJECTIVE

19-22 Compare and contrast congenital and acquired immunodeficiencies.

The absence of a sufficient immune response is called an **immunodeficiency,** which can be either congenital or acquired.

Congenital Immunodeficiencies

Some people are born with a defective immune system. Defects in, or the absence of, a number of inherited genes can result in **congenital immunodeficiencies.** For example, individuals with a certain recessive trait, DiGeorge's syndrome, do not have a thymus gland and therefore lack cell-mediated immunity. An animal equivalent, which is extremely valuable for research in transplantation science, is the nude (hairless) mouse (**Figure 19.12**). These mice have no thymus (the coincidental hairlessness is controlled by the same gene) and therefore do not produce T cells and do not reject transplanted tissue. Even chicken skin, complete with feathers, is readily accepted as a graft.

Acquired Immunodeficiencies

A variety of drugs, cancers, or infectious agents can result in **acquired immunodeficiencies.** For example, Hodgkin's disease (a type of cancer) lowers the cell-mediated response. Many viruses are capable of infecting and killing lymphocytes, lowering the immune response. Removal of the spleen decreases humoral immunity. **Table 19.4** summarizes several of the better known immune deficiency conditions, including AIDS.

CHECK YOUR UNDERSTANDING

- ✓ Is AIDS an acquired or a congenital immunodeficiency? **19-22**

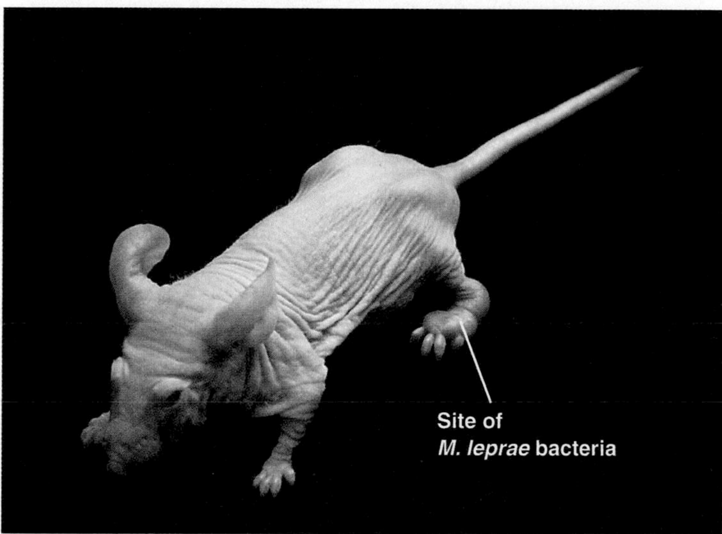

Figure 19.12 A nude (hairless) mouse infected with *Mycobacterium leprae* in the hind foot. Nude mice have no thymus and therefore no cell-mediated immunity. The immune response to infection by *M. leprae* (leprosy pathogen) depends on cell-mediated immunity, so these animals play an important role in leprosy research.

Q **What is the role of the thymus gland in immunity?**

Acquired Immunodeficiency Syndrome (AIDS)

LEARNING OBJECTIVES

19-23 Give two examples of how infectious diseases emerge.

19-24 Explain the attachment of HIV to a host cell.

19-25 List two ways in which HIV avoids the host's antibodies.

19-26 Describe the stages of HIV infection.

19-27 Describe the effects of HIV infection on the immune system.

19-28 Describe how HIV infection is diagnosed.

19-29 List the routes of HIV transmission.

19-30 Identify geographic patterns of HIV transmission.

19-31 List the current methods of preventing and treating HIV infection.

In 1981, a cluster of cases of *Pneumocystis* pneumonia (see page 21) appeared in the Los Angeles area. This extremely rare disease usually occurred only in immunosuppressed individuals. Investigators soon correlated the appearance of this disease with an unusual incidence of a rare form of cancer of the skin and blood vessels called Kaposi's sarcoma. The people affected were all young homosexual men, and all showed a loss of immune function. By 1983, the pathogen causing the loss of immune function

Table 19.4 Immunodeficiencies

Disease	Cells Affected	Comments
Acquired immunodeficiency syndrome (AIDS)	Virus destroys $CD4^+$ T cells	Allows cancer and bacterial, viral, fungal, and protozoan diseases; caused by HIV infection
Selective IgA immunodeficiency	B, T cells	Affects about 1 in 700, causing frequent mucosal infections; specific cause uncertain
Common variable hypogammaglobulinemia	B, T cells (decreased immunoglobulins)	Frequent viral and bacterial infections; second most common immune deficiency, affecting about 1 in 70,000; inherited
Reticular dysgenesis	B, T, and stem cells (a combined immunodeficiency; deficiencies in B and T cells and neutrophils)	Usually fatal in early infancy; very rare; inherited; bone marrow transplant a possible treatment
Severe combined immunodeficiency	B, T, and stem cells (deficiency of both B and T cells)	Affects about 1 in 100,000; allows severe infections; inherited; treated with bone marrow, fetal thymus transplants; gene therapy treatment is promising
Thymic aplasia (DiGeorge syndrome)	T cells (defective thymus causes deficiency of T cells)	Absence of cell-mediated immunity; usually fatal in infancy from *Pneumocystis* pneumonia or viral or fungal infections; due to failure of the thymus to develop in embryo
Wiskott-Aldrich syndrome	B, T cells (few platelets in blood, abnormal T cells)	Frequent infections by viruses, fungi, protozoa; eczema, defective blood clotting; usually causes death in childhood; inherited on X chromosome
X-linked infantile (Bruton's) agammaglobulinemia	B cells (decreased immunoglobulins)	Frequent extracellular bacterial infections; affects about 1 in 200,000; the first immunodeficiency disorder recognized (1952); inherited on X chromosome

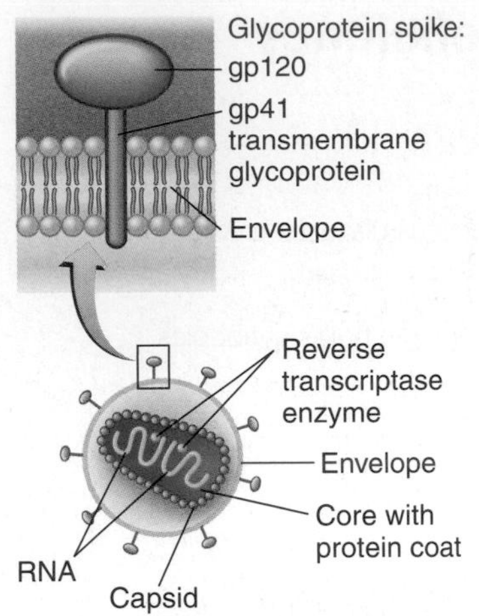

Structure of HIV and infection of a CD4+ T cell. The gp120 glycoprotein spike on the membrane attaches to a receptor on the CD4+ cell. The gp41 transmembrane glycoprotein probably facilitates fusion by attaching to a proposed fusion receptor on the CD4+ cell.

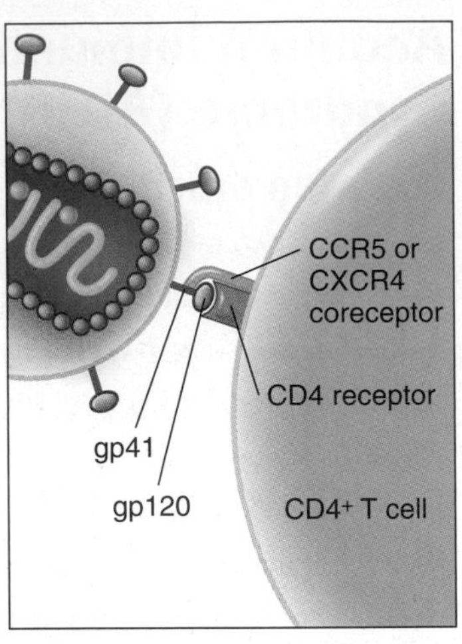

1 Attachment. The gp120 spike attaches to a receptor and to a CCR5 or CXCR4 coreceptor on the cell.

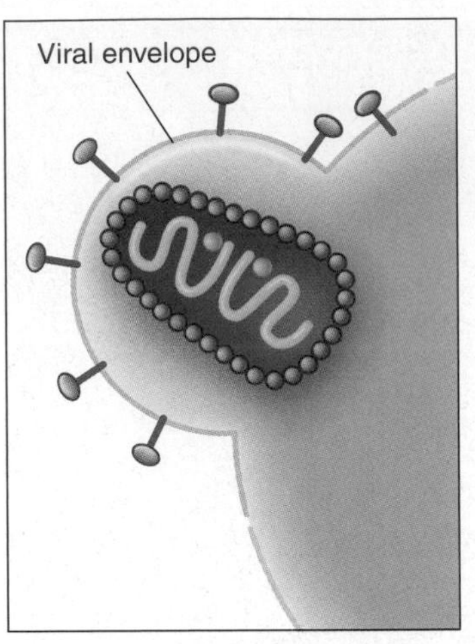

2 Fusion. The gp41 participates in fusion of the HIV with the cell.

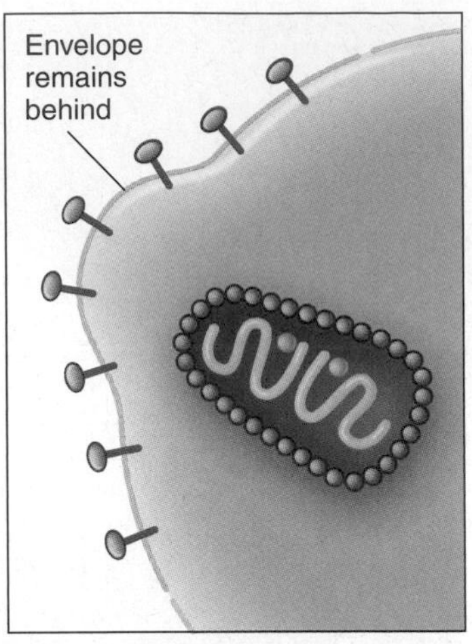

3 Entry. Following fusion with the cell, an entry pore is created. After entry, the viral envelope remains behind and the HIV uncoats, releasing the RNA core (see Figure 19.14b) for directing synthesis of the new viruses.

Figure 19.13 HIV structure and attachment to receptors on target T cell.

Q **Why does HIV preferentially infect $CD4^+$ cells?**

had been identified as a virus that selectively infects T helper cells. This virus is now known as human immunodeficiency virus, or HIV (see Figure 1.1e, page 5).

The Origin of AIDS

HIV is now believed to have arisen by the mutation of a virus that had been endemic in wildlife in some areas of central Africa. Genetic studies of the virus have led to the conclusion that HIV-2 (a type of HIV that is weakly contagious and not often found outside of West Africa) is a mutation of a simian immunodeficiency virus (SIV). Mangabey monkeys in West Africa are naturally and harmlessly infected with this SIV. More recently, studies show that HIV-1 (the primary HIV found worldwide in humans) is genetically related to another SIV that is carried by chimpanzees in Central Africa.

These SIV infections apparently crossed over relatively recently (well into the twentieth century) into the human population, known to eat "bushmeat." Mathematical models of the supposed evolution of HIV, by Bette Korber of the Los Alamos National Laboratory, calculate that the virus probably made the transition to humans around 1930. The disease may have smoldered with little notice as long as transmission was limited to small villages where rates of sexual promiscuity were lower. The virus could not have killed or incapacitated its hosts quickly; otherwise, it could not have been maintained in the village population. With the sudden end of European colonialism, the social structure of sub-Saharan Africa was disrupted. The population became urbanized; the developments that result from urbanization and contribute to an increase in sexual promiscuity, such as an increase in prostitution and the growth of highway transportation, are believed to be responsible for the spread of the disease. The earliest documented case of AIDS is from a patient in Leopoldville, Belgian Congo (now Kinshasa, capital of the Democratic Republic of the Congo). This man died in 1959; preserved samples of his blood contain antibodies to HIV. In the Western world, the first confirmed case of AIDS was the death of a Norwegian sailor in 1976, who probably was infected in 1961 or 1962 by contacts in western Africa.

CHECK YOUR UNDERSTANDING

✓ On what continent did the HIV-1 virus arise? **19-23**

HIV Infection

One of the most common misconceptions is that HIV infection is synonymous with AIDS. AIDS denotes only the final stage of a long infection.

The Structure of HIV

HIV, of the genus *Lentivirus*, is a retrovirus (see Figure 13.19, page 390). It has two identical strands of RNA, the enzyme reverse transcriptase, and an envelope of phospholipid (**Figure 19.13**). The envelope has glycoprotein spikes termed **gp120** (the notation for a glycoprotein with a molecular weight of 120,000).

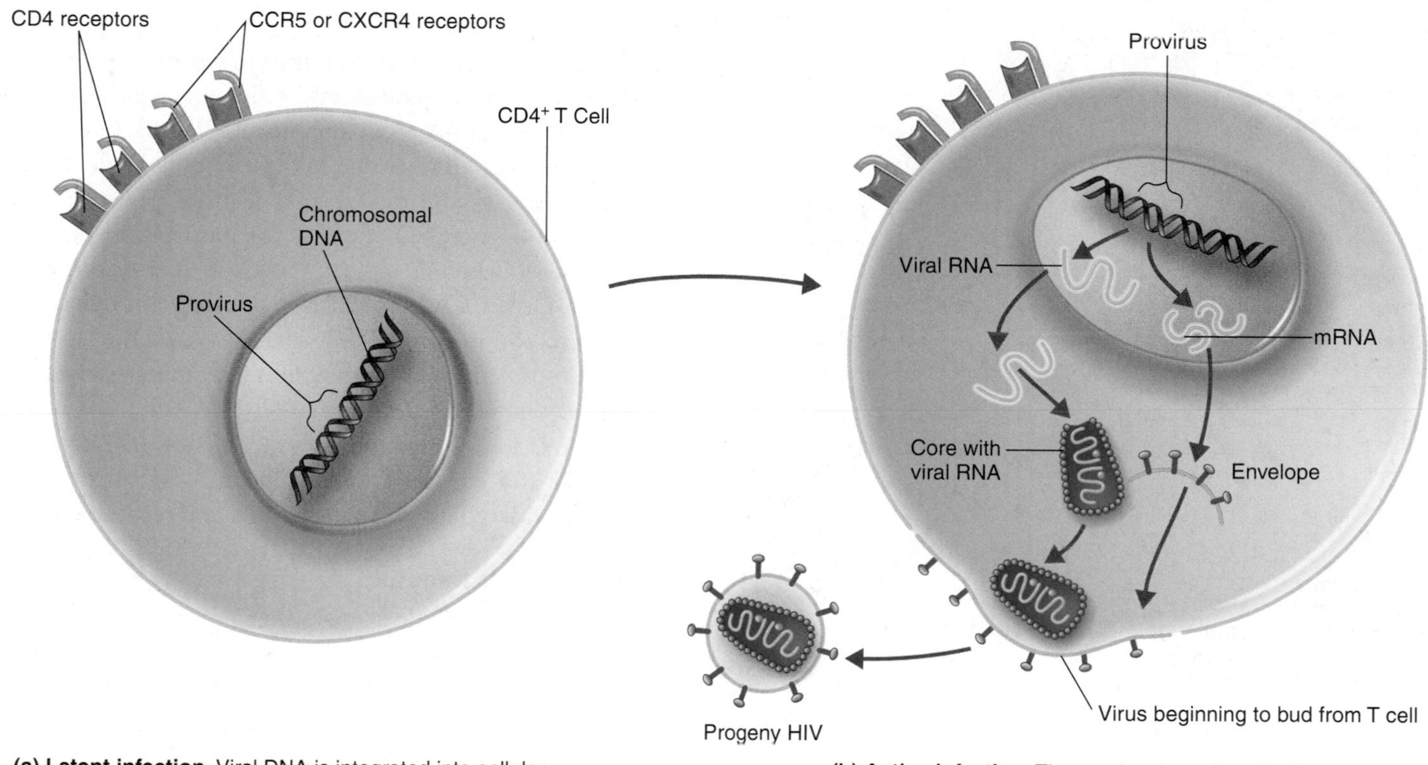

(a) Latent infection. Viral DNA is integrated into cellular DNA and forms a provirus (see page 389) that can later be activated to produce infective viruses.

(b) Active infection. The provirus is activated, allowing it to control the synthesis of new viruses, which bud from the host cell. Final assembly takes place at the cell membrane, taking up the viral envelope proteins as the virus buds from the cell.

Figure 19.14 Latent and active HIV infection in CD4$^+$ T cells.

Q What is a latent infection?

The Infectiveness and Pathogenicity of HIV

There is a strong association between HIV infection and the immune system. HIV is often spread by dendritic cells, which pick up the virus and carry it to the lymphoid organs. There it contacts cells of the immune system, most notably activated T cells, and stimulates an initial strong immune response.

To be infective, HIV must go through the steps of attachment, fusion, and entry in a manner similar to that shown in Figure 13.14 on page 384 and Figure 19.13. Attachment to the target cell depends on the glycoprotein spike (gp120) combining with the CD4$^+$ receptor. Approximately 65,000 of these receptors are found on each CD4$^+$ T helper cell, which is the main target of HIV infection. Certain coreceptors are also required. The two best-known chemokine coreceptors are named CCR5 and CXCR4*. Macrophages and monocytes (page 454) also carry CD4 molecules. (Many cells that do not express the CD4 molecule can also become infected, an indication that other receptors can also serve for infection by HIV).

In the host cell, viral RNA is released and transcribed into DNA by the enzyme reverse transcriptase. This viral DNA then becomes integrated into the chromosomal DNA of the host cell. The DNA may control the production of an active infection in which new viruses bud from the host cell, as shown in **Figure 19.14b**.

Alternatively, this integrated DNA may not produce new HIV but remains hidden in the host cell's chromosome as a *provirus* (**Figure 19.14a** and **Figure 19.15a**). HIV produced by a host cell is not necessarily released from the cell but may remain as *latent virions* in vacuoles within the cell (**Figure 19.15b**). In fact, a subset of the HIV-infected cells, instead of being killed, become long-lived memory T cells in which the reservoir of latent HIV can persist for decades. This ability of the virus to remain as a provirus or latent virus within host cells shelters it from the immune system. Another way HIV evades the immune system is *cell–cell fusion,* by which the virus moves from an infected cell to an adjacent uninfected cell.

The virus also evades immune defenses by undergoing rapid antigenic changes. Retroviruses, with the reverse transcriptase enzyme step, have a high mutation rate compared to DNA viruses. They also lack the corrective "proofreading" capacity of DNA viruses. As a result, a mutation is probably introduced at every

*This nomenclature is based on the beginning amino acid sequence in these proteins. The term CCR5 indicates that the beginning sequence consists of cysteines, thus CC. The letter R is a convention representing the balance of the protein molecule, and the number is for identification. If some other amino acid is located between the first two cysteines, this is shown in the naming—for example, CXCR4.

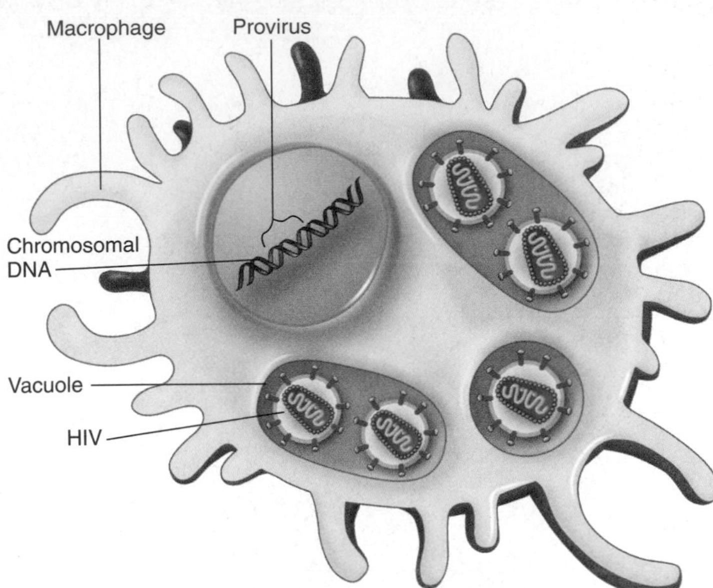

(a) Latently infected macrophage. HIV can persist either as a provirus or as a complete virion in vacuoles.

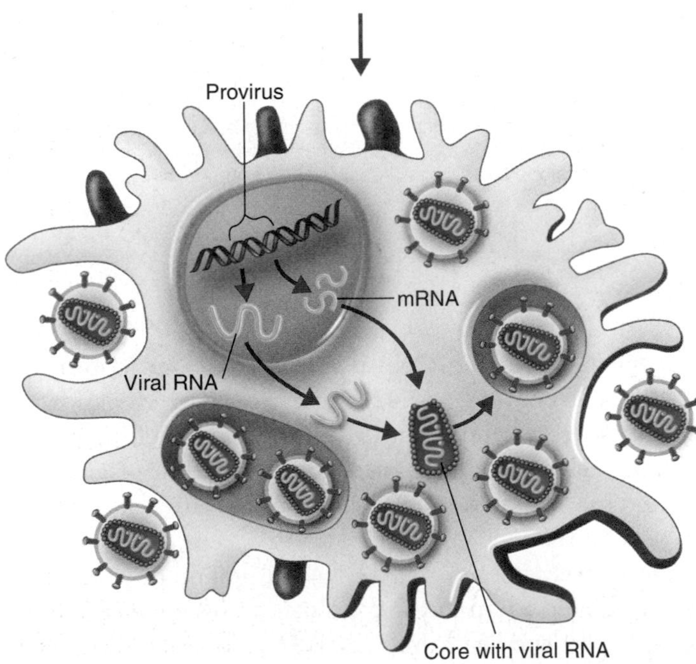

(b) Activated macrophage. New viruses are produced from provirus. Completed virions are either released or persist in the macrophage within vacuoles.

Figure 19.15 **Latent and active HIV infection in macrophages.**

Q How does an active infection differ from a latent infection?

position in the HIV genome many times each day in an infected person. This may amount to an accumulation of 1 million variants of the virus in an asymptomatic person and 100 million variants during the final stages of the infection. These dramatic numbers illustrate the potential problems of drug resistance and obstacles to the development of vaccines and diagnostic tests.

Clades (Subtypes) of HIV

Worldwide, the HIV genome is beginning to separate into distinctive groupings. Based on sequencing of the viral genomes, there are currently three HIV-1 groups named *M (main)*, *O (outlier)*, and *N (non M or O)*. Within group M there are nine clades (Greek for branches) designated A to D, F to H, J, and K. (There are also some subtypes for groups O and N). The most prevalent in the world are clades C and E (actually, clade E is a recombinant renamed CRF-OIAE). Clade C, which is spreading through central Africa down to South Africa, is also the most common in India and southeast Asia and is becoming dominant in parts of China. This clade may now represent half of all HIV infections worldwide. Clade E is found mostly in southeast Asia. In North and South America and Europe, clade B is the most prevalent. These groupings are undergoing constant revision and recombination.

The Stages of HIV Infection

The progress of HIV infection in adults can be divided into three clinical phases (Figure 19.16):

Phase 1. The number of viral RNA molecules per milliliter of blood plasma may reach more than 10 million in the first week or so. Billions of $CD4^+$ T cells may be infected within a couple of weeks. Immune responses and fewer uninfected cells to target deplete viral numbers in blood plasma sharply within a few weeks. The infection may be asymptomatic or cause lymphadenopathy (swollen lymph nodes).

Phase 2. The numbers of $CD4^+$ T cells decline steadily. HIV replication continues but at a relatively low level, probably controlled by $CD8^+$ T cells (see page 489 in Chapter 17) and occurs mainly in lymphatic tissue. Only a relatively few infected cells release HIV, although many may contain viruses in latent or proviral form. There are few serious disease symptoms, but a decline in immune response may become apparent by the appearance of persistent infections by the yeast *Candida albicans*, which can appear in the mouth, throat, or vagina. Other conditions may include fever and persistent diarrhea. Oral leukoplakia (whitish patches on oral mucosa), which results from reactivation of latent Epstein-Barr viruses, shingles, and other indications of declining immunity, may appear.

Phase 3. Clinical AIDS emerges, usually within 10 years of infection. $CD4^+$ T cell counts are below 350 cells/μl (200 cells/μl defines AIDS). Important AIDS indicator conditions appear, such as *C. albicans* infections of bronchi, trachea, or lungs; cytomegalovirus eye infections; tuberculosis; *Pneumocystis* pneumonia; toxoplasmosis of the brain; and Kaposi's sarcoma.

The Centers for Disease Control and Prevention (CDC) classifies the progress of HIV infections based on T cell populations. The purpose is primarily to furnish guidance for treatment, such as when to administer certain drugs. The normal population of a healthy individual is 800 to 1000 $CD4^+$ T cells/μl. In the United States, a count below 200/μl is considered diagnostic for AIDS, regardless of the clinical category observed.

Figure 19.16

FOUNDATION FIGURE The Progression of HIV Infection

Understanding how the HIV infection progresses in a host is integral to understanding the diagnosis, transmission, prevention, and treatment of this pandemic, all of which are discussed in this chapter.

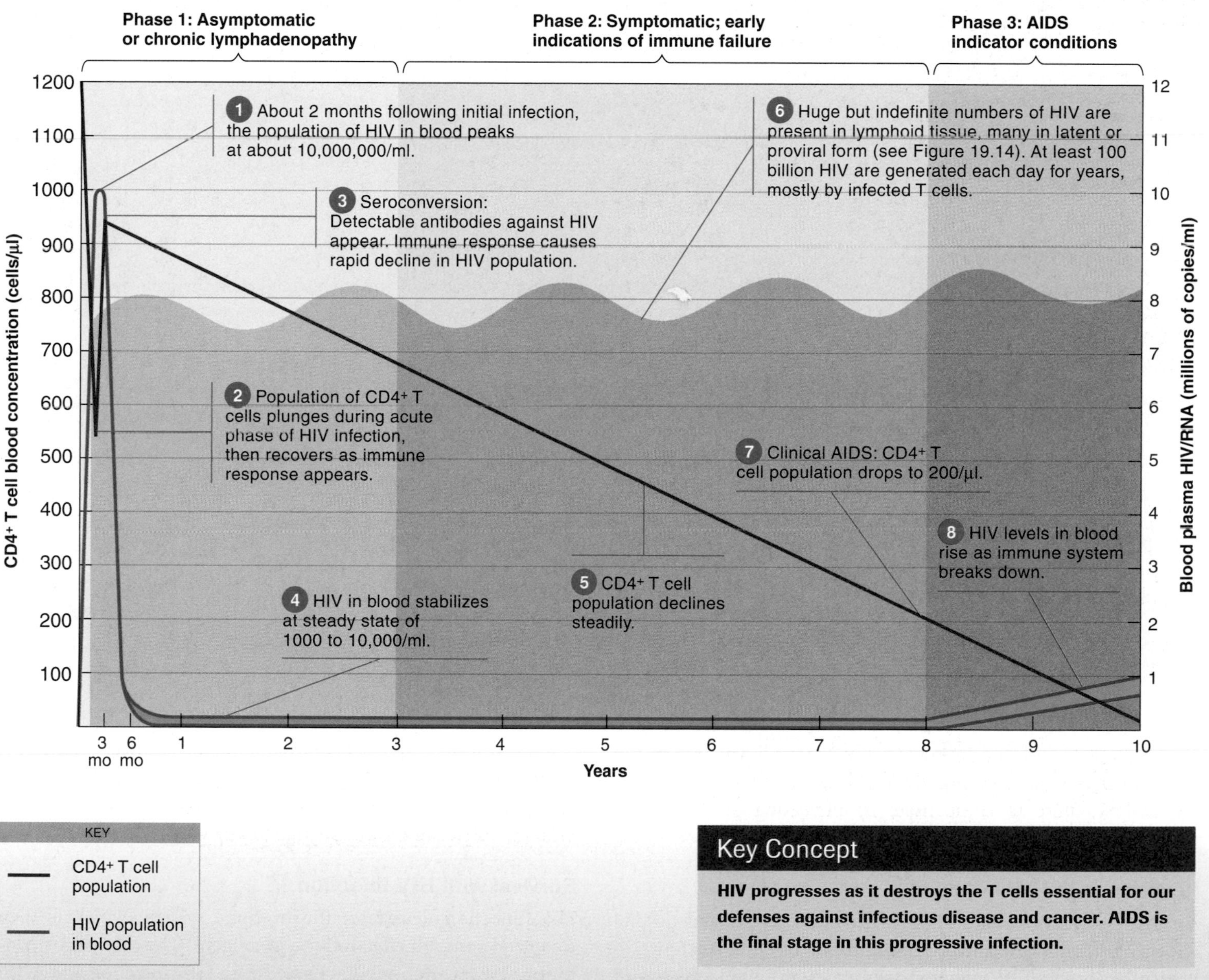

Phase 1: Asymptomatic or chronic lymphadenopathy
Phase 2: Symptomatic; early indications of immune failure
Phase 3: AIDS indicator conditions
1 About 2 months following initial infection, the population of HIV in blood peaks at about 10,000,000/ml.
2 Population of CD4+ T cells plunges during acute phase of HIV infection, then recovers as immune response appears.
3 Seroconversion: Detectable antibodies against HIV appear. Immune response causes rapid decline in HIV population.
4 HIV in blood stabilizes at steady state of 1000 to 10,000/ml.
5 CD4+ T cell population declines steadily.
6 Huge but indefinite numbers of HIV are present in lymphoid tissue, many in latent or proviral form (see Figure 19.14). At least 100 billion HIV are generated each day for years, mostly by infected T cells.
7 Clinical AIDS: CD4+ T cell population drops to 200/µl.
8 HIV levels in blood rise as immune system breaks down.
CD4+ T cell blood concentration (cells/µl)
1200
1100
1000
900
800
700
600
500
400
300
200
100
Blood plasma HIV/RNA (millions of copies/ml)
12
11
10
9
8
7
6
5
4
3
2
1
3 mo
6 mo
1
2
3
4
5
6
7
8
9
10
Years
KEY
CD4+ T cell population
HIV population in blood

Key Concept

HIV progresses as it destroys the T cells essential for our defenses against infectious disease and cancer. AIDS is the final stage in this progressive infection.

The progression from initial HIV infection to AIDS usually takes about 10 years in adults. This figure is typical in industrialized countries; in Africa, it is often about half this. Cellular warfare on an immense scale occurs during this time. At least 100 billion HIVs are generated every day, each with a remarkably short half-life of about 6 hours. These viruses must be cleared by the body's defenses, which include antibodies, cytotoxic T cells, and macrophages. Almost all HIVs, at least 99%, are produced by infected $CD4^+$ T cells, which survive for only about 2 days (T cells normally live for several years). Every day, an average of about 2 billion $CD4^+$ T cells are produced in an attempt to compensate for losses. Over time, however, there is a daily net loss of at least 20 million $CD4^+$ T cells, one of the main markers for the progression of HIV infection. The most recent studies show that the decrease in

Table 19.5 Some Common Diseases Associated with AIDS

Pathogen or Disease	Disease Description
Protozoa	
Cryptosporidium hominis	Persistent diarrhea
Toxoplasma gondii	Encephalitis
Isospora belli	Gastroenteritis
Viruses	
Cytomegalovirus	Fever, encephalitis, blindness
Herpes simplex virus	Vesicles of skin and mucous membranes
Varicella-zoster virus	Shingles
Bacteria	
Mycobacterium tuberculosis	Tuberculosis
M. avium-intracellulare	May infect many organs; gastroenteritis and other highly variable symptoms
Fungi	
Pneumocystis jirovecii	Life-threatening pneumonia
Histoplasma capsulatum	Disseminated infection
Cryptococcus neoformans	Disseminated, but especially meningitis
Candida albicans	Overgrowth on oral and vaginal mucous membranes (category B stage of HIV infection)
C. albicans	Overgrowth in esophagus, lungs (category C stage of HIV infection)
Cancers or Precancerous Conditions	
Kaposi's sarcoma	Cancer of skin and blood vessels (caused by human herpesvirus 8)
Hairy leukoplakia	Whitish patches on mucous membranes; commonly considered precancerous
Cervical dysplasia	Abnormal cervical growth

$CD4^+$ T cells is not due entirely to direct viral destruction of the cells; rather, it is caused primarily by shortened life of the cells and the body's failure to compensate by increasing production of replacement T cells.

Resistance to HIV Infection

A feature of HIV infection is that the virus proliferates despite efforts of the humoral and cellular immune systems. As shown in Figure 19.16, the HIV infection stimulates an initial strong and fairly effective immune response. A few months after infection, virus levels decline greatly. The most important factor in this is probably cytotoxic T cells ($CD8^+$ T cells). Neutralizing antibodies do not emerge until the viremia has passed its peak and rapid genetic changes in the virus lower the effectiveness of antibodies, but $CD8^+$ T cells continue to suppress viral numbers. However, once HIV infection is established, it becomes relentlessly progressive in virtually all patients. This is largely because HIV establishes a pool of latently infected $CD4^+$ T cells early on, and almost no patients completely clear the infection. This reservoir is not eradicated even when antiviral therapy reduces the viremia to undetectable levels (less than 50 molecules per milliliter). The establishment of a latent infection contrasts with almost all other viral infections and is a challenge to any vaccine.

Survival with HIV Infection

HIV infection devastates the immune system, which is then unable to respond effectively to pathogens. The diseases or conditions most commonly associated with HIV infection and AIDS are summarized in **Table 19.5**. Success in treating these conditions has extended the lives of many HIV-infected people.

The age of the infected person can also be an important factor. Older adults are less able to replace $CD4^+$ T cell populations. Infants and younger children have an immune system that is not fully developed. They are much more susceptible to opportunistic infections.

Infants born to HIV-positive mothers are not always infected—in fact, only about 20% are. Infants who are most seriously infected survive less than 18 months.

Exposed, But Not Infected, Population About 1% of the population has a major deletion in CCR5, and these persons (mostly of European ancestry; the mutation is rare in African and Asian populations) are unusually resistant to repeated exposures to HIV. In this population, the CCR5 molecule does not appear on the cell surface, so the infection cannot be completed. (Rare strains of HIV do not require CCR5 and infect the cells anyway).

Another notable population that is resistant to infection by HIV has appeared among certain African prostitutes, who are repeatedly exposed but remain HIV negative. These individuals produce CTLs that are unusually effective in combating HIV.

Long-Term Nonprogressors Even though infected, some individuals remain free of symptoms and do not progress to the stage of AIDS, and their $CD4^+$ Tcell counts remain stable. Survival exceeding 25 years is predicted. The exact mechanism, or mechanisms, by which the patient successfully combats the infection is still uncertain, although there are several hypotheses. In some cases, a genetic factor blocks efficient binding of HIV to the CCR5 coreceptor. There is also evidence that some nonprogressors produce an antiviral factor that inhibits HIV replication.

CHECK YOUR UNDERSTANDING

- ✓ What is the primary receptor on host cells to which HIV attaches? **19-24**
- ✓ Would an antibody against the coat of HIV be able to react with a provirus? **19-25**
- ✓ Would a $CD4^+$ T cell count of 300/μl be diagnostic of AIDS? **19-26**
- ✓ Which cells of the immune system are the main target of an HIV infection? **19-27**

Diagnostic Methods

The CDC now recommends routine screening for HIV infections in several circumstances, especially in patients beginning treatment for tuberculosis and in patients seeking treatment for sexually transmitted infections. The standard procedure for detecting HIV antibodies has been an ELISA test (see Figure 18.14, page 578), which is considered the most sensitive. There are now several relatively inexpensive, rapid tests (10 to 20 minutes) available for HIV screening that are especially useful at urgent care clinics and emergency departments, as well as in developing, resource-poor countries. The tests use urine or fingerstick amounts of blood, and the OraQuick test can even use an oral swab of fluid. Some of these tests can potentially be used for home testing. An estimated 25% of HIV-positive Americans do not realize they are infected; this lack of knowledge fuels the spread of the disease. Inexpensive, rapid routine screening tests should be valuable in changing this.

Positive screening tests for antibodies must be confirmed by additional testing, usually by the Western blot test (see Figure 10.12, page 289).

A problem with antibody-type testing is the window of time between infection and the appearance of detectable antibodies, or **seroconversion.** This interval, which can be as long as 3 months, is illustrated in Figure 19.16, where seroconversion follows the peak number of viruses in circulation. Because of this delay, the recipient of an organ transplant or a blood transfusion can become infected with HIV even though antibody tests did not show the presence of the virus. Improvements in testing have gradually narrowed the window to 21 to 25 days.

An alternative to the Western blot confirmatory test has recently received FDA approval. Instead of antibodies, the APTIMA assay detects the RNA of the HIV-1 virus and should also be easier to read than the Western blot test. This test can also be used to detect early HIV infections, before appearance of antibodies. Its sensitivity is comparable to approved tests used to measure **plasma viral load (PVL)** in the blood of patients and monitor the treatment and progression of AIDS. Conventional PVL tests that detect viral RNA use methods such as PCR (see page 251) or nucleic acid hybridization (see page 291), are costly, and require 2 or 3 days to complete. Viral RNA can be detected in 7 to 10 days and, less reliably, in 2 to 4 days. To ensure safety of the blood supply as much as possible, the American Red Cross has introduced testing for anti-HIV antibody and nucleic acid hybridization testing for viral HIV(see the box on page 727).

Tests that detect viral RNA are the only option during the primary infection, before antibodies appear, and in infants of HIV-infected mothers who have circulating maternal antibodies that interfere with conventional tests to detect antibodies.

A caution to keep in mind in testing for HIV is that current tests may not reliably detect all of the myriad variants of rapidly mutating HIV, especially subtypes that are not normally present in a population. Furthermore, PVL tests sample only the virions circulating in the blood, which is very low compared to the estimated several hundred billion HIV-infected cells.

CHECK YOUR UNDERSTANDING

- ✓ What form of nucleic acid is detected in a PVL test for HIV? **19-28**

HIV Transmission

The transmission of HIV requires the transfer of, or direct contact with, infected body fluids. The most important of these is blood, which contains 1000 to 100,000 infective viruses per milliliter, and semen, which contains about 10 to 50 viruses per milliliter. The viruses are often located within cells in these fluids, especially in macrophages. HIV can survive more than 1.5 days inside a cell but only about 6 hours outside a cell.

Table 19.6	CDC's Universal Precautions for Health Care Personnel
Gloves	Disposable gloves should be used for direct exposure to infected blood, other body fluids, and tissues. Double-gloving is recommended during invasive surgical procedures. Personnel should not work when they have open skin lesions, dermatitis exuding fluids, or cutaneous wounds.
Gowns, Masks, and Goggles	Masks and protective eyewear are recommended when splashes are expected, such as during airway manipulation, endoscopy, and dental procedures, and in the laboratory.
Needles	To minimize the risk of needlesticks, needles should not be resheathed and should be put in a puncture-proof container for sterilization and disposal. More expensive, safer needle devices that minimize the risk of needlesticks are available.
Disinfection	Routine housekeeping in health care settings should include washing floors, walls, and other areas not normally associated with disease transmissions with a 1:100 dilution of household bleach. A 1:10 dilution is recommended for disinfecting a spill.
Preventive Treatment after Exposure	Accidental exposure cannot always be prevented. Studies indicate that exposed individuals can reduce their risk by the prophylactic use of two or three antiviral drugs for 4 weeks.

Source: Adapted from *MMWR*, 2004.

Routes of HIV transmission include intimate sexual contact, breast milk, transplacental infection of a fetus, blood-contaminated needles, organ transplants, artificial insemination, and blood transfusion. The risk to health care workers can be greater. For instance, the risk of infection from needlestick injury is 3 out of 1000, or 0.3%. As a precaution, health care workers should be vaccinated against HBV. Avoiding exposure is the health care worker's first line of defense against HIV. The CDC has developed the strategy of following *universal precautions* in *all* health care settings. These are described in **Table 19.6**. Probably the most dangerous form of sexual contact is anal-receptive intercourse. Vaginal intercourse is much more likely to transmit HIV from man to woman than vice versa, and transmission either way is much greater when genital lesions are present. Although rare, transmission can occur by oral-genital contact.

HIV is not transmitted by insects or casual contact, such as hugging, or sharing household items. Saliva generally contains less than 1 virus per milliliter, and kissing is not known to transmit the virus. In developed countries, transmission by transfusion is unlikely because blood is tested for HIV or HIV antibodies. However, there will always be a slight risk, as discussed earlier.

CHECK YOUR UNDERSTANDING

✓ What is considered to be the most dangerous form of sexual contact for transmission of HIV? **19-29**

AIDS Worldwide

Approximately 25 million people have died from AIDS (**Figure 19.17**). An estimated 5 million are becoming infected every year. It is the leading cause of death in sub-Saharan Africa. As the disease becomes established in the huge populations of Asia, especially China and India, the incidence of HIV could exceed more than a million new cases a year. Eastern Europe, Russia, and central Asia are also areas reporting a steep rise in HIV infections. In Western Europe and the United States, the mortality from AIDS has decreased because of the availability of effective antiviral drugs. It is projected that by 2010 there will be a total of more than 100 million HIV-infected individuals in the world; more than 90% of these will be in developing countries. Deaths from HIV-related causes by 2010 will probably exceed 8 million per year.

Early in the HIV/AIDS pandemic, transmission in the United States and Europe was most commonly by male homosexual activity and with the use of injected drugs. These factors are still very important in the Western world, particularly in North and South America and Europe. Currently, one-third of all HIV infections in eastern Europe and central and southeast Asia are by injected drug use. These infections are also important as a bridge leading to other forms of transmission. Worldwide, heterosexual transmission is predominant (about 85%), especially in the less developed parts of the world such as the center of the pandemic in sub-Saharan Africa. A feature of the pandemic in recent times has been the increasing percentage of women infected (about 42% worldwide, most living in sub-Saharan Africa), with associated mother-to-child transmission. Most of these cases are young women infected by older men.

CHECK YOUR UNDERSTANDING

✓ What is the most common mode, worldwide, by which HIV is transmitted? **19-30**

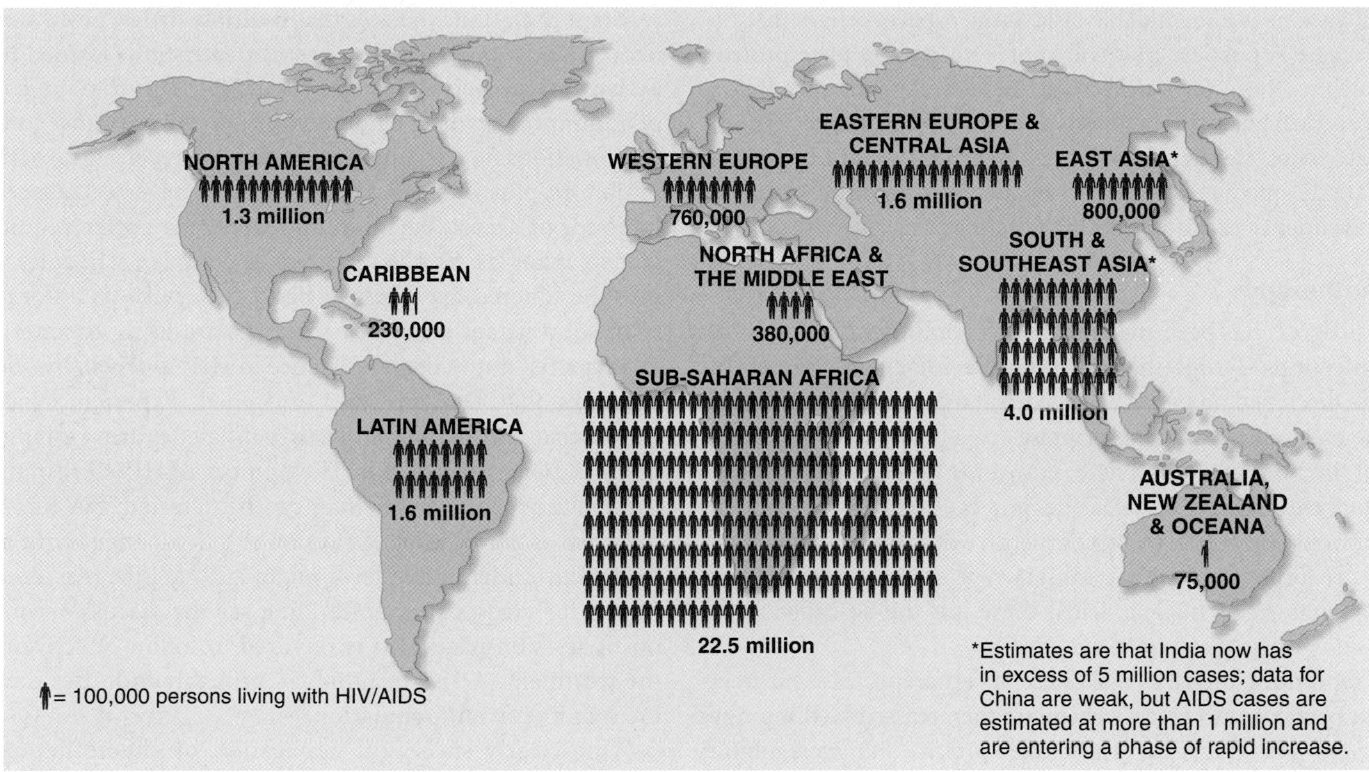

Figure 19.17 Distribution of HIV infection and AIDS in regions of the world. Each figure represents 100,000 persons living with HIV infection or AIDS.
Source: UNAIDS; World Health Organization.

Q **Where do you think the most accurate figures would be available?**

Preventing and Treating AIDS

At present, for most of the world the only practical means of control is to minimize transmission. This requires education programs to promote the use of condoms, as well as discouraging sexual promiscuity. An interesting discovery is that male circumcision lessens the rate of infection by at least 60%. In high-income countries, the availability of medications has made HIV infection no longer a certain death sentence. Unfortunately, improvements in managing HIV infection have resulted in a relaxed attitude toward safer-sex practices. The fact tends to be overlooked that the available drugs only delay the progress of the infection—they are not a cure. Attempts to prevent the use of contaminated needles among injected drug users is also important. To be effective, educational programs often require fundamental social changes that are not easy to accomplish, but they have slowed the infection rate in some areas, such as Thailand.

HIV Vaccines

An important consideration for vaccine development is the fact that the immune system has not shown much capability in coping with natural infections. In some 60 million HIV infections worldwide, there is not a single known case in which the immune system has eradicated the virus. Obstacles to developing a vaccine for HIV are formidable, and the landscape is now littered with unsuccessful vaccine trials. The rapid mutation rate of HIV makes it difficult to develop a vaccine that is effective against all mutational variants of the virus that appear during the course of an infection. Furthermore, the virus has developed clades that differ substantially from one geographic area to another, and each would probably require an appropriate vaccine.

Ideally, a vaccine would produce antibodies that would prevent infection. As we have seen, however, the virus resists antibody binding until a last-second exposure just before attachment and entry into the host cell. For persons already infected, a successful cell-mediated-type vaccine would be necessary to control the progression of the disease. To be considered successful, however, the vaccine would have to stimulate the production of CTLs that are more effective than those produced in response to a natural infection. Cells infected by HIV, it happens, are not very susceptible to attack by CTLs. There is also the problem of a persistent, but immunologically invisible, viral population in the form of proviruses and latent viruses (see Figure 19.15). Finally, a vaccine would have to be affordable in regions of the world where economic subsistence is often marginal. All in all, the development of an HIV vaccine is a formidable task. Some experts believe that no HIV vaccine is possible that will confer nearly complete protection, such as those for smallpox or measles. It is thought that a more practical goal may be to develop a vaccine with more modest goals than "sterilizing immunity." Perhaps a vaccine would stimulate cell-mediated immunity in already infected persons and help the patient's existing immune system to clear the virus.

Chemotherapy

Much progress has been made in using chemotherapy to inhibit HIV infections. Drugs that are available for chemotherapy of HIV are discussed on page 571. The first target of anti-HIV drugs was the enzyme reverse transcriptase (page 255), an enzyme not present in human cells. These *nucleoside reverse transcriptase inhibitors* are analogs of nucleosides and cause the termination of the synthesis of viral DNA by competitive inhibition (page 120). There are other drugs that inhibit reverse transcription but are not analogs of nucleic acids; these are the so-called *non-nucleoside reverse transcriptase inhibitors.*

To reproduce, the virus makes use of certain protease enzymes that cut proteins into pieces, which are then reassembled into the coat of new HIV particles. Drugs called *protease inhibitors* inhibit this enzyme and are now in use.

For infection to take place, the virus must accomplish a series of steps. It must attach to the cell's CD4 receptors, an interplay between the gp120 spike on the virus and the coreceptor (such as CCR5) must occur, and finally there must be a fusion with the cell to allow viral entry into the cell (see Figure 19.13). One of the newer anti-HIV drugs, grouped as *fusion inhibitors*, targets the gp41 region of the viral envelope.

After fusion has been completed, reverse transcription produces a double-stranded cDNA version of HIV, which enters the nucleus. Within the nucleus, the complex containing the cDNA must be integrated into the host chromosome to form the HIV provirus. This step requires an enzyme, HIV integrase, which is a target for drugs called *integrase inhibitors.* Several other points of attack for anti-HIV drugs exist, and drugs such as maturation inhibitors, targeted at these are under investigation or are in clinical trials.

Because of the increasing number of drugs that control reproduction of the virus, at least temporarily, HIV infection is almost at the stage where it can be considered a treatable chronic disease—assuming that the treatment is affordable. The rapid reproductive rate and frequent occurrence of drug-resistant mutations dictates that multiple drugs, given simultaneously, must be used. The current treatment is termed **highly active antiretroviral therapy (HAART).** This therapy consists of administering drug combinations; one of the most common combinations is two nucleoside analog reverse transcriptase inhibitors plus either a non-nucleoside reverse transcriptase inhibitor or a protease inhibitor. Patients are often required to take as many as 40 pills a day on a complex schedule, which must be adhered to rigorously because the virus is unforgiving. Even so, resistant strains of the virus are likely to emerge. As one scientist put it, drug resistance in HIV is driven by selective pressures that Darwin never imagined. Experience has also shown that eliminating all viruses in latent form in lymphoid tissue is especially difficult. The number of HIVs in circulation is often reduced to fewer than can be detected, but this is not the same as eradication. A caution is that a patient with a viral load at an undetectable level might still be infective. Frequent testing for viruses (not antibodies; see the discussion of diagnostic tests on page 545) is required to follow effectiveness of the treatment. A rise in numbers probably indicates development of a resistant population.

One clearly successful application of chemotherapy has been to reduce the chance of HIV transmission from an infected mother to her newborn. The administration of even one nucleoside reverse transcriptase drug alone reduces the incidence.

The AIDS Epidemic and the Importance of Scientific Research

The AIDS epidemic gives clear evidence of the value of basic scientific research. Without the advances in molecular biology of the past century, we would have been unable even to identify the causative agent of AIDS. We would not have been able to develop the tests for screening donated blood, to identify points in the viral life cycle for which selectively toxic drugs could be developed, or even to monitor the course of the infection. In the lifetime of most of us, we will have the opportunity to witness medical history being made as the struggle with this deadly and elusive virus continues.

CHECK YOUR UNDERSTANDING

✓ Does circumcision make a man more or less likely to acquire HIV infection? **19-31**

STUDY OUTLINE

The **MyMicrobiology Place** website (**www.microbiologyplace.com**) will help you get ready for tests with its simple three-step approach: ❶ **take a pre-test** and obtain a personalized study plan, ❷ **learn and practice** with animations, tutorials, and MP3 tutor sessions, and ❸ **test yourself** with quizzes and a chapter post-test.

Introduction (p. 522)

1. Hay fever, transplant rejection, and autoimmunity are examples of harmful immune reactions.
2. Immunosuppression is inhibition of the immune system.
3. Superantigens activate many T-cell receptors that can cause adverse host responses.

Hypersensitivity (pp. 523–531)

1. Hypersensitivity reactions represent immunological responses to an antigen (allergen) that lead to tissue damage rather than immunity.
2. Hypersensitivity reactions occur when a person has been sensitized to an antigen.
3. Hypersensitivity reactions can be divided into four classes: types I, II, and III are immediate reactions based on humoral immunity, and type IV is a delayed reaction based on cell-mediated immunity.

Type I (Anaphylactic) Reactions (pp. 523–526)

4. Anaphylactic reactions involve the production of IgE antibodies that bind to mast cells and basophils to sensitize the host.
5. The binding of two adjacent IgE antibodies to an antigen causes the target cell to release chemical mediators, such as histamine, leukotrienes, and prostaglandins, which cause the observed allergic reactions.
6. Systemic anaphylaxis may develop in minutes after injection or ingestion of the antigen; this may result in circulatory collapse and death.
7. Localized anaphylaxis is exemplified by hives, hay fever, and asthma.
8. Skin testing is useful in determining sensitivity to an antigen.
9. Desensitization to an antigen can be achieved by repeated injections of the antigen, which leads to the formation of blocking (IgG) antibodies.

Type II (Cytotoxic) Reactions (pp. 526–528)

10. Type II reactions are mediated by IgG or IgM antibodies and complement.
11. The antibodies are directed toward foreign cells or host cells. Complement fixation may result in cell lysis. Macrophages and other cells may also damage the antibody-coated cells.

The ABO Blood Group System (pp. 526–527)

12. Human blood may be grouped into four principal types, designated A, B, AB, and O.
13. The presence or absence of two carbohydrate antigens designated A and B on the surface of the red blood cell determines a person's blood type.
14. Naturally occurring antibodies are present in serum against the opposite AB antigen.
15. Incompatible blood transfusions lead to the complement-mediated lysis of the donor red blood cells.

The Rh Blood Group System (pp. 527–528)

16. Approximately 85% of the human population possesses another blood group antigen, designated the Rh antigen; these individuals are designated Rh^+.
17. The absence of this antigen in certain individuals (Rh^-) can lead to sensitization upon exposure to it.
18. An Rh^+ person can receive Rh^+ or Rh^- blood transfusions.
19. When an Rh^- person receives Rh^+ blood, that person will produce anti-Rh antibodies.
20. Subsequent exposure to Rh^+ cells will result in a rapid, serious hemolytic reaction.
21. An Rh^- mother carrying an Rh^+ fetus will produce anti-Rh antibodies.
22. Subsequent pregnancies involving Rh incompatibility may result in hemolytic disease of the newborn.
23. The disease may be prevented by passive immunization of the mother with anti-Rh antibodies.

Drug-Induced Cytotoxic Reactions (p. 528)

24. In the disease thrombocytopenic purpura, platelets are destroyed by antibodies and complement.
25. Agranulocytosis and hemolytic anemia result from antibodies against one's own blood cells coated with drug molecules.

Type III (Immune Complex) Reactions (pp. 528–529)

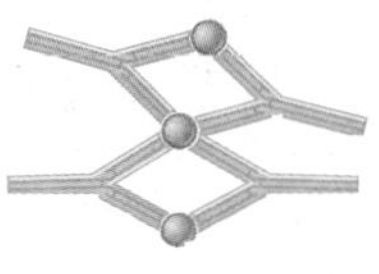

26. Immune complex diseases occur when IgG antibodies and soluble antigen form small complexes that lodge in the basement membranes of cells.
27. Subsequent complement fixation results in inflammation.
28. Glomerulonephritis is an immune complex disease.

Type IV (Delayed Cell-Mediated) Reactions (pp. 529–531)

29. Delayed cell-mediated hypersensitivity reactions are due primarily to T cell proliferation.
30. Sensitized T cells secrete cytokines in response to the appropriate antigen.
31. Cytokines attract and activate macrophages and initiate tissue damage.
32. The tuberculin skin test and allergic contact dermatitis are examples of delayed hypersensitivities.

Autoimmune Diseases (pp. 532–533)

1. Autoimmunity results from a loss of self-tolerance.
2. Self-tolerance occurs during fetal development; T cells that will target host cells are eliminated (clonal deletion) or inactivated.
3. Autoimmunity may be due to antibodies against infectious agents.

4. Graves' disease and myasthenia gravis are cytotoxic autoimmune reactions in which antibodies react to cell-surface antigens.
5. Systemic lupus erythematosus and rheumatoid arthritis are immune complex autoimmune reactions in which the deposition of immune complexes results in tissue damage.
6. Multiple sclerosis, insulin-dependent diabetes mellitus, and psoriasis are cell-mediated autoimmune reactions mediated by T cells.

Reactions Related to the Human Leukocyte Antigen (HLA) Complex

(pp. 533–537)

1. Histocompatibility self molecules located on cell surfaces express genetic differences among individuals; these antigens are coded for by MHC or HLA gene complexes.
2. To prevent the rejection of transplants, HLA and ABO blood group antigens of the donor and recipient are matched as closely as possible.
3. Transplants recognized as foreign antigens may be lysed by T cells and attacked by macrophages and complement-fixing antibodies.
4. Transplantation to a privileged site (such as the cornea) or of a privileged tissue (such as pig heart valves) does not cause an immune response.
5. Pluripotent stem cells differentiate into a variety of tissues that may provide tissues for transplant.
6. Four types of transplants have been defined on the basis of genetic relationships between the donor and the recipient: autografts, isografts, allografts, and xenotransplants.
7. Bone marrow (with immunocompetent cells) can cause graft-versus-host disease.
8. Successful transplant surgery often requires immunosuppressant drugs to prevent an immune response to the transplanted tissue.

The Immune System and Cancer

(pp. 537–538)

1. Cancer cells are normal cells that have undergone transformation, divide uncontrollably, and possess tumor-associated antigens.
2. The response of the immune system to cancer is called immunological surveillance.
3. T_C cells recognize and lyse cancerous cells.
4. Cancer cells can escape detection and destruction by the immune system.
5. Cancer cells may grow faster than the immune system can respond.

Immunotherapy for Cancer (p. 538)

6. Vaccines against liver and cervical cancer are available.
7. Herceptin consists of monoclonal antibodies against a breast cancer growth factor.
8. Immunotoxins are chemical poisons linked to a monoclonal antibody; the antibody selectively locates the cancer cell for release of the poison.

Immunodeficiencies (p. 538)

1. Immunodeficiencies can be congenital or acquired.
2. Congenital immunodeficiencies are due to defective or absent genes.
3. A variety of drugs, cancers, and infectious diseases can cause acquired immunodeficiencies.

Acquired Immunodeficiency Syndrome (AIDS) (pp. 539–548)

The Origin of AIDS (p. 540)

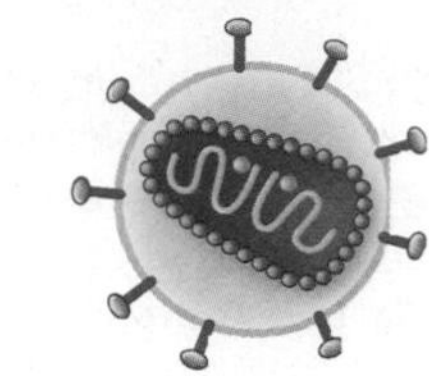

1. HIV is thought to have originated in central Africa and was brought to other countries by modern transportation and unsafe sexual practices.

HIV Infection (pp. 540–545)

2. AIDS is the final stage of HIV infection.
3. HIV is a retrovirus with single-stranded RNA, reverse transcriptase, and a phospholipid envelope with gp120 spikes.
4. HIV spikes attach to $CD4^+$ and coreceptors on host cells; the $CD4^+$ receptor is found on T helper cells, macrophages, and dendritic cells.
5. Viral RNA is transcribed to DNA by reverse transcriptase. The viral DNA becomes integrated into the host chromosome to direct synthesis of new viruses or to remain latent as a provirus.
6. HIV evades the immune system in latency, in vacuoles, by using cell–cell fusion, and by antigenic change.
7. Genetically distinct groups of HIV are classified into clades.
8. HIV infection is categorized by symptoms: Phase 1 (asymptomatic) and Phase 2 (selected symptoms) are reported as AIDS if $CD4^+$ T cells fall below 200 cells/μl; Phase 3 (AIDS indicator conditions) is reported as AIDS.
9. HIV infection is also categorized by $CD4^+$ T cell numbers: 200 $CD4^+$ cells/μl is reported as AIDS.
10. The progression from HIV infection to AIDS takes about 10 years.
11. The life of an AIDS patient can be prolonged by the proper treatment of opportunistic infections.
12. People lacking CCR5 are resistant to HIV infection.

Diagnostic Methods (p. 545)

13. HIV antibodies are detected by ELISA and Western blotting.
14. Plasma viral load tests detect viral nucleic acid and are used to quantify HIV in blood.

HIV Transmission (pp. 545–546)

15. HIV is transmitted by sexual contact, breast milk, contaminated needles, transplacental infection, artificial insemination, and blood transfusion.
16. In developed countries, blood transfusions are not a likely source of infection because blood is tested for HIV antibodies.

AIDS Worldwide (p. 546)

17. Heterosexual intercourse is the primary method of HIV transmission.

Preventing and Treating AIDS (pp. 547–548)

20. The use of condoms and sterile needles prevents the transmission of HIV.
21. Vaccine development is difficult because the virus remains inside host cells.
22. Current chemotherapeutic agents target the virus enzymes, including reverse transcriptase, integrase, and protease.

STUDY QUESTIONS

Answers to the Review and Multiple Choice questions can be found by turning to the blue Answers tab at the back of the textbook.

Review

1. **DRAW IT** Label IgE, antigen, and mast cell, and add an antihistamine to the following figure. What type of cell is this? Singulair stops inflammation by blocking leukotriene receptors. Add this action to the figure.

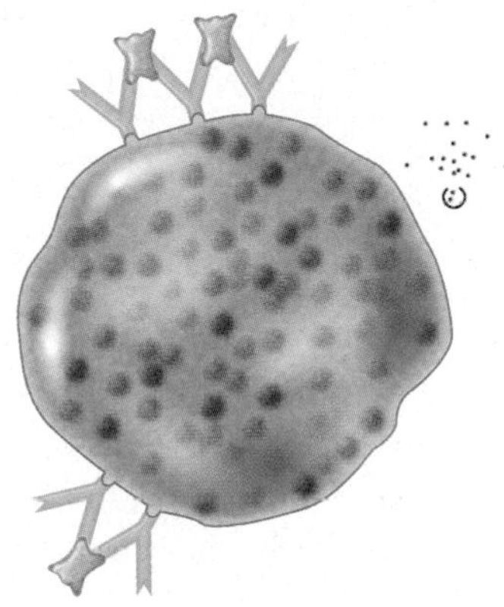

2. In the laboratory, blood is typed by looking for hemagglutination. For example, anti-A antibodies and type A RBCs clump. In a type A person, anti-A antibodies will cause hemolysis. Why?
3. Discuss the roles of antibodies and antigens in an incompatible tissue transplant.
4. Explain what happens when a person develops a contact sensitivity to poison oak.
   a. What causes the observed symptoms?
   b. How did the sensitivity develop?
   c. How might this person be desensitized to poison oak?
5. Why does an ANA (antinuclear antibody) test diagnose lupus?
6. Differentiate the three types of autoimmune diseases. Name an example of each type.
7. Summarize the causes of immunodeficiencies. What is the effect of an immunodeficiency?
8. In what ways do tumor cells differ antigenically from normal cells? Explain how tumor cells may be destroyed by the immune system.
9. If tumor cells can be destroyed by the immune system, how does cancer develop? What does immunotherapy involve?

Multiple Choice

1. Desensitization to prevent an allergic response can be accomplished by injecting small, repeated doses of
   a. IgE antibodies.
   b. the antigen (allergen).
   c. histamine.
   d. IgG antibodies.
   e. antihistamine.
2. What does "pluripotent" mean?
   a. Ability of a single cell to develop into an embryonic or adult stem cell.
   b. Ability of a stem cell to develop into many different cell types.
   c. A cell without MHC I and MHC II antigens.
   d. Ability of a single stem cell to heal different types of diseases.
   e. Ability of an adult cell to become a stem cell.
3. Cytotoxic autoimmunity differs from immune complex autoimmunity in that cytotoxic reactions
   a. involve antibodies.
   b. do not involve complement.
   c. are caused by T cells.
   d. do not involve IgE antibodies.
   e. none of the above
4. Antibodies against HIV are ineffective for all of the following reasons *except*
   a. the fact that antibodies aren't made against HIV.
   b. transmission by cell–cell fusion.
   c. antigenic changes.
   d. latency.
   e. persistence of virus particles in vacuoles.
5. Which of the following is *not* the cause of a natural immunodeficiency?
   a. a recessive gene resulting in lack of a thymus gland
   b. a recessive gene resulting in few B cells
   c. HIV infection
   d. immunosuppressant drugs
   e. none of the above
6. Which antibodies will be found naturally in the serum of a person with blood type A, Rh^+?
   a. anti A, anti B, anti Rh
   b. anti A, anti Rh
   c. anti A
   d. anti B, anti Rh
   e. anti B

Use the following choices to match the type of hypersensitivity to the examples in questions 7 through 10.
   a. type I hypersensitivity
   b. type II hypersensitivity
   c. type III hypersensitivity
   d. type IV hypersensitivity
   e. all of the above

7. Localized anaphylaxis.
8. Allergic contact dermatitis.
9. Due to immune complexes.
10. Reaction to an incompatible blood transfusion.

Critical Thinking

1. When and how does our immune system discriminate between self and nonself antigens?
2. The first preparations used for artificially acquired passive immunity were antibodies in horse serum. A complication that resulted from the therapeutic use of horse serum was immune complex disease. Why did this occur?

3. Do people with AIDS make antibodies? If so, why are they said to have an immune deficiency?

4. What are the methods of action of anti-AIDS drugs?

Clinical Applications

1. Fungal infections such as athlete's foot are chronic. These fungi degrade skin keratin but are not invasive and do not produce toxins. Why do you suppose that many of the symptoms of a fungal infection are due to hypersensitivity to the fungus?

2. After working in a mushroom farm for several months, a worker develops these symptoms: hives, edema, and swelling lymph nodes.

a. What do these symptoms indicate?

b. What mediators cause these symptoms?

c. How may sensitivity to a particular antigen be determined?

d. Other employees do not appear to have any immunological reactions. What could explain this?

(*Hint:* The allergen is conidiospores from molds growing in the mushroom farm.)

3. Physicians administering live, attenuated mumps and measles vaccines prepared in chick embryos are instructed to have epinephrine available. Epinephrine will not treat these viral infections. What is the purpose of keeping this drug on hand?

4. A woman with blood type A+ once received a transfusion of AB+ blood. When she carried a type B+ fetus, the fetus developed hemolytic disease of the newborn. Explain why this fetus developed this condition even though another type B+ fetus in a different type A+ mother was normal.

20 Antimicrobial Drugs

When the body's normal defenses cannot prevent or overcome a disease, it often can be treated by **chemotherapy** with antimicrobial drugs. Like the disinfectants discussed in Chapter 7, **antimicrobial drugs** act by killing or by interfering with the growth of microorganisms. Unlike disinfectants, however, antimicrobial drugs must often act *within* the host without damaging the host. This is the important principle of **selective toxicity.**

Antibiotics were one of the most important discoveries of modern medicine. Within the memory of many people is a time when little could be done to treat many lethal infectious diseases. The introduction of antimicrobials such as penicillin and sulfanilamide to treat conditions such as a ruptured appendix or so-called blood poisoning (sepsis) resulted in cures that seemed almost miraculous.

Today, we are seeing the advances represented by these miracle drugs threatened by the development of **antibiotic resistance.** For example, there are frequent reports of staphylococcal pathogens that are resistant to practically all the available antibiotics. Certain populations of the pathogens causing tuberculosis are now resistant to essentially all of the available antibiotics that were once effective. In some cases, medicine now has only a few more weapons to treat the diseases caused by these pathogens than were available a century ago.

UNDER THE MICROSCOPE

Bacterial lysis caused by an antibiotic.

Q&A

This bacterium is lysing because an antibiotic disrupted its cell wall. Why doesn't the antibiotic lyse human cells?

Look for the answer in the chapter.

The History of Chemotherapy

LEARNING OBJECTIVES

20-1 Identify the contributions of Paul Ehrlich and Alexander Fleming to chemotherapy.

20-2 Name the microbes that produce most antibiotics.

The birth of modern chemotherapy is credited to the efforts of Paul Ehrlich in Germany during the early part of the twentieth century. While attempting to stain bacteria without staining the surrounding tissue, he speculated about some "magic bullet" that would selectively find and destroy pathogens but not harm the host. This idea provided the basis for *chemotherapy,* a term he coined.

In 1928, Alexander Fleming observed that the growth of the bacterium *Staphylococcus aureus* was inhibited in the area surrounding the colony of a mold that had contaminated a Petri plate (Figure 1.5, page 12). The mold was identified as *Penicillium notatum,* and its active compound, which was isolated a short time later, was named penicillin. Similar inhibitory reactions between colonies on solid media are commonly observed in microbiology, and the mechanism of inhibition is called *antibiosis* (Figure 20.1). From this word comes the term **antibiotic,** a substance produced by microorganisms that in small amounts inhibits another microorganism. Therefore, the wholly synthetic sulfa drugs, for example, are technically not antibiotics, a distinction often ignored in practice. The discovery of sulfa drugs arose from a systematic survey of chemicals, looking for any that would be a "magic bullet," by German industrial scientists beginning in 1927. In 1932, a compound termed Prontosil Red (it was a sulfanilamide-containing dye) was found that controlled streptococcal infections in mice. It was later found that the active principle in the compound was the sulfanilamide component, and during World War II the Allied armies made wide use of this. The discovery and use of sulfa drugs made it clear that practical antimicrobials could be effective against systemic bacterial infections and resurrected interest in the earlier reports of penicillin.

In 1940, a group of scientists at Oxford University headed by Howard Florey and Ernst Chain succeeded in the first clinical trials of penicillin. Under wartime conditions in the United Kingdom, research into the development and large-scale production of penicillin was not possible, and this work was transferred to the United States. The original culture of *P. notatum* (later renamed *Penicillium chrysogenum*) was not a very efficient producer of the antibiotic. It was soon replaced by a more prolific strain. This valuable organism was first isolated from a moldy cantaloupe bought at a market in Peoria, Illinois.

Antibiotics are actually rather easy to discover, but few are of medical or commercial value. Some are used commercially other than for treating disease—for example, as a supplement in animal feed (see the box on page 577).

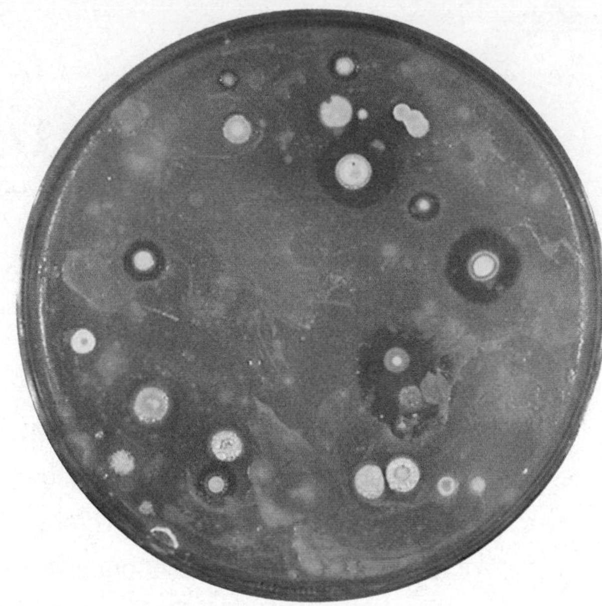

Figure 20.1 Laboratory observation of antibiosis. Anyone plating out microbes from natural environments, especially soil, will frequently see examples of bacterial inhibition by antibiotics produced by bacteria, most commonly *Streptomyces* species.

Q Would there be any advantage to a soil microbe to produce an antibiotic?

More than half of our antibiotics are produced by species of *Streptomyces,* filamentous bacteria that commonly inhabit soil. A few antibiotics are produced by endospore-forming bacteria such as *Bacillus,* and others are produced by molds, mostly of the genera *Penicillium* and *Cephalosporium* (sef-ä-lō-spô'rē-um). See Table 20.1 for the sources of many antibiotics in use today—a surprisingly limited group of organisms. One study screened 400,000 microbial cultures that yielded only three useful drugs. It is especially interesting to note that practically all antibiotic-producing microbes have some sort of sporulation process.

Antibiotic Discovery Today

Most antibiotics in use today were discovered by methods that required identifying and growing colonies of antibiotic-producing organisms, mostly by screening soil samples. It is rather easy to identify microbes in such samples that have antimicrobial activity; however, many are toxic or otherwise not commercially useful. Also, these proved to be examples of "low-hanging fruit," and continued work often resulted in discovery of the same antibiotics. As an example, streptomycin is produced by about 1% of isolates of actinomycetes from soil. To find an antibiotic that is produced by only one soil or sea microbe in 10 million, though, is a daunting task. What is needed are so-called *high-throughput methods* that screen very high numbers of microbes.

Table 20.1 Representative Sources of Antibiotics

Microorganism	Antibiotic
Gram-Positive Rods	
Bacillus subtilis	Bacitracin
Paenibacillus polymyxa	Polymyxin
Actinomycetes	
Streptomyces nodosus	Amphotericin B
Streptomycems venezuelae	Chloramphenicol
Streptomyces aureofaciens	Chlortetracycline and tetracycline
Saccharopolyspora erythraea	Erythromycin
Streptomyces fradiae	Neomycin
Streptomyces griseus	Streptomycin
Micromonospora purpurea	Gentamicin
Fungi	
Cephalosporium spp.	Cephalothin
Penicillium griseofulvum	Griseofulvin
Penicillium chrysogenum	Penicillin

CHECK YOUR UNDERSTANDING

- ✓ Who coined the term *magic bullet*? **20-1**
- ✓ More than half our antibiotics are produced by a certain species of bacteria. What is it? **20-2**

The Spectrum of Antimicrobial Activity

LEARNING OBJECTIVES

20-3 Describe the problems of chemotherapy for viral, fungal, protozoan, and helminthic infections.

20-4 Define the following terms: *spectrum of activity, broad-spectrum antibiotic, superinfection.*

It is comparatively easy to find or develop drugs that are effective against prokaryotic cells and that do not affect the eukaryotic cells of humans. These two cell types differ substantially in many ways, such as in the presence or absence of cell walls, the fine structure of their ribosomes, and details of their metabolism. Thus, selective toxicity has numerous targets. The problem is more difficult when the pathogen is a eukaryotic cell, such as a fungus, protozoan, or helminth. At the cellular level, these organisms resemble the human cell much more closely than a bacterial cell does. We will see that our arsenal against these types of pathogens is much more limited than our arsenal of antibacterial drugs. Viral infections are particularly difficult to treat because the pathogen is within the human host's cells and because the genetic information of the virus is directing the human cell to make viruses rather than to synthesize normal cellular materials.

Some drugs have a **narrow spectrum of microbial activity,** or range of different microbial types they affect. Penicillin G, for example, affects gram-positive bacteria but very few gram-negative bacteria. Antibiotics that affect a broad range of gram-positive or gram-negative bacteria are therefore called **broad-spectrum antibiotics.**

A primary factor involved in the selective toxicity of antibacterial action lies in the lipopolysaccharide outer layer of gram-negative bacteria and the porins that form water-filled channels across this layer (see Figure 4.13c, page 86). Drugs that pass through the porin channels must be relatively small and preferably hydrophilic. Drugs that are lipophilic (having an affinity for lipids) or especially large do not enter gram-negative bacteria readily.

Table 20.2 summarizes the spectrum of activity of a number of chemotherapeutic drugs. Because the identity of the pathogen is not always immediately known, a broad-spectrum drug would seem to have an advantage in treating a disease by saving valuable time. The disadvantage is that these drugs destroy many normal microbiota of the host. The normal microbiota ordinarily compete with and check the growth of pathogens or other microbes. If the antibiotic does not destroy certain organisms in the normal microbiota but does destroy their competitors, the survivors may flourish and become opportunistic pathogens. An example that sometimes occurs is overgrowth by the yeastlike fungus *Candida albicans,* which is not sensitive to bacterial antibiotics. This overgrowth is called a **superinfection,** a term that is also applied to growth of a target pathogen that has developed resistance to the antibiotic. In this situation, such an antibiotic-resistant strain replaces the original sensitive strain, and the infection continues.

CHECK YOUR UNDERSTANDING

- ✓ Identify at least one reason why it is so difficult to target a pathogenic virus without damaging the host's cells. **20-3**
- ✓ Why are antibiotics with a very broad spectrum of activity not as useful as one might first think? **20-4**

The Action of Antimicrobial Drugs

LEARNING OBJECTIVE

20-5 Identify five modes of action of antimicrobial drugs.

Antimicrobial drugs are either **bactericidal** (they kill microbes directly) or **bacteriostatic** (they prevent microbes from growing). In bacteriostasis, the host's own defenses, such as phagocytosis and antibody production, usually destroy the microorganisms. The major modes of action are summarized in Figure 20.2.

Figure 20.2

FOUNDATION FIGURE Major Modes of Action of Antimicrobial Drugs

This figure illustrates the five modes of action of antimicrobials that will be discussed in detail in this chapter. The limited spectrum of activity of individual antibiotics is important to understanding the appropriate applications for different drugs and the mechanisms of antibiotic resistance.

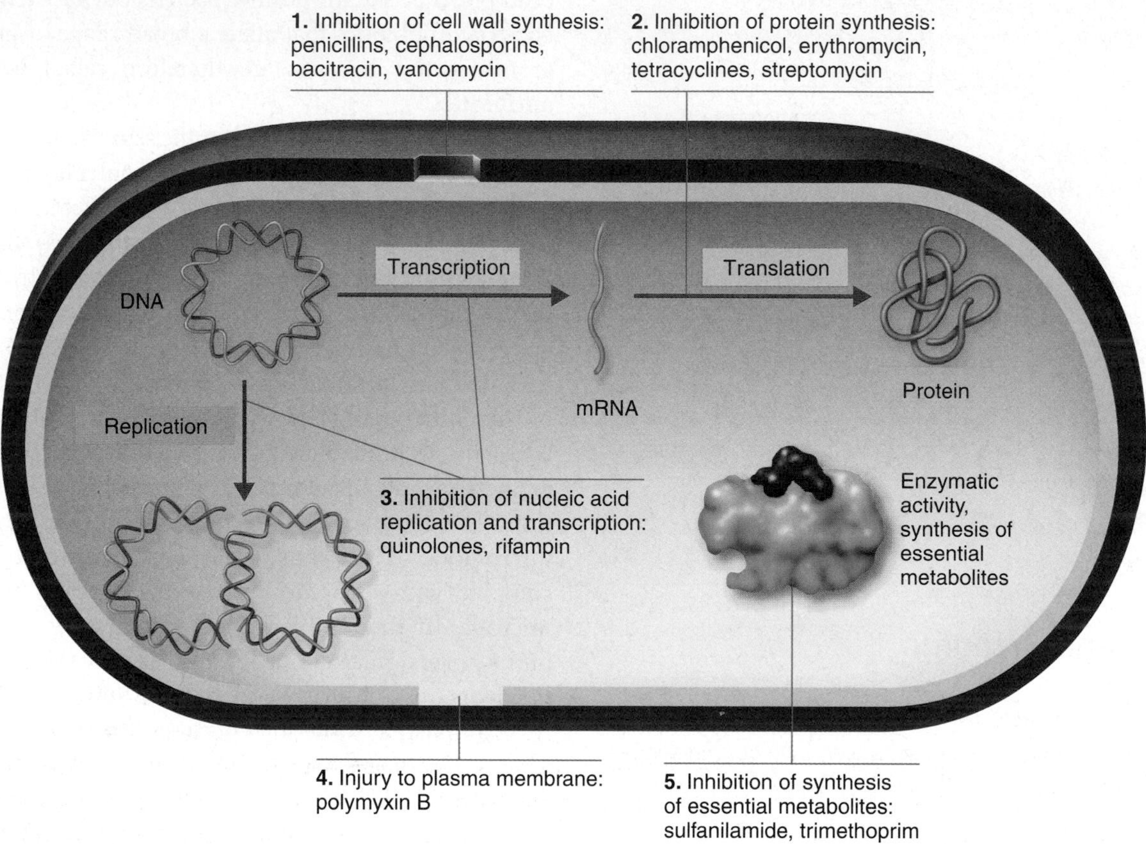

Key Concept

Antimicrobial drugs function in one of the following five ways: inhibiting cell wall synthesis, inhibiting protein synthesis, inhibiting nucleic acid synthesis, injuring the plasma membrane, or inhibiting synthesis of essential metabolites.

Inhibiting Cell Wall Synthesis

Penicillin, the first antibiotic to be discovered and used (if one does not also consider the sulfa drugs), is an example of an inhibitor of cell wall synthesis.

Q&A Recall from Chapter 4 that the cell wall of a bacterium consists of a macromolecular network called peptidoglycan. Peptidoglycan is found only in bacterial cell walls. Penicillin and certain other antibiotics prevent the synthesis of intact peptidoglycan; consequently, the cell wall is greatly weakened, and the cell undergoes lysis (Figure 20.3). Because penicillin targets the synthesis process, only actively growing cells are affected by these antibiotics—and, because human cells do not have peptidoglycan cell walls, penicillin has very little toxicity for host cells.

Inhibiting Protein Synthesis

Because protein synthesis is a common feature of all cells, whether prokaryotic or eukaryotic, it would seem an unlikely target for

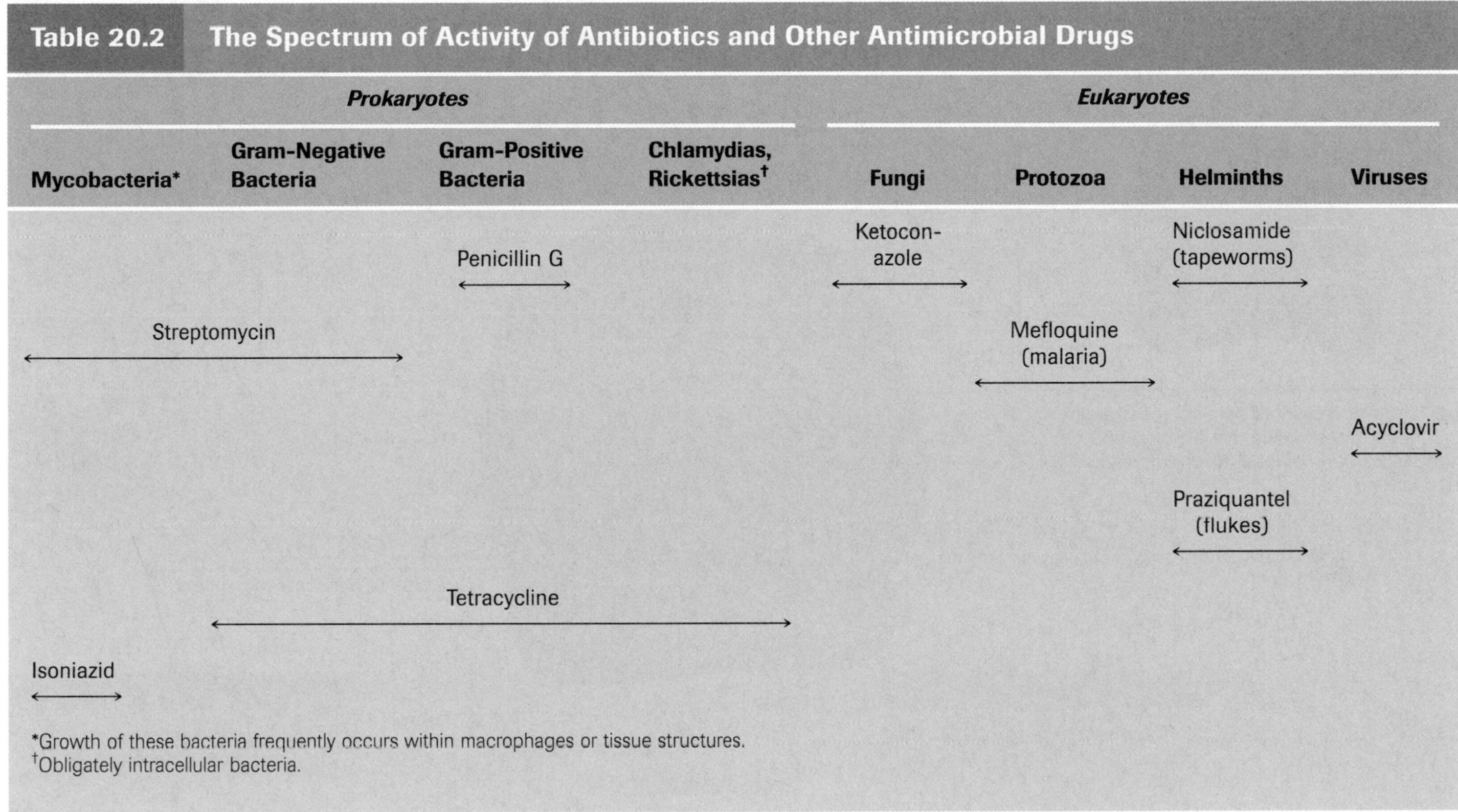

Table 20.2 The Spectrum of Activity of Antibiotics and Other Antimicrobial Drugs

Prokaryotes				*Eukaryotes*			
Mycobacteria*	**Gram-Negative Bacteria**	**Gram-Positive Bacteria**	**Chlamydias, Rickettsias†**	**Fungi**	**Protozoa**	**Helminths**	**Viruses**
		Penicillin G		Ketocon-azole		Niclosamide (tapeworms)	
	Streptomycin				Mefloquine (malaria)		
							Acyclovir
						Praziquantel (flukes)	
		Tetracycline					
Isoniazid							

*Growth of these bacteria frequently occurs within macrophages or tissue structures.
†Obligately intracellular bacteria.

selective toxicity. One notable difference between prokaryotes and eukaryotes, however, is the structure of their ribosomes. As discussed in Chapter 4 (page 95), eukaryotic cells have 80S ribosomes; prokaryotic cells have 70S ribosomes. (The 70S ribosome is made up of a 50S and a 30S unit. The S stands for Svedberg unit, which describes the relative rate of sedimentation in a high-speed centrifuge.) The difference in ribosomal structure accounts for the selective toxicity of antibiotics that affect protein synthesis. However, mitochondria (important eukaryotic organelles) also contain 70S ribosomes similar to those of bacteria. Antibiotics targeting the 70S ribosomes can therefore have adverse effects on the cells of the host. Among the antibiotics that interfere with protein synthesis are chloramphenicol, erythromycin, streptomycin, and the tetracyclines (Figure 20.4).

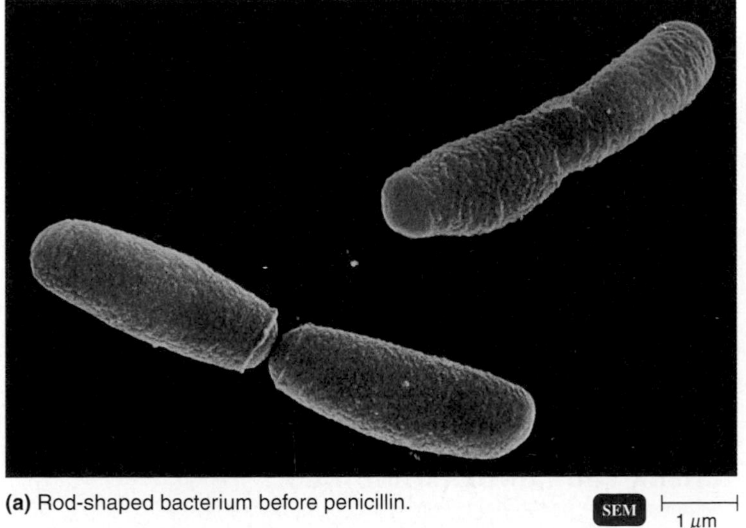

(a) Rod-shaped bacterium before penicillin. SEM 1 μm

(b) The bacterial cell lysing as penicillin weakens the cell wall SEM 1 μm

Figure 20.3 The inhibition of bacterial cell synthesis by penicillin.

Q **Why don't penicillins affect the human cell?**

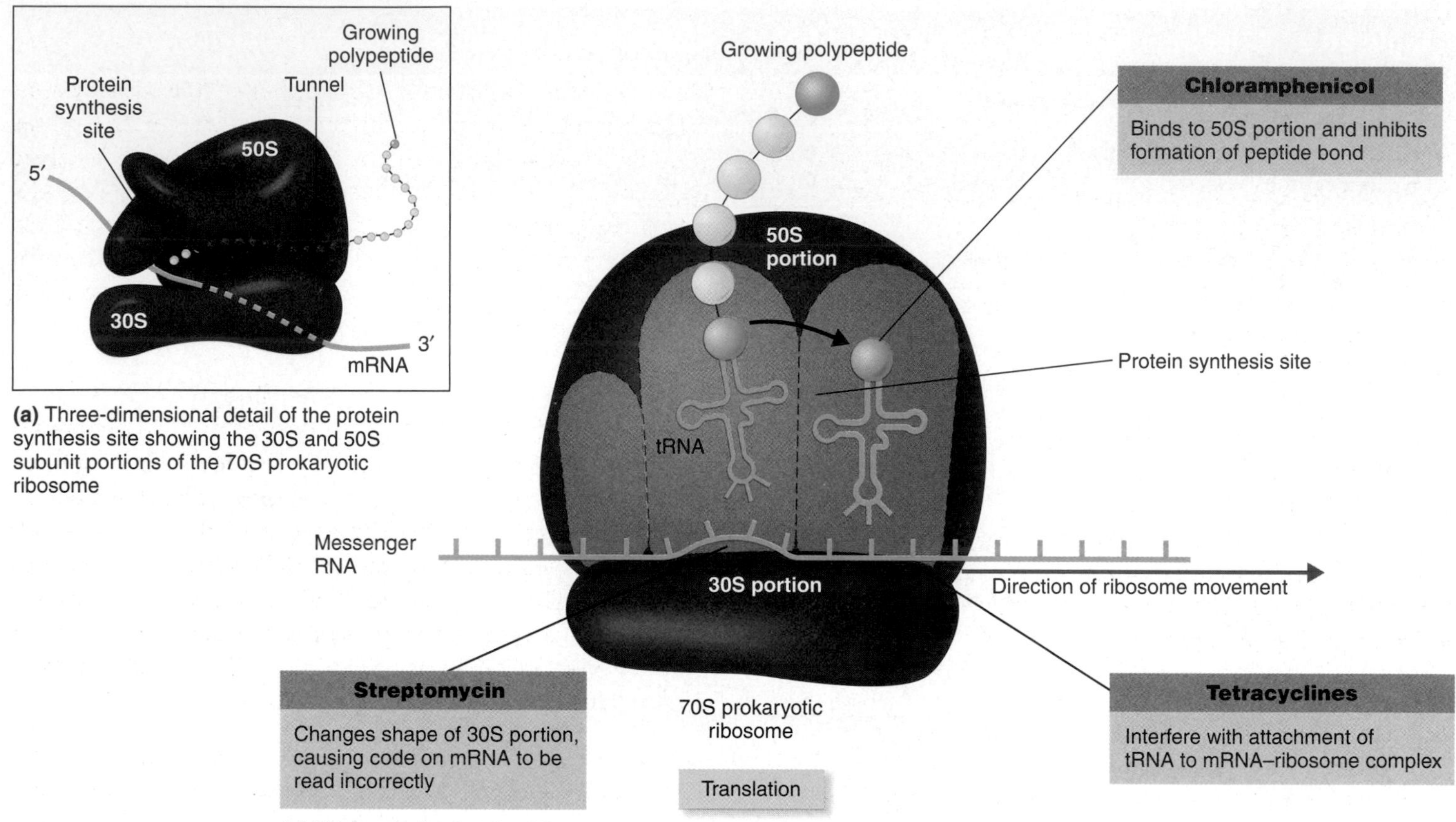
Growing polypeptide
Protein synthesis site
Tunnel
50S
5′
30S
3′
mRNA
(a) Three-dimensional detail of the protein synthesis site showing the 30S and 50S subunit portions of the 70S prokaryotic ribosome
Growing polypeptide
Chloramphenicol
Binds to 50S portion and inhibits formation of peptide bond
50S portion
Protein synthesis site
tRNA
Messenger RNA
30S portion
Direction of ribosome movement
Streptomycin
Changes shape of 30S portion, causing code on mRNA to be read incorrectly
70S prokaryotic ribosome
Translation
Tetracyclines
Interfere with attachment of tRNA to mRNA–ribosome complex
(b) Diagram indicating the different points at which chloramphenicol, the tetracyclines, and streptomycin exert their activities

Figure 20.4 The inhibition of protein synthesis by antibiotics. (**a**) The inset shows how the 70S prokaryotic ribosome is assembled from two subunits, 30S and 50S. Note how the growing peptide chain passes through a tunnel in the 50S subunit from the site of protein synthesis. (**b**) The diagram shows the different points at which chloramphenicol, the tetracyclines, and streptomycin exert their activities.

Q **Why do antibiotics that inhibit protein synthesis affect bacteria and not human cells?**

Injuring the Plasma Membrane

Certain antibiotics, especially polypeptide antibiotics, bring about changes in the permeability of the plasma membrane; these changes result in the loss of important metabolites from the microbial cell.

Some antifungal drugs, such as amphotericin B, miconazole, and ketoconazole, are effective against a considerable range of fungal diseases. Such drugs combine with sterols in the fungal plasma membrane to disrupt the membrane (**Figure 20.5**). Because bacterial plasma membranes generally lack sterols, these antibiotics do not act on bacteria.

Inhibiting Nucleic Acid Synthesis

A number of antibiotics interfere with the processes of DNA replication and transcription in microorganisms. Some drugs with this mode of action have an extremely limited usefulness because they interfere with mammalian DNA and RNA as well.

Inhibiting the Synthesis of Essential Metabolites

In Chapter 5, we mentioned that a particular enzymatic activity of a microorganism can be *competitively inhibited* by a substance (*antimetabolite*) that closely resembles the normal substrate for the enzyme (see Figure 5.7, page 121). An example of competitive inhibition is the relationship between the antimetabolite sulfanilamide (a sulfa drug) and ***para*-aminobenzoic acid (PABA).** In many microorganisms, PABA is the substrate for an enzymatic reaction leading to the synthesis of folic acid, a vitamin that functions as a coenzyme for the synthesis of the purine and pyrimidine bases of nucleic acids and many amino acids. **Animation** Chemotherapeutic Agents: Modes of Action. **www.microbiologyplace.com**

CHECK YOUR UNDERSTANDING

✓ What cellular function is inhibited by tetracyclines? **20-5**

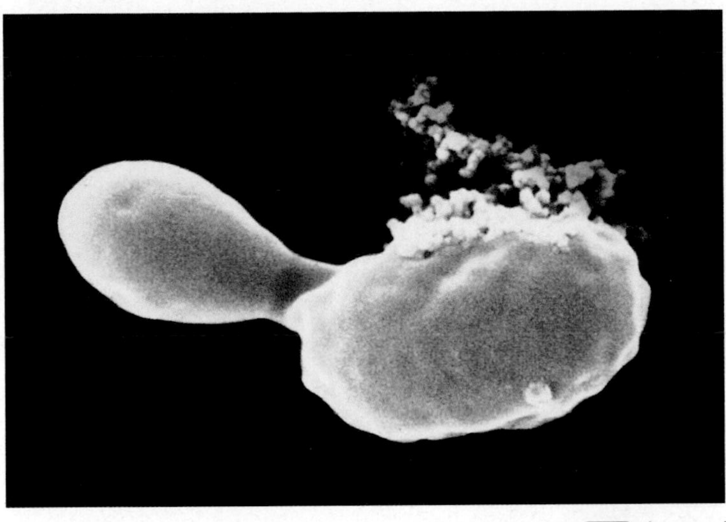

Figure 20.5 **Injury to the plasma membrane of a yeast cell caused by an antifungal drug.** The cell releases its cytoplasmic contents as the plasma membrane is disrupted by the antifungal drug miconazole.

Q Many antifungal drugs combine with sterols in the plasma membrane. Why don't they combine with sterols in human cell membranes?

A Survey of Commonly Used Antimicrobial Drugs

LEARNING OBJECTIVES

20-6 Explain why the drugs described in this section are specific for bacteria.

20-7 List the advantages of each of the following over penicillin: semisynthetic penicillins, cephalosporins, and vancomycin.

20-8 Explain why isoniazid (INH) and ethambutal are antimycobacterial agents.

20-9 Describe how each of the following inhibits protein synthesis: aminoglycosides, tetracyclines, chloramphenicol, macrolides.

20-10 Compare the mode of action of polymyxin B, bacitracin, and neomycin.

20-11 Describe how rifamycins and quinolones kill bacteria.

20-12 Describe how sulfa drugs inhibit microbial growth.

20-13 Explain the modes of action of currently used antifungal drugs.

20-14 Explain the modes of action of currently used antiviral drugs.

20-15 Explain the modes of action of currently used antiprotozoan and antihelminthic drugs.

Table 20.3 (pages 562–563) and Table 20.4 (page 564) summarize commonly used antimicrobial drugs.

Antibacterial Antibiotics: Inhibitors of Cell Wall Synthesis

For antibiotics to function as a "magic bullet," they usually must target microbial structures or functions that are not shared with mammalian structures or functions. As was described in Chapter 4, the eukaryotic mammalian cell usually does not have a cell wall; instead, it has only a plasma membrane. Even this membrane differs in composition from the plasma membrane of prokaryotic cells. For this reason, the microbial cell wall is an attractive target for the action of antibiotics.

Penicillin

The term **penicillin** refers to a group of over 50 chemically related antibiotics (Figure 20.6). All penicillins have a common core structure containing a β-lactam ring called the nucleus. Penicillin molecules are differentiated by the chemical side chains attached to their nuclei. Penicillins can be produced either naturally or semisynthetically. Penicillins prevent the cross-linking of the peptidoglycans, which interferes with the final stages of the construction of the cell walls, primarily of gram-positive bacteria (see Figure 4.13a, page 86).

Natural Penicillins Penicillin extracted from cultures of the mold *Penicillium* exists in several closely related forms. These are the so-called **natural penicillins** (Figure 20.6a). The prototype compound of all the penicillins is *penicillin G.* It has a narrow but useful spectrum of activity and is often the drug of choice against most staphylococci, streptococci, and several spirochetes. When injected intramuscularly, penicillin G is rapidly excreted from the body in 3 to 6 hours (Figure 20.7). When taken orally, the acidity of the digestive fluids in the stomach diminishes its concentration. *Procaine penicillin,* a combination of the drugs procaine and penicillin G, is retained at detectable concentrations for up to 24 hours; the concentration peaks at about 4 hours. Still longer retention times can be achieved with *benzathine penicillin,* a combination of benzathine and penicillin G. Although retention times of as long as 4 months can be obtained, the concentration of the drug is so low that the organisms must be very sensitive to it. Penicillin V, which is stable in stomach acids and can be taken orally, and penicillin G are the natural penicillins most often used.

Natural penicillins have some disadvantages. Chief among them are their narrow spectrum of activity and their susceptibility to penicillinases. *Penicillinases* are enzymes produced by many bacteria, most notably *Staphylococcus* species, that cleave the β-lactam ring of the penicillin molecule (Figure 20.8). Because of this characteristic, penicillinases are sometimes called *β-lactamases.*

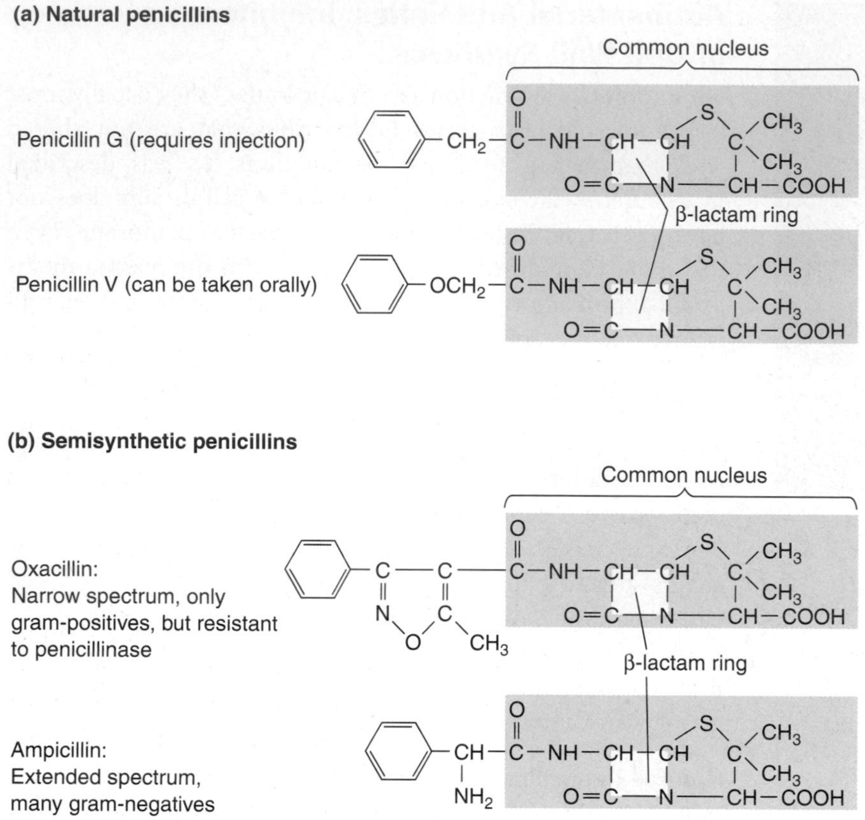

(a) Natural penicillins

Common nucleus

Penicillin G (requires injection)

Penicillin V (can be taken orally)

β-lactam ring

(b) Semisynthetic penicillins

Common nucleus

Oxacillin:
Narrow spectrum, only gram-positives, but resistant to penicillinase

Ampicillin:
Extended spectrum, many gram-negatives

β-lactam ring

Figure 20.6 The structure of penicillins, antibacterial antibiotics. The portion that all penicillins have in common—which contains the β-lactam ring—is shaded in purple. The unshaded portions represent the side chains that distinguish one penicillin from another.

Q What does *semisynthetic* mean?

Semisynthetic Penicillins A large number of **semisynthetic penicillins** have been developed in attempts to overcome the disadvantages of natural penicillins (**Figure 20.6b**). Scientists develop these penicillins in either of two ways. First, they can interrupt synthesis of the molecule by *Penicillium* and obtain only the common penicillin nucleus for use. Second, they can remove the side chains from the completed natural molecules and then chemically add other side chains that make them more resistant to penicillinase, or the scientists can give them an extended spectrum. Thus the term *semisynthetic:* part of the penicillin is produced by the mold, and part is added synthetically.

Penicillinase-Resistant Penicillins Resistance of staphylococcal infections to penicillin soon became a problem because of a plasmid-borne gene for β-lactamase. Antibiotics that were relatively resistant to this enzyme, such as the semisynthetic penicillin, *methicillin,* were introduced, but resistance to them also soon appeared; thus the organisms were termed **methicillin-resistant *Staphylococcus aureus* (MRSA),** usually

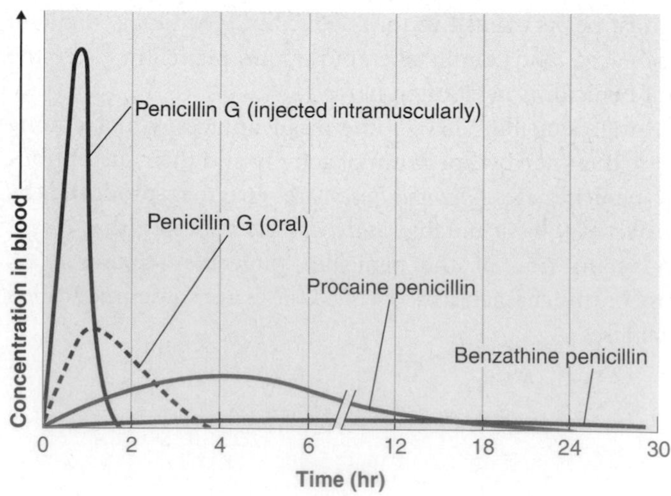

Figure 20.7 Retention of penicillin G. Penicillin G is normally injected (solid red line); when administered by this route, the drug is present in high concentrations in the blood but disappears quickly. Taken orally (dotted red line), penicillin G is destroyed by stomach acids and is not very effective. It is possible to improve retention of penicillin G by combining it with compounds such as procaine and benzathine (blue and purple lines). However, the blood concentration reached is low, and the target bacterium must be extremely sensitive to the antibiotic.

Q How does a low concentration of penicillin G select for penicillin-resistant bacteria?

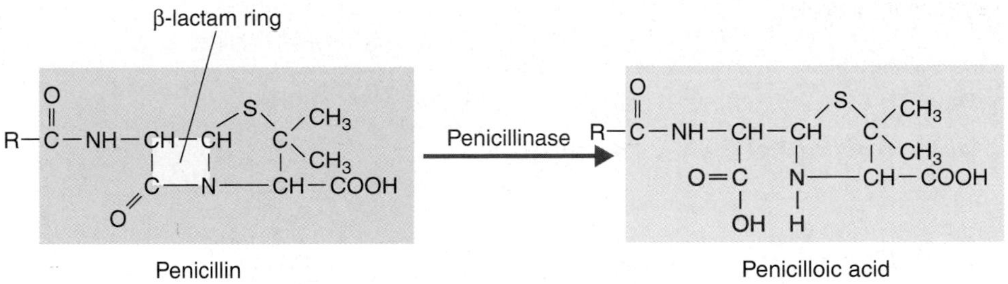

Figure 20.8 The effect of penicillinase on penicillins. Bacterial production of this enzyme, which is shown breaking the β-lactam ring, is by far the most common form of resistance to penicillins. R is an abbreviation for the chemical side groups that differentiate similar or otherwise-identical compounds.

Q What is penicillinase?

β-lactam ring

Penicillin

Penicillinase

Penicilloic acid

pronounced *mersa* (see the box on page 593). Resistance became so prevalent that methicillin has been discontinued in the United States. The term has come to be applied to strains that have developed resistance to a wide range of penicillins and cephalosporins. This includes other penicillinase-resistant antibiotics, such as *oxacillin*, and those combined with β-lactamase inhibitors (discussed below). See the discussion of antibiotic resistance on page 573.

Extended-Spectrum Penicillins To overcome the problem of the narrow spectrum of activity of natural penicillins, broader-spectrum semisynthetic penicillins have been developed. These new penicillins are effective against many gram-negative bacteria as well as gram-positive ones, although they are not resistant to penicillinases. The first such penicillins were the aminopenicillins, such as *ampicillin* and *amoxicillin*. When bacterial resistance to these became more common, the carboxypenicillins were developed. Members of this group, such as *carbenicillin* and *ticarcillin*, have even greater activity against gram-negative bacteria and have the special advantage of activity against *Pseudomonas aeruginosa*.

Among the more recent additions to the penicillin family are the ureidopenicillins, such as *mezlocillin* and *azlocillin*. These broader-spectrum penicillins are modifications of the structure of ampicillin. The search for even more effective modifications of penicillin continues.

Penicillins Plus β-Lactamase Inhibitors A different approach to the proliferation of penicillinase is to combine penicillins with *potassium clavulanate (clavulanic acid)*, a product of a streptomycete. Potassium clavulanate is a noncompetitive inhibitor of penicillinase with essentially no antimicrobial activity of its own. It has been combined with some new broader-spectrum penicillins, such as *amoxicillin* (the combination is best known by its trade name Augmentin).

Carbapenems

The **carbapenems** are a class of β-lactam antibiotics that substitute a carbon atom for a sulfur atom and add a double bond to the penicillin nucleus. These antibiotics, which inhibit cell wall synthesis, have an extremely broad spectrum of activity. Representative of this group is Primaxin, a combination of *imipenem* and *cilastatin*. The cilastatin has no antimicrobial activity but prevents degradation of the combination in the kidneys. Tests have demonstrated that Primaxin is active against 98% of all organisms isolated from hospital patients.

Monobactams

Another method of avoiding the effects of penicillinase is shown by *aztreonam*, which is the first member of a new class of antibiotics. It is a synthetic antibiotic that has only a single ring rather than the conventional β-lactam double ring, and is therefore known as a **monobactam.** Aztreonam's spectrum of activity is remarkable for a penicillin-related compound—this antibiotic, which has unusually low toxicity, affects only certain gram-negative bacteria, including pseudomonads and *E. coli*.

Cephalosporins

In structure, the nuclei of **cephalosporins** resemble those of penicillin (Figure 20.9). Cephalosporins inhibit cell wall synthesis in essentially the same way as do penicillins. They are more widely used than any other β-lactam antibiotics. Their β-lactam ring differs slightly from that of penicillin, but bacteria have developed β-lactamases that inactivate them.

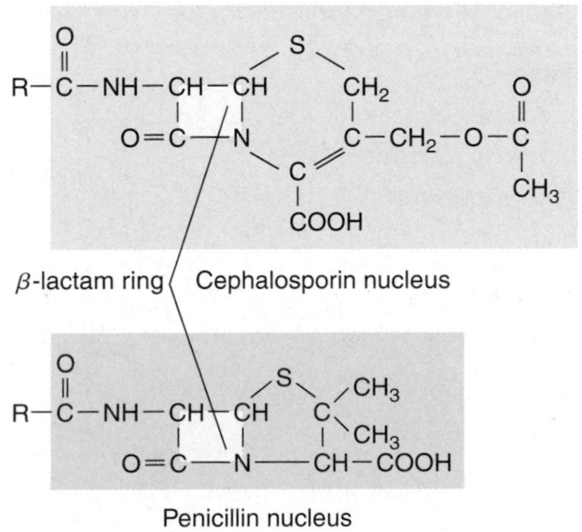

β-lactam ring

Cephalosporin nucleus

Penicillin nucleus

Figure 20.9 The nuclear structures of cephalosporin and penicillin compared.

Q Would a β-lactamase effective against penicillin G be likely to affect cephalosporins?

Table 20.3 Antibacterial Drugs

Drugs by Mode of Action	Comments
INHIBITORS OF CELL WALL SYNTHESIS	
Natural Penicillins	
Penicillin G	Against gram-positive bacteria, requires injection
Penicillin V	Against gram-positive bacteria, oral administration
Semisynthetic Penicillins	
Oxacillin	Resistant to penicillinase
Ampicillin	Broad spectrum
Amoxicillin	Broad spectrum; combined with inhibitor of penicillinase
Aztreonam	A monobactam; effective for gram-negative bacteria, including *Pseudomonas* spp.
Imipenem	A carbapenem; very broad spectrum
Cephalosporins	
Cephalothin	First-generation cephalosporin; activity similar to penicillin; requires injection
Cefixime	Fourth-generation cephalosporin; oral administration
Polypeptide Antibiotics	
Bacitracin	Against gram-positive bacteria; topical application
Vancomycin	A glycopeptide type; penicillinase-resistant; against gram-positive bacteria
Antimycobacterial Antibiotics	
Isoniazid	Inhibits synthesis of mycolic acid component of cell wall of *Mycobacterium* spp.
Ethambutol	Inhibits incorporation of mycolic acid into cell wall of *Mycobacterium* spp.
INHIBITORS OF PROTEIN SYNTHESIS	
Chloramphenicol	Broad spectrum, potentially toxic
Aminoglycosides	
Streptomycin	Broad spectrum, including mycobacteria
Neomycin	Topical use, broad spectrum
Gentamicin	Broad spectrum, including *Pseudomonas* spp.
Tetracyclines	
Tetracycline, oxytetracycline, chlortetracycline	Broad spectrum, including chlamydias and rickettsias; animal feed additives
Macrolides	
Erythromycin	Alternative to penicillin
Azithromycin, clarithromycin	Semisynthetic; broader spectrum and better tissue penetration than erythromycin
Telithromycin (Ketek)	New generation of semisynthetic macrolides; used to cope with resistance to other macrolides
Streptogramins	
Quinupristin and dalfopristin (Synercid)	Alternative for treating vancomycin-resistant gram-positive bacteria
Oxazolidinones	
Linezolid (Zyvox)	Useful primarily against penicillin-resistant gram-positive bacteria
INJURY TO THE PLASMA MEMBRANE	
Polymyxin B	Topical use, gram-negative bacteria, including *Pseudomonas* spp.
INHIBITORS OF NUCLEIC ACID SYNTHESIS	
Rifamycins	
Rifampin (or rifampicin)	Inhibits synthesis of mRNA; treatment of tuberculosis

Table 20.3 *(continued)*

Drugs by Mode of Action	Comments
Quinolones and Fluoroquinolones	
Nalidixic acid, nofloxacin, ciprofloxacin	Inhibit DNA synthesis; broad spectrum; urinary tract infections
Gatifloxacin	Newest generation quinolone; increased potency against gram-positive bacteria
COMPETITIVE INHIBITORS OF THE SYNTHESIS OF ESSENTIAL METABOLITES	
Sulfonamides	
Trimethoprim-sulfamethoxazole	Broad spectrum; combination is widely used

The differential grouping of cephalosporins is most commonly done by reference to generations, reflecting their continued development. The first-generation cephalosporins have a relatively narrow spectrum of activity, primarily against gram-positive bacteria. An example is *cephalothin*. Second-generation cephalosporins have a more extended gram-negative spectrum. Examples are the intravenous drug *cefamandole* and the oral drug *cefaclor*. The third-generation drugs, such as *ceftazidime*, are the most active against gram-negative bacteria, including some pseudomonads, but most must be injected. An oral third-generation cephalosporin is *cefixime*. The fourth-generation drugs, such as *cefepime*, require injections but have the most extended spectrum of activity.

Polypeptide Antibiotics

Bacitracin *Bacitracin* (the name is derived from its source, a *Bacillus* isolated from a wound on a girl named *Tracy*) is a polypeptide antibiotic effective primarily against gram-positive bacteria, such as staphylococci and streptococci. Bacitracin inhibits the synthesis of cell walls at an earlier stage than penicillins and cephalosporins. It interferes with the synthesis of the linear strands of the peptidoglycans (see Figure 4.13a, page 86.) Its use is restricted to topical application for superficial infections.

Vancomycin *Vancomycin* (optimistically named from the word *vanquish*) is one of a small group of glycopeptide antibiotics derived from a species of *Streptomyces* found in the jungles of Borneo. Originally, toxicity of vancomycin was a serious problem, but improved purification procedures in its manufacture have largely corrected this. Although it has a very narrow spectrum of activity, which is based on inhibition of cell wall synthesis, vancomycin has been extremely important in addressing the problem of MRSA (see page 561). Vancomycin has been considered the last line of antibiotic defense for treatment of *Staphylococcus aureus* infections that are resistant to other antibiotics. The widespread use of vancomycin to treat MRSA has led to the appearance of **vancomycin-resistant enterococci (VRE).** These are opportunistic, gram-positive pathogens that are particularly troublesome in hospital settings (see page 319 and the box on page 593). This appearance of vancomycin-resistant pathogens, leaving few effective alternatives, is considered a medical emergency.

Antimycobacterial Antibiotics

The cell wall of members of the genus *Mycobacterium* differs from the cell wall of most other bacteria. It incorporates mycolic acids that are a factor in their staining properties, causing them to stain as acid-fast (page 70). The genus includes important pathogens, such as those that cause leprosy and tuberculosis.

Isoniazid (INH) is a very effective synthetic antimicrobial drug against *Mycobacterium tuberculosis*. The primary effect of INH is to inhibit the synthesis of mycolic acids, which are components of cell walls only of the mycobacteria. It has little effect on nonmycobacteria. When used to treat tuberculosis, INH is usually administered simultaneously with other drugs, such as rifampin (also known as rifampicin) or ethambutol. This minimizes development of drug resistance. Because the tubercle bacillus is usually found only within macrophages or walled off in tissue, any antitubercular drug must be able to penetrate into such sites.

Ethambutol is effective only against mycobacteria. The drug apparently inhibits incorporation of mycolic acid into the cell wall. It is a comparatively weak antitubercular drug; its principal use is as the secondary drug to avoid resistance problems.

CHECK YOUR UNDERSTANDING

- ✓ One of the most successful groups of antibiotics targets the synthesis of bacterial cell walls; why does the antibiotic not affect the mammalian cell? **20-6**
- ✓ What phenomenon prompted the development of the first semisynthetic antibiotics, such as methicillin? **20-7**
- ✓ In what genus of bacteria do we find mycolic acids in the cell wall? **20-8**

Inhibitors of Protein Synthesis

Chloramphenicol

Chloramphenicol inhibits the formation of peptide bonds in the growing polypeptide chain by reacting with the 50S portion of the 70S prokaryotic ribosome. Because of its simple structure (**Figure 20.10**), it is less expensive for the pharmaceutical industry to synthesize it chemically than to isolate it from *Streptomyces*. It is relatively inexpensive, and has a broad

Table 20.4	Antifungal, Antiviral, Antiprotozoan, and Antihelminthic Drugs	
	Mode of Action	**Comments**
ANTIFUNGAL DRUGS		
Agents Affecting Fungal Sterols (Plasma Membrane)		
Polyenes		
Amphotericin B	Injury to plasma membrane	Systemic fungal infections; fungicidal
Azoles		
Clotrimazole, miconazole	Inhibit synthesis of plasma membrane	Topical use
Ketoconazole	Inhibits synthesis of plasma membrane	Can be taken orally for systemic fungal infections
Voriconazole	Inhibits synthesis of plasma membrane	Can penetrate blood-brain barrier to treat aspergillosis of the central nervous system
Allylamines		
Terbinafine, naftifine	Inhibit synthesis of plasma membrane	New class of antifungals frequently used to treat diseases resistant to azoles
Agents Affecting Fungal Cell Walls		
Echinocandins		
Caspofungin (Cancidas)	New class of antifungals that inhibit synthesis of cell wall	
Agents Inhibiting Nucleic Acids		
Flucytosine	Inhibits synthesis of RNA and therefore protein synthesis	
Other Antifungal Drugs		
Griseofulvin	Inhibition of mitotic microtubules	Fungal infections of the skin
Tolnaftate	Unknown	Athlete's foot
ANTIVIRAL DRUGS		
Nucleoside and Nucleotide Analogs		
Acyclovir, ganciclovir, ribavirin, lamivudine	Inhibit DNA or RNA synthesis	Used primarily against herpesviruses
Cidofovir	Inhibits DNA or RNA synthesis	Cytomegalovirus infections; possibly effective against smallpox
Adefovir dipivoxil (Hepsera)	Competitive inhibitor for HBV reverse transcriptase	For resistance against lamivudine
Attachment and Uncoating		
Zanamivir, oseltamivir	Inhibit neuraminidase on influenza virus	Treatment of influenza
Amantadine, zimantadine	Inhibit uncoating	Treatment of influenza
Interferons		
alpha interferon	Inhibits spread of virus to new cells	Viral hepatitis
ANTIPROTOZOAN DRUGS		
Chloroquine	Inhibits DNA synthesis	Malaria; effective against red blood cell stage only
Diiodohydroxyquin	Unknown	Amoebic infections; amoebicidal
Metronidazole, Tinidazole	Interferes with anaerobic metabolisms	Giardiasis, amebiasis, trichomoniasis
Nitazoxanide	Interfere with anaerobic metabolism	Giardiasis; only drug approved for cryptosporidiosis
ANTIHELMINTHIC DRUGS		
Niclosamide	Prevents ATP generation in mitochondria	Tapeworm infections; kills tapeworms
Praziquantel	Alters permeability of plasma membranes	Tapeworm and fluke infections; kills flatworms
Pyantel pamoate	Neuromuscular block	Intestinal roundworms; kills roundworms
Mebendazole, albendazole	Inhibit absorption of nutrients	Intestinal roundworms
Ivermectin	Paralyzes worm	Intestinal roundworms primarily; occasional use for scabies mite and lice

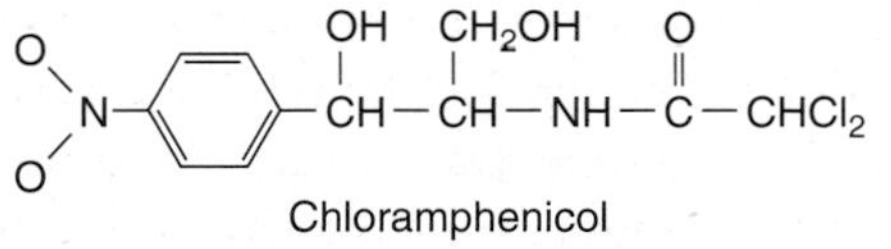

Chloramphenicol

Figure 20.10 The structure of the antibacterial antibiotic chloramphenicol. Notice the simple structure, which makes synthesizing this drug less expensive than isolating it from *Streptomyces.*

Q What effect does the binding of chloramphenicol to the 50S portion of the ribosomes have on a cell?

spectrum, so chloramphenicol is often used where low cost is essential. Its small molecular size promotes its diffusion into areas of the body that are normally inaccessible to many other drugs. However, chloramphenicol has serious adverse effects; most important is the suppression of bone marrow activity. This suppression affects the formation of blood cells. In about 1 in 40,000 users, the drug appears to cause aplastic anemia, a potentially fatal condition; the normal rate for this condition is only about 1 in 500,000 individuals. Physicians are advised not to use the drug for trivial conditions or ones for which suitable alternatives are available.

Other antibiotics inhibiting protein synthesis by binding at the same ribosomal site as chloramphenicol are *clindamycin* and *metronidazole* (see page 471). These three drugs are structurally unrelated, but all have potent anaerobic activity. Clindamycin has a noted association with *Clostridium difficile*–associated diarrhea (see page 720). Its effectiveness against anaerobes has led to its use in the treatment of acne.

Aminoglycosides

Aminoglycosides are a group of antibiotics in which amino sugars are linked by glycoside bonds. Aminoglycoside antibiotics, such as *streptomycin* and *gentamicin*, interfere with the initial steps of protein synthesis by changing the shape of the 30S portion of the 70S prokaryotic ribosome. This interference causes the genetic code of the mRNA to be read incorrectly. They were among the first antibiotics to have significant activity against gram-negative bacteria. Probably the best-known aminoglycoside is *streptomycin,* which was discovered in 1944. Streptomycin is still used as an alternative drug in the treatment of tuberculosis, but rapid development of resistance and serious toxic effects have diminished its usefulness.

Aminoglycosides can affect hearing by causing permanent damage to the auditory nerve, and damage to the kidneys has also been reported. As a result, their use has been declining. *Neomycin* is present in many nonprescription topical preparations. *Gentamicin* (spelled with an "i" to reflect its source, the filamentous bacterium *Micromonospora*) is especially useful against *Pseudomonas* infections. Pseudomonads are a major problem for persons suffering from cystic fibrosis. The aminoglycoside *tobramycin* is administered in an aerosol to help control infections that occur in patients with cystic fibrosis.

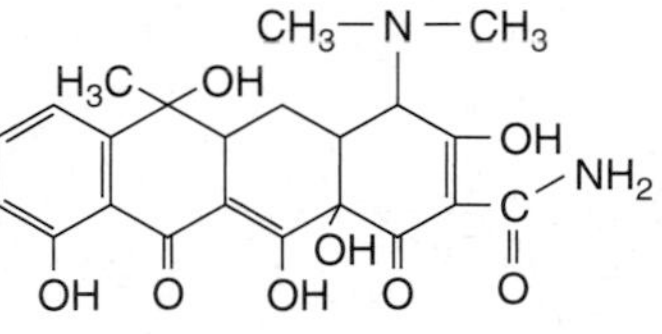

Tetracycline

Figure 20.11 The structure of the antibacterial antibiotic tetracycline. Other tetracycline-type antibiotics share the four-cyclic-ring structure of tetracycline and closely resemble it.

Q How do tetracyclines affect bacteria?

Tetracyclines

Tetracyclines are a group of closely related broad-spectrum antibiotics produced by *Streptomyces* spp. The tetracyclines interfere with the attachment of the tRNA carrying the amino acids to the ribosome at the 30S portion of the 70S ribosome, preventing the addition of amino acids to the growing polypeptide chain. They do not interfere with mammalian ribosomes because they do not penetrate very well into intact mammalian cells. However, at least small amounts are able to enter the host cell, as is apparent from the fact that the intracellular pathogenic rickettsias and chlamydias are sensitive to tetracyclines. The selective toxicity of these drugs is due to a greater sensitivity of the bacteria at the ribosomal level. Tetracyclines not only are effective against gram-positive and gram-negative bacteria but also penetrate body tissues well and are especially valuable against the intracellular rickettsias and chlamydias. Three of the more commonly used tetracyclines are *oxytetracycline* (Terramycin), *chlortetracycline* (Aureomycin), and tetracycline itself (**Figure 20.11**).

Some semisynthetic tetracyclines, such as *doxycycline* and *minocycline,* are available. They have the advantage of longer retention in the body. A derivative of minocycline that has a broad spectrum of activity is a new class of antibiotics, the glycylcyclines. The first of this class is *tigecycline (Tygacil),* which was developed as a response to MRSA.

Tetracyclines are used to treat many urinary tract infections, mycoplasmal pneumonia, and chlamydial and rickettsial infections. They are also frequently used as alternative drugs for such diseases as syphilis and gonorrhea. Tetracyclines often suppress the normal intestinal microbiota because of their broad spectrum, causing gastrointestinal upsets and often leading to superinfections, particularly by the fungus *Candida albicans.* They are not advised for children, who might experience a brownish discoloration of the teeth, or for pregnant women, in whom they might cause liver damage. Tetracyclines are among the most common antibiotics added to animal feeds, where their use results in significantly faster weight gain; however, some human health problems can also result (see the box on page 577).

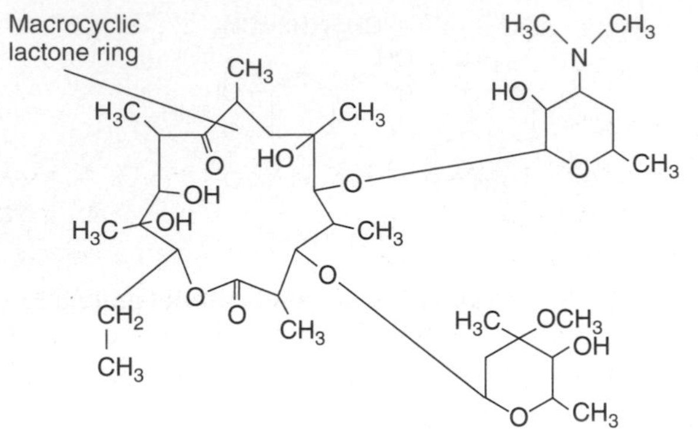

Figure 20.12 The structure of the antibacterial antibiotic erythromycin, a representative macrolide. All macrolides have the macrocyclic lactone ring shown here.

Q How do macrolides affect bacteria?

Macrolides

Macrolides are a group of antibiotics named for the presence of a macrocyclic lactone ring. The best-known macrolide in clinical use is *erythromycin* (**Figure 20.12**). Its mode of action is the inhibition of protein synthesis, apparently by blocking the tunnel shown in Figure 20.4a. However, erythromycin is not able to penetrate the cell walls of most gram-negative bacilli. Its spectrum of activity is therefore similar to that of penicillin G, and it is a frequent alternative drug to penicillin. Because it can be administered orally, an orange-flavored preparation of erythromycin is a frequent penicillin substitute for the treatment of streptococcal and staphylococcal infections in children. Erythromycin is the drug of choice for the treatment of legionellosis, mycoplasmal pneumonia, and several other infections.

Other macrolides now available include *azithromycin* and *clarithromycin.* Compared to erythromycin, they have a broader antimicrobial spectrum and penetrate tissues better. This is especially important in the treatment of conditions caused by intracellular bacteria such as *Chlamydia,* a frequent cause of sexually transmitted infection.

A new generation of semisynthetic macrolides, the **ketolides,** is being developed to cope with increasing resistance to other macrolides. The prototype of this generation is *telithromycin* (Ketek). However, it has a number of significant restrictions related to toxicity.

Streptogramins

We mentioned previously that the appearance of vancomycin-resistant pathogens constitutes a serious medical problem. One answer may be a unique group of antibiotics, the **streptogramins.** The first of these drugs to be released, Synercid, is a combination of two cyclic peptides, *quinupristin* and *dalfopristin,* which are distantly related to the macrolides. They block protein synthesis by attaching to the 50S portion of the ribosome, as do other antibiotics such as chloramphenicol. Synercid, however, acts at uniquely different points on the ribosome. Dalfopristin blocks an early step in protein synthesis, and quinupristin blocks a later step. The combination causes incomplete peptide chains to be released and is synergistic in its action (see page 578). Synercid is effective against a broad range of gram-positive bacteria that are resistant to other antibiotics. This makes Synercid especially valuable, even though it is expensive and has a high incidence of adverse side effects.

Oxazolidinones

The oxazolidinones are another new class of antibiotics developed in response to vancomycin resistance. When the FDA approved this class of antibiotic in 2001, it represented the first new class of antibiotics approved in 25 years. Like several other antibiotics that inhibit protein synthesis, oxazolidinone antibiotics act on the ribosome (see Figure 20.4, page 558). However, they are unique in their target, binding to the 50S ribosomal subunit close to the point where it interfaces with the 30S subunit. These drugs are totally synthetic, which may make resistance slower to develop. Like vancomycin, they have no usefulness against gram-negative bacteria, but they are active against certain enterococci that are not sensitive to Synercid. One member of this antibiotic group is *linezolid* (Zyvox), used mainly to combat MRSA.

CHECK YOUR UNDERSTANDING

✔ Why does erythromycin, a macrolide antibiotic, have a spectrum of activity limited largely to gram-positive bacteria even though its mode of action is similar to that of the broad-spectrum tetracyclines? **20-9**

Injury to the Plasma Membrane

The synthesis of bacterial plasma membranes requires the synthesis of fatty acids as building blocks. Blocking this is the target of several antibiotics and antimicrobials. Examples are the tuberculosis drug *isoniazid* (page 563) and the household antibacterial *triclosan* (page 196). A new antibiotic, *platensimycin,* also exploits the fatty-acid biosynthesis essential for plasma membranes. It is especially significant because it represents another rare example of a new chemical class of antibiotic to appear in the last 40 years. (*Linezolid* and *daptomycin* are other examples). This class of antibiotic may be another weapon to use against the threat of MRSA.

Polymyxin B is a bactericidal antibiotic effective against gram-negative bacteria. For many years, it was one of very few drugs used against infections by gram-negative *Pseudomonas.* Polymyxin B is seldom used today except in the topical treatment of superficial infections.

Both *bacitracin* and *polymyxin B* are available in nonprescription antiseptic ointments, in which they are usually combined with *neomycin,* a broad-spectrum aminoglycoside. In a rare exception to the rule, these antibiotics do not require a prescription.

Many of the antimicrobial peptides discussed on page 578 target the synthesis of the plasma membrane.

CHECK YOUR UNDERSTANDING

✓ Of the three drugs often found in over-the-counter antiseptic creams—polymyxin B, bacitracin, and neomycin—which has a mode of action most similar to that of penicillin? **20-10**

Inhibitors of Nucleic Acid (DNA/RNA) Synthesis

Rifamycins

The best-known derivative of the **rifamycin** family of antibiotics is *rifampin.* These drugs are structurally related to the macrolides and inhibit the synthesis of mRNA. By far the most important use of rifampin is against mycobacteria in the treatment of tuberculosis and leprosy. A valuable characteristic of rifampin is its ability to penetrate tissues and reach therapeutic levels in cerebrospinal fluid and abscesses. This characteristic is probably an important factor in its antitubercular activity, because the tuberculosis pathogen is usually located inside tissues or macrophages. An unusual side effect of rifampin is the appearance of orange-red urine, feces, saliva, sweat, and even tears.

Quinolones and Fluoroquinolones

In the early 1960s, the synthetic drug *nalidixic acid* was developed—the first of the **quinolone** group of antimicrobials. It exerted a unique bactericidal effect by selectively inhibiting an enzyme (DNA gyrase) needed for the replication of DNA. Although nalidixic acid found only limited use (its only application being for urinary tract infections), it led to the development in the 1980s of a prolific group of synthetic quinolones, the **fluoroquinolones.**

The fluoroquinolones are divided into groups, each of which has a progressively broader spectrum of activity. The earliest generations include widely used *norfloxacin* and *ciprofloxacin.* The latter is better known under its trade name of Cipro and has had wide publicity for its use against anthrax infections. A newer group of fluoroquinolones includes *gatifloxacin, gemifloxacin,* and *moxifloxacin.* These antibiotics, with the exception of moxifloxacin, are often the drugs of choice for urinary tract infections and certain types of pneumonia. As a group, the fluoroquinolones are relatively nontoxic. Resistance to them can develop rapidly, even during a course of treatment.

CHECK YOUR UNDERSTANDING

✓ What group of antibiotics interferes with the DNA-replicating enzyme DNA gyrase? **20-11**

Competitive Inhibitors of the Synthesis of Essential Metabolites

Sulfonamides

As noted earlier, **sulfonamides,** or **sulfa drugs,** were among the first synthetic antimicrobial drugs used to treat microbial diseases. Antibiotics have diminished the importance of sulfa drugs in chemotherapy, but they continue to be used to treat certain urinary tract infections and have other specialized uses, as in the combination drug *silver sulfadiazine,* used to control infections in burn patients. Sulfonamides are bacteriostatic; their action is due to their structural similarity to para-aminobenzoic acid (PABA) (see the discussion and formulas on page 121). Microbes sensitive to sulfa drugs must synthesize PABA, whereas humans ingest it with their diet.

Probably the most widely used sulfa drug today is a combination of *trimethoprim* and *sulfamethoxazole* (TMP-SMZ). This combination is an excellent example of drug **synergism.** When used in combination, only 10% of the concentration is needed, compared to concentrations needed when each drug is used alone. The combination also has a broader spectrum of action and greatly reduces the emergence of resistant strains. (Synergism is discussed more fully later in the chapter; see Figure 20.23.)

Figure 20.13 illustrates how the two drugs interfere with different steps of a metabolic sequence leading to the synthesis of precursors of proteins, DNA, and RNA.

CHECK YOUR UNDERSTANDING

✓ Both humans and bacteria need the essential nutrient para-aminobenzoic acid; why, then, are only bacteria affected by sulfa drugs? **20-12**

Antifungal Drugs

Eukaryotes, such as fungi, use the same mechanisms to synthesize proteins and nucleic acids as higher animals. Therefore it is more difficult to find a point of selective toxicity in eukaryotes than in prokaryotes. Moreover, fungal infections are becoming more frequent because of their role as opportunistic infections in immunosuppressed individuals, especially those with AIDS.

Agents Affecting Fungal Sterols

Many antifungal drugs target the sterols in the plasma membrane. In fungal membranes, the principal sterol is ergosterol; in animal membranes, cholesterol. When the biosynthesis of ergosterol in a fungal membrane is interrupted, the membrane becomes excessively permeable, killing the cell. Inhibition of ergosterol biosynthesis is the basis for the selective toxicity of many antifungals, which include members of the polyene, azole, and allylamine groups.

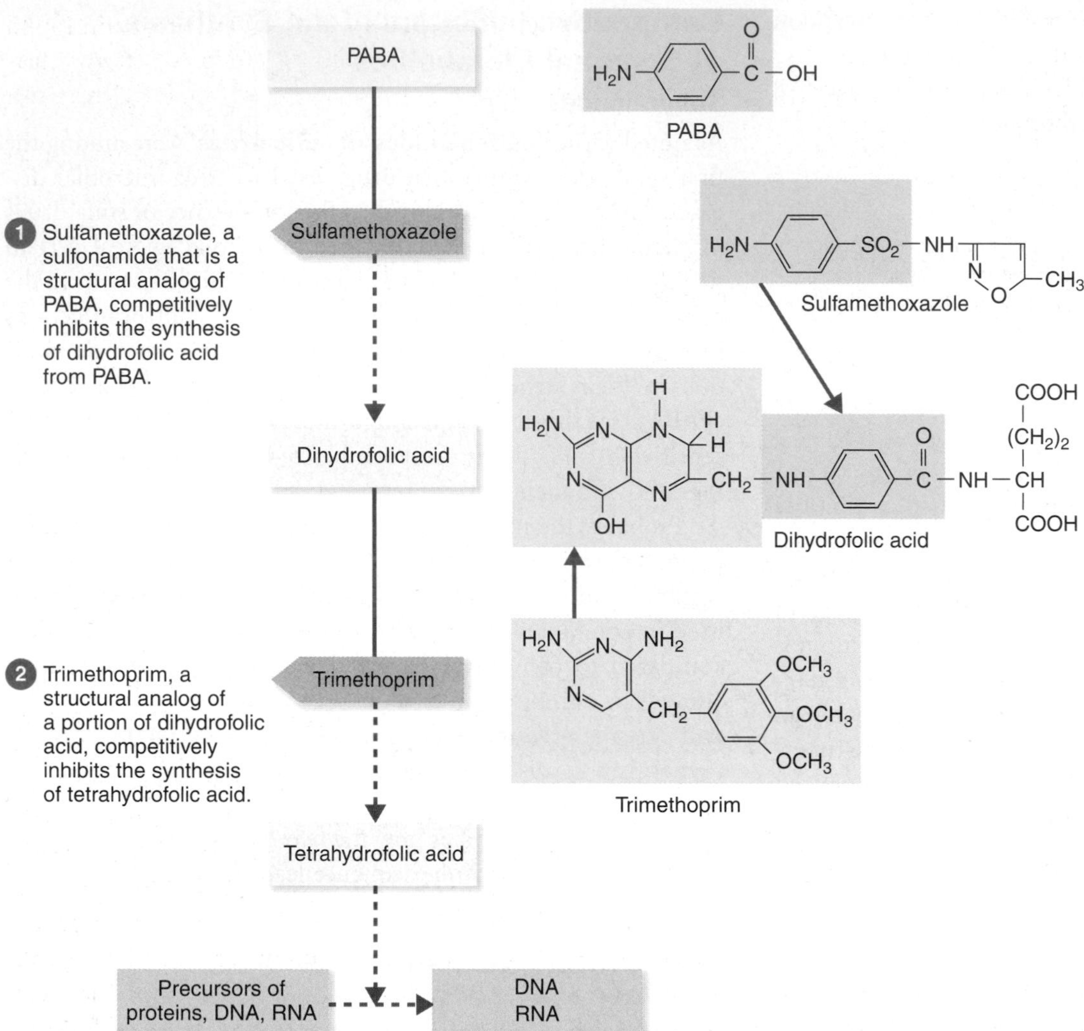
PABA
Sulfamethoxazole
1 Sulfamethoxazole, a sulfonamide that is a structural analog of PABA, competitively inhibits the synthesis of dihydrofolic acid from PABA.
Dihydrofolic acid
2 Trimethoprim, a structural analog of a portion of dihydrofolic acid, competitively inhibits the synthesis of tetrahydrofolic acid.
Trimethoprim
Tetrahydrofolic acid
Precursors of proteins, DNA, RNA
DNA
RNA
PABA
Sulfamethoxazole
Dihydrofolic acid
Trimethoprim

Figure 20.13 Actions of the antibacterial synthetics trimethoprim and sulfamethoxazole. TMP-SMZ works by inhibiting different stages in the synthesis of precursors of DNA, RNA, and proteins. Together the drugs are synergistic.

Q **Define *synergism*.**

Polyenes *Amphotericin B* is the most commonly used member of the antifungal **polyene antibiotics** (Figure 20.14). For many years amphotericin B, produced by *Streptomyces* species of soil bacteria, has been a mainstay of clinical treatment for systemic fungal diseases such as histoplasmosis, coccidioidomycosis, and blastomycosis. The drug's toxicity, particularly to the kidneys, is a strongly limiting factor in these uses. Administering the drug encapsulated in lipids (liposomes) appears to minimize toxicity.

Azoles Some of the most widely used antifungal drugs are represented among the **azole antibiotics.** Before they made their appearance, the only drugs available for systemic fungal infections were amphotericin B and flucytocine (discussed below). The first azoles were **imidazoles,** such as *clotrimazole* and *miconazole* (Figure 20.15), which are now sold without a prescription for topical application for treatment of cutaneous mycoses, such as athlete's foot and vaginal yeast infections. An important addition to this group was *ketoconazole,* which has an unusually broad spectrum of activity among fungi. Ketoconazole, taken orally, proved to be an alternative to amphotericin B for many systemic fungal infections. Ketoconazole topical ointments are used to treat dermatomycoses of the skin.

The use of ketoconazole for systemic infections diminished when the less toxic **triazole** antifungal antibiotics were introduced. The original drugs of this type were *fluconazole* and *itraconazole.* They are much more water soluble, making them easier to use and

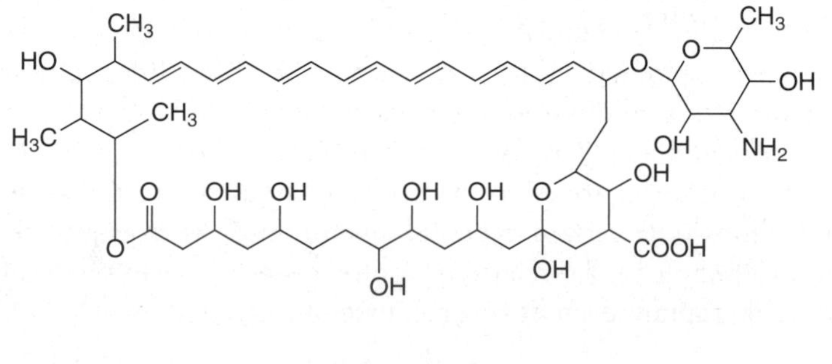

Amphotericin B

Figure 20.14 The structure of the antifungal drug amphotericin B, representative of the polyenes.

Q **Why do polyenes injure fungal plasma membranes and not bacterial membranes?**

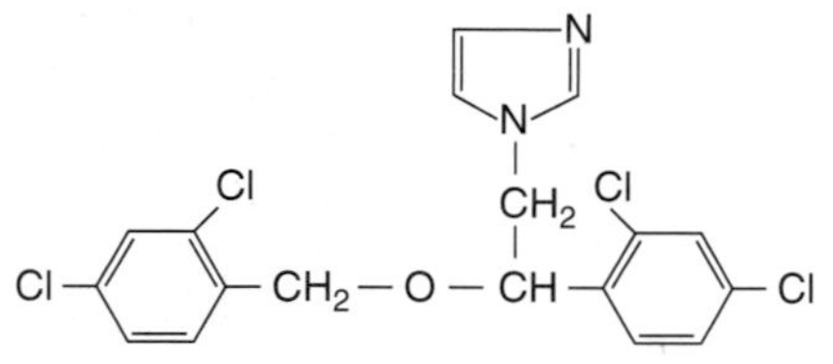

Figure 20.15 The structure of the antifungal drug miconazole, representative of the imidazoles.

Q How do azoles affect fungi?

more effective against systemic infections. The triazole group has recently expanded since the introduction of *voriconazole,* which has become the new standard for treatment of *Aspergillus* infections in immunocompromised patients. The newest triazole drug to be approved is *posaconazole (Noxafil),* which will probably be used to treat a number of systemic fungal infections.

Allylamines The **allylamines** represent a recently developed class of antifungals that inhibit the biosynthesis of ergosterols in a manner that is functionally distinct. *Terbinafine* and *naftifine,* examples of this group, are frequently used when resistance to azole-type antifungals arises.

Agents Affecting Fungal Cell Walls

The fungal cell wall contains compounds that are unique to these organisms. Other than ergosterol, a primary target for selective toxicity among these compounds is β-glucan. The first of a new class of antifungal drugs is the **echinocandins,** which inhibit the biosynthesis of glucans, resulting in an incomplete cell wall and cell lysis. A member of the echinocandin group, *caspofungin (Cancidas)* is now available commercially, and others can be expected. This new antifungal agent is expected to become especially valuable for combating systemic *Aspergillus* infections in persons whose immune system is compromised. It is also effective against important fungi such as *Candida* spp.

Agents Inhibiting Nucleic Acids

Flucytosine, an analog of the pyrimidine cytosine, interferes with the biosynthesis of RNA and therefore protein synthesis. The selective toxicity lies in the ability of the fungal cell to convert flucytocine into 5-fluorouracil, which is incorporated into RNA and eventually disrupts protein synthesis. Mammalian cells lack the enzyme to make this conversion of the drug. Flucytosine has a narrow spectrum of activity, and toxicity to the kidneys and bone marrow further limit its use.

Other Antifungal Drugs

Griseofulvin is an antibiotic produced by a species of *Penicillium.* It has the interesting property of being active against superficial dermatophytic fungal infections of the hair (tinea capitis, or ringworm) and nails, even though its route of administration is oral. The drug apparently binds selectively to the keratin found in the skin, hair follicles, and nails. Its mode of action is primarily to block microtubule assembly, which interferes with mitosis and thereby inhibits fungal reproduction.

Tolnaftate is a common alternative to miconazole as a topical agent for the treatment of athlete's foot. Its mechanism of action is not known. *Undecylenic acid* is a fatty acid that has antifungal activity against athlete's foot, although it is not as effective as tolnaftate or the imidazoles.

Pentamidine isethionate is used in treating *Pneumocystis* pneumonia, a frequent complication of AIDS. The drug's mode of action is unknown, but it appears to bind DNA.

CHECK YOUR UNDERSTANDING

✓ What sterol in the cell membrane of fungi is the most common target for antifungal action? **20-13**

Antiviral Drugs

In developed parts of the world, it is estimated that at least 60% of infectious illnesses are caused by viruses, and about 15% by bacteria. Every year, at least 90% of the U.S. population suffers from a viral disease. Yet relatively few antiviral drugs have been approved in the United States, and they are effective against only an extremely limited group of diseases. Many of the recently developed antiviral drugs are directed against HIV, the pathogen responsible for the pandemic of AIDS. Therefore, as a practical matter the discussion of antivirals is often separated into agents that are directed at chemotherapy of HIV (see page 548) and those with more general (non-HIV) applications (see Table 20.4).

Because viruses replicate within the host's cells, very often using the genetic and metabolic mechanisms of the host's own cells, it is relatively difficult to target the virus without damaging the host's cellular machinery. Many of the antivirals in use today are analogs of components of viral DNA or RNA. However, as more becomes known about the reproduction of viruses, more targets suggest themselves for antiviral action.

Nucleoside and Nucleotide Analogs

Several important antiviral drugs are analogs of nucleosides and nucleotides (page 47). Among the nucleoside analogs, *acyclovir* is the one more widely used (**Figure 20.16**). Although best known for treating genital herpes, it is generally useful for most herpesvirus infections, especially in immunosuppressed individuals. The antiviral drugs *famciclovir,* which can be taken orally, and *ganciclovir* are derivatives of acyclovir and have a similar mode of action. *Ribavirin* resembles the nucleoside guanine and accelerates the already high mutation rate of RNA viruses until the accumulation of errors reaches a crisis point, killing the virus. The nucleoside analog *lamivudine* is used to treat hepatitis B. More recently, a nucleotide analog, *adefovir dipivoxil (Hepsera),* has been introduced for patients resistant to the nucleoside lamivudine. A nucleoside analog, *cidofovir,* is currently used for

Guanine

Deoxyguanosine

Acyclovir

(a) Acyclovir structurally resembles the nucleoside deoxyguanosine.

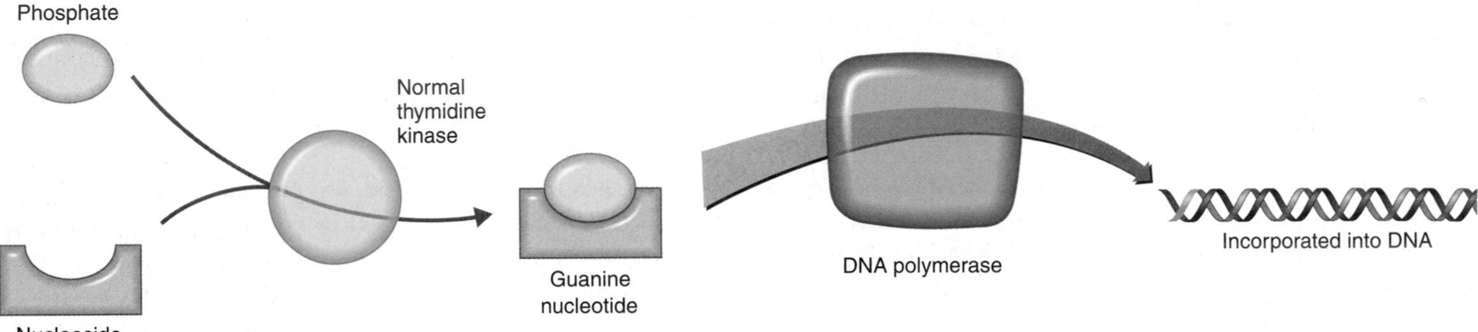

(b) The enzyme thymidine kinase combines phosphates with nucleosides to form nucleotides, which are then incorporated into DNA.

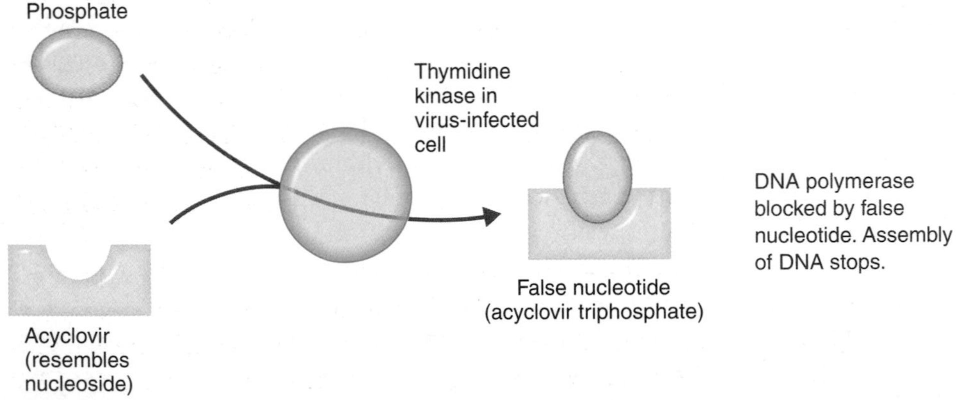

(c) Acyclovir has no effect on a cell not infected by a virus, that is, with normal thymidine kinase. In a virally infected cell, the thymidine kinase is altered and converts the acyclovir (which resembles the nucleoside deoxyguanosine) to a false nucleotide, which blocks DNA synthesis by DNA polymerase.

Figure 20.16 The structure and function of the antiviral drug acyclovir.

Q **Why are viral infections generally difficult to treat with chemotherapeutic agents?**

treatment of cytomegalovirus infections of the eye, but this drug is especially interesting because it shows promise as a possible treatment of smallpox.

Other Enzyme Inhibitors

Two inhibitors of the enzyme neuraminidase (page 694) have been introduced for treatment of influenza. These are *zanamivir* (Relenza) and *oseltamivir* (Tamiflu).

Interferons

Cells infected by a virus often produce interferon, which inhibits further spread of the infection. Interferons are classified as cytokines, discussed in Chapter 17. *Alpha interferon* (see Chapter 16, page 468) is currently a drug of choice for viral hepatitis infections. The production of interferons can be stimulated by a recently introduced antiviral, **imiquimod.** This drug is often prescribed to treat genital warts.

CHECK YOUR UNDERSTANDING

✓ One of the most widely used antivirals, acyclovir, inhibits the synthesis of DNA. Humans also synthesize DNA, so why is the drug still useful in treating viral infections? **20-14**

Antivirals for Treating HIV/AIDS

The interest in effective treatments for the pandemic of HIV infections requires a separate discussion of the many antiviral drugs developed for this. HIV is an RNA virus, and its reproduction depends on the enzyme reverse transcriptase, which controls the synthesis of RNA from DNA (see page 387). Analogs of nucleotides or nucleosides are often a basis for drugs to block this essential step. In fact, the term **antiretroviral** currently implies that a drug is used to treat HIV infections (see the discussion of HAART on page 548). A well-known example of a *nucleoside* analog is *zidovudine*. An example of a *nucleotide* analog is *tenofovir*. In consideration of the large number of drugs required to treat HIV, especially to minimize development of resistant strains, combinations of drugs have been developed. An example is Atripla, which combines tenofovir, emtricitabine, and efavirenz.

Not all drugs that inhibit reverse transcriptase are nucleoside or nucleotide analogs. For example, a few non-nucleoside agents, such as *nevirapine*, block RNA synthesis by other mechanisms.

As the reproduction of HIV became better understood, other approaches to its control became available. When the host cell (at the direction of the infecting HIV) makes a new virus, it must begin by cutting up large proteins with protease enzymes. The resulting fragments are then used to assemble new viruses. Analogs of amino acid sequences in the large proteins can serve as inhibitors of these proteases by competitively interfering with their activity. The **protease inhibitors** *atazanavir, indinavir,* and *saquinavir* have proved especially effective when combined with inhibitors or reverse transcriptase.

Entry of HIV into the cell by fusion can be blocked by **fusion inhibitors** such as *enfuvirtide*—which, however, is dauntingly expensive and must be injected twice daily. It is a synthetic peptide that blocks cell fusion and entry by mimicking a region of the gp41 HIV-1 envelope (see Figure 19.13, page 540).

Drugs that use new targets of HIV reproduction are being considered, and several are undergoing clinical tests. Among these are **integrase inhibitors,** which inhibit an enzyme that integrates viral DNA into the DNA of the infected cell. Blocking entry into the cell by targeting the CCR5 coreceptor (see page 541) is another approach being tested.

Antiprotozoan and Antihelminthic Drugs

For hundreds of years, quinine from the Peruvian cinchona tree was the only drug known to be effective for treating a parasitic infection (malaria). It was first introduced into Europe in the early 1600s and was known as "Jesuit's powder." There are now many antiprotozoan and antihelminthic drugs, although many of them are still considered experimental. This does not preclude their use, however, by qualified physicians. The Centers for Disease Control and Prevention (CDC) provides several of them on request when they are not available commercially.

Antiprotozoan Drugs

Quinine is still used to control the protozoan disease malaria, but synthetic derivatives, such as *chloroquine,* have largely replaced it. For preventing malaria in areas where the disease has developed resistance to chloroquine, the new drug *mefloquine* (Lariam) is recommended, although serious psychiatric side effects have been reported. *Quinacrine* is the drug of choice for treating the protozoan disease giardiasis. *Diiodohydroxyquin (iodoquinol)* is an important drug prescribed for several intestinal amoebic diseases, but its dosage must be carefully controlled to avoid optic nerve damage. Its mode of action is unknown.

Metronidazole (Flagyl) is one of the most widely used antiprotozoan drugs. It is unique in that it acts not only against parasitic protozoa but also against obligately anaerobic bacteria. For example, as an antiprotozoan agent, it is the drug of choice for vaginitis caused by *Trichomonas vaginalis.* It is also used in treating giardiasis and amoebic dysentery. The mode of action is to interfere with anaerobic metabolism, which incidentally these protozoans share with certain obligately anaerobic bacteria, such as *Clostridium.*

Tinidazole, a drug similar to metronidazole, has only recently been approved for use in the United States—although it has long been used elsewhere under the trade name of Fasigyn. It is effective in treating giardiasis, amebiasis, and trichomoniasis. Another new antiprotozoan agent, and the first to be approved for the chemotheraphy of diarrhea caused by *Cryptosporidium hominis,* is *nitazoxanide.* It is active in treating giardiasis and amebiasis. Interestingly, it is also effective in treating several helminthic diseases, as well as having activity against some anaerobic bacteria.

Antihelminthic Drugs

With the increased popularity of sushi, a Japanese specialty often made with raw fish, the CDC began to notice an increased incidence of tapeworm infections. To estimate the incidence, the CDC documents requests for *niclosamide,* which is the usual first choice in treatment. The drug is effective because it inhibits ATP production under aerobic conditions. *Praziquantel* is about equally effective for the treatment of tapeworms; it kills worms by altering the permeability of their plasma membranes. Praziquantel has a broad spectrum of activity and is highly recommended for treating several fluke-caused diseases, especially schistosomiasis. It causes the helminths to undergo muscular spasms and also makes them susceptible to attack by the immune system. Apparently, its action exposes surface antigens, which antibodies can then reach.

Mebendazole and *albendazole* are broad-spectrum antihelminthics that have few side effects and have become the drugs of choice for treating many intestinal helminthic infections. The mode of action of both drugs is to inhibit the formation of microtubules in the cytoplasm, which interferes with the absorption of nutrients by the parasite. These drugs are also widely used

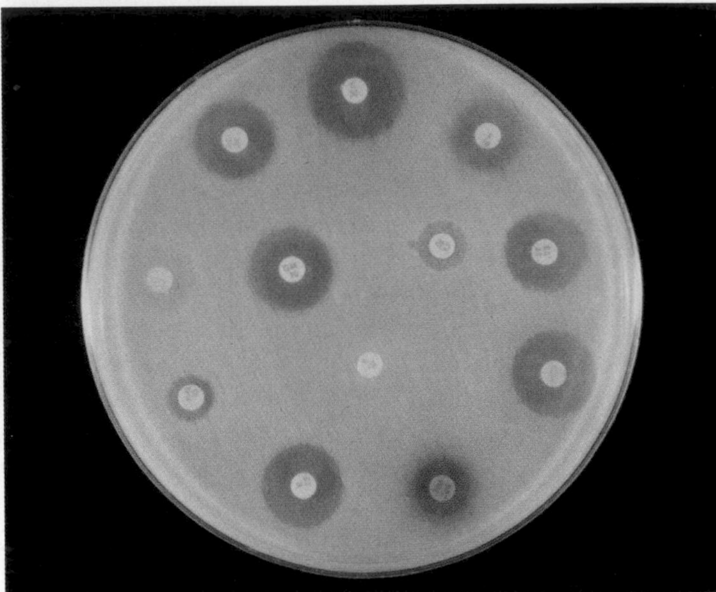

Figure 20.17 The disk-diffusion method for determining the activity of antimicrobials. Each disk contains a different chemotherapeutic agent, which diffuses into the surrounding agar. The clear zones indicate inhibition of growth of the microorganism swabbed onto the agar surface.

Q Which agent is the most effective against the bacterium being tested?

in the livestock industry; for veterinary applications they are relatively more effective in ruminant animals.

Ivermectin is a drug with a wide range of applications. It is known to be produced by only one species of organism, *Streptomyces avermectinius,* which was isolated from the soil near a Japanese golf course. It is effective against many nematodes (roundworms) and several mites (such as scabies), ticks, and insects (such as head lice). (Some mites and insects happen to share certain similar metabolic channels with affected helminths.) Its primary use has been in the livestock industry as a broad-spectrum antihelminthic. Its exact mode of action is uncertain, but the final result is paralysis and death of the helminth without affecting mammalian hosts.

CHECK YOUR UNDERSTANDING

✓ What was the first drug available for use against parasitic infections? **20-15**

Tests to Guide Chemotherapy

LEARNING OBJECTIVE

20-16 Describe two tests for microbial susceptibility to chemotherapeutic agents.

Different microbial species and strains have different degrees of susceptibility to different chemotherapeutic agents. Moreover, the susceptibility of a microorganism can change with time, even during therapy with a specific drug. Thus, a physician must know the sensitivities of the pathogen before treatment can be started. However, physicians often cannot wait for sensitivity tests and must begin treatment based on their "best guess" estimation of the most likely pathogen causing the illness.

Several tests can be used to indicate which chemotherapeutic agent is most likely to combat a specific pathogen. However, if the organisms have been identified—for example, *Pseudomonas aeruginosa,* beta-hemolytic streptococci, or gonococci—certain drugs can be selected without specific testing for susceptibility. Tests are necessary only when susceptibility is not predictable or when antibiotic resistance problems develop.

The Diffusion Methods

Probably the most widely used, although not necessarily the best, method of testing is the **disk-diffusion method,** also known as the *Kirby-Bauer test* (Figure 20.17). A Petri plate containing an agar medium is inoculated ("seeded") uniformly over its entire surface with a standardized amount of a test organism. Next, filter paper disks impregnated with known concentrations of chemotherapeutic agents are placed on the solidified agar surface. During incubation, the chemotherapeutic agents diffuse from the disks into the agar. The farther the agent diffuses from the disk, the lower its concentration. If the chemotherapeutic agent is effective, a **zone of inhibition** forms around the disk after a standardized incubation. The diameter of the zone can be measured; in general, the larger the zone, the more sensitive the microbe is to the antibiotic. The zone diameter is compared to a standard table for that drug and concentration, and the organism is reported as *sensitive, intermediate,* or *resistant.* For a drug with poor solubility, however, the zone of inhibition indicating that the microbe is sensitive will be smaller than for another drug that is more soluble and has diffused more widely. Results obtained by the disk-diffusion method are often inadequate for many clinical purposes. However, the test is simple and inexpensive and is most often used when more sophisticated laboratory facilities are not available.

A more advanced diffusion method, the **E test,** enables a lab technician to estimate the **minimal inhibitory concentration (MIC),** the lowest antibiotic concentration that prevents visible bacterial growth. A plastic-coated strip contains a gradient of antibiotic concentrations, and the MIC can be read from a scale printed on the strip (Figure 20.18).

Broth Dilution Tests

A weakness of the diffusion method is that it does not determine whether a drug is bactericidal and not just bacteriostatic. A **broth dilution test** is often useful in determining the MIC and the **minimal bactericidal concentration (MBC)** of an antimicrobial drug. The MIC is determined by making a sequence of decreasing concentrations of the drug in a broth, which is then inoculated with the test bacteria (Figure 20.19). The wells that do not show growth (higher concentration than the MIC) can be cultured in broth or on agar plates free of the drug. If growth occurs in this broth, the drug was not bactericidal, and the MBC can be determined. Determining the MIC and MBC is important because it avoids the excessive or erroneous use of expensive antibiotics and minimizes the chance of toxic reactions that larger-than-necessary doses might cause.

Dilution tests are often highly automated. The drugs are purchased already diluted into broth in wells formed in a plastic tray. A suspension of the test organism is prepared and inoculated into all the wells simultaneously by a special inoculating device. After incubation, the turbidity may be read visually, although clinical laboratories with high workloads may read the trays with special scanners that enter the data into a computer that provides a printout of the MIC.

Other tests are also useful for the clinician; a determination of the microbe's ability to produce β-lactamase is one example. One popular, rapid method makes use of a cephalosporin that changes color when its β-lactam ring is opened. In addition, a measurement of the *serum concentration* of an antimicrobial is especially important when toxic drugs are used. These assays tend to vary with the drug and may not always be suitable for smaller laboratories.

The hospital personnel responsible for infection control prepare periodic reports called **antibiograms** that record the susceptibility of organisms encountered clinically. These reports are especially useful for detecting the emergence of strains of pathogens resistant to the antibiotics in use in the institution.

CHECK YOUR UNDERSTANDING

✓ In the disk-diffusion (Kirby-Bauer) test, the zone of inhibition indicating sensitivity around the disk varies with the antibiotic. Why? **20-16**

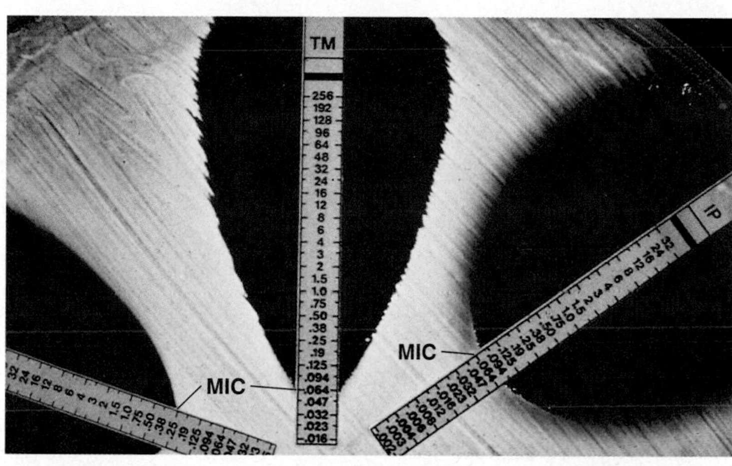

Figure 20.18 The E test (for epsilometer), a gradient diffusion method that determines antibiotic sensitivity and estimates minimal inhibitory concentration (MIC). The plastic strip, which is placed on an agar surface inoculated with test bacteria, contains an increasing gradient of the antibiotic. The MIC is clearly shown.

Q What is the MIC of the central E test?

Resistance to Antimicrobial Drugs

LEARNING OBJECTIVE

20-17 Describe the mechanisms of drug resistance.

One of the triumphs of modern medicine has been the development of antibiotics and other antimicrobials. But the development of resistance to them by the target microbes is an increasing concern. To illustrate this concept, human populations often have a relative resistance to diseases to which they have been exposed for

Figure 20.19 A microdilution, or microtiter, plate used for testing for minimal inhibitory concentration (MIC) of antibiotics. Such plates contain as many as 96 shallow wells that contain measured concentrations of antibiotics. They are usually purchased frozen or freeze dried (page 170). The test microbe is added simultaneously, with a special dispenser, to all the wells in a row of test antibiotics. A button of growth appears if the antibiotic has no effect on the microbe; the microbe is recorded as not sensitive. If there is no growth in a well, the microbe is sensitive to the antibiotic at that concentration. To ensure that the microbe is capable of growth in the absence of the antibiotic, wells that contain no antibiotic are also inoculated (positive control). To ensure against contamination by unwanted microbes, wells that contain nutrient broth but no antibiotics or inoculum are included (negative control).

Q What is *MIC*?

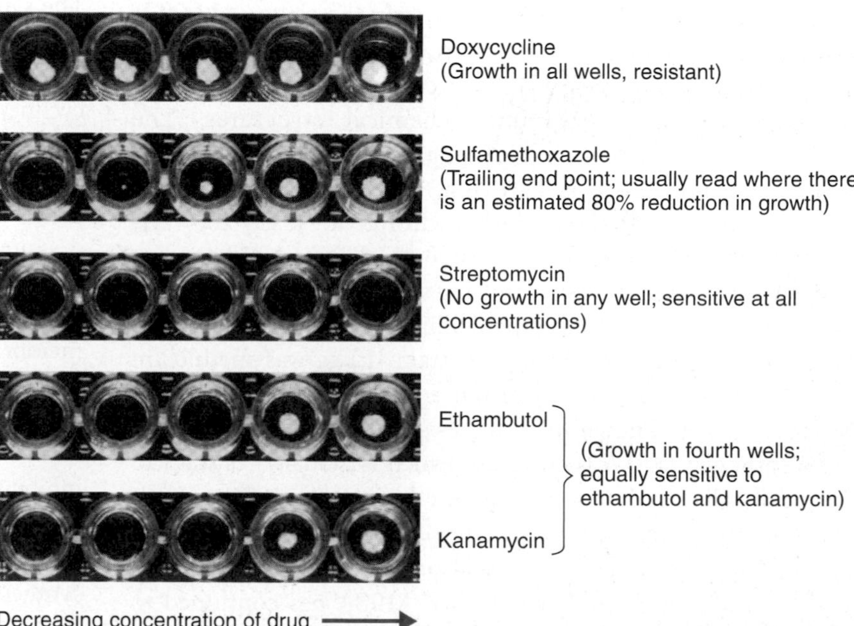

many generations. For example, when Europeans first colonized tropical climes, they proved highly susceptible to diseases to which they had never been exposed, although the local populations were relatively resistant. Antibiotics represent, in a sense, a disease for bacteria. When first exposed to a new antibiotic, the susceptibility of microbes tends to be high, and their mortality rate is also high; there may be only a handful of survivors from a population of billions. The surviving microbes usually have some genetic characteristic that accounts for their survival, and their progeny are similarly resistant.

Such genetic differences arise from random mutations. These mutational differences can be spread *horizontally* among bacteria by processes such as conjugation (page 236) or transduction (page 237). Drug resistance is often carried by plasmids or by small segments of DNA called transposons, which can jump from one piece of DNA to another (Chapter 8, page 240). Some plasmids, including those called resistance (R) factors, can be transferred between bacterial cells in a population and between different but closely related bacterial populations (see Figure 8.29, page 240). R factors often contain genes for resistance to several antibiotics.

Once acquired, however, the mutation is transmitted by normal reproduction, and the progeny carry the genetic characteristics of the parent microbe. Because of the rapid reproductive rate of bacteria, only a short time elapses before practically the entire population is resistant to the new antibiotic.

Mechanisms of Resistance

There are only a few major mechanisms by which bacteria become resistant to chemotherapeutic agents. See **Figure 20.20**.

Enzymatic Destruction or Inactivation of the Drug

Destruction or inactivation by enzymes mainly affects antibiotics that are natural products, such as the penicillins and cephalosporins. Totally synthetic chemical groups of antibiotics such as the fluoroquinolones are less likely to be affected in this manner, although they can be neutralized in other ways. This may simply reflect the fact that the microbes have had fewer years to adapt to these unfamiliar chemical structures. The penicillin/cephalosporin antibiotics, and also the carbapenems, share a structure, the β-lactam ring, which is the target for β-lactamase enzymes that selectively hydrolyze it. Nearly 200 variations of these enzymes are now known, each effective against minor variations in the β-lactam ring structure. When this problem first appeared, the basic penicillin molecule was modified. The first of these penicillinase-resistant drugs was methicillin (see pages 560–561), but resistance to methicillin soon appeared. The best-known of these resistant bacteria is the notorious pathogen MRSA (page 561), which is actually resistant to practically all antibiotics, not just methicillin. Also, *S. aureus* is not the only bacterium of concern; other important pathogens such as *Streptococcus pneumoniae* also have developed resistance to β-lactam antibiotics. Furthermore, MRSA has continued to develop resistance against a succession of new drugs such as vancomycin, even though this antibiotic has a mode of action against cell wall synthesis that is totally different from that of the penicillins. Fortunately, vancomycin-resistant strains have not yet become widespread. These highly adaptable bacteria have even developed resistance against antibiotic combinations that include clavulanic acid, specifically developed as an inhibitor of β-lactamases (see page 561). At first, MRSA was almost exclusively a problem in hospitals and similar health-related settings, accounting for about 20% of bloodstream infections there. However, it is now the cause of frequent outbreaks in the general community, affecting otherwise healthy individuals and causing significant mortality. It has become the most frequent cause of skin and soft-tissue infections presenting at emergency departments in the United States. In consequence, the descriptive terminology now refers to *community-associated MRSA* and *health care–associated MRSA*. Lest one think that this mechanism of resistance is limited to β-lactamases, there are unrelated enzymes that modify and inactivate chloramphenicol and the aminoglycoside group of antibiotics.

Prevention of Penetration to the Target Site within the Microbe

Gram-negative bacteria are relatively more resistant to antibiotics because of the nature of their cell wall, which restricts absorption of many molecules to movements through openings called porins (see page 87). Some bacterial mutants modify the porin opening so that antibiotics are unable to enter the periplasmic space. Perhaps even more important, when β-lactamases are present in the periplasmic space, the antibiotic remains outside the cell, where the enzyme, which is too large to enter even through an unmodified porin, can reach and inactivate it.

Alteration of the Drug's Target Site

The synthesis of proteins involves the movement of a ribosome along a strand of messenger RNA, as shown in Figure 20.4. Several antibiotics, especially those of the aminoglycoside, tetracycline, and macrolide groups, utilize a mode of action that inhibits protein synthesis at this site. Minor modifications at this site can neutralize the effects of antibiotics without significantly affecting cellular function.

Interestingly, the main mechanism by which MRSA gained ascendancy over methicillin was not by a new inactivating enzyme, but by modifying the penicillin-binding protein (PBP) on the cell's membrane. β-Lactam antibiotics act by binding with the PBP, which is required to initiate the cross-linking of peptidoglycan and form the cell wall. MRSA strains become resistant because they have an additional, modified, PBP. The antibiotics continue to inhibit the activity of the normal PBPs, preventing their participation in forming the cell wall. But the additional PBP present on the mutants, although it binds weakly with the antibiotic, still allows synthesis of cell walls that is adequate for survival of MRSA strains.

Figure 20.20

FOUNDATION FIGURE Resistance to Antibiotics

Just as the number of targets for the action of antibiotics is limited, as shown in Figure 20.2, the number of mechanisms by which bacteria are resistant to antibiotics is also limited. This figure illustrates the four major mechanisms of antibiotic resistance that are discussed in detail in this chapter. Knowing how these mechanisms work is the basis for understanding why it is important to prescribe the appropriate antimicrobials to combat specific diseases (like those discussed in the next six chapters).

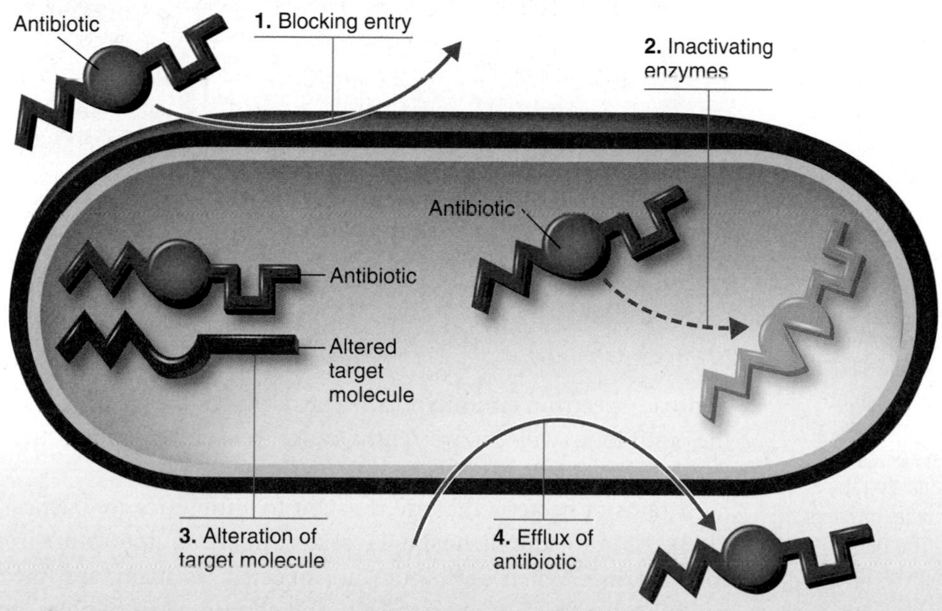

Key Concept

The four main mechanisms of microbial resistance to antimicrobial agents are blocking entry of the drug into the cell, inactivation of the drug by enzymes, alteration of the drug's target sites, and efflux of the drug from the cell.

Rapid Efflux (Ejection) of the Antibiotic

Certain proteins in the plasma membranes of gram-negative bacteria act as pumps that expel antibiotics, preventing them from reaching an effective concentration. This mechanism was originally observed with tetracycline antibiotics, but it confers resistance to practically all major classes of antibiotics. Bacteria normally have many such efflux pumps to eliminate toxic substances.

Variations on these mechanisms also occur. For example, a microbe could become resistant to trimethoprim by synthesizing very large amounts of the enzyme against which the drug is targeted. Conversely, polyene antibiotics can become less effective when resistant organisms produce smaller amounts of the sterols against which the drug is effective. Of particular concern is the possibility that such *resistant mutants* will increasingly replace the susceptible normal populations. Figure 20.21 shows how rapidly bacterial numbers increase as resistance develops.

Antibiotic Misuse

Antibiotics have been much misused, nowhere more so than in the less-developed areas of the world. Well-trained personnel are scarce, especially in rural areas, which is perhaps one reason why antibiotics can almost universally be purchased without prescriptions in these countries. A survey in rural Bangladesh, for example, showed that only 8% of antibiotics had been prescribed by a physician. In much of the world, antibiotics are sold to treat headaches and for other inappropriate uses (Figure 20.22). Even when the use of antibiotics is appropriate, dose regimens are usually shorter than needed to eradicte the infection, thereby encouraging the survival of resistant strains of bacteria. Outdated, adulterated (impure), and even counterfeit antibiotics are common.

The developed world is also contributing to the rise of antibiotic resistance. The CDC estimates that in the United States, 30% of the antibiotic prescriptions for ear infections, 100% of the prescriptions for the common cold, and 50% of prescriptions for sore throats were unnecessary or inappropriate to treat the problem pathogen. At least half of the more than 100,000 tons of antibiotics consumed in the United States each year are not used to treat disease but are used in animal feeds to promote growth—a practice that many people feel should be controlled (see the box on page 577).

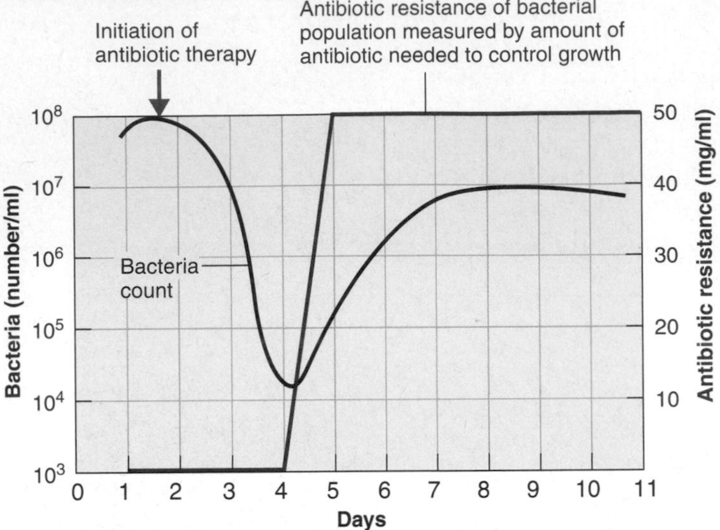

Figure 20.21 The development of an antibiotic-resistant mutant during antibiotic therapy. The patient, suffering from a chronic kidney infection caused by a gram-negative bacterium, was treated with streptomycin. The red line records the antibiotic resistance of the bacterial population. Until about the fourth day, essentially all of the bacterial population is sensitive to the antibiotic. At this time, resistant mutants that require 50,000 μg/ml of antibiotic (a very high amount) to control them appear, and their numbers increase rapidly. The black line records the bacterial population in the patient. After antibiotic therapy is begun, the population declines until the fourth day. At this time, mutants in the population that are resistant to streptomycin appear. The bacterial population in the patient rises as these resistant mutants replace the sensitive population.

 **This test used streptomycin and a gram-negative bacterium. What would the lines have looked like if penicillin G had been the antibiotic?**

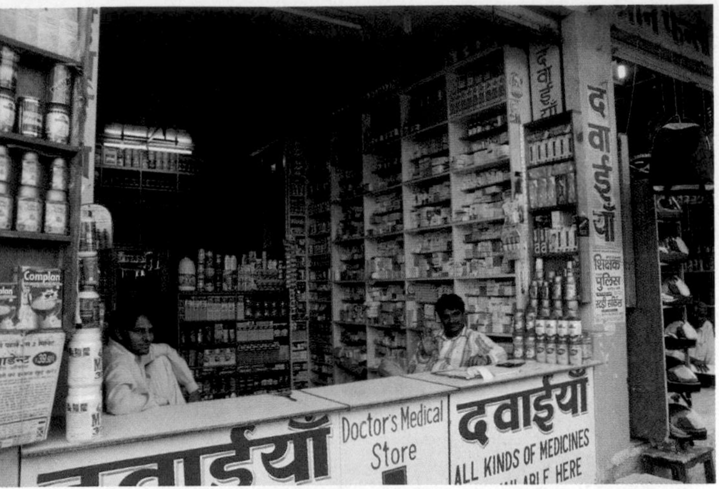

Figure 20.22 Antibiotics have been sold without prescriptions for many decades in much of the world.

 **How does this practice lead to development of resistant strains of pathogens?**

Cost and Prevention of Resistance

Antibiotic resistance is costly in many ways beyond those that are apparent in higher rates of disease and mortality. Developing new drugs to replace those that have lost effectiveness is costly. Almost all of these drugs will be more expensive, sometimes priced in a range that makes them difficult to afford even in highly developed countries. In less-developed parts of the world, the costs are simply unaffordable.

There are many strategies that patients and health care workers can adopt to prevent the development of resistance. Even if they feel they have recovered, patients should always finish the full regimen of their antibiotic prescriptions to discourage the survival and proliferation of the antibiotic-resistant microbes. Patients should never use leftover antibiotics to treat new illnesses or use antibiotics that were prescribed to someone else. Health care workers should avoid unnecessary prescriptions and ensure that the choice and dosages of antimicrobials are appropriate to the situation. Prescribing the most specific antibiotic possible, instead of broad-spectrum antimicrobials, also decreases the chances that the antibiotic will inadvertently cause resistance among the patient's normal flora.

Strains of bacteria that are resistant to antibiotics are particularly common among hospital workers, where antibiotics are in constant use. When antibiotics are injected, as many are, the syringe must first be held vertically and cleared of air bubbles, a practice that causes aerosols of the antibiotic solution to form. When the nurse or physician inhales these aerosols, the microbial inhabitants of the nostrils are exposed to the drug. Many hospitals have special monitoring committees to review the use of antibiotics for effectiveness and cost.

CHECK YOUR UNDERSTANDING

✓ What is the most common mechanism that a bacterium uses to resist the effects of penicillin? **20-17**

Antibiotic Safety

In our discussions of antibiotics, we have occasionally mentioned side effects. These may be potentially serious, such as liver or kidney damage or hearing impairment. Administering almost any drug involves assessing risks against benefits; this is called the *therapeutic index.* Sometimes, the use of another drug can cause toxic effects that do not occur when the drug is taken alone. One drug may also neutralize the intended effects of the other. For example, a few antibiotics have been reported to neutralize the effectiveness of contraceptive pills. Also, some individuals may have hypersensitivity reactions, for example, to penicillins (see the box on page 531).

Antibiotics in Animal Feed Linked to Human Disease

As you read through this box, you will encounter a series of questions that microbiologists ask as they combat antibiotic resistance. Try to answer each question before going on to the next one.

1. Livestock growers use antibiotics in the feed of closely penned animals because the drugs reduce the number of bacterial infections and accelerate the animals' growth. Today, more than half the antibiotics used worldwide are given to farm animals.

   Meat and milk that reach the consumer's table are not heavily laden with antibiotics, so what is the risk of using antibiotics in animal feed?

2. The constant presence of antibiotics in these animals is an example of "survival of the fittest." Antibiotics kill some bacteria, but other bacteria have properties that help them survive.

   How do bacteria acquire resistance genes?

3. Resistance to antimicrobial drugs in bacteria results from mutations. These mutations can be transmitted to other bacteria via horizontal gene transfer (**Figure A**).

   What evidence would show that veterinary use of antibiotics promotes resistance?

4. Vancomycin-resistant *Enterococcus* spp. (VRE) were first isolated in France in 1986 and were found in the United States in 1989. Vancomycin and another glycopeptide, avoparcin, were widely used in animal feed in Europe. In 1996, veterinary use of avoparcin was banned in Germany. After the ban, VRE-positive samples decreased from 100% to 25%, and the human carrier rate dropped from 12% to 3%.

   ***Campylobacter jejuni* is a commensal in the intestines of poultry. What human disease does *C. jejuni* cause?**

5. Annually in the United States, *Campylobacter* causes over 2 million foodborne infections. Fluoroquinolone (FQ)-resistant *C. jejuni* in humans emerged in the 1990s (**Figure B**).

   What FQs are used to treat human infections? (*Hint*: See Table 20.3.)

6. The emergence corresponds with the presence of FQ-resistant *C. jejuni* in grocery store-purchased chicken meat. FQ-resistant *C. jejuni* could be selected for in patients who had previously taken an FQ. However, a study of *Campylobacter* isolates from patients between 1997 and 2001 showed that patients infected with FQ-resistant *C. jejuni* had not taken an FQ prior to their illness and had not traveled out of the United States.

   Suggest a way to decrease emergence of FQ resistance.

7. The use of FQ in chicken feed was banned in 2005, in hope of reducing FQ resistance. A variety of approaches may be necessary to reduce the possibility of illness: (1) prevent colonization in the animals at the farm, (2) reduce fecal contamination of meat during processing at the slaughterhouse, and (3) use proper storage and cooking methods.

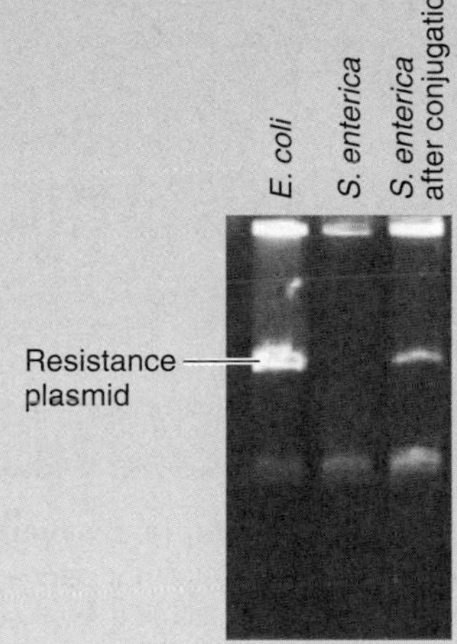

Figure A Cephalosporin-resistance in *E. coli* transferred by conjugation to *Salmonella enterica* in the intestinal tracts of turkeys.

Data sources: CDC and National Antimicrobial Resistance Monitoring System.

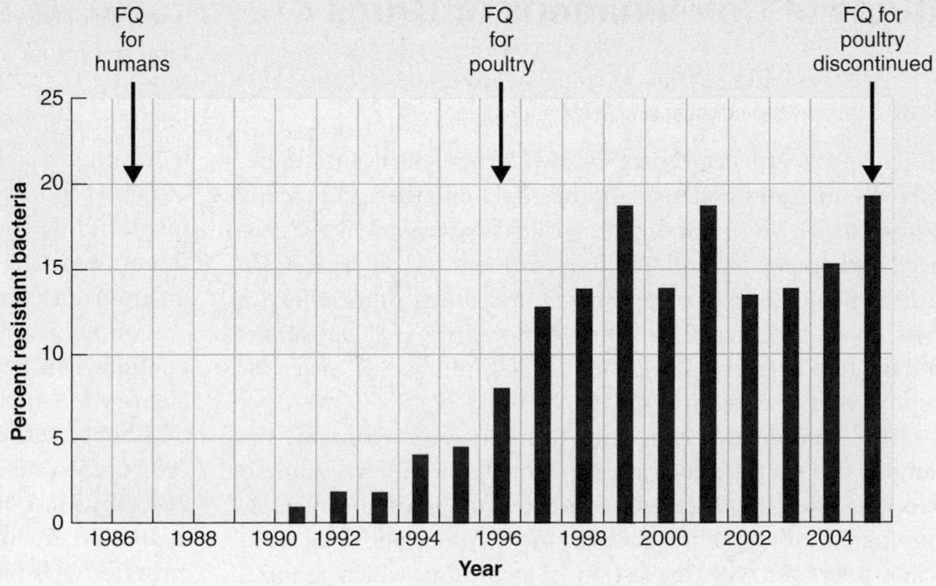

Figure B Flouroquinolone-resistant *Campylobacter jejuni* in the United States, 1982–2005.

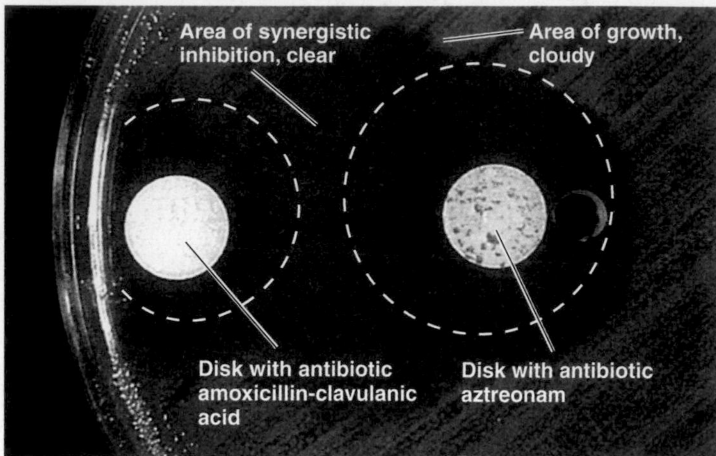

Figure 20.23 An example of synergism between two different antibiotics. The photograph shows the surface of a Petri plate seeded with bacteria. The paper disk at the left contains the antibiotic amoxicillin plus clavulanic acid. The disk on the right contains the antibiotic aztreonam. The dashed circles drawn over the photo show the clear areas surrounding each disk where bacterial growth would have been inhibited if there had been no synergy. The additional clear area between these two areas and outside the drawn circles illustrates inhibition of bacterial growth through the effects of synergy.

What would the plate look like if the two antibiotics had been antagonistic?

A pregnant woman should take only those antibiotics that are classified by the U.S. Food and Drug Administration as presenting no evidence of risk to the fetus.

Effects of Combinations of Drugs

LEARNING OBJECTIVE

20-18 Compare and contrast synergism and antagonism.

The chemotherapeutic effect of two drugs given simultaneously is sometimes greater than the effect of either given alone (Figure 20.23). This phenomenon, called **synergism,** was introduced earlier. For example, in the treatment of bacterial endocarditis, penicillin and streptomycin are much more effective when taken together than when either drug is taken alone. Damage to bacterial cell walls by penicillin makes it easier for streptomycin to enter.

Other combinations of drugs can show **antagonism.** For example, the simultaneous use of penicillin and tetracycline is often less effective than when either drug is used alone. By stopping the growth of the bacteria, the bacteriostatic drug tetracycline interferes with the action of penicillin, which requires bacterial growth.

CHECK YOUR UNDERSTANDING

✓ Tetracycline sometimes interferes with the activity of penicillin. How? **20-18**

The Future of Chemotherapeutic Agents

LEARNING OBJECTIVE

20-19 Identify three areas of research on new chemotherapeutic agents.

As resistance continues to develop against existing antibiotics that have relied on a limited range of targets, research has expanded into investigating new targets for antimicrobial activity other than those summarized in Figure 20.2. Most of our present antibiotics are the products of other microorganisms. Interest is now focusing on antibiotics produced by plants and animals, as well.

Antimicrobial Peptides

Higher organisms, even those that do not have an adaptive immune system, often exhibit extraordinary resistance to microbial infections. This has drawn the attention of scientists seeking a replacement for antibiotics that are losing their effectiveness.

Microorganisms are not the only organisms that produce antimicrobial substances. Essential to the innate immune system of many birds, amphibians, plants, and mammals are **antimicrobial peptides** that they produce. Antimicrobial peptides are, in fact, part of the defense systems of most forms of life. Literally hundreds of such peptides have been identified. They tend to have 100 or fewer amino acids and carry a positive electrical charge, for which they are sometimes termed *cationic peptides.* This is a primary factor in their activity because it allows them to disrupt microbial membranes that are rich in phospholipids with a negative, anionic charge. Enveloped viruses are also affected selectively by similar interaction. In the case of nonenveloped viruses, antimicrobial peptides may interact with viral receptors on the host cell, blocking their attachment. Bacteria are probably affected by the peptides' action against other intracellular targets, as well, including nucleic acid and protein synthesis and even formation of the cell wall. Although many of these peptides are able to directly damage bacteria, viruses, or fungi, mostly by targeting their membranes, they are probably more important in stimulating the host's innate immune and inflammatory responses.

Amphibian skin glands are a rich source of antimicrobial peptides. The best known of these are the *magainins* (from the Hebrew for shield), which have a particularly broad spectrum of activity. *Nisin,* which we have mentioned for its use as a food preservative, is a bacteriocin-like antimicrobial (see pages 200, 401, and 309) that resembles magainin in its mode of action on membranes. It is especially interesting that this antimicrobial has been used for decades without significant development of resistance.

Invertebrates such as insects lack the adaptive immune system that has developed in vertebrate species and must rely on their innate immune systems to combat pathogens. However, the evidence indicates that this response must be very effective. Even more important, these peptides have been around for immense periods of time without having developed microbial resistance to them. The giant silk moth is protected by a peptide, *cecropin*, and bee venom contains another antibacterial peptide, *melittin,* both of which are undergoing testing.

One unexpected fruit of investigation into immune strategies in invertebrates was the discovery of pathogen-associated molecular patterns (see page 450), such as Toll-like receptors (TLRs).* This was a major development in the field of immunology because it showed that TLRs are fundamental components of pathogenic bacteria and serve as an early warning system in adaptive immunity.

The most abundant antimicrobial peptides are the *defensins* that are found in insects, plants, and invertebrates—and in birds and mammals, as well. These are most active against bacteria and fungi. Phagocytized microbes are exposed to defensins in neutrophils, and defensins are released by Paneth cells (specialized secretory cells) in the intestinal tract. They are also found in human skin and mucous membranes.

Antisense Agents

Another promising new approach is the use of short synthetic strands of DNA, called **antisense agents.** (Because antisense agents involve nucleic acids, persons investigating them whimsically call them "nubiotics.") The principle is to identify sites on the DNA or RNA of the pathogen that are responsible for the pathogenic effects. Segments of DNA are then synthesized that will selectively recognize and bind to the target site, a process that blocks the biosynthesis of the target protein. This approach has a great advantage: it prevents a pathogenic protein from being produced, rather than trying to selectively neutralize it once it is made. An antiviral based on antisense principles, *fomivirsen,* has been approved for treating the eye disease cytomegalovirus retinitis.

Mammalian cells have a mechanism that, on occasion, can prevent RNA from giving rise to the proteins for which they are encoded. This has its uses, for example, when an infecting virus tries to take over the cell's metabolism to make viral proteins. The mechanism, called **RNA interference (RNAi),** is proposed as the basis of drugs called **small interfering RNAs (siRNAs)** that could selectively block protein synthesis in pathogens (see page 259). This is similar in principle to antisense agents but is much more effective—at least in theory. No resistance mechanisms currently exist in bacteria. Rather than blocking a single messenger RNA, siRNAs serve as catalysts that would repeat their activity indefinitely. There are many difficulties that will need to be solved before commercially practical drugs appear, however.

*Fruit flies defend themselves from fungal infections by means of a protein called Toll—named for the German word for weird or strange. The term derives from the fact that the Toll protein (see *Toll-like receptors*, page 450) also is involved in development of the embryo, and flies without it have a strange, or weird, appearance.

Other, even more exotic, approaches to solving the problem of antibiotic resistance are being considered. Even before the age of antibiotics, the use of bacteriophages was considered a possibility for disease therapy. These viruses are highly selective in their infective activity, but experiments with them were not very productive. Russian scientists, in particular, have continued to experiment with so-called **phage therapy.** Current thinking is that rather than using intact phage, for which bacteria rapidly develop resistant strains, certain peptides produced by phage that lyse bacteria might be the basis of practical antimicrobials.

Studies of other vertebrates have also proven productive. Compared to mammals, sharks have a rudimentary immune system. The observation that sharks are resistant to infection even in highly contaminated water led to the discovery of an interesting steroid, *squalamine,* that is about as powerful as ampicillin against many pathogens—at least in the laboratory.

A truly novel approach to controlling pathogens has been discovered in laboratory experiments. Antibiotics now in use control pathogens by targeting growth and reproduction factors. The new experiments illustrate an alternative approach: blocking the production of factors that make the pathogen virulent. A screening of small molecules uncovered an experimental compound, *virstatin,* which inhibited the production of toxin and an attachment pilus required for the pathogenicity of the cholera bacillus. Virstatin turned off synthesis of the toxin and also lowered the numbers of the pathogen retained in the intestinal tract—without influencing bacterial growth. The continuing growth of the bacteria may facilitate the development of a normal immune response, which could well be helpful in preventing later infections. This is not the only novel antimicrobial under investigation; some are aimed at, for example, sequestering iron necessary for growth and blocking formation of attachment fimbriae. Most of these new approaches are still in the investigative stages. There is special interest in antiviral drugs, as well as in antifungal and antiparasitic drugs, because our arsenal in these categories is especially limited.

A problem with developing antimicrobial agents is that they are not especially profitable. Like vaccines, they are used only at infrequent occasions. Pharmaceutical companies are understandably more interested in developing drugs that treat chronic conditions, such as high blood pressure or diabetes, for which the patient requires years of regular medication. **Animations** Antibiotic Resistance: Origins of Resistance; Forms of Resistance. **www.microbiologyplace.com**

CHECK YOUR UNDERSTANDING

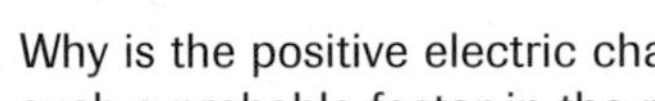

✓ Why is the positive electric charge on antimicrobial peptides such a probable factor in the mode of action? **20-19**

* * *

The past few chapters have shown how profoundly science has changed the effects of infectious disease on human mortality and life span. At the turn of the twentieth century, the most common causes of death were infectious diseases. Most of these, including tuberculosis, typhoid fever, and diphtheria, were caused by bacteria. At the beginning of the twenty-first century, viral diseases such as influenza, viral pneumonias, and AIDS are the only infectious diseases among the ten leading causes of death in the United States. These facts bear testimony to the effectiveness of sanitation, vaccines, and (as discussed in this chapter) the discovery and use of antibiotics. In Chapters 21 through 26, we will see that the struggle against infectious disease continues.

STUDY OUTLINE

The **MyMicrobiologyPlace** website (**www.microbiologyplace.com**) will help you get ready for tests with its simple three-step approach: **1 take a pre-test** and obtain a personalized study plan, **2 learn and practice** with animations, tutorials, and MP3 tutor sessions, and **3 test yourself** with quizzes and a chapter post-test.

Introduction (p. 553)

1. An antimicrobial drug is a chemical substance that destroys pathogenic microorganisms with minimal damage to host tissues.
2. Chemotherapeutic agents include chemicals that combat disease in the body.

The History of Chemotherapy (pp. 554–555)

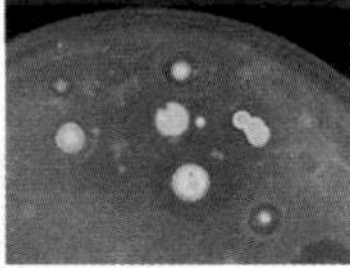

1. Paul Ehrlich developed the concept of chemotherapy to treat microbial diseases; he predicted the development of chemotherapeutic agents, which would kill pathogens without harming the host.
2. Sulfa drugs came into prominence in the late 1930s.
3. Alexander Fleming discovered the first antibiotic, penicillin, in 1929; its first clinical trials were done in 1940.

The Spectrum of Antimicrobial Activity (p. 555)

1. Antibacterial drugs affect many targets in a prokaryotic cell.
2. Fungal, protozoan, and helminthic infections are more difficult to treat because these organisms have eukaryotic cells.
3. Narrow-spectrum drugs affect only a select group of microbes—gram-positive cells, for example; broad-spectrum drugs affect a more diverse range of microbes.
4. Small, hydrophilic drugs can affect gram-negative cells.
5. Antimicrobial agents should not cause excessive harm to normal microbiota.
6. Superinfections occur when a pathogen develops resistance to the drug being used or when normally resistant microbiota multiply excessively.

The Action of Antimicrobial Drugs (pp. 555–559)

1. General action is either by directly killing microorganisms (bactericidal) or by inhibiting their growth (bacteriostatic).
2. Some agents, such as penicillin, inhibit cell wall synthesis in bacteria.
3. Other agents, such as chloramphenicol, tetracyclines, and streptomycin, inhibit protein synthesis by acting on 70S ribosomes.
4. Antifungal agents target plasma membranes.
5. Some agents inhibit nucleic acid synthesis.
6. Agents such as sulfanilamide act as antimetabolites by competitively inhibiting enzyme activity.

A Survey of Commonly Used Antimicrobial Drugs (pp. 559–572)

Antibacterial Antibiotics: Inhibitors of Cell Wall Synthesis (pp. 559–563)

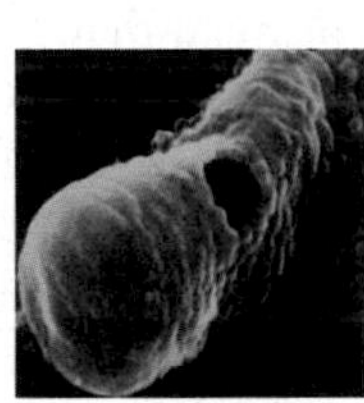

1. All penicillins contain a β-lactam ring.
2. Natural penicillins produced by *Penicillium* are effective against gram-positive cocci and spirochetes.
3. Penicillinases (β-lactamases) are bacterial enzymes that destroy natural penicillins.
4. Semisynthetic penicillins are made in the laboratory by adding different side chains onto the β-lactam ring made by the fungus.
5. Semisynthetic penicillins are resistant to penicillinases and have a broader spectrum of activity than natural penicillins.
6. Carbapenems are broad-spectrum antibiotics that inhibit cell wall synthesis.
7. The monobactam aztreonam affects only gram-negative bacteria.
8. Cephalosporins inhibit cell wall synthesis and are used against penicillin-resistant strains.
9. Polypeptides such as bacitracin inhibit cell wall synthesis primarily in gram-positive bacteria.
10. Vancomycin inhibits cell wall synthesis and may be used to kill penicillinase-producing staphylococci.

Antimycobacterial Antibiotics (p. 563)

11. Isoniazid (INH) and ethambutol inhibit cell wall synthesis in mycobacteria.

Inhibitors of Protein Synthesis (pp. 563–566)

12. Chloramphenicol, aminoglycosides, tetracyclines, macrolides, and streptogramins inhibit protein synthesis at 70S ribosomes.
13. Oxazolidinones prevent formation of 70S ribosomes.

Injury to the Plasma Membrane (pp. 566–567)

14. A new class of antibiotics inhibits fatty-acid synthesis, essential for plasma membranes.
15. Polymyxin B and bacitracin cause damage to plasma membranes.

Inhibitors of Nucleic Acid (DNA/RNA) Synthesis (p. 567)

16. Rifamycin inhibits mRNA synthesis; it is used to treat tuberculosis.
17. Quinolones and fluoroquinolones inhibit DNA gyrase for treating urinary tract infections.

Competitive Inhibitors of the Synthesis of Essential Metabolites (p. 567)

18. Sulfonamides competitively inhibit folic acid synthesis.
19. TMP-SMZ competitively inhibits dihydrofolic acid synthesis.

Antifungal Drugs (pp. 567–569)

20. Polyenes, such as nystatin and amphotericin B, combine with plasma membrane sterols and are fungicidal.
21. Azoles and allylamines interfere with sterol synthesis and are used to treat cutaneous and systemic mycoses.
22. Echinocandins interfere with fungal cell wall synthesis.
23. The antifungal agent flucytosine is an antimetabolite of cytosine.
24. Griseofulvin interferes with eukaryotic cell division and is used primarily to treat skin infections caused by fungi.

Antiviral Drugs (pp. 569–571)

25. Nucleoside and nucleotide analogs, such as acyclovir and zidovudine, inhibit DNA or RNA synthesis.
26. Inhibitors of viral enzymes are used to treat influenza and HIV infection.
27. Alpha interferons inhibit the spread of viruses to new cells.

Antiprotozoan and Antihelminthic Drugs (pp. 571–572)

28. Chloroquine, quinacrine, diiodohydroxyquin, pentamidine, and metronidazole are used to treat protozoan infections.
29. Antihelminthic drugs include mebendazole, praziquantel, and ivermectin.

Tests to Guide Chemotherapy (pp. 572–573)

1. Tests are used to determine which chemotherapeutic agent is most likely to combat a specific pathogen.
2. These tests are used when susceptibility cannot be predicted or when drug resistance arises.

The Diffusion Methods (p. 572)

3. In the disk-diffusion test, also known as the Kirby-Bauer test, a bacterial culture is inoculated on an agar medium, and filter paper disks impregnated with chemotherapeutic agents are overlaid on the culture.

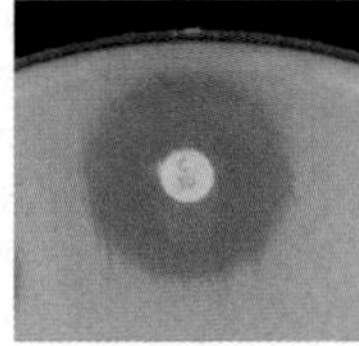

4. After incubation, the diameter of the zone of inhibition is used to determine whether the organism is sensitive, intermediate, or resistant to the drug.
5. MIC is the lowest concentration of drug capable of preventing microbial growth; MIC can be estimated using the E test.

Broth Dilution Tests (pp. 572–573)

6. In a broth dilution test, the microorganism is grown in liquid media containing different concentrations of a chemotherapeutic agent.
7. The lowest concentration of a chemotherapeutic agent that kills bacteria is called the minimum bactericidal concentration (MBC).

Resistance to Antimicrobial Drugs (pp. 573–576)

1. Hereditary drug resistance (R) factors are carried by plasmids and transposons.
2. Resistance may be due to enzymatic destruction of a drug, prevention of penetration of the drug to its target site, cellular or metabolic changes at target sites, or rapid efflux of the antibiotic.
3. Resistance can be minimized by the discriminating use of drugs in appropriate concentrations and dosages.

Antibiotic Safety (pp. 576–578)

1. The risk (e.g., side effects) versus the benefit (e.g., curing an infection) must be evaluated prior to using antibiotics.

Effects of Combinations of Drugs (p. 578)

1. Some combinations of drugs are synergistic; they are more effective when taken together.
2. Some combinations of drugs are antagonistic; when taken together, both drugs become less effective than when taken alone.

The Future of Chemotherapeutic Agents (pp. 578–580)

1. Many bacterial diseases, previously treatable with antibiotics, have become resistant to antibiotics.
2. Chemicals produced by plants and animals are providing new antimicrobial agents called antimicrobial peptides.
3. Protein synthesis in pathogens can be blocked by siRNAs.
4. New agents may inhibit bacterial virulence factors.

STUDY QUESTIONS

Answers to the Review and Multiple Choice questions can be found in the blue Answers tab at the book of the textbook.

Review

1. **DRAW IT** Show where the following antibiotics work: ciprofloxacin, tetracycline, streptomycin, vancomycin, polymyxin B, sulfanilamide, rifampin, erythromycin.

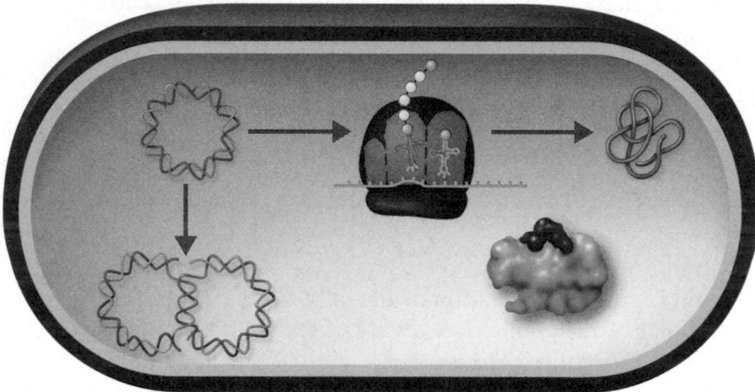

2. List and explain five criteria used to identify an effective antimicrobial agent.
3. What similar problems are encountered with antiviral, antifungal, antiprotozoan, and antihelminthic drugs?
4. Define *drug resistance.* How is it produced? What measures can be taken to minimize drug resistance?
5. List the advantages of using two chemotherapeutic agents simultaneously to treat a disease. What problem can be encountered using two drugs?
6. Why does a cell die from the following antimicrobial actions?
   a. Colistimethate binds to phospholipids.
   b. Kanamycin binds to 70S ribosomes.
7. How does each of the following inhibit translation?
   a. chloroamphenicol
   b. erythromycin
   c. tetracycline
   d. streptomycin
   e. oxazolidinone
   f. streptogramin
8. Dideoxyinosine (ddI) is an antimetabolite of guanine. The –OH is missing from carbon 3 in ddI. How does ddI inhibit DNA synthesis?
9. Compare the method of action of the following pairs:
   a. penicillin and echinocandin
   b. imidazole and polymyxin B

Multiple Choice

1. Which of the following pairs is mismatched?
   a. antihelminthic—inhibition of oxidative phosphorylation
   b. antihelminthic—inhibition of cell wall synthesis
   c. antifungal—injury to plasma membrane
   d. antifungal—inhibition of mitosis
   e. antiviral—inhibition of DNA synthesis
2. All of the following are modes of action of antiviral drugs *except*
   a. inhibition of protein synthesis at 70S ribosomes.
   b. inhibition of DNA synthesis.
   c. inhibition of RNA synthesis.
   d. inhibition of uncoating.
   e. none of the above
3. Which of the following modes of action would not be fungicidal?
   a. inhibition of peptidoglycan synthesis
   b. inhibition of mitosis
   c. injury to the plasma membrane
   d. inhibition of nucleic acid synthesis
   e. none of the above
4. An antimicrobial agent should meet all of the following criteria *except*
   a. selective toxicity.
   b. the production of hypersensitivities.
   c. a narrow spectrum of activity.
   d. no production of drug resistance.
   e. none of the above
5. The most selective antimicrobial activity would be exhibited by a drug that
   a. inhibits cell wall synthesis.
   b. inhibits protein synthesis.
   c. injures the plasma membrane.
   d. inhibits nucleic acid synthesis.
   e. all of the above
6. Antibiotics that inhibit translation have side effects
   a. because all cells have proteins.
   b. only in the few cells that make proteins.
   c. because eukaryotic cells have 80S ribosomes.
   d. at the 70S ribosomes in eukaryotic cells.
   e. none of the above
7. Which of the following will *not* affect eukaryotic cells?
   a. inhibition of the mitotic spindle
   b. binding with sterols
   c. binding to 80S ribosomes
   d. binding to DNA
   e. All of the above will affect them.
8. Cell membrane damage causes death because
   a. the cell undergoes osmotic lysis.
   b. cell contents leak out.
   c. the cell plasmolyzes.
   d. the cell lacks a wall.
   e. none of the above
9. A drug that intercalates into DNA has the following effects. Which one leads to the others?
   a. It disrupts transcription.
   b. It disrupts translation.
   c. It interferes with DNA replication.
   d. It causes mutations.
   e. It alters proteins.
10. Chloramphenicol binds to the 50S portion of a ribosome, which will interfere with
    a. transcription in prokaryotic cells.
    b. transcription in eukaryotic cells.
    c. translation in prokaryotic cells.
    d. translation in eukaryotic cells.
    e. DNA synthesis.

Critical Thinking

1. Which of the following can affect human cells? Explain why or why not.
   a. penicillin
   b. indinavir
   c. erythromycin
   d. polymyxin
2. Why is idoxuridine effective if host cells also contain DNA?
3. Some bacteria become resistant to tetracycline because they don't make porins. Why can a porin-deficient mutant be detected by its inability to grow on a medium containing a single carbon source such as succinic acid?
4. The following data were obtained from a disk-diffusion test.

Antibiotic	Zone of Inhibition
A	15 mm
B	0 mm
C	7 mm
D	15 mm

   a. Which antibiotic was most effective against the bacteria being tested?
   b. Which antibiotic would you recommend for treatment of a disease caused by this bacterium?
   c. Was antibiotic A bactericidal or bacteriostatic? How can you tell?
5. Why do you suppose *Streptomyces griseus* produces an enzyme that inactivates streptomycin? Why is this enzyme produced early in metabolism?
6. The following results were obtained from a broth dilution test for microbial susceptibility.

Antibiotic Concentration	Growth	Growth in Subculture
200 μg/ml	–	–
100 μg/ml	–	+
50 μg/ml	+	+
25 μg/ml	+	+

   a. The MIC of this antibiotic is ___________.
   b. The MBC of this antibiotic is ___________.

Clinical Applications

1. Two patients received corneal transplants of the right eye on the same day, in the same facility, from the same donor. The corneas were kept in a solution of gentamicin and streptomycin, and transplantation was completed within 48 hours of tissue recovery. Both cornea recipients developed eye infections caused by *Clostridium perfringens* and were treated with gentamicin. The eye inflammations persisted. The infections were resolved after treatment was changed to penicillin G. What would you see in a Gram stain of a sample from the infected eyes? Why didn't the first round of treatment cure the infection?
2. A patient with a urinary bladder infection took nalidixic acid, but her condition did not improve. Explain why her infection disappeared when she switched to a sulfonamide.
3. A patient with streptococcal sore throat takes penicillin for 2 days of a prescribed 10-day regimen. Because he feels better, he then saves the remaining penicillin for some other time. After 3 more days, he suffers a relapse of the sore throat. Discuss the probable cause of the relapse.

21 Microbial Diseases of the Skin and Eyes

The skin, which covers and protects the body, is the body's first line of defense against pathogens. As a physical barrier, it is almost impossible for pathogens to penetrate the intact skin. Microbes can, however, enter through skin breaks that are not readily apparent, and the larval forms of a few parasites can penetrate intact skin.

The skin is an inhospitable place for most microorganisms because the secretions of the skin are acidic and most of the skin contains little moisture. Some parts of the body, such as the armpit and the area between the legs, have enough moisture, though, to support relatively large bacterial populations. Drier regions, such as the scalp, support rather small numbers of microorganisms.

Beyond these ecological factors, the skin contains peptide antibiotics called *defensins* that have a wide spectrum of antimicrobial activity (see page 470). These are also found in mucous membranes, especially those lining the gastrointestinal tract.

UNDER THE MICROSCOPE
Candida albicans. A yeastlike fungus.

Q&A

How does the morphology of *Candida albicans* contribute to the microbe's pathogenicity?
Look for the answer in the chapter.

Structure and Function of the Skin

LEARNING OBJECTIVE

21-1 Describe the structure of the skin and mucous membranes and the ways pathogens can invade the skin.

The skin of an average adult occupies a surface area of about 1.9 m^2 and varies in thickness from 0.05 to 3.0 mm. As we mentioned in Chapter 16, skin consists of two principal parts, the epidermis and the dermis (Figure 21.1). The **epidermis** is the thin outer portion, composed of several layers of epithelial cells. The outermost layer of the epidermis, the *stratum corneum,* consists of many rows of dead cells that contain a waterproofing protein called **keratin.** The epidermis, when unbroken, is an effective physical barrier against microorganisms.

The **dermis** is the inner, relatively thick portion of skin, composed mainly of connective tissue. The hair follicles, sweat gland ducts, and oil gland ducts in the dermis provide passageways through which microorganisms can enter the skin and penetrate deeper tissues.

Perspiration provides moisture and some nutrients for microbial growth. However, it contains salt, which inhibits many microorganisms; the enzyme lysozyme, which is capable of breaking down the cell walls of certain bacteria; and antimicrobial peptides.

Sebum, secreted by oil glands, is a mixture of lipids (unsaturated fatty acids), proteins, and salts that prevents skin and hair from drying out. Although the fatty acids inhibit the growth of certain pathogens, sebum, like perspiration, is also nutritive for many microorganisms.

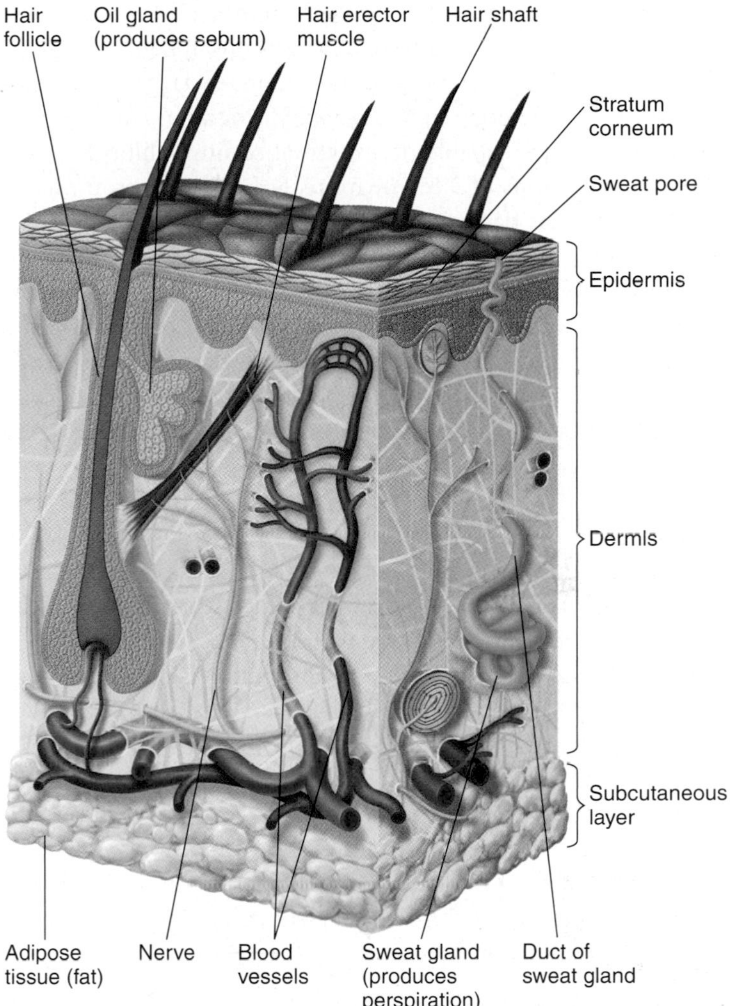

Figure 21.1 The structure of human skin. Notice the passageways between the hair follicle and hair shaft, through which microbes can penetrate the deeper tissues. They can also enter the skin through sweat pores.

Q What do you perceive from this illustration to be the weak points that would allow microbes to reach the underlying tissue by penetrating intact skin?

Mucous Membranes

In the linings of body cavities, such as those associated with the gastrointestinal, respiratory, urinary, and genital tracts, the outer protective barrier differs from the skin. It consists of sheets of tightly packed *epithelial cells.* These cells are attached at their bases to a layer of extracellular material called the *basement membrane.* Many of these cells secrete mucus—hence the name **mucous membrane,** or **mucosa.** Other mucosal cells have cilia; and, in the respiratory system, the mucous layer traps particles, including microorganisms, which the cilia sweep upward out of the body (see Figure 16.4, page 452). Mucous membranes are often acidic, which tends to limit their microbial populations. Also, the membranes of the eyes are mechanically washed by tears, and the lysozyme in tears destroys the cell walls of certain bacteria. Mucous membranes are often folded to maximize surface area; the total surface area in an average human is about 400 m^2, much more than the surface area of the skin.

CHECK YOUR UNDERSTANDING

✔ The moisture provided by perspiration encourages microbial growth on the skin. What factors in perspiration discourage microbial growth? **21-1**

Normal Microbiota of the Skin

LEARNING OBJECTIVE

21-2 Provide examples of normal skin microbiota, and state the general locations and ecological roles of its members.

Although the skin is generally inhospitable to most microorganisms, it supports the growth of certain microbes that are established as part of the normal microbiota. On superficial skin surfaces, certain aerobic bacteria produce fatty acids from sebum. These acids inhibit many microbes and allow better-adapted bacteria to flourish.

Microorganisms that find the skin a satisfactory environment are resistant to drying and to relatively high salt concentrations. The skin's normal microbiota contain relatively large numbers of gram-positive bacteria, such as staphylococci and micrococci. Some of these are capable of growth at sodium chloride (table salt) concentrations of 7.5% or more. Scanning electron micrographs show that bacteria on the skin tend to be grouped into small clumps. Vigorous washing can reduce their numbers but will not eliminate them. Microorganisms remaining in hair follicles and sweat glands after washing will soon reestablish the normal populations. Areas of the body with more moisture, such as the armpits and between the legs, have higher populations of microbes. These metabolize secretions from the sweat glands and are the main contributors to body odor.

Also part of the skin's normal microbiota are gram-positive pleomorphic rods called *diphtheroids.* Some diphtheroids, such as *Propionibacterium acnes,* are typically anaerobic and inhabit hair follicles. Their growth is supported by secretions from the oil glands (sebum), which, as we will see, makes them a factor in acne. These bacteria produce propionic acid, which helps maintain the low pH of skin, generally between 3 and 5. Other diphtheroids, such as *Corynebacterium xerosis* (ze-rō′sis), are aerobic and occupy the skin surface. A yeast, *Malassezia* (mal′as-sēz-ē-ä) *furfur* is capable of growing on oily skin secretions and is thought to be responsible for the scaling skin condition known as *dandruff.* Shampoos for treating dandruff contain the antibiotic ketoconazole or zinc pyrithione or selenium sulfide. All are active against this yeast.

CHECK YOUR UNDERSTANDING

✓ Are skin bacteria more likely to be gram-positive or gram-negative? **21-2**

Microbial Diseases of the Skin

LEARNING OBJECTIVES

21-3 Differentiate staphylococci from streptococci, and name several skin infections caused by each.

21-4 List the causative agent, mode of transmission, and clinical symptoms of *Pseudomonas* dermatitis, otitis externa, acne, and Buruli ulcer.

21-5 List the causative agent, mode of transmission, and clinical symptoms of these skin infections: warts, smallpox, monkeypox, chickenpox, shingles, cold sores, measles, rubella, fifth disease, and roseola.

21-6 Differentiate cutaneous from subcutaneous mycoses, and provide an example of each.

21-7 List the causative agent and predisposing factors for candidiasis.

21-8 List the causative agent, mode of transmission, clinical symptoms, and treatment for scabies and pediculosis.

Rashes and lesions on the skin do not necessarily indicate an infection of the skin; in fact, many diseases manifested by skin lesions are actually systemic diseases affecting internal organs. Variations in these lesions are often useful in describing the symptoms of the disease. For example, small, fluid-filled lesions are **vesicles** (Figure 21.2a). Vesicles larger than about 1 cm in diameter are termed **bullae** (Figure 21.2b). Flat, reddened lesions are known as **macules** (Figure 21.2c). Raised lesions are called **papules** or, when they contain pus, **pustules** (Figure 21.2d). Although the focus of infection is often elsewhere in the body, it is convenient to classify these diseases by the organ most obviously affected: the skin. A skin rash that arises from disease conditions is called an **exanthem;** on mucous membranes, such as the interior of the mouth, such a rash is called an **enanthem.**

Preliminary diagnoses of diseases associated with the skin are often based on the appearance of rashes; these are summarized in Diseases in Focus 21.1, 21.2, and 21.3.

Bacterial Diseases of the Skin

Two genera of bacteria, *Staphylococcus* and *Streptococcus,* are frequent causes of skin-related diseases and merit special discussion. We will also discuss these bacteria in subsequent chapters in relation to other organs and conditions. Superficial staphylococcal and streptococcal infections of the skin are very common. The bacteria frequently come into contact with the skin and have adapted fairly well to the physiological conditions there. Both genera also produce invasive enzymes and damaging toxins.

Staphylococcal Skin Infections

Staphylococci are spherical gram-positive bacteria that form irregular clusters like grapes (see Figure 4.1d, page 37, and Figure 11.18, page 318). For almost all clinical purposes, these bacteria can be divided into those that produce **coagulase,** an enzyme that coagulates (clots) fibrin in blood, and those that do not.

Coagulase-negative strains, such as *Staphylococcus epidermidis,* are very common on the skin, where they may represent 90% of the normal microbiota. They are generally pathogenic only when the skin barrier is broken or is invaded by medical procedures, such as the insertion and removal of catheters into veins. On the surface of the catheter (Figure 21.3), the bacteria are surrounded by a slime layer of capsular material (see discussions of biofilms on pages 18 and 162). This is a primary factor in their importance as a nosocomial pathogen because it protects the bacteria from desiccation and disinfectants.

S. aureus is the most pathogenic of the staphylococci (also see the discussion of MRSA in Chapter 20). It is a permanent resident of the nasal passages of 20% of the population, and an additional 60% carry it there occasionally. Exposed on surfaces, it can survive for months. Typically, it forms golden-yellow colonies. This pigmentation is protective against the antimicrobial effects of sunlight; mutants without it are also more susceptible to

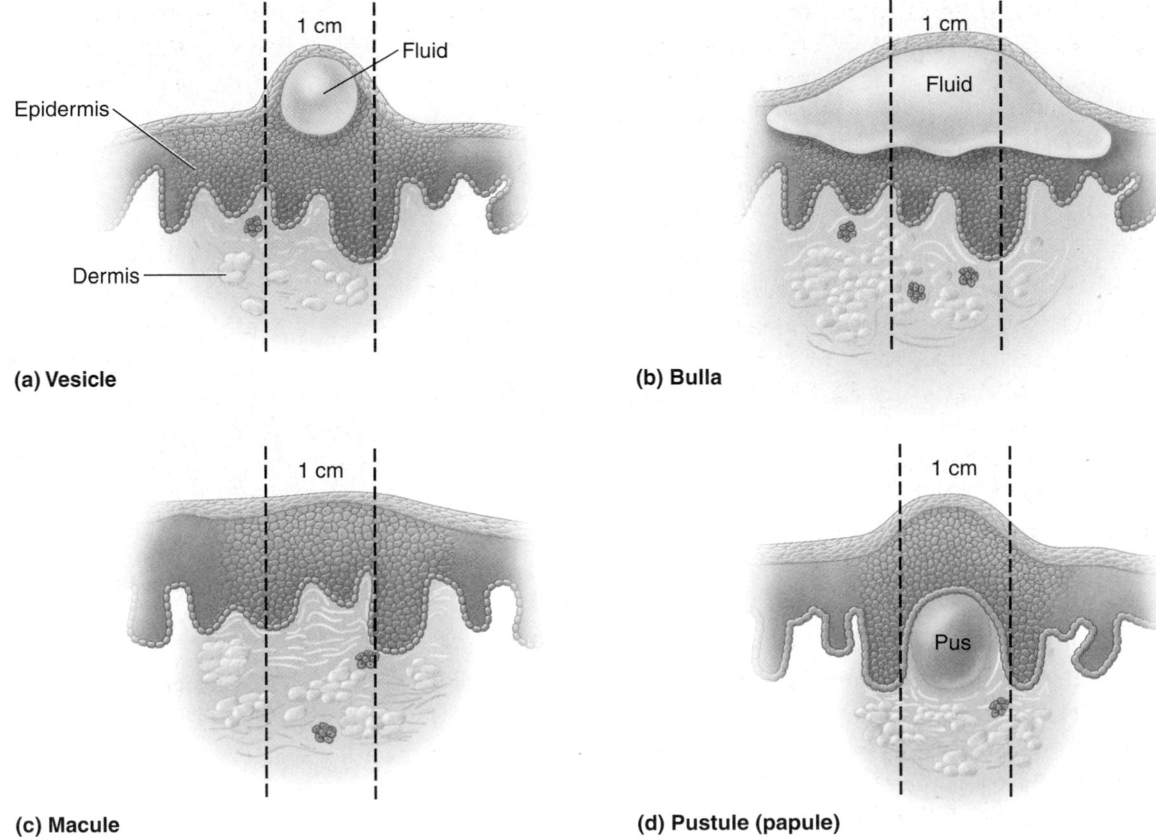

Figure 21.2 Skin lesions. (**a**) Vesicles are small, fluid-filled lesions. (**b**) Bullae are larger fluid-filled lesions. (**c**) Macules are flat lesions that are often reddish. (**d**) Papules are raised lesions; when they contain pus, as shown here, they are called pustules.

Q Are these skin lesions exanthems or enanthems?

killing by neutrophils. Compared to its more innocuous relative *S. epidermidis*, *S. aureus* has about 300,000 more base pairs in its genome—much of it devoted to an impressive array of virulence factors and means of evading host defenses. Almost all pathogenic strains of *S. aureus* are coagulase-positive. There is a high correlation between the bacterium's ability to form coagulase and its production of damaging toxins, several of which facilitate the spread of the organism in tissue, damage tissue, or are lethal to host defenses. In addition, some strains can cause life-threatening sepsis (Chapter 23, page 639), and others produce *enterotoxins* that affect the gastrointestinal tract (see Chapter 25, pages 711 to 712).

Once *S. aureus* infects the skin, it stimulates a vigorous inflammatory response, and macrophages and neutrophils are attracted to the site of infection. However, the bacteria often evade these normal host defenses. Most strains of the pathogen secrete a protein that blocks chemotaxis of neutrophils to the infection site, and if the bacterium does encounter phagocytic cells, it often produces toxins that kills them. It is resistant to opsonization (see page 459) but, failing this, it can survive well within the phagosome. Other proteins it secretes neutralize the antimicrobial peptide defensins on skin, and its cell wall is lysozyme resistant (see page 88). It sometimes responds to the

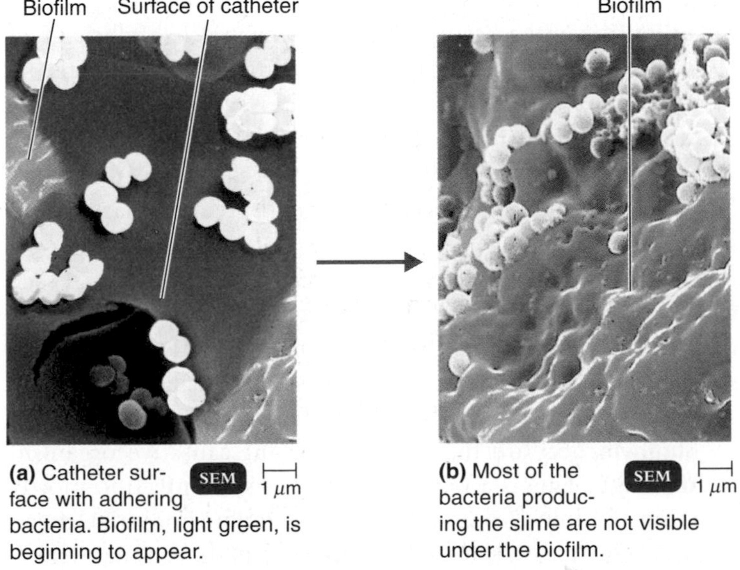

(a) Catheter surface with adhering bacteria. Biofilm, light green, is beginning to appear.

(b) Most of the bacteria producing the slime are not visible under the biofilm.

Figure 21.3 Coagulase-negative staphylococci. These slime-producing bacteria adhere to surfaces such as the plastic catheter in the photos. Once they have adhered to the surface (**a**), they begin to divide. Eventually (**b**), the entire surface is coated with a biofilm containing the organisms.

Q What is the most likely source of the bacteria that grew on the catheter?

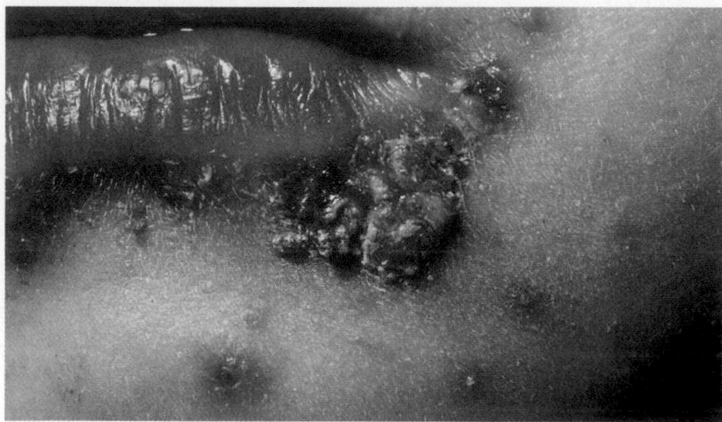

Figure 21.4 Lesions of impetigo. This disease is characterized by isolated pustules that become crusted.

Q What other bacteria also cause impetigo?

immune system as a superantigen (see page 436) but often is able to evade the adaptive immune system entirely. All humans possess antibodies against *S. aureus*, but they do not effectively prevent repeated infections. Antibiotic-resistant strains of *S. aureus* have emerged and are difficult to treat (see pages 563 and 574). These strains are causing infections in hospitals and in the community (see the box on page 593.)

Because this organism is so commonly present in human nasal passages, it is often transported from there to the skin. There it can enter the body through natural openings in the skin barrier, such as the hair follicle; see Figure 21.1. Such infections, or **folliculitis,** often occur as pimples. The infected follicle of an eyelash is called a **sty.** A more serious hair follicle infection is the **furuncle (boil),** which is a type of **abscess,** a localized region of pus surrounded by inflamed tissue. Antibiotics do not penetrate well into abscesses, and the infection is therefore difficult to treat. Draining pus from the abscess is frequently a preliminary step to successful treatment.

When the body fails to wall off a furuncle, neighboring tissue can be progressively invaded. The extensive damage is called a **carbuncle,** a hard, round deep inflammation of tissue under the skin. At this stage of infection, the patient usually exhibits the symptoms of generalized illness with fever.

Staphylococci are the most important causative organism of **impetigo.** This is a highly contagious skin infection mostly affecting children 2 to 5 years of age, among whom it is spread by direct contact. *Streptococcus pyogenes,* a pathogen that we will be discussing shortly, can also cause impetigo, although in fewer cases. Sometimes both *S. aureus* and *S. pyogenes* are involved. The disease takes two forms; *nonbullous impetigo* (see the bulla in Figure 21.2b) is the more common. The pathogen usually enters through some minor break in the skin. The infection can also spread to surrounding areas—a process called *autoinoculation.* Symptoms result from the host's response to the infection. The lesions eventually rupture and form light-colored crusts, as shown in **Figure 21.4**. Topical antibiotics are sometimes applied, but the lesions generally heal without treatment and without scarring.

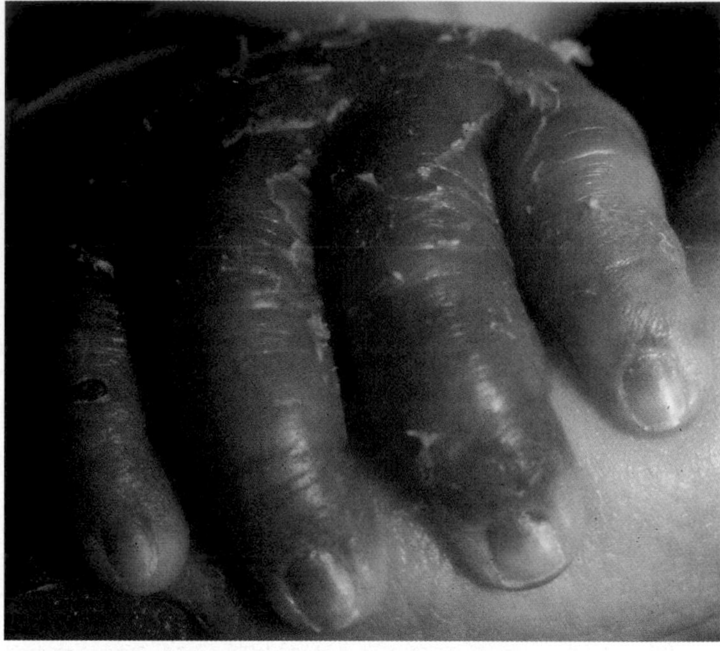

Figure 21.5 Lesions of scalded skin syndrome. Some staphylococci produce a toxin that causes the skin to peel off in sheets, as on the hand of this infant. It is especially likely to occur in children under age 2.

Q What is the name of the toxin that produces this syndrome?

The other type of impetigo, *bullous impetigo,* is caused by a staphylococcal toxin and is a localized form of staphylococcal **scalded skin syndrome.** Actually, there are two serotypes of the toxin; toxin A, which remains localized, causes bullous impetigo, and toxin B, which circulates to distant sites, causes scalded skin syndrome. Both toxins cause a separation of the skin layers, *exfoliation.* See **Figure 21.5**. Outbreaks of bullous impetigo are a frequent problem in hospital nurseries, where the condition is known as **pemphigus neonatorum,** or *impetigo of the newborn.* (See the discussion of hexachlorophene in Chapter 7, page 196).

Scalded skin syndrome is also characteristic of the late stages of **toxic shock syndrome (TSS).** In this potentially life-threatening condition, fever, vomiting, and a sunburnlike rash are followed by shock and sometimes organ failure, especially of the kidneys. TSS originally became known as a result of staphylococcal growth associated with the use of a new type of highly absorbent vaginal tampon; the correlation is especially high for cases in which the tampons remain in place too long. A novel staphylococcal toxin called *toxic shock syndrome toxin 1 (TSST-1)* is formed at the growth site and circulates in the bloodstream. The symptoms are

DISEASES IN FOCUS 21.1

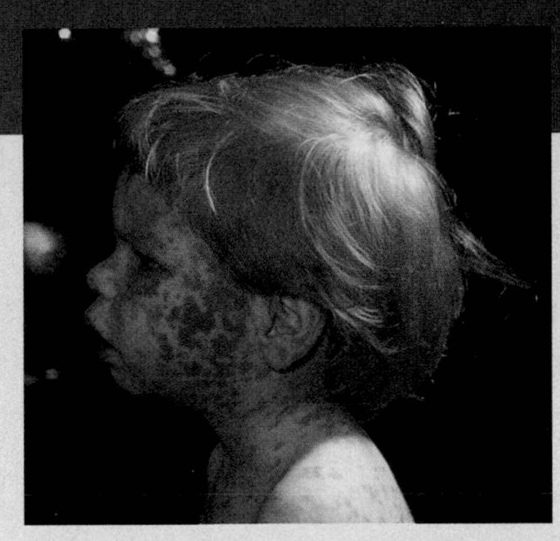

Macular Rashes

Differential diagnosis is the process of identifying the disease from a list of possible diseases that fit the information derived from examining a patient. A differential diagnosis is important for providing initial treatment and for laboratory testing. For example, a 4-year-old boy with a history of cough, conjunctivitis, and fever (38.3°C) now has a macular rash that starts on his face and neck and is spreading to the rest of his body. Use the table below to identify infections that could cause these symptoms. For the solution, go to www.microbiologyplace.com.

Disease	Pathogen	Portal of Entry	Symptoms	Method of Transmission	Treatment
VIRAL DISEASES. Usually diagnosed by clinical signs and symptoms and may be confirmed by serology or PCR.					
Measles (rubeola)	Measles virus	Respiratory tract	Skin rash of reddish macules first appearing on face and spreading to trunk and extremities	Aerosol	No treatment; preexposure vaccine
Rubella (German measles)	Rubella virus	Respiratory tract	Mild macular disease with a rash resembling measles, but less extensive and disappearing in 3 days or less	Aerosol	No treatment; preexposure vaccine
Fifth disease (erythema infectiosum)	Human parvovirus B19	Respiratory tract	Mild disease with a macular facial rash	Aerosol	None
Roseola	Human herpesvirus 6, human herpesvirus 7	Respiratory tract	High fever followed by macular body rash	Aerosol	None
FUNGAL DISEASE. Confirmed by Gram staining of skin scrapings.					
Candidiasis	*Candida albicans*	Skin; mucous membranes	Macular rash	Direct contact; endogenous infection	Miconazole, clotrimozole (topically)

thought to be a result of the superantigenic properties of the toxin (see the discussion of superantigens on page 436).

Today a minority of the cases of TSS are associated with menstruation. Nonmenstrual TSS occurs from staphylococcal infections that follow nasal surgery in which absorbent packing is used, after surgical incisions, and in women who have just given birth.

Streptococcal Skin Infections

Streptococci are gram-positive spherical bacteria. Unlike staphylococci, streptococcal cells usually grow in chains (see Figure 11.19, page 318). Prior to division, the individual cocci elongate on the axis of the chain, and then the cells divide (see Figure 4.1a, page 37). Streptococci cause a wide range of disease conditions beyond those covered in this chapter, including meningitis, pneumonia, sore throats, otitis media, endocarditis, puerperal fever, and even dental caries.

As streptococci grow, they secrete toxins and enzymes, virulence factors that vary with the different streptococcal species. Among these toxins are *hemolysins,* which lyse red blood cells. Depending on the hemolysin they produce, streptococci are categorized as alpha-hemolytic, beta-hemolytic, and gamma-hemolytic (actually nonhemolytic) streptococci (see Figure 6.9, page 168). Hemolysins can lyse not only red blood cells, but almost any type of cell. It is uncertain, though, just what part they play in streptococcal pathogenicity.

Beta-hemolytic streptococci are often associated with human disease. This group is further differentiated into serological groups,

DISEASES IN FOCUS 21.2

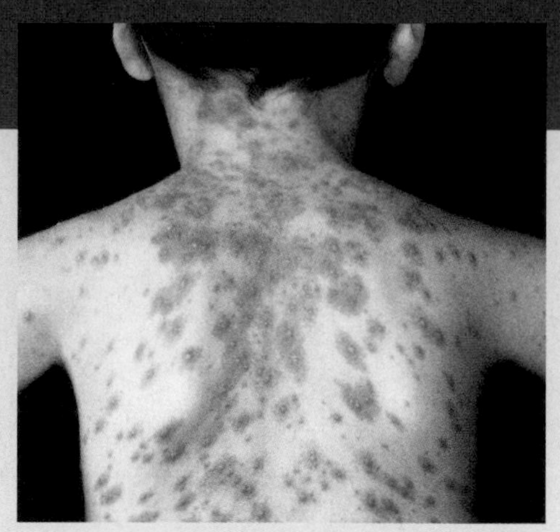

Vesicular and Pustular Rashes

Differential diagnosis is the process of identifying the disease from a list of possible diseases that fit the information derived from examining a patient. A differential diagnosis is important for providing initial treatment and for laboratory testing. For example, an 8-year-old boy has a rash consisting of vesicular lesions of 5 days' duration on his neck and stomach. Within 5 days, 73 students in his elementary school had illness matching the case definition for this disease. Use the table below to identify infections that could cause these symptoms. For the solution, go to www.microbiologyplace.com.

Disease	Pathogen	Portal of Entry	Symptoms	Method of Transmission	Treatment
BACTERIAL DISEASE. Usually diagnosed by culturing the bacteria.					
Impetigo	*Staphylococcus aureus*	Skin	Vesicles on skin	Direct contact; fomites	Topical antibiotics
VIRAL DISEASES. Usually diagnosed by clinical signs and symptoms and may be confirmed by serology or PCR.					
Smallpox (variola)	Smallpox (variola) virus	Respiratory tract	Pustules that may be nearly confluent on skin	Aerosol	None
Monkeypox	Monkeypox virus	Respiratory tract	Pustules, similar to smallpox	Direct contact with or aerosols from infected small mammals	None
Chickenpox (varicella)	Varicella-zoster virus	Respiratory tract	Vesicles in most cases confined to face, throat, and lower back	Aerosol	Acyclovir for immunocompromised patients; preexposure vaccine
Shingles (herpes-zoster)	Varicella-zoster virus	Endogenous* infection of peripheral nerves	Vesicles typically on one side of waist, face and scalp, or upper chest	Recurrence of latent chickenpox infection	Acyclovir preventive vaccine
Herpes simplex	Herpes simplex virus type 1	Skin; mucous membranes	Vesicles around mouth; can also affect other areas of skin and mucous membranes	Initial infection by direct contact; recurring latent infection	Acyclovir

*Endogenous infections are infections caused by microorganisms already part of the host microbiota.

designated A through T, according to antigenic carbohydrates in their cell walls. The group A streptococci (GAS), which are synonymous with the species *Streptococcus pyogenes*, are the most important of the beta-hemolytic streptococci. They are among the most common human pathogens and are responsible for a number of human diseases—some of them deadly. This group of pathogens is divided into over 80 immunological types according to the antigenic properties of the M protein found in some strains (**Figure 21.6**). This protein is external to the cell wall on a fuzzy layer of fibrils. The M protein prevents the activation of complement and allows the microbe to evade phagocytosis and killing by neutrophils (see page 454). It also appears to help the bacteria adhere to and colonize mucous membranes. Another virulence factor of the GAS is their capsule of hyaluronic acid. Exceptionally virulent strains have a mucoid appearance on blood-agar plates from heavy encapsulation and are rich in M protein. Hyaluronic acid is poorly immunogenic (it resembles human connective tissue) and few antibodies against the capsule are produced.

The GAS produce substances that promote the rapid spread of infection through tissue and by liquefying pus. Among these are *streptokinases* (enzymes that dissolve blood clots), *hyaluronidase* (an enzyme that dissolves the hyaluronic acid in the connective tissue, where it serves to cement the cells together), and *deoxyribonucleases* (enzymes that degrade DNA). These streptococci also produce certain enzymes, called *streptolysins*, that lyse red blood cells and are toxic to neutrophils.

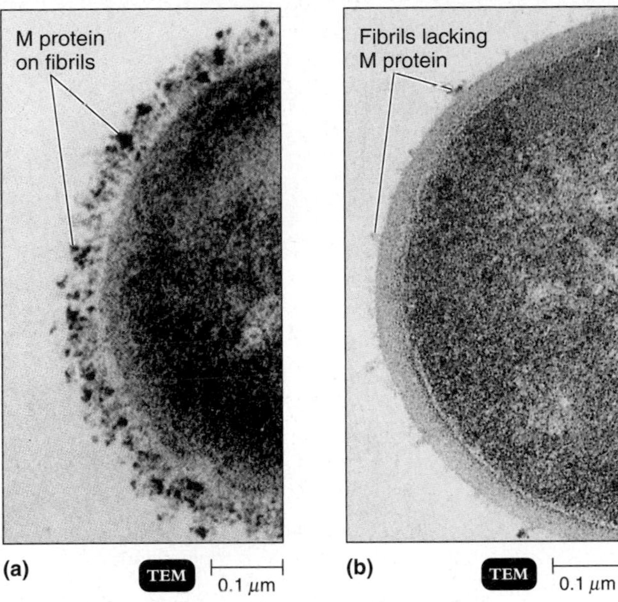

Figure 21.6 The M protein of group A beta-hemolytic streptococci. (**a**) Part of a cell that carries the M protein on a fuzzy layer of surface fibrils. (**b**) Part of a cell that lacks the M protein.

Q **Is the M protein more likely to be antigenic than a polysaccharide capsule?**

Streptococcal skin infections are generally localized, but if the bacteria reach deeper tissue, they can be highly destructive.

When *S. pyogenes* infects the dermal layer of the skin, it causes a serious disease, **erysipelas.** In this disease, the skin erupts into reddish patches with raised margins (Figure 21.7). It can progress to local tissue destruction and even enter the bloodstream, causing sepsis (page 639). The infection usually appears first on the face and often has been preceded by a streptococcal sore throat. High fever is common. Fortunately, *S. pyogenes* has remained sensitive to β-lactam-type antibiotics, especially cephalosporin.

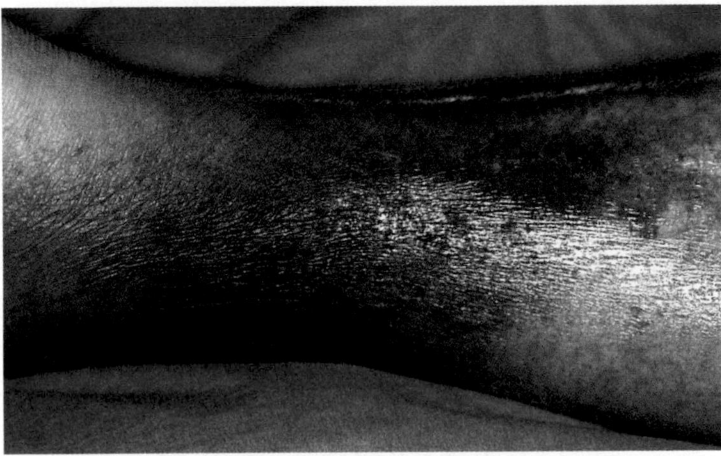

Figure 21.7 Lesions of erysipelas, caused by group A beta-hemolytic streptococcal toxins.

Q **What is the name of the toxin that produces skin reddening?**

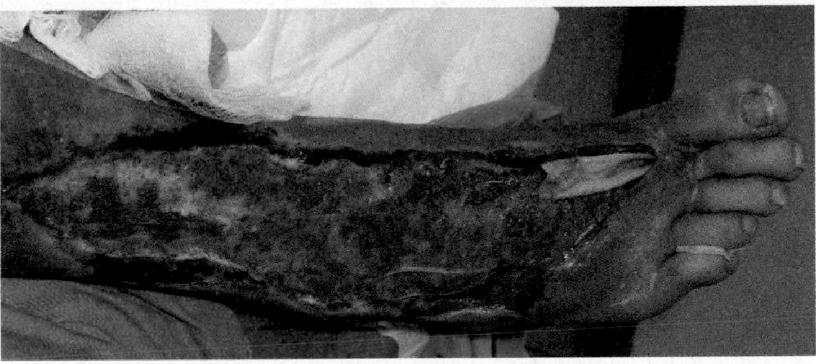

Figure 21.8 Necrotizing fasciitis due to group A streptococci. Extensive damage may require reconstructive surgery or even amputation of limbs.

Q **What is the name of the primary toxin that leads to tissue invasion by the pathogen?**

Some 15,000 cases of invasive group A streptococcal infection, caused by the "flesh-eating bacteria," occur each year in the United States. The infection may be precipitated by minor breaks in the skin, and early symptoms are often unrecognized, delaying diagnosis and treatment—with serious consequences. Once established, **necrotizing fasciitis** (Figure 21.8) may destroy tissue as rapidly as a surgeon can remove it, and mortality rates from systemic toxicity can exceed 40%. Streptococci are considered the most common causative organism, although other bacteria cause similar conditions. An important factor is an exotoxin produced by certain streptococcal M-protein types, *exotoxin A,* which acts as a superantigen, causing the immune system to contribute to the damage. Broad-spectrum antibiotics are usually prescribed because of the possibility that multiple bacterial pathogens are present.

Necrotizing fasciitis is often associated with **streptococcal toxic shock syndrome (streptococcal TSS),** which resembles staphylococcal TSS, described on page 588. In cases of streptococcal TSS, a rash is less likely to be present, but bacteremia is more likely to occur. M proteins shed from the surfaces of these streptococci form a complex with fibrinogen that binds to neutrophils. This causes the activation of the neutrophils, precipitating the release of damaging enzymes and consequent shock and organ damage.

Infections by Pseudomonads

Pseudomonads are aerobic gram-negative rods that are widespread in soil and water. Capable of surviving in any moist environment, they can grow on traces of unusual organic matter, such as soap films or cap liner adhesives, and are resistant to many antibiotics and disinfectants. The most prominent species is *Pseudomonas aeruginosa,* which is considered the model of an opportunistic pathogen.

Pseudomonads frequently cause outbreaks of ***Pseudomonas* dermatitis.** This is a self-limiting rash of about 2 weeks' duration, often associated with swimming pools and

DISEASES IN FOCUS 21.3

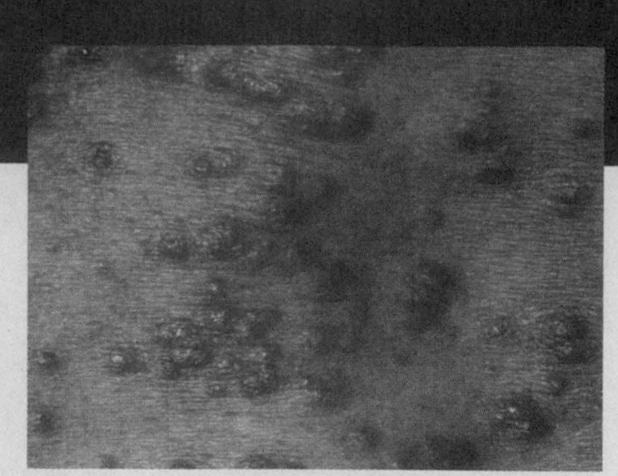

Patchy Redness and Pimple-Like Conditions

Differential diagnosis is the process of identifying a disease from a list of possible diseases that fit the information derived from examining a patient. A differential diagnosis is important for providing initial treatment and lab testing. For example, an 11-month-old boy came to a clinic with a 1-week history of an itchy red rash under his arms. He seemed more bothered at night and had no fever. Use the table below to identify infections that could cause these symptoms. For the solution, go to www.microbiologyplace.com.

Disease	Pathogen	Portal of Entry	Symptoms	Method of Transmission	Treatment
BACTERIAL DISEASES. Usually diagnosed by culturing the bacteria.					
Folliculitis	*Staphylococcus aureus*	Hair follicle	Infection of hair follicle	Direct contact; fomites; endogenous infection*	Draining of pus; topical antibiotics
Toxic shock syndrome	*Staphylococcus aureus*	Surgical incisions	Fever, rash, shock	Endogenous infection*	Antibiotics, depending on sensitivity profile
Necrotizing fasciitis	*Streptococcus pyogenes*	Skin abrasions	Extensive soft-tissue destruction	Direct contact	Surgical tissue removal; broad-spectrum antibiotics
Erysipelas	*Streptococcus pyogenes*	Skin; mucous membranes	Reddish patches on skin; often with high fever	Endogenous infection*	Cephalosporin
***Pseudomonas* dermatitis**	*Pseudomonas aeruginosa*	Skin abrasions	Superficial rash	Swimming water; hot tubs	Usually self-limiting
Otitis externa	*Pseudomonas aeruginosa*	Ear	Superficial infection of external ear canal	Swimming water	Fluoroquinolones
Acne	*Propionibacterium acnes*	Sebum channels	Inflammatory lesions originating with accumulations of sebum that rupture a hair follicle	Direct contact	Benzoyl peroxide, isotretinoin, azelaic acid
Buruli ulcer	*Mycobacterium ulcerans*	Skin	Localized swelling or hardness progressing to deep ulcer	Contaminated water	Antimycobacterial drugs
VIRAL DISEASES. Usually diagnosed by clinical signs and symptoms.					
Warts	*Papillomavirus*	Skin	A horny projection of the skin formed by proliferation of cells	Direct contact	May be removed by liquid nitrogen cryotherapy, electrodesiccation, acids, lasers
FUNGAL DISEASES. Diagnosis is confirmed by microscopic examination.					
Ringworm (tinea)	*Microsporum, Trichophyton, Epidermophyton* spp.	Skin	Skin lesions of highly varied appearance; on scalp may cause local loss of hair	Direct contact; fomites	Griseofulvin (orally); miconazole, clotrimazole (topically)
Sporotrichosis	*Sporothrix schenkii*	Skin abrasions	Ulcer at site of infection spreading into nearby lymphatic vessels	Soil	Potassium iodide solution (orally)
PARASITIC INFESTATIONS. Diagnosis is confirmed by microscopic examination of parasite.					
Scabies	*Sarcoptes scabiei* (mite)	Skin	Papules, itching	Direct contact	Gamma benzene hexachloride, permethrin (topically)
Pediculosis (lice)	*Pediculus humanus capitis*	Skin	Itching	Primarily direct contact; possible fomites such as bedding, combs	Topical insecticide preparations

*Endogenous infections are infections caused by microorganisms already part of the host microbiota.

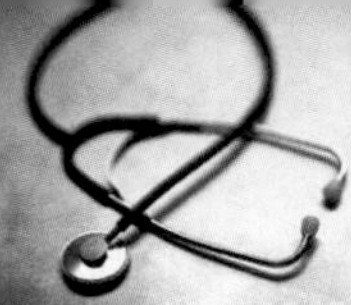

CLINICAL FOCUS From the *Morbidity and Mortality Weekly Report*

Infections in the Gym

As you read through this box you will encounter a series of questions that epidemiologists ask themselves as they try to trace on outbreak to its source. Try to answer each question before going on to the next one.

1. A 21-year-old college football player went to the college health center with an 11 cm × 5 cm area of redness on his right thigh. It was swollen and warm and tender when touched. His temperature was normal. He was given sulfamethoxazole-trimethoprim.
   What is your diagnosis?

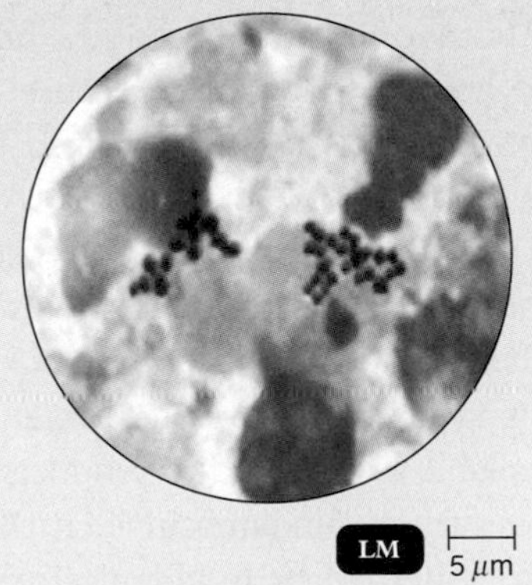

Figure A

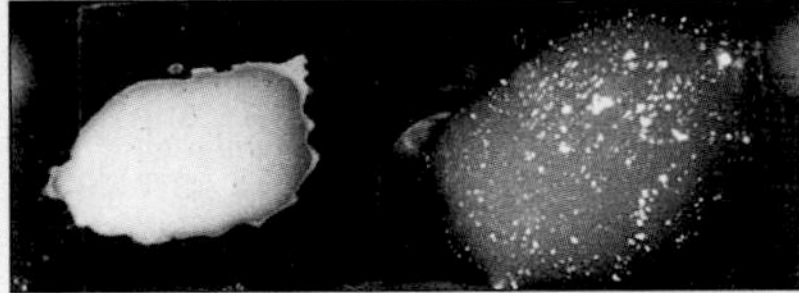

Negative control   Isolate from patient

Figure B

2. After 2 days, he returned saying the area was worse. Examination revealed a broader area of redness. He was diagnosed with cellulitis. The pustule was opened and drained.
   What do you need to know now?
3. Results of Gram stains of the pus are shown in **Figure A**. A coagulase test was also performed on a culture (**Figure B**).
   What is the cause of the infection?
4. The presence of gram-positive, coagulase-positive cocci indicate *Staphylococcus aureus*.
   What treatment is recommended?
5. The results of sensitivity testing are shown in **Figure C**. (P = penicillin, M = methicillin, E = erythromycin, V = vancomycin, X = trimethoprim-sulfamethoxazole.)
   What treatment is appropriate?
6. Over a 3-month period, 10 members of the college football and fencing teams reported to the health center with cellulitis. Seven were hospitalized; one received surgical debridement and skin grafts.
   What is the most likely source of the methicillin-resistant *Staphylococcus aureus* (MRSA)?

Although the investigations described in this report did not determine definitively the roots of MRSA transmission, three factors might have contributed to transmission in these outbreaks. First, abrasions and other skin trauma, which can facilitate entry of

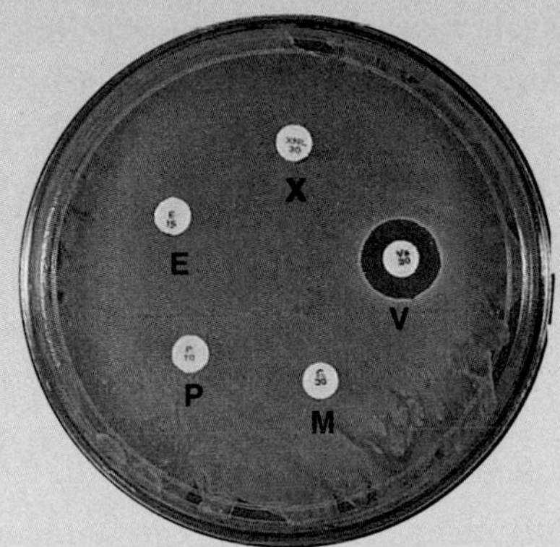

Figure C

pathogens, are likely in some sports. Second, some sports involve frequent physical contact among players. *S. aureus* and other skin microbiota can be transmitted easily from person to person with direct contact. Third, shared equipment or other personal items that are not cleaned or laundered between users could be a vehicle for *S. aureus* transmission.

Investigation of outbreaks of MRSA among professional football (2004) and baseball (2005) players showed that all of the infections occurred at the site of a turf burn and rapidly progressed to large abscesses that required surgery to drain. MRSA was recovered from whirlpools and taping gel and from 35 of the 84 nasal swabs from players and staff members.

Recurrence of infections might be avoided if physicians obtain cultures more routinely when athletes have infected wounds.

Source: Adapted from *MMWR* 52(33):793–795, August 22, 2003; *MMWR* 55(24):677–679, June 23, 2006.

pool-type saunas and hot tubs. When many people use these facilities, the alkalinity rises, and the chlorines become less effective; at the same time, the concentration of nutrients that support the growth of pseudomonads increases. Hot water causes hair follicles to open wider, facilitating the entry of bacteria. Competition swimmers are often troubled with **otitis externa,** or "*swimmer's ear,*" a painful infection of the external ear canal leading to the eardrum that is frequently caused by pseudomonads.

P. aeruginosa produces several exotoxins that account for much of its pathogenicity. It also has an endotoxin. *P. aeruginosa* often grows in dense biofilms (see Figure B, page 57) that contribute to its frequent identification as a cause of nosocomial infections of indwelling medical tubes or devices. This bacterium is also a serious opportunistic pathogen for patients with the genetic lung disease cystic fibrosis; biofilm formation plays a prominent part in this.

P. aeruginosa is also a very common and serious opportunistic pathogen in burn patients, particularly those with second- and third-degree burns. Infection may produce blue-green pus, whose color is caused by the bacterial pigment **pyocyanin.** Of concern in many hospitals is the ease with which *P. aeruginosa* grows in flower vases, mop water, and even dilute disinfectants.

The relative resistance to antibiotics that characterizes pseudomonads is still a problem. However, in recent years, several new antibiotics have been developed, and chemotherapy to treat these infections is not as restricted as it once was. The quinolones and the newer, antipseudomonal β-lactam antibiotics are the usual drugs of choice. Silver sulfadiazine is very useful in the treatment of burn infections by *P. aeruginosa.*

Buruli Ulcer

Buruli ulcer, named for a now-renamed region of Uganda in Africa, is a disease found primarily in western and central Africa. It has also appeared in localized tropical and temperate areas around the globe—including Mexico, Australia, and areas of South America. The disease is caused by *Mycobacterium ulcerans,* which is similar to the mycobacteria that cause tuberculosis and leprosy. When the pathogen is introduced into the skin, it causes a disease that progresses slowly with few serious early signs or symptoms. Eventually, however, the result is a deep ulcer that often becomes massive and seriously damaging. Untreated, this can be so extensive as to require amputation or plastic surgery. This tissue damage is attributed to the production of a toxin, *mycolactone.* Epidemiologically, the infection is associated with contact with swamps and slow-flowing waters.

The incidence of the disease has increased and now exceeds the incidence of leprosy and, in some areas, even tuberculosis. The World Health Organization recently identified it as a global threat to public health.

Buruli ulcer is diagnosed primarily by the appearance of the ulcer, although awareness is higher in endemic areas, and is treated by antimycobacterial drugs such as streptomycin-rifampicin combinations.

Acne

Acne is probably the most common skin disease in humans, affecting an estimated 17 million people in the United States. More than 85% of all teenagers have the problem to some degree. Acne can be classified by type of lesion into three categories: comedonal acne, inflammatory acne, and nodular cystic acne. They require different treatments.

Normally, skin cells that are shed inside the hair follicle are able to leave, but acne develops when cells are shed in higher than normal numbers, they combine with sebum, and the mixture clogs the follicle. As sebum accumulates, whiteheads (comedos) form; if the blockage protrudes through the skin, a blackhead (comedone or open comedo) forms. The dark color of blackheads is due not to dirt, but to lipid oxidation and other causes. Topical agents do not affect sebum formation, which is a root cause of acne and depends on hormones such as estrogens or androgens. Diet has no known effect on sebum production, but pregnancy, some hormone-based contraceptive methods, and hormonal changes with age do reduce sebum formation.

Comedonal (mild) acne is usually treated with topical agents such as azelaic acid (Azelex), salicyclic acid preparations, or retinoids (which are derivatives of vitamin A, such as tretinoin, tazarotene [Tazorac], or adapalene [Differin]). These topical agents do not affect sebum formation.

Inflammatory (moderate) acne arises from bacterial action, especially *Propionibacterium acnes,* an anaerobic diphtheroid commonly found on the skin. *P. acnes* has a nutritional requirement for glycerol in sebum; in metabolizing the sebum, it forms free fatty acids that cause an inflammatory response. Neutrophils that secrete enzymes that damage the wall of the hair follicle are attracted to the site. The resulting inflammation leads to the appearance of pustules and papules. At this stage, therapy is usually focused on preventing formation of sebum; topical agents are not effective for this. A drug that reduces sebum production is isotretinoin (Accutane). Unfortunately, this drug is *teratogenic,* meaning that it can cause serious damage to the developing fetus in a pregnant woman. About one-third of babies born after exposure will have tragic levels of damage. Because of these important side effects, isotretinoin is usually reserved for more severe forms of acne.

Inflammatory acne can also be treated by targeting *P. acnes* with antibiotics. The familiar nonprescription acne treatments containing benzoyl peroxide are effective against some bacteria, especially *P. acnes,* and also cause drying that helps loosen plugged follicles. Benzoyl peroxide is also available as a gel and in products where it is combined with antibiotics such as clindamycin (BenzaClin) and erythromycin (Benzamycin). Alternatives to chemical treatments have been approved by the U.S. Food and Drug Administration (FDA) for treatment of mild to moderate acne. The Clear Light system, which bathes the skin with high-intensity blue light (405–420 μm), and Smoothbeam treatment, which uses laser light, penetrate the skin surface to speed healing and prevent pimples from forming. Also approved recently is a handheld device, ThermaClear, which delivers a brief pulse of heat to the lesions.

Some patients with acne progress to **nodular cystic (severe) acne.** Nodular cystic acne is characterized by nodules or cysts, which are inflamed lesions filled with pus deep within the skin (**Figure 21.9**). These leave prominent scars on the face and upper body, which often leave psychological scars as well. The most important development in treating cystic acne is isotretinoin, which often results in dramatic improvement—but as we have stated, it must be used with extreme caution to avoid administration to a pregnant woman, even for a few days.

CHECK YOUR UNDERSTANDING

- ✓ Which bacterial species features the virulence factor M protein? **21-3**
- ✓ What is the common name for otitis externa? **21-4**

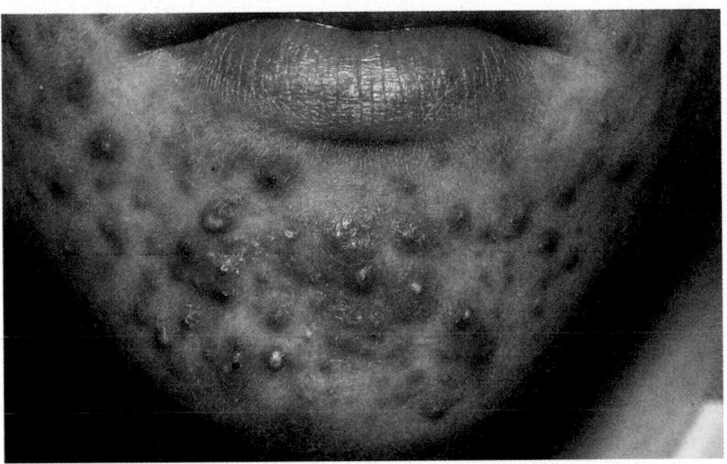

Figure 21.9 Severe acne.

Isotretinoin often leads to dramatic improvement for cases of severe acne, but what precautions must be observed?

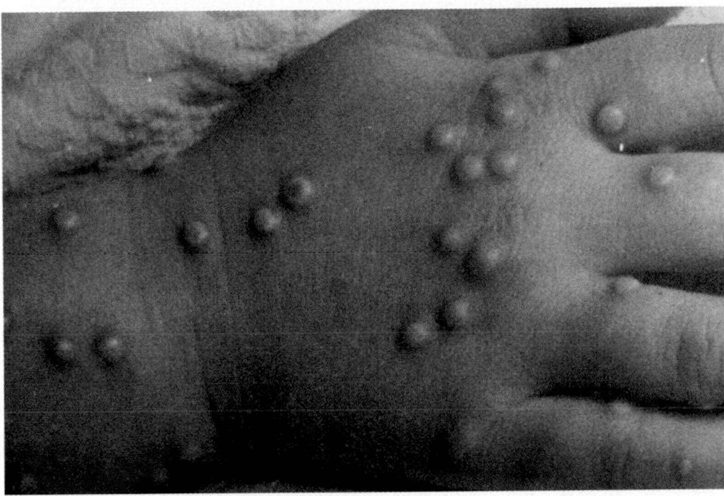

Figure 21.10 Smallpox lesions. In some severe cases, the lesions nearly run together (are confluent).

Q **How do these lesions differ from chickenpox?**

Viral Diseases of the Skin

Many viral diseases, although systemic and transmitted by respiratory or other routes, are most apparent by their effects on the skin.

Warts

Warts, or papillomas, are generally benign skin growths caused by viruses. It was long known that warts can be transmitted from one person to another by contact, even sexually, but it was not until 1949 that viruses were identified in wart tissues. More than 50 types of papillomavirus are now known to cause different kinds of warts, often with greatly varying appearances.

After infection, there is an incubation period of several weeks before the warts appear. The most common medical treatments for warts are to apply extremely cold liquid nitrogen (cryotherapy), dry them with an electrical current (electrodesiccation), or burn them with acids. There is evidence that compounds containing salicylic acids are especially effective. Topical application of prescription drugs such as podofilox or imiquimod (Aldara) is often effective; the latter stimulates production of antiviral interferons. Warts that do not respond to any other treatments can be treated with lasers or injected with bleomycin, an antitumor drug.

Although warts are not a form of cancer, some skin and cervical cancers are associated with papillomaviruses. The incidence of genital warts (see Chapter 26) has reached epidemic proportions.

Smallpox (Variola)

During the Middle Ages, an estimated 80% of the population of Europe contracted **smallpox*** at some time during their lives. Those who recovered from the disease retained disfiguring scars. The disease, introduced by American colonists, was even more devastating to American Indians, who had had no previous exposure and thus little resistance.

Smallpox is caused by an orthopoxvirus known as the smallpox (variola) virus. There are two basic forms of this disease: **variola major,** with a mortality rate of 20% or higher, and **variola minor,** with a mortality rate of less than 1%. (Variola minor first appeared around 1900.)

Transmitted by the respiratory route, the viruses infect many internal organs before they eventually move into the bloodstream, infecting the skin and producing more recognizable symptoms. The growth of the virus in the epidermal layers of the skin causes lesions that become pustular after 10 days or so (**Figure 21.10**).

Smallpox was the first disease to which immunity was artificially induced (see pages 11 and 501) and the first to be eradicated from the human population. The last victim of a natural case of smallpox is believed to be an individual who recovered from variola minor in 1977 in Somalia. (However, 10 months after this case, there was a smallpox fatality in England caused by escape of the virus from a hospital research laboratory.) The eradication of smallpox was possible because an effective vaccine was developed and because there are no animal host reservoirs for the disease. A concerted worldwide vaccination effort was coordinated by the World Health Organization.

Today, only two sites are known to maintain the smallpox virus, one in the United States and one in Russia. Dates for the destruction of these collections have been set and then postponed.

Smallpox would be an especially dangerous agent for bioterrorism. Vaccination in the United States ended in the early 1970s. People who were vaccinated prior to that time have waning immunity; however, they probably have some remaining

*The origin of the name *smallpox* reportedly arose in the late fifteenth century in France, where syphilis had just been introduced. These patients exhibited a severe skin rash called *la grosse verole,* or "the great pox." The rash was compared with that of an endemic disease of the time, which was then referred to as *la petite verole,* or "the small pox." In English, the endemic disease became known as smallpox.

protection that would at least moderate the disease. Stocks of smallpox vaccine are being accumulated as a precaution. No general vaccination program of the entire population is contemplated. However, certain groups, among them military and health care workers, may be an exception. Administered to the general population, the vaccine would cause a significant number of deaths, especially among immunosuppressed individuals.

A search is underway for antiviral drugs that would be effective. An intravenous drug, cidofovir, is one candidate being investigated. The lack of symptomatic smallpox patients prevents really meaningful trials; test tube efficacy does not necessarily transfer to human patients.

With the disappearance of smallpox, there has been some concern with a similar disease, **monkeypox.** This disease first appeared among zoo monkeys that originated in Africa and east Asia and is endemic there in small animals. There are occasional outbreaks among humans in those areas, and one outbreak of more than 50 cases in the United States in 2003 was attributed to contact with pet prairie dogs. These animals apparently were infected by being housed in pet stores with Gambian giant rats imported from western Africa. Monkeypox closely resembles smallpox in symptoms and, while smallpox was endemic, was probably mistaken for it. The mortality rate is typically 1–10% in African adults, highest in children. There were no deaths in the U.S. outbreak. The monkeypox virus, like smallpox virus, is an orthopoxvirus, and vaccination for smallpox has a protective effect. Monkeypox is known to jump from animals to humans, but fortunately its transmission from human to human has been very limited. The World Health Organization is monitoring recent outbreaks, especially to see whether human-to-human transmission increases.

Chickenpox (Varicella) and Shingles (Herpes Zoster)

Chickenpox (*varicella*) is a relatively mild childhood disease. The mortality rate from chickenpox is very low and is usually from complications such as encephalitis (infection of the brain) or pneumonia. Almost half of such deaths occur in adults.

Chickenpox (**Figure 21.11a**) is the result of an initial infection with the herpesvirus varicella-zoster. (The official, but less used, name is human herpesvirus 3; see Chapter 13. We will encounter other herpesviruses for which both the vernacular and the official names will be stated.) The disease is acquired when the virus enters the respiratory system, and the infection localizes in skin cells after about 2 weeks. The infected skin is vesicular for 3 to 4 days. During that time, the vesicles fill with pus, rupture, and form a scab before healing. Lesions are mostly confined to the face, throat, and lower back but can also occur on the chest and shoulders. If varicella infection occurs during early pregnancy, serious fetal damage may occur in about 2% of cases.

Reye syndrome is an occasional severe complication of chickenpox, influenza, and some other viral diseases. A few days after the initial infection has receded, the patient persistently vomits and exhibits signs of brain dysfunction, such as extreme drowsiness or combative behavior. Coma and death can follow. At one time, the death rate of reported cases approached 90%, but this rate has been declining with improved care and is now 30% or lower when the disease is recognized and treated in time. Survivors may show neurological damage, especially if very young. Reye syndrome affects children and teenagers almost exclusively. The use of aspirin to lower fevers in chickenpox and influenza increases the chances of acquiring Reye syndrome.

Like all herpesviruses, a characteristic of varicella-zoster virus is its ability to remain latent within the body. Following a primary infection, the virus enters the peripheral nerves and moves to a central nerve ganglion (a group of nerve cells lying outside the central nervous system), where it persists as viral DNA. Humoral antibodies cannot penetrate into the nerve cell, and because no viral antigens are expressed on the surface of the nerve cell, cytotoxic T cells are not activated. Therefore, neither arm of the specific immune system disturbs the latent virus.

Latent varicella-zoster virus is located in the dorsal root ganglion near the spine. Later, perhaps as long as decades later, the virus may be reactivated (**Figure 21.11b**). The trigger can be stress or simply the lower immune competence associated with aging. The virions produced by the reactivated DNA move along the peripheral nerves to the cutaneous sensory nerves of the skin, where they cause a new outbreak of the virus in the form of **shingles** (*herpes zoster*).

In shingles, vesicles similar to those of chickenpox occur but are localized in distinctive areas. Typically, they are distributed about the waist (the name *shingles* is derived from the Latin *cingulum* for girdle or belt), although facial shingles and infections of the upper chest and back also occur (see Figure 21.11b). The infection follows the distribution of the affected cutaneous sensory nerves and is usually limited to one side of the body at a time because these nerves are unilateral. Occasionally, such nerve infections can result in nerve damage that impairs vision or even causes paralysis. Severe burning or stinging pain is a frequent symptom; occasionally this persists for months or years, a condition called *postherpetic neuralgia.*

Shingles is simply a different expression of the virus that causes chickenpox: different because the patient, having had chickenpox, now has partial immunity to the virus. Exposing children to shingles has led to their contracting chickenpox. Shingles seldom occurs in people under age 20, and by far the highest incidence is among older adults. It is unusual for a patient to develop shingles more than once.

The antiviral drugs acyclovir, valacyclovir, and famciclovir are approved for treatment of shingles. For immunocompromised patients, in which a mortality rate of 17% is reported, and patients with ocular involvement, treatment with antivirals is mandatory.

A live, attenuated varicella vaccine was licensed in 1995. Since then, cases of the disease have declined steadily. There is evidence that the effectiveness of the vaccine, which is about 97% at outset,

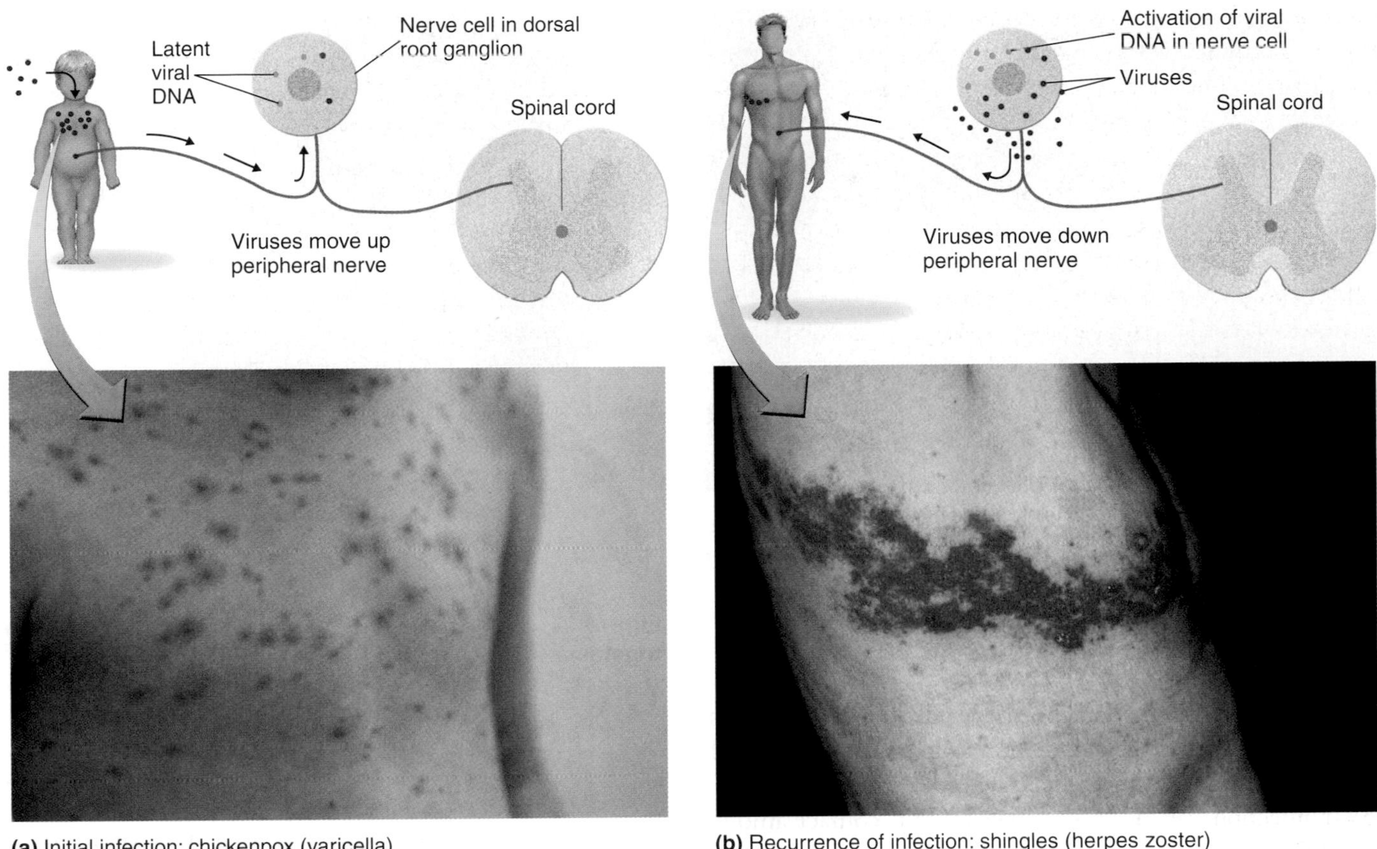

(a) Initial infection: chickenpox (varicella)

(b) Recurrence of infection: shingles (herpes zoster)

Figure 21.11 Chickenpox (varicella) and shingles (herpes zoster). **(a)** Initial infection with the virus, usually during childhood, causes chickenpox. The lesions are vesicles, eventually becoming pustules that rupture and form scabs. The virus then moves to a dorsal root ganglion near the spine, where it remains latent indefinitely. **(b)** Later, usually in late adulthood, the latent virus becomes reactivated, causing shingles. Reactivation can be caused by stress or weakening of the immune system. The skin lesions are vesicles.

Q Does the photo in (a) illustrate an early or late stage of chickenpox?

declines with time. The lack of a booster effect from exposure to new cases of varicella is a factor in this. Therefore, varicella in previously vaccinated persons, called **breakthrough varicella,** is becoming fairly common. Because the vaccine is at least partially effective, it is a relatively mild disease with a rash that does not look much like typical varicella. A booster dose of the vaccine may eventually be needed for complete control of varicella.

Another concern is that the waning effectiveness of the childhood vaccination will lead to a population of susceptible adults, for whom the disease tends to be more severe. Therefore, the current recommendation is that adults 60 years of age or older receive a newly approved *zoster vaccine* even if the subject has had chickenpox or shingles previously.

Herpes Simplex

Herpes simplex viruses (HSV) can be separated into two identifiable groups, HSV-1 and HSV-2. The name *herpes simplex virus,* used here, is the common or vernacular name. The official names are *human herpesvirus 1* and *2.* HSV-1 is transmitted primarily by oral or respiratory routes, and infection usually occurs in infancy. Serological surveys show that about 90% of the U.S. population has been infected. Frequently, this infection is subclinical, but many cases develop lesions known as **cold sores** or **fever blisters.** These are painful, short-lived vesicles that occur near the outer red margin of the lips (Figure 21.12).

Cold sores, caused by herpesvirus infections, are often confused with **canker sores.** The cause of canker sores is unknown, but their occurrence is often related to stress or menstruation. While similar to cold sores in appearance, canker sores usually appear in different areas. They occur as painful sores on movable mucous membranes, such as those on the tongue, cheeks, and inner surface of the lips. They ordinarily heal in a few days but often recur.

HSV-1 usually remains latent in the trigeminal nerve ganglia communicating between the face and the central nervous system

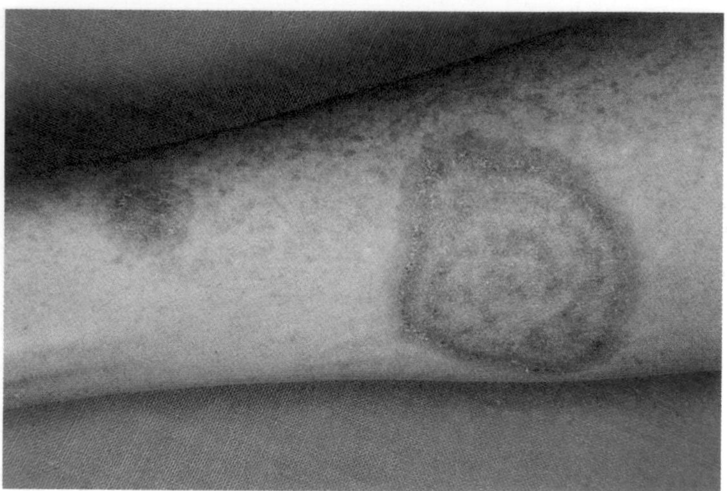

(a) Ringworm

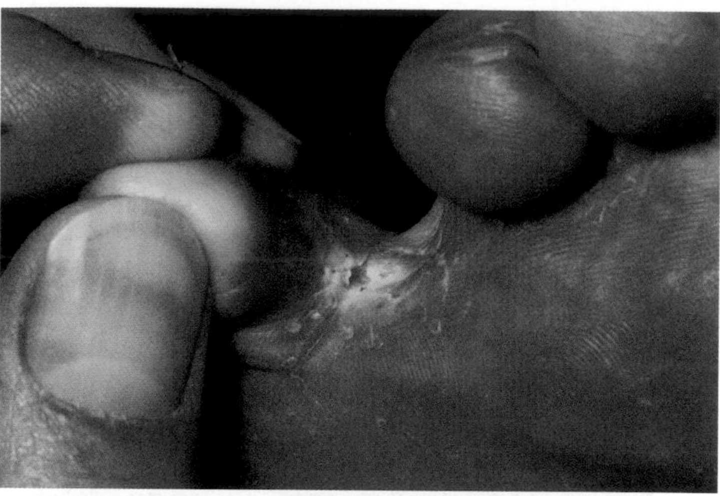

(b) Athlete's foot

Figure 21.16 Dermatomycoses.

 **Is ringworm caused by a helminth?**

occurred. Individuals with an impaired immune system should not receive live vaccine of *any* disease.

Other Viral Rashes

Fifth Disease (Erythema Infectiosum) Parents with young children are often baffled by a diagnosis of fifth disease, which they have never heard of before. The name derives from a 1905 list of skin rash diseases: measles, scarlet fever, rubella, Filatov Dukes' disease (a mild form of scarlet fever), and the fifth disease on the list. This **fifth disease,** or **erythema infectiosum,** produces no symptoms at all in about 20% of individuals infected by the virus (human parvovirus B19, first identified in 1989). Symptoms are similar to a mild case of influenza, but there is a distinctive "slapped-cheek" facial rash that slowly fades. In adults who missed an immunizing infection in childhood, the disease may cause anemia, an episode of arthritis, or, rarely, miscarriage.

Roseola **Roseola** is a mild childhood disease that is very common. The child has a high fever for a few days, which is followed by a rash over much of the body lasting for a day or two. Recovery leads to immunity. The pathogens are human herpesviruses 6 (HHV-6) and 7 (HHV-7)—the latter is responsible for 5–10% of roseola cases. Both viruses are present in the saliva of most adults.

CHECK YOUR UNDERSTANDING

 How did the odd designation of "fifth disease" arise? **21-5**

Fungal Diseases of the Skin and Nails

The skin is most susceptible to microorganisms that can resist high osmotic pressure and low moisture. It is not surprising, therefore, that fungi cause a number of skin disorders. Any fungal infection of the body is called a **mycosis.**

Cutaneous Mycoses

Fungi that colonize the hair, nails, and the outer layer (stratum corneum) of the epidermis (see Figure 21.1) are called **dermatophytes;** they grow on the keratin present in those locations. Termed **dermatomycoses,** these fungal infections are more informally known as *tineas* or *ringworm.* **Tinea capitis,** or ringworm of the scalp, is fairly common among elementary school children and can result in bald patches. This characteristic led the Romans to adopt the name *tinea,* Latin for clothes moth, because the infection resembles the holes left by the wormlike larvae of the moth in wool clothing. The infections tend to expand circularly, hence the term *ringworm* (Figure 21.16a). The infection is usually transmitted by contact with fomites. Dogs and cats are also frequently infected with fungi that cause ringworm in children. Ringworm of the groin, or jock itch, is known as **tinea cruris,** and ringworm of the feet, or athlete's foot, is known as **tinea pedis** (Figure 21.16b). The moisture in such areas favors fungal infections.

Three genera of fungi are involved in cutaneous mycosis. *Trichophyton* (trik-ō-fī′ton) can infect hair, skin, or nails; *Microsporum* (mī-krō-spô′rum) usually involves only the hair or skin; *Epidermophyton* (ep-i-dėr-mō-fī′ton) affects only the skin and nails. The topical drugs available without prescription for tinea infections include miconazole and clotrimazole. Athlete's foot is often difficult to cure. Topical allylamine preparations containing terbinafine or naftifine, as well as another allylamine, butenavine, are recommended and are now available without a prescription. Extended application is usually required. When hair is involved, topical treatment is not very effective. An oral antibiotic, griseofulvin, is often useful in such infections because it can localize in keratinized tissue, such as described on page 569. When nails are infected, called **tinea unguium** or *onychomycosis,* oral itraconazole and terbinafine are the drugs of

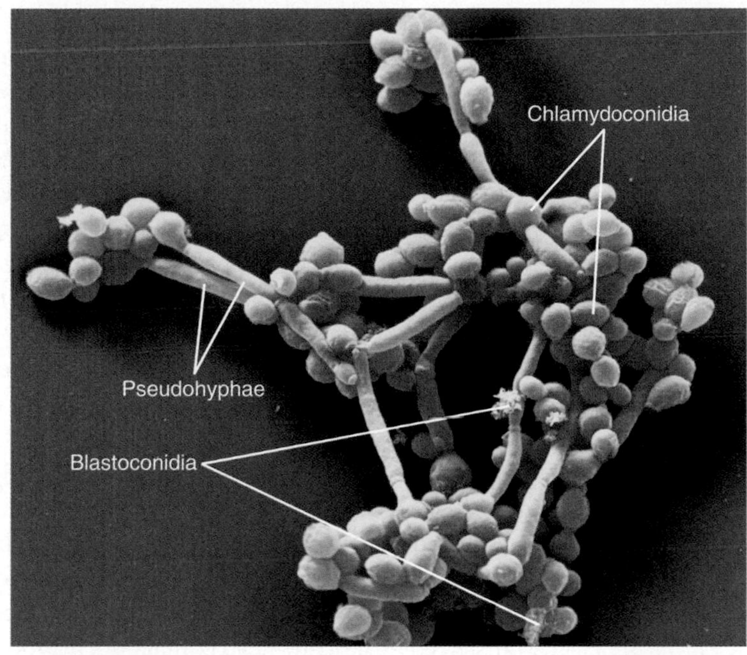

(a) *Candida albicans*

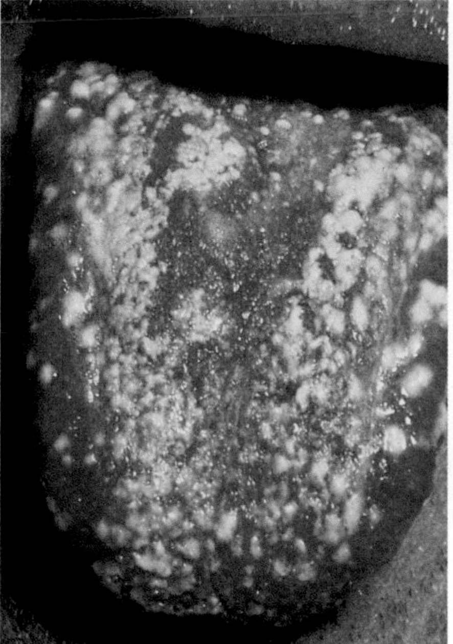

(b) Oral candidiasis, or thrush

Figure 21.17 Candidiasis. (**a**) *Candida albicans.* Notice the spherical chlamydoconidia (resting bodies formed from hyphal cells) and the smaller blastoconidia (asexual spores produced by budding) (see Chapter 12). (**b**) This case of oral candidiasis, or thrush, produced a thick, creamy coating on the tongue.

Q How can antibacterial drugs lead to candidiasis?

choice, but treatment may require weeks and both must be used with caution because of potential severe side effects.

Cutaneous mycosis is usually diagnosed by potassium hydroxide (KOH) microscopic examination of scrapings of the affected areas. The KOH dissolves epithelial tissue, allowing a clear view of fungal hyphae.

Subcutaneous Mycoses

Subcutaneous mycoses are more serious than cutaneous mycoses. Even when the skin is broken, cutaneous fungi do not seem to be able to penetrate past the stratum corneum, perhaps because they cannot obtain sufficient iron for growth in the epidermis and the dermis. Usually subcutaneous mycoses are caused by fungi that inhabit the soil, especially decaying vegetation, and penetrate the skin through a small wound that allows entry into subcutaneous tissues.

In the United States, the most common disease of this type is **sporotrichosis,** caused by the dimorphic fungus *Sporothrix schenkii* (spô-rō′thriks shen′kē-ē). Most cases occur among gardeners or other people working with soil. The infection frequently forms a small ulcer on the hands. The fungus often enters the lymphatic system in the area and there forms similar lesions. The condition is seldom fatal and is effectively treated by ingesting a dilute solution of potassium iodide, even though the organism is not affected in vitro by even a 10% solution of potassium iodide.

Candidiasis

Q&A The bacterial microbiota of the mucous membranes in the genitourinary tract and mouth usually suppress the growth of such fungi as *Candida albicans.* Several other species of *Candida,* for example *C. tropicalis* or *C. krusei* (krūs′ā-ē), may also be involved. The morphology of these organisms is not always yeastlike but can exhibit the formation of pseudohyphae, long cells that resemble hyphae. In this form, *Candida* is resistant to phagocytosis, which may be a factor in its pathogenicity (Figure 21.17a). Because the fungus is not affected by antibacterial drugs, it sometimes overgrows mucosal tissue when antibiotics suppress the normal bacterial microbiota. Changes in the normal mucosal pH may have a similar effect. Such overgrowths by *C. albicans* are called **candidiasis.** Newborn infants, whose normal microbiota have not become established, often suffer from a whitish overgrowth of the oral cavity, called **thrush** (Figure 21.17b). *C. albicans* is also a very common cause of vaginitis (see Chapter 26).

Immunosuppressed individuals, including AIDS patients, are unusually prone to *Candida* infections of the skin and mucous membranes. On people who are obese or diabetic, the areas of the skin with more moisture tend to become infected with this fungus. The infected areas become bright red, with lesions on the borders. Skin and mucosal infections by *C. albicans* are usually

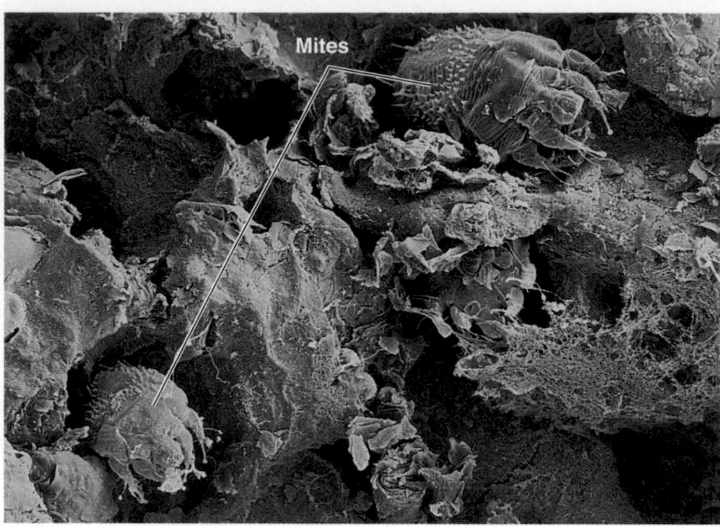

Figure 21.18 Scabies mites in skin.

Q Would it have required a microscope to identify this pathogen?

treated with topical applications of miconazole, clotrimazole, or nystatin. If candidiasis becomes systemic, as can happen in immunosuppressed individuals, *fulminating disease* (one that appears suddenly and severely) and death can result. The usual drug of choice to treat systemic candidiasis is fluconazole. Several new treatments are now also available; for example, some of the new echinocandin class antifungals, such as micafungin and anidulafungin, are now approved for this use.

CHECK YOUR UNDERSTANDING

✓ How do sporotrichosis and athlete's foot differ? In what ways are they similar? **21-6**

✓ How might the use of penicillin result in a case of candidiasis? **21-7**

Parasitic Infestation of the Skin

Parasitic organisms such as some protozoa, helminths, and microscopic arthropods can infest the skin and cause disease conditions. We will describe two examples of common arthropod infestation, scabies and lice.

Scabies

Probably the first documented connection between a microscopic organism (330–450 μm) and a disease in humans was **scabies,** which was described by an Italian physician in 1687. The disease involves intense local itching and is caused by the tiny mite *Sarcoptes scabiei* burrowing under the skin to lay its eggs (**Figure 21.18**). The burrows are often visible as slightly elevated, serpentine lines about 1 mm in width. However, scabies may appear as a variety of inflammatory skin lesions, many of them secondary infections from scratching. The mite is transmitted by intimate contact, including sexual contact, and is most often seen in family members, nursing home residents, and teenagers infected by children for whom they baby-sit.

About 500,000 people seek treatment for scabies in the United States each year; in developing countries, it is even more prevalent. The mite lives about 25 days, but by that time eggs have hatched and produced a dozen or so progeny. Scabies is usually diagnosed by microscopic examination of skin scrapings and usually is treated by topical application of permethrin. Difficult cases are sometimes treated with oral ivermectin.

Pediculosis (Lice)

Infestations by lice, called **pediculosis,** have afflicted humans for thousands of years. Although usually associated in the public mind with poor sanitation, outbreaks of head lice among middle- and upper-class schoolchildren in the United States are common. Parents are usually appalled, but head lice are fairly easily transferred by head-to-head contact, such as occurs among children who know each other well. The head louse, *Pediculus humanus capitis*, is not the same as the body louse, *Pediculus humanus corporis*. These are subspecies of *Pediculus humanus* that have adapted to different areas of the body. Only the body louse spreads diseases, such as epidemic typhus.

Lice (see Figure 12.32a, page 363) require blood from the host and feed several times a day. The victim is often unaware of these silent passengers until itching, which is a result of sensitization to louse saliva, develops several weeks later. Scratching can result in secondary bacterial infections. The head louse has legs especially adapted to grasp scalp hairs (**Figure 21.19a**). During a life span of a little over a month, the female louse produces several eggs (nits) a day. The eggs are attached to hair shafts close to the scalp (**Figure 21.19b**) to benefit from a warmer incubation temperature, and they hatch in about a week. The very young stages of the louse are also called nits. Empty egg cases are whitish and more visible. They do not necessarily indicate the presence of live lice. As the hair grows (at the rate of about 1 cm a month), the attached nit moves away from the scalp.

A point of interest is that the incidence of pediculosis among blacks in the United States is low: in the United States, lice have become adapted to the cylindrical hair shafts found on whites. In Africa, lice have adapted to the noncylindrical hair shafts of blacks.

Treatments of head lice abound, recalling the medical adage that if there are many treatments for a condition it is probably because none of them are really good. Nonprescription medications such as Nix (permethrin insecticide) and Rid (pyrethrin insecticide) are usually the first choice, but resistance has become common. Other topical preparations containing insecticides such as malathion (Ovide) and the more toxic lindane are also available (lindane is banned in California). A single-dose treatment with orally administered ivermectin is occasionally used. A silicone-based product, LiceMD, is effective and nontoxic. The active principle, *dimethicone*, blocks the breathing tubes of the louse. Combing out the nits with fine-toothed louse combs is another treatment option. This is a difficult, time-consuming procedure that has actually led to the appearance of professional removal services in some cities: expensive, but often worth the price to busy mothers.

CHECK YOUR UNDERSTANDING

✓ What diseases, if any, are spread by head lice, such as *Pediculus humanus capitis*? **21-8**

Microbial Diseases of the Eye

LEARNING OBJECTIVES

21-9 Define *conjunctivitis*.

21-10 List the causative agent, mode of transmission, and clinical symptoms of these eye infections: ophthalmia neonatorum, inclusion conjunctivitis, trachoma.

21-11 List the causative agent, mode of transmission, and clinical symptoms of these eye infections: herpetic keratitis, *Acanthamoeba* keratitis.

The epithelial cells covering the eye can be considered a continuation of the skin or mucosa. Many microbes can infect the eye, largely through the *conjunctiva,* the mucous membrane that lines the eyelids and covers the outer white surface of the eyeball. It is a transparent layer of living cells replacing the skin. Diseases of the eye are summarized in Diseases in Focus 21.4.

Inflammation of the Eye Membranes: Conjunctivitis

Conjunctivitis is an inflammation of the conjunctiva, often called by the common name **red eye,** or **pinkeye.** *Haemophilus influenzae* is the most common bacterial cause; viral conjunctivitis is usually caused by adenoviruses. However, a broad group of bacterial and viral pathogens as well as allergies can cause this condition.

The popularity of contact lenses has been accompanied by an increased incidence of infections of the eye. This is especially true of the soft-lens varieties, which are often worn for extended periods. Among the bacterial pathogens that cause conjunctivitis are pseudomonads, which can cause serious eye damage. To prevent infection, contact lens wearers should not use homemade saline solutions, which are a frequent source of infection, and should scrupulously follow the manufacturer's recommendations for cleaning and disinfecting the lenses. The most effective methods for disinfecting contact lenses involve applying heat; lenses that cannot be heated can be disinfected with hydrogen peroxide, which is then neutralized.

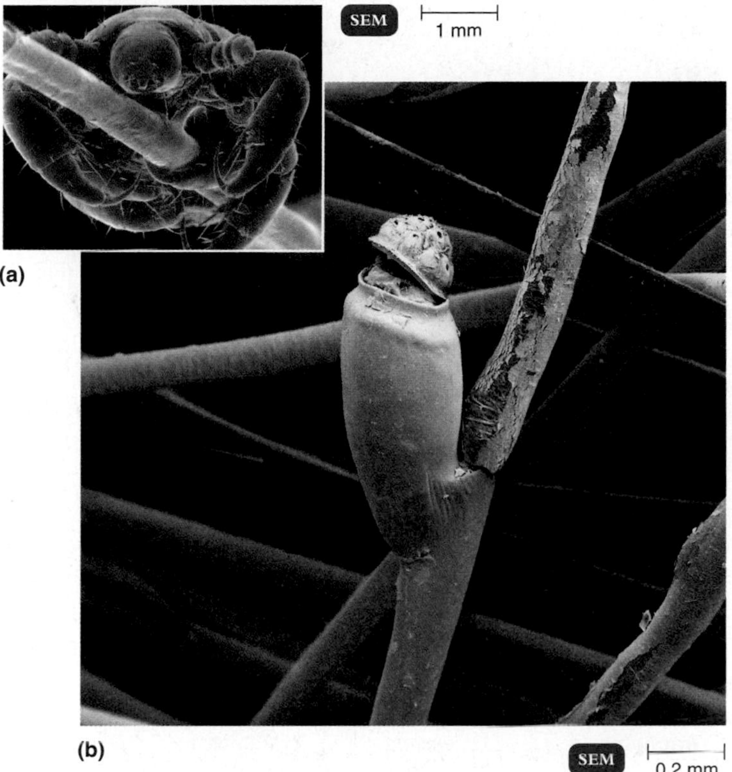

Figure 21.19 Louse and louse egg case. (**a**) Adult louse grasping hair. (**b**) This egg case (nit) contains the nymphal stage of the louse, which is in the process of exiting through the cap (operculum). It does this by gulping air and forcing it out the anus until it pops free, much like a champagne cork.

Q How is pediculosis transmitted?

Bacterial Diseases of the Eye

The bacterial microorganisms most commonly associated with the eye usually originate from the skin and upper respiratory tract.

Ophthalmia Neonatorum

Ophthalmia neonatorum is a serious form of conjunctivitis caused by *Neisseria gonorrhoeae* (the cause of gonorrhea). Large amounts of pus are formed; if treatment is delayed, ulceration of the cornea will usually result. The disease is acquired as the infant passes through the birth canal, and infection carries a high risk of blindness. Early in the twentieth century, legislation required that the eyes of all newborn infants be treated with a 1% solution of silver nitrate, which proved to be a very effective treatment in preventing this eye infection. Between 1906 and 1959, the percentage of admissions to schools for the blind that could be attributed to ophthalmia neonatorum declined from 24% to only 0.3%. Silver nitrate has been almost entirely replaced by antibiotics because of frequent coinfections by gonococci and sexually transmitted chlamydias, and silver nitrate is not effective

DISEASES IN FOCUS 21.4

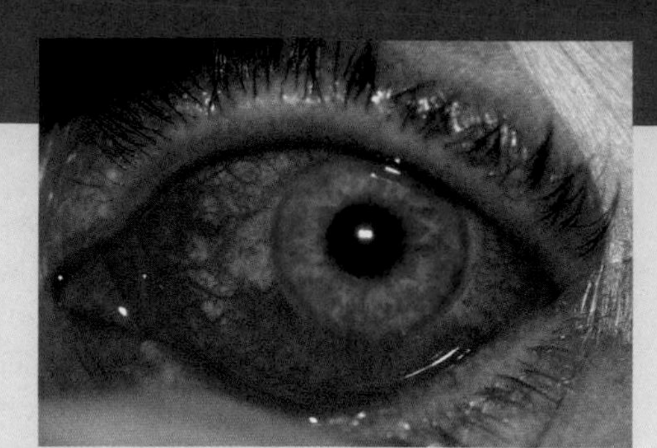

Microbial Diseases of the Eye

Differential diagnosis is the process of identifying the disease from a list of possible diseases that fit the information derived from examining a patient. A differential diagnosis is important for providing initial treatment and for laboratory testing. For example, a 20-year-old man had eye redness with dried mucus crust in the morning. The condition resolved with topical antibiotic treatment. Use the table below to identify infections that could cause these symptoms. For the solution, go to www.microbiologyplace.com.

Disease	Pathogen	Portal of Entry	Symptoms	Method of Transmission	Treatment
BACTERIAL DISEASES					
Conjunctivitis	*Haemophilus influenzae*	Conjunctiva	Redness	Direct contact; fomites	None
Ophthalmia neonatorum	*Neisseria gonorrhoeae*	Conjunctiva	Acute infection with much pus formation	Through birth canal	Prevention: tetracycline, erythromycin, or providone-iodine
Inclusion conjunctivitis	*Chlamydia trachomatis*	Conjunctiva	Swelling of eyelid; mucus and pus formation	Through birth canal; swimming pools	Tetracycline
Trachoma	*Chlamydia trachomatis*	Conjunctiva	Conjunctivitis	Direct contact; fomites; flies	Azithromycin
VIRAL DISEASES					
Conjunctivitis	Adenoviruses	Conjunctiva	Redness	Direct contact	None
Herpetic keratitis	Herpes simplex type 1 virus	Conjunctiva; cornea	Keratitis	Direct contact; recurring latent infection	Trifluridine may be effective
PROTOZOAN DISEASE					
***Acanthamoeba* keratitis**	*Acanthamoeba* spp.	Corneal abrasion; soft contact lenses may prevent removal of amoeba by blinking	Keratitis	Contact with fresh water	Topical propamidine isethionate or miconazole; corneal transplant or eye removal surgery may be required

against chlamydias. In parts of the world where the cost of antibiotics is prohibitive, a dilute solution of povidone-iodine has proven effective.

Inclusion Conjunctivitis

Chlamydial conjunctivitis, or **inclusion conjunctivitis,** is quite common today. It is caused by *Chlamydia trachomatis,* a bacterium that grows only as an obligate intracellular parasite. In infants, who acquire it in the birth canal, the condition tends to resolve spontaneously in a few weeks or months, but in rare cases it can lead to scarring of the cornea. Chlamydial conjunctivitis also appears to spread in the unchlorinated waters of swimming pools; in this context, it is called *swimming pool conjunctivitis.* Tetracycline applied as an ophthalmic ointment is an effective treatment.

Trachoma

A serious eye infection, and probably the greatest single cause of blindness by an infectious disease, is **trachoma**—an ancient name derived from the Greek word for rough. It is caused by certain serotypes of *Chlamydia trachomatis* but not the same ones that cause genital infections (see pages 750, 752 and 755). In the arid parts of Africa and Asia, almost all children are infected early in their lives. Worldwide, there are probably 500 million active cases and 7 million blinded victims. Trachoma also occurs occasionally in the southwestern United States, especially among American Indians.

The disease is a conjunctivitis transmitted largely by hand contact or by sharing such personal objects as towels. Flies may also carry the bacteria. Repeated infections cause inflammation (**Figure 21.20a**), leading to *trichiasis,* an in-turning of the eyelashes (**Figure 21.20b**). Abrasion of the cornea, especially by the eyelashes, eventually causes scarring of the cornea and blindness. Trichiasis can be corrected surgically, a procedure shown in ancient Egyptian papyri. Secondary infections by other bacterial pathogens are also a factor in the disease. Antibiotics to eliminate chlamydia, especially oral azithromycin, are useful in treatment. The disease can be controlled through sanitary practices and health education.

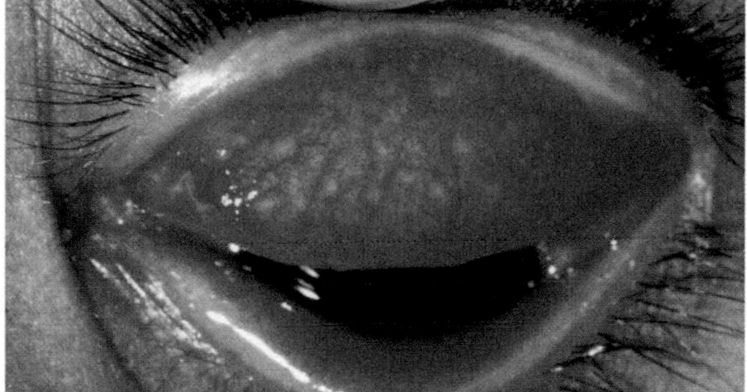

(a) Chronic inflammation of the eyelid

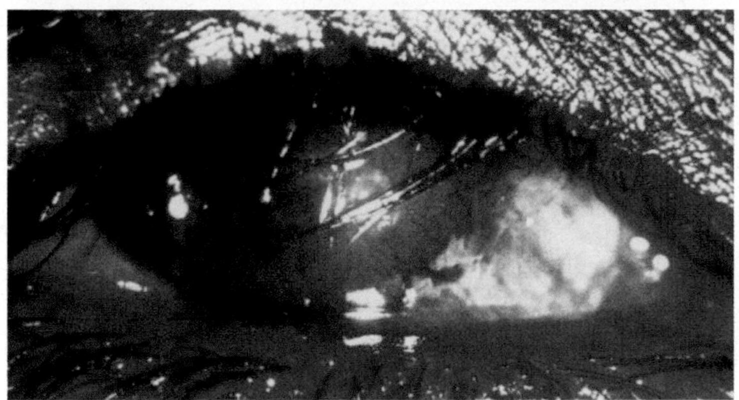

(b) Trichiasis, inturned eyelids, abrading the cornea

Figure 21.20 Trachoma **(a)** Repeated infection with *Chlamydia trachoma* causes chronic inflammation. The eyelid has been pulled back to show the inflammatory nodules that are in contact with the cornea. The abrasion caused by this damages the cornea and makes it susceptible to secondary infections. **(b)** In later stages of trachoma, the eyelashes turn inward (trichiasis) as shown here, further abrading the cornea.

Q How is trachoma transmitted?

CHECK YOUR UNDERSTANDING

- ✓ What is the common name of inclusion conjunctivitis? **21-9**
- ✓ Why have antibiotics almost entirely replaced the less expensive use of silver nitrate for preventing ophthalmia neonatorum? **21-10**

Other Infectious Diseases of the Eye

Microorganisms such as viruses and protozoa can also cause eye diseases. The diseases discussed here are characterized by inflammation of the cornea, which is called *keratitis.* In the United States, keratitis is mostly bacterial in origin; in Africa and Asia, eye infections are mostly caused by fungi, such as *Fusarium* and *Aspergillus.*

Herpetic Keratitis

Herpetic keratitis is caused by the same herpes simplex type 1 virus that causes cold sores and is latent in the trigeminal nerves (see Figure 21.13). The disease is an infection of the cornea, often resulting in deep ulcers, that may be the most common cause of infectious blindness in the United States. The drug trifluridine is often an effective treatment.

Acanthamoeba Keratitis

The first case of ***Acanthamoeba*** (a-kan-thä-mē′bä) **keratitis** was reported in 1973 in a Texas rancher. Since then, well over 4000 cases have been diagnosed in the United States. This amoeba has been found in fresh water, tap water, hot tubs, and soil. Most recent cases have been associated with the wearing of contact lenses, although any cornea damaged by trauma or infection is susceptible. Contributing factors are inadequate, unsanitary, or faulty disinfecting procedures (only heat will reliably kill the cysts), homemade saline solutions, and wearing the contact lenses overnight or while swimming.

In its early stages, the infection consists of only a mild inflammation, but later stages are often accompanied by severe pain. If started early, treatment with propamidine isethionate eye drops and topical neomycin has been successful. Damage is often so severe as to require a corneal transplant or even removal of the eye. Diagnosis is confirmed by the presence of trophozoites and cysts in stained scrapings of the cornea.

CHECK YOUR UNDERSTANDING

- ✓ Of the two eye diseases herpetic keratitis and *Acanthamoeba* keratitis, which is the more likely to be caused by an organism actively reproducing in saline solutions for contact lenses? **21-11**

STUDY OUTLINE

The **MyMicrobiologyPlace** wesbite (**www.microbiologyplace.com**) will help you get ready for tests with its simple three-step approach: ❶ **take a pre-test** and obtain a personalized study plan, ❷ **learn and practice** with animations, tutorials, and MP3 tutor sessions, and ❸ **test yourself** with quizzes and a chapter post-test.

Introduction (p. 584)

1. The skin is a physical barrier against microorganisms.
2. Moist areas of the skin (such as the armpit) support larger populations of bacteria than dry areas (such as the scalp).
3. Human skin produces antibiotics called defensins.

Structure and Function of the Skin (p. 585)

1. The outer portion of the skin (epidermis) contains keratin, a waterproof coating.
2. The inner portion of the skin, the dermis, contains hair follicles, sweat ducts, and oil glands that provide passageways for microorganisms.
3. Sebum and perspiration are secretions of the skin that can inhibit the growth of microorganisms.
4. Sebum and perspiration provide nutrients for some microorganisms.
5. Body cavities are lined with epithelial cells. When these cells secrete mucus, they constitute the mucous membrane.

Normal Microbiota of the Skin (pp. 585–586)

1. Microorganisms that live on skin are resistant to desiccation and high concentrations of salt.
2. Gram-positive cocci predominate on the skin.
3. The normal skin microbiota are not completely removed by washing.
4. Members of the genus *Propionibacterium* metabolize oil from the oil glands and colonize hair follicles.
5. *Malassezia furfur* yeast grows on oily secretions and may be the cause of dandruff.

Microbial Diseases of the Skin (pp. 586–603)

1. Vesicles are small fluid-filled lesions; bullae are vesicles larger than 1 cm; macules are flat, reddened lesions; papules are raised lesions; and pustules are raised lesions containing pus.

Bacterial Diseases of the Skin (pp. 586–598)

Staphylococcal Skin Infections (pp. 586–589)

2. Staphylococci are gram-positive bacteria that often grow in clusters.
3. The majority of skin microbiota consist of coagulase-negative *Staphylococcus epidermidis*.
4. Almost all pathogenic strains of *S. aureus* produce coagulase.
5. Pathogenic *S. aureus* can produce enterotoxins, leukocidins, and exfoliative toxin.

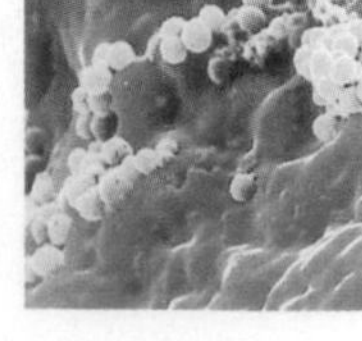

6. Localized infections (sties, pimples, and carbuncles) result from *S. aureus* entering openings in the skin.
7. Impetigo is a highly contagious superficial skin infection caused by *S. aureus*.
8. Toxemia occurs when toxins enter the bloodstream; staphylococcal toxemias include scalded skin syndrome and toxic shock syndrome.

Streptococcal Skin Infections (pp. 589–591)

9. Streptococci are gram-positive cocci that often grow in chains.
10. Streptococci are classified according to their hemolytic enzymes and cell wall antigens.
11. Group A beta-hemolytic streptococci (including *Streptococcus pyogenes*) are the pathogens most important to humans.
12. Group A beta-hemolytic streptococci produce a number of virulence factors: M protein, erythrogenic toxin, deoxyribonuclease, streptokinases, and hyaluronidase.
13. Erysipelas is caused by *S. pyogenes.*
14. Invasive group A beta-hemolytic streptococci cause severe and rapid tissue destruction.

Infections by Pseudomonads (pp. 591–594)

15. Pseudomonads are gram-negative rods. They are aerobes found primarily in soil and water that are resistant to many disinfectants and antibiotics.
16. *Pseudomonas aeruginosa* produces an endotoxin and several exotoxins.
17. Diseases caused by *P. aeruginosa* include otitis externa, respiratory infections, burn infections, and dermatitis.
18. Infections have a characteristic blue-green pus caused by the pigment pyocyanin.
19. Quinolones are useful in treating *P. aeruginosa* infections.

Buruli Ulcer (p. 594)

20. *Mycobacterium ulcerans* causes deep-tissue ulceration.

Acne (p. 594)

21. *Propionibacterium acnes* can metabolize sebum trapped in hair follicles.
22. Metabolic end-products (fatty acids) cause inflammatory acne.
23. Tretinoin, benzoyl peroxide, erythromycin, and light therapy are used to treat acne.

Viral Diseases of the Skin (pp. 595–600)

Warts (p. 595)

24. Papillomaviruses cause skin cells to proliferate and produce a benign growth called a wart or papilloma.
25. Warts are spread by direct contact.
26. Warts may regress spontaneously or be removed chemically or physically.

Smallpox (Variola) (pp. 595–596)

27. Variola virus causes two types of skin infections: variola major and variola minor.
28. Smallpox is transmitted by the respiratory route, and the virus is moved to the skin via the bloodstream.

29. The only host for smallpox is humans.
30. Smallpox has been eradicated as a result of a vaccination effort by the World Health Organization.

Chickenpox (Varicella) and Shingles (Herpes Zoster) (pp. 596–597)

31. Varicella-zoster virus is transmitted by the respiratory route and is localized in skin cells, causing a vesicular rash.
32. Complications of chickenpox include encephalitis and Reye syndrome.
33. After chickenpox, the virus can remain latent in nerve cells and subsequently activate as shingles.
34. Shingles (herpes zoster) is characterized by a vesicular rash along the affected cutaneous sensory nerves.
35. The virus can be treated with acyclovir. An attenuated live vaccine is available.

Herpes Simplex (pp. 597–598)

36. Herpes simplex infection of mucosal cells results in cold sores and occasionally encephalitis.
37. The virus remains latent in nerve cells, and cold sores can recur when the virus is activated.
38. HSV-1 is transmitted primarily by oral and respiratory routes.
39. Herpes encephalitis occurs when herpes simplex viruses infect the brain.
40. Acyclovir has proven successful in treating herpes encephalitis.

Measles (Rubeola) (pp. 598–599)

41. Measles is caused by measles virus and is transmitted by the respiratory route.
42. Vaccination provides effective long-term immunity.
43. After the virus has incubated in the upper respiratory tract, macular lesions appear on the skin, and Koplik's spots appear on the oral mucosa.
44. Complications of measles include middle ear infections, pneumonia, encephalitis, and secondary bacterial infections.

Rubella (pp. 599–600)

45. The rubella virus is transmitted by the respiratory route.
46. An infected individual might experience a red rash and light fever or be asymptomatic.
47. Congenital rubella syndrome can affect a fetus when a woman contracts rubella during the first trimester of her pregnancy.
48. Damage from congenital rubella syndrome includes stillbirth, deafness, eye cataracts, heart defects, and mental retardation.
49. Vaccination with live rubella virus provides immunity of unknown duration.

Other Viral Rashes (p. 600)

50. Human parvovirus B19 causes fifth disease, and HHV-6 causes roseola.

Fungal Diseases of the Skin and Nails (pp. 600–602)

Cutaneous Mycoses (pp. 600–601)

51. Fungi that colonize the outer layer of the epidermis cause dermatomycoses.
52. *Microsporum, Trichophyton,* and *Epidermophyton* cause dermatomycoses called ringworm, or tinea.
53. These fungi grow on keratin-containing epidermis, such as hair, skin, and nails.
54. Ringworm and athlete's foot are usually treated with topical antifungal chemicals.
55. Diagnosis is based on the microscopic examination of skin scrapings or fungal culture.

Subcutaneous Mycoses (p. 601)

56. Sporotrichosis results from a soil fungus that penetrates the skin through a wound.
57. The fungi grow and produce subcutaneous nodules along the lymphatic vessels.

Candidiasis (pp. 601–602)

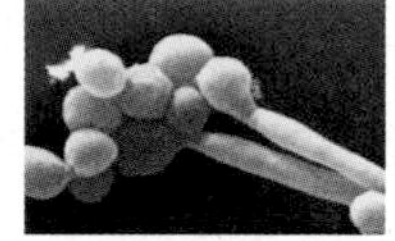

58. *Candida albicans* causes infections of mucous membranes and is a common cause of thrush (in oral mucosa) and vaginitis.
59. *C. albicans* is an opportunistic pathogen that may proliferate when the normal bacterial microbiota are suppressed.
60. Topical antifungal chemicals may be used to treat candidiasis.

Parasitic Infestation of the Skin (pp. 602–603)

61. Scabies is caused by a mite burrowing and laying eggs in the skin.
62. Pediculosis is an infestation by *Pediculus humanus.*

Microbial Diseases of the Eye (pp. 603–605)

1. The mucous membrane lining the eyelid and covering the eyeball is the conjunctiva.

Inflammation of the Eye Membranes: Conjunctivitis (p. 603)

2. Conjunctivitis is caused by several bacteria and can be transmitted by improperly disinfected contact lenses.

Bacterial Diseases of the Eye (pp. 603–605)

3. Bacterial microbiota of the eye usually originate from the skin and upper respiratory tract.
4. Ophthalmia neonatorum is caused by the transmission of *Neisseria gonorrhoeae* from an infected mother to an infant during its passage through the birth canal.
5. All newborn infants are treated with an antibiotic to prevent *Neisseria* and *Chlamydia* infection.
6. Inclusion conjunctivitis is an infection of the conjunctiva caused by *Chlamydia trachomatis.* It is transmitted to infants during birth and is transmitted in unchlorinated swimming water.
7. In trachoma, which is caused by *C. trachomatis,* scar tissue forms on the cornea.
8. Trachoma is transmitted by hands, fomites, and perhaps flies.

Other Infectious Diseases of the Eye (p. 605)

9. *Fusarium* and *Aspergillus* fungi can infect the eye.
10. Herpetic keratitis causes corneal ulcers. The etiology is HSV-1 that invades the central nervous system and can recur.
11. *Acanthamoeba* protozoa, transmitted via water, can cause a serious form of keratitis.

STUDY QUESTIONS

Answers to the Review and Multiple Choice questions can be found by turning to the blue Answers tab at the back of the textbook.

Review

1. Discuss the usual mode of entry of bacteria into the skin. Compare bacterial skin infections with infections caused by fungi and viruses with respect to mode of entry.
2. What bacteria are identified by a positive coagulase test? What bacteria are characterized as group A beta-hemolytic?
3. **DRAW IT** On the figure below, show the sites of the following infections: impetigo, folliculitis, acne, warts, shingles, sporotrichosis, pediculosis.

4. Complete the table of epidemiology below.

Disease	Etiologic Agent	Clinical Symptoms	Mode of Transmission
Acne			
Pimples			
Warts			
Chickenpox			
Fever blisters			
Measles			
Rubella			

5. Why do some states require a test for antibodies against rubella for women before issuing a marriage license?
6. Identify the diseases based on the symptoms in the chart below.

Symptoms	Disease
Koplik's spots	
Macular rash	
Vesicular rash	
Small, spotted rash	
Recurrent "blisters" on oral mucosa	
Corneal ulcer and swelling of lymph nodes	

7. What complications can occur from HSV-1 infections?
8. What is in the MMR vaccine?
9. A patient exhibits inflammatory skin lesions that itch intensely. Microscopic examination of skin scrapings reveals an eight-legged arthropod. What is your diagnosis? How is the disease treated? What would you conclude if you saw a six-legged arthropod?

Multiple Choice

Use the following information to answer questions 1 and 2. A 6-year-old girl was taken to the physician for evaluation of a slowly growing bump on the back of her head. The bump was a raised, scaling lesion 4 cm in diameter. A fungal culture of material from the lesion was positive for a fungus with numerous conidia.

1. The girl's disease was
   a. rubella.
   b. candidiasis.
   c. dermatomycosis.
   d. a cold sore.
   e. none of the above
2. Besides the scalp, this disease can occur on all of the following *except*
   a. feet.
   b. nails.
   c. the groin.
   d. subcutaneous tissue.
   e. none of the above

Use the following information to answer questions 3 and 4. A 12-year-old boy had a fever, rash, headaches, sore throat, and cough. He also had a macular rash on his trunk, face, and arms. A throat culture was negative for *Streptococcus pyogenes.*

3. The boy most likely had
   a. streptococcal sore throat.
   b. measles.
   c. rubella.
   d. smallpox.
   e. none of the above
4. All of the following are complications of this disease *except*
   a. middle ear infections.
   b. pneumonia.
   c. birth defects.
   d. none of the above
   e. encephalitis.
5. A patient has conjunctivitis. If you isolated *Pseudomonas* from the patient's mascara, you would most likely conclude all of the following *except* that
   a. the mascara was the source of the infection.
   b. *Pseudomonas* is causing the infection.
   c. *Pseudomonas* has been growing in the mascara.
   d. the mascara was contaminated by the manufacturer.
   e. none of the above
6. You microscopically examine scrapings from a case of *Acanthamoeba* keratitis. You expect to see
   a. nothing.
   b. viruses.
   c. gram-positive cocci.
   d. eukaryotic cells.
   e. gram-negative cocci.

Use the following choices to answer questions 7 through 9.

a. *Pseudomonas*
b. *S. aureus*
c. scabies
d. *Sporothrix*
e. virus

7. Nothing is seen in microscopic examination of a scraping from the patient's rash.

8. Microscopic examination of the patient's ulcer reveals ovoid cells.

9. Microscopic examination of scrapings from the patient's rash shows gram-negative rods.

10. Which of the following pairs is *mismatched*?

a. leading cause of blindness—*Chlamydia*
b. chickenpox—shingles
c. HSV-1—encephalitis
d. Buruli ulcer—stomach acid
e. none of the above

Critical Thinking

1. A laboratory test used to determine the identity of *Staphylococcus aureus* is its growth on mannitol salt agar. The medium contains 7.5% sodium chloride (NaCl). Why is it considered a selective medium for *S. aureus*?

2. Is it necessary to treat a patient for warts? Explain briefly.

3. Analyses of nine conjunctivitis cases provided the data in the table below. How were these infections transmitted? How could they be prevented?

No.	Etiology	Isolated from Eye Cosmetics or Contact Lenses
5	*S. epidermidis*	+
1	*Acanthamoeba*	+
1	*Candida*	+
1	*P. aeruginosa*	+
1	*S. aureus*	+

4. What factors made the eradication of smallpox possible? What other diseases meet these criteria?

Clinical Applications

1. A hospitalized patient recovering from surgery develops an infection that has blue-green pus and a grapelike odor. What is the probable etiology? How might the patient have acquired this infection?

2. A 12-year-old diabetic girl using continuous subcutaneous insulin infusion to manage her diabetes developed a fever (39.4°C), low blood pressure, abdominal pain, and erythroderma. She was supposed to change the needle-insertion site every 3 days after cleaning the skin with an iodine solution. Frequently she did not change the insertion site more often than every 10 days. Blood culture was negative, and abscesses at insertion sites were not cultured. What is the probable cause of her symptoms?

3. A teenaged male with confirmed influenza was hospitalized when he developed respiratory distress. He had a fever, rash, and low blood pressure. *S. aureus* was isolated from his respiratory secretions. Discuss the relationship between his symptoms and the etiological agent.

22 Microbial Diseases of the Nervous System

Some of the most devastating infectious diseases are those that affect the nervous system, especially the brain and spinal cord. Damage to these areas can lead to deafness, blindness, learning disabilities, paralysis, and death.

Because of the crucial importance of the nervous system, it is strongly protected from accident and infection by bone and other structures. Even pathogens that are circulating in the bloodstream usually cannot enter the brain and spinal cord because of the blood–brain barrier (see Figure 22.2). Occasionally, some trauma will disrupt these defenses with serious consequences. It happens that the fluid (cerebrospinal fluid) of the central nervous system is especially vulnerable because it lacks many of the defenses found in the blood. Pathogens capable of causing diseases of the nervous system often have virulence characteristics of a special nature that enable them to penetrate these defenses. For example, the pathogen can begin replicating in a peripheral nerve and gradually move into the brain and spinal cord.

UNDER THE MICROSCOPE

Naegleria fowleri. This pathogenic amoeba can cause a highly fatal disease.

Q&A

The amoebae in the photo are feeding on a dead amoeba. What tissue do they feed on when they infect humans?

Look for the answer in the chapter.

Structure and Function of the Nervous System

LEARNING OBJECTIVES

22-1 Define *central nervous system* and *blood–brain barrier.*

22-2 Differentiate meningitis from encephalitis.

The human nervous system is organized into two divisions: the central nervous system and the peripheral nervous system (Figure 22.1). The **central nervous system (CNS)** consists of the brain and the spinal cord. As the control center for the entire body, the CNS picks up sensory information from the environment, interprets the information, and sends impulses that coordinate the body's activities. The **peripheral nervous system (PNS)** consists of all the nerves that branch off from the brain and spinal cord. These peripheral nerves are the lines of communication between the central nervous system, the various parts of the body, and the external environment.

Both the brain and the spinal cord are covered and protected by three continuous membranes called *meninges* (Figure 22.2). These are the outermost *dura mater,* the middle *arachnoid mater,* and the innermost *pia mater.* Between the pia mater and arachnoid membranes is a space called the *subarachnoid space,* in which an adult has 100 to 160 ml of *cerebrospinal fluid (CSF)* circulating. Because CSF has low levels of complement or circulating antibodies and few phagocytic cells, bacteria can multiply in it with few checks.

Late in the nineteenth century, experiments in which dyes were injected into the body resulted in the staining of all the organs of the body—with the important exception of the brain. Conversely, when the CSF was injected with dyes, only the brain was stained. These remarkable results were the first evidence of an important feature of anatomy: the **blood–brain barrier.** Certain capillaries permit some substances to pass from the blood into the brain but restrict others. These capillaries are less permeable than others within the body and are therefore more selective in passing materials.

Drugs cannot cross the blood–brain barrier unless they are lipid-soluble. (Glucose and many amino acids are not lipid-soluble, but they can cross the barrier because special transport systems exist for them.) The lipid-soluble antibiotic chloramphenicol enters the brain readily. Penicillin is only slightly lipid-soluble; but, if it is taken in very large doses, enough may cross the barrier to be effective. Inflammations of the brain tend to alter the blood–brain barrier in such a way as to allow antibiotics to cross that would not be able to cross if there were no infection. Probably the most common routes of CNS invasion are the bloodstream and lymphatic system (see Chapter 23), when inflammation alters permeability of the blood–brain barrier.

An inflammation of the meninges is called **meningitis.** An inflammation of the brain itself is called **encephalitis.** If both the brain and the meninges are affected, the inflammation is called **meningoencephalitis.**

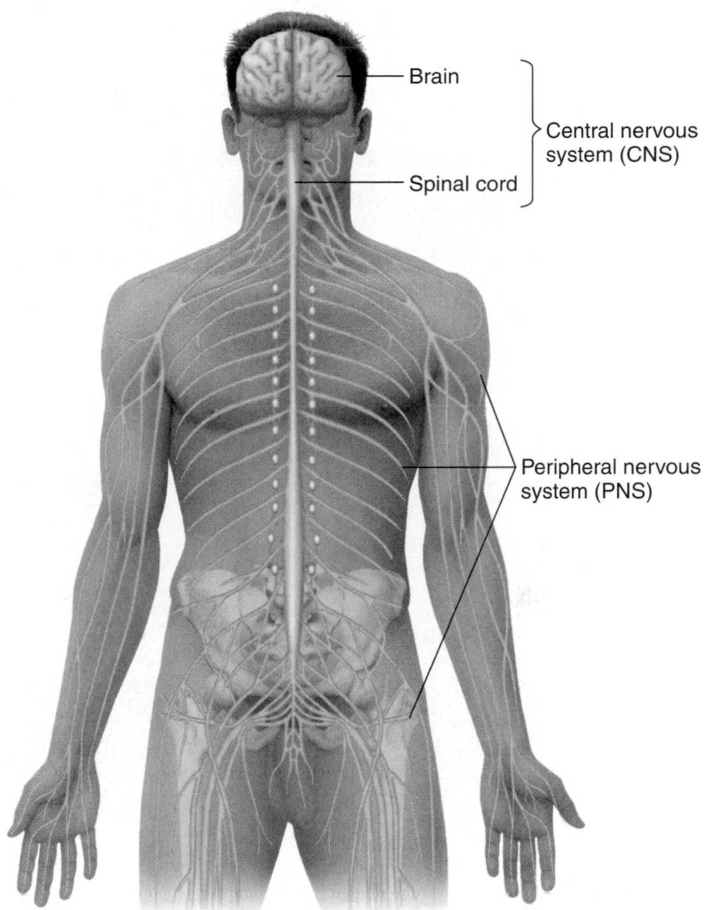

Figure 22.1 The human nervous system. This view shows the central and peripheral nervous systems.

Q **Is meningitis an infection of the CNS or the PNS?**

CHECK YOUR UNDERSTANDING

✓ Why can the antibiotic chloramphenicol readily cross the blood–brain barrier, whereas most other antibiotics cannot? 22-1

✓ Encephalitis is an inflammation of what organ or organ structure? 22-2

Bacterial Diseases of the Nervous System

LEARNING OBJECTIVES

22-3 Discuss the epidemiology of meningitis caused by *Haemophilus influenzae, Neisseria menigitidis, Streptococcus pneumoniae,* and *Listeria monocytogenes.*

22-4 Explain how bacterial meningitis is diagnosed and treated.

22-5 Discuss the epidemiology of tetanus, including mode of transmission, etiology, disease symptoms, and preventive measures.

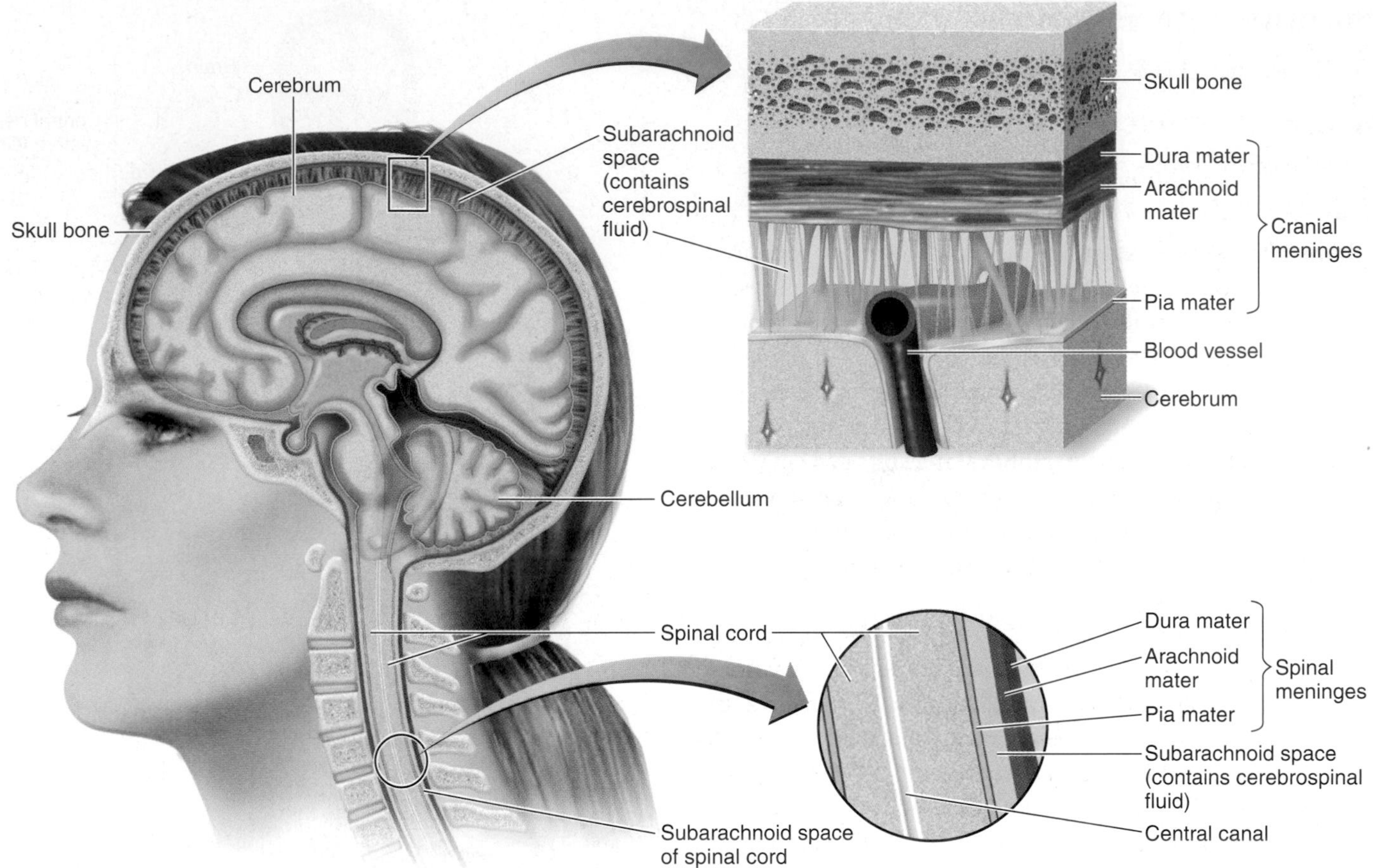

Figure 22.2 The meninges and cerebrospinal fluid. The meninges, whether cranial or spinal, consist of three layers: dura mater, arachnoid mater, and pia mater. Between the arachnoid and the pia mater is the subarachnoid space, in which cerebrospinal fluid circulates. Notice that the CSF is vulnerable to contamination by microbes carried in the blood that are able to penetrate the blood–brain barrier at the walls of the blood vessels.

Q If a patient has meningitis, what barriers would need to be crossed to result in encephalitis?

22-6 State the causative agent, symptoms, suspect foods, and treatment for botulism.

22-7 Discuss the epidemiology of leprosy, including mode of transmission, etiology, disease symptoms, and preventive measures.

Microbial infections of the central nervous system are infrequent but often have serious consequences. In preantibiotic times, they were almost always fatal.

Bacterial Meningitis

The initial symptoms of meningitis are not especially alarming: a triad of fever, headache, and a stiff neck. Nausea and vomiting often follow. Eventually, meningitis may progress to convulsions and coma. The mortality rate varies with the pathogen but is generally high for an infectious disease today. Many people who survive an attack suffer some degree of neurological damage.

Meningitis can be caused by different types of pathogens, including viruses, bacteria, fungi, and protozoa. Viral meningitis (not to be confused with viral encephalitis, page 624), is probably much more common than bacterial meningitis but tends to be a mild disease.

Historically, only three bacterial species have caused most of the cases of meningitis as well as its related mortality. In adult patients, that is, older than 16 years, about 80% of the cases are now caused by *Streptococcus pneumoniae* and *Neisseria meningitidis.* Meningitis caused by *Haemophilus influenzae* type B, once responsible for a majority of cases, has been nearly eliminated in the United States since introduction of an effective vaccine. A conjugate vaccine against *S. pneumoniae* is coming into widespread use and is expected to lower its incidence, especially among children. It may also produce a herd immunity that will benefit the adult population. All three of these pathogens possess a capsule that protects them from phagocytosis as they replicate rapidly in the bloodstream, from which they might enter the cerebrospinal fluid. Death from bacterial meningitis often occurs very quickly, probably from shock and inflammation caused by the release of endotoxins of the gram-negative pathogens or the release of cell wall fragments (peptidoglycans and teichoic acids) of gram-positive bacteria.

Nearly 50 other species of bacteria have been reported to be opportunistic pathogens that occasionally cause meningitis. Especially important are *Listeria monocytogenes*, group B streptococci, staphylococci, and certain gram-negative bacteria.

Haemophilus influenzae Meningitis

Haemophilus influenzae is an aerobic, gram-negative bacterium that is a common member of the normal throat microbiota. Occasionally, however, it enters the bloodstream and causes several invasive diseases. In addition to causing meningitis, it is also frequently a cause of pneumonia (page 688), otitis media (page 679), and epiglottitis. The carbohydrate capsule of the bacterium is important to its pathogenicity, especially those bacteria with capsular antigens of type b. (Strains that lack a capsule are called *nontypable*). Medically, the bacterium is often referred to by the acronym *Hib.*

The name *Haemophilus influenzae* was given because the microorganism was erroneously thought to be the causative agent of the influenza pandemics of 1889 and World War I. *H. influenzae* was probably only a secondary invader during those virus-caused pandemics. *Haemophilus* refers to the need the microorganism has for factors in blood for growth (*hemo* = blood; *philus* = loving).

Hib-caused meningitis occurs mostly in children under age 4, especially at about 6 months, when antibody protection provided by the mother weakens. The incidence is decreasing because of the Hib vaccine, which was introduced in 1988. *H. influenzae* meningitis has accounted for most of the cases of reported bacterial meningitis (45%), with a mortality rate of about 6%.

Neisseria Meningitis (Meningococcal Meningitis)

Meningococcal meningitis is caused by *Neisseria meningitidis* (the **meningococcus**). This is an aerobic, gram-negative bacterium with a polysaccharide capsule that is important to its virulence. Like Hib and the pneumococcus, it is frequently present in the nose and throat of carriers without causing disease symptoms (**Figure 22.3**). These carriers, about 10% of the population, are a reservoir of infection. The symptoms of meningococcal meningitis are mostly caused by an endotoxin that is produced very rapidly and is capable of causing death within just a few hours. The most distinguishing feature is a rash that does not fade when pressed. A case of meningococcal meningitis typically begins with a throat infection, leading to bacteremia and eventually meningitis. It usually occurs in children under 2 years; the maternal immunity weakens at about 6 months and leaves them susceptible. Significant numbers of these children have residual damage, such as deafness.

Death can occur a few hours after the onset of fever; however, antibiotic therapy has helped reduce the mortality rate to about 9–12%. Without chemotherapy, mortality rates approach 80%.

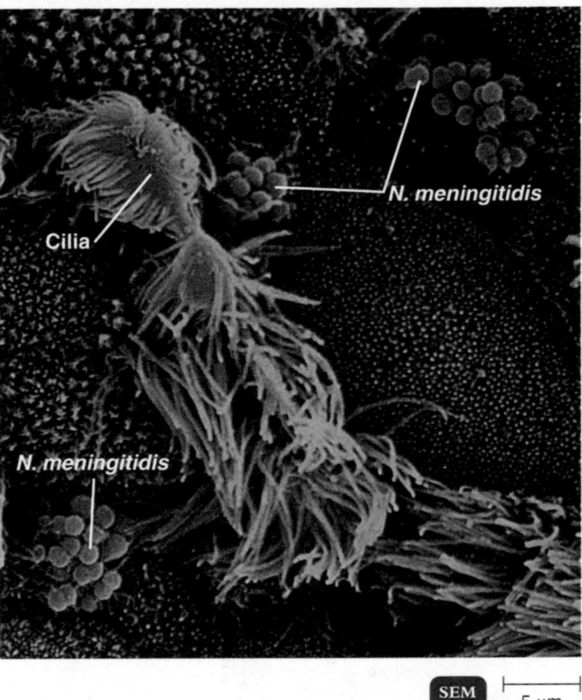

Figure 22.3 ***Neisseria* meningitis.** This scanning electron micrograph shows *Neisseria meningitidis* in clusters attached to cells on the mucous membrane in the pharynx.

Q **What would be the effect if the cilia are inactivated by this infection?**

The meningococcus occurs in five capsular serotypes. In recent years in the United States, meningococcal disease has been caused by several of these serotypes: B (about 35%), C (about 24%), Y (about 34%), and W-135 (about 2%). In Europe, type B predominates, but C also occurs. In arid regions of Africa, China, and the Middle East, type A, and occasionally C and W-135, causes widespread epidemics that coincide with the dry seasons, when the nasal mucous membranes are less resistant to bacterial invasion.

In the United States, sporadic meningococcal outbreaks occur among college students, presumably as a result of crowding of susceptible populations in dormitories. At one time, before vaccination was introduced in 1982, this was a major problem in recruit barracks for the U.S. military. The vaccines are directed at the polysaccharide capsules of serotypes A, C, Y, and W-135; they are recommended for many students entering college and are required by some institutions. No vaccine is available against serotype B, which has a capsule that is not immunogenic in humans. However, an experimental serotype B vaccine is in clinical testing and may be available soon.

The effectiveness of polysaccharide vaccines, which are not effective for very young children, can be much improved by conjugation with protein carriers. Such conjugated vaccines are now available for serotypes A and C in some parts of the world.

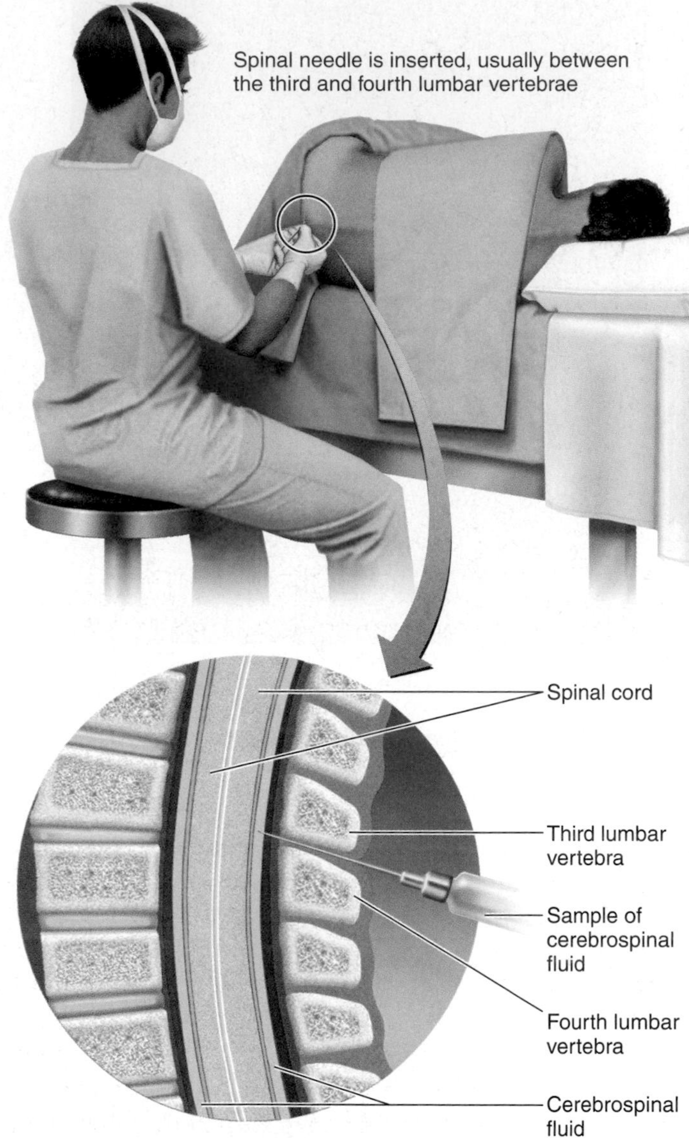

Figure 22.4 Spinal tap (lumbar puncture). Diseases affecting the central nervous system, such as meningitis, often require a spinal tap for diagnosis. A needle is inserted between two vertebrae in the lower spine. A sample of cerebrospinal fluid, which is contained in the subarachnoid space (see Figure 22.2), is withdrawn for laboratory examination.

 **Microscopically, what would you see in CSF from a healthy person? A person with meningococcal meningitis?**

Streptococcus pneumoniae Meningitis (Pneumococcal Meningitis)

Streptococcus pneumoniae, like *H. influenzae,* is a common inhabitant of the nasopharyngeal region. About 70% of the general population are healthy carriers. The pneumococcus, so called because it is best known as a cause of pneumonia (Chapter 24), is a gram-positive, encapsulated diplococcus. It is the leading cause of bacterial meningitis, now that an effective Hib vaccine is in use. In addition to approximately 3000 cases of meningitis, each year *S. pneumoniae* causes 500,000 cases of pneumonia and millions of cases of painful otitis media (earache). Most of the cases of pneumococcal meningitis occur among children between the ages of 1 month and 4 years. For a bacterial disease, the mortality rate is very high: about 30% in children and 80% in the elderly.

A conjugated vaccine, modeled after the Hib vaccine, has been introduced. It is recommended for infants under the age of 2 (see Table 18.3, page 503). One useful side effect of this vaccine is that it results in about a 6–7% decrease in cases of otitis media. The large number of serotypes of the pneumococcus will make it difficult to develop vaccines against all of them.

A serious problem with meningitis and other diseases caused by the pneumococcus is the increasing appearance of antibiotic-resistant strains.

Diagnosis and Treatment of the Most Common Types of Bacterial Meningitis

A diagnosis of bacterial meningitis requires a sample of cerebrospinal fluid obtained by a spinal tap, or lumbar puncture (Figure 22.4). A simple Gram stain is often useful; it will frequently determine the identity of the pathogen with considerable reliability. Cultures are also made from the fluid. For this purpose, prompt and careful handling is required because many of the likely pathogens are very sensitive and will not survive much storage time or even changes in temperature. The most frequently used type of serological tests performed on CSF are latex agglutination tests. Results are available within about 20 minutes. However, a negative result does not eliminate the possibility of less common bacterial pathogens or nonbacterial causes.

Bacterial meningitis is life-threatening and develops rapidly. Therefore, prompt treatment of any type of bacterial meningitis is essential, and chemotherapy of suspected cases is usually initiated before identification of the pathogen is complete. Broad-spectrum third-generation cephalosporins are usually the first choice of antibiotics; some experts recommend including vancomycin. As soon as identification is confirmed, or perhaps when antibiotic sensitivity has been determined from cultures, the antibiotic treatment may be changed. Antibiotics are also valuable in protecting patient contacts against the spread of an outbreak.

Listeriosis

Listeria monocytogenes is a gram-positive rod known to cause stillbirth and neurological disease in animals long before it was recognized as causing human disease. Excreted in animal feces, it is widely distributed in soil and water. The name is derived from the proliferation of monocytes (a type of leukocyte) found in some animals infected by it. In recent years, the disease **listeriosis** has changed from a disease of very limited importance to a major concern for the food industry and health authorities. Since the introduction of Hib vaccination, listeriosis has become the fourth most common cause of bacterial meningitis.

The disease appears in two basic forms: in infected adults and as an infection of the fetus and newborn. In adult humans, it is usually a mild, often symptomless disease, but the microbe sometimes invades the CNS, causing meningitis. This is most likely to happen to persons whose immune system is compromised, such as persons with cancer, diabetes, or AIDS, or who are taking immunosuppressive medications. The mortality rate for CNS infections may be as high as 50%. Occasionally, *L. monocytogenes* invades the bloodstream and causes a wide range of disease conditions, especially sepsis. Recovering or apparently healthy individuals often shed the pathogen indefinitely in their feces. An important factor in its virulence is that when *L. monocytogenes* is ingested by phagocytic cells, it is not destroyed; it even proliferates within them, primarily in the liver. It also has the unusual capability of moving directly from one phagocyte to an adjacent one (Figure 22.5).

L. monocytogenes is especially dangerous when it infects a pregnant woman. She usually suffers no more than mild, flulike symptoms. The fetus, however, can be infected via the placenta, often resulting in an abortion or stillborn infant. In some cases the disease is not manifested until a few weeks after birth, usually as meningitis, which can result in significant brain injury or death. The infant mortality rate associated with this type of infection is about 60%.

In human outbreaks, the organism is mostly foodborne. It is frequently isolated from a wide variety of foods; ready-to-eat deli meats and dairy products have been involved in several outbreaks. *L. monocytogenes* is one of the few pathogens capable of growth at refrigerator temperatures, which can lead to an increase in its numbers during a food's shelf life. The U.S. Food and Drug Administration (FDA) recently approved the use of a bacteriophage-containing spray capable of killing at least 170 *L. monocytogenes* strains for treating ready-to-eat meats. If it meets with consumer approval, it may be a model for similar sprays to control other foodborne pathogens.

Efforts to improve the methods of detecting *L. monocytogenes* in foods are ongoing. Considerable progress has been made with selective growth media and rapid biochemical tests. However, eventually DNA probes and serological tests using monoclonal antibodies are expected to be the most satisfactory (see Chapter 10). Diagnosis in humans depends on isolating and culturing the pathogen, usually from blood or cerebrospinal fluid. Penicillin G is the antibiotic of choice for treatment.

Microbial causes of meningitis and encephalitis are summarized in Diseases in Focus 22.1.

CHECK YOUR UNDERSTANDING

- ✓ Why is meningitis caused by the pathogen *Listeria monocytogenes* frequently associated with ingestion of refrigerated foods? **22-3**
- ✓ What body fluid is sampled to diagnose bacterial meningitis? **22-4**

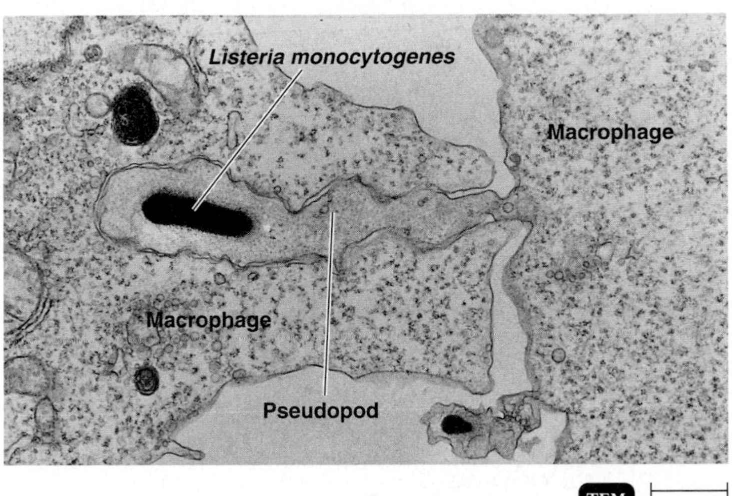

Figure 22.5 Cell-to-cell spread of *Listeria monocytogenes*, the cause of listeriosis. Notice that the bacterium has caused the macrophage on the right, in which it resided, to form a pseudopod that is now engulfed by the macrophage on the left. The pseudopod will soon be pinched off and the microbe transferred to the macrophage on the left.

Q How is listeriosis contracted?

Tetanus

The causative agent of **tetanus,** *Clostridium tetani,* is an obligately anaerobic, endospore-forming, gram positive rod. It is especially common in soil contaminated with animal fecal wastes.

The symptoms of tetanus are caused by an extremely potent neurotoxin, *tetanospasmin,* that is released upon death and lysis of the growing bacteria (see Chapter 15). It enters the CNS via peripheral nerves or the blood. The bacteria themselves do not spread from the infection site, and there is no inflammation.

In a muscle's normal operation, a nerve impulse initiates contraction of the muscle. At the same time, an opposing muscle receives a signal to relax so as not to oppose the contraction. The tetanus neurotoxin blocks the relaxation pathway so that both opposing sets of muscles contract, resulting in the characteristic muscle spasms. The muscles of the jaw are affected early in the disease, preventing the mouth from opening, a condition known as *lockjaw.* In extreme cases, spasms of the back muscles cause the head and heels to bow backward, a condition called *opisthotonos* (Figure 22.6). Gradually, other skeletal muscles become affected, including those involved in swallowing. Death results from spasms of the respiratory muscles.

Because the microbe is an obligate anaerobe, the wound by which it enters the body must provide anaerobic growth conditions—for example, improperly cleaned deep wounds such as those caused by rusty (and therefore presumably dirt-contaminated) nails. Injecting drug users are at high risk: sanitation during injection is not a priority, and the drugs are often contaminated. However, many cases of tetanus arise from

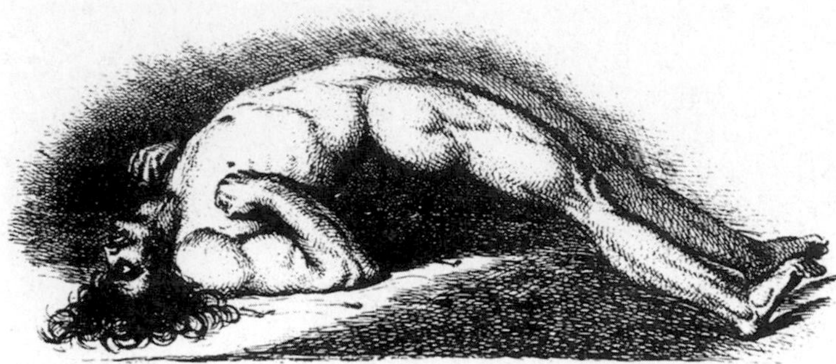

Figure 22.6 An advanced case of tetanus. A drawing of a British soldier during the Napoleonic wars. These spasms, known as opisthotonos, can actually result in a fractured spine. (Drawing by Charles Bell of the Royal College of Surgeons, Edinburgh.)

Q What is the name of the toxin that causes opisthotonos?

trivial injuries, such as sitting on a tack, that are considered too minor to bring to the attention of a physician.

Effective vaccines for tetanus have been available since the 1940s. But vaccination was not always as common as it is today, where it is part of the standard DTaP (diphtheria, tetanus, and acellular pertussis) childhood vaccine. Currently, about 96% of 6-year-olds in the United States have good immunity, but only about 30% of people age 70 do. The tetanus vaccine is a *toxoid,* an inactivated toxin that stimulates the formation of antibodies that neutralize the toxin produced by the bacteria. A booster is required every 10 years to maintain good immunity, but many people do not obtain these vaccinations. Serological surveys show that at least 50% of the U.S. population does not have adequate protection. In fact, 70% of U.S. tetanus cases occur in individuals over age 50. Some were never immunized at all, and others had lost effective antibody levels over time.

Even so, immunization has made tetanus in the United States a rare disease—typically, fewer than 50 cases a year. In 1903, 406 people died of fireworks-related tetanus injuries alone. (Fireworks explosions drive soil particles deep into human tissue.) Worldwide, there are an estimated 1 million cases annually. At least half occur in newborns. In many parts of the world, the severed umbilical cords of infants are dressed with materials such as soil, clay, and even cow dung. Estimates are that the mortality rate from tetanus is about 50% in developing areas; in the United States, it is about 25%.

When a wound is severe enough to need a physician's attention, the doctor must decide whether it is necessary to provide protection against tetanus. Usually there is not enough time to administer toxoid to produce antibodies and block the progression of the infection, even if given as a booster to a patient who has been immunized. However, temporary immunity can be conferred by *tetanus immune globulin (TIG),* prepared from the antibody-containing serum of immunized humans. (Prior to World War I, long before tetanus toxoid became available, similar preparations of preformed antibodies called *antisera* were used. Made by inoculating horses, antisera were very effective in lowering the incidence of tetanus in injured people.)

A physician's decision for treatment depends largely on the extent of the deep injuries and the immunization history of the patient, who may not be conscious. People with extensive injuries who have previously had three or more doses of toxoid within the past 10 years would be considered protected, requiring no action. For extensive wounds in patients with unknown or low immunity, TIG would be given to provide temporary protection. In addition, the first of a toxoid series would be administered to provide more permanent immunity. When TIG and toxoid are both injected, different sites must be used, to prevent the TIG from neutralizing the toxoid. Adults receive a Td (tetanus and diphtheria) vaccine that also boosts immunity to diphtheria. To minimize the production of more toxin, damaged tissue that provides growth conditions for the pathogen should be removed, a procedure called **debridement** (sounds like de-breed-ment), and antibiotics should be administered. However, once the toxin has attached to the nerves, such therapy is of little use.

CHECK YOUR UNDERSTANDING

✓ Is the tetanus vaccine directed at the bacterium or the toxin produced by the bacterium? **22-5**

Botulism

Botulism, a form of food poisoning, is caused by *Clostridium botulinum,* an obligately anaerobic, endospore-forming gram-positive rod found in soil and many aquatic sediments. Ingesting the endospores usually does no harm, as will be explained shortly. However, in anaerobic environments, such as sealed cans, the microorganism produces an exotoxin. This neurotoxin is highly specific for the synaptic end of the nerve, where it blocks the release of acetylcholine, a chemical necessary for transmitting nerve impulses across synapses.

Individuals suffering from botulism undergo a progressive *flaccid paralysis* for 1 to 10 days and may die from respiratory and cardiac failure. Nausea, but no fever, may precede the neurological symptoms. The initial neurological symptoms vary, but nearly all sufferers have double or blurred vision. Other symptoms include difficulty swallowing and general weakness. Incubation time varies, but symptoms typically appear within a day or two. As with tetanus, recovery from the disease does not confer immunity because the toxin is usually not present in amounts large enough to be effectively immunogenic.

Botulism was first described as a clinical disease in the early 1800s, when it was known as the sausage disease (*botulus* is the Latin word for sausage). Blood sausage, the type usually involved, was made by filling a pig stomach with blood and ground meats, tying shut all the openings, boiling it for a short time, and smoking it over a wood fire. The sausage was then stored at room temperature. This attempt at food preservation included most of

DISEASES IN FOCUS 22.1

Meningitis and Encephalitis

Differential diagnosis is the process of identifying the disease from a list of possible diseases that fit the information derived from examining a patient. A differential diagnosis is important for providing initial treatment and for laboratory testing. For example, a worker in a day-care center in eastern North Dakota became ill with fever, rash, headache, and abdominal pain. The patient had a precipitous clinical decline and died on the first day of hospitalization. A Gram stain of the patient's cerebrospinal fluid is shown in the figure. Use the table below to identify infections that could cause these symptoms. For the solution, go to www.microbiologyplace.com.

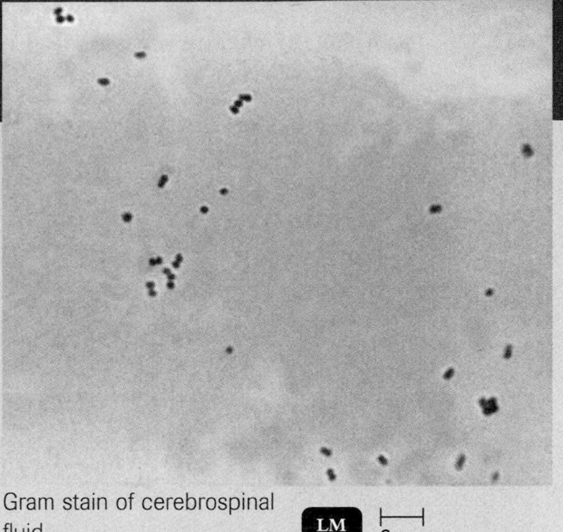

Gram stain of cerebrospinal fluid. LM 6 μm

Disease	Pathogen	Portal of Entry	Method of Transmission	Treatment	Prevention
BACTERIAL DISEASES					
***Haemophilus influenzae* meningitis**	*H. influenzae*	Respiratory tract	Endogenous infection; aerosols	Cephalosporin	Capsular Hib vaccine
Meningococcal meningitis	*Neisseria meningitidis*	Respiratory tract	Aerosols	Cephalosporin	Capsular vaccine against serotypes A, C, Y, W-135
Pneumococcal meningitis	*Streptococcus pneumoniae*	Respiratory tract	Aerosols	Cephalosporin	Polysaccharide vaccine
Listeriosis	*Listeria monocytogenes*	Mouth	Foodborne infection	Penicillin G	Pasteurizing and cooking food
FUNGAL DISEASE					
Cryptococcosis	*Cryptococcus neoformans, C. grubii, C. gattii*	Respiratory tract	Inhaling soil contaminated with spores	Amphotericin B, flucytosine	None
PROTOZOAN DISEASES					
Primary amebic meningoencephalitis	*Naegleria fowleri*	Mucous membrane	Swimming	Amphotericin B	None
Granulomatous amebic encephalitis	*Acanthamoeba* spp.; *Balamuthia mandrillaris*	Mucous membranes	Swimming	Amphotericin B	None

the requirements for an outbreak of botulism. It killed competing bacteria but allowed the more heat-stable *C. botulinum* endospores to survive, and it provided anaerobic conditions and an incubation period for toxin production.

The botulinal toxin will be destroyed by most ordinary cooking methods that bring the food to a boil. Sausage rarely causes botulism today, largely because nitrites are added to it. Nitrites prevent *C. botulinum* from growing after the endospores germinate.

Botulinal toxin is not formed in acidic foods (below pH 4.7). Such foods as tomatoes can therefore be safely preserved without the use of a pressure cooker. There have been cases of botulism from acidic foods that normally would not have supported the growth of the botulism organisms; however, most of these episodes are related to mold growth, which metabolized enough acid to allow *C. botulinum* to begin growing.

Botulinal Types

There are several serological types of the botulinal toxin produced by different strains of the pathogen. These differ considerably in their virulence and other factors.

Type A toxin is probably the most virulent. Deaths have resulted from type A toxin when the food was only tasted but not

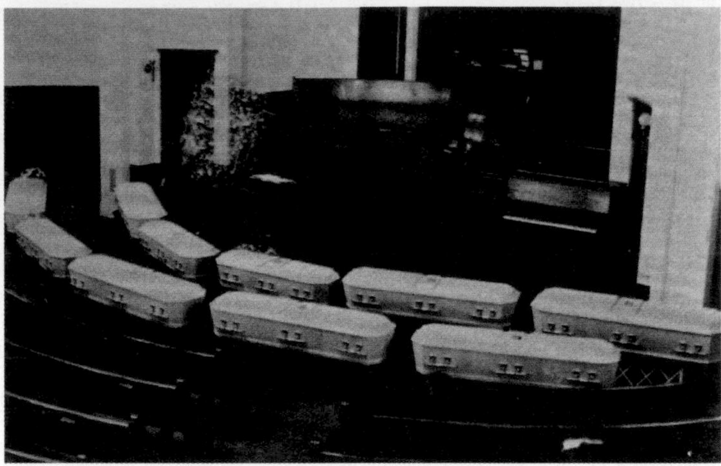

Figure 22.7 Funeral of an Oregon family wiped out by botulism in 1924. The outbreak was caused by home-canned string beans. Altogether there were 12 deaths, but two funerals were held at a different church.

Q Is such a drastic outcome likely today?

swallowed. It is even possible to absorb lethal doses through skin breaks while handling laboratory samples. In untreated cases, the mortality rate is 60–70%. The type A endospore is the most heat-resistant of all *C. botulinum* strains. In the United States, it is found mainly in California, Washington, Colorado, Oregon, and New Mexico. The type A organism is usually proteolytic (the breakdown of proteins by clostridia releases amines with unpleasant odors), but obvious spoilage odor is not always apparent in low-protein foods, such as corn and beans (Figure 22.7).

Type B toxin is responsible for most European outbreaks of botulism and is the most common type in the eastern United States. The mortality rate in cases without treatment is about 25%. Type B botulism organisms occur in both proteolytic and nonproteolytic strains.

Type E toxin is produced by botulism organisms that are often found in marine or lake sediments. Therefore, outbreaks commonly involve seafood and are especially common in the Pacific Northwest, Alaska, and the Great Lakes area. The endospore of type E botulism is less heat-resistant than that of other strains and is usually destroyed by boiling. Type E is nonproteolytic, so the chance of detecting spoilage by odor in high-protein foods such as fish is minimal. The pathogen is also capable of producing toxin at refrigerator temperatures and requires less strictly anaerobic conditions for growth.

Incidence and Treatment of Botulism

Botulism is not a common disease. Only a few cases are reported each year, but outbreaks from social gatherings or restaurants occasionally involve 20 to 30 cases. About half the cases are type A, and types B and E account about equally for the balance. Alaskan native people probably have the highest rate of botulism in the world, mostly of type E. The problem arises from food preparation methods that reflect a cultural tradition of avoiding the use of scarce fuels for heating or cooking. For example, one food involved in Alaskan outbreaks of botulism is *muktuk.* Muktuk is prepared by slicing the flippers of seals or whales into strips and then drying them for a few days. To tenderize them, they are stored anaerobically in a container of seal oil for several weeks until they approach putrefaction. The 40% mortality rate for type E botulism observed in recent years among Alaskan natives reflects the difficulty in getting prompt treatment for isolated ethnic groups.

Botulism organisms do not seem to be able to compete successfully with the normal intestinal microbiota, so the production of toxin by ingested bacteria almost never causes botulism in adults. However, the intestinal microbiota of infants is not well established, and they may suffer from **infant botulism.** Nearly 100 cases occur in the United States annually, several times more than any other form of botulism. Although infants have ample opportunity to ingest soil and other materials contaminated with the endospores of the organism, many reported cases have been associated with honey. Endospores of *C. botulinum* are recovered with some frequency from honey, and a lethal dose may be as few as 2000 bacteria. The recommendation is not to feed honey to infants under 1 year of age; there is no problem with older children or adults who have normal intestinal microbiota. The antitoxin used in adults is derived from horses and has serious side effects, including *serum sickness* (immune complexes formed by reaction with antigens in the antitoxin) and potential anaphylaxis. For infant botulism, safer treatment to neutralize the toxins has been proposed: intravenous human immune globulin (rather than the equine-derived preparation used in adults).

Botulism is diagnosed by inoculating mice with samples from patient serum, stool, or vomitus specimens (Figure 22.8). Different sets of mice are immunized with type A, B, or E antitoxin. All the mice are then inoculated with the test toxin; if, for example, those protected with type A antitoxin are the only survivors, then the toxin is type A. The toxin in food can similarly be identified by mouse inoculation.

The botulism pathogen can also grow in wounds in a manner similar to that of clostridia causing tetanus or gas gangrene (see Chapter 23). Such episodes of **wound botulism** occur occasionally.

The treatment of botulism relies heavily on supportive care. Recovery requires that the nerve endings regenerate; it therefore proceeds slowly. Extended respiratory assistance may be needed, and some neurological impairment may persist for months. Antibiotics are of almost no use because the toxin is preformed. Antitoxins aimed at neutralizing A, B, and E toxins are available and are usually administered together. This trivalent antitoxin will not affect the toxin already attached to the nerve endings and is probably more effective on type E than on types A and B.

The deadly toxin of botulism (Botox) has therapeutic uses for a number of medical conditions, such as chronic headaches. It is also useful for relieving painful muscle contractions in conditions such as cerebral palsy, Parkinson disease, and multiple sclerosis. Injections in the area of facial wounds prevent muscle movement during healing and result in more presentable scar formation. It has been approved to control involuntary eyelid twitching (blepharospasm), crossed eyes (strabismus), and even excessive sweating (hyperhidrosis). This latter use, even though it requires expensive twice-yearly injections, prevents armpit sweating and is favored by professional models to help protect expensive designer clothing. However, the most publicized application has been purely cosmetic: periodic local injections of Botox to eliminate forehead wrinkles (worry lines).

CHECK YOUR UNDERSTANDING

✓ The very name *botulism* is derived from the fact that sausage was the most common food causing the disease. Why is sausage now rarely a cause of botulism? **22-6**

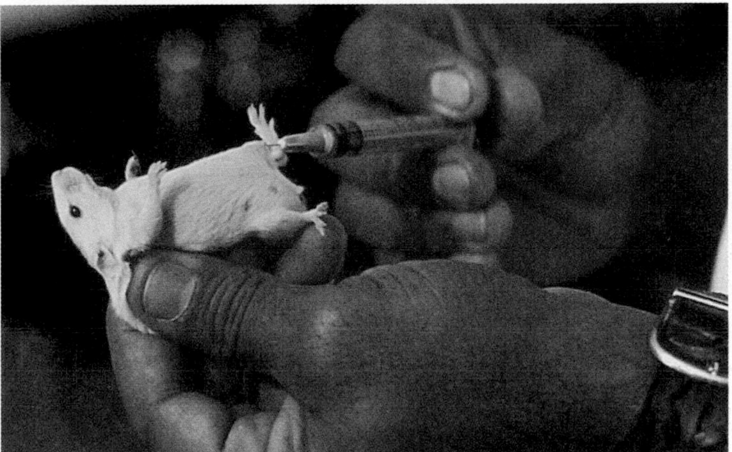

Figure 22.8 Diagnosing botulism by identifying botulinal toxin type. To determine whether botulinal toxin is present, mice are injected with the liquid portion of food extracts or cell-free cultures. If the mice die within 72 hours, toxin is present. To determine the specific type of toxin, groups of mice are passively immunized with antisera specific for *C. botulinum* type A, B, or E. For example, if one group of mice receiving a specific antitoxin lives and the other mice die, the type of toxin in the food or culture has been identified.

Q What are the symptoms of botulism?

Leprosy

Mycobacterium leprae is probably the only bacterium that grows in the peripheral nervous system, although it can also grow in skin cells. It is an acid-fast rod closely related to the tuberculosis pathogen, *Mycobacterium tuberculosis.* The organism was first isolated and identified around 1870 by Gerhard A. Hansen of Norway; his discovery was one of the first links ever made between a specific bacterium and a disease. **Hansen's disease** is the more formal name for **leprosy;** it is sometimes used to avoid the dreaded name.

The organism has an optimum growth temperature of 30°C and shows a preference for the outer, cooler portions of the human body. It survives ingestion by macrophages and eventually invades cells of the myelin sheath of the peripheral nervous system, where its presence causes nerve damage from a cell-mediated immune response. It is estimated that *M. leprae* has a very long generation time, about 12 days. *M. leprae* has never been grown on artificial media. Armadillos have been found to be a useful way to culture the leprosy bacillus; they have a body temperature of 30 to 35°C and are often infected in the wild. Several Texans have actually contracted leprosy from contact with armadillos in that state, where they are common. Probably the most efficient way, however, of culturing *M. leprae* now is the inoculation of the footpads of nude mice (see Figure 19.12, page 539). The ability to grow the bacteria in an animal is invaluable for evaluating chemotherapeutic drugs.

Leprosy occurs in two main forms (although borderline forms are also recognized) that apparently reflect the effectiveness of the host's cell-mediated immune system. The *tuberculoid (neural) form* is characterized by regions of the skin that have lost sensation and are surrounded by a border of nodules (Figure 22.9a). This disease form is roughly the same as *paucibacillary* in the World Health Organization (WHO) leprosy classification system. Tuberculoid disease occurs in people with effective immune reactions. Recovery sometimes occurs spontaneously.

In the *lepromatous (progressive) form* of leprosy (which is much the same as *multibacillary* in the WHO system), skin cells are infected, and disfiguring nodules form all over the body. Patients with this type of leprosy have the least effective cell-mediated immune response, and the disease has progressed from the tuberculoid stage. Mucous membranes of the nose tend to become affected, and a lion-faced appearance is associated with this type of leprosy. Deformation of the hand into a clawed form and considerable necrosis of tissue can also occur (Figure 22.9b). The progression of the disease is unpredictable, and remissions may alternate with rapid deterioration.

The exact means of transfer of the leprosy bacillus is uncertain, but patients with lepromatous leprosy shed large numbers in their nasal secretions and in exudates (oozing matter) of their lesions. Most people probably acquire the infection when secretions containing the pathogen contact their nasal mucosa. However, leprosy is not very contagious and usually is transmitted only between people in fairly intimate and prolonged contact. The time from infection to the appearance of symptoms is usually measured in years, although children can have a much shorter incubation period. Death usually results not from the leprosy itself, but from complications, such as tuberculosis.

Much of the public's fear of leprosy can probably be attributed to biblical and historical references to the disease. In the Middle Ages, people with leprosy were rigidly excluded from normal European society and sometimes even wore bells so that people could avoid them. This isolation might have contributed to the near disappearance of the disease in Europe. But patients with

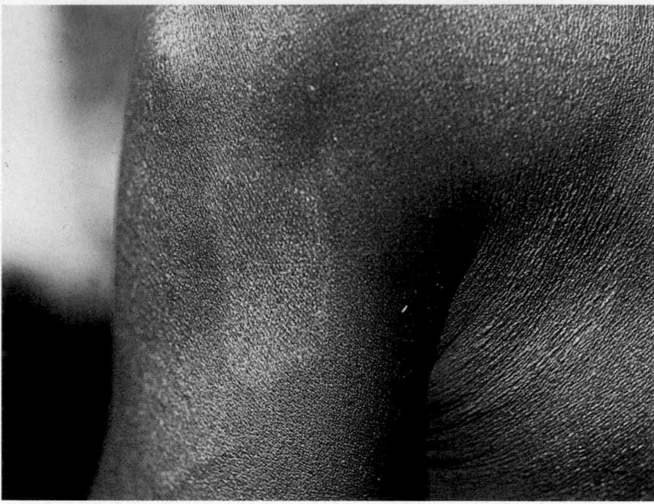

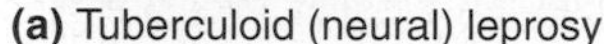

(a) Tuberculoid (neural) leprosy

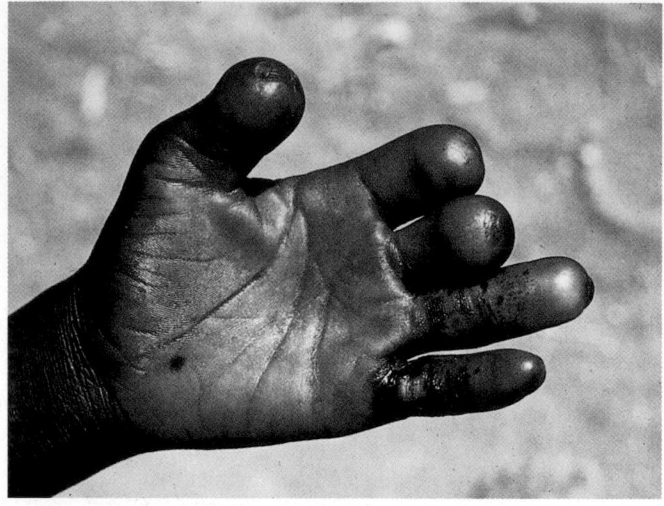

(b) Lepromatous (progressive) leprosy

Figure 22.9 Leprosy lesions. (**a**) The depigmented area of skin surrounded by a border of nodules is typical of tuberculoid (neural) leprosy. (**b**) If the immune system fails to control the disease, the result is lepromatous (progressive) leprosy. This severely deformed hand shows the progressive tissue damage to the cooler parts of the body typical of this later stage.

Q Which form of leprosy is more likely to occur in immunosuppressed individuals?

leprosy are no longer kept in isolation, because they can be made noncontagious within a few days by the administration of sulfone drugs. The National Leprosy Hospital in Carville, Louisiana, once housed several hundred patients but was closed in 1999. Most patients today are treated at centers on an outpatient basis.

The number of leprosy cases in the United States is gradually increasing. Currently, about 100 cases are reported each year. Most are imported; the disease is usually found in tropical climates. Millions of people, most of them in Asia, Africa, and Brazil, suffer from leprosy today, and over half a million new cases are reported each year.

The standard diagnostic test for leprosy is a *skin biopsy sample* taken from the margin of an active lesion. To read this sample reliably, looking for characteristic tissue damage and identifying acid-fast bacilli within nerves, requires an experienced pathologist. Associated procedures, such as the *slit-skin smear*, can be used to enumerate acid-fast bacteria in infected skin. There are no serological tests available.

Dapsone (a sulfone drug), rifampin, and clofazimine, a fat-soluble dye, are the principal drugs used for treatment, usually in combination. A vaccine became commercially available in India in 1998. It is used as an adjunct to chemotherapy. Other vaccines that might be useful in prevention are being tested. One encouraging development is that the Bacillus Calmette-Guérin (BCG) vaccine for tuberculosis (also caused by a *Mycobacterium* species) has been found to be somewhat protective against leprosy.

CHECK YOUR UNDERSTANDING

✓ Why are nude mice and armadillos important in the study of leprosy? **22-7**

Viral Diseases of the Nervous System

LEARNING OBJECTIVES

22-8 Discuss the epidemiology of poliomyelitis, rabies, and arboviral encephalitis, including mode of transmission, etiology, and disease symptoms.

22-9 Compare the Salk and Sabin polio vaccines.

22-10 Compare the preexposure and postexposure treatments for rabies.

22-11 Explain how arboviral encephalitis can be prevented.

Most viruses affecting the nervous system enter it by circulation in the blood or lymph. However, some viruses can enter peripheral nerve axons and move along them toward the CNS.

Poliomyelitis

Poliomyelitis (polio) is best known as a cause of paralysis. However, the paralytic form of poliomyelitis probably affects fewer than 1% of those infected with the poliovirus. The great majority of cases are asymptomatic or exhibit only mild symptoms, such as headache, sore throat, fever, and nausea.

Polio made its first appearance in the United States in an outbreak in Vermont in the summer of 1894. After that, for decades the country was terrified by summertime epidemics. These annual outbreaks increasingly affected adolescents and young adults, and the number of paralytic cases steadily increased. Many victims were killed as their respiratory muscles were paralyzed, and thousands of infants and youths were left with their extremities permanently crippled. Later in the twentieth century, development of the iron lung (**Figure 22.10**) kept alive thousands with paralyzed respiratory systems.

Why did this disease appear so suddenly? The answer is paradoxical—probably because of improved sanitation. Polioviruses can remain infectious for relatively long periods in water and food. The primary mode of transmission is ingestion of water contaminated with feces containing the virus. Improved sanitation delayed exposure to polioviruses in feces until after the protection provided by maternal antibodies had waned. At one time (and today in parts of the world with poor sanitation), exposure to the poliovirus was frequent. Infants were usually exposed to poliovirus while still protected by maternal antibodies. The result was usually an asymptomatic case of the disease and a lifelong immunity. When infection is delayed until adolescence or early adulthood, the paralytic form of the disease appears more frequently.

Because the infection begins when the virus is ingested, its primary areas of multiplication are the throat and small intestine. This accounts for the initial sore throat and nausea. Next, the virus invades the tonsils and the lymph nodes of the neck and ileum (the terminal portion of the small intestine). From the lymph nodes, the virus enters the blood, resulting in *viremia*. In most cases the viremia is only transient, the infection does not progress past the lymphatic system, and clinical disease does not result. If the viremia is persistent, however, the virus penetrates the capillary walls and enters the central nervous system. Once in the CNS, the virus displays a high affinity for nerve cells, particularly motor nerve cells in the upper spinal cord. The virus does not infect the peripheral nerves or the muscles. As the virus multiplies within the cytoplasm of the motor nerve cells, the cells die, and paralysis results. Death can result from respiratory failure.

Polio is usually diagnosed by isolating the virus from feces and throat secretions. Cell cultures can be inoculated, and cytopathic effects on the cells can be observed (see Table 15.4, page 443).

The incidence of polio in the United States has decreased markedly since the availability of the polio vaccines (**Figure 22.11**). The last cases attributed to a wild-type virus were reported in 1979.

There are three different serotypes of the poliovirus, and immunity must be provided for all three. Two vaccines are available. The *Salk vaccine,* developed in 1954, uses viruses that have been inactivated by treatment with formalin. Vaccines of this type, called *inactivated polio vaccines* (IPV), require a series of injections. Their effectiveness rate may be as high as 90% against paralytic polio. The antibody levels decline with time, and booster shots are needed every few years to maintain full immunity. Using only IPV, several European countries have almost eliminated polio from their populations. A newer IPV has been introduced that is produced on human diploid cells. Known as *enhanced inactivated polio vaccine (E-IPV),* it has replaced the original IPV in the United States.

The *Sabin vaccine,* introduced in 1963, contains three living, attenuated strains of the virus (trivalent) and has been more popular in the United States than the Salk vaccine. It is less expensive to administer, and most people prefer taking a sip of orange-flavored drink containing the virus to having a series of injections. The Sabin vaccine is also called *oral polio vaccine,* or OPV. The intestinal immunity achieved with the OPV resembles that acquired by natural infection, and the virus is excreted.

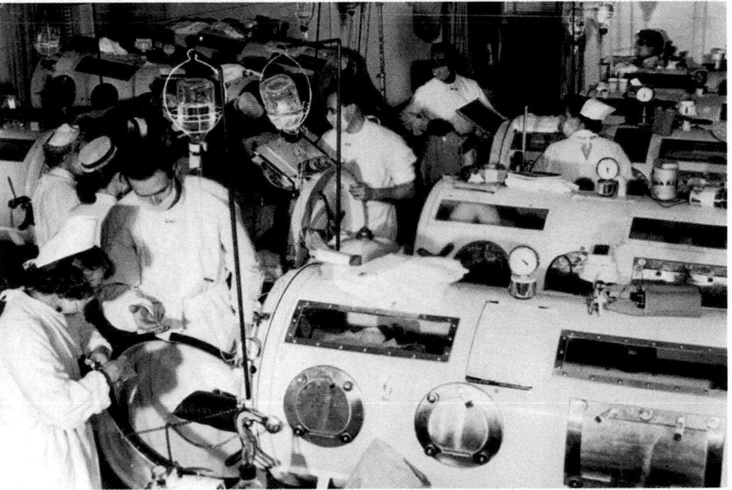

Figure 22.10 Polio patient in an iron lung. Many polio patients were able to breathe only with these mechanical aids. A few survivors from these polio epidemics still use these machines, at least part of the time. Others are able to use portable respiratory aids.

Q **What percentage of polio cases resulted in paralysis?**

One disadvantage is that, on rare occasions—one in 750,000 first doses, one in about 2.4 million on subsequent doses—one of the attenuated strains of the excreted virus (type 3) may revert to virulence and transmit the disease. These cases often occur in secondary contacts, not in the person who received the vaccine. This has caused a few cases a year but also illustrates how recipients of the Sabin vaccine can infect contacts, leading, in most cases, to immunization.

In 2000 the Centers for Disease Control and Prevention (CDC) decided that the advantages of the OPV no longer outweighed the risks. Their recommendation is now to use only IPV for the routine immunization of children. They recommend that the OPV be used only to control widespread outbreaks, to protect children traveling to areas of high risk, or to immunize children who do not receive all four shots of IPV on schedule.

Immunosuppressed individuals should receive E-IPV to reduce the risk of vaccine-related polio from a live virus.

After the successful worldwide eradication of smallpox, polio was targeted. This promised to be more difficult because the fecal–oral transmission route is still common in less-developed parts of the world and because the disease is often not easy to diagnose. For example, only about one child in 200 develops identifiable paralysis after being infected. Launched in 1988, however, a global campaign to eliminate polio has been very successful. The wild-type polio virus is now

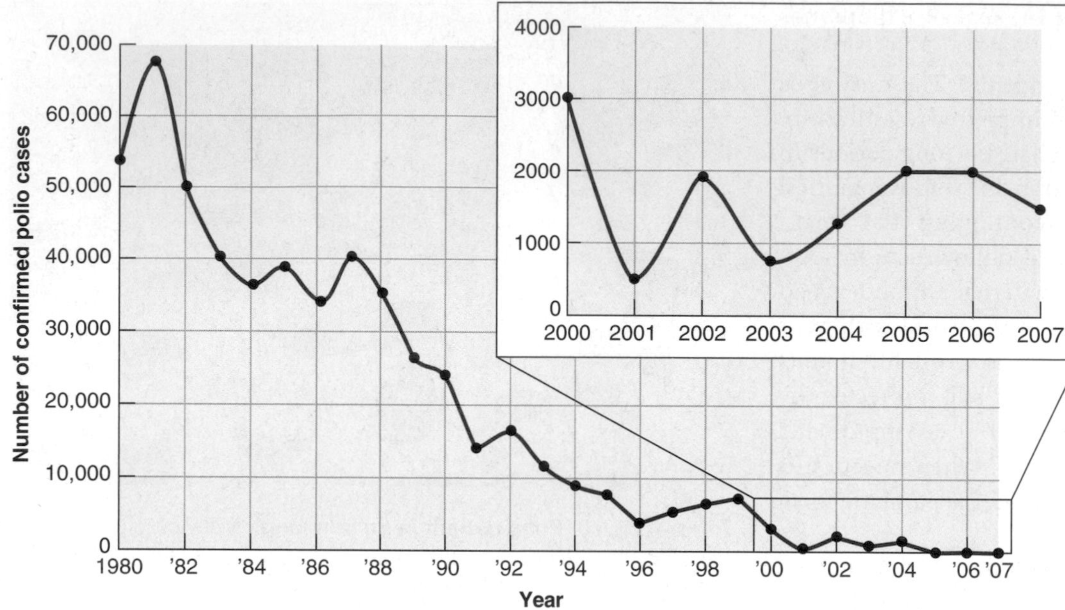

Figure 22.11 Worldwide annual incidence of poliomyelitis. Polio from wild-type viruses has been nearly eliminated in the developed world. A global campaign to eradicate the disease was launched in 1988. Whether it can ever be eradicated completely in less-developed parts of the world is becoming questionable.

Q Why is it possible to eradicate polio but not tetanus?

circulating in only a few countries in Africa and Asia. This has been the result of huge vaccination campaigns in which, in China and India for example, millions of people were immunized with OPV in a single day.

Because of ease of administration, OPV against all three strains of poliovirus (trivalent OPV) is the most practical vaccine in much of the world. A few immunized persons, however, shed virulent, vaccine-derived mutants for long periods. In several areas where vaccination has eliminated the wild-type virus, the disease has reappeared—caused by vaccine-derived viruses. However, discontinuing the vaccination campaigns would soon leave large populations without immunity. For this reason, there may be no choice but to continue immunizing against polio, even in areas where it appears to be nonexistent—probably using IPV in combination with other routine vaccinations.

Stockpiles of the trivalent OPV and the manufacturing facilities will need to be maintained and children routinely vaccinated. The trivalent OPV vaccine is, for several reasons, not as effective in areas with poor sanitation—notably in certain districts of northern India. Even so, of the three types of polio virus, types 2 and 3 are considered to be practically eliminated. Monovalent vaccines directed at type 1 polio, now considered the most likely threat, are being introduced to fight specific outbreaks. They are especially potent against this strain.

During the 1980s, many middle-aged adults who had had polio as children began showing a muscle weakness now called *postpolio syndrome.* It may be that nerve cells that had survived polio originally begin to die. Fortunately, the disease progresses extremely slowly. Exercise therapy is often helpful, and research into drugs that stimulate nerve regeneration is in progress.

CHECK YOUR UNDERSTANDING

- ✓ Why is paralytic polio more likely to occur than a mild or asymptomatic infection in areas with high standards of sanitation? **22-8**
- ✓ Why is the Sabin oral polio vaccine more effective than the injected Salk polio vaccine? **22-9**

Rabies

Rabies (the word is from the Latin for rage or madness) is a disease that almost always results in fatal encephalitis. The causative agent is the rabies virus, a lyssavirus having a characteristic bullet shape (see Figure 13.18a and discussion on page 387). Lyssaviruses are single-stranded RNA viruses with no proofreading capability, and mutant strains develop rapidly. Worldwide, humans usually are infected with the rabies virus from the bite of an infected animal—especially dogs. The virus proliferates in the PNS and moves, fatally, toward the CNS (**Figure 22.12**). In the United States, the most common cause of rabies is a variant of the virus found in silver-haired bats. (Domestic animals have a high rate of vaccination.) This virus has made a unique adaptation and can replicate in epidermal cells and then penetrate to enter a peripheral nerve. Therefore, a lethal dose of the virus can be administered by contact with the unbroken skin. Because deaths from rabies are frequently misdiagnosed, several cases of rabies have been traced to transplanted body tissues, especially corneas.

Rabies is unique in that the incubation period is usually long enough to allow immunity to develop from postexposure vaccination. The immune response is ineffective because the viruses are introduced into the wound in numbers too low to provoke it; also, they do not travel through the bloodstream or lymphatic system, where the immune system could best respond. Initially, the virus multiplies in skeletal muscle and connective tissue, where it remains localized for periods ranging from days to months. Then it enters and travels, at the rate of 15 to 100 mm per day, along peripheral nerves to the CNS, where it causes encephalitis. In some extreme cases, incubation periods of as long as 6 years have been reported, but the average is 30 to 50 days. Bites in areas rich in nerve fibers, such as the hands and face, are especially dangerous, and the resulting incubation period tends to be short.

Once the virus enters the peripheral nerves, it is not accessible to the immune system until cells of the CNS begin to be destroyed, which triggers a belated and ineffective immune response.

Preliminary symptoms are mild and varied, resembling several common infections. When the CNS becomes involved, the patient tends to alternate between periods of agitation and intervals of calm. At this time, a frequent symptom is spasms of the muscles of the mouth and pharynx that occur when the patient feels air drafts or swallows liquids. In fact, even the mere sight or thought of water can set off the spasms—thus the common name *hydrophobia* (fear of water). The final stages of the disease result from extensive damage to the nerve cells of the brain and the spinal cord.

Animals with **furious rabies** are at first restless, then become highly excitable and snap at anything within reach. The biting behavior is essential to maintaining the virus in the animal population. Humans also exhibit similar symptoms of rabies, even biting others. When paralysis sets in, the flow of saliva increases as swallowing becomes difficult, and nervous control is progressively lost. The disease is almost always fatal within a few days.

Some animals suffer from **paralytic rabies,** in which there is only minimal excitability. This form is especially common in cats. The animal remains relatively quiet and even unaware of its surroundings, but it might snap irritably if handled. A similar manifestation of rabies occurs in humans and is often misdiagnosed as Guillain-Barré syndrome, a form of paralysis that is usually transient but sometimes fatal, or other neurological conditions. There is some speculation that the two forms of the disease may be caused by slightly different forms of the virus.

Rabies is usually diagnosed in the laboratory by detection of the viral antigen using the direct fluorescent-antibody test (DFA), which is nearly 100% sensitive and highly specific. These tests can be done on samples of saliva or biopsies of certain external tissues; postmortem samples are usually taken from the brain. For less-developed parts of the world, the CDC has recently developed a *rapid immunohistochemical test (RIT)*. It requires only the use of an ordinary light microscope and has a sensitivity and specificity equivalent to the standard DFA test.

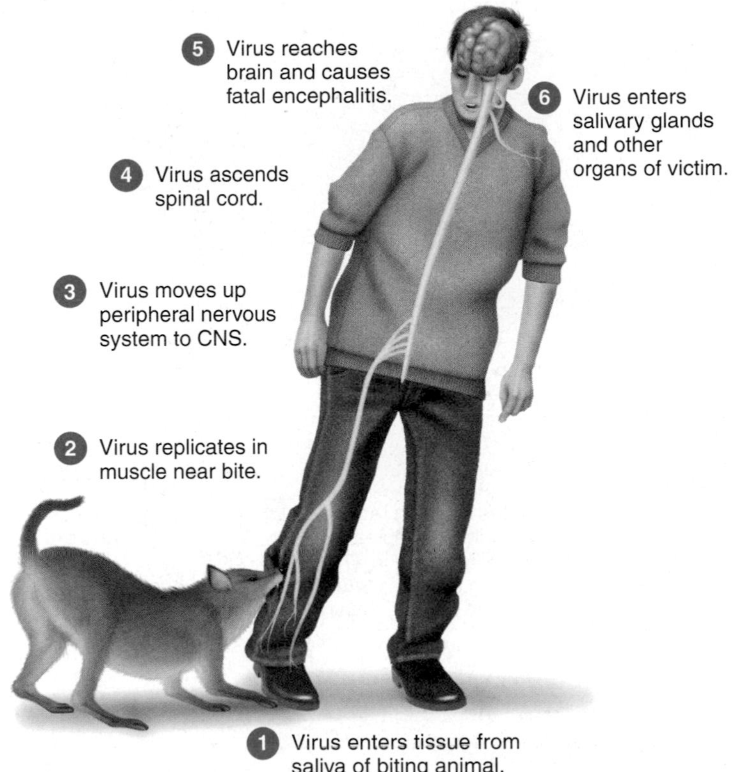
5 Virus reaches brain and causes fatal encephalitis.
6 Virus enters salivary glands and other organs of victim.
4 Virus ascends spinal cord.
3 Virus moves up peripheral nervous system to CNS.
2 Virus replicates in muscle near bite.
1 Virus enters tissue from saliva of biting animal.

Figure 22.12 Pathology of rabies infection.

Q What is the postexposure treatment for rabies?

Prevention of Rabies

Only high-risk individuals, such as laboratory workers, animal control professionals, and veterinarians, are routinely vaccinated against rabies before known exposure. If a person is bitten, the wound should be thoroughly washed with soap and water. If the animal is positive for rabies, the person must undergo *postexposure prophylaxis* (PEP)—meaning a series of antirabies vaccine and immune globulin injections. Another indication for antirabies treatment is any unprovoked bite by a skunk, bat, fox, coyote, bobcat, or raccoon not available for examination. Treatment after a dog or cat bite, if the animal cannot be found, is determined by the prevalence of rabies in the area. The bite of a bat may not be perceptible and may be impossible to rule out in cases where the bat had access to sleeping persons or small children. Therefore, the CDC recommends PEP after any significant encounter with a bat—unless the bat can be tested and shown to be negative for rabies.

The original Pasteur treatment, in which the virus was attenuated by drying in the dissected spinal cords of rabies-infected rabbits, has long been replaced in the United States by *human diploid cell vaccine (HDCV),* or chick embryo–grown vaccines. These vaccines are administered in a series of five to six injections at intervals during a 28-day period. Passive immunization is provided simultaneously by injecting *human rabies immune globulin (RIG)* that has been harvested from people who are immunized against rabies.

Treatment of Rabies

Once the symptoms of rabies appear, there is very little in the way of effective treatment—only a handful of survivors have been reported. Five survivors had received PEP before the

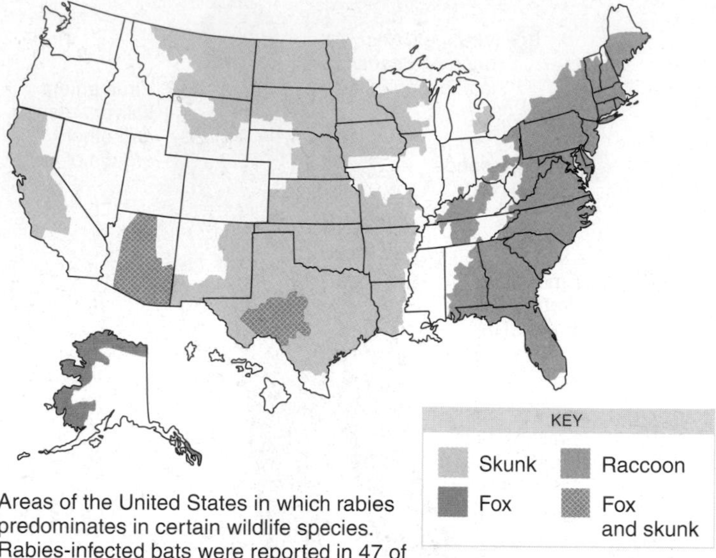

Areas of the United States in which rabies predominates in certain wildlife species. Rabies-infected bats were reported in 47 of the 48 contiguous states. In eastern states in which raccoons are the predominant rabies-infected animal, many cases were also reported in foxes and skunks.

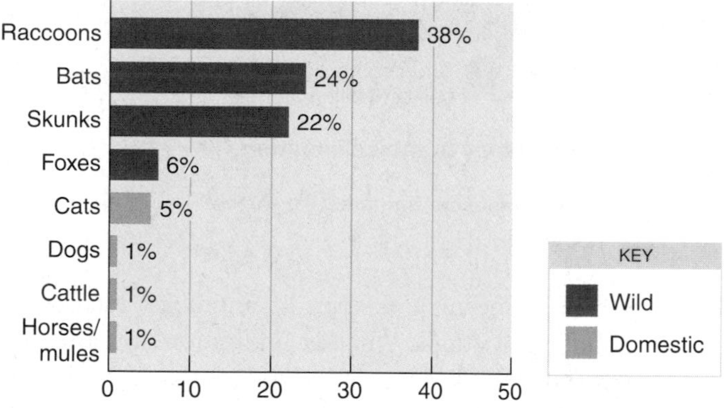

Rabies cases in various wild and domestic animals in the United States. Rabies in domestic animals such as dogs and cats is uncommon because of high vaccination rates. Raccoons, skunks, and bats are the animals most likely to be infected with rabies. Most human cases are caused by bites of bats. Worldwide, most human cases are caused by bites of dogs.

Figure 22.13 Reported cases of rabies in animals. Rabies in foxes include different species in different geographical areas.

Source: CDC 2006.

Q What is the primary reservoir for the rabies virus in your area?

appearance of symptoms. There has been only one reported survival of a patient who had not received PEP. This very recent case was a 15-year-old girl bitten by a rabid bat. The primary treatment was to induce an extended coma to minimize excitability while administering antiviral drugs. She survived with some residual neurological symptoms.

Distribution of Rabies

Rabies occurs all over the world, mostly a result of dog bites. Vaccination of pets is prohibitively expensive in most of Africa, Latin America, and Asia. In these areas, tens of thousands of deaths by rabies occur annually. In the United States, the vaccination of pets is nearly universal, but rabies is widespread among wildlife, predominantly bats, skunks, foxes, and raccoons, although it is also found in domesticated animals (**Figure 22.13**). As many as 40,000 people are administered postexposure rabies vaccine each year, often as a precaution when the rabies status of the biting animal cannot be determined. Rabies is almost never found in squirrels, rabbits, rats, or mice. The disease has long been endemic in vampire bats of South America. In Europe and North America, there are ongoing experiments to immunize wild animals with live rabies vaccine produced in genetically modified vaccinia viruses that are added to food dropped for the animals to find. In the United States, it has been found that gray foxes prefer dog-food bait flavored with vanilla. Coyotes are not at all choosy. In Europe these campaigns have been highly successful, and several countries have been declared free of rabies as a result.

In the United States, 7000 to 8000 cases of rabies are diagnosed in animals each year, but in recent years, only one to six cases have been diagnosed in humans annually (see the box on the facing page).

Related *Lyssavirus* Encephalitis

In recent years a few fatal cases of encephalitis clinically indistinguishable from classic rabies have occurred in Australia and Scotland—countries considered free of rabies. These cases were found to be caused by genotypes of the genus *Lyssavirus* (see page 387) that are closely related to classic rabies virus: the Australian bat lyssavirus (ABLV) and the European bat lyssavirus (EBLV).* Classic rabies is caused by one of seven known genotypes of the genus *Lyssavirus* and is widespread worldwide. Other encephalitis-causing lyssaviruses are indigenous to Europe, Australia, Africa, and Philippines, most commonly in bats.

CHECK YOUR UNDERSTANDING

✓ Why is postexposure vaccination for rabies a practical option? **22-10**

Arboviral Encephalitis

Encephalitis caused by mosquito-borne viruses (called arboviruses) is rather common in the United States. (*Arbovirus* is short for *ar*thropod-*bo*rne virus. This terminology represents a functional grouping; it is not a formal taxonomic term.) The incidence of disease increases in the summer months, coinciding with the proliferation of adult mosquitoes. *Sentinel animals*, such

*A lengthy list of diseases—rabies and similar lyssavirus diseases, the SARS, Ebola, Hendra, and Nipah viruses—are all now known, or strongly suspected, to be transmitted by bats. There are reasons why bats make good disease reservoirs: there are more than a thousand species to occupy various niches; they are long-lived (5 to 50 years), which lends them stability as a reservoir; they tend to roost in close assemblies, which facilitate viral spread; and fly relatively long distances as they forage for food—some are even migratory. Finally, bats seem to be able to carry viruses for long periods without clearing the infection or becoming ill.

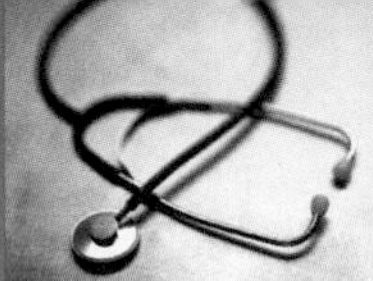

CLINICAL FOCUS From the *Morbidity and Mortality Weekly Report*

A Neurological Disease

As you read through this problem, you will see questions that clinicians ask themselves as they proceed through a diagnosis and treatment. Try to answer each question before going on to the next one.

1. On September 30, a 10-year old girl had pain and weakness in her right arm and temperature of 38.3°C (101°F). On October 3, she began vomiting and had increased arm pain and numbness.
   What could this indicate?
2. A rapid group A streptococcal antigen test was negative. The patient was hospitalized on October 7 when she had difficulty swallowing. Her tongue had a whitish coating and was protruding from her mouth.
   What infections are possible?
3. She was treated with fluconazole for mucosal candidiasis. On October 8, a lumbar puncture showed elevated white blood cell counts.
   What does this indicate?
4. She was treated with vancomycin for meningoencephalitis. She then experienced hypersalivation and lethargy.
   What does this suggest? How would you confirm the disease?
5. Rabies was confirmed by direct fluorescent-antibody staining of a skin biopsy for rabies virus antigens. The patient died on November 2. An abundance of rabies viral inclusions were seen in the brain stem (**Figure A**).
   How would you treat people who had contact with the patient in October and November?
6. Postexposure prophylaxis (PEP) was administered to 66 people, including 31 people in the patient's school.
   Did the delay in diagnosis affect the outcome of the disease?
7. Early diagnosis cannot usually save a patient; however, it may help minimize the number of potential exposures and the need for PEP.
   What else must be determined about this case?
8. In mid-June, the girl had awakened during the night and said a bat had flown into her bedroom window and bitten her. Her mother cleaned a small mark on the girl's arm with an over-the-counter antiseptic but assumed the incident was a nightmare. Two days later, an older sibling took a dead bat away from the yard. The mother did not associate the bat with the previous event and did not seek rabies PEP for the girl.

   The nucleotide sequence of the PCR product was used to identify a rabies virus variant associated with silver-haired bats (**Figure B**).
   Why is rabies surveillance and case reporting important in the United States?
9. During 2000–2007, a total of 20 of the 25 human rabies cases reported in the United States were acquired in the United States. Human rabies is preventable with proper wound care and timely, appropriate administration of human rabies immune globulin and rabies vaccine before onset of clinical symptoms.

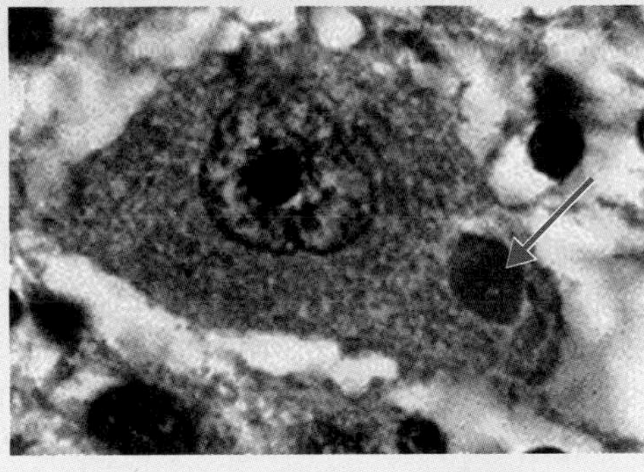

Figure A Negri body at the arrow in an infected neuron.

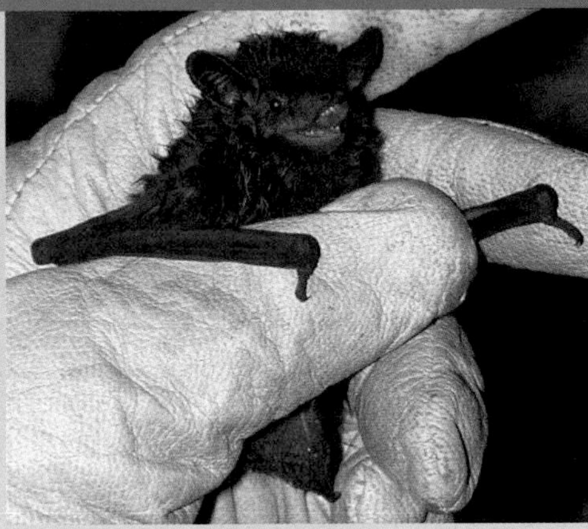

Figure B Silver-haired bat.

Source: Adapted from *MMWR* 56(15):561–565, April 20, 2007, and *MMWR* 57(8):197–200, February 29, 2008.

as caged chickens, are tested periodically for antibodies to arboviruses. This gives health officials information on the incidence and types of viruses in their area.

A number of clinical types of arboviral encephalitis have been identified; all can cause symptoms ranging from subclinical to severe, including rapid death. Active cases of these diseases are characterized by chills, headache, and fever. As the disease progresses, mental confusion and coma occur. Survivors may suffer from permanent neurological problems.

Horses as well as humans are frequently affected by these viruses; thus, there are strains causing *eastern equine encephalitis (EEE)* and *western equine encephalitis (WEE)*. These two viruses are the most likely to cause severe disease in humans. EEE is the most severe; the mortality rate is 30% or more, and survivors experience a high incidence of brain damage, deafness, and other neurological problems. EEE is uncommon (its main mosquito vector prefers to feed on birds); only about 100 cases a year are reported. WEE has been only rarely reported in recent years and has a mortality rate estimated at about 5%.

St. Louis encephalitis (SLE) acquired its name from the location of an early major outbreak (in which it was originally discovered that mosquitoes are involved in the transmission of these

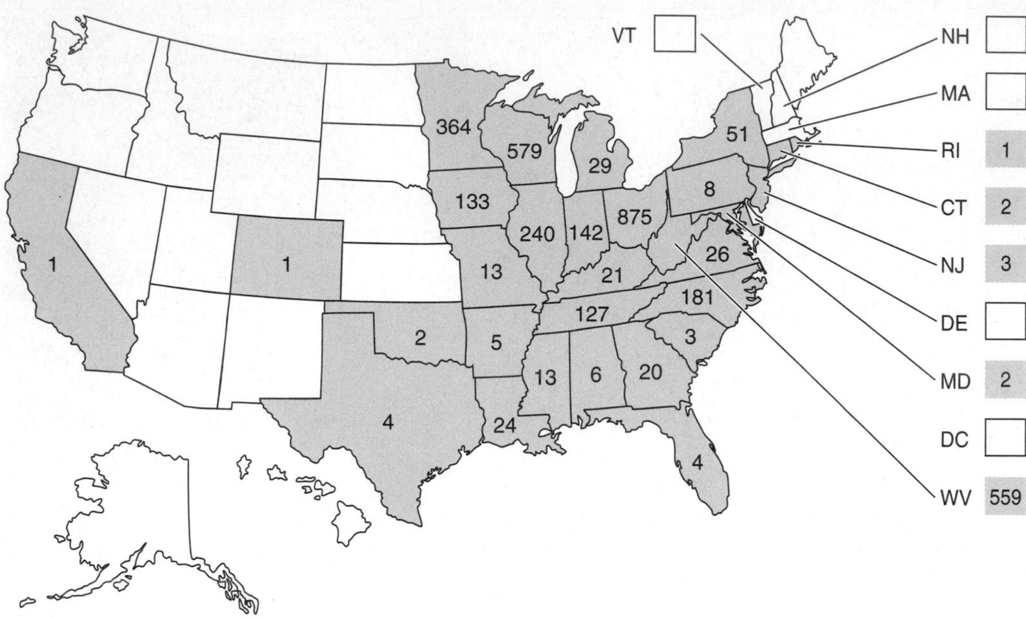

Figure 22.14 California serogroup arbovirus cases: 1964–2006. This is the most common arbovirus encephalitis in the United States. The majority of cases in this serogroup are of the La Crosse virus.

Source: CDC 2007.

Q Why do arboviral infections occur during the summer months?

diseases). SLE is distributed from southern Canada to Argentina, but mostly in the central and eastern United States. Fewer than 1% of people infected exhibit symptoms; it can, however, be a severe disease with a mortality rate in symptomatic patients of about 20%.

California encephalitis (CE) was first identified in that state, but most cases occur elsewhere. The La Crosse strain of CE (first isolated in La Crosse, Wisconsin) is the most commonly encountered arbovirus (Figure 22.14). A relatively mild illness, it is seldom fatal.

A new arbovirus disease, now well known, was introduced into the United States in 1999. First reported in the New York City area, it was quickly identified as being caused by *West Nile virus (WNV)*. The disease is maintained in a bird-mosquito-bird cycle. The primary mosquito is a species of *Culex,* which can overwinter as adults in temperate climates. Birds serve as amplifying hosts; some species, such as house sparrows, can have high levels of viremia without dying. But mortality of infected crows, ravens, and blue jays is high, and public health officials sometimes request reports of dead birds of these species. Most human cases of WNV are subclinical or mild, but the disease can cause a polio-like paralysis or fatal encephalitis, especially in older adults. See Diseases in Focus 22.2 on page 628 for a summary of the arbovirus-caused diseases of the United States.

The Far East also has endemic arboviral encephalitis. **Japanese B encephalitis** is the best known; it is a serious public health problem, especially in Japan, Thailand, Korea, China, and Western India. Vaccines are used to control the disease in these countries and are often recommended for visitors.

Arboviral encephalitis is diagnosed by serological tests, usually ELISA tests to identify IgM antibodies. The most effective preventive measure is local control of the mosquitoes.

CHECK YOUR UNDERSTANDING

✓ When there are serious local outbreaks of arboviral encephalitis, what is the usual response to minimize its transmission? **22-11**

Fungal Disease of the Nervous System

LEARNING OBJECTIVE

22-12 Identify the causative agent, reservoir, symptoms, and treatment for cryptococcosis.

The central nervous system is seldom invaded by fungi. However, one pathogenic fungus in the genus *Cryptococcus* is well adapted to growth in CNS fluids.

Cryptococcus neoformans Meningitis (Cryptococcosis)

The disease **cryptococcosis** is caused by fungi of the genus *Cryptococcus.* They form spherical cells resembling yeasts, reproduce by budding, and produce extremely heavy polysaccharide capsules (Figure 22.15). The primary species pathogenic for humans are *Cryptococcus neoformans* (krip'tō-kok-kus nē-ō-fôr'manz) and *C. grubii* (grub'ē-ē). These organisms are widely distributed, especially in areas contaminated by droppings of birds, most notably pigeons—which excrete an estimated 25 pounds a year. The disease is transmitted mainly by the inhalation of dried, contaminated droppings. The inhaled fungi multiply in persons with compromised immune systems, such as AIDS patients, disseminate to the central nervous system, and cause meningitis that has a high mortality rate. In recent years there have been outbreaks of cryptococcosis in AIDS patients in California caused by *C. gattii* (gat-tē-ē), a species that had previously been reported only in tropical regions (it was thought to have an ecological niche limited to *Eucalyptus* trees but may have a wider distribution). This species has now been isolated in cases of cryptococcosis, even in otherwise healthy individuals, in several areas of western North America as far north as Vancouver Island in Canada.

The best serological diagnostic test is a latex agglutination test to detect cryptococcal antigens in serum or cerebrospinal fluid. The drugs of choice for treatment are amphotericin B and flucytosine in combination. Even so, the mortality rate may approach 30%.

CHECK YOUR UNDERSTANDING

✓ What is the most common source of airborne cryptococcal infections? **22-12**

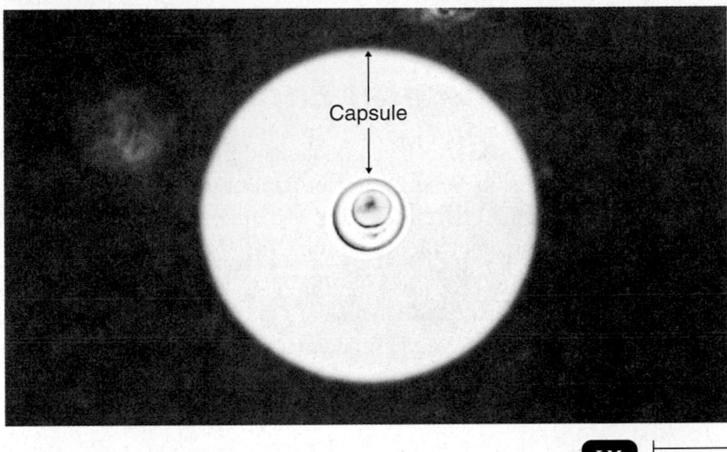

Figure 22.15 *Cryptococcus neoformans.* This yeastlike fungus has an unusually thick capsule. In this photomicrograph, the capsule is made visible by suspending the cells in dilute India ink.

Q What disease does *C. neoformans* cause?

Protozoan Diseases of the Nervous System

LEARNING OBJECTIVE

22-13 Identify the causative agent, vector, symptoms, and treatment for African trypanosomiasis and amebic meningoencephalitis.

Protozoa capable of invading the CNS are rare. However, those that can reach it cause devastating effects.

African Trypanosomiasis

African trypanosomiasis, or sleeping sickness, is a protozoan disease that affects the nervous system. In 1907, Winston Churchill described Uganda during an epidemic of sleeping sickness as a "beautiful garden of death." Even today, estimates are that as many as half a million Africans are infected, and there are about 100,000 new cases yearly.

The disease is caused by two subspecies of *Trypanosoma brucei* that infect humans: *Trypanosoma brucei gambiense* and *Trypanosoma brucei rhodesiense.* They are morphologically indistinguishable but differ significantly in their epidemiology—that is, in their ability to infect nonhuman hosts. Humans are the only significant reservoir for *T.b. gambiense,* whereas *T.b. rhodesiense* is a parasite of domestic livestock and many wild animals. These protozoans are flagellates (see Figure 23.22 on page 661 for the appearance of a similar organism) that are spread by tsetse fly vectors. *T.b. gambiense* is transmitted by a tsetse fly species that inhabits stream vegetation, where there are also concentrations of human populations. It is distributed throughout west and central Africa and is sometimes termed West African trypanosomiasis. More than 97% of reported cases in humans are of this type. Once a person becomes infected, there are few symptoms for weeks or months. Eventually, a chronic form of disease with fever, headaches, and a variety of other symptoms develops that indicates involvement and deterioration of the CNS. Coma and death are inevitable without effective treatment.

In contrast, infections by *T.b. rhodesiense* are transmitted by species of tsetse flies that inhabit savannahs (grasslands with scattered trees) of eastern and southern Africa. Wild animals inhabiting these areas are well adapted to the parasite and are little affected, but humans and domestic animals become acutely ill. This has had a profound effect on sub-Saharan Africa, an area nearly the size of the United States. Agricultural development has been practically prohibited because domestic food and working animals become infected. Infections of humans follow a more acute course than that caused by *T.b. gambiense*; symptoms of illness are apparent within a few days or so of infection. Death occurs within weeks or a few months, sometimes from cardiac problems even before the CNS is affected.

There are some moderately effective chemotherapeutic agents, such as suramin and pentamidine, but these do not alter the course of the disease once the CNS is affected. The drug that does alter the disease's course, however, melarsoprol, is very toxic. In 1992 a new drug, eflornithine, was introduced that crosses the blood–brain barrier and blocks an enzyme required for proliferation of the parasite. It requires an extended series of injections, but it is so dramatically effective against even late stages of *T.b. gambiense* that it has been called the resurrection drug. (Its effectiveness against *T.b. rhodesiense* is variable; melarsoprol is still recommended). The history of this drug provides a valuable illustration of problems in providing health services in poverty-stricken parts of the world. Because the only populations suffering from African trypanosomiasis were unable to afford it, production was soon discontinued. Happily, it was found that the drug had a profitable use in the industrial world:

DISEASES IN FOCUS 22.2

Culex **mosquito engorged with human blood.**

Types of Arboviral Encephalitis

Arboviral encephalitis is usually characterized by fever, headache, and altered mental status ranging from confusion to coma. Vector control to decrease contacts between humans and mosquitoes is the best prevention. Mosquito control includes removing standing water and using insect repellent while outdoors. An 8-year-old girl in rural Wisconsin has chills, headache, and fever and reports having been bitten by mosquitoes. Use the table below to determine which types of encephalitis are most likely. How would you confirm your diagnosis?. For the solution, go to www.microbiologyplace.com.

Disease	Pathogen	Mosquito Vector	Reservoir	U.S. Distribution	Epidemiology	Mortality
Western equine encephalitis	WEE virus (*Togavirus*)	*Culex*	Birds, horses		Severe disease; frequent neurological damage, especially in infants	5%
Eastern equine encephalitis	EEE virus (*Togavirus*)	*Aedes, Culiseta*	Birds, horses		More severe than WEE; affects mostly young children and younger adults; relatively uncommon in humans	>30%
St. Louis encephalitis	SLE virus (*Flavivirus*)	*Culex*	Birds		Mostly urban outbreaks; affects mainly adults over 40	20%
California encephalitis	CE virus (*Bunyavirus*)	*Aedes*	Small mammals		Affects mostly 4- to 18-year age groups in rural or suburban areas; La Crosse strain medically most important. Rarely fatal; about 10% have neurological damage	1% of those hospitalized
West Nile encephalitis	WN virus (*Flavivirus*)	Primarily *Culex*	Primarily birds, assorted rodents, and large mammals		Most cases asymptomatic—otherwise symptoms vary from mild to severe; likelihood of severe neurological symptoms and fatality increases with age	4–18% of those hospitalized

it reduces growth of unwanted facial hair on women. Because of this, the manufacturer has supplied eflornithine at no cost, but for only a limited time, in many African villages.

The current primary approach in combating the disease is to attempt elimination of the vector, the tsetse fly. The use of tentlike, insecticide-treated traps that mimic the color and odor of animal hosts of the insect, combined with large-scale releases of sterile males have eliminated the tsetse fly on the offshore island of Zanzibar. (Female tsetse flies mate only once; the release of artificially reared, radiation-sterilized males in vast numbers

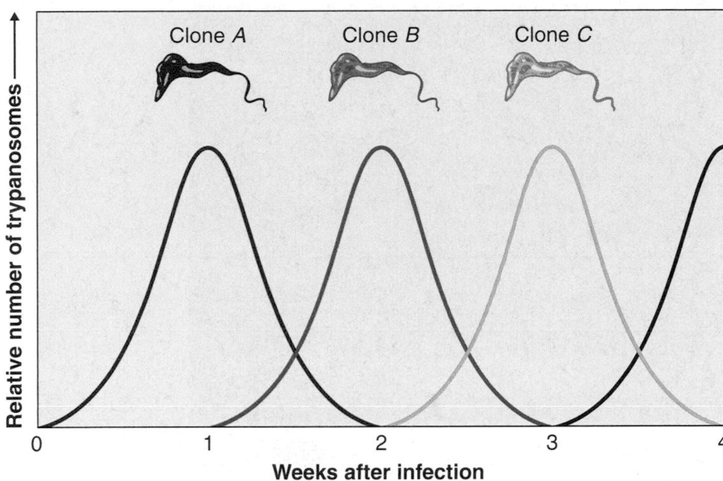

Figure 22.16 How trypanosomes evade the immune system. The population of each trypanosome clone drops nearly to zero as the immune system suppresses its members, but a new clone with a different antigenic surface then replaces the previous clone. The black line represents the population of clone *D*.

Q **Can you think of a viral disease that is causing a worldwide pandemic that would make for a similar figure?**

prevents females that mate with them from producing young). The insect is a weak flyer, and health care officials hope to repeat this eradication on selected areas of the mainland.

A vaccine is being developed, but a major obstacle is that the trypanosome is able to change protein coats at least 100 times and can thus evade antibodies aimed at only one or a few of the proteins. Each time the body's immune system is successful in suppressing the trypanosome, a new clone of parasites appears with a different antigenic coat (Figure 22.16).

Amebic Meningoencephalitis

Q&A There are two species of protozoa that cause amebic meningoencephalitis, a devastating disease of the nervous system. These protozoa are both found in recreational freshwater. Human exposure to them is apparently widespread; many in the population carry antibodies—fortunately, symptomatic disease is rare. *Naegleria fowleri* (nī-glė′rē-ä fou′lėr-ē) is a protozoan (amoeba) that causes a neurological disease, **primary amebic meningoencephalitis (PAM)** (Figure 22.17). Although scattered cases are reported in most parts of the world, only a few cases are reported in the United States annually. The most common victims are children who swim in ponds or streams. The organism initially infects the nasal mucosa and later penetrates to the brain and proliferates, feeding on brain tissue. The fatality rate is nearly 100%, death occurring within a few days after symptoms appear. Diagnosis is typically made at autopsy. The very few known survivors had been treated with the antifungal drug amphotericin B.

A similar neurological disease is **granulomatous amebic encephalitis (GAE).** GAE is caused by a species of

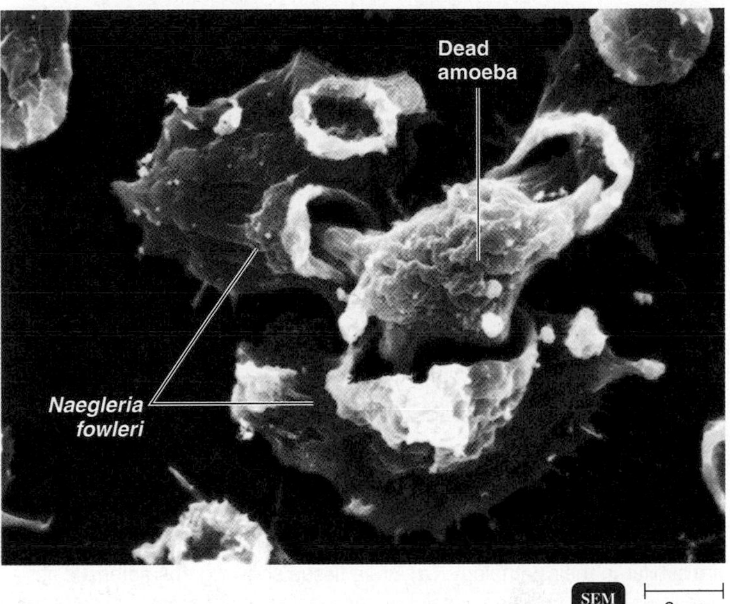

Figure 22.17 *Naegleria fowleri.* This photo shows two vegetative stages of *N. fowleri* beginning to devour a presumably dead amoeba. The suckerlike structures (called amebastomes) function in phagocytic feeding—usually on bacteria or assorted debris that may include host tissue. This protozoan also has a spherical cyst stage and an ovoid flagellated stage (which is most likely to be the infective form) that allows it to swim rapidly in its aquatic habitat.

Q **How is amebic meningoencephalitis transmitted?**

Acanthamoeba, but not the same one that causes *Acanthamoeba* keratitis, a serious disease affecting the eyes. It is chronic, slowly progressive, and fatal in a matter of weeks or months. GAE has an unknown incubation period, and months may elapse before symptoms appear. Granulomas (see Figure 23.28, page 668) form around the organism in response to an immune reaction. The portal of entry is not known but is probably mucous membranes. Multiple lesions are formed in the brain and other organs, especially the lungs. It is probable that many cases of GAE ascribed to *Acanthamoeba* were actually caused by another, similar protozoan, *Balamuthia mandrillaris*, which was first reported in a mandrill baboon in 1989.

CHECK YOUR UNDERSTANDING

✓ What insect is the vector for African trypanosomiasis? **22-13**

Nervous System Diseases Caused by Prions

LEARNING OBJECTIVE

22-14 List the characteristics of diseases caused by prions.

Several fatal diseases affecting the human central nervous system are caused by prions. To explain the term *prion*, we need to recall from the discussion of enzymes in Chapter 5 that the shape of the

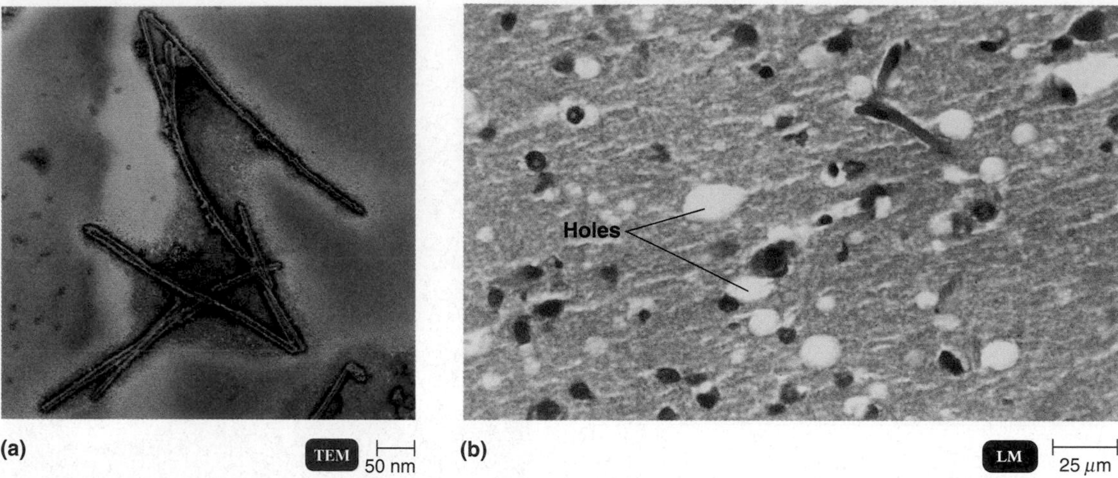

Figure 22.18 Spongiform encephalopathies. These diseases, caused by prions, include bovine spongiform encephalopathy, scrapie in sheep, and Creutzfeldt-Jakob disease in humans. All are similar in their pathology. (**a**) Brain tissue showing characteristic fibrils produced by prion diseases. These fibrils are insoluble aggregates of abnormally folded proteins (prions). Individual prions are not visible by any known technology. (**b**) Brain tissue showing the clear holes that give it a spongiform appearance.

Q What are prions?

enzyme's protein component is essential for its operation. A certain protein is normally found on the surface of brain cell neurons and is even found on the surface of certain stem cells in red bone marrow and cells that become neurons; let us call it *normal protein.* Its function is uncertain, but there is evidence that it may guide maturation of nerve cells. Certainly, the protein's shape causes no damage. But this protein can assume two folded shapes, one normal and the other abnormal (there is no change in the amino acid sequence). If the normal protein encounters an *abnormally folded protein,* a **prion,** the normal protein changes its shape and also becomes abnormally folded—that is, another prion. In fact, a chain reaction of protein misfolding occurs. Therefore, a single infective prion may lead to a cascade of new prions, which then clump together to form the aggregations of fibrils of misfolded proteins that are found in diseased brains. See Figure 22.18a. Autopsies of this infected brain tissue also show that it exhibits a characteristic spongiform degeneration (it is porous, like a sponge), as shown in Figure 22.18b. (Also see the discussion of prions in Chapter 13, page 392, and Figure 13.22). In recent years the study of these diseases, called **transmissible spongiform encephalopathies (TSE),** has been one of the most interesting areas of medical microbiology.

A typical prion disease in animals is **sheep scrapie,** which has been long known in Great Britain and made its first appearance in the United States in 1947. The infected animal scrapes itself against fences and walls until areas of its body are raw. During a period of several weeks or months, the animal gradually loses motor control and dies. The infection can be experimentally passed to other animals by injection of brain tissue from one animal to the next. Similar conditions are seen in mink, possibly resulting from the animals' being fed mutton. A prion disease, *chronic wasting disease,* affects wild deer and elk in the western United States and Canada. It is invariably fatal, and there are concerns that it might infect humans who eat venison and might eventually infect domestic livestock.

Humans suffer from TSE diseases similar to scrapie; **Creutzfeldt-Jakob disease (CJD)** is an example. CJD is rare (about 200 cases per year in the United States). It often occurs in families, an indication of a genetic component. This form of CJD is sometimes referred to as classic CJD to differentiate it from similar variants that have appeared. There is no doubt that an infective agent is involved because transmission via corneal transplants and accidental scalpel nicks of a surgeon during autopsy have been reported. Several cases have been traced to the injection of a growth hormone derived from human tissue. Boiling and irradiation have no effect, and even routine autoclaving is not reliable. This has led to suggestions that surgeons use disposable instruments where there is a risk of exposure to CJD. To sterilize reusable instruments, the World Health Organization currently recommends a strong solution of sodium hydroxide combined with extended autoclaving at 134°C. However, there are reports that applications of a simple cleaning detergent combined with protease enzymes to disrupt the prions may prove an effective solution to the problem. A similar approach to disposing of prion-infected animal carcasses, for which incineration is the current primary method, makes use of digestion. The digester tank shown in Figure 22.19 is in use at the Wisconsin Veterinary Diagnostic Laboratory to dispose of deer carcasses suspected of infection with prion-caused chronic wasting disease (incineration was previously used for this purpose). The animal tissue is subjected to heat and caustic chemicals such as sodium hydroxide or potassium hydroxide. The animal tissue and any microorganisms are reduced to a harmless broth containing only sugars and short peptide chains that can be disposed of in the municipal sanitary sewage system. This process is less expensive than incineration, as well as more environmentally favorable.

Some tribes in New Guinea have suffered from a TSE disease called **kuru** (a native word for shaking or trembling). Transmission of kuru is apparently related to the practice of cannibalistic rituals. Carleton Gajdusek received the Nobel Prize for Physiology and Medicine in 1976 for his investigations of kuru. The disease is disappearing as the practice of ritualistic cannibalism dies out.

Figure 22.19 Tissue digester. (**a**) The inset photo shows the stainless steel tank of the tissue digester that can be used to reduce prion-infected animals into a noninfectious slurry. (**b**) The technician is dropping a deceased prion-infected sheep into the tissue digester.

Q **Will cooking or freezing meat destroy prions?**

Bovine Spongiform Encephalopathy and Variant Creutzfeldt-Jakob Disease

A TSE that is much in the news is **bovine spongiform encephalopathy (BSE).** The disease is better known as *mad cow disease* because of the behavior of the animals. The outbreak that began in 1986 in Great Britain was eventually controlled by drastic culling of herds. The origin of the disease is usually ascribed to feed supplements contaminated with prions from sheep infected with scrapie, a long-endemic neurological disease. As cattle adapted to scrapie, they exhibited the symptoms of BSE. Another hypothesis proposes that BSE resulted from a spontaneous mutation in a cow and that there is no connection with scrapie.

There is an urgent need for reliable tests that will diagnose cases of BSE in early, nonsymptomatic stages in live animals. Currently, the only available tests require postmortem brain tissue and detect only late stages of the disease. In attempts to prevent introduction of BSE into the United States, there are rules prohibiting the use of meat from "downer" animals (fallen and unable to rise and walk) for any purpose and the use of animal protein as a feed supplement. The FDA has banned for human consumption certain portions of the cattle carcass that are most likely to contain a neurological pathogen. Also, only a small percentage of animal carcasses in the United States are tested for BSE—in Europe and Japan, practically all slaughtered animals are tested.

If this disease were to establish itself in domestic cattle in the United States, it would be economically devastating. However, there is another aspect—that the disease could be passed on to humans. In Great Britain and a few other locales around the world, a few cases of apparent classic CJD appeared in relatively young humans. CJD rarely occurs in this age group, and a connection with BSE was feared. Investigation also showed that this variant of CJD (vCJD) differed in significant ways from classic CJD (Table 22.1). Fewer than 200 cases have been identified so far. Considering the long incubation times of prion diseases and that an estimated 1 million cattle had been infected with BSE, it was feared that large numbers of vCJD cases might eventually appear. However, this concern has subsided, especially since the number of cases declined from a small peak in 2000 and after it was shown that the affected patients shared a certain limited genetic profile.

CHECK YOUR UNDERSTANDING

✓ What are the recommendations for sterilizing reusable surgical instruments when prion contamination might be a factor? **22-14**

Table 22.1 Comparative Characteristics of Classic and Variant Creutzfeldt-Jakob Disease

Characteristic	Classic CJD	Variant CJD
Median age at death (yr)	68 (range 23–97)	28 (range 14–74)
Median duration of illness (mo)	4 to 5	13 to 14
Clinical presentation	Dementia; early neurological signs	Prominent psychiatric and behavioral symptoms; delayed neurological signs
Genotype*	Other amino acid combinations	Methionine/methionine

*Victims are homozygous at codon 129, that is, both of their PrP genes (one from each parent) have methionine coded at this position. This is characteristic of only about 37% of Caucasians. Other members of this population have different amino acid combinations at this position—and, although a handful of cases that have appeared were exceptions (homozygous valine/valine, or heterozygous methionine/valine), no one with these genotypes has contracted vCJD to date.

DISEASES IN FOCUS 22.3

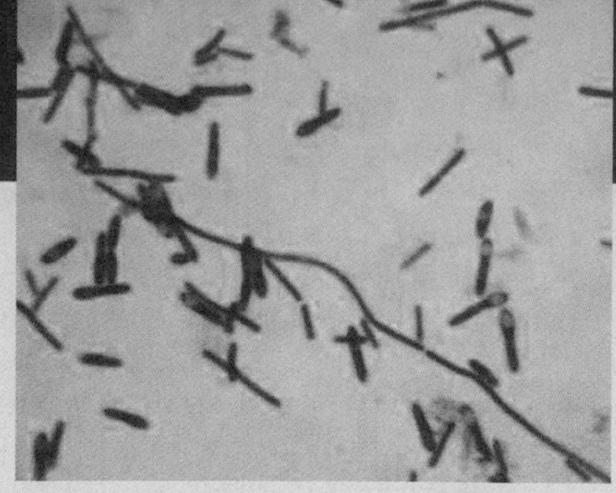

Gram stain from canned chili. LM 10 μm

Microbial Diseases with Neurological Symptoms or Paralysis

Differential diagnosis is the process of identifying the disease from a list of possible diseases that fit the information derived from examining a patient. A differential diagnosis is important for providing initial treatment and for laboratory testing. For example, after eating canned chili two children experienced cranial nerve paralysis followed by descending paralysis. The children are on mechanical ventilation. Leftover canned chili was tested by mouse bioassay. Use the table below to identify infections that could cause these symptoms. For the solution, go to www.microbiologyplace.com.

Disease	Pathogen	Symptoms	Method of Transmission	Treatment	Prevention
BACTERIAL DISEASES					
Tetanus	*Clostridium tetani*	Lockjaw; muscle spasms	Puncture wound	Tetanus immune globulin; antibiotics	Toxoid vaccine (DTaP, Td)
Botulism	*Clostridium botulinum*	Flaccid paralysis	Foodborne intoxication	Antitoxin	Proper canning of foods; infants should not have honey
Leprosy	*Mycobacterium leprae*	Loss of sensation in skin; disfiguring nodules	Probably prolonged contact with contaminated secretions	Dapsone, rifampin, clofaximine	Possibly BCG vaccine
VIRAL DISEASES					
Poliomyelitis	Poliovirus	Headache, sore throat, stiff neck; paralysis if motor nerves infected	Ingesting contaminated water (fecal–oral route)	Mechanical breathing aid	Inactivated polio vaccine (E-IPV)
Rabies	*Lyssavirus*, including rabies virus	Fatal infection; early symptoms include agitation, muscle spasms, difficulty swallowing	Animal bite	Postexposure treatment: rabies immunoglobulin plus vaccine	Human diploid cell vaccine for high-risk individuals; vaccination of domestic animals
PROTOZOAN DISEASE					
African trypanosomiasis	*Trypanosoma brucei rhodesiense, T. b. gambiense*	Fatal infection; early symptoms (headache, fever) progress to coma	Tsetse fly	Suramin; pentamidine	Vector control
PRION DISEASES					
Creutzfeldt-Jakob disease	Prion	Fatal infection; neurologic symptoms include trembling	Inherited; ingested; transplants	None	None
Kuru	Prion	Same as Creutzfeldt-Jakob disease	Contact or ingestion	None	None

Disease Caused by Unidentified Agents

LEARNING OBJECTIVE

22-15 List some possible causes of chronic fatigue syndrome.

Chronic Fatigue Syndrome

The medical community has long been puzzled by patients who complain of persistent fatigue that prevents them from working and has no apparent cause. They often complain as well of multiple allergies. Called **chronic fatigue syndrome (CFS)**, the debilitating condition continues for months or years. For many years, the condition was dismissed as a complaint of people who were depressed or simply complaining about trivial symptoms. Recent research on CFS, however, suggests that it is not "all in the mind" but rather is strongly linked to the immune system and may also have a genetic component. There is now an alternative, more impressive name, *myalgic encephalomyelitis (ME)*. People complaining of CFS often do not adapt well to daily stresses and do not respond strongly to fight infections. CFS often begins with a flulike illness that seems never to go away. Some think that it is triggered by viral illnesses, such as infectious mononucleosis (caused by the Epstein-Barr virus), Q fever, Lyme disease, and others.

The CDC has now developed a diagnostic definition for CFS: a persistent, unexplained fatigue that lasts at least 6 months. The patient should also exhibit at least four of a list of symptoms including sore throat, tender lymph nodes, muscle pain, pain in multiple joints, headaches, unrefreshing sleep, malaise after exercise, and impaired short-term memory or concentration. The condition is not uncommon in the United States; the prevalence is 0.52% in women and 0.29% in men, totaling an estimated 800,000 to 2.5 million persons.

There is no approved treatment for CFS, but an experimental drug, Ampligen, is being tested. It is designed to stimulate production of antiviral interferons.

CHECK YOUR UNDERSTANDING

✓ Name one common disease that may be associated with chronic fatigue syndrome. **22-15**

* * *

Diseases in Focus 22.3 summarizes the main causes of microbial disease involving neurological symptoms and paralysis.

STUDY OUTLINE

The **MyMicrobiologyPlace** website (**www.microbiologyplace.com**) will help you get ready for tests with its simple three-step approach: ❶ **take a pre-test** and obtain a personalized study plan, ❷ **learn and practice** with animations, tutorials, and MP3 tutor sessions, and ❸ **test yourself** with quizzes and a chapter post-test.

Structure and Function of the Nervous System (p. 611)

1. The central nervous system (CNS) consists of the brain, which is protected by the skull bones, and the spinal cord, which is protected by the backbone.
2. The peripheral nervous system (PNS) consists of the nerves that branch from the CNS.
3. The CNS is covered by three layers of membranes called meninges: the dura mater, arachnoid mater, and pia mater. Cerebrospinal fluid (CSF) circulates between the arachnoid mater and the pia mater in the subarachnoid space.
4. The blood–brain barrier normally prevents many substances, including antibiotics, from entering the brain.
5. Microorganisms can enter the CNS through trauma, along peripheral nerves, and through the bloodstream and lymphatic system.
6. An infection of the meninges is called meningitis. An infection of the brain is called encephalitis.

Bacterial Diseases of the Nervous System (pp. 611–620)

Bacterial Meningitis (pp. 612–615)

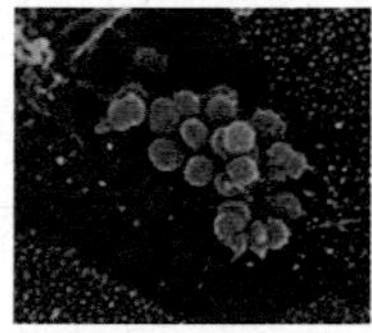

1. Meningitis can be caused by viruses, bacteria, fungi, and protozoa.
2. The three major causes of bacterial meningitis are *Haemophilus influenzae, Streptococcus pneumoniae,* and *Neisseria meningitidis.*
3. Nearly 50 species of opportunistic bacteria can cause meningitis.

Haemophilus influenzae Meningitis (p. 613)

4. *H. influenzae* is part of the normal throat microbiota.
5. *H. influenzae* requires blood factors for growth; serotypes are based on capsules.
6. *H. influenzae* type b is the most common cause of meningitis in children under 4 years old.
7. A conjugated vaccine directed against the capsular polysaccharide antigen is available.

Neisseria Meningitis (Meningococcal Meningitis) (p. 613)

8. *N. meningitidis* causes meningococcal meningitis. This bacterium is found in the throats of healthy carriers.

9. The bacteria probably gain access to the meninges through the bloodstream. The bacteria may be found in leukocytes in CSF.
10. Symptoms are due to endotoxin. The disease occurs most often in young children.
11. Purified capsular polysaccharide vaccine against serotypes A, C, Y, and W-135 is available.

Streptococcus pneumoniae Meningitis (Pneumococcal Meningitis) (p. 614)

12. *S. pneumoniae* is commonly found in the nasopharynx.
13. Young children are most susceptible to *S. pneumoniae* meningitis. Untreated, it has a high mortality rate.
14. A conjugated vaccine is available.

Diagnosis and Treatment of the Most Common Types of Bacterial Meningitis (p. 614)

15. Diagnosis is based on Gram stain, cultures, and serological tests of the bacteria in CSF.
16. Cephalosporins may be administered before the pathogen is identified.

Listeriosis (pp. 614–615)

17. *Listeria monocytogenes* causes meningitis in newborns, the immunosuppressed, pregnant women, and cancer patients.
18. Acquired by ingestion of contaminated food, it may be asymptomatic in healthy adults.
19. *L. monocytogenes* can cross the placenta and cause spontaneous abortion and stillbirth.

Tetanus (pp. 615–616)

20. Tetanus is caused by a localized infection of a wound by *Clostridium tetani.*
21. *C. tetani* produces the neurotoxin tetanospasmin, which causes the symptoms of tetanus: spasms, contraction of muscles controlling the jaw, and death resulting from spasms of respiratory muscles.

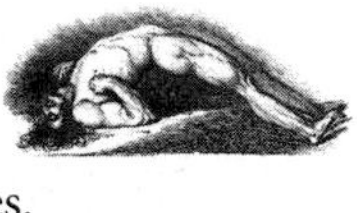

22. *C. tetani* is an anaerobe that will grow in deep, unclean wounds.
23. Acquired immunity results from DTaP immunization.
24. Following an injury, an immunized person may receive a booster of tetanus toxoid. An unimmunized person may receive (human) tetanus immune globulin.
25. Debridement (removal of tissue) and antibiotics may be used to control the infection.

Botulism (pp. 616–619)

26. Botulism is caused by an exotoxin produced by *C. botulinum* growing in foods.
27. Serological types of botulinum toxin vary in virulence; type A is the most virulent.
28. The toxin is a neurotoxin that inhibits the transmission of nerve impulses.
29. Blurred vision occurs in 1 to 2 days; progressive flaccid paralysis follows for 1 to 10 days, possibly resulting in death from respiratory and cardiac failure.
30. *C. botulinum* will not grow in acidic foods or in an aerobic environment.
31. Endospores are killed by proper canning. The addition of nitrites to foods inhibits growth of *C. botulinum.*
32. The toxin is heat labile and is destroyed by boiling (100°C) for 5 minutes.
33. Infant botulism results from the growth of *C. botulinum* in an infant's intestines.
34. Wound botulism occurs when *C. botulinum* grows in anaerobic wounds.
35. For diagnosis, mice protected with antitoxin are inoculated with toxin from the patient or foods.

Leprosy (pp. 619–620)

36. *Mycobacterium leprae* causes leprosy, or Hansen's disease.
37. *M. leprae* has never been cultured on artificial media. It can be cultured in armadillos and mouse footpads.
38. The tuberculoid form of the disease is characterized by loss of sensation in the skin surrounded by nodules.
39. In the lepromatous form, disseminated nodules and tissue necrosis occur.
40. Leprosy is not highly contagious and is spread by prolonged contact with exudates.
41. Untreated individuals often die of secondary bacterial complications, such as tuberculosis.
42. Laboratory diagnosis is based on observations of acid-fast rods in a skin biopsy.
43. Patients with leprosy are treated with sulfone drugs.

Viral Diseases of the Nervous System (pp. 620–626)

Poliomyelitis (pp. 620–622)

1. The symptoms of poliomyelitis are usually sore throat and nausea, and occasionally paralysis (fewer than 1% of cases).
2. Poliovirus is transmitted by the ingestion of water contaminated with feces.
3. Poliovirus first invades lymph nodes of the neck and small intestine. Viremia and spinal cord involvement may follow.
4. Diagnosis is based on isolation of the virus from feces and throat secretions.
5. The Salk vaccine (an inactivated polio vaccine) involves the injection of formalin-inactivated viruses and boosters every few years. The Sabin vaccine (an oral polio vaccine) contains three live, attenuated strains of poliovirus and is administered orally.
6. Polio is a good candidate for elimination through vaccination.

Rabies (pp. 622–624)

7. Rabies virus (*Lyssavirus*) causes an acute, usually fatal, encephalitis called rabies.
8. Rabies may be contracted through the bite of a rabid animal or invasion through skin. The virus multiplies in skeletal muscle and connective tissue.
9. Encephalitis occurs when the virus moves along peripheral nerves to the CNS.
10. Symptoms of rabies include spasms of mouth and throat muscles followed by extensive brain and spinal cord damage and death.
11. Laboratory diagnosis may be made by DFA tests of saliva, serum, and CSF or brain smears.

12. Reservoirs for rabies in the United States include skunks, bats, foxes, and raccoons. Domestic cattle, dogs, and cats may get rabies. Rodents and rabbits seldom get rabies.
13. Postexposure treatment includes administration of human rabies immune globulin (RIG) along with multiple intramuscular injections of vaccine.
14. Preexposure treatment consists of vaccination.
15. Other genotypes of *Lyssavirus* cause rabies-like diseases.

Arboviral Encephalitis (pp. 624–626)

16. Symptoms of encephalitis are chills, headache, fever, and eventually coma.
17. Many types of viruses (called arboviruses) transmitted by mosquitoes cause encephalitis.
18. The incidence of arboviral encephalitis increases in the summer months, when mosquitoes are most numerous.
19. Notifiable arboviral infections are eastern equine encephalitis (EEE), western equine encephalitis (WEE), St. Louis encephalitis (SLE), California encephalitis (CE), and West Nile virus (WNV).
20. Diagnosis is based on serological tests.
21. Control of the mosquito vector is the most effective way to control encephalitis.

Fungal Disease of the Nervous System (pp. 626–627)

Cryptococcus neoformans Meningitis (Cryptococcosis) (pp. 626–627)

1. *Cryptococcus* spp. are encapsulated yeastlike fungi that cause cryptococcosis.
2. The disease may be contracted by inhaling dried infected pigeon or chicken droppings.
3. The disease begins as a lung infection and may spread to the brain and meninges.
4. Immunosuppressed individuals are most susceptible to cryptococcosis.
5. Diagnosis is based on latex agglutination tests for cryptococcal antigens in serum or CSF.

Protozoan Diseases of the Nervous System (pp. 627–629)

African Trypanosomiasis (pp. 627–629)

1. African trypanosomiasis is caused by the protozoa *Trypanosoma brucei gambiense* and *T.b. rhodesiense* and is transmitted by the bite of the tsetse fly.
2. The disease affects the nervous system of the human host, causing lethargy and eventually coma. It is commonly called sleeping sickness.
3. Vaccine development is hindered by the protozoan's ability to change its surface antigens.

Amebic Meningoencephalitis (p. 629)

4. Encephalitis caused by the protozoan *Naegleria fowleri* is almost always fatal.
5. Granulomatous amebic encephalitis, caused by *Acanthamoeba* spp. and *Balamuthia mandrillaris,* is a chronic disease.

Nervous System Diseases Caused by Prions (pp. 629–632)

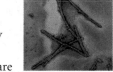

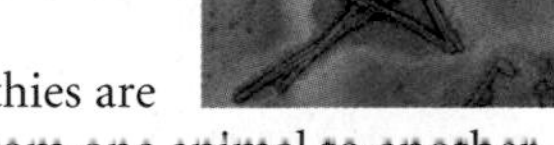

1. Prions are self-replicating proteins with no detectable nucleic acid.
2. Diseases of the CNS that progress slowly and cause spongiform degeneration are caused by prions.
3. Transmissible spongiform encephalopathies are caused by prions that are transferable from one animal to another.
4. Creutzfeldt-Jakob disease and kuru are human diseases similar to scrapie. They are transmitted between humans.

Disease Caused by Unidentified Agents (p. 633)

Chronic Fatigue Syndrome (p. 633)

1. Chronic fatigue syndrome (CFS) may be triggered by a microbial infection.

STUDY QUESTIONS

Answers to the Review and Multiple Choice questions can be found by turning to the blue Answers tab at the back of the textbook.

Review

1. If *Clostridium tetani* is relatively sensitive to penicillin, why doesn't penicillin cure tetanus?
2. What treatment is used against tetanus under the following conditions?
   a. before a person suffers a deep puncture wound
   b. after a person suffers a deep puncture wound
3. Why is the following description used for wounds that are susceptible to *C. tetani* infection: ". . . Improperly cleaned deep puncture wounds . . . ones with little or no bleeding . . ."?
4. Provide the following information on poliomyelitis: etiology, method of transmission, symptoms, prevention. Why aren't the Salk and Sabin vaccines considered treatments for poliomyelitis?
5. Fill in the following table:

Causative Agent of Meningitis	Susceptible Population	Transmission	Treatment
N. meningitidis			
H. influenzae			
S. pneumoniae			
L. monocytogenes			
C. neoformans			

6. **DRAW IT** On the figure below, identify the portal of entry of *H. influenzae, C. tetani*, botulinum toxin, *M. leprae*, poliovirus, *Lyssavirus*, arboviruses, and *Acanthamoeba*.

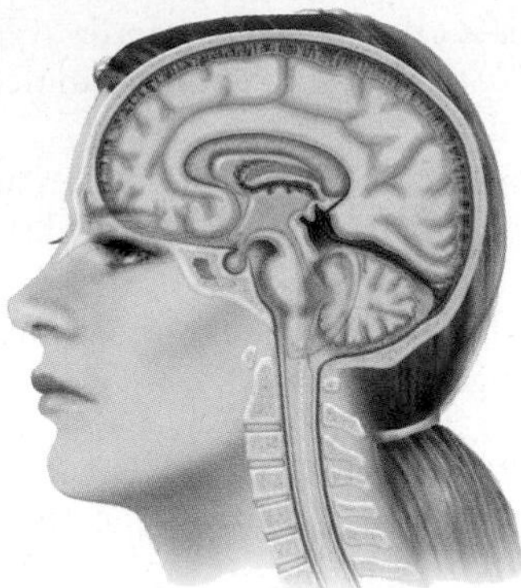

7. Outline the procedures for treating rabies after exposure. Outline the procedures for preventing rabies prior to exposure. What is the reason for the differences in the procedures?
8. Fill in the following table.

Disease	Etiology	Transmission	Symptoms	Treatment
Arboviral-encephalitis				
African-trypanosomiasis				
Botulism				
Leprosy				

9. Provide evidence that Creutzfeldt-Jakob disease is caused by a transmissible agent.

Multiple Choice

1. Which of the following is *not* true?
   a. Only puncture wounds by rusty nails result in tetanus.
   b. Rabies is seldom found in rodents (e.g., rats, mice).
   c. Polio is transmitted by the fecal–oral route.
   d. Arboviral encephalitis is rather common in the United States.
   e. All of the above are true.
2. Which of the following does *not* have an animal reservoir or vector?
   a. listeriosis
   b. cryptococcosis
   c. amebic meningoencephalitis
   d. rabies
   e. African trypanosomiasis
3. A 12-year-old girl hospitalized for Guillain-Barré syndrome had a 4-day history of headache, dizziness, fever, sore throat, and weakness of legs. Seizures began 2 weeks later. Bacterial cultures were negative. She died 3 weeks after hospitalization. An autopsy revealed inclusions in brain cells that tested positive in an immunofluoresence test. She probably had
   a. rabies.
   b. Creutzfeldt-Jakob disease.
   c. botulism.
   d. tetanus.
   e. leprosy.
4. After receiving a corneal transplant, a woman developed dementia and loss of motor function; she then became comatose and died. Cultures were negative. Serological tests were negative. Autopsy revealed spongiform degeneration of her brain. She most likely had
   a. rabies.
   b. Creutzfeldt-Jakob disease.
   c. botulism.
   d. tetanus.
   e. leprosy.
5. Endotoxin is responsible for symptoms caused by which of the following organisms?
   a. *N. meningitidis*
   b. *S. pyogenes*
   c. *L. monocytogenes*
   d. *C. tetani*
   e. *C. botulinum*
6. The increased incidence of encephalitis in the summer months is due to
   a. maturation of the viruses.
   b. increased temperature.
   c. the presence of adult mosquitoes.
   d. an increased population of birds.
   e. an increased population of horses.

Match the following choices to the statements in questions 7 and 8:
   a. antirabies antibodies
   b. HDVC

7. Produces the highest antibody titer.
8. Used for passive immunization.

Use the following choices to answer questions 9 and 10:
   a. *Cryptococcus*
   b. *Haemophilus*
   c. *Listeria*
   d. *Naegleria*
   e. *Neisseria*

9. Microscopic examination of cerebrospinal fluid reveals gram-positive rods.
10. Microscopic examination of cerebrospinal fluid from a person who washes windows on a building in a large city reveals ovoid cells.

Critical Thinking

1. Most of us have been told that a rusty nail causes tetanus. What do you suppose is the origin of this adage?
2. OPV is no longer used for routine vaccination. Provide the rationale for this policy.

Clinical Applications

1. A 1-year-old infant was lethargic and had a fever. When admitted to the hospital, he had multiple brain abscesses with gram-negative coccobacilli. Identify the disease, etiology, and treatment.
2. A 40-year-old bird handler was admitted to the hospital with soreness over his upper jaw, progressive vision loss, and bladder dysfunction. He had been well 2 months earlier. Within weeks he lost reflexes in his lower extremities and subsequently died. Examination of CSF showed lymphocytes. What etiology do you suspect? What further information do you need?
3. One week after bathing in a hot spring, a 9-year-old girl was hospitalized after a 3-day history of progressive, severe headaches, nausea, lethargy, and stupor. Examination of CSF revealed amoeboid organisms. Identify the disease, etiology, and treatment.

23 Microbial Diseases of the Cardiovascular and Lymphatic Systems

The **cardiovascular system** consists of the heart, blood, and blood vessels. The **lymphatic system** consists of the lymph, lymph vessels, lymph nodes, and the lymphoid organs such as the tonsils, appendix, spleen, and thymus. Fluids in both systems circulate throughout the body, intimately contacting many tissues and organs. Physiologically, the blood and lymph distribute nutrients and oxygen to body tissues and carry away wastes. However, these same qualities make the cardiovascular and lymphatic systems vehicles for the spread of pathogens that enter their circulation when an insect bite, needle, or wound penetrates the skin. Because of this, many of the body's innate defensive systems are found in the blood and lymph. Especially important are circulating phagocytic cells; these are also in fixed locations such as the lymph nodes and spleen. The blood is an important part of our adaptive immune system; antibodies and specialized cells circulate to intercept pathogens introduced into the blood. Occasionally the defensive systems found in the blood are overwhelmed, however, and pathogens proliferate explosively with disastrous results.

UNDER THE MICROSCOPE

Ebola virus. A cause of one of the world's emerging viral hemorrhagic fevers.

Q&A

Although the emerging hemorrhagic fevers are almost entirely limited to tropical Africa, they have attracted much attention because of their mortality rate and devastating symptoms. What are these symptoms?

Look for the answer in the chapter.

Structure and Function of the Cardiovascular and Lymphatic Systems

LEARNING OBJECTIVE

23-1 Identify the role of the cardiovascular and lymphatic systems in spreading and eliminating infections.

The center of the cardiovascular system is the heart (**Figure 23.1**). The function of the cardiovascular system is to circulate blood through the body's tissues so it can deliver certain substances to cells and remove other substances from them.

Blood is a mixture of formed elements and a liquid called blood plasma (see the box in Chapter 16, page 467). The lymphatic system is an essential part of the circulation of blood (**Figure 23.2**). As the blood circulates, some blood plasma filters out of the blood capillaries into spaces between tissue cells called *interstitial spaces.* The circulating fluid is called *interstitial fluid.* Microscopic lymphatic vessels that surround tissue cells are called *lymph capillaries.* As the interstitial fluid moves around the tissue cells, it is picked up by the lymph capillaries; the fluid is then called *lymph.*

Because lymph capillaries are very permeable, they readily pick up microorganisms or their products. From lymph capillaries, lymph is transported into larger lymph vessels called *lymphatics,* which contain valves that keep the lymph moving toward the heart. Eventually, all the lymph is returned to the blood just before the blood enters the heart. As a result of this circulation, proteins and fluid that have filtered from the plasma are returned to the blood.

At various points in the lymphatic system are oval structures called *lymph nodes,* through which lymph flows. (Also, see Figure 16.5, page 456.) Within the lymph nodes are fixed macrophages that help clear the lymph of infectious microorganisms. At times the lymph nodes themselves get infected and become visibly swollen and tender; swollen lymph nodes are called **buboes** (see Figure 23.11, page 648).

Lymph nodes are also an important component of the body's immune system. Foreign microbes entering lymph nodes encounter two types of lymphocytes: B cells, which are stimulated to become plasma cells that produce humoral antibodies; and T cells, which then differentiate into effector T cells that are essential to the cell-mediated immune system.

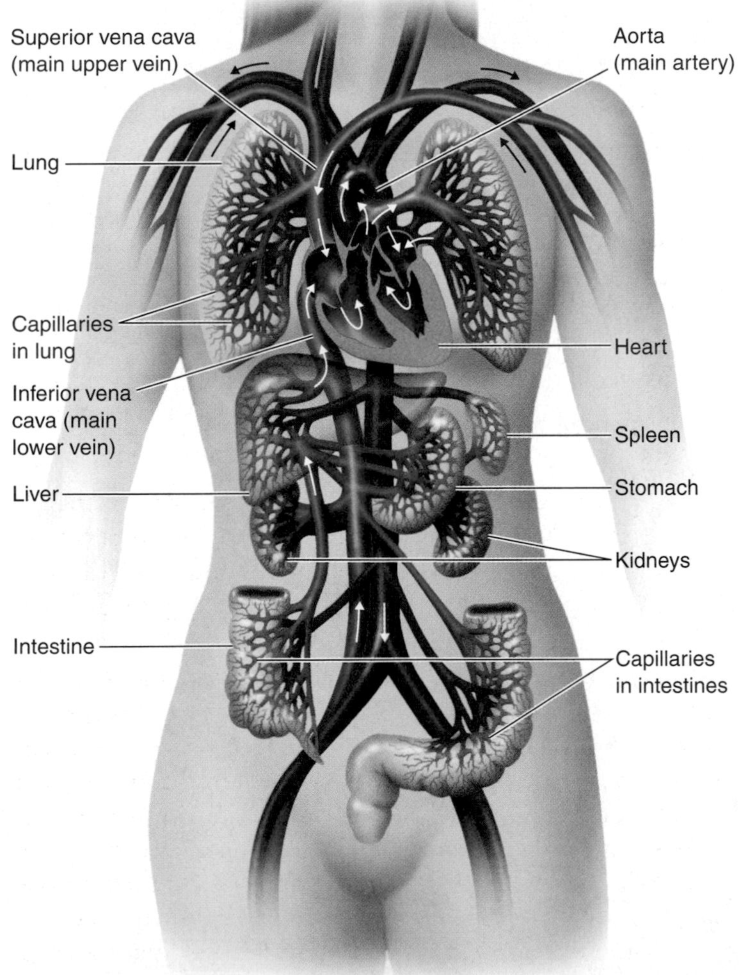

Figure 23.1 The human cardiovascular system and related structures. Details of circulation to the head and extremities are not shown in this simplified diagram. The blood circulates from the heart through the arterial system (red) to the capillaries (purple) in the lungs and other parts of the body. From these capillaries, the blood returns through the venous system (blue) to the heart.

Q **How can a focal infection become systemic?**

CHECK YOUR UNDERSTANDING

✓ Why is the lymphatic system so valuable for the working of the immune system? **23-1**

Bacterial Diseases of the Cardiovascular and Lymphatic Systems

LEARNING OBJECTIVES

23-2 List the signs and symptoms of sepsis, and explain the importance of infections that develop into septic shock.

23-3 Differentiate gram-negative sepsis, gram-positive sepsis, and puerperal sepsis.

23-4 Describe the epidemiologies of endocarditis and rheumatic fever.

23-5 Describe the epidemiology of tularemia.

23-6 Describe the epidemiology of brucellosis.

23-7 Describe the epidemiology of anthrax.

23-8 Describe the epidemiology of gas gangrene.

23-9 List three pathogens that are transmitted by animal bites and scratches.

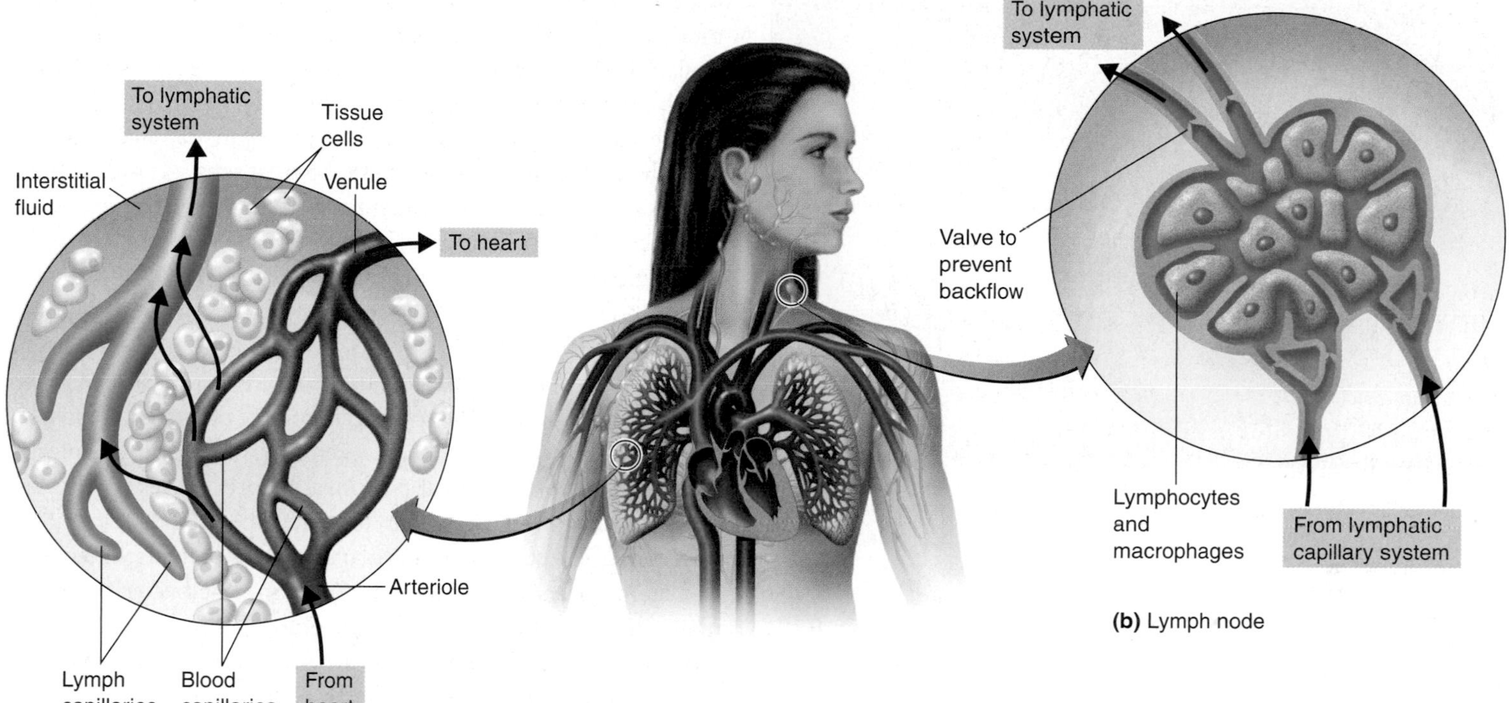

(a) Capillary system in lung

(b) Lymph node

Figure 23.2 The relationship between the cardiovascular and lymphatic systems. (**a**) From the blood capillaries, some blood plasma filters into the surrounding tissue and enters the lymph capillaries. This fluid, now called lymph, returns to the heart through the lymphatic circulatory system (green), which channels the lymph to a vein. (**b**) All lymph returning to the heart must pass through at least one lymph node. (See also Figure 16.5, page 456.)

Q **What is the role of the lymphatic system in defense against infection?**

23-10 Compare and contrast the causative agents, vectors, reservoirs, symptoms, treatments, and preventive measures for plague, Lyme disease, and Rocky Mountain spotted fever.

23-11 Identify the vector, etiology, and symptoms of five diseases transmitted by ticks.

23-12 Describe the epidemiologies of epidemic typhus, endemic murine typhus, and spotted fevers.

Once bacteria gain access to the bloodstream, they become widely disseminated. In some cases, they are also able to reproduce rapidly.

Sepsis and Septic Shock

Although blood is normally sterile, moderate numbers of microorganisms can enter the bloodstream without causing harm. In hospital conditions, the blood frequently is contaminated as a result of invasive procedures, such as insertion of catheters and intravenous feeding tubes. Blood and lymph contain numerous defensive phagocytic cells. Also, blood is low in available iron, which is a requirement for bacterial growth. However, if the defenses of the cardiovascular and lymphatic systems fail, microbes can proliferate in the blood. An acute illness that is associated with the presence and persistence of pathogenic microorganisms or their toxins in the blood is termed **septicemia.** A similar term that is not equated medically with septicemia is sepsis, although there is a tendency to use them interchangeably. **Sepsis** is defined as a *systemic inflammatory response syndrome (SIRS)* caused by a focus of infection that releases mediators of inflammation into the bloodstream. The site of the infection itself is not necessarily the bloodstream, and in about half of the cases no microbes can be found in the blood. The SIRS must exhibit at least two of a set of defined conditions: fever, rapid heart or respiratory rates, and a high count of white blood cells. If the infective bacteria cause red blood cells to lyse, the release of iron-containing hemoglobin can result in accelerated microbial growth. Sepsis and septicemia are often accompanied by the appearance of **lymphangitis,** inflamed lymph vessels visible as red streaks under the skin, running along the arm or leg from the site of the infection (Figure 23.3).

If the body's defenses do not quickly control the infection and the resulting SIRS, the results are progressive and frequently fatal. The first stage of this progression is sepsis. There is evidence of infection and an inflammatory response by the body caused by

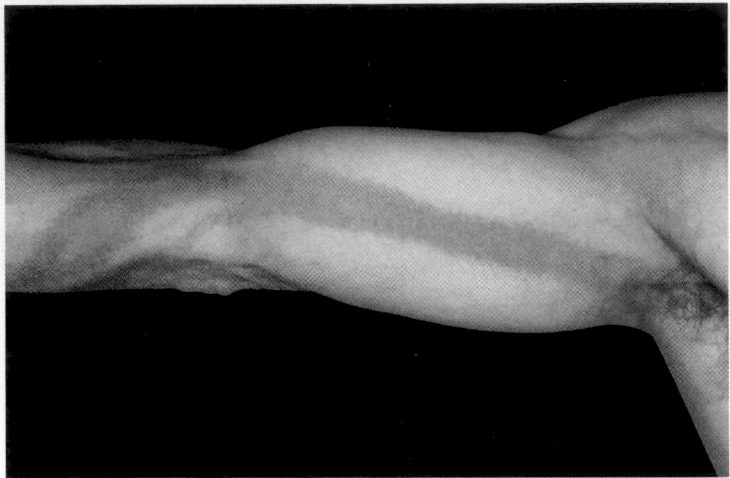

Figure 23.3 Lymphangitis, one sign of sepsis. As the infection spreads from its original site along the lymph vessels, the inflamed walls of the vessels become visible as red streaks.

Q **Why does the red streak sometimes end at a certain point?**

the release and circulation of cytokines. The most obvious signs and symptoms are fever, chills, and accelerated breathing and heart rate. When sepsis results in a drop in blood pressure (*shock*) and dysfunction of at least one organ, it is considered to be **severe sepsis.** Once organs begin to fail, the mortality rate becomes very high. A final stage, when low blood pressure can no longer be controlled by addition of fluids, is **septic shock.**

Gram-Negative Sepsis

Septic shock is most likely to be caused by gram-negative bacteria. Recall that the cell walls of many gram-negative bacteria (LPS; see page 69) contain endotoxins that are released upon lysis of the cell. These endotoxins can cause a severe drop in blood pressure with its associated signs and symptoms. Septic shock is often called by the alternative names *gram-negative sepsis* or *endotoxic shock.* Less than one-millionth of a milligram of endotoxin is enough to cause the symptoms. About 750,000 cases of septic shock occur each year in the United States; at least 225,000 are fatal.

An effective treatment for severe sepsis and septic shock has been a medical priority for many years. The early symptoms of sepsis are relatively nonspecific and not especially alarming. Therefore, the antibiotic treatments that might arrest it then are frequently not administered. The progression to lethal stages is rapid and generally impossible to treat effectively. Administering antibiotics then may even aggravate the condition by causing the lysis of large numbers of bacteria that then release more endotoxins.

In addition to antibiotics, treatment of septic shock involves attempts to neutralize the LPS components and inflammation-causing cytokines. The U.S. Food and Drug Administration (FDA) has approved a drug, drotrecogin alfa (Xigris), which is the first to reduce the death rate of sepsis cases. This drug is a genetically modified version of *human activated protein C*—a natural anticoagulant found at reduced levels in cases of severe sepsis and septic shock (not to be confused with C-reactive protein). The drug reduces clotting, which is a factor in organ damage. Xigris is scarcely the sought-for magic bullet to treat sepsis: it is dauntingly expensive and effective only in a minority of cases. Nonetheless, it is expected to be widely prescribed for treatment of gram-negative sepsis and meningococcal meningitis (see page 613).

Gram-Positive Sepsis

Gram-positive bacteria are now the most common cause of sepsis. Both staphylococci and streptococci produce potent exotoxins that cause toxic shock syndrome, a toxemia discussed in Chapter 21 (page 588). The frequent use of invasive procedures in hospitals allows gram-positive bacteria to enter the bloodstream. Such nosocomial infections are a particular risk for patients who undergo regular dialysis for kidney dysfunction. The bacterial components that lead to septic shock in gram-positive sepsis are not known with certainty. Possible sources are various fractions of the gram-positive cell wall or even bacterial DNA.

An especially important group of gram-positive bacteria are the enterococci, which are responsible for many nosocomial infections. The enterococci are inhabitants of the human colon and frequently contaminate skin. Once considered relatively harmless, two species in particular, *Enterococcus faecium* and *Enterococcus faecalis,* are now recognized as leading causes of nosocomial infections of wounds and the urinary tract. Enterococci have a natural resistance to penicillin and have rapidly acquired resistance to other antibiotics. What has made them something of a medical emergency is the appearance of vancomycin-resistant strains. Vancomycin (see page 563) was often the only remaining antibiotic to which these bacteria, especially *E. faecium,* were still sensitive. Among isolates of *E. faecium* from nosocomial infections of the bloodstream, almost 90% are now resistant.

To this point our discussion of the streptococci has been focused on serologic group A. There is an emerging awareness of group B streptococci (GBS) and of the enterococci. *S. agalactiae* (ā′gal-act-ē-ī) is the only GBS and is the most common cause of life-threatening neonatal sepsis. The CDC recommends that pregnant women be tested for vaginal GBS and that women with GBS be offered antibiotics during labor.

Puerperal Sepsis

Puerperal sepsis, also called **puerperal fever** and **childbirth fever,** is a nosocomial infection. It begins as an infection of the uterus as a result of childbirth or abortion. *Streptococcus pyogenes,* a group A beta-hemolytic streptococcus, is the most frequent cause, although other organisms may cause infections of this type.

Puerperal sepsis progresses from an infection of the uterus to an infection of the abdominal cavity *(peritonitis)* and in many cases to sepsis. At a Paris hospital between 1861 and 1864, of the 9886 women who gave birth, 1226 (12%) died of such infections. These deaths were largely unnecessary. Some 20 years before, Oliver Wendell Holmes in the United States and Ignaz Semmelweis in Austria had clearly demonstrated that the disease was transmitted by the hands and instruments of the attending midwives or physicians and that disinfecting the hands and instruments could prevent such transmission. Antibiotics, especially penicillin, and modern hygienic practices have now made *S. pyogenes* puerperal sepsis an uncommon complication of childbirth.

CHECK YOUR UNDERSTANDING

- ✓ What are two of the conditions that define the systemic inflammatory response syndrome of sepsis? **23-2**
- ✓ Are the endotoxins that cause sepsis from gram-positive or gram-negative bacteria? **23-3**

Bacterial Infections of the Heart

The wall of the heart consists of three layers. The inner layer, called the *endocardium*, lines the heart muscle itself and covers the valves. An inflammation of the endocardium is called **endocarditis.**

One type of bacterial endocarditis, **subacute bacterial endocarditis** (so named because it develops slowly; Figure 23.4), is characterized by fever, general weakness, and heart murmur. It is usually caused by alpha-hemolytic streptococci such as are common in the oral cavity, although enterococci or staphylococci are often involved. The condition probably arises from a focus of infection elsewhere in the body, such as in the teeth or tonsils. Microorganisms are released by tooth extractions or tonsillectomies, enter the blood, and find their way to the heart. A more exotic source of infections that have led to cases of endocarditis has been body piercing, especially of the nose, tongue, and even nipples. Normally, such bacteria would be quickly cleared from the blood by the body's defensive mechanisms. However, in people whose heart valves are abnormal, because of either congenital heart defects or such diseases as rheumatic fever and syphilis, the bacteria lodge in the preexisting lesions. Within the lesions, the bacteria multiply and become entrapped in blood clots that protect them from phagocytes and antibodies. As multiplication progresses and the clot gets larger, pieces of the clot break off and can block blood vessels or lodge in the kidneys. In time, the function of the heart valves is impaired. Left untreated by appropriate antibiotics, subacute bacterial endocarditis is fatal within a few months.

A more rapidly progressive type of bacterial endocarditis is **acute bacterial endocarditis,** which is usually caused by *Staphylococcus aureus.* The organisms find their way from the initial site of infection to normal or abnormal heart valves; the rapid destruction of the heart valves is frequently fatal within a few days or weeks if untreated. Streptococci can also cause **pericarditis,** inflammation of the sac around the heart (the *pericardium*).

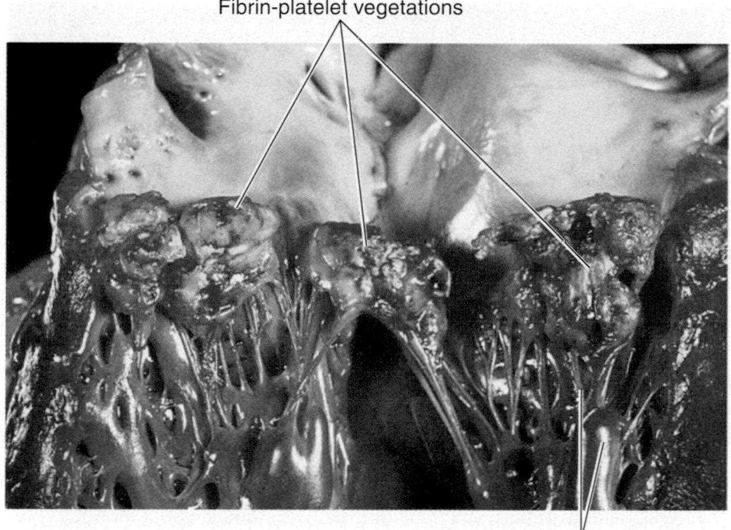

Figure 23.4 Bacterial endocarditis. This is a case of subacute endocarditis, meaning that it developed over a period of weeks or months. The heart has been dissected to expose the mitral valve. The cordlike structures connect the heart valve to the operating muscles.

Endocarditis develops as bacteria attach to the surface and multiply, causing damage that promotes the formation of fibrin-platelet vegetations (shown in photo). These vegetations, a biofilm, bury adherent bacteria and allow them to multiply protected from defenses of the host; further deposition of bacteria cause the vegetation to enlarge in layers.

Symptoms usually include fever and a heart murmur from poor mitral valve function that is detectable by echocardiogram. Treatment with antibiotics in high concentrations is often effective.

Q How is endocarditis contracted?

CHECK YOUR UNDERSTANDING

- ✓ What medical procedures are usually the cause of endocarditis? **23-4**

Rheumatic Fever

Streptococcal infections, such as those caused by *Streptococcus pyogenes,* sometimes lead to **rheumatic fever,** which is generally considered an autoimmune complication. It occurs primarily in people aged 4 to 18 and often follows an episode of streptococcal sore throat. The disease is usually first expressed as a short period of arthritis and fever. Subcutaneous nodules at joints often accompany this stage (Figure 23.5). In about half of persons affected, an inflammation of the heart, probably from a misdirected immune reaction against streptococcal M protein, damages the valves. Reinfection with streptococci renews the immune attack. Damage to heart valves may be serious enough to result in eventual failure and death.

Early in the twentieth century, rheumatic fever killed more school-aged children in the United States than all other diseases combined. The incidence declined steadily in developed

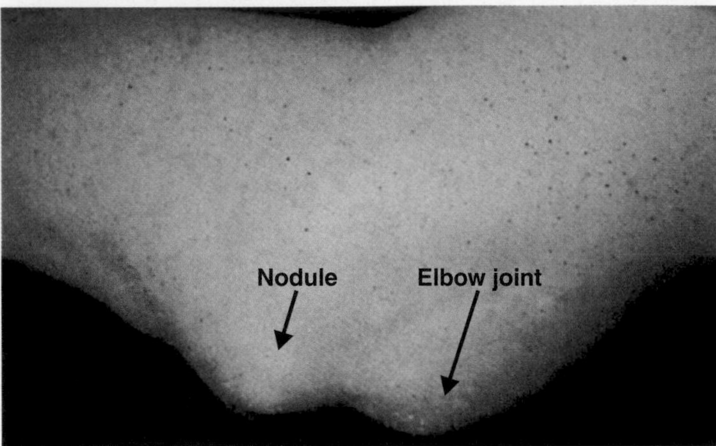

Figure 23.5 A nodule caused by rheumatic fever. Rheumatic fever was named, in part, because of the characteristic subcutaneous nodules that appear at the joints, as shown in this patient's elbow. Infection with group A beta-hemolytic streptococci sometimes leads to this autoimmune complication.

Q Is rheumatic fever a bacterial infection?

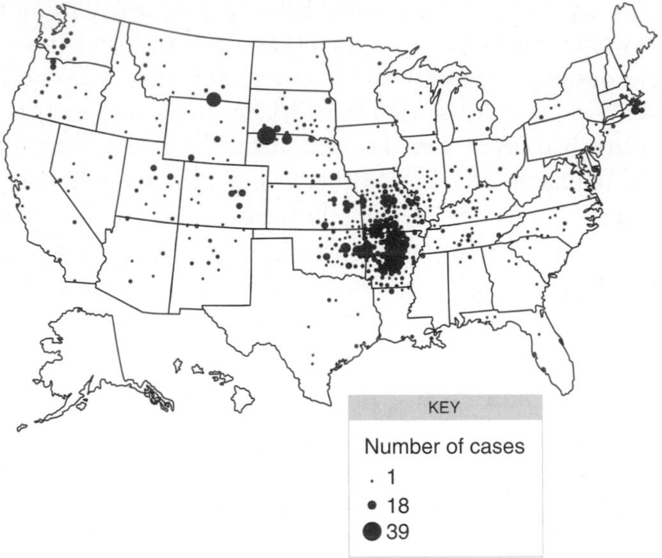

Figure 23.6 Tularemia cases in the United States (1990–2000). There were 1347 cases for which a county of residence was reported. The circle sizes are proportional to the number of cases in a county, ranging from 1 to 39.

SOURCE: CDC, *MMWR* 51(9), March 8, 2002.

Q What area reporting tularemia is closest to you?

countries to the point of becoming rare even before the introduction of effective antimicrobial drugs during the 1930s and 1940s. Many young physicians have never seen a case of the disease, but in much of the underdeveloped world it remains the leading cause of heart disease in the young. The decline of rheumatic fever in the United States is thought to be due to some loss of virulence in the streptococci in circulation. However, since the 1980s there have been a few localized outbreaks of rheumatic fever in the United States that have been related to certain M-protein serotypes. These serotypes had been prevalent during much earlier epidemics of rheumatic fever but had almost disappeared from circulation. People who have had an episode of rheumatic fever are at risk of renewed immunological damage with repeated streptococcal sore throats. The bacteria have remained sensitive to penicillin, and patients at particular risk, such as these, often receive a monthly preventive injection of long-acting penicillin G benzathine.

As many as 10% of people with rheumatic fever develop **Sydenham's chorea,** an unusual complication known in the Middle Ages as Saint Vitus' dance. Several months following an episode of rheumatic fever, the patient (much more likely to be a girl than a boy) exhibits purposeless, involuntary movements during waking hours. Occasionally, sedation is required to prevent self-injury from flailing arms and legs. The condition disappears after a few months.

Sepsis and infections of the heart are summarized in Diseases in Focus 23.1.

Tularemia

Tularemia is an example of a *zoonotic* disease, that is, a disease transmitted by contact with infected animals, most commonly rabbits and ground squirrels. The name derives from Tulare County, California, where the disease was originally observed in ground squirrels in 1911. The pathogen is *Francisella tularensis,* a small gram-negative bacillus. It can enter humans by several routes. The most common is penetration of the skin at a minor abrasion, where it creates an ulcer at the site. About a week after infection, the regional lymph nodes enlarge; many will contain pockets of pus. (See the box on page 644.) The bacterium can multiply in macrophages—as much as a thousand-fold. Mortality is normally less than 3%. If not contained, the proliferation of *F. tularensis* can lead to sepsis and infection of multiple organs.

Almost 90% of cases in the United States are related to contact with rabbits, and the disease is often known locally as *rabbit fever.* Tularemia is also transmitted in some areas by ticks and insects and is known there as *deer fly fever.* Respiratory infection, usually by dust contaminated by urine or feces of infected animals, can cause an acute pneumonia with a mortality rate exceeding 30%. The infective dose is very small, and the organism is dangerous to handle if aerosols are likely to be produced.

At one time, so few cases of tularemia were recorded annually in the United States that it was removed from the list of nationally notifiable diseases, However, concern that it might be used as a biological weapon has recently led to its reinstatement on the list. Figure 23.6 illustrates the geographic distribution of tularemia within the United States. It is also found worldwide in many areas of the Northern Hemisphere.

DISEASES IN FOCUS 23.1

Infections from Human Reservoirs

Differential diagnosis is the process of identifying the disease from a list of possible diseases that fit the information derived from examining a patient. A differential diagnosis is important for providing initial treatment and for laboratory testing. Microorganisms circulating in the blood may reflect a serious, uncontrolled infection. For example, 27-year-old woman had a fever and cough for 5 days. She was hospitalized when her blood pressure dropped. Despite aggressive treatment with fluids and massive doses of antibiotics, she died 5 hours after hospitalization. Catalase-negative, gram-positive cocci were isolated from her blood. Use the table below to identify infections that could cause these symptoms. For the solution, go to www.microbiologyplace.com.

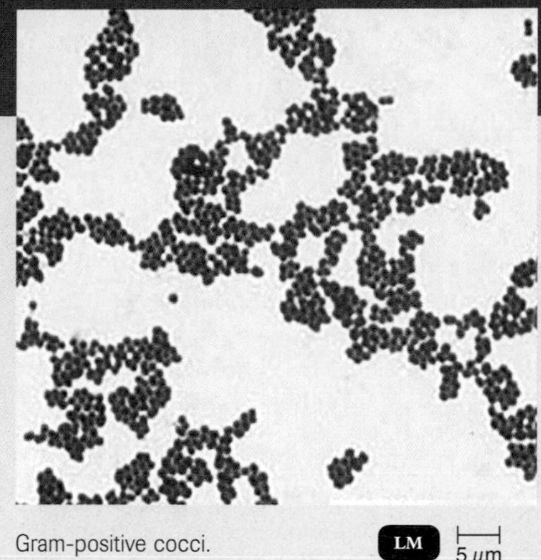

Gram-positive cocci. LM 5 μm

Disease	Pathogen	Symptoms	Reservoir	Method of Transmission	Treatment
BACTERIAL DISEASES					
Septic shock	Gram-negative bacteria, enterococci, group B streptococci	Fever, chills, incresed heart rate; lymphangitis	Human body	Injection; catheterization	Xigris (gram negatives); antibiotics (gram positives)
Puerperal sepsis	*Streptococcus pyogenes*	Peritonitis; sepsis	Human nasopharynx	Nosocomial	Penicillin
Endocarditis **Subacute bacterial** **Acute bacterial**	Mostly alpha-hemolytic streptococci; *Staphylococcus aureus*	Fever, general weakness, heart murmur; damage to heart valves	Human nasopharynx	From focal infection	Antibiotics
Pericarditis	*Streptococcus pyogenes*	Fever; general weakness; heart murmur	Human nasopharynx	From focal infection	Antibiotics
Rheumatic fever	Group A beta-hemolytic streptococci	Arthritis, fever; damage to heart valves	Immune reactions to streptococcal infections		Supportive. Prevention: penicillin to treat streptococcal sore throats
VIRAL DISEASES					
Burkitt's lymphoma	Epstein-Barr virus (EB virus)	Tumor	Unknown	Unknown	Surgery
Infectious mononucleosis	EB virus	Fever, general weakness	Humans	Saliva	None
Cytomegalovirus	Cytomegalovirus	Mostly asymptomatic; initial infection acquired during pregnancy can be damaging to fetus	Humans	Body fluids	Ganciclovir, fomivirsen

The intracellular location of the bacterium is a problem in chemotherapy. Tetracycline is the antibiotic of choice.

CHECK YOUR UNDERSTANDING

✓ What animals are the most common reservoir for tularemia? **23-5**

Brucellosis (Undulant Fever)

With over 500,000 new human cases annually, **brucellosis** is the world's most common bacterial zoonosis. The Middle East is an endemic area, and several countries in the region record the world's highest incidence of the disease. It is widespread around the Mediterranean and in southeastern Europe, Asia, Latin America, and the Caribbean. It is also economically important as a disease of

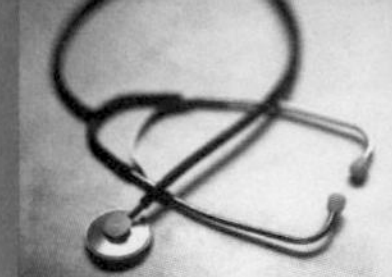

CLINICAL FOCUS From the *Morbidity & Mortality Weekly Report*

A Sick Child

As you read through this problem, you'll encounter questions that primary health care providers ask themselves as they solve a clinical problem. Try to answer each question as a health care provider.

1. On February 15, a 3-year-old boy was seen by his pediatrician for fever, malaise, painful left underarm lymph node, and skin sloughing off his left ring finger. Amoxicillin was prescribed.
   What diseases are possible?
2. The child underwent excisional biopsy of the left axillary lymph node when intermittent fever and enlarged lymph node persisted for 49 days. The excised tissue was cultured; a gram-stain of the bacteria that grew is shown in the figure.
   What additional tests would you do?
3. Serological tests revealed the following results:

Pathogen	Antibody Titer
Bartonella	0
Ehrlichia	0
Francisella	4,096
CMV	0
Toxoplasma gondii	0

   The child improved after treatment with ciprofloxacin.
   What is the cause of the infection? What do you need to know?
4. PCR was used to confirm identification of *Francisella tularensis*. Between January 2 and February 8, the boy's family purchased six hamsters from a pet store. Each hamster died from diarrhea within one week of purchase. One hamster bit the child on the left ring finger.
   Where will you look for the source of the infection?
5. Workers at the pet store reported an unusual number of deaths among hamsters but not other animals during January and February. Eight other customers reported that their hamsters died within 2 weeks of purchase. Available hamsters were negative for *F. tularensis* by serology and culture. One of two cats kept as store pets had a positive serologic test for *F. tularenesis* at a titer of 256. The hamsters came from customers who had pets with unanticipated litters.
   What is the most likely source of infection?
6. The hamsters came from different sources, so they are probably not the origin of the infection. The positive serological test in a pet cat suggests that infected wild rodents infested the store and spread the infection to hamsters by urinating and defecating through the hamster cages. The infected cat might have had an unrecognized illness after catching or eating an infected wild rodent.

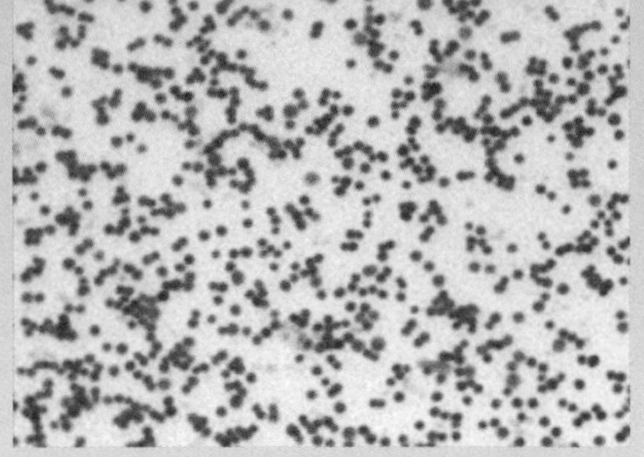

Gram-stained bacteria cultured from lymph node.

Public health officials should be aware that pet rodents might be a source of tularemia. Identifying the organism is important because it is often resistant to antibiotics commonly used for skin and systemic infections and because it is a potential agent of biological terrorism.

Source: Adapted from reports in *MMWR* 53(52):1202, January 7, 2005; and *MMWR* 54(7):170, February 25, 2005.

animals in the developing world. Human cases of brucellosis are usually not fatal, but the disease tends to persist in the reticuloendothelial system (see page 457), where the bacteria evade the host's defenses; they are especially adept at evading phagocytic cells. This ability allows long-term survival and replication. The disease often becomes chronic and is capable of affecting any organ system.

Brucella bacteria are small, aerobic, gram-negative coccoid rods. During laboratory handling, they easily become airborne and are considered dangerous to handle. In fact, they are considered a potential agent of bioterrorism. The three species that are of greatest interest are *Brucella abortus* (brü-sel′ä ä-bōr′tus), which is found primarily in cattle but also infects camels, bison, and several other animals. *Brucella suis* (sü′is) is a species mostly infecting swine but is known to infect cattle when they are kept in contact with swine herds. Abattoir (slaughterhouse) workers who come in contact with swine carcasses are at risk of brucellosis from this species. The most serious pathogen, and the cause of most human cases, is *Brucella melitensis* (me-li-ten′sis). This species is most commonly found today in goats and sheep.

The pattern of brucellosis in the United States has changed over recent decades. At one time, most human cases in the United States were attributed to *B. abortus*. But an intensive campaign of testing and vaccination eliminated brucellosis in cattle (there is a vaccine available for animals but not for humans), and *B. abortus* is now seldom a cause of human disease. At the present time, most cases of brucellosis are caused by *B. melitensis*, predominantly among Hispanic members of the population. The disease is endemic in Mexico and is often imported into the United States in unpasteurized food products such as Mexican soft cheese made from goat milk.

The incubation period is usually 1 to 3 weeks but might be much longer. Symptoms of brucellosis have a wide spectrum, depending on the stage of the disease and the organs affected. Typically they include fever (often rising and falling, which has given the disease an alternative name of *undulant fever*), malaise, night sweats, and muscle aches. Although several serological tests are available, there is still a need for a definitive diagnostic test. The ultimate diagnostic proof is isolation of *Brucella* from the patient's blood or tissue. Because the disease is not common, diagnosis often must start with patient interviews that suggest a contact in endemic areas of the disease.

Antibiotic therapy is possible, and the bacteria have not shown development of resistance. However, treatment must be very long term, usually at least 6 weeks, involving a combination of at least two antibiotics.

CHECK YOUR UNDERSTANDING

✓ What ethnic group in the United States is most commonly affected by brucellosis, and why? **23-6**

Anthrax

In 1877, Robert Koch isolated *Bacillus anthracis,* the bacterium that causes **anthrax** in animals. The endospore-forming bacillus is a large, aerobic, gram-positive microorganism that is apparently able to grow slowly in soil types having specific moisture conditions. The endospores have survived in tests in soil for up to 60 years. The disease strikes primarily grazing animals, such as cattle and sheep. The *B. anthracis* endospores are ingested along with grasses, causing a fulminating, fatal sepsis.

The incidence of human anthrax is now rare in the United States, but occurrences in grazing animals are not uncommon. People at risk are those who handle animals, hides, wool, and other animal products from certain foreign countries.

Infections by *B. anthracis* are initiated by endospores. Once introduced into the body, they are taken up by macrophages, where they germinate into vegetative cells. These are not killed, but multiply, eventually killing the macrophage. The released bacteria then enter the bloodstream, replicate rapidly, and secrete toxins.

The primary virulence factors of *B. anthracis* are two exotoxins. Both toxins share a third toxic component, a cell receptor–binding protein called the *protective antigen,* that binds the toxins to target cells and permits their entry. One toxin, the *edema toxin,* causes local edema (swelling) and interferes with phagocytosis by macrophages. The other toxin, *lethal toxin,* specifically targets and kills macrophages, which disables an essential defense of the host. Furthermore, the capsule of *B. anthracis* is very unusual. It is not a polysaccharide but rather is composed of amino acid residues, which for some reason do not stimulate a protective response by the immune system. Therefore, once the anthrax bacteria enter the bloodstream, they proliferate without any effective inhibition until there are tens of millions per milliliter. These immense populations of toxin-secreting bacteria ultimately kill the host.

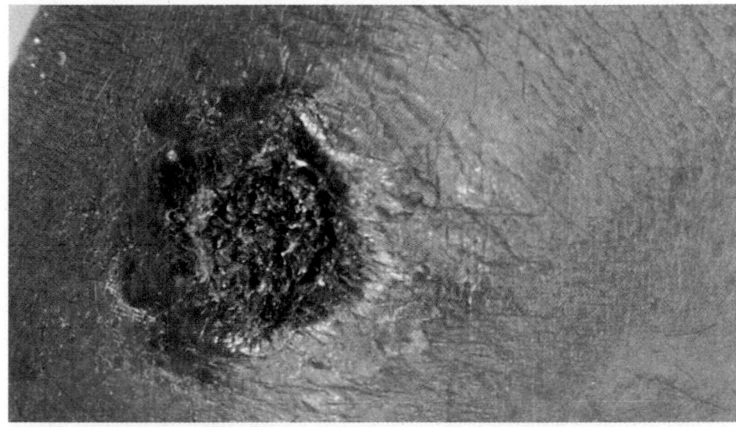

Figure 23.7 Anthrax lesion. The swelling and formation of a black scab that forms around the point of infection is a characteristic of cutaneous anthrax.

Q **What are the other types of anthrax?**

Anthrax affects humans in three forms: cutaneous anthrax, gastrointestinal anthrax, and inhalational (pulmonary) anthrax.

Cutaneous anthrax results from contact with material containing anthrax endospores. Over 90% of naturally occurring cases of anthrax in humans are cutaneous; the endospore enters at some minor skin lesion. A papule appears and then eventually vesicles, which rupture and form a depressed, ulcerated area that is covered by a black eschar (scab) as shown in Figure 23.7. (The name *anthrax* is derived from the Greek word for coal). In most cases the pathogen does not enter the bloodstream, and other symptoms are limited to a low-grade fever and malaise. However, if the bacteria enter the bloodstream, mortality without antibiotic treatment can reach 20%; with antibiotic therapy, mortality is usually less than 1%.

A relatively rare form of anthrax is **gastrointestinal anthrax** caused by ingestion of undercooked food containing anthrax endospores. Symptoms are nausea, abdominal pain, and bloody diarrhea. Ulcerative lesions occur in the gastrointestinal tract ranging from the mouth and throat to, mainly, the intestines. Mortality is usually more than 50%.

The most dangerous form of anthrax in humans is **inhalational (pulmonary) anthrax.** Endospores inhaled into the lungs have a high probability of entering the bloodstream. Symptoms of the first few days of the infection are not especially alarming: mild fever, coughing, and some chest pain. The disease can be arrested at this stage by antibiotics, but unless suspicion of anthrax is high, they are unlikely to be administered. As the bacteria enter the bloodstream and proliferate, the illness progresses in 2 or 3 days into septic shock that usually kills the patient within 24 to 36 hours. The mortality rate is exceptionally high, approaching 100%.

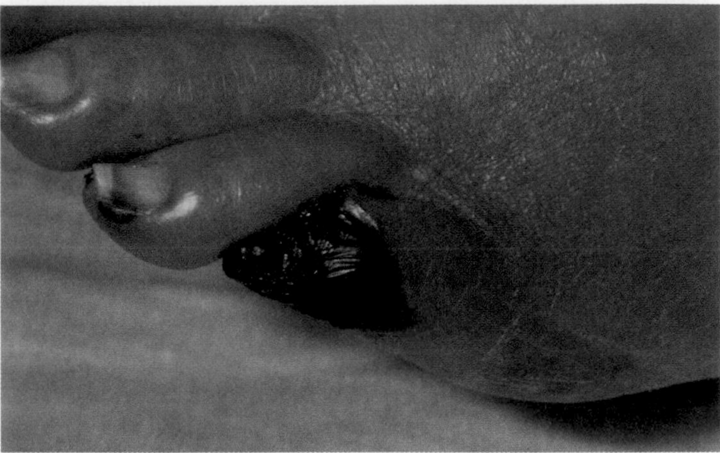

Figure 23.8 The toes of a patient with gangrene. This disease is caused by *Clostridium perfringens* and other clostridia. The black, necrotic tissue, resulting from poor circulation or injury, furnishes anaerobic growth conditions for the bacteria, which then progressively destroy adjoining tissue.

Q **How can gangrene be prevented?**

Antibiotics are effective in treating anthrax if they are administered in time. Currently recommended drugs are ciprofloxacin or doxycycline plus one or two additional agents that are known to be active against the pathogen. People who have been exposed to anthrax endospores can be given preventive doses of antibiotics for a time as a precaution. This time period is usually quite long because experience has shown that up to 60 days can elapse before the inhaled endospores germinate and initiate active disease.

Vaccination of livestock against anthrax is a standard procedure in endemic areas. A single dose of an effective live, attenuated vaccine is used, which is considered unsafe for use in humans. The only vaccine currently approved for use in humans contains an inactivated form of the protective antigen toxin and is designed to prevent entry of the other two toxins into the host's cells. This vaccine requires a series of six injections over a period of 18 months, followed by annual boosters. In view of the recent use of anthrax as a weapon of bioterrorism (see the box on page 649), the need for a more practical human vaccine has become urgent. The target is a vaccine that would require no more than three injections and would work rapidly enough that it could be given *after* exposure to anthrax endospores.

Diagnosis of anthrax has usually consisted of isolation and identification of *B. anthracis* from a clinical specimen—which is too slow for detection of bioterrorism outbreaks. A blood test can detect both inhalational and cutaneous cases of anthrax within an hour. Furthermore, locations such as a few mail-sorting facilities are being equipped with automated electronic sensors that can immediately detect anthrax spores.

CHECK YOUR UNDERSTANDING

✓ How do animals such as cattle become victims of anthrax? **23-7**

Gangrene

If a wound causes the blood supply to be interrupted, a condition known as **ischemia,** the wound becomes anaerobic. Ischemia leads to **necrosis,** or death of the tissue. The death of soft tissue resulting from the loss of blood supply is called **gangrene** (Figure 23.8). These conditions can also occur as a complication of diabetes.

Substances released from dying and dead cells provide nutrients for many bacteria. Various species of the genus *Clostridium,* which are gram-positive, endospore-forming anaerobes widely found in soil and in the intestinal tracts of humans and domesticated animals, grow readily in such conditions. *C. perfringens* is the species most commonly involved in gangrene, but other clostridia and several other bacteria can also grow in such wounds.

Once ischemia and the subsequent necrosis caused by impaired blood supply have developed, **gas gangrene** can develop, especially in muscle tissue. As the *C. perfringens* microorganisms grow, they ferment carbohydrates in the tissue and produce gases (carbon dioxide and hydrogen) that swell the tissue. The bacteria produce toxins that move along muscle bundles, killing cells and producing necrotic tissue that is favorable for further growth. Eventually, these toxins and bacteria enter the bloodstream and cause systemic illness. Enzymes produced by the bacteria degrade collagen and proteinaceous tissue, facilitating the spread of the disease. Without treatment, the condition is fatal.

One complication of improperly performed abortions is the invasion of the uterine wall by *C. perfringens,* which resides in the genital tract of about 5% of all women. This infection can lead to gas gangrene and result in a life-threatening invasion of the bloodstream.

The surgical removal of necrotic tissue and amputation are the most common medical treatments for gas gangrene. When gas gangrene occurs in such regions as the abdominal cavity, the patient can be treated in a **hyperbaric chamber,** which contains a pressurized oxygen-rich atmosphere (Figure 23.9). The oxygen saturates the infected tissues and thereby prevents the growth of the obligately anaerobic clostridia. Small chambers are available that can accommodate a gangrenous limb. The prompt cleaning of serious wounds and precautionary antibiotic treatment are the most effective steps in preventing gas gangrene. Penicillin is effective against *C. perfringens.*

CHECK YOUR UNDERSTANDING

✓ Why are hyperbaric chambers effective in treating gas gangrene? **23-8**

Figure 23.9 Hyperbaric chambers used to treat gas gangrene. A multiplace hyperbaric chamber that can accommodate several patients is shown. Such chambers are usually available at major medical centers. They are also used to treat victims of carbon monoxide poisoning.

Name another bacterial disease that might be treated in a hyperbaric chamber.

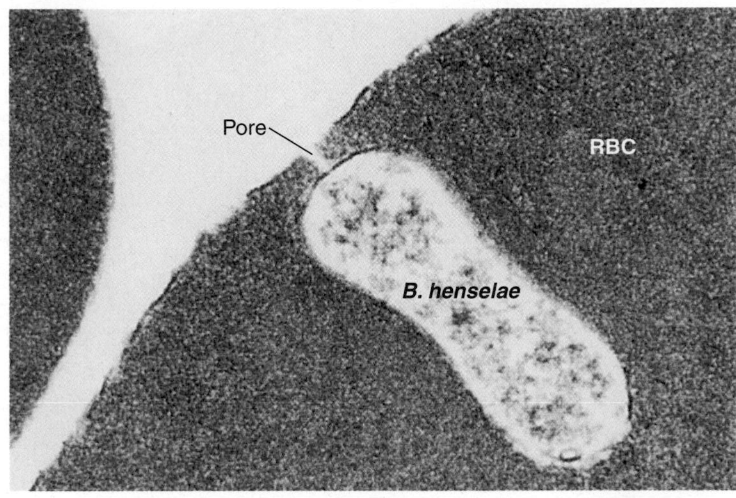

Figure 23.10 Electron micrograph showing the location of *Bartonella henselae* within a red blood cell. Only a pore connects the bacterium with the extracellular fluid.

Q **Why can *B. henselae* infection persist in cats?**

Systemic Diseases Caused by Bites and Scratches

Animal bites can result in serious infections. About 4.4 million animal bites occur in the United States annually, accounting for about 1% of visits to the emergency rooms in hospitals.

Dog bites make up at least 80% of reported bite incidents; cat bites, only about 10%. Cat bites are, however, more penetrating, resulting in a higher infection rate (30–50%), than the bites of dogs (15–20%). Domestic animals often harbor *Pasteurella multocida* (pas-tyėr-el′lä mul-tō′si-dä), a gram-negative rod similar to the *Yersinia* bacterium that causes plague (page 648). *P. multocida* is primarily a pathogen of animals, and it causes sepsis (hence the name *multocida,* meaning many-killing).

Humans infected with *P. multocida* have varied responses. For example, local infections with severe swelling and pain can develop at the site of the wound. Forms of pneumonia and sepsis may develop and are life-threatening. Penicillin and tetracycline are usually effective in treating these infections.

In addition to *P. multocida,* an assortment of anaerobic bacterial species are often found in infected animal bites, as well as species of *Staphylococcus, Streptococcus,* and *Corynebacterium.* Bites by humans, mostly as a result of fighting, are also prone to serious infections. In fact, before antibiotic therapy became available, nearly 20% of victims of human bites on extremities required amputation—currently only about 5% of cases require it.

Cat-Scratch Disease

Cat-scratch disease, although it receives little attention, is surprisingly common. An estimated 22,000 or more cases occur annually in the United States, many more than the well-known Lyme disease. People who own or are closely exposed to cats are at risk. The pathogen is an aerobic, gram-negative bacterium, *Bartonella henselae* (bär′to-nel-lä hen′sel-ī). Microscopy shows that the bacterium can inhabit the interior of some cat red blood cells. It is connected to the exterior of the cell and to the surrounding extracellular fluid by a pore (**Figure 23.10**). Resident there, the bacteria cause a persistent bacteremia in cats; it is estimated that as many as 50% of domestic and feral (wild) cats carry these bacteria in their blood. The primary mode of transmission is by the scratch of a cat; it is uncertain whether bites of cats or of cat fleas transmit the disease to humans. But the presence of cat fleas is definitely a requirement for the infection to be maintained among cats. *B. henselae* multiplies in the digestive system of the cat flea and survives for several days in flea feces. Cat claws then become contaminated from flea feces.

The initial sign is a papule at the infection site, which appears 3 to 10 days after exposure. Swelling of the lymph nodes and usually malaise and fever follow in a couple of weeks. Cat-scratch disease is ordinarily self-limiting, with a duration of a few weeks, but in severe cases antibiotic therapy may be effective.

Rat-Bite Fever

In large urban areas (even in the United States), the rat population is not well controlled, and bites from rats are a fairly common occurrence—and may cause the disease of **rat bite fever.** At one time, the victims of rat bites were small children in substandard housing. Today, rats are popular as laboratory study animals and even as pets; potential patients are now often laboratory technicians who handle rats as well as pet owners and pet store workers.

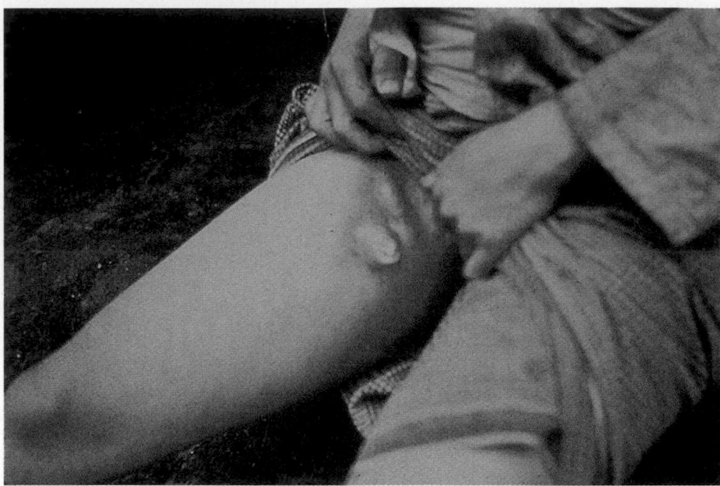

Figure 23.11 A case of bubonic plague. Bubonic plague is caused by infection with *Yersinia pestis.* This photograph shows a bubo (swollen lymph node) on the thigh of a patient. Swollen lymph nodes are a common indication of systemic infection.

Q **In what two ways is plague transmitted?**

Although about half of both wild and laboratory rats are known to carry the bacterial pathogens, only a minority of rat bites (about 10%) result in disease.

There are two similar but distinct diseases. In North America the more common disease, called *streptobacillary rat bite fever,* is caused by *Streptobacillus monilliformis* (when the pathogen is ingested, the disease is termed *Haverhill fever*). This is a filamentous, gram-negative, highly pleomorphic, fastidious bacterium that is difficult to culture, although isolation in culture is the best diagnostic method. The symptoms are initially fever, chills, and muscle and joint pain, followed in a few days by a rash on the extremities. Occasionally there are more serious complications; if untreated, mortality is around 10%.

The other bacterial pathogen causing rat bite fever is *Spirillum minus.* In this case, the disease is called *spirillar fever;* in Asia, where most cases occur, it is known as *sodoku.* It is more likely to occur in bites by wild rodents. The symptoms are similar to those of streptobacillary rat bite fever. Because the pathogen cannot be cultured, diagnosis is made by microscopic observation of the gram-negative, spiral-shaped bacterium. Treatment by penicillin or doxycycline is usually effective for both forms of rat-bite fever.

Cardiovascular infections transmitted to humans by contact with other animals are summarized in Diseases in Focus 23.2.

CHECK YOUR UNDERSTANDING

✓ *Bartonella henselae*, the pathogen of cat-scratch disease, is capable of growth in what insect? **23-9**

Vector-Transmitted Diseases

Vector-borne diseases of the cardiovascular system are summarized in Diseases in Focus 23.3.

Plague

Few diseases have affected human history more dramatically than **plague,** known in the Middle Ages as the Black Death. This term comes from one of its characteristics, the dark blue areas of skin caused by hemorrhages.

The disease is caused by a gram-negative, rod-shaped bacterium, *Yersinia pestis.* Normally a disease of rats, plague is transmitted from one rat to another by the rat flea, *Xenopsylla cheopis* (ze-nop-sil′lä kē-ō′pis) (see Figure 12.32b, page 363). In the far West and Southwest, the disease is endemic in wild rodents, especially ground squirrels and prairie dogs.

If its host dies, the flea seeks a replacement host, which may be another rodent or a human. It can jump about 3½ inches. A plague-infected flea is hungry for a meal because the growth of the bacteria forms a biofilm that blocks the flea's digestive tract, and the blood the flea ingests is quickly regurgitated. An arthropod vector is not always necessary for plague transmission. Contact from the skinning of infected animals; scratches, bites, and licks by domestic cats; and similar incidents have been reported to cause infection.

In the United States, exposure to plague is increasing, as residential areas encroach on areas with infected animals. In parts of the world where human proximity to rats is common, infection from this source still prevails.

From the flea bite, bacteria enter the human's bloodstream and proliferate in the lymph and blood. One factor in the virulence of the plague bacterium is its ability to survive and proliferate inside phagocytic cells rather than being destroyed by them. An increased number of highly virulent organisms eventually emerges, and an overwhelming infection results. The lymph nodes in the groin and armpit become enlarged, and fever develops as the body's defenses react to the infection. Such swellings, called *buboes*, account for the name **bubonic plague** (Figure 23.11). This is the most common form, comprising 80–95% of cases today. The mortality rate of untreated bubonic plague is 50–75%. Death, if it occurs, is usually within less than a week after the appearance of symptoms.

A particularly dangerous condition called **septicemic plague** arises when the bacteria enter the blood and proliferate, causing septic shock. Eventually, the blood carries the bacteria to the lungs, and a form of the disease called **pneumonic plague** results. The mortality rate for this type of plague is nearly 100%. Even today, this disease can rarely be controlled if it is not recognized within 12 to 15 hours of the onset of fever.

Protection against Bioterrorism

The idea of biological weapons, or **bioweapons**—that is, the use of living pathogens for hostile purposes—is not new. The earliest recorded use of biological warfare occurred in 1346. The Tartar army catapulted plague-ridden bodies over the walls of Kaffa (Ukraine). After the fall of Kaffa, survivors escaping the fallen city introduced plague into Europe. Thus began the plague pandemic of 1348–1350. During the Sino-Japanese War (1937–1945), airplanes dropped canisters of fleas carrying *Yersinia pestis* on China.

In 1979, *Bacillus anthracis* was being produced in Sverdlovsk (Soviet Union) when an accidental release of *B. anthracis* resulted in 100 deaths in a 2-week period.

Historically, biological weapons have been associated with military action. The use of biological agents to intimidate civilians and governments, **bioterrorism,** began late in the twentieth century.

- In 1984, a religious cult attacked the people of The Dalles, Oregon, by intentionally contaminating food in restaurants and supermarkets with *Salmonella enterica.*
- In 1996, 15 people developed severe gastroenteritis requiring hospitalization when a laboratory worker intentionally contaminated pastries with *Shigella dysenteriae.*
- In 2001, someone used the U.S. Postal Service to spread *Bacillus anthracis* in New York City and Washington, D.C.

One of the problems with bioweapons is that they contain living organisms (see the table), so their impact is difficult to control or even predict. When the use of biological agents is considered a possibility, military personnel and first-responders (health care personnel and others) are vaccinated if a vaccine for the suspected agent exists. The current plan to protect civilians in the event of an attack with a microbe is illustrated by the smallpox preparedness plan. It is not practical to vaccinate everyone against smallpox. The U.S. government's current strategy following a confirmed smallpox outbreak includes "ring containment and voluntary vaccination." Ring containment consists of identifying people with the infection, vaccinating everyone who has had contact with them, and then vaccinating people in surrounding areas.

It might not be possible to stop all wars, but the public health system is improving its ability to respond to bioweapons. Rapid tests to detect genetic changes in hosts due to bioweapons even before symptoms develop are being investigated. Early warning systems, such as DNA chips or recombinant cells that fluoresce (see the figure) in the presence of a bioweapon, are being developed. New vaccines are being developed, and existing vaccines are being stockpiled for use where needed.

The bioweapons detector called Canary uses B cells specific to a particular bacterium or virus. The B cells are engineered to emit light when they detect their target pathogen.

The "ideal" bioweapon is one that is disseminated by aerosol, is spread efficiently from human to human, causes a debilitating disease, and has no readily available treatment. Lists of potential biological weapons usually contain the following organisms:

Bacteria	Viruses
Bacillus anthracis	"Eradicated" polio and measles
Brucella spp.	Encephalitis viruses
Chlamydophila psittaci	Hemorrhagic fever viruses (Ebola, Marburg, Lassa)
Clostridium botulinum toxin	Influenza A (1918 strain)
Coxiella burnetii	Monkeypox
Francisella tularensis	Nipah virus
Rickettsia prowazekii	Smallpox
Shigella spp.	Yellow fever
Vibrio cholerae	
Yersinia pestis	

DISEASES IN FOCUS 23.2

Infections from Animal Reservoirs Transmitted by Direct Contact

Differential diagnosis is the process of identifying the disease from a list of possible diseases that fit the information derived from examining a patient. A differential diagnosis is important for providing initial treatment and for laboratory testing. The following diseases should be included in the differential diagnosis of patients with exposure to animals. For example, a 10-year-old girl was admitted to a local hospital after having fever (40°C) for 12 days and back pain for 8 days. Bacteria could not be cultured from tissues. She had a recent history of dog and cat scratches. She recovered without treatment. Use the table below to identify infections that could cause these symptoms. For the solution, go to www.microbiologyplace.com.

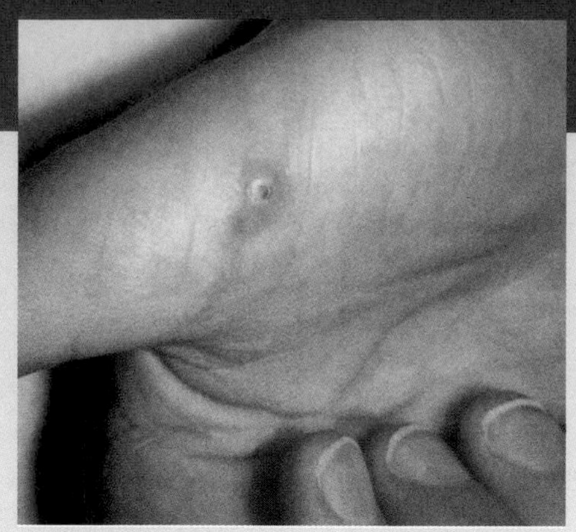

The patient's infected scratch.

Disease	Pathogen	Symptoms	Reservoir	Method of Transmission	Treatment
BACTERIAL DISEASES					
Brucellosis	*Brucella* spp.	Local abscess; undulating fever	Grazing mammals	Direct contact	Tetracycline, streptomycin
Anthrax	*Bacillus anthracis*	Papule (cutaneous); bloody diarrhea (gastrointestinal); septic shock (inhalational)	Soil; large grazing mammals	Direct contact; ingestion; inhalation	Ciprofloxacin; doxycycline
Animal bites	*Pasteurella multocida*	Local infection; sepsis	Animal mouths	Dog/cat bites	Penicillin
Rat-bite fever	*Streptobacillus moniliformis, Spirillum minus*	Sepsis	Rats	Rat bites	Penicillin
Cat-scratch disease	*Bartonella henselae*	Prolonged fever	Domestic cats	Cat bites or scratch, fleas	Antibiotics
PROTOZOAN DISEASE					
Toxoplasmosis	*Toxoplasma gondii*	Mild disease; initial infection acquired during pregnancy can be damaging to fetus; serious illness in AIDS patients	Domestic cats	Ingestion	Pyrimethamine, sulfadiazine and folinic acid

Pneumonic plague is easily spread by airborne droplets from humans or animals. Great care must be taken to prevent airborne infection of people in contact with patients.

Europe was ravaged by repeated pandemics of plague; from the years 542 to 767, outbreaks occurred repeatedly in cycles of a few years. After a lapse of centuries, the disease reappeared in devastating form in the fourteenth and fifteenth centuries. It is estimated to have killed more than 25% of the population, resulting in lasting effects on the social and economic structure of Europe. A nineteenth-century pandemic primarily affected Asiatic countries; 12 million are estimated to have died in India. The last major rat-associated urban outbreak in the United States occurred in Los Angeles in 1924 and 1925. Following this, the disease became a rarity until it reappeared in 1965 on the Navajo reservation in the Southwest. Plague, once established in the ground squirrel and prairie dog communities in this area, has gradually spread over much of the western states (**Figure 23.12**). A peak incidence of 40 cases occurred in 1983. A few cases have also arisen from cats, a new animal reservoir, and one from urban tree squirrels.

Plague has been most commonly diagnosed by isolating the bacterium and then sending it to a laboratory for identification. A rapid diagnostic test, however, can reliably detect the presence of the capsular antigen of *Y. pestis* in blood and other fluids of patients within 15 minutes even under remote field conditions. People exposed to infection can be given prophylactic antibiotic protection. A number of antibiotics, including streptomycin and tetracycline, are effective. Recovery from the disease confers reliable immunity. A vaccine is available for people likely to

DISEASES IN FOCUS 23.3

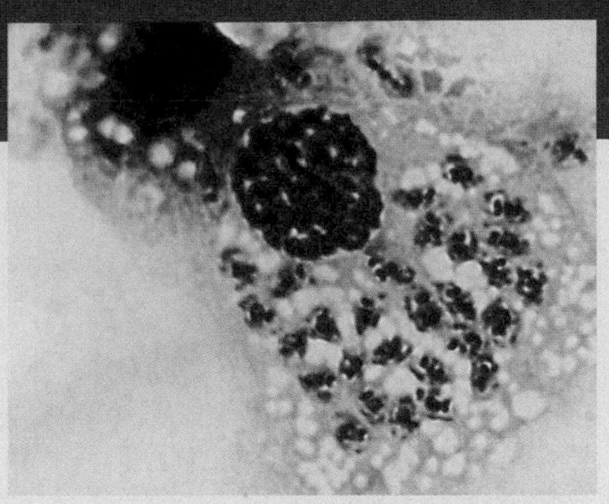

A macrophage practically filled with ovoid cells.

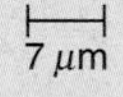

Infections Transmitted by Vectors

Differential diagnosis is the process of identifying the disease from a list of possible diseases that fit the information derived from examining a patient. A differential diagnosis is important for providing initial treatment and for laboratory testing. The following diseases should be considered in the differential diagnosis of patients with a history of tick and insect bites or who have traveled to endemic countries. These diseases are all prevented by controlling exposure to insect and tick bites. For example, a 22-year-old soldier returning from a tour of duty in Iraq had three painless skin ulcers. She reported being bitten by insects every night. Ovoid, protozoa-like bodies were observed within her macrophages by examination with a light microscope. Use the table below to identify infections that could cause these symptoms. For the solution, go to www.microbiologyplace.com.

Disease	Pathogen	Symptoms	Reservoir	Method of Transmission	Treatment
BACTERIAL DISEASES					
Tularemia	*Francisella tularensis*	Local infection; pneumonia	Rabbits; ground squirrels	Direct contact with infected animals, deer fly bite; inhalation	Tetracycline
Plague	*Yersinia pestis*	Enlarged lymph nodes; septic shock	Rodents	Fleas; inhalation	Streptomycin; tetracycline
Relapsing fever	*Borrelia* spp.	Series of fever peaks	Rodents	Soft ticks	Tetracycline
Lyme disease	*Borrelia burgdorferi*	Bull's-eye rash; neurologic symptoms	Field mice; deer	*Ixodes* ticks	Antibiotics
Ehrlichiosis and Anaplasmosis	*Ehrlichia,* spp. *Anaplasma* spp.	Flulike	Deer	Ticks	Tetracycline
Epidemic typhus	*Rickettsia prowazekii*	High fever, stupor, rash	Squirrels	*Pediculus humanus corporis* louse	Tetracycline; chloramphenicol
Endemic murine typhus	*Rickettsia typhi*	Fever; rash	Rodents	*Xenopsylla cheopsis* flea	Tetracycline; chloramphenicol
Rocky mountain spotted fever	*Rickettsia rickettsii*	Macular rash; fever; headache	Ticks; small mammals	*Dermacentor* ticks	Tetracycline, chloramphenicol
VIRAL DISEASE					
Chikungunya fever	Chickungunya virus	Fever; joint pain	Humans	Mosquitoes	Supportive
PROTOZOAN DISEASES					
Chagas' disease (American trypanosomiasis)	*Trypanosoma cruzi*	Damage to heart muscle or peristaltic movement of gastrointestinal tract	Rodents, opossums	Reduviid bug	Nifurtimox
Malaria	*Plasmodium* spp.	Fever and chills at intervals	Humans	*Anopheles* mosquito	Chloroquine
Leishmaniasis	*Leishmania* spp.	*L. donovani*: systemic disease; *L. tropica*: skin sores; *L. braziliensis*: disfiguring damage to mucous membranes	Small mammals	Sandfly	Antimony compounds
Babesiosis	*Babesia microti*	Fever and chills at intervals	Rodents	*Ixodes* ticks	Atovaquone and azithromycin

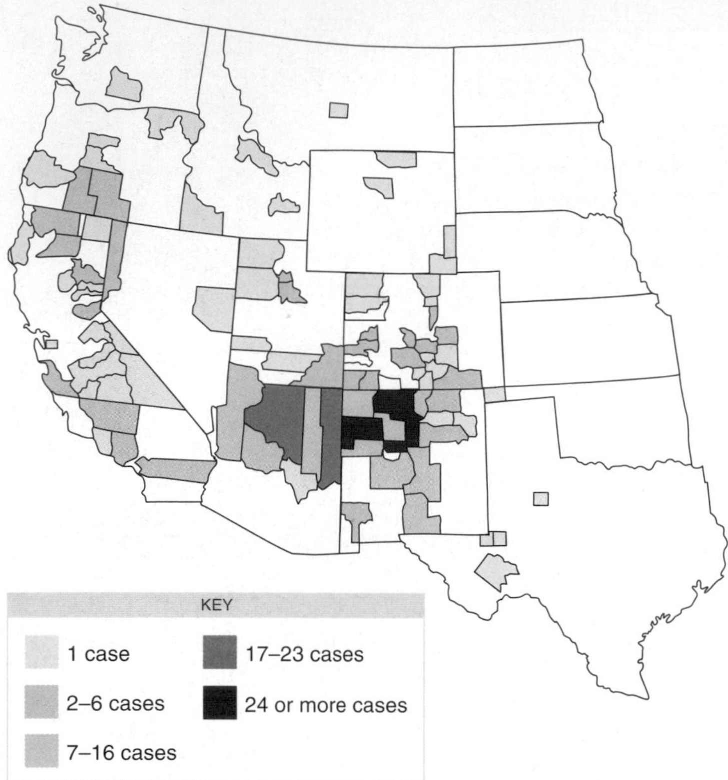

Figure 23.12 The U.S. geographic distribution of human plague, 1970–2004.

Source: CDC, 2007.

Q What area reporting plague is closest to you?

come into contact with infected fleas during field operations or for laboratory workers exposed to the pathogen.

Relapsing Fever

Except for the species that causes Lyme disease (discussed below), all members of the spirochete genus *Borrelia* cause **relapsing fever.** In the United States, the disease is transmitted by soft ticks that feed on rodents. The incidence of relapsing fever increases during the summer months, when the activity of rodents and arthropods increases.

The disease is characterized by fever, sometimes in excess of 40.5°C, jaundice, and rose-colored skin spots. After 3 to 5 days, the fever subsides. Three or four relapses may occur, each shorter and less severe than the initial fever. Each recurrence is caused by a different antigenic type of the spirochete, which evades existing immunity. Diagnosis is made by observing the bacteria in the patient's blood, which is unusual for a spirochete disease. Tetracycline is effective for treatment.

Lyme Disease (Lyme Borreliosis)

In 1975, a cluster of disease cases in young people that was first diagnosed as rheumatoid arthritis was reported near the city of Lyme, Connecticut. The seasonal occurrence (summer months), lack of contagiousness among family members, and descriptions of an unusual skin rash that appeared several weeks before the first symptoms suggested a tickborne disease. In 1983, a spirochete that was later named *Borrelia burgdorferi* was identified as the cause. **Lyme disease** may now be the most common tickborne disease in the United States. In Europe and Asia, the disease is usually known as **Lyme borreliosis.** Often, in these locales the tick and *Borrelia* species differ from those in the United States. Tens of thousands of cases are reported annually. In the United States, Lyme disease is most prevalent on the Atlantic coast (Figure 23.13).

Field mice are the most important animal reservoir. The nymphal stage of the tick feeds on infected mice and is the most likely to infect humans, even though adult ticks are about twice as likely to carry the bacterial pathogen. This is because nymphal ticks are small and less likely to be noticed before the infection is transmitted. Deer are important in maintenance of the disease because the ticks feed and mate on them. However, they are less likely than mice to carry nymphs or to infect them.

The tick (one of two *Ixodes* species) feeds three times during its life cycle (Figure 23.14a). The first and second feedings, as a larva and then as a nymph, are usually on a field mouse. The third feeding, as an adult, is usually on a deer. These feedings are separated by several months, and the ability of the spirochete to

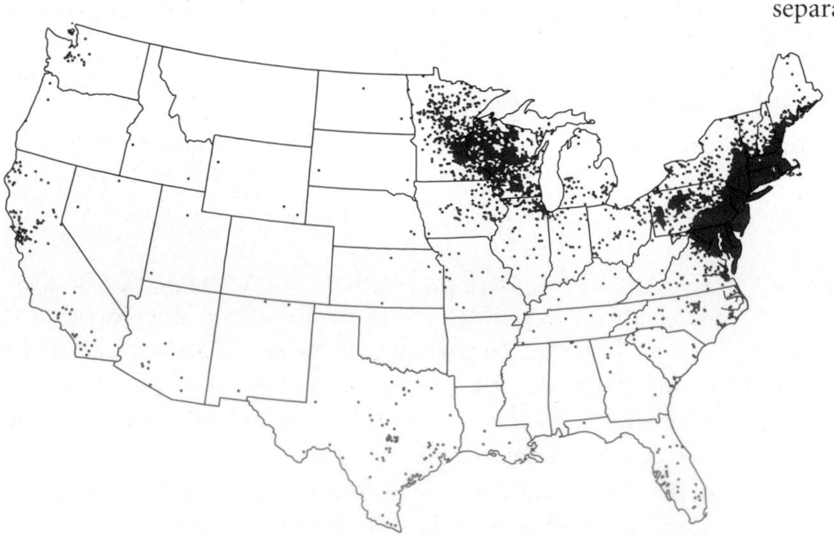

Figure 23.13 Lyme disease in the United States, reported cases by county, 2005.

Source: CDC, *MMWR* 56(23):575, June 15, 2007.

Q What factors are responsible for the geographic distribution of Lyme disease?

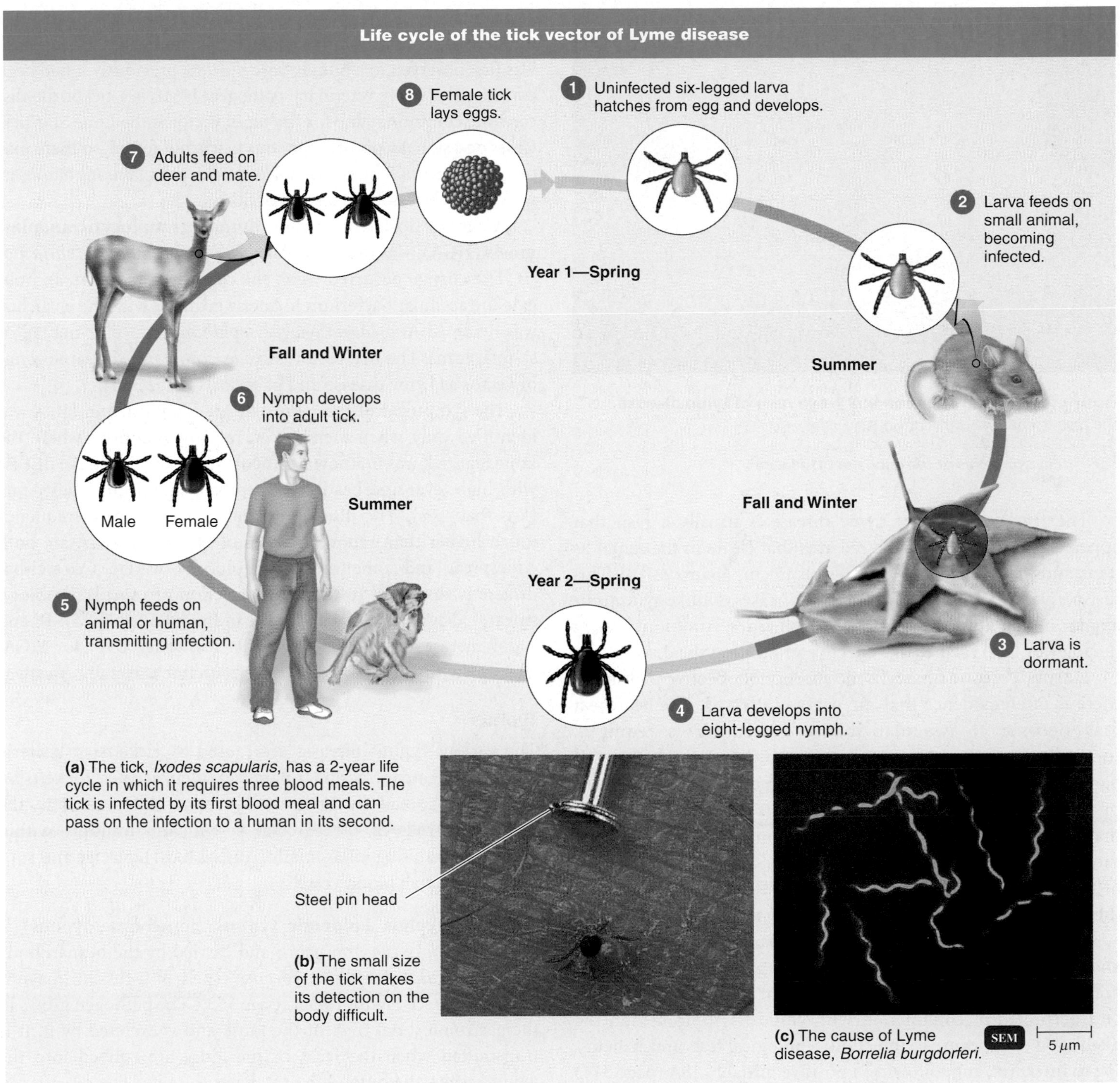
Life cycle of the tick vector of Lyme disease
1 Uninfected six-legged larva hatches from egg and develops.
2 Larva feeds on small animal, becoming infected.
3 Larva is dormant.
4 Larva develops into eight-legged nymph.
5 Nymph feeds on animal or human, transmitting infection.
6 Nymph develops into adult tick.
7 Adults feed on deer and mate.
8 Female tick lays eggs.
Year 1—Spring
Summer
Fall and Winter
Year 2—Spring
Summer
Fall and Winter
Male
Female
Steel pin head
SEM
5 μm

(a) The tick, *Ixodes scapularis*, has a 2-year life cycle in which it requires three blood meals. The tick is infected by its first blood meal and can pass on the infection to a human in its second.

(b) The small size of the tick makes its detection on the body difficult.

(c) The cause of Lyme disease, *Borrelia burgdorferi*.

Figure 23.14 Life cycle of the tick vector of Lyme disease.

Q What other diseases are transmitted by ticks?

remain viable in the disease-tolerant field mice is crucial to maintaining the disease in the wild.

On humans, the ticks usually attach from a perch on shrubs or grass. They do not feed for about 24 hours, and it usually requires 2 or 3 days of attachment before transfer of bacteria and infection occur. Probably only about 1% of tick bites result in Lyme disease.

On the Pacific coast, the tick that transmits Lyme disease is the western black-legged tick *Ixodes pacificus* (iks-ō′dēs pas-i′fi-kus) (see also Figure 12.31, page 362). In the rest of the country, *Ixodes scapularis* (scap-ū-lār′is) is most often responsible. This latter tick is so small that it is often missed (Figure 23.14b). On the Atlantic coast, almost all *Ixodes* ticks carry the spirochete (Figure 23.14c); on the Pacific coast, few are infected because that tick feeds on lizards that do not carry the spirochete effectively.

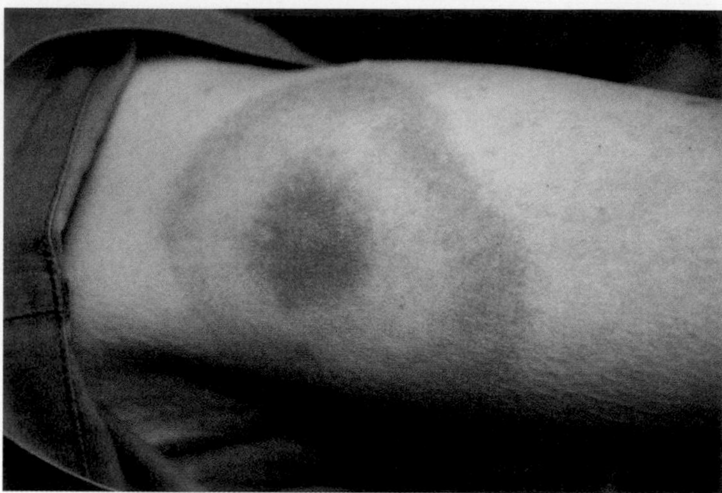

Figure 23.15 The common bull's-eye rash of Lyme disease. The rash is not always this obvious.

Q What symptoms occur once the rash fades?

The first symptom of Lyme disease is usually a rash that appears at the bite site. It is a red area that clears in the center as it expands to a final diameter of about 15 cm (**Figure 23.15**). This distinctive rash occurs in about 75% of cases. Flulike symptoms appear in a couple of weeks as the rash fades. Antibiotics taken during this interval are very effective in limiting the disease.

During a second phase, in the absence of effective treatment, there is often evidence that the heart is affected. The heartbeat may become so irregular that a pacemaker is required. Incapacitating, chronic neurological symptoms, such as facial paralysis, meningitis, and encephalitis, may be seen. In a third phase, months or years later, some patients develop arthritis that may affect them for years. Immune responses to the presence of the bacteria are probably the cause of this joint damage. Many of the symptoms of long-term Lyme disease resemble those of the later stages of syphilis, also caused by a spirochete.

Diagnosis of Lyme disease depends partly on the symptoms and an index of suspicion based on the prevalence in the geographic area. Physicians are cautioned that serological tests must be interpreted in conjunction with clinical symptoms and the likelihood of exposure to infection. Serological tests are challenging to interpret, and following a positive initial ELISA (page 514) or indirect fluorescent-antibody (FA) test (page 513), confirmation should be attempted with a Western immunoblot test (page 516). Also, after effective antibiotic treatment eliminates the bacteria, antibodies—even IgM antibodies—often persist for years and may confuse later attempts at diagnosis.

Several antibiotics are effective in treating the disease, although in the later stages, large amounts may be needed.

Ehrlichiosis and Anaplasmosis

Human monocytotropic ehrlichiosis (HME) is caused by *Ehrlichia chafeensis* (er'lik-ē-ä chaf'ē-en-sis). This is a gram-negative, rickettsia-like, obligately intracellular bacterium. Aggregates of bacteria—called *morulae,* the Latin word for mulberry—form within the cytoplasm of monocytes. *E. chafeensis* was first observed in a human case in 1986; previously it had been considered a solely veterinary pathogen. HME is a tickborne disease; the common name for the usual vector is the Lone Star tick. Cases occasionally occur where this tick is not found, so there may be other vectors. The white-tailed deer is the main animal reservoir, but it does not show signs of illness.

A similar tickborne disease, **human granulocytic anaplasmosis (HGA),** was formerly called *human granulocytic ehrlichiosis.* The change occurred when the causative organism, an obligate intracellular bacterium formerly grouped with the ehrlichia, was renamed *Anaplasma phagocytophilum* (an'ä-plaz-mä fäg'ō-sī-to-fi'lum). The tick vector is *Ixodes scapularis,* the same genus of vector as Lyme disease and babesiosis (page 350).

The symptoms of these diseases are identical, and HGA was identified only when a case occurred in Wisconsin, where the Lone Star tick was unknown. Patients suffer from a flulike disease with high fever and headache; there is a significant fatality rate (less than 5%). The diseases probably occur with a frequency much higher than reported. Cases of HME and HGA are both widespread and sometimes geographically overlap. Once either disease is suspected (often from detection of morulae in blood smears), diagnosis is usually by the indirect FA test for HME and a polymerase chain reaction (PCR) test (page 251) for HGA. Therapy with antibiotics such as tetracycline is usually effective.

Typhus

The various typhus diseases are caused by rickettsias, bacteria that are obligate intracellular parasites of eukaryotes. Rickettsias, which are spread by arthropod vectors, infect mostly the endothelial cells of the vascular system and multiply within them. The resulting inflammation causes local blockage and rupture of the small blood vessels.

Epidemic Typhus Epidemic typhus (louseborne typhus) is caused by *Rickettsia prowazekii* and carried by the human body louse *Pediculus humanus corporis* (ped-ik'ū-lus hü'ma-nus kôr'pô-ris) (see Figure 12.32a, page 363). The pathogen grows in the gastrointestinal tract of the louse and is excreted by it. It is transmitted when the feces of the louse are rubbed into the wound when the bitten host scratches the bite. The disease can flourish only in crowded and unsanitary surroundings, when lice can transfer readily from an infected host to a new host. Anne Frank, the teenaged writer of the famed World War II diary, died of typhus contracted in concentration camp conditions.

Epidemic typhus disease produces a high and prolonged fever that lasts at least 2 weeks. Stupor and a rash of small red spots caused by subcutaneous hemorrhaging are characteristic, as the rickettsias invade blood vessel linings. Mortality rates are very high when the disease is untreated.

Tetracycline and chloramphenicol are usually effective against epidemic typhus, but eliminating conditions in which the disease flourishes is more important. The microbe is considered especially hazardous, and attempts to culture it require extreme care. Vaccines are available for military populations, which historically have been highly susceptible to the disease.

Endemic Murine Typhus **Endemic murine typhus** occurs sporadically rather than in epidemics. The term *murine* (derived from Latin for mouse) refers to the fact that rodents, such as rats and squirrels, are the common hosts for this type of typhus. Endemic murine typhus is transmitted by the rat flea *Xenopsylla cheopis* (see Figure 12.32b, page 363), and the pathogen responsible for the disease is *Rickettsia typhi,* a common inhabitant of rats. With a mortality rate of less than 5%, the disease is considerably less severe than the epidemic form of typhus. Except for the reduced severity of the disease, endemic murine typhus is clinically indistinguishable from epidemic typhus. Tetracycline and chloramphenicol are effective treatments for endemic murine typhus, and rat control is the best preventive measure.

Spotted Fevers Tickborne typhus, or **Rocky Mountain spotted fever,** is probably the best-known rickettsial disease in the United States. It is caused by *Rickettsia rickettsii.* Despite its name (it was first recognized in the Rocky Mountain area), it is most common in the southeastern states and Appalachia (Figure 23.16). This rickettsia is a parasite of ticks and is usually passed from one generation of ticks to another through their eggs, a mechanism called *transovarian passage* (Figure 23.17). Surveys show that in endemic areas, perhaps 1 out of every 1000 ticks is infected. In different parts of the United States, different ticks are involved—in the west, the wood tick *Dermacentor andersoni* (dėr-mä-sen′tôr an-dėr-sōn′ē); in the east, the dog tick *Dermacentor variabilis* (vãr-ē-a′bil-is).

About a week after the tick bites, a macular rash develops that is sometimes mistaken for measles (Figure 23.18); however, it often appears on palms and soles, where viral rashes do not occur. The rash is accompanied by fever and headache. Death, which occurs in about 3% of the approximately 1000 cases reported each year, is usually caused by kidney and heart failure.

Serological tests do not become positive until late in the illness. Diagnosis before the typical rash appears is difficult; symptoms vary widely. Also, on dark-skinned individuals, the rash is difficult to see. A misdiagnosis can be costly; if treatment is not prompt and correct, the mortality rate is about 20%.

Antibiotics such as tetracycline and chloramphenicol are very effective if administered early enough. No vaccine is available.

CHECK YOUR UNDERSTANDING

✓ Why is the plague-infected flea so eager to feed on a mammal? **23-10**

✓ What animal does the infecting tick feed on just before it transmits Lyme disease to a human? **23-11**

✓ Which disease is tickborne: epidemic typhus, endemic murine typhus, or Rocky Mountain spotted fever? **23-12**

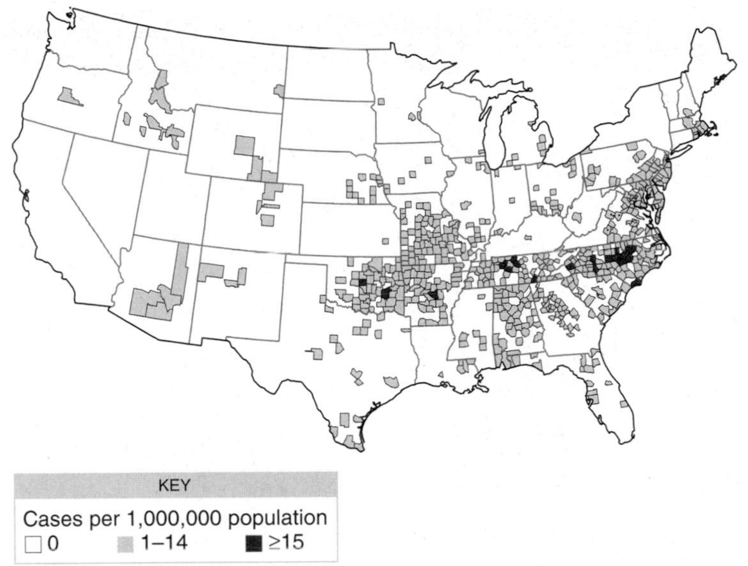

Figure 23.16 The U.S. geographic distribution of Rocky Mountain spotted fever (tickborne typhus) 1997–2002.

Source: CDC, 2007.

Q Geographically, is this a disease of rural or urban areas?

Viral Diseases of the Cardiovascular and Lymphatic Systems

LEARNING OBJECTIVE

23-13 Describe the epidemiologies of Burkitt's lymphoma, infectious mononucleosis, and CMV inclusion disease.

23-14 Compare and contrast the causative agents, vectors, reservoirs, and symptoms of yellow fever, dengue, dengue hemorrhagic fever, and chikungunya fever.

23-15 Compare and contrast the causative agents, reservoirs, and symptoms of Ebola hemorrhagic fever and *Hantavirus* pulmonary syndrome.

Viruses cause a number of cardiovascular and lymphatic diseases, prevalent mostly in tropical areas. However, one viral disease of this type, infectious mononucleosis, is an especially familiar infectious disease among American college-aged individuals.

Burkitt's Lymphoma

In the 1950s, Denis Burkitt, an Irish physician working in eastern Africa, noticed the frequent occurrence in children of a fast-growing tumor of the jaw (Figure 23.19). Known as **Burkitt's lymphoma,** this is the most common childhood cancer in Africa. It has a limited geographic distribution similar to that of malaria in central Africa.

Burkitt suspected a viral cause of the tumor and a mosquito vector. At that time, there was no known virus that caused human cancer, although several viruses were clearly associated with animal cancers. Intrigued by this possibility, in 1964 British

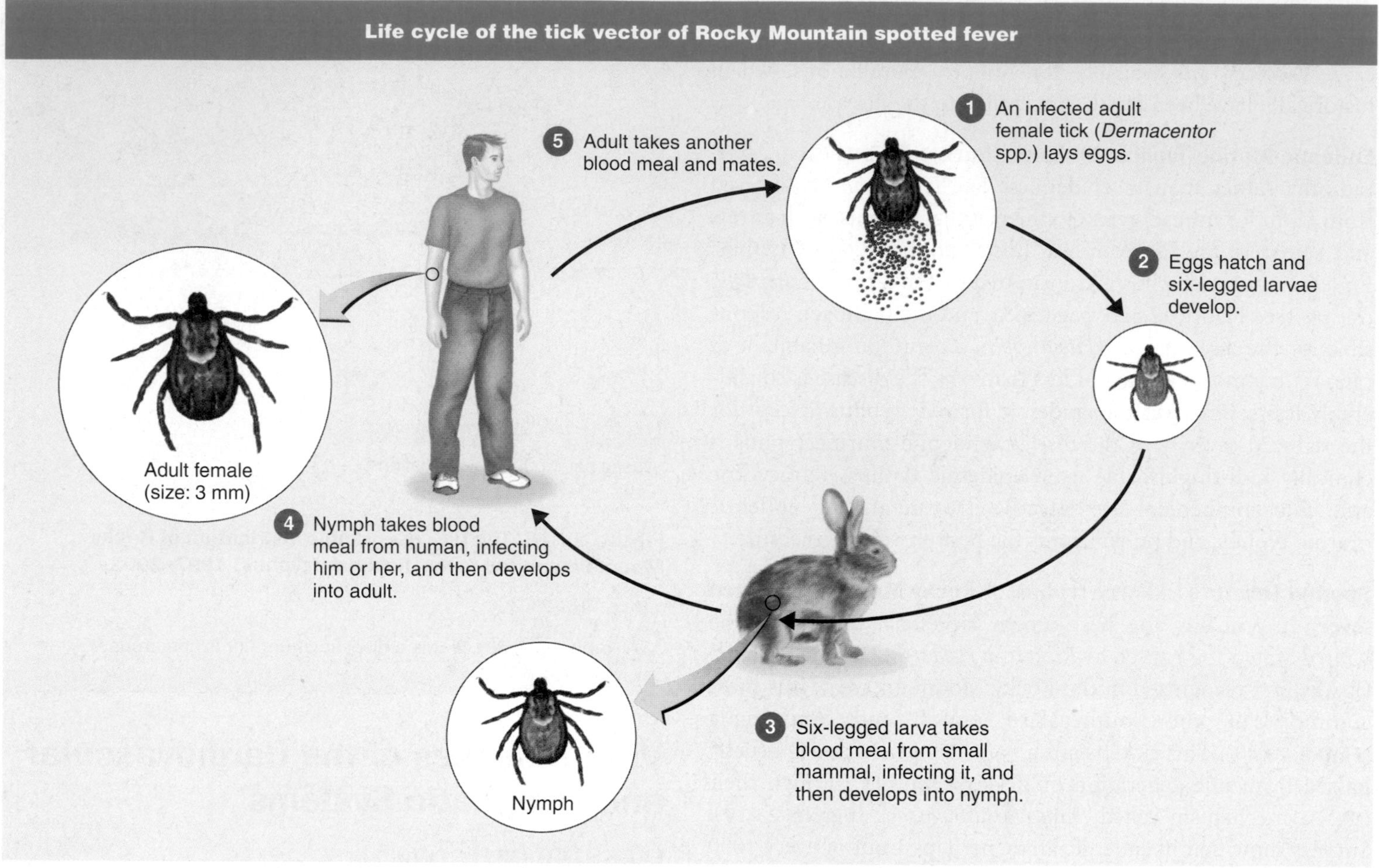

Life cycle of the tick vector of Rocky Mountain spotted fever
1 An infected adult female tick (*Dermacentor* spp.) lays eggs.
2 Eggs hatch and six-legged larvae develop.
3 Six-legged larva takes blood meal from small mammal, infecting it, and then develops into nymph.
4 Nymph takes blood meal from human, infecting him or her, and then develops into adult.
5 Adult takes another blood meal and mates.
Adult female (size: 3 mm)
Nymph

Figure 23.17 The life cycle of the tick vector (*Dermacentor* spp.) of Rocky Mountain spotted fever. Mammals are not essential to survival of the pathogen, *Rickettsia rickettsii,* in the tick population; the bacteria may be passed by transovarian passage, so new ticks are infected upon hatching. A blood meal is required for ticks to advance to the next stage in the life cycle.

Q What is meant by *transovarian passage*?

virologist Tony Epstein and his student, Yvonne Barr, performed biopsies on the tumors. A virus was cultured from this material, and the electron microscope showed a herpeslike virus in the culture cells; it was named the *Epstein-Barr virus (EB virus).* The official name of this virus is human herpesvirus 4.

EB virus is clearly associated with Burkitt's lymphoma, but the mechanism by which it causes the tumor is not understood. Research eventually showed, however, that mosquitoes do not transmit the virus or the disease. Instead, mosquito-borne malarial infections apparently foster the development of Burkitt's lymphoma by impairing the immune response to EB virus, which is almost universally present in human adults worldwide. The virus has, in fact, become so adapted to humans that it is one of our most effective parasites. It establishes a lifelong infection in most people (**Figure 23.20**) that is harmless and rarely causes disease.

In areas without endemic malaria, such as the United States, Burkitt's lymphoma is rare and is usually abdominal. The appearance of the lymphoma in AIDS patients is an indication of the importance of immune surveillance in preventing expression of the disease.

CHECK YOUR UNDERSTANDING

✓ Although not a disease with an insect vector, why is Burkitt's lymphoma most commonly a disease found in malarial areas? **23-13**

Infectious Mononucleosis

The identification of EB virus as the cause of **infectious mononucleosis,** or *mono,* resulted from one of the accidental discoveries that often advance science. A technician in a laboratory investigating EB virus served as a negative control for the virus. While on vacation, she contracted an infection characterized by fever, sore throat, swollen lymph nodes in the neck, and general weakness. The most interesting aspect of the technician's disease was that she now tested serologically positive for EB virus. It was soon

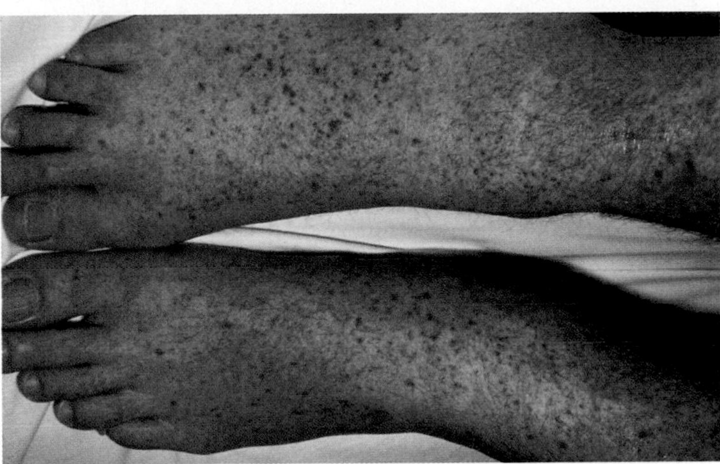

Figure 23.18 The rash caused by Rocky Mountain spotted fever. This rash is often mistaken for measles. People with dark skin have a higher mortality rate because the rash is often not recognized early enough for effective treatment.

Q How can Rocky Mountain spotted fever be prevented?

confirmed that the same virus that is associated with Burkitt's lymphoma also causes almost all cases of infectious mononucleosis.

In developing parts of the world, infection with EB virus occurs in early childhood, and 90% of the children over age 4 have acquired antibodies. Nearly 20% of adults in the United States carry EB virus in oral secretions. Childhood EB virus infections are usually asymptomatic, but if infection is delayed until young adulthood, as is often the case in the United States, the result is more symptomatic probably because of an intense immunological response. The peak U.S. incidence of the disease occurs at about age 15 to 25. A principal cause of the rare deaths is rupture of the enlarged spleen (a common response to a systemic infection) during vigorous activity. Recovery is usually complete in a few weeks, and immunity is permanent.

The usual route of infection is by the transfer of saliva by kissing or, for example, by sharing drinking vessels. It does not spread among casual household contacts, so aerosol transmission is unlikely. The incubation period before appearance of symptoms is 4 to 7 weeks.

EB virus maintains a persistent infection in the oropharynx (mouth and throat), which accounts for its presence in saliva. It is probable that resting memory B cells (see Figure 17.5, page 483) located in lymphoid tissue are the primary site of replication and persistence. Most of the symptoms are attributed to responses of T cells to the infection.

The disease name *mononucleosis* refers to lymphocytes with unusual lobed nuclei that proliferate in the blood during the acute infection. The infected B cells produce nonspecific antibodies called heterophil antibodies. If this test is negative, the symptoms may be caused by cytomegalovirus (see page 658) or several other disease conditions. A fluorescent-antibody test that detects IgM antibodies against EB virus is the most specific diagnostic method. There is no recommended specific therapy for most patients.

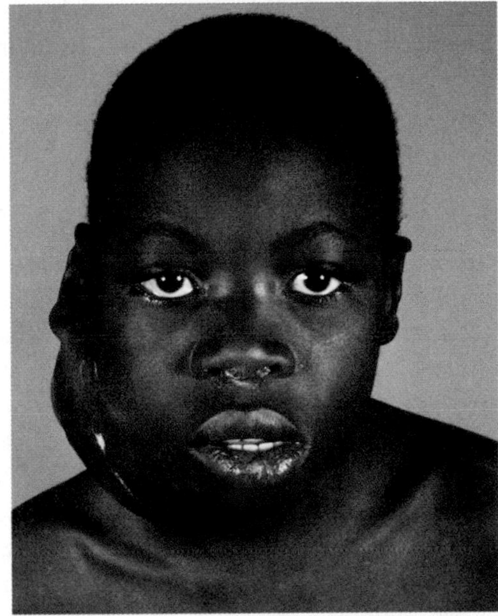

Figure 23.19 A child with Burkitt's lymphoma. Cancerous tumors of the jaw caused by Epstein-Barr virus (EB virus) are seen mainly in children. This child was successfully treated.

Q What is the relationship between malarial areas and areas with Burkitt's lymphoma?

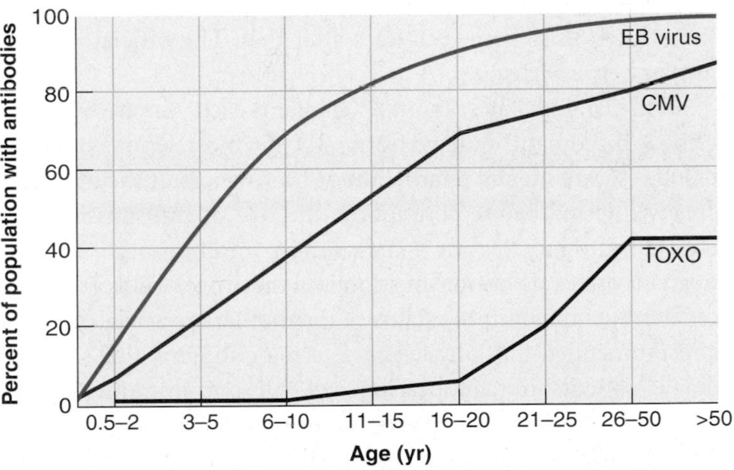

Figure 23.20 The typical U.S. prevalence of antibodies against Epstein-Barr virus (EB virus), cytomegalovirus (CMV), and *Toxoplasma gondii* (TOXO) by age.

Source: Laboratory Management, June 1987, pp. 23ff.

Q To judge from this graph, which of these diseases is more likely to result from early-childhood infections?

Other Diseases and Epstein-Barr Virus

We have just discussed two diseases, Burkitt's lymphoma and infectious mononucleosis, for which there is a clear association with EB virus. There is a lengthy list of diseases for which there is a suspected, but not proven, relationship with EB virus. Some of the more familiar of these include **multiple sclerosis** (autoimmune attack on the nervous system), **Hodgkin's disease** (tumors of the spleen, lymph nodes, or liver), and **nasopharyngeal** (nose and pharynx) **cancer** among certain ethnic groups in southeast Asia and Inuits.

Cytomegalovirus Infections

Almost all of us will become infected with cytomegalovirus (CMV) during our lifetime. The CMV is a very large herpesvirus that, much like the Epstein-Barr virus, remains latent in white blood cells, such as monocytes, neutrophils, and T cells. It is not much affected by the immune system, replicating very slowly and escaping antibody action by moving between cells that are in contact. Carriers of the virus may shed it in body secretions such as saliva, semen, and breast milk. When CMV infects a cell, it causes the formation of distinctive inclusion bodies that are visible by microscopy. When these bodies occur in pairs, they are known as "owl's eyes" and are useful in diagnosis. These inclusion bodies were first reported in 1905 in certain cells of newborn infants affected with congenital abnormalities. The cells were also enlarged, a condition known as *cytomegaly*, from which the virus eventually received its name. This disease of the newborns was given the name of *cytomegalic inclusion disease (CID)*. The inclusion bodies were originally thought to be stages in the life cycle of a protozoan, and a viral cause of the disease was not proposed until 1925. The cytomegalovirus was not isolated until some 30 years after that. The official name is human herpesvirus 5.

In the United States, about 8000 infants each year are born suffering symptomatic damage from CID, the most serious of which includes severe mental retardation or hearing loss. If the mother is already infected before conception, the rate of transmission to a fetus is less than 2%, but if the primary infection occurs during pregnancy, the rate of transmission is in the range of 40–50%. Tests to determine the immune status of the mother are available, and it is recommended that physicians determine the immune status of female patients of childbearing age. All non-immune women should be informed of the risks of infection during pregnancy.

In healthy adults, acquiring a CMV infection causes either no symptoms or those resembling a mild case of infectious mononucleosis. It has been said that if CMV were accompanied by a skin rash, it would be one of the better-known childhood diseases. It is therefore not surprising, given that 80% of the population of the United States is estimated to carry the virus, that CMV is a common opportunistic pathogen in persons whose immune system has become compromised. Figure 23.20 shows the prevalence of antibodies against CMV, Epstein-Barr virus, and *Toxoplasma gondii* (page 661). In developing parts of the world, infection rates of CMV approach 100%. For immunocompromised individuals, CMV is a frequent cause of a life-threatening pneumonia, but almost any organ can be affected. About 85% of AIDS patients exhibit a CMV-caused eye infection, *cytomegalovirus retinitis*. Without treatment, it results in eventual loss of vision. Several antiviral agents, such as ganciclovir and the antisense drug fomivirsen, are effective, but treatment must be lifelong. Such antiviral drugs are also used to treat other CMV-caused diseases.

CMV is transmitted mostly by activities that result in contact with body fluids that contain the virus, such as kissing, and is very common among children in day-care settings. It can also be transmitted sexually, by transfused blood, and by transplanted tissue. Transmission by transfused blood can be eliminated by filtering out the white cells from the blood or by serological testing of the donor for the virus. Transplanted tissue is usually tested for the virus, and products are now available that contain antibodies to neutralize CMV present in donated tissue. Vaccines are under development, but none is currently available.

Chikungunya Fever

The recent introduction of the West Nile virus into the United States has shown that a tropical mosquito-borne disease can spread in temperate climates. Rapid travel and climatic warming, among other factors, are making similar vector-borne diseases a global phenomenon. Another tropical disease now causing concern is **chikungunya fever** (the name sounds like "chicken-gun′ ya," but the disease is often referred to simply as *chik*.) The name comes from an African language and means "that which bends up." The symptoms are a high fever and severe, crippling joint pains—especially in the wrists, fingers, and ankles—that can persist for weeks or months. There is often a rash and even massive blisters. The death rate is very low. The vector is the *Aedes* mosquito, primarily *Aedes aegypti* (ā′e-dēz ē-jip′tē), which spreads the disease widely in Asia and Africa. Recent outbreaks have also been caused by *A. albopictus* (al-bō-pik′tus). A mutation in the virus, which is related to the virus causing western equine encephalitis (WEE) and eastern equine encephalitis (EEE) (page 624), has adapted the virus to multiply in this insect. It is uncertain whether there is an animal reservoir. An outbreak has already occurred in Italy.

A. albopictus is also known as the Asian tiger mosquito because of its bright white stripes. Well adapted to urban settlements, it also survives cold climates and will probably become established eventually even in the northern parts of the United States and the coastal areas of Scandinavia. Because it is an extremely aggressive daytime biter, it is a serious nuisance for outdoor activities. Of greater concern to health officials is that *A. albopictus* is known, so far, to transmit both chikungungya fever and dengue, a disease that will be discussed shortly.

Classic Viral Hemorrhagic Fevers

Most hemorrhagic fevers are zoonotic diseases; they appear in humans only from infectious contact with their normal animal hosts. Some of them have been medically familiar for so long that they are considered "classic" hemorrhagic fevers. First among these is **yellow fever.** The yellow fever virus is injected into the skin by a mosquito, *A. aegypti.*

In the early stages of severe cases of the disease, the person experiences fever, chills, and headache, followed by nausea and vomiting. This stage is followed by jaundice, a yellowing of the skin that gave the disease its name. This coloration reflects liver damage, which results in the deposit of bile pigments in the skin and mucous membranes. The mortality rate for yellow fever is high, about 20%.

Yellow fever is still endemic in many tropical areas, such as Central America, tropical South America, and Central Africa. At one time, the disease was endemic in the United States and occurred as far north as Philadelphia. The last U.S. case of yellow fever occurred in Louisiana in 1905 during an outbreak that resulted in about 1000 deaths. Mosquito eradication campaigns initiated by the U.S. Army surgeon Walter Reed were effective in eliminating yellow fever in the United States.

Monkeys are a natural reservoir for the virus, but human-to-human transmission can maintain the disease. Local control of mosquitoes and immunization of the exposed population are effective controls in urban areas.

Diagnosis is usually by clinical signs, but it can be confirmed by a rise in antibody titer or isolation of the virus from the blood. There is no specific treatment for yellow fever. The vaccine is an attenuated live viral strain and yields a very effective immunity.

Dengue (den′ghee) is a similar but milder viral disease also transmitted by mosquitoes. This disease is endemic in the Caribbean and other tropical environments, where an estimated 100 million cases occur each year. It is characterized by fever, severe muscle and joint pain, and rash. Except for the painful symptoms, which have led to the name **breakbone fever,** classic dengue fever is a relatively mild disease and is rarely fatal.

The countries surrounding the Caribbean are reporting an increasing number of cases of dengue. In most years, more than 100 cases are imported into the United States, mostly by travelers from the Caribbean and South America. The disease does not appear to have an animal reservoir. The mosquito vector for dengue is common in the Gulf states, and there is some worry that the virus will sooner or later be introduced into this region and become endemic. Health officials are concerned about the American introduction of an Asian mosquito, *A. albopictus* (see page 658) an efficient vector for the virus. Control measures are directed at eliminating *Aedes* mosquitoes.

A severe form of dengue, **dengue hemorrhagic fever (DHF),** is probably caused when antibodies from a previous infection combine with the virus. DHF can induce shock in the victim (usually a child) and kill in a few hours; it is a leading cause of death among southeast Asian children. Outbreaks have also occurred in Mexico, South America, and the Caribbean.

CHECK YOUR UNDERSTANDING

✓ Why is the mosquito *Aedes albopictus* a special concern to the populations of temperate climates? 23-14

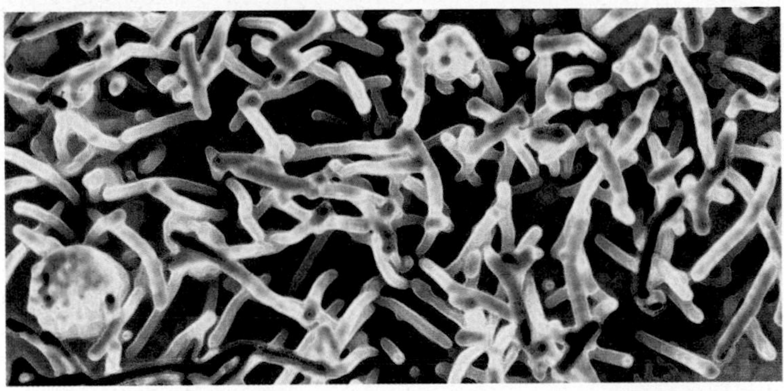

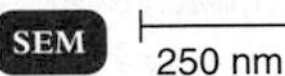

Figure 23.21 Ebola hemorrhagic virus. The viruses causing Ebola hemorrhagic fever are shown here. They disrupt the blood clotting system.

Q **Can you see why the Ebola virus is called a filovirus?**

Emerging Viral Hemorrhagic Fevers

Q&A Certain other hemorrhagic diseases are considered new or "emerging" hemorrhagic fevers. In 1967, 31 people became ill and 7 died after contact with some African monkeys that were imported into Europe. The virus was strangely shaped (in the form of a filament [filoviruses]) and was named for the site of the outbreak in Germany, the **Marburg virus.** The symptoms of infection by hemorrhagic viruses are mild at first; headache and muscle pain. But after a few days the victim suffers from high fever and begins vomiting blood and bleeding profusely, both internally and from external openings such as the nose and eyes. Death comes in a few days from organ failure and shock.

A similar hemorrhagic fever, **Lassa fever,** appeared in Africa in 1969 and was traced to a rodent reservoir. The virus, an arenavirus, is present in the rodent's urine and is the source of human infections. Outbreaks of Lassa fever have killed thousands. Seven years later, outbreaks in Africa of another highly lethal hemorrhagic fever, caused by a filovirus similar to the Marburg virus, caused a disease with a mortality approaching 90%. Named **Ebola virus** for a regional river, this is now a well-publicized disease, the subject of films and books (Figure 23.21).

The natural host reservoir for the Ebola virus is probably a fruit bat, which is used as food and is not acutely affected by the virus it carries. Once a human is infected and shedding blood, the infection is spread by contact with the blood and body fluids and in many cases by the reuse of needles used on patients. The local custom of washing the body before burial often triggers new infections.

South America has several hemorrhagic fevers caused by Lassa-like viruses (arenaviruses) that are maintained in the rodent population. **Argentine** and **Bolivian hemorrhagic fevers** are transmitted in rural areas by contact with rodent excretions. A recent handful of deaths in California have been attributed to the **Whitewater Arroyo virus,** an arenavirus with a reservoir in wood

DISEASES IN FOCUS 23.4

Viral Hemorrhagic Fevers

Viral hemorrhagic fevers are endemic in tropical countries where, except for dengue, they are found in small mammals. However, increasing international travel has resulted in importation of these viruses into the United States. There is no treatment.

Differential diagnosis is the process of identifying the disease from a list of possible diseases that fit the information derived from examining a patient. A differential diagnosis is important for providing initial treatment and for laboratory testing. The CDC's Special Pathogens Branch has specialized containment facilities to confirm diagnosis of viral hemorrhagic fevers by serology, nucleic acids, and virus culture. Use the table below to identify the cause of a rash and severe joint pain in a 20-year-old woman. For the solution, go to www.microbiologyplace.com.

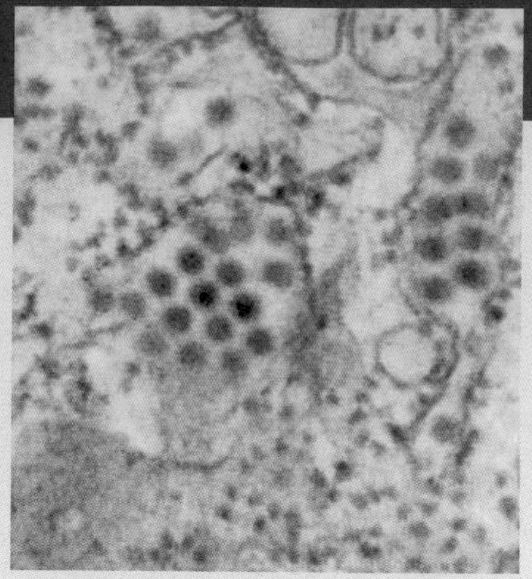

Small viruses seen by electron microscopy in patient's tissues. Upon isolation they were identified as single-stranded RNA viruses in the family Flaviviridae.

Disease	Pathogen	Portal of Entry	Symptoms	Reservoir	Method of Transmission	Prevention
Yellow fever	Flavivirus (yellow fever virus)	Skin	Fever, chills, headache; jaundice	Monkeys	*Aedes aegypti*	Vaccination; mosquito control
Dengue	Flavivirus (dengue fever virus)	Skin	Fever, muscle and joint pain, rash	Humans	*Aedes aegypti; A. albopictus*	Mosquito control
Emerging viral hemorrhagic fevers (Marburg, Ebola, Lassa)	Filovirus, arenavirus	Mucous membranes	Profuse bleeding	Possibly fruit bats and other small mammals	Contract with blood	None
***Hantavirus* pulmonary syndrome**	Bunyavirus (Sin Nombre hantavirus)	Respiratory tract	Pneumonia	Field mice	Inhalation	None

rats. These are the first reports of arenavirus-caused hemorrhagic disease in the Northern Hemisphere.

***Hantavirus* pulmonary syndrome,** caused by the Sin Nombre virus,* a bunyavirus, has become well known in the United States because of several outbreaks, mostly in the western states. It manifests itself as a frequently fatal pulmonary infection, in which the lungs fill with fluids. Actually, diseases of this nature have a long history, especially in Asia and Europe. It is best known there as **hemorrhagic fever with renal syndrome** and primarily affects renal (kidney) function. All these related diseases are transmitted by the inhalation of viruses in dried urine and feces from infected small rodents.

*The virus causing the pulmonary hantavirus outbreak in 1993 in the Four Corners area of the southwestern United States (Arizona, Utah, Colorado, and New Mexico) was originally called the Four Corners virus. Local authorities were concerned about the effect of this name on tourism in the area and complained. The name Sin Nombre, Spanish for no name, was then adopted.

Diseases in Focus 23.4 describes the various viral hemorrhagic fevers.

CHECK YOUR UNDERSTANDING

✓ Which disease does Ebola hemorrhagic fever more closely resemble, Lassa fever or *Hantavirus* pulmonary syndrome? **23-15**

Protozoan Diseases of the Cardiovascular and Lymphatic Systems

LEARNING OBJECTIVES

23-16 Compare and contrast the causative agents, modes of transmission, reservoirs, symptoms, and treatments for Chagas' disease, toxoplasmosis, malaria, leishmaniasis, and babesiosis.

23-17 Discuss the worldwide effects of these diseases on human health.

Protozoa that cause diseases of the cardiovascular and lymphatic systems often have complex life cycles, and their presence may affect human hosts seriously.

Chagas' Disease (American Trypanosomiasis)

Chagas' disease, also known as **American trypanosomiasis,** is a protozoan disease of the cardiovascular system. The causative agent is *Trypanosoma cruzi* (tri-pa-nō-sō′mä kruz′ē), a flagellated protozoan (Figure 23.22). The protozoan was discovered in its insect vector by the Brazilian microbiologist Carlos Chagas in 1910. He named it for the Brazilian epidemiologist Oswaldo Cruz. The disease occurs in Central America and parts of South America, where it chronically infects an estimated 18 million and kills as many as 50,000 annually. It has been introduced into the United States by population migration. In 2006, blood banks began screening for the disease, a practice that will identify many cases.

The reservoir for *T. cruzi* is a wide variety of wild animals, including rodents, opossums, and armadillos. The arthropod vector is the reduviid bug, called the "kissing bug" because it often bites people near the lips (see Figure 12.32d, page 363). The insects live in the cracks and crevices of mud or stone huts with thatched roofs. The trypanosomes, which grow in the gut of the bug, are passed on if the bug defecates while feeding. The bitten human or animal often rubs the feces into the bite wound or other skin abrasions by scratching or into the eye by rubbing. The infection progresses in stages. The acute stage, characterized by fever and swollen glands lasting for a few weeks, may not cause alarm. However, 20–30% of people infected will develop a chronic form of the disease—in some cases, 20 years later. Damage to the nerves controlling the peristaltic contractions of the esophagus or colon can prevent them from transporting food. This causes them to become grossly enlarged, conditions known as *megaesophagus* and *megacolon.* Most deaths are caused by damage to the heart, which occurs in about 40% of chronic cases. Infections of women during the chronic stage can result in congenital infections.

Diagnosis in endemic areas is usually based on symptoms. In the acute phase, the trypanosomes can sometimes be detected in blood samples. During the chronic phase these are undetectable—although patients can transmit the infection by transfusions, transplants, and congenitally. Diagnosis of chronic disease depends on serological tests, which are not very sensitive or specific. Two, or even three, repeated samplings may be required.

Treating Chagas' disease is very difficult when chronic, progressive stages have been reached. The trypanosome multiplies intracellularly and is difficult to reach chemotherapeutically. The only drugs currently available are nifurtimox and benznidazole, which are triazole derivatives (see page 568). Benznidazole therapy was found to eliminate infection in about 60% of infected children. These drugs must be administered for 30 to 60 days, and neither is effective during the chronic stage; both also have serious side effects.

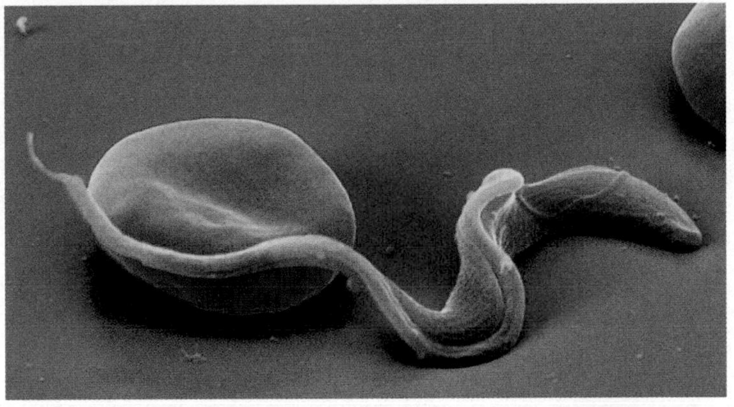

Figure 23.22 ***Trypanosoma cruzi,* the cause of Chagas' disease (American trypanosomiasis).** The trypanosome has an undulating membrane; the flagellum follows the outer margin of the membrane and then projects beyond the body of the trypanosome as a free flagellum. Note the red blood cells in the photo.

Q Name a common trypanosomal disease that occurs in another part of the world. (*Hint*: It was discussed in Chapter 22).

Toxoplasmosis

Toxoplasmosis, a disease of blood and lymphatic vessels, is caused by the protozoan *Toxoplasma gondii. T. gondii* is a spore-forming protozoan, as is the malarial parasite.

Cats are an essential part of the life cycle of *T. gondii* (Figure 23.23). Random tests on urban cats have shown that a large number of them are infected with the organism, which causes no apparent illness in the cat. (A curiosity of the infection in rodents is that it apparently causes them to lose their normal avoidance behavior toward cats, making them more likely to be caught and thus to infect the cat.) The microbe undergoes its only sexual phase in the intestinal tract of the cat. Millions of oocysts are then shed in the cat's feces for 7 to 21 days and contaminate food or water that can be ingested by other animals. The *oocysts* contain *sporozoites* that invade host cells and form trophozoites called *tachyzoites* (about the size of large bacteria, 2 × 7 μm). The intracellular parasite reproduces rapidly (*tachys* is Greek for rapid). The increased numbers cause the rupture of the host cell and the release of more tachyzoites, resulting in a strong inflammatory response.

As the immune system becomes increasingly effective, the disease enters a chronic phase in animals and humans; the infected host cell develops a wall to form a *tissue cyst.* The numerous parasites within such a cyst (in this stage called *bradyzoites; bradys* being Greek for slow) reproduce very slowly, if at all, and persist for years, especially in the brain. These cysts are infective when ingested by intermediate or definitive hosts.

In people with a healthy immune system, toxoplasmosis infection results in only very mild symptoms or none at all. Some surveys have shown that approximately 22–40% of the population,

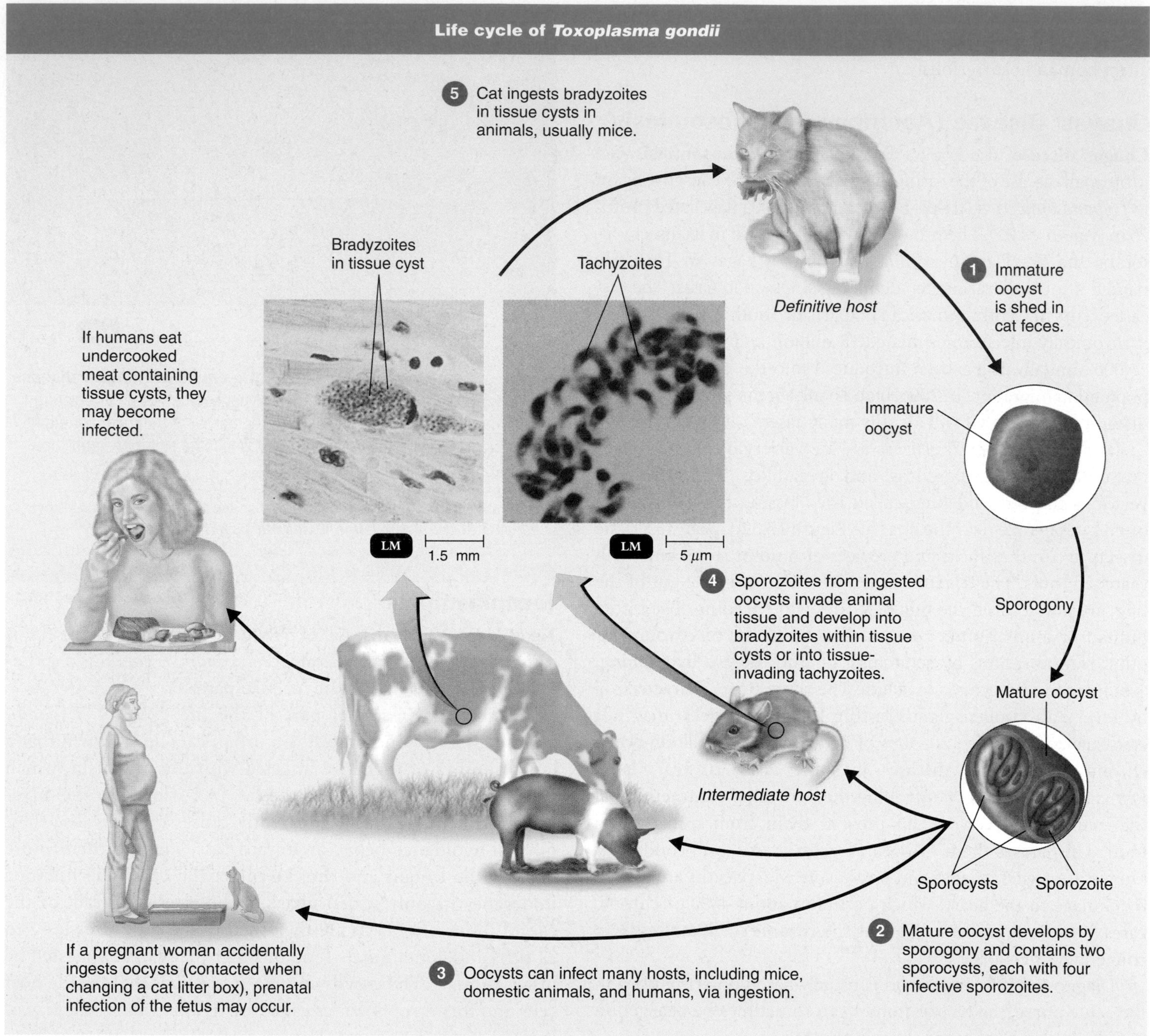

Life cycle of Toxoplasma gondii
5 Cat ingests bradyzoites in tissue cysts in animals, usually mice.
Definitive host
1 Immature oocyst is shed in cat feces.
Immature oocyst
Sporogony
Mature oocyst
Sporocysts
Sporozoite
2 Mature oocyst develops by sporogony and contains two sporocysts, each with four infective sporozoites.
Bradyzoites in tissue cyst
Tachyzoites
LM
1.5 mm
LM
5 μm
4 Sporozoites from ingested oocysts invade animal tissue and develop into bradyzoites within tissue cysts or into tissue-invading tachyzoites.
Intermediate host
If humans eat undercooked meat containing tissue cysts, they may become infected.
If a pregnant woman accidentally ingests oocysts (contacted when changing a cat litter box), prenatal infection of the fetus may occur.
3 Oocysts can infect many hosts, including mice, domestic animals, and humans, via ingestion.

Figure 23.23 The life cycle of *Toxoplasma gondii*, the cause of toxoplasmosis. The domestic cat is the definitive host, in which the protozoa reproduce sexually.

Q How do humans contract toxoplasmosis?

without even being aware of it, eventually develops antibodies to *T. gondii* (see Figure 23.20). Humans generally acquire the infection by ingesting undercooked meats containing tachyzoites or tissue cysts, although there is a possibility of contracting the disease more directly by contact with cat feces. The primary danger is congenital infection of a fetus, resulting in stillbirth or a child with severe brain damage or vision problems. This fetal damage occurs only when the initial infection is acquired during pregnancy. As many as 4000 cases are estimated in the United States annually. The problem also affects wildlife. Off the California coast, a fatal encephalitis of sea otters has appeared, caused by *T. gondii*—apparently, they are being infected by oocysts in waste water contaminated from the flushed contents of cat litter boxes. Loss of immune function, AIDS being the best example, allows the inapparent infection to be reactivated from tissue cysts. It often causes severe neurological impairment and may damage vision from the reactivation of tissue cysts in the eye.

Toxoplasmosis can be detected by serological tests, but interpretation is uncertain. This uncertainty is especially important because in some European countries, a person who becomes toxoplasmosis-positive during pregnancy is encouraged to abort the fetus. Recently, PCR tests have become available. If not contaminated, these tests approach an accuracy of 100%, which has revolutionized prenatal diagnosis. Toxoplasmosis can be treated with pyrimethamine in combination with sulfadiazine and folinic acid. This does not, however, affect the chronic bradyzoite stage and is quite toxic.

Malaria

Malaria is characterized by chills and fever and often by vomiting and severe headache. These symptoms typically appear at intervals of 2 to 3 days, alternating with asymptomatic periods. Malaria occurs wherever the mosquito vector *Anopheles* is found and there are human hosts for the protozoan parasite *Plasmodium*.

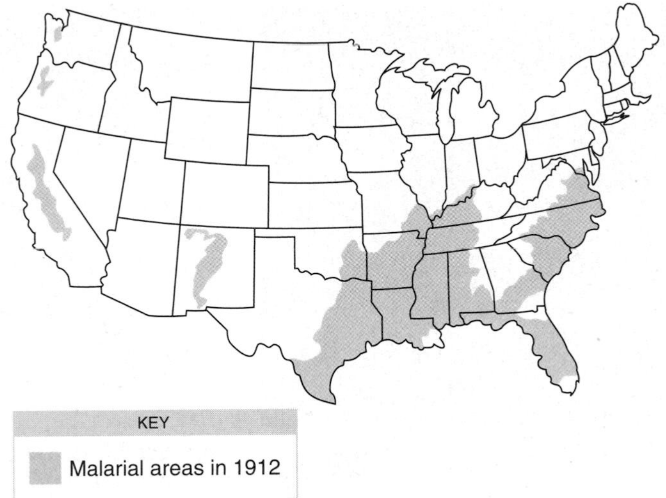

Areas where malaria was endemic as recently as 1912

Figure 23.24 Malaria in the United States.

Q What factors might contribute to the rise in malaria cases since 1990?

The disease was once widespread in the United States (Figure 23.24), but effective mosquito control and a reduction in the number of human carriers caused the reported cases to drop below 100 by 1960. In recent years, however, there has been an upward trend in the number of U.S. cases, reflecting a worldwide resurgence of malaria, increased travel to malarial areas, and an increase in immigration from malarial areas. Occasionally, malaria has been transmitted by unsterilized syringes used by drug addicts. Blood transfusions from people who have been in an endemic area are also a potential risk. In tropical Asia, Africa, and Central and South America, malaria is still a serious problem. It is estimated that malaria affects 300 to 500 million people worldwide and causes 2 to 4 million deaths annually. Actually, there are probably more people dying of malaria today than 30 years ago. It is returning to areas where it had been nearly eradicated, such as eastern Europe and central Asia. Africa, where 90% of the mortality from the disease occurs, suffers the most from malaria. It is estimated that it kills an African child every 30 seconds.

There are four major forms of malaria. *Plasmodium vivax* is widely distributed because it can develop in mosquitoes at a lower temperature and is the cause of the most prevalent form of malaria. Sometimes referred to as "benign" malaria, the cycle of paroxysms occurs every 2 days, and the patients generally survive even without treatment. *P. ovale* and *P. malariae* also cause a relatively benign malaria, but even so, the victims lack energy. These latter two malarial types are lower in incidence and rather restricted geographically.

The most dangerous malaria is that caused by *P. falciparum*. Perhaps one reason for the virulence of this type of malaria is that humans and the parasite have had less time to become adapted to each other. It is believed that humans have been exposed to this parasite (through contact with birds) only in relatively recent history. Referred to as "malignant" malaria, untreated it eventually kills about half of those infected. The highest mortality rates occur in young children. More red blood cells (RBCs) are infected and destroyed than in other forms of malaria. The resulting anemia severely weakens the victim. Furthermore, the RBCs develop surface knobs that cause them to stick to the walls of the capillary vessels, which become clogged. This clogging prevents the infected RBCs from reaching the spleen, where phagocytic cells would eliminate them. The blocked capillaries and subsequent loss of blood supply leads to death of the tissues. Kidney and liver damage is caused in this fashion. The brain is frequently affected, and *P. falciparum* is the usual cause of cerebral malaria.

The disease of malaria and its symptoms are intimately related to its complex reproductive cycle (see Figure 12.18, page 349). Infection is initiated by the bite of a mosquito, which carries the *sporozoite* stage of the *Plasmodium* protozoan in its saliva. The sporozoite enters the bloodstream of the bitten human and within about 30 minutes enters the liver cells. The sporozoites in the liver cells undergo reproductive *schizogony* by a series of steps that finally results in the release of about 30,000 *merozoite* forms into the bloodstream.

The merozoites infect RBCs. Within the RBCs they again undergo schizogony, and after about 48 hours, the RBCs rupture and each releases about 20 new merozoites (Figure 23.25a). Laboratory diagnosis of malaria is usually made by examining a blood smear (Figure 23.25b) for infected RBCs. With the release of the merozoites there is also a simultaneous release of toxic compounds, which is the cause of the paroxysms (recurrent intensifications of symptoms) of chills and fever that are

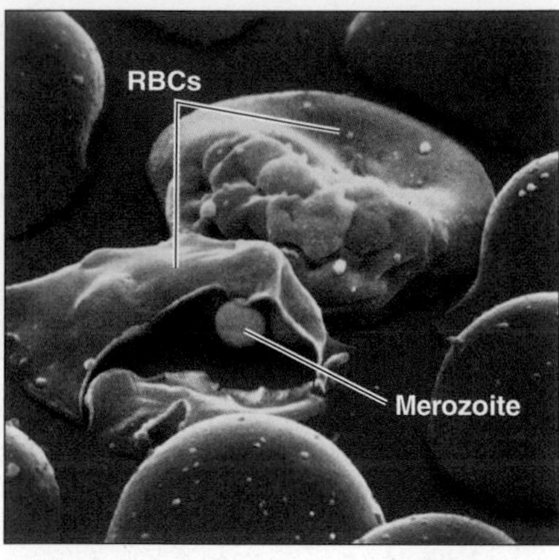

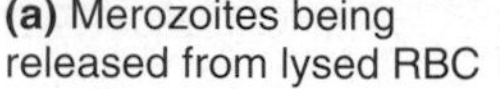

(a) Merozoites being released from lysed RBC

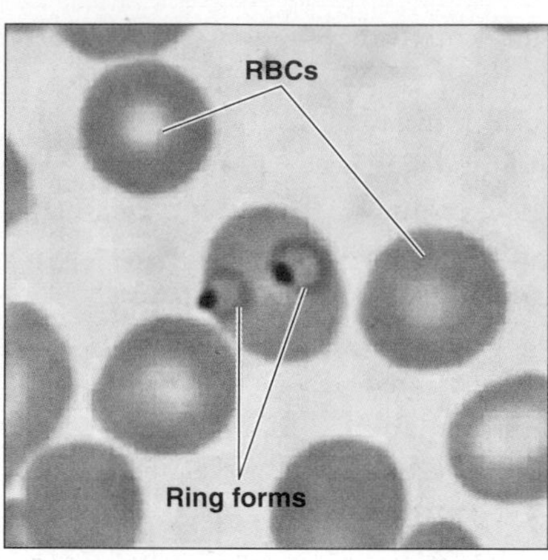

(b) Malarial blood smear; note the ring forms.

Figure 23.25 Malaria. (**a**) Some of the red blood cells (RBCs) are lysing and releasing merozoites that will infect new RBCs. (**b**) Blood smears are used in diagnosing malaria; the protozoa can be detected growing in the RBCs. In the early stages, the feeding protozoan resembles a ring within the RBC. The light central area within the circular ring is the food vacuole of the protozoan, and the dark spot on the ring is the nucleus.

Q **Look at the life cycle of the malarial parasite; in Figure 12.19, which of these stages, (a) or (b), actually occurs first?**

characteristic of malaria. The fever reaches 40°C, and a sweating stage begins as the fever subsides. Between paroxysms, the patient feels normal.

Many of the released merozoites infect other RBCs within a few seconds to renew the cycle in the bloodstream. If only 1% of the RBCs contain parasites, an estimated 100,000,000,000 parasites will be in circulation at one time in a typical malaria patient! Some of the merozoites develop into male or female *gametocytes.* When these enter the digestive tract of a feeding mosquito, they pass through a sexual cycle that produces new infective sporozoites. It took the combined labors of several generations of scientists to unravel this complex life cycle of the malaria parasite.

People who survive malaria acquire a limited immunity. Although they can be reinfected, they tend to have a less severe form of the disease. This relative immunity almost disappears if the person leaves an endemic area with its periodic reinfections. Malaria is especially dangerous during pregnancy because adaptive immunity is suppressed.

Much effort is being expended on the search for an effective vaccine. The sporozoite stage is of primary interest as a target for a vaccine because neutralizing it would prevent the initial infection from becoming well established. A truly global malarial vaccine would have to control not only *P. falciparum* but also the widespread, although milder, *P. vivax.* There are special problems in developing a malarial vaccine. For example, unlike most viral or bacterial pathogens, which are relatively simple genetically and remain much the same during the course of an infection, the malarial parasite has four distinctive stages. In these stages it has as many as 7000 genes that can mutate. The result is that the parasite is very efficient at evading the human immune response. The current goal is to have a vaccine by 2015 that is at least 50% effective and lasts longer than a year, then, by 2025, to have one that is 80% effective and lasts more than 4 years.

The most common diagnostic test for malaria is the blood smear, which requires a microscope. It is also time-consuming and requires skill in interpretation. It is still the "gold standard" for diagnosis when a well-trained staff is available. Rapid, antigen-detecting diagnostic tests that can be performed by staff with minimal training have been developed but are relatively expensive. High-quality rapid diagnostic tests that are affordable and perform reliably under field conditions are urgently needed. In endemic areas malaria is commonly diagnosed by simply observing symptoms, mainly fever, but this frequently leads to misdiagnosis. It has been found that only about half of such patients given prescriptions for antimalarial drugs actually had the disease.

There are two considerations for antimalarial drugs: for prophylaxis (prevention) or for treatment.

Prophylaxis

For travel to the few areas in which the malaria is still sensitive to it, chloroquine is the drug of choice. In chloroquine-resistant areas, the drug malarone, a combination of atovaquone and proguanil, is the best tolerated. Travelers to malarial areas are often prescribed mefloquine (Lariam). It requires only a weekly dosage, but users must be cautioned about possible side effects, which include hallucinations.

Therapy

There is a lengthy list of antimalarial drugs available; recommendations and requirements vary with cost, likelihood of developing resistance, and other factors. In the United States (where there are about 1200 imported cases of malaria annually), if the species cannot be identified it should be assumed that the patient is infected with *P. falciparum.* If the patient is from an area still sensitive to chloroquine, it is the drug of choice; for patients coming from chloroquine-resistant zones, there are several options. The two currently preferred are malarone or oral quinine plus an antibiotic such as tetracycline. Worldwide, key components in malarial treatment are the derivatives of artemisinin, such as artesunate and artemether. These drugs are derived from a plant, wormwood or artemisia, which has long been used as a remedy for fevers in China. Ideally, they are taken in combination with other antimalarials to minimize development of resistance.

As with other tropical diseases, the availability of medications is limited by the very low income of the people affected, which makes their development unprofitable. The most profitable application of antimalarials will probably continue to be prophylaxis of travelers to malarial areas.

Effective control of malaria is not in sight. It will probably require a combination of vector control and chemotherapeutic and immunological approaches. Currently, the most promising control method is the use of insecticide-treated bed nets, because the *Anopheles* mosquito is a night feeder. In malarial areas, a sleeping room often will contain hundreds of mosquitoes, 1–5% of which are infectious. The expense of these efforts and the need for an effective political organization in malarial areas are probably going to be as important in controlling the disease as are advances in medical research.

Leishmaniasis

Leishmaniasis is a widespread and complex disease that exhibits several clinical forms. The protozoan pathogens are of about 20 different species, often categorized into three groups for reasons of simplicity. One group, *Leishmania donovani* (līsh'mā-nē-ä don-ō-van'ē), causes a visceral leishmaniasis in which parasites invade the internal organs. The *L. tropica* (līsh'mā-nē-ä trop'i-ka) and *L. braziliensis* (bra-sil'e-en-sis) groups grow preferentially at cooler temperatures and cause lesions of the skin or mucous membranes. Leishmaniasis is transmitted by the bite of female sandflies, about 30 species of which are found in much of the tropical world and around the Mediterranean. These insects are smaller than mosquitoes and often penetrate the mesh of standard netting. Small mammals are an unaffected reservoir of the protozoans. The infective form, the *promastigote,* is in the saliva of the insect. It loses its flagellum when it penetrates the skin of the mammalian victim, becoming an *amastigote* that proliferates in phagocytic cells, mostly in fixed locations in tissue. These amastigotes are then ingested by feeding sandflies, renewing the cycle. Contact with contaminated blood from transfusions or shared needles can also lead to infection.

A number of cases of leishmaniasis, mostly cutaneous, have occurred among troops fighting the Persian Gulf regions. It was once endemic in countries of southern Europe, such as Spain, Italy, Portugal, and the Balkan peninsula. Occasional cases of leishmaniasis as an opportunistic disease of HIV-infected persons are beginning to reappear in these areas.

Leishmania donovani Infection (Visceral Leishmaniasis)

Leishmania donovani infection occurs in much of the tropical world, although 90% of the cases occur in India, Bangladesh, Sudan, and Brazil. Estimates are that there are about half a million cases per year. Known as *kala azar* in India, visceral leishmaniasis is often fatal. Early symptoms, following infection by as long as a year, resemble the chills and sweating of malaria. As the protozoa proliferate in the liver and spleen, these organs enlarge greatly. Eventually, kidney function is also lost as these organs are invaded. This is a debilitating disease that, if untreated, will lead to death within a year or two.

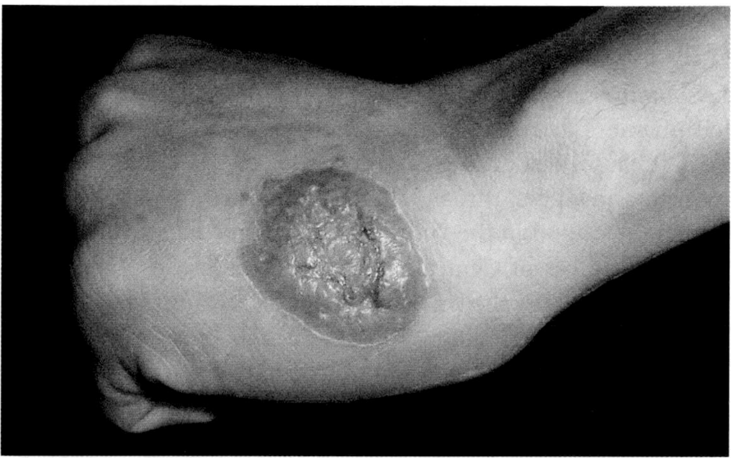

Figure 23.26 Cutaneous leishmaniasis. Lesion on the back of the hand of a patient.

Q Is this case likely to progress to visceral leishmaniasis?

Several inexpensive serological tests that are easy to use have been developed to diagnose visceral leishmaniasis. These have generally replaced microscopic examination of blood and tissues to demonstrate the parasite. PCR tests are very good to confirm diagnosis but usually require a central laboratory.

The primary treatment has long been injected drugs such as sodium stibogluconate that contain the toxic metal antimony. Other drugs are now replacing antimony-based preparations. The first-line treatment in Europe and United States is liposomal amphotericin B, but it is relatively expensive for endemic countries. In many of these areas, conventional formulations of amphotericin B are in use. The first effective oral drug is miltefosine. It has demonstrated a cure rate as high as 82%, but it is teratogenic, resistance develops rapidly, and it is toxic to a significant number of recipients. An inexpensive injectable aminoglycoside antibiotic, paromomycin, has shown good effectiveness, but its availability is limited.

Leishmania tropica Infection (Cutaneous Leishmaniasis)

Leishmania tropica and *L. major* infection causes a *cutaneous* form of leishmaniasis sometimes called *oriental sore.* A papule appears at the bite site after a few weeks of incubation (Figure 23.26). The papule ulcerates and, after healing, leaves a prominent scar. This form of the disease is the most common and is found in much of Asia, Africa, and the Mediterranean region. It has been reported in Mexico, Central America, and the northern part of South America.

DISEASES IN FOCUS 23.5

Infections Transmitted by Soil and Water

Differential diagnosis is the process of identifying the disease from a list of possible diseases that fit the information derived from examining a patient. A differential diagnosis is important for providing initial treatment and for laboratory testing. A minority of systemic infections are acquired by contact with soil and water. The pathogens usually enter through a break in the skin. For example, a 65-year-old man with poor circulation in his legs developed an infection following injury to a toe. Dead tissue further reduced circulation, requiring amputation of two toes. Use the table below to identify infections that could cause these symptoms. For the solution, go to www.microbiologyplace.com.

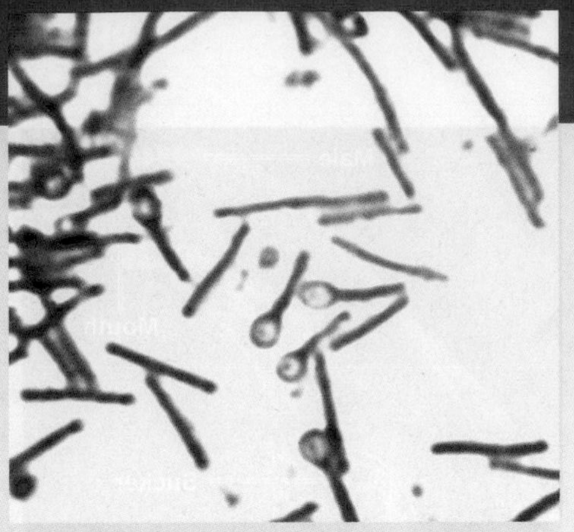

Gram-stained bacteria from the patient's toe.

Disease	Pathogen	Symptoms	Reservoir	Method of Transmission	Treatment
BACTERIAL DISEASE					
Gangrene	*Clostridium perfringens*	Tissue death at infection site	Soil	Puncture wound	Surgical removal of necrotic tissue
HELMINTHIC DISEASES					
Schistosomiasis	*Schistosoma* spp.	Inflammation and tissue damage at site of granulomas (e.g., liver, lungs, bladder)	Definitive host; humans	Cercariae penetrate skin	Proziquantel; oxamniquine Prevention: sanitation; elimination of host snail
Swimmer's itch	Larvae of schistosomes of nonhuman animals	Local inflammation	Wildfowl	Cercariae penetrate skin	None

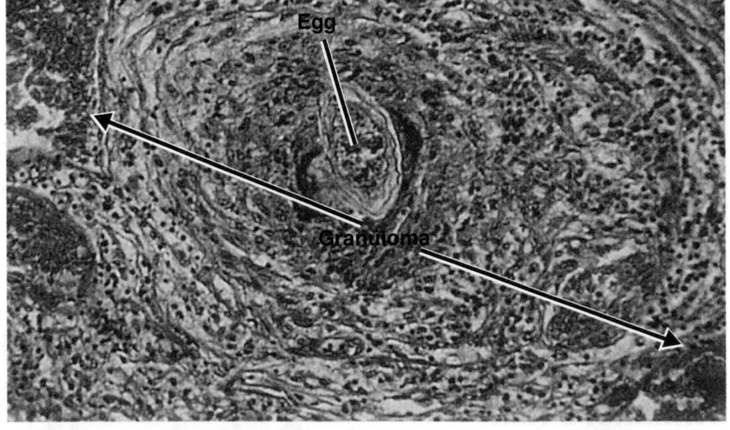

Figure 23.28 A granuloma from a patient with schistosomes. Some of the eggs laid by the adult schistosomes lodge in the tissue, and the body responds to the irritant by surrounding it with scarlike tissue, forming a granuloma.

Q Why is the immune system ineffective against adult schistosomes?

STUDY OUTLINE

The **MyMicrobiologyPlace** website (**www.microbiologyplace.com**) will help you get ready for tests with its simple three-step approach: ❶ **take a pre-test** and obtain a personalized study plan, ❷ **learn and practice** with animations, tutorials, and MP3 tutor sessions, and ❸ **test yourself** with quizzes and a chapter post-test.

Introduction (p. 637)

1. The heart, blood, and blood vessels make up the cardiovascular system.
2. Lymph, lymph vessels, lymph nodes, and lymphoid organs constitute the lymphatic system.

Structure and Function of the Cardiovascular and Lymphatic Systems (p. 638)

1. The heart circulates substances to and from tissue cells.
2. Blood is a mixture of plasma and cells.
3. Plasma transports dissolved substances. Red blood cells carry oxygen. White blood cells are involved in the body's defense against infection.
4. Fluid that filters out of capillaries into spaces between tissue cells is called interstitial fluid.
5. Interstitial fluid enters lymph capillaries and is called lymph; vessels called lymphatics return lymph to the blood.
6. Lymph nodes contain fixed macrophages, B cells, and T cells.

Bacterial Diseases of the Cardiovascular and Lymphatic Systems (pp. 638–655)

Sepsis and Septic Shock (pp. 639–641)

1. Sepsis is an inflammatory response caused by the spread of bacteria or their toxin from a focus of infection. Septicemia is sepsis that involves proliferation of pathogens in the blood.
2. Gram-negative sepsis can lead to septic shock, characterized by decreased blood pressure. Endotoxin causes the symptoms.

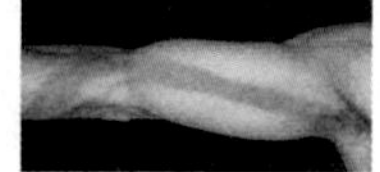

3. Antibiotic-resistant enterococci and group B streptococci cause gram-positive sepsis.
4. Puerperal sepsis begins as an infection of the uterus following childbirth or abortion; it can progress to peritonitis or septicemia.
5. *Streptococcus pyogenes* is the most frequent cause of puerperal sepsis.
6. Oliver Wendell Holmes and Ignaz Semmelweiss demonstrated that puerperal sepsis was transmitted by the hands and instruments of midwives and physicians.

Bacterial Infections of the Heart (p. 641)

7. The inner layer of the heart is the endocardium.
8. Subacute bacterial endocarditis is usually caused by alpha-hemolytic streptococci, staphylococci, or enterococci.
9. The infection arises from a focus of infection, such as a tooth extraction.
10. Preexisting heart abnormalities are predisposing factors.
11. Signs include fever, anemia, and heart murmur.
12. Acute bacterial endocarditis is usually caused by *Staphylococcus aureus.*
13. The bacteria cause rapid destruction of heart valves.

Rheumatic Fever (pp. 641–642)

14. Rheumatic fever is an autoimmune complication of streptococcal infections.
15. Rheumatic fever is expressed as arthritis or inflammation of the heart. It can result in permanent heart damage.
16. Antibodies against group A beta-hemolytic streptococci react with streptococcal antigens deposited in joints or heart valves or cross-react with the heart muscle.
17. Rheumatic fever can follow a streptococcal infection, such as streptococcal sore throat. Streptococci might not be present at the time of rheumatic fever.
18. Prompt treatment of streptococcal infections can reduce the incidence of rheumatic fever.
19. Penicillin is administered as a preventive measure against subsequent streptococcal infections.

Tularemia (pp. 642–643)

20. Tularemia is caused by *Francisella tularensis.* The reservoir is small wild mammals, especially rabbits.
21. Signs include ulceration at the site of entry, followed by septicemia and pneumonia.

Brucellosis (Undulant Fever) (pp. 643–645)

22. Brucellosis can be caused by *Brucella abortus, B. melitensis,* and *B. suis.*
23. The bacteria enter through minute breaks in the mucosa or skin, reproduce in macrophages, and spread via lymphatics to liver, spleen, or bone marrow.
24. Signs include malaise and fever that spikes each evening (undulant fever).
25. Diagnosis is based on serological tests.

Anthrax (pp. 645–646)

26. *Bacillus anthracis* causes anthrax. In soil, endospores can survive for up to 60 years.

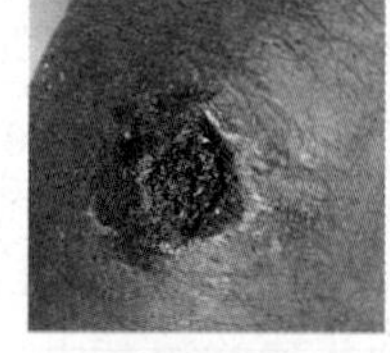

27. Grazing animals acquire an infection after ingesting the endospores.
28. Humans contract anthrax by handling hides from infected animals. The endospores enter through cuts in the skin, respiratory tract, or mouth.
29. Entry through the skin results in a pustule that can progress to sepsis. Entry through the respiratory tract can result in septic shock.
30. Diagnosis is based on isolating and identifying the bacteria.

Gangrene (p. 646)

31. Soft tissue death from ischemia (loss of blood supply) is called gangrene.
32. Microorganisms grow on nutrients released from gangrenous cells.

33. Gangrene is especially susceptible to the growth of anaerobic bacteria such as *Clostridium perfringens,* the causative agent of gas gangrene.
34. *C. perfringens* can invade the wall of the uterus during improperly performed abortions.
35. Surgical removal of necrotic tissue, hyperbaric chambers, and amputation are used to treat gas gangrene.

Systemic Diseases Caused by Bites and Scratches (pp. 647–648)

36. *Pasteurella multocida,* introduced by the bite of a dog or cat, can cause septicemia.
37. Anaerobic bacteria infect deep animal bites.
38. Cat-scratch disease is caused by *Bartonella henselae.*
39. Rat-bite fever is caused by *Streptobacillus moniliformis* and *Spirillum minus.*

Vector-Transmitted Diseases (pp. 648–655)

Plague (pp. 648–650)

40. Plague is caused by *Yersinia pestis.* The vector is usually the rat flea (*Xenopsylla cheopis*).
41. Reservoirs for bubonic plague include European rats and North American rodents.
42. Signs of bubonic plague include bruises on the skin and enlarged lymph nodes (buboes).
43. The bacteria can enter the lungs and cause pneumonic plague.
44. Laboratory diagnosis is based on isolating and identifying the bacteria.
45. Antibiotics are effective in treating plague, but they must be administered promptly after exposure to the disease.

Relapsing Fever (pp. 650–651)

46. Relapsing fever is caused by *Borrelia* species and transmitted by soft ticks.
47. The reservoir for the disease is rodents.
48. Signs include fever, jaundice, and rose-colored spots. Signs recur three or four times after apparent recovery.
49. Laboratory diagnosis is based on the presence of spirochetes in the patient's blood.

Lyme Disease (Lyme Borreliosis) (pp. 651–654)

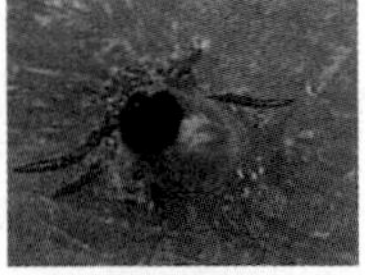

50. Lyme disease is caused by *Borrelia burgdorferi* and is transmitted by a tick *(Ixodes).*
51. Field mice provide the animal reservoir.
52. Diagnosis is based on serological tests and clinical symptoms.

Ehrlichiosis and Anaplasmosis (p. 654)

53. Human ehrlichiosis and anaplasmosis are caused by *Ehrlichia* and *Anaplasma* and are transmitted by *Ixodes* ticks.

Typhus (pp. 654–655)

54. Typhus is caused by rickettsias, obligate intracellular parasites of eukaryotic cells.

Epidemic Typhus (p. 654)

55. The human body louse, *Pediculus humanus corporis,* transmits *Rickettsia prowazekii* in its feces, which are deposited while the louse is feeding.
56. Epidemic typhus is prevalent in crowded and unsanitary living conditions that allow the proliferation of lice.
57. The signs of typhus are rash, prolonged high fever, and stupor.
58. Tetracyclines and chloramphenicol are used in treatment.

Endemic Murine Typhus (p. 655)

59. Endemic murine typhus is a less severe disease caused by *Rickettsia typhi.* It is transmitted from rodents to humans by the rat flea.

Spotted Fevers (p. 655)

60. *Rickettsia rickettsii* is a parasite of ticks (*Dermacentor* spp.) in the southeastern United States, Appalachia, and the Rocky Mountain states.
61. The rickettsia may be transmitted to humans, in whom it causes tickborne typhus fever.
62. Chloramphenicol and tetracyclines effectively treat Rocky Mountain spotted fever, or tickborne typhus.
63. Serological tests are used for laboratory diagnosis.

Viral Diseases of the Cardiovascular and Lymphatic Systems (pp. 655–660)

Burkitt's Lymphoma (pp. 655–656)

1. Epstein-Barr virus (EB virus, HHV-4) causes Burkitt's lymphoma.
2. Burkitt's lymphoma tends to occur in patients whose immune system has been weakened; for example, by malaria or AIDS.

Infectious Mononucleosis (pp. 656–657)

3. Infectious mononucleosis is caused by EB virus.
4. The virus multiplies in the parotid glands and is present in saliva. It causes the proliferation of atypical lymphocytes.
5. The disease is transmitted by the ingestion of saliva from infected individuals.
6. Diagnosis is made by an indirect fluorescent-antibody technique.
7. EB virus may cause other diseases, including cancers and multiple sclerosis.

Cytomegalovirus Infections (p. 658)

8. CMV (HHV-5) causes intranuclear inclusion bodies and cytomegaly of host cells.
9. CMV is transmitted by saliva and other body fluids.
10. CMV inclusion disease can be asymptomatic, a mild disease, or progressive and fatal. Immunosuppressed patients may develop pneumonia.
11. If the virus crosses the placenta, it can cause congenital infection of the fetus, resulting in impaired mental development, neurological damage, and stillbirth.

Chikungunya Fever (p. 658)

12. The chickungunya virus, which causes fever and severe joint pain, is transmitted by *Aedes* mosquitoes.

Classic Viral Hemorrhagic Fevers (pp. 658–659)

13. Yellow fever is caused by the yellow fever virus. The vector is the *Aedes aegypti* mosquito.
14. Signs and symptoms include fever, chills, headache, nausea, and jaundice.
15. Diagnosis is based on the presence of virus-neutralizing antibodies in the host.
16. No treatment is available, but there is an attenuated, live viral vaccine.
17. Dengue is caused by the dengue fever virus and is transmitted by the *Aedes* mosquito.

18. Signs are fever, muscle and joint pain, and rash.
19. Mosquito abatement is necessary to control the disease.
20. Dengue hemorrhagic fever (DHF) can cause shock.

Emerging Viral Hemorrhagic Fevers (pp. 659–660)

21. Human diseases caused by Marburg, Ebola, and Lassa fever viruses were first noticed in the late 1960s.
22. Ebola virus is found in fruit bats; Lassa fever viruses are found in rodents.
23. Rodents are the reservoirs for Argentine and Bolivian hemorrhagic fevers.
24. *Hantavirus* pulmonary syndrome and hemorrhagic fever with renal syndrome are caused by hantavirus. The virus is contracted by inhalation of dried rodent urine and feces.

Protozoan Diseases of the Cardiovascular and Lymphatic Systems (pp. 660–666)

Chagas' Disease (American Trypanosomiasis) (p. 661)

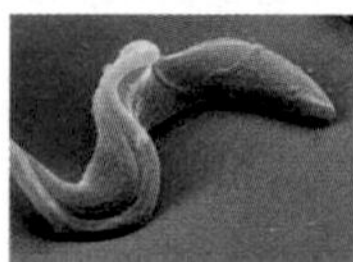

1. *Trypanosoma cruzi* causes Chagas' disease. The reservoir includes many wild animals. The vector is a reduviid, the "kissing bug."

Toxoplasmosis (pp. 661–663)

2. Toxoplasmosis is caused by *Toxoplasma gondii.*
3. *T. gondii* undergoes sexual reproduction in the intestinal tract of domestic cats, and oocysts are eliminated in cat feces.
4. In the host cell, sporozoites reproduce to form either tissue-invading tachyzoites or bradyzoites.
5. Humans contract the infection by ingesting tachyzoites or tissue cysts in undercooked meat from an infected animal or contact with cat feces.
6. Congenital infections can occur. Signs and symptoms include severe brain damage or vision problems.

Malaria (pp. 663–665)

7. The signs and symptoms of malaria are chills, fever, vomiting, and headache, which occur at intervals of 2 to 3 days.
8. Malaria is transmitted by *Anopheles* mosquitoes. The causative agent is any one of four species of *Plasmodium.*
9. Sporozoites reproduce in the liver and release merozoites into the bloodstream, where they infect red blood cells and produce more merozoites.
10. New drugs are being developed as the protozoa develop resistance to drugs such as chloroquine.

Leishmaniasis (pp. 665–666)

11. *Leishmania* spp., which are transmitted by sandflies, cause leishmaniasis.
12. The protozoa reproduce in the liver, spleen, and kidneys.
13. Antimony compounds are used for treatment.

Babesiosis (p. 666)

14. Babesiosis is caused by the protozoan *Babesia microti* and is transmitted to humans by ticks.

Helminthic Diseases of the Cardiovascular and Lymphatic Systems (pp. 666–668)

Schistosomiasis (pp. 666–667)

1. Species of the blood fluke *Schistosoma* cause schistosomiasis.
2. Eggs eliminated with feces hatch into larvae that infect the intermediate host, a snail. Free-swimming cercariae are released from the snail and penetrate the skin of a human.
3. The adult flukes live in the veins of the liver or urinary bladder in humans.
4. Granulomas are from the host's defense to eggs that remain in the body.
5. Observation of eggs or flukes in feces, skin tests, or indirect serological tests may be used for diagnosis.
6. Chemotherapy is used to treat the disease; sanitation and snail eradication are used to prevent it.

Swimmer's Itch (p. 667)

7. Swimmer's itch is a cutaneous allergic reaction to cercariae that penetrate the skin. The definitive hosts for this fluke are wildfowl.

STUDY QUESTIONS

Answers to the Review and Multiple Choice questions can be found by turning to the blue Answers tab at the back of the textbook.

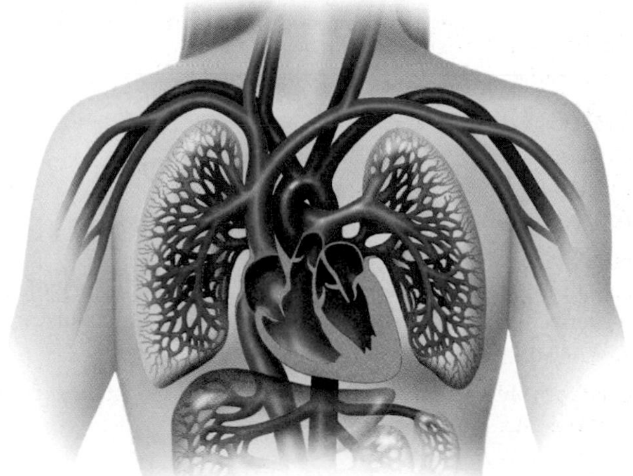

Review

1. **DRAW IT** Show the path of *Streptococcus* from a focal infection to the pericardium. Identify the portals of entry for *Trypanosoma cruzi*, *Hantavirus*, and cytomegalovirus.

2. Complete the following table.

Disease	Frequent Causative Agent	Predisposing Condition(s)
Puerperal sepsis		
Subacute bacterial endocarditis		
Acute bacterial endocarditis		
Rheumatic fever		

3. Compare and contrast epidemic typhus, endemic murine typhus, and tickborne typhus.
4. Complete the following table.

Disease	Causative Agent	Vector	Treatment
Malaria			
Yellow fever			
Dengue			
Relapsing fever			
Leishmaniasis			

5. Complete the following table.

Disease	Causative Agent	Transmission	Reservoir
Tularemia			
Brucellosis			
Anthrax			
Lyme disease			
Ehrlichiosis			
Cytomegalic inclusion disease			
Plague			

6. List the causative agent, method of transmission, and reservoir for schistosomiasis, toxoplasmosis, and Chagas' disease. Which disease are you most likely to get in the United States? Where are the other diseases endemic?
7. Compare and contrast cat-scratch disease and toxoplasmosis.
8. Why is *Clostridium perfringens* likely to grow in gangrenous wounds?
9. List the causative agent and method of transmission of infectious mononucleosis.

Multiple Choice

Use the following choices to answer questions 1 through 4:
a. ehrlichiosis
b. Lyme disease
c. septic shock
d. toxoplasmosis
e. viral hemorrhagic fever

1. A patient presents with vomiting, diarrhea, and a history of fever and headache. Bacterial cultures of blood, CSF, and stool are negative. What is your diagnosis?
2. A patient was hospitalized because of continuing fever and progression of symptoms including headache, fatigue, and back pain. Tests for antibodies to *Borrelia burgdorferi* were negative. What is your diagnosis?
3. A patient complained of headache. A CT (computed tomography) scan revealed cysts of varying size in her brain. What is your diagnosis?
4. A patient presents with mental confusion, rapid breathing and heartbeat, and low blood pressure. What is your diagnosis?
5. A patient has a red circular rash on his arm and fever, malaise, and joint pain. The most appropriate treatment is
a. penicillin.
b. chloroquine.
c. anti-inflammatory drugs.
d. rifampin.
e. no treatment.
6. Which of the following is not a tickborne disease?
a. babesiosis
b. ehrlichiosis
c. Lyme disease
d. relapsing fever
e. tularemia

Use the following choices to answer questions 7 and 8:
a. brucellosis
b. malaria
c. relapsing fever
d. Rocky Mountain spotted fever
e. Ebola hemorrhagic fever

7. The patient's fever spikes each evening. Oxidase-positive, gram-negative cocci were isolated from a lesion on his arm. What is your diagnosis?
8. The patient was hospitalized with fever and headache. Spirochetes were observed in her blood. What is your diagnosis?
9. Which of the following diseases has the highest incidence in the United States?
a. brucellosis
b. Ebola hemorrhagic fever
c. malaria
d. plague
e. Rocky Mountain spotted fever
10. Nineteen workers in a slaughterhouse developed fever and chills, with the fever spiking to 40°C each evening. The most likely method of transmission of this disease is
a. a vector.
b. the respiratory route.
c. a puncture wound.
d. an animal bite.
e. water.

Critical Thinking

1. Indirect fluorescent-antibody (FA) tests on the serum of three 25-year-old women, each of whom is considering pregnancy, provided the information below. Which of these women may have toxoplasmosis? What advice might be given to each woman with regard to toxoplasmosis?

Patient	Antibody Titer Day 1	Day 5	Day 12
Patient A	1024	1024	1024
Patient B	1024	2048	3072
Patient C	0	0	0

2. What is the most effective way to control malaria and dengue?
3. In adults, the second dengue virus infection results in dengue hemorrhagic fever (DHF), which is characterized by bleeding from the skin and mucosa. DHF can be fatal. In infants under 1 year old, the first dengue virus infection results in DHF. Offer an explanation for this.

Clinical Applications

1. A 19-year-old man went deer hunting. While on the trail, he found a partially dismembered dead rabbit. The hunter picked up the front paws for good luck charms and gave them to another hunter in the party. The rabbit had been handled with bare hands that were bruised and scratched from the hunter's work as an automobile mechanic. Festering sores on his hands, legs, and knees were noted 2 days later. What infectious disease do you suspect the hunter has? How would you proceed to prove it?
2. On March 30, a 35-year-old veterinarian experienced fever, chills, and vomiting. On March 31, he was hospitalized with diarrhea, left armpit bubo, and secondary bilateral pneumonia. On March 27, he had treated a cat that had labored respiration; an X-ray image revealed pulmonary infiltrates. The cat died on March 28 and was disposed of. Chloramphenicol was administered to the veterinarian. On April 10, his temperature returned to normal, and on April 20, he was released from the hospital. Sixty human contacts were given tetracycline. Identify the incubation and prodromal periods for this case. Explain why the 60 contacts were treated. What was the etiologic agent? How would you identify the agent?
3. Three of five patients who underwent heart valve replacement surgery developed bacteremia. The causative agent was *Enterobacter cloacae.* What were the patients' signs and symptoms? How would you identify this bacterium? A manometer used in the operations was culture-positive for *E. cloacae.* What is the most likely source of this contaminant? Suggest a way of preventing such occurrences.
4. In August and September, six people who each at different times spent a night in the same cabin developed fever, as shown in the graph below. Three recovered after tetracycline (TET) therapy, two recovered without therapy, and one was hospitalized with septic shock. What is the disease? What is the incubation period of this disease? How do you account for the periodic temperature changes? What caused septic shock in the sixth patient?

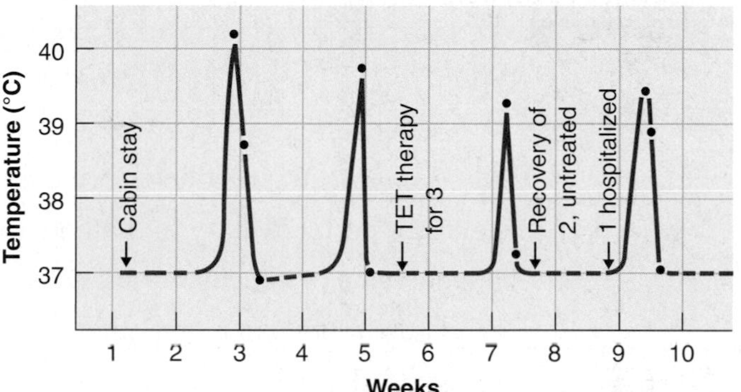

5. A 67-year-old man worked in a textile mill that processed imported goat hair into fabrics. He noticed a painless, slightly swollen pimple on his chin. Two days later he developed a 1-cm ulcer at the pimple site and a temperature of 37.6°C. He was treated with tetracycline. What is the etiology of this disease? Suggest ways to prevent it.

24 Microbial Diseases of the Respiratory System

With every breath, we inhale several microorganisms; therefore, the upper respiratory system is a major portal of entry for pathogens. In fact, respiratory system infections are the most common type of infection—and among the most damaging. Some pathogens that enter via the respiratory route can infect other parts of the body, causing such diseases as measles, mumps, and rubella.

The upper respiratory system has several anatomical defenses against airborne pathogens. Coarse hairs in the nose filter large dust particles from the air. The nose is lined with a mucous membrane that contains numerous mucus-secreting cells and cilia. The upper portion of the throat also contains a ciliated mucous membrane. The mucus moistens inhaled air and traps dust and microorganisms. The cilia help remove these particles by moving them toward the mouth for elimination.

At the junction of the nose and throat are masses of lymphoid tissue, the tonsils, which contribute immunity to certain infections. Because the nose and throat are connected to the sinuses, nasolacrimal apparatus, and middle ear, infections commonly spread from one region to another.

UNDER THE MICROSCOPE

Bordetella pertussis, bacteria (orange) growing on ciliated cells of the upper respiratory system.

Q&A

How do these bacteria, by growing on these particular body cells, cause the disease pertussis? ***Look for the answer in the chapter.***

Structure and Function of the Respiratory System

LEARNING OBJECTIVE

24-1 Describe how microorganisms are prevented from entering the respiratory system.

It is convenient to think of the respiratory system as being composed of two divisions: the upper respiratory system and the lower respiratory system. The **upper respiratory system** consists of the nose, the pharynx (throat), and the structures associated with them, including the middle ear and the auditory (eustachian) tubes (Figure 24.1). Ducts from the sinuses and the nasolacrimal ducts from the lacrimal (tear-forming) apparatus empty into the nasal cavity (see Figure 16.3, page 452). The auditory tubes from the middle ear empty into the upper portion of the throat.

The **lower respiratory system** consists of the larynx (voice box), trachea (windpipe), bronchial tubes, and *alveoli* (Figure 24.2). Alveoli are air sacs that make up the lung tissue; within them, oxygen and carbon dioxide are exchanged between the lungs and blood. Our lungs contain more than 300 million alveoli, with an area for gas exchange of 70 or more square meters in an average adult. The double-layered membrane enclosing the lungs is the *pleura,* or pleural membranes. A ciliated mucous membrane lines the lower respiratory system down to the smaller bronchial tubes and helps prevent microorganisms from reaching the lungs.

As discussed in Chapter 16, particles trapped in the larynx, trachea, and larger bronchial tubes are moved up toward the throat by a ciliary action called the *ciliary escalator* (see Figure 16.4, page 452). If microorganisms actually reach the lungs, phagocytic cells called *alveolar macrophages* usually locate, ingest, and destroy most of them. IgA antibodies in such secretions as respiratory mucus, saliva, and tears also help protect mucosal surfaces of the respiratory system from many pathogens. Thus, the body has several mechanisms for removing the pathogens that cause airborne infections.

CHECK YOUR UNDERSTANDING

✓ What is the function of hairs in the nasal passages? 24-1

Normal Microbiota of the Respiratory System

LEARNING OBJECTIVE

24-2 Characterize the normal microbiota of the upper and lower respiratory systems.

Figure 24.1 Structures of the upper respiratory system.

Q Name the upper respiratory system's defenses against disease.

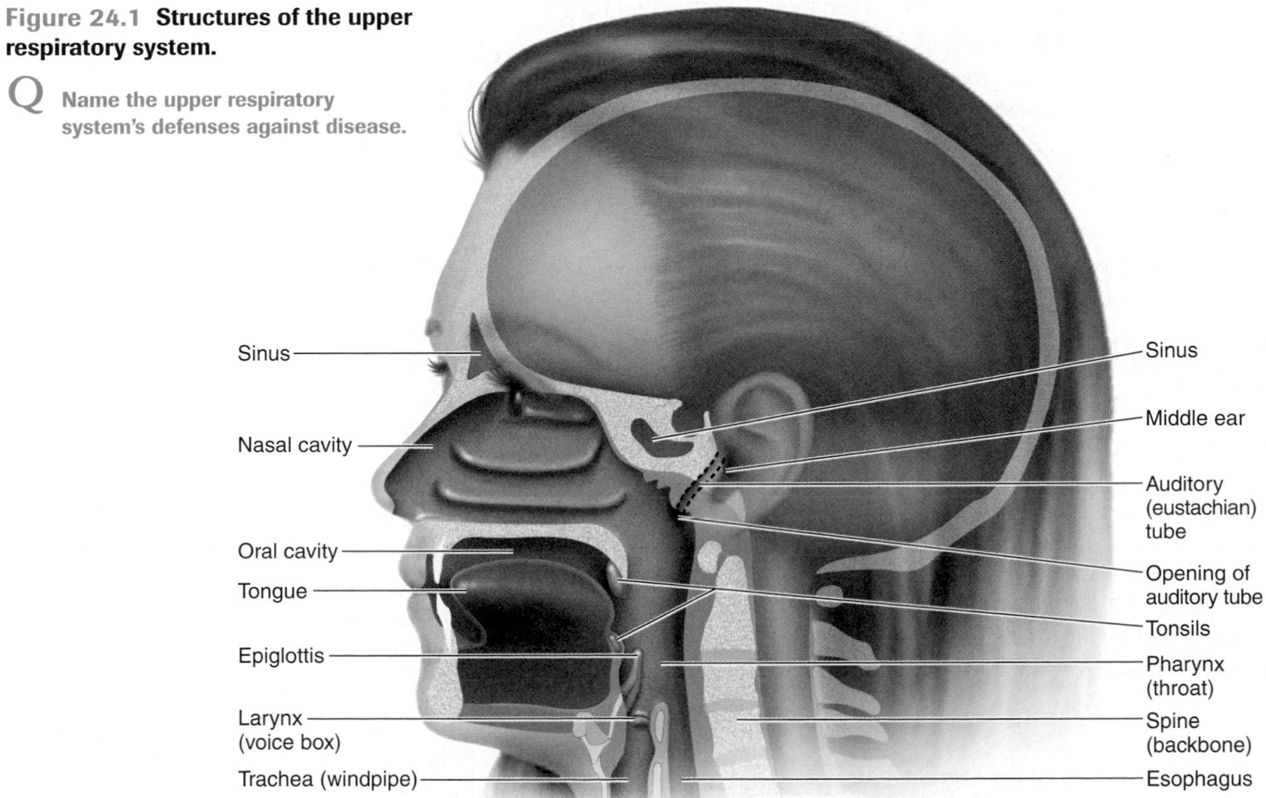

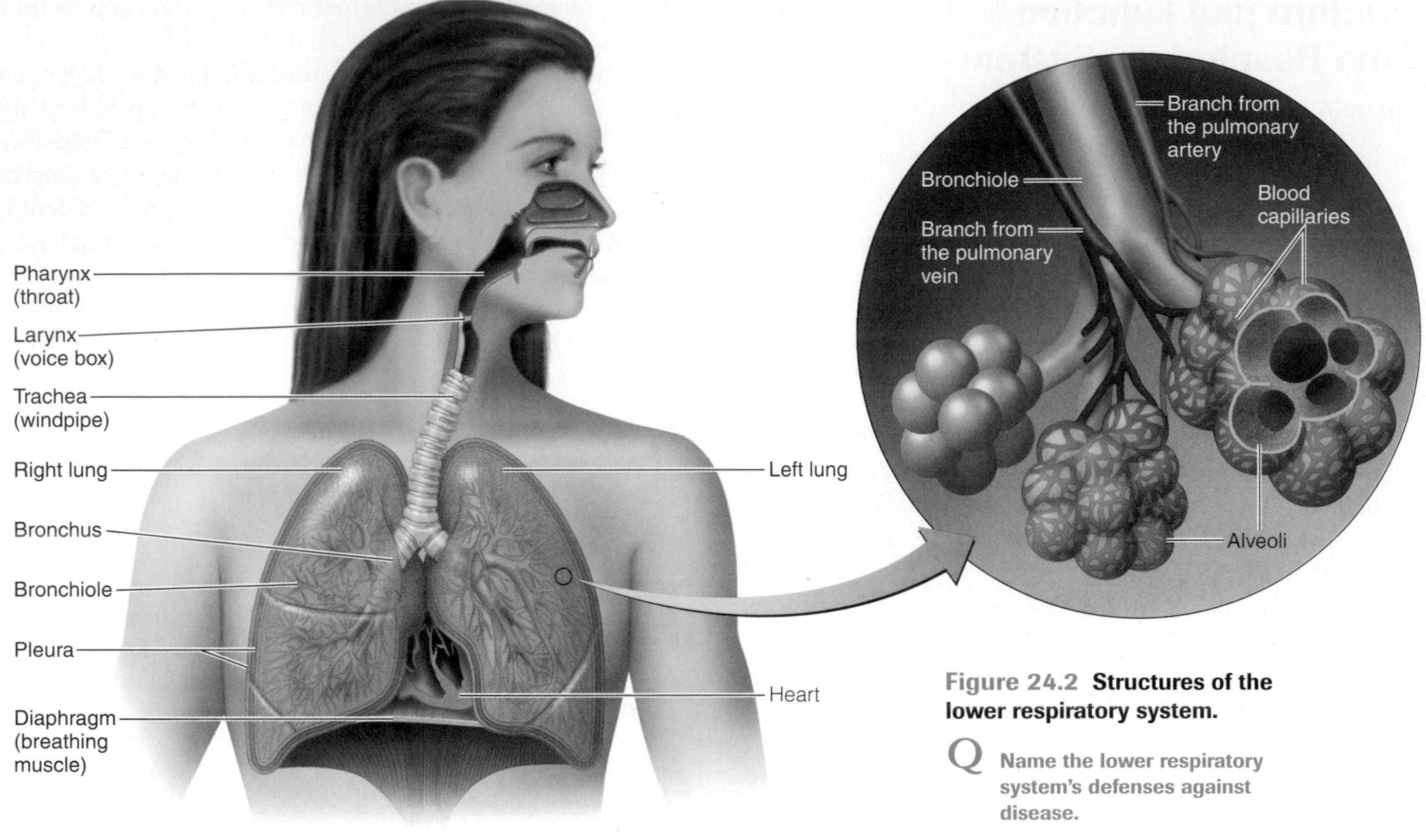

Figure 24.2 Structures of the lower respiratory system.

Q Name the lower respiratory system's defenses against disease.

A number of potentially pathogenic microorganisms are part of the normal microbiota in the upper respiratory system. However, they usually do not cause illness because the predominant microorganisms of the normal microbiota suppress their growth by competing with them for nutrients and producing inhibitory substances.

By contrast, the lower respiratory tract is nearly sterile—although the trachea may contain a few bacteria—because of the normally efficient functioning of the ciliary escalator in the bronchial tubes.

CHECK YOUR UNDERSTANDING

✓ Normally, the lower respiratory tract is nearly sterile. What is the primary mechanism responsible? **24-2**

MICROBIAL DISEASES OF THE UPPER RESPIRATORY SYSTEM

LEARNING OBJECTIVE

24-3 Differentiate pharyngitis, laryngitis, tonsillitis, sinusitis, and epiglottitis.

As most of us know from personal experience, the respiratory system is the site of many common infections. We will soon discuss **pharyngitis,** inflammation of the mucous membranes of the throat, or sore throat. When the larynx is the site of infection, we suffer from **laryngitis,** which affects our ability to speak. The microbes that cause pharyngitis also can cause inflamed tonsils, or **tonsillitis.**

The nasal sinuses are cavities in certain cranial bones that open into the nasal cavity. They have a mucous membrane lining that is continuous with that of the nasal cavity. Infection of a sinus involving heavy nasal discharge of mucus is called **sinusitis.** If the opening by which the mucus leaves the sinus becomes blocked, internal pressure can cause pain or a sinus headache. These diseases are almost always *self-limiting,* meaning that recovery will usually occur even without medical intervention.

Probably the most threatening infectious disease of the upper respiratory system is **epiglottitis,** inflammation of the epiglottis. The epiglottis is a flaplike structure of cartilage that prevents ingested material from entering the larynx (see Figure 24.1). Epiglottitis is a rapidly developing disease that can result in death within a few hours. It is caused by opportunistic pathogens, usually *Haemophilus influenzae* type b. The newly introduced Hib vaccine, although directed primarily at meningitis (see Figure 22.3, page 613), has significantly reduced the incidence of epiglottitis in the vaccinated population.

CHECK YOUR UNDERSTANDING

✓ Which one of the following is most likely to be associated with a headache: pharyngitis, laryngitis, sinusitis, or epiglottitis? 24-3

Bacterial Diseases of the Upper Respiratory System

LEARNING OBJECTIVE

24-4 List the causative agent, symptoms, prevention, preferred treatment, and laboratory identification tests for streptococcal pharyngitis, scarlet fever, diphtheria, cutaneous diphtheria, and otitis media.

Airborne pathogens make their first contact with the body's mucous membranes as they enter the upper respiratory system. Many respiratory or systemic diseases initiate infections here.

Streptococcal Pharyngitis (Strep Throat)

Streptococcal pharyngitis (strep throat) is an upper respiratory infection caused by group A streptococci (GAS). This gram-positive bacterial group consists solely of *Streptococcus pyogenes,* the same bacterium responsible for many skin and soft tissue infections, such as impetigo, erysipelas, and acute bacterial endocarditis.

The pathogenicity of GAS is enhanced by their resistance to phagocytosis. They are also able to produce special enzymes, called *streptokinases,* which lyse fibrin clots, and *streptolysins,* which are cytotoxic to tissue cells, red blood cells, and protective leukocytes.

At one time, the diagnosis of pharyngitis was based on culturing bacteria from a throat swab. Results took overnight or longer, but, beginning in the early 1980s, rapid antigen detection tests that were capable of detecting GAS directly on throat swabs became available. The first rapid tests used latex indirect agglutination methods (see Figure 18.7, page 511). These have been generally replaced by **enzyme immunoassay (EIA)** tests that are more sensitive and easier to read. Currently, there is a wide range of rapid tests commercially available to evaluate cases of pharyngitis, which reflects the fact that millions of patients seek care for it every year. Actually, the majority of patients seen for sore throats do not have a streptococcal infection. Some cases are caused by other bacteria, but many are caused by viruses—for which antibiotic therapy is ineffective. Even the presence of GAS is not a conclusive indication that it is responsible for the sore throat. In areas where acute rheumatic fever occurs the recommendation is to use both bacterial culture and rapid tests.

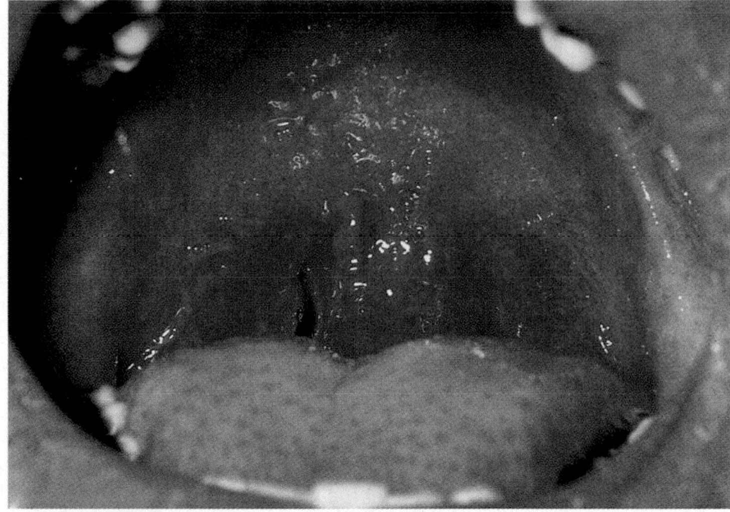

Figure 24.3 Streptococcal pharyngitis. Note the inflammation.

Q **How is strep throat diagnosed?**

Fortunately, GAS have remained sensitive to penicillin, although some resistance to erythromycin has appeared.

Pharyngitis is characterized by local inflammation and a fever (Figure 24.3). Frequently, tonsillitis occurs, and the lymph nodes in the neck become enlarged and tender. Another frequent complication is otitis media (see page 679).

Pharyngitis is now most commonly transmitted by respiratory secretions, but epidemics of streptococcal pharyngitis spread by unpasteurized milk were once frequent.

Scarlet Fever

When the *Streptococcus pyogenes* strain causing streptococcal pharyngitis produces an *erythrogenic* (reddening) *toxin,* the resulting infection is called **scarlet fever.** When the strain produces this toxin, it has been lysogenized by a bacteriophage (see Figure 13.12, page 382). Recall that this means the genetic information of a bacteriophage (bacterial virus) has been incorporated into the chromosome of the bacterium, so the characteristics of the bacterium have been altered. The toxin causes a pinkish red skin rash, which is probably the skin's hypersensitivity reaction to the circulating toxin, and a high fever. The tongue has a spotted, strawberry-like appearance and then, as it loses its upper membrane, becomes very red and enlarged. Classically, scarlet fever has been considered to be associated with streptococcal pharyngitis, but it might accompany a streptococcal skin infection.

The incidence of scarlet fever has varied over time in severity and frequency. Today it is a relatively mild and rare disease.

Diphtheria

Another bacterial infection of the upper respiratory system is **diphtheria.** Until 1935, it was the leading infectious killer of children in the United States. The disease begins with a sore

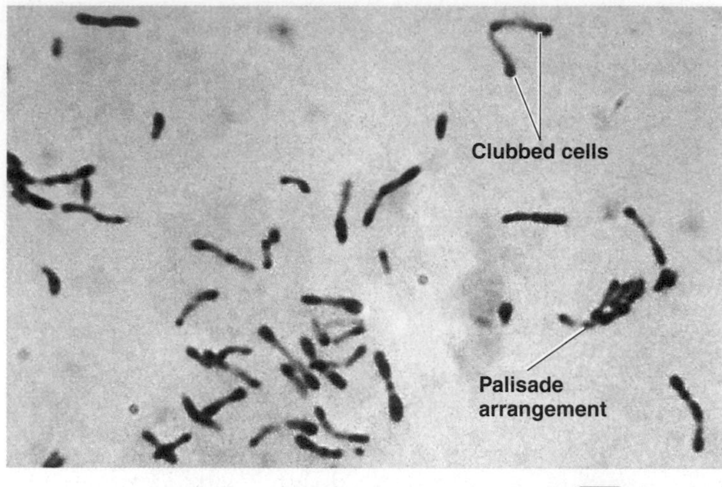

Figure 24.4 *Corynebacterium diphtheriae,* the cause of diphtheria. This Gram stain shows the club-shaped morphology; the dividing cells are often observed to fold together to form V- and Y-shaped figures. Also notice the side-by-side palisade arrangement.

Q Are corynebacteria gram-positive or gram-negative?

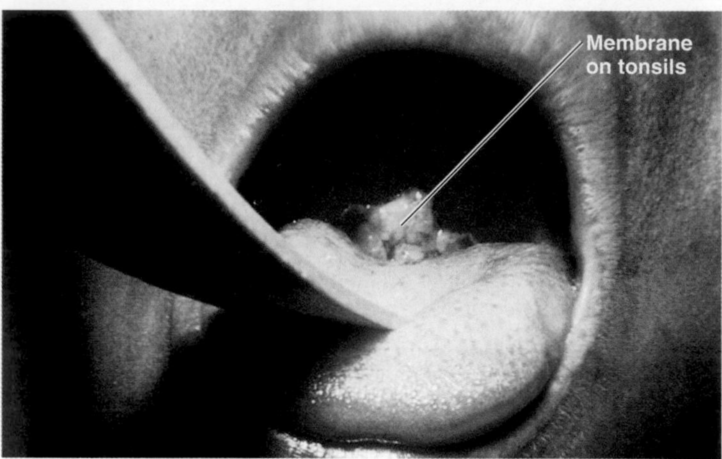

Figure 24.5 A diphtheria membrane. In small children, this leatherlike membrane and accompanying swelling of the breathing passages can block the air supply.

Q What is cutaneous diphtheria?

throat and fever, followed by general malaise and swelling of the neck. The organism responsible is *Corynebacterium diphtheriae,* a gram-positive, non–endospore-forming rod. Its morphology is pleomorphic, frequently club-shaped, and it stains unevenly (Figure 24.4).

Part of the normal immunization program for children in the United States is the **DTaP vaccine.** The D stands for diphtheria toxoid, an inactivated toxin that causes the body to produce antibodies against the diphtheria toxin.

C. diphtheriae has adapted to a generally immunized population, and relatively nonvirulent strains are found in the throats of many symptomless carriers. The bacterium is well suited to airborne transmission and is very resistant to drying.

Characteristic of diphtheria (from the Greek word for leather) is a tough grayish membrane that forms in the throat in response to the infection (Figure 24.5). It contains fibrin, dead tissue, and bacterial cells and can totally block the passage of air to the lungs.

Although the bacteria do not invade tissues, those that have been lysogenized by a phage can produce a powerful exotoxin. Historically, it was the first disease for which a toxic cause was identified. Circulating in the bloodstream, the toxin interferes with protein synthesis. Only 0.01 mg of this highly virulent toxin can be fatal. Thus, if antitoxin therapy is to be effective, it must be administered before the toxin enters the tissue cells. When such organs as the heart and kidneys are affected by the toxin, the disease can rapidly be fatal. In other cases the nerves can be involved, and partial paralysis results.

The number of diphtheria cases reported in the United States each year is currently five or fewer. In young children, the disease occurs mainly in groups that have not been immunized for religious or other reasons. When diphtheria was more common, repeated contacts with toxigenic strains reinforced the immunity, which otherwise weakens with time. Many adults now lack immunity because routine immunization was less available during their childhood. Some surveys indicate effective immune levels in as few as 20% of the adult population. In the United States, when any trauma in adults requires tetanus toxoid, it is usually combined with diphtheria toxoid (Td vaccine).

Diphtheria is also expressed as **cutaneous diphtheria.** In this form of the disease, *C. diphtheriae* infects the skin, usually at a wound or similar skin lesion, and there is minimal systemic circulation of the toxin. In cutaneous infections, the bacteria cause slow-healing ulcerations covered by a gray membrane. Cutaneous diphtheria is fairly common in tropical countries. In the United States, it occurs mostly among American Indians and in adults of low socioeconomic status. It is responsible for most of the reported cases of diphtheria in people over age 30.

In the past, diphtheria was spread mainly to healthy carriers by droplet infection. Respiratory cases have been known to arise from contact with cutaneous diphtheria.

Laboratory diagnosis by bacterial identification is difficult, requiring several selective and differential media. Identification is complicated by the need to differentiate toxin-forming isolates from strains that are not toxigenic; both may be found in the same patient.

Even though antibiotics such as penicillin and erythromycin control the growth of the bacteria, they do not neutralize the diphtheria toxin. Thus antibiotics should be used only in conjunction with antitoxin.

CHECK YOUR UNDERSTANDING

✓ Among streptococcal pharyngitis, scarlet fever, or diphtheria, which two diseases are usually caused by the same genus of bacteria? **24-4**

Otitis Media

One of the more uncomfortable complications of the common cold, or of any infection of the nose or throat, is infection of the middle ear, **otitis media,** or earache. The pathogens cause the formation of pus, which builds up pressure against the eardrum and causes it to become inflamed and painful (Figure 24.6). The condition is most frequent in early childhood because the auditory tube connecting the middle ear to the throat is small and more horizontal than in adults and so is more easily blocked by infection (see Figure 24.1).

A number of bacteria can cause otitis media. The most commonly isolated pathogen is *S. pneumoniae* (about 35% of cases). Other bacteria frequently involved are nonencapsulated *H. influenzae* (20–30%), *Moraxella catarrhalis* (mô-raks-el'lä ka-tär'al-is) (10–15%), *S. pyogenes* (8–10%), and *S. aureus* (1–2%). In about 3–5% of cases, no bacteria can be detected. Viral infections may be responsible in these instances; respiratory syncytial viruses (see page 692) are the most common isolate.

Otitis media affects 85% of children before the age of 3 and accounts for nearly half of office visits to pediatricians—an estimated 8 million cases each year in the United States. Treatment always assumes that bacteria are the cause, and it is estimated that ear infections account for about one-fourth of the prescriptions for antibiotics. Broad-spectrum penicillins, such as amoxicillin, are usually the first choice for children. Many physicians now question the value of antibiotics, uncertain whether these drugs shorten the course of the disease. A conjugate vaccine exists that is intended to prevent pneumonia caused by *S. pneumoniae.* Experience so far has shown that the vaccine has the welcome side effect of reducing the incidence of otitis media by 6–7%. This reduction may not sound like much, but it amounts to over a million fewer cases every year.

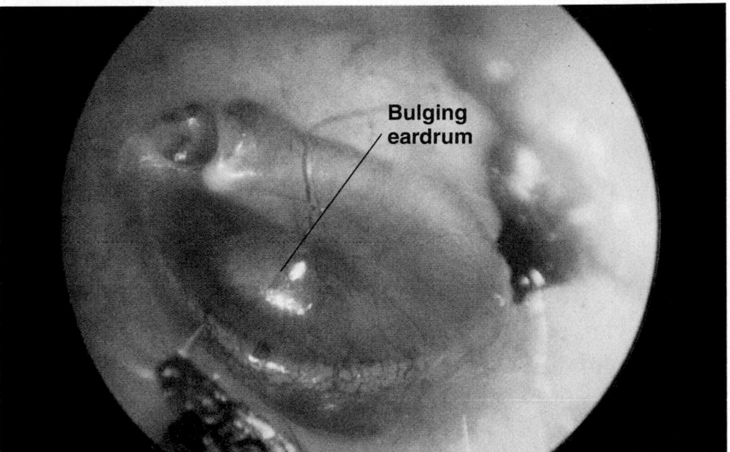

Figure 24.6 Acute otitis media, with bulging eardrum.

Q What is the most common bacterium causing middle ear infections?

Viral Diseases of the Upper Respiratory System

LEARNING OBJECTIVE

24-5 List the causative agents and treatments for the common cold.

Probably the most prevalent disease of humans, at least those living in the temperate zones, is a viral disease affecting the upper respiratory system—the common cold.

The Common Cold

A number of different viruses are involved in the etiology of the **common cold.** About 50% of all colds are caused by rhinoviruses. Coronaviruses probably cause another 15–20%. About 10% of all colds are caused by one of several other viruses. In about 40% of cases, no causative agent can be identified.

We tend to accumulate immunities against cold viruses during our lifetime, which may be a reason why older people tend to get fewer colds. Immunity is based on the ratio of IgA antibodies to single serotypes and has a reasonably high short-term effectiveness. Isolated populations may develop a group immunity, and their colds disappear until a new set of viruses is introduced. Altogether, probably more than 200 agents cause the common cold. There are at least 113 serotypes of rhinoviruses alone, so a vaccine effective against so many different pathogens does not seem practical.

The symptoms of the common cold are familiar to all of us. They include sneezing, excessive nasal secretion, and congestion. The infection can easily spread from the throat to the sinuses, the lower respiratory system, and the middle ear, leading to complications of laryngitis and otitis media. The uncomplicated cold usually is not accompanied by fever.

Rhinoviruses thrive at a temperature slightly below that of normal body temperature, such as might be found in the upper respiratory system, which is open to the outside environment. No one knows exactly why the number of colds seems to increase with colder weather in temperate zones. It is not known whether closer indoor contact promotes epidemic-type transmission or whether physiological changes increase susceptibility.

A single rhinovirus deposited on the nasal mucosa is often sufficient to cause a cold. However, there is surprisingly little agreement on how the cold virus is transmitted to a site in the nose. Experiments with guinea pigs and the influenza virus show that viruses tend to be carried on airborne droplets of water vapor. In the dry air (low humidity) typical of low temperatures, the droplets are smaller and remain airborne longer, facilitating person-to-person transmission. At the same time, the cooler air causes the cilia of the ciliary escalator to work more slowly, allowing inhaled viruses to spread in the upper respiratory system.

A series of experiments involving a group of card players, half with colds and half without colds, supported the hypothesis of airborne transmission. Half of the healthy players were restrained so that they could not use their hands to transfer to their noses viruses picked up from the playing cards, and the other half were not restrained. The players who could not touch their noses came down with as many colds as those who could—an argument for airborne transmission. Even when healthy participants in card games were placed in a room separate from anyone suffering from a cold and isolated from their airborne secretions, but handled cards literally soaked in nasal secretions, none developed colds. In a perhaps less repellent series of experiments, researchers required healthy volunteers to kiss cold sufferers for 60 to 90 seconds; only 8% of the volunteers came down with colds.

Because colds are caused by viruses, antibiotics are of no use in treatment. Recovery time is not reported to be affected by nonprescription drugs, such as zinc lozenges or vitamin C. Symptoms can be relieved by cough suppressants and antihistamines, but these medications do not speed recovery. There is still considerable truth in the medical adage that an untreated cold will run its normal course to recovery in a week, whereas with treatment it will take 7 days.

Promising new approaches to shortening the duration of the common cold are, however, being tested. Almost all rhinoviruses, which are among the more common cold-causing viruses, use the same receptor protein on the host's cells to attach and infect the cells lining the nasal passage. Improved insight into the host–virus attachment mechanisms is considered the most likely key to successful therapy for colds.

The diseases affecting the upper respiratory system are summarized in Diseases in Focus 24.1.

CHECK YOUR UNDERSTANDING

✓ Which viruses, rhinoviruses or coronoviruses, cause about half of the cases of the common cold? **24-5**

MICROBIAL DISEASES OF THE LOWER RESPIRATORY SYSTEM

Many of the same bacteria and viruses that infect the upper respiratory system can also infect the lower respiratory system. As the bronchi become involved, **bronchitis** or **bronchiolitis** develops (see Figure 24.2). A severe complication of bronchitis is **pneumonia,** in which the pulmonary alveoli become involved.

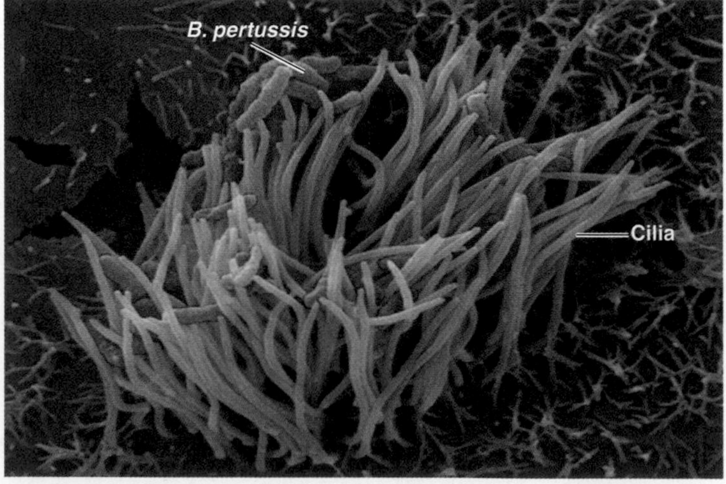

Figure 24.7 Ciliated cells of the respiratory system infected with *Bordetella pertussis.* Cells of *B. pertussis* (orange) can be seen growing on the cilia; they will eventually cause the loss of the ciliated cells.

Q What is the name of the toxin produced by *Bordetella pertussis* that causes the loss of cilia?

Bacterial Diseases of the Lower Respiratory System

LEARNING OBJECTIVE

24-6 List the causative agent, symptoms, prevention, preferred treatment, and laboratory identification tests for pertussis and tuberculosis.

24-7 Compare and contrast the seven bacterial pneumonias discussed in this chapter.

24-8 List the etiology, method of transmission, and symptoms of melioidosis.

Bacterial diseases of the lower respiratory system include tuberculosis and the many types of pneumonia caused by bacteria. Lesser-known diseases such as psittacosis and Q fever also fall into this category.

Pertussis (Whooping Cough)

Q&A Infection by the bacterium *Bordetella pertussis* results in **pertussis,** or **whooping cough.** *B. pertussis* is a small, obligately aerobic, gram-negative coccobacillus. The virulent strains possess a capsule. The bacteria attach specifically to ciliated cells in the trachea, first impeding their ciliary action and then progressively destroying the cells (**Figure 24.7**). This prevents the ciliary escalator system from moving mucus. *B. pertussis* produces several toxins. *Tracheal cytotoxin,* a fixed cell wall fraction of the bacterium, is responsible for damage to the ciliated cells, and *pertussis toxin* enters the bloodstream and is associated with systemic symptoms of the disease.

DISEASES IN FOCUS 24.1

Microbial Diseases of the Upper Respiratory System

Differential diagnosis is the process of identifying the disease from a list of possible diseases that fit the information derived from examining a patient. A differential diagnosis is important for providing initial treatment and for laboratory testing. The differential diagnosis for the following diseases is usually based on clinical symptoms, and throat swabs may be used to culture bacteria. For example, a patient presents with fever and a red, sore throat. Later a grayish membrane appears in the throat. Gram-positive rods were cultured from the membrane. Use the table below to identify infections that could cause these symptoms. For the solution, go to www.microbiologyplace.com.

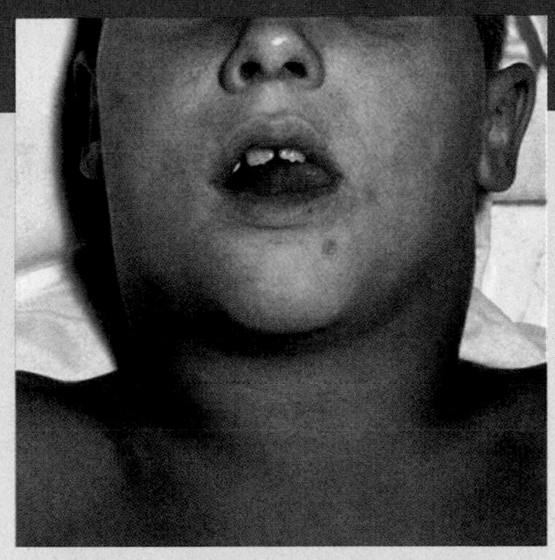

Characteristic swollen lymph nodes of this disease.

Disease	Pathogen	Symptoms	Treatment
BACTERIAL DISEASES			
Epiglottitis	*Haemophilus influenzae*	Inflammation of the epiglottis	Antibiotics; maintain airway Prevention: Hib vaccine
Streptococcal pharyngitis (strep throat)	Streptococci, especially *Streptococcus pyogenes*	Inflamed mucous membranes of the throat	Penicillin
Scarlet fever	Erythrogenic toxin-producing strains of *Streptococcus pyogenes*	Streptococcal exotoxin causes reddening of skin and tongue and peeling of affected skin	Penicillin
Diphtheria	*Corynebacterium diphtheriae*	Membrane forms in throat; cutaneous form also occurs	Penicillin and antitoxin Prevention: DTaP vaccine
Otitis media	Several agents, especially *Staphylococcus aureus, Streptococcus pneumoniae,* and *Haemophilus influenzae*	Accumulations of pus in middle ear build up painful pressure on eardrum	Broad-spectum antibiotics Prevention: pneumococcal vaccine
VIRAL DISEASE			
Common cold	Rhinoviruses, coronaviruses	Familiar symptoms of coughing, sneezing, runny nose	Supportive

Primarily a childhood disease, pertussis can be quite severe. The initial stage, called the *catarrhal stage,* resembles a common cold. Prolonged sieges of coughing characterize the *paroxysmal stage,* or second stage. (The name *pertussis* is derived from the Latin *per,* meaning thoroughly, and *tussis,* meaning cough.) When ciliary action is compromised, mucus accumulates, and the infected person desperately attempts to cough up these mucus accumulations. The violence of the coughing in small children can actually result in broken ribs. Gasping for air between coughs causes a whooping sound, hence the informal name of the disease. Coughing episodes occur several times a day for 1 to 6 weeks. The *convalescence stage,* the third stage, may last for months. Because infants are less capable of coping with the effort of coughing to maintain an airway, irreversible damage to the brain occasionally occurs.

The epidemiology of pertussis has been changing. Before a whole-cell, heat killed vaccine was introduced in the 1940s, pertussis was a major disease affecting almost entirely children under the age of 10. More than half could expect to be infected before beginning school. Introduction of the DTP (diphtheria, tetanus, pertussis) vaccine led to a dramatic decline in cases. However, since about the 1980s the number of cases has significantly increased (peaking in 2004 at about 26,000 cases), but now the adolescent and adult populations are affected.

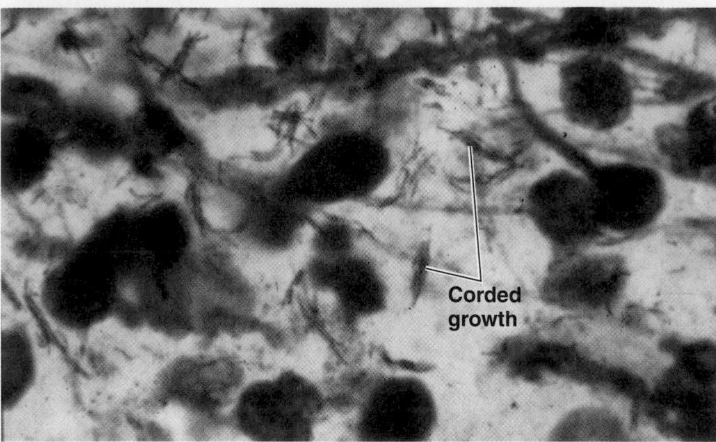

Figure 24.8 *Mycobacterium tuberculosis.* The filamentous, red-stained funguslike growth shown here in a smear from lung tissue is responsible for the organism's name. Under other conditions, it grows as slender, individual bacilli. A waxy component of the cell, cord factor, is responsible for this ropelike arrangement. An injection of cord factor causes pathogenic effects exactly like that caused by tubercle bacilli.

Q What characteristic of this bacterium suggests use of the prefix *myco-*?

New vaccines intended for adolescents and adults (Tdap) are recommended to boost immunity as the effectiveness of childhood vaccination declines. There are several reasons for this decline. The immunity to the DTP vaccine declines after a few years and offers almost no protection by about 12 years. It is uncertain how long the protection offered by the newer acellular vaccine for children (DTaP) will last. In the past, contact with cases of unrecognized or mild pertussis likely provided vaccinated people a booster effect, which helped control the disease. Finally, introduction of PCR diagnostic tests, which are much more sensitive, has probably led to an increase in reported cases in older children and adults. This situation is termed a *pseudoepidemic.*

Diagnosis of pertussis is primarily based on clinical signs and symptoms. The pathogen can be cultured from a throat swab inserted through the nose on a thin wire and held in the throat while the patient coughs. Culture of the fastidious pathogen requires care. As alternatives to culture, PCR methods can also be used to test the swabs for presence of the pathogen, a procedure that is required to diagnose the disease in infants.

Treatment of pertussis with antibiotics, most commonly erythromycin or other macrolides, is not effective after onset of the paroxysmal coughing stage but may reduce transmission.

CHECK YOUR UNDERSTANDING

✓ Another name for pertussis is whooping cough. This symptom is caused by the pathogens' attack on which cells? **24-6**

Tuberculosis

Tuberculosis (TB) is an infectious disease caused by the bacterium *Mycobacterium tuberculosis,* a slender rod and an obligate aerobe. The rods grow slowly (20-hour or longer generation time), sometimes form filaments, and tend to grow in clumps (Figure 24.8). On the surface of liquid media, their growth appears moldlike, which suggested the genus name *Mycobacterium* (*myco* means fungus).

Mycobacteria stained with carbol-fuchsin dye cannot be decolorized with acid-alcohol and are therefore classified as *acid-fast* (see page 70). This characteristic reflects the unusual composition of the cell wall, which contains large amounts of lipids. These lipids might also be responsible for the resistance of mycobacteria to environmental stresses, such as drying. In fact, these bacteria can survive for weeks in dried sputum and are very resistant to chemical antimicrobials used as antiseptics and disinfectants (see Table 7.7 page 203).

Tuberculosis is a particularly good illustration of the ecological balance between host and parasite in infectious disease. A host is not usually aware of pathogens that invade the body and are defeated. If immune defenses fail, however, the host becomes very much aware of the resulting disease. As might be expected, there is a great synergy between HIV infection and TB—many cases are coinfections.

Several factors may affect host resistance levels: the presence of other illness and physiological and environmental factors, such as malnutrition, overcrowding, and stress. A tragic demonstration of individual variation in resistance was the Lübeck disaster in Germany in 1926. By error, 249 babies were inoculated with virulent tuberculosis bacteria instead of the attenuated vaccine strain. Even though all received the same inoculum, there were only 76 deaths, and the remainder did not become seriously ill.

Tuberculosis is most commonly acquired by inhaling the bacillus. Only very fine particles containing one to three bacilli reach the lungs, where they are usually phagocytized by a macrophage in the alveoli (see Figure 24.2). The macrophages of a healthy individual become activated by the presence of the bacilli and usually destroy them.

Pathogenesis of Tuberculosis

Figure 24.9 shows the pathogenesis of TB. An important factor in the pathogenicity of the mycobacteria probably is that the mycolic acids of the cell wall strongly stimulate an inflammatory response in the host. The figure depicts the situation in which the body's defenses fail and the disease progresses to a fatal conclusion. However, most healthy people will defeat a potential infection with activated macrophages, especially if the infecting dose is low.

❶–❷ If the infection progresses, the host isolates the pathogens in a walled-off lesion called a *tubercle* (meaning lump or knob), a characteristic that gives the disease its name.

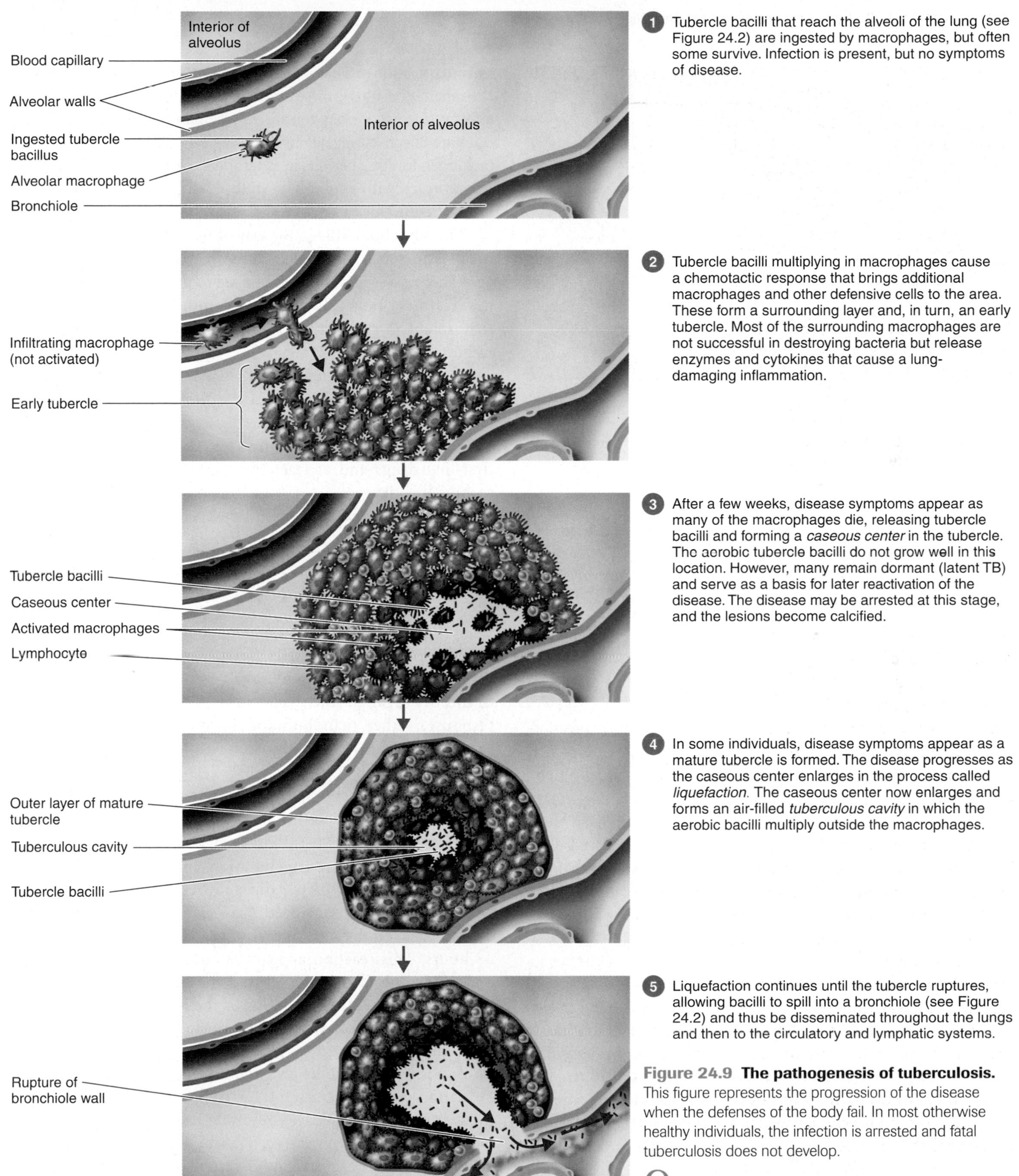

Interior of alveolus
Blood capillary
Alveolar walls
Ingested tubercle bacillus
Alveolar macrophage
Bronchiole
Interior of alveolus
1 Tubercle bacilli that reach the alveoli of the lung (see Figure 24.2) are ingested by macrophages, but often some survive. Infection is present, but no symptoms of disease.
Infiltrating macrophage (not activated)
Early tubercle
2 Tubercle bacilli multiplying in macrophages cause a chemotactic response that brings additional macrophages and other defensive cells to the area. These form a surrounding layer and, in turn, an early tubercle. Most of the surrounding macrophages are not successful in destroying bacteria but release enzymes and cytokines that cause a lung-damaging inflammation.
Tubercle bacilli
Caseous center
Activated macrophages
Lymphocyte
3 After a few weeks, disease symptoms appear as many of the macrophages die, releasing tubercle bacilli and forming a *caseous center* in the tubercle. The aerobic tubercle bacilli do not grow well in this location. However, many remain dormant (latent TB) and serve as a basis for later reactivation of the disease. The disease may be arrested at this stage, and the lesions become calcified.
Outer layer of mature tubercle
Tuberculous cavity
Tubercle bacilli
4 In some individuals, disease symptoms appear as a mature tubercle is formed. The disease progresses as the caseous center enlarges in the process called *liquefaction*. The caseous center now enlarges and forms an air-filled *tuberculous cavity* in which the aerobic bacilli multiply outside the macrophages.
Rupture of bronchiole wall
5 Liquefaction continues until the tubercle ruptures, allowing bacilli to spill into a bronchiole (see Figure 24.2) and thus be disseminated throughout the lungs and then to the circulatory and lymphatic systems.

Figure 24.9 The pathogenesis of tuberculosis. This figure represents the progression of the disease when the defenses of the body fail. In most otherwise healthy individuals, the infection is arrested and fatal tuberculosis does not develop.

Q Almost a third of the earth's population is infected with *Mycobacterium tuberculosis*—does a study of this figure show why this is not the same as a third of the earth's population *having* tuberculosis?

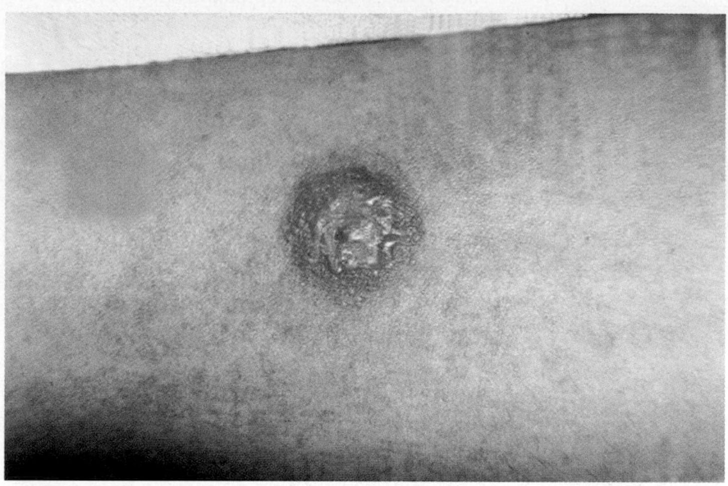

Figure 24.10 A positive tuberculin skin test on an arm.

Q What does a positive tuberculin skin test indicate?

3–4 When the disease is arrested at this point, the lesions slowly heal, becoming calcified. These show up clearly on X-ray films and are called *Ghon's complexes.* (Computed tomography [CT] is more sensitive than X rays in detecting lesions of TB.)

5 If the body's defenses fail at this stage, the tubercle breaks down and releases virulent bacilli into the airways of the lung and then the cardiovascular and lymphatic systems.

Coughing, the more obvious symptom of the lung infection, also spreads the infection by bacterial aerosols. Sputum may become bloodstained as tissues are damaged, and eventually blood vessels may become so eroded that they rupture, resulting in fatal hemorrhaging. The disseminated infection is called *miliary tuberculosis* (the name is derived from the numerous millet seed–sized tubercles formed in the infected tissues). The body's remaining defenses are overwhelmed, and the patient suffers weight loss and a general loss of vigor. At one time, TB was also known as *consumption.*

Treatment of Tuberculosis

The first effective antibiotic for TB treatment was streptomycin, which was introduced in 1944. It is still in use but now considered a second-line drug. The current treatment for TB recommended by the World Health Organization requires the patient to adhere to a minimum of 6 months of antibiotic therapy that includes three or four drugs. That many patients fail to follow such a prolonged regimen faithfully increases the likelihood of resistance developing. The two most powerful anti-TB drugs are isoniazid and rifampin (also known as rifampicin). Other FDA approved first-line drugs include pyrazinamide, rifapentine, and ethambutol—in all, about 10 drugs are approved for treatment of TB, many considered secondary choices. The prolonged treatment is necessary because the tubercle bacillus grows very slowly or is dormant (the only drug effective against dormant bacilli is pyrazinamide), and many antibiotics are effective only against growing cells. Also, the bacillus may be hidden for long periods in macrophages or other locations difficult to reach with antibiotics. Multiple-drug therapy is needed to minimize the emergence of resistant strains.

There has been little in the way of new anti-TB drugs: the last major anti-tubercular drug, rifampin, was introduced decades ago. There is a serious need for a drug or drugs that would reduce the treatment time to less than 3 months, kill persistent bacilli that might later reactivate, and be active against *multidrug resistant (MDR) strains.*

Studies have revealed the presence not only of MDR tuberculosis (defined as resistant to the first-line drugs, isoniazid and rifampin), but also of *extensively drug-resistant (XDR)* tuberculosis, which is defined as resistance to both the first-line drugs and at least three of the six main classes of second-line drugs.

Currently, experimental interest is focused on a novel diarylquinoline drug. In animal testing it has demonstrated a unique specificity for impeding the synthesis of ATP in mycobacteria and is effective in killing both dormant and actively growing bacilli.

Diagnosis of Tuberculosis

People infected with tuberculosis respond with cell-mediated immunity against the bacterium. This form of immune response, rather than humoral immunity, develops because the pathogen is located mostly within macrophages. This immunity, involving sensitized T cells, is the basis for the **tuberculin skin test** (Figure 24.10), a screening test for infection. A positive test does not necessarily indicate active disease. In this test, a purified protein derivative of the tuberculosis bacterium, derived by precipitation from broth cultures, is injected cutaneously. If the injected person has been infected with TB in the past, sensitized T cells react with these proteins, and a delayed hypersensitivity reaction occurs in about 48 hours. This reaction appears as an induration (hardening) and reddening of the area around the injection site. Probably the most accurate tuberculin test is the *Mantoux test,* in which dilutions of 0.1 ml of antigen are injected and the reacting area of the skin is measured.

A positive tuberculin test in the very young is a probable indication of an active case of TB. In older individuals, it might indicate only hypersensitivity resulting from a previous infection or vaccination, not a current active case. Nonetheless, it is an indication that further examination is needed, such as a chest X ray or CT examination to detect lung lesions and attempts to isolate the bacterium.

The initial step in laboratory diagnosis of active cases is a microscopic examination of smears, such as sputum. This may be a conventional acid-fast stain or the more specific fluorescent-antibody microscopy. A confirmation of TB by isolation of the bacterium is complicated by the very slow growth of the pathogen. The formation of a colony might take 3 to 6 weeks, with completion of a reliable identification series adding another 3 to 6 weeks. There has been considerable progress in developing rapid diagnostic tests. DNA probes are available to identify cultured isolates (see Figure 10.16, page 292). Several tests make use of PCR methods that are capable of detecting *M. tuberculosis* directly from sputum or other samples (see page 291). Several new tests detect the presence in blood samples of gamma interferon (IFN-γ), which is released by T cells in response to certain antigens of *M. tuberculosis.* Evidence is that, compared to the skin test, these have higher specificity and less cross-reactivity with BCG vaccination (see the discussion of TB vaccines, following). They do not distinguish latent from active infection. Determination of drug resistance by conventional culture methods is lengthy and labor intensive. However, resistance to rifampin is almost invariably a marker for MDR, and this can be detected more rapidly with certain molecular methods. There is also interest in new *microscopic-observation-drug-susceptibility (MODS)* assays. The bacteria grow faster in liquid than solid medium, and microscopic examination can detect early characteristic cord formation concurrent with exposure to different drugs.

Another mycobacterial species, *Mycobacterium bovis* (bō′vis), is a pathogen mainly of cattle. *M. bovis* is the cause of **bovine tuberculosis,** which is transmitted to humans via contaminated milk or food. Bovine tuberculosis accounts for fewer than 1% of TB cases in the United States. It seldom spreads from human to human, but before the days of pasteurized milk and the development of control methods such as tuberculin testing of cattle herds, this disease was a frequent form of tuberculosis in humans. *M. bovis* infections cause TB that primarily affects the bones or lymphatic system. At one time, a common manifestation of this type of TB was hunchbacked deformation of the spine.

Other mycobacterial diseases also affect people in the late stages of HIV infection. A majority of the isolates are of a related group of organisms known as the *M. avium-intracellulare* (ā′vē-um in′trä-cel-ū-lä-rē) complex. In the general population, infections by these pathogens are uncommon.

Tuberculosis Vaccines

The **BCG vaccine** is a live culture of *M. bovis* that has been made avirulent by long cultivation on artificial media. (BCG stands for bacillus of Calmette and Guérin, the people who originally isolated the strain.) The BCG vaccine has been available since the 1920s and is one of the most widely used vaccines in the world. In 1990, it was estimated that 70% of the world's schoolchildren received it. In the United States, however, the vaccine is currently recommended only for certain children at high risk who have negative skin tests. People who have received the vaccine show a positive reaction to tuberculin skin tests. This has always been one argument against its widespread use in the United States. Another argument against the universal administration of BCG vaccine is its very uneven effectiveness. Experience has shown that it is fairly effective when given to young children, but for adolescents and adults it sometimes has an effectiveness approaching zero. Worse, it has been found that HIV-infected children, who need it most, frequently will develop a fatal infection from the BCG vaccine. Recent work indicates that exposure to members of the *M. avium-intracellulare* complex that is often encountered in the environment may interfere with the effectiveness of the BCG vaccine—which might explain why the vaccine is more effective early in life, before much exposure to such environmental mycobacteria. A number of new vaccines are in the experimental pipeline.

Worldwide Incidence of Tuberculosis

Tuberculosis has emerged as a global epidemic (**Figure 24.11a**). Estimates are that 9 million people develop active tuberculosis every year and that infections result in more than 2 million deaths annually. Probably a third of the world's population is infected. Also, HIV and tuberculosis are almost inseparable, and tuberculosis is the leading direct cause of death in much of the world affected by HIV. The majority of cases in the United States, usually around 14,000 annually, occur among foreign-born individuals (**Figure 24.11b**).

Globally, the number of TB cases is rising at about 2% a year, and the number of MDR cases is rising even faster. To treat MDR patients in the United States can cost tens of thousands of dollars annually and is economically impractical in most of the world. Tuberculosis remains a major worldwide health problem.

Bacterial Pneumonias

The term *pneumonia* is applied to many pulmonary infections, most of which are caused by bacteria. Pneumonia caused by *Streptococcus pneumoniae* is the most common, about two-thirds of cases, and is therefore referred to as *typical pneumonia.* Pneumonias caused by other microorganisms, which can include fungi, protozoa, viruses, and other bacteria, are termed *atypical pneumonias.* This distinction is becoming increasingly blurred in practice.

Pneumonias also are named after the portions of the lower respiratory tract they affect. For example, if the lobes of the lungs are infected, it is called *lobar pneumonia;* pneumonias caused by *S. pneumoniae* are usually of this type. *Bronchopneumonia* indicates that the alveoli of the lungs adjacent to the bronchi are infected. *Pleurisy* is often a complication of various pneumonias, in which the pleural membranes become painfully inflamed. (See Diseases in Focus 24.2.)

Pneumococcal Pneumonia

Pneumonia caused by *S. pneumoniae* is called **pneumococcal pneumonia.** *S. pneumoniae* is a gram-positive, ovoid bacterium (**Figure 24.12**). This microbe is also a common cause of otitis media, meningitis, and sepsis. The cell pairs are surrounded by a

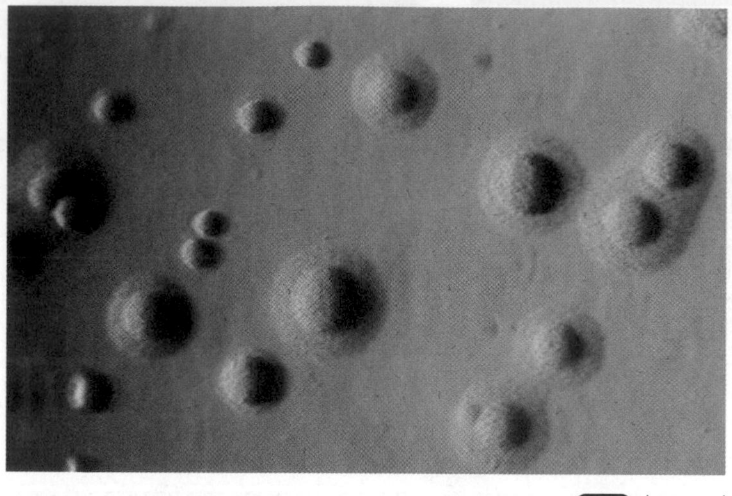

Figure 24.13 Colonies of *Mycoplasma pneumoniae,* the cause of mycoplasmal pneumonia.

Q Could you see these colonies without magnification?

to the infection, alveoli fill with some red blood cells, neutrophils (see Table 16.1, page 455), and fluid from surrounding tissues. The sputum is often rust-colored from blood coughed up from the lungs. Pneumococci can invade the bloodstream, the pleural cavity surrounding the lung, and occasionally the meninges. No bacterial toxin has been clearly related to pathogenicity.

A presumptive diagnosis can be made by isolating the pneumococci from the throat, sputum, and other fluids. Pneumococci can be distinguished from other alpha-hemolytic streptococci by observing the inhibition of growth next to a disk of optochin (ethylhydrocupreine hydrochloride) or by performing a bile solubility test. They can also be serologically typed.

There are many healthy carriers of the pneumococcus. Virulence of the bacteria seems to be based mainly on the carrier's resistance, which can be lowered by stress. Many illnesses of older adults terminate in pneumococcal pneumonia.

A recurrence of pneumococcal pneumonia is not uncommon, but the serological types are usually different. Before chemotherapy was available, the mortality rate was as high as 25%. This has now been lowered to less than 1% for younger patients treated early in the course of their disease. For older patients admitted to a hospital, mortality can approach 20%.

Resistance to penicillin has been an increasing problem, and several other drugs, especially macrolides and fluoroquinolones, are now replacing it.

A conjugated pneumococcal vaccine has recently been introduced that has been effective in preventing infection by the seven serotypes included in it. It has also had an indirect herd effect shown by reduction in other diseases, such as otitis media, attributable to the pneumococcus.

Haemophilus influenzae Pneumonia

Haemophilus influenzae is a gram-negative coccobacillus, and a Gram stain of sputum will differentiate this type of pneumonia from pneumococcal pneumonia. Patients with such conditions as alcoholism, poor nutrition, cancer, or diabetes are especially susceptible. Diagnostic identification of the pathogen uses special media that determine requirements for X and V factors (see page 311). Second-generation cephalosporins are resistant to the β-lactamases produced by many *H. influenzae* strains and are therefore usually the drugs of choice.

Mycoplasmal Pneumonia

The mycoplasmas, which do not have cell walls, do not grow under the conditions normally used to recover most bacterial pathogens. Because of this characteristic, pneumonias caused by mycoplasmas are often confused with viral pneumonias.

The bacterium *Mycoplasma pneumoniae* is the causative agent of **mycoplasmal pneumonia.** This type of pneumonia was first discovered when such atypical infections responded to tetracyclines, indicating that the pathogen was nonviral. Mycoplasmal pneumonia is a common type of pneumonia in young adults and children. It may account for as much as 20% of pneumonias, although it is not a reportable disease. The symptoms, which persist for 3 weeks or longer, are low-grade fever, cough, and headache. Occasionally, they are severe enough to lead to hospitalization. Other terms for the disease are *primary atypical* (that is, the most common pneumonia not caused by the pneumococcus) and *walking pneumonia.*

When isolates from throat swabs and sputum grow on a medium containing horse serum and yeast extract, they form distinctive colonies with a "fried-egg" appearance (Figure 24.13). The colonies are so small that they must be observed with magnification. The mycoplasmas are highly varied in appearance because they lack cell walls (see Figure 11.20, page 319).

Diagnosis based on recovering the pathogens might not be useful in treatment because as long as 3 or more weeks may be required for the slow-growing organisms to develop. Diagnostic tests have improved greatly in recent years, however. They include PCR and serological tests that detect IgM antibodies against *M. pneumoniae.*

Treatment with antibiotics such as tetracycline usually hastens the disappearance of symptoms but does not eliminate the bacteria, which the patient continues to carry for several weeks.

Legionellosis

Legionellosis, or **Legionnaires' disease,** first received public attention in 1976, when a series of deaths occurred among members of the American Legion who had attended a meeting in Philadelphia. Because no obvious bacterial cause could be found, the deaths were attributed to viral pneumonia. Close investigation, mostly with techniques directed at locating a suspected rickettsial agent, eventually identified a previously unknown bacterium, an aerobic gram-negative rod now known as

Legionella pneumophila, which is capable of replication within macrophages. Over 44 species of *Legionella* have now been identified; not all of them cause disease.

The disease is characterized by a high fever of 40.5°C, cough, and general symptoms of pneumonia. No person-to-person transmission seems to be involved. Recent studies have shown that the bacterium can be readily isolated from natural waters. In addition, the microbes can grow in the water of air-conditioning cooling towers, perhaps indicating that some epidemics in hotels, urban business districts, and hospitals were caused by airborne transmission. Recent outbreaks have been traced to whirlpool spas, humidifiers, showers, decorative fountains, and even potting soil.

The organism has also been found to inhabit the water lines of many hospitals. Most hospitals keep the temperature of hot water lines relatively low (43–55°C) as a safety measure, and in cooler parts of the system this inadvertently maintains a good growth temperature for *Legionella.* This bacterium is considerably more resistant to chlorine than most other bacteria and can survive for long periods in water with a low level of chlorine. Evidence indicates *Legionella* exist primarily in biofilms that are highly protective. The bacteria are often ingested by waterborne amoebae when these are present but continue to proliferate and may even survive within encysted amoebae. The most successful method for water disinfection in hospitals with a need to control *Legionella* contamination has been installation of copper-silver ionization systems.

The disease appears to have always been fairly common, if unrecognized. More than 1000 cases are reported each year, but the actual incidence is estimated at over 25,000 annually. Men over 50 are the most likely to contract legionellosis, especially heavy smokers, alcohol abusers, or the chronically ill. (See the box on page 691.)

L. pneumophila is also responsible for **Pontiac fever,** which is essentially another form of legionellosis. Its symptoms include fever, muscular aches, and usually a cough. The condition is mild and self-limiting. During outbreaks of legionellosis, both forms may occur.

The best diagnostic method is culture on a selective charcoal–yeast extract medium. Respiratory specimens can be examined by fluorescent-antibody methods, and a DNA probe test is available. Erythromycin and other macrolide antibiotics, such as azithromycin, are the drugs of choice for treatment.

Psittacosis (Ornithosis)

The term **psittacosis** is derived from the disease's association with psittacine birds, such as parakeets and other parrots. It was later found that the disease can also be contracted from many other birds, such as pigeons, chickens, ducks, and turkeys. Therefore, the more general term **ornithosis** has come into use.

The causative agent is *Chlamydophila psittaci* (sit′tä-sē), a gram-negative, obligate intracellular bacterium. The taxonomy of this organism has recently been revised. The genus name has been changed from *Chlamydia* to *Chlamydophila.* This taxonomic change has also been made with *C. pneumoniae* (see the discussion of chlamydial pneumonia which follows). We will continue to use the generic terms *chlamydial* and *chlamydiae.* One way chlamydiae differ from rickettsias, which are also obligate intracellular bacteria, is that chlamydiae form tiny **elementary bodies** as one part of their life cycle (see Figure 11.24, page 323). Unlike most rickettsias, elementary bodies are resistant to environmental stress; therefore, they can be transmitted through air and do not require a bite to transfer the infective agent directly from one host to another.

Psittacosis is a form of pneumonia that usually causes fever, headache, and chills. Subclinical infections are very common, and stress appears to enhance susceptibility to the disease. Disorientation, or even delirium in some cases, indicates that the nervous system can be involved.

The disease is seldom transmitted from one human to another but is usually spread by contact with the droppings and other exudates of fowl. One of the most common modes of transmission is inhalation of dried particles from droppings. The birds themselves usually have diarrhea, ruffled feathers, respiratory illness, and a generally droopy appearance. Parakeets and other parrots sold commercially are usually (but not always) free of the disease. Many birds carry the pathogen in their spleen without symptoms, becoming ill only when stressed. Pet store employees and people involved in raising turkeys are at greatest risk of contracting the disease.

Diagnosis is made by isolating the bacterium in embryonated eggs or by cell culture. Serological tests can be used to identify the isolated organism. No vaccine is available, but tetracyclines are effective antibiotics in treating humans and animals. Effective immunity does not result from recovery, even when high titers of antibody are present in the serum.

Most years, fewer than 100 cases and very few deaths are reported in the United States. The main danger is late diagnosis. Before antibiotic therapy was available, the mortality rate was about 15–20%.

Chlamydial Pneumonia

Outbreaks of a respiratory illness in populations of college students were found to be caused by a chlamydial organism. Originally the pathogen was considered a strain of *C. psittaci,* but it has been assigned the species name *Chlamydophila pneumoniae,* and the disease is known as **chlamydial pneumonia.** Clinically, it resembles mycoplasmal pneumonia. (There is also strong evidence of association between *C. pneumoniae* and atherosclerosis, the deposition of fatty deposits that block arteries.)

The disease is apparently transmitted from person to person, probably by the respiratory route. Nearly half the U.S. population has antibodies against the organism, an indication that this is a common illness. Several serological tests are useful in diagnosis, but results are complicated by antigenic variation. The most effective antibiotic is tetracycline.

Q Fever

In Australia during the mid-1930s, a previously unreported flulike pneumonia made an appearance. In the absence of an obvious

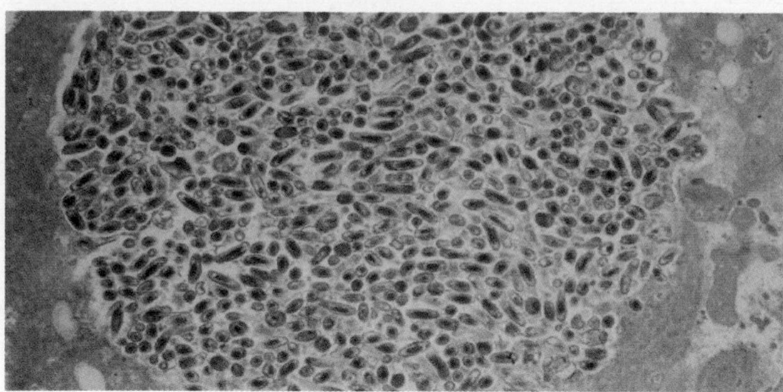

(a) Masses of *Coxiella burnetii* growing in a placental cell

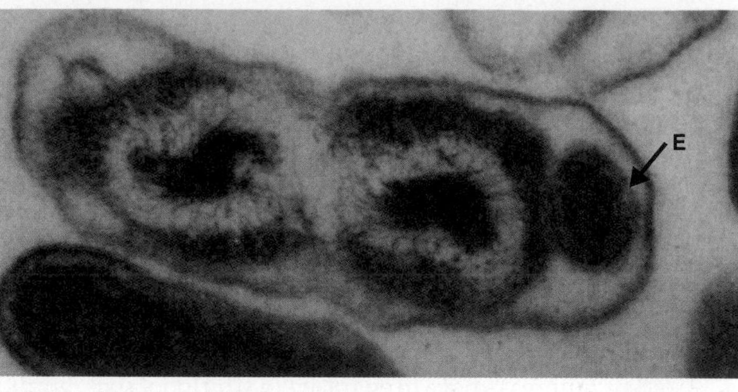

(b) This cell has just divided; notice the endospore-like body (E), which is probably responsible for the relative resistance of the organism.

TEM 0.5 μm

Figure 24.14 ***Coxiella burnetii,*** **the cause of Q fever.**

By what two methods is Q fever transmitted?

cause, the affliction was labeled **Q fever** (for *query*), much as one might say "X fever." The causative agent was subsequently identified as the obligately parasitic, intracellular bacterium *Coxiella burnetii* (käks′ē-el-lä bėr-ne′tē-ē) (Figure 24.14a). Currently, it is classified as a member of the gammaproteobacteria. Along with other bacteria of this group (such as the genera *Franciscella* and *Legionella*), it has the ability to multiply intracellularly. Most intracellular bacteria, such as rickettsia, are not resistant enough to survive airborne transmission, but this microorganism is an exception.

Q fever has a wide range of clinical symptoms, and systematic testing shows that about 60% of cases are not even symptomatic. Cases of *acute Q fever* usually feature symptoms of high fever, headaches, muscle aches, and coughing. A feeling of malaise may persist for months. The heart becomes involved in about 2% of acutely ill patients and is responsible for the rare fatalities. In cases of *chronic Q fever,* the best known manifestation is endocarditis (see page 641). Some 5 to 10 years might elapse between the initial infection and the appearance of endocarditis; and, because these patients show few signs of acute disease, the association with Q fever is often missed. Antibiotic therapy and earlier diagnosis have lowered the mortality rate from chronic Q fever to under 5%.

C. burnetii is a parasite of several arthropods, especially cattle ticks, and it is transmitted among animals by tick bites. Infected animals include cattle, goats, and sheep, as well as most domestic mammalian pets. In animals the infection is usually subclinical. Cattle ticks spread the disease among dairy herds, and the microbes are shed in the feces, milk, and urine of infected cattle. Once the disease is established in a herd, it is maintained by aerosol transmission. The disease is spread to humans by ingesting unpasteurized milk and by inhaling aerosols of microbes generated in dairy barns, especially at calving time from placental material, which contains about a billion bacteria per gram.

Inhaling a single pathogen is enough to cause infection, and many dairy workers have acquired at least subclinical infections. Workers in meat- and hide-processing plants are also at risk. The pasteurization temperature of milk, which was originally aimed at eliminating tuberculosis bacilli, was raised slightly in 1956 to ensure the killing of *C. burnetii.* In 1981, an endosporelike body was discovered, which may account for this heat resistance (Figure 24.14b). This resistant body resembles the elementary body of chlamydiae more than typical bacterial endospores.

The pathogen can be identified by isolation and growth in chick embryos in eggs or in cell culture. Laboratory workers testing for *Coxiella*-specific antibodies in a patient's serum can use serological tests.

A disease found worldwide, most cases of Q fever in the United States occur in the western states. The disease is endemic to California, Arizona, Oregon, and Washington. A vaccine for laboratory workers and other high-risk personnel is available. Doxycycline has been recommended for treatment. When growth within macrophages in chronic infections renders *C. burnetii* resistant, the killing activity can be restored by combining doxycycline with chloroquine, an antimalarial. The chloroquine raises the pH of the phagosome, increasing doxycycline's efficiency.

Melioidosis

In 1911, a new disease was reported among drug addicts in Rangoon, Burma (now Myanmar). The bacterial pathogen, *Burkholderia pseudomallei,* is a gram-negative rod formerly placed in the genus *Pseudomonas.* It closely resembled the bacterium causing glanders, a disease of horses. Therefore, the disease was named **melioidosis** [(mel-ē-öi-dō′-sis) from the Greek *melis* (distemper of asses) and *eidos* (resemblance)]. It is now recognized as a major infectious disease in southeast Asia and northern Australia, where the pathogen is widely distributed in moist soils. It most commonly affects individuals with lower immune capabilities—frequently diabetics. Sporadic cases are

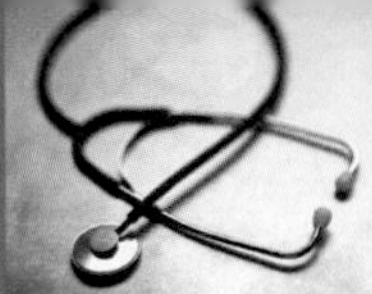

CLINICAL FOCUS From the *Morbidity and Mortality Weekly Report*

Outbreak

As you read through this problem, you will see questions that epidemiologists ask themselves and each other as they solve a clinical problem. Try to answer each question as though you were an epidemiologist.

1. A 64-year-old man saw his primary care physician complaining of fever, malaise, and a cough. His vaccinations were up-to-date, including DTaP. His condition worsened over several days; he had difficulty breathing and his temperature rose to 40.4°C. He was hospitalized, and his lungs showed signs of mild inflammation with thin, watery secretion. A Gram stain of bacteria isolated from the patient is shown in the photo.
   What diseases are possible?
2. The same day, a 37-year-old man went to the emergency department because he had shortness of breath, fatigue, and cough. The day before he had fever and chills, with a maximum body temperature of 38.6°C.
   What additional tests would you do on both patients?
3. Both patients had an antibody titer > 1024 against *Legionella pneumophila* serogroup 1. The local health department was contacted because two patients were hospitalized with legionellosis.
   What do you need to know now?
4. One week before hospitalization, both men stayed in the same hotel within one day of each other. Six additional cases of legionellosis were identified at other hospitals. A follow-up questionnaire was given to all eight patients to ascertain travel that preceded the illness, including location, accommodations, dates, and information about exposures to common sources for infection (see the table).
   What are likely sources of infection?
5. Epidemic legionellosis usually results from exposure of susceptible individuals to an aerosol generated by an environmental source of water contaminated with *Legionella*.
   Why is it important to identify the source?

Patients' Travel History	
Age	37–70 yrs (average: 60)
Gender	6 male; 2 female
Number of Nights at Hotel	1–4 (average: 3)
Diabetes Mellitus	4
Immunocompromised	1
Smoker	5
Showered in Hotel	8
Used Whirlpool Spa at	1
Used Hotel Swimming Pool	6

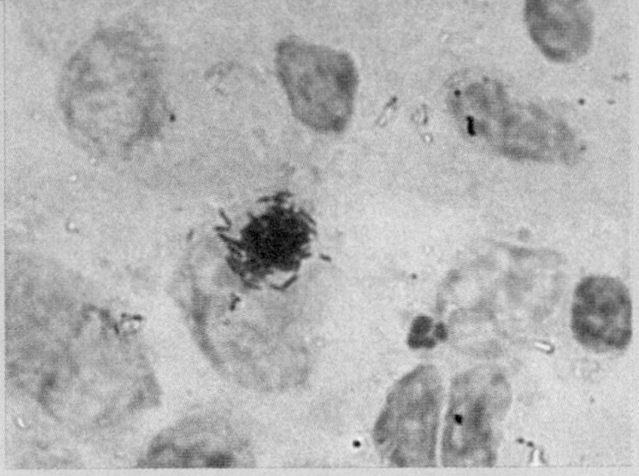

Gram stain shows bacteria within a tissue sample. LM 5 μm

6. Retrospective identification of cases allows control and remediation efforts. *L. pneumophila* of the same monoclonal antibody type was recovered from the hot water storage tanks, cooling tower, and showers and faucets in rooms occupied by patients and well guests.
   Why didn't other hotel guests get sick?
7. During outbreaks, attack rates tend to be highest in specific high-risk groups, including older adults, smokers, and immunocompromised persons.
   What are your recommendations for remediation?

Shower necks and faucets were disinfected with bleach. The spa filter was cleaned, and the potable water system was hyperchlorinated.

Hotels have been common locations for legionellosis outbreaks since the disease was first recognized among hotel guests in Philadelphia in 1976.

Source: Adapted from *MMWR* 56(48):1261–1263, December 7, 2007.

reported in Africa, the Caribbean, Central and South America, and the Middle East. Many animal species are also susceptible.

Clinically, melioidosis is most commonly seen as pneumonia. Mortality arises from dissemination, manifesting itself as septic shock. The mortality rate in southeast Asia is about 50% and in Australia approaches 20%. However, it can also appear as abscesses in various body tissues that resemble necrotizing fasciitis (see Figure 21.8, page 591), as severe sepsis, and even as encephalitis. Transmission is primarily by inhalation, but alternative infective routes are by inoculation through puncture wounds and ingestion. About 7% of American soldiers returning from Vietnam showed serological evidence of exposure, which was highest among helicopter crewmen—probably from inhalation. Incubation periods can be very long, and occasional delayed-onset cases still surface in this population. Most recently, several cases were reported in survivors exposed during the Indian Ocean tsunami disaster of 2004.

Diagnosis is usually by isolation of the pathogen from body fluids. Serological tests in endemic areas are problematic because

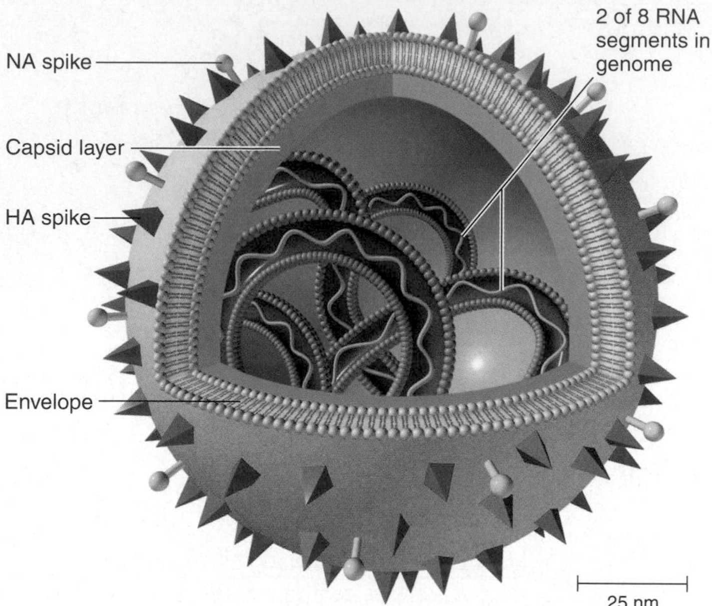

Figure 24.15 Detailed structure of the influenza virus. The virus is composed of a protein coat (capsid) that is covered by a lipid bilayer (envelope) and two types of spikes. The genome is composed of eight segments of RNA. Morphologically, under certain environmental conditions, the influenza virus assumes a filamentous form.

Q What is the primary antigenic structure on the influenza virus?

of widespread exposure to a similar, nonpathogenic bacterium. A rapid PCR test is undergoing clinical testing. Treatment by antibiotic is uncertain in effectiveness; the most commonly used is ceftazidime, a β-lactam antibiotic, but months may be required.

CHECK YOUR UNDERSTANDING

✓ What group of bacterial pathogens causes what is informally called "walking pneumonia"? **24-7**

✓ The bacterium causing melioidosis in humans also causes a disease of horses known as what? **24-8**

Viral Diseases of the Lower Respiratory System

LEARNING OBJECTIVE

24-9 List the causative agent, symptoms, prevention, and preferred treatment for viral pneumonia, RSV, and influenza.

For a virus to reach the lower respiratory system and initiate disease, it must pass numerous host defenses designed to trap and destroy it.

Viral Pneumonia

Viral pneumonia can occur as a complication of influenza, measles, or even chickenpox. A number of enteric and other viruses have been shown to cause viral pneumonia, but viruses are isolated and identified in fewer than 1% of pneumonia-type infections because few laboratories are equipped to test clinical samples properly for viruses. In those cases of pneumonia for which no cause is determined, viral etiology is often assumed if mycoplasmal pneumonia has been ruled out.

Respiratory Syncytial Virus (RSV)

Respiratory syncytial virus (RSV) is probably the most common cause of viral respiratory disease in infants. There are about 4500 deaths from RSV each year in the United States, mostly in infants 2 to 6 months old. It can also cause a life-threatening pneumonia in older adults, where it is easily misdiagnosed as influenza. Epidemics occur during the winter and early spring. Virtually all children become infected by age 2—of whom about 1% require hospitalization. We have previously mentioned that RSV is sometimes implicated in cases of otitis media. The name of the virus is derived from its characteristic of causing cell fusion (*syncytium* formation, Figure 15.7b, page 442) when grown in cell culture. The symptoms are coughing and wheezing that last for more than a week. Fever occurs only when there are bacterial complications. Several rapid serological tests are now available that use samples of respiratory secretions to detect both the virus and its antibodies.

Naturally acquired immunity is very poor. An immune globulin product has been approved to protect infants with lung problems that put them at high risk. Protective vaccines are being clinically tested. For chemotherapy in life-threatening situations, where its cost can be justified, the severity of symptoms can sometimes be reduced by aerosol administration of the antiviral drug ribavirin. The most recent approved treatment, usually reserved for high-risk patients, is the humanized monoclonal antibody, palivizumab (Synagis).

Influenza (Flu)

The developed countries of the world are probably more aware of **influenza (flu)** than any other disease, except the common cold. The flu is characterized by chills, fever, headache, and muscular aches. Recovery normally occurs in a few days, and coldlike symptoms appear as the fever subsides. Still, an estimated 30,000 to 50,000 Americans die annually of flu, even in nonepidemic years. Diarrhea is not a normal symptom of the disease, and the intestinal discomfort attributed to "stomach flu" is probably from some other cause.

The Influenza Virus

Viruses in the genus *Influenzavirus* consist of eight separate RNA segments of differing lengths enclosed by an inner layer of protein and an outer lipid bilayer (Figure 13.3b, page 372, and **Figure 24.15**). Embedded in the lipid bilayer are numerous projections that characterize the virus. There are two types of projections: *hemagglutinin (HA) spikes* and *neuraminidase (NA) spikes.*

The HA spikes, of which there are about 500 on each virus, allow the virus to recognize and attach to body cells before infecting them. Antibodies against the influenza virus are directed mainly at these spikes. The term *hemagglutinin* refers to the agglutination of red blood cells (hemagglutination) that occurs when the viruses are mixed with them. This reaction is important in serological tests, such as the hemagglutination inhibition test often used to identify influenza and some other viruses.

The NA spikes, of which there are about 100 per virus, differ from the HA spikes in appearance and function. Apparently they enzymatically help the virus separate from the infected cell as the virus exits after intracellular reproduction. NA spikes also stimulate the formation of antibodies, but these are less important in the body's resistance to the disease than those produced in response to the HA spikes.

Viral strains are identified by variation in the HA and NA antigens. The different forms of the antigens are assigned numbers—for example, H1, H2, H3, N1, and N2. There are 16 subtypes of HA and 9 of NA. Each number change represents a substantial alteration in the protein makeup of the spike. These changes are called **antigenic shifts,** and they are great enough to evade most of the immunity developed in the human population (see the box in Chapter 13, page 370). This ability is responsible for the outbreaks, including the pandemics of 1918, 1957, and 1968, that are summarized in Table 24.1. Incidentally, influenza viruses were not isolated before 1933, and the antigenic makeup of viruses causing outbreaks before about this time depended on analysis of antibodies taken from persons who had been infected.

Antigenic shifts are probably caused by a major genetic recombination. Because influenza viral RNA occurs as eight segments, recombination is likely in infections caused by more than one strain. Hemagglutinin spikes of the influenza virus bind to sialic acids found on the surface of epithelial cells. The sialic acids of most birds are different from that of humans. This, then, differentiates influenza viruses generally into avian and mammalian strains. Swine and many wild birds can be infected with both strains. Swine are, therefore, good "mixing vessels" in which recombination, *reassortment,* can occur (think of the symbols on a slot machine). Asiatic communities in which humans, domestic chickens, and swine live in close proximity are where reassortment is most likely to occur. Wild ducks and other migratory birds can be infected with both strains and become symptomless carriers that spread the virus over large geographic areas.

In southeast Asia, poultry is now being produced in huge-scale farms that have become a breeding ground for outbreaks of avian influenza such as H5N1, which first appeared in China in 1996. There has been a very limited transmission of the virus from infected birds in these farms to humans. The concern is that mutations could change avian influenza viruses into strains that would allow efficient human-to-human transmission, resulting in a deadly influenza pandemic.

Table 24.1 Human Influenza Viruses*

Type	Antigenic Subtype	Year	Disease Severity
A	H3N2 (the first "modern" pandemic; originated in southern China)	1889	Moderate
	H1N1 (Spanish)	1918	Severe
	H2N2 (Asian)	1957	Severe
	H3N2 (Hong Kong)	1968	Moderate
	H1N1 (Russian)†	1977	Low
B	None	1940	Moderate
C	None	1947	Very mild

*The conventional wisdom is that H1, H2, and H3 are human-infecting strains; H4, H5, H6, and H7 primarily infect animals, especially swine and poultry. (Avian influenza strains H5N1 and H7N7 have caused human fatalities.)

†Probably escaped from a laboratory. At this time persons over age 20 were mostly immune from similar viruses circulating in the 1950s and earlier in the century.

Source: Adapted from C. Mims, J. Playfair, I. Roitt, D. Wakelin, and R. Williams, *Medical Microbiology*, 2nd ed. London: Mosby International, 1998.

Epidemiology of Influenza

Between episodes of such major antigenic shifts, there are minor annual variations in the antigenic makeup called **antigenic drift.** The virus might still be designated as H3N2, for example, but viral strains arise reflecting minor antigenic changes within the antigenic group. These strains are sometimes assigned names related to the locality in which they were first identified. They usually reflect an alteration of only a single amino acid in the protein makeup of the HA or NA spike. Such a minor, one-step mutation is probably a response to selective pressure by antibodies (usually IgA in the mucous membranes) that neutralizes all viruses except for those with new mutations. Such mutations can be expected about once in each million multiplications of the virus. High mutation rates are a characteristic of RNA viruses, which lack much of the "proofreading" ability of DNA viruses.

The usual result of antigenic drift is that a vaccine effective against H3, for example, will be less effective against H3 isolates circulating 10 years after the event. There will have been enough drift in that time that the virus can largely evade the antibodies originally stimulated by the earlier strain.

Influenza viruses are also classified into groups according to the antigens of their protein coats. These groups are A, B, and (rarely) C. The A-type viruses are responsible for the major pandemics. The B-type virus also circulates and mutates, but it is usually responsible for more geographically limited and milder infections.

Almost every year, epidemics of the flu spread rapidly through large populations. The disease is so readily transmissible that epidemics are quickly propagated through populations susceptible to the newly changed strain of virus. The mortality rate from the disease is not high, usually less than 1%, and these deaths are mainly among the very young and the very old.

However, so many people are infected in a major epidemic that the total number of deaths is often large.

Influenza Vaccines

Thus far, it has not been possible to make a vaccine for influenza that gives long-term immunity to the general population. Although it is not difficult to make a vaccine for a particular antigenic strain of virus, each new strain of circulating virus must be identified in time, usually about February, for the useful development and distribution of a new vaccine later that year. Strains of the influenza virus are collected in about 100 centers worldwide, then analyzed in central laboratories. This information is then used to decide on the composition of the vaccines to be offered for the next flu season. The vaccines are usually *multivalent*—directed at the three most important strains in circulation at the time. At present, influenza viruses for manufacturing vaccines are grown in embryonated egg cultures. The vaccines are usually 70–90% effective, but the duration of protection is probably no more than 3 years for that strain.

A vaccine that can be administered as a nasal spray was recently introduced and can be used for children aged 1 to 5. (This vaccine may be extended to other age groups). Research into influenza vaccines that are more effective—and, especially, more rapidly and easily produced—has a high priority. A major problem is that production methods that require growth in egg embryos are clumsy and labor intensive. They also require an unacceptably extended time to respond to the appearance of new viral strains. Furthermore, viruses often grow poorly in egg embryos, and viruses that cause poultry disease usually kill egg embryos. A theoretically promising technology to cope with such problems is to use *reverse genetics.* The viral genome of RNA is converted to DNA and then manipulated to remove the genes causing pathogenicity. The DNA is then converted back to RNA for vaccine production. This procedure results, rather quickly, in a nonpathogenic version of the virus that can be grown in egg embryos or cell culture. It also minimizes the chances of harmful contaminants.

The 1918–1919 Pandemic

In any discussion of influenza, the great pandemic of 1918–1919 must be mentioned.* Worldwide, more than 20 million people died, including an estimated 675,000 in the United States. No one is sure why it was so unusually lethal. Today, the very young and very old are the principal victims, but in 1918–1919, young adults had the highest mortality rate, often dying within a few hours, probably from a "cytokine storm." The infection is usually restricted to the upper respiratory system, but some change in virulence allowed the virus to invade the lungs and cause lethal hemorrhaging.

Evidence also suggests that the virus was able to infect cells in many organs of the body. In 2005, analysis of material preserved from the lungs of U.S. soldiers killed by the flu and from the exhumed body of a victim buried in permanently frozen soil in Alaska led to the complete genetic sequencing of the 1918 virus. The process of reverse genetics was then used to recreate the virus and grow it in chicken embryos and mice. The conclusion is that the pandemic was caused by an H1N1 avian virus with ten changes in amino acids. By contrast, the 1957 and 1968 pandemics were caused by human viruses that had picked up one or two surface proteins from avian viruses by reassortment. However, the human-adapted portion of the virus moderated them enough to keep the virus from matching the lethality of the wholly avian virus of the 1918 pandemic.

Bacterial complications also frequently accompanied the infection and, in those preantibiotic days, were often fatal. The 1918 viral strain apparently became endemic in the U.S. swine population and may have originated there. (See the box in Chapter 13 on page 370.) Occasionally, influenza is still spread to humans from this reservoir, but the disease has not propagated like the virulent disease of 1918 did.

Diagnosis of Influenza

Influenza is difficult to diagnose reliably from clinical symptoms, which numerous respiratory diseases share. However, there are now several commercially available techniques that can diagnose influenza A and B within 20 minutes from a sample taken in a physician's office (from nasal washes or nasal swabs). These rapid tests are usually more than 70% sensitive and 90% specific. A central laboratory with sophisticated equipment is required to identify viral strains.

Treatment of Influenza

The antiviral drugs amantadine and rimantadine significantly reduce the symptoms of influenza A if administered promptly. Two drugs for treatment of influenza have been introduced. They are inhibitors of neuraminidase, which the virus uses to separate itself from the host cell after it replicates. These are zanamivir (Relenza), which is inhaled, and oseltamivir (Tamiflu), which is administered orally. If taken within 30 hours of onset of influenza, these drugs slow replication. This action allows the immune system to be more effective,

*There will always be uncertainty concerning the origin of this most famous pandemic. The best reliable reports place the first well-documented cases among U.S. Army recruits at Camp Funston, Kansas, in March of 1918. The initial wave of influenza was a relatively mild illness that spread rapidly among the crowded troops and reached France as they were dispatched overseas. There the virus underwent a lethal mutation, seriously incapacitating troops on both sides of the front. Military censorship concealed this, and the first newspaper descriptions were published when the outbreak reached the population of neutral Spain, hence the name assigned to the pandemic: the **Spanish flu**. This second wave of influenza, with its high mortality, soon spread throughout the world and reentered the United States in the autumn and winter of 1918.

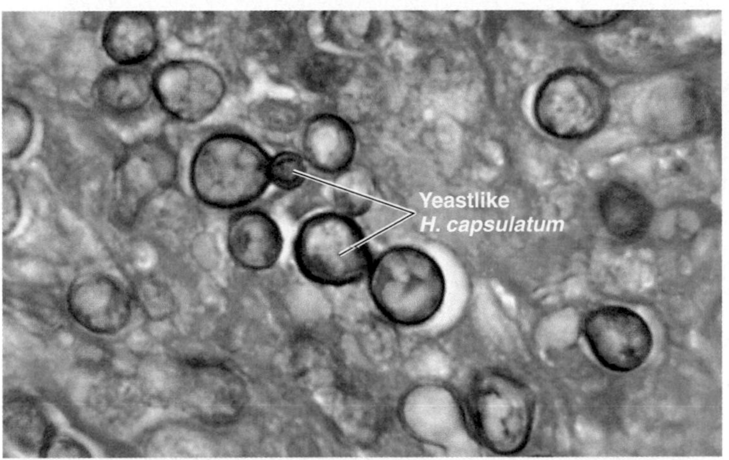

(a) Yeastlike form typical of growth in tissue at 37°C. Notice that one yeastlike cell to the left of center is budding.

LM 5 μm

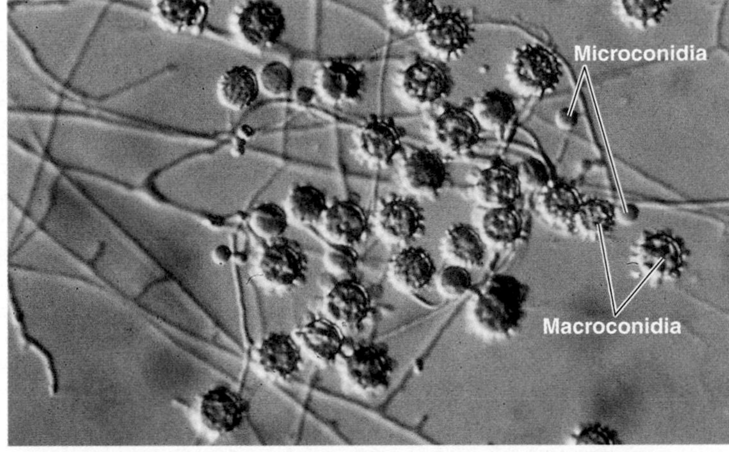

(b) The macroconidia are especially useful for diagnostic purposes. Microconidia bud off from hyphae and are the infectious form. At 37°C in tissues, the organism converts to a yeast phase composed of oval, budding yeasts.

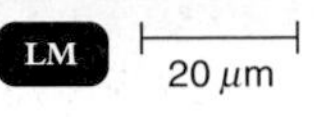
LM 20 μm

Figure 24.16 ***Histoplasma capsulatum,*** **a dimorphic fungus that causes histoplasmosis.**

Q What does the term *dimorphic* mean?

which shortens durations of symptoms and lowers the mortality rate. The bacterial complications of influenza are amenable to treatment with antibiotics.

CHECK YOUR UNDERSTANDING

Is reassortment of the RNA segments of the influenza virus the cause of antigenic shift or antigenic drift? 24-9

Fungal Diseases of the Lower Respiratory System

LEARNING OBJECTIVE

24-10 List the causative agent, mode of transmission, preferred treatment, and laboratory identification tests for four fungal diseases of the respiratory system.

Fungi often produce spores that are disseminated through the air. It is therefore not surprising that several serious fungal diseases affect the lower respiratory system. The rate of fungal infections has been increasing in recent years. Opportunistic fungi are able to grow in immunosuppressed patients, and AIDS, transplant drugs, and anticancer drugs have created more immunosuppressed people than ever before.

Histoplasmosis

Histoplasmosis superficially resembles tuberculosis. In fact, it was first recognized as a widespread disease in the United States when X-ray surveys showed lung lesions in many people who were tuberculin-test negative. Although the lungs are most likely to be initially infected, the pathogens may spread in the blood and lymph, causing lesions in almost all organs of the body.

Symptoms are usually poorly defined and mostly subclinical, and the disease passes for a minor respiratory infection. In a few cases, perhaps fewer than 0.1%, histoplasmosis progresses, and it becomes a severe, generalized disease. This occurs with an unusually heavy inoculum or upon reactivation, when the infected person's immune system is compromised.

The causative organism, *Histoplasma capsulatum* (his-tō-plaz′mä kap-su-lä′tum), is a dimorphic fungus; that is, it has a yeastlike morphology in tissue growth (Figure 24.16a), and, in soil or artificial media, it forms a filamentous mycelium carrying reproductive conidia (Figure 24.16b). In the body, the yeastlike form is found intracellularly in macrophages, where it survives and multiplies.

Although histoplasmosis is rather widespread throughout the world, it has a limited geographic range in the United States (Figure 24.17). In general, the disease is found in the states adjoining the Mississippi and Ohio rivers. More than 75% of the population in some of these states have antibodies against the infection. In other states—Maine, for example—a positive test is a rare event. Approximately 50 deaths are reported in the United States each year from histoplasmosis.

Humans acquire the disease from airborne conidia produced under conditions of appropriate moisture and pH levels. These conditions occur especially where droppings from birds and bats have accumulated. Birds themselves, because of their high body temperature, do not carry the disease, but their droppings provide nutrients, particularly a source of nitrogen, for the

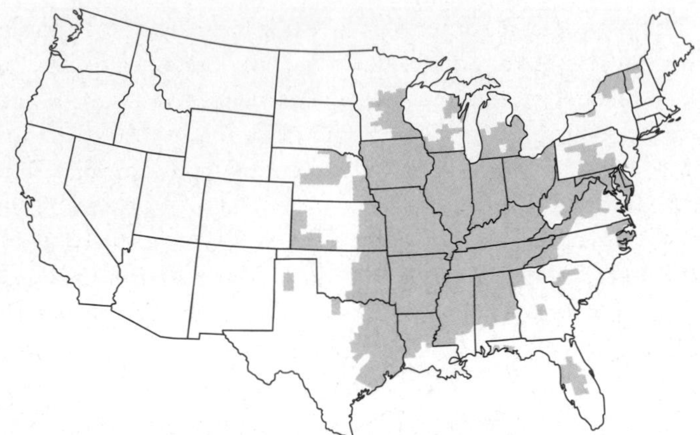

Figure 24.17 Histoplasmosis distribution. Gold indicates the U.S. geographic distribution.

SOURCE: CDC.

Q Compared with the disease distribution shown in the map in Figure 24.19, what can you determine about the moisture requirements in the soil for the two fungi involved?

fungus. Bats, which have a lower body temperature than birds, carry the fungus, shed it in their feces, and infect new soil sites.

Clinical signs and history, serological tests, DNA probes, and, most importantly, either isolation of the pathogen or its identification in tissue specimens are necessary for proper diagnosis. Currently, the most effective chemotherapy is amphotericin B or itraconazole.

Coccidioidomycosis

Another fungal pulmonary disease, also rather restricted geographically, is **coccidioidomycosis.** The causative agent is *Coccidioides immitis,* a dimorphic fungus. The arthroconidia are found in dry, alkaline soils of the American Southwest and in similar soils of South America and northern Mexico. Because of its frequent occurrence in the San Joaquin Valley of California, it is sometimes known as *Valley fever* or *San Joaquin fever.* In tissues, the organism forms a thick-walled body called a *spherule* filled with endospores (**Figure 24.18**). In soil, it forms filaments that reproduce by the formation of arthroconidia. The wind carries the arthroconidia to transmit the infection. Arthroconidia are often so abundant that simply driving through an endemic area can result in infection, especially during a dust storm. An estimated 100,000 infections occur each year.

Most infections are not apparent, and almost all patients recover in a few weeks, even without treatment. The symptoms of coccidioidomycosis include chest pain and perhaps fever, coughing, and weight loss. In less than 1% of cases, a progressive disease resembling tuberculosis disseminates throughout the body. A substantial proportion of adults who are long-time residents of areas where the disease is endemic have evidence of prior infection with *C. immitis* by the skin test.

The incidence of coccidiodomycosis has been increasing recently in California and Arizona (**Figure 24.19**). Contributing factors include an increased number of older residents, an increased prevalence of HIV/AIDS, and a severe drought in

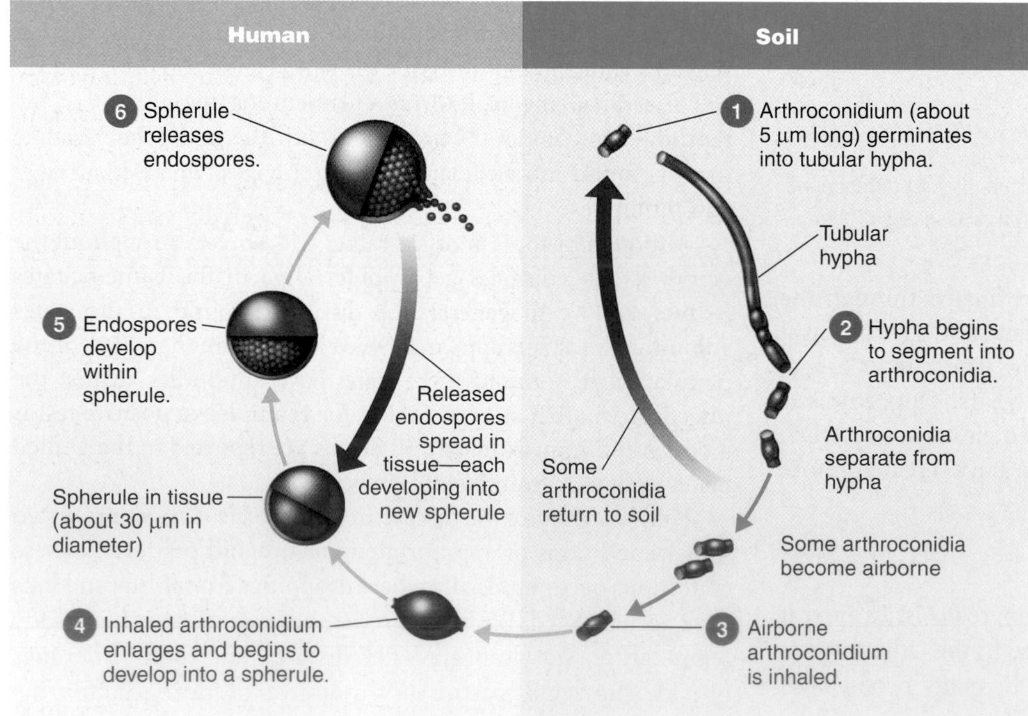

Figure 24.18 The life cycle of *Coccidioides immitis,* the cause of coccidioidomycosis.

Q What is the natural habitat of *Coccidioides*?

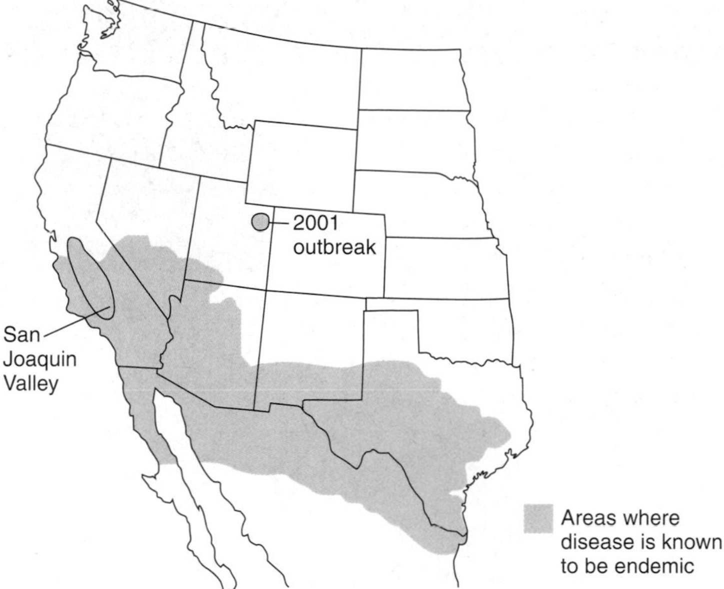

Figure 24.19 The U.S. endemic area for coccidioidomycosis. The outlined area in California is the San Joaquin Valley. Because of the very high incidence there, the disease is also sometimes called Valley fever. The small area on the map in northeastern Utah indicates an outbreak in 2001 in which ten archeologists working on excavations at the Dinosaur National Monument were infected.

SOURCE: CDC, 2004.

Q **Why does the incidence of coccidioidomycosis increase after ecological disturbances, such as earthquakes and construction?**

California that facilitated dustborne transmission. About 50 to 100 deaths occur annually from this disease in the United States.

Diagnosis is most reliably made by identifying the spherules in tissue or fluids. The organism can be cultured from fluids or lesions, but laboratory workers must use great care because of the possibility of infectious aerosols. Several serological tests and DNA probes are available for identifying isolates. A tuberculin-like skin test is used in screening.

Amphotericin B has been used to treat serious cases. However, less toxic imidazole drugs, such as ketoconazole and itraconazole, are useful alternatives.

Pneumocystis Pneumonia

***Pneumocystis* pneumonia** is caused by *Pneumocystis jirovecii* (ye-rō′vet-zē-ē), formerly *P. carinii* (**Figure 24.20**). The taxonomic position of this microbe has been uncertain ever since its discovery in 1909, when it was thought to be a developmental stage of a trypanosome. Since that time, there has been no universal agreement about whether it is a protozoan or a fungus. It has some characteristics of both groups. Analysis of RNA and certain other structural characteristics indicate that it is closely related to certain yeasts and it is usually reported as a fungus.

The pathogen is sometimes found in healthy human lungs. Immunocompetent adults have few or no symptoms, but newly infected infants occasionally show symptoms of a lung infection. Persons with compromised immunity are the most susceptible to symptomatic *Pneumocystis* pneumonia. This population may also be the reservoir of the organism, which is not found in the environment, animals, or very often in healthy humans. This portion of the population has also expanded greatly in recent decades. For example, before the AIDS epidemic, *Pneumocystis* pneumonia was an uncommon disease; perhaps 100 cases occurred each year. By 1993, it had become a primary indicator of AIDS, with more than 20,000 annual reported cases. Presumably, the loss of an effective immune defense allowed the activation of a latent infection. Other groups that are very susceptible to this disease are people whose immunity is depressed because of cancer or who are receiving immunosuppressive drugs to minimize rejection of transplanted tissue.

In the human lung, the microbes are found mostly in the lining of the alveoli. Diagnosis is usually made from sputum samples in which cysts are detected. There, they form a thick-walled cyst in which spherical intracystic bodies successively divide as part of a sexual cycle. The mature cyst contains eight such bodies (see Figure 24.20). Eventually the cyst ruptures and releases them, and each body develops into a trophozoite. The trophozoite cells can reproduce asexually by fission, but they may also enter the encysted sexual stage.

The drug of choice for treatment is currently trimethoprim-sulfamethoxazole, but there are several alternatives.

Blastomycosis (North American Blastomycosis)

Blastomycosis is usually called **North American blastomycosis** to differentiate it from a similar South American blastomycosis. It is caused by the fungus *Blastomyces dermatitidis* (blas-tō-mī′sēz dėr-mä-tit′i-dis), a dimorphic fungus found most often in the Mississippi and Ohio river valleys, where it probably grows in soil. Approximately 30 to 60 deaths are reported each year, although most infections are asymptomatic.

The infection begins in the lungs. It resembles bacterial pneumonia and can spread rapidly. Cutaneous ulcers commonly appear, and there can be extensive abscess formation and tissue destruction. The pathogen can be isolated from pus and biopsy specimens. Amphotericin B or itraconazole is usually an effective treatment.

Other Fungi Involved in Respiratory Disease

Many other opportunistic fungi may cause respiratory disease, particularly in hosts who are immunosuppressed or have been exposed to massive numbers of spores. **Aspergillosis** is an important example; it is airborne by the conidia of *Aspergillus*

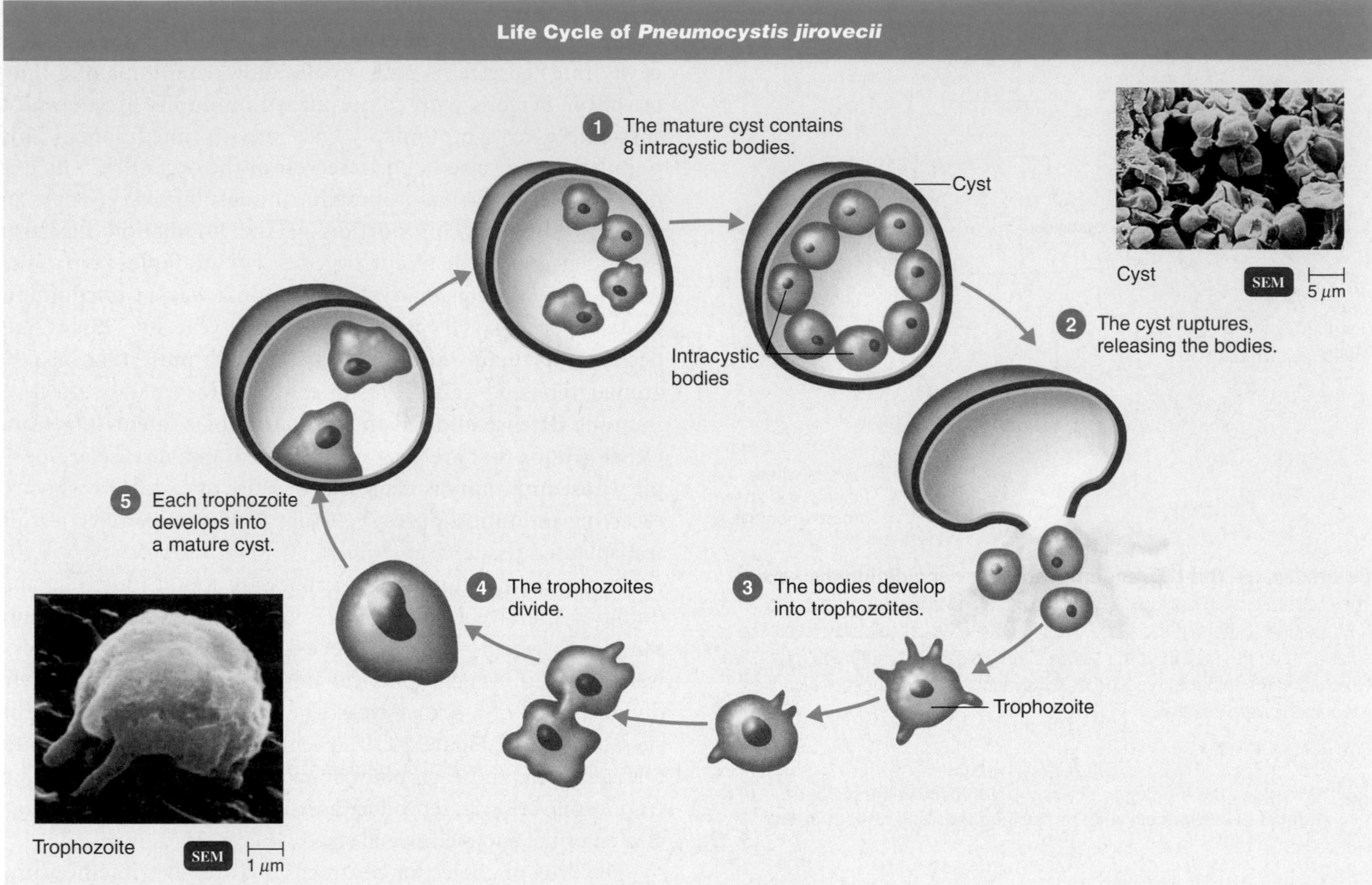
Life Cycle of Pneumocystis jirovecii
1 The mature cyst contains 8 intracystic bodies.
Cyst
Intracystic bodies
Cyst
SEM
5 μm
2 The cyst ruptures, releasing the bodies.
3 The bodies develop into trophozoites.
Trophozoite
4 The trophozoites divide.
5 Each trophozoite develops into a mature cyst.
Trophozoite
SEM
1 μm

Figure 24.20 The life cycle of *Pneumocystis jirovecii,* the cause of *Pneumocystis* pneumonia. Long classified as a protozoan, this organism is now usually considered to be a fungus, but it has characteristics of both groups.

Q Of what value is proper classification of this organism?

fumigatus (fū-mi-gä′tus) and other species of *Aspergillus,* which are widespread in decaying vegetation. Compost piles are ideal sites for growth, and farmers and gardeners are most often exposed to infective amounts of these conidia.

Similar pulmonary infections sometimes result when individuals are exposed to spores of other mold genera, such as *Rhizopus* and *Mucor.* Such diseases can be very dangerous, particularly invasive infections of pulmonary aspergillosis. Predisposing factors include an impaired immune system, cancer, and diabetes. As with most systemic fungal infections, there is only a limited arsenal of antifungal agents available; amphotericin B has proved the most useful.

CHECK YOUR UNDERSTANDING

✓ The droppings of both blackbirds and bats support the growth of *Histoplasma capsulatum;* which of these two animal reservoirs is normally actually infected by the fungus? **24-10**

* * *

Diseases in Focus 24.3 summarizes the microbial respiratory diseases affecting the lower respiratory system discussed in this chapter.

DISEASES IN FOCUS 24.3

Microbial Diseases of the Lower Respiratory System

Differential diagnosis is the process of identifying the disease from a list of possible diseases that fit the information derived from examining a patient. For example, 3 weeks after working on the demolition of an abandoned building in Kentucky, a worker was hospitalized for acute respiratory illness. At the time of demolition, a colony of bats inhabited the building. An X-ray examination revealed a lung mass. A purified protein derivative test was negative; a cytological examination for cancer was also negative. The mass was surgically removed. Microscopic examination of the mass revealed ovoid yeast cells. Use the table below to identify infections that could cause these symptoms. For the solution, go to www.microbiologyplace.com.

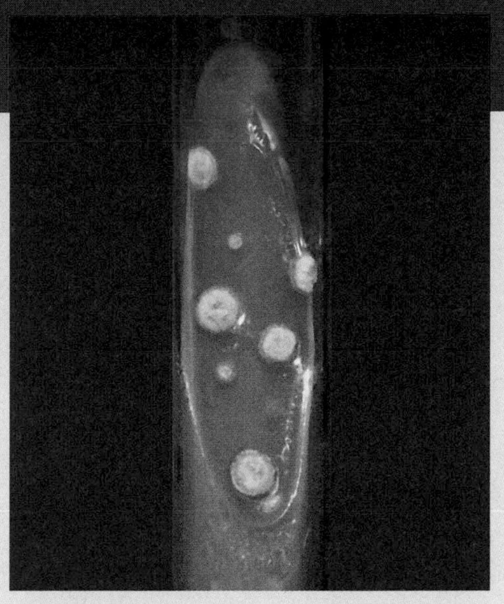

Mycelial culture grown from the patient's lung mass.

Disease	Pathogen	Symptoms	Reservoir	Diagnosis	Treatment
BACTERIAL DISEASES					
Bacterial pneumonia (see Diseases in Focus 24.2, page 687)					
Pertussis (whooping cough)	*Bordetella pertussis*	Spasms of intense coughing to clear mucus	Humans	Bacterial culture	Erythromycin Prevention: DTaP vaccine
Tuberculosis	*Mycobacterium tuberculosis* *Mycobacterium bovis*	Cough, blood in mucus	Humans, cows; can be transmitted via unpasteurized milk	X-ray imaging; presence of acid-fast bacilli in sputum; tests for IFN-γ; bacterial culture	Multiple-antimycobacterial drugs Prevention: pasteurizing milk; BCG vaccine
Melioidosis	*Burkholderia pseudomallei*	Pneumonia, or as tissue abscesses and severe sepsis	Moist soil	Bacterial culture	Ceftazidime
VIRAL DISEASES					
Respiratory syncytial virus (RSV) disease	Respiratory syncytial virus	Pneumonia in infants	Humans	Serological tests	Palivizumab (if life-threatening)
Influenza	*Influenzavirus*; several serotypes	Chills, fever, headache, and muscular aches	Humans, pigs, birds	Serological EIA tests	Amantadine, oseltamivir phosphate (Tamiflu)
FUNGAL DISEASES					
Histoplasmosis	*Histoplasma capsulatum*	Resembles tuberculosis	Soil; widespread in Ohio and Mississippi river valleys	Serological tests	Amphotericin B
Coccidioidomycosis	*Coccidioides immitis*	Fever, coughing, weight loss	Desert soils of U.S. Southwest	Serological tests	Amphotericin B
***Pneumocystis* pneumonia**	*Pneumocystis jirovecii*	Pneumonia	Unknown; possibly humans or soil	Microscopy	Trimethoprim-sulfamethoxazole
Blastomycosis	*Blastomyces dermatitidis*	Abscesses; extensive tissue damage	Soil in Mississippi Valley area	Isolation of pathogen	Amphotericin B

STUDY OUTLINE

The **MyMicrobiologyPlace** website (**www.microbiologyplace.com**) will help you get ready for tests with its simple three-step approach: ❶ **take a pre-test** and obtain a personalized study plan, ❷ **learn and practice** with animations, tutorials, and MP3 tutor sessions, and ❸ **test yourself** with quizzes and a chapter post-test.

Introduction (p. 674)

1. Infections of the upper respiratory system are the most common type of infection.
2. Pathogens that enter the respiratory system can infect other parts of the body.

Structure and Function of the Respiratory System (p. 675)

1. The upper respiratory system consists of the nose, pharynx, and associated structures, such as the middle ear and auditory tubes.
2. Coarse hairs in the nose filter large particles from air entering the respiratory tract.
3. The ciliated mucous membranes of the nose and throat trap airborne particles and remove them from the body.
4. Lymphoid tissue, tonsils, and adenoids provide immunity to certain infections.
5. The lower respiratory system consists of the larynx, trachea, bronchial tubes, and alveoli.
6. The ciliary escalator of the lower respiratory system helps prevent microorganisms from reaching the lungs.
7. Microbes in the lungs can be phagocytized by alveolar macrophages.
8. Respiratory mucus contains IgA antibodies.

Normal Microbiota of the Respiratory System (pp. 675–676)

1. The normal microbiota of the nasal cavity and throat can include pathogenic microorganisms.
2. The lower respiratory system is usually sterile because of the action of the ciliary escalator.

MICROBIAL DISEASES OF THE UPPER RESPIRATORY SYSTEM (pp. 676–680)

1. Specific areas of the upper respiratory system can become infected to produce pharyngitis, laryngitis, tonsillitis, sinusitis, and epiglottitis.
2. These infections may be caused by several bacteria and viruses, often in combination.
3. Most respiratory tract infections are self-limiting.
4. *H. influenzae* type b can cause epiglottitis.

Bacterial Diseases of the Upper Respiratory System (pp. 677–679)

Streptococcal Pharyngitis (Strep Throat) (p. 677)

1. This infection is caused by group A beta-hemolytic streptococci, the group that consists of *Streptococcus pyogenes.*
2. Symptoms of this infection are inflammation of the mucous membrane and fever; tonsillitis and otitis media may also occur.
3. Rapid diagnosis is made by enzyme immunoassays.
4. Immunity to streptococcal infections is type-specific.

Scarlet Fever (p. 677)

5. Strep throat, caused by an erythrogenic toxin-producing *S. pyogenes,* results in scarlet fever.
6. *S. pyogenes* produces erythrogenic toxin when lysogenized by a phage.
7. Symptoms include a red rash, high fever, and a red, enlarged tongue.

Diphtheria (pp. 677–679)

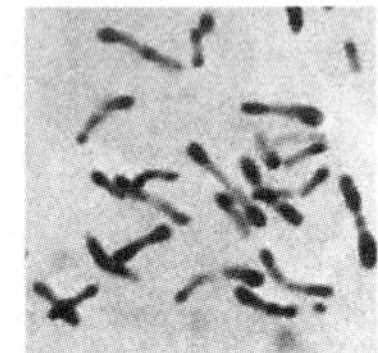

8. Diphtheria is caused by exotoxin-producing *Corynebacterium diphtheriae.*
9. Exotoxin is produced when the bacteria are lysogenized by a phage.
10. A membrane, containing fibrin and dead human and bacterial cells, forms in the throat and can block the passage of air.
11. The exotoxin inhibits protein synthesis, and heart, kidney, or nerve damage may result.
12. Laboratory diagnosis is based on isolation of the bacteria and the appearance of growth on differential media.
13. Routine immunization in the United States includes diphtheria toxoid in the DTaP vaccine.
14. Slow-healing skin ulcerations are characteristic of cutaneous diphtheria.
15. There is minimal dissemination of the exotoxin in the bloodstream.

Otitis Media (p. 679)

16. Earache, or otitis media, can occur as a complication of nose and throat infections.
17. Pus accumulation causes pressure on the eardrum.
18. Bacterial causes include *Streptococcus pneumoniae,* nonencapsulated *Haemophilus influenzae, Moraxella catarrhalis, Streptococcus pyogenes,* and *Staphylococcus aureus.*

Viral Diseases of the Upper Respiratory System (pp. 679–680)

The Common Cold (pp. 679–680)

1. Any one of approximately 200 different viruses can cause the common cold; rhinoviruses cause about 50% of all colds.
2. Symptoms include sneezing, nasal secretions, and congestion.
3. Sinus infections, lower respiratory tract infections, laryngitis, and otitis media can occur as complications of a cold.
4. Rhinoviruses grow best slightly below body temperature.
5. The incidence of colds increases during cold weather, possibly because of increased interpersonal indoor contact or physiological changes.
6. Antibodies are produced against the specific viruses.

MICROBIAL DISEASES OF THE LOWER RESPIRATORY SYSTEM (pp. 680–699)

1. Many of the same microorganisms that infect the upper respiratory system also infect the lower respiratory system.
2. Diseases of the lower respiratory system include bronchitis and pneumonia.

Bacterial Diseases of the Lower Respiratory System (pp. 680–692)

Pertussis (Whooping Cough) (pp. 680–682)

1. Pertussis is caused by *Bordetella pertussis.*
2. The initial stage of pertussis resembles a cold and is called the catarrhal stage.
3. The accumulation of mucus in the trachea and bronchi causes deep coughs characteristic of the paroxysmal (second) stage.
4. The convalescence (third) stage can last for months.
5. Regular immunization for children has decreased the incidence of pertussis.

Tuberculosis (pp. 682–685)

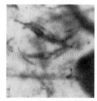

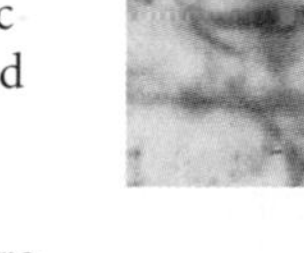

6. Tuberculosis is caused by *Mycobacterium tuberculosis.*
7. Large amounts of lipids in the cell wall account for the bacterium's acid-fast characteristic as well as its resistance to drying and disinfectants.
8. *M. tuberculosis* may be ingested by alveolar macrophages; if not killed, the bacteria reproduce in the macrophages.
9. Lesions formed by *M. tuberculosis* are called tubercles; dead macrophages and bacteria form the caseous lesion that might calcify and appear in an X-ray image as a Ghon's complex.
10. Liquefaction of the caseous lesion results in a tuberculous cavity in which *M. tuberculosis* can grow.
11. New foci of infection can develop when a caseous lesion ruptures and releases bacteria into blood or lymph vessels; this is called miliary tuberculosis.
12. Miliary tuberculosis is characterized by weight loss, coughing, and loss of vigor.
13. Chemotherapy usually involves three or four drugs taken for at least 6 months; multidrug-resistant *M. tuberculosis* is becoming prevalent.
14. A positive tuberculin skin test can indicate either an active case of TB, prior infection, or vaccination and immunity to the disease.
15. *Mycobacterium bovis* causes bovine tuberculosis and can be transmitted to humans by unpasteurized milk.
16. *M. bovis* infections usually affect the bones or lymphatic system.
17. BCG vaccine for tuberculosis consists of a live, avirulent culture of *M. bovis.*
18. *M. avium-intracellulare* complex infects patients in the late stages of HIV infection.

Bacterial Pneumonias (pp. 685–692)

19. Typical pneumonia is caused by *S. pneumoniae.*
20. Atypical pneumonias are caused by other microorganisms.

Pneumococcal Pneumonia (pp. 685–688)

21. Pneumococcal pneumonia is caused by encapsulated *Streptococcus pneumoniae.*
22. Symptoms are fever, breathing difficulty, chest pain, and rust-colored sputum.
23. A vaccine consists of purified capsular material from 23 serotypes of *S. pneumoniae.*

Haemophilus influenzae Pneumonia (p. 688)

24. Alcoholism, poor nutrition, cancer, and diabetes are predisposing factors for *H. influenzae* pneumonia.
25. *H. influenzae* is a gram-negative coccobacillus.

Mycoplasmal Pneumonia (p. 688)

26. *Mycoplasma pneumoniae* causes mycoplasmal pneumonia; it is an endemic disease.
27. *M. pneumoniae* produces small "fried-egg" colonies after 2 weeks' incubation on enriched media containing horse serum and yeast extract.

Legionellosis (pp. 688–689)

28. The disease is caused by the aerobic gram-negative rod *Legionella pneumophila.*
29. The bacterium can grow in water, such as air-conditioning cooling towers, and then be disseminated in the air.
30. This pneumonia does not appear to be transmitted from person to person.

Psittacosis (Ornithosis) (p. 689)

31. *Chlamydophila psittaci* is transmitted by contact with contaminated droppings and exudates of fowl.
32. Elementary bodies allow the bacteria to survive outside a host.
33. Commercial bird handlers are most susceptible to this disease.

Chlamydial Pneumonia (p. 689)

34. *Chlamydophila pneumoniae* causes pneumonia; it is transmitted from person to person.

Q Fever (pp. 689–690)

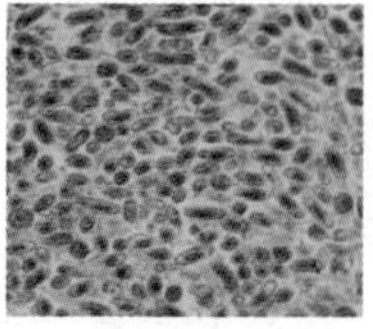

35. Obligately parasitic, intracellular *Coxiella burnetii* causes Q fever.
36. The disease is usually transmitted to humans through unpasteurized milk or inhalation of aerosols in dairy barns.

Melioidosis (pp. 690–692)

37. Melioidosis, caused by *Burkholderia pseudomallei,* is transmitted by inhalation, ingestion, or through puncture wounds. Symptoms include pneumonia, sepsis, and encephalitis.

Viral Diseases of the Lower Respiratory System (pp. 692–695)

Viral Pneumonia (p. 692)

1. A number of viruses can cause pneumonia as a complication of infections such as influenza.

2. The etiologies are not usually identified in a clinical laboratory because of the difficulty in isolating and identifying viruses.

Respiratory Syncytial Virus (RSV) (p. 692)

3. RSV is the most common cause of pneumonia in infants.

Influenza (Flu) (pp. 692–695)

4. Influenza is caused by *Influenzavirus* and is characterized by chills, fever, headache, and general muscular aches.
5. Hemagglutinin (HA) and neuraminidase (NA) spikes project from the outer lipid bilayer of the virus.
6. Viral strains are identified by antigenic differences in the HA and NA spikes; they are also divided by antigenic differences in their protein coats (A, B, and C).
7. Viral isolates are identified by hemagglutination-inhibition tests and immunofluorescence testing with monoclonal antibodies.
8. Antigenic shifts that alter the antigenic nature of the HA and NA spikes make natural immunity and vaccination of questionable value. Minor antigenic changes are caused by antigenic drift.
9. Deaths during an influenza epidemic are usually from secondary bacterial infections.
10. Multivalent vaccines are available for older adults and other high-risk groups.
11. Amantadine and rimantadine are effective prophylactic and curative drugs against influenza A virus.

Fungal Diseases of the Lower Respiratory System (pp. 695–698)

1. Fungal spores are easily inhaled; they may germinate in the lower respiratory tract.
2. The incidence of fungal diseases has been increasing in recent years.
3. The mycoses in the following sections can be treated with amphotericin B.

Histoplasmosis (pp. 695–696)

4. *Histoplasma capsulatum* causes a subclinical respiratory infection that only occasionally progresses to a severe, generalized disease.
5. The disease is acquired by inhaling airborne conidia.
6. Isolating or identifying the fungus in tissue samples is necessary for diagnosis.

Coccidioidomycosis (pp. 696–697)

7. Inhaling the airborne arthroconidia of *Coccidioides immitis* can result in coccidioidomycosis.
8. Most cases are subclinical, but when there are predisposing factors such as fatigue and poor nutrition, a progressive disease resembling tuberculosis can result.

Pneumocystis Pneumonia (p. 697)

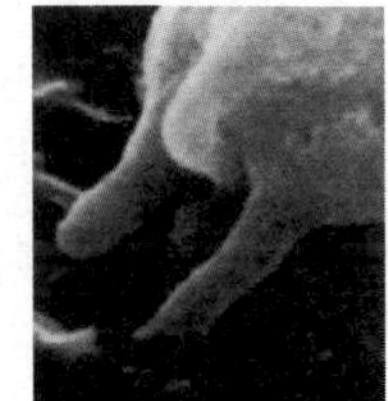

9. *Pneumocystis jirovecii* is found in healthy human lungs.
10. *P. jirovecii* causes disease in immunosuppressed patients.

Blastomycosis (North American Blastomycosis) (p. 697)

11. *Blastomyces dermatitidis* is the causative agent of blastomycosis.
12. The infection begins in the lungs and can spread to cause extensive abscesses.

Other Fungi Involved in Respiratory Disease (pp. 697–698)

13. Opportunistic fungi can cause respiratory disease in immunosuppressed hosts, especially when large numbers of spores are inhaled.
14. Among these fungi are *Aspergillus, Rhizopus,* and *Mucor.*

STUDY QUESTIONS

Answers to the Review and Multiple Choice questions can be found by turning to the blue Answers tab at the back of the textbook.

Review

1. **DRAW IT** Show the locations of the following diseases: common cold, diphtheria, coccidioidomycosis, influenza, pneumonia, scarlet fever, tuberculosis, whooping cough

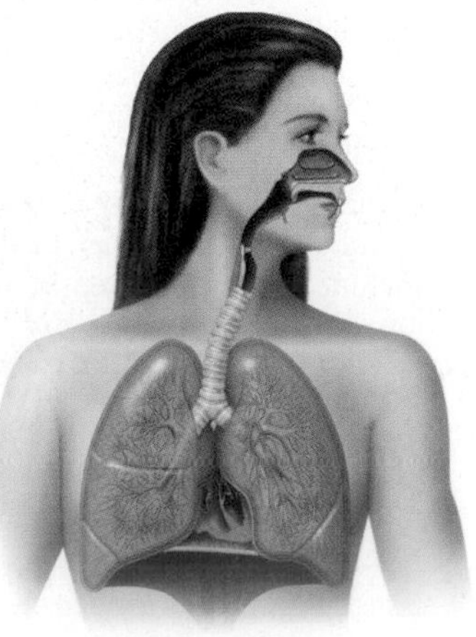

2. Compare and contrast mycoplasmal pneumonia and viral pneumonia.
3. List the causative agent, symptoms, and treatment for four viral diseases of the respiratory system. Separate the diseases according to whether they infect the upper or lower respiratory system.

4. Complete the following table.

Disease	Causative Agent	Symptoms	Treatment
Streptococcal pharyngitis			
Scarlet fever			
Diphtheria			
Whooping cough			
Tuberculosis			
Pneumococcal pneumonia			
H. influenzae pneumonia			
Chlamydial pneumonia			
Otitis media			
Legionellosis			
Psittacosis			
Q fever			
Epiglottitis			
Melioidosis			

5. Under what conditions can the saprophytes *Aspergillus* and *Rhizopus* cause infections?

6. A patient has been diagnosed as having pneumonia. Is this sufficient information to begin treatment with antimicrobial agents? Briefly discuss why or why not.

7. List the causative agent, mode of transmission, and endemic area for the diseases histoplasmosis, coccidioidomycosis, blastomycosis, and *Pneumocystis* pneumonia.

8. Briefly describe the procedures and positive results of the tuberculin test and what is indicated by a positive test.

9. Match the bacteria involved in respiratory infections with the following laboratory test results:

Gram-positive cocci
- Catalase-positive: a. ____________
- Catalase-negative
  - Beta-hemolytic, bacitracin inhibition: b. ____________
  - Alpha-hemolytic, optochin inhibition: c. ____________

Gram-positive rods
- Not acid-fast: d. ____________
- Acid-fast: e. ____________

Gram-negative cocci: f. ____________

Gram-negative rods
- Aerobes
  - Coccobacilli: g. ____________
  - Rods
    - Grow on nutrient agar: h. ____________
    - Require special media: i. ____________
- Facultative anaerobes
  - Coccobacilli j. ____________

Intracellular parasites
- Form elementary bodies k. ____________
- Do not form elementary bodies l. ____________

Wall-less m. ____________

Multiple Choice

1. A patient has fever, difficulty breathing, chest pains, fluid in the alveoli, and a positive tuberculin skin test. Gram-positive cocci are isolated from the sputum. The recommended treatment is
a. penicillin.
b. antitoxin.
c. isoniazid.
d. tetracyclines.
e. none of the above

2. No bacterial pathogen can be isolated from the sputum of a patient with pneumonia. Antibiotic therapy has not been successful. The next step should be
a. culturing for *Mycobacterium tuberculosis.*
b. culturing for *Mycoplasma pneumoniae.*
c. culturing for fungi.
d. a change in antibiotics.
e. none; nothing more can be done.

Match the following choices to the culture descriptions in questions 3 through 6:
a. *Chlamydophila*
b. *Coccidioides*
c. *Histoplasma*
d. *Mycobacterium*
e. *Mycoplasma*

3. Your culture from a pneumonia patient appears not to have grown. You do see colonies, however, when the plate is viewed at 100×.

4. This pneumonia etiology requires cell culture.

5. Microscopic examination of a lung biopsy shows ovoid cells in macrophages. You suspect these are the cause of the patient's symptoms, but your culture grows a filamentous organism.

6. Microscopic examination of a lung biopsy shows spherules.

7. In San Francisco, ten animal health care technicians developed pneumonia 2 weeks after 130 goats were moved to the animal shelter where they worked. Which of the following is *not* true?
a. Diagnosis is made by a blood agar culture of sputum.
b. The cause is *Coxiella burnetii.*
c. The cause is rickettsia.
d. The disease was transmitted by aerosols.
e. Diagnosis is made by complement-fixation tests for antibodies.

8. Which of the following leads to all the rest?
a. catarrhal stage
b. cough
c. loss of cilia
d. mucus accumulation
e. trachaeal cytotoxin

Match the following choices to the statements in questions 9 and 10:
a. *Bordetella pertussis*
b. *Corynebacterium diphtheriae*
c. *Legionella pneumophila*
d. *Mycobacterium tuberculosis*
e. none of the above

9. Causes the formation of a membrane across the throat.

10. Resistant to destruction by phagocytes.

Critical Thinking

1. Differentiate *S. pyogenes* causing strep throat from *S. pyogenes* causing scarlet fever.
2. Why might the influenza vaccine be less effective than other vaccines?
3. Explain why it would be impractical to include cold and influenza vaccinations in the required childhood vaccinations.

Clinical Applications

1. In August, a 24-year-old man from Virginia developed difficulty breathing and bilateral lobe infiltrates 2 months after driving through California. During initial evaluation, typical pneumonia was suspected, and he was treated with antibiotics. Efforts to diagnose the pneumonia were unsuccessful. In October, a laryngeal mass was detected, and laryngeal cancer was suspected; treatment with steroids and bronchodilators did not result in improvement. Lung biopsy and laryngoscopy detected diffuse granular tissue. He was treated with amphotericin B and discharged after 5 days. What was the disease? What might have been done differently to decrease the patient's recovery time from 3 months to one week?
2. During a 6-month period, 72 clinic staff members became tuberculin-positive. A case-control study was undertaken to determine the most likely source of *M. tuberculosis* infection among the staff. A total of 16 cases and 34 tuberculin-negative controls were compared. Pentamidine isethionate is not used for TB treatment. What disease was probably being treated with this drug? What is the most likely source of infection?

	Cases	Control
Works ≥40 hr/week	100%	62%
In room during aerosolized pentamidine isethionate therapy in TB patients	31	3
Patient contact	94	94
Lunch eaten in staff lounge	38	35
Resident of western Palm Beach	75	65
Female	81	77
Cigarette smoker	6	15
Contact with nurse diagnosed with TB	15	12
In unventilated room during collection of TB-positive sputum samples	13	8

3. In March, six members of a family had fever, anorexia, sore throat, cough, headache, vomiting, and muscle pain. Two people were hospitalized. All six improved after therapy with doxycycline. Convalescent serum samples gave titers of 64 and 32. The family had purchased a cockatiel in mid-February and noticed that the bird was irritable. The bird was euthanized in April. A fluorescent-antibody test for antigens was diagnostic. What is the disease? How was it transmitted?

25 Microbial Diseases of the Digestive System

Microbial diseases of the digestive system are second only to respiratory diseases as causes of illness in the United States. Most such diseases result from the ingestion of food or water contaminated with pathogenic microorganisms or their toxins. These pathogens usually enter the food or water supply after being shed in the feces of people or animals infected with them. Therefore, microbial diseases of the digestive system are typically transmitted by a **fecal–oral cycle.** This cycle is interrupted by effective sanitation practices in food production and handling. Modern methods of sewage treatment and disinfection of water are essential. There is also an increasing awareness of the need for new tests that will rapidly and reliably detect pathogens in foods (a perishable commodity).

The Centers for Disease Control and Prevention (CDC) estimates that about 76 million cases of foodborne disease resulting in about 5000 deaths occur annually in the United States. As more of our food products—especially fruits and vegetables—are grown in countries with poor sanitation standards, outbreaks of foodborne disease from imported pathogens are expected to increase.

UNDER THE MICROSCOPE

***Trichinella spiralis,* larva.** These small worms, about 1 mm in length, produce larvae that become encysted in muscle (as shown in the photo) and cause the disease trichinellosis.

Q&A

The United States frequently imports cheese and wine from France. In return, the United States exports horses to France. What does this have to do with the disease of trichinellosis?

Look for the answer in the chapter.

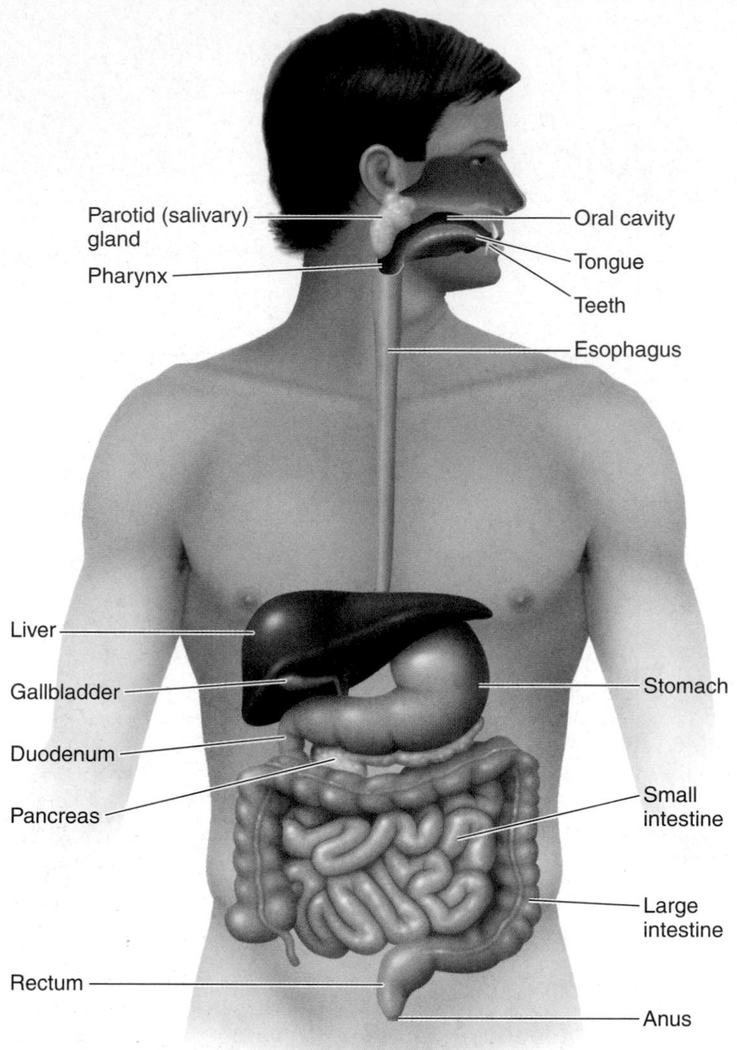

Figure 25.1 The human digestive system.

Q Where are microorganisms normally found in the digestive system?

Structure and Function of the Digestive System

LEARNING OBJECTIVE

25-1 Name the structures of the digestive system that contact food.

The **digestive system** is essentially a tubelike structure, the *gastrointestinal (GI) tract* or *alimentary canal*—mainly the mouth, pharynx (throat), esophagus (food tube leading to the stomach), stomach, and the small and large intestines. It also includes *accessory structures* such as the teeth and tongue. Certain other accessory structures such as the salivary glands, liver, gallbladder, and pancreas lie outside the GI tract and produce secretions that are conveyed by ducts into it (Figure 25.1).

The purpose of the digestive system is to digest foods—that is, to break them down into small molecules that can be taken up and used by body cells. In a process called *absorption,* these end-products of digestion pass from the small intestine into the blood or lymph for distribution to body cells. Then the food moves through the large intestine, where water, vitamins, and nutrients are absorbed from it. Over the course of an average life span, about 25 tons of food pass through the GI tract. The resulting undigested solids, called *feces,* are eliminated from the body through the anus. Intestinal gas, or *flatus,* is a mixture of nitrogen from swallowed air and microbially produced carbon dioxide, hydrogen, and methane. On average, we produce 0.5 to 2.0 liters of flatus every day.

CHECK YOUR UNDERSTANDING

✓ There have been instances in which a surgeon was using spark-producing instruments to remove intestinal polyps and a small explosion occurred. What ignited? **25-1**

Normal Microbiota of the Digestive System

LEARNING OBJECTIVE

25-2 Identify parts of the gastrointestinal tract that normally have microbiota.

Bacteria heavily populate most of the digestive system. In the mouth, each milliliter of saliva can contain millions of bacteria. The stomach and small intestine have relatively few microorganisms because of the hydrochloric acid produced by the stomach and the rapid movement of food through the small intestine. By contrast, the large intestine has enormous microbial populations, exceeding 100 billion bacteria per gram of feces. (Up to 40% of fecal mass is microbial cell material.) The population of the large intestine is composed mostly of anaerobes and facultative anaerobes. Most of these bacteria assist in the enzymatic breakdown of foods, especially many polysaccharides that would otherwise be indigestible. Some of them synthesize useful vitamins.

It is important to understand that food passing through the tubelike GI tract, although it is in contact with the body, remains outside. Unlike the body's exterior, such as the skin, the GI tract is adapted to absorbing nutrients passing through it. However, at the same time that nutrients are absorbed from the GI tract, harmful microbes ingested in food and water must be kept from invading the body. An important factor in this defense is the high acid content of the stomach, which eliminates many potentially harmful ingested microbes.

The small intestine also contains important antimicrobial defenses. Significant among these defenses are millions of specialized, granule-filled cells called *Paneth cells.* These are capable of phagocytizing bacteria, and they also produce antibacterial proteins called *defensins* (see antimicrobial peptides, page 578) and the antibacterial enzyme *lysozyme.*

CHECK YOUR UNDERSTANDING

✓ How are normal microbiota confined to the mouth and large intestine? 25-2

Bacterial Diseases of the Mouth

LEARNING OBJECTIVE

25-3 Describe the events that lead to dental caries and periodontal disease.

The mouth, which is the entrance to the digestive system, provides an environment that supports a large and varied microbial population.

Dental Caries (Tooth Decay)

The teeth are unlike any other exterior surface of the body. They are hard and do not shed surface cells (Figure 25.2). This allows the accumulation of masses of microorganisms and their products. These accumulations, called **dental plaque,** are a type of biofilm (see page 162 in Chapter 6) and are intimately involved in the formation of **dental caries,** or tooth decay.

Oral bacteria convert sucrose and other carbohydrates into lactic acid, which in turn attacks the tooth enamel. The microbial population on and around the teeth is very complex. Based on ribosomal identification methods (see the discussion of FISH on page 292 in Chapter 10), over 700 species of bacteria have been isolated from the oral cavity. Most of these cannot be cultivated by conventional methods. Probably the most important *cariogenic* (caries-causing) bacterium is *Streptococcus mutans,* a gram-positive coccus that is thought to be capable of metabolizing a wider range of carbohydrates than any other gram-positive organism. Some other species of streptococci are also cariogenic but play a lesser role in initiating caries.

The initiation of caries depends on the attachment of *S. mutans* or other streptococci to the tooth (Figure 25.3). These bacteria do not adhere to a clean tooth, but within minutes a freshly brushed tooth will become coated with a pellicle (thin film) of proteins from saliva. Within a couple of hours, cariogenic bacteria become established on this pellicle and begin to produce a gummy polysaccharide of glucose molecules called *dextran* (Figure 25.3b). In the production of dextran, the bacteria first hydrolyze sucrose into its component monosaccharides, fructose and glucose. The enzyme glucosyltransferase then assembles the glucose molecules into dextran. The residual fructose is the primary sugar fermented into lactic acid. Accumulations of bacteria and dextran adhering to the teeth make up dental plaque.

The bacterial population of plaque may harbor over 400 bacterial species but is predominantly streptococci and filamentous

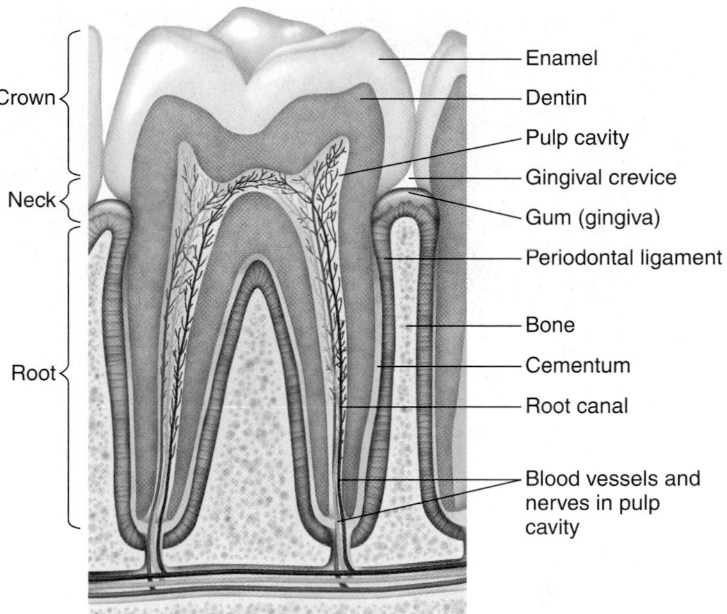

Figure 25.2 A healthy human tooth.

Q Why can a biofilm accumulate on teeth?

members of the genus *Actinomyces.* (Older, calcified deposits of plaque are called *dental calculus,* or *tartar.*) *S. mutans* especially favors crevices or other sites on the teeth protected from the shearing action of chewing or the flushing action of the liter or so of saliva produced in the mouth each day. On protected areas of the teeth, plaque accumulations can be several hundred cells thick. Because plaque is not very permeable to saliva, the lactic acid produced by bacteria is not diluted or neutralized, and it breaks down the enamel of the teeth to which the plaque adheres.

Although saliva contains nutrients that encourage the growth of bacteria, it also contains antimicrobial substances, such as lysozyme, that help protect exposed tooth surfaces. Some protection is also provided by *crevicular fluid,* a tissue exudate that flows into the gingival crevice (see Figure 25.2) and is closer in composition to serum than saliva. It protects teeth by virtue of its flushing action and its phagocytic cells and immunoglobulin content.

Localized acid production within deposits of dental plaque results in a gradual softening of the external *enamel.* Enamel low in fluoride is more susceptible to the effects of the acid. This is the reason for fluoridation of water and toothpastes, which has been a significant factor in the decline in tooth decay in the United States.

Figure 25.4 shows the stages of tooth decay. If the initial penetration of the enamel by caries remains untreated, bacteria can penetrate into the interior of the tooth. The composition of the bacterial population involved in spreading the decayed area from the enamel into the *dentin* is entirely different from that of the population initiating the decay. The dominant microorganisms are gram-positive rods and filamentous bacteria; *S. mutans* is present in small numbers only. Although once considered the cause of dental caries, *Lactobacillus* spp. actually play no role in

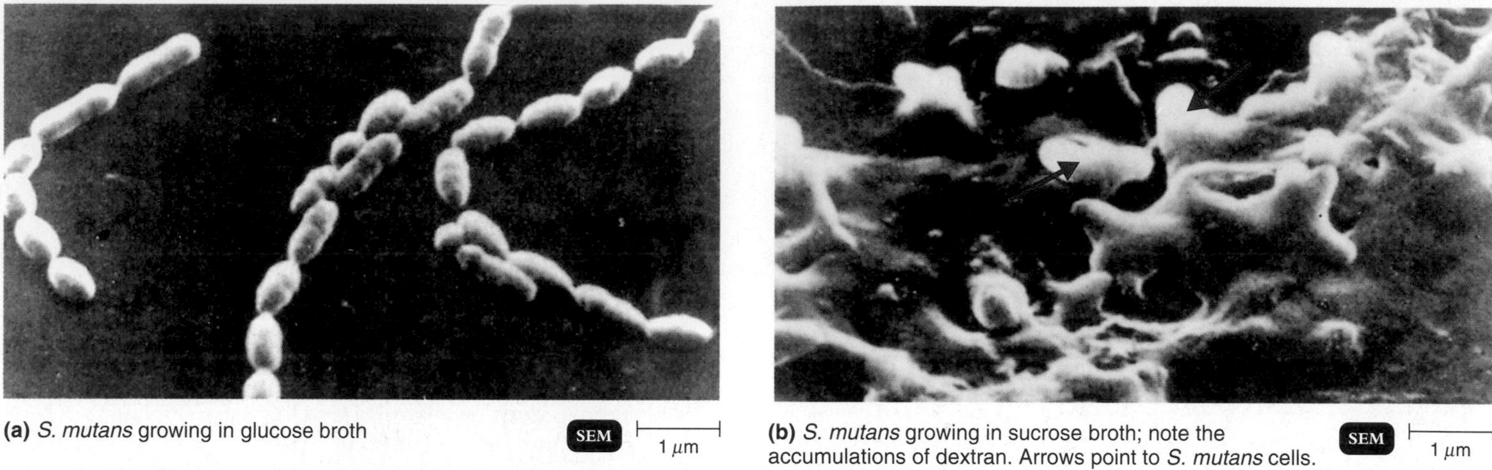

(a) *S. mutans* growing in glucose broth

(b) *S. mutans* growing in sucrose broth; note the accumulations of dextran. Arrows point to *S. mutans* cells.

Figure 25.3 The role of *Streptococcus mutans* and sucrose in dental caries.

Q What is dental plaque?

initiating the process. However, these very prolific lactic acid producers are important in advancing the front of the decay once it has become established.

The decayed area eventually advances to the *pulp* (see Figure 25.4), which connects with the tissues of the jaw and contains the blood supply and the nerve cells. Almost any member of the normal microbiota of the mouth can be isolated from the infected pulp and roots. Once this stage is reached, root canal therapy is required to remove the infected and dead tissue and to provide access for antimicrobial drugs that suppress renewed infection. If untreated, the infection may advance from the tooth to the soft tissues, producing dental abscesses caused by mixed bacterial populations that contain many anaerobes.

Although dental caries is probably one of the more common infectious diseases in humans today, it was scarce in the Western world until about the seventeenth century. In human remains from older times, only about 10% of the teeth contain caries. The introduction of table sugar, or sucrose, into the diet is highly correlated with our present level of caries in the Western world. Studies have shown that sucrose, a disaccharide composed of glucose and fructose, is much more cariogenic than either glucose or fructose individually (see Figure 25.3). People living on high-starch diets (starch is a polysaccharide of glucose) have a low incidence of tooth decay unless sucrose is also a significant part of their diet. The contribution of bacteria to tooth decay has been shown by experiments with germ-free animals. Such animals do not develop caries even when fed a sucrose-rich diet designed to encourage their formation.

Sucrose is pervasive in the modern Western diet. However, if sucrose is ingested only at regular mealtimes, the protective and

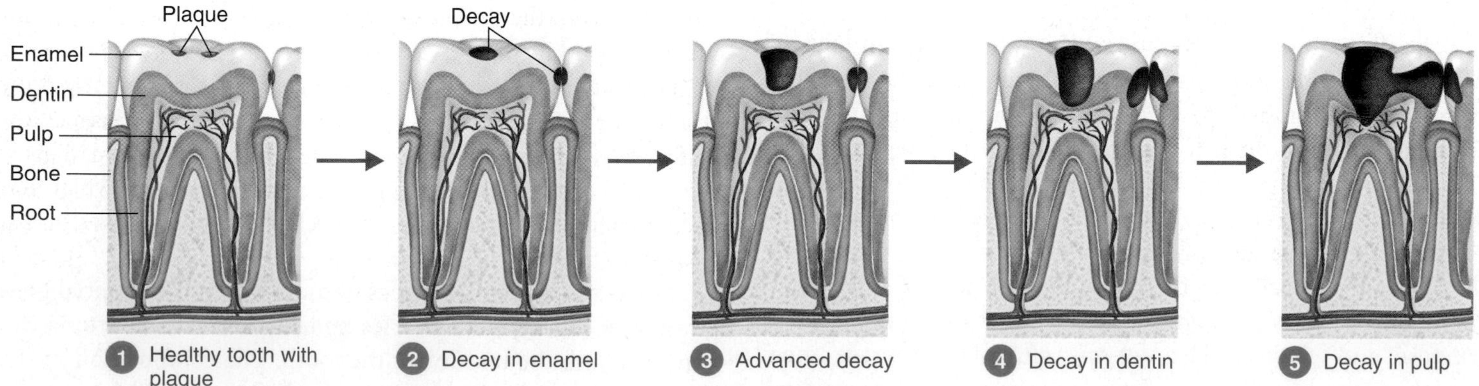

Figure 25.4 The stages of tooth decay. 1 A tooth with plaque accumulation in difficult-to-clean areas. 2 Decay begins as enamel is attacked by acids formed by bacteria. 3 Decay advances through the enamel. 4 Decay advances into the dentin. 5 Decay enters the pulp and may form abscesses in the tissues surrounding the root.

Q Why is the formation of plaque important in tooth decay?

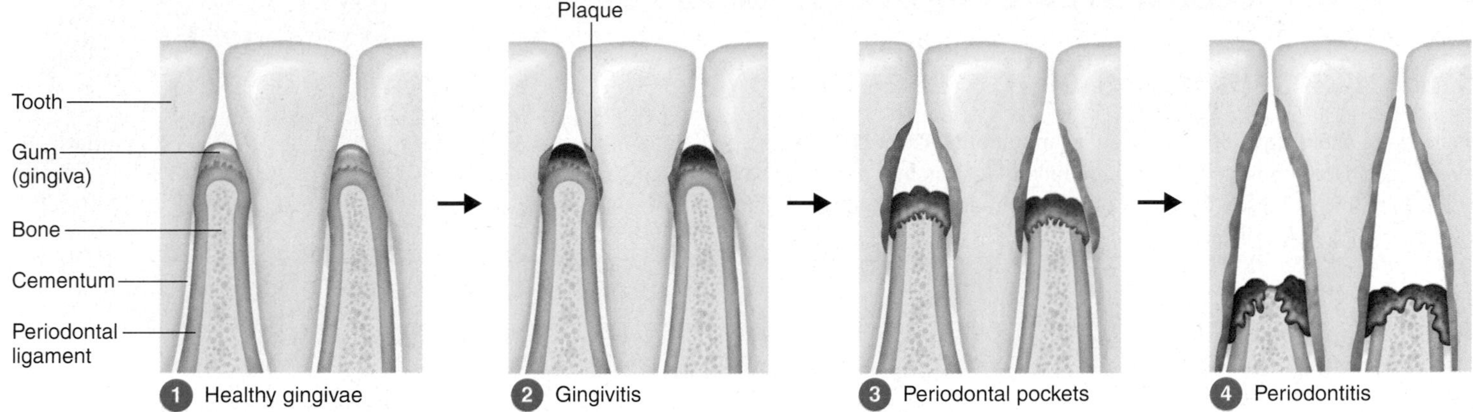

Figure 25.5 The stages of periodontal disease. 1 Teeth firmly anchored by healthy bone and gum tissue (gingiva). 2 Toxins in plaque irritate gums, causing gingivitis. 3 Periodontal pockets form as the tooth separates from the gingiva. 4 Gingivitis progresses to periodontitis. Toxins destroy the gingiva and bone that support the tooth and the cementum that protects the root.

Q What is the cause of "pink tooth brush"?

repair mechanisms of the body are usually not overwhelmed. It is the sucrose that is ingested between meals that is most damaging to teeth. Sugar alcohols, such as mannitol, sorbitol, and xylitol, are not cariogenic; xylitol even appears to inhibit carbohydrate metabolism in *S. mutans.* This is why these sugar alcohols are used to sweeten "sugarless" candies and chewing gum.

The best strategies for preventing dental caries are a minimal ingestion of sucrose; brushing, flossing, and professional cleaning to remove plaque; and the use of fluoride. Professional removal of plaque and tartar at regular intervals lessens the progression to periodontal disease.

Periodontal Disease

Even people who avoid tooth decay might, in later years, lose their teeth to **periodontal disease,** a term for a number of conditions characterized by inflammation and degeneration of structures that support the teeth (Figure 25.5). The roots of the tooth are protected by a covering of specialized connective tissue called *cementum.* As the gums recede with age or with overly aggressive brushing, the formation of caries on the cementum becomes more common.

Gingivitis

In many cases of periodontal disease, the infection is restricted to the gums, or *gingivae.* This resulting inflammation, called **gingivitis,** is characterized by bleeding of the gums while the teeth are being brushed (see Figure 25.5). This is a condition experienced by at least half of the adult population. It has been shown experimentally that gingivitis will appear in a few weeks if brushing is discontinued and plaque is allowed to accumulate. An assortment of streptococci, actinomycetes, and anaerobic gram-negative bacteria predominate in these infections.

Periodontitis

Gingivitis can progress to a chronic condition called **periodontitis.** This is an insidious condition that generally causes little discomfort. About 35% of adults suffer from periodontitis, which is increasing in incidence as more people retain their teeth into old age. The gums are inflamed and bleed easily. Sometimes pus forms in pockets surrounding the teeth (*periodontal pockets;* see Figure 25.5). As the infection continues, it progresses toward the root tips. The bone and tissue that support the teeth are destroyed, leading eventually to loosening and loss of the teeth. Numerous bacteria of many different types, primarily *Porphyromonas* (pôr′fī-rō-mō-nas) species, are found in these infections; the damage to tissue is done by an inflammatory response to the presence of these bacteria. Periodontitis can be treated surgically by eliminating the periodontal pockets.

Acute necrotizing ulcerative gingivitis, also termed **Vincent's disease** or **trench mouth,** is one of the more common serious mouth infections. The disease causes enough pain to make normal chewing difficult. Foul breath (halitosis) also accompanies the infection. Among the bacteria usually associated with this condition is *Prevotella intermedia* (prev′ō-tel-la in′tėr-mē-dē-ä), averaging up to 24% of the isolates. Because these pathogens are usually anaerobic, treatment with oxidizing agents, debridement, and the administration of metronidazole may be temporarily effective. Bacterial diseases of the mouth are summarized in Diseases in Focus 25.1.

CHECK YOUR UNDERSTANDING

✓ Why are "sugarless" candies and gum, which actually contain sugar alcohols, not considered cariogenic (causing caries)? **25-3**

DISEASES IN FOCUS 25.1

Bacterial Diseases of the Mouth

Most adults show signs of gum disease, and about 14% of U.S. adults aged 45 to 54 have a severe case. Use the table below to identify infections that could cause persistent sore, swollen, red, or bleeding gums, as well as tooth pain or sensitivity and bad breath. For the solution, go to www.microbiologyplace.com.

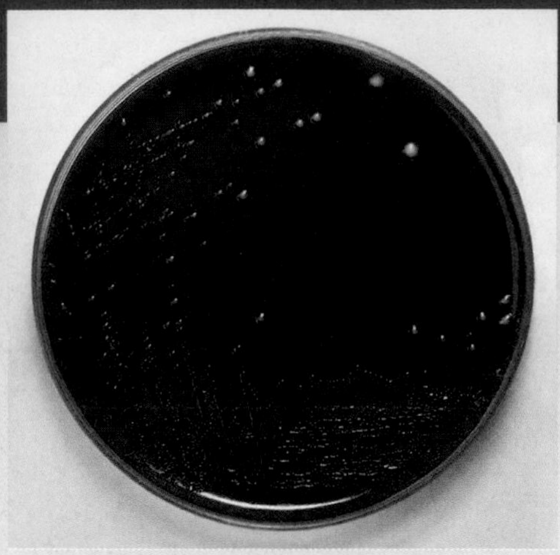

This gram-negative rod accounts for nearly one-quarter of cases.

Disease	Pathogen	Symptoms	Treatment	Prevention
Dental caries	Primarily *Streptococcus mutans*	Discolored or hole in tooth enamel	Remove decayed area	Brushing, flossing, reduction of dietary sucrose
Periodontal disease	Various, primarily *Porphyromonas* spp.	Bleeding gums, pus pockets	Remove damaged area; antibiotics	Plaque removal
Acute necrotizing ulcerative gingivitis	*Prevotella intermedia*	Pain chewing, halitosis	Remove damaged area; metronidazole	Brushing, flossing

Bacterial Diseases of the Lower Digestive System

LEARNING OBJECTIVE

25-4 List the causative agents, suspect foods, signs and symptoms, and treatments for staphylococcal food poisoning, shigellosis, salmonellosis, typhoid fever, cholera, gastroenteritis, and peptic ulcer disease.

Diseases of the digestive system are essentially of two types: infections and intoxications.

An **infection** occurs when a pathogen enters the GI tract and multiplies. Microorganisms can penetrate into the intestinal mucosa and grow there, or they can pass through to other systemic organs. **M cells** (for microfold) are intended to translocate antigens and microorganisms to the other side of the epithelium where they can contact lymphoid tissues (Peyer's patches) to initiate an immune response (see page 486 and Figure 17.9). Infections of the GI tract are characterized by a delay in the appearance of gastrointestinal disturbance while the pathogen increases in numbers or affects invaded tissue. There is also usually a fever, one of the body's general responses to an infective organism.

Some pathogens cause disease by forming toxins that affect the GI tract. An **intoxication** is caused by the ingestion of such a preformed toxin. Most intoxications, such as that caused by *Staphylococcus aureus,* are characterized by a very sudden appearance (usually in only a few hours) of symptoms of a GI disturbance. Fever is less often one of the symptoms.

Both infections and intoxications often cause *diarrhea,* which most of us have experienced. Severe diarrhea accompanied by blood or mucus is called **dysentery.** Both types of digestive system diseases are also frequently accompanied by *abdominal cramps, nausea,* and *vomiting.* Diarrhea and vomiting are both defensive mechanisms designed to rid the body of harmful material.

The general term **gastroenteritis** is applied to diseases causing inflammation of the stomach and intestinal mucosa. Botulism is a special case of intoxication because the ingestion of the preformed toxin affects the nervous system rather than the GI tract (see Chapter 22, page 616).

In developing countries, diarrhea is a major factor in infant mortality. Approximately one child in every four dies of it before the age of 5. It is estimated that mortality from childhood diarrhea could be halved by *oral rehydration therapy* (replacement of lost fluids and electrolytes). This is usually a solution of sodium chloride, potassium chloride, glucose, and sodium bicarbonate to replace lost fluids and electrolytes. These solutions are sold in the infant supply department of many stores. Public health departments often determine the incidence of diarrhea in the population by receiving weekly reports on the sales of oral rehydration preparations.

Diseases of the digestive system are often related to food ingestion.

Staphylococcal Food Poisoning (Staphylococcal Enterotoxicosis)

A leading cause of gastroenteritis is **staphylococcal food poisoning**, an intoxication caused by ingesting an enterotoxin produced by *S. aureus*. Staphylococci are comparatively resistant to environmental stresses, as discussed on page 318. They also have a fairly high resistance to heat; vegetative cells can tolerate 60°C for half an hour. Their resistance to drying and radiation helps them survive on skin surfaces. Resistance to high osmotic pressures helps them grow in foods, such as cured ham, in which the high osmotic pressure of salts inhibits the growth of competitors.

S. aureus is often an inhabitant of the nasal passages, from which it contaminates the hands. It is also a frequent cause of skin lesions on the hands. From these sources, it can readily enter food. If the microbes are allowed to incubate in the food, a situation called **temperature abuse,** they reproduce and release enterotoxin into the food. These events, which lead to outbreaks of staphylococcal intoxication, are illustrated in Figure 25.6.

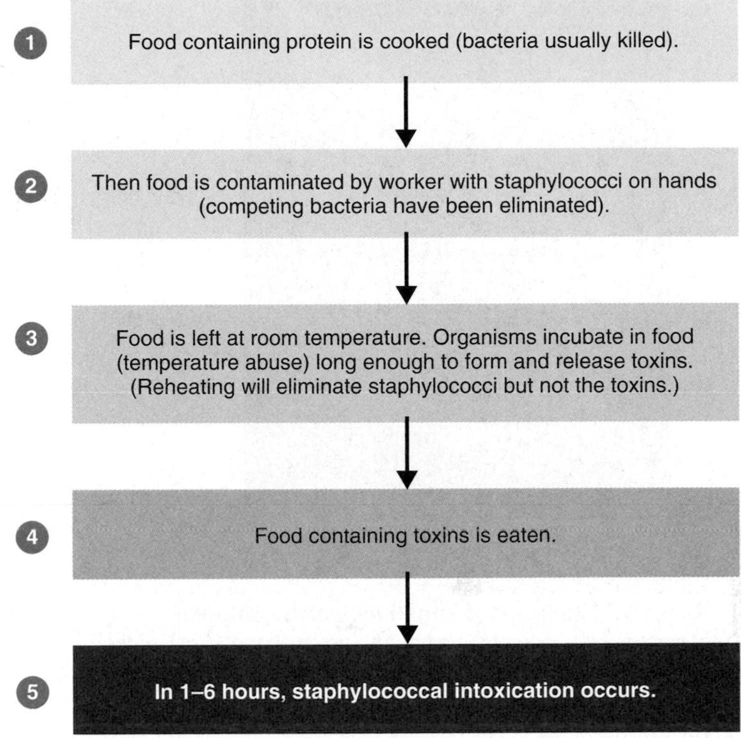

Figure 25.6 The sequence of events in a typical outbreak of staphylococcal food poisoning.

Q How does this differ from foodborne illness caused by a virus?

S. aureus produces several toxins that damage tissues or increase the microorganism's virulence. The production of the toxin of serological type A (which is responsible for most cases) is often correlated with the production of an enzyme that coagulates blood plasma. Such bacteria are described as *coagulase-positive.* No direct pathogenic effect can be attributed to the enzyme, but it is useful in the tentative identification of types that are likely to be virulent.

Generally, a population of about 1 million bacteria per gram of food will produce enough enterotoxin to cause illness. The growth of the microbe is facilitated if the competing microorganisms in the food have been eliminated—by cooking, for example. It is also more likely to grow if competing bacteria are inhibited by a higher-than-normal osmotic pressure or by a relatively low moisture level. *S. aureus* tends to outgrow most competing bacteria under these conditions.

Custards, cream pies, and ham are examples of high-risk foods. Competing microbes are minimized in custards by the high osmotic pressure of sugar and by cooking. In ham they are inhibited by curing agents, such as salts and preservatives. Poultry products can also harbor staphylococci if they are handled and allowed to stand at room temperatures. Because staphylococci do not compete well with the large number of microorganisms hamburger contains, it is rarely a factor in this type of food poisoning. Any foods prepared in advance and not kept chilled are a potential source of staphylococcal food poisoning. Because food contamination by human handlers cannot be avoided completely, the most reliable method of preventing staphylococcal food poisoning is adequate refrigeration during storage to prevent toxin formation.

The toxin itself is heat stable and can survive up to 30 minutes of boiling. Therefore, once the toxin is formed, it is not destroyed when the food is reheated, although the bacteria will be killed.

The toxin quickly triggers the brain's vomiting reflex center; abdominal cramps and usually diarrhea then ensue. This reaction is essentially immunological in character; the staphylococcal enterotoxin is a model example of a superantigen (see page 436). Recovery is usually complete within 24 hours.

The mortality rate of staphylococcal food poisoning is almost zero among otherwise healthy people, but it can be significant in weakened individuals, such as residents of nursing homes. No reliable immunity results from recovery. However, there is a great deal of variation in individual susceptibility to the toxin, and it is suspected that immunity acquired from a previous exposure might account for some of this variation.

The diagnosis of staphylococcal food poisoning is usually based on the symptoms, particularly the short incubation time

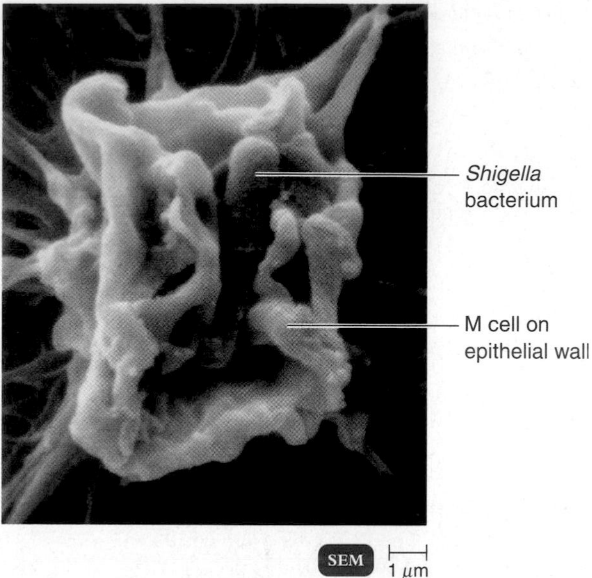

Figure 25.7 Invasion of intestinal wall by *Shigella* bacterium. Note how the membrane ruffles on the M cell of the epithelial wall of the intestine surround the bacterial cell. Invasion by *Salmonella* bacteria is very similar. (Also, see Figure 15.2)

Q If a bacterium succeeds in leaving the interior of the intestine, what elements of the immune system would attempt to deal with it?

characteristic of intoxication. If the food has not been reheated so that the bacteria are not killed, the pathogen can be recovered and grown. *S. aureus* isolates can be tested by *phage typing,* a method used in tracing the source of the contamination (see Figure 10.13, page 290). These bacteria grow well in 7.5% sodium chloride, so this concentration is often used in media for their selective isolation. Pathogenic staphylococci usually ferment mannitol, produce hemolysins and coagulase, and form golden-yellow colonies. They cause no obvious spoilage when growing in foods. Detecting the toxin in food samples has always been a problem; there may be only 1 to 2 nanograms in 100 g of food. Reliable serological methods have become commercially available only recently.

Shigellosis (Bacillary Dysentery)

Bacterial infections, such as salmonellosis and shigellosis, usually have longer incubation periods (12 hours to 2 weeks) than bacterial intoxications, reflecting the time needed for the microorganism to grow in the host. Bacterial infections are often characterized by some fever, indicating the host's response to the infection.

Shigellosis, also known as **bacillary dysentery** to differentiate it from amoebic dysentery (page 731), is a severe form of diarrhea caused by a group of facultatively anaerobic gram-negative rods of the genus *Shigella.* The genus was named for the Japanese microbiologist Kiyoshi Shiga. The bacteria do not have any natural reservoir in animals and spread only from person to person. Outbreaks are most often seen in families, day-care facilities, and similar settings.

There are four species of pathogenic *Shigella: S. sonnei* (sōn′ne-ē), *S. dysenteriae* (dis-en-te′rē-ī), *S. flexneri* (fleks′nėr-ē), and *S. boydii* (boi′dē-ē). These bacteria are residents only of the intestinal tract of humans, apes, and monkeys. They are closely related to the pathogenic *E. coli.*

The most common species in the United States is *S. sonnei;* it causes a relatively mild dysentery. Many cases of so-called traveler's diarrhea might be mild forms of shigellosis. At the other extreme, infection with *S. dysenteriae* often results in a severe dysentery and prostration. The toxin responsible is unusually virulent and is known as the **Shiga toxin** (see enterohemorrhagic *E. coli,* page 717). *S. dysenteriae* is the least common species in the United States.

The infective dose required to cause disease is small; the bacteria are not much affected by stomach acidity. They proliferate to immense numbers in the small intestine, but the primary site of disease is the large intestine. There, the bacteria attach to certain epithelial cells. M cells, membranous cellular ruffles surrounding the cell, take the bacterium into the cell (**Figure 25.7**). The bacteria multiply in the cell and soon spread to neighboring cells, producing Shiga toxin that destroys tissue (**Figure 25.8**). Dysentery is the result of damage to the intestinal wall.

Shigellosis dysentery can cause as many as 20 bowel movements in one day. Additional symptoms of infection are abdominal cramps and fever. *Shigella* bacteria rarely invade the bloodstream. Macrophages not only fail to kill *Shigella* bacteria that they phagocytize, but also are killed by them. Diagnosis is usually based on recovery of the microbes from rectal swabs.

The CDC estimates that about 450,000 cases of shigellosis occur annually; most are caused by *S. sonnei* and chiefly affect children under the age of 5. *S. dysenteriae* has a significant mortality rate, however, and the death rate in tropical areas where it is prevalent can be as high as 20%. Some immunity seems to result from recovery, but a satisfactory vaccine has not yet been developed.

In severe cases of shigellosis, antibiotic therapy and oral rehydration are indicated. At present, fluoroquinolones are the antibiotics of choice.

Salmonellosis (*Salmonella* Gastroenteritis)

The *Salmonella* bacteria (named for their discoverer, Daniel Salmon) are gram-negative, facultatively anaerobic, non–endospore-forming rods. Their normal habitat is the intestinal tracts of humans and many animals. All salmonellae are considered pathogenic to some degree, causing **salmonellosis,** or ***Salmonella* gastroenteritis**. Pathogenically, salmonellae are separated into *typhoidal salmonellae* (see typhoid fever, page 714)

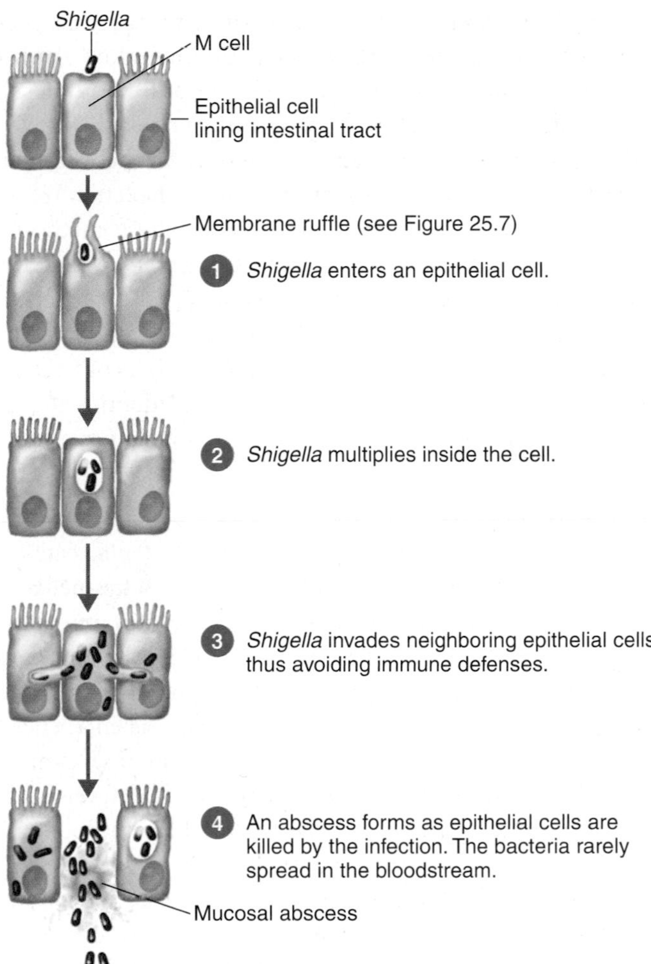

Figure 25.8 Shigellosis. This sequence shows the sequence of infection of the intestinal wall. The bacterium attaches to an M cell (see Figure 25.7) of the epithelial wall located over a Peyer's patch (see page 486, and Figure 17.9). This is a region adapted to facilitate transfer of antigens across the intestinal mucosa.

Q What species of *Shigella* is the most dangerous?

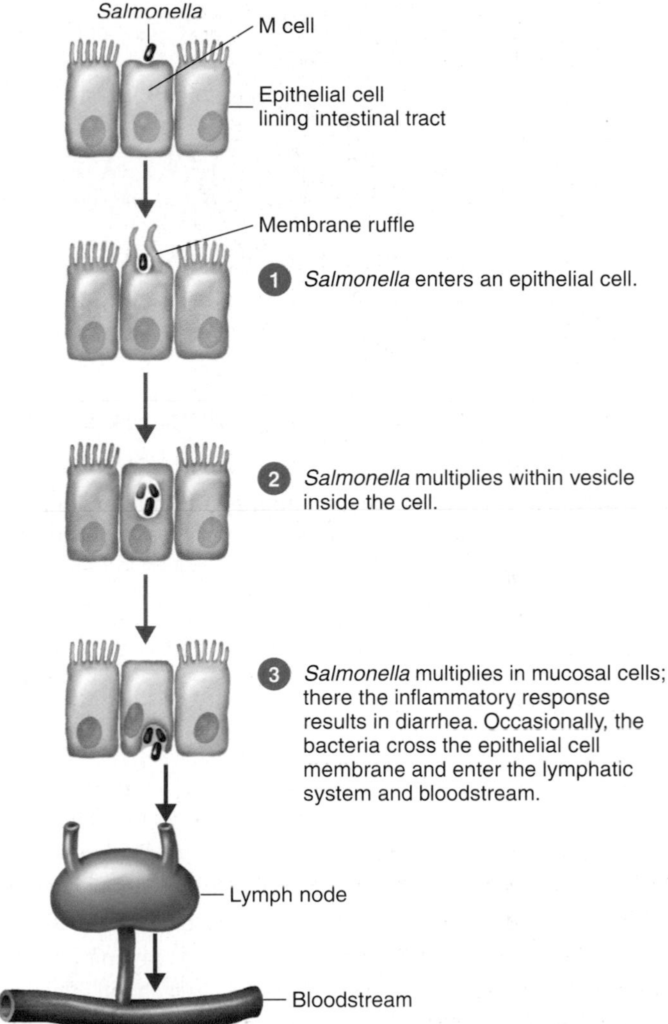

Figure 25.9 Salmonellosis. This sequence shows the sequence of infection of the intestinal wall. Compare with Figure 25.8 showing infection with *Shigella.* Note that invasion of the bloodstream, which happens infrequently, can result in septic shock.

Q Why does salmonellosis have a longer incubation period than a bacterial intoxication?

and the *nontyphoidal salmonellae,* which cause the milder disease of salmonellosis.

The nomenclature of the *Salmonella* microbes differs from the norm. Rather than recognized species, there are more than 2000 serotypes (or serovars), only about 50 of which are isolated with any frequency in the United States. (For a discussion of the nomenclature of the salmonellae, see page 310.) To summarize, many taxonomists consider them to belong to only two species, primarily *Salmonella enterica.* Therefore, you might encounter nomenclature such as *S. enterica* serotype Typhimurium, instead of the conventional name *S. typhimurium.*

The salmonellae first invade the intestinal mucosa and multiply there. Occasionally they manage to pass through the intestinal mucosa at M cells to enter the lymphatic and cardiovascular systems, and from there they may spread to eventually affect many organs (**Figure 25.9**). They replicate readily within macrophages. Salmonellosis has an incubation time of about 12 to 36 hours. There is usually a moderate fever accompanied by nausea, abdominal pain and cramps, and diarrhea. As many as 1 billion salmonellae per gram can be found in an infected person's feces during the acute phase of the illness.

The mortality rate is overall very low, probably less than 1%. However, the death rate is higher in infants and among the very old; death is usually from septic shock. The severity and incubation time can depend on the number of *Salmonella* ingested. Normally, recovery will be complete in a few days, but many patients will continue to shed the organisms in their feces for up to 6 months. Antibiotic therapy is not useful in treating salmonellosis or, indeed, many diarrheal diseases; treatment consists of oral rehydration therapy.

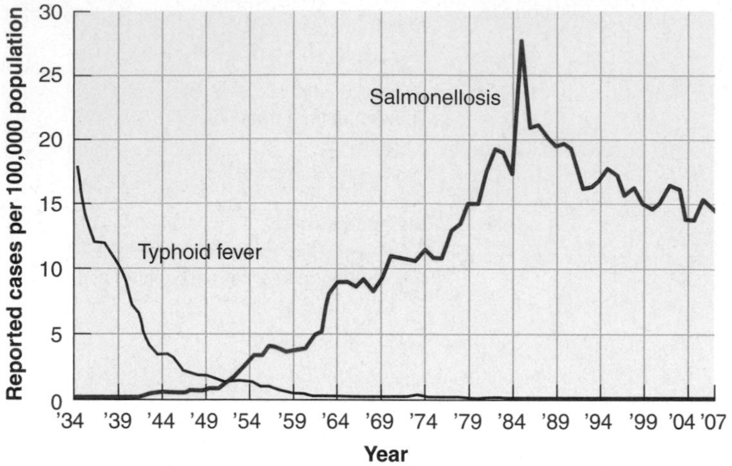

Figure 25.10 The incidence of salmonellosis and typhoid fever. An important factor in comparing the two diseases is that typhoid transmission is almost entirely human, and salmonellosis is transmitted primarily between animal products and humans.

Source: CDC. *MMWR* 56(52), January 4, 2008.

Q Can you suggest reasons for the change in prevalence of those two diseases?

Salmonellosis is probably greatly underreported. There are an estimated 1.4 million cases and 400 deaths annually (Figure 25.10). Meat products are particularly susceptible to contamination by *Salmonella.* The sources of the bacteria are the intestinal tract of many animals. Pet reptiles, such as turtles and iguanas, are also a source; their carriage rate is as high as 90%. In fact, the sale of small turtles ($<$10 cm) as pets is now prohibited by the FDA because of the risk that children may put them in the mouth. *S. enteritidis* and *S. typhimurium* are especially well adapted to commercial chicken production. Hens are highly susceptible to infection, and the bacteria contaminate the eggs. The bacteria have developed the ability to survive in the albumin, which contains natural preservatives such as *lysozyme* (see page 453) and *lactoferrin* (which binds iron the bacteria require). Estimates are that 1 in 20,000 eggs in this country is contaminated by *Salmonella.* Health authorities caution the public to eat only well-cooked eggs. An often unsuspected factor is the presence of inadequately cooked or raw eggs in foods such as hollandaise sauce, cookie batter, and Caesar salad. Surprisingly frequent sources of foodborne illness from ingestion of *Salmonella* and *E. coli* O157:H7 have been raw alfalfa sprouts and tomatoes (see the box on page 715).

Prevention also depends on good sanitation practices to deter contamination and on proper refrigeration to prevent increases in bacterial numbers. Recently, it has been possible to offer eggs in which any *Salmonella* have been killed by a special hot water pasteurization procedure that does not cook the eggs. However, these eggs are more expensive. The microbes are generally destroyed by normal cooking. Chicken, for example, should be cooked at temperatures of 76–82°C and ground beef at 71°C. However, contaminated food can contaminate a surface, such as a cutting board. Although the food first prepared on the board might later be cooked and its bacteria killed, another food subsequently prepared on the board might not be cooked.

Diagnosis usually depends on isolating the pathogen from the patient's stool or from leftover food. Isolation requires specialized selective and differential media; these methods are relatively slow. Also, the small numbers of *Salmonella* generally found in foods present a special problem in detection. The infective dose may be as small as 1000 bacteria. Currently, PCR-based tests are the best for detecting small numbers of *Salmonella* in foods. They require about 5 hours and identify the most common clinical serotypes.

Typhoid Fever

The most virulent serotype of *Salmonella, S. typhi,* causes the bacterial disease **typhoid fever.** Unlike the salmonellae that cause salmonellosis, this pathogen is not found in animals; it is spread only in the feces of other humans. Before the days of proper sewage disposal, water treatment, and food sanitation, typhoid was an extremely common disease. Its incidence has been declining in the United States, whereas that of salmonellosis has been increasing (see Figure 25.10). Typhoid fever is still a frequent cause of death in parts of the world with poor sanitation; an estimated half-million annually.

Instead of being destroyed by phagocytic cells, *S. typhi* multiply within them and are disseminated into multiple organs, especially the spleen and liver. Eventually, the phagocytic cells lyse and release *S. typhi* into the bloodstream. The time required for this explains why the incubation period of typhoid fever (2 or 3 weeks) is much longer than for salmonellosis (12 to 36 hours). The patient with typhoid fever suffers from a high fever of about 40°C and continual headache. Diarrhea appears only during the second or third week, and the fever then tends to decline. In severe cases, which can be fatal, ulceration and perforation of the intestinal wall can occur. Before antibiotic therapy was available, a mortality rate of 20% was common; with the treatments available today, it is less than 1%.

Substantial numbers of recovered patients, about 1–3%, become *chronic carriers.* They harbor the pathogen in the gallbladder and continue to shed bacteria for several months. A number of such carriers continue to shed the organism indefinitely. The classic example of a typhoid carrier was Typhoid Mary. Her name was Mary Mallon; she worked as a cook in New York state in the early part of the twentieth century and was responsible for several outbreaks of typhoid and three deaths. She became well known through the attempts of the state to restrain her from working at her chosen trade.

In recent years there have been about 350 to 400 annual cases of typhoid fever in the United States, of which 70% were acquired during foreign travel. Normally, there are fewer than three deaths each year.

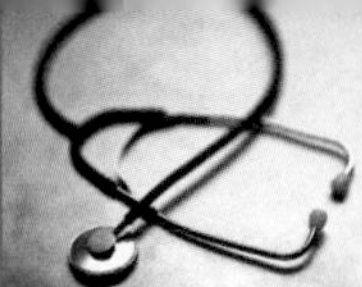

CLINICAL FOCUS From the *Morbidity and Mortality Weekly Report*

A Foodborne Infection

As you read through this problem, you will see questions that epidemiologists ask themselves as they solve a clinical problem. Try to answer each question as an epidemiologist.

1. On June 29, a 36-year-old woman in Ohio was hospitalized with a 3-day history of nausea, vomiting, and diarrhea. She had a temperature of 39.5°C, and she was dehydrated.
   What sample is needed from the patient to determine the cause of her signs and symptoms?
2. Stool culture grew gram-negative, non–lactose-fermenting bacteria.
   Can you identify the bacteria? (See the photograph.)
3. This patient was one of 459 culture-confirmed cases in a 21-state outbreak of salmonellosis.
   What information would you try to obtain from these patients?
4. No single restaurant or restaurant chain was associated with the outbreak.
   How will you determine the source of the infection?
5. Epidemiologists conducted a case-control study to compare 53 case-patients with 53 healthy controls from the same geographic locations. All 106 people were asked to complete a questionnaire about foods eaten. The data collected are shown below.

   Relative risk (RR) is a measure of the probability (risk) that an event will result in disease. RR is calculated in a 2 × 2 table (shown below). The relative risk must be calculated for each source of exposure. The RR for the chicken salad is given.

Salmonella form black colonies on SS agar because of their production of H_2S, which make a black precipitate with iron in the agar.

Exposure	Exposed		Not exposed		Relative risk (RR)
	(a) Ill	*(b)* Not ill	*(c)* Ill	*(d)* Not ill	
Chicken salad	47	40	6	13	**1.71**
Coleslaw	32	20	21	33	
Fruit salad	34	30	19	23	
Potato salad	42	39	11	14	
Tomato salad	47	24	6	29	

Relative Risk Calculation using a 2 × 2 table

	Ill	Not Ill	Relative Risk
Ate ________	*(a)*	*(b)*	$(e) = \frac{a}{a + b}$
Did not eat ________	*(c)*	*(d)*	$(f) = \frac{c}{c + d}$

Relative risk = $\frac{e}{f}$ = ________ = Times more likely to become ill by going to this place

Complete the remaining calculations to determine the probable source of the infection.

6. There is a strong association between illness and consumption of (Roma) tomatoes. The implicated tomatoes had been grown in Florida and diced and packaged at a firm in Kentucky.
   What would you do now?
7. Environmental samples from the farm, drainage ditch water, and animal feces yielded a variety of *Salmonella* strains.
   What factors might contribute to tomatoes as a vehicle of transmission?

In the eastern United States, tomatoes are grown in natural habitats for many known *Salmonella* reservoirs, including birds, amphibians, and reptiles. *Salmonella* can enter tomato plants through roots or flowers and can enter fruit through small cracks in the skin, the stem scar, or the plant itself. Contamination might occur during multiple steps, from the tomato seed nursery to the final kitchen. Eradication of *Salmonella* from the interior of the tomato is difficult without cooking, even if treated with highly concentrated chlorine solution.

Source: Adapted from *MMWR* 56(35):909–911, September 7, 2007.

When the antibiotic chloramphenicol was introduced in 1948, typhoid became a treatable disease. Although mostly replaced by safer (but more expensive) antibiotics, it is still used in endemic areas in the world—but might require 250 capsules during a course of treatment. The most effective antityphoidal drugs are quinolones or third-generation cephalosporins. Treatment of the carrier state might require weeks of antibiotic therapy. Antibiotic resistance is a frequent problem.

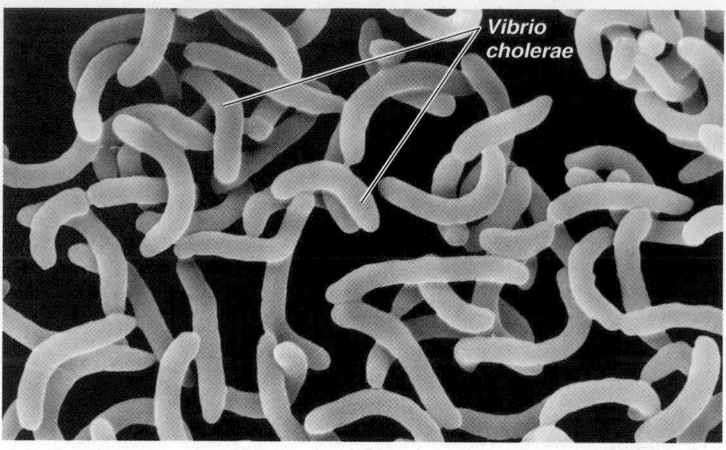

Figure 25.11 ***Vibrio cholerae*, the cause of cholera.** Notice the slightly curved morphology.

Q What are the effects of the sudden loss of fluid and electrolytes during infection with *V. cholerae*?

Recovery from typhoid confers lifelong immunity. Vaccines are seldom used in developed countries except for high-risk laboratory or military personnel. The declining effectiveness of antibiotics has renewed interest in vaccination for less developed parts of the world. The vaccine that has long been in use is a killed-organism type, which must be injected and has high rates of side effects. Newer-generation vaccines have become available that are quite safe and can be used in persons 2 years of age or older. One, a subunit vaccine that requires a single injection, confers good protection for at least 3 years. Another, a live attenuated vaccine that can be taken orally in three or four doses, protects well for as long as 7 years.

CHECK YOUR UNDERSTANDING

✓ Salmonellosis and typhoid fever are caused by closely related organisms. Why was typhoid fever almost entirely eliminated in developed countries by modern sewage treatment whereas salmonellosis has not been? **25-4**

Cholera

The causative agent of **cholera,** one of the most serious gastrointestinal diseases, is *Vibrio cholerae,* a slightly curved, gram-negative rod with a single polar flagellum (**Figure 25.11**). Cholera bacilli grow in the small intestine and produce an exotoxin, *cholera toxin* (see Chapter 15, page 437), that causes host cells to secrete water and electrolytes, especially potassium. The result is watery stools containing masses of intestinal mucus and epithelial cells—called "rice water stools" from their appearance. As much as 12 to 20 liters (3 to 5 gallons) of fluids can be lost in a day, and the sudden loss of these fluids and electrolytes causes shock, collapse, and often death. The blood, lacking fluids, may become so viscous that vital organs are unable to function properly. Violent vomiting generally also occurs. The microbes are not invasive, and a fever is usually not present. The severity of cholera varies greatly, and the number of subclinical cases might be several times the number reported. Untreated cases of cholera may have a mortality rate of 50%, although with proper supportive care it is usually less than 1% today. The diagnosis is based upon symptoms and culturing of *V. cholerae* from feces.

Cholera bacteria, and other members of the genus *Vibrio* in general, are strongly associated with brackish (salty) waters characteristic of estuaries, although they are also readily spread in contaminated fresh water. They form biofilms and colonize copepods (tiny crustaceans), algae, and other aquatic plants and plankton, which aids their survival. It has even been reported that because of this growth habit, straining contaminated water through folded layers of finely-woven cloth (such as saris worn by Indian women) often removes these attached bacteria and makes the water safe to drink. Under unfavorable conditions *V. cholerae* may become dormant; the cell shrinks into a nonculturable, spherical state. A favorable change in the environment causes them to revert rapidly to the culturable form. Both forms are infectious.

Although they survive well in their aquatic environment, cholera bacteria are exceptionally sensitive to stomach acids. Persons with impaired stomach acid secretion or who are taking antacids are at higher risk of infection. Normal individuals may require infective doses on the order of 100 million bacteria to cause severe cholera. Recovery from cholera results in an effective immunity, but only to bacterial strains of the same antigenic characteristics. The serogroup O:1 (see the footnote in Chapter 11, page 310), which caused a pandemic in the 1880s, is known as the classical strain. A later pandemic was caused by a biotype of O:1 named *El Tor* or *eltor* (for the El Tor quarantine camp for pilgrims to Mecca, where it was first isolated). Until the 1990s it was thought that only *V. cholerae* O:1 caused epidemic cholera, but a widespread epidemic in India and Bangladesh by a new serogroup, O:139, changed this view. There are also nonepidemic strains of *V. cholerae,* non-O:1/O:139, that are only infrequently associated with large-scale outbreaks of cholera. They occasionally cause wound infections or sepsis, especially in people with liver disease or who are immunosuppressed.

In the United States there have been occasional cases of cholera caused by the O:1 serogroup. These have all occurred in the Gulf Coast area, and the pathogen may be endemic in these coastal waters. Outbreaks of cholera in this country are limited by high standards of sanitation. This represents the primary means of control and is important because stools may contain 100 million *V. cholerae* per gram. Available oral vaccines provide immunity of relatively short duration and only moderate effectiveness.

Treatment often includes the use of antibiotics such as doxycycline, but the most effective therapy is intravenous replacement of the lost fluids and electrolytes. As much as 10% of the patient's body weight within a few hours may be required. Rehydration therapy is so effective that in Bangladesh, for example, where cholera is common, deaths are considered "unusual."

Noncholera Vibrios

At least 11 species of *Vibrio,* in addition to *V. cholerae,* can cause human illness. Most are adapted to life in salty coastal waters. *Vibrio parahaemolyticus* (pa-rä-hē-mō-li′ti-kus) is found in salt water estuaries in many parts of the world. It is morphologically similar to *V. cholerae* and the most common cause of gastroenteritis by *Vibrio* spp. in humans. The bacterium is present in coastal waters of the continental United States and Hawaii. Raw oysters and crustaceans, such as shrimp and crabs, have been associated with several outbreaks of *gastroenteritis* in the United States in recent years.

Signs and symptoms, which resemble those of cholera, include abdominal pain, vomiting, a burning sensation in the stomach, and watery stools. Treatment by antibiotics and rehydration is usually effective. The incubation time is normally less than 24 hours. Recovery usually follows in a few days.

Because *V. parahaemolyticus* has a requirement for sodium and a high osmotic pressure, isolation media containing 2–4% sodium chloride are used in diagnosing the disease.

Another important pathogenic *Vibrio* is *Vibrio vulnificus,* which is also found in estuaries. It is halophilic and requires 1% sodium chloride in the media used to isolate it. It causes gastroenteritis in only a minority of infections; rather, ingestion can lead to a life-threatening invasion of the bloodstream. People with compromised immune systems are at a higher risk. Anyone suffering from liver disease is also at high risk of sepsis, which in these cases is fatal about 50% of the time. *V. vulnificus* frequently causes very dangerous infections of minor skin lesions incurred in coastal sea waters. Rapidly spreading tissue destruction from these infections may require limb amputation; and, if sepsis occurs, the fatality rate is about 25%. Because these infections are life threatening, they require early antibiotic therapy for successful treatment.

Escherichia coli Gastroenteritis

One of the most prolific microorganisms in the human intestinal tract is *E. coli.* Because it is so common and so easily cultivated, microbiologists often regard it as something of a laboratory pet. *E. coli* are normally harmless, but certain strains can be pathogenic. All pathogenic strains of *E. coli* have specialized fimbriae that allow them to bind to certain intestinal epithelial cells. They also produce toxins that cause gastrointestinal disturbances, collectively termed *E. coli gastroenteritis.*

Traveler's Diarrhea

It has long been observed that travel tends to broaden the mind and to loosen the bowels, the latter being an affliction with the common name of **traveler's diarrhea.** The probable cause of most cases is one of several strains of *E. coli.* **Enterotoxigenic *E. coli* (ETEC)** is not invasive but produces an enterotoxin that causes a watery diarrhea that resembles a mild case of cholera. Another strain that is increasingly recognized as a cause of traveler's diarrhea, possibly second only to ETEC, is **enteroaggregative *E. coli* (EAEC)**. This group of coliforms is named for a growth habit in which the bacteria adhere to each other in a "stacked-brick" configuration. **Enteroinvasive *E. coli* (EIEC)** invade the intestinal wall, resulting in inflammation, fever, and sometimes a *Shigella*-like dysentery.

Traveler's diarrhea can also be caused by gastrointestinal pathogens such as *Salmonella* and *Campylobacter*—as well as by various unidentified bacterial pathogens, viruses, and protozoan parasites. In fact, in most cases the causative agent is never identified. Traveler's diarrhea is usually self-limiting, and chemotherapy is not attempted. Once contracted, the best treatment is the usual oral rehydration recommended for all diarrhea. In severe cases, antimicrobial drugs may be necessary. For preventing traveler's diarrhea, reports indicate that there may be some protective effect from taking prescribed antibiotics. Another option is to take bismuth-containing preparations, such as Pepto-Bismol (two tablets, taken four times a day), if the person does not mind the tongue and stool temporarily turning black. But the best advice in risky areas is to prevent infection, as expressed in the saying, "Boil it, peel it, or don't eat it."

Shiga Toxin–Producing *Escherichia coli*

In recent years, **enterohemorrhagic *E. coli* (EHEC)** strains have become well known in the United States as the cause of several outbreaks of serious disease. The primary virulence factor in these bacteria is the production of Shiga toxin (see page 712), and they are sometimes termed **Shiga-toxin E. coli (STEC)**. (Taxonomists consider *E. coli* to be indistinguishable from members of the genus *Shigella.*) Another virulence factor is their ability to adhere to the intestinal mucosa. The bacteria destroy the microvilli and cause the formation of pedestal-like projections upon which they then rest (Figure 25.12). In the United States, the serotype usually isolated is O157:H7 (see the footnote in Chapter 11, page 310 for an explanation of this nomenclature), but elsewhere other serotypes may predominate. The rearing of livestock has become increasingly industrialized, and masses of cattle are grain-fed in feedlots rather than pastured. One result is that the diet affects the pH of the rumen and promotes the colonization of the animal with STEC, which are relatively acid resistant. Currently, 2–3% of domestic cattle carry STEC, which contaminate the carcass at slaughter. (The

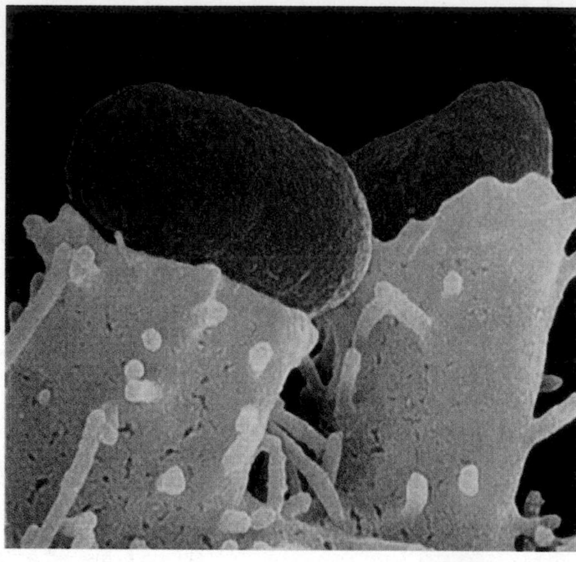

Figure 25.12 Enterohemorrhagic *E. coli* (EHEC) O157:H7. As EHEC bacteria (purple) adhere to the epithelial wall, they destroy the surface microvilli and cause the formation of a pedestal-like projection (yellow) on which they rest. The function of these actin-rich structures is unclear, but they may facilitate the bacteria's spread to adjacent cells.

Q Is adhesion a factor in the pathogenicity of a microbe?

animals do not suffer from obvious symptoms). Furthermore, massive amounts of manure from feedlots tend to contaminate sources of irrigation waters and then leafy vegetables that are eaten raw.

The U.S. Department of Agriculture has found that almost 90% of ground meats are contaminated, although usually at a very low level. Such meats, if not cooked well, are a potential source of infection. Other meats and leafy vegetables may also be contaminated. Ingested food is not the only infection source; some cases have been associated with children's visits to farms or petting zoos. The infective dose is estimated to be very small, probably fewer than 100 bacteria.

In humans the Shiga toxins often cause only self-limiting diarrhea, but in about 6% of people infected, it produces an inflammation of the colon (the large intestine above the rectum) with profuse bleeding, called *hemorrhagic colitis.* Unlike *Shigella* (see Figure 25.8), these *E. coli* do not invade the intestinal wall but release the toxin into the intestinal lumen (space).

Another dangerous complication is *hemolytic uremic syndrome* (HUS). Characterized by blood in the urine, often leading to kidney failure, HUS occurs when the kidneys are affected by the toxin. Some 5–10% of small children who have been infected progress to this stage, which has a mortality rate of about 5%. Management of these patients primarily involves intravenous rehydration and careful monitoring of serum electrolytes. Among survivors of HUS, some may require kidney dialysis or even transplants. An estimated 200 to 500 deaths occur annually.

Because of the attention this pathogen has attracted, researchers have been working, with some success, to develop rapid methods of detecting its presence in food without the need for time-consuming culturing methods. It is recommended that public health laboratories test routinely for STEC O157. A standard method is to use media that differentiate these bacteria by their inability to ferment sorbitol. Any sorbitol-negative colonies should subsequently be tested by a process called *pulsed-field gel electrophoresis (PFGE)*, a technique that subtypes bacteria. The data are entered into a national PulseNet database so that epidemiological information can be compared.

Campylobacter Gastroenteritis

Campylobacter are gram-negative, microaerophilic, spirally curved bacteria that have emerged as the leading cause of foodborne illness in the United States. They adapt well to the intestinal environment of animal hosts, especially poultry. Culturing *Campylobacter* requires conditions of low oxygen and high carbon dioxide developed in special apparatus. The bacteria's optimum growth temperature of about 42°C approximates that of their animal hosts, but the bacteria do not replicate in food. Almost all retail chicken is contaminated with *Campylobacter.* Nearly 60% of cattle excrete the organism in feces and milk, but retail red meats are less likely to be contaminated.

There are more than an estimated 2 million cases of ***Campylobacter* gastroenteritis** in the United States annually, usually caused by *C. jejuni.* The infective dose is fever than a thousand bacteria. Clinically, it is characterized by fever, cramping abdominal pain, and diarrhea or dysentery. Normally, recovery follows within a week.

An unusual complication of campylobacterial infection is that it is linked, in about 1 in 1000 cases, to the neurological disease Guillain-Barré syndrome, a temporary paralysis. Apparently, a surface molecule of the bacteria resembles a lipid component of nervous tissue and provokes an autoimmune attack.

Helicobacter Peptic Ulcer Disease

In 1982, a physician in Australia cultured a spiral-shaped, microaerophilic bacterium observed in the biopsied tissue of stomach ulcer patients. Now named *Helicobacter pylori,* it is accepted that this microbe is responsible for most cases of **peptic ulcer disease.** This syndrome includes gastric and duodenal ulcers. (The duodenum is the first few inches of the small intestine.) About 30–50% of the population in the developed

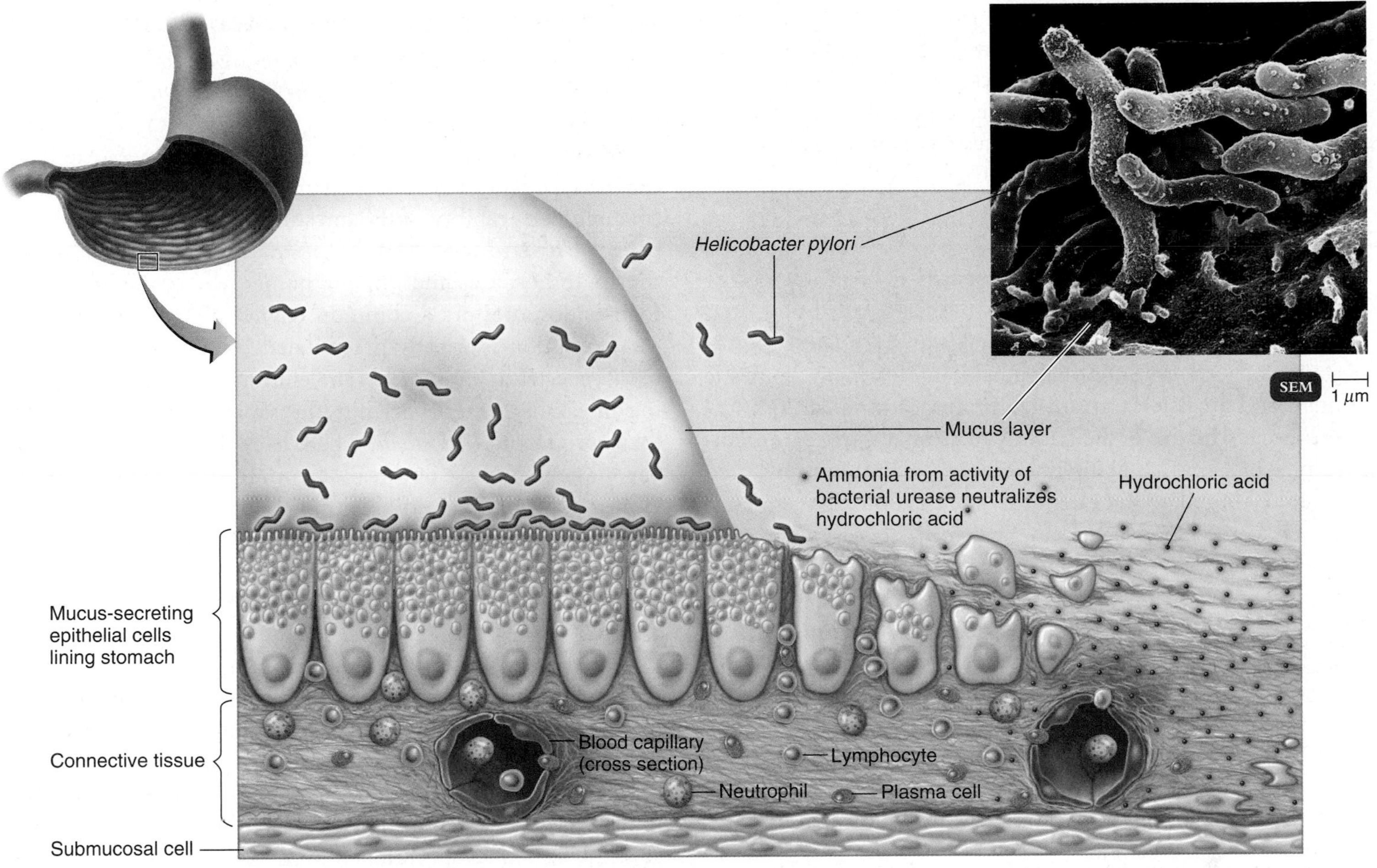

Figure 25.13 *Helicobacter pylori* infection, leading to ulceration of the stomach wall.

Q How does the enzyme urease form ammonia? (*Hint:* Look up the chemical formula of urea.)

world become infected; the infection rate is higher elsewhere. Only about 15% of those infected develop ulcers, so certain host factors are probably involved. For example, people with type O blood are more susceptible, which is also true of cholera. (See page 526.) *H. pylori* is also designated as a carcinogenic bacterium. Gastric cancer develops in about 3% of people infected with these bacteria, but no uninfected persons develop gastric cancer.

The stomach mucosa contains cells that secrete gastric juice containing proteolytic enzymes and hydrochloric acid that activates these enzymes. Other specialized cells produce a layer of mucus that protects the stomach itself from digestion. If this defense is disrupted, an inflammation of the stomach (gastritis) results. This inflammation can then progress to an ulcerated area (Figure 25.13). Through an interesting adaptation, *H. pylori* can grow in the highly acidic environment of the stomach, which is lethal for most microorganisms. *H. pylori* produces large amounts of an especially efficient urease, an enzyme that converts urea to the alkaline compound ammonia, resulting in a locally high pH in the area of growth.

The eradication of *H. pylori* with antimicrobial drugs usually leads to the disappearance of peptic ulcers. Several antibiotics, usually administered in combination, have proven effective. Bismuth subsalicylate (Pepto-Bismol) is also effective and is often part of the drug regimen. When the bacteria are successfully eliminated, the recurrence rate of the ulcer is only about 2–4% a year. Reinfection can result from many environmental sources but is less likely in areas with high standards of sanitation; in fact, there is some evidence that infection by *H. pylori* is slowly disappearing in developed countries.

The most reliable diagnostic test requires a biopsy of tissue and culture of the organism. An interesting diagnostic approach is the urea breath test. The patient swallows radioactively labeled urea; if the test is positive, within about 30 minutes CO_2 labeled with radioactivity can be detected in the breath. This test is most useful for determining the effectiveness of chemotherapy because a positive test is an indication of live *H. pylori.* Diagnostic tests of stools to detect antigens (not antibodies) for *H. pylori* are suitable for follow-up tests following therapy. They are the noninvasive test of choice, especially for children. Serological tests to detect antibodies are inexpensive but not useful in determining eradication.

Yersinia Gastroenteritis

Other enteric pathogens being identified with increasing frequency are *Yersinia enterocolitica* (en'tėr-ō-kōl-it-ik-ä) and *Y. pseudotuberculosis* (sū-dō-tü-bėr-kū-lō'sis). These gram-negative bacteria are intestinal inhabitants of many domestic animals and are often transmitted in meat and milk. Both microbes are distinctive in their ability to grow at refrigerator temperatures of 4°C. This ability increases their numbers in stored refrigerated blood, to the extent that their endotoxins can result in shock to the blood recipient. *Yersinia* has occasionally been the cause of severe reactions when it contaminates transfused blood.

These pathogens cause ***Yersinia*** **gastroenteritis**, or **yersiniosis**. The symptoms are diarrhea, fever, headache, and abdominal pain. The pain is often severe enough to cause a misdiagnosis of appendicitis. Diagnosis requires culturing the organism, which can then be evaluated by serological tests. Adults suffering from yersiniosis usually recover in 1 or 2 weeks; children may take longer. Treatment with antibiotics and oral rehydration may be helpful.

Clostridium perfringens Gastroenteritis

One of the more common, if underrecognized, forms of food poisoning in the United States is caused by *Clostridium perfringens,* a large, gram-positive, endospore-forming, obligately anaerobic rod. This bacterium is also responsible for human gas gangrene (see Chapter 23, page 646).

Most outbreaks of ***Clostridium perfringens*** **gastroenteritis** are associated with meats or meat stews contaminated with intestinal contents of the animal during slaughter. Such foods meet the pathogen's nutritional requirement for amino acids, and when the meats are cooked, the oxygen level is lowered enough for clostridial growth. The endospores survive most routine heatings, and the generation time of the vegetative bacterium is less than 20 minutes under ideal conditions. Large populations can therefore build up rapidly when foods are being held for serving or when inadequate refrigeration leads to slow cooling.

The microbe grows in the intestinal tract and produces an exotoxin that causes the typical symptoms of abdominal pain and diarrhea. Most cases are mild and self-limiting and probably are never clinically diagnosed. If treatment is required, oral rehydration is recommended. The symptoms usually appear 8 to 12 hours after ingestion. Diagnosis is usually based on isolating and identifying the pathogen in stool samples.

Clostridium difficile–Associated Diarrhea

Every year in the United States there are millions of cases of diarrhea associated with infection by *Clostridium difficile,* a gram-positive endospore-forming bacterium. It is found in the stools of many healthy adults. *C. difficile* produces exotoxins that cause inflammation accompanied by increased fluid secretion and permeability of the intestinal mucosa. The condition, called ***Clostridium difficile*–associated diarrhea**, occurs mostly in hospitals and nursing homes, where the bacteria or their endospores are a common environmental contaminant. The condition is usually precipitated by the extended use of broad-spectrum antibiotics. The elimination of most competing intestinal bacteria permits the rapid proliferation of the toxin-producing *C. difficile.* There is evidence that the strains encountered in these outbreaks are more virulent than normal. However, it can cause outbreaks in children in day-care centers that are unrelated to antibiotic use. Caregivers have been known to acquire it from patients. It can be serious; the mortality rate, which is highest in elderly patients, is reported as 1–2.5%.

The disease manifests itself in symptoms ranging from only a mild case of diarrhea to life-threatening colitis (inflammation of the colon). The colitis can result in ulceration of the intestinal wall.

A diagnosis of *C. difficile*–associated diarrhea is often suggested by a history of antibiotic use. It can be confirmed by an immunoassay that detects the responsible exotoxins. A more reliable test, but one that is difficult to perform and requires as long as 48 hours for results to be available, is a cytotoxin assay. Treatment, of course, requires discontinuation of the precipitating antibiotic and oral rehydration therapy. Metronidazole, a drug that targets the metabolism of anaerobes, is also part of the usual therapy. Currently, no antimicrobial treatment, including vancomycin, reliably prevents recurrences. In exceptional cases, enemas containing human stools are used in an attempt to restore the normal microbiota.

Bacillus cereus Gastroenteritis

Bacillus cereus (se'rē-us) is a large, gram-positive, endospore-forming bacterium that is very common in soil and vegetation and is generally considered harmless. It has, however, been identified as the cause of outbreaks of foodborne illness. Heating the food does not always kill the spores, which germinate as the food cools. Because competing microbes have been eliminated in the cooked food, *B. cereus* grows rapidly and produces toxins. Rice dishes served in Asian restaurants seem especially susceptible.

Some cases of ***Bacillus cereus*** **gastroenteritis** resemble *C. perfringens* intoxications and are almost entirely diarrheal in nature (usually appearing 8 to 16 hours after ingestion). Other episodes involve nausea and vomiting (usually 2 to 5 hours after ingestion). It is suspected that different toxins are involved in producing the differing symptoms. Both forms of the disease are self-limiting. The diseases can be differentiated by isolating at least 10^5 *B. cereus* per gram of suspected food.

Bacterial diseases of the GI tract are summarized in Diseases in Focus 25.2.

Viral Diseases of the Digestive System

LEARNING OBJECTIVES

25-5 List the causative agents, modes of transmission, sites of infection, and symptoms for mumps.

25-6 Differentiate hepatitis A, hepatitis B, hepatitis C, hepatitis D, and hepatitis E.

25-7 List the causative agents, mode of transmission, and symptoms of viral gastroenteritis.

Although viruses do not reproduce within the contents of the digestive system like bacteria, they invade many organs associated with the system.

Mumps

The targets of the mumps virus, the parotid glands, are located just below and in front of the ears (see Figure 25.1). Because the parotids are one of the three pairs of salivary glands of the digestive system, it is appropriate to include a discussion of mumps in this chapter.

Mumps typically begins with painful swelling of one or both parotid glands 16 to 18 days after exposure to the virus (**Figure 25.14**). The virus is transmitted in saliva and respiratory secretions, and its portal of entry is the respiratory tract. An infected person is most infective to others during the first 48 hours before clinical symptoms appear. Once the viruses have begun to multiply in the respiratory tract and local lymph nodes in the neck, they reach the salivary glands via the blood. Viremia (the presence of virus in the blood) begins several days before the onset of mumps symptoms and before the virus appears in saliva. The virus is present in the blood and saliva for 3 to 5 days after the onset of the disease and in the urine after about 10 days.

Mumps is characterized by inflammation and swelling of the parotid glands, fever, and pain during swallowing. About 4 to 7 days after the onset of symptoms, the testes can become inflamed, a condition called *orchitis.* This happens in about 20–40% of men past puberty; sterility is a possible but rare consequence. Other possible complications include meningitis, inflammation of the ovaries, and pancreatitis.

An effective attenuated live vaccine is available and is often administered as part of the trivalent measles, mumps, rubella (MMR) vaccine. Second attacks are rare, and cases involving only one parotid gland or subclinical cases (about 15–20% of those infected), are as effective as bilateral mumps in conferring immunity.

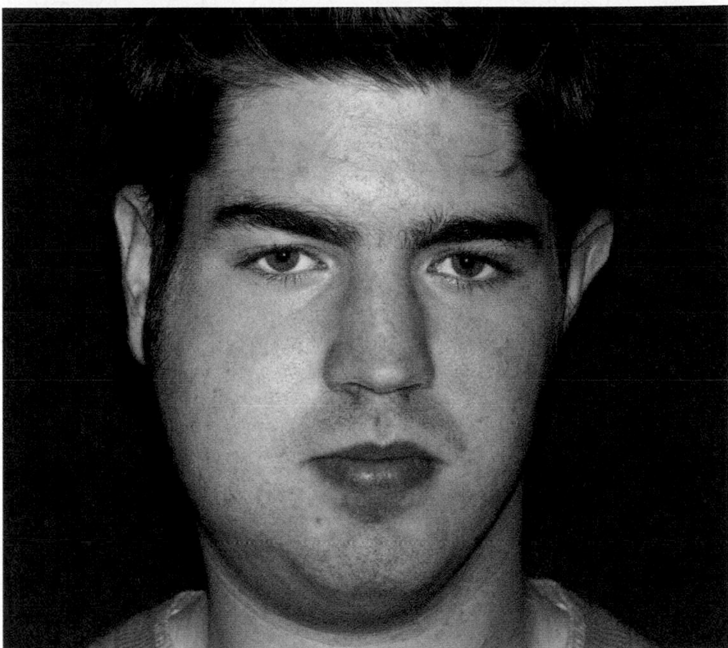

Figure 25.14 A case of mumps. This patient shows the typical swelling of mumps.

Q How is the mumps virus transmitted?

If confirmation of the usual diagnosis based only on symptoms is desired, the virus can be isolated by embryonated egg or cell culture techniques and identified by ELISA tests.

CHECK YOUR UNDERSTANDING

✓ Why is mumps included with the diseases of the digestive system? **25-5**

Hepatitis

Hepatitis is an inflammation of the liver. At least five different viruses cause hepatitis, and probably more remain to be discovered or become better known. Hepatitis is also an occasional result of infections by other viruses such as Epstein-Barr virus (EBV) or cytomegalovirus (CMV). Drug and chemical toxicity can also cause acute hepatitis that is clinically identical to viral hepatitis. The characteristics of the various forms of viral hepatitis are summarized in Diseases in Focus 25.3 on page 724.

Hepatitis A

The *hepatitis A virus (HAV)* is the causative agent of **hepatitis A.** The virus contains single-stranded RNA and lacks an envelope. It can be grown in cell culture.

After a typical entrance via the oral route, HAV multiplies in the epithelial lining of the intestinal tract. Viremia eventually occurs, and the virus spreads to the liver, kidneys, and spleen.

DISEASES IN FOCUS 25.3

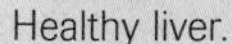

Healthy liver.

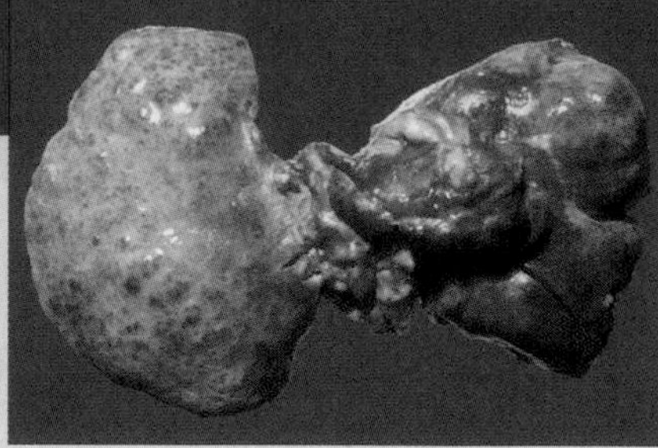

Liver damaged by hepatitis C.

Characteristics of Viral Hepatitis

Hepatitis is an inflammation of the liver. It may be an acute illness with jaundice or elevated serum aminotransferase. Aminotransferases are enzymes found in liver cells and released when the cells are damaged. Chronic hepatitis may be asymptomatic, or there may be evidence of liver disease (including cirrhosis or liver cancer). Hepatitis can be caused by a variety of viruses, alcohol, or drugs; however, it is most often caused by one of the following viruses. For example, after eating at one restaurant, 355 people were diagnosed with the same hepatitis virus. Use the table below to determine which viruses are possible causes of this infection. For the solution, go to www.microbiologyplace.com.

Disease	Pathogen	Symptoms	Incubation Period	Method of Transmission	Diagnostic Test	Antibody Prevalence in U.S.	Vaccine
Hepatitis A	Hepatitis A virus, Picornaviridae	Mostly subclinical; fever, headache; in malaise, jaundice severe cases; no chronic disease	2–6 weeks	Ingestion	IgM antibodies	33%	Inactivated virus. Postexposure immune globulin
Hepatitis B	Hepatitis B virus, Hepadnaviridae	Frequently subclinical; similar to HAV, but no headache; more likely to progress to severe liver damage; chronic disease occurs	4–26 weeks	Parenteral; sexual contact	IgM antibodies	5–10%	Genetically modified vaccine produced in yeast
Hepatitis C	Hepatitis C virus, Flaviviridae	Similar to HBV, more likely to become chronic	2–22 weeks	Parenteral	PCR for viral RNA	1.8%	None
Hepatitis D	Hepatitis D virus, Deltaviridae	Severe liver damage; high mortality rate; chronic disease may occur	6–26 weeks	Parenteral; requires coinfection with hepatitis B	IgM antibodies	Unknown	HBV vaccine is protective
Hepatitis E	Hepatitis E virus, Caliciviridae	Similar to HAV, but pregnant women may have high mortality; no chronic disease	2–6 weeks	Ingestion	IgM antibodies, PCR for viral RNA	0.5%	HAV vaccine is protective

Dane particle, and *filamentous particles,* which are tubular particles similar in diameter to the spherical particles but about ten times as long. The spherical and filamentous particles are unassembled components of Dane particles without nucleic acids; assembly is evidently not very efficient, and large numbers of these unassembled components accumulate. Fortunately, these numerous unassembled particles contain *hepatitis B surface antigen (HB_sAg),* which can be detected with antibodies to them. Such antibody tests make convenient screening of blood for HBV possible.

Physicians, nurses, dentists, medical technologists, and others who are in daily contact with blood have a considerably higher incidence of hepatitis B than members of the general population. It is estimated that thousands of health care workers become infected each year in the United States. Federal regulations require that employees exposed to blood be offered free vaccinations by

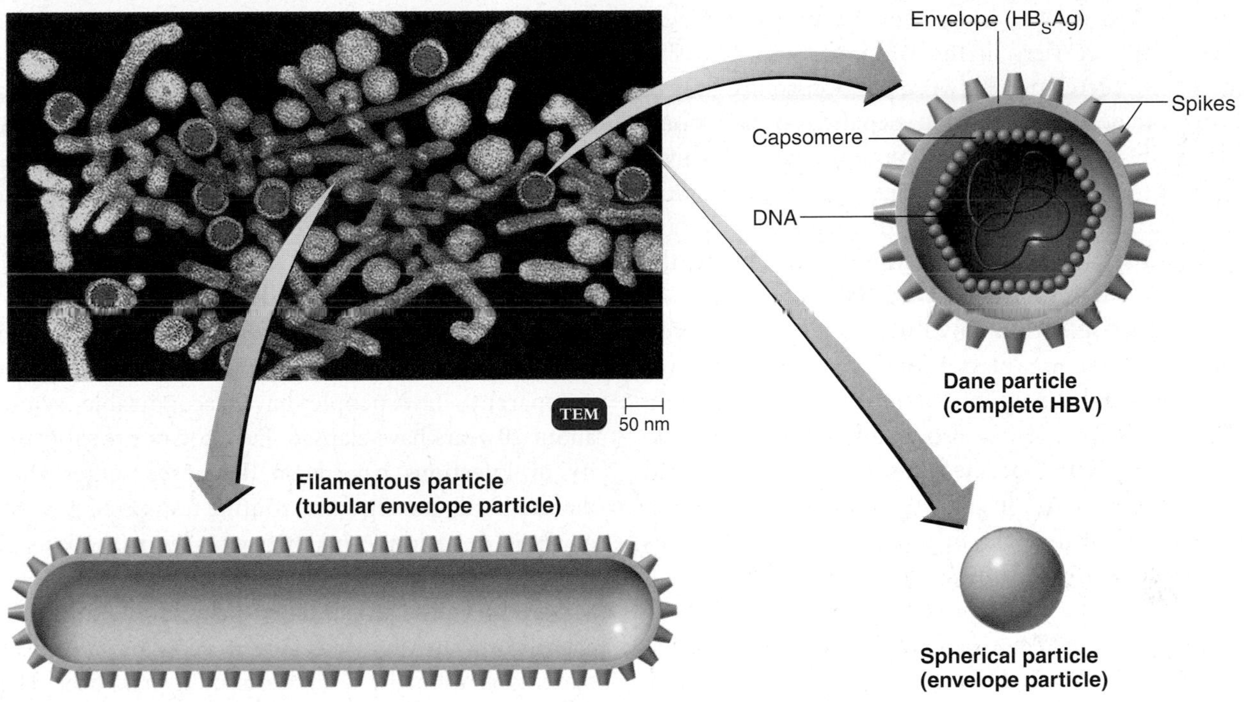

Figure 25.15 Hepatitis B virus (HBV). The micrograph and illustrations depict the distinct types of HBV particles discussed in the text.

Q What are other causes of viral hepatitis?

their employers. There have also been instances of transmission to patients by surgeons and dentists. It is safest to use disposable syringes and needles for each patient. Intravenous drug users often share syringes and needles and fail to sterilize them properly; as a consequence, they also have a high incidence of hepatitis B. Blood may contain up to a billion viruses per milliliter. Therefore, it is not surprising that it is also present in many body fluids, such as saliva, breast milk, and semen, but not in blood-free feces or urine. Transmission by semen donated for artificial insemination has been documented, and semen has been implicated in transmission between heterosexuals with multiple partners and in male homosexuals. Precautions taken to prevent HIV transmission have also had an effect on the incidence of HBV infections. A mother who is positive for HB_sAg, especially if she is a chronic carrier, may transmit the disease to her infant, usually at birth. In most cases, this type of transmission can be prevented by administering hepatitis B immune globulin (HBIG) to the newborn immediately after birth. These babies should also be vaccinated.

A third of the world's population shows serological evidence of past infection—and HBV is estimated to cause a million deaths every year. An estimated 130,000 Americans, mostly young adults, are infected with HBV each year; only about 10,000 cases are actually reported. About 5000 people die each year of HBV-related liver disease, ranging from cirrhosis (hardening and degeneration; see photo on page 724) to cancer. The host's immune response to the virus is primarily responsible for liver damage. The incubation period before the appearance of symptoms averages about 12 weeks: the range is 4 to 26 weeks.

It is important to be able to distinguish between acute and chronic HBV infection. The incubation period for *acute HBV hepatitis* averages about 12 weeks; the range is 4 to 26 weeks. Signs and symptoms are highly variable, and HBV infections cannot be distinguished from other viral hepatitis infections by purely clinical appearance. The patient may have very mild symptoms, such as loss of appetite, low-grade fever, and joint pain. Only a minority of infected infants and small children show any symptoms at all. However, in cases of *fulminant* HBV hepatitis (rapidly increasing in severity), patients might have fever, nausea, and the typical symptoms of jaundice. At least 90% of acute HBV infections end in complete recovery, and the overall mortality rate for HBV infections is less than 1%. However, although fulminating hepatitis occurs in fewer than 2% of infections, it has a very high fatality rate.

If HB_sAg persists for more than about 6 months, it is an indication of *chronic HBV hepatitis;* IgM-type antibodies also will have disappeared at about this time. For individuals who were infected 1 to 5 years previously, the risk of developing chronic disease is highest. The risk for infants is about 90%; in children of 1 to 5 years, about 25–50%. Adolescents and young adults have a much lower risk of developing chronic HBV hepatitis: only 6–10%.

Overall, up to 10% of patients become chronic carriers. These carriers are reservoirs for transmission of the virus, and they also have a high rate of liver disease. It has been estimated that there are 1.25 million HBV carriers in the United States. A special concern is the strong correlation between the occurrence of liver cancer and the incidence of chronic hepatitis B infections. Chronic carriers are about 200 times more likely to get liver cancer than the general population. Liver cancer is the most prevalent form of cancer in sub-Saharan Africa and the Far East, areas where hepatitis B is extremely common. Worldwide, the number of HBV carriers is estimated to be 400 million.

Preventing HBV infection involves several strategies. Important among them are precautions such as disposable needles and syringes and the use of barrier-type contraception. Screening of transfused blood has also greatly reduced the risk. The introduction of HBV vaccines has become widespread worldwide and is now part of the childhood immunization schedule in the United States. The incidence of HBV infections has declined sharply in areas in which the vaccine is in use, and eventual elimination of the disease is conceivable.

It has not been possible to cultivate HBV in cell culture, a step that was necessary for the development of vaccines for polio, mumps, measles, and rubella. The available HBV vaccines use HB_sAg produced by a genetically modified yeast. Vaccination is recommended for high-risk groups; a partial listing would include health care workers exposed to blood and blood products, people undergoing hemodialysis, patients and staff at mental health care institutions, IDUs, and homosexually active men.

There is no specific treatment for acute HBV. Treatments for chronic HBV infection have been limited and are not curative. Lamivudine (a synthetic nucleoside analogue of cytosine) combined with alpha interferon (IFN-α) is expensive but results in improvement in a significant number of patients. Entecavir, recently approved, slows the reproduction of the virus. The first-line drugs are either adefovir dipivoxil (a nucleoside analog) or entecavir. The drug most recently approved for treatment of chronic HBV is telbivudine, and has shown effectiveness similar to that of lamivudine. Liver transplantation is often a final option in treatment.

Hepatitis C

In the 1960s, a previously unsuspected form of transfusion-transmitted hepatitis, now called **hepatitis C,** appeared. This new form of hepatitis soon constituted almost all transfusion-transmitted hepatitis—as testing eliminated HBV in the blood supply. Eventually, serological tests to detect hepatitis C virus (HCV) antibodies were developed that similarly reduced the transmission of HCV to very low levels. However, there is a delay of about 70 to 80 days between infection and the appearance of detectable HCV antibodies. The presence of HCV in contaminated blood cannot be detected during this interval, and about 1 in 100,000 transfusions can still result in infection. Blood-collecting facilities in the United States can now detect HCV-contaminated blood within 25 days of infection. (See the box on safety of the blood supply, page 727). A PCR test can detect viral RNA within 1 to 2 weeks after infection.

HCV has a single strand of RNA and is enveloped. The virus does not kill the infected cell, but it triggers an immune inflammatory response that either clears the infection or slowly destroys the liver. It is capable of rapid genetic variation to evade the immune system. This characteristic, along with the fact that currently it can be cultured only very inefficiently, complicates the search for an effective vaccine.

Hepatitis C has been described as a silent epidemic, killing more people than AIDS in the United States. It is often clinically inapparent—few people have recognizable symptoms until about 20 years have elapsed. Even today, probably only a minority of infections have been diagnosed. Often, hepatitis C is detected only during some routine testing, such as for insurance or blood donation. A majority of cases, perhaps as high as 85%, progress to chronic hepatitis, a much higher rate than with HBV. Surveys indicated an estimated 3.2 million of the U.S. population are chronically infected. About 25% of chronically infected patients develop liver cirrhosis or liver cancer. Hepatitis C is probably the major reason for liver transplantation. Persons infected with HCV should be immunized against both HAV and HBV (a combination vaccine is now available) because they cannot afford the risk of further liver damage.

Preventing HCV is limited to minimizing exposure—even sharing of items such as razors, toothbrushes, or nail clippers is dangerous. A common source of infection is the sharing of injection equipment among IDUs. At least 80% of this group is infected with HCV. In one exceptional case, the disease was transmitted by means of a straw shared for inhaling cocaine. Interestingly, in more than one-third of the cases, a mode of transmission—by contaminated blood, sexual contact, or other means—cannot be identified.

The treatment of choice is a drug combination, peginterferon (the interferon is conjugated with polyethylene glycol, which has a more sustained concentration in the blood) and ribavirin. Its disadvantages are that it is very expensive and requires a regimen of months. It also has many potentially severe side effects. However, complete eradication of HCV is attained in many cases.

Hepatitis D (Delta Hepatitis)

In 1977, a new hepatitis virus, now known as *hepatitis D virus (HDV),* was discovered in carriers of HBV in Italy. People who carried this so-called *delta antigen* and were also infected with HBV had a much higher incidence of severe liver damage and a much higher mortality rate than people who had antibodies against HBV alone. With time, it became clearer that **hepatitis D** can occur as either acute (*coinfection form*) or chronic (*superinfection form*) hepatitis. In people with a case of self-limiting acute hepatitis B, coinfection with HDV disappeared as the HBV was cleared from the system, and the condition resembled a typical case of acute hepatitis B. However, if the HBV infection progressed to the chronic stage, superinfection with HDV was

A Safe Blood Supply

Prior to blood banking, a physician typed the blood of a patient's friends until the proper blood type was found. With the development of blood storage techniques in the 1940s, blood banking became the role of specialists, not the primary care physician (see the photo). The safety of blood products is important for all people, especially those with hemophilia because they regularly receive transfusions of clotting factors. An important advance in protecting the blood supply from infectious agents was the change to an all-volunteer donor system, which occurred in 1979. (Volunteer donors have a lower infection rate than paid donors.)

However, the large number of hemophilia patients who became infected with HIV in the early 1980s raised new questions regarding the safety of the blood supply. More sensitive donor screening was rapidly introduced. Serological tests are now routinely performed on donated blood to detect the presence of T-cell leukemia viruses, HIV-1, HIV- 2, HBV, HCV, *Treponema pallidum* bacteria, and *Trypanosoma cruzi* protozoa.

Unfortunately, contamination in the blood of newly infected donors may not be detected by serological tests because there is a "window" of delay between the time of infection and the appearance of antibodies. Now, virtually all whole blood and plasma donations are screened for HCV, HIV, and West Nile virus by nucleic acid testing (NAT), which detects the virus nucleic acids directly, rather than detecting antibodies. NAT has reduced the window of delay during which a newly acquired infection cannot be detected to approximately 25 days for HCV and 12 days for HIV. However, at present NAT takes several days to complete, so platelets, which become outdated in 5 days, are being released before NAT has been completed.

American physician Charles Drew invented the technique for plasma separation that allowed blood to be stored.

There is also concern over potential contamination of blood by new viruses. One response to the 2003 SARS outbreak was to defer anyone who has traveled in a SARS-affected area from making a donation for a period of 14 days. Technologies are being introduced to clean blood by removing 99.9% of white blood cells, which harbor many viruses. Other new techniques are aimed at inactivating any bacteria or viruses in the blood. The American Red Cross already requires virus-inactivating treatment of blood plasma.

A zero-risk blood supply is probably unattainable, but the goal is to make the blood supply as safe as possible. Synthetic blood substitutes are being developed and may one day replace the need for donor blood.

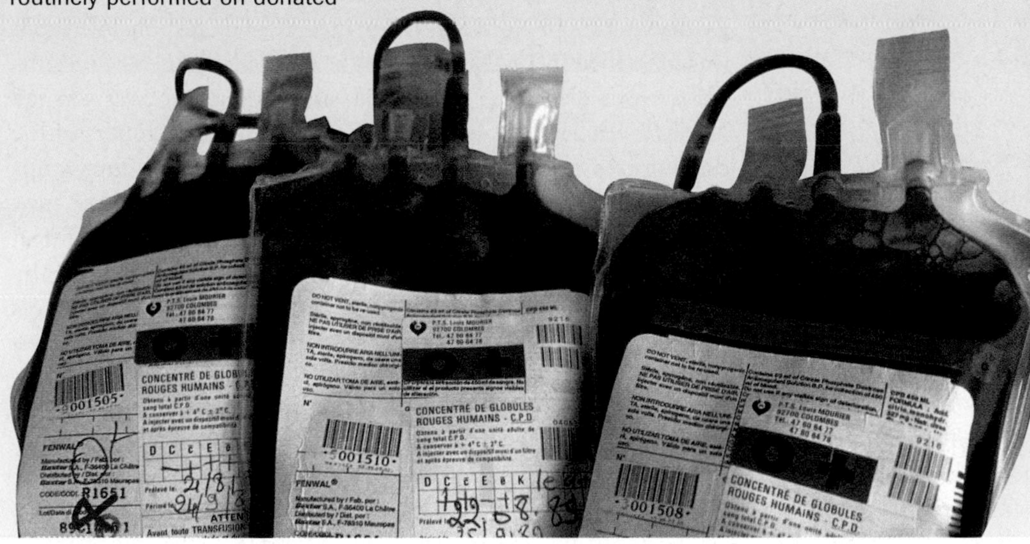

often accompanied by progressive liver damage and a fatality rate several times that of people infected with HBV alone.

Epidemiologically, hepatitis D is linked to the epidemiology of hepatitis B. In the United States and northern Europe, the disease occurs predominantly in high-risk groups, such as IDUs.

Structurally, the HDV contains a single strand of RNA, which is shorter than in any other animal-infecting virus. The particle is not capable of causing an infection. It becomes infectious when an external envelope of HB_sAg, whose formation is controlled by the genome of HBV, covers the HDV protein core (the delta antigen; see Figure 25.15).

Hepatitis E

Hepatitis E is spread by fecal–oral transmission, much like hepatitis A, which it clinically resembles. The pathogen, known as *hepatitis E virus (HEV)*, is endemic in areas of the world with poor sanitation, especially India and southeast Asia. It resembles HAV in being a nonenveloped virus with a single strand of RNA

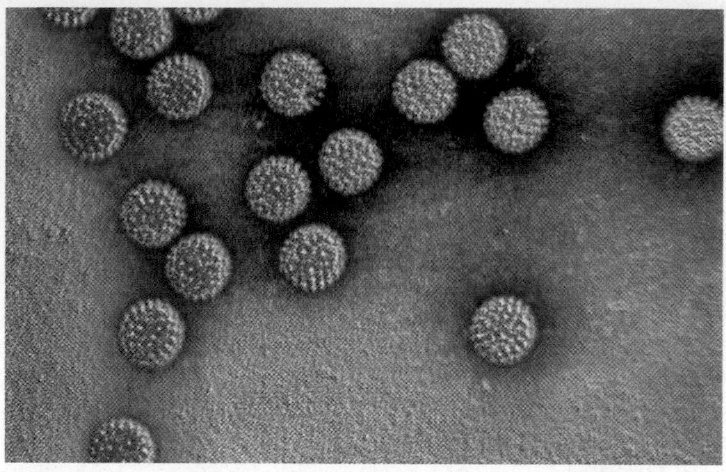
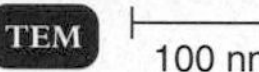

Figure 25.16 Rotavirus. This negatively stained electron micrograph shows the morphology of the rotavirus (*rota* = wheel), which gives the virus its name.

Q What disease does rotavirus cause?

but is not related serologically to it. Like hepatitis A virus, HEV does not cause chronic liver disease, but for some unexplained reason it is responsible for a mortality rate in excess of 20% in pregnant women.

Other Types of Hepatitis

New techniques in molecular biology and serology have provided evidence of blood-transmitted viruses known as *hepatitis F (HFV)* and *hepatitis G (HGV).* The HGV is found worldwide and in the United States is more prevalent than HCV. HGV is closely related to HCV and is sometimes called GB virus C (GBV-C). Apparently, however, it is so well adapted to its human hosts that it causes no significant disease condition. About 5% of cases of chronic liver disease cannot be attributed to any known hepatitis in the series A through E. Whether these will eventually be attributed to HFV, HGV, or to some other member added to this alphabetic explosion is unknown.

CHECK YOUR UNDERSTANDING

✓ Of the several hepatitis diseases, HAV, HBV, HCV, HDV, and HEV, which two now have effective vaccines to prevent them? **25-6**

Viral Gastroenteritis

Acute gastroenteritis is one of the most common diseases of humans, and about 90% of cases of acute viral gastroenteritis are caused by either the rotavirus or the human calciviruses, better known as the Norwalk family of viruses; collectively, the noroviruses.

Rotavirus

Rotavirus (Figure 25.16) is probably the most common cause of viral gastroenteritis, especially in children. It is estimated to cause about 3 million cases, but fewer than 100 deaths, every year in the United States. Mortality is much higher in less developed countries because rehydration therapy is not as available. More than 90% of children in the United States have been infected by the age of 3. In some cases, parents also become infected. Immunity acquired then makes rotavirus infections, except for certain strains, much less common in adults. In most cases, following an incubation period of 2 to 3 days, the patient suffers from low-grade fever, diarrhea, and vomiting, which persists for about a week.

There is usually a peak in cases during the cooler winter months. An infectious dose is estimated to be fewer than 100 viruses, and patients shed billions in every gram of stool. The first vaccine to prevent rotavirus, introduced in 1998, was withdrawn after serious problems developed. In 2006 a live, orally administered vaccine was licensed. Rotavirus infections are routinely diagnosed by several types of commercially available tests, such as enzyme immunoassays. Treatment is usually limited to oral rehydration therapy.

Norovirus

Noroviruses were first identified following an outbreak of gastroenteritis in Norwalk, Ohio, in 1968. The responsible agent was identified in 1972 and called the *Norwalk virus.* Several similar viruses were later identified, and this group was termed *Norwalk-like viruses.* All were determined to be members of the caliciviruses (named for the Latin *calyx,* meaning cup—cup-shaped depressions are visible on the viruses) and are now termed *noroviruses.* They cannot be cultured, nor do they infect the usual laboratory animals. Humans become infected by fecal–oral transmission from food and water and even aerosols from vomiting. The infective dose may be as low as 10 viruses. The viruses continue to be shed for several days after the patient is asymptomatic. More than 20 million cases of norovirus gastroenteritis occur annually in the United States but only about 300 deaths. About half of adult Americans show serological evidence that they have been infected. See the box in Chapter 9, page 266. The currently dominant strain of noroviruses made its appearance around 2002, which is attributed to several possible factors. This strain may be more virulent, or more environmentally stable; also, fewer people may have had resistance to it from previous exposure. Natural resistance to a particular strain may last only a few months—at most about 3 years.

Cleanup and prevention of transmission following an outbreak on a cruise ship or restaurant, for example, has proved to be a challenging problem. The viruses are unusually persistent on environmental surfaces such as door handles or elevator buttons. They are not reliably inactivated by ethanol or detergent-based hand

DISEASES IN FOCUS 25.4

Viral Diseases of the Digestive System

Differential diagnosis is the process of identifying the disease from a list of possible diseases that fit the information derived from examining a patient. A differential diagnosis is important for providing initial treatment and for laboratory testing. For example, an outbreak of diarrhea began in mid-June, peaked in mid-August, and tapered off in September. A clinical case was defined as diarrhea (three loose stools during a 24-hour period) in a person who was a member of a swim club. The virus shown in the photo at right was isolated from one patient. Use the table below to identify infections that could cause these symptoms. For the solution, go to www.microbiologyplace.com.

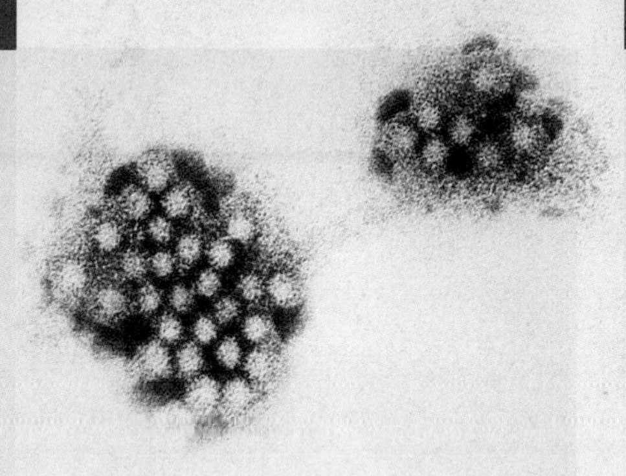

Virus cultured from the patient's stool.

Disease	Pathogen	Symptoms	Incubation Period	Diagnostic Test	Treatment
Mumps	Mumps virus, Paramyxoviridae	Painful swelling of parotid glands	16–18 days	Symptoms; virus culture	Preventive vaccine
Viral gastroenteritis	Rotavirus	Vomiting, diarrhea for 1 week	1–3 days	Enzyme immunoassay for viral antigens in feces	Oral rehydration
	Norovirus	Vomiting, diarrhea for 2–3 days	18–48 hr	PCR	Oral rehydration
Hepatitis (See Diseases in Focus 25.3 on page 724)					

cleaners, although the Centers for Disease Control and Prevention (CDC) recommends the use of sanitizing hand gels such as Purell, which contains >62% ethanol. To decontaminate hard, nonporous surfaces requires solutions containing 1000 to 5000 ppm of hypochlorite (a 1:50 or 1:10 solution of 5.26% bleach, respectively). The EPA recommends use of a peroxygen compound (see page 202) called Virkon-S for surface decontamination.

To detect noroviruses in stool samples, laboratories use sensitive PCR and EIA tests. The availability of such new and sensitive assays has led to recognition of noroviruses as the most common cause (at least half of recent foodborne outbreaks in the United States) of nonbacterial gastroenteritis.

Following an incubation period of 18 to 48 hours, the patient suffers from vomiting and/or diarrhea for 2 or 3 days. Vomiting is the most prevalent symptom in children; most adults experience diarrhea, although many adult patients experience only vomiting. The severity of symptoms often depends upon the size of the infective dose.

The only treatment for viral gastroenteritis is oral rehydration or, in exceptional cases, intravenous rehydration.

Viral diseases of the GI tract are summarized in Diseases in Focus 25.4.

CHECK YOUR UNDERSTANDING

✔ Two very common causes of viral gastroenteritis are caused by rotaviruses and noroviruses. Which of these now can be prevented by a vaccine? **25-7**

Fungal Diseases of the Digestive System

LEARNING OBJECTIVE

25-8 Identify the causes of ergot poisoning and aflatoxin poisoning.

Some fungi produce toxins called *mycotoxins.* When ingested, these toxins cause blood diseases, nervous system disorders, kidney damage, liver damage, and even cancer. Mycotoxin intoxication is considered when multiple patients have similar clinical signs and symptoms. Diagnosis is usually based on finding the fungi or mycotoxins in the suspected food (Diseases in Focus 25.5, page 734).

Ergot Poisoning

Some mycotoxins are produced by *Claviceps purpurea* (kla′vi-seps pur-pu-rēä), a fungus causing smut infections on grain crops. The

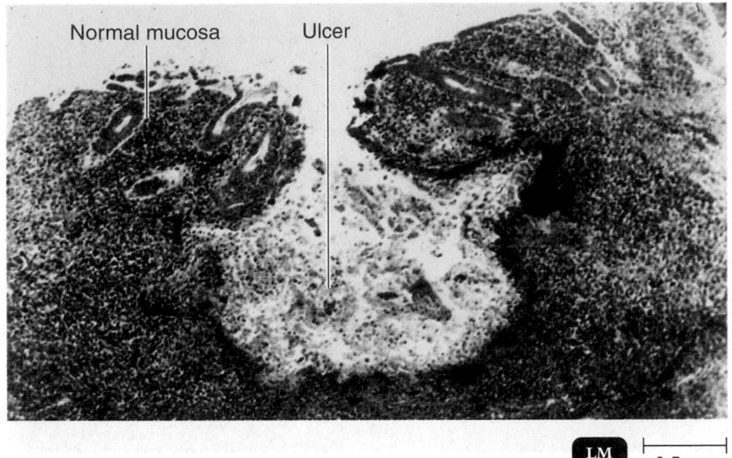

Figure 25.19 Section of intestinal wall showing a typical flask-shaped ulcer caused by *Entamoeba histolytica*.

Q If this lesion progressed far enough, could it be life-threatening?

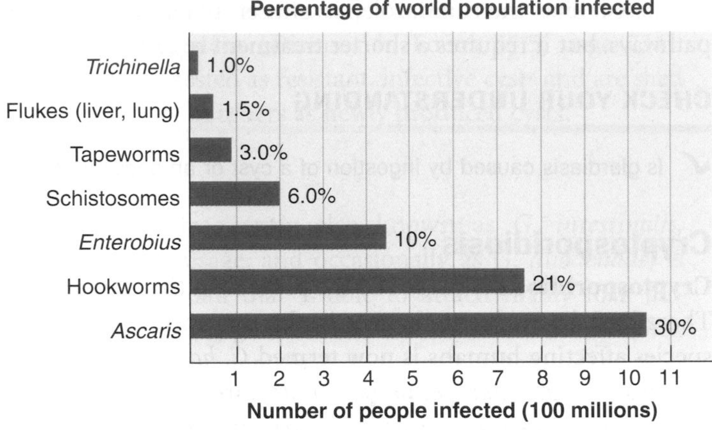

Figure 25.20 The worldwide prevalence of human infections with selected intestinal helminths.
Source: World Health Organization.

Q How is each of these diseases transmitted?

mucus. The trophozoites feed on tissue in the gastrointestinal tract (Figure 25.19).

Severe bacterial infections result if the intestinal wall is perforated. Abscesses might have to be treated surgically, and the invasion of other organs, particularly the liver, is not uncommon. Perhaps 5% of the U.S. population are asymptomatic carriers of *E. histolytica*. Worldwide, one person in ten is estimated to be infected, mostly asymptomatically, and about 10% of these infections progress to the more serious stages.

Diagnosis largely depends on recovering and identifying the pathogens in feces. (Red blood cells, ingested as the parasite feeds on intestinal tissue and observed within the trophozoite stage of an amoeba, help identify *E. histolytica*.) Several serological tests can also be used for diagnosis, including latex agglutination and fluorescent-antibody tests. Such tests are especially useful when the affected areas are outside the intestinal tract and the patient is not passing amoebae.

Metronidazole plus iodoquinol are the drugs of choice in treatment.

Helminthic Diseases of the Digestive System

LEARNING OBJECTIVE

25-10 List the causative agents, modes of transmission, symptoms, and treatments for tapeworms, hydatid disease, pinworms, hookworms, ascariasis, and trichinellosis.

Helminthic parasites are very common in the human intestinal tract, especially under conditions of poor sanitation. Figure 25.20 shows the worldwide estimated incidence of infection with some intestinal helminths. In spite of their size and formidable appearance, they often produce few symptoms. They have become so well adapted to their human hosts, and vice versa, that when their presence is revealed, it is often a surprise.

Diseases in Focus 25.5 summarizes the diseases of the digestive system caused by helminths.

Tapeworms

The life cycle of a typical **tapeworm** extends through three stages. The adult worm lives in the intestine of a human host, where it produces eggs that are excreted in the feces (see Figure 12.26, page 358). The eggs are ingested by animals such as grazing cattle, where the egg hatches into a larval form called a *cysticercus* (plural: *cysticerci*) that lodges in the animal's muscles. Human infections by tapeworms begin with the consumption of undercooked beef, pork, or fish containing cysticerci. The cysticerci develop into adult tapeworms that attach to the intestinal wall by suckers on the scolex (see the photo in Figure 12.26, page 358).

The adult beef tapeworm, *Taenia saginata* (te′-nē-ä sa-ji-nä′tä), can live in the human intestine for 25 years and reaches a length of 6 meters (18 feet) or longer. Even a worm of this size seldom causes significant symptoms beyond a vague abdominal discomfort. There is, however, psychological distress when a meter or more of detached segments (proglottids) break loose and unexpectedly slip out of the anus, which happens occasionally.

Taenia solium (sō′lē-um), the pork tapeworm, has a life cycle similar to that of the beef tapeworm. An important difference is that *T. solium* may produce the larval stage in the human host. **Taeniasis** develops when the adult tapeworm infects the human intestine. This is a generally benign, asymptomatic condition, but the host continuously expels eggs of *T. solium*, which contaminate hands and food under poor sanitary conditions. **Cysticercosis,** infection with the larval stage, can develop when humans or swine ingest *T. solium* eggs. These eggs can

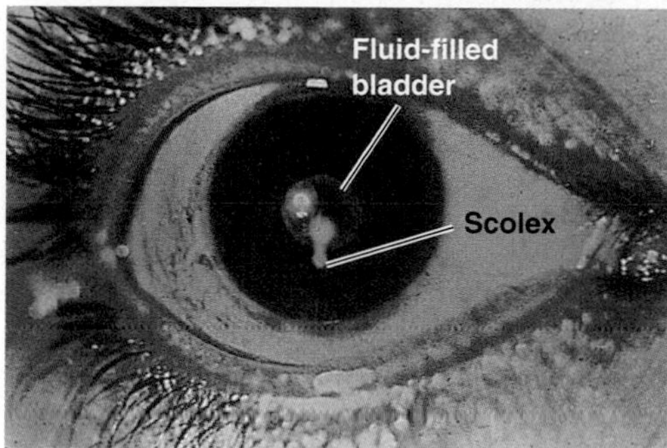

Figure 25.21 Ophthalmic cysticercosis. Some cases of cysticercosis affect the eye.

Q **What organ is most likely to be affected by neurocysticercosis?**

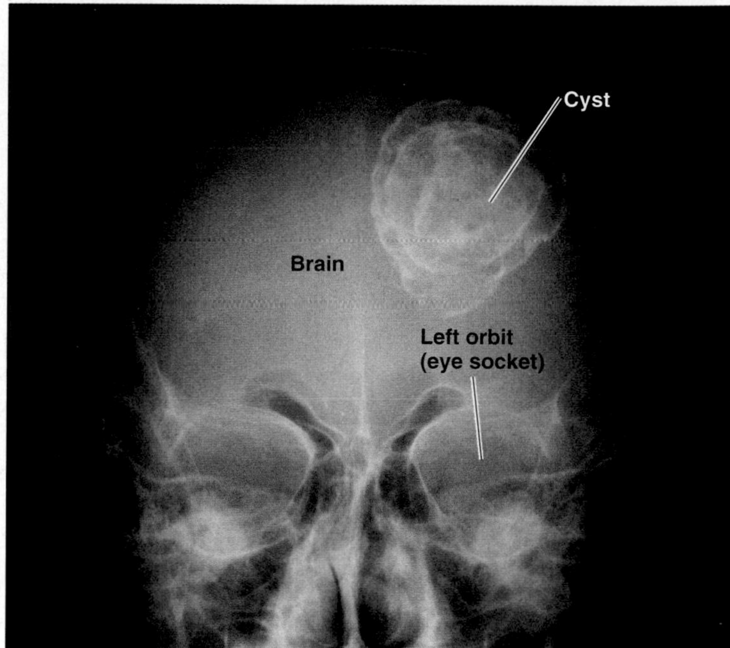

Figure 25.22 A hydatid cyst formed by *Echinococcus granulosus.* A large cyst can be seen in this X-ray image of the brain of an infected individual.

Q **How do hydatid cysts affect the body?**

leave the digestive tract and develop into larvae that lodge in tissue (usually brain or muscles). Cyticerci in muscle tissue are relatively benign and cause few serious symptoms, but the larvae occasionally lodge in an eye, causing **ophthalmic cysticercosis** and affecting vision (Figure 25.21). The most serious, and much more common disease is **neurocysticercosis,** which arises when the larvae develop in the central nervous system such as the brain. Neurocysticercosis, which is endemic in Mexico and Central America, has become a fairly common condition in parts of the United States with large Mexican and Central American immigrant populations.

The symptoms often mimic those of epilepsy or a brain tumor. The number of cases reported reflects, in part, the use of CT (computed tomography) scanning or magnetic resonance imaging (MRI) in diagnosis. In endemic areas, neurological patients can be screened with serological tests for antibodies to *T. solium.*

The fish tapeworm *Diphyllobothrium latum* (dī-fil-lō-bo'thrē-um lā'tum) is found in pike, trout, perch, and salmon. The CDC has issued warnings about the risks of fish tapeworm infection from sashimi and sushi (Japanese dishes prepared from raw fish), foods that have become increasingly popular. To relate a vivid example, about 10 days after eating, one person developed symptoms of abdominal distention, flatulence, belching, intermittent abdominal cramping, and diarrhea. Eight days later, the patient passed a tapeworm 1.2 m (4 ft) long, identified as a species of *Diphyllobothrium.*

Laboratory diagnosis of tapeworms consists of identifying the tapeworm eggs or segments in feces. Adult tapeworms in the intestinal stage can be eliminated with antiparasitic drugs such as praziquantel and albendazole. Cases of neurocysticercosis can sometimes be treated with drugs, but these often worsen the situation and surgery may be required to remove cysticerci.

CHECK YOUR UNDERSTANDING

✓ What species of tapeworm is the cause of cysticercosis? **25-10**

Hydatid Disease

Not all tapeworms are large. One of the most dangerous is *Echinococcus granulosus* (ē-kīn-ō-kok'kus gra-nū-lō'sus), which is only a few millimeters in length (see Figure 12.28, page 359). Humans are not the definitive hosts. The adult form lives in the intestinal tract of carnivorous animals, such as dogs and wolves. Typically, humans become infected from the feces of a dog that has become infected by eating the flesh of a sheep or deer containing the cyst form of the tapeworm. Unfortunately, humans can be an intermediate host, and cysts can develop in the body. The disease occurs most frequently in people who raise sheep or hunt or trap wild animals.

Once ingested by a human, the eggs of *E. granulosus* may migrate to various tissues of the body. The liver and lungs are the most common sites, but the brain and numerous other sites also may be infected. Once in place, the egg develops into a **hydatid cyst** that can grow to a diameter of 1 cm in a few months (Figure 25.22). In some locations, cysts may not be apparent for many years. Some, where they are free to expand, become enormous, containing up to 15 liters (4 gallons) of fluid.

Damage may arise from the size of the cyst in such areas as the brain or the interior of bones. If the cyst ruptures in the host, it can lead to the development of a great many daughter cysts. Another factor in the pathogenicity of such cysts is that the fluid contains proteinaceous material to which the host becomes

DISEASES IN FOCUS 25.5

Fungal, Protozoan, and Helminthic Diseases of the Lower Digestive System

Differential diagnosis is the process of identifying the disease from a list of possible diseases that fit the information derived from examining a patient. A differential diagnosis is important for providing initial treatment and for laboratory testing. For example, public health officials in Pennsylvania were notified of cases of watery diarrhea, with frequent, sometimes explosive, bowel movements among persons associated with a residential facility (e.g., residents, staff, and volunteers). The disease was associated with eating snow peas. Use the table below to identify possible causes of these symptoms. For the solution, go to www.microbiologyplace.com.

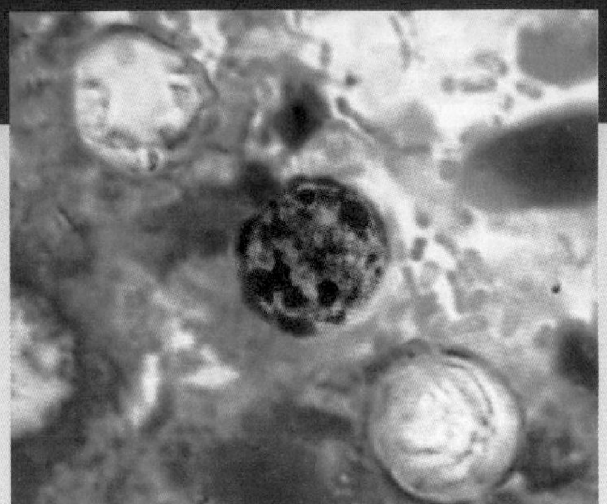

Acid-fast stain from the patient's feces. 6 μm LM

Disease	Pathogen	Symptoms	Reservoir or Host	Diagnostic Test	Treatment
FUNGAL DISEASES					
Ergot poisoning	*Claviceps purpurea*	Restricted blood flow to limbs; hallucinogenic	Mycotoxin produced by fungus growing on grains	Finding fungal sclerotia in food	None
Aflatoxin poisoning	*Aspergillus flavus*	Liver cirrhosis; liver cancer	Mycotoxin produced by fungus growing on food	Immunoassay for toxin in food	None
PROTOZOAN DISEASES					
Giardiasis	*Giardia lamblia*	Protozoan adheres to intestinal wall, may inhibit nutritional absorption; causes diarrhea	Water; mammals	FA	Metronidazole; quinacrine
Cryptosporidiosis	*Cryptosporidium hominis*	Self-limiting diarrhea but may be life-threatening in immunosuppressed patients	Cattle; water	Acid-fast stain; FA; ELISA	Oral rehydration
***Cyclospora* diarrheal infection**	*Cyclospora cayetanensis*	Causes watery diarrhea	Humans; birds; usually ingested with fruits and vegetables	Acid-fast stain	Trimethoprim and sulfamethoxazole
Amoebic dysentery (amoebiasis)	*Entamoeba histolytica*	Amoeba lyses epithelial cells of intestine, causes abscesses; significant mortality rate	Humans	Microscopy; serology	Metronidazole

sensitized. If the cyst suddenly ruptures, the result can be life-threatening anaphylactic shock.

For diagnosis, several serological tests that detect circulating antibodies are useful in screening. If available, physical imaging methods such as X rays, CT, and MRI are best.

Treatment is usually surgical removal, but care must be taken to avoid release of the fluid and the potential spread of infection or anaphylactic shock. If removal is not feasible, the drug albendazole can kill the cysts.

Nematodes

Pinworms

Most of us are familiar with the **pinworm,** *Enterobius vermicularis* (see Figure 12.28, page 360). This tiny worm (females are 8–13 mm in length, males 2–5 mm) migrates out of the anus of the human host to lay its eggs, causing local itching. Whole households may become infected. Diagnosis is usually based on finding eggs around the anus. These can be viewed by pressing

Disease	Pathogen	Symptoms	Reservoir or Host	Diagnostic Test	Treatment
HELMINTHIC DISEASES					
Tapeworms	*Taenia saginata* (beef tapeworm); *T. solium* (pork tapeworm); *Diphyllobothrium latum* (fish tapeworm)	Helminth lives off undigested intestinal contents with few symptoms; pork tapeworm may cause larvae to form in many organs (neurocysticercosis) and cause damage	Intermediate host: cattle, pigs, fish Definitive host: humans	Microscopic exam of feces	Praziquantel; albendazole
Hydatid disease	*Echinococcus granulosus*	Larvae form in body; may be very large and cause damage	Intermediate host: humans Definitive host: dogs	Serology; X-ray exam	Surgical removal; albendazole
Pinworms	*Enterobius vermicularis*	Itching around anus	Intermediate host: humans Definitive host: humans	Microscopic exam	Pyrantel pamoate
Hookworms	*Necator americanus, Ancyclostoma duodenale*	Large infections may result in anemia	Larvae enter skin from soil Definitive host: humans	Microscopic exam	Mebendazole
Ascariasis	*Ascaris lumbricoides*	Helminths live off undigested intestinal contents, causing few symptoms	Intermediate host: humans Definitive host: humans	Microscopic exam	Mebendazole
Trichinellosis	*Trichinella spiralis*	Larvae encyst in striated muscle; usually few symptoms, but large infections may be fatal	Intermediate host: mammals (including humans) Definitive host: mammals (including humans)	Biopsy; ELISA	Mebendazole; corticosteroids

transparent cellulose tape, sticky side down, against the skin, transferring the tape to a microscope slide and viewing the slide under a microscope. Such drugs as pyrantel pamoate (often available without a prescription) and mebendazole are usually effective in treatment.

Hookworms

Hookworm infections were once a very common parasitic disease in the southeastern states. In the United States, the species most often seen is *Necator americanus*. Another species, *Ancyclostoma duodenale*, is widely distributed around the world.

The hookworm attaches to the intestinal wall and feeds on blood and tissue rather than on partially digested food (**Figure 25.23**), so the presence of large numbers of worms can lead to anemia and lethargic behavior. Heavy infections can also lead to a bizarre symptom known as *pica*, a craving for peculiar foods, such as laundry starch or soil containing a certain type of clay. Pica is a symptom of iron deficiency anemia.

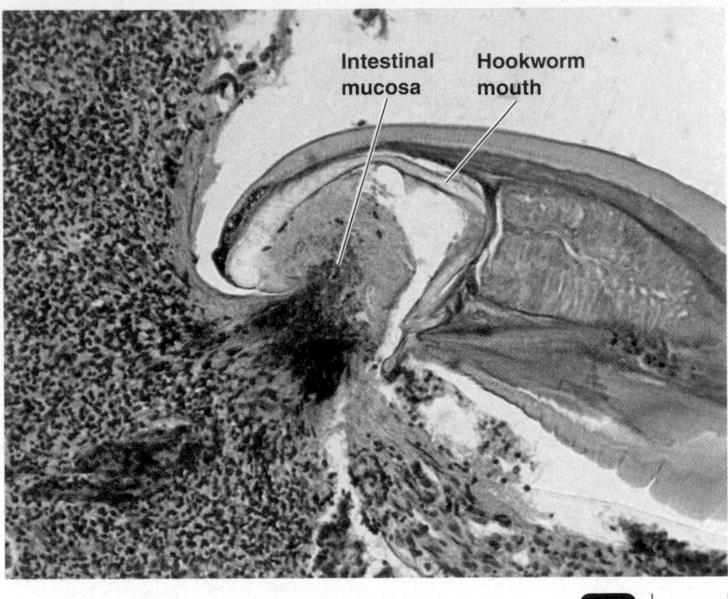

Figure 25.23 An *Ancylostoma* hookworm attached to intestinal mucosa. Notice how the mouth of the worm is adapted to feeding on the tissue.

Q Why can a hookworm infection lead to anemia?

Because the life cycle of the hookworm requires human feces to enter the soil and bare skin to contact contaminated soil, the incidence of the disease has declined greatly with improved sanitation and the practice of wearing shoes. Hookworm infections are diagnosed by finding parasite eggs in feces and can be treated effectively with mebendazole.

Ascariasis

One of the most widespread helminthic infections is **ascariasis,** caused by *Ascaris lumbricoides.* This condition is familiar to many American physicians. As described in Chapter 12 (page 358), diagnosis is often made when an adult worm emerges from the anus, mouth, or nose. These worms can be quite large, up to 30 cm (about 1 ft) in length (Figure 25.24). In the intestinal tract, they live on partially digested food and cause few symptoms.

The worm's life cycle begins when eggs (upwards of 200,000 per day) are shed in a person's feces and, under poor sanitary conditions, are ingested by another person. In the upper intestine, the eggs hatch into small wormlike larvae that pass into the bloodstream and then into the lungs. There they migrate into the throat and are swallowed. The larvae develop into egg-laying adults in the intestines. (All this migration just to return to the place where they started!)

In the lungs, the tiny larvae may cause some pulmonary symptoms. Extremely large numbers may block the intestine, bile duct, or pancreatic duct. The worms do not usually cause severe symptoms, but their presence can be manifested in distressing ways. The most dramatic consequences of infection with *A. lumbricoides* are from the migrations of adult worms. Worms have been known to leave the body of small children through the umbilicus (navel) and to escape through the nostrils of a sleeping person. Microscopic examination of feces for eggs is used for diagnosis. Once ascariasis is diagnosed, it can be effectively treated with mebendazole or albendazole.

Figure 25.24 *Ascaris lumbricoides,* the cause of ascariasis. These intestinal worms are large, the female up to about 30 cm in length.

Q What are the principal features of the life cycle of *A. lumbricoides*?

Trichinellosis

Most infections by the small roundworm *Trichinella spiralis,* called **trichinellosis** (formerly called trichinosis), are insignificant. The larvae, in encysted form, are located in muscles of the host. In 1970, routine autopsies of human diaphragm muscles showed that about 4% of cadavers tested carried this parasite.

Q&A The severity of the disease is generally proportional to the number of larvae ingested. Ingesting undercooked pork is probably the most common mode of infection (Figure 25.25), but eating the flesh of animals that feed on garbage (bears, for example) is an increasing cause of outbreaks. Quite a few human cases of trichinellosis have occurred in France from horsemeat infected in the United States and exported to restaurants. Severe cases can be fatal—sometimes in only a few days.

Any ground meat can be contaminated from machinery previously used to grind contaminated meats. Eating raw sausage or hamburger is a risky habit. One person acquired trichinellosis by chewing the fingernails after handling infected pork. Freezing pork for prolonged periods (for example, −23°C for 10 days) kills *T. spiralis.* However, some species found in wild game, such as *Trichinella nativa,* are not killed by freezing.

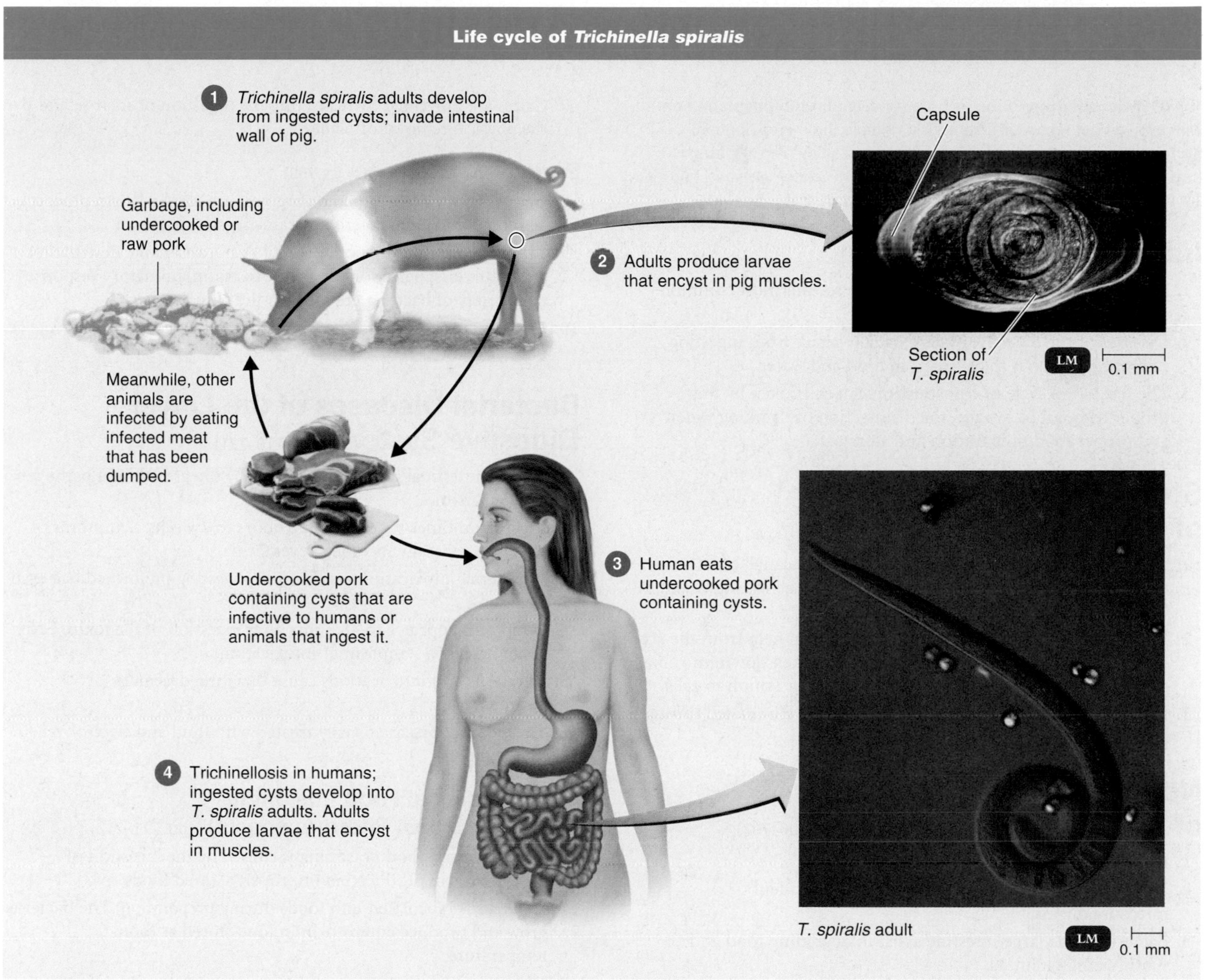
Life cycle of Trichinella spiralis
1 Trichinella spiralis adults develop from ingested cysts; invade intestinal wall of pig.
Garbage, including undercooked or raw pork
Capsule
2 Adults produce larvae that encyst in pig muscles.
Section of T. spiralis
LM 0.1 mm
Meanwhile, other animals are infected by eating infected meat that has been dumped.
Undercooked pork containing cysts that are infective to humans or animals that ingest it.
3 Human eats undercooked pork containing cysts.
4 Trichinellosis in humans; ingested cysts develop into T. spiralis adults. Adults produce larvae that encyst in muscles.
T. spiralis adult
LM 0.1 mm

Figure 25.25 Life cycle of *Trichinella spiralis,* the causative agent of trichinellosis.

Q What is the most common vehicle of infection of *T. spiralis*?

In the muscles of intermediate hosts such as pigs, the *T. spiralis* larvae are encysted in the form of short worms about 1 mm in length. When a human ingests the flesh of an infected animal, digestive action in the intestine removes the cyst wall. The organism then matures into the adult form. The adult worms spend only about a week in the intestinal mucosa and produce larvae that invade tissue. Eventually, the encysted larvae localize in muscle (common sites include the diaphragm and eye muscles), where they are barely visible in biopsied specimens.

Symptoms of trichinellosis include fever, swelling around the eyes, and gastrointestinal upset. Small hemorrhages under the fingernails are often observed. Biopsy specimens, as well as a number of serological tests, can be used in diagnosis. A serological ELISA test that detects the parasite in meats has been developed. Treatment consists of administering mebendazole to kill intestinal worms and corticosteroids to reduce inflammation.

In the past 10 years, the number of cases reported annually in the United States has varied from 16 to 129. Deaths are rare, and in most years there are none.

STUDY OUTLINE

The **MyMicrobiologyPlace** website (**www.microbiologyplace.com**) will help you get ready for tests with its simple three-step approach: ❶ **take a pre-test** and obtain a personalized study plan, ❷ **learn and practice** with animations, tutorials, and MP3 tutor sessions, and ❸ **test yourself** with quizzes and a chapter post-test.

Introduction (p. 705)

1. Diseases of the digestive system are the second most common illnesses in the United States.
2. Diseases of the digestive system usually result from ingesting microorganisms or their toxins in food and water.
3. The fecal–oral cycle of transmission can be broken by the proper disposal of sewage, the disinfection of drinking water, and proper food preparation and storage.

Structure and Function of the Digestive System (p. 706)

1. The gastrointestinal (GI) tract, or alimentary canal, consists of the mouth, pharynx, esophagus, stomach, small intestine, and large intestine.
2. In the GI tract, with mechanical and chemical help from the accessory structures, large food molecules are broken down into smaller molecules that can be transported by blood or lymph to cells.
3. Feces, the solids resulting from digestion, are eliminated through the anus.

Normal Microbiota of the Digestive System (pp. 706–707)

1. Large numbers of bacteria colonize the mouth.
2. The stomach and small intestine have few resident microorganisms.
3. Bacteria in the large intestine assist in degrading food and synthesizing vitamins.
4. Up to 40% of fecal mass is microbial cells.

Bacterial Diseases of the Mouth (pp. 707–710)

Dental Caries (Tooth Decay) (pp. 707–709)

1. Dental caries begin when tooth enamel and dentin are eroded and the pulp is exposed to bacterial infection.
2. *Streptococcus mutans,* found in the mouth, uses sucrose to form dextran from glucose and lactic acid from fructose.
3. Bacteria adhere to teeth by the sticky dextran, forming dental plaque.
4. Acid produced during carbohydrate fermentation destroys tooth enamel at the site of the plaque.
5. Gram-positive rods and filamentous bacteria can penetrate into dentin and pulp.
6. Carbohydrates such as starch, mannitol, sorbitol, and xylitol are not used by cariogenic bacteria to produce dextran and do not promote tooth decay.
7. Caries are prevented by restricting the ingestion of sucrose and by the physical removal of plaque.

Periodontal Disease (p. 709)

8. Caries of the cementum and gingivitis are caused by streptococci, actinomycetes, and anaerobic gram-negative bacteria.
9. Chronic gum disease (periodontitis) can cause bone destruction and tooth loss; periodontitis is due to an inflammatory response to a variety of bacteria growing on the gums.
10. Acute necrotizing ulcerative gingivitis is often caused by *Prevotella intermedia.*

Bacterial Diseases of the Lower Digestive System (pp. 710–721)

1. A gastrointestinal infection is caused by the growth of a pathogen in the intestines.
2. Incubation times range from 12 hours to 2 weeks. Symptoms of infection generally include a fever.
3. A bacterial intoxication results from ingesting preformed bacterial toxins.
4. Symptoms appear 1 to 48 hours after ingestion of the toxin. Fever is not usually a symptom of intoxication.
5. Infections and intoxications cause diarrhea, dysentery, or gastroenteritis.
6. These conditions are usually treated with fluid and electrolyte replacement.

Staphylococcal Food Poisoning (Staphylococcal Enterotoxicosis) (pp. 711–712)

7. Staphylococcal food poisoning is caused by the ingestion of an enterotoxin produced in improperly stored foods.
8. *S. aureus* is inoculated into foods during preparation. The bacteria grow and produce enterotoxin in food stored at room temperature.
9. Boiling for 30 minutes is not sufficient to denature the exotoxin.
10. Foods with high osmotic pressure and those not cooked immediately before consumption are most often the source of staphylococcal enterotoxicosis.
11. Laboratory identification of *S. aureus* isolated from foods is used to trace the source of contamination.

Shigellosis (Bacillary Dysentery) (p. 712)

12. Shigellosis is caused by any of four species of *Shigella.*
13. Symptoms include blood and mucus in stools, abdominal cramps, and fever. Infections by *S. dysenteriae* result in ulceration of the intestinal mucosa.

Salmonellosis (*Salmonella* Gastroenteritis) (pp. 712–714)

14. Salmonellosis, or *Salmonella* gastroenteritis, is caused by many *Salmonella enterica* serovars.
15. Symptoms include nausea, abdominal pain, and diarrhea and begin 12 to 36 hours after eating large numbers of *Salmonella.* Septic shock can occur in infants and in the elderly.

16. Mortality is lower than 1%, and recovery can result in a carrier state.
17. Cooking food will usually kill *Salmonella.*

Typhoid Fever (pp. 714–716)

18. *Salmonella typhi* causes typhoid fever; the bacteria are transmitted by contact with human feces.
19. Fever and malaise occur after a 2-week incubation period. Symptoms last 2 to 3 weeks.
20. *S. typhi* is harbored in the gallbladder of carriers.
21. Typhoid fever is treated with quinolones and cephalosporins; vaccines are available for high-risk people.

Cholera (pp. 716–717)

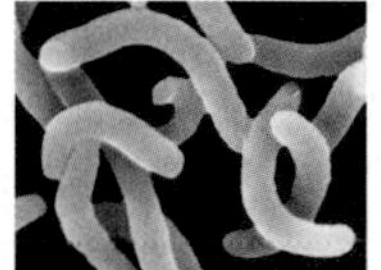

22. *Vibrio cholerae* O:1 and O:139 produce an exotoxin that alters the membrane permeability of the intestinal mucosa; the resulting vomiting and diarrhea cause a loss of body fluids.
23. The symptoms last for a few days. Untreated cholera has a 50% mortality rate.

Noncholera Vibrios (p. 717)

24. Ingestion of other *V. cholerae* serotypes can result in mild diarrhea.
25. *Vibrio* gastroenteritis can be caused by *V. parahaemolyticus* and *V. vulnificus.*
26. These diseases are contracted by eating contaminated crustaceans or contaminated mollusks.

Escherichia coli Gastroenteritis (pp. 717–718)

27. Traveler's diarrhea may be caused by enterotoxigenic or enteroinvasive strains of *E. coli.*
28. The disease is usually self-limiting and does not require chemotherapy.
29. Enterohemorrhagic *E. coli,* such as *E. coli* O157:H7, produces Shiga toxins that cause inflammation and bleeding of the colon, including hemorrhagic colitis and hemolytic uremic syndrome.
30. Shiga toxins can affect the kidneys to cause hemolytic uremic syndrome.

Campylobacter Gastroenteritis (p. 718)

31. *Campylobacter* is the second most common cause of diarrhea in the United States.
32. *Campylobacter* is transmitted in cow's milk.

Helicobacter Peptic Ulcer Disease (pp. 718–720)

33. *Helicobacter pylori* produces ammonia, which neutralizes stomach acid; the bacteria colonize the stomach mucosa and cause peptic ulcer disease.
34. Bismuth and several antibiotics may be useful in treating peptic ulcer disease.

Yersinia Gastroenteritis (p. 720)

35. *Y. enterocolitica* and *Y. pseudotuberculosis* are transmitted in meat and milk.
36. *Yersinia* can grow at refrigeration temperatures.

Clostridium perfringens Gastroenteritis (p. 720)

37. *C. perfringens* causes a self-limiting gastroenteritis.
38. Endospores survive heating and germinate when foods (usually meats) are stored at room temperature.
39. Exotoxin produced when the bacteria grow in the intestines is responsible for the symptoms.
40. Diagnosis is based on isolating and identifying the bacteria in stool samples.

Clostridium difficile–Associated Diarrhea (p. 720)

41. Growth of *C. difficile* following antibiotic therapy can result in mild diarrhea or colitis.
42. The condition is usually associated with hospitalized patients and nursing home residents.

Bacillus cereus Gastroenteritis (pp. 720–721)

43. Ingesting food contaminated with the soil saprophyte *Bacillus cereus* can result in diarrhea, nausea, and vomiting.

Viral Diseases of the Digestive System (pp. 721–729)

Mumps (p. 721)

1. Mumps virus enters and exits the body through the respiratory tract.
2. About 16 to 18 days after exposure, the virus causes inflammation of the parotid glands, fever, and pain during swallowing. About 4 to 7 days later, orchitis may occur.
3. After onset of the symptoms, the virus is found in the blood, saliva, and urine.
4. A measles, mumps, rubella (MMR) vaccine is available.

Hepatitis (pp. 721–728)

5. Inflammation of the liver is called hepatitis. Symptoms include loss of appetite, malaise, fever, and jaundice.
6. Viral causes of hepatitis include hepatitis viruses, Epstein-Barr virus (EBV), and cytomegalovirus (CMV).

Hepatitis A (pp. 721–723)

7. Hepatitis A virus (HAV) causes hepatitis A; at least 50% of all cases are subclinical.
8. HAV is ingested in contaminated food or water, grows in the cells of the intestinal mucosa, and spreads to the liver, kidneys, and spleen in the blood.
9. The virus is eliminated with feces.
10. The incubation period is 2 to 6 weeks; the period of disease is 2 to 21 days, and recovery is complete in 4 to 6 weeks.
11. A vaccine is available; passive immunization can provide temporary protection.

Hepatitis B (pp. 723–726)

12. Hepatitis B virus (HBV) causes hepatitis B, which is frequently serious.
13. HBV is transmitted by blood transfusions, contaminated syringes, saliva, sweat, breast milk, and semen.
14. Blood is tested for HB_sAg before being used in transfusions.
15. The average incubation period is 3 months; recovery is usually complete, but some patients develop a chronic infection or become carriers.
16. A vaccine against HB_sAg is available.

Hepatitis C (p. 726)

17. Hepatitis C virus (HCV) is transmitted via blood.
18. The incubation period is 2 to 22 weeks; the disease is usually mild, but some patients develop chronic hepatitis.
19. Blood is tested for HCV antibodies before being used in transfusions.

Hepatitis D (Delta Hepatitis (pp. 726–727)

20. Hepatitis D virus (HDV) has a circular strand of RNA and uses HB_sAg as a coat.

Hepatitis E (pp. 727–728)

21. Hepatitis E virus (HEV) is spread by the fecal–oral route.

Other Types of Hepatitis (p. 728)

22. There is evidence of the existence of hepatitis types F and G.

Viral Gastroenteritis (pp. 728–729)

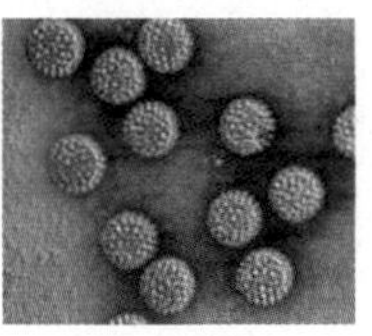

23. Viral gastroenteritis is most often caused by a rotavirus or norovirus.
24. The incubation period is 2 to 3 days; diarrhea lasts up to 1 week.

Fungal Diseases of the Digestive System (pp. 729–730)

1. Mycotoxins are toxins produced by some fungi.
2. Mycotoxins affect the blood, nervous system, kidneys, or liver.

Ergot Poisoning (pp. 729–730)

3. Ergot poisoning, or ergotism, is caused by the mycotoxin produced by *Claviceps purpurea.*
4. Cereal grains are the crop most often contaminated with the *Claviceps* mycotoxin.

Aflatoxin Poisoning (p. 730)

5. Aflatoxin is a mycotoxin produced by *Aspergillus flavus.*
6. Peanuts are the crop most often contaminated with aflatoxin.

Protozoan Diseases of the Digestive System (pp. 730–732)

Giardiasis (pp. 730–731)

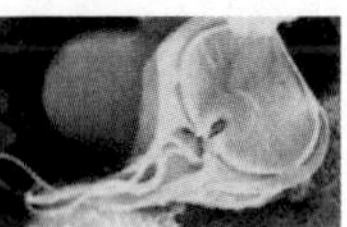

1. *Giardia lamblia* grows in the intestines of humans and wild animals and is transmitted in contaminated water.
2. Symptoms of giardiasis are malaise, nausea, flatulence, weakness, and abdominal cramps that persist for weeks.

Cryptosporidiosis (p. 731)

3. *Crytosporidium hominis* causes diarrhea; in immunosuppressed patients, the disease is prolonged for months.
4. The pathogen is transmitted in contaminated water.

Cyclospora Diarrheal Infection (p. 731)

5. *C. cayetanensis* causes diarrhea; the protozoan was first identified in 1993.
6. It is transmitted in contaminated produce.

Amoebic Dysentery (Amoebiasis) (pp. 731–732)

7. Amoebic dysentery is caused by *Entamoeba histolytica* growing in the large intestine.
8. The amoeba feeds on red blood cells and GI tract tissues. Severe infections result in abscesses.

Helminthic Diseases of the Digestive System (pp. 732–737)

Tapeworms (pp. 732–733)

1. Tapeworms are contracted by the consumption of undercooked beef, pork, or fish containing encysted larvae (cysticerci).
2. The scolex attaches to the intestinal mucosa of humans (the definitive host) and matures into an adult tapeworm.
3. Eggs are shed in the feces and must be ingested by an intermediate host.
4. Adult tapeworms can be undiagnosed in a human.
5. Neurocysticercosis in humans occurs when the pork tapeworm larvae encyst in humans.

Hydatid Disease (pp. 733–734)

6. Humans infected with the tapeworm *Echinococcus granulosus* might have hydatid cysts in their lungs or other organs.
7. Dogs and wolves are usually the definitive hosts, and sheep or deer are the intermediate hosts for *E. granulosus.*

Nematodes (pp. 734–737)

Pinworms (pp. 734–735)

8. Humans are the definitive host for pinworms, *Enterobius vermicularis.*
9. The disease is acquired by ingesting *Enterobius* eggs.

Hookworms (pp. 735–736)

10. Hookworm larvae bore through skin and migrate to the intestine to mature into adults.
11. In the soil, hookworm larvae hatch from eggs shed in feces.

Ascariasis (p. 736)

12. *Ascaris lumbricoides* adults live in human intestines.
13. Ascariasis is acquired by ingesting *Ascaris* eggs.

Trichinellosis (pp. 736–737)

14. *Trichinella spiralis* larvae encyst in muscles of humans and other mammals to cause trichinellosis.
15. The roundworm is contracted by ingesting undercooked meat containing larvae.
16. Adult females mature in the intestine and lay eggs; the new larvae migrate to invade muscles.
17. Symptoms include fever, swelling around the eyes, and gastrointestinal upset.

STUDY QUESTIONS

Answers to the Review and Multiple Choice questions can be found by turning to the blue Answers tab at the back of the textbook.

Review

1. **DRAW IT** Identify the site colonized by the following organisms: *Echinococcus granulosus, Enterobius vermicularis, Giardia, Helicobacter pylori*, hepatitis B virus, mumps virus, rotavirus, *Salmonella, Shigella, Streptococcus mutans, Trichinella spiralis.*

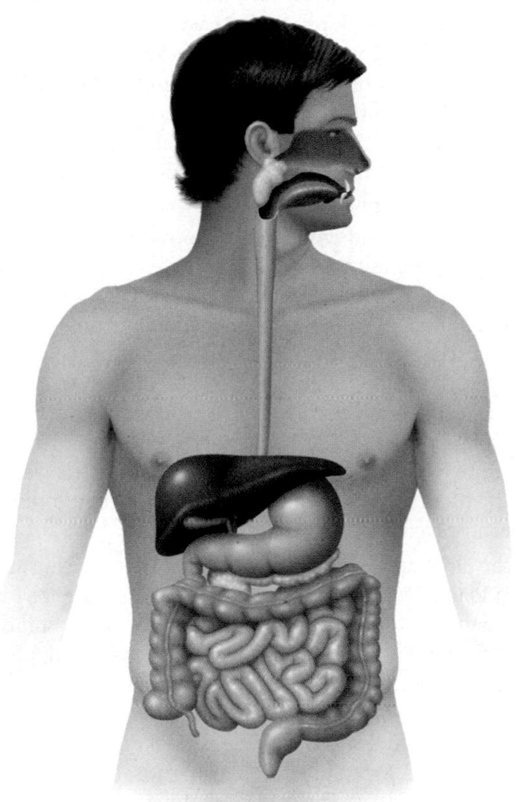

2. Complete the following table:

Disease	Causative Agent	Suspect Foods	Symptoms	Treatment
Staphylococcal food poisoning				
Shigellosis				
Salmonellosis				
Cholera				
Traveler's diarrhea				

3. Complete the following table:

Causative Agent	Suspect Foods	Treatment	Prevention
Vibrio parahaemolyticus			
V. vulnificus			
Enterotoxigenic *E. coli*			
Enteroinvasive *E. coli*			
Enterohemorrhagic *E. coli*			
Campylobacter jejuni			
Yersinia enterocolitica			
Clostridium perfringens			
Bacillus cereus			

4. *E. coli* is part of the normal microbiota of the intestines and can cause gastroenteritis. Explain why this one bacterial species is both beneficial and harmful.
5. Define *mycotoxin.* Give an example of a mycotoxin.
6. Explain how the following diseases differ and how they are similar: giardiasis, amoebic dysentery, *Cyclospora* diarrheal infection, and cryptosporidiosis.
7. Differentiate among the following factors of bacterial intoxication and bacterial infection: prerequisite conditions, causative agents, onset, duration of symptoms, and treatment.
8. Complete the following table:

Disease	Causative Agent	Mode of Transmission	Site of Infection	Symptoms	Prevention
Mumps					
Hepatitis A					
Hepatitis B					
Viral gastro-enteritis					

9. Look at lifecycle diagrams for human tapeworm and trichinellosis. Indicate stages in the life cycles that could be easily broken to prevent these diseases.

Multiple Choice

1. All of the following can be transmitted by recreational (i.e., swimming) water sources *except*
   a. amoebic dysentery.
   b. cholera.
   c. giardiasis.
   d. hepatitis B.
   e. salmonellosis.
2. A patient with nausea, vomiting, and diarrhea within 5 hours after eating most likely has
   a. shigellosis.
   b. cholera.
   c. *E. coli* gastroenteritis.
   d. salmonellosis.
   e. staphylococcal food poisoning.
3. Isolation of *E. coli* from a stool sample is diagnostic proof that the patient has
   a. cholera.
   b. *E. coli* gastroenteritis.
   c. salmonellosis.
   d. typhoid fever.
   e. none of the above

4. Gastric ulcers are caused by
   a. stomach acid.
   b. *Helicobacter pylori.*
   c. spicy food.
   d. acidic food.
   e. stress.
5. Microscopic examination of a patient's fecal culture shows comma-shaped bacteria. These bacteria require 2–4% NaCl to grow. The bacteria probably belong to the genus
   a. *Campylobacter.*
   b. *Escherichia.*
   c. *Salmonella.*
   d. *Shigella.*
   e. *Vibrio.*
6. A cholera epidemic in Peru had all of the following characteristics. Which one *led* to the others?
   a. eating raw fish
   b. sewage contamination of water
   c. catching fish in contaminated water
   d. *Vibrio* in fish intestine
   e. including fish intestines with edibles

Use the following choices to answer questions 7–10:
   a. *Campylobacter*
   b. *Cryptosporidium*
   c. *Escherichia*
   d. *Salmonella*
   e. *Trichinella*

7. Identification is based on the observation of oocysts in feces.
8. A characteristic disease symptom caused by this microorganism is swelling around the eyes.
9. Microscopic observation of a stool sample reveals gram-negative helical cells.
10. This microbe is frequently transmitted to humans via raw eggs.

Critical Thinking

1. Why is a human infection of trichinellosis considered a dead-end for the parasite?
2. Complete the following table:

Disease	Conditions Necessary for Microbial Growth	Basis for Diagnosis	Prevention
Staphylococcal food poisoning			
Salmonellosis			
C. difficile diarrhea			

3. Match the foods in column A with the genus of microorganism (column B) most likely to contaminate each:

Column A	Column B
________ a. Beef	1. *Vibrio*
________ b. Delicatessen meats	2. *Campylobacter*
________ c. Chicken	3. *E. coli* O157:H7
________ d. Milk	4. *Listeria*
________ e. Oysters	5. *Salmonella*
________ f. Pork	6. *Trichinella*

   What disease does each microbe cause? How can these diseases be prevented?

4. Which diseases of the gastrointestinal tract can be acquired by swimming in a pool or lake? Why are these diseases not likely to be acquired while swimming in the ocean?

Clinical Applications

1. In New York on April 26, patient A was hospitalized with a 2-day history of diarrhea. An investigation revealed that patient B had onset of watery diarrhea on April 22. On April 24, three other people (patients C, D, and E) had onset of diarrhea. All three had vibriocidal antibody titers $\geq$ 640. In Ecuador on April 20, B bought crabs that were boiled and shelled. He shared crabmeat with two people (F and G), then froze the remaining crab in a bag. Patient A returned to New York on April 21 with the bag of crabmeat in his suitcase. The bag was placed in a freezer overnight and thawed on April 22 in a double-boiler for 20 minutes. The crab was served 2 hours later in a crab salad. The crab was consumed during a 6-hour period by A, C, D, and E. Individuals F and G did not become ill. What is the etiology of this disease? How was it transmitted, and how could it have been prevented?
2. The 2130 students and employees of a public school system developed diarrheal illness on April 2. The cafeteria served chicken that day. On April 1, part of the chicken was placed in water-filled pans and cooked in an oven for 2 hours at a dial setting of 177°C. The oven was turned off, and the chicken was left overnight in the warm oven. The remainder of the chicken was cooked for 2 hours in a steam cooker and then left in the device overnight at the lowest possible setting (43°C). Two serotypes of a gram-negative, cytochrome oxidase-negative, lactose-negative rods were isolated from 32 patients. What is the pathogen? How could this outbreak have been prevented?
3. A 31-year-old man became feverish 4 days after arriving at a vacation resort in Idaho. During his stay, he ate at two restaurants that were not associated with the resort. At the resort, he drank soft drinks with ice, used the hot tub, and went fishing. The resort is supplied by a well that was dug 3 years ago. He went to the hospital when he developed vomiting and bloody diarrhea. Gram-negative, lactose-negative bacteria were cultured from his stool. The patient recovered after receiving intravenous fluids. What microorganism most likely caused his symptoms? How is this disease transmitted? What is the most likely source of his infection, and how would you verify the source?
4. Three to 5 days after eating Thanksgiving dinner at a restaurant, 112 people developed fever and gastroenteritis. All the food had been consumed except for five "doggie" bags. Bacterial analysis of the mixed contents of the bags (containing roast turkey, giblet gravy, and mashed potatoes) showed the same bacterium that was isolated from the patients. The gravy had been prepared from giblets of 43 turkeys that had been refrigerated for 3 days prior to preparation. The uncooked giblets were ground in a blender and added to a thickened hot stock mixture. The gravy was not reboiled and was stored at room temperature throughout Thanksgiving Day. What was the source of the illness? What was the most likely etiologic agent? Was this an infection or an intoxication?

26 Microbial Diseases of the Urinary and Reproductive Systems

The **urinary system** is composed of organs that regulate the chemical composition and volume of the blood and as a result excrete mostly nitrogenous waste products and water. Because it provides an opening to the outside environment, the urinary system is prone to infections from external contacts. The mucosal membranes that line the urinary system are moist and, compared to skin, more supportive of bacterial growth.

The **reproductive system** shares several of the organs of the urinary system. Its function is to produce gametes to propagate the species and, in the female, to support and nourish the developing embryo and fetus. In the same fashion as the urinary system, it provides openings to the external environment and is therefore prone to infections. This is especially true because intimate sexual contact can promote exchange of microbial pathogens between individuals. It is not surprising, therefore, that certain pathogens have adapted to this environment and a sexual mode of transmission. Often they have done this at the cost of an inability to survive in more rigorous environments.

UNDER THE MICROSCOPE

Leptospira interrogans. This pathogen, which causes a disease, leptospirosis, has a morphology similar to the spirochete that causes syphilis.

Q&A

Both *Leptospira interrogans* and the spirochete that causes syphilis penetrate deeply into the tissue of organs. What is it about their morphology that facilitates this?

Look for the answer in the chapter.

Structure and Function of the Urinary System

LEARNING OBJECTIVE

26-1 List the antimicrobial features of the urinary system.

The **urinary system** consists of two *kidneys,* two *ureters,* a single *urinary bladder,* and a single *urethra* (Figure 26.1). Certain wastes, collectively called *urine,* are removed from the blood as it circulates through the kidneys. The urine passes through the ureters into the urinary bladder, where it is stored prior to elimination from the body through the urethra. In the female, the urethra conveys only urine to the exterior. In the male, the urethra is a common tube for both urine and seminal fluid.

Where the ureters enter the urinary bladder, physiological valves prevent the backflow of urine to the kidneys. This mechanism helps shield the kidneys from lower urinary tract infections. In addition, the acidity of normal urine has some antimicrobial properties. The flushing action of urine during urination also tends to remove potentially infectious microbes.

CHECK YOUR UNDERSTANDING

✓ Does the pH of urine facilitate the growth of most bacteria? 26-1

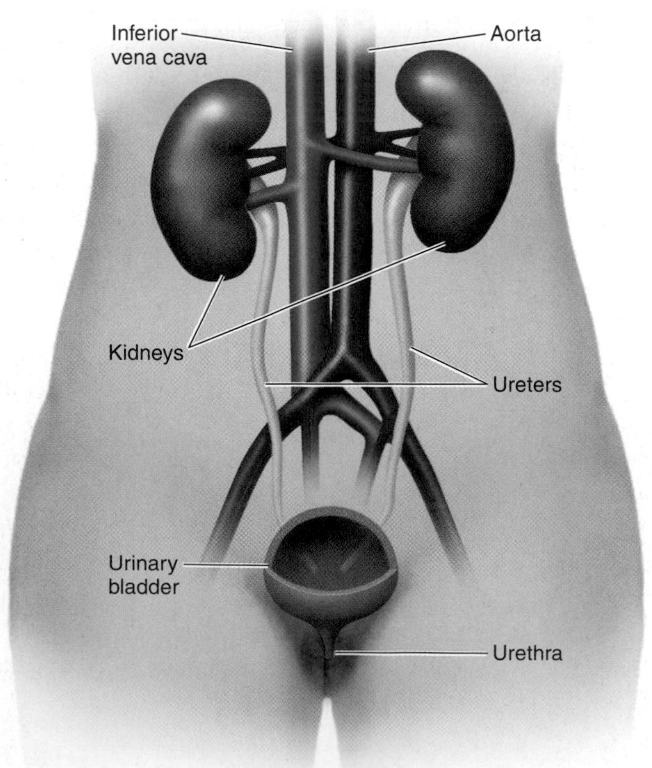

Figure 26.1 Organs of the human urinary system, shown here in the female.

What anatomical features of the urinary system help prevent colonization by microbes?

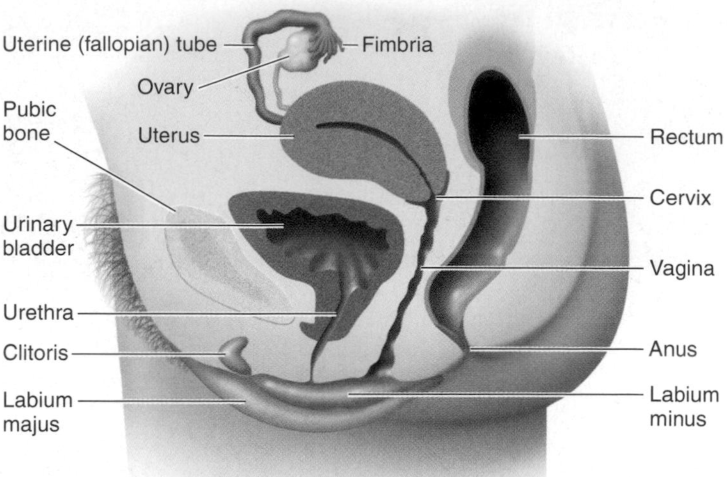

(a) Side view section of female pelvis showing reproductive organs

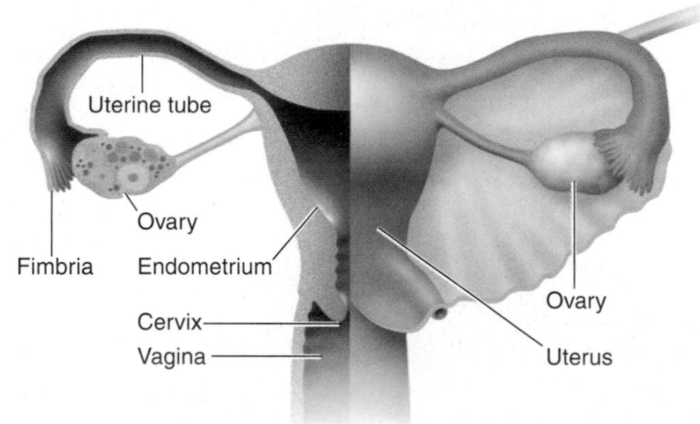

(b) Front view of female reproductive organs, with the uterine tube and ovary to the left in the drawing sectioned. The fimbriae move to create fluid movement that moves the egg into the uterine tube.

Figure 26.2 Female reproductive organs.

Q **Where are normal microbiota found in the female reproductive system?**

Structure and Function of the Reproductive Systems

LEARNING OBJECTIVE

26-2 Identify the portals of entry for microbes into the female and male reproductive systems.

The **female reproductive system** consists of two *ovaries,* two *uterine (fallopian) tubes,* the *uterus,* including the *cervix,* the *vagina,* and *external genitals* (Figure 26.2). The ovaries produce female sex hormones and ova (eggs). When an ovum is released during the process of ovulation, it enters a uterine tube, where fertilization may occur if viable sperm are present. The fertilized ovum (zygote) descends the tube and enters the uterus. It

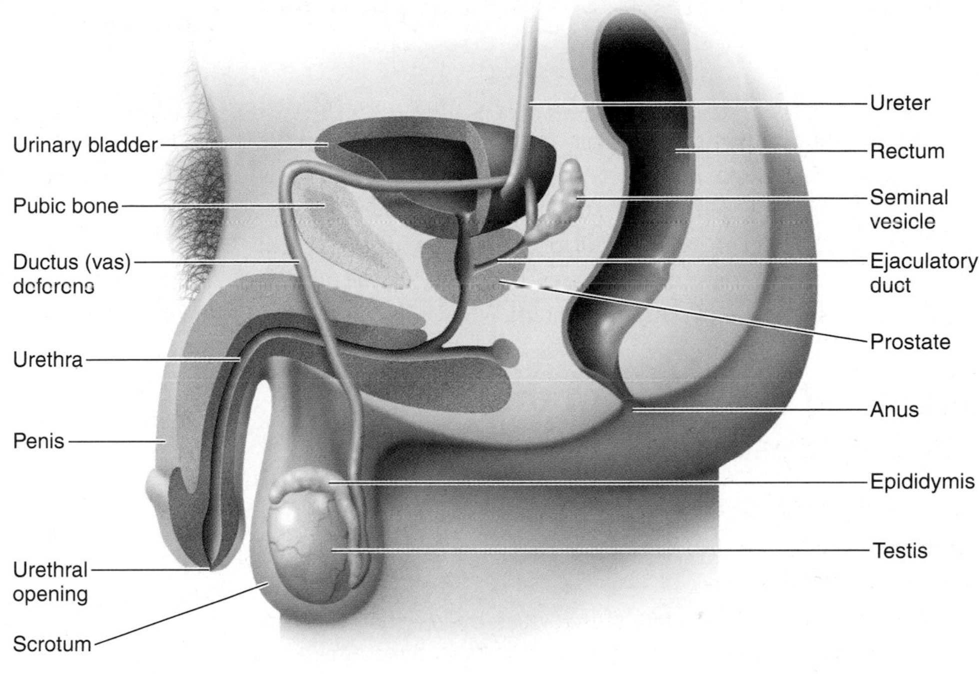

Figure 26.3 Male reproductive and urinary organs. A side view section of a male pelvis.

Q **What factors protect the male urinary and reproductive systems from infection?**

implants in the inner wall of the uterus and remains there while it develops into an embryo and, later, a fetus. The external genitals (*vulva*) include the clitoris, labia, and glands that produce a lubricating secretion during copulation.

The **male reproductive system** consists of two *testes,* a system of *ducts, accessory glands,* and the *penis* (Figure 26.3). The testes produce male sex hormones and sperm. To exit the body, sperm cells pass through a series of ducts: the epididymis, ductus (vas) deferens, ejaculatory duct, and urethra.

CHECK YOUR UNDERSTANDING

✓ Look at Figure 26.2. Is a microbe entering the female reproductive system (the uterus, etc.) necessarily also entering the bladder, causing cystitis? **26-2**

Normal Microbiota of the Urinary and Reproductive Systems

LEARNING OBJECTIVE

26-3 Describe the normal microbiota of the upper urinary tract, the male urethra, and the female urethra and vagina.

Normal urine is sterile, but it may become contaminated with microbiota of the skin near the end of its passage through the urethra. Therefore, urine collected directly from the urinary bladder has fewer microbial contaminants than voided urine.

The predominant bacteria in the vagina are the lactobacilli. These bacteria produce lactic acid, which maintains the acidic pH (3.8 to 4.5) of the vagina, inhibiting the growth of most other microbes. Most vaginal lactobacilli produce hydrogen peroxide, which also inhibits growth of other bacteria. Estrogens (sex hormones) promote the growth of lactobacilli by enhancing the production of glycogen by vaginal epithelial cells. The glycogen quickly breaks down into glucose, which the lactobacilli metabolize into lactic acid.

Other bacteria, such as streptococci, various anaerobes, and some gram-negatives, are also found in the vagina. The yeastlike fungus *Candida albicans* (see pages 758–759) is part of the normal microbiota of 10–25% of women, even when they are asymptomatic.

Pregnancy and menopause are often associated with higher rates of urinary tract infections. The reason is that estrogen levels are lower, resulting in lower populations of lactobacilli and therefore less vaginal acidity.

The male urethra is usually sterile, except for a few contaminating microbes near the external opening.

CHECK YOUR UNDERSTANDING

✓ What is the association between estrogens and the microbiota of the vagina? **26-3**

DISEASES OF THE URINARY SYSTEM

The urinary system normally contains few microbes, but it is subject to opportunistic infections that can be quite troublesome. Almost all such infections are bacterial, although occasional infections by pathogens such as schistosome parasites, protozoa, and fungi occur. In addition, as we will see in this chapter, sexually transmitted diseases often affect the urinary system as well as the reproductive system.

Bacterial Diseases of the Urinary System

LEARNING OBJECTIVES

26-4 Describe the modes of transmission for urinary and reproductive system infections.

26-5 List the microorganisms that cause cystitis, pyelonephritis, and leptospirosis, and name the predisposing factors for these diseases.

Urinary system infections are most frequently initiated by an inflammation of the urethra, or *urethritis.* Infection of the urinary bladder is called *cystitis,* and infection of the ureters is *ureteritis.* The most significant danger from lower urinary tract infections is that they may move up the ureters and affect the kidneys, causing *pyelonephritis.* Occasionally the kidneys are affected by systemic bacterial diseases, such as *leptospirosis.* The pathogens causing these diseases are found in excreted urine.

Bacterial infections of the urinary system are usually caused by microbes that enter the system from external sources. In the United States there are about 7 million urinary tract infections each year. About 900,000 cases are of nosocomial origin, and probably 90% of these are associated with urinary catheters. Because of the proximity of the anus to the urinary opening, intestinal bacteria predominate in urinary tract infections. Most infections of the urinary tract are caused by *Escherichia coli.* Infections by *Pseudomonas,* because of their natural resistance to antibiotics, are especially troublesome.

Cystitis

Cystitis is a common inflammation of the urinary bladder in females. Symptoms often include *dysuria* (difficult, painful, urgent urination) and pyuria.

The female urethra is less than 2 inches long, and microorganisms traverse it readily. It is also closer than the male urethra to the anal opening and its contaminating intestinal bacteria. These considerations are reflected in the fact that the rate of urinary tract infections in women is eight times that of men. In either gender, most cases are due to infection by *E. coli,* which can be identified by cultivation on differential media such as MacConkey's agar: (Interestingly, daily ingestion of cranberry juice prevents *E. coli* from adhering to epithelial cells). Another frequent bacterial cause is the coagulase-negative *Staphylococcus saprophyticus* (sap-rō-fit'i-kus).

As a general rule, a urine sample with more than 100 colony forming units (CFUs) per milliliter of potential pathogens (such as coliforms) from a female patient with cystitis is considered significant. The diagnosis should also include a positive urine test for *leukocyte esterase* (LE), an enzyme produced by neutrophils—which indicates an active infection. Trimethoprim-sulfamethoxazole usually clears cases of cystitis quickly. Fluoroquinolone antibiotics or ampicillin are often successful if drug resistance is encountered.

Pyelonephritis

In 25% of untreated cases, cystitis may progress to **pyelonephritis,** an inflammation of one or both kidneys. Symptoms are fever and flank or back pain. In females, it is often a complication of lower urinary tract infections. The causative agent in about 75% of the cases is *E. coli.* Pyelonephritis generally results in bacteremia; blood cultures and a Gram stain of the urine for bacteria are useful for diagnosis. A urine sample of more than 10,000 CFUs/ml and a positive LE test indicate pyelonephritis. If pyelonephritis becomes chronic, scar tissue forms in the kidneys and severely impairs their function. Because pyelonephritis is a potentially life-threatening condition, treatment usually begins with intravenous, extended-term administration of a broad-spectrum antibiotic, such as a second- or third-generation cephalosporin.

Leptospirosis

Leptospirosis is primarily a disease of domestic or wild animals, but it can be passed to humans and sometimes causes severe kidney or liver disease. The causative agent is the spirochete *Leptospira interrogans* (in-tėr'rä-ganz), shown in Figure 26.4. *Leptospira* has a characteristic shape: an exceedingly fine spiral, only about 0.1 μm in diameter, wound so tightly that it is barely discernible under a darkfield microscope. Like other spirochetes, *L. interrogans* (so named because the hooked ends suggest a question mark) stains poorly and is difficult to see under a normal light microscope. It is an obligate aerobe that can be grown in a variety of artificial media supplemented with rabbit serum.

Animals infected with the spirochete shed the bacteria in their urine for extended periods. Humans become infected by contact with urine-contaminated water or soil or sometimes with animal tissue. People whose occupations expose them to animals or animal products are most at risk. Usually the pathogen enters through minor abrasions in the skin or mucous membranes. When ingested, it enters through the mucosa of the upper digestive system. In the United States, dogs and rats are the most common sources. Domestic dogs have a sizable rate of infection; even when immunized, they may continue to shed leptospira.

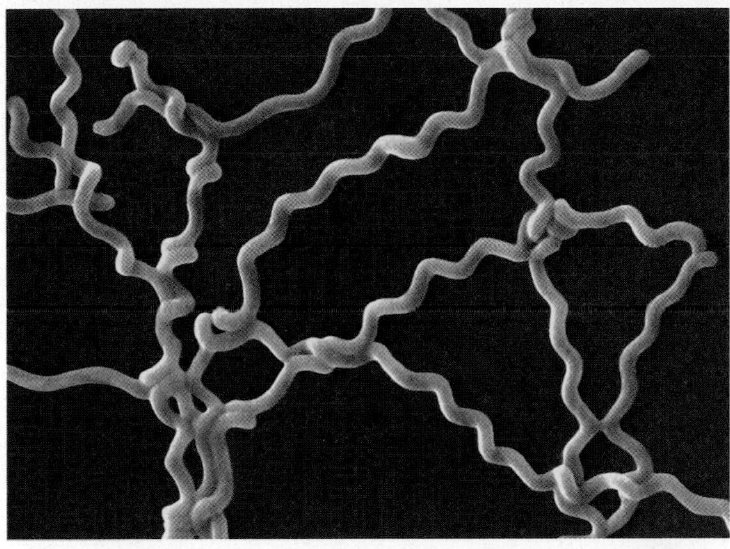

Figure 26.4 ***Leptospira interrogans*, the cause of leptospirosis.** This photo shows several of these tightly coiled spirochetes.

Q On what basis is *L. interrogans* named?

After an incubation period of 1 to 2 weeks, headaches, muscular aches, chills, and fever abruptly appear. Several days later, the acute symptoms disappear, and the temperature returns to normal. A few days later, however, a second episode of fever may occur. In a small number of cases the kidneys and liver become seriously infected (*Weil's disease*); kidney failure is the most common cause of death. Recovery results in a solid immunity, but only to the particular serovar involved. There are usually about 50 cases reported each year in the United States, but because the clinical symptoms are not distinctive, many cases are probably never diagnosed. A recent study in a clinic serving the urban poor in a large eastern U.S. city found that up to 16% of the patients tested positive for infection.

Most cases of leptospirosis are diagnosed by a serological test that is complicated and usually done by central reference laboratories. However, a number of rapid serological tests are available for a preliminary diagnosis. Also, a diagnosis can be made by sampling blood, urine, or other fluids for the organism or its DNA. Doxycyline (a tetracycline) is the recommended antibiotic for treatment; however, administration of antibiotics in later stages is often unsatisfactory. That immune reactions are responsible for pathogenesis in this stage may be an explanation.

Diseases of the urinary system are summarized in Diseases in Focus 26.1.

CHECK YOUR UNDERSTANDING

- ✓ Why is urethritis, an infection of the urethra, frequently preliminary to further infections of the urinary tract? **26-4**
- ✓ Why is *E. coli* the most common cause of cystitis, especially in females? **26-5**

DISEASES OF THE REPRODUCTIVE SYSTEMS

Microbes causing infections of the reproductive systems are usually very sensitive to environmental stresses and require intimate contact for transmission.

Bacterial Diseases of the Reproductive Systems

LEARNING OBJECTIVE

26-6 List the causative agents, symptoms, methods of diagnosis, and treatments for gonorrhea, nongonococcal urethritis (NGU), pelvic inflammatory disease (PID), syphilis, lymphogranuloma venereum (LGV), chancroid, and bacterial vaginosis.

Most diseases of the reproductive systems transmitted by sexual activity have been called **sexually transmitted diseases (STDs).** In recent years there has been a movement to replace this terminology with **sexually transmitted infections (STIs),** a change that is already commonplace in Europe. The reason is that the concept of "disease" implies obvious signs and symptoms, whereas many of the persons infected by the more common sexually transmitted pathogens do not have apparent signs or symptoms. The term *STI* seems more appropriate and is used in this book. More than 30 bacterial, viral, or parasitic infections have been identified as sexually transmitted. In the United States, it is estimated that over 15 million new cases of STIs occur annually. Many of these infections can be successfully treated with antibiotics and can be largely prevented by the use of condoms. However, over 60 million Americans have STIs, mostly viral, for which there is no effective cure.

Gonorrhea

One of the most common reportable, or notifiable, communicable diseases in the United States is **gonorrhea,** an STI caused by the gram-negative diplococcus *Neisseria gonorrhoeae.* An ancient disease, gonorrhea was described and given its present name by the Greek physician Galen in A.D. 150 (*gon* = semen + *rhea* = flow; a flow of semen—apparently, he confused pus with semen). The

DISEASES IN FOCUS 26.1

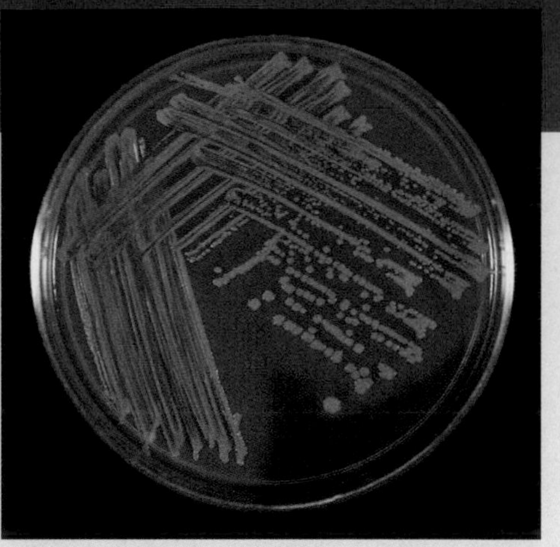

MacConkey agar culture from the patient's urine.

Bacterial Diseases of the Urinary System

Differential diagnosis is the process of identifying the disease from a list of possible diseases that fit the information derived from examining a patient. A differential diagnosis is important for providing initial treatment and for laboratory testing. For example, a 20-year-old woman felt a stinging sensation when urinating and felt an urgent need to urinate, even if very little urine was excreted. Lactose-fermenting, gram-negative rods were cultured from her urine (see the photo). Use the table below to identify infections that could cause these symptoms. For the solution, go to www.microbiologyplace.com.

Disease	Pathogen	Symptoms	Diagnosis	Treatment
Cystitis urinary bladder infection	*Escherichia coli, Staphylococcus saprophyticus*	Difficulty or pain in urination	>100 CFUs/ml potential pathogens and + LE test	Trimethoprim-sulfamethoxazole
Pyelonephritis (kidney infection)	Primarily *E. coli*	Fever; back or flank pain	$>10^4$ CFUs/ml and + LE test	Cephalosporin
Leptospirosis (kidney infection)	*Leptospira interrogans*	Headaches, muscular aches, fever, kidney failure a possible complication	Serological test	Doxycycline

incidence of gonorrhea has tended to decrease in recent years, but more than 300,000 cases are still reported in the United States each year (**Figure 26.5a**). The true number of cases is probably much larger, probably two or three times those reported (**Figure 26.5b**). More than 60% of patients with gonorrhea are aged 15 to 24.

To infect, the gonococcus must attach to the mucosal cells of the epithelial wall by means of fimbriae. The pathogen invades the spaces separating columnar epithelial cells, which are found in the oral-pharyngeal area, the eyes, rectum, urethra, opening of the cervix, and the external genitals of prepubertal females. The invasion sets up an inflammation and, when leukocytes move into the inflamed area, the characteristic pus forms. In men, a single unprotected exposure results in infection with gonorrhea 20–35% of the time. Women become infected 60–90% of the time from a single exposure.

Men become aware of a gonorrheal infection by painful urination and a discharge of pus-containing material from the urethra (**Figure 26.6**). About 80% of infected men show these obvious symptoms after an incubation period of only a few days; most others show symptoms in less than a week. In the days before antibiotic therapy, symptoms persisted for weeks. A common complication is urethritis, although this is more likely to be the result of coinfection with *Chlamydia,* which will be discussed shortly. An uncommon complication is *epididymitis,* an infection of the epididymis. Usually only unilateral, this is a painful condition resulting from the infection ascending along the urethra and the ductus deferens (see Figure 26.3).

In women, the disease is more insidious. Only the cervix, which contains columnar epithelial cells, is infected. The vaginal walls are composed of stratified squamous epithelial cells, which are not colonized. Very few women are aware of the infection. Later in the course of the disease, there might be abdominal pain from complications such as pelvic inflammatory disease (discussed on page 751).

In both men and women, untreated gonorrhea can disseminate and become a serious, systemic infection. Complications of gonorrhea can involve the joints, heart (*gonorrheal endocarditis*), meninges (*gonorrheal meningitis*), eyes, pharynx, or other parts of the body. *Gonorrheal arthritis,* which is caused by the growth of the gonococcus in fluids in joints, occurs in about 1% of gonorrhea cases. Joints commonly affected include the wrist, knee, and ankle.

If the mother is infected with gonorrhea, the eyes of the infant can become infected as it passes through the birth canal. This condition, **ophthalmia neonatorum,** can result in blindness. Because of the seriousness of this condition and the difficulty of being sure the mother is free of gonorrhea, antibiotics are placed in the eyes of all newborn infants. If the mother is known to be

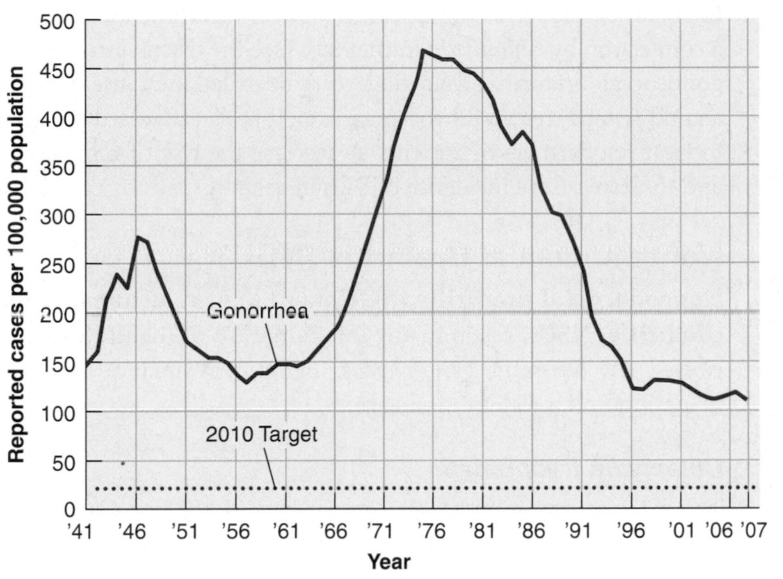

Note: The Healthy People 2010 target for gonorrhea is 19.0 cases per 100,000 population.

(a) Incidence of gonorrhea in the United States, 1941–2007

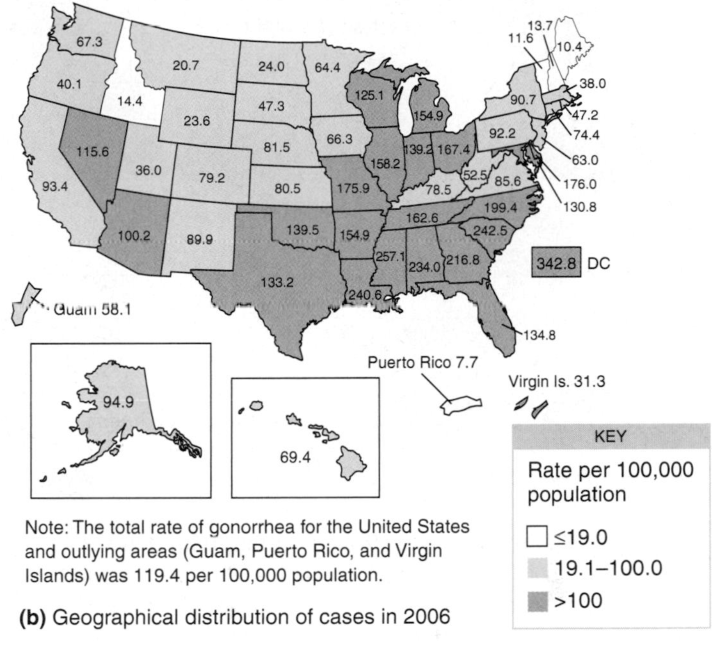

Note: The total rate of gonorrhea for the United States and outlying areas (Guam, Puerto Rico, and Virgin Islands) was 119.4 per 100,000 population.

(b) Geographical distribution of cases in 2006

Figure 26.5 The U.S. incidence and distribution of gonorrhea.
Source: CDC, *STI Surveillance*, November 13, 2007.

Q **How do gonococci attach to mucosal epithelial cells?**

infected, an intramuscular injection of antibiotic is also administered to the infant. Some sort of prophylaxis is required by law in most states. Gonorrheal infections can also be transferred by hand contact from infected sites to the eyes of adults.

Gonorrheal infections can be acquired at any point of sexual contact; pharyngeal and anal gonorrhea are not uncommon. The symptoms of **pharyngeal gonorrhea** often resemble those of the usual septic sore throat. **Anal gonorrhea** can be painful and accompanied by discharges of pus. In most cases, however, the symptoms are limited to itching.

Increased sexual activity with a series of partners, and the fact that in women the disease may go unrecognized, contributed considerably to the increased incidence of gonorrhea and other STIs during the 1960s and 1970s. The widespread use of oral contraceptives also contributed to the increase. Oral contraceptives often replaced condoms and spermicides, which help prevent disease transmission.

There is no effective adaptive immunity to gonorrhea. The conventional explanation is that the gonococcus exhibits extraordinary antigenic variability—which is true. Lately, though, an alternative theory has appeared that provides an additional mechanism. The gonococcus is capable of producing several different opacity (Opa) proteins (see Chapter 15, page 432), which are required for the bacteria to adhere to and infect the cells lining the urinary and reproductive systems. These Opa proteins bind to a family of receptors on host cells; $CD4^+$ lymphocytes (a group including helper T cells and long-lived memory cells) express only one of this family of receptors. When this receptor on the lymphocyte binds to a particular Opa protein on the gonococcus, it prevents activation of the lymphocyte and turns off proliferation. This blocks development of an immunological memory against *N. gonorrhoeae.* (Experiments show that $CD4^+$ lymphocytes lacking this particular Opa receptor are stimulated to a strong immunological response). This mechanism that inhibits the adaptive immune response may also explain why infection with gonorrhea carries an increased risk of acquiring other STIs, including HIV.

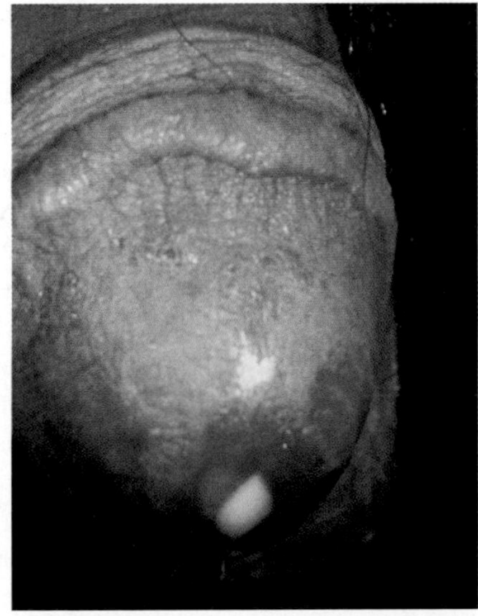

Figure 26.6 Pus-containing discharge from the urethra of a man with an acute case of gonorrhea.

Q **What causes pus formation in gonorrhea?**

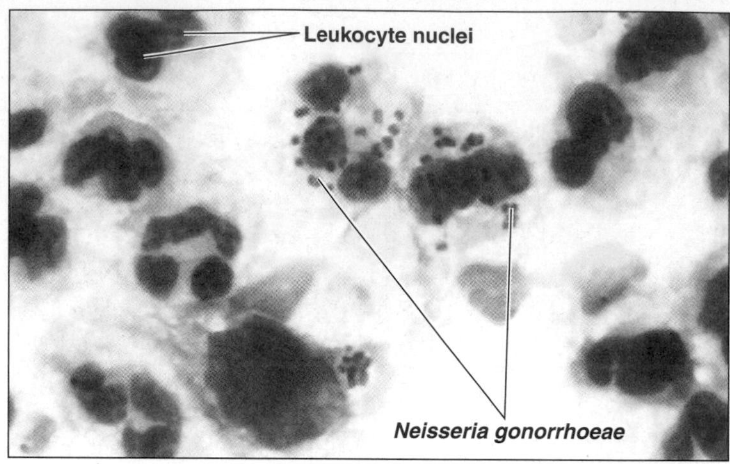

Figure 26.7 A smear of pus from a patient with gonorrhea. The *Neisseria gonorrhoeae* bacteria are contained within phagocytic leukocytes. These gram-negative bacteria are visible here as pairs of cocci. The large stained bodies are the nuclei of the leukocytes.

Q How is gonorrhea diagnosed?

Diagnosis of Gonorrhea

Gonorrhea in men is diagnosed by finding gonococci in a stained smear of pus from the urethra. The typical gram-negative diplococci within the phagocytic leukocytes are readily identified (**Figure 26.7**). It is uncertain whether these intracellular bacteria are in the process of being killed or whether they survive indefinitely. Probably at least a fraction of the bacterial population remains viable. Gram staining of exudates is not as reliable with women. Usually, a culture is taken from within the cervix and grown on special media. Cultivation of the nutritionally fastidious bacterium requires an atmosphere enriched in carbon dioxide. The gonococcus is very sensitive to adverse environmental influences (desiccation and temperature) and survives poorly outside the body. It even requires special transporting media to keep it viable for short intervals before the cultivation is under way. Cultivation has the advantage of allowing determination of antibiotic sensitivity.

Diagnosis of gonorrhea has been aided by the development of an ELISA that detects *N. gonorrhoeae* in urethral pus or on cervical swabs within about 3 hours with high accuracy. Other rapid tests now available use monoclonal antibodies against antigens on the surface of the gonococcus. Nucleic acid amplification tests are very accurate for identifying clinical isolates from suspected cases.

Treatment of Gonorrhea

The guidelines for treating gonorrhea require constant revision as resistance appears (See the Clinical Focus box on the facing page). For gonorrhea affecting the cervical, urethral, or rectal tissues, the current recommendation is first to use cephalosporins, such as ceftriaxone or cefixime. Ceftriaxone is also recommended for cases of pharyngeal infection. Fluoroquinolones, because resistance to them has developed so rapidly, are no longer recommended. Unless a coinfection by *Chlamydia trachomatis* (see the discussion of nongonococcal urethritis, following) can be ruled out, the patient should also be treated for this organism. It is also standard practice to treat sex partners of patients to decrease the risk of reinfection and to decrease the incidence of STIs in general.

Nongonococcal Urethritis (NGU)

Nongonococcal urethritis (NGU), also known as **nonspecific urethritis (NSU),** refers to any inflammation of the urethra not caused by *Neisseria gonorrhoeae.* Symptoms include painful urination and a watery discharge.

Chlamydia trachomatis

The most common pathogen associated with NGU is *Chlamydia trachomatis.* Many people suffering from gonorrhea are coinfected with *C. trachomatis,* which infects the same columnar epithelial cells as the gonococcus. *C. trachomatis* is also responsible for the STI lymphogranuloma venereum (discussed on page 755) and trachoma (see page 605). Of special importance is the fact that five times as many cases are reported in women than men. In women, it is responsible for many cases of pelvic inflammatory disease (discussed on page 751), plus eye infections and pneumonia in infants born to infected mothers. Genital chlamydial infections are also associated with an increased risk of cervical cancer. It is uncertain whether chlamydial infection is an independent factor in this risk or whether it is associated with coinfections with human papillomavirus (page 758).

Because the symptoms are often mild in men and because women are usually asymptomatic, many cases of NGU go untreated. Although complications are not common, they can be serious. Men may develop inflammation of the epididymis. In women, inflammation of the uterine tubes may cause scarring, leading to sterility. As many as 60% of such cases may be from chlamydial rather than gonococcal infection. It is estimated that about 50% of men and 70% of women are unaware of their chlamydial infection.

For diagnosis, culturing is the most reliable method, but this requires specialized cultivation methods, takes 24 to 72 hours, and is not always conveniently available. There are a number of new non–culture-based tests available. Several of them amplify and detect DNA or RNA sequences of *C. trachomatis.* These amplification tests can be done quickly and are very sensitive, in the range of 80–91%; their specificity is close to 100%. They are, however, relatively expensive and require a laboratory with specialized equipment. Urine samples can be used, but the sensitivity is lower than with swabs. The most recent development in amplification testing is to use swab specimens (urethral or vaginal, as the case might be) collected by the patient themselves—which they tend to prefer.

In view of the serious complications often associated with infections by *C. trachomatis,* it is recommended that physicians routinely screen sexually active women 25 years of age and younger for infection. Screening is also recommended for other

CLINICAL FOCUS

Survival of the Fittest

As you read through this problem, you will see questions that health care providers ask themselves and each other as they solve a clinical problem. Try to answer each question as you read through the problem.

1. On May 24, a 35-year-old man was seen at the Denver STD Clinic with a history of painful urination and urethral discharge of approximately 1 month's duration.

   What other information do you need about the patient's history?

2. On March 11, he had returned from a "dating tour" in Thailand, during which he had sexual contact with seven or eight female prostitutes; he denied having had any sexual contact since returning to the United States.

   What sample should be taken, and how should it be tested?

3. *Neisseria gonorrhoeae* was identified by polymerase chain reaction (PCR) of urethral discharge. He was treated with a single 500-mg dose of ciprofloxacin orally.

   What is the advantage of PCR or enzyme immunoassay (EIA) over cultures for diagnosis?

4. PCR and EIA provide results within a few hours, eliminating the need for the patient to return for treatment. In this case, the patient returned to the clinic on June 7 with continuing symptoms. *N. gonorrhoeae* was again detected in urethral discharge. He denied having had any sexual contact since the previous visit. The attending physician requested antimicrobial susceptibility testing of the *N. gonorrhoeae* isolates.

   Why would the physician be interested in antimicrobial susceptibility test results on this patient's specimen?

5. One reason for the patient's failure to respond to ciprofloxacin may be due to infection with a fluoroquinolone-resistant *Neisseria gonorrhoeae.* Susceptibility testing would be helpful to explore this possibility.

   The treatment and control of gonorrhea has been complicated by the ability of *N. gonorrhoeae* to develop resistance to antimicrobial agents. Trends in antibiotic resistance are shown in the figure.

   How does antibiotic resistance emerge?

6. In an environment filled with antibiotics, bacteria that have mutations for antibiotic resistance will have a selective advantage and be the "fittest" to survive.

   How is antibiotic susceptibility determined?

7. *N. gonorrhoeae* must be grown in culture for disk-diffusion or broth dilution tests for antimicrobial susceptibility. The increasing use of nonculture methods for gonorrhea diagnosis such as PCR and EIA is a major challenge to monitoring antimicrobial resistance in *N. gonorrhoeae.*

Source: Data from CDC. *Sexually Transmitted Disease Surveillance 1998* and *2005.*

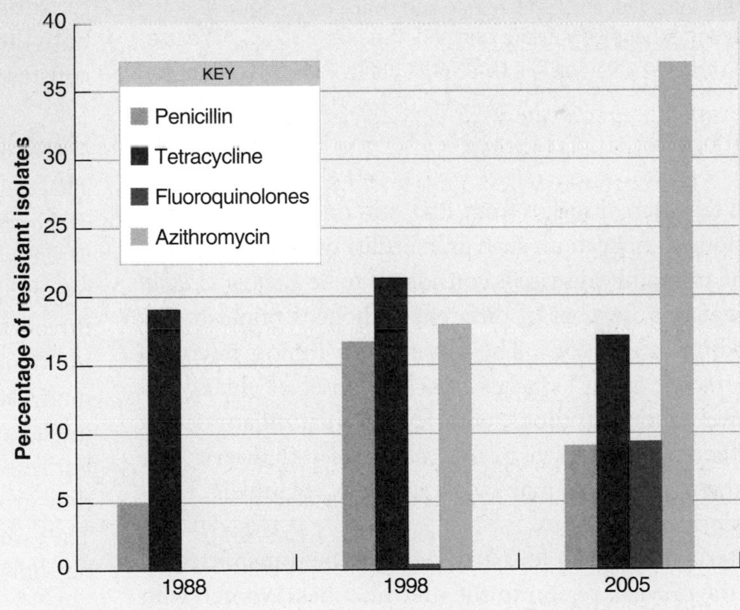

Antibiotic resistance in *N. gonorrhoeae.*

higher-risk groups, such as persons who are unmarried, had a prior risk of STIs, and have multiple sexual partners.

Bacteria other than *C. trachomatis* can also be implicated in NGU. Another cause of urethritis and infertility is *Ureaplasma urealyticum* (ū-rē-ä-lit′i-kum). This pathogen is a member of the mycoplasma (bacteria without a cell wall). Another mycoplasma, *Mycoplasma hominis* (ho′mi-nis), commonly inhabits the normal vagina but can opportunistically cause uterine tube infection.

Both chlamydia and mycoplasma are sensitive to tetracycline-type antibiotics such as doxycycline or to macrolide-type antibiotics such as azithromycin.

Pelvic Inflammatory Disease (PID)

Pelvic inflammatory disease (PID) is a collective term for any extensive bacterial infection of the female pelvic organs, particularly the uterus, cervix, uterine tubes, or ovaries. During their reproductive

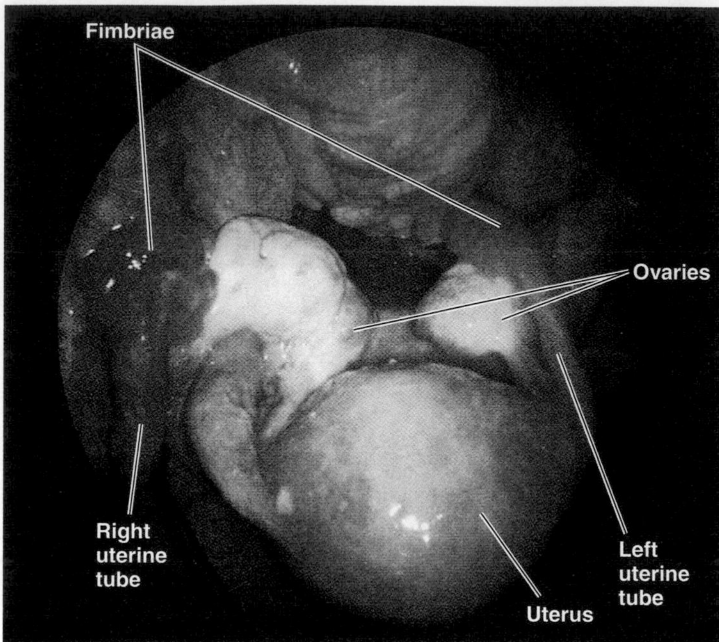

Figure 26.8 Salpingitis. This photograph, taken through a laparoscope (a specialized endoscope), shows an acutely inflamed right uterine tube and inflamed, swollen fimbriae and ovary, caused by salpingitis. The left tube is only mildly inflamed. (See Figure 26.2.) The use of a laparoscope is the most reliable diagnostic method for PID.

Q What is PID?

years, one in ten women suffers from PID, and one in four of these will have serious complications such as infertility or chronic pain.

Pelvic inflammatory disease is considered to be a *polymicrobial infection*—that is, a number of different pathogens might be the cause, including coinfections. The two most common microbes are *N. gonorrhoeae* and *C. trachomatis.* The onset of chlamydial PID is relatively more insidious, with fewer initial inflammatory symptoms than when caused by *N. gonorrhoeae.* However, the damage to the uterine tube may be greater with chlamydia, especially with repeated infections.

The bacteria may attach to sperm cells and be transported by them from the cervical region to the uterine tubes. Women who use barrier contraceptives, especially with spermicides, have a significantly lower rate of PID.

Infection of the uterine tubes, or **salpingitis,** is the most serious form of PID (**Figure 26.8**). Salpingitis can result in scarring that blocks the passage of ova from the ovary to the uterus, possibly causing sterility. One episode of salpingitis causes infertility in 10–15% of women; 50–75% become infertile after three or more such infections.

A blocked uterine tube may cause a fertilized ovum to be implanted in the tube rather than the uterus. This is called an *ectopic* (or *tubal*) *pregnancy,* and it is life-threatening because of the possibility of rupture of the tube and resulting hemorrhage. The reported cases of ectopic pregnancies have been increasing steadily, corresponding to the increasing occurrence of PID.

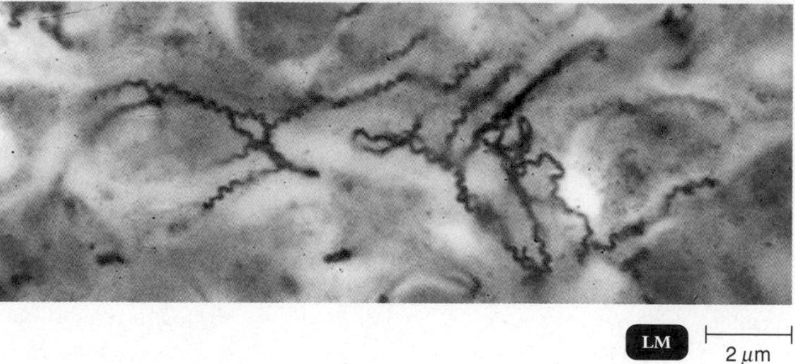

Figure 26.9 *Treponema pallidum,* the cause of syphilis. The microbes are made more visible in this brightfield micrograph by the use of a special silver stain.

Q A diagnostic method for syphilis is the darkfield microscope. Why not use a brightfield microscope?

A diagnosis of PID depends strongly on signs and symptoms, in combination with laboratory indications of a gonorrheal or chlamydial infection of the cervix. The recommended treatment for PID is the simultaneous administration of doxycycline and cefoxitin (a cephalosporin). This combination is active against both the gonococcus and chlamydia. Such recommendations are constantly being reviewed.

Syphilis

The causative agent of **syphilis** is a gram-negative spirochete, *Treponema pallidum* (**Figure 26.9**). Thin and tightly coiled, *T. pallidum* stains poorly with the usual bacterial stains. (The bacterial name is derived from the Greek words for twisted thread and pale.) *T. pallidum* lacks the enzymes necessary to build many complex molecules; therefore, it relies on the host for many of the compounds necessary for life. The organism loses infectiveness outside the mammalian host within a short time. For research purposes they are usually propagated in rabbits, but they grow slowly, with a generation time of 30 hours or more. They can be grown in cell culture, at low oxygen concentrations, but only for a few generations.

Q&A *T. pallidum* has no obvious virulence factors such as toxins, but it produces several lipoproteins that induce an inflammatory immune response. This is what apparently causes the tissue destruction of the disease. Almost immediately on infection, the organisms rapidly enter the bloodstream and invade deeper tissues, easily crossing the junctions between cells. They have a corkscrew-like motility that allows them to swim readily in gel-like tissue fluids.

The earliest reports of syphilis date back to the end of the fifteenth century in Europe, when the return of Columbus from the New World gave rise to a hypothesis that syphilis was introduced to Europe by his men. One English description of the "Morbus Gallicus" (French disease) seems to clearly describe syphilis as early as 1547 and ascribes its transmission in these terms: ". . . It is taken when one pocky person doth synne in lechery one with another."

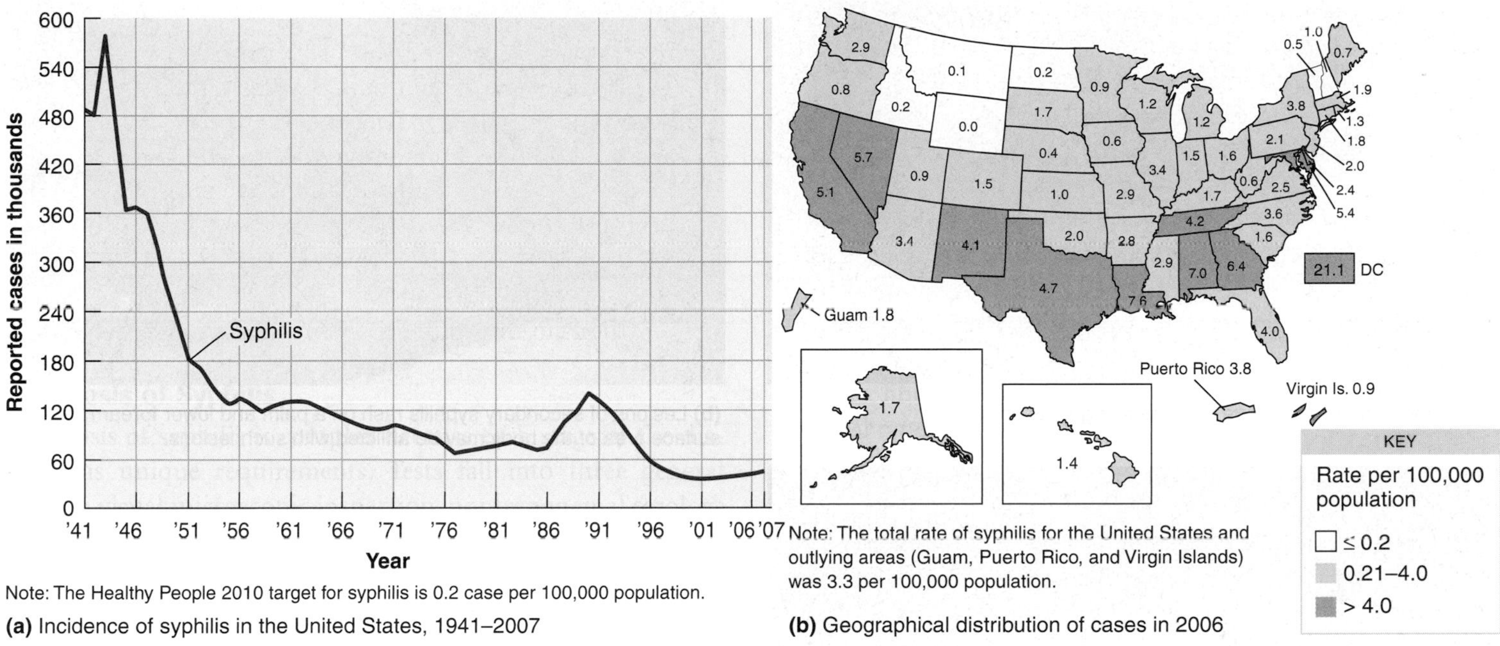

(a) Incidence of syphilis in the United States, 1941–2007

(b) Geographical distribution of cases in 2006

Figure 26.10 The U.S. incidence and distribution of primary and secondary syphilis.

Source: CDC, *STD Surveillance*, November 13, 2007.

Q How is syphilis diagnosed?

Separate strains of *T. pallidum* (subspecies *T.p. pertenue*) are responsible for certain tropically endemic skin diseases such as **yaws.** These cause skin lesions but are not sexually transmitted. However, there is evidence of a probable historical association with syphilis. Current research based on genetic analysis of *Treponema* spp. indicates that a pathogen of yaws found in South America adjacent to the Caribbean mutated into a sexually transmitted disease on contact with European explorers.

The number of new syphilis cases in the United States has remained fairly stable (Figure 26.10) compared with gonorrhea (see Figure 26.5). The relative stability in the incidence of syphilis is remarkable because the epidemiology of the two diseases is quite similar, and concurrent infections are not uncommon. A factor is that syphilis results in a significant, if imperfect, immunity—compared to no conferred immunity from gonorrhea.

Many states discontinued the requirements for premarital syphilis tests because so few cases were detected. At present, the population most at risk is economically disadvantaged inner city residents, especially drug-using prostitutes of both sexes. It is relatively rare in affluent societies.

Syphilis is transmitted by sexual contact of all kinds via syphilitic infections of the genitals or other body parts. The incubation period averages 3 weeks but can range from 2 weeks to several months. The disease progresses through several recognized stages.

Primary Stage Syphilis

In the *primary stage* of the disease, the initial sign is a small, hard-based **chancre,** or sore, which appears at the site of infection 10 to 90 days following exposure—on average, about 3 weeks (Figure 26.11a). The chancre is painless, and an exudate of serum forms in the center. This fluid is highly infectious, and examination with a darkfield microscope shows many spirochetes. In a couple of weeks, this lesion disappears. None of these symptoms causes any distress. In fact, many women are entirely unaware of the chancre, which is often on the cervix. In men, the chancre sometimes forms in the urethra and is not visible. During this stage, bacteria enter the bloodstream and lymphatic system, which distribute them widely in the body.

Secondary Stage Syphilis

Several weeks after the primary stage (the exact length of time varies and may overlap), the disease enters the *secondary stage,* characterized mainly by skin rashes of varying appearance. The rash is widely distributed on the skin and mucous membranes and is especially visible on the palms and on the soles (Figure 26.11b). The damage done to tissues at this stage and the later tertiary stage is caused by an inflammatory response to circulating immune complexes that lodge at various body sites. Other symptoms often observed are the loss of patches of hair, malaise, and mild fever. A few people may exhibit neurological symptoms.

At this stage, the lesions of the rash contain many spirochetes and are very infectious. Transmission from sexual contact occurs during the primary and secondary stages. Dentists and other health care workers coming into contact with fluid from these lesions can become infected by the spirochete entering through

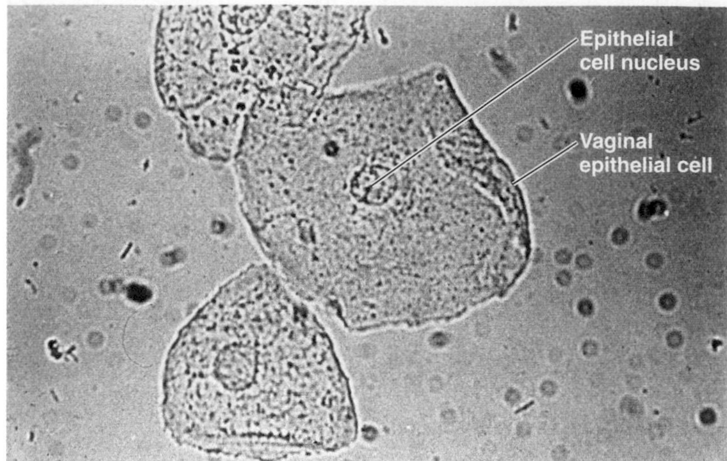

(a) Normal vaginal epithelial cell 12.5 μm

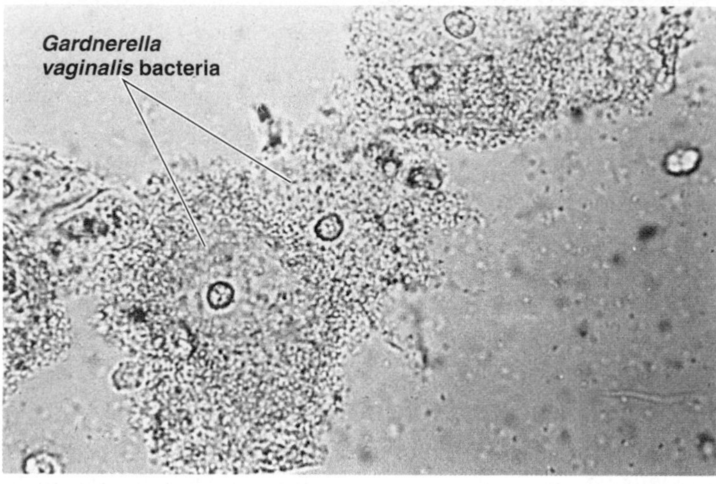

(b) Clue cell LM 12 μm

Figure 26.12 Clue cells. *Gardnerella* bacteria coat the surface of vaginal epithelial cells.

 **What symptoms would cause you to look for clue cells?**

Chancroid (Soft Chancre)

The STI known as **chancroid (soft chancre)** occurs most frequently in tropical areas, where it is seen more often than syphilis. The number of reported cases in the United States has been declining from a peak of 5000 cases in 1988. Almost all occur in New York, Texas, California, Florida, and Georgia. Like syphilis, its incidence is strongly associated with drug use. Because chancroid is so seldom seen by some physicians and is difficult to diagnose, it is probably underreported. It is very common in Africa, Asia, and Latin America.

In chancroid, a swollen, painful ulcer that forms on the genitals involves an infection of the adjacent lymph nodes. Infected lymph nodes in the groin area sometimes even break through and discharge pus to the surface. Such lesions are an important factor in the sexual transmission of HIV, especially in Africa. Lesions might also occur on such diverse areas as the tongue and lips. The causative agent is *Haemophilus ducreyi* (dü-krā'ē), a small gram-negative rod that can be isolated from exudates of lesions. Symptoms and the culture of these bacteria are the primary means of diagnosis. The recommended antibiotics include erythromycin and ceftriaxone.

Bacterial Vaginosis

Inflammation of the vagina due to infection, or **vaginitis,** is most commonly caused by one of several organisms: mainly the fungus *Candida albicans* (kan'did-ä al'bi-kans), the protozoan *Trichomonas vaginalis* (trik-ō-mōn'as va-jin-al'is), or the bacterium *Gardnerella vaginalis,* a small, pleomorphic gram-variable rod (see Diseases in Focus 26.2 on page 759). Most of these cases are attributed to the presence of *G. vaginalis* and are termed **bacterial vaginosis.** (Because there is no sign of inflammation, the term *vaginosis* is preferred to *vaginitis*).

The condition is something of an ecological mystery. It is believed that bacterial vaginosis is precipitated by some event that decreases the number of *Lactobacillus* vaginal bacteria that normally produce hydrogen peroxide. This competitive change allows bacteria, especially *G. vaginalis,* to proliferate, producing amines that contribute to a further rise in pH. These various bacteria, most of which are commonly found in the vaginas of asymptomatic women, are assumed to be metabolically interdependent. This situation does not lend itself to the application of Koch's postulates to determine a specific cause. There is no corresponding disease condition in men, but *G. vaginalis* is often present in their urethras. Therefore, the condition may be sexually transmitted, but it also occurs occasionally in women who have never been sexually active.

Bacterial vaginosis is characterized by a vaginal pH above 4.5 and a copious, frothy vaginal discharge. When tested with a potassium hydroxide solution, these vaginal secretions emit a fishy odor from presence of the amines produced by *G. vaginalis.* Diagnosis is based on the vaginal pH, fishy odor (the *whiff test*), and microscopic observation of *clue cells* in the discharge. These clue cells are sloughed-off vaginal epithelial cells covered with a biofilm of bacteria, mostly *G. vaginalis* (Figure 26.12). The disease has been considered more of a nuisance than a serious infection, but it is now seen as a factor in many premature births and low birth-weight infants.

Treatment is primarily by metronidazole, a drug that eradicates the anaerobes essential to continuation of the disease but allows the normal lactobacilli to repopulate the vagina. Treatments designed to restore the normal population of lactobacilli, such as application of acetic acid gels and even yogurt, have not been shown conclusively to be effective.

CHECK YOUR UNDERSTANDING

 Why is the disease condition of the female reproductive system, principally featuring growth of *Gardnerella vaginalis,* termed *vaginosis* rather than *vaginitis*? **26-6**

Viral Diseases of the Reproductive Systems

LEARNING OBJECTIVE

26-7 Discuss the epidemiology of genital herpes and genital warts.

Viral diseases of the reproductive system are difficult to treat, and so they represent an increasing health problem.

Genital Herpes

A much publicized STI is **genital herpes,** usually caused by *herpes simplex virus type 2 (HSV-2).* (The herpes simplex virus occurs as either type 1 or type 2.) Herpes simplex virus type 1 (HSV-1) is primarily responsible for cold sores or fever blisters (see page 597), but it can also cause genital herpes. The official names are human herpesvirus 1 and 2.

In the United States, one in four persons over the age of 30 is infected with HSV-2—most are unaware they are infected (Figure 26.13). There has been a marked increase in genital HSV-1 infections, which is usually acquired by oral–genital contact, and this now constitutes about half of cases of genital herpes in this country.

Genital herpes lesions appear after an incubation period of up to 1 week and cause a burning sensation. After this, vesicles appear (Figure 26.14). In both men and women, urination can be painful, and walking is quite uncomfortable; the patient is even irritated by clothing. Usually, the vesicles heal in a couple of weeks.

The vesicles contain infectious fluid, but many times the disease is transmitted when no lesions or symptoms are apparent. Semen may contain the virus. Condoms may not provide protection because in women the vesicles are usually on the external genitals (seldom on the cervix or within the vagina) and in men the vesicles may be on the base of the penis.

One of the most distressing characteristics of genital herpes is the possibility of recurrences. There is an element of truth in the medical adage that, unlike love, herpes is forever. As in other herpes infections, such as cold sores or chickenpox-shingles, the virus enters a lifelong latent state in nerve cells. Some people have several recurrences a year; for others, recurrence is rare. Men are more likely to experience recurrences than women. Reactivation appears to be triggered by several factors, including menstruation, emotional stress or illness (especially if accompanied by fever, a factor that is also involved in the appearance of cold sores), and perhaps just scratching the affected area. About 90% of patients with HSV-2 and about 50% of those with HSV-1 will have recurrences. Recurrence rates decrease over time, regardless of treatment.

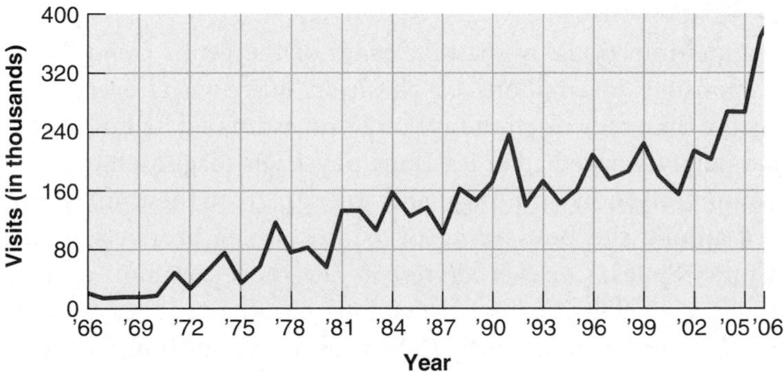

Figure 26.13 **Genital herpes: initial visits to physicians' offices, United States, 1966 to 2006.**

Q What are possible causes for changes in the incidence of a disease, such as shown on this graph?

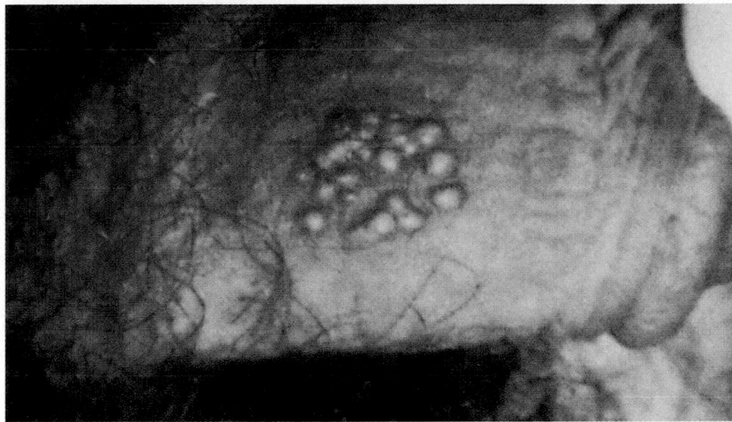

Figure 26.14 **Vesicles of genital herpes on a penis.**

Q What microbe causes genital herpes?

Diagnosis of genital herpes can be done by culture of the virus taken from a vesicle; however, PCR testing of such samples has proven more sensitive and is potentially faster. If there are no lesions to be sampled, serological testing can identify HSV infections or confirm clinical diagnosis by symptoms.

There is no cure for genital herpes, although research on its prevention and treatment is intensive. Discussions of chemotherapy use terms such as *suppression* or *management* rather than *cure.* Currently, the antiviral drugs acyclovir, famciclovir, and valacyclovir are recommended for treatment. They are fairly effective in alleviating the symptoms of a primary outbreak; there is some relief of pain and slightly faster healing. Taken over several months they lower the chances of recurrence during that time. Recent studies have indicated that valacyclovir taken daily can significantly cut sexual transmission of genital herpes. No vaccine is currently available.

Neonatal Herpes

Neonatal herpes is a serious consideration for women of childbearing age. Currently about 1500 cases a year are reported in this country. The virus can cross the placental barrier and affect the fetus. The result can be spontaneous abortion or serious fetal damage, such as mental retardation and defective vision and hearing. Herpes infection of the newborn is most likely to have serious consequences when the mother acquires the initial herpes infection during the pregnancy. Therefore, any pregnant woman without a history of genital herpes should avoid sexual

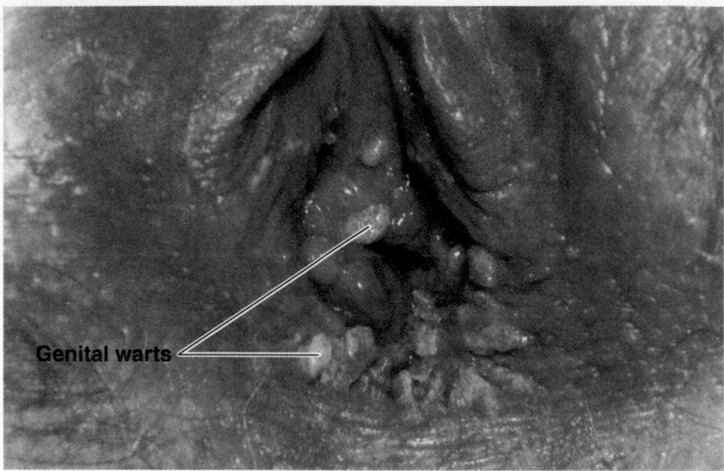

Figure 26.15 Genital warts on a vulva.

What is the relationship between genital warts and cervical cancer?

contact with anyone who might be carrying the virus. Damage to the fetus or newborn is much less likely, by a factor of about ten, from exposure to recurrent or asymptomatic herpes. The reason is that maternal antibodies are apparently somewhat protective.

For practical purposes, the fetus is considered infected if the virus can be grown from amniotic fluid. This procedure is widely available and requires about 5 days. If the fetus is free of the virus, it still needs to be protected from infection during passage through the birth canal. Clinical signs of infection cannot always be seen, and tests to detect asymptomatic shedding of the viruses are not 100% reliable. If there are obvious viral lesions at delivery time, a cesarean section is probably wise. The operation should be done before the fetal membrane ruptures and the viruses spread to the uterus.

Genital Warts

Warts are an infectious disease; since 1907 it has been known that they are caused by viruses known as papillomaviruses. It is probably less well known that warts can be transmitted sexually and that this is an increasing problem. Nearly a million new cases of **genital warts** are estimated to occur in the United States each year.

There are more than 60 serotypes of human papillomaviruses (HPV), and certain serotypes tend to be linked with certain forms of genital warts (technically, *condylomata acuminata*). Morphologically, some warts are extremely large and "warty" in appearance, with multiple fingerlike projections resembling cauliflower; others are relatively smooth or flat (Figure 26.15). The incubation period is usually a matter of a few weeks or months.

Infection by HPV is common. Recent testing found that more than a quarter of women in the United States aged 14 to 59 years were infected—making HPV arguably the most common sexually transmitted infection. Visible genital warts are usually caused by serotypes 6 and 11. These serotypes rarely cause cancer, which is the most serious concern with these infections. The types most likely to cause cancer are types 16 and 18, but these have a relatively low prevalence. Even so, cervical cancer caused by HPV kills at least 4000 women annually in the United States.

As a general rule, warts can be treated but not cured (see the discussion on page 595). However, approximately 90% of cases clear within 2 years. The available methods used for warts (such as surgery or cryotherapy) are not as effective against genital warts. Two patient-applied gels, podofilox and imiquimod, are often useful treatments. Imiquimod (Aldara) stimulates the body to produce interferon (page 468), which appears to account for its antiviral activity. A vaccine has been licensed that is effective against HPV types 6, 11, 16, and 18. It is recommended for routine use in girls aged 11 to 12 years.

AIDS

AIDS, or HIV infection, is a viral disease that is frequently transmitted by sexual contact. However, its pathogenicity is based on damage to the immune system, so it was discussed on pages 539–548. It is important to remember that the lesions resulting from many of the diseases of bacterial and viral origin facilitate the transmission of HIV.

CHECK YOUR UNDERSTANDING

✓ Both genital herpes and genital warts are caused by viruses; which one is the greater danger to a pregnancy? **26-7**

Fungal Disease of the Reproductive Systems

LEARNING OBJECTIVE

26-8 Discuss the epidemiology of candidiasis.

The fungal disease described here is the well-known *yeast infection* for which nonprescription treatments are advertised.

Candidiasis

Vaginal infections by yeastlike fungi of the genus *Candida* are responsible for millions of physician office visits every year. By the time they reach the age of 25, an estimated half of college women will have had at least one physician-diagnosed episode. Nonprescription antifungal therapies to treat these infections are among the best-selling over-the-counter products in the United States. *Candida albicans* is the most common species, causing 85–90% of cases. Infections by other species, such as *C. glabrata*, are more likely to be resistant to antifungals and to be chronic or recurrent.

C. albicans often grows on mucous membranes of the mouth, intestinal tract, and genitourinary tract (see Diseases in Focus 26.2; see also Figure 21.17, page 601). Infections are

DISEASES IN FOCUS 26.2

Characteristics of the Most Common Types of Vaginitis and Vaginosis

Vaginitis, or inflammation of the vagina, often accompanies vaginal infections. Vaginitis may be caused by microbial infections. Some infections are associated with sexual activity, whereas others, such as vaginal candidiasis, are not. *Differential diagnosis* is the process of identifying the disease from a list of possible diseases that fit the information derived from examining a patient. The cause of vaginitis cannot be determined on the basis of symptoms or physical examination alone. Usually, diagnosis involves examining a specimen of vaginal fluid under a microscope (see the photo). Use the table below to identify the infection caused by the organism in the figure. For the solution, go to www.microbiologyplace.com.

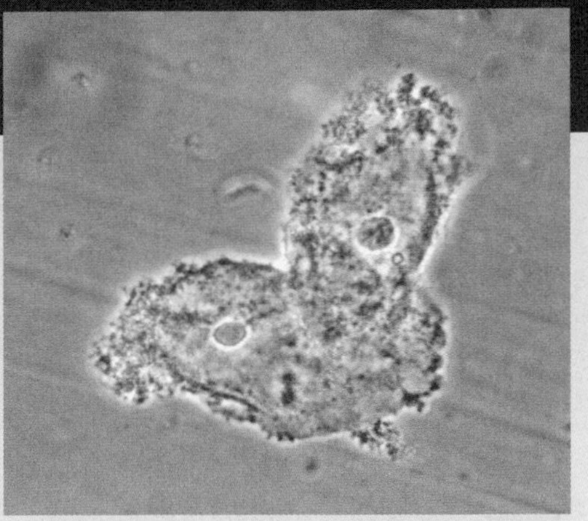

Epithelial cells covered with rod-shaped bacteria from a vaginal swab.

Disease	Pathogen	Symptoms				Diagnosis	Treatment
		Odor, Color, and Consistency of Discharge	*Amount of Discharge*	*Appearance of Vaginal Mucosa*	*pH (normal pH is 3.8–4.2)*		
Candidiasis	Fungus, *Candida albicans*	Yeasty or none, white, curdy	Varies	Dry, red	<4	Microscopic exam	Clotrimazole; fluconazole
Bacterial vaginosis	Bacterium, *Gardnerella vaginalis*	Fishy; gray-white, thin, frothy	Copious	Pink	>4.5	Presence of clue cells	Metronidazole
Trichomoniasis	Protozoan, *Trichomonas vaginalis*	Foul, greenish-yellow; frothy	Copious	Tender, red	5–6	Microscopic exam; DNA probes; monoclonal antibody	Metronidazole

usually a result of opportunistic overgrowth when the competing microbiota are suppressed by antibiotics or other factors. As discussed in Chapter 21, *C. albicans* is the cause of **oral candidiasis,** or thrush. It is also responsible for occasional cases of NGU in men and for **vulvovaginal candidiasis,** which is the most common cause of vaginitis. About 75% of all women experience at least one episode.

The lesions of vulvovaginal candidiasis resemble those of thrush but produce more irritation; severe itching; a thick, yellow, cheesy discharge; and yeasty or no odor. *C. albicans,* the *Candida* species responsible for most cases, is an opportunistic pathogen. Predisposing conditions include the use of oral contraceptives and pregnancy, which cause an increase of glycogen in the vagina (see the discussion of the normal vaginal microbiota earlier in this chapter). It is probable that hormones are a factor; candidiasis is much less common in girls before puberty or in women after menopause. Yeast infections are a frequent symptom in women suffering from uncontrolled diabetes; also, the use of broad-spectrum antibiotics suppresses the normal, competing bacterial microbiota, which leads to opportunistic fungal infections. Thus, diabetes and antibiotic therapy are predisposing factors to *C. albicans* vaginitis.

A yeast infection is diagnosed by microscopic identification of the fungus in scrapings of lesions and by isolation of the fungus in culture. Treatment usually consists of topical application of nonprescription antifungal drugs such as clotrimazole and miconazole. An alternative treatment is a single dose of oral fluconazole or other azole-type antifungal.

CHECK YOUR UNDERSTANDING

✓ What changes in the vaginal bacterial microbiota tend also to favor the growth of the yeast *Candida albicans*? **26-8**

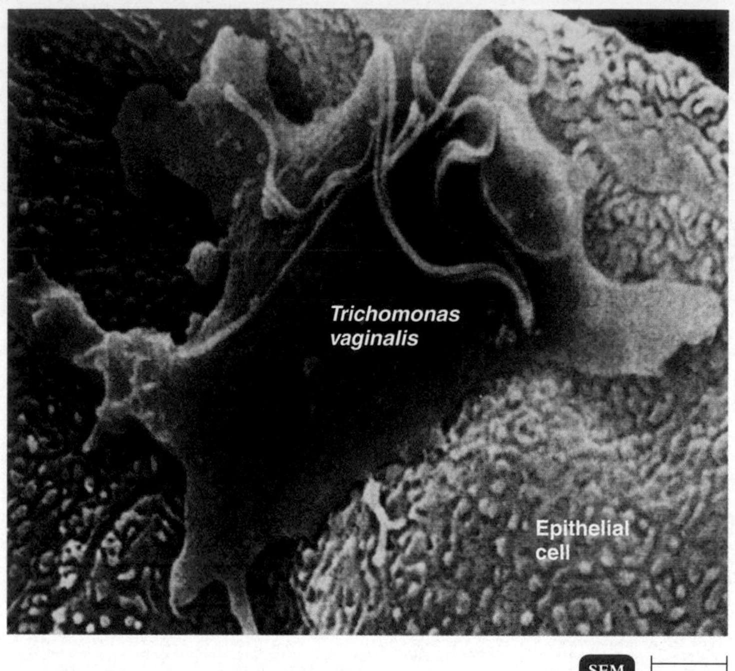

Figure 26.16 ***Trichomonas vaginalis*** **adhering to the surface of an epithelial cell in a cell culture preparation.**
The flagella are clearly visible.
Source: D. Petrin et al., "Clinical and Microbiological Aspects of Trichomonas vaginalis." *ASM Clinical Microbiology Reviews* 11 (1998): 300–317.

Q Are there any harmful effects from infection by this protozoan?

Protozoan Disease of the Reproductive Systems

LEARNING OBJECTIVES

26-9 Discuss the epidemiology of trichomoniasis.

26-10 List reproductive system diseases that can cause congenital and neonatal infections, and explain how these infections can be prevented.

The only STI caused by a protozoan almost entirely affects only women. Although common, it is not widely known.

Trichomoniasis

The anaerobic protozoan *Trichomonas vaginalis* is frequently a normal inhabitant of the vagina in women and of the urethra in many men (Figure 26.16). It is usually sexually transmitted. If the normal acidity of the vagina is disturbed, the protozoan may overgrow the normal microbial population of the genital mucosa and cause **trichomoniasis.** (Men rarely have any symptoms as a result of the presence of the protozoan.) It is often accompanied by a coinfection with gonorrhea. Its prevalence in certain STI clinics is 25% or higher. In response to the protozoan infection, the body accumulates leukocytes at the infection site. The resulting discharge is profuse, greenish yellow, and characterized by a foul odor. This discharge is accompanied by irritation and itching. Up to half the cases, however, are asymptomatic.

The incidence of trichomoniasis is higher than that of gonorrhea or chlamydia, but it is considered relatively benign and is not a reportable disease. It is known, however, to cause preterm delivery and problems associated with this, such as low birth weight.

Diagnosis is usually made by microscopic examination and identification of the organisms in the discharge. They can also be isolated and grown on laboratory media. The pathogen can be found in semen or urine of male carriers. New rapid tests making use of DNA probes and monoclonal antibodies are now available. Treatment is by oral metronidazole, administered to both sex partners, which readily clears the infection.

The TORCH Panel of Tests

We have seen in this and previous chapters that a number of diseases can cause birth defects in newborns when they infect a pregnant woman. TORCH is an acronym for a panel of tests that screens for antibodies to these infections. Confirmation may require additional testing. The panel is made up of the following: **T**oxoplasmosis; **O**ther (such as syphilis, hepatitis B, enterovirus, Epstein-Barr virus, varicella-zoster virus); **R**ubella; **C**ytomegalovirus; **H**erpes simplex virus.

CHECK YOUR UNDERSTANDING

✓ What are the symptoms of the presence of *Trichomonas vaginalis* in the male reproductive system? **26-9**

✓ What is the intent of the TORCH panel of tests? **26-10**

* * *

The major microbial diseases of the urinary and reproductive systems are summarized in Diseases in Focus 26.3.

DISEASES IN FOCUS 26.3

Microbial Diseases of the Reproductive Systems

Differential diagnosis is the process of identifying the disease from a list of possible diseases that fit the information derived from examining a patient. A differential diagnosis is important for providing initial treatment and for laboratory testing. For example, a 26-year-old woman had abdominal pain, painful urination, and a fever. Cultures grown in a high-CO_2 environment revealed gram-negative diplococci. Use the table below to identify infections that could cause these symptoms. For the solution, go to www.microbiologyplace.com.

Gram-negative diplococci on Thayer-Martin agar.

Disease	Pathogen	Symptoms	Treatment
BACTERIAL DISEASES			
Gonorrhea	*Neisseria gonorrhoeae*	Men: painful urination and discharge of pus Women: few symptoms but possible complications, such as PID	Cephalosporins
Nongonococcal urethritis (NGU)	*Chlamydia trachomatis, Mycoplasma hominis, Ureaplasma urealyticum*	Painful urination and watery discharge. In women, possible complications, such as PID.	Doxycycline, azithromycin
Pelvic inflammatory disease (PID)	*N. gonorrhoeae, C. trachomatis*	Chronic abdominal pain; possible infertility	Doxycycline and cefoxitin
Syphilis	*Treponema pallidum*	Initial sore at site of infection, later skin rashes and mild fever; final stages may be severe lesions, damage to cardiovascular and nervous systems.	Benzathine penicillin
Lymphogranuloma venereum (LGV)	*C. trachomatis*	Swelling in lymph nodes in groin	Doxycycline
Chancroid (soft chancre)	*Haemophilus ducreyi*	Painful ulcers of genitals; swollen lymph nodes in groin	Erythromycin; ceftriaxone
Bacterial vaginosis	See Diseases in Focus 26.2, Page 759		
VIRAL DISEASES			
Genital herpes	Herpes simplex virus type 2; HSV type 1	Painful vesicles in genital area	Acyclovir
Genital warts	Human papillomaviruses	Warts in genital area	Podofilox; imiquimod; preventive vaccine
AIDS—see Chapter 19, pp. 539–548			
FUNGAL DISEASE			
Candidiasis	See Diseases in Focus 26.2, Page 759		
PROTOZOAN DISEASE			
Trichomoniasis	See Diseases in Focus 26.2, Page 759		

STUDY OUTLINE

The **MyMicrobiologyPlace** website (**www.microbiologyplace.com**) will help you get ready for tests with its simple three-step approach: ❶ **take a pre-test** and obtain a personalized study plan, ❷ **learn and practice** with animations, tutorials, and MP3 tutor sessions, and ❸ **test yourself** with quizzes and a chapter post-test.

Introduction (p. 743)

1. The urinary system regulates the chemical composition and volume of the blood and excretes nitrogenous waste and water.
2. The reproductive system produces gametes for reproduction and, in the female, supports the growing embryo.
3. Microbial diseases of these systems can result from infection from an outside source or from opportunistic infection by members of the normal microbiota.

Structure and Function of the Urinary System (p. 744)

1. Urine is transported from the kidneys through ureters to the urinary bladder and is eliminated through the urethra.
2. Valves prevent urine from flowing back to the urinary bladder and kidneys.
3. The flushing action of urine and the acidity of normal urine have some antimicrobial value.

Structure and Function of the Reproductive Systems (pp. 744–745)

1. The female reproductive system consists of two ovaries, two uterine tubes, the uterus, the cervix, the vagina, and the external genitals.
2. The male reproductive system consists of two testes, ducts, accessory glands, and the penis; seminal fluid leaves the male body through the urethra.

Normal Microbiota of the Urinary and Reproductive Systems (p. 745)

1. The urinary bladder and upper urinary tract are sterile under normal conditions.
2. Lactobacilli dominate the vaginal microbiota during the reproductive years.
3. The male urethra is normally sterile.

DISEASES OF THE URINARY SYSTEM (pp. 746–747)

Bacterial Diseases of the Urinary System (pp. 746–747)

1. Urethritis, cystitis, and ureteritis are terms describing inflammations of tissues of the lower urinary tract.
2. Pyelonephritis can result from lower urinary tract infections or from systemic bacterial infections.
3. Opportunistic gram-negative bacteria from the intestines often cause urinary tract infections.
4. Nosocomial infections following catheterization occur in the urinary system. *E. coli* causes more than half of these infections.
5. Treatment of urinary tract infections depends on the isolation and antibiotic sensitivity testing of the causative agents.

Cystitis (p. 746)

6. Inflammation of the urinary bladder, or cystitis, is common in females.
7. Microorganisms at the opening of the urethra and along the length of the urethra, careless personal hygiene, and sexual intercourse contribute to the high incidence of cystitis in females.
8. The most common etiologies are *E. coli* and *Staphylococcus saprophyticus.*

Pyelonephritis (p. 746)

9. Inflammation of the kidneys, or pyelonephritis, is usually a complication of lower urinary tract infections.
10. About 75% of pyelonephritis cases are caused by *E. coli.*

Leptospirosis (pp. 746–747)

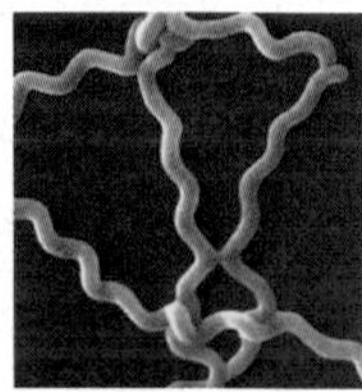

11. The spirochete *Leptospira interrogans* is the cause of leptospirosis.
12. The disease is transmitted to humans by urine-contaminated water.
13. Leptospirosis is characterized by chills, fever, headache, and muscle aches.

DISEASES OF THE REPRODUCTIVE SYSTEMS (p. 747–756)

Bacterial Diseases of the Reproductive Systems (pp. 747–756)

1. Most diseases of the reproductive system are sexually transmitted diseases (STDs), now called sexually transmitted infections (STIs).
2. Most STDs can be prevented by the use of condoms and are treated with antibiotics.

Gonorrhea (pp. 747–750)

3. *Neisseria gonorrhoeae* causes gonorrhea.
4. Gonorrhea is a common reportable communicable disease in the United States.
5. *N. gonorrhoeae* attaches to mucosal cells of the oral-pharyngeal area, genitals, eyes, and rectum by means of fimbriae.
6. Symptoms in men are painful urination and pus discharge. Blockage of the urethra and sterility are complications of untreated cases.
7. Women might be asymptomatic unless the infection spreads to the uterus and uterine tubes (see pelvic inflammatory disease).
8. Gonorrheal endocarditis, gonorrheal meningitis, and gonorrheal arthritis are complications that can affect both sexes if gonorrheal infections are untreated.

9. Ophthalmia neonatorum is an eye infection acquired by infants during passage through the birth canal of an infected mother.
10. Gonorrhea is diagnosed by ELISA or nucleic acid amplification.

Nongonococcal Urethritis (NGU) (pp. 750–751)

11. Nongonococcal urethritis (NGU), or nonspecific urethritis (NSU), is any inflammation of the urethra not caused by *N. gonorrhoeae.*
12. Most cases of NGU are caused by *Chlamydia trachomatis.*
13. *C. trachomatis* infection is the most common STD.
14. Symptoms of NGU are often mild or lacking, although uterine tube inflammation and sterility may occur.
15. *C. trachomatis* can be transmitted to infants' eyes at birth.
16. Diagnosis is based on the detection of chlamydial DNA in urine.
17. *Ureaplasma urealyticum* and *Mycoplasma hominis* also cause NGU.

Pelvic Inflammatory Disease (PID) (pp. 751–752)

18. Extensive bacterial infection of the female pelvic organs, especially of the reproductive system, is called pelvic inflammatory disease (PID).
19. PID is caused by *N. gonorrhoeae, C. trachomatis,* and other bacteria that gain access to the uterine tubes. Infection of the uterine tubes is called salpingitis.
20. PID can result in blockage of the uterine tubes and sterility.

Syphilis (pp. 752–755)

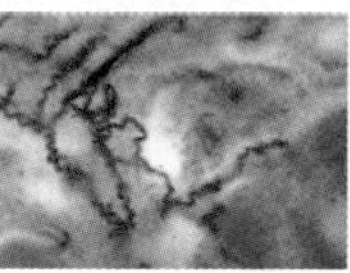

21. Syphilis is caused by *Treponema pallidum,* a spirochete that has not been cultured in vitro. Laboratory cultures are grown in rabbits or cell cultures.
22. The primary lesion is a small, hard-based chancre at the site of infection. The bacteria then invade the blood and lymphatic system, and the chancre spontaneously heals.
23. The appearance of a widely disseminated rash on the skin and mucous membranes marks the secondary stage. Spirochetes are present in the lesions of the rash.
24. The patient enters a latent period after the secondary lesions spontaneously heal.
25. At least 10 years after the secondary lesion, tertiary lesions called gummas can appear on many organs.
26. Congenital syphilis, resulting from *T. pallidum* crossing the placenta during the latent period, can cause neurological damage in the newborn.
27. *T. pallidum* is identifiable through darkfield microscopy of fluid from primary and secondary lesions.
28. Many serological tests, such as VDRL, RPR, and FTA-ABS, can be used to detect the presence of antibodies against *T. pallidum* during any stage of the disease.

Lymphogranuloma Venereum (LGV) (p. 755)

29. *C. trachomatis* causes lymphogranuloma venereum (LGV), which is primarily a disease of tropical and subtropical regions.
30. The initial lesion appears on the genitals and heals without scarring.
31. The bacteria are spread in the lymph system and cause enlargement of the lymph nodes, obstruction of lymph vessels, and swelling of the external genitals.
32. The bacteria are isolated and identified from pus taken from infected lymph nodes.

Chancroid (Soft Chancre) (p. 756)

33. Chancroid, a swollen, painful ulcer on the mucous membranes of the genitals or mouth, is caused by *Haemophilus ducreyi.*

Bacterial Vaginosis (p. 756)

34. Bacterial vaginosis is an infection without inflammation caused by *Gardnerella vaginalis.*
35. Diagnosis of *G. vaginalis* is based on increased vaginal pH, fishy odor, and the presence of clue cells.

Viral Diseases of the Reproductive Systems (pp. 757–758)

Genital Herpes (pp. 757–758)

1. Herpes simplex viruses (HSV-1 and HSV-2) cause genital herpes.
2. Symptoms of the infection are painful urination, genital irritation, and fluid-filled vesicles.
3. The virus might enter a latent stage in nerve cells. Vesicles reappear following trauma and hormonal changes.
4. Neonatal herpes is contracted during fetal development or birth. It can result in neurological damage or infant fatalities.

Genital Warts (p. 758)

5. Human papillomaviruses cause warts.
6. Some human papillomaviruses that cause genital warts have been associated with cancer of the cervix.

AIDS (p. 758)

7. AIDS is a sexually transmitted disease of the immune system (see Chapter 19, pages 539–598).

Fungal Disease of the Reproductive Systems (pp. 758–759)

Candidiasis (pp. 758–759)

1. *Candida albicans* causes NGU in men and vulvovaginal candidiasis, or yeast infection, in women.
2. Vulvovaginal candidiasis is characterized by lesions that produce itching and irritation.
3. Predisposing factors are pregnancy, diabetes, tumors, and broad-spectrum antibacterial chemotherapy.
4. Diagnosis is based on observation of the fungus and its isolation from lesions.

Protozoan Disease of the Reproductive System (pp. 760–761)

Trichomoniasis (p. 760)

1. *Trichomonas vaginalis* causes trichomoniasis when the pH of the vagina increases.
2. Diagnosis is based on observation of the protozoa in purulent discharges from the site of infection.

The TORCH Panel of Tests (p. 760)

3. Antibodies against specific diseases that can infect a fetus are detected by the TORCH tests.

STUDY QUESTIONS

Answers to the Review and Multiple Choice questions can be found by turning to the blue Answers tab at the back of the textbook.

Review

1. **DRAW IT** Diagram the pathway taken by *E. coli* to cause cystitis. Do the same for pyelonephritis. Diagram the pathway taken by *Neisseria gonorrhoeae* to cause PID.

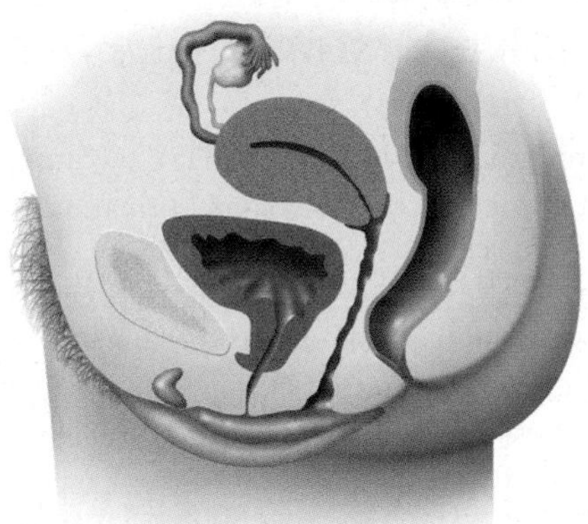

2. How are urinary tract infections transmitted?
3. Explain why *E. coli* is frequently implicated in cystitis in females. List some predisposing factors for cystitis.
4. Name one organism that causes pyelonephritis. What are the portals of entry for microbes that cause pyelonephritis?
5. Complete the following table:

Disease	Causative Agent	Symptoms	Method of Diagnosis	Treatment
Bacterial vaginosis				
Gonorrhea				
Syphilis				
PID				
NGU				
LGV				
Chancroid				

6. Leptospirosis is a kidney infection of humans and other animals. How is this disease transmitted? What types of activities would increase one's exposure to this disease? What is the etiology?
7. Describe the symptoms of genital herpes. What is the causative agent? When is this infection least likely to be transmitted?
8. Name one fungus and one protozoan that can cause reproductive system infections. What symptoms would lead you to suspect these infections?
9. List the genital infections that cause congenital and neonatal infections. How can transmission to a fetus or newborn be prevented?

Multiple Choice

1. Which of the following is usually transmitted by contaminated water?
   a. *Chlamydia*
   b. leptospirosis
   c. syphilis
   d. trichomoniasis
   e. none of the above

Use the following choices to answer questions 2–5:
   a. *Candida*
   b. *Chlamydia*
   c. *Gardnerella*
   d. *Neisseria*
   e. *Trichomonas*

2. Microscopic examination of vaginal smear shows flagellated eukaryotes.
3. Microscopic examination of vaginal smear shows ovoid eukaryotic cell.
4. Microscopic examination of vaginal smear shows epithelial cells covered with bacteria.
5. Microscopic examination of vaginal smear shows gram-negative cocci in phagocytes.

Use the following choices to answer questions 6–8:
   a. candidiasis
   b. bacterial vaginosis
   c. genital herpes
   d. lymphogranuloma venereum
   e. trichomoniasis

6. Difficult to treat with chemotherapy
7. Fluid-filled vesicles
8. Frothy, fishy discharge

Use the following choices to answer questions 9 and 10:
   a. *Chlamydia trachomatis*
   b. *Escherichia coli*
   c. *Mycobacterium hominis*
   d. *Staphylococcus saprophyticus*

9. The most common cause of cystitis.
10. In cases of NGU, diagnosis is made using PCR to detect microbial DNA.

Critical Thinking

1. The tropical skin disease called yaws is transmitted by direct contact. Its causative agent, *Treponema pallidum pertenue*, is indistinguishable from *T. pallidum*. Syphilis epidemics in Europe coincided with the return of Columbus from the New World. How might *T. pallidum pertenue* have evolved into *T. pallidum* in the temperate climate of Europe?
2. Why can frequent douching be a predisposing factor to bacterial vaginosis, vulvovaginal candidiasis, or trichomoniasis?
3. *Neisseria* is cultured on Thayer-Martin media, consisting of chocolate agar and nystatin, incubated in a 5% CO_2 environment. How is this selective for *Neisseria*?

4. The list below is a key to selected microorganisms that cause genitourinary infections. Complete this key by listing genera discussed in this chapter in the blanks that correspond to their respective characteristics.

Gram-negative bacteria
- Spirochete
  - Aerobic ____________________
  - Anaerobic ____________________
- Coccus
  - Oxidase-positive ____________________
- Bacillus, nonmotile
  - Requires X factor ____________________
  - Gram-positive wall ____________________
- Obligate intracellular parasite ____________________
- Lacking cell wall
  - Urease-positive ____________________
  - Urease-negative ____________________
- Fungus
  - Pseudohyphae ____________________
- Protozoa
  - Flagella ____________________
- No organism observed/cultured from patient ____________________

Clinical Applications

1. A previously healthy 19-year-old woman was admitted to a hospital after 2 days of nausea, vomiting, headache, and neck stiffness. Cerebrospinal fluid and cervical cultures showed gram-negative diplococci in leukocytes; a blood culture was negative. What disease did she have? How was it probably acquired?

2. A 28-year-old woman was admitted to a Wisconsin hospital with a 1-week history of arthritis of the left knee. Four days later, a 32-year-old man was examined for a 2-week history of urethritis and a swollen, painful left wrist. A 20-year-old woman seen in a Philadelphia hospital had pain in the right knee, left ankle, and left wrist for 3 days. Pathogens cultured from synovial fluid or urethral culture were gram-negative diplococci that required proline to grow. Antibiotic sensitivity tests gave the following results:

Antibiotic	MIC Tested (μg/ml)	Susceptible MIC (μg/ml)
Cefoxitin	0.5	≤2
Penicillin	8	≤0.06
Spectinomycin	64	≤32
Tetracycline	4	≤0.25

What is the pathogen, and how is this disease transmitted? Which of the antibiotics should be used for treatment? What is the evidence that these cases are related?

3. Using the following information, determine what the disease is and how the infant's illness might have been prevented:

May 11:	A 23-year-old woman has her first prenatal examination. She is $4\frac{1}{2}$ months pregnant. Her VDRL results are negative.
June 6:	The woman returns to her physician complaining of a labial lesion of a few days' duration. A biopsy is negative for malignancy, and herpes test results are negative.
July 1:	The woman returns to her physician because the labial lesion continues to cause some discomfort.
Sept. 15:	The baby's father has multiple penile lesions and a generalized body rash.
Sept. 25:	The woman delivers her baby. Her RPR is 32 and the infant's is 128.
Oct. 1:	The woman takes her infant to a pediatrician because the baby is lethargic. She is told the infant is healthy and not to worry.
Oct. 2:	The baby's father has a persistent body rash and plantar and palmar rashes.
Nov. 8:	The infant becomes acutely ill with pneumonia and is hospitalized. The admitting physician finds signs of osteochondritis.

27 Environmental Microbiology

In previous chapters, we focused primarily on the disease-causing capabilities of microorganisms. In this chapter, you will learn about many of the positive functions microbes perform in the environment. Bacteria and other microorganisms are, in fact, essential to maintaining life on Earth.

Microbes, especially those that belong to the Domains Bacteria and Archaea, live in the most widely varied habitats on Earth. They are found in boiling hot springs, and as many as 5000 bacteria have been isolated from each milliliter of snow at the South Pole. Microbes have been recovered from minute openings in rocks a kilometer (0.62 mile) or more below the planet surface. Explorations of the deepest ocean have revealed large numbers of microbes living there, in eternal darkness and subject to incredible pressures. Microbes are also found in clear mountain streams flowing from a melting glacier and in waters nearly saturated with salts, such as those of the Dead Sea.

UNDER THE MICROSCOPE

Anabaena azollae. These organisms (chains of cells) are nitrogen-fixing cyanobacteria that live symbiotically within a cavity in the leaf of a freshwater plant, *Azolla.*

Q&A

What do these microorganisms have to do with the cultivation of rice in Asia?

Look for the answer in the chapter.

Microbial Diversity and Habitats

LEARNING OBJECTIVES

27-1 Define *extremophile*, and identify two "extreme" habitats.

27-2 Define *symbiosis*.

27-3 Define *mycorrhiza*, differentiate endomycorrhizae from ectomycorrhizae, and give an example of each.

The diversity of microbial populations indicates that they take advantage of any niches found in their environment. Different amounts of oxygen, light, or nutrients may exist within a few millimeters in the soil. As a population of aerobic organisms uses up the available oxygen, anaerobes are able to grow. If the soil is disturbed by plowing, earthworms, or other activity, the aerobes will again be able to grow to repeat this succession.

Microbes that live in extreme conditions of temperature, acidity, alkalinity, or salinity are called **extremophiles.** Most are members of the Archaea. The enzymes (**extremozymes**) that make growth possible under these conditions have been of great interest to industries because they can tolerate extremes of temperature, salinity, and pH that would inactivate other enzymes. The organism *Thermus aquaticus,* found growing in a hot spring in Yellowstone National Park, is the source of the valuable *Taq polymerase* enzyme used in the polymerase chain reaction (PCR) technique (see page 251). The enzyme functions at temperatures as high as 95°C, the boiling point of water in the organism's habitat. At another extreme, bacteria have been found buried deep in the ice sheets of Antarctica and Greenland, somehow surviving at temperatures as low as −40°C within a life-giving film of water only about three molecules thick. In the rainless Atacama Desert of Chile, a species of cyanobacteria lives inside salt crystals. Its only moisture is absorbed from the atmosphere each night, and its energy is derived from sunlight.

Microorganisms live in an intensely competitive environment and must exploit any advantage they can. They may metabolize common nutrients more rapidly or use nutrients that competing organisms cannot metabolize. Some, such as the lactic acid bacteria that are so useful in making dairy products, are able to make an environmental niche inhospitable to competing organisms. The lactic acid bacteria are unable to use oxygen as an electron acceptor and are able to ferment sugars only to lactic acid, leaving most of the energy unused. However, the acidity inhibits the growth of more efficient, competing microbes.

Symbiosis

Recall from Chapter 14 that **symbiosis** is two differing organisms living together in a close association that is beneficial to one or both of them. Economically, the most important example of an animal-microbe symbiosis is that of the ruminants, animals that have a tanklike digestive organ called a *rumen.* Ruminants, such as cattle and sheep, graze on cellulose-rich plants. Bacteria in the rumen ferment the cellulose into compounds that are absorbed into the animal's blood and subsequently are used for carbon and energy. Rumen protozoa keep the bacterial population under control by eating bacteria.

A very important contribution to plant growth is made by **mycorrhizae,** or mycorrhizal symbionts (*myco* = fungus; *rhiza* = root). There are two primary types of these fungi: *endomycorrhizae,* also known as *vesicular-arbuscular mycorrhizae;* and *ectomycorrhizae.* Both types function as do root hairs on plants; that is, they extend the surface area through which the plant can absorb nutrients, especially phosphorus, which is not very mobile in soil.

Vesicular-arbuscular mycorrhizae form large spores that can be isolated easily from soil by sieving. The hyphae from these germinating spores penetrate into the plant root and form two types of structures: vesicles and arbuscules. **Vesicles** are smooth oval bodies that probably function as storage structures. **Arbuscules,** tiny bushlike structures, are formed inside plant cells (**Figure 27.1a**). Nutrients travel from the soil through fungal hyphae to these arbuscules, which gradually break down and release the nutrients to the plants. Most grasses and other plants are surprisingly dependent on these fungi for proper growth, and their presence is nearly universal in the plant kingdom.

Ectomycorrhizae mainly infect trees such as pine and oak. The fungus forms a mycelial *mantle* over the smaller roots of the tree (**Figure 27.1b**). Ectomycorrhizae do not form vesicles or arbuscules. Managers of commercial pine tree farms must ensure that seedlings are inoculated with soil containing effective mycorrhizae (**Figure 27.2a**).

Truffles, known as a food delicacy, are ectomycorrhizae, usually of oak trees (**Figure 27.2b**). In Europe, pigs or trained dogs are used to find them by smell and root them up. To a pig, male or female, the most important component of a truffle's odor is dimethyl sulfide, which is also responsible for the odor of cabbage. In nature, proliferation of the fungus depends on ingestion by an animal, which distributes the undigested spores into new locations. Increasingly, the cultivation of truffles is becoming a farming operation. Oak trees are planted in groves and artificially inoculated with fungal spores that are grown in the laboratory or extracted from ripe truffles.

CHECK YOUR UNDERSTANDING

✓ Can you identify two extreme habitats for extremophile organisms? **27-1**

✓ What is the definition of *symbiosis*? **27-2**

✓ Is a truffle an example of an endomycorrhiza or an ectomycorrhiza? **27-3**

(a) Endomycorrhiza (vesicular-arbuscular mycorrhiza). A fully developed arbuscule of an endomycorrhiza in a plant cell. (The term *arbuscule* means "little bush.") As the arbuscule decomposes, it releases nutrients for the plant. SEM 12 μm

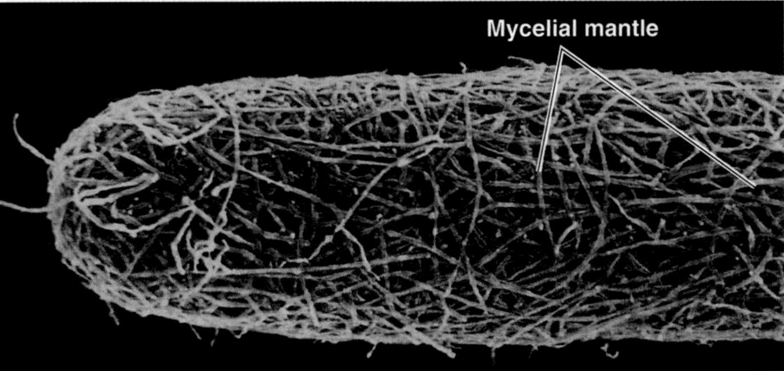

(b) Ectomycorrhiza. The mycelial mantle of a typical ectomycorrhizal fungus surrounding a *Eucalyptus* tree root. 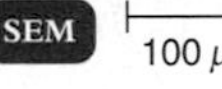 100 μm

Figure 27.1 Mycorrhizae.

 **Of what value is a mycorrhiza to a plant?**

Soil Microbiology and Biogeochemical Cycles

LEARNING OBJECTIVES

27-4 Define *biogeochemical cycle.*

27-5 Outline the carbon cycle, and explain the roles of microoganisms in this cycle.

27-6 Outline the nitrogen cycle, and explain the roles of microorganisms in this cycle.

27-7 Define *ammonification, nitrification, denitrification,* and *nitrogen fixation.*

27-8 Outline the sulfur cycle, and explain the roles of microorganisms in this cycle.

27-9 Describe how the ecological community can exist without light energy.

27-10 Compare and contrast the carbon cycle and the phosphorus cycle.

27-11 Give two examples of the use of bacteria to remove pollutants.

27-12 Define *bioremediation.*

Billions of organisms, including those that are microscopic as well as comparatively huge insects and earthworms, form a vibrant living community in the soil. Typical soil has millions of bacteria in each gram. One gram of soil might seem to be a small sample, but it can yield some startling statistics. Estimates are that it would have 20,000 square meters of surface area. Bacterial numbers in this sample would be about 1 billion (although only about 1% can be cultured), and it might contain more than a kilometer of fungal hyphae. Even so, only a very tiny fraction of the available surface area in that gram of soil is colonized by microbes. The microbial population of soil is largest in the top few centimeters and declines rapidly with depth. The most numerous organisms in soil are bacteria. Although actinomycetes are bacteria, they are usually considered separately.

Bacterial soil populations are usually estimated using plate counts on nutrient media, and the actual numbers are probably greatly underestimated by this method. No single nutrient medium or growth condition can possibly meet all the nutritional and other requirements of soil microorganisms.

We can think of soil as a "biological fire." A leaf falling from a tree is consumed by this fire as microbes in the soil metabolize its organic matter. Elements in the leaf enter the **biogeochemical cycles** for carbon, nitrogen, and sulfur that we will discuss in this chapter. In biogeochemical cycles, elements are oxidized and reduced by microorganisms to meet their metabolic needs. (See the discussion of oxidation-reduction in Chapter 5, page 122.) Without biogeochemical cycles, life on Earth would cease to exist.

The Carbon Cycle

The primary biogeochemical cycle is the **carbon cycle** (Figure 27.3). All organisms, including plants, microbes, and animals, contain large amounts of carbon in the form of organic compounds such as cellulose, starches, fats, and proteins. Let's take a closer look at how these organic compounds are formed.

Recall from Chapter 5 that autotrophs perform an essential role for all life on Earth by reducing carbon dioxide to form organic matter. When you look at a tree, you might think that its mass is from the soil where it grows. In fact, its great mass of cellulose is derived from the 0.03% of carbon dioxide in the atmosphere. This occurs as a result of photosynthesis, the first step of the carbon cycle in which photoautotrophs such as cyanobacteria, green plants, algae, and green and purple sulfur bacteria *fix* (incorporate) carbon dioxide into organic matter using energy from sunlight.

Figure 27.2 Mycorrhizae and their considerable commercial value.

Q Why are mycorrhizae valuable for the uptake of phosphorus?

(a) Infection by mycorrhizae strongly influences the growth of many plants. Shown is the relative growth of two pine seedlings: the seedling on the left was inoculated with mycorrhizae; the seedling on the right was not.

(b) Truffles. Of the three truffles shown, one has been sliced to show the interior.

Figure 27.3 The carbon cycle. On a global scale, the return of CO_2 to the atmosphere by respiration closely balances its removal by fixation. However, the burning of wood and fossil fuels adds more CO_2 to the atmosphere; as a result, the amount of atmospheric CO_2 is steadily increasing.

Q How does the accumulation of carbon dioxide in the atmosphere affect Earth's climate?

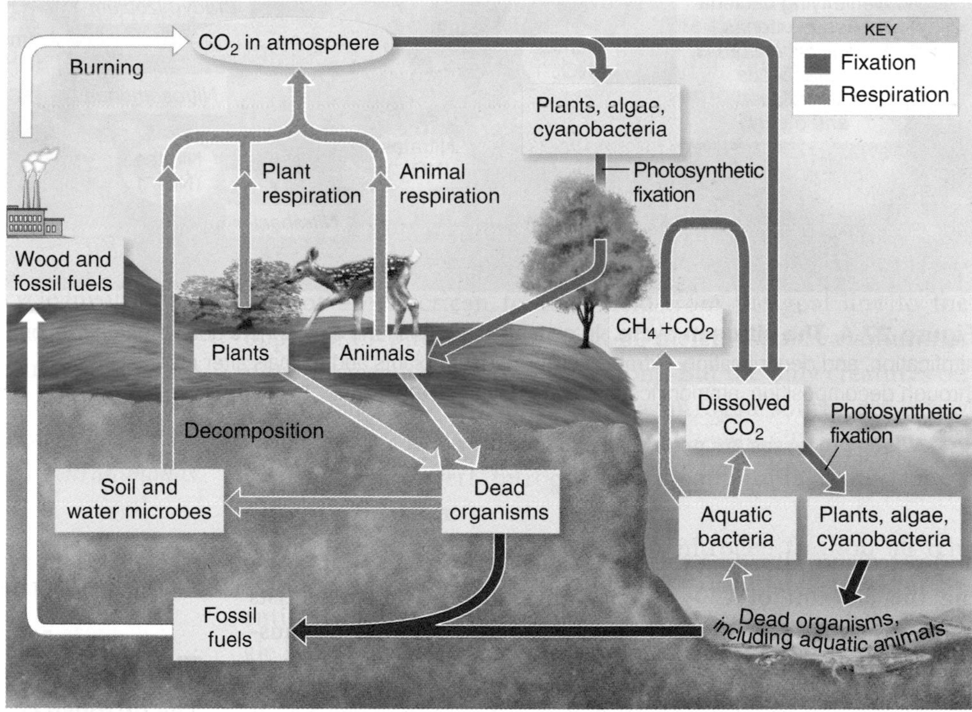

In the next step of the cycle, chemoheterotrophs such as animals and protozoa eat autotrophs and may in turn be eaten by other animals. Thus, as the organic compounds of the autotrophs are digested and resynthesized, the carbon atoms of carbon dioxide are transferred from organism to organism up the food chain.

Chemoheterotrophs, including animals, use some of the organic molecules to satisfy their energy requirements. When this energy is released through respiration, carbon dioxide immediately becomes available to start the cycle over again. Much of the carbon remains within the organisms until they excrete it as wastes or die. When plants and animals die, these organic compounds are decomposed by bacteria and fungi. During decomposition, the organic compounds are oxidized, and CO_2 is returned to the cycle.

Carbon is stored in rocks, such as limestone ($CaCO_3$), and is dissolved as carbonate ions (CO_3^{2-}) in oceans. Vast deposits of fossil organic matter exist in the form of fossil fuels, such as coal

root. The rhizosphere represents something of a nutritional oasis in the soil, especially in grasslands. Among the free-living bacteria that can fix nitrogen are aerobic species such as *Azotobacter.* These aerobic organisms apparently shield the anaerobic nitrogenase enzyme from oxygen by, among other things, having a very high rate of oxygen use that minimizes the diffusion of oxygen into the interior of the cell, where the enzyme is located.

Another free-living obligate aerobe that fixes nitrogen is *Beijerinckia* (bī-yė-rink′ē-ä). Some anaerobic bacteria, such as certain species of *Clostridium,* also fix nitrogen. The bacterium *C. pasteurianum* (pas-tyėr-ē-ā′num), an obligately anaerobic, nitrogen-fixing microorganism, is a prominent example.

There are many species of aerobic, photosynthesizing cyanobacteria that fix nitrogen. Because their energy supply is independent of carbohydrates in soil or water, they are especially useful suppliers of nitrogen to the environment. Cyanobacteria usually carry their nitrogenase enzymes in specialized structures called **heterocysts** that provide anaerobic conditions for fixation (see Figure 11.13a, page 314).

Most of the free-living nitrogen-fixing bacteria are capable of fixing large amounts of nitrogen under laboratory conditions. However, in the soil there is usually a shortage of usable carbohydrates to supply the energy needed to reduce nitrogen to ammonia, which is then incorporated into protein. Nevertheless, these nitrogen-fixing bacteria make important contributions to the nitrogen economy of such areas as grasslands, forests, and the arctic tundra.

Symbiotic Nitrogen-Fixing Bacteria Symbiotic nitrogen-fixing bacteria play an even more important role in plant growth for crop production. Members of the genera *Rhizobium, Bradyrhizobium,* and others infect the roots of leguminous plants, such as soybeans, beans, peas, peanuts, alfalfa, and clover. (These agriculturally important plants are only a few of the thousands of known leguminous species, many of which are bushy plants or small trees found in poor soils in many parts of the world.) Rhizobia, as these bacteria are commonly known, are specially adapted to particular leguminous plant species, on which they form **root nodules** (Figure 27.5). Nitrogen is then fixed by a symbiotic process of the plant and the bacteria. The plant furnishes anaerobic conditions and growth nutrients for the bacteria, and the bacteria fix nitrogen that can be incorporated into plant protein.

There are similar examples of symbiotic nitrogen fixation in nonleguminous plants, such as alder trees. These trees are among the first to appear in forests after fires or glaciation. The alder tree is symbiotically infected with an actinomycete (*Frankia*) and forms nitrogen-fixing root nodules. The growth of 1 acre of alder trees can fix about 50 kg of nitrogen each year; the trees thus make a valuable addition to the forest economy.

Q&A Another important contribution to the nitrogen economy of forests is made by **lichens,** which are a combination of fungus and an alga or a cyanobacterium in a mutualistic relationship (see Figure 12.9, page 340). When one symbiont is a nitrogen-fixing cyanobacterium, the product is fixed nitrogen that eventually enriches the forest soil. Free-living cyanobacteria can fix significant amounts of nitrogen in desert soils after rains and on the surface of arctic tundra soils. Rice paddies can accumulate heavy growths of such nitrogen-fixing organisms. The cyanobacteria also form a symbiosis with a small floating fern, *Azolla,* which grows thickly in rice paddy waters (Figure 27.6). So much nitrogen is fixed by these microbes that other nitrogenous fertilizers are often unnecessary for rice cultivation.

CHECK YOUR UNDERSTANDING

- ✓ What is the common name for the group of microbes that oxidize soil nitrogen into a form that is mobile in soil and likely to be used for nutrition by plants? **27-6**
- ✓ Bacteria of the genus *Pseudomonas,* in the absence of oxygen, will use fully oxidized nitrogen as an electron acceptor, a process in the nitrogen cycle that is given what name? **27-7**

The Sulfur Cycle

The **sulfur cycle** (Figure 27.7) and nitrogen cycle resemble each other in the sense that they represent numerous oxidation states of these elements. The most reduced forms of sulfur are the sulfides, such as the odorous gas hydrogen sulfide (H_2S). Like the ammonium ion of the nitrogen cycle, this is a reduced compound that generally forms under anaerobic conditions. In turn it represents a source of energy for autotrophic bacteria. These bacteria convert the reduced sulfur in H_2S into elemental sulfur granules and fully oxidized sulfates (SO_4^{2-}).

Frequently, elemental sulfur is released from decaying microbes. Elemental sulfur is essentially insoluble in temperate waters, and microbes have difficulty absorbing it. This is probably the origin of huge, prehistoric underground accumulations of sulfur.

Several phototrophic bacteria, such as the green and purple sulfur bacteria, also oxidize H_2S, forming colorful internal sulfur granules (see Figure 11.14, page 315). Like *Beggiatoa,* they can further oxidize the sulfur to sulfate ions. It is important to recognize that these organisms are using light for energy; the hydrogen sulfide is used to reduce CO_2 (see Chapter 5, page 140).

Hydrogen sulfide can be used as an energy source by *Thiobacillus* to produce sulfate ions and sulfuric acid. *Thiobacillus* can grow well at a pH as low as 2 and has practical uses in mining (see Figure 28.14, page 807). Plants and bacteria incorporate sulfates to become part of sulfur-containing amino acids for humans and other animals. There, they form disulfide links that give structure to proteins. As proteins are decomposed,

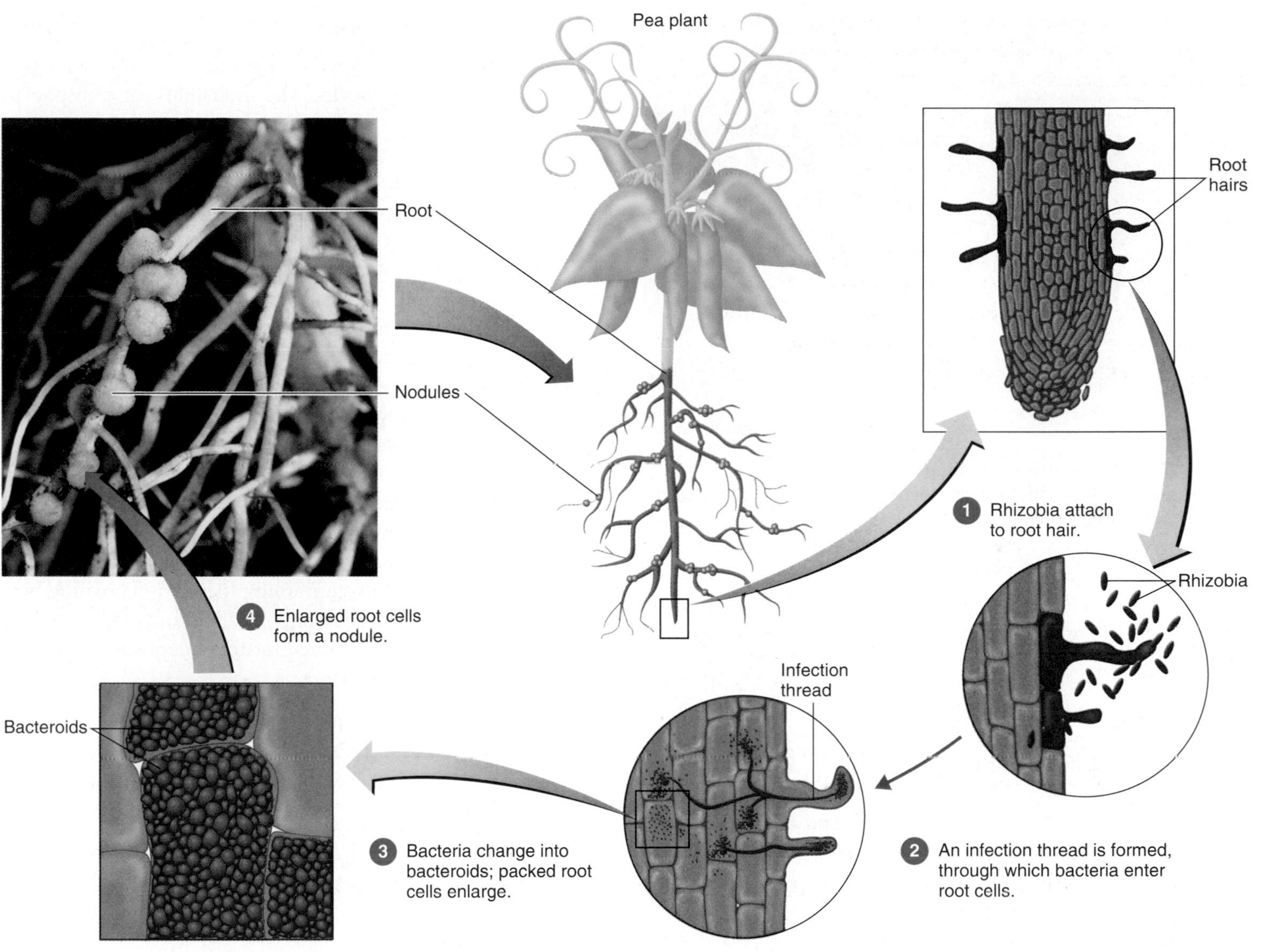
Pea plant
Root
Nodules
Root hairs
1 Rhizobia attach to root hair.
Rhizobia
4 Enlarged root cells form a nodule.
Infection thread
Bacteroids
3 Bacteria change into bacteroids; packed root cells enlarge.
2 An infection thread is formed, through which bacteria enter root cells.

Figure 27.5 The formation of a root nodule. Members of the nitrogen-fixing genera *Rhizobium* and *Bradyrhizobium* form these nodules on legumes. This mutualistic association is beneficial to both the plant and the bacteria.

Q **In nature, are leguminous plants most likely to be valuable in rich agricultural soils or poor desert soils?**

in a process called **dissimilation,** the sulfur is released as hydrogen sulfide to reenter the cycle.

Life without Sunshine

Interestingly, it is possible for entire biological communities to exist without photosynthesis by exploiting the energy in H_2S. Chapter 11 (page 315), presents equations to show that photosynthesis and chemoautotrophic use of H_2S are similar in certain respects. Such communities occur, for example, around deep-sea vents. Deep caves, totally isolated from sunlight, have been discovered that also support entire biological communities. The **primary producers** in these systems are chemoautotrophic bacteria rather than photoautotrophic plants or microbes.

Recently another microbial ecosystem operating far from sunlight has been discovered over 1 km deep within rocks, including shales, granites, and basalts. Such bacteria are called **endoliths** (inside rocks), which must grow in the near absence of oxygen and with minimal nutrient supplies. Also, in these rocks, chemical reactions and radioactivity split H_2O, producing hydrogen, which can be used for energy by autotrophic endolithic bacteria. Carbon dioxide dissolved in the water serves as a carbon source,

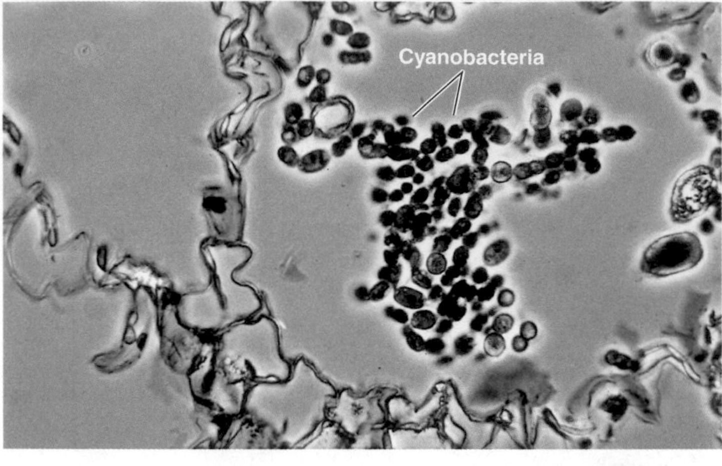

Figure 27.6 The *Azolla*–cyanobacteria symbiosis. A section through the leaf of an Azolla freshwater fern. The cyanobacteria *Anabaena azollae* is visible as chains of cells within the leaf cavity.

Q What is the major contribution of cyanobacteria as symbionts?

and cellular organic matter is produced. Some is excreted, or is released upon the death and lysis of the microbe, and becomes available for the growth of other microbes. Nutrient inputs, especially of nitrogen, are very small in this environment, and generation times may be measured in many years. Various survival strategies have developed for life with minimal nutrition. For example, suspended in a state between life and death, certain of these organisms become dramatically smaller. Ecologists speculating on forms of life that might be found in the harsh environment of Mars are very interested in endoliths.

The Phosphorus Cycle

Another important nutritional element that is part of a biogeochemical cycle is phosphorus. The availability of phosphorus may determine whether plants and other organisms can grow in an area. The problems associated with excess phosphorus (eutrophication) are described later in the chapter.

Phosphorus exists primarily as phosphate ions (PO_4^{3-}) and undergoes very little change in its oxidation state. The **phosphorus cycle** instead involves changes from soluble to insoluble forms and from organic to inorganic phosphate, often in relation to pH. For example, phosphate in rocks can be solubilized by the acid produced by bacteria such as *Thiobacillus.* Unlike the other cycles, there is no volatile phosphorus-containing product to return phosphorus to the atmosphere in the way carbon dioxide, nitrogen gas, and sulfur dioxide are returned. Therefore, phosphorus tends to accumulate in the seas. It can be retrieved by mining the above-ground sediments of ancient seas, mostly as deposits of calcium phosphate. Seabirds also mine phosphorus from the sea by eating phosphorus-containing fish and depositing it as guano (bird droppings). Certain small islands inhabited by such birds have long been mined for these deposits as a source of phosphorus for fertilizers.

CHECK YOUR UNDERSTANDING

- ✓ Certain nonphotosynthetic bacteria accumulated granules of sulfur within the cell; were the bacteria using hydrogen sulfide or sulfates as an energy source? **27-8**
- ✓ What chemical usually serves as an energy source for organisms that survive in darkness? **27-9**
- ✓ Why does phosphorus tend to accumulate in the seas? **27-10**

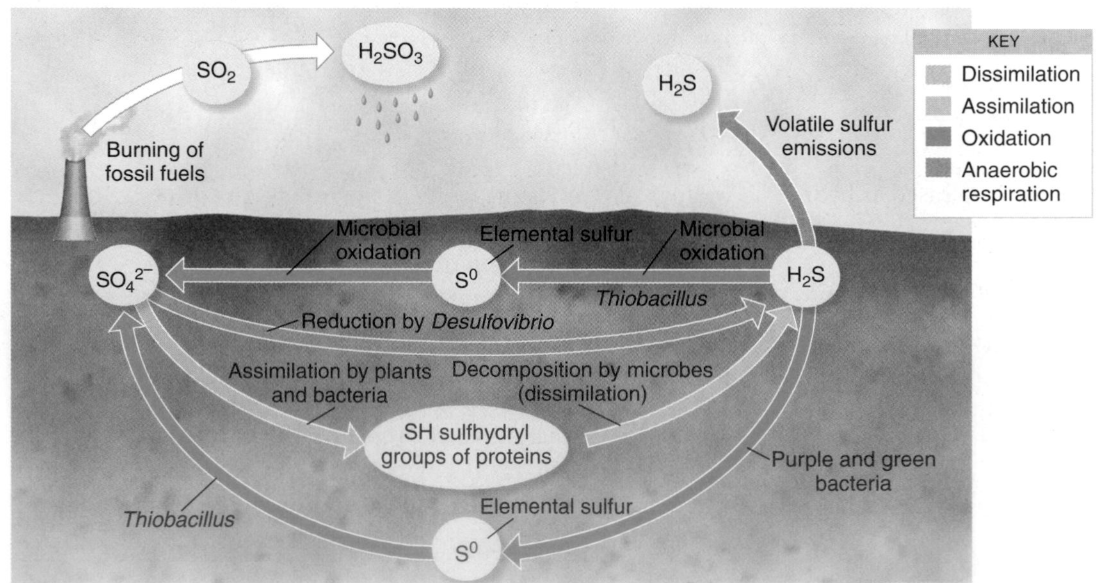

Figure 27.7 The sulfur cycle. Reduced forms of sulfur such as H_2S and elemental sulfur (S°) are energy sources for some microbes under aerobic or anaerobic conditions. Under anaerobic conditions, H_2S can be used as a substitute for H_2O in photosynthesis by purple and green bacteria (see page 777) to produce S°. Oxidized forms of sulfur, such as sulfates (SO_4^{2-}), are used as electron acceptors, as a substitute for oxygen, under anaerobic conditions by certain bacteria. Many organisms assimilate sulfates to make the SH groups of proteins.

Q Why is a source of sulfur necessary for all organisms?

Figure 27.8 2,4-D (black) and 2,4,5-T (red). This graph shows the structures and rates of microbial decomposition of the herbicides 2,4-D (black) and 2,4,5-T (red).

Q Which of these two herbicides is more easily degraded?

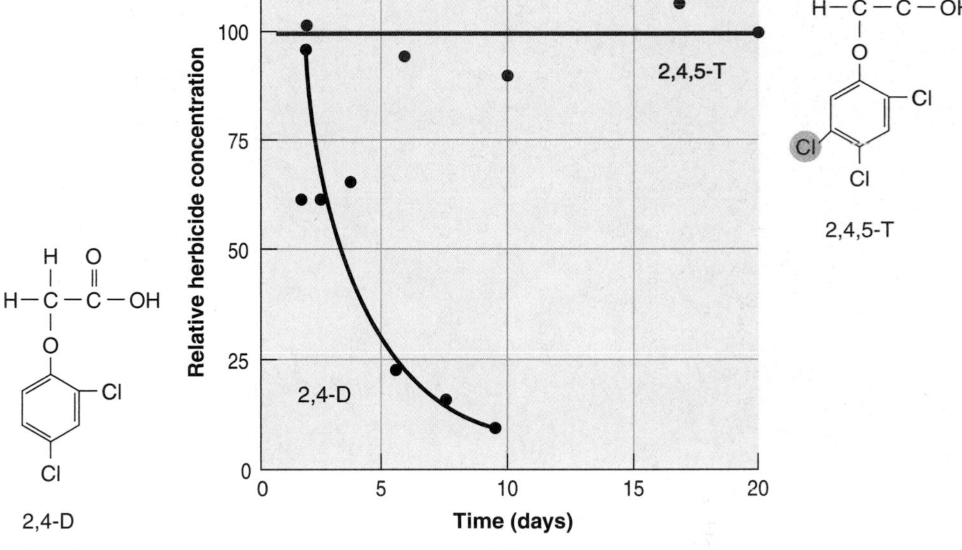

The Degradation of Synthetic Chemicals in Soil and Water

We seem to take for granted that soil microorganisms will degrade materials entering the soil. Natural organic matter, such as falling leaves or animal residues, are in fact readily degraded. However, in this industrial age many chemicals that do not occur in nature (**xenobiotics**), such as plastics, enter the soil in large amounts. In fact, plastics comprise about a fourth of all municipal wastes. A proposed solution to the problem is to develop biodegradable plastics made from polylactide (PLA) produced by lactic acid fermentation. When composted (see Figure 27.10 on page 777), a plastic cup, for example, made of PLA degrades in a few weeks. The barriers are economic, not technological. Many synthetic chemicals, such as pesticides, are highly resistant to degradation by microbial attack. A well-known example is the insecticide DDT, which proved so resistant that it accumulated to damaging levels in the environment.

Some synthetic chemicals are made up of bonds and subunits that are subject to attack by bacterial enzymes. Small differences in chemical structure can make large differences in biodegradability. The classic example is that of two herbicides: 2,4-D (the common chemical used to kill lawn weeds) and 2,4,5-T (used to kill shrubs); both were components of Agent Orange, which was used to defoliate jungles during the Vietnam war. The addition of a single chlorine atom to the structure of 2,4-D extends its life in soil from a few days to an indefinite period (Figure 27.8).

A growing problem is the leaching into groundwaters of toxic materials that are not biodegradable or that degrade very slowly. The sources of these materials may include landfills, illegal industrial dumps, or pesticides applied to agricultural crops.

Bioremediation

The use of microbes to detoxify or degrade pollutants is called **bioremediaton.** Oil spills from wrecked tankers represent some of the most dramatic examples of chemical pollution. The economic losses from contaminated fisheries and beaches can be enormous. To some degree, bioremediation occurs naturally as microbes attack the petroleum if conditions are aerobic. However, microbes usually obtain their nutrients in aqueous solution, and oil-based products are relatively nonsoluble. Also, petroleum hydrocarbons are deficient in essential elements, such as nitrogen and phosphorus. Bioremediation of oil spills is greatly enhanced if the resident bacteria are provided with "fertilizer" containing nitrogen and phosphorus (Figure 27.9). Bioremediaton may also make use of microbes that have been selected for growth on a certain pollutant or of genetically modified bacteria that are specially adapted to metabolize petroleum products. The addition of such specialized microbes is called **bioaugmentation** (see the box in Chapter 2, page 33).

Solid Municipal Waste

Solid municipal waste (garbage) is most frequently placed into large compacted landfills. Conditions are largely anaerobic, and even presumably biodegradable materials such as paper are not very effectively attacked by microorganisms. In fact, recovering a 20-year-old newspaper in readable condition is not at all unusual. But such anaerobic conditions do promote the activity of the same methanogens used in the operation of anaerobic sludge digesters to treat sewage (see page 786). The methane they produce can be tapped with drill holes and burned to generate electricity or purified and introduced into natural gas pipeline systems (see Figure 28.15, page 807). Such systems are part of the design of many large landfills in the United States, some of which

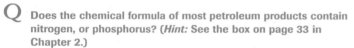

Figure 27.9 Bioremediation at an oil spill in Alaska. The portion of beach on the left is uncleaned, the beach on the right has been treated with applications of carbon-free nutrients (fertilizer). However, below the surface layers, where conditions are anaerobic, oil often remains for much longer periods.

Does the chemical formula of most petroleum products contain nitrogen, or phosphorus? (*Hint:* See the box on page 33 in Chapter 2.)

provide energy for industrial plants and homes. On a smaller scale, studies in India have shown that methane produced from the waste of three cows is enough to supply a family with cooking gas.

The amount of organic matter entering landfills can be considerably reduced if it is first separated from material that is not biodegradable and composted. **Composting** is a process gardeners use to convert plant remains into the equivalent of natural humus (Figure 27.10). A pile of leaves or grass clippings will undergo microbial degradation. Under favorable conditions, thermophilic bacteria will raise the temperature of the compost to 55–60°C in a couple of days. After the temperature declines, the pile can be turned to renew the oxygen supply, and a second temperature rise will occur. Over time, the thermophilic microbial populations are replaced by mesophilic populations that slowly continue the conversion to a stable material similar to humus. Where space is available, municipal wastes are composted in windrows (long, low piles) that are distributed and periodically turned over by specialized machinery. Municipal waste disposal now also makes increasing use of composting methods.

CHECK YOUR UNDERSTANDING

✓ Why are petroleum products naturally resistant to metabolism by most bacteria? **27-11**

✓ What is the definition of the term *bioremediation*? **27-12**

Aquatic Microbiology and Sewage Treatment

LEARNING OBJECTIVES

27-13 Describe the freshwater and seawater habitats of microorganisms.

27-14 Explain how wastewater pollution is a public health problem and an ecological problem.

27-15 Discuss the causes and effects of eutrophication.

27-16 Explain how water is tested for bacteriological purity.

27-17 Describe how pathogens are removed from drinking water.

27-18 Compare primary, secondary, and tertiary sewage treatment.

27-19 List some of the biochemical activities that take place in an anaerobic sludge digester.

27-20 Define *biochemical oxygen demand (BOD), activated sludge system, trickling filter, septic tank,* and *oxidation pond.*

Aquatic microbiology refers to the study of microorganisms and their activities in natural waters, such as lakes, ponds, streams, rivers, estuaries, and oceans. Domestic and industrial wastewater enters lakes and streams, and its degradation and effects on the microbial life are important factors in aquatic microbiology. We will also see that the method of treating wastewater by municipalities mimics a natural filtering process.

Aquatic Microorganisms

Large numbers of microorganisms in a body of water generally indicate high nutrient levels in the water. Water contaminated by inflows from sewage systems or from biodegradable industrial organic wastes is relatively high in bacterial numbers. Similarly, ocean estuaries (fed by rivers) have higher nutrient levels and therefore larger microbial populations than other shoreline waters.

In water, particularly water with low nutrient concentrations, microorganisms tend to grow on stationary surfaces and on particulate matter. In this way, a microorganism has contact with more nutrients than if it were randomly suspended and floating freely with the current. Many bacteria whose main habitat is water often have appendages and holdfasts that attach to various surfaces. One example is *Caulobacter* (see Figure 11.2, page 304).

Freshwater Microbiota

A typical lake or pond serves as an example to represent the various zones and the kinds of microbiota found in a body of fresh water. The **littoral zone** along the shore has considerable rooted vegetation, and light penetrates throughout it. The **limnetic zone** consists of the surface of the open water area away from the shore. The **profundal zone** is the deeper water under the limnetic zone. The **benthic zone** contains the sediment at the bottom.

Microbial populations of freshwater bodies tend to be affected mainly by the availability of oxygen and light. In many

(a) Solid municipal wastes being turned by a specially designed machine

(b) Compost made from municipal wastes awaiting trucks to spread it on agricultural fields

Figure 27.10 **Composting municipal wastes.**

Q **A compost pile of grass and leaves is very high in carbon; does it have much nitrogen?**

ways, light is the more important resource because photosynthetic algae are the main source of organic matter, and hence of energy, for the lake. These organisms are the primary producers of a lake that supports a population of bacteria, protozoa, fish, and other aquatic life. Photosynthetic algae are located in the limnetic zone.

Areas of the limnetic zone with sufficient oxygen contain pseudomonads and species of *Cytophaga, Caulobacter,* and *Hyphomicrobium.* Oxygen does not diffuse into water very well, as any aquarium owner knows. Microorganisms growing on nutrients in stagnant water quickly use up the dissolved oxygen in the water. In the oxygenless water, fish die, and anaerobic activity produces odors. Wave action in shallow layers, or water movement in rivers, tends to increase the amount of oxygen throughout the water and aid in the growth of aerobic populations of bacteria. Movement thus improves the quality of water and aids in the degradation of polluting nutrients.

Deeper waters of the profundal and benthic zones have low oxygen concentrations and less light. Algal growth near the surface often filters the light, and it is not unusual for photosynthetic microbes in deeper zones to use different wavelengths of light from those used by surface-layer photosynthesizers (see Figure 12.10a, page 341).

Purple and green sulfur bacteria are found in the profundal zone. These bacteria are anaerobic photosynthetic organisms that metabolize H_2S to sulfur and sulfate in the bottom sediments of the benthic zone.

The sediment in the benthic zone includes bacteria such as *Desulfovibrio* that use sulfate (SO_4^{2-}) as an electron acceptor and reduce it to H_2S. Methane-producing bacteria are also part of these anaerobic benthic populations. In swamps, marshes, or bottom sediments, they produce methane gas. *Clostridium* species are common in bottom sediments and may include botulism organisms, particularly those causing outbreaks of botulism in waterfowl.

Seawater Microbiota

As knowledge of the microbial life of the oceans expands, largely identified by ribosomal RNA methods (see the discussion of FISH on page 292 in Chapter 10), biologists are becoming more conscious of the importance of oceanic microbes. One conclusion, so far, has been that nearly a third of all life on the planet consists of microbes that live, not in ocean waters, but under the seafloor. These microbes make immense amounts of methane gas that could be environmentally damaging if it were to be released into the atmosphere.

In the upper, relatively sunlit waters of the ocean, photosynthetic cyanobacteria of the genera *Synechococcus* (sin′ē-kō-kok-kus) and *Prochlorococcus* (prō-klôr′ō-kok-kus) are abundant. Populations of different strains vary at different depths according to their adaptations to available sunlight. A drop of seawater might contain 20,000 cells of *Prochlorococcus,* a tiny sphere less than 0.7 μm in diameter. This unseen population of microscopic organisms fills the upper 100 meters of ocean and exerts a profound influence on life on Earth. The support of oceanic life depends largely on such photosynthetic microscopic life, the marine **phytoplankton** (a term derived from the Greek for wandering plants).

Photosynthetic bacteria such as these form the basis of the oceanic food chain. Billions of these in every liter of seawater double in number every few days and are consumed at about the same rate by microscopic predators. They fix carbon dioxide to form organic matter that is eventually released as dissolved organic matter and is used by the ocean's heterotrophic bacteria. A cyanobacterium, *Trichodesmium* (trik′ō-des-mē-um), fixes

Figure 27.11 Bioluminescent bacteria as light organs in fish. This is a deep-sea flashlight fish (*Photoblepharon palpebratus*). The luminous organ under the eye can be covered by a tissue lid.

Q What enzyme is responsible for bioluminescence?

nitrogen and helps replenish the nitrogen that is lost as organisms sink to oceanic depths. Immense populations of another bacterium, *Pelagibacter ubique,* metabolize the waste products of these photosynthetic populations (see the discussion under Microbial Diversity on page 325). Bacteria of many kinds then serve as a particulate food source for a series of increasingly larger consumers. These are first the protozoa, which are in turn prey for multicellular zooplankton (planktonic animal life such as shrimplike krill). These zooplankton are eventually prey for fish. Much of the carbon dioxide and mineral nutrients released by the metabolic activity of bacteria, protozoa, and zooplankton is recycled into the photosynthetic phytoplankton.

In waters below about 100 meters, members of the Archaea begin to dominate microbial life. Planktonic members of this group of the genus *Crenarchaeota* (kren-ärk-e′ō-tä) account for much of the microbial biomass of the oceans. These organisms are well adapted to the cool temperatures and low oxygen levels of oceanic depths. Their carbon is primarily derived from dissolved CO_2.

Microbial **bioluminescence,** or light emission, is an interesting aspect of deep-sea life. Many bacteria are luminescent, and some have established symbiotic relationships with benthic-dwelling fish. These fish sometimes use the glow of their resident bacteria as an aid in attracting and capturing prey in the complete darkness of the ocean depths (Figure 27.11). These bioluminescent organisms have an enzyme called *luciferase* that picks up electrons from flavoproteins in the electron transport chain and then emits some of the electron's energy as a photon of light (see the box on page 780).

CHECK YOUR UNDERSTANDING

✓ Purple and green sulfur bacteria are photosynthetic organisms, but they are generally found deep in freshwater rather than at the surface. Why? **27-13**

The Role of Microorganisms in Water Quality

Water in nature is seldom totally pure. Even rainfall is contaminated as it falls to Earth.

Water Pollution

The form of water pollution that is our primary interest is microbial pollution, especially by pathogenic organisms.

The Transmission of Infectious Diseases Water that moves below the ground's surface undergoes a filtering that removes most microorganisms. For this reason, water from springs and deep wells is generally of good quality. The most dangerous form of water pollution occurs when feces enter the water supply. Many diseases are perpetuated by the fecal–oral route of transmission, in which a pathogen is shed in human or animal feces, contaminates water, and is ingested (see Chapter 25). The Centers for Disease Control and Prevention (CDC) estimates that in the United States 900,000 people become ill each year from waterborne infections. Globally, it is estimated that waterborne diseases are responsible for over 2 million deaths each year, mostly among children under the age of 5. This is the equivalent of 20 jumbo jets crashing every day and represents about 15% of all child deaths in this age group.

Examples of such diseases are typhoid fever and cholera, caused by bacteria that are shed only in human feces. About 100 years ago, the *Journal of the American Medical Association* reported that the typhoid fever mortality rate in Chicago had declined from 159.7 per 100,000 people in 1891 to 31.4 per 100,000 in 1894. This advance in public health had been accomplished by extending the city water supply intake pipes in Lake Michigan to a distance of 4 miles from shore. The medical journal commented that this diluted the sewage contaminating the water supply, which at that time was not treated further. This same article speculated on the need to remove microorganisms that caused specific diseases. They suggested the use of sand filter beds, already widely used in Europe at the time. Sand filtration mimics the natural purification of spring water. Figure 27.12 illustrates the effect of the introduction of such filtration of water supplies on the incidence of typhoid fever in Philadelphia.

Chemical Pollution Preventing chemical contamination of water is a difficult problem. Industrial and agricultural chemicals leached from the land enter water in great amounts and in forms that are resistant to biodegradation. Rural waters often have excessive amounts of nitrate from agricultural fertilizers. When ingested, the nitrate is converted to nitrite by bacteria in the gastrointestinal tract. Nitrite competes for oxygen in the blood and is especially likely to harm infants.

A striking example of industrial water pollution involved mercury in wastewater from paper manufacturing. The metallic mercury was allowed to flow into waterways as waste. It was assumed that the mercury was inert and would remain segregated

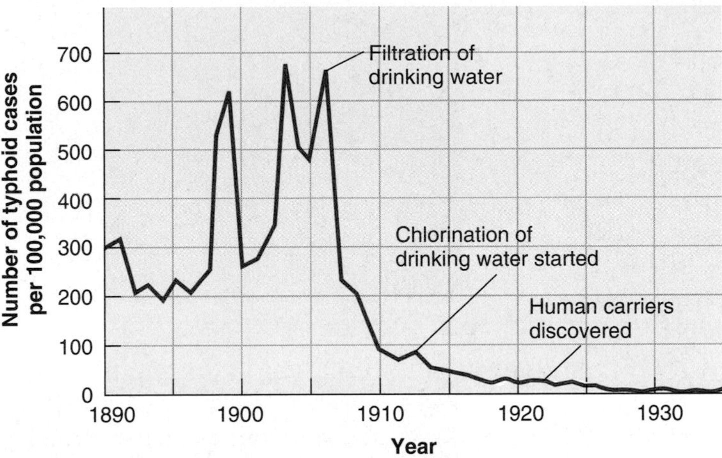

Figure 27.12 The incidence of typhoid fever in Philadelphia, 1890–1935. This graph clearly shows the effect of water treatments on incidence of typhoid.

Source: E. Steel, *Water Supply and Sewerage*, New York: McGraw-Hill, 1953.

Q Why did the incidence of typhoid fever decrease?

Figure 27.13 A red tide. These blooms of aquatic growth are caused by excess nutrients in water. The color is from the pigmentation of the dinoflagellates.

Q What is the primary energy source of the dinoflagellates that cause such aquatic blooms?

in the sediments. However, bacteria in the sediments converted the mercury into a soluble chemical compound, methyl mercury, which was then taken up by fish and invertebrates in the waters. When such seafood is a substantial part of the human diet, the mercury concentrations can accumulate with devastating effects on the nervous system. The U.S. Food and Drug Administration (FDA) advises that pregnant or nursing women not eat certain fish, including swordfish and shark, that are likely to contain high levels of mercury. Bioremediation efforts using bacteria to detoxify mercury in one wildlife refuge are discussed in the box in Chapter 2 (page 33).

Another example of chemical pollution is the synthetic detergents developed immediately after World War II. These rapidly replaced many of the soaps then in use. Because these new detergents were not biodegradable, they rapidly accumulated in the waterways. In some rivers, large rafts of detergent suds could be seen traveling downstream. These detergents were replaced by biodegradable synthetic formulations.

Biogradable detergents, however, still present a major environmental problem because they often contain phosphates. Unfortunately, phosphates pass almost unchanged through sewage systems and can lead to **eutrophication,** which is caused by an overabundance of nutrients in lakes and streams.

To understand the concept of eutrophication, recall that algae and cyanobacteria get their energy from sunlight and their carbon from carbon dioxide dissolved in water. In most waters, only nitrogen and phosphorus supplies, therefore, remain inadequate for algal growth. Both of these nutrients can enter water from domestic, farm, and industrial wastes when waste treatment is absent or inefficient. These additional nutrients cause dense aquatic growths called **algal blooms.** Because many cyanobacteria can fix nitrogen from the atmosphere, these photosynthesizing organisms require only traces of phosphorus to initiate blooms. Once eutrophication results in blooms of algae or cyanobacteria, the eventual effect is the same as adding biodegradable organic matter. In the short run, these algae and cyanobacteria produce oxygen. However, they eventually die and are degraded by bacteria. During the degradation process, the oxygen in the water is used up, killing the fish. Undegraded remnants of organic matter settle to the bottom and hasten the filling of the lake.

Red tides of toxin-producing phytoplankton (Figure 27.13), which were mentioned in Chapter 12, are probably caused by excessive nutrients from oceanic upwellings or terrestrial wastes. In addition to eutrophication effects, this type of biological bloom can affect human health. Seafood, especially clams or similar mollusks, that ingest these plankton become toxic to humans.

Municipal waste containing detergents is likely to be the main source of phosphates in lakes and streams. As a result, phosphate-containing detergents and lawn fertilizers are banned in many places.

Coal-mining wastes, particularly in the eastern United States, are very high in sulfur content, mostly from pyrite (FeS_2). In the process of obtaining energy from the oxidation of the ferrous ion (Fe^{2+}), bacteria such as *Thiobacillus ferrooxidans* convert the FeS_2 into sulfate. The sulfate enters streams as sulfuric acid, which lowers the pH of the water and damages aquatic life. The low pH also promotes the formation of insoluble iron hydroxides, which form the yellow precipitates often seen clouding such polluted waters.

Water Purity Tests

Historically, most of our concern about water purity has been related to the transmission of disease. Therefore, tests have been developed to determine the safety of water; many of these tests are also applicable to foods.

Biosensors: Bacteria That Detect Pollutants and Pathogens

Each year in the United States, industrial plants generate 265 million metric tons of hazardous waste, 80% of which make their way into landfills. Burying these chemicals does not remove them from the ecosystem, however; it just moves them to other places, where they may still find their way into bodies of water. Traditional chemical analyses to locate these chemicals are expensive and cannot distinguish chemicals that affect biological systems from those that lie inert in the environment.

In response to this problem, scientists are developing biosensors, bacteria that can locate biologically active pollutants. Biosensors do not require costly chemicals or equipment, and they work quickly—within minutes.

To work, bacterial biosensors require both a receptor that is activated in the presence of pollutants and a reporter that will make such a change apparent. Biosensors use the *lux* operon from *Vibrio* or *Photobacterium* as a reporter. This operon contains inducer and structural genes for the enzyme luciferase. In the presence of a coenzyme called $FMNH_2$, luciferase reacts with the molecule in such a way that the enzyme–substrate complex emits blue-green light, which then oxidizes the $FMNH_2$ to produce FMN. Therefore, a bacterium containing the *lux* operon will emit visible light when the receptor is activated (see the photographs).

The *lux* operon is readily transferred to many bacteria. Scientists in several countries are investigating the use of *E. coli* containing the *lux* operon to detect hazardous chemicals in soil and water. The soil or water sample is placed in a tube containing the genetically modified *E. coli* bacteria. The bacteria will emit light as long as they are healthy but will stop emitting light if they have been killed by toxic pollutants.

In another application, *Lactococcus* bacteria containing the *lux* operon are being used to detect the presence of antibiotics in milk that is to be used for cheese production. Because the emission of light requires a living cell, the presence of antibiotics is measured as a decrease in light output by the recombinant *Lactococcus* bacteria.

Other biosensors use recombinant microbes carrying a jellyfish gene for green fluorescent protein (GFP) and genes that are induced by pollutants or antibiotics. For example, yeast containing genes encoding mammalian odor receptors and GFP will fluoresce in the presence of TNT. After pollutants are detected, bioremediation processes are still required to remove them.

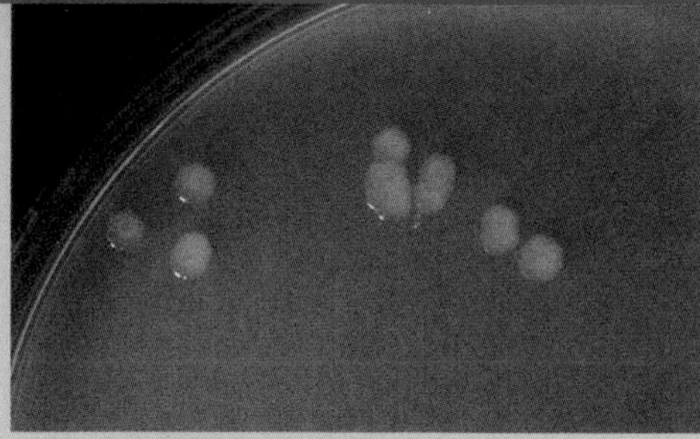

(a)

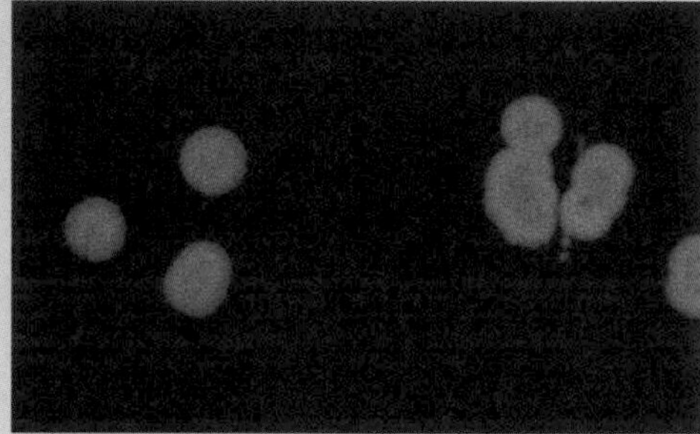

(b)

Vibrio fischeri emits light when energy is released by the transport of electrons to luciferase. Shown here are colonies of *V. fischeri* photographed (**a**) in daylight and (**b**) in the dark, illuminated by their own light.

It is not practical, however, to look only for pathogens in water supplies. For one thing, if we were to find the pathogens causing typhoid or cholera in the water system, the discovery would already be too late to prevent an outbreak of the disease. Moreover, such pathogens would probably be present only in small numbers and might not be included in tested samples.

The tests for water purity in use today are aimed instead at detecting particular **indicator organisms.** There are several criteria for an indicator organism. The most important criterion is that the microbe be consistently present in human feces in substantial numbers so that its detection is a good indication that human wastes are entering the water. The indicator organisms should also survive in the water at least as well as the pathogens would. The indicator organisms must be detectable by simple tests that can be carried out by people with relatively little training in microbiology.

In the United States, the usual indicator organisms in freshwater are the *coliform bacteria.** **Coliforms** are defined as aerobic or facultatively anaerobic, gram-negative, non–endospore-forming, rod-shaped bacteria that ferment lactose to form gas within 48 hours of being placed in lactose broth at 35°C. Because some coliforms are not solely enteric bacteria but are more commonly found in plant and soil samples, many standards for food and water specify the identification of *fecal coliforms.* The predominant fecal coliform is *E. coli,* which constitutes a large proportion of the human intestinal population. There are specialized tests to distinguish fecal coliforms from nonfecal coliforms. Note that coliforms are not themselves pathogenic under normal conditions, although certain strains can cause diarrhea (see Chapter 25, page 717) and opportunistic urinary tract infections (see Chapter 26, page 746).

The methods for determining the presence of coliforms in water are based largely on the lactose-fermenting ability of coliform bacteria. The multiple-tube method can be used to estimate coliform numbers by the most probable number (MPN) method (see Figure 6.19, page 177). The membrane filtration method is a more direct method of determining the presence and numbers of coliforms. This is possibly the most widely used method in North America and Europe. It makes use of a filtration apparatus similar to that shown in Figure 7.4 (page 191). In this application, though, the bacteria collected on the surface of a removable membrane filter are placed on an appropriate medium and incubated. Coliform colonies have a distinctive appearance and are counted. This method is suitable for low-turbidity waters that do not clog the filter and have relatively few noncoliform bacteria that would mask the results.

A newer and more convenient method of detecting coliforms, specifically the fecal coliform *E. coli,* makes use of media containing the two substrates *o*-nitrophenyl-β-D-galactopyranoside (ONPG) and 4-methylumbelliferyl-β-D-glucuronide (MUG). Coliforms produce the enzyme β-galactosidase, which acts on ONPG and forms a yellow color, indicating their presence in the sample. *E. coli* is unique among coliforms in almost always producing the enzyme β-glucuronidase, which acts on MUG to form a fluorescent compound that glows blue when illuminated by long-wave UV light (Figure 27.14). These simple tests, or variants of them, can detect the presence or absence of coliforms or *E. coli* and can be combined with the multiple-tube method to enumerate them. It can also be applied to solid media, such as in the membrane filtration method. The colonies fluoresce under UV light.

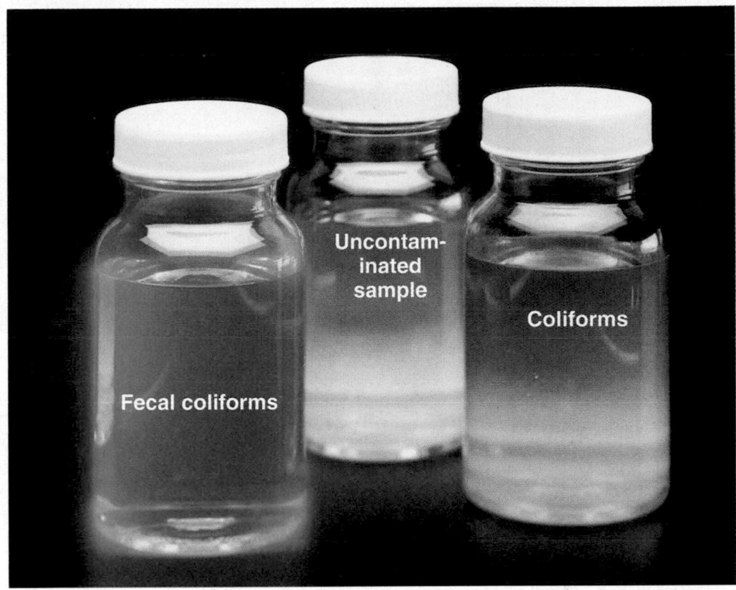

Figure 27.14 The ONPG and MUG coliform test. A yellow color (positive ONPG) indicates the presence of coliforms. Blue fluorescence (positive MUG) indicates the presence of the fecal coliform *E. coli.* The clear medium indicates an uncontaminated sample.

Q What causes the formation of the fluorescent compound in a positive MUG test?

Coliforms have been very useful as indicator organisms in water sanitation, but they have limitations. One problem is the growth of coliform bacteria embedded in biofilms on the inner surfaces of water pipes. These coliforms do not, then, represent external fecal contamination of the water, and they are not considered a threat to public health. Standards governing the presence of coliforms in drinking water require that any positive water sample be reported, and occasionally these indigenous coliforms have been detected. This has led to unnecessary community orders to boil water.

A more serious problem is that some pathogens, especially viruses and protozoan cysts and oocysts, are more resistant than coliforms to chemical disinfection. Through the use of sophisticated methods of detecting viruses, it has been found that chemically disinfected water samples that are free of coliforms are often still contaminated with enteric viruses. The cysts of *Giardia lamblia* and oocysts of *Cryptosporidium* are so resistant to chlorination that completely eliminating them by this method is probably impractical; mechanical methods such as filtration are necessary. A general rule for chlorination is that viruses are more resistant to treatment than is *E. coli* and that the cysts of *Cryptosporidium* and *Giardia* are 100 times more resistant than viruses.

*The U.S. Environmental Protection Agency (EPA) recommends the use of *Enterococcus* bacteria as a safety indicator for waters in oceans and bays. Populations of the enterococci decrease more uniformly than coliforms in both freshwater and seawater.

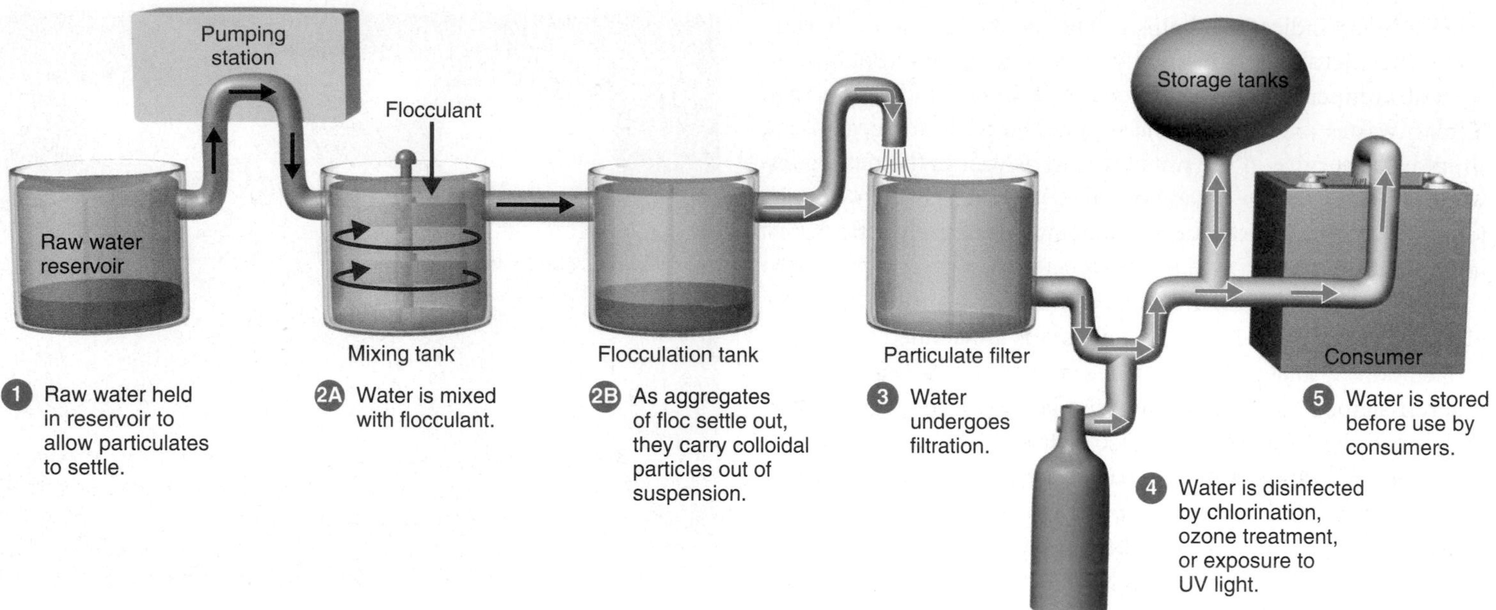

Figure 27.15 The steps involved in water treatment in a typical municipal water purification plant.

Q Does removal of "colloidal particles" by flocculation involve living organisms?

CHECK YOUR UNDERSTANDING

- ✓ Which disease is more likely to be transmitted by polluted water, cholera or influenza? **27-14**
- ✓ Name a microorganism that will grow in water even if there is no source of organic matter for energy or a nitrogen source—but does require small inputs of phosphorus. **27-15**
- ✓ Coliforms are the most common bacterial indicator of health-threatening water pollution in the United States. Why is it usually necessary to specify the term *fecal* coliform? **27-16**

Water Treatment

When water is obtained from uncontaminated reservoirs fed by clear mountain streams or from deep wells, it requires minimal treatment to make it safe to drink. Many cities, however, obtain their water from badly polluted sources, such as rivers that have received municipal and industrial wastes upstream. The steps used to purify this water are shown in Figure 27.15. Water treatment is not intended to produce sterile water, but rather water that is free of disease-causing microbes.

Coagulation and Filtration

Very turbid (cloudy) water is allowed to stand in a holding reservoir for a time to allow as much particulate suspended matter as possible to settle out. The water then undergoes **flocculation,** the removal of colloidal materials such as clay, which is so small (smaller than 10 μm) that it would otherwise remain in suspension indefinitely. A flocculant chemical, such as aluminum potassium sulfate (alum), forms aggregations of fine suspended particles called *floc.* As these aggregations slowly settle out, they entrap colloidal material and carry it to the bottom. Large numbers of viruses and bacteria are also removed this way. Alum was used to clear muddy river water during the first half of the nineteenth century in the military forts of the American West, long before the germ theory of disease was developed.

After flocculation, the water is treated by **filtration**—that is, passing it through beds of 2 to 4 feet of fine sand or crushed anthracite coal. As mentioned previously, some protozoan cysts and oocysts are removed from water only by such filtration treatment. The microorganisms are trapped mostly by surface adsorption onto the sand particles. They do not penetrate the tortuous routing between the particles, even though the openings might be larger than the microbes that are filtered out. These filters are periodically backflushed to clear them of accumulations. Water systems of cities that have an exceptional concern for toxic chemicals supplement sand filtration with filters of activated charcoal (carbon). Charcoal removes not only particulate matter but also most dissolved organic chemical pollutants. A properly operated water treatment plant will remove viruses (which are harder to remove than bacteria and protozoa) with an efficiency of about 99.5%. Low-pressure *membrane filtration systems* are now coming into use. These systems have pore openings as small as 0.2 μm and are more reliable for removal of *Giardia* and *Cryptosporidium.*

Disinfection

Before entering the municipal distribution system, the filtered water is chlorinated. Because organic matter neutralizes chlorine, the plant operators must pay constant attention to maintaining effective levels of chlorine. There has been some concern that chlorine itself might be a health hazard because it could react with organic contaminants of the water to form carcinogenic compounds. At present, this possibility is considered an acceptable risk when compared with the proven usefulness of chlorinating of water.

As noted in Chapter 7 (page 202), another disinfectant for water is ozone treatment. Ozone (O_3) is a highly reactive form of oxygen that is formed by electrical spark discharges and UV light. (The fresh odor of air following an electrical storm or around a UV light bulb is from ozone.) Ozone for water treatment is generated electrically at the site of treatment (Figure 27.16). Ozone treatment is also valued because it leaves no taste or odor. Because it has little residual effect, ozone is usually used as a primary disinfectant treatment and is followed by chlorination. The use of UV light is also a supplement or alternative to chemical disinfection. Ultraviolet tube lamps are arranged so that water flows close to them. This is necessary because of the low penetrating power of UV radiation.

Figure 27.16 Ozone generation. Water treatment plants produce ozone by passing dry air between high-voltage electrodes in tanks called ozonators such as the two shown here.

Q What is a major disadvantage of ozonation of water?

CHECK YOUR UNDERSTANDING

✓ How do flocculants such as alum remove colloidal impurities, including microorganisms, from water? **27-17**

Sewage (Wastewater) Treatment

Sewage, or wastewater, includes all the water from a household that is used for washing and toilet wastes. Rainwater flowing into street drains and some industrial wastes enter the sewage system in many cities. Sewage is mostly water and contains little particulate matter, perhaps only 0.03%. Even so, in large cities the solid portion of sewage can total more than 1000 tons of solid material per day.

Until environmental awareness intensified, a surprising number of large American cities had only a rudimentary sewage treatment system or no system at all. Raw sewage, untreated or nearly so, was simply discharged into rivers or oceans. A flowing, well-aerated stream is capable of considerable self-purification. Therefore, until expanding populations and their wastes exceeded this capability, this casual treatment of municipal wastes did not cause problems. In the United States, most cases of simple discharge have been improved. But this is not true in much of the world. Many of the communities bordering the Mediterranean dump their unprocessed sewage into the sea. At one Asiatic tourist resort, a hotel posted instructions that toilet paper was not to be flushed in the toilets—presumably because floating paper would make it clear that the sewage outlets were near the beach. In areas of Europe and South Africa where tourism is essential to the economy, local administrations are attempting to reassure visitors about bathing water quality with the Blue Flag campaign. The presence of the flag (Figure 27.17) shows that the coastal waters meet certain minimal standards of sanitation.

Primary Sewage Treatment

The usual first step in sewage treatment is called **primary sewage treatment** (Figure 27.18a). In this process, large floating materials in incoming wastewater are screened out, the sewage is allowed to flow through settling chambers to remove sand and similar gritty material, skimmers remove floating oil and grease, and floating debris is shredded and ground. After this step, the sewage passes through sedimentation tanks, where more solid matter settles out. Sewage solids collecting on the bottom are called **sludge**—at this stage, *primary sludge.* About 40–60% of suspended solids are removed from sewage by this settling treatment, and flocculating chemicals that increase the removal of solids are sometimes added at this stage. Biological activity is not particularly important in primary treatment, although some digestion of sludge and dissolved organic matter can occur during long holding times. The sludge is removed on either a continuous or an intermittent basis, and the effluent (the liquid flowing out) then undergoes secondary treatment.

Biochemical Oxygen Demand

An important concept in sewage treatment and in the general ecology of waste management, **biochemical oxygen demand (BOD)** is a measure of the biologically degradable organic matter in water. Primary treatment removes about 25–35% of the BOD of sewage.

BOD is determined by the amount of oxygen required by bacteria to metabolize the organic matter. The classic method of

Figure 27.17 A beach displaying a blue flag.

Q **What sort of bacterial populations would need to be quantified to set standards for beach-area waters?**

measurement is the use of special bottles with airtight stoppers. Each bottle is first filled with test water or dilutions. The water is initially aerated to provide a relatively high level of dissolved oxygen and is seeded with bacteria if necessary. The filled bottles are incubated in the dark for 5 days at 20°C, and the decrease in dissolved oxygen is determined by a chemical or electronic testing method. The more oxygen that is used up as the bacteria degrade the organic matter in the sample, the greater the BOD, which is usually expressed in milligrams of oxygen per liter of water. The amount of oxygen that normally can be dissolved in water is only about 10 mg/liter; typical BOD values of wastewater may be 20 times this amount. If this wastewater enters a lake, for example, bacteria in the lake begin to consume the organic matter responsible for the high BOD, rapidly depleting the oxygen in the lake water. (See the discussion of eutrophication earlier in the chapter, page 779.)

Secondary Sewage Treatment

After primary treatment, the greater part of the BOD remaining in the sewage is in the form of dissolved organic matter. **Secondary sewage treatment,** which is predominantly biological, is designed to remove most of this organic matter and reduce the BOD (Figure 27.18b). In this process, the sewage undergoes strong aeration to encourage the growth of aerobic bacteria and other microorganisms that oxidize the dissolved organic matter to carbon dioxide and water. Two commonly used methods of secondary treatment are activated sludge systems and trickling filters.

In the aeration tanks of an **activated sludge system,** air or pure oxygen is passed through the effluent from primary treatment (Figure 27.19). The name is derived from the practice of adding some of the sludge from a previous batch to the incoming sewage. This inoculum is termed *activated sludge* because it contains large numbers of sewage-metabolizing microbes. The activity of these aerobic microorganisms oxidizes much of the sewage organic matter into carbon dioxide and water. Especially important members of this microbial community are species of *Zoogloea* bacteria, which form bacteria-containing masses in the aeration tanks called floc, or *sludge granules* (Figure 27.20). (See the discussion of floc, earlier in the chapter.) Soluble organic matter in the sewage is incorporated into the floc and its microorganisms. Aeration is discontinued after 4 to 8 hours, and the contents of the tank are transferred to a settling tank, where the floc settles out, removing much of the organic matter. These solids are subsequently treated in an anaerobic sludge digester, which will be described shortly. Probably more organic matter is removed by this settling-out process than by the relatively short-term aerobic oxidation by microbes. The clear effluent is disinfected and discharged.

Occasionally, the sludge will float rather than settle out; this phenomenon is called **bulking.** When this happens, the organic matter in the floc flows out with the discharge effluent, resulting in local pollution. Bulking is caused by the growth of filamentous bacteria of various types; *Sphaerotilus natans* and *Nocardia* species are frequent offenders. Activated sludge systems are quite efficient: they remove 75–95% of the BOD from sewage.

Trickling filters are the other commonly used method of secondary treatment. In this method, the sewage is sprayed over a bed of rocks or molded plastic (Figure 27.21a). The components of the bed must be large enough so that air penetrates to the bottom but small enough to maximize the surface area available for microbial activity. A biofilm (see page 162) of aerobic microbes grows on the rock or plastic surfaces (Figure 27.21b). Because air circulates throughout the rock bed, these aerobic microorganisms in the slime layer can oxidize much of the organic matter trickling over the surfaces into carbon dioxide and water. Trickling filters remove 80–85% of the BOD, so they are generally less efficient than activated sludge systems. However, they are usually less troublesome to operate and have fewer problems from overloads or toxic sewage. Note that sludge is also a product of trickling filter systems.

Another biofilm-based design for secondary sewage treatment is the **rotating biological contactor** system. This is a series of disks several feet in diameter, mounted on a shaft. The disks rotate slowly, with their lower 40% submerged in wastewater. Rotation provides aeration and contact between the biofilm on the disks and the wastewater. The rotation also tends to cause the accumulated biofilm to slough off when it becomes too thick. This is about the equivalent of floc accumulation in activated sludge systems.

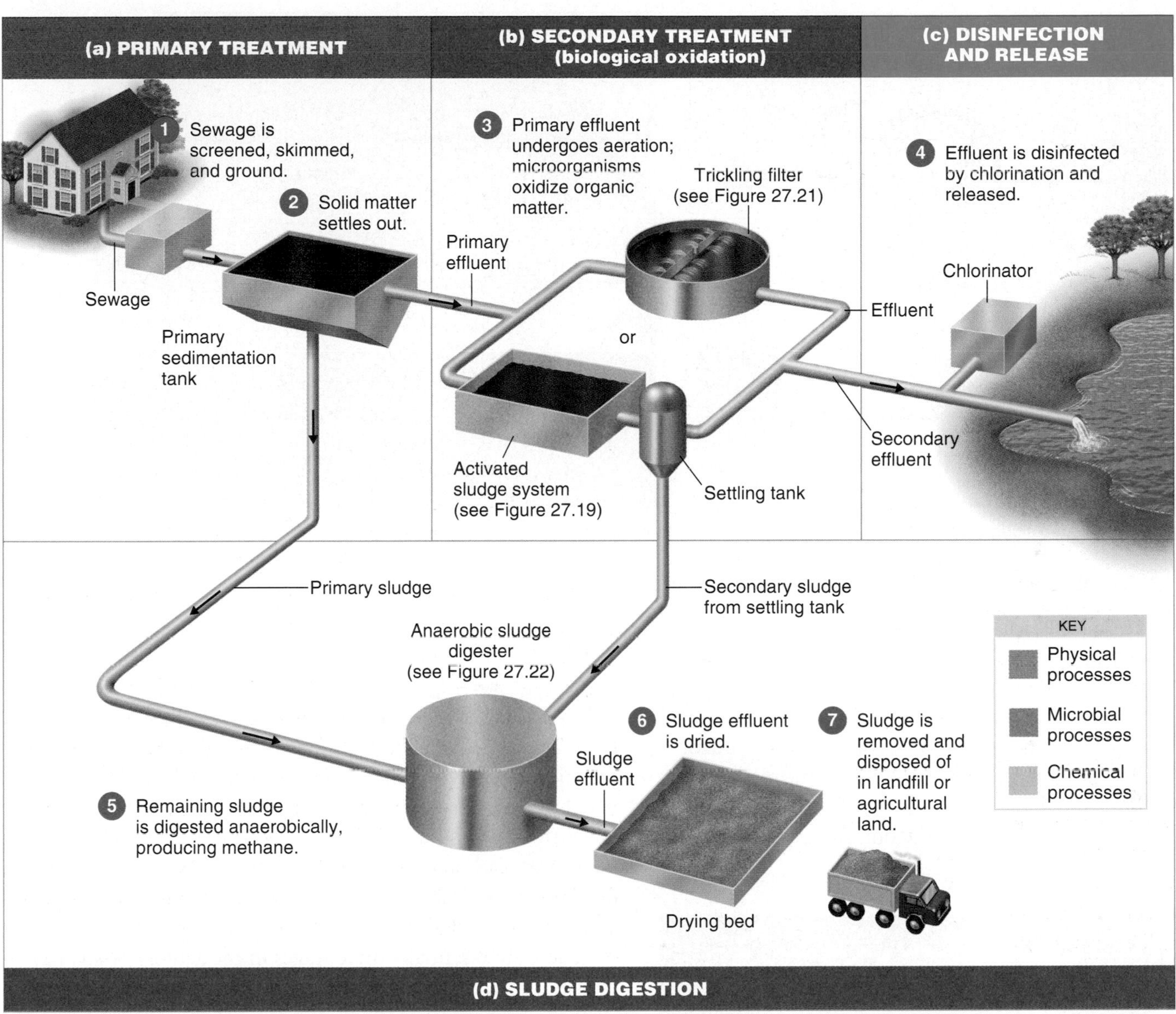

Figure 27.18 The stages in typical sewage treatment. Microbial activity occurs aerobically in trickling filters or activated sludge aeration tanks and anaerobically in the anaerobic sludge digester. A particular system would use either activated sludge aeration tanks or trickling filters, not both, as shown in this figure. Methane produced by sludge digestion is burned off or used to power heaters or pump motors.

Q **Which processes require oxygen?**

Disinfection and Release

Treated sewage is disinfected, usually by chlorination, before being discharged (Figure 27.18c). The discharge is usually into an ocean or into flowing streams, although spray-irrigation fields are sometimes used to avoid phosphorus and heavy metal contamination of waterways.

Sewage can be treated to a level of purity that allows its use as drinking water. This is the practice now in some arid-area cities in the United States and will probably be expanded. In a typical system, the treated sewage is filtered to remove microscopic suspended particles, then passed through a reverse osmosis purification system to remove microorganisms. Any remaining microorganisms are killed by exposure to UV light or other disinfectants.

Sludge Digestion

Primary sludge accumulates in primary sedimentation tanks; sludge also accumulates in activated sludge and in trickling filter secondary treatments. For further treatment, these sludges are often pumped to **anaerobic sludge digesters** (Figure 27.18d and

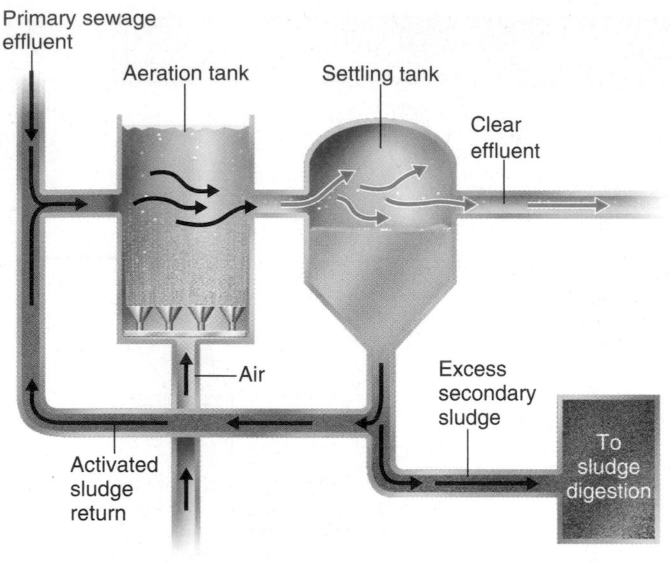

(a) Diagram of an activated sludge system

(b) An aeration tank, showing surface that is frothing from aeration

Figure 27.19 An activated sludge system of secondary sewage treatment.

Q What are the similarities between winemaking and activated sludge sewage treatment?

Figure 27.22). The process of sludge digestion is carried out in large tanks from which oxygen is almost completely excluded.

In secondary treatment, emphasis is placed on the maintenance of aerobic conditions so that organic matter is converted to carbon dioxide, water, and solids that can settle out. An anaerobic sludge digester, however, is designed to encourage the growth of anaerobic bacteria, especially methane-producing bacteria that decrease these organic solids by degrading them to soluble substances and gases, mostly methane (60–70%) and carbon dioxide (20–30%). Methane and carbon dioxide are relatively innocuous end-products, comparable to the carbon dioxide and water from aerobic treatment. The methane is routinely used as a fuel for heating the digester and is also frequently used to run power equipment in the plant.

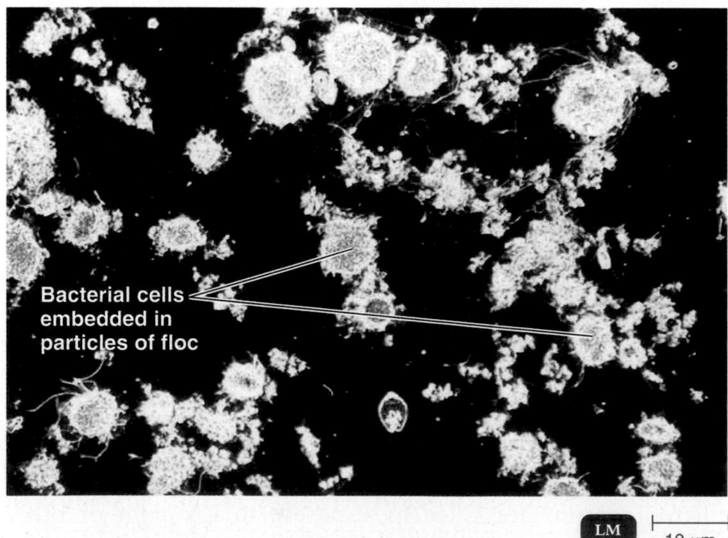

Figure 27.20 Floc formed by an activated sludge system. Gelatinous masses of floc are formed by a species of *Zoogloea* bacteria. If the filamentous bacteria visible in the photo predominate, the floc floats, called bulking—which is undesirable.

Q What happens to the suspended floc when aeration is ended in an activated sludge tank?

There are essentially three stages in the activity of an anaerobic sludge digester. The first stage is the production of carbon dioxide and organic acids from anaerobic fermentation of the sludge by various anaerobic and facultatively anaerobic microorganisms. In the second stage, the organic acids are metabolized to form hydrogen and carbon dioxide, as well as organic acids such as acetic acid. These products are the raw materials for a third stage, in which the methane-producing bacteria produce methane (CH_4). Most of the methane is derived from the energy-yielding reduction of carbon dioxide by hydrogen gas:

$$CO_2 + 4H_2 \longrightarrow CH_4 + 2H_2O$$

Other methane-producing microbes split acetic acid (CH_3COOH) to yield methane and carbon dioxide:

$$CH_3CFOOH \longrightarrow CH_4 + CO_2$$

After anaerobic digestion is completed, large amounts of undigested sludge still remain, although it is relatively stable and inert. To reduce its volume, this sludge is pumped to shallow drying beds or water-extracting filters. Following this step, the sludge can be used for landfill or as a soil conditioner, sometimes

(a) Rotating spray arm of a trickling filter system

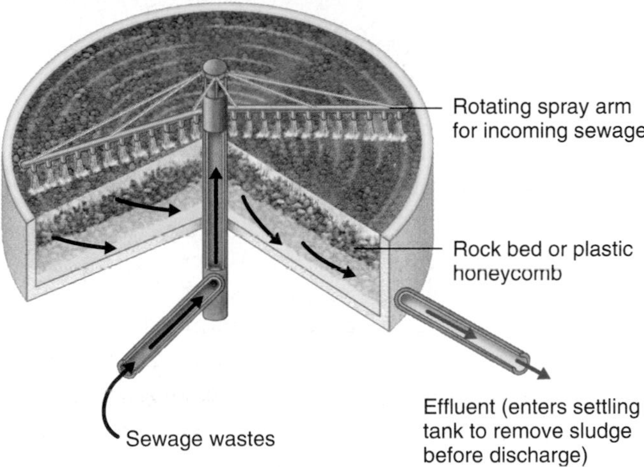

(b) A cutaway view of a trickling filter system

Figure 27.21 A trickling filter of secondary sewage treatment. The sewage is sprayed from the system of rotating pipes onto a bed of rocks or plastic honeycomb designed to have a maximum surface area and to allow oxygen to penetrate deeply into the bed.

Q Which would make the most efficient bed for a trickling filter system, fine sand or golf balls?

(a) An anaerobic sludge digester at a California sewage treatment plant. Much or all of a typical digester is below ground level, especially in cold climates. Methane from such a digester is often used to run pumps or heaters in the treatment plant. Excess methane is being burned off in the flame shown at the top of the digester.

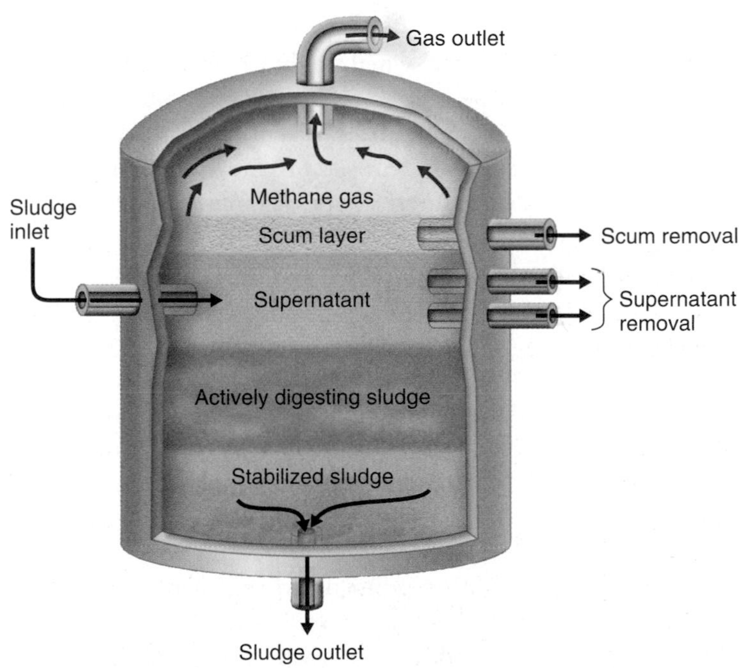

(b) Section of a sludge digester. The scum and supernatant layers are low in solids and are recirculated through secondary treatment.

Figure 27.22 Sludge digestion.

Q What might be some uses for the stabilized sludge?

under the name *biosolids.* Sludge is assigned to two classes: class A sludge contains no detectable pathogens, and class B sludge is treated only to reduce numbers of pathogens below certain levels. Most sludge is class B, and public access to application sites is limited. Sludge has about one-fifth the growth-enhancing value of normal commercial lawn fertilizers but has desirable soil-conditioning qualities, much as do humus and mulch. A potential problem is contamination with heavy metals that are toxic to plants.

Septic Tanks

Homes and businesses in areas of low population density that are not connected to municipal sewage systems often use a **septic tank,** a device whose operation is similar in principle to primary treatment (**Figure 27.23**). Sewage enters a holding tank, and suspended solids settle out. The sludge in the tank must be pumped out periodically and disposed of. The effluent flows through a system of perforated piping into a leaching (soil drainage) field. The effluent entering the soil is decomposed by soil microorganisms. The microbial action necessary for proper functioning of a septic tank can be impaired by excessive amounts of products such as antibacterial soaps, drain cleaners, medications, "every flush" toilet bowl cleaners, and bleach.

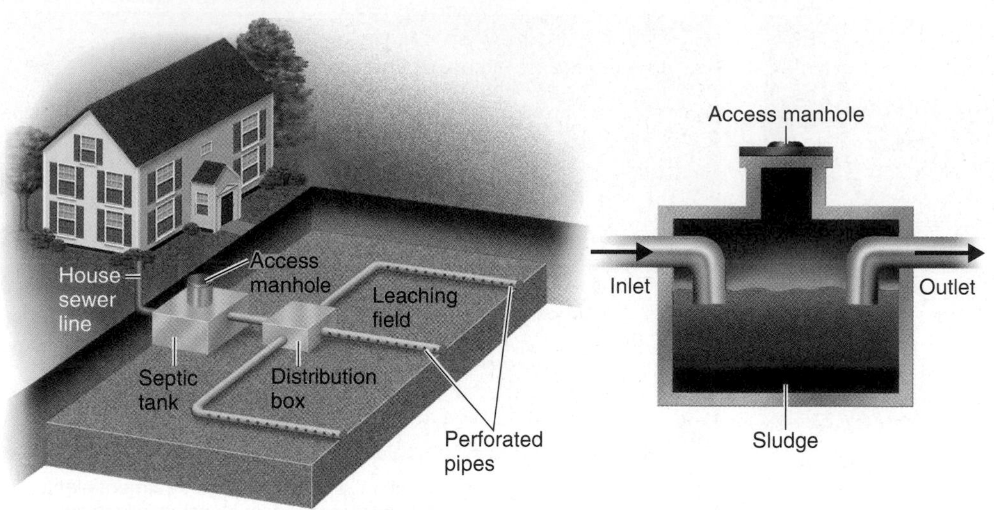

(a) Overall plan. Most soluble organic matter is disposed of by percolation into the soil.

(b) A section of a septic tank

Figure 27.23 A septic tank system.

Q Which type of soil would require the larger drainage area, clay or sandy?

These systems work well when not overloaded and when the drainage system is properly sized to the load and soil type. Heavy clay soils require extensive drainage systems because of the soil's poor permeability. The high porosity of sandy soils can result in chemical or bacterial pollution of nearby water supplies.

Oxidation Ponds

Many industries and small communities use **oxidation ponds,** also called *lagoons* or *stabilization ponds,* for water treatment. These are inexpensive to build and operate but require large areas of land. Designs vary, but most incorporate two stages. The first stage is analogous to primary treatment; the sewage pond is deep enough that conditions are almost entirely anaerobic. Sludge settles out in this stage. In the second stage, which roughly corresponds to secondary treatment, effluent is pumped into an adjoining pond or system of ponds shallow enough to be aerated by wave action. Because it is difficult to maintain aerobic conditions for bacterial growth in ponds with so much organic matter, the growth of algae is encouraged to produce oxygen. Bacterial action in decomposing the organic matter in the wastes generates carbon dioxide. Algae, which use carbon dioxide in their photosynthetic metabolism, grow and produce oxygen, which in turn encourages the activity of aerobic microbes in the sewage. Large amounts of organic matter in the form of algae accumulate, but this is not a problem because the oxidation pond, unlike a lake, already has a large nutrient load.

Some small sewage-producing operations, such as isolated campgrounds and highway rest stop areas, use an *oxidation ditch* for sewage treatment. In this method, a small oval channel in the shape of a racetrack is filled with sewage water. A paddle wheel similar to that on an old-time Mississippi steamboat, but in a fixed location, propels the water in a self-contained flowing stream aerated enough to oxidize the wastes.

Tertiary Sewage Treatment

As we have seen, primary and secondary treatments of sewage do not remove all the biologically degradable organic matter. Amounts of organic matter that are not excessive can be released into a flowing stream without causing a serious problem. Eventually, however, the pressures of increased population might increase wastes beyond a body of water's carrying capacity, and additional treatments might be required. Even now, primary and secondary treatments are inadequate in certain situations, such as when the effluent is discharged into small streams or recreational lakes. Some communities have therefore developed **tertiary sewage treatment** plants. Lake Tahoe in the Sierra Nevada Mountains, surrounded by extensive development, is the site of one of the best-known tertiary sewage treatment systems. Similar systems are used to treat wastes entering the southern portion of San Francisco Bay.

The effluent from secondary treatment plants contains some residual BOD. It also contains about 50% of the original nitrogen and 70% of the original phosphorus, which can greatly affect a lake's ecosystem. Tertiary treatment is designed to remove essentially all the BOD, nitrogen, and phosphorus. Tertiary treatment depends less on biological treatment than on physical and chemical treatments. Phosphorus is precipitated out by combining with such chemicals as lime, alum, and ferric chloride. Filters of fine sands and activated charcoal remove small particulate matter and dissolved chemicals. Nitrogen is converted to ammonia and discharged into the air in stripping towers. Some systems encourage denitrifying bacteria to form volatile nitrogen gas. Finally, the purified water is chlorinated.

Tertiary treatment provides water that is suitable for drinking, but the process is extremely costly. Secondary treatment is less costly, but water that has undergone only secondary treatment still contains many water pollutants. Much work is being done to design secondary treatment plants in which the effluent can be used for irrigation. This design would eliminate a source of water pollution, provide nutrients for plant growth, and reduce the demand on already scarce water supplies. The soil to which this water is applied would act as a trickling filter to remove chemicals and microorganisms before the water reaches groundwater and surface water supplies.

CHECK YOUR UNDERSTANDING

✓ Which type of sewage treatment is designed to remove almost all phosphorus from sewage? **27-18**

✓ What metabolic group of anaerobic bacteria is especially encouraged by operation of a sludge digestion system? **27-19**

✓ What is the relationship between BOD and the welfare of fish? **27-20**

* * *

We hope this chapter on environmental microbiology, as well as previous chapters in the book, has left you with a greater appreciation of the microbial influences around us. Without the natural and human-directed applications of microbes, life would be very different—and perhaps could not sustain itself at all.

STUDY OUTLINE

The **MyMicrobiologyPlace** website (**www.microbiologyplace.com**) will help you get ready for tests with its simple three-step approach: ❶ **take a pre-test** and obtain a personalized study plan, ❷ **learn and practice** with animations, tutorials, and MP3 tutor sessions, and ❸ **test yourself** with quizzes and a chapter post-test.

Microbial Diversity and Habitats (p. 767)

1. Microorganisms live in a wide variety of habitats because of their metabolic diversity and their ability to use a variety of carbon and energy sources and to grow under different physical conditions.
2. Extremophiles live in extreme conditions of temperature, acidity, alkalinity, or salinity.

Symbiosis (p. 767)

3. Symbiosis is a relationship between two different organisms or populations.
4. Symbiotic fungi called mycorrhizae live in and on plant roots; they increase the surface area and nutrient absorption of the plant.

Soil Microbiology and Biogeochemical Cycles (pp. 768–776)

1. In biogeochemical cycles, certain chemical elements are recycled.
2. Microorganisms in the soil decompose organic matter and transform carbon-, nitrogen-, and sulfur-containing compounds into usable forms.
3. Microbes are essential to the continuation of biogeochemical cycles.
4. Elements are oxidized and reduced by microorganisms during these cycles.

The Carbon Cycle (pp. 768–770)

5. Carbon dioxide is incorporated, or fixed, into organic compounds by photoautotrophs and chemoautotrophs.
6. These organic compounds provide nutrients for chemoheterotrophs.
7. Chemoheterotrophs release CO_2 that is then used by photoautotrophs.
8. Carbon is removed from the cycle when it is in $CaCO_3$ and fossil fuels.

The Nitrogen Cycle (pp. 770–772)

9. Microorganisms decompose proteins from dead cells and release amino acids.
10. Ammonia is liberated by microbial ammonification of the amino acids.
11. The nitrogen in ammonia is oxidized to produce nitrates for energy by nitrifying bacteria.
12. Denitrifying bacteria reduce the nitrogen in nitrates to molecular nitrogen (N_2).
13. N_2 is converted into ammonia by nitrogen-fixing bacteria.
14. Nitrogen-fixing bacteria include free-living genera such as *Azotobacter,* cyanobacteria, and the symbiotic bacteria *Rhizobium* and *Frankia.*
15. Ammonium and nitrate are used by bacteria and plants to synthesize amino acids that are assembled into proteins.

The Sulfur Cycle (pp. 772–773)

16. Hydrogen sulfide (H_2S) is used by autotrophic bacteria; the sulfur is oxidized to form S^0 or SO_4^{2-}.
17. Plants and other microorganisms can reduce SO_4^{2-} to make certain amino acids. These amino acids are in turn used by animals.
18. H_2S is released by decay or dissimilation of these amino acids.

Life without Sunshine (pp. 773–774)

19. Chemoautotrophs are the primary producers in deep-sea vents and within deep rocks.

The Phosphorus Cycle (p. 774)

20. Phosphorus (as PO_4^{3-}) is found in rocks and bird guano.
21. When solubilized by microbial acids, the PO_4^{3-} is available for plants and microorganisms.
22. Endolithic bacteria live in solid rock; these autotrophic bacteria use hydrogen as an energy source.

The Degradation of Synthetic Chemicals in Soil and Water (pp. 775–776)

23. Many synthetic chemicals, such as pesticides, are resistant to degradation by microbes.
24. The use of microorganisms to remove pollutants is called bioremediation.
25. The growth of oil-degrading bacteria can be enhanced by the addition of nitrogen and phosphorus fertilizer.
26. Municipal landfills prevent decomposition of solid wastes because they are dry and anaerobic.
27. In some landfills, methane produced by methanogens can be recovered for an energy source.
28. Composting can be used to promote biodegradation of organic matter.

Aquatic Microbiology and Sewage Treatment (pp. 776–789)

Aquatic Microorganisms (pp. 776–778)

1. The study of microorganisms and their activities in natural waters is called aquatic microbiology.
2. Natural waters include lakes, ponds, streams, rivers, estuaries, and the oceans.
3. The concentration of bacteria in water is proportional to the amount of organic material in the water.
4. Most aquatic bacteria tend to grow on surfaces rather than in a free-floating state.
5. The number and location of freshwater microbiota depend on the availability of oxygen and light.
6. Photosynthetic algae are the primary producers of a lake; they are found in the limnetic zone.
7. Pseudomonads, *Cytophaga, Caulobacter,* and *Hyphomicrobium* are found in the limnetic zone, where oxygen is abundant.
8. Microbes in stagnant water use available oxygen and can cause odors and the death of fish.
9. Wave action increases the amount of dissolved oxygen.
10. Purple and green sulfur bacteria are found in the profundal zone, which contains light and H_2S but no oxygen.
11. *Desulfovibrio* reduces SO_4^{2-} to H_2S in benthic mud.
12. Methane-producing bacteria are also found in the benthic zone.
13. Phytoplankton are the primary producers of the open ocean.
14. *Pelagibacter ubique* is a decomposer in ocean waters.
15. Archaea predominate below 100 m.
16. Some algae and bacteria are bioluminescent. They possess the enzyme luciferase, which can emit light.

The Role of Microorganisms in Water Quality (pp. 778–782)

17. Microorganisms are filtered from water that percolates into groundwater supplies.
18. Some pathogenic microorganisms are transmitted to humans in drinking and recreational waters.
19. Resistant chemical pollutants may be concentrated in animals in an aquatic food chain.
20. Mercury is metabolized by certain bacteria into a soluble compound that is concentrated in animals.
21. Nutrients such as phosphates cause algal blooms, which can lead to eutrophication of aquatic ecosystems.
22. Eutrophication is the result of the addition of pollutants or natural nutrients.
23. *Thiobacillus ferrooxidans* produces sulfuric acid at coal-mining sites.
24. Tests for the bacteriological quality of water are based on the presence of indicator organisms, the most common of which are coliforms.
25. Coliforms are aerobic or facultatively anaerobic, gram-negative, non–endospore-forming rods that ferment lactose with the production of acid and gas within 48 hours of being placed in a medium at 35°C.
26. Fecal coliforms, predominantly *E. coli,* are used to indicate the presence of human feces.

Water Treatment (p. 782)

27. Drinking water is held in a holding reservoir long enough that suspended matter settles.
28. Flocculation treatment uses a chemical such as alum to coalesce and then settle colloidal material.
29. Filtration removes protozoan cysts and other microorganisms.
30. Drinking water is disinfected with chlorine to kill remaining pathogenic bacteria.

Sewage (Wastewater) Treatment (pp. 783–789)

31. Domestic wastewater is called sewage; it includes household water, toilet wastes, and rainwater.
32. Primary sewage treatment is the removal of solid matter called sludge.
33. Biological activity is not very important in primary treatment.
34. Biochemical oxygen demand (BOD) is a measure of the biologically degradable organic matter in water.
35. Primary treatment removes about 25–35% of the BOD of sewage.
36. BOD is determined by measuring the amount of oxygen bacteria require to degrade the organic matter.
37. Secondary sewage treatment is the biological degradation of organic matter after primary treatment.
38. Activated sludge systems, trickling filters, and rotating biological contactors are methods of secondary treatment.
39. Microorganisms degrade the organic matter aerobically.
40. Secondary treatment removes up to 95% of the BOD.
41. Treated sewage is disinfected, usually by chlorination, before discharge onto land or into water.
42. Sludge is placed in an anaerobic sludge digester; bacteria degrade organic matter and produce simpler organic compounds, methane, and CO_2.
43. The methane produced in the digester is used to heat the digester and operate other equipment.

44. Excess sludge is periodically removed from the digester, dried, and disposed of (as landfill or soil conditioner) or incinerated.
45. Septic tanks can be used in rural areas to provide primary treatment of sewage.
46. Small communities can use oxidation ponds for secondary treatment.
47. These require a large area in which to build an artificial lake.
48. Tertiary sewage treatment uses physical filtration and chemical precipitation to remove all the BOD, nitrogen, and phosphorus from water.
49. Tertiary treatment provides drinkable water, whereas secondary treatment provides water usable only for irrigation.

STUDY QUESTIONS

Answers to the Review and Multiple Choice questions can be found by turning to the blue Answers tab at the back of the textbook.

Review

1. The koala is a leaf-eating animal. What can you infer about its digestive system?
2. Give one possible explanation of why *Penicillium* would make penicillin, given that the fungus does not get bacterial infections.
3. In the sulfur cycle, microbes degrade organic sulfur compounds, such as (a) __________, to release H_2S, which can be oxidized by *Thiobacillus* to (b) __________. This ion can be assimilated into amino acids by (c) __________ or reduced by *Desulfovibrio* to (d) __________. H_2S is used by photoautotrophic bacteria as an electron donor to synthesize (e) __________. The sulfur-containing by-product of this metabolism is (f) __________.
4. Why is the phosphorus cycle important?
5. **DRAW IT** Identify where the following processes occur: ammonification, decomposition, denitrification, nitrification, nitrogen fixation. Name at least one organism responsible for each process.

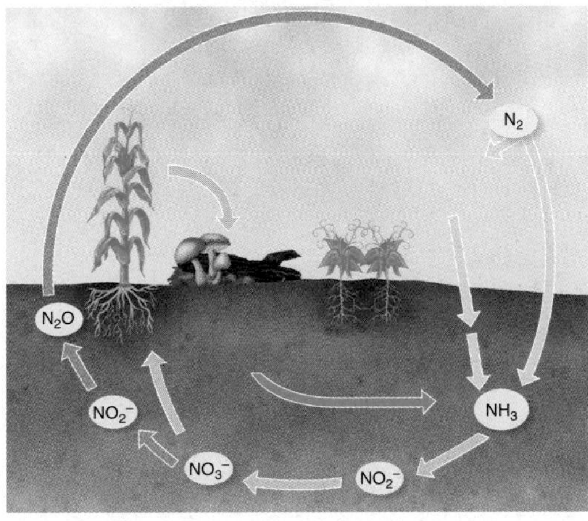

6. The following organisms have important roles as symbionts with plants and fungi; describe the symbiotic relationship of each organism with its host: cyanobacteria, mycorrhizae, *Rhizobium*, *Frankia*.
7. Outline the treatment process for drinking water.
8. The following processes are used in wastewater treatment. Match the stage of treatment with the processes. Each choice can be used once, more than once, or not at all.

Processes	Treatment Stage
__________ a. Leaching field	1. Primary
__________ b. Removal of solids	2. Secondary
__________ c. Biological degradation	3. Tertiary
__________ d. Activated sludge system	
__________ e. Chemical precipitation of phosphorus	
__________ f. Trickling filter	
__________ g. Results in drinking water	

9. Bioremediation refers to the use of living organisms to remove pollutants. Describe three examples of bioremediation.

Multiple Choice

For questions 1–4, answer whether
a. the process takes place under aerobic conditions.
b. the process takes place under anaerobic conditions.
c. the amount of oxygen doesn't make any difference.

1. Activated sludge system
2. Denitrification
3. Nitrogen fixation
4. Methane production
5. The water used to prepare intravenous solutions in a hospital contained endotoxins. Infection control personnel performed plate counts to find the source of the bacteria. Their results:

	Bacteria/100 ml
Municipal water pipes	0
Boiler	0
Hot water line	300

All of the following conclusions about the bacteria can be drawn *except* which one?
a. It was present as a biofilm in the pipes.
b. It is gram-negative.
c. It comes from fecal contamination.
d. It comes from the city water supply.
e. none of the above

Use the following choices to answer questions 6–8:
a. aerobic respiration
b. anaerobic respiration
c. anoxygenic photoautotroph
d. oxygenic photoautotroph

6. $CO_2 + H_2S \xrightarrow{\text{Light}} C_6H_{12}O_6 + S^0$

7. $SO_4^{2-} + 10H^+ + 10e^- \rightarrow H_2S + 4H_2O$

8. $CO_2 + 8H^+ + 8e^- \rightarrow CH_4 + 2H_2O$

9. All of the following are effects of water pollution *except*
a. the spread of infectious diseases.
b. increased eutrophication.
c. increased BOD.
d. increased growth of algae.
e. none of the above

10. Coliforms are used as indicator organisms of sewage pollution because
a. they are pathogens.
b. they ferment lactose.
c. they are abundant in human intestines.
d. they grow within 48 hours.
e. all of the above

Critical Thinking

1. Here are the formulas of two detergents that have been manufactured:

```
C—C—C—C—C—C—C—C—C—C. . .

                C
                |
                C       C
                |       |
    C—C—C—C—C—C. . .
       |
       C
```

Which of these would be resistant, and which would be readily degraded by microorganisms? (*Hint:* Refer to the degradation of fatty acids in Chapter 5.)

2. Explain the effect of dumping untreated sewage into a pond on the eutrophication of the pond. The effect of sewage that has primary treatment? The effect of sewage that has secondary treatment? Contrast your previous answers with the effect of each type of sewage on a fast-moving river.

Clinical Applications

1. Flooding after two weeks of heavy rainfall in Tooele, Utah, preceded a high rate of diarrheal illness. *G. lamblia* was isolated from 25% of the patients. A comparison study of a town 65 miles away revealed that were was diarrheal illness in 2.9% of the 103 people interviewed. Tooele has a municipal water system and a municipal sewage treatment plant. Explain the probable cause of this epidemic and method(s) of stopping it. What would a fecal coliform test have shown?

2. The bioremediation process shown in the photograph is used to remove benzene and other hydrocarbons from soil contaminated by petroleum. The pipes are used to add nitrates, phosphates, oxygen, or water. Why are each of these added? Why is it not always necessary to add bacteria?

28 Applied and Industrial Microbiology

In the previous chapter on environmental microbiology, we saw that microbes are an essential factor in many natural phenomena that make life possible on Earth. In this chapter we will look at how microorganisms are harnessed in such useful applications as the making of food and industrial products. Many of these processes—especially baking, winemaking, brewing, and cheesemaking—have origins long lost in history.

Modern civilization, with its large urban populations, could not be supported without methods of preserving food. In fact, civilization arose only after agriculture produced a year-round stable food supply so that people were able to give up a nomadic hunting-and-gathering way of life. It is also a fact that advances in microbiology, with its insight into spoilage processes and the possibility of disseminating diseases in preserved food, later became an essential element of this.

In Chapter 9, we discussed industrial applications of genetically modified microorganisms that are at the cutting edge of our knowledge of molecular biology. Many of these applications are now essential to modern industry. (See the box in Chapter 1, page 3.)

UNDER THE MICROSCOPE

Saccharomyces cerevisiae, a yeast that is widely used for industrial purposes.

Q&A

To produce ethanol, yeasts require anaerobic conditions. In what widely used industrial process does the growth of *Saccharomyces cerevisiae* require aerobic conditions?

Look for the answer in the chapter.

Food Microbiology

LEARNING OBJECTIVES

28-1 Describe thermophilic anaerobic spoilage and flat sour spoilage by mesophilic bacteria.

28-2 Compare and contrast food preservation by industrial food canning, aseptic packaging, radiation, and high pressure.

28-3 Name four beneficial activities of microorganisms.

Many of the methods of food preservation used today were probably discovered by chance in centuries past. People in early cultures observed that dried meat and salted fish resisted decay. Nomads must have noticed that soured animal milk resisted further decomposition and was still palatable. Moreover, if the curd of the soured milk was pressed to remove moisture and allowed to ripen (in effect, cheesemaking), it was even more effectively preserved and tasted better. Farmers soon learned that if grains were kept dry, they did not become moldy.

Foods and Disease

As more food products are being prepared at central facilities and widely distributed, it is becoming more likely that food, like municipal water supplies, might be a source of widespread disease outbreaks. To minimize the potential for disease outbreaks, communities have established local agencies whose role is to inspect dairies and restaurants. The United States Food and Drug Administration (FDA) and Department of Agriculture (USDA) also maintain a system of inspectors at ports and central processing locations. A recent development in this field has been the introduction of the **Hazard Analysis and Critical Control Point (HACCP)** system, which is intended to safeguard food "from farm to fork." Before the introduction of the HACCP system, the primary role of governmental agencies was to conduct sampling to identify contaminated foods. Such sampling to identify contamination will still have its place, but the HACCP system is designed to prevent contamination by identifying points at which foods are most likely to be contaminated with harmful microbes. Monitoring of these control points can prevent such microbes from being introduced or, if they are present, arrest their proliferation. For example, the HACCP system can identify steps during processing at which meats are likely to become contaminated by the animal's intestinal contents. The HACCP system also requires monitoring of adequate temperatures to kill pathogens during processing and adequate storage temperatures to prevent their reproduction.

Industrial Food Canning

In Chapter 7, you learned that preserving foods by heating a properly sealed container, as in home canning, is not difficult. The challenge in commercial canning is to use the right amount of heat necessary to kill spoilage organisms and dangerous microbes, such as the endospore-forming *Clostridium botulinum*, without degrading the appearance and palatability of food. Thus, much research is applied to determining the exact minimum heat treatment that will accomplish both these goals.

Industrial food canning is much more technically sophisticated than home canning (**Figure 28.1**). Industrially canned goods undergo what is called **commercial sterilization** by steam

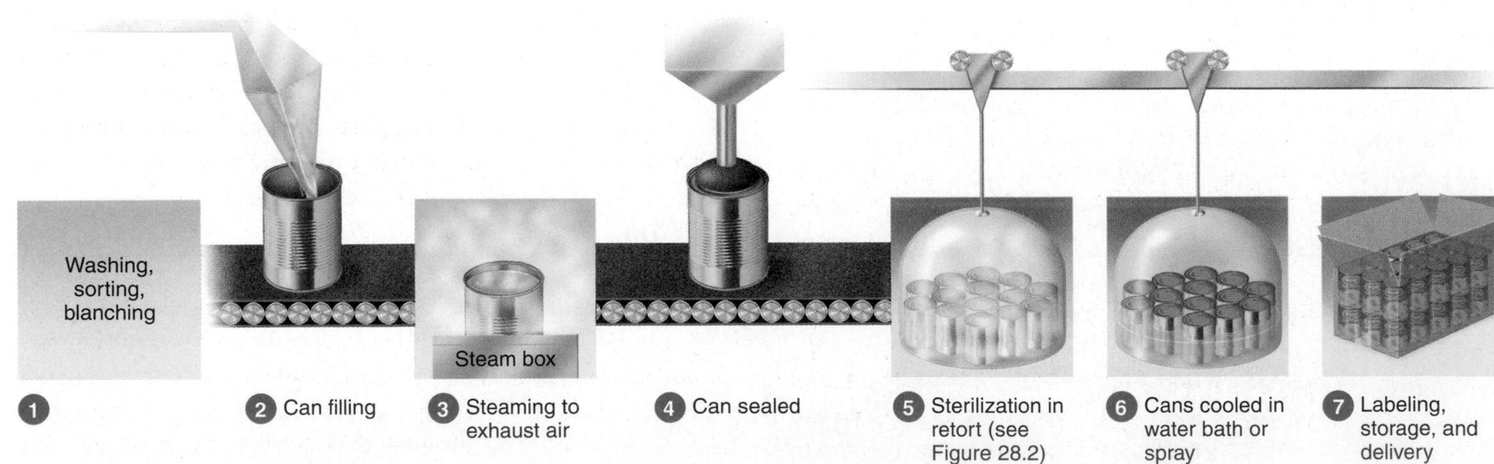

Figure 28.1 The commercial sterilization process in industrial canning. 1 Blanching is a treatment with hot water or steam intended to soften the product so the can will fill better. It also destroys enzymes that might alter the color, flavor, or texture of the product and lower the microbial population. 2 Cans are filled to capacity, leaving as little dead space as possible. 3 To exhaust (drive out) most dissolved air, cans are heated in a steam box. 4 The cans are sealed. 5 Cans are sterilized by steam under pressure. 6 Cans are cooled by submerging them or spraying them with water. 7 Cans are labeled for sale.

Q How does commercial sterilization differ from complete sterilization?

under pressure in a large **retort** (Figure 28.2), which operates on the same principle as an autoclave (see Figure 7.2, page 189). Commercial sterilization is intended to destroy *C. botulinum* endospores and is not as rigorous as complete sterilization. The reasoning is that if *C. botulinum* endospores are destroyed, then any other significant spoilage or pathogenic bacteria will also be destroyed.

To ensure commercial sterilization, enough heat is applied for the **12D treatment** (12-decimal reductions, or *botulinal cook*), by which a theoretical population of *C. botulinum* endospores would be decreased by 12 logarithmic cycles. (See Figure 7.1 and Table 7.2, pages 186–187.) What this means is that if there were 10^{12} (1,000,000,000,000) endospores in a can, after treatment there would be only one survivor. Because 10^{12} is an improbably large population, this treatment is considered quite safe. Certain thermophilic endospore-forming bacteria have endospores that are more resistant to heat treatment than those of *C. botulinum*. However, these bacteria are obligate thermophiles and generally remain dormant at temperatures lower than about 45°C. Therefore, they are not a spoilage problem at normal storage temperatures.

Figure 28.2 Commercial canning retorts. These are much larger then the sterilizing autoclaves used in most microbiology laboratories or hospitals.

Q Is there any difference in principle between a canning retort and a hospital autoclave?

Spoilage of Canned Food

If canned foods are incubated at high temperatures, such as in a truck in the hot sun or next to a steam radiator, the thermophilic bacteria that often survive commercial sterilization can germinate and grow. **Thermophilic anaerobic spoilage** is therefore a fairly common cause of spoilage in low-acid canned foods. The can usually swells from gas, and the contents have a lowered pH and a sour odor. A number of thermophilic species of *Clostridium* can cause this type of spoilage. When thermophilic spoilage occurs but the can is not swollen by gas production, the spoilage is termed **flat sour spoilage.** This type of spoilage is caused by thermophilic organisms such as *Geobacillus stearothermophilus* (ste-rō-thėr-mä′fil-us), which is found in the starch and sugars used in food preparation. Many industries have standards for the numbers of such thermophilic bacteria permitted in raw materials. Both types of spoilage occur only when the cans are stored at higher than normal temperatures, which permits the growth of bacteria whose endospores are not destroyed by normal processing.

Mesophilic bacteria can spoil canned foods if the food is underprocessed or if the can leaks. Underprocessing is more likely to result in spoilage by endospore formers; the presence of non–endospore-forming bacteria strongly suggests that the can leaks. Leaking cans are often contaminated during the cooling of cans after processing by heat. The hot cans are sprayed with cooling water or passed through a trough filled with water. As the can cools, a vacuum is formed inside, and external water can be sucked through a leak past the heat-softened sealant in the crimped lid (Figure 28.3). Contaminating bacteria in the cooling water are drawn into the can with the water. Spoilage from underprocessing or can leakage is likely to produce odors of putrefaction, at least in high-protein foods, and occurs at normal storage temperatures. In such types of spoilage, there is always the potential that botulinal bacteria will be present.

Some acidic foods, such as tomatoes or preserved fruits, are preserved by processing temperatures of 100°C or lower. The reasoning is that the only spoilage organisms that will grow in such acidic foods are easily killed by even 100°C temperatures. Primarily, these would be molds, yeasts, and certain vegetative bacteria.

Occasional problems in acidic foods develop from a few microorganisms that are both heat-resistant and acid-tolerant. Examples of heat-resistant fungi are the mold *Byssochlamys fulva* (bis-sō-klam′is fůl′vä), which produces a *heat-resistant ascospore,* and a few molds, especially species of *Aspergillus,* that sometimes produce specialized resistant bodies called *sclerotia.* A spore-forming bacterium, *Bacillus coagulans* (kō-ag′ū-lanz), is unusual in that it is capable of growth at a pH of almost 4.0. Table 28.1 lists types of spoilage in low- and medium-acid foods.

Aseptic Packaging

The use of **aseptic packaging** to preserve food has been increasing recently. Packages are usually made of some material that cannot tolerate conventional heat treatment, such as laminated paper or plastic. The packaging materials come in continuous rolls that are fed into a machine that sterilizes the material with a hot hydrogen peroxide solution, sometimes aided by ultraviolet (UV) light (Figure 28.4). Metal containers can be sterilized with superheated steam or other high-temperature methods. High-energy electron beams can also be used to sterilize the packaging materials. While still in the sterile environment, the material is formed into packages, which are then filled with

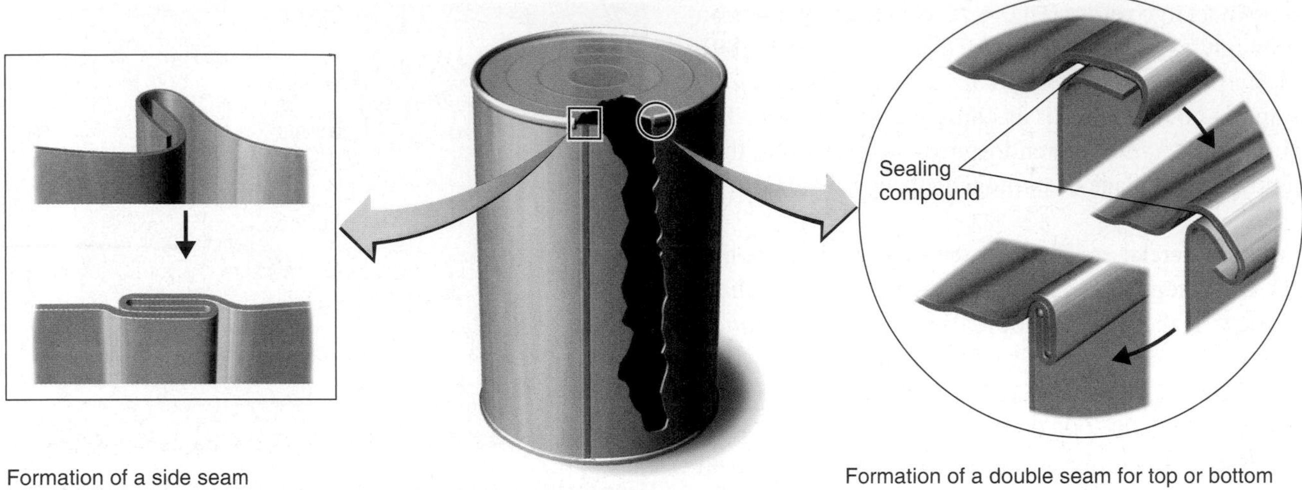

Figure 28.3 The construction of a metal can. Notice the seam construction, which was introduced about 1904. During cooling after sterilization (see Figure 28.1, step 6), the vacuum formed in the can may actually force contaminating organisms into the can along with water.

Q Why isn't the can sealed before it is placed in the steam box?

liquid foods that have been conventionally sterilized by heat. The filled package is not sterilized after it is sealed.

Radiation and Industrial Food Preservation

It has long been recognized that irradiation is lethal to microorganisms; in fact, a patent was issued in Great Britain in 1905 for the use of ionizing radiation to improve the condition of foodstuffs. X rays were specifically suggested in 1921 as a way to inactivate the larvae in pork that are the cause of trichinellosis. Ionizing irradiation inhibits DNA synthesis and effectively prevents microorganisms, insects, and plants from reproducing. The ionizing irradiation is usually X rays or the gamma rays produced by radioactive cobalt-60. Up to certain energy levels, high energy electrons produced by electron accelerators are also used. The main practical difference is in penetration capabilities. These sources inactivate the target organisms and do *not* induce radioactivity in the food or packaging material. The relative doses of radiation needed to kill various organisms are presented in Table 28.2. Radiation is measured in Grays, named for an early radiologist—often in terms of thousands of Grays, abbreviated as kGy.

- *Low doses of irradiation (less than 1 kGy)* are used for killing insects (disinfestation) and inhibiting sprouting, as in stored potatoes. Similarly, it can delay ripening of fruits during storage.
- *Pasteurizing doses (1 to 10 kGy)* can be used on meats and poultry to eliminate or critically reduce the numbers of specific bacterial pathogens.
- *High doses (more than 10 kGy)* are used to sterilize, or at least greatly lower, the bacterial populations in many spices. Spices are often contaminated with 1 million or more bacteria per gram, although these are not considered to be normally hazardous to health.

Table 28.1 Common Types of Spoilage in Low-Acid and Medium-Acid Canned Foods (pH above 4.5)

	Indications of Spoilage	
Type of Spoilage	**Appearance of Can**	**Contents of Can**
Flat sour (*Geobacillus stearothermophilus*)	Can not swollen	Appearance not usually altered; pH markedly lowered; sour; may have slightly abnormal odor; sometimes cloudy liquid
Thermophilic anaerobic (*Thermoanaerobacterium thermosaccharolyticum*)	Swollen	Fermented, sour, cheesy, or butyric acid odor
Putrefactive anaerobic (*Clostridium sporogenes*; possibly *C. botulinum*)	Swollen	May be partially digested; pH slightly above normal; typical putrid odor

Figure 28.4 Aseptic packaging. Rolls of packaging material in foreground, filled packages at right center.

Q Why has the use of this procedure been increasing in recent years?

Table 28.2 Approximate Doses of Radiation Needed to Kill Various Organisms (Prions Are Not Affected)

Organisms	Dose (kGy)*
Higher animals (whole body)	0.005–0.1
Insects	0.01–1
Non-endospore-forming bacteria	0.5–10
Bacterial spores	10–50
Viruses	10–200

*Gray is a measure of ionizing irradiation; kGy is 1000 Grays.
Source: J. Farkas, "Physical Methods of Food Preservation," in *Food Microbiology: Fundamentals and Frontiers,* 2d ed., M.P. Doyle et al. (eds) (Washington, DC: ASM Press, 2001).

A specialized use of irradiation has been to sterilize meats eaten by American astronauts, and a few health facilities have selectively used irradiation to sterilize foods ingested by immunocompromised patients. Millions of implanted medical devices, such as pacemakers, have been irradiated. Irradiated food is marked in the United States with a radura symbol (Figure 28.5) and a printed notice. Unfortunately, this symbol has often been interpreted as a warning rather than the description of an approved processing treatment or preservative. In fact, irradiated foods are not radioactive; consider that the

Figure 28.5 Irradiation logo. This logo, the international radura symbol, indicates that a food has received irradiation treatment.

Q Is irradiation the same as a chemical additive?

X-ray table in a hospital does not become radioactive from repeated daily exposure to ionizing radiation. Recently, the FDA has allowed, upon special approval, substitution of language such as "pasteurization" rather than "irradiation."

When deep penetration is a requirement, the preferred method for irradiation is gamma rays produced by cobalt-60. However, this type of treatment requires several hours of exposure in isolation behind protective walls (Figure 28.6).

High-energy electron accelerators (Figure 28.7) are much faster and sterilize in a few seconds, but this treatment has low penetrating power and is suitable only for sliced meats, bacon, or similar thin products. Also, plasticware used in microbiology is usually sterilized in this way. Another recent application is to irradiate mail to kill possible bioterrorism agents that it might contain, such as anthrax endospores.

High-Pressure Food Preservation

A recent development in food preservation has been the use of a high-pressure processing technique. Prewrapped foods such as fruits, deli meats, and precooked chicken strips are submerged into tanks of pressurized water. The pressure can reach 87,000 pounds per square inch (psi)—which has been compared to the equivalent of about three elephants standing on a dime. This process kills many bacteria, such as *Salmonella, Listeria,* and pathogenic strains of *E. coli,* by disrupting many cellular functions. It also kills nonpathogenic microorganisms that tend to shorten the shelf life of such products.

Because the process does not require additives, it does not require regulatory approval. It has the advantage of preserving colors and tastes of foods better than many other methods and does not provoke the concerns of irradiation.

The Role of Microorganisms in Food Production

In the latter part of the nineteenth century, microbes used in food production were grown in pure culture for the first time. This development quickly led to an improved understanding of the relationships between specific microbes and their products

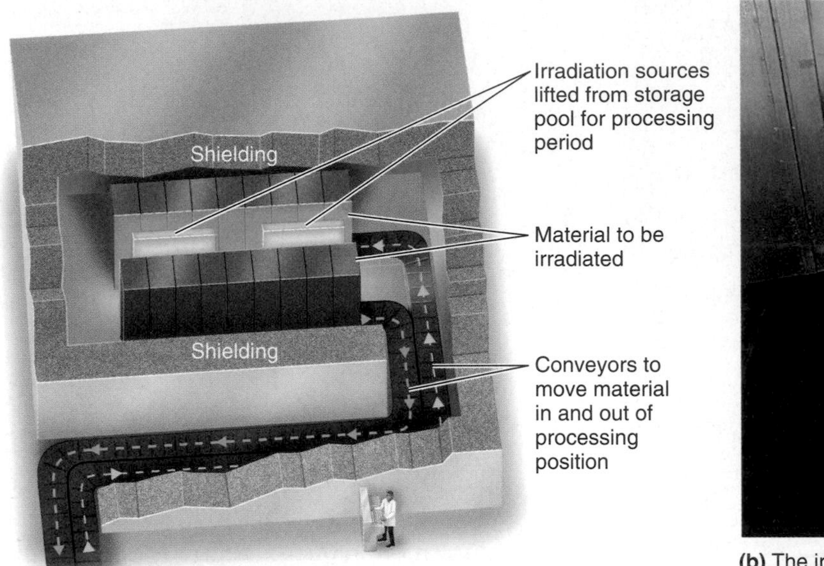

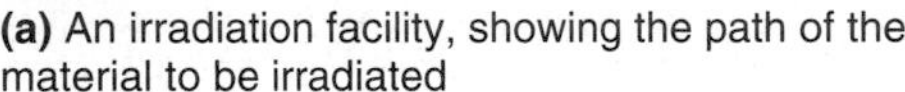

(a) An irradiation facility, showing the path of the material to be irradiated

(b) The irradiation source is in the lowered position in the storage pool. The blue glow is Cerenkov radiation caused by charged particles exceeding the speed of light in water.

Figure 28.6 A gamma-ray irradiation facility.

Q Can microwaves be used to sterilize foods?

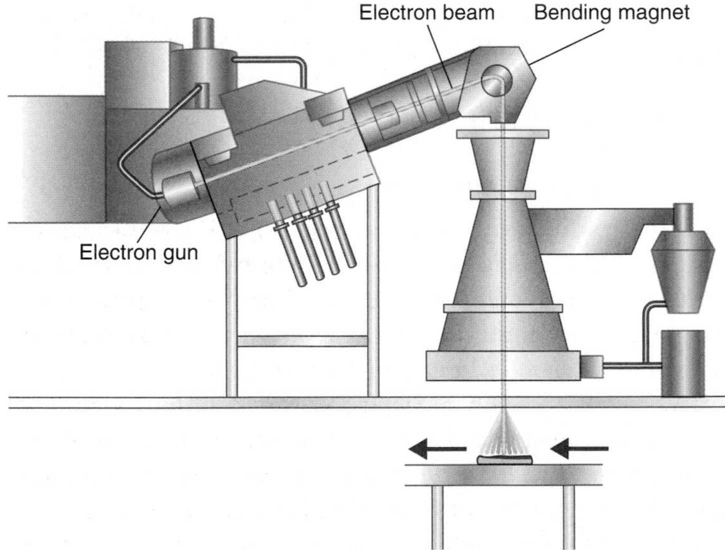

Figure 28.7 Electron-beam accelerator. These machines generate an electron stream that is accelerated down a long tube by electromagnets of the opposite charge. In the drawing, the electron beam is bent by a "bending magnet." This serves to filter out electrons of unwanted energy levels, providing a beam of uniform energy. The vertical beam is swept back and forth over the target as it is moved past the beam. The penetrating power of the beam is limited: if the target substance is expressed as an equivalent thickness of water, the maximum is about 3.9 cm (1.5 in.). In contrast, X rays will penetrate about 23 cm (9 in.).

Q Are high-energy electrons ionizing radiation?

and activities. This period can be considered the beginning of industrial food microbiology. For example, once it was understood that a certain yeast grown under certain conditions produced beer and that certain bacteria could spoil the beer, brewers were better able to control the quality of their products. Specific industries became active in microbiological research and selected certain microbes for their special qualities. The brewing industry extensively investigated the isolation and identification of yeasts and selected those that could produce more alcohol. In this section, we will discuss the role of microorganisms in the production of several common foods.

Cheese

The United States leads the world in the making of cheese, producing millions of tons every year. Although there are many types of cheeses, all require the formation of a **curd,** which can be separated from the main liquid fraction, or **whey** (Figure 28.8). The curd is made up of a protein, **casein,** and is usually formed by the action of an enzyme, **rennin** (or chymosin), which is aided by acidic conditions provided by certain lactic acid–producing bacteria. These inoculated lactic acid bacteria also provide the characteristic flavors and aromas of fermented dairy products during the ripening process. The curd undergoes a microbial ripening process, except for a few unripened cheeses, such as ricotta and cottage cheese.

Cheeses are generally classified by their hardness, which is produced in the ripening process. The more moisture lost from the curd and the more the curd is compressed, the harder the

(a) The milk has been coagulated by the action of rennin (forming curd) and is inoculated with ripening bacteria for flavor and acidity. Here the workers are cutting the curd into slabs.

(b) The curd is chopped into small cubes to facilitate efficient draining of whey.

(c) The curd is milled to allow even more drainage of whey and is compressed into blocks for extended ripening. The longer the ripening period, the more acidic (sharper) the cheese.

Figure 28.8 Making cheddar cheese.

Q Are there living bacteria in the final cheese product?

cheese. Romano and Parmesan cheeses, for example, are classified as very hard cheeses; cheddar and Swiss are hard cheeses. Limburger, blue, and Roquefort cheeses are classified as semisoft; Camembert is an example of a soft cheese.

The hard cheddar and Swiss cheeses are ripened by lactic acid bacteria growing anaerobically in the interior. Such hard, interior-ripened cheeses can be quite large. The longer the incubation time, the higher the acidity and the sharper the taste of the cheese. A *Propionibacterium* (prō-pē-on-ē-bak-ti′rē-um) species produces carbon dioxide, which forms the holes in Swiss cheese. Semisoft cheeses, such as Limburger, are ripened by bacteria and other contaminating organisms growing on the surface. Blue and Roquefort cheeses are ripened by *Penicillium* molds inoculated into the cheese. The texture of the cheese is loose enough that adequate oxygen can reach the aerobic molds. The growth of the *Penicillium* molds is visible as blue-green clumps in the cheese. Camembert cheese is ripened in small packets so that the enzymes of *Penicillium* mold growing aerobically on the surface will diffuse into the cheese for ripening. The box on page 801 describes one use of the whey produced as a by-product by the dairy industry.

Other Dairy Products

Butter is made by churning cream until the fatty globules of butter separate from the liquid *buttermilk.* The typical flavor and aroma of butter and buttermilk are from *diacetyls,* a combination of two acetic acid molecules that is a metabolic end-product of fermentation by some lactic acid bacteria. Today, commerically sold buttermilk is usually not a by-product of buttermaking but is made by inoculating skim milk with bacteria that form lactic acid and the diacetyls. *Cultured sour cream* is made from cream inoculated with microorganisms similar to those used to make buttermilk.

A wide variety of slightly acidic dairy products—probably a heritage of a nomadic past—is found around the world. Many of them are part of the daily diet in the Balkans, eastern Europe, and Russia. One such product is *yogurt,* which is also popular in the United States. Commercial yogurt is made from milk, from which at least one-fourth of the water has been evaporated in a vacuum pan. The resulting thickened milk is inoculated with a mixed culture of *Streptococcus thermophilus,* primarily for acid production, and *Lactobacillus delbrueckii bulgaricus* (bul-gā′ri-kus), to contribute flavor and aroma. The temperature of the fermentation is about 45°C for several hours, during which time *S. thermophilus* outgrows *L. d. bulgaricus.* Maintaining the proper balance between the flavor-producing and the acid-producing microbes is the secret of making yogurt.

Kefir and *kumiss* are fermented milk beverages that are popular in eastern Europe. The usual lactic acid–producing bacteria are supplemented with a lactose-fermenting yeast to give these drinks an alcohol content of 1–2%.

Nondairy Fermentations

Historically, milk fermentation allowed dairy products to be stored and then consumed much later. Other microbial fermentations were used to make certain plants edible. For example, pre-Columbian people in Central and South America learned to ferment chocolate seeds before consumption. The microbial products released during fermentation produce the chocolate flavor.

Microorganisms are also used in baking, especially for bread. The sugars in bread dough are fermented by yeasts. The species of yeast used in baking is *Saccharomyces cerevisiae.* This same species of yeast is also used in the brewing of beer from grains and the fermentation of wines from grapes. (At one time *S. cerevisiae* was classified as multiple species, such as *S. carlsbergensis, S. uvarum,* and *S. ellipsoideus;* these and a few other species names are often encountered in older literature). *S. cerevisiae* will grow readily under both aerobic and anaerobic conditions, although, unlike facultatively anaerobic bacteria such as *E. coli,* it cannot grow anaerobically indefinitely. Various strains of *S. cerevisiae* have been developed over the centuries and are highly adapted to certain fermentation uses.

Q&A Anaerobic conditions for producing ethanol by the yeasts are mandatory for producing alcoholic beverages. In baking, carbon dioxide forms the typical bubbles of leavened bread. Aerobic conditions favor carbon dioxide production and are encouraged as much as possible. This is the reason bread dough is kneaded repeatedly. Whatever ethanol is produced evaporates during baking. In some breads, such as rye or sourdough, the growth of lactic acid bacteria produce the typical tart flavor.

Fermentation is also used in the production of such foods as *sauerkraut, pickles, olives,* and even cocoa and coffee, in which the beans undergo a fermentation step.

Alcoholic Beverages and Vinegar

Microorganisms are involved in the production of almost all alcoholic beverages. Beer and ale are products of grain starches fermented by yeast. **Beer** is fermented slowly with yeast strains that remain on the bottom (*bottom yeasts*). **Ale** is fermented relatively rapidly, at a higher temperature, with yeast strains that usually form clumps that are buoyed to the top by CO_2 (*top yeasts*). Because yeasts are unable to use starch directly, the starch from grain must be converted to glucose and maltose, which the yeasts can ferment into ethanol and carbon dioxide. In this conversion, called **malting,** starch-containing grains, such as malting barley, are allowed to sprout and then are dried and ground. This product, called **malt,** contains starch-degrading enzymes (amylases) that convert cereal starches into carbohydrates that can be fermented by yeasts. Light beers use amylases or selected strains of yeast to convert more of the starch to fermentable glucose and maltose, resulting in fewer carbohydrates and more alcohol. The beer is then diluted to arrive at an alcohol percentage in the usual range. **Sake,** the Japanese rice wine, is made from rice without malting because the mold *Aspergillus* is first used to convert the rice's starch to sugars that can be fermented. (See the discussion of koji, page 805.) For *distilled spirits,* such as *whiskey, vodka,* and *rum,* carbohydrates from cereal grains, potatoes, and molasses are fermented to alcohol. The alcohol is then distilled to make a concentrated alcoholic beverage.

Wines are made from fruits, typically grapes, that contain sugars that yeasts can use directly for fermentation; malting is unnecessary in winemaking. Grapes usually need no additional sugars, but other fruits might be supplemented with sugars to ensure enough alcohol production. The steps of winemaking are shown in Figure 28.9. Lactic acid bacteria are important when wine is made from grapes that are especially acidic from high concentrations of malic acid. These bacteria convert the malic acid to the weaker lactic acid in a process called **malolactic fermentation.** The result is a less acidic, better-tasting wine than would otherwise be produced.

Wine producers who allowed wine to be exposed to air found that it soured from the growth of aerobic bacteria that converted the ethanol in the wine to acetic acid. The result was *vinegar* (*vin* = wine; *aigre* = sour). The process is now used deliberately to make vinegar. Ethanol is first produced by anaerobic fermentation of carbohydrates by yeasts. The ethanol is then aerobically oxidized to acetic acid by acetic acid–producing bacteria of the genera *Acetobacter* and *Gluconobacter.*

CHECK YOUR UNDERSTANDING

- ✓ Is botulism a greater danger in spoilage of canned goods under thermophilic or under mesophilic conditions? **28-1**
- ✓ Canned foods are usually in metal cans. What sorts of containers are used for aseptically packaged foods? **28-2**
- ✓ Roquefort and blue cheeses are characterized by blue-green clumps. What are these? **28-3**

Industrial Microbiology

LEARNING OBJECTIVES

28-4 Define *industrial fermentation* and *bioreactor.*

28-5 Differentiate primary from secondary metabolites.

28-6 Describe the role of microorganisms in the production of industrial chemicals and pharmaceuticals.

28-7 Define *bioconversion,* and list its advantages.

28-8 List biofuels that can be made by microorganisms.

The industrial uses of microbiology had their beginnings in large-scale food fermentations that produced lactic acid from dairy products and ethanol from brewing. These two chemicals also proved to have many industrial uses unrelated to foods. During World Wars I and II, microbial fermentation and similar technologies were used in the production of armament-related chemical compounds such as glycerol and acetone. Present industrial microbiology dates largely from the technology developed to produce antibiotics following World War II. There

From Plant Disease to Shampoo and Salad Dressing

***Xanthomonas campestris* is a gram-negative rod** that causes a disease called black rot in plants. After gaining access to a plant's vascular tissues, the bacteria use the glucose transported in those tissues to produce a sticky, gumlike substance. This substance builds up to form gumlike masses, which eventually block the plant's transport of nutrients. The gum that makes up these masses, xanthan, is composed of a high-molecular-weight polymer of mannose (see the photograph).

In contrast to its effects in plants, xanthan has no adverse effects when ingested by humans. Consequently, xanthan can be used as a thickener in foods, such as dairy products and salad dressings, and in cosmetics such as cold creams and shampoos.

The average American consumes more than 30 pounds of cheese annually, and every pound of cheese creates 9 pounds of the liquid by-product called whey. When researchers at the U.S. Department of Agriculture (USDA) wanted to find some useful product that could be made out of whey, a liquid waste produced in abundance by the dairy industry, they thought of turning it into xanthan. However, because whey is mostly water and lactose, researchers had to figure out how to get *X. campestris* to produce xanthan using lactose rather than glucose.

A research team working with the USDA at Stauffer Chemical Company used an enrichment based on satisfying only two requirements: that the bacteria grow on whey and make xanthan. First, they inoculated a whey medium with *X. campestris* and incubated it for 24 hours. Then they transferred an inoculum of this culture to a flask of lactose broth, to select a lactose-utilizing cell. The strain did not have to make xanthan from this broth; it only had to grow and use lactose.

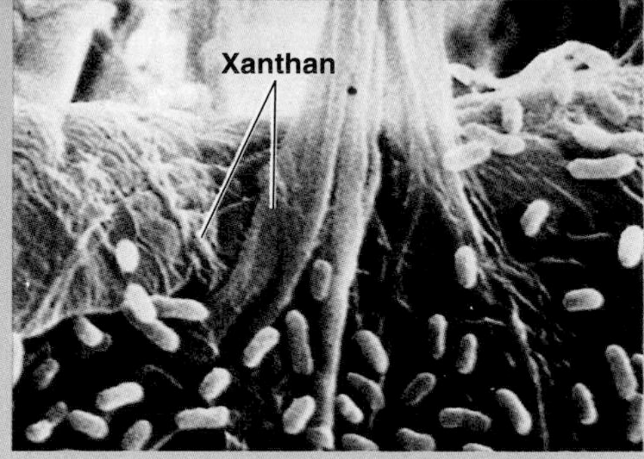

Xanthomonas campestris producing gooey xanthan.

A lactose-utilizing strain was isolated through serial transfers, selecting for the strain with the best ability to grow. After incubation for 10 days, an inoculum was transferred to another flask of lactose broth, and the procedure was repeated two more times. When transferred to a flask of whey medium, the final lactose-utilizing bacteria grew in the whey, and the culture became extremely viscous—xanthan was being produced.

The final result was a process in which 40 g/L of whey powder is converted into 30 g/L of xanthan gum. A quick survey of labels in your neighborhood supermarket will demonstrate just how successful this project was.

is now renewed interest in some of these classic microbial fermentations, especially if they can be used as feedstocks, products that are renewable, or, ideally, products that would otherwise be wasted.

In recent years, industrial microbiology has been revolutionized by the application of genetically modified organisms. An example of a genetically engineered *biosensor* to detect pollution is explained in the box on page 780. In Chapter 9, we discussed the methods for making these modified organisms using recombinant DNA technology and described some of the products derived from them; this technology is now known as **biotechnology.**

Fermentation Technology

The industrial production of microbial products usually involves fermentation. *Industrial fermentation* is the large-scale cultivation

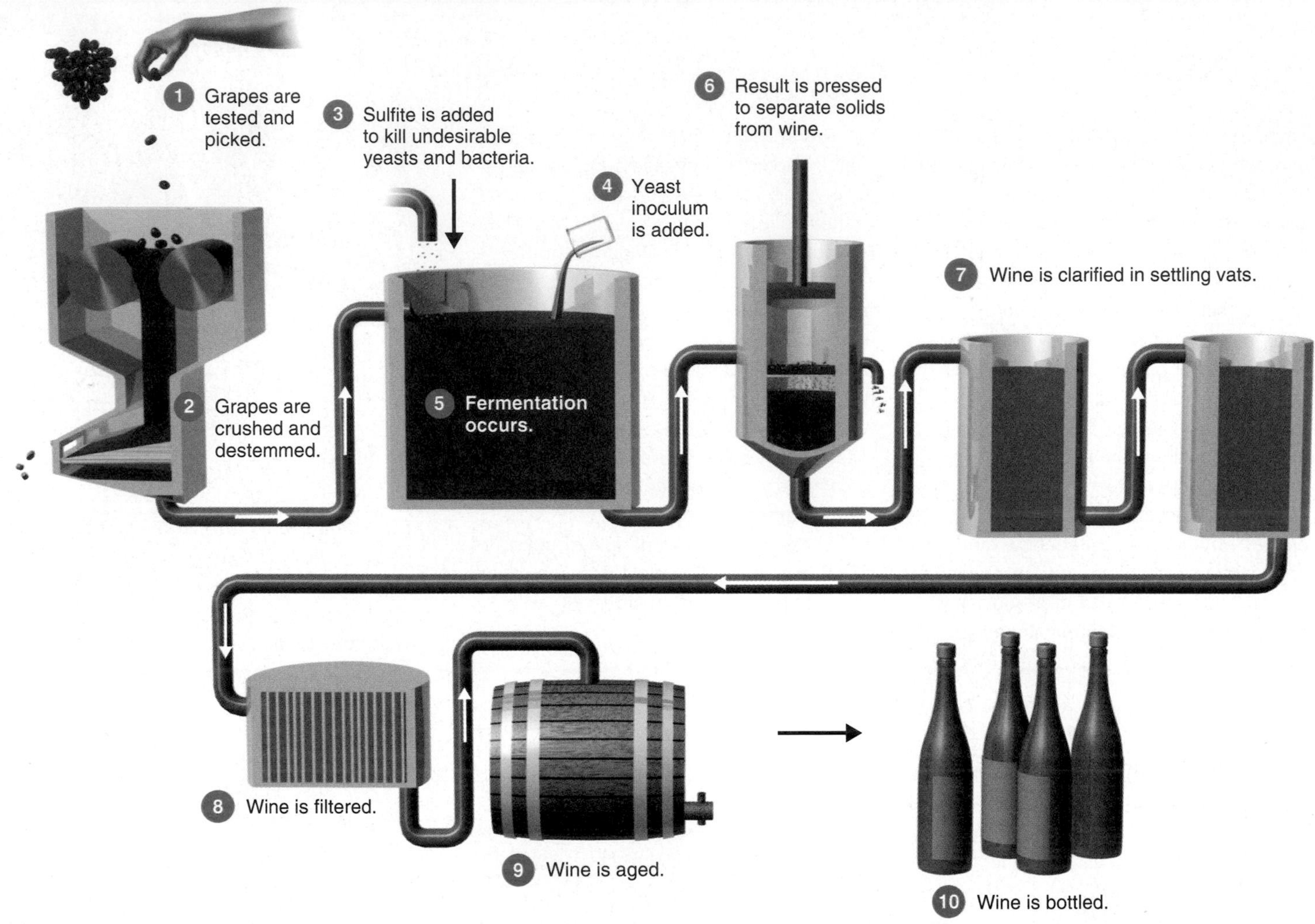

1 Grapes are tested and picked.
2 Grapes are crushed and destemmed.
3 Sulfite is added to kill undesirable yeasts and bacteria.
4 Yeast inoculum is added.
5 Fermentation occurs.
6 Result is pressed to separate solids from wine.
7 Wine is clarified in settling vats.
8 Wine is filtered.
9 Wine is aged.
10 Wine is bottled.

Figure 28.9 The basic steps in making red wine. For white wines, the pressing precedes fermentation so that the color is not extracted from the solid matter.

Q What is the purpose of adding yeast in step 4?

of microbes or other single cells to produce a commercially valuable substance. (See the box in Chapter 5, page 135, for other definitions of *fermentation*). We have just discussed the most familiar examples: the anaerobic food fermentations used in the dairy, brewing, and winemaking industries. Much of the same technology, with the frequent addition of aeration, has been adapted to make other industrial products, such as insulin and human growth hormone, from genetically modified microorganisms. Industrial fermentation is also used in biotechnology to obtain useful products from genetically modified plant and animal cells (see Chapter 9). For example, animal cells are used to make monoclonal antibodies (see Chapter 18, page 507).

Vessels for industrial fermentation are called **bioreactors;** they are designed with close attention to aeration, pH control, and temperature control. There are many different designs, but the most widely used bioreactors are of the continuously stirred type (Figure 28.10). The air is introduced through a diffuser at the bottom (which breaks up the incoming airstream to maximize aeration), and a series of impeller paddles and stationary wall baffles keep the microbial suspension agitated. Oxygen is not very soluble in water, and keeping the heavy microbial suspension well aerated is difficult. Highly sophisticated designs have been developed to achieve maximum efficiency in aeration and other growth requirements, including medium formulation. The high value of the products of genetically modified microorganisms and eukaryotic cells has stimulated the development of newer types of bioreactors and computerized controls for them.

Bioreactors are sometimes very large, holding as much as 500,000 liters. When the product is harvested at the completion of the fermentation, this is known as *batch production.* There are other designs of fermentors. For *continuous flow production,* in which the substrates (usually a carbon source) are fed continuously past

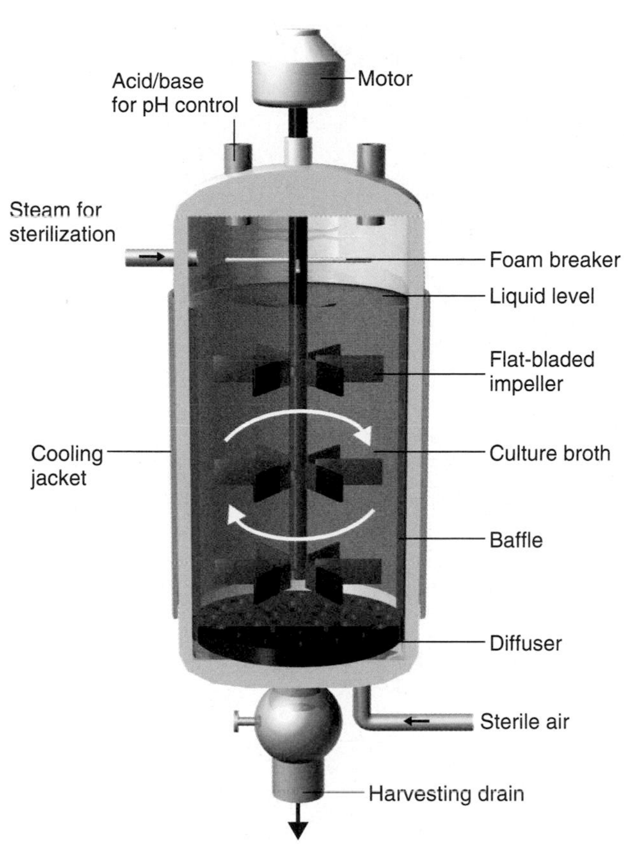

(a) Section of a continuously stirred bioreactor

(b) Bioreactor tank, at left

Figure 28.10 Bioreactors for industrial fermentations.

Identify one essential difference between the bioreactor illustrated and a vat for brewing beer.

immobilized enzymes or into a culture of growing cells, spent medium and desired product are continuously removed.

Generally speaking, the microbes in industrial fermentation produce either primary metabolites, such as ethanol, or secondary metabolites, such as penicillin. A **primary metabolite** is formed essentially at the same time as the new cells, and the production curve follows the cell population curve almost in parallel, with only minimal lag (Figure 28.11a). **Secondary metabolites** are not produced until the microbe has largely completed its logarithmic growth phase, known as the **trophophase,** and has entered the stationary phase of the growth cycle (Figure 28.11b). The following period, during which most of the secondary metabolite is produced, is known as the **idiophase.** The secondary metabolite may be a microbial conversion of a primary metabolite. Alternatively, it may be a metabolic product of the original growth medium that the microbe makes only after considerable numbers of cells and a primary metabolite have accumulated.

Strain improvement is also an ongoing activity in industrial microbiology. (A microbial **strain** differs physiologically in some significant way. For example, it has an enzyme to carry out some additional activity or lacks such an ability, but this difference is not enough to change its species identity). A well-known example is that of the mold used for penicillin production. The original culture of *Penicillium* did not produce penicillin in large enough quantities for commercial use. A more efficient culture was isolated from a moldy cantaloupe from a Peoria, Illinois, supermarket. This strain was treated variously with UV light, X rays, and nitrogen mustard (a chemical mutagen). Selections of mutants, including some that arose spontaneously, quickly increased the production rates by a factor of more than 100. Today, the original penicillin-producing molds produce, not the original 5 mg/L, but 60,000 mg/L. Improvements in fermentation techniques have nearly tripled even this yield. An example of a strain that was developed by enrichment and selection is described in the box on page 801.

Immobilized Enzymes and Microorganisms

In many ways, microbes are packages of enzymes. Industries are increasing their use of free enzymes isolated from microbes to manufacture many products, such as high-fructose syrups, paper, and textiles. The demand for such enzymes is high because they are specific and do not produce costly or toxic waste products. And, unlike traditional chemical processes that require heat or acids, enzymes work under moderate conditions and are safe and biodegradable. For most industrial

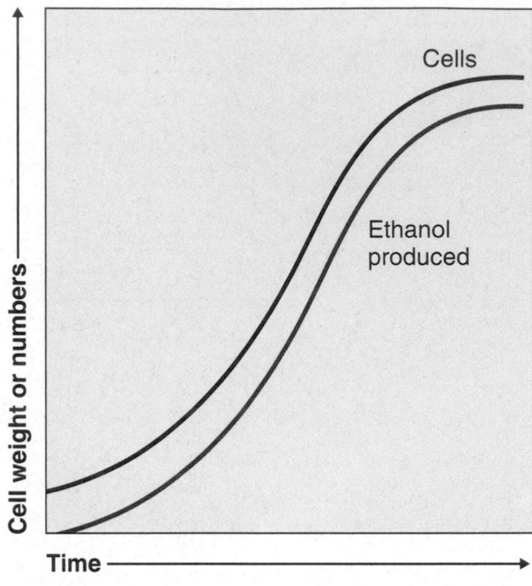

(a) A primary metabolite, such as ethanol from yeast, has a production curve that lags only slightly behind the line showing cell growth.

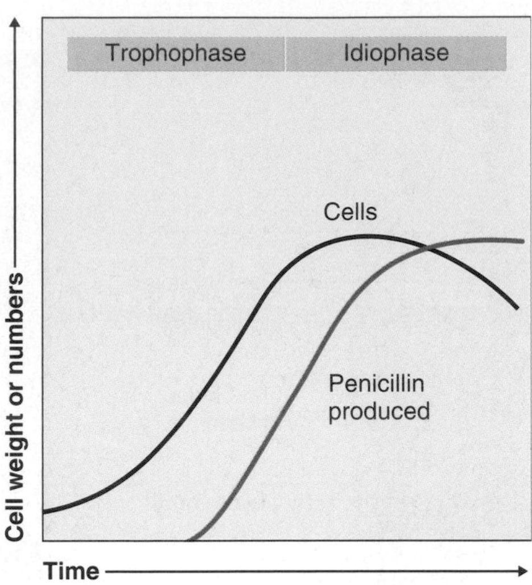

(b) A secondary metabolite, such as penicillin from mold, begins to be produced only after the logarithmic growth phase of the cell (trophophase) is completed. The main production of the secondary metabolite occurs during the stationary phase of cell growth (idiophase).

Figure 28.11 Primary and secondary fermentation.

Q **What is the origin of a secondary metabolite?**

purposes, the enzyme must be immobilized on the surface of some solid support or otherwise manipulated so that it can convert a continuous flow of substrate to product without being lost.

Continuous flow techniques have also been adapted to live whole cells, and sometimes even to dead cells (Figure 28.12). Whole-cell systems are difficult to aerate, and they lack the single-enzyme specificity of immobilized enzymes. However, whole cells are advantageous if the process requires a series of steps that can be carried out by one microbe's enzymes. They also have the advantage of allowing continuous flow processes with large cell populations operating at high reaction rates. Immobilized cells, which are usually anchored to microscopically small spheres or fibers, are currently used to make high-fructose syrup, aspartic acid, and numerous other products of biotechnology.

CHECK YOUR UNDERSTANDING

- ✓ Are bioreactors designed to operate aerobically or anaerobically? **28-4**
- ✓ Penicillin is produced in its greatest quantities after the trophophase of fermentation. Does that make it a primary or secondary metabolite? **28-5**

Industrial Products

As mentioned earlier, cheesemaking produces an organic waste called whey. The whey must be disposed of as sewage or dried and burned as solid waste. Both of these processes are costly and ecologically problematic. However, microbiologists have discovered an alternative use for whey, as discussed in the box on page 801. In this way, microbiologists are devising uses for old products

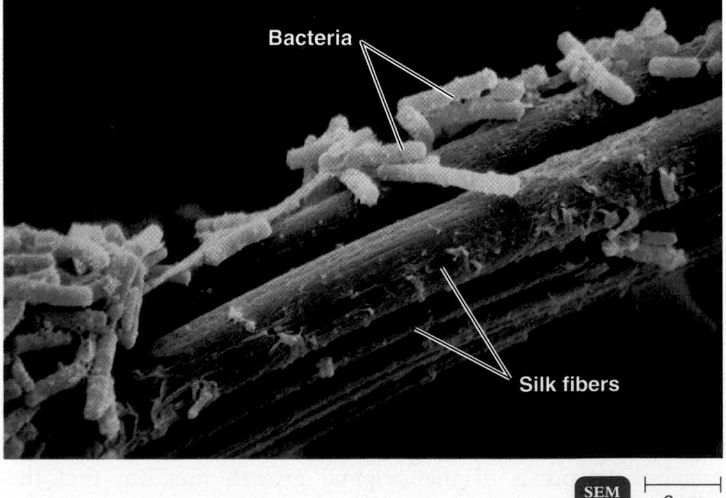

Figure 28.12 Immobilized cells. In some industrial processes, the cells are immobilized on surfaces such as the silk fibers shown here. The substrate flows past the immobilized cells.

Q **How does this process resemble the action of a trickling filter in sewage treatment?**

and creating new ones. In this section, we will discuss some of the more important commercial microbial products and the growing alternative energy industry.

Amino Acids

Amino acids have become a major industrial product from microorganisms. For example, over 600,000 tons of *glutamic acid* (L-glutamate), used to make the flavor enhancer monosodium glutamate, are produced every year. Certain amino acids, such as *lysine* and *methionine,* cannot be synthesized by animals and are present only at low levels in the normal diet. Therefore, the commercial synthesis of lysine and some of the other essential amino acids as cereal food supplements is an important industry. More than 70,000 tons each of lysine and methionine are produced every year.

Two microbially synthesized amino acids, *phenylalanine* and *aspartic acid* (L-aspartate), have become important as ingredients in the sugar-free sweetener aspartame (NutraSweet). Some 3000 to 4000 tons of each of these amino acids are produced annually in the United States.

In nature, microbes rarely produce amino acids in excess of their own needs because feedback inhibition prevents wasteful production of primary metabolites (see Chapter 5, page 120). Commercial microbial production of amino acids depends on specially selected mutants and sometimes on ingenious manipulations of metabolic pathways. For example, in applications in which only the L-isomer of an amino acid is desired, microbial production, which forms only the L-isomer, has an advantage over chemical production, which forms both the **D-isomer** and the **L-isomer** (see Figure 2.13, page 43).

Citric Acid

Citric acid is a constituent of citrus fruits, such as oranges and lemons, and at one time these were its only industrial source. However, over 100 years ago, citric acid was identified as a product of mold metabolism. This discovery was first used as an industrial process when World War I interfered with the picking of the Italian lemon crop. Citric acid has an extraordinary range of uses beyond the obvious ones of giving tartness and flavor to foods. It is an antioxidant and pH adjuster in many foods, and in dairy products it often serves as an emulsifier. Well over 550,000 tons of citric acid are produced every year in the United States. Much of it is produced by a mold, *Aspergillus niger* (nī′jėr), using molasses as a substrate.

Enzymes

Enzymes are widely used in different industries. For example, *amylases* are used in the production of syrups from corn starch, in the production of paper sizing (a coating for smoothness, as on this page), and in the production of glucose from starch. The microbiological production of amylase is considered to be the first biotechnology patent issued in the United States, which was to the Japanese scientist Jokichi Takamine. The basic process by which molds were used to make an enzyme preparation known as **koji** had been used for centuries in Japan to make fermented soy products. Koji is an abbreviation of a Japanese word meaning bloom of mold, reflecting the infiltration of a cereal substrate, either rice or a wheat-soybean mixture, with a filamentous fungus (*Aspergillus*). Primarily, the amylases in koji change starch into sugars, but koji preparations also contain proteolytic enzymes that convert the protein in soybeans into a more digestible and flavorful form. It is the basis of soybean fermentations that are staples of the Japanese diet, such as *soy sauce* and *miso* (a fermented paste of soybeans with a meaty flavor). *Sake,* the well-known Japanese rice wine, makes use of amylases of koji to change the carbohydrates of rice into a form that yeasts can use to produce alcohol. This is roughly the equivalent of the barley malt (page 800) used in beer brewing.

Glucose isomerase is an important enzyme; it converts the glucose that amylases form from starches into fructose, which is used in place of sucrose as a sweetener in many foods. Probably half of the bread baked in this country is made with the aid of *proteases,* which adjust the amount of glutens (protein) in wheat so that baked goods are improved or made uniform. Other proteolytic enzymes are used as meat tenderizers or in detergents as an additive to remove proteinaceous stains. About a third of all industrial enzyme production is for this purpose. *Rennin,* an enzyme used to form curds in milk, is usually produced commercially by fungi but more recently by genetically modified bacteria. An example of a popular clothing product produced with enzymes is described in the box in Chapter 1, page 3.

Vitamins

Vitamins are sold in large quantities combined in tablet form and are used as individual food supplements. Microbes can provide an inexpensive source of some vitamins. *Vitamin* B_{12} is produced by *Pseudomonas* and *Propionibacterium* species. *Riboflavin* (B_2) is another vitamin produced by fermentation, mostly by fungi such as *Ashbya gossypii* (ash′bē-ä gos-sip′ē-ē). *Vitamin C* (ascorbic acid) is produced at the rate of 20,000 tons per year by a complicated modification of glucose by *Acetobacter* species.

Pharmaceuticals

Modern pharmaceutical microbiology developed after World War II, when production of antibiotics was introduced.

All antibiotics were originally the products of microbial metabolism. Many are still produced by microbial fermentations, and work continues on the selection of more productive mutants by nutritional and genetic manipulations. At least 6000 antibiotics have been described. One organism, *Streptomyces hygroscopius,* has different strains that make almost 200 different antibiotics. Antibiotics are typically made industrially by inoculating a solution of growth medium with spores of the appropriate mold or streptomycete and vigorously aerating it.

Vaccines are a product of industrial microbiology. Many antiviral vaccines are mass-produced in chicken eggs or cell cultures. The

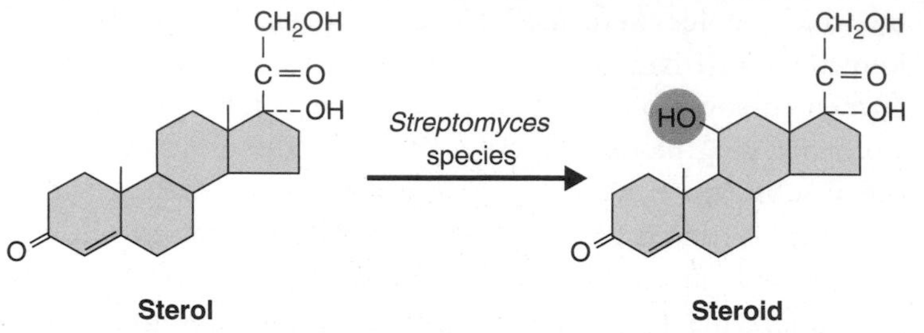

Sterol — Streptomyces species → Steroid

Figure 28.13 The production of steroids. Shown here is the conversion of a precursor compound such as a sterol into a steroid by *Streptomyces.* The addition of a hydroxyl group to carbon number 11 (highlighted in purple on the steroid) might require more than 30 steps by chemical means, but the microorganism can add it in only one step.

Q Name a commercial product that is a steroid.

production of vaccines against bacterial diseases usually requires the growth of large amounts of the bacteria. Recombinant DNA technology is increasingly important in the development and production of subunit vaccines (see Chapter 18, page 502).

Steroids are a very important group of chemicals that include *cortisone,* which is used as an anti-inflammatory drug, and *estrogens* and *progesterone,* which are used in oral contraceptives. Recovering steroids from animal sources or chemically synthesizing them is difficult, but microorganisms can synthesize steroids from sterols or from related, easily obtained compounds. For example, Figure 28.13 illustrates the conversion of a sterol into a valuable steroid.

Copper Extraction by Leaching

Thiobacillus ferrooxidans is used in recovering otherwise unprofitable grades of copper ore, which sometimes contain as little as 0.1% copper. At least 25% of the world's copper is produced this way. *Thiobacillus* bacteria get their energy from the oxidation of a reduced form of iron (Fe^{2+}) in ferrous sulfide to an oxidized form (Fe^{3+}) in ferric sulfate. Sulfuric acid (H_2SO_4) is also a product of the reaction. The acidic solution of Fe^{3+}-containing water is applied by sprinklers and allowed to percolate downslope through the ore body (Figure 28.14). The ferrous iron, Fe^{2+}, and *T. ferrooxidans* are normally present in the ore and continue to contribute to the reactions. The Fe^{3+} in the sprinkling water reacts with insoluble copper (Cu^{+}) in *copper sulfides* in the ore to form soluble copper (Cu^{2+}), which takes the form of *copper sulfates.* To maintain a low enough pH, more sulfuric acid can be added. The soluble copper sulfate moves downslope to collection tanks, where it contacts metallic scrap iron. The copper sulfates react chemically with the iron and precipitate out as metallic copper (Cu^{0}). In this reaction, the metallic iron (Fe^{0}) is converted into ferrous iron (Fe^{2+}) that is recycled to an aerated oxidation pond, where *Thiobacillus* bacteria use it for energy to renew the cycle. This process, although very time consuming, is economical and can recover as much as 70% of the copper in the ore. Uranium, gold, and cobalt ores are processed in a similar manner. The entire arrangement resembles a continuous flow bioreactor.

Microorganisms as Industrial Products

Microorganisms themselves sometimes constitute an industrial product. *Baker's yeast (S. cerevisiae)* is produced in large aerated fermentation tanks. At the end of the fermentation, the contents of the tank are about 4% yeast solids. The cells are harvested by continuous centrifuges and are pressed into the familiar yeast cakes or packets sold for home baking. Wholesale bakers purchase yeast in 50-lb boxes.

Other important microbes that are sold industrially are the symbiotic nitrogen-fixing bacteria *Rhizobium* and *Bradyrhizobium.* These organisms are usually mixed with peat moss to preserve moisture; the farmer mixes the peat moss and bacterial inoculum with the seeds of legumes to ensure infection of the plants with efficient nitrogen-fixing strains (see Chapter 27). For many years, gardeners have used the insect pathogen *Bacillus thuringiensis* to control leaf-eating insect larvae. This bacterium produces a toxin (Bt-toxin) that kills certain moths, beetles, and flies when ingested by their larvae. *B. thuringiensis* subspecies *israelensis* produces Bt-toxin that is especially active against mosquito larvae and is widely used in municipal control programs. Commercial preparations containing Bt-toxin and endospores of *B. thuringiensis* are available at almost any gardening supply store. For an example of microbes being developed to detect chemicals, see the box on page 801.

CHECK YOUR UNDERSTANDING

✓ At one time, citric acid was extracted on an industrial scale from lemons and other citrus fruits. What organism is used to produce it today? **28-6**

Alternative Energy Sources Using Microorganisms

As our supplies of fossil fuels diminish or become more expensive, interest in the use of renewable energy resources will increase. Prominent among these is **biomass,** the collective organic matter produced by living organisms, including crops, trees, and municipal wastes. Microbes can be used for **bioconversion,** the process of converting biomass into alternative energy sources. Bioconversion can also decrease the amount of waste materials requiring disposal.

Methane is one of the most convenient energy sources produced from bioconversion. Many communities produce useful amounts of methane from wastes in landfill sites (Figure 28.15). Large cattle-feeding lots must dispose of immense amounts of animal manure, and much effort has been devoted to devising practical methods for producing methane from these wastes. A major problem with any scheme for large-scale methane production is the need to economically concentrate the widespread

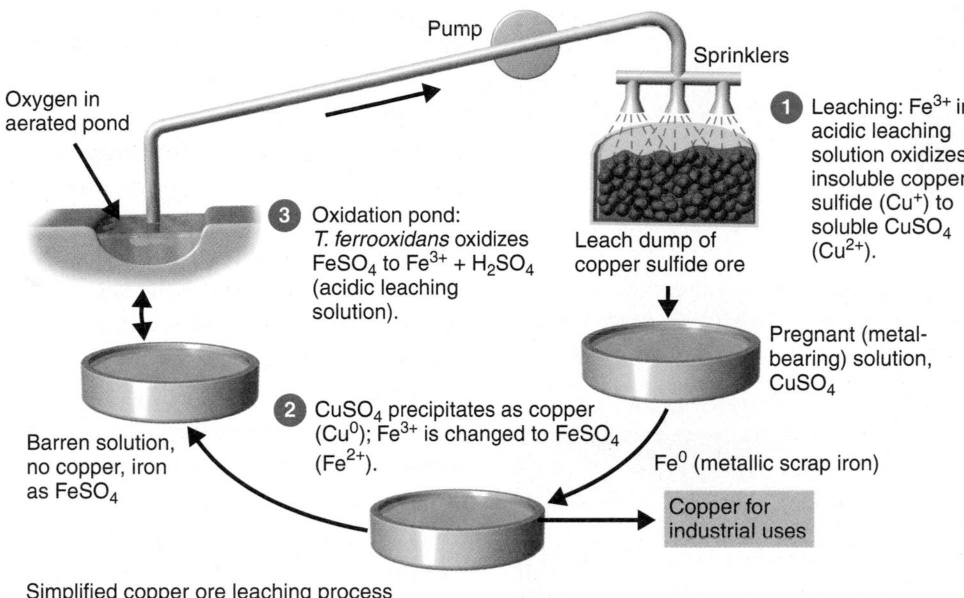
Pump
Sprinklers
Oxygen in aerated pond
1 Leaching: Fe^{3+} in acidic leaching solution oxidizes insoluble copper sulfide (Cu^{+}) to soluble $CuSO_4$ (Cu^{2+}).
3 Oxidation pond: *T. ferrooxidans* oxidizes $FeSO_4$ to Fe^{3+} + H_2SO_4 (acidic leaching solution).
Leach dump of copper sulfide ore
Pregnant (metal-bearing) solution, $CuSO_4$
2 $CuSO_4$ precipitates as copper (Cu^0); Fe^{3+} is changed to $FeSO_4$ (Fe^{2+}).
Barren solution, no copper, iron as $FeSO_4$
Fe^0 (metallic scrap iron)
Copper for industrial uses
Simplified copper ore leaching process

Figure 28.14 Biological leaching of copper ores. The chemistry of the process is much more complicated than shown here. Essentially, *Thiobacillus ferrooxidans* bacteria are used in a biological/chemical process that changes insoluble copper in the ore into soluble copper that leaches out and is precipitated as metallic copper. The solutions are continuously recirculated.

Q **Name another metal that is recovered by a similar process.**

biomass material. If it could be economically concentrated, the animal and human wastes in the United States could provide much of our energy now supplied by fossil fuels and natural gas.

CHECK YOUR UNDERSTANDING

✓ Landfills are the site of a major form of bioconversion—what is the product? **28-7**

Biofuels

As the supplies of fossil petroleum-based fuels become more expensive, and sometimes uncertain, interest in renewable replacement fuels, **biofuels,** is increasing. The initial interest has focused on **ethanol,** which is already widely used as a supplement to gasoline (90% gasoline + 10% ethanol), and the technology is well established. Brazil, for example, produces large amounts of ethanol from sugar cane, about a third of its transportation fuel. In the United States, a limited number of automobiles are adapted to use E85 (15% gasoline + 85% ethanol). Ethanol has, however, a number of deficiencies: it cannot be transported by conventional pipelines (because it absorbs water so avidly), and it has 30% less energy content than gasoline. Also, to produce ethanol from corn creates pressures on the supply and price of a valuable foodstuff.

Gas flaring stacks
Microturbines produce electricity from methane

Figure 28.15 Methane production from solid wastes in landfills. Methane accumulates in landfills and can be used for energy. This installation near Los Angeles has 50 microturbines that produce electricity from methane produced by the landfill. Immediately behind the microturbines are five gas flaring stacks that mask the flames from excess flared methane—a requirement so that aircraft will not confuse it with airport lighting.

Q **How is methane produced in a landfill?**

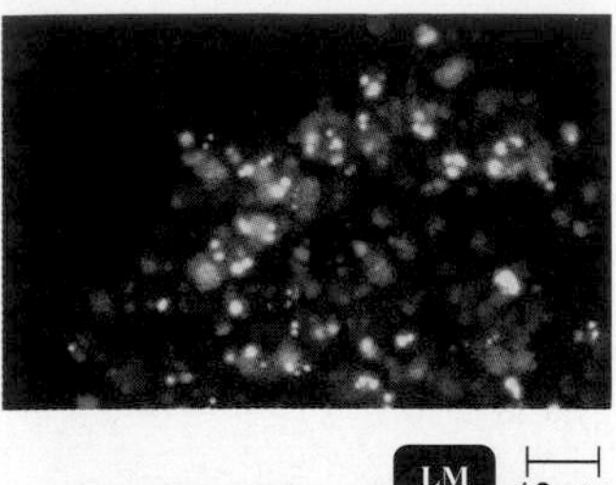

Figure 28.16 Algal bioreactors. An artist's concept of a field of algal bioreactors that could produce biofuels on an industrial scale. Micrograph: This stained culture of green algae shows oil droplets as yellow areas; the red areas hold chlorophyll.

Q Would such a field of algae-growing bioreactors be more likely to be found in Arizona or Iowa?

These drawbacks have increased interest in biofuels derived from cellulosic materials, such as cornstalks, wood, and wastepaper, and from nonfood crops, such as switchgrass—which once carpeted the prairies of the Midwest. Such grasses are perennials and require little more attention than harvesting. The technology for producing ethanol from cellulose is less well known and more expensive than that from corn or sugarcane. The sugar molecules that make up cellulose may be broken apart by enzymes—a possibility that is the focus of intensifying research. Sources of such enzymes might be termites or the fungi that attacked cotton army tenting in World War II. *Cellulose* sources also contain significant amounts of a similar component, *hemicellulose*, which will require organisms capable of digesting it—probably genetically modified microbes. The digestibly resistant cellulosic component, *lignin*, could be burned to heat early steps in fermentation processing.

"Higher" alcohols with longer carbon chains, and especially 'branched' alcohols, would have advantages over conventional ethanol. They would have a lower capacity to absorb water and would have higher energy content. Currently, there is at least one genetically modified bacterium capable of producing several forms of higher alcohols from glucose. Higher alcohols might also be stitched together by chemical processing from short-chained hydrocarbons.

A theoretically attractive organism for producing biofuels is algae. Algae offer a number of advantages; for one, they do not take up valuable farmland needed for food production. Also, algae produce 40 times the energy per acre that corn produces—and the land the algae grow on can be agriculturally nonproductive as long as it has abundant sunlight (**Figure 28.16**). Experimental algal production sites have even used the carbon dioxide emissions from power plants to accelerate growth. The algae can be harvested on an almost daily basis. Oils squeezed from them can be turned into biodiesel fuel and possibly jet fuel: typical algae yield 20% of their weight in oil, and some even more. After oil extraction, the remainder, rich in carbohydrates and proteins, can be used to produce ethanol or as animal feed.

Hydrogen is an attractive candidate as a replacement for fossil fuels, especially if it can be produced by splitting water. It can be used in fuel cells to generate electricity and, if burned to generate energy, does not produce harmful residues. Most research into the production of hydrogen has concentrated on physical and chemical methods, but it is also potentially possible to use bacteria or algae to produce hydrogen from the fermentation of various waste products or by modifications of photosynthesis.

The technologies outlined above will require time to reach their potential. Currently, science is in the early phases of the learning curves that all new technologies face.

CHECK YOUR UNDERSTANDING

✓ How can microbes provide fuels for cars and electricity? **28-8**

Industrial Microbiology and the Future

Microbes have always been exceedingly useful to humankind, even when their existence was unknown. They will remain an essential part of many basic food-processing technologies. The development of recombinant DNA technology has further intensified interest in industrial microbiology by expanding the potential for new products and applications (see the box in Chapter 1, page 3). As the supplies of fossil energy become more scarce, interest in renewable energy sources, such as hydrogen and ethanol, will increase. The use of specialized microbes to produce such products on an industrial scale will probably become more important. As new biotechnology applications and products enter the marketplace, they will affect our lives and well-being in ways that we can only speculate about today.

STUDY OUTLINE

The **MyMicrobiologyPlace** website (**www.microbiologyplace.com**) will help you get ready for tests with its simple three-step approach: ❶ **take a pre-test** and obtain a personalized study plan, ❷ **learn and practice** with animations, tutorials, and MP3 tutor sessions, and ❸ **test yourself** with quizzes and a chapter post-test.

Food Microbiology (pp. 794–800)

1. The earliest methods of preserving foods were drying, the addition of salt or sugar, and fermentation.

Foods and Disease (p. 794)

2. Food safety is monitored by the FDA and USDA and also by use of the HACCP system.

Industrial Food Canning (pp. 794–795)

3. Commercial sterilization of food is accomplished by steam under pressure in a retort.
4. Commercial sterilization heats canned foods to the minimum temperature necessary to destroy *Clostridium botulinum* endospores while minimizing alteration of the food.
5. The commercial sterilization process uses sufficient heat to reduce a population of *C. botulinum* by 12 logarithmic cycles (12D treatment).
6. Endospores of thermophiles can survive commercial sterilization.
7. Canned foods stored above 45°C can be spoiled by thermophilic anaerobes.
8. Thermophilic anaerobic spoilage is sometimes accompanied by gas production; if no gas is formed, the spoilage is called flat sour spoilage.
9. Spoilage by mesophilic bacteria is usually from improper heating procedures or leakage.
10. Acidic foods can be preserved by heat of 100°C because microorganisms that survive are not capable of growth in a low pH.
11. *Byssochlamys, Aspergillus,* and *Bacillus coagulans* are acid-tolerant and heat-resistant microbes that can spoil acidic foods.

Aseptic Packaging (pp. 795–796)

12. Presterilized materials are assembled into packages and aseptically filled with heat-sterilized liquid foods.

Radiation and Industrial Food Preservation (pp. 796–797)

13. Gamma and X-ray radiation can be used to sterilize food, kill insects and parasitic worms, and prevent the sprouting of fruits and vegetables.

High-Pressure Food Preservation (p. 797)

14. Pressurized water is used to kill bacteria in fruit and meat.

The Role of Microorganisms in Food Production (pp. 797–800)

Cheese (pp. 798–799)

15. The milk protein casein curdles because of the action by lactic acid bacteria or the enzyme rennin.
16. Cheese is the curd separated from the liquid portion of milk, called whey.
17. Hard cheeses are produced by lactic acid bacteria growing in the interior of the curd.
18. The growth of microbes in cheese is called ripening.
19. Semisoft cheeses are ripened by bacteria growing on the surface; soft cheeses are ripened by *Penicillium* growing on the surface.

Other Dairy Products (p. 799)

20. Old-fashioned buttermilk was produced by lactic acid bacteria growing during the butter-making process.
21. Commercial buttermilk is made by letting lactic acid bacteria grow in skim milk for 12 hours.
22. Sour cream, yogurt, kefir, and kumiss are produced by lactobacilli, streptococci, or yeasts growing in low-fat milk.

Nondairy Fermentations (pp. 799–800)

23. Sugars in bread dough are fermented by yeast to ethanol and CO_2; the CO_2 causes the bread to rise.
24. Sauerkraut, pickles, olives, soy sauce, and even cocoa and coffee, are products of microbial fermentations.

Alcoholic Beverages and Vinegar (p. 800)

25. Carbohydrates obtained from grains, potatoes, or molasses are fermented by yeasts to produce ethanol in the production of beer, ale, sake, and distilled spirits.
26. The sugars in fruits such as grapes are fermented by yeasts to produce wines.
27. In winemaking, lactic acid bacteria convert malic acid into lactic acid in malolactic fermentation.
28. *Acetobacter* and *Gluconobacter* oxidize ethanol in wine to acetic acid (vinegar).

Industrial Microbiology (pp. 800–808)

1. Microorganisms produce alcohols and acetone that are used in industrial processes.
2. Industrial microbiology has been revolutionized by the ability of genetically modified cells to make many new products.
3. Biotechnology is a way of making commercial products by using living organisms.

Fermentation Technology (pp. 801–804)

4. The growth of cells on a large scale is called industrial fermentation.
5. Industrial fermentation is carried on in bioreactors, which control aeration, pH, and temperature.
6. Primary metabolites such as ethanol are formed as the cells grow (during the trophophase).
7. Secondary metabolites such as penicillin are produced during the stationary phase (idiophase).
8. Mutant strains that produce a desired product can be selected.

Immobilized Enzymes and Microorganisms (pp. 803–804)

9. Enzymes or whole cells can be bound to solid spheres or fibers. When substrate passes over the surface, enzymatic reactions change the substrate to the desired product.
10. They are used to make paper, textiles, and leather and are environmentally safe.

Industrial Products (pp. 804–806)

11. Most amino acids used in foods and medicine are produced by bacteria.
12. Microbial production of amino acids can be used to produce L-isomers; chemical production results in both D- and L-isomers.
13. Lysine and glutamic acid are produced by *Corynebacterium glutamicum.*
14. Citric acid, used in foods, is produced by *Aspergillus niger.*
15. Enzymes used in manufacturing foods, medicines, and other goods are produced by microbes.
16. Some vitamins used as food supplements are made by microorganisms.
17. Vaccines, antibiotics, and steroids are products of microbial growth.
18. The metabolic activities of *Thiobacillus ferrooxidans* can be used to recover uranium and copper ores.
19. Yeasts are grown for wine- and breadmaking; other microbes (*Rhizobium, Bradyrhizobium,* and *Bacillus thuringiensis*) are grown for agricultural use.

Alternative Energy Sources Using Microorganisms (pp. 806–807)

20. Organic waste, called biomass, can be converted by microorganisms into the alternative fuel methane, a process called bioconversion.
21. Fuels produced by microbial fermentation are methane, ethanol, and hydrogen.

Biofuels (pp. 807–808)

22. Biofuels include alcohols and hydrogen (from microbial fermentation) and oils (from algae).

Industrial Microbiology and the Future (p. 808)

23. Recombinant DNA technology will continue to enhance the ability of industrial microbiology to produce medicines and other useful products.

STUDY QUESTIONS

Answers to the Review and Multiple Choice questions can be found by turning to the blue Answers tab at the back of the textbook.

Review

1. What is industrial microbiology? Why is it important?
2. How does commercial sterilization differ from sterilization procedures used in a hospital or laboratory?
3. Why is a can of blackberries preserved by commercial sterilization typically heated to 100°C instead of at least 116°C?
4. Outline the steps in the production of cheese, and compare the production of hard and soft cheeses.
5. Beer is made with water, malt, and yeast; hops are added for flavor.What is the purpose of the water, malt, and yeast? What is malt?
6. Why is a bioreactor better than a large flask for industrial production of an antibiotic?
7. The manufacture of paper includes the use of bleach and formaldehyde-based glue. The microbial enzyme xylanase whitens paper by digesting dark lignins. Oxidase causes the fibers to stick together, and cellulase will remove ink. List three advantages of using these microbial enzymes over traditional chemical methods for making paper.
8. Describe an example of bioconversion. What metabolic processes can result in fuels?
9. **DRAW IT** Label the trophophase and idiophase in this graph. Indicate when primary and secondary metabolites are formed.

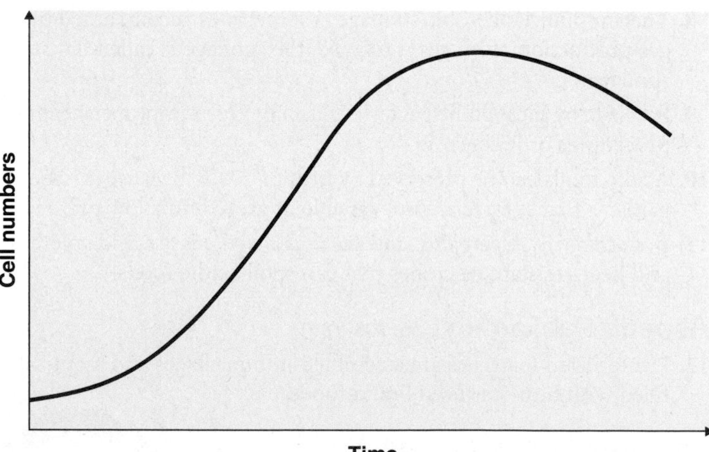

Multiple Choice

1. Foods packed in plastic for microwaving are
   a. dehydrated.
   b. freeze-dried.
   c. packaged aseptically.
   d. commercially sterilized.
   e. autoclaved.
2. *Acetobacter* is necessary for only one of the steps of vitamin C manufacture. The easiest way to accomplish this step would be to
   a. add substrate and *Acetobacter* to a test tube.
   b. affix *Acetobacter* to a surface and run substrate over it.
   c. add substrate and *Acetobacter* to a bioreactor.
   d. find an alternative to this step.
   e. none of the above

Use the following choices to answer questions 3–5:
a. *Bacillus coagulans*
b. *Byssochlamys*
c. flat sour spoilage
d. *Lactobacillus*
e. thermophilic anaerobic spoilage

3. The spoilage of canned foods due to inadequate processing, accompanied by gas production.
4. The spoilage of canned foods caused by *Geobacillus stearothermophilus.*
5. A heat-resistant fungus that causes spoilage in acidic foods.
6. The term *12D treatment* refers to
   a. heat treatment sufficient to kill 12 bacteria.
   b. the use of 12 different treatments to preserve food.
   c. a 10^{12} reduction in *C. botulinum* endospores.
   d. any process that destroys thermophilic bacteria.
7. Which one of the following is *not* a fuel produced by microorganisms?
   a. algal oil
   b. ethanol
   c. hydrogen
   d. methane
   e. uranium
8. Which type of radiation is used to preserve foods?
   a. ionizing
   b. nonionizing
   c. radiowaves
   d. microwaves
   e. all of the above
9. Which of the following reactions is undesirable in winemaking?
   a. Sucrose → ethanol
   b. Ethanol → acetic acid
   c. Malic acid → lactic acid
   d. Glucose → pyruvic acid
10. Which of the following reactions is an oxidation carried out by *Thiobacillus ferrooxidans*?
   a. $Fe^{2+} \rightarrow Fe^{3+}$
   b. $Fe^{3+} \rightarrow Fe^{2+}$
   c. $CuS \rightarrow CuSO_4$
   d. $Fe^0 \rightarrow Cu^0$
   e. none of the above

Critical Thinking

1. Which bacteria seem to be most frequently used in the production of food? Propose an explanation for this.
2. *Methylophilus methylotrophus* can convert methane (CH_4) into proteins. Amino acids are represented by this structure:

$$H_2N-\underset{\substack{|\\R}}{\overset{\substack{H\\|}}{C}}-C\begin{matrix}\nearrow O\\ \searrow OH\end{matrix}$$

   Diagram a pathway illustrating the production of at least one amino acid.

3. "Stone-washed" denim is produced with cellulase. How does cellulase accomplish the look and feel of stone-washing? What is the source of the cellulase?

Clinical Applications

1. Suppose you are culturing a microorganism that produces enough lactic acid to kill itself in a few days.
   a. How can the use of a bioreactor help you maintain the culture for weeks or months? The graph below shows conditions in the bioreactor:

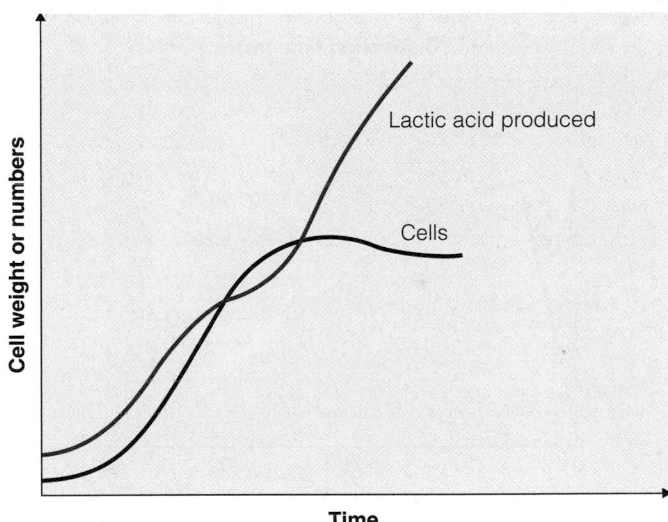

   b. If your desired product is a secondary metabolite, when can you begin collecting it?
   c. If your desired product is the cells themselves and you want to maintain a continuous culture, when can you begin harvesting?
2. Researchers at the CDC inoculated apple cider with 10^5 *E. coli* O157:H7 cells/ml to determine the fate of the bacteria in apple cider (pH 3.7). They obtained the following results:

	Number of *E. coli* O157:H7 cells/ml after 25 days
Apple cider at 25°C	10^4 (mold growth evident by 10 days)
Apple cider with potassium sorbate at 25°C	10^3
Apple cider at 8°C	10^2

   What conclusions can you reach from these data? What disease is caused by *E. coli* O157:H7? (*Hint:* See Chapter 25.)
3. The antibiotic efrotomycin is produced by *Streptomyces lactamdurans. S. lactamdurans* was grown in 40,000 liters of medium. The medium consisted of glucose, maltose, soybean oil, $(NH_4)_2SO_4$, NaCl, KH_2PO_4, and Na_2HPO_4. The culture was aerated and maintained at 28°C. The following results were obtained from analyses of the culture medium during cell growth:

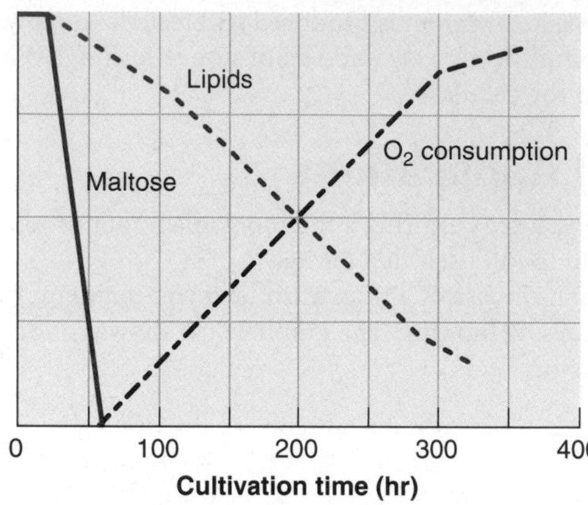

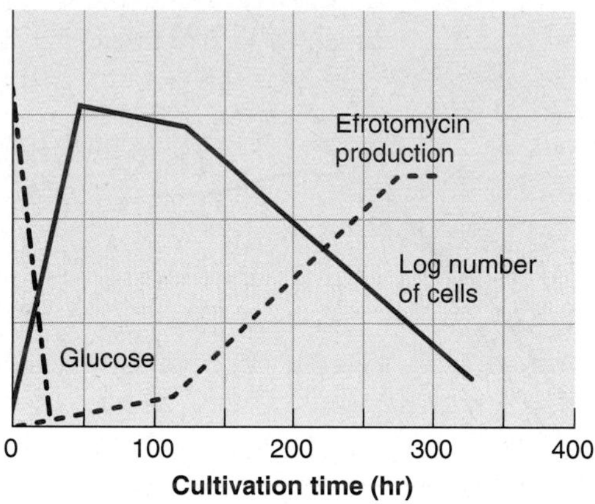

a. Under what conditions is the most efrotomycin produced? Is it a primary or secondary metabolite?

b. Which is used first, maltose or glucose? Suggest a reason for this.

c. What is the purpose of each ingredient in the growth medium? (*Hint:* See Chapter 6.)

d. What is *Streptomyces?* (*Hint:* See Chapter 11.)

ANSWERS TO REVIEW AND MULTIPLE CHOICE STUDY QUESTIONS

Chapter 1

Review

1. People came to believe that living organisms arise from nonliving matter because they would see flies coming out of manure and maggots coming out of dead animals, and see microorganisms appear in liquids after a day or two.
2. **a.** Certain microorganisms cause diseases in insects. Microorganisms that kill insects can be effective biological control agents because they are specific for the pest and do not persist in the environment.
   b. Carbon, oxygen, nitrogen, sulfur, and phosphorus are required for all living organisms. Microorganisms convert these elements into forms that are useful for other organisms. Many bacteria decompose material and release carbon dioxide into the atmosphere that plants use. Some bacteria can take nitrogen from the atmosphere and convert it into a form that plants and other microorganisms can use.
   c. Normal microbiota are microorganisms that are found in and on the human body. They do not usually cause disease and can be beneficial.
   d. Organic matter in sewage is decomposed by bacteria into carbon dioxide, nitrates, phosphates, sulfate, and other inorganic compounds in a wastewater treatment plant.
   e. Recombinant DNA techniques have resulted in insertion of the gene for insulin production into bacteria. These bacteria can produce human insulin inexpensively.
   f. Microorganisms can be used as vaccines. Some microbes can be genetically engineered to produce components of vaccines.
   g. Biofilms are aggregated bacteria adhering to each other and to a solid surface.
3. **a.** 1, 3 **b.** 8 **c.** 1, 4, 5 **d.** 2 **e.** 5 **f.** 3 **g.** 4 **h.** 7
4. **a.** 7 **b.** 4 **c.** 3 **d.** 2 **e.** 6 **f.** 5 **g.** 1
5. **a.** 11 **b.** 14 **c.** 15 **d.** 17 **e.** 3 **f.** 9 **g.** 10 **h.** 2 **i.** 1 **j.** 12 **k.** 18 **l.** 4 **m.** 7 **n.** 5 **o.** 6 **p.** 8 **q.** 13 **r.** 16
6. *Erwinia amylovora* is the correct way to write this scientific name. Scientific names can be derived from the names of scientists. In this case, *Erwinia* is derived from Erwin F. Smith, an American plant pathologist. Scientific names also can describe the organism, its habitat, or its niche. *E. amylovora* is a pathogen of plants (*amylo-* = starch; *vora* = eat).
7. **a.** *B. thuringiensis* is sold as a biological insecticide.
   b. *Saccharomyces* is the yeast sold for making bread, wine, and beer.
8. 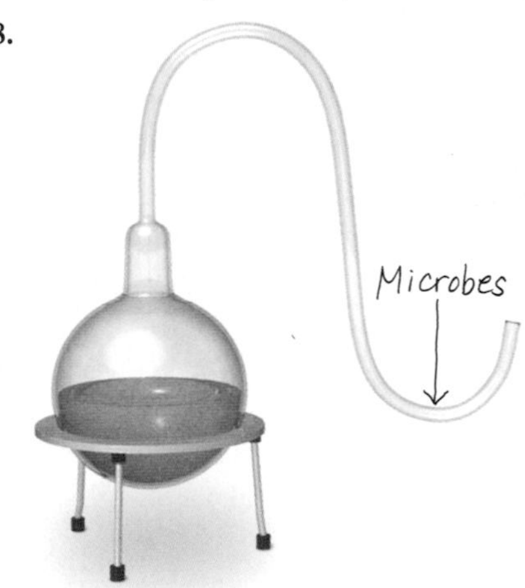

Multiple Choice

1. a
2. c
3. d
4. c
5. b
6. e
7. c
8. a
9. c
10. a

Chapter 2

Review

1. Atoms with the same atomic number and chemical behavior are classified as chemical elements.
2. 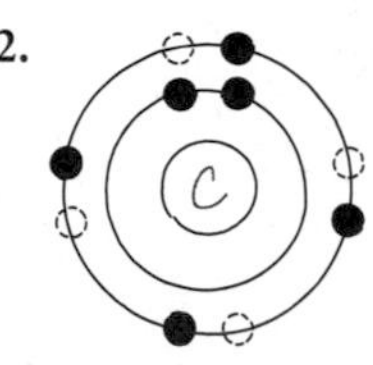

3. **a.** Ionic
   b. Single covalent bond
   c. Double covalent bonds
   d. Hydrogen bond
4. **a.** Synthesis reaction, condensation, or dehydration
   b. Decomposition reaction, digestion, or hydrolysis
   c. Exchange reaction
   d. Reversible reaction

5. The enzyme lowers the activation energy required for the reaction, and therefore speeds up this decomposition reaction.

6. a. Lipid
b. Protein
c. Carbohydrate
d. Nucleic acid

7. a. Amino acids
b. Right to left
c. Left to right

$$H_2N-CH(CH_2-COOH)-C(=O)OH + H-N(H)-CH(CH_2-C_6H_5)-C(=O)O-CH_3 \rightleftharpoons$$

$$H_2N-CH(CH_2-COOH)-C(=O)-N(H)-CH(CH_2-C_6H_5)-C(=O)O-CH_3 + H_2O$$

8. The entire protein shows tertiary structure, held by disulfide bonds. No quaternary structure.

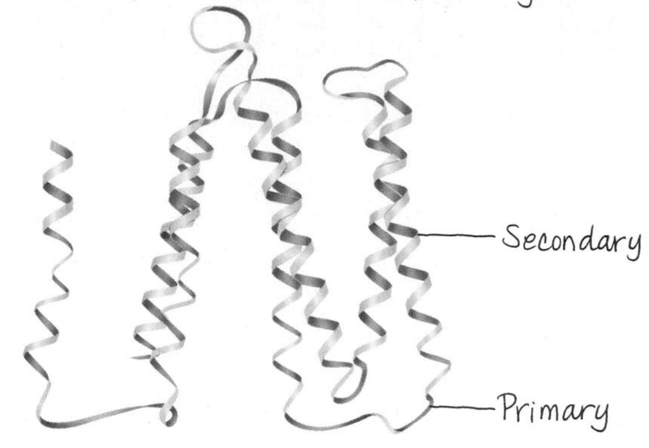

9.

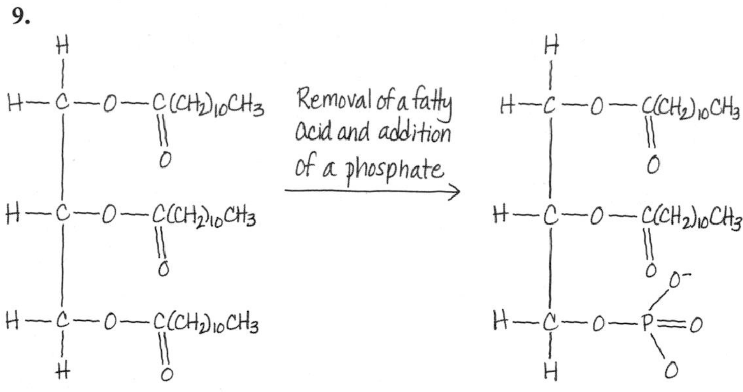

Multiple Choice

1. c	**6.** c
2. b	**7.** a
3. b	**8.** a
4. e	**9.** b
5. b	**10.** c

Chapter 3

Review

1. a. 10^{-6} m; **b.** 1nm; **c.** 10^3 nm

2. a. Compound light microscope
b. Darkfield microscope
c. Phase-contrast microscope
d. Fluorescence microscope
e. Electron microscope
f. Differential interference contrast microscope

3.

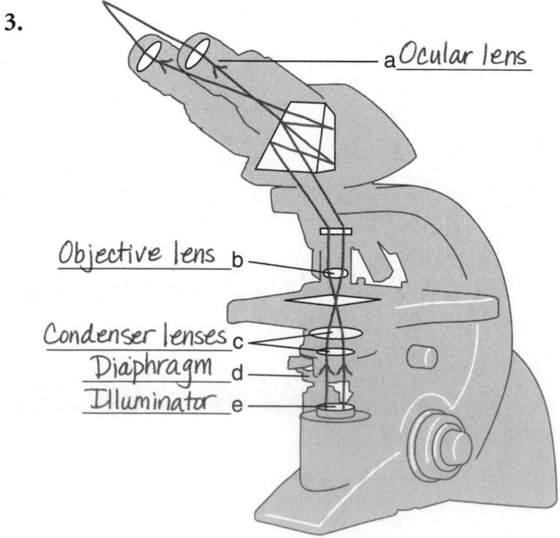

4.

Ocular lens magnification	×	**Oil Immersion lens magnification**	=	**Total magnification of speciman**
10×		100×		1000×

5. **a.** 2,000×
   b. 100,000×
   c. 0.2 μm
   d. 0.0025 μm
   e. Seeing three-dimensional detail.
6. In a Gram stain, the mordant combines with the basic dye to form a complex that will not wash out of gram-positive cells. In a flagella stain, the mordant accumulates on the flagella so that they can be seen with a light microscope.
7. A counterstain stains the colorless non–acid-fast cells so that they are easily seen through a microscope.
8. In the Gram stain, the decolorizer removes the color from gram-negative cells. In the acid-fast stain, the decolorizer removes the color from non–acid-fast cells.
9. **a.** Purple **b.** Purple **c.** Purple **d.** Purple **e.** Purple **f.** Purple **g.** Colorless **h.** Red

Multiple Choice

1. e **2.** d **3.** b **4.** a **5.** a **6.** e **7.** d **8.** b **9.** a **10.** c

Answers

Chapter 4

Review

1. a. and e.

b. and e.

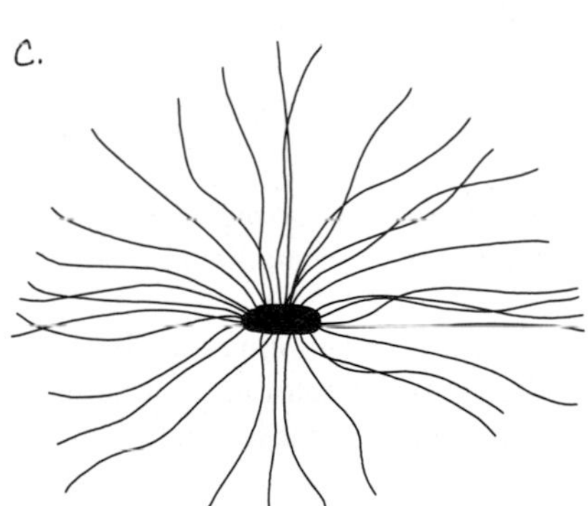

2. **a.** sporogenesis
   b. certain adverse environmental conditions
   c. germination
   d. favorable growth conditions
3.

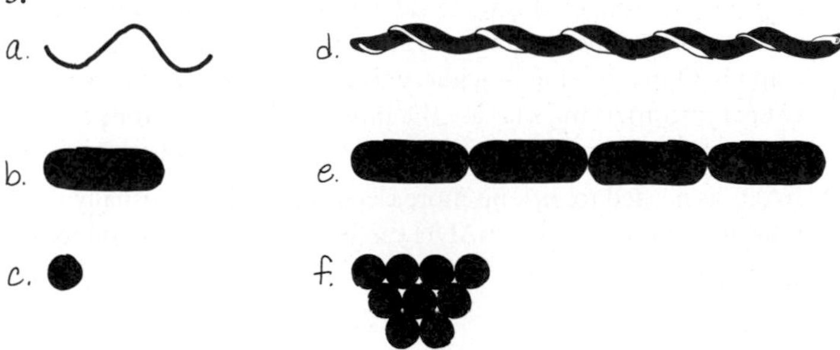

4. **a.** 4
   b. 6
   c. 1
   d. 3
   e. 1, 5
   f. 3, 9
   g. 2, 8
   h. 7
5. An endospore is called a resting structure because it provides a method for one cell to "rest," or survive, as opposed to grow and reproduce. The protective endospore wall allows a bacterium to withstand adverse conditions in the environment.
6. **a.** Both allow materials to cross the plasma membrane from a high concentration to a low concentration without expending energy. Facilitated diffusion requires carrier proteins.
   b. Both require enzymes to move materials across the plasma membrane. In active transport, energy is expended.
   c. Both move materials across the plasma membrane with an expenditure of energy. In group translocation, the substrate is changed after it crosses the membrane.
7. **a.** Diagram (a) refers to a gram-positive bacterium because the lipopolysaccharide–phospholipids–lipoprotein layer is absent.
   b. The gram-negative bacterium initially retains the violet stain, but it is released when the outer membrane is dissolved by the decolorizing agent. After the dye–iodine complex enters, it becomes trapped by the peptidoglycan of gram-positive cells.
   c. The outer layer of the gram-negative cells prevents penicillin from entering the cells.
   d. Essential molecules diffuse through the gram-positive wall. Porins and specific channel proteins in the gram-negative outer membrane allow passage of small water-soluble molecules.
   e. Gram-negative.
8. An extracellular enzyme (amylase) hydrolyzes starch into disaccharides (maltose) and monosaccharides (glucose). A carrier enzyme (maltase) hydrolyzes maltose and moves one glucose into the cell. Glucose can be transported by group translocation as glucose-6-phosphate.

Answers

9. a. 3
   b. 4
   c. 7
   d. 1
   e. 6
   f. 2
   g. 5

Multiple Choice

1. e
2. d
3. b
4. a
5. d
6. e
7. b
8. e
9. a
10. b

Chapter 5

Review

1. (a) is the Calvin-Benson cycle, (b) is glycolysis, and (c) is the Krebs cycle.
   a. Glycerol is catabolized by pathway (b) as dihydroxyacetone phosphate. Fatty acids by pathway (c) as acetyl groups.
   b. In pathway (c) at α-ketoglutaric acid.
   c. Glyceraldehyde-3-phosphate from the Calvin-Benson cycle enters glycolysis. Pyruvic acid from glycolysis is decarboxylated to produce acetyl for the Krebs cycle.
   d. In (a), between glucose and glyceraldehyde-3-phosphate.
   e. The conversion of pyruvic acid to acetyl, isocitric acid to α-ketoglutaric acid, and α-ketoglutaric acid to succinyl~CoA.
   f. By pathway (c) as acetyl groups.
   g.

	Uses	Produces
Calvin-Benson cycle	6 NADPH	
Glycolysis		2 NADH
Pyruvic acid → acetyl		1 NADH
Isocitric acid → α-ketoglutaric acid		1 NADH
α-ketoglutaric acid → Succinyl~CoA		1 NADH
Succinic acid → Fumaric acid		1 FADH2
Malic acid → Oxaloacetic acid		1 NADH

   h. Dihydroxyacetone phosphate; acetyl; oxaloacetic acid; α-ketoglutaric acid.

2.

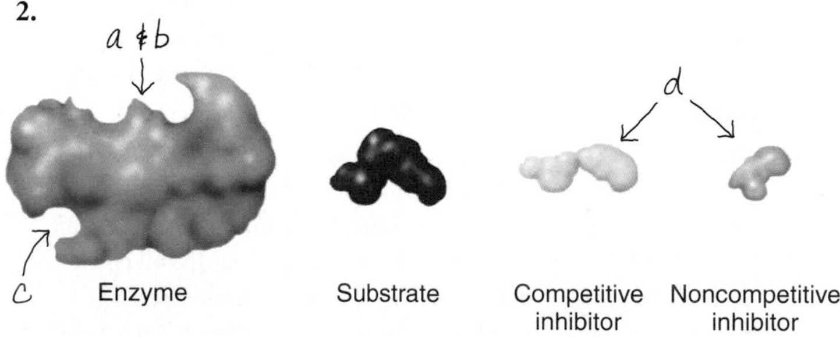

e. When the enzyme and substrate combine, the substrate molecule will be transformed.
When the competitive inhibitor binds to the enzyme, the enzyme will not be able to bind with the substrate.
When the noncompetitive inhibitor binds to the enzyme, the active site of the enzyme will be changed so the enzyme cannot bind with the substrate.

3.

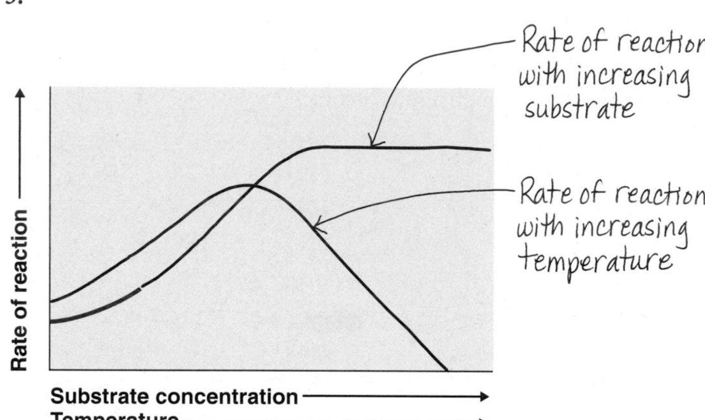

4. Oxidation-reduction: A coupled reaction in which one substance loses electrons and another gains electrons.
   a. The final electron acceptor in aerobic respiration is molecular oxygen; in anaerobic respiration, it is another inorganic molecule.
   b. An electron transport chain is used in respiration but not in fermentation. The final electron acceptor in respiration is usually inorganic; in fermentation it is usually organic.
   c. In cyclic photophosphorylation, electrons are returned to chlorophyll. In noncyclic photophosphorylation, chlorophyll receives electrons from hydrogen atoms.
5. a. Photophosphorylation
   b. Oxidative phosphorylation
   c. Substrate-level phosphorylation
6. oxidation
7. a. CO_2
   b. Light
   c. Organic molecules
   d. Light
   e. CO_2
   f. Inorganic molecules
   g. Organic molecules
   h. Organic molecules
8. Protons are pumped from one side of the membrane to the other; transfer of protons back across the membrane generates ATP.
   a and **b.** Outer portion is acidic, and has a positive electrical charge **c.** Energy-conserving sites are the three loci where protons are pumped out **d.** Kinetic energy is realized at ATP synthase
9. NAD^+ is needed to pick up more electrons. NADH is usually reoxidized in respiration. NADH can be reoxidized in fermentation.

Multiple Choice

1. a
2. d
3. b
4. c
5. c
6. b
7. b
8. a
9. c
10. b

Chapter 6

Review

1. In binary fission, the cell elongates, and the chromosome replicates. Next, the nuclear material is evenly divided. The plasma membrane invaginates toward the center of the cell. The cell wall thickens and grows inward between the membrane invaginations; two new cells result.
2. Carbon: synthesis of molecules that make up a living cell. **H**ydrogen: source of electrons and component of organic molecules. **O**xygen: component of organic molecules; electron acceptor in aerobes. **N**itrogen: component of amino acids. **P**hosphorous: in phospholipids and nucleic acids. **S**ulfur: In some amino acids.
3. **a.** Catalyzes the breakdown of H_2O_2 to O_2 and H_2O.
   b. H_2O_2; peroxide ion is O_2^{2-}.
   c. Catalyzes the breakdown of H_2O_2;

   $$\text{NADH} + \text{H}^+ + \text{H}_2\text{O}_2 \xrightarrow{\text{Peroxidase}} \text{NAD}^+ + \text{H}_2\text{O}$$

   d. O_2^-; this anion has one unpaired electron.
   e. Converts superoxide to O_2 and H_2O_2;

   $$2\text{O}_2^- + 2\text{H}^+ \xrightarrow{\text{Superoxide dismutase}} \text{O}_2 + \text{H}_2\text{O}_2$$

   The enzymes are important in protecting the cell from the strong oxidizing agents, peroxide and superoxide, that form during respiration.
4. Direct methods are those in which the microorganisms are seen and counted. Direct methods are direct microscopic count, plate count, filtration, and most probable number.
5. The growth rate of bacteria slows down with decreasing temperatures. Mesophilic bacteria will grow slowly at refrigeration temperatures and will remain dormant in a freezer. Bacteria will not spoil food quickly in a refrigerator.
6. Number of cells × 2^n generations = Total number of cells

   $$6 \quad \times \quad 2^7 \quad = \quad 768$$
7. Petroleum can meet the carbon and energy requirements for an oil-degrading bacterium; however, nitrogen and phosphate are usually not available in large quantities. Nitrogen and phosphate are essential for making proteins, phospholipids, nucleic acids, and ATP.
8. A chemically defined medium is one in which the exact chemical composition is known. A complex medium is one in which the exact chemical composition is not known.
9.

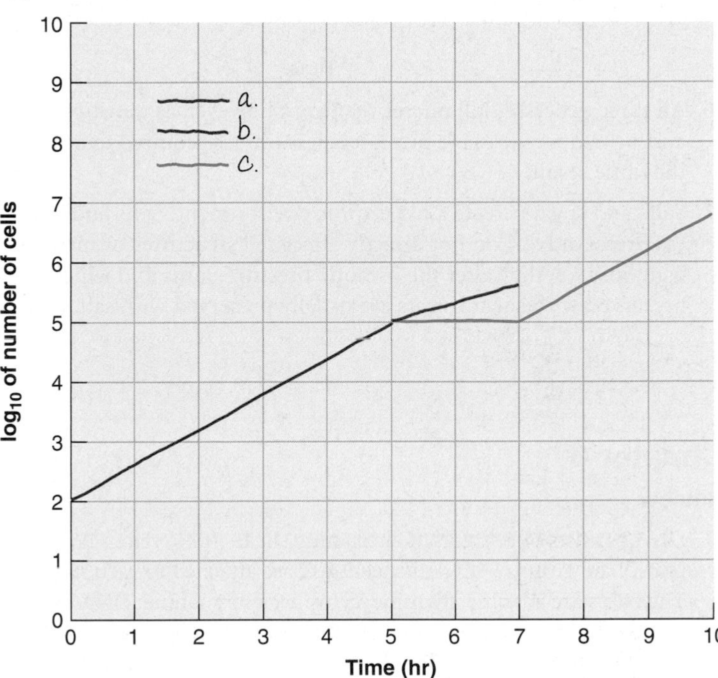

Multiple Choice

1. c
2. a
3. c
4. a
5. c
6. d
7. e
8. c
9. b
10. b

Chapter 7

Review

1. Autoclave. Because of the high specific heat of water, moist heat is readily transferred to cells.
2. Pasteurization destroys most organisms that cause disease or rapid spoilage of food.
3. Variables that affect determination of the thermal death point are
   - The innate heat resistance of the strain of bacteria
   - The past history of the culture, whether it was freeze-dried, wetted, etc.
   - The clumping of the cells during the test
   - The amount of water present
   - The organic matter present
   - Media and incubation temperature used to determine viability of the culture after heating
4. **a.** the ability of ionizing radiation to break DNA directly. However, because of the high water content of cells, free radicals (H· and OH·) that break DNA strands are likely to form.
   b. formation of thymine dimers.

5.

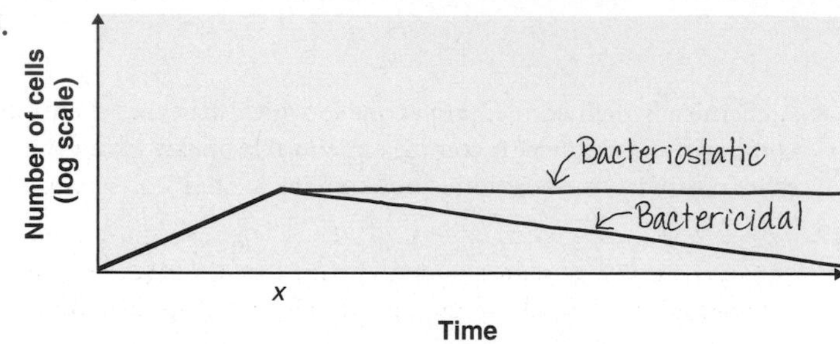

6. All three processes kill microorganisms; however, as moisture and/or temperatures are increased, less time is required to achieve the same result.
7. Salts and sugars create a hypertonic environment. Salts and sugars (as preservatives) do not directly affect cell structures or metabolism; instead, they alter the osmotic pressure. Jams and jellies are preserved with sugar; meats are usually preserved with salt. Molds are more capable of growth in high osmotic pressure than are bacteria.
8. Disinfectant B is preferable because it can be diluted more and still be effective.
9. Quaternary ammonium compounds are most effective against gram-positive bacteria. Gram-negative bacteria that were stuck in cracks or around the drain of the tub would not have been washed away when the tub was cleaned. These gram-negative bacteria could survive the washing procedure. Some pseudomonads can grow on quats that have accumulated.

Multiple Choice

1. d	6. b
2. b	7. b
3. d	8. a
4. d	9. a
5. b	10. b

Chapter 8

Review

1. DNA consists of a strand of alternating sugars (deoxyribose) and phosphate groups with a nitrogenous base attached to each sugar. The bases are adenine, thymine, cytosine, and guanine. DNA exists in a cell as two strands twisted together to form a double helix. The two strands are held together by hydrogen bonds between their nitrogenous bases. The bases are paired in a specific, complementary way: A-T and C-G. The information held in the sequence of nucleotides in DNA is the basis for synthesis of RNA and proteins in a cell.
2.

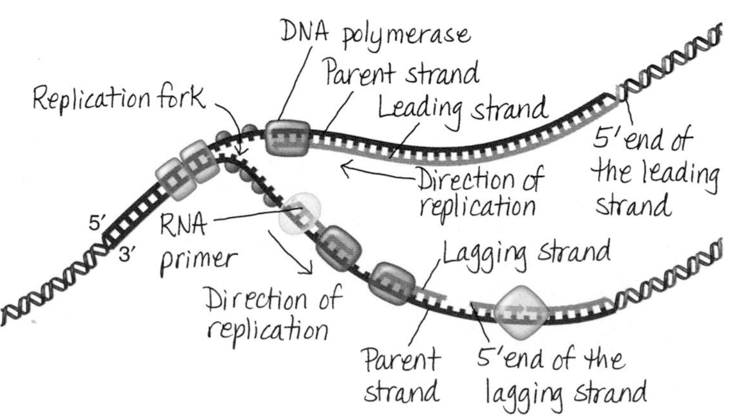

3. a. ATAT<u>TACTTTGCATGGACT</u>.
   b. met-lys-arg-thr-(end).
   c. TATAATGAAACGTTCCTGA.
   d. No change.
   e. Cysteine substituted for arginine.
   f. Proline substituted for threonine (missense mutation).
   g. Frameshift mutation.
   h. Adjacent thymines might polymerize.
   i. ACT.
4. a. 2
   b. 4
   c. 3
   d. 1
   e. 5
5. a. (1) repressor
   (2) operator
   (3) repressor
   (4) transcription
   b. (1) corepressor
   (2) repressor
   (3) operator
   Derepression occurs when the corepressor is needed
   c. None; constitutive enzymes are produced at certain necessary levels regardless of the amount of substrate or end-product.
6. CTTTGA. Endospores and pigments offer protection against UV radiation. Additionally, repair mechanisms can remove and replace thymine polymers.
7. a. Culture 1 will remain the same. Culture 2 will convert to F^+ but will have its original genotype.
   b. The donor and recipient cells' DNA can recombine to form combinations of $A^+B^+C^+$ and $A^-B^-C^-$. If the F plasmid also is transferred, the recipient cell may become F^+.
8. Semiconservative replication ensures the offspring cell will have one correct strand of DNA. Any mutations that may have occurred during DNA replication have a greater chance of being correctly repaired.
9. Mutation and recombination provide genetic diversity. Environmental factors select for the survival of organisms through natural selection. Genetic diversity is necessary for the survival of some organisms through the processes of natural selection. Organisms that survive may undergo further genetic change, resulting in the evolution of the species.

Multiple Choice

1. c	3. c	5. d	7. a	9. d
2. d	4. d	6. b	8. c	10. a

Answers

Chapter 9

Review

1. **a.** Both are DNA. cDNA is a segment of DNA made by RNA-dependent DNA polymerase. It is not necessarily a gene; a gene is a transcribable unit of DNA that codes for protein or RNA.
   b. Both are DNA. A restriction fragment is a segment of DNA produced when a restriction endonuclease hydrolyzes DNA. It is not usually a gene; a gene is a transcribable unit of DNA that codes for protein or RNA.
   c. Both are DNA. A DNA probe is a short, single-stranded piece of DNA. It is not a gene; a gene is a transcribable unit of DNA that codes for protein or RNA.
   d. Both are enzymes. DNA polymerase synthesizes DNA one nucleotide at a time using a DNA template; DNA ligase joins pieces (strands of nucleotides) together.
   e. Both are DNA. Recombinant DNA results from joining DNA from two different sources; cDNA results from copying a strand of RNA.
   f. The proteome is the expression of the genome. An organism's genome is one complete copy of its genetic information. The proteins encoded by this genetic material comprise the proteome.
2. In protoplast fusion, two wall-less cells fuse together to combine their DNA. A variety of genotypes can result from this process. In b, c, and d, specific genes are inserted directly into the cell.
3.

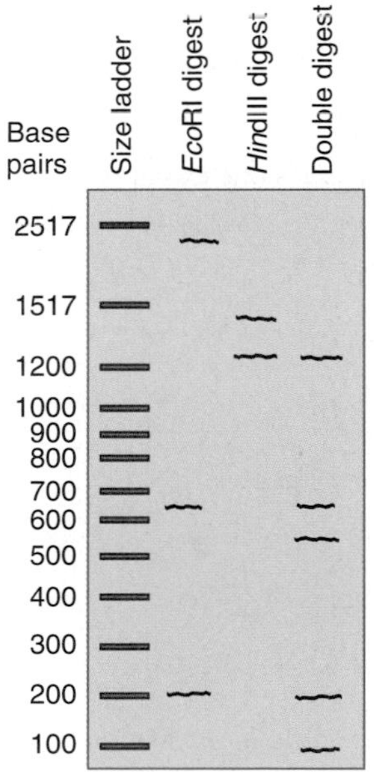

Smallest tet^R fragment with *Hind*III

4. **a.** *Bam*HI, *Eco*RI, and *Hin*dIII make sticky ends.
   b. Fragments of DNA produced with the same restriction enzyme will spontaneously anneal to each other at their sticky ends.
5. The gene can be spliced into a plasmid and inserted into a bacterial cell. As the cell grows, the number of plasmids will increase. The polymerase chain reaction can make copies of a gene using DNA polymerase in vitro.
6. In a eukaryotic cell, RNA polymerase copies DNA; RNA processing removes the introns, leaving the exons in the mRNA. cDNA can be made from the mRNA by reverse transcriptase.
7. See Tables 9.2 and 9.3.
8. You probably used a few plant cells in a Petri plate for your experiment. How will you select the plant cells that actually have the new Ti plasmid? You can grow these cells on plant-cell culture media with tetracycline. Only the cells with the new plasmid will grow.
9. In RNAi, siRNA binds mRNA creating double-stranded RNA, which is enzymatically destroyed.

Multiple Choice

1. b **6.** d
2. b **7.** c
3. b **8.** b
4. b **9.** e
5. c **10.** a

Chapter 10

Review

1. A and D appear to be most closely related because they have similar G-C moles %. No two are the same species.
2. A and D are most closely related.
3. One possible key is shown below. Alternative keys could be made starting with morphology or glucose fermentation.

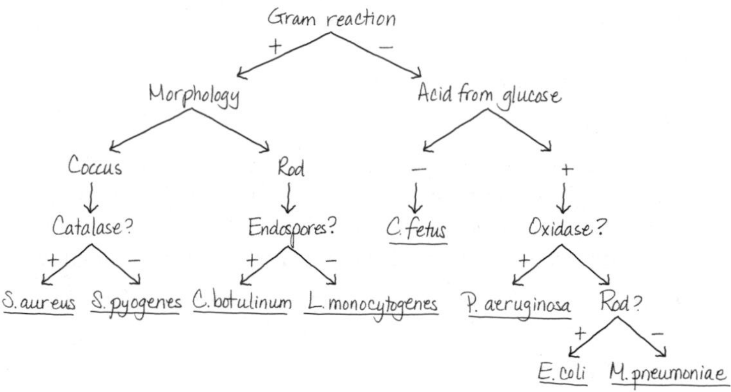

4. The purpose of a cladogram is to show the degree of relatedness between organisms. A dichotomous key can be used for identification but doesn't show relatedness like the cladogram. Mycoplasma and Escherichia are on one branch in the key, but the cladogram indicates Mycoplasma is more closely related to Clostridium.

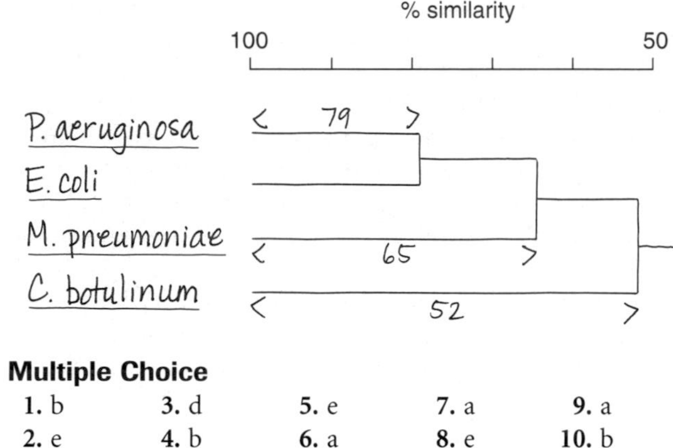

Multiple Choice

1. b	3. d	5. e	7. a	9. a
2. e	4. b	6. a	8. e	10. b

Chapter 11

Review

1. a. *Clostridium*
   b. *Bacillus*
   c. *Streptomyces*
   d. *Mycobacterium*
   e. *Streptococcus*
   f. *Staphylococcus*
   g. *Treponema*
   h. *Spirillum*
   i. *Pseudomonas*
   j. *Escherichia*
   k. *Mycoplasma*
   l. *Rickettsia*
   m. *Chlamydia*
2. a. Both are oxygenic photoautotrophs. Cyanobacteria are prokaryotes; algae are eukaryotes.
   b. Both are chemoheterotrophs capable of forming mycelia; some form conidia. Actinomycetes are prokaryotes; fungi are eukaryotes.
   c. Both are large rod-shaped bacteria. *Bacillus* forms endospores, *Lactobacillus* is a fermentative non–endospore-forming rod.
   d. Both are small rod-shaped bacteria. *Pseudomonas* has an oxidative metabolism; *Escherichia* is fermentative. *Pseudomonas* has polar flagella; *Escherichia* has peritrichous flagella.
   e. Both are helical bacteria. *Leptospira* (a spirochete) has an axial filament. *Spirillum* has flagella.
   f. Both are gram-negative, rod-shaped bacteria. *Escherichia* are facultative anaerobes, and *Bacteroides* are anaerobes.
   g. Both are obligatory intracellular parasites. *Rickettsia* are transmitted by ticks; *Chlamydia* have a unique developmental cycle.
   h. Both lack peptidoglycan cell walls. *Ureaplasma* are archaea; *Mycoplasma* are bacteria (see Table 10.2).
3. There are many ways to draw a key. Here is one example.

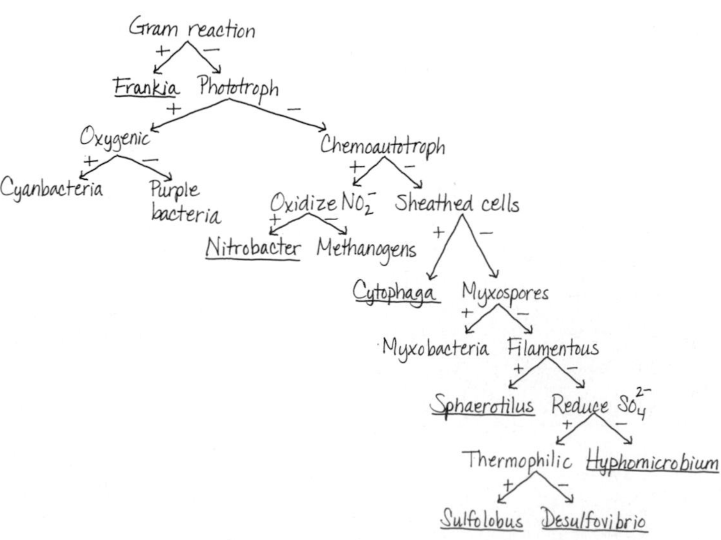

Multiple Choice

1. d	6. c
2. b	7. e
3. e	8. b
4. a	9. b
5. b	10. a

Chapter 12

Review

1. a. Systemic
   b. Subcutaneous
   c. Cutaneous
   d. Superficial
   e. Systemic
2. a. *E. coli*
   b. *P. chrysogenum*
3. As the first colonizers on newly exposed rock or soil, lichens are responsible for the chemical weathering of large inorganic particles and the consequent accumulation of soil.
4. Cellular slime molds exist as individual amoeboid cells. Plasmodial slime molds are multinucleate masses of protoplasm. Both survive adverse environmental conditions by forming spores.
5. a. Flagella
   b. *Giardia*
   c. None
   d. *Nosema*
   e. Pseudopods
   f. *Entamoeba*
   g. None
   h. *Plasmodium*
   i. Cilia
   j. *Balantidium*
   k. Flagella
   l. *Trypanosoma*
6. *Trichomonas* cannot survive for long outside a host because it does not form a protective cyst. *Trichomonas* must be transferred from host to host quickly.
7. Ingestion.
8. 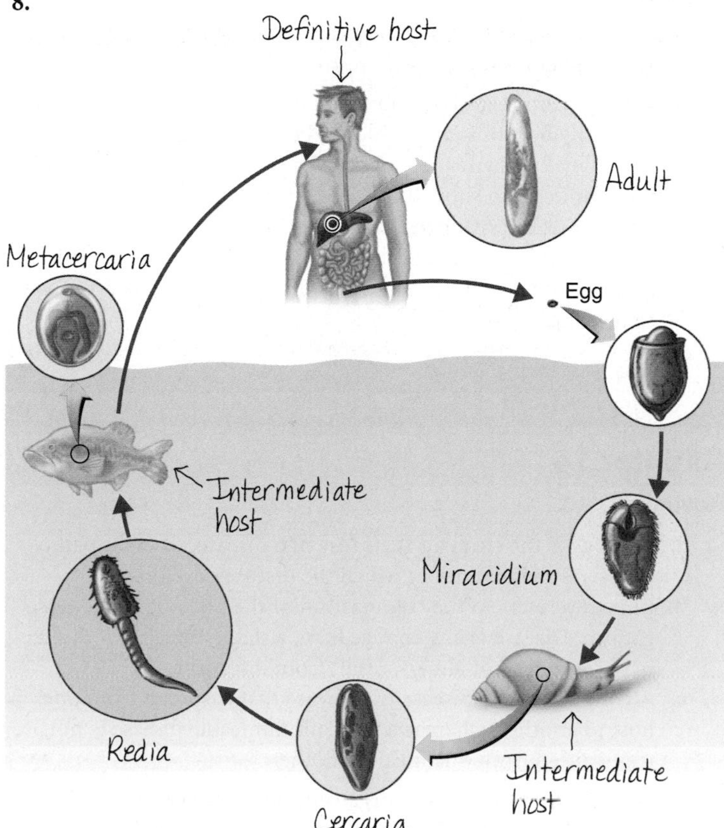

   Phylum: Platyhelminthes
   Class: Trematode
9. The male reproductive organs are in one individual, and the female reproductive organs in another. Nematodes belong to the Phylum Aschelminthes.

Multiple Choice

1. c	3. b	5. d	7. a	9. a
2. b	4. a	6. b	8. c	10. d

Chapter 13

Review

1. Viruses absolutely require living host cells to multiply.
2. A virus
   a. contains DNA or RNA;
   b. has a protein coat surrounding the nucleic acid;
   c. multiplies inside a living cell using the synthetic machinery of the cell; and
   d. causes the synthesis of virions.
   A virion is a fully developed virus particle that transfers the viral nucleic acid to other cells and initiates multiplication.
3. Polyhedral (Figure 13.2); helical (Figure 13.4); enveloped (Figure 13.3); complex (Figure 13.5).
4. Attachment
   Entry
   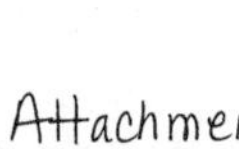
   Uncoating
   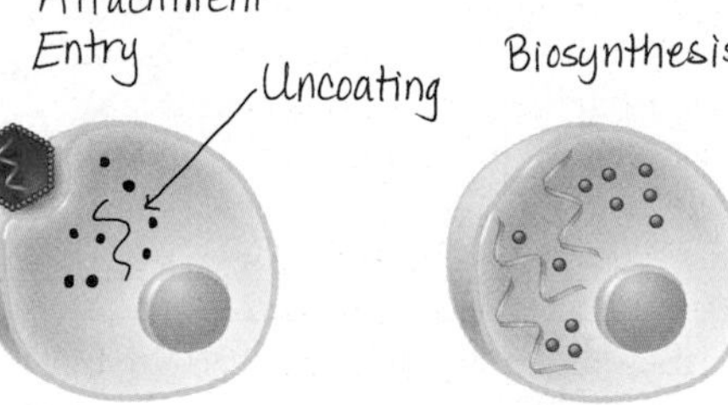
   Biosynthesis
   Maturation
   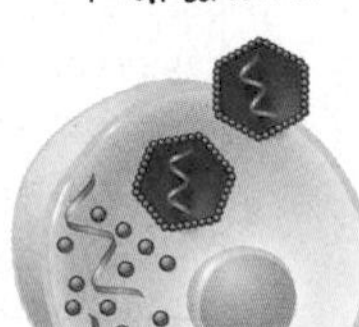

5. Both produce double-stranded RNA with the – strand being the template for more + strands. + strands act as mRNA in both virus groups.
6. Antibiotic treatment of *S. aureus* can activate phage genes that encode P-V leukocidin.
7. **a.** Viruses cannot easily be observed in host tissues. Viruses cannot easily be cultured in order to be inoculated into a new host. Additionally, viruses are specific for their hosts and cells, making it difficult to substitute a laboratory animal for the third step of Koch's postulates.
   b. Some viruses can infect cells without inducing cancer. Cancer may not develop until long after infection. Cancers do not seem to be contagious.
8. **a.** Subacute sclerosing panencephalitis
   b. Common viruses
   c. Answers will vary. One example of a possible mechanism is latent, in an abnormal tissue.
9. **a.** of the rigid cell walls
   b. vectors such as sap-sucking insects
   c. plant protoplasts and insect cell cultures

Multiple Choice

1. e
2. c
3. b
4. c
5. b
6. e
7. c
8. d
9. d
10. c

Chapter 14

Review

1. **a.** Etiology is the study of the cause of a disease, whereas pathogenesis is the manner in which the disease develops.
   b. Infection refers to the colonization of the body by a microorganism. Disease is any change from a state of health. A disease may, but does not always, result from infection.
   c. A communicable disease is a disease that is spread from one host to another, whereas a noncommunicable disease is not transmitted from one host to another.
2. Symbiosis refers to different organisms living together. Commensalism—one of the organisms benefits and the other is unaffected; e.g., corynebacteria living on the surface of the eye. Mutualism—both organisms benefit; e.g., *E. coli* receives nutrients and a constant temperature in the large intestine and produces vitamin K and certain B vitamins that are useful for the human host. Parasitism—one organism benefits while the other is harmed; e.g., *Salmonella enterica* receives nutrients and warmth in the large intestine, and the human host experiences gastroenteritis or typhoid fever.
3.

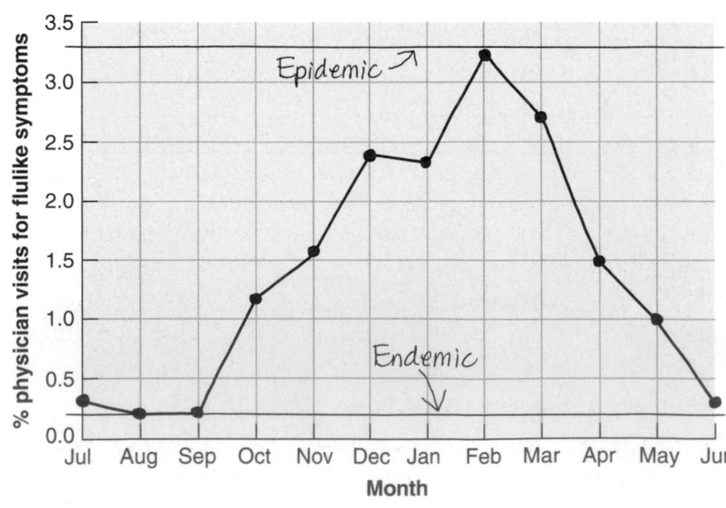

4. **a.** Acute
   b. Chronic
   c. Subacute
5. Hospital patients may be in a weakened condition and therefore predisposed to infection. Pathogenic microorganisms are generally transmitted to patients by contact and airborne transmission. The reservoirs of infection are the hospital staff, visitors, and other patients.
6. Changes in body function that the patient feels are called *symptoms*. Symptoms such as weakness or pain are not measurable by a physician. Objective changes that the physician can observe and measure are called *signs*.
7. When microorganisms causing a local infection enter a blood or lymph vessel and are spread throughout the body, a systemic infection can result.
8. Mutualistic microorganisms are providing a chemical or environment that is essential for the host. Commensal organisms are not essential; another microorganism might serve the function as well.
9. Incubation period, prodromal period, period of illness, period of decline (may be crisis), period of convalescence.

Multiple Choice

1. a
2. b
3. a
4. d
5. a
6. b
7. c
8. a
9. c
10. d

Answers

Chapter 15

Review

1. The ability of a microorganism to produce a disease is called *pathogenicity*. The degree of pathogenicity is *virulence*.
2. Encapsulated bacteria can resist phagocytosis and continue growing. *Streptococcus pneumoniae* and *Klebsiella pneumoniae* produce capsules that are related to their virulence. M protein found in the cell walls of *Streptococcus pyogenes* and A protein in the cell walls of *Staphylococcus aureus* help these bacteria resist phagocytosis.
3. Hemolysins lyse red blood cells; hemolysis might supply nutrients for bacterial growth. Leukocidins destroy neutrophils and macrophages that are active in phagocytosis; this decreases host resistance to infection. Coagulase causes fibrinogen in blood to clot; the clot may protect the bacterium from phagocytosis and other host defenses. Bacterial kinases break down fibrin; kinases can destroy a clot that was made to isolate the bacteria, thus allowing the bacteria to spread. Hyaluronidase hydrolyzes the hyaluronic acid that binds cells together; this could allow the bacteria to spread through tissues. Siderophores take iron from host iron-transport proteins, thus allowing bacteria to get iron for growth. IgA proteases destroy IgA antibodies; IgA antibodies protect mucosal surfaces.
4. **a.** Would inhibit bacteria.
   b. Would prevent adherence of *N. gonorrhoeae*.
   c. *S. pyogenes* would not be able to attach to host cells and would be more susceptible to phagocytosis.
5.

	Exotoxin	Endotoxin
Bacterial source	Gram +	Gram −
Chemistry	Proteins	Lipid A
Toxigenicity	High	Low
Pharmacology	Destroy certain cell parts or physiological functions	Systemic, fever, weakness, aches and shock
Example	Botulinum toxin	Salmonellosis

6. 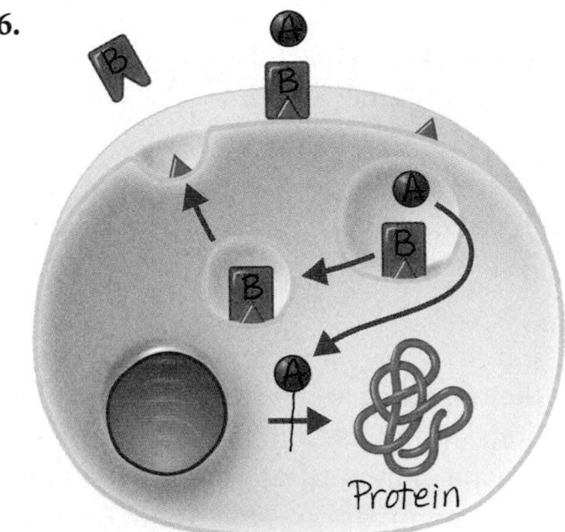

7. Pathogenic fungi do not have specific virulence factors; capsules, metabolic products, toxins, and allergic responses contribute to the virulence of pathogenic fungi. Some fungi produce toxins that, when ingested, produce disease. Protozoa and helminths elicit symptoms by destroying host tissues and producing toxic metabolic wastes.
8. *Legionella*.
9. Viruses avoid the host's immune response by growing inside host cells; some can remain latent in a host cell for prolonged periods. Some protozoa avoid the immune response by mutations that change their antigens.

Multiple Choice

1. e	**6.** a
2. c	**7.** b
3. d	**8.** a
4. d	**9.** d
5. c	**10.** c

Chapter 16

Review

1. **a.** Mechanical: movement out; Chemical: lysozome; acids
   b. Mechanical: movement out; Chemical: acidic environment in female
2. 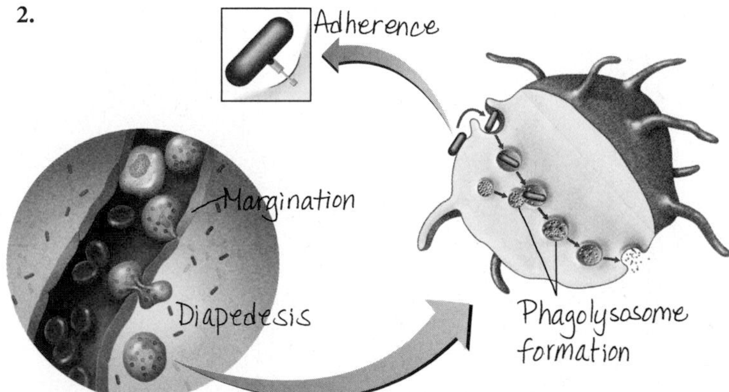

3. Inflammation is the body's response to tissue damage. The characteristic symptoms of inflammation are redness, pain, heat, and swelling.
4. Interferons are antiviral proteins produced by infected cells in response to viral infections. Alpha interferon and beta interferon induce uninfected cells to produce antiviral proteins. Gamma interferon is produced by lymphocytes and activates neutrophils to kill bacteria.
5. Endotoxin binds C3b, which activates C5–C9 to cause cell lysis. This can result in free cell wall fragments, which bind more C3b, resulting in C5–C9 damage to host cell membranes.
6. Toxic oxygen products can kill pathogens.
7. The recipient's antibodies combine with donor antigens and fix complement; the activated complement causes hemolysis.
8. Inhibit formation of C3b; prevent MAC formation; hydrolyze C5a.

9. a. Innate. Facilitate adherence of phagocyte and pathogen.
b. Innate. Bind iron.
c. Innate. Kill or inhibit bacteria.

Multiple Choice

1. a	**6.** a
2. d	**7.** c
3. c	**8.** b
4. d	**9.** d
5. b	**10.** e

Chapter 17

Review

1. a. Adaptive immunity is the resistance to infection obtained during the life of the individual; it results from the production of antibodies and T cells. Innate immunity refers to the resistance of species or individuals to certain diseases that is not dependent on antigen-specific immunity.
b. Humoral immunity is due to antibodies (and B cells). Cellular immunity is due to T cells.
c. Active immunity refers to antibodies produced by the individual who carries them. Passive immunity refers to antibodies produced by another source and then transferred to the individual who needs the antibodies.
d. T_H1 cells produce cytokines that activate T cells. Cytokines produced by T_H2 cells activate B cells.
e. Natural immunity is acquired naturally, i.e., from mother to newborn, or following an infection. Artificial immunity is acquired from medical treatment, i.e., by injection of antibodies or by vaccination.
f. T-dependent antigens: Certain antigens must combine with self-antigens to be recognized by T_H cells and then by B cells. T-independent antigens can elicit an antibody response without T cells.
g. T cells can be classified by their surface antigens: T_H cells possess the CD4 antigen; T_C cells have the CD8 antigen.
h. Immunoglobins = antibodies; TCRs = antigen-receptors on T cells.

2. The major histocompatability complex (MHC) are self-antigens. T_H cells react with MHC II; T_C cells react with MHC I.

3.

Variable; antigen-binding sites
Light chain
-S-S-
-S-S-
Light chain
Fc region
Heavy Chains
IgM

4. See Figure 17.19.

5. Activated T_C cells (CTLs) destroy target cells on contact. T_H cells interact with an antigen to "present" it to a B cell for antibody formation. T_R cells suppress the immune response. Cytokines are chemicals released by cells that initiate a response by other cells.

6.

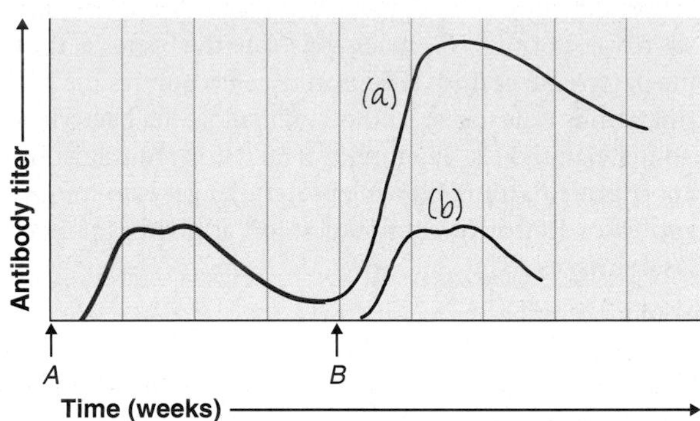

7. Both would prevent attachment of the pathogen; (a) interfere with the attachment site on the pathogen and (b) interfere with the pathogen's receptor site.

8. Rearrangement of the V region genes during embryonic development produces B cells with different antibody genes.

9. The person recovered because he or she produced antibodies against the pathogen. The memory response will continue to protect the person against that pathogen.

Multiple Choice

1. d	**6.** e
2. e	**7.** c
3. b	**8.** d
4. c	**9.** c
5. d	**10.** d

Answers

Chapter 18

Review

1. a. Whole-agent. Live, avirulent virus that can cause the disease if it mutates back to its virulent state.
   b. Whole-agent; (heat-) killed bacteria.
   c. Subunit; (heat- or formalin-) inactivated toxin.
   d. Subunit
   e. Subunit
   f. Conjugated
   g. Nucleic acid
2. If excess antibody is present, an antigen will combine with several antibody molecules. If excess antigen is present, an antibody will combine with several antigens. Refer to Figure 18.3.
3. Particulate antigens react in agglutination reactions. The antigens can be cells or soluble antigens bound to synthetic particles. Soluble antigens take part in precipitation reactions.
4. a. Some viruses are able to agglutinate red blood cells. This reaction is used to detect the presence of large numbers of virions capable of causing hemagglutination (e.g., *Influenzavirus*).
   b. Antibodies produced against viruses that are capable of agglutinating red blood cells will inhibit the agglutination. Hemagglutination inhibition can be used to detect the presence of antibodies against these viruses.
   c. This is a procedure to detect antibodies that react with soluble antigens by first attaching the antigens to insoluble latex spheres. This procedure may be used to detect the presence of antibodies that develop during certain mycotic or helminthic infections.
5.

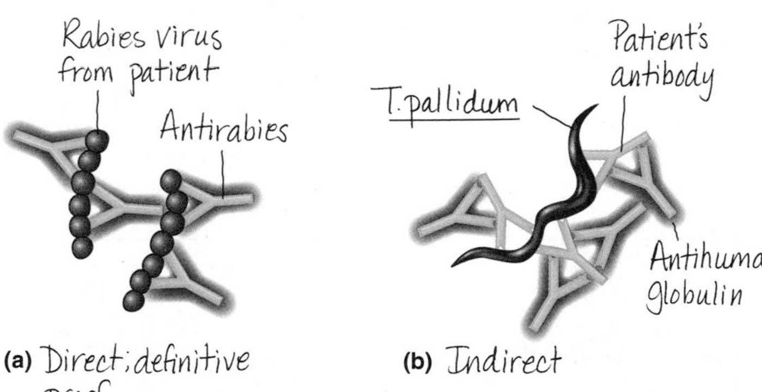

(a) Direct; definitive proof (b) Indirect

6.

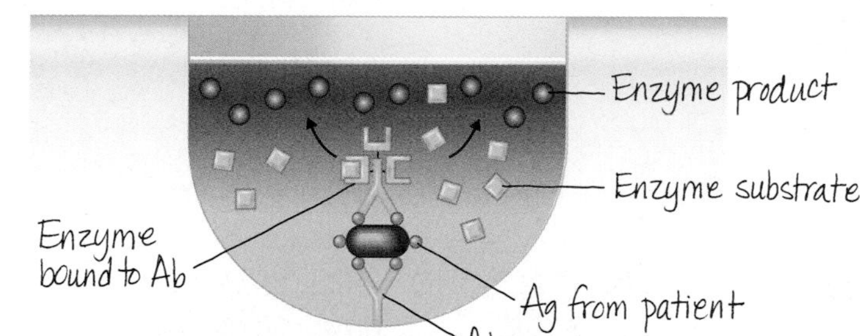

(a) Direct; definitive proof

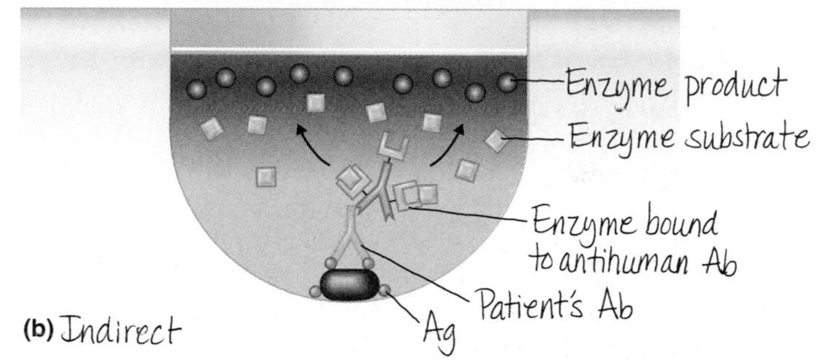

(b) Indirect

7. See Figure 18.2.
8. a. 5 b. 1, 6 c. 1 d. 3 e. 6 f. 2, 4
9. a. 5 b. 3 c. 1 d. 6 e. 2 f. 4

Multiple Choice

1. c
2. d
3. b
4. a
5. a
6. b
7. c
8. a
9. b
10. c

Chapter 19

Review

1.

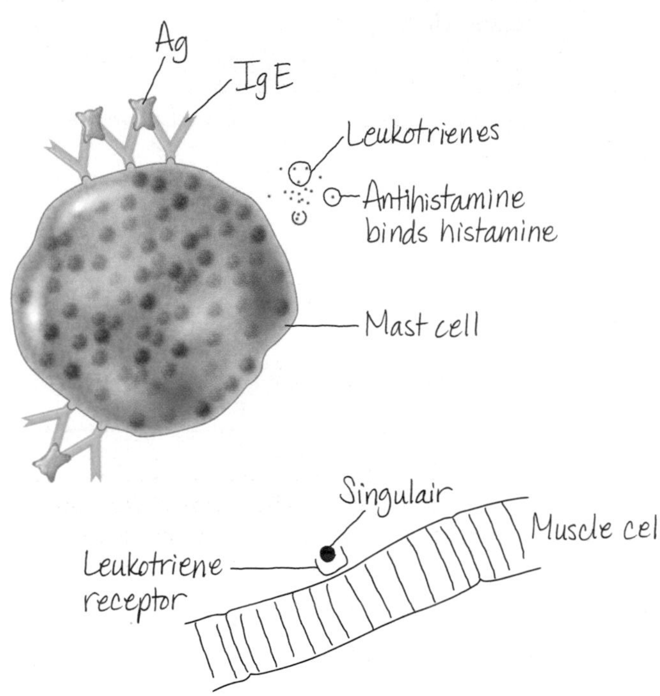

2. Recipient's serum contains complement; activated complement causes hemolysis.
3. Recipient's antibodies will react with donor's tissues.
4. Refer to Figure 19.7.
   a. The observed symptoms are due to lymphokines.
   b. When a person contacts poison oak initially, the antigen (catechols on the leaves) binds to tissue cells, is phagocytized by macrophages, and is presented to receptors on the surface of T cells. Contact between the antigen and the appropriate T cell stimulates the T cell to proliferate and become sensitized. With subsequent exposure to the antigen, sensitized T cells release lymphokines, and a delayed hypersensitivity occurs.
   c. Small repeated doses of the antigen are believed to cause the production of IgG (blocking) antibodies.
5. Lupus patients have antibodies directed at their own DNA.
6. Cytotoxic: Antibodies react with cell-surface antigens.
   Immune complex: Antibody–complement complexes deposit in tissues.
   Cell-mediated: T cells destroy self cells. See Table 19.1.
7. Congenital
   Acquired
   Viral infections, most notably HIV
   Induced by immunosupression drugs
   Result: Increased susceptibility to various infections depending on the type of immune deficiency.
8. Tumor cells have tumor-specific antigens such as TSTA and T antigen. Sensitized T_C cells may react with tumor-specific antigens, initiating lysis of the tumor cells.
9. Some malignant cells can escape the immune system by antigen modulation or immunological enhancement. Immunotherapy might trigger immunological enhancement. The body's defense against cancer is cell-mediated and not humoral. Transfer of lymphocytes could cause graft-versus-host disease.

Multiple Choice

1. b	**6.** e
2. b	**7.** a
3. b	**8.** d
4. a	**9.** c
5. d	**10.** b

Chapter 20

Review

1.

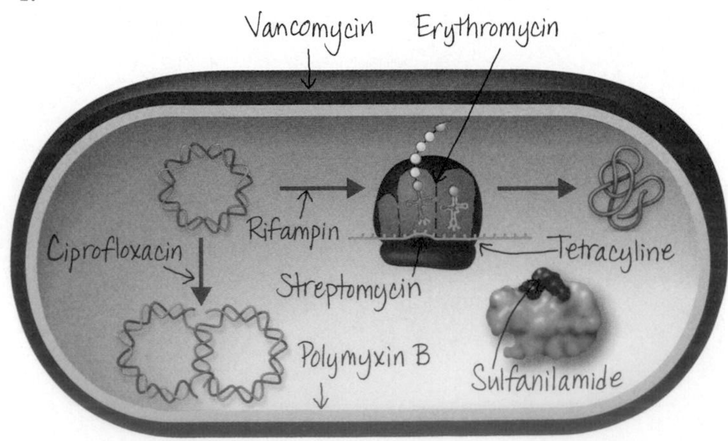

2. The drug (1) should exhibit selective toxicity; (2) should have a broad spectrum; (3) should not produce hypersensitivity in the host; (4) should not produce drug resistance; and (5) should not harm normal microbiota.
3. Because a virus uses the host cell's metabolic machinery, it is difficult to damage the virus without damaging the host. Fungi, protozoa, and helminths possess eukaryotic cells. Therefore, antiviral, antifungal, antiprotozoan, and antihelminthic drugs must also affect eukaryotic cells.
4. Drug resistance is the lack of susceptibility of a microorganism to a chemotherapeutic agent. Drug resistance may develop when microorganisms are constantly exposed to an antimicrobial agent. Ways to minimize the development of drug-resistant microorganisms include judicious use of antimicrobial agents; following directions on the prescription; or administering two or more drugs simultaneously.

5. Simultaneous use of two agents can prevent the development of resistant strains of microorganisms, take advantage of the synergistic effect, provide therapy until a diagnosis is made, and lessen the toxicity of individual drugs by reducing the dosage of each in combination. One problem that can result from simultaneous use of two agents is the antagonistic effect.
6. **a.** Like polymyxin B, causes leaks in the plasma membrane.
   b. Interferes with translation.
7. **a.** Inhibits formation of peptide bond.
   b. Prevents translocation of ribosome along mRNA.
   c. Interferes with attachment of tRNA to mRNA-ribosome complex.
   d. Changes shape of 30S portion of ribosome, resulting in misreading mRNA.
   e. Prevents 70S ribosomal subunits from forming.
   f. Prevents release of peptide from ribosome.
8. DNA polymerase adds bases to the 3′ – OH.
9. **a.** Penicillin inhibits bacterial cell wall synthesis. Echinocandin inhibits fungal cell wall synthesis.
   b. Imidazole interferes with fungal plasma membrane synthesis. Polymyxin B disrupts any plasma membrane.

Multiple Choice

1. b	**6.** d
2. a	**7.** e
3. a	**8.** b
4. b	**9.** c
5. a	**10.** d

Chapter 21

Review

1. Bacteria usually enter through inapparent openings in the skin. Fungal pathogens (except subcutaneous) often grow on the skin itself. Viral infections of the skin (except warts and herpes simplex) most often gain access to the body through the respiratory tract.
2. *Staphylococcus aureus; Streptococcus pyogenes.*
3.

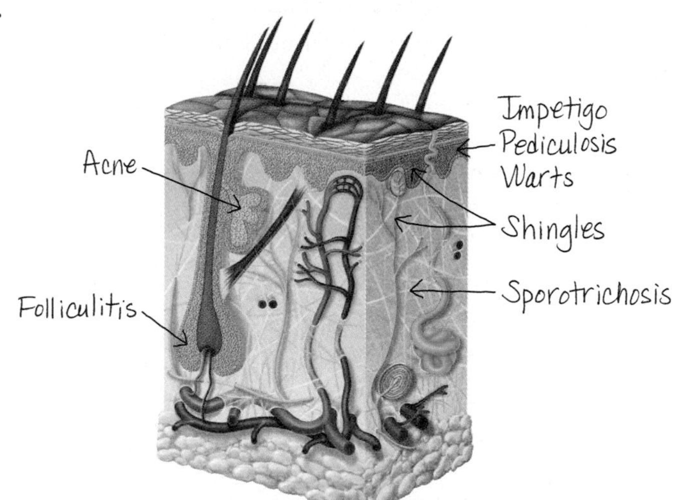

4.

Etiological Agent	Clinical Symptoms	Mode of Transmission
P. acnes	Infected oil glands	Direct contact
S. aureus	Infected hair follicles	Direct contact
Papovavirus	Benign tumor	Direct contact
Herpesvirus	Vesicular rash	Respiratory route
Herpesvirus	Recurrent "blisters"	Direct contact
Paramyxovirus	Papular rash, Koplik's spots	Respiratory route
Togavirus	Macular rash	Respiratory route

5. The test determines the woman's susceptibility to rubella. If the test is negative, she is susceptible to the disease. If she acquires the disease during pregnancy, the fetus could become infected. A susceptible woman should be vaccinated.
6.

Symptoms	Disease
Koplik's spots	Measles
Macular rash	Measles
Vesicular rash	Chickenpox
Small, spotted rash	German measles
"Blisters"	Cold sore
Corneal ulcer	Keratoconjunctivitis

7. The central nervous system can be invaded following keratoconjunctivitis; this results in encephalitis.
8. Attenuated measles, mumps, and rubella viruses.
9. The patient has scabies, an infestation of mites in the skin. It is treated with permethrin insecticide or gamma benzene hexachloride. The presence of a six-legged arthropod (insect) indicates pediculosis (lice).

Multiple Choice

1. c	**6.** d
2. d	**7.** e
3. b	**8.** d
4. c	**9.** a
5. d	**10.** d

Answers

Chapter 22

Review

1. The symptoms of tetanus are due not to bacterial growth (infection and inflammation) but to neurotoxin.
2. **a.** Vaccination with tetanus toxoid.
   b. Immunization with antitetanus toxin antibodies.
3. "Improperly cleaned" because *C. tetani* is found in soil that might contaminate a wound. "Deep puncture" because it is likely to be anaerobic. "No bleeding" because a flow of blood ensures an aerobic environment and some cleansing.
4. Etiology—Picornavirus (poliovirus).
   Transmission—Ingestion of contaminated water.
   Symptoms—Headache, sore throat, fever, nausea; rarely paralysis.
   Prevention—Sewage treatment.
   These vaccinations provide artificially acquired active immunity because they cause the production of antibodies, but they do not prevent or reverse damage to nerves.
5.

Causative Agent	Susceptible Population	Transmission	Treatment
N. meningitidis	Children; military recruits	Respiratory	Penicillin
H. influenzae	Children	Respiratory	Rifampin
S. pneumoniae	Children; elderly	Respiratory	Penicillin
L. monocytogenes	Anyone	Foodborne	Penicillin
C. neoformans	Immunosuppressed individuals	Respiratory	Amphotericin B

6.

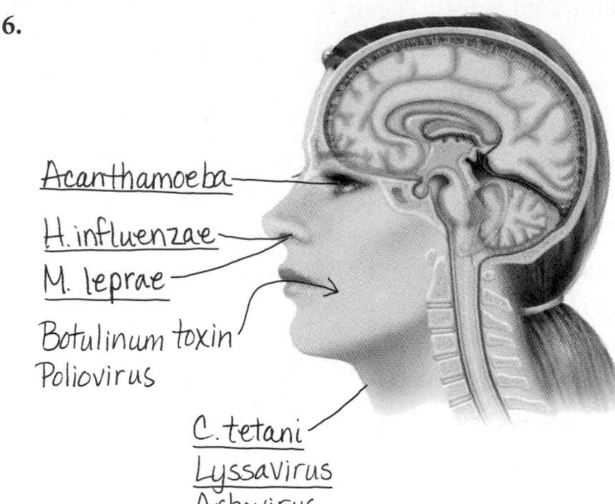

7. Postexposure treatment—Passive immunization with antibodies followed by active immunization with HDCV. Preexposure treatment—Active immunization with HDCV.
   Following exposure to rabies, antibodies are needed immediately to inactivate the virus. Passive immunization provides these antibodies. Active immunization will provide antibodies over a longer period of time, but they are not formed immediately.

8.

Disease	Etiology	Transmission	Symptoms	Treatment
Arboviral encephalitis	Togaviruses, Arboviruses	Mosquitoes (*Culex*)	Headache, fever, coma	Immune serum
African trypanosomiasis	*T. b. gambiense, T. b. rhodesiense*	Tsetse fly	Decreased physical activity and mental acuity	Suramm; melarsoprol
Botulism	*C. botulinum*	Ingestion	Flaccid paralysis	Antitoxin
Leprosy	*M. leprae*	Direct contact	Areas of sensation loss in skin	Dapsone

9. The causative agent of Creutzfeldt–Jakob disease (CJD) is transmissible. Although there is some evidence for an inherited form of the disease, it has been transmitted by transplants. Similarities with viruses are (1) the prion cannot be cultured by conventional bacteriological techniques and (2) the prion is not readily seen in patients with CJD.

Multiple Choice

1. a	**6.** c
2. c	**7.** b
3. a	**8.** a
4. b	**9.** c
5. a	**10.** a

Chapter 23

Review

1.

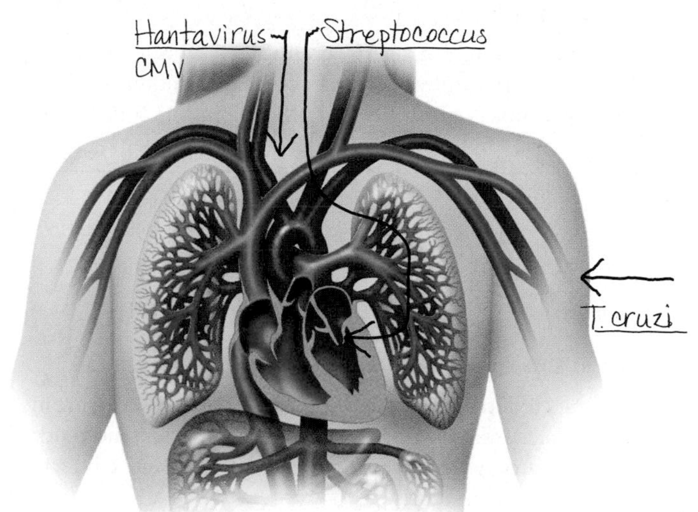

2.

Disease	Causative Agent	Predisposing Conditions
p.s.	*Str. pyogenes*	Abortion or childbirth
s.b.e.	alpha-hemolytic strep.	Preexisting lesions
a.b.e.	*Sta. aureus*	Abnormal heart valves
r.f.	*Str. pyogenes*	Autoimmune

3. All are vectorborne rickettsial diseases. They differ from each other in (1) etiologic agent, (2) vector, (3) severity and mortality, and (4) incidence (e.g., epidemic, sporadic).

4.

Causative Agent	Vector	Treatment
Plasmodium	*Anopheles*	Quinine derivative
Flavivirus	*Aedes aegypti*	None
Flavivirus	*Aedes aegypti*	None
Borrelia	Soft ticks	Tetracycline
Leishmania	Sandflies	Antimony

5.

Disease	Causative Agent	Transmission	Reservoir
Tularemia	*Francisella tularensis*	Skin abrasions, ingestion, inhalation, bites	Rabbits
Brucellosis	*Brucella* spp.	Ingestion of milk, direct contact	Cattle
Anthrax	*Bacillus anthracis*	Skin abrasions, inhalation, ingestion	Soil, cattle
Lyme Disease	*Borrelia burgdorferi*	Tick bites	Deer, mice
Ehrlichiosis	*Ehrlichia*	Tick bites	Deer
Cytomegalic inclusion disease	HHV-5	Saliva, blood	Humans
Plague	*Yersinia pestis*	Flea bites, inhalation	Rodents

6.

Disease	Causative Agent	Transmission	Reservoir	Endemic Area
Schistosomiasis	*Schistosoma* spp.	Penetrate skin	Aquatic snail	Asia, South America
Toxoplasmosis	*Toxoplasma gondii*	Ingestion, inhalation	Cats	United States
Chagas' disease	*Trypanosoma cruzi*	"Kissing bug"	Rodents	Central America

7.

	Reservoir	Etiology	Transmission	Symptoms
Cat-scratch disease	Cats	*Bartonella henselae*	Scratch; touching eyes, fleas	Swollen lymph nodes, fever, malaise
Toxoplasmosis	Cats	*Toxoplasma gondii*	Ingestion	None, congenital infections, neurologic damage

8. Gangrenous tissue is anaerobic and has suitable nutrients for *C. perfringens*.

9. Infectious mononucleosis is caused by EB virus and is transmitted in oral secretions.

Multiple Choice

1. e
2. b
3. d
4. c
5. a
6. e
7. a
8. c
9. c
10. c

Chapter 24

Review

1.

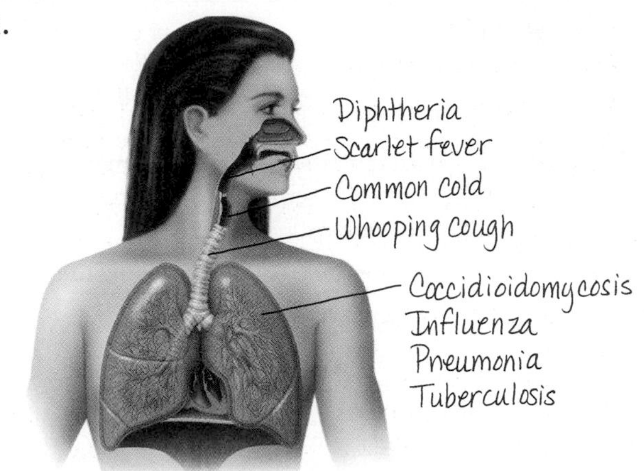

2. Mycoplasmal pneumonia is caused by *Mycoplasma pneumoniae* bacteria. Viral pneumonia can be caused by several different viruses. Mycoplasmal pneumonia can be treated with tetracyclines, whereas viral pneumonia cannot.

3.

Disease	Causative Agent	Symptoms
Upper Respiratory System		
Common cold	Coronaviruses	Sneezing, excessive nasal secretions, congestion
Lower Respiratory System		
Viral pneumonia	Several viruses	Fever, shortness of breath, chest pains
Influenza	Influenzavirus	Chills, fever, headache, muscular pains
RSV	Respiratory syncytial virus	Coughing, wheezing

Amantadine is used to treat influenza; palivizumab, for life-threatening RSV.

4.

Disease	Symptoms
Streptococcal pharyngitis	Pharyngitis and tonsillitis
Scarlet fever	Rash and fever
Diphtheria	Membrane across throat
Whooping cough	Paroxysmal coughing
Tuberculosis	Tubercles, coughing
Pneumococcal pneumonia	Reddish lungs, fever
H. influenzae pneumonia	Similar to pneumococcal pneumonia
Chamydial pneumonia	Low fever, cough, and headache
Otitis media	Earache
Legionellosis	Fever and cough
Psittacosis	Fever and headache
Q fever	Chills and chest pain
Epiglottitis	Inflamed, abscessed epiglottis
Melioidosis	Pneumonia

Refer to Diseases in Focus 24.1, 24.2, and 24.3 to complete the table.

5. Inhalation of large numbers of spores from *Aspergillus* or *Rhizopus* can cause infections in individuals with impaired immune systems, cancer, and diabetes.

6. No. Many different organisms (gram-positive bacteria, gram-negative bacteria, and viruses) can cause pneumonia. Each of these organisms is susceptible to different antimicrobial agents.

7.

Disease	Endemic Areas in the United States
Histoplasmosis	States adjoining the Mississippi and Ohio rivers
Coccidioidomycosis	American Southwest
Blastomycosis	Mississippi
Pneumocystis	Ubiquitous

Refer to Diseases in Focus 24.3 to complete the table.

8. In the tuberculin test, purified protein derivative (PPD) from *M. tuberculosis* is injected into the skin. Induration and reddening of the area around the injection site indicate an active infection or immunity to tuberculosis.

9. **a.** *Staphylococcus aureus*
   b. *Streptococcus pyogenes*
   c. *S. pneumoniae*
   d. *C. diphtheriae*
   e. *Mycobacterium tuberculosis*
   f. *Moraxella catarrhalis*
   g. *Bordetella pertussis*
   h. *Burkholderia pseudomallei*
   i. *Legionella pneumophila*
   j. *Haemophilus influenzae*
   k. *Chlamydophila psittaci*
   l. *Coxiella burnetii*
   m. *Mycoplasma pneumoniae*

Multiple Choice

1. a
2. c
3. e
4. a
5. c
6. b
7. a
8. e
9. b
10. d

Chapter 25

Review

1.

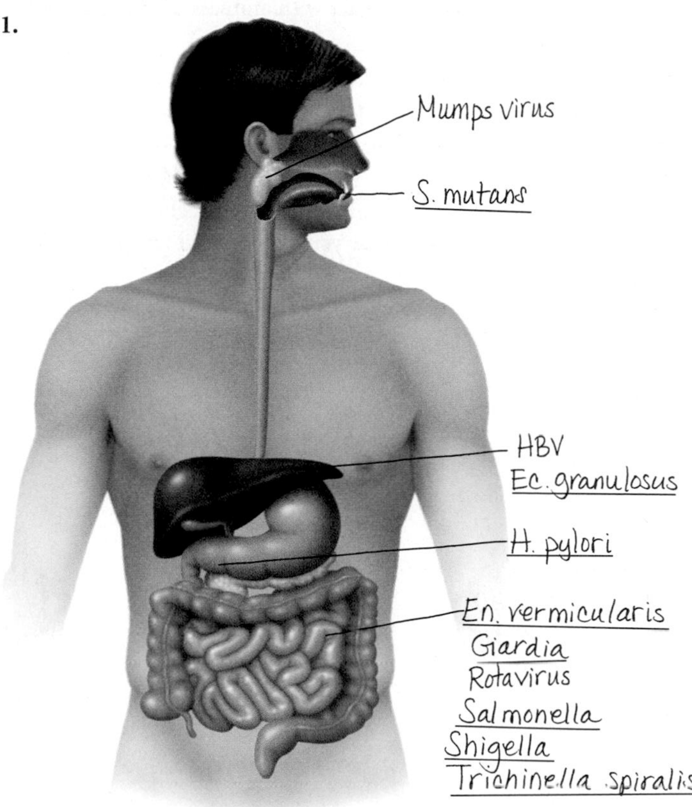

2.

Disease	Suspect Foods	Symptoms
Staph. food poisoning	Not cooked prior to eating	Vomiting and diarrhea
Shigellosis	Contaminated water	Mucus and blood in stools
Salmonellosis	Poultry; contaminated water	Fever, vomiting, and diarrhea
Cholera	Contaminated water	Rice water stools
Traveler's diarrhea	Contaminated water	Vomiting and diarrhea

Refer to Diseases in Focus 25.2 to complete the table.

3.

Causative Agent	Suspect Foods	Prevention
V. parahaemolyticus	Oysters, shrimp	Cooking
V. vulnificus	Contact with coastal waters	
Enterotoxigenic *E. coli*	Water, vegetables	Cooking
Enteroinvasive *E. coli*	Water, vegetables	Cooking
Enterohemorrhagic *E. coli*	Alfalfa sprouts, tomatoes	Cooking
C. jejuni	Chicken	Cooking
Y. enterocolitica	Meat, milk	Cooking
C. perfringens	Meat	Refrigeration after cooking
B. cereus	Rice dishes	Refrigeration after cooking

Refer to Diseases in Focus 25.2 to complete the table.

4. Certain strains of *E. coli* may produce an enterotoxin or invade the epithelium of the large intestine.
5. Toxin produced by fungi; see pp. 729–730.
6. All four are caused by protozoa. The infections are acquired by ingesting protozoa in contaminated water. Giardiasis is a prolonged diarrhea. Amoebic dysentery is the most severe dysentery, with blood and mucus in the stools. *Cryptosporidium* and *Cyclospora* produce severe diseases in persons with immune deficiencies.
7. **Food intoxication:** Microorganisms must be allowed to grow in food from the time of preparation to the time of ingestion. This usually occurs when foods are stored unrefrigerated or improperly canned. The etiologic agents (*Staphylococcus aureus* or *Clostridium botulinum*) must produce an exotoxin. Onset: 1 to 48 hours. Duration: A few days. Treatment: Antimicrobial agents are ineffective. The patient's symptoms may be treated.

   Food infection: Viable microorganisms must be ingested with food or water. The organisms could be present during preparation and survive cooking or be inoculated during later handling. The etiologic agents are usually gram-negative organisms (*Salmonella, Shigella, Vibrio,* and *Escherichia*) that produce endotoxins. *Clostridium perfringens* is a gram-positive bacterium that causes food infection. Onset: 12 hours to 2 weeks. Duration: Longer than intoxication because the microorganisms are growing in the patient. Treatment: Rehydration.

8.

Disease	Site	Symptoms
Mumps	Parotid glands	Inflammation of the parotid glands and fever
Hepatitis A	Liver	Anorexia, fever, diarrhea
Hepatitis B	Liver	Anorexia, fever, joint pains, jaundice
Viral gastroenteritis	Lower GI tract	Nausea, diarrhea, vomiting

Refer to Diseases in Focus 25.3 and 25.4 to complete this question.

9. Cook meat thoroughly. Eliminate the source of contamination to cattle and pigs.

Multiple Choice

1. d
2. e
3. e
4. c
5. e
6. b
7. b
8. e
9. a
10. d

Chapter 26

Review

1.

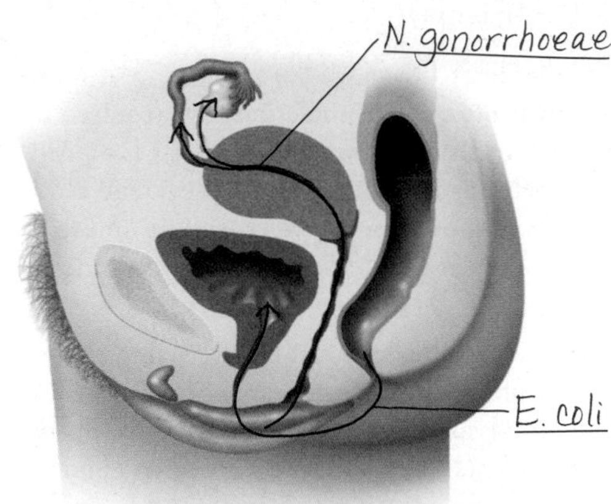

2. Urinary tract infections may be transmitted by improper personal hygiene and contamination during medical procedures. They are often caused by opportunistic pathogens.
3. The proximity of the anus to the urethra and the relatively short length of the urethra can allow contamination of the urinary bladder in females. Predisposing factors for cystitis in females are gastrointestinal infections and vaginal and urinary tract infections.
4. *Escherichia coli* causes about 75% of the cases. Portals of entry are from the lower urinary tract or systemic infections.
5.

Disease	Symptoms	Diagnosis
Bacterial vaginosis	Fishy odor	Odor, pH, clue cells
Gonorrhea	Painful urination	Isolation of *Neisseria*
Syphilis	Chancre	FTA-ABS
PID	Abdominal pain	Culture of pathogen
NGU	Urethritis	Absence of *Neisseria*
LGV	Lesion, lymph node enlargement	Observation of *Chlamydia* in cells
Chancroid	Swollen ulcer	Isolation of *Haemophilus*

Refer to Diseases in Focus 26.2 and 26.3 to complete table.

6. Transmission—Water; enters via wounds.
   Activities—Water contact; contact with animals or rodent-infested places.
   Etiology—*Leptospira interrogans.*
7. Symptoms—Burning sensation, vesicles, painful urination.
   Etiology—Herpes simplex virus type 2 (sometimes type 1). When the lesions are not present, the virus is latent and noncommunicable.
8. *Candida albicans*—Severe itching; thick, yellow, cheesy discharge.
   Trichomonas vaginalis—Profuse yellow discharge with disagreeable odor.
9.

Disease	Prevention of Congenital Disease
Gonorrhea	Treatment of newborn's eyes
Syphilis	Prevention and treatment of mother's disease
NGU	Treatment of newborn's eyes
Genital herpes	Cesarean delivery during active infection

Multiple Choice

1. b
2. e
3. a
4. c
5. d
6. c
7. c
8. e
9. b
10. a

Chapter 27

Review

1. The koala should have an organ housing a large population of cellulose-degrading microorganisms.
2. *Penicillium* might make penicillin to reduce competition from faster-growing bacteria.
3. **a.** Amino acids
   b. SO_4^{2-}
   c. plants and bacteria
   d. H_2S
   e. carbohydrates
   f. S^0
4. Phosphorus must be available for all organisms.
5.

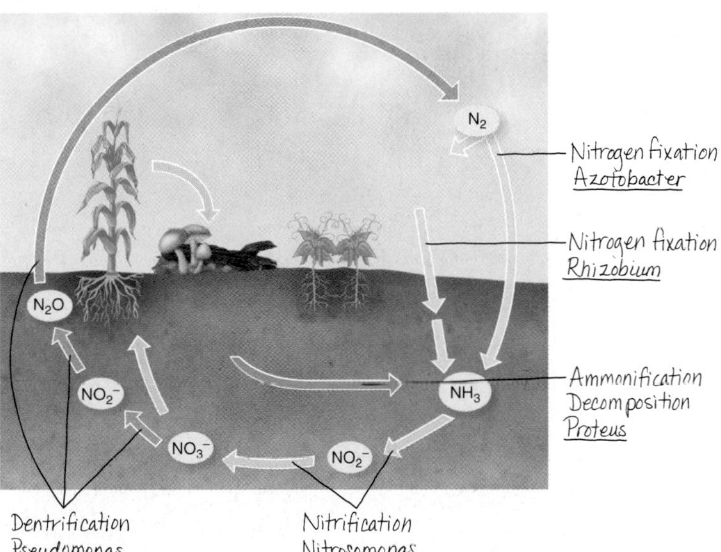

6. Cyanobacteria: With fungi, cyanobacteria act as the photoautotrophic partner in a lichen; they may also fix nitrogen in the lichen. With *Azolla*, they fix nitrogen.
   Mycorrhizae: Fungi that grow in and on the roots of higher plants; increase absorption of nutrients.
   Rhizobium: In root nodules of legumes; fix nitrogen.
   Frankia: In root nodules of alders, roses, and other plants; fix nitrogen.
7. Settling
   Flocculation treatment
   Sand filtration (or activated charcoal filtration)
   Chlorination
   The amount of treatment prior to chlorination depends on the amount of inorganic and organic matter in the water.
8. **a.** 2 **b.** 1 **c.** 2 **d.** 2 **e.** 3 **f.** 2 **g.** 3
9. Biodegradation of sewage, herbicides, oil, or PCBs.

Multiple Choice

1. a	**6.** c
2. b	**7.** b
3. b	**8.** b
4. b	**9.** e
5. c	**10.** c

Chapter 28

Review

1. Industrial microbiology is the science of using microorganisms to produce products or accomplish a process. Industrial microbiology provides (1) chemicals such as antibiotics that would not otherwise be available, (2) processes to remove or detoxify pollutants, (3) fermented foods that have desirable flavors or enhanced shelf life, and (4) enzymes for manufacturing a variety of goods.
2. The goal of commercial sterilization is to eliminate spoilage and disease-causing organisms. The goal of hospital sterilization is complete sterilization.
3. The acid in the berries will prevent the growth of some microbes.
4. Milk  Curd + Whey

   ↓ ↓
   Cheese Waste

   Hard cheese is ripened by lactic acid bacteria growing anaerobically in the interior of the curd. Soft cheese is ripened by molds growing aerobically on the outside of the curd.
5. Nutrients must be dissolved in water; water is also needed for hydrolysis. Malt is the carbon and energy source that the yeast will ferment to make alcohol. Malt contains glucose and maltose from the action of amylases on starch in seeds (barley).
6. A bioreactor provides the following advantages over simple flask containers:
   - Larger culture volumes can be grown.
   - Process instrumentation for monitoring and controlling critical environmental conditions such as pH, temperature, dissolved oxygen, and aeration can be used.
   - Sterilization and cleaning systems are designed in place.
   - It offers aseptic sampling and harvest systems for in-process sampling.
   - Improved aeration and mixing characteristics result in improved cell growth and high final cell densities.
   - A high degree of automation is possible.
   - Process reproducibility is improved.

7. (1) Enzymes don't produce hazardous wastes. (2) Enzymes work under reasonable conditions, e.g., they don't require high temperatures or acidity. (3) Use of enzymes eliminates the need to use petroleum in chemical syntheses of solvents such as alcohol and acetone. (4) Enzymes are biodegradable. (5) Enzymes are not toxic.
8. The production of ethyl alcohol from corn; or methane from sewage. Alcohols and hydrogen are produced by fermentation; methane is produced by anaerobic respiration.
9.

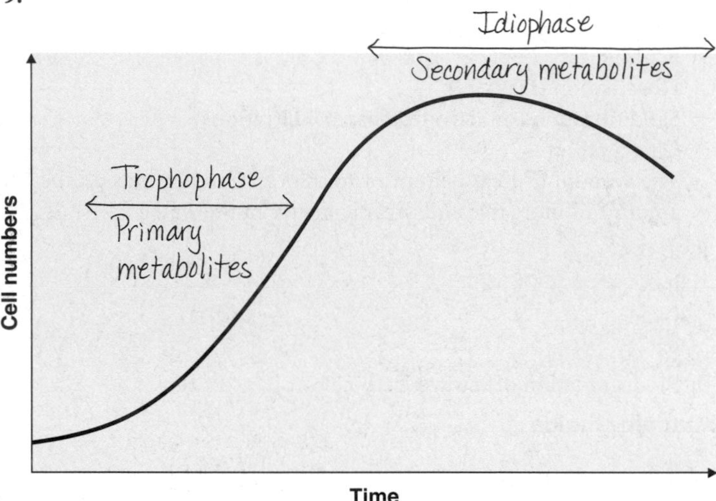

Multiple Choice

1. c
2. b
3. e
4. c
5. b
6. c
7. e
8. a
9. b
10. a

APPENDIX A

Metabolic Pathways

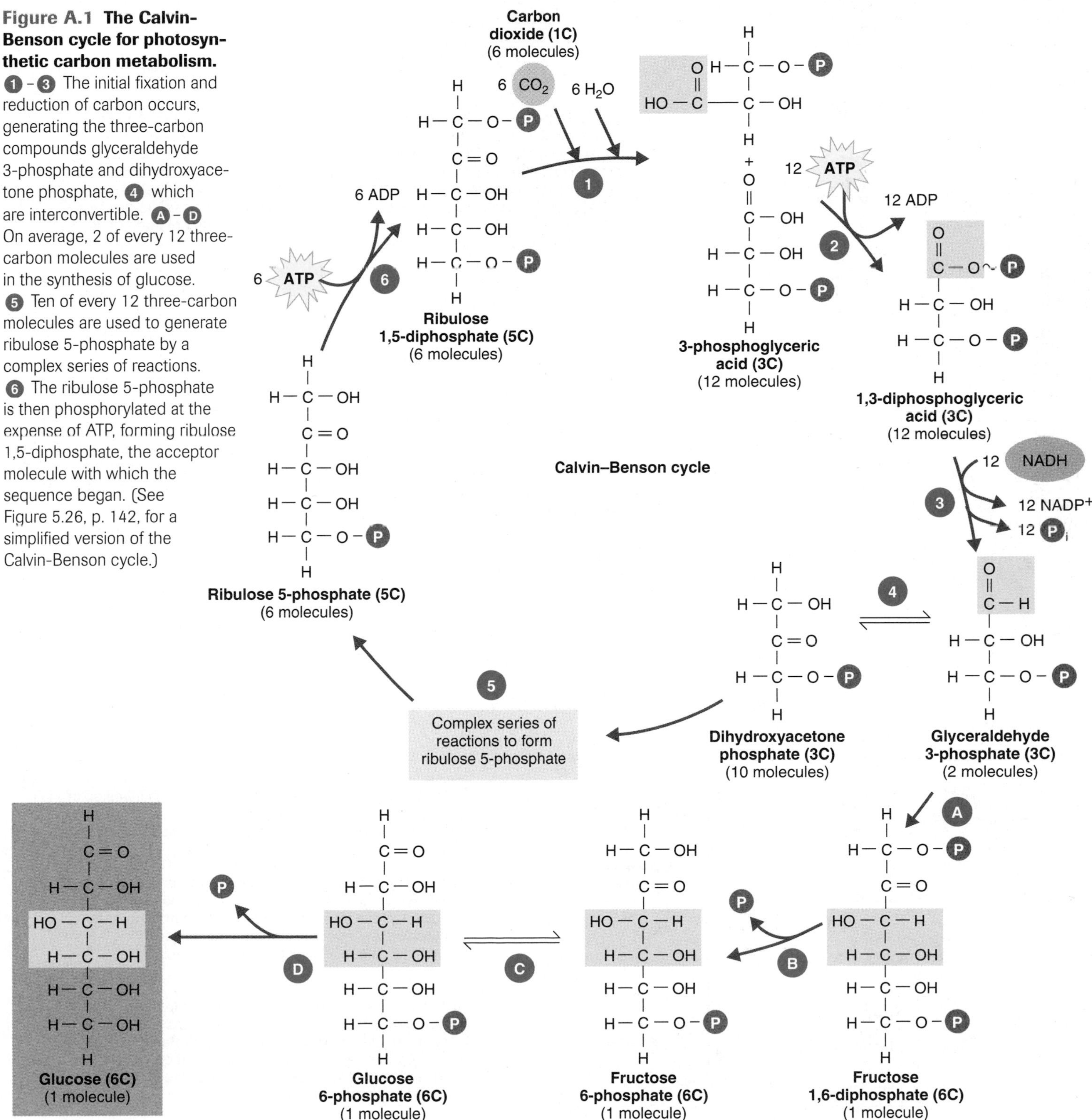

Figure A.1 The Calvin-Benson cycle for photosynthetic carbon metabolism. (1)-(3) The initial fixation and reduction of carbon occurs, generating the three-carbon compounds glyceraldehyde 3-phosphate and dihydroxyacetone phosphate, (4) which are interconvertible. (A)-(D) On average, 2 of every 12 three-carbon molecules are used in the synthesis of glucose. (5) Ten of every 12 three-carbon molecules are used to generate ribulose 5-phosphate by a complex series of reactions. (6) The ribulose 5-phosphate is then phosphorylated at the expense of ATP, forming ribulose 1,5-diphosphate, the acceptor molecule with which the sequence began. (See Figure 5.26, p. 142, for a simplified version of the Calvin-Benson cycle.)

Figure A.2 Glycolysis (Embden-Meyerhof pathway). Each of the ten steps of glycolysis is catalyzed by a specific enzyme, which is named under each step number. (See Figure 5.12, p. 126, for a simplified version of glycolysis.)

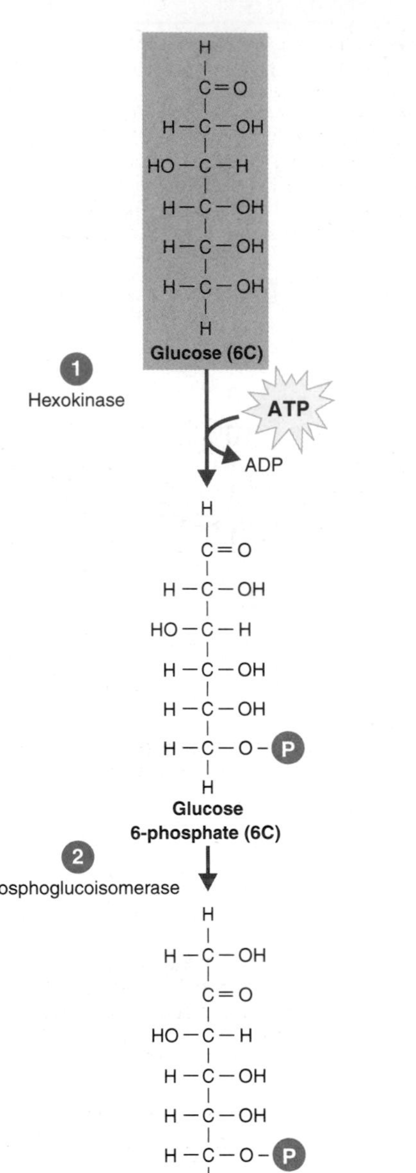

1 Glucose enters the cell and is phosphorylated by the enzyme hexokinase, which transfers a phosphate group from ATP to the number 6 carbon of the sugar. The product of the reaction is glucose 6-phosphate. The electrical charge of the phosphate group traps the sugar in the cell because of the impermeability of the plasma membrane to ions. Phosphorylation of glucose also makes the molecule more chemically reactive. Although glycolysis is supposed to *produce* ATP, in step 1, ATP is actually consumed—an energy investment that will be repaid with dividends later in glycolysis.

2 Glucose 6-phosphate is rearranged to convert it to its isomer, fructose 6-phosphate. Isomers have the same number and types of atoms but in different structural arrangements.

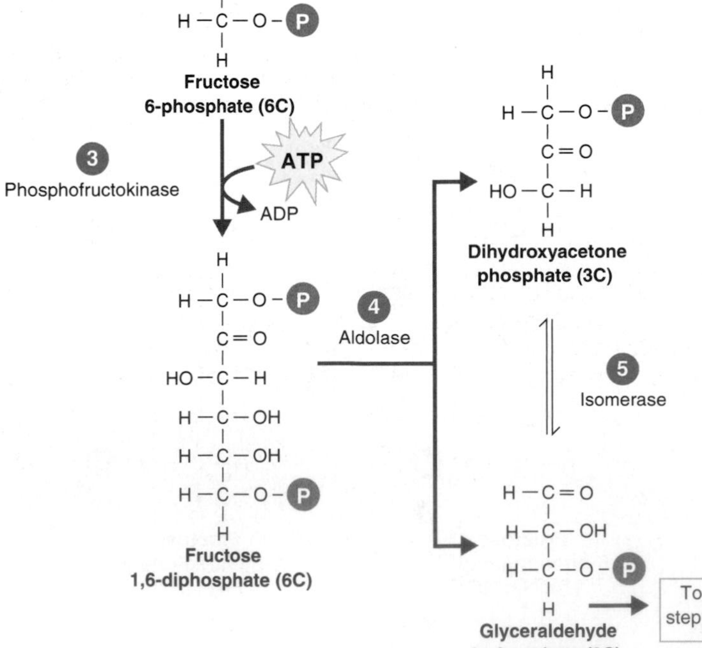

3 In this step, still another molecule of ATP is invested in glycolysis. An enzyme transfers a phosphate group from ATP to the sugar, producing fructose 1,6-diphosphate.

4 This is the reaction from which glycolysis gets its name ("sugar splitting"). An enzyme cleaves fructose 1,6-diphosphate into two different three-carbon sugars: glyceraldehyde 3-phosphate and dihydroxyacetone phosphate. These two sugars are isomers.

5 The enzyme isomerase interconverts the three-carbon sugars. The next enzyme in glycolysis uses only glyceraldehyde 3-phosphate as its substrate. This pulls the equilibrium between the two three-carbon sugars in the direction of glyceraldehyde 3-phosphate, which is removed as fast as it forms.

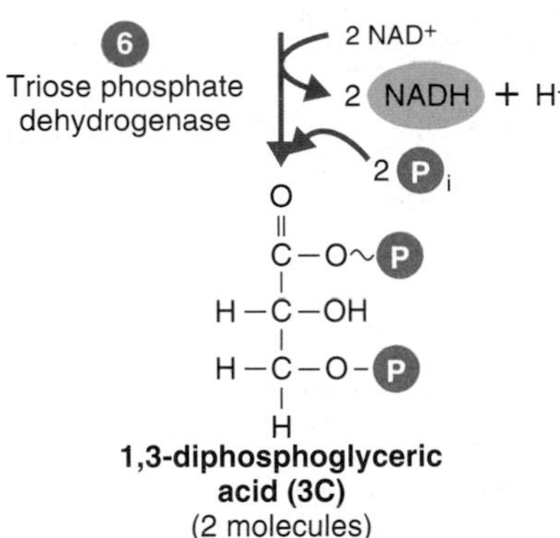

⑥ An enzyme now catalyzes two sequential reactions while it holds glyceraldehyde 3-phosphate in its active site. First, the sugar is oxidized at the number 1 carbon and NAD^+ is reduced, resulting in the formation of NADH + H^+. Second, the enzyme couples this reaction to the creation of a high-energy phosphate bond at the number 1 carbon of the oxidized substrate. The source of the phosphate is inorganic phosphate, which is always present in the cell. As products, the enzyme releases NADH + H^+ and 1,3-diphosphoglyceric acid. Notice in the figure that the new phosphate bond is symbolized with a high-energy bond (~), which indicates that the bond is at least as energetic as the phosphate bonds of ATP.

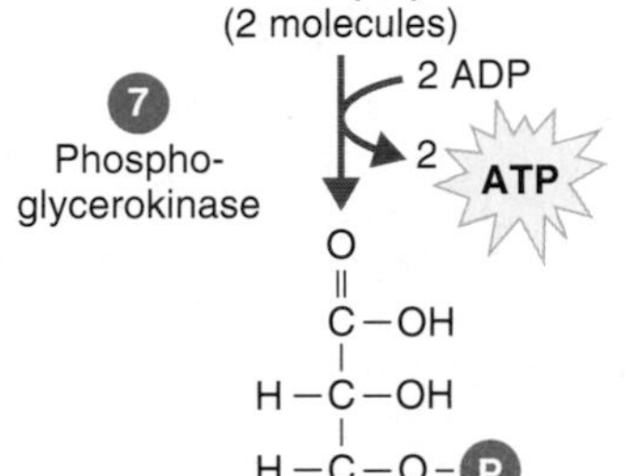

⑦ At this step, glycolysis produces ATP. The phosphate group, with its high-energy bond, is transferred from 1,3-diphosphoglyceric acid to ADP. For each glucose molecule that began glycolysis, step ⑦ produces two molecules of ATP, because every product after the sugar-splitting step (step ④) is doubled. Of course, two ATPs were invested to get sugar ready for splitting. The ATP ledger now stands at zero. By the end of step ⑦, glucose has been converted to two molecules of 3-phosphoglyceric acid.

8
Phospho-glyceromutase

O
C–OH
H–C–O–P
H–C–OH
H

2-phosphoglyceric acid (3C)
(2 molecules)

⑧ Next, an enzyme relocates the remaining phosphate group of 3-phosphoglyceric acid to form 2-phosphoglyceric acid. This prepares the substrate for the next reaction.

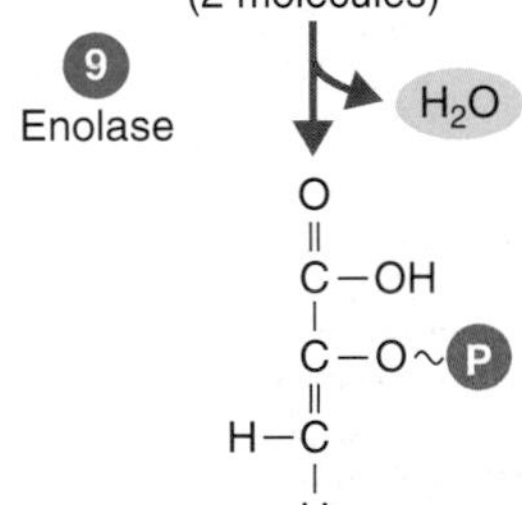

⑨ An enzyme forms a double bond in the substrate by extracting a water molecule from 2-phosphoglyceric acid to form phosphoenolpyruvic acid. This results in the electrons of the substrate being arranged in such a way that the remaining phosphate bond becomes very unstable.

⑩ The last reaction of glycolysis produces another molecule of ATP by transferring the phosphate group from phosphoenolpyruvic acid to ADP. Because this step occurs twice for each glucose molecule, the ATP ledger now shows a net gain of two ATPs. Thus, the glycolysis of one molecule of glucose results in two molecules of pyruvic acid, two molecules of NADH + H^+, and two molecules of ATP. Each molecule of pyruvic acid can now undergo respiration or fermentation.

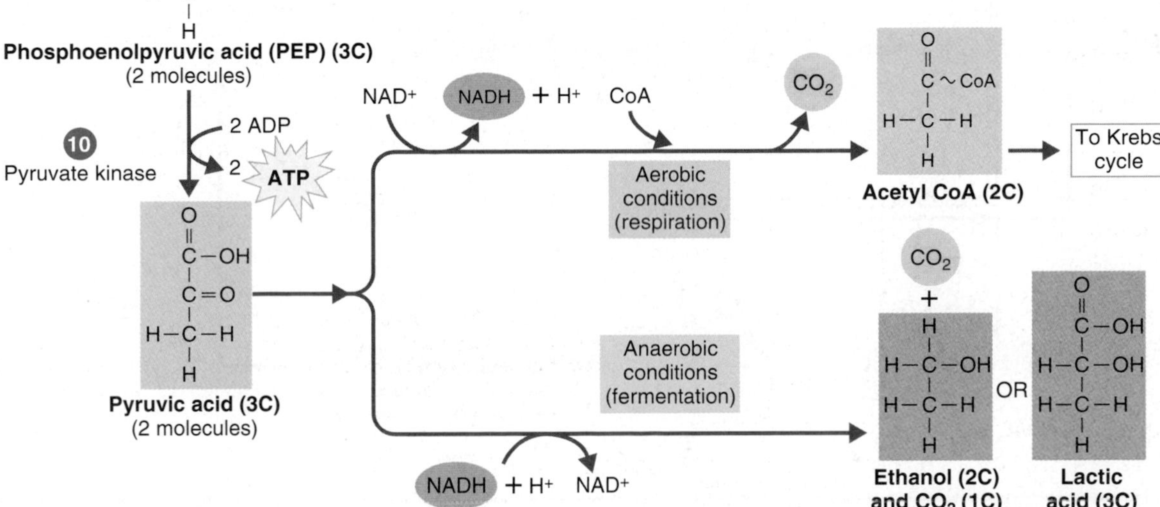

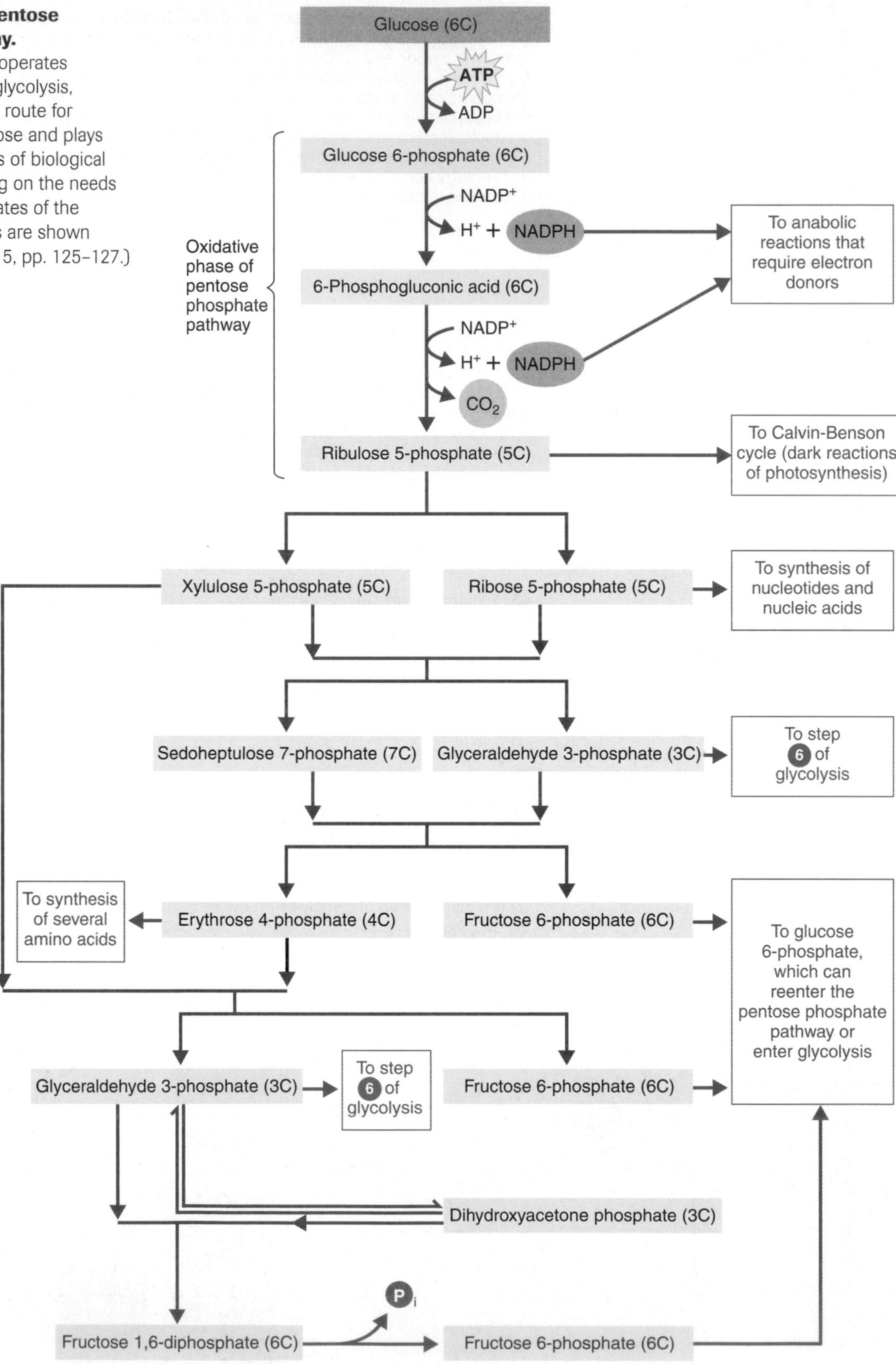

Figure A.3 The pentose phosphate pathway. This pathway, which operates simultaneously with glycolysis, provides an alternate route for the oxidation of glucose and plays a role in the synthesis of biological molecules, depending on the needs of the cell. Possible fates of the various intermediates are shown in blue. (See Chapter 5, pp. 125–127.)

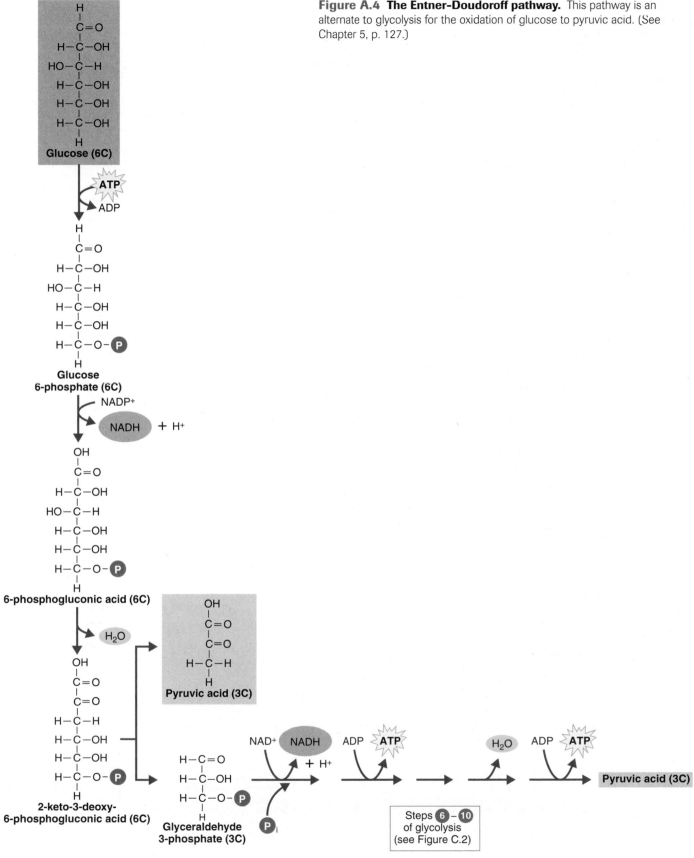

Figure A.4 The Entner-Doudoroff pathway. This pathway is an alternate to glycolysis for the oxidation of glucose to pyruvic acid. (See Chapter 5, p. 127.)

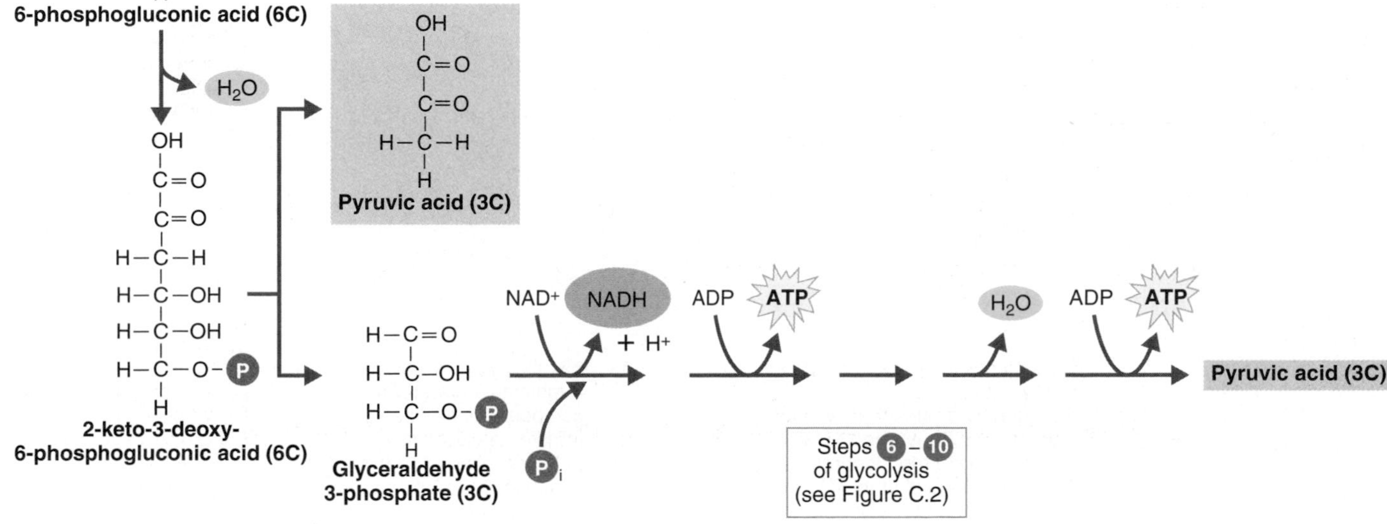

Figure A.5 The Krebs cycle. See Figure 5.13, p. 128, for a simplified version.

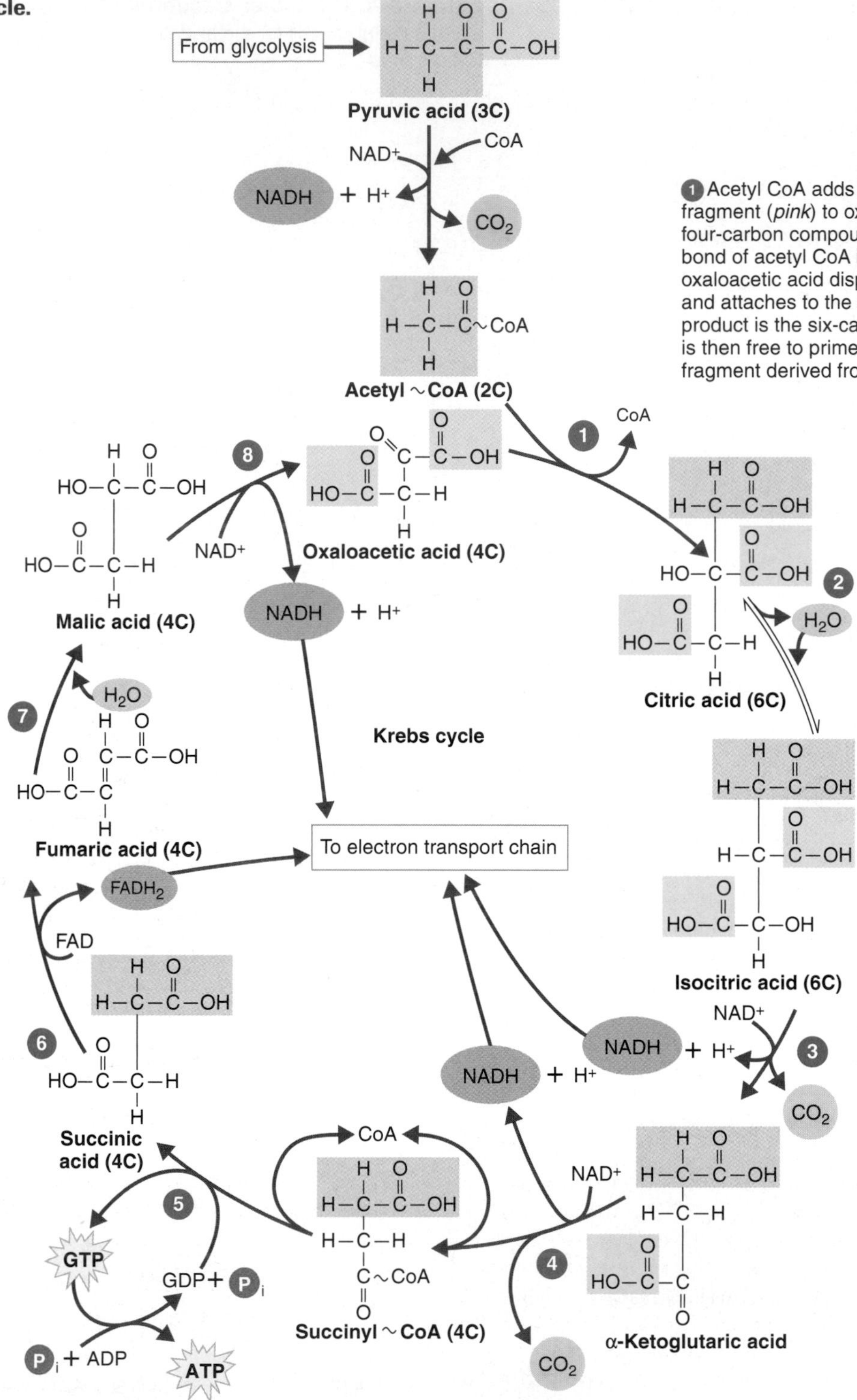

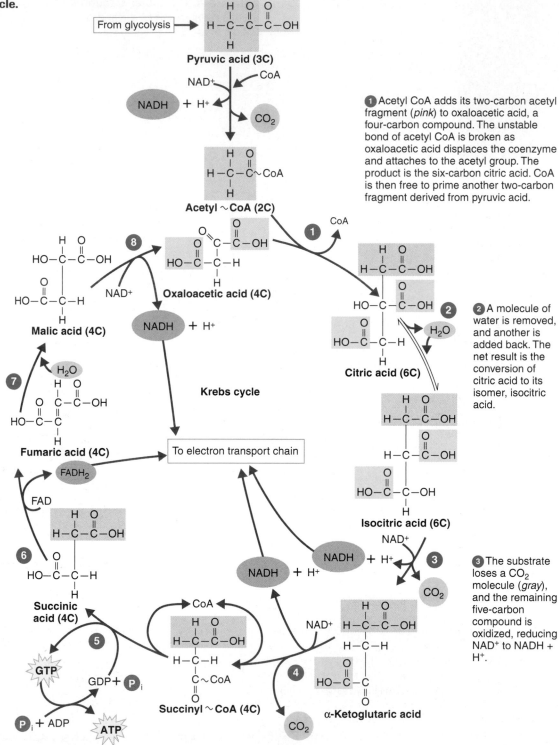

1 Acetyl CoA adds its two-carbon acetyl fragment (*pink*) to oxaloacetic acid, a four-carbon compound. The unstable bond of acetyl CoA is broken as oxaloacetic acid displaces the coenzyme and attaches to the acetyl group. The product is the six-carbon citric acid. CoA is then free to prime another two-carbon fragment derived from pyruvic acid.

2 A molecule of water is removed, and another is added back. The net result is the conversion of citric acid to its isomer, isocitric acid.

3 The substrate loses a CO_2 molecule (*gray*), and the remaining five-carbon compound is oxidized, reducing NAD^+ to NADH + H^+.

4 CO_2 (*gray*) is lost; the remaining four-carbon compound is oxidized by the transfer of electrons to NAD^+ to form NADH + H^+ and is then attached to CoA by an unstable bond.

5 Substrate-level phosphorylation occurs in this step. CoA is displaced by a phosphate group, which is then transferred to GDP to form guanosine triphosphate (GTP). GTP is similar to ATP, which is formed when GTP donates a phosphate group to ADP.

6 In another oxidative step, two hydrogens are transferred to FAD to form $FADH_2$. The function of this coenzyme is similar to that of NADH + H^+, but $FADH_2$ stores less energy.

7 Bonds in the substrate are rearranged in this step by the addition of a water molecule.

8 The last oxidative step produces another molecule of NADH + H^+ and regenerates oxaloacetic acid, which accepts a two-carbon fragment from acetyl CoA for another turn of the cycle.

APPENDIX B

Exponents, Exponential Notation, Logarithms, and Generation Time

Exponents and Exponential Notation

Very large and very small numbers, such as 4,650,000,000 and 0.00000032, are cumbersome to work with. It is more convenient to express such numbers in exponential notation—that is, as a power of 10. For example, 4.65×10^9 is in standard exponential notation, or **scientific notation**: 4.65 is the *coefficient*, and 9 is the power or *exponent*. In standard exponential notation, the coefficient is always a number between 1 and 10, and the exponent can be positive or negative.

To change a number into exponential notation, follow two steps. First, determine the coefficient by moving the decimal point so there is only one nonzero digit to the left of it. For example,

0.0000003 2

The coefficient is 3.2. Second, determine the exponent by counting the number of places you moved the decimal point. If you moved it to the left, the exponent is positive. If you moved it to the right, the exponent is negative. In the example, you moved the decimal point seven places to the right, so the exponent is −7. Thus

$$0.00000032 = 3.2 \times 10^{-7}$$

Now suppose you are working with a larger number instead of a very small number. The same rules apply, but the exponential value will be positive rather than negative. For example,

$$4{,}650{,}000{,}000 = 4.65 \times 10^9$$

To multiply numbers written in exponential notation, multiply the coefficients and *add* the exponents. For example,

$$(3 \times 10^4) \times (2 \times 10^3) = (3 \times 2) \times (10^{4+3}) = 6 \times 10^7$$

To divide, divide the coefficient and *subtract* the exponents. For example,

$$\frac{3 \times 10^4}{2 \times 10^3} = \frac{3}{2} \times 10^{4-3} = 1.5 \times 10^1$$

Microbiologists use exponential notation in many situations. For instance, exponential notation is used to describe the number of microorganisms in a population. Such numbers are often very large (see Chapter 6). Another application of exponential notation is to express concentrations of chemicals in a solution—chemicals such as media components (Chapter 6), disinfectants (Chapter 7), or antibiotics (Chapter 20). Such numbers are often very small. Converting from one unit of measurement to another in the metric system requires multiplying or dividing by a power of 10, which is easiest to carry out in exponential notation.

Logarithms

A **logarithm (log)** is the power to which a base number is raised to produce a given number. Usually we work with logarithms to the base 10, abbreviated $\log_{10}$. The first step in finding the $\log_{10}$ of a number is to write the number in standard exponential notation. If the coefficient is exactly 1, the $\log_{10}$ is simply equal to the exponent. For example

$$\log_{10} 0.00001 = \log_{10}(1 \times 10^{-5}) = -5$$

If the coefficient is not 1, as is often the case, the logarithm function on a calculator must be used to determine the logarithm.

Microbiologists use logs for calculating pH levels and for graphing the growth of microbial populations in culture (see Chapter 6).

Calculating Generation Time

As a cell divides, the population increases exponentially. Numerically this is equal to 2 (because one cell divides into two) raised to the number of times the cell divided (generations);

$$2^{\text{number of generations}}$$

To calculate the final concentration of cells:

$$\text{Initial number of cells} \times 2^{\text{number of generations}} = \text{Number of cells}$$

For example, if 5 cells were allowed to divide 9 times, this would result in

$$5 \times 2^9 = 2560 \text{ cells}$$

To calculate the number of generations a culture has undergone, cell numbers must be converted to logarithms. Standard logarithm values are based on 10. The log of 2 (0.301) is used because one cell divides into two.

$$\text{Number of generations} = \frac{\text{log number of cells (end)} - \text{log number of cells (beginning)}}{0.301}$$

To calculate the generation of time for a population:

$$\frac{60 \text{ min/hr} \times \text{hours}}{\text{number of generations}} = \text{minutes/generation}$$

As an example, we will calculate the generation time if 100 bacterial cells growing for 5 hours produced 1,720,320 cells:

$$\frac{\log 1{,}720{,}320 - \log 100}{0.301} = 14 \text{ generations}$$

$$\frac{60 \text{ min/hr} \times 5 \text{ hours}}{14 \text{ generations}} = 21 \text{ minutes/generation}$$

A practical application of the calculation is determining the effect of a newly developed food preservative on the culture. Suppose 900 of the same species were grown under the same conditions as the previous example, except that the preservative was added to the culture medium. After 15 hours, there were 3,276,800 cells. Calculate the generation time, and decide whether the preservative inhibited growth.

Answer: 75 minutes/generation. The preservative did inhibit growth.

APPENDIX C

Methods for Taking Clinical Samples

To diagnose a disease, it is often necessary to obtain a sample of material that may contain the pathogenic microorganism. Samples must be taken aseptically. The sample container should be labeled with the patient's name, room number (if hospitalized), date, time, and medications being taken. Samples must be transported to the laboratory immediately for culture. Delay in transport may result in the growth of some organisms, and their toxic products may kill other organisms. Pathogens tend to be fastidious and die if not kept in optimum environmental conditions.

In the laboratory, samples from infected tissues are cultured on differential and selective media in an attempt to isolate and identify any pathogens or organisms that are not normally found in association with that tissue.

Appendices

Universal Precautions*

The following procedures should be used by all health care workers, including students, whose activities involve contact with patients or with blood or other body fluids. These procedures were developed to minimize the risk of transmitting HIV or AIDS in a health care environment, but adherence to these guidelines will minimize the transmission of *all* nosocomial infections.

1. Gloves should be worn when touching blood and body fluids, mucous membranes, and nonintact skin and when handling items or surfaces soiled with blood or body fluids. Gloves should be changed after contact with each patient.
2. Hands and other skin surfaces should be washed immediately and thoroughly if contaminated with blood or other body fluids. Hands should be washed immediately after gloves are removed.
3. Masks and protective eyewear or face shields should be worn during procedures that are likely to generate droplets of blood or other body fluids.
4. Gowns or aprons should be worn during procedures that are likely to generate splashes of blood or other body fluids.
5. To prevent needlestick injuries, needles should not be recapped, purposely bent or broken, or otherwise manipulated by hand. After disposable syringes and needles, scalpel blades, and other sharp items are used, they should be placed in puncture-resistant containers for disposal.
6. Although saliva has not been implicated in HIV transmission, mouthpieces, resuscitation bags, and other ventilation devices should be available for use in areas in which the need for resuscitation is predictable. Emergency mouth-to-mouth resuscitation should be minimized.
7. Health care workers who have exudative lesions or weeping dermatitis should refrain from all direct patient care and from handling patient-care equipment.
8. Pregnant health care workers are not known to have a greater risk of contracting HIV infection than health care workers who are not pregnant; however, if a health care worker develops HIV infection during pregnancy, the infant is at risk of infection. Because of this risk, pregnant health care workers should be especially familiar with, and strictly adhere to, precautions to minimize the risk of HIV transmission.

Instructions for Specific Sampling Procedures

Wound or Abscess Culture

1. Cleanse the area with a sterile swab moistened in sterile saline.
2. Disinfect the area with 70% ethanol or iodine solution.
3. If the abscess has not ruptured spontaneously, a physician will open it with a sterile scalpel.
4. Wipe the first pus away.
5. Touch a sterile swab to the pus, taking care not to contaminate the surrounding tissue.
6. Replace the swab in its container, and properly label the container.

Ear Culture

1. Clean the skin and auditory canal with 1% tincture of iodine.
2. Touch the infected area with a sterile cotton swab.
3. Replace the swab in its container.

Eye Culture

This procedure is often performed by an ophthalmologist.

1. Anesthetize the eye with topical application of a sterile anesthetic solution.
2. Wash the eye with sterile saline solution.
3. Collect material from the infected area with a sterile cotton swab. Return the swab to its container.

Blood Culture

1. Close the room's windows to avoid contamination.
2. Clean the skin around the selected vein with 2% tincture of iodine on a cotton swab.
3. Remove dried iodine with gauze moistened with 80% isopropyl alcohol.
4. Draw a few milliliters of venous blood.
5. Aseptically bandage the puncture.

Urine Culture

1. Provide the patient with a sterile container.
2. Instruct the patient to first void a small volume from the urinary bladder before collection (to wash away extraneous bacteria of the skin microbiota) then to collect a midstream sample.
3. A urine sample may be stored under refrigeration (4°–6°C) for up to 24 hours.

Fecal Culture

For bacteriological examination, only a small sample is needed. This may be obtained by inserting a sterile swab into the rectum or feces. The swab is then placed in a tube of sterile enrichment broth for transport to the laboratory. For examination for parasites, a small sample may be taken from a morning stool. The sample is placed in a preservative (polyvinyl alcohol, buffered glycerol, saline, or formalin) for microscopic examination for eggs and adult parasites.

Sputum Culture

1. A morning sample is best because microorganisms will have accumulated while the patient is sleeping.
2. The patient should rinse his or her mouth thoroughly to remove food and normal microbiota.
3. The patient should cough deeply from the lungs and expectorate into a sterile glass wide-mouth jar.
4. Care should be taken to avoid contaminating health care workers.
5. In cases such as tuberculosis in which there is little sputum, stomach aspiration may be necessary.
6. Infants and children tend to swallow sputum. A fecal sample may be of some value in these cases.

**Source:* Centers for Disease Control and Prevention and National Institutes of Health. *Biosafety in Microbiological and Biomedical Laboratories.* Available from www.cdc.gov.

APPENDIX D

Pronunciation of Scientific Names

Rules of Pronunciation

The easiest way to learn new material is to talk about it, and that requires saying scientific names. Scientific names may look difficult at first glance, but keep in mind that generally every *syllable* is pronounced. The primary requirement in saying a scientific name is to communicate it.

The rules for the pronunciation of scientific names depend, in part, on the derivation of the root word and its vowel sounds. We have provided some general guidelines here. Pronunciations frequently do not follow the rules because a common usage has become "accepted" or the derivation of the name cannot be determined. For many scientific names there are alternative correct pronunciations.

Vowels

Pronounce all the vowels in scientific names. Two vowels written together and pronounced as one sound are called a *diphthong* (for example, the *ou* in *sound*). A special comment is needed about the pronunciation of the vowel endings *-i* and *-ae:* There are two alternative ways to pronounce each of these. In this book, we usually give the pronunciation of a long *e* (*ē*) to the *-i* ending, and a long *i* (*ī*) to the *-ae* ending. However, the reverse pronunciations are also correct and in some cases are preferred. For example, *coli* is usually pronounced kō′lī.

Consonants

When *c* or *g* is followed by *ae, e, oe, i,* or *y,* it has a soft sound. When *c* or *g* is followed by *a, o, oi,* or *u*, it has a hard sound. When a double *c* is followed by *e, i,* or *y,* it is pronounced as *ks* (e.g., cocci).

Accent

The accented syllable is usually the next-to-last or third-to-last syllable.

1. The accent is on the next-to-last syllable:
   a. When the name contains only two syllables. Example: pes′tis.
   b. When the next-to-last syllable is a diphthong. Example: a-kan-thä-mē′bä.
   c. When the vowel of the next-to-last syllable is long. Example: tre-pō-nē′mä. The vowel in the next-to-last syllable is long in words ending in the following suffixes:

Suffix	Example
-ales	Orders such as Eubacteriales
-ina	*Sarcina*
-anus,-anum	*pasteurianum*
-uta	*diminuta*

   d. When the word ends in one of the following suffixes:

Suffix	Example
-atus,-atum	*caudatum*
-ella	*Salmonella*

2. The accent is on the third-to-last syllable in family names. Families end in *-aceae*, which is always pronounced *-ā′sē-ē*.

Pronunciation of Microorganisms in this Text

Pronunciation key:

a	hat	ē	see	o	hot	th	thin
ā	age	ė	term	ō	go	u	cup
ã	care	g	go	ô	order	ů	put
ä	father	i	sit	oi	oil	ü	rule
ch	child	ī	ice	ou	out	ū	use
e	let	ng	long	sh	she	zh	seizure

Acanthamoeba polyphaga a-kan-thä-mē′bä pol′if-ä-gä
Acetobacter a-sē′tō-bak-tėr
Acinetobacter baumanii a-si-ne′tō-bak-tėr bou′man-ē-ē
Actinomyces israelii ak-tin-ō-mī′sēs is-rā′lē-ē
Aedes aegypti ā′ē-dēz ē-jip′tē
A. albopictus al-bō-pik′tus
Aeromonas hydrophilia ãr′ō-mō-nas hī′dro-fil-ē-ä
Agrobacterium tumefaciens ag′rō-bak-ti′rē-um tü′me-fãsh-enz
Ajellomyces ä-jel-lō-mī′sēs
Alcaligenes al′kä-li-gen-ēs
Alexandrium äl-eg-zan′drē-um
Amanita phalloides am-an-ī′ta fal-loi′dēz
Amoeba proteus ä′mē-bä prō′tē-us
Anabaena azollae an-ä-bē′-nä ā′zō-lī
Anaplasma phagocytophilum an′ä-plaz-mä fäg′o sī-to-fil-um
Ancylostoma duodenale an-sil-os′tō-mä dü′o-den-al-ē
Anopheles an-of′e-lēz
Aquaspirillum ä-kwä-spī-ril′lum
A. serpens sėr′pens
Arcanobacterium phocae är′kā-no-bak-ti-rē-um fō′sī
Arthroderma är-thrō-dėr′mä
Ascaris lumbricoides as′kar-is lum-bri-koi′dēz
Ashbya gossypii ash′bē-ä gos-sip′ē-ē
Aspergillus flavus a-spėr-jil′lus flā′vus
A. niger nī′jėr
A. rouxii rō′ē-ē
Azolla ā-zō′lä
Azomonas ā-zō-mō′nas
Azospirillum ā-zō-spī′ril-lum
Azotobacter ä-zo′tō-bak-tėr
Babesia microti ba-bē′sē-ä mī-krō′tē
Bacillus amyloliquefaciens bä-sil′lus a′mil-ō-li-kwi-fa-shens
B. anthracis an-thrā′sis
B. cereus se′rē-us
B. circulans sėr′ku-lans
B. coagulans kō-ag′ū-lanz
B. licheniformis lī-ken-i-fôr′mis
B. sphaericus sfe′ri-kus
B. subtilis su′til-us
B. thuringiensis thůr-in-jē-en′sis
Bacteroides fragilis bak-tė-roi′dēz fra′jil-is
Balamuthia bal′am-üth-ē-ä
Balantidium coli bal-an-tid′ē-um kō′lī (or kō′lē)
Bartonella henselae bär′tō-nel-lä hen′sel-ī
Baylisascaris procyonis bā′lis-as-kar-is prō′sē-on-is
Bdellovibrio bacteriovorus del-lō-vib′rē-ō bak-tė-rē-o′vô-rus
Beauveria bō-vãr′ē-a
Beggiatoa alba bej′jē-ä-tō-ä al′bä
Beijerinckia bī-yė-rink′ē-ä
Bifidobacterium bī-fi-dō-bak-ti′rē-um
Blastomyces dermatitidis blas-tō-mī′sēz dėr-mä-tit′i-dis

Appendices

Blattabacterium blat-tä-bak-ti′rē-um
Bordetella pertussis bôr′de-tel-lä pėr-tus′sis
Borrelia burgdorferi bôr′-rel-ē-ä burg-dôr′fėr-ē
Bradyrhizobium brad-ē-rī-zō′bē-um
Brevibacterium bre′vē-bak-ti-rē′um
Brucella abortus brü′sel-lä ä-bôr′tus
B. melitensis me-li-ten′sis
B. suis sü′is
Burkholderia bėrk′hōld-ėr-ē-ä
B. cepacia se-pā′sē-ä
B. pseudomallei sū-dō-mal′le-ē
Byssochlamys fulva bis-sō-klam′is fül′vä
Campylobacter fetus kam′pi-lō-bak-tėr fē′tus
C. jejuni jē-jū′nē
Candida albicans kan′did-ä al′bi-kanz
C. glabrata gla′brä-tä
C. utilis ū′til-is
Canis familiaris kānis fa-mil′ē-ār-is
Carsonella rudii kar′son-el-lä ru′dē-ē
Caulobacter kô-lō-bak′tėr
Cephalosporium sef-ä-lō-spô′rē-um
Ceratocystis ulmi sē-rä-tō-sis′tis ul′mē
Chilomastix kē′lō-ma-sticks
Chlamydia trachomatis kla-mi′dē-ä trä-kō′mä-tis
Chlamydomonas klam-i-dō-mō′näs
Chlamydophila pneumoniae kla-mi-do′fil-ä nü-mō′nē-ī
C. psittaci sit′tä-sē
Chlorobium klō-rō′bē-um
Chloroflexus klô-rō-flex′us
Chromatium krō-mā′tē-um
Chrysops krī′sops
Citrobacter sit′rō-bak-tėr
Claviceps purpurea kla′vi-seps pür-pū-rē′ä
Clonorchis sinensis klo-nôr′kis si-nen′sis
Clostridium acetobutylicum klôs-tri′dē-um a-sē-tō-bū-til′li-kum
C. botulinum bo-tū-lī′num
C. butyricum bü-ti′ri-kum
C. difficile dif′fi-sil
C. pasteurianum pas-tyėr-ē-ā′num
C. perfringens pėr-frin′jens
C. sporogenes spô-rä′jen-ēz
C. tetani te′tan-ē
Coccidioides immitis kok-sid-ē-oi′dēz im′mi-tis
Coniothyrium minitans kon′ē-ō-ther-ē-um mi′ni-tanz
Corynebacterium diphtheriae kôr′ī-nē-bak-ti-rē-um dif-thi′rē-ī
C. glutamicum glü-tam′i-kum
C. xerosis ze-rō′sis
Coxiella burnetii käks′ē-el-lä bėr-ne′tē-ē
Crenarchaeota kren-ärk-e′ ō-tä
Crucibulum vulgare krü-si-bū′lum vul′gār-ē
Cryphonectria parasitica kri-fō-nek′trē-ä par-ä-si′ti-kä
Cryptococcus gattii krip′tō-kok-kus gat′tē-ē
C. grubii grub′ē-ē
C. neoformans nē-ō-fôr′manz
Cryptosporidium hominis krip′tō-spô-ri-dē-um ho′min-is
C. parvum pär′vum
Culex kū′leks
Culiseta kū-li′se-tä
Cyclospora cayetanensis sī′klō-spô-rä kī′ē-tan-en-sis
Cytophaga sī-täf′äg-ä
Dermacentor andersoni dėr-mä-sen′tôr an-dėr-sōn′ē
D. variabilis vār-ē-a′bil-is
Desulfotomaculum nigrificans dē′sul-fō-to-ma-kū-lum nī′gri-fi-kans
Desulfovibrio desulfuricans dē′sul-fō-vib-rē-ō dē-sul-fėr′i-kans
Dictyostelium dik′tē-ō-stel-ē-um
Diphyllobothrium latum dī-fil-lō-bo′thrē-um lā′tum
Dipylidium caninum dī′pil-i-dē-um kan-i-num
Dirofilaria immitis di′rō-fi-lār-ē-ä im′mi-tis
Dracunculus medininsis dra-kun′ku-lus med-in′in-sis
Echinococcus granulosus ē-kīn-ō-kok′kus gra-nū-lō′sus
Ectothiorhodospira mobilis ek′tō-thī-ō-rō-dō-spī-rä mō′bil-is
Ehrlichia chafeensis ėr′lik-ē-ä chaf′ē-en-sis
Entamoeba coli en-tä-mē′bä kō′lē
E. dispar dis′par
E. histolytica his-tō-li′ti-kä
Enterobacter aerogenes en-te-rō-bak′tėr ā-rä′jen-ēz
E. cloacae klō-ā′kē
Enterobius vermicularis en-te-rō′bē-us ver-mi-kū-lar′is
Enterococcus faecalis en-te-rō-kok′kus fē-kā′lis
E. faecium fē′sē-um
Entomophaga en′tō-mo-fäg-ä
Epidermophyton ep-i-dėr-mō-fī′ton
Epulopiscium fishelsoni ep′ū-lō-pis-sē-um fish′el-sō-nē
Erwinia ėr-wi′nē-ä
Erysipelothrix rhusiopathiae ār-i-si-pel′ō-thrix rus-ē-ō-path′ē-ī
Escherichia coli esh-ė-rik′ē-ä kō′lī (or kō′lē)
Eucalyptus ū′kal-ip-tus
Euglena ū-glē′nä
Filobasidiella fi-lō-ba-si-dē-el′lä
Fonsecaea pedrosoi fon-se′kē-ä pe-drō′sō-ē
Francisella tularensis fran′sis-el-lä tü′lä-ren-sis
Frankia frank′ē-ä
Fusarium fu′sār-ē-um
Fusobacterium fū-sō-bak-ti′rē-um
Gambierdiscus toxicus gam′bē-ėr-dis-kus toks′i-kus
Gardnerella vaginalis gärd-nė-rel′lä va-jin-al′is
Gemmata obscuriglobus jem′mä-tä ob′skėr-ē-glob-us
Geobacillus stearothermophilus gē′ō-bä-sil-lus ste-rō-thėr-mä′fil-us
Giardia duodenalis jē-är′dē-ä dü′ō-den-al-is
G. intestinalis in′tes-tin-al-is
G. lamblia lam′lē-ä
Glaucocystis nostochinearum glou′ko-sis-tis no′stok-in-ē-ār-um
Gloeocapsa glē-ō-kap′sä
Glossina gläs-sē′nä
Gluconacetobacter xylinus glü′kon-a-sē-tō-bak-tėr zy′lin-us
Gluconobacter glü′kon-ō-bak-tėr
Gracilaria gra′sil-ār-ē-ä
Gymnoascus jim-nō-as′kus
Gymnodynium breve jim-nō-din′ē-um brev′ē
Haemophilus aegyptius hē-mä′fil-us e′jip-tē-us
H. ducreyi dü-krā′ē
H. influenzae in-flü-en′zī
Haloarcula hā′lō-är-kū-lä
Halobacterium ha-lō-bak-ti′rē-um
Helicobacter pylori hē′lik-ō-bak-tėr pī′lô-rē
Histoplasma capsulatum his-tō-plaz′mä kap-su-lä′tum
Homo sapiens hō′mō sā′pē-ens
Hydrogenomonas hī-drō-je-nō-mō′näs
Hyphomicrobium hī-fō-mī-krō′bē-um
Isospora ī-so′spô-rä
Isthmia nervosa isth′mē-ä nėr′vō-sä
Ixodes scapularis iks-ō′dēs skap-ū-lār′is
I. pacificus pas-i′fi-kus
Klebsiella pneumoniae kleb-sē-el′lä nü-mō′nē-ī
Lactobacillus acidophilus lak-tō-bä-sil′lus a′sid-o-fil-us
L. brevis brev′is
L. delbrueckii bulgaricus del-brük′ē-ē bul′gā-ri-kus
L. leichmannii līk-man′nē-ē
L. plantarum plan-tä′rum
L. sanfranciscensis san-fran-si′sken-sis
Lactococcus lak-tō-kok′kus
Laminaria japonica lam′i-när-e-ä ja-pon′i-kä
Legionella pneumophila lē-jä-nel′lä nü-mō′fi-lä
Leishmania braziliensis lish′mā-nē-ä brä-sil′ē-en-sis
L. donovani don′ō-van-ē
L. tropica trop′i-kä
Leptospira interrogans lep-tō-spī′rä in-tėr′rä-ganz
Leuconostoc mesenteroides lü-kō-nos′tok mes-en-ter-oi′dēz
Limulus polyphemus lim′ū-lus pol-if′i-mus
Listeria monocytogenes lis-te′rē-ä mo-nō-sī-tô′je-nēz

Macrocystis porifera ma'krō-sis-tis pōr'i-fė-rä
Magnetospirillum magnetotacticum mag-nē-tō-spī-ril-lum mag-ne-tō-tak'ti-kum
Malassezia furfur mal'as-sēz-ē-a fur'fur
Mannheimia haemolytica man-hī'me-ä hē'mō-li-ti-kä
Metarrhizium me'tär-rī-zē-um
Methylophilus methylotrophus meth-i-lo'fi-lus meth-i-lō-trōf'us
Microcladia mī-krō-klād'ē-ä
Micrococcus luteus mī-krō-kok'kus lü'tē-us
Micromonospora purpurea mī-krō-mo-nä'spô-rä pėr-pū-rē'ä
Microsporum g mī-krō-spô'rum ji
Mixotricha mix-ō-trik'ä
Moraxella catarrhalis mô-raks-el'lä ka-tär'al-is
M. lacunata la-kü-nä'tä
Mucor indicus mū'kôr in'di-kus
Mycobacterium abscessus mī-kō-bak-ti'rē-um ab'ses-sus
M. avium-intracellulare ā'vē-um-in'trä-cel-ū-lä-rē
M. bovis bō'vis
M. leprae lep'rī
M. smegmatis smeg-ma'tis
M. tuberculosis tü-bėr-kū-lō'sis
M. ulcerans ul'sėr-anz
Mycoplasma hominis mī-kō-plaz'mä ho'mi-nis
M. pneumoniae nu-mō'nē-ī
Myxococcus fulvus micks-ō-kok'kus ful'vus
M. xanthus zan'thus
Naegleria fowleri nī-gle'rē-ä fou'lėr-ē
Necator americanus ne-kā'tôr ä-me-ri-ka'nus
Neisseria gonorrhoeae nī-se'rē-ä go-nôr-rē'ī
N. meningitidis me-nin-ji'ti-dis
Nitrobacter nī-trō-bak'tėr
Nitrosomonas nī-trō-sō-mō'näs
Nocardia asteroides nō-kär'dē-ä as-tėr-oi'dēz
Nosema locustae nō'sē-mä lō'kus-tē
Oocystis ō-ō-sis'tis
Ornithodorus ôr-nith-ō'dô-rus
Paecilomyces fumosoroseus pī'sil-ō-mī-cēs fū'mō-sō-rō-sē-us
Paenibacillus polymixa pi'nē-bä-sil-lus po-lē-miks'ä
Pantoea agglomerans pan'tō-ē-ā äg'glom-ér-anz
Paracoccus denitrificans pār-ä-kok'kus dē-nī-tri'fi-kanz
Paragonimus westermani pār-ä-gōn'e-mus we-stėr-ma'nē
Paramecium pār-ä-mē'sē-um
Pasteurella multocida pas-tyėr-el'lä mul-tō'si-dä
Pediculus humanus capitis ped-ik'ū-lus hü'ma-nus kap'-i tis
P. humanus corporis hü'ma-nus kôr'pô-ris
Pediococcus pe-dē-ō-kok'kus
Pelagibacter ubique pel-aj'ē-bak-tėr ū'bēk
Penicillium chrysogenum pen-i-sil'lē-um krī-so'jen-um
P. griseofulvum gri-sē-ō-fůl'vum
P. notatum nō'ta-tum
Peridinium per-i-din'ē-um
Petriellidium pet-rē-el-li'dē-um
Pfiesteria fē'ster-ē-ä
Phlebotomus fle'-bo-to-mus
Photobacterium fō'tō-bak-ti-rē-um
Physarum fī'sār-um
Phytophthora cinnamoni fī-tof'thô-rä cin'nä-mō-nē
P. infestans in-fes'tans
P. ramorum ra'môr-um
Pityrosporum ovale pit-i-ros'pô-rum ovale
Plasmodium falciparum plaz-mō'dē-um fal-sip'är-um
P. malariae mä-lā'rē-ī
P. ovale ō-vä'lē
P. vivax vī'vaks
Plesiomonas shigelloides ple-sē-ō-mō'nas shi-gel-loi'des
Pneumocystis jirovecii nü-mō-sis'tis ye-rō'vet-zē-ē
Porphyromonas pôr'fī-rō-mō-nas
Prevotella intermedia prev'ō-tel-la in'tėr-mē-dē-ä
Prochlorococcus prō-klôr'ō-kok-kus
Propionibacterium acnes prō-pē-on'ē-bak-ti-rē-um ak'nēz
P. freudenreichii froi-den-rīk'ē-ē
Proteus mirabilis prō'tē-us mi-ra'bi-lis
Pseudomonas aeruginosa sū-dō-mō'nas ā-rü-ji-nō'sä
P. carboxydohydrogena kär'boks-i-dō-hī-drō-je-nä
P. fluorescens flôr-es'ens
P. putida pü'tē-dä
P. syringae sėr-in'jī
Pyridictum abyssi pir-i'dik-tum a-bis'sē
Quercus kwer'kus
Ralstonia räl'stō-nē-ä
Rhizobium rī-zō'bē-um
R. meliloti mel-i-lot'ē
Rhizopus stolonifer rī'zō-půs stō'lon-i-fėr
Rhodococcus bronchialis rō-dō-kok'kus bron-kē'al-is
R. erythropolis er-i-throp'ō-lis
Rhodopseudomonas rō-dō-su-dō-mō'nas
Rhodospirillum rubrum rō-dō-spī-ril'um rūb'rum
Rickettsia prowazekii ri-ket'sē-ä prou-wä-ze'kē-ē
R. rickettsii ri-ket'sē-ē
R. typhi tī'fē
Rosa pratincula rō-sä pra'tin-kü-lä
Saccharomyces sak-ä-rō-mī'sēs
S. carlsbergensis kärls'bėrg-en-sis
S. cerevisiae ellipsoides se-ri-vis'ē-ī ē'lip-soi-dez
S. exiguus egz-ij'ū-us
Salmonella bongori sal'mön-el-lä bon'gôr-ē
S. choleraesuis kol-ėr-ä-sü'is
S. enterica en-ter'i-kä
S. enteritidis en-tėr-ī'ti-dis
S. typhi tī'fē
S. typhimurium tī-fi-mėr'ē-um
Saprolegnia ferax sa'-prō-leg-nē-ä fe'raks
Sarcoptes scabiei sär-kop'tēs skā'bē-ē
Sargassum sär-gas'sum
Schistosoma haemotobium shis-tō-sō'mä (or skis-tō-sō'mä) hē'mō-tō-bē-um
Schizosaccharomyces skiz-ō-sak-ä-rō-mī'sēs
Serratia marcescens ser-rä'tē-ä mär-ses'sens
Shigella boydii shi-gel'lä boi'dē-ē
S. dysenteriae dis-en-te'rē-ī
S. flexneri fleks'nėr-ē
S. sonnei sōn'ne-ē
Sphaerotilus natans sfe-rä'ti-lus nā'tans
Spirillum minus spī'ril-lum mī'nus
S. volutans vō'lū-tans
Spiroplasma spī-rō-plaz'mä
Spirulina spī-rü-lī'nä
Spongomorphora spon'jō-môr-fô-rä
Sporothrix schenkii spô-rō'thriks shen'kē-ē
Stachybotrys stak'ē-bo-tris
Staphylococcus aureus staf-i-lō-kok'kus ô'rē-us
S. epidermidis e-pi-der'mi-dis
Stella stel'lä
Streptobacillus moniliformis strep-tō-bä-sil'lus mon'il-i-fôr-mis
Streptococcus agalactiae strep-tō-kok'kus ā'gal-act-ē-ī
S. mutans mū'tans
S. pneumoniae nü-mō'nē-ī
S. pyogenes pī-äj'en-ēz
S. thermophilus thėr-mo'fil-us
Streptomyces aureofaciens strep-tō-mī'sēs ô-rē-ō-fa'si-ens
S. erythraeus ā-rith'rē-us
S. fradiae frā'dē-ī
S. griseus gri'sē-us
S. hygroscopius hī'grō-skō-pē-us
S. lactamdurans lak'tam-dėr-anz
S. nodosus nō-dō'sus
S. venezuelae ve-ne-zü-e'lī
Sulfolobus sul'fō-lō-bus
Synechococcus sin'ē-kō-kok-kus
Taenia saginata te'nē-ä sa-ji-nä'tä
T. solium sō'lē-um
Talaromyces macrosporus ta-lä-rō-mī'sēs ma'-krō-spôr-us

Taxomyces tacks′ō-mī-sēs
Tetrahymena tet-rä-hī′me-nä
Thermoactinomyces vulgaris thėr-mō-ak-tin-ō-mī′sēs vul-ga′ris
Thermoanaerobium thermosaccharolyticum thėr′mō-an-e-rō-bē-um thėr-mō-sak-kär-ō-li′ti-kum
Thermoplasma thėr-mō-plaz′mä
Thermotoga maritima thėr′mō-tō-gä mar-it′ē-mä
Thermus aquaticus thėr′mus ä′kwä-ti-kus
Thiobacillus ferrooxidans thī-ō-bä-sil′lus fer-rō-oks′i-danz
T. thiooxidans thī-ō-oks′i-danz
Thiocapsa floridana thī-ō-kap′sä flôr′i-dä-nä
Thiomargarita namibiensis thī′ō-mär-gär-ē-tä na′mi-be-en-sis
Toxoplasma gondii toks-ō-plaz′mä gon′dē-ē
Treponema pallidum tre-pō-nē′mä pal′li-dum
Triatoma trī-ä-tō′ma
Tribonema vulgare trī′bō-nē-mä vul′gār-ē
Triceratium latus trī-ser′ä-te-um lā-tus
Trichinella nativa trik-in-el′lä na′tē-vä
T. spiralis spī-ra′lis
Trichoderma viride trik′ō-dėr-mä vir′-i-dä
Trichodesmium trik′-ō-des-mē-um
Trichomonas vaginalis trik-ō-mōn′as va-jin-al′is
Trichonympha sphaerica trik-ō-nimf′ä sfe′ri-kä
Trichophyton trik-ō-fī′ton
Trichosporon trik-ō-spôr′on
Trichuris trichiura trik′ėr-is trik-ē-yėr′a
Tridacna trī-dak′nä
Tropheryma whipplei trō-fer-ē′mä whip′plē-ī
Trypanosoma brucei gambiense tri-pa′nō-sō-mä brüs′ē gam-bē-ens′
T. brucei rhodesiense rō-dē-sē-ens′
T. cruzi kruz′ē
Ulva ul′vä
Ureaplasma urealyticum ū-rē-ä-plaz′mä ū-rē-ä-lit′i-kum
Usnea üs′nē-ä
Veillonella vī-lo-nel′lä
Vibrio cholerae vib′rē-ō kol′ėr-ī
V. fischeri fish′ėr-ē
V. parahaemolyticus pa-rä-hē-mō-li′ti-kus
V. vulnificus vul′ni-fi-kus
Vorticella vôr′ti-sel-lä
Wolbachia wol-ba′kē-ä
Wuchereria bancrofti vū-kėr-ār′ē-ä ban-krof′tē
Xanthomonas campestris zan′thō-mō-nas kam′pe-stris
Xenopsylla cheopis ze-nop′sil-lä kē-ō′pis
Yersinia enterocolitica yėr-sin′ē-ä en′tėr-ō-kōl-it-ik-ä
Y. pestis pes′tis
Y. pseudotuberculosis sū-dō-tü-bėr-kū-lō′sis
Zoogloea zō′ō-glē-ä

APPENDIX E

Word Roots Used in Microbiology

The Latin rules of grammar pertain to singular and plural forms of scientific names.

	Gender		
	Feminine	**Masculine**	**Neuter**
Singular	-a	-us	-um
Plural	-ae	-i	-a
Examples	alga, algae	fungus, fungi	bacterium, bacteria

a-, an- absence, lack. Examples: abiotic, in the absence of life; anaerobic, in the absence of air.
-able able to, capable of. Example: viable, having the ability to live or exist.
actino- ray. Example: actinomycetes, bacteria that form star-shaped (with rays) colonies.
aer- air. Examples: aerobic, in the presence of air; aerate, to add air.
albo- white. Example: *Streptomyces albus* produces white colonies.
ameb- change. Example: ameboid, movement involving changing shapes.
amphi- around. Example: amphitrichous, tufts of flagella at both ends of a cell.
amyl- starch. Example: amylase, an enzyme that degrades starch.
ana- up. Example: anabolism, building up.
ant-, anti- opposed to, preventing. Example: antimicrobial, a substance that prevents microbial growth.
archae- ancient. Example: archaeobacteria, "ancient" bacteria, thought to be like the first form of life.
asco- bag. Example: ascus, a baglike structure holding spores.
aur- gold. Example: *Staphylococcus aureus,* gold-pigmented colonies.
aut-, auto- self. Example: autotroph, self-feeder.
bacillo- a little stick. Example: bacillus, rod-shaped.
basid- base, pedestal. Example: basidium, a cell that bears spores.
bdell- leech. Example: *Bdellovibrio,* a predatory bacterium.
bio- life. Example: biology, the study of life and living organisms.
blast- bud. Example: blastospore, spores formed by budding.
bovi- cattle. Example: *Mycobacterium bovis,* a bacterium found in cattle.
brevi- short. Example: *Lactobacillus brevis,* a bacterium with short cells.
butyr- butter. Example: butyric acid, formed in butter, responsible for rancid odor.
campylo- curved. Example: *Campylobacter,* curved rod.
carcin- cancer. Example: carcinogen, a cancer-causing agent.
caseo- cheese. Example: caseous, cheeselike.
caul- a stalk. Example: *Caulobacter,* appendaged or stalked bacteria.
cerato- horn. Example: keratin, the horny substance making up skin and nails.
chlamydo- covering. Example: chlamydoconidia, conidia formed inside hypha.
chloro- green. Example: chlorophyll, green-pigmented molecule.
chrom- color. Examples: chromosome, readily stained structure; metachromatic, intracellular colored granules.
chryso- golden. Example: *Streptomyces chryseus,* golden colonies.
-cide killing. Example: bactericide, an agent that kills bacteria.
cili- eyelash. Example: cilia, a hairlike organelle.
cleisto- closed. Example: cleistothecium, completely closed ascus.
co-, con- together. Example: concentric, having a common center, together in the center.
cocci- a berry. Example: coccus, a spherical cell.
coeno- shared. Example: coenocyte, a cell with many nuclei not separated by septa.
col-, colo- colon. Examples: colon, large intestine; *Escherichia coli,* a bacterium found in the large intestine.
conidio- dust. Example, conidia, spores developed at the end of aerial hypha, never enclosed.
coryne- club. Example: *Corynebacterium,* club-shaped cells.
-cul small form. Example: particle, a small part.
-cut the skin. Example: Firmicutes, bacteria with a firm cell wall, gram-positive.
cyano- blue. Example: cyanobacteria, blue-green pigmented organisms.
cyst- bladder. Example: cystitis, inflammation of the urinary bladder.
cyt- cell. Example: cytology, the study of cells.
de- undoing, reversal, loss, removal. Example: deactivation, becoming inactive.
di-, diplo- twice, double. Example: diphlococci, pairs of cocci.
dia- through, between. Example: diaphragm, the wall through or between two areas.
dys- difficult, faulty, painful. Example: dysfunction, disturbed function.
ec-, ex-, ecto out, outside, away from. Example: excrete, to remove materials from the body.
en-, em- in, inside. Example: encysted, enclosed in a cyst.
entero- intestine. Example: *Enterobacter,* a bacterium found in the intestine.
eo- dawn, early. Example: *Eobacterium,* a 3.4-billion-year-old fossilized bacterium.
epi- upon, over. Example: epidemic, number of cases of a disease over the normally expected number.
erythro- red. Example: erythema, redness of the skin.
eu- well, proper. Example: eukaryote, a proper cell.
exo- outside, outer layer. Example: exogenous, from outside the body.
extra- outside, beyond. Example: extracellular, outside the cells of an organism.
firmi- strong. Example: *Bacillus firmus* forms resistant endospores.
flagell- a whip. Example: flagellum, a projection from a cell; in eukaryotic cells, it pulls cells in a whiplike fashion.
flav- yellow. Example: *Flavobacterium* cells produce yellow pigment.
fruct- fruit. Example: fructose, fruit sugar.
-fy to make. Example: magnify, to make larger.
galacto- milk. Example: galactose, monosaccharide from milk sugar.
gamet- to marry. Example: gamete, a reproductive cell.
gastr- stomach. Example: gastritis, inflammation of the stomach.
gel- to stiffen. Example: gel, a solidified colloid.
-gen an agent that initiates. Example: pathogen, any agent that produces disease.
-genesis formation. Example: pathogenesis, production of disease.
germ, germin- bud. Example: germ, part of an organism capable of developing.
-gony reproduction. Example: schizogony, multiple fission producing many new cells.
gracili- thin. Example: *Aquaspirillum gracile,* a thin cell.
halo- salt. Example: halophile, an organism that can live in high salt concentrations.
haplo- one, single. Example: haploid, half the number of chromosomes or one set.
hema-, hemato-, hemo- blood. Example: *Haemophilus,* a bacterium that requires nutrients from red blood cells.
hepat- liver. Example: hepatitis, inflammation of the liver.
herpes creeping. Example: herpes, or shingles, lesions appear to creep along the skin.
hetero- different, other. Example: heterotroph, obtains organic nutrients from other organisms; other feeder.
hist- tissue. Example: histology, the study of tissues.
hom-, homo- same. Example: homofermenter, an organism that produces only lactic acid from fermentation of a carbohydrate.
hydr-, hydro- water. Example: dehydration, loss of body water.

hyper- excess. Example: hypertonic, having a greater osmotic pressure in comparison with another.

hypo- below, deficient. Example: hypotonic, having a lesser osmotic pressure in comparison with another.

im- not, in. Example: impermeable, not permitting passage.

inter- between. Example: intercellular, between the cells.

intra- within, inside. Example: intracellular, inside the cell.

io- violet. Example: iodine, a chemical element that produces a violet vapor.

iso- equal, same. Example: isotonic, having the same osmotic pressure when compared with another.

-itis inflammation of. Example: colitis, inflammation of the large intestine.

-karyo, -caryo a nut. Example: eukaryote, a cell with a membrane-enclosed nucleus.

kin- movement. Example: streptokinase, an enzyme that lyses or moves fibrin.

lacti- milk. Example: lactose, the sugar in milk.

lepis- scaly. Example: leprosy, disease characterized by skin lesions.

lepto- thin. Example: *Leptospira,* thin spirochete.

leuko- whiteness. Example: leukocyte, a white blood cell.

lip-, lipo- fat, lipid. Example: lipase, an enzyme that breaks down fats.

-logy the study of. Example: pathology, the study of changes in structure and function brought on by disease.

lopho- tuft. Example: lophotrichous, having a group of flagella on one side of a cell.

luc-, luci- light. Example: luciferin, a substance in certain organisms that emits light when acted upon by the enzyme luciferase.

lute-, luteo- yellow. Example: *Micrococcus luteus,* yellow colonies.

-lysis loosening, to break down. Example: hydrolysis, chemical decomposition of a compound into other compounds as a result of taking up water.

macro- largeness. Example: macromolecules, large molecules.

mendosi- faculty. Example: mendosicutes, archaeobacteria lacking peptidoglycan.

meningo- membrane. Example: meningitis, inflammation of the membranes of the brain.

meso- middle. Example: mesophile, an organism whose optimum temperature is in the middle range.

meta- beyond, between, transition. Example: metabolism, chemical changes occurring within a living organism.

micro- smallness. Example: microscope, an instrument used to make small objects appear larger.

-mnesia memory. Examples: amnesia, loss of memory; anamnesia, return of memory.

molli- soft. Example: Mollicutes, a class of wall-less eubacteria.

-monas a unit. Example: *Methylomonas,* a unit (bacterium) that utilizes methane as its carbon source.

mono- singleness. Example: monotrichous, having one flagellum.

morpho- form. Example: morphology, the study of the form and structure of organisms.

multi- many. Example: multinuclear, having several nuclei.

mur- wall. Example: murein, a component of bacterial cell walls.

mus-, muri- mouse. Example: murine typhus, a form of typhus endemic in mice.

mut- to change. Example: mutation, a sudden change in characteristics.

myco-, -mycetoma, -myces a fungus. Example: *Saccharomyces,* sugar fungus, a genus of yeast.

myxo- slime, mucus. Example: Myxobacteriales, an order of slime-producing bacteria.

necro- a corpse. Example: necrosis, cell death or death of a portion of tissue.

-nema a thread. Example: *Treponema* has long, threadlike cells.

nigr- black. Example: *Aspergillus niger,* a fungus that produces black conidia.

ob- before, against. Example: obstruction, impeding or blocking up.

oculo- eye. Example: monocular, pertaining to one eye.

-oecium, -ecium a house. Examples: perithecium, an ascus with an opening that encloses spores; ecology, the study of the relationships among organisms and between an organism and its environment (household).

-oid like, resembling. Example: coccoid, resembling a coccus.

oligo- small, few. Example: oligiosaccharide, a carbohydrate composed of a few (7–10) monosaccharides.

-oma tumor. Example: lymphoma, a tumor of the lymphatic tissues.

-ont being, existing. Example: schizont, a cell existing as a result of schizogony.

ortho- straight, direct. Example: orthomyxovirus, a virus with a straight, tubular capsid.

-osis, -sis condition of. Examples: lysis, the condition of loosening; symbiosis, the condition of living together.

pan- all, universal. Example: pandemic, an epidemic affecting a large region.

para- beside, near. Example: parasite, an organism that "feeds beside" another.

peri- around. Example: peritrichous, projections from all sides.

phaeo- brown. Example: Phaeophyta, brown algae.

phago- eat. Example: phagocyte, a cell that engulfs and digests particles or cells.

philo-, -phil liking, preferring. Example: thermophile, an organism that prefers high temperatures.

-phore bears, carries. Example: conidiophore, a hypha that bears conidia.

-phyll leaf. Example: chlorophyll, the green pigment in leaves.

-phyte plant. Example: saprophyte, a plant that obtains nutrients from decomposing organic matter.

pil- a hair. Example: pilus, a hairlike projection from a cell.

plankto- wandering, roaming. Example: plankton, organisms drifting or wandering in water.

plast- formed. Example: plastid, a formed body within a cell.

-pnoea, -pnea breathing. Example: dyspnea, difficulty in breathing.

pod- foot. Example: pseudopod, a footlike structure.

poly- many. Example: polymorphism, many forms.

post- after, behind. Example: posterior, a place behind a (specific) part.

pre-, pro- before, ahead of. Examples: prokaryote, a cell with the first nucleus; pregnant, before birth.

pseudo- false. Example: pseudopod, false foot.

psychro- cold. Example: psychrophile, an organism that grows best at low temperatures.

-ptera wing. Example: Diptera, the order of true flies, insects with two wings.

pyo- pus. Example: pyogenic, pus-forming.

rhabdo- stick, rod. Example: rhabdovirus, an elongated, bullet-shaped virus.

rhin- nose. Example: rhinitis, inflammation of mucous membranes in the nose.

rhizo- root. Examples: *Rhizobium,* a bacterium that grows in plant roots; mycorrhiza, a fungus that grows in or on plant roots.

rhodo- red. Example: *Rhodospirillum,* a red-pigmented, spiral-shaped bacterium.

rod- gnaws. Example: rodents, the class of mammals with gnawing teeth.

rubri- red. Example: *Clostridiium rubrum,* red-pigmented colonies.

rumin- throat. Example: *Ruminococcus,* a bacterium associated with a rumen (modified esophagus).

saccharo- sugar. Example: disaccharide, a sugar consisting of two simple sugars.

sapr- rotten. Example: *Saprolegnia,* a fungus that lives on dead animals.

sarco- flesh. Example: sarcoma, a tumor of muscle or connective tissues.

schizo- split. Example: schizomycetes, organisms that reproduce by splitting and an early name for bacteria.

scolec- worm. Example: scolex, the head of a tapeworm.

-scope, -scopic watcher. Example: microscope, an instrument used to watch small things.

semi- half. Example: semicircular, having the form of half a circle.

sept- rotting. Example: septic, presence of bacteria that could cause decomposition.

septo- partition. Example: septum, a cross-wall in a fungal hypha.

serr- notched. Example: serrate, with a notched edge.

sidero- iron. Example: *Siderococcus,* a bacterium capable of oxidizing iron.

siphon- tube. Example: Siphonaptera, the order of fleas, insects with tubular mouths.

soma- body. Example: somatic cells, cells of the body other than gametes.

speci- particular things. Examples: species, the smallest group of organisms with similar properties; specify, to indicate exactly.

spiro- coil. Example: spirochete, a bacterium with a coiled cell.

sporo- spore. Example: sporangium, a structure that holds spores.

staphylo- grapelike cluster. Example: *Staphylococcus,* a bacterium that forms clusters of cells.

-stasis arrest, fixation. Example: bacteriostasis, cessation of bacterial growth.

strepto- twisted. Example: *Streptococcus,* a bacterium that forms twisted chains of cells.

sub- beneath, under. Example: subcutaneous, just under the skin.

super- above, upon. Example: superior, the quality or state of being above others.

sym-, syn- together, with. Examples: synapse, the region of communication between two neurons; synthesis, putting together.

-taxi to touch. Example: chemotaxis, response to the presence (touch) of chemicals.

taxis- orderly arrangement. Example: taxonomy, the science dealing with arranging organisms into groups.

tener- tender. Example: Tenericutes, the phylum containing wall-less eubacteria.

thallo- plant body. Example: thallus, an entire macroscopic fungus.

therm- heat. Example: *Thermus,* a bacterium that grows in hot springs (to 75°C).

thio- sulfur. Example: *Thiobacillus,* a bacterium capable of oxidizing sulfur-containing compounds.

-thrix See trich-.

-tome, -tomy to cut. Example: appendectomy, surgical removal of the appendix.

-tone, -tonic strength. Example: hypotonic, having less strength (osmotic pressure).

tox- poison. Example: antitoxin, effective against poison.

trans- across, through. Example: transport, movement of substances.

tri- three. Example: trimester, three-month period.

trich- a hair. Example: peritrichous, hairlike projections from cells.

-trope turning. Example: geotropic, turning toward the Earth (pull of gravity).

-troph food, nourishment. Example: trophic, pertaining to nutrition.

-ty condition of, state. Example: immunity, the condition of being resistant to disease or infection.

undul- wavy. Example: undulating, rising and falling, presenting a wavy appearance.

uni- one. Example: unicellular, pertaining to one cell.

vaccin- cow. Example: vaccination, injection of a vaccine (originally pertained to cows).

vacu- empty. Example: vacuoles, an intracellular space that appears empty.

vesic- bladder. Example: vesicle, a bubble.

vitr- glass. Example: in vitro, in culture media in a glass (or plastic) container.

-vorous eat. Example: carnivore, an animal that eats other animals.

xantho- yellow. Example: *Xanthomonas,* produces yellow colonies.

xeno- strange. Example: axenic, sterile, free of strange organisms.

xero- dry. Example: xerophyte, any plant that tolerates dry conditions.

xylo- wood. Example: xylose, a sugar obtained from wood.

zoo- animal. Example: zoology, the study of animals.

zygo- yoke, joining. Example: zygospore, a spore formed from the fusion of two cells.

-zyme ferment. Example: enzyme, any protein in living cells that catalyzes chemical reactions.

APPENDIX F

Classification of Bacteria According to *Bergey's Manual**

Domain: Bacteria
- **Phylum Aquificae**
  - Class I: Aquificae
    - Order I: Aquificales
      - Family I: Aquificaceae
        - *Aquifex*
        - *Calderobacterium*
        - *Hydrogenobacter*
        - *Hydrogenobaculum*
        - *Hydrogenothermus*
        - *Persephonella*
        - *Sulfurihydrogenibium*
        - *Thermocrinis*
      - Genera incertae sedis
        - *Balnearium*
        - *Desulfurobacterium*
        - *Thermovibrio*
- **Phylum Thermotogae**
  - Class I: Thermotogae
    - Order I: Thermotogales
      - Family I: Thermotogaceae
        - *Fervidobacterium*
        - *Geotoga*
        - *Marinitoga*
        - *Petrotoga*
        - *Thermosipho*
        - *Thermotoga*
- **Phylum Thermodesulfobacteria**
    - Order I: Thermodesulfobacteriales
      - Family I: Thermodesulfobacteriaceae
        - *Thermodesulfobacterium*
- **Phylum Deinococcus-Thermus**
  - Class I: Deinococci
    - Order I: Deinococcales
      - Family I: Deinococcaceae
        - *Deinococcus*
    - Order II: Thermales
      - Family I: Thermaceae
        - *Marinithermus*
        - *Meiothermus*
        - *Oceanithermus*
        - *Thermus*
        - *Vulcanithermus*
- **Phylum Chrysiogenetes**
  - Class I: Chrysiogenetes
    - Order I: Chrysiogenales
      - Family I: Chrysiogenaceae
        - *Chrysiogenes*
- **Phylum Chloroflexi**
  - Class I: Chloroflexi
    - Order I: Chloroflexales
      - Family I: Chloroflexaceae
        - *Chloroflexus*
        - *Chloronema*
        - *Heliothrix*
        - *Roseiflexus*
      - Family II: Oscillochloridaceae
        - *Oscillochloris*
    - Order II: Herpetosiphonales
      - Family I: Herpetosiphonaceae
        - *Herpetosiphon*
  - Class II: Anaerolineae
    - Order I: Anaerolinaeles
      - Family I: Anaerolinaceae
        - *Anaerolineae*
- **Phylum Thermomicrobia**
  - Class I: Thermomicrobia
    - Order I: Thermomicrobiales
      - Family I: Thermomicrobiaceae
        - *Thermomicrobium*
- **Phylum Nitrospira**
  - Class I: Nitrospira
    - Order I: Nitrospirales
      - Family I: Nitrospiraceae
        - *Leptospirillum*
        - *Magnetobacterium*
        - *Nitrospira*
        - *Thermodesulfovibrio*
- **Phylum Deferribacteres**
  - Class I: Deferribacteres
    - Order I: Deferribacterales
      - Family I: Deferribacteraceae
        - *Deferribacter*
        - *Denitrovibrio*
        - *Flexistipes*
        - *Geovibrio*
      - Genera incertae sedis
        - *Caldithrix*
        - *Synergistes*
- **Phylum Cyanobacteria**
  - Class I: Cyanobacteria
    - Subsection I
        - *Chamaesiphon*
        - *Chroococcus*
        - *Cyanobacterium*
        - *Cyanobium*
        - *Cyanothece*
        - *Dactylococcopsis*
        - *Gloeobacter*
        - *Gloeocapsa*
        - *Gloeothece*
        - *Microcystis*
        - *Prochlorococcus*
        - *Prochloron*
        - *Synechococcus*
        - *Synechocystis*
    - Subsection II
        - *Chroococcidiopsis*
        - *Cyanocystis*
        - *Dermocarpella*
        - *Myxosarcina*
        - *Pleurocapsa*
        - *Stanieria*
        - *Xenococcus*
    - Subsection III
        - *Arthrospira*
        - *Borzia*
        - *Crinalium*
        - *Geitlerinema*
        - *Halospirulina*
        - *Leptolyngbya*
        - *Limnothrix*
        - *Lyngbya*
        - *Microcoleus*
        - *Oscillatoria*
        - *Planktothrix*
        - *Prochlorothrix*
        - *Pseudoanabaena*
        - *Spirulina*
        - *Starria*
        - *Symploca*
        - *Trichodesmium*
        - *Tychonema*
    - Subsection IV
        - *Anabaena*
        - *Anabaenopsis*
        - *Aphanizomenon*
        - *Calothrix*
        - *Cyanospira*
        - *Cylindrospermopsis*
        - *Cylindrospermum*
        - *Nodularia*
        - *Nostoc*
        - *Rivularia*
        - *Scytonema*
        - *Tolypothrix*
    - Subsection V
        - *Chlorogloeopsis*
        - *Fischerella*
        - *Geitleria*
        - *Iyengariella*
        - *Nostochopsis*
        - *Stigonema*
- **Phylum Chlorobi**
  - Class I: Chlorobia
    - Order I: Chlorobiales
      - Family I: Chlorobiaceae
        - *Ancalochloris*
        - *Chlorobaculum*
        - *Chlorobium*
        - *Chloroherpeton*
        - *Pelodictyon*
        - *Prosthecochloris*
- **Phylum Proteobacteria**
  - Class I: Alphaproteobacteria
    - Order I: Rhodospirillales
      - Family I: Rhodospirillaceae
        - *Azospirillum*
        - *Inquilinus*
        - *Magnetospirillum*
        - *Phaeospirillum*
        - *Rhodocista*
        - *Rhodospira*
        - *Rhodospirillum*
        - *Rhodovibrio*
        - *Roseospira*
        - *Skermanella*
        - *Thallassospira*
        - *Tistrella*
      - Family II: Acetobacteraceae
        - *Acetobacter*
        - *Acidiphilium*
        - *Acidisphaera*

**Bergey's Manual of Systematic Bacteriology*, 2nd ed., 5 vols. (2004), is the reference for classification. *Bergey's Manual of Determinative Bacteriology*, 9th ed. (1994), should be used for identifying culturable bacteria and archaea.

Acidocella
Acidomonas
Asaia
Craurococcus
Gluconacetobacter
Gluconobacter
Kozakia
Muricoccus
Paracraurococcus
Rhodopila
Roseococcus
Rubritepida
Stella
Teichococcus
Zavarzinia
Order II: Rickettsiales
Family I: Rickettsiaceae
Orientia
Rickettsia
Family II: *Anaplasmataceae*
Aegyptianella
Anaplasma
Cowdria
Ehrlichia
Neorickettsia
Wolbachia
Xenohaliotis
Family III: Holosporaceae
Holospora
Genera incertae sedis
Caedibacter
Lyticum
Odyssella
Pseudocaedibacter
Symbiotes
Tectibacter
Order III: Rhodobacterales
Family I: Rhodobacteraceae
Ahrensia
Albidovulum
Amaricoccus
Antarctobacter
Gemmobacter
Hirschia
Hyphomonas
Jannaschia
Ketogulonicigenium
Leisingera
Maricaulis
Methylarcula
Oceanicaulis
Octadecabacter
Pannonibacter
Paracoccus
Pseudorhodobacter
Rhodobaca
Rhodobacter
Rhodothalassium
Rhodovulum
Roseibium
Roseinatronobacter
Roseivivax
Roseobacter
Roseovarius
Roseovivax
Rubrimonas
Ruegeria
Sagittula
Silicibacter
Staleya
Stappia
Sulfitobacter
Order IV: Sphingomonadales
Family I: Sphingomonodaceae
Blastomonas
Erythrobacter
Erythromicrobium
Erythromonas
Novosphingobium
Porphyrobacter
Rhizomonas
Sandaracinobacter
Sphingobium
Sphingomonas
Sphingopyxis
Zymomonas
Order V: Caulobacterales
Family I: Caulobacteraceae
Asticcacaulis
Brevundimonas
Caulobacter
Phenylobacterium
Order VI: Rhizobiales
Family I: Rhizobiaceae
Agrobacterium
Allorhizobium
Carbophilus
Chelatobacter
Ensifer
Rhizobium
Sinorhizobium
Family II: Aurantimonadaceae
Aurantimonas
Fulvimarina
Family III: Bartonellaceae
Bartonella
Family IV: Brucellaceae
Brucella
Mycoplana
Ochrobactrum
Family V: Phyllobacteriaceae
Aminobacter
Aquamicrobium
Defluvibacter
Mesorhizobium
Nitratireductor
Phyllobacterium
Paseuaminobacter
Family VI: Methylocystaceae
Albibacter
Methylocystis
Methylopila
Methylosinus
Terasakiella
Family VII: Beijerinckiaceae
Beijerinckia
Chelatococcus
Methylocapsa
Methylocella
Family VIII: Bradyrhizobiaceae
Afipia
Agromonas
Blastobacter
Bosea
Bradyrhizobium
Nitrobacter
Oligotropha
Rhodoblastus
Rhodopseudomonas
Family IX: Hyphomicrobiaceae
Ancalomicrobium
Ancylobacter
Angulomicrobium
Aquabacter
Azorhizobium
Blastochloris
Devosia
Dichotomicrobium
Filomicrobium
Gemmiger
Hyphomicrobium
Labrys
Methylorhabdus
Pedomicrobium
Prosthecomicrobium
Rhodomicrobium
Rhodoplanes
Seliberia
Starkeya
Xanthobacter
Family X: Methylobacteriaceae
Methylobacterium
Microvirga
Protomonas
Roseomonas
Family XI: Rhodobiaceae
Rhodobium
Roseospirillum
Order VII: Parvularculales
Family I: Parvularculaceae
Parvularcula
Class II: Betaproteobacteria
Order I: Burkholderiales
Family I: Burkholderiaceae
Burkholderia
Cupriavidus
Lautropia
Limnobacter
Pandoraea
Paucimonas
Polynucleobacter
Ralstonia
Thermothrix
Wautersia
Family II: Oxalobacteraceae
Duganella
Herbaspirillum
Janthinobacterium
Massilia
Oxalicibacterium
Oxalobacter
Telluria
Family III: Alcaligenaceae
Achromobacter
Alcaligenes
Bordetella
Brackiella
Oligella
Pelistega
Pigmentiphaga
Sutterella
Taylorella
Family IV: Comamonadaceae
Acidovorax
Alicycliphilus
Brachymonas
Caldimonas
Comamonas
Delftia
Diaphorobacter
Hydrogenophaga
Hylemonella
Lampropedia
Macromonas
Ottowia
Polaromonas

Ramlibacter
Rhodoferax
Variovorax
Xenophilus
Genera incertae sedis
Aquabacterium
Ideonella
Leptothrix
Roseateles
Rubrivivax
Schlegelella
Sphaerotilus
Tepidimonas
Thiomonas
Xylophilus
Order II: Hydrogenophilales
Family I: Hydrogenophilaceae
Hydrogenophilus
Thiobacillus
Order III: Methylophilales
Family I: Methylophilaceae
Methylobacillus
Methylophilus
Methylovorus
Order IV: Neisseriales
Family I: Neisseriaceae
Alysiella
Aquaspirillum
Chromobacterium
Eikenella
Formivibrio
Iodobacter
Kingella
Laribacter
Microvirgula
Morococcus
Neisseria
Prolinoborus
Simonsiella
Vitreoscilla
Vogesella
Order V: Nitrosomonadales
Family I: Nitrosomonadaceae
Nitrosolobus
Nitrosomonas
Nitrosospira
Family II: Spirillaceae
Spirillum
Family III: Gallionellaceae
Gallionella
Order VI: Rhodocyclales
Family I: Rhodocyclaceae
Azoarcus
Azonexus
Azospira
Azovibrio
Dechloromonas
Dechlorosoma
Ferribacterium
Propionibacter
Propionivibrio
Quadricoccus
Rhodocyclus
Sterolibacterium
Thauera
Zoogloea
Order VII: Procabacteriales
Family I: Procabacteriaceae
Procabacter
Class III: Gammaproteobacteria
Order I: Chromatiales
Family I: Chromatiaceae
Allochromatium
Amoebobacter
Chromatium
Halochromatium
Isochromatium
Lamprobacter
Lamprocystis
Marichromatium
Nitrosococcus
Pfennigia
Rhabdochromatium
Rheinheimera
Thermochromatium
Thioalkalicoccus
Thiobaca
Thiocapsa
Thiococcus
Thiocystis
Thiodictyon
Thioflavicoccus
Thiohalocapsa
Thiolamprovum
Thiopedia
Thiorhodococcus
Thiorhodovibrio
Thiospirillum
Family II: Ectothiorhodospiraceae
Alcalilimnicola
Alkalispirillum
Arhodomonas
Ectothiorhodospira
Halorhodospira
Nitrococcus
Thioalkalispira
Thioalkalivibrio
Thiorhodospira
Order II: Acidithiobacillales
Family 1: Acidithiobacillaceae
Acidithiobacillus
Family II: Thermithiobacillaceae
Thermithiobacillus
Order III: Xanthomonadales
Family I: Xanthomonadaceae
Frateuria
Fulvimonas
Luteimonas
Lysobacter
Nevskia
Pseudoxanthomonas
Rhodanobacer
Schineria
Stenotrophomonas
Thermomonas
Xanthomonas
Xylella
Order IV: Cardiobacteriales
Family I: Cardiobacteriaceae
Cardiobacterium
Dichelobacter
Suttonella
Order V: Thiotrichales
Family I: Thiotrichaceae
Achromatium
Beggiatoa
Leucothrix
Thiobacterium
Thiomargarita
Thioploca
Thiospira
Thiothrix
Family II: Francisellaceae
Francisella
Family III: Piscirickettsiaceae
Cycloclasticus
Hydrogenovibrio
Methylophaga
Piscirickettsia
Thioalkalimicrobium
Thiomicrospira
Order VI: Legionellales
Family I: Legionellaceae
Legionella
Family II: Coxiellaceae
Aquicella
Coxiella
Rickettsiella
Order VII: Methylococcales
Family I: Methylococcaceae
Methylobacter
Methylocaldum
Methylococcus
Methylomicrobium
Methylomonas
Methylosarcina
Methylosphaera
Order VIII: Oceanospirillales
Family I: Oceanospirillaceae
Balneatrix
Marinomonas
Marinospirillum
Neptunomonas
Oceanobacter
Oceanospirillum
Oleispira
Pseudospirillum
Thalassolituus
Family II: Alcanivoraceae
Alcanivorax
Fundibacter
Family III: Hahellaceae
Hahella
Zooshikella
Family IV: Halomonadaceae
Halomonas
Carnimonas
Chromohalobacter
Cobetia
Deleya
Zymobacter
Family V: Oleiphilaceae
Oleiphilus
Family VI: Saccharospirillaceae
Saccharospirillum
Order IX: Pseudomonadales
Family I: Pseudomonadaceae
Azomonas
Azotobacter
Cellvibrio
Chryseomonas
Flavimonas
Mesophilobacter
Pseudomonas
Rhizobacter
Rugamonas
Serpens
Family II: Moraxellaceae
Acinetobacter
Moraxella
Psychrobacter
Family III: Incertae sedis
Enhydrobacter
Order X: Alteromonadales
Family I: Alteromonadaceae
Aestuariibacter
Alishewanella
Alteromonas
Colwellia
Ferrimonas
Glaciecola
Idiomarina

Marinobacter
Marinobacterium
Microbulbifer
Moritella
Pseudoalteromonas
Psychromonas
Shewanella
Thalassomonas
Family II: Incerta sedis
Teredinibacter
Order XI: Vibrionales
Family I: Vibrionaceae
Allomonas
Catenococcus
Enterovibrio
Grimontia
Listonella
Photobacterium
Salinivibrio
Vibrio
Order XII: Aeromonadales
Family I: Aeromonadaceae
Aeromonas
Oceanimonas
Oceanisphaera
Tolumonas
Family II: Succinivibrionaceae
Anaerobiospirillum
Ruminobacter
Succinomonas
Succinivibrio
Order XIII: Enterobacteriales
Family I: Enterobacteriaceae
Alterococcus
Arsenophonus
Brenneria
Buchnera
Budvicia
Buttiauxella
Calymmatobacterium
Cedecea
Citrobacter
Edwardsiella
Enterobacter
Erwinia
Escherichia
Ewingella
Hafnia
Klebsiella
Kluyvera
Leclercia
Leminorella
Moellerella
Morganella
Obesumbacterium
Pantoea
Pectobacterium
Phlomobacter
Photorhabdus
Plesiomonas
Pragia
Proteus
Providencia
Rahnella
Raoultella
Saccharobacter
Salmonella
Samsonia
Serratia
Shigella
Sodalis
Tatumella
Trabulsiella
Wigglesworthia
Xenorhabdus
Yersinia
Yokenella
Order XIV: Pasteurellales
Family I: Pasteurellaceae
Actinobacillus
Gallibacterium
Haemophilus
Lonepinella
Pasteurella
Mannheimia
Phocoenobacter
Class IV: Deltaproteobacteria
Order I: Desulfurellales
Family I: Desulfurellaceae
Desulfurella
Hippea
Order II: Desulfovibrionales
Family I: Desulfovibrionaceae
Bilophila
Desulfovibrio
Lawsonia
Family II: Desulfomicrobiaceae
Desulfomicrobium
Family III: Desulfohalobiaceae
Desulfohalobium
Desulfomonaas
Desulfonatronovibrio
Desulfothermus
Family IV: Desulfonatronumaceae
Desulfonatronum
Order III: Desulfobacterales
Family I: Desulfobacteraceae
Desulfatibacillum
Desulfobacter
Desulfobacterium
Desulfobacula
Desulfobotulus
Desulfocella
Desulfococcus
Desulfofaba
Desulfofrigus
Desulfomusa
Desulfonema
Desulforegula
Desulfosarcina
Desulfospira
Desulfotignum
Family II: Desulfobulbaceae
Desulfobulbus
Desulfocapsa
Desulfofustis
Desulforhopalus
Desulfotalea
Family III: Nitrospinaceae
Nitrospina
Order IV: Desulfarcales
Family I. Desulfarculaceae
Desulfarculus
Order V: Desulfuromonales
Family I: Desulfuromonaceae
Desulfuromonas
Desulfuromusa
Malonomonas
Pelobacter
Family II: Geobacteraceae
Geobacter
Trichlorobacter
Order VI: Syntrophobacterales
Family I: Syntrophobacteraceae
Desulfacinum
Syntrophobacter
Desulforhabdus
Desulfovirga
Thermodesulforhabdus
Family II: Syntrophaceae
Desulfobacca
Smithella
Syntrophus
Order VII: Bdellovibrionales
Family I: Bdellovibrionaceae
Bacteriovorax
Bdellovibrio
Micavibrio
Vampirovibrio
Order VIII: Myxococcales
Family I: Cystobacteraceae
Anaeromyxobacter
Archangium
Cystobacter
Hyalangium
Melittangium
Stigmatella
Family II: Myxococcaceae
Corallococcus
Myxococcus
Pyxicoccus
Family III: Polyangiaceae
Byssophaga
Chondromyces
Haploangium
Jahnia
Polyangium
Sorangium
Family IV: Nannocystaceae
Nannocystis
Plesiocystis
Family V: Haliangiaceae
Haliangium
Family VI: Kofleriaceae
Kofleria
Class V: Epsilonproteobacteria
Order I: Campylobacterales
Family I: Campylobacteraceae
Arcobacter
Campylobacter
Dehalospirillum
Sulfurospirillum
Family II: Helicobacteraceae
Helicobacter
Sulfurimonas
Thiovulum
Wolinella
Phylum Firmicutes
Class I: Bacilli
Order I: Bacillales
Family I: Bacillaceae
Alkalibacillus
Amphibacillus
Anoxybacillus
Bacillus
Cerasibacillus
Filobacillus
Geobacillus
Gracilibacillus
Halobacillus
Haloactibacillus
Lentibacillus
Marinicoccus
Oceanobacillus
Paraliobacillus
Saccharococcus
Tenuibacillus
Thalassobacillus
Virgibacillus

Family II: Alicyclobacillaceae
Alicyclobacillus
Family III: Listeriaceae
Brochothrix
Listeria
Family IV: Paenibacillaceae
Ammoniphilus
Aneurinibacillus
Brevibacillus
Cohnella
Oxalophagus
Paenibacillus
Thermobacillus
Family VI: Planococcaceae
Caryophanon
Filibacter
Jeotgalibacillus
Kurthia
Marinibacillus
Planomicrobium
Planococcus
Planomicrobium
Sporosarcina
Ureibacillus
Family VII: Sporolactobacillaceae
Sporolactobacillus
Family VIII: Staphylococcaceae
Jeotgalicoccus
Macrococcus
Salinicoccus
Staphylococcus
Family IX: Thermoactinomycetaceae
Thermoactinomyces
Laceyella
Mechercharimyces
Planifilum
Seinonella
Shimazuella
Thermoflavimicrobium
Incertae Sedis
Exiguobacterium
Gemella
Thermicanus
Order II: Lactobacillales
Family I: Lactobacillaceae
Lactobacillus
Paralactobacillus
Pediococcus
Family II: Aerococcaceae
Abiotrophia
Aerococcus
Dolosicoccus
Eremococcus
Facklamia
Globicatella
Ignavigranum
Family III: Carnobacteriaceae
Alkalibacterium
Allofustis
Alloiococcus
Atopobacter
Atopococcus
Atopostipes
Carnobacterium
Desemzia
Dolosigranulum
Granulicatella
Isobaculum
Marinilactibacillus
Trichococcus
Family IV: Enterococcaceae
Enterococcus
Melissococcus
Tetragenococcus
Vagococcus
Family V: Leuconostocaceae
Leuconostoc
Oenococcus
Weissella
Family VI: Streptococcaceae
Lactococcus
Lactovum
Streptococcus
Class II: Clostridia
Order I: Clostridiales
Family I: Clostridiaceae
Alkaliphilus
Anaerobacter
Anoxynatronum
Caloramator
Caloranaerobacter
Caminicella
Clostridium
Natronincola
Oxobacter
Sarcina
Thermobrachium
Thermohalobacter
Tindallia
Family V: Lachnospiraceae
Acetitomaculum
Anaerostipes
Bryantella
Butyrivibrio
Catonella
Dorea
Hespellia
Johnsonella
Lachnobacterium
Moryella
Oribacterium
Parasporobacterium
Lachnospira
Pseudobutyrivibrio
Roseburia
Shuttleworthia
Sporobacterium
Syntrophococcus
Family VII: Peptostreptococcaceae
Filifactor
Peptostreptococcus
Tepidibacter
Family II: Eubacteriaceae
Acetobacterium
Acetobacterium
Alkalibacter
Anaerofustis
Eubacterium
Garciella
Pseudoramibacter
Family VI: Peptococcaceae
Cryptanaerobacter
Dehalobacter
Desulfitobacterium
Desulfonispora
Desulfosporosinus
Desulfotomaculum
Pelotomaculum
Peptococcus
Syntrophobotulus
Thermincola
Family III: *Gracilibacteraceae*
Gracilibacter
Family IV: Heliobacteriaceae
Heliobacterium
Heliobacillus
Heliophilum
Heliorestis
Family VIII: Ruminococcaceae
Ruminococcus
Family X: Veillonellaceac
Acetonema
Acidaminococcus
Allisonella
Anaeroarcus
Anaeroglobus
Anaeromusa
Anaerosinus
Anaerovibrio
Centipeda
Dendrosporobacter
Dialister
Megasphaera
Mitsuokella
Pectinatus
Phascolarctobacterium
Propionispira
Propionispora
Quinella
Schwartzia
Selenomonas
Sporomusa
Succiniclasticum
Succinispira
Thermosinus
Veillonella
Zymophilus
Incertae Sedis
Anaerococcus
Finegoldia
Gallicola
Fusibacter
Family IX: Syntrophomonadaceae
Pelospora
Syntrophomonas
Syntrophospora
Syntrophothermus
Thermosyntropha
Order III: Thermoanaerobacteriales
Family I: Thermoanaerobacteriaceae
Ammonifex
Caldanaerobacter
Carboxydibrachium
Coprothermobacter
Gelria
Moorella
Thermacetogenium
Thermanaeromonas
Thermoanaerobacter
Incertae Sedis
Caldicellulosiruptor
Mahell
Thermoanaerobacterium
Thermosediminibacter
Thermovenabulum
Order II: Halanaerobiales
Family I: Halanaerobiaceae
Halanaerobium
Halocella
Halothermothrix
Family II: Halobacteroidaceae
Acetohalobium
Halanaerobacter
Halonatronum
Natroniella
Orenia
Selenihalanaerobacter
Sporohalobacter

Appendices

Phylum Tenericutes
Order I: Mycoplasmatales
Family I: Mycoplasmataceae
Eperythrozoon
Haemobartonella
Mycoplasma
Ureaplasma
Order II: Entomoplasmatales
Family I: Entomoplasmataceae
Entomoplasma
Mesoplasma
Family II: Spiroplasmataceae
Spiroplasma
Order III: Acholeplasmatales
Family I: Acholeplasmataceae
Acholeplasma
Phytoplasma
Order IV: Anaeroplasmatales
Family I: Anaeroplasmataceae
Anaeroplasma
Asteroleplasma
Phylum Actinobacteria
Class I: Actinobacteria
Order I: Acidimicrobiales
Family I: Acidimicrobiaceae
Acidimicrobium
Order II: Rubrobacterales
Conexibacter
Rubrobacter
Solirubrobacter
Thermoleophilum
Order III: Coriobacteriales
Family I: Coriobacteriaceae
Atopobium
Collinsella
Coriobacterium
Cryptobacterium
Denitrobacterium
Eggerthella
Olsenella
Slackia
Order IV: Sphaerobacterales
Family I: Sphaerobacteraceae
Sphaerobacter
Order V: Actinomycetales
Suborder: Actinomycineae
Family I: Actinomycetaceae
Actinobaculum
Actinomyces
Arcanobacterium
Mobiluncus
Varibaculum
Suborder: Micrococcineae
Family I: Micrococcaceae
Arthrobacter
Citricoccus
Kocuria
Micrococcus
Nesterenkonia
Renibacterium
Rothia
Stomatococcus
Yania
Family II: Bogoriellaceae
Bogoriella
Family III: Rarobacteraceae
Rarobacter
Family IV: Sanguibacteraceae
Sanguibacter
Family V: Brevibacteriaceae
Brevibacterium
Family VI: Cellulomonadaceae
Cellulomonas
Oerskovia
Tropheryma
Family VII: Dermabacteraceae
Brachybacterium
Dermabacter
Family VIII: Dermatophilaceae
Dermatophilus
Kineosphaera
Family IX: Dermacoccaceae
Dermacoccus
Demetria
Kytococcus
Family X: Intrasporangiaceae
Arsenicicoccus
Intrasporangium
Janibacter
Nostocoidia
Ornithinicoccus
Ornithinimicrobium
Terrabacter
Terracoccus
Tetrasphaera
Family XI: Jonesiaceae
Jonesia
Family XII: Microbacteriaceae
Agrococcus
Agromyces
Aureobacterium
Clavibacter
Cryobacterium
Curtobacterium
Frigoribacterium
Leifsonia
Leucobacter
Microbacterium
Rathayibacter
Subtercola
Family XIII: Beutenbergiaceae
Beutenbergia
Georgenia
Salana
Family XIV: Promicromonosporaceae
Cellulosimicrobium
Promicromonospora
Xylanibacterium
Xylanimonas
Suborder: Corynebacterineae
Family I: Corynebacteriaceae
Corynebacterium
Family II: Dietziaceae
Dietzia
Family III: Gordoniaceae
Gordonia
Skermania
Family IV: Mycobacteriaceae
Mycobacterium
Family V: Nocardiaceae
Nocardia
Rhodococcus
Family VI: Tsukamurellaceae
Tsukamurella
Family VII: Williamsiaceae
Williamsia
Suborder: Micromonosporineae
Family I: Micromonosporaceae
Actinoplanes
Asanoa
Catellatospora
Catenuloplanes
Couchioplanes
Dactylosporangium
Micromonospora
Pilimelia
Spirilliplanes
Verrucosispora
Virgisporangium
Suborder: Propionibacterineae
Family I: Propionibacteriaceae
Luteococcus
Microlunatus
Propionibacterium
Propioniferax
Propionimicrobium
Tessaracoccus
Family II: Nocardioidaceae
Aeromicrobium
Actinopolymorpha
Friedmanniella
Hongia
Kribbella
Micropruina
Marmoricola
Nocardioides
Propionicimonas
Suborder: Pseudonocardineae
Family I: Pseudonocardiaceae
Actinoalloteichus
Actinopolyspora
Amycolatopsis
Crossiella
Kibdelosporangium
Kutzneria
Prauserella
Pseudonocardia
Saccharomonospora
Saccharopolyspora
Streptoalloteichus
Thermobispora
Thermocrispum
Family II: Actinosynnemataceae
Actinokineospora
Actinosynnema
Lechevalieria
Lentzea
Saccharothrix
Suborder: Streptomycineae
Family I: Streptomycetaceae
Kitasatospora
Streptomyces
Streptoverticillium
Suborder: Streptosporangineae
Family I: Streptosporangiaceae
Acrocarpospora
Herbidospora
Microbispora
Microtetraspora
Nonomuraea
Planobispora
Planomonospora
Planopolyspora
Planotetraspora
Streptosporangium
Family II: Nocardiopsaceae
Nocardiopsis
Streptomonospora
Thermobifida
Family III: Thermomonosporaceae
Actinomadura
Spirillospora
Thermomonospora
Suborder: Frankineae
Family I: Frankiaceae
Frankia
Family II: Geodermatophilaceae
Blastococcus
Geodermatophilus
Modestobacter
Family III: Microsphaeraceae
Microsphaera

Family IV: Sporichthyaceae
Sporichthya
Family V: Acidothermaceae
Acidothermus
Family VI: Kineosporiaceae
Cryptosporangium
Kineococcus
Kineosporia
Suborder: Glycomycineae
Family I: Glycomycetaceae
Glycomyces
Order VI: Bifidobacteriales
Family I: Bifidobacteriaceae
Aeriscardovia
Bifidobacterium
Falcivibrio
Gardnerella
Parascardovia
Scardovia
Family II: Unknown Affiliation
Actinobispora
Actinocorallia
Excellospora
Pelczaria
Turicella

Phylum Planctomycetes
Order I: Planctomycetales
Family I: Planctomycetaceae
Gemmata
Isosphaera
Pirellula
Planctomyces

Phylum Chlamydiae
Order I: Chlamydiales
Family I: Chlamydiaceae
Chlamydia
Chlamydophila
Family II: Parachlamydiaceae
Neochlamydia
Parachlamydia
Family III: Simkaniaceae
Rhabdochlamydia
Simkania
Family IV: Waddliaceae
Waddlia

Phylum Spirochaetes
Class I: Spirochaetes
Order I: Spirochaetales
Family I: Spirochaetaceae
Borrelia
Brevinema
Clevelandina
Cristispira
Diplocalyx
Hollandina
Pillotina
Spirochaeta
Treponema
Family II: Serpulinaceae
Brachyspira
Serpulina
Family III: Leptospiraceae
Leptonema
Leptospira

Phylum Fibrobacteres
Class I: Fibrobacteres
Family I: Fibrobacteraceae
Fibrobacter

Phylum Acidobacteria
Family I: Acidobacteriaceae
Acidobacterium
Geothrix
Holophaga

Phylum Bacteroidetes
Class I: Bacteroidetes
Order I: Bacteroidales
Family I: Bacteroidaceae
Acetofilamentum
Acetomicrobium
Acetothermus
Anaerophaga
Anaerorhabdus
Bacteroides
Megamonas
Family II: Rikenellaceae
Alistipes
Marinilabilia
Rikenella
Family III: Porphyromonadaceae
Dysgonomonas
Porphyromonas
Tannerella
Family IV: Prevotellaceae
Prevotella
Class II: Flavobacteria
Order I: Flavobacteriales
Family I: Flavobacteriaceae
Aequorivita
Arenibacter
Bergeyella
Capnocytophaga
Cellulophaga
Chryseobacterium
Coenonia
Croceibacter
Empedobacter
Flavobacterium
Gelidibacter
Gillisia
Mesonia
Muricauda
Myroides
Ornithobacterium
Polaribacter
Psychroflexus
Psychroserpens
Riemerella
Saligentibacter
Tenacibaculum
Weeksella
Zobellia
Family II: Blattabacteriaceae
Blattabacterium
Class III: Sphingobacteria
Order I: Sphingobacteriales
Family I: Sphingobacteriaceae
Pedobacter
Sphingobacterium
Family II: Saprospiraceae
Haliscomenobacter
Lewinella
Saprospira
Family III: Flexibacteraceae
Belliella
Cyclobacterium
Cytophaga
Dyadobacter
Flectobacillus
Flexibacter
Hongiella
Hymenobacter
Meniscus
Microscilla
Reichenbachia
Runella
Spirosoma
Sporocytophaga
Family IV: Flammeovirgaceae
Flammeovirga
Flexithrix
Persicobacter
Thermonema
Family V: Crenotrichaceae
Chitinophaga
Crenothrix
Rhodothermus
Salinibacter
Toxothrix

Phylum Fusobacteria
Class I: Fusobacteria
Order I: Fusobacteriales
Family I: Fusobacteriaceae
Fusobacterium
Ilyobacter
Leptotrichia
Propionigenium
Sebaldella
Sneathia
Streptobacillus
Family II: Incertae sedis
Cetobacterium

Phylum Verrucomicrobia
Class I: Verrucomicrobiae
Opitutus
Prosthecobacter
Verrucomicrobium
Victivallis
Xiphinematobacter

Phylum Dictyoglomi
Class I: Dictyoglomi
Order I: Dictyoglomales
Family I: Dictyoglomaceae
Dictyoglomus

Phylum Gemmatimonadetes
Class I: Gemmatimonadetes
Order I: Gemmatimonadales
Gemmatimonas

GLOSSARY

9 + 2 array Attachment of microtubules in eukaryotic flagella and cilia; 9 pairs of microtubules plus two microtubules.

12D treatment A sterilization process that would result in a decrease of the number of *Clostridium botulinum* endospores by 12 logarithmic cycles.

ABO blood group system The classification of red blood cells based on the presence or absence of A and B carbohydrate antigens.

abscess A localized accumulation of pus.

A-B toxin Bacterial exotoxins consisting of two polypeptides.

acellular vaccine A vaccine consisting of antigenic parts of cells.

acetyl group

$$H_3C-\overset{\overset{\displaystyle O}{\|}}{C}-$$

acid A substance that dissociates into one or more hydrogen ions (H^+) and one or more negative ions.

acid-fast stain A differential stain used to identify bacteria that are not decolorized by acid-alcohol.

acidic dye A salt in which the color is in the negative ion; used for negative staining.

acidophile A bacterium that grows below pH 4.

acquired immunodeficiency The inability, obtained during the life of an individual, to produce specific antibodies or T cells, due to drugs or disease.

activated macrophage A macrophage that has increased phagocytic ability and other functions after exposure to mediators released by T cells after stimulation by antigens.

activated sludge system A process used in secondary sewage treatment in which batches of sewage are held in highly aerated tanks; to ensure the presence of microbes efficient in degrading sewage, each batch is inoculated with portions of sludge from a precious batch.

activation energy The minimum collision energy required for a chemical reaction to occur.

active site A region on an enzyme that interacts with the substrate.

active transport Net movement of a substance across a membrane against a concentration gradient; requires the cell to expend energy.

acute disease A disease in which symptoms develop rapidly but last for only a short time.

acute-phase proteins Serum proteins whose concentration changes by at least 25% during inflammation.

adaptive immunity The ability, obtained during the life of the individual, to produce specific antibodies and T cells.

adenosarcoma Cancer of glandular epithelial tissue.

adenosine diphosphate (ADP) The substance formed when ATP is hydrolyzed and energy is released.

adenosine triphosphate (ATP) An important intracellular energy source.

adherence Attachment of a microbe or phagocyte to another's plasma membrane or other surface.

adhesin A carbohydrate-specific binding protein that projects from prokaryotic cells; used for adherence, also called a ligand.

adjuvant A substance added to a vaccine to increase its effectiveness.

aerobe An organism requiring molecular oxygen (O_2) for growth.

aerobic respiration Respiration in which the final electron acceptor in the electron transport chain is molecular oxygen (O_2).

aerotolerant anaerobe An organism that does not use molecular oxygen (O_2) but is not affected by its presence.

aflatoxin A carcinogenic toxin produced by *Aspergillus flavus*.

agar A complex polysaccharide derived from a marine alga and used as a solidifying agent in culture media.

agglutination A joining together or clumping of cells.

agranulocyte A leukocyte without visible granules in the cytoplasm; includes monocytes and lymphocytes.

alarmone A chemical signal that promotes a cell's response to environmental stress.

alcohol An organic molecule with the functional group—OH.

alcohol fermentation A catabolic process, beginning with glycolysis, that produces ethyl alcohol to reoxidize NADH.

aldehyde An organic molecule with the functional group

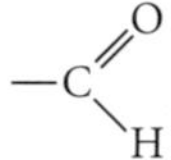

alga (plural: **algae**) A photosynthetic eukaryote; may be unicellular, filamentous, or multicellular but lack the tissues found in plants.

algal bloom An abundant growth of microscopic algae producing visible colonies in nature.

algin A sodium salt of mannuronic acid ($C_6H_8O_6$); found in brown algae.

allergen An antigen that evokes a hypersensitivity response.

allergy *See* hypersensitivity.

allograft A tissue graft that is not from a genetically identical donor (i.e., not from self or an identical twin).

allosteric inhibition The process in which an enzyme's activity is changed because of binding to the allosteric site.

allosteric site The site on an enzyme at which a noncompetitive inhibitor binds.

allylamines Antifungal agents that interfere with sterol synthesis.

amanitin A polypeptide toxin produced by *Amanita* spp., inhibits RNA polymerase.

Ames test A procedure using bacteria to identify potential carcinogens.

amination The addition of an amino group.

amino acid An organic acid containing an amino group and a carboxyl group. In alpha-amino acids the amino and carboxyl groups are attached to the same carbon atom called the alpha-carbon.

aminoglycoside An antibiotic consisting of amino sugars and an aminocyclitol ring; for example, streptomycin.

amino group $-NH_2$.

ammonification The release of ammonia from nitrogen-containing organic matter by the action of microorganisms.

amphibolic pathway A pathway that is both anabolic and catabolic.

amphitrichous Having flagella at both ends of a cell.

anabolism All synthesis reactions in a living organism; the building of complex organic molecules from simpler ones.

anaerobe An organism that does not require molecular oxygen (O_2) for growth.

anaerobic respiration Respiration in which the final electron acceptor in the electron transport chain is an inorganic molecule other than molecular oxygen (O_2); for example, a nitrate ion or CO_2.

anaerobic sludge digester Anaerobic digestion used in secondary sewage treatment.

anal pore A site in certain protozoa for elimination of waste.

analytical epidemiology Comparison of a diseased group and a healthy group to determine the cause of the disease.

anamnestic response *See* memory response.

anamorph Ascomycete fungi that have lost the ability to reproduce sexually; the asexual stage of a fungus.

anaphylaxis A hypersensitivity reaction involving IgE antibodies, mast cells, and basophils.

Angstrom (Å) A unit of measurement equal to 10^{-10} m, or 0.1 nm.

animalia The kingdom composed of multicellular eukaryotes lacking cell walls.

anion An ion with a negative charge.

anoxygenic Not producing molecular oxygen; typical of cyclic photophosphorylation.
antagonism Active opposition; (1) When two drugs are less effective than either one alone. (2) Competition among microbes.
antibiogram Report of antibiotic susceptibility of a bacterium.
antibiotic An antimicrobial agent, usually produced naturally by a bacterium or fungus.
antibody A protein produced by the body in response to an antigen, and capable of combining specifically with that antigen.
antibody-dependent cell-mediated cytotoxicity (ADCC) The killing of antibody-coated cells by natural killer cells and leukocytes.
antibody titer The amount of antibody in serum.
anticodon The three nucleotides by which a tRNA recognizes an mRNA codon.
antigen Any substance that causes antibody formation; also called immunogen.
antigen-antibody complex The combination of an antigen with the antibody that is specific for it; the basis of immune protection and many diagnostic tests.
antigen-binding sites A site on an antibody that binds to an antigenic determinant.
antigenic determinant A specific region on the surface of an antigen against which antibodies are formed; also called epitope.
antigenic drift A minor variation in the antigenic makeup of influenza viruses that occurs with time.
antigenic shift A major genetic change in influenza viruses causing changes in H and N antigens.
antigenic variation Changes in surface antigens that occur in a microbial population.
antigen-presenting cell (APC) A macrophage, dendritic cell, or B cell that engulfs an antigen and presents fragments to T cells.
antihuman immune serum globulin (anti-HISG) An antibody that reacts specifically with human antibodies.
antimetabolite A competitive inhibitor.
antimicrobial peptide An antibiotic that is bactericidal and has a broad spectrum of activity; *see* bacteriocin.
antisense DNA DNA that is complementary to the DNA encoding a protein; the antisense RNA transcript will hybridize with the mRNA encoding the protein and inhibit synthesis of the protein.
antisense strand (– strand) Viral RNA that cannot act as mRNA.
antisepsis A chemical method for disinfection of the skin or mucous membranes; the chemical is called an antiseptic.
antiserum A blood-derived fluid containing antibodies.
antitoxin A specific antibody produced by the body in response to a bacterial exotoxin or its toxoid.
antiviral protein (AVP) A protein made in response to interferon that blocks viral multiplication.
apoenzyme The protein portion of an enzyme, which requires activation by a coenzyme.
apoptosis The natural programmed death of a cell; the residual fragments are disposed of by phagocytosis.
aquatic microbiology The study of microorganisms and their activities in natural waters.
arbuscule Fungal mycelia in plant root cells.
archaea Domain of prokaryotic cells lacking peptidoglycan; one of the three domains.
arthroconidia An asexual fungal spore formed by fragmentation of a septate hypha.
Arthus reaction Inflammation and necrosis at the site of injection of foreign serum, due to immune complex formation.
artificially acquired active immunity The production of antibodies by the body in response to a vaccination.
artificially acquired passive immunity The transfer of humoral antibodies formed by one individual to a susceptible individual, accomplished by the injection of antiserum.
artificial selection Choosing one organism from a population to grow because of its desirable traits.
ascospore A sexual fungal spore produced in an ascus, formed by the ascomycetes.
ascus A saclike structure containing ascospores; found in the ascomycetes.
asepsis The absence of contamination by unwanted organisms.
aseptic packaging Commercial food preservation by filling sterile containers with sterile food.
aseptic surgery Techniques used in surgery to prevent microbial contamination of the patient.
aseptic techniques Laboratory techniques used to minimize contamination.
asexual spore A reproductive cell produced by mitosis and cell division (eukaryotes) or binary fission (actinomycetes).
atom The smallest unit of matter that can enter into a chemical reaction.
atomic force microscopy *See* scanned-probe microscopy.
atomic number The number of protons in the nucleus of an atom.
atomic weight The total number of protons and neutrons in the nucleus of an atom.
atrichous Bacteria that lack flagella.
attenuated whole-agent vaccine A vaccine containing live, attenuated (weakened) microorganisms.
autoclave Equipment for sterilization by steam under pressure, usually operated at 15 psi and 121°C.
autograph A tissue graft from one's self.
autoimmune disease Damage to one's own organs due to action of the immune system.
autotroph An organism that uses carbon dioxide (CO_2) as its principal carbon source. chemoautotroph, photoautotroph.
auxotroph A mutant microorganism with a nutritional requirement that is absent in the parent.
axial filament The structure for motility found in spirochetes; also called endoflagellum.
azole Antifungal agents that interfere with sterol synthesis.

bacillus (plural: **bacilli**) (1) Any rod-shaped bacterium. (2) When written as a genus *(Bacillus)* refers to rod-shaped, endospore-forming, facultatively anaerobic, gram-positive bacteria.
bacteremia A condition in which there are bacteria in the blood.
bacteria Domain of prokaryotic organisms, characterized by peptidoglycan cell walls; **bacterium** (singular) when referring to a single organism.
bacterial growth curve A graph indicating the growth of a bacterial population over time.
bactericide A substance capable of killing bacteria.
bacteriocin An antimicrobial peptide produced by bacteria that kills other bacteria.
bacteriochlorophyll A photosynthetic pigment that transfers electrons for photophosphorylation; found in anoxygenic photosynthetic bacteria.
bacteriology The scientific study of prokaryotes, including bacteria and archaea.
bacteriophage (phage) A virus that infects bacterial cells.
bacteriostasis A treatment capable of inhibiting bacterial growth.
base A substance that dissociates into one or more hydroxide ions (OH^-) and one or more positive ions.
base pairs The arrangement of nitrogenous bases in nucleic acids based on hydrogen bonding; in DNA, base pairs are A-T and G-C; in RNA, base pairs are A-U and G-C.
base substitution The replacement of a single base in DNA by another base, causing a mutation; also called point mutation.
basic dye A salt in which the color is in the positive ion; used for bacterial stains.
basidiospore A sexual fungal spore produced in a basidium, characteristic of the basidiomycetes.
basidium A pedestal that produces basidiospores; found in the basidiomycetes.
basophil A granulocyte (leukocyte) that readily takes up basic dye and is not phagocytic; has receptors for IgE Fc regions.
batch production An industrial process in which cells are grown for a period of time after which the product is collected.
B cell A type of lymphocyte; differentiates into antibody-secreting plasma cells and memory cells.
BCG vaccine A live, attenuated strain of *Mycobacterium bovis* used to provide immunity to tuberculosis.
beer Alcoholic beverage produced by fermentation of starch.

benthic zone The sediment at the bottom of a body of water.

Bergey's Manual *Bergey's Manual of Systematic Bacteriology,* the standard taxonomic reference on bacteria; also refers to *Bergey's Manual of Determinative Bacteriology,* the standard laboratory identification reference on bacteria.

β-lactam Core structure of penicillins.

beta oxidation The removal of two carbon units from a fatty acid to form acetyl CoA.

binary fission Prokaryotic cell reproduction by division into two daughter cells.

binomial nomenclature The system of having two names (genus and specific epithet) for each organism; also called scientific nomenclature.

bioaugmentation The use of pollutant-acclimated microbes or genetically engineered microbes for bioremediation.

biochemical oxygen demand (BOD) A measure of the biologically degradable organic matter in water.

biocide A substance capable of killing microorganisms.

bioconversion Changes in organic matter brought about by the growth of microorganisms.

bioenhancer Nutrients such as nitrate and phosphate that promote microbial growth.

biofilm A microbial community that usually forms as a slimy layer on a surface.

biogenesis The theory that living cells arise only from preexisting cells.

biogeochemical cycle The recycling of chemical elements by microorganisms for use by other organisms.

bioinformatics The science of determining the function of genes through computer-assisted analysis.

biological transmission The transmission of a pathogen from one host to another when the pathogen reproduces in the vector.

bioluminescence The emission of light from the electron transport chain; requires the enzyme luciferase.

biomass Organic matter produced by living organisms and measured by weight.

bioreactor A fermentation vessel with controls for environmental conditions, e.g., temperature and pH.

bioremediation The use of microbes to remove an environmental pollutant.

Biosafety Level (BSL) Safety guidelines for working with live microorganisms in a laboratory, four levels called BSL-1 through BSL-4.

biosynthetic *See* anabolism.

biotechnology The industrial application of microorganisms, cells, or cell components to make a useful product.

biotype *See* biovar.

biovar A subgroup of a serovar based on biochemical or physiological properties; also called biotype.

blade A flat leaflike structure of multicellular algae.

blastoconidium An asexual fungal spore produced by budding from the parent cell.

blebbing Bulging of plasma membrane as a cell dies.

blood-brain barrier Cell membranes that allow some substances to pass from the blood to the brain but restrict others.

brightfield microscope A microscope that uses visible light for illumination; the specimens are viewed against a white background.

broad-spectrum antibiotic An antibiotic that is effective against a wide range of both gram-positive and gram-negative bacteria.

broth dilution test A method of determining the minimal inhibitory concentration by using serial dilutions of an antimicrobial drug.

bubo An enlarged lymph node caused by inflammation.

budding (1) Asexual reproduction beginning as a protuberance from the parent cell that grows to become a daughter cell. (2) Release of an enveloped virus through the plasma membrane of an animal cell.

budding yeast Following mitosis, a yeast cell that divides unevenly to produce a small cell (bud) from the parent cell.

buffer A substance that tends to stabilize the pH of a solution.

bulking A condition arising when sludge floats rather than settles in secondary sewage treatment.

bullae (singular: **bulla**) Large serum-filled vesicles in the skin.

bursa of Fabricus An organ in chickens responsible for maturation of the immune system.

Calvin-Benson cycle The fixation of CO_2 into reduced organic compounds; used by autotrophs.

capnophile A microorganism that grows best at relatively high CO_2 concentrations.

capsid The protein coat of a virus that surrounds the nucleic acid.

capsomere A protein subunit of a viral capsid.

capsule An outer, viscous covering on some bacteria composed of a polysaccharide or polypeptide.

carbapenems Antibiotics that contain a β-lactam antibiotic and cilastatin.

carbohydrate An organic compound composed of carbon, hydrogen, and oxygen, with the hydrogen and oxygen present in a 2:1 ratio; carbohydrates include starches, sugars, and cellulose.

carbon cycle The series of processes that converts CO_2 to organic substances and back to CO_2 in nature.

carbon fixation The synthesis of sugars by using carbons from CO_2. *See also* Calvin-Benson cycle.

carbon skeleton The basic chain or ring of carbon atoms in a molecule; for example,

```
  |   |   |
—C—C—C—
  |   |   |
```

carboxyl group

```
      O
     //
  —C
     \
      OH
```

carboxysome A prokaryotic inclusion containing ribulose 1,5-diphosphate carboxylase.

carcinogen Any cancer-causing substance.

carrier Organisms (usually refers to humans) that harbor pathogens and transmit them to others.

casein Milk protein.

catabolism All decomposition reactions in a living organism; the breakdown of complex organic compounds into simpler ones.

catabolite repression Inhibition of the metabolism of alternate carbon sources by glucose.

catalase An enzyme that breaks down hydrogen peroxide: $2H_2O_2 \rightarrow 2\,H_2O + O_2$

catalyst A substance that increases the rate of a chemical reaction but is not altered itself.

cation A positively charged ion.

CD (cluster of determination) Number assigned to an epitope on a single antigen, for example, CD4 protein, which is found on T helper cells.

cDNA (complementary DNA) DNA made in vitro from an mRNA template.

cell culture Eukaryotic cells grown in culture media; also called tissue culture.

cell theory All living organisms are composed of cells and arise from preexisting cells.

cellular immunity An immune response that involves T cells binding to antigens presented on antigen-presenting cells; T cells then differentiate into several types of effector T cells.

cellular respiration *See* respiration.

cell wall The outer covering of most bacterial, fungal, algal, and plant cells; in bacteria, it consists of peptidoglycan.

Centers for Disease Control and Prevention (CDC) A branch of the U.S. Public Health Service that serves a central source of epidemiological information.

central nervous system (CNS) The brain and the spinal cord. *See also* peripheral nervous system.

centriole A structure consisting of nine microtubule triplets, found in eukaryotic cells.

centrosome Region in a eukaryotic cell consisting of a pericentriolar area (protein fibers) and a pair of centrioles; involved in formation of the mitotic spindle.

cercaria A free-swimming larva of trematodes.

cfu (colony-forming unit) Visible bacterial colonies on solid media.

chancre A hard sore, the center of which ulcerates.

chemical bond An attractive force between atoms forming a molecule.

chemical element A fundamental substance composed of atoms that have the same atomic number and behave the same way chemically.

chemical energy The energy of a chemical reaction.
chemical reaction The process of making or breaking bonds between atoms.
chemically defined medium A culture medium in which the exact chemical composition is known.
chemiosmosis A mechanism that uses a proton gradient across a cytoplasmic membrane to generate ATP.
chemistry The science of the interactions between atoms and molecules.
chemoautotroph An organism that uses an inorganic chemical as an energy source and CO_2 as a carbon source.
chemoheterotroph An organism that uses organic molecules as a source of carbon and energy.
chemokine A cytokine that induces, by chemotaxis, the migration of leukocytes into infected areas.
chemotaxis Movement in response to the presence of a chemical.
chemotherapy Treatment of disease with chemical substances.
chemotroph An organism that uses oxidation-reduction reactions as its primary energy source.
chimeric monoclonal antibody A genetically engineered antibody made of human constant regions and mouse variable regions.
chlamydoconidium An asexual fungal spore formed within a hypha.
chlorophyll *a* A photosynthetic pigment that transfers electrons for photophosphorylation; found in plant, algae, and cyanobacteria.
chloroplast The organelle that performs photosynthesis in photoautotrophic eukaryotes.
chlorosome Plasma membrane folds in green sulfur bacteria containing bacteriochlorophylls.
chromatin Threadlike, uncondensed DNA in an interphase eukaryotic cell.
chromatophore An infolding in the plasma membrane where bacteriochlorophyll is located in photoautotrophic bacteria; also known as thylakoids.
chromosome The structure that carries hereditary information, chromosomes contain genes.
chronic infection An illness that develops slowly and is likely to continue or recur for long periods.
ciliary escalator Ciliated mucosal cells of the lower respiratory tract that move inhaled particulates away from the lungs.
cilium (plural: **cilia**) A relatively short cellular projection from some eukaryotic cells, composed of nine pairs plus two microtubules. *See* flagellum.
cis Hydrogen atoms on the same side across a double bond in a fatty acid. *See* trans.
cistern A flattened membranous sac in endoplasmic reticulum and the Golgi complex.
clade A group of organisms that share a particular common ancestor; a branch on a cladogram.
cladogram A dichotomous phylogenetic tree that branches repeatedly, suggesting the classification of organisms based on the time sequence in which evolutionary branches arose.
class A taxonomic group between phylum and order.
clonal deletion The elimination of B and T cells that react with self.
clonal selection The development of clones of B and T cells against a specific antigen.
clone A population of cells arising from a single parent cell.
clue cells Sloughed-off vaginal cells covered with *Gardnerella vaginalis*.
coagulase A bacterial enzyme that causes blood plasma to clot.
coccobacillus (plural: **coccobacilli**) A bacterium that is an oval rod.
coccus (plural: **cocci**) A spherical or ovoid bacterium.
codon A sequence of three nucleotides in mRNA that specifies the insertion of an amino acid into a polypeptide.
coenocytic hypha A fungal filament that is not divided into uninucleate cell-like units because it lacks septa.
coenzyme A nonprotein substance that is associated with and that activates an enzyme.
coenzyme A (CoA) A coenzyme that functions in decarboxylation.
coenzyme Q *See* ubiquinone.
cofactor (1) The nonprotein component of an enzyme. (2) A microorganism or molecule that acts with others to synergistically enhance or cause disease.
coliforms Aerobic or facultatively anaerobic, gram-negative, nonendospore-forming, rod-shaped bacteria that ferment lactose with acid and gas formation within 48 hours at 35°C.
collagenase An enzyme that hydrolyzes collagen.
collision theory The principle that chemical reactions occur because energy is gained as particles collide.
colony A visible mass of microbial cells arising from one cell or from a group of the same microbes.
colony hybridization The identification of a colony containing a desired gene by using a DNA probe that is complementary to that gene.
colony-stimulating factor (CSF) A substance that induces certain cells to proliferate or differentiate.
commensalism A symbiotic relationship in which two organisms live in association and one is benefited while the other is neither benefited nor harmed.
commercial sterilization A process of treating canned goods aimed at destroying the endospores of *Clostridium botulinum*.
communicable disease Any disease that can be spread from one host to another.
competence The physiological state in which a recipient cell can take and incorporate a large piece of donor DNA.
competitive exclusion Growth of some microbes prevents the growth of other microbes.
competitive inhibitor A chemical that competes with the normal substrate for the active site of an enzyme. *See also* noncompetitive inhibitor.
complement A group of serum proteins involved in phagocytosis and lysis of bacteria.
complementary DNA (cDNA) DNA made in vitro from an mRNA template.
complement fixation The process in which complement combines with an antigen–antibody complex.
complex medium A culture medium in which the exact chemical composition is not known.
complex virus A virus with a complicated structure, such as a bacteriophage.
composting A method of solid waste disposal, usually plant material, by encouraging its decomposition by microbes.
compound A substance composed of two or more different chemical elements.
compound light microscope (LM) An instrument with two sets of lenses that uses visible light as the source of illumination.
compromised host A host whose resistance to infection is impaired.
condensation reaction A chemical reaction in which a molecule of water is released; also called dehydration synthesis.
condenser A lens system located below the microscope stage that directs light rays through the specimen.
confocal microscopy A light microscope that uses fluorescent stains and laser to make two- and three-dimensional images.
congenital Refers to a condition existing at birth; may be inherited or acquired in utero.
congenital immunodeficiency The inability, due to an individual's genotype, to produce specific antibodies or T cells.
conidiophore An aerial hypha bearing conidiospores.
conidiospore *See* conidium.
conidium An asexual spore produced in a chain from a conidiophore.
conjugated monoclonal antibody *See* immunotoxin.
conjugated vaccine A vaccine consisting of the desired antigen and other proteins.
conjugation The transfer of genetic material from one cell to another involving cell-to-cell contact.
conjugative plasmid A prokaryotic plasmid that carries genes for sex pili and for transfer of the plasmid to another cell.
constitutive enzyme An enzyme that is produced continuously.
contact inhibition The cessation of animal cell movement and division as a result of contact with other cells.
contact transmission The spread of disease by direct or indirect contact or via droplets.
contagious disease A disease that is easily spread from one person to another.
continuous cell line Animal cells that can be maintained through an indefinite number of generations in vitro.
continuous flow An industrial fermentation in which cells are grown indefinitely with continual addition of nutrients and removal of waste and products.
corepressor A molecule that binds to a repressor protein, enabling the repressor to bind to an operator.
cortex The protective fungal covering of a lichen.

counterstain A second stain applied to a smear, provides contrast to the primary stain.
covalent bond A chemical bond in which the electrons of one atom are shared with another atom.
crisis The phase of a fever characterized by vasodilation and sweating.
crista (plural: **cristae**) Folding of the inner membrane of a mitochondrion.
crossing over The process by which a portion of one chromosome is exchanged with a portion of another chromosome.
CTL (cytotoxic T lymphocytes) An activated T_C cell; kills cells presenting endogenous antigens.
culture Microorganisms that grow and multiply in a container of culture medium.
culture medium The nutrient material prepared for growth of microorganisms in a laboratory.
curd The solid part of milk that separates from the liquid (whey) in the making of cheese, for example.
cutaneous mycosis A fungal infection of the epidermis, nails, or hair.
cuticle The outer covering of helminths.
cyanobacteria Oxygen-producing photoautotrophic prokaryotes.
cyclic AMP (cAMP) A molecule derived from ATP, in which the phosphate group has a cyclic structure; acts as a cellular messenger.
cyclic photophosphorylation The movement of an electron from chlorophyll through a series of electron acceptors and back to chlorophyll; anoxygenic; purple and green bacterial photophosphorylation.
cyst A sac with a distinct wall containing fluid or other material; also, a protective capsule of some protozoa.
cysticercus An encysted tapeworm larva.
cytochrome A protein that functions as an electron carrier in cellular respiration and photosynthesis.
cytochrome c oxidase An enzyme that oxidizes cytochrome *c*.
cytokine A small protein released from human cells that regulates the immune response; directly or indirectly may induce fever, pain, or T-cell proliferation.
cytokine storm Overproduction of cytokines; can cause damage to the human body.
cytolysis The destruction of cells, resulting from damage to their cell membrane, that causes cellular contents to leak out.
cytopathic effect (CPE) A visible effect on a host cell, caused by a virus, that may result in host cell damage or death.
cytoplasm In a prokaryotic cell, everything inside the plasma membrane; in a eukaryotic cell, everything inside the plasma membrane and external to the nucleus.
cytoplasmic streaming The movement of cytoplasm in a eukaryotic cell.
cytoskeleton Microfilaments, intermediate filaments, and microtubules that provide support and movement for eukaryotic cytoplasm.
cytosol The fluid portion of cytoplasm.
cytostome The mouthlike opening in some protozoa.
cytotoxin A bacterial toxin that kills host cells or alters their functions.

darkfield microscope A microscope that has a device to scatter light from the illuminator so that the specimen appears white against a black background.
deamination The removal of an amino group from an amino acid to form ammonia. *See also* ammonification.
death phase The period of logarithmic decrease in a bacterial population; also called logarithmic decline phase.
debridement Surgical removal of necrotic tissue.
decarboxylation The removal of CO_2 from an amino acid.
decimal reduction time (DRT) The time (in minutes) required to kill 90% of a bacterial population at a given temperature; also called D value.
decolorizing agent A solution used in the process of removing a stain.
decomposition reaction A chemical reaction in which bonds are broken to produce smaller parts from a large molecule.
deep-freezing Preservation of bacterial cultures at −50°C to −95°C.
defensins Small peptide antibiotics made by human cells.
definitive host An organism that harbors the adult, sexually mature form of a parasite.
degeneracy Redundancy of the genetic code; that is, most amino acids are encoded by several codons.
degerming The removal of microorganisms in an area; also called degermation.
degranulation The release of contents of secretory granules from mast cells or basophils during anaphylaxis.
dehydration synthesis *See* condensation reaction.
dehydrogenation The loss of hydrogen atoms from a substrate.
delayed-type hypersensitivity Cell-mediated hypersensitivity.
denaturation A change in the molecular structure of a protein, usually making it nonfunctional.
dendritic cell A type of antigen-presenting cell characterized by long fingerlike extensions; found in lymphatic tissue and skin.
denitrification The reduction of nitrogen in nitrate to nitrite or nitrogen gas.
dental plaque A combination of bacterial cells, dextran, and debris adhering to the teeth.
deoxyribonucleic acid (DNA) The nucleic acid of genetic material in all cells and some viruses.
deoxyribose A five-carbon sugar contained in DNA nucleotides.
dermatomycosis A fungal infection of the skin; also known as tinea or ringworm.
dermatophyte A fungus that causes a cutaneous mycosis.
dermis The inner portion of the skin.
descriptive epidemiology The collection and analysis of all data regarding the occurrence of a disease to determine its cause.
desensitization The prevention of allergic inflammatory responses.
desiccation The removal of water.
diapedesis The process by which phagocytes move out of blood vessels.
dichotomous key An identification scheme based on successive paired questions; answering one question leads to another pair of questions, until an organism is identified.
differential interference contrast (DIC) microscope An instrument that provides a three-dimensional, magnified image.
differential medium A solid culture medium that makes it easier to distinguish colonies of the desired organism.
differential stain A stain that distinguishes objects on the basis of reactions to the staining procedure.
differential white blood cell count The number of each kind of leukocyte in a sample of 100 leukocytes.
diffusion The net movement of molecules or ions from an area of higher concentration to an area of lower concentration.
dimorphism The property of having two forms of growth. *See also* sexual dimorphism.
dioecious Referring to organisms in which organs of different sexes are located in different individuals.
diplobacilli (singular: **diplobacillus**) Rods that divide and remain attached in pairs.
diplococci (singular: **diplococcus**) Cocci that divide and remain attached in pairs.
diploid cell A cell having two sets of chromosomes; diploid is the normal state of a eukaryotic cell.
diploid cell line Eukaryotic cells grown in vitro.
direct agglutination test The use of known antibodies to identify an unknown cell-bound antigen.
direct contact transmission A method of spreading infection from one host to another through some kind of close association between the hosts.
direct FA test A fluorescent-antibody test to detect the presence of an antigen.
direct microscopic count Enumeration of cells by observation through a microscope.
disaccharide A sugar consisting of two simple sugars, or monosaccharides.
disease An abnormal state in which part or all of the body is not properly adjusted or is incapable of performing normal functions; any change from a state of health.
disinfection Any treatment used on inanimate objects to kill or inhibit the growth of microorganisms; a chemical used is called a disinfectant.
disk-diffusion method An agar-diffusion test to determine microbial susceptibility to chemotherapeutic agents; also called Kirby-Bauer test.
D-isomer Arrangement of four different atoms or groups around a carbon atom. *See* L-isomer.
dissimilation A metabolic process in which nutrients are not assimilated but are excreted as ammonia, hydrogen sulfide, and so on.

dissimilation plasmid A plasmid containing genes encoding production of enzymes that trigger the catabolism of certain unusual sugars and hydrocarbons.
dissociation The separation of a compound into positive and negative ions in solution. *See also* ionization.
disulfide bond A covalent bond that holds together two atoms of sulfur.
DNA base composition The moles-percentage of guanine plus cytosine in an organism's DNA.
DNA chip A silica wafer that holds DNA probes; used to recognize DNA in samples being tested.
DNA fingerprinting Analysis of DNA by electrophoresis of restriction enzyme fragments of the DNA.
DNA gyrase *See* topoisomerase.
DNA ligase An enzyme that covalently bonds a carbon atom of one nucleotide with the phosphate of another nucleotide.
DNA polymerase Enzyme that synthesizes DNA by copying a DNA template.
DNA probe A short, labeled, single strand of DNA or RNA used to locate its complementary strand in a quantity of DNA.
DNA sequencing A process by which the nucleotide sequence of DNA is determined.
DNA vaccine A vaccine made up of DNA, usually in the form of a plasmid.
domain A taxonomic classification based on rRNA sequences; above the kingdom level.
donor cell A cell that gives DNA to a recipient cell during genetic recombination.
droplet transmission The transmission of infection by small liquid droplets carrying microorganisms.
DTaP vaccine A combined vaccine used to provide active immunity, containing diphtheria and tetanus toxoids and *Bordetella pertussis* cell fragments.
D value *See* decimal reduction time.
dysentery A disease characterized by frequent, watery stools containing blood and mucus.

eclipse period The time during viral multiplication when complete, infective virions are not present.
ecology The study of the interrelationships between organisms and their environment.
edema An abnormal accumulation of interstitial fluid in body parts or tissues, causing swelling.
electron A negatively charged particle in motion around the nucleus of an atom.
electron acceptor An ion that picks up an electron that has been lost from another atom.
electron donor An ion that gives up an electron to another atom.
electronic configuration The arrangement of electrons in shells or energy levels in an atom.
electron microscope A microscope that uses electrons instead of light to produce an image.
electron shell A region of an atom where electrons orbit the nucleus, corresponding to an energy level.
electron transport chain, electron transport system A series of compounds that transfer electrons from one compound to another, generating ATP by oxidative phosphorylation.
electroporation A technique by which DNA is inserted into a cell using an electrical current.
elementary body The infectious form of chlamydiae.
ELISA (enzyme-linked immunosorbent assay) A group of serological tests that use enzyme reactions as indicators.
embryonic stem cell A cell from an embryo that has the potential to become a wide variety of specialized cell types.
emerging infectious disease (EID) A new or changing disease that is increasing or has the potential to increase in incidence in the near future.
Embden-Meyerhof pathway *See* glycolysis.
enanthem Rash on mucous membranes. *See also* exanthem.
encephalitis Infection of the brain.
encystment Formation of a cyst.
endemic disease A disease that is constantly present in a certain population.
endergonic reaction A chemical reaction that requires energy.
endocarditis Infection of the lining of the heart (endocardium).
endocytosis The process by which material is moved into a eukaryotic cell.
endoflagellum *See* axial filament.
endogenous (1) Infection caused by an opportunistic pathogen from an individual's own normal microbiota. (2) Antigens, usually of viral origin and degraded into fragments, generated within a cell.
endolith An organism that lives inside rock.
endoplasmic reticulum (ER) A membranous network in eukaryotic cells connecting the plasma membrane with the nuclear membrane.
endospore A resting structure formed inside some bacteria.
endosymbiotic theory A model for the evolution of eukaryotes which states that organelles arose from prokaryotic cells living inside a host prokaryote.
endotoxic shock *See* gram-negative sepsis.
endotoxin Part of the outer portion of the cell wall (lipid A) of most gram-negative bacteria; released on destruction of the cell.
end-product inhibition *See* feedback inhibition.
energy level Potential energy of an electron in an atom. *See also* electron shell.
enrichment culture A culture medium used for preliminary isolation that favors the growth of a particular microorganism.
enteric The common name for a bacterium in the family Enterobacteriaceae.
enterotoxin An exotoxin that causes gastroenteritis, such as those produced by *Staphylococcus, Vibrio,* and *Escherichia.*
Entner-Doudoroff pathway An alternate pathway for the oxidation of glucose to pyruvic acid.
envelope An outer covering surrounding the capsid of some viruses.
enzyme A molecule that catalyzes biochemical reactions in a living organism, usually a protein. *See also* ribozyme.
enzyme-linked immunosorbent assay *See* ELISA.
enzyme–substrate complex A temporary union of an enzyme and its substrate.
eosinophil A granulocyte whose granules take up the stain eosin.
epidemic disease A disease acquired by many hosts in a given area in a short time.
epidemiology The science that studies when and where diseases occur and how they are transmitted.
epidermis The outer portion of the skin.
epitope *See* antigenic determinant.
equilibrium The point of even distribution.
equivalent treatments Different methods that have the same effect on controlling microbial growth.
ergot A toxin produced in sclerotia by the fungus *Claviceps purpurea* that causes ergotism.
ester linkage Bonding between fatty acids and glycerol in bacterial and eukaryotic phospholipids:

$$\text{----C—O—}\overset{\overset{\text{O}}{\|}}{\text{C}}\text{----}$$

E test An agar diffusion test to determine antibiotic sensitivity using a plastic strip impregnated with varying concentrations of an antibiotic.
ethambutol A synthetic antimicrobial agent that interferes with the synthesis of RNA.
ethanol

$$\begin{array}{ccccc} & \text{H} & & \text{H} & \\ & | & & | & \\ \text{H—} & \text{C} & \text{—} & \text{C} & \text{—OH} \\ & | & & | & \\ & \text{H} & & \text{H} & \end{array}$$

ether linkage Bonding between fatty acids and glycerol in archaeal phospholipids: -----C—O—C-----
etiology The study of the cause of a disease.
eukarya All eukaryotes (animals, plants, fungi, and protists); members of the Domain Eukarya.
eukaryote A cell having DNA inside a distinct membrane-enclosed nucleus.
eukaryotic species A group of closely related organisms that can interbreed.
eutrophication The addition of organic matter and subsequent removal of oxygen from a body of water.
exanthem Skin rash. *See also* enanthem.
exchange reaction A chemical reaction that has both synthesis and decomposition components.
exergonic reaction A chemical reaction that releases energy.
exon A region of a eukaryotic chromosome that encodes a protein.

Glossary

exotoxin A protein toxin released from living, mostly gram-positive bacterial cells.

experimental epidemiology The study of a disease using controlled experiments.

exponential growth phase *See* log phase.

extracellular polymeric substance (EPS) A glycocalyx that permits bacteria to attach to various surfaces.

extreme thermophile *See* hyperthermophile.

extremophile A microorganism that lives in environmental extremes of temperature, acidity, alkalinity, salinity, or pressure.

extremozymes Enzymes produced by extremophiles.

facilitated diffusion The movement of a substance across a plasma membrane from an area of higher concentration to an area of lower concentration, mediated by transporter proteins.

facultative anaerobe An organism that can grow with or without molecular oxygen (O_2).

facultative halophile An organism capable of growth in, but not requiring, 1–2% salt.

FAD Flavin adenine dinucleotide; a coenzyme that functions in the removal and transfer of hydrogen ions (H^+) and electrons from substrate molecules.

FAME Fatty acid methyl ester; identification of microbes by the presence of specific fatty acids.

family A taxonomic group between order and genus.

feedback inhibition Inhibition of an enzyme in a particular pathway by the accumulation of the end-product of the pathway; also called end-product inhibition.

fermentation The enzymatic degradation of carbohydrates in which the final electron acceptor is an organic molecule, ATP is synthesized by substrate-level phosphorylation, and O_2 is not required.

fermentation test Method used to determine whether a bacterium or yeast ferments a specific carbohydrate; usually performed in a peptone broth containing the carbohydrate, a pH indicator, and an inverted tube to trap gas.

fever An abnormally high body temperature.

F factor (fertility factor) A plasmid found in the donor cell in bacterial conjugation.

fibrinolysin A kinase produced by streptococci.

filtration The passage of a liquid or gas through a screenlike material; a 0.45–μm filter removes most bacteria.

fimbria (plural: **fimbriae**) An appendage on a bacterial cell used for attachment.

FISH Fluorescent in situ hybridization; use of rRNA probes to identify microbes without cutturing.

fission yeast Following mitosis, a yeast cell that divides evenly to produce two new cells.

fixed macrophage A macrophage that is located in a certain organ or tissue (e.g., liver, lungs, spleen, or lymph nodes); also called a histiocyte.

fixing (1) In slide preparation, the process of attaching a specimen to a slide. (2) Regarding chemical elements, combining elements so that a critical element can enter the food chain. *See also* Calvin-Benson cycle; nitrogen fixation.

flaccid paralysis Loss of muscle movement, loss of muscle tone.

flagellum (plural: **flagella**) A thin appendage from the surface of a cell; used for cellular locomotion; composed of flagellin in prokaryotic cells, composed of 9 + 2 microtubules in eukaryotic cells.

flaming The process of sterilizing an inoculating loop by holding it in an open flame.

flat sour spoilage Thermophilic spoilage of canned goods not accompanied by gas production.

flatworm An animal belonging to the phylum Platyhelminthes.

flavoprotein A protein containing the coenzyme flavin; functions as an electron carrier in electron transport chains.

flocculation The removal of colloidal material during water purification by adding a chemical that causes colloidal particles to coalesce.

flow cytometry A method of counting cells using a flow cytometer, which detects cells by the presence of a fluorescent tag on the cell surface.

fluid mosaic model A way of describing the dynamic arrangement of phospholipids and proteins comprising the plasma membrane.

fluke A flatworm belonging to the class Trematoda.

fluorescence The ability of a substance to give off light of one color when exposed to light of another color.

fluorescence-activated cell sorter (FACS) A modification of a flow cytometer that counts and sorts cells labeled with fluorescent antibodies.

fluorescence microscope A microscope that uses an ultraviolet light source to illuminate specimens that will fluoresce.

fluorescent-antibody (FA) technique A diagnostic tool using antibodies labeled with fluorochromes and viewed through a fluorescence microscope; also called immunofluorescence.

FMN Flavin mononucleotide; a coenzyme that functions in the transfer of electrons in the electron transport chain.

focal infection A systemic infection that began as an infection in one place.

folliculitis An infection of hair follicles, often occurring as pimples.

fomite A nonliving object that can spread infection.

forespore A structure consisting of chromosome, cytoplasm, and endospore membrane inside a bacterial cell.

frameshift mutation A mutation caused by the addition or deletion of one or more bases in DNA.

free radical A compound with an unpaired electron. *See* superoxide.

free wandering macrophage A macrophage that leaves the blood and migrates to infected tissue.

freeze-drying *See* lyophilization.

FTA-ABS test An indirect fluorescent-antibody test used to detect syphilis.

fulminating A condition that develops quickly and rapidly increases in severity.

functional group An arrangement of atoms in an organic molecule that is responsible for most of the chemical properties of that molecule.

fungus (plural: **fungi**) An organism that belongs to the Kingdom Fungi; a eukaryotic absorptive chemoheterotroph.

furuncle An infection of a hair follicle.

fusion The merging of plasma membranes of two different cells, resulting in one cell containing cytoplasm from both original cells.

gamete A male or female reproductive cell.

gametocyte A male or female protozoan cell.

gamma globulin The serum fraction containing immunoglobulins (antibodies); also called immune serum globulin.

gastroenteritis Inflammation of the stomach and intestine.

gas vacuole A prokaryotic inclusion for buoyancy compensation.

gel electrophoresis The separation of substances (such as serum proteins or DNA) by their rate of movement through an electrical field.

gene A segment of DNA (a sequence of nucleotides in DNA) encoding a functional product.

gene silencing A mechanism to inhibit gene expression. *See* RNAi.

gene therapy Treating a disease by replacing abnormal genes.

generalized transduction The transfer of bacterial chromosome fragments from one cell to another by a bacteriophage.

generation time The time required for a cell or population to double in number.

genetic code The mRNA codons and the amino acids they encode.

genetic engineering *See* recombinant DNA technology.

genetic recombination The process of joining pieces of DNA from different sources.

genetics The science of heredity and gene function.

genetic screening Techniques for determining which genes are in a cell's genome.

genome One complete copy of the genetic information in a cell.

genomic library A collection of cloned DNA fragments created by inserting restriction enzyme fragments in a bacterium, yeast, or phage.

genomics The study of genes and their function.

genotype The genetic makeup of an organism.

genus (plural: **genera**) The first name of the scientific name (binomial); the taxon between family and species.

germicide *See* biocide.

germination The process of starting to grow from a spore or endospore.

germ theory of disease The principle that microorganisms cause disease.

global warming Retention of solar heat by gases in the atmosphere.

globulin The class of globular proteins that includes antibodies. *See also* immunoglobulin.

glycocalyx A gelatinous polymer surrounding a cell.

glycolysis The main pathway for the oxidation of glucose to pyruvic acid; also called Embden-Meyerhof pathway.

Golgi complex An organelle involved in the secretion of certain proteins.
graft-versus-host (GVH) disease A condition that occurs when a transplanted tissue has an immune response to the tissue recipient.
gram-negative bacteria Bacteria that lose the crystal violet color after decolorizing by alcohol; they stain red after treatment with safranin.
gram-negative sepsis Septic shock caused by gram-negative endotoxins.
gram-positive bacteria Bacteria that retain the crystal violet color after decolorizing by alcohol; they stain dark purple.
gram-positive sepsis Septic shock caused by gram-positive bacteria.
Gram stain A differential stain that classifies bacteria into two groups, gram-positive and gram-negative.
granulocyte A leukocyte with visible granules in the cytoplasm; includes neutrophils, basophils, and eosinophils.
granzymes Proteases that induce apoptosis.
green nonsulfur bacteria Gram-negative, nonproteobacteria; anaerobic and phototrophic; use reduced organic compounds as electron donors for CO_2 fixation.
green sulfur bacteria Gram-negative, nonproteobacteria; strictly anaerobic and phototrophic; no growth in dark; use reduced sulfur compounds as electron donors for CO_2 fixation.
group translocation In prokaryotes, active transport in which a substance is chemically altered during transport across the plasma membrane.
gumma A rubbery mass of tissue characteristic of tertiary syphilis.

HAART (highly active antiretroviral therapy) A combination of drugs used to treat HIV infection.
halogen One of the following elements: fluorine, chlorine, bromine, iodine, or astatine.
halophile An organism that requires a high salt concentration for growth.
H antigen Flagella antigens of enterics, identified by serological testing.
haploid cell A eukaryotic cell or organism with one of each type of chromosome.
hapten A substance of low molecular weight that does not cause the formation of antibodies by itself but does so when combined with a carrier molecule.
HA (hemagglutinin) spike Antigenic projections from the outer lipid bilayer of *Influenzavirus*.
Hazard Analysis and Critical Control Point (HACCP) System of prevention of hazards, for food safety.
helminth A parasitic roundworm or flatworm.
hemagglutination The clumping of red blood cells.
hemoflagellate A parasitic flagellate found in the circulatory system of its host.
hemolysin An enzyme that lyses red blood cells.
herd immunity The presence of immunity in most of a population.
hermaphroditic Having both male and female reproductive capacities.
heterocyst A large cell in certain cyanobacteria; the site of nitrogen fixation.
heterolactic Describing an organism that produces lactic acid and other acids or alcohols as end-products of fermentation; e.g., *Escherichia*.
heterotroph An organism that requires an organic carbon source; also called organotroph.
Hfr cell A bacterial cell in which the F factor has become integrated into the chromosome; Hfr stands for high frequency of recombination.
high-efficiency particulate air (HEPA) filter A screenlike material that removes particles larger than 0.3 μm from air.
high-temperature short-time (HTST) pasteurization Pasteurizing at 72°C for 15 seconds.
histamine A substance released by tissue cells that causes vasodilation, capillary permeability, and smooth muscle contraction.
histocompatibility antigen An antigen on the surface of human cells.
histone A protein associated with DNA in eukaryotic chromosomes.
holdfast The branched base of an algal stipe.
holoenzyme An enzyme consisting of an apoenzyme and a cofactor.
homolactic Describing an organism that produces only lactic acid from fermentation; e.g., *Streptococcus*.
horizontal gene transfer Transfer of genes between two organisms in the same generation. *See also* vertical gene transfer.
host An organism infected by a pathogen. *See also* definitive host; intermediate host.
host range The spectrum of species, strains, or cell types that a pathogen can infect.
hot-air sterilization Sterilization by the use of an oven at 170°C for approximately 2 hours.
human leukocyte antigen (HLA) complex Human cell surface antigens. *See also* major histocompatibility complex.
humanized antibody Monoclonal antibodies that are partly or fully human proteins.
humoral immunity Immunity produced by antibodies dissolved in body fluids, mediated by B cells; also called antibody-mediated immunity.
hyaluronidase An enzyme secreted by certain bacteria that hydrolyzes hyaluronic acid and helps spread microorganisms from their initial site of infection.
hybridoma A cell made by fusing an antibody-producing B cell with a cancer cell.
hydrogen bond A bond between a hydrogen atom covalently bonded to oxygen or nitrogen and another covalently bonded oxygen or nitrogen atom.
hydrolysis A decomposition reaction in which chemicals react with the H^+ and OH^- of a water molecule.
hydroxide OH^-; the anion that forms a base.
hydroxyl —OH; covalently bonded to a molecule forms an alcohol.
hydroxyl radical A toxic form of oxygen (OH•) formed in cytoplasm by ionizing radiation and aerobic respiration.
hyperacute rejection Very rapid rejection of transplanted tissue, usually in the case of tissue from nonhuman sources.
hyperbaric chamber An apparatus to hold materials at pressures greater than 1 atmosphere.
hypersensitivity An altered, enhanced immune reaction leading to pathological changes; also called allergy.
hyperthermophile An organism whose optimum growth temperature is at least 80°C; also called extreme thermophile.
hypertonic solution A solution that has a higher concentration of solutes than an isotonic solution.
hypha A long filament of cells in fungi or actinomycetes.
hypotonic solution A solution that has a lower concentration of solutes than an isotonic solution.

ID_{50} The number of microorganisms required to produce a demonstrable infection in 50% of the test host population.
idiophase The period in the production curve of an industrial cell population in which secondary metabolites are produced; a period of stationary growth following the phase of rapid growth. *See also* trophophase.
IgA The class of antibodies found in secretions.
IgD The class of antibodies found on B cells.
IgE The class of antibodies involved in hypersensitivities.
IgG The most abundant class of antibodies in serum.
IgM The first class of antibodies to appear after exposure to an antigen.
immune complex A circulating antigen-antibody aggregate capable of fixing complement.
immune serum globulin *See* gamma globulin.
immune surveillance The body's immune response to cancer.
immunity *See* adaptive immunity, innate immunity.
immunization *See* vaccination.
immunodeficiency The absence of an adequate immune response; may be congenital or acquired.
immunodiffusion test A test consisting of precipitation reactions carried out in an agar gel medium.
immunoelectrophoresis The identification of proteins by electrophoretic separation followed by serological testing.
immunofluorescence *See* fluorescent-antibody technique.
immunogen *See* antigen.
immunoglobulin (Ig) A protein (antibody) formed in response to an antigen and can react with that antigen. *See also* globulin.
immunology The study of a host's defenses to a pathogen.
immunosuppression Inhibition of the immune response.
immunotherapy Making use of the immune system to attack tumor cells, either by enhancing the normal immune response or by using toxin-bearing specific antibodies. *See also* immunotoxin.
immunotoxin An immunotherapeutic agent consisting of a poison bound to a monoclonal antibody.

inapparent infection *See* subclinical infection.
incidence The fraction of the population that contracts a disease during a particular period of time.
inclusion Material held inside a cell, often consisting of reserve deposits.
inclusion body A granule or viral particle in the cytoplasm or nucleus of some infected cells; important in the identification of viruses that cause infection.
incubation period The time interval between the actual infection and first appearance of any signs or symptoms of disease.
indicator organism A microorganism, such as a coliform, whose presence indicates conditions such as fecal contamination of food or water.
indirect (passive) agglutination test An agglutination test using soluble antigens attached to latex or other small particles.
indirect contact transmission The spread of pathogens by fomites (nonliving objects).
indirect FA test A fluorescent-antibody test to detect the presence of specific antibodies.
inducer A chemical or environmental stimulus that causes transcription of specific genes.
induction The process that turns on the transcription of a gene.
infection The growth of microorganisms in the body.
infectious disease A disease in which pathogens invade a susceptible host and carry out at least part of their life cycle in the host.
inflammation A host response to tissue damage characterized by redness, pain, heat, and swelling; and sometimes loss of function.
innate immunity Host defenses that afford protection against any kind of pathogen. *See also* adaptive immunity.
inoculum A culture medium in which microorganisms are implanted.
inorganic compound A small molecule that does not contain carbon and hydrogen.
insertion sequence (IS) The simplest kind of transposon.
integrase An enzyme produced by HIV that allows the integration of HIV DNA into the host cell's DNA.
interferon (IFN) A specific group of cytokines. Alpha- and beta-IFNs are antiviral proteins produced by certain animal cells in response to a viral infection. Gamma-IFN stimulates macrophage activity.
interleukin (IL) A chemical that causes T-cell proliferation. *See also* cytokine.
intermediate host An organism that harbors the larval or asexual stage of a helminth or protozoan.
intoxication A condition resulting from the ingestion of a microbially produced toxin.
intron A region in a eukaryotic gene that does not code for a protein or mRNA.
intubation Placing a tube into the body; tracheal intubation provides access for air to the lungs.
invasin A surface protein produced by *Salmonella typhimurium* and *Escherichia coli* that rearranges nearby actin filaments in the cytoskeleton of a host cell.
iodophor A complex of iodine and a detergent.
ion A negatively or positively charged atom or group of atoms.
ionic bond A chemical bond formed when atoms gain or lose electrons in the outer energy levels.
ionization The separation (dissociation) of a molecule into ions.
ionizing radiation High-energy radiation with a wavelength less than 1nm; causes ionization. X rays and gamma rays are examples.
ischemia Localized decreased blood flow.
isograft A tissue graft from a genetically identical source (i.e., from an identical twin).
isomer One or two molecules with the same chemical formula but different structures.
isotonic solution A solution in which, after immersion of a cell, osmotic pressure is equal across the cell's membrane.
isotope A form of a chemical element in which the number of neutrons in the nucleus is different from the other forms of that element.

karyogamy Fusion of the nuclei of two cells; occurs in the sexual stage of a fungal life cycle.
kelp A multicellular brown alga.
keratin A protein found in epidermis, hair, and nails.
ketolide Semi-synthetic macrolide antibiodies; effective against macrolide-resistant bacteria.
kinase (1) An enzyme that removes a Ⓟ from ATP and attaches it to another molecule. (2) A bacterial enzyme that breaks down fibrin (blood clots).
kingdom A taxonomic classification between domain and phylum.
kinin A substance released from tissue cells that causes vasodilation.
Kirby-Bauer test *See* disk-diffusion method.
Koch's postulates Criteria used to determine the causative agent of infectious diseases.
koji A microbial fermentation on rice; usually *Aspergillus oryzae;* used to produce amylase.
Krebs cycle A pathway that converts two-carbon compounds to CO_2, transferring electrons to NAD^+ and other carriers; also called tricarboxylic acid (TCA) cycle or critic acid cycle.

lactic acid fermentation A catabolic process, beginning with glycolysis, that produces lactic acid to reoxidize NADH.
lagging strand During DNA replication, the daughter strand that is synthesized discontinuously.
lag phase The time interval in a bacterial growth curve during which there is no growth.
larva The sexually immature stage of a helminth or arthropod.
latent disease A disease characterized by a period of no symptoms when the pathogen is inactive.
latent infection A condition in which a pathogen remains in the host for long periods without producing disease.
LD_{50} The lethal dose for 50% of the inoculated hosts within a given period.
leading strand During DNA replication, the daughter strand that is synthesized continuously.
lectin Carbohydrate-binding proteins on a cell.
lepromin test A skin test to determine the presence of antibodies to *Mycobacterium leprae,* the cause of leprosy.
leukocidins Substances produced by some bacteria that can destroy neutrophils and macrophages.
leukocyte A white blood cell.
leukotriene A substance produced by mast cells and basophils that causes increased permeability of blood vessels and helps phagocytes attach to pathogens.
L-form Prokaryotic cells that lack a cell wall; can return to walled state.
lichen A mutualistic relationship between a fungus and an alga or a cyanobacterium.
ligand *See* adhesin.
light-dependent reaction The process by which light energy is used to convert ADP and phosphate to ATP. *See also* photophosphorylation.
light-independent reactions The process by which electrons and energy from ATP are used to reduce CO_2 to sugar. *See also* Calvin-Benson cycle.
light-repair enzyme *See* photolyase.
limnetic zone The surface zone of an inland body of water away from the shore.
***Limulus* amoebocyte lysate (LAL) assay** A test to detect the presence of bacterial endotoxins.
lipase An enzyme that breaks down triglycerides into their component glycerol and fatty acids.
lipid A non–water soluble organic molecule, including triglycerides, phospholipids, and sterols.
lipid A A component of the gram-negative outer membrane; endotoxin.
lipid inclusion *See* inclusion.
lipopolysaccharide (LPS) A molecule consisting of a lipid and a polysaccharide, forming the outer membrane of gram-negative cell walls.
L-isomer Arrangement of four different atoms or groups around a carbon atom. *See* D-isomer.
lithotroph *See* autotroph.
littoral zone The region along the shore of the ocean or a large lake where there is considerable vegetation and where light penetrates to the bottom.
local infection An infection in which pathogens are limited to a small area of the body.
localized anaphylaxis An immediate hypersensitivity reaction that is restricted to a limited area of skin or mucous membrane; for example, hayfever, a skin rash, or asthma. *See also* systemic anaphylaxis.

logarithmic decline phase *See* death phase.
log phase The period of bacterial growth or logarithmic increase in cell numbers; also called exponential growth phase.
lophotrichous Having two or more flagella at one end of a cell.
luciferase An enzyme that accepts electrons from flavoproteins and emits a photon of light in bioluminescence.
lymphangitis Inflammation of lymph vessels.
lymphocyte A leukocyte involved in specific immune responses.
lyophilization Freezing a substance and sublimating the ice in a vacuum; also called freeze-drying.
lysis (1) Destruction of a cell by the rupture of the plasma membrane, resulting in a loss of cytoplasm. (2) In disease, a gradual period of decline.
lysogenic conversion The acquisition of new properties by a host cell infected by a lysogenic phage.
lysogenic cycle Stages in viral development that result in the incorporation of viral DNA into host DNA.
lysogeny A state in which phage DNA is incorporated into the host cell without lysis.
lysosome An organelle containing digestive enzymes.
lysozyme An enzyme capable of hydrolyzing bacterial cell walls.
lytic cycle A mechanism of phage multiplication that results in host cell lysis.

macrolide An antibiotic that inhibits protein synthesis; for example, erythromycin.
macromolecule A large organic molecule.
macrophage A phagocytic cell; a mature monocyte. *See* fixed macrophage, free wandering macrophage.
macule A flat, reddened skin lesion.
maculopapular A rash with macules and papules.
magnetosome An iron oxide inclusion, produced by some gram-negative bacteria, that acts like a magnet.
major histocompatibility complex (MHC) The genes that code for histocompatibility antigens; also known as human leukocyte antigen (HLA) complex.
malolactic fermentation The conversion of malic acid to lactic acid by lactic acid bacteria.
malt Germinated barley grains containing maltose, glucose, and amylase.
malting The germination of starchy grains resulting in glucose and maltose production.
margination The process by which phagocytes stick to the lining of blood vessels.
mast cell A type of cell found throughout the body that contains histamine and other substances that stimulate vasodilation.
matrix Fluid in mitochondria.
maximum growth temperature The highest temperature at which a species can grow.
M (microfold) cell Intestinal cells that take up and transfer antigens to lymphocytes.
mechanical transmission The process by which arthropods transmit infections by carrying pathogens on their feet and other body parts.
medulla A lichen body consisting of algae (or cyanobacteria) and fungi.
meiosis A eukaryotic cell replication process that results in cells with half the chromosome number of the original cell.
membrane attack complex (MAC) Complement proteins C5–C9, which together make lesions in cell membranes that lead to cell death.
membrane filter A screenlike material with pores small enough to retain microorganisms; a 0.45-μm filter retains most bacteria.
memory cells A long-lived B or T cell responsible for the memory, or secondary, response.
memory response A rapid rise in antibody titer following exposure to an antigen after the primary response to that antigen; also called anamestic response or secondary response.
meningitis Inflammation of the meninges, the three membranes covering the brain and spinal cord.
merozoite A trophozoite of *Plasmodium* found in red blood cells or liver cells.
mesophile An organism that grows between about 10°C and 50°C; a moderate-temperature–loving microbe.
mesosome An irregular fold in the plasma membrane of a prokaryotic cell that is an artifact of preparation for microscopy.
messenger RNA (mRNA) The type of RNA molecule that directs the incorporation of amino acids into proteins.
metabolic pathway A sequence of enzymatically catalyzed reactions occurring in a cell.
metabolism The sum of all the chemical reactions that occur in a living cell.
metacercaria The encysted stage of a fluke in its final intermediate host.
metachromatic granule A granule that stores inorganic phosphate and stains red with certain blue dyes; characteristic of *Corynebacterium diphtheriae*. Collectively known as volutin.
methane The hydrocarbon CH_4, a flammable gas formed by the microbial decomposition of organic matter; natural gas.
methylate Addition of a methyl group ($—CH_3$) to a molecule; methylated cytosine is protected from digestion by restriction enzymes.
microaerophile An organism that grows best in an environment with less molecular oxygen (O_2) than is normally found in air.
micrometer (μm) A unit of measurement equal to 10^{-6}m.
microorganism A living organism too small to be seen with the naked eye; includes bacteria, fungi, protozoa, and microscopic algae; also includes viruses.
microtubule A hollow tube made of the protein tubulin; the structural unit of eukaryotic flagella and centrioles.
microwave Electromagnetic radiation with wavelength between 10^{-1} and 10^{-3} m.
minimal bactericidal concentration (MBC) The lowest concentration of chemotherapeutic agent that will kill test microorganisms.
minimal inhibitory concentration (MIC) The lowest concentration of a chemotherapeutic agent that will prevent growth of the test microorganisms.
minimum growth temperature The lowest temperature at which a species will grow.
miracidium The free-swimming, ciliated larva of a fluke that hatches from the egg.
missense mutation A mutation that results in the substitution of an amino acid in a protein.
mitochondrion (plural: **mitochondria**) An organelle containing Krebs cycle enzymes and the electron transport chain.
mitosis A eukaryotic cell replication process in which the chromosomes are duplicated; usually followed by division of the cytoplasm of the cell.
mitosome Eukaryotic organelle derived form degenrate mitochondria, found in *Trichomonas* and *Giardia*.
MMWR *Morbidity and Mortality Weekly Report;* a CDC publication containing data on notifiable diseases and topics of special interest.
mole An amount of a chemical equal to the atomic weights of all the atoms in a molecule of the chemical.
molecular biology The science dealing with DNA and protein synthesis of living organisms.
molecular weight The sum of the atomic weights of all atoms making up a molecule.
molecule A combination of atoms forming a specific chemical compound.
monoclonal antibody (Mab) A specific antibody produced in vitro by a clone of B cells hybridized with cancerous cells.
monocyte A leukocyte that is the precursor of a macrophage.
monoecious Having both male and female reproductive capacities.
monomer A small molecule that collectively combines to form polymers.
mononuclear phagocytic system A system of fixed macrophages located in the spleen, liver, lymph nodes, and red bone marrow.
monosaccharide A simple sugar consisting of 3–7 carbon atoms.
monotrichous Having a single flagellum.
morbidity (1) The incidence of a specific disease. (2) The condition of being diseased.
morbidity rate The number of people affected by a disease in a given period of time in relation to the total population.
mordant A substance added to a staining solution to make it stain more intensely.
mortality The number of deaths from a specific notifiable disease.
mortality rate The number of deaths resulting from a disease in a given period of time in relation to the total population.
most probable number (MPN) method A statistical determination of the number of coliforms per 100 ml of water or 100 g of food.

motility The ability of an organism to move by itself.

M protein A heat- and acid-resistant protein of streptococcal cell walls and fibrils.

mucous membranes Membranes that line body openings, including the intestinal tract, open to the exterior; also called mucosa.

mutagen An agent in the environment that brings about mutations.

mutation Any change in the nitrogenous base sequence of DNA.

mutation rate The probability that a gene will mutate each time a cell divides.

mutualism A type of symbiosis in which both organisms or populations are benefited.

mycelium A mass of long filaments of cells that branch and intertwine, typically found in molds.

mycolic acid Long-chained, branched fatty acids characteristic of members of the genus *Mycobacterium.*

mycology The scientific study of fungi.

mycorrhiza A fungus growing in symbiosis with plant roots.

mycosis A fungal infection.

mycotoxin A toxin produced by a fungus.

NAD^+ A coenzyme that functions in the removal and transfer of hydrogen ion (H^+) and electrons from substrate molecules.

$NADP^+$ A coenzyme similar to NAD^+.

nanobacteria Bacteria well below the generally accepted lower limit diameter (about 200 nanometres) for bacteria.

nanometer (nm) A unit of measurement equal to 10^{-9} m, 10^{-3} μm.

NA (neuraminidase) spikes Antigenic projections from the outer lipid bilayer of *Influenzavirus.*

natural killer (NK) cell A lymphoid cell that destroys tumor cells and virus-infected cells.

naturally acquired active immunity Antibody production in response to an infectious disease.

naturally acquired passive immunity The natural transfer of humoral antibodies, for example, transplacental transfer.

natural selection Process by which organisms with certain inherited characteristics are more likely to survive and reproduce than organisms with other characteristics.

necrosis Tissue death.

negative (indirect) selection The process of identifying mutations by selecting cells that do not grow using replica plating.

negative staining A procedure that results in colorless bacteria against a stained background.

neurotoxin An exotoxin that interferes with normal nerve impulse conduction.

neutralization An antigen–antibody reaction that inactivates a bacterial exotoxin or virus.

neutron An uncharged particle in the nucleus of an atom.

neutrophil A highly phagocytic granulocyte; also called polymorphonuclear leukocyte (PMN) or polymorph.

nitrification The oxidation of nitrogen in ammonia to produce nitrate.

nitrogen cycle The series of processes that converts nitrogen (N_2) to organic substances and back to nitrogen in nature.

nitrogen fixation The conversion of nitrogen (N_2) into ammonia.

nitrosamine A carcinogen formed by the combination of nitrite and amino acids.

noncommunicable disease A disease that is not transmitted from one person to another.

noncompetitive inhibitor An inhibitory chemical that does not compete with the substrate for an enzyme's active site. *See also* allosteric inhibition; competitive inhibitor.

noncyclic photophosphorylation The movement of an electron from chlorophyll to NAD^+; plant and cyanobacterial photophosphorylation.

nonionizing radiation Short-wavelength radiation that does not cause ionization; ultraviolet (UV) radiation is an example.

non-nucleoside reverse transcriptase inhibitor A drug that binds with and inhibits the action of the HIV reverse transcriptase enzyme.

nonsense codon A codon that does not encode any amino acid.

nonsense mutation A base substitution in DNA that results in a nonsense codon.

normal microbiota The microorganisms that colonize a host without causing disease; also called normal flora.

nosocomial infection An infection that develops during the course of a hospital stay and was not present at the time the patient was admitted.

notifiable infectious disease A disease that physicians must report to the U.S. Public Health Service; also called reportable disease.

nuclear envelope The double membrane that separates the nucleus from the cytoplasm in a eukaryotic cell.

nuclear pore An opening in the nuclear envelope through which materials enter and exit the nucleus.

nucleic acid A macromolecule consisting of nucleotides; DNA and RNA are nucleic acids.

nucleic acid hybridization The process of combining single complementary strands of DNA.

nucleoid The region in a bacterial cell containing the chromosome.

nucleolus (plural: **nucleoli**) An area in a eukaryotic nucleus where rRNA is synthesized.

nucleoside A compound consisting of a purine or pyrimidine base and a pentose sugar.

nucleoside reverse transcriptase inhibitor A nucleoside analog antiretroviral drug.

nucleotide A compound consisting of a purine or pyrimidine base, a five-carbon sugar, and a phosphate.

nucleotide analog A chemical that is structurally similar to the normal nucleosides in nucleic acids but with altered base-pairing properties.

nucleotide excision repair The repair of DNA involving removal of defective nucleotides and replacement with functional ones.

nucleus (1) The part of an atom consisting of the protons and neutrons. (2) The part of a eukaryotic cell that contains the genetic material.

numerical identification Bacterial identification schemes in which test values are assigned a number.

nutrient agar Nutrient broth containing agar.

nutrient broth A complex medium made of beef extract and peptone.

O antigen Polysaccharide antigens in the outer membrane of gram-negative bacteria, identified by serological testing.

objective lenses In a compound light microscope, the lenses closest to the specimen.

obligate aerobe An organism that requires molecular oxygen (O_2) to live.

obligate anaerobe An organism that does not use molecular oxygen (O_2) and is killed in the presence of O_2.

obligate halophile An organism that requires high osmotic pressures such as high concentrations of NaCl.

ocular lens In a compound light microscope, the lens closest to the viewer; also called the eyepiece.

oligodynamic action The ability of small amounts of a heavy metal compound to exert antimicrobial activity.

oligosaccharide A carbohydrate consisting of 2 to approximately 20 monosaccharides.

oncogene A gene that can bring about malignant transformation.

oncogenic virus A virus that is capable of producing tumors; also called oncovirus.

oocyst An encysted apicomplexan zygote in which cell division occurs to form the next infectious stage.

Opa A bacterial outer membrane protein; cells with Opa form opaque colonies.

operator The region of DNA adjacent to structural genes that controls their transcription.

operon The operator and promoter sites and structural genes they control.

opportunistic pathogen A microorganism that does not ordinarily cause a disease but can become pathogenic under certain circumstances.

opsonization The enhancement of phagocytosis by coating microorganisms with certain serum proteins (opsonins); also called immune adherence.

optimum growth temperature The temperature at which a species grows best.

order A taxonomic classification between class and family.

organotroph *See* heterotroph.

organelle A membrane-enclosed structure within eukaryotic cells.

organic compound A molecule that contains carbon and hydrogen.

organic growth factor An essential organic compound that an organism is unable to synthesize.

osmosis The net movement of solvent molecules across a selectively permeable membrane from an area of lower solute concentration to an area of higher solute concentration.
osmotic lysis Rupture of the plasma membrane resulting from movement of water into the cell.
osmotic pressure The force with which a solvent moves from a solution of lower solute concentration to a solution of higher solute concentration.
oxidation The removal of electrons from a molecule.
oxidation pond A method of secondary sewage treatment by microbial activity in a shallow standing pond of water.
oxidation-reduction A coupled reaction in which one substance is oxidized and one is reduced; also called redox reaction.
oxidative phosphorylation The synthesis of ATP coupled with electron transport.
oxygenic Producing oxygen, as in plant and cyanobacterial photosynthesis.
ozone O_3.

PAMP (pathogen-associated molecular patterns) Molecules present on pathogens and not self.
pandemic disease An epidemic that occurs worldwide.
papule Small, solid elevation of the skin.
parasite An organism that derives nutrients from a living host.
parasitism A symbiotic relationship in which one organism (the parasite) exploits another (the host) without providing any benefit in return.
parasitology The scientific study of parasitic protozoa and worms.
parenteral route A portal of entry for pathogens by deposition directly into tissues beneath the skin and mucous membranes.
pasteurization The process of mild heating to kill particular spoilage microorganisms or pathogens.
pathogen A disease-causing organism.
pathogenesis The manner in which a disease develops.
pathogenicity The ability of a microorganism to cause disease by overcoming the defenses of a host.
pathology The scientific study of disease.
pellicle (1) The flexible covering of some protozoa. (2) Scum on the surface of a liquid medium.
penicillins A group of antibiotics produced either by *Penicillium* (natural penicillins) or by adding side chains to the β-lactam ring (semisynthetic penicillins).
pentose phosphate pathway A metabolic pathway that can occur simultaneously with glycolysis to produce pentoses and NADH without ATP production; also called hexose monophosphate shunt.
peptide bond A bond joining the amino group of one amino acid to the carboxyl group of a second amino acid with the loss of a water molecule.
peptidoglycan The structural molecule of bacterial cell walls consisting of the molecules N-acetylglucosamine, N-acetylmuramic acid, tetrapeptide side chain, and peptide side chain.
perforin Protein that makes a pore in a target cell membrane, released by T_C cells.
pericarditis Inflammation of the pericardium, the sac around the heart.
period of convalescence The recovery period, when the body returns to its predisease state.
peripheral nervous system (PNS) The nerves that connect the outlying parts of the body with the central nervous system.
periplasm The region of a gram-negative cell wall between the outer membrane and the cytoplasmic membrane.
peritrichous Having flagella distributed over the entire cell.
peroxidase An enzyme that destroys hydrogen peroxide: $H_2O_2 + 2\,H^+ \rightarrow 2\,H_2O$
peroxide anion An oxygen anion consisting of two atoms of oxygen (O_2^{2-}).
peroxisome Organelle that oxidizes amino acids, fatty acids, and alcohol.
peroxygen A class of oxidizing-type sterilizing disinfectants.
persistent viral infection A disease process that occurs gradually over a long period.
pfu (plaque-forming units) Visible viral plaques counted.
pH The symbol for hydrogen ion (H^+) concentration; a measure of the relative acidity or alkalinity of a solution.
phage *See* bacteriophage.
phage conversion Genetic change in the host cell resulting from infection by a bacteriophage.
phage typing A method of identifying bacteria using specific strains of bacteriophages.
phagocyte A cell capable of engulfing and digesting particles that are harmful to the body.
phagocytosis The ingestion of solids by eukaryotic cells.
phagolysosome A digestive vacuole.
phagosome A food vacuole of a phagocyte; also called a phagocytic vesicle.
phalloidin A peptide toxin produced by *Amanita phalloides,* affects plasma membrane function.
phase-contrast microscope A compound light microscope that allows examination of structures inside cells through the use of a special condenser.
phenol OH Also called carbolic acid.
phenolic A synthetic derivative of phenol used as a disinfectant.
phenotype The external manifestations of an organism's genotype, or genetic makeup.
phosphate group A portion of a phosphoric acid molecule attached to some other molecule, Ⓟ,

$$PO_4^{3-},\ {}^{-}O-\overset{\overset{\large O}{\|}}{\underset{\underset{\large O^-}{|}}{P}}-O^-$$

phospholipid A complex lipid composed of glycerol, two fatty acids, and a phosphate group.
phosphorous cycle The various solubility stages of phosphorus in the environment.
phosphorylation The addition of a phosphate group to an organic molecule.
photoautotroph An organism that uses light as its energy source and carbon dioxide (CO_2) as its carbon source.
photoheterotroph An organism that uses light as its energy source and an organic carbon source.
photolyase An enzyme that splits thymine dimers in the presence of visible light.
photophosphorylation The production of ATP in a series of redox reactions; electrons from chlorophyll initiate the reactions.
photosynthesis The conversion of light energy from the sun into chemical energy; the light-fueled synthesis of carbohydrate from carbon dioxide (CO_2).
phototaxis Movement in response to the presence of light.
phototroph An organism that uses light at its primary energy source.
phylogeny The evolutionary history of a group of organisms; phylogenetic relationships are evolutionary relationships.
phylum A taxonomic classification between kingdom and class.
phytoplankton Free-floating photoautotrophs.
pilus (plural: **pili**) An appendage on a bacterial cell used for conjugation and gliding motility.
pinocytosis The engulfing of fluid by infolding of the plasma membrane, in eukaryotes.
plankton Free-floating aquatic organisms.
plantae The kingdom composed of multicellular eukaryotes with cellulose cell walls.
plaque A clearing in a bacterial lawn resulting from lysis by phages. *See also* dental plaque.
plasma (1) The liquid portion of blood in which the formed elements are suspended. (2) Excited gases used for sterilizing.
plasma cell A cell that an activated B cell differentiates into; plasma cells manufacture specific antibodies.
plasma (cytoplasmic) membrane The selectively permeable membrane enclosing the cytoplasm of a cell; the outer layer in animal cells, internal to the cell wall in other organisms.
plasmid A small circular DNA molecule that replicates independently of the chromosome.
plasmodium (1) A multinucleated mass of protoplasm, as in plasmodial slime molds. (2) When written as a genus, refers to the causative agent of malaria.

plasmogamy Fusion of the cytoplasm of two cells; occurs in the sexual stage of a fungal life cycle.
plasmolysis Loss of water from a cell in a hypertonic environment.
plate count A method of determining the number of bacteria in a sample by counting the number of colony-forming units on a solid culture medium.
pleomorphic Having many shapes, characteristic of certain bacteria.
pluripotent A cell that can differentiate into a many different types of tissue cells.
pneumonia Inflammation of the lungs.
point mutation *See* base substitution.
polar flagella Having flagella at one or both ends of a cell.
polar molecule A molecule with an unequal distribution of charges.
polymer A molecule consisting of a sequence of similar molecules, or monomers.
polymerase chain reaction (PCR) A technique using DNA polymerase to make multiple copies of a DNA template in vitro. *See also* cDNA.
polymorphonuclear leukocyte (PMN) *See* neutrophil.
polypeptide (1) A chain of amino acids. (2) A group of antibiotics.
polysaccharide A carbohydrate consisting of 8 or more monosaccharides joined through dehydration synthesis.
porins A type of protein in the outer membrane of gram-negative cell walls that permits the passage of small molecules.
portal of entry The avenue by which a pathogen gains access to the body.
portal of exit The route by which a pathogen leaves the body.
positive (direct) selection A procedure for picking out mutant cells by growing them.
pour plate method A method of inoculating a solid nutrient medium by mixing bacteria in the melted medium and pouring the medium into a Petri dish to solidify.
prebiotics Chemicals that promote growth of beneficial bacteria in the body.
precipitation reaction A reaction between soluble antigens and multivalent antibodies to form visible aggregates.
precipitin ring test A precipitation test performed in a capillary tube.
predisposing factor Anything that makes the body more susceptible to a disease or alters the course of a disease.
prevalence The fraction of a population having a specific disease at a given time.
primary cell line Human tissue cells that grow for only a few generations in vitro.
primary infection An acute infection that causes the initial illness.
primary metabolite A product of an industrial cell population produced during the time of rapid logarithmic growth. *See also* secondary metabolite.
primary producer An autotrophic organism, either chemotroph or phototroph, that converts carbon dioxide into organic compounds.
primary response Antibody production in response to the first contact with an antigen. *See also* memory response.
primary sewage treatment The removal of solids from sewage by allowing them to settle out and be held temporarily in tanks or ponds.
prion An infectious agent consisting of a self-replicating protein, with no detectable nucleic acids.
privileged site (tissue) An area of the body (or a tissue) that does not elicit an immune response.
probiotics Microbes inoculated into a host to occupy a niche and prevent growth of pathogens.
prodromal period The time following the incubation period when the first symptoms of illness appear.
profundal zone The deeper water under the limnetic zone in an inland body of water.
proglottid A body segment of a tapeworm containing both male and female organs.
prokaryote A cell whose genetic material is not enclosed in a nuclear envelope.
prokaryotic species A population of cells that share certain rRNA sequences; in conventional biochemical testing, it is a population of cells with similar characteristics.
promoter The starting site on a DNA strand for transcription of RNA by RNA polymerase.
prophage Phage DNA inserted into the host cell's DNA.
prophylactic Anything used to prevent disease.
prostaglandin A hormonelike substance that is released by damaged cells, intensifies inflammation.
prostheca A stalk or bud protruding from a prokaryotic cell.
protease An enzyme that digests protein (proteolytic enzymes).
protein A large molecule containing carbon, hydrogen, oxygen, and nitrogen (and sulfur); some proteins have a helical structure and others are pleated sheets.
protein kinase An enzyme that activates another protein by adding a Ⓟ from ATP.
proteobacteria Gram-negative, chemoheterotrophic bacteria that possess a signature rRNA sequence.
proteomics The science of determining all of the proteins expressed in a cell.
protist Term used for unicellular and simple multicellular eukaryotes; usually protozoa and algae.
proton A positively charged particle in the nucleus of an atom.
protoplast A gram-positive bacterium or plant cell treated to remove the cell wall.
protoplast fusion A method of joining two cells by first removing their cell walls; used in genetic engineering.
protozoan (plural: **protozoa**) Unicellular eukaryotic organisms; usually chemoheterotrophic.
provirus Viral DNA that is integrated into the host cell's DNA.
pseudohypha A short chain of fungal cells that results from the lack of separation of daughter cells after budding.
pseudopod An extension of a eukaryotic cell that aids in locomotion and feeding.
psychrophile An organism that grows best at about 15°C and does not grow above 20°C; a cold-loving microbe.
pscyhrotroph An organism that is capable of growth between about 0°C and 30°C.
purines The class of nucleic acid bases that includes adenine and guanine.
purple nonsulfur bacteria Alphaproteobacteria; strictly anaerobic and phototrophic; grow on yeast extract in dark; use reduced organic compounds as electron donors for CO_2 fixation.
purple sulfur bacteria Gammaproteobacteria; strictly anaerobic and phototrophic; use reduced sulfur compounds as electron donors for CO_2 fixation.
pus An accumulation of dead phagocytes, dead bacterial cells, and fluid.
pustule A small pus-filled elevation of skin.
pyocyanin A blue-green pigment produced by *Pseudomonas aeruginosa.*
pyrimidines The class of nucleic acid bases that includes uracil, thymine, and cytosine.

quaternary ammonium compound (quat) A cationic detergent with four organic groups attached to a central nitrogen atom; used as a disinfectant.
quorum sensing The ability of bacteria to communicate and coordinate behavior via signaling molecules.

R Used to represent nonfunctional groups of a molecule. *See also* resistance factor.
random shotgun sequencing A technique for determining the nucleotide sequence in an organism's genome.
rapid plasma reagin (RPR) test A serological test for syphilis.
r-determinant A group of genes for antibiotic resistance carried on R factors.
RecA Catalyzes joining of DNA strands, facilitates recombination of DNA.
receptor An attachment for a pathogen on a host cell.
recipient cell A cell that receives DNA from a donor cell during genetic recombination.
recombinant DNA (rDNA) A DNA molecule produced by combining DNA from two different sources.
recombinant DNA (rDNA) technology Manufacturing and manipulating genetic material in vitro; also called genetic engineering.
recombinant vaccine A vaccine made by recombinant DNA techniques.
redia A trematode larval stage that reproduces asexually to produce cercariae.
redox reaction *See* oxidation-reduction.
red tide A bloom of planktonic dinoflagellates.
reducing medium A culture medium containing ingredients that will remove dissolved oxygen from the medium to allow the growth of anaerobes.
reduction The addition of electrons to a molecule.
refractive index The relative velocity with which light passes through a substance.

relative risk A comparison of the risk of disease in two groups.
rennin An enzyme that forms curds as part of any dairy fermentation product; originally from calves' stomachs, now produced by molds and bacteria.
replica plating A method of inoculating a number of solid minimal culture media from an original plate to produce the same pattern of colonies on each plate.
replication fork The point where DNA strands separate and new strands will be synthesized.
repression The process by which a repressor protein can stop the synthesis of a protein.
repressor A protein that binds to the operator site to prevent transcription.
reservoir of infection A continual source of infection.
resistance The ability to ward off diseases through innate and adaptive immunity.
resistance (R) factor A bacterial plasmid carrying genes that determine resistance to antibiotics.
resistance transfer factor (RTF) A group of genes for replication and conjugation on the R factor.
resolution The ability to distinguish fine detail with a magnifying instrument; also called resolving power.
respiration A series of redox reactions in a membrane that generates ATP; the final electron acceptor is usually an inorganic molecule.
restriction enzyme An enzyme that cuts double-stranded DNA at specific sites between nucleotides.
reticulate body The intracellular growing stage of chlamydiae.
retort A device for commercially sterilizing canned food by using steam under pressure; operates on the same principle as an autoclave but is much larger.
reverse genetics Genetic analysis that begins with a piece of DNA and proceeds to find out what it does.
reverse transcriptase An RNA-dependent DNA polymerase; an enzyme that synthesizes a complementary DNA from an RNA template.
reversible reaction A chemical reaction in which the end-products can readily revert to the original molecules.
RFLP Restriction fragment length polymorphism; a fragment resulting from restriction-enzyme digestion of DNA.
Rh factor An antigen on red blood cells of rhesus monkeys and most humans; possession makes the cells Rh^+.
rhizine A rootlike hypha that anchors a fungus to a surface.
ribonucleic acid (RNA) The class of nucleic acids that comprises messenger RNA, ribosomal RNA, and transfer RNA.
ribose A five-carbon sugar that is part of ribonucleotide molecules and RNA.
ribosomal RNA (rRNA) The type of RNA molecule that forms ribosomes.
ribosomal RNA (rRNA) sequencing Determination of the order of nucleotide bases in rRNA.
ribosome The site of protein synthesis in a cell, composed of RNA and protein.
ribozyme An enzyme consisting of RNA that specifically acts on strands of RNA to remove introns and splice together the remaining exons.
ring stage A young *Plasmodium* trophozoite that looks like a ring in a red blood cell.
RNAi RNA interference; stops gene expression at transcription by using a short interfering RNA to make double-stranded RNA.
RNA primer A short strand of RNA used to start synthesis of the lagging strand of DNA, and to start the polymerase chain reaction.
root nodule A tumorlike growth on the roots of certain plants containing symbiotic nitrogen-fixing bacteria.
rotating biological contactor A method of secondary sewage treatment in which large disks are rotated while partially submerged in a sewage tank exposing sewage to microorganisms and aerobic conditions.
rough ER Endoplasmic reticulum with ribosomes on its surface.
roundworm An animal belonging to the phylum Nematoda.

S (Svedberg unit) Notes the relative rate of sedimentation during ultra-high-speed centrifugation.
salt A substance that dissolves in water to cations and anions, neither of which is H^+ or OH^-.
sanitization The removal of microbes from eating utensils and food preparation areas.
saprophyte An organism that obtains its nutrients from dead organic matter.
sarcina (plural: **sarcinae**) (1) A group of eight bacteria that remain in a packet after dividing. (2) When written as a genus, refers to gram-positive, anaerobic cocci.
saturation (1) The condition in which the active site on an enzyme is occupied by the substrate or product at all times. (2) In a fatty acid, having no double bonds.
saxitoxin A neurotoxin produced by some dinoflagellates.
scanned probe microscopy Microscopic technique used to obtain images of molecular shapes, to characterize chemical properties, and to determine temperature variations within a specimen.
scanning acoustic microscope (SAM) A microscope that uses high-frequency ultrasound waves to penetrate surfaces.
scanning electron microscope (SEM) An electron microscope that provides three-dimensional views of the specimen magnified 1000–10,000×.
scanning tunneling microscopy *See* scanned-probe microscopy.
schizogony The process of multiple fission, in which one organism divides to produce many daughter cells.
scientific nomenclature *See* binomial nomenclature.
sclerotia The compact mass of hardened mycelia of the fungus *Claviceps purpurea* that fills infected rye flowers; produces the toxin ergot.
scolex The head of a tapeworm, containing suckers and possibly hooks.
secondary infection An infection caused by an opportunistic microbe after a primary infection has weakened the host's defenses.
secondary metabolite A product of an industrial cell population produced after the microorganism has largely completed its period of rapid growth and is in a stationary phase of the growth cycle. *See also* primary metabolite.
secondary response *See* memory response.
secondary sewage treatment Biological degradation of the organic matter in wastewater following primary treatment.
secretory vesicle A membrane-enclosed sac produced by the ER; transports synthesized material into cytoplasm.
selective medium A culture medium designed to suppress the growth of unwanted microorganisms and encourage the growth of desired ones.
selective permeability The property of a plasma membrane to allow certain molecules and ions to move through the membrane while restricting others.
selective toxicity The property of some antimicrobial agents to be toxic for a microorganism and nontoxic for the host.
self Host tissue.
semiconservative replication The process of DNA replication in which each double-stranded DNA molecule contains one original strand and one new strand.
sense codon A codon that codes for an amino acid.
sense strand (+ strand) Viral RNA that can act as mRNA.
sensitivity Percentage of positive samples correctly detected by a diagnostic test.
sentinel animal An organism in which changes can be measured to assess the extent of environmental contamination and its implication for human health.
sepsis The presence of a toxin or pathogenic organism in blood and tissue.
septate hypha A hypha consisting of uninucleate cell-like units.
septicemia The proliferation of pathogens in the blood, accompanied by fever; sometimes causes organ damage.
septic shock A sudden drop in blood pressure induced by bacterial toxins.
septum A cross-wall in a fungal hypha.
serial dilution The process of diluting a sample several times.
seroconversion A change in a person's response to an antigen in a serological test.
serological testing Techniques for identifying a microorganism based on its reaction with antibodies.
serology The branch of immunology that studies blood serum and antigen–antibody reactions in vitro.
serotype *See* serovar.
serovar A variation within a species; also called serotype.
serum The liquid remaining after blood plasma is clotted; contains antibodies (immunoglobulins).
sexual dimorphism The distinctly different appearance of adult male and female organisms.
sexual spore A spore formed by sexual reproduction.
Shiga toxin An exotoxin produced by *Shigella dysenteriae* and enterohemorrhagic *E. coli*.
shock Any life-threatening loss of blood pressure. *See also* septic shock.

Glossary

shuttle vector A plasmid that can exist in several different species; used in genetic engineering.
siderophore Bacterial iron-binding proteins.
sign A change due to a disease that a person can observe and measure.
simple stain A method of staining microorganisms with a single basic dye.
singlet oxygen Highly reactive molecular oxygen (O_2^-).
siRNA Short interfering RNA; An intermediate in the RNAi process in which the long double-stranded RNA has been cut up into short (~21 nucleotides) double-stranded RNA.
site-directed mutagenesis Techniques used to modify a gene in a specific location to produce the desired polypeptide.
slide agglutination test A method of identifying an antigen by combining it with a specific antibody on a slide.
slime layer A glycocalyx that is unorganized and loosely attached to the cell wall.
sludge Solid matter obtained from sewage.
smear A thin film of material containing microorganisms, spread over the surface of a slide.
smooth ER Endoplasmic reticulum without ribosomes.
solute A substance dissolved in another substance.
solvent A dissolving medium.
Southern blotting A technique that uses DNA probes to detect the presence of specific DNA in restriction fragments separated by electrophoresis.
specialized transduction The process of transferring a piece of cell DNA adjacent to a prophage to another cell.
species The most specific level in the taxonomic hierarchy. *See also* bacterial species; eukaryotic species; viral species.
specific epithet The second or species name in a scientific binomial. *See also* species.
specificity Percentage of false positive results given by a diagnostic test.
spectrum of microbial activity The range of distinctly different types of microorganisms affected by an antimicrobial drug; a wide range is referred to as a broad spectrum of activity.
spheroplast A gram-negative bacterium treated to damage the cell wall, resulting in a spherical cell.
spicule One of two external structures on the male roundworm used to guide sperm.
spike A carbohydrate-protein complex that projects from the surface of certain viruses.
spiral *See* spirillum and spirochete.
spirillum (plural: **spirilla**) (1) A helical or corkscrew-shaped bacterium. (2) When written as a genus, refers to aerobic, helical bacteria with clumps of polar flagella.
spirochete A corkscrew-shaped bacterium with axial filaments.
spontaneous generation The idea that life could arise spontaneously from nonliving matter.
spontaneous mutation A mutation that occurs without a mutagen.
sporadic disease A disease that occurs occasionally in a population.
sporangiophore An aerial hypha supporting a sporangium.
sporangiospore An asexual fungal spore formed within a sporangium.
sporangium A sac containing one or more spores.
spore A reproductive structure formed by fungi and actinomycetes. *See also* endospore.
sporogenesis *See* sporulation.
sporozoite A trophozoite of *Plasmodium* found in mosquitoes, infective for humans.
sporulation The process of spore and endospore formation; also called sporogenesis.
spread plate method A plate count method in which inoculum is spread over the surface of a solid culture medium.
staining Colorizing a sample with a dye to view through a microscope or to visualize specific structures.
staphylococci (singular: **staphylococcus**) Cocci in a grapelike cluster or broad sheet.
stationary phase The period in a bacterial growth curve when the number of cells dividing equals the number dying.
stem cell An undifferentiated cell that gives rise to a variety of specialized cells.
stereoisomers Two molecules consisting of the same atoms, arranged in the same manner but differing in their relative positions; mirror images; also called D-isomer and L-isomer.
sterile Free of microorganisms.
sterilization The removal of all microorganisms, including endospores.
steroid A specific group of lipids, including cholesterol and hormones.
stipe A stemlike supporting structure of multicellular algae and basidiomycetes.
storage vesicle Organelles that form from the Golgi complex; contain proteins made in the rough ER and processed in the Golgi complex.
strain Genetically different cells within a clone. *See* serovar.
streak plate method A method of isolating a culture by spreading microorganisms over the surface of a solid culture medium.
streptobacilli (singular: **streptobacillus**) Rods that remain attached in chains after cell division.
streptococci (singular: **streptococcus**) (1) Cocci that remain attached in chains after cell division. (2) When written as a genus, refers to gram-positive, catalase-negative bacteria.
streptokinase A blood-clot dissolving enzyme, produced by beta-hemolytic streptococci.
streptolysin A hemolytic enzyme, produced by streptococci.
structural gene A gene that determines the amino acid sequence of a protein.
subacute disease A disease with symptoms that are intermediate between acute and chronic.
subclinical infection An infection that does not cause a noticeable illness; also called inapparent infection.
subcutaneous mycosis A fungal infection of tissue beneath the skin.
substrate Any compound with which an enzyme reacts.
substrate-level phosphorylation The synthesis of ATP by direct transfer of a high-energy phosphate group from an intermediate metabolic compound to ADP.
subunit vaccine A vaccine consisting of an antigenetic fragment.
sulfhydryl group —SH.
sulfur cycle The various oxidation and reduction stages of sulfur in the environment, mostly due to the action of microorganisms.
sulfur granule *See* inclusion.
superantigen An antigen that activates many different T cells, thereby eliciting a large immune response.
superficial mycosis A fungal infection localized in surface epidermal cells and along hair shafts.
superinfection The growth of a pathogen that has developed resistance to an antimicrobial drug being used; the growth of an opportunistic pathogen.
superoxide dismutase (SOD) An enzyme that destroys superoxide: $O_2^- + O_2^- + 2\,H^+ \rightarrow H_2O_2 + O_2$
superoxide radical A toxic anion (O_2^-) with an unpaired electron.
surface-active agent Any compound that decreases the tension between molecules lying on the surface of a liquid; also called surfactant.
susceptibility The lack of resistance to a disease.
symbiosis The living together of two different organisms or populations.
symptom A change in body function that is felt by a patient as a result of a disease.
syncytium A multinucleated giant cell resulting from certain viral infections.
syndrome A specific group of signs or symptoms that accompany a disease.
synergism The principle whereby the effectiveness of two drugs used simultaneously is greater than that of either drug used alone.
synthesis reaction A chemical reaction in which two or more atoms combine to form a new, larger molecule.
synthetic drug A chemotherapeutic agent that is prepared from chemicals in a laboratory.
systematics The science organizing groups of organisms into a hierarchy.
systemic anaphylaxis A hypersensitivity reaction causing vasodilation and resulting in shock; also called anaphylactic shock.
systemic (generalized) infection An infection throughout the body.
systemic mycosis A fungal infection in deep tissues.

tachyzoite A rapidly growing trophozoite form of a protozoan.
T antigen An antigen in the nucleus of a tumor cell.
tapeworm A flatworm belonging to the class Cestoda.

Glossary

target cell An infected body cell to which defensive cells of the immune system bind.
taxa Subdivisions used to classify organisms, e.g., domain, kingdom, phylum.
taxis Movement in response to an environmental stimulus.
taxonomy The science of the classification of organisms.
T cell A type of lymphocyte, which develops from a stem cell processed in the thymus gland, that is responsible for cell-mediated immunity. *See also* T cytotoxic cells, T helper cells, T regulatory cells.
TCRs (T-cell receptors) Molecules on T cells that recognize antigens.
T cytotoxic (T_C) cells A specialized T cell that destroys infected cells presenting antigens.
T helper (T_H) cell A specialized T cell that often interacts with an antigen before B cells interact with the antigen.
T regulatory (T_{reg}) cells Lymphocytes that appear to suppress other T cells.
T-dependent antigen An antigen that will stimulate the formation of antibodies only with the assistance of T helper cells. *See also* T-independent antigen.
teichoic acid A polysaccharide found in gram-positive cell walls.
telomere Noncoding regions of DNA at the ends of eukaryotic chromosomes.
teleomorph The sexual stage in the life cycle of a fungus; also refers to a fungus that produces both sexual and asexual spores.
temperate phage A phage capable of lysogeny.
temperature abuse Improper food storage at a temperature that allows bacteria to grow.
terminator The site on a DNA strand at which transcription ends.
tertiary sewage treatment A method of waste treatment that follows conventional secondary sewage treatment; nonbiodegradable pollutants and mineral nutrients are removed, usually by chemical or physical means.
tetrad A group of four cocci.
thallus The entire vegetative structure or body of a fungus, lichen, or alga.
thermal death point (TDP) The temperature required to kill all the bacteria in a liquid culture in 10 minutes.
thermal death time (TDT) The length of time required to kill all bacteria in a liquid culture at a given temperature.
thermoduric Heat resistant.
thermophile An organism whose optimum growth temperature is between 50°C and 60°C; a heat loving microbe.
thermophilic anaerobic spoilage Spoilage of canned foods due to the growth of thermophilic bacteria.
thylakoid A chlorophyll-containing membrane in a chloroplast. A bacterial thylakoid is also known as a chromatophore.
thymus A mammalian organ responsible for maturation of the immune system.
tincture A solution in aqueous alcohol.
T-independent antigen An antigen that will stimulate the formation of antibodies without the assistance of T helper cells. *See also* T-dependent antigen.
tinea Fungal infection of hair, skin, or nails.
titer An estimate of the amount of antibodies or viruses in a solution; determined by serial dilution and expressed as the reciprocal of the dilution.
TLRs (toll-like rceptors) Transmembrane proteins of immune cells that recognize pathogens and activate an immune responses directed against those pathogens.
topoisomerase Enzyme that relaxes supercoiling of DNA ahead of replication form; separates DNA circles at the end of DNA replication.
total magnification The magnification of a microscopic specimen, determined by multiplying the ocular lens magnification by the objective lens magnification.
toxemia The presence of toxins in the blood.
toxigenicity The capacity of a microorganism to produce a toxin.
toxin Any poisonous substance produced by a microorganism.
toxoid An inactivated toxin.
T plasmid An *Agrobacterium* plasmid carrying genes for tumor induction in plants.
trace element A chemical element required in small amounts for growth.
trans Hydrogen atoms on opposite side across a double bond in a fatty acid. *See* cis.
transamination The transfer of an amino group from an amino acid to another organic acid.
transcription The process of synthesizing RNA from a DNA template.
transduction The transfer of DNA from one cell to another by a bacteriophage. *See also* generalized transduction; specialized transduction.
transferrin A human iron-binding protein that reduces iron available to a pathogen.
transfer RNA (tRNA) The type of RNA molecule that brings amino acids to the ribosomal site where they are incorporated into proteins.
transfer vesicle Membrane-bound sacs that move proteins from the Golgi complex to specific areas in the cell.
transformation (1) The process in which genes are transferred from one bacterium to another as "naked" DNA in solution. (2) The changing of a normal cell into a cancerous cell.
transient microbiota The microorganisms that are present in an animal for a short time without causing a disease.
translation The use of mRNA as a template in the synthesis of protein.
transmission electron microscope (TEM) An electron microscope that provides high magnifications (10,000–100,000×) of thin sections of a specimen.
transport vesicle Membrane-bound sacs that move proteins from the rough ER to the Golgi complex.
transporter protein A carrier protein in the plasma membrane.
transposon A small piece of DNA that can move from one DNA molecule to another.
trickling filter A method of secondary sewage treatment in which sewage is sprayed out of rotating arms onto a bed of rocks or similar materials, exposing the sewage to highly aerobic conditions and microorganisms.
triglyceride A simple lipid consisting of glycerol and three fatty acids.
triplex agent A short segment of DNA that binds to a target area on a double strand of DNA blocking transcription.
trophophase The period in the production curve of an industrial cell population in which the primary metabolites are formed; a period of rapid, logarithmic growth. *See also* idiophase.
trophozoite The vegetative form of a protozoan.
tuberculin skin test A skin test used to detect the presence of antibodies to *Mycobacterium tuberculosis*.
tumor necrosis factor (TNF) A polypeptide released by phagocytes in response to bacterial endotoxins.
tumor-specific transplantation antigen (TSTA) A viral antigen on the surface of a transformed cell.
turbidity The cloudiness of a suspension.
turnover number The number of substrate molecules acted on per enzyme molecule per second.
two-photon microscope A light microscope that uses fluorescent stains and long wavelength light.

ubiquinone A low–molecular weight, nonprotein carrier in an electron transport chain; also called coenzyme Q.
ultra-high-temperature (UHT) treatment A method of treating food with high temperatures (140–150°C) for very short times to make the food sterile so that it can be stored at room temperature.
uncoating The separation of viral nucleic acid from its protein coat.
undulating membrane A highly modified flagellum on some protozoa.
unsaturated A fatty acid with one or more double bonds.
use-dilution test A method of determining the effectiveness of a disinfectant using serial dilutions.

vaccination The process of conferring immunity by administering a vaccine; also called immunization.
vaccine A preparation of killed, inactivated, or attenuated microorganisms or toxoids to induce artificially acquired active immunity.
vacuole An intracellular inclusion, in eukaryotic cells, surrounded by a plasma membrane; in prokaryotic cells, surrounded by a proteinaceous membrane.
valence The combining capacity of an atom or a molecule.
vancomycin An antibiotic that inhibits cell wall synthesis.
variolation An early method of vaccination using infected material from a patient.
vasodilation Dilation or enlargement of blood vessels.

VDRL test A rapid screening test to detect the presence of antibodies against Treponema pallidum. (VDRL stands for Venereal Disease Research Laboratory.)
vector (1) A plasmid or virus used in genetic engineering to insert genes into a cell. (2) An arthropod that carries disease-causing organisms from one host to another.
vegetative Referring to cells involved with obtaining nutrients, as opposed to reproduction.
vehicle transmission The transmission of a pathogen by an inanimate reservoir.
vertical gene transfer Transfer of genes from an organism or cell to its offspring.
vesicle (1) A small serum-filled elevation of the skin. (2) Smooth oval bodies formed in plant roots by mycorrhizae.
V factor NAD^+ or $NADP^+$.
vibrio (1) A curved or comma-shaped bacterium. (2) When written as a genus (*Vibrio*), a gram-negative, motile, facultatively anaerobic curved rod.
viral hemagglutination The ability of certain viruses to cause the clumping of red blood cells in vitro.
viral hemagglutination inhibition test A neutralization test in which antibodies against particular viruses prevent the viruses from clumping red blood cells in vitro.
viral species A group of viruses sharing the same genetic information and ecological niche.
viremia The presence of viruses in the blood.
virion A complete, fully developed viral particle.
viroid Infectious RNA.
virology The scientific study of viruses.
virulence The degree of pathogenicity of a microorganism.
virus A submicroscopic, parasitic, filterable agent consisting of a nucleic acid surrounded by a protein coat.
volutin Stored inorganic phosphate in a prokaryotic cell. *See also* metachromatic granule.

Western blotting A technique that uses antibodies to detect the presence of specific proteins separated by electrophoresis.
whey The fluid portion of milk that separates from curd.

xenobiotics Synthetic chemicals that are not readily degraded by microorganisms.
xenodiagnosis A method of diagnosis based on exposing a parasite-free normal host to the parasite and then examining the host for parasites.
xenotransplantation product A tissue graft from another species; also called xenotransplant.
X factor Substances from the heme fraction of blood hemoglobin.

yeast Nonfilamentous, unicellular fungi.
yeast infection Disease caused by growth of certain yeasts in a susceptible host.

zone of inhibition The area of no bacterial growth around an antimicrobial agent in the disk-diffusion method.
zoonosis A disease that occurs primarily in wild and domestic animals but can be transmitted to humans.
zoospore An asexual algal spore; has two flagella.
zygospore A sexual fungal spore characteristic of the zygomycetes.
zygote A diploid cell produced by the fusion of two haploid gametes.

CREDITS

Illustration Credits

All illustrations by Precision Graphics unless otherwise noted.

2.1, 2.3, 2.9, 2.12, 2.14, 2.15: G. J. Tortora and S. R. Grabowski, *Principles of Anatomy and Physiology*, 8th ed. F2.1, 2.4, 2.6, 2.9, 2.12, 2.13, 2.14. © 1996 Biological Sciences Textbooks. Reproduced by permission of Pearson Science. Illustrations, Jared Schneidman Design.

4.22, 4.24, 4.25, 4.26: Adapted from G. J. Tortora and S. R. Grabowski, *Principles of Anatomy and Physiology*, 9th ed. F3.1, 3.25, 3.20, 3.21. © 2000 Biological Sciences Textbooks. Reproduced by permission of Wiley.

5.1: G. J. Tortora and S. R. Grabowski, *Principles of Anatomy and Physiology*, 8th ed. F25.1. © 1996 Biological Sciences Textbooks. Reproduced by permission of Pearson Science. Illustration, Page Two Associates.

8.7: P. Berg and M. Singer, *Dealing With Genes: The Language of Heredity* (Mill Valley, CA: University Science Books, 1992) F13.1.

9.16: From J. D. Watson et al., *Recombinant DNA*, 2nd ed. (New York: W. H. Freeman, 1992).

13.T2: From R. I. B. Francki et al., eds., "Classification and Nomenclature of Viruses, Fifth Report of the Intl. Comm. on Taxonomy of Viruses," *Archives of Virology: Supplementum 2* (Vienna: Springer-Verlag, 1991).

14.4: CDC

16.5: G. J. Tortora and S. R. Grabowski, *Principles of Anatomy and Physiology*, 7th ed., F22.1, 22.10. © 1993 Biological Sciences Textbooks. Reproduced by permission of Pearson Science.

19.14, 19.15: P. D. Greenberg, "Immunopathogenesis of HIV Infection," *Hospital Practice* 27:109. Illustrations, Ilil Arbel. Reproduced with permission.

19.17: Adapted from UNAIDS data by the Map Design Unit of the World Bank. Reproduced by permission of Confronting AIDS: Public Priorities in a Global Epidemic and the U. of California San Francisco data from HIVInSite, 2005.

20.21: T. D. Brock et al., *Biology of Microorganisms*, 7th ed. © Prentice-Hall, 1993, p. 410. Reproduced with permission.

21.13, 22.12: Adapted from P. R. Murphy et al., *Medical Microbiology*, 3rd ed. (Philadelphia: Lippincott Williams & Wilkins, 1993). Reproduced with permission.

23.14: A. C. Steere, "Current Understanding of Lyme Disease," *Hospital Practice* 28: 4, p. 37. Illustration, Nancy Lou Makris Riccio. Reproduced with permission.

24.9: A. M. Dannenberg, Jr., "Pulmonary Tuberculosis," *Hospital Practice* 28:51. Illustrations, Seward Hung and Laura Pardi Duprey. Reproduced with permission.

24.19: CDC

25.8, 25.9: M. Schaechter et al., eds., *Mechanisms of Microbial Disease*, 2nd ed. (Baltimore, MD: Lippincott Williams & Wilkins, 1993) F18.1.

25.13: From "Helicobacter Pylori Infection," *Hospital Practice* 26:F1. Illustration, Laura Pardi Duprey. Reproduced with permission.

Photo Credits

Chapter 1. Opener: SciMAT/Photo Researchers. **1.1 (a):** CNRI/SPL/Photo Researchers. **1.1 (b):** Biophoto Associates/Photo Researchers. **1.1 (c):** K. W. Jean/ Visuals Unlimited. **1.1 (d):** Stephen Durr. **1.1 (e):** NIBSC/Photo Researchers. **1.2 (a):** Pfizer. **1.2 (b):** Christine Case. **1.3:** Charles O'Rear/Bettmann/Corbis. **1.4 (1, 2):** Bettmann/Corbis. **1.4 (3):** Rockefeller Archive Center. **1.5:** Bettmann/ Corbis. **1.6:** Michael M. Kliks/Donald Heyneman. **1.7:** SciMAT/Photo Researchers. **1.8:** Rodney M. Donlan and Janice Carr, CDC. **AM (top, middle):** Sascha Drewlo. **AM (bottom):** Digital Vision/Alamy.

Chapter 2. Opener: C. Ginocchio, S. Olmstead, C. Wells, and J. E. Galan. Contact with epithelial cells induces the formation of surface appendages on Salmonella typhimurium. *Cell.* 1994 Feb 25; 76(4):717–24. Copyright © 1994 by Elsevier Science Ltd. Reproduced with permission. **AM:** Accent Alaska/Ken Graham Agency.

Chapter 3. Opener: Biophoto Associates/Photo Researchers. **3.1:** Leica Microsystems. **3.2 (1):** The Scanning Probe Microscopy Unit/University of Bristol, UK. **3.2 (2):** Eye of Science/Photo Researchers. **3.2 (3):** Scimat/Photo Researchers. **3.2 (4):** Mae Melvin, CDC. **3.2 (5):** Tom Murray/*BugGuide.net*. **3.4:** Specimens prepared and photographed by L. Brent Selinger, Department of Biological Sciences, University of Lethbridge, Alberta, Canada; Pearson Science. **3.5:** Mike Abbey/Visuals Unlimited. **3.6:** CDC. **3.7:** Dennis Kunkel/Phototake. **3.8:** A. Diaspro, P. Fronteb, M. Raimondoa, M. Fatoc, G. DeLeoc, F. Beltramec, F. Cannoned, G. Chiricod, and P. Ramoinob. Functional imaging of living paramecium by means of confocal and two-photon excitation fluorescence microscopy. *Proceedings of SPIE*. 2002; 4622. © 2002 SPIE. From the Diaspro Lab, *www.lambs.it*, Department of Physics, University of Genoa. **3.9:** M. S. Good, C. F. Wend, L. J. Bond, J. S. McLean, P. D. Panetta, S. Ahmed, S. L. Crawford, and D. S. Daly. An estimate of biofilm properties using an acoustic microscope. *IEEE Transactions on Ultrasonics, Ferroelectrics and Frequency Control.* 2006 Sept; 53(9):1637–48. Fig. 5B. © 2006 IEEE. **3.10 (a):** Phototake Electra/ Phototake. **3.10 (b):** Karl Aufderheide/Visuals Unlimited. **3.11 (a):** M. Amrein et al. Scanning tunneling microscopy of recA-DNA complexes coated with a conducting film. *Science.* 1988 Apr 22; 240(4851):514–16. Reprinted with permission. ©1988 AAAS. **3.11 (b):** D. M. Czajkowsky et al. Vertical collapse of a cytolysin prepore moves its transmembrane beta-hairpins to the membrane. *EMBO.* 2004 Aug 18; 23(16):3206–15. Reprinted by permission from Macmillan Publishers. Image provided by Zhifeng Shao, University of Virginia. **3.12:** Jack Bostrack/Visuals Unlimited. **3.13:** E. C. S. Chan/Visuals Unlimited. **3.14 (a):** Jack M. Bostrack/ Visuals Unlimited. **3.14 (b):** Joseph W. Duris and Silvia Rossbach, Western Michigan University. **3.14 (c):** Eric Graves/Photo Researchers. **AM (top):** Sebastien Vilain. **AM (middle left):** Eshel Ben-Jacob, School of Physics and Astronomy, Tel Aviv University, Israel. **AM (middle right):** Heinrich Lünsdorf, Helmholtz Center for Infection Research, Germany. **AM (bottom):** Digital Vision/Alamy. **T3.2 (1, 2, 3):** Specimens prepared and photographed by L. Brent Selinger, Department of Biological Sciences, University of Lethbridge, Alberta, Canada; Pearson Science. **T3.2 (4):** Mike Abbey/Visuals Unlimited. **T3.2 (5):** CDC. **T3.2 (6):** Dennis Kunkel/Phototake. **T3.2 (7):** M. S. Good, C. F. Wend, L. J. Bond, J. S. McLean, P. D. Panetta, S. Ahmed, S. L. Crawford, and D. S. Daly. An estimate of biofilm properties using an acoustic microscope. *IEEE Transactions on Ultrasonics, Ferroelectrics and Frequency Control.* 2006 Sept; 53(9):1637–48. Fig. 5B. © 2006 IEEE. **T3.2 (8):** Phototake Electra/Phototake. **T3.2 (9):** Karl Aufderheide/Visuals Unlimited. **T3.2 (10):** M. Amrein et al. Scanning tunneling microscopy of recA-DNA complexes coated with a conducting film. *Science.* 1988 Apr 22; 240(4851):514–16. Reprinted with permission. ©1988 AAAS. **T3.2 (11):** A. Diaspro, P. Fronteb, M. Raimondoa, M. Fatoc, G. DeLeoc, F. Beltramec, F. Cannoned, G. Chiricod, and P. Ramoinob. Functional imaging of living paramecium by means of confocal and two-photon excitation fluorescence microscopy. *Proceedings of SPIE.* 2002; 4622. © 2002 SPIE. From the Diaspro Lab, *www.lambs.it*, Department of Physics, University of Genoa. **T3.2 (12):** D. M. Czajkowsky et al. Vertical collapse of a cytolysin prepore moves its transmembrane beta-hairpins to the membrane. *EMBO.* 2004 Aug 18; 23(16):3206–15. Reprinted by permission from Macmillan Publishers. Image provided by Zhifeng Shao, University of Virginia. **CA:** Biophoto Associates/ Science Source/Photo Researchers.

Chapter 4. Opener: NIAID/RML. **4.1 (a top):** David M. Phillips/Visuals Unlimited. **4.1 (a bottom):** Oliver Meckes and Nicole Ottawa/Photo Researchers. **4.1 (b, c):** G. Shih and R. Kessel/Visuals Unlimited. **4.1 (d):** David Scharf/Peter Arnold. **4.2 (a, b):** Manfred Kage/Peter Arnold. **4.2 (c):** Dennis Kunkel/Phototake. **4.2 (d):** Microworks Color/Phototake. **4.3:** N. H. Mendelson and J. J. Thwaites, *ASM News* 1993; 59:25. Fig. 2. Reprinted by permission. **4.4 (a):** London School of Hygiene/Photo Researchers. **4.4 (b):** Stanley Flegler/Visuals Unlimited. **4.4 (c):** Charles Stratton/Visuals Unlimited. **4.5 (a):** Horst Volker and Heinz Schlesner, Institut fur Allgemeine Mikrobiologie, Kiel/Michael Thomm. **4.5 (b):** H. W. Jannasch, Woods Hole Oceanographic Institution. **4.6:** Ralph A. Slepecky/Visuals Unlimited. **4.7 (a):** Science Source/Photo Researchers. **4.7 (b):** David M. Phillips/Visuals Unlimited. **4.7 (c):** Biomedical Imaging Unit, Southampton General Hospital/SPL/ Photo Researchers. **4.7 (d):** Ed Reschke/Peter Arnold. **4.9:** Lee D. Simon/Science Source/Photo Researchers. **4.10:** Custom Medical Stock Photo. **4.11:** Kwangshin Kim/Photo Researchers. **4.14:** T. J. Beveridge/Biological Photo Service. **4.15:** H. S. Pankratz and R. L. Uffen, Michigan State University/Biological Photo Service. **4.16:** Christine Case. **4.20:** D. Balkwill and D. Maratea/Visuals Unlimited. **4.21:** Visuals Unlimited. **4.22 (left):** Biophoto Associates/Photo Researchers. **4.22 (right):** D.W. Fawcett/Photo Researchers. **4.23:** David M. Phillips/Visuals Unlimited. **4.24:** CNRI/SPL/Photo Researchers. **4.25:** R. Bolender and D. Fawcett/ Photo Researchers. **4.26:** M. Powell/Visuals Unlimited. **4.27:** Keith Porter/Photo Researchers. **4.28:** E. H. Newcomb and W. P. Wergin/Biological Photo Service. **AM (top):** Dean Soulia and Lynn Margulis/University of Massachusetts.

AM (bottom): Nigel Cattlin/FLPA. **T4.1:** Manfred Kage/Peter Arnold. **T4.2:** Experiment performed and photographed by L. Brent Selinger, Department of Biological Sciences, University of Lethbridge, Alberta, Canada; Pearson Science.

Chapter 5. Opener: I. Rosenshine et al. A pathogenic bacterium triggers epithelial signals to form a functional bacterial receptor that mediates actin pseudopod formation. *The EMBO Journal.* 1996 June 3; 15(11):2613–24. Cover. **5.4:** T.A. Steitz, Yale University. **5.22, 5.23, 5.24:** Christine Case. **AM:** Bryan Reinhart/Masterfile. **CF:** Christine Case.

Chapter 6. Opener: Scimat/Photo Researchers. **6.7:** Lester Lefkowitz/Corbis. **6.8:** Jim Gathany, CDC. **6.9:** Christine Case. **6.10:** Courtesy of and © of Becton, Dickinson and Company. **6.11:** Christine Case. **6.12:** Lee D. Simon/Photo Researchers. **6.18 (a):** Pall/Visuals Unlimited. **6.18 (b):** K. Taiaro/Visuals Unlimited. **CF (b):** Christine Case.

Chapter 7. Opener: Dennis Kunkel/Phototake. **7.3, 7.6:** Christine Case. **7.8:** CDC. **CF:** Christine Case.

Chapter 8. Opener, 8.1: Gopal Murti/Photo Researchers. **8.6:** NIH Kakefuda/ Science Source/Photo Researchers. **8.7:** M. Guthold, Wake Forest University, and C. Bustamante, University of California, Berkeley. **8.10:** Visuals Unlimited. **8.26 (a):** Charles C. Brinton, Jr., University of Pittsburgh. **8.26 (b):** Omikron/Photo Researchers. **8.29:** Gopal Murti/Phototake. **CF:** Linda Stannard, UCT/Photo Researchers.

Chapter 9. Opener: Secchi-Lecaque, Roussel-Uclaf/SPL/Photo Researchers. **9.5:** From D. Cheney and B. Metz, unpublished. **9.6:** Matt Meadows/Peter Arnold. **9.7:** Institute Pasteur/Phototake. **9.10:** Matt Meadows/Peter Arnold. **9.13:** Secchi-Lecaque, Roussel-Uclaf/SPL/Photo Researchers. **9.17:** CDC. **9.18:** R. S. Oremland et al. Structural and spectral features of selenium nanospheres produced by Se-respiring bacteria. *Applied Environmental Microbiology.* 2004 Jan; 70(1):52–60. Fig. 1A. © 2004, American Society for Microbiology. **9.19:** Holt Studios International Ltd./Alamy. **CA:** Mike Zeller, Office of Biotechnology, Iowa State University. **CF (b):** S. U. Parshionikar, S. Willian-True, G. S. Fout, D. E. Robbins, S. A. Seys, J. D. Cassady, and R. Harris. Waterborne outbreak of gastroenteritis associated with a norovirus. *Applied and Environmental Microbiology.* Sept 2003; 69(9):5263–68. Fig. 2. **CT:** Christine Case.

Chapter 10. Opener: L. L. Pifer, W. T. Hughes, M. J. Murphy Jr. Propagation of Pneumocystis carinii in vitro. *Pediatric Research.* 1977 Apr; 11(4):305–16. **10.3:** Jerome Pickett-Heaps/Photo Researchers. **10.4 (a):** Georgette Douwma/Nature Picture Library. **10.4 (b):** Christine Case. **10.4 (c):** S. M. Awramik, University of California/Biological Photo Service. **10.9, 10.10:** Christine Case. **10.11 (a):** Sinclair Stammers/Photo Researchers. **10.11 (b):** Colin Cuthbert/Photo Researchers. **10.12:** CDC. **10.13:** Chris Jones, University of Leeds. **10.14:** Pascal Goetgheluck/SPL/ Photo Researchers. **10.17 (a):** Volker Steger/SPL/Photo Researchers. **10.17 (d):** James Liao's Lab, UCLA DNA Microarray Core Facility. **10.18:** V. A. Kempf, K. Trebesius, I. B. Autenrieth. Fluorescent in situ hybridization allows rapid identification of microorganisms in blood cultures. *Journal of Clinical Microbiology.* 2000 Feb; 38(2):830–38. Fig 1. © 2000 American Society for Microbiology. **AM (top):** Darryl W Bush, Marine World/Africa USA, Vallejo, CA. **AM (bottom):** Terry Whittaker/ FLPA. **T10.1 (1):** Oliver Meckes/Nicole Ottawa/Photo Researchers. **T10.1 (2):** Scimat/Photo Researchers. **T10.1 (3):** Stanley Flegler/Visuals Unlimited/Getty Images.

Chapter 11. Opener: Heide Schulz/Max Planck Institute. **11.1:** Animal and Plant Health Inspection Service (APHIS), USDA. **11.2:** Yves V. Brun. **11.3:** Biological Photo Service. **11.4:** John D. Cunningham/Visuals Unlimited. **11.5:** Science VU/Visuals Unlimited. **11.6:** David M. Phillips/Visuals Unlimited. **11.7:** Linda Stannard, University of Cape Town/Photo Researchers. **11.8:** Dennis Kunkel/Visuals Unlimited. **11.9 (a):** Institute Pasteur/CNRI/Phototake. **11.9 (b):** Pablo Zunino, Laboratorio de Microbiología, Instituto de Investigaciones Biológicas Clemente Estable. **11.10:** S. Rendulic, J. Berger, and S. Schuster, Max Planck Institute. **11.11:** Jonathan Eisenback/Phototake. **11.12:** B. Dowsett, CAMR/Photo Researchers. **11.13 (a):** Robert Calentine/Visuals Unlimited. **11.13 (b):** Ron Dengler/Visuals Unlimited. **11.14:** Paul Johnson/Biological Photo Service. **11.15:** Cabisco/Visuals Unlimited. **11.16:** Esther R. Angert. **11.17 (a):** From Hannay and Fitz James, *Canadian Journal of Microbiology* 1, 1955/National Researcher Council of Canada. **11.17 (b):** R. E. Strange and J. R. Hunter, in G. W. Gould and A. Hurst, eds., *The Bacterial Spore.* 1969, p.461, Fig. 4. (Orlando, FL: Academic Press, 1969). **11.18:** Eye of Science/Photo Researchers. **11.19:** SciMAT/Photo Researchers. **11.20 (a):** D. C. Krause and D. Taylor-Robinson, 1992. Mycroplasmas which infect humans. Fig. 1. in J. Maniloff et al., eds., *Mycoplasmas: Molecular Biology and Pathogenesis.* Reproduced by permission of the American Society for Microbiology. **11.20 (b):** Michael Gabridge/Visuals Unlimited. **11.21:** Kitasato University. **11.22:** David M. Phillips/Visuals Unlimited. **11.23:** J. Wang, C. Jenkins, R. I. Webb, and J. A. Fuerst. Isolation of gemmata-like and isosphaera-like planctomycete bacteria from soil and freshwater. *Applied and Environmental Microbiology.* January 2002; 68(1):417–22. Fig. 1c. **11.24:** Kurt Reed, Marshfield Medical Research Foundation. **11.25:** J. A. Breznak and H. S. Pankratz/Biological Photo Service. **11.26:** Chris Bjornberg/Photo Researchers. **11.27:** Karl O. Stetter. **11.28:** Heide Schulz/Max Planck Institute. **AM (top):** University of Bath. **AM (bottom):** Arco Images GmbH/Alamy.

Chapter 12. Opener: Yuuji Tsukii, Hosei University, Japan. **12.2:** Christine Case. **12.3:** David Scharf/Peter Arnold. **12.4:** Christine Case. **12.5 (a):** Dennis Kunkel/Phototake. **12.5 (b):** M. F. Brown/Visuals Unlimited. **12.5 (c, d):** David M. Phillips/Visuals Unlimited. **12.5 (e), 12.6 (1):** G. Shih and R. Kessel/Visuals Unlimited. **12.6 (2):** M. F. Brown/Visuals Unlimited. **12.7:** J. Dijksterhuis, J. Nijsse, F. A. Hoekstra, E. A. Golovina. High viscosity and anisotropy characterize the cytoplasm of fungal dormant stress-resistant spores. *Eukaryot Cell.* 2007 Feb; 6(2):157–70. Fig. 7. **12.8 (1):** Biophoto Associates/Photo Researchers. **12.8 (2), 12.9, 12.10:** Christine Case. **12.11:** Jacqui Hurst/Corbis. **12.12:** Manfred Kage/Peter Arnold. **12.14 (a):** Fred Rhoades/Mycena Consulting. **12.14 (b):** Yuuji Tsukii, Hosei University, Japan. **12.15:** M. Abbey/Visuals Unlimited. **12.16 (b):** David M. Phillips/Visuals Unlimited. **12.16 (c):** Mary Anne Harrington, TML/MSH Shared Microbiology Service. **12.16 (d):** Melanie Moser, DPDx-CDC. **12.17:** Manfred Kage/Peter Arnold. **12.19:** E. R. Degginger/Photo Researchers. **12.20:** Michael Abbey/Visuals Unlimited. **12.21:** Carolina Biological Supply Company/ Phototake. **12.22:** Christine Case. **12.23:** D. O. Rosenberry. Malformed frogs in Minnesota: An update. USGS Water Fact Sheet. 2001; FS-043–01. Photo by David Hoppe. **12.24:** Steve J. Upton, Parasitology Research, Division of Biology, Kansas State University **12.26:** Stanley Flegler/Visuals Unlimited. **12.27 (a):** M. B. Hildreth, M. D. Johnson, K. R. Kazacos. Echinococcus multilocularis: A zoonosis of increasing concern in the United States. *Compendium on Continuing Education for the Practicing Veterinarian.* 1991 May; 13(5): 727–41. **12.27 (b), 12.28:** Robert Calentine/Visuals Unlimited. **12.29:** C. E. Atkins. Heartworm Caval Syndrome. *Sem. Vet. Med. Surg.* 1987; 2:64–71. **12.30:** Hans Pfletschinger/Peter Arnold. **12.31:** Tom Murray/*BugGuide.Net.* **CF:** Melanie Moser, DPDx-CDC.

Chapter 13. Opener: Eye of Science/Photo Researchers. **13.2:** R. C. Valentine and H. G. Pereira, *Journal of Molecular Biology*/Biological Photo Service. **13.3:** K. G. Murti/Visuals Unlimited. **13.4:** Frederick A. Murphy, CDC. **13.5 (a):** Eye of Science/Photo Researchers. **13.5 (b):** Hans Gelderblom/Visuals Unlimited. **13.6:** Christine Case. **13.9:** G. Steven Martin/Visuals Unlimited. **13.14 (a):** D. O. White and F. J. Fenner, *Medical Virology*, 4th ed., Academic Press California (1994). **13.14 (b):** Chris Bjornberg/Photo Researchers. **13.16 (a):** C. Garon and J. Rose, CDC. **13.16 (b):** Linda Stannard, University of Cape Town/Photo Researchers. **13.18 (a):** Frederick A. Murphy, University of Texas Medical Branch, Galveston. **13.18 (b):** Linda Stannard, University of Cape Town/Photo Researchers. **13.20:** Visuals Unlimited/Corbis. **13.23:** T. O. Diener, USDA/Agricultural Research Service Honey Breeding.

Chapter 14. Opener: Eye of Science/Photo Researchers. **14.1 (a):** Juergen Berger/Photo Researchers. **14.1 (b):** SPL/Photo Researchers. **14.1 (c):** David Scharf/Science Faction/Getty Images. **14.6 (a):** Stone/Getty Images. **14.6 (b):** Helen King/Corbis. **14.6 (c):** Stockbyte Platinum/Alamy. **14.6 (d):** Andrew Davidhazy, Photo Arts and Sciences at Rochester Institute of Technology. **14.7:** iStockphoto. **14.8:** Imagebroker/Alamy. **CF (a):** Christine Case. **CF (b):** iStockphoto.

Chapter 15. Opener: Gillette Corporation/Photo Researchers. **15.1 (b):** SPL/Photo Researchers. **15.1(c):** Gillette Corporation/Photo Researchers. **15.2:** C. Ginocchio, S. Olmstead, C. Wells and J. E. Galan. Contact with epithelial cells induces the formation of surface appendages on Salmonella typhimurium. *Cell.* 1994 Feb 25; 76(4):717–24. Copyright © 1994 by Elsevier Science. Reproduced by permission. **15.7 (a):** Frederick A. Murphy, School of Veterinary Medicine, University of California Davis. **15.7 (b):** Diana Hardie, University of Cape Town Medical School, South Africa. **15.8:** John P. Bader/Biological Photo Service. **CF (a):** P. Marazzi/Photo Researchers. **CF (b):** Janice Carr, CDC. **CF (b inset):** Jason D. Pimentel.

Chapter 16. Opener: Janice Carr, CDC. **16.2:** Ed Reschke/Peter Arnold. **16.4:** Anatomical Travelogue/Photo Researchers. **16.6:** Eye of Science/Photo Researchers. **16.10:** RD Schreiber, DC Morrison, ER Podack and HJ Muller-Eberhard. Bactericidal activity of the alternative complement pathway generated from 11 isolated plasma proteins. *Journal of Experimental Medicine.* 1979; 149:870–82, Copyright © 1979 by Rockefeller University Press. **AM:** Susumu Nishinaga/Photo Researchers. **T16.1 (1, 2, 3, 4, 6, 8, 10):** Barbara Safiejko-Mroczka and Paul B. Bell Jr./Dept. of Zoology, University of Oklahoma. **T16.1 (5):** Giuseppe Bigi, *www.giuseppebigi.it.* **T16.1 (6):** Nivaldo Medeiros, *www.hematologyatlas.com.* **T16.1 (7, 9):** Copyright American Society of Hematology. All rights reserved.

Chapter 17. Opener: David Scharf/Peter Arnold. **17.3:** Zhifeng Shao, University of Virginia. **17.9:** T. Kato and R. L. Owen in *Mucosal Immunology* (2005), eds. J. Mestecky, M. E. Lamm, W. Strober, J. Bienenstock, J. R. McGhee, and L. Mayer. (Elsevier Academic Press, San Diego), pp. 131–51. **17.12:** Gopal Murti/Photo Researchers. **17.13:** David Scharf/Peter Arnold. **17.14:** Lennart Nilsson, Albert Bonniers Forlag AB. **17.15:** Anthony Butterworth. **AM (top):** Crystal structure of IL-12 as published in the Protein Data Bank (PDB: 1F45). Rendered with Pymol by Ramin Herati. **AM (bottom):** AGStockUSA/Alamy.

Chapter 18. Opener: F. Marsik/Visuals Unlimited. **18.1:** Maine Biological. **18.4:** Christine Case. **18.6:** Biological Photo Service. **18.11:** F. Marsik/Visuals Unlimited. **18.13:** Laurent/Photo Researchers. **CF:** P. Marazzi/Photo Researchers.

Chapter 19. Opener: David Scharf/Peter Arnold. **19.1:** Lennart Nilsson, Albert Bonniers Forlag AB. **19.2:** David Scharf/Peter Arnold. **19.3:** James King-Holmes/ Photo Researchers. **19.7:** Jim W. Grace/Photo Researchers. **19.8:** Harvard Medical School. **19.11:** Lennart Nilsson, Albert Bonniers Forlag AB. **19.12:** J. Krahenbuhl, National Hansen's Disease Laboratory, Baton Rouge, LA. **CF:** P. Marazzi/Photo Researchers.

Chapter 20. Opener: Lennart Nilsson, Albert Bonniers Forlag AB. **20.1:** Michael T. Madigan. **20.3:** Lennart Nilsson, Albert Bonniers Forlag AB. **20.5:** Madeline Bastide, Laboratoire D'Immunologie et Parasitologie, Universite' de Montpellier, France. **20.17:** Christine Case. **20.18:** AB Biodisk. **20.19:** National Library of Medicine. **20.22:** Arco Images/Alamy. **20.23:** Eddy Vercauteren. **CF:** C. Poppe, L. C. Martin, C. L. Gyles, R. Reid-Smith, P. Boerlin, S. A. McEwen, J. F. Prescott, and K. R. Forward. Acquisition of resistance to extended-spectrum cephalosporins by Salmonella enterica subsp. enterica, serovar Newport and Escherichia coli in the turkey poultry intestinal tract. *Applied and Environmental Microbiology*. March 2005; 71(3): 1184–92. Fig. 1.

Chapter 21. Opener: Eye of Science/Photo Researchers. **21.3:** M. E. Olson, I. Ruseska, and J. W. Costerton. Colonization of n-bityl-2-cyanoacrylate tissue adhesive by Staphylococcus epidermis. *Journal of Biomedical Materials*. 22:485–95. ©1988 by Wiley. Reproduced by permission. **21.4:** Pulse Picture Library/CMP Images/Phototake. **21.5:** Charles Stoer/Camera M. D. Studios. **21.6:** P. P. Cleary, University of Minnesota School of Medicine/Biological Photo Service. **21.7:** Cavallini/Custom Medical Stock Photo. **21.8:** Ben Barankin, Dermatlas, *www.dermatlas.org*. **21.9:** Pulse Picture Library/ CMP Images/Phototake. **21.10:** Custom Medical Stock Photo. **21.11 (a):** Peter Usbeck/ Alamy. **21.11 (b), 21.12:** P. Marazzi/Photo Researchers. **21.14:** Lowell Georgia/Photo Researchers. **21.15:** Franceschini/CNRI/Photo Researchers. **21.16 (a):** Medical-on-Line/ Alamy. **21.16 (b):** Jane Shemilt/Photo Researchers. **21.17 (a):** Eye of Science/Photo Researchers. **21.17 (b):** Biophoto Associates/Photo Researchers. **21.18:** Eye of Science/Photo Researchers. **21.19 (a):** BSIP/PIR/Photo Researchers. **21.19 (b):** Eye of Science/Photo Researchers. **21.20:** World Health Organization—Prevention of Blindness and Deafness. **CF (a):** Stephen Tristram, School of Human Life Science, Tasmania. **CF (b, c):** Christine Case. **DIF 21.1:** CDC. **DIF 21.2:** Medical-on-Line/ Alamy. **DIF 21.3:** P. Marazzi/Photo Researchers. **DIF 21.4:** SPL/Photo Researchers.

Chapter 22. Opener: D. T. John, T. B. Cole Jr., and F. M. Marciano-Cabral. Sucker-like structures on the pathogenic amoeba Naegleria fowleri. *Applied Environmental Microbiology*. 1984 Jan; 47(1):12–14. Fig. 7.14. **22.3:** D. S. Stephens, Emory University School of Medicine. **22.5:** L. Tilney, P. S. Connelly, and D. A. Portnoy. **22.6:** Bettmann/Corbis. **22.7:** C. E. Dolman. Botulism as a world health problem. *Botulism: Proceedings of a Symposium*, edited by K. H. Lewis and K. Cassel, American Public Health Service Publication No. 999-FP-1, 1964. **22.8:** FDA. **22.9 (a):** Biophoto Associates/Photo Researchers. **22.9 (b):** C. James Webb/Phototake. **22.10:** March of Dimes Birth Defects Foundation. **22.15:** Edward J. Bottone, Mount Sinai School of Medicine. **22.17:** D. T. John, T. B. Cole Jr., and F. M. Marciano-Cabral. Sucker-like structures on the pathogenic amoeba Naegleria fowleri. *Applied Environmental Microbiology*. 1984 Jan; 47(1):12–14. Fig. 7.14. **22.18 (a):** Vla/Photo Researchers. **22.18 (b):** Ralph Eagle Jr./Photo Researchers. **22.19:** Waste Reduction by Waste Reduction, *www.wr2 .net*. **CF (a):** Frederick A Murphy/University of Texas Medical Branch, Galveston. **CF (b):** Bureau of Land Management. **DIF 22.1:** Brodsky, CDC. **DIF 22.2:** Jim Gathany, CDC. **DIF 22.3:** CDC.

Chapter 23. Opener: T. Geisbert, USAMRIID. **23.3:** Bart's Medical Library/Phototake. **23.4:** Edward P. Ewing, CDC. **23.5:** National Museum of Health and Medicine/Armed Forces Institute of Pathology. **23.7:** Science Source/Photo Researchers. **23.8:** P. Marazzi/ Photo Researchers. **23.9:** Gregory G. Dimijian/Photo Researchers. **23.10:** D. L. Kordick and E. B. Breitschwerdt. Intraerythrocytic presence of Bartonella henselae. *J Clin Microbiol.* 1995 June; 33(6): 1655–56. Fig. 3. **23.11:** CDC. **23.14 (b):** Bernard Furnival/ Fran Heyl Associates. **23.14(c):** Scott Camazine/Photo Researchers. **23.15:** James Gathany, CDC. **23.18:** Custom Medical Stock Photo. **23.19:** M. A. Ansary/Photo Researchers. **23.21:** T. Geisbert, USAMRIID. **23.22:** Oliver Meckes/Photo Researchers. **23.23 (1):** Original photo by A. Kimbal from *A Pictorial Presentation of Parasites*, edited by Herman Zaiman. **23.23 (2):** SFI Program. **23.25 (a):** Lennart Nilsson, Albert Bonniers Forlag AB. **23.25 (b):** DPDx-CDC. **23.26:** Walter Reed Army Institute of Research. **23.27:** NIH/Science Source/Photo Researchers. **23.28:** Original photo by M. Voge from *A Pictorial Presentation of Parasites*, edited by Herman Zaiman. **AM (top):** Georgia Tech Photo by Gary Meek. **AM (bottom):** Michael Abbey/Photo Researchers. **CF:** Larry Stauffer, Oregon State Public Health Laboratory, CDC. **DIF 23.1:** Ann Smith/ Reproduced with permission of the ASM MicrobeLibrary, *www.microbelibrary.org*. **DIF 23.2:** Sellers, Emory University, CDC. **DIF 23.3:** A.J. Sulzer, CDC. **DIF 23.4:** CDC. **DIF 23.5:** Visuals Unlimited/Corbis.

Chapter 24. Opener: J. A. Edwards, N. A. Groathouse, and S. Boitano, Bordetella bronchiseptica adherence to cilia is mediated by multiple adhesin factors and blocked by surfactant protein A. *Infectious Immunology*. 2005 Jun; 73(6):3618–26. Cover. © 2005, American Society for Microbiology. **24.3:** P. Marazzi/Photo Researchers. **24.4:** P. B. Smith, CDC. **24.5:** Science VU/Visuals Unlimited. **24.6:** Tony Wright, Institute of Laryngology and Otology/SPL/Photo Researchers. **24.7:** J. A. Edwards, N. A. Groathouse, and S. Boitano, Bordetella bronchiseptica adherence to cilia is mediated by multiple adhesin factors and blocked by surfactant protein A. *Infectious Immunology*. 2005 Jun; 73(6):3618–26. Cover. © 2005, American Society for Microbiology. **24.8:** Biophoto Associates/Photo Researchers. **24.10:** Mediscan/Visuals Unlimited. **24.12:** Raymond B. Otero/Visuals Unlimited. **24.13:** Michael Gabridge/Visuals Unlimited. **24.14 (a):** Moredun Animal Health Ltd./SPL/Photo Researchers. **24.14 (b):** National Museum of Health and Medicine/ Armed Forces Institute of Pathology. **24.16 (a):** Libero Ajello, CDC. **24.16 (b):** Arthur M. Siegelman/Visuals Unlimited. **24.20 (top):** A. B. Dowsett/ Photo Researchers. **24.20 (bottom):** L. L. Pifer, W. T. Hughes, and M. J. Murphy Jr. Propagation of Pneumocystis carinii in vitrio. *Pediatric Research*. 1977 Apr; 11(4):305–16. **CF:** CDC. **DIF 24.1:** Edgar O. Ledbetter, *Visual Red Book on CD-ROM* 2001, ©AAP. Used with permission of the American Academy of Pediatrics. **DIF 24.2:** Christine Case. **DIF 24.3:** Lenore Haley, CDC.

Chapter 25. Opener: Eric Grave/Photo Researchers. **25.3:** Hutton D. Slade. *Microbiology Review*. 1980; 44:331–84, Fig. 5. ASM News. **25.7:** B. B. Finlay and P. Cossart, *Science*. 1997; 276:718–25. Reproduced by permission of the American Assoc. for the Advancement of Science. **25.11:** Dennis Kunkel/Visuals Unlimited. **25.12:** I. Rosenshine et al. A pathogenic bacterium triggers epithelial signals to form a functional bacterial receptor that mediates actin pseudopod formation. *The EMBO Journal*. 1996 June 3; 15(11):2613–24. Cover. **25.13:** SPL/Photo Researchers. **25.14:** P. Marazzi/Photo Researchers. **25.15:** Linda Stannard, UCT/Photo Researchers. **25.16:** Gopal Murti/Phototake. **25.17:** Robert Owen et al. Ultrastructural observations of giardiasis in a murin model. *Gastroenterology*. 76:757–69. ©1979 by the American Gastroenterological Assoc. **25.18:** EM Unit, London School of Hygiene and Tropical Medicine, London, UK. **25.19:** Armed Force Institute of Pathology. **25.21:** Original photo by A. Kimball from *A Pictorial Presentation of Parasites* edited by Herman Zaiman. **25.22:** Armed Forces Institute of Pathology. **25.23:** Robert Calentine/Visuals Unlimited. **25.24:** Sinclair Stammers/Photo Researchers. **25.25 (a):** Eric Grave/Photo Researchers. **25.25 (b):** Daniel Snyder/Visuals Unlimited/Getty Images. **AM (top):** Addison N. Scurlock Collection, Archives Center, National Museum of American History, Smithsonian Institution. **AM (bottom):** Eureka Slide/SuperStock. **CF:** Christine Case. **DIF 25.1:** Laura Ahonen. **DIF 25.2:** Mauro Rodrigues/ Shutterstock. **DIF 25.3 (left):** Southern Illinois University/Photo Researchers. **DIF 25.3 (right):** Custom Medical Stock Photo. **DIF 25.4:** E. L. Palmer, CDC. **DIF 25.5:** Melanie Moser, DPDx-CDC.

Chapter 26. Opener, 26.4: Dennis Kunkel/Visuals Unlimited. **26.6:** CDC. **26.7:** Gary E. Kaiser, *http://student.ccbcmd.edu/~gkaiser/*. **26.8:** David Soper. **26.9:** Michael Abbey/Photo Researchers. **26.11 (a):** Biophoto Associates/Photo Researchers. **26.11 (b):** Collection CNRI/Phototake. **26.11 (c):** CDC. **26.12:** Seattle STD/HIV Prevention Training Center at the University of Washington. **26.14:** Michael Remington, University of Washington Viral Disease Clinic. **26.15:** Bart's Medical Library/Phototake. **26.16:** D. Petrin et al. Clinical and microbiological aspects of Trichomonas vaginalis. *Clinical Microbiology Review*. 1998 Apr; 11(2):300–17, Fig. 1. **DIF 26.1:** Christine Case. **DIF 26.2:** M. Rein, CDC. **DIF 26.3:** Renelle Woodall, CDC.

Chapter 27. Opener: George J. Wilder/Visuals Unlimited. **27.1 (a):** M. F. Brown/ Biological Photo Service. **27.1 (b):** R. L. Peterson/Biological Photo Service. **27.2 (a):** Mycorrhizal Application, *www.mycorrhizae.com*. **27.2 (b):** Michel Viard/Peter Arnold. **27.5:** Holt Studios Int'l. Ltd./Alamy. **27.6:** George J. Wilder/Visuals Unlimited. **27.9:** Visuals Unlimited. **27.10:** Nancy Pierce/Photo Researchers. **27.11:** Kenneth Lucas. **27.13:** Pete Atkinson/ANT Photo Library. **27.14:** IDEXX Laboratories. **27.16:** Ron Hendricksen, Water Treatment Plants, City of Fargo, ND. **27.17:** Blue Flag, *www.blueflag.org*. **27.19:** Virgil Paulson/Biological Photo Service. **27.20:** Environmental Leverage, *www.EnvironmentalLeverage.com*. **27.21 (a):** Douglas Munnecke/Biological Photo Service. **27.22:** Christine Case. **AM (a, b):** Christine Case. **AM (bottom):** Pedro Armestre/AFP/Getty Images. **CA:** Randall Von Wedel, CytoCulture Int'l.

Chapter 28. Opener: David Scharf/Peter Arnold. **28.2:** Packaging Technologies & Inspection, Tuckahoe, NY. **28.4:** International Paper. **28.6:** Hank Morgan/Rainbow. **28.8 (a, b):** David M. Frazier/Photo Researchers. **28.8(c):** Junebug Clark/Photo Researchers. **28.10:** From the collection of the North Carolina Biotechnology Center. **28.12:** Manfred Kage/Peter Arnold. **28.15:** Capstone Microturbines, *www.microturbine.com*. **28.16 (a):** Solix Biofuels. **28.16 (b):** Keith Cooksey/Montana State University. **AM (top):** Reproduced by permission of Kelco Biopolymers, San Diego. **AM (middle):** *Seelevel.com*. **AM (left):** StockFood/Getty Images.

INDEX

Note: a *t* following a page number indicates tabular material, an *f* following a page number indicates a figure or illustration, a *b* indicates a boxed feature, and a page number in **boldface** indicates a definition.

A-B toxins, **435**, 436*f*, 436*t*, 437, 438*t*
ABO blood group system, **526**–527, 527*t*
  IgM antibody and, 480, 526
abortions, gas gangrene and, 646
abscess, **588**
abscess formation, **462**
  in inflammatory response, 461*f*, 462
absorbance (optical density/OD), 178, 179*f*
Acanthamoeba, 348, 354*t*, 629
Acanthamoeba keratitis, **605**
accelerators (chemical), allergic reactions and, 531
accessory glands, of male reproductive system, 745, 745*f*
accidental inoculation, 406
Accutane (isotretinoin), 453, 594
acellular vaccines, **503**
acetaldehyde, in alcohol fermentation, 135*b*, 136*f*
acetaminophen, 437
acetate kinase, 116*t*
acetic acid
  acetobacter and, 137*t*, 300*t*, 303
  bacteria that produce, 300*t*
  fermentation and, 134*f*, 137*t*
  industrial/commercial use, 137*t*
Acetobacter genus/spp., **303**
  as acetic acid producers, 300*t*, 303
  biochemical tests for, 139
  fermentation and, 137*t*
  in taxonomic hierarchy, 300*t*
  industrial importance of, 303
  used in production of vinegar, 800
acetoin, as fermentation end-product, 134*f*
acetone, 2
  as fermentation end-product, 134*f*, 137*t*
  biotechnology and, 246
acetyl CoA (acetyl coenzyme A), 124, 125*f*
  in biosynthesis of amino acids, 146–147, 148*f*
  in biosynthesis of lipids, 146, 147*f*
  in lipid catabolism, 136*f*
  Krebs cycle and, 127–129, 128*f*
acetyl group, 127
acetyl-CoA synthetase, 116*t*
acid fuchsin dye, 69
acid precipitation, lichens and, 340
acid-anionic sanitizers, **199**, 204*t*
acid-base balance, 35–37, 36*f*
acid-fast bacteria, **70**–71, 71*f*
  cell walls and staining of, 87–88
  mycolic acid in cell walls, 87–88
acid-fast stain, **70**–71, 71*f*
  mycolic acid and, 87–88
  procedures, 70–71, 71*f*
acid-tolerant microbes, 37, 325
acidic dyes, **68**–69
acidic solutions, 35
  alkaline *vs.,* 35–36, 36*f*
  microbial growth and, 37, 158–159
acidophiles, **159**
  archaea, 325
  bacteria, 159
acidophilic inclusion bodies, 442, 443*t*
acidosis, fever and, 463
acids, 34, **35**–36, 36*f*
Acinetobacter baumanii, antibiotic resistance and, 309
Acinetobacter genus/spp., **309**
  as normal microbiota of skin, 402*t*
  genetic transformation natural in, 236
  nosocomial infections and, 414, 414*t*
acne, 453, **594**, 595*f*
  bacterial, 320
  clindamycin to treat, 565
  lesions caused by, 592*b*
acoustic microscopy, scanning (SAM), **63**, 63*f*, 67*t*
aquatic microbiology, **776–789**
aquatic microorganisms, 776–778
  freshwater, **776**–777
  seawater, **777–778**
acquired immunity. *See also* adaptive immunity
  active *vs.* passive, 494–495, 494*f*
  natural *vs.* artificial, 494–495, 494*f*
acquired immunodeficiencies, **538**, 539*t*
acquired immunodeficiency syndrome (AIDS). *See* AIDS
acridine dyes, 230
Actimmune (alpha interferon), to treat osteoporosis, 470
actin, rearranged by pathogen-produced invasins, 433
Actinobacteria, 301*t*, **320**–321, 321*f*
  as high G + C gram-positive bacteria, 281*f*, 301*t*, 315–316
  important genera/special features, 301*t*
  pathogenic genera of, 320
Actinomyces genus/spp., 301*t*, **321**, 321*f*
  actinomycetes informal name for, 320
  as filamentous, branching bacteria, 301*t*
  as human pathogens, 301*t*
  as normal microbiota of mouth, 403*t*
  Streptococcus mutans, dextran, and dental plaque, 431
Actinomyces israelii, actinomycosis caused by, 321
Actinomycetales, 301*t*
actinomycetes, 320
  antibiotics produced by, 554, 555*t*
  G + C ratio of, 316
  morphology of, 320
  reproductive methods, 171
actinomycosis, 321
Activase (tissue plasminogen activator), genetically engineered, 260*t*
activated sludge system, **784**, 785*f*, 786*f*
activation energy, **115**, 115*f*
active immunity, **494**, 494*f*
  artificially acquired, 494*f*, **495**
  naturally acquired, **494**, 494*f*
active site of enzymes, 115, **117**, 118*f*, 120*f*
active transport processes, 91, **94**, 146
acute bacterial endocarditis, **641**, **643*b***
acute disease, **406**
acute inflammation, 460
acute necrotizing ulcerative gingivitis (trench mouth), **709**, 710*f*
acute-phase proteins, **460**
acyclovir, **569**, 570*f*
  mode of action/uses, 564*t*
  spectrum of activity, 557*t*
  structure and function, 570*f*
  to treat shingles, 596
ADA (adenosine deaminase) deficiency, 17–18
adaptive immunity, 433, **450**, 450*f*, **476**–499
  active
    artificially acquired, 494*f*, **495**
    naturally acquired, **494**, 494*f*
  antigens, **478**–479, 479*f*
  as third line of defense, 450, 450*f*
  blood's role in, 637–638, 639*f*
  cellular immunity, 477–**478**, 486–489
  dual nature of, 477–478, 496*f*
  humoral immunity and, **477**, 496*f*
    B cells in, 482–486
  lymph's role in, 637–638, 639*f*
  memory component of, 450, 493–494, 496*f*
  nonself *vs.* self and, 477, 492–493, 494, 496*f*
  passive
    artificially acquired, 494*f*, **495**
    naturally acquired, **494**, 494*f*
  specificity of, 450
  summary, 496*f*
  types of, 494–495, 494*f*
ADCC (antibody-dependent cell-mediated cytotoxicity), 484, **485**, 485*f*, **491**, 492*f*
Addison's disease, 534*t*
adefovir dipivoxil (Hepsera), 564*t*
  to treat hepatitis B, 569
adenine (A), 47, 48*f*, 49*f*, 211
  exposed to mutagenic nitrous acid, 229, 229*f*
  in DNA replication, 212–215, 214*f*, 215*f*, 216*f*
  in translation phase of protein synthesis, 216, 218*f*
adenine nucleotide (adenosine monophosphate/AMP), 47, 48*f*
  difference between ATP and, 214
adenocarcinomas, virus-induced (in mice), **389**
adenosine, 47, 49*f*
adenosine deaminase (ADA) deficiency, 17–18
adenosine diphosphate (ADP), **47**–49, 49*f*
  anabolic reactions and, 114, 114*f*
adenosine diphosphoglucose (ADPG), 146, 146*f*
adenosine monophosphate/AMP (adenine nucleotide), 47, 48*f*
  difference between ATP and, 214, 215*f*
adenosine triphosphate (ATP), **47**
  chemical structure, 47–49, 49*f*
  role in coupling anabolic and catabolic reactions, 114, 114*f*
Adenoviridae, **385**, 387*f*
  characteristics/important genera/clinical features, 375*t*
adenoviruses, **385**, 387*f*
  as opportunistic pathogens, 404
  cytopathic effects of, 443*t*
  Mastadenovirus, 372*f*, 373, 387*f*
  size of, 369*f*
  used in gene therapy, 251, 259
adherence, in second phase of phagocytosis, **458**–459, 458*f*
adherence (adhesion) of pathogens, 428*f*, **431**–432, 431*f*, 445*f*
adhesins (ligands) of pathogens, **431**, 431*f*
  diversity of, 431
adhesion (adherence) of pathogens, 428*f*, **431**–432, 431*f*
adjuvants to antigens, **506**
ADP
  in Calvin-Benson cycle, 142*f*
  in generation of ATP, 122, 123
  in photosynthesis, 140, 141*f*
ADP (adenosine diphosphate), **47**–49, 49*f*
ADPG (adenosine diphosphoglucose), 146, 146*f*
adsorption. *See* attachment (adsorption)
adult stem cells (ASCs), 535
Aedes aegypti (mosquito), 413*t*, 658
Aedes albopictus (mosquito), 658
Aedes (mosquito)
  California encephalitis transmitted by, 628*b*
  chikungunya fever transmitted by, 658
  dengue fever/yellow fever/heartworm transmitted by, 362*t*, 413*t*, 658
  eastern equine encephalitis transmitted by, 628*b*
  heartworm disease and, 360
aerial hyphae, 331, **331***f*, 332–333, 335*f*
aerobes
  culture media for, 167–168
  obligate, **161**, 161*t*
  vs. anaerobes, **127**
    energy production from nutrients and, 161
    growth rates for, 132
aerobic respiration, **127–132**, 133*f*
  anaerobic respiration *vs.*, 137*t*
  ATP yield, 132*t*
  chemiosmosis, 130–131, 130*f*
  effect of oxygen on bacterial growth, 161, 161*t*
  electron transport chain, 129–130, 129*f*
  fermentation *vs.*, 137*t*
  Krebs cycle, 127–129, 128*f*
  summary, 131–132, 133*f*
Aeromonas hydrophilia, 283*b*
aerotolerant anaerobes, 161*t*, **162**
affinity, in antigen-antibody complex, **484**
aflatoxin, **443**
  as frameshift mutagen, 230
  poisoning, **730**, **734*b***
  produced by *Aspergillus flavus* mold, 443
AFM (atomic force microscope), **65**, 65*f*, 68*t*
  DNA double helix image, 58*f*
  perfringolysin O toxin image, 65*f*, 68*t*
  specimen preparation and, 65, 68*t*
  specimen size and, 58*f*
African trypanosomiasis (sleeping sickness), 222, 329, 351, 354*t*, 362*t*, 363, 413*t*, 444, **627**–629, **632*b***
agar, **165**
  algae-derived, 165, 343
  bismuth sulfite, 168
  blood, 168, 168*f*
  MacConkey's, 746, 748*f*
  mannitol salt, 168–169, 169*f*
  nutrient, **165**
  peptone ion, 139, 139*f*
  properties of, 165
  Sabouraud's dextrose, 168
  salt concentration and, 159
  temperature and, 165
agarose gel, 262
Agent Orange, decomposition rate, 775, 775*f*
agglutination, **484**–485, 485*f*, **510**, 510*f*
  epitopes of antigens and, 478, 479*f*
  IgG antibodies and, 481*t*, 484–485
  IgM antibody and, 480
  slide agglutination test, **287**, 287*f*
agglutination reactions, **510**–512, 510*f*, 511*f*
  direct, **510**–511, 511*f*
  hemagglutination, **512**, 512*f*
  indirect (passive), **511**–512, 511*f*
  latex, **511**–512, 511*f*
aging
  immune system's gradual decline and, 462
  phagocyte's progressive inefficiency and, 462

agranulocytes, **454**, 455t
  dendritic cells, **454**, 455t
  lymphocytes, **454**, 455t
  monocytes, **454**, 455f
agranulocytosis, **528**
Agre, Peter, 15t
agriculturally important bacteria
  Agrobacterium, 300t, **304–305**
  Azospirillum, 300t, **303**
  Bradyrhizobium, 300t, **304**
  Nitrobacter, 145, 300t, **305**
  Nitrosomonas, 145, 300t, **305**
  Rhizobium (rhizobia), 300t, **304**
agriculture
  fungi's desirable/undesirable effects, 339
  microbial insect control, 17, 265, 265f, 267, 267t
  rDNA technology applications, 264–267, 265f
  wastes of, fermentation and, 137t
  widespread use of antibiotics and, 239–240
Agrobacterium genus/spp., 300t, **304–305**
  as vehicle to introduce rDNA into plant cells, 240, 264, 265f, 304–305
  Entner-Doudoroff pathway and, 127
Agrobacterium tumefaciens
  crown gall disease and, 264, 265f, 304
  inserts plasmid with genetic data into plant DNA, 264, 265f, 304–305
AIDS, **21**, **539–548**, 539t. *See also* HIV; HIV infection
  animal models for treatments/vaccine development, 377
  antiviral drugs to treat, 571
  as an epidemic disease, 546, 547f, 548
  as emerging infectious disease, 417t
  as final stage of HIV infection, 540
  as notifiable infectious disease, 421t
  as persistent viral infection, 394t
  candidiasis and, 339
  CD4+ T cells and, 5f, 415, 541–545, 541f, 543f
  chemotherapy for, 548
  chimpanzees and, 377
  clinical, definition of, 542
  deaths from, worldwide, 546, 547f
  diagnostic methods, 545
  diseases commonly associated with, 544t
  distribution of cases, by world region, 547f
  earliest documented case of, 540
  ELISA test to detect HIV antibodies, 287, 288f, 514, 516, 518f, 545
  feline, 377
  first confirmed case of, 540
  historical aspects, 539–540
  incubation period, 430t
  Microspora causal agent for many infections in, 348
  opportunistic fungal infections in, 337, 339
  origins of, 540
  Pneumocystis pneumonia as leading cause of death, 329, 337
  portals of entry, 430t
  prevention of, 547
  progression from initial HIV infection to, 542–544, 543f
  reported cases in United States 1979–2006, 407f
  simian, 377
  toxoplasmosis and, 542, 544t, 662–663
  vaccine development and, 259, 547–548
  worldwide aspects of, 546
air pollutants, lichens used to determine, 340
airborne infections, chlamydials, 322
airborne microbes
  HEPA filters and, 168, 191
  UV light to control, 193
airborne microorganisms, 2
airborne pathogens
  early aseptic techniques and, 8
  of systemic mycoses, 336
airborne radioactive cesium-137, lichen and, 340
airborne transmission of disease agents, 412
  of nosocomial infections, 415–416
alanine, **44t**, 45f
alanine deaminase, 116t
alanine racemase, 116t
alarm signals (chemical)
  alarmone, 226
  cyclic AMP as, 225–226
alarmone, 226
  cyclic AMP as, 225–226
albendazole
  mode of action/uses, 564t
  to treat intestinal helminthic infections, 571–572
alcohol
  enzymes in peroxisomes oxidize, 105
  in Gram stain method, 69, 87
alcohol fermentation, **135b**, 136f
alcohol functional group, 37, 38t
alcohol-acetone solution, in Gram stain method, 69
alcoholic beverages
  fermentation and, 9, 135b
  microbes used in production of, 800
alcohols, 37, 38t
  as antimicrobial agents, **197**–198, 198t, 200f, 203t, 204t
  bacterial plasma membrane damaged by, 91, 197
aldehyde group, 38t
aldehydes, as antimicrobial agents, **200**, 205t
ale, microbes used in production of, **800**
allergic reactions, as type of immune response, 478–479
Alexandrium
  neurotoxin produced by, 344, 354t, 444
  red tide and, **344**
Alexidine, 197
alfalfa plants, symbiotic relationships with microbes and, 267
algae/alga, 2, 5f, **6**, 329, **340–345**
  agar derived from, 165, 343
  as a biofuel, 808, 808f
  as eukarya/eukaryotes, 6, 76, 341
  as photoautotrophs, 143–145, 143f, 330t, 341
  as plankton, 341
  atmospheric nitrogen converted by, 17
  brown, 341, 341f, **342**, **343t**
  cell structure, 5f, 6, 98, 99f
  cell wall of, 98, 99f, 253
  cellulose and, 40, 98
  characteristics, 341–342, 341f
  characteristics of selected phyla, 343t
  chemoheterotrophic, 342
  chloroplasts of, **105**, 106f, 140
  classification and, 342
  copper sulfate as algicide, 199, 204t
  cyanobacteria once called, 313–314
  death of and dissolved oxygen levels, 344
  diatoms, 341f, **343**, 343f, **343t**
  dinoflagellates (plankton), 341f, **343–344**, 344f
  Earth's molecular oxygen produced by, 345
  filamentous, 341, 341f
  fungal-like (oomycotes/water molds), 342, 343t
  green, 342f, **343**, **343t**
  habitats of, 341, 341f
  identification of, 341
  in lichens, 339–340, 340f
  increase in poisonings due to, 329
  inserting foreign DNA into cells of, 253, 253f
  kelp (brown algae), 341, 341f, **342**, **343t**
  life chyle of, 342, 342f
  morphology of, 341
  multicellular, 341–342, 341f, 342f
  neurotoxins produced by, 444
  nutrition and, 6, 342, 343t
  pathogenic, 444
  photosynthesis and, 140, 141f, 145t, 341f, 342, 343t
  plankton, 341f, **343–344**, 344f
  pond, 5f
  red, 341f, **343**, **343t**
  reproductive methods, 6
  role in nature, 344–345
  rules for naming, 279
  shapes of, 5f, 6
  thalli of, 341–342, 341f, 343f
  unicellular, 341, 341f
  vegetative structures, 341–342
  water molds, **343t**, **344**, 345f
algal blooms, **344**, **779**
Algin, **342**
alginic acid, 343t
alkaline habitats, cyanobacteria and, 37
alkaline solutions, 35
  acidic *vs.*, 35–36, 36f
  microbial growth and, 159
alkylation, 201
All Species Inventory project, 274
allergen, **523**
allergic contact dermatitis, **530**–531, 530f, 531b
allergic reactions, **523**–531. *See also* hypersensitivity reactions
  IgE antibodies and, 481, 523, 524f
  to *Bacillus thuringiensis* (BT) toxin insecticide, 268
allergy, 523
allografts, **536**
allolactose, 224, 225f
allosteric enzyme inhibitors, **120**, 120f
  feedback (end-product) inhibition and, 120–121, 121f
allosteric site, **120**, 120f
allylamines, mode of action/uses, 564t
alpha interferon, 469–470, 469f
  as rDNA product, used to treat leukemia/melanoma/hepatitis, 260t
  Intron A to treat virus-associated disorders, 470
  mode of action/uses, 564t
alpha-amino acids, 43
alpha-hemolytic streptococci, 319
alpha-ketoglutaric acid, 127, 128f, 148f, 149f
alphaproteobacteria, 300t, 303–305, 304f, 305f
  important genera/special features, 300t
Alphavirus, 413t
  causing dengue fever, 413t
alphaviruses, 375t, 387
alternative pathway of complement activation, 464f, **466**–467, 466f
alum, as adjuvant to antigen effectiveness, 506
alveolar macrophages, 457
Alveolata, 350
alveoli, 675, 676f
Amanita phalloides (deathcap mushroom), 443
amanitin, **443**
amantadine, mode of action/uses, 564t
amebic encephalitis, granulomatous, **617b**
amebic meningoencephalitis, **617b**, **629**
  primary, **617b**
American Academy of Microbiology, 262
American leishmaniasis, 666
American trypanosomiasis. *See* Chagas' disease
Ames test, **232**–233, 233f
amination, **147**, 148f
amino acids, 38, **43**, 43f, 44t
  amphibolic pathways and, 147, 149f
  biosynthesis of, 146–147, 148f
  detection test for presence of enzymes that catabolize, 138, 139f
  distinguishing feature, 43
  found in proteins (structural formula/characteristic R group), 44t
  in protein biosynthesis, 146–147, 148f
  in protein catabolism, 136, 138f
  in translation (protein synthesis), 219–220, 219f, 220–221f
  metabolism, coenzyme in, 117t
  peptide bonds of, 43, 45f
amino group, 37, 38t
  in amino acids, 38, 43, 43f, 44t
  in deamination conversion, 136
aminoglycosides, **565**
  mode of action/spectrum of activity, 562t
ammonia
  as an energy source, 141, 143f
  as energy source, 145
  in chloramines, 197
ammonification, 770–**771**, 770f
ammonium ion, 136
  in quats, 199, 199f
Amoeba proteus, 348f
amoebas (Amoebozoans), 5f, **348**, 348f
  amoebicidal agents, quats as, 199
  food acquisition methods, 346
  movement methods, 5f, 6, 348, 348f
  position in evolutionary tree, 275f
  pseudopods of, **348**, 348f
  slime molds and, 4, 351–352, 352f
amoebiasis. *See* amoebic dysentery
amoebic diseases, diiodohydroxyquin (iodoquinol) to treat, 571
amoebic dysentery, 329, 348, 348f, 354t, **731–732**, 732f, **734b**
  metronidazole to treat, 571
  portal of entry for, 429
Amoebozoa/Amoebozoans, 5f, **348**, 348f. *See also* amoebas
amoxicillin, 561
  mode of action/spectrum of activity, 562t
amp (ampicillin-resistance gene), 250, 251f, 256, 256f
AMP/adenosine monophosphate (adenine nucleotide), 47, 48f
amphibolic pathways, **147**, 149f
amphitrichous flagella, **81**, 81f
amphotericin B, 568, 568f
  damages plasma membrane, 558
  mode of action/comments, 564t
  produced by *Streptomyces nodosus,* 555t
ampicillin, 561
  mode of action/spectrum of activity, 560f, 562t
ampicillin resistance gene, 250, 251f, 256, 256f
amplified DNA, 247
  by polymerase chain reaction (PCR) process, 251, 252f
Ampligen, 633
AMPs. *See* antimicrobial peptides
amylases, 40
Anabaena azollae, 776f
Anabaena genus/spp., 301t
  as oxygenic photosynthetic bacteria, 301t
  characteristics, compared, 314t
anabolic chemical reactions. *See* anabolism
anabolism, **32**, 113, **114**, 114f, 146–147, 146f, 147f, 148f
  amphibolic pathways and, 147, 149f, 150
anaerobes
  aerotolerant, 161t, **162**
  facultative, **161**, 161t
  growth media for, 166–167, 167f
  obligate, culture media for, 169t

anaerobes *vs.* aerobes, **127**
in energy production from nutrients, 161
in growth rates, 132
anaerobic bacteria, *Clostridium,* 301*t*
anaerobic chambers, 167, 167*f*
anaerobic growth media, 166–167, 167*f*
anaerobic respiration, **127**, 132, 161
aerobic respiration *vs.*, 137*t*
fermentation *vs.*, 137*t*
anaerobic sludge digesters, **785**–786, 785*f*, 787*f*
anal gonorrhea, **749**
anal pore, **346**
of *Paramecium,* 350*f*
of protozoa digestive system, 346
analytical epidemiology, **419**–420
anamnestic response. *See* secondary response
Anamorphorhs, pathogenic fungi of, 338*t*
anamorphs (asexual fungi), **335**
Pneumocystis as, 337, 338*t*
Anopheles mosquito, malaria and, 348–350, 349*f*
anaphylactic reactions, **523**–**526**, 523*t*, 524*f*
as Type I hypersensitivity, 523*t*
IgE antibodies and, 481, 523–526, 524*f*
inherited complement deficiencies and, 468
localized, 523, **525**–526
preventing, 526
skin tests to identify antigens, 526, 526*f*
systemic, 523, **524**
anaphylactic shock, **524**
anaphylaxis, **523**
localized, **525**–526, 525*f*
systemic, **524**, 524*f*
Anaplasma phagocytophilum, 654
anaplasmosis caused by, 654
Ixodes scapularis as tick vector, 654
anaplasmosis, 291, **651***b*, **654***b*
as notifiable infectious disease, 421*t*
ancestor, universal, 275, 275*f*
ancestor DNA, 275*f*
ancestral relationships, classification systems and, 274, 275*f*
Ancylostoma duodenale, 359–360, 361*t*
anemia
Babesia microti causing, 350
genetically engineered erythropoietin to treat, 260*t*
human parvovirus B19, 375*t*
angiospermae, position in taxonomic hierarchy, 280*f*
Angstrom (Å), 55
animal bites
bat, 624*f*, 624 *footnote*, **625***b*
cat, 311
dog. *See* dog bites
infections transmitted by, 647–648, 647*f*, **650***b*
rat, 647–648, 650*t*
animal bites and scratches, infections transmitted by, 647–648, 647*f*, **650***b*
animal dander, allergic reactions to, 525
animal feed antibiotics, 554, 562*t*, 565, 575, **577***b*
avoparcin, 577*b*
fluoroquinolones, 577*b*
human disease linked to, and safety of, 575, **577***b*
tetracyclines, 562*t*, 565
vancomycin, 577*b*
animal husbandry
animal feed antibiotics and, 554, 562*t*, 565, 575, 577*b*
bovine growth hormone (bGH) and, 267, 267*t*
porcine growth hormone (pGH), 267*t*
rDNA products important to, 267, 267*t*
animal pathogens, *Saprolegnia ferax,* 329*f*
animal reservoirs, **409**
animal viruses
cultivation of, 377–379
in cell cultures, **378**–379, 378*f*
in embryonated eggs, 377–378, 377*f*, 404–405, 504
in living animals, 377
genetic modification of, 259
genetically modified, 259
latent, 382
multiplication stages, 382–389
attachment, **383**, 384*t*, 386*f*
biosynthesis
of DNA viruses, 384*t*, **385**–386, 385*t*
of RNA viruses, 384*t*, 385*t*, **387**–389, 388*f*, 389*f*
entry by fusion, **383**, 384*f*, 384*t*
entry by pinocytosis/endocytosis, **383**, 384*f*, 384*t*
maturation, **389**, 391*f*
release, **389**
uncoating, **384**–385, 384*t*
vs. bacteriophages, 384*t*
"animalcules," 7, 7*f*
Animalia (kingdom)
energy source, 282
in Linnaeus's classification system, 274
major differences among eukaryotic microbes and, 330*t*
organisms included in, 282
position in taxonomic hierarchy, 280*f*
animals
as kingdom in Domain Eukarya, 6, 274, 275*f*, 280*f*
as reservoirs, **409**
cell structure, 98–106, 99*f*
cells used to produce viral vaccines, 247
DNA vaccines approved for, 503
eukaryotic cells composing, 76
introducing foreign DNA into by microinjection, 254, 254*f*
mud-dwelling, 13
nutritional classification of, 143, 143*f*
position in evolutionary tree, 275*f*
spontaneous abortion in domestic caused by *Campylobacter fetus,* 312
wild, veterinary microbiologists and, 283*b*
antimicrobial agents, usnic acid from *Usnea* lichen, 340
anionic detergents, 88*t*, 199
susceptibility of gram-negative *vs.* gram-positive bacteria to, 88*t*
anions, **30**, 35
superoxide, **161**–**162**
anisakiasis (sashimi worms), 361*t*
anisakines, 361
Anopheles (mosquito), as malaria vector, 348–350, 349*f*, 362–363, 362*t*, 410*t*, 413*t*, 663
anoxygenic photosynthetic bacteria, 314*t*, **315**, 315*f*
characteristics, compared, 314*t*
gas vacuoles and, 96
processes, **144**, 145*t*
antagonism
in combination antibiotics, **578**
normal microbiota and, **401**
antheridiol hyphae, 345*f*
anthrax, **645**–646, 645*f*, **650***b*
as biological weapon, 317, 646, 649*b*
as emerging infectious disease, 417*t*
as notifiable infectious disease, 421*t*
as zoonotic disease, 410*t*
causative agent discovered, 11, 404
caused by *Bacillus anthracis,* 11, 61, 80, 239, **317**, 404, 410*t*, 417*t*
ciprofloxacin (Cipro) to treat, 567, 646
contamination, chlorine dioxide gas to fumigate, 201
cutaneous, 430, **645**, 645*f*, 650*b*
endospores of, 96, 97*f*, 645
gastrointestinal, 430
inhalational (pulmonary), **645**–646, 649*b*, 650*b*
virulence of, 430, 645–646
stain used to diagnose, 61
vaccination of livestock and, 646
vaccine for humans, 646
virulence, 430
capsules of bacteria and, 432
portals of entry and, 430
antibiosis, 554, 554*f*
antibiotic resistance, 12–13, 20, **553**, 573–576, **575** *f*
animal feed antibiotics and, 554, 562*t*, 565, 575, 577*b*
approaches to solving, 578–579
as global health crisis, 19–20, 575, 576*f*
biofilms and, 19, 163, 186, 189, 431
cost of, 576
developed during antibiotic therapy, 575, 576*f*
Enterobacter genus/spp. and, 319
genetic mutation and, 210, 228, 231, 574
horizontal gene transfer, 577, 577*b*
resistant mutant, 575, 576*f*
in hospital workers, 576
infectious disease reemergence and, 21
mechanisms of, 574–575, 575*f*
altering target molecules, 574, 575*f*
blocking entry (porins modified), 574, 575*f*
inactivating enzymes, 574, 575*f*
rapid ejection, 575, 576*f*
misuse/overuse of antibiotics and, 19, 416, **575**–578, 576*f*, 577*b*
MRSA, **20**, 422*b*, **560**–561, **593** *b*. *See also* MRSA
of gonorrhea pathogen, 751*b*
of *Neisseria gonorrhoeae,* **751***b*
of pseudomonads, 594
of *Staphylococcus aureus*, 318. *See also* MRSA
plasmids and, 95, 250, 251*f*, 439–441, 574. *See also* plasmids
pneumococcal diseases and, 614
prevention of, 576
R factors and, **239**–240, 240*f*, 308, 308*f*, 414, 439–441, 574, 577*b*
sex pili, enterics and, 309
transposons and, 574
triclosan and, 196
VISA, **20**, 417*t*, 421*t*, 422*b*
VRSA, 12–13, 19–**20**, 210, 241, 417*t*, 421*t*, 422*b*, 563
antibiotics, **12**, 553*f*, **554**. *See also* antimicrobial drugs
antagonism in combinations, **578**
antifungal, 564*t*, 567–569, 568*f*, 569*f*
antihelminthic, 564*t*, 571–572
antimycobacterial, 562*t*, **563**
antiprotozoan, 12, 528, 529*f*, 564*t*, **571**
antiviral, 564*t*, 569–571, 570*f*
as antimicrobial agents, 200
azoles, **568**, 569*f*
bacterial inactivation of, 70
bacterial inhibition in laboratory cultures (antibiosis), 554, 554*f*
blood-brain barrier and, 611
broad-spectrum, **555**, 557*t*
opportunistic fungal infections and, 337, 339
Clostridium difficile and, 316
damage to cell walls by, 88–89, 94
derived from microbes, 247, 249, 301*t*, 317, 321, 339, 554, 555*t*, 563
destruction of plasma membrane by, 91
discovery of, 12, 12*f*, 246, 553, 554
endotoxins and, 437
eukaryotic cell wall and, 98
future of, 578–579
gastrointestinal infections following, 401
gram-negative bacteria and, 87
gram-positive bacteria and, 70
in animal feeds, 554, 562*t*, 565, 575, 577*b*
investigative studies in promising approaches, 579
misuse/overuse, 19, 416, **575**–**578**, 576*f*, 577*b*
as factor in emerging infectious diseases, 416
widespread use in medicine/agriculture, 239–240
modes of action
of commonly used, 562–563*t*, 564*t*
overview, 556*f*
polyenes, **568**, 568*f*
protein synthesis inhibition and, 95, 106
rashes induced by, 531*b*
resistance to. *See* antibiotic resistance
safety issues, 576–578, 577*b*
sensitivity tests, 195, 196*f*, **572**–573, 751*b*
solutions of, filtration used to sterilize, 191
Streptomyces species produce, 321, 554
superinfections and, **555**
susceptibility in Archaea/Bacteria/Eukarya compared, 276*t*
susceptibility tests, 195, 196*f*, 572–573, 751*b*
synergism in combinations, 566, 567, 568*f*, **578**, 578*f*
therapeutic index and, 576
triclosan-resistant bacteria and, 196
with ribosomal activity, 563, 565–566
antibodies (immunoglobulins), **61**–**62**, 61*f*, **479**–**482**, 481*f*, 481*t*
against pathogen capsules, 432
antibody titer, **493**
antigen-antibody binding results, **484**–**486**, 485*f*
agglutination, **484**–485, 485*f*
complement activation, 484, **485**, 485*f*
cytotoxicity, 484, **485**, 485*f*
neutralization, 484, **485**, 485*f*
opsonization, **485**, 485*f*
antitoxins (against exotoxins) produced by, 435, 439*t*
as globulin proteins, 42, 479
as third line of defense, 450*f*
B cells (B lymphocytes) and, 482–486, 482*f*, 483*f*
blood's role in, 637–638, 639*f*
can differentiate between amino acids/isomers, 484
diversity of, 484
early discoveries about, 477–478
endotoxins and, 439, 439*t*
first ones produced in response to infection, 481*t*, 493
fully human antibodies, **509**
half-life of an injected antibody, 495
humanized, **509**
humoral immunity carried out by, **477**, 482
IgA proteases enzymes and, 433
immunoglobulin classes, 479–481, 480*f*, 481*t*
in artificially acquired passive immunity, 494*f*, 495
intracellular antigens and, 486
ionic bonding's role when combating infection, 30
monoclonal, **507**–**509**, 508*f*
rDNA products in medical therapies, 260*t*

Index

placental transfer of, 494–495
primary response to an antigen, **493**
solutions of (antiserum) to identify bacteria, 287
specificity of, **484**
structure of, 479, 479*f*, 480*f*, 481*t*
T-dependent antigens and, **482**, 482*f*
to differentiate *Salmonella* serovars, 310
toxoplasma gondii and, 350
viruses and, 373, 379
antibody titer, **493**, 494*f*, **510**
measuring with direct agglutination test, **510**, 511*f*
antibody-dependent cell-mediated cytotoxicity (ADCC), 484, **485**, 485*f*, **491**, 492*f*, 529
anticancer drugs
nucleoside analogs and, 229–230, 229*f*
taxol produced by *Taxomyces* fungus, 339
anticodon, **219**, 220*f*
antifungal drugs, 443, 567–569, 568*f*, 569*f*
ineffective against bacteria, 279
summary, by mode of action/uses, 564*t*
that damage plasma membrane, 558, 559*f*, 567–569, 568*f*
antigen, virus-specific, on tumor cells, 391
antigen-antibody complex, **484–486**, 485*f*
antigen-antibody reactions
classical pathway of complement activation and, 464*f*, 466, 466*f*
fluorescent-antibody (FA) technique to identify, 61–62, 61*f*
ionic bonding in, 30
antigen-binding sites, **479**, 479*f*, 480*f*
antigen-presenting cells (APCs), 482, **486**, **489**–490, 490*f*, 496*f*
activated macrophages as, **490**, 490*f*
dendritic cells as, 476*f*, **490**, 490*f*
antigenic determinants (epitopes), **478**, 479*f*, 480*f*, 484
antigenic drift, **693–694**
antigenic shift, **370**–371*b*, **693**
bird flu and, 370*b*, 693, 693*t*
influenza virus and, **370***b*, 371*f*, **693**, 693*t*
antigenic variation, **433**, 442
examples of pathogens capable of, 433
gonorrhea and, 433, 749
HIV and, 541–542
Opa-encoding gene and, 433
used by *Giardia* protozoa, 444
used by trypanosomes, 433, 444, 629, 629*f*
antigens, **61–62**, 61*f*, **478**–482
allergens and, **523**
antibody-antigen binding results
agglutination, **484**–485, 485*f*
complement activated, **485**, 485*f*
cytotoxicity, **485**, 485*f*
neutralization, 484, **485**
opsonization, 484, **485**, 485*f*
antibody-dependent cell-mediated cytotoxicity and, **485**, 485*f*
as vaccines, **495**
binding sites, **479**, 479*f*, 480*f*
early discoveries about, 477
endogenous, **488**
epitopes and, **478**, 479*f*, 480*f*, 484
extracellular (free), B cell activation and, 482, 482*f*
free (extracellular), 482
H antigen, **82**
haptens and, **478**, 479*f*
histocompatibility complex and, **482**, 482*f*, 496*f*, **533–534**
microscope to observe, in real time, 62
nature of, 478–479, 479*f*
neutralization by antibodies, 484, **485**, 485*f*
number recognized by human immune system, 484
O polysaccharide functioning as, 87
opsonization by antibodies, 484, **485**, 485*f*
primary immune response to, **493–494**, 494*f*
Salmonella serovars and, 310
secondary immune response to, **493–494**, 494*f*
superantigens, **436**, 438*t*, 492, 522
T antigen, **391**
T-dependent, **482**, 482*f*
T-independent, **484**, 484*f*, 503
antigenic variation, **433**
antihelminthic drugs, 564*t*, 571–572
antihuman immune serum globulin (antiHISG), **513**–514, 515*f*
antimicrobial agents
added to household cleaning products, 20
alcohols, **197**–198, 198*t*, 200*f*, 203*t*, 204*t*
aldehydes, **200**, 204*t*
as therapy, 12–13, 12*f*. *See also* antibiotics
biguanides, **196**–197, 204*t*
biofilms and, 163
bisphenols, **196**, 196*f*, 203*t*, 204*t*
Cepacol, 199, 204*t*
chemical food preservatives, 199–200
chemical sterilization, 200–201, 205*t*
chlorhexidine, 197, 203*t*, 204*t*
considerations in choosing, 195
copper, 198–199, 198*f*, 204*t*
destruction of plasma membrane by, 91
detergents, 199
disk-diffusion test to evaluate, **195**, 196*f*, 201*b*
drawbacks of, 12
environmental influences, 186
evaluating disinfectants, 195
factors influencing effectiveness, 186, 187*f*
glutaraldehyde, 200, 203*t*, 205*t*
halogens, **197**, 204*t*
heavy metals, 198–199, 198*f*
hydrogen peroxide, 202
mechanisms of action, 186–187
mercury, 199, 203*t*
microbial exponential death rate and, 186, 186*t*, 187*f*
peroxygens, 202
phenolics, **195**, 196*f*, 203*t*, 204*t*
phenols, **195**, 196*f*, 203*t*, 204*t*
plasma sterilization, 201
resistance
biofilms and, 19, 163, 431
emerging infectious diseases (EIDs) and, 19
hemodialysis-associated infections and, 422*b*
misuse/overuse of, 19, 416, **575–578**, 576*f*, 577*b*
porins and, 202
to biocides, 202–203, 203*f*, 203*t*
silver, 198–199, 198*f*
silver nitrate, 198, 204*t*
silver sulfadiazine, 198, 204*t*
soaps, 199
supercritical fluids, 201–202
Surfacine, 198–199
terminology of, 185–186, 185*t*
time of exposure and, 186
use-dilution test to evaluate, 195
Zephiran, 198, 199, 199*f*, 200*f*, 201*b*
zinc, 199
antimicrobial drugs, 553–583. *See also* antibiotics
bactericidal *vs.* bacteriostatic, **555**
future of, 578–579
history of, 554
microbes that produce, 247, 249, 301*t*, 317, 321, 339, 554, 555*t*, 563
microbial susceptibility/sensitivity tests, 572, 572–573, 573*f*
modes of action, 555–559, 556*f*, 562–564*t*
cell-wall synthesis inhibitors, 556, 556*f*, 557*f*
essential-metabolite synthesis inhibitors, 558, 558*f*
nucleic acid synthesis inhibitors, 558, 558*f*
plasma membrane injury, 558, 559*f*
protein synthesis inhibitors, 556–557, 556*f*, 558*f*
resistance to. *See* antibiotic resistance
spectrum of activity, 555, 557*t*
antimicrobial peptides (AMPs), **470**–471, 472*t*, **578**–579
antimicrobial resistance not developed by, 471, 578–579
septic shock and, 471
synergy shown by, 470
Toll-like receptors and, 470, 579
antimicrobial resistance. *See* antibiotic resistance
antimicrobial substances of innate immunity, 463–472
antimicrobial peptides, **470**–471, 472*t*, **578–579**
as second line of defense, 450, 450*f*, 463
complement system, **463–468**
interferons, 468–470, **469**, 469*f*
iron-binding proteins, **470**
antimycobacterial antibiotics, **563**
modes of action/spectrum of activity, 562*t*
antiprotozoan drugs, 12, 528, 529*f*, 564*t*, **571**
antiretroviral drugs, 548, **571**
antisense agents, **579**
antisense DNA
explored as gene therapy, 259
pectin degradation and, 267, 267*t*
antisense DNA technology, MacGregor tomatoes, **267**, 267*t*
antisense drugs, 658
antisense strand (- strand), **387**, 388*f*
antisepsis, **185**, 185*t*
antiseptics, 195–199
alcohols, 197–198, 198*t*, 200*f*, 203*t*, 204*t*
alexidine, 197
bacitracin, 566–567
bacteria able to grow in, 196*f*, 202
biguanides, **196**–197, 204*t*
bisphenols, **196**, 196*f*, 203*t*, 204*t*
Cepacol, 199, 204*t*
chloramines, 197
chlorhexidine, 197, 203*t*, 204*t*
chlorine dioxide, 197
chlorine gas, 197
copper, 198, 199
effectiveness of various, 199, 200*f*
halogens, **197**, 204*t*
heavy metals, 198–199, 198*f*
hexachlorophene, 196, 196*f*
iodine, 197, 200*f*, 201*b*, 203*t*, 204*t*
iodophors, 197
isopropanol, 198
Lysol, 195
mercuric chloride, 199
mercury, 199, 203*t*
phenols/phenolics, 195–196, 196*f*, 203*t*, 204*t*
pHisoHex, 196
Purell, 198
quaternary ammonium compounds, 199, 199*f*
silver, 198–199, 198*f*
silver sulfadiazine, 198, 204*t*
soap and, 199, 200*f*
Surfacine, 198–199
triclosan, 196, 196*f*
vs. disinfectants, 185
zinc, 199
antisera/antiserum, **287**, **495**, 495*f*, 616
antitoxins, **435**, 439*t*, 477
neutralization tests and, 512, 513*f*
antitrypsin, produced by genetically modified sheep, 260*t*
antitumor activity of oncolytic viruses, 369
antitumor drugs, nucleoside analogs as, 229–230, 229*f*
antiviral drugs, 564*t*, **569**–571, 570*f*, 579
acyclovir, 557*t*, **569**, 570*f*
AZT, 230
enzyme inhibitors, 570
HIV/AIDS, 571
interferons, 246, 468–470, 469*f*, 570
nucleoside analogs and, 229–230, 229*f*
nucleoside/nucleotide analogs, 569–570, 570*f*
summary, by mode of action, 564*t*
antiviral proteins (AVPs), **469**–470, 469*f*
ants
fire, 346
fungi-farming, 330
APCs. *See* antigen-presenting cells
aphids
cauliflower mosaic virus transmitted by, 394*t*
pea, *Wolbachia* bacteria and, 307*b*
potato yellow dwarf virus transmitted by, 394*t*
Apicomplexa, **348**–350, 349*f*
as obligate intracellular parasites, 348
life cycle of, 348–350, 349*f*
oocyst of, **346**
apicomplexans, 348–350, 349*f*
aplastic anemia, chloramphenicol causing, 565
apoenzyme portion of enzymes, **116**, 116*f*
apoptosis, 455*t*, **489**, 489*f*, 496*f*
apple juice, contaminated, 264*f*
Approved Lists of Bacterial Names, 282
APTIMA assay, to detect early HIV infections, 545
aquaporins, 92, 92*f*
Aquaspirillum serpens, plasma membrane of, 90*f*
aquatic environments
algae's need for water, 341
low-nutrient, bacteria found in, 304, 304*f*, 305*f*
aquatic prokaryotes
gas vacuoles and, 96, 314
Planctomycetes as, 322, 322*f*
aqueous solutions, *vs.* tinctures as antiseptics, 198, 200*f*
Arachnida, 362, 362*t*
vectors of/diseases caused by, 362*t*
arachnoid mater, 611, 612*f*
Arber, Werner, 10*f*, 14*t*
arboviral disease, as nationally notifiable infectious disease, 421*t*
arboviral encephalitis, **624–626**, 626*f*
Culex mosquito as vector, 362*t*, 413*t*
eastern equine encephalitis in humans, 625
horses affected by, 625
St. Louis encephalitis (SLE), 625–626
symptoms, 625
western equine encephalitis in humans, 625
arboviruses, 223*b*, 387, 624–626
arboviral encephalitis, 624–626, 626*f*, 628*b*
types of, 628*b*
sentinel animals tested for antibodies to, 624–625
West Nile virus, 20, 212, 223*b*, 223*f*, 626
arbuscules, **767**, 768*f*
Arcanobacterium phocae, found in wounded seals, 283*b*
archaea, **4**, **274–275**, 275*f*, 276*t*
acidophilic, 325
cell walls of, 87, 274, 276*t*, 325

Index

  Gram staining and, 87
  morphology, 87, 325, 325f
  nitrifying, 325
  nutritional supply, 325
  optimum growth temperature and, 158, 325
  origins of, 275, 275f, 277, 277f, 281f
  thermophilic, 325
Archaea (domain), **4**, 6, **274–275**, 275f, **302t**
  Bacteria Domain compared to, 276t
  characteristics of, 276t
  Eukarya Domain compared to, 276t
  evolution and, 275, 275f, 277, 277f, 281f
  extremophiles of, **325**
  halophiles (extreme)of, 275, 275f, **325**
  important genera/special features of, 302t
  methanogens of, 4, 275, 275f, 281f, 302t, 325
  phyla of, 302t
  phylogenetic relationships, 275f, 281f
  thermophiles (hyperthermophiles), 4, **157**, 157f, 158, 275, 275f, 302t, **325**, 325f
Archaezoa, **347–348**, 347f
  as symbionts in animal digestive tracts, 347
  distinguishing features, 347, 354t
  parasitic species, 347–348, 354t
  position in evolutionary tree, 275f
  sources of human infections, 354t
archaezoans, 347–348, 347t, 354t
Arenaviridae, characteristics/important genera/clinical features, 376t
Arenavirus, 376t, 659–660
Argentine hemorrhagic fever, **659**
Arginine (Arg) amino acid, structural formula/characteristic R group, 44t
arithmetic death curves, *vs.* logarithmetic calculations, 187f
arm, of compound light microscope, **56**, 56f
armadillos
  as disease reservoir, 661
  used to culture leprosy bacillus, 167, 619
arrangement of cells (microbes), 76, 77–79, 78f, 79f, 101t, 330t
arsenic, combined with enzymes to prevent cell functioning, 120
arsenic derivative, first used to treat syphilis, 12
artery plaque, scanning acoustic microscopy (SAM) to study, 63, 63f, 67t
arthritis
  gonorrheal, 748
  psoriatic, 533
  rheumatoid, 460, 492, 493b, 509, **532**, 533
  septic, *Haemophilus influenzae* causing, 311
arthroconidia, **333**, 334f, 338t
  in *Coccidioides immitis,* 334f
arthropods, 361
  Alphavirus transmitted by, 375t
  as vectors, 329, 361–363, 362f, 362t, 363f
    diseases they transmit/causative agent, 413t
    transmission methods, 412–413, 413t
  classes of, 362
    Arachnida, 362, 362t
  mosquitos and West Nile virus, 20, 212, 223b, 223f, 626
  viruses that can replicate in, 376t
arthroscopic surgical instruments, 201
artifacts, 64
  mesosomes as, **91**
artificial blood, 259
artificially acquired immunity
  active, 494f, **495**. *See also* vaccination
  passive, 494f, **495**
ascariasis, 361t, 735b, **736**, 736f
Ascaris lumbricoides, 358–359, 361t, 736, 736f
ascomycetes, 334, 336f
Ascomycota (sac fungi), 280f, **334**, 336f
  pathogenic fungi of, 338t
  position in taxonomic hierarchy, 280f
ascorbic acid (vitamin C), fermentation and, 137t
ascospores, **334**, 336f
ASCs (adult stem cells), 535
ascus, **334**, 336f
asepsis, **186**
aseptic encephalitis, 223b
aseptic packaging, to preserve food, 202, 795–796, 797f
aseptic surgery, **184**
aseptic techniques, **8**, 186
  in hospitals, 416
  nosocomial infections and, 413, 416
asexual reproduction
  in algae, 342, 342f
  in diatoms, 343f
  in fungi, **332**, **333**, 335f-337f
  in *Plasmodium vivax,* 348–349, 349f
  in protozoa, 346
asexual spores, **332–333**
  of Ascomycota fungi, 334, 334f, 336f
  of Basidiomycota fungi, 335, 337f
  of pathogenic fungi, 334f, 338t
  of prokaryotic actinomycetes, 320
  of Zygomycota fungi, 333, 334f, 335f
Asian liver fluke (*Clonorchis sinensis*), 356, 356f
Asparagine (Asn), structural formula/characteristic R group, 44t
aspartic acid (Asp)
  in transamination, 148f
  structural formula/characteristic R group, 44t
aspergillosis, 338, 388t, 569, **697–698**
Aspergillus flavus
  aflatoxin produced by, 230, 443
  conidia and conidiophore of, 333, 334f
Aspergillus fumigatus, 698
Aspergillus genus/spp./Ascomycota, **334**, 336, 336f
  as opportunistic pathogen, 338t, 388
  conidiospores produced by, 333, 334f
  fermentation and, 137t
  produces sclerotia to resist food processing temperatures, 795
  used in production of sake, 800
Aspergillus niger
  genetically engineered rennin produced by, 267t
  used to produce citric acid for food/beverages, 339
  vegetative and aerial hyphae of, 331f
aspirin, reduce fever by inhibiting prostaglandin synthesis, 437
asthma, 523, 523t
  as an allergic reaction, 525
  leukotrienes and, 524
  near epidemic in children, 525
atazanavir, 571
atherosclerosis, 18
athlete's foot (tinea pedis), 338t, 409, 410t, 568, **600**, 600f
Atlantic coast horseshoe crab, 439
atmospheric oxygen levels, photosynthetic cyanobacteria and, 314–315
atomic force microscope (AFM), **65**, 65f, 68t
  DNA double helix image, 58f
  perfringoglysin O toxin image, 65f, 68t
  specimen preparation and, 65, 68t
  specimen size and, 58f
atomic number, **27**
  notation of, 28
  of common elements, 27t
atomic weight, **27**
  notation of, 28
  of common chemical elements, 27t
atoms, **27–28**, 27f, 27t
  chemical elements and, 27–28, 27t
  inert, 28
  molecule formation by, 28–32
  structure, 27, 27f
ATP
  active transport and, 91, **94**
  breakdown of, 114, 114f
  chemiosmotic generation of, **130–131**, 130f
  confocal microscopy to observe distributions/concentration of, 62
  generation of, 122–123
  in active membrane transport processes, 91
  in Calvin-Benson cycle, 140, 142f
  in glycolysis, 124, 125f
  in photophosphorylation, 123
  in photosynthesis, 140, 141f
  microbial uses for, 146
  mitochondria's role in production of, 105
  oxidation-reduction reactions and, 122, 122f, 123f
  requirements for production of, 141, 143f
  synthesis, volutin and, 95
  synthesis of, 114, 114f
  unstable bonds of, 121
  yield in aerobic respiration, 132t, 137t
  yield in anaerobic respiration, 132, 137t
  yield in fermentation, 134f, 137t
ATP (adenosine triphosphate), chemical structure, **47–49**, 49f
ATP synthase, 130, 130f, 131f
ATP synthesis
  nitrogen requirements, 160
  phosphorus requirements, 160
atrichous bacteria, **81**
Atripla, 571
attachment (adsorption) in viral multiplication, **380**
  in animal viruses, 383, 384t, 386f
  in bacteriophages, 380, 381f, 384t
attenuated whole agent vaccines, **501**–502
attractants, types of bacterial movements toward, 82
AUG codon as start codon, 219
Augmentin, 561
auramine O, 61
Aureomycin (chlortetracycline), 565
  mode of action/spectrum of activity, 562t
  produced by produced by *Streptomyces aureofaciens,* 555t
autism, MMR vaccine and, 506
autoclaves/autoclaving, **188**–190, 189f, 189t, 194t
  endotoxins released by bacteria killed by, 439t, 440b
autografts, **536**
autoimmune diseases, **532**–533
  cell-mediated, 532–533
  cytotoxic, **532**
  Graves' disease, **532**
  immune complex reactions, **532**
  insulin-dependent diabetes mellitus, **533**
  multiple sclerosis, **532**–533
  myasthenia gravis, **532**
  psoriasis, **533**
  rheumatoid arthritis, 460, 492, 493b, 509, **532**, 533
  self-tolerance loss and, 532
autoinoculation, 588
autotrophs (lithotrophs), **142–143**, 143f, 146
auxotrophs, **232**
Avery, Oswald T., 10f, 16, 47, 235
avian influenza A H5N1 (bird flu), **19**, 370–371b
  emerging infectious diseases and, 19, 416, 417t
  genetic recombination and, 416, 693
  recent human cases, by subtype/location, 370t
  vaccines and, 19
avian influenza A (H5N1) virus
  emerging infectious diseases and, 19, 416, 417t
  genetic recombination and, 416, 693
avian sarcoma viruses, derived from normal part of chicken genes, 391
avirulent microbial strains
  defined, 11
  vaccines produced from, immunity and, 11
avoparcin, 577b
AVPs (antiviral proteins), **469–470**, 469f
axial filaments (endoflagella), 82–**83**, 84f
  of spirochetes, **322**, 324f
axostyle, of *Trichomonas vaginalis,* 347f
azidothymidine (AZT), as nucleoside analog, 230
azithromycin, 566
  mode of action/spectrum of activity, 562t
azlocillin, 561
azole antibiotics, **568**, 569f
azoles (antifungals), **568**
  mode of action/uses, 564t
Azolla-cyanobacteria symbiosis, 772, 774f
Azomonas genus/spp., 300t, **308**
  as free-living nitrogen fixers, 300t
Azospirillum genus/spp., 300t, **303**
  agricultural importance as nitrogen fixer, 303
Azotobacter genus/spp., 300t, **308**
  as free-living nitrogen fixers, 300t
  lipid inclusions of, 96
AZT (azidothymidine), as nucleoside analog, 230
aztreonam, mode of action/spectrum of activity, 562t

B cells, **454**, **478**
  activation of, 482–484, 482f
  as memory cells, 483, 483f, 486–487, 493–494
  as plasma cells, 482f, **483**, 483f, 494
  as third line of defense, 450f
  cancerous, 507, 508f
  clonal selection of, **483**–484, 483f
  differentiation of, 483–484, 483f
  function of, 478
  IgD antibody and, 481t
  in compromised hosts, 415
  in humoral immunity, 482–486, 482f, 483f
  lymph node location of, 456, 638, 639f
  monoclonal antibodies and, 507–509, 508f
  T-dependent, 482, 496f
B lymphocytes. *See* B cells
B vitamins, in complex culture media, 165, 165t
β-1, 4 linkage, 85f
β-galactosidase, 224, 224f, 226f
β-galactosidase (*lacZ*) gene, 256, 256f
  encoding as marker gene, 251f
β-lactam antibiotics, gram-negative bacteria susceptibility and, 89
Babesia microti, 350, 354t, 666
babesiosis, 354t, **651b**, **666**
BAC (bacterial artificial chromosome), 261f
Baccillariophyta, 343t
Bacillales, **317–318**, 317f, 318f
  important genera/special features, 301t
bacillary dysentery (shigellosis), 310, 459, **712**, 712f, 713f, **722b**
  as notifiable infectious disease, 421t
  incubation period, 430t
  portal of entry for, 429, 430t
  Shigella bacteria causing, 310. *See also* *Shigella*
  waterborne transmission and, 411

*Bacillus amyloliquefaciens, Bam*HI restriction enzyme used in rDNA technology, 249*t*
Bacillus anthracis, **317**
as biological weapon, 317, 646, 649*b*, **649*b***
capsule of, 80, 645
virulence and, 432, 645
emerging infectious diseases and, 417*t*
encapsulated, phagocytosis and, 80
fluorochrome used to stain, 61
Koch's experiments with, 11, 404
portals of entry, 429
reservoirs/transmission method, 410*t*
toxins of, 239
Bacillus cereus, 317, 317*f*
endospore staining and, 72*f*
gastroenteritis, 317, **720**–721, 723*b*
temperature and growth of, 159*f*
Bacillus coagulans, capable of growth in canned foods, 795
Bacillus genus/spp., **77–78**, 78*f*, **317**–318, 317*f*, 318*f*
anaerobic respiration and, 132
antibiotics derived from, 317
as gram-variable bacteria, 87
as human pathogens, 301*t*
calcium propionate active against, 200
endospores and, **96**–98, 97*f*, 301*t*
enzymes of
bioremediation and, 17
in household detergents, 17
fermentation and its end-products, 134*f*
genetic transformation naturally occurs in, 236
in taxonomic hierarchy, 301*t*
lipid inclusions of, 96
selenium toxicity and nanotechnology, 264, 264*f*
Bacillus sphaericus, survived in fossilized amber for millions of years, 278
Bacillus subtilis
bacitracin derived from, 555*t*
double-stranded helix formed by, 78*f*
pentose phosphate pathway and, 127
secreting-product capability of, and genetic modification, 258
Bacillus thuringiensis
as best-known microbial insect pathogen, 317, 317*f*
human allergies to, 268
insecticidal toxin (Bt toxin) derived from, 265, 267, 267*t*, 317
monarch butterflies and, 268
Pseudomonas fluorescens engineered to produce toxin normally produced by, 267, 267*t*
sold industrially, 806
used in pest control, 17, 317
bacillus/bacilli (rod-shaped bacteria), **77**, 78, 78*f*
as chain of rods (streptobacilli), **78**, 78*f*
as oval rod (coccobacilli), **78**, 78*f*
as pair of rods (diplobacilli), **78**, 78*f*
as single rod, 78, 78*f*
bacitracin, 555*t*, **563**, 566–567
inhibits cell wall synthesis, 556, 556*f*, 557*f*
mode of action/spectrum of activity, 562*t*
backmutation of microbes, attenuated vaccines, 502
bactercidal antimicrobial drugs, **555**
bacteremia, **407**
as emerging infectious disease, 417*t*
methicillin-resistant *Staphylococcus aureus* and, 417*t*
nosocomial, 415*t*
epidemiological analysis of, 422*b*
vancomycin-resistant enterococci and, 417*t*
vancomycin-resistant *Staphylococcus aureus* and, 417*t*
Bacteria (domain)
characteristics, compared to Archaea, Eukarya, 276*t*
phylogenetic relationships, 281, 281*f*
selected prokaryotes of, 300–302*t*
bacteria/bacterium, 2, **3–4**, 5*f*, **274**, 275*f*, 276*t*
acetic-acid producers, 300*t*
acid-fast, 70–71, 71*f*
anatomy, 3–4, 79–98, 80*f*
antibiotic resistance and. *See* antibiotic resistance
as atmospheric nitrogen converters, 17
as carbon recyclers, 16–17
as domain in three-domain system, 6, 274, 275*f*, 276*t*
as pest controllers, 17
beneficial activities of, 16–18, 20
biofilms, percentage existing in, 77. *See also* biofilms
bioremediation and, 17, 33*b*
cell walls, 39, 80*f*, **84**–89, 86*f*
acid-fast, 87–88
atypical, 87–88
damage to, 88–89
gram stain mechanism and, 87
gram-negative, 85, 86*f*, 87
gram-positive, 85, 86*f*, 87
structures inside, 80*f*, 89–98
structures outside, 79–84, 80*f*
classification of, 279–281, 280*f*
differences between fungi and, 330*t*
diseases caused by. *See* bacterial diseases
early representations, 7, 7*f*
emerging infectious diseases caused by, 417*t*
evolution of, 275, 275*f*, 276*t*, 277–278, 277*f*
fermentation process and, 9
first used in genetic research, 16
flagella of, 4, 81–82, 81*f*, 82*f*
food spoiled by, *vs.* by molds, 339
fungi compared to, 330*t*
genetic recombination in, 234–241
genetic transformation in, 234–236, 235*f*
germ theory of disease and, 9, 11, 477
glycocalyx of, 79–81
gram-negative, 69–70, 70*f*
gram-positive, 69–70, 70*f*
identification methods, 282–294
by biochemical testing, 137–139, 139*f*, 285–287
by cell morphology, 284–285
by differential staining, 285
by rapid identification methods, 285–286, 291, 292*f*
by serology, 287, 287*f*
in foods, radiation doses needed to kill, 797*t*
in repair of UV-induced damage, 230–231, 230*f*
industrial/commercial fermentation uses, 137*t*
L forms of, **89**
lactic-acid bacteria, 135*b*
mercury cleanup by, 33*b*
metabolism, 113–155
monomorphic, **79**
movements of, 82, 83*f*
nutritional classification of, 143, 143*f*
nutritional requirements, 4
oil-eating, 33*b*
origin of, 275, 275*f*, 276*t*, 277–278, 277*f*
osmotic solutions subjected to, 93–94, 93*f*
overview, 3–4
parasitic, 403
pasteurization process and, 9
pH and growth of, 158–159
pH of, 68
photosynthesis by, 4
phylogenetic relationships, 274–278, 278*f*
plasma (cytoplasmic) membrane, 89–91, 90*f*
pleomorphic, **79**
quorum sensing and, 57*b*, 163
rapid identification tests for, 286, 286*f*, 291, 292*f*
reproductive methods, 4, 171, 171*f*
"resting" cells formed by, 96
scientific nomenclature and, 278–279
shapes of, 4, 5*f*, 77–79, 78*f*, 79*f*
cell wall and, 80*f*, 84–89, 86*f*
shrinkage/collapse of, 94
sizes of, 4, 5*f*, 77–79, 78*f*, 79*f*
large (seen by unaided eye), 13
specimen preparation for microscopy, 54, 68–72
staining of, 54, 68–72
star-shaped, 79, 79*f*
symbiotic, 106, 107*b*, 267, 300*t*
taxonomic hierarchy of, 279, 280*f*
used as carcinogen indicators, 232–233, 233*f*
used in cotton production, 3*b*
virulence of, **71**, 80
viruses compared to, 368, 368*t*
bacterial artificial chromosome (BAC), 261*f*
bacterial biosensors, **780*b***
bacterial chromosome, 210*f*
bacterial chromosomes, **94**–95, 97*f*
maps, 212, 212*f*
bacterial cultures
bacterial division, 171, 171*f*
generation times for, **171**
growth curves, **172**–174, 173*f*
obtaining pure, 170, 170*f*
phases of growth, 172–174, 173*f*
preserving, 170
bacterial diseases
of cardiovascular system, 636–655, **643*b***, **650*b***, **651*b***, **668*b***
of digestive system, 707–721, **710*b***, **722*b***
of eyes, 603–605, **604*b***
of lymphatic system, 636–655, **643*b***, **650*b***, **651*b***, **668*b***
of nervous system, 611–620, **617*b***, **632*b***
of reproductive system, 747–756, **759*b***, **761*b***
of respiratory system, 677–692, **681*b***, **687*b***, **699*b***
of skin, 586–598, **589*b***, **590*b***, **592*b***
of urinary system, 746–747, **748*b***
bacterial endospores, *vs.* other spores, 97
bacterial enzymes
as restriction enzymes in rDNA technology, 249, 249*t*
bioremediation and, 17
bacterial growth curve, **172**–174, 172*f*, 173*f*
bacterial heat resistance, 188
bacterial meningitis, **612–615**, **617*b***
diagnosis and treatment, 614, 614*f*
Hib vaccine and, 613, 614
leading cause of, 614
bacterial pneumonias, **685–692**, **687*b***
bacterial populations, logarithmic representations, 171–172, 172*f*
bacterial species
members defined taxonomically, 279, 281
strains of, 279, **281**
bacterial vaginosis, **756**, 756*f*, **759*b***
bacterial viruses. *See* bacteriophages
bacterial zoonoses, 410*t*
bacteriochlorophyll, 140, 144, 145*t*, 315
bacteriocins, 42, **239**, 310
produced by *Escherichia coli,* 401
Bacteriodetes, **324**
Bacteriological Code, 279
bacteriology, **13**
bacteriophage f2, size of, 369*f*
bacteriophage lambda, 377*f*, 380, 382, 382*f*
bacteriophage M13, size of, 369*f*
bacteriophage MS2, size of, 369*f*
bacteriophage T4, size of, 369*f*
bacteriophages (phages), **237**, 239*f*, 288, 290*f*, **368–369**
as complex virus, 373, 374*f*
cultivation of, 374, 377, 377*f*
plaque-forming units (PFU), **377**
plaques, **374**, 377, 377*f*
entry/penetration into host cells, 380, 383–384, 384*f*, 384*t*
genes of that contribute to pathogenicity, 441
lysogeny, prophages and, 441
multiplication of, 379–382
lysogenic cycle, **379**, 380, 382, 382*f*
lytic cycle, **379**–380, 381*f*
vs. animal viruses, 384*t*
phage DNA, 237, 249, 380
phage library, 255*f*
phage lysozyme, **380**
phage therapy, 368–369, **579**
virus-host interactions research, 369, 579
phage typing, **288**, 290*f*
protein coat of, 237
reproduction and, 237
restriction enzymes and, 249
sizes of selected, 369*f*
T-even bacteriophages, 374*f*, 379–380, 381*f*
viral plaques formed by, 374, 377, 377*f*
bacteriostasis, **186**
bacteriostatic antimicrobial drugs, **555–559**
Bacteroidales, important genera/special features, 302*t*
Bacteroides fragilis, plasmids from that encode resistance to clindamycin (antibiotic), 240*f*
Bacteroides genus/spp., 302*t*, **324**
as normal microbiota of large intestine, 403*t*
as normal microbiota of mouth, 403*t*
as normal microbiota of urethra, 403*t*
deep tissue infections and, 324
gingival crevices inhabitants, 324
inhabit human intestinal tract, 302*t*, 324
phylogenetic relationships, 281*f*
Bacteroidetes, 302*t*
baker's yeast. *See Saccharomyces cerevisiae*
bakery products, microbes used in, 800
Balamuthia, granulomatous amoebic encephalitis caused by, 348, 354*t*
Balantidium coli, 350, 354*t*
Baltimore, David, 10*f*, 14*t*
Bam HI restriction enzyme, 249*t*, 251
bandages, quat antiseptics neutralized by, 199, 201*b*
Bang, Olaf, 389
Barré-Sinoussi, Françoise, 15*t*
Barr, Yvonne, 10*f*, 391
Bartonella genus/spp., 300*t*, **305**
as human pathogens, 300*t*
Bartonella henselae
cat-scratch disease and, 305, 410*t*, 417*f*, **647**, 647*f*, **650*b***
reservoirs/transmission method, 410*t*
basal body
eukaryotic cell, 99*f*
of flagella, 81, 82*f*
base, of compound light microscope, **56**, 56*f*
base pairs, **211**
base substitution (point mutation), **227**, 227*f*
baseplate, of a T-even bacteriophage, 374*f*, 381*f*
bases, 34, **35**–36, 36*f*
complementary, 47, 48*f*, 211
in nucleotides, 47, 48*f*
basic dyes, **68**–69
basidiomycetes, **335**, 337*f*

Basidiomycota (club fungi), **335**, 337*f*
life cycle, 337*f*
pathogenic fungi of, 338*t*
Cryptococcus neoformans (Filobasidiella), 338*t*
Malassezia, 338*t*
basidiospores, **335**, 337*f*
basidium, **335**, 337*f*
basophilic inclusion bodies, 442, 443*t*
basophils, **454**, 455*t*
histamine present in, 460
IgE antibodies and, 481
in hypersensitivity reactions, **523**, 524*f*
staining and, 454
Bassi, Agostino, 10*f*, 11
bathrooms, fungi capable of growing in, 333
bats
as disease reservoirs, 410*t*, 624 *footnote*
bites
incidence/diseases transmitted by, 624*f*, 624 *footnote*
rabies case report (Clinical Focus), **625*b***, 625*f*
fruit, possibly transmitting hemorrhagic fevers, 660*b*
histoplasmosis and, 695–696
rabies virus variant found in, 622, 625*b*, 625*f*
Baylisascaris procyonis (helminth), causing raccoon roundworm encephalitis, in humans, 417*t*
BCG vaccine, **685**
Bdellovibrio genus/spp., 301*t*, **312**, 312*f*
other bacteria attacked by, 301*t*, 312, 312*f*
Bdellovibrionales, important genera/special features, 301*t*
Beadle, George W., 10*f*, 14*t*, 16
bears, as disease reservoirs, 410*t*
Becton Dickinson's Enterotube II, 286*f*
bee stings
anaphylaxis and, 523–524
desensitization success and, 526
beef, "measly" appearance, tapeworm infection and, 357
beef products, infected with mad cow disease, **20**
beef tapeworm, humans as definitive hosts for, 357
beer, **800**
fermentation and, 135*b*, 137*t*
microbes used in production of, 800
pasteurization of, 190–191
souring and spoilage, 9
bees, Israeli acute paralysis virus and, 367
Beggiatoa alba, as gliding, sulfur-oxidizer, **307**–308, 772
Beggiatoa genus/spp., 145, 300*t*, **307**–308, 312, 772
Beijerinck, Martinus, 16–17
Beijerinckia genus/spp.
as free-living nitrogen fixers, 300*t*
in taxonomic hierarchy, 300*t*
benthic zone, **776**
benzalkonium chloride. *See* Zephiran
benzathine penicillin, 559, 560*f*
benzoic acid, 204*t*
benzopyrene, as frameshift mutagen, 230
benzoyl peroxide, 202
Berg, Paul, 15*t*, 16
Bergey, David, 282
Bergey's Manual
description of strains and, 286
earlier editions and morphologic bacteria groupings, 299
latest edition based on phylogenetic system, 299
Bergey's Manual of Determinative Bacteriology, 1st ed., 282
Bergey's Manual of Determinative Bacteriology, 9th ed., 282
Bergey's Manual of Systematic Bacteriology, 279, 282 *footnote*, 321
beta interferon (human), 469–470, 469*f*
as rDNA product in medical therapies, 260*t*
to treat multiple sclerosis (Betaferon), 470
to treat osteoporosis (Actimmune), 470
beta-hemolysis, 168, 168*f*
beta hemolytic group A streptococcus, 319
beta-hemolytic group B streptococcus, 319
beta-hemolytic streptococci, 168*f*, 319
beta-lactam antibiotics, 559–561, 560*f*, 561*f*
beta-lactam ring in penicillins, 559, 560*f*, 574
beta-lactamase enzymes, that inactivate antibiotics, 574, 575*f*
beta-lactamase inhibitors, 561
beta-lactamases, 559
beta-oxidation
in lipid catabolism, 136*f*
of petroleum/oil spills, 33*b*, 136
Betadine, 197
Betaferon (beta-interferon), to treat multiple sclerosis, 470
betamethasone, 201*b*
betaproteobacteria, 300*t*, 305–306, 306*f*
important genera/special features, 300*t*
beverage industry, *Aspergillus niger* fungus used to produce citric acid for, 339
bGH (bovine growth hormone), 267, 267*t*
bidirectional DNA replication, bacterial, 214, 217*f*
Bifidobacterium genus/spp., as normal microbiota of large intestine, 403*t*
biguanides, **196**–197, 204*t*
bile, most microbes destroyed by, 429
bile salts, gram-negative bacteria and, 87
binary fission, 4, **77**, 101*t*, **171**, 171*f*, 276*t*
of cyanobacteria, 314, 314*f*
viruses and, 368*t*
binomial nomenclature, **278–279**
bioaugmentation, **775**
biochemical oxygen demand (BOD), in sewage treatment, **783–785**
biochemical reactions, metabolic, 113–155. *See also* chemical reactions
biochemical tests, 137–139, 139*f*, 144*b*, 169
identify human pathogens isolated from marine mammals, 283*b*
importance with enterics group of bacteria, 309
to identify microorganisms, 285–287, 285*f*
biocides, **186**. *See also* antimicrobial agents
bioconversion, **806**–807
biodegradable alternatives, to plastic, 3*b*
bioenhancers, plant fertilizers used in oil-spill bioremediation, 33*b*
biofilms, **18**–19, 19*f*, 77, 81, 162–**163**, 163*f*
adherence and pathogenicity of, 431
algae on swimming pool walls as, 431
antimicrobial resistance and, 19, 163, 186, 189, 431
as bacterial packs exhibiting group behavior, **57*b***
bacteria existing in (percentage), 77
bacterial endocarditis and, 641*f*
bacterial fimbriae's role in forming, 83
catheters and, 19*f*, 163, 164*b*, 431, 586, 587*f*
confocal microscopy and, 163
dental plaque as, 431
evading phagocytosis, 459–460
examples of, 431
gliding motility and, 84
glycocalyx as important component of, 81, 162, 163*f*, 431
heart valves and, 163, 431, 641, 641*f*
in sewage treatment, 784, 787*f*
inducer (signaling chemical) and, 57*b*
laboratory containers, endotoxins, and autoclaving, 440*b*
Legionella, hospital water lines and, 689
pathogenicity and, 431
Pseudomonas aeruginosa often grows in, 593
quorum sensing and, 57*b*
scanning acoustic microscope and, 63, 63*f*, 67*t*
shower door scum as, 431
that lead to disease, 57*b*, 431
virulence factors, 431
biofuels, **807**–808, 808*f*
biogenesis theory, **8**
biogeochemical cycles, **768**–776
carbon cycle, **768**–770, 769*f*
microbial benefits to, 16–17
nitrogen cycle, **770–772**, 770*f*
phyosphorus cycle, **774**
sulfur cycle, **772**–774, 774*f*
bioinformatics, **261**–262
biological oxidations, 122, 122*f*, 123*f*
biological transmission of disease, by arthropods, **412–413**
biological weapons, 21, 262, **649*b***, 649*f*
bioweapons detectors, 193, 649*b*
list of potential bioweapons (bacteria/viruses), 649*b*
biology, molecular, **16**
bioluminescence, **778**, 778*f*
chemical pathway of, 57*b*
bioluminescent bacteria, *Vibrio fischeri,* 57*b*
biomass, **806**–807
bioreactors
algal, that could produce biofuels, 808
in industrial fermentation, 303*f*, **802**–803
bioremediation, **17**, **775**, 776*f*
mercury contamination, 33*b*
oil spills, 33*b*
biosafety level 4 (BSL-4), 168, 168*f*
biosafety levels of laboratories, 168, 168*f*
biosensors, 801
biosensors (bacterial) to detect pollutants/pathogens, **780*b***
biosolids, 787
biosynthesis, 146
in bacteriophage viral multiplication cycle, **380**, 381*f*, 384*t*
of DNA viruses, 384*t*, **385**–386, 385*t*, 386*f*
of polysaccharides, 146, 146*f*
of RNA viruses, 384*t*, 385*t*, **387**–389, 388*f*, 389*f*
speed of in eukaryotic *vs.* prokaryotic cells, 146
biosynthetic chemical reactions. *See* anabolism
biotechnology, **17**–18, 246–272, **247**, 249*t*
ethical issues, 268
restriction enzymes used in, 249, 249*t*
safety issues, 268
tools, 247–252
polymerase chain reaction (PCR), **251**, 252*f*
restriction enzymes, **249**–250, 249*t*, 250*f*
selection (artificial), 247, **249**
site-directed mutagenesis, **249**
vectors, 250–251, 250*f*, 251*f*. *See also* vectors
bioterrorism, **649*b***
bioweapons detectors, 193, 649*b*
list of potential biological weapons, 649*b*
biotin, coenzymatic functions, 117*t*, 160
biotypes, **310**
biovars, **287**, **310**
bioweapons, 21, 262, **649*b***, 649*f*
detectors, 193, 649*b*, 649*f*
potential pathogens list (bacteria/viruses), 649*b*
bird flu (avian influenza A H5N1), **19**, 370–371*b*
as emerging infectious disease, 19, 416, 417*t*
genetic recombination and, 416
vaccines and, 19
birds
as disease reservoirs, 410*t*
as West Nile virus disease reservoirs, 20, 223*b*, 410*t*
influenza A viruses and, **19**, 370–371*b*
Bishop, J. Michael, 15*t*, 391
bismuth sulfite agar, 168
bisphenols, **196**, 196*f*, 203*t*, 204*t*
bites
animal. *See* animal bites
by Komodo dragon, *Pasteurella multocida* and, 311
insect. *See* insect bites
blades of algae, 341*f*, **342**
blastoconidia, **333**, 334*f*
Blastomyces (Ajellomyces) dermatitidis, 338*t*
Blastomyces dermatitidis, blastomycosis caused by, 697
blastomycosis
airborne transmission and, 412
amphotericin B to treat, 568
blastomycosis (North American blastomycosis), **697**, **699*b***
bleaching agents
as disinfectants, 197
safety improved by microbiology applications, 3*b*
blebs/blebbing, 489, 489*f*
blindness
Acanthamoeba causing, 348
herpetic keratitis causing, 605
ophthalmia neonatorum causing, 603, 748
trachoma causing, 322, 604–605, 605*f*
blood, **454**, 638, 638*f*
components
formed elements, **454**–456, 455*t*, 638
erythrocytes, 455*t*
leukocytes, **454**, 455*t*
platelets, **455*t***
plasma, **454**
filtration by kidney glomeruli, 529
blood agar, 168, 168*f*
blood banking, safe blood supplies, 727*b*
blood capillaries, relation to lymphatic capillaries, tissue cells, 456*f*
blood clotting
fibrinogen and, 460
in inflammatory response, 461*f*
platelets' function as, 455*t*
blood diseases, 407
blood flukes, 356
Schistosoma, 356, 361*t*, 666, 667*f*
blood parasites (hemoflagellates), **351**, 667*f*, 668*b*
blood plasma, **201**, **454**, 457, **467*b***, 638
blood plasma substitute, dextran as, 40
blood platelets
histamine present in, 460
quinine and, 528, 529*f*
thrombocytopenic purpura and, 528, 529*f*
blood poisoning. *See* septicemia
blood transfusions
HIV transmission and, 546
reactions, 523*t*, 526–528, 527*t*
Rh incompatibility, 527–528, 528*f*
blood types, 526–528, 527*t*
blood vessels, in inflammatory response, 461*f*
blood-brain barrier, 610, **611**, 627
blood-clotting proteins, activated by endotoxins, 437
blood-feeding arthropods, 20, 223*b*
Blue cheese, ripened by *Penicillium* molds, 799
blue-green algae, cyanobacteria misnomer, 313–314
blue-white screening, 256, 256*f*
blunt ends of cut DNA strands, **249**, 250*f*

- body defenses, 18. *See also* host defenses; immunity
  - adaptive immunity, 433, **450**, 450f, **476–499**
  - adaptive *vs.* innate, 450, 450f
  - complement system, 463–468
  - first line of defense, 450–453, 450f, **471t**
  - innate immunity, 449–475, **450**, 450f, **476**
  - overview, 450f
  - second line of defense, 450f, 454–472
    - antimicrobial substances, 463–472
    - fever, 463
    - inflammation, 460–463
    - phagocytes, 457–460
  - third line of defense, 450f
- body piercing, bacterial endocarditis and, 641
- body temperature
  - fever and, 463
  - high, intensifies antiviral interferon's effects, 463
- body tube, of compound light microscope, **56**, 56f
- boil (furuncle), **588**
- boiling water, to control microbial growth, 188, 194t
- boils, 462
  - acute inflammation of, 460
- Bolivian hemorrhagic fevers, **659**
- bonds, chemical. *See* chemical bonds
- bone marrow, red, 456, 456f
- bone marrow transplants, 535, **536**
- bone morphogenic proteins, genetically engineered; useful in healing fractures/reconstructive surgery, 260t
- booster immunizations, 418, 501, 502, 616
  - recommendations for, 502t, 503t, 616
- *Bordetella* genus/spp., 300t, **306**
  - as human pathogens, 300t
- *Bordetella pertussis,* 674f
  - complement system evasion by, 468
  - emerging infectious diseases and, 417t
  - incubation period, 430t
  - pertussis (whooping cough) caused by, 13, 306, 411, 417t, 421t, 430t, 680–682, 680f, 699b
  - portals of entry, 430t
  - vaccine against, 13, 502, 502t, 503t, 504t, 682
- *Borrelia burgdorferi*
  - Lyme disease caused by, 287, 389f, 413t, 417t
  - reservoirs/transmission method, 410t
- *Borrelia* genus/spp., 302t, **323**
  - as human pathogens, 302t, 323
  - causing relapsing fever, 413t
  - transmitted by *Ornithodorus* (tick), 413t
- Botox, 618–619
- bottlenose dolphins, 283b
- botulinum toxin, **437**, 616–617
  - as A-B neurotoxin, 437, 441
  - as an exotoxin, 439t
  - different types of, with different potencies, 437
  - glycoproteins, plasma membranes and, 90
  - mechanism of, 438t, 616–617
  - naming of, 435
  - potency of, 430–431, 435
  - produced by *Clostridium botulinum,* 435, 437
  - serotypes of, 617–618
  - symptoms induced by, 437, 438t
  - therapeutic uses (Botox), 618–619
- botulism, **616–619**, **632** *b*. *See also* *Clostridium botulinum*
  - as nationally notifiable infectious disease, 421t
  - as special case of intoxication, 710
  - *Clostridium botulinum* causing, **316**
  - diagnosis of, 618, 619f
  - from bacteria in soil, 409
  - from improper home canning methods, 188, 190
  - incidence of, 618
  - nitrites active against, 200, 204t
  - refrigeration and, 618
  - symptoms, 437, 438t, 616–617
  - treatment of, 618
  - wound, **618**
- bovine growth hormone (bGH), 267, 267t
- bovine spongiform encephalopathy (BSE), **20**, 203, 393, 417t, **631–632**
- bovine tuberculosis, **685**
- *Bradyrhizobium* genus/spp.
  - as symbiotic nitrogen fixers, 300t
  - in taxonomic hierarchy, 300t
- bradyzoites, in toxoplasmosis, 661, 662f
- brain, 611, 611f
  - as immunologically privileged site, 534
  - inflammation of. *See* encephalitis
  - inflammations of, blood-brain barrier and, 611
  - parasitic helminths and, 361t
  - pathogenic invasion routes to, 611
  - prions and, 630–632, 630f
- brain abscess, caused by amoeba *Balamuthia,* 348
- bread, rye, fermentation and, 137t
- bread dough, what makes it rise, 135b
- bread molds, 5f
- breakbone fever, **659**
- breakthrough varicella, **597**
- breast cancer
  - genetic screening and, 262
  - monoclonal antibodies (Herceptin) to treat, 509, 538
- breast milk, IgA antibodies in, 480, 481
- *Brevibacterium,* as normal microbiota of skin, 402t
- brightfield illumination microscope, **59**, 60f, 67t
- broad-spectrum antibiotics, **555**, 557t
  - commonly used, by mode of action, 562t, 563t
  - newer penicillins as, 561
  - normal microbiota destroyed by, 555
- bronchopneumonia, streptococcal, 407
- bronchitis, *Haemophilus influenzae* as cause of, 311
- broth dilution tests, **572–573**, 573f
- brown algae (kelp), 341, 341f, **342**, **343t**
- *Brucella abortus,* 644
- *Brucella* genus/spp., 300t, **305**
  - adept at evading phagocytes, 459, 644
  - as human pathogens, 300t, 305
  - as potential biological weapon, 644, **649b**
  - brucellosis caused by, 305, 457, 644
  - portals of entry, 430t
  - reservoirs/transmission method, 410t
- *Brucella melitensis,* 644
- *Brucella suis,* 644
- brucellosis (undulant fever), 283b, 305, **643–645**, **650b**
  - as notifiable infectious disease, 421t
  - as zoonotic disease, 410t, 643
  - direct agglutination test to diagnose, 510
  - incubation period, 430t
  - portals of entry, 430t
  - vaccine for animals, 644
- BSE (bovine spongiform encephalopathy), **20**
- BSL-1 to BSL-3 (biosafety level 1 to 3) labs, 168
- BSL-4 (biosafety level 4) labs, 168, 168f
- Bt corn plants, 267t
- Bt cotton plants, 267t
- Bt toxin, 265, 317, 806
- buboes, **638**, 648f
  - of bubonic plague, 648, 648f
- bubonic plague, **648**, 648f, **650b**
- budding bacteria, **171**
  - *Hyphomicrobium* as, 300t, 304, 305f
- budding of enveloped viruses, **389**, 391f
- budding yeasts, **332**, 332f
- buffers
  - chemical, in laboratory growth mediums, 159
  - pH, **36–37**
  - temperature, 35
- bulking in sewage treatment, **784**
  - *Sphaeroltilus* bacteria and, 305, 306f, 784
- bullae (lesions), **586**, 587f
- bullous impetigo, 588, 588f
- Bunyaviridae, characteristics/important genera/clinical features, 376t
- *Bunyavirus,* 376t, 660, 660b
- *Bunyavirus* /CE virus (California encephalitis), 376t, 626, 626f
- *Burkholderia cepacia,* hospital equipment, disinfectants and, 305–306, 440b, 440f, 449f
- *Burkholderia* genus/spp., 300t, **305**–306
  - as opportunistic pathogens, 300t
  - biocide resistance and, 202, 306
  - biofilms formed, 440b, 440f
  - cystic fibrosis and, 306, 308
  - formerly grouped with *Pseudomonas,* 279, 305
  - grow in disinfectants, 202, 306, 440b, 440f
- *Burkholderia pseudomallei* causes melioidosis, 279, 306, 690
- Burkholderiales (order), important genera/special features, 300t
- Burkitt, Denis, 655
- Burkitt's lymphoma, 375t, 391, **643b**, **655–656**, **657f**
- burn patients
  - nosocomial infection susceptibility and, 415
  - *Pseudomonas* bacterial infections and, 308
  - silver sulfadiazine to treat, 567
- Burnet, Frank Macfarlane, 14t
- burning, as method of microbial control, 191, 194t
- burns, genetically engineered epidermal growth factor to heal, 260t
- bursa of Fabricius, **477**
- Buruli ulcer, **594**
  - identified as global health threat, 594
  - rash caused by, 592b
- butanediol, as fermentation end-product, 134f
- butanol, 2
  - as fermentation end-product, 134f, 137t
- by-products, metabolic pathway, 123
- *Byssochlamys fulva,* produces heat-resistant ascospores, 795

- C-reactive protein, 460
- C1 to C9 complement proteins, 464–468, 464f, 466f, 468f
- cabbage
  - fermentation and, 137t
  - lactic acid fermentation and, 135b
- cachectin, 437. *See also* tumor necrosis factor
- cadherin, used by pathogens to move from cell to cell, 433
- calcium, as cofactor, 117
- calcium (Ca)
  - atomic number/atomic weight, 27t
  - microbial requirements, 160
- calcium chloride solution, to make cells competent to take up external DNA, 253
- calcium hypochlorite, 197. *See also* chloride of lime
- calcium ion
  - as a cation, 30
  - confocal microscopy to observe distributions/concentration of, 62
- calcium propionate, 199, 200, 204t
- Caliciviridae, characteristics/important genera/clinical features, 375t
- California encephalitis (CE virus/ *Bunyavirus*), 376t, 626, 626f, **628b**
- California sea lions, leptospirosis deaths, 283b
- California sea otters, toxoplasmosis deaths, 283b, 662
- calves, colostrum and, 494–495
- Calvin-Benson cycle, 140, 142f, 145, 146
- cAMP (cyclic AMP), **225–226**, 226f
- camphor, bacteria that use as energy/carbon source, 239
- *Campylobacter fetus,* causes spontaneous abortion in domestic animals, 312
- *Campylobacter* genus/spp., 301t, **312**
  - as human pathogens, 301t, 312
  - culturing, 168
- *Campylobacter jejuni,* foodborne infections caused by, 312, 577b, 577f
- Campylobacterales, important genera/special features, 301t
- cancer. *See also* carcinogens
  - activated macrophages destroy, 533t
  - AIDS-associated, 544t
  - antisense DNA explored as gene therapy, 259
  - breast, genetic screening and, 262
  - carcinogenic mutagens and, 232
  - cell transformation and proliferation, 391, 537–538
  - cervical. *See* cervical cancers
  - cytotoxic T lymphocytes (CTLs) destroy, 537, 537f
  - Epstein-Barr (EB) virus causing, 391
  - hepatitis B virus (HBV) causing, 391
  - immune system response to, 537–538, 537t
  - immunotherapy for, **538**
  - interferons' discovery and, 16
  - interferons to treat, 470
  - interleukin-12 and, 493b
  - interleukins as possible therapy, 260t
  - Kaposi's sarcoma, 21, 375t, 385, 470, 539, 542, 544t
  - liver, 391
  - monoclonal antibodies to treat, 509
  - ovarian, 260t
  - *p53 gene* and, 259
  - *Papillomavirus* and, 385–386
  - percentage known to be virus-induced, 391
  - RNA interference (RNAi) as promising therapy, 259
  - scanning acoustic microscopy (SAM) to study, 63, 63f, 67t
  - skin, exposure to UV light and, 231
  - stomach, 312, 719
  - tumor cell transformation, 391, 537
  - vaccines, 538
  - viruses and, 375t, 382, 389–391
- Cancidas (caspofungin), 564t, 569
- *Candida albicans*
  - antibiotics and overgrowth of, 555
  - as budding yeast, 332, 601f
  - as normal microbiota of vagina, 403t, 745
  - as pathogenic, 338t, 443
  - blastoconidia spores found in, 333, 334f
  - candidiasis caused by, **339**, 601–602, 601f, **758–759**, **759b**
  - in diabetics, 601–602
  - in HIV/AIDS patients, 542, 544t, 601–602
  - incubation period, 430t
  - microbial antagonism phenomenon and, 401
  - nosocomial infections and, 414, 414t
  - portals of entry, 430t
  - skin infections caused by, 443
- *Candida* genus/spp.
  - as normal microbiota of large intestine, 403t, 758
  - as normal microbiota of mouth, 403t, 758

  - as normal microbiota of skin, 402*t*
  - as normal microbiota of vagina, 403*t*, 745
  - biofilms and, 163
- candidiasis (yeast infection), **339**, **601–602**, 601*f*, **758–759**, **759*b***
  - *Candida albicans* causing, 339, 601, 601*f*, 758
  - incubation period, 430*t*
  - oral (thrush), 339, **601**, 601*f*, **759**
  - portals of entry, 430*t*
  - rash caused by, 589*b*
  - vulvovaginal, 339, **759**
- candle jars, 167–168
- Canidae, position in taxonomic hierarchy, 280*f*
- *Canis familiaris* (species), position in taxonomic hierarchy, 280*f*
- Canis genus/spp., position in taxonomic hierarchy, 280*f*
- canker sores, **597**
- canned foods
  - commercial sterilization and, **185**, 185*t*, 188, 190, 439, 440*b*, 594*f*, **794–795**, 794*f*, 795*f*
  - heat-preserved, 188
  - home "canning," 188, 190
  - metal can construction, 796*f*
  - types of spoilage in, 795, 796*t*
- Cano, Raul, 278, 290–291
- CAP (catabolic activator protein), 225–226, 226*f*
- capnophiles, **167**
- capsids, viral, **372**, 372*f*, 373*f*, 374*f*, 381*f*
  - classification of viruses and, 373
- capsomeres, viral, **372**, 372*f*, 373*f*
- capsules of bacteria, **79–81**, 80*f*, 101*t*
  - antibodies against produced by human immune system, 432
  - as examples of T-independent antigens, 484, 484*f*
  - chemical nature impairs phagocytosis (host defense), 432
  - complement activation prevented by, 468
  - microbial evasion of phagocytosis and, 459
  - of *Neisseria gonorrhoeae*, 306*f*, 459
  - of *Streptococcus pneumoniae*, 234–236, 235*f*, 432, 459
  - staining, 71, 72*f*, 72*t*, 79
  - virulence of pathogens increased by, 80–81, 234, 432, 459
- carbapenems, **561**
  - mode of action/spectrum of activity, 562*t*
  - penicillin allergy and, 524
- carbenicillin, 561
- carbohydrate catabolism, **124–135**, 125*f*
  - fermentation test and, 138–139, 139*f*
  - gas formation and, 138, 139*f*
- carbohydrates, **39–40**
  - amphibolic pathways and, 147, 149*f*
  - functions of, 39
  - microbes, photosynthesis and, 17
  - simple, 39
- carbolfuchsin stain, 69, 71, 72*t*
  - mechanism of action, 87
- carbolic acid. *See* phenol
- carbon (C)
  - atomic number/atomic weight, 27*t*
  - bacterial recyclers of, 17
  - electronic configuration, 29*t*
  - in methane formation, 31, 31*f*
  - in organic compounds, 37
  - organisms classified by their source of, 142–145, 143*f*
  - requirements for microbial growth, 160
  - simplified diagram, 27*f*
  - uniqueness of, 34
- carbon cycle, **768–770**, 769*f*
- carbon dioxide, 34
  - as by-product of Krebs cycle, 128, 128*f*, 139
  - as fermentation end-product, 134*f*, 135*b*, 137*t*, 139
  - crosses plasma membrane by simple diffusion process, 91
  - in alcohol fermentation, 135*b*, 136*f*
  - in Calvin-Benson cycle, 140, 142*f*
  - in photosynthesis, 140, 141*f*
  - incubators for growing bacteria, 167
  - made by yeasts, 135*b*
  - photosynthetic bacteria and, 96
  - supercritical, 201–202, 205*t*
- carbon fixation, 117*t*, 140, **140**, 142*f*, 145, 146
- carbon monoxide, as energy source, 145
- carbon skeleton, **37–38**
- carbonate, anaerobic respiration and, 132
- carboxyl group, 37, 38*t*
  - in amino acids, 43, 43*f*, 44*t*
  - in fatty acids, 40, 41*f*
- carboxysomes, **96**
- carbuncle, **588**
- carcinogens, **232**
  - Ames test and, 232–233, 233*f*
  - chemical, identifying, 232
  - frameshift mutagens as, 230
  - nitrosamines, **200**
- cardiac muscle, regeneration capacity of, 462
- cardiotoxins, 435
- cardiovascular syphilis, 754
- cardiovascular system, **637**, 638*f*
  - lymphatic system in relationship with, 637–638, 639*f*
  - microbial diseases of, 637–673
    - bacterial, 638–643, **643*b***, **650*b***, **651*b***, **668*b***
    - helminthic, 666–667, **668*b***
    - protozoan, **650*b***, **651*b***, 660–666
    - vector-borne, 648, **651*b***, 652–655
    - viral, **643*b***, 655–660
  - structure/function, 637–638, 638*f*
- carcinogens, *Helicobacter* as, 301*t*
- caribou, lichens and, 340
- Carnivora (order), position in taxonomic hierarchy, 280*f*
- carotene, 343*t*
- carotenoids, 146
- carrageenan, 343
- carriers of infectious disease, **409**
- *Carsonella ruddii*, extremely small genome of, 326
- case control method, in analytical epidemiology, 419–420
- case reporting
  - CDC's *MMWR* and, 420
  - uses in establishing chain of transmission, 420
- casein, **798**
- caspofungin (Cancidas), 564*t*, 569
- catabolic activator protein (CAP), 225–226, 226*f*
- catabolic chemical reactions. *See* catabolism
- catabolism, **32**, 113, **114**, 114*f*
  - amphibolic pathways and, 147, 149*f*, 150
  - carbohydrate, **124–135**, 125*f*
  - lipid, **136–137**, 136*f*, 138*f*
  - of various organic molecules, 138*f*
  - protein, 136–137, 136*f*, 138*f*
- catabolite repression (glucose effect), 226
- catalase, 105, 161*t*, **162**
  - hydrogen peroxide and, 202
- catalysts, **115**
- cataract surgery, 440*b*, 440*f*
- cathelicidins, produced by neutrophils/macrophages/epithelium, 470
- catheterization
  - intravenous, 415*t*
  - urinary, 415*t*
- catheters
  - biofilms and, 19*f*, 163, 164*b*, 431, 586, 587*f*
  - infections (Clinical Focus), 164*b*, 442*b*
  - intravenous, nosocomial bacteremia and, 415*t*, 422*b*
  - nosocomial infections and, 415*t*
  - *Staphylococcus epidermidis* and, 586, 587*f*
- cationic detergents, as antimicrobial agents, 199, 204*t*
- cationic peptides, 578. *See also* antimicrobial peptides
- cations, **30–31**, 31*f*, 35
- cats
  - as disease reservoirs, 410*t*, 650*b*
  - bites and *Pasteurella* bacteria, 311
  - cat-scratch disease, 305, 410*t*, 417*f*, 417*t*, **647**, 647*f*, **650*b***
  - feline AIDS and, 377
  - feline leukemia virus (FeLV), 391
  - heartworm in, 360
  - litter box contents flushed, sea otter deaths and, 662
  - plague transmitted by, 650
  - reported cases of rabies in, 624*f*
  - sarcoma viruses in, 391
  - testing positive for tularemia pathogen, 644*b*
  - toxoplasmosis and, 350, 661–662, 662*f*
  - vaccinated against leptospirosis, 324
- cattle
  - bovine tuberculosis, **685**
  - reported cases of rabies in, 624*f*
  - *Salmonella* bacteria common intestinal inhabitants of, 310
  - sepsis caused by *Pasteurella* bacteria, 311
  - ticks, 690
- cauliflower mosaic virus, 394*t*
- *Caulobacter* genus/spp., 300*t*, **304**, 304*f*, 776, 777
  - stalked cell arrangement of, 300*t*, 304*f*
- Caulobacterales, important genera/special features, 300*t*
- CCR5 (chemokine coreceptors), 541, 548
- CCR5 coreceptors, 571
  - HIV resistance and, 545
- CD (clusters of differentiation) of T cells, **487**
- CD4+ T cells, 5*f*, 441, **487–488**, 488*f*
  - in gonorrhea, 749
  - in HIV infection, 540–542, 540*f*, 541*f*
  - normal count *vs.* in AIDS patients, 542
- CD8+ T cells, **488–489**, 489*f*, 496*f*
- CDC (Centers for Disease Control and Prevention), **420**
  - nosocomial infection estimates by, 413, 414*t*, 415*t*
- cDNA (complementary DNA), **254–255**, 255*f*
- cDNA library, 255
- CE virus/ *Bunyavirus* (California encephalitis), 376*t*, 626, 626*f*
- cefaclor, 563
- cefamandole, 563
- cefixime, mode of action/spectrum of activity, 562*t*
- ceftriaxone, 422*b*
- cell arrangements
  - microbial, 76, 77–79, 78*f*, 79*f*, 101*t*, 330*t*
  - of algae, 343*t*
- cell counters, 176, 178, 178*f*
- cell cultures, **378–379**, 378*f*
  - for vaccine development, 504, 506
- cell division
  - bacterial growth curves, **172–174**, 172*f*, 173*f*
  - DNA complementary structure and, 211
  - in eukaryotes *vs.* prokaryotes, 77
  - in prokaryotic *vs.* eukaryotic cells, 101*t*
- cell growth, biosynthetic reactions that generate materials for, 114, 114*f*
- cell lines, **378**
  - continuous, **378**
  - diploid, **378**
  - immortal, 378
  - maintenance of, 379
  - primary, **378**
- cell metabolism, carbohydrate catabolism, **124–135**, 125*f*
- cell structure
  - confocal microscopy to reconstruct three-dimensional images of, 62
  - proteins as integral part of, 42
- cell theory, **7**
- cell wall
  - in prokaryotic *vs.* eukaryotic cells, 101*t*
  - inserting foreign DNA through, 253–254, 253*f*
  - of algae, 343*t*
  - of archaea, 276*t*
  - of bacteria, 69, 276*t*, 330*t*
  - of eukaryotes, 77, 99*f*, 101*t*, 330*t*
  - of fungi, 330*t*
    - antifungal drugs affecting, 564*t*, 569
  - of prokaryotes, 77, 80*f*, 82, **84–89**, 99*f*, 101*t*
  - of T-even bacteriophage, 381*f*
  - synthesis, antibiotics that inhibit, 556, 556*f*, 557*f*, 559–561, 562*t*, 563
- cell-cell fusion of HIV to evade immune system, 541
- cell to cell chemical communication. *See* quorum sensing
- cell-to-cell interactions
  - glycocalyx's role in, 98
  - proteins involved in, 90
- cellular immunity, 477–**478**, 486–489
  - activated macrophages, **490**, 490*f*, 491*t*
  - antibody-dependent cell-mediated cytotoxicity, **491**, 492*f*
  - antigen-presenting cells, **489–490**
  - congenitally absent thymus gland and, 538
  - cytokines and, **491–492**
  - dendritic cells, **490**, 490*f*
  - interleukin-12 activates, 493*b*
  - intracellular antigens, 486, 496*f*
  - natural killer (NK) cells, 491
  - principal cells that function in, 491*t*
  - T cells, 486–489
    - cytotoxic cells, 488–489, 489*f*
    - helper cells, 487–488, 488*f*
    - regulatory cells, 489
- cellular metabolism, rate of, 147
- cellular oxidations, 122, 122*f*, 123*f*
- cellular prion protein, 393, 393*f*
- cellular respiration (respiration), 124, **127–132**
  - aerobic, 127–132
    - electron transport chain, 124, **129–130**, 129*f*
    - Krebs cycle, 124, **127–129**, 128*f*
  - anaerobic, **127**
  - glycolysis in, **124**, 125, 126*f*
  - location of, 105
  - overview figure, **125 *f***
- cellular slime molds, **351–352**, 352*f*
  - life cycle, 351, 352*f*
  - position in evolutionary tree, 275*f*
- cellulases, 40
  - genetically engineered, 267*t*
  - used in producing "stone-washed" denim, 3*b*, 40
- cellulitis, MRSA causing, **593*b***
- cellulose, 2, 3*b*, 40, 101
  - in green algae cell walls, 342*f*, 343
  - of algae cell walls, 6, 98
  - of plant cell walls, gene guns and, 253, 254*f*
  - peptidoglycan *vs.*, 4
  - termites and, 107*b*
- Centers for Disease Control and Prevention (CDC), **420**
  - nosocomial infection estimates by, 413, 414*t*, 415*t*
  - priorities for emerging infectious diseases, 418
  - Universal Precautions for Health Care Personnel, 546*t*

centimeter (cm), metric/U.S. equivalent, 55*t*
central nervous system (CNS), **611**, 611*f*
centrifugation, in serum collection, 467*b*
centriole, 99*f*, 105
centrosome, 99*f*, **105**
Cepacol, 199, 204*t*
cephalosporins, **561**, 561*f*, 563, 617*b*
  cell wall synthesis inhibited by, 556, 556*f*, 557*f*
  gram-positive bacteria and, 70
  mode of action/spectrum of activity, 562*t*
  penicillin allergy history and, 531*b*
  peptidoglycan and, 98
  structure of, compared to penicillin, 561*f*
Cephalosporium, 554, 555*t*
cephalothin, 531*b*, 562*t*, 563
  produced by *Cephalosporium,* 554, 555*t*
Ceratocystis ulmi, Dutch elm disease caused by, 339
cercariae, swimmer's itch in reaction to, 667
cerebrospinal fluid (CSF), 610, 611, 612*f*
  has low levels of defensive cells, 611
  spinal tap (lumbar puncture), 614, 614*f*
cervical cancers
  caused by human papillomavirus (HPV), 391, 394*t*
  HPV vaccine (Gardasil) to prevent, 260*t*, 391, 503*t*, 538, 758
cervical dysplasia, in AIDS patients, 544*t*
cervical mucus, antimicrobial activity of, 453
cervix, 744, 744*f*
Cestodes (tapeworms), **356**–358, 358*f*, **361***t*
  food acquisition methods, 356
cetacean Morbillivirus (CM), marine mammal deaths and, 283*b*
cetylpyridinium chloride (Cepacol), 199, 204*t*
CF. *See* cystic fibrosis
CF (confocal microscopy), **62**, 62*f*, 67*t*
  advantages of, 62
  Paramecium multimicronucleatum image, 62*f*, 67*t*
  specimen preparation and, 62
CFS (chronic fatigue syndrome), **633**
CFU (colony-forming units), **174**, 175*f*
CGD (chronic granulomatous disease), recombinant gamma interferon to treat, 469
Chagas, Carlos, 10*f*, 285, 661
Chagas' disease (American trypanosomiasis), 351, 354*t*, 413*t*, 459, **651***b*, **661**–663, 662*f*
  as emerging infectious disease, 417*t*
Chain, Ernst, 10*f*, 14*t*
Chain, Ernst, 554
chain of transmission, case reporting procedure and, 420
chancre, **753**, 754*f*
chancroid (soft chancre), **756**, **761***b*
  as notifiable infectious disease, 421*t*
  Haemophilus ducreyi cause of, 311, 756
charge (property of subatomic particles), **27**
Chatton, Edouard, 274
cheese
  chemical food preservatives added to, 199–200
  fermentation and, 137*t*
  microbes used in making of, 798–799, 799*f*
  nisin added to inhibit bacteria, 200
  pH and spoilage, 159
chemical agents
  carcinogens, 232–233, 233*f*
  mutagenic, 228
chemical agents of microbial control, 195–202. *See also* antimicrobial agents
  summary, by agent/mechanism of action/preferred use, 204–205*t*
chemical bonds, **28**–32
  covalent, **30**–31, 31*f*
  ionic, 28, **30**, 30*f*
chemical elements, **27**–28, 27*t*
  most abundant in living matter, 27, 27*t*
  number commonly found in living things, 27, 27*t*
chemical energy, **32**
  ATP and, 47–49, 49*f*
chemical food preservatives, 199–200, 204*t*
chemical messengers, 478
chemical methods of microbial control, 195–202. *See also* antimicrobial agents
chemical mutagens, 228–230, 229*f*
  causing frameshift mutations, 230
chemical pesticides, safety issues, 268
chemical principles, importance to microbiologists, 26
chemical products, 2
chemical reactions, **32**–34, 33*b*
  anabolic, 32
  catabolic, 32, 113, **114**, 114*f*
  collision theory and, **115**
  condensation, 38
  coupled, 114, 114*f*
  decomposition, 32, 33*b*
  dehydration, 38
  endergonic, 32
  energy absorption/release in, 32
  energy of, 32
  enzymes and, 115, 115*f*
  exchange, 33–34, 39
  exergonic, 32
  reaction rate, **115**
  reversible, 34, 39*f*
  synthesis, 32
chemical signals
  alarm signals (alarmone), 225–226
  biofilms and, 57*b*, 163
chemical spills, bioremediation and, 17
chemical sterilization, 200–201, 205*t*
  by ethylene oxide, 200–201, 205*t*
  by plasmas, 201, 205*t*
  by supercritical fluids, 201–202, 205*t*
chemically defined culture media, **165**, 165*t*
chemiosmosis, 123, 125*f*, **130**–131, 130*f*
chemistry, 26–53, **27**
  atoms, 27–28, 27*f*, 27*t*
  chemical bonds, 28–32
  chemical reactions, 32–34, 33*b*
  elements, 27–28, 27*t*
  importance to microbiologists, 26
  molecules, **27**, 28–32
chemoautotrophs, 143, 143*f*, **145**, 159, 305
  carbon requirements for growth, 160
  culture media for, 169*t*
  Nitrobacter, 305
  Nitrosomonas, 305
  Thiobacillus, 305
chemoheterotrophs, 143*f*, **145**
  carbon requirements for growth, 160
  chemically defined medium for growing, 165*t*
  culture media for, 169*t*
  fungi as, 330, 330*t*, 333
  green sulfur bacteria as, 314*t*
  proteobacteria as, 302–312, 313*f*, 314*t*
  protozoa as, 346
chemokine coreceptors, CCR5 and CXCR4, 541
chemokines, 462, **492**
chemosterilants (gaseous), 200–201
chemotaxis, **82**
  as first phase in phagocytosis, **458**, 458*f*
  kinins and, 462
  neutrophils attracted to, 462
chemotherapeutic drugs. *See also* antibiotics; antimicrobial drugs
  future of, 578–579
  major modes of action (overview), 556*f*
  salvarsan (antisyphilitic), 12
  spectrum of activity of, 555, 557*t*
  synthetic drugs, **12**
  toxicity to humans and, 12
chemotherapy, **12**, 260*t*, **553**
  history of, 554
  selective toxicity and, 553
  tests for microbial susceptibility/sensitivity, 572–573, 572*f*, 573*f*
chemotrophs, **142**, 143*f*
Chernobyl nuclear disaster, lichens and, 340
chestnut trees, fungal blight by *Cryphonectria parasitica,* 339
chick embryos, viruses for vaccines grown in, 377*f*, 504
chickenpox (varicella), 375*t*, 385, **596**–597, 597*f*
  as notifiable infectious disease, 421*t*
  breakthrough varicella, **597**
  incubation period, 430*t*, 596
  portals of entry, 430*f*, 596
  rash caused by, 590*b*
  Reye syndrome complication of, **596**
  vaccine, 13, 503*t*
chickens
  antibiotics in chicken feed, 577*b*
  avian sarcoma virus derived from normal chicken genes, 391
  cholera in (fowl cholera), 311
  influenza A viruses and, 19
  leukemia in, 389
  viral-caused sarcoma in, **389**
chikungunya fever, **651***b*, **658**
childbirth, genetically engineered relaxin to assist, 260*t*
childbirth fever (puerperal sepsis), 11, 197, 418, **640**–641, 643*b*
childhood immunizations, 502, 504*t*
chills and fever, 463
Chilomastix, 347*f*
chimeric monoclonal antibodies, **509**
  as immunosuppressives, 537
chitin, 4, 40, 101*t*
  in cell walls of algae, 98
  in cell walls of fungi, 330*t*
Chlamydia genus/spp./Chlamydiales, 302*t*, **322**, 323*f*
  antibiotics effective against, 565, 566, 751
  as gram-negative coccoid bacteria, 322
  as intracellular parasites, human pathogens, 302*t*
  can survive in phagocytes, 459
  culture media and, 167, 322
  elementary bodies, **322**, 323*f*
    size of, 369*f*
  life cycle, 322, 323*f*
  no longer grouped with rickettsial bacteria, 299
  pathogenic species of, 322
  phylogenetic relationships, 281*f*
  pneumonia caused by, 322, **687***b*, **689**
  portals of entry, 429, 430*t*
  transmission routes, 322
  viruses compared to, 368*t*
Chlamydia trachomatis
  as notifiable infectious disease, 421*t*
  gonorrhea coinfection and, 750
  inclusion conjunctivitis caused by, 604, **604***b*
  incubation period, 430*t*
  lymphogranuloma venereum caused by, 322, 459, 755
  portals of entry, 429, 430*t*
  toxin produced by, 262
  trachoma caused by, 322, **604**–605, 605*f*
  urethritis (nonspecific) caused by, 430*t*, **750**–751, **761***b*
Chlamydiae, **322**
  important genera/special features, 302*t*
chlamydoconidia, **333**, 334*f*, 338*t*
Chlamydomonas (green alga), 342*f*
Chlamydophila genus/spp., 302*t*, **322**, 323*f*
  classification and, 279
  intracellular parasites, human pathogens, 302*t*
Chlamydophila psittaci, 322, 323*f*
  as potential biological weapon, 649*b*
  psittacosis (ornithosis) caused by, **687***b*, **689**
  reservoirs/transmission method, 410*t*
Chlamydphila pneumoniae, 322
Chlor-Floc tablets, to clarify water, 197
chloramines, 197
  to sanitize glassware/eating utensils/dairy equipment, 197, 204*t*
chloramphenicol, **563**, 565, 565*f*
  adverse effects, 565
  bacterial resistance to, 239, 240*f*
  blood-brain barrier and, 611
  mode of action/spectrum of activity, 562*t*
  produced by *Streptomyces venzuelae,* 555*t*
  protein synthesis inhibited by, 95, 556–557, 556*f*, 558*f*, 563, 565–566
  susceptibility of gram-negative *vs.* gram-positive bacteria to, 88*t*
chlorhexidine, 197, 203*t*, 204*t*
chloride ion
  as an anion, 30
  in table salt, dissolved in water, 35, 35*f*, 36*f*
chloride of lime, 184, 197. *See also* calcium hypochlorite
chlorination
  chlorine dioxide gas and, 201
  of drinking water, 197
  ozone as supplement to, 202, 205*t*
chlorine, as antimicrobial agent, 196*f*, 197, 201, 203*t*, 204*t*
chlorine (Cl)
  as bleaching agent, peroxide *vs.*, 3*b*
  atomic number/atomic weight, 27*t*
chlorine gas, used to disinfect water, 197, 204*t*
chlorines, as antimicrobial agent, 196*f*, 197, 201, 203*t*, 204*t*
Chlorobi, 301*t*, 314*t*
  important genera/special features, 301*t*
Chlorobium genus/spp., 301*t*, 314*t*
  as anoxygenic photoautotrophs, 144, 301*t*
  characteristics, compared, 314*t*
chlorobium vesicles (chlorosomes), 144
Chloroflexi, 301*t*, 314*t*
  important genera/special features, 301*t*
Chloroflexus genus/spp., 145, 301*t*, 314*t*
  as photosynthetic, anoxygenic bacteria, 301*t*
  characteristics, compared, 314*t*
chlorophyll a, 140, 141*f*, 144, 145*t*
  in algae, 343, 343*t*
chlorophyll b, in green algae, 343, 343*t*
chlorophyll c, in brown algae, 343*t*
chlorophyll d, in red algae, 343*t*
chlorophylls, 105, 140, 141*f*, 144, 145*t*, 146
Chlorophyta, characteristics of green algae, 343*t*
chloroplasts, 99*f*, 102, **105**, 106*f*, 140, 145*t*
  of *Euglena,* 350, 351*f*
  origins of, 275*f*, 277, 277*f*
chloroquine
  mode of action/uses, 564*t*
  to treat malaria, 571
chlorosomes (chlorobrium vesicles), 144, 145*t*
chlortetracycline (Aureomycin), 565
  mode of action/spectrum of activity, 562*t*
  produced by produced by *Streptomyces aureofaciens,* 555*t*
chocolate seeds, fermented before eating, 800
cholera, 19, 309, **716**–717, 716*f*, **722** *b*. *See also Vibrio cholerae*
  as emerging infectious disease, 417*t*
  as notifiable infectious disease, 421*t*
  convalescing patient and disease spread, 409
  epidemic of 1848 (London) and discovery of source, 418

in chickens (fowl cholera), 311
incubation period, 430*t*
modern transportation and spread of, 416
new strains of, 717. *See Vibrio cholerae*
noncholera vibrios, 717
portals of entry, 429, 430*f*
symptoms, 438*t*
toxins (A-B enterotoxins) causing, 437, 438*t*
glycoproteins, plasma membranes and, 90
vaccine, 502
waterborne transmission and, 411
cholesterol
structure of, 41, 42*f*
synthesis of, 146, 147*f*
Chordata, position in taxonomic hierarchy, 280*f*
Chromatiales, important genera of, 300*t*
chromatin, 102*f*, **103**
Chromatium genus/spp., 300*t*
as anoxygenic photoautotrophs, 144, 145*t*
characteristics, compared, 314*t*
chromatophores, 144, 145*t*
chromatophores (thylakoids), **91**, 91*f*
Chromista, position in evolutionary tree, 275*f*
chromophore, 68
chromosomes, **94**, **211**
bacterial, **94**–95, 101*t*, 103
DNA and, 211–212, 212*f*
eukaryotic, 101*t*, **103**
maps of, 212, 212*f*
of *Escherichia coli*, 211–212, 212*f*
prokaryotic, 211, 212*f*
chronic fatigue syndrome (CFS), **633**
chronic granulomatous disease (CGD), recombinant gamma interferon to treat, 469
chronic inflammation/inflammatory response, 460, 463
chronic (persistent) viral infections, **392**, 392*f*
examples (disease/primary effect/causative virus), 394*t*
chronic wasting disease, prion disease affecting wild deer/elk, 630
Chrysops (deer fly), as vector transmitting tularemia, 362*t*, 363*f*
Cidex, 200, 205*t*
cidofovir
mode of action/uses, 564*t*
to treat cytomegalovirus eye infections, 569–570
Ciechanover, Aaron, 15*t*
ciguatera, **344**, 354*t*
cilastatin, 561
cilia/cilium
of eukaryotic cells, **98**, 100*f*
origins of, 106
of human respiratory tract, 98, **452**, 452*f*
as defense against pathogens, 452, 471*t*
of mucous membranes, 585
of *Paramecium,* 350*f*
of protozoa, 6, 100*f*
ciliary escalator, **452**, 452*f*, 675, 676
ciliated cells, 452*f*
ciliates (Ciliophora), **350**, 350*f*
conjugation in, 346, 346*f*
food acquisition methods, 346
position in evolutionary tree, 275*f*
Ciliophora. *See* ciliates
Cipro (ciprofloxacin), 563*t*, 567
ciprofloxacin (Cipro), 567
mode of action/spectrum of activity, 563*t*
cis fatty acid, 40, 41*f*
cisterns, **103**, 103*f*
citric acid, 127–128, 128*f*, 149*f*
as fermentation end-product, 137*t*
Aspergillus niger fungus used to produce, 339
biotechnology and, 246
citric acid cycle. *See* Krebs cycle
Citrobacter genus/spp.
as enteric bacteria, 285
as normal microbiota of large intestine, 403*t*
as opportunistic pathogens, 300*t*
in taxonomic hierarchy, 300*t*
nosocomial infections and, 414, 414*t*
CJD (Creutzfeldt-Jakob disease), **20**, 393, **630**–631, 631*f*, 631*t*
clade, 223*b*, **281**
of HIV genome, 542
cladograms, 275*f*, 281*f*, **293**, 294*f*
steps for constructing, 294*f*
clams
paralytic shellfish poisoning (PSP) and, 344, 354*t*, 444
unicellular algae symbionts in giant *Tridacna,* 345
clarithromycin, 566
mode of action/spectrum of activity, 562*t*
class, in taxonomic hierarchy, **279**, 280*f*
classical pathway of complement activation, 464*f*, **466**, 466*f*
classification of organisms, 273–298
binomial nomenclature used in, 278–279
by nutritional patterns, 142–145, 143*f*
methods, 282–294
natural, reflecting phylogenetic relationships, 274, 277
of eukaryotes, 275*f*, 280*f*, 281–282
of prokaryotes, 275*f*, 279–281, 280*f*
of viruses, 373–374, 375–376*t*
study of phylogenetic relationships, 274–278
hierarchies, 277–278, 278*f*
taxonomic hierarchy and, 279, 280*f*
three-domain system, 274–278, 275*f*
Claviceps purpurea (fungus), 443
clavulanic acid (potassium clavulanate), 561
climate, incidence of infectious diseases and, 408
clindamycin, 565
Bacteroides fragilis resistant to, 240*f*
microbial resistance to, 240*f*
mode of action, 565
clofazimine, to treat leprosy, 620, 632*b*
clonal deletion, **484**
clonal expansion of B cells, 482, 482*f*
clonal selection of B cells, **483**–484, 483*f*
clone, **279**, 281
clones/cloning, **247**
applications, 258–267
agricultural, 264–267
scientific, 261–264
therapeutic, 258–259
of plant cells, 264–265, 267
selecting, 256–257, 256*f*, 257*f*
types of cells used for rDNA products, 257–258, 257*f*
uses for, 258
vectors and, 250–251, 250*f*, 251*f*
cloning vectors, 15*t*, 247, 248*f*, 250–251, 250*f*, 251*f*
Clonorchis sinensis (Asian liver fluke), 356, 356*f*
Clorox, 196*f*, 197
clostridia. *See Clostridium*
Clostridium acetobutylicum, fermentation and, 137*t*
Clostridium botulinum, **316**. *See also* botulism
as obligate anaerobe, 161, 316, 616
as potential biological weapon, 649*b*
botulism caused by, **616**. *See also* botulism
commercial sterilization to destroy, 185, 193, **794**–795, 794*f*, 795*f*
gastric juice unable to destroy, 453
grows at refrigerator temperatures, 618
in soil, 409, 616
lysogenic phages and, 382
neurotoxin produced by, **437**. *See also* botulinum toxin
nitrites active against, 200, 204*t*
Clostridium difficile
antibiotic therapy and, 316
diarrhea-associated, 316, **720**, 720*b*
clindamycin and, 565
normal microbiota, antibiotic therapy and, 401
nosocomial infections and, 414, 414*t*
toxin of similar to *Chlamydia trachomatis* toxin, 262
Clostridium genus/spp./Clostridiales, 316–317, **316**, 316*f*, 777
as anaerobic human pathogen, 161, 301*t*
as gram-variable bacteria, 87
as normal microbiota of vagina, 403*t*
canned food spoilage by, 795, 796*t*
endospores of, **96**–98, 97*f*, 185, 301*t*, 795
fermentation and, 134*f*
important genera/special features, 301*t*
in taxonomic hierarchy, 301*t*
low G + C content of, 316
Clostridium perfringens
gas gangrene caused by, 316, 430*t*
gastroenteritis, **720**, 723*t*
incubation period, 430*t*
penfringoglycin toxin produced by, 65*f*, 68*t*
portals of entry, 430*t*
Clostridium tetani, 316*f*
in soil, 409
incubation period, 430*t*
neurotoxin of, 239, 437, 438*t*, 615
portals of entry, 429, 430*t*
tetanus caused by, 316, 406, **615**–616, 616*f*, **632***b*
vaccine against, 13, 435, 502*t*
clotrimazole, 564*t*, 568, 600
club fungi. *See* Basidiomycota
clue cells, of bacterial vaginosis, 756, 756*f*
clumping of cells/viruses, IgM antibodies and, 480, 484
clusters of differentiation (CD) in T cells, **487**
cm (centimeter), metric/U.S. equivalent, 55*t*
CM (cetacean morbillivirus) virus, marine mammal deaths and, 283*b*
CMV. *See* cytomegalovirus
CoA (coenzyme A), **117**
coagulase, **432**, **586**
coagulase-negative staphylococci, 414, 414*t*, 586–589
coagulase-positive, gram-positive cocci, 422*b*, 422*f*, 587
pathogenicity and, 587
coal mines, 159
coal tar, phenolics derived from, 195, 196*f*
coarse focusing knob, of compound light microscope, **56**, 56*f*
cobalt, as cofactor, 117
cocarboxylase, 117*t*
Coccidioides immitis (fungus)
arthroconidia formed by, 333, 334*f*, 696, 696*f*
characteristics as pathogenic fungi, 338*t*
coccidioidomycosis caused by, 13, 416, 696–697
emerging infectious diseases and, 417*t*
infections caused by, increasing rates of, 13, 416
coccidioidomycosis, **696**–697, 696*f*, 697*f*, **699***b*
airborne transmission and, 412
amphotericin B effective against, 568
as a systemic mycosis, 336
as emerging infectious disease, 417*t*
as notifiable infectious disease, 421*t*
epidemic area for, 696, 697*f*
incidence increase following natural disaster, 416
sometimes known as Valley fever/San Joaquin fever, 696
coccobacilli, **78**, 78*f*, 303, 305, 308
coccus/cocci (spherical-shaped bacteria), **77**–78, 78*f*
cell arrangements of, 77–78, 78*f*
cocoa
fermentation used in production of, 800
infected by *Phytophythora infestans,* 344
codeine, as genetically engineered product, 258
codons, 212, **219**–220, 219*f*, 220–221*f*
nonsense, **219**
sense, **219**
start, 212, 219–220, 220–221*f*
stop, 212, 219–220, 220–221*f*
coenocytic hyphae, **331**, 331*f*, 333
coenzyme A (CoA), **117**, 117*t*
coenzyme Q (ubiquinones), **129**, 129*f*
coenzymes, **116**, 116*f*
derived from vitamins (selected), 117*t*
vitamins functioning as, 162
cofactors for enzymes, **116**, 116*f*, 160
coffee, fermentation used in production of, 800
cohort groups/cohort method in analytical epidemiology, 420
cold sores (fever blisters), 385, **597**, 598*f*, 757
herpes simplex virus type 1 (HSV-1) causing, 385, 597, 598*f*, 757
latent state in nerve cells, 392, 394*t*, 598*f*
cold temperatures, to control microbial growth, 157–158, 158*f*, 170, 191–192, 194*t*
cold virus. *See* colds
cold-loving microbes (psychrophiles), **157**, 157*f*
colds, common
Coronavirus and, 376*t*
Rhinovirus and, 375*t*
Coley, William B., 538
Coley's toxins, 538
coliform bacteria
as indicator organisms, **780**, 780*f*
counting methods, 175, 177*f*, 780
colitis
fatal, 401
hemorrhagic, 718
collagen vascular disorders, caused by inherited complement deficiencies, 468
collagenase, **433**
collision theory, **115**
colonies (microbial), 156, **170**, 170*f*
colony-forming units (CFU), **174**, 175*f*
Escherichia coli, fimbriae, colonization and, 83, 84*f*
mutant, replica plating to identify, 231–232, 232*f*
colony hybridization, **256**–257, 257*f*
colony-forming units (CFU), **174**, 175*f*
colony-stimulating factor (CSF), 492
genetically engineered, therapeutic uses, 258, 260*t*
Colorado tick fever, 376*t*
colorimeter, to measure turbidity, 178, 179*f*
colostrum, 494–495
gastrointestinal infections and, 481
IgA's presence in, 481
comedonal acne, **594**
commensalism, **401**, 402*f*
commercial applications, of microbes, 2
commercial sterilization, **185**, 185*t*, 188, 190, 439, 440*b*, 594*f*, **794**–795
12D treatment (*botulinal cook*), **795**
canning retorts, **795**, 795*f*
in industrial canning, **794**–795, 794*f*, 795*f*
common arm of tRNA, of Archaea/Bacteria/Eukarya compared, 276*t*
common cold, **679**–680, **681***b*
adenoviruses causing, 385
antibody protection against, 480–481

  - *Coronavirus* causing, 376t, 679
  - portal of entry for, 429
  - *Rhinovirus* causing, 375t, 679
  - transmission of, 679–680
  - treatments for, 680
- common variable hypogammaglobulinemia, 539t
- communicable diseases, **406**
  - control methods, 501
- competence (genetic), **236**, 253
  - transformation and, 234–236, 235f, 253
    - making *Escherichia coli* competent, 253
- competitive exclusion, normal microbiota and, **401**
- competitive inhibitors
  - of enzymes, **120**, 120f
  - of essential metabolite synthesis, 556f, 558, 562t, 567, 568f
- complement. *See also* complement system
  - cytolysis caused by, 465f, 479
  - deficiency of, 467b
  - early discoveries about, 477
  - Fc regions of antibodies and, 479
  - testing serum for presence of, 467b
- complement fixation, 465, 467b, **512**–513, 514f
  - by immunoglobulin class, 481t
- complement system, **463**–468, 472t
  - activation of, 465–468, **466**
    - alternative pathway, 464f, **466**–467, 466f
    - by antibodies, 480, 484, **485**, 485f
    - classical pathway, 464f, **466**, 466f
    - in transfusion reactions, 526–527, 529
    - lectin pathway, 464f, **467**–468, 468f
  - cascading action of, 465
  - diseases may play role in, 468
  - evasion by microbes, 468
  - functions of, 464
  - in second line of host's defenses, 450f, 472t
  - inherited deficiencies and resulting disorders, 468
  - outcomes of activation (overview), 464f
    - cytolysis, 464, 464f, 465, 465f
    - inflammation, 464, 464f, 465, 465f
    - phagocytosis, 464, 464f
    - phagocytosis (via) opsonization, 464f
  - protein C3 activation, 464f, 465
  - protein designations, 464–465
  - regulation of, 468
  - testing for in serum (lab test), 467b
- complement-fixation reactions, **512**–513, 514f
- complementary base pairs, 47, 48f, 211
  - DNA replication and, 212–215, 214f, 216f
  - sticky ends of DNA strands and, 249, 250f
- complementary DNA (cDNA), **254**–255, 255f
- complex culture media, **165**, 166t
- complex lipids (phospholipids), **40**–41, 42f
- complex viruses, **373**, 374f
- composting, **776**
  - thermophiles important to, 158, 776, 777f
- compound light microscope, 7, 7f, 55, **56**, 56f, 58–59, 59f
  - path of light in, 56, 56f
  - principle parts and functions, 56, 56f
  - specimen sizes and, 58, 58f
- compound microscope, 7, 7f. *See also* compound light microscope
  - first, 55
- compounds, **28**
  - inorganic, **34**–37
  - organic, **34**, **37**–**49**
- compromised host, 414f, **415**
- concentration gradient, 91–94, 92f, 93f
- condensation reaction, **38**, 39f
- condenser lenses of compound light microscope, **56**, 56f, 59f
- conidiospore (conidium/conidia), **333**, 334f
- *condylomata acuminata*. *See* genital warts
- confocal microscopy (CF), **62**, 62f, 67t
  - advantages of, 62
  - biofilms and, 163
  - *Paramecium multimicronucleatum* image, 62f, 67t
  - specimen preparation and, 62
- congenital immunodeficiencies, **538**, 539t
- congenital rubella syndrome, **599**
- congenital syphilis, **755**
  - as notifiable infectious disease, 421t
- congenital toxoplasmosis, 350, 354t
- congestive heart failure, from heartworm disease, 360
- conidia/conidium (conidiospore), 171, **333**, 334, 334f, 336f, 338t
- conidiophores, **333**, 334f, 336f
  - of *Aspergillus flavus*, 333, 334f
- conidiospores (conidium/conidia), 171, **333**, 334, 334f, 336f, 338t
  - of *Streptomyces*, 321, 321f
- conidium. *See* conidia/conidium (conidiospore)
- conifers, as eukarya, 6
- *Coniothyrium minitans*, 339
- conjugated proteins, 45
- conjugated vaccines, **503**
- conjugation
  - in bacteria, 16, **236**–237, 237f, 238f
    - biofilms and, 163
    - sex pili and, 84, 236–237, 237f
  - in protozoa, **346**, 346f
- conjugation fungi. *See* Zygomycota
- conjugation (sex) pili, **84**, 236–237, 237f
- conjugative plasmids, **238**–**239**
- conjunctiva of eyes
  - as portal of entry for pathogens, 429, 430t, 445f, 603
  - normal microbiota of, 402t
- conjunctivitis, 354t, 429, **603**, **604b**
  - inclusion, **604**, 604b
  - swimming pool, 604
- connective tissue, histamine present in, 460
- constant (C) regions, of antibodies, 479, 480f
- constitutive genes, 222
- contact dermatitis, allergic, **530**–531, 530f, 531b
  - lichens or their acids causing, 340
- contact inhibition, **442**, 443f
  - loss of, and unregulated cell growth, 442, 443f
  - viruses and, 442, 443f
- contact lens, hydrogen peroxide as disinfectant, 202
- contact lenses
  - biofilms colonizing, 431
  - conjunctivitis and, 603
- contact transmission, **410**, 411f
- contagious diseases, **406**
- *contagium vivum fluidum* ("contagious fluid"), virus first described as, 368
- contaminated intravenous heparin solution, bloodstream infection and, 164b
- continuous cell lines, **378**
- control of microbial growth, 184–209
  - by altering plasma membrane, 186
  - by damaging proteins/nucleic acids, 187
  - chemical methods, 195–202
  - death rate graphs, 186, 186t, 187f
  - in nosocomial infections, 416
  - microbial characteristics and, 202–205
  - physical methods, 187–194, 194t
    - cold, 157–158, 158f, 170, 191–192, 194t
    - desiccation, 192, 194t
    - dry heat, 191, 194t
    - filtration, 177f, 184f, 191, 191f, 194t
    - heat, 188–191
    - high pressure, 192
    - incineration, 191, 194t
    - low temperatures, 191–192
    - moist heat, 188–190, 189f, 194t
    - osmotic pressure, 192
    - radiation, 192–193
    - summary by method/mechanism of action/preferred use, 194t
  - terminology of, 185–186, 185t
- convalescent home infections. *See* nosocomial infections
- convalescent period (recovery) stage, **409**
- convulsive symptoms of tetanus, 437
- copper, 37
- copper 8–hydroxyquinoline, 199
- copper
  - antimicrobial activity of, 198–199, 198f, 204t
  - as cofactor, 117
- copper ore extraction by leaching, 806, 807f
- copper sulfate, as an algicide, 199, 204t
- cordite, 2
- core polysaccharide, 86f, **87**
- corepressors, **225**, 225f
- corn borer, European, 267
- corn plants, transposons discovered in, 240
- cornea
  - *Acanthamoeba* keratitis of, **605**
  - herpetic keratitis of, **605**
- coronary artery disease, 18
  - antisense DNA explored as gene therapy, 259
- Coronaviridae, characteristics/important genera/clinical features, 376t
- *Coronavirus*
  - common cold caused by, 376t, 679
  - SARS-associated, 367, 376t, 417t
- cortex, of lichen, **339**, 340f
- corticosteroids, to treat psoriasis, 533
- *Corynebacterium*, as normal microbiota of eye, 402t
- *Corynebacterium diphtheriae*, 678f
  - diphtheria toxin produced by, 237, 320, 382, 436, 436f, 438t, 677, 678f
  - emerging infectious diseases and, 417t
  - metachromatic granules of, 95
- *Corynebacterium* genus/spp., 301t, **320**
  - as human pathogens, 301t, 320
  - as normal microbiota of mouth, 403t
  - as normal microbiota of skin, 402t, 586
  - as pleomorphic bacteria, 79
  - G + C ratio, 316
- *Corynebacterium xerosis*, as normal microbiota of skin, 586
- cosmetics and allergic contact dermatitis, 530
- cotton balls, quat antiseptics neutralized by, 199, 201b
- cotton plants, insect toxin genetically engineered into, 264–265
- cotton production, microbes used in, 3b, 40
- Coulter counters, 176, 178, 178f
- counterstains, **69**, 71
- coupled chemical reactions, 114, 114f, 122, 122f
- covalent bonds, **30**–31, 31f
  - double, 31
  - ionic bonds *vs.*, 31
  - of different elements, 31, 31f
  - single, 31, 31f
  - triple, 31
- cowpox vaccine, 501
- cowpox virus, 11, 501
  - caused by Poxviridae, 385
- cows
  - bovine tuberculosis, **685**
  - dairy, bovine growth hormone (bGH) in milk production, 267, 267t
  - livestock
    - animal feed antibiotics, 554, 562t, 565, 575, 577b
    - *Pasteurella* -caused sepsis in cattle, 311
  - *Salmonella* common intestinal inhabitants of, 310
- *Coxiella burnetii*, 309
  - as potential biological weapon, 649b
  - endospore-like structures formed by, 96
  - Q fever caused by, 309, 690, 690f
  - replicates inside phagolysosomes, 459
- *Coxiella* genus/spp., 300t, **309**
  - as obligately intracellular human pathogens, 300t
- coxsackievirus, 375t
- coyotes, tapeworm *Echinococcus granulosus* in, 358, 359f
- CPE (cytopathic effect), in cell cultures, **378**, 378f
- cranberry juice, prevents *E. coli* from adhering to cells, 746
- crayfish, lung flukes and, 356, 357f, 361t
- Crenarchaeota, **302** f
- *Crenarchaeota*, 778
- Crenarchaeota, gram-negative archaea, 302t
- cresols, 195, 196f
- Creutzfeldt-Jakob disease (CJD), **20**, 393, **630**–631, 631f, **632b**
  - variant CJD, compared, 631t
- Crick, Frances H. C., 10f, 14t, 16, 47
- crisis phase of pyrogenic response (fever), **463**
- cristae/crista, **104**, 105f
- Crohn's disease, 460
  - interleukin-12 and, 493b
  - monoclonal antibodies to treat, 509
- crop plants
  - insect toxin genetically engineered into, 264–265
  - MacGregor tomatoes, 267, 267t
- cross-bridge amino acid, 85, 86f
- crossing over (genetic recombination), **234**, 234f
- crown gall disease, 264, 265f, 304
- Crustacea, 362
- crustaceans, chitin exoskeleton of, 98
- crustose lichens, 339, 340f
- Cruz, Oswaldo, 4t
- *Cryphonectria parasitica*, chestnut tree blight caused by, 339
- *Cryptococcus gattii* (fungus), **617b**, 627
- *Cryptococcus grubii* (fungus), **617b**, 626–627
- cryptococcosis, **617b**, **626**–627, 627f
- *Cryptococcus* (fungus)
  - as opportunistic fungus, 339
  - blastoconidia spores found in, 333
- *Cryptococcus neoformans* (fungus), **626**–627, 627f
  - in AIDS patients, 544t
  - pathogenic properties, 338t, 443
- cryptosporidiosis, **731**, **734b**
  - as an emerging infectious disease, **20**–21, 417t
  - as notifiable infectious disease, 421t
- *Cryptosporidium hominis* (protozoa)
  - causing persistent diarrhea in AIDS patients, 544t
  - diarrhea caused by, nitazoxanide to treat, 571
- *Cryptosporidium* (protozoa), 350, 354t
  - causing waterborne diarrhea outbreaks, 329, 355b
  - chlorine-resistance and, 355b
  - emerging infectious diseases and, 417t
  - interleukin-12 and, 493b
  - preventing outbreaks, 355b
  - transmission routes, 355b
- crystal violet stain, 68, 69
  - in Gram stain, 69, 70f, 87
- crystal violet-iodine (CV-1) complex, 69, 87
- CSF (cerebrospinal fluid), 610, 611, 612f
  - spinal tap (lumbar puncture), 614, 614f
- CSF (colony-stimulating factor), 492
  - genetically engineered, therapeutic uses, 258, 260t

Index

CTL (cytotoxic T lymphocyte), **487**, 489*f*, 496*f*, 529, 545
  cancer cell and, 537, 537*f*
cucumbers, lactic acid fermentation and, 135*b*
Culex (mosquito)
  as vector for arboviral encephalitis, 362*t*, 413*t*, 628*b*
  as vector for St. Louis encephalitis, 628*b*
Culiseta (mosquito), as vector for eastern equine encephalitis, 628*b*
culture, **164**
culture media, **164**–169
  agar, 139, 159, 165, 343
  alternative methods to, 404–405
  chemically defined, **165**, 165*t*
  complex, **165**, 166*t*
  criteria for, 164–165
  differential, 139, 139*f*, **168**–169, 168*f*, 169*f*
  enrichment, **169**, 169*t*
  filtration and, 191
  for aerobes, 167–168, 168*f*
  for anaerobes, 166–167, 167*f*
  for growing bacteriophages, 374, 377, 377*f*
  Haemophilus bacteria and, 311
  nutrient broth/nutrient agar, **165**
  reducing media, **166**
  salt concentration and, 159
  selective, **168**, 169*f*, 169*t*, 285–286
  solidifying agents, 165. *See also* agar
  special techniques, 167–168, 168*f*
  sterilization and, 164–165, 191
  summary, by type/purpose, 169*t*
  trace elements and, 160
  transport, **284**
  viruses and, 374, 377–379
curd, in cheese production, **798**–799, 799*f*
cutaneous anthrax, **645**, 645*f*, 650*b*
  virulence of, 430
cutaneous diphtheria, **678**
cutaneous mycoses (dermatomycoses), **337**, 338*t*, **600–601**, 600*f*
  ketoconazole to treat, 568
  potassium hydroxide (KOH) to diagnose, 601
cuticles
  of flukes, **356**, 356*f*
  of tapeworms, 356
CXCR4 (chemokine coreceptors), 541
cyanide
  combined with enzymes to prevent cell functioning, 120
  iron and, 120
cyanobacteria, 301*f*, 301*t*, **313**–315, 314*f*, 314*t*
  alkaline habitats and, 37
  as nitrogen fixers, 17, 160, 314
  as photoautotrophs, 143–145, 143*f*, 301*t*
  environmental role of, 314
  evolutionary contributions to life on Earth, 314
  fossils of, 277
  gas vacuoles and, 96, 314
  habitat and, 341*f*
  important genera of/special features, 301*t*
  in taxonomic hierarchy, 301*t*
  lichens and, 339
  pH ranges and, 37
  photosynthesis and, 140, 143–145, 143*f*, 145*t*
  phylogenetic relationships, 281*f*, 313–314
  position in evolutionary tree, 275*f*
  selected characteristics, compared, 314*t*
cyanocobalamin (vitamin B12), 117*t*
Cyanophora paradoxa, 277, 277*f*
cyclic AMP (cAMP), **225**–226, 226*f*
  amoeba-produced, 351, 352*f*
cyclic compounds, 43, 43*f*, 47
cyclic photophosphorylation, **140**, 141*f*
cyclic R group, characteristic of various amino acids, 43, 43*f*, 44*t*
cyclic structure, 43
Cyclospora, 329, 350, 354*t*
Cyclospora cayetanensis (protozoan), 350, 354*t*, 417*t*, 731, 734*b*
Cyclospora diarrheal infection, **731**, **734*b***
cyclosporine, discovery and successful transplants, 536
cysteine (cys)
  disulfide bridges of, 45, 46*f*
  structural formula/characteristic R group, 44*t*
cystic acne, 453
cystic fibrosis, 18
  biofilm-forming *P. aeruginosa* in, 57*b*, 163
  biofilms and, 163
  Burkholderia infections and, 306, 308
  DNA sequencing to identify cause, 262
  genetically engineered enzyme used to treat, 260*t*
  Pseudomonas infections and, 308
  tobramycin to control *Pseudomonas* infections common to, 565
cysticerci, **357**
cysticercosis, 358, **732–733**
cystitis, **746**, **748*b***
cysts of protozoa, 330*t*, **346**
  chlorine dioxide activity against, 197
  of *Chilomastix*, 347*f*
  of *Giardia*, 347*f*
  oocyst of Apicomplexa and, 346
  resistance to chemical biocides, 203, 203*f*
cyt (cytochromes), **129**–130, 129*f*
cytochrome c oxidase, 139
cytochrome oxidase, 116*t*
cytochromes (cyt), **129**–130, 129*f*
cytocidal effects, 441
cytokine storm, **492**, 522
  of 1918 influenza pandemic, 694
cytokines, 436, **450**, **462**, **491**–492
  as chemical messengers of immune cells, 491–492
  as therapeutic agents, 492, 493*b*
  cellular immunity and, 491–492
  chemokines, **492**
  hematopoietic, **492**
  in B cell activation, 482, 482*f*
  in cellular immunity, 491–492, 496*f*
  in humoral immunity, 469, 482–484, 482*f*, 492, 496*f*
  in tissue repair stage of inflammatory response, 463
  inflammatory response and, 461*f*, 462
  interferons as, 469, 492
  interleukin-1 (IL-1), **437**, 438*t*
  interleukin-12 (IL-12) as "magic bullet," 493*b*
  overproduction (cytokine storm), 492
  phagocytosis and, 459
  roles of, 450
  secreted by T cells, 478
  symptoms induced by, 436
  toxic at high concentrations, 437
  tumor necrosis factor alpha (TNF-α), **437**, 438*f*, 492
cytolysis, **454**, 455*t*
  by complement activation, 464, 464*f*, 465, 465*f*, 466, 466*f*
cytomegalovirus, **643*b***, **658**
  as AIDS-associated disease, 544*t*
  cytopathic effects of, 443*t*
  eye infections in AIDS patients, 542, 658
  inclusion bodies of, 385
  pregnancy and, 760
  typical U.S. prevalence of antibodies against, 567, 567*f*, 568
Cytomegalovirus (HHV-5), 375*t*, 643*b*, **643*b***, **658**
cytomegalovirus retinitis, 542, 658
cytopathic effects (CPE), **441**–442, 442*f*, 443*t*
  in cell cultures, **378**, 378*f*
Cytophaga genus/spp., **324**, 777
cytoplasm
  eukaryotic cells, 99*f*, **100**–101
  in prokaryotic *vs.* eukaryotic cells, 101*t*
  prokaryotic cell, 80*f*, 82*f*, **94**
cytoplasmic membrane. *See* plasma membrane
cytoplasmic streaming, **101**, 101*t*, **352**, 353*f*
cytosine (C), 47, 48*f*, 211
  in DNA replication, 212–215, 214*f*–216*f*
  in translation phase of protein synthesis, 216, 218*f*
cytoskeleton, **101**, 101*t*
  invasins and, 433
  pathogens use of actin in, 433
cytosol, **101**
cytostome, 330*t*, **346**
  of euglenoids, 351*f*
  of *Paramecium*, 350*f*
cytotoxic reactions, 523*t*, **526**–528
  as Type II hypersensitivity reaction, 523*t*, 526–528
  drug-induced, **528**, 529*f*
cytotoxic T lymphocyte (CTL), **487**, 489*f*, 496*f*, 529, 545
  cancer cell and, 537, 537*f*
cytotoxins, 435
  diseases caused by, 438*t*

D value/DRT (decimal reduction time), **188**
D-glucose, 43
daclizumab, 537
dairies, disinfectants used in, 197
dairy cows, benefits of bGH to milk production, 267, 267*t*
dairy equipment, chloramines to sanitize, 197, 204*t*
dairy products
  butter/buttermilk, 779
  cheese. *See* cheese
  cultured sour cream, 799
  estimating bacterial populations in, 176, 178*f*
  genetically engineered rennin and, 267*t*
  microbes used in production of, 798–799, 799*f*
  pasteurization of, 190–191
  phosphatase test and, 190
  yogurt, 799
dalfopristin, 566
dalfopristin (Synercid), mode of action/spectrum of activity, 562*t*
dandruff, 586
  Malassezia causing, 338*t*
dapsone, to treat leprosy, 620, 632*b*
daptomycin, 566
darkfield microscope, **59**, 60*f*, 66*t*
Darwin, Charles, 274
daughter cells
  daughter DNA strands, 2, 211, 214, 214*f*
  in flow of genetic information, 211, 213*f*, 214*f*
DDT, 17, 19
deamination, **136**, **771**, 771*f*
death, fever and, 463
death (logarithmic decline) phase of bacterial growth curve, 173*f*, **174**
death rate of microbes, antimicrobial agents and exponential death rates, 186, 186*t*, 187*f*
deathcap mushroom (*Amanita phalloides*), 443
debridement, **616**
decarboxylation, **127**, 128*f*, **136**
  biochemical test for, 138, 139*f*
decimal reduction time (DRT/D value), **188**
decimeter (dm), metric/U.S. equivalent, 55*t*
decolorizing agent in gram staining, **69**
decomposers
  fungi as, 330
  oomycotes as, 345*f*
  water molds as, 344
decomposition reactions, **32**, 33*b*
deep-freezing
  to control microbial growth, 194*t*
  to preserve bacterial cultures, **170**
deep-sea hydrothermal vents, microbes associated with, 158
deer, as disease reservoirs, 410*t*, 651*b*
deer fly (*Chrysops*), as vector for tularemia, 362*t*, 363*f*, 642
deer fly fever, 642
deer mice, as disease reservoirs, 410*t*
defecation, **452**, 471*t*
defenses of human body. *See* immunity
defensins, 579, 584
  produced by neutrophils/macrophages/epithelium, 470
defensive cells of innate immunity, 472*t*
  natural killer (NK) cells, **454**, 455*t*, 472*t*
  phagocytes, 450*f*, **457**–460, 457*f*, 472*t*
definitive host, **349**
  of *Echinococcus granulosus*, 358, 359*f*, 361*t*
  of *Paragonimus westermani*, 356, 357*f*
  of *Plasmodium vivax*, 349–350, 349*f*
  of selected parasitic helminths, 361*t*
  of *Taenia saginata*, 357, 361*t*
  of *Taenia solium*, 357–358, 361*t*
degeneracy (of genetic code), **219**, 226–227, 255
degenerative evolution, 320
degerming/degermation, **185**, 185*t*, 197, 204*t*
  alcohol swabs as, 197, 204*t*
  soaps as, 199, 204*t*
degradation of synthetic chemicals
  bioremediation, 775, 776*f*
  composting, **776**
  solid municipal waste, 775–776
degradative chemical reactions. *See* catabolism
degranulation, **524**, 524*f*
dehydration, fever and, 463
dehydration synthesis, **38**, 39*f*, 114
  peptide bonds formed by, 43, 45*f*
dehydrogenases, 116
dehydrogenation, **122**, **136**. *See also* Oxidation reaction
  biochemical test for, 138, 139*f*
Deisenhofer, Johann, 15*t*
delayed hypersensitivity reactions, **529**–**531**, 530*f*-532*f*, 531*b*
  as Type IV delayed cell-mediated reactions, 523*t*, 529
  T cells and, 529–530, 530*f*
  transplant rejection and, 529, 531*b*
delayed-onset bloodstream infections, 164*b*
Delbrück, Max, 10*f*, 14*t*
delirium, fever and, 463
delta hepatitis. *See* hepatitis D
deltaproteobacteria, 301*t*, **312**
  important genera/special features, 301*t*
Deltaviridae, characteristics/important genera/clinical features, 376*t*
denaturation, **45**, **119**, 119*f*
  antimicrobial agents and, 187
  by moist heat, 188–190, 189*f*
  by pasteurization, 9, 190–191, 194*t*
dendritic cells, **454**, 455*t*, **490**
  antimicrobial proteins (AMPs) and, 471
  as antigen-presenting cells, 476*f*, **490**, 490*f*
  in second line of defense, 450, 450*f*
dengue, 376*t*, 417*t*, **659**, **660*b***
  Aedes mosquito as vector, 362*t*, 413*t*
  as emerging infectious disease, 417*t*

dengue hemorrhagic fever (DHF), **659**, 660*b*
as emerging infectious disease, 417*t*
denim blue jeans, made by microbes, 3*b*
dental caries (tooth decay), **707–709**, 707*f*, 708*f*, **710*b***
caused by *Streptococcus mutans*, 319, 707–708, 708*f*
dental plaque, **707**
as biofilm, 431
dextran, *Actinomyces*, *Streptococcus mutans* and, 431, 440
dentification, 770*f*, **771**
deoxyribonucleases, 590
deoxyribonucleic acid. *See* DNA
deoxyribose, 39, **47**, 48*f*, 211
Dermacentor andersoni (wood tick)
as vector of *Rickettsia rickettsii*, 413*t*
Rocky Mountain spotted fever transmitted by, 362*t*, 655
Dermacentor variabilis (dog tick), Rocky Mountain spotted fever transmitted by, 655
dermatitis
Malassezia causing, 338*t*
Pseudomonas, **591**, 593–594
dermatomycoses (cutaneous mycoses), **337**, 338*t*, **600**–601, 600*f*
ketoconazole to treat, 568
potassium hydroxide (KOH) to diagnose, 601
dermatophytes, **337**, **600**
dermatomycoses (cutaneous mycoses) caused by, 337, 338*t*
keratinase enzyme secreted by, 337, 600
dermicidin, 470
dermis, **451**, 451*f*, **585**, 585*f*
descriptive epidemiology, **419**
desensitization, **526**
desiccation, to control microbial growth, **192**, 194*t*
designer jeans, made by microbes, 3*b*
Desulfovibrio desulfuricans, mercury contamination and, 33*b*
Desulfovibrio genus/spp., 301*t*, **312**, 777
anaerobic respiration and, 132
as sulfate reducers, 301*t*
found in intestinal tracts of human/animals, 312
Desulfovibrionales, 301*t*, **312**
as sulfur reducing bacteria, 312
important genera/special features, 301*t*
Desulfurococcales
important genera/special features, 302*t*
in taxonomic hierarchy, 302*t*
detergent (SDS), 257*f*
detergents, 199
anionic, susceptibility of gram-negative *vs.* gram-positive bacteria to, 88*t*
as antimicrobial agents, 199, 204*t*
cationic, 199, 204*t*
gram-negative bacteria and, 87, 88*t*
deuteromycetes, unclassified, 335
Deuteromycota, 335
developing countries, parasitic diseases and, 329
devescovinids, 107*b*
dextran, 40
Actinomyces, *Streptococcus mutans* and dental plaque, 431, 440
dextransucrase, produced by *Streptococcus mutans*, 440
DHAP (dihydroxyacetone phosphate), 124, 125*f*
diabetes
gene therapy and, 18
insulin produced by bacteria and rDNA technology, 247
diabetes mellitus
insulin-dependent, **533**
mucormycosis and, 338
diagnostic immunology, 507–518. *See also* diagnostic tools
future of, 516–517
diagnostic tools
agglutination reactions, **510**–512, 510*f*, 511*f*
complement-fixation reactions, **512**–513, 514*f*
DNA probes and, 517. *See also* DNA probes
enzyme-linked immunosorbent assay (ELISA), **514**–516, 518*f*
fluorescent-antibody (FA) tests, **513**–514, 515*f*
for HIV detection, 545
for viral RNA, 545
monoclonal antibodies, **507**–509, 508*f*
neutralization reactions, **512**, 513*f*
precipitation reactions, **509**–510, 509*f*, 510*f*
rDNA technology and, 262
sensitivity and, **507**
specificity and, **507**
Western blotting, **287**, 289*f*, 379, 510, **516**
dialysis patients, at risk for gram-positive sepsis, 640
diapedesis, 461*f*, **462**
diaphragm, of compound light microscope, **56**, 56*f*
diarrhea, 710
antibiotic associated, 438*t*
cholera and, 309, 437, 438*t*
Clostridium difficile -associated, 438*t*, **720**, 720*b*
cryptosporidiosis and, 20–21, 354*t*
Cryptosporidium causing, 354*t*
Cyclospora cayetanensis causing, 354*t*, 417*t*
Escherichia coli 0157:H7 and, 20, 83, 84*f*, 113, 417*t*
hemorrhagic, 417*t*
infant, 239
infant mortality and, 710–711
microspora causing, 348, 354*t*
nosocomial, 414*t*
persistent, in HIV/AIDS patients, 542, 544*t*
traveler's, 239, 310
waterborne (recreational), 350, 355*b*
diarylquinoline, experimental anti-TB drug, 684
diatoms, 341*f*, **343**, 343*f*, **343*t***
identifying, 343
in kingdom Stramenopila, 343
neurological disease caused by, 343
oil storage by, 343, 343*t*
DIC (differential interference contrast) microscope, **60**, 61*f*, 66*t*
DIC (disseminated intravascular coagulation), 437
dichotomous keys, **293**
examples of, 283*b*, 285*f*, 303
Dictyostelium, 352*f*
differential culture media, **168**–169, 168*f*, 169*f*, 169*t*
to identify pathogenic *Escherichia coli*, 139, 139*f*, 169
differential interference contrast (DIC) microscope, **60**, 61*f*, 66*t*
differential stains, **69**–71
acid-fast stain, **70**–71, 71*f*
gram stain, **69**–70, 70*f*
used to identify microorganisms, 285
differential white blood cell count, **454**, 455*t*, 456, 457
diffraction of light rays, 60, 60*f*
diffusion
chemiosmosis and, 130–131, 130*f*, 131*f*
facilitated, **91**–92, 92*f*
simple, **91**, 92*f*
diffusion methods (to evaluate antibiotic sensitivity)
disk-diffusion method, **195**, 196*f*, 201*b*, **572**, 572*f*
E test, **572**, 573*f*
DiGeorge's syndrome, 538, 539*t*
digestion, as phase of phagocytosis, 457, 458*f*, **459**
digestive aids, 2
digestive system, 705–742
fecal-oral cycle, **705**
infection *vs.* intoxication, **710**
microbial diseases
bacterial, 707–721, **710*b***, **722*b***
fungal, 729–730, **734*b***
helminthic, 732–737, **735*b***
protozoan, 730–732, **734*b***
viral, 721–729, **724*b***, **729*b***
normal microbiota of, **706**–707
ruminant, microbes in biofilms and, 163
structure/function, 706, 706*f*
digestive system infections, Reoviridae and, 387
dihydroxyacetone phosphate
in biosynthesis of lipids, 147*f*
in lipid catabolism, 136*f*
dihydroxyacetone phosphate (DHAP), 124, 125*f*
diiodohydroxyquin (iodoquinol), to treat intestinal amoebic diseases, 571
diiodohydroxyquin, mode of action/uses, 564*t*
dilation, of blood vessels (vasodilation), 460, 461*f*
dilution tests of antibiotics, 572–573, 573*f*
dimers
secretory IgA, 480
unrepaired, and skin cancers, 231
dimorphic fungi, **332**, 332*f*, 338*t*
dimorphism, **332**, 332*f*
sexual, 358–359
Dinoflagellata
blooms of certain species indicate polluted water, 344
characteristics of, 343*t*
in kingdom Stramenopila, 343
dinoflagellates, as protozoa, but often studied with algae, 346
dinoflagellates (plankton) (algae), 341*f*, **343**–344, 343*t*, 344*f*
dioecious parasitic helminths, 354–355
dipeptide, 43, 45*f*
diphtheria, 19, 95, 237, **677**–679, 678*f*, **681*b***
1990's epidemic, and adult vaccination booster, 418
as emerging infectious disease, 417*t*
as notifiable infectious disease, 421*t*
Corynebacterium diphtheriae causing, 320, 471*t*, 678
cutaneous diphtheria, **678**
cytotoxin's mechanism of action, 438*t*
membrane in throat characteristic of, 678, 678*f*
toxin causing. *See* diphtheria toxin
vaccine, 435, 502, 502*t*, 504*t*, 678
diphtheria toxin, 435, **436**, 438*t*, 439*t*, 441
as example of A-B toxin, 435, 436*f*
mechanism of action (model), 436*f*
produced by *Corynebacterium diphtheriae*, 237, 438*t*
if infected by lysogenic phage carrying *tox* gene, 436, 436*f*
vaccine produced from purified, 502*t*
diphtheroids
as normal microbiota of eye, 402*t*
as normal microbiota of nose, 403*t*
as normal microbiota of skin, 586
as normal microbiota of urethra, 403*t*
dipicolinic acid, 97
diplobacilli, **78**, 78*f*
diplococci, **77**, 78*f*
diploid cell lines, **378**
direct agglutination tests, **510**–511, 511*f*
direct contact transmission, **410**, 410*t*, 411*f*
in nosocomial infections, 415–416
direct ELISA tests, 514, **515**–516, 518*f*
direct FA tests, **513**, 515*f*
direct flaming sterilization, **191**, 194*t*
direct microscopic count of bacteria, **176**, 177*f*, 178
direct (positive) selection method to identify mutations, **231**
Dirofilaria immitis (nematode), 360, 360*f*
Wolbachia bacteria essential to, 360
disaccharides, **39**, 39*f*
disease, **400**. *See also* infectious diseases; microbial diseases
acute, **406**
chronic, **407**
communicable, **406**
contagious, **406**
degenerative, *vs.* infectious, 404
diagnosis, antibody presence (IgM) and, 480
diagnosis of, 406
duration or severity of, 406–407
endemic, **406**
epidemic, **406**
from cooperation among microbes, 404
fulminating, 602
general principles of, 399–418
classification, 406–407
emerging, 416–418
etiology, **400**, 404–406
hospital-acquired, 413–416
normal microbiota and, 400–404
patterns of, 408–409
spread of infection, 409–413, 444
germ theory and, 9, 11, 404, 477
incidence, **406**
infectious, **19**, **404**. *See also* infectious diseases
inherited (genetic), *vs.* infectious, 404
microbial. *See* microbial diseases
noncommunicable diseases, **406**
occurrence of, 406
pandemic, **406**
pathogenesis of, **400**
pathogens and, **399**
pathology as study of, **400**
patterns of, 408–409, 408*f*
predisposing factors, **408**
prevalence, **406**
self-limiting, 676
severity or duration of, 406–407
signs and symptoms, **406**
sporadic, **406**
spread of, 409–413, 444
stages/sequence of events during, 408–409, 408*f*
syndromes, **406**
vs. infection, **400**
disease reservoirs, **409**, 410*t*
animal and human, 409
nonliving (soil/water), 409
of zoonoses/with transmission methods, 410*t*
disinfectants
alcohols, **197**–198, 198*t*, 203*t*, 204*t*
aldehydes, **200**, 204*t*
antibiotics as, 200
bacteria that can grow in, 196*f*, 202, 449*f*
bacterial plasma membrane damaged by, 91
biguanides, **196**–197, 204*t*
bisphenols, **196**, 196*f*, 203*t*, 204*t*
Cepacol, 199, 204*t*
chemical food preservatives, 199–200
chemical sterilization, 200–201, 205*t*
chlorhexidine, 197, 203*t*, 204*t*
considerations in choosing, 195
copper, 198–199, 198*f*
detergents, 199
early uses of, 11
evaluating effectiveness, 195

Index

formaldehyde, 200
glutaraldehyde, 200, 203*t*, 205*t*
halogens, **197**, 204*t*
hydrogen peroxide, 202
mercury, 199, 203*t*
peroxygens, 202
phenolics, **195**, 196*f*, 203*t*
phenols, **195**, 196*f*
plasma sterilization, 201
silver, 198–199, 198*f*
silver sulfadiazine, 198, 204*t*
soaps, 199
sulfur dioxide, 199
supercritical fluids, 201–202
surface-active agents, 199
Surfacine, 198–199
temperature and effectiveness of, 186
types of, 195–202
use-dilution tests, 195
vs. antiseptics, 185
zinc, 199
disinfection, **185**, 185*t*
evaluating effectiveness of, 195, 196*f*
principles of, 195
water treatment, 782*f*, **783**
disinfection and release in sewage treatment, **785**, 785*f*
disk-diffusion method, **195**, 196*f*, 201*b*, **572**, 572*f*
disseminated intravascular coagulation (DIC), 437
dissemination of disease, 409–413, 444
dissimilation, **773**, 774*f*
dissimilation plasmids, **238–239**
dissimilatory metabolism, 312
dissociation (ionization), **35**, 35*f*, 36*f*
distilled water, microbial growth and, 159
disulfide bridges, 45, 46*f*
antimicrobial agents and, 187
of antibodies, 479, 480*f*
diversity, genetic, 231, 241
dm (decimeter), metric/U.S. equivalent, 55*t*
DNA, 39, **47**
amplification of, 247
antimicrobial agents and, 187
bacterial chromosome, 94–95, 97*f*
base pairs, **211**
bullets via gene guns, 253, 254*f*
complementary (cDNA), **254**–255, 255*f*
complementary structure and duplication of, 211
conjugation, 16, **84**
damage/destruction by radiation, 193, 194*t*
double helix, **47**, **48***f*, 58*f*, 211
enzymes of replication process, 212–215, 214*f*, 215*f*, 215*t*, 216*f*
experiments with bacteria and science of, 16
for identification purposes, 288–294
from mummies/extinct plants/animals, 262, 264
genetic transformation and hereditary information, 235–236
in bacterial binary fission, 171*f*
in prokaryotic cell/eukaryotic cell/eukaryotic organelles, 276*t*
"junk," 261
location in eukaryotic cells, 102–103
location in prokaryotic/bacterial cell, 80*f*, 94–95
mitochondrial, 105
mutagenic agents and, 228–233
mutation and, 226–233
"naked," and transformation process, 234, 253
of viruses, 371–372
probes, **256**–257, 257*f*
protein involved in repair of, 65*f*
protein synthesis and, 147
recombinant, **16**
STM microscopes to view, 65, 65*f*
structure, 47, **48** *f*
as supercoiled (twisted), 212
complementary nature allowing for precise duplication, 211
linear sequence of bases of, 211
paired DNA strands and, 213–215, 214*f*
supercoiled strands in replication process, 212–213, 214*f*
synthetic, **255**, 256*f*
ultraviolet light damage to, 193, 194*t*
vaccines, **503–504**
DNA base composition, to identify microorganisms, **288–289**
DNA chips, 262, 292, 293*f*, 517
DNA fingerprinting, **262**, **290**
forensic microbiology and, 262, 264*f*
to identify microbes, 289–**290**, 290*f*
to track infectious disease, 262, 264*f*
DNA gyrase, 215*t*
DNA hybridization studies
by colony hybridization, **256–257**, 257*f*
in classification of microbes, 279
using DNA chip technology, 292, 293*f*
value to understanding evolutionary relationships, 278
DNA ligase, 116*t*, 216*f*
in making recombinant DNA, 250, 250*f*
DNA oncogenic viruses, 391
viral families found within, 391
DNA polymerase, **213**, 215*t*, 216*f*
in polymerase chain reaction (PCR) process, 251, 552*f*
proofreading capability of, 214–215
DNA probes, **256**, **291**, 517
by colony hybridization, **256**–257, 257*f*
by Southern blotting, 263*f*, **291**, 292*f*
DNA chip technology and, **292**, 293*f*
to identify pathogens, **256**–257, 257*f*, 262, **291**–292, 292*f*, 293*f*
DNA repair
by excision repair, 230–231, 230*f*
by light-repair enzymes, 230
important enzymes in, 215*t*, 230
radiation causing errors in, 230
DNA replication, 212–215, 214 *f* -216*f*, 215*t*
5′ to 3′ direction of DNA strands, 214, 214*f*, 215*f*
bidirectional replication in bacteria, 214, 217*f*
energy supply for, 214, 215*f*
enzymes important in, 212–214, 215*t*
genetic information flow and, 212, 213*f*
in DNA viruses, 385–386, 385*t*, 386*f*
mistakes in, 214–215, 226–233
rates for spontaneous errors, 231
mistakes in (mutations), 226–233
nucleoside analogs and, 220*f*, 229–230
radiation causing errors in, 230
replication fork, 213–214, 214*f*
events at (summary), 216*f*
in *Escherichia coli* bacteria, 214, 217*f*
semiconservative, **213**
DNA reverse transcriptase viruses, 385*t*
DNA sequencing, **261**–262, 261*f*
fungi and, 274
DNA strands
blunt ends, **249**, 250*f*
complementary, 211
sticky ends, 215*t*, 241*f*, **249**, 250*f*
DNA synthesis
antibiotics that inhibit, 567
from nucleosides with deoxyribose, 214
nitrogen requirements, 160
DNA technology
commercial successes of, 258
recombinant, **16**. *See also* recombinant DNA (rDNA) technology
DNA template strand, 215*t*, 216, 218*f*
DNA transfer, pili's role in, 83–84
DNA vaccines, **259**
advantages of, 506
DNA vectors (gene-cloning vectors/cloning vectors), 247, 248*f*, 250–251, 250*f*, 251*f*
required properties, 250
DNA viruses
biosynthesis of, 384*t*, **385**–386, 385*t*, 386*f*
that infect humans (summary), 375*t*
dogs
as disease reservoirs, 410*t*
bites of, *Pasteurella multocida* transmitted by, 311
heartworm in, 360
reported cases of rabies in, 624*f*
tapeworm *Echinococcus granulosus* in, 358, 359*f*
vaccinated against leptospirosis, 324
Doherty, Peter C., 15*t*
dolphins
bottlenose, 283*b*, 283*f*
Maui, 283*b*
Domain Archaea, 76, **274**–275, 275*f*, 276*t*, 300, 325
members of, 275, 275*f*
position in evolutionary tree, 275*f*
Domain Bacteria, **274**, 275*f*, 276*t*, **300–325**. *See also* bacteria; prokaryotes
position in evolutionary tree, 275*f*
position in taxonomic hierarchy, 280*f*
Domain Eukarya, **274**, 275*f*, 276*t*. *See also* eukaryotes
Kingdoms in, 275*f*
position in evolutionary tree, 275*f*
position in taxonomic hierarchy, 280*f*
domain name, defined, **279**, 280*f*
domains
Domain Archaea, 76, **274**–275, 275*f*, 276*t*, **300**, 325
Domain Bacteria, **300**, 300–302*t*, 302–325
of the three-domain system, 6, 274–277, 275*f*, 276*t*
domoic acid intoxication, **343**
donor cells, in gene transfers, **234**, 234*f*
double covalent bond, 31
double helix, DNA, **47**, 48*f*, 211
double-stranded DNA viruses, 385*t*
enveloped viruses, 375*t*, 385*f*
noneveloped viruses, 375*t*
double-stranded RNA viruses, 385*t*, 388*f*
nonenveloped viruses, 376*t*
doxycycline, 565, 646
Dracunculus medinensis (guinea worm), 13, 13*f*
drain cleaners, 2, 17
drinking water
liquid form of compressed chlorine gas used to disinfect, 197
parasitic protozoa and, 354*t*
droplet transmission, **411**, 411*f*
diseases spread by, 411
drotrecogin alfa (Xigris), 640
DRT/D value (decimal reduction time), **188**
drug resistance, 12–13. *See also* antibiotic resistance
drug-induced cytotoxic reactions, 528, 529*f*
drugs
antibiotics, **12**–13, 12*f*. *See also* antibiotics
antimicrobial, **553–583**. *See also* antimicrobial drugs
synthetic, **12**–13
dry heat sterilization, 191
dry weight, as measure of microbial numbers, 178–179
DTaP vaccine, 616, **678**
recommended schedule, 504*t*
Duchenne's muscular dystrophy, 18
ducks, influenza A viruses and, 19
ducts, of male reproductive system, 745, 745*f*
Dulbecco, Renato, 10*f*, 14*t*
dura mater, 611, 612*f*
dust mites, 525, 525*f*
Dutch elm disease, 339
dye derivatives, as antimicrobial agents, 12
dyes
acidic, **68**–69
basic, **68**–69
gram-negative bacteria and, 87
dysentery, **710**
amoebic. *See* amoebic dysentery
bacillary. *See* shigellosis
Balantidium coli causing, 350, 354*t*
epidemics, antibiotic resistance and, 239
life-threatening, *Shigella* and, 310
dysuria, 746

EAEC (enteroaggregative *E. coli*), **717**, **723*b***
earaches, *Haemophilus influenzae* causing, 311
Earth's carbon cycle, *Pelagibacter ubique* role in, 303
eastern equine encephalitis (EEE/ *Togavirus*), 375*t*, 625, **628*b***
eating utensils (restaurant), calcium hypochlorite (chloride of lime) to disinfect, 197, 204*t*
EB virus. *See* Epstein-Barr (EB) virus
Ebola hemorrhagic fever (EHF), **20**
as an emerging infectious disease, 20, 417*t*
Ebola virus, 637*f*, **659**, 659*f*, **660*b***
as an emerging infectious diseases, 20, 417*t*, 659
as filovirus, 373*f*, 376*t*
as helical virus, 373, 373*f*
as potential biological weapon, 649*b*
size of, 369*f*
echinocandins, **569**
echinocandins (antifungals), 564*t*
Echinococcus granulosus (tapeworm), 358, 359*f*, 361*t*
life cycle, 358, 359*f*
Echinococcus multilocularis, 359*f*
echoviruses, 375*t*, 394*t*
as opportunistic pathogens, 404
eclipse period in viral multiplication cycle, **380**
ecological niche (host range), of viral species, 374
ecology, microbial, **17**
Eco RI restriction enzyme, 249*t*, 251*f*
ecosystems, without sunlight, 773–774
ectomycorrhizae, 767, 768*f*, 769*f*
ectopic pregnancies, pelvic inflammatory disease and, 752
ectosymbiosis, 107*b*
Edelman, Gerald M., 10*f*, 14*t*
edema, of inflammation, **460**
edema toxin, of *Bacillus anthracis*, 645
EDTA (ethylenediaminetetraacetic acid), 89
EEE/ *Togavirus* (eastern equine encephalitis), 375*t*, 625, **628*b***
efavirenz, 571
eflornithine, to treat African sleeping sickness, 627–628
EGF (epidermal growth factor), genetically engineered, heals burns/wounds/ulcers, 260*t*
eggs
embryonated, to grow viruses, 377–378, 377*f*, 504
food allergies and, 525
EHEC (enterohemorrhagic *E. coli*), **717–718**, 718*f*, **723*b***
EHF (Ebola hemorrhagic fever), **20**
Ehrlich, Paul, 10*f*, 12, 14*t*, 477, 478, 554
Ehrlichia chaffeensis
ehrlichiosis caused by, 654

Lone Star tick as vector, 654
PCR used to identify, 291, 654
Ehrlichia genus/spp., 300t, **303**
arthropod vectors that transmit, 413t
as obligately intracellular human pathogens, 300t, 303
ehrlichiosis and, 291, 303, 352t, 362t, 410t, 413t
reservoirs/transmission method, 410t
ehrlichiosis, 291, 303, 352t, 362t, 410t, 413t, **651b**, **654**
as notifiable infectious disease, 421t
causative agent/arthropod vector, 413t
human granulocytic anaplasmosis, 291, **651b**, **654**
EIA (enzyme immunoassay), 514, **677**, **755**
EIDs. *See* emerging infectious diseases
EIEC (enteroinvasive *E. coli*), **717**, **723b**
electrolyte imbalances, fever and, 463
electromagnetic fields, in plasma sterilization, 201, 205t
electromagnetic lenses, used in electron microscopes, 63–65, 64f
electron acceptors, 30, 30f
final, in energy-producing processes, 133, 134f, 137t, 141, 143f
electron carriers
in energy production, 141, 143f
used in oxidative phosphorylation, 123, 123f
electron donors, 30, 30f
in energy production, 141, 143f
electron microscopes/microscopy, 16, **63**–65, 64f, 67t
classification of microbes and, 174
scanning electron microscope (SEM), **64**–65, 64f, 67t
transmission electron microscope (TEM), **63**–64, 64f, 67t
viral size and, 369, 369f
viruses and the invention of, 368, 379
electron shells, 27f, **28**, 29t
electron transfers, coenzymes important in, 117t, 122
electron transport, vitamin K as coenzyme used in, 117t
electron transport chain (system), **123**, 124, 125f, **129**–130, 129f
as mechanism of ATP synthesis (chemiosmosis), 130–131, 130f, 131f
ATP yield, 132t
in aerobic respiration, 129–132, 129f, 133f
in eukaryotic cells, 129, 131f
in photosynthesis, 140, 141f
in prokaryotic cells, 129, 131f
electronic cell counters, 176, 178, 178f
electronic configurations, 27, 27f, **28**, 29t
electrons, **27**, 27f
chemical bonds and, 28–32
in cellular oxidations, 122, 122f, 123f
in ionizing radiation, mutagens and, 230, 230f
used instead of light in microscopes, 63–65, 64f
wavelength size, *vs.* that of visible light, 63
electrophoresis. *See* gel electrophoresis
electroporation, **253**
elementary bodies, of *Chlamydophila psittaci*, **322**, 323f, **689**
elements (chemical), 27–28, 27t
isotopes, **27**–28
trace, 117, 160
elephantiasis, 444
ELISA (enzyme-linked immunosorbent assay), **287**, 288f, **514**–516, 518f
HIV antibodies detected by, 516, 518f, 545
syphilis, 755
Toxoplasma gondii detected by, 350
Ellerman, Wilhelm, 389
elm trees, Dutch elm disease and *Ceratocystis ulmi* fungus, 339
embalming chemicals, 200
Embden-Meyerhof pathway (glycolysis), 124
embryo formation among eukaryotic microbes, 330t
embryonated eggs
influenza viruses grown in to make vaccine, 502f
to culture animal viruses, 377–378, 377f
embryonic stem cells (ESCs), **535**, 535f
emerging eukaryotic pathogens, 329
emerging infectious diseases (EIDs), **19**–21, 223b, **416**–418
criteria for identifying, 416
examples of, by microbe/year/disease, 417t
factors contributing to, 19, 416, 418
genetics important to understanding, 210
nosocomial, 422b
vaccine development and, 506
viral hemorrhagic fevers, 637, 659–660, **660b**
Emerging Infectious Diseases (scientific journal), 418
emphysema, 260t
emtricitabine, 571
emulsification, 199
enanthem rashes, **586**, 587f
Enbrel (etanercept), 509
encephalitis, **611**
arboviral, **624**–626, 626f
Culex mosquito as vector, 362t, 413t
aseptic, 223b
Balamuthia causing, 348, 354t
California encephalitis serogroup, 376t, 626, 626f, **628b**
fatal, from rabies, 622
granulomatous amebic, 348, 617b, **617b**, **629**
Hendra virus causing, 417t
Japanese B, **626**
Nipah virus causing, 417t
progressive, 394t
raccoon roundworm causing (*Baylisascaris procyonis*), 417t
subacute sclerosing panencephalitis (SSPE), 392, 394t
West Nile, 19–20, 212, 223b, 223f, 626
encephalitis viruses, as potential biological weapon, 649b
encystment of protozoa, **346**
end-product, 120, 123
end-product inhibition/feedback inhibition, **120**–121, 121f
end-products of fermentation, 133, 134f, 135b
industrial or commercial uses, 137t
endemic disease, **406**
endemic murine typhus, 303, 410t
causative agent/arthropod vector, 413t
Rickettsia typhi as causal agent, 303
Xenopsylla (rat flea) as vector transmitting, 362t
endergonic chemical reactions, **32**, 114
Enders, John F., 14t
endo medium, for enumerating coliforms, 177f
endocarditis, **641**, 641f, **643b**
acute bacterial, **641**, **643b**
gonorrheal, 748
subacute bacterial, **641**, 641f, **643b**
vancomycin-resistant enterococci and, 417t
endocytosis, **100**, 384t
endoflagella (axial filaments), 82–**83**, 84f, **322**, 324f
endogenous antigens, **488**
endogenous pyrogen. *See* interleukin-1
endoliths, **773**–**774**
endomycorrhizae (vesicular-arbuscular mycorrhizae), 767, 768f, 769f
endonucleases, 215t, 231
endoplasmic reticulum (ER), **103**, 103f
rough, 99f, **103**, 103f
smooth, **103**, 103f
endoscopes, peracetic acid and, 201
endospore suspensions, to test for successful sterilization, 190
endospores, **71**, 72f, **96**–98, 97f
alcohols and, 197, 203t, 204t
bacterial, *vs.* spores of other prokaryotes/eukaryotes, 97, 332
boiling water survival time and, 98, 188
chemical antimicrobials activity against, 203, 203t
chlorine dioxide activity against, 197
desiccation and, 192
ethylene oxide and, 201
formation process, 96–98, 97f
fungal spores compared to, 332
heating to destroy, 185, 188
high pressure techniques to control, 192
in foodstuffs, radiation doses needed to kill, 797t
iodine and, 197
microbial death curves and, 186, 187f
of *Bacillus,* 317–318, 317f
of *Clostridium botulinum,* 301t, 316–317, 316f
of thermophilic bacteria, 98, 158
resistance to chemical biocides, 203f, 203t
staining, **71**, 72f, 72t, 96
endosymbionts, 275
endosymbiosis, 106
endosymbiotic bacteria, 106, 277, 277f
endosymbiotic theory, **106**, 275
prokaryotic cells/eukaryotic organelles compared, 276t, 277
endothelial cells, 451
endotoxic shock (gram-negative sepsis), 437, **640**
endotoxins, 435f, **437**, 438f, 439, 439t
antitoxins and, 439, 439t
as immunotherapy for cancer patients, 538
as lipopolysaccharides, 437, 439t
autoclaving and, 439t, 440b
blood-clotting proteins activated by, 437
examples of microbes that produce, 439
exotoxins *vs.*, **439t**
fever (pyrogenic response) and, 438f, 439t
gram-negative bacteria and, 437, 439t
lethal dose and, 439t
lipid A as, **437**, 439t
properties of, 437, 439
compared to exotoxins, 439t
symptoms induced by, 437, 439t
toxicity of, 439t
energy
activation, **115**, 115f
cellular
active transport, 94
carbohydrates as source of, 39
chemical, ATP and, 47–49, 49f, 114, 114f
chemical reactions that release, 113, 114, 114f. *See also* catabolism
chemical reactions that require, 113, 114, 114f. *See also* anabolism
coupling anabolic and catabolic reactions, 114, 114f
group translocation requirements, 94
high-energy bonds, 122
organisms classified by their source of, 142–145, 143f
PEP and, 94
potential, in glucose, 122
radiant, 193, 193f
required for a chemical reaction, 115, 115f
supply in DNA replication, 214
energy levels of electrons, **28**
energy production mechanisms, 121–123
aerobic respiration, 127–132, 133f, 141, 143f
aerobic/anaerobic respiration, fermentation compared, 137t
anaerobic respiration, 127, 132, 137t, 141, 143f
by carbohydrate catabolism, 124–135, 125f
by lipid catabolism, 136–137, 136f, 138f
by protein catabolism, 136–137, 136f, 138f
fermentation, 9, 124, 125f, 132–134, 134f, 135b, 136f, 143f
metabolic pathways and, 123
number of ATP molecules produced/glucose molecule, 137t
oxidation-reduction reactions, 116t, **122**, 122f, 123, 123f, 141, 143f
photosynthesis, 140, 141f
sources for, 141, 143f
summary, 141
energy storage, lipids function in, 146
energy stores, affecting rate of biochemical reactions, 150
enfuvirtide, 571
enrichment culture media, **169**, 169t
Entamoeba dispar, 348, 354t
Entamoeba histolytica, amoebic dysentery caused by, 348, 348f, 354t, 731–732, 732f, 734b
etanercept (Enbrel), 509
enterics, **309**–311. *See also* Enterobacteriales
bacteriocins produced by and ecological balance in intestines, 310
biochemical tests to identify, 285, 285f, 286–287, 286f, 309
clinical importance of, 309
enteroaggregative *E. coli* (EAEC), **717**, **723b**
Enterobacter aerogenes, hospital-acquired infections caused by, 311
Enterobacter cloacae, hospital-acquired infections caused by, 311
Enterobacter genus/spp., 300t, **311**
antibiotic resistance high in, 319
as enteric bacteria, 285, 311
as normal microbiota of large intestine, 403t
as opportunistic pathogens, 300t
fermentation and, 134f, 309
nosocomial infections and, 319, 414, 414t
Enterobacteriaceae, 280f, 285
Enterobacteriales (order), 280f, **309**–**311**
important genera/special features, 300t
enterobactin, bacterial siderophores and, 434, 434f
Enterobius vermicularis, 358, 360f, 361t
enterococci, 319, 640
causing septic shock, **643b**
natural resistance to penicillin, 640
vancomycin-resistant (VRE), 417t, **563**, 577b, 640
Enterococcus faecalis, **319**
classification and, 279
indwelling catheters and, 319
leading cause of nosocomial infections, 319, 640
pentose phosphate pathway and, 127
surgical wound infections and, 319
urinary tract infections and, 319
vancomycin-resistant, 12–13
resistance transferred to *Staphylococcus aureus* via Tn1546 transposon, 241
Enterococcus faecium, classification and, 279
Enterococcus genus/spp., 301t, **319**
as causing nosocomial infections, 414, 414t
as normal microbiota of large intestine, 403t
as normal microbiota of urethra, 403t
as opportunistic pathogens, 301t, 319
classification and, 279

enterohemorrhagic *E. coli* (EHEC), **717–718**, **723*b***
enteroinvasive *E. coli* (EIEC), **717**, **723*b***
enterotoxicosis, staphylococcal, **711–712**, 711*f*, **722*b***
enterotoxigenic *E. coli* (ETEC), **717**, **723*b***
enterotoxins, 435
 diseases caused by, 438*t*
 produced by *Clostridium difficile,* 438*t*
 produced by *E. coli* strains, 310
 produced by *Staphylococcus aureus,* **318**, 438*t*
 produced by *Vibrio cholerae,* 437, 438*t*
 staphylococcal, 430, **437**, 439*t*, 440
 traveler's diarrhea and, 438*t*
Enterotube II, 286*f*
Enterovirus genus/spp., 375*t*
 cytopathic effects of, 443*t*
 pregnancy and, 760
Entner-Doudoroff pathway, **127**
 in purine/pyrimidine biosynthesis, 147, 148*f*
Entomophaga fungus, as pest control, 339
entry stage, in animal virus multiplication cycle, **383**, 384*t*, 386*f*
envelope, viral, **372**, 372*f*
enveloped helical viruses, 373
 Influenzavirus as example, 372*f*, 373
enveloped polyhedral viruses, 373
 herpes simplex virus as example, 373, 387*f*
enveloped viruses, 372*f*, **373**
 biguanide disinfectants and, 197
 biocidal resistance and, 203, 203*f*
 helical, 372*f*, 373
 polyhedral, 373, 387*f*
 quats active against, 199, 204*t*
environmental microbiology, 766–792
 issues in biotechnology, 268
 oil-degrading bacteria and oil spills, 33*b*
environmental niches
 giant bacteria and, 326
 microbial diversity and, 325–326
environments, extreme, 4, 275
enzyme immunoassay (EIA), 514, **677**, **755**
enzyme poisons, 120
enzyme-linked immunosorbent assay (ELISA), **287**, 288*f*, **514–516**, 518*f*
 direct ELISA tests, 514, **515**–516, 518*f*
 indirect ELISA tests, 514, **516**, 518*f*
enzyme-substrate complex, **115**, 115*f*, **117**–118, 118*f*
enzymes, 42, **115–121**
 amylases, 40
 as biological catalysts, **115**
 bacterial virulence and, 432–433
 biochemical tests to detect presence of, 138–139, 139*f*
 cellulases used in "stone-washed"-jeans production, 3*b*
 classification, by type of chemical reaction catalyzed, 116, 116*t*
 coagulases, **432**
 cofactors, **116**, 116*f*, 160
 collagenase, **433**
 collision theory and, 115
 components of, 116–117, 116*t*
 denaturation of, **119**, 119*f*, 194*t*
 efficiency of, 116
 extracellular, 92, 92*f*, 321
 extracellular (coenzymes), 432
 filtration used to sterilize, 191
 genes' relationship to, 16
 hyaluronidase, **432**–433
 in bacterial plasma membranes, 91
 in cytoplasmic fluid of prokaryotes, *vs.* in organelles of eukaryotes, 101
 in DNA replication, 212–213, 215*t*
 inhibitors of, **120**–121, 120*f*, 121*f*
 kinases, **432**
 light-repair (photolyases), 215*t*, **230**
 mechanism of action, 117–118, 118*f*
 mechanism that prevent synthesis of, 221–226
 metabolic pathways and, 123
 metabolic pathways of cells and, 114
 microbial, manipulated to synthesize new substances, 2
 naming of, 116
 phage lysozyme, **380**
 photolyases, 215*t*, **230**
 produced by bacteria that break down large molecules for crossing plasma membrane, 92
 regulation mechanisms
  induction, **224**, 224*f*
  repression, **224**, 224*f*
 restriction. *See* restriction enzymes
 role in coordinating anabolic/catabolic reactions, 147
 shape/molecular weight/structure of, 45, 46*f*, 116, 187
 specificity of, 116
 streptococci-produced, and tissue destruction, 318
 synthesis, factors influencing, 118
 synthesis of, 222
 temperature and, 115
 turnover number and, **116**
 viral, 379, 380, 385
eosin dye, 69
eosinophils, **454**, 455*t*
 adhering to parasitic fluke larvae, 492*f*
 as second line of defense, 450*f*
 histamine released by, 454
 produce toxins against parasites, 454
 staining and, 454
epidemic disease, **406**
epidemics, emerging infectious diseases and, **19**–21
epidemiologists, role in hospital infection control, 416
epidemiology, **418–422**, 419*f*, 421*t*
 analytical, **419**–420
 case reporting, 420
 descriptive, **419**
 early efforts of Nightingale, Semmelweis, Snow, 418
 examples of epidemiological graphs, 419, 419*f*
 experimental, **420**
 information sources in, 420
 MMWR's importance to, 420
 morbidity rate/mortality rate and, **420**
 notifiable infectious diseases reports, **420**
 public health departments, state and federal, 420
 topics of study, 418–419
 types of investigations, 419
epidermal growth factor (EGF), as rDNA product in medical therapy, 260*t*
epidermis, **451**, 451*f*, **585**, 585*f*
 as physical barrier to microbes, 451, **471*t***, 584
 cutaneous mycoses and, 337, 338*t*
 fungal infections, 337
Epidermophyton
 as pathogenic fungi, 338*t*
 reservoirs/transmission method, 410*t*
Epidermophyton (fungus), 600
epiglottis, 675, 675*f*
epiglottitis, **452**, 471*t*, **676**–677, **681*b***
 Haemophilus influenzae causing, 311, 613
epinephrine, anaphylactic shock and, 524
epithelial cells
 of mucous membranes, 451
 of skin, 451
epithelium
 cathelicidins produced by, 470
 defensins produced by, 470
epitopes (antigenic determinants), **478**, 479*f*, 480, 484
EPO (erythropoietin), genetically engineered to treat anemia, 260*t*
EPS (extracellular polymeric substance), **81**
epsilonproteobacteria, 301, **312**
 important genera/special features, 301*t*
Epstein, Michael, 10*f*, 391
Epstein-Barr (EB) virus (*Lymphocryptovirus*)
 Burkitt's lymphoma associated with, **643*b***, 655–656, 657*f*
 cancer and, 391
 incubation period, 430*t*
 infectious mononucleosis caused by, 430*t*, **643*b***
 portals of entry, 430*t*
 pregnancy and, 760
 reactivated in HIV/AIDS patients, 542
 typical U.S. prevalence of antibodies against, 567, 567*f*
Epulopiscium genus/spp., 301*t*, **316–317**, 316*f*
 as giant bacteria, 301*t*, 316–317, 326
equilibrium, in simple diffusion process, 91, 92*f*
equivalent treatments, **191**
ER (endoplasmic reticulum), **103**, 103*f*
ergot poisoning, 729–**730**, **734*b***
ergot toxin, **443**
 as natural source LSD (lysergic acid diethylamide), 443
ergotism, 443
Erwinia genus/spp., 300*t*, **311**
 as plant pathogens, 300*t*, 311
erysipelas, 319, **591**, 591*f*
 caused by *Streptococcus pyogenes,* 406
 rash caused by, 592*b*
Erysipelothrix rhusiopathiae, 283*b*
erythema infectiosum. *See* fifth disease
erythroblastosis fetalis. *See* hemolytic disease of newborn
erythrocytes (red blood cells), **455*t***
 agglutination by envelope spikes of influenza viruses, 376*t*
 blood agar and, 168, 168*f*
 functions of, 455*t*
 in inflammatory response, 461*f*
 parasites
  Babesia microti, 350
  Plasmodium vivax, 348–350, 349*f*
erythrogenic toxins, **437**, 438*t*, 441, 677
 Streptococcus pyogenes, 237, 437, 438*t*
erythrolitmin, dye used in litmus paper extracted from lichens, 340
erythromycin, 566, 566*t*
 inhibits protein synthesis, 95, 556–557, 556*f*, 558*f*, 566
 mode of action/spectrum of activity, 562*t*
 produced by *Saccharopolyspora erythraea,* 555*t*
erythropoietin (EPO), genetically engineered; used to treat anemia, 260*t*
Escherich, Theodor, 3, 10*f*
Escherichia coli, 3*f*
 adhesins on fimbriae, *Shigella* and, 431, 431*f*
 agriculturally important products genetically engineered in, 267*t*
 as causing nosocomial infections, 414, 414*t*
 as facultative anaerobe, 161
 as important biological research tool, 310
 as normal microbiota of large intestine, 403*t*
 bacteriocins produced by, 401
 beneficial activities of, 20
 biochemical tests to identify pathogenic, 139, 139*f*, 169
 cephalosporin resistance transferred to *Salmonella enterica* by, 577*b*
 chemically defined medium for growing, 165*t*
 colony-stimulating factor and, 260*t*
 competence and, 236, 253
 conjugation in, 236–237, 238*f*
 cystitis caused by, 746
 disinfectants evaluated by disk-diffusion method, 196*f*
 DNA replication by, 214, 217*f*
 E. coli 0157:H7 strain
  and laboratory tests, 87
  as an emerging infectious disease, **20**, 417*t*
  as serovars, 82
  culture media and, 169
  emerging infectious diseases and, **20**, 416
  genetic recombination and, 416
  hemolytic uremic syndrome (HUS) and, 113*f*, 210, 718
  naming of, 310 *footnote*
  outbreak tracked by DNA fingerprinting, 264*f*
  Shiga toxin gene and, 210
  tomatoes and, 310, 714, 715*b*
 Eco RI restriction enzyme used in rDNA technology, 249*t*
 endotoxins produced by and genetic engineering and, 257
 enteropathogenic strains, 440
 epidermal growth factor (EGF) and, 260*t*
 feedback inhibition in, 121, 121*f*
 gastroenteritis, **717–718**, 718*f*, 723*b*
 genetically engineered pharmaceutical products produced by, 260*t*
 genetically engineered to produce gene products (e.g., gamma interferon), 257–258, 257*f*
 genetically engineered to produce pharmaceutical products, 259, 260*t*
 genetically modified, 246*f*
 genetically modified to produce gamma interferon, 246*f*
 genetically modified, to produce human growth hormone, 247, 260*t*
 genome has been mapped, 261
 growth rate of *E. coli* on glucose, 225–226, 226*f*
 interferons produced by, 260*t*
 lactose metabolism in, 224–225, 224*f*, 225*f*
 leukemia therapy and, 260*t*
 nosocomial infections and, 414, 414*t*
 number of H antigens for, 82
 oxygen availability and, 156*f*
 pathogenic strains, Shiga toxin and, 382
 pathogenic, TaqMan system uses PCR to identify, 291
 pentose phosphate pathway and, 127
 plasmid vector pUC19 used for cloning, 251*f*
 position in taxonomic hierarchy, 280*f*
 pyelonephritis caused by, 746
 RecA protein (STM image), 65*f*, 68*t*
 Salmonella strains and, effects on host's plasma membrane, 433, 433*f*
 size of, compared to selected viruses, 369*f*
 that causes diarrhea in infants and travelers, 239
 transduction in, 237, 239*f*
 traveler's diarrhea enterotoxins and, 438*t*
 twitching motility in, 83
 used in indigo production, 3*b*
 using to synthesize gene products, 257–258, 257*f*
Escherichia genus/spp., **310**
 as enterics, 285, 300*t*, 310
 as pathogens, 300*t*
 fermentation and, 134*f*
 in taxonomic hierarchy, 280*f*, 300*t*
 oxidase test and, 139
 resistance plasmid R100 and, 239, 240*f*
ESCs (embryonic stem cells), **535**, 535*f*

ester group, 38t
ester linkage, 40, 41f
ETEC (enterotoxigenic *E. coli*), **717**, **723b**
ethambutol, **563**
  mode of action/spectrum of activity, 562t
ethanol, 37
  Acetobacter and, 139, 303
  as a biofuel, **807–808**
  as antimicrobial agent, 198, 198t, 200f, 204t
  as fermentation end-product, 134f, 135b, 136f, 137t, 332
  as primary metabolite of industrial fermentation, **803**, 804f
  biotechnology and, 246
  Gluconobacter and, 303
  industrial uses for, 137t
ether group, 38t
ethical issues, of genetic engineering, 268
ethylene oxide gas, 200–201
  vs. hydrogen peroxide gas, 202
ethylenediaminetetraacetic acid (EDTA), 89
etiology, **400**
Eucalyptus species, infected by *Phytophthora cinnamoni,* 344
Euglena (alga), flagellum of, 98, 100f
Euglenoids, **350**–351, 351f
  algae and, 346, 351
  as photoautotrophs, 350, 351
  habitat, 341f
  red eyespot of, 350, 351f
Euglenozoa, **350**–351, 351f
  as photoautotrophs, 350–351, 351f
  position in evolutionary tree, 275f
Eukarya (domain), 6, **274**, 275f, 276t
  algae, 340–345. *See also* algae
  Archaea compared to, 276t
  arthropods as vectors and, 361–363
  Bacteria compared to, 276t
  fungi, 330–339. *See also* fungi
  helminths, 352–361. *See also* helminths
  kingdoms in, 275f, 281–282
  protozoa, 345–351. *See also* protozoa
eukaryotes/eukaryotic cells, **4**, 76, 99f
  active transport processes used by, 94
  anatomy
    cell wall, **98**, 99f
    cilia and flagella, **98**, 99f, 100f
    flagella and cilia, **98**, 99f, 100f
  as vehicles for expressing genetically engineered genes, 258
  characteristics that distinguish, 77
  classification of, 275f, 281–282
  cloning genes from, 254–255, 255f
  DNA arrangement of, 76
  evolution, 106, 275–278, 275f, 276t
    Cyanophora paradoxa as modern example of, 277, 277f
  genetic recombination in, 234
  major differences among, 330t
  mutation identification and, 231
  nucleus, and prokaryotic *Gemmata obscuriglobus,* 322, 322f
  organelles of, compared to prokaryotic cells, 275, 276t, 277
  origin of, 10f, 106, 275–278, 275f, 276t, 277f, 322
  photosynthesis in, compared to prokaryotes, 145t
  photosynthetic, 6
  plasmids and, 238
  principal differences between prokaryotic cells and, 101t, 106
  prokaryotic cells compared to, 77, 81, 82, 95, 101t
  protein synthesis in, 220, 222f
  ribosomal differences, 95
  Saccharomyces cerevisiae as best understood genome, 258
  size of, *vs.* prokaryotic cells, 101t
  typical cell structure, 99f
eukaryotic pathogens, over half world's population infected with, 329
eukaryotic species, **279**
  taxonomic definition of, *vs.* prokaryotic species, 279, 281
European corn borer, 267
Euryarchaeota, **302t**
  gram-positive to variable archaea, 302t
eutrophication, **779**
evaporated milk, algae-produced thickeners used in, 343
evaporating ponds, solar, extreme halophiles (archaea) found in, 325
evolution
  Carsonella ruddii's small genome and, 326
  cyanobacteria fossil evidence, 314–315
  definition of, 274
  degenerative, 320
  EIDs and, 19
  eukaryotic, 106, 107b
  genetically modified crops and, 268
  microbial pathogenicity, virulence and, 428
  mutation rates and, 231
  natural selection and, 274
  nucleoplasm/nucleoplasmic cells as universal ancestor, 275, 275f, 277–278
  of the three domains, 275–277, 275f, 276t, 277t
  phylogeny and, **274**
  species adaptation and genetic mutation, 231
  systematics and, **274**
  Wolbachia's unique biology and its influence on, 307b
evolutionary relationships
  cladograms to map, 293, 294f
  ribotyping to determine, 292
  study of, 274
evolutionary tree
  the three-domain system, 274, 275f
  Thermotoga maritima near origin or "root" of, 277
exanthem rashes, **586**, 587f
exchange reactions, 38
exergonic chemical reactions, **32**, 114, 214
exergonic hydrolysis, 214
exfoliation, 588, 588f
exfoliative toxins, 239, 588
exoenzymes (extracellular enzymes), virulence and, 432
exons, 215t, **220**, 222f, **254**–255, 255f
exonucleases, 215t
exotoxins, 42, **434**–437, 435f
  altered (inactivated) as toxoids, 435, 439t
  as enzymes, 434
  as one of most lethal substances known, 435
  as proteins, 434, 437
  diseases caused by, 438t
  endotoxins *vs.*, **439t**
  genes of carried on bacterial plasmids or phages, 434
  lethal dose and, 439t
  naming of, 435
  properties of, 434–435, 435f
    compared to endotoxins, 439t
  representative examples of, 436–437
  symptoms induced by, 435, 439t
  toxicity of, 439t
  types of, 435–436
    A-B toxins, **435**, 436f, 438t
    exotoxin A and streptococcal M proteins, 591
    membrane-disrupting toxins, **436**, 438t
    superantigens, **436**, 438t
experimental epidemiology, **420**
exponential growth (log) phase, in bacterial growth curve, **173**, 173f
expression, gene, 211, 213f, 220–221
extracellular antigens, in humoral immunity, 482, 482f, 496f
extracellular enzymes, in facilitated diffusion, 92, 92f
extracellular enzymes (exoenzymes), virulence and, 432
extracellular polymeric substance (EPS), **81**
extrachromosomal genetic elements (plasmids), 95
extreme acidophiles, **325**
extreme halophiles, 4, **159**, 275, 275f, 325
  phylogenetic relationships, 281f
  position in evolutionary tree, 275f
extreme thermophiles. *See* hyperthermophiles
extremophiles, **325**, **767**
extremozymes, **767**
Exxon Valdez oil spill (1989), 17
  bacterial cleanup of, 33b
eyelids, 451, 452f
eyepiece (ocular lens), of compound light microscope, **56**, 56f
eyes
  infections
    inflammation of (Clinical Focus), 440b, 440f
    Moraxella bacteria and, 308
    toxic anterior segment syndrome (TASS), 440b
  lacrimal apparatus and tears produced by, 451–452, 452f
  microbial diseases of, 603–605, **604b**
  normal microbiota of, 402t
eyespot
  of euglenoids, 350, 351f
  of green algae, 342f

F cells. *See* F factor
F factor (fertility factor), **236–237**, 238f
  as conjugative plasmid, 238
F factor (fertility factor) cells, 84, 95, 236–237, 238f
FA tests. *See* fluorescent-antibody (FA) tests
fabric, produced using microbiological methods, 3b
facilitated diffusion, **91**–92, 92f
FACS (fluorescence-activated cell sorter), **514**, 516f
factor B complement protein, 466, 466f
factor D complement protein, 466, 466f
factor P (properdin) complement protein, 466, 466f
Factor VII, genetically engineered; used to treat hemorrhagic stroke, 260t
Factor VIII, genetically engineered; used to treat hemophilia/improve clotting, 260t
facultative anaerobes, **161**, 161t
  fungi as, 330t
facultative halophiles, **159**
FAD, in Krebs cycle, 127, 128f
FAD (flavin adenine dinucleotide)
  enzymatic functions, **117**
  in electron transport chain, 129–130, 129f
  oxidative phosphorylation and, 122
FADH
  in electron transport chain, 129–130, 129f
  in Krebs cycle, 124, 125f, 127, 128, 128f
fallopian (uterine) tubes, 744, 744f
famciclovir, 569
FAME (fatty acid methyl ester) test, to identify microorganisms, 288
family name (taxonomic), defined, **279**, 280f
farm animals
  antibiotics in animal feed, 554, 562t, 565, 575, 577b
    antibiotic resistance and, 577b
    linked to human disease, 577b
  antihelminthic (ivermectin) to treat, 571
  as disease reservoirs, 410t
Fasigyn (tinidazole), 571
fastidious microorganisms, 165, 166t
  chemically defined medium for growing, 165, 166t
  transport media for pathogenic, **284**
fatal familial insomnia, 393
fats (triglycerides), 40, 41f, 136
  fat molecule formation, 40, 41f
  in lipid catabolism, 136, 136f, 138f
  synthesis of, 146, 147f
fatty acid profiles (FAME), **288**
fatty acids, 40, 41f
  bacteria, petroleum products and, 33b, 136
  cis fatty acids, 40, 41f
  in biosynthesis of lipids, 146, 147f
  in lipid catabolism, 136, 136f, 138f
  saturated, 40, 41f, 42f
  synthesis, biotin and, 117t
  trans fatty acids, 40
  unsaturated, 40, 41f, 42f
Fc (stem) region of antibodies, 479, 480f, 481, 523–524, 524f
fecal-oral cycle, **705**
feces, 706
  phenolics to disinfect, 195
feedback inhibition/end-product inhibition, **120**–121, 121f
  in regulation of amino acid production, 121
  in regulation of bacterial gene expression, 221–226
feline AIDS, 377
feline leukemia virus (FeLV), 391
FeLV (feline leukemia virus), 391
fermentation, **9**, 124, 125f, **132–134**, 134f, 135b, 136f
  ability, as method of identifying bacteria, 285, 285f
  aerobic respiration *vs.*, 137t
  alcohol, **135b**, 136f
  anaerobic respiration *vs.*, 137t
  end-products, 133, 134f, 135b, 136f, 137t
  foundation figure, **125f**
  industrial uses for different types of, 137t
  lactic acid, **135b**, 136f
  of mannitol, 168–169, 169f
  of milk products, 798–799, 799f
fermentation test, **138**–139, 139f
ferns, as eukarya, 6
ferritin, 434, **470**
ferrous iron, as energy source, 145
fertility factor (F factor), **236**–237, 238f
  as conjugative plasmid, 238
fertility factor (F factor cells), 84, 95, 236–237, 237f, 238f
fetal calf serum, 495
fetal genetic screening, 262
fetus
  IgG antibodies and, 481t
  immune system tolerance of, 534–535
  rejection as nonself and, 534–535
fever, 450f, **463**, **472t**
  as pyrogenic response to endotoxins, 437, 438f
  as second line of defense, 450f, 463, 472t
  Babesia microti causing, 350
  chills of, 463
  complications of, 463
  cytokines and, 437, 438f, 463
  death and, 463
  endotoxins and, 437, 438f, 441, 463
  Plasmodium vivax causing, 349–350
  prostaglandin synthesis and, 437, 438f
  shivering and, 463
  Streptococcus pyogenes causing, 406
  tumor necrosis factor alpha (TNF-α) and, 463
fever blisters. *See* cold sores
fibrinogen, 460

fibrinolysin (streptokinase), 432
fibrosis, in forming scar tissue, 463
fibrous proteins, shape/structure of, 45, 46f
fifth disease (erythema infectiosum), 375t, 383, **600**
 human parvovirus B19 causing, 589b, 600
 macular rash caused by, 589b
50S ribosomes, 95, 95f
filament of flagella, 81, 82f
filamentous algae, 341, 341f
filamentous bacteria
 reproductive methods, 171, 320
 soil inhabitants, actinomycetes and, 320
filamentous fungi, advantages of, 320
filamentous streamers, biofilms and, 163
Filoviridae, characteristics/important genera/clinical features, 376t
Filovirus, 373f, 376t
 Ebola virus as, 373, 373f
filterable agents, 367, 368
filterable viruses, 191
filters
 HEPA, **191**
 membrane, **191**, 191f
filtration
 sterilizing liquids or gases by, 185
 to control microbial growth, 177f, 184f, 191, 191f, 194t
 to count bacteria, **175**, 177f
 water treatment, **782–783**, 782f
fimbriae/fimbria, **83–84**, 84f, **432**
 of enterics, 309
 of *Neisseria gonorrhoeae* increase its virulence, 432
 of prokaryotic cells, 80f, 306f
 of uterine (fallopian) tubes, 744f
final electron acceptors during fermentation/aerobic respiration/anaerobic respiration, 133, 134f, 137t
fine focusing knob, of compound light microscope, **56**, 56f
fingernails, cutaneous mycoses and, 337, 539
Fire, Andrew, 15t
Firmicutes (low G + C ratios), 281f, 301t, 315–316, **316–320**
 important genera/special features, 301t
fish
 food allergies and, 525
 killed by toxic marine algae, 344
 transmitting parasitic worms, 361
FISH (fluorescent in situ hybridization), **292–293**, 294f
 to identify new bacteria in marine mammals, 283b
 to identify tooth bacteria, 707
Fisher, Edmond H., 15t
fission yeasts, **332**
FITC (fluorescein isothiocyante), 61
five-kingdom system, Whittaker's proposal of, 274
5′→3′ direction, 214, 214f, 215f, 216, 218f, 219
5-bromouracil, 229–230, 229f
fixed macrophages (histiocytes), **457**, 638, 639f
fixed/fixing specimens, **68**
 electron microscopes and, 63–64
flaccid paralysis, caused by botulinum toxin, 437, 616
flagella staining, **71**, 72f, 72t
 negative staining, 63
flagella/flagellum, **71**, **81**
 bacterial, 4, **71**, 80f, **81–82**, 81f, 82f
  arrangements of, 81, 81f
  in gram-negative *vs.* gram-positive bacteria, 88t
  staining for, 63, 71, 72f, 72t
  structure, 81–82, 82f
 energy use and, 82
 eukaryotic cell, 99f
 eukaryotic cell, **98**, 99f, 100f
 evolutionary aspects, 106, 107b
 in prokaryotic *vs.* eukaryotic cells, 101t
 motility and, 82–83, 83
 movement in eukaryotic *vs.* prokaryotic cells, 98, 100f
 of alga, 98, 100f
 of archaezoans, 347–348, 347f
 of *Campylobacter,* 312
 of *Chilomastix* protozoa, 347f
 of dinoflagellates, 344f
 of *Euglena,* 350, 351f
 of giant bacteria *Epulopiscium fishelsoni,* 316
 of *Helicobacter,* 312, 313f
 of oomycote spores (zoospores), 344, 345f
 of *Proteus mirabilis,* 310, 311f
 of protozoa, 6, 98, 107b
 of *T. sphaerica,* 107b
 of *Trichomonas vaginalis,* 347f
 of unicellular green algae, 342, 342f
 origin of in eukaryotes, 106
 preemergent, of *Euglena,* 351, 351f
flagellin, 81
Flagyl (metronidazole), to treat vaginitis caused by *Trichomonas vaginalis,* 571
flaming (dry heat sterilization), **191**, 194t
flat sour spoilage, of canned foods, **795**, 796t
flatworms (Platyhelminthes), 6, 353, **356–358**, 356f, 357f, 358f
 cestodes (tapeworms), **356–358**, 358f, **361t**
 trematodes (flukes), **356**, 356f, **361t**
flavin, 129
flavin adenine dinucleotide (FAD), **117**
flavin mononucleotide (FMN), **117**
 in cellular respiration, 129, 129f
Flaviviridae, characteristics/important genera/clinical features, 376t
Flavivirus, **376t**, 660b
 reservoirs/transmission method, 410t
 St. Louis encephalitis caused by, 376t, 625–626, **628b**
 West Nile virus epidemiological tracking and, 223b
flavoproteins, 117t, **129**, 129f
flavoring uses, for fermentation end-products, 137t
flea-transmitted diseases
 endemic murine typhus, 362t
 plague and *Yersinia pestis,* 311
 typhus and *Rickettsia typhi,* 303
fleas
 as vectors, 362t, 410t, 413t, 648
 diseases transmitted by, 303, 311, 362t
Fleming, Alexander, 10f, 12, 14f, 453, 554
"flesh-eating" bacteria (necrotizing fasciitis), **20**, 287, 318, 321, 422b, 591, 591f
flies
 diseases transmitted by, 354t, 362, 362t
 that are vectors, 362, 362t
floc formation in activated sludge systems, 784, 785f
flocculating agents, 197
flocculation, **782**
flora. See normal microbiota
Florey, Howard, 10f, 14f, 554
flow cytometers, 288, 514, 516f
flow cytometry, 288
fluconazole, 568–569
flucytosine, 564t
fluid mosaic model, **90**
flukes (trematodes), **356**, 356f, **361t**
 immune system attack on, 491, 492f
 praziquantel to treat, 557t
fluorescein isothiocyanate (FITC), 61
fluorescence, **61–62**, 61f
fluorescence microscopy, **61–62**, 61f, 66t
fluorescence-activated cell sorter (FACS), **514**, 516f
fluorescent dyes (fluorochromes), 61
fluorescent in situ hybridization (FISH), **292–293**, 294f
 to identify new bacteria in marine mammals, 283b
fluorescent treponemal antibody absorption (FTA-ABS) test, 61f, 62, **755**
fluorescent-antibody (FA) tests, **61–62**, 62f, 66t, 350, **513–514**, 515f
fluoride
 calcium and, 120
 magnesium and, 120
fluorochromes (fluorescent dyes), 61, 61f
fluoroquinolones (FQ), 422b, **567**, 577b
 in chicken feed, 577b
 mode of action/spectrum of activity, 563t
 Neisseria gonorrhoeae resistant, 750, 751b
 resistant *Campylobacter jejuni* and, 577b
FMN (flavin mononucleotide)
 enzymatic functions, **117**
 in electron transport chain, 129–130, 129f
focal infection, **407**
folic acid
 coenzymatic functions, 117t
 synthesis of, 120
foliose lichens, 339, 340f
folliculitis, **588**, **592b**
fomites, 410t, **411**, 411f, 415
fomivirsen, 579, 658
food acquisition methods
 by absorption *vs.* ingestion, 330t, 333
 of algae, 330t, 342
 of amoebas, 346
 of animals, 282, 333
 of archaea, 325
 of bacteria, 330t
 of flukes, 356
 of fungi, 281, 330, 330t, 333
 of helminths, 330t
  tapeworms, 356
 of parasitic helminths, 354
 of plants, 282
 of protozoa, 330t, 346
 of viruses, 282
food allergies, 525–526
food canning
 home, 188, 190
 industrial, 794–795, 794f, 795f
food poisoning. *See also* gastroenteritis
 algae-associated, 329, 341
 botulism. *See* botulism
 endospores and, 96
 exotoxins causing, 438t
 mushrooms, 729–730, 734b
 Salmonella and. *See* salmonellosis
 shellfish, 344, 354t, 444
 staphylococcal, 318, 438t, **711–712**, 711f, 722b, **722b**
 symptoms, 438t
 vectors transmitting bacteria that cause, 412
food preservation
 by adding antibiotics, 200
 by aseptic packaging, 795–796, 795–797, 797f
 by chemical additives, 199–200, 204t
 by commercial sterilization, **185**, 185t, 188, 594f, **794–**795, 794f, 795f
 by heat, 185, 185t, 188
 by high-pressure processing, 797
 by industrial canning processes, 794–795, 794f, 795f
 by irradiation, 796–797, 797t, 798f
 HACCP system to prevent contamination in, 794
 osmotic pressure and, 159–160, 160f, 192
 temperatures and, 157–158, 157f, 158f, 159f
food production, 2
 Aspergillus niger fungus in citric acid production, 339
 disinfectants used in, 197
 endospore-forming bacteria problems for, 98
 genetically engineered products, 267t
 inspection agencies, 794
 microbes used in, 797–800
  cheese making, 798–799, 799f
food spoilage
 acidic foods and, 795
 bacterial *vs.* mold damage, 339
 Clostridium bacteria and, 618, 795
 commercial sterilization to prevent, **185**, 185t, **794–**795, 794f, 795f
 fermentation and, 9, 135b
 of canned foods, **795**, 796t
  flat sour, **795**, 796t
  thermophilic anaerobic, **795**, 796t
 pasteurization to prevent, **9**, 185, 185t, **190–**191, 194t
 pH and, 159
 Pseudomonas bacteria and, 308
 refrigeration and, 191–192, 308, 319, 615, 618
 relationship between microbes and, 9
 Salmonella bacteria and, 310
 temperature and, 157–158, 157f–159f
 thermophilic anaerobic spoilage, **795**, 796t
food thickeners
 algin (from brown algae), 342
 carrageenan (from red algae), 342
 xanthan (from *Xanthomonas campestris*), 801b
food vacuoles
 in digestive system of protozoa, 346
 of *Amoeba proteus,* 348f
 of *Chilomastix* protozoa, 347f
 of *Paramecium,* 350f
food-associated infections, phage typing to trace, 288, 290f
foodborne illness
 Listeria monocytogenes and, 615
 salmonellosis, 310, 715b
foodborne infections
 Clostridium perfringens and, 316
 E. coli enterotoxins causing, 310
 epidemics, *E. coli* O157:H7, 20, 82
 hemolytic uremic syndrome (HUS), 113f
 incidence in U.S., 705
foodborne transmission of disease agents, 411–412
foods
 freeze-dried, 192
 microbes used in production of, 247
forensic medicine, DNA fingerprinting and, 262, 264f
forensic microbiology, **262**, 264
forensics, uses of cloned DNA in, 258
forespore, 96, 97f
formaldehyde, 200, 205t
formalin, 200
formic acid, as fermentation end-product, 134f
formylmethionine, 219, 276t
fossilized materials, DNA studies of and science of taxonomy, 262, 264
fossils
 Bacillus sphaericus survived embedded in, 278
 cyanobacteria and atmospheric oxygen, 314–315
 cyanobacteria-like, 277, 278f
 of prokaryotes, 275, 277, 278f
 oldest known, 275, 277
 phylogenetic relationships, classification and, 277–278, 278f
fowl cholera, 311
 caused by *Pasteurella,* 311

foxes
  as disease reservoirs, 410t
  reported cases of rabies in, 624f
FQ. *See* fluoroquinolones
fractures, genetic engineering product to heal, 260t
frameshift mutagens, 230
  carcinogens and, 230
frameshift mutations, **228**, 228f
Francisella genus/spp., 300t, **308**
Francisella tularensis, 308
  as potential biological weapon, 642, 642f, 649b
  can remain dormant within phagocytes, 459, 643
  tularemia caused by, 308, 642–643, **643b**
Frankia genus/spp., 301t, **320**
  actinomycetes informal name for, 320
  alder tree root pathogen, 320
  as symbiotic nitrogen-fixers, 301t, 320
Franklin, Rosalind, 47
free (extracellular) antigens
  B cell activation and, 482, 482f
  in humoral immunity, 482, 482f, 496f
free nucleotides in DNA replication, 213–214, 213f, 214f
free radicals, 201, 230, 260t
free ribosomes, 102
free (wandering) macrophages, **457**
freeze-drying (lyophilization)
  to control microbial growth, 194t
  to preserve bacterial cultures, **170**
freezing temperatures
  bacteria and, 192
  food spoilage and, 158, 158f, 192
freshwater microbiota, 2, 300t, 304, 305, **776–777**
frogs, deformed, 356f
fructose, 39, 39f, 146, 146f
  how it crosses plasma membrane, 92, 92f
  in dehydration synthesis, 39f
  in hydrolysis, 39f
  microbes in manufacture of, 246
fruit flies, *Wolbachia* bacteria and, 307b
fruit juices
  fermentation and, 137t
  preserved by high pressure techniques, 192
fruiting bacteria, 312, 313f
fruits, genetically engineered MacGregor tomatoes, 267, 267t
fruits and vegetables, PAA for washing/disinfecting, 202
FTA-ABS tests, 61f, 62, **755**
fuel products, fermentation and, 137t
fugal spores, localized anaphylaxis and, 525
fully human antibodies, **509**
fulminating disease, 602
fumaric acid, 128f, 149f
functional groups, **37–38**, 38t
fungal blights on trees, 339
fungal diseases. *See also* fungal infections
  antifungal drugs to treat, 558
  of digestive system, 729–730, **734b**
  of nervous system, **617b**, 626–627
  of reproductive systems, 758–759, **759b**
  of respiratory system, 695–698, **699b**
  of skin, **600–602**, 600f
    rashes caused by, 589b
fungal infections (mycoses), **335–339**, 338t
  by *Coccidioides immitis,* 13
  of skin and nails, **600–602**, 600f
fungal zoonoses, 410t
Fungi (kingdom), **4**, 6, 274, 275f, **281**, **330–339**
  characteristics of, 330t, 331–335, 331f, 332f
  develop from spores or hyphae, 281, 331f
  energy sources of, 281, 330
  lichens and, **339–340**, 340f
  nutritional needs of, 281, 330, 330t
  organisms included in, 281
  position in evolutionary tree, 275f
  position in taxonomic hierarchy, 280f
fungi-farming ants, 330
fungi/fungus, 2, **4**, **281**, 329–339, 330t, 338t
  alcohols and antimicrobial activity against, 197, 204t
  anamorphic, **335**
  anamorphs (asexual fungi), 337
  antibiotics derived from, 12, 12f, 249
  antibiotics produced by, 555t
  as biological controls of pests, 339
  as carbon recyclers, 17
  as chemoheterotrophs, 330, 330t
  as decomposers of plant matter, 330
  as eukaryotic cells, 6, 76, **281**, 330t
  as kingdom in Domain Eukarya, **4**, 6, 274, 275f, 330–339
  as teleomorphs, **335**
  as unicellular/multicellular microbes, **4**
  asexual (anamorphs), **335**
  asexual spores of, 330t, 331f, **332–333**, 334 f-337f, 338t
  bacteria compared to, 330t, 333
  beneficial activities of, 17, 330
  beneficial uses of, 339
  biofilms and, 163
  biotechnology uses for, 339
  branch of microbiology that studies, 13, **330**
  cell walls of, 98
  cellular structure of, 4, 5f
  cellulases produced by, 40
  characteristics of, 330t, 331–335, 331f, 332f
  chitin in cell wall, 40, 330t
  conjugation (Zygomycota), **333**, 335f
  dimorphic, **332**, 332f, 338t
  diseases caused by, 335–339
  economic effects of, 339
  emerging infectious diseases caused by, 417t
  filamentous, 320
  fleshy, 331, 331f
  human uses of, 330
  hyphae of, **331–332**, 331f
  identification methods for, 331
  identification of, 282
  in lichens, 339–340, 340f
  in soil, 409
  iodine active against, 197
  ketoconazole to treat, 557t
  life cycle of, 332–333, 334f
  low-moisture environments and growth capability, 333
  medically important, 333–335, 338t
  metabolism of, 330, 330t
  moist heat sterilization to kill, 188
  mucor, 5f
  mycology as study of, **330**
  nutrition of, 4
  nutritional adaptations, 333
  nutritional classification of, 143, 143f
  osmotic pressure resistance of, 333
  pathogenic, 330, 339, 443
    dimorphism and, 332
  penicillin produced by, 12, 12f
  Penicillium, rDNA technology and, 249
  pH ranges tolerated by, 37, 333
  Pneumocystis classification and, 285
  rapid identification tests for, 286
  reproduction in, 4
    aerial hyphae and, 331, 331f
    asexual spores, 330t, **332–333**, 334 f-337f
    filamentous fungi, 332
    sexual spores, 330t, **333**, 335 f-337f
  resistance to chemical biocides, 203f
  rules for naming and, 279
  sexual spores of, 330t, **333**, 335 f-337f
  silkworm disease and, 11
  skin's keratin no obstacle to, 429
  spores of, 331f, **332–333**, 334f
    asexual, **332–333**, 334f
    resistance to chemical biocides, 203f
  toxin-producing, 339, 443
  vegetative structures of, 331–332, 332f
fungicides, 186
fungistats, 200
fungus. *See* fungi/fungus
furious rabies (in animals), **623**
furuncle (boil), **588**
Fusarium (fungus), toxin of, 443
fusiform bacteria, 324, 324f
fusion, in viral multiplication, **384**, 384t
fusion inhibitors, **571**
  to treat HIV infection, 548
Fusobacteria, 302t, **324**
Fusobacteriales, important genera/special features, 302t
Fusobacterium genus/spp., 302t, **324**, 324f
  as normal microbiota of large intestine, 302t, 324f, 403t
  as normal microbiota of mouth, 403t
  in gingival crevices, 324

G + C ratios, high, 281f, 301t, 315–316. *See also* Actinobacteria
G-CSF (granulocyte-colony stimulating factor), 492
GAE (granulomatous amebic encephalitis), 617b, **629**
Gajdusek, Carleton, 631
gal gene, in specialized transduction, 382, 383f
galactose, 39
  how it crosses plasma membrane, 92, 92f
galactose fermentation, 382
galactose-binding lectins, 348
Gambierdiscus toxicus (dinoflagellate), ciguatera disease and, 344
gametes (gametocytes), **346**
  in life cycle of *Rhizopus,* 335
  of plasmodial slime mold, 353f
gamma globulin, **495**
gamma interferon (human), 469
  as rDNA product in medical therapies, 260t, 469
  E. coli genetically engineered to produce, 246f, 257f
  induces neutrophils/macrophages to kill bacteria, 469
gamma rays, 192, 193f
  as mutagens, 228, 230–231, 230f
  in food irradiation, 797–798, 798f
gammaproteobacteria, 280f, 300–301t, **306–312**
  important genera/special features, 300–301t
ganciclovir, 569
  mode of action/uses, 564t
gangrene, 96, **646**, 646f, **668b**
  Clostridium perfringens causing, 646, 646f, 668b
  gas, 316, 430t, 433, 438t, 439t, **646**, 647f
  portals of entry, 429, 430t
Gardasil (HPV vaccine), 260t, 391, 503t, 538, 758
Gardnerella genus/spp., 301t, **320**
  as gram-variable, pleomorphic bacteria, 320
  as human pathogens, 301t, 320
Gardnerella vaginalis, vaginitis caused by, 320, 756, 756f
gas formation, in carbohydrate catabolism, 138, 139f
gas gangrene, 316, **646**
  Clostridium perfringens causing, 316, 430t
  collagenase enzyme of *Clostridium* help spread, 433
  exotoxin causing, 438t, 439t
  hyperbaric chambers to treat, **646**, 647f
  incubation period, 430t
  symptoms, 438t
GAS (group A streptococci), 590–591, 591f, 640
gas vacuoles, **96**, 314
gas vesicles, 96
gaseous chemosterilants, 200–201
gastric juice
  as chemical defense against pathogens, **453**, 471t
  pH of, 453
  toxins not destroyed by, 453
gastritis, *Helicobacter pylori* and, 453
gastroenteritis, **710**
  Bacillus cereus, **720**–721, 723b
  Campylobacter, **718**, 723b
  Clostridium perfringens, **720**, 723t
  Escherichia coli, **717–718**, 718f, 723b
    traveler's diarrhea, **717**, 723b
  genomics used to trace outbreaks, 262, 266b
  hepatitis E virus and, 375t
  norovirus-associated, **728–729**, **729b**
  rotovirus-associated, **728**, **729b**
  Salmonella, 310, 410t, **712–714**, 713f, 714f, **722b**
  Vibrio parahaemolyticus and, 309
  viral, **728–729**, **729b**
  Yersinia, **720**, **723b**
gastrointestinal anthrax, **645**
  virulence of, 430
gastrointestinal (GI) tract. *See* digestive system
gastrointestinal infections, following antibiotic therapy, 401
gastrointestinal tract
  as portal of entry, 429, 430t, 445f
  parasitic helminths and, 361t
  physical forms of defense against microbes, 452
gatifloxacin, 567
  mode of action/spectrum of activity, 563t
gauze, quat antiseptics neutralized by, 199
GB virus, 386, 728
GBS (group B streptococci), neonatal sepsis caused by, 640
gel electrophoresis
  in Southern blotting, 262, 263f
  pulsed-field (PFGE), 718
  to separate serum proteins, 495, 495f
  to view amplified DNA, 251, 290
gemifloxacin, 567
Gemmata genus/spp., 302t, **322**
  cell walls of, 302t
  observations of true nucleus in, 277, 302t, 322, 322f
Gemmata obscuriglobus, 322, 322f
  as model for origin of eukaryotic nucleus, 322
GenBank, 262
gene expression, 211, 213f, 220–221. *See also* transcription; translation
  enzymes important in, 215t
  regulation of, 221–226
    induction, **224**, 224f
    operon model, 224–225, 224f
    positive regulation, 225–226, 226f
    repression, **224**, 225f
  silencing of, 259, 259f
gene gun, **253**, 254f
  to inject vaccines, 503
gene library. *See* genomic library
gene mapping
  conjugation and location of genes on bacterial chromosome, 237
  Human Genome Project and, 261
  Human Proteome project and, 261

Index

gene silencing, **259**, 259*f*
as natural process occurring in organisms, 259
reverse genetics and, 262
gene therapy, **17–18**, **259**
viral DNA as vectors, 251, 259
gene transfers
between bacteria, 233–241
by conjugation, **236**–237, 237*f*, 238*f*
by crossing over, **234**, 234*f*
by transduction, **237**, 239*f*
by transformation, **234**–236, 235*f*, 236*f*
by transposition (transposons), 240–241, 241*f*
horizontal, 213*f*, **234**, 577*b*
vertical, 213*f*, **234**
gene-cloning vectors, 15*t*, 247, 248*f*, 250–251, 250*f*, 251*f*
gene-cloning vectors/cloning vectors, 247, 248*f*
Genencor, 3*b*
genera. *See* Genus/genera
generalized (systemic) infection, **407**
generalized transduction in bacteria, **237**, 239*f*
generation time, **171**
genes, 16, 47, **211**. *See also* DNA
altered or rearranged by mutation, transposition, recombination, 241
artificial, 254–255, 255*f*
as products, 257–258, 257*f*. *See also* genetic engineering
chemically synthesized, 255
cloning and, 247, 254–255, 255*f*
constitutive, 222
eukaryotic, transcription in, 220
evolution and, 241
genetic transfers (transformation) and, 234–236
in plasmids, 95
inducible, 224, 224*f*
libraries of, 254, 255*f*
minimum necessary for free-living existence, 320, 326
mutation and, 226–233
mutation rates, 231
number needed by immune cells in antigen recognition, 484
prokaryotic
in protein synthesis, 216–221
transcription, 213*f*, **216**–217, 218*f*, 221
translation, 217, 219–220, 219*f*, 220–221*f*
location on bacterial chromosomes, 212
repressible, 224, 224*f*
sources for rDNA technology, 254–255, 255*f*
structural, 224*f*, 225, 225*f*
synthetic, 255, 256*f*
genetic change, plasmids and transposons as mechanisms of, 237
genetic code, **219**, 219*f*
degeneracy and, 219, 226–227, 255
genetic counseling, ethical issues, 268
genetic diseases
gene therapy and, 17–18, **259**
screening for, 262
genetic diversity
evolution and, 241
random, low-frequency mutations and species adaptation, 231
genetic engineering, 247, 253–258. *See also* recombinant DNA (rDNA) technology
agricultural products, 264–267, 267*t*
animal husbandry products, 267, 267*t*
food production products, 267*t*
genomic libraries, 254–255, 255*f*
in plants, 264–267, 265*f*
inserting foreign DNA into cells, 253–254, 264, 265
methods
electroporation, **253**
gene gun, 253, 254*f*
microinjection, **254**, 254*f*
producing protoplasts, 89, 253, 253*f*
protoplast fusion, **253**, 253*f*
transformation, **234**–236, 235*f*, 236*f*, **253**
obtaining DNA for, 254–255, 255*f*
overview of typical procedure, 248*f*
pharmaceutical products, 258–259, 260*t*
pharmaceutical products of, 260*t*
shuttle vectors and multicellular organisms, 251
techniques of, 253–258
therapeutic products, 258–259, 260*t*
transgenic animals, 259, 260*t*, 267*t*
genetic information
flow of from one generation to next, 212, 213*f*
location in bacterial cell, 80*f*, 94–95
transcription of, **216**–217, 218*f*, 221
translation of, 217, 219–221, 219*f*, 220–221*f*
genetic makeup of cell, relationship of metabolic pathways and enzymes, 114
genetic mapping
Human Genome Project and, 261–262
of resistance plasmid R100, 239, 240*f*
genetic material
changes in (mutation), 226–233
DNA and chromosomes, 211–212
DNA replication processes, 212–215, 216*f*
genotype and, 211
information flow and, 212, 213*f*
phenotype and, 211
protein synthesis and, 216–221
recombination processes, 233–241. *See also* genetic recombination
RNA and protein synthesis, 216–221
structure/function of, 211–221
genetic modification techniques. *See* genetic engineering
genetic recombination, 233, **234**–241. *See also* recombinant DNA (rDNA) technology
beneficial aspects of, 234
between organisms, avian influenza (H5N1) and, 416, 693
by crossing over, **234**
by gene transfers, 234–236
conjugation, 236–237
plasmids, 238–240
reassortment and antigenic shifts of flu virus, 693
transduction, 237
transformation, 234–236
transposons, 240–241
genetic screening, **262**
genetic transformation, **234**–236, 235*f*, 236*f*, **253**
genetically modified plants, 258, 264–267, 265*f*, 267*t*
genetics, **211**
microbial, **16**, 210–245. *See also* genetic material
molecular, cloning procedures of, 249
of bacterial morphology, 79
reverse, **262**, 694
genital herpes (herpes simplex virus type 2/HSV-2), **757**, 757*f*, **761***b*
acyclovir to treat, 569, 570*f*, 757
alpha interferon to treat, 470
incidence, 567*f*
latent state in nerve cells, 757
genital infections
Chlamydia trachomatis causing, 421*t*
Trichomonas vaginalis causing, 347, 347*f*
genital warts, 429, **757**, 758*f*, **761***b*
human papillomavirus causing, 375*t*, 385–386, 386*f*
imiquimod to treat, 570
genitourinary tract
as portal of entry, 429, 430*t*, 445*f*
pathogenic diseases, 429, 430*t*
genome, **211**
eukaryotic, baker's yeast as best understood, 258
genome sequencing, 261–262, 261*f*
genomes
minimum genetic requirements, 320, 326
of flavivirus, 223*b*
scientific applications, 261–262
viral, availability of, 261
genomic library, **254**–255, 255*f*
genomics, **13**
in West Nile virus tracking, 223*b*, 223*f*
to trace a norovirus infection outbreak, 262, 266*b*
genomics of pathogens, infectious diseases and, 262
genotypes, **211**
changes in, 226. *See also* mutations
ways bacteria acquire new, 237
gentamicin, 565
mode of action/spectrum of activity, 562*t*
produced by *Micromonospora purpurea*, 555*t*, 565
protein synthesis inhibited by, 95, 562*t*, 565
genus name, defined, **2**, **278**–279, 280*f*
Geobacillus stearothermophilus, causing food spoilage, 795, 796*t*
geosmin, produced by *Streptomyces*, 321
germ theory of disease, **9**, 11, 404–406, 477
German measles. *See* rubella
germfree mammals (without microbiota), used in research, 401
germicidal lamps, 193
germicides, **186**
germination, **97**
germs, 2. *See also* microbes/microorganisms
Gerstmann-Sträussler-Scheinker syndrome, 393
giant bacteria
Epulopiscium, 301*t*, **316**–317, 316*f*, 326
Thiomargarita namibiensis, 299*f*, 300*t*, 326
giant clam (*Tridacna*), symbiotic host to dinoflagellate algae, 345
Giardia duodenalis. *See Giardia lamblia*
Giardia intestinalis. *See Giardia lamblia*
Giardia lamblia, 347*f*, 348, 354*t*, 730–731, 734*b*
pathogenic mechanisms of, 443–444
Giardia (protozoa), 347*f*, 348
antigenic variation used by to evade immune response, 444
lack of mitochondria in, 104
parasitic species, 347*f*, 348, 354*t*
trophozoites of, 347*f*
giardial enteritis, 354*t*
giardiasis, 348, **730**–731, 730*f*, **734***b*
as notifiable infectious disease, 421*t*
metronidazole to treat, 571
portal of entry for, 429
quinacrine to treat, 571
gingival bacteria
Bacteroides, 324
Fusobacterium, 324, 324*f*
gingivitis, **709**, 710*b*
gliding motility, **83**–84
of cyanobacteria, 314
of *Cytophaga*, 324
of *Myxococcus*, 312
global warming
carbon cycle and, **770**
emerging infectious diseases, 416
globular proteins
enzymes as, 116
flagellin, 81
shape/structure of, 45, 46*f*
globulin proteins, antibodies as, 479
Gloeocapsa, binary fission of, 314*f*
Gloeocapsa genus/spp.
as photosynthetic bacteria, 301*t*
in taxonomic hierarchy, 301*t*
glomerulonephritis, **529**
Glossina (tsetse fly), African trypanosomiasis transmitted by, 351, 354*t*, 362*t*, 413*t*, 627–628
glucans, in cell walls of fungi, 330*t*
Gluconacetobacter xylinus, used in fabric production, 3*b*
Gluconobacter genus/spp., **303**
as acetic acid producers, 300*t*
fermentation and, 137*t*
in taxonomic hierarchy, 300*t*
industrial importance of, 303
used in production of vinegar, 800
glucose 6–phosphate
enzyme specificity and, 118
in glycogen synthesis, 146, 146*f*
glucose
as an energy source, 141, 143*f*
as main energy-supplying molecule of living cells, 39, 39*f*
cyclic AMP and positive regulation, 225–226, 226*f*
growth rate of *E. coli* on, 225–226, 226*f*
how it crosses plasma membrane, 92, 92*f*
in biosynthesis of lipids, 146, 147*f*
in biosynthesis of polysaccharides, 146, 146*f*
in Calvin-Benson cycle, 142*f*
in chemically defined mediums, 165, 165*t*
in dehydration synthesis, 39*f*
in energy production, 122, 122*f*, 123*f*, 124, 125*f*
glycolysis, 124, 125*f*
oxidation reactions, 122, 122*f*, 123*f*
in hydrolysis, 39*f*
number of ATP molecules produced per molecule of, in eukaryotes/prokaryotes, 137*t*
synthesis of, 146, 146*f*
transport by group translocation, 94
glucose effect (catabolite repression), **226**
glucose-phosphate isomerase, 116*t*
glucosyltransferase, produced by *Streptococcus mutans*, 431
glutamic acid (Glu)
in transamination, 148*f*
structural formula/characteristic R group, 44*t*
glutamine (Gln)
in biosynthesis of purine/pyrimidine nucleotides, 148*f*
structural formula/characteristic R group, 44*t*
glutaraldehyde, 200, 203*t*, 205*t*
glyceraldehyde 3–phosphate (GP), 124
in biosynthesis of lipids, 147*f*
in Calvin-Benson cycle, 142*f*
in lipid catabolism, 136*f*
glycerol, 41*f*
as fermentation end-product, 137*t*
in biosynthesis of lipids, 146, 147*f*
in complex lipids (phospholipids), 40–41, 42*f*
in fat molecule formation, 40, 41*f*
in lipid catabolism, 136, 136*f*, 138*f*
in simple lipids (fats/triglycerides), 40, 41*f*
glycine (Gly), 38, 45*f*
structural formula/characteristic R group, **44***t*
glycocalyx, **79–81**
as bacterial capsule, 79–81

as slime layer, 80, 81
biofilms and, 81
eukaryotic cell, **98**
in prokaryotic *vs.* eukaryotic cells, 101t
glycogen, 40
synthesis of, 146, 146f
glycogen granules, in presence of iodine, 95
glycolipids, **90**
glycolysis (Embden-Meyerhof pathway), **124**
alternatives to, 125, 127
ATP yield, 124, 132t
chemical reactions of (outline), **126f**
energy-conserving stage, 124, 126f
fermentation and, 125f, 132–134, 134f, 135b, 136f
in lipid catabolism, 136f
in synthesis of new cell components, 146–147, 146f, 147f, 148f
preparatory stage, 124, 126f
glycoproteins, 45, **90**
as adhesins (ligands) of pathogens, 431
glycylalanine, 45f
glycylclines, 565
glyphosate (herbicide)
insecticidal toxin (Bt toxin) and, 265
plants genetically engineered to resist, 265, 267t
goblet cells, of ciliary escalator, 452f
gold
used in staining of specimens, 63
used with gene guns, 253, 254f
Golgi complex, 99f, **104**, 104f
gonorrhea. *See also Neisseria gonorrhoeae*
antigenic variability and host defenses, 433, 749
arthritis as complication of, 748
as epidemic disease, 406, 749f
as notifiable infectious disease, 421t
Chlamydia trachomatis and, 750
diagnosis of, 750, 750f
endocarditis as complication of, 748
incidence and distribution, 748, 749f
incubation period, 430t, 749
meningitis as complication, 748
ophthalmia neonatorum and, 198, 204t, 429, **603**–604, **604b**, **748**–749
pelvic inflammatory disease and, 748
portals of entry, 429, 430t, 749
pregnancy and, 748–749
tetracyclines to treat, 565
treatment, 750
GP (glyceraldehyde 3–phosphate), 124
gp120 spike on HIV, 540, 540f, 541, 548
Gracilaria (red algae)
some species produce lethal toxin, 343
used by humans for food, 343
graft-versus-host (GVH) disease, 492, **536**
grafts, **535–536**
hyperacute rejection and, **536**
Graham sticky-tape method, 358
grains
aflatoxin and, 230
ergot toxin and, 443
fermentation and, 137t
molds and spoilage of, 192, 230
Gram, Hans Christian, 10f, 69
Gram stain, **69**–70, 70f, 88t
Archaea and, 87
importance of, 69–70
mechanism, and bacterial cell wall structure, 87
procedures for, 69, 70f
reaction, in Gram-negative *vs.* Gram-positive bacteria, 88t
gram-negative bacteria, **69**–70, 70f
antibacterial drugs used against, 562t
cell walls, 85, 86f
Gram stain mechanism and, 87
hypotonic solutions and, 94
characteristics of, 88t, 202
colonizing water pipes, laboratory containers, 440b, 440f
conjugation in, 236
cytolysis susceptibility and, 465
disinfectants effective against, 196, 196f
endotoxic shock caused by, 437, 439
Entner-Doudoroff pathway and, 127
fimbriae of, 83–84, 84f
flagella of, 81–82, 82f
Gram stain mechanism and cell walls of, 87
important genera/special features, 301t
lipid A inducing symptoms of infection by, 87
lipopolysaccharide (LPS) of, 69, 86f, **87**, 437, 450
nonproteobacteria, 301t, **313**–315, 314f, 315f
nosocomial infections and, 414, 414t
position in evolutionary tree, 275f
proteobacteria, 300–301t, 302–312, **303**, 304f-306f, 308f, 309f, 311f-313f
resistance to chemical biocides, 202–203, 203f
resistance to physical disruption, 88t
Rickettsia genus as, 303
vs. gram-positive bacteria, 69, 81, 87, 88t
gram-negative sepsis (endotoxic shock), 437, **640**
gram-positive bacteria, **69**–70, 70f
actinobacteria, 301t
antibacterial drugs used against, 562t
cell walls, 85, 86f
Gram stain mechanism and, 87
characteristics of, 88t
conjugation in, 236
cytolysis resistance and, 465
disinfectants effective against, 196, 196f
endospores and, **96**–98, 97f
firmicutes, 301t
flagella of, 81–82, 82f
high G + C ratio, 281f, 301t, 315–316. *See also* Actinobacteria
low G + C ratio, 281f, 301t, 315–316. *See also* Firmicutes
nosocomial infections and, 414, 414t
phylogenetic relationships, 281, 281f
position in evolutionary tree, 275f
resistance to chemical biocides, 202–203, 203f
resistance to physical disruption, 88t
vs. gram-negative bacteria, 69, 81, 87, 88t
gram-positive sepsis, **640**
gram-variable bacteria, 87
grana/granum, **105**, 106f
granddaughter DNA, altered, 229, 229f
granules
metachromatic, **95**
polysaccharide, **95**
granulocyte-colony stimulating factor (G-CSF), 492
granulocytes, **454**
granulomas, of schistosomiasis, **666**, 668f
granulomatous amebic encephalitis (GAE), **617b**, **629**
caused by *Acanthamoeba,* 617b, 629
caused by *Balamuthia,* 348, 617b, 629
granulomatous disease, chronic, gamma-interferon to treat, 260t
granum, **105**, 106f
granzymes, **454**, **489**
grapes, fermentation and, 137t
graphs, microbial death curve, 186, 187f
grappling hook model of twitching motility, 83
grasshoppers, protozoa *Nosema locustae* as insecticide against, 346
Graves' disease, **532**
HLA typing to determine susceptibility, 534t
Great Salt Lake, extreme halophiles (archaea) found in, 325
green algae, 342f, **343**, **343t**
cellulose cell walls of, 343
life cycle of, 342, 342f
multicellular, 341f, 342f
terrestrial plants origins and, 343
unicellular, 341f, 342f
green bacteria, 143, 143f, **144**, 145t
photosynthesis in, 145t
green monkeys, AIDS in, 377
green nonsulfur bacteria, **145**, 314t
characteristics, compared, 314t
nutritional classification of, 143, 143f
phylogenetic relationships, 281f
green plants, as photoautotrophs, 143–145, 143f
green scum in ponds, formed by filamentous green algae, 343
green sulfur bacteria, 144, 314t, **315**, 777
characteristics, compared, 314t
chlorobium vesicles found in, 144
phylogenetic relationships, 281f
Griffith, Frederick, 10f, 234–236, 235f
griseofulvin, 569, 600
mode of action, 564t
produced by *Penicillium griseofulvum,* 555t, 569
group A streptococci (GAS), 590–591, 591f, 640
group B streptococci (GBS), neonatal sepsis caused by, 640
group translocation, **94**
growth characteristics of pathogenic fungi, 338t
growth conditions for energy-producing processes, 137t
growth deficiencies in children, genetically engineered human growth hormone to treat, 260t
growth media
salt concentration and, 159
trace elements and, 160
growth (microbial), 156–179
chemical requirements, 156, 160–162, 161t
in cultures, 170–179
cell division, 156, 171–173, 171f-173f
generation time, **171**
measurement techniques, 156, 174–179, 175f-179f
media for, 156, 164–169
in prokaryotic cell/eukaryotic cell/eukaryotic organelles, 276t
of bacterial cultures, 171–179
phases of, 156, 172–174, 172f, 173f
physical requirements, 157–160
refrigeration and, 157–158, 158f, 159f
requirements for, 156, 157–163
osmotic pressure, 159–160, 160f
pH, 158–159
temperature, 157–158, 157f, 158f, 159f
growth temperatures (minimum/optimum/maximum) for microbes, **157**, 157f
guanine (G), 47, 48f, 211
in DNA replication, 212–215, 214f, 215f, 216f
in transcription phase of protein synthesis, 216, 218f
guinea worm (*Dracunculus medinensis*) infection, 13, 13f
gummatous syphilis, 754, 754f
gunpowder, 2
GVH (graft-versus-host) disease, 492, **536**
Gymnodinium breve (dinoflagellate), neurotoxin (saxitoxins) produced by, 344
gyrase, DNA, 212, 215t

H antigen, **82**
for *E. coli,* 82
HA spikes of influenzavirus, 692–693, 692f
HAART (highly active antiretroviral therapy), **548**
habitats, of pathogenic fungi, 338t
HACCP (Hazard Analysis and Critical Control Point) system, **794**
Haeckel, Ernst, 274
Hae III restriction enzyme, 249t
Haemophilus aegyptius, in rDNA technology, 249t
Haemophilus ducreyi, chancroid caused by, 311, 756, 761b
Haemophilus genus/spp., 301t, **311**
as human pathogens, 301t
as normal microbiota of mouth, 403t
genetic transformation natural occurrence in, 236
inhabit mucous membranes, 311
nosocomial infections and, 414, 414t
require blood in culture medium, 311
Haemophilus influenzae, 5f, **311**
as normal microbiota of throat, 403t
as notifiable infectious disease, 421t
complement system evasion by, 468
Hind III restriction enzyme used in rDNA technology, 249t
meningitis and, 432, 612, **613**, **617b**
otitis media caused by, 679
phagocytes and, 228, 432
pneumonia caused by, 311, 432, 613, **687b**, **688**
typeb
meningitis caused by, 432, 612, **613**, **617b**
septic shock and, 439
vaccine against, 502t, 503, 504t, 612
hair
cutaneous mycoses and, 337, 338t
sebum and, 453
hair follicles, 585, 585f
hairs, of nasal mucous membrane, **452**, 471t
hairy leukoplakia, in AIDS patients, 544t
half-life, of injected antibodies, 495
Haloarcula genus, shape of, 79, 79f
halobacteria, gas vacuoles and, 96
Halobacteriales, important genera/special features, 302t
Halobacterium genus/spp., 302t, **325**
Haloccoccus genus/spp., 302t, **325**
halogens, **197**, 204t
halophiles
extreme, 4, **159**, 275, 275f, 325
facultative, **159**
obligate, **159**
halophilic archaea, shape of, 79, 79f
hamsters, tularemia case study, 644b
hand lotion, brown algae used in production, 342
hand-washing procedures, as most important means of infection control, 416
Hansen's disease. *See* leprosy
Hantavirus, 376t
causing *Hantavirus* pulmonary syndrome, 417t
emerging infectious diseases and, 417t
PCR used to identify as cause of hemorrhagic fever outbreak, 290
reservoirs/transmission method, 410t
Hantavirus pulmonary syndrome, 376t, 410t, **660**, **660b**
as emerging infectious disease, 417t
as notifiable infectious disease, 421t
global warming and, 416
hapten-carrier conjugate, 479f
haptens, **478**, 479f
allergic contact dermatitis and, 530
Hartmut, Michel, 10f, 15t
hasiliximab, 537
Haverhill fever, 648
hay fever, 523, 523t, 525
IgA antibodies and, 481

Hazard Analysis and Critical Control Point (HACCP) system, **794**
HDNB (hemolytic disease of newborn), **527–528**, 528*f*
head lice, ivermectin effective against, 572
health care facilities, infections. *See* nosocomial infections
health care personnel
antibiotic resistance and, 576
hospital-acquired infections and. *See* nosocomial infections
Universal Precautions for (CDC), 546*t*
hearing loss, caused by aminoglycoside antibiotics, 565
heart attacks, genetically engineered products used in, 260*t*
heart infections (bacterial), 641–642, 641*f*, **643*b***
endocarditis, **641**, 641*f*
pericarditis, **641**, 643*b*
rheumatic fever, 319, **641–642**, 643*b*
heart transplant patients, impaired innate defenses of, 462
heart valves
abnormal, endocarditis risks and, 641
as privileged tissue, 535
biofilms colonizing, 163, 431, 641, 641*f*
rheumatic fever and, 641–642, 643*b*
heartworm (*Dirofilaria immitis*), **360**, 360*f*
Aedes mosquito as vector, 360, 362*t*
Wolbachia bacteria essential to, 360
heat
of inflammation, 460
of microbial energy production, 146
released from anabolic and catabolic reactions, 114, 114*f*
heat absorption by molecules, 35
heat treatments, 188–191, 194*t*
dry heat sterilization, 191, 194*t*
enzyme denaturation and, 188, 194*t*
equivalent treatments and, 191
factors influencing effectiveness, 186
flaming, 191, 194*t*
hot-air sterilization, 191, 194*t*
how it kills microbes, 188
moist heat treatments, 188–190, 189*f*, 194*t*
resistance to, 188
to remove *Clostridium botulinum* endospores, 185, 185*t*
heat-labile enterotoxin, produced by *E. coli* strains, 437
heat-labile enterotoxins, 440
heat-loving microbes (thermophiles), **157**, 157*f*
heat-resistant (thermoduric) bacteria, pasteurization and, 190
heavy chains of antibodies, 479, 480*f*
heavy metals
as biocidal or antiseptic agents, 198–199, 198*f*
gram-negative bacteria and, 87
R factors that confer resistance to, 239
used in staining of specimens, 63
HeLa cell line, 378
helical virus, **373**, 373*f*
enveloped, 372*f*
helical viruses, **373**, 373*f*
helicase, 212–213, 215*t*
helices of protein structure, 45, 46*f*
Helicobacter genus/spp., 301*t*, **312**
as carcinogenic bacteria, 301*t*
as human pathogens, 301*t*
Helicobacter pylori
neutralizes stomach acid, so it can grow, 453
peptic ulcer disease, 312, 313*f*, **718–721**, 719*t*, 723
helium, used with gene guns, 253, 254*f*
helminthic diseases
of cardiovascular/lymphatic systems, 666–667, **668*b***
of digestive system, 732–737, 732*f*, **735*b***
helminthic zoonoses, 410*t*
helminths, **6**, 352–361, **353**, **361*t***
antihelminthic drugs, 564*t*, 571–572
as multicellular eukaryotic animals, 353
characteristics of, 353–354
emerging infectious diseases caused by, 417*t*
parasitic, 329, 353–361, **361*t***
habitat, 354
life cycle, 354–355
nutrition, 354
reproductive methods, 354
pathogenic, 361*t*, 444
helper T cells, **487–488**, 488*f*, 496*f*
CD4+ T cells and, **487–488**, 488*f*
in antibody production, 482–483, 482*f*
hemagglutination, **512**
influenza viruses and, 372*f*, 373
viral, **512**, 512*f*
hemagglutinin (H) proteins, influenza A virus subtypes and, 370*b*, 371*t*
hemagglutinin (HA) spikes of *Influenzavirus*, **692–693**, 692*f*
hematologic disorders, sickle cell disease, 228
hematopoietic cytokines, **492**
Hemiascomycetes, position in taxonomic hierarchy, 280*f*
hemodialysis
antibiotic resistance developing from, 422*b*
disinfectants used in, 197
patients at risk for gram-positive sepsis, 640
hemoflagellates (blood parasites), **351**
hemoglobin, 434, **470**
hemolysins, 470, 589
hemolysis, in complement testing, 467*b*
hemolytic anemia, **528**
hemolytic disease of newborn (HDNB), **527–528**, 528*f*
hemolytic streptococci, 319, 589–590
hemolytic uremic syndrome (HUS)
as notifiable infectious disease, 421*t*
E. coli 0157:H7 and, 113*f*, 210, 718
hemophilia, 18
hemophilia B, gene therapy to treat, 259
hemorrhagic colitis, 718
hemorrhagic fever viruses, 20, 290, 376*t*, 658–659, **660*b***
as potential biological weapon, 649*b*
emerging, 637, 659–660, **660*b***
hemorrhagic fever with renal syndrome, **660**
Hendra virus, emerging infectious diseases and, 417*t*
HEPA (high-efficiency particulate air) filters, 168, **191**
Hepadnaviridae, **386**
as DNA virus, 385*t*
biosynthesis of, 385*t*
characteristics/important genera/clinical features, 375*t*
synthesize DNA using reverse transcriptase, 386
Hepadnavirus
hepatitis B and, 375*t*, 430*t*
hepatitis D and, 376*t*
incubation period, 430*t*
portals of entry, 430*t*
hepatitis, 721–728
alpha interferon to treat, 260*t*, 570
antisense DNA explored as gene therapy, 259
as emerging infectious disease, 417*t*
blood banking supplies and, 727*b*
caused by hepatitis C virus, 417*t*
caused by hepatitis E virus, 417*t*
genetically engineered interferons to treat, 260*t*
other types of, 728
hepatitis A, **721–723**, **724*b***
as notifiable infectious disease, 421*t*
incubation period, 430*t*
hepatitis A virus (HAV), 375*t*, 407, **721–723**, **724*b***
as RNA virus, 386
portals of entry, 429, 430*t*
vaccine, 503*t*, 504*t*
hepatitis B, **723–726**, **724*b***, 725*f*
adefovir dipivoxil (Hepsera) to treat, 569
alpha interferon to treat, 470
as chronic disease, 407
as notifiable infectious disease, 421*t*
incubation period, 430*t*
lamivudine to treat, 569
portals of entry, 430*t*
pregnancy and, 760
hepatitis B vaccine, 13, 502, 503*t*, 504, 726
genetically engineered, 260*t*
genetically modifying yeasts and, 247, 259
made from genetically modified *Saccharomyces cerevisiae*, 339
recommended schedule, 504*t*
hepatitis B virus (HBV), **723–726**, **724*b***, 725*f*
as cancer-causing virus, 391
gene silencing and, 259, 259*f*
Hepadnaviridae and, 386
Hepadnavirus, 375*t*, 385, 430*t*
incubation period, 430*t*
portals of entry, 430*t*
hepatitis C, **724*b***, **726**
alpha interferon to treat, 470
as notifiable infectious disease, 421*t*
hepatitis C virus (HCV), 376*t*, **724*b***, **726**
as RNA virus, 386
hepatitis D (delta hepatitis), 376*t*, **724*b***, **726**
as RNA virus, 386
depends on coinfection with hepadnavirus, 376*t*
hepatitis E, **724*b***, **727–729**
hepatitis E virus (HEV), 375*t*, 417*t*, **724*b***, **727–729**
as RNA virus, 386
hepatitis F virus (HFV), 728
as RNA virus, 386
hepatitis G virus (HGV), 728
as RNA virus, 386
hepatotoxins, 435
Hepsera (adefovir dipivoxil), mode of action/uses, 564*t*
Hepsera (adefovir dipivoxil) to treat hepatitis B, 569
heptoses, 39
herbicide resistance, genetically engineered into crop plants, 265, 267*t*
herbicides
decomposition rate of Agent Orange, 775, 775*f*
RoundUp, 265, 267*t*
Herceptin (trastuzumab), 509, 538
herd immunity, **407**, 501, 598, 612
hereditary traits, determination of, 16, 47
heredity, science of. *See* genetics
hermaphroditic helminths, **355**
herpes encephalitis, **598**
herpes gladiatorum, **598**
herpes simplex viruses
as AIDS-associated, 544*t*
as latent infections, 392, 394*t*
portals of entry/incubation period, 430*t*
pregnancy and, 760
type 1 (HSV-1), 385, 387*f*, 590*b*, **597–598**, 598*f*, 757
type 2 (HSV-2), 385, 598, **757**, 757*f*, **761*b***
herpes-zoster (shingles), 375*t*, 394*t*, 407, **596–597**
as a latent varicella-zoster virus disease, 407, 596
in HIV/AIDS patients, 542, 544*t*
rash caused by, 590*b*, 597*f*
vaccine, 503*f*, 596–597
Herpesviridae, **385**, 387*f*
as DNA virus, 385
biosynthesis of, 385*t*
characteristics/important genera/clinical features, 375*t*
portals of entry, 429, 430*t*
vaccine, 503*t*
herpesviruses (HHV), **385**, 387*f*. *See also specific herpes virus*
acridine dyes and, 230
contaminated red bone marrow transplant and, 406
incubation period, 430*t*
infections, acyclovir to treat, 569, 570*f*
latent infections and, 392
portals of entry, 429, 430*t*
species (HHV-1 to HHV-8), **385**
used to insert corrective genes into human cells, 251
herpetic keratitis, **605**
herpetic whitlow, **598**
Hershey, Alfred D., 10*f*, 14*t*
Hershko, Avram, 15*t*
heterocyclic R group, of various amino acids, 43, 44*t*
heterocysts, **314**, 314*f*, **772**
heterofermentative (heterolactic) microbes, **135*b***
heterolactic (heterofermentative) microbes, **135*b***
heterotrophs (organotrophs), **142–143**, 143*f*, 146
complex medium for growing, 166*t*
fungi and, 330*t*
hexachlorophene, 196, 196*f*
hexose monophosphate shunt (pentose phosphate pathway), **125**, 127
hexoses, 39
Hfr cell (high frequency of recombination), 236–237, 237*f*, 238*f*
HGA (human granulocytic anaplasmosis), 291, 421*t*, **651*b***, **654*b***
hGH (human growth hormone), produced by genetically modified *E. coli*, 247
HHV (human herpes virus), **385**, 387*f*. *See also* herpesviruses
HHV-1 *simplexvirus*, 385, 387*f*, **597–598**, 598*f*, 757
HHV-2 *simplexvirus*, 385, 387*f*, **597–598**, 598*f*, **757**, 757*f*, 761*b*
HHV-3 *Varicellovirus*, 385. *See also* varicella-zoster virus
HHV-4 (*Lymphocryptovirus*), 375*t*, 385. *See also Lymphocryptovirus*
HHV-6 *Roseolovirus*, 385, 600
HHV-7 infant measles-like rashes, 385
HHV-8 Kaposi's sarcoma. *See* Kaposi's sarcoma
Hib. *See Haemophilus influenzae*, typeb
high frequency of recombination (Hfr) cell, 236–237, 237*f*, 238*f*
high G + C gram-positive bacteria, 281*f*, 301*t*, 315–316. *See also* Actinobacteria
high pressure treatments, to control microbial growth, 192, 194*t*
high-efficiency particulate air (HEPA) filters, 168, **191**
high-energy bond, 122
symbol for, 122
high-energy electron beams, 192, 193
high-temperature short time (HTST) pasteurization, **190–191**
high-throughput screening methods of soil samples, 554
highly active antiretroviral therapy (HAART), **548**

Hind III restriction enzyme, 249*t*, 251*f*
hinge region of antibodies, 479, 480*f*
hip replacement components, biofilms colonizing, 431
histamine, **424**, **460**, 461*f*, 465*f*, **524**
  complement system and, 460, 461*f*, 464*f*, 465*f*
  in allergic reactions, 481, **524**, 524*f*
  released by eosinophils, 454
histidine (his)
  auxotrophs, Ames test and, 232–233, 233*f*
  replica plating technique and, 231–232, 232*f*
  structural formula/characteristic R group, 44*t*
histiocytes, 457. *See also* fixed macrophages
histocompatibility antigens, 482, **533**
  major histocompatibility complex (MHC) and, **482**, 482*f*, 496*f*, **533**–534
  tissue rejection and, 482
histones
  in prokaryotic cell/eukaryotic cell/eukaryotic organelles, 276*t*
  prokaryotic *vs.* eukaryotic DNA and, 77, 101*t*, **103**
Histoplasma (Ajellomyces) capsulatum, 338*t*
  as pathogenic fungi, 338*t*
Histoplasma (Ajellomyces) dermatitidis, as pathogenic fungi, 338*t*
Histoplasma capsulatum (fungus)
  AIDS-related, 544*t*
  histoplasmosis caused by, 430*t*, **695**–696, 695*f*, 696*f*
Histoplasma (fungus), interleukin-12 and, 493*b*
histoplasmosis, **695**–696, 695*f*, 696*f*, **699*b***
  airborne transmission and, 412
  amphotericin B effective against, 568
  as a systemic mycosis, 336
  caused by *Histoplasma capsulatum,* 338*t*, 430*t*
  incubation period, 430*t*
  portals of entry, 430*t*
HIV, 5*f*, **21**, 540–542
  antigenic variation undergone by, 541–542
  as a provirus, 389, 541, 541*f*, 542*f*
  as a retrovirus, **387**, 389, 390*f*
  as mutation of simian immunodeficiency virus, 540
  can survive in phagocytes, 459
  CD4+ T cell targets and, 540, 540*f*, 541, 541*f*
  clads of genome, 542
  cytopathic effects of, 443*t*
  early recognition of, 367, 539–540
  ELISA test to detect, 287, 288*f*, **516**, 518*f*, 545
  emerging infectious diseases and, 417*t*
  evading immune defenses, 441–442, 443*t*, 459, 541–542
  gp120 and, 540, 540*f*, 541
  HIV-1, HIV-2 subspecies and, 376*t*, 387, 540, 571
  incubation period, 430*t*
  infection. *See* HIV infection
  macrophages as targets, 541, 542*f*
  mechanisms for attacking immune system directly, 441
  of genus *Lentivirus*, viral family Retroviridae, 376*t*, 540
  pathogenicity of, 384*f*, 540*f*, 541–542, 541*f*, 542*f*
  portals of entry, 429, 430*t*
  resistance to, 544
  reverse transcriptase enzyme and, 387, 390*f*, 392, 540, 540*f*
  structure, 540, 540*f*
  transmission of, **545**–546
  vaccine development and, 259, 547–548
  Western blot test to confirm, 545
HIV infection, 540–545
  active, 541, 541*f*, 542*f*
  APTIMA assay to detect, 545
  as notifiable infectious disease, 421*t*
  as persistent viral infection, 394*t*
  blood banking and, 727*b*
  CD4+ T cells and, 5*f*, 415, 541–545, 541*f*, 543*f*
    cell counts during stages of, 542, 543*f*
  chemotherapy, 548. *See also* HIV infection, treatment regimens
  clinical phases of, 542–544, 543*f*
  diagnostic methods, 545
  distribution of cases, by world region, 547*f*
  ELISA test to detect, 287, 288*f*, 514, 516, 518*f*, 545
  estimated number of new cases per year, 546
  HIV structure, 540, 540*f*
  infants born to HIV-positive mothers and, 544
  latent, 541, 541*f*, 542*f*
  long-term nonprogressors and, 545
  progression stages to AIDS, 542–544, 543*f*
  resistance to, 544
  retrovirus *Lentivirus* HIV causing, 376*t*, 387, 389, 390*f*
  survival with, 544–545, 544*t*
  transmission of, 545–546, 546*t*
  treatment regimens, 548, 571
    alpha interferon, 470
    antivirals, 571
    atazanavir, 571
    Atripla, 571
    chemotherapies, 548
    efavirenz, 571
    emtricitabine, 571
    enfuvirtide, 571
    fusion inhibitors, 548, **571**
    genetically engineered colony-stimulating factor, 260*t*
    indinavir, 571
    integrase inhibitors, 548, 571
    interleukin-12 (IL-12) and, 493*b*
    lactic acid bacteria probiotic therapy, 403
    nevirapine, 571
    nucleoside reverse transcriptase inhibitors, 548
    protease inhibitors, 548, **571**
    saquinavir, 571
    tenofovir, 571
    zidovudine, 571
  vaccine development and, 259, 547–548
  Western blotting to confirm, 287, 289*f*
hives, 523, 525
HLA (human leukocyte antigen) complex, 482, **533**–537, 533*t*, 534*t*
HLA tissue typing, 533–534, 533*f*
HME (human monocytotrophic ehrlichiosis). *See* ehrlichiosis
Hodgkin's disease
  as acquired immunodeficiency, 538
  Epstein-Barr virus and, 658
  HLA typing to determine susceptibility, 534*t*
holdfasts of algae, **342**
holdfasts of multicellular algae, **342**
Holmes, Oliver Wendell, 641
holoenzyme, **116**, 116*f*
home canning of foods, 188, 190
home pregnancy test, 515, 517*f*
homofermentative (homolactic) bacteria, 135*b*
homolactic (homofermentative) microbes, 135*b*
hook of flagella, 81, 82*f*
Hooke, Robert, 7, 10*f*, 55
hookworms, 329, 359–360, 361*t*, **735**–736, 735*b*, 736*f*
  larvae bore through intact skin, 429
horizontal gene transfer, 213*f*, **234**, **577*b***
  antibiotic resistance and, 575, 577*b*
hormones, proteins as, 42
hormones, genetically engineered
  bovine growth hormone (bGH), 267, 267*t*
  human growth hormone (hGH), 247, 260*t*
  insulin, 2, 247, 255, 258
  porcine growth hormone (pGH), 267*t*
  somatostatin, 259
horsepox (extinct), 501
horses
  DNA vaccine against West Nile virus approved for, 503
  eastern equine encephalitis in, 625, **628*b***
  influenza A viruses and, 19, 370*b*
  reported cases of rabies in, 624*f*
  western equine encephalitis in, 375*t*, 625, **628*b***
hospital nurseries, outbreaks of impetigo (pemphigus neonatorum) in, 588
hospital-acquired infections. *See* nosocomial infections
hospitals
  control of nosocomial infections in, 416
  Universal Precautions for health care workers (CDC), 546*t*
  UV lamps to control microbes, 193
  ventilation systems, nosocomial infections and, 415
  water pipes in, *Legionella* in biofilms and, 689
  workers, resistance to antibiotics, 576
host cells
  complementary surface receptors for pathogenic adhesins, 431, 431*f*
  how bacterial pathogens damage, 434–441
host defenses
  how pathogens penetrate, 432–433, 433*f*
  how viruses evade, 441–442, 442*f*, 443*t*
  IgA antibodies and, 433
  non-specific (innate immunity), **449**–475, **476**. *See also* innate immunity
  phagocytosis, bacterial capsules and, 432
  virulence and, 428, 432–433
host environments, for parasitic helminths, 354
host interactions
  emerging infectious diseases and, 418
  viral, phage therapy research and, 369, **579**
host range (viral), **368**
  ecological niches and, 374
  species barrier crossings, 368, 370–371*b*
hosts
  compromised, 414*f*, **415**
  definitive, **349**
  how pathogens damage cells, 434–441
  how pathogens enter, 429–432, 430*t*
  how pathogens penetrate defenses, 432–433
  intermediate, **349**
  symbiotic, 345
  viral (mammalian cells in culture), 258
hot environments, archaea found growing in, 275, 275*f*, 325, 325*f*
hot springs, microbes associated with, 158
hot tubs/saunas, rashes and, 592–593
hot zone labs, 168, 168*f*
hot-air sterilization, **191**, 194*t*
houseflies, as vectors, 362
HPV (human papillomavirus), 391
  cervical cancers caused by, 391
  HPV-16, 391
  vaccine, 260*t*, 391, 503*t*
HPV vaccine (Gardasil), 260*t*, 391, 503*t*, 538, 758
HSV-1. *See* herpes simplex viruses
HSV-2. *See* genital herpes
HTLV-1 and HTLV-2 (human T-cell leukemia virus), 391, 394*t*
HTST (high-temperature short-time) pasteurization, **190**–191
HTST pasteurization, **190**–191
Huber, Robert, 10*f*, 15*t*
human activated protein C, Xigris genetically engineered from to treat sepsis, 640
human cells, as eukaryotes/eukaryotic cells, 76
human diploid cell vaccine (HDCV), 623
human disease
  biofilms and, 18–19, 19*f*
  emerging infectious diseases (EIDs), **19**–21
  infectious, **19**
  normal microbiota and, 18, 18*f*
human DNA sequencing, 262
human eye, specimen sizes resolved by, 58, 58*f*
human genome, mapping of, 261
Human Genome Project, 261
human granulocytic anaplasmosis/HGA, 291, **651*b***, **654**
human granulocytic ehrlichiosis, 291
human growth hormone (hGH)
  as rDNA product in medical therapy, 260*t*
  industrial fermentation used to produce, 802
  produced by genetically modified *E. coli,* 247
human herpesviruses (HHV), **385**, 589*b*, 596, 597–598, 598*f*
  latent infections and, 392
human immunodeficiency virus. *See* HIV
human insulin. *See* insulin (human)
human intestinal bacteria
  Escherichia, 300*t*
  Proteus, 300*t*
human leukocyte antigen (HLA) complex, 482, **533**–537, 533*t*, 534*t*
  bone marrow transplants, 536–537
  diseases related to, 534*t*
  grafts, **535**–536
  reactions to transplantation, 534–535
  stem cells and, 535, 535*f*
  tissue typing, 533–534, 533*f*
  using PCR in matching donors in transplant surgery, 534
human monocytotrophic ehrlichiosis (HME). *See* ehrlichiosis
human papillomavirus (HPV), 375*t*, 391
  cervical cancers caused by, 391
  vaccine, 260*t*, 391, 503*t*, 758
  warts caused by, 375*t*, 385–386, 386*f*
    genital warts, **758**, 758*f*, 761*b*
human parasites
  Acanthamoebea, 348
  Babesia microti, 350
  Balamuthia, 348
  Balantidium coli, 350
  Cryptosporidium, 350
  Cyclospora cayetanensis, 350
  Entamoeba histolytica, 348
  Giardia lamblia, 347*f*, 348
  microsporidial protozoa, 348
  Plasmodium vivax, 348–349, 349*f*
  Toxoplasma gondii, 350
  Trichomonas vaginalis, 347, 347*f*
human parvovirus B19, 375*t*
human pathogens (bacteria)
  Bordetella, 300*t*
  Borrelia, 302*t*
  Brucella, 300*t*
  Campylobacter, 301*t*
  Chlamydia, 302*t*
  Chlamydophila, 302*t*
  Citrobacter, 300*t*
  Coxiella, 300*t*
  Ehrlichia, 300*t*
  Enterobacter, 300*t*
  Escherichia, 300*t*
  Francisella, 300*t*
  Haemophilus, 301*t*
  Helicobacter, 301*t*

Legionella, 300*t*
Leptospira, 302*t*
Moraxella, 300*t*
Neisseria, 300*t*
Proteus, 300*t*
Pseudomonas, 300*t*
Rickettsia, 300*t*
Salmonella, 301*t*
Shigella, 301*t*
Streptobacillus, 302*t*
Treponema, 302*t*
Vibrio, 300*t*
human pathogens (fungal), 335–339, 338*t*
Human Proteome Project, 261
human rabies immune globulin (RIG), 623
human reservoirs, **409**
human T-cell leukemia viruses (HTLV-1 and HTLV-2), 391
humanized antibodies, **509**
humans, bacterial infections in and biofilms, 163
humidifiers, as disease reservoirs, 416
humoral immunity, **477–486**, 496*f*
antibody titer and, 493, 494*f*, **510**, 511*f*
B cells and, 482–486, 482*f*, 483*f*
effective against freely circulating pathogens, 486
immulogical memory and, 493–494
primary response, **493–**494, 494*f*
secondary response, **493–**494, 494*f*
spleen removal decreases, 538
humors, health and, 477
Huntington's disease, genetic mutation that causes, 228
HUS. *See* hemolytic uremic syndrome
hyaluronidase, **432–**433, 590
produced by some clostridia, 433
therapeutic uses, 433
hybridization reactions
colony, **256–**257, 257*f*
fluorescent in situ (FISH), **292–**293, 294*f*
forensic microbiology and, 262, 263*f*
nucleic acid, 291–293, 291*f*
hybridomas, **507**, 508*f*
hydatid cyst, **358**, 359*f*
hydatid disease (hydatid cyst), **733–**734, **735***b*
hydatidosis, 361*t*
hydrocarbons
petroleum and bacteria that use as energy/carbon source, 239
petroleum, natural gas formed by early planktonic algae, 345
hydrochloric acid (HCL)
as a base, 35–36, 36*f*
most microbes destroyed by, 429
hydrogel, biofilm as, 163
hydrogen, 34
an energy source, 141, 143*f*, 145*t*
as a biofuel, microbes and, 808
as energy source, 145
as fermentation end-product, 134*f*
atomic number/atomic weight, 27*t*
electronic configuration, 29*t*
formation, 30, 30*f*
green bacteria and, 144, 145*t*
in biological oxidations, 122, 123*f*
in methane formation, 31, 31*f*
in organic compounds, 37
hydrogen bonds, **31–**32
of amino acids in protein's structural levels, 46*f*
of water molecules, 32*f*, 35
hydrogen ions, acid-base balance and, 35–36, 36*f*
hydrogen peroxide
as antiseptic, 202, 205*t*
as disinfectant, 202, 205*t*
as toxic oxygen product of lysosomal enzymes, 459
catalase and, 162, 202
enzymes in peroxisomes decompose, 105
for aseptic packaging, 202
in plasma sterilization, 201, 205*t*
magnetosomes can decompose, 96
hydrogen sulfide
anaerobic respiration and, 132
as energy source, 145
bacteria that consume, 13
biochemical tests to identify, 139, 139*f*
green bacteria and, 144, 145*t*
Hydrogenomonas, 145
Hydrogenophilales, important genera of, 300*t*
hydrolase enzyme, 116*t*
hydrolysis, **39**, 39*f*, 116*t*
in DNA replication, 214, 215*f*
hydrolytic reactions, 114
hydrophilic heads of phospholipids, 41, 42*f*, 89, 90*f*
hydrophilic molecules, 41, 42*f*
hydrophobia, as sign of rabies, 623
hydrophobic molecules, 41, 42*f*
hydrophobic portion of phospholipids, 41, 42*f*, 89, 90*f*
hydrophobic side groups of proteins, 45, 46*f*
hydrothermal vents, deep-sea, microbes associated with, 158, 325
hydroxide ion, 35–36, 36*f*
hydroxyl group
in amino acids, 43, 44*t*
in fatty acids, 40, 41*f*
of alcohols, 37
when two monomers join, 38
hydroxyl radicals, **162**
ionizing radiation and, 192–193
hygiene hypothesis, 525
hyperacute rejection, **536**
hyperbaric chambers, to treat gas gangrene, **646**, 647*f*
hypercholesterolemia, gene therapy, 18
hypersensitivity reactions, **523–**531, 523*t*
anaphylactic (Type I), **523–**526, 523*t*, 524*f*
inherited complement deficiencies and, 468
cytotoxic (Type II), **526–**528
delayed (Type IV), **529–**531
eosinophils increase during, 454
IgE antibodies and, 481, 523, 524*f*
immune complex (Type III), **528–**529
hyperthermophiles (extreme thermophiles), 4, **157**, 157*f*, 158, **158**, 275, 275*f*, 302*t*, **325**, 325*f*
phylogenetic relationships, 281*f*
position in evolutionary tree, 275*f*
hypertonic environments, microbial growth and, 159
hypertonic solution, 93*f*, **94**
hyphae, **331–**332, 331*f*
coenocytic, **331**, 331*f*, 333
fragmentation and, 331, 334*f*
fungal, of lichen, 339, 340*f*
fungi develop from, 4, 281, 331, 331*f*
of Basidiomycota, 335, 337*f*
of *Candida albicans,* 334*f*
of *Mucor,* 5*f*
of *Talaromyces,* 336*f*
septate, **331**, 331*f*
vegetative, 331, **331** *f*
Hyphomicrobium genus/spp., 300*t*, **304**, 305*f*, 777
hypochlorous acid
antimicrobial activity of, 459
germicidal action of, 197
hypotension, endotoxic shock and, 437, 439
hypothalamus, as body's thermostat, 463
hypotonic environments, microbial growth and, 159
hypotonic solution, 93*f*, **94**
I gene, 225, 225*f*
ibritumomab (Zevalin), 509
ice cream, algae-produced thickeners used in, 342
ice formation, *Pseudomonas syringae* and genetically engineered plants, 267*t*
icosahedron-shaped viruses, 372*f*, 373
ICTV (International Committee on Taxonomy of Viruses), 282, 374
ID50, virulence of pathogens and, **429–**430
identification of microorganisms, 282–294
by biochemical tests, 285–287, 285*f*
by differential staining, 285
by DNA base composition, 288–289
by DNA fingerprinting, 289–**290**, 290*f*
by enzymatic activity tests, 285, 285*f*
by fatty acid profiles (FAME tests), 288
by flow cytometry, 288
by metabolic characteristics, 285–287, 285*f*
by morphological characteristics, 284–285, 285
by nucleic acid hybridization, 291–293
by phage typing, 288, 290*f*
by polymerase chain reaction (PCR), 290–291
by rapid identification methods, 285–286, 286*f*
by serological testing, 287, 287*f*
by Western blotting, **287**, 289*f*
cladograms and, 275*f*, 281*f*, 293, 294*f*
criteria and methods for, 282, 284–294
dichotomous keys and, 283*b*, 285*f*, 293
identification schemes in *Bergey's Manual,* 9th ed., 282
lab report form example, 284, 284*f*
microscopic, 282
of prokaryotes, 282, 284
relationship of taxonomy to, 273
source and habitat considerations, 282, 284
idiophase, **803**
idoquinol (diiodohydroxyquin), to treat amoebic diseases, 571
IFNs (interferons), **468–**472. *See also* interferons
IgA, 433, 479, **480–**481, 481*t*, 486
proteases enzymes, **433**
serum IgA, 480
IGAS (invasive group *A Streptococcus*), 20
IgD, 479, 480*f*, **481**, 481*t*
IgE, 479, 480*f*, **481**, 481*t*
anaphylactic reactions and, 523–526, 523*t*, 524*f*
IgG, 479, **480**, 480*f*, 481*t*, 484–485, 493, 494*f*, 509
desensitization process and, 526
immune complex reactions and, 528–529, 529*f*
IgM, 479, **480**, 481*t*, 484, 493, 494*f*, 509
IL-1. *See* interleukin-1
IL-12. *See* interleukin-12
illuminator, of compound light microscope, **56**, 56*f*
imidazoles, **568**, 569*f*
imipenem, 561, 562*t*
imiquimod, **570**
immersion oil, 59, 59*f*
"immortal" cell lines, 378
immune adherence (opsonization), complement activation and, 464*f*, 465
immune complex autoimmune diseases, **532**
complement deficiency and, 467*b*
immune complex reactions, **528–**529, 529*f*
as Type III hypersensitivity, 523*t*, 528–529, 529*f*
immune deficiency diseases, **539***t*
immune surveillance, **537**
immune system, **523–**531
aging and decline of, 462, 522
antibody-mediated humoral response, 477, 482–486, 493
biofilms and, 163
cancer and, 537–538, 537*f*
complement system's role in, 463–468
disorders, 522–552
AIDS, 539–548
autoimmune diseases, 532–533, **532–**533
HLA complex reactions, 533–537
hypersensitivity, **523–**531
immunodeficiencies (congenital), **538**
extracellular killing by, 491
number of antigens recognized by, 474
self *vs.* nonself recognition and, 477, 482, 485, 486, 492–493, 494, 496*f*, 532–536
suppressed
susceptibility to nosocomial infections, 415, 415*t*
to prevent transplant rejection, 522
immunity
activation mechanisms, 450
active, naturally or artificially acquired, 494–495, 494*f*
adaptive, **450**, 450*f*, **476–**499
vs. innate, 450, 450*f*
as something that can be acquired, 433, 477
cellular, **477–478**
discovery of, 11
first line of defense, 450–453, 450*f*, **471***t*
chemical factors, 453, **471***t*
normal microbiota, 453
physical factors, 451–452, 451*f*, **471***t*
skin and mucous membranes, 450–453, 450*f*, **471***t*
herd, **407**, 501, 598, 612
humoral, **477**, 482–486
innate, 449–475, **450**, 450*f*
vs. adaptive, 450, 450*f*
non-specific host defenses, **449–**475
of population, disease spread and, 407
overview, 450, 450*f*
passive, naturally or artificially acquired, 494–495, 494*f*
second line of defense, 450*f*, 454–472
antimicrobial substances, 463–472
fever, 463
inflammation, 460–463
phagocytes, 457–460
third line of defense, 450*f*
vaccination rates and, 407, 505*b*
immunization, 494*f*, **495**. *See also* vaccination
adult booster, 418, 501, 502*t*, 503*t*
childhood, 502
recommended schedule for, 504*t*
immunocompromised patients
human parvovirus B19 and, 375*t*
nosocomial infection susceptibility and, 415
immunodeficiencies, **538**, 539*f*, 539*t*
acquired, **538**, 539*t*
congenital, **538**, 539*t*
immunodiffusion tests, **509–**510
immunoelectrophoresis, **510**
immunofluorescence. *See* fluorescent-antibody (FA) tests
immunoglobulins (Ig), **479–**482, 480*f*. *See also* antibodies
classes of, 479–481
complement fixation of, 481*t*
functions of, by class, 481*t*
IgA, 479, **480**, 481*t*
IgD, 479, 480*f*, **481**, 481*t*
IgE, 479, 480*f*, **481**, 481*t*
IgG, 479, **480**, 480*f*, 481*t*, 484–485, 493, 494*f*, 509
IgM, 479, **480**, 481*t*, 484, 493, 494*f*, 509
location in body, 481*t*
molecular weight of, 481*t*
placental transfer of, 494–495
summary table, **481***t*

immunological memory, **493**–494, 494f
immunologically privileged sites/tissues, transplant rejection and, 534–535
immunology, **13**, 16
  diagnostic tests based on, 507–518. *See also* diagnostic tools
  early history, 477–478, 507
  golden age of, 506
  practical applications
    diagnostic tools, 507–518
    vaccines, 501–506
  therapeutic, future and, 516–517
immunosuppression in transplant surgery, 536–537
immunosuppressive drugs
  basiliximab, 537
  chimeric monoclonal antibodies as, 537
  cyclosporine, 536
  daclizumab, 537
  opportunistic mycoses and, 337–339
  sirolimus (Rapamune), 536–537
  Tacrolimus (FK506), 536
immunotherapy, **538**
  for allergies, 526, 526f
  for cancers, 538
immunotoxin, **538**
impetigo, 319, **588**, 588f
  vesicular rash caused by, 590b
impetigo of newborn (pemphigus neonatorum), **588**
implants (medical)
  bacterial colonization on, 531b
  decontaminated with supercritical carbon dioxide, 202
in-phase light rays, 59
inactivated whole-agent vaccines, **502**
  bacteria, 502
  viruses, 502, 502f
inapparent (subclinical) infections, 494
incidence of disease, **406**
incineration, sterilization and, 191
inclusion bodies (viral), **441**–442, 442f
  cytomegalic inclusion disease, 385, 658
inclusion conjunctivitis, **604**, 604b
inclusions, of prokaryotic cells, 80f, **95**–96
incubation period in infectious diseases, **408**, 408f, 430t
India ink, in capsule staining, 71, 72f
indicator organisms in water purity tests, **780**–781
indicators, sterilization, 190, 190f
indigo, produced by bacteria, 3b
indinavir, 571
indirect contact transmission, **410**, 411f
  in nosocomial infections, 415–416
indirect ELISA tests, 514, **516**, 518f
indirect FA tests, **513**–514, 515f
indirect (negative) selection method to identify mutations, **231**
indirect (passive) agglutination tests, **511**–512, 511f
indole, 3b
induced pluripotent stem cells (iPS), 535
inducer (signaling chemical), quorum sensing and, 57b, 163
inducers, **224**
inducible enzymes, 224
inducible operons, 224f, **225**
inducible promoters, 257
induction, **224**–225, 224f
industrial applications of microbiology, 800–808
  alternative energy sources, 806–807
  amino acids products, 805
  antibiotics, 800
    microbes used to produce, 247, 249, 301t, 317, 321, 339, 554, 555t, 563, 805
  biofuels, **807**–808, 808f
  biotechnology, 801. *See also* biotechnology
  chemical detection microbes, 801b, 806
  citric acid products, 805
  commercial microbial products, 804–806
  enzyme products, 805
  fermentation technology, **801**–804
  food preservation, **794**–795, 794f, 795f
  future of, 808
  in copper production, 806
  microbes as industrial products, 806
  pharmaceuticals, 805–806, 806f
  renewal energy sources, 806–807
  vaccines, 805–806
  vitamins, 805
industrial fermentation
  in human growth hormone, 802
  in insulin production, 802
  in making monoclonal antibodies, 802
  primary metabolite produced by, **803**–804, 804f
  secondary metabolite produced by, **803**–804, 804f
industrially important bacteria, 40, 137t, 303
  lactobacilli, 318
  mining industry microbes, 247
indwelling catheters
  biofilms and, 19f, 163, 164b, 431, 586, 587f
  bloodstream infections and, 164b
  silver incorporated in for antimicrobial effects, 198
infant diarrhea, pathogenic *E. coli* and, 239
infants born to HIV-positive mothers, 544
infection control
  hand-washing as single most important activity, 416
  in hospitals, 416
infections, **400**
  drug-resistant, 12–13
  fungal, **335**–**339**, 338t
  germ theory of disease and, 9, 11, 404–406, 477
  hospital-acquired. *See* nosocomial infections
  in digestive tract, *vs.* intoxication, **710**
  local, **407**
  nosocomial, **413**, 414f. *See also* nosocomial infections
    biofilms and, 19f, 163
    catheterization and, 164b
  primary, **407**
  secondary, **407**
  spread of, 409–413, 444
    reservoirs, 409
    transmission, 409–413
  subclinical (inapparent), **407**, 494
  WBC types during initial/middle/late stages of, 457
infectious diseases, **19**, **404**. *See also* microbial diseases
  acute, **406**
  carriers of, **409**
  chronic, **407**
  classifying, 406–407
  climate and, 408
  communicable diseases, **406**
  contagious diseases, **406**
  control methods, 501. *See also* vaccines
  DNA fingerprinting to track, 262, 264f, 290, 290f
  duration or severity of, 406–407
  emerging (EIDs), **19**–**21**. *See also* emerging infectious diseases
  endemic disease, **406**
  epidemic disease, **406**, 407f
  etiology of, 404–406, 405f
  extent of host involvement and, 407
  frequency of occurrence and, **406**
  genomics of pathogens and, 262
  human reservoirs, **409**
  incidence of, **406**
  incubation periods, **408**, 408f, 430t
  Koch's postulates and, **404**–406, 405f
  noncommunicable diseases, **406**
  norovirus infections, 266b
  occurrence of, 406
  pandemic disease, **406**
  patterns of, 408–409, 408f
  period of convalescence stage, 408f, 409
  period of decline stage, 408f, **409**
  period of illness stage, **408**, 408f
  predisposing factors, **408**
  prevalence of, **406**
  prions causing, 392–393
  prodromal period, **408**, 408f
  related to human leukocyte antigens (HLAs), 534t
  severity or duration of, 406–407
  signs and symptoms, **406**
  sporadic diseases, **406**
  spotted fever group, *Rickettsias* and, 303
  spread of infection
    disease reservoirs, 409
    transmission of disease, 409–413
  stages/sequence of events during, 408–409, 408f
  syndromes and, **406**
  transmission
    by contact (direct or indirect), **410**–411, 411f
    by droplets, **411**, 411f
    by vehicle, **411**–412, 412f
  vaccination rates, herd immunity and, **407**, 501, 598, 612
  weather and, 408
  zoonoses, **409**
infectious mononucleosis, 375t, 385, **643b**, **656**–657
  as chronic disease, **407**
  caused by Epstein-Barr virus, 430t, 656–657
  hemagglutination test to diagnose, 512
  incubation period, 430t
  portals of entry, 430t
infectious proteins (prions), 20
infertility, from pelvic inflammatory disease, 752, 761b
inflammation, 450f, **460**–463, 461f
  acute, 460
  antibody-complement complex and, 464, 464f, 465, 465f, 485, 485f
  as second line of defense, 450f, 460, 461f
  blood clot formation in, 461f
  chemokines important in, 492
  chronic, 460
  functions of, 460
  monoclonal antibodies to treat, 509
  permeability of blood vessels increases in, 460, 461f, 462
  phagocyte migration/phagocytosis in, 461f, 462
  scar tissue and, 463
  signs/symptoms, 460
  stages of, 460–463, 461f
  tissue repair stage of, 461f, 462–463
  vasodilation stage, 460, 461f, 462
inflammatory acne, **594**
inflammatory response, 461f
inflammatory responses
  associated with acne, 453
  of autoimmune diseases, 492
infliximab (Remicade), 509
influenza (flu), **692**–**695**, 692f, 693t, **699b**
  1918–1919 pandemic, 694
  antigenic drift and, **693**–694, 693t
  antigenic shift and, **370b**, 371f, **693**
  antigenic variation and, 433
  as pandemic disease, 406, 693
  as zoonotic disease, 410t
  cytokine storm and, 492, 694
  diagnosis of, 694
  epidemiology of, 693–694
  portals of entry, 429, 430t
  transmission methods, 410t, 411
  treatment of, 694–695
  vaccine, 13, 502, 502f, 503t, 504t, 694
influenza viruses, 376t, **692**–693, 692f, 693t
  antigenic drift and, **693**–694
  antigenic shift and, **370b**, 371f, **693**
  as potential biological weapon, 649b
  bird flu (avian influenza A H5N1), **19**, 370–371b, 693
    recent human cases, 370t
  dimensions/viral family/clinical features, 376t
  glycoproteins, plasma membranes and, 90
  hemagglutination and, 372f, 373
  incubation period, 430t
  influenza A viruses, 376t
    as potential biological weapon, 649b
    avian influenza A H5N1 (bird flu), **19**, 370–371b, 370t, 693
    crossing species barrier, 370–371b
    different animal species found in, 19, 370b
    Influenzavirus A2 spikes and hemagglutination, 372f, 373
    pandemics during Twentieth Century, 371t
    subtypes of, 370b, 372f, 376t
  portals of entry, 429, 430t
  vaccines, 13
    avian influenza virus, 19
    DNA vaccine being tested, 259
    genetically engineered, 260t
influenza-associated pediatric mortality, as notifiable infectious disease, 421t
Influenzavirus, **692**–693, 692f, 693t
  antigenic variation capabilities of, 433
  hemagglutinin (HA) spikes, **692**–693, 692f
  incubation period, 430t
  neuraminidase (NA) spikes of *Influenzavirus*, **692**–693, 692f
  portals of entry, 430t
  reservoirs/transmission method, 410t
Influenzavirus A2, spikes, hemagglutination and, 372f, 373
information storage, biological, 211. *See also* genetics
ingestion phase of phagocytosis, 458, 458f, **459**
INH. *See* isoniazid
inhalation anthrax, virulence of, 430
inhalation fungal pathogens, 336, 338, 338t
inhalational (pulmonary) anthrax, **645**–646, 649b, 650b
inherited disorders
  complement deficiencies, 468
  Huntington's disease, 228
  sickle cell disease, 228
  xeroderma pigmentosum, 231
inherited traits of microbes, 210–245. *See also* genetic material
inhibition by basic dyes, of gram-negative *vs.* gram-positive bacteria, 88t
inhibitors of enzymes, **120**, 120f
injection site, microbial controls and, 185, 185t
innate immunity, 449–475, **450**, 450f, **476**. *See also* immunity
  antimicrobial substances, 463–472
    antimicrobial peptides, 470–471, **578**
    complement system, 463–468
    interferons, 468–470, 469f
    iron-binding proteins, 470
  blood's role in, 454–456, 455t, 637–638, 639f
  chemical factors, 450f, 453
  fever, 463
  first line of defense, 450–453, 450f, **471t**
  inflammation, 460–463, 461f

lymphatic system's role in, 456–457, 456*f*
lymph's role in, 637–638, 639*f*
mucous membranes and, 450–453, 452*f*
normal microbiota and, 450*f*, 453*f*
overview, 450, 450*f*
phagocytes, 457–460, 637–638, 639*f*
physical factors, 451–542, 451*f*
second line of defense, 450*f*, 454–471, **472*t***
skin, 450–453, 450*f*, 451*f*
summary, by component/functions, **471–472*t***
inner membrane. *See* plasma (cytoplasmic) membrane
inoculating loop sterilization, 191, 194*t*
inoculation of embryonated eggs with animal viruses, 377–378, 377*f*, 504
inoculum, **164**
inorganic compounds, **34**–37
water, 32*f*, 34–35, 35*f*
insect bites
flea, 303, 311, 362*t*, 410*t*, 648
rickettsias bacteria transmitted to humans by, 303
sand fly, leishmaniasis and, 354*t*, 665
insect venom
anaphylaxis and, 523–524
desensitization success and, 526
Insecta, 362, 362*t*
insecticides
allergic reactions to *Bacillus thuringiensis* (BT) toxin, 268
apicomplexan protozoan to reduce fire ant egg production, 346
protozoa *Nosema locustae* to kill grasshoppers, 346
insects
as eukarya, 6
as vectors, 362, 362*t*
chitin exoskeleton of, 98
diseases transmitted by, 362, 362*t*
evolutionary influence of *Wolbachia* bacteria, 307*b*
in foodstuffs, radiation doses needed to kill, 797*t*
plant resistance to, and genetic engineering, 17
symbiotic relationships, 106, 107*b*
that are vectors, 362*t*
Wolbachia as symbionts of, 300*t*, 305, 307*b*
insertion sequences (IS), 240, 241*f*
insulin (human), 2
chemically synthesized genes and, 255
E. coli bacteria used to produce, 247, 260*t*
genetically engineered, 258, 260*t*
industrial fermentation used to produce, 802
insulin-dependent diabetes mellitus, **533**
integral proteins
of prokaryotic plasma membrane, 89–90, 90*f*
role in facilitated diffusion, 91–92, 92*f*
integrase inhibitors, **571**
to treat HIV infection, 548, 571
interference (relative darkness)
in phase-contrast microscopy, 59–60, 60*f*
of light rays, 59
interferons (IFNs), **442**, **468**–472, 469*f*, **472*t***, **492**, **570**
alpha interferon, 469–470, 469*f*, **570**
as rDNA product in medical therapies, 260*t*
to treat viral hepatitis, 564*t*, 570
to treat virus-associated disorders (Intron A), 470
as anticancer agents, 470
as antimicrobial chemicals, 18
as antiviral drugs, 246, 469, 469*f*, 564*t*, **570**
to stimulate natural interferon production, 570
as cytokines, 469, 492
as natural defenses against disease, 18
beta interferon, 469–470, 469*f*
as rDNA product in medical therapies, 260*t*
to treat multiple sclerosis (Betaferon), 470
to treat osteoporosis (Actimmune), 470
disadvantages of, 470
discovery of, 13, 16
gamma interferon, 469
as rDNA product in medical therapies, 260*t*, 469
E. coli genetically engineered to produce, 246*f*, 257*f*
induces neutrophils/macrophages to kill bacteria, 469
host-cell-specific, but not virus-specific feature, 469, 469*f*
human types of, 469
alpha interferon (IFN-α), 469, 469*f*
beta interferon (IFN-β), 469, 469*f*
gamma interferon (IFN-γ), 469, 469*f*
in second line of host's defenses, 472*t*
principal function of, 469
produced from chemically synthesized genes, 255
species-specific nature of, 246
viruses and, 368*t*
interleukin-1 (IL-1), **437**, 438*f*, 463
interleukin-12 (IL-12)
as promising "magic bullet" therapy, 493*b*
HIV and, 493*b*
humoral response and, 493*b*
measles virus and, 442, 493*b*
psoriasis treatment success and, 493*b*
interleukins, **491–492**, **493*b***
genetically engineered, 260*t*
intermediate bodies, *Chlamydophila psittaci* and, **323*f***
intermediate filaments, 101
intermediate host, **349**, 361*t*
lung flukes and, 357*f*
of *Paragonimus westermani*, 356, 357*f*
of *Plasmodium vivax*, 349, 349*f*
of selected parasitic helminths, 361*t*
of tapeworm *Echinococcus granulosus*, 358, 359*f*, 361*t*
International Code of Botanical Nomenclature, 279
International Code of Zoological Nomenclature, 279
International Committee on Systematics of Prokaryotes, 279
International Committee on Taxonomy of Viruses (ICTV), 282, 374
International Journal of Systematic and Evolutionary Microbiology, 279
interstitial fluid, 456*f*, 457, 638, 639*f*
intestinal bacteria
ecological balance and, 310
enterics as, **309–311**
Escherichia, 300*t*
Proteus, 300*t*
intestinal parasites, 329, **361*t***
flatworms, 356, 356*f*, 361*t*
protozoa, 354*t*
roundworms, 358–361, 361*t*
tapeworms, 356–358, 358*f*, 361*t*
intestinal protozoa, capable of anaerobic growth, 346
intestines, normal microbiota of, 400*f*, 403*t*
intoxication
botulism as special case of, 710
in digestive tract, **710**
staphylococcal, 711–712, 711*f*
vs. infection, 435
intracellular antigens
cellular immunity and, 486, 496*f*
humoral immunity and, 486
intracellular parasites
Chlamydia, 302*t*
Chlamydophila, 302*t*
Rickettsia, 300*t*, 303
viruses as, 282, **368**, 368*t*
intracellular pathogens, obligate, 300*t*
intravenous (IV) catheters
contaminated heparin solution, infection, 164*b*
nosocomial bacteremia and, 415*t*, 422*b*
Intron A (alpha interferon), 470
introns, 215*t*, **220**, 222*f*, **254**–255, 255*f*
as "junk DNA," 261
viroids and, 394–395
intubation devices, as disease reservoirs, 416
invasins, **433**
invasive group A *Streptococcus* (IGAS), 20
iodide ion, as an anion, 30
iodine, as antimicrobial agent, **197**, 200*f*, 201*b*, 203*t*, 204*t*
iodine (I)
as mordant, 69, 87
atomic number/atomic weight, 27*t*
glycogen/starch granules and, 95
in Gram stain mechanism, 87
in water treatment, 197, 203*t*
iodophors, **197**, 204*t*
ionic bonds, 28, **30**, 30*f*
covalent bonds *vs.*, 31
strong *vs.* weak, 30
ionization (dissociation), **35**, 36*f*
ionizing radiation, **192**–193, 193*f*, 194*t*
as mutagenic, 230
ions, **28**, 35, 35*f*, 36*f*
in ionizing radiation, mutagens and, 230, 230*f*
Iospora belli (protozoa), causing gastroenteritis in AIDS patients, 544*t*
iPS (induced pluripotent stem cells), 535
Irish Famine, *Phytophthora infestans* infecting potato crops and, 329, 344
Irish moss, 343
iron (Fe)
as cofactor, 117
as requirement for bacterial growth, 434, 463, 470, 639
atomic number/atomic weight, 27*t*
biofilms and, 163
cyanide and, 120
lactoferrin and, 163, 434
siderophores and, 434, 434*f*
iron oxide, in magnetosomes, 96
iron-binding proteins, **470**, 472*t*
siderophores of pathogenic bacteria and, **434**, 434*f*, 445*f*, **470**
irradiation of foodstuffs, **796–797**
doses needed to kill various organisms, 796*t*
electron-beam accelerators used in, 797, 798*f*
gamma ray processing, 797, 798*f*
irradiation logo, 797*f*
IS (insertion sequences), 240, 241*f*
ischemia, **646**
isocitrate lyase, 116*t*
isocitric acid, 127, 128*f*, 149*f*
isografts, **536**
isoleucine, synthesis, 121, 121*f*
Isoleucine (Ile), structural formula/characteristic R group, 44*t*
isomerase enzymes, type of chemical reaction catalyzed/examples, 116*t*
isomers, **39**
of amino acids, 43, 43*f*
isoniazid (INH), 562*t*, **563**, 566, 684
mode of action/spectrum of activity, 562*t*
isoprenoids, as genetically engineered product, 258
isopropanol (rubbing alcohol), 37
as disinfectant, 198, 204*t*
isopropyl alcohol, as fermentation end-product, 134*f*
isotonic solutions, **93**, 93*f*, 159, 160*f*
isotopes, **27–28**
isotretinoin (Accutane), 453, 594
Israeli acute paralysis virus, pollinating bees and, 367
Isthmia nervosa (diatom), 343*f*
itraconazole, 568–569
ivermectin, **572**
mode of action/uses, 564*t*
produced by *Streptomyces avermectinius*, 572
to treat lice, 603
veterinary applications, 571
Iwanowski, Dimitri, 16, 367
Ixodes pacificus (tick), Lyme disease vector on Pacific coast, 362*f*, 413*t*, 653
Ixodes scapularis (tick)
as vector for babesiosis, 350, 362*t*
as vector for Lyme disease, 653
Ixodes (tick)
as vector for babesiosis, 350, 362*t*
as vector for ehrlichiosis, 362*t*, 413*t*
as vector of Lyme disease, 362*t*, 413*t*
life cycle of, 653*f*

j (joining) chain, 480, 481*t*
Jacob, François, 10*f*, 14*t*, 16, 224
Janssen, Zaccharias, 55
Japanese B encephalitis, **626**
Jenner, Edward, 10*f*
smallpox vaccine and, 11, 501
Jerne, Niels Kai, 15*t*
jock itch (tinea cruris), 338*t*, **600**
"junk DNA," 261

kanamycin resistance, 241*f*
Kaposi's sarcoma, 21, 375*t*, 385, 544*t*
alpha interferon to treat, 470
early recognition of HIV connection, 21, 539
in AIDS patients, 542, 544*t*
karyogamy, **333**, 335*f*, 336*f*
Kauffmann-White scheme, 310
Kefir (fermented milk beverage), 799
kelp (brown algae), 341, 341*f*, **342**, **343*t***
keratin, **337**, 338*t*, **451**, 451*f*, **585**
as resistant barrier of skin, 402*t*, 451, 451*f*, 584, **585**
dermatophytes degrade, 337, 338*t*
fungi and, 429
keratinase, 337
keratitis, 354*t*, 605
Acanthamoeba, **605**
herpetic, **605**
keratoconjunctivitis, 348, 354*t*
Ketek (telithromycin), 566
mode of action/spectrum of activity, 562*t*
ketoconazole, 568
damages plasma membrane, 558
mode of action/comments, 564*t*
spectrum of activity, 557*t*
ketolides, **566**
ketone group, 38*t*
kidney dialysis
antibiotic resistance developing from, 422*b*
disinfectants used in, 197
patients at risk for gram-positive sepsis, 640
kidney diseases
hemolytic uremic syndrome, 113*f*, 210, 421*t*, 718
leptospirosis, **746**–747, 747*f*, **748*b***
pyelonephritis, **746**, 748*b*
kidney transplant patients
impaired innate body defenses and, 462
monoclonal antibody (muromonab-CD3) to minimize rejection, 508, 509

kidneys, 744, 745*f*
  glomeruli, 529
killing curve for microbial populations, 186, 187*f*
kilometer (km), metric/U.S. equivalent, 55*t*
kinases, **432**
kinetic energy, heat absorption by molecules and, 35
Kingdom Monera (Prokaryotae), 274
kingdom name, defined, **279**, 280*f*
Kingdom Prokaryotae (Monera), 274
  first proposed, 274
Kingdom Protista, Haeckel's proposal, 274
kinins, 460, 461*f*, **462**
"kissing bug" (*Triatoma* /reduviid bug), Chagas' disease transmitted by, 351, 354*t*, 362*t*, 363*f*, 413*t*, 661
Kitasato, Shibasaburo, 10*f*
Klebsiella genus/spp., 300*t*, **310**
  as nitrogen fixers, 310
  as normal microbiota of large intestine, 403*t*
  as normal microbiota of urethra, 403*t*
  as opportunistic pathogens, 300*t*
  polysaccharide capsule of and virulence, 80
  resistance plasmid R100 and, 239, 240*f*
Klebsiella pneumoniae, 283*b*, 310
  capsule staining to identify, 72*f*
  capsule-producing, virulence and, 432
  nosocomial infections and, 414, 414*t*
Klug, Aaron, 10, 15*t*
km (kilometer), metric/U.S. equivalent, 55*t*
Koch, Robert, 9, 10*f*, 11, 14*t*, 404, 507
Koch's postulates, **11**, **404**–406, 405*f*
KOH (potassium hydroxide), to diagnose cutaneous mycoses, 601
Köhler, Georges J. F., 15*t*
Komagataelia pastoris (yeast), genetically engineered superoxide dismutase produced by, 260*f*
Komodo dragon bites, *Pasteurella multocida* and, 311
Koplik's spots, 599
Korean hemorrhagic fever, 376*t*
Krebs cycle, 124, 125*f*, **127**–129, 128*f*
  anaerobic conditions and, 132
  as source for amino acid synthesis, 147
  ATP yield, 132*t*
  in lipid catabolism, 136, 136*f*
  in protein catabolism, 136
  lipid biosynthesis and, 146, 147*f*
Krebs, Edwin G., 15*t*
Krebs, Hans A., 14*t*
kumiss (fermented milk beverage), 799
Kupffer's cells, 457
kuru disease, 393, **631**, **632*b***

L forms of bacteria, **89**
lab report form example, 284, 284*f*
laboratory tests, 137–139, 139*f*, 144*b*
  serum collection, 467*b*
lac operon, 224*f*, 225, 225*f*, 226*f*, 257, 380
lac repressor, 380
lac structural genes, 225
lacrimal apparatus, **451**, 452*f*
  tears and innate immunity defenses, 451, 471*t*
lacrimal canals, 451, 452*f*
lacrimal glands, 451–452, 452*f*
lactate dehydrogenase, 116*t*
lactic acid
  aerotolerant anaerobes and, 162
  as fermentation end-product, 134*f*
  bacteria used in winemaking, 800
  in amphibolic pathways, 149*f*
  industrial/commercial uses for, 137*t*
  Streptococcus and, 137*t*
lactic acid bacteria. *See Lactobacillus*
lactic acid fermentation, 134*f*, **135*b***, 136*f*
lactobacilli
  as normal microbiota of vagina, 403*t*
  used in acidic-fermented foods, 162
Lactobacillus acidophilus, in vaginal secretions, 453
Lactobacillus delbrueckii, used to make rye bread, 137*t*
Lactobacillus delbrueckii bulgaricus, used to make yogurt, 799
Lactobacillus genus/spp./Lactobacillales, 301*t*, **318**, **318**–319, 318*f*
  antibiotic therapy and, 402
  as lactic acid producers, 135*b*, 137*t*, 301*t*, 318
  as normal microbiota of large intestine, 403*t*
  as normal microbiota of mouth, 403*t*
  as normal microbiota of urethra, 403*t*
  chemically defined medium to culture, 165
  fermentation and, 134*f*, 137*t*
  important genera/special features, 301*t*
  industrial importance of, 137*t*, 318
Lactobacillus plantarum, sauerkraut and, 137*t*
lactoferrin, 163, 434, **470**
lactose, 39, 224, 224*f*
lactose in milk, allergy to, 525
lactose metabolism, in *E. coli,* 224–225, 224*f*, 225*f*
lactose operon regulation, 224*f*, 225, 225*f*
  by positive regulation, 225–226, 226*f*
lacZ, 251*f*
lag phase, in bacterial growth curve, **173**, 173*f*
lagging strand in DNA replication, 216*f*
lake bacteria. *See also* freshwater microbiota
  Caulobacter genus and, 304, 776
LAL (limulus amoebocyte lysate) assay, **439**, 440*b*
Laminaria japonica (algae), 342
lamivudine, to treat hepatitis B, 569
Lancefield, Rebecca C., 10*f*, 16, 287
landfills
  bacterial biosensors to detect pathogens/pollutants, **780*b***
  degradation of synthetic chemicals in, 775–776
Langerhans cells/Langerhans DC, 490
laparoscopic surgical instruments, 201
large intestine, 456*f*
  microbial antagonism in, 401
  normal microbiota of, 400*f*, 403*t*
  parasitic helminths and, 361*t*
Lariam (mefloquine) to prevent malaria, 571, 664
Lariam (mefloquine) to treat malaria, 557*t*
larval stage of parasitic helminths, **354**
laryngitis, **676**
Lassa fever, 376*t*, **659**, **660*b***
  as potential biological weapon, 649*b*
latency, in lysogenic cycle, 380, 382
latent disease, **407**
latent infections (viral), **392**, 392*f*, 394*t*
  examples (disease/primary effect/causative virus), 394*t*
  HIV infection, 541, 541*f*, 542*f*
  provirus and, 389, 541, 541*f*, 542*f*
latent virions, 541, 541*f*
latex agglutination tests, **511**–512, 511*f*, 677
latex allergy, 530–531, 532*f*
lattices, 509
Lavoisier, Anton Laurent, 8
LD50, to express potency of toxins, **430**–431
LDL-receptor deficiency, 18
lead, used in staining of specimens, 63
leading strand in DNA replication, 216*f*
leafhoppers
  potato yellow dwarf virus transmitted by, 394*t*
  wound tumor virus transmitted by, 394*t*
lectin pathway of complement activation, 464*f*, **467**–468, 468*f*
lectins, **467**, 468*f*
  binding of, 348
  mannose-binding lectin, **468**, 468*f*
Lederberg, Joshua, 10*f*, 14*t*, 16
Legionella genus/spp., 300*t*, **309**
  colonize hospital warm-water lines/air conditioning systems, 309
Legionella pneumophila
  Legionnaires' disease caused by, 309, 404, 417*t*, 689
  phosphoprotein synthesis by bacteria and, 45
Legionellales, 300*t*, **309**
  important genera/special features, 300*t*
legionellosis (Legionnaires' disease), 309, 404, 417*t*, **687*b***, **688**–689, **691*b***
  as notifiable infectious disease, 421*t*
  erythromycin effective against, 566
  outbreak (case study), 691*b*
Leishmania braziliensi, 665, 666
Leishmania donovani, visceral leishmaniasis caused by, **651*b***, **665**
Leishmania (protozoa), 354*t*, 665
  can survive in phagocytes, 459
  interleukin-12 and, 493*b*
leishmaniasis, 354*t*, 459, **651*b***, **665**–666, 665*f*
  American, 666
  cutaneous, **665**–666, 665*b*, 665*f*
  mucocutaneous, 651*b*, **666**
  visceral, 651*b*, **665**
length, basic metric units of, 55, 55*t*
lenses of microscopes
  early, 7, 7*f*, 55
  electromagnetic, 63–65, 64*f*
  multiple, in compound light microscope, 55, 56, 56*f*, 59*f*
Lentivirus HIV
  as retrovirus, 376*t*, **387**, 389, 390*f*
  characteristics/viral family/clinical features, 376*t*
lepromatous (progressive) form of leprosy, 619, 620*f*
leprosy (Hansen's disease), 320, 404, **619**–620, 620*f*, **632*b***
  acid-fast stains to diagnose, 70, 71*f*
  as notifiable infectious disease, 421*t*
  culturing leprosy bacillus, 539*f*, 619
  diagnosis of, 620
  treatments for, 563, 567, 620, 632*b*
  types of, 619, 620*f*
  vaccines useful for, 620
Leptospira genus/spp., 302*t*, **323**–**324**
  as human pathogen, 302*t*, 323, 748*f*
  reservoirs/transmission method, 410*t*
Leptospira interrogans, 746–747, **748*b***, 748*f*
leptospirosis, 323–324, 410*t*, **746**–**747**, 747*f*, **748*b***
  California sea lion deaths and, 283*b*
  waterborne transmission and, 411
lesions, skin, **586**, 587*f*
lethal dose, **430**–431, 439*t*
lethal toxin, of *Bacillus anthracis,* 645
lettuce, norovirus infection outbreak, 266*b*
Leucine (Leu), structural formula/characteristic R group, 44*t*
Leuconostoc mesenteroides, pentose phosphate pathway and, 127
leukemia
  as latent viral infection, 394*t*
  caused by oncoviruses, 376*t*, 391
  chicken, 389
  genetically engineered colony-stimulating factor in therapy for, 260*t*
  in cats, 391
  patients, mucormycosis and, 338
  viruses, 376*t*, 391
leukocidin toxin, 422*b*
  produced by *S. aureus,* 76*f*, 422*b*
leukocyte esterase, 746
leukocytes (white blood cells), **454**, 455*t*
  agranulocytes, **454**, 455*t*
  basophils, **454**, 455*t*
  decreases in during infections (leukopenia), 454
  differential white blood cell count, **454**, 455*t*, 456, 457
  eosinophils, **454**, 455*t*
  granulocytes, **454**, 455*t*
  increases during infections (leukocytosis), 454
  polymorphs, 454
leukocytosis, 454
leukopenia, 454
leukoplakia, oral/hairy, 542, 544*t*
leukotoxins, 435
leukotrienes, 461*f*, **462**, **524**
Level 4 labs, 168, 168*f*
Levi Strauss denim blue jeans, made by microbes, 3*b*
LGV (lymphogranuloma venereum), 322, 459, 755, **761*t***
libraries
  cDNA, 255
  genomic, **254**–255, 255*f*
  phage library, 255*f*
  plasmid, 255*f*
lice (pediculosis), 362, 362*t*, **602**–603, 603*f*
  head, ivermectin effective against, 572
  Lyme disease and, 323
  Pediculus humanus (parasite) species and, 362*t*, 363*f*, 602
  rash caused by, 592*b*
  sucking, 362*t*
  treatments for, 602–603
  typhus transmitted by, 303
lichens, **339**–340, 340*f*, **772**
  as air quality testers, 340
  as major food for tundra herbivores, 340
  morphologic types, 339, 340*f*
lidocaine, 201*b*
life, definition of, 368
life cycles, of helminths, 330*t*
life-support processes. *See* metabolism
ligands (adhesins) of pathogens, **431**, 431*f*
  diversity of, 431
  location of, 431
ligase enzymes
  DNA ligase, 215*t*
  type of chemical reaction catalyzed/examples, 116*t*
light chains of antibodies, 479, 480*f*
light energy, converting to chemical energy (photosynthesis), 123, 140, 141*f*
light microscope/microscopy (LM), **56**, 56*f*, 58–62, 66*t*, 67*t*
  brightfield, **59**, 60*f*, 66*t*
  confocal, **62**, 62*f*, 67*t*
  darkfield, **59**, 60*f*, 66*t*
  differential interference contrast, **60**, 61*f*, 66*t*
  fluorescence, **61**–62, 61*f*, 66*t*
  phase-contrast, **59**–60, 60*f*, 66*t*
  resolution and, 56, 58–59
  specimen preparation for, 68–72
  specimen sizes and, 58*f*
  summary of (features/typical image/uses), 66*t*, 67*t*
light (photic) zone of bodies of water, algae and, 341*f*, 342
light (ultraviolet/UV light), to control microbes, 193, 193*f*
light (visible)
  diffraction of, 60, 60*f*
  in repair of thymine dimers, 230
  microscopes and, 55–62
    path of light in, 56, 56*f*, 58–59, 59*f*, 60*f*
  reflection of, 59, 60*f*
  refractive index of specimens and, 58–59

reinforcement of, 59
staining specimens and, 58–59
ultraviolet light and, 61
wave nature of, 59–60, 60*f*
wavelength size, *vs.* that of electrons, 63
light-dependent (light) reactions, **140**, 141*f*
light-independent (dark) reactions, **140**, 141*f*
light-repair enzymes (photolyases), 215*t*, **230**
lignin (component of wood), most fungi capable of metabolizing, 333
limnetic zone, **776–**777
limulus amoebocyte lysate (LAL) assay, **439**, 440*b*
Limulus polyphemus (crab), endotoxin testing and, 439
lindane, 602, 603
linezolid (Zyvox), 566
mode of action/spectrum of activity, 562*t*
Linnaeus, Carolus, 2, 10*f*, 274, 279
lipases, 116*t*
in lipid catabolism, 136, 136*f*
lipid A, 86*f*, **87**, **437**, 468
antimicrobial proteins (AMPs) and, 471
lipid bilayer, 89, 90*f*
osmosis through, 92*f*
simple diffusion through, 91, 92*f*
lipid catabolism, 136–137, 136*f*, 138*f*
lipid inclusions, **96**
lipid-carbohydrate complex, alternative pathway of complement activation and, 466–467, 466*f*
lipids, **40–**41, 41*f*, 42*f*
as nonpolar molecules, 40
biosynthesis of, 146, 147*f*
complex, **40–**41, 42*f*
fats (triglycerides), 40, 41*f*
functions of, 40–41
in gram-negative *vs.* gram-positive bacteria, 88*t*
in lipoproteins, 45
metabolism, coenzymes involved in, 117*t*
phospholipids, 40–41, 42*f*
simple, **40–**41, 41*f*
lipids (fats/triglycerides), catabolism of, 136, 136*f*, 138*f*
lipophilic viruses, biocidal resistance and, 203, 203*f*
lipopolysaccharide (LPS), 69, 86*f*, **87**, 437, 450
in gram-negative *vs.* gram-positive bacteria, 88*t*
selective toxicity of antibiotics and, 555
lipoproteins, 45
as adhesins (ligands) of pathogens, 431
in gram-negative *vs.* gram-positive bacteria, 88*t*, 437
lipoteichoic acid, 85, 86*f*
Lister, Joseph, 10*f*, 11, 184, 195, 413
Lister, Joseph Jackson, compound microscope and, 55
Listeria genus/spp., 301*t*, **319**
as human pathogens, 301*t*
dairy product/food contaminant, 319
in milk, flow cytometry to detect, 288
use actin of host to propel themselves, 433
Listeria monocytogenes
adhesin production in, 431
can grow at refrigerator temperatures, 319, 615
can survive in phagocytes, 459
membrane attack complexes produced by, 459
meningitis caused by, 613, 614–615, 615*f*, 617*b*
pregnancy dangers and, 319, 615
sepsis caused by, 615
listeriosis, 192, 459, **614–**615, 615*f*, 617*b*
as foodborne infection, 615, 617*b*
as notifiable infectious disease, 421*t*
cell-to-cell spread of, 615, 615*f*
lithotrophs (autotrophs), **142–**143, 143*f*
littoral zone, **776**
algal habitats, 341*f*
liver, parasitic helminths and, 361*t*
liver cancer, hepatitis B virus and, 391, 394*t*
liver flukes, 356
liver transplants, 536
liver tumors, caused by hepatitis B virus, 375*t*
livestock
animal feed antibiotics, 554, 562*t*, 565, 575, 577*b*
antihelminthics (ivermectin) to treat, 571
as disease reservoirs, 410*t*
Pasteurella-caused sepsis in cattle, 311
LM (light microscope), **56**, 56*f*, 58–59, 59*f*
local infection, **407**
localized anaphylaxis, **525–**526, 525*f*
lockjaw, 437, 615. *See also* tetanus toxin
log (exponential growth) phase, in bacterial growth curve, **173**, 173*f*
logarithmic representations of bacterial populations, 171–172, 172*f*
microbial death curve, 186*f*, 187*f*
lophotrichous flagella, **81**, 81*f*
low G + C gram-positive bacteria, 281*f*, 301*f*, 315–316. *See also* Firmicutes
low-density lipoprotein (LDL) deficiency, 18
LPS, in evasion of complement system, 468. *See* lipopolysaccharide (LPS)
LSD (lysergic acid diethylamide), ergot toxin of fungus natural source of, 443
luciferase enzyme, bioluminescence and, 57*b*, 778
lumbar puncture (spinal tap), 614, 614*f*
luminescent microorganisms, 61, 61*f*
lung diseases
cystic fibrosis, 18, 57*b*
pneumonia. *See* pneumonia
tuberculosis, 13, 54*f*
lung flukes, 356, 357*f*, 361*t*
lungs, parasitic helminths and, 361*t*
Luria, Salvador E., 10*f*, 14*t*
lux operon, bacterial biosensors and, **780***b*
Lwoff, André, 14*t*
lyase enzymes, type of chemical reaction catalyzed/examples, 116*t*
Lyme borreliosis. *See* Lyme disease
Lyme disease, 409, 410*t*, **651***b*, **652–**654, 652*f*, 654*f*
as emerging infectious disease, 417*t*
as notifiable infectious disease, 421*t*
bacterial axial filaments and, 83
Borrelia burgdorferi causing, 323, 652, 653*f*
causative agent/arthropod vector, 413*t*
diagnosed by Western blotting, 287, 289*f*
field mice as animal reservoir, 652–653, 653*f*
increases in, and animal control measures, 416
Ixodes tick as vector, 362, 362*f*, 362*t*, 652–653, 653*f*
reported cases 1992–2007, by year, 419*f*
reported cases 2007, by month, 419*t*
reported cases by county, 2005, 652*f*
symptoms, 651*b*, 654, 654*f*
lymph, 456, 456*f*, 638
lymph nodes, 456, 456*f*, 638, 639*f*
reticular fibers of, 456–457
site of activation of T cells, B cells, 456, 638
swollen (buboes), **638**, 648*f*
lymphangitis, **639**, 640*f*
lymphatic capillaries, 456, 456*f*, 638, 639*f*
relation to tissue cells, blood capillaries, 456*f*
lymphatic ducts, 456*f*, 457
lymphatic system, **456–**457, 456*f*, **637**, 638, 639*f*
cardiovascular system's relationship with, 637–638, 639*f*
microbial diseases of, 637–673
bacterial, 638–655, **643***b*, **650***b*, **651***b*
helminthic, 666–667, **668***b*
protozoan, **650***b*, **651***b*, 660–666
vector-borne, 648, **651***b*, 652–655
viral, **643***b*, 655–660
structure/function, 456*f*, 637–638, 639*f*
lymphatic vessels/lymphatics, 456, 456*f*, 638, 639*f*
Lymphocryptovirus (HHV-4/Epstein-Barr virus), 375*t*, 385
cancer and, 391
incubation period, 430*t*
infectious mononucleosis caused by, 430*t*
portals of entry, 430*t*
lymphocytes, **454**, 455*t*
as third line of defense, 450*f*
functions of, 478
natural killer (NK) cells, **454**, 455*t*, 472*t*
lymphocytic choriomeningitis, 376*t*
lymphogranuloma venereum (LGV), 322, 459, 755, **761***t*
lymphoid organs, 455*t*, 457
lymphoid tissue, 456, 456*f*, 457
lymphocytes of, 456
lymphoma
Burkitt's, 375*t*
human, 391
lyophilization (freeze-drying)
desiccation and, 192
to control microbial growth, 194*t*
to preserve bacterial cultures, **170**
lysergic acid diethylamide (LSD), ergot toxin of fungus natural source of, 443
lysine (lys)
allergic contact dermatitis and, 530
structural formula/characteristic R group, 44*t*
lysis, **85**, **380**, 381*f*
osmotic, **89**
lysogenic conversion, **440**, 445*f*
lysogenic cycle
animal viruses and, 382
of viral multiplication, **379**, 380, 382, 382*f*
lysogenic (temperate) phages, 380
of *Vibrio cholerae,* 441
lysogeny, **380**, 382, 382*f*
pathogenicity and, 440–441
phage conversion and, **382**, 441
prophage genes and, 380, 382, 382*f*, 440–441
specialized transduction and, **382**, 383*f*
Lysol, main ingredient in, 195, 196*f*
lysosomes, 99*f*, **104**
in phagocytosis, 458*f*, 459
toxic oxygen products produced by, 459
lysozyme, **453**
cell wall damage done by, 88–89, 88*t*, 94, 453
gram-negative bacteria and, 87–89, 88*t*
gram-positive bacteria and, 88–89, 88*t*
immunity functions of, 453, 471*t*
in perspiration, 585
in perspiration, 453
in phagocytosis, 459
in tears, 88, 453
in viral multiplication cycle, **380**
phage, **380**
Lyssavirus (rabies virus), 376*t*
bats as good disease reservoirs for, 624 *footnote*
encephalitis cases caused by closely related genotypes, 624
incubation period, 430*t*
portals of entry, 430*t*
reservoirs/transmission method, 410*t*
lytic cycle of bacteriophage multiplication, **379–**380, 381*f*

M cells, **486**, 487*f*, **710**
Shiga toxin and, 712, 712*f*, 713, 713*f*
m (meter), metric/U.S. equivalent, 55*t*
M protein, **432**
increased virulence of *Streptococcus pyogenes* and, 319, 432, 459, 590–591, 591*f*
microbial evasion of phagocytosis and, 459
rheumatic fever and, 641–642, 643*b*
mabs. *See* monoclonal antibodies
MAC (membrane attack complex), 459, 464*f*, **465**
complement fixation test based on, 465
MAC-resistant bacteria, 465
MacConkey's agar, 746, 748*f*
MacGregor tomatoes, 267, 267*t*
MacKirron, Roderick, 15*t*
MacLeod, Colin M., 10*f*, 16, 47, 235
Macrocystis (brown algae), 341*f*
macrolides, **566**, 566*f*
mode of action/spectrum of activity, 562*t*
macromolecules, 34, **38**
polysaccharides as, 39
macronucleus, of *Paramecium,* 350*f*
macrophages, **454**, 455*t*, **490**, 490*f*
activated, **490**, 490*f*
as antigen-presenting cells, 490, 490*f*
as phagocytes, 455*t*, 457, 457*f*, 490
as second line of defense, 450*f*
cathelicidins produced by, 470
defensins produced by, 470
fixed, **457**, 638, 639*f*
free (wandering), **457**
HIV in, 541, 542*f*
in adaptive cellular immunity, 487, 490, 490*f*
in inflammatory response, 461*f*
in innate immunity, 490
latent and active HIV infections in, 541, 542*f*
mononuclear phagocytic (reticuloendothelial) system and, **457**
macular rashes, diseases that cause, 589*b*
mad cow disease, **20**, 203, 393, 417*t*, 631–632
magainins, 578–579
"magic bullet" chemotherapies, 12–13, 493*b*, 554, 559
magnesium
as bridge between enzyme and ATP, 117
as cofactor, 117
fluoride and, 120
magnesium (Mg)
atomic number/atomic weight, 27*t*
electronic configuration, 29*t*
microbial requirements, 160
magnet-like inclusions formed by gram-negative bacteria (magnetosomes), 96
magnetosomes, **96**
magnetospirillum magnetobacterium, magnetosomes of, 96, 96*f*
major histocompatibility complex (MHC), **482**, 482*f*, 496*f*, **533–**534
malachite green stain, 68, 71, 72*t*
malaria, 19, 329, 348–349, 349*f*, 410*t*, 459, **651***b*, **663–**665, 663*f*, 664*f*
Anopheles mosquito as vector, 348–350, 349*f*, 362–363, 362*t*, 410*t*, 413*t*, 663
as fourth leading cause of death worldwide, 346
as notifiable infectious disease, 421*t*
chloroquine to treat, 571, 664
DNA vaccine being tested, 259
global warming and, 416
incidence in U.S., 663, 663*f*
incubation period, 430*t*
"malignant", *P. falciparum* and, 663
mefloquine (Lariam) to prevent, 571, 664
mefloquine (Lariam) to treat, 557*t*
Plasmodium (protozoan parasite) causing, 348–349, 349*f*, 354*t*, 651*b*, 663
portals of entry, 430*t*

Index

quinine to treat, 12, 571
red blood cells in, 663–664, 664f
sickle cell disease and, 408
treatments for, 664–665
Malassezia (fungus)
as normal microbiota of skin, 402t
as pathogenic, 338t
Malassezia furfur, as normal microbiota of skin, 586
malic acid, 149f
malignant melanoma, alpha interferon to treat, 470
malolactic fermentation, **800**
malt, **800**
Malt extract, fermentation and, 137t
Mammalia (class), position in taxonomic hierarchy, 280f
mammalian cells in culture
advantages for making foreign gene products, 258
genetically engineered erythropoietin (EPO), 260t
genetically engineered interferons, 260t
genetically engineered pulmozyme to treat cystic fibrosis, 260t
genetically modified to host viruses, 258
used to produce monoclonal antibodies, 260t
mammals, domestic or wild, as disease reservoirs, 410t
manganese, as cofactor, 117
mannans, in cell walls of fungi, 330t
Mannheimia haemolytica, 283b
mannitol, biochemical tests and, 139, 139f
mannitol-salt agar, 168–169, 169f, 422b
mannose, as receptor on host cells, 431
mannose-binding lectin (MBL), 460, **468**, 468f
Mantoux test for tuberculosis, 684
mapping (genetic), by conjugation, 237
Marburg virus, **20**, **659**, **660**b
as filovirus, 376t
as potential biological weapon, 649b
margination, 461f, **462**
Margulis, Lynn, 10f, 106
marine algae, toxic, 344
marine mammals
cetacean morbillivirus (CM) and, 283b
killed by toxic algae, 344
mortality rates and veterinary microbiology, 283b
marine microbes, 2
discovered by using FISH technique, 292, 303
marker genes, 256, 256f
common, 250, 251f
Marshall, Barry, 15t
mast cells
histamine present in, 460
IgE antibodies and, 481
in complement activation, 464f, 465, 465f
in hypersensitivity reactions, **523**, 524f
recruited by antimicrobial proteins (AMPs), 471
Mastadenovirus, 375t, 387f
as an icosahedral adenovirus, 372f, 373
cytopathic effects of, 443t
matrix, mitochondrial, **104**, 105f
mattress sterilization, 201
maturation stage
in animal virus multiplication cycle, 385, 386f, **389**, 391f
in bacteriophage multiplication cycle, **380**, 381f
Maui dolphin, brucellosis and, 283b
maximum growth temperature, **157**, 157f
Mayer, Aldolf, 367
MBC (minimal bactericidal concentration) of antibiotics, **572–573**, 573f
McCarty, Maclyn, 10f, 16, 47, 235
McClintock, Barbara, 10f, 15t, 240
MDR tuberculosis, 684
ME (myalgic encephalomyelitis), 633
measles, German. *See* rubella
Measles Initiative, 505b
measles (rubeola), **598–599**, 599f
as a world health problem, 505b
as contagious viral illness, 505b
as notifiable infectious disease, 421t
as persistent viral infection, 392, 394t
incubation period, 430t
macular rash caused by, 589b
mortality rates, vaccination and, 505b
portals of entry, 429, 430t
vaccine, 13, 501, 503t, 504t, 505b
measles virus (*Morbillivirus*), 376t
airborne transmission and, 412, 429, 430t
as potential biological weapon, 649b
cytopathic effects of, 443t
incubation period, 430t
portals of entry, 429, 430t
vaccine, 13, 501, 503t, 504t, 505b
"measly" beef, 357
measurement of microorganisms, 55, 55t
metric units of length/U.S. equivalent, 55t
meat extracts, in complex culture media, 165, 166t
meat products, fermentation and, 137t
mebendazole
mode of action/uses, 564t
to treat intestinal helminthic infections, 571–572
mechanical transmission of disease, by arthropods, **412**, 412f
Medawar, Peter Brian, 14t
mediators (chemical), in allergic reactions, 523–524, 524f
medical discoveries, accidental, 12–13
medical implants
bacterial colonization on, 531b
decontaminated with supercritical carbon dioxide, 202
medical microbiology, 282
medical research, importance of rDNA technology to, 259
medicine
applications of rDNA in, 258–259, 259f
widespread use of antibiotics and, 239–240
medium, light-bending ability of, 58
medulla, of lichen, **339**, 340f
mefloquine (Lariam)
spectrum of activity, 557t
to prevent/treat malaria, 557t, 571, 664
megacolon, 661
megaesophagus, 661
meiosis, 101t, 103
fungal, **333**, 335f, 336f, 337f
in algae, 342f
in plasmodial slime mold, 353f
melanin, genetically engineered, 258
melanoma
genetically engineered interferons to treat, 260t
malignant, alpha interferon to treat, 470
melioidosis, 279, 306, **690–692**, **699** f
Mello, Craig, 15t
membrane attack complex (MAC), 459, 464f, **465**
membrane filters, **191**, 191f
membrane ruffling, 433, 433f
membrane-bound ribosomes, 102
membrane-disrupting toxins, **436**, 438t
membrane-enclosed organelles, in eukaryotes/eukaryotic cells *vs.* prokaryotes, 77
memory cells, **483**, 483f
immunological, **493–494**, 494f, 496f
of B cells, **483**, 483f, 486–487, 493–494, 494f, 496f
of T cells, 486–487, 494
meninges, 611, 612f
inflammation of. *See* meningitis
meningitis, **611**, **612–615**, **617**b
bacterial, **612–615**, **617**b
cephalosporin to treat, 617b
cryptococcosis and, 626–627
Cryptococcus neoformans (fungus) causing, 443
diagnosis and treatment, 405, 614, 614f
gonorrheal, 748
Haemophilus influenzae causing, 311, 432, 502t, 612, **613**, **617**b
in AIDS patients, 626–627
meningococcal. *See* meningococcal meningitis
method of transmission, 617b
Neisseria meningitidis causing. *See* meningococcal meningitis
respiratory tract as portal of entry, 617b
Streptococcus pneumoniae causing, **614**, 617b
vaccines for, 502t, 503, 504t, 612, 617b
viral, 612
meningococcal meningitis, **613**, 613f, **617**b
as notifiable infectious disease, 421t
endotoxins and, 439, 439t, 613
Neisseria meningitidis causing, 306, 404, 421t, 433, 439, 439t, 502t, 612, **613**, 613f, **617**b
vaccine, 502t, 504t
Xigris to treat, 640
meningococcus, **613**
serotypes of, 613
meningoencephalitis, 354t, **611**
primary amebic, **617**b, **629**, 629f
menopause, normal microbiota of reproductive tract and, 745
mercuric chloride, 199
mercury
combined with enzymes to prevent cell functioning, 120
genes for resistance to, 239, 240f
methyl, 33b
mercury contamination, bioremediation of, 33b
mercury poisoning, 33b
mercury water pollution, 778–779
merozoites, **349**, 349f
mesophiles, **157**, 157f, 158
mesosomes, **91**
messenger RNA (mRNA), 16, **47**, 211, 212
discovery of, 16
metabolic activity, measuring to estimate bacterial numbers, 178
metabolic pathways, **114**
amphibolic, **147**
Calvin-Benson cycle, **140**, 142f
energy production and, 123
Entner-Doudoroff pathway, 127
enzymes' role in, 114, 115–121
Krebs cycle, 124, 125f, **127–129**, 128f, 147
microbial diversity among, 142–145, 143f, 145t
pentose phosphate pathway, 127
metabolic rate, increased, with fever, 463
metabolism (microbial), **114–155**
anabolic reactions, 114
biochemical tests and, 137–139
biosynthetic processes, 146–147, 148f
carbohydrate catabolism, 124–135
catabolic reactions, 114
diversity among organisms, 142–145
energy production, 121–123, 141
energy use, 146–147
enzymes' role in, 115–121
integration of, 147, 149f, 150
lipid biosynthesis, 146
lipid catabolism, 136–137
photosynthesis, 140
polysaccharide biosynthesis, 146
protein biosynthesis, 146–147
protein catabolism, 136–137
metachromatic granules, **95**
metal atoms, in metalloproteins, 45
metal ions, as cofactors, 117
metalloproteins, 45
metals, heavy, used in staining of specimens, 63
Metarrhizium fungus, as pest control (parasitic weevils), 339
Metchnikoff, Elie, 10f, 14f
meter (m), metric/U.S. equivalent, 55t
methane
anaerobic respiration and, 132
as energy source produced from bioconversion, 806–807, 807f
as fermentation end-product, 137t
formation, 31, 31f
landfills and, 775–776
methanogens and, 4, 275
produced by methanogens, 4
Methanobacteriales, **302**t
important genera/special features, 302t
Methanobacterium genus/spp. 302t
methanogens, 4, 275, 275f, 302t, 325
no known bacterial examples, 325
phylogenetic relationships, 275f, 281f
Methanogens genus/spp., 302t
as human microbiota, 325
as strict anaerobic archaea, 325
economic importance of, 325
methanol, 37
Methanosarcina, fermentation and, 137t
methicillin, 19, 560, 561
methicillin-resistant *Staphylococcus aureus.* *See* MRSA
methionine (Met)
in protein synthesis, 219, 219f, 220–221f, 276t
structural formula/characteristic R group, 44t
methods of microbial control
considerations when selecting, 188
physical, 187–194, 194t
methotrexate, to treat psoriasis, 533
methyl cyanocobalamide, 117t
methyl group, 38t, 215t, 231, 249
methyl mercury, 33b
methylases, 215t, **231**
methylates (cellular process), **249**
methylene blue dye, 68
methylene blue stain, 69
metric measurements, 55
units of length and U.S. equivalent, 55t
metronidazole, 565
mode of action/uses, 564t
metronidazole (Flagyl), to treat vaginitis caused by *Trichomonas vaginalis,* 571
mezlocillin, 561
MF59 adjuvant, 506
MHC (major histocompatibility complex), **482**, 482f, 496f, **533–534**
MIC (minimal inhibitory concentration) of antibiotics, 284f, **572–573**, 573f
mice
as model for studying viral replication, 377
deer, as disease reservoirs, 410t
field, as disease reservoirs, 410t
genetically modified to make human-murine hybrid, 509
monoclonal antibodies and, 508f, 509
nude (hairless), transplant research and, 538, 539f
Michel, Hartmut, 15t
miconazole, 568, 569f, 600
damages plasma membrane, 558, 559f
mode of action/comments, 564t

microaerophiles, 161*t*, **162**
culturing, 168
microaerophilic bacteria, 312
microbes/microorganisms, **2**
antagonism (competitive exclusion) and, 401
antibiotic-producing, 247, 249, 301*t*, 317, 321, 339, 554, 555*t*, 563
as biofilms, 18–19, 19*f*
as biological weapons, 21, 193, 262, **649*b***, 649*f*
as recyclers of vital elements, 16–17
beneficial activities, 16–18
against human disease, 18–21
biotechnology and, 246–272
chemistry of, 26–53
classification of, 6, 273–298. *See also* classification of organisms
by nutritional patterns, 142–145, 143*f*
classified by nutritional patterns, 142–145, 143*f*
control of, 184–209. *See also* control of microbial growth
cooperation among, 404
culture media for growing, 164–169
disease principles and, 399–418
fastidious, 165, 166*t*
germ theory of disease and, **9**, 11, 404–406, 477
growth in, 156–179. *See also* growth (microbial)
infectious diseases caused by, 404
Koch's postulates and, 11, **404**–406, 405*f*
mechanisms of pathogenicity, 428–448
metabolic classes of, 142–145, 143*f*
metabolic diversity among, 142–145, 143*f*, 145*t*
metabolism, 113–155
naming (nomenclature), 2–3, 4*t*
normal microbiota in humans, 18, 18*f*, 400–404, 402*f*
nosocomial infections and, 414–416, 414*t*
opportunistic, 403–404
pH ranges and, 37
portals of entry, 429, 430*t*
preferred temperature ranges, 157–158, 157*f*
preparing specimens of, 54, 68–72
recombinant DNA technology and, 246–272
symbiosis and, 401–403, 402*f*
types of, 3–6, 5*f*
used as genetic engineering "factories," 247
used in production of foodstuffs, 797–800
virulence of pathogenic, **428**
virulence determination, 71
with ability to survive inside phagocytes, 459
microbial diseases
of cardiovascular system, 637–673, **643*b***, **650*b***, **651*b***, **668*b***
of digestive system, 705–742, **710*b***, **722*b***, **724*b***, **729*b***, **734*b***, **735*b***
of eyes, 603–605, **604*b***
of lymphatic system, 637–673, **643*b***, **650*b***, **651*b***, **668*b***
of nervous system, 610–636, **617*b***, **628*b***, **632*b***
of reproductive systems, 747–761, **759*b***, **761*b***
of respiratory system, 676–698, **681*b***, **687*b***, **699*b***
of skin, 584–603, 586–598, **589*b***, **590*b***, **592*b***
of urinary system, 743, 746–747, **748*b***
microbial diversity
environmental niches and, 325
in habitats, 767
reflected by extremes in size, genomes, 325–326
symbiosis and, **767**
microbial ecology, **17**
microbial enzymes, 2
microbial genetics, **16**, 210–245. *See also* genetic material
microbial growth measurements, 174–179, 175*f*-179*f*
microbiological assays, 165, 169, 178
microbiology
agricultural, 303
branches of, 13, 16
forensic, **262**
history, 6–16
biogenesis theory, 8
first observations, 7, 7*f*
Golden Age of, 9–11, 9*f*, 10*f*
milestones, 9, 10*f*
modern achievements, 13–16, 14–15*t*
spontaneous generation debates, 8, 9*f*
medical, 282
Nobel prizes awarded in, 14–15*t*
soil, 768–776
veterinary, 223*b*, 283*b*
microbiota, normal, **18**, 18*f*, **400–404**, 400*f*, 402–403*t*
animal, protozoa as part of, 346
Microcladia (red algae), 341*f*
Micrococcus
as normal microbiota of eye, 402*t*
as normal microbiota of skin, 402*t*
Micrococcus genus/spp., as normal microbiota of urethra, 403*t*
microfilaments, 99*f*, 101
microfold cells. *See* M cells
microglial cells, 457
microinjection (of foreign DNA), **254**, 254*f*
micrometer (μm), **55**
metric/U.S. equivalent, 55*t*
Micromonospora purpurea, gentamicin derived from, 555*t*
micronucleus, of *Paramecium*, 350*f*
microorganisms. *See* microbes/microorganisms
microscopes/microscopy, 2, 54–75
compound light, 7, 7*f*, 55, **56**, 56*f*, 58–59, 58*f*, 59*f*, 60*f*
electron, 16, **63–65**, 64*f*, 67*t*
scanning (SEM), **64**–65, 64*f*, 67*t*
transmission (TEM), **63**–64, 64*f*, 67*t*
first used to see bacteria, 7*f*, 55
in identification of microorganisms, 282
light (LM), 7, 7*f*, 55, **56**, 56*f*, 58–59, 58*f*, 59*f*, 60*f*
brightfield, **59**, 60*f*, 66*t*
confocal, **62**, 62*f*, 67*t*
darkfield, **59**, 60*f*, 66*t*
differential interference contrast, **60**, 61*f*, 66*t*
fluorescence, **61**–62, 61*f*, 66*t*
phase-contrast, **59**–60, 60*f*, 66*t*
relationship between specimen size and resolving power, 58, 58*f*
scanned-probe, **65**, 65*f*, 68*t*
atomic force (AFM), 58*f*, **65**, 65*f*, 68*t*
scanning tunneling (STM), **65**, 65*f*, 68*t*
scanning acoustic (SAM), **63**, 63*f*, 67*t*
summary of (features/typical image/uses), 66–68*t*
three-dimensional images by, 61*f*, 62*f*, 64*f*, 65*f*, 66–68*t*
to determine temperature variations inside cells, 65
to study living cellular activities, 62, 62*f*, 63, 63*f*, 65
two-photon (TPM), **62**, 62*f*, 67*t*
units of measurements for, 55, 55*t*
microscopic count of bacteria, **176**, 177*f*, 178
Microspora (fungus), **348**
human diseases caused by, 348
position in evolutionary tree, 275*f*
Microsporum, 338*t*
Microsporum (fungus)
as pathogenic fungi, 338*t*
cutaneous mycosis and, 600
reservoirs/transmission method, 410*t*
microtiter plates, 510, 511*f*
microtubules, **98**, 99*f*, 100*f*, 101, 101*t*
microsporidial protozoa and, 348
of centrioles, 105
microwaves, **193**
mildew
copper compounds to prevent, 199
damp shower curtains and, 192
mercurials to control in paints, 199
milk
breast, IgA antibodies in, 480, 481
counting number of bacteria in, 176, 178*f*
fermentation and, 137*t*
food allergies and, 525
lactic acid fermentation and, 135*b*
pasteurization and, 9
pasteurization of, 190–191
milk production (dairy cow), bovine growth hormone and, 267, 267*t*
milk products, *Listeria* in, flow cytometry to detect, 288
millimeter (mm), metric/U.S. equivalent, 55*t*
Milstein, César, 15*t*
minimal bactericidal concentration (MBC) of antibiotics, **572–573**, **573*f***
minimal inhibitory concentration (MIC) of antibiotics, 284*f*, **572**–573, 573*f*
minimum growth temperature, **157**, 157*f*
mining industry, microbes used in, 247
minocycline, 565
miscarriage, induced by endotoxins, 437
missense mutation, **227**–228, 228*f*
Mitchell, Peter, 10*f*, 14*t*
mites, 362*t*
ivermectin effective against, 572
mitochondria/mitochondrion, 99*f*, 102, **104**–105, 105*f*
Archaezoa as eukaryotes that lack, 347
electron transport chain (system) and, 129
origin of, 275*f*, 277, 277*f*, 326
mitosis, 101*t*, 103, 276*t*
fungal, 336*f*
in algae, 342, 342*f*
in diatoms, 343*f*
mitosome, organelle of archaezoans, 347
mitotic spindle, 105
Mixotricha (protozoan), that lives in termite hindgut, 107*b*
μm (micrometer), **55**
metric/U.S. equivalent, 55*t*
mm (millimeter), metric/U.S. equivalent, 55*t*
MMR vaccine, 504*t*, 598
MMWR (*Morbidity and Mortality Weekly Report*), **420**
moderate-temperature-loving microbes (mesophiles), **157**, 157*f*
moist heat sterilization, 188–190, 189*f*, 194*t*
molasses, fermentation and, 137*t*
mold-related illness, increases in, and safe exposure laws, 329
moldlike forms of dimorphic fungi, **332**, 332*f*
molds, 2, 4, 5*f*
as aerobic organisms, 333
as eukarya, 6
as eukaryotes/eukaryotic cells, 76
bacterial food spoilage *vs.* damage done by, 339
bread, 5*f*
chemical food preservatives active against, 199–200, 204*t*
growing in homes, allergic responses and, 443
high osmotic pressure and growth of, 192
hyphae of, 331, **331**–332, 331*f*
included among ascomycota, 334
included in Kingdom Fungi, 281
low moisture conditions and growth of, 192
mucor, 5*f*
penicillin discovery and, 12
pH and growth of, 159
prokaryotic actinomycetes resembling, 320
saprophytic, 333
slime, 4
spore formation and, 331*f*, 332–333, 334*f*
thallus (body) of, 331
molecular biology, **16**
molecular genetics
cloning procedures of, 249
ethical issues and, 268
molecular weight, **32**
molecules, **27**
covalent bonds and, 30–31, 31*f*
heat absorption by, 35
how atoms form, 28–32
hydrogen bonds and, 31–32, 32*f*
important biological,, 34–49. *See also* *specific molecules*
inorganic, 34–37
ionic bonds and, 28, 30, 30*f*
macromolecules, 34, **38**
organic, 37–49
polar, **34**–35
mole(s) (unit of measure), **32**
Molluscipoxvirus, 375*t*
mollusks, paralytic shellfish poisoning (PSP) and, 344, 354*t*, 444
monkeypox, **596**
as orthopoxvirus, 596
as potential biological weapons, 649*b*
human-human transmission being monitored, 596
rash caused by, 590*b*
monkeys
as disease reservoirs, 410*t*, 659, 660*b*
green, AIDS in, 377
monobactams, **561**
mode of action/spectrum of activity, 562*t*
monoclonal antibodies (Mabs), **507–509**, 508*f*
as tool for delivering cancer therapies, 538
chimeric, **509**
discovery of, 507
humanized antibodies, **509**
hybridoma and, **507**
importance in diagnostics and medical therapy, 260*t*, 507, 508
in home pregnancy tests, 515, 517*f*
industrial fermentation used in making, 802
to treat psoriatic arthritis, 533
to treat viral infection, 383
monocytes, **454**, 455*t*
developing into phagocytic macrophages, 457
in inflammatory response, 461*f*
Monod, Jacques, 10*f*, 14*t*, 16, 224
monoecious parasitic helminths, **355**
monomer, **38**
antibody structure as, **479**, 480*f*
monomorphic bacteria, **79**
mononuclear phagocytic (reticuloendothelial) system, **457**, 644
mononucleosis, infectious, 375*t*
monosaccharides, **39**
monotrichous flagella, **81**, 81*f*
Montagnier, Luc, 15*t*
Montagu, Mary, 501
Moraxella catarrhalis, otitis media caused by, 679
Moraxella genus/spp., 300*t*, **308**
as human pathogens, 300*t*
Moraxella lacunata, conjunctivitis and, 308
Morbidity and Mortality Weekly Report (*MMWR*), **420**
morbidity rate, **420**

Morbillivirus (measles virus), 376*t*
as a persistent viral infection, 392
mordant, **69**, 72*t*, 87
morphological characteristics, to classify/identify organisms, 284–285
morphology of bacteria, 77–79, 78*f*, 79*f*
mortality, influenza-associated pediatric, as notifiable infectious disease, 421*t*
mortality rate, **420**
mosaic disease of cauliflower, 394*t*
mosaic disease of tobacco, 16, 367, 368, 369*f*
mosquitos, 362*f*, 362*t*
Culex, transmitting West Nile virus, 626
diseases transmitted by, **362*t***
that are vectors, **362*t***
viruses transmitted by. *See* arboviruses
mosses, as eukarya, 6
most probable number (MPN) method, **175**, 177*f*
motility, **82**, 83*f*
characteristic of protozoa, 330*t*
gliding, **83–84**
of spirochetes, 82–83, 84*f*, 322–323, 324*f*
pili's role in, 83–84
twitching, **83**
mouse cells, genetically engineered products produced in, 260*t*
mouse mammary tumor virus, 389*f*
mouth
dental caries (tooth decay), **707–709**, 707*f*, **710*b***
gingivitis, **709**, 710*b*
normal microbiota of, 1*f*, 18*f*, 403*t*
periodontal disease, **709**, 709*f*, 710*b*
periodontitis, **709**
mouthwashes, 199
movement
of bacteria, 82, 83*f*
proteins and, 42
moxifloxacin, 567
MPN (most probable number) method, **175**, 177*f*
mRNA, in RNA viruses, 387, 388*f*
mRNA (messenger RNA), 16, **47**, 211, 212, **216**, 218*f*
codons and, **219**, 219*f*, 220–221*f*
in eukaryotic RNA processing, 220, 222*f*
in induction, 224, 224*f*
in transcription, 216–217, 218*f*, 221
in translation, 217, 219–221, 220–221*f*
positive RNA acting as in translation process, 223*b*
viral RNA and reverse-transcription PCR, 251
MRSA (methicillin-resistant *Staphylococcus aureus*), **20**, **560–561**
cellulitis caused by, 593*b*
community outbreaks *vs.* health-care-associated, 574
emerging infectious diseases and, **20**, 417*t*
hemodialysis patients and, 422*b*
nosocomial infections and, 422*b*
outbreaks among professional athletes, 593*b*
platensimycin developed in response to, 566
resistant to new drugs, 574
tigecycline (Tygacil) developed in response to, 565
USA100, USA 300 strains, 422*b*
MS (multiple sclerosis), genetically engineered interferons to treat, 260*t*, 270
mucocutaneous mycoses, 338*t*
Mucor
as opportunistic fungi, 338
as pathogenic fungi, 338*t*
Mucor (fungi), 5*f*
Mucor indicus, as dimorphic fungi, 332, 332*f*
mucormycosis, 338
mucous membranes, **451**, **585**
as barrier to pathogens, 451, **471*t***, 584
as first line of defense, 450*f*, **471*t***
as portals of entry, 429, 430*t*, 445*f*
broken, susceptibility to infections, 415, 415*t*, 451
ciliated, of lower respiratory tract, 675, 676*f*
IgA antibodies and, 481*t*
of gastrointestinal tract, 452
of genitourinary tract, 452
of nose, 452
of respiratory tract, 452
structure of, 585
mucus, **451**, **452**
cervical, antimicrobial activity of, 453
ciliary escalator and, 452, 452*f*
IgA antibodies in, 480
lysozyme in, 88, 453
mules, reported cases of rabies in, 624*f*
Mullis, Kary B., 15*t*
multicellular animal parasites, 6
multicellularity
algae and, 341–342, 341*f*
among eukaryotic microbes, 330*t*
multiple sclerosis (MS)
as cell-mediated autoimmune disease, **532–533**
Epstein-Barr virus and, 533, 658
genetically engineered interferons to treat, 260*t*, 470, 533
HLA typing to determine susceptibility, 534*t*
interleukin-12 and, 493*b*
monoclonal antibodies to treat, 517
multiplication of viruses, 379–389
of animal viruses, 382–389. *See also* animal viruses
of bacteriophages, 379–382. *See also* bacteriophages
one-step growth curve, **379**, 379*f*
mumps, **721**, 721*f*, **729*b***
mumps virus (*Rubulavirus*), 376*t*
as notifiable infectious disease, 421*t*
incubation period, 430*t*
portals of entry, 430*t*
vaccine, 13, 502, 503*t*, 504*t*, 721
municipal chlorination, household bleach equivalent in emergencies, 197
municipal water treatment systems, chloramines to disinfect, 197
munitions, 2
murein. *See* peptidoglycan
murine (mouse) cells, 508*f*, 509
muromonab-CD3, 260*t*, 508, 509
Murray, Joseph E., 15*t*
Murray, Robert G.E., 274
muscle contraction, proteins and, 42
muscle contractions, uncontrollable, tetanus toxin causing, 437
muscles, parasitic helminths and, 361*t*
muscular dystrophy, Duchenne's, 18
mushrooms, 4
as eukarya, 6
macroscopic species included in Kingdom Fungi, 281
produced by Basidiomycota fungi, 337
toxins produced by, 443
mussels, paralytic shellfish poisoning (PSP) and, 344, 354*t*
mutagenesis, site-directed, **249**
mutagens, **228**, 229–231
Ames test and, 232–233, 233*f*
as carcinogens, 232–233
chemical, 228–230, 229*f*, 232
experimental uses of, 231
radiation, 228, 230–231, 230*f*
spontaneous rate of mutation and, 231
mutation rate, **231**
mutations, **226**–233
acquired by West Nile virus, 223*b*
antibiotic resistance and, 574, 577*t*
beneficial, 231
evolution and, 231
frameshift, **228**, 228*f*
frequency of, 231
HIV and, 541–542
horizontal gene transfers to other bacteria, 574, 577*b*, 577*f*
identifying, 231–232, 232*f*
chemical carcinogens, 232–233, 233*f*
missense, **228**, 228*f*
nonsense, **228**, 228*f*
point (base substitution), **227**, 227*f*
positive (direct) selection to identify, 231
random, 231, 428
rate of, **231**
repair of, 230–231, 230*f*
retroviruses and, 541–542
selection methods to identify mutations, **231**
silent (neutral), 226–227
spontaneous, **228**
that result in resistance to antibiotics, 228
types, 227–228, 228*f*
mutualism, **402–403**, 402*f*
in lichens, 339
myalgic encephalomyelitis (ME), 633
myasthenia gravis, **532**
mycelia/mycelium, 4, 331*f*, **332**, 335*f*, 336*f*, 337*f*
mycelium. *See* mycelia/mycelium
mycetoma, 321
mycobacteria
antibiotics that inhibit, **563**
antimicrobial resistance and, 201*b*, 203, 203*f*, 203*t*, 320
as aerobic, non-endospore-forming rods, 320
filamentous growth and, 320
mycolic acids in cell walls of, 87–88, 320, 563
pathogenicity of, 320
rapidly-growing, 201*b*
slow-growing, 201*b*, 320
identification tests, 144*b*, 203
Mycobacterium abscessus infection, from Zephiran-soaked cotton balls, 201*b*
Mycobacterium avium, 144*b*
interleukin-12 (IL-12) and, 493*b*
Mycobacterium avium-intracellulare, 544*t*, 685
Mycobacterium bovis, 144*b*, 685
Mycobacterium genus/spp., 301*t*, **320**
acid-fast stains and, 70, 71*f*, 87–88
as acid-fast bacteria, 301*t*
as human pathogens, 301*t*
cell walls distinctive to, 41, 85, 86*f*, 320
disinfectants susceptible to, 195
G + C ratio of, 316
lipid inclusions of, 96
mycolic acids in cell walls of, 87–88, 320
pathogenic species of, 320
Mycobacterium leprae
acid-fast stain identifies, 70, 71*f*
armadillos used to culture, 619
as slow growing mycobacteria, 201*b*
cultivation and, 404
culture media for, 167
grows in peripheral nervous system, skin cells, 619
leprosy caused by, 619–620, 632*b*
Mycobacterium tuberculosis, 54*f*, 682–685, 682*f*, 683*f*
acid-fast stain to identify, 70
antibiotics to treat, 563
as disease associated with AIDS, 544*t*
as slow growing mycobacteria, 201*b*
can survive in phagocytes, 459
desiccation resistance and, 192
disinfectants and, 202
fluorochrome used to stain, 61
found in Egyptian mummies, 6
incubation period, 430*t*
Koch's experiments with, 404
lipid-rich cell wall of, 41
pathogenesis of, 682, 683*f*, 684
portals of entry for, 429, 430*t*
urease test to identify, 144*b*
virulence and, 432
Mycobacterium ulcerans, Buruli ulcer caused by, 594
mycolactone, 594
mycolic acid, **87–88**, **432**
antibiotics that inhibit synthesis of, 563
of *Mycobacterium tuberculosis*, virulence and, 432
mycology, **13**, **330**
branches included in, 13
mycophenolate mofetil, 537
Mycoplasma genus/spp., 87, 301*t*, 319, **320**
appearance of, 320
as human pathogens, 301*t*
atypical (no) cell walls of, 87, 301*t*, 319*f*
culture media for, 320
degenerative evolution and, 320
originally considered to be viruses, 320
plasma membrane uniqueness, 87, 89
small genomes of, 320
Mycoplasma pneumoniae, 319*f*, 320
mycoplasmal pneumonia, **687*b***, **688**, 688*f*
tetracyclines to treat, 565
mycoplasmas, 319–320, 319*f*
G + C ratio of, 316
sterols in plasma membrane of, 41, 87, 89
Mycoplasmatales (order), **319–320**
important genera/special features, 301*t*
mycorrhizae (symbiotic fungi), **330**, **767**, 768*f*, 769*f*
mycoses (fungal infections), **335–339**, 338*t*
cutaneous, **337**, 568, **600–601**, 600*f*
drugs affecting and effect on animal cells, 336
of skin and nails, **600–602**, 600*f*
opportunistic, 336, **337**–339
subcutaneous, 336, **336**–337, **601**
superficial, 336, **337**
systemic, 336, **336**
mycosis, **335**, **600**. *See also* mycoses
mycotoxins, **443**
myeloma cells, in monoclonal antibody production, 508*f*
myelomas, 507
Myxobacteria, **312**
fruiting body of, 57*b*, 57*f*, 312, 313*f*
gliding motility of, 83–84, 312, 313*f*
nonexistent as individual cells, 57*b*, 57*f*
Myxococcales, **312**, 313*f*
important genera/special features, 301*t*
life cycle of, 313*f*
predatory features, 312
Myxococcus fulvus, nutrition source is bacteria encountered, 312, 313*f*
Myxococcus genus/spp., 301*t*, **312**, 313*f*
as gliding, fruiting bacteria, 301*t*, 312, 313*f*
Myxococcus xanthus
exhibiting pack behavior, 57*b*, 57*f*
nutrition source is bacteria encountered, 312, 313*f*
myxospores, of *Myxococcus* bacteria, 312, 313*f*

N-acetylglucosamine (NAG), 85, 85*f*, 86*f*
chitin and, 98
N-acetylmuramic acid (NAM), 85, 85*f*, 86*f*
N-acetyltalosaminuronic acid, 87
NA spikes of influenzavirus, 692–693, 692*f*
NAD+ (nicotinamide adenine dinucleotide)
in cellular oxidations, 122, 123, 123*f*, 124, 125*f*

  in electron transfers, 117*t*
  in electron transport chain, 129–130, 129*f*
  in fermentation, 125*f*, 132, 134, 134*f*, 136*f*
  in Krebs Cycle, 127–128, 128*f*
NADH
  in alcohol fermentation, 135*b*, 136*f*
  in cellular oxidations, 122, 123, 123*f*, 124, 125*f*
  in electron transport chain (system) in eukaryotic cells, 129–130, 129*f*
  in Krebs cycle, 127–128, 128*f*
  in photosynthesis, 140, 141*f*
NADP+ (nicotinamide adenine dinucleotide phosphate)
  as coenzyme in cellular metabolism, 117
  in Calvin-Benson cycle, 142*f*
  in fermentation, 125*f*, 132, 134, 134*f*, 136*f*
  in photosynthesis, 140, 141*f*
NADPH
  in Calvin-Benson cycle, 142*f*
  in photophosphorylation, 123
  in photosynthesis, 140, 141*f*
Naegleria fowleri (amoeba), 354*t*, 610*f*, **629**, 629*f*
naftifine, 564*t*
NAG (N-acetylglucosamine), 85, 85*f*, 86*f*
  chitin and, 98
nails (finger/toe), cutaneous mycoses and, 337, 569
naked DNA
  transformation process and, 234, 253
  vaccines and, 503
naked RNA, viroids and, 394
nalidixic acid, 567
  mode of action/spectrum of activity, 563*t*
NAM (N-acetylmuramic acid), 85, 85*f*, 86*f*
names for living organisms (scientific nomenclature), 2–3, 4*t*, 278–279
nanobacteria, **326**
nanometer (nm), **55**
  metric/U.S. equivalent, 55*t*
nanons, **326**
nanotechnology, **264**, 264*f*
naphthoquinones, 117*t*
narrow spectrum antibiotics, **555**, 557*t*
nasal epithelium, normal microbiota of, 400*f*
nasal passages, bacteria found in, 168, 588
nasal secretions, staphylococci in, 318, 588
nasal spray for influenza vaccine, 506
nasolacrimal ducts, 451*f*
nasopharyngeal cancer, Epstein-Barr virus and, 658
Natamycin (pimaricin), antifungal antibiotic used in foods, 200
Nathans, Daniel, 10*f*, 14*t*
National Institute of Allergy and Infectious Diseases (NIAID), interleukin-12 research by, 493*b*
natural classification systems, 274, 277
natural gene silencing, 259
natural killer (NK) cells, **454**, 455*t*, 472*t*, **491**, 491*t*
natural penicillins, **559**, 560*f*
natural recombination of DNA in microbes
  competence and, 236, 253
  conjugation and, 236, 253
  occurrence of, 247
  transformation and, **234**–236, 235*f*
    in genetic engineering, **253**
natural selection, 241
  antibiotic resistance and, 577*b*
  artificial selection *vs.*, 249
  Charles Darwin and, 274
  coevolution and, 428
  definition of, 428
  evolution and, 241, 274, 428
  horizontal gene transfer and, 213*f*, **234**, 577*b*
  resistance factors of bacteria and, 239–240, 240*f*
Necator americanus (hookworm), 359–360, 361*t*
necrosis, **646**
necrotizing fasciitis, 287, 318, 422*b*, **591**, 591*f*
  rash caused by, 592*b*
necrotizing illness, due to leukocidin toxin MRSA-infection, 422*b*
Needham, John, 8
negative (indirect) selection method to identify mutations, **231**
negative staining, 63, **69**, 72*f*, 72*t*
  bacterial capsules and, 72*f*, 72*t*, 79
  bacterial flagella and, 72*f*, 72*t*
  of *Mastadenovirus*, 387*f*
Neisser, Max, 10*f*
Neisseria genus/spp., 300*t*, **306**
  antibiotic resistance and, 751*b*
  as human pathogens, 300*t*
  as normal microbiota of mouth, throat, 403*t*
  genetic transformation natural in, 236
  penicillinase-producing plasmid and *Streptococcus*, 240
Neisseria gonorrhoeae, 306, 306*f*, 747–750, 750*f*
  ability to destroy IgA antibodies with IgA proteases, 433
  antibiotic resistance and, **751*b***
  antigenic variation and, 433, 749
  chemically defined medium used to cultivate, 165, 166*t*
  complement system evasion by, 468
  desiccation and, 192
  fimbriae contain adhesins that attach to host cells, 431–432
  fluoroquinolone-resistant, 750, 751*b*
  gonorrhea caused by, 306, **747**. *See also* gonorrhea
  grows inside human epithelial cells, leukocytes, 432
  incubation period, 430*t*, 748
  ophthalmia neonatorum and, **603–604**, **748**–749
  oxidase test for, 139
  portals of entry, 429, 430*t*, 749
  twitching motility in, 83
Neisseria meningitidis, 306
  as endotoxin producer, 439
  as opportunistic pathogen, 404
  can produced IGA proteaes that destroy host's antibodies, 433
  iron source for, 470
  meningitis caused by, 306, 404, 421*t*, 433, 439, 439*t*, 502*t*, 612, **613**, 613*f*, **617*b***
  phagocytes and, 228
  vaccine against, 502*t*
Neisseria meningitis. *See* meningococcal meningitis
Neisseriales, important genera of, 300*t*
Nelson, Karen, 277
Nematoda, 192, 353, **358–361**, 361*t*. *See also* nematodes
  Ancylostoma duodenale, 359–360, 361*t*
  Anisakines, 361, 361*t*
  Ascaris lumbricoides, 358–359, 361*t*
  Dirofilaria immitis, 360, 360*f*
  Enterobius vermicularis, 358, 360*f*, 361*t*
  Necator americanus, 359–360, 361*t*
  Trichinella spiralis, 361*t*
nematodes (roundworms), 192, **358**–361, 360*f*
  aniskines, 361
  characteristics, 358
  eggs infective for humans, 358–359, 360*f*
  freezing temperatures and, 192
  habitat, 358
  heartworm, 360, 360*f*
  hookworm, 359–360
  ivermectin effective against, 572
  larvae infective for humans, 359–360
  pinworms, 358, 360*f*
  sexual dimorphism in, **358**–359
  Wolbachia bacteria and, 307*b*, 360
neomycin, 565
  mode of action/spectrum of activity, 562*t*
  produced by *Streptomyces fradiae*, 555*t*
neonatal herpes, **757–758**
neonatal sepsis, caused by *Streptococcus agalactiae*, 640
nephritis, 405
nervous system
  microbial diseases, 610–636
    bacterial, 611–620, **617*b***
    fungal, **617*b***, 626–627
    prions, 629–632, **632*b***
    protozoan, **617*b***, 627–629, 629*f*
    viral, **617*b***, 620–626, **628*b***, **632*b***
  pathogenic invasion routes, 611
  structure/function of, 611, 611*f*
neuraminidase (N) proteins
  influenza A virus subtypes and, 370*b*, 371*t*
  to treat influenza, 570
neuraminidase (NA) spikes of *Influenzavirus*, 692–693, 692*f*
neurocysticercosis, 361*t*, **733**
neurological diseases, caused by prions, 393
neurological disorders
  caused by diatoms, 343
  Huntington's disease, 228
neurosyphilis, 754
neurotoxins, 239, 435, 437, 440, 443
  algae produced, 343, 344, 444
  fungi produced, 443
  plasmids and *Clostridium tetani*, 239
neutral (silent) mutations, 226–227
neutralization reactions, 484, **485**, 485*f*, **512**, 513*f*
  cytopathic effects of viruses and, 441, 512
  viral hemagglutination inhibition test, **512**, 513*f*
neutrons, 27, 27*f*
neutrophils, **454**, 455*t*
  as second line of defense, 450*f*
  cathelicidins produced by, 470
  defensins produced by, 470
  in inflammatory response, 461*f*
  staining and, 454
nevirapine, 571
newborn diseases, neonatal sepsis caused by *Streptococcus agalactiae*, 319
newborns
  candidiasis occurring in, 339
  IgG antibodies and, 481*t*
  silver nitrate solutions and, 198, 204*t*
  skin infections in, 196, 196*f*
Newcastle disease in chickens, 376*t*
newspapers, fungi capable of growing in, 333
NGU. *See* nongonococcal urethritis
niacin (nicotinic acid), coenzymatic functions, 117*t*
NIAID (National Institute of Allergy and Infectious Diseases), interleukin-12 research by, 493*b*
nickel allergy, 530
niclosamide
  mode of action/uses, 564*t*
  spectrum of activity, 557*t*
  to treat tapeworms, 571
nicotinamide adenine dinucleotide (NAD+), **117**
  as coenzyme in cellular metabolism, 117
  in electron transfers, 117*t*
nicotinamide adenine dinucleotide phosphate (NADP+), **117**
  as coenzyme in cellular metabolism, 117
nicotinic acid (niacin), coenzymatic functions, 117*t*
Nightingale, Florence, 418
nigrosin dye, 69
NIH (National Institutes of Health), priorities for issues relating to emerging infectious diseases, 418
9 + 0 array (microtubules), 105
9 + 2 array (microtubules), 98, 100*f*
Nipah virus
  as potential biological weapon, 649*b*
  emerging infectious diseases and, 417*t*
nisin, 200
  as a bacteriocin, 310, 401, 578
nitazoxanide, 571
  mode of action/uses, 564*t*
nitrate reduction test, 144*b*
nitrates
  anaerobic respiration and, 132
  importance to agriculture, 305
  Pseudomonas bacteria and nitrogen fertilizers, 308
nitrates/nitrites, as food preservatives, 200, 204*t*
nitric oxide, 459
nitrification, 305
nitrifying bacteria, 96, 300*t*, **305**, 771
nitrile gloves, 531
nitrites
  anaerobic respiration and, 132, 137*t*
  as energy source, 145, 305
nitrites/nitrates, as food preservatives, 200, 204*t*
Nitrobacter genus/spp., 145, 300*t*, **305**, 770*f*, 771
nitrocellulose filters, 257*f*, 263*f*
nitrogen, 2
  anaerobic respiration and, 132, 137*t*
  cyanobacteria use of, 17, 160
  electronic configuration, 29*t*
  hydrogen bonding and, 31–32
  in organic compounds, 37
  requirements for microbial growth, 160
  sources of, 160
  symbol/atomic number/atomic weight, 27*t*
nitrogen cycle, **770–772**, 770*f*
  anaerobic respiration and, 132
nitrogen fixation, **160**, 770*f*, **771–772**
  alphaproteobacteria genera and, 303
  Azotobacter and *Azomonas* used to demonstrate, 308
  genetically modified plants and, 267, 267*t*
nitrogen-fixing bacteria
  Azomonas, 300*t*, 308
  Azospirillum, 300*t*
  Azotobacter, 300*t*, 308, 772
  Beijerinckia, 300*t*, 772
  Bradyrhizobium, 300*t*
  Clostridium, 772
  cyanobacteria, 314, 314*f*, 772
  free-living, 772–773
  Rhizobium, 300*t*
  symbiotic, **772**
    Bradyrhizobium, 300*t*, 772
    Frankia, 301*t*, 772
    Rhizobium, 300*t*, 772
nitrogenous bases, normal, *vs.* nucleoside analogs, 229–230
nitrosamines, **200**
Nitrosomonadales, important genera of, 300*t*
Nitrosomonas genus/spp., 145, 300*t*, **305**
nitrous acid, as mutagenic chemical, 229, 229*f*
nitrous oxide, anaerobic respiration and, 132
NK cells. *See* natural killer (NK) cells
nm (nanometer), **55**
  metric/U.S. equivalent, 55*t*
Nobel prizes awarded in microbiology, 14–15*t*
  first prize awarded, 477
Nocardia asteroides, pulmonary infection caused by, 321
Nocardia genus/spp., 301*t*, **321**
  actinomycetes informal name for, 320
  acid-fast stains identify pathogenic species of, 70, 87

Index

as filamentous, branching aerobic bacteria, 301*t*, 321
as opportunistic pathogens, 301*t*
mycolic acid in cell walls of, 87–88
nodular cystic acne, **594**, 595*f*
norfloxacin, mode of action/spectrum of activity, 563*t*
nomenclature
binomial, **278**–279
scientific, 2–3, 4*t*, 278–279
rules for name assignment, 279
non nucleoside reverse transcriptase inhibitors, 548
non-Hodgkin's lymphoma, monoclonal antibodies to treat, 509
nonbullous impetigo, 587*f*, 588
noncommunicable diseases, **406**
noncompetitive inhibitors of enzymes, **120**, 120*f*
noncyclic photophosphorylation, **140**, 141*f*
noncytocidal effects, 441
nonenveloped polyhedral virus, 372*f*
nonenveloped viruses, 372*f*, **373**, 389
biocidal resistance and, 203, 204*t*
nongonococcal urethritis (NGU), 322, 459, **750**–751, **761*b***
nonionizing radiation, **193**, 193*f*, 194*t*
nonpolar tails of phospholipids, 41, 42*f*
of plasma membrane, 89, 90*f*
nonproteobacteria gram-negative bacteria, 301*t*, **313**–315
important genera/special features, 301*t*
nonself *vs.* self recognition, 477, 492–493, 494, 496*f*
autoimmune diseases and, 532–533
hyperacute rejection and, 536
immune system tolerance of fetus and, 534–535
major histocompatibility complex (MHC) and, **482**, 486, **533**–534
thymic selection and, 486, 532
transplant rejection and, 534–535
nonsense codons (stop codons), 212, **219**, 219*f*
nonsense mutation, **228**, 228*f*
nonspecific urethritis (NSU), **750**–751, **761** *f*
nontyphoidal salmonellae, 713
norfloxacin, 567
normal flora. *See* normal microbiota
normal microbiota, of skin, 402*t*, 450*f*, 453, **585**–586
normal microbiota (flora), **18**, 18*f*, **400**–404, **453**
body's defenses and, 401, 450*f*, 453
broad-spectrum antibiotics destruction of, 555
by body region, 400*f*, 402–403*t*
competitive exclusion and, **401**
composition of, 401
distribution of, 401
factors affecting, 401
in animals
absent in germfree research mammals, 401
protozoa part of, 346
infectious disease *vs.*, 18
innate immunity and, 450*f*, 453
microbial antagonism and, **401**
of digestive system, 706–707
of reproductive system, 745
of respiratory system, 675–676
of skin, **585**–586
of urinary system, 745
symbiotic relationships with host, **401**–403, 402*f*
commensalism, **401**, 402*f*
mutualism, **402**–403, 402*f*
parasitism, 402*f*, **403**
transient microbiota and, **400**
Norovirus, 375*t*
noroviruses, **728**–729, **729*b***
outbreak, traced using genomics of pathogens, 262, 266*b*
North American blastomycosis (blastomycosis), **697**, **699*b***
nose, normal microbiota of, 403*t*
Nosema, 354*t*
Nosema locustae, 346
nosocomial infections, **413**–416, 414*f*, 422*b*
antibiotic Primaxin active against, 561
antibiotic-resistant pathogens and, 414
before aseptic surgery, 184
biofilms and, 18, 19*f*, 163
case history reports (Clinical Focus)
infection following steroid injection, 201*b*
nosocomial infections (bacteremia), 422*b*
catheterization and, 164*b*
causes of, 413, 414*f*
chain of transmission and, 414*f*, 415–416
childbirth fever of mid-1800's, 11, 418
compromised hosts and, 414*f*, **415**, 415*t*
control of, 416
delayed bloodstream infection following catheterization, 164*b*
DNA fingerprinting to determine source, 290, 290*f*
gram-negative microbes and, 414, 640
gram-positive microbes and, 414
immune systems responses to, 415
invasive procedures/devices and risks of, 19*f*, 414
microbes in hospital settings, 414, 414*t*
opportunistic pathogens and, 403–404, 414
principal sites in body, 415*t*
Pseudomonas bacteria responsible for one in ten, 308
sepsis as, **639**–**641**, **643*b***
Staphylococcus aureus and, 19–20, 318
vancomycin-resistant enterococci (VRE) and, 417*t*, **563**, 577*b*, 640
Notifiable Infectious Diseases (U.S. Public Health Service), **420**
Novo Nordisk Biotech, 3*b*
Noxafil (posaconazole), 569
NSU (nonspecific urethritis), **750**–751, **761*b***
"nubiotics," 579
nuclear envelope, 102*f*, **103**
of *Gemmata obscuriglobus,* 322*f*
nuclear membrane, 101*t*
nuclear pores, 102*f*, **103**
nucleic acid hybridization, 291–293, 291*f*
by DNA chip technology, 292, 293*f*
by DNA probes, **291**, 292*f*
by fluorescent in situ hybridization (FISH), **292**–293, 294*f*
by ribotyping and rRNA sequencing, **292**
by Southern blotting, **291**, 292*f*
in HIV testing, 545
nucleic acids, **47**–49, 48*f*
antimicrobial agents and, 187, 556*f*, 558, 558*f*, 562*t*, 567
gram-positive bacteria and, 315–316
hydrogen bonds of, 32
in definition of life, 368
in nucleoproteins, 45
synthesis inhibition
by antibiotics, 556*f*, 558, 558*f*, 562*t*, 567
by antifungal drugs, 564*t*
by antimicrobial agents, 187
viral, 368, 372–373
West Nile virus and sequencing, 223*b*
nucleobases (adenine/thymine/cytosine/guanine), 47, 48*f*, 49*f*, 211
nucleoid (prokaryotic cell), 80*f*, **94**–95
of *Gemmata obscuriglobus,* 322*f*
nucleoli/nucleolus, 99*f*, 101*t*, 102*f*, **103**
nucleoplasm, 275, 275*f*, 277
nucleoplasmic cell, original, 277, 277*f*
nucleoproteins, 45
nucleoside analog, **229**–230, 229*f*
nucleoside reverse transcriptase inhibitors, 548
nucleosides, **47**
nucleoside analogs and, 229–230, 229*f*
nucleoside triphosphates, 214, 215*f*
nucleosome, 103
nucleotide excision repair, **230**–231, 230*f*
defect, and inherited xeroderma pigmentosum, 231
nucleotides, **47**, 48*f*
adding to DNA, 215*f*
biosynthesis of, 147, 148*f*
in DNA replication, 212–215, 213*f*, 214*f*, 215*f*, 215*t*
mutations and, 226–233. *See also* mutations
nucleobases (adenine/thymine/cytosine/guanine) and, 211
nucleoside analogs and, 229–230, 229*f*
nucleus
eukaryotic, 90*f*, 99*f*, **102**, 102*f*
as site of transcription, 220
eukaryotic *vs.* in prokaryotes, 101*t*
in *Gemmata* bacteria, 277
nucleus, atomic, **27**, 27*f*
nude mice
to culture leprosy bacillus, 539*f*, 619
transplant research and, 538, 539*f*
numerical identification, **286**, 286*f*
nurseries (hospital), effective disinfectants for, 196, 196*f*
nursing home infections. *See* nosocomial infections
nutrient agar, **165**
composition, 166*t*
nutrient broth (complex medium), **165**
nutrients, glucose's value as, 122
nutrition
in parasitic helminths, 354
of bacteria, 4
of protozoa, 346
organisms classified by patterns of, 142–145, 143*f*
nuts (tree-grown), food allergies and, 525

O polysaccharide, 86*f*, **87**, 468
O-phenylphenol, 195, 196*f*
oak tree deaths, caused by *Phytophthora ramorum,* 344
objective lenses, of compound light microscope, **56**, 56*f*
obligate aerobes, **161**, 161*t*
obligate anaerobes, **161**, 161*t*, 162
culture media for, 169*t*
obligate halophiles, **159**
obligate intracellular bacteria, culture media and, 167
obligate intracellular parasites, viruses as, 282
obligately intracellular human pathogens
Coxiella, 300*t*
Ehrlichia, 300*t*
Rickettsia, 300*t*
obligatory intracellular parasites, viruses as, 282, 368
oceans, most algae species found in, 341, 341*f*
ocular lens (eyepiece), of compound light microscope, **56**, 56*f*
OD (optical density)/absorbance, 178, 179*f*
oil
bioremediation of, 17
stored by diatoms, 343*t*
oil glands of skin
antimicrobial properties of, 402*t*
sebum secreted by, 453, 471*t*, 585
oil immersion lens
magnification and, 56
refractive index and, 59, 59*f*
oil immersion objective lens, 59, 59*f*
oil spills, bacteria that degrade, 33*b*, 136
Okazaki fragments, 215*t*, 216*f*
Old World flavivirus, introduced into New World, 223*b*
oleic acid, 41*f*
oligoadenylate synthetase, 470
oligodynamic action, **198**, 198*f*
oligosaccharides, **39**
olives, fermentation used in production of, 800
omalizumab (Xolair), 525
oncogenes, **390**–**391**
activated by virus, 442
oncogenic viruses (oncoviruses), 376*t*, **391**
among DNA viruses, 391
among RNA viruses, 391–392
latent infection and, **392**
retroviruses and, 389, 391–392
oncolytic viruses, 369
oncoviruses. *See* oncogenic viruses
onychomycosis (tinea unguium), **600**–601
oocysts
of Apicomplexa protozoa, **346**
of *Cryptosporidium,* 355*b*, 355*f*
of *Toxoplasma gondii,* 350
Oomycota (water molds), 343*t*, **343*t***, **344**, 345*f*
oomycotes
as decomposers in fresh water, 345*f*
position in evolutionary tree, 275*f*
terrestrial, as plant parasites, 344
OPA (ortho-phthaladehyde), 200
Opa protein, **432**
antigenic variation and, 433
gonococcal bacteria and, 749
open-reading frames, DNA regions likely to encode proteins, 212
operator, **225**, 225*f*
operon, 224*f*, **225**, 225*f*
operon model of gene expression, 224–225, 224*f*, 225*f*
operons
inducible, 224*f*, **225**
repressible, **225**, 225*f*
ophthalmia neonatorum, 198, 204*t*, 429, **603**–604, **604*b***, **748**–749
opisthotonos, **615**, 616*f*
opossums, as disease reservoirs, 651*b*, 661
opportunistic fungal infections, **337**–339, 338*t*
opportunistic pathogens, 300*t*, **337**, 338*t*, **403**–404
found in dolphins, 283*b*
opsonization (immune adherence)
in antigen-antibody binding, **485**, 485*f*
in complement activation pathways, 464*f*, 465, 466, 466*f*, 468, 468*f*
microbial evasion of, 468
ophthalmic cysticercosis, **733**, 733*f*
optical density/OD (absorbance), 178, 179*f*
optimum growth temperature, **157**, 157*f*
oral candidiasis (thrush), 339, **601**, 601*f*, **759**
oral cavity bacteria
gingival crevice *Bacteroides,* 324
Prevotella, 302*t*, 324
spirochetes, 322–323, 324*f*
Streptococcus mutans, 319
oral groove, of *Chilomastix* protozoa, 347*f*
oral rehydration therapy, for diarrhea, 710–711
order name, defined, **279**, 280*f*
ore, 37
bacteria used to extract, 247
organelles
eukaryotic, 99*f*, **102**–105, 102*f*, 106*f*
prokaryotic cells compared to, 101*t*, 276*t*
mitosome of Archaezoa, **347**
of apicomplexans, 348
of euglenoids, 350, 351*f*

Index

organic compounds, **34**, 37–49
  chemistry of, 37–38, 38*t*
  most common elements found in, 27*t*, 37
  structure, 37–38, 38*t*
organic growth factors, **162**
organic molecules, 37–49. *See also* organic compounds
organisms. *See also* microbes/microorganisms
  classification of, 278–282
    methods, 282–294
  evolutionary relationships among, 274–277, 275*f*, 281*f*
  identification of, 282–294
  scientific names for, 2–3, 4*t*, 278–279
organotrophs (heterotrophs), **142–143**, 143*f*
  complex medium for growing, 166*t*
Ornithodorus (tick), as vector for relapsing fever, 362*t*
ornithosis (psittacosis), 322, 410*t*
  as notifiable infectious disease, 421*t*
orphan viruses, 387
ortho-phthaladehyde (OPA), 200
orthoclone OKT3, important rDNA products in medical therapy, 260*t*
Orthomyxoviridae, characteristics/important genera/clinical features, 376*t*, 389
orthomyxoviruses, 389
Orthopoxvirus, 375*t*
Orthopoxvirus genus/spp., 374*f*
oseltamivir (Tamiflu)
  mode of action/uses, 564*t*
  to treat influenza, 570
osmium, used in staining of specimens, 63
osmosis, **92**–94, 93*f*
osmotic lysis, **89**, 94
  hypotonic solutions and, 94
osmotic pressure, **93**, 93*f*
  microbial growth requirements, 159–160, 160*f*
  most fungi resistant to, 333
  to control microbial growth, 192, 194*t*
  to preserve foods, 192
osteoporosis, beta interferon (Actimmune) to treat osteoporosis, 470
otitis externa, **593**
  rash caused by, 592*b*
otitis media, **679**, 679*f*, **681***b*
  Haemophilus influenzae causing, 613, 679
  Moraxella catarrhalis causing, 679
  Streptococcal pneumoniae causing, 614, 679
  Streptococcal pyogenes caused by, 679
out-of-phase light rays, 59
outer membrane, 86*f*, 87, 90*f*
  in gram-negative *vs.* gram-positive bacteria, 88*t*
ovarian cancer, genetically engineered Taxol used to treat, 260*t*
ovaries, 744, 744*f*
oxacillin, 561
  mode of action/spectrum of activity, 560*f*, 562*t*
oxalate decarboxylase, 116*t*
oxaloacetic acid, 148*f*, 149*f*
  in transamination, 148*f*
oxazolidinones, **566**
  developed in response to vancomycin resistance, 566
  inhibits protein synthesis, 558*f*, 566
  mode of action/spectrum of activity, 562*t*
oxidase test, 139
oxidases, 116
oxidation ponds, in sewage treatment, **788**
oxidation reaction, **122**, 122*f*, 123*f*
  in hot-air sterilization, 191, 194*t*
oxidation-reduction (redox) reactions, 116*t*, **122**, 122*f*, 123, 123*f*
  in Krebs cycle, 127–128, 128*f*
oxidative burst response
  Pseudomonas aeruginosa and, 460
  toxic oxygen products produced by, 459
oxidative phosphorylation, **122–123**
  aerobic respiration and, 137*t*
  anaerobic respiration and, 137*t*
  ATP yield, 132*t*, 137*t*
oxidoreductase enzyme, 116
  type of chemical reaction catalyzed/examples, 116*t*
oxygen, 34
  as poisonous gas, 161
  bacterial growth and, 161–162, 161*t*
  molecular, Earth's supply produced mostly by planktonic algae, 344
  photosynthetic algae provide most of Earth's supply, 344
  photosynthetic processes and, 140, 141*f*, 143, 145*t*, 344
  reducing media to grow anaerobes, 166–167, 167*f*
  requirements for microbial growth, 161–162, 161*t*
  singlet, 62, **161**
  toxic forms of, 161–162, 459
  transported across plasma membrane by simple diffusion process, 91
oxygen (atmospheric), photosynthetic cyanobacteria and, 314–315
oxygen (molecular), Earth's supply and planktonic algae, 344
oxygen (O)
  atomic number/atomic weight, 27*t*
  electronic configuration, 29*t*
  hydrogen bonding and, 31–32, 32*f*
  in organic compounds, 37
oxygenic photosynthetic bacteria, **313–315**, 314*f*, 314*t*
  environmental importance of, 314
  evolutionary contributions to life on Earth, 314
  processes of, **143–144**, 145*t*
oxytetracycline, mode of action/spectrum of activity, 562*t*
oxytetracycline (Terramycin), 565
ozone, **162**
  as disinfectant, 202, 205*t*, 783, 783*f*
  ozonators in water treatment plants, 783, 783*f*
ozone layer of atmosphere, UV light and, 230

P antigen, 383
p53 gene, 259
PAA/peroxyacetic acid (peracetic acid), 201, 202, 205*t*
PABA (*para* -aminobenzoic acid), 120, **558**
  sulfonamides and, 558, 567
packaging material, made by microbes, 3*b*
Paecilomyces fumosoroseus, to kill termites, 339
Paenibacillus, exhibiting bacterial pack behavior, 57*b*, 57*f*
Paenibacillus polymyxa, Polymyxin derived from, 555*t*
pain, of inflammation, 460
paints
  copper added to prevent mildew, 199
  mercurials to control mildew in, 199
palivizumab (Synagis), 692
palmitic acid, 41*f*
PAMPs. *See* pathogen-associated molecular patterns
pandemic disease, **406**
Paneth cells, 706–707
  defensins released by, 579
pantothenic acid, coenzymatic functions, 117*t*
paper products, microbes and their enzymes in manufacture of, 246
Papillomavirus, 375*t*
  vaccine, 503*t*
Papovaviridae, **385–386**, 386*f*
  as DNA virus, 385
  as plant virus, 394*t*
  biosynthesis of, 385*t*
  characteristics/important genera/clinical features, 375*t*, 443*t*
papovavirus
  cytopathic effects of, 443*t*
  replication of DNA in, 386*f*
papules (lesions), **586**, 587*f*
para-aminobenzoic acid (PABA), 120, **558**
  sulfonamides and, 558, 567
paragonimiasis (lung flukes), 356*f*, 357*f*, 361*t*
Paragonimus westermani (lung fluke), life cycle, 356, 357*f*, 361*t*
parainfluenza disease, 376*t*
paralysis
  flaccid, caused by botulinum toxin, 437, 616
  polio as best known infectious cause, 620, 632*b*
paralytic rabies (in animals), **623**
paralytic shellfish poisoning (PSP), **344**, 354*t*, 444
Paramecium, 61*f*, 62*f*, 63*f*, 64*f*, 66*t*
  cell structure, 350*f*
Paramecium multimicronucleatum, 62*f*
Paramyxoviridae, characteristics/important genera/clinical features, 376*t*
Paramyxovirus, 376*t*
paramyxoviruses, 389
parasites, 6, **145**
  animal, 6
  biological transmission of disease and, 412–413
  coevolution between host and, 428
  human
    Giardia lamblia, 347*f*, 348, 443–444
    Trichomonas vaginalis, 347, 347*f*
  natural killer (NK) cells can attack, 491
  obligatory intracellular, viruses as, 282, 368
  of bacteria (*Bdellovibrio*), 301*t*
  vectors and, 362–363
parasitic archaezoans, 347, 347*f*, 348
parasitic bacteria, *Rickettsias,* 303
parasitic helminths, 6, 13, 13*f*, 192, 329, 353–361, **361***t*
  antibody-dependent cell-mediated cytotoxicity and, 491, 492*f*
  flatworms, 6, 353, **356–358**, 356*f*–359*f*, **361***t*
  flukes, **356**, 356*f*, 357*f*, **361***t*
  identification of, 282
  roundworms, 6, 192, 353, **358**–361, 360*f*, **361***t*
  rules for naming and, 279
parasitic infections
  as top 20 microbial causes of death, 329
  IgE increases during, 481
  of skin, **602–603**, 603*f*
parasitic protozoa, 346
  antibody-dependent cell-mediated cytotoxicity and, 491, 492*f*
  Cyclospora, 329
  encystment and survival outside host, 346
  features/diseases caused by/source of infection, 354*t*
  Giardia lamblia, 347*f*, 348
  Plasmodium vivax, 348–349, 349*f*
  Trichomonas vaginalis, 347, 347*f*
parasitic weevils, controlled by *Metarrhizium* fungus, 339
parasitic worm infections, eosinophils increase during, 454
parasitic worms
  IgE antibodies and, 481
  immune system attacks on, 491, 492*f*
parasitism, in symbiosis, 402*f*, **403**
parasitology, **13**
parenchyma, in tissue repair of inflammatory response, 462
parent cells
  in flow of genetic information, 212, 213*f*, 214*f*
  parental DNA strands, 214*f*, 216*f*
parenteral route, **429**, 430*t*
  as portal of entry, 429, 430*t*, 445*f*
  diseases caused by, 429, 430*t*
parrots, as disease reservoirs, 410*t*
parthenogenesis, 307*b*
Parvoviridae
  as DNA virus, 385
  biosynthesis of, 385*t*
  characteristics/important genera/clinical features, 375*t*
parvovirus B19, P antigen and, 383
passive immunity
  acquired, 494*f*, **495**
  gamma globulin most often used to transfer, 495
  natural (at birth), **494**, 494*f*
passive natural, 480
passive transport processes (of moving materials across membranes), **91**
  facilitated diffusion, **91–92**, 92*f*
  osmosis, **92–94**, 93*f*
  simple diffusion, **91**–92, 92*f*
Pasteur, Louis, 8, 9*f*, 10*f*, 11, 184, 190, 477
Pasteurella genus/spp., 301*t*, **311**
  as domestic animal pathogens, 311
  as human pathogens, 301*t*
Pasteurella multocida, 283*b*
  transmitted by dog and cat bites, 311
Pasteurellales, **311**
  important genera/special features, 301*t*
pasteurization, **9**, **190–191**, 194*t*
  commercial sterilization and, **185**, 185*t*, **794–795**, 794*f*, 795*f*
  high-temperature short-time (HTST), **190–191**
  ultra-high-temperature (UHT) treatments and, 190–191
patch test to determine cause of dermatitis, 531
paternity determination, DNA fingerprinting and, 262, 264*f*
pathogen-associated molecular patterns (PAMPs), **450**, 458–459, 458*f*
pathogenesis, disease, **400**
pathogenic amoeba, 348, 348*f*
pathogenic bacteria
  at refrigerator temperatures, 192
  plasmids coding for proteins that enhance, 239
pathogenic fungi, **335–339**
  dimorphism in, temperatures and, 332
  summary of, 338*t*
pathogenic microbes, 2
  modern chemotherapy genesis and, 12–13, 12*f*
  vegetative, disinfection to control, 185, 185*t*
  virulence determination, **71**, 72*f*
pathogenic prokaryotes, included in Domain Bacteria, 274
pathogenicity, **428–448**, 445*f*
  altered, 228. *See also* mutations
  cytopathic effects and, 441–442, 442*f*, 443*t*, 445*f*
  damaging host cells, 434–441, 445*f*
    by direct damage, 434, 445*f*
    by producing toxins, 434–439, 438*t*, 439*t*, 445*f*
    by using host's nutrients, 434, 434*f*
  entering the host, 429–432, 430*t*
  lysogeny and, 440–441, 445*f*
  number of invading microbes and, 429–431, 445*f*
  of algae, 444
  of fungi, 443

of helminths, 444
of protozoa, 443–444
of viruses, 441–443, 443t
penetration of host defenses
by antigenic variation, 433
by capsule presence, 432
by cell wall components, 432
by enzymes, 432–433
by invasins, 433, 433f
plasmids and, 439–441
portals of entry, **429**, 430t
portals of exit, **444**
prophages and, 441
siderophores and, **434**, 434f
virulence and, **429**
numbers of invading microbes and, 429–431, 445f
pathogens, **399**
bacterial biosensors to detect, **780b**
first line of defense against, 449–453, 450f, **471** t. *See also* immunity
how they enter a host, 429–432
how they harm host cells, 434–441
how they penetrate host defenses, 432–433
human. *See* human pathogens
plant. *See* plant pathogens
second line of host defenses, 450f, 454–471, **472** t. *See also* immunity
that can cause multiple diseases, 406
third line of defense against, 450f
pathology, science of, **400**
paucibacillary leprosy, 619
PCR microarrays, 262, 517
PCR (polymerase chain reaction), **251**, 252f
1st and 2nd cycle procedures, 252f
as diagnostic tool, 251
DNA probes and, 262
real-time PCR, 251
reverse-transcription PCR, 251, 266b
to identify microorganisms, **290**–291
from ancient *Bacillus* bacteria, 290–291
Hantavirus hemorrhagic fever outbreak, 291
in norovirus infection, 266b
in Whipple's disease, 290
of tickborne disease, 291
to identify source of rabies virus, 291
West Nile virus, 379
to study genetic material from extinct plants/animals, 264
pea aphids, *Wolbachia* bacteria and, 307b
peanut butter, aflatoxin and, 443
peanuts
aflatoxin and, 230, 443
food allergies and, 525–526
peas, food allergies and, 525
pectin, 267, 267t
in cell walls of diatoms, 343, 343t
pediculosis (lice), 362, 362t, **602**–603, 603f
head, ivermectin effective against, 572
Lyme disease and, 323
Pediculus humanus (parasite) species and, 362t, 363f, 602
rash caused by, 592b
sucking, 362t
treatments for, 602–603
typhus transmitted by, 303
Pediculus humanus capitis (head louse), **602**–603, 603f
Pediculus humanus corporis (body louse), **602**
transmits typhus, relapsing fever, 362t, 363f, 413t
Pediococcus, summer sausage and, 137t
Pelagibacter genus/spp., **303**
FISH study showed relatedness to rickettsias, 292
most abundant living organism in oceans, 303
small genome of, 303
Pelagibacter ubique, 778
FISH studies and, 292, 303
pellicles
of euglenoids, 351f
of *Paramecium,* 350f
of protozoa, 98, 346, 350f, 351f
pelvic inflammatory disease (PID), **751–752**, 752f, **761b**
Chlamydia trachomatis causing, 750
ectopic pregnancies and, 752
Neisseria gonorrhoeae causing, 752
possible infertility resulting from, 752
pemphigus neonatorum (bulleous impetigo of newborn), **588**
penetration stage in viral multiplication cycle, **380**, 381f
penfringoglycin O toxin, 65f, 68t
penicillin, 76, **559**–561, 560f, 563
allergic reactions to, 478–479, 524
increase in, from overuse 40 years ago, 531b
as a hapten, 478, 524
as secondary metabolite of industrial fermentation, **803**, 804f
blood-brain barrier and, 611
broader-spectrum, 561
discovery of, 12–13, 12f, 553
gram-negative bacteria and, 87, 88t, 89
gram-positive bacteria and, 70, 87, 88f, 89, 559
mode of action, 85, 86f, 89, 556, 556f, 557f
natural, **559**, 560f
penicillinase-resistant, **560**–561, 561f
peptidoglycan and, 98, 556
produced by *Penicillium* mold, yield increased by rDNA techniques, 249
resistance to, 559–561, 561f
semisynthetic, **560**, 560f
mode of action/spectrum of activity, 562t
structure, 560f
spectrum of activity, 562t
structure of, compared to cephalosporin, 561f
susceptibility of gram-negative *vs.* gram-positive bacteria to, 88t
penicillin G
mode of action/spectrum of activity, 557t, 562t
retention of, 559, 560f
structure, 559, 560f
penicillin V
mode of action/spectrum of activity, 562t
structure, 559, 560f
penicillin-resistant *S. aureus,* 19–20
penicillinase-resistant penicillins, **560**–561, 561f
penicillinases, 559, 561f
monobactams to inhibit, 561
potassium clavulanate to inhibit, 561
Penicillium chrysogenum, antibiotic penicillin derived from, 4t, 12, 12f, 554, 555t, 559, 560
Penicillium genus/spp.
as opportunistic fungus, 339
example of anamorph arising from mutation in teleomorph, 335
used to ripen cheeses, 799
Penicillium griseofulvum, antibiotic griseofulvin derived from, 555t
Penicillium notatum, 12, 12f, 554
penis, 745, 745f
pentamidine, to treat African sleeping sickness, 627
pentamidine isethionate, to treat *Pneumocystis* pneumonia, 569
pentose phosphate pathway (hexose monophosphate shunt), **125**, 127
bacteria that use, 127, 135b
in purine/pyrimidine biosynthesis, 147, 148f
pentoses, 39
PEP (phosphoenolpyruvic acid), 94
peptic ulcer disease, *Helicobacter,* 312, **718**–721, 719t, 723
peptidases, 136
peptide antibiotics. *See* antimicrobial peptides
peptide bonds, **43**, 45f, 46f, 219, 220f
peptide cross-bridge, 85, 86f
peptides, 43
peptidoglycan (murein), 4, 82f, **85**, 86f, 90f
archaea cell walls and, 275, 325
biosynthesis of, 146, 146f
fungal cell walls and, 330t
in bacterial cell walls
gram-negative, 85, 86f, 87, 88t, 437
gram-positive, 69, 85, 86f, 88t, 450
in eukaryotes *vs.* prokaryotes, 77, 98, 101t
lysozyme damage to, 88–89, 88t, 453
peptone iron agar, to detect hydrogen sulfide production, 139, 139f
peptones, complex culture media and, 165
peracetic acid (peroxyacetic acid/PAA), 201, 202, 205t
perforin, **454**, **488**
pericarditis, **641**, **643b**
pericentriolar material, 99f, 105
Peridinium (dinoflagellate), 344f
periodontal disease, **709**, 709f, 710b
period of convalescence in infectious disease, 408f, **409**
period of decline in infectious disease, 408f, **409**
period of illness in infectious disease, **408**, 408f
periodontitis, **709**
peripheral nervous system (PNS), **611**, 611f
leprosy pathogen and, 619–620, 620f
rabies virus and, 622
peripheral proteins, of prokaryotic plasma membrane, 89, 90f
periplasm, 85
periplasmic space, in gram-negative *vs.* gram-positive bacteria, 88t
peristalsis, **452**, 471t
in response to microbial toxins, 452
peritoneal macrophages, 457
peritoneal tuberculosis, 144b
peritonitis, 324, 418
etiology determination and, 405
peritrichous flagella, **81**, 81f
permeability
of blood vessels in inflammatory response, 460, 461f
selective, **90**
permease, 224, 224f
permeases, in facilitated diffusion, 91–92, 92f
peroxidase, 3b, **162**
peroxide
as bleaching agent, chlorine *vs.,* 3b
microbes (yeasts) in production of, 3b
peroxide anion, **162**
peroxisomes, 99f, **105**
peroxyacetic acid/PAA (peracetic acid), 201, 202, 205t
peroxygens, **202**, 205t
persistent (chronic) viral infections, **392**, 392f
examples (disease/primary effect/causative virus), 394t
persistent enterovirus infection, 394t
person-to-person transmission, of avian influenza versus, 19
perspiration, **453**
of skin, 585, 585f
pertussis (whooping cough), 306, **680**–682, 680f, **699b**
as emerging infectious disease, 417t
as notifiable infectious disease, 421t
incubation period, 430t
portals of entry, 430t
spread by droplet transmission, 411
vaccine, 13, 502, 503t, 504t, 682
pest control
fungi used for, 339
microorganisms used in, 17
pesticide resistance, engineered into crop plants, 265
Pestivirus, 376t
Petri dishes, 165
Petri, Julius, 10f
Petroff-Hausser cell counter, 176, 178f
petroleum, formed from diatoms/planktonic organisms living millions of years ago, 344–345
petroleum hydrocarbons, bacteria that use as energy/carbon source, 239
petroleum products, beta-oxidation of, 33b, 136
Peyer's patches, 456f, 457, 710
M cells and, **486**, 487f, 710
Pfiesteria (dinoflagellate), 344, 354t
PFU (plaque-forming units), **377**
PG (polygalacturonase), 267
pGH (porcine growth hormone), 267t
pH buffers, **36–37**
pH scale, 35–36, **36** f
pH values, 35–37, **36**, 36f
disinfectant's activity and, 195
enzymatic activity and, **119**, 119f
extreme, acidophilic archaea and, 325
growth of bacteria and, 37, 158–159
pH scale, **36** f
PHA (polyhydrooxyalkanoate), as biodegradable alternative to plastic, 3b
Phaeophyta, characteristics of brown algae, 343t
phage conversion, as result of lysogeny, **382**
phage DNA, 237, 249, 380
as prophage, **380**
phage library, **255** f
phage lysozyme, **380**
phage therapy, 368–369, **579**
phage typing, **288**, 290f, 712
phages. *See* bacteriophages (phages)
phagocytes, **457–460**, 457f
actions of, 457, 457f
aging and progressive inefficiency of, 462
as second line of defense, 450f, 457, **472t**
fixed macrophages, **457**
inability to produce and, 462
macrophages as, **454**, 455t, 457, 457f, 490
microbes that survive inside, 459
phagocytic vesicle (phagosome), 458f, **459**
phagocytosis, 94, 100, **457–460**, 457f, 458f
adaptive immunity's role in, 460, 487, 489–490, 490f, 496f
biofilms and, 459–460
Brucella able to survive, 305
capsule presence and, 80
capsules of pathogens impairs, 432
cells that perform, 455t, 457, 472t
IgG antibody and, 481t
in inflammatory response, 461f, 462
lysosomes, and toxic forms of oxygen, 162, 458f
mechanism of, 458–459, 458f
phases of
chemotaxis, **458**, 458f
adherence, 458, 458f
ingestion, 458, 458f, **459**
digestion, 458, 458f
Streptococcus pneumoniae and, 234–236, 235f, 432
Streptococcus pyogenes avoidance feature, 319
phagolysosomes, 458f, **459**
phagosome (phagocytic vesicle), 458f, **459**
phalloidin, **443**
pharmaceutical agents, algae-produced thickeners used in, 343

pharmaceutical products, genetically engineered, 258–259, 260*f*
pharmaceutical uses for fermentation end-products, 137*t*
pharyngeal gonorrhea, **749**
pharyngitis, streptococcal (strep throat), 319, **676**, **677**, 677*f*, **681***b*
phase-contrast microscope, **59**–60, 60*f*, 66*t*
phenol (carbolic acid), **195**, 196*f*, 204*t*
  bacteria capable of metabolizing, enrichment mediums for, 169
  early uses in surgery, 11, 195
  local anesthetic effects, 195
  mechanism of action, 195, 204*t*
  molecular structure, 196*f*
  preferred use, 204*t*
phenolics, **195**, 196*f*, 203*t*, 204*t*
  mechanism of action, 195, 204*t*
  preferred use, 204*t*
phenotypes, **211**
  changes in, 226. *See also* mutations
    identifying mutants and, 231–232, 232*f*
    reversions and, 232
phenylalanine (phe), structural formula/characteristic R group, 44*t*
pHisoHex, 196
Phlebotomus (sand fly), leishmaniasis and, 354*t*, 665
phloem, 342
phocid distemper virus, found in seals, 283*b*
phosphatase test, pasteurization and, 190
phosphate
  in DNA structure, 47, 48*f*
  in RNA structure, 49*f*
phosphate functional group, 38*t*
  in phosphoproteins, 45
phosphate group, in nucleotides, 211
phosphate salts, culture media and, 159
phosphoenolpyruvic acid (PEP), in group translocation, 94
phosphoglyceric acid, 149*f*
phospholipids (complex lipids), **40**–41, 42*f*, 89, 90*f*
  in lipid bilayer of prokaryotic plasma membranes, 89, 90*f*
phosphoproteins, 45
phosphorus (P)
  atomic number/atomic weight, 27*t*
  electronic configuration, 29*t*
  in organic compounds, 37
  requirements for microbial growth, 160
  sources of, 160
phosphorylation, **122**
  type used to generate ATP, compared, 137*t*
photic (light) zone of bodies of water, 341*f*, 342
photic zones for algal habitats, 341–342, 341*f*
  nutrition and, 342
photoautotrophic prokaryotes, included in Domain Bacteria, 274
photoautotrophs, **143**–145, 143*f*, 145*t*
  algae as, 330*t*, 341, 341*f*
  carbon requirements for growth, 160
  culture media for, 169*t*
photoheterotrophs, 143, 143*f*, **145**
photolyases (light-repair enzymes), 215*t*, **230**
photophosphorylation, **123**
  noncyclic, **140**, 141*f*
photophosphorylation, cyclic, **140**, 141*f*
photosynthesis, 2, 106*f*, 123, **140**, 141*f*
  algae and, 6
  anoxygenic, **144**, 145*t*, 315
  bacterial, 4, 144, 145*t*, 313–315, 314*t*
  bacterial plasma membrane enzymes and, 91
  carbon dioxide, carbohydrates and, 17
  chloroplasts and, 105, 106*f*
  in selected eukaryotes and prokaryotes, compared, 145*t*
  lichens and, 340
  life without sunlight, endoliths and, 773–774
  light-dependent (light) reaction stage, **140**, 141*f*
  light-independent (dark) reaction stage, **140**, 141*f*
  oxygenic, **143**–144, 145*t*, 313–315
photosynthetic bacteria, 143–145, 143*f*, 145*t*
  Anabaena, 301*t*
  Chlorobium, 301*t*
  Chloroflexus, 301*t*
  Chromatium, 144, 145*t*, 300*t*, 314*t*
  cyanobacteria, 313–315, 314*f*, 314*t*
  enzyme required by, 96
  Gloeocapsa, 301*t*
  Rhodospirillum, 301*t*
  selected characteristics, compared, 314*t*
photosynthetic pigments, of algae, 343*t*
phototaxis, **82**
phototrophs, **142**, 143*f*
phycobiliproteins, 343*t*
phylogenetic relationships, 274–278
  phylogenic hierarchies, 277–278, 278*f*
  ribotyping to determine, 292
  the three Domains, 275–277, 275*f*, 276*t*
phylogeny (systematics), **274**
phylum, in taxonomic hierarchy, **279**, 280*f*
phyosphorus cycle, **774**
Physarum, 353*f*
physical methods of microbial control, 187–194, 194*t*
Phytophthora cinnamoni, *Eucalyptus* species infected by, 344
Phytophthora infestans, potato crops infected by, 329, 344
phytoplankton, **777**
pia mater, 611, 612*f*
pickles
  lactic acid fermentation and, 135*b*, 800
  pH and, 159
Picornaviridae, **387**, 388*f*
  as RNA virus, 385*t*
  biosynthesis of, 385*t*, 387
  characteristics/important genera/clinical features, 375*t*
  sense strand (+ strand) in, **387**
PID. *See* pelvic inflammatory disease
pig influenza viruses, 19
pigments
  bacterial, protection from sunlight and, 193
  photosynthetic, 141, 143*f*
    of algae, 343, 343*t*
  photosynthetic, of algae, 343*t*
pigs
  as disease reservoirs, 410*t*
  genetically modifying as organ donors, 536
  heart valves of, 535
  influenza A viruses and, 19, 370–371*b*
pili/pilus, 80*f*, **83**–84
  conjugation (sex) pili, **84**, 236–237, 237*f*, 238
pilin subunits, 83
pilot whales, CM virus and, 283*b*
pimaricin (Natamycin), antifungal antibiotic used in foods, 200
pin, of a T-even bacteriophage, 374*f*, 381*f*
pink eye/red eye (conjunctivitis), **603**, **604***b*
pinocytosis, 94, 100, **383**, 384*f*
pinworms (nematode), 358, 360*f*, 361*t*, **734**–735, 735*b*
Pityrosporum (fungus), as normal microbiota of skin, 402*t*
placebo, in experimental epidemiology, 420
placental transfer of immunoglobulins, 481*t*, 494
plague, **648**, 648*f*, 650, **651***b*, 652, 652*f*
  as notifiable infectious disease, 421*t*
  as zoonotic disease, 410*t*
  bacteria capsules and virulence, 432
  bubonic, **648**, 648*f*, **650***b*
  causative agent/arthropod vector, 410*t*, 413*t*
  portals of entry, 429
  rat flea (*Xenopsylla*) as vector, 362*t*, 410*t*, 413*t*, 648
  septicemic, **648**, **650***b*
  vaccine, 650, 652
  Yersinia pestis causing, 311, 410*t*, 413*t*, 432, 648
Planctomyces genus/spp., 302*t*, **322**
  as aquatic, stalk-producing bacteria, 302*t*, 322
  Gemmata obscuriglobus and origin of eukaryotic nucleus, 322, 322*f*
Planctomycetales, 302*t*
Planctomycetes, 302*t*, **322**, 322*f*
  important genera/special features, 302*t*
plankton (dinoflagellates), 341*f*, **343**–**344**, 344*f*
  photosynthesis of and Earth's oxygen supply, 344
planktonic bacteria, biofilms and, 163, 163*f*
plant alkaloids, genetically engineered, 258
plant breeding, 264
plant cells
  genetic engineered to produce valuable products, 258
  Ti plasmids and, 264, 265*f*
plant diseases, caused by viroids, **394**–395
plant pathogens
  Agrobacterium, 300*t*, **304**–305
  Erwinia, 300*t*
  rhizobia, **304**–305
  Saprolegnia ferax, 329*f*
plant pollens, allergic reactions and, 523–526
plant rot disease, 311
plant viruses, 393–395, 394*t*
  classification (viral family/genus/morphology/transmission), 394*t*
Plantae (kingdom)
  energy source, 282
  in Linnaeus' classification system, 274
  organisms included in, 282
  position in evolutionary tree, 275*f*
  position in taxonomic hierarchy, 280*f*
plants
  advantages of genetically engineered for production of human therapeutics, 258
  applications of rDNA technology, 264–267, 265*f*
  as eukarya, 6
  as kingdom in Domain Eukarya, 6, 274, 275*f*, 280*f*
  as potential source for vaccines, 506
  dependency on symbiotic fungi and, 330
  eukaryotic cells composing, 76
  genetically engineered, 264–267, 265*f*
    introducing foreign DNA into, 253–254, 253*f*, 254*f*, 264, 265*f*
    Ti plasmid and, 264, 265*f*
    to act as "factories" for producing desirable chemicals, 247
    uses of bacteria in, 258, 264–265, 265*f*
  green
    as photoautotrophs, 143–145, 143*f*
    carbon fixation, photosynthesis and, 140
  oxygen-producing and cyanobacteria, 314–315
  photosynthesis in, 145*t*
  typical cell structure, 99*f*
plaque method for detecting, counting viruses, **374**, 377, 377*f*
plaque (tooth/dental), biofilms and, 163
plaque-forming units (PFU), **377**
plaques, viral, **374**, 377, 377*f*
plasma, blood, **201**, **454**, 457, 467*b*
plasma cells, 99*f*, 482*f*, **483**, 483*f*, 494, 496*f*
plasma (cytoplasmic) membrane
  electron transport chain (system) and, 129
  in prokaryotic *vs.* eukaryotic cells, 101*t*
plasma (cytoplasmic) membrane (eukaryotes), 99*f*, **100**
plasma membrane, 40, 80*f*, 82*f*, 86*f*, **89**–91, 90*f*
  antifungal drugs that damage, 564*t*
  antimicrobial drugs that damage, 91, 186, 195, 204–205*t*, 556*f*, 558, 559*f*, 562*t*, 566–567
  functions, 90–91
  membrane ruffling and, 433*f*, 443
  movement of substances across, 91–94, 92*f*, 93*f*
    energy sources for, 94
  of *Mycoplasma* genus, 41, 42*f*, 87
  of T-even bacteriophage, 381*f*
  penetration by pathogenic invasins, 433, 433*f*
  proteins of, 95
  selective permeability of, 91, 186
  sterols and, 41, 42*f*, 87, 89, 558
  structure, 89–90, 90*f*
plasma sterilization, **201**
plasma viral load (PVL), **545**
  tests of, 545
plasmid library, 255*f*
plasmid vectors, used to deliver therapeutic genes, 259
plasmids, 80*f*, **95**, 236, **237**–240, 240*f*
  Agrobacterium tumefaciens as vehicle for, 264, 265*f*, 304–305
  as genetic engineering tool, 240, 250
  as vectors, 250–251, 250*f*, 251*f*
    acting as shuttle vectors, 251
    as primary vector used for cloning, 250, 251*f*, 256–257, 256*f*, 257*f*, 305
  circular DNA as protective property, 250
  conjugative, **238**–239
  dissimilation, **238**–239
  F factor and, 95, 236–237, 237*f*, 239*f*
  genes determining pathogenicity and, 439–441
  in bacterial conjugation, 236–237, 237*f*, 239*f*
  in synthesis of bacteriocins to kill other bacteria, 239
  in transfer of resistance to antibiotics, 239–240, 240*f*
  in typical genetic modification procedure, 247, 248*f*
  pathogenicity and, 439–441
  R factor and, 239–240, 240*f*, 250, 250*f*, 439–441
  recombinant, 248*f*, 258
  Ti plasmids, **264**, 265*f*
  transfer to other species and, 239–240
  virulence factors and, 439–440
  yeasts and expression of foreign eukaryotic genes, 258
plasmodial slime molds, 351, **352**
  cytoplasmic streaming and, **352**, 353*f*
  life cycle of, 353*f*
  plasmodium and, **352**, 353*f*
  position in evolutionary tree, 275*f*
Plasmodium falciparum, 663
Plasmodium malariae, 663
Plasmodium ovale, 663
Plasmodium (protozoa)
  can survive in phagocytes, 459
  vectors and, 362–363*f*
Plasmodium vivax (protozoan), **348**–350, 349*f*, 354*t*. *See also* malaria
  Anopheles mosquito as vector, 348–350, 349*f*, 362–363, 362*t*, 410*t*, 413*t*, 663
  incubation period, 430*t*
  life cycle, 348–349, 349*f*
  pathogenic mechanisms of, 443

  portals of entry, 430*t*
  reservoirs for, 410*t*, 413*t*
  sporozoite as infective stage of, 348–349, 349*f*
plasmogamy, **333**, 335*f*, 336*f*, 337*f*
plasmolysis, 94, **159**, 160*f*, 192, 194*t*
plastic
  biodegradable alternative to, 3*b*
  made by microbes, 3*b*
plate counts, **174**–175, 175*f*, 176*f*
platelets, **455*t***
  functions of, 455*t*
  histamine present in, 460
  thrombocidin produced by, 470
platensimycin, 566
platinum, used in staining of specimens, 63
Platyhelminthes (flatworms), 353, **356**–358, 356*f*, 357*f*, 358*f*, **361*t***
  cestodes (tapeworms), 356–358, 358*f*, **361*t***
  trematodes (flukes), **356**, 356*f*, 357*f*, **361*t***
pleated sheets of protein structure, 45, 46*f*
pleomorphic bacteria, **79**
  actinobacteria and, 320–321
Plesiomonas shigelloides, 283*b*
pleura, 675, 676*f*
pleurisy, 685
pluripotent stem cells, 535
PMNs/polymorphs (polymorphonuclear leukocytes), common name for neutrophils, 454
pneumatocyst of algae, 341*f*, 342
pneumococcal meningitis, **614**, 617*b*
pneumococcal pneumonia vaccine, 13
Pneumocystis carinii. See Pneumocystis jirovecii
Pneumocystis (fungus)
  as emerging eukaryotic pathogen, 329
  as leading cause of death in AIDS patients, 329, 337
  as opportunistic pathogen, 338*t*
  formerly classed as protozoan, now as fungus, 337
Pneumocystis jirovecii (fungus), 285, 403, 417*t*, 544*t*
  life cycle of, 698*f*
Pneumocystis pneumonia, 273*f*, **697**, 698, **699*b***
  in AIDS patients, 21, 285, 329, 337, 403, 417*t*, 542, 544*t*, 697
  pentamidine isethionate to treat, 569
  trimethoprim-sulfamethoxazole to treat, 697
pneumonia, 405
  antibiotic resistant, as an emerging infectious disease, 417*f*
  atypical *vs.* typical, 685
  bacterial, **685–692**, **687*b***
  bronchopneumonia, 685
  chlamydial, 322, **687*b***, **689**
  etiology determination and, 405
  fluoroquinolones to treat, 567
  Haemophilus influenzae, 311, 432, 613, **687*b***, **688**
  in animals, *Pasteurella* causing, 311
  incubation period, 430*t*
  Klebsiella pneumoniae causing, 5*f*, 283*b*, 310, 414, 414*t*, 432
  legionellosis, **687*b***, **688**–689, **691*b***
  lobar, 685
  methicillin-resistant *Staphylococcus aureus* and, 417*t*
  mycoplasmal, 319*f*, 320, 565, **687*b***, **688**, 688*f*
  nosocomial, 414*t*, 415*t*
  Pneumocystis jirovecii causing, 273*f*, 285, 403, 417*t*, 544*t*, 697, 698*f*
  pneumonococcal. *See* pneumonococcal pneumonia
  portals of entry, 429, 430*t*
  psittacosis (ornithosis), **687*b***, **689**
  Q fever, 96, 309, 459, **687*b***, **689–690**, 690*f*
  spread by droplet transmission, 411
  Staphylococcus aureus causing, 414*t*
  streptococcal bronchopneumonia post-influenza, 407
  Streptococcus pneumoniae causing. *See* pneumonococcal pneumonia
  typical *vs.* atypical, 685
  vaccine, 13, 688
  vancomycin-resistant *Staphylococcus aureus* and, 417*t*
  viral, **692**
  walking, 688
pneumonic plague, **648**, **650*b***, 652
pneumonococcal meningitis, 612, **614**, 617*b*
pneumonococcal pneumonia, 13, 319, 430*t*, 432, 502, 502*t*, 504*t*, **685–688**, 686*f*, **687*t***
PNS (peripheral nervous system), **611**, 611*f*
point mutation (base substitution), **227**, 227*f*
poison ivy reactions, 530, 530*f*
poisoning, due to algae, 329
poisonous gases, oxygen as, 161
poisons, enzyme, 120
polar bears, algae living in hair of, 341
polar flagella, **81**, 81*f*
polar head of phospholipids, 41, 42*f*
  of plasma membrane, 89, 90*f*
polar molecule, **34**
  water as, 34–35
polio. *See* poliomyelitis
poliomyelitis (polio), **620–622**, 621*f*, 622*f*, 632*b*
  as notifiable infectious disease, 421*t*
  diagnosis of, 443*t*, 621
  incidence, worldwide, 621, 622*f*
  iron lung developed for, 620–621, 621*f*
  portals of entry, 429, 632*b*
  postpolio syndrome, 622
  vaccine, 13, 419, 502, 503*t*, 504*t*, 506, 621, 632*b*
poliovirus, 375*t*, 407
  as an icosahedral virus, 373
  as potential biological weapon, 649*b*
  cytopathic effects of, 443*t*
  GI tract as portal of entry for, 429, 632*b*
  size of, 369*f*
  uncoating in, 385
  vaccine, 13, 419, 502, 503*t*, 504*t*, 506, 621, 632*b*
pollens, plant
  allergic reactions and, 523–526
  as antigens, IgE antibodies and, 481, 523, 524
  localized anaphylaxis and, 525, 525*f*
pollution
  bacterial biosensors to detect, **780*b***
  bioremediation and, 17, 33*b*
  using bacteria to degrade, 33*b*
  water, 17, 33*b*, **778**–779
poly-beta-hydroxybutyric acid, 96
polyene antibiotics, **568**, 568*f*
polyenes, **568**, 568*f*
  mode of action/uses, 564*t*
polyester manufacture, bacteria used in, 3*b*
polyethylene glycol, 253, 253*f*
polygalacturonase (PG), 267
polyhedral (icosahedral) virus, 372*f*
polyhedral viruses, 372*f*, **373**
polyhydroxyalkanoates (PHAs), as biodegradable alternative to plastic, 3*b*
polymerase chain reaction (PCR), **251**, 252*f*
  1st and 2nd cycle procedures, 252*f*
  as diagnostic tool, 251
  real-time PCR, 251
  reverse transcription, 251, 266*b*
  to distinguish MRSA strains, 422*b*
  to estimate the different bacterial species in soil sample, 326
  to identify microorganisms, **290**–291
  to identify viruses, 379
polymers, **38**
polymorphonuclear leukocytes (PMNs/polymorphs), 454
polymorphs/PMNs (polymorphonuclear leukocytes), common name for neutrophils, 454
polymyxin B, 91, 555*t*, 556*f*, 566–567
  mode of action/spectrum of activity, 562*t*
Polyomavirus, 375*t*
  cytopathic effects of, 443*t*
polypeptide antibiotics, 562*t*, **563**
polypeptides, 43
  in bacterial cell walls, **85**, 86*f*
polyphosphate reserves, volutin as, 95
polyribosomes, 102, 222*f*
polysaccharide
  core, 86*f*, **87**
  O, 86*f*, **87**
polysaccharide capsules, phagocytosis prevention and resulting virulence, 80, 234
polysaccharide granules, **95**
polysaccharides, **39**–40
  biosynthesis of, 146, 146*f*
pond alga, *Volvox,* 5*f*
pond scum, formed by filamentous green algae, 343
Pontiac fever, **689**
"popcorn" strain of *Wolbachia,* 307*b*
populations (bacterial), 156
  logarithmic representations, 171–172, 172*f*
porcine growth hormone (pGH), 267*t*
pores of integral proteins, 90*f*, 91
porins, **87**, 202, 308
  as factor in selective toxicity of antibiotics, 555
pork tapeworm, 357–358, 361*t*, 410*t*
  humans as definitive host, 357–358
Porphyromonas, periodontitis and, 709
portals of entry, **429**, 430*t*, 445*f*
portals of exit, **444**, 445*f*
Porter, Rodney R., 10*f*, 14*t*
posaconazole (Noxafil), 569
positive (direct) selection to detect mutant cells, **231**
positive regulation, 225–226, 226*f*
  of *lac* operon, 225–226, 226*f*
positive RNA, 223*b*
positive staining, 63
postherpetic neuralgia, 596
postoperative infections, principal sites of, 415*t*
postpolio syndrome, 622
potassium clavulanate (clavulanic acid), 561
potassium hydroxide (KOH), to diagnose cutaneous mycoses, 601
potassium ion (K+), as a cation, 30
potassium (K)
  atomic number/atomic weight, 27*t*
  microbial requirements for, 160
potassium sorbate, 199–200
potato crops
  insect toxin genetically engineered into, 264–265
  Phytophthora infestans infecting, 329, 344
potato spindle tuber viroid (PSTV), 394, 395*f*
potatoes, genetically engineered to produce antigenic proteins, 506
potential chemical energy, Krebs cycle and, 127–129, 128*f*
potential energy, in glucose, 122
Potyviridae, causing watermelon wilt, 394*t*
poultry
  as disease reservoirs, 410*t*
  cephalosporin-resistance in *E. coli* transferred to *Salmonella enterica* in, 577*b*
  fowl cholera in caused by *Pasteurella,* 311
  influenza A viruses and, 19, 370*t*
  Salmonella bacteria common intestinal inhabitants of, 310
pour plate method, **174**, 176*f*
povidone-iodine, 197
Poxviridae, 385, **385**
  as DNA virus, 385
  biosynthesis of, 385*t*
  characteristics/important genera/clinical features, 375*t*
poxviruses
  as example of complex virus, 373, 374*f*
  uncoating in, 384
prairie dogs, plague endemic to, 648, 650
praziquantel
  mode of action/uses, 564*t*
  spectrum of activity, 557*t*
  to treat schistosomiasis, 571
  to treat tapeworms, 571
prebiotics, 402
precipitation curve, 509*f*
precipitation reactions, **509**–510, 509*f*, 510*f*
precipitin ring test, **509**, 510*f*
precursors, used in amino acid synthesis, 147
pregnancy
  chlamydial infections and, 750
  cytomegalovirus and, 760
  gonorrhea infection and, 748–749
  immune system tolerance of fetus and, 534–535
  Listeria monocytogenes and, 319, 615
  neonatal herpes and, 757–758
  normal microbiota of reproductive tract and, 745
  pelvic inflammatory disease and, 752
  rubella and, 421*t*, 599, 760
  Toxoplasma gondii dangers to, 350
premergent flagellum, of euglenoids, 351, 351*f*
preparation of specimens, 54, 68–72. *See also* stains/staining
pressure cookers, 188, 190
prevalence of disease, **406**
Prevotella genus/spp., 302*t*, 324
  in human oral cavity, 302*t*, 324
Prevotella intermedia, trench mouth and, 709
primary amebic meningoencephalitis (PAM), **617*b***, **629**, 629*f*
primary cell lines, **378**
primary immune response, **493–494**, 494*f*
  vaccines provoke, 501
primary infection, **407**
primary sewage treatment, **783**, 785*f*
primary stain, **69**, 71, 87
primary structure of proteins, 45, 46*f*
primase, 215*t*
Primaxin, 561
primers
  in PCR microarrays, 262
  nucleic acid, 251, 251*f*
  PCR process and, 252*f*
  RNA, 215*t*, 216*f*
prions, **392**–393, 393*f*, **630**, 630*f*, **632*b***
  antimicrobial resistance and, 185, 203, 203*f*, 630
  bovine spongiform encephalopathy (BSE), **20**, 203, 393, 417*t*, 630*f*, **631–632**
  Creutzfeldt-Jakob disease (CJD), **20**, 393, **630**–631, 631*f*, 631*t*
  emerging infectious diseases caused by, 20, 417*t*
  how proteins become infectious, 393, 393*f*
  irradiation of foods does not affect, 797*t*
  mad cow disease and, 20, 203, 393, 417*t*, 631–632
  nervous system diseases caused by, 629–632, **632*b***
  sheep scrapie, **630**
  size of, 369*f*
privileged sites/tissues, transplant rejection and, 534–535
probes, DNA, **256**–257, 257*f*
  to identify pathogens, 257

probiotics, **402**–403
procaine penicillin, 559, 560f
Prochlorococcus, 777
prodromal period in infectious diseases, **408**, 408f
produce DNA viruses, 376t
products, of chemical reactions, 32, 34, 115, 115f, 118f
profundal zone, **776**
proglottids, **356**–357, 358f
programmed cell death (apoptosis), **489**, 489f
progressive encephalitis, 394t
prokaryotes/prokaryotic cells, **4**, 76, 77–98, **77**, 80f, 299–328
- Archaea Domain as, 76, 325
- Bacteria Domain as, 76, 302–325
- cell structure, 79–98
  - external to cell wall, 79–84
  - internal to cell wall, 89–98
  - typical, 80f
- distinguishing characteristics of, 77
- DNA arrangement of, 76, 77
- eukaryotic cells/organelles *vs.*, 77, 81, 82, 95, 275, 276t, 277
- evolution of, 275, 275f, 277–278
  - phylogenetic relationships, 281, 281f
- flagella of, 81–82, 81f, 82f
- historical and current definitions of, 274
- mutation and, 226–233
  - identification techniques, 231
- origins of, 275, 275f, 277–278, 277f
- pH ranges and, 37
- photosynthetic, 143–145, 145t, 314f
- plasma (cytoplasmic) membrane, 89–91, 90f
- principal differences between eukaryotic cells and, 101t
- protein synthesis in, 216–221
- ribosomal differences, 95, 95f
- RNA types in bacterial cells, 216
- rules for naming and, 279
- sizes/shapes/arrangement of, 77–79, 78f, 79f
  - *vs.* eukaryotic cells, 101t
- term introduced by Chatton, 274

prokaryotic species, *vs.* eukaryotic species, **279**, 281
proline (pro), structural formula/characteristic R group, 44t
promoter (region on DNA strand), **216**, 218f, 225, 225f
promoters, inducible, 257
proofreading capability of DNA polymerase, 214–215
properdin (factor P) complement protein, 466, 466f
prophage, **380**, 382, 382f, 383f
- lysogenic conversion and, **441**
- *vs.* provirus, 389

Propionibacterium
- as normal microbiota of eye, 402t
- as normal microbiota of skin, 402t

Propionibacterium acnes
- as normal microbiota of skin, 586, 594
- bacterial acne caused by, 320, 594
- pH ranges and, 37

Propionibacterium freudenreichii, Swiss cheese and, 137t, 320
Propionibacterium genus/spp., 301t, **320**
- added to cheese in ripening process, 799
- as propionic acid producers, 301t, 320
- fermentation and, 134f, 139

propionic acid
- as fermentation end-product, 134f, 137t, 139
- *Propionibacterium* genera able to produce, 320

prospective studies, 419
prostaglandins, 437, 438f, 461f, **462**, **524**
- as mediators in allergic reactions, **524**
- aspirin, acetaminophen inhibit synthesis of, 437
- fever and, 437, 438f, 463

prosthecae, **303**
- of *Caulobacter,* 304
- of *Hyphomicrobium,* 304

protease enzymes, to inactivate prions, 203
protease inhibitors, 548, **571**
proteases, 136
- granzymes, **454**, 489, **489**

protein denaturation, **45**, **119**, 119f, 188–190, 189f, 194t
- by antimicrobial agents, 187, 197, 204–205t
- by moist heat, 188–190, 189f, 194t
- by pasteurization, 9, 190–191, 194t

protein kinase, 470
protein synthesis, 216–221
- early discoveries about, 16
- evolutionary aspects, 106
- genetic code for, 211, 219f
- inhibitors
  - antimicrobial agents, 197
  - antimicrobial drugs, 556–557, 556f, 558f, 562t, **563**, 565–566
- nitrogen requirements, 160
- prokaryotic cell
  - site of, 95, 101, 216, 219
  - transcription, **216**–217, 218f, 221
  - translation, 213f, **217**, 219–220, 219f, 220–221f
  - *vs.* eukaryotic cell, 106
- regulation of, 222
- ribosomes and, 102, 216
- RNA and, 216–221
- transcription, **216**–217, 218f, 221
- translation, 213f, **217**, 219–220, 219f, 220–221f

proteinaceous infectious particle (prion), **392**–**393**
proteins, **42**–**46**
- amino acids found in, **43**, 43f, 44t
- antimicrobial agents and, 187
- biosynthesis of, 146–147, 148f
- catabolism of, 136–137, 138f
- conjugated, 45
- denaturation of, **45**, **119**, 119f
- DNA as blueprint for, 212–213, 213f
- enzymatic activity of. *See* enzymes
- flagellin, 81
- functions, 42
- globular, flagellin, 81
- Human Proteome Project and, 261
- hydrogen bonds of, 32
- in complex culture media, 165, 165t
- infectious (prions), 392–393, 393f
- iron-binding, **470**
- mapping all expressed in human cells, 261
- negative staining in study of, 63
- of complement system, 463–468
- phenotypes and, 211
- proteomics science and, 262
- simple, 45
- structural levels, 45, 46f
- three-dimensional shape of, 45, 46f, 187
- transporter, 42
- viral, biosynthesis of in multiplication cycle, 380, 381f

proteobacteria, 280f, **300t**, 302–312, **303**, 313f
- alphaproteobacteria, 300t, 303–305, 304f, 305f
- betaproteobacteria, 300t, 303
- deltaproteobacteria, 301t, 303
- epsilonproteobacteria, 301t, 303
- gammaproteobacteria, 300t, 303
- important genera/special features, 300t
- photosynthetic bacteria of, 314t
- phylogenetic relationships, 281f, 303
- position in taxonomic hierarchy, 280f, 300t

proteomics, **262**
Proteus genus/spp., 300t, **310**, 311f
- as endotoxin producer, 439
- as normal microbiota of large intestine, 403t
- as normal microbiota of urethra, 403t
- as pathogens, 300t, 310
- L forms of, **89**
- swarming movement of, 82, 310, 311f

Proteus mirabilis, 310, 311f
- rapid identification methods, 286, 286f

Protista (kingdom), 6, 274, 275f, **281**
- Haeckel's proposal of, 274

protists, 6, 274, 277
- algae as, 330t
- as host cell for *Cyanophora paradoxa,* 277, 277f
- as kingdom in Domain Eukarya, 6, 274, 275f, **281**
- clads and, **281**
- protozoa phyla now classified as, 346–351

proton acceptors, bases as, 35
proton donors, acids as, 35
proton motive force, 130
proton pumps, 130–131, 130f, 131f
protons, **27**, 27f
- in cellular oxidations, 122, 123f

protoplast fusion, **253**, 253f
- for plant cells, 264

protoplasts, **89**, **253**
- protoplast fusion, **253**, 253f

protozoa/protozoan, 2, **4**, 5f, 6, 329, **345**–**351**
- AIDS-related diseases caused by, 544t
- antiprotozoan drugs, 12, 528, 529f, 564t, **571**
- as aerobic heterotrophs, 346
- as eukarya, 6, 346
- as eukaryotes/eukaryotic cells, 76, 98, 99f, 346
- as insecticides, 346
- cell structure, 5f, 6, 99f
- cell wall and, 98
- characteristics of, 346
- classification and, 346–347
- conjugation in, **346**
- *Cryptosporidium* causing diarrhea outbreaks, 21
- cysts of, antimicrobial agents and, 203, 203f
- cysts/oocysts and biocidal effectiveness, 203
- digestion occurs in food vacuoles, 346
- early representations, 7, 7f
- emerging infectious diseases caused by, 417t
- habitat of, 346
- identification by microscope, 282
- immune system attacks on, 491, 492f
- life cycle of, 346
- locomotion and, 5f, 6
- medically important phyla, 346–351
- nutrition of, 6, 346
  - as animal-like, 346
  - classification and, 143, 143f
- parasitic, 346–351, 354t
- Pasteur's research on, 11
- pathogenicity of, 443–444
- phyla now classified as protists, 346–351
- reproduction methods, 6, 346
- rules for naming and, 279
- silkworm disease and, 11
- trophozoite as vegetative form of, 346
- vegetative, resistance to chemical biocides, 203f

protozoan diseases, 354t, 443–444
- of cardiovascular system, **650b**, **651b**, 660–666
- of digestive system, 730–732, **734b**
- of eyes, **604b**, 605
- of lymphatic system, **650b**, **651b**, 660–666
- of nervous system, **617b**, 627–629, 629f
- of reproductive system, 759b, 760–761, 761b
- zoonotic, 410t

prourokinase, genetically engineered, used in anticoagulant therapy, 260t
provirus, **389**, 390f
- HIV as, 541, 541f, 542f

Prusiner, Stanley B., 10f, 15t, 392
pseudohyphae, **332**, 334f
- pathogenic fungi with, 338t

Pseudomonadales, 300t, **308**
- important genera/special features, 300t

pseudomonads. *See Pseudomonas*
Pseudomonas aeruginosa
- as nosocomial infection, 414, 414t
- biofilm-forming, 57b, 57f
- carboxypenicillins effective against, 561
- disinfectants evaluated by disk-diffusion method, 196f
- neutrophil response slower when in biofilms, 459
- R factors and genes determining antibiotic resistance, 414
- skin infections caused by, 591, 592t, 593–594
- triclosan resistance and, 196
- twitching motility in, 83
- weakened hosts susceptible to, 308

Pseudomonas carboxydohydrogena, 145
Pseudomonas dermatitis, **591**, 592b, 593–594
Pseudomonas fluorescens, genetically engineered to produce *Bacillus* toxin, 267
Pseudomonas fluorescens infection, indwelling catheters and, 164b, 164f, 308
Pseudomonas genus/spp., 300t, **308**, 308f
- ability to degrade/detoxify, plasmids and, 239
- able to grow in some antiseptics, 199, 201b, 308
- anaerobic respiration and, 132
- antibiotic resistance and, 308
- antibiotics effective against, 565
- as normal microbiota of urethra, 403t
- as oil degraders, 33b
- as opportunistic pathogens, 300t
- bioremediation uses, 17, 33b
- can grow at refrigerator temperatures, 308
- cystic fibrosis patients and, 308
- dissimilation plasmids and, 238–239
- Entner-Doudoroff pathway and, 127
- hospital-acquired infections and, 164b, 164f, 308
- in taxonomic hierarchy, 300t
- oxidase test and, 139
- quat compounds and, 199
- quaternary ammonium compounds resistance and, 199, 201b, 308
- resistance to biocides, 196f, 199, 202, 308
- skin infections caused by, 591, 592t, 593–594
- soil and other natural environments common habitat, 308
- some species transferred to genus *Burkholderia,* 279, 305, 308
- Zephiran resistance and, 199, 201b

Pseudomonas putida, 3b
Pseudomonas syringae
- as occasional plant pathogen, 308
- genetically engineered plants and, 267t

pseudomurein, 87
pseudopods, 5f, 6, 100, **348**, **459**
- of *Amoeba proteus,* 348f
- of amoebas, **348**, 348f
- of phagocytes, 457f, 458f, **459**

psittacosis, as notifiable infectious disease, 421t

Index

psittacosis (ornithosis), 322, 410t
psoriasis, **533**
  interleukin-12 therapy and, 493b
psoriatic arthritis, **533**
PSP (paralytic shellfish poisoning), **344**, 354t, 444
PSTV (potato spindle tuber viroid), 394, 395f
psychotrophs, **157**, 157f, 158
  growth at refrigerator temperatures, 191–192
psychrophiles, **157**, 157f
psychrotrophs, **157**, 157f
public health
  E. coli 0157:H7 outbreaks, **20**
  emerging infectious diseases and, 19–21, 418
public health issues
  measles vaccination, 505b
  West Nile virus, 223b, 626
puerperal sepsis (childbirth fever), 11, 197, 418, **640–641**, **643b**
pulmonary (inhalational) anthrax, **645–646**, 649b, 650b
pulmonary syndrome, *Hantavirus*, 376t, 410t, 416, 417t
pulmonary tuberculosis, 144b
Pulmozyme (rhDNase), genetically engineered, 260t
pulsed-field gel electrophoresis (PFGE), 718
puncture wounds, fungal infections and, 337, 338t
pure bacterial cultures, streak plate method for obtaining, 170, 170f
Purell hand cleaner, 198
purine nucleotides, **47**
  biosynthesis of, 117t, 147, 148f
purple bacteria, 143, 143f, **144**, 145t
  photosynthesis in, 145t
purple nonsulfur bacteria, **145**, 314t, **315**
  characteristics of, compared, 314t
  nutritional classification of, 143, 143f
  Rhodospirillum, 300t, 314t
  Rhodospirillum rubrum, 91, 91f
purple sulfur bacteria, 144, 314t, **315**, 315f
  characteristics of, compared, 314t
  Chromatium, 144, 145t, 300t, 314t
purpura, 428
pus, **462**
  formation in inflammatory response, 462
  phenolics to disinfect, 195
pustules (lesions), **586**, 587f
  formation in inflammatory response, 462
putrefaction spoilage, of canned foods, 795, 796t
PVL (plasma viral load), **545**
  tests of, 545
pyrantel pamoate, mode of action/uses, 564t
pyelonephritis, **746**, 748b
pyocyanin, **593**
pyrimidine dimers, 215t
pyrimidine nucleotides, **47**
  biosynthesis of, 117t, 147, 148f
Pyrodictium abyssi (archaea), unusual morphology of, 325f
Pyrodictium (archaea), 302t
  as hyperthermophiles, 302t
pyrogenic response (fever), 450f, **463**. *See also* fever
  endotoxins causing, 437, 438f
pyruvic acid
  alcohol fermentation and, 135b, 136f
  coenzymes and, 117t
  fermentation and, 125f, 132–135, 134f
  glycolysis and, 124, 125f
  in biosynthesis of lipids, 147f
  in lipid catabolism, 136f
  Krebs cycle and, 127, 128f
  lactic acid fermentation and, 135b, 136f

Q fever, 96, 309, 459, **687b**, 689–**690**, 690f
  as notifiable infectious disease, 421t
quaternary ammonium compounds (quats), 91, 196f, **199**, 199f, 201b, 202, 203t, 204t
  pseudomonads resistance to, 199, 201b
quaternary structure of proteins, 45, 46f
quats (quaternary ammonium compounds), 91, 196f, **199**, 199f, 202, 203t, 204t
quinacrine, to treat giardiasis, 571
quinine, 12
  as antiprotozoan drug, **571**
  inducing cytotoxic reaction, 528, 529f
  to control malaria, 571
quinolones, **567**
  mode of action/spectrum of activity, 556f, 563t
quinones, 117t
quinupristin, 566
  mode of action/spectrum of activity, 562t
quorum sensing, biofilms and, 57b, 163

R factors (resistance factors), **239**–240, 240f
  antibiotic resistance and, 239–240, 240f, 308, 308f, 414, 439–441, 574, 577b
  plasmids as vectors and, 250
  r-determinant genes and, **239**, 240f
  resistance transfer factor (RTF) genes and, **239**, 240f
  transposons and, 240, 241f
R groups of organic compounds, 37, 38t
R groups (side groups) of amino acids, 43, 43f, 44t
r-determinant part of R factor, **239**, 240f
R100 (resistance plasmid R100), 239, 240f
  insertion of transposon Tn5 into, 241f
rabbits
  as disease reservoirs, 651b
  rabbit fever (tularemia) contact transmission, 642–643
rabies, **622–624**, 623f, 624f, **625b**, **632b**
  as notifiable infectious disease (animal/human), 421t
  as zoonotic disease, 410t, 622, 625b
  bat bites and, 624f, **625b**
  diagnosis of, 62, 623, 625b
  distribution in wildlife, 624, 624f
  incidence, by animal species, 624, 624f
  incubation period, 430t, 622–623
  portals of entry, 430t, 623, 623f
  postexposure prophylaxis for, 623
  prevention of, 623
  signs in animals, 623
  symptoms in humans, 623
  treatment for, 623–624, 632b
  vaccines for, 623
rabies virus (*Lyssavirus*), 376t, **387**, 389f
  as a rhabdovirus, 387, 389f
  as helical virus, 373
  can mimic neurotransmitter acetylcholine, 441
  inclusion bodies produced by, 442, 442f
  incubation period, 430t, 622–623
  PCR used to identify source, 291
  portals of entry, 430t
  size of, 369f
  vaccine for animals, 502
  vaccine for humans, 502, 503t
raccoons
  as disease reservoirs, 410t, 417t
  reported cases of rabies in, 624f
radiant energy spectrum, 192–193, 193f
radiation, 162, **192**–193, 193f
  ionizing, **192**–193, 193f
  mutagenic, 228, 230–231, 230f
  nonionizing, **193**, 193f
  to control microbial growth, 192–193, 193f, 194t
  to kill microbes in foods, 796–797, 797t, 798f
radiation therapy, impaired innate defenses and, 462
radicals
  hydroxyl, **162**
  superoxide, **161**–162
random mutations, 428
random sequencing. *See* random shotgun sequencing
random shotgun sequencing, **261**, 261f
Rapamune (sirolimus), 536–537
rapid diagnostic tests (RDTs) for syphilis, **755**
rapid identification methods, 285–286, 286f, 291, 292f
rapid plasma reagin (RPR) test, for syphilis, **755**
rapidly growing mycobacteria, 201b
rashes, **586**, 587f
  antibiotic-induced, 531b
  delayed (Clinical Focus), 531b
  diseases that cause, macular rashes, 589b
  enanthem, **586**, 587f
  exanthem, **586**, 587f
  of scarlet fever, 437
  of syphilis, 753, 754f
  of yaws, 753
  patchy redness, 592b
  pimple-like conditions, 592b
  pustular, 590b
  vesicular, 590b
rat liver extract, 232–233, 233f
rats
  rat bite fever, **647–648**, **650b**
  rat flea (*Xenopsylla*) transmitting plague, typhus, 362t, 363f, 410t, 413t, 648
  Yersinia bacteria carried by, 311, 410t
rDNA. *See* recombinant DNA (rDNA) technology
RDTs (rapid diagnostic tests), for syphilis, **755**
reactants, in chemical reactions, 32, 34, 115, 115f
reading frames
  open, 212
  translational, frameshift mutations and, 228
reagents in Gram staining, 87
real-time PCR, 251
RecA protein
  from *E. coli*, 65f, 68t
  in crossing over (genetic recombination), 234, 234f
  in genetic transformation, 236, 236f
receptor sites, in viral multiplication, 383
receptors, surface, on host cells, **431**, 431f
recipient cells, in gene transfers, **234**, 234f
recognition sites
  in making recombinant DNA, 250f
  in transposition, 240, 241f
recombinant DNA, **16**
recombinant DNA (rDNA) technology, **16**, **247**, 248f
  advantages, 247
  applications, 258–267
    agricultural, 264–267
    scientific, 261–264
    therapeutic, 258–259
  artificial selection and, **249**
  biotechnology and, 17–18, **246**. *See also* biotechnology
  enzymes produced by, 17
  ethical issues, 268
  gene therapy and, 17–18
  genetic modification techniques, 253–258
  Human Genome Project and, 261
  Human Proteome Project and, 261
  pharmaceutical products of, 260t
  safety issues, 268
  tools/procedures, 247–252
    artificial selection, 247, **249**
    inserting foreign DNA into cells, 253–254, 254f, 264
    making a gene product, 257–258
    making synthetic DNA, 255, 256f
    mutation, 249
    obtaining DNA, 254–255
    overview, 247, 248f
    restriction enzymes, **249**–250, 249t, 250f
    selecting a clone, 256–257
    selection and, 247, **249**
    site-directed mutagenesis, **249**
  vaccines produced by, 17, 247
recombinant interferons (rIFNs), 470
recombinant plasmids, 248f, 258
recombinant vaccines, **502–503**, 506
recombinants/recombinant cells, 213f, 234, 236, 236f
recombination, genetic, 233–241. *See also* genetic recombination
reconstructive surgery, genetically engineered bone morphogenic proteins helps induce new bone formation, 260t
rectangular-shaped bacteria, 79, 79f
red algae, 341f, **343**, **343t**
red blood cells (RBCs), **455** t. *See also* erythrocytes
  ABO blood type and, 526–528, 527t
red bone marrow, 456, 456f
  radiation therapy damage to, 462
red eye/pink eye (conjunctivitis), **603**, **604b**
red tides, **344**, 444, 779, 779f
Redi, Francesco, 8
redness, of inflammation, 460
redox (oxidation-reduction) reaction, **122**, 122f
reducing culture media, **166**, 169t
reduction reactions, **122**, 122f
redwood trees, infected by *Phytophthora ramorum*, 344
Reed, Walter, 659
refractive index, **58**–59, 59f
refrigeration
  for preserving bacterial cultures, 170
  Listeria species can grow in, 319
  temperature and microbial growth in, 157–158, 158f
  to control microbial growth, 191–192, 194t
regulatory genes, I gene, 225, 225f
regulatory proteins
  CD59 of complement system, 468
  repressors, **224**, 224f
regulatory T cells, **489**
rehydration therapy, oral, 710–711
reindeer, lichens and, 340
reinforcement (relative brightness), in phase-contrast microscopy, 59–60, 60f
relapsing fever, 323, **651b**, **652**
  Borrelia species causing, 652
  causative agent/arthropod vector, 413t
  Ornithodorus (tick) as vector, 362t
relative brightness (reinforcement), in phase-contrast microscopy, 59–60, 60f
relative darkness (interference), in phase-contrast microscopy, 59–60, 60f
relaxation pathway, 437
relaxin, genetically engineered, 260t
release stage
  in animal virus multiplication cycle, 385, 386f, **389**
  in bacteriophage multiplication cycle, **380**, 381f
Relenza (zanamivir), 565t, 570
Remicade (infliximab), 509
rennin
  genetically engineered, 267t
  in cheese making, **798**

Reoviridae, **387**, 388*f*
as RNA virus, 385*t*
biosynthesis of, 385*t*
characteristics/important genera/clinical features, 376*t*
Reovirus, 376*t*
wound tumor virus (in plants), 394*t*
repellants, types of bacterial movements toward, 82
replica plating, to identify mutation, **231**–232, 232*f*
replication, semiconservative, **213**
replication enzymes (DNA), 212–213, 214*f*-216*f*
replication fork (DNA), 213
enzymes important to, 212–213, 215*t*
events at (summary), 216*f*
in *E. coli* bacteria, 214, 217*f*
replication of DNA. *See* DNA replication
repressible operons, **225**, 225*f*
repression, **224**, 224*f*
repressor proteins, 224*f*, **225**
repressors, **224**, 224*f*
reproduction
fungal sexual spores and, 333, 335*f*
parthenogenesis as type of, 307*b*
sporulation in bacteria and, 97
reproductive choices, genetic screening, ethics involved, 268
reproductive methods
algal, 342, 342*f*, 343*t*
bacterial, 4, 171, 171*f*
fungal, 330*t*, 331, 331*f*, 332–333, 334*f*–337*f*
of dioecious parasitic helminths, 354–355
reproductive systems
bacterial diseases of, 747–756, **759***b*, **761***b*
female, **744**–745, 744*f*
fungal diseases of, 758–759, **759***b*, **761***b*
male, **745**, 745*f*
normal microbiota of, 403*t*
protozoan diseases of, 759*b*, 760–761, 761*b*
viral diseases of, 757–758, **761***b*
reptiles, as disease reservoirs, 410*t*
research, medical, importance of rDNA technology to, 259
reservoirs of disease, **409**, 410*t*
animal and human, 409
bats as especially good, 624, 624 *footnote*
nonliving (soil and water), 409
of zoonoses/with transmission methods, 410*t*
residual body formation in phagocytosis, 458*f*, 459
resistance, **18**, **449**. *See also* immunity
innate immunity, **449–475**, **476**
resistance factors in bacteria. *See* R factors
resistance plasmid R100, genetic map of, 239, 240*f*
resistance to antimicrobial drugs. *See* antibiotic resistance
resistance to drought, engineered into crop plants, 265
resistance to drying, in gram-negative *vs.* gram-positive bacteria, 88*t*
resistance to microbes, antimicrobial peptides do not appear to develop, 471, **578**
resistance to sodium azide, by gram-negative *vs.* gram-positive bacteria, 88*t*
resistance transfer factor (RTF), **239**–240, 240*f*
resistant bacteria
triclosan and, 196
widespread antibiotic use and, 239–240
resolution (resolving power) of microscopes, **56**
resolving power (resolution), of microscopes, **56**
respiration, cellular. *See* cellular respiration
respiration (cellular respiration), **127**–132
aerobic, **127**
anaerobic, **127**
respirators, as disease reservoirs, 416
respiratory syncytial virus (RSV), 679, **692**, **699***b*
respiratory system
as portals of entry, 429, 430*t*, 445*f*
bacterial diseases, 677–692, **681***b*, **699***b*
fungal diseases, 695–698, **699***b*
lower respiratory tract, **675**, 676*f*
bacterial diseases, 680–692, **687***b*, **699***b*
microbial diseases of, 18, 57*b*, 80, 674–704
commonly contracted via, 429, 430*t*
nosocomial, 414*t*, 415*t*
Reoviridae and, 376*t*, 387
normal microbiota of, 403*t*, 675–676
physical defenses against microbes, 452, 452*f*, 674–675, 675*f*, 676*f*
structure/function, 675–676, 675*f*, 676*f*
upper respiratory tract, **675**, 675*f*
bacterial diseases, 677–679, **681***b*
common cold caused by adenoviruses, 385
IgA antibody protection and, 480–481
Mastadenovirus and, 375*t*
viral diseases, 679–680, **681***b*
restaurant eating utensils, calcium hypochlorite to disinfect, 197
restriction enzymes, **249**–250, 249*t*, 250*f*
blunt ends/sticky ends, **249**, 250*f*
used in rDNA technology, 249*t*
restriction fragment length polymorphisms (RFLPs), **262**, 290
to identify viruses, 379
reticular dysgenesis, 539*t*
reticulate bodies, *Chlamydophila psittaci* and, **323**, 323*f*
reticuloendothelial system
brucellosis persists in, 644
macrophages and, **457**
retorts, 188, **795** *f*
retrospective studies, 419
retroviruses (Retroviridae), 376*t*, **387**, 389, 390*f*
ability to induce tumors and reverse transcriptase, 390*f*, 392
as RNA virus, 385*t*
biosynthesis of, 385*t*
characteristics/important genera/clinical features, 376*t*
HIV as, 376*t*, 387, 540, 541–542
human T-cell leukemia (HTLV-1, HTLV-2) as, 391
multiplication and inheritance processes in, 387, 389, 390*f*
mutation rate high in, 541–542
oncogenic, 389, 391–392
provirus and, **389**
reverse transcriptase and, 387, 389, 390*f*, 392
used as vectors in gene therapy, 251, 259
reverse genetics, **262**, 694
reverse transcriptase, **254**–255, 255*f*, **387**
anti-HIV drugs that inhibit, 548
Hepadnaviridae and, 386
Retroviridae and, 387, 389, 390*f*
retroviruses and, 387, 389, 390*f*, 392
reverse-transcription PCR (RT-PCR), 251
used to confirm norovirus infection, 266*b*
reversible chemical reactions, **34**, 39*f*
reversion rate, spontaneous, 232–233
reversions/revertant bacteria, 232–233, 233*f*
Reye syndrome, **596**
RFLPs (restriction fragment length polymorphisms), 262, **262**, 290
Rh blood group system, **527**–528, 528*f*
Rh factor, **527**–528, 528*f*
Rhabdoviridae, **387**, 388*f*, 389*f*
as RNA virus, 385*t*
biosynthesis of, 385*t*
characteristics/important genera/clinical features, 376*t*
cytopathic effects of, 443*t*
potato yellow dwarf virus caused by, 394*t*
rhabdoviruses, **387**, 388*f*, 389*f*
cytopathic effects of, 443*t*
rhDNase (Pulmozyme), genetically engineered, 260*t*
rheumatic fever, 319, **641**–642, 642*f*, **643***t*
HLA typing to determine susceptibility, 534*t*
rheumatoid arthritis, 460, 467*b*, 492, 493*b*, 509, **532**, 533
interleukin-12 and, 493*b*
monoclonal antibodies to treat, 509, 533
tumor necrosis factor and, 492, 509
rheumatoid factors, 532
Rhinovirus
common cold caused by, 375*t*, 679
size of, 369*f*
rhizines, **339**, 340*f*
rhizobia, **304**–305
Rhizobium genus/spp./Rhizobiales
as pleomorphic bacteria, 79
as symbiotic nitrogen fixers, 300*t*
Entner-Doudoroff pathway and, 127
in taxonomic hierarchy, 300*t*
sold industrially, 806
Rhizobium meliloti, genetically engineered to enhance nitrogen fixation, 267, 267*t*
Rhizopus
as pathogenic fungi, 338*t*
life cycle of, 335*f*
spores produced by, 333, 334*f*, 335*f*
Rhizopus stolonifer, common black bread mold, 333, 334*f*, 335*f*
Rhodococcus bronchialis, DNA fingerprinting and, 290
Rhodococcus erythropolis, desulfurized oil (petroleum) and, 145
Rhodocyclales, important genera of, 300*t*
Rhodophyta, characteristics of red algae, 343*t*
Rhodopseudomonas, 145
Rhodospirillales, important genera/special features, 300*t*
Rhodospirillum genus/spp., 300*t*
as photosynthetic, anoxygenic bacteria, 300*t*, 314*t*
characteristics, compared, 314*t*
Rhodospirillum rubrum, chromatophores of, 91, 91*f*
RhoGAM, 528
ribavirin, 564*t*, 569
Ribeiroia (trematode), 356*f*
riboflavin
coenzymatic functions, 117*t*
in cellular respiration, 129
ribonucleic acid (RNA), **47**
ribose, **47**, 48*f*
ribosomal RNA (rRNA), **47**, 95, 103, 211, 274, 292, 299
sequencing. *See* rRNA sequencing
ribosomes, **95**, 95*f*, 99*f*, **101**–102, 102*f*, 103*f*
antibiotics that inhibit, 563, 565–567
as site of translation, 219, 220–221*f*
eukaryotic
40S subunit, 101–102
60S subunit, 101–102
80S ribosomes, 95, 101–102
free, 102
importance in study of phylogenetic relationships, 274
in prokaryotic cell/eukaryotic cell/eukaryotic organelles, 276*t*
in prokaryotic *vs.* eukaryotic cells, 101*t*
membrane-bound, 102
polyribosomes, 102, 222*f*
prokaryotic, 80*f*, **95**, 95*f*
complete 70S, 95, 95*f*, 557, 558*f*
subunit 30S, 95, 95*f*
subunit 50S, 95, 95*f*
viruses and, 369*t*
Ribotyping, **292**
ribozymes, **121**, 215*t*, 220
ribulose 1, 5–diphosphate carboxylase, 96
ribulose diphosphate, in Calvin-Benson cycle, 142*f*
Rickettsia genus/spp., 300*t*, **303**, 304*f*
as human pathogen, 300*t*
as intracellular parasites, 300*t*, 303, 565
can survive in phagocytes, 459
cultivation and, 404
culture media and, 167
diseases caused by, 303
no longer grouped with *Chlamydia,* 299, 303
ocean bacterium *Pelagibacter* related to, 292
tetracyclines effective against, 565
transmitted to humans by tick and insect bites, 303
viruses compared to, 368, 368*t*
Rickettsia prowazekii, 300*t*, 303, 654–655
as potential biological weapon, 649*b*
considered hazardous to culture, 655
epidemic typhus and, 303, 413*t*, 651*b*, **654**–655
Rickettsia rickettsii
incubation period, 430*t*
portals of entry, 430*t*
reservoirs/transmission method, 410*t*
Rocky Mountain spotted fever and, 303, 413*t*, 430*t*
Rickettsia typhi
endemic murine typhus and, 303, 413*t*
reservoirs/transmission method, 410*t*
rickettsial infections, tetracyclines to treat, 565
Rickettsiales, 300*t*
rifampicin. *See* rifampin
rifampin, 563, **567**
mode of action/spectrum of activity, 556*f*, 562*t*
to treat leprosy, 620, 632*b*
to treat tuberculosis, 684
rifamycins, 556*f*, 562*t*, 563, **567**
rIFNs (recombinant interferons), 470
RIG (human rabies immune globulin), 623
right lymphatic duct, 456*f*, 457
ring stage, **349**, 349*f*
ringworm
athlete's foot (tinea pedis), 338*t*, 409, 410*t*, **600**, 600*f*
jock itch (tinea cruris), **600**
nails (tinea unguium), **600**–601
of scalp/skin (tinea capitis), 592*b*, **600**, 600*f*
griseofulvin to treat, 569, 600
rituximab (Rituxan), 509
rizosphere, 771–772
RNA
naked, viroids and, 394–395
viral, 372–373
RNA interference (RNAi), **259**, 259*f*, **579**
RNA nucleotides, required in transcription phase of protein synthesis, 216–217, 218*f*
RNA oncogenic viruses, 391–392
RNA polymerase, 215*t*
in eukaryotic transcription, 220, 222*f*
in prokaryotic transcription, 215*t*, 216–217, 216*f*, 218*f*, 221, 222*f*
repressor regulatory proteins and, 224
RNA primers, 215*t*, 216*f*
RNA reverse transcriptase viruses, 385*t*
RNA (ribonucleic acid), **47**, 49*f*
antibiotics that inhibit, 563, 565–567

antimicrobial agents and, 187
in protein synthesis, 147, 211, 216–221
messenger, 16, **47**, **216**, 218*f*
processing in eukaryotic cells, 220, 222*f*
ribosomal, **47**. *See also* ribosomal RNA
ribozymes and, **121**
structure, 211
transfer, **47**
RNA synthesis
antibiotics that inhibit, 567
from nucleoside triphosphates with ribose, 214
nitrogen requirements, 160
phosphorus requirements, 160
RNA transcript, 220, 222*f*
RNA tumor viruses, 376*t*
RNA viruses, 375–376*t*
biosynthesis, 385*t*, 387–389, 388*f*, 389*f*
multiplication of, 385*t*, 387
pathways, 388*f*
RNA-dependent RNA polymerase, 387, 388*f*
RNA-RNA hybridization reactions, 291
RNAi (RNA interference), **259**, 259*f*, **579**
Robbins, Frederick C., 14*t*
Roberts, Richard J., 15*t*
rock-eating microorganisms, 143
Rocky Mountain spotted fever, 410*t*, 459, **651*b***, **655**
as notifiable infectious disease, 421*t*
as tickborne typhus, 655
caused by *Rickettsia rickettsii*, 303, 413*t*, 655, 656*f*
Dermacentor spp. as tick vector, 655
life cycle of, 656*f*
incubation period, 430*t*
portals of entry, 430*t*
rash caused by, 655, 657*f*
transovarian passage of bacteria and, 655, 656*f*
U.S. geographic distribution, 655, 655*f*
rod-shaped bacteria (bacillus/bacilli), **77**, 78, 78*f*, 107*b*, 303, 316, 316*f*
rodents
as disease reservoirs, 410*t*, 651*b*, 661
as pets
rat bite fever and, 647–648, 650*b*
tularemia and, 644*b*
ground squirrels
plague and, 648, 650
tularemia carried by, 642–643
Hantavirus pulmonary syndrome associated with, 376*t*
prairie dogs and plague, 648, 650
rats. *See* rats
sarcoma viruses in, 391
toxoplasmosis-infected, cats and, 661
root nodules, **772**, 773*f*
Roquefort cheeses, ripened by *Penicillium* molds, 799
Rosa genus/spp., 280*f*
Rosa pratincula, 280*f*
Rosaceae, 280*f*
Rosales, 280*f*
Rose, Irwin, 15*t*
Roseolovirus (HHV-6), 375*t*
roseola, 385, **600**
herpesviruses 6 and 7 causing, 600
rash caused by, 589*b*
Ross, Ronald, 14*t*
rot, plant, 311
rotating biological contactor system, **784**
Rotavirus, 376*t*
vaccine, recommended schedule, 504*t*, 506
rotovirus, **728**, **729*b***
rough ER, 99*f*, **103**, 103*f*
RoundUp herbicide, 265, 267*t*
roundworms (nematodes), 6, 192, 353, **358**–361, 360*f*
aniskines, 361
characteristics, 358
eggs infective for humans, 358–359, 360*f*
freezing temperatures and, 192
heartworm, 360, 360*f*
hookworm, 359–360
larvae infective for humans, 359–360
pinworm, 358, 360*f*
sexual dimorphism in, **358**, 359
Wolbachia bacteria and, 307*b*, 360
Rous, F. Peyton, 10*f*, 14*t*, 389
RPR (rapid plasma reagin) test for syphilis, **755**
rRNA loop, of Archaea/Bacteria/Eukarya compared, 276*t*
rRNA (ribosomal RNA), **47**, 95, 103, 211, 274, 292, 299
sequencing. *See* rRNA sequencing
rRNA sequencing
Chlamydia species moved to new genus as result of, 279, 299
"signature" sequences within domains, phylums, 292
to show evolutionary relationships, 274, 278, 290–292
cladograms, 275*f*, 281*f*, 293, 294*f*
in fossilized materials, 278, 290–291
RSV (respiratory syncytial virus), 679, **692**, **699*b***
RT-PCR (reverse transcription PCR), 266
RTF (resistance transfer factor), **239**–240, 240*f*
rubber, synthetic, 258
rubber tires, 145
brown algae used in production, 342
rubbing alcohol (isopropanol), 37
as antiseptic/disinfectant, 198
rubella (German measles), **599**–**600**, 599*f*
as notifiable infectious disease, 421*t*
congenital syndrome as notifiable infectious disease, 421*t*
incubation period, 430*t*
macular rash caused by, 589*b*
portals of entry, 430*t*
pregnancy and, 421*t*, 760
Rubivirus as cause, 375*t*, 394*t*, 430*t*
vaccine, 13, 502, 503*t*, 504*f*, 599–600
rubella virus. *See Rubivirus*
rubeola. *See* measles
Rubivirus (rubella virus), 375*t*
as persistent viral infection, 394*t*
incubation period, 430*t*
portals of entry, 430*t*
transmission route, 375*t*
vaccine, 13, 502, 503*t*, 504*t*
Rubulavirus (mumps virus)
as notifiable infectious disease, 421*t*
incubation period, 430*t*
portals of entry, 430*t*
vaccine, 13, 502, 503*t*, 504*t*
"run" movement, in bacterial motility, 82, 83*f*
rusts, as basidiomycocytes, 338*t*
rye bread, fermentation and, 137*t*

Sabin polio vaccine, 502, 621
Sabouraud's dextrose agar, 168
sac fungi (Ascomycota), 280*f*, **334**, 336*f*
Talaromyces life cycle, 336*f*
Saccharomyces carlsbergensis, 800
Saccharomyces cerevisiae, sold industrially as baker's yeast, 806
Saccharomyces cerevisiae (baker's yeast), 4*t*, 793*f*
as budding yeast, 332, 332*f*
as vehicle for expressing genetically engineered genes, 258
colony-stimulating factor produced by, 260*t*
fermentation end-products and, 134*f*, 135*b*, 137*t*
genetically modified is used to make hepatitis B vaccine, 339
genome has been mapped, 261
in taxonomic hierarchy, 280*f*
influenza vaccine and, 260*t*
interferons produced by, 260*t*
plasmids found in, 238
strains developed over centuries, 800
used to make bread, beer, wine, 339, 800
vaccine for cervical cancer produced by, 260*t*
Saccharomyces ellipsoideus, 800
Saccharomyces genus/spp.
ethanol produced by for brewed beverages, 332
in taxonomic hierarchy, 280*f*
Saccharomyces uvarum, 800
Saccharomycetaceae, in taxonomic hierarchy, 280*f*
Saccharomycetales, in taxonomic hierarchy, 280*f*
Saccharopolyspora erythraea, erythromycin derived from, 555*t*
safety issues, in biotechnology, 268
safranin stain, 68, 69, 71, 72*t*
as counterstain, 69, 71
in capsule staining, 71, 72*f*, 72*t*
in Gram stain method, 69, 70*f*, 87
Saint Vitus' dance (Sydenham's chorea), **642**
sake, microbes used in production of, 800
saliva, **452**, 453
as defense against pathogens, 453, 471*t*
IgA antibodies in, 480
lysozyme in, 88, 453
pH of, 453
phenolics to disinfect, 195
salivary amylase enzyme of, 453
spirochete bacteria and, 322
substances in that inhibit microbial growth, 453
salivary amylase, of saliva, starch digestion and, 453
salivary glands, 452
Salk polio vaccine, 419, 502, 621
salmon, DNA vaccine approved for, 503
Salmon, Daniel, 4*t*
Salmonella bongori, 310
Salmonella enterica
antibiotic therapy, lactic acid bacteria and, 402
cephalosporin-resistance transferred by *E. coli*, 577*b*
incubation period, 430*t*
phage typing to identify strain of, 290
portals of entry, 430*t*
reservoirs/transmission method, 410*t*
salmonellosis caused by, 430*t*, **712**–714, 714*f*, **722*b***
serovars (serotypes), 310, 517
surviving in phagocytes, altered pathogenicity and, 228
Salmonella genus/spp., 301*t*, **310**
as common inhabitants of animal intestinal tracts, 310
as enteric bacteria, 285, 309–310
as human pathogens, 301*t*, 310
biochemical tests that identify, 139, 139*f*
Bt toxin and, 265
complement system evasion by, 468
E. coli and, effects on host's plasma membrane, 433, 433*f*
fermentation and its end-products, 134*f*
genetic recombination and flagellar proteins, 234
hybridization reactions to identify strains of, 291, 292*f*
in Ames test to identify carcinogens, 232–233, 233*f*
nomenclature unusual for, 310
resistance plasmid R100 and, 239, 240*f*
salmonella release (regulatory molecule), 26
Salmonella typhi
as endotoxin producer, 439
culture medium and, 168
portals of entry, 429, 430*t*
typhoid fever caused by, 310
typhus caused by, 430*t*
Salmonella typhimurium, 4*t*, 26*f*
as serovar of *Salmonella* genus, 310
membrane ruffling by invasins of, 433, 433*f*
salmonellosis, 310, 410*t*, **712**–714, 713*f*, 714*f*, **722*b***
as notifiable infectious disease, 421*t*
incidence of, 714*f*
incubation period, 430*t*
outbreak (tomatoes), 715*b*
portals of entry, 430*t*
salpingitis, **752**, 752*f*
salt. *See also* sodium chloride
to preserve foods, 192
salt crystals, formation of, 30, 30*f*
salts, 34, **35**–36, 36*f*
in food preservation, 159, 192
salty environments
extreme halophiles (archaea) and, 4, **159**, 275, 275*f*, 325
microbial growth and, 159, 169
Staphylococcus aureus and, 168–169, 169*f*
salvarsan, 12
SAM (scanning acoustic microscopy), **63**, 63*f*, 67*t*
San Joaquin fever. *See* coccidioidomycosis
sand fly bites, leishmaniasis and, 354*t*, 665
sanitation, **185**, 185*t*
sanitizers, acid-anionic, **199**, 204*t*
Saprolegnia ferax, 345*f*
pathogenic to plants/animals, 329*f*
saprophytes, **145**
saprophytic fungi, 333, 336
saquinavir, 571
SAR 11, 303
Sarcina genus/spp.
in taxonomic hierarchy, 301*t*
occur in cubical packets, 301*t*
sarcinae, **78**, 78*f*
sarcoma, **389**
sarcoma viruses
as oncogenic retroviruses, 391–392
chicken/avian, 389, 391
feline, 391
Sarcoptes scabiei (mite), 60*f*, 602
Sargasso Sea, *Pelagibacter ubique* discovered in by FISH technique, 303
Sargassum (brown alga), found in subtropical Sargasso Sea, 341
SARS (severe acute respiratory syndrome)
as emerging infectious disease, 417*t*
Coronavirus and, 367, 376*t*, 421*t*
vaccine against, 259
SARS-CoV (severe acute respiratory syndrome-associated coronavirus), 421*t*
sashimi worms (anisakiasis), 361*t*
saturated fatty acids, 40, 41*f*, 42*f*
saturation condition of enzymes, **119**, 119*f*
sauerkraut
lactic acid fermentation and, 135*b*, 137*t*, 800
pH and, 159
saunas/hot tubs, rashes and, 592–593
sausage, fermentation and, 137*t*
saxitoxins, **344**, **444**
scab formation, in inflammatory response, 461*f*
scabies, 362, **602**, 602*f*
ivermectin effective against, 572
rash caused by, 592*b*
scalded skin syndrome, 438*t*, **588**, 588*f*
scanned-probe microscopy, **65**, 65*f*, 68*t*
atomic force microscope (AFM), 58*f*, **65**, 65*f*, 68*t*
scanning tunneling (STM), **65**, 65*f*, 68*t*

scanning acoustic microscopy (SAM), **63**, 63f, 67t
scanning electron micrograph, 64f, 65
scanning electron microscope (SEM), **64–65**, 64f, 67t
  Paramecium image, 64f, 67t
  specimen sizes and, 58f
scanning tunneling microscopy (STM), **65**, 65f, 68t
  RecA protein from *E. coli* image, 65f, 68t
  specimen preparation and, 65
scar tissue formation, **463**
  normal tissue function and, 463
scarlet fever, 319, **677**, **681b**
  caused by *Streptococcus pyogenes*, 406
  exotoxin causing, 439t, 677
  Streptococcus pyogenes causing, 437, 677
Schaeffer-Fulton endospore stain, **71**, 72f, 72t
Schistosoma (blood fluke), 356, 361t, 666–668, **668b**
  life cycle of, 667f
Schistosoma haematobium, 666
Schistosoma japonicum, 666
Schistosoma mansoni, 666
schistosomiasis, 329, 356, 361t, **666–668**, 667f, **668b**
  as major world health problem, 356
  praziquantel to treat, 571, 666–667
schizogony, **346**
  in *Plasmodium*, 348–349, 349f, 663
  trypanosomes and, 351, 661
Schizosaccharomyces, as fission yeasts, **332**
Schulz, Heide, 13
SCID (severe combined immunodeficiency disease), 18
scientific applications, of rDNA technology, 261–264
scientific nomenclature, 2–3, 4t, 278–279
sclerotia, **443**
scolex of tapeworms, **356**, 358f
scrapie disease in sheep, 392
  mad cow disease and, 393
screening procedures for clone selection, 256–257, 256f
scum, green, 341
sea lions, leptospirosis deaths in, 283b
sea otters, toxoplasmosis deaths, 283b, 662
seafood allergies, 525
seals
  Arcanobacterium phocae and veterinary microbiology, 283b
  influenza A viruses and, 19, 370b
  phocid distemper virus caused deaths in, 283b
seawater microbiota, **777–778**
seaweeds, 341–342
sebaceous (oil) glands of skin, 453
sebum, **453**, 471t
  protective functions of, 453, 471t, 585
secondary immune response, **493–494**, 494f
  vaccines and subsequent antigen encounters, 501
secondary infection, **407**, 409
secondary infections, difficulty in treating in hospitalized patients, 414
secondary sewage treatment, **784**, 785f
secondary structure of proteins, 45, 46f
secretion of gene products, advantages to genetic engineering, 258
secretory component, IgA antibody and, 480–481
secretory IgA, 480
secretory vesicles, **104**, 104f
seizures, fever and, 463
selection, 249
  artificial, **249**
  marker genes within vectors and, 250
  natural. *See* natural selection
  of genetically desirable plants, 264
selection methods to identify mutations, **231**
selection (preferential survival) of bacteria with resistance factors, 239–240, 240f
selective culture media, **168**, 169t
  enrichment culture *vs.*, 169, 169t
  identification of microorganisms and, 285–286
selective IgA immunodeficiency, 539t
selective permeability, **90**
selective toxicity, **553**
  of antibiotics, 553, 555, 557, 558f
  tetracyclines, 565
selenium, toxicity reduced in anaerobic bacteria cultures, 264, 264f
self molecules of MHC, 482, 486, 533
self *vs.* nonself recognition, 477, 485, 492–493, 494, 496f
  autoimmune diseases and, 532–533
  hyperacute rejection and, **536**
  immune system tolerance of fetus and, 534–535
  major histocompatibility complex (MHC) and, **482**, 486, **533**–534
  thymic selection and, 486, 532
  transplant rejection and, 534–535
self-replication capability, DNA vectors and, 250
self-tolerance loss in autoimmune diseases, 532
SEM (scanning electron microscope), **64–65**, 64f, 67t
  Paramecium image, 64f, 67t
  specimen sizes and, 58f
semiconservative replication, **213**
semipermeability, **90**
Semmelweis, Ignaz, 10f, 11, 184, 197, 413, 418, 641
sense codons, **219**
sense strand (+ strand), **387**, 388f
sensitivity of diagnostic tests, **507**
sensitized individuals, 523
sentinel animals, tested for arbovirus antibodies, 624–625
sepsis, **186**, **407**, 414t, **639**–641
  cytokine storm and, 492
  endotoxin release with antibiotic therapy for, 640
  gram-negative (endotoxic shock), **640**
  gram-positive, **640**
  Listeria monocytogenes causing, 615
  lymphangitis and, 639, 640f
  neonatal, 640
  puerperal (childbirth fever), **640**–641, **643b**
  severe, **640**
  Staphylococcus aureus causing, 587. *See also* nosocomial infections
sepsis in cattle, *Pasteurella* bacteria causing, 311
septa, **331**
septate hyphae, **331**, 331f
  conidia formed by, 333, 334f
  pathogenic fungi with, 338t
septic arthritis, *Haemophilus influenzae* causing, 311
septic shock, **437**, 439, 639–641, **640**, **643b**
  antimicrobial peptides (AMPs) and, 471
septic tanks, **787**–788, 788f
septicemia, **407**, 418, **639**
  indwelling venous catheters and, 164b
  lymphangitis and, 639, 640f
septicemic plague, **648**, **650b**
sequencing, DNA, **261–262**, 261f
  random shotgun sequencing, **261**, 261f
serial dilution, **174**, 175f
serine (Ser), structural formula/characteristic R group, 44t
seroconversion, 543f, **545**
serological testing, **287**, 287f, 310
  tissue typing, 533–534, 533f
  virus typing, 512
serology, **287**, 310, **495**
serotypes, **287**, **310**
  of meningococcus, 613
  of *Salmonella enterica*, **310**
serovars, **82**, **287**, **310**
  direct agglutination tests and, 510
  of *Salmonella enterica*, **310**
  of *Vibrio cholerae* 0139, evolution and, 416
Serratia genus/spp., 301t, 310, **310**
  as opportunistic pathogens, 301t
  found in catheters/sterile solutions, 310
  hospital respiratory/urinary tract infections and, 310
Serratia marcescens, red pigment produced by, 301t, 310
serum, **467b**
  antibody percentages, 479–481, 481t
  antibody titer, **493**, 494f, **510**, 511f
  antiserum and, **287**, **495**, 495f
  fetal calf, 495
  laboratory collection of, 467b
  separation of proteins by gel electrophoresis, 495, 495f
  testing for chemicals or enzymes in blood, 467b
serum IgA, 480
70S ribosomes, 95, 95f, 101t
  in mitochondria, 105
severe acute respiratory syndrome (SARS)
  as emerging infectious disease, 417t
  Coronavirus and, 367, 376t
  vaccine against, 259
severe acute respiratory syndrome-associated coronavirus (SARS CoV), 421t
severe combined immunodeficiency, 539t
severe combined immunodeficiency disease (SCID), 18
  gene therapy to treat, 259
severe sepsis, **640**
sewage
  bacteria found in, 305, 306f
  Enterobacter common to, 311
  liquid form of compressed chlorine gas used to disinfect, 197
sewage treatment, **783–789**
  aquatic microorganisms and, 776–778
  archaea methanogens used in, 325, 787f
  biochemical oxygen demand (BOD), **783–785**
  biofilms and, 163, 787f
  disinfection and release, **785**, 785f
  oxidation ponds, **788**
  primary, **783**, 785f
  secondary, **784**, 785f
  septic tanks, **787–788**, 788f
  sludge digestion, **785**–787, 785f, 786f
  tertiary, **788**–789
  Zoogloea found in, 300t
sewage treatment plants
  beneficial microbes used in, 17
  by-products of bacterial conversions, 17
sex pili, 84, 236, 237f, 238f
  of enterics, 309
sexual dimorphism, **358**–359
  in nematode *Ascaris lumbricoides*, 358–359
sexual recombination, in prokaryotic *vs.* eukaryotic cells, 101t
sexual reproduction
  fungal, **333**, 335f
  in algae, 342, 342f
  in *Plasmodium vivax*, 348–349, 349f
  of protozoa, 346, 346f
sexual spores, **333**
  of Ascomycota fungi, 334, 336f
  of Basidiomycota fungi, 335, 337f
  of Zygomycota fungi, 333, 335f
sexually transmitted diseases (STDs), 322, **747**. *See also* sexually transmitted infections
sexually transmitted infections (STIs), 322, **747–761**
  AIDS. *See* AIDS
  bacterial, 746–756, **759b**, **761b**
  chancroid (soft chancre), 311, **756**, **761b**
  clamydias, 322, 429, 430t, 750–751, 761b
  epidemics, 21
  genital herpes, 569, 570f, 740, **757**, 757f, **761b**
  genital warts, 375t, 385–386, 386f, 429, **758**, 758f, **761b**
  gonorrhea, 306, **747**. *See also* gonorrhea
  HIV infection. *See* HIV infection
  lymphogranuloma venereum, 322, 459, **755**, **761b**
  pelvic inflammatory disease, **751**–752, 752f, **761b**
  portals of entry for, 429, 430t
  syphilis, **323**, **752**. *See also* syphilis
  trichomoniasis, **759b**, **760**, 760f
  urethritis, nongonococcal, 322, **750**–751, 761b
  vaginitis, **756**, 756f, **759b**
  vaginosis, **756**, 756f, **759b**
shadow casting, 63
  TEM image, 80
Sharp, Phillip A., 15t
sheath, of a T-even bacteriophage, 374f, 381f
sheathed bacteria, 301t, 305
sheep
  genetically modified to produce therapeutic drugs in milk, 259, 260t
  scrapie disease in, 392, **630**
sheep scrapie, 392, **630**
  mad cow disease and, 631
Shiga, Kiyoshi, 10f
Shiga toxin
  potency of, 430
  prophage gene and, 382
  shigellosis and, **712**, **722b**
Shiga-toxin *E. coli*, 210, 237, 382, 441, **717**–718, 718f, 723b
  as notifiable infectious disease, 421t
Shigella genus/spp., 301t, **310**, 712f, 713f
  as enteric bacteria, 285, 309–310
  as human pathogens, 301t
  as potential biological weapon, 649b
  biochemical tests that detect, 139
  can survive in phagocytes, 459
  capable of using actin to advantage, 433
  E. coli 0157:H7
    adherence and pathogenicity, 431
    Shiga toxin and, 210, 237, 382, 441, 717–718, 718f, 723b
  portals of entry, 430t
  shigellosis caused by, 310, 411, 421t, 429, 430t, **712**
  traveler's diarrhea and, 438t
shigellosis (bacillary dysentery), 310, 459, **712**, 712f, 713f, **722b**
  as notifiable infectious disease, 421t
  incubation period, 430t
  portals of entry, 429, 430t
  Shigella bacteria causing, 310. *See also* *Shigella*
  waterborne transmission and, 411
shingles (herpes-zoster), 375t, 394t, 407, **596**–597
  as a latent varicella-zoster virus disease, 407, 596
  in HIV/AIDS patients, 542, 544t
  rash caused by, 590b, 597f
  vaccine, 503t, 596–597
shivering, **463**
shock, **437**, 640
  anaphylactic, **524**
  endotoxic, 437
  septic, **437**, 439, 471, 639–641, **640**, **643b**
shoe leather, fungi capable of growing in, 333

Index

shuttle vectors, **251**
sialic acid, 468
sickle cell disease, 408
  gene therapy and, 18
  missense mutation and, 228
side chain amino acid (tetrapeptide side chain), 85, 86f
side groups (R groups) of amino acids, 43, 43f, 44t, 45
siderophores, **434**, 434f, 445f
  enterobactin, 434f
  iron-binding proteins and, **470**
signals (chemical)
  alarm signals (alarmone), 224, 225–226
  biofilms and, 57b, 163
signs, *vs.* symptoms, **406**
silencing, gene, **259**, 259f
silent (neutral) mutations, 226–227
silica, in cell walls of diatoms, 343, 343f
silkworm disease, early research on cause, 11
silver, antimicrobial activity of, 198–199, 198f
silver nitrate, 198, 204t, 603–604, 604b
silver sulfadiazine, 198, 204t, 567, 594
silver-haired bats, rabies virus variant associated with, 625b, 625f
silver-impregnated dressings, for antibiotic-resistant bacteria, 198
simian AIDS, 377
simian immunodeficiency virus (SIV), 540
simple carbohydrates, 39
simple diffusion, **91**–92, 92f
simple lipids, 40–41, 41f
simple proteins, 45
simple stains, **69**, 72t
simple sugars, 39
Simplexvirus (HHV-1, HHV-2), 375t, 385, 392
Sin Nombre hantavirus, 660, 660b
single covalent bond, 31, 31f
single-stranded DNA nonenveloped viruses, 375t
single-stranded DNA viruses, 385t
single-stranded RNA, + strand enveloped viruses, 375–376t, 388f
single-stranded RNA, + strand nonenveloped viruses, 375t, 388t
singlet oxygen, 62, **161**
  as toxic oxygen product produced by lysosomal enzymes, 459
  sunlight and, 193
sinusitis, **676**
siRNAs (small interfering RNAs), **259**, 259f, **579**
sirolimus (Rapamune), 536–537
SIRS (systemic inflammatory response syndrome), 639
site-directed mutagenesis, **249**
SIV (simian immunodeficiency virus), 540
sizes, of viruses, 369f
skin, **451**, 451f
  acidity of, 453
  as first line of defense, 450, 450f, 471t
  as physical barrier to pathogens, 450f, 451–452, 451f, **471t**, 584
  as portal of entry, 429, 430t, 445f
  bacteria most likely to cause infections of, 451
  broken, susceptibility to infections, 415, 415t, 451
  chemicals that defend, 453, 471t, 584
  commensal microbes of, 453
  cutaneous mycoses and, 337, 338t
  delayed cell-mediated hypersensitivity reactions to, 530–531, 530f, 531b, 532f
  dermis of, **451**, 451f, 471t, **585**, 585f
  epidermis of, **451**, 451f, 471t, **585**, 585f
  function of, 584, 585
  fungal diseases of, **600**–602, 600f
  hookworm larvae and, 429
  immune system and, 450–453, 471t
  keratin and, **337**, 338t, 402t, **451**, 451f
  lesions, **586**, 587f
  microbial diseases, 584–603
    bacterial, 586–598
    nosocomial, 415t
    viral, 595–600
  normal microbiota of, 402t, **585**–586
    innate immunity and, 450f, 453
  parasitic infestations of, **602**–603, 603f
  perspiration flushes microbes from surface, 453
  pH, 453
  pH of, 453, 586
  physical factors that defend, 451–452, 451f
  Propionibacterium bacteria on, 320
  rashes. *See* rashes
  regeneration capacity of, 462
  sebum and, 453, 585, 585f
  staphylococcal infections, 586–589
  staphylococci on, 318
  streptococcal infections, **589**–591
  structure of, 585, 585f
  sweat glands and perspiration, 453, 585f
skin cancers, unrepaired dimers and, 231
skin infections
  caused by *Streptococcus pyogenes*, 406
  staphylococcal, 586–589
skin tests
  for antigen sensitivities, 526, 526f
  for food allergies, 525
  for leprosy, 620
  for penicillin sensitivity, 524
  for tuberculosis, 507, 530
skunks
  as disease reservoirs, 410t
  reported cases of rabies in, 624f
slants, 165
SLE (St. Louis encephalitis), 376t, 625–626, **628b**
sleeping sickness. *See* African trypanosomiasis
slide agglutination test, **287**, 287f
slime, biofilms and, 162–163, 163f
slime layers, **80**, 81, 101t. *See also* biofilms
  catheters and, 586, 587f
slime molds, 4, 6, **351**–352, 352f
  cellular, **351**, 352f
  plasmodial, 351, **352**
slime trails
  Myxococcus bacteria, 312, 313f
  produced by *M. xanthus*, 57b
sloth, South American, algae living in hair of, 341
slow-growing mycobacteria, identification tests for, 144b
sludge, **783**, 785f
  primary, 783
sludge digestion in sewage treatment, **785**–787, 785f, 786f
small interfering RNAs (siRNAs), **259**, 259f, **579**
small intestine, 456f
small intestine enzymes, most microbes destroyed by, 429
small intestines, parasitic helminths and, 361t
small nuclear ribonucleoproteins. *See* snRNPs
smallpox vaccine, 503t
  as first vaccine developed, 477
  cowpox virus and, 11, 501
  early experiments to develop, 11, 501
  early testings of, 406, 501
  importance to science of immunology, 501
  variolation procedure and, 501
smallpox (variola), 375t, **595**–596, 596f
  as notifiable infectious disease, 421t
  as potential biological weapon, 649b
  bioterrorism dangers as immunity declines, 596
  cidofovir may be effective against, 570, 596
  early epidemics, 11
  first disease for which vaccine was developed, 477
  mortality rate in eighteenth century, 501
  orthopoxvirus causing, 374f, 375t, 595
  Poxviridae causing, 385
  rash caused by, 590b
  respiratory tract as portal of entry, 429
smallpox (variola) virus. *See* smallpox (variola)
smear (specimen), **68**
Smith, Hamilton, 10f, 14t, 231
smooth ER, 99f, **103**, 103f
smuts, as basidiomycocytes, 338t
Snow, John, 418
snRNPs (small nuclear ribonucleoproteins), 215t, **220**, 222f
soaps, 199, 200f
  antiseptic effectiveness, 200f, 204t
SOD (superoxide dismutase), 161t, **162**
  genetically engineered, 260t
sodium benzoate, 199, 199–200
sodium chloride
  dissociation of, 35, 36f
  water acting as solvent for, 35, 35f
sodium chloride (NaCl), *Staphylococcus aureus* and selective culture media, 168–169, 169f
sodium chloride (NaCl) formation, 30, 30f
sodium dichloroisocyanurate, 197
sodium hydroxide (NaOH)
  as a base, 35, 36f
  colony hybridization and, 257f
sodium hypochlorite, as disinfectant, 196f
sodium hypochlorite (NAOCl), 196f, 197
sodium ion, **28**, 30
  as a cation, 30
  in table salt, dissolved in water, 35, 35f, 36f
sodium (Na), atomic number/atomic weight, 27t
sodium nitrate/nitrite
  as food preservatives, 200, 204t
  as meat preservative, 200
sodium thioglycolate, in reducing media, 166
sodoku (rat bite fever), 648
soft chancre (chancroid), **756**, **761b**
soft-rot diseases of plants, *Erwinia* bacteria as cause, 311
soil
  as disease reservoirs, 2, 3b, 306, 308, 310, 311, 317–318, 320, 321, 321f, 324, 409, 646, 668b
  DNA probes to identify specific, 262, 293
  enrichment mediums and, 169
  pathogenic fungi in, 336–337
  screening for antibiotic-producing microbes, 554
soil microbiology
  biogeochemical cycles and, **768**–**776**
  degrading synthetic chemicals, 775–776
solar evaporating ponds, extreme halophiles (archaea) found in, 325
solid municipal waste (garbage), 775–776
solutes, **35**
solutions
  acidic *vs.* alkaline, 35–36, 36f
  hypertonic, 93f, **94**, 159, 160f
  hypotonic, 93f, **94**, 159
  isotonic, **93**, 93f, 159, 160f
solvents, **35**, 35f
somatostatin, genetically modified *E. coli* culture and production of, 259
sorbic acid, 199–200, 204t
sorbitol
  fermentation and, 137t
  fermentation by *E. coli* and biochemical test for pathogenic strain, 138, 139f
sorbose, as fermentation end-product, 137t
sore throat
  caused by *Streptococcus pyogenes*, 406
  Streptococcus pyogenes and, 319
sound waves, scanning acoustic microscopy uses of, 63, 63f, 67t
souring and spoilage, alcoholic beverages and, 9
Southern blotting, **262**, 263f, **291**, 292f
soy products, food allergies and, 525
soybeans, infected by *Phytophythora infestans*, 344
Spallanzani, Lazzaro, 9
spasmodic contractions, of tetanus, 437
special stains/staining, **71**–72, 72f
  summary, 72t
specialized transduction, **237**, 239f
  lysogeny and, **382**
species
  eukaryotic, **278**–279, 280f
  name (specific epithet), **2**, **278**–279, 280f
  prokaryotic, **278**–279, 280f
  viral, **282**
species barrier
  antigenic shift and, 370–371b, 371f
  influenza A viruses crossing, 370–371b
specific defenses of hosts against pathogens, **476**–**499**. *See also* adaptive immunity
specific epithet (species) name, **2**, **278**–279, 280f
specificity
  of antibodies, **484**
  of enzymes, 116
specificity and diagnostic tests, **507**
specimen preparation, 54, 68–72
  artifacts and, 63
  fixing to slide, 68
  heat involved in, 68
  microscope resolution and size, 58f
  smears, 68
spectrophotometers
  endotoxing testing and, 439
  to measure turbidity, 178, 179f
Sphaeroltilus genus/spp., 300t, **305**, 306f
  as sheathed bacteria, 300t
  found in fresh water, sewage, 305
Sphaeroltilus natans, 305, 306f
spherical-shaped bacteria, **77**–78, 78f
spheroplasts, **89**
spicules of nematodes, **358**, 360f
spikes
  gp120 (on HIV), 540, 540f, 541, 548
  viral, **372**–373, 372f
spinal cord, 611, 611f
spinal tap (lumbar puncture), 614, 614f
spiral bacteria, **77**, 79, 79f
spirilla/spirillum, **79**, 79f
spirillar fever, 648
Spirillum genus/spp.
  lipid inclusions of, 96
  position in taxonomic hierarchy, 300t
Spirillum minus, causing rat bite fever (spirillar fever), 648
Spirillum volutans, 305, 306f
  flagella staining of, 72, 72t
Spirochaetales, 302t
  important genera/special features, 302t
Spirochaetes, 302t
spirochete that causes Lyme disease, 362
spirochetes, **79**, 79f, 107b, 322–323, 324f
  motility of, 82–83, 84f, 322–323, 324f
  phylogenetic relationships, 281f
Spiroplasma genus/spp., 301t, **320**
  as plant pathogens, 301t
  as pleomorphic bacteria with no cell wall, 301t
  parasites of plant-feeding insects, 320
  plant pathogens, 320
spleen, 456f, 457
  in monoclonal antibody production, 508f
  macrophages of, 457
spoilage, food. *See* food spoilage
sponges, as eukarya, 6

spongiform encephalopathies, prions and, 203, 393, 630*f*
spontaneous generation theory, **8**, 9*f*
spontaneous mutations, **228**
frequency in bacteria, 240
sporadic disease, **406**
sporangia
of *mucor,* 5*f*
of plasmodial slime mold, 353*f*
sporangioles, 313*f*
sporangiophores, **333**
of *Rhizopus,* 334*f*, 335*f*
sporangiospores, **333**, 334*f*
of *Rhizopus,* 333, 334*f*, 335*f*, 338*t*
pathogenic fungi and, 338*t*
sporangium (spore sac), **333**, 334*f*, 335*f*
spore caps, of cellular slime molds, 352, 352*f*
spore clusters, of *M. xanthus* cells, 57*b*, 57*f*
spore coat, 96, 97*f*
spore (endospore) staining, **71**, 72*f*, 96
spore sac (sporangium), 333, 334*f*
spore septum, 96, 97*f*
spores, **332**–333, 334*f*
asexual, 330*t*, 331*f*, **332**, **333**, 334*f*, 335*f*
bacterial endospores compared to, 332
fungal, 281, 330*t*, 331*f*, 332–333, 334*f*, 335*f*
importance in identifying fungi, 333
resistance to chemical biocides, 203*f*
fungal, airborne transmission and, 412
growth of hyphae from, 331, 331*f*
in slime molds, 351–352, 352*f*, 353*f*
inhalation of, systemic mycoses and, 336
of amoeboid cells, 351, 352*f*
reproductive, 331*f*, 332
sexual, 330*t*, **333**, 335*f*
zygospores, **333**, 335*f*
sporicidal agents, peracetic acid, 202
sporicides
glutaraldehyde, 200, 203*t*, 205*t*
hydrogen peroxide, 202
Sporothrix schenckii, as pathogenic fungi, 338*t*, 601
sporotrichosis, **601**
as subcutaneous mycosis, 337
rash caused by, 592*b*
transmission route, 337
sporozoite, **348**–349, 349*f*
in toxoplasmosis, 661, 662*f*
sporulation/sporogenesis, **96**–97, 97*f*
evolutionary development, 317
reproduction and, 97
spotted fever diseases, *Rickettsias* and, 303
spread plate method, **175**, 176*f*
squalamine, 579
squid, transmitting parasitic worms, 361
squirrels, ground
plague carried by, 311, 648, 650, 651*b*
tularemia carried by, 642–643
src gene, cancer-causing, 391
SSPE. *See* subacute sclerosing panencephalitis
St. Louis encephalitis (SLE), 376*t*, 625–626, **628***b*
as an arbovirus, 625, 628*b*
Culex mosquito as vector, 628*b*
Stachybotrys (fungus), pathogenic properties of, 337–338, 338*t*, 443
stage, of compound light microscope, **56**, 56*f*
stains/staining
composition of, 68–69
counterstains, **69**
decolorizing agents, **69**
desirability of, 59–60
differential, **69**–71, 70*f*
distortion from, 59–60
electron microscopes and, 63–64
fixing specimens before, 68
Gram stain, **69**70, 70*f*, 87
negative, 63, **69**, 79
of basophils, 454
of eosinophils, 454
of neutrophils, 454
positive, 63
preparing smears for, 68–69
primary stain, **69**
refractive index and, 58–59
simple, **69**
special stains, **71**–72, 72*f*, 72*t*
techniques
differential, **69**–71, 70*f*
simple, **69**
special, **71**–72, 72*f*, 72*t*
TEM microscope and, 63–64
stalked-cell bacteria, 300*t*, 303, 304*f*
Stanier, Roger, 274
Stanley, Wendell, 16, 368
staphylococcal food poisoning, 318, 438*t*, **711**–712, 711*f*, **722***b*
staphylococcal enterotoxicosis, **711**–712, 711*f*, **722***b*
staphylococcal enterotoxin, 430, **437**, 439*t*, 440, 441
staphylococcal skin infections, 586–589
staphylococci, **78**, 78*f*. *See also Staphylococcus* genus/spp.
characteristics that account for their pathogenicity, 318
disinfectants effective against, 196, 196*f*
most likely to cause skin infections, 451
nosocomial infections and, 414, 414*t*, 422*b*
Staphylococcus aureus, 2–3, 76*f*, **318**, 318*f*, 399*f*
acute inflammation caused by, 460
adherence mechanism resembles viral attachment, 432
as coagulase-positive, 587
as most pathogenic staphylococci, 586–589
as normal microbiota of eye, 402*t*
as normal microbiota of nose, throat, 403*t*
biochemical tests and, 283*b*
biofilms and catheters, 19*f*
culture media to identify, 168–169, 169*f*
destroying a phagocyte, 76*f*
disk-diffusion method evaluating disinfectants, 196*f*
endocarditis caused by, 641, **643***b*
enterotoxins produced by, 318, 437, 439*t*
food poisoning caused by, 318, 438*t*, **711**–712, 711*f*, **722***b*
found on healthy skin, 399*f*, 402*t*
gastric juice unable to destroy, 453
impetigo and, 588, 588*f*
in skin infections, 586–589
lactic acid bacteria to prevent surgical wound infections, 402–403
methicillin-resistant, **20**, 417*t*, 422*b*, **560**–561
nosocomial infections and, 414, 414*t*, 422*b*
otitis media caused by, 679
penicillin resistance, 19–20, 318
produces superantigen that affects intestines, 437
scalded skin syndrome caused by, 438*t*
staphylokinase produced by, 432
toxic shock syndrome and, 437
toxins produced by
enterotoxins, 318, 437, 439*t*
exfoliative, 239
vancomycin-intermediate resistant (VISA) and, **20**, 417*t*, 421*t*, 422*b*
vancomycin-resistant (VRSA), 12–13, 19–**20**, 210, 241, 417*t*, 421*t*, 422*b*, 563
Staphylococcus epidermidis
as a nosocomial pathogen, 586, 587*f*
as normal microbiota of eye, 402*t*
as normal microbiota of nose, throat, 403*t*
catheters, biofilms, other medical procedures, 586, 587*f*
fermentation test to detect, 139, 139*f*
in differential culture media, 169, 169*f*
Staphylococcus genus/spp., 301*t*, **318**, 318*f*
as human pathogens, 19, 301*t*, 318
as normal microbiota of mouth, 403*t*
as normal microbiota of skin, 402*t*
as normal microbiota of urethra, 403*t*
fermentation test to detect, 139, 139*f*
genetic transformation natural in, 236
produces leukocidins that kill phagocytes, 459
Staphylococcus saprophyticus, cystitis caused by, 746
staphylokinase, produced by *Staphylococcus aureus,* 432
star-shaped bacteria, 79, 79*f*
starch granules, in presence of iodine, 95
starch storage, by green algae, 342*f*, 343, 343*t*
starches, 40
as carbohydrates, 39–40
start codons, 212
stationary phase, in bacterial growth curve, **173**, 173*f*
STDs. *See* sexually transmitted diseases
steam heat, to control microbial growth, 188–190, 189*f*, 189*t*, 194*t*
stearic acid, 41*f*
STEC. *See* Shiga-toxin *E. coli*
Stella genus, star-shaped cells of, 79, 79*f*
stem cells
adult, 535
as part of lymphatic system, 456, 456*f*
bone marrow, B cells, T cells originate from, 486, 486*f*
embryonic stem cells (ESCs), **535**, 535*f*
transplantation medicine and, 535, 535*f*
umbilical cord blood cells, 535, 536
stents, cardiac, immunosuppressive drug to treat, 436–537
stents, cardiovascular, biofilms colonizing, 431
stereoisomers, **43**, 43*f*
sterilants, **185**
gaseous chemosterilants, 200–201
glutaraldehyde, 200, 203*t*, 205*t*
peracetic acid, 202
sterile culture media, **164**–165
sterilization, **185**, 185*t*
autoclaves and, **188**–190, 189*f*, 189*t*, 439, 440*b*
boiling water, 188
calculating time necessary for, 186, 187*f*
chemical, 200–201, 205*t*
commercial, **185**, 185*t*, 190, **794**–795, 794*f*, 795*f*
dry heat (flaming), **191**
endotoxins survival despite, 439
fungi and, 188
hot-air, **191**
indicators of successful, 190, 190*f*
moist heat, **188**–190, 189*f*, 189*t*
of gases, 185
of liquids, 185
of milk, by UHT treatments, 190
plasma, **201**
reliable temperatures for, 188
viruses and, 188
sterilizing agents, glutaraldehyde, 200, 203*t*, 205*t*
steroids, **41**, 42*f*
synthesized from microbes, 806, 806*f*
sterols, 101*t*, 330*t*
antifungal drugs affecting, 564*t*, 567–569, 568*f*
in plasma membranes of fungi, 330*t*, 558
in plasma membranes of *Mycoplasma,* 41, 42*f*, 87, 89, 330*t*
Stewart, Sarah, 10*f*, 390
sticky ends of cut DNA strands, **249**, 250*f*
replication and, 215*t*
transposase and, 241*f*
Stigmatella genus/spp.
as gliding, fruiting bacteria, 301*t*
in taxonomic hierarchy, 301*t*
stipe, algae, 341*f*
stipes of algae, 341*f*, **342**
STIs. *See* sexually transmitted infections
STM (scanning tunneling microscopy), **65**, 65*f*, 68*t*
RecA protein from *E. coli* image, 65*f*, 68*t*
specimen preparation and, 65
stomach
enzymes destroy most microbes (except some toxins), 429, 453
gastric juice, **453**
stomach cancer, *Helicobacter pylori* and, 719
stomach lining, normal microbiota of, 400*f*
"stone-washed" denim jeans, applications of microbiology and, 3*b*, 40
stop codons (nonsense codons), 212, 219, 219*f*
storage materials, of algae, 343*t*
storage vesicles, **104**
strains
Bergey's Manual and, 286
improvements, industrial microbiology active in, **803**
of bacterial species, 279, **281**
serological testing to identify, 287
Stramenopila (kingdom), includes diatoms, dinoflagellates, water molds, 343
- strand (antisense strand), **387**, 388*f*
-strand, multiple strands of RNA viruses, 376*t*
-strand, one strand of RNA viruses, 376*t*
- strand RNA viruses, 385*t*
+ strand RNA viruses, 385*t*
+ strand (sense strand), **387**, 388*f*
stratum corneum, 585, 585*f*
streak plate method, **170**, 170*f*
strep throat (streptococcal pharyngitis), 168, **677**, 677*f*, 681*b*
streptobacillary rat bite fever, 648
streptobacilli, **78**, 78*f*
Streptobacillus genus/spp., in taxonomic hierarchy, 302*t*
Streptobacillus moniliformis, streptobacillary rat bite fever caused by, 648
streptococcal infections
as notifiable infectious disease, 421*t*
sulfa drugs effective against during WWII, 554
streptococcal pharyngitis (strep throat), 168, **677**, 677*f*, 681*b*
streptococcal skin infections, **589**–591, 591*f*
streptococci, **77**–78, 78*f*
as normal microbiota of eye, 402*t*
beta-hemolytic (group A, B), 319
carrying lysogenic phage, and toxic shock syndrome, 382
disinfectants effective against, 196, 196*f*
enzymes produced by and tissue destruction, 287, 318
group A (GAS). *See Streptococcus pyogenes*
identification of, 16, 287
immunological techniques, 16
M protein and, 319, 590–591, 591*f*
non-beta-hemolytic, 319
nonpathogenic species of, dairy industry and, 319
serotypes of, 16, 287
streptolysin released by kills phagocytes, 459
viridans streptococci, 319
Streptococcus agalactiae, neonatal sepsis caused by, 319
"Streptococcus faecalis," 279
"Streptococcus faecium," 279
Streptococcus genus/spp., **318**–319, 318*f*
as chemoheterotroph, 143*f*

  as human pathogens, 301t, 318–319
  as lactic acid bacteria, 135b, 137t
  as normal microbiota of mouth, 403t
  as normal microbiota of vagina, 403t
  fermentation and, 134f, 137t
  genetic transformation natural in, 236
  in taxonomic hierarchy, 301t
  low G + C content ratio, 316
  metabolic characteristics, 318
  penicillinase-producing plasmid and *Neisseria*, 240
Streptococcus mutans
  Actinomyces, dextran, and dental plaque, 431, 707
  dental caries caused by, 319, 707–709, 708f, 710b
  glucosyltransferase produced by, 431
Streptococcus pneumoniae
  as normal microbiota of nose, throat, 403t
  as notifiable infectious disease, 421t
  as opportunistic pathogen, 404
  avirulent strains, 234–236, 235f, 432
  capsule/encapsulated, virulent strain, 80, 234–236, 235f, 441
  classification and, 279
  DNA transformation process and, 234–236, 235f
  drug-resistant, as notifiable infectious disease, 421t
  emerging infectious diseases and, 417t
  Griffith's experiments with, 234–236, 235f
  incubation period for, 430t
  meningitis (pneumonococcal) caused by, 612, **614**, 617b
  nonencapsulated, avirulent strain, 234–236, 235f, 432
  otitis media caused by, 679
  phagocytes and, 228
  pneumonia caused by, 13, 319, 430t, 432, 502, 502t, 504t, **685**, 686f, 687–688, **687**b
  polysaccharide capsule and virulence of, 80
  portals of entry for, 430t
  post-influenza bronchopneumonia caused by, 407
  resistance to beta-lactam antibiotics, 574
  vaccine against, 502t, 614
Streptococcus pyogenes, 4t, 319, 590–591, 591f
  as "flesh-eating" bacteria, **20**, 287, 318, 321, 422b, 591, 591f
  causing toxic shock syndrome, 382, 417t
  differential media to identify, 168, 168f, 319
  diseases caused by, 406
  erythrogenic toxin and, 237
  ethanol activity against, 198t
  has genetic material to synthesize three types of cytotoxins, 437
  impetigo and, 588, 588f
  iron source for, 470
  M protein produced by increases virulence, 319, 432, 459, 590–591, 591f
  otitis media caused by, 679
  pericarditis caused by, **641**, **643**b
  puerperal sepsis caused by, 418, 640–641, 643b
  strep throat caused by, 677, 677f, 681b
  streptokinase (fibrinolysin) produced by, 432
Streptococcus thermophilus, used to make yogurt, 799
streptogramins, **566**
  mode of action/spectrum of activity, 562t
streptokinase (fibrinolysin), 432
streptokinases, 590, 677
streptolysins, 590, 677
Streptomyces venezuelae, chloramphenicol derived from, 555t
Streptomyces, vancomycin derived from, 563
Streptomyces aureofaciens, chlortetracycline, tetracycline derived from, 555t
Streptomyces fradiae, neomycin derived from, 555t
Streptomyces genus/spp., **321**, 321f
  actinomycetes informal name for, 320
  antibiotics derived from, 301t, 555t, 563
    vancomycin, 563
  as filamentous branching bacteria, 301t
  as pleomorphic bacteria, 320
  G + C content of, 316
  in taxonomic hierarchy, 301t
  reproductive asexual spores of, 321
  used in production of steroids, 806, 806f
Streptomyces griseus, streptomycin derived from, 555t
Streptomyces nodosus, amphotericin B derived from, 555t
streptomycin, **565**
  derived from *Streptomyces griseus*, 555
  mode of action/spectrum of activity, 557t, 562t
  protein synthesis inhibited by, 95, 556–557, 556f, 558f, 562t, 565
  resistance, 239, 240f
  spectrum of activity, 557t
  susceptibility of gram-negative *vs.* gram-positive bacteria to, 88t
stroke, hemorrhagic, genetically engineered Factor VII to treat, 260t
stroma, in tissue repair of inflammatory response, 462
stromatolites, 277, 278f
sty, **588**
subacute bacterial endocarditis, **641**, 641f, **643**b
subacute disease, **407**
subacute sclerosing panencephalitis (SSPE), 392, 394t, 407, **599**
subarachnoid space, 611, 612f
subclavian veins, 456f
subclinical (inapparent) infection, **407**
subcutaneous mycoses, 336, **336–337**, **601**
  caused by saprophytic fungi, 336
  sporotrichosis acquired by farmers/gardeners, 337
sublimation, in preserving bacterial cultures, 170
sublittoral zone, algal habitats, 341f
substrate, **115**, 116f, 117–118, 118f, 120f
substrate concentration, **119**
  as factor influencing enzymatic activity, 119, 119f
substrate-level phosphorylation, **122**, 124, 137t
  aerobic respiration and, 137t
  ATP yield, 132t
  in Krebs cycle, 128, 128f
subunit vaccines, **259**, **502–503**
subunits of ribosomes, 95, 95f, 101–102
succinic acid, 149f
  as fermentation end-product, 134f
succinyl CoA, 149f
suckers, oral and ventral (flukes), 356, 356f
sucking lice, 362t
sucrase, 116t
sucrose (table sugar), 39, 39f
Sudan dyes, 96
sudden oak death, caused by *Phytophthora ramorum*, 344
sugar (table), fermentation and, 137t
sugar-phosphate backbone of DNA, 211, 214f, 215t, 230, 250, 250f
sugars
  as carbohydrates, 39–40
  carbon dioxide used to synthesize, 140, 141f
  deoxyribose, 47, 48f
  in glycoproteins, 45
  milk (lactose), 39
  simple, 39
  table (sucrose), 39, 39f
sulfa drugs. *See* sulfonamides
sulfamethoxazole, 567, 568f
sulfanilamide, 120, 554, 556f
  as antimetabolite to PABA, 558
sulfate ion, 160
  anaerobic respiration and, 132
sulfate-reducing bacteria, *Desulovibrio*, 301t
sulfhydryl group, 37, 38t, 43
sulfide ion, as an anion, 30
sulfites, allergic reactions to, 526
Sulfolobales, important genera/special features, 302t
Sulfolobus genus/spp., 302t, **325**
  as extreme acidophiles, 325
  as hyperthermophiles, 302t
sulfonamides (sulfa drugs), 12, 553, 554, **567**, 568f
  bacterial resistance to, 239, 240f
  mode of action/spectrum of activity, 563t
  susceptibility of gram-negative *vs.* gram-positive bacteria to, 88t
sulfone drugs, to treat leprosy, 620
sulfur
  as energy source, 145
  atomic number/atomic weight, 27t
  chemoautotrophic bacteria and, 159
  electronic configuration, 29t
  extreme thermophiles and, 158
  green bacteria and, 144, 145t
  in cysteine (amino acid), 44t, 45
  in methionine (amino acid), 44t
  in organic compounds, 37
  requirements for microbial growth, 160
  sources of, 160
  Thiobacillus ferroxidans and, 37
sulfur cycle, **772–774**, 774f
  anaerobic respiration and, 132
  bacteria important to, 305, 774f
  deltaproteobacteria and, 312
sulfur dioxide
  as disinfectant, 199
  as food additive, 199
sulfur granules, **96**, 314t
sulfur-oxidizing bacteria, 300t
  Beggiatoa, 300t
sulfuric acid
  chemoautotrophic bacteria and, 159
  Thiobacillus ferroxidans and, 37
summer sausage, fermentation and, 137t
sunlight, antimicrobial effect of, 193
sunscreens, genetically engineered melanin in, 258
suntanning, excessive, skin cancers and, 231
superantigens, **436**, 438t, 492, 522, 589
  erythrogenic toxins as, 437
  mechanism of action, 436
supercoiled DNA strands, 212
  enzymes that unwind, 212–213, 215t
supercritical carbon dioxide, 202
superficial mycoses, **337**
superinfection, **555**
superinfections, tetracyclines use often leads to, 565
superoxide anions, **161–162**
superoxide dismutase (SOD), 161t, **162**
  genetically engineered, 260t
superoxide radicals, **161–162**, 459
suppressor T cells. *See* T regulatory cells
suramin, 627
surface disinfectants, phenolics, 195
surface-active agents (surfactants), as antimicrobial agents, **199**
Surfacine, 198–199
surfactants (surface-active agents), as antimicrobial agents, **199**
surgery, nosocomial infections and, 422b
surgical dressings, in nosocomial infections, 416
surgical gloves, latex allergy and, 530–531
surgical hand scrubs, 197, 204t
surgical infections
  Bacteroides and, 324
  phage typing to trace, 288, 290f
surgical instruments, 201
  endotoxins in ultrasonic baths to clean, 440b
  prion contamination and, 203
surgical wounds
  aseptic techniques and, 184
  body's normal defenses, sterilization and, 185
  infections
    at surgical site, 415t
    lactic acid bacteria to prevent, 402–403
    microbes causing, 414t
    phage typing to tract, 288, 290f
    Staphylococcus aureus and, 318, 403
  MRSA-infected patients post-surgery, 422b
sulfur, as an energy source, 141, 143f, 145t
susceptibility, **449**
susceptibility testing, antibiotics, 572–573, 751b
Svedberg units, 95
"swarm" movement, in bacterial motility, 82, 83f, 310, 311f
swarming bacteria, *Proteus*, 82, 310, 311f
sweat glands, 585, 585f
  dermicidin produced by, 470
sweat (perspiration), **453**, 585, 585f
  antimicrobial properties, 402t
  fever and, 463
  glands in skin, perspiration and, 453
swelling (edema), of inflammation, 460
"swim" movement, in bacterial motility, 82, 83f
swimmer's itch, **667**, **668**b
swimming pools
  conjunctivitis, 604
  liquid form of compressed chlorine gas used to disinfect, 197
  otitis externa and, **593**
  rashes and, 592–593
swine, as disease reservoirs, 410t
Swiss cheese, bacteria important in ripening/fermentation of, 137t, 320, 799
Sydenham's chorea (Saint Vitus' dance), **642**
symbionts
  of animals, unicellular algae as, 345
  of insects, *Wolbachia* as, 300t
symbiosis, 106, 107b, 267, 402f, **403**, **767**
  between normal microbiota and host, **401–**403, 402f
  mycorrhizae fungi and, **330**, **767**, 768f, 769f
  of *Azolla* -cyanobacteria, 772, 774f
  ruminants and, 767
  truffles and, 767, 769f
symbiotic bacteria
  Bradyrhizobium, 300t
  Carsonella ruddii, 326
  Frankia, 301t
  genetically modified *Rhizobium* and, 267
  nitrogen fixers, 300t
  Rhizobium, 267, 300t
  Wolbachia, 300t
symbiotic fungi (mycorrhizae), **330**
symptoms, *vs.* signs, **406**
Synagis (palivizumab), 692
syncytium, **442**, 442f, 692
syndrome, **406**
Synechococcus, 777
Synercid (dalfopristin), mode of action/spectrum of activity, 562t
Synercid, 566
synergism, **567**, **578**, 578f
  in combination antibiotics, **578**, 578f
    Synercid, 566
    TMP-SMZ, 567, 568f
  of antimicrobial peptides (AMPs), 471, **578**

synthesis reactions, **32**
as anabolic reactions, **32**
synthetic DNA, **255**, 256*f*
used to produce human insulin, 258
synthetic drugs, **12**
history of, 12–13
synthetic genes. *See* synthetic DNA
synthetic rubber, 258
syphilis, 323, 324*f*, **752**–755, 752*f*, 753*f*, 754*f*, **761** *b*. *See also Treponema pallidum*
as a once fatal epidemic disease, 21
as notifiable infectious disease, 421*t*
central nervous system affected by late-stage, 754
congenital, 421*t*, **755**
culturing media and, 167, 404
diagnosis of, 755
darkfield microscope, 59, 66*t*
immunofluorescence, 61*f*, 62, 66*t*, 755
gummas of, 754, 754*f*
incidence and distribution, 753, 753*f*
incubation period, 430*t*, 753
latent period, 754
portals of entry, 429, 430*t*, 753
pregnancy and, 760
progression of
primary stage, **753**, 753*f*
secondary stage, 753, 754*f*
rashes of, 753, 754*f*
tertiary (late-stage), 754, 754*f*
treatments for
benzathine penicillin, 559, 755
salvarsan first used to treat, 12
tetracyclines, 565, 755
systematics (phylogeny), **274**
systemic anaphylaxis (anaphylactic shock), **524**
systemic (generalized) infection, **407**
systemic inflammatory response syndrome (SIRS), 639
systemic lupus erythematosus, as immune complex disease, 467*b*
systemic mycoses, **336**, 338*t*

T antigen, **391**
T cells, **454**, **478**, **486**–489
as third line of defense, 450*f*
cellular immunity and, 477–**478**, 486–489
classes of, 487–489
cytotoxic, **487**, 488–489, 489*f*
CD8+ T cells and, **488**–489, 489*f*
dendritic cells importance to, 490
differentiation from stem cells in red bone marrow, 486, 486*f*
evolved to combat intracellular antigens, 486
function of, 478
HIV (*Lentivirus*) and destruction of, 443*t*. *See also* HIV
in compromised hosts, 415
in delayed hypersensitivity reactions, 529–530, 530*f*, 531*b*
in HIV/AIDS, 539–548, 540 *f* -**543** *f*
leukemia viruses and, 391
lymph node location of, 456, 638, 639*f*
memory cells, 486–487, 494
regulatory, **489**
superantigens stimulate proliferation of, 436
thymus as site for maturation, 457
T helper cells, **487**–488, 488*f*, 496*f*
activation of, 487–488, 488*f*, 496*f*
CD4 T cells and, 487–488, 488*f*
in antibody production, 482–483, 482*f*
T lymphocytes. *See* T cells
T regulatory cells, **489**
T suppressor cells. *See* T regulatory cells
T-cell receptors (TCRs), **478**
T-dependent antigen, **482**, 482*f*
T-DNA, 264, 265*f*
T-even bacteriophages, 374*f*
multiplication of in host *E. coli*, 379, 380, 381*f*
TEM image, 58*t*
T-independent antigens, **484**, 484*f*, 503
table sugar (sucrose), 39, 39*f*
tachycardia, as complication of fever, 463
tachyzoites, **350**
in toxoplasmosis, 661, 662*f*
Tacrolimus (FK506), 536
Taenia saginata (beef tapeworm), 357, 361*t*
Taenia solium (pork tapeworm), 357–358, 361*t*
reservoirs/transmission method, 410*t*
Taeniasis, **732**
tail, of T-even bacteriophage, 374*f*, 381*f*
tail fiber, of a T-even bacteriophage, 374*f*, 381*f*
Talaromyces, life cycle, 336*f*
Talaromyces life cycle, 336*f*
Tamiflu (oseltamivir), to treat influenza, 570
Tamm, Sid, 107*b*
tap water, *Acanthamoeba* grows in, 348
tapeworms (Cestodes), **356**–358, 358*f*, 361*t*, **732**–733, 735*b*
anatomy of adult, 356–357, 358*f*
beef (*Taenia saginata*), 357, 361*t*
foodborne transmission of, 412
life cycle of *Echinococcus granulosus*, 358, 359*f*
niclosamide to treat, 557*t*
pork (*Taenia solium*), 357–358, 361*t*, 410*t*
Taq polymerase enzyme, 767
TaqMan, 291
TASS (toxic anterior segment syndrome), 440*b*
Tatum, Edward L., 10*f*, 14*t*, 16
taxa/taxon, **274**
taxis, **82**
taxol
genetically engineered; used to treat ovarian cancer, 260*t*
produced by *Taxomyces* fungus, 339
Taxomyces fungus, produces anticancer drug taxol, 339
taxonomic hierarchy of organisms, 279, **280***f*
taxonomy, 273, **274**
advances in via DNA studies of fossilized materials, 262
as tool for natural classification system, 277
importance to understanding microbial susceptibilities, 275*f*
of microbes, 273–298. *See also* classification
of viruses, 373–374
TB. *See* tuberculosis
TCRs (T-cell receptors), **478**
TDP (thermal death point), **188**
TDT (thermal death time), **188**
tears
as protective mechanism, 451–452, 452*f*, 471*t*
IgA antibodies in, 480
lysozyme in, 88, 453
teeth, biofilm formation as plaque, 163
"teflon pathogen" *Treponema pallidum*, 754
teichoic acids, 85, 86*f*, 87
in gram-negative *vs.* gram-positive bacteria, 88*t*
telithromycin (Ketek), 566
mode of action/spectrum of activity, 562*t*
telomeres, as "junk DNA," 261
TEM (transmission electron microscope), **63**–64, 64*f*
advantages/disadvantages, 63–64
Paramecium image, 64*f*, 67*t*
specimen preparation and, 63–64
specimen size and, 58*f*
T-even bacteriophages image, 58*f*
Temin, Howard, 10*f*, 14*t*
temperate/lysogenic phages, 380
temperature
agar and, 165
as factor influencing enzymatic activity, 118–119, 119*f*
autoclaves and, 188–190, 189*f*, 189*t*, 439, 440*t*
disinfectants' effectiveness and, 186
enzymes increase chemical reaction rate without raising, 115
high, known record for bacterial growth, 158
maximum for microbial growth, 157, 157*f*
microbial growth requirements, 157–158, 157*f*, 158*f*, 159*f*
minimum for microbial growth, **157**, 157*f*
optimal for most disease-producing bacteria in human body, 119
optimum for microbial growth, 157, 157*f*
optimum for pathogenic bacteria, 158
pasteurization and, 190–191
steam heat microbial control and, 188–191, 189*t*
thermal death point and, 188
thermal death time and, 188
temperature abuse, **711**
temperature buffer, water as, 35
temperature (environmental)
extremes, archaea found in, 4, 275, 275*f*
water protects cells from fluctuations in, 35
template stand, DNA, 215*t*, 216, 218*f*
tenofovir, 571
teratogenic drugs, isotretinoin (Accutane), 594
terbinafine, mode of action/comments, 564*t*
terminator site on DNA strand, **216**, 218*f*
terminology
of microbial control, 185, 185*t*
scientific nomenclature, 2–3, 4*t*, 278–279
termites
as example of endosymbiosis, 107*b*
Paecilomyces fumosoroseus fungus as biocontrol, 339
spirochetes bacteria and, 322–323
Terramycin (oxytetracycline), 565
terrorism, biological weapons and, 21, 193, 262, **649***b*
tertiary sewage treatment, **788**–789
tertiary structure of proteins, 45, 46*f*
test tubes, culture media and, 165
testes, 745, 745*f*
tests for water purity, 779–782
tetanospasmin. *See* tetanus toxin
tetanus, 96, 161, **615**–616, 616*f*, **632***b*
as notifiable infectious disease, 421*t*
caused by *Clostridium tetani*, 316, 316*f*, 438*t*, 615, **632***b*
from bacteria in soil, 409
incubation period, 430*t*
neurotoxin (tetanospasmin) causing, 435, **437**, 438*t*, 439*t*
portals of entry, 429, 430*t*
symptoms, 438*t*
vaccine, 13, 502*t*, 504*t*
as a toxoid, 435, **502**, 502*t*
tetanus immune globulin (TIG), 616
tetanus toxin (tetanospasmin), 435, **437**, 438*t*, 440, 615
produced by *Clostridium tetani*, 239, 437, 438*t*, 615
vaccine made from purified, 13, 435, 502*t*, 504*t*
tetracyclines, **565**, 565*f*
bacterial resistance to, 239, 240*f*
inhibits protein synthesis, 95, 556–557, 556*f*, 558*f*
mode of action/spectrum of activity, 562*t*
produced by *Streptomyces aureofaciens*, 555*t*
selective toxicity of, 565
spectrum of activity, 557*t*
structure, 565*t*
susceptibility of gram-negative *vs.* gram-positive bacteria to, 88*t*
tetrads, **78**, 78*f*
Tetrahymena (protozoan), cilia of, 98, 100*f*
tetrapeptide side chain (side chain amino acid), in bacterial cell walls, 85, 86*f*
Tetraviridae, 394*t*
tetroses, 39
textiles, microbes and their enzymes in manufacture of, 246
thallus (body)
fleshy fungi or molds, **331**
of algae, 341–342, 341*f*
of lichen, 339, 340*f*
thawing, of freeze-thaw cycle, 192
therapeutic index, antibiotics and, 576
therapeutic substances, 2
thermal death point (TDP), **188**
thermal death time (TDT), **188**
thermalcycler, 251
Thermoactinomyces vulgaris, regerminated 7500–year-old endospores of, 96
thermoduric bacteria
acid-anionic sanitizers and, 199
pasteurization and, 190
thermophiles, extreme (hyperthermophiles), 4, **157**, 157*f*, 158, 275, 275*f*, **325**, 325*f*
thermophilic anaerobic spoilage, of canned foods, **795**, 796*t*
thermophilic archaea, optimal growth temperatures, 158, 325
thermophilic bacteria
food spoilage and, 185
Thermus aquaticus as, 251
Thermotoga
phylogenetic relationships, 281*f*
position in evolutionary tree, 275*f*
Thermotoga maritima, as one of earliest cells, 277
Thermus aquaticus, 767
used in thermalcycler to amplify DNA, 251
thiamine (vitamin B1), 117*t*, 160
thickening agents
algin (from brown algae), 342
from brown algae (algin), 342
from red algae (carrageenan), 342
Thiobacillus ferroxidans
pH ranges and, 37
used in copper ore recovery, 145, 806
Thiobacillus genus/spp., 300*t*, **305**, 772
as sulfur oxidizers, 300*t*, 772
sulfur granules of, 96
Thiobacillus thiooxidans, 145
Thiomargarita genus/spp., as giant bacteria, 300*t*, 317
Thiomargarita namibiensis, 13, 299*f*, 326, 326*f*
Thiotrichales, important genera of, 300*t*
30S ribosomes, 95, 95*f*
Thomas, E. Donnall, 15*t*
thoracic (left lymphatic) duct, 456*f*, 457
three-dimensional images
AFM microscope and, 65, 65*f*, 68*t*
confocal microscope and, 62, 62*f*
DIC microscope and, 60, 61*f*, 66*t*
SEM microscopes and, 64–65, 64*f*
three-domain system, 274–277, 275*f*, 276*t*
evolutionary relationship, 275–278, 275*f*, 276*t*
threonine (Thr)
in feedback inhibition and isoleucine synthesis, 121
structural formula/characteristic R group, 44*t*
throat, normal microbiota of, 403*t*
thrombocidin, produced by platelets, 470

thrombocytes. *See* blood platelets
thrombocytopenic purpura, **528**, 529*f*
thrush (oral candidiasis), 339, **601**, 601*f*, **759**
thylakoids (chromatophores) of bacteria, **91**, 91*f*, 140, 145*t*
thylakoids of eukaryotic cells, **105**, 106*f*
thymic aplasia (DiGeorge syndrome), 538, 539*t*
thymic selection, **486**, 532
thymine dimers, 193
  nucleotide excision repair and, **230**–231, 230*f*
  unrepaired, skin cancers and, 231
  UV light exposure and, 230–231, 230*f*
thymine nucleotide, 47, 48*f*
thymine (T), 47, 48*f*, 211
  exposure to UV light and, 230–231, 230*f*
  in DNA replication, 212–215, 214*f*, 215*f*, 216*f*
  in translation phase of protein synthesis, 216, 218*f*
thymus, 456*f*, 457, **477**
  T cells mature in, 457, 486
Ti plasmid, **264**, 265*f*
ticarcillin, 561
tickborne diseases, 323, 362, **362***t*
ticks
  as disease reservoirs, 651*b*
  as vectors, 362*t*
  cattle, 690
  Dermacentor species, 655, 656*f*
  ivermectin effective against, 572
  Ixodes species, 350, 353*f*, 362*f*, 362*t*, 413*t*, 653
  Lone Star, 654
  Ornithrodorus species, 291, 362*t*
TIG (tetanus immune globulin), 616
tigecycline (Tygacil), 565
time factors
  antimicrobial agents and, 186, 195
  ethylene oxide's antimicrobial action and, 201
  resistant microbes and, 186
tincture of iodine, **197**, 200*f*, 204*t*
tincture of Zephiran, 200*f*
tinctures, **197**
  vs. aqueous solutions, 198, 200*f*
tinea capitis (ringworm), **600**, 600*f*
  griseofulvin to treat, 569
tinea cruris (jock itch), 338*t*, **600**
tinea pedis (athlete's foot), 338*t*, 409, 410*t*, 568, **600**, 600*f*
tinea unguium (onychomycosis), **600**–**601**
tinidazole, 571
  mode of action/uses, 564*t*
tires, rubber, 145
tissue cells, relation to lymphatic capillaries, blood capillaries, 456*f*
tissue cysts, 661–663, 662*f*
tissue digester, to dispose of prion-infected animals, 631, 631*f*
tissue fluids, lysozyme in, 453
tissue plasminogen activator (Activase), genetically engineered; used to dissolve blood clots, 260*t*
tissue rejection
  histocompatibility antigens and, 482
  surgery-damaged cells and, 534
tissue repair, in inflammation, 461*f*, 462
tissue typing, 533–534, 533*f*
tissue-destroying diseases, 20, 287, 318
  actinomycosis, 321
  mycetoma, 321
  necrotizing fasciitis, 20, 287, 318
  necrotizing fasciitis ("flesh-eating" bacteria), **20**, 287, 318, 321, 422*b*, 591, 591*f*
TLRs. *See* Toll-like receptors
TMD (tobacco mosaic disease). *See* tobacco mosaic virus
TMV (tobacco mosaic virus), 16, 367, 368, 369*f*
Tn5 transposon, 241*f*
TNF. *See* tumor necrosis factor
tobacco mosaic virus (TMV), 16, 367, 368, 369*f*
Tobamovirus, 394*t*
tobramycin, to treat *Pseudomonas* infections of cystic fibrosis, 565
toenails, cutaneous mycoses and, 337, 539
Togaviridae, **387**
  as RNA virus, 385*t*
  biosynthesis of, 385*t*, 387
  characteristics/important genera/clinical features, 375*t*
Togavirus /EEE virus (eastern equine encephalitis), 375*t*, 625, **628***b*
Togavirus /WEE virus (western equine encephalitis), 375*t*, 625, **628***b*
tolerance to toxic metals, plasmids and, 95
Toll-like receptors (TLRs), **450**, 458–459, 458*f*
  antimicrobial proteins (AMPs) and, 470
  as early warning system in adaptive immunity, 579
  in inflammatory response, 460
tolnaftate, to treat athlete's foot, 564*t*, 569
toluene, bacteria that use as energy/carbon source, 239
tomato-associated diarrhea, DNA fingerprinting and, 290
tomatoes, salmonella outbreak, 715*b*
tomatoes (MacGregor variety), 267, 267*t*
Tonegawa, Susumu, 15*t*, 484
tongue, normal microbiota of, 1*f*, 18*f*
tonoplast, 104
tonsillectomies, bacterial endocarditis and, 641
tonsillitis, **676**
tonsils, 456*f*, 457
tooth decay (dental caries), **707**–**709**, 707*f*, 708*f*, **710***b*
  Streptococcus mutans and, 319, 431
tooth extractions, bacterial endocarditis and, 641
topoisomerase, 212, 215*t*
TORCH panel of tests, **760**
total magnification of specimen, **56**
toxemias, **407**, **434**, 588, 640
toxic oxygen products, 161–162, 459
toxic shock syndrome (TSS), **588**–**589**
  as notifiable infectious disease, 421*t*
  gram-positive sepsis and, 650
  mechanisms of exotoxins causing, 438*t*
  rash caused by, 592*b*
  Staphylococcus aureus causing, 318, 417*t*, 438*t*, 589
    toxic shock syndrome toxin 1 (TSST-1) strain, 588–589
  streptococcal, 588, **591**
  Streptococcus pyogenes causing, 5, 382, 417*t*, 588
  symptoms, 438*t*
toxic waste sites, bioremediation and, 17
toxic wastes, minimized by applying microbiology to commercial production methods, 3*b*
toxigenicity, **434**
toxins, **434**–439, 438*t*, 439*t*
  A-B toxins, **435**, 436*f*, 438*t*
  algal, *Gracilaria* and, 343
  amanitin, 443
  antitoxins and, **435**, 439*t*, 477
  as proteins, 42
  bacteriocins, 42, **239**
  cardiotoxins, 435
  cellular, R factor plasmids that confer resistance to, 239
  commercial sterilization and, **185**, 185*t*, 439, 440*b*, **794**–795, 794*f*, 795*f*
  cytotoxins, 435
  domoic acid produced by diatoms, 343
  early research with, 477
  enterotoxins, 435
    of *E. coli*, 310
  environmental, microbial ecology and, 17
  ergot, 443
  erythrogenic, 237, 677
  exfoliative, 239
  exotoxins, **434**–437, 435*f*
  fungal, 443
  gastric juice ineffective against, 453
  hepatoxins, 435
  IgG antibodies and, 481*t*
  intoxication *vs.* infections, 435
  leukocidin, 422*b*
  leukotoxins, 435
  lysogenic bacteriophages and, 382
  membrane-disrupting toxins, **436**, 438*t*
  neurotoxins, 239, 435, 437, 440, 443
  of *Clostridium botulinum*, 185, 185*t*, 382, 435, 437
  of *Corynebacterium diphtheriae*, 237, 382
  of streptococci, 382
  phalloidin, 443
  plasmids and, 95
  pore-forming, secreted by intracellular pathogens, 459
  potency expression, **430**–431
  produced by gram-negative *vs.* gram-positive bacteria, 88*t*
  prophage genes and, 380, 382, 382*f*
  saxitoxins, **344**
  Shiga, 210, 237, 382
  trichothecenes, 443
toxoids (inactivated toxins), **435**, 439*t*
  as vaccines, 435, **502**, 616
Toxoplasma gondii (protozoa), 350, 354*t*
  causing encephalitis in AIDS patients, 544*t*
  interleukin-12 and, 493*b*
  reservoirs/transmission method, 410*t*
  typical U.S. prevalence of antibodies against, 657*f*, 662
Toxoplasma (protozoa), pathogenic mechanisms of, 443
toxoplasmosis, 410*t*, **650***b*, 661–663, 662*f*
  California sea otter deaths and, 283*b*, 662
  cats infected with, 661, 662*f*
  of brain, in AIDS patients, 542, 544*t*, 662–663
  pregnancy and, 760
  Toxoplasma gondii causing, 354*t*, 661, 662*f*
  typical U.S. prevalence of antibodies against, 657*f*, 662
TPM (two-photon microscopy), **62**, 62*f*, 67*t*
  advantages, 62, 62*f*
  Paramecium image, 62*f*, 67*t*
trace elements
  activating enzymes and, 117
  microbial requirements, 160
Tracheophyla, position in taxonomic hierarchy, 280*f*
Tracheophyta, position in taxonomic hierarchy, 280*f*
trachoma, 322, 429, 459, **604**–**605**, **604***b*, 605*f*
  blindness, 322
trans fatty acid, 40
transacetylase, 224, 224*f*
transamination, **147**, 148*f*
transcription, 213*f*, **216**–217, 218*f*, 221
  DNA viruses and, 385, 385*t*
  enzymes required, 216–217, 218*f*
  in eukaryotic cells, 220, 222*f*
  RNA viruses and, 387–389, 388*f*
  simultaneous with translation in bacteria, 220, 222*f*
transduction (bacterial), **237**, 239*f*
  generalized, **237**, 239*f*, 382
  specialized, **237**, 239*f*, 382, 383*f*
transfer RNA (tRNA), **47**, 211, **219**
  in translation, 219, 220–221*f*, 221
transfer vesicles, **104**, 104*f*
transferase enzymes, type of chemical reaction catalyzed/examples, 116*t*
transferrin, **470**
transferrin, 434
transferrins, high body temperature increases production of, 463
transformation, **234**–236, 235*f*, 236*f*, **253**
  as genetic engineering technique, 253
  in continuous cell lines, 378
  in tumor cells, 391, 537–538, 537*f*
  naturally occurring, 236, 253
  of polyomavirus, 443*t*
  viruses causing, 442
transfusion reactions, 523*t*, 526–528
translation, 213*f*, 217, **219**–221, 219*f*, 220–221*f*
  DNA viruses and, 385, 385*t*
  in eukaryotic cells, 220, 222*f*
  simultaneous with transcription in bacteria, 220, 222*f*
  site of, 219
translational reading frame, frameshift mutations and, 228
transmembrane proteins, of prokaryotic plasma membrane, 89–90, 90*f*
transmissible spongiform encephalopathies (TSE), **630**–631, 631*f*
  kuru, 631
transmission electron micrograph, 63, 64*f*
transmission electron microscope (TEM), **63**–64, 64*f*
  advantages/disadvantages, 63–64
  Paramecium image, 64*f*, 67*t*
  specimen preparation and, 63–64
  specimen size and, 58*f*
  T-even bacteriophages image, 58*f*
transmission of disease
  by direct contact, **410**, 410*t*, 411*f*
  by droplet transmission, **411**, 411*f*
  by flea bites, 410*t*
  by indirect contact, **410**–411, 410*t*, 411*f*
  by ingestion, 410*t*
  by mosquito bites, 410*t*
  by nonliving fomites, 410*t*
  by tick bites, 410*t*
  by vectors, **412**–413, 413*t*
  by vehicle transmission, **411**–412, 412*f*
  in nosocomial infections, 414*f*, 415–416
transplacental transfer of immunoglobulins, 481*t*, 494
transplant rejection
  cytotoxic T lymphocytes (CTLs) and, 487, 529
  delayed hypersensitivity in, 529–530, 531*b*
  drugs that prevent, and resulting impaired innate defenses, 462
  genetically engineered products to minimize, 260*t*
  immunologically privileged sites/privileged tissue, 534–535
  mechanisms of, 534–535
  monoclonal antibodies to minimize, 508
transplant surgery
  HLA tissue typing, 533–534, 533*f*
  immunosuppression and, 536–537
  using PCR in matching donors, 534
transplantation
  reactions to, 534–535
  stem cells and, 535, 535*f*
transplants
  bone marrow, **536**
  liver, 536
transport media, **284**
transport vesicle, **104**, 104*f*
transporters (integral proteins), 42
  in active transport processes, 92*f*, 94

Index

in facilitated diffusion, 91–92, 92*f*
nonspecific, 91*f*, 92*f*
specific, 91, 92*f*
transposase, 215*t*, 240, 241*f*
transposition, 240–241, 241*f*
frequency of, 240
transposons, 237–238, **240**–241, 241*f*
antibiotic resistance and, 240–241, 241*f*, 574
as "junk DNA," 261
as mediator of evolution in organisms, 241
complex, 240, 241*f*
gene silencing as defense against, 259
trastuzumab (Herceptin), 509, 538
traveler's diarrhea, **717**, 723*b*
exotoxins causing, 239, 310, 438*t*
pathogenic *E. coli* and, 239, 310, **717**, 723*b*
trees
ascomycete *Cryphonectria parasitica* and chestnut trees, 339
Ceratocystis ulmi causing Dutch elm disease, 339
Trematodes (flukes), **356**, 356*f*, **361***t*
trench mouth (acute necrotizing ulcerative gingivitis), **709**, 710*f*
Treponema genus/spp., 302*t*, **323**, 324*f*
as human pathogens, 302*t*
as normal microbiota of mouth, 403*t*
portals of entry for, 429, 430*t*
Treponema palladium pertenue, yaws caused by, 753
Treponema pallidum, **323**, 324*f*, 752–754, 752*f*. *See also* syphilis
adherence method of, 431, 752
as "teflon pathogen," 754
axial filaments of, 82–83, 324*f*
blood banks screen for, 727*b*
cultivation of virulent strains and, 404
darkfield microscopy to detect, 59, 66*t*
FTA-ABS test, 61*f*, 62
LM image, 66*t*
mucous membrane penetrated by, 451
portals of entry, 429, 430*t*
syphilis caused by, 323, 324*f*. *See also* syphilis
yaws caused by subspecies strains, **753**
triangular-shaped bacteria, 79
Triatoma (kissing bug), as vector for Chagas' disease, 351, 354*t*, 362*t*, 363*f*, 413*t*
triazoles, **568**–569, 661
Treponema vulgare (algal cell), 99*f*
tricarboxylic acid (TCA) cycle. *See* Krebs cycle
trichiasis, 605, 605*f*
Trichinella spiralis (helminth), 361*t*, 705*f*
life cycle of, 736–737, 737*f*
portals of entry, 430*t*
reservoirs/transmission method, 410*t*
trichinellosis, 360, 361*t*, 410*t*, 705*f*, **735***b*, **736**–737, 737*f*
as notifiable infectious disease, 421*t*
freezing temperatures and, 192
incubation period, 430*t*
pork cooked in microwave ovens and, 193
Trichoderma (fungus)
cellulases produced by used for industrial purposes, 40
in "stone-washed" denim jean production, 3*b*
used commercially to produce enzyme cellulase for clear fruit juice, 339
Trichodesmium, 777–778
Trichomonas vaginalis (protozoan), 347, 347*f*, 354*t*
as normal microbiota of vagina, 403*t*, 760
metronidazole effective against, 571
vaginitis caused by, 347, 756, 759*b*
trichomoniasis, **759***b*, **760**, 760*f*
Trichonympha sphaerica, 107*b*
Trichophyton (Arthroderma), 338*t*
Trichophyton (fungus)
as pathogenic fungi, 338*t*, 443
cutaneous mycosis and, 600
reservoirs/transmission method, 410*t*
trichothecenes, 443
trickling filters, in sewage treatment, **784**, 785*f*, 786*f*
triclosan, 196, 196*f*, 566
Tridacna (giant clam), symbiotic host to dinoflagellate algae, 345
trigeminal nerve ganglia, herpes simplex virus and, 597–598, 598*f*
triglycerides (fats), 40, 41*f*
trimethoprim, 567, 568*f*
trimethoprim-sulfamethoxazole (TMP-SMZ), 567, 568*f*
mode of action/spectrum of activity, 556*f*, 563*t*
trioses, 39
tripeptide, 43
triple covalent bond, 31
tRNA (transfer RNA), **47**, 211, **219**
in translation, 219, 220–221*f*, 221
Tropheryma whipplei, Whipple's disease caused by, 290
trophophase, **803**, 804*f*
trophozoites, **346**
of *Giardia*, 347*f*
of *Toxoplasma gondii* and, 350
vegetative form of protozoa, 346
true flies, as vectors of human diseases, 362*t*
truffles, 767, 769*f*
Trypanosoma brucei gambiense (protozoa), 351, 354*t*, 413*t*
African trypanosomiasis caused by, **627**–**629**
antigenic variation in, 433, 444, 629, 629*f*
tsetse fly as vector, 351, 354*t*, 362*t*, 413*t*, 627–628
Trypanosoma brucei rhodesiense (protozoa), 354*t*, 413*t*, 627
Trypanosoma cruzi, blood banks screen for, 727*b*
Trypanosoma cruzi (protozoa), 4*t*, 351, 354*t*, 413*t*, 417*t*, 459, 650*b*, 661, 661*f*
Trypanosoma (protozoa)
antigenic variation in, 433, 444, 629, 629*f*
surface glycoproteins and gene expression, 222
trypanosomes
antigenic variation in, 433, 444, 629, 629*f*
in Chagas' disease, 661, 661*f*
schizogony and, 351
trypanosomiasis
African, 329, 351, 354*t*, 362*t*, 363, 413*t*
American. *See* Chagas' disease
tryptophan synthesis, repression of, 225, 225*f*
tryptophan (trp), structural formula/characteristic R group, 44*t*
trytophan, in indigo production, 3*b*
TSE (transmissible spongiform encephalopathies), **630**
tsetse fly, as vector for African trypanosomiasis, 351, 354*t*, 362*t*, 413*t*, 627–629
TSS. *See* toxic shock syndrome
TSTA (tumor-specific transplantation antigen, **391**
tuberculin skin test, 507, 530, **684**, 684*f*
tuberculocidal agents
glutaraldehyde, 200
instruction labels and, 202
tests for effectiveness, 203
tuberculoid (neural) form of leprosy, 619, 620*f*
tuberculosis (TB), **682**–**685**, 686*f*, **699***b*
acid-fast stain to identify, 70
airborne transmission and, 412
antibiotics to treat, 562*t*, 563, 567
as chronic disease, 407
as notifiable infectious disease, 421*t*
biochemical tests to detect, 144*b*
bovine (*Mycobacterium bovis*), **685**
causative agent of, 54*f*, 144*b*, 682*f*. *See also* *Mycobacterium tuberculosis*
chronic inflammation of, 460
diagnosis, 684–685
fluorescent dyes to identify, 61
in AIDS patients, 542, 544*f*
incubation period, 430*t*
multidrug resistant (MDR) strains, 684
pathogenesis, 682–684, 683*f*
peritoneal, 144*b*
portals of entry for, 429, 430*t*
pulmonary, 144*b*
reported cases, 1948–2007, 419*t*
skin test, 507, 530, **684**, 684*f*
treatments for, 684
vaccine, 13, 502, **685**
worldwide incidence, 685, 686*f*
tubulin, 98
tularemia, **642**–643, 642*f*, **651***b*
as notifiable infectious disease, 421*t*
as potential biological weapon, 642, 649*b*
as zoonotic disease, 642
caused by *Francisella tularensis*, 308, 459, 644*b*
Chrysops (deer fly) as vector transmitting, 362*t*
number of cases in U.S. (1990–2000), 642, 642*f*
pet rodents case study, **644***b*
"tumbles" (bacterial motility), 82, 83*f*
tumor cells
natural killer (NK) cells can destroy, 491
transformation and, **391**, 537–538, 537*f*
tumor necrosis factor alpha (TNF-α), **437**, 438*f*
disorders leading from excessive production of, 460
in fever, 463
in inflammatory response, 460
psoriasis and, 533
tumor necrosis factor (TNF), **492**
as cytokines, 492
endotoxic shock and, 437, 439
genetically engineered, 260*t*
tumor-destroying viruses, 369
tumor-specific transplantation antigen (TSTA), **391**
tumors
interleukin-12 (IL-12) and, 493*b*
Mastadenovirus and, 375*t*
Papillomavirus and, 375*t*
tungsten
used in staining of specimens, 63
used with gene guns, 253, 254*f*
turbidity, measuring, to estimate bacterial growth, **178**, 179*f*
turkey farming, animal feed antibiotics and, 577*b*
turnover number of enzymes, **116**
12D treatment (*botulinal cook*), in commercial sterilization, **795**
twitching motility, **83**
two-kingdom classification system, 274
two-photon microscopy (TPM), **62**, 62*f*, 67*t*
advantages, 62, 62*f*
Paramecium image, 62*f*, 67*t*
2–aminopurine, 229–230, 229*f*
Tygacil (tigecycline), 565
Type I hypersensitivity, **523**–**526**, 523*t*
Type II hypersensitivity, 523*t*, **526**–528
Type III hypersensitivity, 523*t*, **528**–529
Type IV hypersensitivity, 523*t*, **529**–531
typhoid fever, 429, **714**–**716**, 714*f*, **722***b*
as notifiable infectious disease, 421*t*
culture medium and, 168
endotoxin causing, 439*t*
incidence of, 714, 714*f*
incubation period, 430*t*
infection still spread by convalescing person, 409
portals of entry, 429, 430*t*
preferred portals and, 429
Salmonella typhi as cause of, 310, 430*t*
vaccine, 502, 716
Typhoid Mary, 409
typhoidal salmonellae, 712–713
typhus, 459, **651***b*, **654**–655
causative agent/arthropod vector, 413*t*
endemic murine, 310, 410*t*, 413*t*, **651***b*, **655**
caused by *R. typhi*, 303, 413*t*
Xenopsylla (rat flea) as vector, 362*t*, 413*t*
epidemic, **651***b*, **654**–655
caused by *Rickettsia prowazekii*, 303, 413*t*, 654, 655
Nightingale's epidemiologic analysis of, 418
Pediculus humanus corporis (body louse) as vector, 362*t*, 363*f*, 413*t*, 654
tickborne, **655**. *See also* Rocky Mountain spotted fever
vaccine, 655
tyrosine (tyr), 43, 43*f*
structural formula/characteristic R group, 44*t*

ubiquinones (coenzyme Q), **129**, 129*f*
UDP-N-acetylglucosamine (UDPNAc), 146, 146*f*
UDPG (uridine diphosphoglucose), 146, 146*f*
UDPNAc (UDP-N-acetylglucosamine), 146, 146*f*
UHT (ultra-high-temperature) treatments, **190**–191
ulcers
genetically engineered epidermal growth factor to heal, 260*t*
Helicobacter pylori and, 453
ultra-high-temperature (UHT) treatments, **190**–191
ultrasonic baths, endotoxins and, 440*b*
ultraviolet light, as mutagen, 228
ultraviolet (UV) light, 61, 61*f*, 66*t*
mutagenic, 230, 230*f*
to control microbes, 193
Ulva (multicellular green alga), 342*f*
umbilical cords, stem cells harvested from, 535, 536
uncoating
in animal virus multiplication cycle, **384**–385, 384*t*, 385*t*, 386*f*
in bacteriophage multiplication cycle, **384**, 384*t*
undecylenic acid, antifungal activity of, 569
undulant fever. *See* brucellosis
undulating membrane, of *Trichomonas vaginalis*, **347**, 347*f*
unicellular algae, 341–342, 341*f*, 342*f*
universal ancestor split into three lineages, 275, 275*f*
Universal Precautions for Health Care Personnel (CDC), 546*t*
unsaturated fatty acids, 40, 41*f*, 42*f*
uracil (U), 47, 49*f*
in translation phase of protein synthesis, 216, 218*f*
uranium, 37
used in staining of specimens, 63
Ureaplasma genus/spp., 301*t*, **320**
no cell wall, 301*t*
urinary tract infections and, 320
urease test, 144*b*, 144*f*
ureteritis, 746
ureters, 744, 744*f*
urethra, 744, 744*f*

urethritis, 746
  Chlamydia trachomatis causing, 430*t*, **750**–751, **761***b*
  nongonococcal/nonspecific, 322, 430*t*, 459, **750**–751, **761***b*
  Trichomonas vaginalis causing, 354*t*
uridine diphosphoglucose (UDPG), 146, 146*f*
uridine triphosphate (UTP), 146, 146*f*
urinary bladder, 744, 744*f*
urinary catheters, number of MRSA-infected patients related to, 422*b*
urinary system, **743**–745, 744*f*
  normal microbiota of, 403*t*, 745
  structure/function, 744, 744*f*
urinary tract infections
  as emerging infectious disease, 417*t*
  caused by *E. coli*, 310
  caused by *Proteus* bacteria, 310
  endotoxin causing, 439*t*
  fluoroquinolones to treat, 567
  nosocomial, 414*t*, 415*t*, 422*b*
  sulfa drugs to treat, 567
  tetracyclines to treat, 565
  Trichomonas vaginalis causing, 347, 347*f*
  vancomycin-resistant enterococci and, 417*t*
urine, **453**
  lysozyme in, antimicrobial activity and, 453, 471*t*
  normal microbiota of urinary tract and, 745
  pH of, 453
  urinary catheters altering flow, infections and, 452
  washes microbes from urethra, **452**, 471*t*
U.S. Geological Survey research, using nanotechnology to reduce toxic selenium, 264
U.S. Postal Service, anthrax bioterrorism and, 646, 649*b*
U.S. Public Health Service, 420
USA100 MRSA strain, 422*b*
USA300 MRSA strain, 422*b*
use-dilution tests, **195**
usnic acid, from *Usnea* lichen, 340
uterine (fallopian) tubes, 477*f*, 744
  infection (salpingitis), **752**
uterus, 744, 744*f*
UTP (uridine triphosphate), 146, 146*f*
UV (ultraviolet) light, to control microbes, 193
UV wavelengths, those most effective for killing microbes, 193

V factor, *Haemophilus* bacteria and, **311**
V-P (Voges-Proskauer) test, 286*f*, 287
vaccination, 477, **495**
  adult booster, 418, 501, 502*t*, 503*t*, 616
  and disease carriers, 407
  as artificially acquired active immunity, **495**
  development of new vaccines, 504, 506
  discovery of why it works, 11, 477
  emerging infectious diseases and, 418
  herd immunity and, **407**, 501, 598, 612
  most effective method of disease control, 501
  principles and effects of, 501
  rates of, 407
  term coined from Jenner's research, 501
vaccines, **495**, **501**–**506**
  acellular, **503**
  adjuvants and, 506
  against bacterial diseases, **502***t*
  against viral diseases, **503***t*
  as nasal sprays, 503*t*, 506
  attenuated, **501**–502
  available (selected list), 13
  Bacillus Calmette-Guérin (BCG), 620
  BCG, 685
  boosters, 418, 501, 502, 502*t*, 503*t*, 616
  cervical cancer, 260*t*, 391
  chickenpox, 13, 503*t*
  childhood, 502, **504***t*, 616
  conjugated, **503**
  cowpox, 501
  development of new, 504, 506
  diphtheria, 435, 502, 502*t*
  DNA, **259**, **503**–504
  DTaP, 616, **678**
  filtration used to sterilize, 191
  first, 11
  flu, 13, 502, 502*f*, 503*t*, 504*t*, 694
  for travelers, 501
  Gardasil (HPV vaccine), 260*t*, 391, 503*t*, 538, 758
  gene guns to inject, 503
  genetically engineered yeast and hepatitis virus, 247
  German measles (rubella), 13, 502, 503*t*, 504*t*
  hepatitis A, 503*t*, 504*t*
  hepatitis B, 260*t*, 503*t*, 504*t*
  herpes zoster, 503*t*
  HIV, 547–548
  HPV, 260*t*, 391, 503*t*
  HPV (Gardasil), 260*t*, 391, 503*t*, 538, 758
  human papillomavirus (HPV), 260*t*, 391, 503*t*
  inactivated whole-agent, **502**
  influenza, 13, 260*t*, 502, 502*f*, 503*t*, 504*t*, 506, 694
  injection sites, dendritic cells and, 490
  liver cancer, 538
  malaria, difficulties with, 348
  measles, 13, 501, 503*t*, 504*t*
  meningitis, 502*t*, 503, 504*t*, 612, 614, 617*b*
  microbes used in commercial production of, 247
  MMR vaccine, 502, 504*t*, 598
    autism possible connection and, 506
  mumps virus, 13, 502, 503*t*, 504*t*
  nucleic acid (DNA vaccines), **503**–504
  oral, 506
  pertussis, 13, 502, 503, 503*t*, 504*t*, 682
  pneumonia, 13, 502*t*, 504*t*
  pneumonococcal meningitis, 502*t*, 614
  polio, 13, 419, 502, 503*t*, 504*t*, 621, 632*b*
  primary immune response provoked by, 501
  principles of vaccination, 501
  rabies, 502, 503*t*, 623
  recombinant, **502**–503
  recommendations, 502*t*, 503*t*
  rotavirus, 504*t*, 506
  rubella, 13, 502, 503*t*, 504*t*
  safety of, 506
  secondary immune response and, 501
  shingles, 503*t*
  smallpox, 11, 406, 477, 501, 503*t*
  sources for recommended immunizations, 501
  subunit, **259**, **502**–503
  tetanus, 13, 502, 502*t*, 504*t*, 616
  toxoids (inactivated toxins), 435, **502**, 616
  tuberculosis, 13
  types of, 11, 501–504
  UV light to disinfect, 193
  vaccinia virus, 501
  varicella, 504*t*, 596–597
  viral, animal cells used to produce, 247
  yellow fever, 659
  zoster, 597
vaccinia virus, 375*t*
  confers immunity to smallpox, 501
  genetically modified, 259
  size of, 369, 369*f*
vaccinia virus vaccine, 501
vacuoles, 99*f*, **104**
  contractile, 350*f*
  food, 346, 350*f*
  gas, **96**
  of protozoa, **346**
vagina, 744, 744*f*
  normal microbiota of, 403*t*, 745
  pH of, 745
vaginal infections, caused by *E. coli*, lactic acid bacteria probiotics to prevent, 403
vaginal secretions
  as defense against pathogens, **452**, **453**, 471*t*
  pH of, 453
vaginal yeast infections, miconazole to treat, 568, 569*f*
vaginitis, 320, 401, **756**, 756*f*, **759***b*
  Candida albicans causing, 756, 759, 759*b*
  Gardnerella vaginalis causing, 756, 759*b*
  Trichomonas vaginalis causing, 347, 354*t*, 756, 759*b*
vaginosis, bacterial, **756**, 756*f*, **759***b*
valence, **28**, 29*t*
  of antibodies, **479**
valine (Val), structural formula/characteristic R group, **44***t*
Valley fever. *See* coccidioidomycosis
van Leeuwenhoek, Anton, 7, 7*f*, 10*f*, 13, 54, 55, 322
vancomycin, 422*b*, **563**
  inhibits cell wall synthesis, 556, 556*f*, 557*f*
  mode of action/spectrum of activity, 562*t*
  MRSA problem and its importance to, 563
  resistance
    antibiotics developed in response to, 566
    by *S. aureus* (VISA), **20**, 417*t*, 421*t*, 422*b*
    by *S. aureus* (VRSA), 12–13, 19–**20**, 210, 241, 417*t*, 421*t*, 422*b*, 563
    transposons and, 241
vancomycin-intermediate resistant *Staphylococcus aureus* (VISA), 421*t*, 422*b*
vancomycin-intermediate *Staphylococcus aureus* (VISA), **20**, 421*t*, 422*b*
vancomycin-resistant enterococci (VRE), 417*t*, **563**, 577*b*, 640
vancomycin-resistant *Staphylococcus aureus* (VRSA), 12–13, 19–**20**, 210, 241, 417*t*, 421*t*, 422*b*, 563
variable (V) regions, of antibodies, 479, 480*f*
varicella (chickenpox), 375*t*, 385, **596**–597, 597*f*
  as notifiable infectious disease, 421*t*
  breakthrough varicella, **597**
  incubation period, 430*t*, 596
  portals of entry, 430*t*, 596
  rash caused by, 590*b*
  Reye syndrome complication of, **596**
  vaccine, 13, 503*t*, 504*t*, 596–597
varicella-zoster virus (*Varicellovirus* /HHV-3), 375*t*
  as AIDS-associated, 544*t*
  causing chickenpox, 375*t*, **596**–597. *See also* chickenpox
  causing shingles, 375*t*, **596**–597. *See also* shingles
  incubation period, 430*t*
  portals of entry, 430*t*
  pregnancy and, 760
  vaccine, 13, 503*t*, 504*t*, 596–597
Varicellovirus /HHV-3. *See* varicella-zoster virus
variola major, **595**
variola minor, **595**
variola virus, 374*f*
variolation, **501**
Varmus, Harold E., 15*t*, 391
vasodilation, **460**
  in inflammatory response, **460**, 461*f*, 462
  kinins and, 460
VDRL test, for syphilis, **755**
vectorborne diseases
  by arthropod vector/disease, 362*t*
  elimination methods, 363
vectors, **361**
  arthropods as, 361–363, 362*f*, 362*t*, 363*f*, **412**–413, 413*t*
  DNA molecules as, 247, 248*f*, 250–251, 250*f*, 251*f*
  shuttle, **251**
  viral DNA as, 251
vegetables and fruits, PAA for washing/disinfecting, 202
vegetative bacteria
  desiccation and, 192
  freezing temperatures and, 192
  high pressure to control, 192
vegetative cells, temperatures that kill, 98
vegetative cells of endospore-forming bacteria, 96–97, 97*f*, 332
vegetative hyphae, 331, **331***f*
vegetative (non-endospore-forming) pathogens, disinfection to control, 185, 185*t*
vegetative pathogens
  disinfection to control, 185, 185*t*
  microwave ovens and, 193
  moist heat sterilization to kill, 188–190, 189*f*, 193
vegetative structures
  of algae, 341–342
  of fungi, 331–332, 331*f*
  of protozoa, 346
vehicle transmission of disease agents, 411, **411**–412, 412*f*
Veillonella genus/spp., as normal microbiota of mouth, 403*t*
veins, parasitic helminths and, 361*t*
Venezuelan hemorrhagic fever, 376*t*, 416
  as emerging infectious disease, 417*t*
Venezuelan hemorrhagic virus, 19, 417*t*
ventilator-related procedures, MRSA-infected patients and, 422*b*
vertebrates, as eukarya, 6
vertical gene transfers, 213*f*, **234**
vesicles, **767**, 768*f*
vesicles (lesions), **586**, 587*f*
vesicular stomatitis virus (VSV), 376*t*, 378*f*, 389*f*
vesicular-arbuscular mycorrhizae (endomycorrhizae), 767, 768*f*, 769*f*
Vesiculovirus, 376*t*
veterinary microbiology
  marine mammal deaths, 283*b*
  West Nile virus, 223*b*, 223*f*
Vibrio cholerae, 77, **309**
  A-B enterotoxin (cholera toxin) produced by, 437
  as potential biological weapon, 649*b*
  coevolution and, 428
  incubation period, 430*t*
  lysogenic phages and, 441
  noncholera vibrios, **717**, 722*b*
  portals of entry, 429, 430*t*
  strain 0:139, 716
  strain 0:139
    emerging infectious diseases and, 417*t*
    new serovar and evolutionary changes, 19, 416, 417*t*
    terminology used in naming, 310 *footnote*
  virulence and, 430
Vibrio enterotoxin (cholera toxin), **437**
  produced by *Vibrio cholerae*, 435, 437
Vibrio fischeri, producing enzyme luciferase, 57*b*

Vibrio genus/spp., **79**, 79*f*, 300*t*, **309**, 309*f*
as human pathogens, 300*t*
found in dolphins, 283*b*
Vibrio parahaemolyticus, **309**
Vibrio vulnificus, 717, 723*b*
Vibrionales, **309**
important genera/special features, 300*t*
vibriosis, as notifiable infectious disease, 421*t*
Vincent's disease (trench mouth), **709**, 710*f*
vinegar
fermentation and, 137*t*
microbes used in production of, 800
viral agents, first used to produce immunity, 11
viral coat protein, 247
viral diseases
development of drugs to treat and, 12
interferons' discovery and, 16
of cardiovascular system, **643***b*, 655–660
of eyes, **604***b*, 605
of lymphatic system, **643***b*
of nervous system, 620–626, **632***b*
of reproductive system, 757–758, 761*b*
of respiratory system
lower, 692–695, **699***b*
upper, 679–680, **681***b*
of skin, 595–600
rashes caused by, 589*b*, 590*b*, 592*b*
suspected to be transmitted by bats, 624 *footnote*
viral DNA, as a vector, 251
viral genomes, 261
directing biosynthesis inside host cell and, 282
viral hemagglutination inhibition test, **512**, 513*f*
viral hemorrhagic fevers, 658–659, **660***b*
emerging, 659–660, **660***b*
viral infections
attachment sites and drug development, 383
engineered into crop plants, 265
gene silencing as promising therapy for, 259
latent, **392**, 392*f*, 394*t*
persistent, 392, 392*f*, 394*t*
persistent (chronic), **392**, 392*f*, 394*t*
viral meningitis, 612
viral multiplication, 379–389
animal viruses, 382–389
compared to bacteriophages, 374*t*
fusion, 383, 384*f*
pinocytosis, 383, 384*f*
uncoating, 384–385
bacteriophages, 379–382
compared to animal viruses, 384*t*
lysogenic cycle, **379**, 380, 382, 382*f*
lytic cycle, **379**–380, 381*f*
drugs that interfere with, 368
one-step growth curve, **379**, 379*f*
requirements for, host range and, 368
stages of, **380**, 381*f*, 382–385, 384*t*
differences in, 384*t*
viral pneumonia, **692**
viral protein, DNA vaccines and, 259
viral proteins, produced by *S. cerevisiae* and cervical cancer vaccine, 260*t*
viral RNA, reverse-transcription PCR and, 251
viral RNA testing, 545
viral species, **282**, **374**
three-domain system and, 282
viral therapy, safety of, 369
viral vaccine preparations, egg proteins and allergies to, 378
viral zoonoses, 410*t*
Virchow, Rudolf, 8
viremia, poliovirus causing, 621
viridans streptococci, 319
virions, **370**. *See also* viruses
latent, 541, 541*f*
one-step growth curve and, 379, 379*f*
viroids, **394**–395, 395*f*
causing plant diseases, 394, 395*f*
introns and, 394–395
size of, 369*f*
virology, **16**
virosomes, 506
virstatin, 579
virucides, 186
virulence, **71**, **428**, 429–431
antigenic variation and, 433
bacterial capsules' role in, 80, 432
cell wall components and, 432
early experiments in, 11
extracellular enzyme production and, 432–433
fungal, 443
genetic transformation and, 234–236, 235*f*
host cell cytoskeleton and, 433, 433*f*
ID50 and, **429**–430
LD50 and, **430**–431
lysogeny and, 441
of algae, 444
of helminths, 444
of protozoa, 443–444
of viruses, 441–442, 442*f*, 443*t*
plasmid genes encoding for, 440–441
virus-host interactions, phage therapy research and, 369, **579**
viruses, 5*f*, **6**, 76, 367–398, **368**. *See also* antiviral drugs
advantages of electron microscopes to view, 64–65, 64*f*
animal. *See* animal viruses
antigenic changes induced by, 442
antiviral drugs to combat, 569–571. *See also* antiviral drugs
as acellular microbes, 6
as emerging infectious diseases (EIDs), 19–21, 417*t*
as obligatory intracellular parasites, 368
attachment (adsorption) in, **380**, 381*f*, 383, 384*t*, 386*f*
adherence in *S. aureus* adherence similar to, 432
bacteria compared to, 368, 368*t*
bacterial. *See* bacteriophages
biosynthesis stage in multiplication cycle, **380**, 381*f*
cancer and, 389–392
as unrecognized etiology, 390
DNA oncogenic viruses, 391
first demonstrated, 389
RNA oncogenic viruses, 391–392
transformation of normal cells into tumor cells, 390–391
tumor cells naturally infected by, 369
capsid of, **372**–373, 372*f*, 373*f*
capsomeres of, **372**, 372*f*, 373*f*
cell cultures, 378–379, 378*f*
characteristics of, 368–369, 368*t*, 369*f*
chromosomal changes induced by, 442
classification of, 282, 370
complex viruses, **373**, 374*f*
cultivation of, 374, 377–379
animal viruses
in cell cultures, **378**–379, 378*f*
in embryonated eggs, 377–378, 377*f*
in living animals, 377
bacteriophages, 374, 377, 377*f*
cytopathic effects of, **441**–442, 442*f*, 443*t*, 445*f*
distinguishing features, 368, 368*t*
early descriptions of, 367, 368
eclipse period in viral multiplication cycle, **380**
enveloped, **372**, 372*f*, **373**, 373*f*
alcohols and, 197, 204*t*
biguanide disinfectants and, 197
Influenzavirus as example, 372*f*, 373
quats active against, 199, 204*t*
evolution of, 282
filterable, 191, 367, 368
gene silencing as defense against, 259
genetically modified to infect tumor cells, 369
helical, **373**, 373*f*
host range of, **368**–369
identification of, 379
IgG antibodies and, 481*t*
in foodstuffs, radiation doses needed to kill, 797*t*
interferons to counter, 468–472, 469*f*, 472*t*
isolation of, 374, 377, 377*f*
latent viral infections, 382, 392, 392*f*, 394*t*
Latin meaning of term, 368
lipophilic, biocidal resistance and, 203, 203*f*
mammalian cells in culture as hosts for, 258, 378*f*
maturation stage in multiplication cycle, **380**, 381*f*
mechanisms for evading host defenses, 441–442, 442*f*, 443*t*
moist heat sterilization and, 188
molecular methods of identifying, 379
multiplication, 379–389
in animal viruses, 382–389
in bacteriophages, 379–382
natural killer (NK) cells can destroy, 491
negative staining of, 63, 387*f*
nonenveloped, 372*f*, **373**
alcohols and, 197, 204*t*
resistance to chemical biocides, 203, 203*f*
Old World viruses introduced into New World viruses, 223*b*
oncogenic (oncoviruses), 376*t*, 389, **391**–392
oncolytic, 369
origins of, 282
orphan, 387
pathogenic properties of, 441–442, 442*f*, 443*t*
penetration stage in multiplication cycle, **380**, 381*f*
peracetic acid effective against, 202
plant, 393–395
plaques formed by bacteriophages, 374, 377, 377*f*
release stage in viral multiplication, **380**, 381*f*
reproduction and, 6, 12
rickettsias/chlamydias compared to, 368, 368*t*
size of, 369, 369*f*
vs. bacterium (*E. coli*), 369*f*
vs. red blood cell, 369*f*
spikes covering envelope of, **372**–373, 372*f*
structure
capsomeres, **372**, 372*f*, 373*f*
enveloped, **372**, 373*f*
nucleic acid of, 371–372, 372*f*, 373*f*
protein coat (capsid), **372**–373, 372*f*, 373*f*
size, 6, 369, 369*f*
spikes, **372**–373, 372*f*
virions, **370**
survival time in boiling water, 188
taxonomy of, 373–374
summary of viruses that infect humans, 375–376*t*
viral species and, **374**
three-domain system and, 282
vaccines and, 502–503, 502*f*, 503*t*
animal cells used to produce, 247
viral enzymes and host enzymes, 379
virions and, 370
with lipid envelopes, resistance to chemical biocides, 203, 203*f*
VISA (vancomycin-resistant *S. aureus*), **20**, 417*t*, 421*t*, 422*b*
visible light
compound light microscopes and, 56, 56*f*, 58–59
energy of, photolyase and, 215*t*
vitamin B1 (thiamine), 117*t*
vitamin B2 (riboflavin), 117*t*
vitamin B6 (pyridoxine), 117*t*
vitamin B12, porins and, 87
vitamin B12 (cyanocobalamin), 117*t*
vitamin C (ascorbic acid), fermentation and, 137*t*
vitamin E, coenzymatic functions, 117*t*
vitamin K, coenzymatic functions, 117*t*
vitamins
as organic growth factors, 162
coenzymatic functions of selected, 117*t*
how they cross plasma membrane, 92, 92*f*
in complex culture media, 165, 166*t*
microbes used in commercial production of, 247
microbiological assays, 165
Voges-Proskauer (V-P) test, 286*f*, 287
volutin, **95**
vomiting, to expel microbes, **452**, 471*t*
von Behring, Emil A., 10*f*, 14*t*, 477
von Nägeli, Carl, 274
voriconazole, 569
mode of action/comments, 564*t*
Vorticella, 350*f*
VRE (vancomycin-resistant enterococci), 417*t*, **563**, 577*b*, 640
VRSA (vancomycin-resistant *Staphylococcus aureus*), 12–13, 19–**20**, 210, 241, 417*t*, 421*t*, 422*b*, 563
VSV (vesicular stomatitis virus), 376*t*, 378*f*, 389*f*
vulnerability to disease. *See* susceptibility
vulva, 744*f*, 745
vulvovaginal candidiasis, 339, **759**

Waksman, Selman A., 14*t*
wandering (free) macrophages, **457**
Warren, J. Robin, 15*t*
warts (papillomas), **595**
genital, 429, **757**, 758*f*, **761***b*
imiquimod to treat, 570
Papillomavirus causing, 375*t*, 385–386, 386*f*
symptoms of, 592*b*
treatments for, 595
Wassermann test, 513
waste products, metabolic pathway, 123
wasting syndrome, caused by *Cyclospora cayetanensis*, 417*t*
water, 32*f*, 34–35, 35*f*
as a polar molecule, **34**–35, 35*f*
as a reactant/product in chemical reactions, 35
as a solvent, **35**, 35*f*
as nonliving reservoirs of infectious disease, 409
as temperature buffer, 35
boiling point, 35
dissociation of common compounds in (examples), 35, 36*f*
distilled, microbial growth and, 159
hydrogen bond formation in, 31, 32*f*
in dehydration synthesis, 38, 39*f*
in hydrolysis, 38, 39*f*
methods for crossing plasma membrane, 92–94, 92*f*, 93*f*
microbial growth and, 159
mole of, 32
molecular weight, 32

passing through plasma membrane via aquaporins, 92*f*
properties, 34–35, 35*f*
structure, 32*f*, 34
water (drinking)
chloramines to disinfect, 197
household bleach to disinfectant in emergencies, 197
water molds (Oomycota), **343***t*, **344**, 345*f*
as decomposers of dead algae, animals, 344
in kingdom Stramenopila, 343
water pipes, *Burkholderia* form biofilms in, 440*b*, 440*f*, 689
water pollution
bioremediation and, 17, 33*b*
blooms of dinoflagellate species as indicators, 344
chemicals, 778–779
detergents, 779
microbial ecology and, 17
pathogenic organisms in, 411, 778
typhoid fever and, 778, 779*f*
water quality
chemicals in, 778–779
pathogenic organisms in, 778, 779*f*
purity tests, 779–782
water treatment, 782–783, 782*f*
chloramines to disinfect, 197
disinfection, 782*f*, **783**
filtration, **782**–783, 782*f*
flocculation, **782**, 782*f*
ozonators, 783, 783*f*
water-damaged homes, *Stachybotrys* fungus found growing in, 338
waterborne diarrhea
Cryptosporidium causing, 355*b*
Cyclospora cayetanensis causing, 350
Watson, James D., 10*f*, 14*t*, 16, 47
wavelengths of light
microscopes and, 56, 56*f*, 58–59
photic zone of bodies of water, algae and, 341*f*, 342
waxes, 146
waxy lipid (mycolic acid), 432
weapons, microbes as. *See* bioweapons
weather, infectious disease incidence and, 408
WEE virus/ *Togavirus* (western equine encephalitis), 375*t*, 625, **628***b*
Weil's disease, 747
Weizmann, Chaim, 2
Weller, Thomas H., 14*t*
West Nile encephalitis (WNE), **20**, 212, 223*b*, 223*f*, **628***b*
as an arbovirus, **628***b*
as an emerging infectious disease, 20, 417*t*
as *Flavirus*, 376*t*, 628*b*
as zoonotic disease, 410*t*
West Nile virus (WNV), 19–**20**, 212, 223*b*, 223*f*, 376*t*
as an arbovirus, 626
birds as disease reservoirs, 20, 223*b*, 410*t*, 626, 628*b*
DNA vaccine for horses, 503
emerging infectious diseases and, 19–20, 417*t*
modern transportation and spread of, 416
mosquitos (*Culex*) as vectors for, 362, 626, 628*b*
PCR used to identify, 379
SEM micrograph, 223*f*
Western blotting, **287**, 289*f*, 510, **516**
to identify viruses, 379
western equine encephalitis (WEE virus/ *Togavirus*), 375*t*, 625, **628***b*
whales
influenza A viruses and, 19, 370*b*
pilot, cetacean morbillivirus (CM) and, 283*b*
wheat, food allergies and, 525
whey, **798**
as liquid waste by-product of dairy industry, 801*b*
in cheese production, **798**–799, 799*f*
used for xanthan production, 801*b*
Whipple's disease, PCR used to identify cause, 290
white blood cells, **454**, 455*t*. *See also* leukocytes
White Cliffs of Dover, as fossilized colonies of marine protist, 277
whiteflies
biocontrol by fungi, 339
watermelon wilt transmitted by, 394*t*
Whitewater Arroyo virus, **569**–660
Whittaker, Robert H., 274
WHO. *See* World Health Organization
whooping cough (pertussis), 306, **680–682**, 680*f*, **699***b*
as emerging infectious disease, 417*t*
as notifiable infectious disease, 421*t*
incubation period, 430*t*
portals of entry, 430*t*
spread by droplet transmission, 411
vaccine, 13, 502, 503*t*, 504*t*, 682
wildlife management, veterinary microbiologists and, 283*b*
Wilkins, Maurice A. F., 14*t*, 47
wine
fermentation and, 135*b*, 137*t*, 800, 802*f*
souring and spoilage, 9, 800
steps in winemaking, 800, 802*f*
sulfur dioxide used as disinfectant, 199, 802*f*
Winogradsky, Sergei, 10*f*, 16–17
Wiskott-Aldrich syndrome, 539*t*
WNE. *See* West Nile encephalitis
WNV. *See* West Nile virus
Woese, Carl R., 6, 10*f*, 274
Wolbachia genus/spp., 300*t*, **305**, 307*b*, 307*f*
as endosymbionts of insects/invertebrates, 300*t*, 305, 307*b*, 307*f*
evolutionary implications of, 307*b*
heartworm life cycle and, 360
wood-eating termites, 107*b*
World Health Organization (WHO)
disease rankings, 329
priorities for emerging infectious diseases, 418
worms. *See* helminths
wound tumor virus (plant virus), 394*t*
wounds. *See also* surgical wounds
botulism, **618**
dressings, silver used in, 198–199
genetically engineered epidermal growth factor to heal, 260*t*
Wuchereria bancrofti (roundworm), elephantiasis caused by, 444

X factor, *Haemophilus* bacteria and, **311**
X ray crystallography, 373
X rays, 192, 193*f*. *See also* radiation
as mutagens, 228, 230–231, 230*f*
X-gal (culture medium), 256, 256*f*
X-linked infantile (Bruton's) agammaglobulinemia, 539*t*
xanthan (thickening agent), produced from whey, 801*b*
xanthins, 343*t*
xanthophylis, 343*t*
xenobiotics, **775**
xenografts. *See* xenotransplantation products
Xenopsylla (rat flea), as vector for typhus, plague, 362*t*, 363*t*, 410*t*, 413*t*, 648
xenotransplantation products, **536**
hyperacute rejection and, **536**
xeroderma pigmentosum, 231
Xigris (drotrecogin alfa), 640
Xolair (omalizumab), 525
xylem, 342

yaws, **753**
yeast extracts, in complex culture media, 165, 166*t*
yeast infection. *See* candidiasis
yeastlike forms of dimorphic fungi, **332**, 332*f*
yeasts, 2, 4, 5*f*, **332**, 332*f*
advantages of using in biotechnology, 258
as ascomycota, 334
as eukaryotes/eukaryotic cells, 6, 76
as facultative anaerobes, 333
as nonfilamentous, unicellular fungi, 332
budding, **332**, 332*f*
carbon dioxide produced by, 332
cell wall of, 98
ethanol produced by, 332
facultative anaerobic growth capabilities of, 332
fermentation and, 9. *See also* fermentation
fission, **332**
genetically modified to produce subunit vaccines, 259
high osmotic pressures and growth of, 192
identification by biochemical testing, 331
in Kingdom Fungi, 281
living, that are millions of years old, 278
peroxidase production and, 3*b*
pH and growth of, 159
rapid identification tests for, 286
reproductive processes, 304, 332–333, 334*f*
yellow fever, **659**, **660***b*
Aedes mosquito as vector, 362*t*, 413*t*, 659, 660*b*
as notifiable infectious disease, 421*t*
as potential biological weapon, 649*b*
filterable agents and, 367
vaccine, 659
viral family that causes, 376*t*, 413*t*, 660*b*
Yersinia enterocolitica, 283*b*, 720
Yersinia gastroenteritis (yersiniosis), **720**, **723***b*
Yersinia genus/spp., 301*t*, **311**
as human pathogens, 301*t*
Yersinia pestis
as potential biological weapon, 649*b*
capsule of, virulence and, 432
plague caused by, 311. *See also* plague
portals of entry, 429
reservoirs/transmission method, 410*t*
yersiniosis (*Yersinia* gastroenteritis), **720**, **723***b*
yogurt
fermentation and, 135*b*, 137*t*
microbes used to make, 799

zanamivir (Relenza), 564*t*, 570
Zephiran (benzalkonium chloride), 198, 199, 199*f*, 200*f*, 201*b*, 204*t*
Zevalin (ibritumomab), 509
zidovudine, 571
zimantadine, 564*t*
zinc
as antimicrobial agent, 199
as cofactor, 117
zinc chloride, 199
zinc pyrithione, 199
Zinkernagel, Rolf. M., 15*t*
zippers, made by microbes, 3*b*
Zoogloea genus/spp., 300*t*, **306**
in sewage treatment, 306, 784, 786*f*
zoonoses/zoönosis, **409**, 410*t*
zoospores, **344**, 345*f*
zur Hausen, Harald, 15*t*
zygomycetes, 333, 334*f*, 335*f*
Zygomycota, **333**, 335*f*
as conjugation fungi, 333
pathogenic fungi of, 338*t*
Mucor, 338, 338*t*
Rhizopus, 333, 334, 335*f*, 338, 338*t*
zygosporangium, 335*f*
zygospore, **333**, 335*f*
zygote
in life cycle of *Rhizopus*, 335*f*
in protozoan reproduction, 346
Zyvox (linezolid), 566
mode of action/spectrum of activity, 562*t*

TAXONOMIC GUIDE TO DISEASES

BACTERIA AND THE DISEASES THEY CAUSE

Alphaproteobacteria

Anaplasmosis	*Anaplasma phagocytophilum*	p. 654
Brucellosis	*Brucella* spp.	pp. 643–645
Cat-scratch disease	*Bartonella henselae*	p. 647
Ehrlichiosis	*Ehrlichia* spp.	p. 654
Endemic murine typhus	*Rickettsia typhi*	p. 655
Epidemic typhus	*R. prowazekii*	p. 654
Rocky Mountain spotted fever	*R. rickettsii*	p. 655

Betaproteobacteria

Gonorrhea	*Neisseria gonorrhoeae*	pp. 747–750
Melioidosis	*Burkholderia pseudomallei*	pp. 690–691
Meningitis	*N. meningitidis*	pp. 613–614
Nosocomial infections	*Burkholderia* spp.	p. 440
Ophthalmia neonatorum	*N. gonorrhoeae*	pp. 603, 748
Pelvic inflammatory disease	*N. gonorrhoeae*	pp. 751–752
Whooping cough	*Bordetella pertussis*	pp. 680–682
Rat-bite fever	*Spirillum minor*	p. 647

Gammaproteobacteria

Animal bites	*Pasteurella multocida*	pp. 647–648
Bacillary dysentery	*Shigella* spp.	p. 712
Chancroid	*H. ducreyi*	p. 756
Cholera	*Vibrio cholerae*	pp. 716–717
Conjunctivitis	*H. influenzae*	p. 603
Cystitis	*Escherichia coli*	p. 746
Dermatitis	*Pseudomonas aeruginosa*	p. 591
Epiglottitis	*Haemophilus influenzae*	pp. 676–677
Gastroenteritis	*E. coli*	pp. 717–718
Gastroenteritis	*V. parahaemolyticus*	p. 717
Gastroenteritis	*V. vulnificus*	p. 717
Gastroenteritis	*Y. enterocolitica*	p. 720
Legionellosis	*Legionella pneumophila*	pp. 688–689, 691
Meningitis	*H. influenzae*	p. 613
Otitis externa	*P. aeruginosa*	p. 593
Otitis media	*H. influenzae*	p. 679
Otitis media	*Moraxella catarrhalis*	p. 679
Plague	*Yersinia pestis*	pp. 648, 650
Pneumonia	*H. influenzae*	p. 688
Pyelonephritis	*E. coli*	p. 746
Q fever	*Coxiella burnetti*	pp. 689–690
Salmonellosis	*Salmonella enterica*	pp. 712–714, 715
Septicemia	*P. fluorescens*	p. 164
Tularemia	*Francisella tularensis*	pp. 642–643
Typhoid fever	*S. typhi*	pp. 714–716

Epsilonproteobacteria

Gastroenteritis	*Campylobacter jejuni*	p. 718
Gastritis	*Helicobacter pylori*	pp. 718–720
Peptic ulcers	*H. pylori*	pp. 718–720

Clostridia

Botulism	*C. botulinum*	pp. 616–619
Gangrene	*C. perfringens*	p. 646
Gastroenteritis	*C. difficile*	p. 720
Gastroenteritis	*C. perfringens*	p. 720
Tetanus	*Clostridium tetani*	pp. 615–616

Mollicutes

Pneumonia	*Mycoplasma pneumoniae*	p. 688
Urethritis	*Mycoplasma, Ureaplasma*	p. 751

Bacilli

Anthrax	*Bacillus anthracis*	p. 645
Bacterial endocarditis	*Staphylococcus aureus*	p. 641
Cystitis	*S. saprophyticus*	p. 746
Dental caries	*S. mutans*	pp. 707–709
Endocarditis	Alpha-hemolytic streptococci	p. 641
Erysipelas	*Streptococcus pyogenes*	p. 591
Folliculitis	*S. aureus*	p. 588
Food poisoning	*S. aureus*	pp. 711–712
Gastroenteritis	*B. cereus*	p. 720
Impetigo	*S. aureus*	p. 588
Listeriosis	*Listeria monocytogenes*	pp. 614–615
Meningitis	*S. pneumoniae*	p. 614
MRSA	*S. aureus*	pp. 422, 586–587
Necrotizing fasciitis	*S. pyogenes*	p. 591
Otitis media	*S. pneumoniae*	p. 679
Pneumonia	*S. pneumoniae*	pp. 685–688

Guide continues

Puerperal sepsis	*S. pyogenes*	pp. 640–641
Rheumatic fever	*S. pyogenes*	p. 641
Scalded skin syndrome	*S. aureus*	p. 588
Scarlet fever	*S. pyogenes*	p. 677
Sepsis	*Enterococcus* spp.	p. 640
Sepsis	*S. agalactiae*	p. 640
Strep throat	*S. pyogenes*	p. 677
Toxic shock syndrome	*S. aureus*	p. 588
Toxic shock syndrome	*S. pyogenes*	p. 591

Actinobacteria

Acne	*Propionibacterium acnes*	p. 594
Buruli ulcer	*Mycobacterium ulcerans*	p. 594
Diphtheria	*Corynebacterium diphtheriae*	pp. 677–678
Leprosy	*M. leprae*	pp. 619–620
Mycetoma	*Nocardia asteroides*	p. 321
Rapidly growing mycobacteria	*Mycobacterium* spp.	p. 201
Tuberculosis	*M. tuberculosis*	pp. 682–685
Tuberculosis	*M. bovis*	p. 144
Vaginosis	*Gardnerella vaginalis*	p. 756

Chlamydiae

Inclusion conjunctivitis	*Chlamydia trachomatis*	p. 604
Lymphogranuloma venereum	*C. trachomatis*	p. 755
Pelvic inflammatory disease	*C. trachomatis*	pp. 751–752
Pneumonia	*Chlamydophila pneumoniae*	p. 689
Psittacosis	*C. psittaci*	p. 689
Trachoma	*C. trachomatis*	pp. 604–605
Urethritis	*C. trachomatis*	pp. 750–751

Spirochetes

Leptospirosis	*Leptospira interrogans*	pp. 746–747
Lyme disease	*Borrelia burgdorferi*	pp. 651–654
Relapsing fever	*B.* spp.	pp. 650–651
Syphilis	*Treponema pallidum*	pp. 752–755

Bacteroidetes

Acute necrotizing gingivitis	*Prevotella intermedia*	p. 709
Periodontal disease	*Porphyromonas* spp.	p. 709

Fusobacteria

Rat-bite fever	*Streptobacillus moniliformis*	p. 647

FUNGI AND THE DISEASES THEY CAUSE

Ascomycetes

Aspergillosis	*Aspergillus fumigatus*	p. 697–698
Blastomycosis	*Blastomyces dermatitidis*	p. 697
Histoplasmosis	*Histoplasma capsulatum*	pp. 695–696
Ringworm, Athlete's foot	*Microsporum, Trichophyton*	pp. 600–601

Anamorphs

Candidiasis	*Candida albicans*	pp. 601–602, 758–759
Coccidioidomycosis	*Coccidioides immitis*	pp. 696–697
Pneumonia	*Pneumocystis jirovecii*	p. 697
Sporotrichosis	*Sporothrix schenckii*	p. 601

Basidiomycetes

Dandruff	*Malassezia furfur*	p. 586
Meningitis	*Cryptococcus neoformans*	pp. 626–627
Mycotoxins		pp. 443, 729–730

PROTOZOA AND THE DISEASES THEY CAUSE

Archaezoa

Giardiasis	*Giardia lamblia*	p. 730
Trichomoniasis	*Trichomonas vaginalis*	p. 760

Apicomplexa

Babesiosis	*Babesia microti*	p. 666
Cryptosporidiosis	*Cryptosporidium* spp.	pp. 355, 731
Cyclospora infection	*Cyclospora cayetanensis*	p. 731
Malaria	*Plasmodium* spp.	pp. 663–665
Toxoplasmosis	*Toxoplasma gondii*	pp. 661–663

Amoebozoa

Amoebic dysentery	*Entamoeba histolytica*	pp. 731–732
Encephalitis	*Acanthamoeba* spp.	p. 605
Keratitis	*Acanthamoeba, Balamuthria mandrillaris*	p. 629

Euglenzoa

African trypanosomiasis	*Trypanosoma brucei*	pp. 627, 629
Chagas' disease	*T. cruzi*	p. 661
Leishmaniasis	*Leishmania* spp.	pp. 665–666
Meningoencephalitis	*Naegleria fowleri*	p. 629

HELMINTHS AND THE DISEASES THEY CAUSE

Platyhelminths

Hydatid disease	*Echinococcus granulosus*	pp. 733–734
Schistosomiasis	*Schistosoma* spp.	pp. 666–667
Swimmer's itch	Schistosomes	p. 667
Tapeworm infections	*Taenia* spp.	pp. 732–733

Nematodes

Ascariasis	*Ascaris lumbricoides*	p. 736
Hookworm disease	*Necator americanus, Ancyclostoma*	pp. 735–736
Pinworms	*Enterobius vermicularis*	p. 734–738
Trichinellosis	*Trichinella spiralis*	pp. 736–737

ALGAE AND THE DISEASES THEY CAUSE

Red Algae, Diatoms, and Dinoflagellates	*Alexandrium, Pfiesteria*	pp. 343–344
Oomycota	*Phytophthora*	p. 344

ARTHROPODS AND THE DISEASES THEY CAUSE

Pediculosis	*Pediculus humanus*	pp. 602–603
Scabies	*Sarcoptes scabiei*	p. 602

VIRUSES AND THE DISEASES THEY CAUSE

DNA Viruses

Burkitt's lymphoma	Herpesvirus	pp. 655–656
Chickenpox	Herpesvirus	pp. 596–597
Cold sores	Herpesvirus	pp. 597–598
Cytomegalic inclusion disease	Herpesvirus	p. 658
Fifth disease	Parvovirus	p. 600
Genital herpes	Herpesvirus	pp. 757–758
Genital warts	Papovavirus	p. 758
Hepatitis B	Hepadnavirus	pp. 723–726
Infectious mononucleosis	Herpesvirus	pp. 656–657
Keratitis	Herpesvirus	p. 605
Monkeypox	Poxvirus	p. 596
Roseola	Herpesvirus	p. 600
Shingles	Herpesvirus	p. 596
Smallpox	Poxvirus	pp. 595–596
Warts	Papovavirus	p. 595

RNA Viruses

AIDS	Retroviruses	pp. 539–548
Chikungunya fever	Togavirus	p. 658
Common cold	Picornavirus	p. 679
Dengue	Flavivirus	p. 659
Encephalitis	Bunyavirus	pp. 624–626
Encephalitis	Flavivirus	pp. 223, 624–626
Encephalitis	Rhabdovirus	p. 624
Encephalitis	Togavirus	pp. 624–626
Gastroenteritis	Calcivirus	pp. 266, 728
Gastroenteritis	Reovirus	p. 728
Hantavirus pulmonary syndrome	Bunyavirus	p. 660
Hemorrhagic fever	Filovirus, Arenavirus	p. 659
Hepatitis A	Picornavirus	pp. 721–723
Hepatitis C	Flavivirus	p. 726
Hepatitis D	Deltavirus	pp. 726–727
Hepatitis E	Calcivirus	pp. 727–728
Influenza	Orthomyxovirus	pp. 370, 692–694
Measles	Paramyxovirus	pp. 504, 598–599
Mumps	Paramyxovirus	p. 721
Poliomyelitis	Picornavirus	pp. 620–622
Rabies	Rhabdovirus	pp. 622–624
RSV infection	Paramyxovirus	p. 692
Rubella	Togavirus	p. 599
Yellow fever	Flavivirus	p. 659

PRIONS AND THE DISEASES THEY CAUSE

Transmissible spongiform encephalopathies		pp. 392–393, 630–631